T0396227

Hydraulic Engineering of Dams

Hydraulic Engineering of Dams

Willi H. Hager
Laboratory of Hydraulics, Hydrology and Glaciology (VAW), ETH Zürich, Zürich, Switzerland

Anton J. Schleiss
Laboratoire de Constructions Hydrauliques (LCH), Ecole Polytechnique Fédérale de Lausanne, Lausanne, Switzerland

Robert M. Boes
Laboratory of Hydraulics, Hydrology and Glaciology (VAW), ETH Zürich, Zürich, Switzerland

Michael Pfister
Haute école d'ingénierie et d'architecture (HES-SO), Fribourg, Switzerland

CRC Press
Taylor & Francis Group
Boca Raton London New York Leiden

CRC Press is an imprint of the
Taylor & Francis Group, an **informa** business

A BALKEMA BOOK

CRC Press/Balkema is an imprint of the Taylor & Francis Group, an informa business

© 2021 Taylor & Francis Group, London, UK

Original edition: Dam Hydraulics, D.L. Vischer & W.H. Hager, June 1998, John Wiley & Sons Ltd.,

ISBN 978-0-47197289-1

Typeset by Apex CoVantage, LLC

Library of Congress Cataloging-in-Publication Data
Names: Hager, Willi H., author.
Title: Hydraulic engineering of dams / Willi H. Hager, Laboratory of
    Hydraulics, Hydrology and Glaciology, Department of Civil, Environmental
    and Geomatic Engineering, ETH Zürich, Zürich, Switzerland, Anton J.
    Schleiss, Laboratoire de Constructions Hydrauliques, Ecole Polytechnique
    Fédérale de Lausanne, Lausanne, Switzerland, Robert M. Boes, Laboratory of
    Hydraulics, Hydrology and Glaciology, Department of Civil, Environmental
    and Geomatic Engineering, ETH Zürich, Zürich, Switzerland, Michael Pfister,
    Haute école d'ingénierie et d'architecture, Fribourg, Switzerland.
Description: London, UK : CRC Press/Balkema is an imprint of the Taylor &
    Francis Group, an Informa Business, [2019] | Series: Technology—hydraulic
    engineering | Includes bibliographical references.
Identifiers: LCCN 2019020884 (print) | ISBN 9780415621533 (hbk : alk.
    paper)
Subjects: LCSH: Dams. | Hydraulic engineering. | Hydraulics.
Classification: LCC TC540 .H246 2019 (print) | LCC TC540 (ebook) |
    DDC 627/.8—dc23
LC record available at https://lccn.loc.gov/2019020884
LC ebook record available at https://lccn.loc.gov/2019981064

Published by: CRC Press/Balkema
        Schipholweg 107c, 2316 XC Leiden, The Netherlands
        e-mail: Pub.NL@taylorandfrancis.com
        www.crcpress.com – www.taylorandfrancis.com

ISBN: 978-0-415-62153-3 (hbk)
ISBN: 978-0-203-77143-3 (eBook)

DOI: 10.1201/9780203771433
DOI: https://doi.org/10.1201/9780203771433

Dedication

*We are proud and happy to dedicate this book to our former
teacher, tutor and colleague
Emeritus Prof. Dr. Dr. h.c. Daniel L. Vischer*

Contents

Preface

Hydraulics is the engineering science of physical fluid flow, typically encountered both in pressurized and free-surface flows. In contrast, hydraulic engineering includes hydraulics applied to hydraulic structures usually designed by civil and environmental engineers. The book *Dam Hydraulics* of Vischer and Hager, published in 1998, was mainly directed to basic hydraulic problems dealing with dams, whereas the present book includes also questions of hydraulic engineering, so that the chosen book title appears adequate.

During the past two decades, a large number of relevant topics have been developed both at universities and in engineering practice. To mention just a few, these include embankment weirs mainly under submerged flow conditions, scale effects associated with weir flow (Chapter 2), spatial features of side channels, labyrinth weirs, piano key weirs (Chapter 3), air transport on chutes, chute aerators, aerator spacing (Chapter 4), undular hydraulic jumps, free fall outlets including questions of scour (Chapter 5), design of ski jumps and questions relating to granular and rock scour (Chapter 6), diversion tunnels including problems with choking flow and scour features (Chapter 7), intakes and problems with air entrainment, gate flow and low-level outlets whose main features have only been poorly understood until then (Chapter 8), reservoir sedimentation with the hydraulic performance of sediment bypass tunnels, the characteristics of turbidity currents, and sedimentation control in general (Chapter 9), both 2D and 3D impulse waves in reservoirs with a description of the relevant features during wave generation and propagation, wave run-up including overland flow, plus the dam overtopping features for both rigid and erodible crests (Chapter 10), and finally hydraulic breaches including both progressive and instantaneous cases along with the main features of fuse plugs (Chapter 11). In addition to these mostly new topics hardly encountered in engineering books, the more standard issues are of course also retained in the book. The entire work thus aims at providing modern knowledge in hydraulic engineering related to dams presented in an attractive way, so that readers will feel comfortable with its contents.

The book is one of the few currently available on the market on dam hydraulics and engineering. The authors of the current book have all been (partially) educated at ETH Zurich, having a close relation to its former VAW-Director, Em. Prof. Dr. Dr. h.c. Daniel L. Vischer, to whom this book is dedicated. The authors have been involved in the past, or are still active at either EPF Lausanne or at ETH Zurich, the two Swiss Federal Institutes of Technology. Notably, the authors have either a background in hydraulics or in hydraulic engineering, or even both, so that the topics considered in this book are adequately covered. Note that the book contents are mainly based on laboratory experimentation, at places on prototype observations, but only sporadically on computational hydraulics. This reflects the origin of the authorship, whose main message should be the collection of scientific data, their interpretation and analyses, serving in addition as a data basis for numerical studies. The book also includes few computational examples by which a certain procedure is highlighted. The entire book volume is so large that the addition of more examples would be hardly possible in one volume, so that this task should be covered in other works, possibly based on the present book.

Each chapter is completed by a notation section, in which the main parameters are again defined. Note that identical notation with a different meaning may occur in the various chapters. The notation is followed by the references, and a bibliography section is also provided to detail mainly works published ahead of the year 2000, which are often not found on the Internet. Readers interested in additional works may therein find adequate information not detailed in the main text.

The authors have not only written a book; they were eager to show also the technical beauties of water flow related to dam engineering with selected figures and photographs, given that few have seen hydraulic engineering from this perspective. All figures were prepared by Mr. Andreas Schlumpf, VAW, ETH Zurich, to whom we would like to express our sincere thanks for the excellent result.

Zürich/Lausanne/Fribourg, December 2019
WHH, AJS, RMB, MP

Authors' CVs

Willi H. **Hager** (*1951) was educated at ETH Zurich, including the civil engineering degree (1976), the PhD degree (1981), the habilitation degree (1994), and professorship of hydraulics (1998–2016). He was research associate at ETH Lausanne (1983–1988), returning in 1989 to *Versuchsanstalt für Wasserbau, Hydrologie und Glaziologie* (VAW) as scientific head. He was interested in hydraulic structures, wastewater hydraulics, high-speed flows, impulse waves, scour and erosion, and in the history of hydraulics. He has published papers and books on these topics, including Dam hydraulics (1998), *Constructions hydrauliques* (2009), Wastewater hydraulics (2010) or Hydraulicians. He served the  *Journal of Hydraulic Engineering* (ASCE) as Associate Editor, and the *Journal of Hydraulic Research* (IAHR) as Editor. He was awarded the 1997 Ippen Lecture (IAHR) becoming its Honorary Member in 2013, and was the recipient of the Hydraulic Structures Medal (ASCE), among other distinctions.
Homepage: http://www.vaw.ethz.ch/people/hy/hagerw

Anton J. **Schleiss** (*1953) graduated in Civil Engineering from ETH Zurich, obtaining a PhD on the Design of pressure tunnels. He worked for 11 years with an International Engineering Consulting Company in Zurich, and was involved in the design of hydro projects as an expert in hydraulic engineering and as project manager. In 1997, he was appointed full professor and director of the Laboratory of Hydraulic Constructions (LCH) at the Ecole polytechnique fédérale de Lausanne (EPFL). Until having become honorary professor in 2018, he supervised more than 50 PhD and Postdoc researches in the field of engineering of hydraulic systems and schemes. He was listed in 2011 among the 20 international personalities that "have made the biggest difference  to the sector Water Power & Dam Construction over the last 10 years". For his outstanding contributions to advance the art and science of hydraulic structures engineering, he obtained in 2015 the ASCE-EWRI Hydraulic Structures Medal. The French Hydro Society (SHF) awarded him with the Grand Prix SHF 2018. After having served as vice-president from 2012 to 2015, he was president of the International Commission on Large Dams (ICOLD) until 2018. With more than 40 years of experience, he is regularly involved as an expert in large water infrastructure projects including hydropower and dams all over the world. More details including publications see: http://people.epfl.ch/anton.schleiss?lang=en

Robert M. **Boes** (*1969) studied civil engineering in Germany and France, graduating from the Technical University of Munich in 1996. He obtained a Doctorate in hydraulic engineering from ETH Zurich in 2000. In 2002, he joined the Hydro Engineering Department of the Tyrolean Utility TIWAG in Innsbruck, Austria, where he was involved in projects relating to hydropower schemes, hydraulics and flood protection and became head of the Dam Construction Group. In 2009, he was appointed Professor of Hydraulic Structures at ETH Zurich, where he directs the Laboratory of Hydraulics, Hydrology and Glaciology (VAW). He is vice-president of the Swiss National Committee on Dams, board member of the Swiss Association for Water Resources Management, and the Energy Science Council at ETH Zurich. He is work package leader "Hydropower" of the Swiss Competence Center for Energy Research – Supply of Electricity, and steering committee member of the Joint Programme Hydropower in the framework of the European Energy Research Alliance, and is involved as an expert in hydropower, dam and flood protection projects. His research interests include dam hydraulics and dam safety, reservoir sedimentation, impulse waves, hydropower & environment, sediment monitoring, hydro-abrasive wear of hydraulic structures and machinery, and flood protection.

Homepage: https://vaw.ethz.ch/en/people/person-detail.NjM3MTQ=.TGlzdC8xOTYxLDE1MTczNjI1ODA=.html

Michael **Pfister** (*1976) graduated in Civil Engineering from ETH Zurich in 2002. He then joined the Laboratory of Hydraulics, Hydrology and Glaciology (VAW), ETH Zurich, obtaining in 2007 a Doctorate in Sciences. In 2010, he moved to the Laboratory of Hydraulic Constructions (LCH), EPF Lausanne, as a Research and Teaching Associate. There, he initiated research axes and advised PhD students, was a Lecturer and a project manager for hydropower-related model studies. The *Haute école d'ingénierie et d'architecture Fribourg* (HEIA-FR, HES-SO), Switzerland, appointed him in 2016 as Professor for Hydraulic Engineering. During his activities, he was responsible for research projects on high-speed two-phase flows, Piano Key weirs, urban drainage structures, and for wave dynamics. Physical modelling was the main approach for many of his researches. Michael has published peer-reviewed journal papers and served as book editor in the above-mentioned topics. He is a member of national and international committees and associations.

Chapter 1 Frontispiece (a) Buttress dam Al-Massira in Morocco with spillway (Type 2133 according to Figure 1.6) with bottom outlets located underneath and powerhouse on the left side (Courtesy Roland Bischof), (b) 220 m high Contra Arch Dam in Switzerland with spillway in operation with about 300 m^3/s (ungated ogee crest followed by chutes and ski jumps: Type 1133 according to Figure 1.6; design discharge 2150 m^3/s) (Courtesy Verzasca SA, Gordola, Switzerland)

Chapter 1

Introduction

1.1. Definition and purposes of dams

Large dams are civil engineering structures closing a valley well above the maximum flood-water stage. According to the International Commission on Large Dams (ICOLD), the denomination of *Large dams* comprises dams exceeding 15 m in structural height above the foundation or, for 10–15 m high dams, impounding a reservoir volume exceeding $3 \times 10^6 \text{ m}^3 = 3 \text{ hm}^3$. Dams belong to the largest man-made structures and are often called *useful pyramids* (Schnitter, 1994). The *reservoir* created by water impoundment typically stores a large part of the inflow, thus creating a balance between water supply and demand. Usually, water is stored during flood periods and used during dry seasons. Depending on the storage and inflow volumes, as well as the utilization concept, a distinction is made between daily, weekly, monthly, seasonal, and multi-seasonal reservoirs.

Besides storing water, dams are also erected to create a hydraulic head or a water surface. A *hydraulic head* increases the net pressure on hydropower turbines and may improve river navigation by creating sufficient backwater. It also facilitates to withdraw water from a river and to convey it by gravity to headrace channels for drinking water or irrigation water supply. A *water surface* enables, among others, navigation, lake recreation, groundwater replenishment, and fishery as well as flood peak damping.

Dams and reservoirs are classified according to their main purpose, namely (rounded % according to ICOLD Register):

- Irrigation (50%)
- Hydropower (20%)
- Water supply (11%)
- Flood control (9%)
- Recreation and tourism (5%)
- Navigation and fish farming (<1%)
- Others (5%).

Most of these dams are single-purposed (75%), but many dams (25%) are also multi-purposed, simultaneously satisfying two or more of the aforementioned purposes. This particularly pertains to the combinations of irrigation and water supply, hydropower and flood control, as well as flood protection and water replenishment for navigation. Often, besides their main purpose, there are additional benefits from dams such as serving as recreation areas, nature reserves, and tourist attractions. Today, large water infrastructures projects like reservoirs and dams should be designed as multipurpose schemes to benefit from synergies between different objectives of use and protection, and to gain wide acceptance from all stakeholders (Schleiss, 2016a, 2017a, 2018).

Regarding flood protection, it should be noted that reservoirs inherently have positive effects on the flood risk attenuation in their tailwater reaches by retaining large volumes of the inflow, so that the peak outflow is damped. Provided there is an adequately designed unregulated spillway, this holds true even for reservoirs completely filled up to the spillway crest due to the retention effect created by the water surface. Regarding droughts, existing dams and reservoirs as well as future projects will have to play a key role in mitigating the effects of climate change worldwide, particularly in semi-arid and arid regions.

1.2. Worldwide importance of dams and reservoirs

According to the ICOLD Register (2018), today almost 59,000 large dams are satisfying the worldwide vital needs for water, energy, food, and flood protection (Figure 1.1). The total reservoir volume of all dams registered by ICOLD is about 7500 km^3, of which about 4000 km^3 can be used directly. If this useful storage is compared with the entire water stored at a certain instant in all rivers worldwide, i.e. 1000–2000 km^3, it is clearly recognized that these reservoirs can significantly influence the global water cycle. This is, above all, of highest importance for the vital worldwide food production.

Figure 1.2 shows the number of large dams under construction as reported yearly by the World Atlas & Industry Guide (Schleiss, 2017b). The dam construction activity since the year 2000 is almost constant, with between 320 and 370 dams higher than 60 m and between 30 and 60 dams higher than 150 m under construction every year. Note that, besides a slight fluctuation, no direct effect of the world economic crisis on dam construction is visible. This

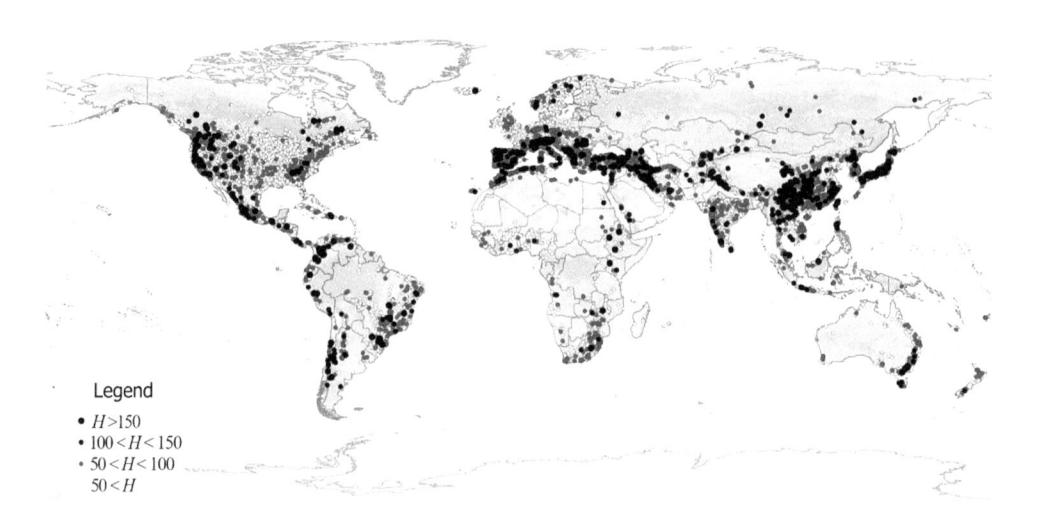

Legend
- $H > 150$
- $100 < H < 150$
- $50 < H < 100$
- $50 < H$

Figure 1.1 Location of the almost 59,000 large dams of height H [m] according to the ICOLD Register (Courtesy Patrick Le Delliou)

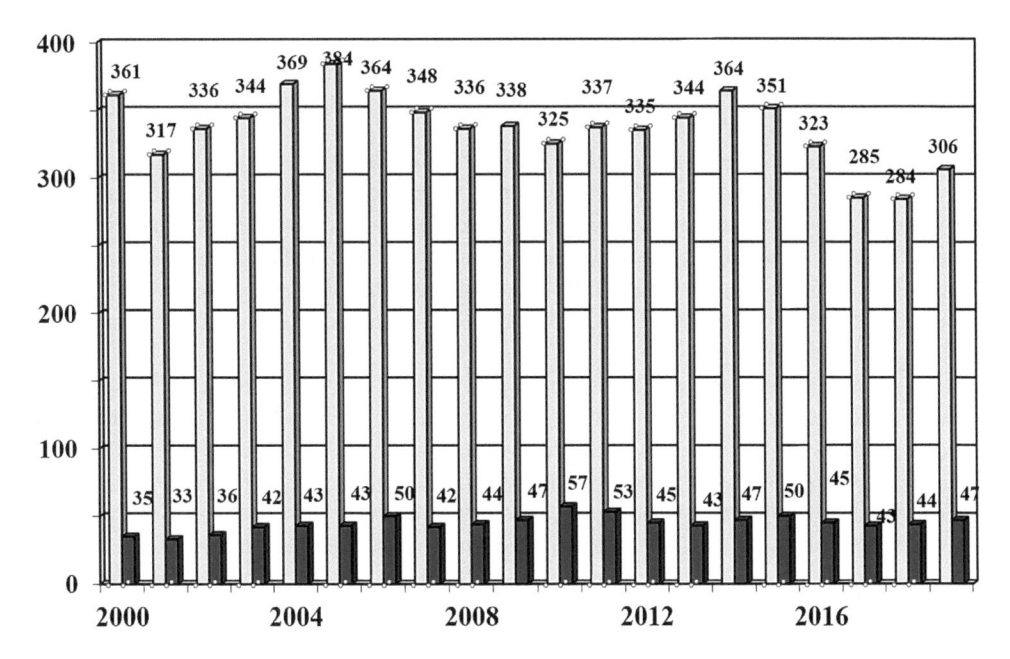

Figure 1.2 Dams under construction worldwide from 2000 to 2019 of height ($\square$) > 60 m, ($\blacksquare$) > 150 m
(Schleiss, 2018)

reveals that dams are still and will always be needed as vital infrastructures on which the soundness of the economy depends.

Dams and hydropower are strongly related (Schleiss, 2018). Hydropower is still the cheapest and most flexible renewable energy with a significant potential worldwide. It is a fact that only about one-quarter of the technically feasible hydro potential of 16,000 TWh have been developed worldwide so far. In South America, only one-third of the economically feasible potential is in operation. In Asia (including Russia and Turkey), where the economy in many countries is growing fast with a tremendous increase of the energy demand, only one-fifth of the hydropower potential is used until today. In Africa only about 12% of the economic and environmentally feasible hydropower potential is used on average on the continent, and only some 17 GW is under construction. The remaining economic and environmental feasible hydropower potential per year in Africa corresponds to 150% of the yearly generation of all hydropower plants under operation in Europe.

This reveals again the high potential for dams and reservoir development in Africa as a trigger for welfare. Nevertheless, for hydropower including dams and reservoirs, this development has to be prepared in a sustainable way in order to protect the rich ecosystems such as flood plains and wildlife reserves. Such ecological electricity generation will be a selling argument in the interconnected intercontinental electricity market in the future. In Europe and in North America, already 70% and 75% of this potential is used, respectively. Some countries in Europe, as Switzerland (90%) and France (97%), are using most of their technically and environmentally feasible potential. Nevertheless, important investments have to be made in the near future to upgrade existing hydropower to increase flexibility in view of the strong increase of new renewable energies such as solar and wind, whose production is highly volatile.

The major problems of the world population in this century will be without doubt the safe supply of ecological and renewable energy as well as the supply of water of good quality and sufficient quantity to fight against famine, poverty, and disease. Still today, water supply and sanitation services leave much to be desired; two-thirds of the world population suffers from the lack of safe water (insufficient quantity) or from the lack of safe sanitation (inappropriate quality). Furthermore, an important part of the world population is threatened by famine. This risk could be considerably lessened by the irrigation of arid areas not cultivable today. Thus, in many countries, especially in Africa, there is still an urgent need for increased development of water and energy resources as a basis for the economic prosperity and cultural wealth of the societies.

Furthermore, regarding the life-cycle analysis, hydropower is by far the best option in view of sustainability. When looking at the so-called recovery factor or energy payback ratio of primary energy, which is obtained by the ratio of the total energy produced to the total expense of non-renewable energy (direct and indirect) during a lifetime to operate an installation, hydropower is unbeatable (Schleiss, 2000). For storage hydropower plants with reservoirs created by dams, the energy payback ratio is set between 205 and 280, and for run-off-river power plants it is between 170 and 270. In fact, these numbers exceed definitely those attained by other renewable energies such as solar photovoltaic (3 to 6) or wind (18 to 34). Their recovery factors are rather small today, but important technical progress can be expected in the future. Thermal power plants producing electricity with non-renewable fuels have, as expected, an even lower energy payback ratio. Furthermore, recent life-cycle analyses also confirm that hydropower can reduce greenhouse gas (GHG) emissions significantly even by developing only a part of the remaining economically feasible hydro potential.

Recent studies in Switzerland revealed that for hydropower, the equivalent CO_2 emission regarding greenhouse gases is extremely small compared to other electricity generation technologies; it is caused mainly by the acquisition of material required for construction and maintenance (Frischknecht *et al.*, 2012; Bauer *et al.*, 2017). Swiss run-off-river power plants produce between 4 and 5 g CO_2-equivalent/kWh whereas storage power plants between 6 and 7 g CO_2eq/kWh. For nuclear energy, these emissions are between 10 and 20 g CO_2eq/kWh, depending on the type of reactor. For new renewable energies like wind and solar (photovoltaic, or PV), average values of 15 g CO_2eq/kWh and between 38 and 95 g CO_2eq/kWh are obtained, respectively, the latter depending on the PV technology. Swiss biogas power plants with thermal-power coupling produce between 150 and 450 g CO_2eq/kWh. These values are still relatively small compared to the emission of gas power plants (480 to 640 g CO_2eq/kWh) and hard coal and lignite power plants (820 to 980 g CO_2eq/kWh).

The safe and sustainable development of dams and reservoirs can significantly contribute to the 16 Sustainable Development Goals (SDGs) declared by Agenda 2030 of the United Nations. The following goals are directly related to dam and reservoir development (Schleiss, 2017b):

- Goal 2. End hunger, achieve food security and improved nutrition and promote sustainable agriculture
- Goal 7. Ensure access to affordable, reliable, sustainable and modern energy for all
- Goal 8. Promote sustained, inclusive and sustainable economic growth, full and productive employment and decent work for all

- Goal 9. Build resilient infrastructure, promote inclusive and sustainable industrialization and foster innovation
- Goal 13. Take urgent action to combat climate change and its impacts.

Hydropower and dam projects often provoke controversial discussions (WCD, 2000). In order to gain wide acceptance and to obtain a win-win situation between all stakeholders, large water infrastructure projects have to be designed as multipurpose projects by multi-disciplinary teams with a complex system approach. This needs excellence in engineering sciences and management.

Figure 1.3 shows where the new large dams have been built since the year 2000 (Schleiss, 2016a, 2018). Thanks to these new water infrastructures, a kind of a security belt around the world can be recognized ensuring water, food, and energy. There is a high new dam density from Southern Europe over the Middle East and Central Asia to East Asia. It covers the area with high water stress in arid and semi-arid regions as well as monsoon climate–exposed regions of extremely high population density. The belt is less visible across North America over the world's most productive crop growing region, where only a few dams have been built in this century, but significant development there took place in the last century. It has to be noticed that the regions along this belt are already touched perceptibly today by climate change, whose effects will become even more dramatic in the future according to estimates. The actual dams and reservoirs as well as the future projects will have to play a key role in the mitigation of the effects of climate change. The belt of dams and reservoirs which covers these threatened and very vulnerable regions around the world will give safety for food, water, and energy, thus it can be called a security belt.

Figure 1.3 highlights another worldwide problem due to the construction of new dams, namely the huge economical gradient from north to south originating from developed countries to emerging and developing countries. Overcoming this challenge is a millennium goal.

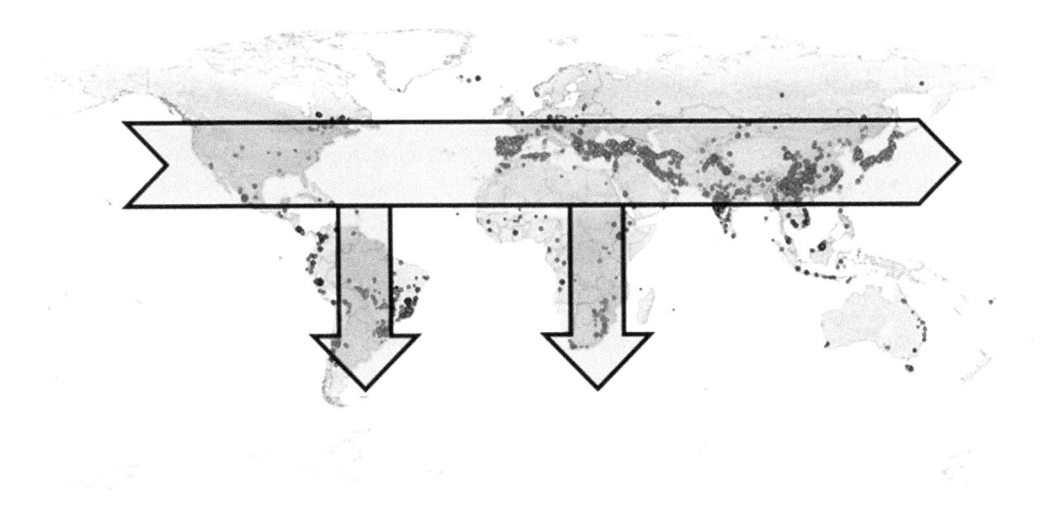

Figure 1.3 New dams and reservoirs commissioned since 2000 creating a security belt around the world to ensure water, food and energy (NEXUS and SDGs) (Schleiss, 2016a, 2017b)

In South America, the north-south spreading of the new dams attached to the aforementioned security belt can be clearly seen. In Africa, unfortunately, it can be hardly recognized. In a translated way, it can be said that the security belt has to be fixed with suspenders to the south in order not only to secure food, water, and energy but also equitable worldwide wealth for all countries. Dams and reservoirs are the water infrastructures or the main elements which give the required strength to the security belt and its suspenders.

1.3. Historical overview and challenges of dam engineering

Dam construction and dam engineering has a long tradition of several thousands of years. While ancient dams and reservoirs mainly served for drinking and irrigation water supply as well as flood protection purposes, many larger dams were built from the 16th century to drive water mills and later hydropower plants. A short description of the history of dam engineering is given below, mainly based on Garbrecht (1987) and Schnitter (1994).

The first noteworthy ancient dams were constructed by the earlier civilizations of antiquity, particularly in the Nile Valley, Mesopotamia, China, and Southern Asia. The oldest known remains stem from the Sadd-el Kafara Dam in Egypt built between 2950 and 2750 BC. This 14 m high and 113 m long fill dam was composed of a central core made of fines (silt and sand) and rockfill shells, the faces of which were protected by rubble masonry. The water from the reservoir of 0.5×10^6 m^3 capacity was used for irrigation purposes. This dam was destroyed by breaching, probably due to flood overtopping after a relatively short period of operation (Novak *et al.*, 2007).

An impressive historical example of vital water infrastructures is Marib Dam in Yemen (Schleiss, 2015, 2016b). The construction of a 20 m high and 700 m long embankment dam was started in 519 BC during the Kingdom of Saba, providing a reservoir volume of 30×10^6 m^3 allowing for the storage of 15% of the annual runoff. This dam was the economical basis of the success of the Kingdom of Saba, providing irrigation water for about 110 km^2 of agriculture land, not only for food production but also for the cultivation of spices and aromatics such as frankincense and myrrh, and for trading, thus strongly contributing to the kingdom's wealth and prosperity. The dam, with its impressive irrigation system including tunnels, guaranteed the success of the kingdoms in the region for almost 1300 years. The dam was the largest structure built in the ancient times. Due to well-organized maintenance works, the dam survived several large floods and the damages could be repaired successfully. When the dam failed due to a large flood around AD 570, probably due to neglected maintenance works, it resulted also in an economic disaster of the prospering region, provoking the migration of more than 50,000 persons. This historical example demonstrates that the prosperity of a society is strongly linked to water infrastructure. Based on the support of Abu Dhabi, a new 39 m high embankment dam was inaugurated 3 km upstream of the ruins of the historical dam in 1986 (Figure 1.4).

The first major rubble masonry dam of 10 m height was built in Turkey in the same period as the Marib Dam. The Romans were the first to use cementitious material in large quantities for dam construction. Their concrete consisted of sand, gravel, burnt lime, water, and volcanic ash. To today's knowledge, they also first adopted the arch principle in dam engineering.

Masonry dams were realized in other parts of the world, particularly in Iran. The best known are the Kebar (completed around AD 1300, 26 m high, crest length 55 m, base thickness 6 m) and the Kurit masonry dams of cylindrical shape, with decreasing structural width from the base to the crest. Numerous homogeneous earthfill dams were constructed in Japan from the 6th century AD. All these ancient dams were equipped with a spillway, either cut

Figure 1.4 New Marib Dam in Yemen built in 1986 (Schleiss, 2016b)

into a rock block, installed at the flank or taking advantage of a natural depression of the surrounding terrain.

In the Harz region of northern Germany, various earthfill dams were built from AD 1298 (Mittlerer Pfauen Dam, 10 m high), culminating in large dam construction activity in the 16th century to exploit the hydropower potential for mining. The first fill dams with central core were realized there after 1715. During the same era, dams for drinking water, irrigation, and navigation were erected in the United Kingdom (UK) and France.

While the ancient dam engineering relied on proven experience and empiricism, the art of designing and building dams developed rapidly with the industrialization in the 19th century. The demographic growth and the resulting need for water and energy supply boosted dam construction and the technological advance of rational design methods. Related scientific domains such as structural analysis and material science progressed remarkably from the 19th century. It was not until the 20th century, however, that numerical methods along with modern soil and rock mechanic theories as well as engineering geology emerged, advancing modern dam construction.

Today, many scientific disciplines are involved in dam construction projects. Besides technical professions including civil, mechanical, and electrical engineers, hydrologists, geologists, and surveyors and land use planners, the involvement of ecologists of various disciplines, particularly on fauna as well as fish and aquatic habitats, has become standard. Needless to say, the disciplines of economy and financial engineering as well as law and politics play important roles in the public participation processes of large dam projects. As mentioned, large water infrastructures projects like dams and reservoirs have to be designed more and more as multipurpose projects by multidisciplinary teams with a complex system approach requiring excellence in engineering sciences and management.

1.4. Dams as critical water infrastructures

Dam and reservoirs are among critical water infrastructures whose failure may have catastrophic consequences with risk of fatalities and high economic losses. Dams are individually unique structures that differ from all other major civil engineering structures, and not only because the safety requirements are extremely high over all stages from the planning and design over the construction and commissioning to the operation. Every dam has to be considered as a prototype due to its unique and site-specific features including foundation geology, material properties, catchment hydrology, meteorology, seismic load, and reservoir operation mode. Moreover, both the loading conditions and the resistance of the structure may change over the functional life span of a dam due to environmental changes and aging. This is particularly of concern as the life span of a dam is typically on the order of a century.

Dam engineering includes a wide range of technical disciplines with many critical interfaces between them. Nevertheless, aspects of dam hydraulics are of high relevance since the majority of past dam incidents and failures have been caused by inadequate design or malfunction of appurtenant hydraulic structures including river diversions and spillways. The main causes of dam failures are (Singh, 1996):

- Floods exceeding spillway capacity: 30% (±5%)
- Foundation problems (seepage, internal erosion, excessive pore pressures, fault movement, settlement): 37% (±8%)
- Slides into reservoir (earth, rock, glaciers, avalanches): 10% (±5%)
- Improper design and construction, inferior quality of material, acts of war, lack of operation and maintenance: 23% (±12%).

Floods exceeding the spillway capacity and foundation problems are the most important causes of dam failures resulting in dam-break waves. According to Schnitter (1993), almost half of the failures after 1900 were due to overtopping. In 41% of those cases, the spillway was underdesigned, whereas overtopping was the result of operational problems with spillway gates in 21%. Dam overtopping may also be caused by tsunami-like waves, the so-called impulse waves originating from mass movements at the reservoir hill slopes and banks and their impact into the water body.

The requirement for a reservoir is a certain storage volume maintained over the expected life span of the dam. Climate change will also significantly increase reservoir sedimentation, one of the most serious problems endangering the sustainable use of the worldwide reservoir volume. Therefore, measures against excessive sedimentation have to be applied from the very beginning of a dam project.

1.5. Safe operation of dams and reservoirs through advanced dam safety concepts: example of Switzerland

Dams are considered to be safe if risks are kept under control through appropriate measures (Darbre, 2011). To achieve a high level of safety, a dam needs to be designed and constructed in a manner that ensures stability under all conceivable loading and operational conditions. The resulting minimization of risk is sustained by a surveillance that permits the early identification of any form of deterioration of the dam or its surroundings and any unanticipated occurrences or abnormal behavior, and by a maintenance program aimed at preventing these occurrences. However, it is not possible to eliminate all risks. Thus, an emergency planning concept has to be defined in which surveillance plays a significant role in ensuring that any imminent uncontrolled release of reservoir water can be identified at the earliest possible stage. The Swiss dam safety concept, illustrated in Figure 1.5, is based on these considerations and comprises three components: structural safety, surveillance and maintenance, and emergency planning.

Structural safety calls for sound engineering and construction practices plus compliance with minimum performance requirements (e.g. dam stability and prevention of uncontrolled release of reservoir water during the safety assessment earthquake or safe passage of the probable maximum flood). It is implemented at all dams subject to the provisions of the Swiss federal legislation, irrespective of the population downstream or the potential extent of damage in the event of a breach. Every person who could be affected in the event of a dam incident or accident is thus assured of a minimum level of protection, which may be regarded as the enforcement of a limit on individual risk.

The structural requirements are kept in line with the development of scientific and technical knowledge. They are formulated in the form of performance targets with certain quantitative requirements. Two of the latter relate to flood and earthquake hazards. The safe passage of a 1000-year design flood has to be guaranteed with sufficient freeboard and under the assumption that one gate of the spillway or outlet with the largest discharge capacity is blocked (the $n - 1$ rule). Furthermore, a safety check flood, e.g. the probable maximum flood (PMF), must safely pass the dam without surpassing its critical water level (the $n - 1$ rule still applies to embankment dams). The earthquake safety of the largest dams (defined

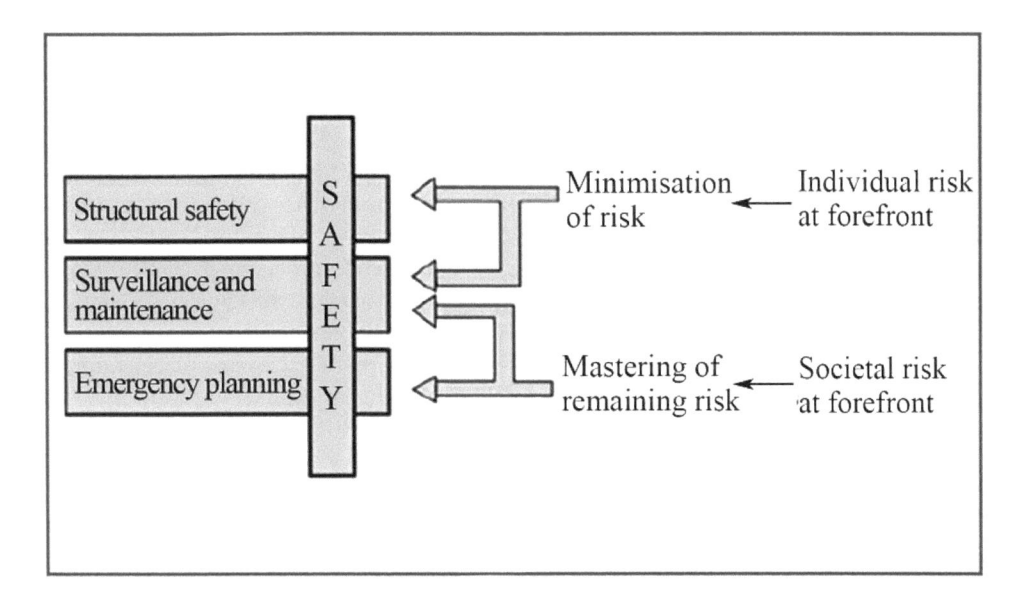

Figure 1.5 Swiss dam safety concept (Adapted from Darbre, 2011)

as having the same minimal geometrical dimensions as for performing the 5-year in-depth safety review, see below) is assessed on the basis of a 10,000-year return period, while a return period of 5000 years is used for the other dams under direct federal supervision, and a 1000-year period applies for the smaller dams under direct supervision of the cantons.

Surveillance and maintenance activities are divided into three to four levels. The first level calls for periodical visual inspections and measurements as well as maintenance by the operator's personnel on site. This includes an annual functional test of the gated discharge works at full reservoir, though with limited water discharge (full opening as dry tests only).

The second level calls for an experienced engineer to carry out an initial assessment of the inspection and measurement results as soon as they are obtained, to conduct an annual inspection and prepare an annual safety report in which the findings of the first level are integrated. Predictive behavior models are being used to an ever-increasing extent for assessing the degree of deformation of concrete dams. This allows for identifying irreversible developments (due to creep, swelling, valley deformation, etc.) at an early stage and separate them from reversible issues resulting from changes in the reservoir water level and temperature (or season).

The third level concerns the performance of an in-depth safety review every 5 years for dams with a height of at least 40 m, or of at least 10 m if the reservoir contains more than 10^6 m^3. This review is carried out by an experienced civil engineer and a geologist for the purpose of evaluating the findings of the past 5 years and assessing them over a long-term perspective as well as addressing any specific safety issues that may arise.

The content and structure of these three levels for each dam are specified in a document that has to be approved by the supervisory authority. The fourth level involves the supervisory authority itself. Here its specialists have to ensure that all safety requirements are complied with, carry out their own site inspections, and evaluate the annual and 5-year safety reports (including results of the functional tests of water release devices) as well as other special reports (e.g. earthquake analyses). In this process, they check the content of the technical documents for plausibility and carry out their own assessments. If necessary, they ask the owners, engineers, or experts to carry out special additional studies (e.g. concerning concrete swelling).

Emergency planning. If a threatening situation develops that could culminate in the uncontrolled release of large masses of water, emergency countermeasures need to be taken both to prevent such an uncontrolled release (as far as possible), as well as to warn, alarm, and if necessary evacuate the population at risk before the arrival of the flood wave. Surveillance supports this process by permitting the early identification of a threat, thus providing additional time for taking the necessary countermeasures or ultimately for evacuating the population at risk. Emergency planning addresses the key issues of *whom to alarm* (based on dam breach analyses), *when to alarm* (as soon as an uncontrolled release of large masses of water is detected or feared), *how to alarm* (with the aid of suitable systems), and *how to evacuate* (on the basis of evacuation maps). This requires the involvement of several institutions at the planning stage, a clear assignment of responsibilities and duties during an event, and effective alarm systems. With respect to the latter, dedicated sirens (so-called water-alarm sirens) have been installed downstream of reservoirs in Switzerland containing a volume greater than 2×10^6 m^3, i.e. those facilities that would cause the most severe harm (in terms of loss of human lives) in the event of a breach. This is clearly a consideration of the societal risk.

1.6. Appurtenant structures of dams

1.6.1 Overview

Besides the retaining structure itself, be it a concrete or an embankment dam, so-called appurtenant or ancillary structures are important for the economic and safe operation of the dam and its reservoir during its implementation, commissioning, and service life. These structures may even largely determine the selection of the dam type and its layout. For instance, the safe passage of significant flood discharges may be determining for the selection of a gravity dam versus an embankment dam, provided the other boundary conditions do not prohibit the construction of this type of concrete dam, because massive spillway concrete structures should not be located on embankments due to differential settlements.

In the following, a short overview of appurtenant dam structures is given, whereas their interaction with the respective dam type and the specific hydraulic design principles relating to each structure is detailed in the subsequent chapters. Basically, one may distinguish between discharge facilities such as spillways, which are crucial for dam safety, and intake structures for the withdrawal of water to satisfy the purpose of the reservoir, e.g. to supply drinking water to treatment plants or irrigation water directly to agricultural land as well as to deliver storage headwater to the turbines of a hydropower plant.

1.6.2 Spillways including overflow and dissipation structures

A spillway is a safety structure against overflow. It should inhibit water overflow over the dam at locations which were not considered for overtopping. The spillway is thus the main element for *overflow safety* and especially the safety structure against dam overtopping. The spillway protects not only the dam itself but also the population downstream against any uncontrolled flood release resulting in damages.

Spillways can be characterized by the three main elements forming them, namely inlet structures, transport or conveyance structures, and outlet or energy dissipation structures. A further criterion is whether the spillway is controlled by regulated devices like gates or semi-regulated fuse elements such as gates or plugs. Figure 1.6 represents all theoretical combinations in a so-called morphological box.

Based on these elements there are theoretically 300 combinations ($4 \times 5 \times 5 \times 3$) of possible spillway types. Nevertheless, only one-third of these are technically feasible and only one-quarter of these are really reasonable combinations. This underlines that the best choice

Control	① without gates	② with gates	③ with fuse gates	④ fuse plugs	
Inlet structure	① frontal weir	② side weir	③ circular weir	④ siphon	⑤ orifice
Transport structure	① free-falling jet	② cascade, stepped chute	③ smooth chute	④ free surface tunnel	⑤ pressurized tunnel or shaft
Outlet structure	① missing	② stilling basin	③ ski jump		

Figure 1.6 Morphological box of structural elements characterizing a spillway

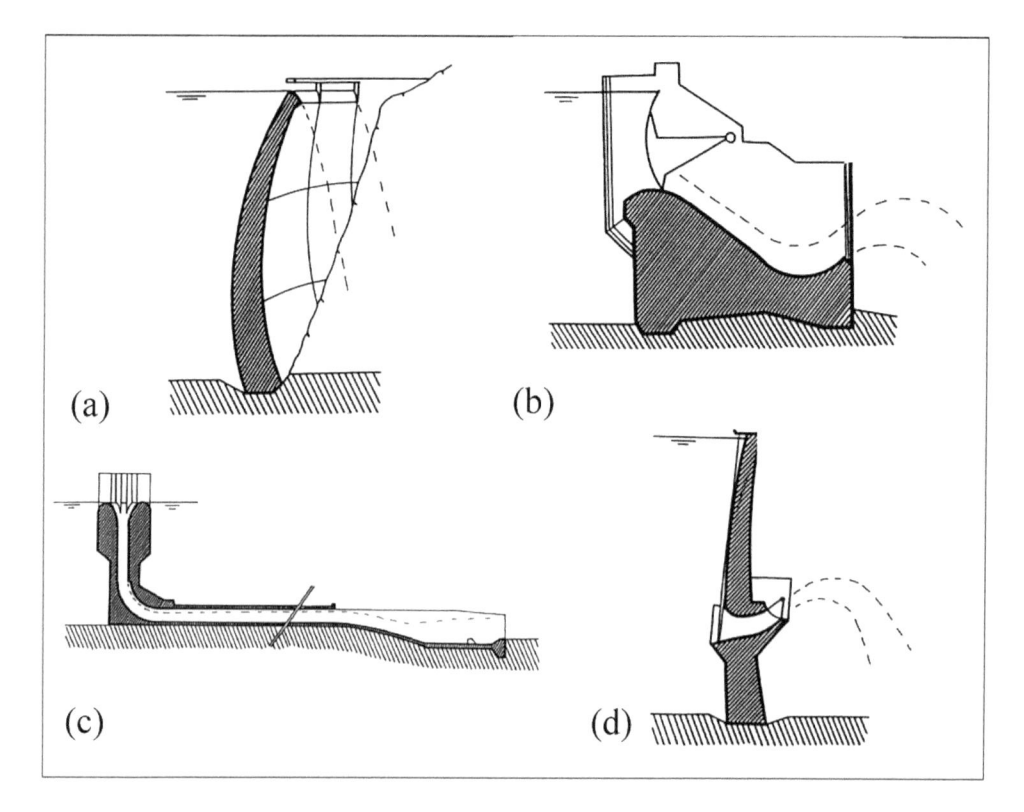

Figure 1.7 Examples of combinations of structural elements according to Figure 1.6 forming spillway types:

    (a) 1111: Ungated frontal weir on arch dam crest with free-falling jet (Gebidem Dam, Switzerland)

    (b) 2133: Gated frontal weir on gravity dam crest followed by smooth chute and ski jump (Jaguara Dam, Brazil)

    (c) 1342: Circular weir followed by vertical shaft, free-surface tunnel and a stilling basin (Heart-Butte Dam, USA)

    (d) 2513: Gated orifice in arch dam followed by ski jump (Cabora Bassa Dam, Mozambique)

of a spillway type in combination with a certain dam type is highly challenging and rather a design art than a simple engineering task (Figure 1.7).

For the design of a spillway, in addition to hydraulic issues, one has to consider also problems related to floating debris, ice formation, and abrasion due to suspended sediment. In general, hydraulic model tests are required for validating the safe operation under design and safety check floods.

1.6.3 Bottom outlets

In many countries, the design of a bottom or low-level outlet is mandatory, and its inclusion is strongly recommended in all other instances. Primarily, a bottom outlet is a *safety structure*, and it applies secondarily to flush *sediment deposits* or for discharging surplus water during floods. The load test of a dam occurs during the *first filling* of the reservoir. Accordingly, this test has to be conducted with extreme care, because a failure has serious consequences. The filling must be made progressively by accounting for the stability and the water tightness of both the dam and its surroundings. A dam is thus typically filled to the first third of its height, and observations are conducted during several days or weeks. If the results are satisfactory, the dam is filled further, typically to two-thirds of its height. Particular attention will be directed to the response of the dam under complete initial filling. This *filling procedure* is only amenable if the reservoir level can be controlled, i.e. if a bottom outlet is available. The bottom outlet must thus be designed so that the reservoir level can be kept constantly under arbitrary levels during the first filling.

In case of danger, the bottom outlet may also allow for a sufficiently fast reservoir drawdown. The hydraulic pressure force on a dam increases with the second or third power of the water depth, depending on the dam type. A drawdown of the reservoir level thus induces a significant stress relief. If, for example, a full reservoir were drawn down to two-thirds of its height, then the hydraulic pressure on the dam would be reduced by 56% to 70%, depending on the dam type. Accordingly, the drawdown of the reservoir level of an endangered dam is highly efficient.

The *drawdown time* is a significant design element. Under emergency, one would like to draw down the reservoir level of say, the top third of the reservoir height, typically over a few days or weeks. The design of the bottom outlet combined with other water release devices below the target level should be checked also in this regard. The drawdown time can be integrated directly as a design condition for a suitable domain of solution. For reservoirs with a large storage volume, the drawdown time is dictated normally by restrictions. For reservoirs whose storage volume is a multiple of the mean annual inflow volume, drawdown times of months or even years may result. Under these conditions, a rapid reservoir drawdown is obviously not feasible. Nevertheless, a reservoir has to be emptied partially or entirely under normal operation, such as when revisions on the waterside of the dam or at the inlet of the bottom outlet or the intake are needed.

In certain cases, bottom outlets are used for sediment flushing or turbidity current venting. For the design of the outlet structure, the continuity of the sediment transport downstream of the dam is important. Deposits in the tailwater may lead to submerged flow conditions at the bottom outlet and thus endanger its safe operation. The ratio between water and sediment quantities has to be such that it corresponds to the transport capacity of the tailwater. The clogging of a bottom outlet is another problem. In certain countries, a minimum diameter of 1.8–2 m is prescribed.

1.6.4 Intakes

Intakes are vital hydraulic structures to satisfy the goals of the reservoir purpose and the economical operation of the dam. The design discharge of an intake results from an overall optimization of the hydraulic scheme considering often multipurpose criteria. Vibration of structures, the management of floating debris, and air entrainment are often critical design issues of intake structures.

1.6.5 River diversion

The construction of a dam across a valley with its river, including a river diversion, is a major engineering challenge involving considerable risk. The purpose of the river diversion is to put the foundation site of the future dam with its appurtenant structures into dry conditions before starting dam construction. A dry construction pit is required to perform the following works:

- Final geological and geotechnical investigations
- Excavation of the foundation
- Improvement of the underground by injections
- Realization of the dam with its appurtenant structures.

The notion *river diversion* comprises all structural measures for controlling a river during construction. The task of an appropriate river diversion is to divert probable floods during dam execution works to protect the construction site against any damages. The river diversion also has to ensure that the presence of the construction site does not increase the flooding risk both upstream and downstream of the dam. Floods during construction are particularly dangerous because the dam with its appurtenant structures, as well as other scheme structures such as the powerhouse, are not yet finalized and thus highly vulnerable. The appropriate selection of the design discharge for a river diversion is crucial in order to limit the risk of a catastrophic cofferdam failure. Although the river diversion is often composed of temporary structures, they have to be designed correctly to ensure proper operation during any flood event considering the cost and risks involved.

In principal, a river diversion is achieved by three different methods:

- Integral river diversion
- Diversion of river through construction site
- Balancing of river and its constriction during construction.

The first method consists of diverting fully or integrally the river by placing a diversion tunnel or channel around the construction site, which allows to fully dry the river at the future dam site. Diversion structures as tunnels or channels can be built independently from the river and its discharge. Nevertheless, important construction works besides the diversion tunnel itself, like cofferdams and closure structures, have to be realized for the integral river diversions.

The diversion of a river directly across the construction site may be feasible in wide valleys by the help of a diversion through concrete culverts or tunnels excavated in the rock located below the foundation of the future dam. The diversion structure has to be built during the dry season with low discharge under the protection of temporary upstream and downstream cofferdams.

In the case of river balancing, the latter remains in the valley but is constricted locally by a cofferdam, protecting a certain phase of the construction works. This method is feasible in sufficiently wide valleys and allows for minimizing the works related to the river diversion. Furthermore, the perturbation of the river flow regime is limited, which is a must in cases where ship navigation has to be maintained during construction. The closure of the river and of the temporary diversion is often a critical operation, which can strongly influence the planning and duration of the construction works.

1.7 Hydraulic engineering of dams: structure of the book

Within the scope of this book on hydraulic engineering aspects of dams, the stress is put on the term 'hydraulic'. Dam hydraulics comprises all hydraulic questions relating to the design, construction, operation, management, and safety of dams. The hydraulic engineering of dams is thus directed to the *hydraulic design* of the following appurtenant dam structures:

- Spillways including overfalls and gated weirs (Chapters 2 and 3) or gated inlet/orifice structures (Chapter 8.2), transport or conveyance structures (Chapter 4), and dissipation structures (Chapters 5 and 6);
- Diversion facilities during construction (Chapter 7);
- Intake structures including gates (Chapters 8.3–8.5);
- Gated bottom and low-level outlets (Chapter 8.7).

Particular hydraulic problems for the above appurtenant structures include vortex formation at intakes, floating debris, air entrainment, cavitation damage and structural vibration, energy dissipation, and scour and erosion.

Further special topics of hydraulic engineering of dams to be considered include:

- Reservoir sedimentation (Chapter 9)
- Impulse waves due to slope instabilities (Chapter 10)
- Dam breaches (Chapter 11).

References

Bauer, C., Hirschberg, S. (eds), Bäuerle, Y., Biollaz, S., Calbry-Musyka, A., Cox, B., Heck, T., Lehnert, M., Meier, A., Prasser, H.-M., Schenler, W., Treyer, K., Vogel, F., Wieckert, H.C., Zhang, X., Zimmermann, M., Burg, V., Bowman, G., Erni, M., Saar, M. & Tran, M.Q. (2017) *Potentials, Costs and Environmental Assessment of Electricity Generation Technologies*. PSI, WSL, ETHZ, EPFL. Paul Scherrer Institut PSI, Villigen, Switzerland.

Darbre, G. (2011) Dam safety in Switzerland. In: *Dams in Switzerland: Source for Worldwide Swiss Dam Engineering*. Swiss Committee on Dams, Switzerland, pp. 11–13.

Frischknecht, R., Itten, R. & Flury, K. (2012) *Treibhausgas-Emissionen der Schweizer Strommixe (Greenhouse gas emissions of the Swiss electricity mixes). Studie im Auftrag des Bundesamtes für Umwelt (BAFU)*. ESU-Services Ltd., Uster, Switzerland (in German).

Garbrecht, G. (ed) (1987) *Historische Talsperren* (Historical Dams). Wittwer, Stuttgart (in German).

International Commission on Large Dams (2018) *World Register of Dams*. ICOLD, Paris.

Novak, P., Moffat, A.I.B., Nalluri, C. & Narayanan, R. (2007) *Hydraulic Structures*. Taylor & Francis, London.

Schleiss, A.J. (2000) The importance of hydraulic schemes for sustainable development in the 21st century. *Hydropower & Dams*, 7(1), 19–24.

Schleiss, A.J. (2015) Hydropower and dams: Vital water infrastructures as a basis for worldwide economic prosperity. *World Atlas & Industry Guide*, 2015, 8–9.

Schleiss, A.J. (2016a) Dams and reservoirs as security belt around the world to ensure water, food and energy. *World Atlas & Industry Guide*, 2016, 12–13.

Schleiss, A. (2016b) Talsperren und Speicher als lebenswichtige Infrastrukturanlagen für den weltweiten Wohlstand (Dams and reservoirs as vital infrastructure for the global welfare). *Wasserwirtschaft*, 106(6), 12–15 (in German).

Schleiss, A.J. (2017a) ICOLD welcomes his 100th member country supporting his mission "Better Dams for a Better World". *World Atlas & Industry Guide*, 2017, 10–11.

Schleiss, A.J. (2017b) Better water infrastructures for a better world – the important role of water associations. *Hydrolink*, 2017, 86–87.

Schleiss, A.J. (2018) Sustainable and safe development of dams and reservoirs as vital water infrastructures in this century – The important role of ICOLD. *Proceedings of International Dam Safety Conference, 23 & 24 January 2018 Thiruvananthapuram, Kerala*, pp. 3–16.

Schnitter, N.J. (1993) Dam failures due to overtopping. *Proceedings of Workshop Dam Safety Evaluation Grindelwald*, 1, 13–19.

Schnitter, N.J. (1994) *A History of Dams: The Useful Pyramids*. Balkema, Rotterdam.

Singh, V.P. (1996) *Dam Break Modelling Technology*. Kluwer, Dordrecht.

WCD World Commission on Dams (2000) *Dams and Development: A New Framework. Report of the World Commission on Dams*. Earthscan, London.

Chapter 2 Frontispiece (a) Itaipu Buttress Dam with gated spillway of discharge capacity of 62,200 m³/s, Brazil (Courtesy VAW), (b) Karakaya Dam, Turkey, view from downstream onto dam (Courtesy Pöyry Engineering, Switzerland)

Chapter 2

Frontal crest overflow

2.1 Introduction

2.1.1 Overflow structures

The hydraulic behavior of spillway overflow structures is such that the discharge increases significantly with the head on the overflow crest. Nevertheless, the overflow height is usually a small portion of the dam height. Two types of overflow structures exist: gated structures and non-controlled free overflow structures. Gates positioned on the spillway crest control the overflow. During floods, and if the reservoir is filled, gates are completely opened to promote the overflow. A large number of reservoirs with a relatively small design discharge are ungated (Figures 2.1, 2.2).

Currently, most large dams are equipped with gates to allow for a flexible reservoir operation and flood management. The cost of the gates increases mainly with the magnitude of the flood, i.e. with the overflow area. Improper operation or malfunction of gates is a major concern leading to serious dam overtopping or unintentional flood release. To avoid undesirable floods in the dam tailwater, gates of any water release structure are to be moved according to strict gate operation rules. Gates should be checked against resonance due to flow-induced vibrations.

The advantages of gated overflow structures are:

- Flexible variation of reservoir level
- Flood control with possible preventive gate operation
- Benefit from higher storage level.

Disadvantages to be considered include:

- Potential risk of malfunction
- Reduction of discharge capacity if the $n - 1$ rule is required for the design flood
- Additional construction and equipment cost
- Maintenance.

Depending on the size of the reservoir and its location, gates are preferred for:

- Large reservoirs
- Large floods
- Easy access for gate operation and maintenance.

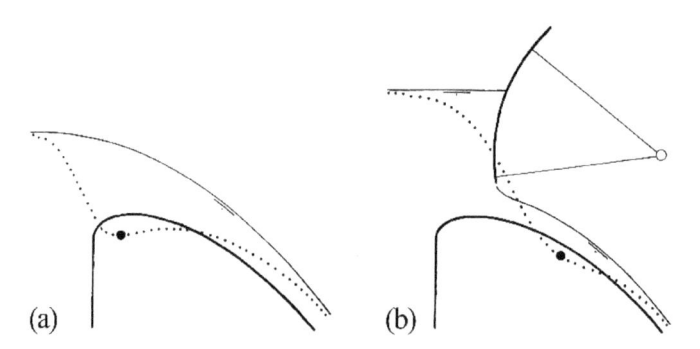

Figure 2.1 Overflow structures (a) ungated, (b) gated with (. . .) bottom pressure head profile and (•) minimum bottom pressure head

Figure 2.2 Photographs of gated and ungated spillways (a) Dinh Binh Dam, Vietnam, (b) Wyaralong Dam, Australia (Courtesy Michael Pfister)

Three gate types are currently favored (Hartung, 1973; ICOLD, 1987; Erbisti, 2014), including hinged flap gates, vertical lift gates, and radial gates (Figure 2.3a–c). In addition to these classical hydraulic steel structures, inflatable rubber tubes are frequently used at small dams and reservoirs (Figure 2.3d). Flaps are normally used for heads of only some meters but may span over a considerable width. The vertical gate can have a considerable height but requires substantial slots, a heavy lifting device, and a significant superstructure endangered by earthquakes. These gates are equipped with fixed wheels and may have flaps at the top. Radial gates are most frequently used for medium and large overflow structures because of their simple construction, modest force required for operation, and the absence of gate slots. They may be up to 20 m wide and 25 m high, or for example also 12 m high and up to 40 m wide. This gate is limited by the strength of the trunnion bearings.

Rubber dams are typically used for large widths up to 50 m or even more without intermediate pillar, but limited to small to moderate heights of up to 6–8 m. Like for steel gates, the stated maximum widths and heights may not simultaneously be combined.

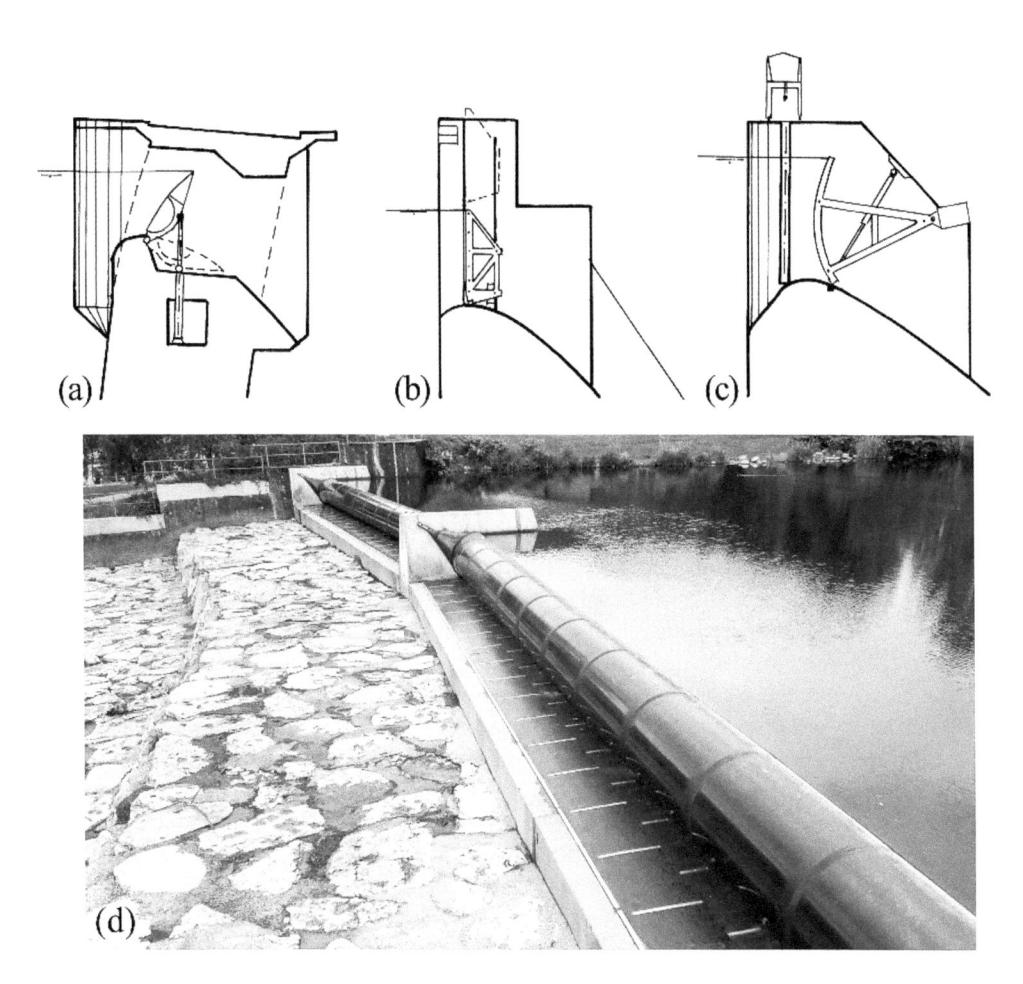

Figure 2.3 Gate types (a) flap gate, (b) vertical gate, (c) radial gate, (d) inflatable rubber dam at Hausen hydropower plant on Wiese River, Germany (Courtesy Robert M. Boes)

Erbisti (2014) distinguishes between gate types and their specific applications:

- A *flap gate* involves a straight or curved surface, pivoted on a fixed axis from below. In its fully raised position, the angle with the horizontal is up to 70°, whereas it can be lowered to the weir bottom at the fully opened position, thereby avoiding any obstacle to the overflow. Seals are provided along the bottom and the sidewalls. Partially opened flap gates may be subjected by severe vibrations due to sub-pressure below the overflowing nappe, which are removed by an adequate air supply.
- A *slide gate* is a basic type of flap gate, consisting of a leaf gate sliding alongside guides. The leaf is provided with sliding surfaces, usually metallic, which act as seals under tight contact at the bearing surfaces. This gate type is used in applications because of the

simple and safe operation and little maintenance. Note that no vibrations occur under any gate opening degree – a prime advantage for bottom outlets (Chapter 8). These gates usually do not close under gravity only, given the large friction exerted along the side guides.

- A *caterpillar gate* consists of a leaf supported by vertical girders at the sides. Continuous roller trains are mounted around the vertical girders traveling with the gate. The low friction on the rollers and their high load capacity recommend their use in high-head installations requiring closure by its own weight. The main disadvantages include high cost, considerable maintenance because of the large number of movable parts, and precise adjustment between roller tread and tracks for adequate gate operation. These gates are usually employed at high-head intakes, but only either in the fully opened or closed positions (Chapter 8).

- A *segment gate* consists of a curved skin plate formed to a cylinder segment, supported by radial compressed arms transferring the hydraulic forces to fixed bearings, so that it counts as the simplest gate type. It rotates about the horizontal axis passing through the bearing center coinciding with the center of the skin plate curvature radius. The resultant thrust from the water pressure therefore passes through the point of rotation without a tendency to close or open the gate. The segment gate must be lifted to discharge water. A large discharge results for a small gate opening because of the orifice-like arrangement. These gates are often equipped with a small flap gate at their top to pass ice or debris without loss of large water volumes.

- A *sector gate* has a curved skin plate as the segment gate, but continues along its upper portion by a full surface in the radial direction up to the bearings allowing for overflow. This gate is hinged at the downstream side with its leaf shaped of an open body on its lower radial side. In the raised position, the sector gate is kept open by water pressure on the inner face of the upper radial side. Gate operation is fully hydraulic so that no hoists are required. Its movable structure is placed in the gate chamber built in the crest structure. When water is allowed to flow into the recess chamber, the gate rotates upward due to pressure on the gate bottom. The gate is lowered by opening the outlet valves, draining the water in the chamber. This gate type provides safe and accurate automatic control without resort to an external power supply. Its length is as long as required, but its height is limited to some 8 m.

- A *roller gate* involves a horizontal steel cylinder with toothed gears provided at each end. Racks are placed in the piers along the track recesses. This gate type was and still is used in the cold season when ice may block the operation of conventional gates. This gate applies to low-head dams or if a wide opening between the piers is relevant for ice or debris passage. The gate is sealed at its end and at the bottom. Maximum dimensions reach 50 m in width and 8 m in height. It is often the heaviest and the most expensive of all gates.

- *Stoplogs* are primarily employed for maintenance of gates or other hydraulic elements located in the tailwater, as stilling basins. These devices are typically found upstream of intakes, spillway gates, and bottom outlets, and downstream of turbines or draft tube emergency gates. Stop logs include either one or several members, depending on the height. Criteria include the lifting capacity of the crane, crane height, storage capacity, and transport limitations. A maximum height of 3 m is normally not exceeded. Weir stoplogs are provided with seals along all sides to become watertight. These elements are flexible, quick in operation, and safe.

Other gate types include the Stoney gate, the drum gate, the bear-trap gate, the fixed-wheel gate, and the visor gate. Note the extensive description of details by Erbisti (2014), including questions relating to the selection of the gate type.

The risk of gate blocking at seismic sites is relatively small if the gate is set inside a stiff one-piece frame. For safety reasons, there should be a number of moderately sized gates rather than a few large gates. It is customary for the overflow design to assume that the largest gate is out of operation. The regulation is ensured by hoists or by hydraulic jacks driven by electric motors. Standby diesel-electric generators should be provided if power failures are likely (ICOLD, 1987).

2.1.2 Overflow types

Depending on the site conditions and hydraulic particularities, a spillway overflow structure involves various designs. Figure 2.4 shows the basic types, namely:

- Frontal overflow
- Side-channel overflow
- Shaft overflow.

Nonlinear overflow structures such as the *labyrinth spillway* or the *piano key weir* involve frontal overflow but with a crest consisting of successive triangles or trapezoids in plan view (Chapter 3). Non-frontal overflow structures apply for small to intermediate discharges up to design floods of, say, 2500 m^3s^{-1}.

Still another type is the *orifice spillway* of arch dams as the Kariba Dam (Figure 2.5) or the Cabora Bassa Dam on the Zambezi River in Africa. Its design discharge must be accurately defined because of dangerous overtopping otherwise. Orifice spillways are thus normally combined with overflow structures at the dam crest for safety reasons. The orifice is governed by an insensitive discharge-head equation as compared with the overflow structure. The orifice structure demands a high degree of reliability and control from the hydromechanical equipment. The material properties used must be detailed under high-velocity flow. Additional provision for flood safety is required.

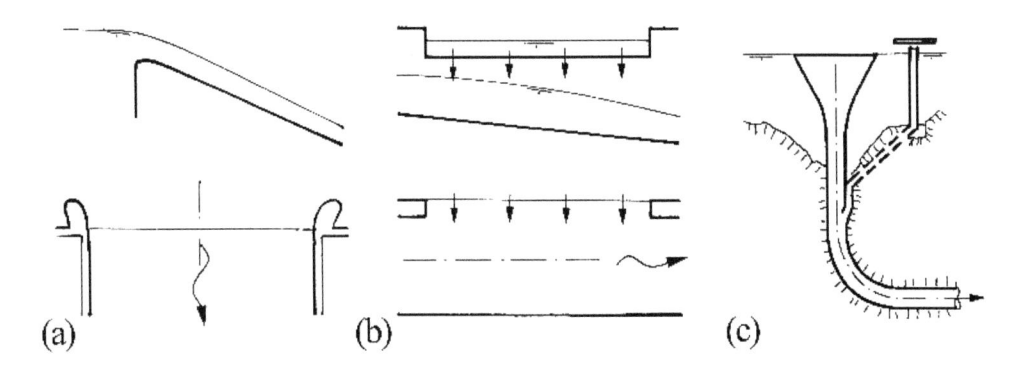

Figure 2.4 Main types of overflow structures in section and plan (a) frontal, (b) side, (c) shaft overflow

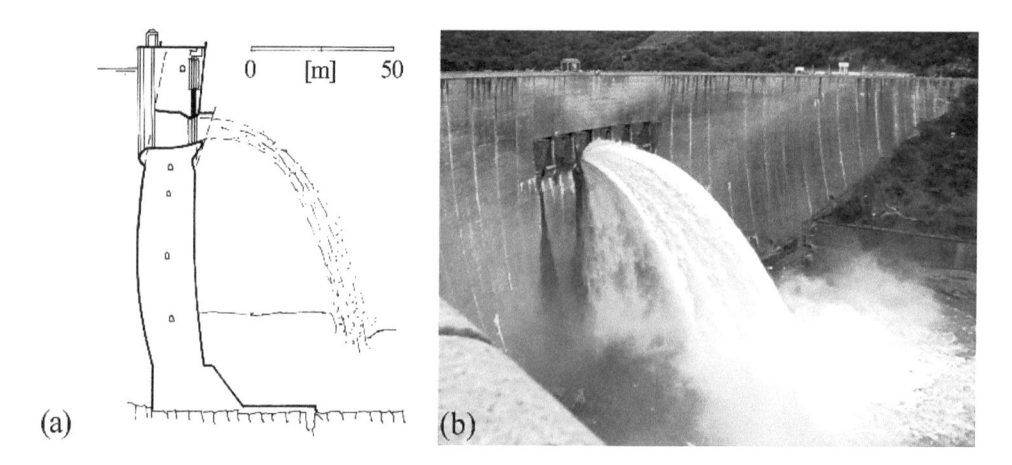

Figure 2.5 Orifice-type dam (a) section (b) photo of Kariba Dam with one orifice partially open (Courtesy Anton J. Schleiss)

The *shaft type spillway* was developed in the 1930s and has proved to be economical if the diversion tunnel is used as a tailrace. The structure consists of three main elements, namely the intake, the vertical shaft with a 90° bend at its base, and the almost horizontal tunnel. Air is supplied by aeration conduits to prevent pulsations and cavitation damage at the transition between the shaft and the tunnel. To account for flood safety, only non-submerged flow is tolerated for the design flood with a certain safety margin so that free surface flow occurs along the entire structure from the intake to the dissipator. The hydraulic capacity of both the shaft and the tunnel is thus larger than that of the intake structure. The system intake shaft is also referred to as 'morning glory overflow' due to the similarity with a flower of cup shape. Figure 2.6a, b refers to the Hungry Horse spillway with a design discharge of 1500 m³/s. The shaft has a non-standard 45° inclination. Cavitation damage occurred at the transition from the sloping to the horizontal tailrace tunnel. Figure 2.6c shows an intake in Morocco.

The *side channel overflow* was successfully used at Hoover Dam (USA) in the late 1930s (Figure 2.7). This arrangement is advantageous at locations where a frontal overflow is not feasible, as for earth and rockfill dams, or if a different location at the dam side yields a better and simpler connection to the stilling basin or a flip bucket. Side channels include a frontal type overflow structure and a spillway with axis parallel to the overfall crest (Fig. 2.7c). Its specific discharge is normally limited to 10 m³/sm for lengths in excess of 100 m. The side channel of Hoover Dam has the particularity that the discharge is directed to the dissipation structure via a spillway tunnel. Many other side channels exist worldwide, yet this overflow type is limited to small and medium discharges (Chapter 3).

The *frontal type overflow* corresponds to the most widely used spillway overflow structure, both due to simplicity in design and construction, and a direct reservoir-to-tailwater connection. It is normally located at the crest of concrete dams. Also, earth and rockfill dams and frontal overflows excavated in the rock abutments combine well if particular attention is focused on overtopping. The frontal overflow is easily extended with gates and piers to control the reservoir level and to improve the approach flow to the spillway. Gated overflows

Figure 2.6 Shaft type spillway (a) section and plan with ① aeration conduits, ② ring gate, (b) photograph (*Engineering News-Record* 1953, Nov. 5: 46), (c) Shaft spillway of El Makkhazine Dam, Morocco, connected to one of the diversion tunnels. Free overflow structure with piers to avoid swirling flow. Shaft equipped with orifices for frequent floods. On the left submerged power intake connected to the second diversion tunnel (Courtesy Electrowatt Engineering, Zurich)

of 25 m gate height and more have been successfully designed, with a unit width capacity of up to 250 m^2/s. These overflows apply thus for medium to large discharges. Particular attention has to be directed to cavitation damage due to the immense heads generating pressure

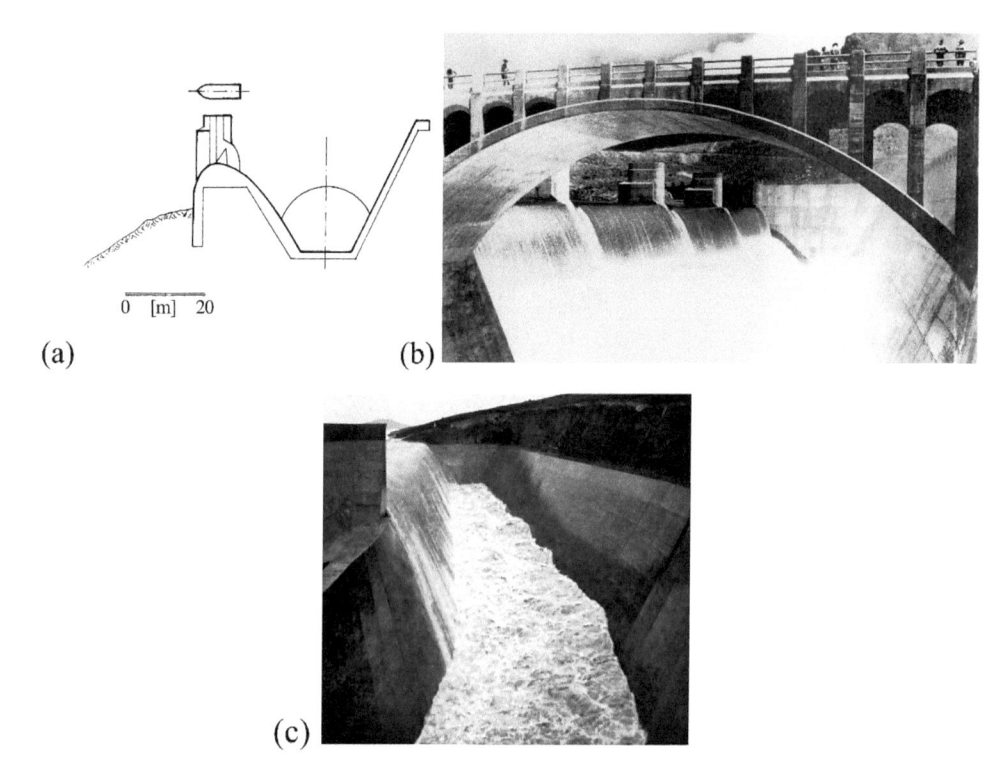

(a)   (b)

(c)

Figure 2.7 Side-channel type overflow of Hoover Dam, USA (a) section, (b) photograph (US Dept. of the Interior, Bureau of Reclamation, The Reclamation Engineering Center: Denver, 1950), (c) Karahnjukar Dam, Iceland (Courtesy Sigurdur M. Gardarsson)

below vapor pressure in the crest domain. Also, gate piers and sidewall noses have to be shaped carefully to generate symmetrical approach flow. Figure 2.8 shows a typical frontal overflow.

The downstream portion of a frontal overflow may have various designs. Usually, a spillway chute is connected to the overfall crest as transition between the overflow section and the energy dissipator (Figure 2.9a). The crest may also abruptly end for arch dams to induce a falling nappe impacting onto the tailwater (Figure 2.9b). Another design uses a *stepped chute* to dissipate energy right away from the crest end to the tailwater, so that only a reduced stilling basin is required (Figure 2.9c). The standard design involves a smooth chute conveying high-velocity flow either directly to the stilling basin or to a trajectory bucket where the jet is lifted into the air to elongate the impact zone far from the dam and its abutments but also to promote jet dispersion, thereby reducing the impact action on the downstream riverbed. Figure 2.10 refers to these two types of energy dissipators, considered in Chapters 5 and 6, respectively.

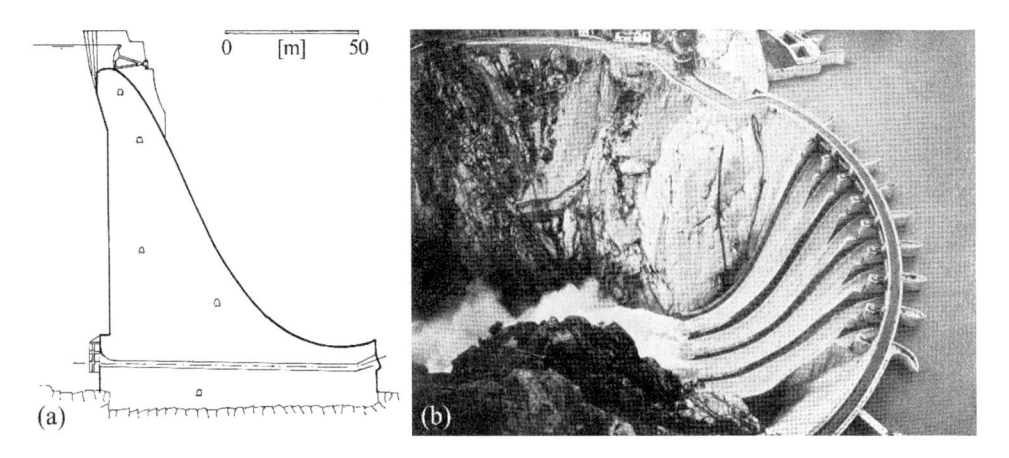

Figure 2.8 Frontal type overflow (a) section, (b) photo of Aldeadavila Dam, Spain (ICOLD Q33, R22)

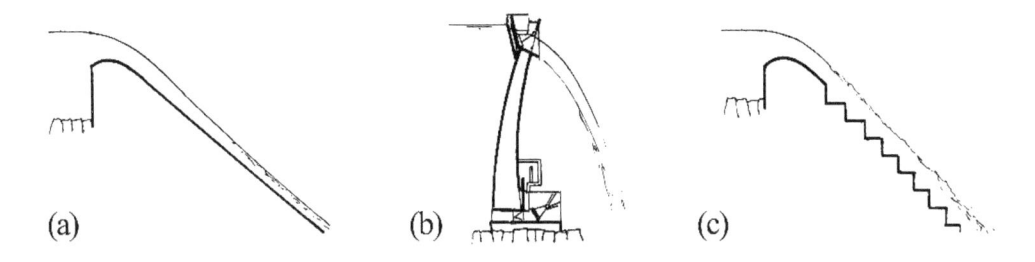

Figure 2.9 Connection between frontal overflow and dissipator (a) smooth chute, (b) free fall, (c) stepped chute

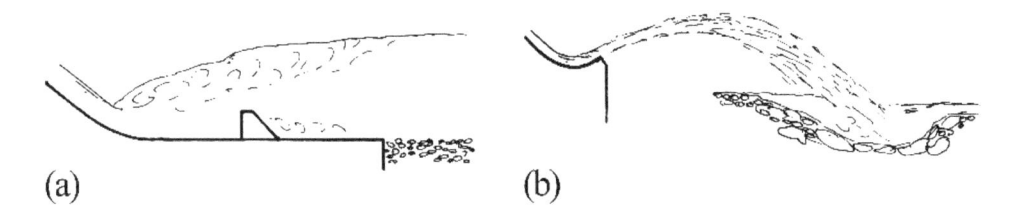

Figure 2.10 Dissipation structures (a) stilling basin, (b) trajectory basin

2.1.3 Significance of overflow structure

According to ICOLD (1987, 2017), overflow structure and design discharge have a strong effect on dam safety. Scale models of overfall structures are definitely required in the following cases (ICOLD, 1987, 2017):

* The valley is narrow and approach flow velocity is large, resulting in *asymmetric* flow pattern and vorticity formation;

- Asymmetric gate operation and presence of significant floating debris;
- The overflow and pier geometries are not of standard shapes;
- The chute is not aligned (contraction and/or direction change) with overflow structure;
- Structures at either overflow side may disturb the spilling process.

For all other cases, the design of the overflow structure is standardized so that no model study is required, except for details departing from the recommendations.

The following deals with frontal overflow structures as the basis of all weir-type structures, whereas Chapter 3 highlights particularities of spatial overflow structures, in which the flow pattern is no more as simple as for the two-dimensional (2D) flow. Both chapters form the basis of spillway hydraulics, given the fundamental relevance of weirs in the field of dam engineering. At the end of Chapter 2, the round-crested weir is detailed to address scale effects.

2.2 Frontal overflow

2.2.1 Crest shapes and standard crest

Overflow structures of different shapes are shown in Figure 2.11. In plan view, the crest is straight (standard), curved, polygonal, or of labyrinth shape. The latter structure has an increased overflow capacity with regard to the structural width.

The usual transverse section of a prototype overflow structure is rectangular, trapezoidal, or triangular (Figure 2.12). The rectangular cross section corresponds to the standard design to enhance a symmetric downstream flow and to accommodate gates.

The longitudinal (streamwise) section of the overflow is typically broad-crested (Hager and Schleiss, 2009), circular-crested (Hager, 2010), or of standard crest shape (Figure 2.13). For heads larger than 3 m, say, the standard overflow crest shape is recommended. Although its cost is higher than for the other crest shapes, advantages result both in discharge capacity and safety against cavitation damage.

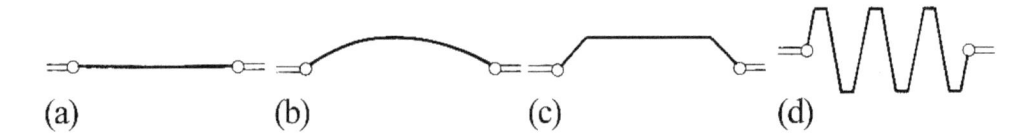

Figure 2.11 Plan views of overflow structures (a) straight, (b) curved, (c) polygonal, (d) labyrinth

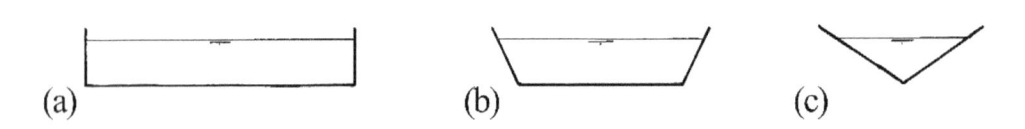

Figure 2.12 Transverse sections of overflow structure (a) rectangular, (b) trapezoidal, (c) triangular

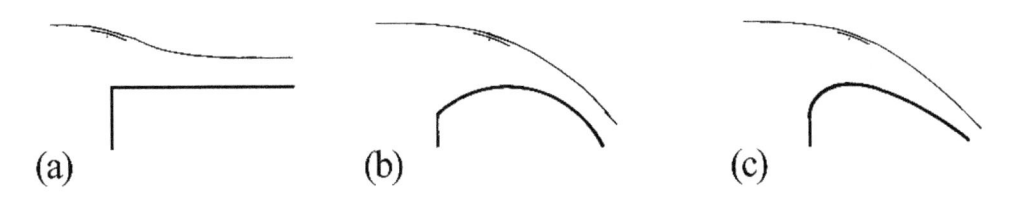

Figure 2.13 Longitudinal section of overflow structure (a) broad-shaped, (b) circular-shaped, (c) standard-shaped

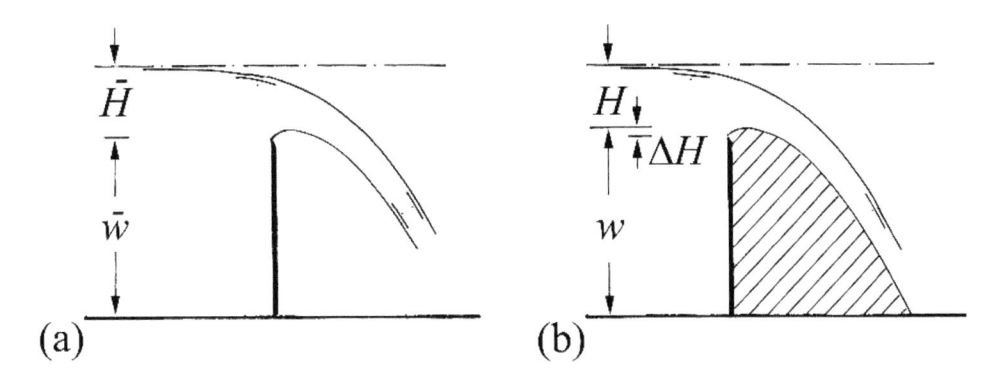

Figure 2.14 (a) Thin-plate, sharp-crested weir fully aerated, (b) corresponding standard crest, also referred to as Creager or Ogee weir crest (Hager and Castro-Orgaz, 2017)

The flow over a structure involves curved streamlines with the curvature origin below the flow. The gravity component of a fluid element is thus reduced by the centrifugal force. If the curvature effect is large, the internal pressure drops below atmospheric pressure, attaining values below the vapor pressure for large hydraulic heads. Cavitation (Chapter 8) then occurs with a potential for *cavitation damage*; these conditions are unacceptable given the importance of the spillway overflow structure.

For medium and large overflow structures, the crest shape corresponding to the lower nappe of a fully aerated *thin-plate weir* (sharp-crested weir) is adopted, because the resulting overflow jet composes the 'natural' shape involving atmospheric pressure *both* along the lower and the upper nappes. The basis of the crest shape is well-defined and involves a smooth thin-plate weir of a certain angle relative to the vertical. Of particular interest is the vertical thin-plate weir shown in Figure 2.14. The standard crest is knife sharp, with 2 mm horizontal crest thickness and a 45° downstream beveling. To inhibit scale effects due to viscosity and surface tension under laboratory conditions, the head $\bar{H}$ on the weir should be at least 50 mm, and the weir height $\bar{w}$ has to be at least twice as large as the maximum head, i.e. $\bar{H}_{max}/\bar{w} \leq 0.5$. Then, effects of the approach flow velocity remain insignificant.

Figure 2.15 refers to thin-plate weir flow and the corresponding overflow structure at design discharge. The lower nappe separates at the upstream crest edge from the weir plate,

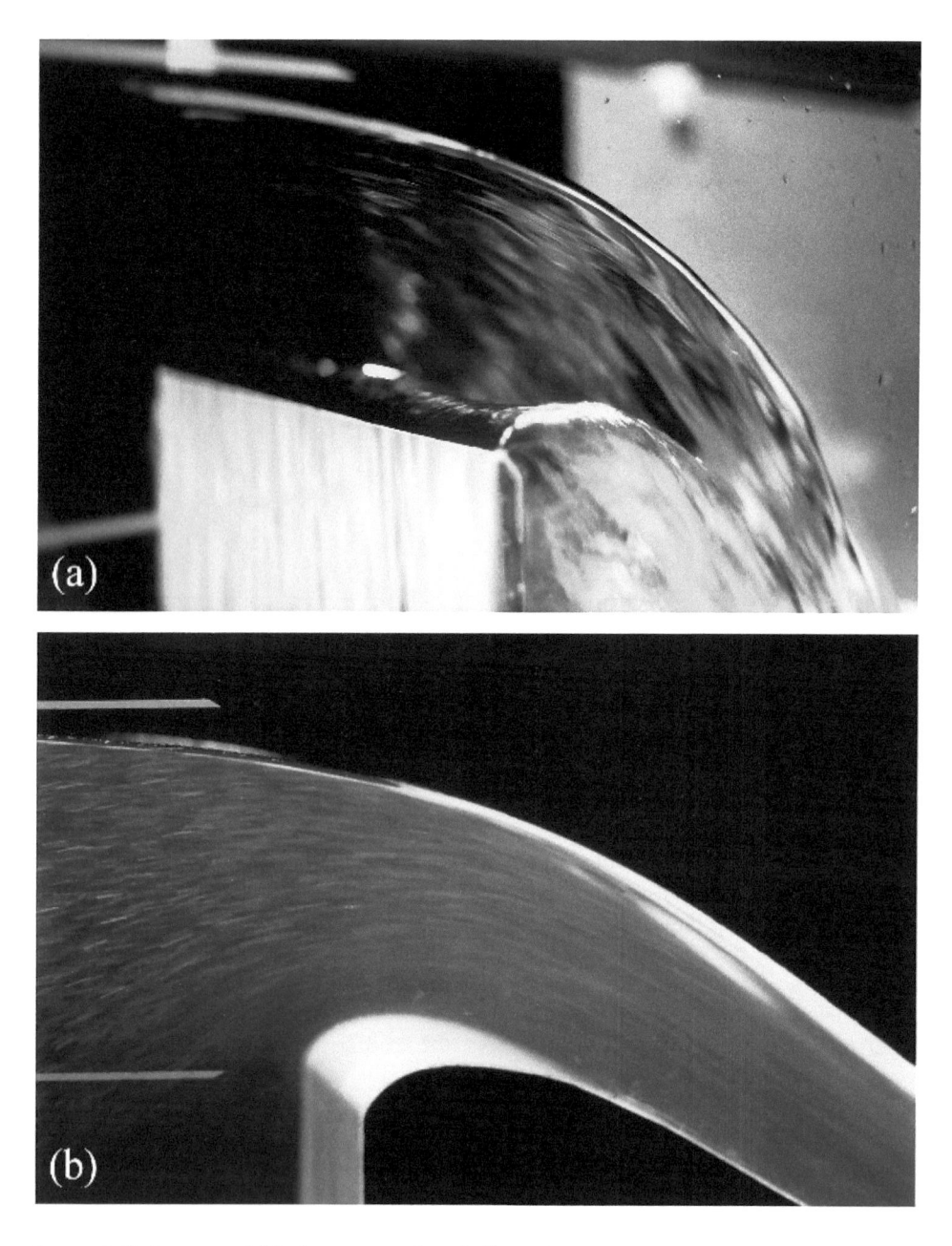

Figure 2.15 Images of (a) sharp-crested weir flow, (b) standard-crested overflow at design discharge (Courtesy Willi H. Hager)

rises to the maximum trajectory point, and then falls into the tailwater. The falling lower nappe portion was considered more important until it was realized that the dominant low pressures on the corresponding standard overflow occur along the rising nappe portion.

For overflow depths larger than 50 mm, say, for which viscous and surface tension effects are absent, weir flow and overflow over a standard crest are fully equivalent. For a weir crest having its highest point by an amount ΔH above the sharp crest, equal discharge passes for identical upstream water elevation. The crest shape is important regarding the bottom pressure distribution. Slight modifications have a significant effect on it, while the discharge characteristics remain practically unaffected. The geometry of the lower nappe cannot simply be expressed analytically. The best-known approximation for the crest shape geometry is due to the US Army Corps of Engineers (USACE, 1970). A three-arc profile for the upstream quadrant was proposed whereas a power function describes the downstream quadrant, with the crest as origin of the Cartesian coordinate system $(x; z)$. The crest shape is plotted in Figure 2.16.

The governing scaling length of the standard overflow structure is the so-called design head H_D (subscript D for design). All other lengths are normalized with H_D, including the radii of the upstream crest profile $R_1/H_D = 0.50$, $R_2/H_D = 0.20$, and $R_3/H_D = 0.04$. The curvature origins O_1, O_2, and O_3 and the transition points P_1, P_2, and P_3 of the *upstream quadrant* are detailed in Table 2.1.

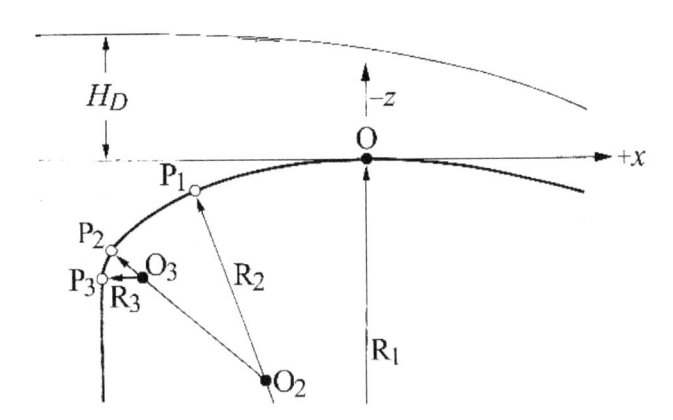

Figure 2.16 USACE crest shape for vertical upstream weir face, and zero velocity of approach

Table 2.1 Coordinates of curvature origins O_i, and transition points P_i for standard crest shape according to Figure 2.16 (USACE, 1970)

Point	O_1	O_2	O_3	P_1	P_2	P_3
x/H_D	0.000	−0.105	−0.242	−0.175	−0.276	−0.2818
z/H_D	0.500	0.219	0.136	0.032	0.115	0.1360

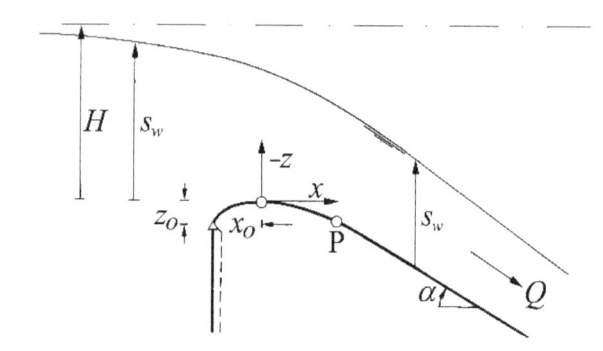

Figure 2.17 Continuous crest profile with transition to straight chute bottom at tangency point P

The crest shape of the *downstream quadrant* was originally proposed by Creager (for a later version, see Creager *et al.*, 1945; Hager and Castro-Orgaz, 2017 provide a historical background to this issue) as

$$z / H_D = 0.50(x / H_D)^{1.85}, \quad \text{for } x > 0. \tag{2.1}$$

This shape is employed up to the so-called tangency point P at the transition to the constantly sloped spillway chute (Figure 2.17). The disadvantage of the USACE crest shape is the abrupt change of curvature at points P_1 to P_3 and at the crest, so that it cannot be applied for computational methods due to the curvature discontinuities. An alternative approach with a smooth curvature development was proposed by Hager (1987). Based on the results of observations, the crest shape proposed is (Figure 2.17)

$$Z^* = -X^* \ln X^*, \quad \text{for } X^* > -0.2818. \tag{2.2}$$

Here $(X^*; Z^*)$ are transformed coordinates based on Table 2.1 as $X^* = 1.3055(X + 0.2818)$ and $Z^* = 2.7050(Z + 0.1360)$, with $X = x/H_D$ and $Z = z/H_D$. Equation (2.2) has the property that the second derivative is $d^2Z^*/dX^{*2} = -1/X^*$, e.g. the inverse curvature varies linearly with X^*. The differences between the two crest geometries of USACE (1970) and according to Eq. (2.2) are usually negligible for crest construction.

2.2.2 Free surface profile and discharge characteristics

The free surface profile over a standard overflow structure is important in relation to free-board design and for gated flow. Figure 2.18a refers to the USACE curves both for 2D flow without presence of piers and for the axial profiles between two piers. Along the crest piers (see below), the free surface is lower than for plane flow due to the transverse acceleration.

Hager (1991a) provided a generalized approach for plane flow over the standard-shaped overflow crest (Figure 2.19). According to Figure 2.17, the surface elevation s_w is referred to the crest level upstream from the crest origin O, and to the bottom elevation downstream from point O (Figure 2.16). The dimensionless free surface profile $S(X)$ with $S = s_w/H_D$

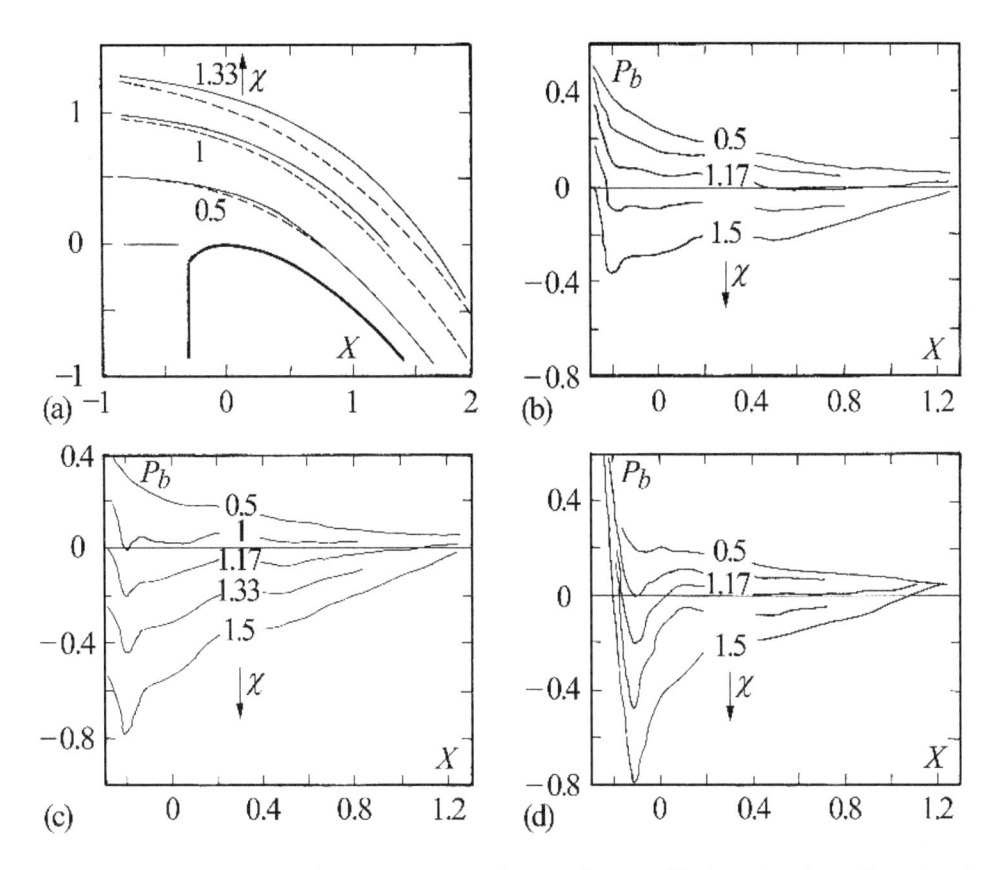

Figure 2.18 Standard overflow structure (a) free surface profile (—) for plane flow, (- - -) between piers; bottom pressure distribution for (b) plane flow, (c) axially between piers, (d) along piers (USACE, 1970)

decreases almost linearly for $-2 < X/\chi^{1.1} < +2$ with $\chi = H/H_D$ as the head ratio. Its approximation is (Figure 2.19)

$$S = 0.75[\chi^{1.1} - (1/6)X].\tag{2.3}$$

The discharge Q over an overflow structure is expressed with C_d as the discharge coefficient, b the overflow width, and g the gravity acceleration as

$$Q = C_d b(2gH^3)^{1/2}.\tag{2.4}$$

As all the other parameters, C_d varies only with the relative head $\chi = H/H_D$. The experimental data plotted in Figure 2.20b follow independently of the chute bottom angle α (Figure 2.17) up to $\chi = 3$

$$C_d = \frac{2}{3\sqrt{3}}\left[1 + \frac{4\chi}{9 + 5\chi}\right].\tag{2.5}$$

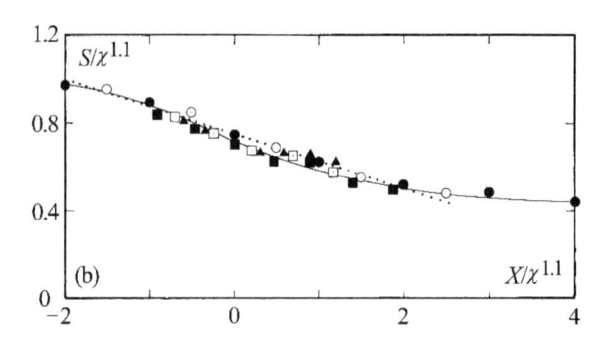

Figure 2.19 Free surface profile $S(X, \chi)$, experiments and (-) average data curve, ($\cdots$) Eq. (2.3)

For $\chi \to 0$, the overflow is shallow so that nearly hydrostatic pressure occurs. Then, the overflow depth is equal to the critical flow depth, with the discharge coefficient of $C_d = 2/(3)^{3/2} = 0.385$. For design discharge, $\chi = 1$, so that $C_d = 0.495$ from Eq. (2.5). The discharge coefficient is thus increased by 30% as compared with parallel-streamlined flow. For $\chi > 2$, the increase is only moderate, however, i.e. $C_d(\chi = 2) = 0.55$.

2.2.3 Bottom pressure characteristics

The bottom (subscript b) pressure distribution $p_b(x)$ is important because it yields

- Index for potential danger of cavitation damage
- Location where piers should end without inducing flow separation.

Figure 2.18(b–d) provide bottom pressure head data $p_b/(\rho g)$ non-dimensionalized by the design head H_D as $P_b = p_b/(\rho g H_D)$ versus location $X = x/H_D$ for various values of χ. Note that the *minimum* (subscript m) pressure p_m occurs on the upstream quadrant throughout, and that bottom pressures are positive for $\chi = H/H_D \leq 1$. The most severe bottom pressure minima occur along the piers due to significant streamline curvature.

A generalized data analysis was conducted by Hager (1991a) by accounting for all published observations relative to plane overflow. These data are plotted as $\bar{P}_m = p_m/(\rho g H)$ over the relative head χ in Figure 2.20a, following the approximation

$$\bar{P}_m = (1 - \chi). \tag{2.6}$$

The minimum bottom pressure head thus is positive as compared with atmospheric pressure if $\chi < 1$. The minimum pressure head $p_m/(\rho g)$ is proportional to the effective head H, and $(1 - \chi)$. The location of minimum bottom pressure is $X_m = -0.15$ for $\chi < 1.5$, and $X_m = -0.27$ for $\chi > 1.5$, i.e. just at the transition of the crest to the vertical weir face.

The crest (subscript c) bottom pressure index $\bar{P}_c = p_c/(\rho g H)$ versus χ is seen to be significantly above the minimum bottom pressure (Figure 2.21a). Model data suggest $\bar{P}_c = (2/3)\bar{P}_m$. Figure 2.21b refers to the location of zero bottom (subscript 0) pressure head, i.e. the location of atmospheric bottom pressure. The data do not only vary with the

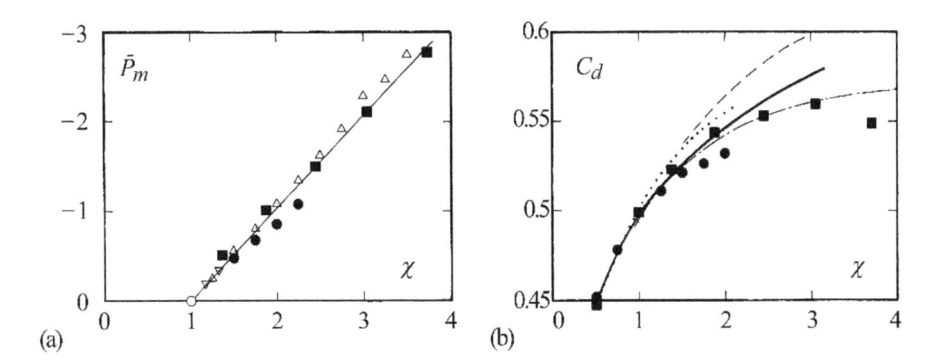

Figure 2.20 (a) Minimum bottom pressure index $\bar{P}_m = p_m/(\rho g H)[\chi]$, (b) discharge coefficient $C_d[\chi]$ with $\chi = H/H_D$ (Hager, 1991a)

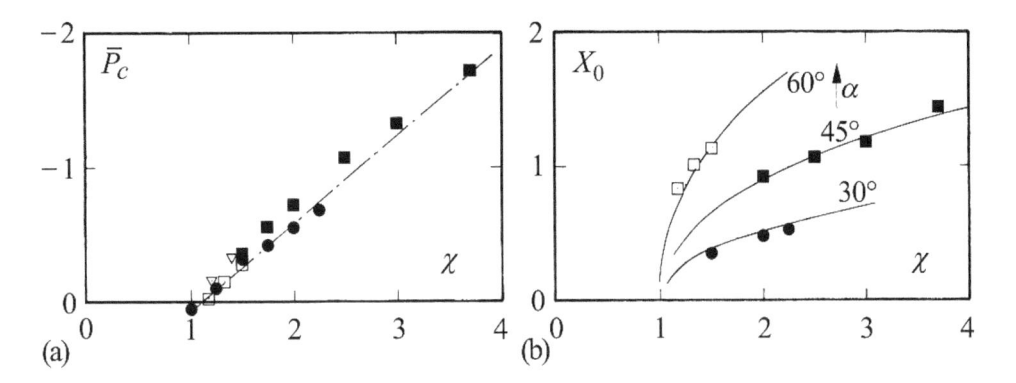

Figure 2.21 (a) Crest pressure $\bar{P}_c(\chi)$, (b) location of atmospheric bottom pressure $X_0(\chi)$ (Hager, 1991a)

relative head χ but also with the tailwater chute angle α (Figure 2.17). The relative position $X_0 = x_0/H_D$ is (Hager, 1991a)

$$X_0 = 0.9 \tan \alpha (\chi - 1)^{0.43}. \tag{2.7}$$

2.2.4 Velocity distribution

Velocity distributions were recorded by Hager (1991a, b) for chute angles of $\alpha = 30°$ and 45°. Figure 2.22 shows normalized plots of the relative velocity $\mu = V/(2gH_D)^{1/2}$ versus X and Z for various relative heads χ. Away from the crest, the velocity distribution is almost uniform and the pressure distribution thus almost hydrostatic. However, along the crest, there is a significant velocity increase in the streamwise and depthwise directions due to the free surface gradient and streamline curvature effects, respectively. From these generalized plots result velocities at any location $(X; Z)$.

Figure 2.23 refers to typical 2D flow over the standard overflow structure. The flow is seen to be absolutely smooth, with small air bubbles contained in the approach flow revealing the

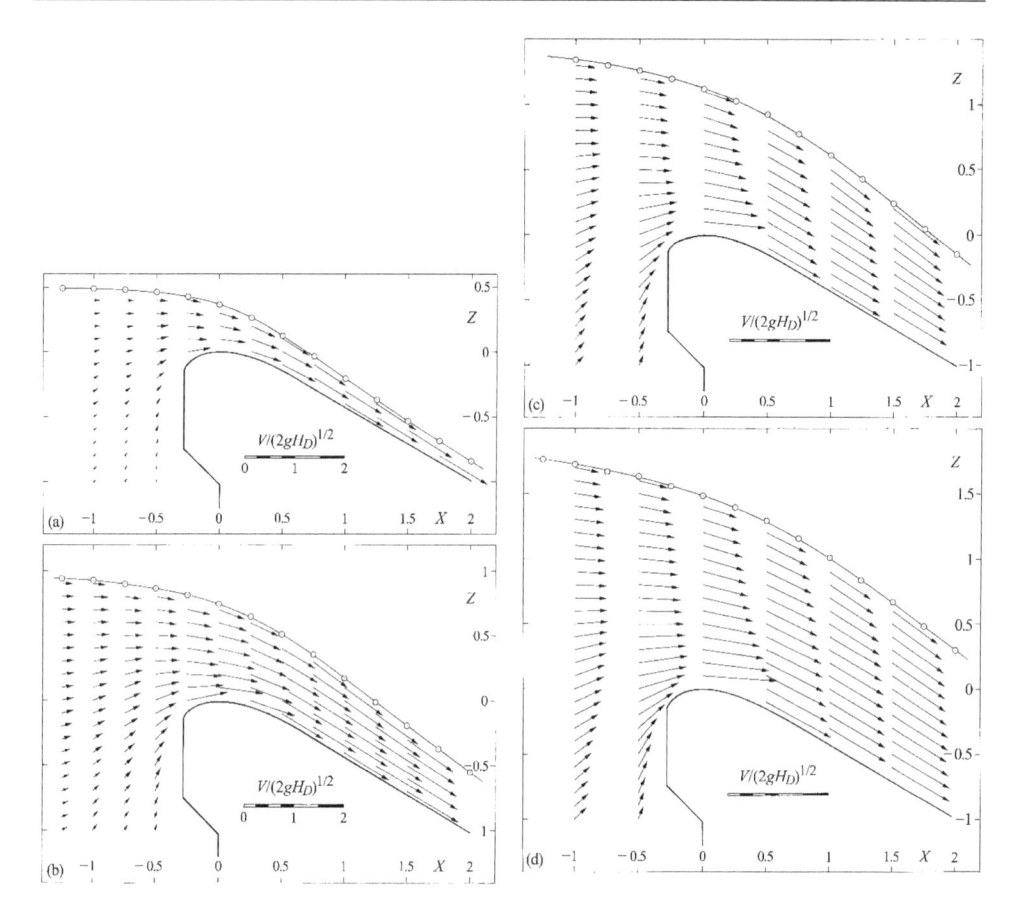

Figure 2.22 Velocity distribution $V/(2gH_D)^{1/2}[X; Z]$ for χ = (a) 0.5, (b) 1, (c) 1.5, (d) 2 (Hager, 1991a, b)

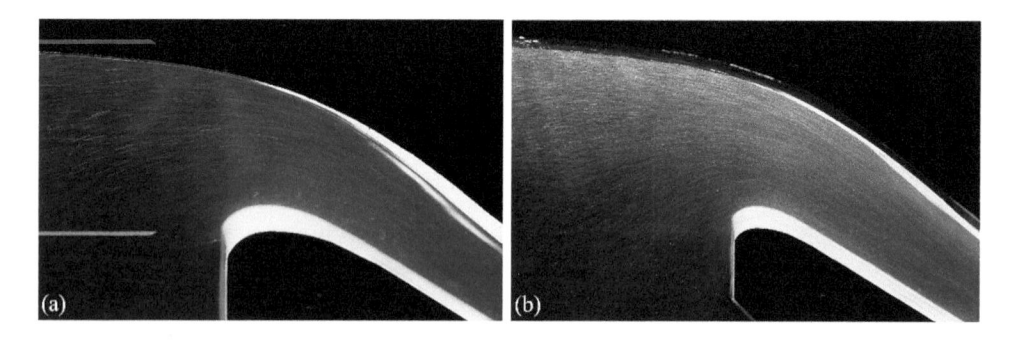

Figure 2.23 Flow pattern over standard overflow structure for χ = (a) 1, (b) 2 (Courtesy Willi H. Hager)

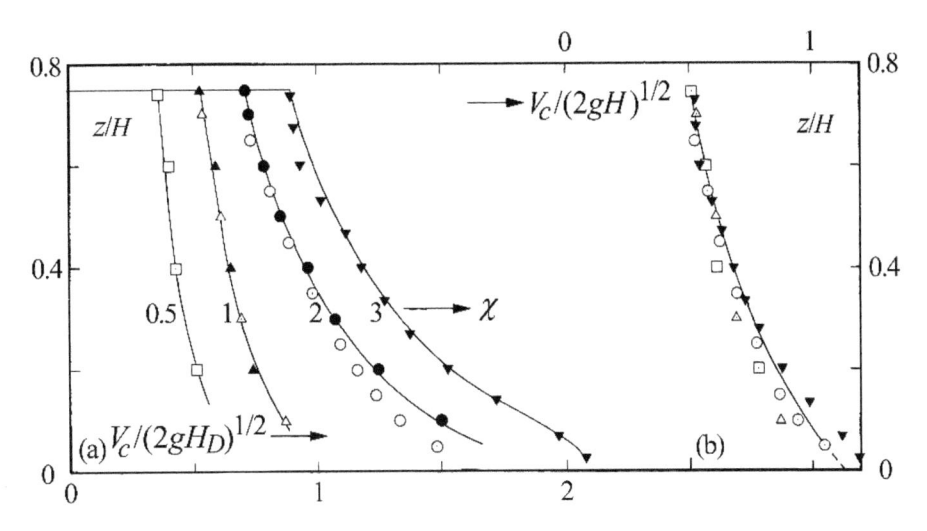

Figure 2.24 Crest velocity distribution (a) $V_c/(2gH_D)^{1/2}$, (b) $V_c/(2gH)^{1/2}$ versus relative elevation over crest z/H for various χ

streamlines. The design head was $H_D = 0.20$ m, as indicated with white horizontal lines, and the chute bottom angle $\alpha = 30°$ (Hager, 1991a).

The *crest velocity distribution* $V_c(z)$ is of particular relevance; Figure 2.24a shows the distributions normalized by H_D as $\mu_c = V_c/(2gH_D)^{1/2}$. The increase of crest velocity with both depth z and χ is obvious, except for the thin bottom boundary layer. An even simpler plot yields a normalization by the effective head H, instead of the design head H_D (Fig. 2.24b). Computations indicate that the *free vortex* model approximates the crest section favorably, i.e. that the product of tangential velocity times the radius of curvature remains constant. The resulting velocity distribution thus is (Hager, 1991b)

$$\frac{V_c}{(2gH)^{1/2}} = \frac{1}{2}\left[\frac{r+0.75}{r+z/H}\right]. \tag{2.8}$$

Here $r = d^2Z/dX^2 = 0.584$ is the dimensionless crest curvature from Eq. (2.2). A complex plane flow problem is thus governed by a fundamental physical law.

2.2.5 Cavitation design

Standard crest overflows with $\chi < 1$ are referred to as underdesigned, while the *overdesign* involves $\chi > 1$, resulting in sub-atmospheric bottom pressures. Overdesign of dam overflows was originally associated with advantages in discharge capacity. However, as follows from Figure 2.20b, the increase of C_d for $\chi > 1$ is relatively small, whereas the decrease of minimum pressure p_m from Figure 2.20a is significant. Overdesigning does thus add to the cavitation potential. *Incipient cavitation* is known to be a statistical process depending mainly on the water quality and the local turbulence pattern. The incipient (subscript i) pressure head is

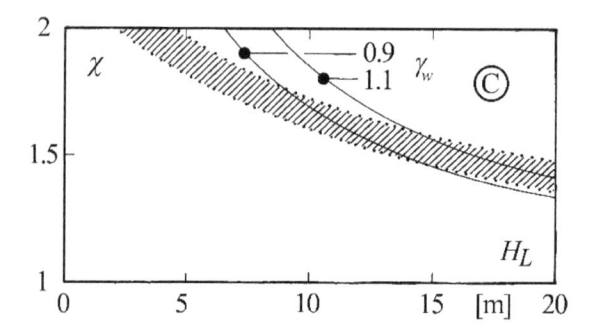

Figure 2.25 Limit head H_L (m) versus relative head $\chi = H/H_D$ on plane overflow structure,
(−−) Eq. (2.9), ($\cdot\cdot\cdot$) domain of Abecasis (1970), © domain of definite cavitation

assumed to be $p_{vi}/(\rho g) = -7.6$ m as compared with -10 m roughly for static and clean water (Abecasis, 1970). The limit (subscript L) head H_L for incipient cavitation to occur thus is from Eq. (2.6)

$$H_L = [\gamma_w(1-\chi)]^{-1}[p_{vi}/(\rho g)].\tag{2.9}$$

Because the incipient cavitation head is a length, H_L is also expressed in meters. Figure 2.25 compares Eq. (2.9) with data of Abecasis (1970), indicating general agreement. The constant γ_w accounts for additional effects as the variability of p_{vi} with χ.

2.2.6 Crest piers

Piers on overflow structures are provided to:

* Improve approach flow conditions
* Mount overflow gates and stoplogs
* Divide spillway in various chute portions
* Support a weir bridge
* Aerate chute flow at pier ends.

The front pier shape was studied by USACE (1970), with two typical designs shown in Figure 2.26a, b. Other front shapes include an almost rectangular pier nose with rounded edges, and a triangular nose, yet the designs shown are standard. Currently, parabolic or elliptical pier shapes close to USACE Type 2 are often used to have a resistant rounded nose in view of floating debris impact.

A pier modifies the overflow from plane to spatial. In terms of discharge, this effect is accounted for by the *effective width* b_e (subscript e) instead of the geometrical width b; with K_p as pier coefficient, and H as head on the overflow structure

$$b_e = b - 2K_p H.\tag{2.10}$$

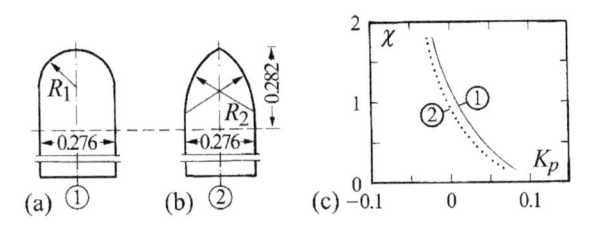

Figure 2.26 Typical pier front shapes (a) circular, (b) circular-arced, with numbers to be multiplied with design head H_D, (c) $K_p(\chi)$ values for designs ① and ② according to USACE (1970)

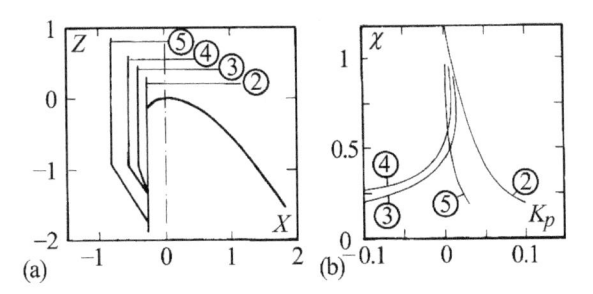

Figure 2.27 Upstream front pier position (a) definition, (b) pier coefficient $K_p(\chi)$ (USACE, 1970)

According to Figure 2.26c the parameter K_p decreases slightly from 0.05 for $\chi = 0.2$ to $K_p = -0.05$ for $\chi = 1.5$; for preliminary design purposes $K_p \cong 0$, so that the recommended pier shapes do not significantly perturb the plane overflow pattern.

The *upstream front position* of the pier relative to the crest is subject to variation, as shown in Figure 2.27a, to save dam material, provided static considerations allow for this economic design. The effect on the pier coefficient K_p is again noted to be small, except for small relative heads χ (Figure 2.27b).

Additional information on the pier effect provide Webster (1959) and Indlekofer (1976). Figure 2.28 shows views of overflow piers from an upstream reservoir. Note the formation of intake vortices and a wavy water surface close to the gates, due to the flow impact on the gates, and the formation of capillary waves.

The tailwater end of an overflow pier corresponds to an abrupt flow expansion. Because chute flow is supercritical, standing shock waves have their origins at the pier ends, propagating all along the chute (Chapter 4). To suppress these *pier shock waves*, two designs are available:

- Sharpening pier end both in width and height
- Continue with pier as dividing wall along chute.

Both designs are not ideal, because even a slim pier end perturbs the flow, whereas dividing walls are costly especially for long chutes.

Figure 2.28 Approach flow to piers located at overflow structures (a) overall view, (b) detail with bow wave (Courtesy Willi H. Hager)

Reinauer and Hager (1994) proposed an alternative design sketched in Figure 2.29. It involves the so-called pier extension (Figure 2.29c), corresponding to a one-sided extension wall of the abruptly terminating pier. The shock wave is reduced by the interference principle. The location of the pier extension (subscript E) x_E in horizontal chutes with b_p as the pier width, $F_o = V_o/(gh_o)^{1/2}$ the approach flow Froude number with V_o as approach flow velocity at the pier end, and h_o as the corresponding flow depth is

$$X_E = \frac{x_E}{b_p F_o} = 0.41(2h_o/b_p)^{2/3}. \tag{2.11}$$

For spillway flow without pier extension, pier wave 1 is located along the pier axis, and a reflection wave 2 is generated from the adjacent pier, or by the presence of a chute wall. Both waves depend significantly on the ratio $(h_o/b_p)^{1/2}$, as found by Reinauer and Hager (1994). The pier extension is considered a simple means to save cost for high chute walls, and to avoid uncontrolled chute flow aeration at pier ends.

Bridge piers manifest an increased risk for driftwood or debris blockage. A sufficiently large weir bay and an adequate freeboard up to the weir bridge are required to avoid a debris accumulation related to a reduced discharge capacity (STK, 2017). As for the Palagnedra Dam, Switzerland, the piers and the crest crossing bridge were removed (Figure 2.29e) after a catastrophic driftwood accumulation (Bruschin *et al.*, 1982). A new bridge in the tailwater of the dam was set up.

2.2.7 Overflow crest gates

Gated overflows are regulated to a desired or prescribed reservoir level. Floods entering a reservoir are modified at the outlet due to flood regulation maneuvers. The head on the turbines is increased as compared with ungated overflow structures, but any gate opening may pose a problem under extreme flood conditions, mainly due to debris, power

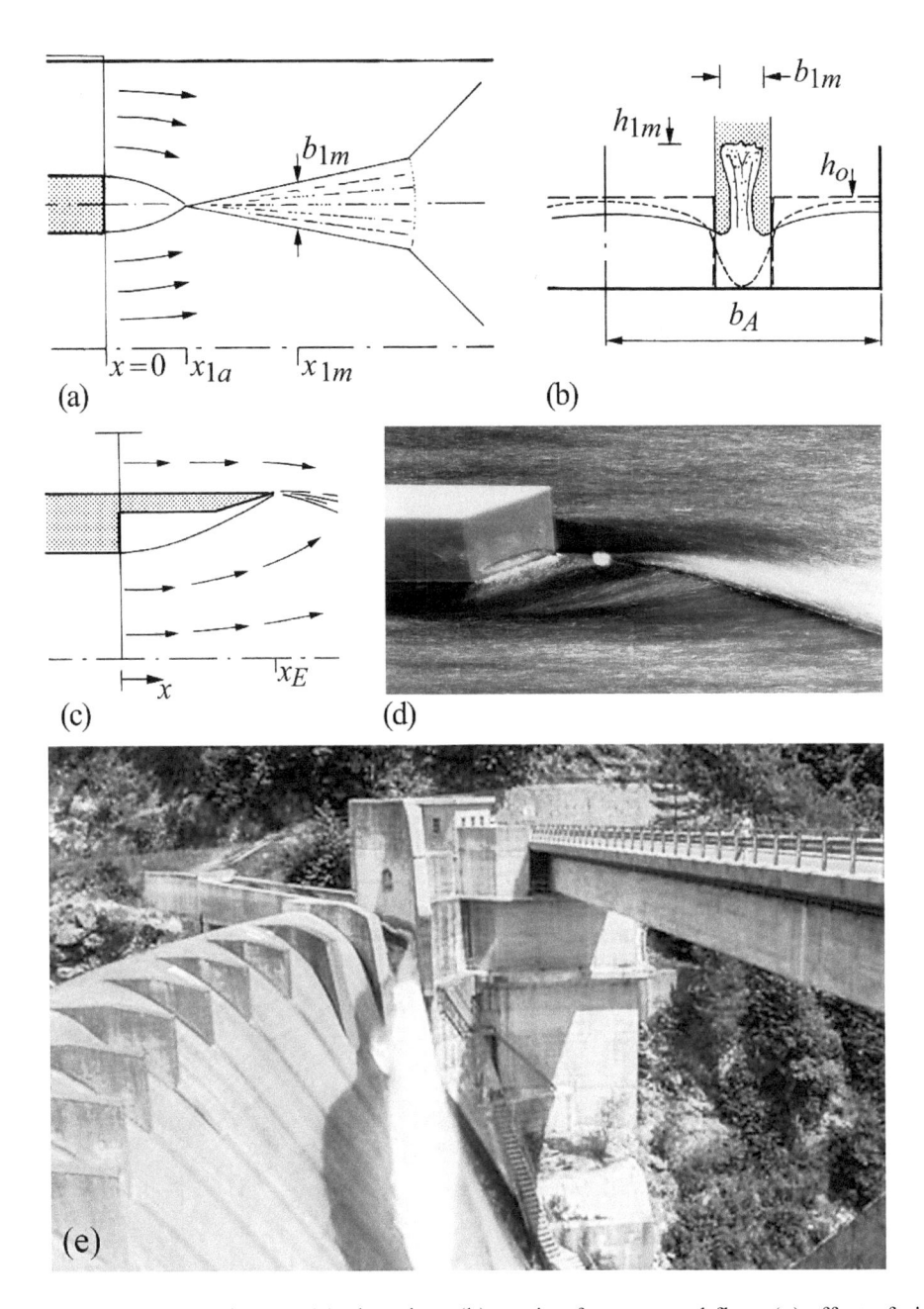

Figure 2.29 Pier shock wave (a) plan view, (b) section for untreated flow, (c) effect of pier extension, (d) laboratory photograph (Reinauer and Hager, 1994), (e) piers removed to reduce probability of driftwood blockage at Palagnedra Dam, Switzerland (Photo: polier.ch)

breakdown, and unexpected flood generation in the tailwater. Gates on dams during large floods must be especially secured against failure because their safe operation is of major importance.

Gates must be simple and reliable in operation, and easy in maintenance. Usually, either vertical plane gates or cylindrical sector gates are provided on the crest of the overflow structure, eventually extended by flap gates to control small overflows. Figure 2.30 shows a typical segment or radial gate on an overflow structure.

The hydraulics of gates on overflow structures includes three major problems:

- Discharge characteristics
- Crest pressure distribution
- Gate vibrations.

The *vertical plane gate* located at the overflow crest section as shown in Figure 2.31a was analyzed by Hager and Bremen (1988). The discharge Q depends on the head H_o on the gate, the corresponding head H without gate, the gate opening a, and the crest design head H_D. The design discharge Q_D is defined with $C_{dD} = C_d(\chi = 1) = 0.495$ from Eq. (2.5) as

$$Q_D = C_{dD}b(2gH_D^3)^{1/2}. \tag{2.12}$$

The discharge Q_g under a gated (subscript g) overflow is with $\chi_o = H_o/H_D$ as relative head of the gated flow and $A = a/H_D$ as the relative gate opening

$$Q_g/Q_D = [\chi_o^{3/2} - (\chi_o - A)^{3/2}][(1/6) + A]^{1/9}. \tag{2.13}$$

Equation (2.13) holds for $0 < A < 2$ and $\chi_o > (4/3)A$ because the gate is not submerged as $\chi_o < (4/3)A$. For ungated overflow, i.e. $\chi_o \to \chi$, and $Q_g \to Q$, Eq. (2.13) simplifies to

$$Q/Q_D = \frac{2}{3\sqrt{3}}\frac{\chi^{3/2}}{C_{dD}}\left[1 + \frac{4\chi}{9 + 5\chi}\right]. \tag{2.14}$$

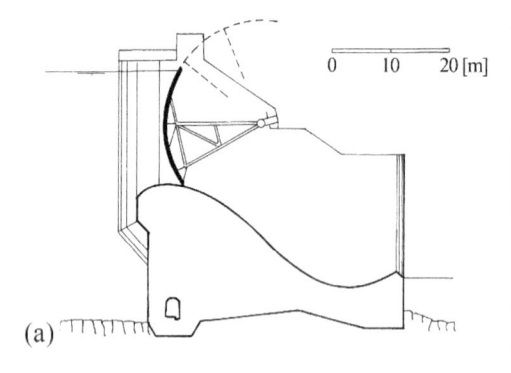

Figure 2.30 Sector gate on standard crest overflow structure (a) schematic view, (b) Jirau Dam, Brazil (Courtesy Michael Pfister)

Equations (2.13) and (2.14) are plotted in Figure 2.31b as the generalized discharge-head diagram for vertical overflow gates. The effects of gate location relative to the overflow crest and the corresponding crest pressures were studied by Lemos (1981).

The *radial gate* or *Tainter gate* positioned on the standard overflow structure is influenced by a large number of parameters, including (Figure 2.32):

- Relative radius of gate curvature $R = r/H_D$
- Position of gate lip $(X_l = x_l/H_D; Z_l = z_l/H_D)$
- Gate seat coordinate $X_s = x_s/H_D$
- Gate trunnion coordinates $(X_t = x_t/H_D; Z_t = z_t/H_D)$
- Gate lip angle α_g
- Shortest distance G normalized by H_D from overflow profile to gate lip
- Corresponding horizontal coordinate $X_w = x_w/H_D$
- Profile angle γ_G at this point.

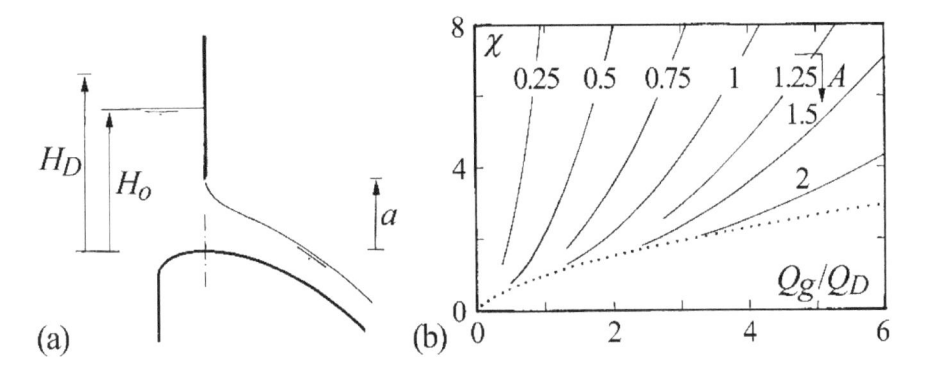

(a)   (b)

Figure 2.31 Vertical gate at crest of overflow structure (a) definition, (b) discharge-head equation with (---) Eq. (2.13), (· · ·) Eq. (2.14) for ungated flow (Hager and Bremen, 1988)

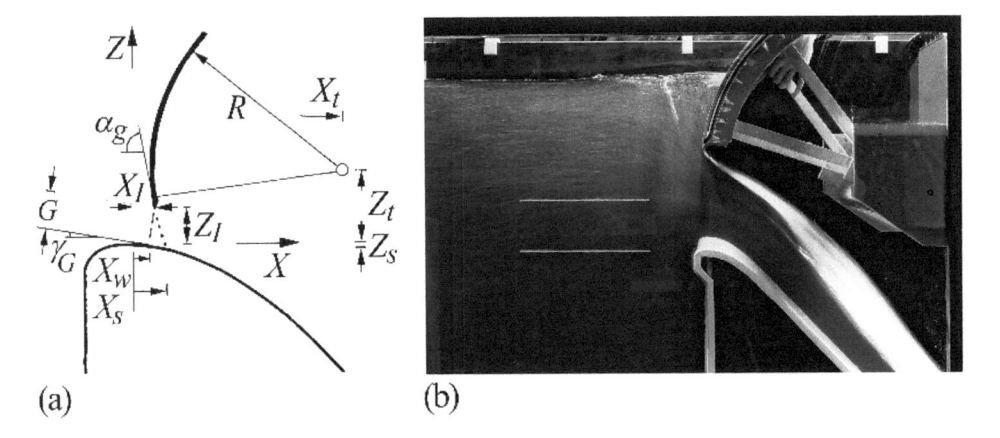

(a)   (b)

Figure 2.32 Radial gate on standard overflow structure (a) parameter definition, (b) photo (Courtesy Willi H. Hager)

The independent parameters include the design head H_D of the overflow crest, the operational head H, while the effective (subscript e) head H_e corresponds to the approach flow head over the crest for $x_l < 0$, and to the approach flow head plus the height of the standard profile for $x_l \geq 0$, respectively. The discharge follows with C_{dg} as the discharge coefficient of gated (subscript g) flow from

$$Q = C_{dg} b(GH_D)(2gH_e)^{1/2}. \tag{2.15}$$

Sinniger and Hager (1989) and Hager and Schleiss (2009) presented a complete analysis from which the discharge characteristics result.

According to Rhone (1959), the gate seat coordinate should be positive to inhibit significant sub-pressure, but x_s should be confined to small values to limit the gate height. Values range typically within $0 \leq X_s \leq + 0.2$, for which $Z_s < 0.025$.

The *bottom pressure distribution* over the gated standard overflow structure was studied by, e.g. Lemos (1981). The location of minimum pressure was determined to $\Delta X \cong +0.20$ downstream from the gate seat coordinate. These sub-pressures define both the cavitation potential and the separation features of the overflow profile.

The *minimum bottom pressures* $P_m = p_m/(\rho g H_D)$ are practically uninfluenced by the relative gate radius R, whereas the effect of χ_o is significant. The curves $P_m(Z_l)$ increase to a maximum value at $Z_l = 0.4$, equal to $P_m \cong -0.2$ for $\chi = 1$, and $P_m \cong -0.4$ for $\chi = 1.25$, depending on the trunnion elevation. For a detailed analysis, refer to Lemos (1981) or to Hager and Schleiss (2009).

2.3 Additional weir effects

2.3.1 Influence of weir face slopes

Introduction

Circular-crested weirs are used for discharge measurement and serve as an overflow structure in hydraulic applications. Their advantages include stable hydraulic overflow condition, simplicity in design, and associated low cost. The weir is designed with a circular crest of radius R_w, an upstream face perpendicular to the approach flow direction, and a downstream slope angle of around 45°. In low-head dam applications, the two angles may be much smaller. For free overflow, the discharge coefficient C_d from Eq. (2.4) determines the overflow discharge Q with $H_o = h_o + Q^2/[2gb^2(h_o + b)^2]$ as the approach flow energy head relative to the weir crest, with h_o as the approach flow overflow depth. The effects of weir height w and weir geometry are contained in C_d. A proposal for C_d was derived by Montes (1970) and adapted by Hager (1994) to

$$C_d = \frac{2}{3\sqrt{3}}\left(1 + \frac{3\rho_k}{11 + \Omega\rho_k}\right). \tag{2.16}$$

Here $\rho_k = H_o/R_w$ is the relative crest curvature and $\Omega = 4.5$. An overview on empirical C_d formulas for these weirs is provided in, e.g. Chanson and Montes (1998). However, their equation does not account for the effect of upstream and downstream weir face slopes. Note that sloping weirs have a higher discharge capacity than the standard broad-crested weir with vertical faces. The study of Schmocker *et al.* (2011) provides information on circular-crested weirs and their discharge behavior.

The most notable study carried out to investigate the discharge over weirs include Bazin (1898), the first analyzing systematically the discharge coefficient of embankment weirs. His classification includes three flow types: (1) free flow, (2) plunging nappe flow, and (3) submerged flow. Jaeger (1933) used the Boussinesq equation to determine the discharge coefficient of circular-crested weirs, stating that the C_d depends on the relative crest curvature R_w/H_o and the relative overflow depth. Both Hégly (1939) and Jaeger (1940) presented formulas for C_d depending on H_o/R_w. The effect of small overflow depths on the free surface profile was studied by Bretschneider (1961) for weirs sloping 1:1 (V:H), 1:1.5, and 1:2.

An equation for C_d including the effects of streamline curvature, viscosity, and surface tension was published by Matthew (1963) thereby accounting for scale effects in hydraulic modeling of overflow structures. The velocity distribution at the crest section along with an equation for the discharge coefficient was derived by Montes (1970). Rao and Rao (1973) and Rao (1975) investigated the coefficient of discharge of hydrofoil weirs taking especially into account the submergence limit. Bos (1976) presented a standard work on discharge measurement structures including circular-crested weirs with vertical upstream faces and downstream angles of 45°. Four flow types downstream of a weir were distinguished, depending on the tailwater depth: (1) hydraulic jump, with the toe located at or downstream of the weir; (2) plunging jet; (3) surface wave flow; and (4) surface jet flow. This classification was reconsidered by Schmocker *et al.* (2011) (Figure 2.33). The effects of viscosity and surface tension on embankment weir flow were studied by Ranga Raju *et al.* (1990). They identified the characteristic parameter $\Phi = \mathsf{R}^{0.2}\mathsf{W}^{0.6}$, with $\mathsf{R} = \left(gh_o^3\right)^{1/2} v^{-1}$ as Reynolds number, and $\mathsf{W} = \rho gh_o^2\sigma^{-1}$ as Weber number, with v as fluid viscosity, ρ as fluid density, and σ as surface tension. Scale effects due to surface tension and viscosity are absent if $\Phi > 10^3$, almost independently of embankment slope, weir height, and discharge.

The effect of weir faces on circular-crested weir flow was studied by Ramamurthy and Vo (1993a). They tested upstream angles of 60°, 75°, and 90° and downstream angles of 45°, 60°, and 75° for a range of H_o/R_w, finding that the upstream slope had no effect on C_d for a certain downstream slope. For a fixed upstream weir slope, they observed an increase of C_d with increasing downstream slope. Ramamurthy and Vo (1993b) determined the weir discharge coefficient versus the total approach flow head and the crest radius by applying the theory of Dressler (1978). Chanson and Montes (1998) presented tests on circular-crested weir overflow in terms of weir radii and weir heights including an upstream ramp, stating that the latter presence had no effect on C_d but a major influence on the upstream flow conditions. The general work on the hydraulics of embankment weirs by Fritz and Hager (1998) includes the coefficient of discharge in terms of relative crest length for trapezoidal-crested weirs. Heidarpour *et al.* (2008) applied the potential flow theory to the circular-crested weir. The crest velocity distribution, normalized crest pressure and the pressure correction coefficient were determined validating their resulting equations with laboratory data. A 2D model for critical flow developed by Castro-Orgaz (2008) was applied to the circular-crested weir. The discharge coefficient was found to increase with relative head E/R_w, with E as specific energy by including streamline curvature effects. The researches of Fritz and Hager (1998), and

Figure 2.33 Main flow patterns downstream of circular-crested weir flow (a) hydraulic jump, (b) plunging jet, (c) surface wave flow, (d) surface jet flow (Schmocker *et al.*, 2011)

Castro-Orgaz and Hager (2014) are presented below because lack of material currently available on the tailwater features of weir flow, and a computational approach to determine scale effects relating to weir flow.

The study of Schmocker *et al.* (2011) deals with the hydraulic features of the circular-crested weir of various upstream and downstream face slopes. The main parameters include (1) free surface profile, (2) coefficient of discharge, (3) modular limit and transition range, (4) discharge reduction, and (5) velocity distribution. The results complete the hydraulic knowledge of circular-crested weirs including a novel formula for the discharge coefficient by accounting for both the upstream and downstream weir face angles. The results apply to hydraulic structures in general, and particularly to the plane dike break problem (Chapter 10), because the circular-crested weir shape is similar to that of an embankment during erosion by overtopping. The results apply therefore to the discharge features during a plane breach of an embankment dike.

Experimentation

Tests were conducted in a rectangular horizontal channel of 150 l/s discharge capacity. It was 0.50 m wide, 0.70 m high, and 7.0 m long. The uniformity of the approach flow was improved with a flow straightener so that the approach flow was free of surface waves, turbulence, and flow concentrations. The tailwater level was adjusted with a flap gate. The horizontal velocity component was determined to ±5%.

Figure 2.34 shows the test setup for both free and submerged overflow conditions, with h as flow depth, α_o as upstream (subscript o), and α_u as downstream (subscript u) weir face angles, x as streamwise coordinate measured from the weir crest, h_u as downstream depth, and h_t as tailwater (subscript t) depth measured positively up from the weir crest. Circular-crested weir models of radii $R_w = 0.15$ m and 0.30 m and upstream and downstream weir face angles of 20°, 30°, 45°, and 90° were tested. The approach overflow depths h_o ranged from 0.05 to 0.20 m.

As to the test procedure, the depth h_o was set and the hydraulic jump positioned at the downstream weir toe (Figure 2.34a). The axial free surface profile $h(x)$ was then measured. The point gage was placed 1 mm above the approach flow water surface; the downstream flow depth was then increased until the aforementioned elevation was reached to determine the modular limit. The downstream water level was then further increased until transition from plunging jet to surface wave flows occurred to determine the transition ranges. Both the upstream and downstream water levels were again measured. The discharge was determined for several submergence ratios to account for the discharge reduction due to tailwater increase.

Figure 2.35a shows typical free surface profiles. The upstream flow is subcritical. Transitional flow occurs on the weir crest whereas the flow along the downstream slope is supercritical. The critical (subscript c) flow depth $h_c = [Q^2/(gb^2)]^{1/3}$ is located nearly at the crest section. The roller (subscript r) length L_r is roughly equal to that of the classical hydraulic

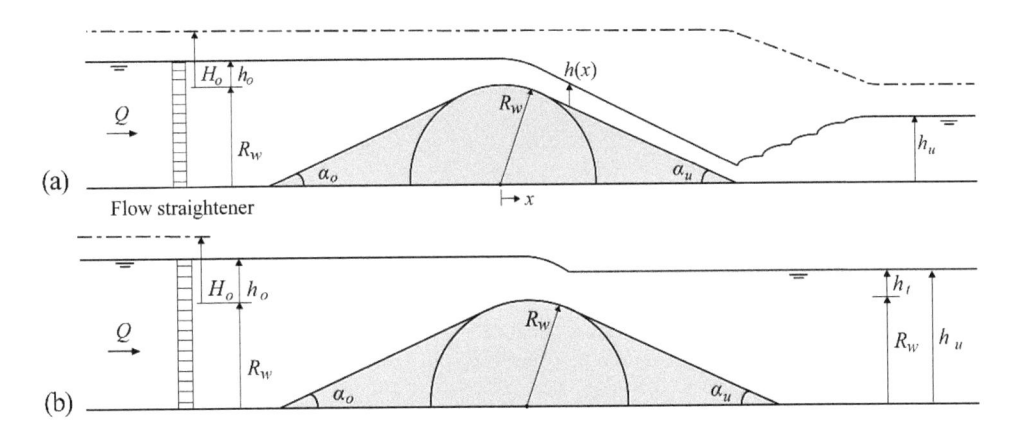

Figure 2.34 Circular-crested weir with sloping weir faces, definition of flow geometry for (a) free, (b) submerged overflow (Schmocker et al., 2011)

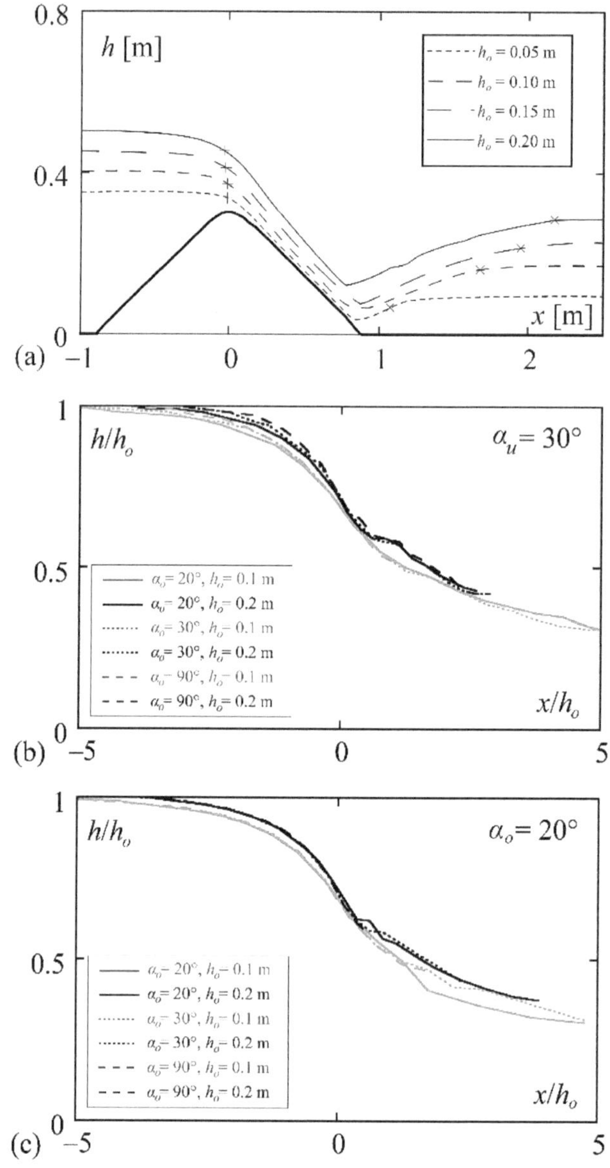

Figure 2.35 Free surface profiles (a) $h(x)$ with (+) critical flow depth, (×) end of roller; normalized free surface profiles $h/h_o[x/h_o]$ for (b) $\alpha_u = 30°$, $R_w = 0.30$ m and various values of α_o, (c) $\alpha_o = 20°$, $R_w = 0.30$ m and various values of α_u for (−−) $h_o = 0.10$ m and (▬) $h_o = 0.20$ m (Schmocker *et al.*, 2011)

jump (Hager, 1992). Figure 2.35b compares the dimensionless free surface profiles $h/h_o[x/h_o]$ for various values of α_o for $\alpha_u = 30°$ and $R_w = 0.30$ m. The surface profiles almost collapse for two approach flow depths and all tested upstream slopes, so that the upstream slope has no effect on the free surface profile. Figure 2.35c compares the dimensionless free surface profiles $h/h_o[x/h_o]$ for various values of α_u for $\alpha_o = 20°$ and $R_w = 0.30$ m. Given the identical upstream weir angle, the surface profiles again almost collapse along this reach. Differences occur for $x/h_o > 1$ as the effect of α_o on the free surface profile is directly linked to the downstream weir slope.

For each overflow depth, the discharge coefficient was determined with Eq. (2.4). The function $C_d(\rho_k)$ along with Eq. (2.16) is shown for $\alpha_u = 90°$ and $\alpha_o = 20°, 30°, 45°$, and $90°$ in Figure 2.36a, whereas Figure 2.36b relates to $\alpha_o = 90°$ and $\alpha_u = 20°, 30°, 45°$, and $90°$. The measured data are lower than from Eq. (2.16) because of the absence of the weir slope effect. Note that C_d of the circular-crested weir with upstream and downstream slopes is lower than of the standard circular-crested weir. The data indicate that for fixed values of ρ_k and α_u, C_d hardly changes with α_o (Figure 2.36a), in agreement with Ramamurthy and Vo (1993a). The

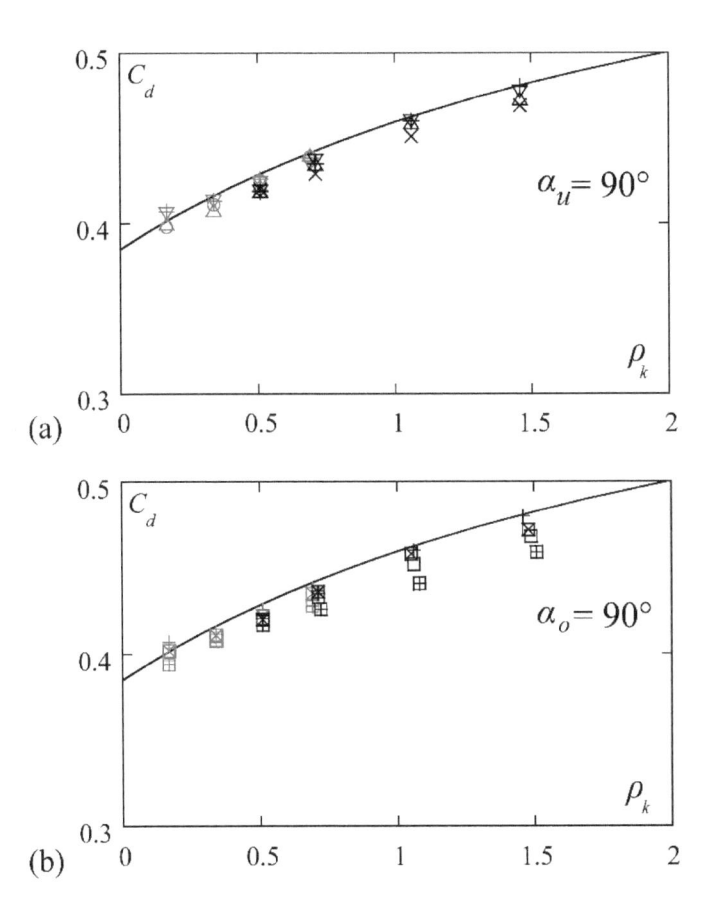

Figure 2.36 Measured discharge coefficient $C_d(\rho_k)$ for (a) $\alpha_u = 90°$ and $\alpha_o = 20°, 30°, 45°, 90°$; (b) $\alpha_o = 90°$ and $\alpha_u = 20°, 30°, 45°, 90°$ with (–) Eq. (2.16) (Schmocker *et al.* 2011)

data for $\alpha_o = \alpha_u = 90°$ nearly collapse with Eq. (2.16); the data for $\alpha_u = 20°$ and $\alpha_u = 30°$ under variable α_o indicate the same trend.

The data for fixed ρ_k and α_o under variable α_u indicate that C_d increases with α_u (Figure 2.36b). This effect increases with ρ_k. The C_d coefficient is smallest for $\alpha_u = 20°$. A steeper downstream face angle for a fixed overflow depth improves the performance of discharge. Compared with the upstream angle, the effect of the downstream angle on the discharge is larger. The data for $\alpha_o = \alpha_u = 90°$ nearly collapse with Eq. (2.16).

To account for the weir face angles, the C_d values were normalized with H_o/R_w as the relative weir crest curvature and $[(\alpha_o + 2\alpha_u)/270]^{1/3}$ as the weir angle ratio as

$$\rho_k' = \frac{H_o}{R_w}\left(\frac{\alpha_o + 2\alpha_u}{270}\right)^{\frac{1}{3}}.$$

$$(2.17)$$

Given the dominant effect of the downstream angle, it is multiplied by two, resulting thereby in the best data fit. For the standard circular-crested weir $\alpha_o = \alpha_u = 90°$, this ratio equals 1, so that $\rho_k' = \rho_k = H_o/R_w$. The discharge coefficient is therefore expressed according to Hager (1994) with ρ_k' instead of ρ_k and $\Omega = 4.5$ as

$$C_d = \frac{2}{3\sqrt{3}}\left(1 + \frac{3\rho_k'}{11 + \Omega\rho_k'}\right).$$

$$(2.18)$$

Figure 2.37 compares $C_d(\rho_k')$ for all test data with Eq. (2.18). The weir angle ratio increases the data fit strongly as compared with Figure 2.36. Equation (2.18) therefore determines the discharge coefficient for circular-crested weirs of arbitrary upstream and downstream weir face angles if $0.1 \le \rho_k' \le 1.46$.

Ramamurthy and Vo (1993a) investigated the discharge coefficient for $\alpha_o = 90°$, $75°$, and $60°$, $\alpha_u = 75°$, $60°$, and $45°$, and $0 \le H/R_w \le 25$. A maximum discharge coefficient was stated for $H/R_w = 5.5$ followed by a reduction if $H/R_w > 5.5$. Figure 2.38 compares their data $C_d(\rho_k')$ with Eq. (2.18) for $0 \le \rho_k' \le 6$. For $\rho_k' < 1$, Eq. (2.18) overestimates C_d, whereas Eq. (2.18) underestimates C_d otherwise. Compared with Ramamurthy and Vo (1993a), the dimensionless weir

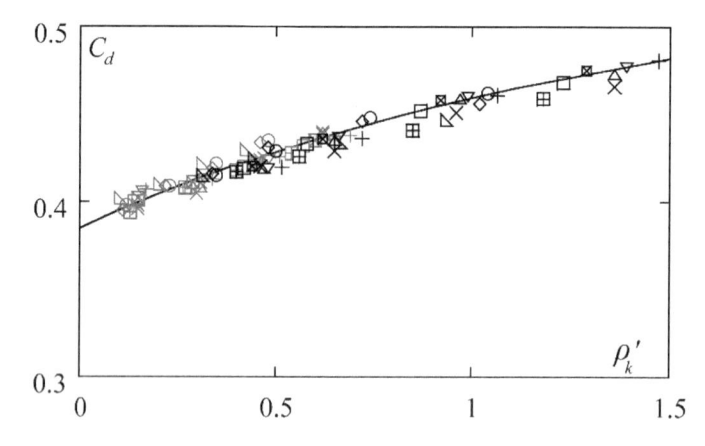

Figure 2.37 Comparison of measured discharge coefficients $C_d(\rho_k')$ with (−) Eq. (2.18) (Schmocker *et al.*, 2011)

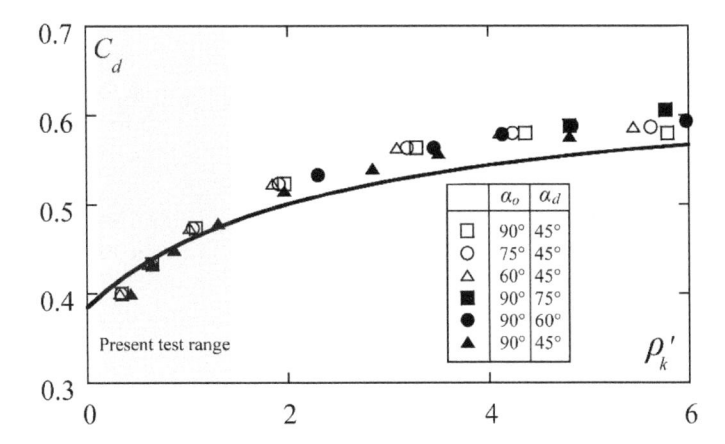

Figure 2.38 Discharge coefficients $C_d(\rho_k')$ of Ramamurthy and Vo (1993a) and from (−) Eq. (2.18)

height w/H is comparatively low in the present study. The weir height may therefore have an effect on C_d. Ramamurthy and Vo (1993a) used weir model radii from 0.0095 m up to 0.1516 m along with $w/H \geq 3$. Their data indicate that C_d for $\rho_k' \to 0$ is below the theoretical value of $2/(3 \cdot 3^{0.5}) \approx 0.385$, pointing at scale effects due to extremely small overflow depths. Schmocker *et al.* (2011) noted scale effects for $h_o < 0.05$ m due to a reduction of the discharge coefficient.

Modular limit and submergence effect

The modular limit (subscript L) separates free from submerged overflows. The degree of submergence is described by $y_t = h_t/h_o$. For free flow, the discharge is independent of y_t. Submerged laboratory overflow is characterized by an increase of +1 mm overflow depth h_o due to the tailwater elevation. Figure 2.39a shows the modular limit ratio $y_L = h_{tL}/h_o$ versus ρ_k', expressed as

$$y_L = 0.57 + 0.12\rho_k'. \tag{2.19}$$

The modular limit thus increases linearly with ρ_k'. The data scatter is considerable, without definite effects of H_o/R_w, α_o and α_u. For $\rho_k' < 0.5$, small scale effects contribute to the data scatter, despite all overflow energy heads were $H_o > 0.05$ m. The modular limit ratio is therefore more or less independent of both weir face angles. Circular-crested weirs are less prone to submergence than sharp-crested weirs. Note that free flow is generated for circular-crested weirs under 50% up to 80% of submergence, similar to trapezoidal-crested weirs (Fritz and Hager, 1998).

Increasing the tailwater level results in the transition (subscript T) from plunging jet to surface wave flow at a certain tailwater depth h_{tT}. Reducing then the tailwater level, the transition from surface wave to plunging jet flow occurs at another tailwater depth. Figure 2.39b shows the submergence ratio $y_T = h_{tT}/h_o$ versus ρ_k' under increasing tailwater level. The best-fit equation is

$$y_T = 0.97 + 0.039 \cdot \ln\left(\rho_k'\right). \tag{2.20}$$

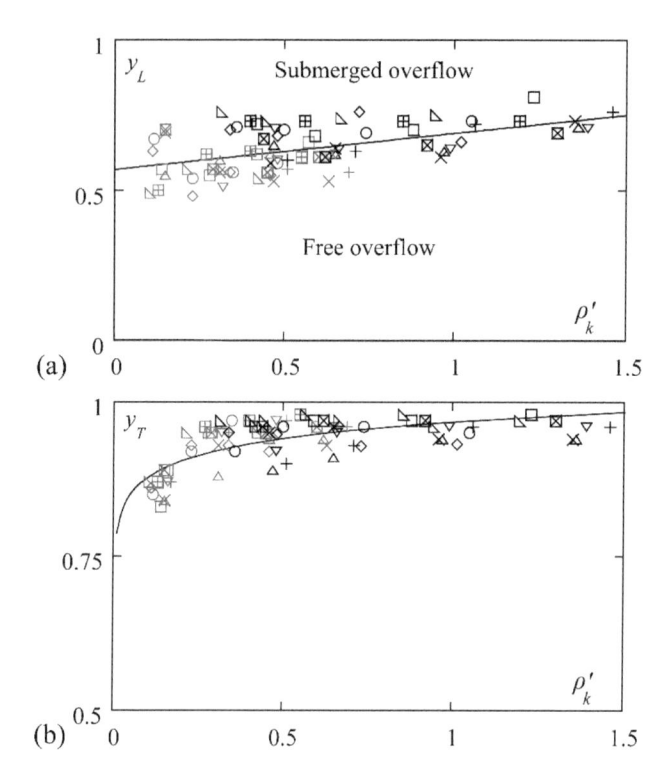

Figure 2.39 (a) Modular limit $y_L(\rho_k')$ with (−) Eq. (2.19), (b) transition submergence $y_T(\rho_k')$ with (−) Eq. (2.20) (Schmocker *et al.*, 2011)

A high submergence ratio is therefore reached as compared with trapezoidal-crested weirs, with y_T increasing with ρ_k'. The transition submergence has a lower value of $y_T = 0.84$ for $\rho_k' = 0.1$, increases to $y_T = 0.95$ for $\rho_k' = 0.45$, and then remains constant for $\rho_k' \geq 0.75$. Plunging flow results for $y_T < 0.84$, whereas surface wave flow occurs if $y_T > 0.95$. Between these, both flow types may occur. No effect of weir face slopes on y_T was found.

The discharge Q_s of submerged (subscript s) weir flow is smaller than the discharge Q of free overflow. The discharge reduction factor $\psi_s = Q_s/Q$ depends only on the submergence ratio $y_t = h_t/h_o$, as all other parameters are contained in Q. The factor ψ_s varies between $\psi_s = 0$ for $y_t(h_t = h_o) = 1$ and $\psi_s = 1$ for $y_t(h_t = h_{tL}) = y_L$. To generalize results, the relative submergence $Y_t = (y_t - y_L)/(1 - y_L)$ with $0 < Y_t < 1$ according to Fritz and Hager (1998) was used. Figure 2.40 indicates that $\psi_s(Y_t)$ follows

$$\psi_s = \left(1 - Y_t^3\right)^{1/6}. \tag{2.21}$$

No decisive effects of H_o/R_w, α_o, or α_u are visible because all data collapse more or less on a single curve. Note that the discharge reduction for a certain value of Y_t is slightly smaller for flat than for steep downstream weir face angles. The main results of the effect of weir face angles on the discharge features are thus available. Additional information details the effects of tailwater submergence and the modular limit.

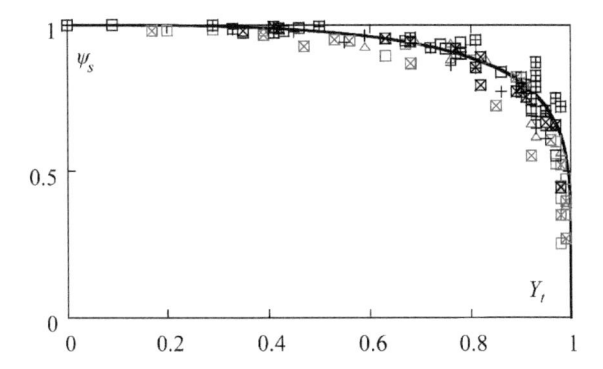

Figure 2.40 Discharge reduction factor ψ_s versus relative submergence Y_t (Schmocker *et al.* 2011)

2.3.2 Embankment weir

Introduction

Weirs of trapezoidal shape constitute a standard engineering structure applied for dikes, embankments, or cofferdams (Chapter 7). From geotechnical concerns, the embankment slope is normally 1V:2H (±10%) to satisfy bank stability and seepage control. The study of Fritz and Hager (1998) investigates embankments of variable crest length in the flow direction. The limit crest length of zero (triangular section) is interesting from the hydraulic point of view, whereas the maximum crest length studied was equal to the dam height. For free embankment overflow, the coefficient of discharge determines the overflow capacity. Sloping embankments have a higher discharge capacity than the standard broad-crested weir with vertical weir faces.

Submerged embankments as found in flood plains may undergo various regimes, including plunging jet flow with a surface roller, and both the surface wave and the surface jet flows, both with a bottom recirculation. The type of embankment overflow cannot yet be predicted despite its effects on the tailwater protection, embankment erosion, and storage considerations. The complete velocity field for 2D flow was determined, including forward and backward flow portions and velocity distributions. These results exhibit surprises regarding the flow features of embankment dams.

Bazin (1898) initiated the study of embankment weirs. Among broad-crested weirs, experiments included trapezoidal weirs with both upstream and downstream dam sides sloping <1:1 (45°). Embankment profiles with 1V:2H and crest lengths of $L_w = 0, 0.10$, and 0.20 m, respectively, were also considered. The discharge coefficient was determined for free overflow and the effect of submergence was investigated.

Horton (1907) summarized the then available data. In contrast to Bazin, no details on the flow structures were described. Yarnell and Nagler (1930) conducted a study on the flow over railway and highway embankments. Their laboratory channel was 58 m long, 3 m wide, and 3 m deep. The embankment was 1.2 m high and the crest had a length of 4 m. The study mainly aimed at the definition of submergence curves.

Govinda Rao and Muralidhar (1961) reanalyzed the Bazin data and presented an equation for the discharge coefficient by including the effects of relative crest length, upstream and downstream weir face slopes, and relative embankment height. Kikkawa *et al.* (1961) presented velocity and pressure distributions on inclined sharp- and broad-crested weirs. The effect of low overflow depths on the free surface configuration was studied by Bretschneider (1961) for embankments sloping 1:1, 1:1.5, and 1:2. He concentrated on the undular surface regime resulting in flows with a small overflow depth relative to the crest length. A comparable study was conducted by Kindsvater (1964), including observations of the crest velocity distribution and the boundary layer development along the crest. A summary of these results is provided in Hulsing (1968).

Ranga Raju *et al.* (1990) studied the effects of viscosity and surface tension on the flow over embankment dams. The Reynolds and the Weber numbers were identified as the governing parameters. Overflow depths larger than 50 mm do not exhibit these effects, so they were excluded by Fritz and Hager (1998). The flow features downstream of any submerged weir have received scarce attention. In the following, studies relating to discharge characteristics are excluded, given the standard works of Lakshmana Rao (1975), Bos (1976), or Miller (1994). An early classification of flow types is due to Escande (1939). Depending on the tailwater depth, the four flow types in the order of raising tailwater level are as follows (Figure 2.41):

- Type A, *hydraulic jump*, with toe located at or downstream of dam structure;
- Type B, *plunging flow*, with main flow along tailwater bottom and a surface roller;
- Type C, *surface wave flow*, with main flow along free surface, and a bottom recirculation zone;
- Type D, *surface jet flow*, analogous to Type C, but with a nearly horizontal flow surface.

The first hydrodynamic contribution to Type D flow was due to Rajaratnam and Muralidhar (1969) for the vertical thin-plate weir. The tailwater depth relative to the weir crest was always larger than 93% of the overflow depth. The velocity field in the surface jet was described as a fully developed flow region of which the potential core corresponds to a half-turbulent free jet. The length of the bottom recirculation zone is eight times the weir height, independent of the Reynolds number and the normalized overflow height. Pressure distributions on a 1:1 downstream sloping submerged embankment have been determined by Pinto and Ota (1980).

Leutheusser and Birk (1991) stated that 'little is known about the currents that may be produced in the downstream pools of low overflow structures. Under unfavorable circumstances, these secondary flows . . . can transform a seemingly innocuous low dam into a veritable drowning machine'. Further, 'Even competent swimmers have difficulties in trying to escape from the downstream pool of a low dam harboring a submerged hydraulic jump'. Several accidents are reported.

Hager *et al.* (1994) described accidents in Switzerland, along with a discussion of the main flow features. Wu and Rajaratnam (1996) conducted the first experimental work on submerged flow patterns downstream of sharp-crested vertical weirs. Four flow types were distinguished: (1) impinging jet, also referred to as plunging jet; (2) breaking wave; (3) surface wave; and (4) surface jet. The transitions between these are governed by hysteresis effects. A discharge equation was developed based on the ratio of discharges under submerged and free flows versus the submergence rate.

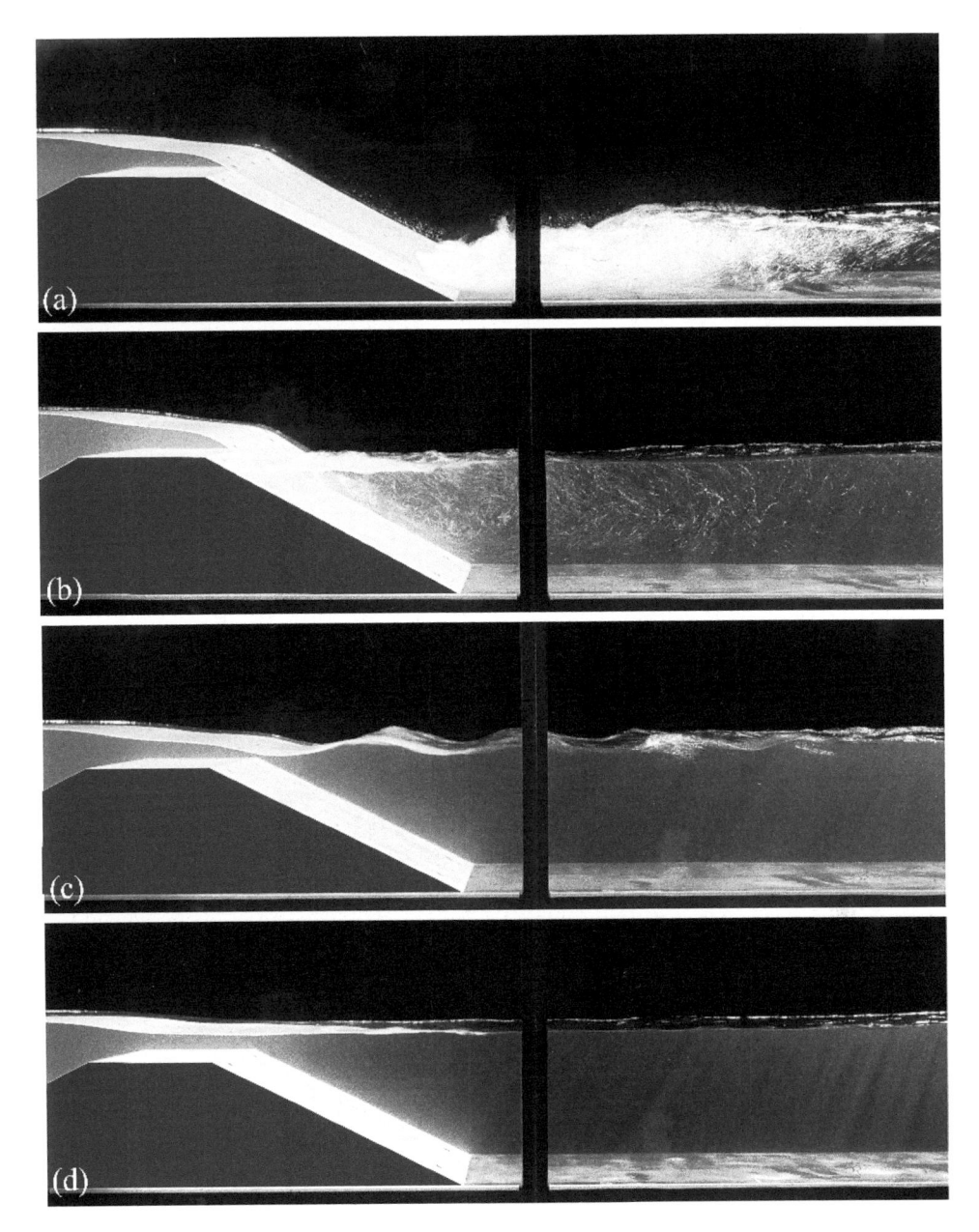

Figure 2.41 Main flow patterns at embankment weirs: (a) A-jump (Type A), (b) plunging jet (Type B), (c) surface wave flow (Type C), (d) surface jet flow (Type D) with flow parameters as in Figure 2.47 (Fritz and Hager, 1998)

Free overflow

Experiments were conducted using a rectangular horizontal channel 499 mm wide and 700 mm high. The symmetrical weir was 300 mm high and 1.5 m long with crest lengths of $L_w = 0$, 50, and 300 mm. Weir face slopes were 1V:2H both for the upstream and downstream faces. The crest was horizontal and had sharp-crested corners. Figure 2.42 shows the free overflow, with h_o as overflow depth measured from the crest, $H_o = h_o + Q^2[2gb^2(h_o + w)^2]$ as the total overflow head, w as embankment height, h_1 as flow depth at the embankment toe, x' as the streamwise coordinate measured from the dam toe, L_r' as the roller length, z_r as height of the forward flow, and h_d as the downstream flow depth.

Figure 2.43 shows surface profiles across embankments for $L_w = 0$ and 300 mm. For $L_w = 0$ (Figure 2.43a), the flow separates at the crest and accelerates down the dam face. Hydraulic

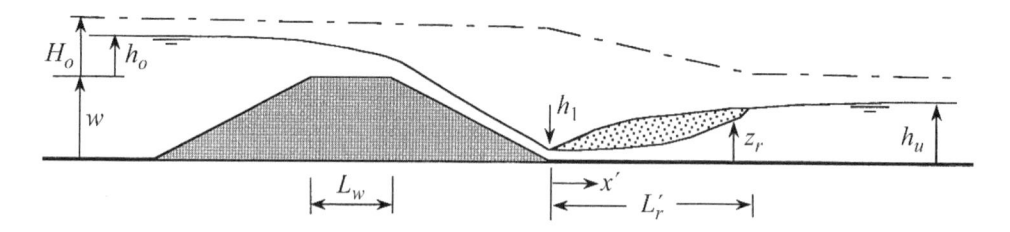

Figure 2.42 Definition sketch of free flow over embankment dam (Fritz and Hager, 1998)

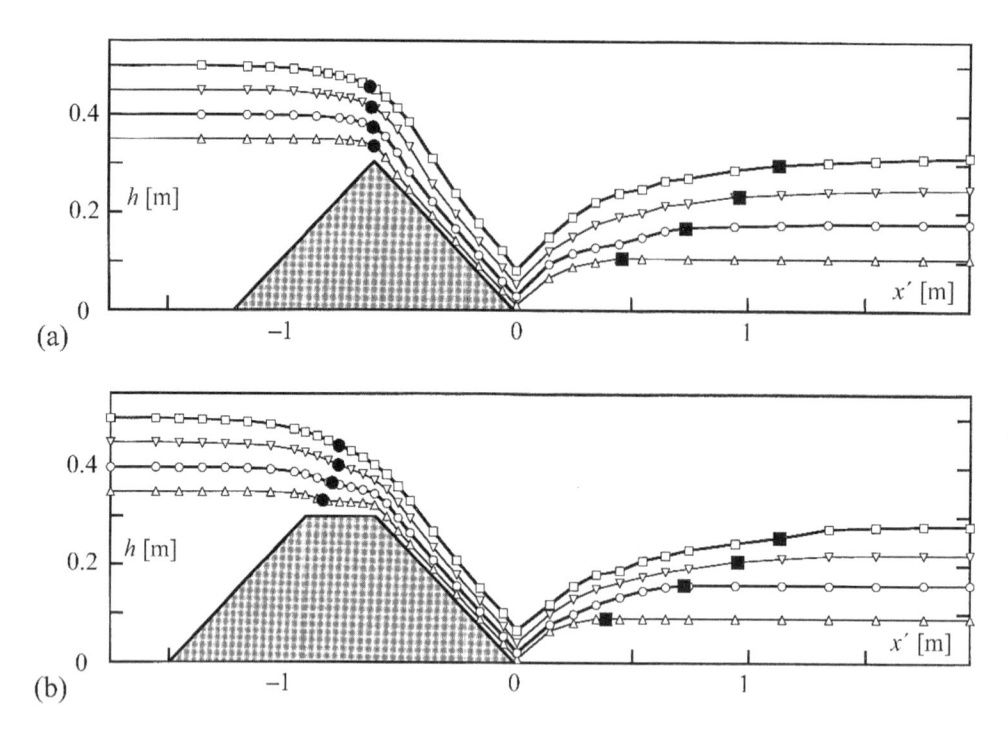

(a)

(b)

Figure 2.43 Free-surface profiles $h(x')$ for L_w = (a) 0, (b) 300 mm for h_o [mm] = ($\triangle$) 50, ($\circ$) 100, ($\triangledown$) 150, ($\square$) 200; ($\bullet$) critical flow depth h_c, ($\blacksquare$) end of roller (Fritz and Hager, 1998)

jumps were located with the toe at the embankment foot (A-jump). The critical depth $h_c =$ $[Q^2/(gb^2)]^{1/3}$ is located at the crest section. For $L_w = 300$ mm (Figure 2.43b), the critical section is slightly downstream of the upper crest section. The lengths of roller L_r' agree with these of the classical hydraulic jump (Hager, 1992). It was concluded that the latter and the A-jump had essentially identical flow features.

The discharge was related to the overflow energy head H_o with C_d as discharge coefficient as

$$Q = C_d b \left(2 g H_o^3 \right)^{1/2}. \tag{2.22}$$

Compared with conventional formulations involving the overflow depth h_o, Eq. (2.22) includes the effect of the approach flow velocity, at least up to $H_o/w < 1/6$. Figure 2.44 shows C_d versus the relative crest length $\xi = H_o/(H_o + L_w)$, with $0 < \xi < 1$. All data from Bazin (1898), Kindsvater (1964), and Fritz and Hager (1998) collapse on the curve

$$C_d = 0.43 + 0.06\sin[\pi(\xi - 0.55)]. \tag{2.23}$$

Surface undulations typical for 'long broad-crested weirs' with $\xi < 0.10$ (Hager and Schwalt, 1994) occur also in the present configuration. This domain was not tested further because the amplitudes are much smaller and the discharge capacity remains comparably small. For $0.1 < \xi < 0.3$, weir flows are referred to as 'broad-crested', for $0.3 < \xi < 0.6$ reference is made to 'short-crested weirs', and $\xi = 1$ corresponds to 'sharp-crested weirs'. For long broad-crested weirs, $C_d = 0.37$ remains almost constant; for short-crested weirs, C_d linearly increases with ξ reaching the value $C_d = 0.485$ for $\xi = 1$. Figure 2.44 also includes the curve $C_d(\xi)$ for the standard broad-crested weir with vertical weir faces (Hager and Schwalt, 1994), indicating differences of some 10% due to flow separation at the leading weir crest face.

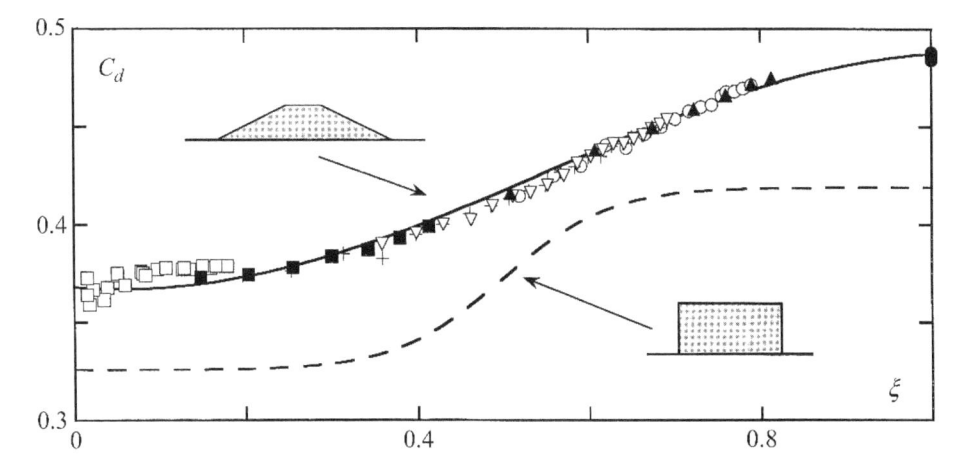

Figure 2.44 Discharge coefficient $C_d(\xi)$ according to Bazin (1898), series (o) 170, (+) 178 with zero correction by −3 mm, (▽) 179; (□) Kindsvater (1964) for L_w [mm] = (•) 0, (▲) 50, and (■) 300. (—) Eq. (2.23), and (-·- -) Hager and Schwalt (1994) for standard broad-crested weir

Submerged overflow

Figure 2.45 defines the parameters of submerged overflow. In addition to Figure 2.42, subscript R refers to the recirculation zone and h_t is the tailwater depth measured positively up from the weir crest. The coordinate origin of x is located at the downstream embankment crest. The values of L_r and L_R are the associated lengths of surface roller and bottom recirculation.

Figure 2.46 refers to typical observations for flow with $\xi = 0.25$ and various degrees of submergence $y_t = h_t/h_o$. For free overflow (Figure 2.46a), an A-jump is generated at the embankment toe. For $y_t = 1\%$, plunging jet flow is established with a concentration of forward flow along the bottom and a surface roller up to $x = 1.9$ m (Figure 2.46b). Increasing the tailwater to $y_t = 56\%$ results in surface wave flow (Figure 2.46c), with marked standing waves downstream of the embankment crest. The forward flow is now along the surface, and recirculation up to $x = 2.3$ m along the bottom and the lower embankment face. For submergence rates up to $y_t = 75\%$, the approach flow is not submerged. Figure 2.46d refers to $y_t = 98\%$ with an almost horizontal free surface, i.e. surface jet regime. The forward flow is again along the surface whereas the bottom recirculation extends up to $x = 1.5$ m. Both the roller and bottom recirculation regions start under a specific slope, from the lower crest, and tend to plateau. The mixing layer between the forward and backward flow regions decreases as y_t increases.

An illustration of the four regimes of embankment overflow is provided in Figure 2.41, in the order of increasing tailwater level. These images represent the data shown in Figure 2.46. For given discharge, the air entrainment increases as y_t decreases. For plunging jets and A-jumps, the air entrainment is concentrated at the leading edge of the mixing layer. For surface jets, the mixing layer is submerged by the main flow zone, and air cannot be entrained except for local zones due to wave breaking.

The modular limit (subscript L) separates free and submerged overflows. The latter is characterized in the laboratory setup by an increase of +1 mm overflow depth when increasing

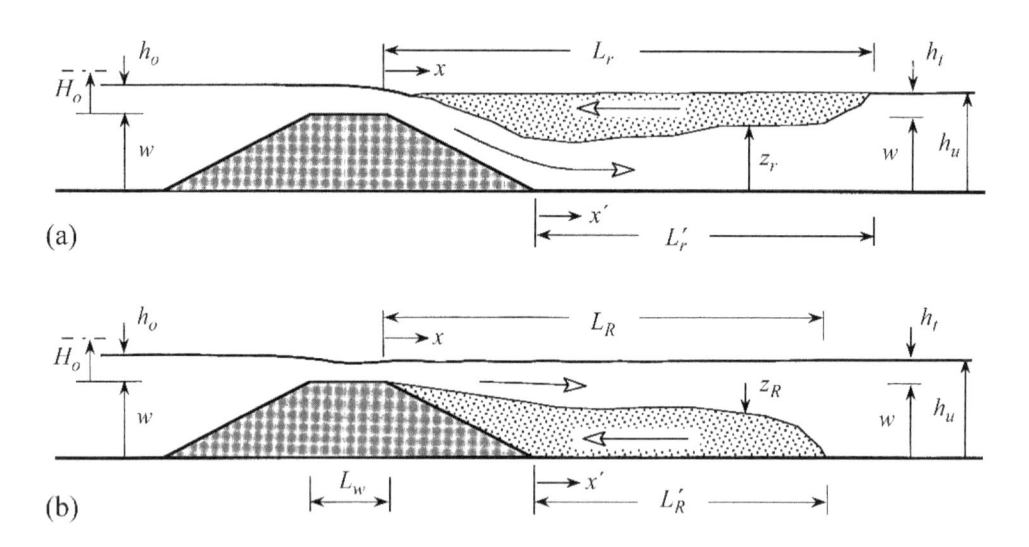

(a)

(b)

Figure 2.45 Submerged embankment flow, notation: (a) plunging jet with surface roller, (b) surface jet with bottom recirculation (Fritz and Hager, 1998)

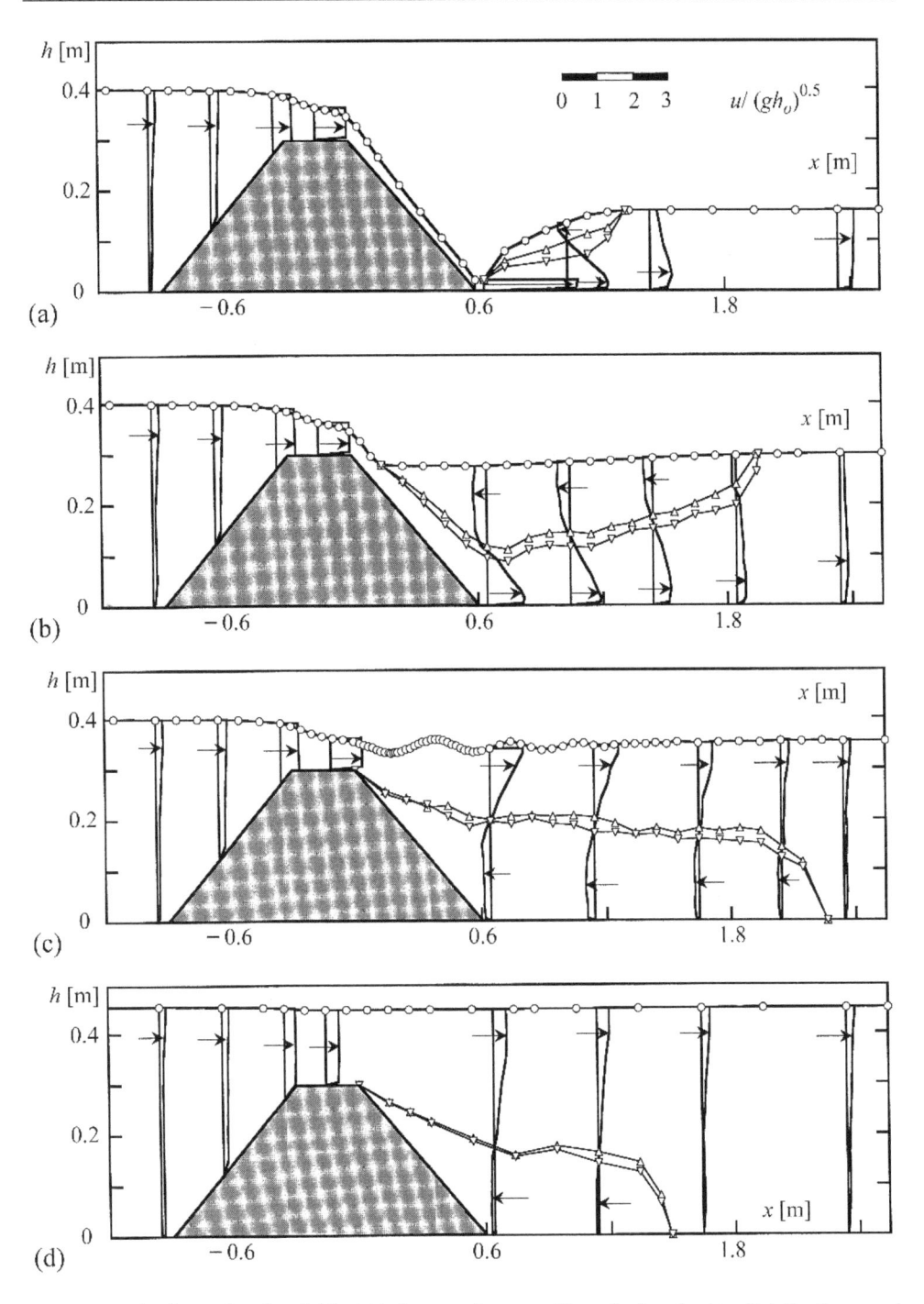

Figure 2.46 Flows for $\xi = 0.25$, (O) free-surface profiles, ($\rightarrow$) velocity distributions and mixing layers for $y_t =$ (a) A-jump, (b) 1%; (c) 56%, (d) 98%. Velocity scale is free-flow velocity $(gh_o)^{1/2}$. Mixing layer ($\triangledown$) lower and ($\triangle$) upper limits (Fritz and Hager, 1998)

the tailwater level. For free flow, the discharge is independent of y_t. Figure 2.47a shows the modular limit ratio $y_L = h_{tL}/h_o$ versus the relative crest length $\xi = H_o/(H_o + L_w)$. For $\xi \to 0$, i.e. the very long broad-crested weir, a maximum $y_L = 0.85$ is reached (Kindsvater, 1964). As the crest length decreases, i.e. ξ increases, the modular limit decreases almost linearly as (Figure 2.47a)

$$y_L = 0.85 - 0.50\xi. \tag{2.24}$$

A broad-crested weir is much less prone to submergence than a sharp-crested weir, so that free flow equations apply up to 80 or even 85% of submergence.

Increasing the tailwater level results at a certain level y_{t1} in the transition (subscript T) from the plunging jet to the surface jet regime. Decreasing the tailwater level, this transition is normally at another level y_{t2}. The prediction of the flow type is imperative for engineering purposes. Figure 2.47b shows a generalized diagram. With $y_T = h_T/h_o$ as transition submergence and for given ξ, the governing regime is determined, namely plunging flow, transition range, or surface wave. The transition range separates the diagram, starting at $0 < y_T < 0.60$ for $\xi = 0$, increasing to $0.65 < y_T < 0.78$ for $\xi = 1$. Clearly, flows are always plunging for $y_T < 0$ (i.e., tailwater level below weir crest), whereas surface waves occur as $y_T > 0.88$. Between these limits, either flow regime may occur. This switching mechanism adds to the complexity of submerged weir flow. Note that Figure 2.47b relates to the 1:2 sloping embankment, and the embankment slope has an additional effect. Figure 2.47b also contains the modular limit according to Eq. (2.24).

The discharge of a submerged (subscript s) embankment weir Q_s is smaller than the discharge Q over the same weir under free overflow. The discharge reduction factor $\psi_s = Q_s/Q$ depends exclusively on the submergence ratio y_t, because all other parameters are contained in Q. The factor ψ_s varies between 0 and 1, with $\psi_s = 0$ for $y_t = 1$ and $\psi_s = 1$ for $y_t = y_L$. The parameter $Y_t = (y_t - y_L)/(1 - y_L)$, with $0 < Y_t < 1$, allows for a data presentation $\psi_s(Y_t)$ as (Figure 2.48a)

$$\psi_s = (1 - Y_t)^{1/n}. \tag{2.25}$$

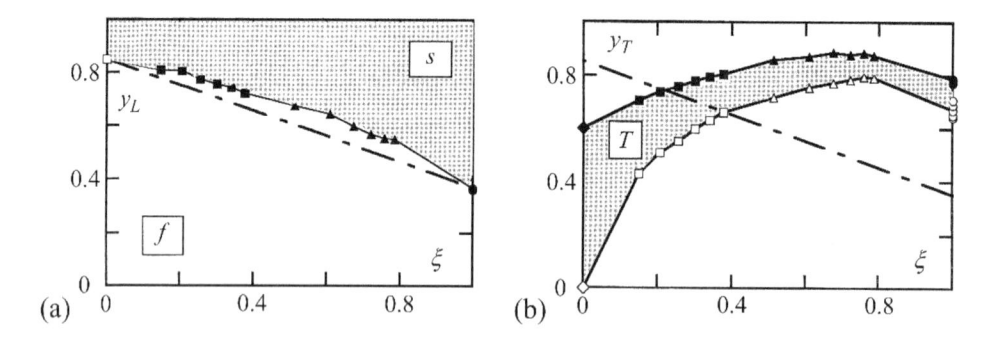

Figure 2.47   (a) Modular limit $y_L(\xi)$ for L_w [m] = ($\bullet$) 0, ($\blacktriangle$) 0.05, ($\blacksquare$) 0.3; (--) Eq. (2.24), (f) free and (s) submerged overflow, (b) (T) Transition range $y_T(\xi)$, ($\square$) Kindsvater (1964) (Fritz and Hager, 1998)

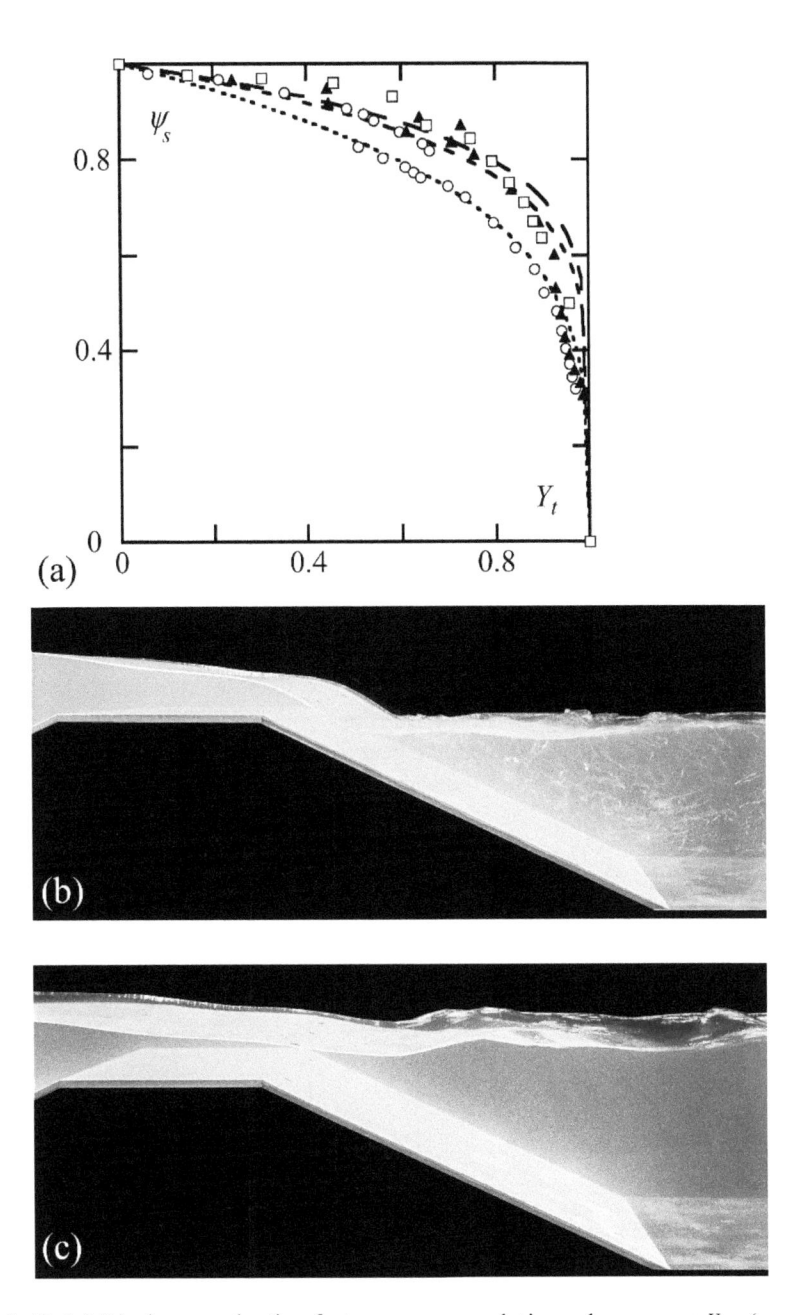

Figure 2.48 (a) Discharge reduction factor ψ_s versus relative submergence $Y_t = (y_t - y_L)/(1 - y_L)$ for $\xi =$ (□) 0.25 and $n = 7$, (▲) 0.67 and $n = 6$, (●) 1 and $n = 4$, (---) Eq. (2.25). Flow details for (b) plunging jet, (c) surface jet (Fritz and Hager, 1998)

The exponent n varies with the relative crest length $\xi = H_o/(H_o + L_w)$ as $n(\xi = 0.25) = 7$, $n(\xi = 0.67) = 6$, and $n(\xi = 1) = 4$. Accordingly, the discharge reduction for a certain value of Y_t is larger for sharp-crested than for broad-crested embankments. If an accuracy of $\pm10\%$ in ψ_s is sufficient, then $n = 6$. Figure 2.48(b–c) relates to details of plunging and surface flows, respectively.

Tailwater flow field

For a low tailwater level, the flow from the crest follows the embankment face and the tailwater bottom. Over the forward flow zone, a recirculation zone is established. With $Z_r = z/h_u$ as relative depth of the forward flow zone (Figure 2.49a), the development of $Z_r(X_r)$ depends on $X_r = x/L_r$, with L_r as the length of surface roller, measured from the lower embankment crest, as (Figure 2.49a)

$$Z_r = 1 - A_r X_r + AX_r^2. \tag{2.26}$$

The function $Z_r(X_r)$ thus starts and ends at $Z_r = 1$, is nearly symmetric about $X_r = 0.5$, with a minimum of $1 - (1/4)A_r$. The effect of A_r is relatively small with an average of $A_r = 2.4$. The relative roller length measured from the embankment toe to the surface stagnation point (Figure 2.49a) for all plunging jets, including the A-jump, is $L'_r/h_u = 4.3$ ($\pm10\%$), in agreement with the classical hydraulic jump.

Figure 2.49b relates to the surface regime (wave and jet flows) with a recirculation length L_R measured from the lower embankment crest. The shape of the recirculation zone has the following features. Just downstream of the crest with $z_R(x = 0) = w$, the zone decreases linearly with x to reach a portion of nearly constant depth. Further downstream, the recirculation zone decreases rapidly toward the channel bottom, with $z_R(x = L_R) = 0$. Accordingly,

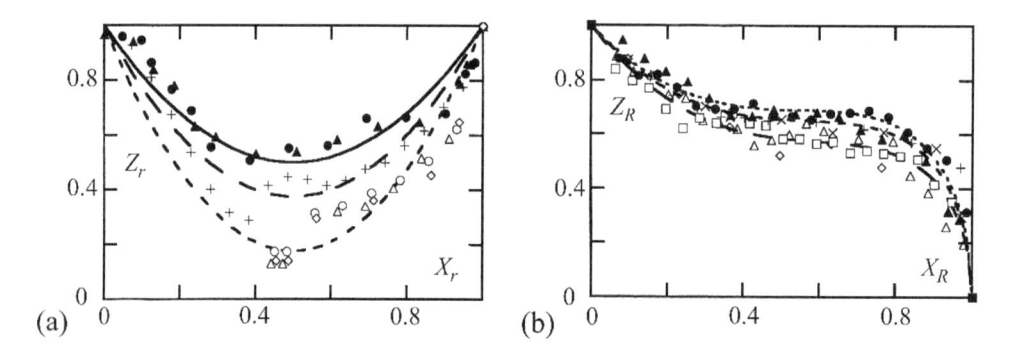

Figure 2.49 Interface between forward and backward flows for (a) plunging jet $Z_r(X_r)$ with $y_t = (\bullet,\blacktriangle)$ 76% and 84%, (+) 1%, (o, △, ◊) A-jump with L_w [m] = (0, 0.05, 0.3). A_r values in Eq. (2.26) are, respectively, $A_r = 2.3, 2.5, 3.3$; (b) Surface jet $Z_R(X_R)$ with $y_t = (\bullet,\blacktriangle)$ 70% and 94%, $B_R = 1$, $y_t = (\times, +)$ 84%, 96%, $B_R = 1.1$, $y_t = (\square, △, ◊)$ 56%, 82%, 98%, $B_R = 1.3$ (Fritz and Hager, 1998)

with $Z_R = z_R/w$ and $X_R = x_R/L_R$, the conditions are $Z_R(0) = 1$ and $Z_R(1) = 0$, respectively. All data follow the relation (Figure 2.49b)

$$Z_R = (1 - B_R X_R + 1.9 X_R^2)(1 - X_R)^{1/2}. \tag{2.27}$$

The first term gives a linear reduction for small X_R and the plateau value for $X_R = 0.5$, whereas the second term yields a rapid decrease of Z_R as $X_R \to 1$. The effect of y_t is small so that the average value is $B_R = 1.1$. The length L_R of the recirculation zone measured from the lower embankment crest varies essentially with y_t so that the effect of crest length remains small. The present data follow the fit (Fritz and Hager, 1998)

$$L_R / h_u = 6.8(1 - y_t)^{1/6}. \tag{2.28}$$

Because surface jets occur for $y_t > y_T$, typically $y_t > 0.6$ (Figure 2.47b). For $0.6 < y_t < 0.9$, Eq. (2.28) yields $L_R/h_u = 5.3$ ($\pm 10\%$), comparable with the length of the classical hydraulic jump.

For a known flow field in the embankment tailwater in terms of y_t, the velocity field is defined. Based on Rajaratnam (1965) and Hager (1992) for the classical hydraulic jump, a generalization of their results was attempted. Figure 2.50 defines the parameters and relates to the minimum (subscript m) and maximum (subscript M) velocities in the two flow regimes. The boundary layer thickness δ_0 is also included.

The maximum streamwise velocity u_M depends on the approach flow velocity u_A at the lower embankment crest, and the nominal tailwater velocity $u_u = Q/(bh_u)$. With $U_M = (u_M - u_u)/(u_A - u_u)$ as the velocity scale, the domain of U_M ranges between 1 at the lower crest and 0 in the tailwater. The distribution of U_M with the length scale $\chi_r = x/L_r$ for plunging jets, and $\chi_r = x'/L_r'$ for A-jumps is shown in Figure 2.51a. This plot indicates no effects of relative crest length ξ and tailwater submergence y_t; the data follow

$$U_M = \exp(-2.3\chi_r^2). \tag{2.29}$$

This is comparable to the expression of Hager (1992) for classical hydraulic jumps. For $\chi_r = 1$, i.e. at the roller end, the maximum velocity is 10% and the flow has almost reached the tailwater velocity.

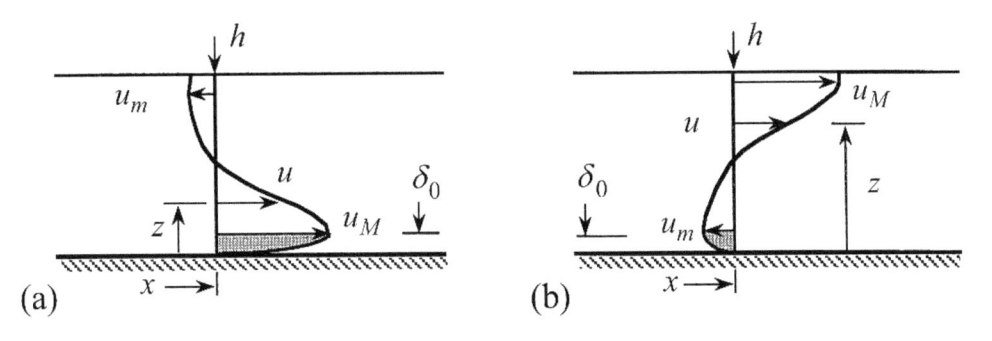

(a)   (b)

Figure 2.50 Notation for velocity distribution in (a) plunging jet regime, (b) surface jet regime

Figure 2.51b relates to the surface jet regime. The velocity decay also follows Eq. (2.29). The length definitions are here $\chi_R = x/L_R$ for $y_t < 85\%$, and $\chi_R = (1/2)x/L_R$ for $y_t > 95\%$ to include both surface waves and surface jets. The decay of their maximum forward velocities is thus governed by the same physical processes applying to the classical hydraulic jump, and nearly to the classical wall jet (Rajaratnam, 1965). The absolute (subscript A) maximum velocity u_A, or $U_A = u_A/(gh_u)^{1/2}$, as introduced in the definition of U_M, is plotted versus y_t in Figure 2.52. Again, both jets follow the relation

$$U_A = (1-y_t)^{1/2}. \tag{2.30}$$

The highest velocity of a plunging jet just downstream of the lower embankment crest is $(gh_u)^{1/2}$ for $y_t = 0$, independent of crest length and discharge.

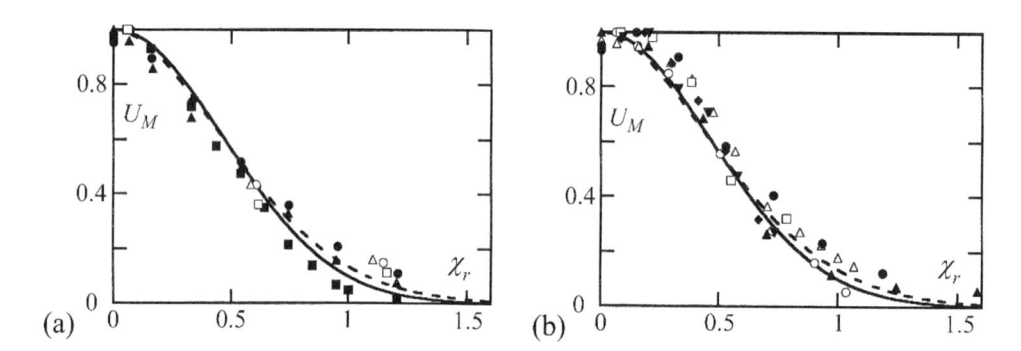

Figure 2.51 Decay of maximum velocity $U_M(\chi_r)$ for (a) plunging jet with (L_w [m]; y_t) = (•) (0; 76%), (▲) (0.05; 84%), (■) (0.3; 1%) and A-jumps with L_w [m] = (○) 0, (△) 0.05, (□) 0.30; (−) Eq. (2.29), (---) Hager (1992); (b) surface jet with (L_w [m]; y_t) = (•, ♦) (0; 70%, 94%), (▲, ▼) (0.05; 84%, 96%), (○, △, □) (0.30; 56%, 82%, 98%. (−) Eq. (2.29), (---) Hager (1992)

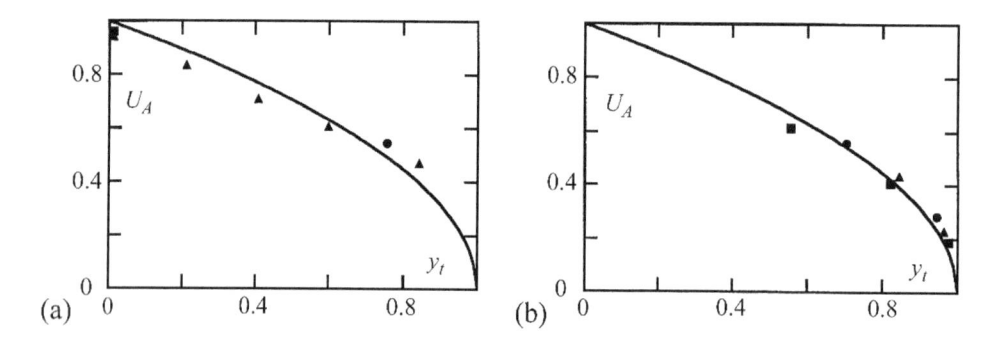

Figure 2.52 Absolute maximum velocity $U_A = u_A/(gh_u)^{1/2}$ versus y_t for (a) plunging jet regime, (b) surface jet regime for L_w [m] = (•) 0, (▲) 0.05, (■) 0.30; (−) Eq. (2.30) (Fritz and Hager, 1998)

The maximum backward (subscript m) velocity $U_m = u_m/u_a$ with u_a as the absolute maximum backward velocity for plunging jets varies with location $\chi_r = x/L_r$ for $y_t > 0$, and $\chi_r = x'/L_r'$ for A-jumps. The data follow the classical hydraulic jump as (Figure 2.53a)

$$U_m = \sin\chi_r. \tag{2.31}$$

Figure 2.53b relates to the surface jet regime with $\chi_R = x/L_R$, for which the data follow

$$U_m = 2[\chi_R(1-\chi_R)]^{1/2}. \tag{2.32}$$

The backward velocity for plunging jets is $U_m = 0$ at $\chi_R = 0$, attaining the maximum $U_m = 1$ at $\chi_R = 0.5$, decreasing to $U_m = 0$ at the roller end at $\chi_R = 1$, to finally increase to the tailwater velocity. For surface jets, the increase and decrease of the function $U_m(\chi_r)$ close to $\chi_r = 0$ and 1 is steeper because of the confined recirculation zone.

The absolute (subscript a) maximum of the backward velocity $U_a = u_a/(gh_u)^{1/2}$ for both the plunging and surface regimes is (Figure 2.54)

$$U_a = -0.25(1-y_t)^{1/2}. \tag{2.33}$$

The ratio between the absolute maximum forward and backward velocities according to Eq. (2.30) is $U_a/U_A = 0.25$. The maximum forward velocities for both regimes are thus four times larger than the corresponding backward velocities.

Figure 2.50 defines the notation, with x as location, z as distance from the bottom, $\delta_0(x)$ as boundary layer thickness, and $h(x)$ as the local flow depth. By introducing the relative elevation $Z_\delta = (z - \delta_0)/(h - \delta_0)$ between the maximum and minimum velocities u_M and u_m, respectively, and $U = (u - u_m)/(u_M - u_m)$ as normalized velocity in a plunging jet, all profiles follow (Figure 2.55a)

$$U = [\cos(100Z_\delta^{0.8})]^2. \tag{2.34}$$

The data of Hager (1992) follow essentially also this curve.

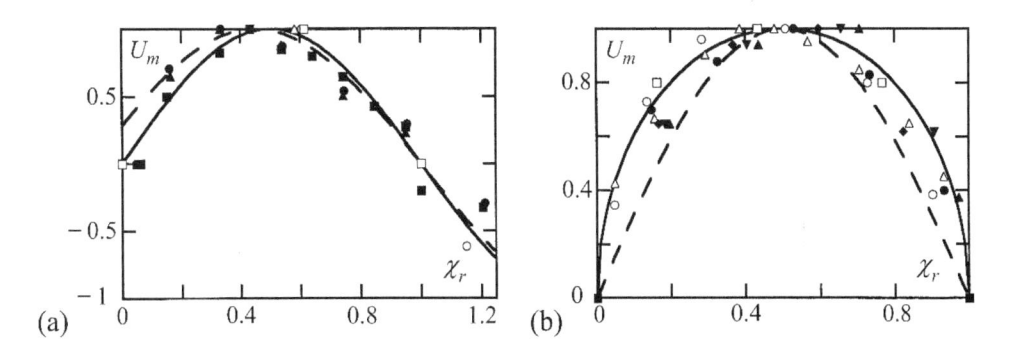

Figure 2.53 Maximum backward velocity $U_m(\chi_r)$ for (a) plunging jet regime with (–) Eq. (2.31) and (---) Hager (1992), (b) surface jet regime with (–) Eq. (2.32) and (---) Eq. (2.31). Notation as in Figure 2.51 (Fritz and Hager, 1998)

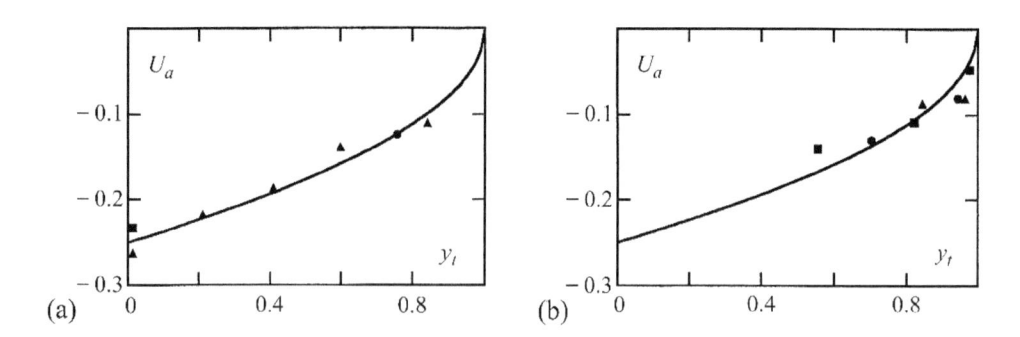

Figure 2.54 Absolute maximum backward velocity $U_a = u_a/(gh_u)^{1/2}$ versus y_t for (a) plunging jet regime, (b) surface jet regime, L_w [m] = (•) 0, (▲) 0.05, (■) 0.3, (−) Eq. (2.33) (Fritz and Hager, 1998)

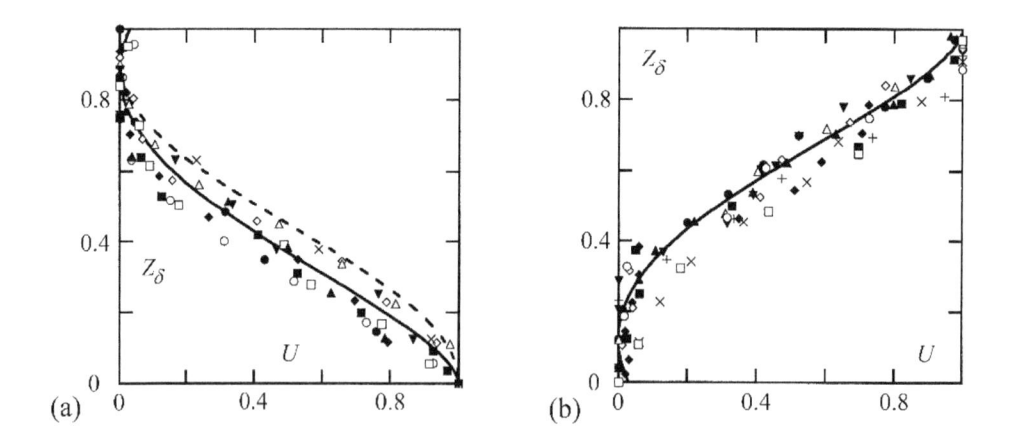

Figure 2.55 Velocity profile $U(Z_\delta)$ for $L_w = 0.3$ m (a) plunging jet regime with $y_t = 1\%$ at x [m] = (•) 3.95, (o) 4.3, (■) 4.5, (□) 4.7, (♦) 4.9, (◊) 5.1,(▲) 5.3, (△) 5.5, (▼) 5.6, and A-jump at (×) 4.7 m; (b) surface jet regime with $y_t = 56\%$ at x [m] = (□) 3.8, (o) 4.0, (△) 4.3, (◊) 4.8, (+) 5.3, (×) 5.7, and $y_t = 82\%$ at x [m] = (■) 4.0, (•) 4.3, (▲) 4.7, (♦) 5.2, (▼) 5.5, (−) Eq. (2.35)

By transforming $Z_\delta \rightarrow (1 - Z_\delta)$, the data of the surface jet regime are described by

$$U = \{\cos[100(1 - Z_\delta)^{0.8}]\}^2. \tag{2.35}$$

Therefore, these and all other results for different crest lengths are generalized. The velocity profiles of both the plunging and surface jet regimes are described with the same Eqs. (2.34) and (2.35) when inverting the sign of Z_δ. The effects of crest length, submergence ratio, and weir height are small. A flow in the surface regime is nothing else but an inverted plunging jet with identical physical characteristics. This study is one of the few dealing with submerged weir flow in which the discharge coefficient, the governing regimes, the modular limit, the submergence characteristics, the lengths of roller and recirculation, the velocity fields, and the scaling lengths and velocities are defined.

2.4 Scale effects

2.4.1 Real fluid effects in weir flow

Introduction

Scale effects are defined as features of normally laboratory-tested hydraulic models, which do not correctly upscale to prototype scale, because not all physical issues are correctly reproduced with only one scaling number. For free surface flows, the Froude similitude normally governs these flows, exclusively considering inertia and gravity effects. Additional effects stemming from fluid viscosity or surface tension are not included. These effects, i.e. the related neglected viscous and surface tension forces, can influence the flow behavior particularly under small weir flow depths. If identical fluids are employed both in the scale model as in the prototype, then these scale effects do not allow for correct model similitude with the prototype. In the following, scale effects will be considered based on the work of Castro-Orgaz and Hager (2014).

Scale effects distort the model-prototype relationship under certain conditions. Heller (2011) presented a comprehensive review of scale effects in hydraulic engineering. This section deals with scale effects in relation to the discharge characteristics of round-crested weirs, an important overflow structure of high dams (Cassidy, 1965; Naudascher, 1987; Montes, 1998), or as flow measuring device (Hager, 1993; Ramamurthy and Vo, 1993a, 1993b). Fuentes-Aguilar and Acuña (1971), and Ramamurthy et al. (1994) reported extremely thin boundary layers over round-crested weirs and excellent performance of the ideal fluid flow theory if the crest curvature radius R_c is large enough to avoid significant scale effects. The two major sources of scale effects originate from surface tension and viscosity (Sarginson, 1984; Naudascher, 1987). Whereas the head losses related to the boundary layer development are small in the discharge characteristics of prototype structures, laboratory models are of smaller scale so that these may result in major alterations of the head-discharge relation (Varshney, 1977; Isaacs, 1981; Naudascher, 1987). Scale effects in control structures were tackled in an International Association for Hydraulic Research (IAHR) symposium, and addressed in an IAHR monograph (Kolkman, 1984, 1994). However, neither of these two proposed generalized equations for scale effects. The test data of Lakshmana Rao and Jagannadha Rao (1973), or Varshney (1977) reveal that the boundary layer of round-crested weir flow is laminar at laboratory scale. Maxwell and Weggel (1969) and Sarginson (1972, 1984) observed the relevance of viscosity at low heads. Matthew (1963) found an analytical solution for the laminar boundary layer thickness profile, yet his solution was never verified. Isaacs (1981) conducted the only computation of laminar boundary layers in weir models. He numerically solved the integral laminar boundary layer equations using Thwaites' method. Yet, neither 2D nor 1D solutions of the laminar boundary layer development for round-crested weirs are available, despite their relevance for the discharge features, thereby avoiding a general approach for scale effects due to viscosity. A major contradiction from the current approach stems by using the laminar boundary layer equations of the flat plate (Naudascher, 1987, 1991). This method contradicts the basic weir flow features encompassing accelerating flow and favorable pressure gradients.

A general round-crested weir flow equation accounting for real fluid flow effects due to viscosity and surface tension is developed first, accounting for surface tension effects at high heads, given its dependence on streamline curvature. Second, the laminar boundary layer development of round-crested weir flow is solved by using 2D and integral 1D laminar boundary layer solutions. These allow for an approximate analytical model used as predictor of viscosity effects in the weir flow equation. The discharge prediction is verified with a set of experimental data. The theory proposed is also used to predict the discharge features of broad-crested weirs using a separation bubble. A flow analogy between the separation bubble and a round-crested weir is developed. The theory is further employed to evaluate the minimum overflow head necessary to avoid significant scale effects.

Effect of surface tension

Consider the flow of a partially developed real fluid over a weir of arbitrary bottom profile $z_b(x)$ (Figure 2.56). The free surface energy head is

$$H = z_b + h + \frac{p_s}{\gamma} + \frac{V_s^2}{2g} = \text{const.} \tag{2.36}$$

Here H is the total energy head, V_s the free surface velocity, p_s/γ the free surface pressure head with γ as specific fluid weight, and h the flow depth. The value of $p_s \neq 0$ due to surface tension, producing with the surface tension coefficient σ and the free surface radius of curvature R_s the boundary condition (Liggett, 1994)

$$p_s = \frac{\sigma}{R_s}. \tag{2.37}$$

To unveil surface tension, the boundary layer displacement thickness δ^* is first overlooked. The effects of curvilinear flow are accounted for by the coefficient $\chi_s = V_s/(q/h)$, with q as unit discharge (Wilkinson, 1974). The weir discharge coefficient C_d is defined with E as specific energy head at the weir crest as (Montes, 1998)

$$q = C_d (gE^3)^{1/2}. \tag{2.38}$$

From Eqs. (2.36) to (2.38) results with $C_o = \chi_s^{-1/2}$ in the common weir flow equation

$$C_d = \left(\frac{2}{3}\right)^{\frac{3}{2}} C_o C_\sigma. \tag{2.39}$$

With subscripts indicating ordinary differentiation with respect to x (Fig. 2.56), and $h/E = 2/3$,

$$C_\sigma = \left(1 + \frac{3\sigma}{\gamma E R_s}\right)^{1/2} = \left[1 + \frac{3\sigma}{\gamma E} \frac{h_{xx} + z_{xx}}{(1 + h_x^2)^{3/2}}\right]. \tag{2.40}$$

To apply it to weir flow, knowledge of both C_o and C_σ is required. The coefficient C_o accounts for curvilinear flow of ideal fluid under irrotational motion. It is determined numerically by solving the Laplace equation (Cassidy, 1965; Fuentes-Aguilar and Acuña, 1971; Montes, 1992) or by approximating curvilinear flow (Fawer, 1937; Jaeger, 1956; Matthew, 1963,

1991; Lenau, 1967; Montes, 1970; Hager, 1985; Montes, 1998). Matthew (1991) obtained from a third-order Boussinesq-type equation

$$C_o = 1 + 0.271 \frac{E}{R_w} - 0.045 \left(\frac{E}{R_w} \right)^2.$$ (2.41)

The term C_o is a surface tension correction coefficient. Its magnitude depends on the boundary radius of curvature, $z_{xx} = -1/R_w$, as well as on the crest derivatives of the flow depth, h_x and h_{xx}. An equation for C_o results from $h_x^2 = h/(3R_w), (1 + h_x^2)^{3/2} \approx 1$ and $h_{xx} = (4/9)/R_w$ (Matthew, 1963; Hager, 1985). However, a small weir under high head may have surface tension effects in which nonlinear contributions and higher-order terms in h_x and h_{xx} become relevant. Therefore, more accurate predictors of similar mathematical validity to Eq. (2.41) are (Matthew, 1991)

$$h_x^2 = \frac{2}{9} \frac{E}{R_w} \left(1 - \frac{236}{729} \frac{E}{R_w} \right), \quad h_{xx} = \frac{4}{9R_w} \left(1 + \frac{4783}{16038} \frac{E}{R_w} \right).$$ (2.42)

Combining Eqs. (2.41) and (2.42) with Eq. (2.40), the discharge characteristics of weir flow subjected to scale effects due to surface tension are modeled.

Effect of viscosity

The viscous effect of weir flow is accounted for by the boundary layer theory. Its theoretical inclusion is described, e.g. by Ackers *et al.* (1978). The weir discharge q is reduced as compared to potential flow by accounting for a correction coefficient C_v with δ_c^* as the boundary layer displacement thickness at the weir crest as

$$C_v = \left(1 - \frac{\delta_c^*}{E} \right).$$ (2.43)

The general weir flow equation accounting for scale effects is thus

$$C_d = \left(\frac{2}{3} \right)^{\frac{3}{2}} C_o C_\sigma C_v.$$ (2.44)

This is similar to that of the round-nosed broad-crested weir, for which $C_o = C_\sigma = 1$. To determine C_v requires for computing δ_c^* by the boundary layer theory (Craya and Delleur, 1952; Delleur, 1955; 1957, Hall, 1962; Harrison, 1967a, 1967b; Vierhout, 1973). For broad-crested weirs, the boundary layer is turbulent so that the initial laminar flow portion is neglected due to weak flow acceleration that allows for parallel-streamlined flow ($C_o = C_\sigma = 1$). For round-crested weir flow, this is different.

Consider accelerating flow over the weir shown in Figure 2.56. At the upstream weir portion, the free surface only slightly converges, so that the flow acceleration is weak and the bottom velocity U_e remains essentially constant. An increasing boundary layer thickness profile δ^* similar to the laminar flat plate is therefore expected. As the flow approaches the weir crest, it accelerates within a short distance from the upstream to the crest velocity, thereby increasing U_e, and inhibiting boundary layer development, producing a decreasing thickness of $\delta^*(x)$. Therefore, the laminar upstream boundary layer of round-crested weirs is likely to

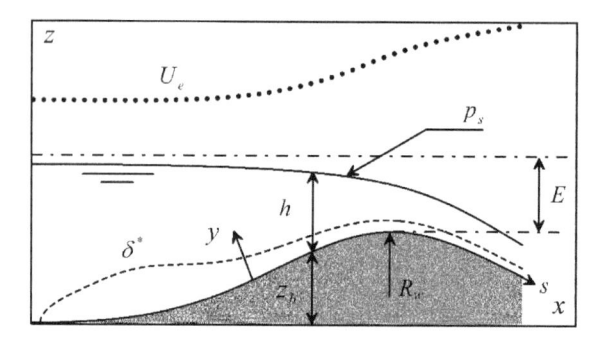

Figure 2.56 Definition sketch for real fluid flow over round-crested weir

remain laminar along the crest, as verified by Lakshmana Rao and Jagannadha Rao (1973). Matthew (1963) assumed laminar flow and produced an expression for $\delta^*(x)$, which was not verified with 1D or 2D laminar boundary layer flow computations, however. Whereas Matthew (1963) advocated the relevance of accelerating flow toward a round-crested weir, Naudascher (1987, 1991) proposed the laminar flat-plate equations to account for the boundary layer correction at a weir crest. Both proposals neglect the effect of flow acceleration, therefore.

The viscous effect in Froude models with curvilinear flow appears mainly at low weir heads (Maxwell and Weggel, 1969), whereas surface tension may also play a role at higher heads. Sarginson (1972) used an empirical equation to account for viscous effects at low heads. The relevance of viscosity in Froude models was experimentally verified by Varshney (1977), who found a dependence of C_d on the Reynolds number. Viscosity in weir flow models at low heads is thus relevant, with the boundary layer remaining laminar as the flow passes along the crest. However, no systematic study predicted $\delta^*(x)$, so that a study of the laminar boundary layer development of round-crested weirs is required.

2.4.2 Boundary layer development

2D Potential flow solution

The flow over a round-crested weir model is treated by using Prandtl's boundary layer approximation (Lemos, 1965). The flow is divided into an outer potential flow zone and a thin viscous layer attached to the solid wall (White, 1991). The potential flow solution provides the outer flow velocity $U_e(s)$, with s as curvilinear coordinate along the solid wall. For the potential flow computations, the x-ψ method (Montes, 1992, 1994) is used, with ψ as the stream function. The Laplace equation for the free surface elevation z is expressed with (ψ, x) as

$$\frac{\partial^2 z}{\partial x^2}\left(\frac{\partial z}{\partial \psi}\right)+\frac{\partial^2 z}{\partial \psi^2}\left[1+\left(\frac{\partial z}{\partial x}\right)^2\right]-2\frac{\partial^2 z}{\partial x \partial \psi}\frac{\partial z}{\partial x}\frac{\partial z}{\partial \psi}=0. \tag{2.45}$$

Its solution for flow over control structures is described by Montes (1992, 1994), Dey (2002), or Castro-Orgaz (2013). Note that other methods for computations of potential flow apply (Cassidy, 1965, or Fuentes-Aguilar and Acuña, 1971).

2D Laminar boundary layer solution

The viscous flow features within the boundary layer are modeled using the system of continuity and momentum equations (White, 1991; Schlichting and Gersten, 2000)

$$\frac{\partial u}{\partial s} + \frac{\partial v}{\partial y} = 0,$$ (2.46)

$$u\frac{\partial u}{\partial s} + v\frac{\partial u}{\partial y} = U_e\frac{\mathrm{d}U_e}{\mathrm{d}s} + v\frac{\partial^2 u}{\partial y^2}.$$ (2.47)

Here v is the kinematic viscosity, u and v are velocities in the s- and y-directions, and y is the curvilinear coordinate normal to s (Figure 2.56). The boundary conditions for no wall slip are $u = 0$ and $v = 0$, whereas $u = U_e$ along the boundary layer thickness to match the outer potential flow to viscous flow close to the wall. Equations (2.46–2.47) imply constant pressure within the boundary layer in the y-direction prescribed by the outer potential flow. The ratio of boundary layer thickness to boundary radius of curvature δ/R_w must be small, therefore, as is typical for this flow (Lemos, 1965). The full Navier-Stokes equations are elliptic and must be solved simultaneously in the entire computational domain. In contrast, Eqs. (2.46–2.47) are parabolic so that their numerical solution is space-marching. Computations start at a given position s for which the profiles (u, v) along y are prescribed, and computations progress to a new position s to compute the velocity profiles. The implicit finite-difference model of White (1991) was adopted with details given by Castro-Orgaz and Hager (2014).

Integral solution

This method was used by Isaacs (1981) to estimate the boundary layer development in accelerated flow at the exit drop of a small-scale broad-crested weir. He found that the method produced acceptable results, yet the necessity for further verification of his method was stated. Given the good results of Isaacs (1981), it was applied to round-crested weir flow, testing its accuracy for the 2D solution. The integral form of the boundary layer Eqs. (2.46–2.47) is given by the von Kármán equation, reduced for a laminar boundary layer by Thwaites to (White, 1991; Schlichting and Gersten, 2000)

$$\frac{\theta^2}{v} = \frac{\theta_o^2}{v} + aU_e^{-b}\int_{s_o}^{s} U_e^{b-1}\mathrm{d}s .$$ (2.48)

Here subscript o refers to the upstream section at which computations start, with the empirical parameters $a = 0.45$ and $b = 6$. Using the outer potential velocity distribution $U_e(s)$, the integral in Eq. (2.48) is evaluated numerically and the profile $\theta(s)$ thereby computed. The profile $\delta^*(s)$ is then determined using the shape factor S_c involving a correlation polynomial developed by White (1991).

Matthew's theory

Matthew (1963) assumed an exponential function for the potential bottom velocity distribution in the upstream weir portion as

$$U_e = U_c \exp\left[s\left(2R_w E\right)^{-1/2}\right].$$ (2.49)

For parallel streamlines, the crest velocity at $s = 0$ is $U_c = [(2/3)E/g]^{1/2}$ for critical flow. Using Eq. (2.49), the exact Goldstein solution of the laminar boundary layer applies. An exponential type profile for the boundary layer thickness is (Matthew, 1963)

$$\delta^* = 0.7\nu^{1/2}\left(\frac{3R_w}{g}\right)^{1/4}\exp\left[-\frac{s}{2(2R_wE)^{1/2}}\right]. \tag{2.50}$$

At the crest, Eq. (2.50) yields as the relevant expression for discharge computations

$$\delta_c^* = 0.7\nu^{1/2}\left(\frac{3R_w}{g}\right)^{1/4}. \tag{2.51}$$

Analytical solution

An analytical solution for C_d results by inserting Eq. (2.51) into Eq. (2.43). Although the exponential behavior of U_e may be reasonable close at the weir crest, the value used for U_c may not be accurate for high E/R_w. The crest bottom velocity is thus given by potential flow using $h_{xx} = (4/9)/R_w$ as (Castro-Orgaz et al., 2008)

$$U_c = V_s\exp\left[-hz_{xx} - \frac{hh_{xx}}{2}\right] \approx \left(\frac{2}{3}gE\right)^{1/2}\exp\left(\frac{14}{27}\frac{E}{R_w}\right). \tag{2.52}$$

Inserting Eq. (2.49) into Eq. (2.48) and using Eq. (2.52) provides for the momentum thickness profile

$$\frac{\theta^2}{\nu} = \frac{\theta_o^2}{\nu} + \frac{a(2R_wE)^{1/2}}{U_c(b-1)}\exp\left[-(2R_wE)^{-1/2}s\right]\left[1 - \exp\left(\frac{b-1}{(2R_wE)^{1/2}}(s_o - s)\right)\right]. \tag{2.53}$$

The last term containing the flow development length $(s_o - s)$ is of small magnitude and neglected, therefore. Taking values at the crest $s = 0$, and using the shape factor S_c at that position to transform θ_c to δ_c^*, the crest displacement thickness finally is

$$\delta_c^* = \left(\delta_o^{*2} + \delta_a^{*2}\right)^{1/2}. \tag{2.54}$$

Here $\delta_o^* = S_c\theta_o$, and the acceleration contribution is

$$\delta_a^* = \left(\frac{aS_c^2}{b-1}\right)^{1/2}\nu^{1/2}\left(\frac{3R_w}{g}\right)^{1/4}\exp\left(-\frac{7}{27}\frac{E}{R_w}\right) = \lambda\nu^{1/2}\left(\frac{3R_w}{g}\right)^{1/4}\exp\left(-\frac{7}{27}\frac{E}{R_w}\right). \tag{2.55}$$

For $a = 0.45$, $b = 6$, and $S_c = 2.3$ follows $\lambda = 0.69$ in Eq. (2.55), close to 0.7 as in Eq. (2.51).

Results

Potential flow over a round-crested weir of bottom profile $z_b = 20\exp[-0.5(x/24)^2]$ is considered as test case for operational heads of $E/R_w = 0.516$ and 0.253, respectively (Figure 2.57). Free surface and bottom pressure predictions are successfully compared with the test data of Sivakumaran *et al.* (1983). Details of the potential flow solution are described by Montes (1992, 1994) or Castro-Orgaz (2013).

The potential flow simulations were used to obtain the potential bottom velocity U_e plotted in Figure 2.58 for comparative purposes with Eq. (2.49) using $U_c = [(2/3)E/g]^{1/2}$ (Matthew, 1963), and Eq. (2.52) for improved crest velocity predictions. Note that the exponential function provides excellent bottom velocity predictions close to the crest. Deviations from Matthew's theory are attributed to the use of $U_c = [(2/3)E/g]^{1/2}$. Far from the crest, in a zone of low-flow velocity, the exponential function deviates from the 2D potential velocity distribution. Given the agreement near the crest, good δ^* data are expected there if Eqs. (2.49) and (2.52) are used to predict $U_e(s)$.

The laminar boundary layer displacement thickness profiles from the 2D and 1D models are shown in Figure 2.59 for $R_w = 0.288$ m. The agreement of the 1D method with the 2D solution supports its use for computing laminar boundary layers in weir models. The figure includes also Matthew's Eq. (2.50), which is close to neither 1D nor 2D models, except for near the weir crest. Computations for $E/R_w = 0.516$ and $R_w = 0.025$ m are shown in Figure 2.58b, resulting in a larger boundary layer thickness, further supporting the accuracy of the method.

The results for $E/R_w = 0.253$ and weir scales of $R_w = 0.288$ m and 0.025 m are shown in Figure 2.60. The boundary layer is thicker under this smaller head for identical R_w, confirming the excellence of Thwaites' method for the entire domain, although the peaks are slightly over-predicted. Matthew's method is only reliable near the crest.

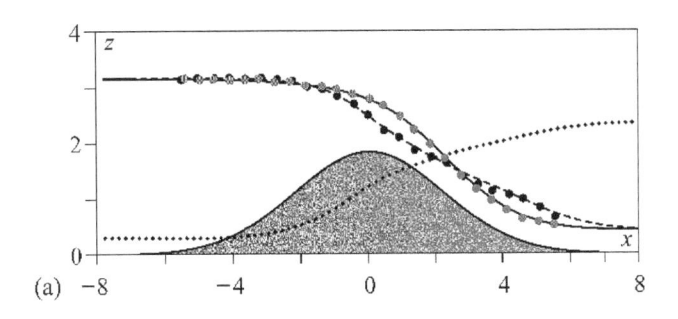

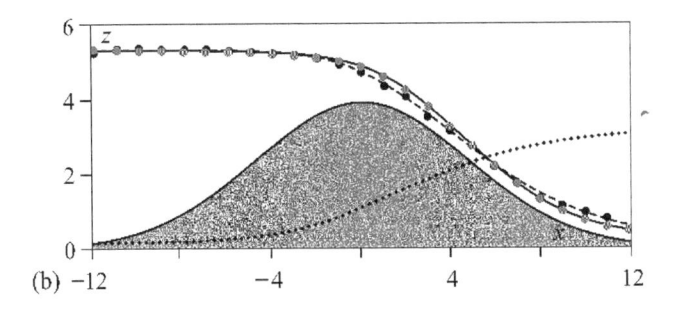

Figure 2.57 Potential flow solution for round-crested weirs with $E/R_w =$ (a) 0.516, (b) 0.253. Variables are normalized using critical flow condition (z/h_c, z_b/h_c, $p_b/(\gamma h_c)$, U_e/U_c, x/h_c), with $h_c = (q^2/g)^{1/3}$ (−−) computed and (•) measured free surfaces; (---) computed and (•) measured bottom pressure heads; (·) computed bottom velocity (Castro-Orgaz and Hager, 2014)

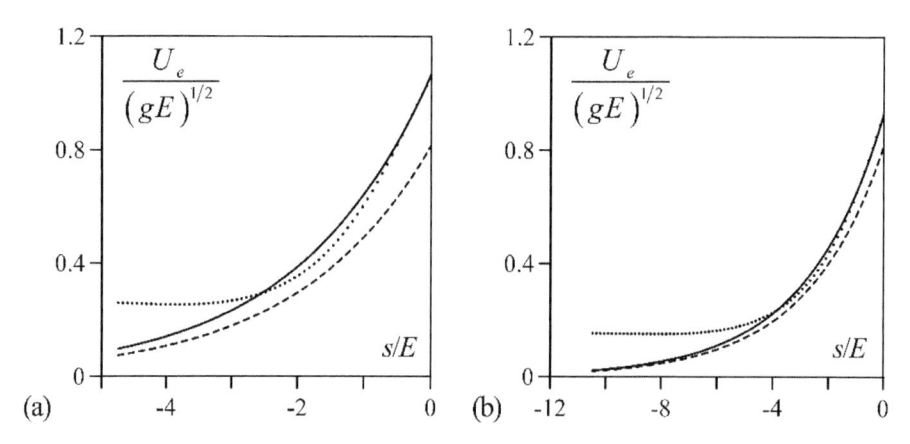

Figure 2.58 Comparison of relative potential bottom velocity $U_e/(gE)^{1/2}[s/E]$ for E/R_w = (a) 0.516, (b) 0.253, (•••) 2D solution, (—) analytical solution, (---) Matthew (1963) (Castro-Orgaz and Hager, 2014)

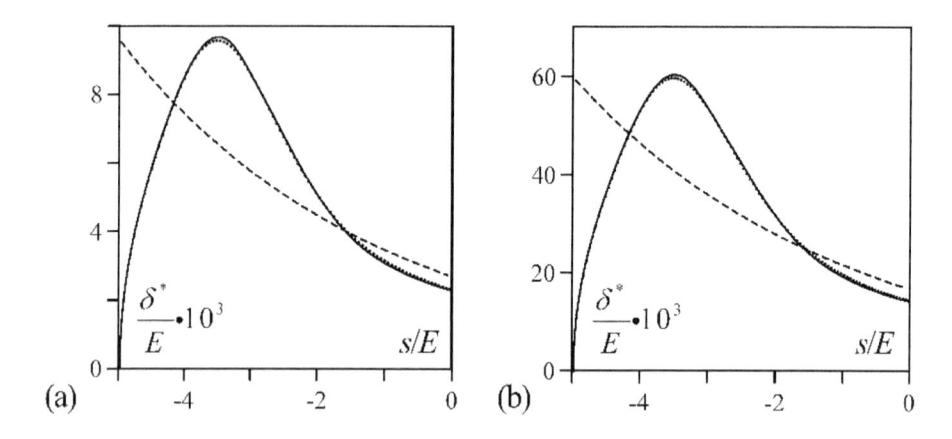

Figure 2.59 Comparison of boundary layer displacement thickness profiles $(\delta^*/E) \cdot 10^3[s/E]$ for E/R_w = 0.516 and R_w = (a) 0.288 m, (b) 0.025 m with (•••) 2D solution, (—) integral solution, (---) Matthew (1963) (Castro-Orgaz and Hager, 2014)

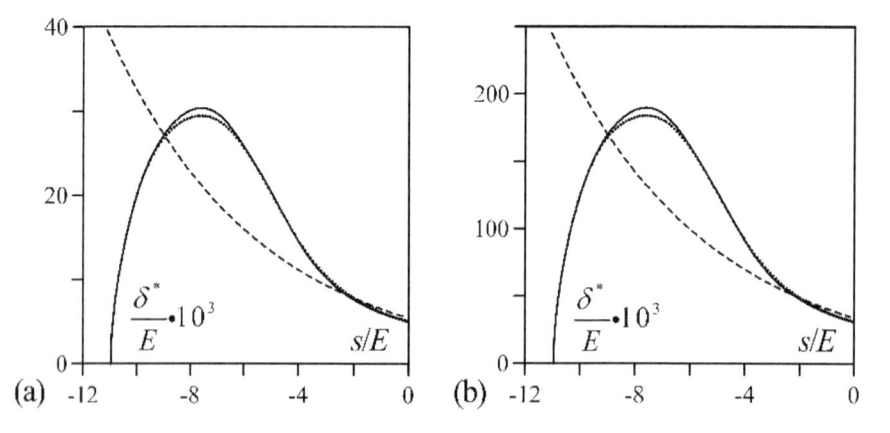

Figure 2.60 Comparison of boundary layer displacement thickness profiles $(\delta^*/E) \cdot 10^3$ for E/R_w = 0.253 and R_w = (a) 0.288 m, (b) 0.025 m, notation see Figure 2.59 (Castro-Orgaz and Hager, 2014)

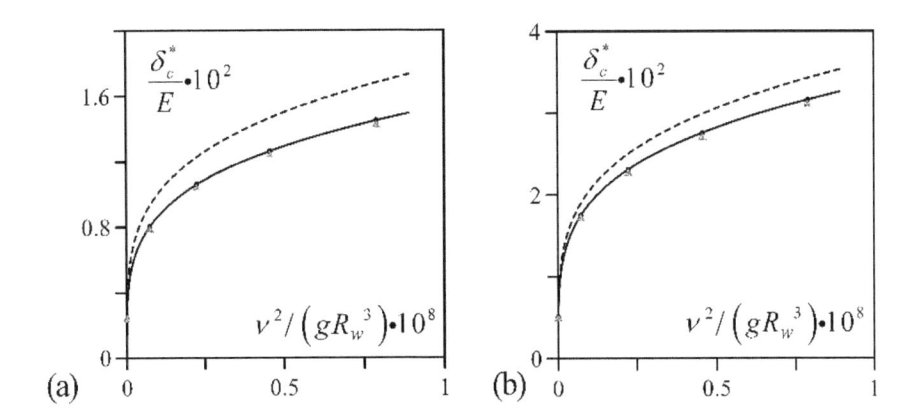

Figure 2.61 Comparison of crest boundary layer displacement thickness $(\delta_c^*/E)\cdot 10^2$ for E/R_w = (a) 0.516, (b) 0.253 with (•) 2D solution, (▲) integral solution, (—) analytical solution, (---) Matthew (1963) (Castro-Orgaz and Hager, 2014)

For discharge predictions, the displacement thickness at the weir crest is relevant. Simulations for $E/R_w = 0.516$ and 0.253 and different weir scales using the 1D and 2D models result for δ_c^*/E in Figure 2.61. The prediction using the analytical solution given by Eq. (2.55) is also shown. It agrees well with both the 1D and 2D models, thereby confirming its usefulness as predictor.

2.4.3 Discharge coefficient

The discharge coefficient C_d of Eq. (2.44) is evaluated by using Eq. (2.41) for C_o, Eq. (2.40) for C_σ combined with Eqs. (2.42–2.43) for C_v using Eq. (2.54) for δ_c^*/E combined with Eq. (2.55) for δ_a^*/E. Figure 2.62 compares the present approach with the data of Matthew (1963) for a circular-crested weir of $R_w = 0.0254$ m and 10°C fluid temperature. Matthew's experimental setup consisted of a half cylinder mounted with vertical walls (inset of Figure 2.62), taking the boundary layer displacement thickness at the starting point 'a' of the boundary layer zero. Hence, its effect on the crest point 'b' originates from the flow acceleration from 'a' to 'b'. Figure 2.62 indicates that the predictions agree well with the test data. The ideal fluid flow line is given by $C_d = (2/3)^{3/2}C_o$. Note the small deviations in this test from Matthew's equation

$$C_d = \left(\frac{2}{3}\right)^{\frac{3}{2}} \left[\begin{array}{l} 1 + 0.271\dfrac{E}{R_w} - 0.045\left(\dfrac{E}{R_w}\right)^2 - 1.05\left(\dfrac{3}{g}\right)^{1/4} \times \\[2ex] \times \nu^{1/2}R_w^{-3/4}\left(\dfrac{R_w}{E}\right) - 0.833\left(\dfrac{\sigma}{\gamma R_w^2}\right)\left(\dfrac{R_w}{E}\right) \end{array} \right]. \tag{2.56}$$

Therefore, the estimation of δ_c^*/E in Matthew's equation is adequate for discharge computation in this test case. The good prediction confirms that $\delta_o^* = 0$ at point 'a', so that the boundary layer development along the vertical weir face is avoided.

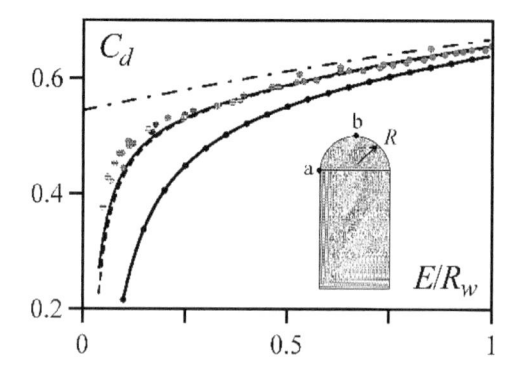

Figure 2.62 Discharge coefficient $C_d(E/R_w)$ of circular weir of $R_w = 0.0254$ m with (—) theory, (---) Matthew (1963), (—·—) ideal fluid flow, (•••) flat plate (Naudascher, 1987), (•) experiments of Matthew (1963) (Castro-Orgaz and Hager, 2014)

If the prototype operated at $R_w \to \infty$, then the difference between the ideal curve and the present theory would represent the scale effect. Since the prototype suffers viscous effects as well, this corresponds to the largest possible amount if viscous effects were negligibly small in the prototype.

To check the flat-plate approach for viscous effects, the method of Naudascher (1987, 1991) was further tested with

$$\delta_c^* = 1.73L\left(\frac{U_c L}{v}\right)^{-1/2}. \tag{2.57}$$

Here the flow development length L between points 'a' and 'b' is $L = \pi R_w/2$, and $U_c = [(2/3)E/g]^{1/2}$. Results using Eq. (2.57) are plotted in Figure 2.62, indicating an overestimation of scale effects.

The present theory is positively compared in Figure 2.63a with the 3D numerical simulations by Pfister *et al.* (2013) for $R_w = 0.005$ m. Predictions from Eq. (2.56) are also included, in which scale effects are overpredicted, as noted by Pfister *et al.* (2013). The enhanced present predictions stem from the improved treatment of surface tension effects at high normalized heads E/R_w. Predictions are again compared with the 3D simulations of Pfister *et al.* (2013) for $R_w = 0.010$ m (Figure 2.63b). Again, the theory agrees well with the numerical data, improving predictions as compared with Eq. (2.56).

The experiments of Chanson and Montes (1997) are considered in Figure 2.64. The setup consisted of full cylinders mounted on a thin plate (inset of Figure 2.64a). Only the circular weir with $R_w = 0.029$ m is retained because of scale effect presence. A data check indicates smaller values of C_d at high normalized heads ($E/R_w = 2.5$ to 3) than predicted by the ideal fluid flow theory. A comparison of $C_d(E/R_w)$ with another data set of Montes (see Chanson and Montes 1997) using the same R_w and an upstream ramp indicates the same trend at high E/R_w. The latter case was also tested by Ramamurthy and Vo (1993a, b). They found that the flow was essentially irrotational, with a very thin boundary layer. Their discharge data are in excellent agreement with ideal fluid flow computations. Montes (1970) states that the crest flow conditions at a circular weir are essentially irrotational under high E/R_w. For these,

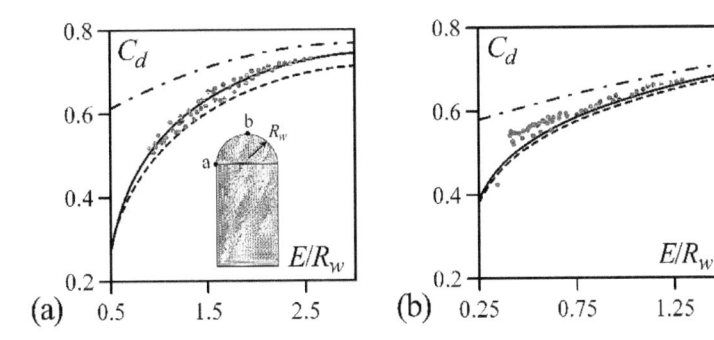

Figure 2.63 Discharge coefficient of circular weir $C_d(E/R_w)$ with R_w = (a) 0.005 m, (b) 0.010 m (Castro-Orgaz and Hager, 2014)

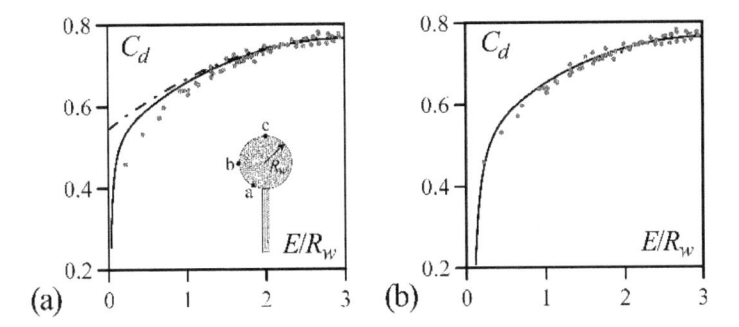

Figure 2.64 Discharge coefficient $C_d(E/R_w)$ of circular weir flow with R_w = 0.029 m for predictions (a) without [(—) only acceleration effects, (–·–) ideal fluid flow, (•) Montes' (1992) data], (b) with initial boundary layer [(—) boundary layer included, (•) Montes' (1992) data] (Castro-Orgaz and Hager, 2014)

a close agreement between the ideal fluid flow theory and experiments is reported (Fawer, 1937; Ramamurthy and Vo, 1993a, 1993b, Ramamurthy *et al.*, 1994). The corrected data in Figure 2.64a agree excellently with the ideal fluid flow theory for high normalized heads. This correction hardly affects the low-head data, however, for which the kinetic energy head is small, so that this is the flow zone of interest for scale effects. Note the appreciable drop in C_d originating from real fluid flow features.

Castro-Orgaz and Hager (2014) thus compared their predictions with observations in Figure 2.64a, by which scale effects for 20°C fluid temperature are underpredicted. Why did the theory work so well in the former test cases, and so poorly in this case? A reexamination of the experimental setup suggests that the particular design may induce additional scale effects. For the half cylinder mounted with vertical walls, the flow accelerates toward the weir crest, where the boundary layer starts. For a full cylinder mounted on a thin plate, the flow in the half-submerged cylinder portion is close to stagnation. However, the flow development there may provoke a boundary layer development from 'a' to 'b' (inset of Figure 2.64a), so that $\delta_o^* \neq 0$ at point 'b', in contrast to the former setup. Thus, the additionally induced scale effects include the boundary layer development from points 'a' to 'b'.

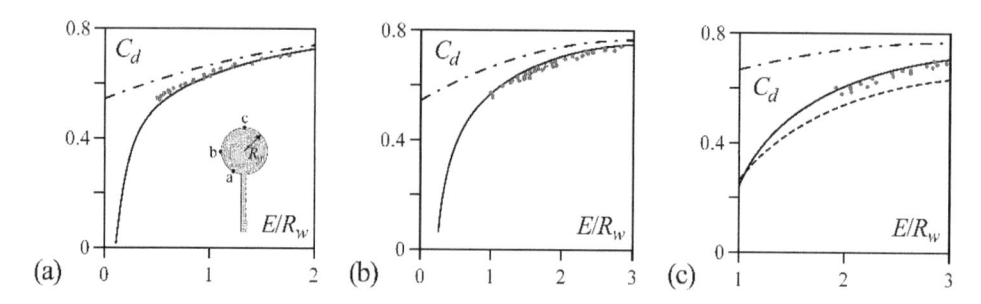

Figure 2.65 Discharge coefficient of circular weir $C_d(E/R_w)$ for R_w = (a) 0.0125 m, (b) 0.0065 m, (c) 0.003 m with (—) present theory, (−·−) ideal fluid flow, (---) Matthew, 1963), (•) experiments Sarginson (1972) (Castro-Orgaz and Hager, 2014)

Neglecting acceleration effects between these, Thwaites' equation yields for the momentum thickness at point 'b'

$$\frac{\theta_o^2}{v} = aU_e^{-b}\int_0^{s_o} U_e^{b-1}ds \approx aU_o^{-1}s_o .$$ (2.58)

Inserting it in Eq. (2.53), the displacement thickness at point 'c' is predicted for prescribed s_o and U_o. The exponential bottom velocity $U_o = U_c\exp[-R_w(2R_wE)^{-1/2}]$ was adopted. The starting point 'a' of the boundary layer was assumed $s_o = 0.5R_w$, resulting in Figure 2.64b, producing an almost perfect prediction of C_d for all heads E/R_w.

Sarginson (1972, 1984) conducted tests using full cylinders mounted on thin plates, as previously. His data set was used to further check the present theory. Predictions are successfully compared with experimental data for R_w = 0.0125, 0.0065, and 0.003 m in Figure 2.65 at fluid temperature of 20°C. Note the large deviations of the ideal fluid flow theory for the smallest R_w = 0.003 m operating at high head.

2.4.4 Round-crested weir flow analogy

Separation bubble

Weir flow close to the upstream corner of a broad-crested weir is related to the free streamline separation from the crest, provoking a recirculating flow zone. According to Moss (1972), the pocket flow is essentially recirculating, with a shear layer above it. He proposed a simplified model, involving a stagnant bubble and an irrotational stream passing above it. The shear layer is approximated by a thin boundary layer, which was overlooked by Moss. His treatment is based on an idea of Hunter Rouse in the 1930s, essentially assuming that a small spillway profile is added to the upstream broad-crested weir corner. Its discharge characteristics are determined using the elements of the round-crested weir, with s_M as the maximum separation thickness of the corner bubble, and H_o as the upstream head over the broad-crested weir (inset of Figure 2.66b).

With C_D as discharge coefficient, the discharge over the broad-crested weir is (Hager and Schwalt, 1994)

$$q = C_D\left(gH_o^3\right)^{1/2} .$$ (2.59)

The separation bubble is assumed to be solid, guiding an external irrotational stream (Moss, 1972). Using the flow analogy, the specific energy at the virtual round-crested weir is $E = H_o - s_M$. Combining Eqs. (2.59) and (2.38) with C_D as function of the discharge coefficient of the round-crested weir C_d results in

$$C_D = C_d \left(\frac{H_o - s_M}{H_o} \right)^{3/2}. \tag{2.60}$$

For irrotational flow without surface tension at the water surface and a shear layer above the bubble, the experimental data indicate that $R_w = 1.2H_o$ (Hager and Schwalt, 1994). Moss (1972) found $s_M = 0.15H_o$, whereas Hager and Schwalt (1994) measured $s_M = 0.2H_o$. With $s_M = 0.185H_o$, $E/R_w = 0.679$, $C_d = (2/3)^{3/2}C_o = 0.633$, so that $C_D = 0.466$, in agreement with the accepted value of 0.463. The round-crested weir flow analogy thus allows for a realistic prediction of C_D. The broad-crested weir coefficient for ideal fluid flow without separation bubble originating from Bélanger's critical flow condition is $(2/3)^{3/2} = 0.544$; the reduction of 0.544 to 0.466 is due to flow separation.

The data of Hager and Schwalt (1994) are plotted in Figure 2.66a versus H_o/L. The effects of surface tension and viscosity were accounted for by Eq. (2.44). This admits the existence of the shear layer above the bubble, a feature so far overlooked. The C_D curve of the present theory using approximate values of R_w and s_M varies only with the absolute head H_o. The weir length $L = 0.5$ m was used to scale the results, with the prediction compared to the data in Figure 2.66a. Note that the present theory predicts a drop in C_D as H_o/L reduces due to inclusion of viscosity and surface tension effects related to the existence of a shear layer above the separation bubble and surface tension at the weir surface. For $0.1 < H_o/L < 0.4$, Figure 2.66a states a good agreement with data. The data sets of Bazin (1896) and Tison (1950) are compared with the theory in Figure 2.66b, resulting in acceptable agreement for the broad-crested weir range. In the long-crested weir range $H_o/L < 0.1$, the experimental data are overpredicted, however.

Round-crested weirs without significant scale effects

Given a weir model of crest radius R_w, what is the minimum overflow head E_{min} to avoid significant scale effects? This basic question is answered by applying the above theory. For

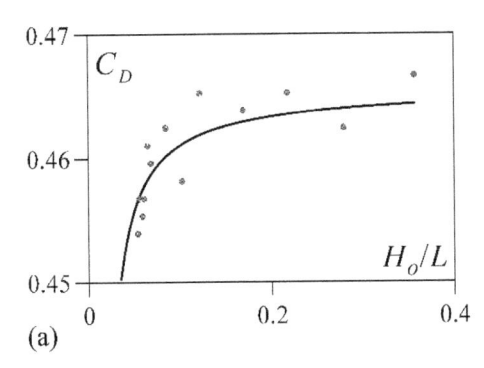
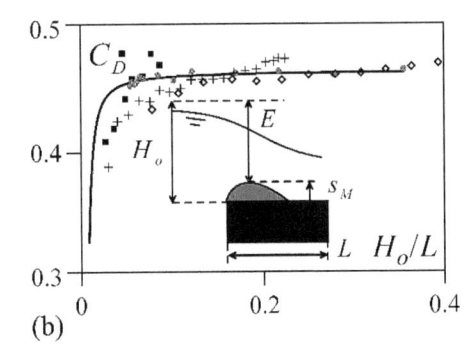

(a)   (b)

Figure 2.66 Broad-crested weir flow using round-crested weir flow analogy, comparison of theory with (a) data of Hager and Schwalt (1994), (b) (—) theory, data of (●) Hager and Schwalt (1994), (■) Tison (1950), Bazin (1896) (+) series 115, (◊) series 114 (Castro-Orgaz and Hager, 2014)

a given R_w, the curve $C_d(E/R_w)$ is determined from the present theory. Let the ideal fluid flow curve be $C_{di} = (2/3)^{3/2}C_o$. Accepting a drop in C_d due to scale effects, say $C_d/C_{di} = 0.98$, the corresponding value of E/R_w is determined. Multiplying it by R_w results in E_{min}. Figure 2.67a indicates that E_{min}/R_w is high for small R_w, the domain where Matthew's Eq. (2.56) is inaccurate. For comparative purposes, E_{min} was then determined using Eq. (2.56), with the results shown in Figure 2.67b. Note the disagreement between Matthew's and the present theory for small R_w (Figure 2.67c). The 3D data of Pfister *et al.* (2013) are also plotted, confirming values of E_{min} below computations from Eq. (2.56). Given the uncertainty of E_{min} for $R_w < 0.01$ m, this is the smallest weir radius proposed for laboratory application. For practical applications within $0.01 < R_w < 0.30$ m, the discharge curve is free from significant scale effects if $E_{min} > 0.04$ m. Selected experiments conducted at ETH Zurich confirm a design free of significant scale effects. A circular weir made up of a half-cylindrical crest of $R_w = 0.30$ m and downstream face slopes of 30° and 90° was inserted in a 0.50 m wide flume. From Figure 2.67b results $E_{min} = 0.0325$ m, corresponding to $E/R_w = 0.108$. This value served as minimum overflow head for experimentation. The discharge curve was experimentally determined from an overflow head of 0.05 m. The experimental results are plotted in Figure 2.67d, along with the theoretical C_d and C_{di} curves. Note that the discharge curve is free from significant scale effects.

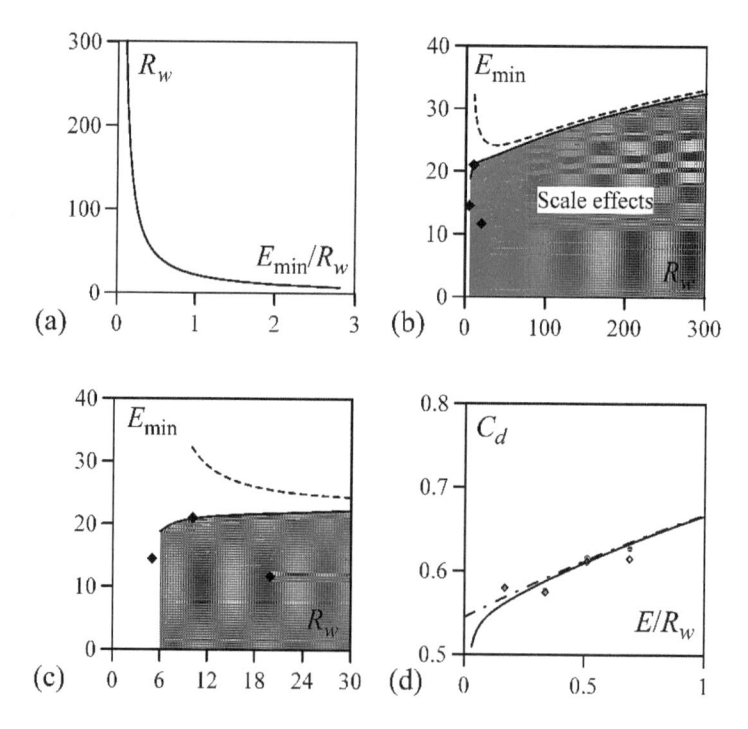

Figure 2.67 Scale effects at round-crested weir (a) limiting value R_w [mm] versus (E_{min}/R_w) for 2% C_d reduction, (b) minimum overflow head E_{min} [mm] versus R_w [mm] to avoid significant scale effects with (—) theory, (---) Matthew, 1963, (♦) Pfister *et al.*, 2013, (c) detail of (b) for $R_w < 30$ mm, (d) $C_d(E/R_w)$ with (—) theory, (—·—) ideal fluid flow, VAW data (•) 90° and (◊) 30° downstream weir face angle (Castro-Orgaz and Hager, 2014)

Boundary layer presence does not necessarily result in significant scale effects (e.g. if it is turbulent in both prototype and model). They are likely to result if the boundary layer is turbulent in the prototype and laminar in the model, or if the boundary layer is laminar in both the prototype and the model.

Notation

A	Relative gate opening $= a/H_D$ (-)
A_r	Constant of embankment weir flow (-)
a	Gate opening (m)
B_R	Dimensionless recirculation fit value (-)
b	Weir width (m)
b_{1m}	Width of shock at maximum pier wave height (m)
b_A	Width of weir overflow and pier (m)
b_e	Effective overflow width (m)
b_p	Pier width (m)
C_D	Discharge coefficient of broad-crested weir (-)
C_d	Discharge coefficient (-)
C_{dD}	Discharge coefficient under design head (-)
C_{dg}	Gated-weir discharge coefficient (-)
C_{di}	Discharge coefficient for ideal fluid flow (-)
C_o	Discharge correction coefficient for curvilinear flow (-)
C_σ	Discharge correction coefficient due to surface tension (-)
C_v	Discharge correction coefficient due to viscosity (-)
E	Specific energy head (m)
F_o	Approach flow Froude number $= V_o/(gh_o)^{1/2}$ (-)
G	Normalized shortest distance between overflow profile and gate lip (-)
g	Gravity acceleration (m/s^2)
H	Head on weir (m)
H_D	Weir design head (m)
H_e	Effective head on gated weir (m)
H_L	Limit head to avoid cavitation damage (m)
H_o	Energy head upstream of broad-crested weir (m)
h	Flow depth (m)
h_1	Flow depth at embankment toe (m)
h_{1m}	Maximum height of pier wave (m)
h_c	Critical flow depth $= [Q^2/(gb^2)]^{1/3}$ (m)
h_d	Downstream flow depth of embankment weir (m)
h_o	Approach flow depth (m)
h_t	Tailwater flow depth (m)
h_{tT}	Transition flow depth (m)
h_u	Downstream flow depth (m)
K_p	Weir pier coefficient (-)
L	Length of broad-crested weir (m)
L_r	Roller length (m)

L_R	Length of bottom recirculation (m)
L_w	Embankment crest weir length (m)
n	Exponent (-)
P_b	Relative bottom pressure head $= p_b/(\rho g H_D)$ (-)
$\bar{P}_c$	Minimum relative crest pressure head $= p_c/(\rho g H)$ (-)
P_m	Relative minimum bottom pressure head of gated flow $= p_m/(\rho g H_D)$ (-)
$\bar{P}_m$	Minimum relative bottom pressure head $= p_m/(\rho g H)$ (-)
p_b	Bottom pressure (N/m²)
p_s	Free surface pressure (N/m²)
$p_{vi}/(\rho g)$	Incipient vapor pressure head (m)
Q	Discharge (m³/s)
Q_g	Discharge below gate (m³/s)
Q_s	Submerged discharge (m³/s)
q	Unit discharge (m²/s)
R	Relative gate curvature radius $= r/H_D$ (-)
R	Weir flow Reynolds number $= \left(gh_o^3\right)^{1/2}/n$ (-)
R$_s$	Local Reynolds number based on momentum thickness $= U\theta/v$ (-)
R_i	Radius of crest profile (m)
R_w	Weir crest radius (m)
R_s	Curvature radius of free surface (m)
r	Relative crest radius (-)
S	Relative value $= s_w/H_D$ (-)
S_c	Crest shape factor (-)
s	Curvilinear coordinate along weir bottom profile (m)
s_M	Maximum separation bubble thickness (m)
s_o	S-coordinate at starting point (m)
s_w	Free surface elevation above weir crest (m)
U_A	Relative absolute maximum surface velocity $= u_A/(gh_u)^{1/2}$ (-)
U_a	Relative absolute maximum backward velocity $= u_a/(gh_u)^{1/2}$ (-)
U_c	Critical bottom velocity (m/s)
U_e	Potential bottom velocity (m/s)
U_M	Relative tailwater velocity (-)
U_m	Relative maximum backward velocity $= u_m/u_a$ (-)
U_o	Approach flow velocity (m/s)
u	Velocity in s-direction (m/s)
u_A	Approach flow velocity to tailwater (m/s)
u_a	Absolute maximum backward velocity for plunging jet (m/s)
u_M	Maximum streamwise velocity (m/s)
u_u	Nominal tailwater velocity (m/s)
V	Absolute velocity (m/s)
V_c	Crest velocity (m/s)
V_o	Approach flow velocity (m/s)
V_s	Free surface velocity (m/s)
v	Velocity in vertical direction (m/s)

W	Weir flow Weber number $= \rho g h_o^2 / \sigma$ (-)
w	Weir height (m)
X	Dimensionless horizontal crest coordinate $= x/H_D$ (-)
X_0	Relative position of zero bottom pressure $= x_0/H_D$ (-)
X^*	Transformed dimensionless crest coordinate (-)
X_E	Dimensionless pier extension length $= x_E/(b_p F_o)$ (-)
X_l	Relative horizontal gate lip position $= x_l/H_D$ (-)
X_m	Relative location of minimum bottom pressure $= x_m/H_D$ (-)
X_R	Relative recirculation length $= x_R/L_R$ (-)
X_r	Length relative to roller length $= x_r/L_r$ (-)
X_s	Gate seat position $= x_s/H_D$ (-)
X_t	Relative horizontal gate trunnion position $= x_t/H_D$ (-)
X_w	Relative location of $G = x_w/H_D$ (-)
x	Streamwise coordinate (m)
x'	Distance measured from embankment toe (m)
x_{1m}	Location of maximum pier wave height (m)
x_E	Pier extension length (m)
x_R	Streamwise location within recirculation zone (m)
x_r	Length from jump toe (m)
Y_t	Relative submergence factor (-)
y	Coordinate normal to bottom (m)
y_L	Modular limit ratio $= h_{tL}/h_o$ (-)
y_t	Weir submergence degree $= h_t/h_o$ (-)
y_T	Weir submergence ratio $= h_{tT}/h_o$ (-)
Z	Dimensionless vertical crest coordinate $= z/H_D$ (-)
Z^*	Transformed dimensionless crest coordinate (-)
Z_l	Relative vertical gate lip position $= z_l/H_D$ (-)
Z_R	Relative recirculation height $= z_R/w$ (-)
Z_r	Relative forward flow depth $= z_r/h_u$ (-)
Z_t	Relative vertical gate trunnion position $= z_t/H_D$ (-)
Z_δ	Relative elevation in tailwater jet $= (z-\delta_0)/(h-\delta_0)$ (-)
z	Vertical coordinate (m)
z_b	Bottom elevation (m)
z_o	Maximum elevation of lower weir flow trajectory above crest (m)
z_R	Recirculation height (m)
z_r	Height of forward flow (m)
α	Spillway face angle (-)
α_g	Gate lip angle (-)
α_o	Upstream weir angle (-)
α_u	Downstream weir angle (-)
γ	Specific fluid weight (N/m³)
γ_G	Profile angle at location of G (-)
δ	Boundary layer thickness of weir flow (m)
δ_0	Jet boundary layer thickness (m)
δ^*	Boundary layer displacement thickness (m)

δ_c^* Boundary layer displacement thickness at weir crest (m)
ψ Stream function (m^2/s)
ψ_s Discharge reduction factor $= Q_s/Q$ (-)
Φ Characteristic scale effect parameter $= \mathsf{R}^{0.2}\mathsf{W}^{0.6}$ (-)
χ Relative head on weir $= H/H_D$ (-)
χ_o Relative head on gated weir $= H_o/H_D$ (-)
χ_R Relative bottom recirculation distance $= x/L_R$ (-)
χ_r Relative surface jet distance $= x/L_r$ (-)
χ_s Surface velocity coefficient $= V_s/(q/h)$ (-)
μ Relative velocity $= V/(2gH_D)^{1/2}$ (-)
μ_c Relative crest velocity $= V_c/(2gH_D)^{1/2}$ (-)
v Kinematic fluid viscosity (m^2/s)
ρ_k Relative crest curvature $= H_o/R_w$ (-)
ρ_k' Expanded relative crest curvature (-)
σ Surface tension (N/m)
θ Boundary layer momentum thickness (m)
θ_c Boundary layer momentum thickness at weir crest (m)
ζ Crest length parameter $= H_o/(H_o + L_w)$ (-)

References

Abecasis, F.M. (1970) Discussion of Designing spillway crests for high-head operation. *Journal of Hydraulics Division ASCE*, 96(12), 2654–2658.

Ackers, P., White, W.R., Perkins, J.A. & Harrison, A.J.M. (1978) *Weirs and Flumes for Flow Measurement*. Wiley, Chichester.

Bazin, H. (1896) Expériences nouvelles sur l'écoulement par déversoir [Recent experiments on water flow of over weirs]. *Annales des Ponts et Chaussées*, 66, 645–731 (in French).

Bazin, H. (1898) Expériences nouvelles sur l'écoulement en déversoir [New experiments on weir discharge]. *Annales des Ponts et Chaussées*, 68(2), 151–265 (in French).

Bos, M.G. (1976) Discharge measurement structures. *Rapport* 4. Laboratorium voor Hydraulica an Afvoerhydrologie. Landbouwhogeschool, Wageningen NL.

Bretschneider, H. (1961) Abflussvorgänge bei Wehren mit breiter Krone [Hydraulics of broad-crested embankment weirs]. *Mitteilung* 53. Institut für Wasserbau und Wasserwirtschaft, Technische Universität, Berlin (in German).

Bruschin, J., Bauer, S., Delley, P. & Trucco, G. (1982) The overtopping of the Palagnedra dam. *Water Power & Dam Construction*, 34(1), 13–19.

Castro-Orgaz, O. (2008) Curvilinear flow over round-crested weirs. *Journal of Hydraulic Research*, 46(4), 543–547.

Castro-Orgaz, O. (2013) Potential flow solution for open channel flows and weir-crest overflow. *Journal of Irrigation Drainage Engineering*, 139(7), 551–559.

Castro-Orgaz, O., Hager, W.H. (2014) Scale effects of round-crested weir flow. *Journal of Hydraulic Research*, 52(5), 653–665.

Castro-Orgaz, O., Giraldez, J.V. & Ayuso, J.L. (2008) Higher order critical flow condition in curved streamline flow. *Journal of Hydraulic Research*, 46(6), 849–853.

Cassidy, J.J. (1965) Irrotational flow over spillways of finite height. *Journal of Engineering Mechanics Division ASCE*, 91(EM6), 155–173.

Chanson, H. & Montes, J.S. (1997) Overflow characteristics of cylindrical weirs. *Research Report CE 154*. Department of Civil Engineering, University of Queensland, Brisbane.

Chanson, H. & Montes, J.S. (1998) Overflow characteristics of circular crested weirs: effects of inflow conditions. *Journal of Irrigation and Drainage Engineering*, 124(3), 152–162.

Craya, A.E. & Delleur, J.W. (1952) An analysis of boundary layer growth in open conduits near critical regime. *Technical Report 1*. Department of Civil Engineering, Columbia University, New York.

Creager, W.P., Justin, J.D. & Hinds, J. (1945) *Engineering for dams*. 2nd ed. Wiley, Chapman & Hall, London.

Delleur, J.W. (1955) The boundary layer development on a broad crested weir. *Proceedings of 4th Midwestern Conference on Fluid Mechanics*, Purdue University, pp. 183–193.

Delleur, J.W. (1957) The boundary layer development in open channels. *Journal of Engineering Mechanics Division ASCE*, 83(1), 1–24.

Dey, S. (2002) Free overfall in open channels: state-of-the-art review. *Flow Measurement and Instrumentation*, 13(5–6), 247–264.

Dressler, R.F. (1978) New nonlinear shallow flow equations with curvature. *Journal of Hydraulic Research*, 16(3), 205–222.

Erbisti, P.C.F. (2014) *Design of Hydraulic Gates*. 2nd ed. CRC Press, Boca Raton, USA.

Escande, L. (1939) Recherches nouvelles sur les barrages déversoirs noyés [New experiments on submerged overflow dams]. *La Technique Moderne*, 31(18), 617–620 (in French).

Fawer, C. (1937) Étude de quelques écoulements permanents à filets courbes [Study of some steady flows with curved streamlines]. *Thesis*, Université de Lausanne. La Concorde, Lausanne (in French).

Fritz, H.M. & Hager, W.H. (1998) Hydraulics of embankment weirs. *Journal of Hydraulic Engineering*, 124(9), 963–971.

Fuentes-Aguilar, R. & Acuña, E. (1971) Estudio de algunas características del escurrimiento sobre vertederos de arista redondeada mediante la solución numérica de la ecuación de Laplace [Study of some flow features of round-crested weir flow using a numerical solution of the Laplace equation]. *Publicación SH5*. Departamento de Obras Civiles, Facultad de Ciencias Físicas y Matemáticas, Universidad de Chile (in Spanish).

Govinda Rao, N.S. & Muralidhar, D. (1961) A new discharge formula for waste-weirs (bye washes). *Irrigation and Power*, 18(10), 908–917; 19(7), 546–549.

Hager, W.H. (1985) Critical flow condition in open channel hydraulics. *Acta Mechanica*, 54(3/4), 157–179.

Hager, W.H. (1987) Continuous crest profile for standard spillway. *Journal of Hydraulic Engineering*, 113(11), 1453–1457.

Hager, W.H. (1991a) Experiments on standard spillway flow. *Proceedings of the Institution Civil Engineers*, 91(2), 399–416.

Hager, W.H. (1991b) Flow features over modified standard spillway. *Proceedings of 24th IAHR Congress Madrid D*, 243–250.

Hager, W.H. (1992) *Energy Dissipators and Hydraulics Jumps*. Kluwer, Dordrecht NL.

Hager, W.H. (1993) Abfluss über Zylinderwehr [Flow over cylindrical weir]. *Wasser und Boden*, 44(1), 9–14 (in German).

Hager, W.H. (1994) Discussion of 'Momentum model for flow past weir', by Ramamurthy, A.S., Vo., N.-D., Vera, G. *Journal of Irrigation and Drainage Engineering*, 120(3), 684–685.

Hager, W.H. (2010) *Wastewater hydraulics*, 2nd ed. Springer, Berlin.

Hager, W.H. & Bremen, R. (1988) Plane gate on standard spillway. *Journal of Hydraulic Engineering*, 114(11), 1390–1397.

Hager, W.H. & Castro-Orgaz, O. (2017) Ogee weir crest definition: historical advance. *Proceedings of 37th IAHR World Congress, Kuala Lumpur*, pp. 1937–1948.

Hager, W.H., Reinauer, R. & Lauber, G. (1994) Wenn die Absturzströmung zur Todesfalle wird [When drop flow becomes a death trap]. *Schweizer Ingenieur und Architekt*, 112(38), 738–741 (in German).

Hager, W.H. & Schwalt, M. (1994) Broad-crested weir. *Journal of Irrigation and Drainage Engineering*, 120(1), 13–26.

Hager, W.H. & Schleiss, A.J. (2009) *Constructions hydrauliques: ecoulements stationnaires* [Hydraulic structures: steady flows]. PPUR, Lausanne (in French).

Hall, G.W. (1962) Analytical determination of the discharge characteristics of broad-crested weirs using boundary layer theory. *Proceeding of ICE*, 22(2), 177–190.

Harrison, A.J.M. (1967a) Boundary layer displacement thickness on flat plates. *Journal of Hydraulics Division ASCE*, 93(4), 79–91.

Harrison, A.J.M. (1967b) The streamlined broad crested weir. *Proceeding of ICE*, 38, 657–678.

Hartung, F. (1973) Gates in spillways of large dams. Proceedings of 11th *ICOLD Congress Madrid*, Q41(R72), 1361–1374.

Hégly, V.M. (1939) Expériences sur l'écoulement de l'eau au-dessus et en dessous des barrages cylindriques [Discharge over and below cylindric weirs]. *Annales des Ponts et Chaussées*, 109(9), 235–281 (in French).

Heidarpour, M., Habili, J.M. & Haghiabi, A.H. (2008) Application of potential flow to circular-crested weir. *Journal of Hydraulic Research*, 46(5), 699–702.

Heller, V. (2011) Scale effects in physical hydraulic engineering models. *Journal of Hydraulic Research*, 49(3), 293–306; 50(2), 244–246.

Horton, R.E. (1907) Weir experiments, coefficients, and formulas. *Water Supply and Irrigation Paper 200*. U.S. Department of the Interior, Government Printing Office, Washington DC.

Hulsing, H. (1968) Measurement of peak discharge at dams by indirect methods. Techniques of Water Resources Investigations, U.S. Geological Survey, *Book* 3(A5). U.S. Government Printing Office, Washington DC.

ICOLD (1987) Spillways for dams. *Bulletin 58*. International Commission for Large Dams, Paris.

ICOLD (2017) Technical advancements in spillway design: progress and innovations from 1985 to 2015. *Bulletin 172*. International Commission for Large Dams, Paris.

Indlekofer, H. (1976) Zum hydraulischen Einfluss von Pfeileraufbauten bei Überfall-Entlastungsanlagen [Hydraulic effect of piers on overfall dams]. *Mitteilung 13*. Institut Wasserbau und Wasserwirtschaft, Rheinisch-Westfälische Technische Hochschule, Aachen (in German).

Isaacs, L.T. (1981) Effects of laminar boundary layer on a model broad-crested weir. *Research Report CE28*, Department of Civil Engineering, University of Queensland, Brisbane.

Jaeger, C. (1933) Notes sur le calcul des déversoirs et seuils [Discharge calculations for weirs and sills]. *Bulletin Technique de la Suisse Romande*, 59(13), 153–156; 59(14), 166–169 (in French).

Jaeger, C. (1940) Erweiterung der Boussinesqschen Theorie des Abflusses in offenen Gerinnen: Abflüsse über abgerundete Wehre [Extension of the Boussinesq theory for open channel flow: flow over round-crested weirs]. *Wasserkraft und Wasserwirtschaft*, 35(4), 83–86 (in German).

Jaeger, C. (1956) *Engineering Fluid Mechanics*. Blackie & Son, Glasgow.

Kikkawa, H., Ashida, K. & Tsuchiya, A. (1961) Study on the discharge coefficient of broad-crested weirs. *Journal of Research*, 5(4), 1–20.

Kindsvater, C.E. (1964) Discharge characteristics of embankment shaped weirs. *Geological Survey Water Supply Paper* 1617-A, U.S. Government Printing Office, Washington, DC.

Kolkman, P.A. (1984) Considerations about the accuracy of discharge relations of hydraulic structures and the use of scale models for their calibration. *Symposium on Scale Effects in Modelling Hydraulic Structures, Esslingen, Germany*, 2(1), 1–12.

Kolkman, P.A. (1994) Discharge relationships and component head losses for hydraulic structures. Discharge characteristics. *IAHR Hydraulic Structures Design Manual*, 8(3), 55–151.

Lakshmana Rao, N.S. (1975) Theory of weirs. *Advances in Hydroscience*, 10, 309–406.

Lakshmana Rao, N.S. & Jagannadha Rao, M.V. (1973) Characteristics of hydrofoil weirs. *Journal of Hydraulics Division ASCE*, 99(HY2), 259–283; 100(HY7), 1076–1079; 100(HY12), 1836–1837.

Lemos, F. de Oliveira (1965) A instabilidade da camada limite: Sua influencia na concepcao dos descarregadores das barragens [Instability of boundary layer: influence on design of dam spillways]. *Thesis*, Laboratorio Nacional de Engenharia Civil LNEC, Lisboa (in Portuguese).

Lemos, F. de Oliveira (1981) Criterios para o dimensionamento hidraulico de barragens descarregadioras [Criteria for the hydraulic design of gated spillways]. *Memoria 556*. Laboratorio Nacional de Engenharia Civil LNEC, Lisboa (in Portuguese).

Lenau, C.W. (1967) Potential flow over spillways at low heads. *Journal of Engineering Mechanics Division ASCE*, 93(EM3), 95–107; 94(EM1), 354–359.

Leutheusser, H.J. & Birk, W.M. (1991) Drownproofing of low overflow structures. *Journal of Hydraulic Engineering*, 117(2), 205–213; 118(11), 1586–1589.

Liggett, J.A. (1994) *Fluid Mechanics*. McGraw-Hill, New York.

Matthew, G.D. (1963) On the influence of curvature, surface tension and viscosity on flow over round crested weirs. *Proceedings of the Institution Civil Engineers*, 25(4), 511–524; 28(4), 557–569.

Matthew, G.D. (1991) Higher order one-dimensional equations of potential flow in open channels. *Proceeding of ICE*, 91(3), 187–201.

Maxwell, W.H.C. & Weggel, J.R. (1969) Surface tension in Froude models. *Journal of Hydraulics Division ASCE*, 95(HY2), 677–701; 96(HY3), 845.

Miller, D.S. (1994) Discharge characteristics. In: Miller, D.S. (ed) *IAHR Hydraulic Structures Design Manual 8*, Balkema, Rotterdam NL.

Montes, S. (1970) Flow over round crested weirs. *L'Energia Elettrica*, 47(3), 155–164.

Montes, J.S. (1992) Potential flow analysis of flow over a curved broad crested weir. *Proceedings of 11th Australasian Fluid Mechanics Conference, Auckland*, pp. 1293–1296.

Montes, J.S. (1994) Potential flow solution to the 2D transition from mild to steep slope. *Journal of Hydraulic Engineering*, 120(5), 601–621, 121(9), 681–682.

Montes, J.S. (1998) *Hydraulics of Open Channel Flow*. ASCE Press, Reston VA.

Moss, W.D. (1972) Flow separation at the upstream edge of a square-edged broad-crested weir. *Journal of Fluid Mechanics*, 52, 307–320.

Naudascher, E. (1987) *Hydraulik der Gerinne und Gerinnebauwerke* [Hydraulics of Channels and Canal Structures]. Springer-Verlag, Wien (in German).

Naudascher, E. (1991) Hydrodynamic forces. *IAHR Hydraulic Structures Design Manual 3*. Balkema, Rotterdam.

Pfister, M., Battisacco, E., De Cesare, G. & Schleiss, A.J. (2013) Scale effects related to the rating curve of cylindrically crested piano key weirs. Proceedings of 2nd International Workshop on *Labyrinth and Piano Key Weirs – PKW 2013*, Paris, France.

Pinto, N.L. de S. & Ota, J.J. (1980) Distribuicao das pressoes na face de jusante das barragens de enrocarnento submersas: investigatio experimental [Pressure distribution at submerged overflows]. Proceedings of 9th *Congresso Latinoamericano Hidraulica* Merida E(3), 1–12 (in Portuguese).

Rajaratnam, N. (1965) The hydraulic jump as a wall jet. *Journal of Hydraulics Division of ASCE*, 91(3), 107–132.

Rajaratnam, N. & Muralidhar, D. (1969) Flow below deeply submerged rectangular weirs. *Journal of Hydraulic Research*, 7(3), 355–374.

Ramamurthy, A.S. & Vo, N.-D. (1993a) Characteristics of circular crested weir. *Journal of Hydraulic Engineering*, 119(9), 1055–1062; 120(12), 1494–1495.

Ramamurthy, A.S. & Vo, N.-D. (1993b) Application of Dressler theory to weir flow. *Journal of Applied Mechanics*, 60(1), 163–166.

Ramamurthy, A.S., Vo, N.D. & Balachandar, R. (1994) A note on irrotational curvilinear flow past a weir. *Journal of Fluids Engineering*, 116(2), 378–381.

Ranga Raju, K.G., Srivastava, R. & Porey, P.D. (1990) Scale effects in modelling flows over broad-crested weirs. *Irrigation and Power India*, 47(3), 101–106.

Rao, N.S.L. (1975) Theory of weirs. *Advances in Hydroscience*, 10, 309–406.

Rao, N.S.L. & Rao, M.V.J. (1973) Characteristics of hydrofoil weirs. *Journal of Hydraulics Division ASCE*, 99(2), 259–283.

Reinauer, R. & Hager, W.H. (1994) Supercritical flow behind chute piers. *Journal of Hydraulic Engineering*, 120(11), 1292–1308.

Rhone, T.J. (1959) Problems concerning use of low head radial gates. *Journal of Hydraulics Division ASCE*, 85(HY2), 35–65; 85(HY7), 151–154; 85(HY9), 113–117; 86(HY3), 31–36.

Sarginson, E.J. (1972) The influence of surface tension on weir flow. *Journal of Hydraulic Research*, 10(4), 431–446; 11(3), 299–306.

Sarginson, E.J. (1984) Scale effects in model tests on weir. *Symposium on Scale Effects in Modelling Hydraulic Structures, Esslingen, Germany*, 3(3), 1–4.

Schlichting, H. & Gersten, K. (2000) *Boundary Layer Theory*. Springer, Berlin, Germany.

Schmocker, L., Halldórsdóttir, B.R. & Hager, W.H. (2011) Effect of weir face angles on circular-crested weir flow. *Journal of Hydraulic Engineering*, 137(6), 637–643.

Sinniger, R.O. & Hager, W.H. (1989) *Constructions Hydrauliques* [Hydraulic structures]. Presses Poly techniques Fédérales, Lausanne (in French).

Sivakumaran, N.S., Tingsanchali, T. & Hosking, R.J. (1983) Steady shallow flow over curved beds. *Journal of Fluid Mechanics*, 128, 469–487.

STK (2017) Schwemmgut an Hochwasserentlastungsanlagen von Stauanlagen [Floating debris at dam spillways]. *Report of the Working Group on Floating Debris*. Swiss Committee on Dams, Lucerne, Switzerland (in German).

Tison, L.J. (1950) Le déversoir epais [Broad-crested weir]. *La Houille Blanche*, 5(4), 426–439 (in French).

USACE (1970) *Hydraulic Design Criteria*. US Army Waterways Experiment Station, Vicksburg MI.

Varshney, D.V. (1977) Model scale and the discharge coefficient. *Water Power and Dam Construction*, 29(4), 48–52.

Vierhout, M.M. (1973) On the boundary layer development in rounded broad-crested weirs with a rectangular control section. *Research Report* 3. Laboratory of Hydraulics and Catchment Hydrology, Agricultural University of Wageningen NL.

Webster, M.J. (1959) Spillway design for Pacific Northwest projects. *Journal of Hydraulics Division ASCE*, 85(HY8), 63–85.

White, F.M. (1991) *Viscous Fluid Flow*. McGraw-Hill, New York.

Wilkinson, D.L. (1974) Free surface slopes at controls in channel flow. *Journal of Hydraulics Division ASCE*, 100(HY8), 1107–1117.

Wu, S. & Rajaratnam, N. (1996) Submerged flow regimes of rectangular sharp-crested weirs. *Journal of Hydraulic Engineering*, 122(7), 412–414.

Yarnell, D.L. & Nagler, F.A. (1930) Flow of flood water over railway and highway embankments. *Public Roads*, 11(2), 30–34.

Bibliography

Various references are not cited in the main text but are still considered either historically interesting or technically relevant, so that they are mentioned below.

Aeration of weirs

Albrecht, D. (1969) Schätzung der Sauerstoffzufuhr durch Wehre und Kaskaden [Estimation of oxygen supply by weirs and cascades]. *Die Wasserwirtschaft*, 59(11), 321–323 (in German).

Albrecht, D. (1971) Belüftungsversuche mit frei abstürzendem Wasser [Aeration tests with freely falling water jets]. *gwf-wasser/abwasser*, 112(1), 29–32 (in German).

Apted, R.W. & Novak, P. (1973) Some studies of oxygen uptake at weirs. *Proceedings of 15th IAHR Congress* Istanbul, 2(B23), 177–186.

Avery, S.T. & Novak, P. (1978) Oxygen transfer at hydraulic structures. *Journal of Hydraulics Division*, ASCE, 104(HY11), 1521–1540; 105(HY9), 1211–1217.

Barrett, M.J., Gameson, A.L.H. & Ogden, C.G. (1960) Aeration studies at four weir systems. *Water and Water Engineering*, 64(9), 407–413.

Bugliarello, G. (1951) L'aerazione dello stramazzo tipo [Aeration of typical weir]. *L'Energia Elettrica*, 28(12), 688–694 (in Italian).

Buttes, T.A. & Evans, R.L. (1983) Small stream channel dam aeration characteristics. *Journal of Environmental Engineering*, 109(3), 555–573; 110(3), 728–735.

Chanson, H. (1995) Predicting oxygen content downstream of weirs, spillways and waterways. *Proceedings of ICE Water, Maritime & Energy*, 112(1), 20–30; 130(2), 115–116.

Cummings, P.D. & Chanson, H. (1997) Air entrainment in the developing flow region of plunging jets. *Journal of Fluids Engineering*, 119(9), 597–608.

Ervine, D.A. (1976) The entrainment of air in water. *Water Power & Dam Construction*, 28(12), 27–30.

Ervine, D.A. & Elsawy, E.M. (1975) The effect of a falling nappe on river aeration. *Proceedings of 16th IAHR Congress* Sao Paulo, 3(C45), 390–397; 6(C45), 347–349.

Gameson, A.L.H. (1957) Weirs and the aeration of rivers. *Journal of the Institution of Water Engineers*, 11, 477–490.

Gameson, A.L.H., Vandyke, K.G. & Ogden, C.G. (1958) The effect of temperature on aeration at weirs. *Water and Water Engineering*, 62(11), 489–492.

Gulliver, J.S. & Rindels, A.J. (1993) Measurement of air-water oxygen transfer at hydraulic structures. *Journal of Hydraulic Engineering*, 119(3), 327–349.

Hibbs, D.E. & Gulliver, J.S. (1997) Prediction of effective saturation concentration at spillway plunge pools. *Journal of Hydraulic Engineering*, 123(11), 940–949.

Hutarew, A. (1978) Möglichkeiten zur Verbesserung des Sauerstoffeintrags an Absturzbauwerken [Possibilities to improve oxygen uptake at drop structures]. *Wasserwirtschaft*, 68(3), 76–79 (in German).

Hutarew, A. & Minor, H.-E. (1975) Oxygen uptake of a free overfalling water nappe. *Proceedings of 16th IAHR Congress* Sao Paulo, 3(C39), 338–345; 6(C39), 341–343.

Johnson, J.W. (1935) The aeration of sharp-crested weirs. *Civil Engineering* 5(3), 177–178.

Kusabiraki, D., Niki, H., Yamagiwa, K. & Ohkawa, A. (1990) Gas entrainment rate and flow pattern of vertical plunging liquid jets. *Canadian Journal of Chemical Engineering*, 68(18), 893–903.

Markofsky, M. & Kobus, H. (1978) Unified presentation of weir-aeration data. *Journal of Hydraulics Division* ASCE, 104(HY4), 562–568.

Nakasone, H. (1987) Study of aeration at weirs and cascades. *Journal of Environmental Engineering*, 113(1), 64–81, 115(1), 267–271.

Rindels, A.J. & Gulliver, J.S. (1986) Air-water oxygen transfer at spillways and hydraulic jumps. *Water Forum*, 86(1), 1041–1048.

Rogola, R. (1981) Entraînement d'air par lame déversante [Air entrainment by overflow jet]. *La Houille Blanche*, 36(1), 15–21 (in French).

Sananes, F. (1959) Entraînement d'air sous une lame déversante [Air entrainment below an overflow jet]. *Proceedings of 8th IAHR Congress* Montreal, 13(D1), 1–15 (in French).

Sakthivadivel, R. & Seetharaman, S. (1971) Aeration of weirs. *Irrigation and Power*, 29(4), 177–186; 32(1), 103–106.

van der Kroon, G.T.M., Schram, A.H. (1969) Weir aeration. H_2O, 2(22), 528–545.

Watson, C.C., Walters, R.W. & Hogan, S.A. (1998) Aeration performance of low drop weirs. *Journal of Hydraulic Engineering*, 124(1), 65–71; 125(6), 666–668.

Cylindrical weir

Anwar, H.O. (1967) Inflatable dams. *Journal of the Hydraulics Division* ASCE, 93(HY3), 99–119; 94(HY1), 321–325; 94(HY6), 1521–1523.

Binnie, A.M. (1955) The effect of viscosity upon the critical flow of a liquid through a constriction. *Quarterly Journal of Mechanics and Applied Mathematics*, 8(4), 394–414.

Blau, E. (1963) Der Abfluss und die hydraulische Energieverteilung über einer parabelförmigen Wehrschwelle [The flow and the hydraulic energy distribution over a parabolic weir]. *Wasser- und Grundbau*, Heft 7, 5–212. Forschungsanstalt Schiffahrt, Wasser- und Grundbau, Berlin (in German).

Chanson, H. (1998) Overflow characteristics of circular weirs: effects of inflow conditions. *Journal of Irrigation and Drainage Engineering*, 124(3), 152–162.

Davidson, P.A. & Matthew, G.D. (1982) Potential flow over weirs of moderate curvature. *Journal of the Engineering Mechanics Division* ASCE, 108(EM5), 689–707.

De Marchi, G. (1950) Sul cambiamento di regime di una corrente lineare a pelo libero, in un alveo di sezione costante [On the transition of regime of a free surface flow in a channel of constant cross-section]. *L'Energia Elettrica*, 27(3), 125–132 (in Italian).

Eisner, F. (1931) Überfallversuche in verschiedener Modellgrösse [Overflow tests at various scale models]. *Zeitschrift für Angewandte Mathematik und Mechanik*, 11(6), 416–422 (in German).

Golaz, M. (1931) Contribution à l'étude rationnelle des ouvrages de décharge [Contribution to the rational study of waste weirs]. *Mémoires et Comptes Rendus des Ingénieurs Civils de France*, 84, 417–453 (in French).

Hégly, V.-M. (1939) Expériences sur l'écoulement de l'eau au-dessus et en dessous des barrages cylindriques [Experiences on the water flow over and below cylindrical weirs]. *Annales des Ponts et Chaussées*, 109(2), 235–281 (in French).

Jaeger, C. (1933) Notes sur le calcul des déversoirs et seuils [Notes on the computation of weirs and sills]. *Bulletin Technique de la Suisse Romande*, 59(13), 153–156; 59(14), 166–169 (in French).

Keutner, C. (1929) Neues Berechnungsverfahren für den Abfluss an Wehren aus der Geschwindigkeitsverteilung des Wassers über der Wehrkrone [Novel computational approach for the weir discharge based on the velocity distribution of water over the weir crest]. *Die Bautechnik*, 7(37), 575–582; 7(40), 636 (in German).

Keutner, C. (1933) Der Einfluss der Krümmung der Wasserfäden auf die Energiebilanz und das Wasserabführungsvermögen von abgerundeten und scharfkantigen Wehrkörpern [The effect of streamline curvature on the energy balance and the discharge capacity of rounded and sharp-crested weirs]. *Wasserkraft und Wasserwirtschaft*, 28(3), 25–29; 28(4), 43–47; 28(14), 165–166 (in German).

Kirschmer, O. (1928) Untersuchungen der Überfallkoeffizienten und der Kolkbildungen am Absturzbauwerk I im Semptflutkanal der Mittleren Isar [Study of discharge coefficient and scour formation at drop structure I in the Sempt Canal of Mittlere Isar]. *Mitteilung* 1. Forschungsinstitut für Wasserbau und Wasserkraft e.V. München. Oldenbourg, München (in German).

Lauffer, H. (1936) Strömung in Kanälen mit gekrümmter Sohle [Flows in channels with curved bottom]. *Wasserkraft und Wasserwirtschaft*, 31(19), 245–249; 31(20), 260–264 (in German).

Morris, R.H. & Houston, A.J.R. (1922) An investigation of the Herschel type of weir. *Mechanical Engineering*, 44(10), 651–654.

Ramamurthy, A.S., Vo, N.-D. & Vera, G. (1992) Momentum model of flow past weir. *Journal of Irrigation and Drainage Engineering*, 118(6), 988–994; 120(4), 684–686.

Rouvé, G., Indlekofer, H. (1974) Abfluss über geradlinige Wehre mit halbkreisförmigem Überfallprofil [Flow over straight weirs of semi-circular crest]. *Der Bauingenieur*, 49(7), 250–256 (in German).

Shen, S.S.P. & Shen, M.C. (1990) On the limit of subcritical free-surface flow over an obstruction. *Acta Mechanica*, 82, 225–230.

Vanden-Broeck, J.-M. (1987) Free-surface flow over an obstruction in a channel. *Physics of Fluids*, 30(8), 2315–2317.

Nappe characteristics

Bazin, H. (1898) *Expériences nouvelles sur l'écoulement en déversoir exécutées à Dijon de 1886 à 1895* [New experiments on weir flow conducted at Dijon from 1886 to 1895]. Dunod, Paris (in French).

Binnie, A.M. (1972) The stability of a falling sheet of water. *Proceedings of Royal Society London*, A, 326, 149–163.

Blaisdell, F.W. (1954) Equation of the free-falling nappe. *Proceedings of ASCE*, 80(482), 1–16; 81(624), 15–19; 81(794), 1.

Boyer, P., Castex, L. & Nougaro, J. (1971) Pressions et vitesses dans une lame déversante sur un seuil en mince paroi [Pressures and velocities in a jet overflowing a sharp-crested weir]. *La Houille Blanche*, 26(1), 49–58 (in French).

D'Alpaos, L. (1986) Effetti scala nei moti di efflusso fortemente accelerati [Scale effects of strongly accelerated outflows]. Proceedings of 20th *Convegno di Idraulica e Costruzioni idrauliche* Padova, pp. 763–786 (in Italian).

D'Alpaos, L. & Ghetti, A. (1984) Some new experiments on surface tension and viscosity effects on the trajectory of a falling jet. Symposium *Scale Effects in Modelling Hydraulic Structures*, 2(5), 1–8.

De Marchi, G. (1928) Ricerche sperimentali sulle dighe tracimanti [Experimental researches on overflow dams]. *Annali dei Lavori Pubblici*, 66(7), 581–620 (in Italian).

Dias, F. & Tuck, E.O. (1991) Weir flows and waterfalls. *Journal of Fluid Mechanics*, 230, 525–539.

Gaion, A. (1984) Studio sperimentale e modello teorico del moto di vene tracimanti [Experimental model and theoretical study of overflowing jets]. *Studi e Ricerche* 366. Istituto di Idraulica Giovanni Poleni. Università degli Studi, Padova (in Italian).

Ghetti, A. & D'Alpaos, L. (1977) Effets des forces de capillarité et de viscosité dans les écoulements permanents examinées en modèle physique [Effects of capillarity and viscous forces in permanent flows observed at physical models]. Proceedings of 17th *IAHR Congress* Baden-Baden, 2(A124), 389–396 (in French).

Hay, N. & Markland, E. (1958) The determination of the discharge over weirs by the electrolytic tank. *Proceeding of ICE*, 10(1), 59–86; 11(3), 381–382.

Kandaswamy, P.K. & Rouse, H. (1957) Characteristics of flow over terminal weirs and sills. *Journal of the Hydraulics Division* ASCE, 83(HY4, 1345), 1–13; 84(HY1, 1558), 59–60; 84(HY5, 1832), 51–54; 85(HY2), 85–86.

Kistler, S.F. & Scriven, L.E. (1994) The teapot effect: sheet-forming flows with deflection, wetting and hysteresis. *Journal of Fluid Mechanics*, 263, 19–62.

Rajaratnam, N., Subramanya, K. & Muralidhar, D. (1968) Flow profiles over sharp-crested weirs. *Journal of the Hydraulics Division* ASCE, 94(HY3), 843–847.

Scimemi, E. (1930) Sulla forma delle vene tracimanti [On the form of overflow jets]. *L'Energia Elettrica*, 7(4), 293–305 (in Italian).

Straus, M.W. & Young, W.R., eds. (1948) *Studies of Crests for Overfall Dams*. Boulder Canyon Project, Final Reports 6: Hydraulic investigations. Denver CO.

Strelkoff, T.S. (1964) Solution of highly curvilinear gravity flows. *Journal of Engineering Mechanics Division* ASCE, 90(EM3), 195–221; 90(EM5), 467–470; 91(EM1), 172–178; 92(EM3), 95–96.

Tuck, E.O. (1976) The shape of free jets of water under gravity. *Journal of Fluid Mechanics*, 76(4), 625–640.

Vanden-Broeck, J.-M. & Keller, J.B. (1986) Pouring flows. *Physics of Fluids*, 29(12), 3958–3961.

Sharp-crested weir flow

Barbe, A., Boyer, P., Coulomb, R., Moreau de Saint-Martin, J. & Nougaro, J. (1967) Loi hauteur-débit d'un déversoir en mince paroi fonction de la charge totale *H* et valable dans une large plage d'utilisation [The head/discharge relationship for a sharp-crested weir based on total head *H* and with a wide range of application]. *La Houille Blanche*, 23(3), 249–256 (in French).

Benedini, M. (1966) Lo stramazzo Bazin in canali di grandi dimensioni [Bazin weir in large channels]. *L'Energia Elettrica*, 43(7), 412–423 (in Italian).

D'Alpaos, L. (1976) Sull'efflusso a stramazzo al di sopra di un bordo in parete sottile per piccolo valori del carico [On the flow over sharp-crested weirs with small discharges]. *Atti dell'Istituto Veneto di Scienze, Lettere ed Arti*, 145, 169–190 (in Italian).

Jameson, A.H. (1948) Flow over sharp-crested weirs: effect of thickness of crest. *Journal of the Institution of Civil Engineers*, 31(1), 36–55.

Lazzari, E. (1961) Ricerca sperimentale sull'influenza dello spessore della cresta nell'efflusso da stramazzi in parete sottile [Experimental research on the effect of crest thickness on the overflow of sharp-crested weirs]. *L'Energia Elettrica*, 39(6), 559–568 (in Italian).

Levi, E. (1983) A universal Strouhal number. *Journal of Engineering Mechanics*, 109(3), 718–727; 110(5), 839–845.

Milano, V. (1981) Ricerca sperimentale sull'efflusso di correnti lente su stramazzi in parete sottile a bassa soglia [Experimental research on the flow of a slow stream over sharp-crested weirs of low weir height]. *Idrotecnica*, 7(6), 263–274 (in Italian).

Milano, V. (1983) Ricerca sperimentale sull'efflusso di correnti veloci su stramazzi in parete sottile a bassa soglia [Experimental research on the flow of a fast stream over sharp-crested weirs of low weir height]. *Idrotecnica*, 9(2), 263–274 (in Italian).

Rajaratnam, N. & Muralidhar, D. (1971) Pressure and velocity distribution for sharp-crested weirs. *Journal of Hydraulic Research*, 9(2), 241–248.

Ramamurthy, A.S., Tim, U.S. & Rao, M.V.J. (1987) Flow over sharp-crested plate weirs. *Journal of Irrigation and Drainage Engineering*, 113(2), 163–172.

Ranga Raju, K.G., Ali, J. & Ahmad, I. (1972) Discharge relationship for suppressed and contracted thin-plate weirs. *Journal of the Institution of Engineers,* India CI, 52(11), 286–293.

Ranga Raju, K.G. & Asawa, G.L. (1977) Viscosity and surface tension effects on weir flow. *Journal of the Hydraulics Division* ASCE, 103(HY10), 1227–1231; 104(HY7), 1114–1116; 105(HY4), 426.

Salih, A.M.A. & Francis, J.R.D. (1971) The rounding of the upstream edge of a broad-crested weir. *Proceeding of ICE*, 50(2), 169–172; 53(1), 117–121.

Unser, K. & Holzke, H. (1975) Abfluss an scharfkantigen Wehren ohne Seitenkontraktion bei anliegendem, unbelüftetem Überfallstrahl [Discharge of sharp-crested overflow-weirs without side contraction with adjacent, non-aerated nappe]. *Wasser und Boden*, 27(12), 314–317 (in German).

Vanden-Broeck, J.-M. & Keller, J.B. (1987) Weir flows. *Journal of Fluid Mechanics*, 176, 283–293.

White, W.R. (1977) Thin plate weirs. *Proceeding of ICE*, 63(2), 255–269.

Standard spillway overflow

Abecasis, F.M. (1961) Soleiras descarregadoreas: alguns problemas especiais [Overflow spillways: some special problems]. *Memoria* 175. LNEC, Lisboa (in Portuguese).

Abecasis, F.M. (1977) The behavior of spillway crests under flows higher than the design flow. Proceedings of 17th *IAHR Congress* Baden-Baden, 4(C70), 559–566.

Cassidy, J.J. (1970) Designing spillway crests for high-head operation. *Journal of the Hydraulics Division* ASCE, 96(HY3), 745–753; 96(HY8), 1778–1779; 96(HY9), 1916–1917; 96(HY11), 2395–2396; 96(HY12), 2654–2658; 97(HY1), 180–181; 97(HY10), 1755–1756.

Cassidy, J.J. & Elder, R.A. (1984) Spillways of high dams. *Developments in Hydraulic Engineering*, 2, 159–182, P. Novak, ed. Elsevier, London.

Creager, W.P. (1917) *Engineering for Masonry Dams*. Wiley, New York.

Creager, W.P. & Justin, J.D. (1927) *Hydro-electric Handbook*. Wiley, New York.

Diersch, H.-J., Schirmer, A. & Busch, K.-F. (1977) Analysis of flows with initially unknown discharge. *Journal of the Hydraulics Division* ASCE, 103(HY3), 213–232.

Escande, L. (1933) Détermination pratique du profil optimum d'un barrage-déversoir: Tracé des piles par des méthodes aérodynamiques [Practical determination of the optimum overflow dam profile: Pier design by aerodynamic methods]. *Science et Industrie*, 17(9), 430–433; 17(10), 467–474 (in French).

Escande, L. (1952) Dighe tracimanti con fessura aspirante [Overflow dams with aeration slit]. *L'Energia Elettrica*, 29(6), 363–374 (in Italian).

Escande, L. (1953) *Nouveaux compléments d'hydraulique*. Publications Scientifiques et Techniques du Ministère de l'Air 280, Paris (in French).

Ferroglio, L. (1941) Alcune osservazioni sul deflusso sopra le dighe tracimanti [Various observations on dam overflows]. *L'Energia Elettrica*, 18(7), 455–459 (in Italian).

Golaz, M. (1928) Recherches sur la dynamique des courants déversants en régime hydraulique permanent [Researches on the dynamics of steady overflows]. *Bulletin Techniques de la Suisse Romande*, 54(21), 245–250; 54(22), 257–261 (in French).

Grzywienski, A. (1950) Über die Wahl des Profils bei vollkommenen Überfällen [On the profile selection of perfect overfalls]. *Österreichische Bauzeitschrift*, 5(7), 111–114; 5(8), 130–136 (in German).

Grzywienski, A. (1951) Anti-vacuum profiles for spillways of large dams. Proceedings of 4th *ICOLD Congress* New Delhi, Q12(R19), 105–124.

Lemos, F. de Oliveira (1975) Directivas para a colocaçao das comportas nos descarregadores das barragen [Directives for the placement of gates on dams]. *Memoria* 469. LNEC, Lisboa (in Portuguese).

Lemos, F. de Oliveira (1979) Criteria for the hydraulic design of overflow dams with 2:3 upstream face slope. *Memoria* 518. LNEC, Lisboa.

Levi, E. (1965) Longitudinal streakings in liquid currents. *Journal of Hydraulic Research*, 3(2), 25–39.

Muller, R. (1908) Development of a practical type of concrete spillway dam. *Engineering Record*, 58(17), 461–462; 59(3), 83–84.

Offitzeroff, A.S. (1940) Model studies of overflow spillway sections. *Civil Engineering*, 10(8), 523–526; 10(10), 660; 10(11), 727; 10(12), 789–790.

O'Shaughnessy, P.S. (1941) Conformity between model and prototype tests: Madden Dam spillway. *Civil Engineering*, 11(8), 491–493; 11(10), 614.

Randolph, R.R. (1938) Hydraulic tests on the spillway of the Madden Dam. *Transactions of ASCE*, 103, 1080–1132.

Reese, A.J. & Maynord, S.T. (1987) Design of spillway crests. *Journal of Hydraulic Engineering*, 113(4), 476–490.

Rouse, H. & Reid, L. (1935) Model research on spillway crests. *Civil Engineering*, 5(1), 10–14.

Scimemi, E. (1937) Il profilo delle dighe sfioranti [The profile of overflow dams]. *L'Energia Elettrica*, 14(12), 937–940 (in Italian).

Smetana, J. (1948) Etude de la surface d'écoulement des grands barrages [Study of flow surface of large dams]. *Revue Générale de l'Hydraulique*, 14(7/8), 185–194; 15(1/2), 19–32 (in French).

Song, C.C.S. & Zhou, F. (1999) Simulation of free surface flow over spillway. *Journal of Hydraulic Engineering*, 125(9), 959–967.

Various (1944) Conformity between model and prototype: a symposium. *Transactions of ASCE*, 109, 1–193.

Chapter 3 Frontispiece (a) Van Phong Dam piano key weir, Vietnam (Courtesy Michael Pfister), (b) labyrinth weir of Linville Land Harbor Dam, located in North Carolina (Courtesy Brian Crookston, Utah State University, Logan UT)

Chapter 3

Spatial crest overflow

3.1 Introduction

Overflow structures are arranged in plan view either two-dimensionally or three-dimensionally, corresponding then to frontal and spatial overflows, respectively. The former are considered in Chapter 2, whereas the latter are highlighted in this chapter. The spatial overflow of weir structures is of course more complex than frontal overflow, so that these are less mathematically developed. However, spatial overflows represent an important class of hydraulic structures often applied in practice, so that their description is relevant. Of the many designs developed in the past, the selection presented here includes:

- The *side channel* as a weir resembling a drop structure discharging water into a collector channel, referred to as side channel. These were introduced in hydraulic engineering in the 1920s and are often employed in connection with rockfill dams, given that these normally do not include a crest overflow due to safety aspects with failure following overtopping. Instead, the spillway is arranged sideways of the dam so as not to endanger the dam. A historical overview is presented by Hager and Pfister (2011).
- *Morning glory overfall,* typically arranged in the reservoir with a circular plan allowing for water overflow that is radially collected and conveyed to a vertical shaft below the bottom elevation of the reservoir. A $90°$ bend then discharges the overflowing water to an energy dissipator (Chapter 5). The causes for this structure are similar to the side channel, thereby also avoiding a direct overflow of the dam crest. These hydraulic structures were introduced in the 1930s.
- *Labyrinth weirs* of relatively small weir height, located at the dam crest zone. As compared to the fundamental frontal weir overflow, these weirs have an increased overflow length by their particular plan design, made up of weir cycles arranged at a certain angle as compared with the basic frontal overflow structure. The zigzag weir arrangement may hydraulically increase the discharge capacity of this structure, particularly if the crest width is limited for one reason or the other. Note that the increase of discharge capacity is limited, particularly under relatively high relative overflow depths because of spatial interference of the various weir cycles. This type of overflow structure was introduced in the late 1920s (Hager *et al.*, 2015).
- *Piano key weirs* as a development of the labyrinth weir, in which not only the weir plan but also the weir section is adopted so as to increase the discharge capacity further and to reduce the structural footprint. These hydraulic structures have experienced a rapid improvement and count to the novelties in hydraulic engineering. Their benefit is

considered large if the crest width dimensions are limited since the structural footprint is small. The head on the weir should be kept below certain limits.

- The *siphon* as an old hydraulic structure, combining the hydraulics of overflow and pressurized bend flow to allow for a large discharge capacity. The original siphon has produced in several cases dangerous flow phenomena mainly due to excessively low pressure, so that the 'black-water siphon' was replaced by the 'white-water siphon' decades ago by which under-pressures are controlled. Both types of siphons will be presented, and their particular ads and pros are highlighted.

As for the other chapters, the purpose of Chapter 3 is to first describe the structure considered, then to present the hydraulic theory, and then to finally note the typical fields of applications in current hydraulic engineering. The descriptions are supported by figures and photos allowing for an overall account of what is considered relevant.

3.2 Side channel

3.2.1 Typology

The side channel or the side-channel spillway is a common structure employed as dam spillway headworks. Its axis is parallel to the overflow crest, whereas the axis of frontal overflow structures is perpendicular to the crest, as discussed in 2.2. Whereas a frontal overflow is located at the dam structure, the side channel is separated from the dam, so that the discharge is conveyed along the valley down to the tailwater. Figure 3.1 shows the side channel, the spillway, and the stilling basin as dissipator.

Side channels are often considered adequate at sites where:

- A narrow gorge does not allow for sufficient width for frontal overflow
- Sufficient overflow length is available
- Impact forces and scour are a concern for arch dams
- A dam spillway is not feasible, as for an earth dam
- Topographic conditions are favorable for side channel.

Side channels are avoided at embankment dams of large design discharge due to the limited overflow head of, say, 3 m. Also, side channels are normally not equipped with gates; however, for cases in which the reservoir level control is important, flap gates or drum gates are used. The length of the overflow structure is increased by discharging the flow from both sides into the side channel. Figure 3.2 refers to a design with multiple inlets.

The cross-sectional shape of side channels is either rectangular (Figure 3.3) or trapezoidal to save excavation cost for large overflow structures. Side channels should not submerge the reservoir outflow due to complex interactions between the overflow and the downstream channel. Also, no blocks or other appurtenances should be placed along the side channel invert to reduce the lateral flow velocity, or to direct the overflow along its axis because of axial flow disturbance (USBR 1938). The bottom slope of the side channel is usually constant, with a possible exception at the dead-end where a transition from the overflow to the side channel invert may be considered. The side channel wall roughness is comparable with that of a chute, i.e. it consists of smooth concrete. The side channel width is often constant or slightly diverging in the streamwise direction, yet there are no hydraulic advantages from the latter design regarding the head losses.

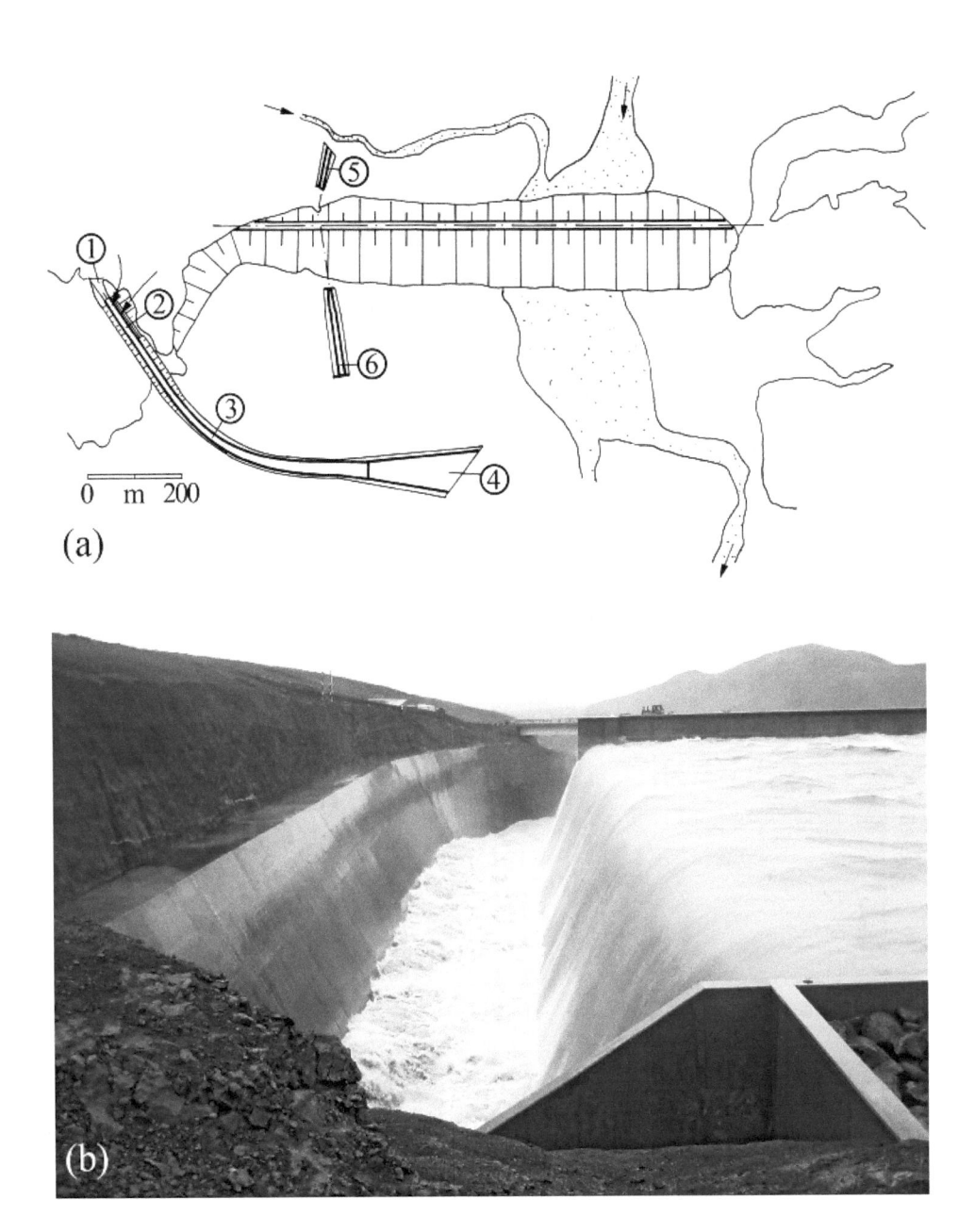

Figure 3.1 (a) General arrangement of side channel with ① frontal gated inlet, ② side channel, ③ chute tunnel, ④ stilling basin/flip bucket, ⑤ intake structure and ⑥ outlet structure of bottom outlet, (b) Karahnjukar Dam (Iceland) side channel in operation (Photo: Sigurður M. Garðarsson)

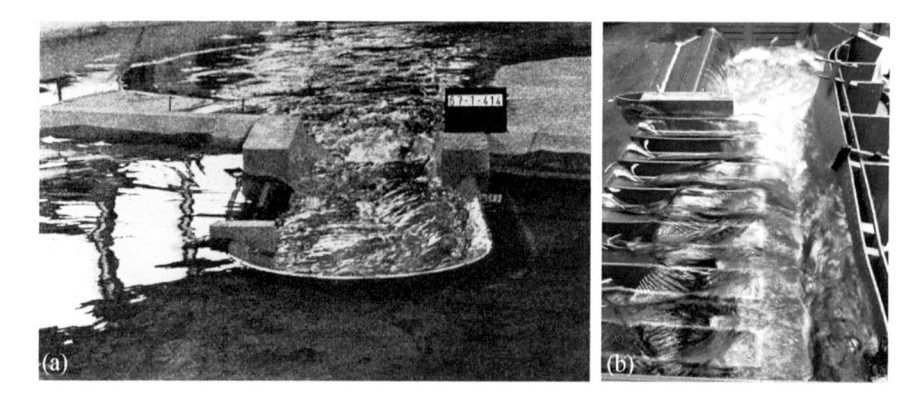

Figure 3.2 Side channel (a) with multiple discharge inlets (Bretschneider *Wasserwirtschaft* 61(5), 143), (b) hydraulic model of spillway of Gage Dam, France: lateral spillway with piano key weir combined with flap weir entering a side channel ending in a tunnel (Courtesy LCH-EPFL)

Figure 3.3 Typical views at laboratory side channel (a) upstream, (b) downstream portions, (c) plan with two-cell vorticity flow (Courtesy Willi H. Hager)

3.2.2 Hydraulic design

The governing one-dimensional equation of the free-surface profile follows from momentum considerations (Chow, 1959). Let h be the local average flow depth varying with the longitudinal coordinate x, S_o the bottom slope, S_f the friction slope, $U\cos\varphi$ the lateral inflow component in the streamwise direction, V the average side-channel velocity, Q the locally varied discharge, g the gravity acceleration, A the cross-sectional area with $\partial A/\partial x$ as surface width change, and $\mathsf{F}^2 = Q^2(\partial A/\partial h)/(gA^3)$ as the square of the local Froude number; then (Figure 3.4)

$$\frac{\mathrm{d}h}{\mathrm{d}x} = \frac{S_o - S_f - \left[2 - \dfrac{U\cos\phi}{V}\right]\dfrac{Q(\mathrm{d}Q/\mathrm{d}x)}{gA^2} + \dfrac{Q^2(\partial A/\partial x)}{gA^3}}{1 - \mathsf{F}^2}. \tag{3.1}$$

For $\mathrm{d}Q/\mathrm{d}x = 0$, i.e. no lateral inflow, and $\partial A/\partial x = 0$, i.e. a prismatic channel, Eq. (3.1) reduces to the standard backwater equation. Equation (3.1) follows also from energy considerations to yield (Hager, 2010)

$$H = h + z + \frac{V^2}{2g}, \quad \frac{\mathrm{d}H}{\mathrm{d}x} = -\left[S_f + \left(1 - \frac{U\cos\phi}{V}\right)\frac{Q(\mathrm{d}Q/\mathrm{d}x)}{gA^2}\right]. \tag{3.2}$$

Equation (3.2) states that the change of total head H with x (where z is the elevation above a reference datum) is equal to the friction slope S_f plus a term proportional to the local velocity $V = Q/A$ divided by velocity $V_d = gA/(\mathrm{d}Q/\mathrm{d}x)$, times a reduction factor depending on the lateral inflow characteristics. For $90°$ inflow, the latter term is equal to unity, and the term $V/V_d \geq 0$ is only added to the friction slope. This term is large for either large side-channel velocity or large lateral inflow.

   If the side-channel geometry, the lateral flow, the wall roughness, and a boundary condition are specified, one may solve Eq. (3.2) numerically. However, this method is simplified for practical purposes as follows. Assuming:

- A prismatic side channel, $\partial A/\partial x = 0$;
- An average value $S_{fa} = (S_{fu} + S_{fd})/2$ for the friction slope with subscripts u and d for the upstream and downstream ends, instead of a spatially variable friction slope $S_f(x)$; and
- Constant lateral inflow intensity $p_s = \mathrm{d}Q/\mathrm{d}x$,

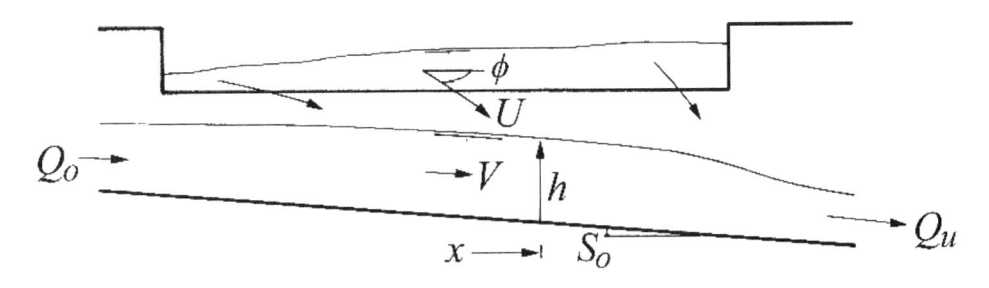

Figure 3.4 Flow in side channel, notation

the difference $J = S_o - S_{fa}$ is independent of x and corresponds to a substitute slope larger than zero. Let the coordinates for location x_s, and the typical flow depth h_s be (Bremen and Hager, 1989)

$$x_s = \frac{8p_s^2}{gb^2J^3}, \quad h_s = \frac{4p_s^2}{gb^2J^2} = Jx_s/2. \tag{3.3}$$

Non-dimensionalising as $X = x/x_s$, $y = h/h_s$ yields the governing equation for the free-surface profile $y(X)$ instead of Eq. (3.3) as

$$\frac{dy}{dx} = 2\frac{y^3 - Xy}{y^3 - X^2}. \tag{3.4}$$

The scaling values $(x_s; h_s)$ have so been selected so that Eq. (3.4) becomes *singular* at $(X; y) = (1; 1)$, i.e. both its numerator and denominator tend to zero. In other words, this singularity (subscript s) occurs for a flow which is simultaneously uniform because $dy/dX = 0$, and critical because $dX/dy = 0$. Physically this flow occurs for side channels in which the tailwater flow is supercritical along with a *transition* from subcritical upstream (typically at dead-end) to supercritical downstream flow.

For all side channels of length L_s with a supercritical tailwater flow, i.e. where the bottom slope of the downstream channel is larger than the critical slope, two cases have to be distinguished (Figure 3.5):

- If length L_s is larger than the location of the singular point x_s, then a transition from sub- to supercritical flow occurs at the singular point $x = x_s$;
- If length L_s is smaller than x_s, the latter has no physical relevance so that critical flow is forced at the end of the side channel, at the *critical point* $x = L_s$.

Though Eq. (3.4) can be solved numerically (Hager, 2010), two questions of design relevance in connection with dam side channels remain for which the tailwater flow is *supercritical* throughout:

- What is the maximum (subscript max) flow depth h_{max}, and where is its location x_{max} along the side channel?
- What is the dead-end flow depth h_o?

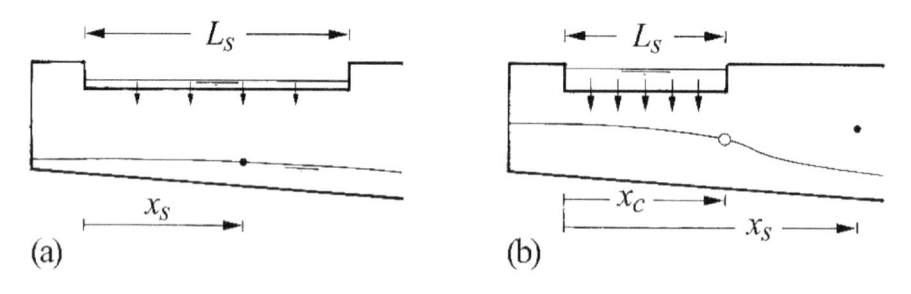

(a)   (b)

Figure 3.5 Transitional flow in side channel (a) ($\bullet$) singular point x_s, (b) ($\circ$) critical point x_c

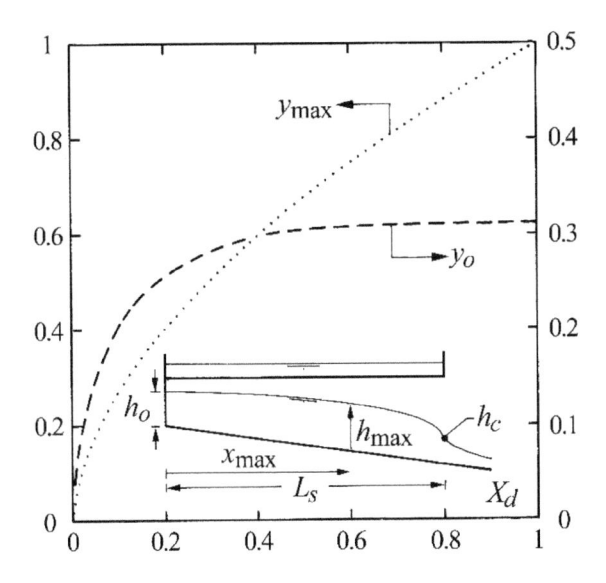

Figure 3.6 Side channel with *critical flow* at downstream end, relative upstream flow depth y_o and relative maximum flow depth y_{max} versus $X_d = L_s/x_s$

Figure 3.6 shows both $y_{max} = h_{max}/h_c$ and $y_o = h_o/h_c$ versus $X_d = L_s/x_s \leq 1$ based on numerical computations. For a given critical flow depth $h_c = [Q_d^2/(gb^2)]^{1/3}$ result the upstream flow depth h_o, and the maximum flow depth h_{max}, allowing to sketch the free-surface profile $y(X)$. The location of the maximum flow depth thereby is $X_{max} = y_{max}^2$ from Eq. (3.4). Flows with subcritical tailwater are not typical in dam hydraulics, and are therefore not further considered (Hager, 2010).

3.2.3 Spatial flow features

Side channels have a tendency to develop spiral flow due to the interaction of the lateral inflow and the axial side-channel flow components. The previous approach yields the one-dimensional (1D) surface profile $h(x)$ that nearly coincides with the flow depth at the lateral jet impact point. However, as shown in Figure 3.7, for various side-channel shapes, the wall flow depth t_s, or the corresponding axial flow depth for twin inlets may become considerably larger than $h(x)$ due to secondary flow presence. According to limited observations, and with $P_h = b + 2h$ as wetted perimeter, z_s as lateral fall depth, and for a trapezoidal side channel of transverse slope 1 (horizontal): 0.6 (vertical) (Hager, 2010)

$$\frac{t_s}{h} = 1 + 5.5 \left[\frac{p_s^2}{ghP_h} (z_s/h)^{\frac{1}{2}} \right]^{1/2}. \tag{3.5}$$

For rectangular side channels with $U_s = p_s/h_s$ as lateral inflow velocity results

$$\frac{t_s}{h} = 1 + \gamma_s \frac{p_s U_s}{gh^2}. \tag{3.6}$$

Here the average proportionality factor is $\gamma_s = 1$, and the maximum increases to $\gamma_s = 1.5$. The local flow depth $h(x)$ is computed from the 1D approach, as previously outlined.

Side channels often discharge into tunnel spillways (Chapter 8). To satisfy the basic requirements of dam overflow structures, these tunnels must always have *free surface* flow, because submergence easily leads to dam overtopping. Of particular concern is the tunnel inlet (Figure 3.8). If the flow touches the inlet vertex, gate type flow forms with a rapid increase of the upstream head. Then, the overflow structure may get submerged, depending on the crest elevation relative to the vertex. The free overflow thus changes to submerged overflow, with a strong increase in reservoir elevation.

Gate type flow at the tunnel inlet can lead to a serious *vortex breakdown* action. If the submergence of the overflow crest is sufficient, a transition from two-cell vortex flow to one-cell vortex flow (Figure 3.7) develops, associated with a strong longitudinal circulation. The pressure decreases toward the vortex center, so that air may be sucked against the flow direction up to the dead-end, where the line vortex attaches to the wall. Vortex breakdown may seriously decrease the discharge capacity and lead to an additional reservoir level rise (Figure 3.8). It is thus important to verify that free-surface flow forms not only along the side channel, but also in the tailwater tunnel.

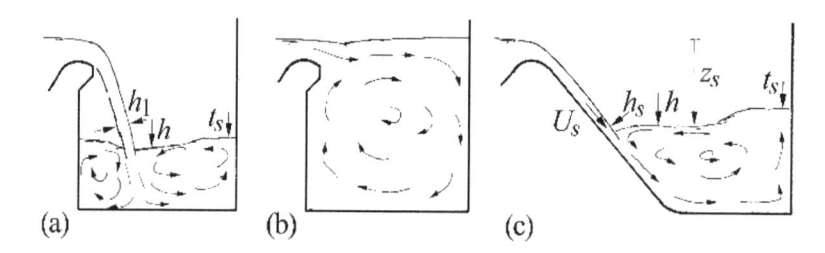

Figure 3.7 Lateral flow configurations in side channel (a) free, (b) submerged flow in rectangular cross section, (c) free flow in trapezoidal cross section

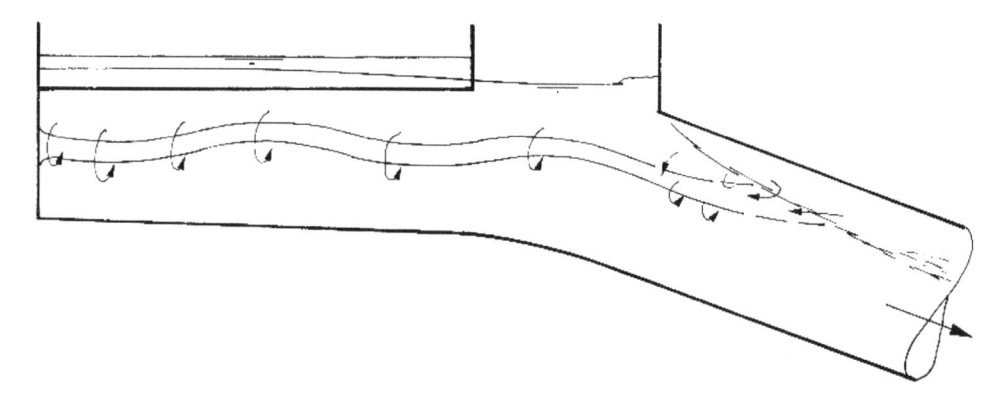

Figure 3.8 Side channel connected with tailwater tunnel under surging flow causing submerged side-channel flow. Note pressurized tunnel intake by which flow is no more fully aerated, and tornado vortex along side channel

3.2.4 Examples of physical model studies

Introduction

Side channels constitute hydraulic structures collecting lateral inflow and directing it toward a downstream channel, chute, or tunnel (Hager, 2010). The hydraulic theory of side channels is based on the conservation of momentum, assuming that the lateral inflow has no effect on the streamwise side-channel velocity. The hydraulic energy over the weir crest is partially dissipated at impact onto the side channel (USBR, 1960). This assumption allows for a 1D approach of the free-surface flow with a spatially increasing discharge (USBR, 1960; Hager, 2010). Despite a considerable interest in the past, limited experimental data on side channels are available. Lucas *et al.* (2015) therefore computed streamwise free-surface profiles of three side channels and then compared the results with experimental laboratory data. Additionally, cross-sectional profiles of the model investigations were gathered with a focus on the elevation of the water surface at the sidewall opposite to the overflow weir. The flow patterns and the formation of air entraining vortices are also discussed.

Notable studies on the hydraulics of side channels have been furnished from the 1920s in the USA relating mainly to the large dam construction activities. The first complete study was presented by Hinds (1926), who both conducted an experimental account and introduced the equations of spatially varied flow. A similar work is due to Favre (1932), who derived the identical governing equations using a more rigorous approach, however. McConaughy (1933) was the first applying the previous approach to the largest side channel then under construction, namely the Nevada Spillway of the Boulder (today Hoover) Dam. His results were discussed by Meyer-Peter and Favre (1934) based on a model erected at ETH Zurich. McConaughy responded in his closure to the basic research direction of his 1933 approach. A large experimental campaign was then undertaken by the US Bureau of Reclamation (USBR, 1938) also relating to the Boulder Dam, resulting in numerous flow details of the hydraulics of this structure.

Another notable paper was presented by Camp (1940), relating to side channels as employed in final sedimentation tanks of wastewater treatment stations. De Marchi (1941) and his collaborator Citrini (1942, 1948) at the Hydraulic Laboratory of Milan, Italy, investigated both experimentally and computationally side-channel flow. They also considered the effect of a gradual width increase of the side channel.

Keulegan (1950) introduced the concept of the singular point analysis to determine the critical flow features. This allows for the initiation of the free-surface computation with the generalized spatially varied flow equation. Li (1955) described the side-channel hydraulics with generalized backwater curves, and determined the effects of both sloping sidewalls and the bottom slope on the solution. Frictional effects were also accounted for. The large number of discussers of his paper reflects the general interest in this fascinating topic.

Sassoli (1959) conducted a large and detailed experimental campaign on side-channel flow providing a rich data basis mainly for computational purposes. Hager *et al.* (1988) employed these to validate their approach using the Boussinesq equation. Liggett (1961) provided an insight into the numerical procedure available for the solution of the free-surface equation of spatially varied flows, thereby again attracting various discussers. Yen (1973) revisited these equations based on both the energy and the momentum concepts, discussing their fundamental issues. Based on laboratory experimentation, Bremen and Hager (1990) studied the spatial flow development and the air entrainment features of side-channel flow, including a description of the so-called tornado vortex. A similar study was also published by Guercio and Magini (1994). Unami *et al.* (1999) aimed at a description of side-channel flow by employing a computational approach. Note that few additions have been presented in the past years, mainly relating to guidelines for purposes in hydraulic practice. Therefore, the following refers to a number of projects in which these issues have been dealt with.

Laboratory experimentation

VAW research dealt with the side channels of three projects analyzed in physical model studies at ETH Zurich. The Trängslet and Kárahnjúkar physical models were built in 1:45 scale, whereas the Lyssbach side channel was reproduced at a scale of 1:16. For each side channel, three to four load cases (LC) were investigated (Table 3.1), with discharges Q or unit discharges q increasing from LC1 to LC3 or LC4. The cross-sectional and longitudinal water levels were measured by point gages for both the Trängslet and Lyssbach side channels, and by ultrasonic sensors for the Kárahnjúkar project. The three side channels are described below, and a comparison of their features is given.

TRÄNGSLET DAM

Within the upgrade of Trängslet Dam, Sweden, and the increase of the design flood, the spillway capacity was increased (Boes *et al.*, 2013). A new flap gate and an unregulated overflow weir were planned next to the existing Tainter-gated spillway, both discharging into a new side channel. It is followed by a new spillway canal, and a stepped spillway merging the discharges from the side channel and the existing spillway (Figures 3.9, 3.12a–b). The trapezoidal side channel has a bottom width expanding from $b = 20$ m to 30 m (initial design) in the streamwise direction with a bottom slope of $S_o = 3\%$. The left-hand sidewall has a slope of 10:1 (V:H), whereas the right-hand sidewall toward the ogee weir and the flap gate has a slope of 1:1. The horizontal ogee weir crest lies at the full reservoir supply level of total length $L = 125$ m. The side-channel height between its bottom and the weir crest is $t = 9.8$ m at the mid-section. The ogee weir height relative to the upstream reservoir bottom elevation is $w = 1.55$ m; its design (subscript d) head is $H_d = 2.86$ m relative to the weir design discharge $Q_d = 1284$ m³/s. The 30 m wide flap gate is separated from the ogee weir by a 3 m wide pier. Downstream of the flap gate, the side channel is extended with a vertical right-hand sidewall and a constant bottom width of 30 m of 0.3% bottom slope.

Table 3.1 Total discharges Q and unit discharges q over overflow weir of investigated side channels and Q_{fg} over flap gate for Trängslet project for different LC (Lucas *et al.* 2015)

			LC1	LC2	LC3	LC4
Trängslet	Q	[m³/s]	224	482	945	1,340
	q	[m²/s]	1.8	3.9	7.6	10.7
	Q_{fg}	[m³/s]	348	433	562	680
Kárahnjúkar	Q	[m³/s]	800	1,350	2,250	–
	q	[m²/s]	5.7	9.6	16.1	–
Lyssbach	Q	[m³/s]	25.2	42.6	63.9	–
	q	[m²/s]	0.7	1.2	1.8	–

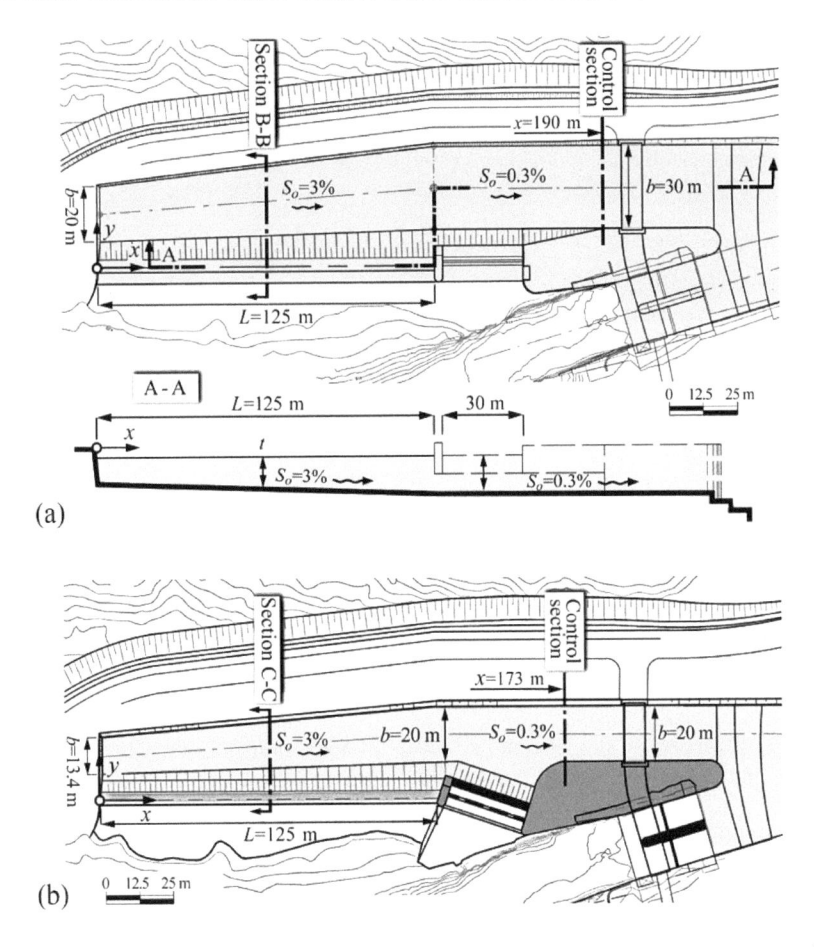

Figure 3.9 Trängslet side channel, (a) plan and (−·−) longitudinal section A-A of initial design, (b) plan of final design with overflow weir, flap gate, existing spillway, new spillway canal and start of stepped spillway, including measurement Sections B-B and C-C (Lucas *et al.* 2015)

To reduce excavation work and thus increase project economy, the side-channel width was reduced by one-third from the initial to the final design. Additionally, the flap gate was rotated by 18° to the side-channel axis to generate a favorable approach flow from the reservoir to the new spillway canal. This improvement also added to a better passage of driftwood through the flap gate. Note the slightly modified plan of the triangular pier-shaped element downstream of the flap gate.

KÁRAHNJÚKAR DAM

The spillway of the newly built Kárahnjúkar Dam, Iceland, consists of a side channel followed by an open chute (Tomasson *et al.*, 2006; Pfister *et al.*, 2008). The trapezoidal side channel has an increasing bottom width b from 6 m to 12 m in the streamwise direction and a bottom slope of $S_o = 0.26\%$ (Figure 3.10). The sidewalls have a slope of 2:1 (Figure 3.12c).

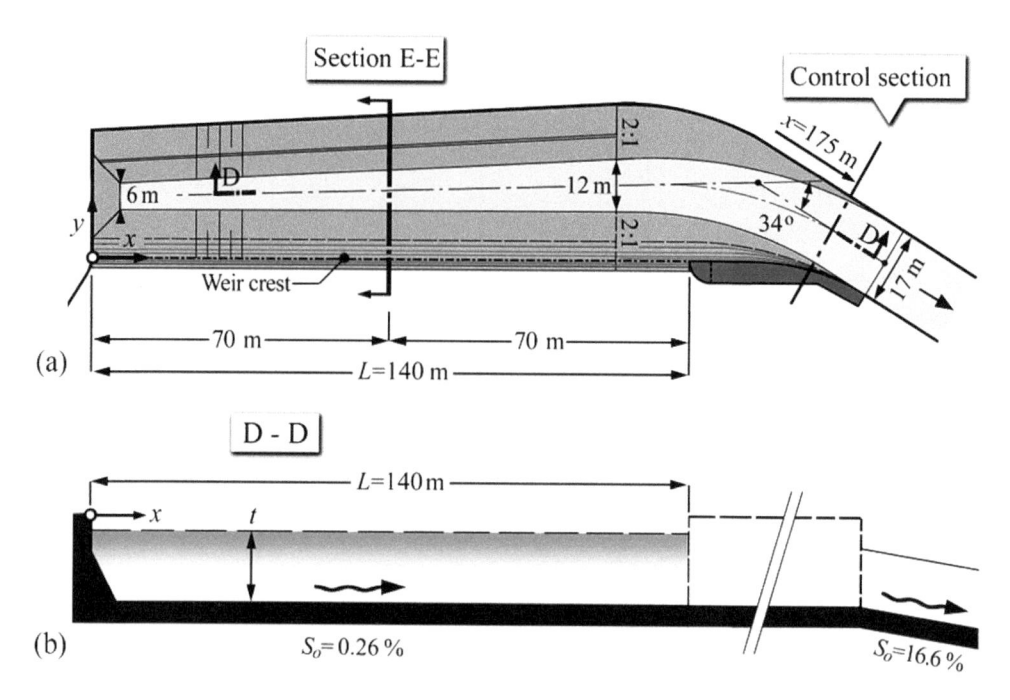

Figure 3.10 Kárahnjúkar side channel and start of open chute (a) plan, (b) (−·−) longitudinal section D-D (Lucas *et al.* 2015)

The crest length of the horizontal standard weir profile is $L = 140$ m and the height is $w = 3$ m. The weir design discharge is $Q_d = 710$ m³/s with a design overflow head of $H_d = 1.67$ m. Note that the weir design discharge Q_d is lower than the total spillway (subscript spw) design discharge $Q_{d,spw}$, so that for all load cases $Q > Q_d$ (Table 3.1). A transition bend of 34° deviation angle and a cross-sectional transition from trapezoidal to rectangular shape leads to the open chute. The bottom width increases gradually from 12 to 17 m along the bend.

LYSSBACH DIVERSION TUNNEL

A discharge portion of the Lyssbach was routed through a diversion tunnel during floods to guarantee flood safety to the town of Lyss, Switzerland (Pfister and Rühli, 2011). The flow was conveyed from a small head pond into a rectangular side channel made up of a standard profile weir (Vischer and Hager, 1998) of crest length $L = 35.8$ m and $Q_d = 42.6$ m³/s. The weir height is $w = 1.75$ m. The side channel has a constant bottom width of $b = 4.68$ m with a transverse (subscript t) bottom slope of $S_t = 3\%$ (Figure 3.11, 3.12d). The streamwise slope is $S_o = 0.80\%$ with an average side-channel depth of $t = 4.5$ m. A transition bend with 17° deviation angle and a slope increase to 1.06% follows in the tailwater channel. An acceleration reach of variable slope and a change from the rectangular to circular cross section leads to the diversion tunnel.

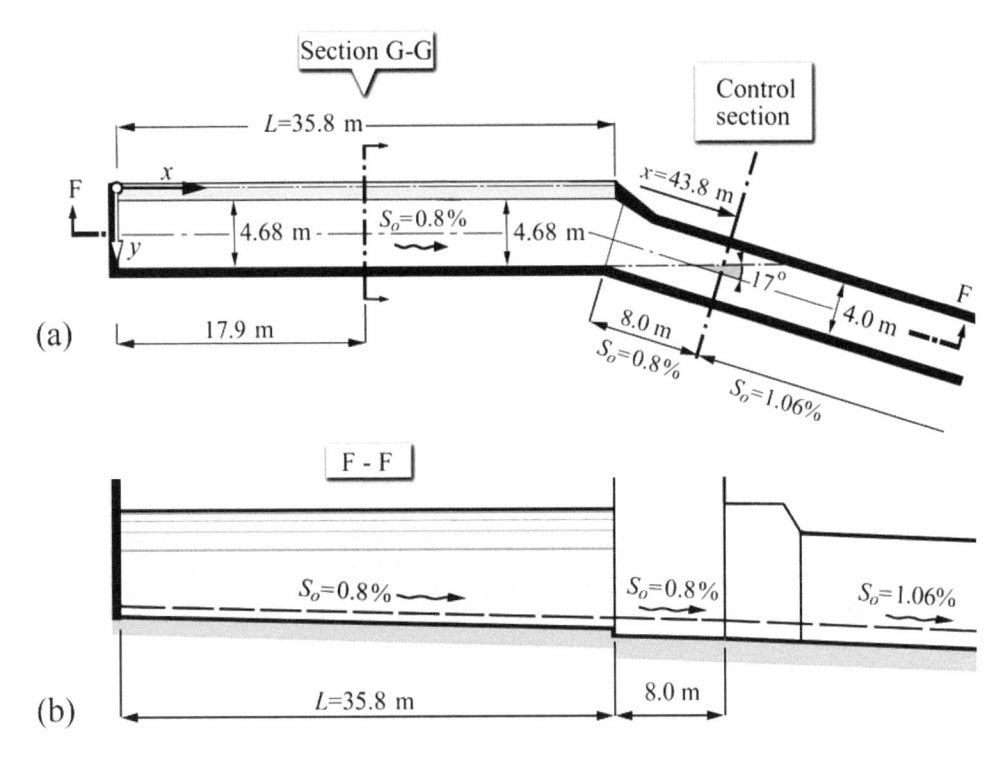

Figure 3.11 Lyssbach side channel and acceleration reach (a) plan, (b) (−·−) section F-F (Lucas *et al.* 2015)

Comparison of side-channel features

The main parameters of the side channels shown in Figures 3.9–3.13 are compared in Table 3.2. The overflow length L is similar for Trängslet and Kárahnjúkar, but at Trängslet there is a 30 m long flap gate discharging into the same side channel next to the ogee weir. For spillway design discharge conditions, the head ratio $\chi = H/H_d = 1.5$ at Kárahnjúkar, with H_d as the scaling length for the crest shape, whereas $\chi = 0.5$ and 1 for Tränsglet and Lyssbach, respectively. Sub-atmospheric bottom pressure occurs at the Kárahnjúkar upstream weir quadrant, associated with increased spillway capacity.

With a higher t and a smaller b, Kárahnjúkar side channel is much more compact than the wide Trängslet side channel. Both have a trapezoidal cross section with a streamwise-varied channel bottom width b. The dimensions of the Lyssbach side channel are considerably smaller than of the two others, and the cross section is rectangular (Figure 3.12). The streamwise bottom slope is $S_o = 3\%$ for the Trängslet side channel and less than 1% for the Kárahnjúkar and Lyssbach projects. The Lyssbach side channel is the only having a transverse bottom slope S_t. Its effect on the flow pattern is considered negligible. The Trängslet and Kárahnjúkar side channels discharge into open chutes, whereas the Lyssbach side channel leads to a diversion tunnel of circular cross section. Both the Lyssbach and Kárahnjúkar projects have bended transition sections following the side channels.

Table 3.2 Side-channel parameters: Comparison of Trängslet original and final designs with Kárahnjúkar and Lyssbach side channels

		Trängslet original design	Trängslet optimized design	Kárahnjúkar	Lyssbach
Overflow weir length L	[m]	120	120	140	35.8
Average bottom width b	[m]	25	16.7	9.0	4.68
Average height t	[m]	9.8		18.0	4.5
Weir height w	[m]	1.55		3.0	1.75
Bottom slope S_o	[%]	3		0.26	0.8
Transverse bottom slope S_t	[%]	0		0	3.0
Side-channel cross section		trapezoidal		trapezoidal	rectangular
Wall slope weir side	V:H	1:1		2:1	∞
Wall slope shore side	V:H	10:1		2:1	∞
Weir design discharge Q_d	[m³/s]	1284		710	42.6
Weir design head H_d	[m]	2.86		1.67	0.68
Design discharge $Q_{d,spw}$ (LC2)	[m³/s]	1975[1]		1350	42.6
Head ratio χ for $Q_{d,spw}$	[-]	0.54		1.51	1.0
Model scale	[-]	1:45		1:45	1:16

[1] Including discharge of 1060 m³/s through existing spillway.

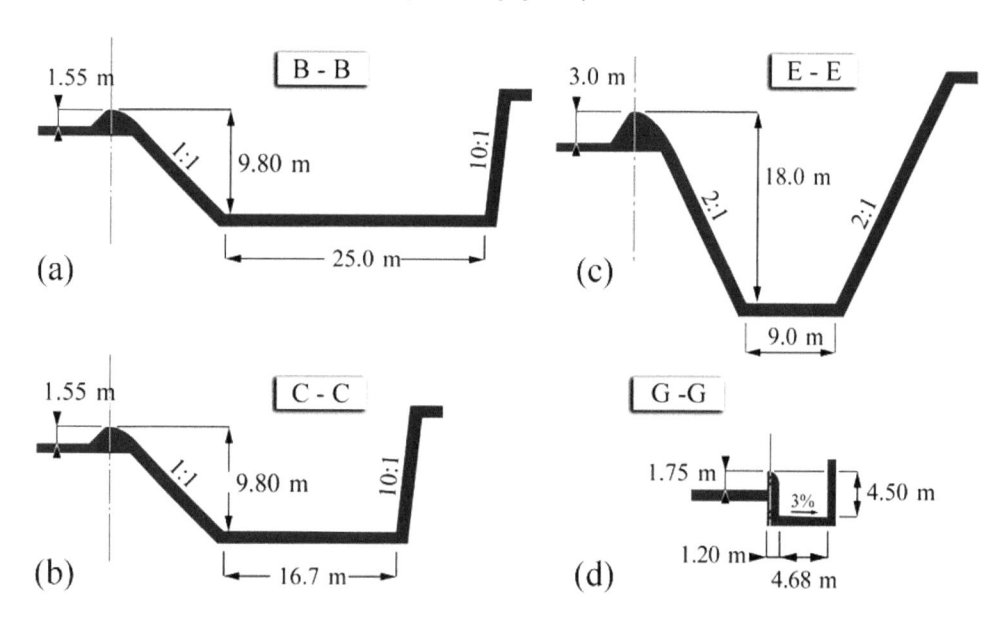

Figure 3.12 Side-channel cross sections for Trängslet (a) original, (b) final design, (c) Kárahnjúkar, (d) Lyssbach, numbers in [m]. For plans and streamwise sections see Figures 3.9–3.11 (Lucas *et al.* 2015)

Flow patterns

The flow in a side channel is characterized by the lateral inflow from the overflow weir on the one hand, and by the axial channel flow characteristics on the other hand. For side channels of mild weir face slope, two different flow patterns are distinguished (Figure 3.13):

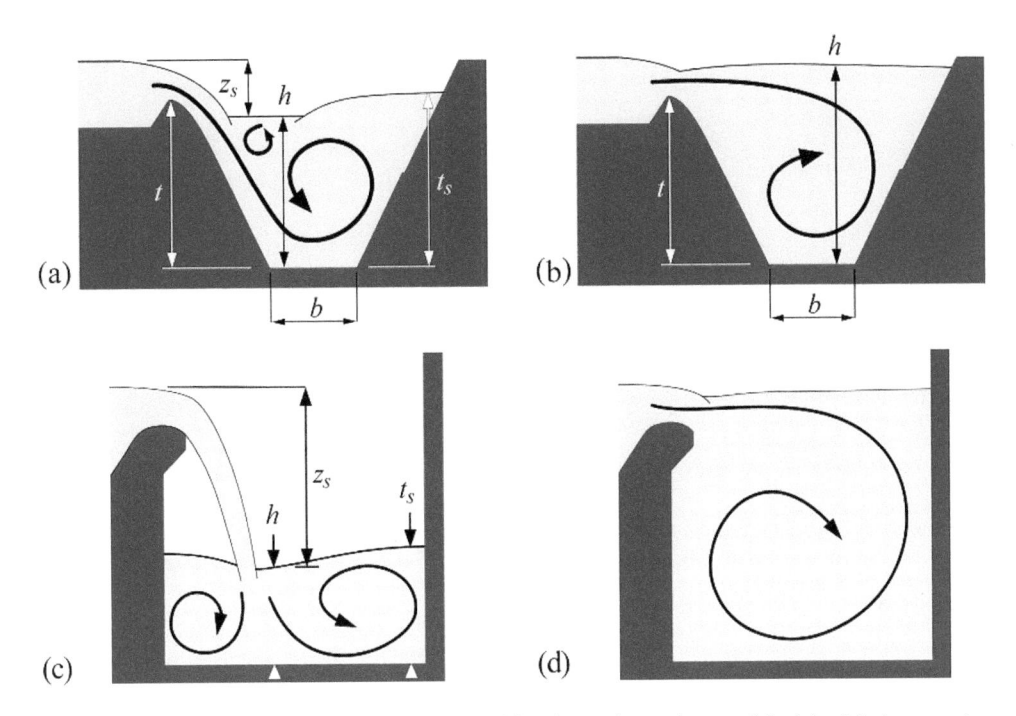

Figure 3.13 Trapezoidal and rectangular side-channel sections with (a), (c) two-vortex flow, (b), (d) single-vortex flow (Lucas *et al.* 2015)

(a) At relatively small discharge or small flow depth in the side channel, the interaction of the lateral plunging jet and the axial side-channel flow results in two-vortex spiral flow, referred to as plunging jet in frontal weir flow. The overflow plunges onto the side channel, follows its bottom and emerges along the opposite side, inducing a strong longitudinal vortex. A secondary vortex develops between the weir overflow and the dominant spiral flow. (b) At high discharge or high flow depth in the side channel, the weir crest is submerged, resulting in a transition from the plunging to surface jet regime, associated with a change to a single-vortex structure, indicating the discharge capacity limit. Whereas the flow surface for single-vortex flow is relatively smooth, the two-vortex spiral flow is characterized by a turbulent flow surface. The water level along the opposite sidewall is higher than the water elevation at the impact point due to the spiral flow (Figure 3.13a).

For rectangular side channels with vertical sidewalls (Figure 3.13), the free-falling plunging jet of the lateral inflow generates (c) two vortices at relatively low discharge and low water elevation inducing high water levels along both sidewalls associated with a lower axial flow depth. A large amount of air is entrained into the side-channel flow. (d) At higher discharge or high flow depth, the weir is submerged, generating a surface jet that induces a single vortex as for the trapezoidal side channel.

For the initial design of the Trängslet side channel, a weak two-vortex flow was observed for all LC because the large side-channel width did not allow for full spiral flow development. Given the large freeboard, the discharge capacity of the side channel was not exploited. Therefore, the effect of a width reduction was studied. Figure 3.14 compares the final Trängslet side-channel flow at reduced channel width for two LCs. The flap gate and the dividing

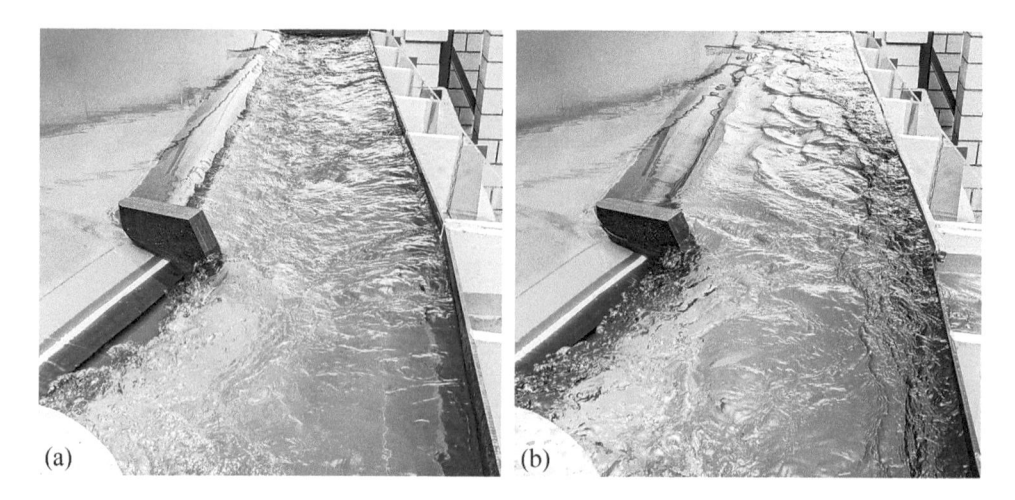

Figure 3.14 Tailwater views of Trängslet side channel for final design with reduced width (a) LC3 with two-vortex structure, (b) LC4 with single-vortex structure (Lucas *et al.* 2015)

Figure 3.15 Kárahnjúkar side channel (a) side, (b) tailwater views for LC3 with pronounced two-vortex flow and visible tornado vortex with air entrainment (Lucas *et al.* 2015)

pier to the overflow weir are also visible. The two-vortex flow in Figure 3.14a occurs for LC3 with a smaller flow depth at the line of jet impact h than along the sidewall t_s. Figure 3.14b shows the transition to the single-vortex pattern for LC4 with a relatively smooth flow surface. The side-channel discharge capacity is fully exploited without any remaining freeboard. Note the absence of air entrainment for both flow patterns.

The two-vortex flow of the Kárahnjúkar side channel is more pronounced than for the Trängslet side channel (Figure 3.15). The difference between the flow depths h and t_s is distinctive. In the relatively narrow side channel, the emergent jet has a higher energy component from the plunging jet as in the flat and wide Trängslet project. At high discharges, a winding line vortex develops along the side-channel axis, referred to as tornado vortex (Hager, 2010). Its formation is due to the high tangential velocity component at the spiral flow core, inducing a sub-atmospheric pressure strong enough to entrain air from the tailwater. At the upstream end of the side channel, a transition to single-vortex flow occurs for LC3, simultaneously to the persistent two-vortex flow in the center and downstream channel portions. For smaller discharges the dominant spiral current following the left side dies off in the transition bend (LC1), whereas for LC3 it propagates into the bend and disturbs the upstream chute flow (Figure 3.15).

In contrast to the prismatic Trängslet and Kárahnjúkar side channels, the Lyssbach side channel has a rectangular cross section. Due to the free lateral inflow, the impact of the plunging jet onto the side-channel flow is higher, entraining a large air quantity. The two-vortex spiral flow is strongly pronounced, similar to the Kárahnjúkar side channel; at high discharges (LC3) a tornado vortex is formed (Figures 3.16, 3.17). The vortex is unstable, however, alternating with a chain of aligned air bubbles and air clusters rotating around the vortex core. Despite the smaller dimensions, the aspect ratio of average width to depth of the side channel is similar for both Kárahnjúkar and Lyssbach projects with nearly 1, whereas the average aspect ratio of the Trängslet side channel is 3 and 2 for the initial and reduced channel widths, respectively.

Streamwise surface profile

The side-channel inflow over the weir for the projects investigated is perpendicular to its longitudinal axis allowing for neglect the streamwise momentum component of the lateral inflow. Steady, 1D open-channel flow with a streamwise increasing discharge, i.e. spatially varied flow, is considered. The origin of the streamwise coordinate x is located at the upstream end of the side channel. For a channel of cross section $A(x, h)$, Eq. (3.1) describes

Figure 3.16 Upstream views of Lyssbach side channel (a) LC2, (b) LC3 (Lucas *et al.* 2015)

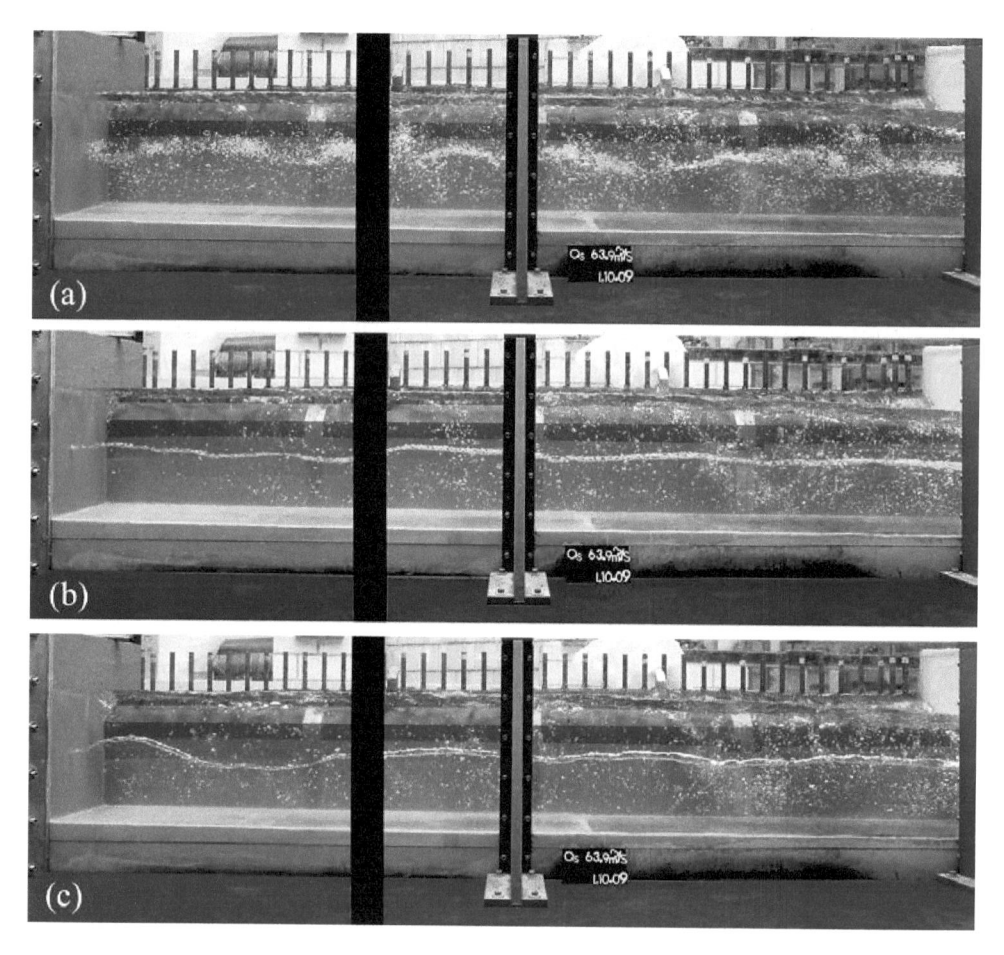

Figure 3.17 Lyssbach side channel, side views for LC3 (a) rotating bubble cluster, (b) aligned air bubble chain, (c) tornado vortex (Lucas *et al.* 2015)

the longitudinal surface profile $h(x)$ (Chow, 1959). Given that $\varphi = 90°$ in the simplified version, the term $[2 - (U\cos\varphi/V] \rightarrow 2$. According to Bremen and Hager (1989), the average flow depth $h(x)$ coincides with that at the lateral jet impact point (Figure 3.13). Note that the momentum correction factor was ignored, given its complexity in terms of general specification.

In all three cases considered, not only the discharge is spatially varied but also the cross-sectional profile $A(x)$ and the bottom slope $S_o(x)$. These three properties were mathematically formulated as continuous functions to allow for the definition of the singular points. The locations so determined are shown as 'Control section' in Figures 3.9–3.11. The slope of the free-surface profile was then determined following Bremen and Hager (1989), allowing for the application of a Runge-Kutta approach to find $h(x)$ both upstream and downstream of the critical sections. Note that the singular point was always located downstream of the side channel because of (1) the presence of the spillway canal with additional lateral discharge (Trängslet, Figure 3.9), and (2) both a contraction and slope increase toward the tailwater chute (Kárahnjúkar, Figure 3.10; Lyssbach, Figure 3.11). Computations of the free-surface profiles $h(x)$ started at the 'Critical section' against and in the flow directions for sub- and supercritical flows, respectively.

The computed free-surface profiles $h(x)$ from Eq. (3.1) or its simplified version for gradually varied flow for which $dQ/dx = 0$, are compared with the measured profiles $t_s(x)$ from the Trängslet model. The nearly horizontal transverse cross-sectional free-surface profiles allows for the simplification $h(x) \approx t_s(x)$ (Figure 3.21). The critical flow section is located at the start of the prismatic new spillway canal at $x = 190$ m for the initial design and at $x = 173$ m for the final design (Figure 3.9). The flap gate of lateral discharge intensity $dQ/dx = 348/30$ m²/s $= 11.6$ m²/s for LC1 is located between $x = 125$ m and 155 m. Given this relatively large value, the free-surface slope is there relatively steep. Further upstream, $dQ/dx = 224/125 = 1.8$ m²/s, resulting in a much smaller free-surface slope. Figure 3.18 relates to the load cases LC1 and LC3 both over length L of the Trängslet side channel, and the new spillway canal toward the stepped spillway (Table 3.2). For LC1 of the initial design, the computed flow depth along the side channel practically coincides with the data, whereas deviations increase along the canal and further downstream due to local flow perturbations of the flap gate and flow deflection. A sudden peak in the data downstream of the overflow weir is noted for both LC1, and the final design, due to wave run-up of the flap-gate inflow at the opposite sidewall. This roughly agrees with the corresponding energy head $H(x = 162$ m$) = 5.3 + (572/(20 \cdot 5.3))^2/2g = 6.7$ m. For LC3, the measured and computed data along the side channel down to the critical point fit better for the final than the initial design,

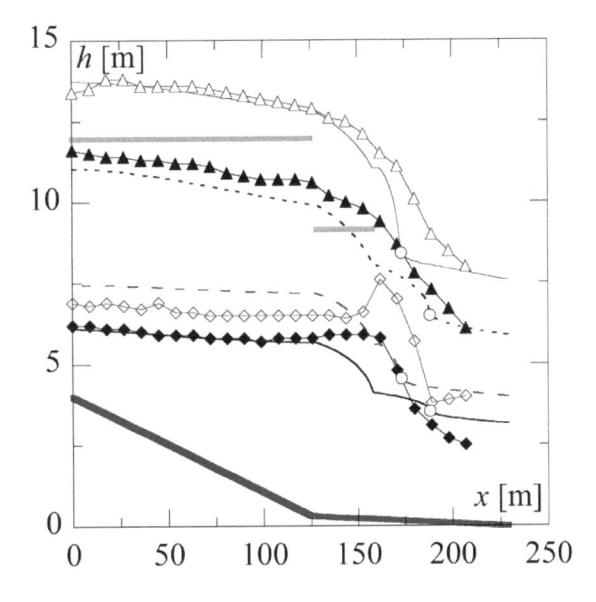

Figure 3.18 Trängslet side channel: Computed $h(x)$ and measured $t_s(x) \approx h(x)$ streamwise free-surface profiles. Initial design, LC1, (♦) measured, (–) computed; LC3, (▲) measured, (⋯) computed; final design with reduced side-channel width, LC1, (◊) measured, (--) computed; LC3 (△) measured, (–) computed. (—) Channel bottom and (–) crests of weir and lowered flap gate. ☉ Critical points (computed) (Lucas *et al.* 2015)

with deviations further downstream due to the start of the stepped chute. The local inflow from the flap gate is relatively small for LC3, so that no peak in $h(x)$ is observed.

The Kárahnjúkar side channel is characterized by a transition bend at its downstream end, which is overlooked in the 1D approach. The critical flow section is located at $x = 175$ m in the bend portion (Figure 3.10). The data were collected along the smallest flow depths of the side channel close to its axis. For LC1, the experimental and computational data agree well (Figure 3.19). With increasing discharge and thus flow depths, deviations between the two increase, with the measured flow depths higher than these computed. From safety arguments, the computation lies on the unsafe side, so that additional safety provisions need to be formulated. The deviations are explained by the bend flow structure for LC3 where the prevailing lateral flow component causes a water level rise (Figure 3.19). The resulting 3D flow pattern corroborates the use of model experimentation to determine these details.

As to the Lyssbach side channel, the measured streamwise flow depths at the line of submergence and the computed data based on Eq. (3.1) coincide well (Figure 3.20). The critical flow section is located at $x = 43.8$ m at the start of the acceleration reach (Figure 3.20). Note the simple plan and streamwise geometries of the Lyssbach project as compared with the other two. From these three selected cases, it becomes obvious that the data fit with the measurements does not improve for large discharge intensities or a specific Froude number range.

Transverse surface profiles

The streamwise profiles presented were recorded along the central overflow section (Figures 3.9–3.11). The origin of the transverse coordinate y was set at the weir crest. Figures 3.21 and 3.22 show the cross sections of the side channels. The Trängslet and Kárahnjúkar channels are presented at identical scale, whereas the Lyssbach channel is reproduced at a larger scale due to the small prototype dimensions. As described, the two-vortex flow pattern is

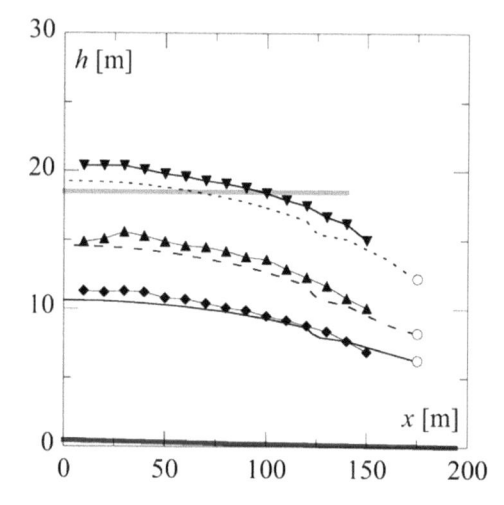

Figure 3.19 Kárahnjúkar side channel: Computed and measured streamwise free-surface profiles $h(x)$. LC1, ($\blacklozenge$) measured, (−) computed; LC2, ($\blacktriangle$) measured, (--) computed; LC3, ($\blacktriangledown$) measured, ($\cdots$) computed. Notation otherwise as in Figure 3.18. $\odot$ Critical points (computed) (Lucas *et al.* 2015)

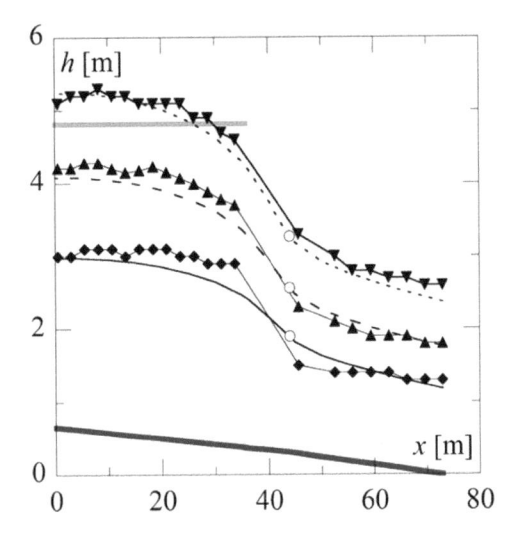

Figure 3.20 Lyssbach side channel: Computed and measured streamwise free-surface pro-
files $h(x)$. LC1, (♦) measured, (−−) computed; LC2, (▲) measured, (--) computed;
LC3, (▼) measured, (···) computed. Notation otherwise as in Figure 3.18.
⊙ Critical points (computed) (Lucas *et al.* 2015)

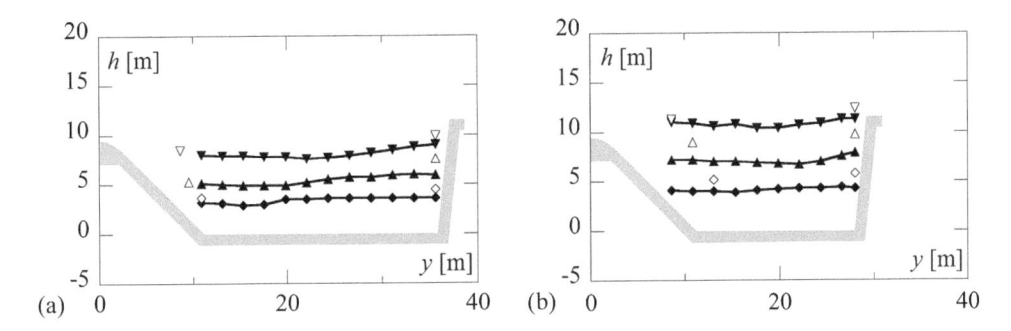

Figure 3.21 Trängslet side channel: Transverse free-surface profiles $h(y)$ at mid-section
for (a) initial design, (b) final design with reduced side-channel width, for (♦)
LC1, (▲) LC2, (▼) LC3 measured; (◊) LC1, (△) LC2, (▽) LC3 computed by Eq.
(3.1) (Lucas *et al.* 2015)

less pronounced for both versions of the Trängslet side channel than for the Lyssbach and
Kárahnjúkar projects. For the initial Trängslet design, the water level does not reach the
level of the weir crest up to LC3, so that the side-channel capacity is not exploited. For the
final design with the reduced side-channel width, the weir is submerged under LC3, however,
along with a transition to the single-vortex flow reached only at the higher discharge of LC4.

The observations of pronounced two-vortex flow in the Kárahnjúkar side channel are confirmed by the transverse free-surface data. The difference between the axial flow depth at the
line of submergence and that along the sidewall is distinctive for the three load cases. For
LC3, the water level h is at about the same elevation as the weir crest, whereas the water

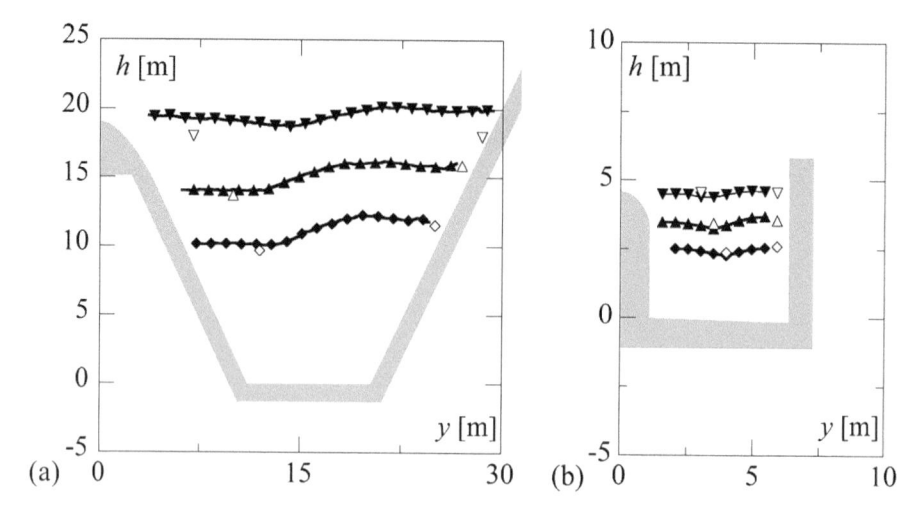

Figure 3.22 Transverse free surface profiles $h(y)$ at mid-section for (a) Kárahnjúkar, (b) Lyssbach side-channels. ($\blacklozenge$) LC1, ($\blacktriangle$) LC2, ($\blacktriangledown$) LC3 measured; ($\lozenge$) LC1, ($\triangle$) LC2, ($\triangledown$) LC3 computed by Eqs. (3.7) and (3.8), respectively (Lucas *et al.* 2015)

level along the sidewall t_s exceeds the weir crest level. The transition to single-vortex flow is observed at the upstream end of the side channel for LC3 simultaneously to the persistent and pronounced two-vortex flow in the central and downstream channel portions (Figure 3.15b). The transverse free-surface profiles of the Lyssbach side channel indicate a lower axial water level h as compared with t_s along both sidewalls. Due to the free-falling plunging jet, two vortices are generated on both sides of the impact point, resulting in a flat V-shaped free-surface profile (Figures 3.13c, 3.16).

Vischer and Hager (1998) relate the flow depth h to the maximum flow depth t_s and the fall depth z_s between the reservoir level and the average local side-channel flow depth h for a trapezoidal side channel as given in Eq. (3.5) (Figure 3.13). Equation (3.5) is based on limited data relating to a trapezoidal side channel of transverse slope 0.6:1 (V:H) and wetted perimeter P_h. (Note the printing error in Bremen and Hager, 1990.)

For a rectangular cross section as in the Lyssbach project, Eq. (3.5) reduces with U_s as the lateral inflow velocity and $\gamma_s \approx 1$ to 1.5 as a factor of proportionality to Eq. (3.6). For the Lyssbach side channel, the measured data follow Eq. (3.6) well if the computed flow depth $h(x)$ and $\gamma_s = 1$ are used (Figure 3.22b). Deviations between predicted and measured t_s values are in the range of ± 1.5–3.8%. For the trapezoidal Trängslet and Kárahnjúkar side channels Eq. (3.5) was applied, despite deviations in the transverse slopes from that underlying Eq. (3.5). The computational agreement with the data for LC1 and LC2 of the Kárahnjúkar project with a transverse slope of 2:1 compared to 0.6:1 underlying Eq. (3.5) is in the range of ± 1.4–2.2% (Figure 3.22a), whereas the correlation for LC3 (flow structure in the bend) and for the data of the Trängslet project (Figure 3.21) is poor, with deviations up to 26%. The reasons for the latter lie in the (1) large deviation in transverse slope (10:1 vs. 0.6:1) and (2) relatively wide asymmetrical cross section (Figure 3.12), leading to flat cross-sectional free-surface profiles with $t_s \approx h$ (Figure 3.21).

The conclusions are:

- Agreement between computation and measurements is good for nearly prismatic side channels, straight in plan and unaffected by local perturbations as the presence of gates or side slope changes;
- Computed free-surface profiles agree well with flow depth at line of jet impact;
- Deviations between computation and measurements are neither related to discharge intensity nor to cross-sectional shape;
- Transition from two-vortex to single-vortex flow structures is not predictable based on 1D approach;
- Presence or absence of tornado vortex along side channel is not solely explained with 1D approach.

A number of additional features of side-channel flow are highlighted, including a verification of the increase of lateral flow depth along the outer side-channel wall relative to the computed streamwise free-surface profile, or the minor effect of the momentum correction factor on the free-surface profile. Using additional safety provisions, side-channel flows thus are approximated with this procedure. In all other cases, hydraulic laboratory modeling is proposed to account for unexpected features, as described.

3.3 Morning glory overfall

3.3.1 Hydraulic concept

A shaft spillway or morning glory overflow structure is composed of three elements, namely the cup-shaped overflow portion, the vertical shaft, and the nearly horizontal diversion tunnel spillway. The design discharge has to produce *free surface flow*, i.e. both the shaft and the diversion tunnel must have a larger capacity than the overflow structure. To promote atmospheric pressure all along the spillway up to the dissipator, an aeration conduit is normally inserted at the shaft base. A representative section of the structure is shown in Figure 2.4c.

Morning glory spillways are typically used for dams of small to medium design discharges, with a maximum of some 1000 m^3s^{-1}. The structural height is up to 100 m, although 50 m is more relevant. The structure has a circular standard-crested overfall geometry, a vertical shaft, a bottom bend including the aeration device, and a diversion tunnel discharging to the energy dissipator. The structure is advantageous if:

- Seismic action is moderate
- A horizontal spillway may be connected to the existing diversion used during construction
- Floating debris is insignificant
- Space for overflow structure is limited
- Geologic conditions are excellent against settlement
- A short diversion channel is sought.

Debris may not be a concern if crest and shaft radii are sufficiently large (see below). Figure 3.23 shows a typical arrangement in a project involving an earthfill dam.

Freestanding morning glory spillways have a vertical shaft. Shafts excavated into rocky reservoir banks may be inclined. The overfall structure is normally circular-shaped but may

Figure 3.23 (a) Morning glory structure with ① intake for river diversion during dam construction, ② overflow structure, ③ inspection access, ④ tunnel spillway used
as river diversion during construction, ⑤ stilling basin; (b) El Makhazine Dam
on Oued Loukkos River (Morocco) 50 m high morning glory spillway designed
for 1450 m^{3}s^{-1} combined with intakes for bottom outlets using identical tunnel.
Four piers on circular crest avoid spiral approach flow and are extended along
shaft to give more structural strength under seismic action. Second diversion
tunnel is used for placing penstock connected to power intake (left of morning glory spillway), (c) vertical cross section and plan view of morning glory
structure with elements mentioned above (Courtesy Elektrowatt Engineering,
Zurich)

also be of so-called duckbill shape to increase the crest length (Figure 3.2). It can have a semi-circular crest if attached directly to the rock at the reservoir bank or to the upstream concrete face of a gravity dam (Hager and Schleiss, 2009).

The design of the morning glory spillway involves the overflow structure, shaft structure, tunnel spillway, and aeration conduit. The latter has not been systematically studied until now, so that a hydraulic model study is recommended for this element. The approach flow currents to the structure should be considered using a general model to account for the spatial flow features. The structure is prone to rotational approach flow, to be inhibited with a selected location of the shaft relative to the reservoir topography and the dam axis. The *radial* flow pattern is improved with piers placed on the overfall crest. Figure 3.24 relates to a typical morning glory overflow.

The morning glory overflow structure should be located so that the structural height in the reservoir becomes small and the approach flow remains nearly free of circulation. The connection to an existing diversion tunnel should involve least cost. The *asymmetry* of approach flow is improved by excavating protruding soil and rock formations, and by providing the overflow with *crest piers*. Figure 3.25a refers to an improved approach flow current of the Lower Shing Mun Reservoir in Hong Kong (Smith, 1966). According to detailed model observations, the overflow was improved with a relatively modest excavation. Figure 3.25b shows the overflow at 425 m^3s^{-1}. The approach flow was improved by the addition of two piers.

Figure 3.24 Morning glory overflow structure of Paradela Dam (Portugal), with aeration chimney on right side (Courtesy Michael Pfister)

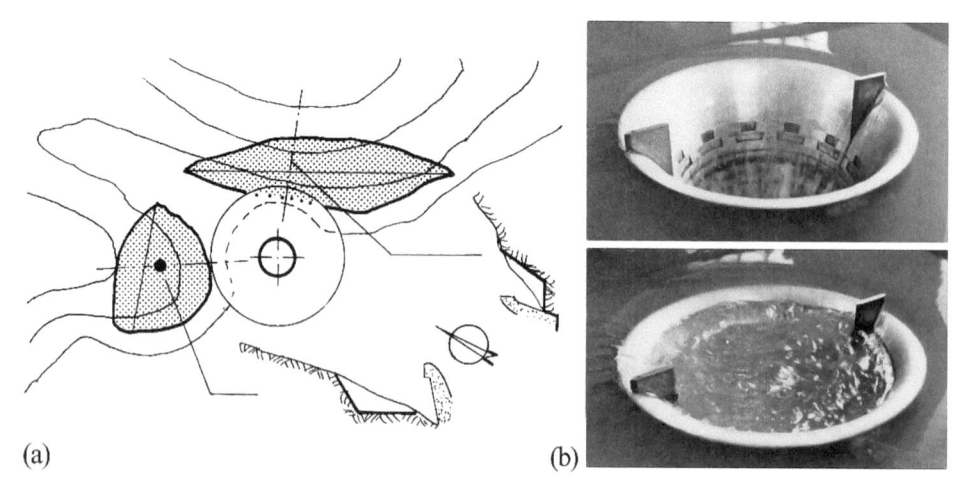

(a)   (b)

Figure 3.25 Lower Shing Mun Reservoir (Hong Kong) (a) Excavation work to improve approach flow, (b) overflow patterns (Smith, 1966). Submerged flow as shown in lower right photo is definitely not recommended

The shape of the reservoir banks around the shaft structure to improve the radial approach flow was also studied by Novak and Cabelka (1981). The height of the protruding shaft portion from the excavation level to the crest should be at least 1.6 times the design head. The plan shape of excavation was also specified. If the shaft and circular overfall crest have to be located close to the reservoir banks due to local constraints, the approach flow is considerably improved by piers placed on the circular crest (Emami and Schleiss, 2016).

The effects of pier shape and pier geometry was analyzed by Indlekofer (1976), proposing parallel-sided or conical piers with a circular-shaped front face. The pier effect is normally small for free overfall, so that the information relevant for straight-crested overflow structures may be used also for the morning glory overflow.

3.3.2 Crest shape

The shape of the morning glory overfall is a logic extension of the standard overfall crest. To control the crest pressures, sharp-crested circular weirs were model-tested allowing for the determination of both the discharge characteristics and the lower crest geometry. A basic study conducted by Wagner (1956) extended the 1948 USBR project. His experimental arrangement is shown in Figure 3.26. All quantities referring to the circular sharp-crested weir are over-barred to indicate the difference with the standard frontal overflow structure. The pipe radius is $\bar{R}$, the overflow head relative to the sharp crest is $\bar{H}$ and the coordinate system $(\bar{x}; \bar{z})$ is located at the weir crest.

As for the overflow with a straight crest in plan view, the circular weir overflow has a lower nappe increasing to the *maximum* (subscript m) point P with coordinates $(x_m; z_m)$. The coordinate system of the standard morning glory overflow is located at P, the corresponding crest radius is $R = \bar{R} - x_m$, and the head is $H = \bar{H} - z_m$. With $\bar{X} = \bar{x}/\bar{H}$ and $\bar{Z} = \bar{z}/\bar{H}$ as

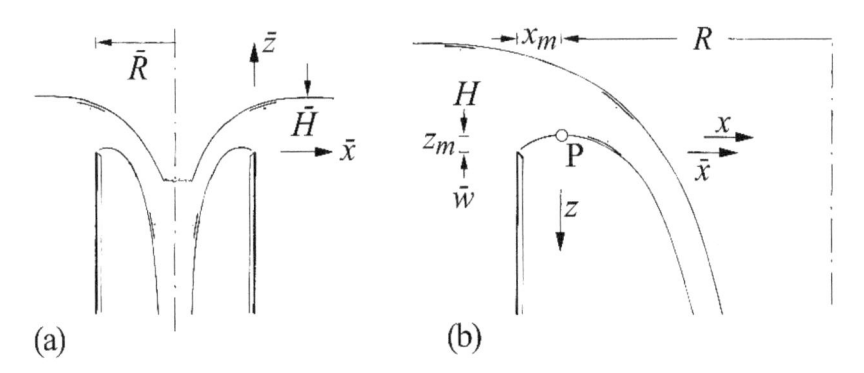

Figure 3.26 Crest geometry of sharp-crested circular pipe overflow (a) overall view, (b) crest detail

dimensionless nappe coordinates, the location of point P is for $0.1 < \bar{H}/\bar{R} < 0.5$ (Hager, 1990)

$$\alpha = \bar{X}_m / \bar{X}_{mo} = [1.04 - 1.055(\bar{H}/\bar{R})], \tag{3.7}$$

$$\beta = \bar{Z}_m / \bar{Z}_{mo} = [1.04 - 1.020(\bar{H}/\bar{R})]. \tag{3.8}$$

Note the differences of 4% as $\bar{H}/\bar{R} \to 0$ from observations $(\bar{X}_{mo}; \bar{Z}_{mo}) = (0.250; 0.112)$.

The nappe profiles are shown in Figure 3.27a for various relative heads $\bar{H}/\bar{R}$, limited to $\bar{H}/\bar{R} < 0.40$. For larger relative heads, the vertical pipe is submerged so that the overflow changes to orifice flow. According to Wagner (1956), the effect of approach flow velocity is negligible if $\bar{w} > 2\bar{R}$ with $\bar{w}$ as the weir height. Scale effects remain insignificant for heads $\bar{H} > 40$ mm.

According to Figure 3.27a, a distinct nappe profile is attributed to each value of $\bar{H}/\bar{R}$. Using the transformations $\bar{X}^* = \bar{X}/\alpha$ and $\bar{Z}^* = \bar{Z}/\beta$ with α and β defined in Eqs. (3.7) and (3.8), a single generalized nappe profile for all relevant relative radii results (Figure 3.27b). The *lower* nappe equation thus reads for $\bar{X}^* < 1.6$ in analogy to Eq. (2.2)

$$\bar{Z}^* = -\bar{X}^* \ln \bar{X}^*. \tag{3.9}$$

The *upper* nappe is approximated as

$$\bar{Z} = 1 - 0.26\left(\frac{\bar{X}}{1 - 1.2(\bar{H}/\bar{R})^{2.3}} + 0.6\right)^2. \tag{3.10}$$

These relations could be transformed to the $(X; Z)$ coordinate system of the overfall structure, but it is easier to refer to the coordinates $(\bar{X}; \bar{Z})$ of the corresponding sharp-crested weir. This task is easily accomplished once the basic parameters $\bar{H}$ and $\bar{R}$ are determined.

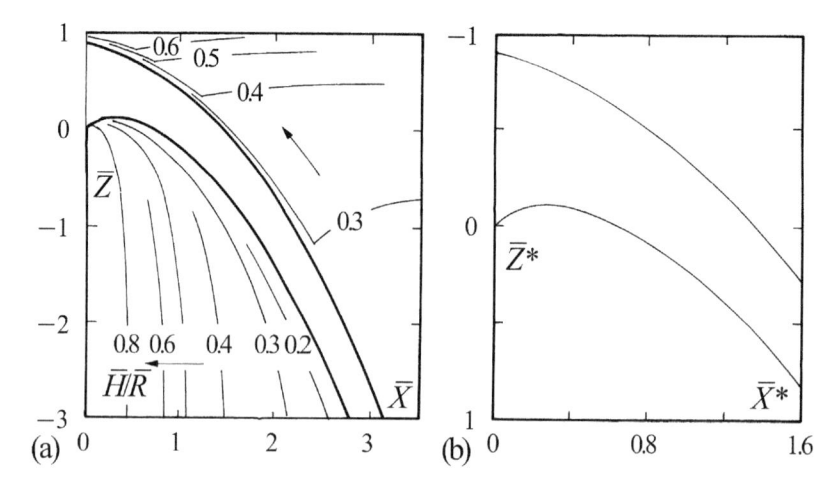

Figure 3.27 Nappe profiles for various relative pipe radii $\bar{H}/\bar{R}$ (a) $\bar{Z}(\bar{X})$, (b) generalization $\bar{Z}^*(\bar{X}^*)$ for $0.2 < \bar{H}/\bar{R} < 0.6$ with (—) $\bar{H}/\bar{R} = 0$ (Hager, 1990)

3.3.3 Discharge and pressure characteristics

The discharge over a morning glory overfall is, in analogy to the straight-crested overfall, with C_d as discharge coefficient relative to parameters R and H (Figure 3.26b)

$$Q = C_d 2\pi R (2gH^3)^{1/2}. \tag{3.11}$$

According to Indlekofer (1978) and for values $0.2 < H/R < 0.5$ (Figure 3.28b)

$$C_d = 0.515[1 - 0.20(H/R)]. \tag{3.12}$$

The discharge coefficient decreases as the relative head increases due to obstruction of flow (Figure 3.27a). For $H/R = 0.2$, the discharge coefficient is equal to the base value $C_d = 0.495$ of the straight-crested overflow. The range $0 < H/R < 0.2$ is hydraulically uninteresting and therefore should be avoided.

For a given design discharge, the pair H and R for $0.2 \leq H/R \leq 0.5$ is thus determined. An approximation for the weir parameter $\bar{H}/\bar{R}$ is

$$H/R = 1.06(\bar{H}/\bar{R})^{1.07}. \tag{3.13}$$

The crest coordinates are then determined from Eqs. (3.7) and (3.8), and the nappe profiles result from Figure 3.27.

The crest pressure was determined for a slightly different crest shape by Lazzari in 1955 (Hager and Schleiss, 2009). For overflow heads below the design head, no sub-pressures were noted. Attention should be paid to the transition from the crest to the shaft profile so that the curvature change remains small, because the nappe may separate from the crest otherwise. According to Bradley (1956), the crest and the transition to the shaft as well as the

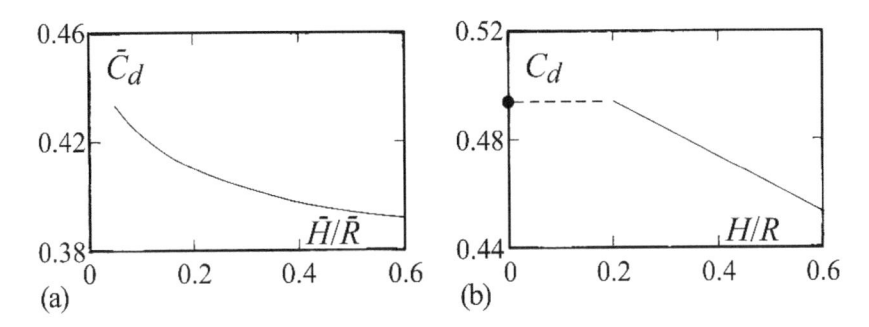

Figure 3.28 Discharge coefficient (a) $\bar{C}_d(\bar{H}/\bar{R})$, (b) $C_d(H/R)$

difficult transitions along the bottom bend should have a smooth lining without offsets and fins to counter cavitation damage.

3.3.4 Vertical shaft structure

The morning glory overflow structure is connected with the vertical shaft to the diversion tunnel. The shaft (subscript s) has a vertical axis, a constant diameter D_s, and conveys the water along the shaft wall with an internal air core to guarantee *free surface flow*. If the air-flow is cut, the air transport is choked resulting in vibrations associated with air backflow, and sub-pressure becomes a concern (Peterka, 1956). To enhance and stabilize the airflow, a separate *air supply conduit* is normally inserted just upstream of the 90° bottom bend. Free-surface flow is thus forced also across the bend into the diversion tunnel. Figure 3.29 shows a typical morning glory overflow structure with details of the air supply arrangement. Note the offset at the junction between the shaft and the aeration conduit to obtain a stable wall separation. This device is particularly suited for overflows for which vibrations are a concern. A less sophisticated flow aeration uses an air supply conduit not integrated into the overflow structure and discharging sideways into the flow, as shown in Figure 2.4c.

For a preliminary design, the shaft radius R_s is correlated to the crest radius R as (Hager and Schleiss, 2009)

$$R_s\,[\mathrm{m}] = 1\,[\mathrm{m}] + 0.1R\,[\mathrm{m}] \cdot \tag{3.14}$$

The minimum shaft diameter should be $R_s = 1.5$ m if no surface debris is expected. As surface racks should not be mounted, the shaft has to be sufficiently large to receive logs of wood. If large surface debris is expected, the morning glory overflow is a poor design. The bend radius R_b should at least be equal to $6R_s$ for small fall heights of up to 20 m, and $10R_s$ for larger fall heights of up to 50 m. A number of further structural details are discussed by Bretschneider (1980). Currently, no standard shaft design is available, so that the shaft bend is arranged depending on site conditions. It is recommended that the air supply conduit should be arranged at the shaft bend end, as shown in Figures 3.30a and 2.4c. The design shown in Figure 3.30b is not recommended due to problems with continuous free-surface flow along the bend. Furthermore, the bend cross section just upstream of the air supply

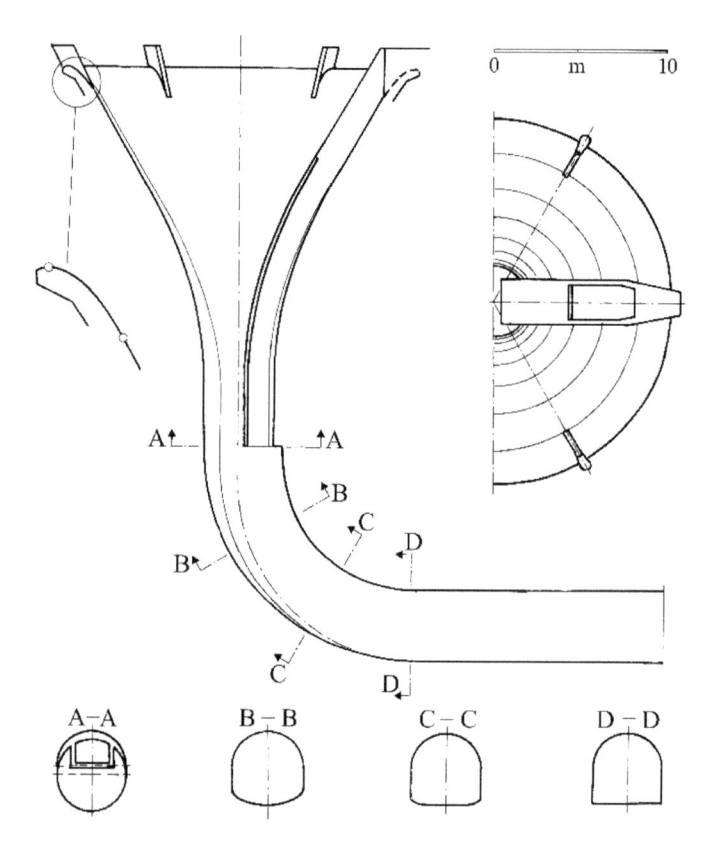

Figure 3.29 Morning glory overflow at Innerste-Talsperre, Germany (Bretschneider and Krause, 1965)

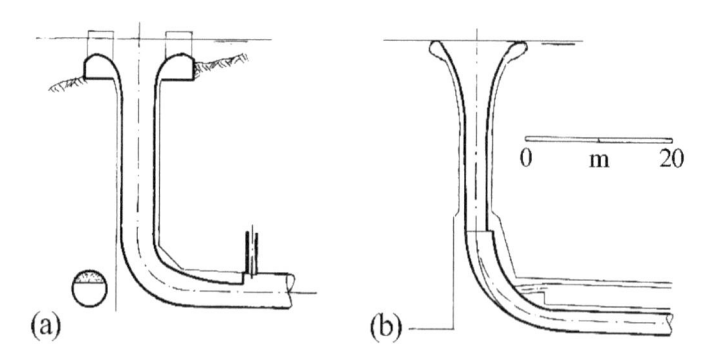

Figure 3.30 Detail of shaft bend, abrupt expansion at (a) bend start (Aabach Dam, Germany), (b) bend end according to Gardel (1949), not recommended

should be a flow control section for the extreme case to be avoided of fully submerged and pressurized flow to exclude cavitation damage along the shaft.

3.3.5 Shaft air supply

The aeration of a shaft associated with free-surface flow is an important concept in the design of the morning glory overflow structure. It promotes a relatively smooth flow of nearly atmospheric pressure distribution inhibiting vibrations, cavitation and air backflow. The design of the aeration device is thus of significance.

The air (subscript a) discharge Q_a is related to the water discharge Q and the shaft geometry. For small water discharges, the air discharge increases proportionally, whereas for nearly pipe-full water discharge, the available section is the limiting factor for air discharge, i.e. it decreases almost linearly up to pipe-full flow for which the air discharge ceases. Figure 3.31 shows both the discharge-head relation $H(Q)$, and the air-water-discharge relation $Q_a(Q)$. For small discharges Q, the discharge relation is of overflow type so that $Q \sim H^{3/2}$, whereas for large water discharge orifice-type flow occurs with $Q \sim H^{1/2}$. The transition between these two is plotted with a dot and corresponds to the incipient pipe full-discharge (Figure 3.31a).

The *maximum air discharge* (2 in Figure 3.31b) results at the transition between the free and submerged regimes. Although standard morning glory overflow structures are not designed for these flow conditions, this limiting value is of interest for the design of the air supply system. According to Novak and Cabelka (1981), the aeration characteristics cannot yet be generally predicted. Based on a literature review, it was found that the:

- Aeration of morning glory overflow depends mainly on the shaft geometry and bend shape at transition from shaft to nearly horizontal tailwater tunnel;
- Maximum air discharge occurs for supercritical tunnel flow, i.e. without a hydraulic jump downstream of shaft bend;

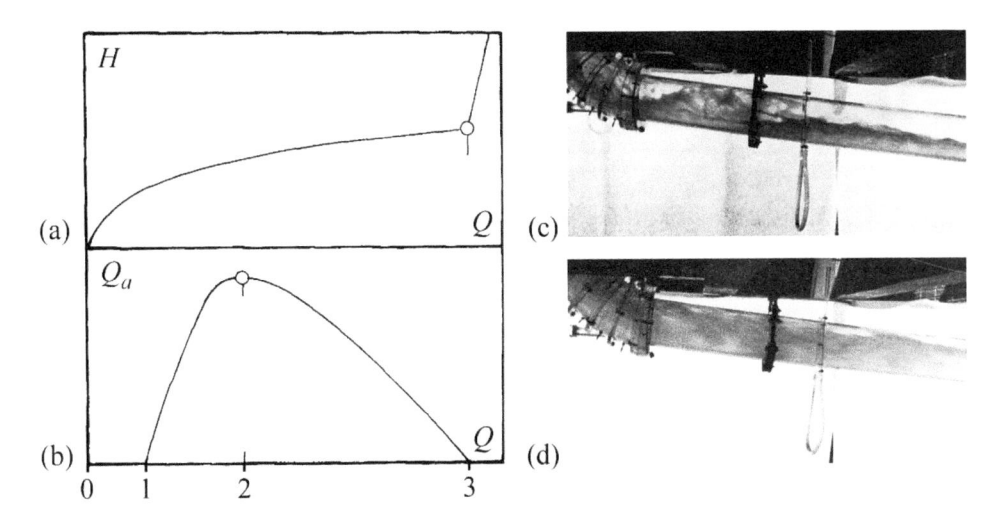

Figure 3.31 Discharge characteristics of shaft spillways (a) head-discharge relation $Q(H)$, (b) air-water discharge relation $Q_a(Q)$, (c) unsubmerged, (d) submerged shaft flow for final design of Lower Shing Mun Reservoir, Hong Kong (Smith, 1966)

- Maximum air discharge from geometrical considerations occurs by assuming co-current air-water tunnel (subscript t) flow, so that with A_t as tunnel section and A_w as the section of water flow

$$Q_{aM} = [(A_t - A_w)/A_w]Q. \tag{3.15}$$

This expression has a maximum value because A_w decreases with increasing water discharge Q. The following issues are of relevance:

- If Δp is sub-pressure downstream of the aeration pipe and V_o the approach flow velocity to the junction with the air supply, i.e. $\mathsf{E} = [\Delta p/\rho g)]/(V_o^2/2g)$ as Euler number of the aeration device, the air-water ratio $\beta_a = Q_a/Q$ depends on E, Froude number F, and relative tunnel length L_t/D_t with D_t as tunnel diameter.
- Aeration system should be designed for maximum air velocity of $V_a = 50$ m/s so that compressibility effects remain small.

The aeration of air-water flow in vertical shafts without the 90° bend and vortex intake were studied by Viparelli (1954). For unsubmerged flow in a shaft of diameter D_s, length L_s, and with K_s as the Strickler roughness coefficient, his results follow (Hager, 1994)

$$\beta_a = \frac{1}{\pi} \left[\frac{Q}{\pi K_s D_s^{8/3}} [1 + (66 D_s/L_s)]^{1/2} \right]^{-1/2} - 1. \tag{3.16}$$

Figure 3.32 shows this relation and defines the shaft outlet geometry. Accordingly, β_a depends significantly on the relative discharge $q_s = Q/(\pi K_s D_s^{8/3})$ and slightly on the relative shaft length L_s/D_s. The maximum (subscript M) air discharge $Q_{aM} = 0.41Q$ results from differentiation and reads (Hager, 2010)

$$Q_{aM} = 0.011 K_s L_s^{1/3} D_s^{7/3}. \tag{3.17}$$

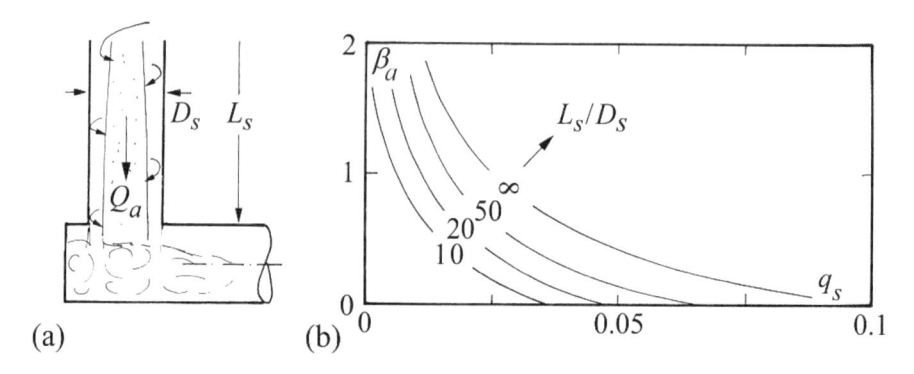

(a)   (b)

Figure 3.32 Air entrainment in vertical shaft (a) outlet geometry considered, (b) air-water ratio $\beta_a = Q_a/Q$ versus relative discharge $q_s = Q/(\pi K_s D_s^{8/3})$ for various relative shaft lengths L_s/D_s (Hager, 2010)

Ervine and Himmo (1984) considered the flow arrangement shown in Figure 3.33. A vertical shaft was connected to a horizontal tunnel by a sharp-edged miter bend. The air pockets located at the tunnel inlet are either transported to the tailwater, or rise back through to the shaft inlet. The contracted flow depth h_1 downstream of the sharp-edged miter bend of cross section A_1 and Froude number $F_s = V_s/(gD_s)^{1/2}$ is

$$h_1/D_s = 1 - \frac{1}{2}F_s^2\left[\left(\frac{\pi D_s^2}{4A_1}\right) - 1\right]. \qquad (3.18)$$

With $F_f = (Q_a + Q)/[\pi D_s^2/4)(gD_s)^{1/2}]$ as the total discharge Froude number, the relevant flow types include (Figure 3.33b):

① Air backflow to shaft for $F_f < 0.3$, development of long air pockets in tunnel,
② No air backflow if $0.25 < F_f < 0.5$ and $\beta_a < 0.3$, resulting in small air entrainment, but air pockets increase in length,
③ Supercritical tunnel flow with transition to pressurized flow if $F_f > 0.3$ and $\beta_a > 0.7(F_f - 0.3)$, and
④ No separation from inner bend if $\beta_a < 0.7(F_f - 0.3)$, air bubbles moving along tunnel vertex.

The maximum relative separation height $T_1 = 1 - h_1/D$ varies with F_f and the air ratio β_a as shown in Figure 3.34. A maximum value establishes for $F_{fmax} \cong 0.3$. Ervine and Himmo (1984) pointed at the particular character of their results with regard to the geometry and tunnel length $L_s/D_s = 53$. For large tunnel dimensions, hydraulic model tests are recommended involving a model shaft diameter of $D_s \geq 150$ mm to exclude scale effects (Heller, 2011).

A project description in which the air entrainment and detrainment are important aspects was provided by Stephenson and Metcalf (1991). Their model studies of air entrainment in the Muela dropshaft indicates that hydraulic modeling combined with the analysis of scale effects are currently the only design approach. In general, Froude similitude governs aeration

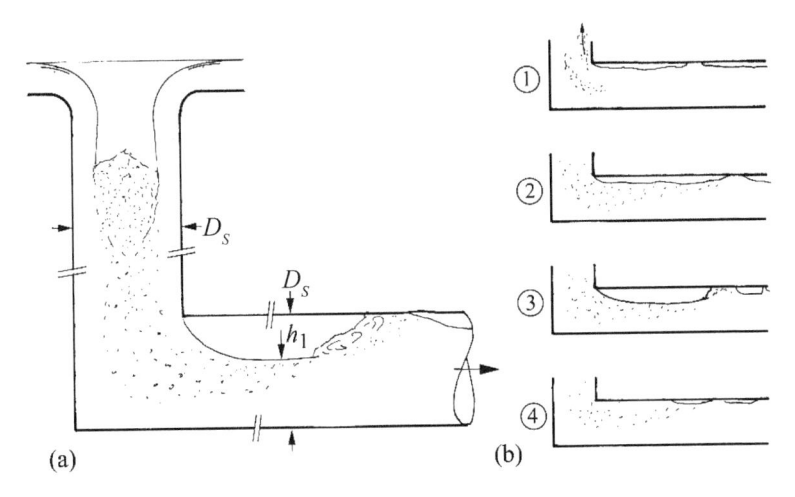

Figure 3.33 System shaft-tunnel (a) definition of geometry, (b) flow types in tunnel spillway

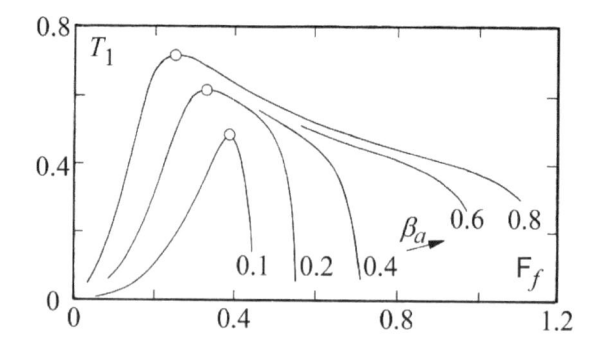

Figure 3.34 Maximum thickness of relative separation height T_1 versus shaft Froude number F_f and air ratio β_a (Adapted from Ervine and Himmo, 1984)

processes in free-surface flows. A contribution to the scaling of the tunnel flow downstream of a 90° bend was presented by Mussalli (1978). Again, a particular case was investigated; a generalization to standard design is currently impossible. Further details on morning glory spillways including complicated intake arrangements, pressure distributions along the shaft, transition from shaft to tunnel spillway, and air entrainment along the shaft are described by Hager and Schleiss (2009).

Further information on the morning glory overfall structure was provided by Bollrich (1968, 1971). Both the design bases and the hydraulic characteristics are highlighted. A particular attention is directed toward the air entrainment features.

3.3.6 Case study

Pinotti and Lais (2010) presented a case study involving hydraulic model tests (Figure 3.35). The reservoir is controlled by a 56 m high gravity-arch dam creating 3.3×10^6 m³ live storage volume. Three spillway structures including a gated chute spillway and a middle and a bottom outlet with a total spilling capacity of 240 m³/s serve for flood evacuation. The design flood hydrograph is based on a maximum reservoir inflow of 280 m³/s for a return period of 1000 years, a discharge exceeding the available total spilling capacity. The operator thus planned to erect an additional morning glory spillway (MGS) of 75 m³/s discharge capacity. Its concept included the existing middle outlet tunnel for a time- and cost-saving design. The new structure is located at the right valley flank, close to the gated chute spillway. The MGS tunnel connects the existing middle outlet tunnel with a curve-tangent junction. The energy dissipation at the tunnel outlet is assured by a stilling basin equipped with chute and baffle blocks.

The routing of the design flood hydrograph through the reservoir with the gated chute spillway, the bottom outlet and the planned MGS in operation (middle outlet excluded) predicted a peak discharge of 75 m³/s released through the MGS at a reservoir water level of 538.9 m a.s.l. at design discharge. The operation sequence of the spillway varies with the reservoir water level. As the water level reaches 538.2 m a.s.l., the radial gate of the chute spillway is fully opened, then discharging 150 m³/s. The purpose of the hydraulic model tests

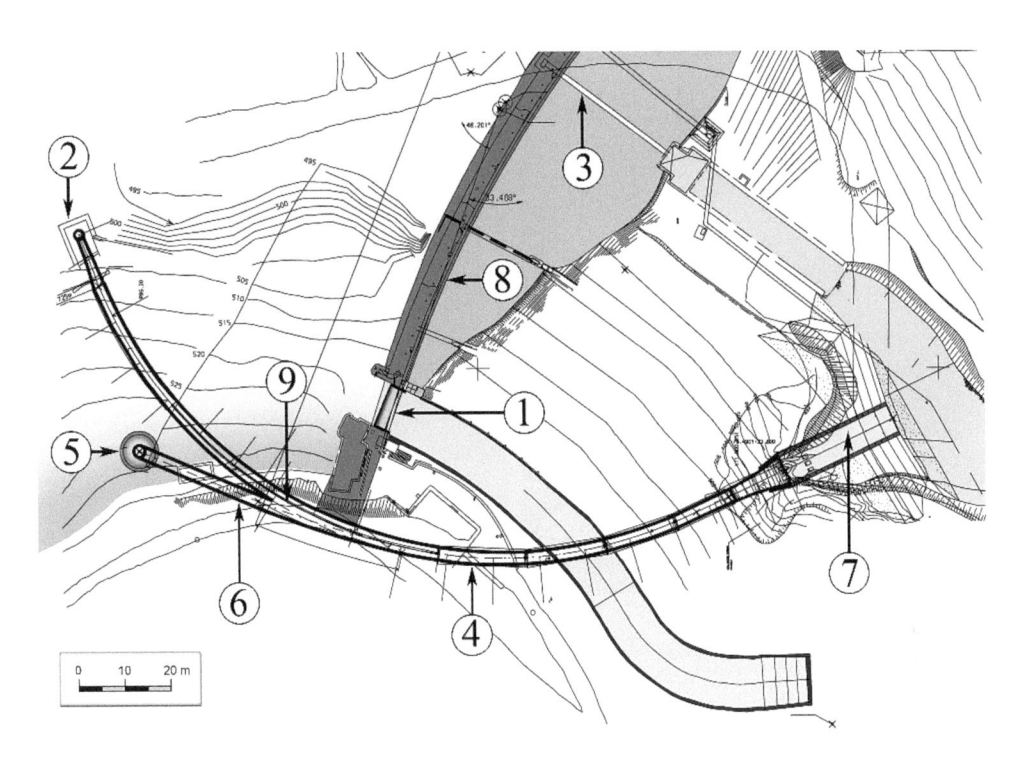

Figure 3.35 Plan view with existing spillway structures ① gated chute spillway, ② middle and ③ bottom outlets, ④ middle outlet tunnel, ⑤ projected morning glory spillway, ⑥ spillway tunnel, ⑦ stilling basin, ⑧ gravity-arch dam, ⑨ junction structure (Adapted from Pinotti and Lais, 2010)

was to evaluate the overall hydraulic behavior of the MGS and to improve its efficiency and safety.

The hydraulic model included the prototype reservoir area of 10,000 m², a portion of the right valley flank and all the spillway structures of the water supply facility in a 1:25 physical model using Froude similitude (Figure 3.36). This model scale allowed for an adequate reproduction of the prototype conditions, despite the accurate modeling of all its hydraulic parameters is impossible. Scale effects due to water viscosity and surface tension are significant on the discharge capacity at small overflow depths; these were considered negligible for prototype overflow depths above 0.55 m, i.e. for discharges exceeding 22 m³/s.

Both an initial and the final design were investigated. The initial design included (Figure 3.37) a cup-shaped overflow structure (transition tube) with a crest elevation at 537.5 m a.s.l. and crest diameter of 8.0 m, designed for an overflow head of 1.2 m and roughly representing the Creager crest profile; a 32.04 m high vertical shaft, 2.0 m in diameter, connected to the overflow structure at 530.0 m a.s.l. (throat of transition tube); a circular bottom bend of 4.0 m curvature radius; and a spillway tunnel 26.84 m long, 2.0 m in diameter and with a slope of 1.6%, joining the existing middle outlet tunnel at elevation 492.52 m a.s.l. This design did not include aeration devices.

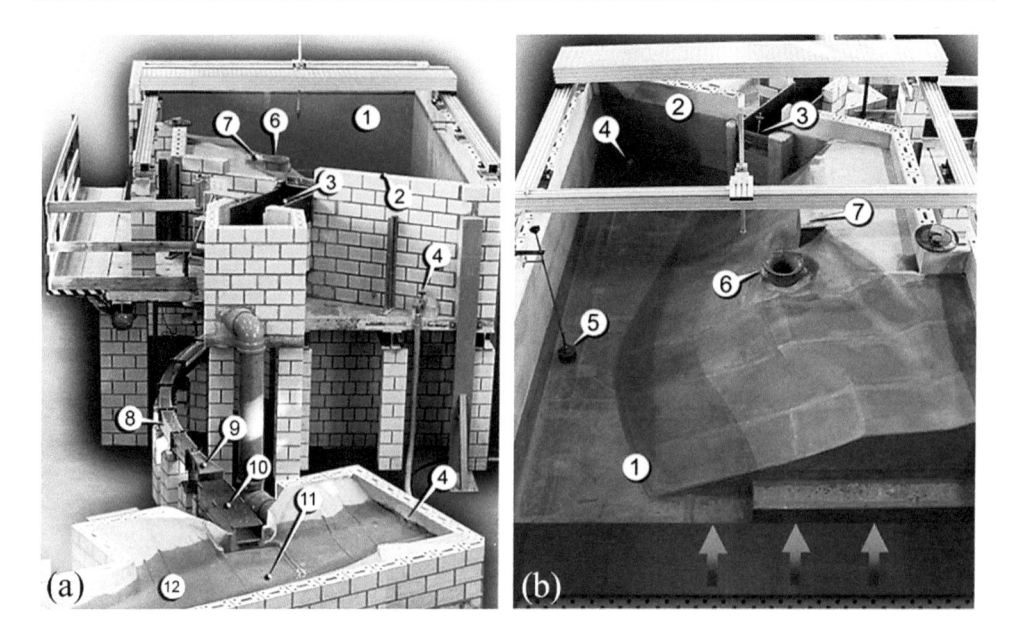

Figure 3.36 Physical model at scale 1:25 (a) downstream view, (b) upper tank with (1) reservoir, (2) dam portion, (3) chute spillway with radial gate, (4) bottom outlet, (5) inlet valve of middle outlet, (6) cup-shaped overflow structure, (7) service building, (8) middle outlet tunnel, (9) transition chute, (10) stilling basin, (11) riprap apron, (12) portion of valley (Adapted from Pinotti and Lais, 2010)

The flow is governed by the crest discharge characteristics for reservoir levels up to 538.75 m a.s.l., so that the discharge follows Eq. (3.11). According to the service regulation, the chute spillway is fully opened for reservoir levels above 538.2 m a.s.l. Then, the strong current toward the open gated chute spillway affects the approach flow to the MGS, resulting in a slightly reduced discharge capacity and a reduced value of $C_d = 0.468$, as compared with the hypothetical scenario in which the chute spillway is out of operation (Figures 3.38, 3.40).

For reservoir levels above 538.75 m a.s.l., the overflow crest is submerged and the entire MGS operates under pressure. The control section is then shifted from the crest to the downstream end of the spillway tunnel. The large increase of reservoir water level under pressurized flow only leads to a small gain in spilling capacity (Figure 3.38).

Annular flow occurs in the overflow structure up to $Q = 40$ m³/s. The overflow nappe clings to the cup-shaped overflow structure with the free air core remaining in the center of the vertical shaft. As the discharge increases, the overflowing annular nappe becomes thicker and converges to a solid vertical jet at the crotch point. The air-water mixture flow chokes the vertical shaft, thereby preventing free air supply. The flow regime changes from free to pressurized flow forming a so-called boil above the crotch point. Both the boil and the crotch rise progressively as the discharge increases. Due to the high net head and the absence of free-surface aeration, sub-atmospheric pressures occur at the throat of the transition tube (Figure 3.37). Due to throttle absence in the hydraulic system, the vertical shaft is exposed to negative pressures close to vapor pressure for $Q \geq 50$ m³/s. The upscaled pressures at the

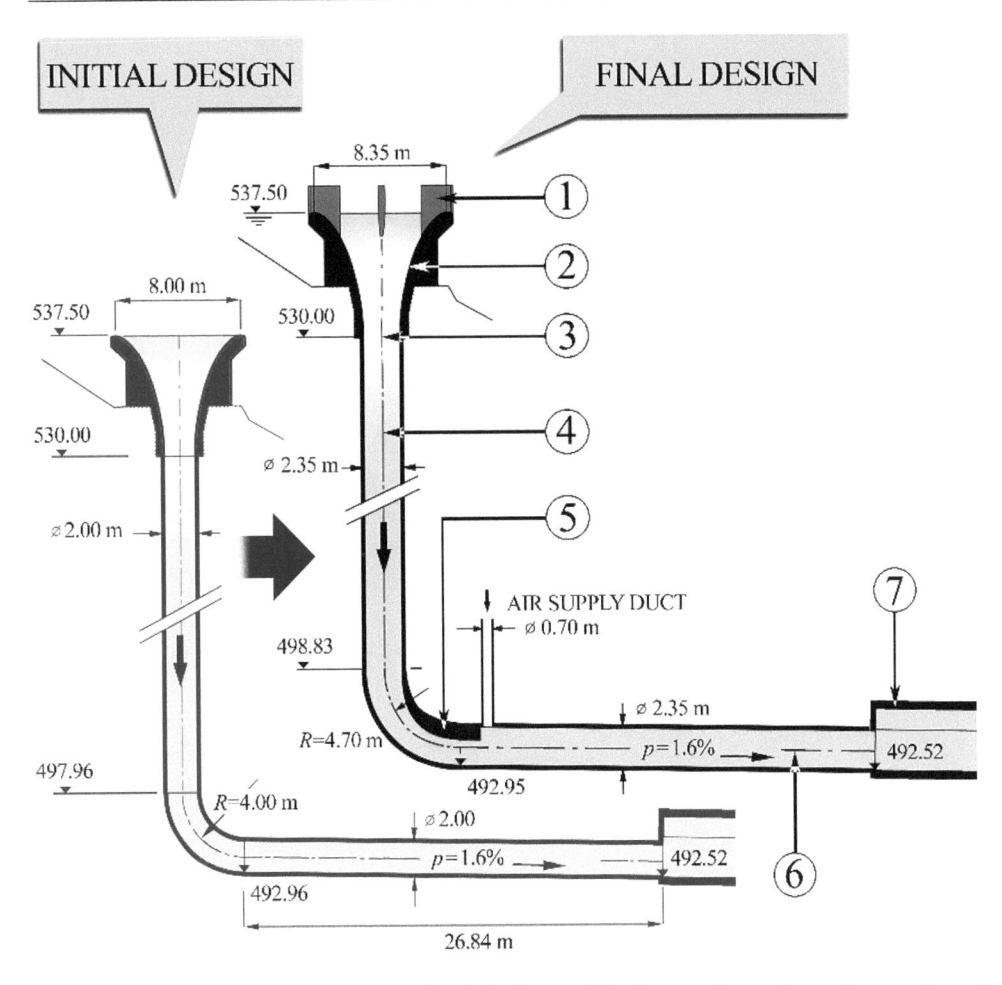

Figure 3.37 Layouts of initial and final designs with ① overflow piers, ② cup-shaped overflow structure, ③ throat of transition tube, ④ vertical shaft, ⑤ throttled circular bottom bend, ⑥ spillway tunnel, ⑦ junction with middle outlet tunnel (Adapted from Pinotti and Lais, 2010)

throat of the transition tube drop below vapor pressure, resulting in a phase change from water to vapor, i.e. the water column separates. The flow is then controlled by this section and the available net head is strongly reduced, as is the discharge capacity.

For a theoretical net head of 45.23 m between the reservoir water level and the tunnel axis elevation just upstream of the junction with the middle outlet tunnel, the discharge from Bernoulli's equation is 72 m³/s. The corresponding energy and pressure gradients are shown in Figure 3.39. The predicted relative minimum pressure of -19.4 m water column (WC) at the throat of the transition tube marginally deviates from the measured and upscaled value of -19.5 m WC.

The phenomenon of water column separation was not reproduced in the hydraulic model, as Froude similarity would theoretically require a reduction of the atmospheric model

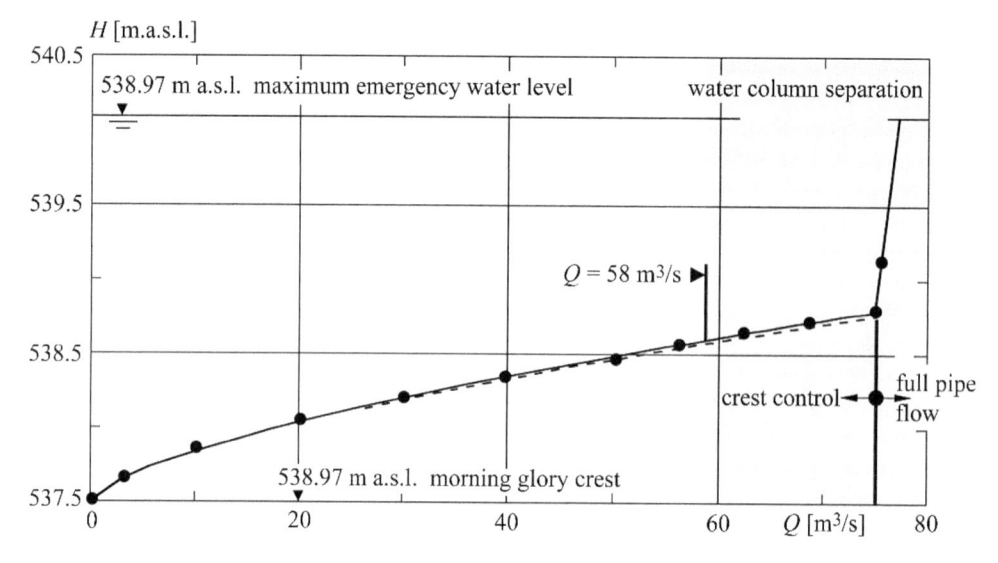

Figure 3.38 Initial design, measured reservoir water level *H* versus discharge *Q* for (hypothetical) (---) closed and (•) fully opened chute spillway, (−) Eq. (3.11) (Adapted from Pinotti and Lais, 2010)

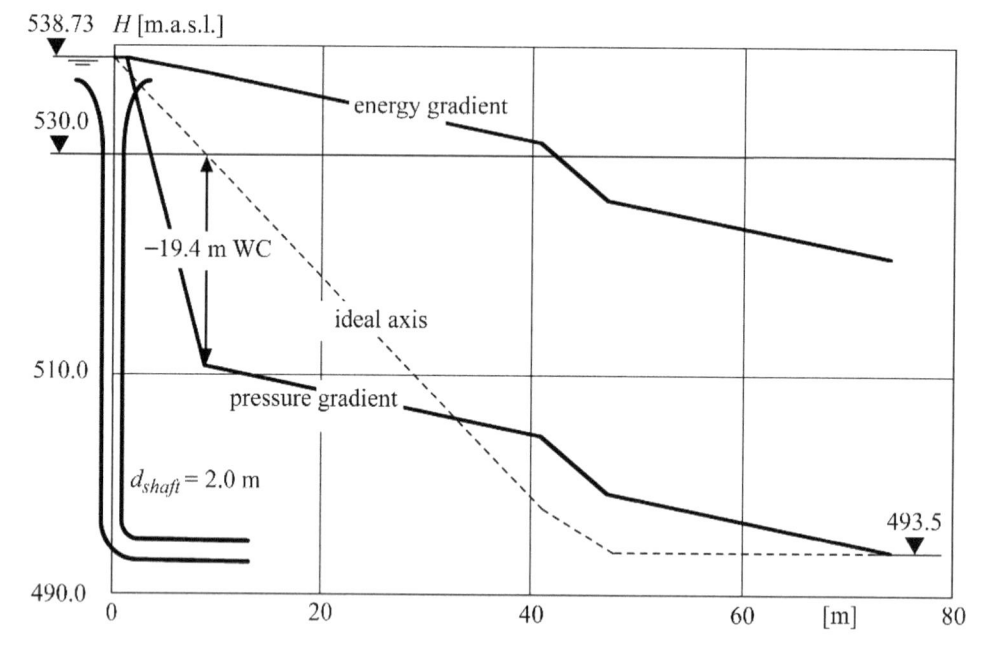

Figure 3.39 Predicted energy and pressure gradients for *Q* = 72 m³/s, initial design. Note minimum pressures below absolute vacuum (−10 m WC) (Adapted from Pinotti and Lais, 2010)

pressure. The maximum prototype discharge of the initial design was computed theoretically. Assuming the throat pressure at the transition tube equal to vapor pressure and full-pipe flow conditions in the vertical shaft results in $Q = 58$ m^3/s, for which the water column would separate, so that the discharge capacity of the initial design is limited to 58 m^3/s (Figure 3.38).

A modification of the initial design aimed at improving the flow conditions to ensure safe operation avoiding water column separation. The modification further improved the discharge capacity to 75 m^3/s. The final design consisted of:

- Increased shaft diameter from 2.00 to 2.35 m resulting in a crest diameter of 8.35 m, without altering the drop inlet location and shape;
- Redesign of bottom bend, including a throttle at its end, to raise the pressure gradient ensuring pressures higher than the maximum tolerable negative pressure head of $-(7 \div 8)$ m WC at the throat of the transition tube. Note that the minimum pressure head there should not drop below this limit to avoid water column separation;
- Increased spillway tunnel diameter of 2.35 m;
- Addition of air supply duct of at least 0.70 m in diameter downstream of throttled bottom bend to ensure sufficient air supply and free-surface flow conditions in spillway tunnel.

The approach flow toward the MGS is affected both by the protruding slope of the right valley flank and the nearby service building foundation. For small discharges, radial flow occurred (Figure 3.40a). At higher discharges, however, the mainstream toward the open chute spillway circulated around the intake, developing a strong current between the overflow structure and the right valley flank. This flow is deflected by the foundation, negatively influencing the velocity distribution near the MG intake (Figure 3.40b). For reservoir water levels above 538.9 m a.s.l., a bathtub vortex developed. A considerable amount of kinetic energy is dissipated by this strong rotational swirl reducing the discharge capacity, i.e. a discharge reduction along with a vortex growth under increasing reservoir water level (Figure 3.41). The maximum discharge of the modified design reached 72 m^3/s at 538.9 m a.s.l., so that it did not provide the required discharge capacity of 75 m^3/s.

The final design included four overflow piers 0.40 m wide and 2 m long to ensure radial inflow (Figures 3.37, 3.42). The MGS is submerged for $Q = 75$ m^3/s at a reservoir level of 538.8 m a.s.l. The overflow piers impose radial inflow and suppress the vortex formation enhancing the hydraulic performance (Figure 3.42). However, the swirl along the right valley flank toward the open chute spillway is still strongly deflected by the service building foundation. A 3D flow field develops, resulting in a single vortex-jet close to the MG crest (Figure 3.42), eventually leading to erosion at the intake vicinity and entraining debris into the MGS. For $Q > 75$ m^3/s and for a reservoir level above 538.8 m a.s.l., the MGS operates under pressure. The vertical shaft chokes for discharges $Q > 56$ m^3/s. At the maximum emergency water level of 538.97 m a.s.l., a discharge of 76.7 m^3/s was measured (Figure 3.41).

For discharges $Q < 40$ m^3/s, the mean pressure heads along the cup-shaped overflow structure and the vertical shaft are nearly atmospheric along with small pressure fluctuations. For $Q > 56$ m^3/s, however, full-pipe flow conditions occur along with sub-atmospheric pressures in the upper portion of the vertical shaft. A minimum mean value of -6.53 m WC was there detected at $Q = 75$ m^3/s (final design), in agreement with the design criteria. Nevertheless, the upscaled local sub-pressures drop below vapor pressure, leading to flow pulsations with considerable pressure peaks, including the formation of localized cavitation bubbles. These

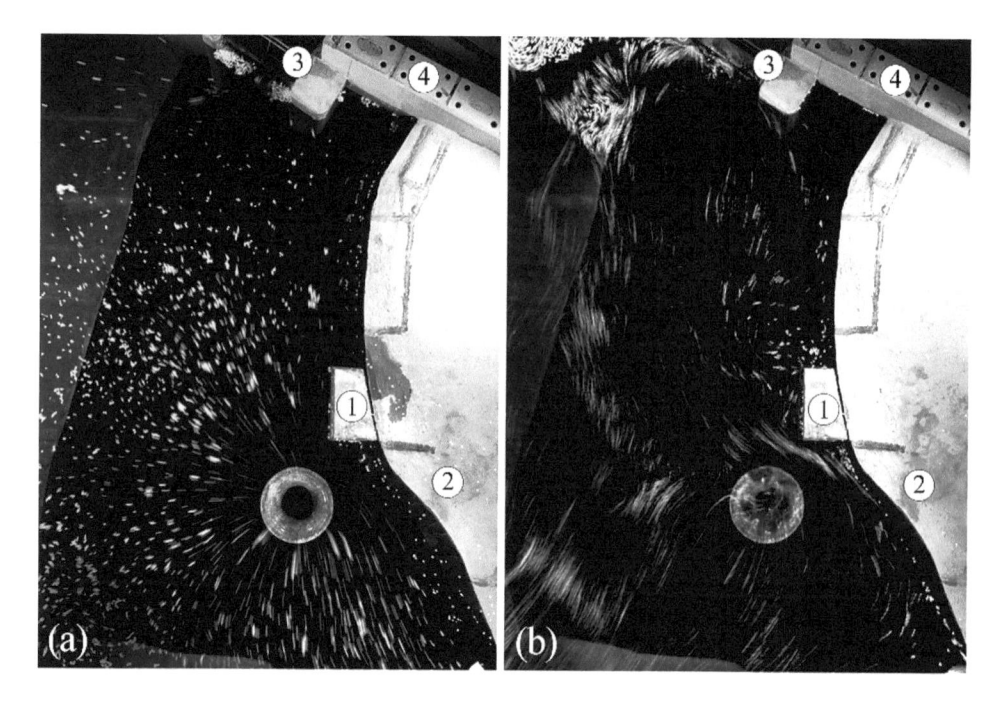

Figure 3.40 Reservoir plan view of modified design ① service building, ② protruding slope of right valley flank, ③ chute spillway, ④ dam. At (a) $Q = 10$ m³/s (chute spillway out of operation) radial flow occurs, whereas at (b) 72 m³/s main flow toward fully opened chute spillway causing vorticity. Bathtub vortex forms under submerged overflow conditions, limiting discharge capacity to $Q = 259$ m³/s at inflow, and $Q = 187$ m³/s at chute intake (Adapted from Pinotti and Lais, 2010)

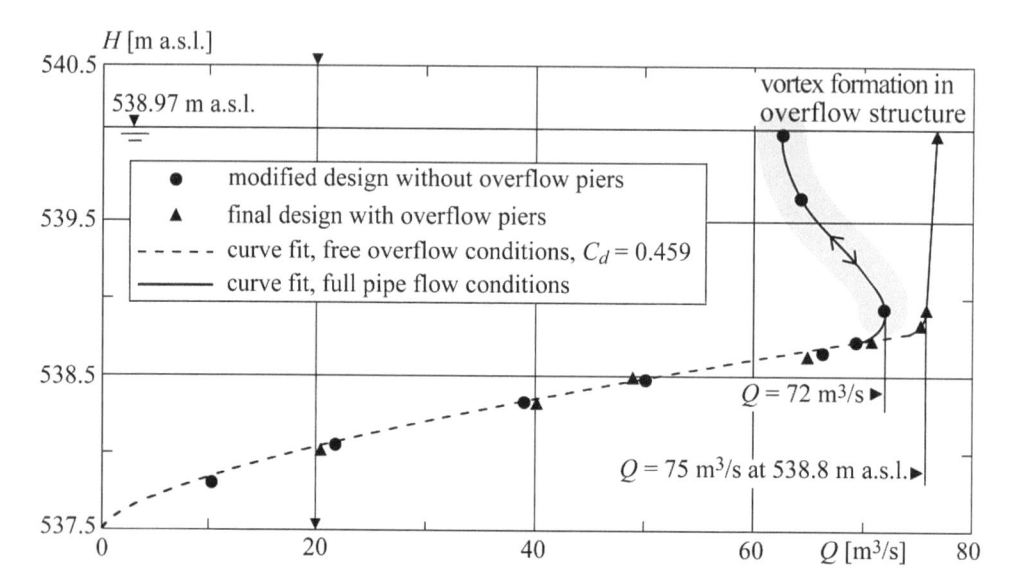

Figure 3.41 Final design, measured reservoir water level H versus discharge Q, with maximum discharge capacity of 72 m³/s for modified design at 538.9 m a.s.l. For higher reservoir water levels, vortex formation reduces discharge capacity. Final design with overflow piers inhibits vortex formation ensuring required discharge capacity of 75 m³/s (Adapted from Pinotti and Lais, 2010)

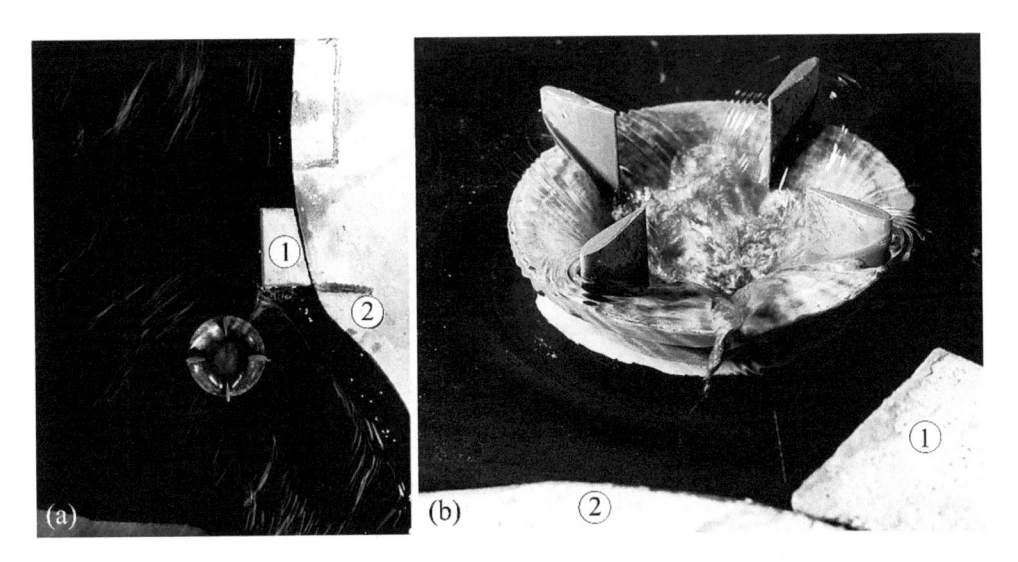

Figure 3.42 Final design with overflow piers at Q = 75 m³/s (a) plan, (b) side view with ① service building, ② protruding right valley flank. Overflow piers ensure radial inflow. Note spiral flow with vortex in its core between overflow piers (Adapted from Pinotti and Lais, 2010)

phenomena may damage the concrete surface. If a certain amount of air is entrained by the flow, these negative consequences are reduced. Prototype experience with similar structures indicated no cavitation damages if the mean pressure is above vapor pressure. The occurrence of extreme minimum pressures below vapor pressure was avoided by raising the pressure gradient to obtain a higher mean pressure at the throat of the transition tube. This was achieved by enlarging the shaft diameter and by an adequate adaption of the throttling orifice.

Pressure fluctuations in the vertical shaft due to the formation of an oscillating boil and two-phase flow phenomena during the transition phase from free to full-pipe flow have to be particularly considered. These fluctuations were analyzed by the spectral density of a time series, i.e. the frequency of the measured pressures. The spectral density at Q = 65 m³/s indicated one distinct peak at 0.03 Hz, corresponding to the boil oscillation frequency causing a pressure amplitude of 1.8 m WC for the modified design, and 0.7 m WC for the final design with overflow piers. At submerged inflow, the asymmetrical approach flow toward the MGS without overflow piers led to the formation of an air-core vortex within the overflow structure. This phenomenon caused additional intensive pressure fluctuations due to unstable flow conditions as the air-core bathtub vortex intermittently collapsed in the inlet structure (Figure 3.43). Instantaneous fluctuations of up to 20 m WC were detected. The final design with overflow piers added suppressed the formation of these vortices, resulting in reduced pressure fluctuations.

This case study highlights the merits of hydraulic modeling for a problem which currently can hardly be assessed otherwise, due to the presence of two-phase flow, complicated low-pressure flow zones, and cavitation damage to be expected. Starting with the initial design of the MG spillway, the upscaled minimum pressures were noted to be below vapor pressure

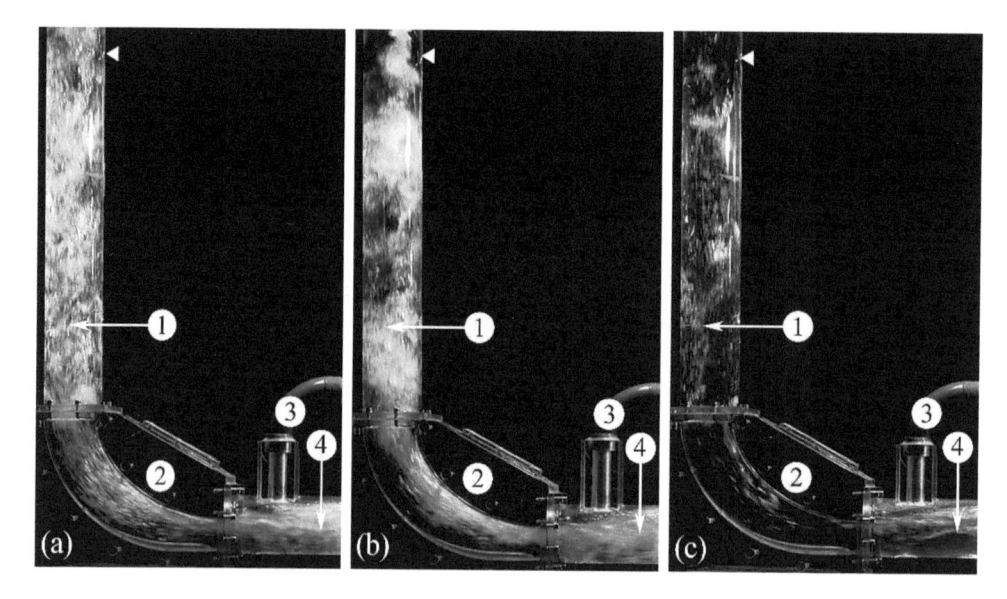

Figure 3.43 Side views of ① vertical shaft, ② throttled bottom bend, ③ aeration duct, ④ spillway tunnel for (a) transition phase at 65 m³/s, final design; (b) modified, (c) final design at submerged inflow. Without overflow piers, air entrainment is batch-wise due to intermittent collapse of air-core bathtub vortex, resulting in strong pressure fluctuations. Note that entrained air considerably reduces with presence of overflow piers; (◁) pressure measuring position (Adapted from Pinotti and Lais, 2010)

at the transition from free to pressurized flow conditions in the vertical shaft, resulting in a phase change from water to vapor, i.e. water column separation. This scenario would cause strong vibrations in a prototype structure probably generating structural damage and limited spillway capacity. The initial design was improved by increasing the vertical shaft diameter and installing a throttled bottom bend. Furthermore, overflow piers at the MG crest were provided to ensure radial inflow avoiding air entraining vortex formation. The required discharge capacity of 75 m³/s was confirmed and extreme pressure fluctuations were limited.

3.4 Labyrinth weir

3.4.1 Historical evolution

The discharge of rectangular weirs is overproportional, so that weirs are particularly apt for discharge retention up to the weir crest, generating high discharge once it is surpassed, as typically during floods. The weir is a fundamental hydraulic structure, by which discharge is measured, rivers are made navigable, or flooding is prevented, particularly if gates are added. Given that the weir overflow zone is often laterally restricted, the overflow depths can become excessive. As a result, nonlinear weirs in plan as the labyrinth weir (LW) or the piano key weir (PKW) have been developed to increase both the weir length and the discharge capacity within a fixed structural width. The LW and the PKW (3.5) have an increased weir length by adoption of a polygonal weir plan.

Hager *et al.* (2015) reviewed the current knowledge on LW up to 1985. Murphy (1909) appears to be the first who highlighted its advantages. Figure 3.44 shows an example of the weir shape adopted. The effective crest length was more than three times that of a straight weir, discharging 17 m^3/s. Given the narrow space at the overflow section, as shown in Figure 3.45, seven bays or cycles were arranged in plan view. These were 3 m long and 1.5 m wide, resulting in a total crest length of 60 m in the 18 m wide overflow channel. The inlet bays were designed with reducing cross-sectional areas along their length to generate nearly uniform overflow. Tests indicated that the discharge was practically identical with that of the standard straight-crested weir of equivalent length (Chapter 2).

The first hydraulic study on the LW was conducted by the Italian Gentilini. First, the plan of the weir shape is explained: let L be the straight structural length, and α the intended angle. Figure 3.46 shows a number of weir arrangements tested, all of which feature the sharp standard weir crest shape. Gentilini (1941) found that the head-discharge relationship per cycle was independent of the number of cycles, e.g., weir types 1, 2, and 3, or types 5 and 6 in Figure 3.46. It was further observed that the data of weir types 1–4 and 5–8 both with $\alpha = 90°$, and 9 and 10 with slightly different plan geometries, respectively, had identical discharge-head relations. Note that the notation is based on that of the various papers discussed, so that an identical abbreviation may apply to various parameters.

Figure 3.44 Photographs of reinforced-concrete spillway of Keno Canal, near Klamath Falls, OR, USA (Murphy, 1909)

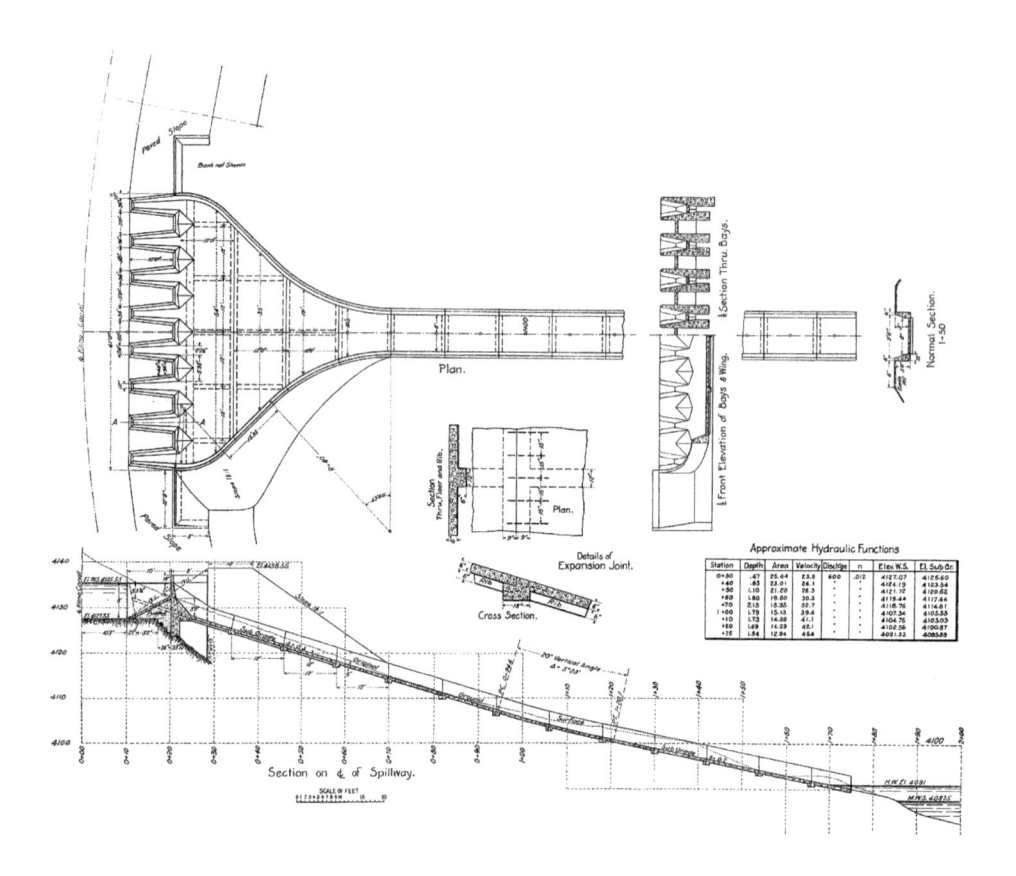

Figure 3.45 Keno Canal Spillways with concentrated crest length (Murphy, 1909)

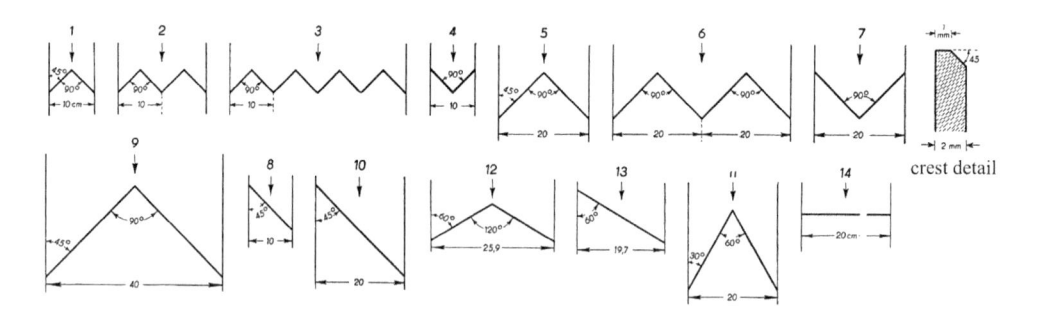

Figure 3.46 Plan geometries of LWs tested (Adapted from Gentilini, 1941)

As to the approach flow direction perpendicular (subscript n) to the weir, the discharge coefficient μ_n in the free weir flow equation generally varies with the relative head on the weir $[h/(h+p)]$ with h as the overflow depth and p as the weir height, and the Reynolds and Weber numbers accounting for viscosity and surface tension effects. If chemically clean water is employed, and absolute overflow depths are larger than $h = 0.05$ m, say, then the latter two effects remain insignificant. For oblique approach flow, the corresponding discharge coefficient μ depends in addition on the relative overflow depth h/L and α, with L as the projected weir length perpendicular to the approach flow direction. Gentilini (1941) shows a plot μ/μ_n versus h/L for $\alpha = 30°, 45°$, and $60°$ (Figure 3.47), indicating that $\mu/\mu_n = 1$ for $h/L = 0$, from where all curves drop exponentially versus a final plateau value. The drop rate changes inversely with α; no drop at all occurs for $\alpha = 180°$. Note also from Figure 3.47 that for e.g. $h/L = 1$, $\mu/\mu_n(\alpha = 60°) = 0.90$, $\mu/\mu_n(\alpha = 45°) = 0.79$, and $\mu/\mu_n(\alpha = 30°) = 0.65$, indicating a massive loss of discharge capacity as α reduces on the one hand, which is made up on the other hand by the increase in crest length, resulting in sum in an absolute increase of discharge capacity.

The effect of LW discharge modification is finally addressed by Gentilini (1941). Figure 3.48 shows the discharge ratio of the LW and the normal weir versus the variable h/L, with L as the LW length across the channel for various numbers n of singular oblique weirs and $\alpha = 45°$. As in Figure 3.47 all curves start at $Q/Q_n(h/L = 0) = 1.41$ for all n, reducing as h/L increases to attain $Q/Q_n(h/L \rightarrow \infty) = 1$. The reduction of the discharge ratio remains small for $n = 1$, but increases with n. A slightly modified result applies as the angle α is changed. The work of Gentilini (1941) results in a definite increase of LW discharge capacity, particularly for relative small overflow depths h/L.

Hay and Taylor (1970) state that the only relevant hydraulic criterion for the weir performance is the ratio between the discharges of the labyrinth (subscript L) and the normal (subscript N) weir configurations Q_L/Q_N in a channel of common width. Figure 3.49 shows a scheme of the LW and the surface profiles generated. Note the flow contraction in

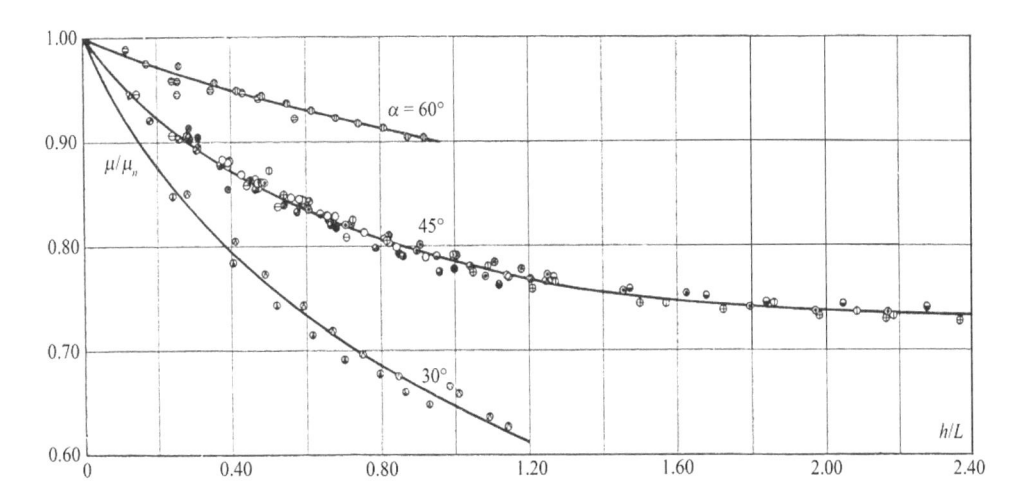

Figure 3.47 Effects of relative overflow depth h/L, and α on discharge coefficient ratio μ/μ_n (Adapted from Gentilini, 1941)

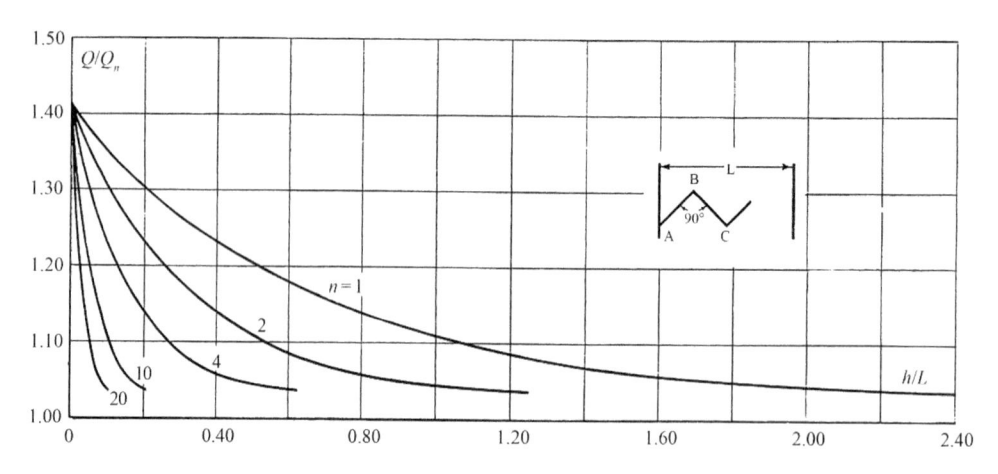

Figure 3.48 Effects of relative overflow depth h/L, and obliquity angle α on discharge ratio Q/Q_n (Adapted from Gentilini, 1941)

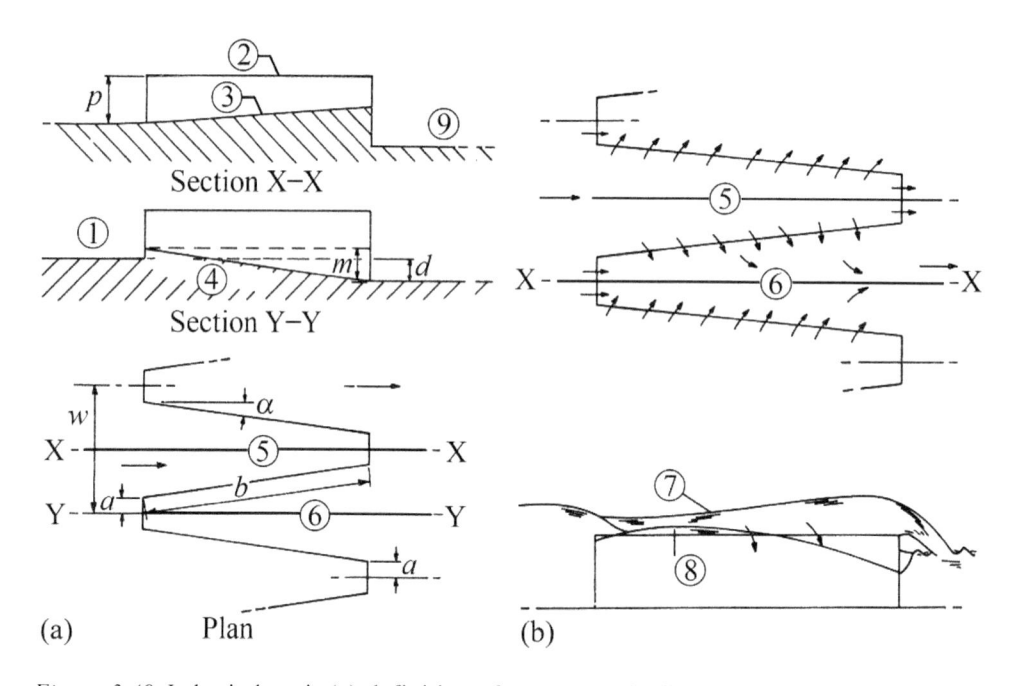

Figure 3.49 Labyrinth weir (a) definition of geometry, (b) flow features with ① approach flow channel, ② crest, ③ upstream apron, ④ downstream, apron, ⑤ upstream channel, ⑥ downstream channel, ⑦ upstream and ⑧ downstream surface profiles, ⑨ runoff channel (Adapted from Hay and Taylor, 1970)

Figure 3.49b due to water flowing from the upstream to the downstream channel. In contrast, along the upstream channel up to the crest region, the surface profile increases, and only then falls into the downstream channel. The application of the usual weir discharge formula was thus questioned. Figure 3.50 shows further details, namely the plan geometry of the models tested, varying from the trapezoidal, to the rectangular, and the triangular shapes, and the parameter ranges tested in terms of the relative weir cycle width w/p, the length magnification l/w with l here as the length of one weir cycle, the number of weir cycles n, and the weir plan shapes.

Hay and Taylor (1970) concluded that the weir discharge performance Q_L/Q_N depends exclusively on the parameters h/p, w/p, l/w, α, and n. Note that all weirs tested were made of Perspex of standard crest shape. The normal discharge Q_N was expressed with the equation of Kindsvater and Carter (1957), whereas submerged flow was normalized with the proposal of Villemonte (1947).

As to the head-to-crest height ratio $h/p \to 0$ for all weir types, $Q_L/Q_N \to l/w$, i.e., the relative increase in labyrinth weir discharge, relative to the linear weir becomes equal to the weir length magnification factor at small upstream heads. As the relative overflow depth h/p increases, then $Q_L/Q_N \to 1$ (in theory) because of the interference of the overflow jets, head losses, and possibly tailwater effects. As to the sidewall angle α, varying from 0 for the rectangular to the maximum for the triangular plan shapes, the data indicate that the discharge performance increases with increasing α. Thus, the triangular plan shape renders the highest weir discharge performance. Note the limits of this statement, because effects of

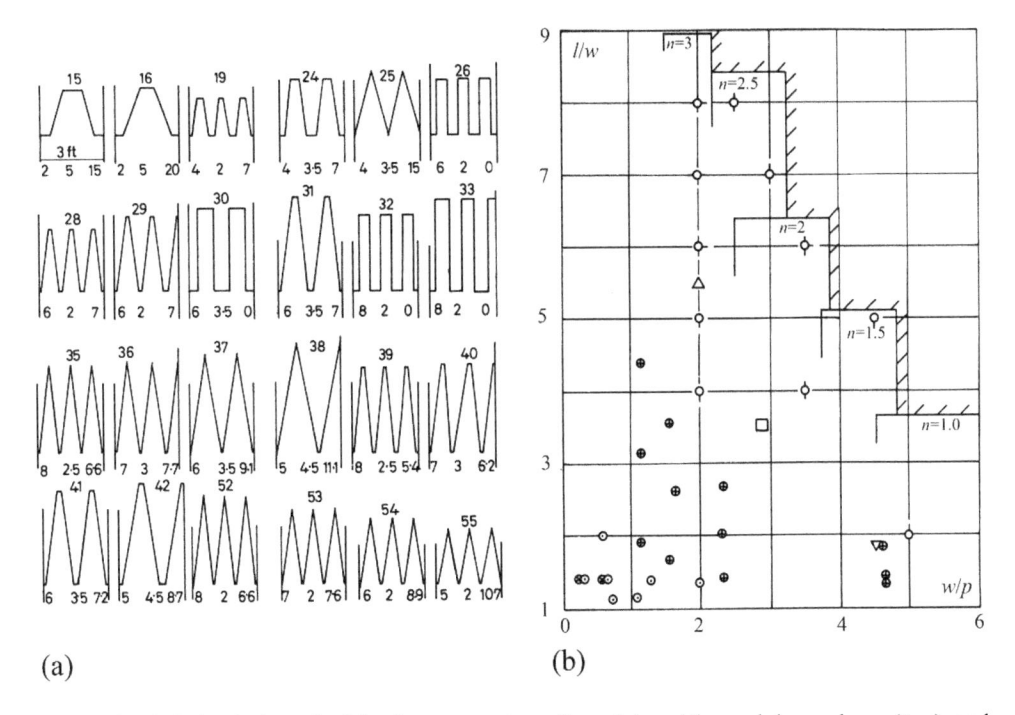

(a)   (b)

Figure 3.50 Labyrinth weir (a) plan geometry of models with model numbers (top) and values of (l/w, w/p, α) (bottom), (b) tested parameter ranges (Adapted from Hay and Taylor, 1970)

flow interference increase as the angle α decreases. No effect of the vertical aspect ratio w/p was noted. The effect of nappe interference is less pronounced for trapezoidal plan shape. As to the number of weir cycles n, no effect was observed. The effect of the downstream interference due to problems with nappe aeration was noted, so that an adequate air supply should be considered. As to the weir submergence effect, the ratio Q_L/Q_N reasonably follows Villemonte's equation. Hay and Taylor (1970) state that the LW should not be submerged with a tailwater elevation higher than the weir crest elevation, in analogy to normal weir flow conditions. As to crest shapes other than sharp-crested, no effect was found if the normal discharge is determined with the appropriate overflow relation, including e.g. broad-crested or round-crested weirs.

As to the LW design guidelines, Hay and Taylor provide discharge performance diagrams $Q_L/Q_N[h/p, l/w]$ for the triangular and the trapezoidal weir plan shapes for a sidewall angle of $\alpha = 0.75\alpha_{max}$ with and without downstream interferences. Note from Figure 3.51 that the weir discharge performance remains excellent if $h/p < 0.1$, whereas it is too small for $h/p > 0.5$. The design guidelines further include comments relating to the crest length magnification, the vertical aspect ratio, the sidewall angle, the apron elevation difference, the crest geometry, the submergence, and other limitations.

A notable research on the LW, referred to by the authors as the polygonal weir, was conducted by Indlekofer and Rouvé (1975). Figure 3.52 shows weir plan shape arrangements including curved and straight weir crests, of which only the latter were considered under fully aerated nappe. A so-called corner weir consists of two or a multiple of straight vertical weirs

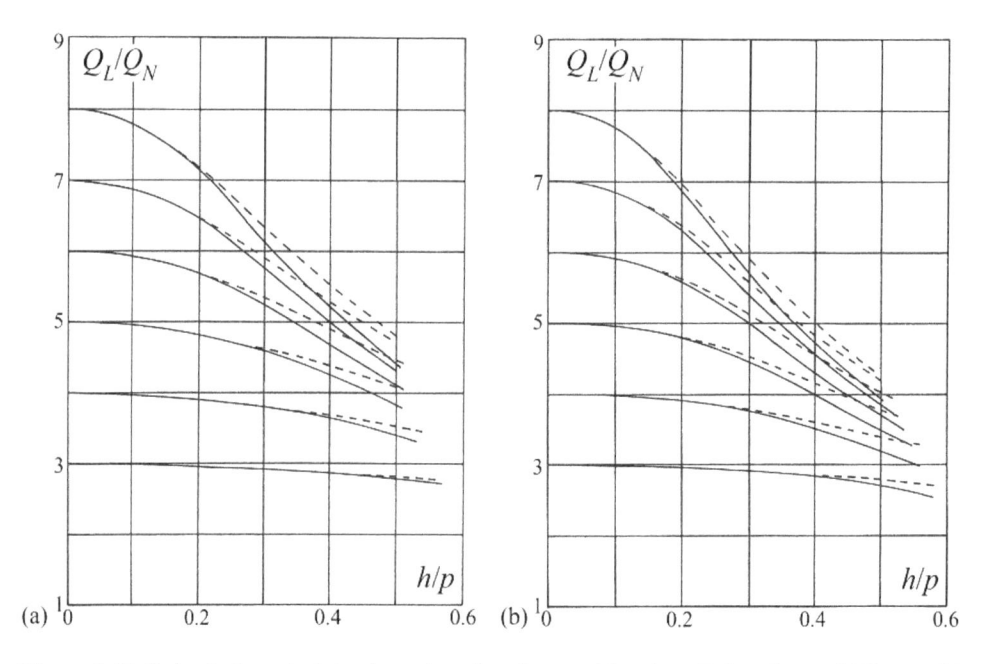

Figure 3.51 Labyrinth weir (a) triangular plan form with $w/p \geq 2.5$ and $\alpha = 0.75\alpha_{max}$, (b) trapezoidal plan form with $w/p \geq 2.0$ and $\alpha = 0.75\alpha_{max}$ under (---) no downstream interference, (—) downstream interference (Adapted from Hay and Taylor, 1970)

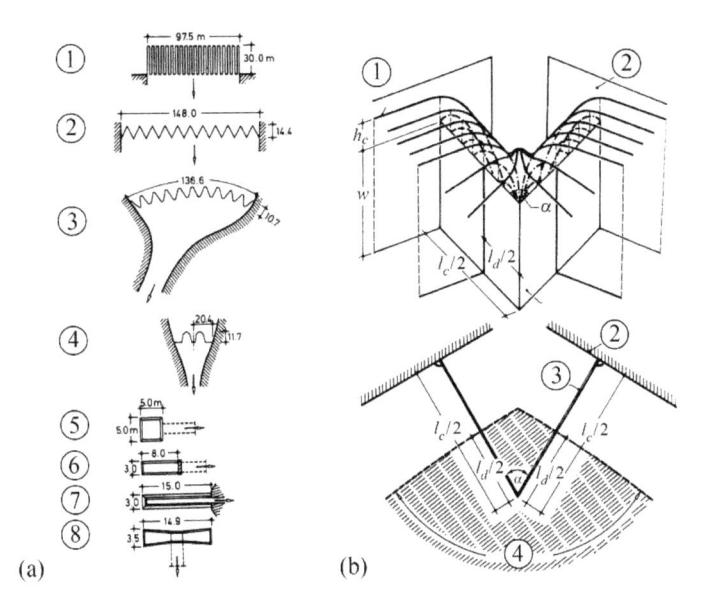

Figure 3.52 (a) Polygonal weir crests ① duck-bill overfall, ② and ③ labyrinth weirs, ④ polygonal weir, ⑤ square intake, ⑥ rectangular intake, ⑦ rectangular spillway, ⑧ polygonal intake tower. (b) Corner weir from upstream (top), and plan (bottom) with ① streamline, ② sidewall, ③ weir crest, ④ disturbed zone (Adapted from Indlekofer and Rouvé, 1975)

with intended angle α of $0 < \alpha \leq 180°$. Model tests for $\alpha = 47°$, $62°$, $90°$, and $123°$ for the standard crest shape were made. The resulting discharge was related to the Rehbock (1929) equation. Based on laboratory observations, various overflow zones affect the discharge. A computational approach was presented including the overflow length, and the related disturbance coefficients, resulting in a complex generalized Rehbock discharge equation.

Cassidy *et al.* (1985) describe a spillway providing protection against the possible maximum flood of an earth dam and reservoir in north central Oregon. The zoned-filled earth dam is 31 m high, across Six Mile Canyon. To counter a dam failure under maximum flood conditions, a labyrinth-type spillway was integrated into the dam crest. After a description of the spillway requirements for two stages of reservoir development, various spillway types are discussed. Among the straight crest, the gate-controlled crest, and the labyrinth-crest types, the latter was selected for the final design for reasons stated previously. Figure 3.53 shows a plan of the overflow reach, including pressure tap locations. The straight crest length was 110 m, so that two symmetrical triangular plan-shape labyrinth portions of 39° intended angle were selected along the 36.6 m wide overflow section. To increase the discharge coefficient, the crest shape adopted was semi-circular of diameter 0.46 m. The weir height varied from 2.8 m upstream to 4.3 m downstream of the structure, with an average of $p = 3.5$ m. Based on Hay and Taylor (1970), the discharge magnification ratio varied from 2.7 to 3, which followed also from model tests. The free-surface profiles were determined along the weir crest to estimate the freeboard requested along the weir section.

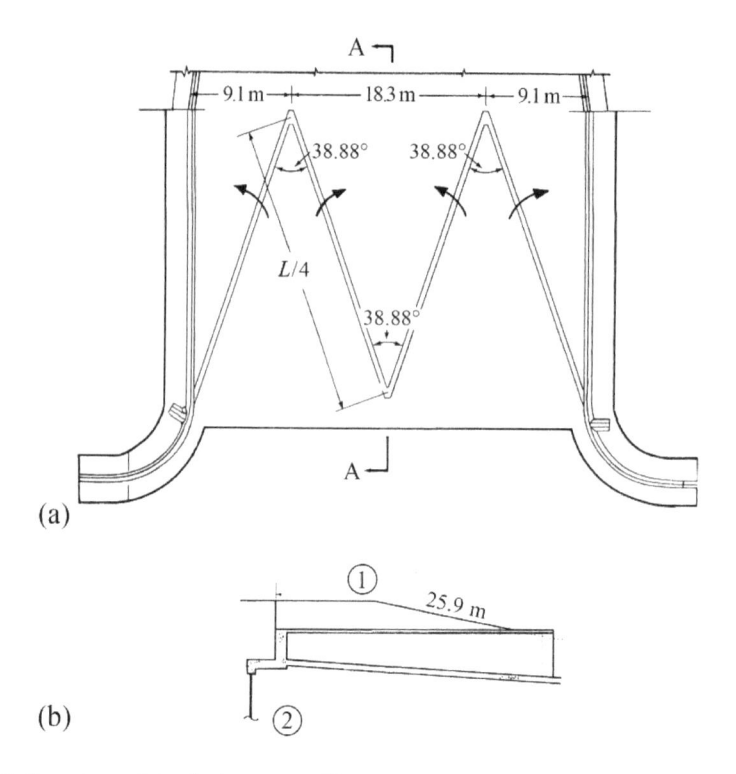

Figure 3.53 Boardman labyrinth-crest spillway (a) plan, (b) section A-A with ① crest elevation 206.5 m, ② grout curtain (Adapted from Cassidy *et al.*, 1985)

3.4.2 Design criteria

Introduction

Tullis *et al.* (1995) presented a comprehensive work on the design aspects of labyrinth
weirs (LW). Figure 3.54 shows a typical layout. A LW has advantages as compared with
the straight overflow weir and the standard ogee crest. The total LW length is typically
three to five times the spillway width. Its capacity varies with head and is typically about
twice that of a standard weir or overflow crest of the same width. LW apply to increase
the discharge capacity for a given weir crest elevation and length, or to increase storage
by raising the crest while maintaining the spillway capacity. The design variables include
its length and width, the crest height, the labyrinth angle, the number of cycles, and
other less important variables such as the wall thickness, the crest shape, and the apex
configuration.

Apart from Hay and Taylor (1970), a number of hydraulic models have been tested to
improve the design of LW. Darvas (1971) used the experimental data of the model studies
of the Woronora and Avon Weirs in Australia developing a family of design curves. Mayer
(1980) used a 1:20 scale model to study the effect on discharge of a proposed labyrinth weir
spillway to be added to the Bartlett's Ferry project. The conceptual design of the structure

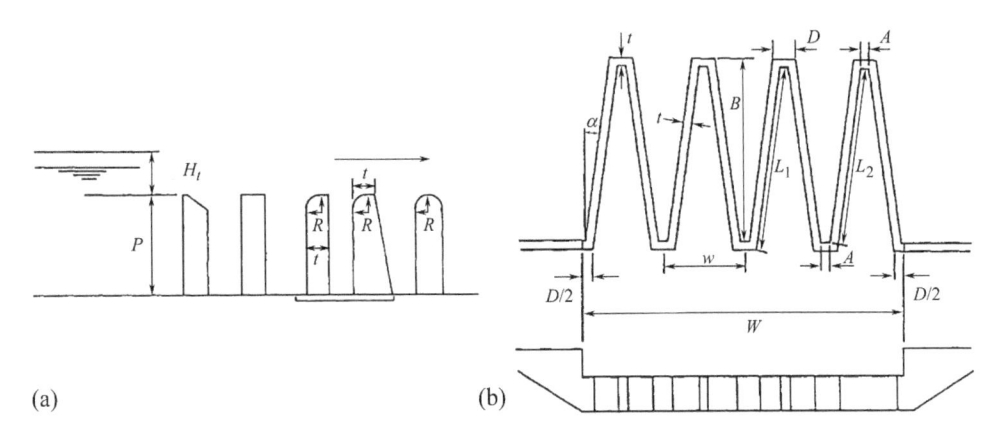

(a)   (b)

Figure 3.54 (a) Layout and crest shapes, (b) details of labyrinth weir with effective crest
length $L = 2N(A + L_2)$, total crest length $N(2L_1 + A + D)$, actual length of side
leg L_1, effective length of side leg L_2, length of labyrinth B, width of labyrinth
W, and number of cycles N (Adapted from Tullis *et al.*, 1995)

based on the approach of Hay and Taylor (1970) was found inadequate, as the structure
would not pass the required discharge. Lux (1984) assessed the hydraulic performance of
LW using data obtained from flume studies and site-specific models. He developed a discharge-head equation.

The US Bureau of Reclamation (USBR) tested models for the labyrinth spillways of the
Ute Dam and the Hyrum Dam (Houston, 1982, 1983; Hinchliff and Houston, 1984). Testing of the originally designed 10-cycle model of the Ute Dam labyrinth spillway was based
on Hay and Taylor (1970), indicating that the design discharge could not be passed at the
maximum reservoir elevation. The discrepancy between their results and these of Hay and
Taylor (1970) was partly due to the difference in head definition. Houston (1982, 1983) and
Lux (1984) used the total head instead of the piezometric head. The latter does not allow for
differences in the approach flow velocity, thereby introducing significant errors. USBR also
completed a model of the Ritschard Dam LW (Vermeyen, 1991). The model results were
used to design a LW for Standley Lake (Tullis, 1993).

Researchers have been involved in developing a procedure to improve the design of LW.
Experimentation was conducted at Utah Water Research Laboratory (UWRL) to evaluate the crest coefficient (Amanian; 1987; Waldron, 1994). Each used the same 1.0 m-wide
flume. Linear weirs with flat, quarter-round, and half-round crest shapes were tested over a
range of H/P from 0.05 to 1.0. The model weirs were 152 and 229 mm high and tested at
angles from 6° and 35°. Most of these were 152 mm high and 25.4 mm thick with quarter-
round and half-round crest shapes (Figure 3.54). A point gage, located 1 m upstream from
the LW, was used to measure flow depths. The data of these studies were used to develop a
design procedure, based on the specific crest geometry employed. The procedure allows for
flexibility in selecting the number of weir cycles and the angle of the side legs. Limitations
are placed on the design variables. The hydraulic design of labyrinth weirs has also been

summarized and described in detail by Falvey (2003). Figure 3.55(b–c) show top and side views of the Bospoort Dam gravity dam equipped with a labyrinth weir.

Discharge equation

The fundamental weir discharge equation determines the discharge coefficient C_d as (Tullis et al., 1995)

$$Q = (2/3)C_d L(2gH_t^3)^{1/2}. \tag{3.19}$$

Here, C_d is the dimensionless crest coefficient, g gravity acceleration, L the effective weir length, and H_t the total head on the crest. For a weir with a short approach length involving negligible inlet losses, H_t is the elevation difference between the reservoir water and the weir crest elevations. For a linear weir without side contractions and with normal approach flow conditions, the effective length L is the actual measured weir length (Chapter 2). The crest coefficient depends on H_t/P, the wall thickness t, the crest configuration, and nappe aeration. The variation of C_d with H_t/P for the aerated linear weir with $t/P = 1/6$ and the crest rounded at $R = P/12$ radius on the upstream corner indicates an increase for small H_t/P, and a constant value of $C_d = 0.75$ if $H_t/P > 0.40$. This behavior represents the upper limit of C_d for LW. Their effective length L is defined in the caption of Figure 3.54; the discharge coefficient depends on the same variables plus the LW configuration at its apex, and the labyrinth angle.

Four basic options for the crest shape, namely sharp-crested, flat, quarter-round on the upstream side, and half-round were tested (Figure 3.54). The wall thickness is determined from structural analysis and depends on the crest height, hydraulic forces, ice loading, and specific site conditions. For economy and strength, it is preferable to taper the downstream wall, which has no effect on the discharge coefficient. Sharp-crest and flat-crest weirs are not preferred for the LW because their discharge coefficients are below these of round-crested weirs. The most efficient and practical shape is attributed to the quarter-round crest because it has a large C_d value at higher values of H_t/P and is easy to construct. The following is therefore based on this crest shape with a top wall thickness of $t = (1/6)P$, or a tapered wall with $R = (1/12)P$.

As to the discharge coefficient, C_d of the frontal weir increases with H_t/P reaching $C_d = 0.75$ if $H_t/P > 0.40$. In contrast, C_d increases for small values of H_t/P, reaching the maximum at $H_t/P \cong 0.2$ to 0.3, then reducing as H_t/P is further increased due to flow interaction. Under these conditions, the discharge capacities of equal frontal and labyrinth weirs are nearly identical. Figure 3.55 shows this fact and the limitation of H_t/P to 0.90. For higher values, the LW still works, but its advantages diminish, so that there is no more economic use. Given that the apex width A affects the discharge capacity significantly, this value should be a minimum, typically one or two times the wall thickness. Further, the LW should never be submerged so that there is no reduction of discharge capacity, as for the frontal weir. A slight discharge reduction results for non-perpendicular approach flow to the labyrinth of angle β_{af}, with $\beta_{af} = 0$ for perpendicular approach flow. According to Tullis et al. (1995), the reductions of C_d for $\beta_{af} = 15°$, 30°, and 45° are 1%, 4%, and 6%, respectively. Below, the standard perpendicular approach flow direction is only considered. As for frontal overflow of sharp-crested weirs, the nappes of a LW should be fully aerated to reduce vibrations.

Figure 3.55 shows the discharge coefficient C_d of the standard LW versus the relative head H_t/P on the weir for various labyrinth angles α from 6° to 90° (frontal overflow). These values

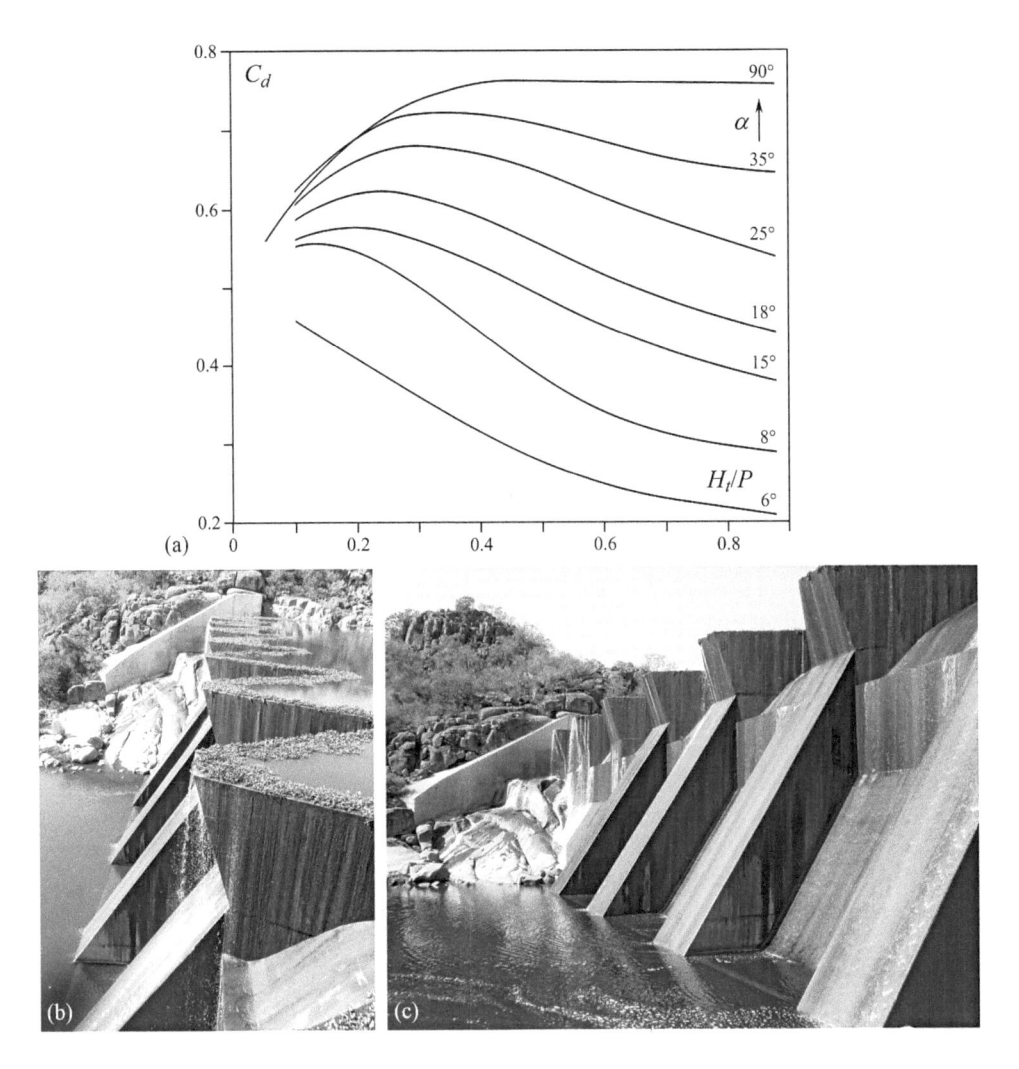

Figure 3.55 (a) Discharge coefficient of standard labyrinth weir $C_d(H/P)$ for various laby-rinth angles α (Adapted from Tullis *et al.*, 1995), (b, c) Labyrinth weir installed at crest of 28 m high concrete gravity Bospoort Dam, South Africa (Courtesy Anton J. Schleiss)

are limited to $1 \leq A/t \leq 2$, $H_t/P < 0.90$, $t/P = 1/6$, quarter-rounded weir crest, and $R/P = 1/12$ (Tullis *et al.*, 1995). This plot illustrates that all data appear to start at $C_d(H_t/P \to 0) = 0.5$, i.e. some 10% below the accepted value of $3^{-1/2} = 0.577$. For $\alpha > 6°$, C_d increases to a maximum below the previously mentioned plateau value 0.75 for frontal overflow ($\alpha = 90°$), from where it reduces as H_t/P is increased.

Tullis *et al.* (1995) also provide a detailed design procedure along with examples. Their design was verified at nine different prototype sites, from which less than 10% discharge deviations resulted as compared with these determined from Figure 3.55. It is concluded that important labyrinth weirs should still be laboratory-tested given the uncertainty under slightly modified structural and hydraulic conditions. Further indications to the LW are given by Tullis *et al.* (2005, 2007), Falvey (2003), or Schleiss (2011).

3.5 Piano key weir

3.5.1 Historical evolution

Several labyrinth weirs have been built so far and they prove good performance. They are a standard structure in flow control due to their pronounced hydraulic performance and simple geometry, mainly including a base slab and vertical walls. Drawbacks include their considerable reinforced-concrete volumes, a reduced hydraulic capacity for high discharges, and their substantial foundation structure (Lempérière *et al.*, 2011). Particularly the latter limits their application in rivers.

The piano key weir (PKW, Figure 3.56) is a recent further development of the LW. It maintains the main hydraulic advantage of the labyrinth weir, namely high discharge capacity due to the developed crest length. Its folded geometry provides a long overflow crest as compared with the width, with related high discharges per transverse unit width. In parallel, the substantial foundation is omitted. The footprint is reduced to the statically required minimum in the streamwise direction. The most distant parts of the developed crest thus become overhanging keys, allowing for its application on top of spillways located on concrete dams, and not only in rivers. Further, the concrete volumes reduce due to the much shorter foundation. To facilitate its construction, the plan view of the crest is squared (instead of trapezoidal as for labyrinth weirs). It consists thus of repeating cycles, made up of vertical walls connected by alternatingly inclined bottom plates. This geometry provides less flow resistance near the key extremes, i.e. at the most distant parts of the overhangs. Compared with a labyrinth weir, the flow has more space in the transverse direction and is orientated upward according to the key slope. This increases the unit discharge capacity as compared to the labyrinth weir.

To summarize, the advantages of the PKW as compared with the labyrinth weir are:

1   Reduced structural footprint allowing for the installation on top of (existing) gravity dams (Lempérière and Ouamane, 2003).
2   High discharge capacity, because the developed crest length corresponds several times the transverse weir width. The inclined key bottom instead of the horizontal-vertical arrangement of labyrinth weirs improves the hydraulic features efficiently (Laugier *et al.*, 2009; Anderson and Tullis, 2011, 2012).
3   Construction cost is relatively moderate, despite of the complex geometry. The repetitive cycle character allows for serial production using prefabricated formwork. Further, construction time is relatively short (Laugier *et al.*, 2013).

Early tests including hybrids between labyrinth weirs and PKWs were conducted at the LNH Laboratory in Chatou (France) in 1998, initiated by Hydrocoop and Electricité de France (EDF). The squared shape was not yet fully adopted, but the principle to include overhanging keys was included. Blanc and Lempérière (2001) suggested two types of modified labyrinth weirs, one of which was close to a PKW. They noted that such an adapted labyrinth weir applies to increase the discharge capacity of spillways on existing dams. Then, test series of PKWs were conducted at Roorkee University (India) and at Biskra University (Algeria).

Figure 3.56 Examples of piano key weirs at (a) Charmines Dam (F) under construction (December 2014; Courtesy Michael Pfister), (b) Gloriettes Dam (F) under construction (October 2010; Photo EDF, France), (c) Malarce Dam (F) (Courtesy Michael Pfister)

The results were published by Lempérière and Ouamane (2003), particularly introducing the name 'PKW'. The tests included a single geometry of a so-called A- and B-type PKW (Figure 3.57), as defined below. No systematic parameter variation was conducted at this stage. A graph with the rating curve is provided, comparing the two tested PKW geometries with the performance of the ogee crest. Results highlight the relevance of the ratio between the developed crest length L and the transverse width W. The authors point at the efficient PKW rating curve, and that construction cost is presumably substantially lower than for ogees.

Since 2006, PKW prototypes were built by EDF to update the discharge capacity of spillways on existing dams. Ogees were replaced by PKWs (Laugier et al., 2013, 2017). Early PKWs were also installed on rivers in Vietnam (Ho Ta Khanh et al., 2011; Ho Ta Khanh, 2017).

As stated above, PKWs are an attractive alternative as inlet element and discharge control of spillways because (1) they are non-gated inlets without risk of failure or false operation; (2) the $n - 1$ rule does not apply, so that the full weir width applies to spill a flood; (3) generation of high unit discharges under small heads, reducing thus the water level increase in the reservoir during a flood; (4) they are free of inserts as bridge piers or piers for gate fixation sensitive to block under driftwood presence; and (5) generation of free-surface flow in subsequent chute. These aspects are relevant for new spillways or to increase the capacity of existing spillways. The replacement of an ogee by a PKW can significantly increase the unit discharge, asking only for a small lowering of the maximum reservoir operation level or a relatively moderate dam heightening (e.g. Phillips and Lesleighter, 2013; Pinchard et al., 2013; Erpicum et al., 2017; Chapuis et al., 2017).

3.5.2 PKW types and notation

Four types of PKWs (A to D) are distinguished according to their geometry, including:

- A: Upstream and downstream overhanging keys (Figure 3.57a), possibly of different lengths on each side. The hydraulic efficiency is excellent, particularly if optimizing the geometrical parameters.
- B: Exclusively overhanging keys on the upstream side (Figure 3.57b), rendering this type also hydraulically efficient.

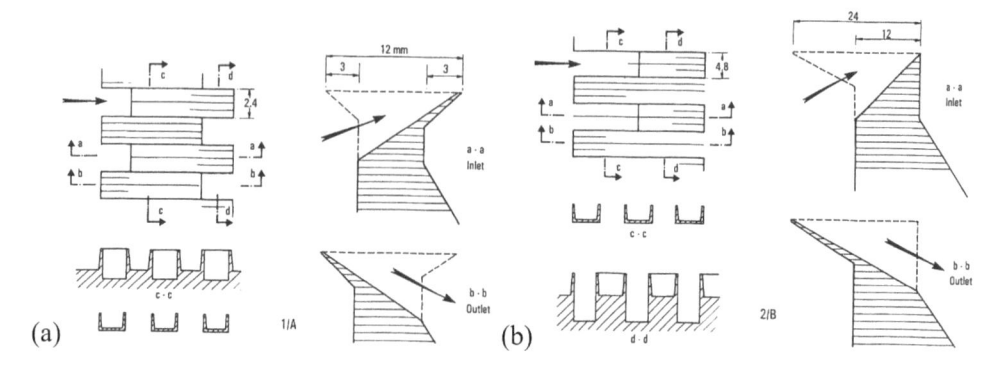

Figure 3.57 Early PKWs tested by Lempérière and Ouamane (2003) (a) A-type with overhanging keys up- and downstream, (b) B-type with overhanging keys upstream only

- C: Overhanging keys exclusively on the downstream weir side, yet hydraulically less efficient. However, no objects are trapped on the upstream side given the absence of overhangs.
- D: Without overhangs, thus close to a labyrinth weir, but with two differences: Plan view is squared, and keys include slopes.

Besides the selection of the PKW type and its individual geometrical dimensions, the crest shape has to be defined. Common are cylindrical and broad-crested shapes; few applications include quadrants oriented toward the flow or away from it. An effect of the crest shape is observed only for small overflow heads $H/P < 0.3$, with P as weir height. The cylindrical and quadrant crests are hydraulically more efficient than the broad-crested shape, as for frontal weir flow. Differences are 10% to 20% for small heads. Above these, the crest effect becomes negligible given that a discharge coefficient close to that of a sharp-crested weir is approached, independent of the crest types (Cicero and Delisle, 2013).

The standard notation for the PKW geometry follows Pralong et $al.$ (2011) (Figure 3.58), with B as the streamwise crest length, P as the vertical height, T_s as the wall thickness, and R_p as the parapet wall height. Subscript i refers to the inlet key, i.e. the key filled with water for a reservoir surface at the PKW crest elevation, whereas subscript o refers to the outlet key, i.e. the 'dry' key for the latter reservoir level.

3.5.3 Rating curve

The head-discharge function (rating curve) of a non-gated weir is one of its most important hydraulic features. It ranges between the (1) reservoir full supply level as maximum normal operation level, identical with the weir crest elevation, and (2) maximum reservoir level during floods and spillway operation under PMF. The latter range defines a reservoir head or volume needed for flood release but not exploitable for water storage. The rating curve depends on the (a) selected crest type (and its critical section), e.g. an ogee or a board-crested weir, and (b) crest length. As to item (a), a cylindrical or similar crest type is most efficient on a PKW. As to item (b), an extended crest length is required. Generally speaking, this is

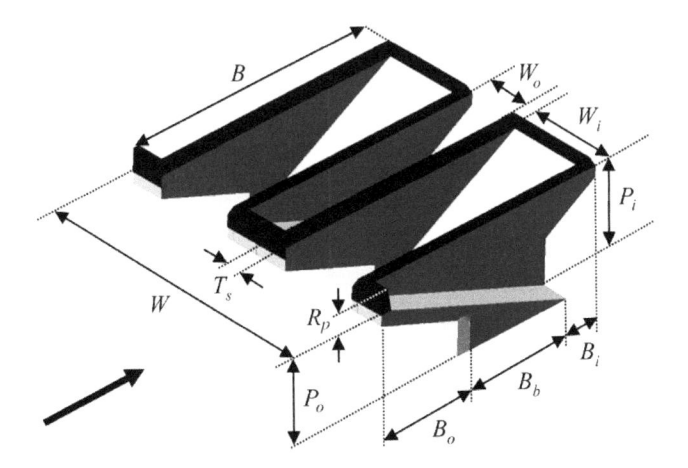

Figure 3.58 Notation of PKWs according to Pralong *et al.* (2011)

mostly in conflict with the topography and the cost. It would technically be possible to provide a long crest, for example if designing a side channel. The latter is, however, only efficient and economical under particular conditions. For an increase of the spilling capacity at existing dams, these structures are often difficult to add since no space is available, whereas a PKW is effective because it adds a developed crest length L of up to some seven times the transverse chute width W.

Before comparing the relative efficiency of PKWs illustrated by a particular case, the so far published PKW rating curves are presented, namely that of Kabiri-Samani and Javaheri (2012), Leite Ribeiro *et al.* (2012), and Machiels *et al.* (2014, 2015). The experimentally obtained rating curves are limited to effects of physical modeling and its related parameter variation (Table 3.3). These limits have to be respected when deriving reliable rating curves. The equation of Kabiri-Samani and Javaheri (2012) appears particularly sensitive in this regard, as suggested by Pfister *et al.* (2012).

Study of Kabiri-Samani and Javaheri

This study was conducted in a 12 m long and 0.40 m wide channel, including a total of 600 tests resulting in some individual 3000 pieces of data. The tests included PKW types A, B, C, and D. The PKW performance was investigated under free and submerged flow conditions. Here, only free weir flow is discussed. The Poleni equation served as basis to derive the rating curve via the PKW discharge Q_{PKW}, defined as

$$Q_{PKW} = \frac{2}{3} C_d W \sqrt{2gH^3} . \tag{3.20}$$

Based on a dimensional analysis, the discharge coefficient C_d was derived versus the geometrical (Figure 3.58) and hydraulic parameters as

$$C_d = \left[0.212 \left(\frac{H}{P} \right)^{-0.675} \left(\frac{L}{W} \right)^{0.377} \left(\frac{W_i}{W_o} \right)^{0.426} \left(\frac{B}{P} \right)^{0.306} \exp\left(1.504 \frac{B_o}{B} + 0.093 \frac{B_i}{B} \right) \right]$$
$$+0.606. \tag{3.21}$$

Study of Leite Ribeiro et al.

Systematic model tests were conducted in a channel 3 m long and 0.50 m wide. Of the total of 380 tests, 49 different PKW geometries of type A were considered, relating exclusively to

Table 3.3 Application limits of PKW rating curves (Figure 3.58)

	L/W	H/P	W_i/W_o	B/P	$B_i/B, B_o/B$	Crest type
Kabiri-Samani and Javaheri (2012)	2.5–7.0	0.1–0.6	0.33–1.22	1.0–2.5	0.00–0.26	Sharp-crested
Leite Ribeiro *et al.* (2012)	3.0–7.0	0.1–2.8	0.50–2.00	1.5–4.6	0.20–0.40	Cylindrical
Machiels *et al.* (2014, 2015)	4.2–5.0	0.1–5.0	0.50–2.00	1.0–6.0	0.29–0.33	Broad-crested

free overflow. The rating curve relates the PKW discharge Q_{PKW} to the reference discharge of a linear sharp-crested weir Q_s as

$$r = \frac{Q_{PKW}}{Q_S} = \frac{Q_{PKW}}{0.42 W \sqrt{2gH^3}}. \tag{3.22}$$

The discharge increase ratio r is a function of primary and secondary parameters. The primary effects dominating the discharge capacity are developed length L, the total transverse width W, the vertical height P_i, and the head H. The secondary parameters influence the discharge capacity marginally, including the ratio of inlet to outlet key widths W_i/W_o, the ratio of inlet to outlet heights P_i/P_o, the relative overhang length $(B_i + B_o)/B$, and the relative parapet wall height R_o/P_o. Their effects are expressed with individual correction (C) factors, namely w_C, p_C, b_C, and a_C. The estimation of r is then

$$r = 1 + 0.24 \left(\frac{(L-W)\,P_i}{WH} \right)^{0.9} \cdot \left(w_C p_C b_C a_C \right). \tag{3.23}$$

The values of the individual correction factors range between 0.92 and 1.20 for the geometries tested. The effect of the ratio between inlet and outlet key widths is

$$w_C = \left(\frac{W_i}{W_o} \right)^{0.05}, \tag{3.24}$$

that of the ratio between outlet and inlet key heights is

$$p_C = \left(\frac{P_o}{P_i} \right)^{0.25}, \tag{3.25}$$

that of the ratio between key overhangs to the base length is

$$b_C = \left(0.3 + \frac{B_o + B_i}{B} \right)^{-0.50}, \tag{3.26}$$

and that of the relative parapet wall height is

$$a_C = 1 + \left(\frac{R_o}{P_o} \right)^2. \tag{3.27}$$

Study of Machiels et al.

Model tests were conducted in a 7.2 m long and 1.20 m wide channel. Again, the proposed design equation is only valid for free overflow, involving PKW types A, B, and C. The unit PKW discharge q includes the sum of the specific discharges on the downstream (subscript d), the upstream (subscript u), and the side crests (subscript s_C), so that

$$q = \frac{Q_{PKW}}{W} = q_u \frac{W_o}{W_u} + q_d \frac{W_i}{W_u} + q_{sC} \frac{2B}{W_u} K_{Wi} K_{Wo}. \tag{3.28}$$

With W_u as cycle width, defined as $W_u = W_i + W_o + 2T$, the specific discharges are with $P_T = P + P_d$, with P_d as dam height below the PKW,

$$q_u = 0.374 \left(1 + \frac{1}{1000H + 1.6}\right)\left(1 + 0.5\left(\frac{H}{H + P_T}\right)^2\right)\sqrt{2gH^3},$$ (3.29)

$$q_d = 0.445 \left(1 + \frac{1}{1000H + 1.6}\right)\left(1 + 0.5\left(\frac{H}{H + P}\right)^2\right)\sqrt{2gH^3},$$ (3.30)

$$q_{sC} = 0.41 \left(1 + \frac{1}{0.833H + 1.6}\right)\left(1 + 0.5\left(\frac{0.833H}{0.833H + P_e}\right)^2\right) \times$$

$$\left(\frac{P_e^{\alpha} + \beta_{sC}}{(0.833H + P_e)^{\alpha} + \beta_{sC}}\right)\sqrt{2gH^3}.$$ (3.31)

Furthermore, P_e is the mean weir height along the sidewall, given as

$$P_e = P_T \frac{B_o}{B} + \frac{P}{2}\left(1 - \frac{B_o}{B}\right).$$ (3.32)

Parameters α_i and β_{sC} characterize the effect of the inlet key slope on the side crest discharge efficiency. With $S_i = P/(B - B_o)$ as slope of the inlet key,

$$\alpha_i = \frac{0.7}{S_i^2} - \frac{3.58}{S_i} + 7.55,$$ (3.33)

$$\beta_{sC} = 0.029 \exp\left(-\frac{1.446}{S_i}\right).$$ (3.34)

Let K_{Wi} describe the effect of flow velocity variation along the side crest on its efficiency as

$$K_{Wi} = 1 - \frac{\gamma}{\gamma + W_i^2},$$ (3.35)

$$\gamma = 0.0037 \left(1 - \frac{W_i}{W_o}\right),$$ (3.36)

The parameter K_{Wo} accounts for the side crest length decrease induced by the outlet key flow and the side nappe interference. It depends on H/W_o as

$$K_{Wo} = 1, \quad \text{for } H/W_o \leq L_1,$$ (3.37)

$$K_{Wo} = \frac{2}{(L_2 - L_1)^3}\left(\frac{H}{W_o}\right)^3 - \frac{3(L_2 + L_1)}{(L_2 - L_1)^3}\left(\frac{H}{W_o}\right)^2 + \frac{6L_2L_1}{(L_2 - L_1)^3}\left(\frac{H}{W_o}\right) + \frac{L_2^2(L_2 - 3L_1)}{(L_2 - L_1)^3}$$ (3.38)

$$\text{for } L_1 \leq H/W_o \leq L_2,$$

$$K_{Wo} = 0, \text{ for } H / W_o \geq L_2. \tag{3.39}$$

To apply Eqs. (3.37) to (3.39), the limits L_1 and L_2 include the outlet key slope $S_{oK} = P/(B - B_i)$, defined as

$$L_1 = -0.788 \, S_{oK}^{-1.88} + 5, \tag{3.40}$$

$$L_2 = 0.236 \, S_{oK}^{-1.94} + 5. \tag{3.41}$$

Comparison of results at a prototype case study

A comparison of the above equations relating to PKW rating curves is given in the frame of a virtual prototype. The latter has to be in accordance with the limits of the individual studies (Table 3.3). The selected prototype remains somehow arbitrary in terms of its characteristics. The following virtual site is considered:

1   Roller-compacted concrete gravity dam with a $W = 100$ m wide chute on its downstream face, a dam height of $P_d = 20$ m below the PKW foundation, free weir overfall, without contraction effects of the distal weir ends.
2   Design discharge of $Q_D = 2500$ m³/s.
3   Symmetrical A-type PKW, mounted at the dam crest as unregulated control structure, with $B = 8$ m as total streamwise length, $P = P_i = P_o = 5$ m as vertical height, $T_s = 0.35$ m as wall thickness, $R_p = 0$ m (without parapet walls), $W_i = 1.80$ m as inlet and $W_o = 1.50$ m as outlet key widths, and $B_i = B_o = 2$ m as overhang lengths (Figure 3.58). The crest shape is broad-crested.

The PKW characteristics are cycle width $W_u = W_i + W_o + 2T_s = 4$ m, number of cycles $N = W/W_u = 25$, developed crest length $L = W + (2NB) = 500$ m, $L/W = 5.00$, $B/P = 1.60$, $W_i/W_o = 1.20$, $B_i/B = B_o/B = 0.25$, and $S_i = S_o = 0.83$. The latter contradicts the limitations of Leite Ribeiro et al. (2012), requiring S_i and $S_{oK} < 0.70$. The three compared approaches are based on model studies using different crest shapes (Table 3.3). The discharges were first computed according to the above equations (Figure 3.59a), and transformed to the broad-crested weir then (Figure 3.59b). The ratios of discharge coefficients C_d were used as reference for the transformation. The C_d values per crest type involve:

- Hager and Schwalt (1994) for broad-crested weir, where C_d varies from $C_d = 0.33$ for $H < 0.2$ m to $C_d = 0.42$ for $H > 1$ m.
- Hager and Schleiss (2009) for sharp-crested weir, where $C_d = 0.42$ for all H.
- Castro-Orgaz (2012) for cylindrical weir crests if $H/(0.5T_s) < 2$ (up to $C_d = 0.53$ at $H = 0.35$ m), then linearly tending to $C_d = 0.42$ until $H/(0.5T_s) < 25$, equivalent to $H = 4.35$ m (Ramamurthy and Vo, 1993).

The resulting rating curves are shown in Figure 3.59a, yet without including the crest type. The rating curve of the linear ogee is also shown (Vischer and Hager, 1998) for the design head of $H = 5$ m. The rating curves of the three PKW studies are similar. Differences occur for small ($H < 1$ m) and large ($H > 3$ m) heads. Given that particularly small heads are of interest, the effect of the crest type was added via C_d in Figure 3.59b. Then, the predictions collapse for small heads, whereas a slightly larger spread occurs for $H > 1.5$ m. Therefore,

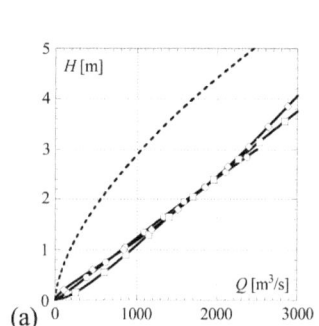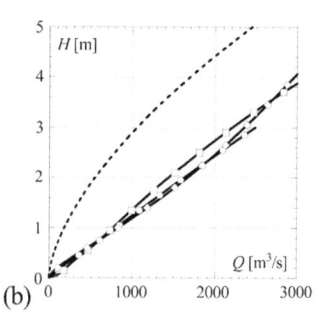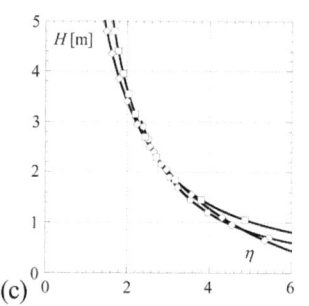

Figure 3.59 PKW rating curves $Q(H)$ of case study, (a) according to literature, (b) correcting crest-type effect, (c) discharge ratio comparing PKW and ogee according to (○) Kabiri-Samani and Javaheri (2012), (□) Leite Ribeiro *et al.* (2012), (◊) Machiels *et al.* (2014, 2015), (---) Ogee weir, Vischer and Hager (1998)

the effect of the crest type should be considered, given that PKWs are particularly efficient and often applied under small heads. Figure 3.59c shows the discharge ratio $\eta_{PKW} = Q_{PKW}/Q_O$ versus H, with Q_O as ogee and Q_{PKW} as PKW discharges. The efficiency of PKWs is particularly high for small heads, because the developed length L is then fully active, similar to a long linear weir. For a head of $H = 1$ m, a PKW spills 4.5 times the ogee discharge. For higher discharges, the flow interaction in the PKW crest wedges and corners (in plan view) becomes important, reducing the capacity to approach the ogee discharge ($\eta_{PKW} \to 1$). For heads $H > 3.5$ m a discharge ratio of $\eta_{PKW} \leq 2$ occurs.

Considering the design discharge of $Q_D = 2500$ m³/s and ignoring the crest shape effect (Figure 3.59a), the prediction of Leite Ribeiro *et al.* (2012) results in $H = 3.12$ m whereas Machiels *et al.* (2014, 2015) gives $H = 3.23$ m. Interesting is the comparison of the PKW discharge capacity with that of the ogee. For the latter, the head is $H = 5$ m for $Q_D = 2500$ m³/s according to its design criterion. Its required hydraulic head is thus around 157% of the PKW head, for an identical discharge. The related absolute head difference is around 1.83 m. The dam height may thus be reduced by this difference, resulting in lower construction cost, or the height difference is available to generate additional storage volume by rising the crest elevation. If the PKW replaces an existing ogee by keeping the same crest elevation and the same maximum reservoir level during a flood (so that $H = 5$ m), then the discharge capacity is upgraded from 2500 m³/s to some 3500 to 4000 m³/s, depending of the considered estimation.

Tailwater submergence

The non-submergence range is defined as condition under which the tailwater head H_d exceeds the weir crest but the free-flow rating curve still applies. Note that the tailwater head H_d involves the weir crest as reference level. Kabiri-Samani and Javaheri (2012) state that the non-submergence range is $H_d/H \leq 0.6$ for A, B, C, and D types. Small L/W values up to 2.5–4 are more sensitive and the submergence starts earlier, so that the limit is then $H_d/H \leq 0.5$. The reduced, submergence discharge Q_{SUB} relative to the non-submerged, free discharge Q_{PKW} is

$$C_{SUB} = \frac{Q_{SUB}}{Q_{PKW}}, \tag{3.42}$$

$$C_{SUB} = \left[1 - 0.858\frac{H_d}{H} + 2.628\left(\frac{H_d}{H}\right)^2 - 2.489\left(\frac{H_d}{H}\right)^3\right]\left(\frac{L}{W}\right)^{0.055}.$$ (3.43)

Dabling and Tullis (2012) conducted model tests involving A-type PKWs, concluding that the non-submergence range is $H_d/H \leq 0.48$. Belaabed and Ouamane (2013) state $H_d/H \leq 0.8$ for an A-type PKW as non-submergence limit. The cited studies suggest that the effect of submergence, namely the increase of the upstream head to spill a certain discharge, or vice versa the discharge reduction for a given head, starts only for comparatively high tailwater heads exceeding the elevation to some $0.5H$.

3.5.4 Further design aspects

Geometry and cost

The selection of the individual geometry affects the hydraulic efficiency and cost of a PKW. Machiels (2012), Laugier *et al.* (2013), and Machiels *et al.* (2014) recommend:

- If a PKW is built on top of an existing gravity dam to enhance the spillway flood capacity (e.g. replacement of ogee), then a part of the dam crest (or of ogee) has to be removed for the PKW foundation. Besides, the reservoir level is lowered during construction. To limit both constrictions, a small PKW height P is beneficial (Figure 3.58). Low PKWs with $P_i/(W_o + W_i + 2T_s) = P_i/W_u \approx 0.5$ satisfy this requirement, with $P = 4$ to 5 m. However, the remaining parameters affecting the hydraulic efficiency are then to be optimized, including $6 \leq L/W \leq 7$, B_o/B_i up to 2, and $1.2 \leq W_i/W_o \leq 1.5$. Accordingly, an unequal width results between inlet and outlet keys. Two separate formworks for the inclined key bottoms have to be provided, generating additional cost but accelerating in parallel the construction process.
- If a PKW is added to a new spillway, then fewer constrictions regarding its height P exist. Vice versa, a larger liberty regarding the arrangement of the PKW remains, particularly regarding the transverse chute width W. A high PKW with $1 \leq P_i/(W_o + W_i + 2T_s) = P_i/W_u \leq 1.3$ is preferable, typically with $P = 5$ to 10 m. Further, $4 \leq L/W \leq 6$, $B_o/B_i = 1$, and $W_i/W_o = 1$. A single (and thus economic) standard formwork for the slopes applies on both inlet and outlet keys. Additional weir width can compensate the not optimal (in the sense of most performant) hydraulic layout.
- Crest shape should be cylindrical as the hydraulic optimum. The increased cost as compared with the broad-crested shape is negligible if building multiple cycles with the same formwork set. For few cycles and a new PKW with liberty regarding its layout, the broad-crested shape appears reasonable.

Laugier *et al.* (2013, 2017) list the cost of six projects in France involving a PKW. For all, the existing spillway capacity was insufficient and an auxiliary structure including a PKW was added. The total project cost are reported for these six projects between 4000 and 4500 k€, of which only some 230–800 k€ refer to the PKW itself. Typical construction times are around six months only.

Driftwood blockage risk

A retention of floating woody debris at overflow control structures potentially results in reduced discharge efficiency (higher upstream head for given weir discharge). If the catchment is forested, the potential effects of driftwood on the PKW rating curve should be considered as part of the hydraulic design.

There are few studies on the effect of debris at PKWs. Ouamane and Lempérière (2006) conducted preliminary model tests, observing that no debris was collected under the over-hangs during reservoir filling. During operation, debris was stuck in the inlet key at small discharges, lowering the discharge coefficient by some 10% (for $H/P < 0.5$). As the dis-charge increased, the debris was eventually washed downstream. Laugier (2007) reports of systematic debris model tests on the PKW of Goulours Dam (France). He concludes that (1) increasing head tends to remove previously blocked debris; (2) principal gated spillways attracts debris before PKW operation; (3) most debris passes PKW if the flow depths at the PKW exceed 1 m; (4) flow streamlines typically pass below debris of a fully blocked PKW, entering the inlet key unhindered (i.e. the water flows 'around' block-age); and (5) residual discharge of a blocked PKW is more than 80% of the debris-free condition.

Pfister *et al.* (2013a, b) conducted model tests on driftwood blocking at PKWs for reser-voir conditions upstream of the weir. They conducted trunk and rootstock tests to determine the blocking probability of an individual element, and accumulative debris tests to evaluate the reservoir head increase for large debris volumes.

The dimensionless trunk diameter D/H was found to indicate whether an individual trunk is trapped by or passes a PKW. A 50% blocking probability was observed if the trunk diam-eter D equaled the theoretical critical flow depth $(2/3)H$ at the PKW crest. Individual trunks of smaller diameter normally pass, but trunks of larger diameter are typically trapped. If $D \geq H$, then a blocking probability of 100% was observed (all trunks blocked), whereas for $D \leq (1/3)H$, 0% blocking probability results, i.e. all trunks passed the PKW. The trunk length, though relevant, was less significant. The debris elements used featured lengths T that were proportional to their diameter D ($T/D = 20$). Individually arriving rootstocks have a higher blocking probability than trunks. A 50% blocking probability was observed for $D \leq (1/5)H$, indicating that already small trunks with roots block, as compared with trunks without roots.

The data indicate a less clear trend for accumulation tests, i.e. driftwood batches or carpets as frequently occur at prototype weirs. In general, a notable increase in head is recognizable if (1) high debris volumes arrive at PKW, (2) reference heads H_r are small, and (3) chute widths W are narrow, increasing the 'unit debris volume'. The reference head H_r corresponds to the head for the considered discharge without driftwood effect, i.e. before trunks arrive at the PKW. It is essential that the relative head increase is only significant if H_r is small, i.e. for smaller discharges. For larger heads, typically used as PKW design values, the relative head increase reduces, as some debris is washed over the crest by the increasing hydraulic load. Pfister *et al.* (2013a) propose for the relative head increase H/H_r, with V_D as arriving driftwood volume (pure wood mass) and W as weir width (Figure 3.58)

$$\frac{H}{H_r} \cong 1 + \tanh\left(0.007\frac{V_D}{H_r^2 W}\right). \tag{3.44}$$

For 'low' heads ($H_r/P < 0.2$), the debris accumulations increased H up to 70%. For 'higher' heads ($0.2 < H_r/P < 5.0$), its increase in H was below 20%. For $H_r/P > 0.2$, above certain debris volume limit, the addition of more floating debris resulted in a marginal increase of

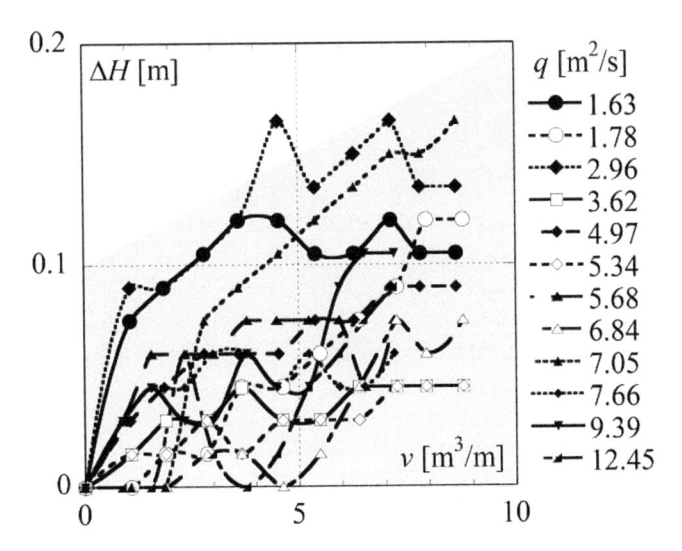

Figure 3.60 Effect of unit driftwood volume v and unit discharge q on absolute ΔH reservoir head increase (Pfister *et al.*, 2013b)

H. An explicit approach is shown in Figure 3.60 (Pfister *et al.*, 2013b). It relates the absolute additional head ΔH generated by the driftwood blockage to the unit discharge $q = Q/W$ and the unit approach flow wood volume $v = V_D/W$. If no driftwood arrives at a PKW (both small v and q values), then the initial blockage generates an additional head of some 0.1 m. If more wood arrives under this discharge ($v \leq 9$ m³/m and small q values), the additional head reaches up to some 0.2 m. 'High' unit discharges generate generally smaller additional heads than 'small' discharges.

Overall, PKWs with an upstream reservoir appear less sensitive to the driftwood effect on the discharge capacity than classical weirs, as ogees. The plain blockage is often limited to the reach upstream of the PKW, ending close to the upstream crest of the outlet key (Figure 3.61a). The PWK itself remains mostly clear, and the water passes below the debris entering the inlet key unhindered by blockage. From numerical simulations (e.g. Ackers *et al.*, 2013) and physical model tests (e.g. Machiels *et al.*, 2011), a significant part of the spilled discharge travels into the inlet key from below, without reaching the free surface upstream of the PKW where blockage occurs.

Prototype observations at EDF Malarce Dam (France) confirm these observations (Pfister *et al.*, 2015). During a flood in November 2014 of about 4 m²/s unit discharge (near 30% of the design discharge), significant debris volumes arrived. Most trunks passed the PKW, and few blocked at the upstream crest of the outlet key, mostly at the right PKW portion. Consequently, the PKW itself remained practically clear and unaffected, even at its right portion where debris blocked. Flow surfaces and depths in the inlet keys appeared visually to be similar between the right (blocked) and left (debris-free) PKW sides, because the inlet keys are mainly fed from below the water column. The hydraulic head associated to the measured discharge was slightly below $H = 0.5$ m, with the related critical flow depth $(2/3)H \cong 0.3$ m.

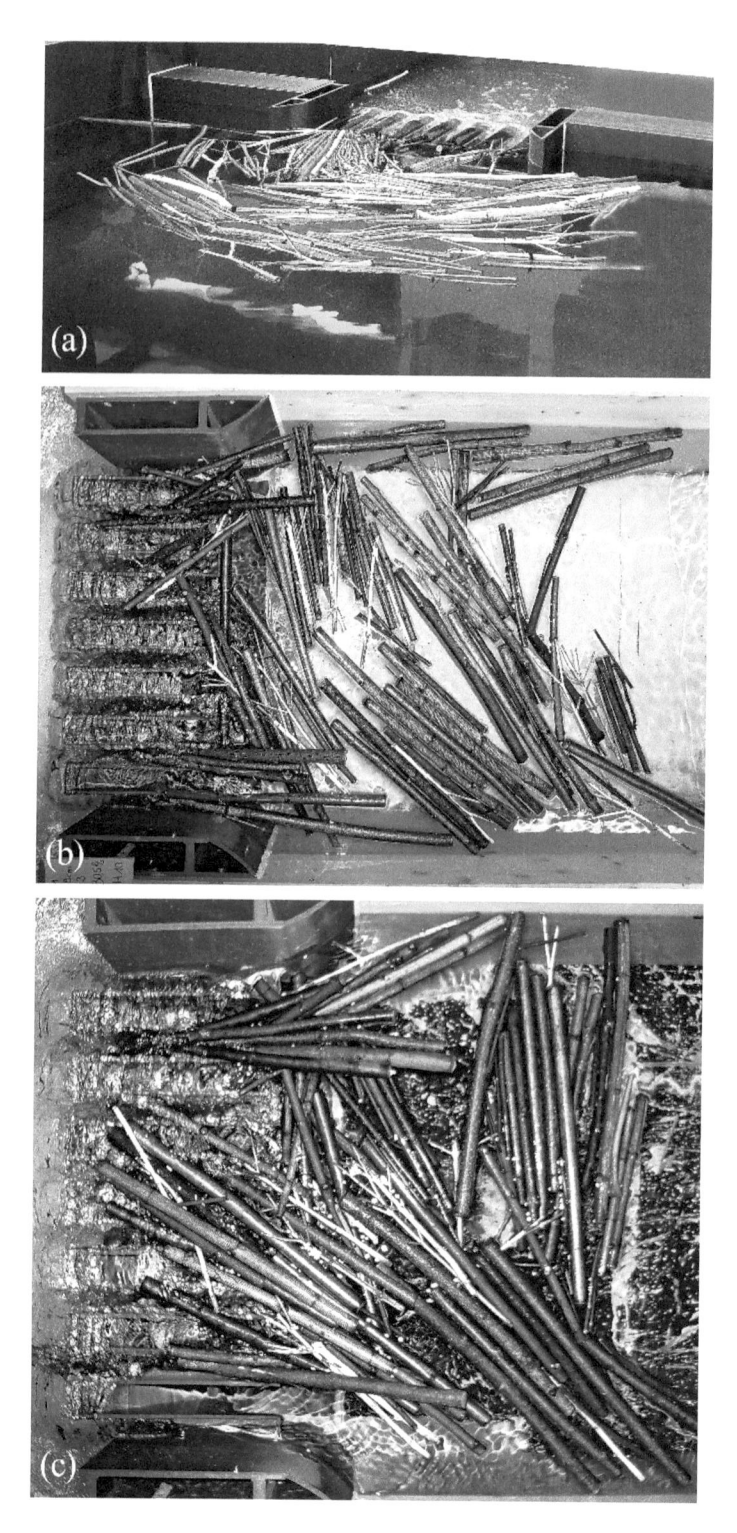

Figure 3.61 Density and shape of blockage on PKW versus approach flow Froude number, with $F_o =$ (a) 0 (reservoir), (b) 0.05, (c) 0.10, under otherwise similar conditions (Venetz, 2014)

Observations indicated that mainly trunks of diameters $D \geq 0.3$ m initiated the blockage, in agreement with the above described individual trunk tests.

The application of these results is, among others, limited to the following conditions:

- PKWs installed on dam crests with an upstream reservoir: The approach flow velocity of the trunks is negligible. The debris blocking process is affected if the trunks arrive fast (e.g. Schmocker and Hager, 2011), e.g. if a PKW is placed on a river. As noted from Figure 3.61, the blocking becomes dense if the approach flow Froude number, involving the approach flow velocity, increases from $F_o = 0$ for the reservoir to $F_o = 0.1$ for an intake channel. Second, the wood is pushed onto the PKW so that the flow cannot enter the inlet keys below the blockage.
- Debris arriving at the crest if the PKW is already in operation: The possibility of driftwood blockage under the overhanging keys during reservoir filling was not considered. Debris accumulations could arrive before the PKW spills, mainly due to a reservoir current induced by an operating principal spillway or due to wind.
- Steady discharge during debris arrival: Effective hydrographs during a flood event may influence the behavior of the debris at the weir.
- Effects of additional appurtenant structures as bridge piers, along with the PKW if debris passage were not considered.

Structural aspects

The water jets falling on the downstream face of the inlet keys enclose each a trapped air volume below the key, if the discharge exceeds a certain value. This nappe starts to fluctuate due to sub-pressures, and because of a fluctuating flow the detachment point on the crest. As for sub-pressures and the related vibrations, these are reduced if the nappe is sufficiently aerated. The aeration ducts are inserted in the PKW concrete walls (Vermeulen *et al.*, 2017), a challenging task due to the high degree of reinforcement and the limited wall thickness (Figure 3.56c). Alternatively, the aeration ducts are fixed on the concrete surface (Figure 3.56b). As to the fluctuation of the flow detachment point on the crest, a fixation of this point by adding a discontinuity or flow splitters at the crest is efficient. Sensitive are particularly cylindrical weir crests (Crookston and Tullis, 2013; Lodomez *et al.*, 2017).

Laugier *et al.* (2013) report of the EDF experience in building PKWs. They point at the fact that PKWs are hyper-static and stiff structures, subjected by gravity, hydrostatic, hydrodynamic, seismic, and thermal loads. Because their wall thickness is small, 0.25 m $\leq T_s \leq$ 0.40 m, more steel reinforcement is required for the thermal loads than for classic structural forces. Structural computations conducted by EDF indicated that the center of gravity is usually located within the PKW base, even for a water level at the maximum reservoir level during floods. PKWs are thus not as unbalanced as they look, and a slight anchoring into the foundation is generally sufficient.

Denys *et al.* (2017) discuss the fluid-structure interaction at PKWs. They found, for the cases considered, that resonance behavior may occur, caused by upstream flow separation and vortices shed from it.

3.5.5 Downstream toe scour on riverbed

In parallel to their most frequent application as auxiliary dam spillway, PKWs are also installed on low head barrages in rivers in combination with run-off-river power plants

(Ho Ta Khanh *et al.*, 2011). However, scour occurs at their downstream sides on a loose river-bed if no technical protection is placed. Such a protection measure can include a stilling basin (Truong Chi and Ho Ta Khanh, 2017) or a plunge pool (Jüstrich *et al.*, 2016; Pfister *et al.*, 2017).

Model tests of stilling basins downstream of a PKW under various hydraulic conditions were executed by Truong Chi and Ho Ta Khanh (2017). They tested stilling basins with a stepped chute between the PKW toe and the basin, and basins with and without end sills or baffles, for which relatively long basins resulted. A basin length of 82.75 m resulted as an example for a unit discharge of 28.5 m²/s, still inducing a weak scour downstream of the end sill. The experiments of Jüstrich *et al.* (2016) serve as reference for the plunge pool type (in a loose bed) and are detailed hereafter, with the notation given in Figure 3.62.

The conditions at PKWs are considered as clear-water scour, given that a PKW with its high crest traps incoming sediment. The question if sediment transport occurs on the loose bed downstream of the PKW is answered by combining the Meyer-Peter and Müller transport equation with the Manning-Strickler uniform flow equation. The scour (sub-script S) depth z_F at the downstream PKW foundation varies with the maximum scour depth z_{SM} as

$$\frac{z_F}{z_{SM}} \cong 0.36. \tag{3.45}$$

The sediment directly at the PKW toe is eroded to roughly 1/3 of the maximum scour depth z_{SM}, i.e. a considerable value. This is due to inclined jets issued by the outlet keys touching the sediment at the end of the concrete structure. The maximum scour hole depth is nor-malized with the characteristic sediment diameter d_{50}, and is expressed versus the relative discharge h_c/d_{50} with $h_c = (q^2/g)^{1/3}$ as critical flow depth (with $q = Q/W$), and the relative head difference $\Delta H/h_d$ (Figure 3.62). Note that the tailwater head H_d as used here to derive ΔH considers the sediment bed as reference, and not the crest level as for submergence effects. Significant discharges under high heads generate thus extended scour holes, mainly for fine sediment. Within the test limits $0.8 < Z_{SM} < 133$, the relative maximum scour depth is

$$Z_{SM} = \frac{z_{SM}}{d_{50}} = 0.42 \left(\frac{h_c}{d_{50}} \right)^{1.7} \left(\frac{\Delta H}{h_d} \right)^{0.3}. \tag{3.46}$$

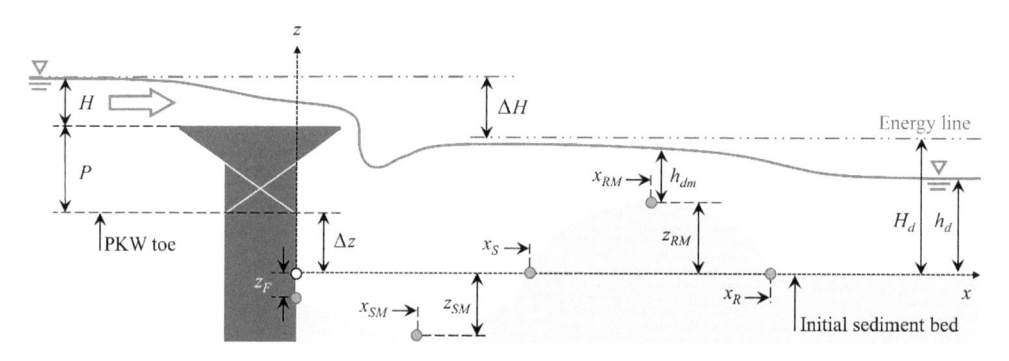

Figure 3.62 Notation for bed scour hole and ridge downstream of PKW (Jüstrich *et al.*, 2016)

The streamwise location x_{SM} of the maximum scour depth occurs at

$$x_{SM} = 1.20z_{SM} + \frac{B_i}{2}. \tag{3.47}$$

The maximum scour depth is generated slightly downstream of the PKW foundation because the inclined jets issued from the outlet keys require a certain length to pass the inflection point and reorientate upwards. The scour hole works like a submerged flip bucket. Further, a nearly vertical jet falls along B_i. The maximum scour hole depth occurs thus with a streamwise shift of $B_i/2$ from the PKW foundation. Further, the maximum depth occurs at a distance of only 1.2 times the maximum depth, resulting in a steep upstream slope of the scour hole. The scour hole extends up to x_S (Figure 3.62), defined where the sediment surface crosses the initial bed level, i.e.

$$x_S = 2.70z_{SM} + B_i. \tag{3.48}$$

The dominant parameter of the scour hole geometry is thus its maximum vertical depth z_{SM}, with all other dimensions varying with it. The unit scour hole volume V_S per width W is

$$\frac{V_S}{z_{SM}^2} = 2. \tag{3.49}$$

An identical relation was proposed by Stein and Julien (1994) for scour hole volumes below free overfalls. They remark that the maximum observed scour hole angles tend to the angle of repose of the bed material, as also observed for the PKW scour hole.

If a sufficient sediment volume is eroded in the scour hole, the ridge (subscript R) increases until the incipient flow velocity v_H is reached to activate sediment transport on its top. The latter velocity is given by Hjulström (1935, subscript H). A related incipient flow depth follows as $h_H = Q/(Wv_H)$. Smaller flow depths erode the ridge, whereas larger depths allow for ridge height increase, or stable conditions. If h_H is reached and the scour process supplies further sediment, then the ridge length increases. The maximum ridge height under sufficient sediment transport is thus

$$z_{RM} = h_d - h_H = h_d - \frac{Q}{Wv_H}. \tag{3.50}$$

It is independent of the maximum scour hole depth z_{SM}, but linked to the sediment characteristics and the tailwater depth h_d. The streamwise location of the maximum ridge height is roughly at

$$x_{RM} = 6z_{SM} + B_i. \tag{3.51}$$

A maximum height was identified for all ridges, equivalent to the plateau height of 'flat' ridges. Its streamwise location is thus again dominated by the maximum scour hole depth (or its related volume). The streamwise ridge end location x_R is at

$$x_R = 9z_{SM} + B_i. \tag{3.52}$$

The ridge is thus typically twice as long as the scour hole, if sufficiently fed.

Figure 3.63 shows an example of a PKW scour hole and its ridge as observed in a physical model. A comparison of the PKW scour hole depths with literature suggests that PKW conditions are described as a jet-induced scour, issued by the outlet keys. Almost the entire discharge per cycle is concentrated at its outlet key. The above estimates are helpful to define an appropriate foundation depth of a PKW installed on a movable riverbed. If the requested depth is significant, then the estimate provides a riprap block size to limit the scour depth. A scour protection is reasonable for PKWs not founded on rock. The scour downstream of a PKW is reduced by installing a small deflector at the end of the outlet keys, orientating the issued jets horizontally. According to Pagliara *et al.* (2006), this measure significantly reduces the maximum scour depth. Noui and Ouamane (2013) suggest that a classical flip bucket in the outlet key could reduce the PKW discharge capacity. Alternatively, a concrete apron combined with an end sill (e.g. a stilling basin) is also a measure to avoid scour.

3.5.6 Upstream riverbed

Some PKWs are implemented in low-head run-on-the-river plants. If PKWs could replace gates, then the sediment passage potentially deposited in the reservoir during floods must be assured. The level of the upstream sediment bed related to the hydraulic conditions at the PKW, combined with its sediment passage capacity, is essential to avoid an overtopping of the upstream river levees.

The model study conducted by Noseda (2017) gives first indications on the mobile riverbed development in the near-field upstream of a PKW. It was demonstrated that sediments

Figure 3.63 Example of PKW scour hole and ridge in physical model (Jüstrich *et al.*, 2016)

arriving at the PKW pass the latter. The upstream flow depth is dominated by the sediment transport capacity of the flow, as given for instance by van Rijn (1984) or Hjulström (1935). Subtracting from the latter the PKW-generated head (or the related flow depth) results in the sediment bed level upstream of the PKW.

Note that PKW-generated heads do *not* correspond to the rating curves presented in 3.5.3, given that the sediments potentially lift the bed level into the PKW. Then, the conditions that the kinetic approach flow head does not affect the weir head is not respected anymore. Noseda, et al. (2019) gives an adapted rating curve for these conditions.

3.6 Siphon

3.6.1 Description

A siphon is a ducted overflow structure with either free-surface or pressurized flow. Siphons are used as spillway either in parallel or in addition to other flood releases, or as intake of small power plants, as described by Xian-Huan (1989). In the following, the latter are excluded. Siphons either correspond to a saddle siphon or a shaft siphon (Figure 3.64). Under increasing discharge, both behave hydraulically similar to a weir. At a certain discharge, priming occurs so that the flow becomes pressurized under higher discharges. The transition between free-surface and pressurized flows depends mainly on the aeration and de-aeration features of the siphon crest.

A siphon looks like a covered round-crested weir. Its advantages compared with a weir are the priming action and a much increased discharge capacity under low heads. For larger heads, the weir has a larger capacity reserve in discharge, however (Figure 3.65). Therefore, siphons apply at locations where the head is relatively small and no discharge control is available. The disadvantages of a siphon are:

- Tailwater surges at rapid priming, as for sudden gate opening
- Cavitation damage induced by low pressure
- Clogging by ice or wood.

The first two disadvantages are mitigated by constructional means.

The *saddle siphon* is commonly used and is considered below. Siphons were incorporated both into dams and weirs. There are few modern designs and a dearth of literature. Detailed

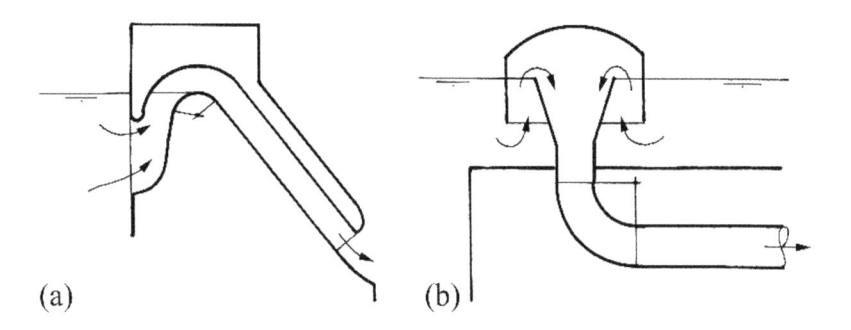

Figure 3.64 (a) Saddle siphon, (b) shaft siphon

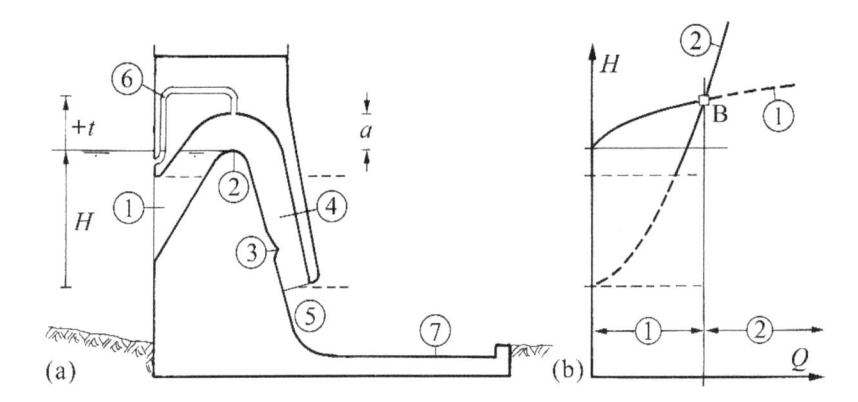

Figure 3.65 (a) Conventional siphon with ① intake, ② crest, ③ priming nose, ④ siphon barrel, ⑤ outlet, ⑥ aeration pipe, ⑦ stilling basin, (b) discharge-head relation with ① weir regime, ② orifice regime. For equal width, weir discharges less than siphon in domain ①; both structures have equal discharge at point B

descriptions of siphons are found, e.g. by Govinda Rao (1956), Press (1959), Samarin *et al.* (1960), and Preissler and Bollrich (1985). The rehabilitation of a siphon with a relatively large head of 16 m is described by Bollrich (1994).

According to Head (1975) a distinction is made between the *black-water siphon*, only operating on an on-off basis (i.e. full-bore, or no discharge at all), and the *air-regulated siphon*, which automatically adjusts its discharge over the full range of flow to maintain a virtually constant headwater level. The latter inhibits flooding due to abrupt priming, and vibration problems. The feature of the air-regulated siphon is that it sucks in both water *and* air from where the white-water originates. In the following, both types of siphons are described.

3.6.2 Black-water siphon

The transition between free-surface and pressurized flow occurs relatively abruptly for this arrangement. As the reservoir level increases over the siphon crest, overflow starts by entraining air due to the large velocities. As soon as the inflow primes, the air is not substituted from the intake. If the downstream air access is also cut, as by a jet deflector or priming nose (Figure 3.65) or tailwater submergence, the siphon primes. An observer then sees black-water at the outlet. The section of a saddle siphon is often rectangular, of typical height to width ratio 1:1.5 to 1:2.5. To reduce oscillations in the reservoir elevation and for an adequate hydraulic inlet shape, an aeration conduit is added, with a typical cross section 3% to 5% of the siphon crest section.

The *hydraulic design* of a siphon refers mainly to the maximum discharge, i.e. pressurized flow. According to the generalized Bernoulli equation, the discharge Q is

$$Q = A_d V_d = ab\eta(2gH_o)^{1/2}. \tag{3.53}$$

Here A_d and V_d are cross section and average velocity of the siphon outlet, a and b are height and width at the siphon crest, H_o is the head on the siphon, and η is a siphon efficiency coefficient. For $\eta = 1$, the siphon flow would involve no hydraulic losses. In practice, typical values of η range between 0.7 and 0.9, with $\eta = (1 + \Sigma\xi_i)^{-1/2}$ in which ξ_i is the ith loss coefficient. The sum over all losses involves the intake, the bend, the outlet, and friction losses. For the siphon according to Figure 3.66, Bollrich (1994) states $\eta = 0.86$ for a head of $H_o = 16$ m and a discharge of $Q = 51.7$ m³s⁻¹.

In a second step, the sub-pressure at the crest section has to be checked against *cavitation damage*. By analogy to the standard spillway, the minimum pressure occurs at the siphon crest. Assuming concentric streamlines over the crest (subscript C) region yields an expression for discharge Q_C with incipient cavitation as

$$Q_C = br_i \ln\left(\frac{r_a}{r_i}\right)\left[\frac{2g(h_A - h_{cA} + t_S)}{1 + \Sigma\xi_i}\right]^{1/2}. \tag{3.54}$$

Here, r_i and r_a are inner and outer crest radii, respectively, h_A [m] $= 10.3 - (A/900)$ is the atmospheric pressure head at elevation A [m] above sea level, h_{cA} is the pressure head at incipient cavitation, t_S is the reservoir elevation above the siphon crest, and $\Sigma\xi_i$ is the sum of all head-loss coefficients, as discussed above. In general, a conservative value of $h_{cA} = 2$ m is assumed. If the maximum discharge Q_M is larger than the cavitation discharge Q_C, then the siphon has to be modified until the condition $Q_M < Q_C$ is met.

For heads larger than, say, 8 m on the siphon, the barrel has to be contracted from the crest section toward the outlet, to improve the pressure conditions at the crest section. To shorten the barrel, the outlet can also be located above the tailwater level. The flow is discharged over a flip bucket into a plunge pool, or over a spillway into a stilling basin (Chapter 5).

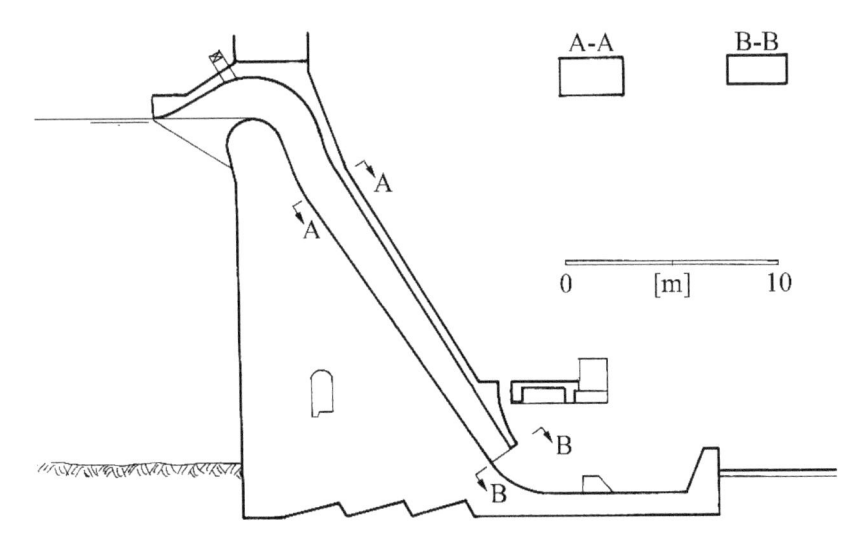

Figure 3.66 Siphon spillway of Burgkhammer Dam (Germany) according to Bollrich (1994). Pair of siphons have equal crest but different inlet nose elevations

To reduce the risk of tailwater flooding during siphon priming, various siphons in parallel are arranged. The siphon crests are located at increasing elevations with successive priming. Figure 3.67 refers to an example of six individual siphons, two of which have identical priming conditions. The difference of head to the next pair of siphons is 0.1 m. The resulting discharge-head curve is shown in Figure 3.67b. Note the typical differences between the curves for increasing, and decreasing discharges. Under adverse discharge conditions, siphons may vibrate (Govinda Rao, 1956). A general discussion on the topic follows in 8.4.

3.6.3 White-water siphon

Figure 3.68 shows a typical structure, with an upward sloping converging inlet, a covered crest of almost prismatic geometry, and a downward sloping outlet. The priming nose seals the siphon pipe against the atmosphere. The head on white-water siphons varies between 0.5 m and 2 m, and their unit discharge capacity is up to 4 m^2/s. The manometer tapping is not essential but provides a useful means of monitoring performance.

The *discharge-head relation* of Head (1975) includes three ranges, namely weir flow, air-regulated flow, and drowned flow. Their hydraulic features are (Figure 3.69):

1. *Weir flow* behaves like overflow over a cylindrical crest under atmospheric air pressure in pipe.

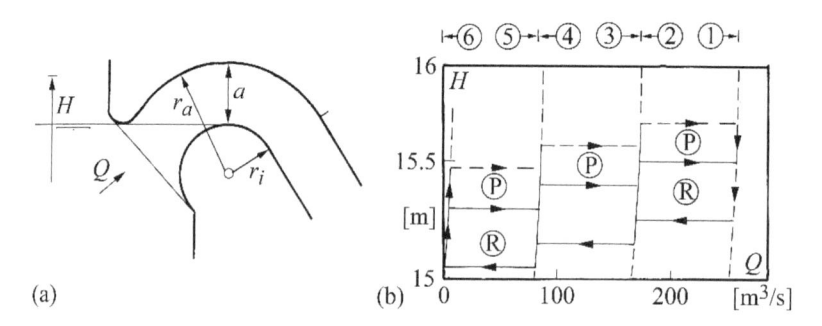

(a)   (b)

Figure 3.67 (a) Definition of siphon crest geometry, (b) siphon operation cycle with P priming and R reduction of flow for Burgkhammer siphon with total discharge of 250 m^3s^{-1} (Bollrich, 1994)

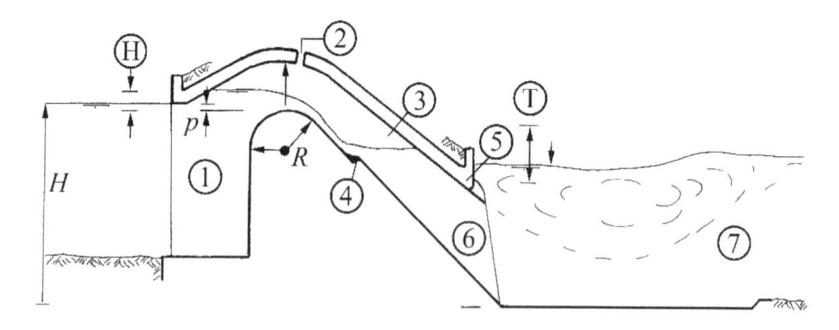

Figure 3.68 Type siphon structure with ① entry, ② tapping, ③ hood, ④ deflector, ⑤ downstream lip, ⑥ outlet, ⑦ stilling basin. ⑭ headwater range, ⑪ tailwater range

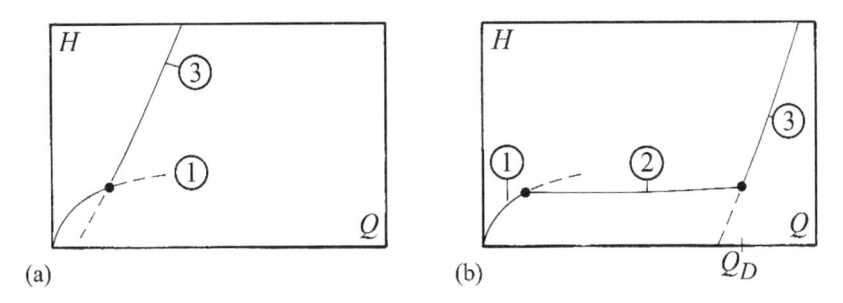

Figure 3.69 Typical stage-discharge curves for (a) conventional, (b) air-regulated siphon

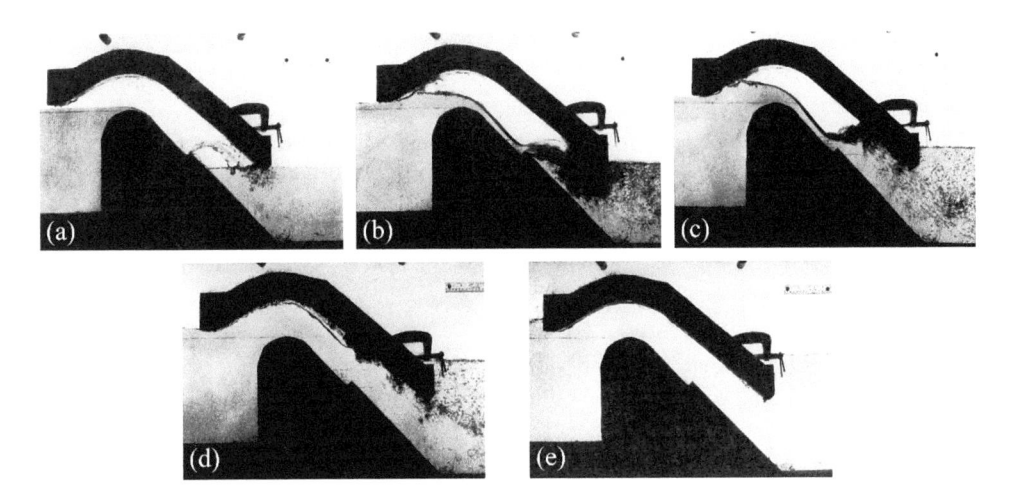

Figure 3.70 Flow phases in air-regulated siphon (a) weir flow, (b) deflected nappe, (c) drowned nappe, (d) air-partialized flow, (e) drowned flow (Head, 1975)

2. *Air-regulated flow* starts when barrel is sealed from atmosphere due to jet deflection. Air evacuation continues creating partial vacuum that increases discharge without significant headwater level rise. Air thus is continuously sucked into siphon to compensate for extraction at siphon outlet. Headwater rise yields reduced airflow, and fall leads to increased airflow thus reducing discharge. Accordingly, the white-water siphon is *self-regulating*.

3. *Drowned flow*, including transition from air-partialized range with air-water mixture in barrel, to full-pipe flow. Discharge is computed from generalized Bernoulli equation involving the difference level between headwater and tailwater elevations (2.3.2). Air regulation has ceased so that this range seldom occurs, e.g. under exceptional flood. Design has to account for transition discharge between air regulation and drowning flow.

Figure 3.70 shows the five stages of an air-regulated siphon.

Figure 3.71 Syphon spillway at Räterichsboden gravity dam, Switzerland (a) frontal, (b) side view (Courtesy KWO, Switzerland)

The crest position of a siphon spillway depends on particular site criteria. Once a siphon starts to operate in the air-regulating range, it will flatten the stage-discharge curve. Both Head (1975) and Ervine and Oliver (1980) provide design guidelines. Figure 3.71 refers to a typical siphon structure.

Notation

A	Cross-sectional area (m^2); also LW apex width (m)
A_d	Cross-sectional area of siphon outlet (m^2)
A_t	Tunnel cross-sectional area (m^2)
A_w	Cross-sectional section of water flow (m^2)
A_1	Miter bend cross-sectional area (m^2)
$\partial A/\partial x$	Streamwise channel width increase (m)
a	Height of siphon crest zone (m); also efflux width (m)
a_C	Parapet wall height correction for PKW (-)
B	Length of labyrinth weir/PKW (m)
B_i	PKW inlet overhang length (m)
B_o	PKW outlet overhang length (m)
b	Side channel or siphon width (m)
b_C	Key overhang correction coefficient for PKW (-)
C_d	Discharge coefficient (-)
C_{SUB}	Submerged discharge ratio (-)
D	Trunk diameter (m)
D_s	Shaft diameter (m)
D_t	Tunnel diameter (m)
d_{50}	Mean grain size (m)
E	Euler number of aeration device $=[\Delta p/\rho g)]/\left(V_o^2/2g\right)(-)$
F	Local Froude number $= Q(\partial A/\partial h)^{1/2}/(gA^3)^{1/2}$ (-)
F$_f$	Total discharge Froude number $=(Q_a+Q)/[(\pi D_s^2/4)(gD_s)^{1/2}](-)$

F_o	Approach flow Froude number (-)
F_s	Shaft Froude number $= V_s/(gD_s)^{1/2}$ (-)
g	Gravity acceleration (m/s^2)
H	Energy head (m); also morning glory head on weir $= \bar{H} - z_m$ (m)
ΔH	Additional head generated by driftwood presence (m)
$\bar{H}$	Morning glory overflow head (m)
H_d	Design head of overflow weir, also downstream energy head of PKW (m)
H_o	Head on siphon crest (m)
H_r	Reference head (m)
H_t	Total head on labyrinth weir crest (m)
h	Local flow depth (m)
h_1	Contracted flow depth (m)
h_A	Atmospheric pressure head (m)
h_c	Critical flow depth $= [Q_d^2/(gb^2)]^{1/3}$ or $(q^2/g)^{1/3}$ (m)
h_{cA}	Pressure head at incipient cavitation (m)
h_d	Downstream flow depth (m)
h_H	Incipient flow depth for ridge erosion (m)
h_{max}	Maximum flow depth (m)
h_o	Dead-end flow depth (m)
h_s	Singular point flow depth (m)
J	Slope difference $= S_o - S_{fa}$ (-)
K_s	Strickler roughness coefficient (m$^{1/3}$/s)
K_{Wi}	Effect of PKW velocity variation (-)
K_{Wo}	Effect of PKW key outlet (-)
L	Developed PKW length; also crest length (m)
LW	Labyrinth weir
L_1, L_2	Length of side leg (m), also PKW limitations (-)
L_i	Intake channel width (m)
L_s	Side-channel length; also dropshaft length (m)
L_t	Tunnel length (m)
N	Flow depth perpendicular to bottom profile (m)
N_T	Tullis' number of weir cycles, also PKW cycles (-)
n	Number of labyrinth weir cycles (-)
P	Labyrinth/PKW weir height (m)
PKW	Piano key weir
P_d	Dam height (m)
P_e	Mean PKW weir height (m)
P_i	Inlet PKW height (m)
P_o	Outlet PKW height (m)
P_h	Wetted perimeter $= b + 2h$ (m)
p	Labyrinth weir height (m)
p_C	Inlet height correction coefficient for PKW (-)
p_s	Lateral discharge intensity $= dQ/dx$ (m^2/s)
Δp	Sub-pressure downstream of aeration pipe (N/m^2)
Q	Water discharge (m^3/s)
ΔQ	Lateral outflow (m^3/s)

Q_a	Air discharge (m³/s)
Q_{aM}	Maximum air discharge (m³/s)
Q_c	Siphon cavitation discharge (m³/s)
Q_C	Siphon discharge subjected by incipient cavitation (m³/s)
Q_d	Design discharge (m³/s)
Q_{fg}	Discharge over flap gate (m³/s)
Q_L	Labyrinth weir discharge (m³/s)
Q_M	Maximum siphon discharge (m³/s)
Q_N	Normal weir discharge (m³/s)
Q_n	Frontal discharge (m³/s)
Q_O	Ogee crest discharge (m³/s)
Q_{PKW}	PKW discharge (m³/s)
Q_S	Reference discharge of PKW (m³/s)
Q_{SUB}	Submerged PKW discharge (m³/s)
q	Unit width discharge (m²/s)
q_D	$= Q/(gb^2H_o^3)^{1/2}$ (-)
q_d	Unit downstream PKW discharge (m²/s)
q_s	Relative shaft discharge $= Q/(\pi K_s D_s^{8/3})$ (-)
q_{sc}	Unit side crest PKW discharge (m²/s)
q_u	Unit upstream PKW discharge (m²/s)
R	Morning glory crest radius $= \bar{R} - x_m$ (m); also upstream crest radius of LW (m)
$\bar{R}$	Morning glory shaft radius (m)
R_b	Bend radius (m)
R_c	Critical Reynolds number $= 4R_{hc}V_c/v$ (-)
R_{hc}	Critical hydraulic radius (m)
R_O	Orifice Reynolds number $= Q/(v\Delta L)$ (-)
R_p	Parapet wall height (m)
R_s	Shaft radius (m)
r	Ratio of PKW to reference discharges (m³/s)
r_a	Outer siphon crest radius (m)
r_i	Inner siphon crest radius (m)
S_f	Friction slope (-)
S_{fa}	Average friction slope (-)
S_i	Slope of PKW inlet key (-)
S_o	Bottom slope (-)
S_{oK}	PKW outlet key slope (-)
S_t	Transverse bottom slope (-)
T	Length of debris element (m)
T_1	Maximum separation height $= 1 - h_1/D$ (-)
T_d	Relative downstream intake channel depth $= t_d/t$ (-)
T_s	PKW wall thickness (m)
T_u	Relative upstream intake channel depth $= t_u/t$ (-)
t	Side channel depth, intake channel depth, also labyrinth weir crest width (m)
t_d	Downstream intake channel depth (m)
t_S	Reservoir elevation above siphon crest (m)
t_s	Lateral flow depth (m)
t_u	Upstream intake channel depth (m)

U	Transverse velocity component (m/s)
U_s	Lateral inflow velocity (m/s)
V	Streamwise average velocity (m/s)
V_D	Arriving driftwood volume (m³)
V_a	Airflow velocity (m/s)
V_c	Critical flow velocity (m/s)
V_d	Reference velocity = $gA/(dQ/dx)$ (m/s), also siphon outlet velocity (m/s)
V_o	Approach flow velocity (m/s)
V_s	Surface velocity (m/s)
V_S	Unit scour hole volume (m²)
v	Unit wood volume (m²)
v_H	Incipient velocity to erode ridge (m/s)
W	Width of labyrinth weir/PKW (m)
WC	Water column (m)
W_i	Inlet PKW key width (m)
W_o	Outlet PKW key width (m)
W_u	PKW cycle width (m)
w	Approach flow weir height, or weir cycle width (m)
$\overline{w}$	Approach flow weir height relative to weir crest elevation (m)
w_C	Key width correction coefficient for PKW (-)
X	Non-dimensional streamwise location = x/x_s, also relative location x/h_c (-)
$\overline{X}$	Morning glory relative location = $\overline{x}/\overline{H}$ (-)
$\overline{X}_m$	Location of maximum lower nappe elevation (-)
$\overline{X}_{mo}$	Location of maximum lower nappe elevation for straight crest (-)
$\overline{X}^*$	Relative value of $\overline{X}/\alpha$ (-)
X_d	Relative side-channel length = L_s/x_s (-)
X_{max}	Dimensionless location of maximum flow depth (-)
X_o	= x/H_o (-)
x	Streamwise coordinate (m)
$\overline{x}$	Morning glory radial coordinate (m)
x_m	Morning glory location of maximum lower napper elevation (m)
x_{max}	Location of maximum flow depth (m)
x_R	Location of ridge end (m)
x_{RM}	Location of maximum ridge height (m)
x_s	Singular point coordinate (m)
x_S	Scour hole extension (m)
x_{SM}	Streamwise location of maximum scour hole depth (m)
y	Non-dimensional flow depth = h/h_s (-); also transverse coordinate (m)
y_{max}	Relative maximum flow depth = h_{max}/h_c (-)
y_o	Relative dead-end flow depth = h_o/h_c (-)
Z	Relative flow depth = N/h_c (-)
$\overline{Z}$	Morning glory relative elevation from crest = $\overline{z}/\overline{H}$ (-)
$\overline{Z}_m$	Maximum relative elevation of lower nappe (-)
$\overline{Z}_{mo}$	Maximum relative elevation of lower nappe for straight-crested weir (-)
$\overline{Z}^*$	Relative value of = $\overline{Z}/\beta$ (-)
Z_o	N/H_o (-)

Z_{SM} Relative maximum scour depth $= z_{SM}/d_{50}$ (-)
z Vertical bottom elevation (m)
$\overline{z}$ Morning glory vertical coordinate (m)
z_F scour depth downstream of PKW (m)
z_m Morning glory maximum lower nappe elevation (m)
z_{RM} Ridge height (m)
z_s Lateral fall depth (m)
z_{SM} Maximum PKW scour depth (m)
α Horizontal relative location of morning glory nappe maximum $= \overline{X}_m/\overline{X}_{mo}$ (-) also intended LW angle (-)
α_i Effect of inlet key slope (-)
β Vertical relative location of morning glory nappe maximum $\overline{Z}_m/\overline{Z}_{mo}$ (-)
β_a Air-water discharge ratio $= Q_a/Q$ (-)
β_{af} Approach flow LW angle (-)
β_{sC} Effect of side crest PKW discharge (-)
γ PKW width effect (-)
γ_s Proportionality factor (-)
η Siphon efficiency coefficient (-)
η_{PKW} PKW discharge efficiency (-)
χ Relative head on weir $= H/H_d$ (-)
μ Discharge coefficient of labyrinth weir (-)
μ_n Discharge coefficient of flow perpendicular to weir (-)
v Kinematic viscosity (m²/s)
φ Lateral outflow angle (-)
X Transformed streamwise coordinate $= C_d AX$ (-)
ξ Siphon loss coefficient (-)

References

Ackers, J.C., Bennett, F.C.J., Scott, T.A. & Karunaratne, G. (2013) Raising the bellmouth spillway at Black Esk reservoir using Piano Key weirs. *Labyrinth and Piano Key Weirs – PKW*, 2013, 235–242.

Amanian, N. (1987) Performance and design of labyrinth spillway. *MSc Thesis*. Utah State University, Logan UT.

Anderson, R.M. & Tullis, B. (2011) Influence of Piano Key Weir geometry on discharge. *Labyrinth and Piano Key Weirs – PKW*, 2011, 75–80.

Anderson, R.M. & Tullis, B. (2012) Comparison of Piano Key and rectangular Labyrinth Weir hydraulics. *Journal of Hydraulic Engineering*, 138(4), 358–361.

Belaabed, F. & Ouamane, A. (2013) Submerged flow regimes of piano key weirs. *Labyrinth and Piano Key Weirs – PKW*, 2013, 85–92.

Blanc, P. & Lempérière, F. (2001) Labyrinth spillways have a promising future. *International Journal of Hydropower & Dams*, 8(4), 129–131.

Boes, R.M., Lutz, N., Lais, A., Lucas, J. (2013) Hydraulic modelling of floating debris conveyance for a spillway upgrade at a large rockfill dam. Proceedings of 9th *ICOLD European Club Symposium*, Venice, Italy, Paper A8.

Bollrich, G. (1968) Gestaltung und hydraulische Berechnung senkrechter Fallschächte (Design and hydraulic computation of vertical shaft spillways). *Wissenschaftliche Zeitschrift der TU Dresden*, 17(4), 833–841 (in German).

Bollrich, G. (1971) Combined function of hydraulic structures with shaft spillways. *Water Power*, 23(10), 363–367.

Bollrich, G. (1994) Hydraulic investigations of the high-head siphon spillway of Burgkhammer. Proceedings of the 18th *ICOLD Congress* Durban, Q71(R2), 11–20.

Bradley, J.N. (1956) Shaft spillways: prototype behavior. *Transactions of ASCE*, 121, 312–344.

Bremen, R. & Hager, W.H. (1989) Experiments in side-channel spillways. *Journal of Hydraulic Engineering*, 115(5), 617–635.

Bremen, R. & Hager, W.H. (1990) Flow features in side-channel spillways. Proceedings of the 12th *Convegno di Idraulica e Costruzioni Idrauliche* Cosenza, pp. 91–105.

Bretschneider, H. (1961) Abflussvorgänge bei Wehren mit breiter Krone (Hydraulics of broad-crested embankment weirs). *Mitteilung 53*. Institut für Wasserbau und Wasserwirtschaft, Technische Universität, Berlin (in German).

Bretschneider, H. (1980) Kreisförmige Fallschächte für die Hochwasserentlastung bei Talsperren (Morning glory shaft for spillways on dams). *Wasserwirtschaft*, 70(3), 88–93 (in German).

Bretschneider, H. & Krause, D. (1965) Die Modellversuche für die Hochwasserentlastungsanlage der Innerste-Talsperre im Harz (The scale experiments for the spillway of the Innerste Dam in the Harz). *Mitteilung 62*. Institut für Wasserbau und Wasserwirtschaft, Technische Universität, Berlin (in German).

Camp, T.R. (1940) Lateral spillway channels. *Transactions of ASCE*, 105, 606–637.

Cassidy, J.J., Gardner, C.A. & Peacock, R.T. (1985) Boardman labyrinth-crest spillway. *Journal of Hydraulic Engineering*, 111(3), 398–416; 113(6), 808–819.

Castro-Orgaz, O. (2012) Discussion of Overflow characteristics of circular-crested weirs. *Journal of Hydraulic Research*, 50(2), 241–243.

Chapuis, A., Thomas, M., Deroo, L., Richit, C. & Touzet, C. (2017) Piano Key Weir on a lateral spillway: from an analytical preliminary design to a detailed design validated by a physical model. *Labyrinth and Piano Key Weirs – PKW*, 2017, 205–212.

Chow, V.T. (1959) *Open Channel Hydraulics*. McGraw-Hill, New York.

Cicero, G.M. & Delisle, J.R. (2013) Effects of the crest shape on the discharge efficiency of a type A Piano Key Weir. *Labyrinth and Piano Key Weirs – PKW*, 2013, 41–48.

Citrini, D. (1942) Canali rettangolari con portata e larghezza gradualmente variabili (Rectangular channels with linearly varied discharge and width). *L'Energia Elettrica*, 19(5), 254–262; 19(6), 297–301 (in Italian).

Citrini, D. (1948) Canali rettangolari con apporto laterale di portata (Rectangular channels with spatially-varied discharge). *L'Energia Elettrica*, 25(4), 155–166 (in Italian).

Crookston, B. & Tullis, B. (2013) Hydraulic design and analysis of Labyrinth Weirs. II: nappe aeration, instability, and vibration. *Journal of Irrigation and Drainage Engineering*, 139(5), 371–377.

Dabling, M.R. & Tullis, B.P. (2012) Piano Key Weir submergence in channel applications. *Journal of Hydraulic Engineering*, 138(7), 661–666.

Darvas, L.A. (1971) Discussion of Performance and design of labyrinth weirs. *Journal of Hydraulics Division* ASCE, 97(HY8), 1246–1251.

De Marchi, G. (1941) Canali con portata progressivamente crescente (Channels with progressively increasing discharge). *L'Energia Elettrica*, 18(6), 351–360 (in Italian).

Denys, F.J.M., Basson, G.R. & Strasheim, J.A.B. (2017) Fluid structure interaction of Piano Key Weir. *Labyrinth and Piano Key Weirs – PKW*, 2017, 119–126.

Emami, S. & Schleiss, A.J. (2016) Hydraulic scaled model tests for the optimization of approach channel excavation and approach flow conditions of Haraz morning glory spillway. International Symp. on *Appropriate Technology* to ensure proper development, operation and maintenance of dams in developing countries. Proc. 84th ICOLD Annual Meeting, Johannesburg, SA (2a), 87–95.

Erpicum, S., Archambeau, P., Dewals, B., Pirotton, M, Tralli, H. & Alende, J. (2017) A Piano Key Weir to improve the discharge capacity of the Oule Dam spillway (France). *Labyrinth and Piano Key Weirs – PKW*, 2017, 195–204.

Ervine, D.A. & Himmo, S.K. (1984) Modelling the behaviour of air pockets in closed conduit hydraulic systems. *Scale Effects in Modelling Hydraulic Structures*, Esslingen, 4(15), 1–12.

Ervine, D.A. & Olivier, G.C.S. (1980) The full-scale behaviour of air-regulated siphon spillways. *Proceedings of the Institution of Civil Engineers*, 69(2), 687–706.

Favre, H. (1932) *Contribution à l'étude des courants liquides* [Contribution to the study of liquid currents]. Rascher, Zurich, Switzerland (in French).

Falvey, H.T. (2003) *Hydraulic Design of Labyrinth Weirs*. ASCE Press, Reston VA.

Gardel, A. (1949) Les évacuateurs de crues en déversoirs circulaires (The spillways for circular overfalls). *Bulletin Technique de la Suisse Romande*, 75(27), 341–349 (in French).

Gentilini, B. (1941) Stramazzi con cresta a pianta obliqua e a zig-zag (Weirs with oblique and zig-zag crests). *L'Energia Elettrica*, 18(10), 653–664 (in Italian).

Govinda Rao, N.S. (1956) Design of siphons. *Publication* 59. Central Board of Irrigation and Power, New Delhi.

Guercio, R. & Magini, R. (1994) Effetti dell'immissione sul deflusso nei canali di gronda: indagine sperimentale e modellazione numerica (Effect of the lateral discharge in side-channels: experimental and numerical modelling). *Giornale del Genio Civile*, 132(1–3), 47–67 (in Italian).

Hager, W.H. (1990) Vom Schachtüberfall zum Schachtwehr (From shaft weir to shaft spillway). *Wasserwirtschaft*, 80(4), 182–188 (in German).

Hager, W.H. (1994) Wasser-Luftgemische in Vertikalrohren (Air-water mixtures in vertical pipes). *gwf Wasser-Abwasser*, 135(7), 391–397 (in German).

Hager, W.H. (2010) *Wastewater Hydraulics*, 2nd ed. Springer, Berlin.

Hager, W.H., Edder, O. & Rappaz, J. (1988) Streamline curvature effects in side-channel spillway flow. *Acta Mechanica*, 72, 95–110.

Hager, W.H. & Pfister, M. (2011) Historical development of side-channel spillway in hydraulic engineering. Proceedings of 34th *IAHR World Congress*, Brisbane (1636), 3906–3913.

Hager, W.H., Pfister, M. & Tullis, B.P. (2015) Labyrinth weirs: developments until 1985. Proceedings of 36th *IAHR World Congress* Den Haag, 79454, 1–9 (CD-Rom).

Hager, W.H. & Schleiss, A.J. (2009) *Constructions Hydrauliques: Ecoulements stationnaires* (Hydraulic structures: Steady flows). PPUR, Lausanne (in French).

Hager, W.H. & Schwalt, M. (1994) Broad-crested weir. *Journal of Irrigation and Drainage Engineering*, 120(1), 13–26.

Hay, N. & Taylor, G. (1970) Performance and design of labyrinth weirs. *Journal of Hydraulics Division* ASCE, 96(HY11), 2337–2357; 97(HY8), 1246–1251; 98(HY4), 708–711.

Head, C.R. (1975) Low-head air-regulated siphons. *Journal of Hydraulics Division* ASCE, 101(HY3), 329–345; 102(HY1), 102–105; 102(HY3), 422–425; 102(HY10), 1597.

Heller, V. (2011) Scale effects in physical hydraulic engineering models. *Journal of Hydraulic Research*, 49(3), 293–306; 50(2), 244–246.

Hinchliff, D.L. & Houston, K.L. (1984) Hydraulic design and application of labyrinth spillways. Proceedings of 4th *USCOLD Lecture Dam Safety and Rehabilitation*. USCOLD, Denver CO.

Hinds, J. (1926) Side channel spillways: hydraulic theory, economic features, and experimental determination of losses. *Transactions of ASCE*, 89, 881–939.

Hjulström, F. (1935) Studies of the morphological activity of rivers as illustrated by the River Fyris. *Ph.D. Thesis*, University of Uppsala. Almqvist & Wiksells, Sweden.

Ho Ta Khanh, M. (2017) History and development of Piano Key Weirs in Vietnam. *Labyrinth and Piano Key Weirs – PKW*, 2017, 3–16.

Ho Ta Khanh, M., Truong, C.H. & Nguyen, T.H. (2011) Main results of the P.K. weir model tests in Vietnam (2004 to 2010). *Labyrinth and Piano Key Weirs – PKW*, 2011, 191–198.

Houston, K.L. (1982) Hydraulic model study of the Ute Dam labyrinth spillway. *Report* GR-82–13. USBR, Denver CO.

Houston, K.L. (1983) Hydraulic model study of Hyrum auxiliary labyrinth spillway. *Report* GR-82–13. USBR, Denver CO.

Indlekofer, H. (1976) Zum hydraulischen Einfluss von Pfeileraufbauten bei Überfall-Entlastungsanlagen (Hydraulic effect of piers on overfall dams). *Mitteilung* 13. Institut Wasserbau und Wasserwirtschaft, Rheinisch-Westfälische Technische Hochschule, Aachen (in German).

Indlekofer, H. (1978) Zur Berechnung der Überfallprofile von kelchförmigen Hochwasserentlastungen (Computation of crest profiles of cup-shaped spillways). *Die Bautechnik*, 55(11), 368–371 (in German).

Indlekofer, H. & Rouvé, G. (1975) Discharge over polygonal weirs. *Journal of the Hydraulics Division* ASCE, 101(HY3), 385–401; 102(HY1), 105.

Jüstrich, S., Pfister, M. & Schleiss, A.J. (2016) Mobile riverbed scour downstream of a Piano Key weir. *Journal of Hydraulic Engineering*, 142(11), 04016043, 1–12.

Kabiri-Samani, A. & Javaheri, A. (2012) Discharge coefficient for free and submerged flow over Piano Key weirs. *Journal of Hydraulic Research*, 50(1), 114–120.

Keulegan, G.H. (1950) Determination of critical depth in spatially-variable flow. Proceedings of the 1st *Midwestern Conference on Fluid Mechanics*, pp. 68–80.

Kindsvater, C.E. & Carter, R.W. (1957) Discharge characteristics of rectangular thin-plate weirs. *Journal of the Hydraulics Division* ASCE, 83(HY6, Paper 1453), 1–35; 84(HY2), Paper 1616), 93–99; 84(HY3, Paper 1690), 21–39; 84(HY6, Paper 1856), 39–41; 85(HY3), 46–49.

Laugier, F. (2007) Design and construction of the first Piano Key Weir spillway at Goulours Dam. *International Journal of Hydropower & Dams*, 13(5), 94–100.

Laugier, F., Lochu, A., Gille, C., Leite Ribeiro, M. & Boillat, J.-L. (2009) Design and construction of a labyrinth PKW spillway at St-Marc Dam, France. *International Journal of Hydropower & Dams*, 15(5), 100–107.

Laugier, F., Vermeulen, J. & Blancher, B. (2017) Overview of design and construction of 11 Piano Key Weirs spillways developed in France by EDF from 2003 to 2016. *Labyrinth and Piano Key Weirs – PKW*, 2017, 37–51.

Laugier, F., Vermeulen, J. & Lefebvre, V. (2013) Overview of Piano Key Weirs experience developed at EDF during the past few years. *Labyrinth and Piano Key Weirs – PKW*, 2013, 213–226.

Leite Ribeiro, M., Pfister, M., Schleiss, A.J. & Boillat, J.-L. (2012) Hydraulic design of A-type Piano Key Weirs. *Journal of Hydraulic Research*, 50(4), 400–408.

Lempérière, F. & Ouamane, A. (2003) The Piano Keys weir: a new cost-effective solution for spillways. *International Journal of Hydropower & Dams*, 10(5), 144–149.

Lempérière, F., Vigny, J.-P. & Ouamane, A. (2011) General comments on Labyrinth and Piano Key Weirs: the past and present. *Labyrinth and Piano Key Weirs – PKW*, 2011, 17–24. CRC Press, Leiden NL.

Li, W.-H. (1955) Open channels with nonuniform discharge. *Transactions of ASCE*, 120, 255–280.

Liggett, J.A. (1961) General solution for open channel profiles. *Journal of the Hydraulics Division* ASCE, 87(HY6), 89–107; 88(HY2), 193–194; 88(HY3), 255–257; 88(HY6), 175–176.

Lodomez, M., Pirotton, M., Dewals, B., Archambeau, P. & Erpicum, S. (2017) Could Piano Key Weirs be subjected to nappe oscillations? *Labyrinth and Piano Key Weirs – PKW*, 2017, 135–144.

Lucas, J., Lutz, N., Lais, A., Hager, W.H., Boes, R.M. (2015) Side channel flow: Physical model studies. *Journal of Hydraulic Engineering*, 141(9) 10.1061/(ASCE) HY, 1943-7900. 0001029, 05015003.

Lux III, F. (1984) Discharge characteristics of labyrinth weirs. In: Schreiber, D.L. (ed) *Water for Resource Development*. ASCE, New York, pp. 385–389.

Machiels, O. (2012) Experimental study of the hydraulic behavior of Piano Key Weirs. *Ph.D. Thesis*, ULgetd-09252012–224610, University of Liège, Liège.

Machiels, O., Erpicum, S., Dewals, B., Archambeau, P. & Pirotton, M. (2011) Experimental observation of flow characteristics over a Piano Key Weir. *Journal of Hydraulic Research*, 49(3), 359–366.

Machiels, O., Pirotton, M., Archambeau, P., Dewals, B. & Erpicum, S. (2014) Experimental parametric study and design of Piano Key Weirs. *Journal of Hydraulic Research*, 52(3), 326–335; 53(4), 545.

Machiels, O., Pirotton, M., Archambeau, P., Dewals, B. & Erpicum, S. (2015) Closure of experimental parametric study and design of piano key weirs. *Journal of Hydraulic Research*, 53(4), 545.

Mayer, P.G. (1980) Bartletts Ferry project labyrinth weir model studies. *Project* E-20–610. Georgia Institute of Technology, Atlanta GA.

McConaughy, D.C. (1933) Spillways in canyon walls to handle floodwaters. *Engineering News-Record*, 111(Dec.21), 754–756; 114(Apr.4), 480–482.

Meyer-Peter, E. & Favre, H. (1934) Analysis of Boulder Dam spillways made by Swiss laboratory. *Engineering News-Record*, 113(Oct. 25), 520–522.

Murphy, D.W. (1909) A reinforced-concrete spillway with concentrated crest length. *Engineering News*, 62(11), 278–279.

Mussalli, Y.G. (1978) Size determination of partly full conduits. *Journal of the Hydraulics Division* ASCE, 104(HY7), 959–974; 105(HY8), 1039–1041.

Noseda, M. (2017) Upstream erosion at Piano Key weirs. *MSc Thesis*, Laboratory of Hydraulic Constructions (LCH), Ecole Polytechnique Fédérale de Lausanne (EPFL), Lausanne, Switzerland.

Noseda, M., Stojnic, I., Pfister, M. & Schleiss, A.J. (2019) Upstream erosion and sediment passage at piano key weirs. *Journal of Hydraulic Engineering*, 145(8), 04019029.

Noui, A. & Ouamane, A. (2013) Study of optimization of the Piano Key Weir. *Labyrinth and Piano Key Weirs – PKW*, 2013, 175–182.

Novak, P. & Cabelka, J. (1981) *Models in Hydraulic Engineering*. Pitman, Boston and London.

Ouamane, A. & Lempérière, F. (2006) Design of a new economic shape of weir. Proceedings of International Symposium of *Dams in the Societies of the 21st Century*, Barcelona, pp. 463–470.

Pagliara, S., Hager, W.H. & Minor, H.-E. (2006) Hydraulics of plane plunge pool scour. *Journal of Hydraulic Engineering*, 132(5), 450–461.

Peterka, A.J. (1956) Shaft spillways: performance tests on prototype and model. *Transactions of ASCE*, 121, 385–409.

Pfister, M., Berchtold, T. & Lais, A. (2008) Kárahnjúkar dam spillway: optimization by hydraulic model tests. Proceedings of 3rd IAHR International Symposium of *Hydraulic Structures*, Hohai University, Nanjing, 6, 2106–2111.

Pfister, M., Capobianco, D., Tullis, B. & Schleiss, A.J. (2013a) Debris blocking sensitivity of Piano Key Weirs under reservoir-type approach flow. *Journal of Hydraulic Engineering*, 139(11), 1134–1141.

Pfister, M., Erpicum, S., Machiels, O., Schleiss, A. & Pirotton, M. (2012) Discussion of Discharge coefficients for free and submerged flow over Piano Key Weirs. *Journal of Hydraulic Research*, 50(6), 642–645.

Pfister, M., Jüstrich, S. & Schleiss, A.J. (2017) Toe-scour formation at Piano Key Weirs. *Labyrinth and Piano Key Weirs – PKW*, 2017, 147–156.

Pfister, M. & Rühli, E. (2011) Junction flow between drop shaft and diversion tunnel in Lyss, Switzerland. *Journal of Hydraulic Engineering*, 137(8), 836–842.

Pfister, M., Schleiss, A.J. & Tullis, B. (2013b) Effect of driftwood on hydraulic head of Piano Key Weirs. *Labyrinth and Piano Key Weirs – PKW*, 2013, 255–264.

Pfister, M., Tullis, B. & Schleiss, A.J. (2015) Closure of Debris blocking sensitivity of Piano Key weirs under reservoir type approach flow. *Journal of Hydraulic Engineering*, 141(10), 07015013.

Phillips, M.A. & Lesleighter, E.J. (2013) Piano Key Weir spillway: upgrade option for a major dam. *Labyrinth and Piano Key Weirs – PKW*, 2013, 159–168.

Pinchard, T., Farges, J.L., Boutet, J.M., Lochu, A. & Laugier, F. (2013) Spillway capacity upgrade at Malarce Dam: construction of an additional Piano Key Weir spillway. *Labyrinth and Piano Key Weirs – PKW*, 2013, 243–252.

Pinotti, M. & Lais, A. (2010) Investigating Val Noci. *Water Power & Dam Construction*, 62(10), 30–33.

Pralong, J., Vermeulen, J., Blancher, B., Laugier, F., Erpicum, S., Machiels, O., Pirotton, M., Boillat, J.-L., Leite Ribeiro, M. & Schleiss, A.J. (2011) A naming convention for the piano key weirs geometrical parameters. *Labyrinth and Piano Key Weirs* – PKW, 2011, 271–278.

Preissler, G. & Bollrich, G. (1985) *Technische Hydromechanik* (Technical hydromechanics) 1. VEB Verlag für Bauwesen, Berlin (in German).

Press, H. (1959) *Wehre* (Weirs). Ernst & Sohn, Berlin (in German).

Ramamurthy, A.R. & Vo, N.D. (1993) Characteristics of circular-crested weir. *Journal of Hydraulic Engineering*, 119(9), 1055–1062.

Rehbock, T. (1929) Wassermessung mit scharfkantigen Überfällen (Discharge measurement with sharp-crested weirs). *Zeitschrift des Vereines Deutscher Ingenieure*, 73(24), 817–823 (in German).

Samarin, E.A., Popow, K.W. & Fandejew, W.W. (1960) *Wasserbau* (Translation from Russian). VEB Verlag für Bauwesen, Berlin (in German).

Sassoli, F. (1959) Canali collettori laterali a forte pendenza (Side-channels with a strong bottom slope). *L'Energia Elettrica*, 36(1), 26–39 (in Italian).

Schleiss, A.J. (2011) From labyrinth to Piano Key weirs: a historical review. Proceedings of International Conference of *Labyrinth and Piano Key weirs,* Liège. CRC Press, Boca Raton FL, pp. 3–15.

Schmocker, L. & Hager, W.H. (2011) Probability of drift blockage at bridge decks. *Journal of Hydraulic Engineering*, 137(4), 470–479.

Smith, W. (1966) Bellmouth spillway and stilling basin: hydraulic model tests for Lower Shing Mun Reservoir. *Civil Engineering and Public Works Review*, 61(3), 302–305; 61(4), 499–501.

Stein, O.R. & Julien, P.Y. (1994) Sediment concentration below free overfall. *Journal of Hydraulic Engineering*, 120(9), 1043–1059.

Stephenson, D. & Metcalf, J.R. (1991) Model studies of air entrainment in the Muela drop shaft. *Proceedings of the Institution of Civil Engineers*, 91(2), 417–434.

Tomasson, G.G., Gardarsson, S.M., Petry, B. & Stefansson, B. (2006) Design challenges and solutions for the Kárahnjúkar spillway. *International Journal of Hydropower and Dams*, 13(5), 84–88.

Truong Chi, H. & Ho Ta Khanh, M. (2017) Experimental study for energy dissipation using stilling basin downstream of Piano Key Weir type A. *Labyrinth and Piano Key Weirs* – PKW, 2017, 157–164.

Tullis, B.P., Willmore, C.M. & Wolfhope, J.S. (2005) Improving performance of low-head labyrinth weirs. In: *Impacts of Global Climate Change*. ASCE, Reston VA, pp. 418–426.

Tullis, B.P., Young, J. & Chandler, M. (2007) Head-discharge relationships for submerged labyrinth weirs. *Journal of Hydraulic Engineering*, 133(3), 248–254; 133(9), 1097.

Tullis, J.P. (1993) Standley Lake service spillway model study. *Hydraulic Report* R 341. Utah State University Foundation. Utah Water Resources Laboratory, Logan UT.

Tullis, J.P., Amanian, N. & Waldron, D. (1995) Design of labyrinth spillways. *Journal of Hydraulic Engineering*, 121(3), 247–255.

Unami, K., Kawachi, T., Munir Babar, M. & Itagaki, H. (1999) Two-dimensional numerical model of spillway flow. *Journal of Hydraulic Engineering*, 125(4), 369–375.

USBR (1938) Model studies of spillways. *Bulletin* 1. Boulder Canyon Projects, Final Reports. Part VI: Hydraulic investigations. US Department of the Interior, Bureau of Reclamation, Denver, CO.

USBR (1960) *Design of Small Dams*. US Bureau of Reclamation USBR, Denver, CO.

van Rijn, L.C. (1984) Sediment transport I: bed load transport. *Journal of Hydraulic Engineering*, 110(10), 1431–1456.

Venetz, P. (2014) Einfluss von Schwemmholz auf die Abflusscharakteristik von Klaviertastenwehren (Effect of driftwood on the flow characteristics of Piano Key Weirs). *MSc Thesis*, Laboratory of Hydraulic Constructions (LCH), Ecole Polytechnique Fédérale de Lausanne (EPFL), Lausanne, Switzerland (in German).

Vermeulen, J., Lassus, C. & Pinchard, T. (2017) Design of a Piano Key Weir aeration network. *Labyrinth and Piano Key Weirs* – PKW, 2017, 127–133.

Vermeyen, T. (1991) Hydraulic model study of Ritschard Dam spillways. *Report* R-91–08. USBR, Denver CO.

Villemonte, J.R. (1947) Submerged-weir discharge studies. *Engineering News-Record*, 131(Nov.18), 748–750.

Viparelli, M. (1954) Trasporto di aria da parte de correnti idriche in condotti chiusi (Air transport as a part of a hydraulic current in a closed conduit). *L'Energia Elettrica*, 31(11), 813–826 (in Italian).

Vischer, D.L. & Hager, W.H. (1998) *Dam Hydraulics*. Wiley, Chichester UK.

Wagner, W.E. (1956) Determination of pressure-controlled profiles. *Transactions of ASCE*, 121, 345–384.

Waldron, D.R. (1994) Design of labyrinth weirs. *MSc Thesis*. Utah State University, Logan UT.

Xian-Huan, W. (1989) Siphon intakes for small hydro plants in China. *Water Power and Dam Construction*, 41(8), 44–53.

Yen, B.C. (1973) Open-channel flow equations revisited. *Journal of Engineering Mechanics Division* ASCE, 99(EM5), 979–1009; 100(EM5), 1055–1060; 101(EM4), 485–488.

Bibliography

Various references are not mentioned in the main text but are still considered either historically interesting or technically relevant, so that they are mentioned below.

Labyrinth weirs

Afshar, A. (1988) The development of labyrinth spillway designs. *Water Power & Dam Construction*, 40(5), 36–39.

Aichel, O.G. (1908) Experimentelle Untersuchungen über den Abfluss des Wassers bei vollkommenen schiefen Überfallwehren [Experimental investigation of the water flow over free oblique weirs]. *Zeitschrift des Vereines Deutscher Ingenieure*, 52(44), 1752–1755; also *Forschungsarbeiten auf dem Gebiete des Ingenieurwesens*, 80. Springer, Berlin (in German).

Armanious, S. (1978) Abfluss über Schräg- und Rundwehre [Flow over oblique and round-crested weirs]. *Mitteilung*, 59. Leichtweiss-Institut für Wasserbau, Technische Universität, Braunschweig (in German).

Asthana, K.C., Husain, S.T. & Yousuf, S. (1961) Flow over curved weirs. *Journal of Irrigation and Power*, 18(8), 744–761.

Cassidy, J.J., Gardner, C. & Peacock, R. (1983) Labyrinth-crest spillway: planning, design and construction. *Hydraulic Aspects of Floods & Flood Control* (C1), 59–80.

Deo, V., Modi, P.N. & Dandekar, M.M. (1976) Dentated end sill as a depth controlling device in channels. *Journal of Irrigation and Drainage*, 33(2), 217–224.

Escande, L. & Sabathé, G. (1937) Déversoir incline par rapport à l'axe d'un canal [Inclined weir relative to the channel axis]. *Comptes Rendus de l'Académie des Sciences*, 204, 1547–1549 (in French).

Houston, K.L. & DeAngelis, C.S. (1982) A site specific study of a labyrinth spillway. In: Smith, P.E. (ed) *Applying Research to Hydraulic Practice*. ASCE, New York, pp. 86–95.

Istomina, V.S. (1937) Stramazzi oblique in parete sottile e grossa [Oblique weirs with sharp and broad crests]. *L'Energia Elettrica*, 14(2), 178–181 (in Italian).

Lux III, F. (1989) Design and application of labyrinth weirs. In: Albertson, M.L. & Kia, R.A. (eds) *Design of Hydraulic Structures '89*. Balkema, Rotterdam NL, pp. 205–215.

Lux III, F. & Hinchliff, D.L. (1985) Design and construction of labyrinth spillways. Proceedings of 15th *ICOLD Congress*, Lausanne, Q59(R15), 249–274.

Parrett, N. (1986) New approaches to rehabbing old dams. *Civil Engineering*, 56(6), 74–76.

Pinto Magalhaes, A. (1983) Descarregadores em labirinto [Labyrinth overflow]. *Mémoria* 605. Laboratorio Nacional de Engenharia Civil LNEC, Lisboa (in Portuguese).

Pinto de Magalhaes, A. (1985) Labyrinth-weir spillways. Proceedings of 15th *ICOLD Congress,* Lausanne, Q59(R24), 395–407.

Pinto Magalhaes, A. & Lorena, M. (1989) Hydraulic design of labyrinth weirs. *Memoria* 736. Laboratorio Nacional de Engenharia Civil LNEC, Lisboa.

Ricceri, G. (1968) Esame teorico e sperimentale del comportamento idraulico di un particolare tipo di sfioratore [Theoretical and experimental examination of the hydraulic behavior of a particular weir type]. *L'Acqua*, 46(5), 125–134 (in Italian).

Schlag, A. (1962) Formule de débit de déversoirs dont le seuil est constitué par une série de dents situées dans un plan horizontal [Discharge equation of weirs whose crest is constituted with a series of teeth located along a horizontal plane]. *La Tribune de CeBeDeau*, 15(2), 180–183 (in French).

Tacail, F., Evans, B. & Babb, A. (1989) Case study of a labyrinth weir spillway. *Annual Conference Canadian Society for Civil Engineers*, St. John's, pp. 79–93.

Tacail, F., Evans, B. & Babb, A. (1989) Case study of a labyrinth weir spillway. *Canadian Journal of Civil Engineering*, 17(1), 1–7.

Tison, G. & Fransen, T. (1963) Essais sur déversoirs de forme polygonale en plan [Experiments on weirs of polygonal plan shape]. *Revue C Tijdschrift*, 3(3), 38–51 (in French).

Piano Key Weirs

Labyrinth and Piano Key Weirs (2011) Sébastien Erpicum, Frédéric Laugier, Jean-Louis Boillat, Michel Pirotton, Bernard Reverchon & Anton J. Schleiss (eds). CRC Press, Leiden NL, ISBN 978-0-415-68282-4.

Labyrinth and Piano Key Weirs II (2013) Sébastien Erpicum, Frédéric Laugier, Michael Pfister, Michel Pirotton, Guy-Michel Cicero & Anton J. Schleiss (eds). CRC Press, Leiden NL, ISBN 978-1-138-00085-8.

Side channels

Guercio, R. & Magini, R. (1996) Influenza dell'immissione sul deflusso nei canali di gronda [Effect of lateral discharge on the flow in side channels]. Proceedings of 25th *Convegno di Idraulica e Costruzioni Idrauliche,* Genova, 3, pp. 166–177 (in Italian).

Hajdin, G. (1967) Canal collecteur à déversoir lateral, destiné à évacuer des crues [Side channel for flood evacuation]. Proceedings of 19th *ICOLD Congress,* Istamboul, C(5), 359–366 (in French).

Kao, D.T.Y. & Perry, S.D. (1976) Spatially varied flow in channel transitions. Proceedings of *Rivers '76*, Fort Collins. ASCE, New York, pp. 1551–1571.

Kawagoshi, N., Kudo, A. & Sasanabe, S. (1974a) Hydraulic model study on the spillway design of the Sakuda Dam. *Bulletin*, 22, 85–98. Faculty of Agriculture, Hirosaki University (in Japanese with English Summary).

Kawagoshi, N., Kudo, A. & Sasanabe, S. (1974b) Hydraulic model study on the spillway design of the Matakido Dam. *Bulletin*, 22, 99–127. Faculty of Agriculture, Hirosaki University (in Japanese with English Summary).

Kebelmann, G. (1964) Verfahren zur Bemessung der Sammelkanäle von Hangüberfällen mit seitlicher Schussrinne [Procedure for designing side channels of overflows with lateral chute]. *Heft* 10, 36–207. Mitteilungen der Forschungsanstalt für Schiffahrt, Wasser- und Grundbau, Berlin (in German).

Kim, C. & Roccas, S.M. (1966) Hydraulik des Abflusses mit zunehmender Wassermenge [Hydraulics of flows with increasing discharge]. *Wasser- und Energiewirtschaft*, 58(6), 173–177 (in German).

Kudo, A. & Kawagoshi, N. (1980) Studies on side channel spillways. *Bulletin*, 33, 28–40. Faculty of Agriculture, Hirosaki University (in Japanese with English Summary).

Lopes, V.L. & Shirley, E.D. (1993) Computation of flow transitions in open channels with steady uniform lateral inflow. *Journal of Irrigation and Drainage Engineering*, 119(1), 187–200.

Maione, U. (1963) Contributo sperimentale allo studio delle correnti con portata uniformemente crescente [Experimental contribution to the study of flows with uniformly increasing discharge]. Proceedings of 8th *Convegno di Idraulica*, Pisa, A(5), 1–14 (in Italian).

Reinauer, R., Müller, D. & Filippini, L. (1994) Zur Hydraulik eines Entlastungsbauwerkes [To the hydraulics of a relief structure]. *Wasser, Energie, Luft*, 86(11/12), 354–358 (in German).

Morning glory overfalls

Anderson, C.L. & Blaisdell, F.W. (1979) Prototype spillway performance verification. *Journal of the Hydraulics Division* ASCE, 105(HY6), 707–720.

Anonymous (1898) A funnel-shaped spillway from a pond at Providence RI. *Engineering News*, 39(June 9), 373.

Billore, J., Jaoui, A., Kolkman, P.A., Radu, M.T. & de Vries, A.H. (1979) Recherches hydrauliques pour la dérivation provisoire les déversoirs en puits et la vidange de fond du barrage de M'Dez au Maroc [Hydraulic research on the temporary diversion, shaft spillways and bottom outlet of M'Dez Dam, Morocco]. Proceedings of 13th *ICOLD Congress*, New Delhi, Q50(R62), 1085–1106 (in French).

Binnie, W.J.E. (1938) Bellmouthed weirs and tunnel outlets for the disposal of flood water. *Transactions of Institution of Water Engineers*, 42(1), 103–145.

Binnie, W.J.M. (1938) Model-experiments on bellmouth and siphon-bellmouth overflow spillways. *Proceedings of the Institution of Civil Engineers*, 10(1), 65–90; 12(8), 277–278.

Binnie, W.J.E. & Gourley, H.J.F. (1938) The Gorge Dam. *Proceedings of the Institution of Civil Engineers*, 11(5), 179–222; 12(8), 429–449.

Blau, H. (1970) Die Berechnung flachkroniger Einläufe von Schachtüberfällen [Computation of flat-crested intakes of shaft spillways]. *Wasserwirtschaft-Wassertechnik*, 20(6), 209–212(in German).

Bovoli, V., Marone, V. (1992) Sul proporzionamento degli scaricatori di superficie a calice prossimi alla saturazione [On the design of surface shaft overflows close to choking]. *Idrotecnica*, 19(1), 31–43 (in Italian).

Cabelka, J. (1971) Contribution of hydraulic research to solution of shaft spillways. *Vodni hospodarstvi*, 21(5), 123135 (in Czech).

Caufourier, P. (1924) Le puits-déversoir pour l'évacuation souterraine des crues du barrage-réservoir de Davis Bridge, Vermont [The overflow shaft to evacuate floods of Davis Bridge Dam, Vermont]. *Le Génie Civil*, 85(13), 269–272 (in French).

Dallwig, H.-J. (1982) Zur Leistungsfähigkeit von Kelchüberfällen [On the efficiency of cup-shaped overflows]. *Technischer Bericht* 28. Institut für Wasserbau, TH Darmstadt, Darmstadt (in German).

Faure, J. & Pugnet, L. (1959) Etude de l'alimentation d'un évacuateur en puits [Study on the discharge characteristics of a shaft spillway]. Proceedings of 6th *Convegno di Idraulica*, Padova, B(6), 1–7 (in French).

Goldring, B.T. (1983) Air voids at downshaft-tunnel bends. *Journal of Hydraulic Engineering*, 109(2), 189–198.

Hajdin, G. (1979) Two contributions to spillway designing based on experimental studies. Proceedings of 13th *ICOLD Congress*, New Delhi, Q50(R45), 781–788.

Indlekofer, H. (1977a) Abflusscharakteristik und hydraulische Leistungsfähigkeit von kelchförmigen Überfallbauwerken [Discharge characteristics and hydraulic efficiency of cup-shaped overfall structures]. *Bauingenieur*, 52(2), 67–75 (in German).

Indlekofer, H. (1977b) Zur Frage der Profilform kelchförmiger Überfallbauwerke [On the question of cup-shaped overfall structures]. *Die Bautechnik*, 54(6), 203–207 (in German).

Indlekofer, H. (2006a) Die geometrische Analyse der Überlaufkronen von in Deutschland existierenden Kelchüberfallbauwerken [Geometrical analysis of in Germany existing cup-shaped overflow structures]. *Wasserbau-Heft* 8, Fach-Hochschule Aachen, Aachen (in German).

Indlekofer, H. (2006b) Die Profilgeometrie und hydraulische Leistungsfähigkeit von Kelchüberfall-bauwerken [Profile geometry and hydraulic efficiency of cup-shaped overflow structures]. *Wasser und Abfall*, 8(6), 44–52 (in German).

Indri, E. (1959) Esperienze su modelli di scaricatori a pozzo a costanza di livello nel serbatoio [Experiments with models of constant head overflow structures]. *L'Energia Elettrica*, 36(4), 332–343 (in Italian).

Kurtz, F. (1925) The hydraulic design of the shaft spillway for the Davis Bridge Dam, and hydraulic tests on working models. *Transactions of ASCE*, 88, 1–86.

Lazzari, E. (1959) Ricerca sperimentale sugli sfioratori a calice [Experimental research on morning glory overfalls]. *L'Energia Elettrica*, 36(7), 641–651 (in Italian).

Leopardi, M. (1995) Limite di saturazione degli sfioratori a calice [Choking limits of shaft spillways]. *Idrotecnica*, 22(5), 273–291 (in Italian).

Mussali, Y.G. & Carstens, M.R. (1969) A study of flow conditions in shaft spillways. *Completion Report* OWRR Project No. B-022-GA. School of Civil Engineering, Georgia Institute of Technology, Atlanta GA.

Sastry, P.G. (1962) Hochwasserentlastung durch Schachtüberfälle [Flood evacuation by shaft spillways]. *Wissenschaftliche Zeitschrift der TU Dresden*, 11(4), 713–721 (in German).

Savic, L., Kapor, R., Kuzmanovic, V. & Milovanovic, B. (2014) Shaft spillway with deflector downstream of vertical bend. *Proceedings of the Institution of Civil Engineers, Water Management*, 167(5), 269–278.

Schmidt, M. (1965) Die Innerste-Talsperre im Harz [The Innerste Dam in the Harz Region]. *Wasser und Boden*, 17(5), 154–160 (in German).

Schmidt, M. (1969) Überlauftürme als Hochwasserentlastungsanlagen [Overflow towers as spillways]. *Die Wasserwirtschaft*, 59(1), 8–12 (in German).

Smith, C.D. (1981) Simplified elbow design for low head drop inlet spillways. Proceedings of 5th *Canadian Hydrotechnical Conference*, Fredericton NB, pp. 879–893.

Wickert, G. (1955) Hochwasserentlastungsanlagen mit kreisförmiger Überlaufkrone [Overflow structures of circular-crested shape]. *Der Bauingenieur*, 30(7), 263–267 (in German).

Chapter 4 Frontispiece (a) LG2 cascade spillway, Hydro-Quebec, Canada (Courtesy Hydro-Quebec), (b) stepped chute along embankment dam abutments of Upper Dam of Siah-Bishe pumped-storage power plant in Iran (Courtesy IWPC, Tehran)

Chapter 4

Spillway chute

4.1 Introduction

A dam spillway consists of three distinctly different portions, namely the:

- *Overflow zone* or headworks, involving a weir-type structure, eventually combined with gates, by which the upstream water elevation is fully controlled. Depending on the width of the overfall structure, various weir types apply, as discussed in Chapter 2. Frontal and side weirs, and low-level openings combine the fundamental weir and gate flow features with an approach flow resulting in spatial flow. These latter hydraulic structures are treated in Chapter 3.
- *Chute zone* or transport structure consisting essentially of a straight, wide conveyance structure to convey the water flow from the upstream to the downstream portions. Chutes are often straight, of constant width and bottom slope, and may be provided with additional elements as aerators, bends in plan view or slope changes, junctions, contractions or expansions, by which complicated aspects of shock wave formation are generated. These are also considered in this chapter. Further, stepped chutes are detailed herein as an alternative to smooth chutes.
- *Terminal zone* or outlet and dissipation structure made up either of a stilling basin, a ski jump combined with a plunge pool, or a drop. Whereas stilling basins are based on the hydraulic jump phenomena, involving mainly turbulence generation and energy dissipation (Chapter 5), trajectory basins deflect the chute flow into the atmosphere, thereby generating high-speed jet flow impacting onto the tailwater in a plunge pool. Concerns then are both scour and sediment aggradation, as treated in Chapter 6.

The chute has become an important dam appurtenant structure once dam heights increased over several tens of meters. Originally, spillways were often cut into the adjacent rock in the form of a cascade, an early form of stepped chutes. The latter was introduced in the early 1980s based on the advent of the roller-compacted concrete (RCC) technology. It is currently considered a standard structure of concrete gravity dams but more recently often used in combination with rock- or earthfill dams when excavated in the rocky dam abutments.

The word 'chute' has its origin in French, meaning fall. In the present context, chute describes a water conveyance or transport structure along which the water is guided by a normally rectangular cross section from the dam crest zone to the terminal structure zone. Three fundamental chute flow features have to be considered, namely:

- *Transition* from black-water to aerated chute flow due to the increase of boundary layer thickness from the weir crest to the free surface further downstream. The typical feature of developed chute flow is the 'white-water' generation due to air entrainment downstream from the point of air inception. This particular process is highlighted below, and expressions are provided to detail the free-surface depth and the velocity distribution.

- *Developing chute flow*, from the air inception point to the zone along which there is equilibrium between the driving and the retaining forces, as for uniform open-channel flow, but here relating to the uniform-aerated chute flow. Along this flow zone, the air entrained at the flow surface is transported to the chute invert by the action of turbulence, thereby progressively aerating the flow. Given the combined presence of air and water, instead of water alone as in a typical canal, the air-water flow undergoes *flow bulking*, thereby generating a higher free-surface elevation along the chute as compared with water flow alone. To avoid dangerous chute overtopping, the chute sidewalls or training walls thus have to be higher than for black-water flow. This and other aspects are dealt with in this chapter.
- *Uniform-aerated chute flow* is simpler to describe than the two other aforementioned phenomena, and often retains, at least approximately, the essence of chute flow dynamics. As for uniform channel flow, the uniform-aerated chute flow follows essentially the laws established by the fundamental open-channel hydraulics. In addition to bottom slope and boundary roughness, the degree of flow aeration has to be added. Often, the average cross-sectional air concentration is employed for its description.

Cavitation damage is the main concern in chute flow. Given the high flow velocities in chutes, even minute boundary imperfections may lead to local pressure below the atmospheric pressure, tending eventually to vapor pressure, so that the fluid undergoes a phase change from fluid to gas. Cavitation, namely boiling of water at temperatures well below the static boiling point, is per se not a problem in hydraulic structures, but becomes a challenge if the pressure increases shortly downstream, resulting in vapor bubble implosion accompanied with extreme pressure action close to the chute boundary, so that the boundary material may be seriously damaged. The action of cavitation damage is prevented by so-called chute aerators, by which air from the atmosphere is supplied to the flow close at the chute bottom, thereby significantly reducing the cavitation damage potential. These aspects will be dealt with also in this chapter.

Open chutes as previously described are simple as compared with tunnel chutes. In addition to 'open-air' chutes, tunnel chutes have a limited supply of air and a high risk to choke, given that the chute is covered. Relatively few research has been conducted for tunnel chutes, whereas there is more research available from sewer flow, involving a smaller diameter but with otherwise similar hydraulic features. The straight tunnel chute is considered in Chapter 8. Given the lack of knowledge in additional tunnel flow features as presence of bends or aerators, it is proposed to employ these appurtenances based on open chute flow, thereby providing a conservative safety margin, given their importance, and their high damage potential in case of structural failure.

4.2 Smooth chute

4.2.1 Hydraulic design

A chute is a sloping open channel made of high-quality concrete by which excess discharge is conveyed from the overflow structure to the outlet structure and tailwater. Its particular feature is its high-velocity flow between 20 and 50 m/s. Problems with air entrainment, shock waves, cavitation, highly dynamic bottom pressures, and abrasion are frequent. To guarantee reliability under extreme flood conditions, the chute must be designed and constructed carefully. Endangering a chute often means a danger to the dam, as experienced in 2017 at Oroville Dam, California (Goguel, 2017).

The chute is one of the most spectacular portions of a dam spillway, representing a particular challenge for each designer and hydraulic engineer. Chutes are often separated from the dam structure. Outstanding examples are located at Itaipu Dam (Brazil/Paraguay) or Tarbela Dam (Pakistan). A gated spillway chute may have a design unit discharge of up to 300 m^2/s, with bottom slope angles ranging from 20° to 60°. Usually, the chute sidewalls are vertical or have a slope of 1:2, and the chute surface is smooth to reduce cavitation damage. Particular attention must be paid to concrete joints because even a small offset of less than 10 mm may generate cavitation (Falvey, 1990).

The concept of the smooth chute is different from the stepped chute, in which the discharge is tumbling over a series of steps. In smooth chutes, even small flow perturbations are amplified, easily visualized by the presence of crosswaves. As an example, the perturbations of gate piers generate so-called rooster tails, namely high standing waves (Figure 4.1). Due to the low pressure at the rear of the pier, air is locally entrained in the flow, tracing the way downstream from the origin of perturbation. These waves spread diagonally over the chute and eventually reach the sidewalls, where wave reflection occurs, associated with a local increase of the near-wall flow depth. Crosswaves should thus be reduced in height for an economic chute design.

Figure 4.1 Pier waves and chute flow at Foz do Areia Dam (Brazil) (a) view from upstream, (b) side view with rooster tails, (c) view from downstream with ski jump for $Q = 2500$ m^3s^{-1} (Courtesy Nelson Pinto)

Forced air supply to high-speed chute flow is effective and economic to counter cavitation damage. Special aeration devices, so-called chute aerators, are provided at locations where the natural free-surface air entrainment is insufficient for concrete protection (see 4.4). In contrast to the air entrainment at chute perturbations (including the rooster tails previously discussed), the chute aerator is designed for a certain air discharge, which is distributed over the entire chute width. One should distinguish thus between local perturbations and chute aerators for controlling the air entrainment.

The freeboard required remains relatively small and the energy dissipator, located downstream of the chute, often experiences no adverse effect. The hydraulic design of a chute thus has to account for:

- Drawdown curve
- Surface air entrainment and bulking of the air-water mixture
- The effect of chute aerators
- The effect of shock waves and roll waves.

These items are discussed in the following sections.

Mason (2017) considers practical design and construction details of chutes, based on his longtime experience. These include typical chute cross sections, lateral invert joint and drain details, invert slab design and anchorage requirements, sidewall design, wall joint details, end drainage, and inspections, operations, and maintenance.

4.2.2 Surface air entrainment

General

Fast-flowing water is known to become white because air is entrained into the flow (Figure 4.2). This air entrainment or self-aeration is also referred to as natural, as opposed to so-called forced air entrainment with aerators. For air entrainment to occur, two conditions must be satisfied:

1 Flow must be fully turbulent, i.e. boundary layer thickness must be equal to flow depth
2 Kinetic energy of surface eddies must be higher than surface tension energy.

The latter condition is usually satisfied in prototype structures. Depending on the degree of turbulence, or the turbulence number, chute flow is naturally aerated in prototype structures if the velocity exceeds 10–15 m/s.

Uniform aerated flow is similar to conventional uniform flow in channels, except for the equilibrium between tractive and retarding forces for aerated flow. Using the concept of uniform-aerated flow, one may distinguish between equilibrium and non-equilibrium reaches along a chute (Figure 4.3).

The growth of the turbulent boundary layer downstream of an overflow structure follows (Wood *et al.*, 1983)

$$\frac{\delta}{x} = 0.0212 \left(\frac{x}{H_s} \right)^{0.11} \left(\frac{x}{k_s} \right)^{-0.10}. \tag{4.1}$$

Figure 4.2 (a) Pier waves visible down to first aerator, (b) high concentration air-water flow up to trajectory basin at Itaipu Dam (Brazil/Paraguay) (Courtesy Nelson Pinto)

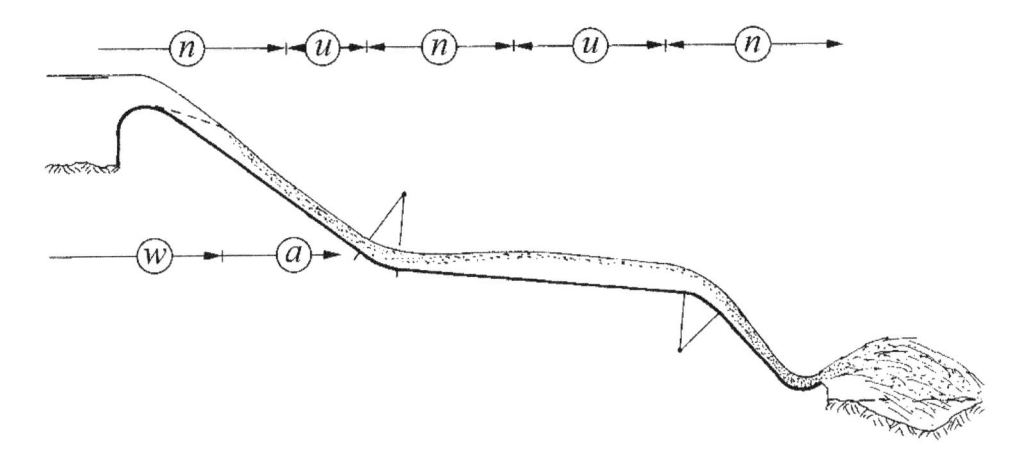

Figure 4.3 Chute flow from overflow structure to energy dissipator, with *n* non-equilibrium and *u* uniform or equilibrium flow reaches, *w* water, and *a* air-water flow

Here δ is the boundary layer thickness, x the longitudinal coordinate measured from the weir crest, H_s the difference of water elevations between the air inception point (subscript i) and the reservoir level, and k_s the equivalent roughness height according to Nikuradse (Figure 4.4). Approximately, this simplifies to

$$\delta / x = 0.02 (k_s / H_s)^{0.10}.$$ (4.2)

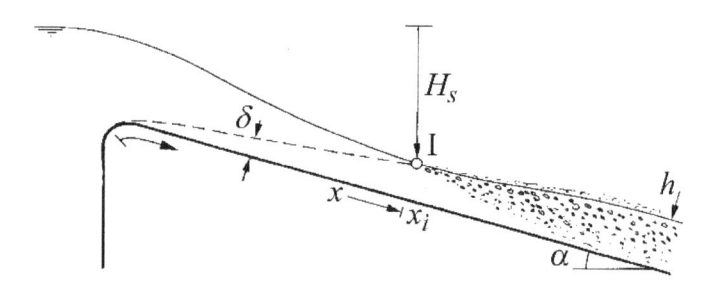

Figure 4.4 Determination of air inception point I

The air inception (subscript i) point $x = x_i$ is located where the local flow depth $h = h(x)$ is equal to the boundary layer thickness $\delta = \delta(x)$, i.e. $\delta_i = h_i$. Note from Eq. (4.2) that increasing the equivalent roughness has the same effect as decreasing the elevation difference. The point of inception is determined with a drawdown curve. Reinauer and Hager (1996) presented a simplified analysis that was checked with laboratory data.

Incipient air entrainment

Castro-Orgaz and Hager (2010) considered air inception in chute flow in relation to the turbulent boundary layer development. This flow is characterized by a Froude number largely in excess of unity, e.g. $F > 3$, referred to as hyper-critical flow (Reinauer and Hager, 1996; Hager and Blaser, 1998). Chute flow becomes aerated if the turbulent boundary layer generated by bottom friction reaches the free surface at the so-called inception point (Cain and Wood, 1981; Wood *et al.*, 1983; Wood, 1991), resulting in developing aerated flow and finally in fully developed chute flow (Montes, 1998; Chanson, 2004). Depending on the hydraulic head and the flow depth, the boundary layer may not reach the free surface up to the energy dissipator, so that the chute flow energy loss needs to be determined to deduce its inflow conditions (US, Army Corps of Engineers, 1995). The chute design up to incipient-aerated flow involves a developing boundary layer above which is irrotational flow (Chanson, 1996) (Figure 4.5). Design questions relating to the freeboard of mixture flow, shock waves, and safety against cavitation damage depend significantly on the local Froude number $F = V/(gh)^{1/2}$ (Falvey, 1980, 1990; Hager, 1992), to be determined with a drawdown curve $h = h(x)$. Here, V is cross-sectional average flow velocity, g gravity acceleration, h flow depth, and x streamwise coordinate measured from the spillway crest. Further, mixture-flow computations can only be made once the inception point is determined, based on the boundary layer profile $\delta = \delta(x)$, with δ as the boundary layer thickness measured perpendicularly to the chute profile. Thus, non-aerated chute flow dictates both predictions of $h = h(x)$ and $\delta = \delta(x)$.

Isolated work yet with unrelated proposals is available in literature to determine both of these profiles (Bauer, 1954; Campbell *et al.*, 1965; Wood *et al.*, 1983; Falvey, 1990; Wood, 1991; Hager and Blaser, 1998; Castro-Orgaz, 2009). Some considered a developing boundary layer flow different from the standard backwater approach for $h = h(x)$ (Silberman, 1980), whereas the backwater equation based on the Darcy-Weisbach or the

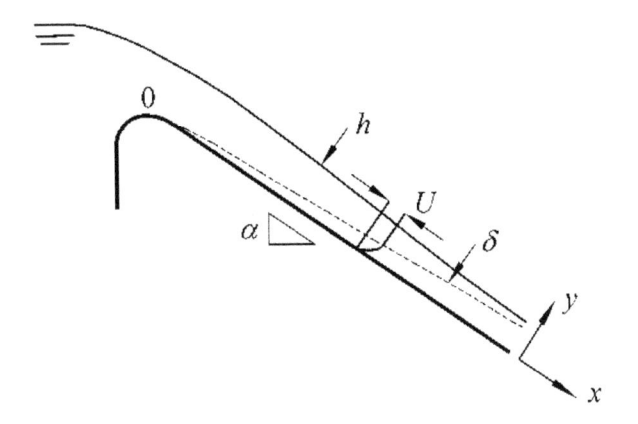

Figure 4.5 Developing chute flow, definition sketch

Gauckler-Manning-Strickler roughness coefficients was proposed in other cases. Empirical power law type equations were proposed as

$$\frac{\delta}{x} = \gamma \left(\frac{x}{k_s} \right)^{-\beta_B}.$$ (4.3)

Here, k_s is the equivalent roughness height, γ the coefficient of the boundary layer profile, and β_B its exponent. Different values for γ and β_B were proposed (Halbronn, 1952, 1955; Bauer, 1951, 1954; Campbell *et al.*, 1965; Wood *et al.*, 1983; Hager and Blaser, 1998), and the application domain of each approach remained unspecified. Scientific and technical organizations including ICOLD (Hager, 1992), IAHR (Wood, 1991), the US Bureau of Reclamation (Falvey, 1980, 1990), and the US Army Corps of Engineers (Campbell *et al.*, 1965) adopted various approximations in their design recommendations. Below, a unified approach for engineering practice is presented.

The only numerical solution to this problem is the approach of Keller and Rastogi (1975, 1977), presenting an approach for the developing flow zone by integrating the Reynolds equations with a *k-ε* model. However, no numerical solution based on the integral boundary layer type flow equations is available (White, 1991).

Models for developing flow

The complete numerical solution for the developing flow zone involves a developing boundary layer below irrotational flow. The energy head of the free surface remains constant, so that the free-surface profile follows (Silberman, 1980; Castro-Orgaz, 2009; Castro-Orgaz and Hager, 2017)

$$\frac{dh}{dx} = \frac{\sin\alpha - \dfrac{q^2}{g(h-\delta^*)^3}\dfrac{d\delta^*}{dx}}{\cos\alpha - \dfrac{q^2}{g(h-\delta^*)^3}}.$$ (4.4)

Here, α is the chute bottom angle relative to the horizontal, q the unit discharge, and δ^* the boundary layer displacement thickness (Figure 4.5). Equation (4.4) describes the irrotational flow zone linked to the turbulent flow inside the boundary layer by δ^*. Conservation of momentum for the boundary layer results in the von Karman integral equation (Bauer, 1954; White, 1991)

$$\frac{d\theta_m}{dx} = \frac{1}{2}C_f - \left(1 + \frac{1}{2}S\right)\frac{\theta_m}{U^2}\frac{dU^2}{dx}. \tag{4.5}$$

Here, $C_f = \tau_o/(\rho U^2/2)$ is the skin friction coefficient, τ_o the boundary shear stress, ρ the water density, U the potential flow velocity, θ_m the boundary layer momentum thickness, and $S = \delta^*/\theta_m$ the boundary layer shape factor. Equations (4.4) and (4.5) involve the unknowns $h = h(x)$, $\delta^* = \delta^*(x)$, $\theta_m = \theta_m(x)$, and $C_f = C_f(x)$, so that two additional closures are required. Further, the term $U^{-2}dU^2/dx$ in Eq. (4.5) is determined by requiring energy conservation of the potential flow zone.

The time-averaged velocity profile of a turbulent boundary layer is approximated by a wall-wake model as (Montes, 1998)

$$\frac{u}{u^*} = \frac{1}{\kappa}\ln\left(\frac{y}{k_s}\right) + B + \frac{1}{\kappa}(1+6\Pi)\left(\frac{y}{\delta}\right)^2 - \frac{1}{\kappa}(1+4\Pi)\left(\frac{y}{\delta}\right)^3. \tag{4.6}$$

Here, y is the distance normal to chute bottom measured from the top plane of the roughness elements plus the distance to the virtual origin of the law of the wall, u is the velocity parallel to chute bottom at distance y, $u^* = (\tau_o/\rho)^{1/2}$ the shear velocity, κ the von Karman constant, B the constant of integration for the law of the wall, and Π the wake strength parameter. The first two terms of Eq. (4.6) are referred to as the law of the wall, whereas the terms involving Π correspond to the law of the wake. Normally, high supercritical velocity corresponds to $(hu^*)/v > 70$ in the turbulent rough flow regime. However, a polished concrete surface has a micro-roughness height resulting in $(hu^*)/v < 70$, for which Eq. (4.6) should be modified allowing for smooth-wall conditions. From Eq. (4.6) evaluated at $u = U$ for $y = \delta$, the relation $C_f = C_f(\delta)$ reads

$$\frac{U}{u^*} = \left(\frac{2}{C_f}\right)^{1/2} = \frac{1}{\kappa}\ln\left(\frac{\delta}{k_s}\right) + B + \frac{2\Pi}{\kappa}. \tag{4.7}$$

In turn, the integral parameters are approximated as (White, 1991)

$$\frac{\delta^*}{\delta} = \frac{1}{\delta}\int_0^\delta \left(1 - \frac{u}{U}\right)dy \approx \frac{1+\Pi}{\left(\kappa U/u^*\right)}, \tag{4.8}$$

$$\frac{\theta_m}{\delta} = \frac{1}{\delta}\int_0^\delta \frac{u}{U}\left(1 - \frac{u}{U}\right)dy = \frac{\delta^*}{\delta} - \frac{2+3.2\Pi+1.5\Pi^2}{\left(\kappa U/u^*\right)^2}. \tag{4.9}$$

From Eqs. (4.8) and (4.9) follows $\delta^* = \delta^*(\theta)$. The acceleration term $U^{-2}dU^2/dx$ in Eq. (4.5) is from energy conservation in irrotational flow with H_o as the energy head at the spillway crest

$$U^{-2}\frac{\mathrm{d}U^2}{\mathrm{d}x}=\left(\sin\alpha-\frac{\mathrm{d}h}{\mathrm{d}x}\right)\left(H_o+x\sin\alpha-h\right)^{-1}.$$ (4.10)

Substituting Eqs. (4.7) to (4.10) into Eqs. (4.4) and (4.5), a system of two ordinary differential equations for $\mathrm{d}h/\mathrm{d}x$ and $\mathrm{d}\delta/\mathrm{d}x$ results, subject to the boundary conditions $h/h_c(x=0)=1$ and $\delta/h_c(x=0)=10^{-3}$ at the spillway crest, with $h_c=(q^2/g)^{1/3}$ as the critical depth of parallel-streamlined flow. The last condition was adopted for computational purposes when using the logarithmic law of the wall implicit in Eq. (4.6). The system of ordinary differential equations was solved using the fourth-order Runge-Kutta method. Further, the constants in Eq. (4.6) used were $\kappa=0.41$, $B=8.5$, and $\Pi=0.2$ (Castro-Orgaz, 2009). For typical chute flow with $\alpha>30°$, streamline curvature at the crest zone has a small effect so that $H_o/h_c=1.5$. The numerical model was solved for $\alpha=33.5°$ and $r=k_s/h_c=0.0023$. Figure 4.6 compares predictions with Bormann's (1968) data, resulting in a reasonable agreement for $Y(\chi)$, $D(\chi)$, and the turbulent velocity profiles $u/u^*(y/\delta)$ at various sections, with $Y=h/h_c$, $D=\delta/h_c$, and $\chi=x/h_c$.

Given that no analytical solutions are available for Eqs. (4.4) and (4.5) considering the wall-wake model given by Eq. (4.6) (White, 1991), Castro-Orgaz (2009) proposed with Λ and n as coefficients of the *power-law velocity profile* (Chen, 1991)

$$\frac{u}{u^*}=\Lambda\left(\frac{y}{k_s}\right)^{1/n}.$$ (4.11)

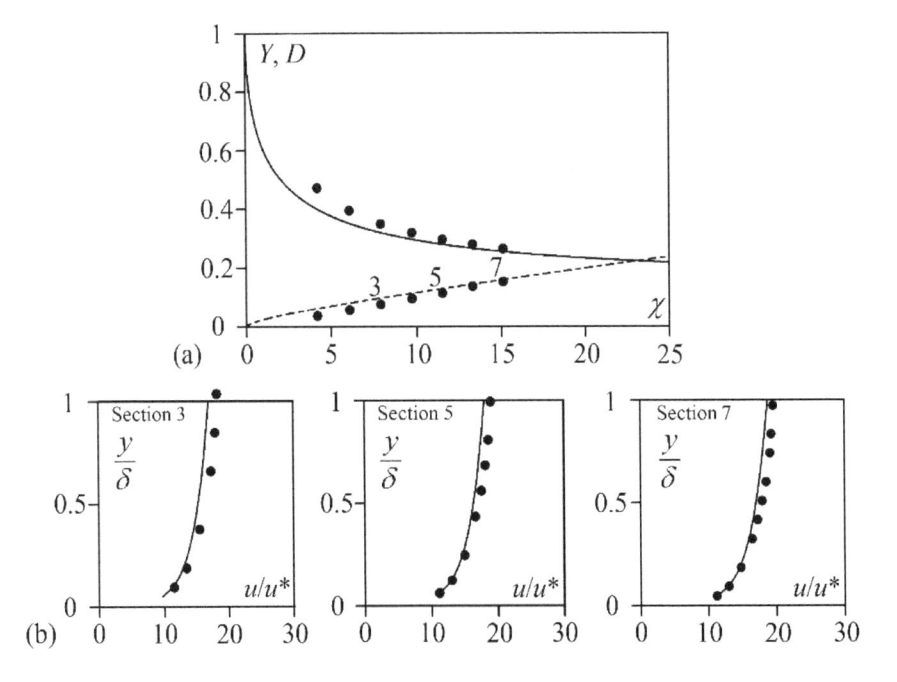

Figure 4.6 (a) (—) Drawdown curve $Y(\chi)$, (---) turbulent boundary layer development $D(\chi)$, (b) (—) velocity profiles $u/u^*(y/\delta)$ from numerical solution and Bormann's (1968) data at sections 3, 5, 7 for (•) $\alpha=33.5°$ and $r=0.0023$ (Castro-Orgaz and Hager, 2010)

The analytical solution for hypercritical flow $F > 3$ is (Castro-Orgaz and Hager, 2010)

$$Y = \frac{D}{1+n} + \left[1 + 2\chi \sin\alpha\right]^{-1/2}, \tag{4.12}$$

$$\frac{D}{\chi} = \left[\frac{2}{\Lambda^2} \frac{(n+1)(n+2)^2}{(5n^2+8n+4)}\right]^{\frac{n}{n+2}} \left(\frac{\chi}{r}\right)^{-\frac{2}{n+2}}. \tag{4.13}$$

The power law approach is simpler for chute flow than the wall-wake law and results in no-slip. The coefficients Λ and n were determined from an equivalent boundary layer in terms of C_f and δ^*, resulting in

$$n = \kappa \left(\frac{2}{C_f}\right)^{1/2} \left(\frac{11}{12} + \Pi\right)^{-1} - 1, \tag{4.14}$$

$$\Lambda = \left(\frac{2}{C_f}\right)^{1/2} \left(\frac{\delta}{k_s}\right)^{-1/n}. \tag{4.15}$$

The profiles $Y(\chi)$ and $D(\chi)$ follow from Eqs. (4.12) and (4.13) using Eqs. (4.14) and (4.15) for $\Lambda = \Lambda(D = \delta/h_c, r = k_s/h_c, C_f)$ and $n = n(C_f)$. Equation (4.7) serves for closure of C_f, whereas $D/\chi \cong 0.08(\chi/r)^{-0.233}$ is used in Eqs. (4.14) and (4.15) to avoid numerical iterations, rendering the system of governing equations explicit.

Wood *et al.* (1983) assumed a boundary layer growth equation as given by Eq. (4.3) and the power-law velocity distribution

$$\frac{u}{U} = \left(\frac{y}{\delta}\right)^{1/n}. \tag{4.16}$$

Considering $U^2/(2g) \approx x \sin\alpha$ and imposing incipient aeration at $h = \delta$, the coefficients of Eq. (4.3) were obtained by fitting their formulae at the inception point to the theoretical results of Keller and Rastogi (1975, 1977). The final solution for the boundary layer profile was (Wood *et al.*, 1983; Wood, 1991)

$$\frac{\delta}{x} = 0.0212 \sin\alpha^{-0.11} \left(\frac{x}{k_s}\right)^{-0.1}. \tag{4.17}$$

This equation is strictly valid only at the inception point, and for the domain of Keller and Rastogi's (1975, 1977) simulations. However, the chute design involves the entire profiles $h = h(x)$ and $\delta = \delta(x)$. Chanson (2004) proposed Eq. (4.17) as the complete boundary layer profile, and conservation of the free-surface energy head with a power law profile as Eq. (4.16) assuming $n = 6$ to compute the free surface profile. This is equivalent to Eq. (4.4) with $\delta^* = \delta/7$ and Eq. (4.17) to obtain $h = h(x)$ and $\delta = \delta(x)$.

Hager and Blaser (1998) reduced the backwater equation to

$$\frac{dh}{dx} = \frac{\sin\alpha - S_f}{\cos\alpha - \dfrac{q^2}{gh^3}}. \tag{4.18}$$

For $F > 3$ and with S_f as friction slope expressed with the Manning-Strickler formula, the analytical solution reads

$$Y = \left[Y_N^{-1} - \left(Y_N^{-1} - 1\right)\exp\left(-\frac{1}{3}\sin\alpha^{2/5} r^{1/5} \chi\right)\right]^{-1}. \tag{4.19}$$

Here, $Y = h/h_c$ and $Y_N = h_N/h_c$ are the dimensionless flow depths based on the critical flow depth h_c, the latter given by

$$Y_N = 0.32\left(\frac{r}{\sin\alpha^3}\right)^{1/10}. \tag{4.20}$$

Reanalyzing the data of Bauer (1954), the boundary layer growth is approximated to

$$\frac{D}{\chi} = 0.029\sin\alpha^{2/7}\left(\frac{\chi}{r}\right)^{-1/7}. \tag{4.21}$$

Equations (4.20) and (4.21) are simple analytical tools to obtain the profiles $Y(\chi)$ and $D(\chi)$. Using an equation for fully developed flows in a developing flow problem, the free-surface profile $Y(\chi)$ is independent of the turbulent boundary layer. This results from the independence of the friction formulae for fully developed flows on δ, as for $\delta = h$. Thus, $Y(\chi)$ follows without knowledge of $D(\chi)$. Note that $Y(\chi)$ and $D(\chi)$ in Eqs. (4.12) and (4.13) are coupled.

Other approximations based on the standard backwater approach were suggested. ICOLD (Hager, 1992) proposed the backwater approach using Eq. (4.18) to obtain $h = h(x)$, whereas Eq. (4.17) served for $\delta = \delta(x)$. Falvey (1990) used a backwater approach with the standard-step-method combined with the Colebrook-White equation, and the profile $\delta = \delta(x)$ was based on flat-plate equations. An approach based on the numerical solution of Eq. (4.18) and the Colebrook-White equation for fully developed rough hydraulic flows with f as the friction factor reads (Montes, 1998; Chanson, 2004)

$$f = \left[2\log\left(\frac{14.8h}{k_s}\right)\right]^{-2}. \tag{4.22}$$

For smooth-wall or transitional flows there is an effect of the Reynolds number on f. To avoid iterative computations, Haaland's equation was employed (White, 1991). Computational results using Eq. (4.22) are comparable to these if f is improved by accounting for boundary layer development as (Castro-Orgaz, 2009)

$$f = 8\left[\frac{1}{\kappa}\ln\left(\frac{\delta}{k_s}\right) + B + \frac{2\Pi}{\kappa} - \frac{1}{\kappa}\frac{\delta}{h}\left(\frac{11}{12} + \Pi\right)\right]^{-2}. \tag{4.23}$$

The ASCE Task Force on Friction Factors (1963) indicates an increase of f associated with developing boundary layers. Castro-Orgaz (2009) obtained f for these flows, verifying that f is too small if Eq. (4.22) approximates Eq. (4.23). Thus, Eq. (4.18) is considered either using Eqs. (4.22) or (4.23) along with Eq. (4.17) to estimate $\delta = \delta(x)$.

Campbell *et al.* (1965) reanalyzed the boundary layer development data of Bauer (1951, 1954) and proposed as adopted by the US Army Corps of Engineers (1995)

$$\frac{\delta}{x} = 0.08\left(\frac{x}{k_s}\right)^{-0.233}. \tag{4.24}$$

The computation proposed for $h = h(x)$ is thus essentially equivalent to the numerical solution of Eq. (4.4) using $n = 4.5$.

Bauer (1951) made the first systematic analysis of boundary layer development on chutes. Chow (1959) approximated his design chart by the fit equation

$$\frac{\delta}{x} = 0.024\left(\frac{x}{k_s}\right)^{-0.13}. \tag{4.25}$$

This was adopted by Montes (1998) using $n = 8$ in an integral form of Eq. (4.4). The results of the numerical solution of Eqs. (4.4) and (4.5) are compared in Figure 4.7 with these of

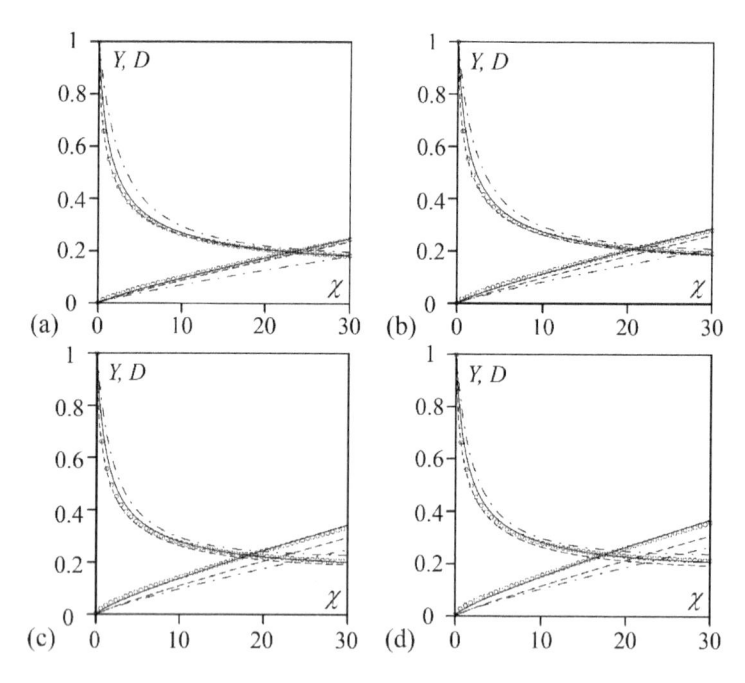

Figure 4.7 Drawdown curve $Y(\chi)$ and turbulent boundary layer development $D(\chi)$ from (○) present numerical solution, (—) Castro-Orgaz (2009), (— · —) Hager and Blaser (1998), (---) Wood (1991) for $\alpha = 45°$ and $r = $ (a) 0.001, (b) 0.003, (c) 0.010, (d) 0.015. Note: $Y = h/h_c$, $D = \delta/h_c$, and $\chi = x/h_c$

Wood (1991), Hager and Blaser (1998), and Castro-Orgaz (2009) for $\alpha = 45°$ and $r = 0.001$, 0.003, 0.01, and 0.015, to highlight the effect of relative roughness. Note that the variable power-law velocity approach given by Eqs. (4.12) and (4.13) generally agrees with the numerical solution of both $Y(\chi)$ and $D(\chi)$, regardless of r.

The method of Wood (1991) yields a free-surface profile $Y(\chi)$ in agreement with the numerical solution. Both his approach and the numerical solution based on Eq. (4.4) indicate that Eq. (4.17) yields a good estimate of δ^* and $d\delta^*/dx$ for $Y(\chi)$. The estimation of $D(\chi)$ from Eq. (4.17) is in excellent agreement with the numerical solution $D(\chi)$ for low r values, whereas it significantly deviates as r increases. Thus, Wood's approach yields a good estimate of $Y(\chi)$, but $D(\chi)$ is limited to relatively smooth beds. The fully developed flow approach of Hager and Blaser (1998) is seen to yield reasonable estimates of $Y(\chi)$ regardless of r, but the predictions are above the numerical solution. According to Campbell et al. (1965), the roughness as determined by Bauer (1951, 1954) was exceedingly high, resulting in a low gradient $D/\chi(\chi/r)$ if scaled with the corresponding value of k_s. Thus, Eq. (4.19) yields reasonable results, but Eq. (4.21) is improved by using a different k_s value. As to the effect of chute slope, all models yield good estimates for $Y(\chi)$. Note again that the variable power-law approach generally fits the numerical results for $D(\chi)$ (Castro-Orgaz and Hager, 2010). The roughness effect under variable chute slope was also detailed.

The results for $D(\chi)$ shown in Figure 4.8 are reproduced for clarity again in Figure 4.9, to compare with the classical approaches of Bauer (1954) and Campbell et al. (1965). Note that Eq. (4.21) generally fits Eq. (4.25) of Bauer for $\alpha = 45°$, for any value of r, whereas Eq. (4.24) generally agrees with the numerical solution only for low r values, but deviates as r increases. This deviation is opposed to both Eqs. (4.17) and (4.21), resulting in higher $D(\chi)$. As further shown in Figure 4.8, as expected from Eqs. (4.24) and (4.25), there is no effect of chute slope, because only Eqs. (4.17) and (4.21) are affected by α. Therefore, based on this comparative analysis, the effect of α on $D(\chi)$ is small, whereas deviations between this approach and the complete numerical solution are based on the assumed relation between $D(\chi)$ and r. Thus, an improved approach relies on fitting formulae of the type $D/\chi(\chi/r)$, as α may be neglected.

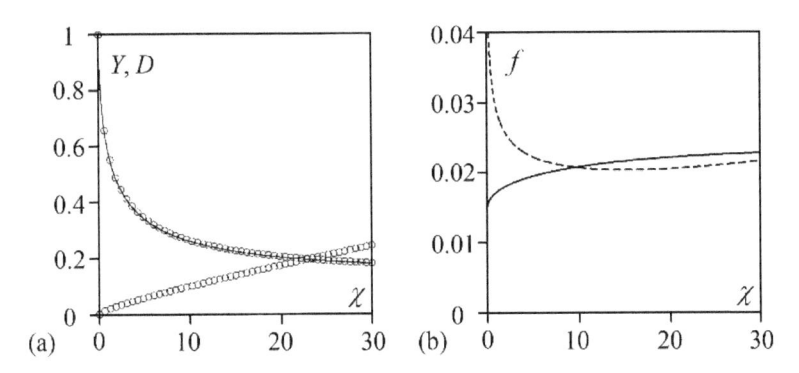

Figure 4.8 (a) $Y(\chi)$, $D(\chi)$ from ($\circ$) present numerical solution, (—) backwater approach with Colebrook-White $f = f(Y/r)$, (---) backwater approach with $f = f(D/r, D/Y)$ according to boundary layer theory for $\alpha = 45°$ and $r = 0.001$, (b) results for (—) $f = f(Y/r)$ and (---) $f = f(D/r, D/Y)$ (Castro-Orgaz and Hager, 2010)

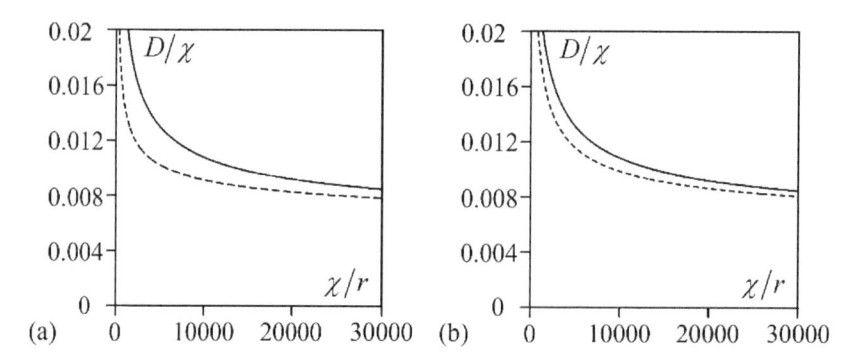

Figure 4.9 Dimensionless boundary layer thickness $D/\chi(\chi/r)$ from Eqs. (4.4) and (4.5) for (a) $\alpha = 45°$ and $r = $ (---) 0.01, (—) 0.0005, (b) $r = 0.001$ and $\alpha = $ (---) 60°, (—) 20° (Castro-Orgaz and Hager, 2010)

The backwater approach by Eq. (4.18) combined with Eq. (4.22) to estimate $Y(\chi)$ needs to be clarified, as proposed by ICOLD (Hager, 1992) or Falvey (1990), without verification against the boundary layer methods. Note that Eq. (4.18) neglects the effect of the Boussinesq velocity coefficient and its variation with x, whereas this effect is included in Eqs. (4.4) and (4.5) by the terms $d\delta*/dx$ and $d\theta_m/dx$. However, this effect is small, as verified below. Equation (4.18) was solved numerically for the typical case $\alpha = 45°$ and $r = 0.001$ using Eq. (4.22) for $f = f(Y/r)$. The equation was solved again using Eq. (4.23) for $f = f(D/r, D/Y)$ using Eq. (4.17) to estimate δ/h. The results are plotted in Figure 4.9a, thereby including the results of the numerical solution using Eqs. (4.4) and (4.5). Surprisingly, there is no appreciable effect between Eq. (4.18) neither using Eqs. (4.22) nor (4.23), despite the differences both in magnitude and behavior observed in f (Figure 4.9b). Figure 4.9b agrees with the f increase outlined by ASCE (1963) for developing chute flow. However, the effect of the improved formulation on the free-surface profile $Y(\chi)$ is insignificant. More surprisingly, the backwater solution of Eqs. (4.18) and (4.23) yields almost identical results as Eqs. (4.4) and (4.5). Accordingly, a simple backwater solution applies to predict the free-surface profile. A disadvantage of Eq. (4.18) is that it does not provide information on the development of $D(\chi)$, and an empirical equation is thus needed to determine the inception point. Further, Eq. (4.18) requires a numerical solution, whereas engineering practice would benefit from simple, explicit design computations. From this point of view, Eqs. (4.12) and (4.13) are a reasonable choice, as the solution is analytical, explicit, and both $Y(\chi)$ and $D(\chi)$ are easily determined simultaneously. A simplified form of these equations is presented below to facilitate practice.

A simplified design approach is proposed. Equations (4.12) and (4.13) generally result in accurate predictions as compared with Eqs. (4.4) and (4.5). The relation $D/\chi(\chi/r, r, \alpha)$ from Eqs. (4.4) and (4.5) is examined in Figure 4.10. The numerical results for $\alpha = 45°$, and $r = 0.01$ or $r = 0.0005$ are plotted in Figure 4.10a, from which no similarity in terms of a unique function $D/\chi(\chi/r)$ is observed, as also for $r = 0.001$ and both $\alpha = 20°$ and 60° (Figure 4.10b). Despite the small deviations, this numerical model is not a good choice to define a simplified design equation $D/\chi(\chi/r)$.

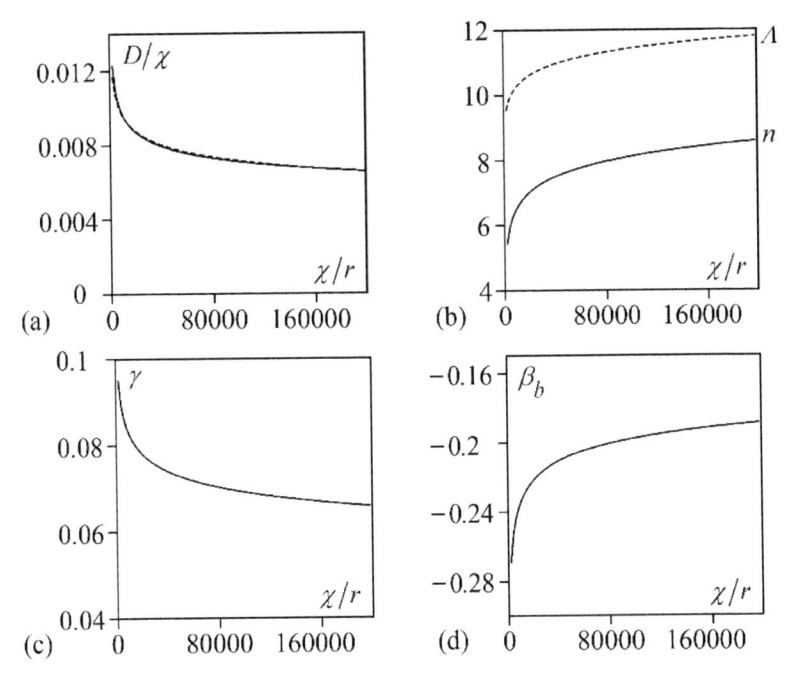

Figure 4.10 Design model (a) $D/\chi(\chi/r)$ from (—) Eq. (4.12), (---) Eq. (4.28), (b) (---) $\Lambda = \Lambda(\chi/r)$, (—) $n = n(\chi/r)$, (c) (—) $\gamma = \gamma(\chi/r)$, (d) (—) $\beta_b = \beta_b(\chi/r)$ (Castro-Orgaz and Hager, 2010)

In contrast, Eq. (4.13) results in a complete similarity function $D/\chi(\chi/r)$ as seen in Figure 4.11a if variable coefficients $\Lambda = \Lambda(\chi/r)$ and $n = n(\chi/r)$ from Eqs. (4.14) and (4.15) are used (Figure 4.11b). The power-law parameters of Eq. (4.3) are then (Figure 4.11c, d)

$$\gamma = \left[\frac{2}{\Lambda^2} \frac{(n+1)(n+2)}{(5n^2 + 8n + 4)} \right]^{\frac{n}{n+2}}, \tag{4.26}$$

$$\beta_b = -\frac{2}{n+2}. \tag{4.27}$$

For practical purposes, the curve $D/\chi(\chi/r)$ in Figure 4.11a is approximated in the range $2 \times 10^3 \leq \chi/r \leq 2 \times 10^5$ by the fit ($R^2 = 0.992$)

$$\frac{D}{\chi} = 0.0302 \left(\frac{\chi}{r} \right)^{-1/8}. \tag{4.28}$$

Computations are simplified by using Eq. (4.28). For design purposes, the variations of Λ, n, γ, and β_b along the chute can be overlooked. For more accurate predictions, however, use

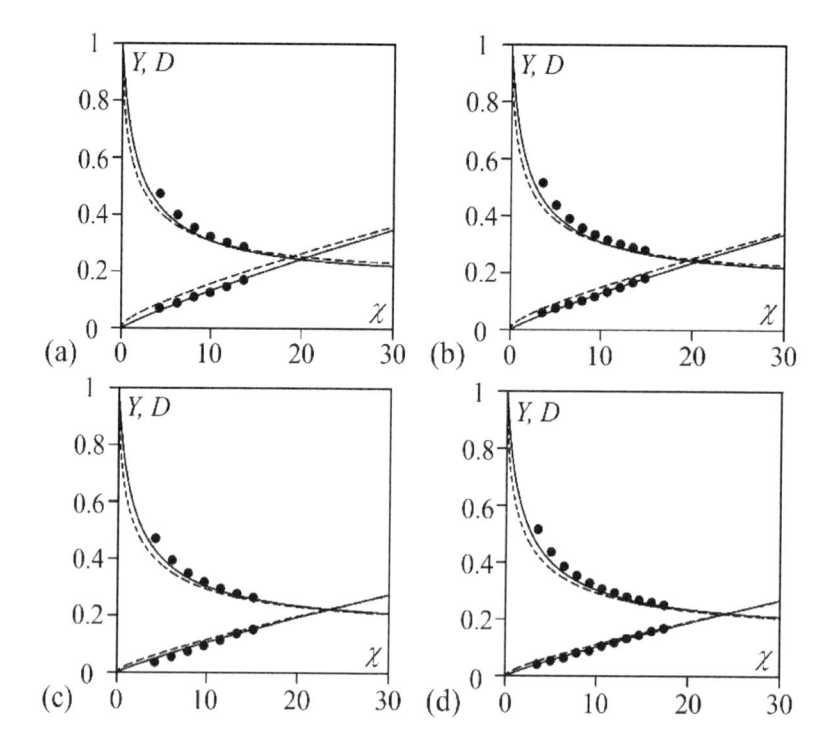

Figure 4.11 $Y(\chi)$ and $D(\chi)$ from (—) Eqs. (4.12), (4.28), (---) Eqs. (4.4), (4.5), (•) Bormann's data (1968) for α = 33.5°, r = (a) 0.0144, (b) 0.0109, (c) 0.0023, (d) 0.0017 (Castro-Orgaz and Hager, 2010)

the complete analytical model. The accuracy of Eq. (4.28) is tested in Figure 4.11 against the numerical solution of Eqs. (4.4) and (4.5) for α = 33.5° and $0.0017 \leq r \leq 0.0144$ (Bormann 1968). For n = 6 the simplified free-surface profile $Y(\chi)$ is

$$Y = \frac{D}{7} + \left[1 + 2\chi\sin\alpha\right]^{-1/2}.$$
(4.29)

Note that the agreement of the simplified approach given by Eqs. (4.28) and (4.29) with both the numerical solution and the experimental data is reasonable, supporting its use for practical purposes, so that the effect of α in Eq. (4.4) remains small. No numerical computations are therefore necessary.

In conclusion, the chute design up to incipient-aerated flow involving a developing boundary layer and irrotational flow above it was considered. Both the drawdown curve $h = h(x)$ and the boundary layer development profile $\delta = \delta(x)$ are analyzed, either using the boundary layer theory, empirical relations, or backwater approximations. The results serve as a practical guide. All models tested result in accurate $h = h(x)$ predictions, whereas estimations of $\delta = \delta(x)$ differ significantly.

The backwater approach combined with Colebrook-White's equation is discussed. Based on a comparative analysis, the variable power-law velocity profile approach agrees well with the complete numerical solution of a boundary layer equation. Its solution is simpler than the backwater approach, resulting in explicit expressions. A simplified design is verified against test data and the numerical solution, so that chute flow computations reduce to two explicit equations, namely Eqs. (4.28) and (4.29).

4.2.3 Development of aerated chute flow

Introduction

Spillway chutes and tunnels represent important appurtenant hydraulic structures at dams regarding their safety. Due to high-speed flow associated with a low-pressure potential, dynamic uplift and cavitation may occur on the bottom and sidewalls of these conveyance structures, causing major damage, as observed e.g. on the Karun I Dam in Iran in 1977, or the Glen Canyon Dam in Colorado, USA, in 1983. Peterka (1953) evidenced that an average cross-sectional air concentration of $C_a \cong 5\%$ reduces the cavitation risk almost completely. The elastic properties of water strongly change under the presence of air bubbles. Even if the amount of air needed for cavitation protection was questioned in the past, there is general agreement that a small air concentration close to the chute or tunnel boundaries (subscript b), such as $C_b = 0.01$, reduces the risk of cavitation damage significantly.

In chute flow, air is either entrained at the free surface or at the chute bottom with chute aerators. For a long time in the past, no reliable design guideline was available for the distance required between two aerators to produce sufficient bottom air concentration, although aerators have been proposed from the 1970s. The guidelines elaborated in the 20th century for aerator spacing were scientifically weakly founded because no detailed laboratory experiments focusing on the air detrainment processes have been conducted. Based on prototype observations on Russian high-head dams, Semenkov and Lantyaev (1973) proposed an average (subscript a) air detrainment of $\Delta C_a = 0.40\%$ to 0.80% per chute meter length, whereas Prusza (1983) suggested $\Delta C_a = 0.15\%$ to 0.20%. These numbers were used for prototype chutes such as *Foz de Areia* to determine the required aerator spacing inhibiting cavitation damage. It was later noted that too much air was entrained so that some aerators had to be eliminated (Pinto *et al.*, 1982). Falvey (1990) summarized studies relating to the air detrainment downstream of an aerator versus chute distance. Minor (2000) mentioned a smaller air detrainment rate of only $\Delta C_a = 0.10\%$ to 0.15% based on photometrical data of the Restitution Chute. May (1987) introduced a computational model including the air concentration downstream of an aerator and the point of air inception. Chanson (1989a) suggested to account for chute slope, specific discharge, equilibrium air concentration, average air concentration, inflow water depth, and bubble rise velocity.

Cavitation protection on chutes by forced air addition is separated into three zones, including (1) the aerator zone where the air is detrained from the flow, (2) a zone of minimum air concentration where a second aerator is possibly needed, and (3) a zone downstream of it affected by free-surface aeration due to stream turbulence (see 4.4).

For the flow region downstream of the air inception point, several methods were developed, including terms containing the friction factor, a Boussinesq number, an Eötvös number, and chute slope (Yevdjevich and Levin, 1953; Straub and Anderson, 1960; Lakshmana Rao and Kobus, 1973; Falvey, 1980; Volkart, 1982; Hager and Blaser, 1998). Due to incomplete model data including all hydraulic parameters, and limited instrumentation for two-phase flows, additional model data on the behavior of the streamwise air detrainment downstream of aerators are required. Three books are available, namely these of Wood (1991), Chanson (1996) and Vischer and Hager (1998). Kramer *et al.* (2006) focused on the streamwise development of the bottom air concentration along chutes, of whom the results are reported below.

Experiments

The hydraulic experiments were conducted in a 14 m long prismatic rectangular chute model of width $b = 0.50$ m, variable bottom slope $0\% \leq S_o \leq 50\%$, approach (subscript o) flow depths up to $h_o = 0.12$ m, and water (subscript w) discharges of up to $Q_w = 250$ l/s (Kramer, 2004). The chute bottom was made of PVC, whose sand-roughness height under uniform flow was $k_s = 0.00023$ mm. All quantities, namely (1) water discharge Q_w, (2) air discharge Q_A, (3) positioning in streamwise x and perpendicular z directions, and (4) automized data collection, were controlled by a LabView-based program. The automatic data acquisition included air concentration profiles, velocity fields, and the characteristic mixture-flow depth h_{90} (Figure 4.12).

Two different air supply devices for air-water flow generation were employed: (1) deflector similar to a prototype aerator, and (2) pre-aerated flow by adding air to the pressurized supply pipe. Both setups were investigated separately to determine the effect of the remaining hydraulic parameters and that of the aerator. The effects of Froude number, chute slope, and inflow depth on the streamwise development of chute air concentration are studied below.

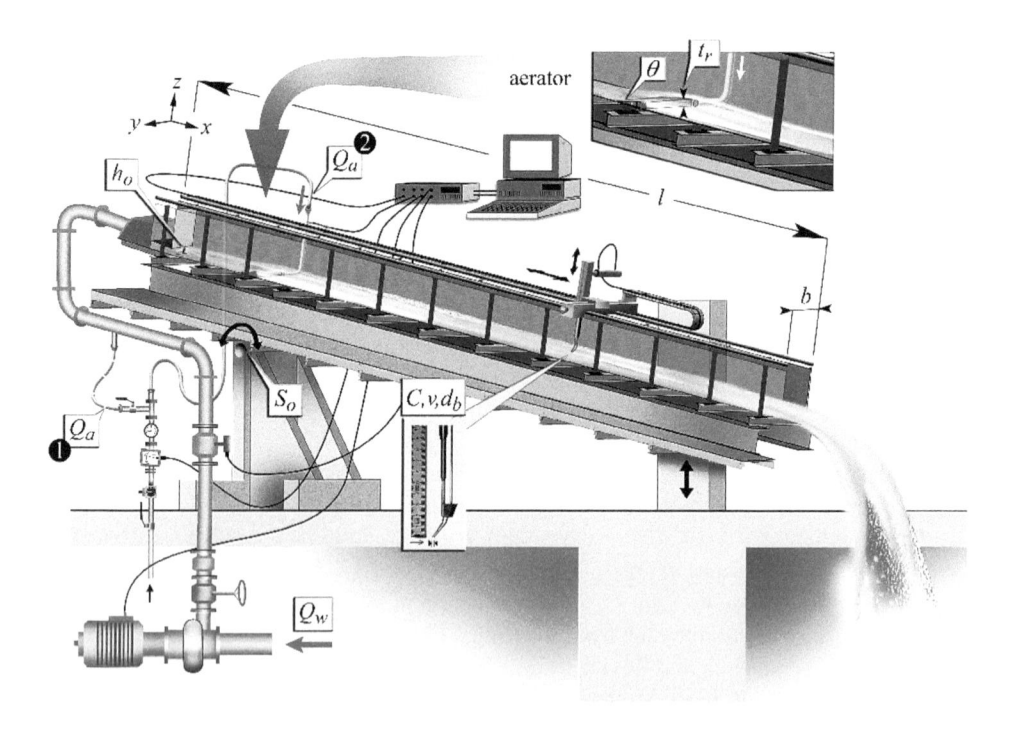

Figure 4.12 Hydraulic laboratory model (Kramer, 2004)

Air concentration profiles

Local air concentration, flow velocity, and air bubble size are evidenced by contour lines and air concentration profiles. The mixture-flow depth h_{90} and the average air concentration C_{a90} are based on air concentration $C = 90\%$, the currently established value for the free surface definition. For pre-aerated flow, a jet-box provided well-defined flow conditions of variable approach (subscript o) flow depth h_o. Figure 4.12 shows the coordinate system (x, y, z). The approach flow velocity was defined by the specific water q and the specific air q_A discharges as $u_o = (q_A + q)/h_o$, so that the approach flow Froude number is $F_o = u_o/(gh_o)^{0.5}$ and the aeration rate $\beta = Q_A/Q$.

The scaling of two-phase chute flow follows Froude similitude if (1) model scale is large enough to suppress viscous effects and (2) model turbulence intensity has adequate kinetic energy to overcome the restraining surface tension between air and water and bubble uplift forces. Therefore, all measurements accounted for fully turbulent flow, with approach flow Reynolds numbers $5.8 \times 10^5 \leq R_o \leq 2.3 \times 10^6$ where $R_o = u_o h_o/v$ and approach flow Weber numbers $88 \leq W_o \leq 210$ with $W_o = u_o/[\sigma/(\rho h_o)]^{0.5}$, where ρ is water density, v kinematic fluid viscosity, and σ_t surface tension. Most of the scale limitations available in the literature, e.g. Boes and Hager (2003a), were met.

Figure 4.13a shows the streamwise development of the air concentration profiles $C(X_{90u})$ over the non-dimensional mixture-flow depth $Z_{90} = z/h_{90}$ for typical pre-aerated flow, with h_{90} as the local mixture-flow depth and h_{90u} as the uniform mixture-flow depth. The first profile at $X_{90u} = x/h_{90u} = 0.4$ ($x = 0.022$ m, $h_{90u} = 0.054$ m) is influenced by the jet box and thus not representative. All profiles further downstream are subject to continuous air detrainment with distance. Beyond $X_{90u} = 47.0$, the air concentration profiles remain practically constant indicating conditions close to uniform mixture flow. Figure 4.13b is the equivalent to Figure 4.13a for aerator flow. The first profile at the downstream aerator end at $X_{90u} = 0.3$ ($x = 0.019$ m) involves almost 100% air up to $Z_{90} = 0.2$, followed by black-water for $0.20 \leq Z_{90} \leq 0.85$. At the top region, black-water forms a sharp boundary to the free surface. For $Z_{90} \geq 1$, the air concentration ranges within $0.9 \leq C \leq 1.0$, as proposed by the definition of Z_{90}. At $X_{90u} \cong 9.6$ ($x = 0.614$ m), the lower jet nappe impacts the chute bottom. Further downstream,

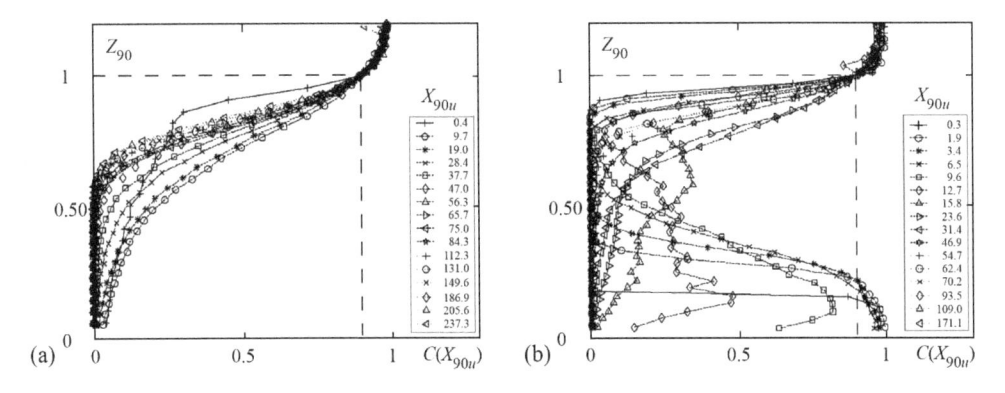

Figure 4.13 Air concentration profiles $C(X_{90u})[Z_{90}]$ for $\beta = 0.21$, $S_o = 10\%$ (a) pre-aerated flow with $F_o = 10$, $h_o = 0.05$ m, $h_{90u} = 0.054$ m, (b) aerator flow with $F_o = 8$, $h_o = 0.06$ m, $h_{90u} = 0.064$ m (Kramer, 2004)

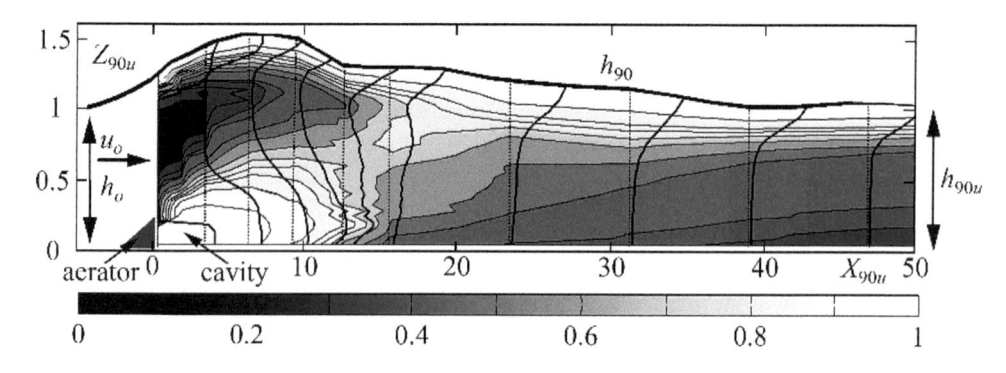

Figure 4.14 Air contour plot for aerator flow at $\mathsf{F}_o = 8$, $h_o = 0.06$ m, $\beta = 0.21$, $S_o = 10\%$, $h_{90u} = 0.064$ m, with (—) air concentration contour lines $C(X_{90u})$, ($\cdots$) origin of air concentration lines (Kramer, 2004)

the bottom air concentration reduces drastically to attain almost zero at $X_{90u} \cong 15.8$. In parallel, the air concentration increases in the center region $0.20 \leq Z_{90} \leq 0.80$, and the profiles develop into a fluid body in the top layer with detrainment further downstream. These two examples for $S_o = 10\%$ describe the principle behavior of air transport in high-speed water flow by means of air concentration profiles for pre-aerated and aerator flows, respectively. From the 150 tests conducted, selected data illustrate the basic flow behavior, whereas the fit equations presented below are based on all data.

Figure 4.14 shows axial air concentration profiles in the downstream aerator region based on Figure 4.13b. Shading indicates air concentration contours linearly interpolated between the observational grid. For the bottom (subscript b) air concentration C_b, the minimum distance from the chute bottom ranged within 3 mm $\leq z \leq$ 5 mm. The uniform flow depth h_{90u} was assumed to occur at the chute end. The plot shows for $X_{90u} \leq 50$ the development of air concentration in the streamwise direction. The air moves upwards through the flow, from the aerator downstream cavity to the free surface while it is carried with the main flow. The light shade stands for a high amount of air, whereas dark areas indicate regions of small air concentration. The latter relates to the bottom region downstream of jet impact onto the chute bottom where air is detrained, and to the central inflow region where no air is yet entrained from the aerator or the free surface. The transitions between the grayscales in Figure 4.14 correspond to equal air concentration lines (contour lines), marked with black fine lines. Air detrainment is determined as the gradient of the contour lines det$C = C$dz/dx, thereby assuming that the air bubble rise velocity depends on the local air concentration and turbulence. The bubble rise velocity for these flow conditions is detailed in 4.2.5. The bold black lines in Figure 4.14 represent vertical air concentration profiles.

Velocity profiles

Figure 4.15 shows typical velocity profiles for pre-aerated and aerator flows. The local mixture (subscript m) velocity u_m was scaled with the characteristic mixture velocity u_{90} involving the mixture-flow depth h_{90} so that the non-dimensional velocity is $U_{90} = u/u_{90}$. No velocity information was available for $Z_{90} \geq 1$ at the chute end and in the cavity region due to the lack

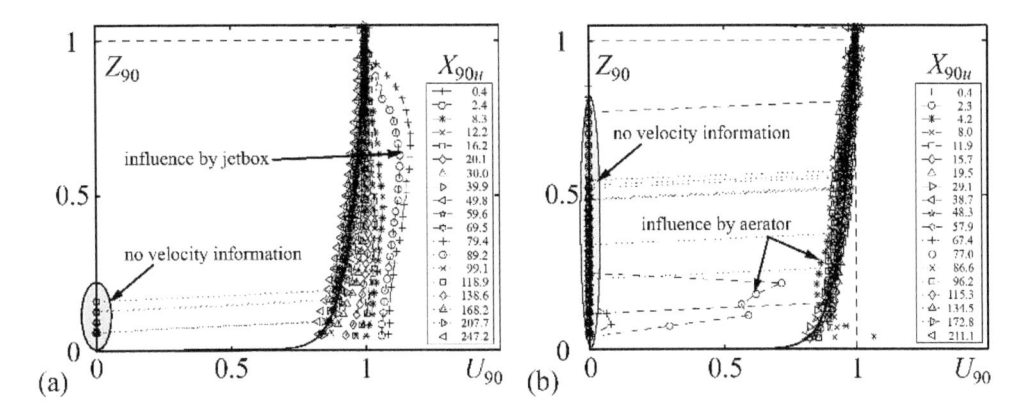

Figure 4.15 Velocity profiles $U_{90}(Z_{90})$ at various X_{90u} and $S_o = 50\%$ for (a) pre-aerated flow
with $F_o = 10$, $h_o = 0.06$ m, $\beta = 0.21$, $h_{90u} = 0.051$ m, (b) aerator flow with $F_o = 8$,
$h_o = 0.06$ m, $\beta = 0.21$, $h_{90u} = 0.052$ m (Kramer, 2004)

of air bubbles, as indicated with the dotted curves. The fully developed velocity profiles follow the power law $U_{90} = aZ_{90}^{1/n}$, with a and n as empirical constants. It is reached much faster
for aerator flow (Figure 4.15b) than for pre-aerated flow (Figure 4.15a). A deflector thus acts
as turbulence generator (Ervine and Falvey, 1987). The best fit for the velocity profiles in
Figure 4.15 is $n = 16$ and $a = 1$, of which $n = 16$ is high as compared with e.g. Cain (1978)
proposing $n = 6.3$ for prototypes, or Boes and Hager (2003a) for stepped chutes with $n = 4.3$
and $a = 1.05$ for $0.004 \leq Z_{90} \leq 0.8$. The value $n = 16$ originates from air presence in the bottom
zone, reducing the wall friction in the boundary layer (Wood, 1985).

Air concentration development analysis

Figure 4.16a shows the schematic development of bottom air concentration $C_b(x)$ for aerator
flow. It starts downstream of the cavity involving a characteristic approach flow air concentration C_{bo} between zones (1) and (2). The air concentrations of pre-aerated and aerator flows
are similar. Further downstream, the air concentration $C_{b,det}$ decreases to a minimum $C_{b,min}$.
Kramer *et al.* (2006) demonstrated that the point of air inception $x = x_i$ occurs at the minimum average air concentration $C_{a,min}$, where the local flow depth $h = h_i$ equals the thickness
of turbulent boundary layer $\delta = \delta(x)$ (4.2.2). The minimum bottom air concentration $C_{b,min}$
is reached downstream of x_i, where the air bubbles have advanced to the chute bottom from
the free surface. Region (3) in Figure 4.16a spans therefore between the point of air inception and the point where the air bubbles reach the chute bottom at the transition to zone (4).

In the region downstream of $C_{b,min}$, air is entrained from the free surface resulting in
increasing $C_{b,ent}$ up to uniform air concentration C_{bu}. The hydraulic effects on this typical
pattern are analyzed based on five zones: (1) approach flow air concentration C_{bo}, (2) air
concentration in detrainment region $C_{b,det}$, (3) minimum air concentration $C_{b,min}$, (4) air concentration in entrainment region $C_{b,ent}$, and (5) uniform air concentration C_{bu}. Figure 4.16d
highlights that a definite air concentration $C_{b,min}$ is reached for $S_o = 50\%$, whereas no minimum occurred for chute slopes $S_o \leq 10\%$ (Figure 4.16b). Turbulence generation on flat

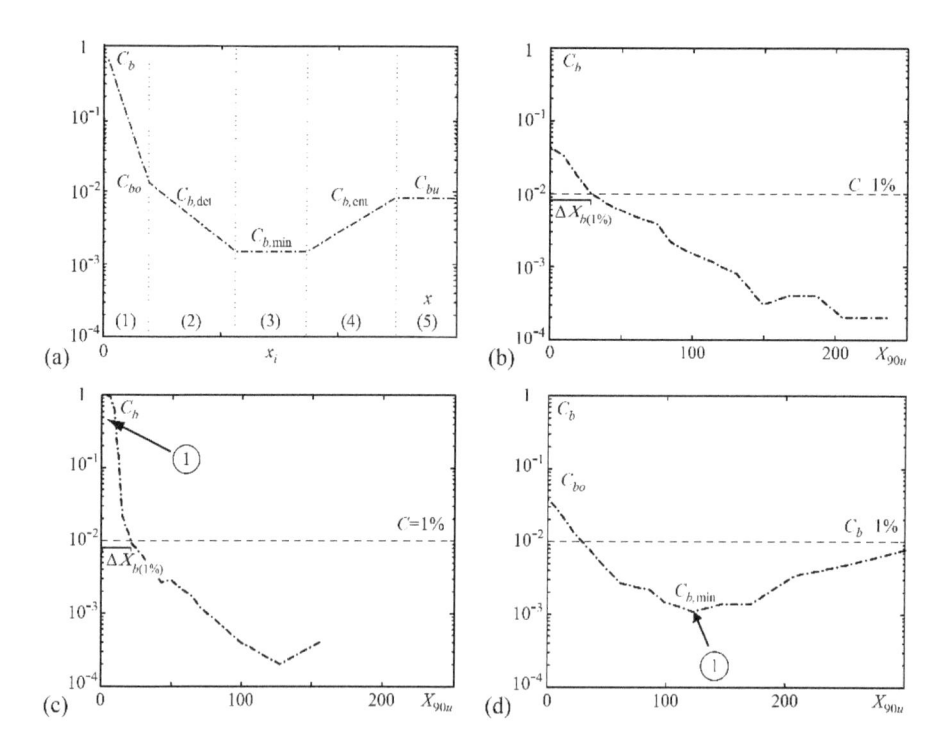

Figure 4.16 Streamwise development of bottom air concentration (a) air concentration development with (1) cavity zone, (2) air detrainment zone, (3) minimum air concentration zone, (4) air entrainment zone, (5) uniform flow zone, (b) pre-aerated flow $F_o = 10$, $h_o = 0.05$ m, $h_{90u} = 0.054$ m, $S_o = 10\%$, (c) aerator flow $F_o = 8$, $h_o = 0.06$ m, $h_{90u} = 0.064$ m, $S_o = 10\%$ with ① cavity downstream of aerator, (d) pre-aerated flow $F_o = 10.28$, $h_o = 0.06$ m, $h_{90u} = 0.051$ m, $S_o = 50\%$ with ① point of air inception (Kramer, 2004)

slopes is insufficient to overcome surface tension and entraining air down to the chute bottom. Therefore, only data on chute slopes $S_o \geq 10\%$ are considered for the analysis of the minimum air concentration.

Bottom air concentration

The bottom air concentration is relevant for cavitation protection. Figure 4.16b–c shows examples of the streamwise development of the bottom air concentration C_b for the tests in Figure 4.15. Figure 4.16d refers to pre-aerated flow with the boundary conditions as in Figure 4.16b, and $S_o = 50\%$. Note the increasing value of C_b for $X_{90u} \geq 120$ for the steep slope, the point where the air reaches the chute bottom from the free surface.

Figure 4.17 shows C_b for pre-aerated flow versus $X_{90u} = x/h_{90u}$ in a semi-logarithmic plot for $S_o = 0$, 10%, 30%, and 50%, and $4.5 \leq F_o \leq 14.4$. Close to $X_{90u} \approx 0$, $C_{bo} \approx 5\%$, a typical value of bottom air concentration generated by a jet box under pre-aerated flow. For an improved data analysis, only the detrainment region between the characteristic approach

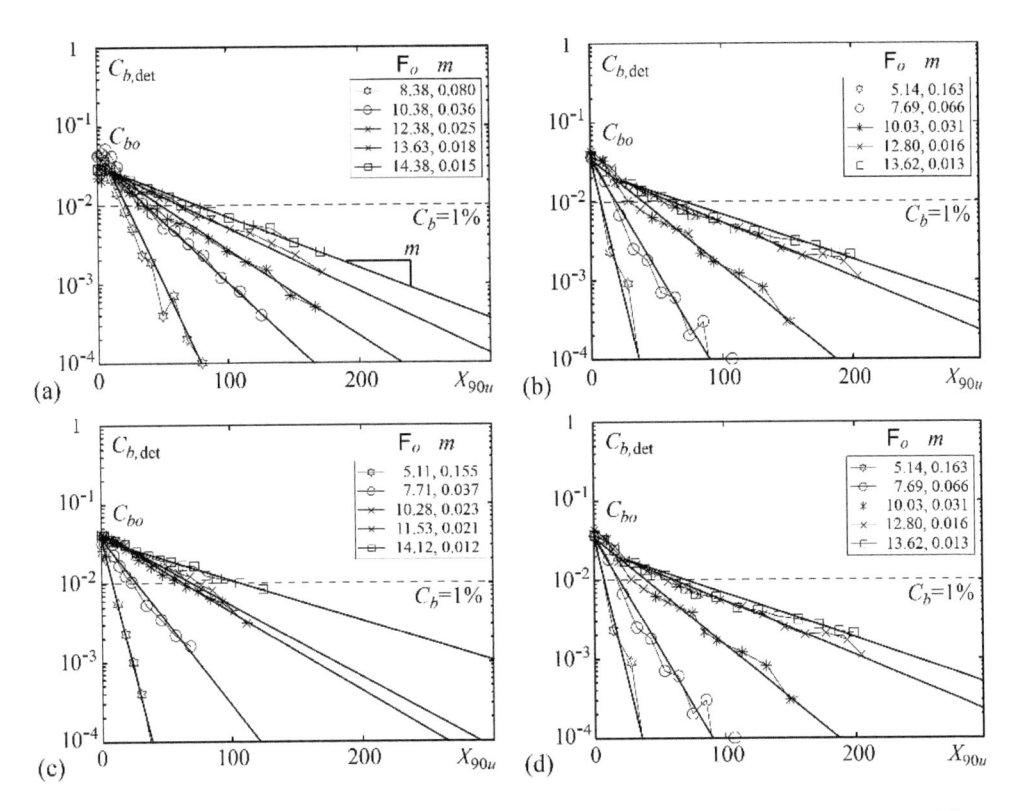

Figure 4.17 Bottom air concentration $C_{b,det}(X_{90u})$ in detrainment region for pre-aerated flow for S_o = (a) 0, (b) 10%, (c) 30%, (d) 50%, (——) Eq. (4.30) (Kramer, 2004)

flow air concentration C_{bo} and the minimum air concentration at the point of air inception $C_{b,min}$ are shown (Zone (2) in Figure 4.16a). The approach flow Froude number F_o represents the main parameter for the streamwise development of the bottom air concentration. All curves start at the characteristic value C_{bo}, to decrease fan-shaped in the semi-logarithmic plot, depending on F_o. The bottom slope has two effects: (1) gradient of decreasing curves $C_{b,det}(X_{90u})$ and (2) minimum air concentration $C_{b,min}$. For almost horizontal chutes, the minimum air concentration falls to zero further downstream, whereas there is a clear minimum value at the point where the air concentration increases for steeper slopes.

The semi-logarithmic, fan-shaped data of Figure 4.17 are described by the gradients of each line, depending on $F_o = u_o/(gh_o)^{1/2}$, C_{bo} as the characteristic upstream bottom air concentration and m as the slope of the detrainment lines as

$$C_{b,det} = C_{bo}e^{-mX_{90u}}. \tag{4.30}$$

The best fit for the air detrainment gradient $m = f(F_o)$ is the power function

$$m = nF_o^{-2.5}, \text{ for } 5 \leq F_o \leq 12. \tag{4.31}$$

For $F_o \geq 10$ the bottom slope has almost no effect on the detrainment gradient; the best fit for n versus chute slope is ($R^2 = 0.94$)

$$n = 7.2 \cdot 0.006^{-S_o} 6.6, \text{ for } S_o \leq 50\%. \tag{4.32}$$

Combining Eqs. (4.30) to (4.32) results in the general equation for the bottom air detrainment region as (Kramer *et al.*, 2006)

$$C_{b,det} = C_{bo}\exp[-(7.2 \cdot 0.006^{S_o} + 6.6)F_o^{-2.50}X_{90u}], \text{ for } 5 \leq F_o \leq 12, S_o \leq 50\%. \tag{4.33}$$

The bottom air concentration in the detrainment region thus reduces exponentially. Therefore, air detrainment depends directly on the upstream air concentration C_{bo}, the approach flow Froude number, and chute bottom slope. A high air concentration on a relatively flat chute for small F_o results in maximum air detrainment, and vice versa. The a priori hypothesis was thus systematically proven in this research.

According to Kramer *et al.* (2006), the average minimum air concentration $C_{a,min}$ depends on the streamwise development of the Froude number $F(x)$. The minimum bottom air concentration is related as $C_{b,min} = f[S_o, F(x)]$, with the flow depth contained in both the chute slope and the Froude number. Because the Froude number depends on S_o, this is rewritten as $C_{b,min} = f[F(x)]$. The best fit for the minimum bottom air concentration involves the streamwise Froude number $F_{o-min} = [F_o + F(C_{b,min})]/2$ as the arithmetic average of F_o and the Froude number $F(C_{b,min})$ at the location of minimum bottom air concentration. The relation $F_{o-min}(C_{b,min})$ is shown in Figure 4.18a with $R^2 = 0.85$ as (Kramer *et al.*, 2006)

$$C_{b,min} = 3.6 \cdot 10^{-7} \exp[0.70F_{o-min}], \text{ for } F_{o-min} > 11, S_o \geq 10\%. \tag{4.34}$$

The limit $F_{o-min} > 11$ avoids that $C_{b,min}$ falls below $C_b = 0.01\%$ imposed by measuring accuracy. Equation (4.34) depends indirectly on the chute slope S_o because of the drawdown curve up to the point of inception, and thus the Froude number at the point of minimum bottom air concentration $F(C_{b,min})$. For $F_{o-min} \geq 14.6$ the bottom air concentration never falls below $C_b = 1\%$, from Eq. (4.34).

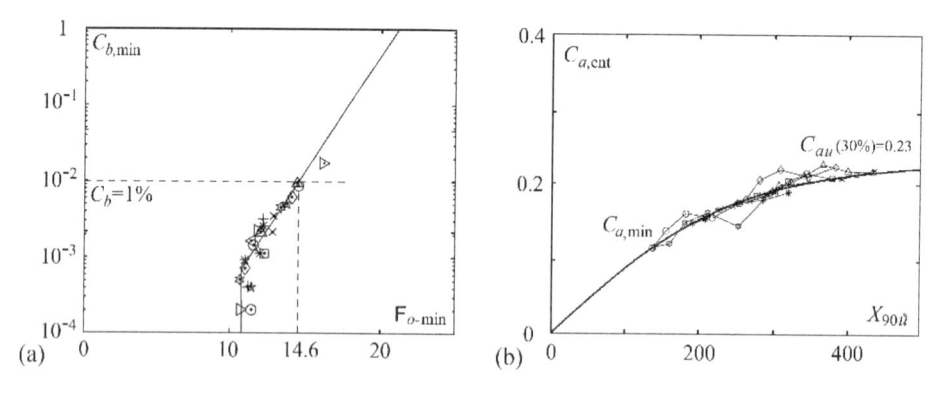

Figure 4.18 (a) Minimum bottom air concentration $C_{b,min}(F_{o-min})$ for $S_o \geq 10\%$ (−) Eq. (4.34), $F_o \geq 7$, (b) air entrainment $C_{a,ent}(X_{90u})$ for $S_o = 30\%$, (−) Eq. (4.35)

Straub and Anderson (1960) established that the average air concentration C_a increases noticeably downstream of the point of inception to reach equilibrium far downstream. This is an important feature for aerator spacing because no additional aerator is normally needed beyond the inception point. Figure 4.18b shows the average air entrainment (subscript ent) on a 30% chute. For air entrainment on chute slopes up to 50%, the best fit is with $b = 0.004$ for both pre-aerated and aerator flow, and $b = 0.006$ for no upstream air supply (Kramer et al., 2006)

$$C_{a,\text{ent}} = C_{au}\tanh(bX_{90u}).$$ (4.35)

Accordingly, $C_{a,\text{ent}}$ depends on C_{au} and the relative flow distance only. The uniform air concentration C_{au} increases with the bottom slope as (Kramer et al., 2006)

$$C_{au} = 0.11S_o^{0.24}, \text{ for } S_o \le 50\%.$$ (4.36)

Only a small effect of the free-surface air entrainment on the bottom air concentration C_b was noted (Hager, 1991; Chanson, 1996).

4.2.4 Spacing of chute aerators

Cavitation protection

Cavitation occurs if the local pressure p_o falls below the vapor (subscript v) pressure of water p_v. It also occurs if the average (subscript a) flow velocity u_a increases so that the cavitation number $\sigma = \left(p_o - p_v\right)/\left(\rho u_a^2 / 2\right)$ falls below the critical (subscript c) value σ_c. Here the total local pressure $p_o = p_{\text{atm}} + p_g$ includes the atmospheric pressure p_{atm} and the piezometric pressure p_g. The number σ is expressed in terms of absolute pressure as above, or as local pressure head $h_o = h_{\text{atm}} + h_g$ involving the atmospheric pressure head h_{atm} [m], the piezometric pressure head h_g [m], and the vapor pressure head h_v [m] as

$$\sigma = \frac{h_g + h_{atm} - h_v}{u_a^2 / 2g}.$$ (4.37)

Substituting $h_{\text{atm}} - h_v = (p_{\text{atm}} - p_v)/(\rho g) = 10.0$ m by assuming $p_{\text{atm}} = 1000$ hPa $= 0.100$ N/mm^2 and constant vapor pressure of $p_v = 16.0$ hPa $= 0.0016$ N/mm^2 for water of 14°C yields

$$\sigma = \frac{h_g + 10\,[\text{m}]}{u_{ac}^2 / 2g}.$$ (4.38)

Figure 4.19 shows the critical average velocity u_{ac} for various piezometric heads h_g and five critical cavitation numbers $0.1 \le \sigma_c \le 0.3$ according to Eq. (4.38). For a flow depth of $h_g = 3.0$ m and a critical cavitation number of $\sigma_c = 0.2$, cavitation occurs for $u_{ac} \ge 35.8$ m/s. Figure 4.19 also shows that h_g has a small effect on u_{ac}, as compared with the critical cavitation number σ_c. The typical number presently adopted for cavitation damage as observed on prototype chutes is $\sigma_c = 0.2$ (Falvey (1983).

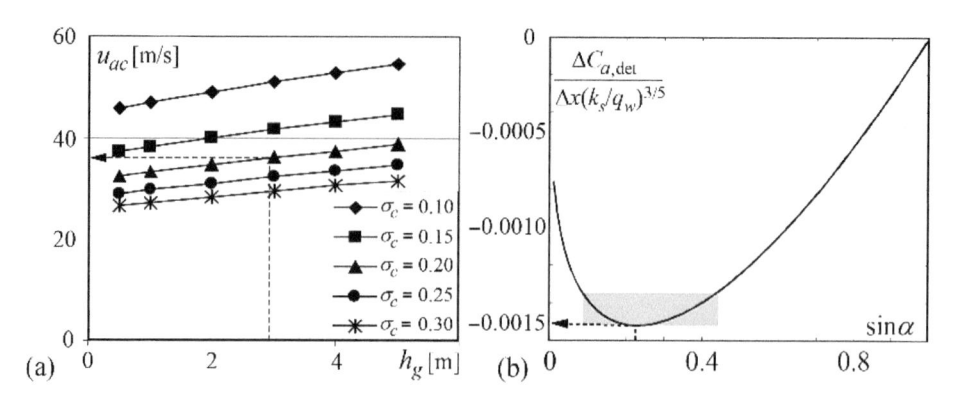

Figure 4.19 (a) Critical average flow velocity u_{ac} [m/s] against piezometric head h_g [m] for various critical cavitation numbers σ_c depending on Eq. (4.38), (b) relative air detrainment $\Delta C_{a,det}/[\Delta x(k_s/q_w)^{3/5}]$ versus chute angle $\sin\alpha$, (---) Eq. (4.43), ($\cdots$) almost constant maximum detrainment zone

Air detrainment process

The detrainment process appears to be influenced by high pressures along the impact region downstream of aerators. If the flow transport capacity is low, the air entrained escapes quickly, whereas the air is captured in turbulent flows of large transport capacity (Wood, 1991). Furthermore, the air detrainment process is directly linked to the free-surface entrainment since it reaches an equilibrium state at a point where the air entrained by turbulence is equal to the air detrained (4.2.3).

In the past, either aerators or uniform chute flow were investigated, whereas the development of air concentration between these two chute regions remained excluded. New model data from a hydraulic chute model of Kramer *et al.* (2005) were based on dimensional analysis. It was intended to address the following items:

- Basic behavior of air-water flow downstream of an aerator (Kramer, 2004);
- Effects of chute slope, Froude number and approach flow depth on average air detrainment, minimum air concentration, and air entrainment process;
- Bottom air concentration development over chute length;
- Bubble rise velocity in highly turbulent flow;
- Practical application and design procedure for spacing of chute aerators.

The following presents the equations of average air concentration for a deflector. A similar behavior applies for pre-aerated flow (Kramer, 2004). The flow downstream of an air supply device is divided into the reaches (Figure 4.16a):

1   Approach flow air concentration reach C_{ao}
2   Air detrainment region $C_{a,det}$
3   Minimum air concentration region $C_{a,min}$
4   Air entrainment region $C_{a,ent}$
5   Uniform air concentration reach C_{au}.

The air detrainment $C_{a,det}$ and entrainment $C_{a,ent}$ functions describe the streamwise development of average air concentration along the chute.

The approach air concentration C_{ao} depends on the aerator efficiency. A maximum average inflow air concentration of 30% appears realistic (Rutschmann, 1988), with $C_{a90} = (1/h_{90})\int C(z)dz$, and h_{90} as the depth with a local air concentration of $C = 90\%$, z as the vertical coordinate measured perpendicularly to the chute bottom, and the integral taken from $z = 0$ to $z = h_{90}$. The air detrainment varies with the non-dimensional flow distance $X_{90u} = x/h_{90u}$ as (Kramer et al., 2005)

$$C_{a90,det} = aX_{90u} + C_{ao}. \tag{4.39}$$

Here, x is the distance from the air supply device and h_{90u} the uniform air-water flow depth. The parameter a depends on the chute angle as $a = -0.0085(1 - \sin\alpha)$, so that

$$C_{a,det} = -0.0085(1 - \sin\alpha)X_{90u} + C_{ao}, \text{ for } \alpha \leq 30°, X_{90u} \leq 200. \tag{4.40}$$

According to Eq. (4.40), the air detrainment depends on the relative chute distance from the origin $x = 0$ where $C_a = C_{ao}$, and the chute angle α. The detrainment rate is large for small chute angles, and vice versa.

The air detrainment per chute length $\Delta C_{a,det}/\Delta x$ from Eq. (4.40) is (Kramer et al., 2005)

$$\frac{\Delta C_{a,det}}{\Delta x} = -\frac{0.0085(1 - \sin\alpha)}{h_{90u}}, \text{ for } C_{ao} \geq C_{a,det} \geq C_{a,min}. \tag{4.41}$$

Equation (4.41) is a function of the chute angle α and the uniform mixture-flow depth h_{90u}, which depends on the chute angle too. The uniform flow depth in Eq. (4.41) is substituted by the Manning-Strickler formula $u_a = KR_h^{2/3}(\sin\alpha)^{1/2}$, including thereby the average flow velocity u_a [m/s], the wall roughness coefficient K [m$^{1/3}$/s], the hydraulic radius R_h [m], and the chute angle α [°]. For a wide chute $R_h \approx h_w$, the flow velocity is $u_a = Q_w/(bh_w)$; with the clear-water depth h_w, the chute width b, and the water discharge Q_w, the transposed Manning-Strickler formula gives

$$h_{90u} = \left(\frac{Q_w^2}{K^2 b^2 \sin\alpha}\right)^{\frac{3}{10}}. \tag{4.42}$$

Substituting Eq. (4.42) into Eq. (4.41) leads with the specific discharge $q_w = Q_w/b$ to

$$\frac{\Delta C_{a,det}}{\Delta x} = -0.0085\left(\frac{K}{q_w}\right)^{\frac{3}{5}}(\sin\alpha)^{\frac{3}{10}}(1 - \sin\alpha). \tag{4.43}$$

The discussion of Eq. (4.43) reveals two zeros for $\sin\alpha = 0$ and 1, so that no air is detrained for both the horizontal and vertical chutes. Note that the 'horizontal chute' is academic because the uniform flow depth would be infinite. Further, Eq. (4.43) has a minimum for $\sin\alpha = 0.23$, corresponding to $\alpha = 13.3°$, for which the air detrainment is maximum. According to Figure 4.19b, the maximum air detrainment is in the region where the curve has its minimum, i.e. $0.10 \leq \sin\alpha \leq 0.45$. The maximum air detrainment per chute meter length therefore depends on K and q_w only, with (Kramer $et\ al.$, 2005)

$$\left(\frac{\Delta C_{a,det}}{\Delta x}\right)_{max} = -0.0015\left(\frac{K}{q_w}\right)^{\frac{3}{5}}, \text{ for } 0.10 \leq \sin\alpha \leq 0.45. \tag{4.44}$$

This suggests that the average unit air detainment decreases with increasing specific discharge q_w due to relative reduction of surface roughness as the flow depth increases. Turbulence reduces with higher discharges, causing a smaller effect on air bubbles, therefore.

According to Kramer $et\ al.$ (2005) the minimum average air concentration is with $F_{o-min} = (1/2)[F_o + F(C_{min})]$, as stated previously,

$$C_{a,min} = 0.015 F_{o-min}. \tag{4.45}$$

Note that the minimum average air concentration falls never below $C_{a,min} = 0.15$ for $F_{o-min} \geq 10$. The air entrainment downstream of the point of air inception is given by Eq. (4.35). Its general form depends with $b = 0.004$ only on the uniform air concentration C_{au}, either according to Eq. (4.40), or (Kramer, 2004)

$$C_{au} = \frac{1}{3}(\sin\alpha)^{1/4}. \tag{4.46}$$

Combining Eqs. (4.35) and (4.46) thus gives the general air entrainment function as

$$C_{a,ent} = \frac{1}{3}(\sin\alpha)^{0.25}\tanh(0.004X_{90u}), \text{ for } 0 \leq S_o \leq 50\%. \tag{4.47}$$

Equation (4.47) applies for small chute roughness, typically $0.0001 \leq k_s/h_w \leq 0.0005$.

Computational procedure

The design procedure for a chute is presented to apply the equations developed. The first steps follow the design procedure for aerator design on a chute as mentioned, e.g. by Rutschmann (1988) or Falvey (1990). A critical average air concentration of $C_{ac} = 5\%$ along with a critical bottom air concentration of $C_{b,c} = 1.0\%$ is recommended. The general procedure for aerator spacing is (Kramer $et\ al.$, 2005):

- Determine black-water flow depth $h_w(x)$, flow velocity $u_w(x)$, Froude number $F_w(x) = u_w/(gh_w)^{0.5}$ and cavitation number $\sigma(x)$ as defined in Eq. (4.38) based on topography, dam structure and maximum discharge Q_{max}.
- Cavitation characteristics depend on the critical cavitation number $\sigma_c = 0.2$. Follow Falvey (1990) for surface offset tolerances to determine the minimum value σ_{min}.
- If $\sigma_{min} \leq \sigma_c$, an aerator is required. The location of the inception point x_i is where the boundary layer thickness equals the flow depth $\delta = h_w$; it is from Eq. (4.17) (Wood $et\ al.$, 1983)

$$\frac{\delta}{x_s} = 0.0212\left(\frac{x_s}{H_s}\right)^{0.11}\left(\frac{k_s}{x_s}\right)^{0.10}.$$

Here, x_s is streamwise coordinate measured from the spillway crest, k_s the equivalent roughness height and H_s the difference of water elevations between inception point and reservoir level.

- Average air entrainment downstream of the point of inception follows from Eq. (4.47). If the entrained air exceeds the critical air concentration $C_{a,ent} \geq C_{ac} \approx 5\%$, no aerator is required. Otherwise, an aerator must be placed at the point where the cavitation number falls below $\sigma = 0.2$.
- Follow a standard aerator design procedure, e.g. Koschitzky (1987), Rutschmann (1988), or Falvey (1990). The optimum aerator design must not necessarily be an aerator associated with maximum air concentration C_{ao} but a design resulting in minimum shock waves and pressure peaks along jet impact.
- Streamwise Froude number is $F_{o-min} = [F_o + F(C_{min})]/2$. If minimum average air concentration $C_{a,min}$ from Eq. (4.45), and minimum bottom air concentration from Eq. (4.34) (Kramer et al., 2005)

$$C_{b,min} = 3.6 \cdot 10^{-7} \exp(0.70 F_{o-min})$$

satisfy $C_{a,min} \geq C_{ac} \approx 5\%$ and $C_{b,min} \geq C_{b,c} = 1\%$, no second aerator is required.
- Streamwise average air detrainment $C_{a,det}$ follows Eq. (4.40) and the bottom air detrainment (Kramer et al., 2005)

$$C_{b,det} = C_{bo} \exp[-(30 \times 0.006^{So}) \times F_o^{-2.50} \times X_{90u}]. \tag{4.48}$$

A second aerator is required once the bottom or the average air concentrations fall below their critical values.
- Repeat procedure beyond the reach of the second aerator.

Since only one aerator geometry was investigated, an optimized aerator design may increase the protected chute distance. Note the design example of Kramer et al. (2005), from which the detailed computational procedure is highlighted.

4.2.5 Air transport phenomena

Air bubble rise velocity

According to Falvey and Ervine (1988), four types of forces affect the transport of a single air bubble, namely (1) inertia, (2) drag, (3) buoyancy, and (4) turbulent eddy transport. Rutschmann et al. (1986) determined in addition an effect of non-hydrostatic pressure gradient on the bubble rise velocity, and Kobus (1991) mentioned an influence of bubble clouds on the single bubble transport process. Figure 4.20 shows schematically the various forces acting on air bubbles in chute flow.

The bubble transport processes were developed in fluid mechanics, heat transfer, biochemistry, and nuclear engineering (Soo, 1967; Braeske et al., 1997). A basic contribution to the rise velocity of a single bubble in still water is due to Haberman and Morton (1956) (Figure 4.21). Stokes' law of drag on rigid spherical bubbles was improved. The features of the so-called Morton number are described by Pfister and Hager (2014). Comolet (1979) provided a theory based on the drag, the weight, and the buoyant forces. The motion of a single

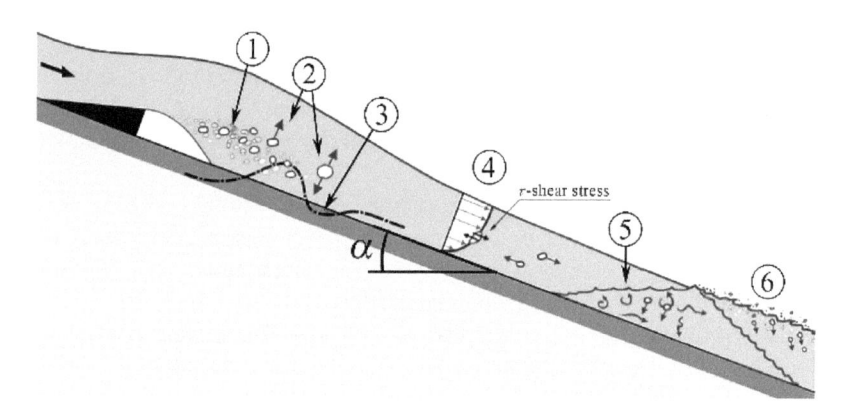

Figure 4.20 Forces acting on air bubbles downstream of chute aerator with effects of ① bubble cloud, ② buoyancy, ③ non-hydrostatic pressure gradient, ④ velocity profile, ⑤ turbulence transport, ⑥ free-surface entrainment (Adapted from Rutschmann *et al.*, 1986)

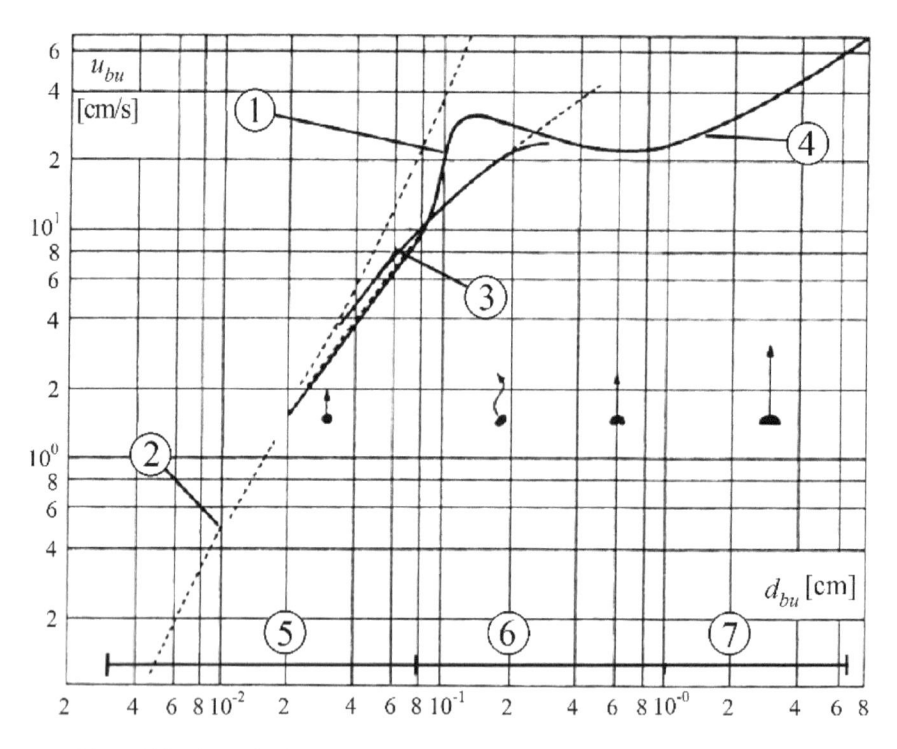

Figure 4.21 Bubble rise velocity u_{bu} in still water versus bubble size d_{bu} with ① experimental data for distilled water, ② Stokes' equation (4.49), ③ Eq. (4.50), ④ Eq. (4.51) for ⑤ spherical, ⑥ ellipsoidal, ⑦ spherical cap regimes (Adapted from Haberman and Morton, 1956)

bubble is often expressed in dimensional terms with the bubble (subscript bu) rise velocity u_{bu} versus the equivalent bubble diameter d_{bu}. The drag resistance coefficient was related to the bubble Reynolds number $R_{bu} = u_{bu}d_{bu}/v$, with v as the kinematic viscosity of water.

Expressions of the dimensional rise velocity are plotted in Figure 4.21. The transport of small rigid spherical bubbles depends on the surface tension. With g as the gravity acceleration, ρ_A density of air, and ρ_w density of water, air bubbles in water with $d_{bu} < 0.068$ mm follow according to Stokes (Falvey, 1980)

$$u_{bu} = \frac{2}{9}\frac{d_{bu}^2 g}{v}\left(1 - \frac{\rho_A}{\rho_w}\right). \tag{4.49}$$

For 0.068 mm $< d_{bu} < 0.80$ mm, the bubble rise velocity is (Comolet, 1979)

$$u_{bu} = \frac{1}{18}\frac{d_{bu}^2 g}{v}\left(1 - \frac{\rho_A}{\rho_w}\right). \tag{4.50}$$

With increasing bubble size, the bubble shape changes from spherical to the oblate spheroid with the bubbles rising along a spiral trajectory (Kobus, 1991). Comolet (1979) observed that both surface tension and buoyancy are relevant; for bubbles with 0.80 mm $\leq d_{bu} < 10$ mm and σ_t as the interfacial surface tension between air and water

$$u_{bu} = \sqrt{0.52 g d_{bu} + 2.14\frac{\sigma_t}{\rho w d_{bu}}}. \tag{4.51}$$

For $d_{bu} \geq 10$ mm, surface tension effects are small so that then simply (Falvey, 1980)

$$u_{bu} = \sqrt{g d_{bu}}. \tag{4.52}$$

Turbulence reduces the terminal bubble rise velocity, because turbulent shear fractures larger bubbles into smaller. Falvey (1980) estimates the critical bubble size $d_{bu(95)}$ for which less than 95% of the air is contained in this bubble diameter from the loss of energy according to Hinze (1975). According to Rouse (1950), the critical bubble size $d_{bu(95)}$ depends on the energy slope S_E and the average flow velocity u_a as

$$d_{bu(95)} = 0.725\left[\left(\frac{\sigma_t}{\rho_w}\right)^3\left(\frac{1}{g S_E u_a}\right)^2\right]^{\frac{1}{5}}. \tag{4.53}$$

Let u_{bu} be the terminal bubble rise velocity in still water, and $u_{bu,t}$ the terminal rise velocity in turbulent flow. With $u_{bu,t} = au_{bu}$ and a as coefficient versus dimensionless discharge $Q_w^2/\left(gh_w^5\right)$, with Q_w as water discharge and h_w as the water flow depth, only rising bubbles may be considered. For $Q_w^2/\left(gh_w^5\right) > 5$ follows $a \approx 0.25$; air bubbles then rise four times slower than in still water (Chanson and Toombes, 2002a). The presence of finer bubbles is attributed to bubble breakup processes by turbulent shear.

A bubble cloud in stagnant water moves with a considerably larger mean velocity than a single bubble because of the upward current. This effect is counteracted in the same order of magnitude by bubbles displaced laterally along their paths and a less induced vertical water flow (Kobus, 1991). Volkart (1985) explains the reduced rise velocity of bubble clouds by the large number of bubbles in mutual contact and interference, resulting in collisions, deformations, bubble collapse, and bubble coalescence, causing a loss of kinetic energy and a higher drag of air bubbles.

According to Rutschmann *et al.* (1986), the bubble rise velocity in the jet impact zone onto a chute bottom is affected by a non-hydrostatic pressure gradient. The effect of the vertical pressure gradient dp/dz on the bubble rise velocity results on the balance between the upward turbulent diffusion and the downward pressure gradient. Although their equation proposed was dimensionally inhomogeneous, the effect of non-hydrostatic pressure gradient on the bubble rise velocity is retained (Figure 4.20). Wood (1988) predicted similarly the bubble rise velocity under non-hydrostatic pressure conditions. Using the results of Rutschmann *et al.* (1986) and Chanson (1988), he found that the pressure at the impact zone may become much larger than the hydrostatic pressure. This overpressure leads to pressure gradients resulting in massive air detrainment in the impact region. According to Chanson (1988), the non-hydrostatic pressure gradient is responsible for the bubble rise velocity, i.e. with u_{bu} as the bubble rise velocity in stagnant water (Kramer and Hager, 2005)

$$u_{bu,t}^2 = u_{bu}^2 \frac{1}{\rho_w g} \frac{dp}{dz}.$$

(4.54)

The bubble transport direction depends on the algebraic sign of the pressure gradient; bubbles thus rise toward the free surface for $dp/dz > 0$, whereas they are captured by the flow if $dp/dz < 0$. According to Volkart (1985), the bubble rise velocity in turbulent flow may be 10 times slower than in still water.

Air concentration contours

The local air concentration, flow velocity, and air bubble size were determined by interpolated contour lines and air concentration profiles. The mixture-flow depth h_{90} and the value C_{a90} were defined in terms of the standard air concentration $C = 90\%$. The non-dimensional coordinates $X_{90u} = x/h_{90u}$ and $Z_{90u} = z/h_{90u}$ were adopted, with the uniform (subscript u) mixture-flow depth h_{90u} measured at the chute end. Figure 4.22 shows air concentration contours for flow in a horizontal chute in which the approach flow Froude number $F_o = (Q_w + Q_A)/(gb^2 h_o^3)^{1/2}$, the approach flow depth h_o and surface roughness were kept constant. The local air bubble rise velocity was indirectly determined using the gradient $dz(C)/dx[C]$, thereby assuming that air bubbles do not rise with a steady velocity u_{bu} but depend on local air concentration and turbulence. The contour lines involved $C = 5\%$, $C = 1\%$, and $C = 0.1\%$, respectively. They rise steadily along the chute model until levelling off at a point determined individually for each run. The detrainment gradient is defined between the lower starting point of the contour line and the kink with respect to the selected air concentration as

$$\det{}_{(C)} = \frac{dZ_{90u}(C)}{dX_{90u}(C)} = \frac{\dfrac{C}{X_{90u}}}{\dfrac{C}{Z_{90u}}}.$$

(4.55)

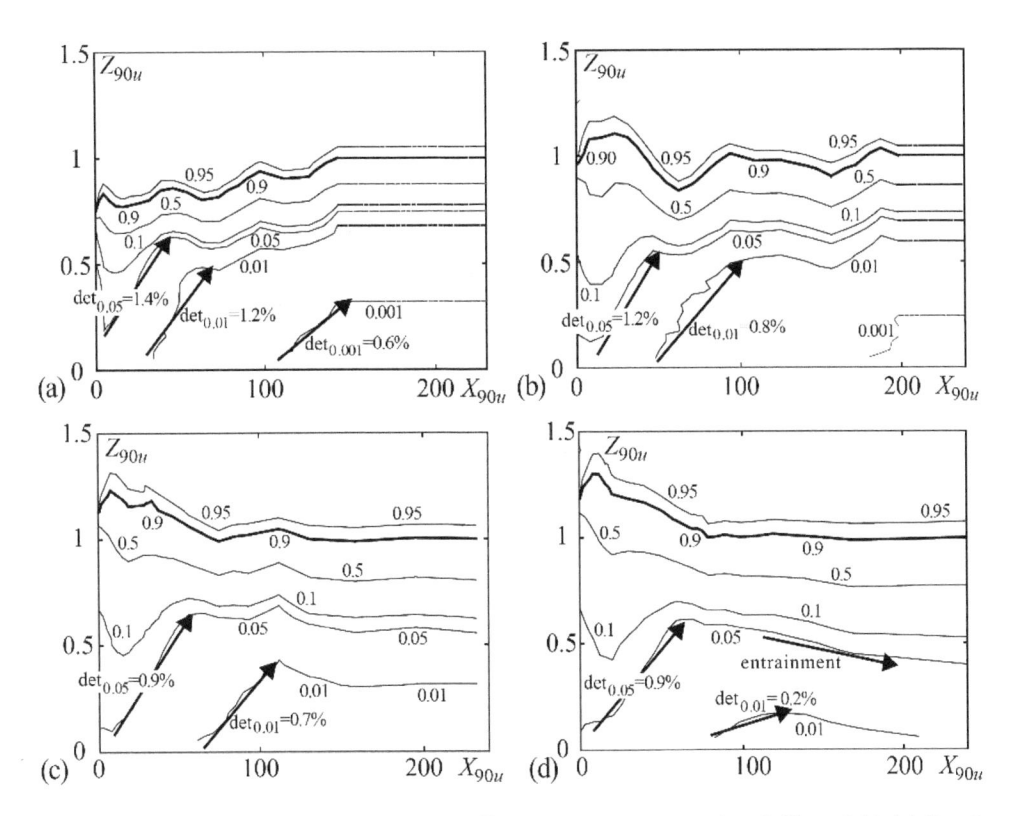

Figure 4.22 Air concentration contours for F_o = 10.3, h_o = 0.06 m, $\beta = Q_A/Q_w$ = 0.21 (a) S_o = 0, h_{90u} = 0.087 m, (b) S_o = 10%, h_{90u} = 0.064 m, (c) S_o = 30%, h_{90u} = 0.054 m, (d) S_o = 50%, h_{90u} = 0.051 m (Kramer, 2004)

Small air concentrations of C = 1.0% and C = 0.1% were correlated with small bubble rise velocities or detrainment gradients, therefore. Downstream of the kink, the contour lines either remain parallel to the chute bottom or decrease almost linearly. The 5% contour line appears to be unaffected by the bottom slope whereas the contour $det_{0.01}$ flattens for steeper chutes. For F_o = 10.3, the air concentration falls below C = 0.1% only at the chute end. Kramer (2004) details tests for other F_o.

The bubble size d_{bu} [mm] is relevant for the bubble rise velocity u_{bu} (Figure 4.21). The fiber-optical measuring system detected the local bubble sauter cord size by means of flow velocity, frequency, and time-averaged air concentration using a statistic algorithm, relying on all three parameters, except for flow regions where bubbles remained undetected. Figure 4.23a shows the distribution of bubble sizes d_{bu} [mm] over the non-dimensional chute length X_{90u} and the flow depth Z_{90u} using contour plots. The bold line represents the mixture-flow depth h_{90}. In the lower flow zone $0 \leq Z_{90u} \leq 0.2$, the bubble size is small with $d_{bu} <$ 1.0 mm due to high turbulence. The zone $0.2 \leq Z_{90u} \leq 0.7$ includes 1.0 mm $\leq d_{bu} \leq$ 5.0 mm, similar to the size quoted by e.g. Haberman and Morton (1956). For $Z_{90u} \geq 0.7$, air bubbles increase rapidly in size due to air pockets ejected from the flow into the atmosphere. This finding is supported by Figure 4.23b, 4.23c, showing bubble size profiles $d_{bu}/d_{bu90}(Z_{90u})$ for

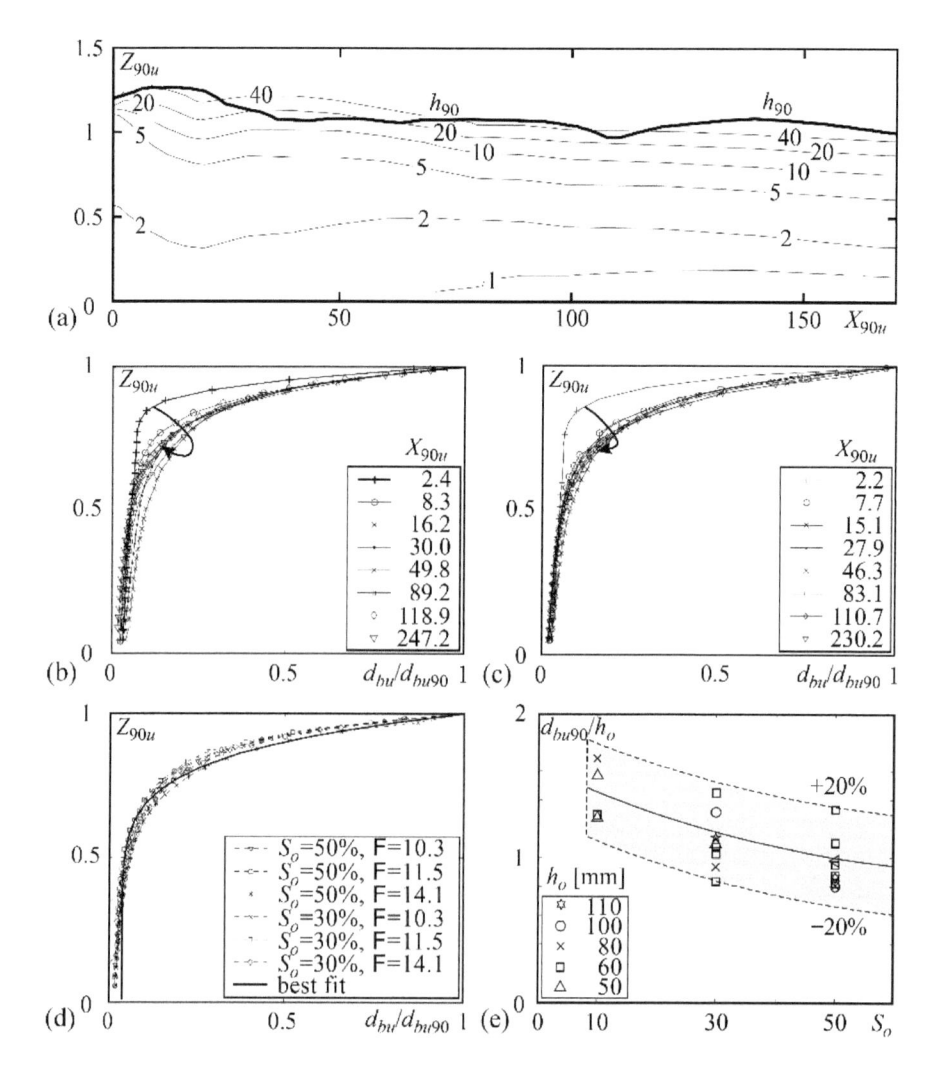

Figure 4.23 Bubble size distribution $d_{bu}/d_{bu90}(Z_{90u})$ for $h_o = 0.06$ m, $\beta = 0.21$, $S_o = 50\%$, (a) contour lines $d_{bu}(Z_{90u})$ [mm] for $F_o = 10.3$, $h_{90u} = 0.051$ m, (b) bubble size profiles d_{bu}/d_{bu90} against Z_{90u} for $F_o = 10.3$, $h_{90u} = 0.051$ m and (c) $F_o = 14.1$, $h_{90u} = 0.054$ m, (d) typical uniform profiles, (——) best fit Eq. (4.56), (e) bubble size in top region d_{bu90}/h_o versus chute slope S_o for various h_o (Kramer and Hager, 2005)

selected locations X_{90u}. The first profile of Figure 4.23b at $X_{90u} = 2.4$ ($x = 0.122$ m) was influenced by the jet-box presence. All others increase up to $X_{90u} = 49.8$ ($x = 2.540$ m) prior to decreasing to uniform flow conditions. Figure 4.23c shows a similar trend for a higher approach flow Froude number; higher turbulence accelerates the bubble development so that uniform bubble distribution is attained faster. Measurements in the uniform flow region suggest for the characteristic bubble size distribution (Figure 4.23d)

$$d_{bu}/d_{bu90u} = \left(Z_{90u} - 0.005\right)^7 + 0.04, \text{ for } S_o \geq 30\%. \tag{4.56}$$

The bubble size in the uniform flow region d_{bu90u} divided by the approach flow depth h_o against the chute slope S_o is shown in Figure 4.23e. The bubble size is close to the approach flow depth $0.8 \leq d_{bu90u}/h_o \leq 1.6$, with a decreasing trend for steeper chute slopes. Air pockets instead of air bubbles thus represent the mixing region with an air concentration of $C = 90\%$ (Falvey (1980). Figure 4.23 suggests that a constant bubble size distribution in chute flow is unrealistic due to highly turbulent flow. Assuming a constant bubble rise velocity as proposed in the literature appears to be questionable in view of u_{bu} as a function of bubble size, therefore.

Effects of Froude number and chute slope

The increase of the air concentration isoline is defined by Eq. (4.55), with typical examples shown in Figure 4.22. Air detrainment gradients $\det_{(C)}$ are influenced by the local air concentration C and the local air bubble rise velocity u_{bu}. Rise velocities are low in regions of small air bubbles $d_{bu} \leq 2$ mm (Figure 4.23).

Figure 4.24 shows air detrainment gradients $\det_{(C)}$ against F_o for the four chute slopes investigated relative to $C = 5\%$, 1%, and 0.1%. Linear trend lines correlate with the gradients $\det_{(0.05)}$, $\det_{(0.01)}$, and $\det_{(0.001)}$. The trend of $\det_{(0.05)}$ is relatively weak, although a larger air concentration seems to give smaller gradients for high Froude numbers, whereas the detrainment is larger for smaller F_o. Despite the data scatter, a relation between the approach flow Froude

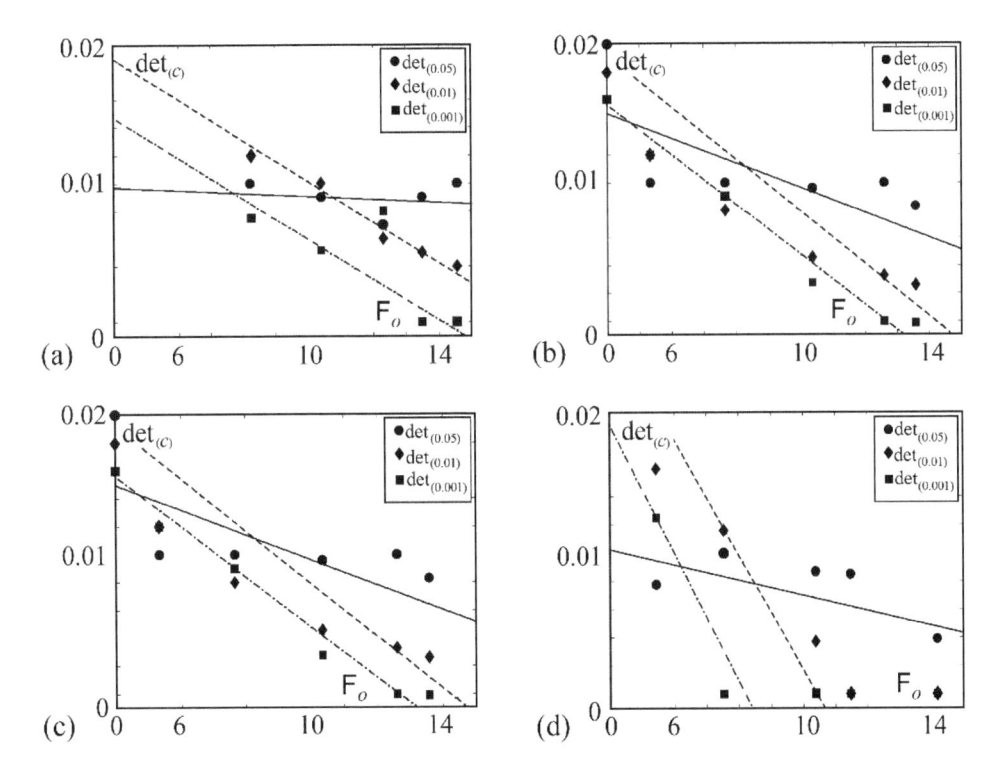

Figure 4.24 Air detrainment gradient $\det_{(C)}$ versus F_o with linear trend lines (——) $\det_{(0.05)}$, (- - -) $\det_{(0.01)}$ and (- · - · -) $\det_{(0.001)}$ for S_o = (a) 0, (b) 10%, (c) 30%, (d) 50% (Kramer and Hager, 2005)

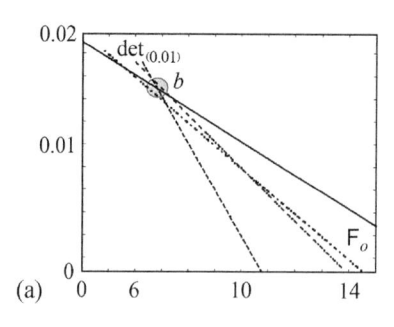 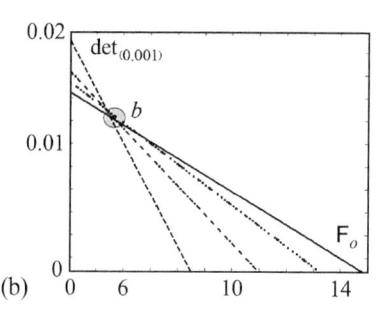

Figure 4.25 Air detrainment gradients $\det_{(C)}$ versus F_o for $S_o = (—)\ 0$, $(\cdots)\ 10\%$, $(-\cdot-\cdot-)\ 30\%$, $(- - -)\ 50\%$, (a) $\det_{(0.01)}$, (b) $\det_{(0.001)}$ (Kramer and Hager, 2005)

number F_o and the air detrainment gradient $\det_{(C)}$ follows for the $C = 1\%$ and $C = 0.1\%$ contour lines. The $\det_{(0.001)}$ trend line is below $\det_{(0.01)}$, indicating that a small air presence remains longer in the flow.

For gradients $\det_{(0.01)}$ and $\det_{(0.001)}$ equal to zero, the air concentration $C = 1\%$ or $C = 0.1\%$ never fell below these values along the entire chute. The air concentration thus has a minimum of $C = 0.1\%$ for $F_o \geq 14.5$ in the horizontal chute (Figure 4.24a) and $F_o \geq 8.3$ for $S_o = 50\%$ (Figure 4.24d). Likewise, the air concentration always falls below $C = 1\%$ in horizontal chutes but never below this value for $F_o \geq 10.7$ and $S_o = 50\%$. The data for chute slopes $10\% \leq S_o \leq 30\%$ are intermediate to the extremes investigated. Both the Froude number *and* the chute slope influence the air detrainment gradients.

Figure 4.25 shows trend lines for $C = 1\%$ and 0.1%; these for $\det_{(0.05)}$ were dropped because the large data scatter and irrelevance in practice. The trend line for the horizontal chute $S_o = 0$ is at the top with the largest detrainment gradients for $\det_{(0.01)}$ and $\det_{(0.001)}$, respectively. For larger chute slopes, the trend lines steepen, indicating smaller air detrainment gradients for a given F_o. The gradient $\det_{(C)}$ is thus a function of air concentration, approach flow Froude number and chute slope with $a = f(S_o)$ as

$$\det_{(C)} = -aF_o + b. \tag{4.57}$$

The common point $b = aF_b + \det_b$ depends on the air concentration C, with $F_{b(C)}$ as the approach flow Froude number F_o at point b and $\det_{b(C)}$ as the corresponding detrainment gradient $\det_{(C)}$. The data follow with the two parameters $\det_{b(0.01)} = 0.015$ and $F_{b(0.01)} = 6.9$ for $C = 1\%$, and $\det_{b(0.001)} = 0.012$ and $F_{b(0.001)} = 5.8$ for $C = 0.1\%$

$$\det_{(C)} = -a(F_o - F_{b(C)}) + \det_{b(C)}. \tag{4.58}$$

The variable a for the four chute slopes, as well as for $\det_{(0.01)}$ and $\det_{(0.001)}$, is almost independent of the air concentration, which only influences the location of point b in Figure 4.25, whereas slope a is a function of chute slope only. For $0 \leq \sin\alpha \leq 0.5$, the best fit is (Kramer and Hager, 2005)

$$\det_{(C)} = -1.14 \cdot 10^{-3}[1 + 0.5\tan(2.7\sin a)] \cdot (F_o - F_{b(C)}) + \det_{b(C)}. \tag{4.59}$$

This relation is based on a large number of test conditions in the turbulent smooth regime. It may serve for an estimate of the flow features in prototype chutes where experimentation is difficult.

Dimensionless air bubble rise velocity

Although the bubble size d_{bu} is determined above, a relation between it and the bubble rise velocity u_{bu} appears to be uncertain in highly turbulent flow. Moreover, the bubble size varies with turbulence and air concentration. A novel approach between the air concentration C and the bubble rise velocity u_{bu} was considered, therefore. It is based on the air detrainment gradient $det_{(C)}$ according to Eq. (4.55) describing the increase of air concentration contours with distance, thereby accounting indirectly for the average approach flow velocity $u_o = (Q_w + Q_A)/(bh_o)$. The time-averaged bubble rise velocity with respect to the air concentration C is thus defined as

$$u_{bu(C)} = det_{(C)}u_o. \tag{4.60}$$

The air concentration contour lines $C = 1\%$ and $C = 0.1\%$ referring to $det_{(0.01)}$ and $det_{(0.001)}$ were used to determine the bubble rise velocity $u_{bu(0.01)}$ and $u_{bu(0.001)}$, where $u_{bu(0.01)}$ refers to the rise velocity for the $C = 1\%$ air concentration contour line. For $det_{(0.01)} = dZ_{90u}/dX_{90u} = 0.32/40 = 0.008$ relating to the $C = 1\%$ contour and $u_o = 9$ m/s, the bubble rise velocity is $u_{bu(0.01)} = det_{(0.01)}u_o = 0.008 \times 9 = 0.072$ m/s.

Chanson (1989b) assumed a constant bubble rise velocity for air detrainment processes, namely $u_{bu} = 0.16$ m/s for scale models and $u_{bu} = 0.40$ m/s for prototypes. However, Figure 4.24 shows that the air concentration contours related to small air concentrations decay slower than these relating to larger air concentrations because: (1) small air concentration contours refer to small air bubbles whose rise velocity is reduced according to Figure 4.21, or (2) turbulent and diffusive effects act as a random generator and distribute air bubbles in regions with smaller air concentrations. Figure 4.26 shows the non-dimensional bubble rise velocity

$$U_{bu} = \frac{u_{bu(C)}}{\sqrt{gh_o}} \tag{4.61}$$

relating to $C = 1\%$ versus the streamwise average Froude number

$$F_{o-P} = \frac{F_o + F_P}{2}. \tag{4.62}$$

This involves the arithmetic average of Froude numbers of the approach flow F_o and at the kink of the contour line F_P thereby incorporating the average flow behavior in the region where the air bubbles actually rise, including the drawdown effect and thus indirectly the increase of turbulence and shear. The data of Figure 4.26 refer to the four investigated chute slopes. Despite data scatter, U_{bu} reduces as F_{o-P} increases. Further, the bubble rise velocity falls to zero if $F_{o-P} \geq 13.7$, indicating that these associated with $C = 1\%$ are then kept at the chute bottom. With F_o as Froude number where $U_{bu} = 0$, a constant slope of $t_b = 0.029$ results for $F_0 = 13.7$ and $C = 1\%$ so that (Figure 4.26)

$$U_{bu} = -t_b(F_{o-P} - F_0), \quad F_{o-P} \geq 7.5. \tag{4.63}$$

Equation (4.63) is limited to $U_{bu} \leq 0.15$ for $C = 1\%$, and to $U_{bu} \leq 0.10$ for $C = 0.1\%$.

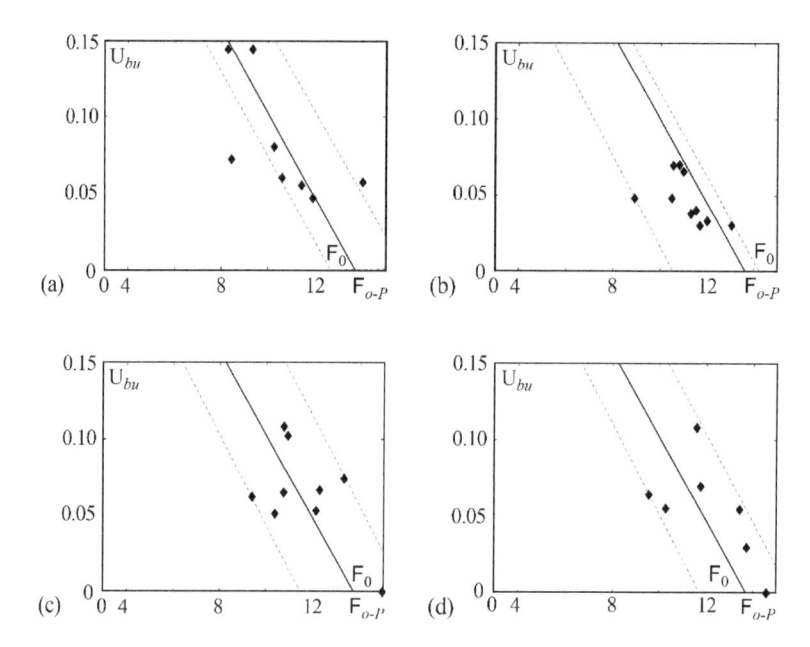

Figure 4.26 Non-dimensional bubble rise velocity U_{bu} for $C = 1.0\%$ versus streamwise Froude number F_{o-P} (——) trend line with $t_b = 0.029$ from Eq. (4.63) for $S_o =$ (a) 0, (b) 10%, (c) 30%, (d) 50% (Kramer and Hager, 2005)

From a physical point of view, the non-dimensional bubble rise velocity reduces for steeper chute slopes due to increasing turbulence so that slope t_b of the trend lines in Figure 4.26 increases for steeper chutes. Note that F_{o-P} accounts for the effect of slope, incorporating the gross flow features with a reducing flow depth $h(x)$ in the streamwise direction. Note also the simplicity of this analysis for the bubble rise velocity in highly turbulent flow. For $C = 0.1\%$, a similar trend was observed, satisfying also $t_b = 0.029$ but $F_0 = 11.8$ instead of 13.7 for $C = 1\%$; the bubble rise velocity is thus zero for a higher streamwise Froude number.

The non-dimensional bubble rise velocity U_{bu} depends only on the streamwise Froude number F_{o-P}. From a summary plot of all dimensional velocities u_{bu} in [m/s] according to Eq. (4.60) including all chute slopes against F_o results for $C = 1\%$ in 0.02 m/s $\leq u_{bu} \leq$ 0.12 m/s, and in $0 \leq u_{bu} \leq 0.08$ m/s for $C = 0.1\%$. The bubble rise velocity in model chute flow based on Froude similitude is thus of the order $u_{bu} \approx 0.08$ m/s.

4.3 Uniform-aerated chute flow

4.3.1 Experimental approach

Based on the data set of Straub and Anderson (1960) for chutes of bottom angles α from 7.5° to 75°, Wood (1985) plotted generalized concentration profiles. With h_{90} as flow depth over the chute bottom up to 90% air concentration, and y as the coordinate perpendicular to it,

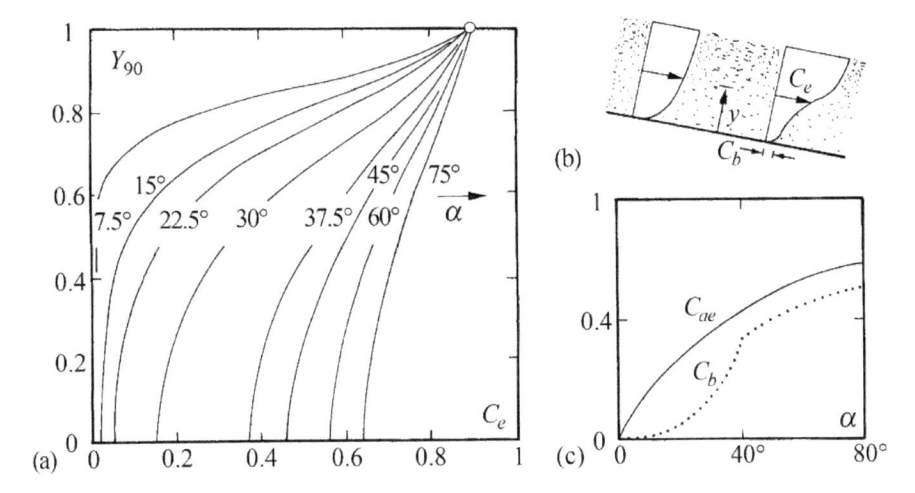

Figure 4.27 (a) Concentration profiles $C_e(Y_{90})$ for uniform-aerated chute flow and various chute inclinations α (Wood, 1985), (b) definition plot, (c) bottom and average air concentrations C_b and C_{ae} according to Hager (1991)

the equilibrium (subscript e) concentration distribution C_e varies only with non-dimensional depth $Y_{90} = y/h_{90}$, and the chute inclination α. Figure 4.27a shows the corresponding curves currently adopted for concrete prototype chutes.

Hager (1991) reanalyzed the data thereby proposing for the bottom (subscript b) concentration $C_b = C_e(Y_{90} = 0)$ when overlooking boundary layer effects (Figure 4.27b, 4.27c)

$$C_b = 1.25\alpha^3, \qquad 0 < \alpha < 0.70\,(40°), \tag{4.64}$$

$$C_b = 0.65\sin\alpha, \qquad 0.70 < \alpha < 1.40\,(80°). \tag{4.65}$$

All concentration curves are scaled with $c = (C_e - C_b)/(0.90 - C_b)$ varying only with $y^* = 2(1 - Y_{90})(\sin\alpha)^{-1/2}$. Figure 4.28b shows the universal concentration profile $c(y^*)$.

The cross-sectional average (subscript a) equilibrium concentration C_{ae} results from integrating the concentration curves over the flow thickness as (Figure 4.27c)

$$C_{ae} = 0.75(\sin\alpha)^{0.75}. \tag{4.66}$$

Accordingly, the concentration increases almost linearly for usual chute slopes. A generalized approach is presented in Eq. (4.68) below.

The Darcy-Weisbach friction factor f for chute flow depends on the relative sand-roughness height and the aspect ratio. If f_w is the factor for water flow and f_e the corresponding factor for equilibrium air-water flow, then the ratio f_e/f_w varies with the average air concentration C_{ae}, as shown in Figure 4.28a. For $C_{ae} < 20\%$, i.e. for $\alpha < 10°$, this effect is negligible from Eq. (4.68), whereas the resistance of an air-water mixture is smaller than of the corresponding black-water flow for $\alpha \geq 10°$.

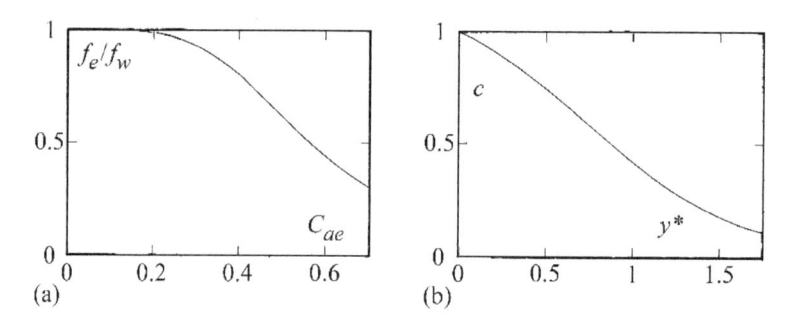

Figure 4.28 Uniform or equilibrium aerated chute flow (a) ratio of friction coefficients $f_e/f_w(C_{ae})$ for aerated and non-aerated flows, (b) universal concentration profile $c(y^*)$ (Hager, 1991)

The increase of flow depth (flow bulking) due to air-water mixture presence varies mainly with the chute roughness characteristics $\eta_R = [h_w \sin^3\alpha/(n_M^2 g^3)]^{1/4}$, with h_w as the corresponding black-water flow depth and $n_M = 1/K$ as the Manning, or the inverse as Strickler friction coefficient. Let $Y_{99} = h_{99}/h_w$ be the relative flow depth at 99% air concentration, and b the chute width, then (ICOLD, 1992)

$$Y_{99} = 1 + \frac{1.35\eta_R}{1 + 2(h_w/b)}. \tag{4.67}$$

The freeboard $(Y_{99} - 1)$ due to the presence of air-water flow depends on η_R, i.e. mainly on the chute slope and on the wall roughness, but less on the black-water flow depth.

Equation (4.66) is based on Straub and Anderson's data collected in a channel of definite boundary roughness of roughness value n_M. For shallow chute flow $(h_w/b \ll 1)$, Eq. (4.67) reduces to a generalization of Eq. (4.68) involving both n_M and h_w besides the chute angle α

$$C_{ae} = 1.35\eta_R. \tag{4.68}$$

For gradually varied flow, the effects of backwater curve and air entrainment may be separated, at least to the lowest order of approximation. From ICOLD (1992), the backwater curve of the black-water flow is computed first, to which the effect of air-water flow is added according to Eq. (4.67). An alternative is due to Falvey (1990).

Generalized approach

PROPERTIES OF AIR-WATER MIXTURES

Aerated water flows faster than black-water under otherwise identical conditions (e.g. Eddington, 1970; Bogdevich *et al.*, 1977; Madavan *et al.*, 1984). Drag reduction was analyzed by Lumley (1977), Legner (1984), Marié (1987), Ebadian *et al.* (2003), Ferrante and Elghobashi (2004), Lu *et al.* (2005), or Ortiz-Villafuerte and Hassan (2006). In hydraulic engineering, Straub and Anderson (1960) furnished the most-cited data set collected in a 15 m long and 0.46 m wide channel. Both smooth and rough boundary conditions were studied, but equilibrium conditions resulted only for the latter. In contrast to turbulence research, the causes of self-aeration received limited attention, resulting in uncertainty concerning the chute design: neither model tests nor computations revealed that the velocities at a ski jump would be notably higher than predicted, thus causing scour at the opposite river bank. A reliable velocity prediction at a chute end requires physical understanding of the phenomenon.

Based on the Weber number, Hinze (1955) states a criterion for bubble break-up involving dynamic pressure fluctuations, the black-water density ρ_w, and fluid surface tension σ_t. Relating these fluctuations to turbulence production P_t, the limit (subscript L) break-up diameter d_L is

$$d_L = \frac{0.725}{P_t^{0.4}} \left(\frac{\sigma_t}{\rho_w} \right)^{0.6}. \tag{4.69}$$

Falvey (1980) replaced P_t by gS_EV (Rouse, 1950), with S_E as the energy line gradient and V as the mean velocity, respectively, under nearly uniform flow. The complete expression for P_t under steady 2D uniform flow reads, however, $\rho_w P_t = \tau_t(\partial \bar{u}/\partial y)$ (e.g. Schlichting and Gersten, 1996), with τ_t as the turbulent shear stress and $\bar{u}$ as the time-averaged velocity. Inserting into Eq. (4.69) gives with l_p as Prandtl's mixing length and $\tau_0(1 - \eta_0)$ with as τ_0 as the local shear stress, and η_0 as the relative wall distance, the expression (Zünd, 2008)

$$\frac{d_L}{l_P} = \frac{0.725}{r_P^{0.2}} \cdot \left(\frac{\sigma_t/l_P}{\tau_0(1 - \eta_0)} \right)^{0.6}. \tag{4.70}$$

Thomas (1976) obtained an expression for bubble coalescence, yet his criterion applies only to two individual bubbles. Thus, Eq. (4.70) just states a sufficient condition.

Wood (1991) concludes from a dimensional analysis that the concentration profile $C(z)$ normalized with the flow depth only depends on chute slope S_o as $C(z)/h_{90} = f(S_o)$, applying also to the bottom air concentration, yet excluding the effect of surface tension. Straub and Anderson (1960) described the air distribution by a diffusion equation. Using the cosine of the chute slope to reduce the bubble rise velocity, the result is with ε_c as the exchange coefficient and u_{bu} as the bubble rise velocity $\varepsilon_c(dC/dz) = Cu_{bu}\cos\alpha$, or normalized with the approach flow black-water depth h_o,

$$\frac{(1/C)dC}{d(z/h_o)} = \frac{h_o u_b \cos\alpha}{\varepsilon_c}. \tag{4.71}$$

Note that $h_o u_{bu}/\varepsilon_c$ reduces with increasing shear, so that the air distribution normalized with h_o does not only depend on the chute slope, but also on h_o. Another consequence is that S_o has an effect on the bubble rise velocity u_{bu}, but that the energy line gradient S_E acts on the turbulence and hence on both the exchange and the bubble size and thus on their rise velocity. For developing flows, S_o and S_E have thus to be distinguished.

RESISTANCE WITHOUT AIR PRESENCE

Consider aerated chute flow on a slope $S_o = \sin\alpha$. Given an air distribution $C = C(z)$, the theoretical black-water depth $h_w = h_w(z)$ is defined as $h_w = \int(1 - C)dz'$ integrated from $z' = 0$

to ∞. Here, the integration variable z' is distinguished from the lower integration limit z. The total shear at this level is $\tau = \int \rho g S_o dz' = \int \rho_w (1 - C) g S_o dz' = \rho_w g h_o S_o$. Note that τ does not depend on the air distribution above z. The clear-water depth at the invert is $h_o = h(z = 0)$. With $u_\tau = (\tau_0/\rho_w)^{1/2}$ as the shear velocity and subscript '0' referring to $z = 0$,

$$\tau_0 = \rho_w g h_o S_o = \rho_w u_\tau^2. \tag{4.72}$$

Introducing the 'air-cleaned' non-dimensional wall distance $\eta_0 = 1 - h(z)/h_o$, the total shear at elevation z is $\tau = \tau(z) = (1 - \eta_0)\tau_0$.

For steady uniform flow, the 2D Reynolds equations state as shear balance

$$\tau = \tau_v + \tau_t = \mu(\partial \overline{u} / \partial z) - \rho \overline{u'v'}. \tag{4.73}$$

Here, μ is dynamic viscosity, and u' and v' are the velocity fluctuations parallel and perpendicular to the chute invert, respectively. Since their signs are opposite, the product $-\overline{u'v'}$ is positive. In turbulent flow the viscous portion τ_v is small, so that

$$\tau = \tau_t = (1-\eta)\rho u_\tau^2 = -\overline{u'v'}. \tag{4.74}$$

Let the inner variables be $\tau^+ = \tau/\tau_0$, $u^+ = u/u_\tau$, and $z^+ = z \cdot u_\tau/v$ with v/u_τ as the hydraulic length scale. Dividing Eq. (4.75) by $\tau_0 = \rho u_\tau^2$ leads to the dimensionless expression

$$\tau^+ = \tau_v^+ + \tau_t^+ = \frac{\partial \overline{u}^+}{\partial y^+} - \frac{\overline{u'v'}}{u_\tau^2}. \tag{4.75}$$

MIXING LENGTH CONCEPTS

Consider non-aerated chute flow (Figure 4.29): A portion of water at level z_1 flowing at velocity $\overline{u}_1$ is seized by a turbulent fluctuation v' perpendicular to the flow. It is transported

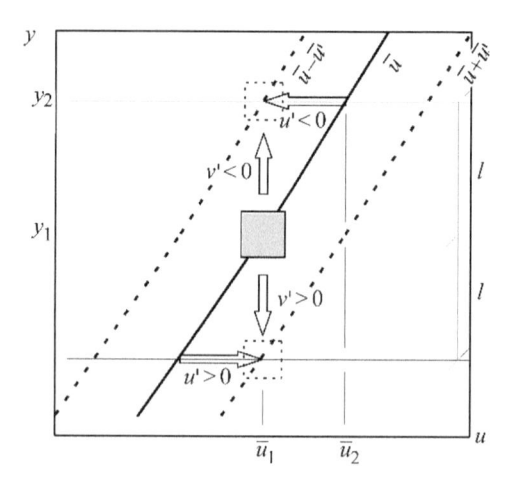

Figure 4.29 Prandtl's mixing length concept according to Bezzola (2002)

across the flow and brought to a higher level $z_2 = z_1 + l$, where it is mixed with the surrounding fluid. However, the 'correct' velocity there is $\bar{u}_2$, resulting in the difference $u' = \bar{u}_2 - \bar{u}_1$ noted by a local observer as fluctuation in flow direction. This model leads to Prandtl's basic relation for the velocity gradient $\partial\bar{u}/\partial z = \bar{u}'/l$. It reveals that the Reynolds stresses describe a momentum exchange: A momentum $\rho u'$ per unit volume is transported across the flow at a velocity $|v'|$. Assuming $\overline{u;'} \approx -\overline{v;'}$ and $\overline{\bar{u}'\bar{v}'} \approx \overline{u'v'}$, Eq. (4.74) yields

$$\tau_t = \rho l^2 \left(\frac{\partial\bar{u}}{\partial z}\right)^2. \tag{4.76}$$

The vertical non-dimensional velocity distribution for a smooth boundary is with B_i as integration constant (Schlichting and Gersten, 1997)

$$\bar{u}/u_\tau = \bar{u}^+ = \frac{1}{\kappa}\ln z^+ + B_i = \frac{1}{\kappa}\ln\frac{zu_\tau}{v} + B_i. \tag{4.77}$$

In turn, for rough boundary conditions, z scales with the roughness length k rather than with v/u_τ, resulting in

$$\bar{u}/u_\tau = \frac{1}{\kappa}\ln\frac{z}{k} + B_i. \tag{4.78}$$

The standard constants are $\kappa = 0.41$, and $B_i = 5$ and 8 for hydraulic smooth and rough conditions, respectively. The logarithmic velocity profile starts at $(z; \bar{u}/u_\tau) = (k; B_i)$. A second integration across the flow depth yields the specific discharge q, from which the mean flow velocity V for smooth and rough chute boundaries results in

$$V/u_\tau = \frac{1}{\kappa}\left(\ln h_o^+ - 1\right) + B_i, \tag{4.79a}$$

$$V/u_\tau = \frac{1}{\kappa}\left(\ln\frac{h_o}{k} - 1\right) + B_i. \tag{4.79b}$$

Many models were proposed to adequately simulate the mixing layer of turbulent flow, including e.g. Escudier (1966). Bezzola (2002) refers to studies concerning the near-wall region (e.g. Dittrich and Koll, 1997). They noted the existence of a layer above the reference level, where the measured data of $-\rho\,\overline{u'v'}$ remain constant and no longer cover the total shear $\tau_0(1 - \eta_0)$. It is referred to as the roughness sublayer of thickness z_R, of the order of k. The data also indicate that turbulent conditions persist far in between the roughness elements. Bezzola's model is based on two hypotheses:

1   Roughness elements produce secondary flow structures. Below z_R these take over a portion of the total shear layer. The region affected by this effect is referred to as the roughness sublayer, whose size is of the order of the roughness layer itself. Within this layer the turbulent shear stress is reduced to $\tau_0(1 - \eta_R)$, with $\eta_R = 1 - h(y_R)/h_o$.
2   As in the log-region, the velocity gradient in the turbulent sublayer is also solely determined by the reduced turbulent shear stress. Secondary flow structures only have an indirect effect by reducing the turbulent shear stress, resulting in a reduction of the turbulent shear stress close to the roughness elements.

Numerous drag reduction mechanisms are available in turbulence research. They are divided based on (1) turbulence suppression or modification and (2) changes in the mean fluid properties involving density and viscosity. Flow over smooth surfaces is usually considered, focusing on the near-wall region, thereby discussing the increase of viscosity. Therefore, the viscous forces are thought to dominate inertia over a wide range, thus 'thickening' the viscous sublayer (e.g. Gyr and Bewersdorff, 1995). Below, only fluid property mechanisms, referred to as the density effect and the viscosity effect, are considered.

Wood et al. (1983) postulate that Darcy's friction coefficient f be a function of the air concentration, whereas the fundamental property is that f depends exclusively on the wall properties. The wall shear is related to velocity according to Darcy-Weisbach as (Schlichting and Gersten, 1996)

$$\tau_0 = \left(f/8\right)\rho g V^2, \text{ or } \left(8/f\right)^{1/2} = \left(V/u_\tau\right). \tag{4.80}$$

For a fixed pressure gradient, the wall-shear stress τ_0 in pipe flow follows this law, stating that the invariability of f does not contradict observations, since the velocity V increases or decreases inversely to the root of fluid density ρ.

The question in relation with chute flow is: how does air act upon the momentum exchange model in free surface flow? The mixing length itself is assumed not to undergo changes from pure water to air-water mixture flows. However, the particle density is reduced, and so will be the momentum. From Prandtl's theory, the clear-water density on the right-hand-side of Eq. (4.76) has to be replaced by (Zünd, 2008)

$$\tau_t = \left(1-C\right)\rho_w l_P^2 \left(\frac{\partial \bar{u}}{\partial z}\right)^2. \tag{4.81}$$

The velocity gradient then becomes

$$\frac{\partial \bar{u}}{\partial z} = \sqrt{\frac{\tau_t}{\left(1-C\right)\rho_w}} \frac{1}{l_P} = \sqrt{\frac{1-\eta_0}{1-C}} \frac{u_\tau}{l_P}. \tag{4.82}$$

To obtain the velocity profile, this gradient has to be integrated across the flow depth by accounting for the local values of η_0, C, and l_P. A second integration yields the cross-sectional average velocity. Zünd (2008) describes the density effect; for the reference example selected, approximate agreement with the data of Straub and Anderson (1960) is noted. The density effect is further described to capture the drag reduction effect in air-water mixture chute flows. As to the viscous effect, Zünd (2008) concludes that for clear-water depths >0.10 m, hence including prototype chutes, no viscous effect is to be expected. For scale models, however, this effect cannot be fully excluded if the model flow depth is below this limit.

To obtain the velocity profile of chute flows, the gradient expressed in Eq. (4.82) is considered. The mixing length as proposed by Bezzola (2002) was admitted, whereas the bubble diameter given in Eq. (4.70) was adopted. The computational results aimed at a prediction of the ratio of air-water mixture velocity to the average black-water velocity V_m/V_a versus the average cross-sectional air concentration C_a. The agreement between the two is almost perfect for all data of Straub and Anderson (1960), thus including the relevant slope and roughness effects in hydraulic practice.

The major conclusions concerning dam engineering include:

1   Effect of flow aeration interacts with the development of the air distribution due to gravity and turbulence in high-speed free-surface flow. A physically based analysis of the air concentration profile does not exist yet, so that the concentration profile $C(z)$ follows from Figure 4.27a, based on Straub and Anderson (1960). A check computation based on the absence of any friction often determines the upper velocity limit.
2   Bubble size plays no major role in chute hydraulics, but may be different at rougher conditions. The viscous effect is limited to small flow depths as in scale models.

4.4 Chute aerator

4.4.1 Motivation and historical development

Floods are evacuated from reservoirs by the spillway chute, so it is an important dam safety element. The high-speed flow along the chute is due to the significant flood discharge conveyed, and due to the high-level difference between the reservoir surface and the tailwater. Spillways are a key element for most dams. As heads and unit discharges increased significantly after 1900, the maximum chute velocity often is above 20–25 m/s, resulting in cavitation damage. Engineers were perplexed at first, but realized that similar damages had almost put an end to the career of the turbine designer Victor Kaplan (1876–1934). Cavitation was investigated from the 19th century, including Lord Rayleigh (1842–1919) and Osborne Reynolds (1842–1912).

The physical cavitation phenomenon corresponds to pitting of a solid surface along with its consecutive loss of integrity and mechanical resistance. The damage was sometimes so intense that the spillway as a dam safety element was lost and the chute no longer operational. Bradley (1945) investigated devices to artificially entrain air into the chute flow to reduce its cavitation potential at the Boulder Dam spillway tunnels. To avoid damages due to cavitation, a smooth lining and 'means to supply air into the spillway flow' were proposed to act as a cushion between the high-speed flow and the concrete lining. Bradley tested sills, orifices, and short tubes. Deflectors provided the best results in terms of prototype feasibility and economy (Figure 4.30 ③, ⑬).

A first symposium on cavitation damage of chutes was held in 1945. Several papers were presented, authored by:

Vennard (1947) relating to the nature of cavitation with examples on diffusors, blades, turbines, and pumps. Two techniques were proposed to investigate cavitation processes, namely model tests in vacuum tanks to determine cavitation damage directly on the model, and near atmospheric conditions in the standard environment to measure dynamic pressure fields, to indirectly determine the cavitation potential.

Harrold (1947) reports of cavitation problems of the Madden Dam tunnel spillway in the Panama Canal due to a sharp inlet in the conduit roof. Other problems associated with cavitation were found at gate slots and piers of stilling basins. Model tests on chutes revealed cavitation damage at a break in chute alignment. Additional tests related to the 'pitting' of concrete surfaces, yet without definite results.

Warnock (1947) reviews problems of cavitation damage at the Arizona Spillway of Boulder Dam, and the spillway pier faces of Parker Dam, both on the Colorado River. The surface of new concrete in the eroded area was carefully finished, producing a sound, continuous and

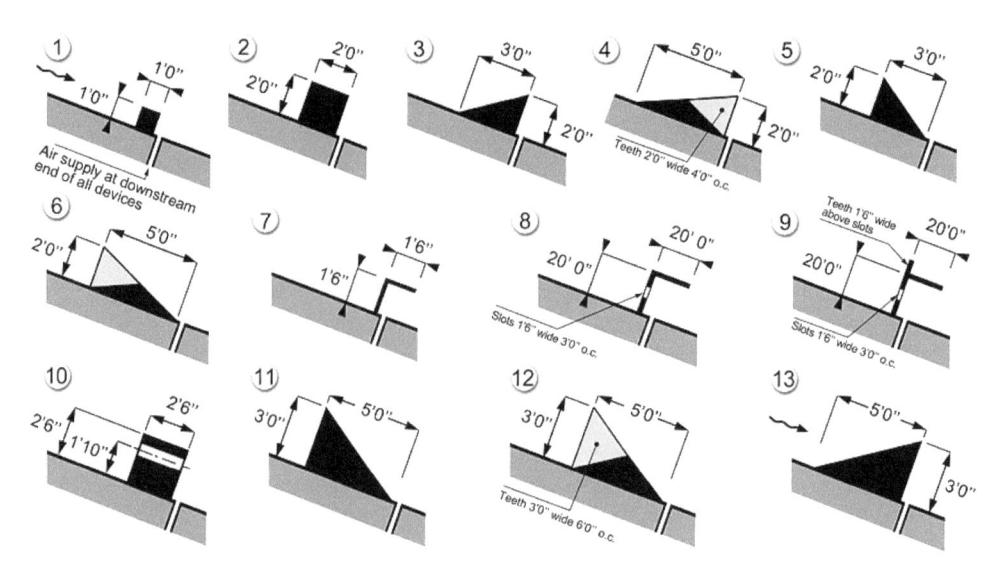

Figure 4.30 Aerator devices tested by Bradley (1945), sketch modified by Hager and Pfister (2009)

uninterrupted boundary. This surface treatment only received attention, yet no provisions were proposed to reduce the origins of cavitation damage.

Hickox (1947) considers cavitation damage at the sluice gates of the Norris Dam. Repairs of the sluice inlet included grinding and welding of a metal liner, using pitting-resistant materials. It was also proposed to modify the inlet curvature until pressures were sufficiently high to avoid cavitation.

Peterka (1953) presents a benchmark paper on the relation between cavitation pitting and the ambient air concentration. He stated that the presence of air bubbles has a positive effect on the reduction of cavitation damage. Yet, up to 1953, no information was available on the amount of air required for practical purposes. Peterka thus conducted a model study with concrete samples subjected to a high-speed air-water mixture of well-defined speed and average air concentration. The test data were plotted as either the weight loss of concrete or the cavitation index, both versus the average air concentration. For an air concentration of some 2%, the loss of weight was small but increased significantly if less air was supplied to the flow. The corresponding value in terms of cavitation index is $\sigma = 0.12$. Peterka thus demonstrated that a small amount of air concentration between 1% and 2% may suffice to reduce cavitation damage significantly, as expressed by the loss of concrete.

Semenkov and Lentyaev (1973) considered chute cavitation, supporting Peterka's (1953) proposal that flow aeration substantially reduces the cavitation risk. Upstream of the natural self-aerated reach, technical aerators have to be added to artificially supply air close to the chute invert. The loss of air downstream of such an aerator was specified to some 0.5% per meter chute length, allowing for the definition of the position of a subsequent aerator. The latter is necessary to increase air concentration again to a level at which the cavitation risk is removed. Galperin *et al.* (1979) published a noteworthy book on cavitation in hydraulic structures, probably the first entirely on this subject in dam hydraulics. Both laboratory and field tests are reported, including high dams and tunnel spillways. The causes of various damages are analyzed and means to prevent them are discussed.

Between 1960 and 1980, several chute damages occurred (Falvey, 1990), destroying their concrete invert surfaces. Pinto *et al.* (1982) initiated works on chute aerators that became a standard. They focused on the limit flow velocity above which aeration starts, on the air discharge entrained by an aerator, and on the aerator spacing to maintain a sufficient protection level. Deflectors and invert drops (also referred to as offsets) were considered, or combinations of the two. Hydraulic model data related the air entrainment coefficient $\beta = Q_A/Q$ at the aerator, with Q_A as the air (subscript A) discharge entrained via the air duct and Q as the water discharge, respectively, to the relative jet length from take-off at the aerator lip to jet impact onto the chute invert. The associated equation was tested with model and prototype data, in which also the sub-pressure in the jet cavity was varied. These works and those of Pinto and Neidert (1983) and Pinto (1984) led to the design basis for chute aerators.

Extensive experiments were then conducted on invert chute aerators, including Volkart and Chervet (1983), proposing various aerator designs. Further studies were conducted by Galperin et al. (1971), Tan (1984), Low (1986), Rutschmann (1988), Chanson (1988), Rutschmann and Hager (1990, a, b), and Skripalle (1994). They describe the efficiency of chute aerators based on the local air entrainment coefficient $\beta = Q_A/Q$. However, these studies do not account for the spatial streamwise air transport and air detrainment, respectively, as β specifies just the local overall air supply to the flow. The streamwise development of the invert (subscript b) concentration C_b and the average (subscript a) air concentration C_a downstream of an aerator was particularly considered by Pfister (2011), and Pfister et al. (2011).

4.4.2 Cavitation potential

According to the phase diagram, water changes phase to vapor either if the ambient temperature increases over 100°C or if the ambient pressure decreases below vapor pressure. For a temperature of 10°C, for instance, vapor pressure is 1.23 kPa (Table 4.1). If the absolute pressure reaches these values locally, then water changes its phase and includes numerous small vapor bubbles. These are transported with the flow toward zones of pressures above vapor pressure and there implode, marking the return to the liquid phase. The implosion is accompanied by an acoustic crackle and a deformation of the imploding bubble to a 'pancake'-type geometry with a distinctive micro-jet toward the concrete surface.

Table 4.1 Characteristic parameters of water

Temperature [°C]	Density [kg/m³]	Vapor pressure [kPa]	Vapor pressure h_v [m WC]
0	999.9	0.61	0.06
5	1000.0	0.87	0.09
10	999.7	1.23	0.13
15	999.1	1.70	0.17
20	998.2	2.33	0.24
25	997.0	3.16	0.32
30	995.7	4.23	0.43
35	994.0	5.62	0.58
40	992.2	7.38	0.76

Isselin *et al.* (1998) estimate the pressure peak associated with the implosion process to 210 MPa, whereas Peterka (1953) mentions a value of 760 MPa. Suslick (1989) and Tomita and Shima (1990) suggest a maximum velocity of the micro-jet of 134 and 111 m/s, respectively. Note that the distinctive pressure peak and the hyper-fast micro-jet per bubble harm a concrete surface or other types of lining material such as steel. The countless 'hammering' over hours can remove the superficial material layer locally, thereby exposing the aggregates. This increases the damage, since a rough surface represents a higher potential for pressure fluctuations and thus phase changes. The phenomenon is thus self-intensifying, similar to hydro-abrasion.

Besides vapor pressure, the water quality influences the phase change process from liquid to vapor. The cohesion of pure water is higher than of water of high suspended sediment load. Particles represent breaking points and allow for a phase change under reduced vapor pressure (Keller, 1988). High-speed flows with their related turbulence level are particularly prone to cavitation damage. A high volatility in velocity is associated to pressure fluctuations on the chute invert, and thus to phase changes. Apart from this feature, the concrete surface pattern is essential. Micro-roughness affects turbulence, and macro roughness or irregularities generate flow separation zones from the surface (Figure 4.31). These zones include at its start low-pressure areas below vapor pressure, and pressures above vapor pressure at the flow reattachment. The exposure duration of the surface to cavitation affect in addition the damage extent (Falvey, 1990).

In 1930, Jakob Ackeret (1898–1981) wrote one of the first PhD theses on the formation of cavitation bubbles in water, introducing thereby the cavitation index. It allows for estimating the cavitation potential (Falvey, 1990), based on average flow parameters as velocity V [m/s], hydrostatic pressure depth $h_p = h \cos\alpha$ [m], chute angle α from horizontal, flow depth h [m] perpendicular to chute invert, atmospheric pressure head h_{atm} [m], and vapor pressure head h_v [m] as (Tables 4.1 and 4.2)

$$\sigma = \frac{h_p + h_{atm} - h_v}{V^2 / 2g}. \tag{4.83}$$

(a) Positive step   (b) Offset   (c) Convex invert

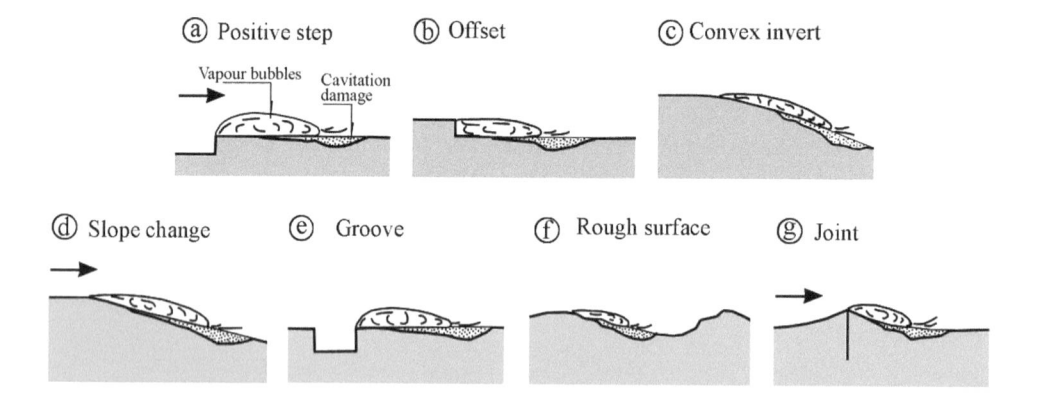

(d) Slope change   (e) Groove   (f) Rough surface   (g) Joint

Figure 4.31 Frequent irregularities on chutes generating flow separation zones with increased cavitation risk (Modified from Ball, 1976)

Table 4.2 International Standard Atmosphere (ISO, 1975)

Altitude [m a.s.l.]	*Atmospheric pressure* [kPa]	*Atmospheric pressure* h_{atm} [m WC]	*Temperature* [°C]
0	101.3	10.34	15.0
500	95.5	9.73	11.8
1000	89.9	9.16	8.5
1500	84.6	8.62	5.3
2000	79.5	8.11	2.0
2500	74.7	7.61	−1.2
3000	70.1	7.15	−4.5

The atmospheric pressure head h_{atm} is based on the International Standard Atmosphere (ISA; ISO, 1975). At mean sea level, the standard gives a pressure of 101.33 kPa (10.34 m WC), a temperature of 15°C, and an air density of 1.225 kg/m³. Further atmospheric pressure values relative to the elevation above mean sea level are given in Table 4.2.

The cavitation index is computed along the spillway. The streamwise flow depth and velocity required in Eq. (4.83) follow from drawdown curves. Falvey (1990) states a critical (subscript c) value of $\sigma_c = 0.2$ below which cavitation damage is probable for smooth concrete chutes. This value is conservative and independent of spillway operation duration, given that the latter is unknown at design stage.

4.4.3 Cavitation protection

Cavitation damage is in principal limited if the:

1   Discharge and chute alignment are designed so that no low-pressure zones exist
2   Chute surface is smooth without irregularities and/or is made particularly resistant to endure the pitting phenomenon
3   Flow is aerated along chute invert.

Limited flow velocity V and a sufficient pressure head h_p lead to a higher cavitation index σ. A manipulation of the latter parameters is possible to a limited extent with design measures and an adapted operation regime as the regulated spillway. The reservoir management depends on several factors, however, so that cavitation protection would represent another restriction. Constructional measures include flip buckets. Those remove the flow from the chute before critical velocities occur, generating a free jet. Alternatively, σ is controlled via the longitudinal chute invert profile, i.e. the invert pressure head h_p is increased by providing a concave chute alignment (Falvey, 1990). This is partially feasible on spillways at the air face of gravity and arch gravity dams.

Cavitation damage often occurs shortly downstream of construction joints or voids in the concrete, and near irregular surfaces due to negative local invert kinks (Figure 4.31). For high flow velocities, even a small surface roughness of few millimeters is sufficient to generate a cavitation damage (Drewes, 1988). Bradley (1945) already recommended smooth concrete surfaces for the Hoover Dam spillway. Russell and Sheehan (1974) investigated concrete elements of different surface characteristics and found that the cavitation damage extent was larger and more intense for a relatively rough surface. In practice, acceptable tolerances are difficult to respect. Schrader (1983) mentions a best feasible geometrical accuracy of 3 mm per 3 m chute length. Minor (1988) states a few millimeters as an acceptable irregularity to avoid cavitation.

If cavitation cannot be avoided, then its damaging effect has to be limited, e.g. by resistant surfaces. As to concrete resistance against cavitation damage, the bending tensile strength and the fractural energy are relevant (Jacobs *et al.*, 2001). Russell and Sheehan (1974) correlate the concrete cylinder strength with its resistance against cavitation. According to Schrader (1983), 0.5–1.5 volume percent of steel fibers have a positive effect. The concrete resistance against cavitation damage increased roughly by a factor of three if the invert layer was reinforced by fibers. A surface treatment with epoxy or polymer products can also have a positive effect. However, questions relating to interconnection, material aging, and variable temperature deformation should be addressed. Steel lining is known to be efficient, but is limited to bottom outlets and pressure shafts, but not applied on chutes due to high cost. Note that any overlaying resistant surfaces are prone to dynamic uplift due to the high-velocity flow if not carefully bonded or anchored to the underlying chute concrete.

Flow aeration by chute aerators of reaches with cavitation indices below the critical value is the most reliable and economic measure to protect chutes from cavitation damage. Modern spillways are therefore usually equipped with chute aerators (Vischer and Hager, 1998; Khatsuria, 2004). Air bubbles are 'elastic' and thus damp pressure fluctuations, particularly at locations of local pressure reduction below vapor pressure. Further, they absorb pressure peaks associated with imploding vapor pressures. Air thus significantly reduces the formation of cavitation damage and damps its effects. Volkart (1988) suggests that air further disperses the local effects of cavitation. According to the experiments of Peterka (1953) and Rasmussen (1956), some 1% to 2% of average air concentration is required to limit the damaging effect of cavitation. All sources refer to average air concentrations but do not specify the value at the boundary surface where the damage is generated. Air is entrained either via self-entrainment or technically by means of chute bottom aerators.

The flow close to the aerator is divided into four zones as shown in Figure 4.32 (Volkart and Rutschmann, 1984):

1. Approach flow usually with surface aeration
2. Transition zone deflected by aerator
3. Bottom aeration downstream offset by aeration device
4. Bottom de-aeration with gradually varied aerated flow.

A highly turbulent spray develops in the cavity downstream of the aerator due to sub-atmospheric pressure Δp. The air is supplied from the atmosphere by an air intake and supply channels. The air is mixed in the chute flow mainly along the lower nappe, from the offtake at the ramp to the jet impact point P, along with a significant bottom pressure increase (Figure 4.32). An aerator increases the local turbulence that enhances the nappe air entrainment (Ervine *et al.*, 1995).

According to Vischer *et al.* (1982), a chute aerator should include either a *ramp*, an *offset*, or even a combination of the two. Grooves are to be avoided due to aerator submergence. An aerator should provide a nearly uniform air distribution over the chute width. Figure 4.33 shows two types of chute aerators without and with an air manifold. The latter applies for wide chutes resulting in a nearly perfect air distribution with an almost plane chute flow free of shock waves.

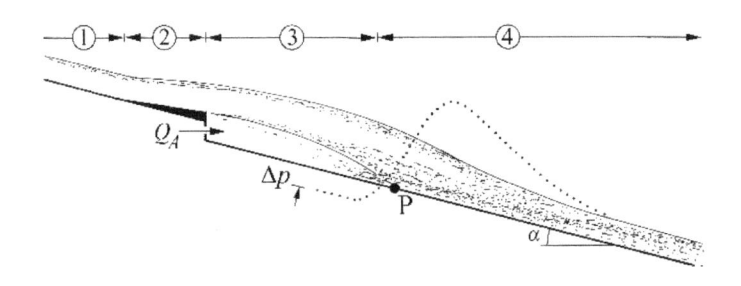

Figure 4.32 Flow zones in vicinity of chute aerator with ① approach flow region, ② chute deflector with air supply Q_A, ③ flow aeration reach, ④ air detrainment region, (. . .) bottom pressure curve

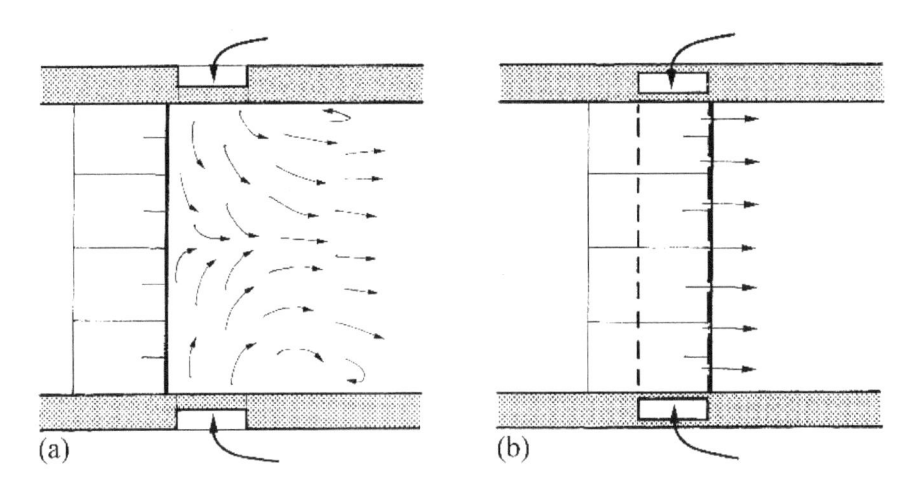

Figure 4.33 Air supply for chute aeration (a) without, (b) with distribution duct (Vischer *et al.*, 1982)

4.4.4 Aerator geometry and air supply system

Aerators have to satisfy the following requirements:

- Entrainment of sufficient air into the flow under variable discharge, to guarantee a reliable value for cavitation protection near the chute invert along a certain reach;
- No generation of significant sub-pressures in air cavity (4.4.12) to avoid cavitation risk;
- Autonomous and self-entraining operation mode without energy supply;
- Drained air supply duct unable to choke by chute water, or block due to ice or snow;
- Simple and economic construction;
- Exclusion of shock waves or a transverse heterogenic unit discharge distribution.

All chute aerators locally separate the high-speed chute flow from the invert, producing a short free jet with an air cavity below connected via the air duct to the atmosphere. The jet disintegrates due to flow turbulence generated upstream, resulting from the turbulent boundary layer emerging along the chute and eventually from a deflector. The disintegration of the lower jet trajectory is supported by gravity, pulling down turbulent cells ejected from the jet, and by the abrupt pressure drop at the aerator take-off lip. The pressure drop is even amplified by deflector presence. The jet surface thus becomes 'rough' and spreads, thereby promoting flow aeration (Figures 4.34 and 4.35, Pfister and Hager, 2009, 2012). The average air concentration increases with the jet length. For offset aerators, the jet length is reduced, as is the level of jet disintegration (Figure 4.35 as compared with Figure 4.34). At the jet reattachment point P on the chute invert (Figure 4.32), a portion of the jet air concentration is transported with the chute flow. The air duct compensates this air removed from the jet cavity.

Various types of chute aerators are applied, depending on the specific needs. The local flow separation from the invert is generated with grooves (slots), deflectors, offsets (drops), or combinations of these elements (Figure 4.36). The most commonly applied aerator is the chute deflector, deviating the flow locally upwards, thereby increasing the turbulence level and the pressure gradient at jet start. Typical deflector heights t measured perpendicularly

Figure 4.34 Flow over model jet deflector combined with offset (Pfister, 2008) (a) black-water approach flow and jet, (b) jet reattachment and aerated downstream flow. Note: air duct is not visible

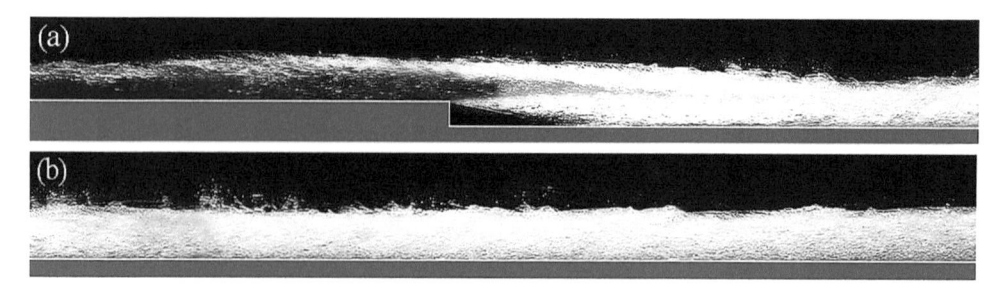

Figure 4.35 Flow over model offset only (Pfister, 2008) (a) black-water approach flow, jet, and reattachment, (b) aerated downstream flow. Note: air duct is not visible

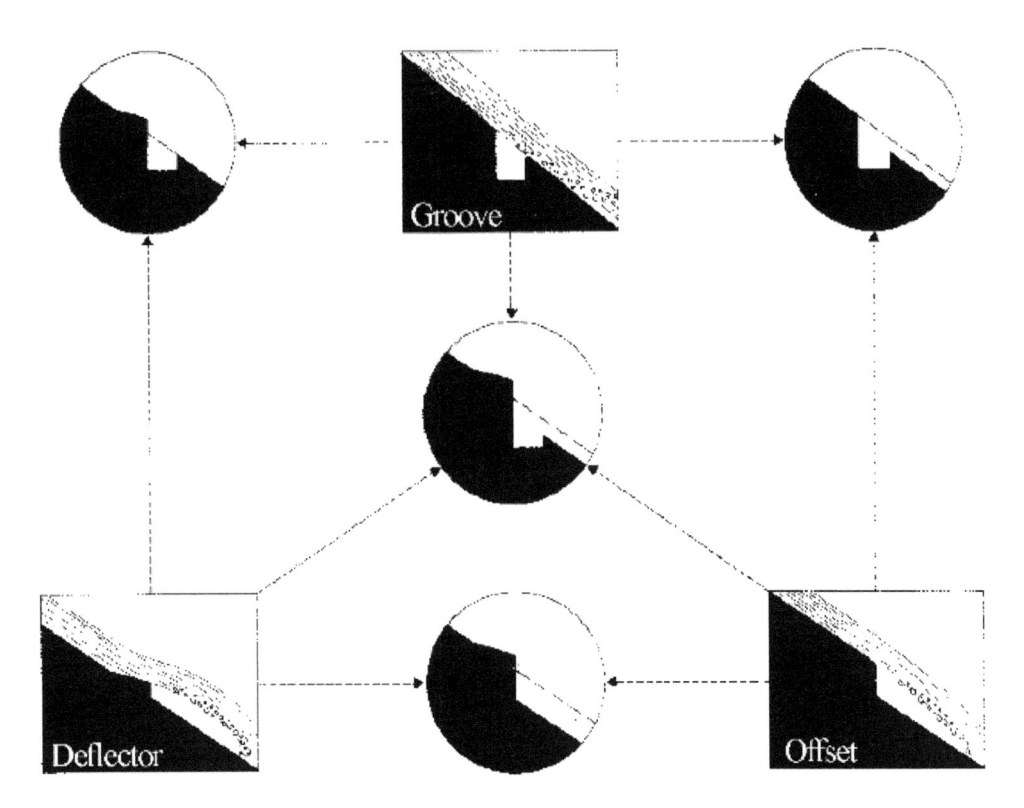

Figure 4.36 Basic chute aerator types with groove, offset, and deflector (Modified from Volkart and Rutschmann, 1984)

to the chute invert (Figure 4.37) range from 0.1 to 0.5 m, and deflector angles ϕ from 5° to 12° (corresponding to 1:10–1:15, Kells and Smith, 1991; Volkart and Rutschmann, 1984). Deflectors entrain relatively high air concentrations, even under comparably small approach flow Froude numbers $F_o = V/(gh_o)^{0.5}$ (Figure 4.37). The deflector efficiency depends on its relative height t/h_o. 'High' deflectors act as a kind of ski jump, with the flow following the deflector at the take-off, whereas 'low' deflectors represent an element by which the flow is hardly deviated. The transverse cross section of the air cavity induced subsequent to deflectors is relatively small, so that the crosswise airflow is restricted. To avoid local cavity subpressures with a limited air entrainment into the flow, the air supply should thus be arranged laterally across both sidewalls as well as through the deflector for wide chutes (4.4.5).

Second, offsets (drops) are used as chute aerators. In contrast to deflectors, they do not include an active flow deviation element generating a longer jet, additional turbulence or an increased pressure gradient. Offsets are therefore less efficient than deflectors and apply mainly for relatively large F_o as occur on terminal chute portions or in bottom outlets. Typical offset heights are of the order of $s < 1.0$ m (Figure 4.37, Kells and Smith, 1991; Volkart and Rutschmann, 1984).

Third, grooves (slots) apply for widths up to 2.5 m and depths from 1 to 2.5 m. Their main restriction relates to the short separation length, and thus a limited jet disintegration level

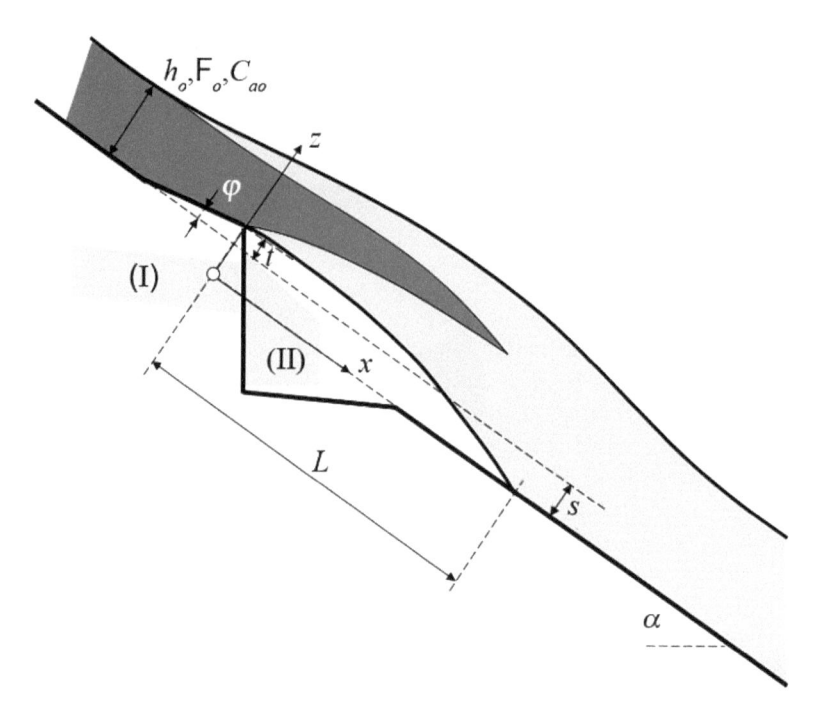

Figure 4.37 Geometry and parameter definition sketch of chute aerator combing offset and deflector, and including an enlarged transverse air cavity surface. (I) represents frontal, transversally distributed connection, (II) lateral connection to atmosphere

combined with small air entrainment as well as the choking risk. The latter is pronounced for small discharges on non-gated spillways, for which the starting discharge enters the groove. A powerful drainage is thus mandatory. Further, the impact load at their downstream nose is significant. To limit these drawbacks, grooves must be combined with small offsets, as applied on bottom outlets.

The optimum aerator configuration depends on the particular application (Pfister 2011; Pfister and Hager 2010a, b; Pfister *et al.* 2011). Generally, a combination of the three above elements seems adequate (Figure 4.37). Note that all aerator types require a specific minimum F_o to entrain air. Fortunately, small F_o typically involve a combination of small V and large h, so that the cavitation index σ is usually not critical along these reaches. The following guidelines may provide a first draft of the aerator geometry:

- Flat chutes of invert angles $\alpha \leq 30°$ (Figure 4.37) and moderate discharges should include a steep and high deflector combined with an offset;
- Steep chutes with invert angles $\alpha > 30°$ and high-speed approach flows can include a small deflector combined with an offset;
- Steep deflectors ($\phi = 11°$) entrain air if $F_o > 4$, flat deflectors ($\phi = 6°$) if $F_o > 5$;
- Offset solution without deflector appears adequate if $F_o > 6$, as occur in bottom outlets;

- Bottom outlets usually include aerators consisting of small offsets ($s < 0.5$ m), often combined with a groove to increase the transverse airflow cross section.

On spillway chutes, cavitation damage typically occurs on the invert but rarely along the sidewalls, so that invert aeration is often sufficient. In bottom outlets, the flow velocity and thus the cavitation potential is much higher than on chutes, so that sidewall aeration is provided, serving also as air duct for the invert aeration. Note that aerators are combined with streamwise slope changes of the chute invert; transitions from flat to steep are prone to cavitation damage, however (Figure 4.31c–d).

The air supply system (also referred to as air duct) connects the air cavity below the jet to the free atmosphere. It assures the compensation of the air entrained into the flow at the chute aerator. The air is supplied into the cavity either (I) frontally, i.e. via openings in the vertical aerator wall across the chute, or (II) laterally, i.e. across openings in the chute sidewalls (Figure 4.37), depending mainly on the chute width and the cavity dimensions. Narrow chutes with 'large' cavities are fed laterally (II), whereas broad chutes and aerators with small cavities apply type (I).

A proper transverse air flux in the cavity is essential to uniformly distribute the air across the cavity, even to the most remote zones. Areas which are not adequately fed with air will provoke local sub-pressures (4.4.9), reducing the aerator efficiency and thus the cavitation protection. In the worst case, choked aerators act as cavitation generators. Choking occurs for under-designed cavity cross sections generating cavity sub-pressures, large energy losses in the air supply system, or from inflowing water. A proper drainage of the air supply system is essential. Snowfall or ice formation can also affect the proper operation of the air supply system (Rutschmann, 1988). As for bottom outlet aerators, the cross section of the air supply system is included in the design process (e.g., Rabben, 1984; Sharma, 1976). On chutes, this is not the case since the supply systems are shorter and sufficient space is generally available.

Obviously, the air supply system should provide a sufficient air discharge to supply the aerator avoiding sub-pressures, but its dimensions should be economic in parallel. A simple criterion to estimate the adequate cross section (for *short* supply systems only!) concerns the limitation of the maximum air velocity, combined with the air intake discharge at the aerator (air entrainment coefficient, 4.4.6) and continuity. According to Pugh and Rhone (1988), the maximum air velocity in the supply system should not exceed some 60–90 m/s. Volkart (1988) gives a limit of 50–60 m/s for standard air systems and 80–100 m/s for hydro-dynamically optimized systems. This limitation results from air compressibility, influencing the airflow unfavorably (Blevins, 1984). *Long* air supply systems should respect this criterion, but the pressure drop due to energy losses should also be restricted. In practice, air velocities in aeration ducts connected to chimneys at the chute sidewall should never exceed 30 m/s to guarantee stable aerator operation without extreme noise development.

Note that all equations below, except for those in 4.4.9, relate to negligible air cavity sub-pressure. For these, the air supply is not restricting or even influencing the aerator operation at any transverse location, so that the latter 'breathes' freely.

4.4.5 Air transport downstream of aerator

The flow on a representative model-aerator consisting of a deflector and an offset is shown in Figure 4.34. Along the approach flow section (Figure 4.34a), the water appears black without air transport, except at the turbulent free surface. The deflector deviates the flow and produces a free jet subjected to the disintegration process. The roughness of the jet surfaces increases considerably in the streamwise direction, while the black-water core thickness reduces. Downstream of a certain location, the jet is fully aerated and appears white (Pfister and Hager, 2009). The lower jet trajectory impacts the chute invert to the left of Figure 4.34b. The impinging flow is partially deflected by the chute invert generating spray at the water surface. The turbulent flow downstream of the reattachment zone is fully aerated, as is typical for two-phase air-water flows. The aerator effect is visible when comparing the flows at the approach flow section (Figure 4.34a, left) and at the model end (Figure 4.34b, right). These phenomena occur less pronounced for an offset aerator without deflector (Figure 4.35).

Figure 4.38 shows the characteristic air concentration C contour plot of two model tests (Pfister, 2008). Note that the streamwise coordinate x starting at the deflector lip is normalized with the jet length L (Figure 4.37) so that $x/L = 1$ at jet impact on the invert. Light gray shades and thus a high degree of aeration appear close to the jet impact location, indicating a considerable air transport at the jet end. Shortly downstream of the reattachment point, the flow close to the chute invert turns quickly into dark gray, because of local air detrainment and the remaining small air concentration. Note also the irregular contour levels: air concentrations below $C = 0.10$ have a smaller contour spacing, as they typically appear near

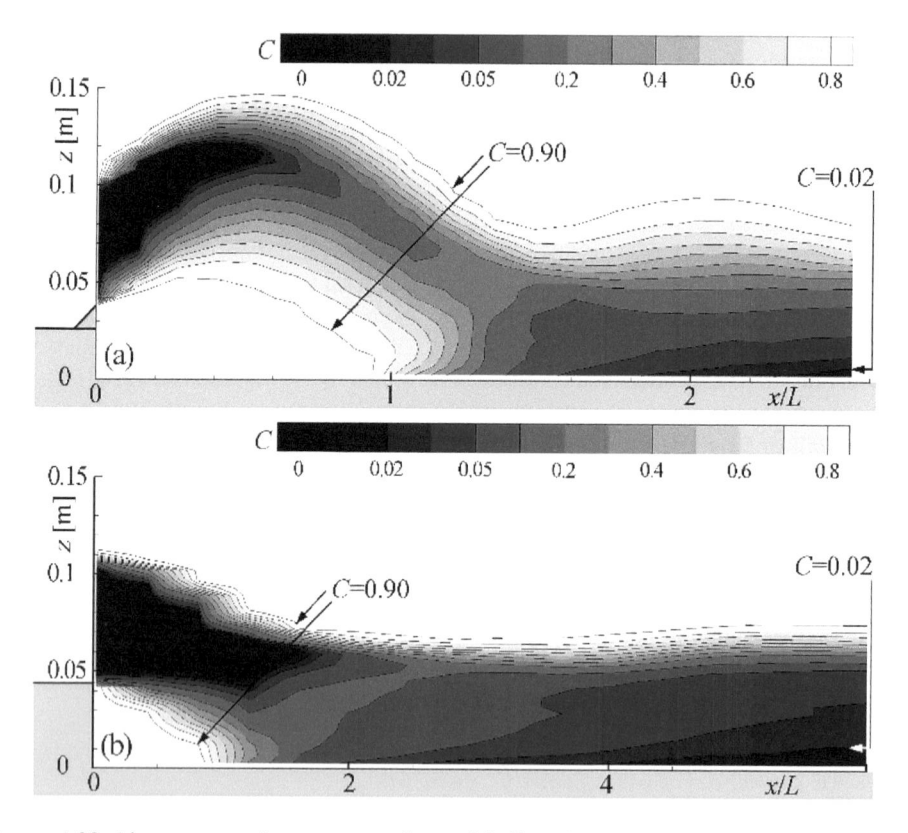

Figure 4.38 Air concentration contour plots with $F_o = 9.1$, $h_o = 66$ mm for (a) combined defector and offset $\varphi = 8.1°$, $t = 13.3$ mm, $s = 25$ mm, $\alpha = 30°$, (b) offset with $\varphi = 0°$, $t = 0$ mm, $s = 44$ mm, $\alpha = 12°$ (Pfister, 2008)

the chute invert where cavitation damage is a concern. The water surfaces defined along the $C = 0.90$ isoline are also shown.

Chanson (1989b) classified the air concentration distribution downstream of chute aerators into six regions: (1) approach flow, (2) transition flow, (3) aerator flow, (4) impact flow, (5) downstream flow, and (6) equilibrium flow. Kramer (2004) studied the air detrainment gradients in the far field of chute aerators (4.2.3). Both found an intensive invert de-aeration near jet reattachment and a significant aerator effect on the average air concentration shortly downstream of the latter.

Pfister and Hager (2010a, b, Pfister 2011, Pfister *et al.* 2011) stated that the air transport downstream of an offset, a deflector or a combination of both follows three general flow zones, namely:

I   Jet zone
II   Reattachment and spray zone
III   Far-field zone.

These zones depend on the relative jet length x/L (Figure 4.37) describing flow reaches of similar hydraulic characteristic. This classification is based on the (a) relative upper (subscript u) water surface $z_u/h_o[x/L]$, (b) average air concentration $C_a[x/L]$, and (c) invert air concentration $C_b[x/L]$ (Figure 4.39).

The average air concentration C_a along the jet ($0 \leq x/L \leq 1$) is defined between the upper z_u (at $C = 0.90$) and the lower (subscript l) z_l (at $C = 0.90$) jet surfaces as

$$C_a = \frac{1}{z_u - z_l} \int_{z_l}^{z_u} C(z)\, \mathrm{d}z. \tag{4.84}$$

Downstream of the reattachment point, C_a is integrated between the chute invert and the flow surface z_u ($C = 0.90$) as

$$C_a = \frac{1}{z_u} \int_0^{z_u} C(z)\, \mathrm{d}z. \tag{4.85}$$

The invert air concentration C_b describes the lowest value close to the chute invert.

The *jet zone* (I) starts at the aerator lip (jet take-off point) $x/L = 0$ (Figure 4.37) and ends at the jet reattachment at $x/L = 1$; it is characterized by the jet disintegration process. The *upper* air-water surface follows the parabolic jet trajectory and remains locally unaffected at $x/L = 1$ because the reattachment process involves the lower jet trajectory (Figure 4.39a). The average air concentration C_a (Figure 4.39b) is close to zero at jet take-off $x/L = 0$. Along the jet, C_a increases to relatively high values of up to $C_a = 0.5$ (Pfister and Hager, 2009). Note that C_a represents mainly entrapped air in this zone, due to jet disintegration initiated by turbulence, aerodynamic effects and relaxation of the velocity and pressure profiles, respectively. The invert air concentration is $C_b = 1$ along the jet zone, because of the air cavity below the jet (Figure 4.39c). Shortly upstream of the jet reattachment point, C_b decreases due to the invert roller near the jet impact and buoyancy further downstream.

The *reattachment and spray zone* (II) is a rapidly varied flow zone including the reach $1 < x/L < 3$, characterized by two local phenomena: (1) flow compression within the reattachment region along $1 < x/L \leq 1.25$, and (2) surface turbulence within the spray region along $1.25 < x/L < 3$. Both phenomena affect the water surface profile $z_u/h_o[x/L]$ (Figure 4.39a), the

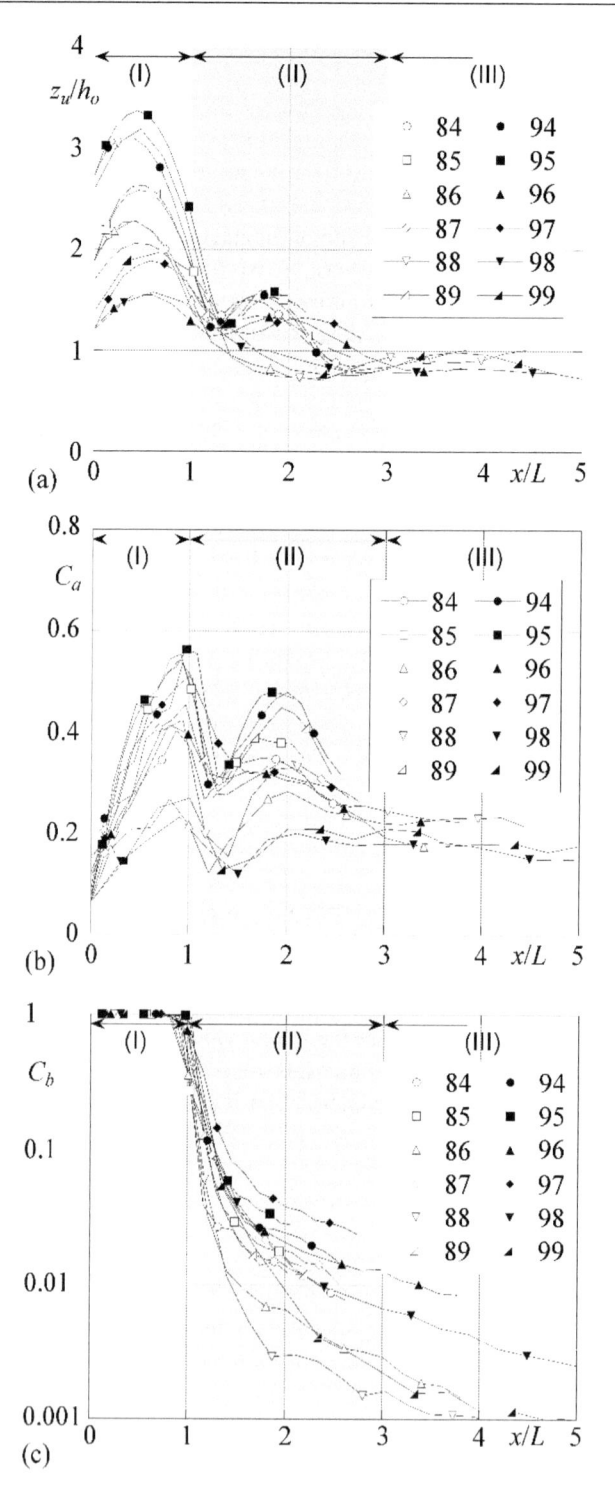

Figure 4.39 Characteristic concentration profiles for representative model tests (nos. in legend, for details see Pfister and Hager, 2010a) for $\alpha = 30°$ (a) upper free surface $z_u/h_o[x/L]$, (b) average air concentration $C_a[x/L]$, (c) semi-logarithmic invert air concentration $C_b[x/L]$. Air transport zones: (I) Jet zone, (II) reattachment and spray zone (in grey), and (III) far-field zone

average air concentration distribution $C_a[x/L]$ (Figure 4.39b), and the invert air concentration $C_b[x/L]$ (Figure 4.39c). The two subzones feature:

- Upper water surface along reattachment region initially following the parabolic jet trajectory, while the lower flow boundary is represented by the chute invert at $z = 0$, so that the flow depth reduces (Figure 4.39a). As water is nearly incompressible, a flow depth reduction is related to flow de-aeration and acceleration, affecting the flow portion close to the chute invert. The intense transport of entrapped air along the lower jet trajectory does thus only partially continue downstream of the jet reattachment point. In contrast, the air transport along the upper water surface remains unaffected within the jet reattachment region. Consequently, C_a decreases along this relatively short flow region (Figure 4.39b): Chanson (1994c) reports a de-aeration rate of up to 80% and correlates this rate with the jet impact angle on the chute invert (Pfister, 2012). Due to the flow reorganization near the chute invert, C_b drastically reduces along the reattachment region (Figure 4.39c).
- The spray region is dominated by flow deflection on the chute invert, so that water particles are deflected and ejected from the flow body generating an intense surface spray whose height can increase to several approach flow depths. The average air concentration C_a also increases as spray presence raises locally the $C = 0.90$ isoline (Figure 4.38a). The spray disappears once the ejected water particles have returned to the flow. The local C_a increase is therefore mainly caused by the air entrapped near the flow surface. Spray influences the flow only near the free surface, and hardly affects the chute invert region. An intense de-aeration along the chute invert up to the end of zone (II) reduces the invert concentrations to around $C_b = 0.01$, a small value for cavitation protection as compared with $0.01 < C_a < 0.06$ (Peterka, 1953). Note that (1) the required values of Peterka are averages of C_a, (2) C_b is always lower than the C_a value due to the typical mixture-flow air-profile, (3) a major portion of C_a is transported in the surface zone and so unavailable for cavitation protection (Pfister, 2008), (4) prototype experiences indicate no serious cavitation damage after aerator addition, and (5) critical reach for cavitation damage starts at $x \cong 3L$. This length corresponds to 40–50 m downstream of prototype aerators, consistent with common aerator spacing (Chanson, 1989a, Kramer, 2004). The described spray is missing in some profiles of Figure 4.39a because these represent tests with short jets generated by small values of F_o and h_o, for which turbulence at the jet reattachment point is insufficient to eject water clusters out of the flow.

The *far-field zone* (III) is a gradually varied flow reach and includes $x/L \geq 3$ up to incipient self-aeration (4.3). This zone is characterized by a continuous development of all hydraulic parameters toward the equilibrium conditions. The free-surface profile follows a drawdown curve (Figure 4.39a), and C_a remains nearly constant (Figure 4.39b). In contrast, C_b further decreases, yet with a reduced detrainment rate (Figure 4.39c) as compared with zone (II). Kramer (2004) describes the far-field zone (4.2.3).

The efficiency of a chute aerator becomes apparent along the flow zones (II) and (III) mainly in terms of the streamwise invert air concentration $C_b[x]$ development. Therefore, the air entrainment ratio $\beta = Q_A/Q$ of an aerator is irrelevant, because the streamwise air detrainment is overlooked. Nevertheless, β serves for designing the air system supply (4.4.4), providing the maximum air discharge to be supplied without generating sub-pressures in the nappe below the jet (4.4.9). Note that the maximum air discharge often occurs for discharges below the design value.

4.4.6 Jet length and air entrainment coefficient

The jet (or cavity) length L between take-off at the aerator lip ($x = 0$, Figure 4.37) and jet reattachment represents a basic parameter to specify the general air transport zones as described in 4.4.5. The jet length L is defined in terms of flow and aerator parameters, or via a classical jet trajectory computation (Steiner et al., 2008; Pfister, 2009, 2012; Chapter 6). For the trajectory approach flow, a corrected take-off angle and velocity are assumed. The jet length L downstream of deflectors, offsets or combinations is approximated as (Pfister and Hager, 2010a, Figure 4.37)

$$\frac{L}{h_o} = 0.77 \mathsf{F}_o (1 + \sin \alpha)^{1.5} \left[\left(\frac{s+t}{h_o} \right)^{0.5} + \mathsf{F}_o \tan \varphi \right]. \tag{4.86}$$

This empirical relation includes three effects: (1) approach flow condition, (2) chute invert angle, and (3) aerator geometry. The latter term indicates that offsets with $\phi = t = 0$ generate a shorter L than deflectors. In Eq. (4.86), h_o can be dropped and does therefore not affect L, as expected for the lower jet trajectory with a take-off point fixed by the aerator lip. Equation (4.86) was validated with model data of Chanson (1988), Rutschmann (1988), and Pfister (2008). Moreover, prototype data sets were considered, namely these from the Foz do Areia Dam, Brazil (Pinto et al., 1982; Wood, 1991); Guri Dam, Venezuela (Marcano and Castillejo, 1984); and Grande Dixence Dam, Switzerland (Rutschmann, 1988). Note that Eq. (4.86) only applies for small cavity sub-pressure of the order of $P < 0.1$ (4.4.9).

The coefficient $\beta = Q_A/Q$ is a parameter describing the local air entrainment capacity of an aerator. Although β represents only the *local* air entrainment without specifying the downstream air concentration development, it applies to design the air supply system (4.4.4). It is known that β is primarily a function of F_o (Rutschmann, 1988; Chanson, 1988; Rutschmann and Hager, 1990a; Kökpinar and Göğüş, 2002) increasing significantly with F_o. A minimum value of $\mathsf{F}_o \geq 4$ is required to initiate air entrainment at an aerator. Offsets without deflectors are less sensitive, while deflector aerators are more sensitive toward changes of F_o. Furthermore, deflectors with relatively large ϕ induce a higher dependence on F_o than do flat deflectors. Particularly steep deflectors entrain thus more air than offsets under otherwise identical conditions. The effects of the parameters h_o, t, s, and α on β were found to be insignificant. According to Pfister and Hager (2010b), the air entrainment coefficient follows (Figure 4.37)

$$\beta = 0.0028 \, \mathsf{F}_o^2 \left[1 + \mathsf{F}_o \tan \alpha \right] - 0.1. \tag{4.87}$$

Equation (4.87) applies to aerators consisting of deflectors ($\phi > 0$), offsets ($\phi = 0$), or combinations. The air duct should consist of a sufficiently large cross section to avoid relevant sub-pressures causing reduced β values.

According to Pfister and Hager (2010b), a minimum F_o is required to generate air entrainment. The minimum values are $\mathsf{F}_o = 6$ for offsets, $\mathsf{F}_o = 5$ for aerators with flat deflectors ($\phi = 5.7°$), and $\mathsf{F}_o = 4$ for these with steep deflectors ($\phi = 11.3°$). Rutschmann (1988) stated similar values. Equation (4.87) suggests that the air entrainment depends primarily on F_o

and only slightly on the aerator geometry. Furthermore, aerators with deflectors are more efficient in terms of air entrainment than those with only offsets.

Equation (4.87) includes model data from Koschitzky (1987), Rutschmann (1988), Skripalle (1994), and Pfister and Hager (2010b). Prototype data of Wood (1991) are also considered, namely these of Foz do Areia Dam (Brazil), Emborcação Dam (Brazil), Amaluza Dam (Ecuador), Colbun Dam (Chile), and Tarbela Dam (Pakistan). Note that Eq. (4.87) applies strictly for insignificant cavity sub-pressure (4.4.9).

4.4.7 Downstream air concentration development

Average air concentration

Downstream of the jet reattachment point, i.e. along $x/L \geq 1$ (Figures 4.37 and 4.39b), the average air concentration C_a is defined according to Eq. (4.87). In the context of cavitation prevention, the values of $C_a(x/L)$ are of interest in the reattachment and the spray zone (II) $(1 < x/L < 3)$, as well as in the far-field zone (III) $(x/L \geq 3)$. The jet zone (I) is irrelevant, because no cavitation occurs if the flow separates from the chute invert.

The values of C_a strongly decrease in the reattachment region $(1 < x/L < 1.25)$ due to flow compression. A local maximum of C_a occurs along the spray region within $1.5 < x/L < 2$, generated by water particles ejected from the flow surface due to jet reattachment. Therefore, the transport of entrapped air near the surface increases, while the air concentration close to the chute invert decreases. These local flow phenomena influence C_a along zone (II), complicating its quantification. However, the value of C_a at $x/L = 3$ (subscript $3L$) is estimated as (Pfister and Hager, 2010b)

$$C_{a(3L)} = 0.008 \frac{L}{h_o} + 0.1. \tag{4.88}$$

The average approach (subscript o) flow air concentration (Figure 4.37) is typically close to $C_{ao} = 0$ under non-aerated approach flow to the aerator. For significantly pre-aerated approach flow, consider Eq. (4.97).

The development of $C_a(x/L)$ along the far-field zone (III) indicates that the values tend to decrease in the streamwise direction for $\alpha = 12°$, remain constant for $\alpha = 30°$, and slightly increase for $\alpha = 50°$. For given α, all respective curves similarly decrease or increase for $x/L > 3$, indicating an insignificant effect of the other parameters. The concentration curves were therefore defined as a function of α with respect to $C_{a(3L)}$ at the end of zone (II), identical with the start of zone (III) according to Eq. (4.88). The values of $C_a(x/L \geq 3)$ follow the trend (Pfister and Hager, 2010b)

$$C_a = C_{a(3L)} + 0.02 \left(\frac{x}{L} - 3 \right) \sin(\alpha - 30°). \tag{4.89}$$

Flows on 'steep' chutes with $\alpha \geq 30°$ have sufficient air transport capacity to maintain the average air concentration, while the flow de-aerates otherwise.

Invert air concentration

The invert, i.e. the bottom air concentration profile $C_b(x/L)$ downstream of the jet reattachment point (Figures 4.37 and 4.39c) was evaluated by Pfister and Hager (2010b). These include zone (II) for (1) offset aerators without deflectors ($s > 0$ m and $\phi = 0$), and (2) deflector aerators with or without offsets ($s \geq 0$ m and $\phi > 0$), while the far-field zone (III) is described independently of the aerator type.

Consider first the invert air concertation along zone (II). Aerators consisting of *offsets* without a deflector are characterized by s only, as $t = \phi = 0$. Combining the individual effects of all parameters influencing C_b along $1 \leq x/L \leq 3$ results in (Pfister and Hager, 2010b, Figure 4.37)

$$C_b = 1 - \tanh\left(4.8\left(\frac{x}{L} - 1\right)^{0.2} \mathsf{F}_o^{-0.4}\left(\frac{h_o}{s + h_o}\right)^{-0.26}(\sin\alpha)^{0.08}\right). \tag{4.90}$$

Equation (4.90) applies for offset aerators without deflector along the reach $1 \leq x/L \leq 3$. The range of validity starts thus at the jet reattachment point where $C_b = 1$. At the end of zone (II), the values of C_b reduce, unless self-aeration occurs (Kramer, 2004). A sensitivity analysis indicates that $C_b(x)$ mainly increases with F_o and h_o, while the effects of s and α are marginal. Parameters F_o, h_o, and partially α are usually given for a certain spillway by design and economic considerations. The efficiency of an offset aerator can therefore only marginally be increased by augmenting s, so that it should exclusively be used for large $\mathsf{F}_o > 6$, as is typical in bottom outlets or at chute ends. Offsets operate generally less effective in terms of invert air entrainment than deflector aerators, because the turbulence-generating deflector is absent (Ervine *et al.*, 1995), and L is relatively short. The invert air concentration $C_{b(3L)}$ at the end of zone (II) is obtained by inserting $x/L = 3$ in Eq. (4.90) as (Pfister and Hager, 2010b)

$$C_{b(3L)} = 1 - \tanh\left(5.5\mathsf{F}_o^{-0.4}\left(\frac{h_o}{s + h_o}\right)^{-0.26}(\sin\alpha)^{0.08}\right). \tag{4.91}$$

Aerator configurations with *deflectors* (and eventually offsets, $s \geq 0$ m) generate more complex two-phase flow features than only offsets. The effect of the dominant parameters affecting the air transport process downstream of these aerators is thus described differently. According to Pfister and Hager (2010b), α influences indirectly the exponent n attributed to F_o, and the exponent m attributed to the effect of ϕ. These two exponents vary with α as

$$n = -1 - (1.5\sin\alpha)^3, \tag{4.92}$$

$$m = 0.5 - (1.5\sin\alpha)^3. \tag{4.93}$$

The negative values of n increase with α, indicating an amplified effect of F_o on the invert air transport particularly for 'steep' chutes, i.e. $\alpha > 30°$. In contrast, the effect of m is dual: the exponent is positive for $\alpha \leq 30°$ and negative for $\alpha > 30°$. The invert air transport is reduced on 'flat' chutes when applying steep deflectors and increases on 'steep' chutes with steep deflectors. This phenomenon results from the jet impact features: Steep deflectors on flat chutes generate a large impact angle of the aerated jet on the chute invert, known to cause significant de-aeration (Chanson, 1988; Ervine *et al.*, 1995; Pfister, 2012). In contrast, the effect of ϕ on the impact angle is reduced on 'steep' chutes while the disintegration process is amplified.

The invert air concentration is again predicted starting from the jet reattachment point at $x/L = 1$ with $C_b = 1$. The values tend to $C_b \rightarrow 0$ at the end of zone (II) if self-aeration is absent.

The general trend of C_b along $1 \leq x/L \leq 3$ for deflector aerators combined with or without offsets is expressed as (Pfister and Hager, 2010b)

$$C_b = 1 - \tanh\left[4.8\left(\frac{x}{L}-1\right)^{0.25} \mathsf{F}_o^{0.25n} (\tan\varphi)^{0.25m}\left(\frac{h_o}{s+h_o}\right)^{-0.05}\right]. \tag{4.94}$$

Considering the combined effects of all relevant parameters on $C_b(x)$ from Eq. (4.94) and from L/h_o (Eq. 4.86) indicates again a distinction between 'flat' ($\alpha \leq 30°$) and 'steep' chutes ($\alpha > 30°$). An increase of the parameters F_o, ϕ, α, h_o, t, and s causes an increase of the invert air concentration. Their effect is small on 'flat' chutes, whereas large values of F_o and ϕ strongly increases the invert air transport on 'steep' chutes.

The approach flow conditions are usually determined by the spillway geometry and the discharge, which cannot be varied for unregulated chutes. Given that the chute angle follows the valley topography and economic considerations, these parameters may generally not be varied to optimize an aerator. Hence, only the aerator geometry characterized by ϕ, s, and t can be selected. An optimization may be necessary for low F_o and should involve ϕ. For $\alpha \leq 30°$, a combination of steep ϕ and high s and t values is effective, while for $\alpha > 30°$ mainly ϕ influences $C_b(x)$.

At the end of zone (II), $C_{b(3L)}$ is given by Eq. (4.94) by substituting $x/L = 3$, so that

$$C_{b(3L)} = 1 - \tanh\left[5.7\mathsf{F}_o^{0.25n}(\tan\varphi)^{0.25m}\left(\frac{h_o}{s+h_o}\right)^{-0.05}\right]. \tag{4.95}$$

In the far-field zone (III), i.e. between $x/L = 3$ and the self-aeration point, continuous de-aeration occurs with reduced rates as compared with zone (II). The concentration curves start with their initial values $C_{b(3L)}$ from Eq. (4.91) for offset aerators, or Eq. (4.95) for deflectors, resulting in (Pfister and Hager, 2010b)

$$C_b = C_{b(3L)} \exp\left[-8.5\left(\frac{x}{L}-3\right)\mathsf{F}_o^{-1.5}\right]. \tag{4.96}$$

Equation (4.96) applies for aerators consisting of deflectors, offsets and combinations. The $C_b(x)$ values derived from Eq. (4.96) and those provided by Kramer (2004) are almost identical, given that Kramer's equation was applied (1) in the far-field zone (III), and (2) if the initial value $C_{b(3L)}$ is derived from Eqs. (4.91) or (4.95). Again, F_o is the governing parameter regarding the invert air transport, besides a long L and a high initial value of $C_{b(3L)}$.

4.4.8 Effect of pre-aerated approach flow

Air is detrained from air-water mixture flow under the effect of buoyancy, except for strong turbulence, so that the invert air concentration generally reduces in the streamwise direction. On 'long' or relatively 'flat' chutes, more than one aerator may be required for cavitation protection due to the limited influence length of the first aerator (Chanson, 1989a). The first aerator is therefore operated with unaerated black-water approach flow, whereas the following aerator(s) involve pre-aerated approach flow.

The air entrainment coefficient β describes the air discharge entrained through the air duct relative to the water discharge (4.4.4). Experiments conducted by Pfister et al.

(2011) indicate that the average approach flow air concentration just upstream of the aerator C_{ao} (computed according to Eq. (4.85), Figure 4.37) has a limited effect, because Eq. (4.87) remains reliable even for pre-aerated approach flow. The discharge of air Q_A to be supplied does not increase, therefore, because pre-aerated approach flow mainly affects the flow surface, whereas the flow close to the chute invert is hardly influenced. The turbulence level close to the chute invert appears to remain also unaffected, as otherwise β would increase (Ervine et al., 1995; Terrier, 2016).

The jet length L, defined between $x = 0$ and the jet reattachment point (Figure 4.37), is also not strongly affected by flow pre-aeration. Because the invert concentration is hardly affected by pre-aeration, the lower jet trajectory also remains unaffected. As L is unaffected by C_{ao}, Eq. (4.86) also applies for pre-aerated approach flow. Again, the air transport resulting from pre-aeration mainly affects the free-surface flow region, whereas the invert flow region is hardly influenced, resulting in similar values of L.

Experiments by Pfister et al. (2011) indicate that the streamwise change of the average air concentration C_a downstream of an aerator (4.4.7) is slightly affected by the presence of pre-aerated approach flow, since the flow close to the surface transports an increased amount of air. As to the effect on the invert air concentration C_b development, slightly reduced values were recorded for pre-aerated flow.

Effect on average air concentration

As stated in 4.4.7, no general streamwise development of $C_a(x)$ was derived by Pfister and Hager (2010b) along zone (II). However, $x/L = 3$ was selected as relevant location to define the starting value of $C_{a(3L)}$ for the streamwise development of the average air concentration further dowstream. Equation (4.90) states the average air concentration at the end of zone (II) (at $x/L = 3$), imposing a priori $C_{ao} = 0$ as the average approach flow air concentration resulting for unaerated black-water approach flow. However, when including the approach flow air concentration in the latter equation, $C_{a(3L)}$ increases under pre-aerated approach flow. Hence, C_{ao} is inserted in Eq. (4.88) using the term $0.1[1 + (C_{ao}/C_{au})]^{0.8}$, so that (Pfister et al. 2011)

$$C_{a(3L)} = 0.008\frac{L}{h_o} + 0.1\left(1 + \frac{C_{ao}}{C_{au}}\right)^{0.8}. \tag{4.97}$$

The term $0.1[1 + (C_{ao}/C_{au})]^{0.8} = 0.10$ for $C_{ao} = 0$, i.e. for unaerated approach flow from (Eq. 4.88), and $0.1[1 + (C_{ao}/C_{au})]^{0.8} = 0.17$ for $C_{ao} = C_{au}$. Note that C_{au} is the average air concentration of fully aerated uniform flow (4.3), for which from Eq. (4.66)

$$C_{au} = 0.75(\sin\alpha)^{0.75}. \tag{4.98}$$

Chanson (1996) provides the relation

$$C_{au} = 0.9\sin\alpha, \tag{4.99}$$

whereas Kramer (2004) derived

$$C_{au} = 0.33(\sin\alpha)^{0.25}. \tag{4.100}$$

Pfister et al. (2011) observed no particular effect of C_{ao} on the development of $C_a[x/L]$ along the far-field zone (III), so that Eq. (4.89) applies. Accordingly, the detrainment gradient of $C_a[x/L]$ is not affected along this zone. The effect of pre-aeration is included in Eq. (4.97), implicitly affecting Eq. (4.89).

Effect on invert air concentration

The data of Pfister et al. (2011) indicate that pre-aerated approach flow slightly reduces the streamwise invert air concentrations downstream of chute aerators, as compared with unaerated approach flow. Despite of the small effect of C_{ao} on $C_b(x/L)$ along the spray and reattachment zone (II) for aerators with deflectors and offsets, Eq. (4.94) was completed with an additional term for pre-aeration, so that

$$C_b = 1 - \tanh\left(4.8\left(\frac{x}{L}-1\right)^{0.25} \mathsf{F}_o^{0.25n}(\tan\varphi)^{0.25m}\left(\frac{h_o}{s+h_o}\right)^{-0.05}\left(1+\frac{C_{ao}}{C_{au}}\right)^{0.13}\right). \tag{4.101}$$

Without pre-aeration ($C_{ao}=0$), the additional term reduces to $[1+(C_{ao}/C_{au})]^{0.13}=1$, whereas it is $=1.09$ for $C_{ao}=C_{au}$, so that pre-aerated approach flow slightly lowers the invert air concentrations. Inserting $x/L=3$ in Eq. (4.101) provides $C_{b(3L)}$ as starting value of zone (III). Under pre-aerated approach flow, Eq. (4.95) modifies to

$$C_{b(3L)} = 1 - \tanh\left(5.7\mathsf{F}_o^{0.25n}(\tan\varphi)^{0.25m}\left(\frac{h_o}{s+h_o}\right)^{-0.05}\left(1+\frac{C_{ao}}{C_{au}}\right)^{0.13}\right). \tag{4.102}$$

The far-field flow zone (III) is only indirectly affected by pre-aerated approach flow via Eq. (4.102), whereas the streamwise detrainment gradient for $x/L>3$ is independent of C_{ao}. The original detrainment gradient of the basic study is thus given in Eq. (4.96).

4.4.9 Steep deflectors and cavity sub-pressure

Common chute aerators involve deflector angles ϕ (Figure 4.37) ranging from 5° to 12° (DeFazio and Wei, 1983; Marcano and Castillejo, 1984; Pinto, 1984, 1989; Koschitzky, 1987; Wood, 1991). Steeper deflectors are rarely applied as they generate shock waves, spray, and steep jets requiring high chute sidewalls. However, steep deflectors operate generally more efficient in terms of air entrainment than flat deflectors. Pfister (2011) conducted model tests to compare the aerator efficiency of common deflector angles of 0°, 5.7° (1:10, denoted as flat), 8.1° (1:7), 11.3° (1:5, denoted as steep) with excessively steep angles of $\phi=18.4°$ (1:3) and 26.6° (1:2).

The jet length L (Figure 4.37) under nearly atmospheric cavity pressure increases for steep deflectors as compared with flat, given that also the jet take-off angle increases. Equation (4.86) gives reliable values of L up to deflector angles of roughly 18° (1:3). Its application range includes thus $0° \leq \phi \leq 18°$. For steeper deflectors, Eq. (4.86) overestimates L. Similarly, Eq. (4.87) for the air entrainment coefficient β applies for $0° \leq \phi \leq 18°$. Steeper deflectors (e.g. $\phi=26.6°$) entrain less air than predicted by Eq. (4.87). The optimum angle ϕ in terms of aerator performance appears to be $\phi \leq 12°$ (1:5); steeper aerators do not perform significantly better.

According to model observations, a deflector aerator with a steep deflector of $\phi > 12°$ generates a higher air transport at the jet end than defectors with $\phi \leq 12°$. This is due to the longer and thus further disintegrated jet. The additional air is detrained rapidly from the chute invert due to the steeper impact angle, however (Pfister, 2012), combined with a higher invert pressure, as indicated by the $C_b(x)$ profiles. Yet, the reference length L is longer and thus also the normalized streamwise distance x/L, so that the local air concentration is slightly higher for a steep deflector at a certain location x than for a flat deflector. Steep deflectors do thus slightly increase the invert air concentration C_b as compared with 'standard' deflectors. The concentrations C_a along $x/L \geq 3$ are derived for $\phi \leq 18°$ from Eqs. (4.88) and (4.89), as well as from Eqs. (4.92) to (4.96) for C_b.

To summarize, a deflector with $\phi = 12°$ (1:5) generates high air concentration in the streamwise direction with limited shock wave height, spray, and jet height thereby not requiring excessively high sidewalls. Steeper deflectors ($\phi \geq 18°$) entrain locally slightly more air, yet without significantly affecting the streamwise air concentration. For standard cases, deflector angles of $5° \leq \phi \leq 12°$ appear adequate. Particularly, 'fast' flows with large F_o can include $0° \leq \phi \leq 5°$, whereas rather 'slow' flows $11° \leq \phi \leq 15°$.

Sub-pressures limit the aerator performance in the air supply system. They control the equilibrium between the air demand of the flow at the aerator, given by the air entrainment coefficient β according to Eq. (4.87), and the capacity of the air supply system. Notable sub-pressures provoke poor flow features at the aerator, including: (1) Insufficient protection against cavitation damage along the chute due to reduced air entrainment, (2) pulsations in the air supply system, and (3) transforming aerators to 'cavitation generators'. The effect of cavity sub-pressure on β was investigated e.g. by Tan (1984), Chanson (1988), Rutschmann (1988), and Pfister and Hager (2011). The relation between cavity sub-pressure and streamwise air concentration field was described by Chanson (1989b).

The air supply system connects the cavity below the jet with the free atmosphere. It consists typically of two lateral (or several frontal) air ducts connected to the transverse jet cavity, i.e. the air space below the jet. The frontal, transversally distributed connection (I) and the lateral connection (II) are shown in Figure 4.37. This cavity is generated by the offset or the deflector (Figure 4.36), optionally combined with a lowered chute invert downstream of the aerator (as in Figure 4.37). This setup prevails especially for wide chutes to avoid extensive cavity sub-pressure and to assure uniform cavity airflow (Marcano and Castillejo, 1984).

The air entrained at an aerator provokes airflow velocities in the supply system related to energy losses determined by the Bernoulli equation (Rutschmann et al., 1986). Losses reduce the airflow, limiting the air entrainment capacity of the aerator. Therefore, a steady equilibrium emerges between the aerator air demand and the air supply capacity of the air supply duct. For large sub-pressures, this equilibrium results in small air entrainment rates with a poor aerator performance. A thorough design of the air supply system involving large cross sections and an aerodynamic shape is therefore at least as important as the aerator design. The air supply system should also self-drain to avoid aerator choking.

The air cavity sub-pressure head h_s expressed in [m] water column, i.e. the relative sub-pressure in the transverse air cavity below the jet normalized with h_o is (Figure 4.37)

$$P = \frac{h_s}{h_o}. \tag{4.103}$$

As proposed by Rutschmann and Hager (1990a), the effect of P on L and β is normalized with their maximum values under negligible cavity sub-pressures close to $P \approx 0$. These maxima represent the optimum aerator performance without restrictions from the air supply system. As soon as the free airflow is limited, a sub-pressure $P > 0$ occurs and both L and β reduce. Non-influenced values of L for $P \approx 0$ follow from Eq. (4.88), whereas these of β for $P \approx 0$ follow from Eq. (4.87).

An exponential relation between the shortened and the non-affected jet length L relative to P was proposed by Pfister (2011), including also data of Rutschmann (1988) and Tan (1984). These test series involve cavity sub-pressure up to $P = 3$. The shortened jet length L_s affected by cavity sub-pressure is expressed as a function of non-influenced value L (according to Eq. 4.86) as

$$\frac{L_s}{L} = \exp\left[-0.85P\right]. \tag{4.104}$$

Note that even a small value of P drastically reduces L. For $P > 1$, the cavity typically collapses and chokes, as observed in model tests. Then, a two-phase air-water flow occurs in the residual cavity rotating with high intensity. This regime must be avoided in any case, as previously noted.

To derive the effect of P on the air entrainment coefficient β (Eq. 4.87), Pfister (2011) combined his data with these of Tan (1984), Rutschmann (1988) and Chanson (1988). A data analysis indicated that both P and F_o affect the reduction of β. It was found that this reduction is smaller for large than for small F_o, under otherwise identical conditions, so that

$$\frac{\beta_S}{\beta} = -6(P/\mathsf{F}_o)^{1/2} + 1.3. \tag{4.105}$$

Here, β_S is the reduced coefficient due to cavity sub-pressure, while β is the non-affected coefficient for nearly atmospheric cavity pressures (according to Eq. 4.87). Equation (4.105) is valid for roughly $\beta_S/\beta \geq 0.5$, and the choking limit may be set to $P \approx \mathsf{F}_o/10$. This also correlates with the expression $P \approx \mathsf{F}_o/9.4$ derived from the data of Chanson (1995a), who linked the choking with intense spray along the cavity trajectory mentioning also P and F_o as the relevant parameters. The effect of cavity sub-pressure is relevant for $P > \mathsf{F}_o/400$, i.e. by setting $\beta_S/\beta = 1$ in Eq. (4.105).

To ensure optimum aerator performance, the sub-pressure coefficient should be as small as possible, i.e. $P \leq 0.1$ for the relevant discharges with a cavitation potential. Then, the cavity sub-pressure hardly affects the air entrainment. For $0.1 < P < 1$, the aerator air entrainment is reduced by the cavity sub-pressure, whereas the aerator will choke for $P > 1$, in agreement with Chanson (1995a).

The tests for $P > 0$ indicate that the invert streamwise air concentration C_b reduces as P increases, mainly because L over-proportionally decreases with P (Pfister, 2011). The effect of reduced L is not adequately contained in Eq. (4.94), so that a correction term respecting P was added as (along $1 \leq x/L \leq 3$)

$$C_b = 1 - \tanh\left(4.8\left(\frac{x}{L} - 1\right)^{0.25} \mathsf{F}_o^{0.25n}(\tan\varphi)^{0.25m}\left(\frac{h_o}{s + h_o}\right)^{-0.05}\left(\frac{1}{1 + P}\right)^{0.18}\right). \tag{4.106}$$

For $P = 0$ results $[1/(1 + P)]^{0.18} = 1$, whereas the term is 0.8 for $P = 1$ to adjust the effect from Eq. (4.104). Note that if $P > 0.1$, the 'shortened' value L_s has to be considered in Eq. (4.106) instead of L.

4.4.10 Design procedure

The design of chute aerators includes the following steps for all relevant discharges:

1. Define spillway type and its dimensions.
2. Assure that concrete surface is smooth and no distinctive joints or irregularities exist.
3. Compute drawdown curves, resulting in the black-water profile $h(x)$ (and the related hydrostatic invert pressure depths $h_p(x)$) and the depth-averaged flow velocities V. The streamwise Froude number then is $\mathsf{F} = V/(gh)^{1/2}$.
4. Determine streamwise development of cavitation index σ from Eq. (4.83).
5. Estimate the location of the natural self-inception point (4.2.2).
6. Check for reaches upstream of the natural self-aeration point with a cavitation index below the critical value of typically 0.2. Consider also invert slope changes or invert irregularities to define the endangered reaches.
7. Extract h and F at the most upstream point where cavitation could occur for typically the maximum discharge. These values define the approach flow characteristics h_o and F_o to design the (first) aerator, for all considered discharges. Ensure that the approach flow is *a priori* non-aerated with an average air concentration $C_{ao} \cong 0$ (Figure 4.37).
8. Select aerator type, mainly based on F_o and α. As for a bottom outlet aerator, take the offset solution if $\mathsf{F}_o > 6$. On chutes, combine an offset with a deflector. Note that steep deflectors ($\phi = 11°$) entrain air only if $\mathsf{F}_o > 4$, and flat deflectors ($\phi = 6°$) if $\mathsf{F}_o > 5$. For chute invert angles $\alpha \leq 30°$, provide a steep ($\phi = 11°$) and high ($t = 0.5$ m) deflector; if $\alpha \geq 50°$ a pure offset is sufficient, possibly combined with a small deflector ($\phi = 5°$, $s = 0.2$ m).
9. Compute jet length L with Eq. (4.86), assuming nearly atmospheric cavity pressure.
10. Compute streamwise decrease of invert air concentration $C_b(x/L)$ along reattachment and spay zone (II, $1 \leq x/L \leq 3$) from Eq. (4.90) for offset aerators, or Eq. (4.94) for deflectors (possibly combined with offsets).
11. Compute streamwise decrease of invert air concentration $C_b(x/L)$ along far-field zone (III, $x/L > 3$) from Eq. (4.91) for offset aerators, or Eq. (4.95) for deflectors (possibly combined with offsets) as start values, and Eq. (4.96) as streamwise C_b decrease.
12. Define a minimum invert air concertation of typically $C_b = 0.01$ required for cavitation protection. Evaluate aerator influence reach versus the aerator parameters t, ϕ, and s. Optimize these parameters if necessary to increase the influence reach.
13. Check if any chute portion downstream of aerator reach remains non-protected. If yes, provide second aerator and reapply procedure from step 6 on. Consider that the second aerator is subjected to pre-aerated approach flow. Derive the degree of pre-aeration C_{ao} with Eqs. (4.88) and (4.89) referring to first aerator, and use Eq. (4.101) to compute $C_b(x/L)$ of second aerator.
14. Determine air entrainment coefficient β with Eq. (4.87), and air supply discharge Q_A. Note that its maximum may not occur for the design, but for a smaller flood!
15. Sketch air supply system. Provide a large cavity cross section by lowering locally the invert (Figure 4.37). For wide chutes and transversally 'small' air cavities, select a linear air supply across vertical aerator wall plus sidewall openings (I and II); for 'narrow' chutes, provide two lateral openings in chute sidewalls (II in Figure 4.37).
16. Based on the maximum air discharge to provide Q_A, chose a minimum air supply cross section limiting maximum airflow velocities to 30 m/s. Check pressure (energy losses) in the supply system and increase cross section if the relative air cavity pressure head drops below $0.1h_o$. If impossible, consider reduced aerator operation mode from 4.4.9.

17 Provide a powerful drainage system keeping the air supply system free from water.
18 Assure that the maximum jet height locally has sufficient freeboard along sidewalls, so that significant shock waves generated at the aerator are absent and aerator-generated air-water chute flow does not overtop the sidewalls.

After the dangerous cavitation damages having occurred in the 1980s on spillways all over the world, engineers tended to add many and powerful aerators. The effect was that a surplus of air was entrained, increasing flow velocities (Minor, 1987) thus affecting the operation mode of the stilling basin or the flip bucket. Additionally, the freeboard was reduced and the chutes appeared 'full' with air-water flow even for discharges far below the design value. Spray and the related saturation of terrain and ice formation are further potential concerns of unnecessarily high flow aeration. Aerators should thus not be 'overdesigned' to achieve the maximum air entrainment.

Another remark relates to the basic assumption of a straight, constantly sloping, prismatic chute without neither lateral discharge addition nor reduction. Whenever the standard chute is of more general shape, the above indications are subject to changes. Given that chutes are hardly designed with the complications mentioned, the proposed design normally applies for design purposes. However, if these complications prevail in important chute structures, their effects have to be considered for safety reasons.

Figure 4.40 shows the chute of the Foz do Areia Dam. Note the well visible section of the first aerator, whereas that of the second aerator is slightly downstream, causing overaeration

Figure 4.40 Chute flow with aerators at Foz do Areia Dam, Brazil (Courtesy Nelson Pinto)

of the chute flow. Farther away is even a third aerator location and the trajectory bucket is noted at the chute end. At this discharge, the chute almost overtops, endangering the lateral zone with erosion, by which the entire spillway may be lost if this effect keeps over a sufficiently long duration. Note also the air ducts arranged along the chute, by which air is supplied to the flow from the atmosphere.

4.5 Shock waves

4.5.1 Introduction

Shock waves (shocks in short) or crosswaves are a surface pattern in supercritical flow along which perturbations are propagated. These waves are steady under constant flow conditions. They are easily noted because of large-scale flow perturbations often accompanied with air entrainment; they point at the source of flow perturbation when looking in the upstream direction. Crosswaves are running obliquely toward the chute walls causing local maxima of the wall flow profile. These flows become non-uniform with local flow extrema due to the shocks. Figure 4.41 shows an example.

Any perturbation of a supercritical flow generates a shock, i.e. any deviation from uniform chute flow. Typical examples of perturbations are:

- Chute curve
- Change of chute slope

Figure 4.41 Shock wave at rear of gate piers, Gardiner Dam, Canada (Toth, 1968). Note truck for comparison of scale

- Chute contraction and expansion
- Chute junction.

A trajectory bucket induces shocks due to the change of slope along the bucket. The shocks are often visible for small discharges but they are blurred for larger due to the turbulence effect. A shock is not a real concern for the chute, except that the freeboard must be increased and that a stilling basin may experience asymmetric approach flow. A discussion on the basic phenomena and means to reduce shocks therefore appears appropriate. Note that shocks are generated for any supercritical flow. However, as for hydraulic jumps corresponding to a limit case of shock waves, undular flow pattern occurs for transitional flow with Froude numbers below, say, 2.5. Whereas shocks for larger Froude numbers are abrupt in terms of free surface elevation, undular shocks are more complicated and also tend to a slight backwater effect. Undular shocks are excluded here, given the complications of description, the flow instabilities involved, and the relatively small shock heights generated as compared with direct shocks.

The basic phenomenon of a shock wave occurs in the horizontal, smooth, wide and rectangular channel of approach flow (subscript 1) depth h_1, and approach flow velocity V_1. The approach flow Froude number is thus equal to the ratio of approach flow velocity to the approach flow wave celerity $F_1 = V_1/(gh_1)^{1/2}$. The approach flow is referred to as supercritical if $F_1 > 1$, and subcritical otherwise. A subdivision into transcritical ($0.7 < F_1 < 1.5$), and hypercritical flows ($F_1 > 3$) is made. The following discussion includes super- and mainly hypercritical flows. Transcritical flows are excluded because of their irrelevance in chute flows, the effect of streamline curvature, and the formation of weak jumps, as mentioned above.

Figure 4.42 shows the abrupt wall deflection. At the flow deflection point P, the wall is turned by the angle θ. Because perturbations in supercritical flow are carried downstream only, a perturbation front of shock (subscript s) angle β_s is generated. The shock originates at point P, and all streamlines follow the walls upstream and downstream of the shock front. Note the abrupt free surface elevation from the undisturbed approach flow ① to the perturbed tailwater flow ②, indicating an approach flow Froude number in excess of ③. This example corresponds to the fundamental flow feature in disturbed hypercritical flow, in analogy to the

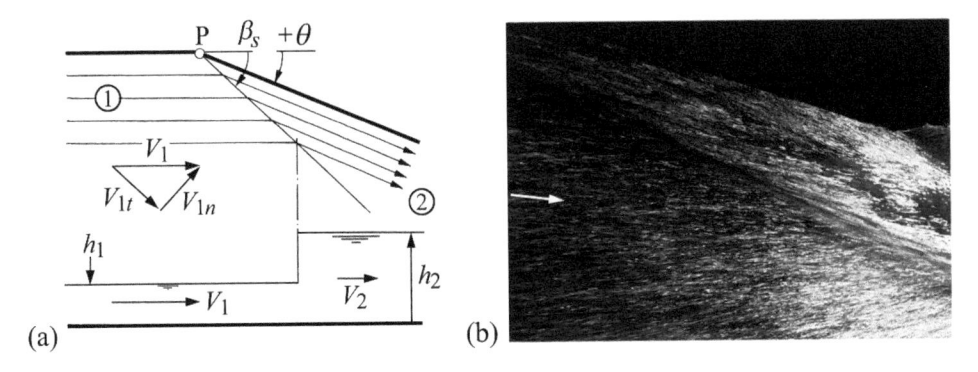

(a)   (b)

Figure 4.42 Abrupt wall deflection (a) plan and side views, (b) photograph of shock front
(Courtesy Willi H. Hager)

classical hydraulic jump as the particular flow feature of energy dissipation in free surface flows.

Of engineering relevance are the ratio of flow depths h_1 and h_2 in front and in the tailwater of the shock, the shock angle β_s, and the tailwater Froude number F_2. Based on the momentum equations applied parallel and perpendicular to the shock front yields (e.g. Chow, 1959)

$$Y = \frac{1}{2}[(1+8F_1^2 \sin^2 \beta_s)^{1/2} - 1], \tag{4.107}$$

$$Y = \frac{\tan \beta_s}{\tan(\beta_s - \theta)}, \tag{4.108}$$

$$F_2^2 = Y^{-1}[F_1^2 - (2Y)^{-1}(Y-1)(Y+1)^2]. \tag{4.109}$$

Here $Y = h_2/h_1$ is the depth ratio with subscript '1' referring to the unperturbed, and '2' to the perturbed flow (Figure 4.42a). For given values of h_1, V_1, and θ the unknowns h_2, V_2, and β_s follow. An explicit approach for $F_1 \sin \beta_s > 1$ yields (ICOLD, 1992)

$$Y = \sqrt{2}F_1 \sin \beta_s - \frac{1}{2}, \tag{4.110}$$

$$\beta_s = \theta + F_1^{-1}, \tag{4.111}$$

$$F_2 / F_1 = [1 + (F_1 \sin \beta_s / \sqrt{2})]^{-1}. \tag{4.112}$$

If the sine is replaced by the arcus function for the small values of θ considered, all three parameters depend exclusively on the so-called shock number $S = \theta \cdot F$ as

$$Y = 1 + \sqrt{2}S_1, \tag{4.113}$$

$$\beta_s F_1 = 1 + S_1, \tag{4.114}$$

$$F_2 / F_1 = (1 + S_1 / \sqrt{2})^{-1}. \tag{4.115}$$

According to Figure 4.43, the wall flow profile $h_w(x)$ increases abruptly at point P from h_1 to h_2. Schwalt and Hager (1992) analyzed the flow around an abrupt wall deflection to establish

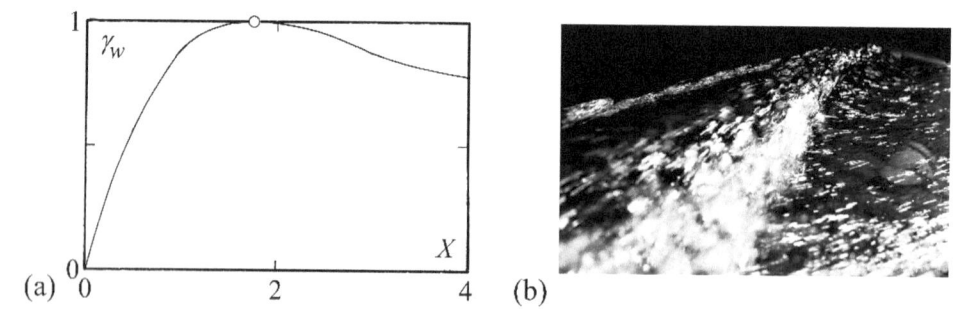

(a) (b)

Figure 4.43 Abrupt wall deflection (a) normalized wall profile $\gamma_w(X)$ according to Schwalt and Hager (1992), (b) shock wave for $S_1 = 0.8$, view against flow direction (Courtesy Willi H. Hager)

a generalized surface geometry. Figure 4.43a shows the generalized wall (subscript w) profile $\gamma_w(X)$ with $X = x/(h_1 F_1)$ as the longitudinal coordinate measured from P and $\gamma_w = (h_w - h_1)/(h_M - h_1)$ as the dimensionless wall flow depth. The maximum wall flow depth h_M depends exclusively on the approach flow shock number S_1 as

$$Y_M = \frac{h_M}{h_1} = 1 + \sqrt{2} S_1 [1 + (1/4) S_1]. \tag{4.116}$$

Compared with Eq. (4.113), the latter has the second order correction $[1 + (1/4) S_1]$ becoming significant as $S_1 > 0.5$, so that the linear theory based on the hydrostatic pressure distribution becomes invalid.

The channel contraction is an application of the abrupt wall flow deflection. Figure 4.44a defines the geometry of the straight-walled structure, with point A as shock origin, point B as shock crossing, point C as wall impingement, and point D as contraction end. At the latter point, a so-called negative wall deflection is generated, with a smooth formation of expansion waves. The following applies to a tailwater width b_3 larger than 25% of b_1.

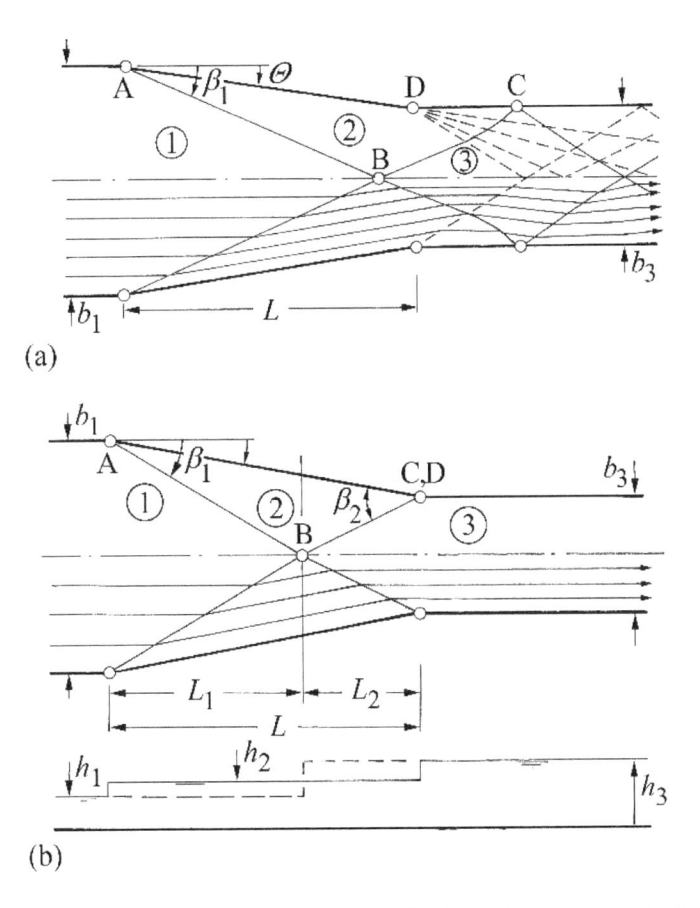

(a)

(b)

Figure 4.44 Straight-wall chute contraction (a) definition of geometry and off-design flow, (b) design according to Ippen and Dawson (1951) with plan (top) and section (bottom)

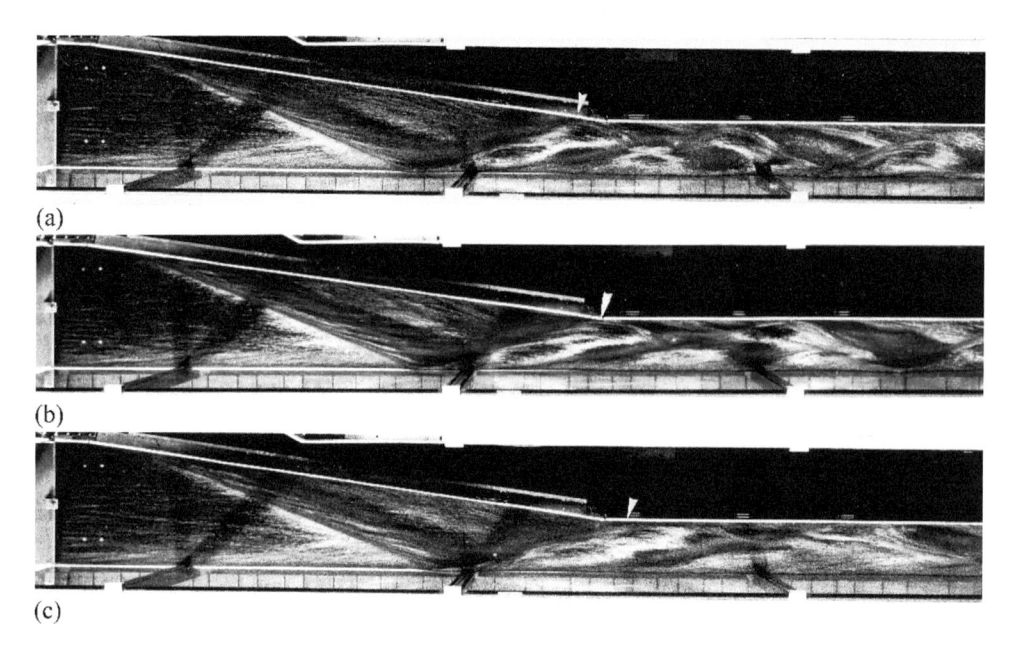

(a)

(b)

(c)

Figure 4.45 Interference principle in chute contraction with shock front directed (a) slightly upstream, (b) exactly at, (c) slightly downstream of contraction end point (Reinauer and Hager, 1998)

Ippen and Dawson (1951) considered supercritical contraction flow, introducing a design procedure involving the wave interference principle as known from optics (Figure 4.44b). The positive wave originating at point A is supposed to interfere with the wave originating from point D, of which the height is equal but of negative sign. Accordingly, for wave interference to occur, point C must be shifted to point D. The hydraulic condition reads for $\theta < 10°$ (ICOLD, 1992)

$$\arctan \theta = \left(\frac{b_1}{b_3} - 1\right)\frac{1}{2F_1}.\qquad(4.117)$$

The disadvantage of wave interference is that it may be satisfied for one discharge only. Also, Ippen and Dawson's design principle was proved to be in error (Figure 4.45).

4.5.2 Chute expansion

A channel expansion typically occurs at locations where chutes merge, as in the tailwater of gate piers, or at stilling basins with various parallel chutes subdivided with separation walls. Such flow may also occur at junction structures of small junction angle with only one branch in operation, as well as in bottom outlets. Expanding chutes are geometrically designed either with an abrupt expansion, or with a Rouse-type expansion.

Rouse *et al.* (1951) proposed a continuous chute wall expansion geometry, with a reversed transition to inhibit shock waves in the tailwater chute. The proposal was model tested by

Mazumder and Hager (1993) finding acceptable results even for a 50% transition length as compared with Rouse *et al.*'s original design.

The abrupt chute expansion performs so well that it is often applicable. Figure 4.46 shows the expansion geometry in the horizontal rectangular chute, including the definitions of the axial (subscript *a*) and wall (subscript *w*) free surface profiles. The scaling lengths are the

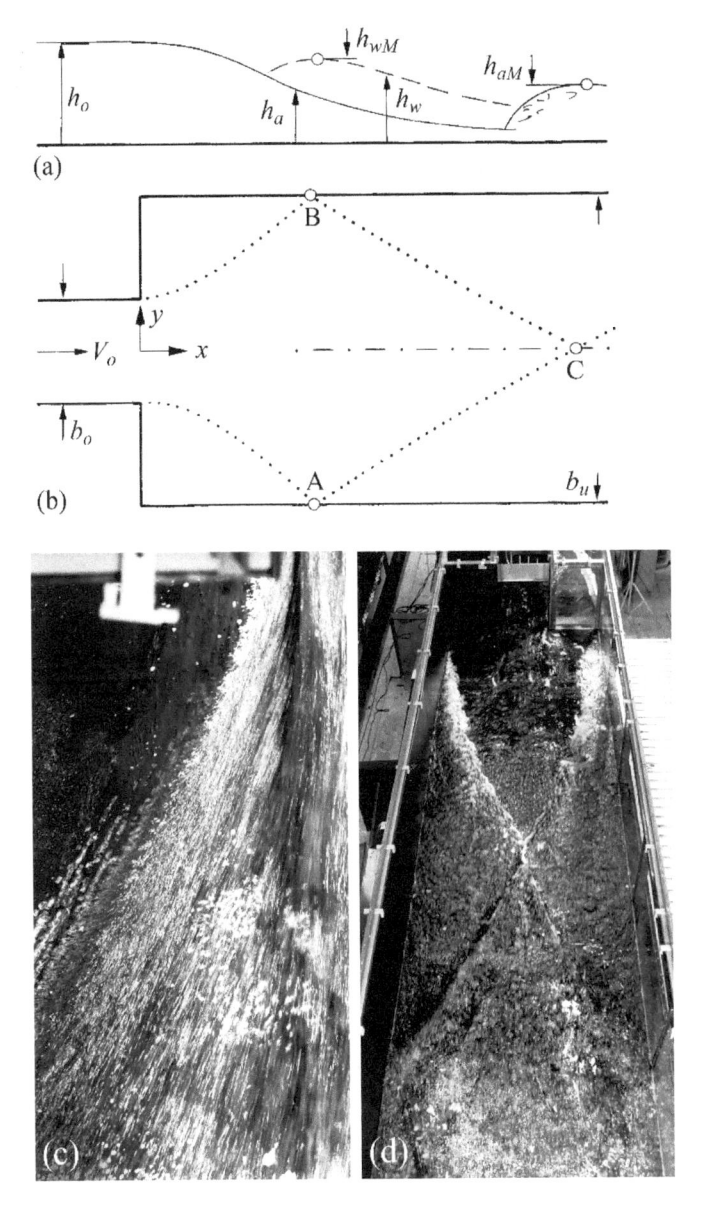

Figure 4.46 Definition of abrupt chute expansion (a) section with (—) axial, (- - -) wall free surface profiles, (b) plan view with (. . .) shocks, (c, d) photos (Hager and Mazumder, 1992)

approach (subscript o) flow width and height b_o and h_o, and $\beta_e = b_u/b_o$, $Y_a = h_a/h_o$, $Y_w = h_w/h_o$, $X = x/(b_o F_o)$ are the parameters for the width ratio, the axial and the wall free surface profiles versus the relative streamwise coordinate. The free surface profiles determined from model studies read (Hager and Mazumder, 1992)

$$Y_a = 0.2 + 0.8 \exp(-X^2), \tag{4.118}$$

$$Y_w = Y_{wM} \tau \exp(1 - \tau). \tag{4.119}$$

The maximum (subscript M) wall flow depth for $1.8 < \beta_e < 6$ thereby is

$$Y_{wM} = 1.27 \beta_e^{-0.4}. \tag{4.120}$$

According to Eq. (4.120), h_{wM} depends exclusively on the wall expansion ratio β_e remaining always smaller than h_o. Coordinate τ depends on X and β_e as

$$\tau = \frac{X - (1/6)(\beta_e - 1)}{0.52\beta_e^{0.86} - (1/6)(\beta_e - 1)}. \tag{4.121}$$

The formation of a hydraulic jump downstream from the abrupt chute expansion is discussed in Chapter 5.

4.5.3 Chute bend

Supercritical bend flow should be avoided in chutes because of shock generation. A definition plot provides Figure 4.47a. According to Knapp (1951), the shock angle β_s depends on the relative chute curvature $\rho_a = b/R_a$ with b as chute width and R_a as axial radius of curvature, and the elementary shock angle $\sin\theta \cong \tan\theta = 1/F_o^{-1}$ as

$$\tan\beta_s = \frac{(b/R_a)F_o}{(1 + 2b/R_a)} \cong (b/R_a)F_o. \tag{4.122}$$

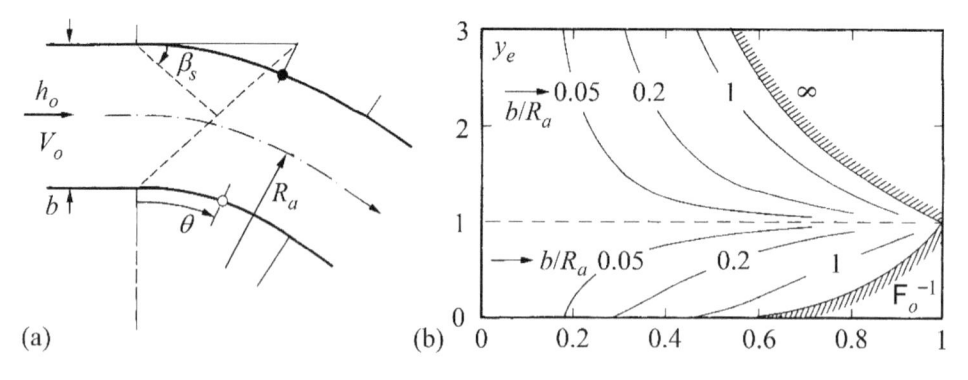

(a)   (b)

Figure 4.47 Supercritical flow in chute bend (a) definition of flow, (b) extreme flow depths $y_e(F_o)$

Because the energy loss across a weak shock is small, the energy equation applies for the extreme (subscript e) wave heights $y_e = h_e/h_o$ along the outer and inner bend walls, resulting in (Figure 4.47b)

$$y_e = \left[1 \pm \frac{1}{2}(b/R_a)F_o^2\right]^2. \tag{4.123}$$

Reinauer and Hager (1997) analyzed strongly deflected bend flow by selected tests. The approach of Knapp (1951) was confirmed for small relative curvature b/R_a, and extended to large values. The study deals with circular bends in rectangular horizontal and smooth chutes. The governing parameter of supercritical bend flow is the bend number $B_o = (b/R_a)^{1/2}F_o$. Figure 4.48 shows typical free surfaces across a bend flow for $B_o = 1.5$ and 2.3. In Figure 4.48a

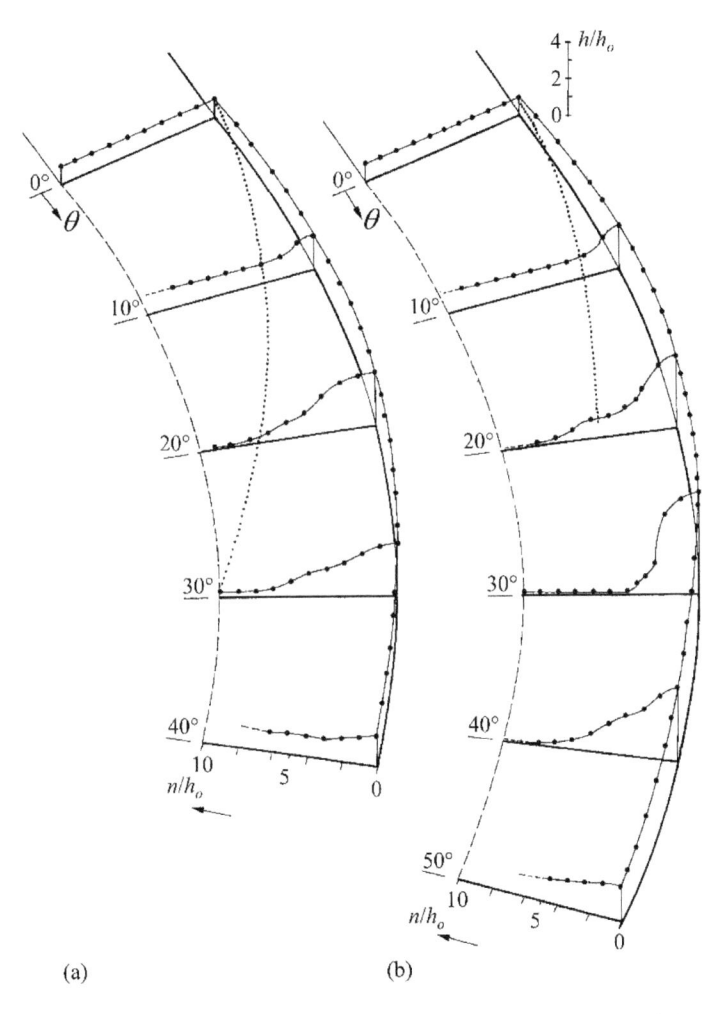

(a) (b)

Figure 4.48 Surface $h/h_o[\theta, n/h_o]$ of chute bend flow for $B_o =$ (a) 1.5, (b) 2.3 (Reinauer and Hager, 1997)

the bend flow is transitional between weak and strong flows, whereas the case with $B_o = 2.3$ corresponds to strong bend flow (Figure 4.48b). For weak bend flow, the surface is continuous and the transverse sections are nearly trapezoidal. Flow separation from the inner bend wall is locally developed. For strong bend flow, the flow is fully separated from the inner bend wall, with the transverse surface profile becoming nearly triangular. Note that shock wave may also break.

Of particular design interest are the maximum (subscript M) wall flow depth h_M, and the corresponding location θ_M. Based on laboratory tests, the results for weak and strong bend flows are with $Z_M = (h_M/h_o)^{1/2} - 1$, respectively, (Figure 4.49a)

$$Z_M = 0.40B_o^2, \quad B_o \leq 1.5, \tag{4.124}$$

$$Z_M = 0.60B_o, \quad B_o > 1.5. \tag{4.125}$$

Note that the free surface remains undisturbed in a straight chute, i.e. $Z_M(B_o = 0) = 0$. For $B_o \leq 1.5$ the result is basically in agreement with that of Knapp (1951), whereas a linear dependence exists for strong bend flow.

The bend angle θ_M is more difficult to determine, given the larger data scatter. Figure 4.49b shows that $\tan\theta_M$ varies with $(b/R_a)F_o = (b/R_a)^{1/2}B_o$, slightly different from the bend number. The experimental data follow the fits (Reinauer and Hager, 1997)

$$\tan\theta_M = (b/R_a)F_o, \qquad \text{for } (b/R_a)F_o \leq 0.35, \tag{4.126}$$

$$\tan\theta_M = 0.60[(b/R_a)F_o]^{1/2}, \quad \text{for } (b/R_a)F_o > 0.35. \tag{4.127}$$

Reinauer and Hager (1997) also analyzed both the minimum wave characteristics, and the wall profiles. The velocity distribution is almost tangential with the absolute value close to the approach flow velocity V_o throughout the main body of the flow and practically without

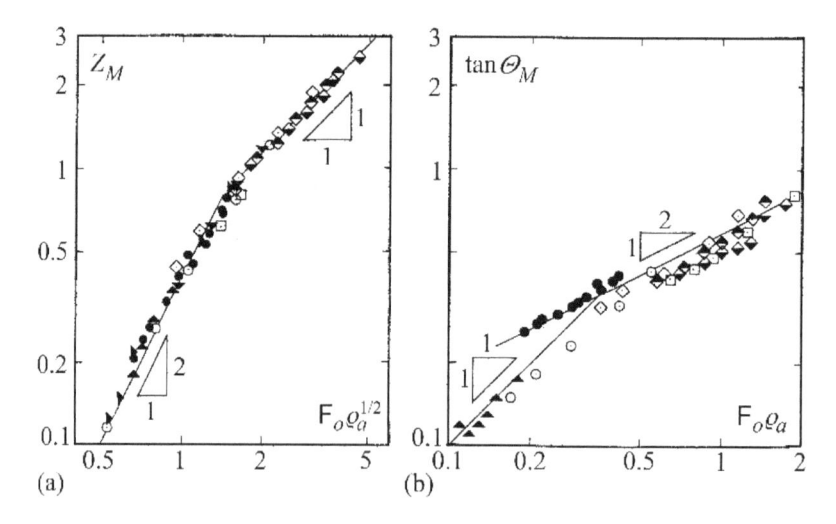

Figure 4.49 Chute bend flow (a) maximum wave height $Z_M(B_o)$, (b) location $\tan\theta_M$ for $\rho_a = b/R_a = (\blacktriangle) 0.04, (\blacktriangleright) 0.06, (\circ) 0.07, (\bullet) 0.08, (\blacktriangledown) 0.10, (\lozenge) 0.14, (\square) 0.31$ (Reinauer and Hager, 1997)

any radial component. Figure 4.50 shows the development of a curved shock wave, whereas Figure 4.51 refers to views at separated bend flow.

Currently, two aspects of supercritical bend flow have not yet received attention: (1) effect of bottom slope and (2) choking conditions. The latter item is extremely important and has to be carefully analyzed in sufficiently large hydraulic models. If a chute bend would have been

Figure 4.50 Development of curved shock wave for $b/R_a = 0.07$ and $\mathsf{F}_o = $ (a) 2, (b) 8 (Reinauer and Hager, 1997)

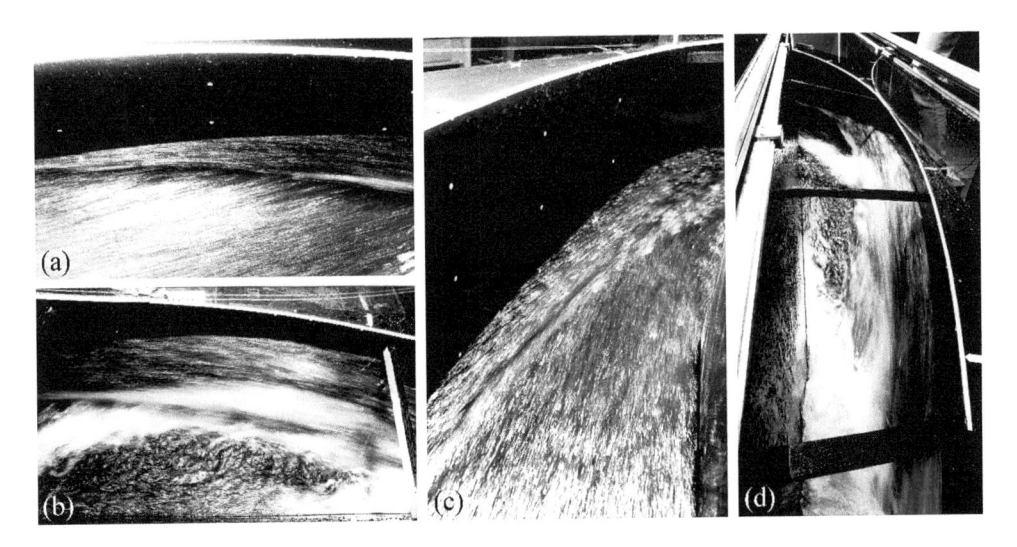

Figure 4.51 Views at separated bend flows for $b/R_a = 0.14$ and $\mathsf{F}_o = 4.5$. Plan views (a) upper and (b) lower portions, views (c) in and (d) against flow direction (Reinauer and Hager, 1997)

designed for supercritical flow yet choking occurs, the flow breaks down leading to dangerous and highly unstable hydraulic jumps along with massive air entrainment. Typically, choking in horizontal chute bends occurs for Froude numbers $F_o < 2$ to 4, depending on the relative curvature and the bottom slope. The effect of bottom slope on the wave maximum is small as for chute contractions.

Shocks in tunnel spillways may lead to dangerous surging phenomena with abrupt transitions from free surface to pressurized duct flow. Tunnels with bends have to be designed sufficiently large to guarantee free surface flow for all discharges along with co-flowing air above water flow. The two-phase air-water flow must be stratified. Figure 4.52 shows aerated flow in a sloping tunnel indicating the large transverse free surface slopes that may provoke choking and surging. The bottom slope of a tunnel containing supercritical flow, and the tailwater elevation at the tunnel outlet should be selected so that no hydraulic jumps occur along the entire tunnel at any discharges.

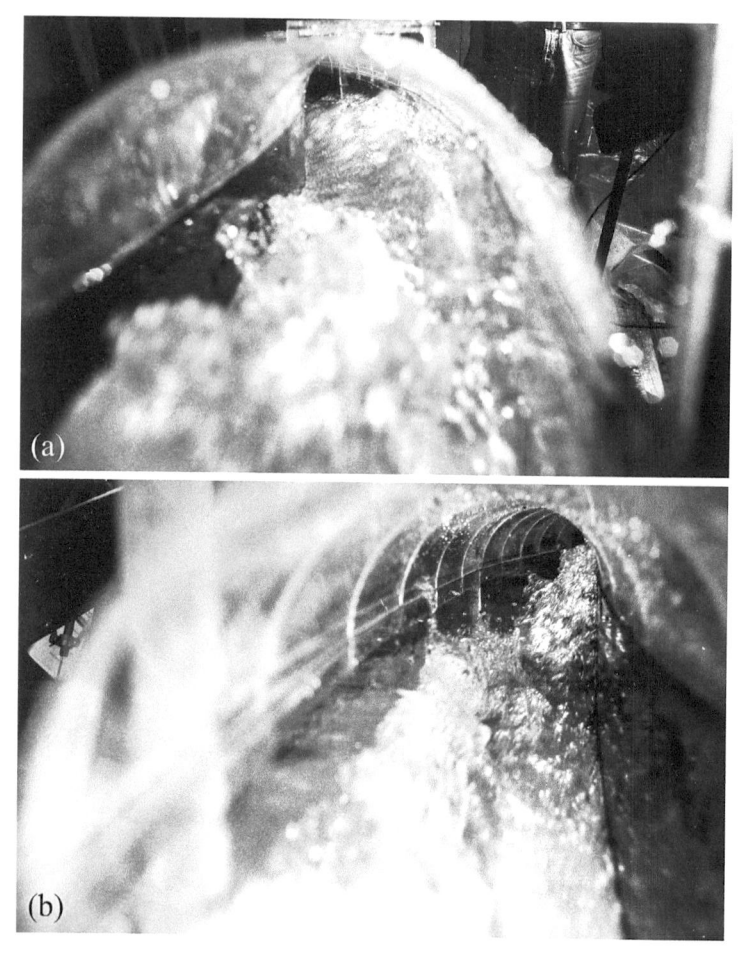

Figure 4.52 Supercritical tunnel flow, views from (a) upstream, (b) downstream with shock along outer tunnel wall (Courtesy Willi H. Hager)

4.5.4 Chute contraction

A chute contraction is not a common design mainly because of problems with shocks, and the resulting poor approach flow to the energy dissipator. However, the basic approach of Ippen and Dawson (1951) was extended by Reinauer and Hager (1998) for funnel-shaped contractions, resulting in an improved design. Three contraction types exist (Figure 4.53): funnel, fan, and nozzle. The funnel is simple for construction and involves abrupt wall deflections; the fan applies for chute inlets, typically below small overflow structures; the nozzle was shown to perform poorly, because of strong shocks at the inlet. The following refers to the fan-shaped contraction with reference to Anastasi (1980), Vischer (1988a, b), whereas ICOLD (1992) applies for funnel contractions.

Ippen and Dawson (1951) provided a rational design for funnel contractions. The procedure involves the principle of wave interference at the contraction end. Figure 4.54 shows

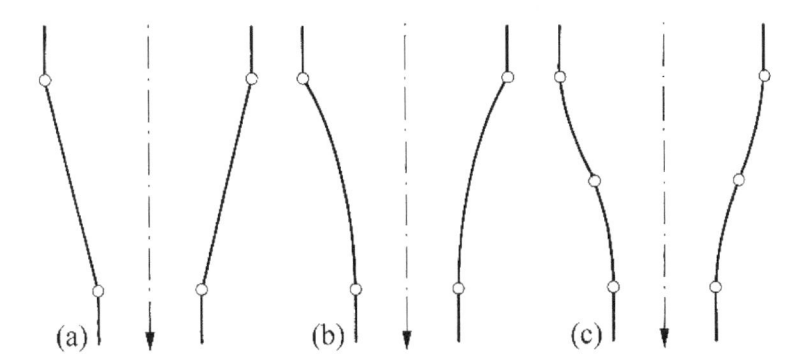

Figure 4.53 Chute contraction shapes (a) funnel, (b) fan, (c) nozzle

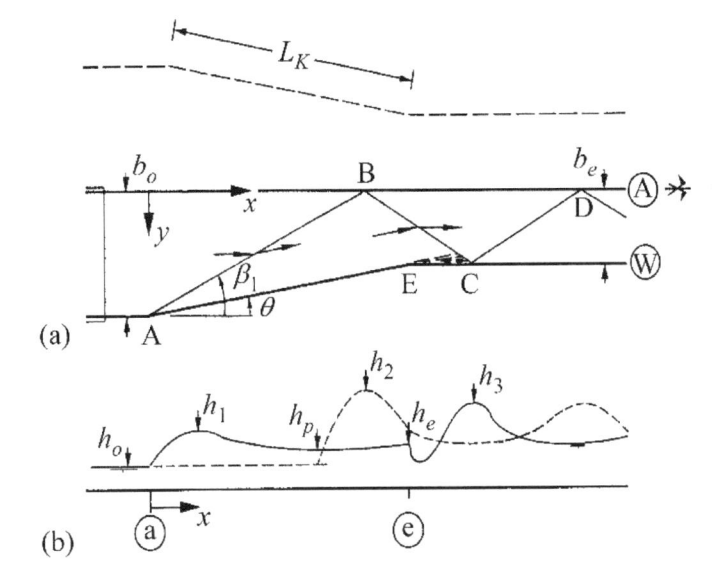

Figure 4.54 Chute contraction (a) plan, (b) side view with (—) wall, (- - -) axial free surface profiles

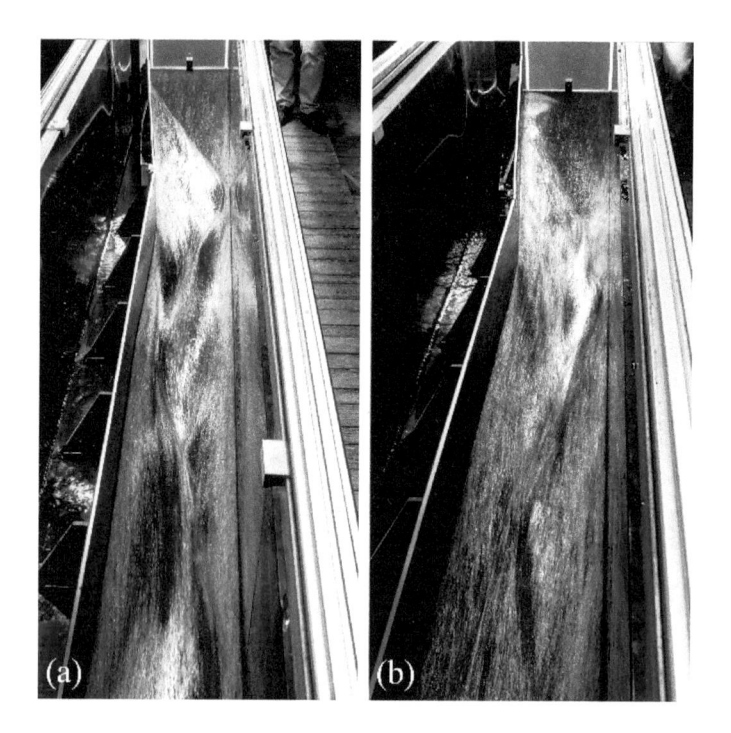

Figure 4.55 Effect of shock diffractor in chute contraction (a) without, (b) with diffractor element (Reinauer and Hager, 1998)

a sketch involving the contraction angle θ and the contraction rate b_3/b_1 so that the positive shock originating at contraction point A is directed to the contraction end point D, from which a negative shock of equal intensity is generated, resulting in undisturbed tailwater flow. Based on detailed observations, this interference principle adopted from optics was demonstrated to fail in chute flow because shock waves are of finite width and streamline curvature effects are significant (Figure 4.55).

Figure 4.54 shows the flow pattern in a plane funnel-shaped chute contraction of approach flow width b_o and end width b_e. At the contraction point A, a shock of angle β_1 is generated due to the abrupt wall deflection. The shock is propagated to the channel axis (point B) continuing to the wall to impinge at point C. At the contraction end point E, a negative wave forms along with a complicated tailwater wave pattern. The axial (subscript a) and the wall (subscript w) free surface profiles are relevant.

Three shock waves are distinguished:

- Wave 1 of height h_1 downstream of point A
- Wave 2 of height h_2 in the chute axis close to point B
- Wave 3 of height h_3 along chute wall usually downstream of point E.

The locations of the wave maxima are x_1, x_2, and x_3. Downstream of wave 1, the wall flow depth h_p is nearly constant and the flow depth at point E is h_e. The velocity across the shocks along the contraction remains equal to the approach flow velocity V_o.

The significant parameters of supercritical contraction flow are the heights and locations of waves 1–3, because all further tailwater waves are lower. For waves 1 and 2, the width ratio $\beta_e = b_e/b_o$ is insignificant and the governing parameters are the approach flow shock number $S_o = \theta F_o$ in analogy to the abrupt wall deflection (4.5.1), and the chute bottom inclination $\alpha[°]$. Based on extensive model tests, the wave heights are (Reinauer and Hager, 1998)

$$\text{Wave 1:} \quad Y_1 = h_1/h_o = (1 + 2^{-1/2}S_o)^2, \tag{4.128}$$

$$\text{Wave 2:} \quad Y_2 = h_2/h_o = (1 + 2^{1/2}S_o)^2, \tag{4.129}$$

$$\text{Wave 3:} \quad Y_3 = h_3/h_o = \beta_e^{-1} - 0.2\alpha^{0.6} + 1.8S_o^{1/2}. \tag{4.130}$$

Note that the first order approximation of Eq. (4.128) is equal to Eq. (4.113) based on shallow-water assumptions. Also, both waves 1 and 2 are independent of chute slope, at least up to 45°. For wave 3, an increasing slope reduces the wave height, a general result applying to supercritical chute flow. The wave positions were also determined.

To reduce a shock wave along an abrupt wall deflection, Reinauer and Hager (1998) proposed the shock diffractor. The purpose of this pyramid-shaped element is to break up the compact shock wave and thus reduce its height. Figure 4.56 compares the shock waves due to contraction without, and with shock diffractor highlighting the positive effect in the tailwater. The optimum diffractor shape was experimentally found as shown in Figure 4.56. The triangular-shaped base of length $6h_o$ and width $4h_o$ increases to the point of abrupt wall deflection, where its height is $0.9h_o$, independent of F_o. The back of the element is vertical, and may be aerated from the side to inhibit local cavitation damage to the diffractor. The element is small and easily inserted to existing chutes. A second diffractor may be added in the contraction axis to further reduce tailwater waves. A thorough design is given by Reinauer and Hager (1998).

The hydraulic effect of a shock diffractor involves generation of a negative wave at its rear resulting in wave diffraction, i.e. the shock is transversally stretched and thus reduced in height (Figure 4.55). To improve the flow, Diffractor 1 is positioned with its vertical face at the contraction point A, whereas Diffractor 2 in the contraction axis. Diffractor 2 then directs wave 1 into its rear wave, so that only a small wave 2 is generated. Off-design was found to be insignificant; the diffractor should be designed for the flow with the largest shock number.

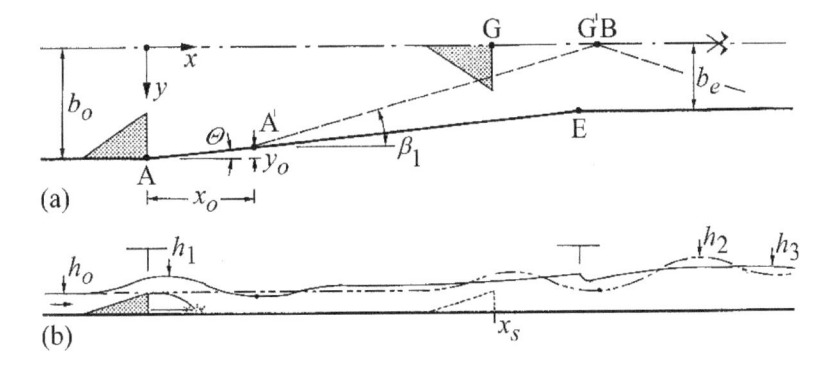

Figure 4.56 Definition sketch of shock diffractors in chute contraction (a) plan, (b) side view with reduced wave pattern. (——) Wall and (- - -) axial free-surface profiles (Reinauer and Hager, 1998)

Chute contraction flow with bottom elements includes three waves:

- Wall wave 1 downstream of contraction inlet
- Axial wave 2 at contraction center
- Wall wave 3 downstream of contraction end.

Without shock development, the increase of flow depth across a contraction would be $h_3/h_1 = b_1/b_3$, for constant velocity. The heights of waves 1–3 with shock diffractors thus are (Reinauer and Hager, 1998)

$$Y_1 = h_1 / h_o = (1 + 1.7\mathsf{S}_o + 0.011\mathsf{F}_o^2) / \cos\alpha, \qquad (4.131)$$

$$Y_2 = (1 + \mathsf{S}_o)^2, \qquad (4.132)$$

$$Y_3 = \beta_e^{-1} - 0.2\alpha^{0.6} + 1.2\mathsf{S}_o^{1/2}. \qquad (4.133)$$

The term containing F_o in Eq. (4.131) stems from the diffractor self-perturbation, and Y_1 with diffractor may be smaller or larger than without diffractor. Waves 2 and 3 are always much lower with a diffractor than without. The decision whether a diffractor should be inserted depends mainly on an economical consideration involving cost and reduction of wall height. Diffractors are easily added to existing chutes, and an aerator is provided if cavitation damage is a concern. The chute contraction as presented here has its hydraulic optimum for width reductions between 50% and 80%. For $\beta_e < 0.5$ fan-shaped contractions should be considered (ICOLD, 1992).

Contractions under supercritical approach flow are prone to choking, i.e. a breakdown of supercritical chute flow associated with the formation of a hydraulic jump. For such a contraction, overtopping may occur because it was designed for a much smaller flow depth. A safety check against choking is thus required. Choking flow was also analyzed by Henderson (1966). For incipient choking such as due to a discharge reduction, the flow at the contraction end section is critical. Applying the Bernoulli equation over the contraction relates the approach flow Froude number F_o to the width ratio β_e as (Reinauer and Hager, 1998)

$$\beta_e = \mathsf{F}_o \left(\frac{3}{2 + \mathsf{F}_o^2} \right)^{3/2}. \qquad (4.134)$$

From extended model observations it is noted that the choking phenomenon is stable in the sense that the toe of the hydraulic jump is shifted always slightly upstream of the contraction point A. Figure 4.57 shows a series of photographs illustrating choking sequences in a horizontal chute contraction. Choking starts as the critical flow depth is established at the downstream end of the contraction (Figure 4.57a). A moving hydraulic jump of small Froude number then travels upstream, whose height increases because the upstream Froude number at the toe of the jump increases (Figure 4.57b–d). Shortly downstream of the gate, which was used for these particular tests, the hydraulic jump is fully developed, and reaches the gate section in Figure 4.57f. Obviously, the flow depth along the jump is much higher than for supercritical flow.

For a choked chute contraction, a larger approach flow Froude number is required for blowout, i.e. for returning to supercritical flow. Figure 4.58 refers to the choking control. In a horizontal chute, incipient choking F_o^- from Eq. (4.134) is distinguished from blowout at

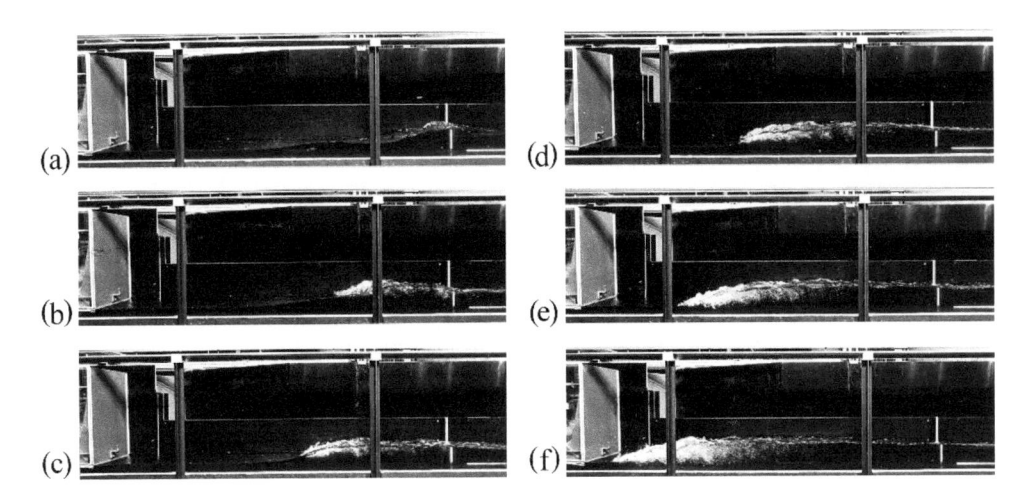

Figure 4.57 Choking flow in horizontal chute contraction with $\beta_e = 0.6$ for incipient choking at $F_o = 2.4$ (Reinauer and Hager, 1998)

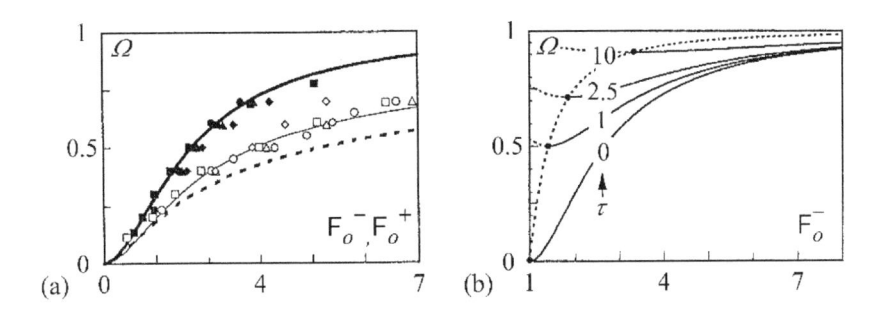

Figure 4.58 (a) Choking in *horizontal* chute contraction with (—) incipient choking F_o^-, (. . .) blowout for F_o^+ (b) incipient choking in *sloping* contraction for various slope parameters τ. Note: choking occurs *above* a specific curve (Reinauer and Hager, 1998)

F_o^+. For sloping contractions, the parameter $\tau = (S_o - S_f)(L_k/h_o)$ controls F_o^- as shown in Figure 4.58b, with S_o as bottom slope, S_f as average friction slope, and L_k as contraction length. For $0.5 < \beta_e < 1$, or $0 < \Omega = 1 - \beta_e < 0.5$, choking occurs only if $\tau > 1$. Choking is normally not a concern for chute slopes in excess of 5°, therefore.

4.6 Roll waves

4.6.1 Definition and early advances

Depending on the generation mechanisms, the topographic environment and their relative scale, water waves have a wide variety of appearance. One type of water waves is referred to as shallow-water waves, whose wave amplitude is large as compared with the still water depth. These waves are opposite to deep-water waves as are typically generated by wind, with practically zero mass transfer. Shallow-water waves involve almost hydrostatic pressure and uniform velocity distributions. The governing equations were derived by Adhémar Barré de Saint-Venant (1797–1886), so that this set of partial differential equations is also referred to as Saint-Venant equations. They read in one-dimensional (1D) form (e.g. Stoker, 1957, Chow, 1959)

$$\partial h / \partial t + \partial (Vh) / \partial x = 0, \tag{4.135}$$

$$\partial V / \partial t + V \partial V / \partial x + g \partial h / \partial x = g(S_o - S_f). \tag{4.136}$$

The continuity equation (4.135) relates the temporal change of flow depth h to the local change of unit discharge $q = Vh$, with V as cross-sectional velocity. The energy equation (4.136) accounts for the total acceleration $d(Vh)/dt$, with the two source terms S_o bottom slope and S_f friction slope. With appropriate initial and boundary conditions, this system is solved with a standard numerical approach (Chapter 11).

Roll waves constitute a periodic wave train generated on steep slopes, provided that the (1) so-called Vedernikov number $V > 1$, and (2) the channel is longer than required for wave initiation. The latter condition is accounted for by the Montuori number M, constituting the sufficient requirement for roll wave generation. Both the required and the sufficient conditions have to be satisfied, therefore.

Roll waves exist because of an instability inherent to the Saint-Venant equations. This phenomenon is not easily appreciated because it does not follow obvious hydraulic arguments. For a long prismatic channel of constant bottom slope, boundary roughness and discharge, uniform flow will eventually establish, i.e. the flow velocity and the flow depth become constant. However, if both the necessary and sufficient conditions for roll wave generation are satisfied, this most simple state of flow breaks down. As demonstrated in the 1980s, the energy transfer of roll waves is smaller than of uniform supercritical flow, so that nature tends rather to wave formation instead of uniform flow. Given that roll waves exist only for

Figure 4.59 Roll waves (a) 2D on Santa Anita Wash, Arcadia, California (Brock, 1969), (b) 3D on chute of 91 m high Llyn Brianne rockfill Dam in Wales, UK, discharge capacity 850 m³/s (Courtesy Aled Hughes, Binnie & Partners Engineering)

large Froude numbers, they are hardly experienced in natural watercourses, and they normally occur in man-made hydraulic structures, including chutes or relatively smooth roads (Figure 4.59a).

The implications of roll waves in hydraulic engineering are widespread. Designing a chute for uniform flow instead of for roll waves, the freeboard required is insufficient, so that water overtops the chute causing erosion, such as on earth dams. If bridges cross a chute carrying roll waves, flow choking may occur generating hydraulic jumps upstream, and again cause lateral overflow. If roll waves develop in tunnels, the flow may choke because wave crests become high enough to touch the tunnel ceiling causing complicated two-phase flow instead of supercritical free surface tunnel flow. Noting the large energy of hypercritical flow, these phenomena may cause massive damage, associated with complicated hydraulic flow features and extreme pressure fluctuations resulting eventually in the loss of a complete structure. Therefore, and especially under flood conditions, the presence of roll waves has to be included in a hydraulic design, to account for the determining flow conditions and to prevent damages due to unexpected flow features.

The history of roll waves is more than 100 years old (Hager, 2002). Starting with the first hydraulic description of roll waves, the main findings are reviewed up to 1961 when Montuori proposed his sufficient condition for roll wave generation.

Forchheimer (1903) was the first to physically describe roll waves. He correctly stated that roll waves are generated only under large bottom slope, and noted the surprising observation of periodic wave trains. Consider uniform flow in a rectangular channel: For subcritical flow, a small perturbation dies out within a comparably short reach, due to frictional effects. However, for a flow with a Froude number F in excess of, say, 2, a perfect uniform flow eventually develops into a wave pattern if the flow depth is sufficiently large, to exclude viscous effects. This phenomenon is observed on steeper roads during rainfall, with the formation of wave trains down the road. Although these are referred to as slug waves due to the small flow depth, their basic features are similar to roll waves. Forchheimer observed that the wave celerity c was larger than the average flow velocity V, and that the waves become steady relative to an observer moving with celerity c. The continuity equation applied at the wave front (subscript f) and at the wave peak (subscript M) then gives (Forchheimer, 1903)

$$(V_f - c)h_f = (V_M - c)h_M = (V - c)h. \qquad (4.137)$$

At locations where the free surface profile is parallel to the channel bottom, so-called pseudo-uniform flow establishes. Accounting for the Chézy (subscript C) uniform (subscript u) flow relation according to which $V_u = C_c(S_o h)^{1/2}$ for a wide channel, with C as the Chézy roughness coefficient, Eq. (4.137) then relates the two velocities at the wave front and the wave maximum $\mu = V_f/V_M$ as

$$c/V_M = 1 + \mu - \frac{\mu}{1+\mu}. \qquad (4.138)$$

The physical domain of μ is confined between 0 and 1, so that from Eq. (4.138) the ratio of velocities is $1 < c/V_M < 1.5$. For small perturbations, the wave celerity tends to 1.5 times the average flow velocity, which is close to observation. Forchheimer (1903) applied his proposal also to uniform flow in which V varies with flow depth as $h^{0.7}$ (Forchheimer, 1914), instead of the exponent 0.5 according to Chézy, expanding the range of c/V_M to 1.7, instead of 1.5. He computed the discharge below a roll wave and thus provided insight in the phenomenon.

Cornish dedicated his early life mainly to waves in various environment including sand waves, surge waves in tidal rivers, sea waves, and roll waves. Cornish (1907) introduced the English notion of roll waves. He conducted observations in England and Switzerland, comparing roll waves with tidal bores as occur for instance at the Severn River, UK (Cornish, 1914). Figure 4.59b refers to small roll waves on the chute of the 91 m high Llyn Brianne rockfill dam in Wales, UK. Given their smallness, they could also be referred to as slug waves observed on steep roads during rainfall.

The first mathematical approach to roll waves is due to Jeffreys (1925). He distinguished between laminar and turbulent plane flows aiming to determine fluid viscosity inversely by measuring discharge and flow depth in a small channel. These observations were troubled by the presence of progressive waves, however, which were nothing else than slug waves. Jeffreys stipulated that these waves 'might be due to an essential instability of uniform flow of the stream in certain conditions'. Assuming fully turbulent flow, and denoting uniform flow of velocity V_u and depth h_u, the actual values are expressed as $V = V_u + v$, and $h = h_u + \eta$. Performing a standard perturbation approach with Eqs. (4.135) and (4.136) results in a first order differential equation whose solutions are exponentials. Roll waves are demonstrated to occur only for Froude numbers $\mathsf{F} = V/(gh)^{1/2} > 2$, as noted by Forchheimer (1903). Jeffreys was the first to state that each shallow-water flow with $\mathsf{F} > 2$ eventually develops an instability, resulting in a train of roll waves rather than in uniform flow.

Vedernikov (1945, 1946) extended Jeffreys' analysis and proposed the so-called Vedernikov number. Using the generalized Saint-Venant equations for arbitrary cross-sectional shape, and the generalized equation for the friction slope $S_f = CV^p / R_h^r$, he introduced the cross-sectional parameter

$$M = 1 - R_h \left(dP / dF \right).$$
(4.139)

Here R_h is the hydraulic radius, P the wetted perimeter, and F the cross-sectional area. With the shallow-water celerity (e.g. Chow, 1959) $c = V \pm (gF/b)^{1/2}$, the Froude number is

$$\mathsf{F} = V / \left(gF / b \right)^{1/2}.$$
(4.140)

Vedernikov computed derivatives of all parameters in the governing system of equations, proposing the Vedernikov number as

$$\mathsf{V} = \left(r / p \right) M \mathsf{F}.$$
(4.141)

For $\mathsf{V} > 1$, roll waves may occur. Consider a rectangular wide and rough channel with $p = 2$ and $r = 4/3$, and $P = h$ and $R_h = h$, so that $M = 1$ resulting in $\mathsf{V} = 3/2$. For Froude numbers close to 1, therefore, supercritical flow does not result in roll waves.

Powell (1948) became aware of Vedernikov's approach and published a summary of results. The Vedernikov number V was defined as

$$\mathsf{V} = [(1 + \beta) MV / p(c - V)].$$
(4.142)

Here β and p are exponents of the uniform flow equation $S_f = kV^p / R_h^{1+\beta}$. From Manning $p = 2$ and $\beta = 1/3$, yet Vedernikov suggested $p = 2$ only. Powell stated $p + \beta = 2$, resulting in

$$\mathsf{V} = [(1 + \beta) / (2 - \beta)] M \mathsf{F}.$$
(4.143)

The Vedernikov number may thus be described as a Froude number extended with the purely geometrical parameter M and the hydraulic parameter β. Note that $V > 1$ does not imply the formation of roll waves because this is only the necessary condition.

Dressler (1949) presented the most complete mathematical work on roll waves. Starting also with the Saint-Venant equations, he sought solutions for a moving observer deriving an equation for the free surface profile, similar to the equation of backwater curves. Because roll waves for the moving observer involve a subcritical portion close to the wave crests, and a supercritical portion close to the wave fronts, so-called transitional flow must be accounted for, in which the relative (subscript r) Froude number $F_r = (V-c)/(gh)^{1/2}$ tends to 1. Dressler demonstrated that roll waves, always for the moving observer, involve a singularity, in which simultaneously $F_r = 1$ and $[S_o-(rV_c)^2/(gh_c)] = 0$, with r as roughness coefficient in the Chézy formula and subscript c denoting critical flow for the observer. This relation demonstrates that roll waves are coupled with friction, and do not exist in inviscid fluid. The necessary condition for roll waves to occur was found again as $4r^2 < \tan\alpha$, indicating that the roughness is related to the chute bottom slope.

Dressler then provided a discontinuous solution for roll waves. Using the singular point approach, of which the two real roots are h_A and h_B so that $0 < h_B < h_A < h_c$ (Figure 4.60), a so-called shock separates the wave packets. Let h_f and h_b be the flow depths just ahead and behind the shock S. By applying the standard shock equations gives for $Y_f = h_f/h_b$, in analogy to the classical hydraulic jump, with $F_b = (c-V_b)/(gh_b)^{1/2} > 0$ as the relative Froude number related to section x_s,

$$Y_f = (1/2)[(1 + 8F_b^2)^{1/2} -1].\tag{4.144}$$

These results relate the wave length Λ to the wave height h_b, independent of wave celerity c. Dressler plotted the entire features of roll waves, including the rising wave portion from h_f to h_b, and the shock separating the wave packets. For given discharge, the wave celerity c is determined for given h_o and V_o. A classification of all possible surface profiles is also presented, in analogy to backwater curves classified by Chow (1959). Apart from a comparison with observations, the approach of Dressler thus may be regarded complete. An extension was presented by Dressler and Pohle (1953).

Mayer (1961) conducted an experimental study on roll waves with a distinction between slug waves for small flow depths (where viscous effects are significant), and roll waves for larger flow depths where gravity plays the key role. His paper was discussed, of which emerged additional knowledge. Ishihara et al. (1962) related the dimensionless wave velocity

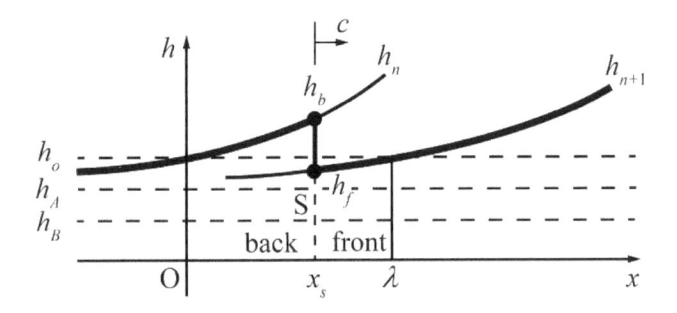

Figure 4.60 Plot of roll waves as seen by moving observer (Dressler, 1949)

$\mu_w = V_w/(gR_o\cos\alpha)^{1/3}$ to the Froude number defined as $\mathsf{F}_o = V_o/(gR_o\cos\alpha)^{1/2}$ that reads simplified as (Hager, 2002)

$$\mu_w = 2 + 0.5(\mathsf{F}_o - 1). \tag{4.145}$$

The relative wave length $\lambda = L_w S_o (g\cos\alpha / V_w^2)$ and wave period $\tau_w = \Lambda/\mu_w$ practically do not vary with the Froude number and are expressed for turbulent flow as $\Lambda = 0.7$, and $\tau_w = 0.15$, respectively. In contrast to Mayer, Ishihara *et al.* (1962) verified no effect of surface tension on the roll wave development.

Escoffier and Boyd (1962) considered arbitrary cross-sectional shapes, determining the necessary criterion for roll wave inception. Their analysis was directed toward the circular, trapezoidal and rectangular chute profiles. Design charts were prepared to distinguish between tranquil flow ($\mathsf{F} < 1$), steady rapid flow ($\mathsf{V} < 1$) and 'pulsating' rapid flow ($\mathsf{V} \geq 1$). These diagrams relate the relative discharge $Q/(gD^5)^{1/2}$ to relative bottom slope $j = S_o^{1/2} D^{1/6} / g^{1/2}$, with D as conduit diameter and K as the Manning-Strickler roughness coefficient. Using simplified expressions for the circular conduit (Hager, 2010), i.e. $\mathsf{V} = 0.56 K S_o^{1/2} D^{1/6} h^{1/2}$ for uniform velocity, and $A = D^{0.6} h^{1.4}$ for cross-sectional area, the stage variable obtains $\omega = 2(2gh)^{1/2}$, and the Vedernikov number is independent of stage $\mathsf{V} = (9/16) K S_o^{1/2} D^{1/6} / g^{1/2}$. This compares well with Escoffier and Boyd (1962). To the same order of approximation, the Froude number of uniform conduit flow may be expressed as $\mathsf{F} = (9/16) K S_o^{1/2} D^{1/6} / g^{1/2}$, so that $\mathsf{V} = (80/9)\mathsf{F} = 8.9\mathsf{F}$, i.e. much larger than in the rectangular channel. Therefore, roll waves may be hardly seen in conduit flow. Roll wave inception (subscript i) thus depends mainly on the conduit bottom slope and the roughness coefficient and is computed in circular conduits by

$$[K S_o^{1/2} D^{1/6} / g^{1/2}]_i = 0.2. \tag{4.146}$$

4.6.2 Advances from Montuori

Montuori added to the knowledge of roll waves. His first paper (1961) involves a rederivation of known facts, and observations relating to the rectangular, trapezoidal, triangular and semicircular chutes. He then set up a diagram relating the so-called Montuori number $\mathsf{M} = V^2/(gLS_o)$ versus the Vedernikov number V, with V as uniform velocity and L the distance required to initiate roll waves. He defined two regimes, depending on whether M is larger or smaller than the inception (subscript i) number M_i, with (Figure 4.61, Montuori, 1984)

$$\mathsf{M}_i = (1/12)(\mathsf{V} - 1). \tag{4.147}$$

This criterion corresponds to the sufficient condition for roll wave formation. His benchmark paper was also translated in English (Montuori, 1965).

Brock (1969) conducted experimental work to roll waves investigating mainly the wave development for channel bottom slopes up to $S_o = 0.12$. The study aimed at the determination of the minimum flow depths during wave passage, the wave period, the wave celerity c, and the maximum wave heights h_M. It was found that the chute inlet conditions were important: if the bottom of the inlet was smooth as the entire chute, roll waves developed further upstream as for a rough inlet portion. This difference was associated with differing transition developments of the laminar to the turbulent boundary layer along the chute bottom. Accordingly,

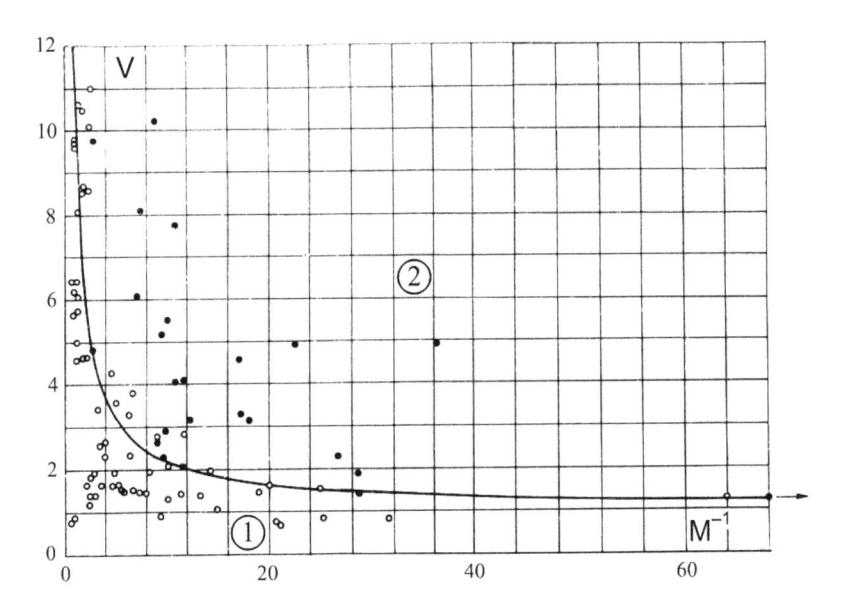

Figure 4.61 Delimitation V(M⁻¹) between flows ① with, and ② without roll waves (Montuori, 1984)

all tests were run with the rough inlet, for which the boundary layer was turbulent right from the chute inlet.

Let h_N be the uniform flow depth of undisturbed chute flow, so that the mean dimensionless (subscript *av*) wave period is $T_{av} = S_o t_{av} (g/h_N)^{1/2}$, the mean wave propagation velocity $C_{av} = c_{av}/(gh_N)^{1/2}$, with $X = x/h_N$ as the dimensionless location, $Y_M = h_M/h_N$, and $Y_m = h_m/h_N$ as maximum (subscript *M*), and minimum (subscript *m*) wave heights. Figure 4.62 shows the development of these four governing wave parameters, indicating an increase of both $Y_M(X)$ and $T_{av}(X)$ for a bottom slope of $S_o = 0.084$, whereas $Y_m(X)$ reduces and $C_{av}(X)$ depends on the bottom slope. Brock mentions that these curves look similar for the other bottom slopes investigated. Note from Figure 4.62a the small wave growth rate up to $X = 1500$, followed by a nearly linear growth up to $X = 9000$. Figure 4.62b indicates that T_{av} remains constant up to $X = 2500$, from where it also increases almost linearly until the chute end due to the wave overtaking process. As to the development of the wave minimum, Figure 4.62c shows a rapid decay up to $X = 4000$, from where $Y_m \cong 0.57$. A plot $Y_M(T)$ reveals that roll waves become higher at smaller as compared with larger bottom slopes.

The data of the wave maximum development $Y_M(X)$ collapse for various uniform flow depths at equal bottom slopes. Within the slope range $0.05 \leq S_o \leq 0.12$ the data are made up by the upward convex portion, followed beyond the deflection point by a downward concave portion. According to Brock (1969), waves typically break at the point of inflection, i.e. at $Y_M \cong 1.3$. It is noted that Brock's three curves are similar, and that the effect of the uniform flow Froude number $F_N = V_N/(gh_N)^{1/2}$ is easily accounted for by using the transformation $X^* = XF_N^2$. The result is

$$Y_M = 1 + 1.75[\tanh(10^{-5} X^*)]^{2.5}, \text{ if } 3.5 < F_N < 5.6, X < 10^4. \tag{4.148}$$

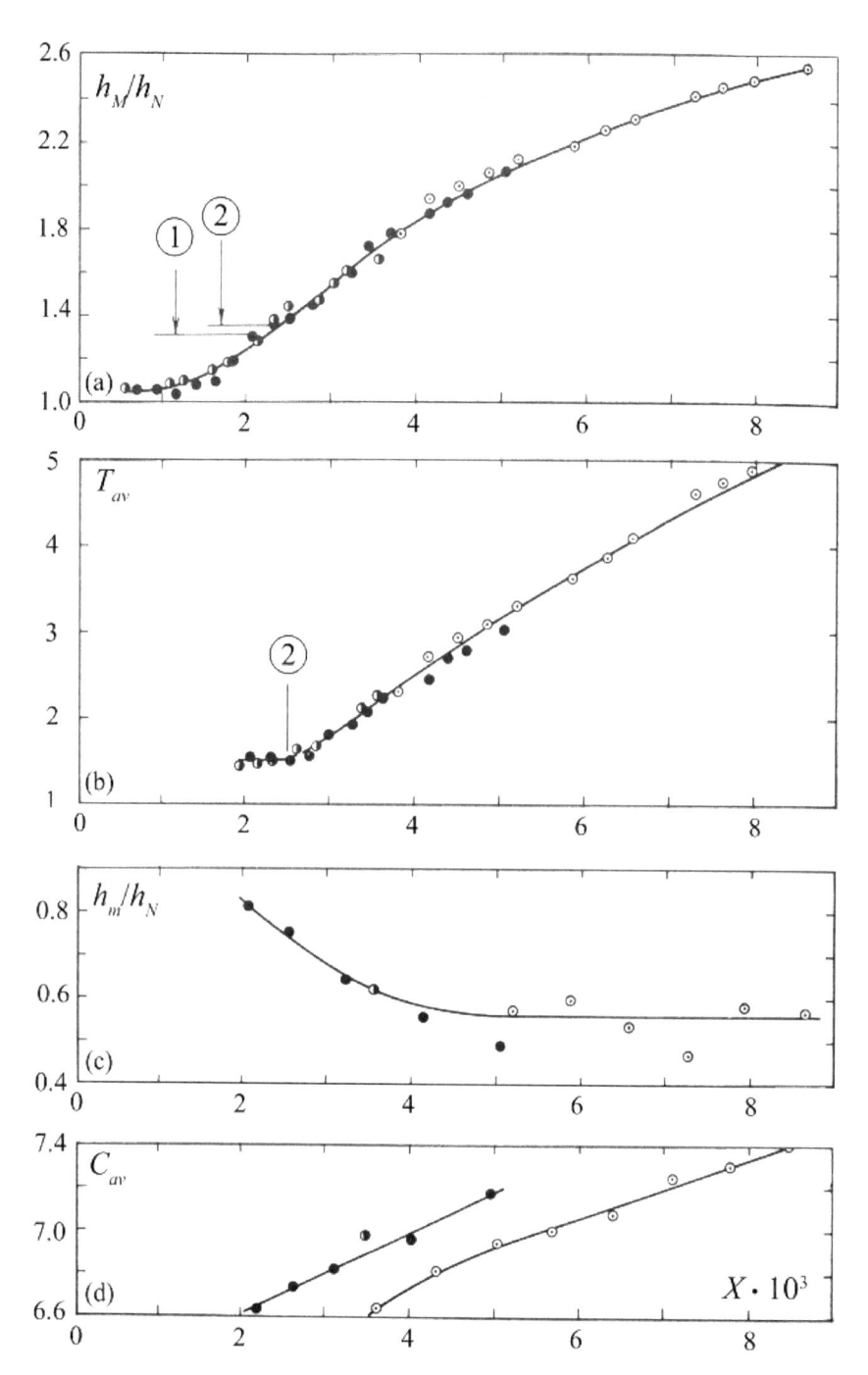

Figure 4.62 Development of roll waves according to Brock (1969) for $S_o = 0.084$ with (a) $h_M/h_N(X)$, (b) $T_{av}(X)$, (c) $h_m/h_N(X)$, (d) $C_{av}(X)$ with ① wave formation, ② start of wave overtaking. Test parameters $[h_N \, [\text{m}]; F_N] = (\odot) \, [0.06; 4.63], (\bullet) \, [0.09; 4.96],$ (◗) [0.12; 5.05]

From Eq. (4.148), the absolute maximum roll wave height tends to a value of almost $Y_{Mabs} = 3$, indicating its significant effect on the freeboard design, and the damages to be expected when overlooking their presence. Note that definite wave breaking occurs at $Y_M \cong 2$, i.e. once the wave height is double of the uniform flow depth.

In the discussion, an additional relation of the Montuori chart is given for prototype chutes as $V_i = 0.07F_N^2/(S_o X)$; if $V > V_i$ then roll waves are generated, whereas they are absent otherwise. Additional information was provided by Brock (1970). A review of results was provided by Berlamont (1986). Rosso *et al.* (1990) studied the frictional effect on roll waves, whereas Cassidy (1990) presented practical aspects in dam hydraulics, pointing at the implications of roll waves. The evolution of roll waves from a mathematical point of view is discussed by Kranenburg (1992).

4.7 Stepped chute

4.7.1 Introduction

Definitions

Whereas a 'smooth' standard chute (4.2) directs the overflow to the outlet structure, followed by concentrated energy dissipation, the stepped chute (cascade) corresponds to a hybrid structure. It combines, at least up to a certain extent, the transport function with an anticipated energy dissipation. Accordingly, the terminal structure has potentially a reduced rate of energy to dissipate, and is thus smaller. Figure 4.63 compares the conventional system chute-stilling basin with the stepped chute.

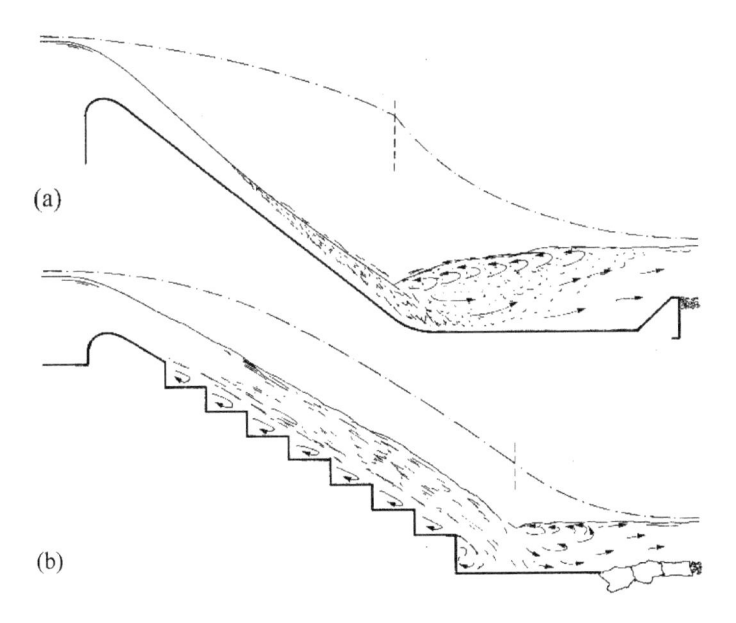

(a)

(b)

Figure 4.63 Comparison of (a) smooth chute with a limited energy dissipation and (b) stepped chute with a comparably enhanced energy dissipation, (- - -) energy headline

Stepped chutes are mainly built with the roller-compacted concrete (RCC) construction technique. They are frequently used on the downstream face of RCC dams, but also gain importance as auxiliary chutes beside or on embankments dams. The former application usually implies a steep stepped chute with inclination angles of around 50° relative to the horizontal (~1V:0.8H) (Figure 4.64), whereas embankment dam chutes are typically flat with angles of some 15° to 30° (~1:4 to 1:1.7) (Figure 4.65).

(a)

(b)

(c)

Figure 4.64 Stepped chute of (a) Hinze Dam, (b) Wyaralong Dam, both Australia (Courtesy Michael Pfister), (c) Murum Dam, Borneo, Malaysia, spilling at a unit discharge of $\cong 10$ m²/s (Courtesy Brian Forbes)

(a)   (b)

Figure 4.65 Stepped chute of (a) Big Beaver Dam (CO), and (b) Upper Las Vegas Wash Dam (NV), both USA (Courtesy Robert M. Boes)

The flow features along a stepped chute differ significantly from those of a smooth chute due to the pronounced step-induced surface roughness:

- Three flow regimes in terms of the unit discharge, namely nappe flow for small, transition flow for intermediate, and skimming flow for high unit discharges (4.7.4, Figures 4.63b and 4.64c);
- Steps increase flow turbulence, dissipating energy (4.7.4) and causing natural air entrainment of the flow for relatively small discharges, as compared with smooth chutes (Figure 4.64c);
- Flow depth is larger than on smooth chutes due to highly aerated flow, thus requiring higher side or training walls;
- Small discharges tend to spray formation, due to flow impact onto a series of horizontal step treads (4.7.4).

Stepped chutes are described with the following geometrical parameters: vertical step height s, horizontal step length l, and chute angle relative to the horizontal $\phi = \tan(s/l)$. The latter is identical with the pseudo-bottom angle, a virtual plane formed if connecting all step edges. Usually, the pseudo-bottom is a plane, and exceptionally only curved near a possible ogee at the stepped chute start (4.7.3).

Selected historical aspects

Chanson (1995b, 1996, 2002) reports of early stepped chutes erected in Iran around 700 BC. The 'rediscovery' of the concept of stepped chutes dates from the 19th century, with the New Croton Dam as most prominent application (Hager and Pfister, 2013). The New Croton Aqueduct was built from 1885 to 1890, satisfying the increasing demand of drinking water of New York City. The dam should have originally consisted of a 185 m long straight wall, combined with a 300 m long and bent section over which the spillway should be erected. The final design was slightly modified, with the spillway discharging into a side channel that collects and then discharges the flood water. The spillway consists of a stepped chute and counts to the largest erected at this time, with a maximum design discharge of 1550 m^3/s (Wegmann, 1911).

An early hydraulic study on cascades, the so-called 'step spillway,' was conducted analytically by Armani (1894). The water passed a succession of either free or submerged pool-sill elements, described by applying the basic weir flow formula. One of the plots shows a wavy surface structure which is reported to have occurred during a thunderstorm in the Vistula River Basin in 1893. A similar laboratory study was conducted by Koch (1923), yet without answering the main questions on the detailed hydraulic features of these flows.

Stepped spillways as currently employed became a design variation of the standard spillway from 1900. Typical examples include the Urft Dam, Germany, then the largest dam in Europe (Küppers, 1905), or the Eschbach Dam, or the cascade of the Los Angeles Aqueduct involving the currently used stepped spillway (Heinly, 1913).

The first thorough study of stepped spillways was conducted by Gausmann and Madden (1923) on the Gilboa Dam (USA) by the Board of Water Supply of New York City for the Catskill System. The dam consists of a 400 m long masonry section and a 300 m long earthen dam. Its downstream face is formed of 6 m high steps. The 50 m high masonry portion was then the highest spillway of the world. Laboratory experiments were conducted on scales 1:50, 1:20, and 1:8, under a prototype approach flow head of 0.90 m.

Risks

Major problems have rarely occurred on stepped chutes so far, despite of the large number of existing RCC dams. This might, however, be related to the relatively small unit discharges so far released. Typically, 20–40 m^2/s are conveyed, requesting wide chutes to transport high absolute discharges. These are financially unattractive in terms of safety-relevant auxiliary dam structure (not included on the downstream dam face), and may require a chute contraction at the dam toe. If a less conservative design is chosen with higher unit discharges, even up to the values of smooth chutes, then particular phenomena are expected, which have to be considered during design. These are, among others, cavitation (4.7.3) and fluctuating (negative) bottom pressures (4.7.4) acting on the concrete upstream of the incipient point of natural self-aeration.

A specific analysis of accidents and failures on stepped chutes was conducted by Chanson (2002). He concludes that the following issues caused failure: improper hydrological assessment, foundation failure and instability, flow instabilities associated with the transition flow regime (4.7.4), lack of maintenance, and poor quality of construction. From a hydraulic design standpoint, the transition flow regime on a stepped chute seems unfavorable, since the latter generates rapid longitudinal flow variations and fluctuations. The related hydrodynamic instabilities cause fluctuating hydraulic loads and vibrations. Design engineers should avoid the transition regime, particularly for floods close to the design discharge. The transition to skimming flow should take place under relatively small absolute discharges, or be skipped by an adequate gate operation for regulated inlets. High unit discharges generate a drastically reduced relative energy dissipation, as compared to small flows on stepped chutes. This is linked to the relatively reduced step roughness. A flood exceeding the design value requires thus an overproportionally large dissipation structure, as compared with the design flood.

4.7.2 Main application

Stepped spillways on gravity dams

Stepped chutes are frequently applied at RCC gravity dams (Figure 4.64), because most have stepped downstream faces as a result of the construction method. Beginning in 1982 with the first major dam constructed of RCC, Willow Creek in the USA (Hansen and Reinhard, 1991; Boes, 1999a), the use of stepped chutes has since greatly increased worldwide as it complements the RCC construction method and because of the high energy dissipation rate, reducing the cost of stilling basins. RCC has gained a wide application in dam engineering because of a number of advantages, resulting in more time- and cost-efficient dam design and construction (Hansen and Reinhard, 1991; Hansen, 1991, 1996; Boes, 1999a). At the initial stage of the RCC technique, many designs were conservative, with conventional vibrated concrete faces and smooth chutes used. Nowadays, most RCC gravity dams have stepped downstream faces.

The layout of stepped chutes is similar to that of smooth chutes on concrete gravity dams, consisting of control structure, spillway chute and terminal structure. However, the hydraulic performance of the stepped chute requires design modifications (4.7.4): (1) crest-to-chute transition has to be designed carefully to avoid flow deflection and spray; (2) training wall height has to account for flow bulking; (3) cavitation risk has to be carefully analyzed and aeration provided if needed; and (4) terminal structure accounting for air-water flow characteristics and energy dissipation along the chute.

Overflow RCC stepped spillways on embankment dams

Stepped chutes are also frequently employed at embankment dams in the form of an over-topping protection with an RCC overlay. Typically, existing embankment dams with limited hydrological safety are in need of overflow emergency spillways supplementing the service spillway to discharge large floods (Berga, 1995; PCA, 2002; FEMA, 2014; Toledo *et al.*, 2015; Boes *et al.*, 2015). A concrete slab armoring the downstream slope and rendering over-flowing possible is an attractive design option (Boes, 1999b). The RCC technique is often used for constructing these slabs (Hansen and France, 1986). As RCC is typically placed in horizontal layers, the downstream face constituting the spillway becomes stepped (Figure 4.66). For small dams (<15 m high), this alternative is cost-effective when compared to other measures, such as enlarging the existing service spillway, constructing a new spillway separated from the dam body, or heightening the dam (Hansen, 2003; Bass *et al.*, 2012). The statistics of 109 overflow RCC spillways in the USA (FEMA, 2014) show that these range in height from 5 to 20 m, and that the maximum height to date is 35 m; regarding the design unit discharge, the average is below 3 m^2/s and the maximum around 10 m^2/s (ICOLD, 2017).

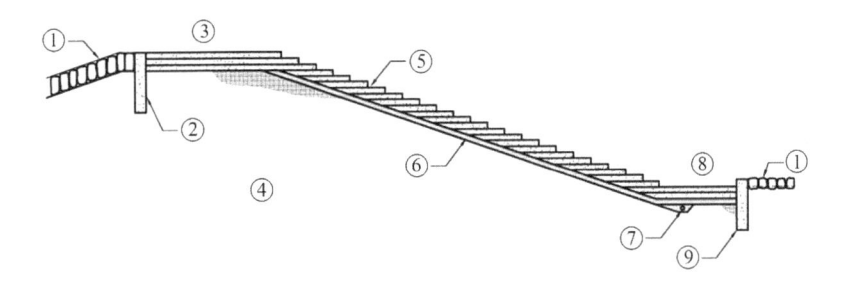

Figure 4.66 Typical cross section of embankment dam with RCC overtopping spillway, ① riprap, ② upstream cutoff wall, ③ broad-crested weir, ④ homogeneous embankment, ⑤ RCC, ⑥ filter drain, ⑦ drain collector pipe, ⑧ downstream apron, ⑨ downstream cutoff wall (Adapted from Bass *et al.*, 2012)

Despite the common use of RCC, the design and construction conditions of overtopping protection stepped chutes are different from those on gravity dams, as the size and the finishing of the steps vary. The usual step height is 0.30 m (1 lift), the typical downstream slope (V:H) between 1:2 (26.6°) and 1:5 (11.3°) and the facing may be formed or unformed (Figure 4.65). When unformed, the step face is not compacted and it has to be considered as a sacrificial concrete (ICOLD, 2017). Additionally, the hydraulic performance of the unformed protections differs from the formed and less energy dissipation occurs (FEMA, 2014). The lift width is determined by construction and structural requirements. Special care has to be taken with the uplift pressures that could occur beneath the slab, and a drainage system has to be provided. The minimum lift width is around 2.5 m, limited by the vibrating roller size (ICOLD, 2017). Larger width is required for large unit discharges and for flat slopes, where a minimum slab thickness of 0.6 m is recommended (PCA, 2002).

Stepped spillways along embankment dam abutments

Stepped chutes have been applied in spillways located along embankment dam abutments excavated into rock. It is a promising use in future projects aiming at increased energy dissipation (ICOLD, 2017), benefitting from research on the hydraulic performance of stepped chutes on moderate and flat slopes (Ohtsu *et al.*, 2004; Boes and Hager, 2003a, b; André, 2004; González and Chanson, 2007; Meireles and Matos, 2009; Lutz *et al.*, 2015). These chutes need to be adapted to the prevailing topography in terms of chute slope and step height (ICOLD, 2017). To keep the latter constant over distances as long as possible, the use of transition channels may be an option (Baumann *et al.*, 2006; Boes *et al.*, 2015; Lutz *et al.*, 2015).

Stepped masonry spillways

Prior to the era of modern concrete, stepped chutes made of masonry with and without pooled steps with vertical end sills were often applied as service spillways next to both gravity and earth embankment dams along natural terrain. The slopes (V:H) of these spillways are typically in the range of 1:3 (18.4°) to 1:10 (5.7°) (ICOLD, 2017). Unstable flow phenomena or even failures of two spillways in the UK at Boltby Dam in 2005 and Ulley Dam in 2007

(a)   (b)

Figure 4.67 (a) Robina Gabion Weir, Australia, with $s = 0.6$ m and $l = 1.1$ to 2.0 m (Courtesy Davide Wüthrich), (b) nappe flow down a Gabion stepped cascade (Boes and Minor, 2002)

have led to investigations on their possible causes (Herrmann, 2003; Winter *et al.*, 2010). Variations of the hydrodynamic pressures on the sidewalls were highlighted as a significant vulnerability and the need for good continued maintenance of both the masonry and the mortar was highlighted as one essential aspect for these spillways (ICOLD, 2017).

Gabion structures

Stepped chutes made up with gabions are mostly applied as temporary structures (Figure 4.67). Gabion steps are flexible and adapt to settlements, are sustainable, have a reduced environmental impact, are easy to construct and to remove, and are relatively cheap. Gabions also proved to reduce the noise of the flow (Boes and Schmid, 2003). Some disadvantages are a possible deterioration and a more complex flow behavior because of the interaction between seepage and free-surface flows. Similar to 'classical' stepped chutes, experimental studies including gabions have shown that they feature enhanced energy dissipation rates as compared with 'smooth standard' structures. However, their energy dissipation rate is smaller than of the 'classical' stepped spillway (impermeable, made for instance of RCC).

The effect of gabion step roughness on the flow properties was studied by Gonzalez *et al.* (2008) and Bung and Schlenkhoff (2010). Both studies showed that rough configurations were characterized by faster flow associated with less energy dissipation. The major difference between 'classical' stepped chutes and gabions is the gabion porosity, allowing for a portion of the total discharge to seep through the gabions made up of stones. Stephenson (1979) was one of the first to present results of both flow regimes and energy dissipation on stepped gabion weirs. Peyras *et al.* (1992) studied the flow patterns and energy dissipation of gabion stepped weirs, while Kells (1993) discussed the interactions between seepage and free-surface flows. A comprehensive study on the hydraulics of gabion stepped spillways was performed by Wüthrich and Chanson (2014a, b) for a wide range of discharges, covering the nappe, transition and skimming flow regimes. For the latter study, a gabion configuration was installed over an impervious structure.

Visually, the flow over gabion steps is less aerated and the inception point is located further downstream as compared to a 'classical' stepped chute. Higher flow velocities were measured on all gabion steps, resulting in a lower energy dissipation rate as compared to a 'classical' stepped chute. Although it might appear counter-intuitive, the global friction factor of gabion structures was lower than for 'classical' stepped chute. The former's friction factor was on average half of that of the 'classical' stepped chute for otherwise identical conditions.

An altered step niche recirculation due to the interaction between seepage and free surface flows was noted. Flow observations at step niche level were conducted by Wüthrich and Chanson (2014a, b) by dye injection and colored strings positioned into the cavity. The presence of a clear-water core in the niche center was noted, limiting the recirculation to the downstream end of the niche. A bubbly motion was observed inside the gabions downstream of the air inception point. Zhang and Chanson (2016a, b) performed bubble tracking in the gabions, showing large bubble velocities particularly next to the step edge. Large bubble velocity fluctuations were recorded throughout the gabions. Altogether, the interactions between seepage and free-surface flows were significant, leading to a complex cavity recirculation flow pattern.

Adapted configurations of gabion steps with capped structures were tested by Wüthrich and Chanson (2015) by introducing horizontal impervious step cappings. Results showed that the air-water properties of capped gabion stepped chutes were in between the 'classical' and gabion stepped chute flow properties.

Stepped chute flows are known for their natural self-aeration performance. Wüthrich and Chanson (2015) showed that, under a small discharge, the aeration efficiency is higher on the gabion stepped chute, whereas for large discharges, the 'classical' stepped chute is more efficient in terms of aeration and mass transfer.

4.7.3 General considerations

Cavitation potential

Steps represent a series of bottom irregularities protruding into the flow, generating thus flow separation. Consequently, the local average pressure reduces locally, as measurements on physical models confirm (4.7.4). Considering in addition the dynamic pressure fluctuations in these separation zones, it becomes evident that the absolute pressure can approach values leading to a phase change from water to vapor, potentially associated to cavitation occurrence at the end of the separation zones (downstream end of horizontal and upper end of vertical step face, respectively). As shown in (4.4.2), bottom irregularities and abrupt slope changes are sensitive in terms of cavitation development and should be avoided. It is consequently appropriate to discuss the cavitation potential of stepped chute flow.

LIMIT PRESSURE HEAD FOR CAVITATION INCEPTION

The cavitation occurrence of stepped chutes is assessed by considering hydrodynamic pressures. According to Lopardo (2002), the minimum pressure arising with 0.1% probability, $p_{0.1\%}$, is the representative negative pressure for cavitation occurrence in macro-turbulent flow. Figure 4.68 depicts dimensionless pressure coefficients $C_{p0.1\%}$ along a spillway chute

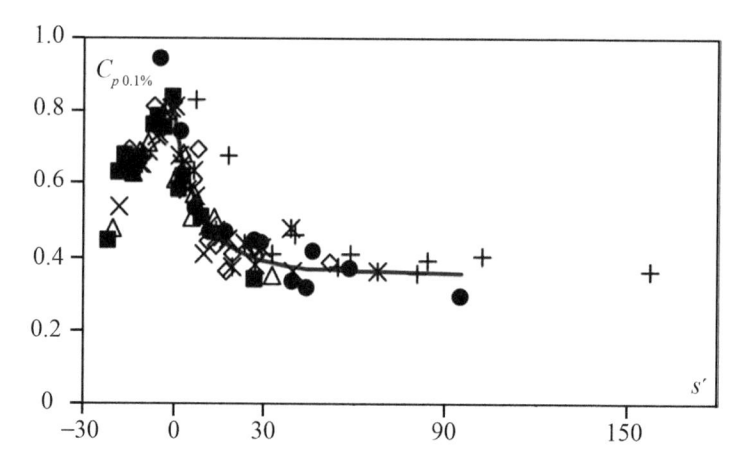

Figure 4.68 Pressure coefficient of minimum pressures with 0.1% probability $C_{p0.1\%}$ on vertical step faces near outer step edge at $z/l = 0.07$ measured for skimming flow along a stepped chute scale model of slope 1:0.8 with s' = relative distance to inception point of air entrainment for different relative flow depths h_c/s, where h_c is critical flow depth, with h_c/s = (■) 3.21, (△) 2.93, (×) 2.65, (◊) 2.25, (•) 1.85, (•) 1.41, (+) 0.89, (−) $C_{p0.1\%}$ = 0.358/[1−0.543exp(−0.062s')] (Adapted from Amador *et al.*, 2009)

with $\phi = 51.3°$ (1:0.8) at the critical point of a step at a relative vertical distance $z/l = 0.07$ from the outer step edge, where the minimum measured pressures occur (4.7.4). Note that the pressure coefficient is defined as the ratio of pressure head to velocity head, i.e. $C_p = p/(\rho g)/[V^2/(2g)]$. The maximum $C_{p0.1\%}$ values of 0.8–0.9 occur near the inception point of air entrainment.

Pressure measurements allow for a direct comparison of the local minimum pressure of 0.1% probability with the limit pressure head for cavitation inception of around -10 m water column (w.c.) for pure water. The presence of particles in suspension and dissolved gas can cause cavitation at higher pressures. Thus, a pressure head of about -7 to -8 m w.c. is typically considered as limit pressure head in hydraulic engineering design (USACE, 1995). Cavitation may also occur due to pressure fluctuations falling below the limit pressure head even if the mean pressure is well above the mentioned limit. Based on the 0.1% percentile of pressure fluctuations at the natural inception point (Lopardo, 2002), Amador *et al.* (2005) suggest to limit the unit discharge on stepped chutes to 13.9, 11.5, and 8.4 m²/s for step heights of $s = 1.2, 0.9$, and 0.6 m, respectively, for a pseudo-bottom angle $\phi = 51.3°$ (1:0.80).

CRITICAL VELOCITY FOR CAVITATION INCEPTION

The critical velocity V_c for which the minimum pressure of 0.1% probability reaches the critical pressure p_c is with ρ_w as water density

$$V_c = \sqrt{\frac{-2p_c/\rho_w}{C_{p0.1\%}}} . \tag{4.149}$$

Example: With a critical pressure $p_c = p_v - p_{atm} = (1.23 - 95.46)$ kPa $= -94.23$ kPa, where $p_v = 1.23$ kPa is the vapor pressure at 10°C and 95.46 kPa atmospheric pressure p_{atm} at 500 m altitude a.s.l. based on the International Standard Atmosphere, and $C_{p0.1\%}$ is in the range of 0.45–0.95 upstream of the inception point of air entrainment (Figure 4.68 for $s' < 0$), the resulting critical velocity is $V_c = 14$ to 21 m/s. Amador *et al.* (2009) recommend to limit the flow velocity to 15 m/s at the inception point, thereby considering $C_{p0.1\%} \approx 0.84$ relevant for the minimum pressures with 0.1% probability.

LIMIT BOTTOM AIR CONCENTRATION FOR CAVITATION INCEPTION

As mentioned, cavitation inception has to be expected for high-velocity flows above ~15–20 m/s. A bottom air concentration of 1%–2% significantly reduces the cavitation risk (Chanson, 1989a); values of 5%–8% are sufficient to avoid cavitation damages (Peterka, 1953) (4.7.4). The point just upstream of the air inception point is thus the most critical for cavitation inception. Using the approximation of Boes and Hager (2003a) for the general streamwise development of bottom air concentration for $26° < \phi < 55°$ (4.7.4) and considering $C_b = 0.05$ and $C_b = 0.08$ as minimum values to avoid cavitation damage to spillway concrete, respectively (Peterka, 1953), the required non-dimensional distances $X_{i,\text{crit}}$ from the inception point are

$$X_{i,\text{crit}} (C_{b,\text{crit}} = 0.05) = 5(\sin\phi)^{-2.3}, \tag{4.150a}$$

$$X_{i,\text{crit}} (C_{b,\text{crit}} = 0.08) = 10(\sin\phi)^{-3.0}. \tag{4.150b}$$

This suggests that the distance required to attain a sufficiently high bottom air concentration depends significantly on the value $C_{b,crit}$. Because the aeration tends to be more pronounced in the prototype than on spillway models underlying Eq. (4.150a, b), $C_{b,crit} = 0.05$ is suggested for design purposes. For typical gravity dams with a downstream slope of 1:0.8, the critical distance from the inception point amounts to about $9h_i$, with h_i = flow depth at inception point (4.7.4). Similar to the black-water region upstream of the inception point, the downstream developing region (4.7.4) down to $X_{i,crit}$ might also be prone to cavitation damage for velocities larger than the critical cavitation inception velocity mentioned above. Because potentially critical velocities of 20 m/s are attained at $X_{i,crit}$ for unit discharges close to 25 m²/s, depending on the chute angle and the step height, an aerator placed in the unaerated chute region downstream from the crest (4.7.4) might become necessary for larger unit discharges.

Critical cavitation index

Cavitation occurs as long as the air concentration of the near-bottom flow is below a few percent (4.4.3). Accordingly, the black-water reach between the chute top and the point of natural self-aeration is potentially endangered. Given that the cavitation index σ lowers with increasing velocity and thus increasing chute length, the region immediately upstream of the air inception point of natural self-aeration is most critical. The σ value at the point of bottom air inception along a stepped chute may be predicted based on Boes and Hager (2003a). For a given unit discharge, the mixture-flow depth at the inception point follows from 4.7.4. Using the average value $C_{a,i} = 0.228$ at the inception point for $\phi = 50°$ results in the corresponding black-water flow depth, hydrostatic pressure, and black-water velocity. These data allow for the computation of σ at the bottom inception point in analogy to 4.4.3. The cavitation index σ is related in Figure 4.69 to the unit discharge q for a pseudo-bottom angle of $\phi = 50°$ and various step heights. The cavitation index at the inception point for $s = 1.2$ m and $q = 30$ m²/s for instance is $\sigma = 0.42$, whereas it is $\sigma = 0.17$ for $q = 100$ m²/s. The distances from the

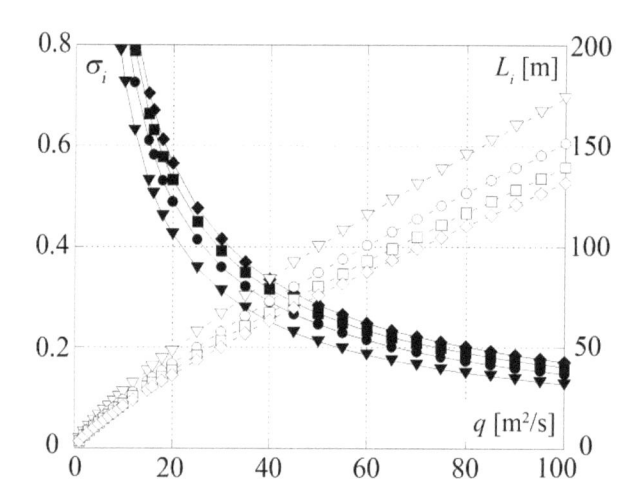

Figure 4.69 Cavitation index σ_i (solid symbols) at inception point L_i (open symbols) for $\phi = 50°$ and step height s [m] = ($\triangledown$) 0.30, ($\circ$) 0.60, ($\square$) 0.90, ($\diamond$) 1.20 (Pfister and Boes, 2014)

spillway crest to the location of the bottom air inception point as derived from Boes and Hager (2003a) (4.7.4) are also plotted in Figure 4.69.

The bottom air inception point is the most critical location for cavitation damage. Further upstream, σ is larger, whereas the flow close to the chute bottom is aerated further downstream. The cavitation indices at the most endangered section can thus be estimated, but the critical value for cavitation inception remains to be defined.

Following Drewes (1988) and Falvey (1990), the critical cavitation index σ_c for an individual step protruding into the flow is of the order of $\sigma_c = 1.0$. The critical index for a sequence of steps is not discussed, but is most likely lower because (i) upstream steps orient the flow, (ii) a uniformly rough surface has a lower cavitation potential than an isolated roughness of identical geometry due to reduced velocities and wake effects (Arndt and Ippen, 1968; Frizell and Mefford, 1991), and (iii) steps form large offsets away from the flow direction, preventing cavitation from residing on the boundary (Frizell and Mefford, 1991). The effective cavitation index computation along a smooth chute usually is based on hydrostatic pressure, as detailed in 4.4.3. For stepped chutes, this appears inadequate, as the local pressure below the step edge is much smaller than the hydrostatic value. Based on effectively measured pressures, up-scaled to steps of $s = 1.2$ m, Pfister *et al.* (2006b) conclude that a critical index of $\sigma_c = 0.9$ is reached at the top of the stepped chute (first vertical step face below a standard crest, Figure 4.70a) if $q > 30$ m²/s. Choosing lower critical σ_c values allows for higher unit discharges. According to Boes (2012), the limit cavitation index is around 0.5 for stepped chutes in the common slope range of 1:2.9 ($\phi = 19°$) to 1:0.7 ($\phi = 55°$).

Frizell *et al.* (2013) conducted a comprehensive cavitation study on stepped chutes in a reduced ambient pressure conduit. In analogy to Arndt and Ippen (1968), with f_b as Darcy-Weisbach friction factor (4.7.4), they suggest a critical cavitation index of

$$\sigma_c = 4 f_b. \tag{4.151}$$

The stepped chute data collapsed with 'rough' surface data suffering from cavitation. They found values in the range of $0.31 < \sigma_c < 0.65$. Frizell *et al.* (2015) performed additional tests on a high-speed facility. Damages were observed at $\sigma = 0.3$ for a flow velocity of 22 m/s. Pitting was detected near the downstream third of the horizontal step face (jet impact location) at all steps for $\phi = 21.8°$, and less damages for $\phi = 68.2°$.

Chanson (2015) discusses the work of Frizell *et al.* (2013), pointing at no relevant cavitation damages so far reported from prototype stepped chutes, although few chutes were operated under relatively large unit discharges up to $q = 72$ m²/s. It remains a fact, however,

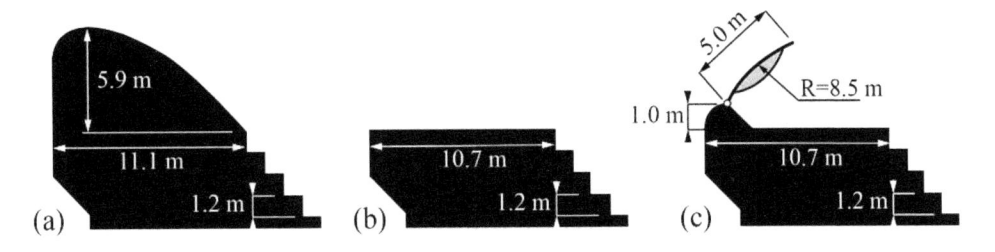

Figure 4.70 Possible crest types in combination with stepped chutes (Pfister, 2009) in prototype dimensions (a) standard-crested, (b) broad-crested, (c) flap gate

that most of today's 700 RCC dams were built during the past 20–30 years. They are thus relatively young structures. In addition, spillways are designed to convey rarely occurring but extreme floods. A statistically low operation probability is, at least for the moment, combined with a short lifetime. Moreover, the flood duration is another important parameter for cavitation damage to occur. The longer the cavitation indices fall below critical values, the higher the likeliness to observe damages. Furthermore, most of these stepped chutes were designed for moderate unit discharges far below the values mentioned by Chanson (2015). The lack of cavitation damages can therefore also be related to the rare operation under significant unit discharges, besides unfavorable flow characteristics for cavitation occurrence (Chanson, 2015). The consideration of the flow cavitation potential on a stepped chute is nevertheless mandatory during the design process.

Inlet control structure and transition to stepped chute

The inlet and control structures are located at the upstream end of the stepped chute. The following control structures are found in combination with stepped spillways (Chanson, 2006; Pfister, 2009; Boes, 2012):

1 Ogee-crested weir with or without steps (Figure 4.70a)
2 Broad-crested weir with or without gates (Figure 4.70b, 4.70c)
3 Piano key weir.

For an uncontrolled inlet, the crest is designed to optimize the discharge capacity combined with a minimum reservoir head and low cost.

The probably most common unregulated inlet structure is the *standard crested* weir (2.2.1, Figure 4.70a). The latter follows the lower jet trajectory of a sharp-crested weir under design head. The flow is then tangentially orientated along the profile and the pressures on the invert are atmospheric. The stepped chute of constant step height starts at the tangency point T (Figure 4.71), where the ogee profile tangentially joins the pseudo-bottom angle. Given that

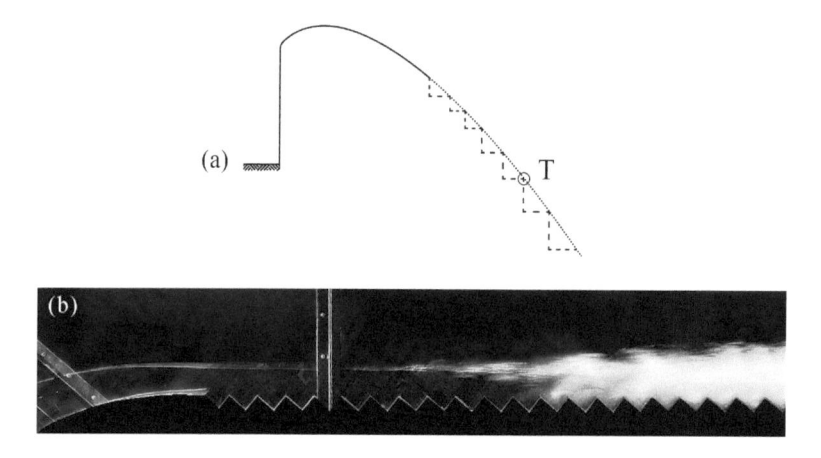

Figure 4.71 Stepped spillway (a) transition from crest to uniformly stepped spillway, (b) model flow attached to chute bottom at transition from standard profile to chute (Pfister, 2009, photo rotated to horizontal)

the standard crest follows the jet trajectory, an insertion of steps upstream of the tangency point T is *a priori* unsuitable as the steps interfere with the flow. However, for discharges far below the design flow, the thin water film impacts onto the first horizontal step face and is then horizontally deflected. A local spray zone just downstream of the transition between the crest and stepped chute is the consequence, intensifying with increasing jet velocity, i.e. with increasing distance from the crest (4.7.4). To counter spray generation at the sudden transition for small discharges, a transition zone including a few small steps upstream of T (downstream of the standard crest) is foreseen (Figure 4.72). A detailed geometry of such a standard crest including steps is proposed by Elviro and Mateos (1995).

The standard-crested weir combined with a stepped chute guarantees smooth flow conditions and modest spray development for small discharges. This crest design involves excellent conditions even for large discharges, optimizing energy dissipation along the chute since the flow is always in contact with the steps (Figure 4.72).

An alternative inlet structure is the *broad-crested* weir (Figure 4.70b), whose rating curve is less effective than the standard-crested profile. The flow passing a broad-crested weir is not oriented parallel to the chute pseudo-bottom when entering the stepped reach, but almost horizontally. The transition involves an abrupt slope change (particularly on an RCC dam with a steep chute). A free jet emerges for intermediate discharges, whereas the flow remains attached for the smallest and maximum values, yet with pressure and free surface

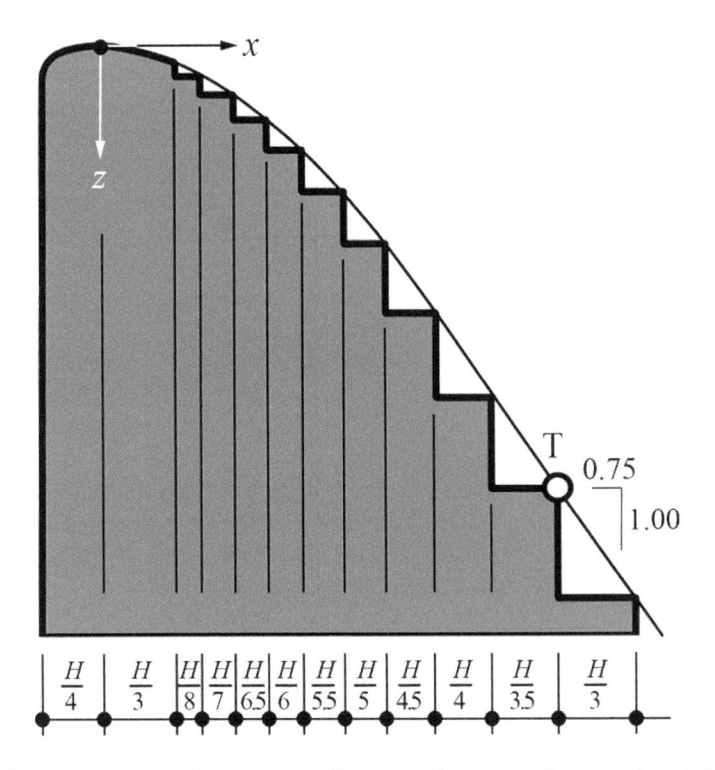

Figure 4.72 Crest profile of stepped spillway with steps of increasing height fitted to standard ogee-crested weir profile $(-)$ $z/H_D = 0.50(x/H_D)^{1.85}$ down to point of tangency T (Elviro and Mateos, 1995)

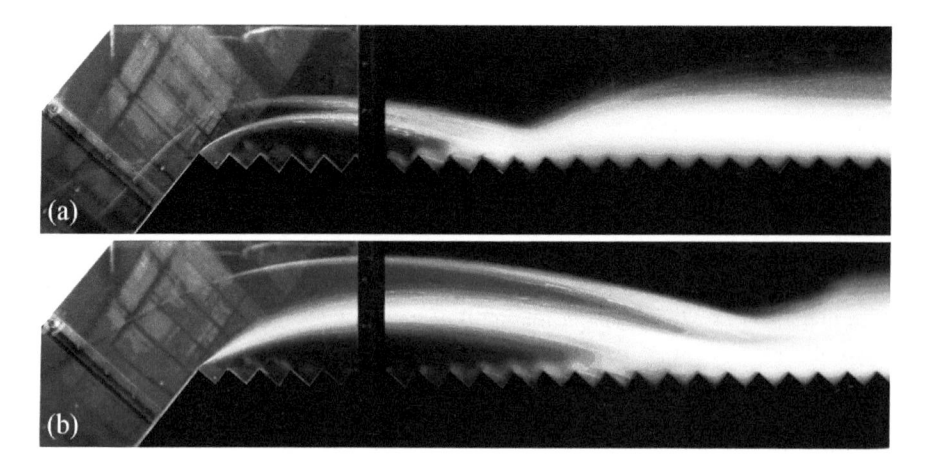

Figure 4.73 Flow on stepped chute model with broad-crested weir as inlet structure and jet cavity aeration for unit discharges q = (a) 10, (b) 40 m²/s (Pfister, 2009, photos rotated to horizontal)

fluctuations. Negative mean pressures occur below the jet at least on the first vertical step face, increasing with discharge and reaching values up to around $h_p = -5s$ for attached chute flow with $q = 40$ m²/s (Pfister, 2009). Furthermore, the energy dissipation is suspended along the jet reach. At the impact location of the free jet on the stepped chute, several steps downstream of the upstream chute end, an intense spray occurs due to the significant jet impact velocity. Consequently, the stepped chute cannot exhaust its energy dissipation capacity. To stabilize the flow downstream of a broad-crested weir and to prevent excessive sub-pressures and their fluctuation, an air supply system can be provided laterally in a step niche. With an air supply below the jet, the flow detaches for all discharges (Figure 4.73). The jet lengths are then comparably longer than those of the closed air supply system. The mean pressure on the steps in the jet reach is close to zero.

The broad-crested weir causes a priori fluctuating flow with sub-pressures in the first step niches, with a certain potential for cavitation damage. An air supply should be adopted, operating efficiently and stabilizing the flow. However, a considerable hydrodynamic impact accompanied by spray occurs at the reattachment point.

A short *flap-gate* (Figure 4.70c) as inflow control generates pronounced fluctuating flow, with long and high jets, a considerable hydrodynamic impact on the chute at the reattachment point combined with intense spray. The installation of flaps is delicate and requests a thorough study, combined with mitigation measures. An optimization of the impact area on the weir body combined with an inclined crest end at the jet take-off may improve the situation (Pfister, 2009). If a broad-crested weir is regulated by radial or flap gates, it should be followed by a sufficiently long and rather mildly sloped smooth chute connected to a constantly sloped stepped chute by a parabolic transition, as e.g. performed at Folsom Dam in California (Frizell and Renna, 2011).

Experiments conducted by Silvestri *et al.* (2013) give indications related to a piano key weir (PKW, 3.5) as inlet structure upstream of a stepped chute. The residual energy on a stepped chute following a PKW was compared to that generated by a standard-crested weir.

It is concluded that uniform flow is achieved at a slightly higher chute elevation for the PKW than for the standard crest. In parallel, the energy dissipated in uniform flow below a PKW seems *smaller* than that downstream of an ogee. This phenomenon is explained with the flow concentration below PKWs, where water emerges concentrated through the outlet keys, whereas the zones downstream of the inlet keys are hardly fed. For the zones below the outlet keys, higher unit discharges occur, explaining the relatively reduced energy dissipation.

Control structures and the transition from the latter to the stepped chute strongly affect the flow feature along the stepped chute, a fact confirmed by Chanson (2006) for skimming flow. This observation is linked to the different modes of turbulent boundary layer development, related to the subsequent energy dissipation in the zone between the inlet and the uniform flow reach of the chute. Note that most design recommendations concerning stepped chutes are based on physical model tests with a standard-crested weir as inlet, or a structure generating similar flow features at the start of the stepped chute. Other control structures can produce flow features that differ from the characteristics detailed herein.

Selection of step height

The step height should primarily be based on the construction procedure. RCC dams are often erected in layers of 0.3 m and formwork heights of 0.6–1.2 m, so that step heights from 0.3 to 1.2 m are convenient. Higher steps of 2–3 m may be favorable for stepped chutes cut into rock such as in the valley flanks of embankment dams to facilitate construction by drilling and blasting (Lutz *et al.*, 2015). Regarding overtopping protection on embankment dams, the usual step height is 0.30 m (1 lift), whereas it should be 0.6 m for large unit discharge and mild slopes (PCA, 2002).

Results of model tests indicate that large step heights are preferable in terms of hydraulic behavior. First, the location of the inception point of air entrainment moves slightly towards the spillway crest with increasing step height, so that the unaerated spillway portion prone to cavitation damage is shorter (see 4.7.4). Second, energy dissipation slightly increases with increasing step height (Boes and Minor, 2002; Boes and Hager, 2003b) (see also 4.7.4).

The step height should be selected in such a way that the chute flow regime is either distinctively in the skimming flow or nappe flow regimes for both the design and the safety check floods (see 4.7.4). Otherwise the maximum hydraulic load would occur in the transition regime with potential hydrodynamic instabilities resulting from a change from aerated to unaerated nappe flows, or vice versa (Chanson, 2002). Obviously, for ungated spillways, the transition regime cannot be avoided if the chute is designed for skimming flow.

A non-uniform step height distribution along the chute has a minor effect on the flow features as compared to constant step height (Felder and Chanson, 2011). Yet, non-uniform configurations can induce flow instabilities for small discharges.

Slope change

Apart from a few exceptions, such as the Upper Stillwater Dam in the USA (Houston, 1987), and New Victoria Dam in Australia, most stepped chutes along downstream dam faces include a constant pseudo-bottom chute slope. However, the variation of the slope along the stepped chute is an interesting feature to implement the latter on valley flanks, such as at the Lower Siah-Bishe Dam in Iran (Baumann *et al.*, 2006) and Trängslet Dam in Sweden (Boes

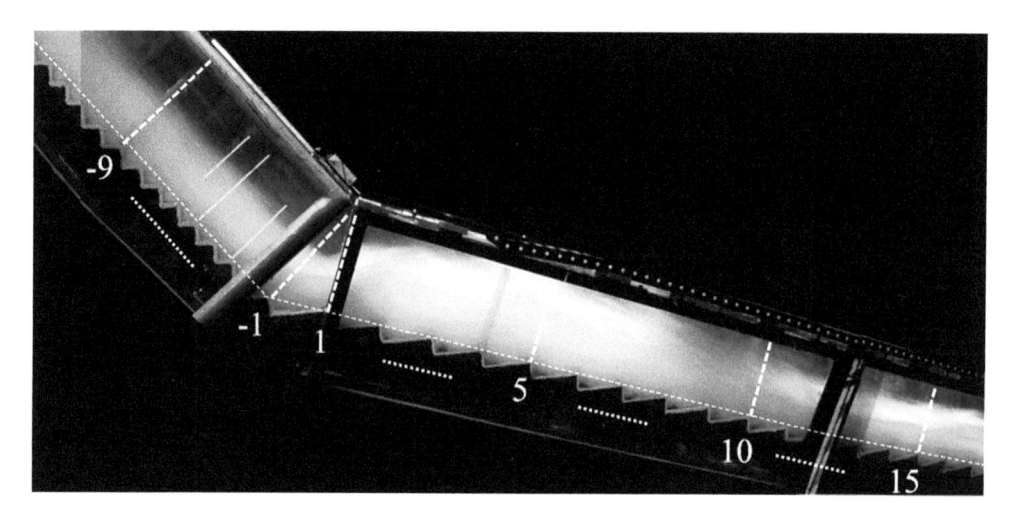

Figure 4.74 Stepped model chute with abrupt pseudo-bottom slope change from 50° to 18.6°, with flow features, spray formation, and step numbers from origin (Ostad Mirza *et al.*, 2017). Copyright © International Association for Hydro-Environment Engineering and Research, reprinted by permission of Taylor & Francis Ltd, www.tandfonline.com on behalf of International Association for Hydro-Environment Engineering and Research

et al., 2015; Lutz *et al.*, 2015), or to gain flexibility for the section of a RCC gravity dam. The effect of an abrupt slope change from steep to flat on selected skimming flow properties is described by Ostad Mirza *et al.* (2017). They conducted systematic model tests on a large-scale model (Figure 4.74) including 50° to 30° and 50° to 18.6° abrupt slope changes, considered fully aerated uniform flow approaching the slope change, and normalized discharges between $2.6 \leq h_c/s \leq 9.2$.

Ostad Mirza *et al.* (2017) observed that:

- Abrupt slope changes locally affect the air concentration characteristics and the flow depths;
- Influence range of slope change is limited from $0.5h_c$ upstream to $9h_c$ downstream of slope change;
- Mean air concentration C_a decreases immediately upstream of slope change, as compared to the uniform flow value initiated by upstream chute. Then, C_a significantly increases due to spray and flow deflection (Figure 4.74) with a peak value at $5h_c$. Further, C_a reduces after the peak to reach an almost constant value at the end of the influence reach at $9h_c$;
- Advective diffusion model of Chanson and Toombes (2002a) describes well the air concentration profiles in the spray region;
- Despite flow compression along with de-aeration near slope change, bottom air concentrations at step edges near the pseudo-bottom typically are $C_b > 0.1\%$ (with fully aerated uniform flow upstream of slope change), expected to assure safety against cavitation damage;
- Characteristic flow depths remain unaffected upstream of slope change, whereas a significant increase occurs further downstream due to spray (similar to a jet trajectory). In general, peak values of mixture-flow depth h_{90}, h_{95}, and h_{99} relevant for the sidewall height design are observed at approximately $5h_c$;

- Maximum bulk flow depth is generally lower than the critical flow depth h_c. However, the 50° to 18.6° configuration combined with minimum discharge generated a spray up to $h_{99}/h_c = 1.6$;
- Streamwise development of equivalent clear-water depth is not significantly influenced by slope change;
- Minimum discharges in the lower skimming flow regime or even the transition flow regime could perform differently, as suggested by tests with $h_c/s \geq 2.6$. Such flows would generate higher local C_a and C_b values, related to a more pronounced spray occurrence;
- Tested slope change was abrupt; a gradual change could reduce flow bulking and de-aeration up to a certain degree.

Terminal energy dissipation structure

Typically, the design discharge of a spillway is high, ranging up to almost 100,000 m³/s for rare cases, combined with considerable heads of 50–200 m or even more. It is evident that the dominant portion of the flow energy is not dissipated on the chute, even if it is stepped (4.7.4). Particular structures are provided at its toe to dissipate – or govern – the remaining flow energy. The flow then enters the natural downstream watercourse in balanced conditions, i.e. avoiding erosion and guaranteeing stability of the dam foundation.

Various types of technical energy dissipaters are applied below smooth chutes, namely stilling basins (Figures 4.63b, 4.64a), plunge pools, or submerged buckets (Vischer and Hager, 1995). The US Bureau of Reclamation (USBR) conducted extensive model studies in the 1960s and 1970s on classical stilling basins, considered as reference today (e.g. Peterka, 1963). The type II and the shorter type III Basins, developed in the former study, are frequently applied. The main mechanism of stilling basins' energy dissipation is the installation of a stable hydraulic jump on a rigid horizontal bottom. Hydraulic jumps are particularly efficient to dissipate flow energy, with a dissipation rate up to 85% for high inflow Froude numbers (Peterka, 1963). The basic features of a hydraulic jump, namely its sequent depths and energy dissipation, follow by using the concept of momentum conservation. Both, the Froude number and the momentum equation apply for clear-water flow conditions with a hydrostatic pressure distribution and a characteristic velocity profile.

These conditions are rarely given for a stilling basin downstream of a stepped chute. The flow there is typically both highly aerated and turbulent. The pressure and velocity distributions also differ from smooth chute approach flow conditions. Further, unit discharges supplied to stepped chutes are increased up to some 160 m²/s in recent projects (Lueker et al., 2008). Consequently, the basic design assumptions for the type II and III basins do not apply or are questionable for stepped chutes.

Energy dissipation at the dam toe was often neglected in the past since stepped chutes are known to dissipate overproportional energy rates, as compared to smooth chutes, for small unit discharges. For stepped chutes, the following investigations refer to energy dissipation at their toe, among others:

- Houston (1987) conducted a model study of the stepped chute and dissipation structure (scale factors from 1:5 to 1:15) of the Upper Stillwater Dam (Utah, USA). It focused on the hydraulic efficiency of the stepped chute and the size of the stilling basin. Wave heights and impact pressures at the basin invert were measured, as well as the debris-handling capacity of the basin. The key objective was to reduce the stilling basin length.

Data indicated that the inflow velocities were much lower downstream of the stepped chute as compared to a smooth chute, so that the stilling basin length was reduced to 10 m, because the hydraulic jump occurred in the upper basin portion. However, the unit discharge was limited to a small value of 12 m^2/s.

- Frizell (1990) studied the converging stepped chute for McClure Dam, finally providing a unit design discharge of almost 31 m^2/s at the dam toe. Down to the entrance of the stilling basin, some 53% of the flow energy was already dissipated on the chute. It was concluded that a 40 m long stilling basin with end sill 'produced an excellent energy dissipation for all discharges'.

- Based on systematic model tests on the hydraulic characteristics of stepped chutes, Boes (2000), Boes and Minor (2002) and Boes and Hager (2003b) suggested to determine the energy head at the toe of a stepped chute based on equivalent clear-water velocity V_w and flow depths h_w, respectively, i.e. to use $F_1 = V_w/(gh_w)^{0.5}$ as entrance Froude number for stilling basin design.

- Baumann et al. (2006) describe the two stepped spillways with stilling basins installed at the upper and lower reservoirs of the Siah-Bishe pumped-storage scheme in Iran. The arrangement was chosen mainly as a result of instabilities at the toe of the lower dam, requesting for an increased energy dissipation along the spillway and a thorough dissipation of the residual energy within the basin. The hydraulics was optimized using physical model tests with scale factors of 1:15 and 1:20. Rather small stilling basin lengths were sufficient.

- Cardoso et al. (2007) investigated a type III stilling basin at the end of a stepped chute of 51° bottom angle, focusing on the basin invert pressures. They varied the model step heights between 2 and 8 cm, and the unit model discharges between 0.08 and 0.20 m^2/s. The measured pressure heads on the basin invert were almost independent of the step height.

- Lueker et al. (2008) conducted a model study of scale factor 1:26 of the auxiliary stepped spillway of Folsom Dam near Sacramento, CA, including a large stilling basin. The initial basin length was 52 m with a double row of 2.7 m high baffle blocks and a 1.4 m high end sill. However, this configuration was found to be insufficient to keep the hydraulic jump inside the basin under design flow conditions. The basin length was increased to 76 m, and the two baffle rows were replaced by one single row of seven 4.8 m high baffle blocks. The 1.4 m high solid end sill was kept. A flow energy consideration indicated that some 80% of the total head was dissipated at the basin exit for a particular discharge, of which roughly half on the stepped chute.

- Frizell (2009) discussed factors affecting the design of standard stilling basins for stepped chutes, based on a data reanalysis. The chute angle upstream of the basin was demonstrated to affect its performance, an aspect neglected by Peterka (1963). Mildly sloped chutes would require longer basins. This uncertainty is even more pronounced for mildly sloping stepped chutes. A comparison of the velocity profiles at the upstream basin end for a smooth and a stepped chute, under otherwise similar conditions, indicated that a stepped chute significantly decreases the average velocity and smoothens the velocity profile. An entrance flow depth h_1 resulted in $F_1 = 4.6$ for the stepped and $F_1 = 7.8$ for the smooth chute. Accordingly, the type III length would be 39 m for the stepped, and 58 m for the smooth chute.

- Meireles et al. (2010) and Meireles (2011) measured the flow characteristics within a type III basin downstream of stepped chutes (Figure 4.63b). The results indicated (1) at the basin entrance, the pressure head is higher than the respective hydrostatic pressure due to streamline curvature, (2) the agreement between water levels along the studied basin and type III basin is acceptable, (3) pressure profiles at the basin entrance indicated considerably higher values with an antecedent stepped than a smooth chute, (4) the hydraulic jump stabilizes faster downstream of a stepped chute, and (5) differences in flow characteristics for the basin with or without baffle blocks are negligible.

- Frizell and Svoboda (2012) studied a standard type III basin designed for $F_1 = 8$. The model inlet consisted of a jet box and an adjustable flap gate at the model exit to set the tailwater depth. The chute consisted of a stepped chute of 38.1 mm step height, as well as a smooth chute, and of pseudo-bottom angles between 14° and 51°. The supplied unit discharges ranged from 0.25 m²/s to 0.62 m²/s. A stepped chute and a smooth invert were investigated, to compare the basin performance. Parameters as average air concentrations at the inflow section were measured. As for the stepped chute, the flow at the basin entrance was close to uniform conditions and thus fully aerated. The main conclusion was that the use of the clear-water parameters of stepped spillway flow allows for a consistent application of the USBR design principles for the type III basin, thus corroborating the results of Boes and Hager (2003b). This observation is not justified physically, but pointing at flow depth measurement inaccuracies of the initial USBR study implementing unintentionally some reserve in the design principle. For stepped chute flows with $F_1 < 6$, even a significantly lower tailwater was required to maintain a stable hydraulic jump.
- Boes et al. (2015) report from model investigations on a 1:45 model of a 550 m long mildly sloped stepped chute of varying step height, width and bottom slope along the abutment of a large Swedish rockfill dam. The measured energy dissipation rates varied between 73% and 81% of the upstream head for unit discharges from 30.4 to 56 m²/s in the lower chute section of bottom slope 26.6° just upstream of the stilling basin. The latter could thus be designed with compact dimensions.
- The performance of a classical stilling basin downstream of stepped chutes was studied in detail by Stojnic (2020) and Stojnic et al. (2020). It was found that hydraulic jumps downstream of stepped chutes have longer dimensionless lengths being normalized with the downstream sequent depth, namely $L_J = 6.7h_2$, as compared to $L_J = 5.75h_2$ downstream of smooth chutes.

There is, at least for the moment, no general information available on the operation mode and efficiency of flip buckets below stepped chutes. Single design approaches are available, however, e.g. by Rice and Kadavy (1997). Indications on the jet behavior of highly pre-aerated flow approaching a flip bucket are provided by Schmocker et al. (2008), or Pfister and Hager (2012).

4.7.4 Hydraulic design

Flow regime

Steps strongly affect the flow along a chute, particularly if the flow depth is small, the steps high, and the chute slope mild. Intuitively for small flow depths and high steps, the flow along a cascade appears differently than for large flow depths and small steps. The ratio between flow depth and step height is of relevance, representing a specific roughness. Consequently, three basic flow regimes exist (Chanson, 2002):

1 *Nappe flow regime* (Figure 4.75a) appearing at low flow depths combined with high steps. The flow from each step hits the step below as a free-falling jet, generating a succession of jets with air nappes below them. A fully or partially developed hydraulic jump

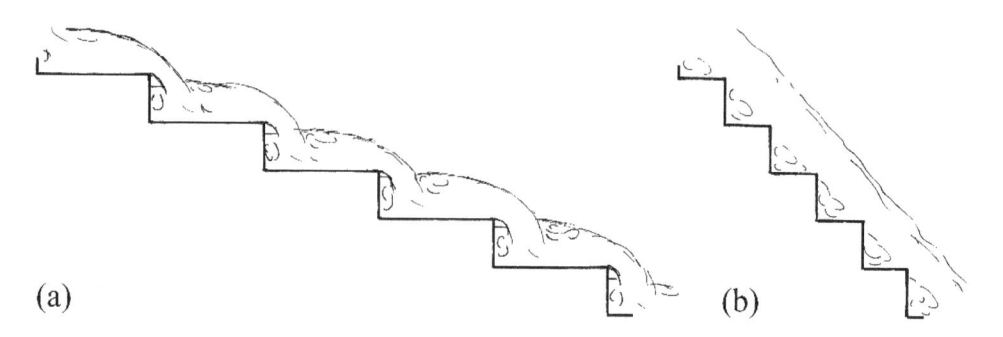

Figure 4.75 Stepped chute with (a) nappe flow, (b) skimming flow

Figure 4.76 Stepped model chute with (a) transition flow and air cavities in step niches for $q = 19$ m²/s, (b) skimming flow for $q = 35$ m²/s (prototype values) (Courtesy VAW, ETH Zurich)

forms on the step following the jet impact location. For larger discharges, the hydraulic jump interferes with the downstream step edge or disappears. Then, supercritical flow appears with shock and roll waves. Nappe flow is effective in terms of energy dissipation, due to the repeated jet dispersion, impact and the (at least partial) presence of a hydraulic jump. These processes typically generate a highly aerated flow, with potential nappe oscillation due to local sub-pressures. Nappe splitters open the entrapped air volume to the atmosphere, reducing vibrations.

2   *Transition flow regime* (Figure 4.76a) occurring for intermediate flow depths and characterized by a chaotic flow motion associated with intense splashing and irregular flow depths. Flow velocities increase, so that the jet impact location on the subsequent horizontal step is close to the next step edge. The jet is deflected partially, causing surface spray. This flow condition should be avoided, or this regime occurs at relatively low discharges only (4.7.1).

3   *Skimming flow regime* (Figures 4.75b and 4.76b) including a coherent stream 'skimming' over the steps, with rollers in the step niches and a relatively smooth flow body above the pseudo-bottom. Skimming flow appears for large flow depths combined with small steps. As visible from Figures 4.64c and 4.71b, first unaerated flow occurs along the top chute portion. As the turbulent boundary layer reaches the flow surface (at the inception point), natural self-aeration (4.7.4) starts and develops until uniform

flow is reached. The flow there is homogenously aerated and looks similar to fully aerated smooth chute flow with turbulent but steady characteristics.

The stepped chute flow regime varies with the normalized critical flow depth h_c/s and the channel slope expressed with $s/l = \tan\phi$, with $h_c = (q^2/g)^{1/3}$. Implicitly, the unit discharge $q = Q/b$ and thus a potential chute convergence affect the flow regime. With varying discharges as are typical for unregulated inlets, the flow regime consequently changes with discharge. For increasing discharge, the first regime appearing is nappe flow, then transition flow, and finally skimming flow for large discharge.

The upper and lower limits of the transition flow regime were defined by Chanson and Toombes (2004) based on numerous literature data and own experiments. The lower limit of the transition flow regime is

$$\frac{h_c}{s} = 0.92 - 0.38\frac{s}{l} \qquad \text{for } 0 < s/l < 1.7 \ (0 < \phi < 59.5°). \tag{4.152}$$

The upper limit of the transition flow regime is

$$\frac{h_c}{s} = \frac{0.98}{\left(\dfrac{s}{l} + 0.39\right)^{0.38}}, \qquad \text{for } 0 < s/l < 1.5 \ (0 < \phi < 56.3°). \tag{4.153}$$

Accordingly, nappe flow occurs as long as the normalized discharge h_c/s is smaller than the value predicted by Eq. (4.152), the transition regime appears if h_c/s is between the predictions of Eqs. (4.152) and (4.153), and the skimming flow regime occurs if h_c/s is larger than the limit given by Eq. (4.153).

Self-aeration and air inception point location

While the flow typically becomes aerated from chute start due to air entrainment by hydraulic jumps and intense splashing for nappe and transition flows, two regions differing in aeration are distinguished for skimming flow (Figure 4.77). In the upstream spillway portion downstream of the crest, the flow is unaerated. This region is also denominated as black-water or clear-water reach. Here, the flow features a compact, transparent look, without air bubbles or pockets. The turbulent boundary layer starts growing from the crest. As the boundary layer reaches the water surface, the degree of turbulence is high enough to entrain air into the clear-water flow, provided effects of surface tension are small. The air-water friction is large enough to cause surface irregularities, creating air pockets incepted and rapidly distributed within the flow (Amador et al., 2009). The location of natural air entrainment from the crest is termed inception point, downstream of which the flow is thus aerated. This region is also named the white-water region.

The inception (subscript i) point location is expressed either in terms of length L_i of the black-water reach (Figure 4.77), or by the vertical distance $w_i \approx L_i \sin\phi$ from the crest. According to Boes and Minor (2002) and Boes and Hager (2003a), L_i is defined as

$$L_i = \frac{5.90 h_c^{1.2}}{(\sin\phi)^{1.4} s^{0.2}}, \qquad \text{for } 26° < \phi < 75°. \tag{4.154}$$

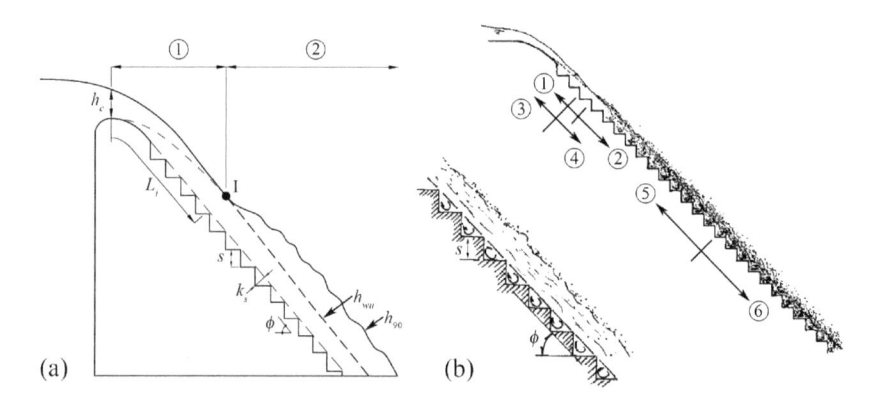

Figure 4.77 Skimming flow development with (a) ① unaerated and ② aerated flow regions,
(---) boundary layer and inception point I (ICOLD, 2017), (b) sub-regions with
③ developing flow, ④ developed flow, ⑤ gradually varied flow, ⑥ quasi-
uniform flow (Ohtsu *et al.*, 2001)

Eq. (4.154) indicates that the critical depth h_c or the unit discharge q govern the value of L_i,
whereas the effect of s is small. By doubling s, L_i is reduced by only 13%, whereas doubling
q leads to an increase of 74%. The steeper the spillway slope, the more upstream the flow
becomes aerated. An increase of bottom slope from 1:2 to 1:0.8 ($\phi = 26.6°$ to $51.3°$) reduces
L_i by about 54%.

While Eq. (4.154) is useful for steep stepped chutes, the inception point location follows
a different relation for mild slopes. According to Chanson and Toombes (2002b), the non-
dimensional length L_i/s reads with the roughness Froude number $\mathsf{F}_* = q/[g \cdot \sin\phi(k_s)^3]^{1/2} =$
$q/[g \cdot \sin\phi(s \cdot \cos\phi)^3]^{1/2}$ and $k_s = s \cdot \cos\phi$ as the step (roughness) height perpendicular to the
pseudo-bottom (Figure 4.77)

$$\frac{L_i}{s} = 12.34 \frac{\cos\phi}{\sin^{0.08}\phi} \mathsf{F}_*^{0.47} \quad \text{for } 16° < \phi < 22° \text{ and } 0.85 < \mathsf{F}_* < 3.4. \tag{4.155}$$

In turn, Meireles and Matos (2009), propose

$$\frac{L_i}{s} = 5.25 \cos\phi \mathsf{F}_*^{0.95} \qquad \text{for } 16° < \phi < 26.6°. \tag{4.156}$$

For stepped chutes on a broad-crested embankment dam, the non-dimensional length from
the spillway crest is given by (Hunt and Kadavy, 2013)

$$\frac{L_i}{s} = 5.19 \cos\phi \mathsf{F}_*^{0.89} \qquad \text{for } \phi < 26.6° \text{ and } 0.1 < \mathsf{F}_* \leq 28; \tag{4.157a}$$

$$\frac{L_i}{s} = 7.48 \cos\phi \mathsf{F}_*^{0.78} \qquad \text{for } \phi < 26.6° \text{ and } 28 < \mathsf{F}_* < 10^5. \tag{4.157b}$$

Note the increasing values of L_i/s with $\cos\phi\mathsf{F}_*$. Another relation for the location of the air
inception point was proposed by Chanson (1994a, 1994b) for $\phi > 27°$.

The flow depth at the inception point h_i has also been analyzed e.g. by Boes and Hager (2003a) or Boes *et al.* (2015) to yield

$$h_i = \frac{0.40 h_c^{0.9} s^{0.1}}{(\sin \phi)^{0.3}}, \qquad \text{for } 18° < \phi < 55°. \tag{4.158}$$

Here again, h_c has a greater effect on h_i than s. For spillways of embankment dams, the non-dimensional flow depth at the inception point is (Meireles and Matos, 2009)

$$\frac{h_i}{s} = 0.28 \cos \phi \mathsf{F}_*^{0.68}, \qquad \text{for } 16° < \phi < 26.6°. \tag{4.159}$$

The depth-averaged air concentration $C_{a,i}$ at the inception point is given by (Boes and Hager, 2003a)

$$C_{a,i} = 1.2 \cdot 10^{-3}\,(240° - \phi), \qquad \text{for } 26° < \phi < 55°. \tag{4.160}$$

Uniform flow characteristics

Three different regions occur within the aerated skimming flow zone regarding flow aeration and development (Figure 4.77b). The flow depth increases significantly due to the highly turbulent aeration process. In the first region in streamwise direction, located immediately downstream of the inception point, the air entrained at the free surface is rapidly distributed by turbulence and the flow varies gradually up to the stabilization of the aeration process. The two-phase flow emulsion of air and water, also termed mixture flow, is gradually varied in the second aerated region until quasi-uniform flow is attained, constituting the third aerated flow region. The mixture-flow depth increases until reaching a constant value under uniform flow conditions.

ATTAINMENT OF UNIFORM FLOW

The normalized vertical distance from the spillway crest needed for uniform flow (subscript u) to be attained increases almost linearly with ϕ and follows the power formula (Boes and Minor, 2002; Boes and Hager, 2003b)

$$\frac{z_u}{h_c} \approx 24(\sin \phi)^{2/3}, \qquad \text{for } 26° < \phi < 55°. \tag{4.161}$$

UNIFORM FLOW DEPTHS

Once the flow on stepped chutes becomes aerated, a distinction is made between the equivalent clear-water (subscript w) depth h_w and the mixture-flow depth h_{90}, the latter defined as the flow depth with an air concentration of 90% (Figure 4.78). The term 'equivalent' implies that $h_w = h_{90}(1-C_a)$ is a computed value in two-phase flow determined analytically with the characteristic mixture-flow depth h_{90} and the depth-averaged (subscript a) air concentration C_a.

The uniform equivalent clear-water depth $h_{w,u}$ is given by (Boes and Minor, 2002; Boes and Hager, 2003b)

$$\frac{h_{90,u}}{h_c} = 0.215(\sin \phi)^{-1/3}, \qquad \text{for } 26° < \phi < 55°. \tag{4.162}$$

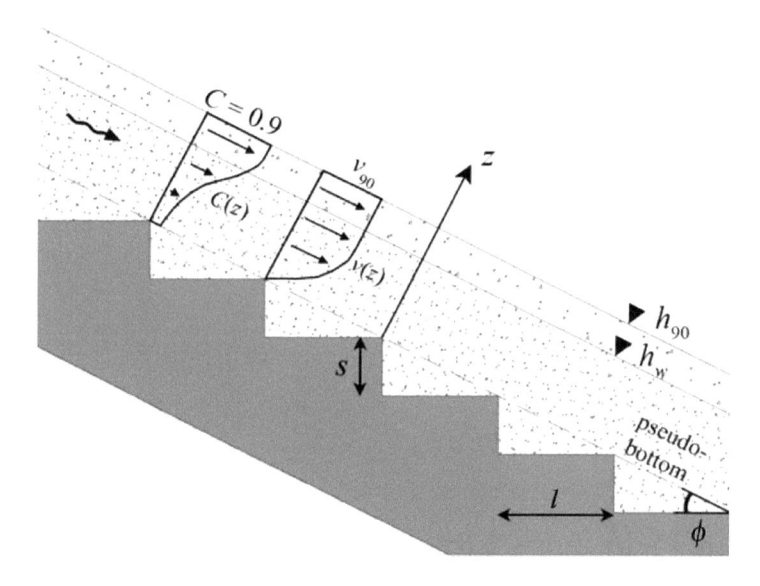

Figure 4.78 Definition sketch of stepped chute flow (Lutz *et al.*, 2015)

This ratio thus varies only with chute angle ϕ, independent from s and q. The uniform mixture-flow depth $h_{90,u}$ is described with $F_{s*} = q/(g\sin\phi\, s^3)^{1/2}$ as the characteristic roughness Froude number by (Boes and Minor, 2002; Boes and Hager, 2003b)

$$\frac{h_{90,u}}{s} = 0.50 F_{s*}^{(0.1\tan\phi+0.5)}, \quad \text{for } 26° < \phi < 55°. \tag{4.163}$$

For given relative discharge h_c/s, both $h_{w,u}$ and $h_{90,u}$ decrease with the chute slope.

Air entrainment and air concentration

The depth-averaged air concentration C_a is of interest to compute the equivalent clear-water depth h_w to determine the energy dissipation rate. The characteristic air concentration distribution across the flow depth and particularly the near-bottom air concentration are also important to assess the cavitation potential (4.7.3).

DEPTH-AVERAGED AIR CONCENTRATION

Boes and Hager (2003a) express the depth-averaged air concentration C_a in skimming flow versus the relative vertical distance $Z_i = (z - z_i)/h_c$ from the inception point (Figure 4.77). By normalizing the difference between the mean air concentration in a given cross section $C_a(Z_i)$ and that at the inception point $C_a(Z_i = 0) = C_{a,i}$ from Eq. (4.160) with $C_{a,u} - C_{a,i}$, where $C_{a,u} = 0.75(\sin\phi)^{0.75}$ is the uniform (subscript u) value for smooth chutes versus chute slope only (4.3), the normalized mean air concentration is $c_i = [C_a(Z_i) - C_{a,i}]/[C_{a,u} - C_{a,i}]$.

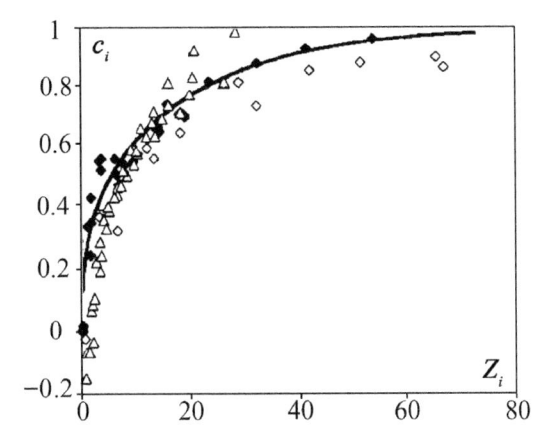

Figure 4.79 Normalized depth-averaged air concentration $c_i = [C_a(Z_i)-C_{a,i}]/[C_{a,u}-C_{a,i}]$ versus relative vertical distance from inception point $Z_i = (z-z_i)/h_c$, (—) Eq.(4.164), ($\diamond$) $s = 20$ mm, ($\blacklozenge$) $s = 60$ mm (Boes, 2000), ($\triangle$) (Wahrheit-Lensing, 1996) for $\phi = 50°$

This parameter tends to unity with increasing distance from the inception point and follows (Figure 4.79)

$$c_i(Z_i) = \frac{C_a(Z_i)-C_{a,i}}{C_{a,u}-C_{a,i}} = \left[\tanh\left(5\cdot10^{-4}\left(100° - \phi\right)Z_i\right)\right]^{1/3}. \tag{4.164}$$

Here the chute angle ϕ with $26° \leq \phi \leq 55°$ is expressed in degrees, and the non-dimensional coordinate Z_i originates at the inception point (Figure 4.77). The uniform average air concentration for smooth chutes was taken as normalizing parameter, because spillway roughness is irrelevant for $C_{a,u}$ (Chanson, 1994d; Boes, 2000; Matos, 2000). The uniform depth-averaged air concentration for skimming flow on stepped chutes therefore agrees well with the uniform value of smooth chutes of identical slope. This is demonstrated by the asymptotic trend of the curve towards unity in Figure 4.79, comparing experimental data for $\phi \approx 51°$ with Eq. (4.164).

Takahashi and Ohtsu (2012) proposed for $19° \leq \phi \leq 55°$ and $h_c/s < 5$ in quasi-uniform skimming flow

$$C_{a,u} = \left(\frac{6.9}{\phi}-0.12\right)\frac{s}{h_c}+0.656(1-e^{-0.0356(\phi-10.9)})+0.073. \tag{4.165}$$

The mean air concentration increases generally with a regime change from skimming, to transition, and to nappe flows. For a given flow regime, an increasing specific discharge q or higher ratio of h_c/s, e.g. due to a reduction of chute width, leads to lower C_a. Figure 4.80 compares the measured C_a values in a 1:2 sloped ($\phi = 26.6°$) stepped chute model for varying q-values with Eq. (4.165). The overall agreement between the two is surprisingly good, given that Eq. (4.165) applies for quasi-uniform skimming flow, whereas the data cover the entire range from nappe to skimming flows.

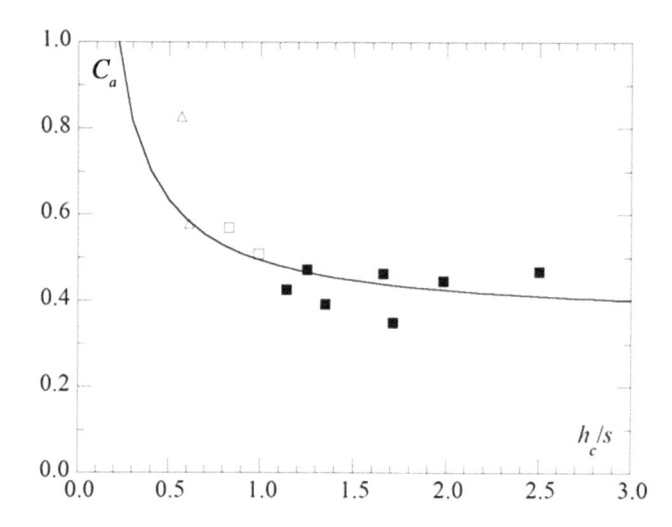

Figure 4.80 Measured mean air concentrations $C_a(h_c/s)$ versus relative discharge for (■) skimming flow, (□) transition flow, (△) nappe flow, compared with (—) Eq. (4.165) (Lutz *et al.*, 2015)

CHARACTERISTIC AIR CONCENTRATION DISTRIBUTION

The air concentration distribution over the flow depth of uniform-aerated flow is given by the air bubble advective-diffusion equation as (Chanson, 2000)

$$C(z) = 1 - \tanh^2\left[K' - \frac{z/h_{90}}{2D'} \right],$$

(4.166)

with

$$K' = \operatorname{arctanh}(0.1^{0.5}) + \frac{1}{2D'}$$

(4.167)

and

$$C_{a,u} = 2D'\left(\tanh\left(\operatorname{arctanh}\left(0.1^{0.5}\right) + \frac{1}{2D'} \right) - 0.1^{0.5} \right).$$

(4.168)

Here, K' is an integration constant and D' the turbulent diffusivity normal to the flow direction. Assuming a constant diffusivity over the flow depth, D' is determined from Eq. (4.168) for a known $C_{a,u}$ value and is typically $D' = 0.3–0.4$. This approach was validated for smooth-invert open channel flow, and is also in agreement for stepped chute flows (Chanson, 2000).

An advanced air concentration distribution model for skimming flow involves a non-constant diffusivity (Chanson and Toombes, 2002b)

$$C(z) = 1 - \tanh^2\left[K' - \frac{z/h_{90}}{2D_o} + \frac{(z/h_{90} - 1/3)^3}{3D_o} \right],$$

(4.169)

with

$$K' = \operatorname{arctanh}^2(0.1^{0.5}) + \frac{1}{2D_0} - \frac{8}{81D_0}.$$

(4.170)

The relation between the dimensionless coefficient D_0 and the average air concentration C_a is

$$C_a = 0.7622(1.0434 - e^{-3.614D_0}).$$

(4.171)

According to Bung (2009), Eq. (4.169) represents well the air concentration distribution at the step edges if $C_a < 0.3$. For higher C_a values the air concentration is underestimated, especially close to the pseudo-bottom, i.e. for $z/h_{90} < 0.40$. For $C_a \geq 0.3$ and $z/h_{90} < 0.4$, Eq. (4.166) gives better results than Eq. (4.169).

The measured axial and sidewall air concentration profiles in a 1:2 ($\phi = 26.6°$) sloped stepped chute model are compared in Figure 4.81 with Eqs. (4.166) and (4.169) based on the measured mean air concentrations C_a. For $0.4 < z/h_{90} < 1.0$, the values at the sidewall

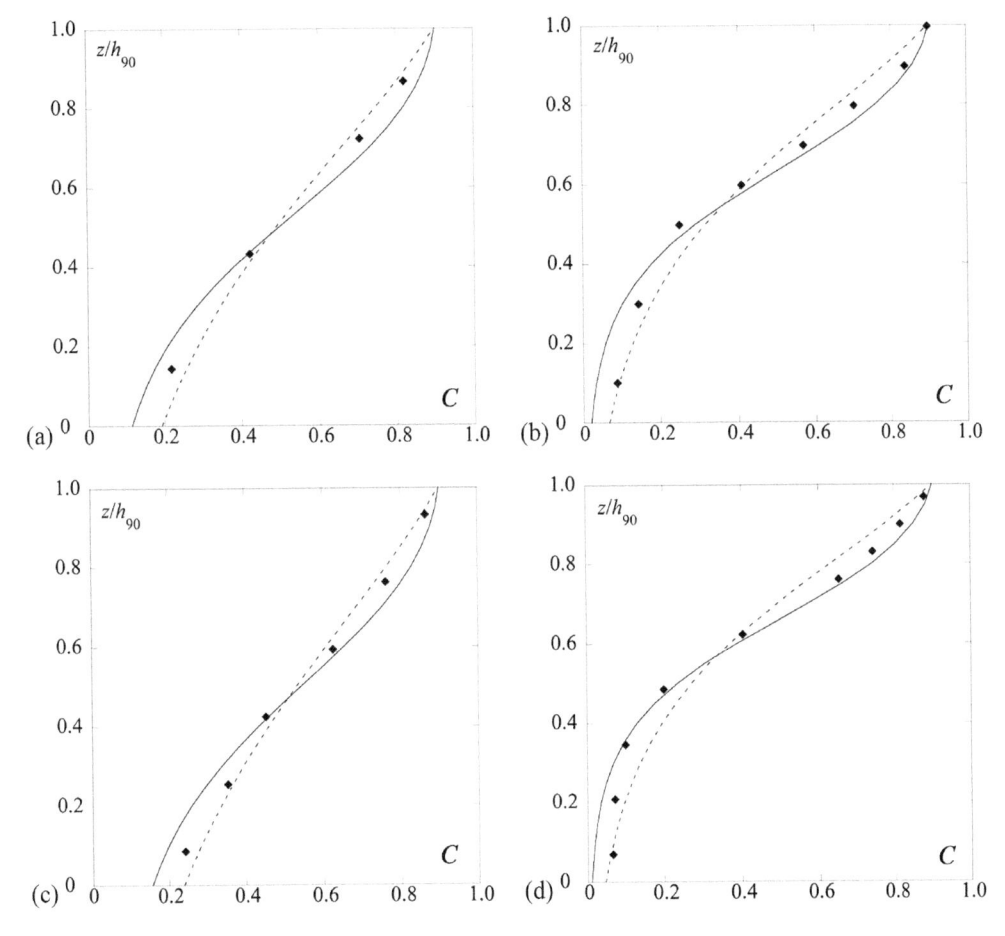

Figure 4.81 Air concentration profiles $C(z/h_{90})$ for skimming flow on 1:2 sloped stepped chute along (a), (c) axis, (b), (d) sidewall for $[q \ (\text{m}^2/\text{s}); \ C_a (\text{-})] = $ (a) [19; 0.51], (b) [19; 0.38], (c) [35; 0.55], (d) [35; 0.35], with ($\blacklozenge$) model data, (--) Eq. (4.166), (—), Eq. (4.169) (Lutz *et al.*, 2015)

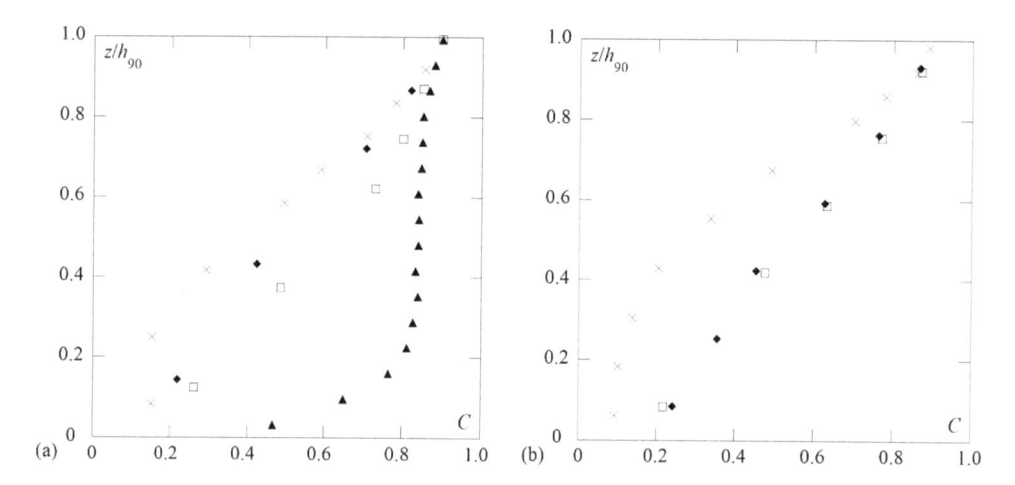

Figure 4.82 Axial air concentration profiles $C(z/h_{90})$ on 1:2 sloped stepped chute for (a) skimming flow with $q = (\blacklozenge, \square)$ 19 m²/s, and $(\times)$ 30.4 m²/s, $(\blacktriangle)$ nappe flow with $q = 30.4$ m²/s, $(\square)$ transition flow with $q = 19$ m²/s, (b) skimming flow with $q = (\blacklozenge, \square)$ 35 m²/s and $(\times)$ 56 m²/s, $(\blacktriangle)$ transition flow with $q = 56$ m²/s (Lutz *et al.*, 2015). Step heights $s = (\blacklozenge)$ 2 m, $(\square, \times)$ 4 m, $(\blacktriangle)$ 8 m

with $C_a < 0.4$ fit better Eq. (4.169), while those at the chute axis with $C_a > 0.5$ better follow Eq. (4.166), in agreement with Bung (2009). In contrast, the air concentrations near the step edges $z/h_{90} < 0.2$ are underestimated with Eq. (4.169) and fit better with Eq. (4.166) for the entire range of C_a.

The model data near the pseudo-bottom (subscript b) originating from the 1:2 ($\phi = 26.6°$) sloping stepped chute model indicate $C_b > 0.2$ for transition flow, and $C_b > 0.4$ for nappe flow (Figure 4.82), while $C_b > 0.05$ for skimming flow (Figures 4.81, 4.82). The shapes of the air concentration profiles for nappe flow differ from those for skimming and transition flows, indicating higher and nearly constant concentrations over the entire flow depth, except for close to the pseudo-bottom. The measured air concentration profiles for both transition and skimming flows fit Eqs. (4.166) and (4.169) well. Note that Eq. (4.166) generally compares better with the data for $z/h_{90} < 0.2$, whereas Eq. (4.169) applies better for $z/h_{90} \geq 0.2$.

PSEUDO-BOTTOM AIR CONCENTRATION

The growth of the pseudo-bottom air concentration $C(z = 0) = C_b$ can be expressed as a function of non-dimensional distance from the inception point $X_i = (x - L_i)/h_i$, with x as the streamwise coordinate originating at the spillway crest (Figure 4.83), and L_i and h_i computed from Eqs. (4.154) and (4.158), respectively. According to Boes and Hager (2003a), the air concentration at the pseudo-bottom follows for $26° < \phi < 55°$

$$C_b(X_i) = 0.015 X_i^{\sqrt{\tan \phi/2}}. \tag{4.172}$$

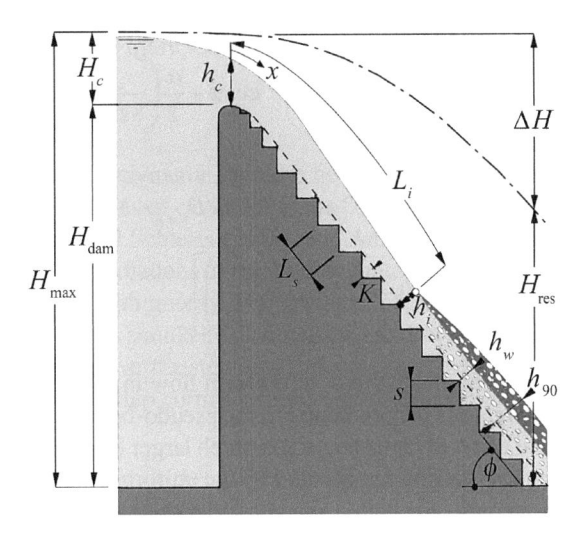

Figure 4.83 Longitudinal section of stepped chute with origin of streamwise coordinate $x = 0$ at spillway crest, (---) pseudo-bottom, flow region with equivalent clear-water depth h_w (lightly shaded) and mixture-flow depth h_{90} (darkly shaded), (·—·—·) energy head line, ($\circ$) air inception point I (Boes and Hager, 2003b)

Energy dissipation

The energy dissipation efficiency η of a spillway is the ratio of the difference between the energy head H_{max} at the spillway crest, i.e. the reservoir water level and the residual energy head H_{res} at the chute end to H_{max} (Figure 4.83)

$$\eta = \frac{H_{max} - H_{res}}{H_{max}} = \frac{\Delta H}{H_{max}}. \tag{4.173}$$

Here H_{max} is estimated to $H_{max} = H_{dam} + 1.5h_c$; ΔH is the dissipated energy head. The residual head H_{res} above the pseudo-bottom at any section along a stepped chute (Figure 4.83), regardless of uniform or non-uniform flow conditions, is expressed by

$$H_{res} = z' + h_w \cos\phi + \alpha' \frac{q^2}{2gh_w^2}. \tag{4.174}$$

Here z' = vertical distance above reference plain and $\alpha' \approx 1.1$ (Boes, 2000) is the velocity head correction coefficient.

To compute the relative residual energy head of stepped chutes, the relative spillway height is important, i.e. whether the flow at the chute end is still developing or if uniform flow has established (see gradually varied vs. uniform flow in Figure 4.83 and Eq. (4.161). If the chute is too short to attain uniform flow ($H_{chute} < z_u$), the following approximation of Boes and Hager (2003b) applies

$$\frac{H_{res}}{H_{max}} = \exp\left[\left(-0.045\left(\frac{k_s}{D_{h,w}}\right)^{0.1}(\sin\phi)^{-0.8}\right)\frac{H_{chute}}{h_c}\right]. \tag{4.175}$$

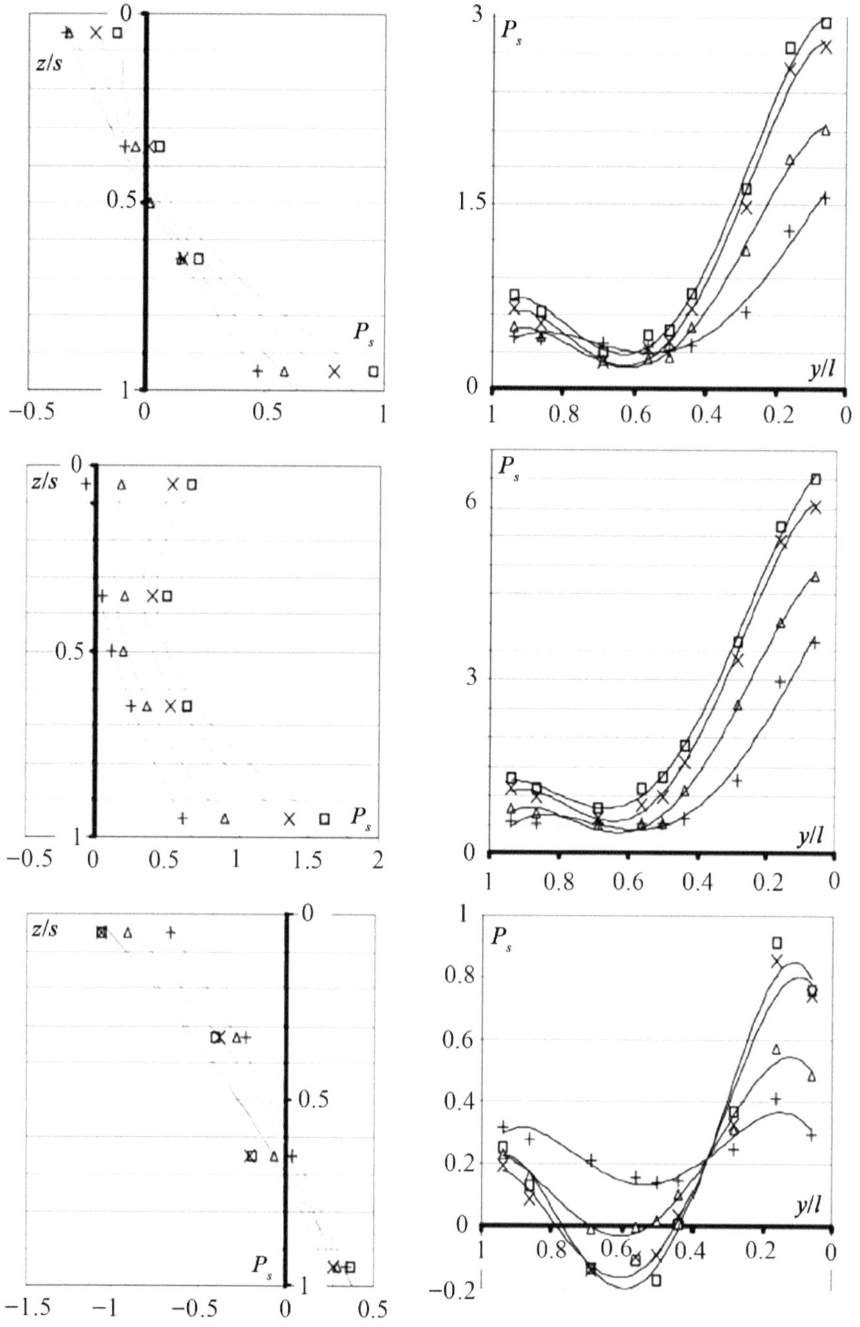

Figure 4.85 Profiles of mean (top), maximum (center) and minimum (bottom) normalized pressures $P_s = p/(\rho g s)$ on vertical (left) and horizontal (right) step faces of stepped chute model of slope 1:0.8 far downstream of air inception point $(L/K\sim60)$ for different relative flow depths h_c/s = (+) 0.891, ($\triangle$) 1.415, ($\times$) 1.854, ($\square$) 2.246 (Sánchez-Juny *et al.*, 2000)

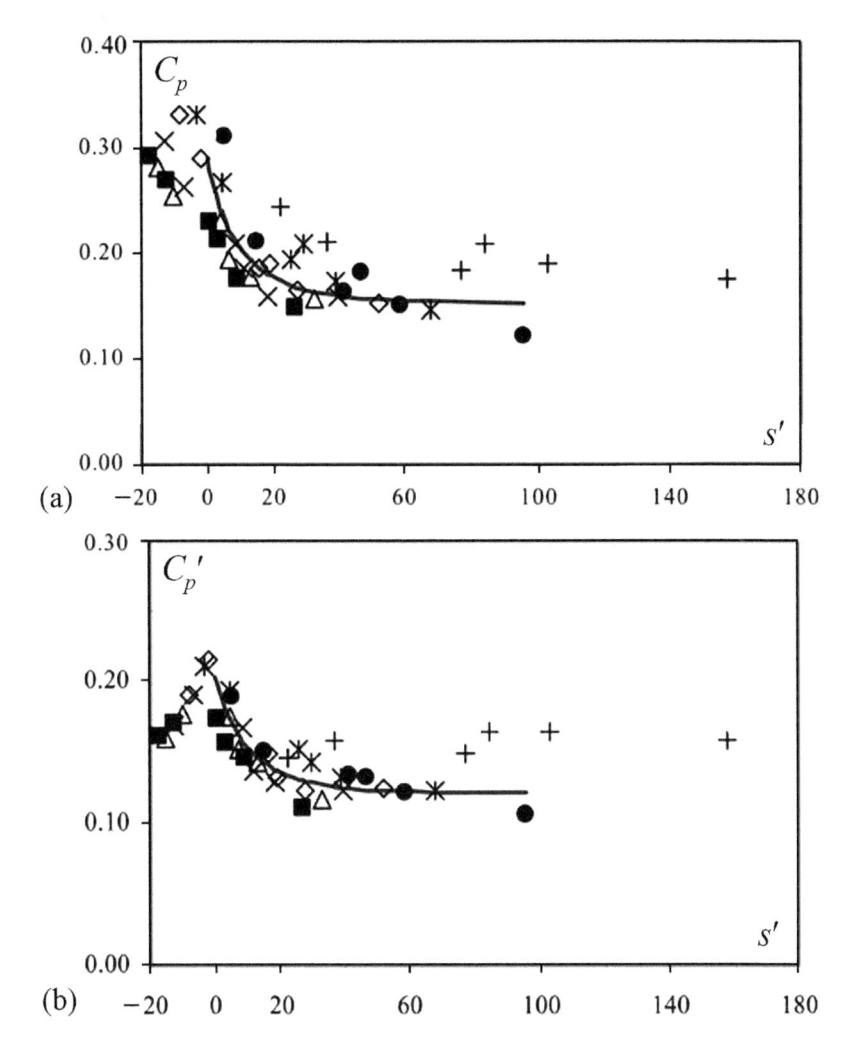

Figure 4.86 (a) Mean pressure coefficient $C_p(s')$, (b) coefficient of fluctuating pressures $C_p'(s')$ on horizontal step faces near outer step edge at $y/l = 0.14$ measured along chute with $s' = (L - L_i)/h_i$ as relative distance to air inception point for skimming flow on chute of slope 1:0.8 and different relative flow depths $h_c/s = (+) 0.89$, ($\bullet$) 1.41, ($*$) 1.85, ($\diamond$) 2.25, ($\times$) 2.65, ($\triangle$) 2.93, ($\blacksquare$) 3.21 (Amador *et al.*, 2009)

large pressure fluctuations, this region is identified as potentially critical for cavitation risk in the developing flow region. Fluctuations are more pronounced near the step edges and for higher discharges. Pressure has a positive gradient from the step edge to the inner corner, where its value is similar to that of the horizontal face (Figure 4.85, left). The mean pressures near the outer step edge (small values z/s from the step edge) are always negative. Positive mean pressures near the outer step edge occur at steps closer to the air inception point, where flow velocities are lower. The highest pressure fluctuations along the vertical step face are

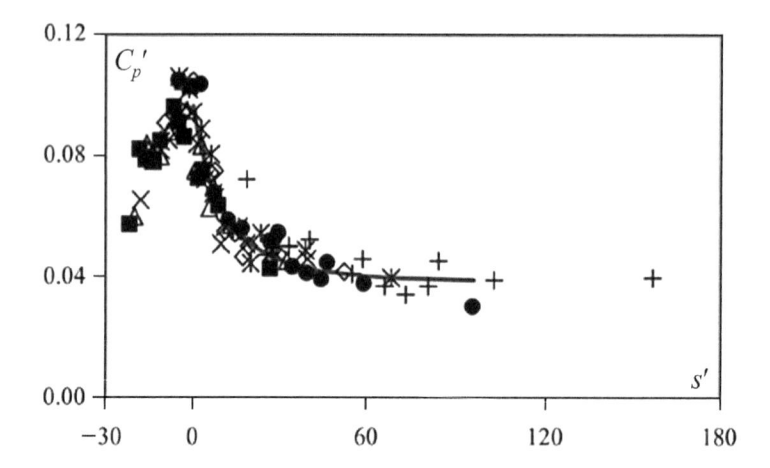

Figure 4.87 Coefficient of fluctuating pressures $C'(s')$ on vertical step faces near outer step edge at $z/l = 0.07$ along chute with $s' = (L - L_i)/h_i$ as relative distance to air inception point for skimming flow on chute of slope 1:0.8 and different relative flow depths h_c/s as in Figure 4.86 (Amador *et al.*, 2009)

near the outer step edge where also the highest negative pressures are measured (Amador *et al.*, 2009).

As observed at the horizontal step face, coefficient C_p' decreases downstream of the air inception point (Figure 4.87). The pressure fluctuations near the outer step edge at the vertical step faces are smaller than those acting on the horizontal faces.

The above empirical pressure coefficients were derived by Amador *et al.* (2009) from model tests on a stepped chute of slope 1V: 0.8H, i.e. 51.2°. André *et al.* (2004) provide information for milder chute angles of 30°.

PRESSURE FIELD ALONG CHUTE

As to the evolution of the pressure field along a stepped chute, the most delicate part is the non-aerated flow region (Figures 4.86, 4.87 near $s' = 0$), while the aerated region is not prone to cavitation damage (Zhang *et al.*, 2012). Extreme pressures were recorded close to the air inception point; further downstream, the presence of air bubbles produces a cushion effect reducing the pressure and its fluctuation.

Pressure field on sidewalls

The collapse of two old stepped masonry spillways in the UK, at Boltby Dam in 2005 and at Ulley Dam in 2007, led to a research program on the hydrodynamic effects of stepped chutes (Winter *et al.*, 2010; Winter, 2010). These studies revealed that at high skimming flows the center of the horizontal vortices at each step is able to generate high negative pressures which, in turn, act directly on the associated local zones of the sidewall. Moreover this occurs adjacent to other wall zones subject to high positive pressures due to step impact (Figure 4.88). Poor maintenance and cracks in the mortar may lead to a high back-pressurization

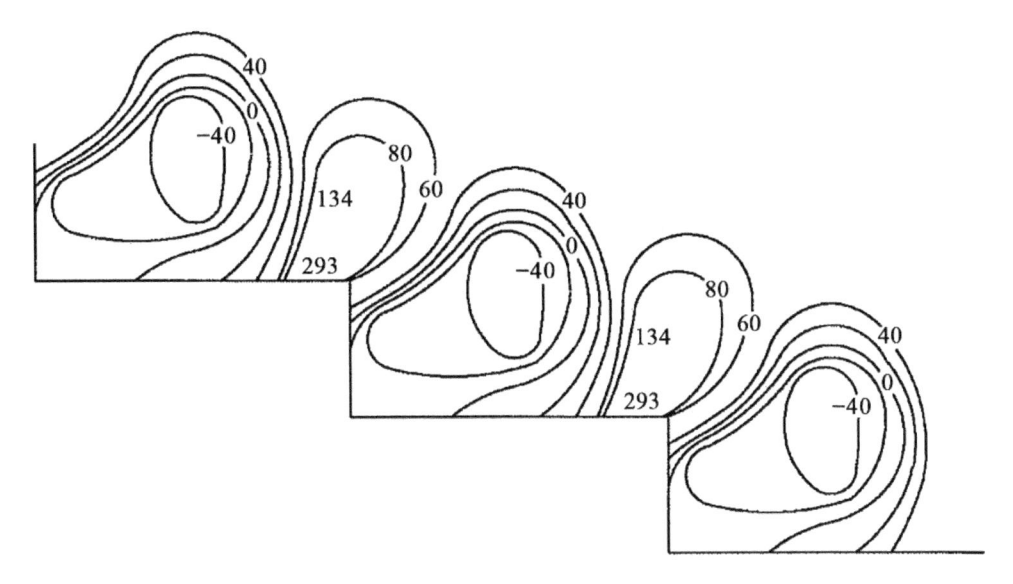

Figure 4.88 Mean pressure contours [Pa] on sidewall of stepped chute model under skimming flow for flow over a set of steps with rough approach chute (Winter *et al.*, 2010)

of masonry in zones subjected to external negative pressures. This results in blocks being 'plucked' from the wall, a local turbulence increase, and in extreme cases, complete wall collapse (ICOLD, 2017).

Aerator design

The cavitation potential of a stepped chute becomes relevant for large unit discharges, as indicated in 4.7.3. A technical chute aeration is then potentially necessary to avoid damages. To overcome the uncertainties regarding cavitation risk, different prototypes were equipped with aerators, namely:

- Dachaoshan Dam (China) stepped chute is equipped with flaring gate piers (Lin and Han, 2001). The spillway bottom is smooth up to the pier end; different bottom angles between the smooth and the stepped lower part create a cavity entraining air (Deng *et al.*, 2003). The unit discharge is 200 m^2/s for the safety check flood.
- Wadi Dayqah RCC Dam (Oman) spillway includes aerated crest-splitter teeth (Prisk *et al.*, 2009; Al Harthy *et al.*, 2010). Additionally, a continuous horizontal lip protruding 4 m into the flow is located below the splitters. Both the splitter teeth and the lip are equipped with aeration ducts on their downstream face (Figure 4.89). The 1000-year flood corresponds to a unit discharge of almost 50 m^2/s, and the PMF to some 84 m^2/s.
- Boguchany Dam (Russia) combines rockfill and conventional gravity dams, and includes a stepped chute (Toloshinov *et al.*, 2009). Two groove aerators are implemented, one at the transition from the standard profile to the steps and the second further downstream.

Figure 4.89 Wadi Dayqah RCC Dam with aerated crest-splitter teeth above aerated lip and aeration gallery (a) cross section with ① minimum 400 mm grout enriched RCC against formwork, ② training wall, ③ aeration gallery, ④ aerated crest-splitter teeth above aerated lip, ⑤ sidewall, (b) photo of near-completed dam (Adapted from Prisk *et al.*, 2009), (c) view from downstream (Courtesy Davide Wüthrich)

- Enlarged Cotter RCC Dam (Australia) features a spillway on its stepped air face for unit discharges of up to 48 m²/s at PMF (Willey *et al.*, 2010). An aerator is located 25 m vertically below the crest (Figure 4.90) due to the maximum flow velocity of 15 m/s (Amador *et al.*, 2009). Some steps were combined to form a deflector fed by a manifold-type aeration duct with discrete outlets at intervals across the chute width (Figure 4.90) as proposed by Ozturk *et al.* (2008). The air supply for these ducts of 2.5 m diameter (Figure 4.91a) is through intakes located directly outside the primary spillway training walls. The intakes extend above the PMF water surface and were shaped to minimize the disturbance to the flow during extreme events.

These are practical applications previously tested on physical models, yet not designed according to guidelines, given that the latter are rare for stepped chute aerators. First proposals to technically aerate stepped chutes were made by Pfister *et al.* (2006a). A deflector was installed at the end of the standard profile, performing as aerator only optimal at sufficiently high approach flow Froude numbers. A different type of stepped chute aerator, the

Figure 4.90 Enlarged Cotter RCC Dam with aeration duct outlets at 10 m intervals across chute width (Courtesy Brian Forbes)

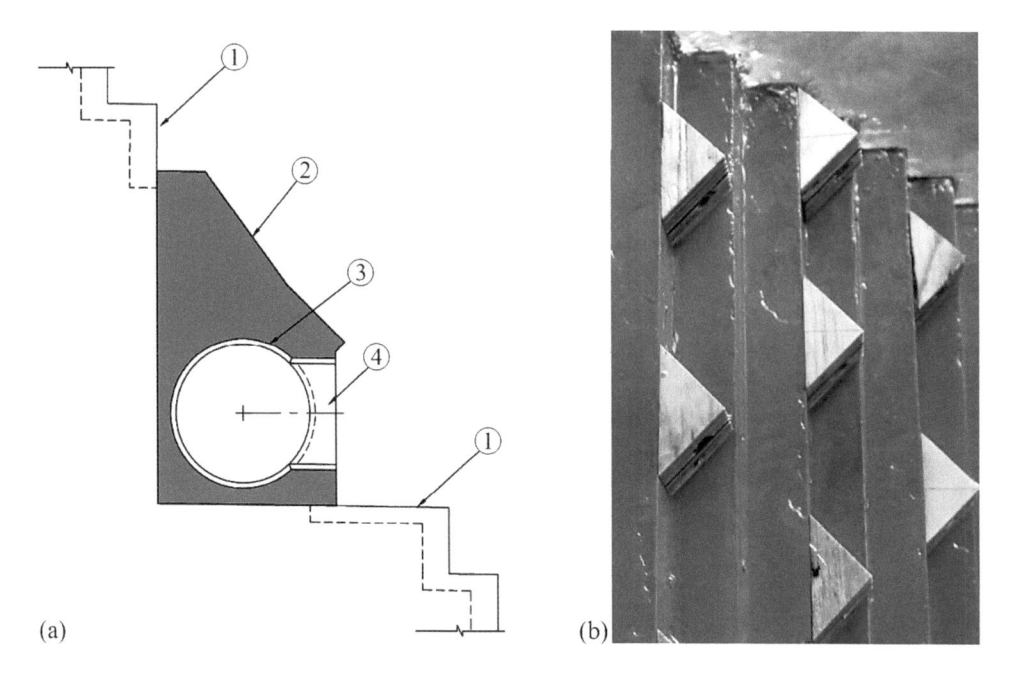

(a)　　　　　　　　　　　　　　　　　(b)

Figure 4.91 (a) Cross section of aeration step detail of Enlarged Cotter RCC Dam with ① chute steps, ② reinforced-concrete aeration step, ③ 2.5 m feeder pipe, ④ 1.8 m aeration outlet pipe (Willey *et al.*, 2010), (b) top view of triangular protrusions on stepped chute scale model (Wright, 2010)

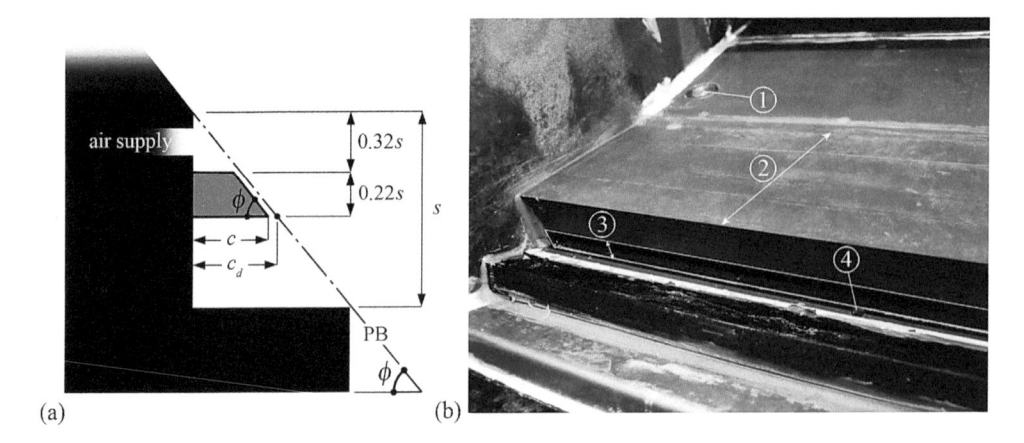

(a)   (b)

Figure 4.92 Definition sketches of (a) step-aerator, with PB = pseudo-bottom (Schiess *et al.*, 2008), (b) deflector aerator with ① pressure sensor, ② deflector, ③ air slot, ④ manometer (Terrier, 2016)

step-aerator, was introduced by Pfister *et al.* (2006b) and developed by Schiess *et al.* (2008). The sub-pressures along the upper part of the vertical step face are isolated and connected to the free atmosphere, so that air is entrained (Figure 4.92a). Aeration is further enhanced by using splitters and protrusions. Based on scale modeling, Wright (2010) evidenced that triangular protrusions located on the upstream spillway portion (Figure 4.91b) shorten the length of the black-water reach to the inception point, thereby enhancing aeration. For an upper velocity limit of 20 m/s at the air inception point, a chute fitted with triangular protrusions will be effective up to $q = 40$ m²/s for steps of $s = 1.2$ m.

A thorough model study was conducted by Terrier (2016), focusing on deflector aerators (Figure 4.92b) located sufficiently *downstream* of the transition between the control structure and the stepped chute, i.e. at a location with a sufficiently high approach flow Froude number. A minimum of $\mathsf{F} = V/(gh)^{0.5} = 2.0$ is required to initiate air entrainment for $\phi = 50°$ along with a deflector angle of some $10°$, and a minimum of $\mathsf{F} = 2.5$ for $\phi = 30°$. The bottom air concentration downstream of the aerator decreases locally at the jet impact zone to a local minimum, increasing then further downstream toward uniform flow conditions. To increase the minimum pseudo-bottom air concentration, the value of F, the pseudo-bottom angle, the deflector angle and the deflector height have to be increased. In practice, only the deflector angle and deflector height can be adapted. For all tested geometries and hydraulic conditions, even for unfavorable cases, a minimum bottom air concentration of $C_b = 0.09$ was noted by Terrier (2016), i.e. sufficient to prevent cavitation damage. Apparently, stepped chutes are capable to maintain the turbulence generated by the deflector at the jet impact point, reducing thus the de-aeration process along the chute.

Spray reduction

Spray is a nuisance in hydraulic engineering, because of soil saturation with related instabilities, surface erosion, or ice formation when freezing. Stepped chutes are known for spray generation, particularly at the transition between the control structure and the stepped chute

(4.7.3) under small discharges. The spray height may even become larger than the flow depth under design discharge. From a practical point of view, such flow is unacceptable because of the large sidewall freeboard required, the small energy dissipation, and the poor hydraulic performance.

Supercritical flow is sensitive to any changes related to wall deflections, drops, or curves because of shock wave generation. The first step of a stepped chute may be considered as a vertical drop, if the approach flow depth is small and the flow is issued by an upstream standard profile and thus aligned to the pseudo-bottom (Figure 4.70a). The jet is directed to the vertical due to gravity and a relatively large jet portion impacts onto the horizontal first step face. The flow is then deflected away from the chute, and the contact with the chute is lost for a number of steps. Upon reimpact onto the stepped chute, the flow is redeflected, resulting in a flow that may be described neither by skimming nor by nappe flow.

There are generally two options to reduce the spray height along the first steps of the constantly sloping stepped chute at small discharges. Mateos Iguácel and Elviro García (2000) proposed a gradual increase of step height from shortly downstream of the standard profile crest to the start of the stepped chute, as illustrated in Figure 4.72. This measure limits further flow separation at the transition, as indicated by tests of Houston (1987) for the Upper Stillwater Dam. An alternative installation for spray reduction at the transition is due to Pfister *et al.* (2006a). To reduce spray, the jet impact angle on the first horizontal step edge(s) should be small. Instead of using a standard step with a horizontal step face, the step edge was cut close to the jet impact angle. To simplify construction, step insets were added in the model to the original steps. Figure 4.93 shows the flow features with and without insets. It was noted that the insets should have a 'nose-angle' smaller than that of the pseudo-bottom. Adding

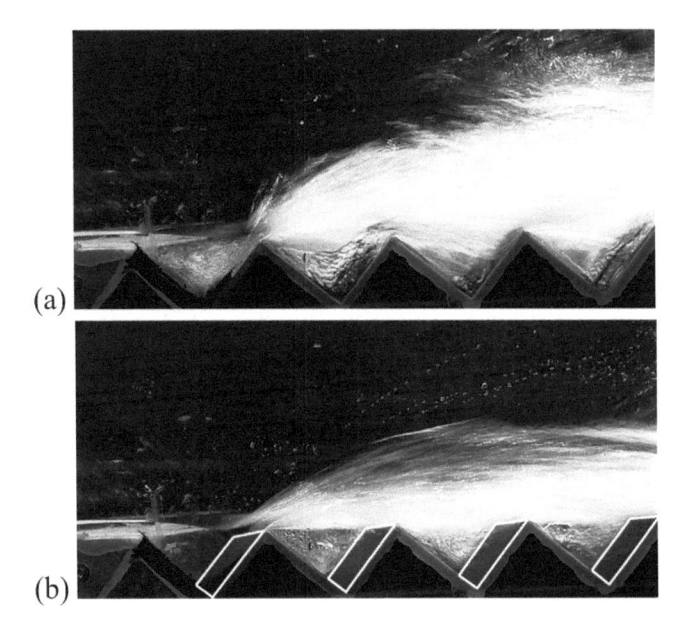

(a)

(b)

Figure 4.93 Spray generation at transition between standard profile and start of stepped chute, (a) without insets, (b) with five insets. (Pfister *et al.*, 2006a). Note photos rotated to horizontal

these insets to two steps reduces spray significantly, and five insets practically eliminate all spray. Without insets, the spray reaches up to almost four times the step height for small discharges.

Training wall design

The design of training walls along stepped chutes should account for the bulking of aerated flow. The flow depth of the two-phase flow is larger than that of an equivalent clear-water flow over a smooth chute. In assessing the required wall height, a safety factor is usually applied depending on the nature of the surrounding topography (Boes and Minor, 2002; Boes and Hager, 2003b). Special care is needed if the chute flow overtopping the training walls could affect an adjacent earthfill embankment dam.

While the local flow depths in the upper chute portion are mainly governed by considerable spray formation for small discharges, the uniform mixture-flow depth $h_{90,u}$ at design discharge may serve as guide for the design of training walls in the aerated region further downstream, where considerable aeration leads to flow bulking.

UPPER SPILLWAY PORTION

The required training wall (subscript t) height along the upper chute portion (subscript 1) $h_{t,1}$, where jet deflection causes spray (subscript s), follows the envelope of spray profiles of Pfister *et al.* (2007) and Schiess *et al.* (2008). It is based on the dimensionless coordinate $X_s = (x - x_s)/(s\mathsf{F}_o)$, where $x_s = 3s/\sin\phi$ is the origin of spray flow and $\mathsf{F}_T = v_T/(gh_T)^{1/2}$ the Froude number at the first step edge, i.e. at the point of tangency (subscript T) for the smooth standard-crested profile (Figure 4.70a), with v_T as local flow velocity and h_T as local flow depth. With $Z_s = (h_{s,1} - h_T)/(h_{s,\max} - h_T)$, where $h_{s,1}$ is spray height due to jet deflection, the spray profile follows with $M = 2$ for a chute *without* aerator and $M = 1$ *with* a step-aerator (Figure 4.92a)

$$Z_s = \left[0.65 M X_s \cdot \exp\left(1 - 0.65 M X_s\right)\right]^{1/2}, \quad \text{for } 0 < X_s < 1.5. \tag{4.181}$$

The maximum spray height $h_{s,\max}$ is given by (Pfister *et al.*, 2007, Schiess *et al.*, 2008)

$$h_{s,\max} / h_T = 6.6\left(h_c / s\right)^{-2\varsigma}, \quad \text{for } 0.35 < \left(h_c / s\right)^\varsigma < 1.5. \tag{4.182}$$

The exponent $\zeta = (1 + n_s)^{-1/3}$ accounts for the number n_s of treated step edges. Equation (4.181) indicates that for chutes below a step-aerator, the jet reattaches further downstream than on standard-stepped chutes without aerator due to the slightly lifted lower jet trajectory beyond its contact with the step-aerator (Schiess *et al.*, 2008). Note that Eqs. (4.181) and (4.182) apply for gravity dam chutes with $\phi \approx 50°$. A safety factor $\eta_S > 1$ applies when computing the design training wall height $h_{t,1}$, so that

$$h_{t,1} = \eta_S \cdot h_{s,1}. \tag{4.183}$$

Depending on the erosion potential along the chute training walls, it is recommended to select $\eta_S = 1.2$ if spray is a nuisance, e.g. for concrete dams, and $\eta_S = 1.5$ for chutes where spray is a concern, e.g. on embankment dams or on valley flanks prone to erosion or for nearby switchyards (Boes and Hager, 2003b).

LOWER SPILLWAY PORTION

The proposed design height $h_{t,2}$ for the training walls along the aerated lower chute portion (subscript 2) reads, again with a safety factor accounting for the erosion potential besides the chute as given above,

$$h_{t,2} = \eta_S \cdot h_{90,u}. \tag{4.184}$$

Note that Eq. (4.184) is based upon the skimming flow regime and valid for the entire range of chute angles $19° \leq \phi \leq 55°$. For nappe flow, the jet impact onto the steps may cause considerable spray, overtopping the training walls designed after Eq. (4.184), so that the sidewalls should then be designed according to Eqs. (4.181) and (4.183).

For chutes where uniform flow is not attained (Eq. 4.161), the training wall height follows the mixture-flow depth $h_{90}(x)$ gradually varying in the streamwise direction, based on the differential equation of the backwater curve (Boes and Minor, 2000)

$$h_{90}(x) = 0.55 \left(\frac{q^2}{g \sin \phi} \right)^{1/4} \tanh \left(\frac{\sqrt{gs \sin \phi}}{3q}(x - L_i) \right) + 0.42 \left(\frac{q^{10} s^3}{(g \sin \phi)^5} \right)^{1/8}. \tag{4.185}$$

Here x is the distance along the chute from the crest (Figure 4.83). Similar to smooth chutes, a minimum freeboard $f_{min}(x) = 0.61 + 0.00513 V_w(x) h_{90}(x)^{1/3}$ [m] according to USBR (1977) can then be added to the characteristic mixture-flow depth $h_{90}(x)$. Note that $V_w = q/h_w(x)$ and $h_w(x) = h_{90}(x)[1 - C_a(x)]$, where $C_a(x)$ is obtained from Eq. (4.164).

CONVERGING CHUTE

The design of a training wall height in converging chutes requires a specific approach (Hunt et al., 2012) due to near-wall flow concentration. These chutes are commonly found in narrow valleys and on embankment dams where the chute overflow protection may also spill over the dam abutments constituting a converging stepped chute (ICOLD, 2017).

NO TRAINING WALLS

The stepped chutes without training walls is a novel design characterized by lateral flow expansion, resulting in lower unit discharges at the dam toe. For low unit discharge and good foundation conditions, omitting training walls is a cost-effective alternative to conventional designs with sidewalls. Model studies by Estrella et al. (2012) indicate that waiving training walls reduce unit discharges from 50% to 70% due to lateral flow expansion.

Step surface roughness and roughness elements

ROUGH STEP SURFACE

Stepped chutes do not mandatorily include smooth concrete surfaces, but can involve rough faces as gabions (4.7.2) or damaged concrete, for instance. The latter influences the flow features in terms of turbulence and energy dissipation. A model study of Gonzalez et al. (2008)

with $\phi = 21.8°$ and $s = 0.008$ m including different surface roughness arrangements resulted in the following observations for skimming flow:

- No effect of surface roughness on flow regime;
- Downstream shift of air inception point location for a rough as compared to a smooth surface;
- Slightly higher flow velocity on rough step surface under otherwise identical conditions;
- Slightly less aerated flow over rough than over smooth steps.

ROUGHNESS ELEMENTS ON STEP EDGES

Steps can be equipped with particular geometrical elements, located on the horizontal step surface near the edge, to affect the flow features. These elements typically consist of a small coherent end sill per step edge of height $s_r = 0.50s$ and thickness $l_r = 0.43s$ (Figure 4.94). They are arranged over the entire chute width, or constitute fragmented sills covering only a certain portion, similar to baffle blocks in stilling basins (Figure 4.94). The most important flow features affected by these elements are the natural air entrainment into the flow, as well as energy dissipation. Both are by trend amplified. Model tests on both a 'standard' stepped chute and one equipped with various types of step edge elements were presented by André et al. (2006, 2008a, b), for $\phi = 18.6°$ and $30°$. The effects of the elements on the basic flow features are:

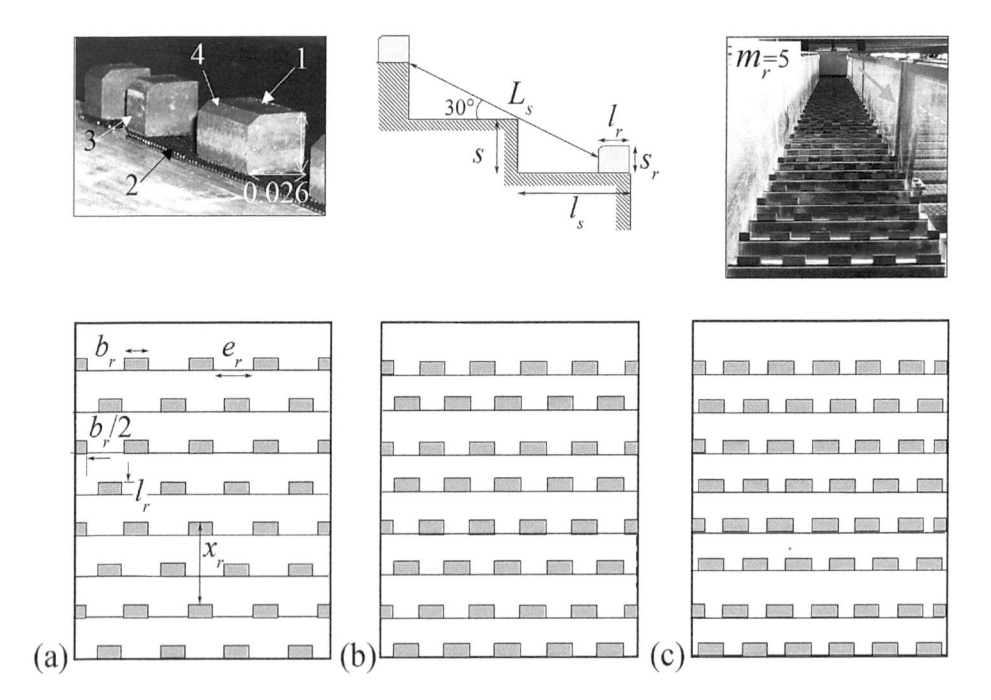

Figure 4.94 Sketch of tested stepped macro-roughness overlays ($L_s = 2l_s/\cos(\phi-30°)$, $s = 0.06$ m, $s_r = 0.03$ m, $l_r = 0.026$ m, $l_s = 0.104$ m, $\phi = 30°$) (a) spaced alternated blocks $m_r = 4$, (b) aligned alternated blocks $m_r = 5$, (c) closely packed alternated blocks $m_r = 6$. ① Block, ② Velcro, ③ half block, ④ chamfer (modified from André et al., 2008a, b)

- A coherent end sill provided over the entire chute width hardly affects the flow features as compared with a standard-stepped chute;
- Fragmented and transversally irregularly placed blocks have the largest effect on flow features;
- Location of air inception point is not significantly influenced by elements. Only well-developed skimming flow over fragmented sills shortens non-aerated chute reach by some 17%;
- Average uniform air concentration for skimming flow remains unchanged under coherent end sills, whereas alternate blocks enhance the values relative to standard steps. Higher air concentrations pertain near the pseudo-bottom;
- Energy dissipation along the chute is hardly affected by laterally coherent end sills. In contrast, the fragmented setup with alternate blocks increases the energy dissipation by 5% to 8%, depending on block spacing.

The block configuration with spaced alternated blocks ($m_r = 4$, Figure 4.94) of relative block width $b_r/s_r = 5/3$, relative block height $s_r/s = 1/2$, and lateral block spacing to block width $e_r/b_r = 1.5$ was the most efficient configuration in view of energy dissipation.

OCCURRENCE OF JUMP WAVES ON POOLED STEPS

For low-gradient pooled stepped chutes, for example if the steps are equipped with end sills (see above), flow instabilities in the form of jump waves may occur. These have to be considered in the design to avoid overtopping of sidewalls and damages of pooled step chutes. Thorwarth et al. (2009) developed a semi-empirical approach allowing for the prediction of the flow condition for occurrence of jump waves depending on bottom slope and geometrical conditions of the pooled steps.

Notation

A_i	Integration constant (-)
a	Relative terminal bubble rise velocity (m/s)
B	Constant of integration (-)
B	Bend number (-)
B_i	Integration constant (-)
b	Chute width (m)
b_e	End width of chute contraction (m)
b_o	Approach flow channel width (m)
b_r	Width of macro-roughness element (m)
b_u	Downstream channel width (m)
C	Air concentration (-)
C_{90}	Air concentration at 90% (-)
ΔC_a	Streamwise air detrainment decay (-)
C_a	Average air concentration (-)
$C_{a(3L)}$	Average air concentration at $x/L = 3$ (-)
C_{ac}	Critical average air concentration (-)
C_{ae}	Average equilibrium air concentration (-)
$C_{a,ent}$	Average air entrainment concentration (-)
$C_{a,i}$	Average air concentration at air inception point (-)
$C_{a,min}$	Minimum average air concentration (-)

C_{ao}	Average approach flow air concentration (-)
C_{au}	Air concentration for fully aerated uniform flow (-)
C_b	Bottom (invert) air concentration (-)
$C_{b,crit}$	Critical bottom concentration (-)
$C_{b,det}$	Bottom air concentration along detrainment zone (-)
$C_{b,ent}$	Bottom air concentration along entrainment zone (-)
$C_{b,min}$	Minimum bottom air concentration (-)
$C_{b(3L)}$	Bottom air concentration at $x/L = 3$ (-)
C_{bo}	Approach flow bottom air concentration (-)
C_C	Chézy roughness coefficient (m$^{1/2}$/s)
C_e	Equilibrium air concentration (-)
C_f	Skin friction coefficient (-)
C_p	Pressure head coefficient (-)
C_p'	Fluctuating pressure head coefficient (-)
c	$= (C_e - C_b)/(0.90 - C_b)$ (-); also wave celerity (m/s)
c_i	$= (C_a - C_{a,i})/(C_{a,u} - C_{a,i})$ (-)
D	Relative boundary layer thickness (-)
D'	Turbulent diffusivity (-)
D_0	Dimensionless coefficient in diffusivity model (-)
$D_{h,w}$	Hydraulic diameter (m)
d_{bu}	Equivalent bubble diameter (m)
d_{bu90}	Bubble size in top flow region (m)
$d_{bu(95)}$	Critical bubble size (m)
d_L	Limit break-up diameter (m)
e_r	Macro-roughness spacing (m)
F	Cross-sectional flow area (m^2)
F	Froude number (-)
F$_*$	Roughness Froude number (-)
F$_0$	Froude number where U$_{bu}$ = 0 (-)
F$_1$	Entrance Froude number for stilling basin design (-)
F$_b$	Froude number at roll wave front (-)
F$_o$	Approach flow Froude number (-)
F$_{o-min}$	Arithmetically averaged Froude number (-)
F$_r$	Relative Froude number of roll waves (-)
F$_T$	Froude number at point of tangency (-)
F$_w$	Water flow Froude number (-)
f	Friction factor (-)
f_e	Friction factor for equilibrium air-water flow (-)
f_b	Darcy-Weisbach friction factor (-)
f_{min}	Minimum freeboard height (m)
f_w	Friction factor for water flow (-)
g	Gravity acceleration (m/s^2)
H_D	Weir crest design head (m)
H_{chute}	Chute height (m)
H_{dam}	Dam height (m)
H_{max}	Energy head at spillway crest (m)
H_o	Energy head at spillway crest (m)

H_{res}	Residual energy head (m)
H_s	Difference of water elevations between inception point and reservoir level (m)
ΔH	Dissipated energy head (m)
h	Local flow depth (m)
h_1	Entrance flow depth at stilling basin (m)
h_{90}	Mixture-flow depth at 90% air concentration (m)
h_{99}	Mixture-flow depth at 99% air concentration (m)
$h_{90,u}$	Uniform mixture-flow depth at 90% air concentration (m)
h_a	Axial flow depth (m)
h_{atm}	Atmospheric pressure head (m)
h_c	Critical flow depth (m)
h_e	Extreme bend wave height (m)
h_f	Roll wave front height (m)
h_g	Piezometric pressure head (m)
h_i	Local flow depth at air inception point (m)
h_M	Maximum wall flow depth (m); also roll wave peak flow depth (m)
h_o	Approach flow depth (m)
h_p	Hydrostatic pressure depth (m)
h_s	Cavity sub-pressure head (m)
$h_{s,1}$	Spray height due to jet deflection (m)
$h_{s,max}$	Maximum spray height (m)
h_T	Flow depth at point of tangency (m)
h_t	Height of training wall (m)
h_v	Vapor pressure head (m)
h_w	Clear-water flow depth (m), also wall flow depth (m)
h_{wM}	Maximum shock height (m)
$h_{w,u}$	Uniform equivalent clear-water depth (m)
K	Strickler's roughness coefficient ($m^{1/3}$/s), also step roughness $= s\cos\phi$ (m)
K'	Integration constant (-)
k	Roughness height (m)
k_s	Equivalent roughness height (m)
L	Jet length (m)
L_i	Distance from spillway crest to air inception point (m)
L_s	Shortened jet length (m)
l	Horizontal step length (m)
l_P	Prandtl's mixing length (m)
l_r	Length of macro-roughness overlay (m)
M	Number accounting for aerator presence (-); also cross-sectional parameter (-)
M	Montuori number (-)
m	Slope of air detrainment line, also exponent (-)
m_r	Macro-roughness type number (-)
N_s	Number of steps (-)
n	Exponent of power-law velocity profile, also coefficient of $m(\mathsf{F}_o)$ (-)
n_M	Manning roughness coefficient ($s/m^{1/3}$)
n_s	Number of steps treated against spray flow (-)
P	Relative sub-pressure (-); also wetted perimeter (m)
P_t	Turbulence production (-)

p	Pressure (N/m²)
$p_{0.1\%}$	Pressure with 0.1% probability (N/m²)
p_a	Time-averaged local pressure (N/m²)
p_{atm}	Atmospheric pressure (N/m²)
p_c	Critical pressure (N/m²)
p_g	Piezometric pressure (N/m²)
p_o	Local pressure (N/m²)
p_v	Vapor pressure (N/m²)
Q	Discharge (m³/s)
Q_A	Air discharge (m³/s)
Q_w	Water discharge (m³/s)
q	Unit discharge (m²/s)
q_A	Unit air discharge (m²/s)
q_w	Unit water discharge (m²/s)
R_h	Hydraulic radius (m)
R_{bu}	Bubble Reynolds number (-)
R_o	Approach flow Reynolds number (-)
r	Relative chute roughness (-)
S	Shape factor of boundary layer (-)
S	Shock number (-)
S_E	Energy line slope (-)
S_f	Friction slope (-)
S_o	Chute bottom slope (-)
s	Offset or step height (m)
s'	Relative distance to air inception point (-)
s_r	Height of macro-roughness overlay (m)
t	Deflector height (m), also time (s)
t_b	Bubble rise velocity coefficient (-)
U	Potential flow velocity (m/s)
U_{90}	$= u/u_{90}$ (-)
U_{bu}	Non-dimensional bubble rise velocity (-)
u	Velocity parallel to chute bottom (m/s)
u_a	Average flow velocity (m/s)
u_{ac}	Critical average velocity (m/s)
u'	Streamwise velocity fluctuation (m/s)
u_{90}	Mixture velocity at $C = 90\%$ (m/s)
u^*	Shear velocity (m/s)
u_{bu}	Air bubble rise velocity (m/s)
$u_{bu,t}$	Terminal air bubble rise velocity (m/s)
u_m	Mixture velocity (m/s)
u_o	Approach flow velocity (m/s)
u_τ	Shear velocity (m/s)
V	Local (mean) flow velocity (m/s)
V	Vedernikov number (-)
V_a	Average black-water velocity (m/s)
V_f	Roll wave front velocity (m/s)
V_M	Roll wave peak velocity (m/s)

V_m	Average mixture velocity (m/s)
V_o	Approach flow velocity (m/s)
V_u	Uniform flow velocity (m/s)
v'	Streamwise velocity fluctuation (m/s)
v_a	Time-averaged local velocity (m/s)
v_c	Critical velocity (m/s)
v_T	Flow velocity at point of tangency (m/s)
v_w	Clear-water flow velocity (m/s)
W_o	Approach flow Weber number (-)
w_i	Vertical distance from weir crest to inception point elevation (m)
X_{90u}	$= x/h_{90u}$ (-)
X	$= x/(h\mathsf{F})$ at wall deflection, $= x/(b\mathsf{F})$ at chute expansion (-)
X_i	$= (x - L_i)/h_i$ (-)
$X_{i,crit}$	Critical relative distance to air inception point (-)
X_s	$= (x - x_s)/(s\mathsf{F}_T)$ (-)
x	Streamwise coordinate (m)
x_i	Location of air inception point from weir crest (m)
x_r	Streamwise distance between macro-roughness overlays (m)
x_s	$= 3s/\sin\phi$ (m)
Y	Relative flow depth (-)
Y_{90}	$= y/h_{90}$ (-)
Y_{99}	$= h_{99}/h_w$ (-)
Y_a	$= h_a/h_o$ (-)
Y_f	Relative roll wave height (-)
Y_M	Relative maximum wall flow depth (-)
Y_N	Dimensionless uniform flow depth (-)
Y_w	$= h_w/h_o$ (-)
Y_{wM}	H_{wM}/h_o (-)
y	Distance from chute invert, also transverse coordinate (m)
y^*	$2(1 - Y_{90})/(\sin\alpha)^{1/2}$ (-)
y_e	Extreme bend wave height ratio (-)
Z_{90}	$= z/h_{90}$ (-)
Z_{90u}	$= z/h_{90u}$ (-)
Z_i	$= (z - z_i)/h_c$ (-)
Z_M	Relative maximum bend shock height (-)
Z_s	$= (h_{s,1} - h_T)/(h_{s,max} - h_T)$ relative spray height (-)
z	Coordinate perpendicular to chute invert (m), also vertical coordinate (m)
z'	Vertical distance above reference plain (m)
z_l	Lower air-water jet elevation (m)
z_R	Roughness sublayer thickness (m)
z_u	Upper air-water jet elevation (m), also vertical distance from weir crest to uniform flow (m)
α	Chute bottom angle (-)
α'	Velocity head correction coefficient (-)
β	Aeration rate, air entrainment coefficient (-)
β_b	Exponent of boundary layer profile (-)
β_e	Ratio of chute expansion/contraction (-)

β_S Air entrainment coefficient under cavity sub-pressure (-)
β_s Shock angle
χ Relative streamwise distance (-)
γ Coefficient of boundary layer profile (-)
γ_w Generalized wall flow profile (-)
δ Boundary layer thickness (m)
δ^* Boundary layer displacement thickness (m)
δ_i Boundary layer thickness at air inception point (m)
ε_c Turbulent exchange coefficient (-)
ϕ Stepped chute invert angle of pseudo-bottom (-)
η Energy dissipation efficiency (-)
η_R Chute roughness characteristics (-)
η_S Safety factor (-)
η_0 Relative wall distance (-)
η_R Relative roughness wall distance (-)
ϕ Deflector angle (-)
κ Von Karman constant (-)
Λ Coefficient of power-law velocity profile (-)
λ Relative roll wave length (-)
μ Dynamic viscosity (Pa·s); also velocity ratio of roll waves (-)
μ_w Relative roll wave velocity (-)
ν Kinematic fluid viscosity (m^2/s)
Π Wake strength parameter (-)
ρ Fluid density (kg/m^3)
ρ_A Air density (kg/m^3)
ρ_a Relative chute curvature (-)
ρ_w Water density (kg/m^3)
σ Cavitation number (-)
σ_c Critical cavitation number (-)
σ_{min} Minimum cavitation number (-)
σ_p Root-mean-square of pressure fluctuations (N/m^2)
σ_t Surface tension (N/m)
θ Wall deflection angle (-)
θ_M Location of maximum bend shock height (-)
θ_m Boundary layer momentum thickness (m)
τ Dimensionless value of chute expansion/contraction (-)
τ_0 Local boundary shear stress (N/m^2)
τ_t Turbulent shear stress (N/m^2)
τ_v Viscous shear stress (N/m^2)
ζ Exponent (-)
Ω $= 1 - \beta_e$ (-)

References

Al Harthy, S.H., Hall, J.K., Hieatt, M.J. & Wheeler, M. (2010) The day Wadi Dayqah roared. *Water Power & Dam Construction*, 62(12), 40–43.

Amador, A., Sánchez-Juny, M. & Dolz, J. (2005) Discussion of two phase flow characteristics of stepped spillways. *Journal of Hydraulic Engineering*, 131(5), 421–423.

Amador, A., Sánchez-Juny, M. & Dolz, J. (2006) Diseño hidráulico de aliviaderos escalonados en presas de HCR [Hydraulic design of stepped spillways for RCC dams]. *Ingeniería del Agua*, 13(4), 289–302 (in Spanish).

Amador, A., Sánchez-Juny, M. & Dolz, J. (2009) Developing flow region and pressure fluctuations on steeply sloping stepped spillways. *Journal of Hydraulic Engineering*, 135(12), 1092–1100.

Anastasi, G. (1980) L'eliminazione delle onde d'urto negli scivoli a contrazione [The elimination of shock waves at chute contractions]. Proceedings of 17th *Convegno di Idraulica e Costruzioni Idrauliche*, Palermo, A2, 1–12 (in Italian).

André, S. (2004) High velocity aerated flows on stepped chutes with macro-roughness elements. In: Schleiss, A. (ed) *Communication* 20. Laboratoire de Constructions Hydrauliques, EPFL, Lausanne, Switzerland.

André, S., Boillat, J.-L. & Schleiss, A.J. (2006) Blocks on stepped overlay: a promising measure to protect the downstream slope and the base of an overtopped embankment dam. Proceedings of 22nd *ICOLD Congress*, Barcelone, Q86(R74), 1225–1237.

André, S., Boillat, J.-L. & Schleiss, A.J. (2008a) Ecoulements aérés sur évacuateurs en marches d'escalier équipées de macro-rugosités 1: caractéristiques hydrauliques [Aerated stepped chute flow equipped with macro roughness elements 1: hydraulic characteristics]. *La Houille Blanche*, 63(1), 91–100 (in French).

André, S., Boillat, J.-L. & Schleiss, A.J. (2008b) Ecoulements aérés sur évacuateurs en marches d'escalier équipées de macro-rugosités 2: dissipation d'énergie hydrauliques [Aerated stepped chute flow equipped with macro roughness elements 2: energy dissipation]. *La Houille Blanche*, 63(1), 101–108 (in French).

André, S., Matos, J., Boillat, J.-L. & Schleiss, A.J. (2004) Energy dissipation and hydro-dynamic forces of aerated flow over macro-roughness linings for overtopped embankment dams. Proceedings of International Conference of *Hydraulics of Dams & River Structures*, Tehran, Iran. Taylor & Francis, London, pp. 189–196.

André, S. & Schleiss, A.J. (2008) Discussion of Pressures on a stepped spillway, by Sánchez-Juny, M., Bladé, E., Dolz, J. *Journal of Hydraulic Research*, 46(4), 574–576.

Armani, A. (1894) Über die Bewegung des Wassers in gestaffelten Gerinnen [On the movement of water in stepped channels]. *Zeitschrift des Österreichischen Ingenieur- und Architekten-Vereines*, 46(52), 589–592 (in German).

Arndt, R.E.A. & Ippen, A.T (1968) Rough surface effects on cavitation inception. *ASME Journal of Basic Engineering*, 9(3), 249–261.

ASCE (1963) Friction factors in open channels. *Journal of the Hydraulics Division* ASCE, 89(HY2), 97–143.

Ball, J.W. (1976) Cavitation from surface irregularities in high velocity. *Journal of the Hydraulics Division* ASCE, 102(9), 1283–1297; 103(HY4), 469–472; 103(HY8), 945–946; 104(HY8), 1199–1200.

Bass, R.P., Fitzgerald, T. & Hansen, K.D. (2012) Lesson learned: more than 100 RCC overtopping spillways in the United States. Proceedings of 6th International Symposium *Roller Compacted Concrete Dams*, Zaragoza, Spain (CD-ROM).

Bauer, W.J. (1951) The development of the turbulent boundary layer on steep slopes. *Ph.D. Thesis*. University of Iowa, Iowa.

Bauer, W.J. (1954) Turbulent boundary layer on steep slopes. *Transactions of ASCE*, 119, 1212–1233.

Baumann, A., Arefi, F. & Schleiss, A.J. (2006) Design of two stepped spillways for a pumped storage scheme in Iran. Proceedings of International Conference *Hydro*, Porto Carras, Greece (CD-ROM).

Berga, L. (1995) Hydrologic safety of existing embankment dams and RCC for overtopping protection. Proceedings of 2nd International Symposium *Roller Compacted Concrete Dams*, Santander, Spain, pp. 639–652.

Berlamont, J. (1986) Unstable turbulent channel flow. *Encyclopedia of Fluid Mechanics*, 2, 98–121, N.P. Cheremisinoff, ed. Gulf: Houston.

Bezzola, G.R. (2002) Fliesswiderstand und Sohlenstabilität natürlicher Gerinne unter besonderer Berücksichtigung des Einflusses der relativen Überdeckung [Flow resistance and bed stability of natural rivers under particular consideration of the relative coverage effect]. *Ph.D. Thesis*, also *VAW Mitteilung*, 173, H.-E. Minor (ed). Versuchsanstalt für Wasserbau, Hydrologie und Glaziologie. ETH Zurich, Zürich (in German).

Blevins, R.D. (1984) *Applied fluid dynamics handbook*. Van Nostand Reinhold, New York.

Boes, R. (1999a) Gewichtsstaumauern aus Walzbeton [Gravity dams of roller compacted concrete]. *Wasser, Energie, Luft*, 91(1/2), 11–15 (in German).

Boes, R. (1999b) Sanierung von Schüttdämmen mit Walzbeton [Rehabilitation of fill dams with roller compacted concrete]. *Wasserwirtschaft*, 89(6), 292–296 (in German).

Boes, R. (2000) Zweiphasenströmung und Energieumsetzung an Grosskaskaden [Two-phase flow and energy dissipation at cascades]. Doctoral Dissertation, *Mitteilung* 166, Laboratory of Hydraulics, Hydrology and Glaciology (VAW), ETH Zurich, Zürich, Switzerland (in German).

Boes, R.M. (2012) Guidelines on the design and hydraulic characteristics of stepped spillways. Proceedings of 24th *ICOLD Congress*, Kyoto, Q94(R15), 203–220.

Boes, R.M. & Hager, W.H. (2003a) Two-phase flow characteristics of stepped spillways. *Journal of Hydraulic Engineering*, 129(9), 661–670.

Boes, R.M. & Hager, W.H. (2003b) Hydraulic design of stepped spillways. *Journal of Hydraulic Engineering*, 129(9), 671–679.

Boes, R.M., Lutz, N. & Lais, A. (2015) Upgrading spillway capacity at large, non-overtoppable embankment dams. Proceedings of 25th *ICOLD Congress*, Stavanger, Norway, Q97(R23), 332–348.

Boes, R.M. & Minor, H.-E. (2000) Guidelines to the hydraulic design of stepped spillways. In: Minor, H.E. & Hager, W.H. (eds) Proceedings of International Workshop *Hydraulics of Stepped Spillways*, VAW, ETH, Zurich. Balkema, Rotterdam, the Netherlands, pp. 163–170.

Boes, R.M. & Minor, H.-E. (2002) Hydraulic design of stepped spillways for RCC Dams. *International Journal of Hydropower & Dams*, 9(3), 87–91.

Boes, R.M. & Schmid, H. (2003) Weir rehabilitation using gabions as a noise abatement option. *HYDRO, Cavtat*, 781–785.

Bogdevich, V.G., Evseev, A.R., Malyuga, A.G. & Migirenko, G.S. (1977) Gas-saturation effect on near-wall turbulence characteristics. Proceedings of 2nd International Conference *Drag Reduction*, D2, 25–37.

Bormann, K. (1968) Der Abfluss in Schussrinnen unter Berücksichtigung der Luftaufnahme [Chute flow under consideration of air entrainment]. In: Hartung, F. (ed) Versuchsanstalt für Wasserbau *Bericht* 13. Oskar v. Miller Institut, Technische Hochschule München, München (in German).

Bradley, J.N. (1945) Study of air injection into the flow in the Boulder Dam spillway tunnels, Boulder Canyon Project. Bureau of Reclamation, *Hydraulic Laboratory Report*, 186. US Dept. Interior, Denver CO.

Braeske, H., Domnick, J. & Brenn, G. (1997) Experimentelle Grundlagenuntersuchungen zur Strömungsmechanik in Blasensäulen [Experimental studies on flow mechanisms of bubble columns]. *Report* LSTM 552/E97. Lehrstuhl für Strömungsmechanik, Technische Fakultät der Friedrich-Alexander-Universität, Erlangen-Nürnberg (in German).

Brock, R.R. (1969) Development of roll-wave trains in open channels. *Journal of the Hydraulics Division* ASCE, 95(HY4), 1401–1427; 96(HY4), 1069–1072.

Brock, R.R. (1970) Periodic permanent roll waves. *Journal of the Hydraulics Division* ASCE, 96(HY12), 2565–2580; 97(HY8), 1251–1252.

Bung, D.B. (2009) Zur selbstbelüfteten Gerinneströmung auf Kaskaden mit gemäßigter Neigung [Self-aerated cascade flow of moderate slope]. *Ph.D. Thesis*, Bergische University, Wuppertal, Germany (in German).

Bung, D.B. & Schlenkhoff, A. (2010) Self-aerated flow on embankment stepped spillways: the effect of additional micro-roughness on energy dissipation and oxygen transfer. *IAHR European Congress* Munich, Germany, (CD-ROM).

Cain, P. (1978) Measurements within self-aerated flow on a large spillway. *Report* 78–18. University of Canterbury, Christchurch NZ.

Cain, P. & Wood, I.R. (1981) Measurements of self-aerated flow on a spillway. *Journal of the Hydraulics Division* ASCE, 107(HY11), 1425–1444; 109(1), 145–146.

Campbell, F.B., Cox, R.G. & Boyd, M.B. (1965) Boundary layer development and spillway energy loss. *Journal of the Hydraulics Division* ASCE, 91(HY3), 149–163.

Cardoso, G., Meireles, I. & Matos, J. (2007) Pressure head along baffle stilling basins downstream of steeply sloping stepped chutes. Proceedings of 32nd *IAHR World Congress*, Venice, Italy (CD-ROM).

Cassidy, J.J. (1990) Fluid mechanics and design of hydraulic structures. *Journal of Hydraulic Engineering*, 116(8), 961–977.

Castro-Orgaz, O. (2009) Hydraulics of developing chute flow. *Journal of Hydraulic Research*, 47(2), 185–194.

Castro-Orgaz, O. & Hager, W.H. (2010) Drawdown curve and turbulent boundary layer development for chute flow. *Journal of Hydraulic Research*, 48(5), 591–602.

Castro-Orgaz, O. & Hager, W.H. (2017) *Non-hydrostatic Free Surface Flows*. Springer, Berlin.

Chanson, H. (1988) Study of air entrainment and aeration devices on spillway model. *Ph.D. Thesis*, University of Canterbury, Christchurch NZ.

Chanson, H. (1989a) Study of air entrainment and aeration devices. *Journal of Hydraulic Research*, 27(3), 301–319.

Chanson, H. (1989b) Flow downstream of an aerator: aerator spacing. *Journal of Hydraulic Research*, 27(4), 519–536.

Chanson, H. (1994a) Hydraulics of skimming flows over stepped channels and spillways. *Journal of Hydraulic Research*, 32(3), 445–460; 33(3), 414–419.

Chanson, H. (1994b) State of the art of the hydraulic design of stepped chute spillways. *Hydropower and Dams*, 1(4), 33–42.

Chanson, H. (1994c) Aeration and de-aeration at bottom aeration devices on spillways. *Canadian Journal of Civil Engineering*, 21(3), 404–409.

Chanson, H. (1994d) *Hydraulic Design of Stepped Cascades, Channels, Weirs and Spillways*. Pergamon, Oxford, UK.

Chanson, H. (1995a) Predicting the filling of ventilated cavities behind spillway aerators. *Journal of Hydraulic Research*, 33(3), 361–372.

Chanson, H. (1995b) History of stepped channels and spillways: a rediscovery of the "wheel". *Canadian Journal of Civil Engineering*, 22(2), 247–259.

Chanson, H. (1996) *Air Bubble Entrainment in Free-surface Turbulent Shear Flows*. Academic Press, San Diego CA.

Chanson, H. (2000) Discussion of Characteristics of skimming flow over stepped spillways by M.R. Chamani and N. Rajaratnam. *Journal of Hydraulic Engineering*, 126(11), 862–865.

Chanson, H. (2002) *The Hydraulics of Stepped Chutes and Spillways*. Balkema, Lisse, the Netherlands.

Chanson, H. (2004) *The Hydraulics of Open Channel Flows: An Introduction*. Butterworth-Heinemann, Oxford UK.

Chanson, H. (2006) Hydraulics of skimming flows on stepped chutes: the effects of inflow conditions? *Journal of Hydraulic Research*, 44(1), 51–60.

Chanson, H. (2015) Discussion of cavitation potential of flow on stepped spillways. *Journal of Hydraulic Engineering*, 141(5), 07014025.

Chanson, H. & Toombes, L. (2002a) Air-water flows down stepped chutes: turbulence and flow structure observations. *International Journal of Multiphase Flow*, 28(11), 1737–1761.

Chanson, H. & Toombes, L. (2002b) Experimental investigations of air entrainment in transition and skimming flows down a stepped chute. *Canadian Journal of Civil Engineering*, 29(1), 145–156.

Chanson, H. & Toombes, L. (2004) Hydraulics of stepped chutes: the transition flow. *Journal of Hydraulic Research*, 42(1), 43–54.

Chen, C.L. (1991) Unified theory on power laws for resistance. *Journal of Hydraulic Engineering*, 117(3), 371–389.

Chow, V.T. (1959) *Open Channel Hydraulics*. McGraw Hill, New York.

Comolet, R. (1979) Sur le mouvement d'une bulle de gaz dans un liquide [On the movement of a gas bubble in a liquid]. *La Houille Blanche*, 34(1), 31–42 (in French).

Cornish, V. (1907) Progressive waves in rivers. *Geographical Journal*, 29(1), 23–31.

Cornish, V. (1914) *Waves of Sand and Snow*. Unwin, London.

DeFazio, F.G. & Wei, C.Y. (1983) Design of aeration devices on hydraulic structures. In: Shen, H.T. (ed) Proceedings of *Frontiers in Hydraulic Engineering*, 426–431. Massachusetts Institute of Technology, Cambridge, MA.

Deng, Y., Lin, K. & Han, L. (2003) Design and prototype test of stepped overflow surface at Dachaoshan hydropower station in China. Proceedings of 4th Symposium *Roller Compacted Concrete Dams*, Madrid, Spain. Balkema, Lisse, pp. 431–432.

Dittrich, A. & Koll, K. (1997) Velocity field and resistance of flow over rough surfaces with large and small relative submergence. *International Journal of Sediment Research*, 12(3), 21–33.

Dressler, R.F. (1949) Mathematical solution of the problem of roll-waves in inclined open channels. *Communications on Pure and Applied Mathematics*, 2(2/3), 149–194.

Dressler, R.F. & Pohle, F.V. (1953) Resistance effects on hydraulic instability. *Communications on Pure and Applied Mathematics*, 6(1), 93–96.

Drewes, U. (1988) Oberflächentoleranzen bei Betonschussrinnen im Hinblick auf Kavitation [Surface tolerances at concrete chutes relative to cavitation]. *VAW Mitteilung*, 99, 11–33 (in German).

Ebadian, M.A., Skudarnov, P.V., Lin, C.X. & Philippidis, G.P. (2003) Analytical assessments related to microbubble drag reduction. *Final Technical Report*. Hemispheric Center for Environmental Technology, Florida International University, Miami FL.

Eddington, R.B. (1970) Investigation of supersonic phenomena in a two-phase (liquid-gas) tunnel. *Journal of AIAA*, 8(1), 65–74.

Elviro, V. & Mateos, C. (1995) Spanish research into stepped spillways. *Hydropower & Dams*, 2(5), 61–65.

Ervine, D.A. & Falvey, H.T. (1987) Behaviour of turbulent water jets in the atmosphere and in plunge pools. *Proceedings of the Institution of Civil Engineers*, 33(2), 295–314.

Ervine, D.A., Falvey, H.T. & Kahn, A.R. (1995) Turbulent flow structure and air uptake at aerators. *Hydropower and Dams*, 2(4), 89–96.

Escoffier, F.F. & Boyd, M.B. (1962) Stability aspects of flow in open channels. *Journal of the Hydraulics Division* ASCE, 88(HY6), 145–166; 89(HY4), 259–273; 90(HY1), 261–263.

Escudier, M.P. (1966) The distribution of mixing length in turbulent flows near walls. Heat Transfer Section, *Report* TWF/TN/1. Imperial College, London UK.

Estrella, S., Sánchez-Juny, M., Pomares, J., Dolz, J., Ibáñez de Aldecoa, R., Domínguez, M., Rodríguez, J. & Balairón, L. (2012) Recent trends in stepped spillways design: behavior without sidewalls. Proceedings of 24th *ICOLD Congress*, Kyoto, Japan, Q94(R28), 433–446.

Falvey, H.T. (1980) Air water flow in hydraulic structures. Water Resources Technical Publication, *Engineering Monograph* 41. U.S. Dept. of Interior, U.S. Printing Office, Denver CO.

Falvey, H.T. (1983) Prevention of cavitation on chutes and spillways. In: Shen, H.T. (ed) Proceedings of *Frontiers in Hydraulic Engineering*, 432–437. Massachusetts Institute of Technology, Cambridge, MA.

Falvey, H.T. (1990) Cavitation in chutes and spillways. *Engineering Monograph* 42. Water Resources Technical Publication. US Printing Office, Bureau of Reclamation, Denver CO.

Falvey, H.T. & Ervine, D.A. (1988) Aeration in jets and high velocity flows. In: Burgi, P.H. (ed) Proceedings of *Model-Prototype Correlation of Hydraulic Structures*, Colorado Springs CO. ASCE, New York, pp. 25–55.

Felder, S. & Chanson, H. (2011) Energy dissipation down a stepped spillway with nonuniform step heights. *Journal of Hydraulic Engineering*, 137(11), 1543–1548.

FEMA (2014) Technical Manual: overtopping protection for dams. *FEMA P-1015*, Federal Emergency Management Agency, Denver CO, USA.

Ferrante, A. & Elghobashi, S. (2004) On the physical mechanisms of drag reduction in a spatially developing turbulent boundary layer laden with microbubbles. *Journal of Fluid Mechanics*, 503, 345–355.

Forchheimer, P. (1903) Wasserbewegung in Wanderwellen [Water flow in roll waves]. *Zeitschrift für Gewässerkunde*, 6(6), 321–339 (in German).

Forchheimer, P. (1914) *Hydraulik* [Hydraulics]. Teubner, Leipzig (in German).

Frizell, K.H. (1990) Hydraulic model study of McClure Dam existing and proposed RCC stepped spillway. *Report* R-90–02. US Bureau of Reclamation, Denver CO.

Frizell, K.H. & Mefford, B.W. (1991) Designing spillways to prevent cavitation damage. *Concrete International*, 13(5), 58–64.

Frizell, K.H. & Renna, F.M. (2011) Laboratory studies on the cavitation potential of stepped spillways. *Proceedings of 34th IAHR Congress*, Brisbane. IAHR, Madrid, pp. 2420–2427.

Frizell, K.W. (2009) Cavitation potential of the Folsom auxiliary spillway basin baffle blocks: laboratory studies. *Hydraulic Laboratory Report* HL-2009–06, USBR, Denver CO.

Frizell, K.W., Renna, F.M. & Matos, J. (2013) Cavitation potential of flow on stepped spillways. *Journal of Hydraulic Engineering*, 139(6), 630–636.

Frizell, K.W., Renna, F.M. & Matos, J. (2015) Closure of Cavitation potential of flow on stepped spillways. *Journal of Hydraulic Engineering*, 141(8), 07015009.

Frizell, K.W. & Svoboda, C.D. (2012) Performance of type III stilling basins: stepped spillway studies. *Hydraulic Laboratory Report* HL-2012–02, USBR, Denver CO.

Galperin, R.S., Kuzmin, K.K., Novikova, I.S., Oskolkov, A.G., Semenkov, V.M. & Tsedrov, G.N. (1971) Cavitation in elements of hydraulic structures and methods of controlling it. *Hydraulic Construction*, 5(8), 726–732.

Galperin, R.S., Oskolkov, A.G., Semenkov, V.M. & Tsedrov, G.N. (1979) *Cavitation in Hydraulic Structures*. Energiya, Moscow.

Gausmann, R.W. & Madden, G.G. (1923) Experiments with models of the Gilboa Dam and Spillway. *Transactions of ASCE*, 86, 280–319.

Goguel, B. (2017) La crise d'Oroville en Février 2017: Eclairage technico-historique [Oroville February 2017 crisis: technical and historical background]. *Proceedings of Colloque CFBR-SHF Hydraulique des barrages et des digues, Chambéry*, pp. 174–185 (in French).

González, C.A. & Chanson, H. (2007) Diseño hidráulico de vertedores escalonados con pendientes moderadas: metodología basada en un estudio experimental [Hydraulic design of stepped spillways with moderate slope: methodology based on an experimental study]. *Ingeniería Hidráulica en México*, 22(2), 5–20 (in Spanish).

Gonzalez, C., Takahashi, M. & Chanson, H. (2008) An experimental study of effects of step roughness in skimming flows on stepped chutes. *Journal of Hydraulic Research*, 46(1), 24–35.

Gyr, A. & Bewersdorff, H.-W. (1995) *Drag Reduction of Turbulent Flows by Additives*. Kluwer, Dordrecht NL.

Haberman, W.L. & Morton, R.K. (1956) An experimental study of bubbles moving in liquids. *Transactions of ASCE*, 121, 227–252.

Hager, W.H. (1991) Uniform aerated chute flow. *Journal of Hydraulic Engineering*, 117(4), 528–533; 118(6), 944–946.

Hager, W.H. (1992) Spillways: shock waves and air entrainment: review and recommendations. *Bulletin* 81. ICOLD, Paris.

Hager, W.H. (2010) *Wastewater Hydraulics* (2nd ed). Springer, Berlin.

Hager, W.H. (2002) History of roll waves. *L'Acqua*, 81(4), 7–12.

Hager, W.H. & Blaser, F. (1998) Drawdown curve and incipient aeration for chute flow. *Canadian Journal of Civil Engineering*, 25(3), 467–473.

Hager, W.H. & Mazumder, S.K. (1992) Supercritical flows at abrupt expansions. *Proceedings of Institution Civil Engineers Water Maritime and Energy*, 96, 153–166.

Hager, W.H. & Pfister, M. (2009) Historical advance of chute aerators. Proceedings of 33rd *IAHR Congress*, Vancouver, 10228, 5827–5834.

Hager, W.H. & Pfister, M. (2013) Stepped spillways: Technical advance from 1900. Proc. 35th *IAHR World Congress*, Chengdu, A10515, 1–8.

Halbronn, G. (1952) Étude de la mise en régime des écoulements sur les ouvrages à forte pente: application au problème de l'entrainement d'air [Analysis of flows on hydraulic structures with a large slope: application to air entrainment problem]. *La Houille Blanche*, 7(1), 21–40; 7(3), 347–371; 7(5), 702–722 (in French).

Halbronn, G. (1955) Discussion of Turbulent boundary layer on steep slopes. *Transactions of ASCE*, 119, 1234–1240.

Hansen, K.D. (1991) Advantages of roller compacted concrete dams. *RCC Newsletter, Portland Cement Association*, 7(1), 1–2.

Hansen, K.D. (1996) Roller compacted concrete: a civil engineering innovation. *Concrete International: Design and Construction*, 18(3), 49–53.

Hansen, K.D. (2003) RCC use in dam rehabilitation projects. In: Berga, L. *et al.* (eds) *Proceedings of 4th International Symposium Roller Compacted Concrete Dams*. Balkema, Rotterdam, the Netherlands, pp. 79–89.

Hansen, K.D. & France, J.W. (1986) RCC: a dam rehab solution unearthed. *Civil Engineering*, 56(9), 60–63.

Hansen, K.D. & Reinhardt, W.G. (1991) *Roller-Compacted Concrete Dams*. McGraw-Hill, New York, USA.

Harrold, J.C. (1947) Experiences of the Corps of Engineers. *Transactions of ASCE*, 112, 16–42; 112, 116–119.

Heinly, B.A. (1913) Completion, testing and dedication of the Los Angeles Aqueduct. *Engineering News*, 70(19), 920–923.

Henderson, F.M. (1966) *Open Channel Flow*. MacMillan, New York.

Herrmann, G. (2003) Abflussinstabilitäten der Hochwasserentlastung Sorpetalsperre [Flow instabilities at Sorpe Dam spillway]. *Master Thesis*, Lehrstuhl für Wasserbau und Wasserwirtschaft, Rheinisch-Westfälischen Hochschule Aachen, Germany (in German).

Hickox, G.H. (1947) Experiences of the Tennessee Valley Authority. *Transactions of ASCE*, 112, 59–67.

Hinze, J.O. (1955) Fundamentals of the hydrodynamic mechanism of splitting in dispersion processes. *Journal of AIChe*, 1(3), 289–295.

Hinze, J.O. (1975) *Turbulence*. McGraw-Hill, New York.

Houston, K.L. (1987) Hydraulic model studies of Upper Stillwater Dam stepped spillway and outlet works. *Report* REC-ERC-87–6, US Bureau of Reclamation, Denver CO.

Hunt, S.L. & Kadavy, K.C. (2013) Inception point for embankment dams stepped spillways. *Journal of Hydraulic Engineering*, 139(1), 60–64.

Hunt, S.L., Temple, D.M., Abt, S.R., Kadavy, K.C. & Hanson, G. (2012) Converging stepped spillways: simplified momentum analysis approach. *Journal of Hydraulic Engineering*, 138(9), 796–802.

ICOLD (1992) Spillways. Shockwaves and air entrainment. *Bulletin* 81, prepared by W.H. Hager. International Commission of Large Dams, Paris.

ICOLD (2017) Technical advancements in spillway design: progress and innovations from 1985 to 2015. *Bulletin* 172, International Commission on Large Dams, Paris.

Ippen, A.T. & Dawson, J.H. (1951) Design of channel contractions. *Transactions of ASCE*, 116, 326–346.

Ishihara, T., Iwagaki, Y. & Iwasa, Y. (1962) Discussion of Roll waves and slug flows in inclined open channels. *Transactions of ASCE*, 126(1), 548–563.

ISO (1975) *Standard atmosphere*. International Organization for Standardization, ISO 2533.

Isselin, J.C., Alloncle, A.P. & Autric, M. (1998) On laser inducted single bubble near a solid boundary: contribution to the understanding of erosion phenomena. *Journal of Applied Physics*, 84(10), 5766–5771.

Jacobs, F., Winkler, K., Hunkeler, F. & Volkart, P. (2001) Betonabrasion im Wasserbau [Concrete abrasion in hydraulic structures]. In: Minor, H.E. (ed) *VAW Mitteilung* 168, VAW, ETH Zürich (in German).

Jeffreys, H. (1925) The flow of water in an inclined channel of rectangular section. *Philosophical Magazine,* Series 6(49), 793–807.

Keller, A.P. (1988) Kavitationsuntersuchungen an Profilmodellen als Beitrag zur Klärung von Fragen über Massstabseffekte bei Kavitationsbeginn [Cavitation studies at profile models to answer questions relating to scale effects at cavitation inception]. *VAW Mitteilung*, 99, 171–194, (in German).

Keller, R.J. & Rastogi, A.K. (1975) Prediction of flow development on spillways. *Journal of Hydraulics Division ASCE*, 101(HY9), 1171–1184; 102(HY9), 1401–1404; 103(HY6), 664.

Keller, R.J. & Rastogi, A.K. (1977) Design chart for prediction critical point on spillways. *Journal of Hydraulics Division ASCE*, 103(HY12), 1417–1429; 104(HY12), 1678.

Kells, J.A. (1993) Spatially varied flow over rockfill embankments. *Canadian Journal of Civil Engineering*, 20(5), 820–827; 21(1), 161–166.

Kells, J.A. & Smith, C.D. (1991) Reduction of cavitation on spillways by induced air entrainment. *Canadian Journal of Civil Engineering*, 18, 358–377; 19, 924–929.

Khatsuria, R.M. (2004) *Hydraulics of Spillways and Energy Dissipators*. Dekker, New York.

Knapp, R.T. (1951) Design of channel curves for supercritical flow. *Transactions of ASCE*, 116, 296–325.

Kobus, H. (1991) Introduction to air-water flows. In: Wood, I.R. (ed) *IAHR Hydraulic Structures Design Manual* 4, Air entrainment in free-surface flows. Balkema, Rotterdam, pp. 1–28.

Koch, L. (1923) Modellversuche über die Wirksamkeit von Wassertreppen (Stufenüberfälle) [Modell tests on the effect of water stairs]. *Der Bauingenieur*, 4(16), 472–474 (in German).

Kökpinar, M.A. & Göğüş, M. (2002) High-speed jet flows over spillway aerators. *Canadian Journal of Civil Engineering*, 29(6), 885–898.

Koschitzky, H.-P. (1987) Dimensionierungskonzept für Sohlenbelüfter in Schussrinnen zur Vermeidung von Kavitationsschäden [Design concept for chute aerators to avoid cavitation damage]. *Mitteilung* 65. Institut für Wasserbau, Technische Universität, Stuttgart (in German).

Kramer, K. (2004) Development of aerated chute flow. *Ph.D. Thesis* 15428. ETH Zurich, Zürich.

Kramer, K. & Hager, W.H. (2005) Air transport in chute flows. *International Journal of Multiphase Flows*, 31(10–11), 1181–1197.

Kramer, K., Hager, W.H. & Minor, H.-E. (2006) Development of air concentration on chute spillways. *Journal of Hydraulic Engineering*, 132(9), 908–915.

Kramer, K., Minor, H.-E. & Hager, W.H. (2005) Spacing of chute aerator for cavitation protection. *Hydropower and Dams*, 12(4), 64–70.

Kranenburg, C. (1992) On the evolution of roll waves. *Journal of Fluid Mechanics*, 245, 249–261.

Küppers, W. (1905) Die grösste Talsperre Europas bei Gmünd (Eifel) und hydraulische Kraftstation [The largest dam of Europe near Gmünd (Eifel) and its power station]. *Die Turbine*, 2(3), 61–64 (in German).

Lakshmana Rao, N.S. & Kobus, H. (1973) *Characteristics of Self-aerated Free-surface Flows*. Water and Waste Water, Current Research and Practice 10. Erich Schmidt, Berlin.

Legner, H.H. (1984) A simple model for gas bubble drag reduction. *Physics of Fluids*, 27(12), 2788–2790.

Lin, K.J. Han, L. (2001) Stepped spillway for Dachaoshan RCC dam. *29th IAHR Congress, Special Seminar on Key Hydraulics Issues of Huge Water Projects, Beijing, China*. IAHR, Madrid, pp. 88–93.

Lopardo, R. (2002) Contribution of hydraulic models on the safe design of large dams stilling basins. *Proceedings of IAHR Symposium Hydraulic and Hydrological Aspects of Reliability and Safety Assessment of Hydraulic Structures*. IAHR, St. Petersburg, Russia.

Low, H.S. (1986) Model studies of Clyde Dam spillway aerators. *Research Report* 86–6, Department of Civil Engineering, University of Canterbury, Christchurch NZ.

Lu, J., Fernández, A. & Tryggvason, G. (2005) The effect of bubbles on the wall drag in a turbulent channel flow. *Physics of Fluids*, 17, 095102.

Lueker, M.L., Mohseni, O., Gulliver, J.S., Schulz, H. & Christopher, R.A. (2008) The physical model study of the Folsom Dam Auxiliary Spillway System. *Project Report* 511, St. Anthony Falls Laboratory, University of Minnesota, Minneapolis.

Lumley, J.L. (1977) Drag reduction in two phase and polymer flows. *Physics of Fluids*, 20(10), 64–71.

Lutz, N., Lucas, J., Lais, A. & Boes, R.M. (2015) Stepped chute of Trängslet Dam: physical model study. *Journal of Applied Water Engineering and Research*, 3(2), 166–176.

Madavan, N.K., Deutsch, S. & Merkle, C.L. (1984) Reduction of turbulent skin friction by microbubbles. *Physics of Fluids*, 27(2), 356–363.

Marcano, A. & Castillejo, N. (1984) Model-prototype comparison of aeration devices of Guri Dam spillway. *Proceedings of Scale Effects in Modelling Hydraulic Structures*, Esslingen, 4(6), 1–5.

Marié, J.L. (1987) A simple analytical formulation for microbubble drag reduction. *Physico-Chemical Hydrodynamics*, 8(2), 213–220.

Mason, P.J. (2017) Spillway chutes: practical design considerations and details. *Hydropower & Dams*, 24(5), 79–86.

Mateos Iguácel, C. & Elviro García, V. (2000) Stepped spillway studies at CEDEX. In: Minor, H.-E. & Hager, W.H. (eds) *Proceedings of Hydraulics of Stepped Spillways*. Balkema, Rotterdam, pp. 87–94.

Matos, J. (2000) Discussion of Characteristics of skimming flow over stepped spillways, by M.R. Chamani and N. Rajaratnam. *Journal of Hydraulic Engineering*, 126(11), 865–869.

May, R.W.P. (1987) Cavitation in hydraulic structures: occurrence and prevention. *Hydraulic Research Report* SR 79. Wallingford, UK.

Mayer, P.G. (1961) Roll waves and slug flows in inclined open channels. *Transactions of ASCE*, 126(1), 505–564.

Mazumder, S.K. & Hager, W.H. (1993) Supercritical expansion flow in Rouse modified and reversed transitions. *Journal of Hydraulic Engineering*, 119(2), 201–219.

Meireles, I. (2011) Hydraulics of stepped chutes: experimental-numerical-theoretical study. *Ph.D. Thesis*, Universidade de Aveiro, Departamento de Engenharia Civil, Portugal.

Meireles, I. & Matos, J. (2009) Skimming flow in the non-aerated region of stepped spillways over embankment dams. *Journal of Hydraulic Engineering*, 135(8), 685–689.

Meireles, I., Matos, J. & Silva Afonso, A. (2010) Flow characteristics along a USBR type III stilling basin downstream of steep stepped spillways. In: Janssen, R. & Chanson, H. (eds) *Proceedings of 3rd International Junior Researcher and Engineer Workshop on Hydraulic Structures*, Edinburgh UK; *Hydraulic Model Report* CH80/10, School of Civil Engineering, The University of Queensland, Brisbane, Australia.

Minor, H.-E. (1987) Erfahrungen mit Schussrinnenbelüftung [Experiences with chute aeration]. *Wasserwirtschaft*, 77(6), 292–295 (in German).

Minor, H.E. (1988) Konstruktive Details zur Vermeidung von Kavitationsschäden [Structural details to avoid cavitation damage]. VAW *Mitteilung*, 99, 367–378 (in German).

Minor, H.-E. (2000) Cavitation damage on Alicura and Restitucion Dams. Personal communication.

Montes, J.S. (1998) *Hydraulics of Open Channel Flow*. ASCE, Reston VA.

Montuori, C. (1961) La formazione spontanea dei treni d'onde su canali a pendenza molto forte [Spontaneous formation of wave trains in channels of steep slope]. *L'Energia Elettrica*, 38(2), 127–141 (in Italian).

Montuori, C. (1965) Spontaneous formation of wave trains in channels with a very steep slope. *WES Translation* 65–12. US Army Engineer Waterways Experiment Station, Vicksburg MS.

Montuori, C. (1984) Sviluppi recenti nello studio delle correnti supercritiche [Recent developments in the study of supercritical flows]. Dipartimento di Idraulica, Gestione delle Risorse Idriche ed Ingegneria Ambientale *Memoria* 508. Università, Napoli (in Italian).

Ohtsu, I., Yasuda, Y. & Takahashi, M. (2001) Discussion of Onset of the skimming flow on stepped spillways, by M.R. Chamani and N. Rajaratnam. *Journal of Hydraulic Engineering*, 127(6), 522–524.

Ohtsu, I., Yasuda, Y. & Takahashi, M. (2004) Flow characteristics of skimming flows in stepped channels. *Journal of Hydraulic Engineering*, 130(9), 860–869.

Ortiz-Villafuerte, J. & Hassan, Y.A. (2006) Investigation of microbubble boundary layer using particle tracking velocimetry. *Journal of Fluids Engineering*, 128(3), 507–519.

Ostad Mirza, M.J., Matos, J., Pfister, M. & Schleiss, A.J. (2017) The effect of an abrupt slope change on air entrainment and flow depths on stepped spillways. *Journal of Hydraulic Research*, 55(3), 362–375.

Ozturk, M., Aydin, M.C. & Aydin, S. (2008) Damage limitation: a new spillway aerator. *International Journal of Water Power Dam Construction*, 60(5), 36–40.

PCA (2002) *Design Manual for RCC Spillways and Overtopping Protection.* URS Greiner Woodward Clyde. Portland Cement Association, Illinois.

Peterka, A.J. (1953) The effect of entrained air on cavitation pitting. Joint Meeting Paper, *Proceedings of IAHR/ASCE Congress*, Minneapolis, USA. IAHR, Delft, pp. 507–518.

Peterka, A.J. (1963) Hydraulic design of stilling basins and energy dissipators. *Engineering Monograph* 25, US Bureau of Reclamation, Denver CO.

Peyras, L.A., Royet, P. & Degoutte, G. (1992) Flow and energy dissipation over stepped gabion weirs. *Journal of Hydraulic Engineering*, 118(5), 707–717.

Pfister, M. (2008) Schussrinnenbelüfter: Lufttransport ausgelöst durch interne Abflussstruktur [Chute aerators: air transport due to internal flow structure]. In: Minor, H.-E. (ed) *VAW Mitteilung* 203, ETH Zurich, Zürich (in German).

Pfister, M. (2009) Effect of control section on stepped spillway flow. *Proceedings of 33rd IAHR Congress, Vancouver*, 10229, 1964–1971.

Pfister, M. (2011) Chute aerators: steep deflectors and cavity sub-pressure. *Journal of Hydraulic Engineering*, 137(10), 1208–1215.

Pfister, M. (2012) Jet impact angle on chute downstream of aerator. *Proceedings of 4th IAHR International Symposium on Hydraulic Structures*, Porto (P), CD 7, 1–9.

Pfister, M. & Boes, R.M. (2014) Discussion of Skimming, non-aerated flow on stepped spillways over roller compacted concrete dams. *Journal of Hydraulic Engineering*, 140(10), 07014012.

Pfister, M. & Hager, W.H. (2009) Deflector-generated jets. *Journal of Hydraulic Research*, 47(4), 466–475.

Pfister, M. & Hager, W.H. (2010a) Chute aerators I: air transport characteristics. *Journal of Hydraulic Engineering*, 136(6), 352–359.

Pfister, M. & Hager, W.H. (2010b) Chute aerators II: hydraulic design. *Journal of Hydraulic Engineering*, 136(6), 360–367.

Pfister, M. & Hager, W.H. (2012) Deflector-jets affected by pre-aerated approach flow. *Journal of Hydraulic Research*, 50(2), 181–191.

Pfister, M. & Hager, W.H. (2014) History and significance of the Morton number in hydraulic engineering. *Journal of Hydraulic Engineering*, 140(5), 02514001.

Pfister, M., Hager, W.H. & Minor, H.-E. (2006a) Bottom aeration of stepped spillways. *Journal of Hydraulic Engineering*, 132(8), 850–853.

Pfister, M., Hager, W.H. & Minor, H.-E. (2006b) Stepped chutes: pre-aeration and spray reduction. *International Journal of Multiphase Flow*, 32(2), 269–284.

Pfister, M., Hager, W.H. & Minor, H.-E. (2007) Step aerator and spray reduction for stepped chutes. *Proceedings of 32nd IAHR Congress, Venice*, 282, 1–10.

Pfister, M., Lucas, J. & Hager, W.H. (2011) Chute aerators: pre-aerated approach flow. *Journal of Hydraulic Engineering*, 137(11), 1452–1461.

Pinto, N.L. (1984) Model evaluation of aerators in shooting flow. *Proceedings of Scale Effects in Modelling Hydraulic Structures*, Esslingen, 4.2, 1–6.

Pinto, N.L. (1989) Designing aerators for high velocity flow. *Water Power and Dam Construction*, 41(7), 44–48.

Pinto, N.L. & Neidert, S.H. (1983) Evaluating entrained air flow through aerators. *Water Power & Dam Construction*, 35(8), 40–42.

Pinto, N.L., Neidert, S.H. & Ota, J.J. (1982) Aeration at high velocity flows. *Water Power & Dam Construction*, 34(2), 34–38; 34(3), 42–44.

Powell, R.W. (1948) Vedernikov's criterion for ultra-rapid flow. *Transactions of AGU*, 29(6), 882–886; 32(4), 603–607.

Prisk, M., Richards, M. & Hieatt, M. (2009) Delivering Wadi Dayqah, Oman's tallest dam. *Proceedings of the Institution of Civil Engineers*, 162(6), 42–50.

Prusza, V. (1983) Remedial measures against spillway cavitation. *Proceedings of 20th IAHR Congress, Moscow*, 3, 468–476.

Pugh, C.A. & Rhone, T.J. (1988) Cavitation in Bureau of Reclamation Tunnel Spillways. *Proceedings of International Symposium on Hydraulics for High Dams*. IAHR China, Beijing, pp. 645–652.

Rabben, S. (1984) Untersuchung der Belüftung an Tiefschützen unter besonderer Berücksichtigung von Massstabseffekten [Study of aeration at bottom outlets with a particular interest on scale effects]. *Mitteilung* 53, Institut für Wasserbau und Wasserwirtschaft, RWTH Aachen (in German).

Rasmussen, R.E.H. (1956) Some experiments on cavitation erosion in water mixed with air. *Proceedings of 1st International Symposium Cavitation in Hydrodynamics*, 20, 1–25.

Reinauer, R. & Hager, W.H. (1996) Generalized drawdown curve for chutes. *Proceedings of Institution Civil Engineers, Water Maritime and Energy*, 118(4), 196–198.

Reinauer, R. & Hager, W.H. (1997) Supercritical bend flow. *Journal of Hydraulic Engineering*, 123(3), 208–218.

Reinauer, R. & Hager, W.H. (1998) Supercritical flow in chute contraction. *Journal of Hydraulic Engineering*, 124(1), 55–64.

Rice, C.E. & Kadavy, K.C. (1997) Physical model study of the proposed spillway for Cedar Run Site 6, Fauquier County, Virginia. *Applied Engineering in Agriculture*, 13(6), 723–729.

Rosso, M., Schiara, M. & Berlamont, J. (1990) Flow stability and friction factor in rough channels. *Journal of Hydraulic Engineering*, 116(9), 1109–1118.

Rouse, H. (1950) *Engineering Hydraulics*. Wiley, New York.

Rouse, H., Bootha, B.V. & Hsu, E.Y. (1951) Design of channel expansions. *Transactions of ASCE*, 116, 1369–1385.

Russell, S.O. & Sheehan, G.J. (1974) Effect of entrained air on cavitation damage. *Canadian Journal of Civil Engineering*, 1(1), 97–107.

Rutschmann, P. (1988) Belüftungseinbauten in Schussrinnen [Aeration devices in chutes]. In: Vischer, D. (ed) *VAW Mitteilung* 97, ETH Zurich, Zürich (in German).

Rutschmann, P. & Hager, W.H. (1990a) Air entrainment by spillway aerators. *Journal of Hydraulic Engineering*, 116(6), 765–782; 117(4), 545; 118(1), 114–117.

Rutschmann, P. & Hager, W.H. (1990b) Design and performance of spillway chute aerators. *Water Power & Dam Construction*, 41(1), 36–42.

Rutschmann, P., Volkart, P. & Wood, I.R. (1986) Air entrainment at spillway aerators. *Proceedings of 9th Australasian Fluid Mechanics Conference, Auckland*, pp. 350–353.

Sánchez-Juny, M., Blade, E. & Dolz, J. (2007) Pressures on stepped spillways. *Journal of Hydraulic Research*, 45(4), 505–511.

Sánchez-Juny, M., Pomares, J. & Dolz, J. (2000) Pressure field in skimming flow over a stepped spillway. In: Minor, H.-E. & Hager, W.H. (eds) Proceedings of International Workshop *Hydraulics of Stepped Spillways*, VAW, ETH Zurich. Balkema, Rotterdam, the Netherlands, pp. 137–145.

Schiess, A., Pfister, M., Hager, W.H. & Minor, H.-E. (2008) Hydraulic performance of step aerator. *Journal of Hydraulic Engineering*, 134(2), 127–134.

Schlichting, H. & Gersten, K. (1996) *Grenzschicht-Theorie* [Boundary layer theory]. Springer, Berlin (in German).

Schmocker, L., Pfister, M., Hager, W.H. & Minor, H.-E. (2008) Aeration characteristics of ski jump jets. *Journal of Hydraulic Engineering*, 134(1), 90–97.

Schrader, E.K. (1983) Cavitation resistance of concrete structures. *Proceedings of Frontiers in Hydraulic Engineering*. ASCE, New York, pp. 419–424.

Schwalt, M. & Hager, W.H. (1992) Shock pattern at abrupt wall deflection. *ASCE Conference Environmental Engineering Water Forum '92 Baltimore*, pp. 231–236.

Semenkov, V.M. & Lentyaev, L.D. (1973) Spillway with nappe aeration. *Gidrotekhnicheskoe Stroitel'stvo*, 43(5), 16–20 (in Russian); *Hydrotechnical Construction*, 7, 437–441.

Sharma, H.R. (1976) Air entrainment of high head gated conduits. *Journal of Hydraulics Division ASCE*, 102(HY11), 1629–1646; 103(HY10), 1254–1255; 103(HY11), 1365–1366; 103(HY12), 1486–1493; 104(HY8), 1200–1202.

Silberman, E. (1980) Boundary layers in developing open channel flow. *Journal of Hydraulics Division ASCE*, 106(HY7), 1237–1241; 1981, 107(HY4), 527–528, 107(HY10), 1275.

Silvestri, A., Archambeau, P., Pirotton, M., Dewals, B. & Erpicum, S. (2013) Comparative analysis of the energy dissipation on a stepped spillway downstream of a Piano Key Weir. *Labyrinth and piano key weirs – PKW*, 2013, 111–120.

Skripalle, J. (1994) Zwangsbelüftung von Hochgeschwindigkeitsströmungen an zurückspringenden Stufen im Wasserbau [Forced aeration of high-speed flows at aerators in hydraulic structures]. *Mitteilung* 124, Technische Universität, Berlin (in German).

Soo, S.L. (1967) *Fluid Dynamics of Multiphase Systems*. Blaisdell, Waltham MA.

Steiner, R., Heller, V., Hager, W.H. & Minor, H.-E. (2008) Deflector ski jump hydraulics. *Journal of Hydraulic Engineering*, 134(5), 562–571.

Stephenson, D. (1979) Rockfill and gabions for erosion control. *The Civil Engineer in South Africa*, 21(9), 203–208.

Stojnic, I., Pfister, M., Matos, J. & Schleiss, A. J. (2020) Bottom pressure characteristics in a stilling basin downstream of a stepped spillway for two different chute slopes. In International Symposium on Hydraulic Structures (ISHS2020). Proceedings of the 8th IAHR International Symposium on *Hydraulic Structures ISHS2020*, Santiago, Chile, R. Janssen and H. Chanson, Editors. The University of Queensland, Brisbane, Australia.

Stojnic, I. (2020) Stilling basin performance downstream of stepped spillways. *Ph.D. Thesis* 7481, École Polytechnique Fédérale de Lausanne (EPFL), Switzerland, and Instituto Superior Técnico (IST), Lisbon, Portugal.

Stoker, J.J. (1957) *Water waves*. Interscience, New York.

Straub, L.G. & Anderson, A.G. (1960) Experiments on self-aerated flow in open channels. *Transactions of ASCE*, 125, 456–486.

Suslick, K.S. (1989) Die chemischen Wirkungen von Ultraschall [Chemical reaction of ultrasound]. *Spektrum der Wissenschaft*, 4, 60–66 (in German).

Takahashi, M. & Ohtsu, I. (2012) Aerated flow characteristics of skimming flow over stepped chutes. *Journal of Hydraulic Research*, 50(4), 427–434.

Tan, T.P. (1984) Model studies of aerators on spillways. *Research Report* 84–6. Department of Civil Engineering, University of Canterbury, Christchurch NZ.

Terrier, S. (2016) Hydraulic performance of stepped spillway aerators and related downstream flow features. *Ph.D. Thesis* 6989, EPFL Lausanne, Lausanne.

Thomas, H.H. (1976) *The Engineering of Large Dams*. Wiley, London, New York.

Thorwarth, J., Schleiss, A.J., Köngeter, J. & Schüttrumpf, H. (2009) Discussion of Flow patterns in nappe flow regime down low-gradient stepped chutes by L. Toombes, H. Chanson. *Journal of Hydraulic Research*, 47(6), 830–832.

Toledo, M.A., Morán, R. & Oñate, E. (eds) (2015) *Dam Protections against Overtopping and Accidental Leakage*. CRC Press, Leiden, the Netherlands.

Toloshinov, A.V., Volynchikov, A.N., Prokof'ev, V.A. & Sudol'skii, G.A. (2009) Development of the design for the No. 2 spillway at the Boguchany hydroproject. *Power Technology and Engineering*, 43(3), 135–142.

Tomita, Y. & Shima, A. (1990) High-speed photographic observations of laser-inducted cavitation bubbles in water. *Acustica*, 71(1), 161–171.

Toth, I. (1968) Le barrage Gardiner sur la riviére South Saskatchewan, Canada [The Gardiner Dam on South Saskatchewan River, Canada]. *Technique des Travaux*, 44, 363–374.

USACE (1995) Hydraulic design criteria. *U.S. Army Corps of Engineers*, Waterways Experiment Station, Vicksburg MI, USA.

US Army Corps of Engineers (1995) *Hydraulic Design of Spillways*. ASCE, Reston VA.

USBR (1977) *Design of Small Dams*. Technical report. US Department of the Interior, Bureau of Reclamation (USBR), Denver CO, USA.

Vedernikov, V.V. (1945) Conditions at the front of a translation wave disturbing a steady motion of a real fluid. *Comptes Rendus de l'Académie des Sciences URSS*, 48(4), 239–242.

Vedernikov, V.V. (1946) Characteristic features of a liquid flow in an open channel. *Comptes Rendus de l'Académie des Sciences URSS*, 52(8), 207–210.

Vennard, J.K. (1947) Nature of cavitation. *Transactions of ASCE*, 112, 2–15.

Vischer, D.L. (1988a) A design principle to avoid shockwaves in chutes. *Proceedings of International Symposium of Hydraulics for High Dams, Beijing*, pp. 391–396.

Vischer, D.L. (1988b) Recent developments in spillway design. *Water Power & Dam Construction*, 40(1), 8–9.

Vischer, D.L. & Hager, W.H. (1995) Energy dissipators. *IAHR Hydraulic Structures Design Manual* 9. Balkema, Rotterdam.

Vischer, D.L. & Hager, W.H. (1998) *Dam Hydraulics*. Wiley, Chichester.

Vischer, D.L., Volkart, P. & Siegenthaler, A. (1982) Hydraulic modelling of air slots in open chute spillways. *Hydraulic Modelling of Civil Engineering Structures, Coventry*, pp. 239–252.

Volkart, P. (1982) Self-aerated flow in steep, partially filled pipes. *Journal of Hydraulics Division ASCE*, 108(9), 1029–1046.

Volkart, P. (1985) Transition from aerated supercritical to subcritical flow and associated bubble de-aeration. *Proceedings of 21st IAHR Congress, Melbourne*, 5, 2–6.

Volkart, P. (1988) Kavitation an Schussrinnen und deren Vermeidung durch Belüftung [Cavitation at chutes and its avoidance]. VAW *Mitteilung*, 99, 125–147 (in German).

Volkart, P. & Chervet, A. (1983) Air slots for flow aeration. In: Vischer, D. (ed) VAW *Mitteilung* 66. ETH, Zurich, Zürich.

Volkart, P. & Rutschmann, P. (1984) Air entrainment devices. In: Vischer, D. (ed) VAW *Mitteilung* 72. ETH Zurich, Zürich.

Volkart, P. & Rutschmann, P. (1986) Aerators on spillway chutes: fundamentals and applications. In: Arndt, R.E.A., Stefan, H.G., Farrell, C. & Peterson, S.M. (eds) *Proceedings of Advancements in Aerodynamics, Fluid Mechanics, and Hydraulics, Minneapolis MN*. ASCE, New York, pp. 1–15.

Wahrheit-Lensing, A. (1996) Selbstbelüftung und Energieumwandlung beim Abfluss über treppen-förmige Entlastungsanlagen [Self-aeration and energy dissipation of stepped spillway flow]. *Ph.D. Thesis*, University of Karlsruhe, Germany (in German).

Warnock, J.E. (1947) Experiences of the Bureau of Reclamation. *Transactions of ASCE*, 112, 43–58.

Wegmann, E. (1911) *The Design and Construction of Dams, Including Masonry, Earth, Rock-Fill, Timber, and Steel Structures*. Wiley, New York.

White, F.M. (1991) *Viscous Fluid Flow*. McGraw-Hill, New York.

Willey, J., Ewing, T., Lesleighter, E. & Dymke, J. (2010) Numerical and physical modelling for a complex stepped spillway. *Hydropower and Dams*, 17(3), 108–113.

Winter, C. (2010) Research into the hydrodynamic forces and pressures acting within stepped masonry spillways. *Dams and Reservoirs*, 20(1), 16–26.

Winter, C., Mason, P.J., Baker, R. & Ferguson, A. (2010) Guidance for the design and maintenance of stepped masonry spillways. *Project SC080015*, UK Environment Agency, Bristol, UK.

Wood, I.R. (1985) Air water flows. *Proceedings of 21st IAHR Congress, Melbourne*, 6, 18–29.

Wood, I.R. (1988) Aerators: the interaction of nappe and duct air entrainment. *Proceedings of International Symposium Hydraulics for High Dams, Beijing*, pp. 611–618.

Wood, I.R. (1991) Free surface air entrainment on spillways. In: Wood, I.R. (ed) IAHR Hydraulic Structures Design Manual 4, *Air Entrainment in Free-surface Flows*. Balkema, Rotterdam, pp. 55–84.

Wood, I.R., Ackers, P. & Loveless, J. (1983) General method for critical point on spillways. *Journal of Hydraulic Engineering*, 109(2), 308–312.

Wright, H.J. (2010) Improved energy dissipation on stepped spillways with the addition of triangular protrusions. *Proceedings of 78th ICOLD Annual Meeting, Hanoi, Vietnam*, pp. 1–10.

Wüthrich, D. & Chanson, H. (2014a) Aeration and energy dissipation over stepped gabion spillways: a physical study. *Hydraulic Model Report* No. CH92/13, School of Civil Engineering, the University of Queensland, Brisbane.

Wüthrich, D. & Chanson, H. (2014b) Hydraulics, air entrainment, and energy dissipation on a gabion stepped weir. *Journal of Hydraulic Engineering*, 140(9), 04014046.

Wüthrich, D. & Chanson, H. (2015) Aeration performances of a gabion stepped weir with and without capping. *Journal of Environmental Fluid Mechanics*, 15(4), 711–730.

Yasuda, Y. & Ohtsu, I. (1999) Flow resistance of skimming flows in stepped channels. In: Bergmann, H., Krainer, R. & Breinhälter, H. (eds) *Proceedings of 28th IAHR Congress, Graz, Austria* (CD-ROM).

Yevdjevich, V. & Levin, L. (1953) Entrainment of air in flowing water and technical problems connected with it. *Proceedings of 5th IAHR Congress, Minneapolis, Minnesota*, pp. 439–454.

Zhang, G. & Chanson, H. (2016a) Gabion stepped spillway: interactions between free-surface, cavity, and seepage flows. *Journal of Hydraulic Engineering*, 142(5), 06016002.

Zhang, G. & Chanson, H. (2016b) Interaction between free-surface aeration and total pressure on a stepped chute. *Experimental Thermal and Fluid Science*, 74(June), 368–381.

Zhang, J., Chen, J. & Wang, Y. (2012) Experimental study on time-averaged pressures in stepped spillways. *Journal of Hydraulic Research*, 50(1), 236–240.

Zünd, B. (2008) Fliesswiderstand eines belüfteten Freispiegelabflusses im Gleichgewichtszustand [Flow resistance of uniform-aerated free surface flow]. *Ph.D. Thesis*, also *VAW Mitteilung* 209, H.-E. Minor (ed). Versuchsanstalt für Wasserbau, Hydrologie und Glaziologie. ETH Zurich, Zürich.

Bibliography

Cavitation damage

Aksoy, S. & Ethembabaoglu, S. (1979) Cavitation damage at the discharge channels of Keban Dam. Proceedings of 13th *ICOLD Congress*, New Delhi, Q50(R21), 369–379.

Brown, F.R. (1963) Cavitation in hydraulic structures: problems created by cavitation phenomena. *Journal of Hydraulics Division ASCE*, 89(HY1), 99–115; 89(HY5), 141–145; 90(HY2), 359.

Colgate, D. (1959) Cavitation damage of roughened concrete surfaces. *Journal of Hydraulics Division ASCE*, 85(HY11), 1–10.

Engelund, F. & Munch-Petersen, J. (1953) Steady flow in contracted and expanded rectangular channels. *La Houille Blanche*, 8(8/9), 464–474; 9(3/4), 179–188.

Falvey, H.T. (1982) Predicting cavitation in tunnel spillways. *Water Power & Dam Construction*, 34(8), 9–15; 35(3), 56–57; 35(9), 54.

Hamilton, W.S. (1983) Preventing cavitation damage to hydraulic structures. *Water Power & Dam Construction*, 35(11), 40–43; 35(12), 48–53; 36(1), 42–45.

Hammitt, F.G. (1986) Cavitation and erosion: monitoring and correlating methods. *Encyclopedia of Fluid Mechanics*, 2(38), 1119–1139, N.P. Cheremisinoff, ed. Gulf: Houston.

Hellstrom, B. (1953) Fan-shaped outlets. *La Houille Blanche*, 8(12), 873–880.

Houston, K.L., Quint, R.J. & Rhone, T.J. (1985) An overview of Hoover Dam tunnel spillway damage. *Waterpower '85*(2), 1421–1430.

Johnson, V.E. (1963) Mechanics of cavitation. *Journal of Hydraulics Division ASCE*, 89(HY3), 251–275.

Katz, J. & O'Hern, T.J. (1986) Cavitation in large scale shear flows. *Journal of Fluids Engineering*, 108(9), 373–376.

Kenn, M.J. (1971) Protection of concrete from cavitation. *Proceedings of ICE*, 49(1), 75–79; 50(4), 581–583.

Kenn, M.J. & Garrod, A.D. (1981) Cavitation damage and the Tarbela Tunnel collapse of 1974. *Proceedings of ICE*, 70(1), 65–89; 70(4), 779–810.

Knapp, R.T., Daily, J.W. & Hammitt, F.G. (1970) *Cavitation*. McGraw-Hill, New York.

Lauterborn, W. & Bolle, H. (1975) Experimental investigations of cavitation-bubble collapse in the neighbourhood of a solid boundary. *Journal of Fluid Mechanics*, 72(2), 391–399.

Lee, W. & Hoopes, J.A. (1996) Prediction of cavitation damage for spillways. *Journal of Hydraulic Engineering*, 122(9), 481–488.

Lowe III, J., Bangash, H.D. & Chao, P.C. (1979) Some experiences with high velocity flow at Tarbela Dam project. Proceedings of 13th *ICOLD Congress, New Delhi*, Q50(R13), 215–247.

Pugh, C.A. (1984) Modeling aeration devices for Glen Canyon Dam. In: Schreiber, D.L. (ed) *Water for Resource Development,* ASCE, New York, pp. 412–416.

Regan, R.P., Munch, A.V. & Schrader, E.K. (1979) Cavitation and erosion damage of sluices and stilling basins at two high-head dams. Proceedings of *13th ICOLD Congress, New Delhi,* Q50(R11), 177–198.

Rouse, H. (1947) Fundamental aspects of cavitation. *Proceedings of 3rd National Conference on Industrial Hydraulics,* Illinois Institute of Technology, pp. 1–8.

Trevena, D.H. (1987) *Cavitation and tension in liquids.* Hilger, Bristol.

Various (1947) Cavitation in hydraulic structures: a symposium. *Transactions of ASCE,* 112, 1–124.

Vennard, J.K. (1947) Nature of cavitation. *Transactions of ASCE,* 112, 2–15.

Warnock, J.E. (1947) Experiences of the Bureau of Reclamation. *Transactions of ASCE,* 112, 43–58.

Free surface aeration

Aki, S. (1969) Field measurements of velocity and pressure on spillway chutes. *Proceedings of 13th IAHR Congress,* Kyoto, 5(1), 167–172.

Anderson, A.G. (1965) Influence of channel roughness on the aeration of high-velocity, open-channel flow. *Proceedings of 11th IAHR Congress, Leningrad,* 1(37), 1–13.

Annemüller, H. (1961) Berechnung der Abflusstiefen in Schussrinnen [Computation of flow depths in chutes]. *Der Bauingenieur,* 36(6), 222–226 (in German).

Ashfar, N.R., Ranga Raju, K.G. & Asawa, G.L. (1994) Air concentration distribution in self-aerated flow. *Journal of Hydraulic Research,* 32(4), 623–631; 33(4), 582–592.

Chanson, H. (1993) Self-aerated flows on chutes and spillways. *Journal of Hydraulic Engineering,* 119(2), 220–243; 120(6), 778–782.

Durand, W.F. (1940) The flow of water in channels under steep gradients. *Transactions of ASME,* 62(1), 9–14.

Ehrenberger, R. (1926) Wasserbewegung in steilen Rinnen (Schusstennen) mit besonderer Berück-sich-tigung der Selbstbelüftung [Water flow in steep channels with a particular consideration of self-aeration]. *Zeitschrift des Österreichischen Ingenieur- und Architekten-Vereines,* 78(15/16), 155–160; 78(17/18), 175–179 (in German).

Ehrenberger, R. (1930) Eine neue Geschwindigkeitsformel für künstliche Gerinne mit starken Neigun-gen und Berechnung des Selbstbelüftung des Wassers [A new velocity formula for chutes and predic-tion of self-aeration of water]. *Die Wasserwirtschaft,* 23(28), 573–575; 23(29), 595–598 (in German).

Falvey, H.T. (1979) Mean air concentration of self-aerated flows. *Journal of Hydraulics Division ASCE,* 105(HY1), 91–96; 105(HY11), 1469; 106(HY8), 1407.

Gangadharaiah, T., Lakshmana Rao, N.S. & Seetharamiah, K. (1970) Inception and entrainment in self-aerated flows. *Journal of Hydraulics Division ASCE,* 96(HY7), 1549–1565.

Hager, W.H. & Kramer, K. (2003) Historical development of free surface chute aeration. *Proceedings of 30th IAHR Congress, Thessaloniki* E, pp. 389–396.

Halbronn, G. (1952) Study of the setting up of the flow regime on high gradient structures. *La Houille Blanche,* 7(5/6), 348–371 (in French).

Halbronn, G., Durand, R. & Cohen de Lara, G. (1953) Air entrainment in steeply sloping flumes. *Proceedings of 5th IAHR Congress, Minneapolis,* pp. 455–466.

Hall, L.S. (1943) Open channel flow at high velocities. *Transactions of ASCE,* 108, 1434.

Harrison, A.J.M. & Owen, M.W. (1967) A new type of structure for flow measurement in steep streams. *Proceedings of ICE,* 36(2), 273–296; 37(4), 811–818.

Hickox, G.H. (1945) Air entrainment on spillway faces. *Civil Engineering,* 15(12), 562–563;

Houk, I.E. (1927) New design features in Willwood Diversion Dam. *Engineering News-Record,* 99(17), 660–664.

Keller, R.J., Lal, M.K. & Wood, I.R. (1974) Developing region in self-aerated flows. *Journal of Hydraulics Division ASCE,* 100(HY4), 553–568; 100(HY11), 1729–1730; 101(HY2), 319–320; 101(HY9), 1278–1280; 102(HY4), 529–530.

Lakshmana Rao, N.S., Seetharamiah, K. & Gangadharaiah, T. (1970) Characteristics of self-aerated flows. *Journal of Hydraulics Division ASCE*, 96(HY2), 331–355; 96(HY12), 2622–2627; 97(HY12), 2077.

Lakshmana Rao, N.S. & Gangadharaiah, T. (1971) Self-aerated flow characteristics in wall region. *Journal of Hydraulics Division ASCE*, 97(HY9), 1285–1303.

Lane, E.W. (1936) Recent studies on flow conditions in steep chutes. *Engineering News-Record*, 116(1), 5–7.

Lane, E.W. (1939) Entrainment of air in swiftly flowing water. *Civil Engineering*, 9(2), 89–91; 9(6), 371.

Straub, L.G., Killen, J.M., Lamb, O.P. (1954) Velocity measurement of air-water mixtures. *Transactions of ASCE*, 119, 207–220.

Straub, L.G. & Lamb, O.P. (1953) Experimental studies of air entrainment in open channel flow. *Proceedings of 5th IAHR Congress, Minneapolis*, pp. 425–437.

Straub, L.G. & Lamb, O.P. (1956) Studies of air entrainment in open-channel flows. *Transactions of ASCE*, 121, 30–44.

Thandaveswara, B.S. & Lakshmana Rao, N.S. (1978) Developing zone characteristics in aerated flows. *Journal of Hydraulics Division ASCE*, 104(HY3), 385–396; 105(HY3), 279–281; 105(HY11), 1451–1452.

Viparelli, M. (1951) Premessa ad una ricerca sulle correnti rapide [Introduction to a study on high-speed flows]. *L'Energia Elettrica*, 28(1), 13–22 (in Italian).

Viparelli, M. (1953) The flow in a flume with 1:1 slope. *Proceedings of 5th IAHR Congress, Minneapolis*, pp. 415–423.

Viparelli, M. (1954) Correnti rapide: risultati in canaletta a 45° [Rapid flows: results in 45° chute]. *L'Energia Elettrica*, 31(6), 393–405 (in Italian).

Viparelli, M. (1958) Correnti rapide [High-speed flows]. *L'Energia Elettrica*, 35(7), 633–649 (in Italian).

Volkart, P. (1980) The mechanism of air bubble entrainment in self-aerated flow. *International Journal of Multiphase Flow*, 8(5), 411–423.

Wilhelms, S.C. & Gulliver, J.S. (1994) Self-aerated flow on Corps of Engineers spillways. *Technical Report* W-94-2. US Corps of Engineers, Washington DC.

Wood, I.R. 81983) Uniform region of self-aerated flow. *Journal of Hydraulic Engineering*, 109(3), 447–461.

General

Ellis, J. (1989) Guide to analysis of open-channel spillway flow. *CIRIA Technical Note 134*. Construction Industry Research and Information Association, London UK.

Novak, P. & Cabelka, J. (1981) *Models in Hydraulic Engineering: Physical Principles and Design Applications*. Pitman, Boston MA.

Subramanya, K. (1982) *Flow in Open Channels*. Tata McGraw-Hill, New Delhi.

Roll waves

Arsenishvili, K.I. (1965) *Effect of Wave Formation on Hydro-engineering Structures*. Israel Program for Scientific Translations, Jerusalem.

Chen, C.-l. (1995) Free-surface stability criterion as affected by velocity distribution. *Journal of Hydraulic Engineering*, 121(10), 736–743; 123(7), 666–668.

Craya, A. (1952) The criterion for the possibility of roll-wave formation. Gravity waves. In: Sawyer, C. (ed) *Circular* 521. National Bureau of Standards, Washington DC, pp. 141–151.

Hedberg, J. (1942) Report on steep-slope flow. *Transactions of AGU*, 23(1), 74–76.

Holmes, W.H. (1936) Traveling waves in steep channels. *Civil Engineering*, 6(7), 467–468.

Keulegan, G.H. & Patterson, G.W. (1940) A criterion for instability of flow in steep channels. *Transactions of AGU*, 21(2), 594–596.

Montuori, C. (1964) L'onda di riempimento di un canale vuoto [Choking of an empty conduit]. *L'Energia Elettrica*, 41(12), 853–865 (in Italian).

Ponce, V.M. & Maisner, M.P. (1993) Verification of theory of roll-wave formation. *Journal of Hydraulic Engineering*, 119(6), 768–773.

Ponce, V.M. & Porras, P.J. (1995) Effect of cross-sectional shape on free-surface instability. *Journal of Hydraulic Engineering*, 121(4), 376–380.

Scobey, F.C. (1931) Unusual flow phenomena. *Civil Engineering*, 1(12), 1101–1103.

Shock waves

Anastasi, G. (1982) Die Aufhebung von Stosswellen in Schussrinnenverengungen [The elimination of shock waves along chute contractions]. In: Vischer (ed) *VAW-Mitteilung* 59, D. Versuchsanstalt für Wasserbau, Hydrologie und Glaziologie. ETH Zürich, Zürich (in German).

Anastasi, G., Bisaz, E., Gerodetti, M., Schaad, F. & Soubrier, G. (1979) Essais sur modèle hydraulique et études d'évacuateurs par rapport aux conditions de restitution [Hydraulic model study on spillway rehabilitation]. *Proceedings of 13th ICOLD Congress, New Delhi*, Q50(R33), 515–558 (in French).

Bagge, G. & Herbich, J.B. (1967) Transitions in supercritical open-channel flow. *Journal of Hydraulics Division ASCE*, 93(HY5), 23–41; 94(HY3), 803–804; 95(HY1), 453–454.

Behlke, C.E. & Pritchett, H.D. (1966) The design of supercritical flow channel junctions. *Highway Research Record*, 123, 17–35.

Berger, R.C. & Stockstill, R.L. (1995) Finite-element model for high-velocity channels. *Journal of Hydraulic Engineering*, 121(10), 710–716; 122(6), 581–582.

Carlyle, W.J. (1969) Llyn Brianne Dam, South Wales. *Civil Engineering and Public Works Review*, 64(12), 1195–1200.

Causon, D.M., Mingham, C.G. & Ingram, D.M. (1999) Advances in calculation methods for supercritical flow in spillway channels. *Journal of Hydraulic Engineering*, 125(10), 1039–1050; 127(4), 328–330.

Christodoulou, G.C. (1993) Incipient hydraulic jump at channel junctions. *Journal of Hydraulic Engineering*, 119(3), 409–421; 119(7), 875; 120(6), 768–771.

Citrini, D. (1940) Sul movimento di una corrente veloce in un canale in curva [On the movement of a high-speed flow in a channel curve]. *L'Energia Elettrica*, 17(9), 509–525 (in Italian).

Courant, R. & Friedrichs, K.O. (1948) *Supersonic Flow and Shock Waves*. Interscience, New York.

Ellis, J. (1985) Numerical analysis of Kielder Dam spillway. *Journal of Institution of Water Engineers*, 39, 254–270.

Ellis, J. 1995) Guide to analysis of open-channel spillway flow. *Technical Note* 134. CIRIA, London UK.

Ellis, J. & Pender, G. (1982) Chute spillway design calculations. *Proceedings of ICE*, 73(2), 299–312.

Gerodetti, M. (1978) Schussrinnen im Strassenbau [Chutes in road construction]. *Strasse und Verkehr*, 64(7), 271–275 (in German).

Greated, C.A. (1968) Supercritical flow through a junction. *La Houille Blanche*, 23(8), 693–695.

Hager, W.H. (1989a) Transitional flow in channel junctions. *Journal of Hydraulic Engineering*, 115(2), 243–259.

Hager, W.H. (1989b) Supercritical flow in channel junctions. *Journal of Hydraulic Engineering*, 115(5), 595–616.

Hager, W.H. & Altinakar, M.S. (1984) Infinitesimal cross-wave analysis. *Journal of Hydraulic Engineering*, 110(8), 1145–1150.

Hager, W.H. & Mazumder, S.K. (1993) Flow choking in an expanding bucket. *Water Power & Dam Construction*, 45(4), 50–52.

Hager, W.H., Schwalt, M., Jiménez, O. & Chaudhry, M.H. (1994) Supercritical flow near an abrupt wall deflection. *Journal of Hydraulic Research*, 32(1), 103–118.

Hager, W.H. & Yasuda, Y. (1997) Unconfined expansion of supercritical water flow. *Journal of Engineering Mechanics*, 123(5), 451–457.

Haindl, K. & Liskovec, L. (1973) *Supercritical streaming in hydraulic engineering. Vyzkumny Ustav Vodohospodarsky v Praze Prace a studie 132*. VUV, Praha (in Czech).

Harrison, A.J.M. (1966) Design of channels for supercritical flow. *Proceedings of ICE*, 35(3), 475–490; 37(3), 557–565.

Hartung, F. (1972) Gestaltung von Hochwasserentlastungsanlagen bei Talsperrendämmen [Shape of flood conveyances at dams]. *Wasserwirtschaft*, 62(1/2), 39–51 (in German).

Hartung, F. & Knauss, J. (1967) Developments to improve economy, capacity, and efficiency of structures controling the passage of flood water through reservoirs. *Proceedings of 9th ICOLD Congress, Istamboul*, Q33(R14), 227–249.

Herbrand, K. (1976) Zusammenführung von Schussstrahlen: Zwei praktische Beispiele konstruktiver Lösungen aus Modellversuchen [Merging of high-speed jets: two examples of design solutions based on laboratory experimentation]. *Bericht*, 32, 183–201, F. Hartung (ed). Versuchsanstalt für Wasserbau. TU München, München (in German).

Herbrand, K. & Scheuerlein, H. (1979) Examples of model tests dealing with special problems and design criteria at large capacity spillways. *Proceedings of 13th ICOLD Congress, New Delhi*, Q50(R10), 161–176.

Ippen, A.T. (1943) Gas-wave analogies in open channel flow. *Proceedings* of 2nd Hydraulics Conference. *Bulletin*, 27, 248–265. University of Iowa, Iowa City IA.

Ippen, A.T. (1951) Mechanics of supercritical flow. *Transactions of ASCE*, 116, 1290–1317.

Ippen, A.T. & Harleman, R.F. (1956) Verification of theory for oblique standing waves. *Transactions of ASCE*, 121, 678–694.

Ippen, A.T. & Knapp, R.T. (1936) A study of high-velocity flow in curved channels of rectangular cross-section. *Transactions of AGU*, 17(2), 516–521.

Jiménez, O.F. & Chaudhry, M.H. (1988) Computation of supercritical free-surface flows. *Journal of Hydraulic Engineering*, 114(4), 377–395.

Jobes, J.G. & Douma, J.H. (1942) Testing theoretical losses in open channel flow. *Civil Engineering*, 12(11), 613–615; 12(12), 667–669.

Killen, J.M. & Anderson, A.G. (1969) A study of the air-water interface in air-entrained flow in open channels. *Proceedings of 13th IAHR Congress, Kyoto*, 2(36), 1–9; 5(2), 241–243.

Knapp, R.T. & Ippen, A.T. (1938) Curvilinear flow of liquids with free surfaces at velocities above that of wave propagation. *Proceedings of 5th International Congress Applied Mechanics, Cambridge MA*, pp. 531–536.

Knauss, J. (1967) Ein Beitrag zur Behandlung von stehenden Wellen, die nicht aus seiner parallelen Anströmung hervorgehen [Contribution to the treatment of standing waves not originating from parallel approach flow]. *Bericht*, 9, 15–118, F. Hartung (ed) Versuchsanstalt für Wasserbau. TH München, München (in German).

Koch, K. (1968) Ein Beitrag zum Verhalten spitzwinklig zusammenprallender Schussstrahlen und die Anwendungsmöglichkeiten im Wasserbau [Sharp-angled impacting chute flows and applications in hydraulic engineering]. *Bericht* 15, F. Hartung (ed) Versuchsanstalt für Wasserbau. TH München, München (in German).

Krause, D. (1970) Einfluss der Trassierungselemente auf den Spiegelverlauf in gekrümmten Schussrinnen [Influence of insets on the free surface of curved chute flows]. *Technischer Bericht* 6. Institut für Hydraulik und Hydrologie, TH Darmstadt, Darmstadt (in German).

Kutija, V. (1993) On the numerical modelling of supercritical flow. *Journal of Hydraulic Research*, 31(6), 841–858; 32(5), 783–791.

Lenau, C.W. (1979) Supercritical flows in bends of trapezoidal section. *Journal of Engineering Mechanics Division ASCE*, 105(EM1), 43–54.

Leopardi, M. & Maggi, A. (2003) Water surface superelevation conditioning in channels of nonlinear alignment. *L'Acqua*, 82(6), 17–23 (in Italian).

Liggett, J.A. & Vasudev, S.U. 81965) Slope and friction effects in two-dimensional, high speed channel flow. *Proceedings of 11th IAHR Congress, Leningrad*, 1(25), 1–12.

Marchi, E. (1988) Correnti veloci in curva a 90° molto strette [High-speed flows in strong 90° bends]. *Idrotecnica*, 15(6), 439–455 (in Italian).

Martin Vide, J.P., Azlor, J. & Viver, L. (1995) The design of converging overflow spillways. *Hydropower & Dams*, 2(11), 87–92.

McLean, F., Charbonneau, A.L. & Peterson, A.W. (1971) The Brazeau spillway. *Engineering Journal*, 54(1/2), 25–29.

Meyer, T. (1908) Über zweidimensionale Bewegungsvorgänge in einem Gas, das mit Überschallgeschwindigkeit strömt [On 2D supercritical gas flow]. Forschungsarbeiten auf dem Gebiete des Ingenieurwesens, *Mitteilung 82*. Schade, Berlin (in German).

Molinas, A. & Marcus, K.B. (1998) Choking in water supply structures and natural channels. *Journal of Hydraulic Research*, 36(4), 675–694.

Murty Bhallamudi, S. & Chaudhry, M.H. (1992) Computation of flows in open-channel transitions. *Journal of Hydraulic Research*, 30(1), 77–93.

Neilson, F.M. (1976) Convex chutes in converging supercritical flow. *Miscellaneous Paper* H-76–19. US Army Waterways Experiment Station, Vicksburg MS.

Pinto, J.F.R. (1941) Escoamento de agua em conduta livre a altas velocidades [High-speed water flow]. *Técnica*, 16(117), 198–215; 16(118), 236–249 (in Portuguese).

Poggi, B. (1956) Correnti veloci nei canali in curva [High-speed flows in curved channels]. *L'Energia Elettrica*, 33(7), 465–480 (in Italian).

Prandtl, L. (1907) Neue Untersuchungen über die strömende Bewegung der Gase und Dämpfe [New studies on the flow of gases and vapors]. *Physikalische Zeitschrift*, 8(1), 23–30 (in German).

Prandtl, L. (1936) Allgemeine Betrachtungen über die Strömung zusammendrückbarer Flüssigkeiten [General considerations on compressible fluid flows]. *ZAMM*, 16(3), 129–142 (in German).

Prandtl, L. & Busemann, A. (1929) Näherungsverfahren zur zeichnerischen Ermittlung von ebenen Strömungen mit Überschallgeschwindigkeit [Graphical method to determine plane supercritical gas flow]. In: Honegger, E. (ed) *Festschrift*, Prof. Dr. A. Stodola, pp. 499–509, Orell Füssli, Zürich (in German).

Preisswerk, E. (1937) Zweidimensionale Strömung schiessenden Wassers [Two-dimensional supercritical water flow]. *Schweizerische Bauzeitung*, 109(20), 237–238 (in German).

Rakotoarivelo, W. & Sananes, F. (1967) Etude de l'écoulement supercritique dans un canal dont la section est en forme de U [Study of supercritical channel flow whose cross-section is U-shaped]. *Proceedings of 12th IAHR Congress, Fort Collins*, 1(A38), 1–7 (in French).

Reinauer, R. & Hager, W.H. (1996a) Shockwave in air-water flows. *International Journal of Multiphase Flow*, 22(6), 1255–1263.

Reinauer, R. & Hager, W.H. (1996b) Shockwave reduction by chute diffractor. *Experiments in Fluids*, 21(3), 209–217.

Rudavsky, A.B. (1966) Model solves supercritical flow conveyance in flood control channels. *Proceedings of 2nd Annual American Water Resources Conference, Chicago*, pp. 306–327.

Schwalt, M. & Hager, W.H. (1994) Shock-wave reduction at bottom drop. *Journal of Hydraulic Engineering*, 120(10), 1222–1227.

Schwalt, M. & Hager, W.H. (1995) Experiments to supercritical junction flow. *Experiments in Fluids*, 18(6), 429–437.

Slopek, R.J. & Nunn, J.O.H. (1989) Freeboard allowances for chute spillways. *International Conference on Hydropower Waterpower '89, Niagara Falls*. ASCE, New York, pp. 518–527.

Stockstill, R.L., Berger, R.C. & Nece, R.E. (1997) Two-dimensional flow model for trapezoidal high-velocity channels. *Journal of Hydraulic Engineering*, 123(10), 844–852.

Strassburger, A.G. & Sias, J (1969) Model and prototype tests of buoyant spillway gates. *Journal of Power Division ASCE*, 95(PO2), 335–349.

Sturm, T.W. (1985) Simplified design of contractions in supercritical flow. *Journal of Hydraulic Engineering*, 111(5), 871–875; 113(3), 422–427.

Täubert, U. (1971) Der Abfluss in Schussrinnen-Verengungen [Flow in chute contractions]. *Der Bauingenieur*, 46(11), 385–392 (in German).

Täubert, U. (1974) The design of spillway contractions using computer simulation. *Water Power & Dam Construction*, 26(8), 282–287.

Tursunov, A.A. (1965) Methods of governing of geometrical forms of supercritical high velocity flows. *Proceedings of 11th IAHR Congress, Leningrad*, 1(23), 1–9.

Vischer, D.L. & Hager, W.H. (1994) Reduction of shockwaves: a typology. *Hydropower & Dams*, 1(7), 25–29.

von Karman, T. (1938) Eine praktische Anwendung der Analogie zwischen Überschallströmung in Gasen und überkritischer Strömung in offenen Gerinnen (A practical application of the analogy between supersonic gas flow and supercritical open channel flow). *ZAMM*, 18(1), 49–56 (in German).

Stepped chutes

Anonymous (1990) Knellport Dam: the world's first rollcrete arch-gravity dam. *The Civil Engineer in South Africa*, 32(2), 47–50; 32(9), 375.

Anonymous (1994) *Roller-compacted Concrete: Technical Engineering and Design Guides as Adapted from the US Army Corps of Engineers*, Nr. 5. ASCE, New York.

Anonymous (1996) International symposium reviews RCC progress. *Hydropower & Dams*, 3(2), 50–58.

Baker, R., Pravdivets, Y. & Hewlett, H. (1994) Design considerations for the use of wedge-shaped precast concrete blocks for dam spillways. *Proceedings of the Institution of Civil Engineers, Water, Maritime & Energy*, 106(12), 317–323.

Baylar, A. & Emiroglu, M.E. (2003) Study of aeration efficiency at stepped channels. *Water & Maritime Engineering*, 156(WM3), 257–263.

Bindo, M., Gautier, J. & Lacroix, F. (1993) The stepped spillway of M'Bali Dam. *Water Power & Dam Construction*, 45(1), 35–36.

Chamani, M.R. & Rajaratnam, N. (1994) Jet flow on stepped spillways. *Journal of Hydraulic Engineering*, 120(2), 254–259; 121(5), 441–448.

Chamani, M.R. & Rajaratnam, N. (1999a) Characteristics of skimming flow over stepped spillways. *Journal of Hydraulic Engineering*, 125(4), 361–368; 126(11), 860–873; 121(1), 90.

Chamani, M.R. & Rajaratnam, N. (1999b) Onset of skimming flow on stepped spillways. *Journal of Hydraulic Engineering*, 125(9), 969–971; 127(6), 519–525.

Chanson, H. (1993) Stepped spillway flows and air entrainment. *Canadian Journal of Civil Engineering*, 20(3), 422–435.

Chanson, H. (1996) Prediction of the transition nappe/skimming flow on a stepped channel. *Journal of Hydraulic Research*, 34(3), 421–429.

Croce, N. & Peruginelli, A. (1993) Scala di stramazzi: Esperienza di laboratorio [Scales of weirs: laboratory experimentation]. *Ingegneria Sanitaria Ambientale*, 22(4), 43–51 (in Italian).

Dunstan, M.R.H. (1999) Recent developments in RCC dams. *Hydropower & Dams*, 6(1), 40–45.

Frizell, K.H. (1992) Hydraulics of stepped spillways for RCC dams and dam rehabilitations. In: Hansen, K.D. & McLean, F.G. (eds) *Proceedings of Conference Roller Compacted Concrete III, San Diego CA*. ASCE, New York, pp. 423–439.

Geringer, J.J. & du Buisson, N.J. (1995) Neusberg: an RCC weir across South Africa's largest river. *Hydropower & Dams*, 2(5), 53–60.

Gerodetti, M. (1981) Model studies of an overtopped rockfill dam. *Water Power & Dam Construction*, 33(9), 25–31.

Hansen, K.D. (1997) Current RCC dam activity in the USA. *Hydropower & Dams*, 4(5), 62–65.

Hewlett, H.W.M. & Baker, R. (1992) The use of stepped blocks for dam spillways. In: *Water Resources and Reservoir Engineering*. Telford, London UK, pp. 183–190.

Houston, K.L. (1987) Stepped spillway design with a RCC dam. In: Ragan, R.M. (ed) *Proceedings of Conference Hydraulic Engineering, Williamsburg VA*. ASCE, New York.

Jansen, R.B. (1988) *Advanced Dam Engineering for Design, Construction and Rehabilitation*. Van Nostrand Reinhold, New York.

Lejeune, M. & Lejeune, A. (1994) About the energy dissipation of skimming flows over stepped pillways. In: Verwey, A. *et al.* (eds) *Hydroinformatics '94*: 595–600. Balkema, Rotterdam.

Lejeune, A., Lejeune, M. & Lacroix, F. (1994) Study of skimming flow over stepped spillways. In: *Modelling, Testing & Monitoring for Hydropower Plants, Budapest*, Aqua-Media International, Sutton, Surrey UK, pp. 285–295.

Mateos Iguacel, C. & Garcia, V.E. (1995) Stepped spillways: design for the transition between the spillway crest and the steps. *Proceedings of IAHR Congress Hydra 2000*, 1, 260–265. IAHR, Delft.

Matsushita, F. (1988) Flow characteristics of large scale roughness on the steep slope channel. *Transactions of JSIDRE*, 135, 107–117 (in Japanese).

Ohtsu, I. & Yasuda, Y. (1997) Characteristics of flow conditions on stepped channels. *Proceedings of 27th IAHR Congress, San Francisco*, 4, 583–588. IAHR, Delft.

Pagliara, S., Peruginelli, A. & Casadidio, A. (2002) Scale di stramazzi con basso numero di gradini [Stepped channel with low steps number]. *L'Acqua*, (3), 15–22 (in Italian).

Pegram, G.G.S., Officer, A.K. & Mottram, S.R. (1999) Hydraulics of skimming flow on modeled stepped spillway. *Journal of Hydraulic Engineering*, 125(5), 500–510; 126(12), 947–954.

Peterson, D.F. & Mohanty, P.K. (1960) Flume studies of flow in steep, rough channels. *Journal of Hydraulics Division ASCE*, 86(HY9), 55–76; 87(HY4), 245–251; 88(HY1), 83–87; 88(HY3), 199–202.

Peyras, L., Royet, P. & Degoutte, G. (1991) Ecoulement et dissipation sur les déversoirs en gradins de gabion [Flow and dissipation on gabion spillways]. *La Houille Blanche*, 46(21), 37–47 (in French).

Pravdivets, Y.P. & Bramley, M.E. (1989) Stepped protection blocks for dam spillways. *Water Power & Dam Construction*, 41(7), 49–56.

Pravdivets, Y.P. & Slissky, S.M. (1981) Passing floodwaters over embankment dams. *Water Power & Dam Construction*, 33(7), 30–32.

Rajaratnam, N. (1990) Skimming flow in stepped spillways. *Journal of Hydraulic Engineering*, 116(4), 587–591; 118(1), 111–113.

Razvan, E. (1966) Solutii modern pentru disipatori de energie la baraje [Modern solutions of energy dissipation on dams]. *Hidrotehnica, Gospodarirea apelor, Meteorologia*, 11(5), 231–240 (in Romanian).

Rhone, T.J. (1977) Baffled apron as spillway energy dissipator. *Journal of Hydraulics Division ASCE*, 103(HY12), 1391–1401.

Rice, C.E. & Kadavy, K.C. (1996) Model study of a roller compacted concrete stepped spillway. *Journal of Hydraulic Engineering*, 122(6), 292–297; 123(10), 931–936.

Sarkaria, G.S. & Dworsky, B.H. (1968) Model studies of an armoured rockfill overflow dam. *Water Power*, 20(11), 455–462.

Schrader, E.K. (1982) World's first all-rollcrete dam. *Civil Engineering ASCE*, 52(4), 45–48.

Schrader, E.K. (1995) RCC: current practices, controversies and options. *Hydropower & Dams*, 2(5), 80–87.

Sorensen, R.M. (1985) Stepped spillway hydraulic model investigation. *Journal of Hydraulic Engineering*, 111(12), 1461–1472; 113(8), 1095–1097.

Stephenson, D. (1979) Gabion energy dissipators. *Proceedings of 13th ICOLD Congress, New Delhi*, Q50(R3), 33–43.

Stephenson, D. (1988) Stepped energy dissipators. *Proceedings of International Symposium Hydraulics for High Dams, Beijing*. IAHR, Delft, 1228–1235.

Stephenson, D. (1991) Energy dissipation down stepped spillways. *Water Power & Dam Construction*, 43(9), 27–30.

Young, M. 1982) Feasibility study of a stepped spillway. In: Smith, P.E. (ed) *Proceedings of Conference Applying Research to Hydraulic Practice*, Jackson MI. ASCE, New York, pp. 96–105.

Chapter 5 Frontispiece (a) Schiffenen Dam, Switzerland, with three gated crest spillways controlled by flap gates and four low level (bottom) outlets; total discharge capacity of 1200 m^3/s (Courtesy groupe e), (b) stilling basin highlighting turbulent action of water flow along with strong air entrainment

Chapter 5

Dissipation structures

5.1 Introduction

The hydraulic jump is the basic feature used to dissipate excess hydromechanical energy below a spillway. It corresponds to the most discontinuous and turbulent flow in an open channel. The prominent features of the hydraulic jump are:

- Highly turbulent flow
- Pulsations in jump body
- Air entrainment at toe and air detrainment at jump end
- Generation of spray and sound
- Energy dissipation due to turbulence production
- Erosive potential and generation of tailwater waves.

A hydraulic jump occurs either (1) as transition from supercritical to subcritical flow in a channel with a varied location or (2) in a stilling basin with a fixed location.

Hydraulic jumps have various appearances, including:

- *Classical hydraulic jump* as fundamental jump type occurring in a prismatic, horizontal and quasi-frictionless rectangular channel;
- *Undular hydraulic jump* for approach flow Froude numbers below, say 2.5;
- *Sloping hydraulic jump* in a sloping, normally rectangular channel;
- *Submerged hydraulic jump* if the jump toe is covered by the tailwater;
- *Expanding hydraulic jump* in nearly horizontal channel of diverging plan.

The main features of a hydraulic jump are its efficiency in energy dissipation, and its stability under varied boundary conditions. Because jumps in channels of non-rectangular cross section are prone to spatial instability, they are rarely applied in dam hydraulics (Elevatorsky, 1959; Rajaratnam, 1967; McCorquodale, 1986; Hager, 1992).

The *stilling basin* uses the hydraulic jump as the hydraulic element for dissipating excess flow energy in a specific structure, installed as a spillway outlet structure. A distinction is made between the hydraulic jump basin and the baffle basin. The hydraulic jump basin corresponds to a structure with a nearly horizontal and smooth bottom to fix the location of dissipation. It is designed to be resistant against spray, scour, pulsations, cavitation, tailwater waves, and jump blowout (Figure 5.1). These basins are recommended for small approach flow energy heads of less than 10 m, or heads between 30 and 50 m (Mason, 1982). For

Figure 5.1 Stilling basin of Manitoba Hydro Limestone (Canada) with one turbine of the adjacent powerhouse in operation (*Water Power and Dam Construction,* July 1991)

higher heads, cavitation may become a concern so that the basin bottom has to be lowered thereby ending with a step.

For small to medium approach flow energy heads of 10–30 m, a stilling basin provided with baffles is favorable to shorten the structure. Elements including blocks, steps, or sills are located at the basin bottom by which the approach flow is deflected to the surface, yet without inducing excessive pulsations.

A general account on both the hydraulic jump and baffle stilling basins is given in 5.2. For the latter, standardized designs are available, to be considered before introducing a novel design. Currently, the number of basin types available is so large that novel designs can only be justified if particular site conditions prevail. For larger basins, extended model studies are recommended because of unexpected features even for small deviations from standardized designs. Plunge pools of drop structures in general are intermediate to hydraulic jumps and trajectory basins, for which both elements play a role. Its main features are described in Chapter 6 and in 5.3. In contrast to stilling basins, the bottom of drop structures may be either of concrete or of rock. Figure 5.2 refers to a typical drop structure as applied with arch dams.

For heads larger than, say 50 m, the *trajectory basin* is the standard design, as described in 6.2. The effect of jet dispersion is thereby used to dissipate a portion of the excess energy (Figure 5.3). The knowledge on lined trajectory basins is still limited as compared with hydraulic jump stilling basins, including aspects of dynamic pressures response and the potential of scour.

The following intends to deal with the various hydraulic aspects of hydraulic jumps and stilling basins. Both the direct and the so-called undular hydraulic jumps are considered, of which the latter should be avoided in engineering structures due to its potential instability and the extensive length. Given that the approach flow Froude number is often fixed, and sometimes is below the limit of 2.5, undular jumps have to be dealt with. In addition, their

Figure 5.2 Drop and dissipation structure of Wagendrift Dam, South Africa (*Journal of Hydropower and Dams*, November 1994)

(b)　　　　　　　　　　　　　　(c)

Figure 5.3 Trajectory basins of Tarbela Dam (Pakistan) (a) auxiliary spillway (*ICOLD* Q69, R24), (b) outlet structure

complexities in terms of hydraulic description are fascinating, so that the main features are addressed.

As to stilling basins, the main types are considered including the baffle-sill and the baffle-block basins, the expanding stilling basin, the slotted-bucket stilling basin and the general basin characteristics. The important issue of scour related to stilling basins has to be considered by a sound engineering approach, given damages that become evident otherwise to the dam foundation, to the tailwater river, and the valley sides. Drop structures are discussed 5.4, whereas free fall jets are considered in Chapter 6.

5.2 Hydraulic jump

5.2.1 Classical hydraulic jump

The basic flow type of a hydraulic jump is referred to as classical. As mentioned, it occurs in a straight, prismatic, and horizontal channel of rectangular shape in which wall friction is negligible. All notation referring to the classical hydraulic jump is denoted with an asterisk (*). Major contributions to the classical hydraulic jump were made by Rouse $et\ al.$ (1959) in presenting the turbulence characteristics, by Schröder (1963), and Rajaratnam (1965) relating it to the wall jet phenomenon, and by McCorquodale and Khalifa (1983) in presenting an early numerical approach.

Figure 5.4 shows a scheme of the classical hydraulic jump, or in short the classical jump. Its body is located between the toe and the end of the jump. Just upstream from the jump toe (subscript 1), the flow depth is h_1, the average velocity V_1, and x is the longitudinal coordinate measured from the toe. At the end of the jump (subscript 2), the flow depth is h_2^*, the velocity V_2^* and the jump length is L_j^*. All quantities are time-averaged because of the significant pulsating action. The approach flow expands vertically along the bottom as a wall jet rising

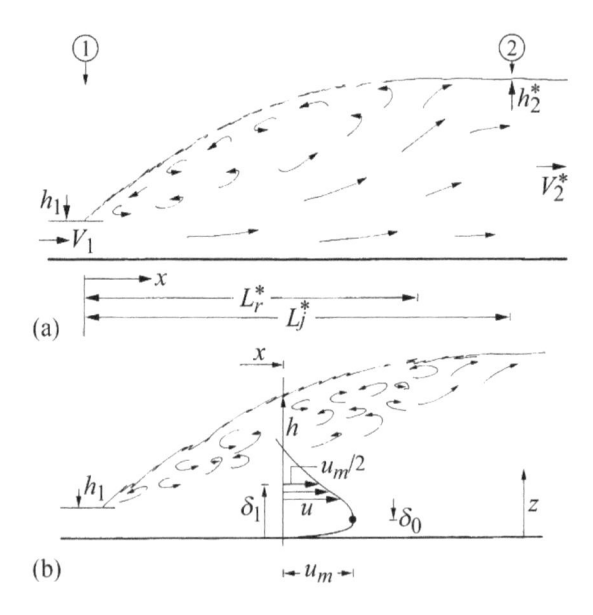

(a)

(b)

$Figure\ 5.4$ Definition sketch of classical hydraulic jump (a) lengths, (b) velocity characteristics

Dissipation structures 351

to the surface, where a stagnation point forms at location L_r^*. A portion of the flow returns to the toe as a *surface roller* that is entrained at the toe zone, whereas the remainder is deflected into the tailwater. Due to the large velocity gradient between the wall jet and roller flow, air is entrained close to the toe section, resulting in a significant turbulence level.

The classical jump is an example for which the basic momentum equation applies successfully, so that the hydraulic energy loss across the jump may be predicted. The following refers to the gross flow pattern first, and then to the internal flow features associated with a classical jump.

The ratio of *sequent depths* $Y^* = h_2^*/h_1$ between sections 1 and 2 is based on:

· Hydrostatic pressure distributions
· Uniform velocity distributions
· Ideal fluid
· Negligible effect of air entrainment
· Time-averaged quantities.

The momentum equation reads with ρ as fluid density, g as gravity acceleration and b as channel width (Figure 5.4a)

$$\frac{1}{2}\rho g b h_1^2 + \rho Q V_1 = \frac{1}{2}\rho g b h_2^{*2} + \rho Q V_2^*. \tag{5.1}$$

The discharge is $Q = b h_1 V_1 = b h_2^* V_2^*$. Let the approach flow Froude number be

$$F_1 = \frac{Q}{(gb^2 h_1^3)^{1/2}}. \tag{5.2}$$

The solution of Eq. (5.1) is then

$$Y^* = \frac{1}{2}[(1+8F_1^2)^{1/2} - 1]. \tag{5.3}$$

For $F_1 > 2$ as considered below, this is approximated as

$$Y^* = \sqrt{2}F_1 - (1/2). \tag{5.4}$$

For a given depth h_1, the sequent depth h_2^* increases thus linearly with F_1.

The *efficiency* $\eta^* = (H_1 - H_2^*)/H_1$ of the classical hydraulic jump with H as energy head follows from the energy equation; for $F_1 > 2$ it reads

$$\eta^* = \left(1 - \frac{\sqrt{2}}{F_1}\right)^2. \tag{5.5}$$

For $F_1 < 3$ the relative loss of head is below 30%, for $F_1 = 5$ it is 50%, whereas $\eta^* > 70\%$ for $F_1 > 9$. Typical Froude numbers F_1 of engineering practice range between 3 and 9.

Classical hydraulic jumps have various appearances, depending on the approach flow Froude number. For $F_1 < 1.6$ to 1.7 jumps are referred to as undular because of the undulating surface pattern. They are inefficient as energy dissipators generating tailwater waves (Reinauer and Hager, 1995). For $1.7 < F_1 < 2.5$, the efficiency is still small and (a) pre-jumps are generated. The (b) transition jump with $2.5 < F_1 < 4.5$ has a pulsating action. These jumps occur often in the tailwater of low-head structures. The best performance has the (c) stabilized jump for which $4.5 < F_1 < 9$. It is compact and stable with a considerable dissipation and a good stilling action. The (d) choppy jump for $F_1 > 9$ is rough and overforced, and is therefore not recommended (Figure 5.5). Note that hydraulic jumps depend exclusively on F_1, so that knowledge of this number is a prerequisite in all related questions. Note also that the Froude similitude is limited to approach flow depths in excess of some 50 mm, as proposed by Hager and Bremen (1989). For smaller flow depths, viscous effects dominate so that Froude scaling is impossible.

The *length of roller* is approximately

$$L_r^* / h_2^* = 4.5. \tag{5.6}$$

The *length of jump* is nearly equal to

$$L_j^* / h_2^* = 6. \tag{5.7}$$

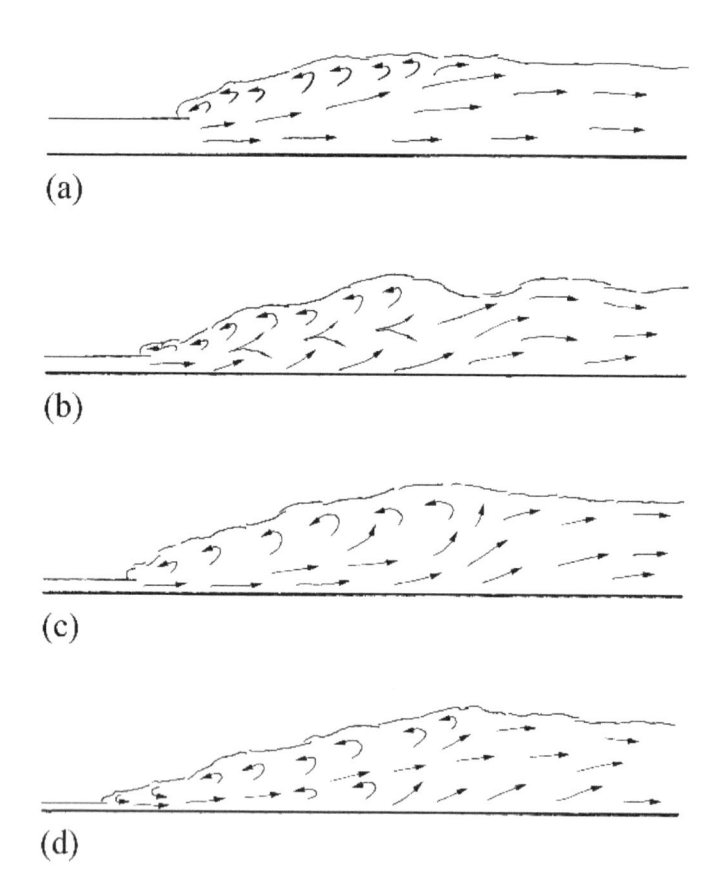

(a)

(b)

(c)

(d)

Figure 5.5 Forms of classical hydraulic jump in terms of F_1 (Peterka, 1958)

Vischer and Hager (1995) present detailed information on the internal flow pattern of the classical hydraulic jump, including the average free surface profile, the velocity distributions, maximum forward and backward velocities, boundary layer growth, pressure fluctuations, and air entrainment characteristics. All quantities exclusively depend on the approach flow Froude number F_1 and on the approach flow depth h_1.

The *turbulent pressure characteristics* are important in designing stilling basins to become resistant against negative pressures, cavitation and dynamic uplift. With p' as the fluctuating pressure component on the basin invert and p_f as the root-mean-square (rms) value, the dimensionless pressure value $P = p_f / (\rho V_1^2 / 2)$ varies only with the dimensionless location $X = x/L_r^*$. The maximum (subscript m) P_m of the function $P(X)$ varies also with F_1 according to Figure 5.6a being 0.08 for $F_1 = 4.5$, and smaller otherwise. The pressure distribution $P(X)$ along the jump follows

$$P / P_m = [3X \exp(1 - 3X)]^2. \tag{5.8}$$

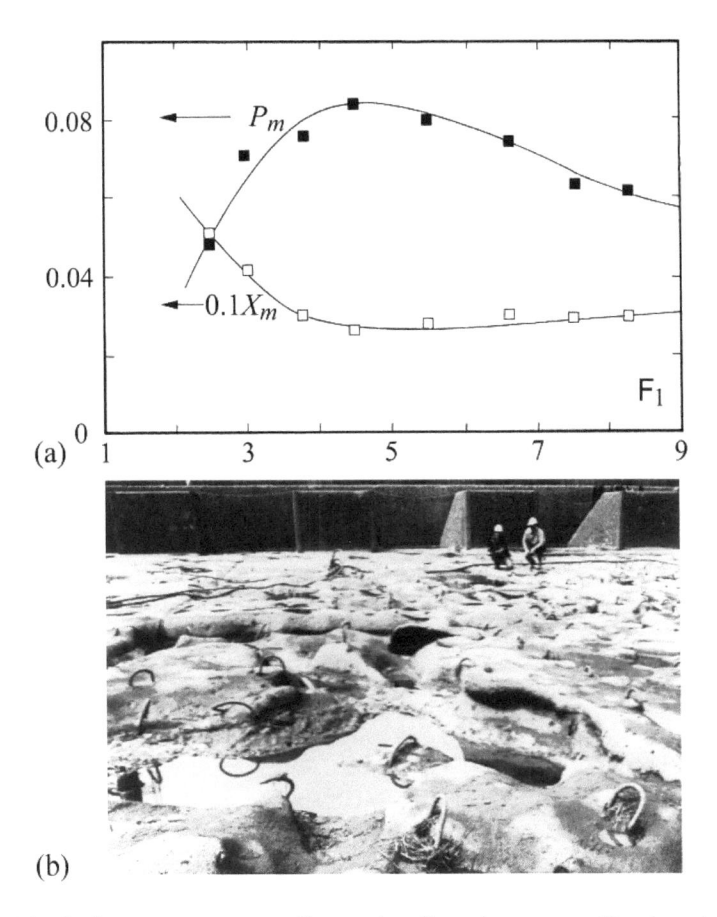

(a)

(b)

Figure 5.6 (a) Maximum rms pressure fluctuation P_m and corresponding location X_m versus F_1 (Vischer and Hager, 1995), (b) damage on stilling basin (*Water Power & Dam Construction* 39(5): 56)

For a given value of F_1 and thus P_m according to Figure 5.6a, the turbulent pressure has a maximum at $X_m = 1/3$, i.e. just behind the jump toe. The extreme turbulent pressures during 24-hour tests can reach values of up to $P_m = 1$, indicating that single pressure peaks are as large as the kinetic head entering the jump, namely $\pm\left(V_1^2 / 2g\right)$.

The *air entrainment* of the classical hydraulic jump is particular, because the air concentration increases from the bottom to the surface, with small bubbles close to the bottom and larger close to the flow surface. The cross-sectional average concentration increases sharply from the toe to a maximum and then decreases until beyond the end of the jump. Expressions for the maximum air concentration, the aeration length, and the location of maximum concentration, among others, involve again the approach flow Froude number F_1 (Wang and Chanson, 2015). The effects of the approach flow bottom slope and submergence are described by Hager (1992).

5.2.2 Hydraulic approach

Governing equations

Figure 5.7a shows the simplified flow structure of the two-dimensional (2D) classical hydraulic jump, composed of the main flow zone of thickness $h = h(x)$, causing a net mass transport in the streamwise direction x of

$$q = \int_0^h u(z)\mathrm{d}z = V_1 h_1, \qquad (5.9)$$

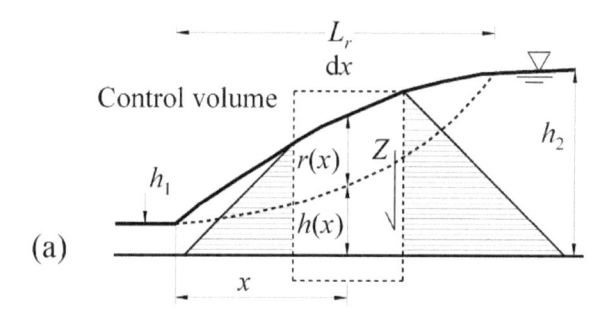

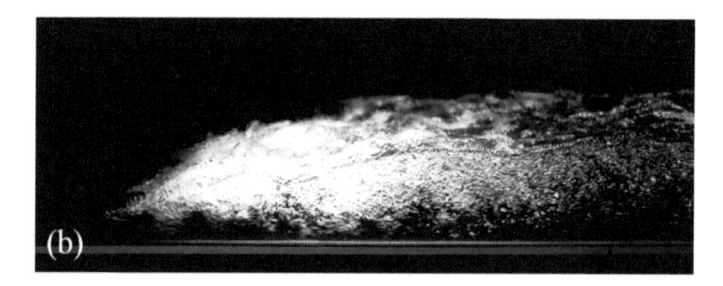

Figure 5.7 2D classical hydraulic jump (a) definition sketch assuming hydrostatic pressure distribution, (b) typical view of hydraulic jump with surface roller

and a surface roller of thickness $r(x)$. Herein, u is the horizontal velocity component at vertical distance z above the horizontal channel bottom, V_1 the mean approach flow velocity, and h_1 the corresponding flow depth. For approach flow Froude numbers $F_1 = q / \left(gh_1^3\right)^{1/2} > 1.5$ to 2 a surface roller is generated. The roller thickness $r(x)$ is characterized from mass conservation as (Resch $et\ al.$, 1976; Hager and Hutter, 1983; Valiani, 1997)

$$\int_h^{h+r} u(z)\mathrm{d}z = 0. \tag{5.10}$$

According to laboratory observations, velocity profiles are highly non-uniform across the flow depth (Rajaratnam, 1965; Murzyn and Chanson, 2007) and the momentum flux is non-zero in the roller (Schröder, 1963; Valiani, 1997). The simplest computational approach involves thus a hydrostatic pressure distribution, a uniform horizontal velocity profile within the main flow zone and negligible momentum flux in the roller as compared with the main flow zone. Neglecting the bottom shear forces, specific momentum conservation $S = (1/2)(h + r)^2 + q^2/(gh)$ in the x-direction for the control volume of Figure 5.7a yields (Woycicki, 1931)

$$\frac{\mathrm{d}}{\mathrm{d}x}\left[\frac{(h+r)^2}{2} + \frac{q^2}{gh}\right] = 0. \tag{5.11}$$

Assuming hydrostatic pressure distribution, the fluid weight in the control volume is in equilibrium with the bottom reaction. To conserve momentum in the vertical direction, the net vertical shear forces arising from tangential Reynolds stresses are equal to the change in the vertical momentum flux. The vertical shear force due to Reynolds stresses τ_{xz} is generally

$$T = \int_0^{h+r} \frac{\tau_{xz}}{\rho_w g}\mathrm{d}z. \tag{5.12}$$

In turn, the vertical momentum flux is with v as the vertical velocity vector component

$$M_z = \int_0^{h+r} \frac{uv}{g}\mathrm{d}z. \tag{5.13}$$

For the control volume shown in Figure 5.7a with two vertical sections distant by $\mathrm{d}x$, the conservation of vertical momentum implies (Valiani, 1997)

$$\mathrm{d}(T + M_z) = 0. \tag{5.14}$$

Equation (5.14) implies that the vertical specific force Z remains constant within the jump for preserving the vertical momentum balance, i.e. (Valiani, 1997)

$$Z = T(x) + M_z(x) = T_1 + M_{z1} = \text{constant}. \tag{5.15}$$

The sequent depths ratio from Eq. (5.3) is

$$\Lambda = \frac{h_2}{h_1} = \frac{1}{2}\left[(1 + 8F_1^2)^{1/2} - 1\right]. \tag{5.16}$$

The equilibrium of moments of momentum (i.e. the angular momentum balance) per unit mass with respect to the channel invert at a longitudinal location x requires

$$T(x) + M_z(x) = \frac{\mathrm{d}}{\mathrm{d}x}\left[\frac{1}{6}(h+r)^3\right] = \text{constant}. \tag{5.17}$$

With $\lambda = \Lambda(\Lambda+1)$, $\varsigma = h/h_1$, $\sigma = r/h_1$, $Y = (h+r)/h_1$ and $\chi = x/L_r$, integration of Eq. (5.17), subject to the boundary conditions $Y(\chi = 0) = 1$ and $Y(\chi = 1) = \Lambda$, defines the total vertical specific force Z as well as the integration constant. The analytical solution for the free surface profile $Y = \varsigma + \sigma = Y(\chi)$ is (Valiani, 1997)

$$Y = \left[1+(\Lambda^3-1)\chi\right]^{1/3}. \tag{5.18}$$

Note that the second boundary condition is not exact, because the sequent depth is not yet reached at $\chi = 1$ (Hager, 1992), resulting in $Y(\chi = 1) = 0.905$ instead of 1. From Eq. (5.11), the free surface profile $\varsigma = \varsigma(\chi)$ is

$$\varsigma = \lambda\left[(1+\lambda)-\left\{(\Lambda^3-1)\chi+1\right\}^{2/3}\right]^{-1}. \tag{5.19}$$

Figure 5.8a shows the dimensionless roller profile $\sigma = \sigma(X)$, with $X = (\Lambda^3-1)\chi$, pointing at the strong effect of F_1 on the roller and the main flow zone features. For practical purposes the roller maximum (subscript M) is approximated in the domain $0 < X < 100$ as $\sigma_M = 0.2X^{5/9}$. Figure 5.8b shows the normalized roller profile $\sigma/\sigma_M(\chi)$, exhibiting a strong similarity. The family of curves for a range of F_1 is described by the semi-ellipse $\sigma/\sigma_M = [1-[2(\chi-0.5)]^2]^{1/2}$. Figure 5.8c relates to the dimensionless free surface profile $Y = Y(X)$, noted to be independent of F_1, whereas Figure 5.8d reveals a weak dependence of $Y/Y_M(\chi)$ on F_1 for $0 < \chi < 0.5$.

Experimental verification

The verification of the above theory indicates an excellent agreement for both the free surface and the forward flow zones $Y(\chi)$ and $\varsigma(\chi)$. The total head profile along the hydraulic jump also follows well the theoretical approach. Castro-Orgaz and Hager (2009) further considered the effects of the pressure coefficient by which the non-hydrostatic pressure distribution is accounted for, as well as for both the kinetic α_k and the momentum β_S correction coefficients, accounting for non-uniform velocity distribution. Its effect on the free surface profile is important, particularly in the central hydraulic jump region. A simple approximation is $\beta_S \approx 1 + r/h$. The effect of non-hydrostatic pressure distribution is significant on the streamwise momentum balance, but it cannot further be detailed with the current knowledge.

Castro-Orgaz and Hager (2009) also investigated the transition from super- to subcritical flow across a classical hydraulic jump using the singular point analysis. Using the energy head equation $H = h + q^2/(2gh^2)$ along with the critical flow condition $\mathsf{F} = 1$, the location of the critical (subscript c) point is from Eq. (5.19)

$$\chi_c = \left\{\left[(1+\lambda)-\lambda\mathsf{F}_1^{-2/3}\right]^{3/2}-1\right\}(\Lambda^3-1)^{-1}. \tag{5.20}$$

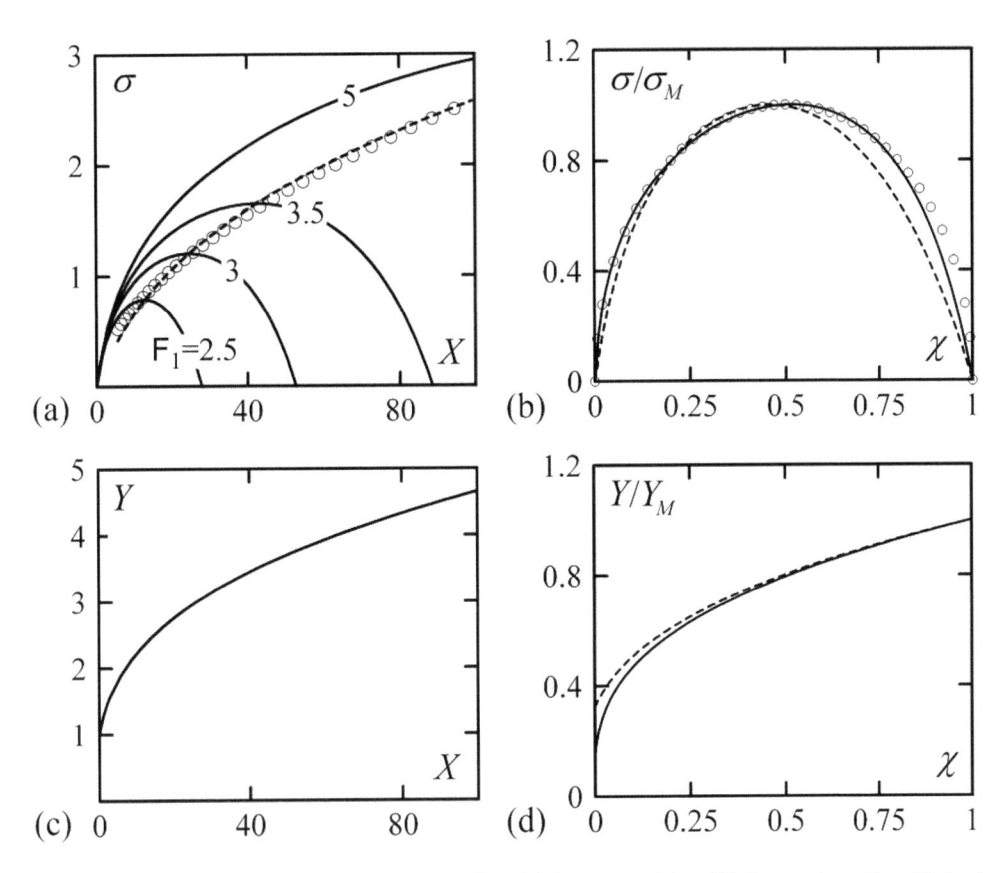

Figure 5.8 Classical hydraulic jump (a) roller thickness profile $\sigma(X)$ for various F_1 with (---) locii of maxima, ($\circ$) approximation for $\sigma_M(X)$, (b) relative roller thickness profile $\sigma/\sigma_M(\chi)$ for (—) $F_1 = 5$, (---) $F_1 = 2.5$, ($\circ$) semi-elliptical approximation, (c) (—) free surface profile $Y(X)$, (d) relative free surface profile $Y/Y_M(\chi)$ for (—) $F_1 = 5$, (---) $F_1 = 2.5$ (Adapted from Castro-Orgaz and Hager, 2009)

Using Eq. (5.11) permits to define the energy line slope S_e as

$$S_e = -\frac{dH^*}{d\chi} = \frac{d\varsigma}{d\chi}F^2\left[\frac{\sigma}{\sigma+\varsigma}\right].$$

(5.21)

This indicates a strong dependence of S_e within jump flow from the slope of the main flow zone $d\varsigma/d\chi$, the roller thickness σ and F_1. As expected, S_e is large where both $d\varsigma/d\chi$ and σ are large, whereas at the extreme jump sections $\sigma = 0$, so that $S_e = 0$.

Singular point analysis

The flow transition from $F > 1$ to $F < 1$ within the main flow zone may also be analyzed using the momentum principle. The energy approach does not allow for information on sucessice derivations of h_c at the critical point ($\mathsf{F} = 1$) because $dH/dx = -S_e \neq 0$ (Hager, 1985). However, the momentum approach applies to develop a singular point analysis at $\mathsf{F} = 1$. According to Eq. (5.11), $dS/dx = d^2S/dx^2 = \ldots = 0$, i.e. all derivatives of $S(x)$ are identical to zero. With $()' = d()/dx$, $()'' = d()/dx^2$, this yields

$$S = \frac{(h+r)^2}{2} + \frac{q^2}{gh} = \frac{h_1^2}{2} + \frac{q^2}{gh_1}, \tag{5.22}$$

$$S' = (h+r)(h'+r') - \frac{q^2}{gh^2}h', \tag{5.23}$$

$$S'' = (h+r)(h''+r'')^2 + (h'+r')^2 - \frac{q^2}{gh^2}h'' + \frac{q^2}{2gh^3}h'^2. \tag{5.24}$$

For a continuous main flow profile $h(x)$, the following properties are obtained in dimensionless form after differentiation of $\varsigma = [1+2\mathsf{F}_1^{-2}(1-Y^2)]^{-1}$

$$\varsigma' = \mathsf{F}_1^{-2}YY'\varsigma^2, \tag{5.25}$$

$$\varsigma'' = \mathsf{F}_1^{-2}\varsigma^2(YY''+Y'^2) + 2\mathsf{F}_1^{-2}YY''\varsigma\varsigma', \tag{5.26}$$

$$\varsigma''' = 4\mathsf{F}_1^{-2}\varsigma\varsigma'(YY''+Y'^2) + 2\mathsf{F}_1^{-2}YY'(\varsigma\varsigma''+\varsigma'^2) + \mathsf{F}_1^{-2}\varsigma^2(YY''+3Y''Y'). \tag{5.27}$$

For critical flow, $\mathsf{F} = 1$ or $q/\left(gh_c^3\right)^{1/2} = 1$, follows $\varsigma = \varsigma_c = \mathsf{F}_1^{2/3}$. With $X_c = (x - x_c)/L_r$ as the dimensionless streamwise distance from the critical (subscript c) point, where $X_c = 0$, the main flow profile is approximated by a Taylor polynomial up to third order as

$$\varsigma = \varsigma_c + \varsigma'_c X_c + \varsigma''_c \frac{X_c^2}{2} + \varsigma'''_c \frac{X_c^2}{6}. \tag{5.28}$$

The supercritical (subcritical) flow profile is computed in the upstream (downstream) direction, in contrast to standard backwater computations (Hager and Hutter, 1983). From Eq. (5.18), the derivatives of $Y(\chi)$ are with $Y_c = [1+2\mathsf{F}_1^2(1-\mathsf{F}_1^{-2/3})]^{1/2}$

$$Y' = \frac{(\varLambda-1)^3}{3}Y^{-2}, \tag{5.29}$$

$$Y'' = \frac{2(\varLambda-1)^3}{3}Y'Y^{-3}, \tag{5.30}$$

$$Y''' = \frac{2(\varLambda-1)^3}{3}(Y''Y^{-3} - 3Y'^2Y^{-4}). \tag{5.31}$$

The Taylor polynomial for the main flow profile computed from the critical point in the upstream and downstream directions is compared in Figure 5.9a with Eq. (5.19) for $\mathsf{F}_1 = 2$. The results are almost identical, confirming that the rules of standard backwater computation are not generally valid. The two approaches are further compared in Figure 5.9b for $\mathsf{F}_1 = 4$, again with excellent results for almost the entire roller lenght. As expected, higher-order terms need to be added to Eq. (5.31) as F_1 increases.

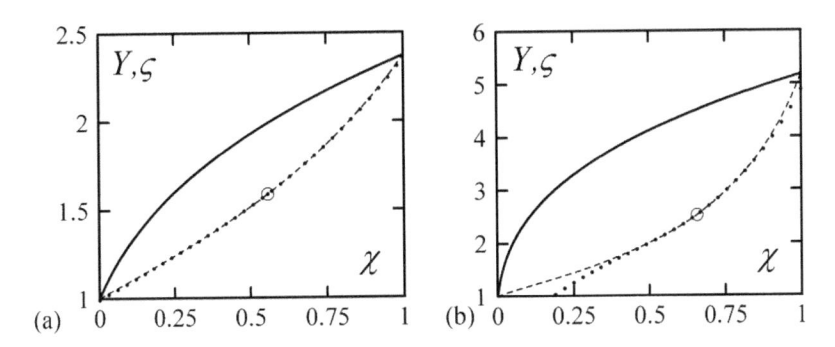

Figure 5.9 Main flow profiles from (○) critical point (—) $Y(\chi)$ from Eq. (5.18), (---) $\varsigma(\chi)$ from Eq. (5.19), (●) $\varsigma(\chi)$ from Eq. (5.34), for F_1 = (a) 2, (b) 4 (Castro-Orgaz and Hager, 2009)

Streamline curvature effects

The transition from F < 1 to F > 1, typically for weir flow, is dominated by streamline curvature and slope effects (Hager, 1985). In the reverse transition from F > 1 to F < 1, as for hydraulic jump flow, the effect of streamline curvature on the flow profiles needs consideration. With $h' = \mathrm{d}h/\mathrm{d}x$ and $h'' = \mathrm{d}^2h/\mathrm{d}x^2$, weakly curvilinear streamline flow follows the generalized specific momentum equation as (Hager and Hutter, 1984a)

$$S = \frac{(h+r)}{2} + \frac{q^2}{gh}\left(1 + \frac{hh''- h'^2}{3}\right). \tag{5.32}$$

With the scaling factor h_1/L_r accounting for the transformation $h' \to \varsigma'$ and $h'' \to \varsigma''$ involving different vertical and horizontal length scales, Eq. (5.32) yields in dimensionless form

$$\frac{1}{2} + F_1^2 = \frac{Y^2}{2} + \frac{F_1^2}{\varsigma}\left[1 + \left(\frac{h_1}{L_r}\right)^2 \frac{(\varsigma\varsigma''-\varsigma'^2)}{3}\right]. \tag{5.33}$$

Streamline curvature effects for the profile $\varsigma(\chi)$, whose zero-order (subscript $_{(0)}$) solution is with $Y_{(0)}$ given by Eq. (5.18)

$$\varsigma_{(0)} = \left[1 + 2F_1^{-2}\left(1 - Y_{(0)}^2\right)\right]^{-1}. \tag{5.34}$$

According to Hager (1985), the first order approximation to $\varsigma(\chi)$ results by estimating the profile derivations of the zero-order approach. From Eq. (5.34) with $\varsigma'_{(0)}$ and $\varsigma''_{(0)}$ given by Eqs. (5.25) and (5.26)

$$\varsigma_{(0)} = \left[1 + 2F_1^{-2}\left(1 - Y_{(0)}^2\right)\right]^{-1}\left[1 + \left(\frac{h_1}{L_r}\right)^2 \frac{(\varsigma\varsigma''-\varsigma'^2)_{(0)}}{3}\right]. \tag{5.35}$$

If streamline curvature effects are inversely set to $Y(\chi)$, one obtains from Eq. (5.33)

$$Y_{(1)} = \left[1 + 2F_1^{-2}\left\{1 - \varsigma_{(0)}^{-1}\left[1 + \left(\frac{h_1}{L_r}\right)^2 \frac{(\varsigma\varsigma'' - \varsigma'^2)_{(0)}}{3}\right]\right\}\right]^{1/2}. \qquad (5.36)$$

Equations (5.35) and (5.36) are plotted in Figure 5.10a together with the zero-order solution for $F_1 = 2.43$ using $h_1/L_r = 0.0074$ from Hager and Hutter (1983). Streamline curvature effects are seen to be small, indicating that both $Y(\chi)$ and $\varsigma(\chi)$ are so weakly curved and sloped that there is practically no effect on these profiles. Figure 5.10b further presents this effect for $F_1 = 5.08$ using $h_1/L_r = 0.0334$ from Schröder (1963). Note that streamline curvature effects increase with F_1, but remain small, except for the extremes of the hydraulic jump. The small value of the scaling $h_1/L_r = 0.0364$ indicates the magnitude of the average free surface slope, namely $h' \approx (h_2 - h_1)/L_r = (h_1/L_r)(0.918\lambda - 1) \approx 0.187$ (10.6°→1V:5.35H). The forward and free surface profiles of a hydraulic jump are thus governed by the hydrostatic pressure distribution, with a minor effect of higher-order Boussinesq terms, thereby simplifying the computational effort for these otherwise complicated two-phase flows.

This research demonstrates the following main issues:

- Free surface and roller flow profiles are predicted based on the conservation of horizontal, vertical, and angular momentums within the jump.
- Model corresponds to the lowest possible order, yet it yields resonable results for practice. The 2D flow features are favorably compared with experimental data.
- Hydraulic jump flow is rationally related to the critical flow condition. The specific energy of the main flow zone reaches its minimum at $F = 1$, whereas total energy head decreases due to energy dissipation.
- Effects of streamline curvature and two-phase flow features on pressure field and momentum flux are small because of small deviations from 2D model.
- Effect of non-uniform velocity distribution, however, impacts flow structure.
- Both the two-phase flow features and turbulence contribution of Reynolds stresses are small only if non-uniform velocity effects are correctly treated.

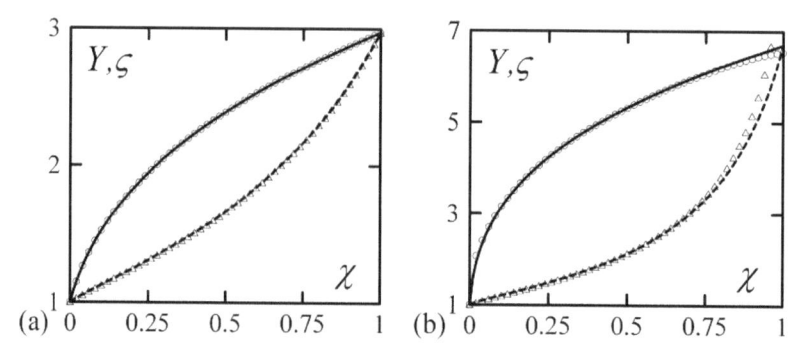

Figure 5.10 Streamline curvature effect (—) $Y(\chi)$ from Eq. (5.18), (---) $\varsigma(\chi)$ from Eq. (5.19), (○) $\varsigma(\chi)$ from Eq. (5.35), (Δ) $Y(\chi)$ from Eq. (5.36) for $F_1 = $ (a), 2.43, (b) 5.08

5.2.3 Undular hydraulic jump

General

The undular jump constitutes a particular transition from super- to subcritical flows, referred also to as weak hydraulic jump. In contrast to the direct hydraulic jump, its effectiveness in terms of energy dissipation is small, its compactness is poor because of extreme length, and its stability is weak, given that it is displaced from a certain location due to extremely small variations of either the discharge or the tailwater elevation. Undular jumps are therefore avoided in hydraulic engineering, but they still can occur so that the main flow features should be known.

Despite these jumps have been studied in the 19th century, the main features have only been described recently. Henry Bazin (1829–1917) was the first to describe the hydraulic characteristics in his *Recherches hydrauliques*, published in 1865. Joseph V. Boussinesq (1842–1929) was the first to theoretically tackle the problem based on the generalized open channel flow equations by accounting for streamline curvature. He demonstrated that undular hydraulic jumps are steady waves, referred to as the cnoidal wave. This simplified approach is restricted to small approach flow Froude numbers F_1 only, for which the phenomenon remains two-dimensional. As F_1 increases, the jump becomes spatial, thereby complicating the theoretical approach.

Figure 5.11 shows a sketch of the undular hydraulic jump, with section '1' again referring to the approach flow, with h_1 as the approach flow depth, $V_1 = Q/(bh_1)$ as the average approach flow velocity in a rectangular channel of width b, so that $F_1 = V_1/(gh_1)^{1/2}$. Given the 3D flow, a distinction is made between the axial and the wall surface profiles; let subscript M describe the wave maxima, and subscript m the wave minima, with h_{1M}, h_{2M}, etc. as axial wave maxima, and h_{3M}, h_{4M}, etc. as wall maxima. In turn, the wave minima are described by h_{1m}, h_{2m}, etc. along the channel axis, whereas these along the channel wall are denoted as h_{3m}, h_{4m}, etc.

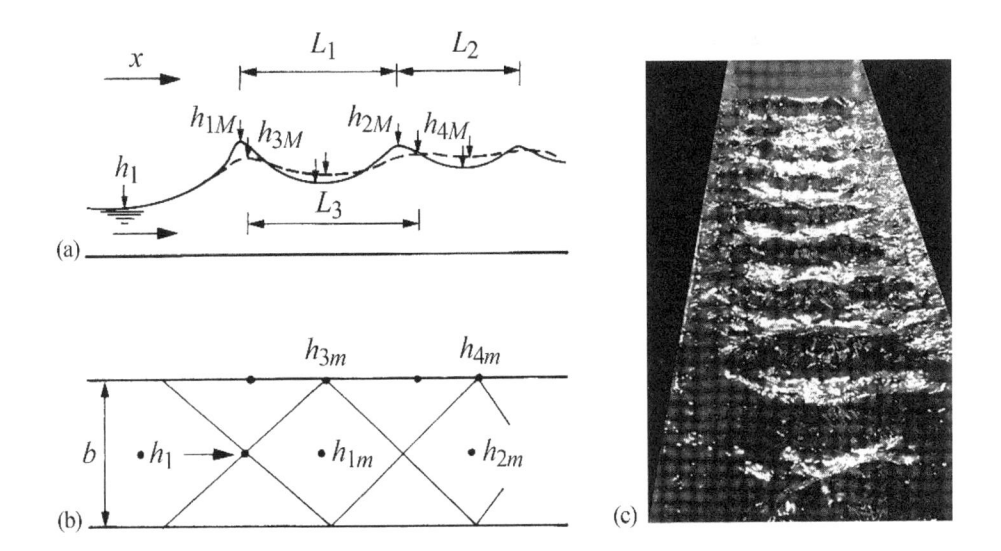

Figure 5.11 Undular hydraulic jump, description of flow pattern (a) streamwise section with (–) axial and (---) wall profiles, (b) plan, (c) photo (Reinauer and Hager, 1995)

A model was developed based on k-ε turbulence closure. Depth-averaged equations were applied by employing a suitable time-averaged velocity distribution based on a composite power-law model, in which both streamline curvature and vorticity are accounted for. The bed-shear stress closure was included by a boundary layer method. Predictions of the Reynolds Averaged Navier-Stokes (RANS) model are close to the 2D RANS solution and the experimental data.

Consider an undular jump along a plane bottom of inclination α. The Cartesian coordinate system (x, y) with velocity components (u, v) is defined in Figure 5.14. The RANS momentum equation in x-direction is (Rodi, 1993; Steffler and Jin, 1993)

$$\frac{\partial u^2}{\partial x} + \frac{\partial}{\partial y}(uv) = -\frac{1}{\rho}\frac{\partial p}{\partial x} + \frac{1}{\rho}\left(\frac{\partial \tau_{yx}}{\partial y} + \frac{\partial \sigma_x}{\partial x}\right) + g\sin\alpha. \tag{5.41}$$

The tangential Reynolds stress is τ_{yx}, the normal turbulent stress in the x-direction is σ_x, and the time-averaged fluid pressure is p. This can be rewritten as

$$\frac{\partial}{\partial x}\left(\frac{u^2}{g} + \frac{p - \sigma_x}{\gamma}\right) = -\frac{1}{g}\frac{\partial}{\partial y}\left(uv - \frac{\tau_{yx}}{\rho}\right) + \sin\alpha. \tag{5.42}$$

To integrate Eq. (5.42) over the time-averaged flow depth $h(x)$, kinematic and dynamic boundary conditions need to be prescribed. At the channel bottom ($y = 0$), the non-slip kinematic boundary conditions are $u = v = 0$. At the time-averaged water surface ($y = h$), the kinematic boundary condition reads (Grillhofer and Schneider, 2003; Jurisits and Schneider, 2012)

$$v = u\frac{dh}{dx}. \tag{5.43}$$

Under identical conditions, the dynamic boundary condition in x-direction at a turbulent free surface is (Grillhofer and Schneider, 2003; Jurisits and Schneider, 2012)

$$\tau_{yx} = (\sigma_x - p)\frac{dh}{dx}. \tag{5.44}$$

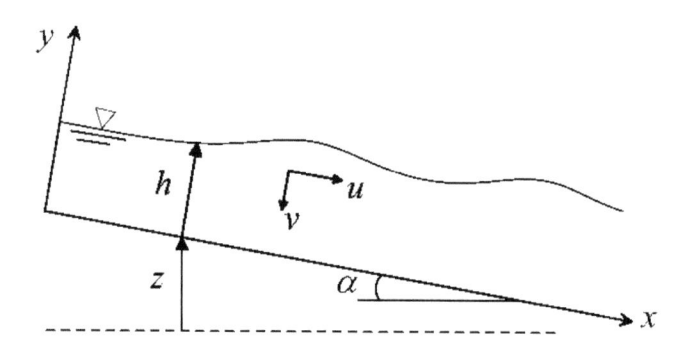

Figure 5.14 Definition sketch of undular hydraulic jump

This follows from integrating the Navier-Stokes equation in x-direction between the instantaneous free surface and the time-averaged free surface h. Integrating Eq. (5.42) from $y = 0$ to $y = h$, and imposing the kinematic and dynamic boundary conditions leads with the bed-shear stress τ_b to the depth-averaged momentum equation

$$\frac{dS}{dx} = h \sin \alpha - \frac{\tau_b}{\gamma}. \tag{5.45}$$

The specific momentum S for turbulent flow is

$$S = \int_0^h \left(\frac{u^2}{g} + \frac{p - \sigma_x}{\gamma} \right) dy. \tag{5.46}$$

A bottom drag model and a closed-form S function from Eq. (5.46) are required to solve Eq. (5.45). The distributions of velocity u and pressure p must be prescribed, in addition to introducing a turbulent closure for σ_x.

The pressure distribution from Leibniz' rule is (Castro-Orgaz and Hager, 2011)

$$\frac{p}{\gamma} = (h - y) \cos \alpha + \frac{\sigma_y}{\gamma} - \frac{v^2}{g} + \frac{1}{g} \frac{\partial}{\partial x} \int_y^h \left(uv - \frac{\tau_{yx}}{\rho} \right) dy. \tag{5.47}$$

Inserting Eq. (5.47) into Eq. (5.46), the variable pressure is eliminated from the governing equations and the problem is reduced to find the time-averaged velocity field (u, v) and the turbulence closure. The latter is used to relate the Reynolds stresses to the time-averaged velocity field.

The turbulence closure follows from the eddy viscosity concept (Rodi, 1993). As the turbulence closure is based on depth-averaged quantities, the computation of the Reynolds stress is further simplified by introducing depth-averaged gradients, i.e. $\partial u/\partial x \approx \partial U/\partial x$. Limiting the free surface contributions to the leading order terms hh_{xx} and h_x^2, thereby dropping higher-order terms from Taylor expansions and products of derivatives originating from $\partial v/\partial x$, Castro-Orgaz and Hager (2011) find

$$S = \frac{h^2}{2} \cos \alpha + 4 \frac{V_t}{g} \frac{q}{h} h_x + \int_0^h \frac{u^2}{g} dy - \int_0^h \frac{v^2}{g} dy + \frac{1}{g} \int_0^h \left\{ \frac{\partial}{\partial x} \int_y^h uv dy \right\} dy. \tag{5.48}$$

The time-averaged velocity field of undular hydraulic jumps is modeled by accounting for both vorticity and streamline curvature effects as

$$u = U (1 + N) \eta^N \left[1 + \left(\frac{hh_{xx}}{2} - h_x^2 \right) \left(\eta^2 - \frac{1}{3} \right) \right]. \tag{5.49}$$

Inserting this in the continuity equation permits to compute the corresponding vertical velocity profile. Considering only the leading terms results in (Bose and Dey, 2007)

$$v = U h_x \eta^{N+1}. \tag{5.50}$$

Inserting Eqs. (5.49) and (5.50) into Eq. (5.48) produces

$$S = \frac{h^2}{2}\cos\alpha + 4\frac{V_t}{g}\frac{q}{h}h_x + \frac{q^2}{gh}\left(\lambda_1 + \lambda_2 h h_{xx} - \lambda_3 h_x^2\right).$$
(5.51)

Here

$$\lambda_1 = \frac{(1+N)^2}{2N+1},$$
(5.52)

$$\lambda_2 = \frac{1+N}{2N+3} + \frac{4N(1+N)^2}{3(2N+3)(2N+1)},$$
(5.53)

$$\lambda_3 = \frac{1}{2N+3} + \frac{8N(1+N)^2}{3(2N+3)(2N+1)}.$$
(5.54)

Based on vorticity transport along the streamlines, Castro-Orgaz and Hager (2011) find a theoretical expression for N as

$$N = N_1\left(\frac{h}{h_1}\right)^2.$$
(5.55)

This simple equation applies for N closure and avoids the need of additional transport equations. Upstream uniform flow conditions are considered at approach flow section 1, from which N_1 follows with $\kappa = 0.41$, the friction factor f_o using Haaland's equation (White, 1991), with $R_1 = q/v$ as Reynolds number and v as kinematic viscosity as

$$N_1 = \frac{1}{\kappa}\left(\frac{f_o}{8}\right)^{1/2},$$
(5.56)

$$f_o = \left[-1.8\log_{10}\left(\frac{6.9}{4R_1}\right)\right]^{-2}.$$
(5.57)

Castro-Orgaz and Hager (2011) also model the bottom shear closure.

The measured free surface profile for $F_1 = 1.11$ and a bottom slope of 1/282 is plotted in Figure 5.15 (Gotoh et al., 2005). This is an adequate test case for the depth-averaged undular jump model, given the absence of crosswaves, leading to a perfect 2D jump structure. The numerical simulations of Schneider et al. (2010) are also included. Herein, the depth-averaged RANS model with the simplified k-ε turbulence model were solved numerically for the unknowns $S(x)$, $k(x)$, $\varepsilon(x)$, $h_x(x)$, and $h(x)$ using the fourth-order Runge-Kutta method to solve a system of ordinary differential equations, taking section 1 as starting point along with $h_1 = 0.0953$ m and the free surface slope $h_x = 0$. The power-law exponent N_1 was computed from Eq. (5.56) using Eq. (5.57) for f_o. The coefficients λ_1, λ_2, and λ_3 were computed from Eqs. (5.52)–(5.54), resulting in $k_1 = h_1^2\left(f_o/8\right)^{-2}\left(q/h_1\right)^{-4}\varepsilon_1^2 c_{2\varepsilon}c_{\varepsilon}^{-1}$. The normalized free surface profile $h/h_c[x/h_c]$ with $h_c = (q^2/g)^{1/3}$ computed from the depth-averaged model is plotted in Figure 5.15.

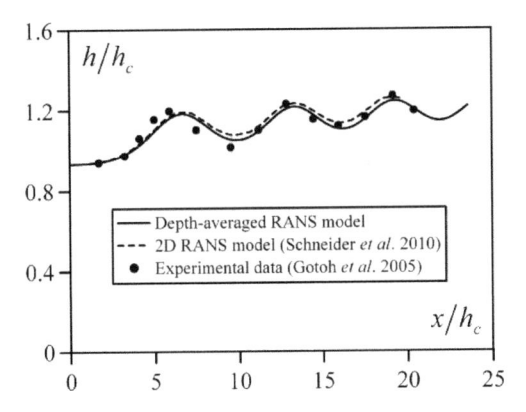

Figure 5.15 Free surface profile $h/h_c[x/h_c]$ of undular hydraulic jump for $F_1 = 1.11$ (Castro-Orgaz and Hager, 2015)

The results of the depth-averaged RANS model are in excellent agreement with the 2D results (Schneider *et al.*, 2010) and with the experimental data (Gotoh *et al.*, 2005). Note how the depth-averaged model follows the undular 2D RANS pattern, keeping a similar wave length and amplitude.

To compare with previous solutions, consider the Serre (1953) model

$$S = \frac{h^2}{2}\cos\alpha + \frac{q^2}{gh}\left(1 + \frac{hh_{xx} - h_x^2}{3}\right), \quad \frac{dS}{dx} = h\sin\alpha - \frac{f_o}{8g}\frac{q^2}{h^2}. \tag{5.58}$$

It uses the depth-averaged specific momentum S of potential flow, and the streamwise momentum balance including bottom friction and a standard friction factor formulae. The simulations corresponding to Eqs. (5.58) are plotted in Figure 5.16 for comparative purposes. Note that Serre's model produces a first wave crest close to the 2D RANS model, which is overpredicted, however. Along the entire wave train, Serre's model produces over- and underpredictions of wave crests and troughs, respectively. The depth-averaged model produces underpredictions of both wave crests and troughs, but the results are closer to 2D data. Serre's model diverges from the 2D RANS model in wave phase, contrary to the new depth-integrated RANS model, producing a wave train almost in phase with the full 2D RANS solution.

Further effort was directed to compare bed-shear stresses and the momentum thickness profile. Velocity profiles at typical sections and pressure distributions were also determined. The following statements apply:

- Undular hydraulic jump is a turbulent flow originally treated by the potential flow theory (Mandrup-Andersen, 1978), or potential flow models, in which the energy loss was included in the streamwise energy balance (Serre, 1953; Marchi, 1963; Hager and Hutter, 1984b). An advanced method relies on depth-averaging of the RANS equations (Steffler and Jin, 1993; Bose and Dey, 2007). Herein, the general depth-integrated RANS equations are solved with a turbulence closure based on eddy viscosity.

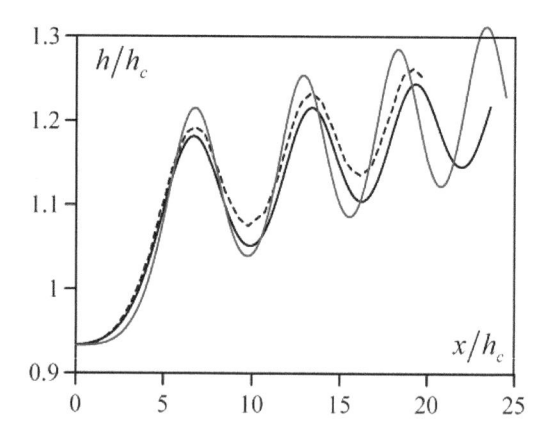

Figure 5.16 Comparison of free surface profile $h/h_c[x/h_c]$ with solutions of undular jump for
$F_1 = 1.11$ with (—) depth-averaged RANS, (—) Serre, (−−) 2D RANS models
(Schneider *et al.*, 2010)

- Present model is based on momentum conservation of a turbulent flow by using the energy equation to solve turbulent flow structure. Yet, the specific energy equation contains additional unknowns so that it is preferable to use a momentum-based model.
- Turbulence inclusion in the undular jump problem was generalized by introducing Reynolds stresses both in horizontal and vertical directions (Khan and Steffler, 1996). Turbulence closure for this problem employs a simplified, depth-averaged k-ε model.
- No specific constrains in terms of F or R were set up in the present model, so that the model equations apply to finite values of R, and to F different from unity.
- Composite velocity power law compares well with data, including the interaction of streamline curvature with frictional effects, and providing improved predictions as compared with a single power law.
- Depth-averaged open channel flow models usually compute boundary shear stress using standard friction formulae. An alternative is proposed by using the N index. An averaged friction factor closure is developed for depth-averaged computations.
- Depth-averaged, turbulent open channel flow models are usually based on the hydrostatic pressure approach. Computational solutions involving shocks thus cannot physically reproduce undulating patterns due to non-hydrostatic pressure distribution. The present model includes this effect and applies it as sub-model into depth-averaged hydrostatic approach. It also applies to generate the initial free surface profile in CFD codes solving the 2D turbulent, non-hydrostatic RANS equations in a vertical plane.
- Laboratory investigations of free surface flow problems usually relates to Froude similitude. Avoiding scale effects originating from real fluid flow effects are a major concern. For given F_1, the approach flow h_1 depth should not fall below a threshold (Reinauer and Hager, 1995), so that viscous effects are neglected in the physical model. The present depth-averaged model is a real fluid flow predictor describing for given F_1 the viscous effect for any model scale, so that no specific limits are required.

5.3 Stilling basins

5.3.1 General

In a stilling basin excess hydromechanical energy is converted mainly into heat, spray, and sound. The stilling basin is a hydraulic structure located between the outlet works of a spill-way and the tailwater, to where it should safely return the excess flow. It is a structure in which a hydraulic jump is generated, economically designed in terms of length, tailwater level, tailwater wave generation, and scour.

The selection of a stilling basin depends on factors as:

- Hydraulic approach flow conditions
- Tailwater characteristics
- Scour potential
- Site-specific constraints.

The approach flow energy head should range between 10 and 30 m, allowing for a successful basin performance. A number of standard basins, tested extensively both in the laboratory and at prototypes, is available (Peterka, 1958). Problems with stilling basins typically occur under high approach flow velocity >25 m/s, approach flow Froude numbers <2.5, asymmet-ric approach flow conditions, curved inflow or outflow, or low tailwater level. In general, the minimum tailwater level should at least be equal to the sequent depth given by Eq. (5.4). If this is not the case, the basin has to be deepened and a positive step has to be inserted. Baffles are provided mainly to shorten the hydraulic jump without gaining additional tailwater level.

Compared with the hydraulic jump basin, in which the approach flow momentum is bal-anced by an adequate tailwater level, the stilling basin has in addition baffle elements located on the basin bottom, involving steps, sills, or blocks. The dissipation can be increased with a diverging basin. Increasing the wall roughness or adding lateral discharge has not received attention in practice, mainly because of problems with negative pressures, dynamic uplift, and jump stability. Figure 5.17 shows a number of elements proposed to affect one or the other aspect of stilling basin flow.

Among the large variety of basin shapes proposed (Vischer and Hager, 1995), both the baffle-sill and the baffle-block basins are certainly the most popular designs. These elements are prone to approach flow heads >30 m due to cavitation damage. In contrast, abrasion is not a concern for basins downstream of overflow structures but for bottom outlets. For stilling basins sub-jected to flows with a high-abrasion potential, baffles should have steel armoring, or a simple basin should be provided. Below, baffle basins together with the abruptly expanding basin are discussed. The USBR standard basins as introduced by Peterka (1958) are not redescribed here.

5.3.2 Baffle-sill basin

The baffle-sill basin involves a transverse sill of height s of minimum width for structural resistance. For given approach flow depth h_1 and Froude number F_1, various flow types result, depending on the relative sill height $S_s = s/h_1$ and tailwater level h_2:

- *A-jump* with roller end above sill
- *B-jump* with a lower tailwater level, a surface boil on sill, and roller extending into tailwater

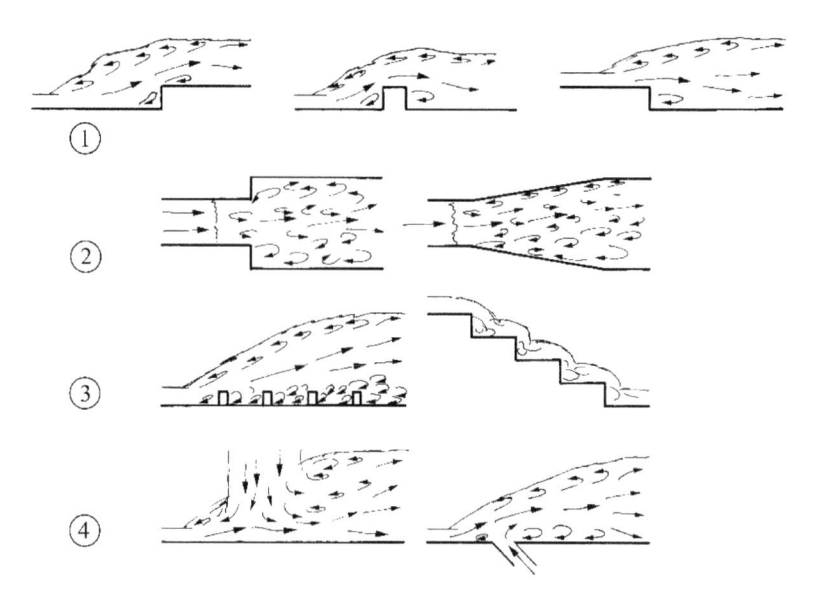

Figure 5.17 Basic elements of stilling basins (Vischer and Hager, 1995) involving ① bottom geometry, ② plan geometry, ③ boundary roughness, ④ discharge addition

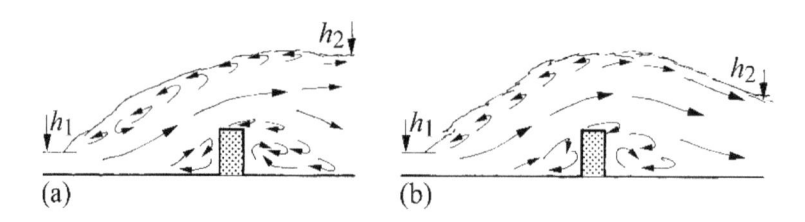

Figure 5.18 Stilling basin with (a) sufficient submergence and effective dissipation (B-jump), (b) insufficient tailwater submergence (C-jump) (Adapted from Vischer and Hager, 1998)

- *Minimum B-jump* with a secondary roller, and plunging flow beyond sill not reaching basin bottom
- *C-jump* with plunging flow causing inappropriate tailwater flow, and adding to scour potential
- *Wave type flow* with supercritical flow over sill and inacceptable dissipation.

Both the A- and B-jumps are effective for stilling basins because of sufficient tailwater submergence, whereas the C-jump and the wave type flow are inadequate. Figure 5.18 shows the significance of tailwater submergence in stilling basin design. This example illustrates that a slight reduction of tailwater level below the sequent depth from Eq. (5.4) has dramatic consequences on the basin performance. The purpose of any baffle element should thus involve a *length reduction* but no significant reduction of the tailwater level.

Figure 5.19 compares the baffle-sill basin with the hydraulic jump basin. The sill is defined by its relative height $S_s = s/h_1$ and relative sill location $\Lambda_s = L_s/L_r^*$. The sequent depth ratio required $Y_s = Y^* - \Delta Y_s$ includes the effects of classical jump, and of the sill as (Hager, 1992)

$$\Delta Y_s = 0.7 S_s^{0.7} + 3 S_s (1 - \Lambda_s)^2. \tag{5.59}$$

For any sill height S_s, a minimum (subscript m) approach flow Froude number F_{1m} is required for jump formation, with the corresponding maximum (subscript M) sill height S_M for any approach flow Froude number

$$S_M = \frac{1}{6} F_1^{5/3}. \tag{5.60}$$

The relative sill height should be limited in practice to $S_M = 2$. Also, the sill should neither be too small nor too large to inhibit ineffectiveness or jump overforcing. The optimum (subscript opt) relative sill height S_{opt} is

$$S_{opt} = 1 + \frac{1}{200} F_1^{2.5}. \tag{5.61}$$

Depending mainly on the relative sill position Λ_s, three jump types occur:

1. $\Lambda_s > 0.8$ (to 1): *A-jump* suitable for easily erodible beds
2. $0.65 < \Lambda_s < 0.8$: *B-jump* with small erosion mainly along tailwater walls
3. $0.55 < \Lambda_s < 0.65$: *minimum B-jump* only suitable for rocky tailwater channels.

The jump length L_j from the toe to the end of the bottom roller relative to the length of the classical jump L_j^* is

$$L_j / L_j^* = 1 - 0.6 S^{1/3} (1 - \Lambda_s). \tag{5.62}$$

The basin length L_B (Figure 5.19b) for all three flow types is slightly shorter than that of the classical jump. A sill basin stabilizes a hydraulic jump under variable tailwater. The effect

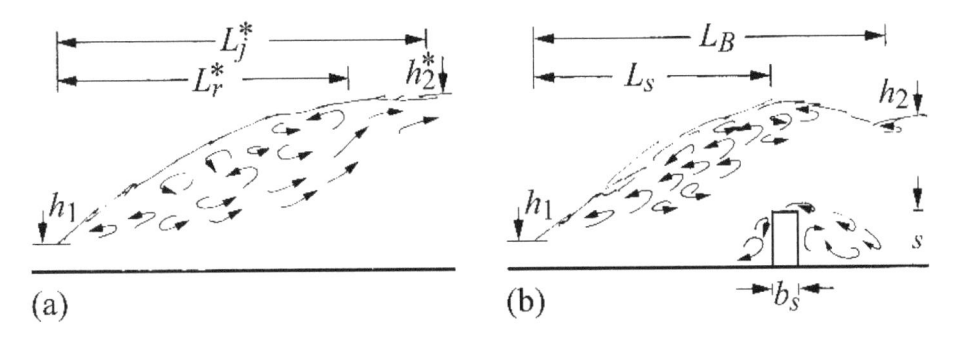

Figure 5.19 Comparison of (a) classical hydraulic jump with (b) baffle-sill basin

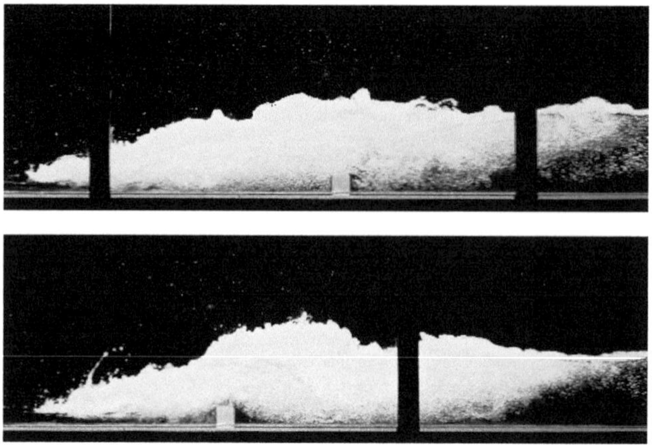

Figure 5.20 Photographs of (a) A-jump, (b) B-jump for $F_1 = 5$ (Adapted from Vischer and Hager, 1998)

of additional appurtenances is discussed below. Figure 5.20 shows photos of the A- and B-jumps.

5.3.3 Baffle-block basin

As for the baffle-sill basin, a baffle-block basin involves various flow types. For optimum basin flow, the blocks must have an adequate location and height to counter ineffective or overforced flow. According to Basco (1971), the optimum (subscript opt) height $S_{opt} = s_{opt}/h_1$ and the optimum basin length are, respectively,

$$S_{opt} = 1 + \frac{1}{40}(F_1 - 2)^2,$$ (5.63)

$$(L_B / h)_{opt} = 1.6 + 7.5F_1^{-2}.$$ (5.64)

Figure 5.21 shows the basin provided with standard USBR blocks, of spacing equal to the block width $e = b_B$ and $e/s = 0.75$. The force F_B on the blocks involves the force coefficient $\Phi = F_B/[\rho g b h_2^{*2}/2]$; for optimum basin performance

$$\Phi_{opt} = \frac{1}{7} + \frac{F_1}{100}.$$ 5.65)

The sequent depth ratio obtains

$$Y_B = \left(\frac{2}{1+\Phi}\right)^{1/2} F_1 - \frac{1}{2}.$$ (5.66)

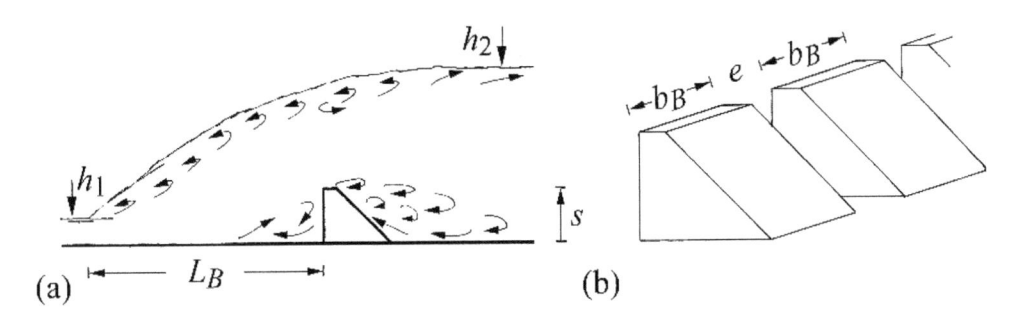

Figure 5.21 Baffle-block basin (a) side view, (b) standard block geometry

The tailwater reduction is thus over 10% as compared with the classical jump. A staggered block row is insignificant in terms of basin performance. Details on the pressure character- istics are given by Hager (1992). Figure 5.22a–c refers to a typical laboratory-tested stilling basin. Note that the blocks are exposed to high-speed flow so that they are prone to cavitation formation. Instream-oriented surfaces generating positive pressure are favorable, as well as rounded or chamfered edges, eventually reinforced by steel profiles, for edges oriented in the streamwise direction (Figure 5.22d).

5.3.4 Abruptly expanding stilling basin

Stilling basins expand behind outlets due to partial operation, or because of a width increase from the approach flow channel to the tailwater. In practice, the abrupt expansion is of inter- est due to structural chute compactness. This design is considered here with reference to Hager (1992) for gradually expanding basins.

At an abrupt expansion with h_1, b_1, F_1 as the approach flow conditions and b_2 as the tailwa- ter width, various types of flows occur (Figure 5.23):

- *R-jump* with supercritical flow in expansion and hydraulic jump in tailwater
- *S-jump* with toe of jump more upstream but still in expanding reach and generation of an oscillating or even asymmetric jet flow
- *T-jump* with jump toe in approach flow channel and jump body in expansion.

Although the R-jump is a stable configuration, it may develop into the S-jump under slightly increasing tailwater. The S-jump is highly spatial, unstable, and excessively long. For an expanding basin, only the T-jump is acceptable.

The sequent depth ratio $Y = h_2/h_1$ of the T-jump depends on the relative toe location $X_1 = x_1/L_r^*$ with L_r^* as roller length of the classical jump, the width ratio $\beta = b_2/b_1$ and the classical sequent depth ratio Y^* (Figure 5.24a). According to Hager (1992)

$$\frac{Y^*-Y}{Y^*-1} = (1 - \beta^{-1/2})[1 - \tanh(1.9X_1)]. \tag{5.67}$$

Figure 5.22 Stilling basin with chute blocks, baffle blocks and end sill at design discharge
(a) side view, (b) approach flow detail, (c) tailwater view (Vischer and Hager,
1998), (d) stilling basin downstream of gated weir, with chute and reinforced
baffle blocks (Courtesy Anton J. Schleiss)

For $\beta = 1$ the asymptotic result is $Y = Y^*$. For $X_1 > 1.3$, the jump end is located upstream from
the expansion section, so that $Y = Y^*$ (Figure 5.23d).

The efficiency of the T-jump increases as X_1 decreases, but its performance decreases in
parallel so that the T-jump is not an effective dissipator. The performance of expanding stilling basins is improved with a baffle sill. The comments below refer also to basins with an
expansion angle larger than 30° (Bremen and Hager, 1994).

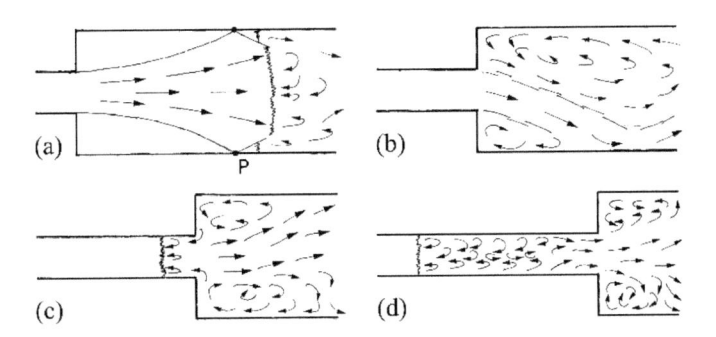

Figure 5.23 Flow types in abruptly expanding stilling basin without bottom elements (a) R-jump, (b) S-jump, (c) T-jump, (d) classical jump in approach flow channel

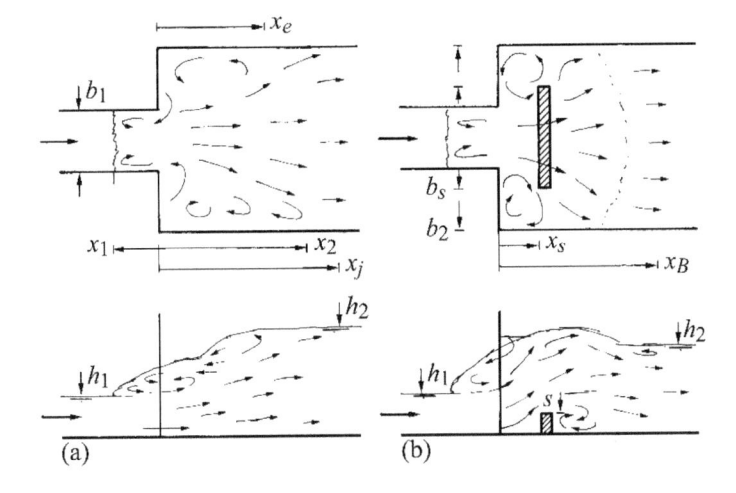

Figure 5.24 Abruptly expanding stilling basin (a) without, (b) with central sill. Plan (top), section (bottom)

The *optimized* expanding stilling basin (Figure 5.24b) involves a transverse central sill of width

$$b_s / b_1 = 1 + (1/4)(\beta - 1). \tag{5.68}$$

The sill forces the approach flow jet to expand and induces two corner vortices (Figure 5.25). The sill has a relative position $X_s = x_s / L_r^*$ and a relative height $S_s = s/h_1$ correlated to the approach flow Froude number $\mathsf{F}_1 = Q/\left(gb_1^2 h_1^3\right)^{1/2}$ as

$$X_s = \frac{3}{4}(S_s / \mathsf{F}_1)^{3/4}. \tag{5.69}$$

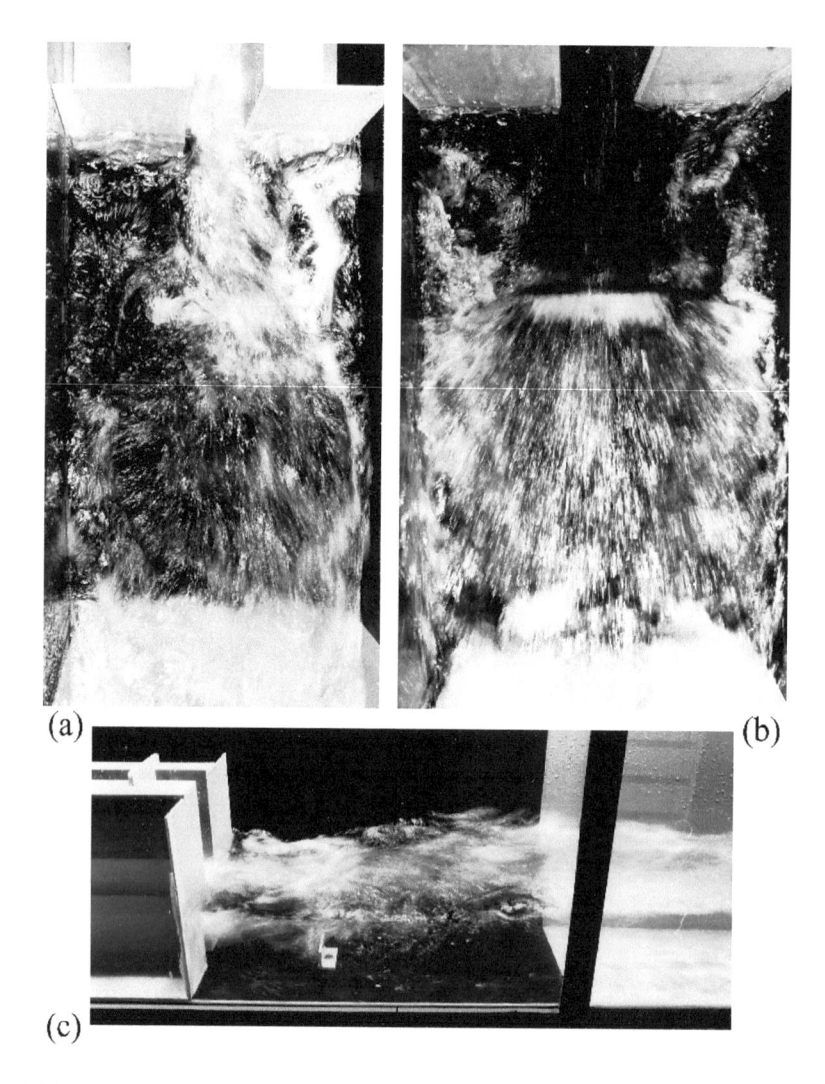

Figure 5.25 Expanding stilling basin (a) without, (b) with corner vorticity well-developed, (c) side view ($F_1 = 5$, $\beta = 5$) (Bremen and Hager, 1994)

The sill height increases with increasing X_s, and F_1, therefore. The sill position should be contained within

$$0.4 < \frac{x_s / b_1}{(\beta - 1)} < 0.6. \tag{5.70}$$

A reasonable relative sill height is

$$S_s < (1/2)(F_1 - 1). \tag{5.71}$$

The basin length x_B is equal to the roller length L_r^* if a conventional end sill is added. Conditions to be satisfied include (Hager, 1992): (1) $3 < \mathsf{F}_1 < 10$ for the approach flow, (2) $1 < \beta < 5$ for width ratio, and (3) $0.1 \leq X_s \leq 0.6$ for relative sill position, next to those for stilling basins in general. Figure 5.25 shows the performance of the expanding stilling basin in a laboratory channel.

5.3.5 Slotted-bucket stilling basin

Beichley and Peterka (1959) developed the slotted-bucket stilling basin with an 8° sloping apron on which teeth of 45° are mounted (Figure 5.26). These generate stable flow and little boil action. Three flow types are distinguished:

- Sweep-out due to too low tailwater level
- Minimum tailwater level because of excessive surface waves and scour
- Maximum tailwater level above which dividing flow results.

The slotted-bucket basin has a lower and an upper *limit of operation*. These depend on the approach flow Froude number $\mathsf{F}_1 = V_1/(gN_1)^{1/2}$ and the relative bucket radius $\rho_b = (R_b/N_1)$ $[1 + (1/2)\mathsf{F}_1^2]$, with V_1 as approach flow velocity, N_1 flow depth measured perpendicularly to the chute bottom, and R_b as bucket radius (Figure 5.26c).

The *minimum* bucket radius should be $R_{bm}/N_1 = 2.2\mathsf{F}_1^{1/2}$; the extreme tailwater levels t_b/N_1 are given in Figure 5.27 versus ρ_b and F_1. Material entering the bucket may cause abrasion damage. Figure 5.28 refers to a basin with optimum roller action.

5.3.6 Basin characteristics

Tailwater level

Any stilling basin has to perform satisfactorily under various approach flow and tailwater characteristics. If the tailwater depth is too low, *sweep-out* results along with a significant tailwater scour. It is imperative that such poor flow never occurs under any flow scenario, because of large-scale scour in the tailwater. For all basins in which the tailwater depth is lower than 90% of the sequent depth of the classical hydraulic jump, the design should be verified in sufficiently large models. Damage of a stilling basin has to be countered under

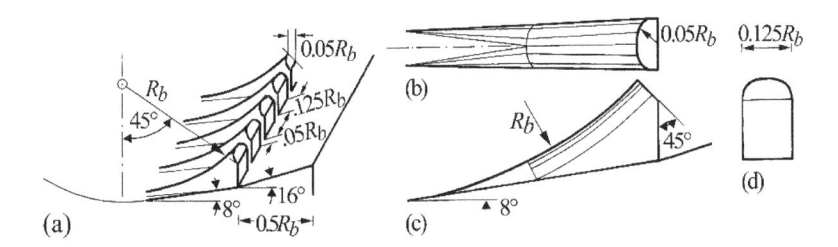

Figure 5.26 Slotted-bucket stilling basin, notation and geometry (a, c, d) side views, (b) plan view

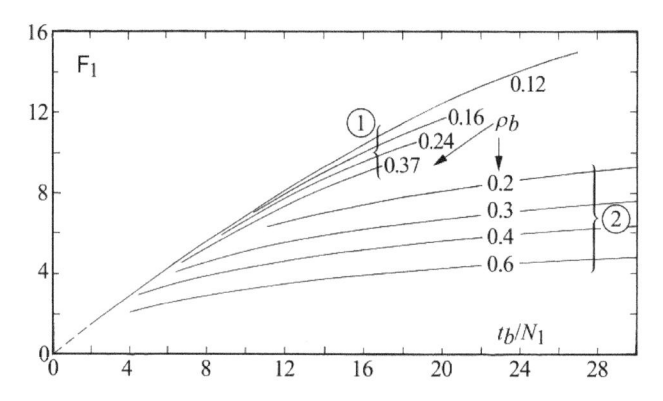

Figure 5.27 Extreme tailwater levels for *slotted* stilling basin $t_b/N_1(F_1;\ \rho_b = (R_b/N_1)\ [1 + (1/2)F_1^2])$ with ① minimum and ② maximum tailwater depths

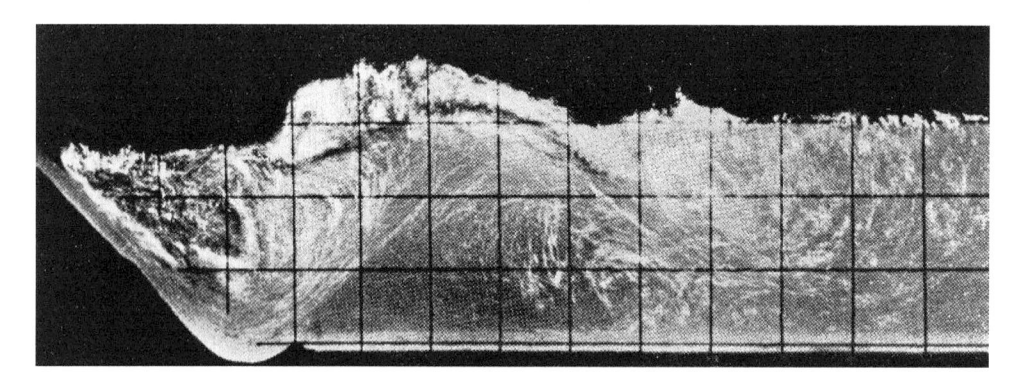

Figure 5.28 Slotted-bucket stilling basin in laboratory model (*L'Acqua* 1970, 48(5), 125)

any circumstances given the implications to the safety of the entire dam. The tailwater depth-discharge relation has thus to be known before the selection of the stilling basin. Other aspects to be satisfied are cavitation and dynamic uplift resistance, scour control, and tailwater waves.

Control of cavitation and dynamic uplift

Cavitation damage occurs due to turbulent pressure fluctuations with instantaneous values below vapor pressure. Two cases are of concern:

1   Zone of peak turbulence in front basin portion ($x/L_r^* < 0.4$)
2   Zone of appurtenances, i.e. at the rear of baffle sills and baffle blocks.

The *lining* of a stilling basin is related to the pressure fluctuations; a minimum slab thickness should be $0.27(V_1^2/2g)$. The two design conditions are (Hager, 1992):

1   Design discharge including unbalanced uplift force
2   Maximum reservoir level with empty stilling basin.

Detailed guidelines for the design of stilling basin linings with both sealed and unsealed joints considering dynamic uplift provide Barjastehmaleki *et al.* (2016a–c).

Tailwater scour control

The transition from a stilling basin to the unprotected tailwater is important relative to scour. Its potential is greatly reduced by providing suitable end sills deflecting bottom currents to the water surface thereby inducing a bottom return current (Figure 5.29). According to Novak (1955) and denoting by z_t the tailwater bed elevation relative to the basin bottom, and by H_1 the dam height above the stilling basin:

- *Apron angle $\alpha_a = 20°$* minimizes scour area and location of maximum scour
- *Sill width t_s* should correspond to structural minimum
- *Sill height s* should satisfy conditions $z/s \geq 0.4$ and $(s - z_t)/H_1 < 0.14$
- *Basin length L_B* should be equal to roller length L_r^* of classical jump
- *Submergence degree* $(h_2 + z_t)/h_2 = 1.05$ to 1.1 is optimum.

The optimum size of *riprap* was studied by Peterka (1958).

Tailwater waves

Stilling basins at $2.5 < F_1 < 4.5$ are prone to tailwater waves (Figure 5.30a). Abou-Seida (1963) studied the relative wave height h_w/h_2, and wave steepness $\sigma_w = h_w/(gt_w^2)$ with t_w as the wave period. The inverse *wave Froude number* $F_w^{-2} = (gh_w)/V_1^2$ and the dimensionless wave period $T_w = (gt_w)/V_1$ vary with the sequent depth ratio Y and F_1. The effect of submergence $Y_w = h_2/h_1$ is again significant (Figure 5.30b–c).

Field experience

Berryhill (1963) studied various stilling basins and arrived at:

- Tailwater depth should at least be equal to sequent depth of classical jump
- Adequate tailwater submergence shortens basin

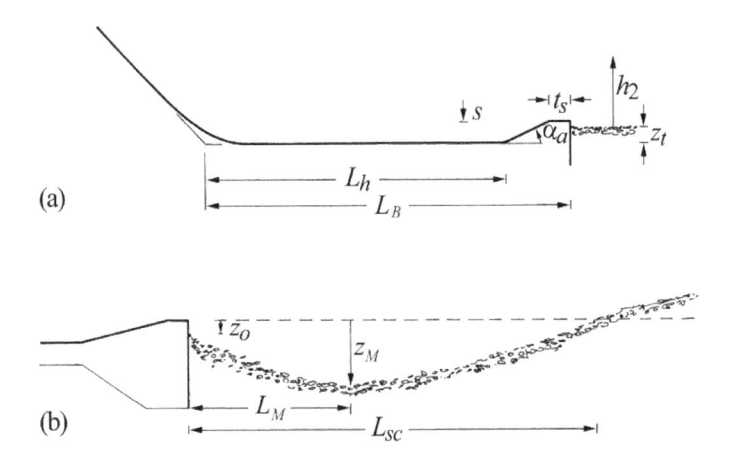

Figure 5.29 Scour control for stilling basins (a) basin geometry, (b) tailwater scour

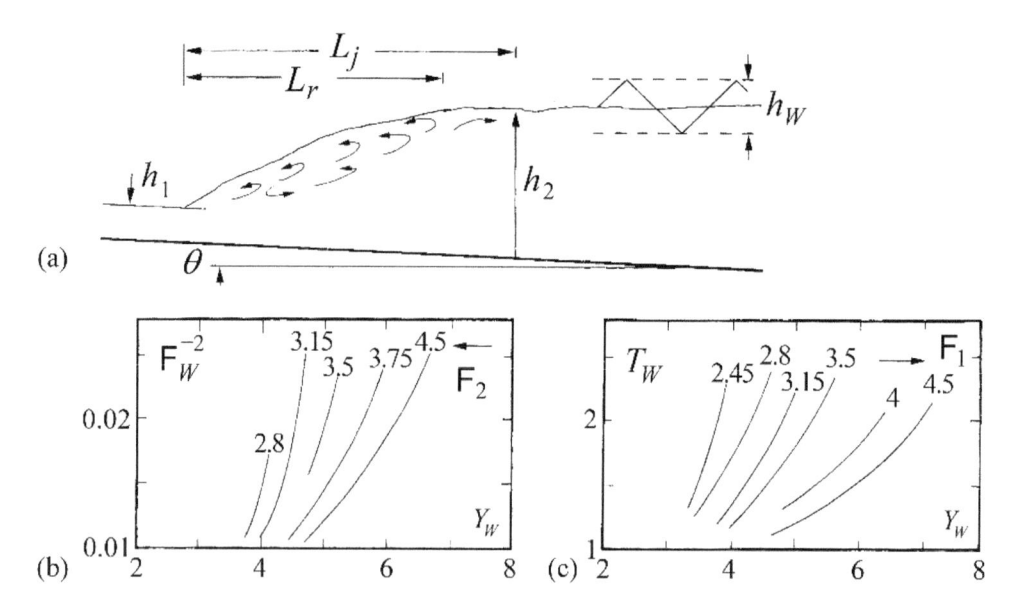

Figure 5.30 Tailwater waves in stilling basin (a) geometry, (b) wave Froude number F_w and F_2 versus Y_w, (c) wave period $T_w(Y_w)$ in horizontal stilling basin (Adapted from Vischer and Hager, 1998)

- Dividing walls contribute to stilling action and reduce flow concentrations
- Cavitation damage increases under high approach flow velocity and low tailwater levels
- End sills reduce scour significantly.

Difficulties with stilling basins occur due to (ICOLD, 1987):

- Unacceptable scour
- Long periods of floods damaging appurtenances
- Uplift pressure due to dynamic action
- Absence of appropriate basin lining
- Overforcing of appurtenances and cavitation damage
- Insufficient self-cleaning of basin and resulting basin abrasion.

Based on a review on damages and field experiences, Cassidy *et al.* (1994) conclude:

- Stilling basins contain no chute blocks for approach flow velocities >30 m/s; baffle blocks for approach flow velocities >30 m/s are carefully model tested;
- High, narrow baffle blocks are subjected to large fluctuating lateral forces and should be avoided;
- Interference effects from adjacent blocks often prevent organization of vortex shedding from blocks.

Cases of lining damage have been reported for Malpaso Dam (Sanchez Bribiesca and Capella Viscaino, 1973), Tarbela Dam, and Karnafuli Dam (Bowers and Toso, 1988), as well as to the hydroelectric power plants of Liu Jia-Xia (in 1966) and Wu Qiang-Xi (in 1996) (Liu and Li, 2007) highlighting the aforementioned experiences.

Stilling basins are popular and among the designer's favorite choice for energy dissipation, certainly because of knowledge and experience acquired over the years. They have proved to be a reliable hydraulic structure if the approach flow conditions and the tailwater elevation are within certain limits.

5.4 Drop structures

5.4.1 Basic flow features

Drop structures are used if the tailwater required for a stilling basin is not available. The approach flow direction is nearly vertical onto a water pool so that impact forces become significant. The scour potential in addition is relevant, whereas the basin length is short and thus not of real concern.

The basic flow types in a prismatic drop structure depend on the approach flow and tailwater depths h_o and h_u measured from the drop elevation; these include (Figure 5.31):

- Free-falling jet and supercritical tailwater flow
- Hydraulic jump if tailwater depth is smaller than drop height w
- Plunging jet flow for flow depth ratio $h_o/h_u \geq 1.17$
- Undulating surface jet flow for $1 < h_o/h_u < 1.17$.

Plunging jet flow is more effective because of confined length and no tailwater waves. However, a strong surface return current is thereby generated of intensity comparable with the roller of a classical hydraulic jump. Various basin types were developed involving a drop structure, including these of Rand in 1955, the Inlet Drop Spillway of Blaisdell and Donnelly

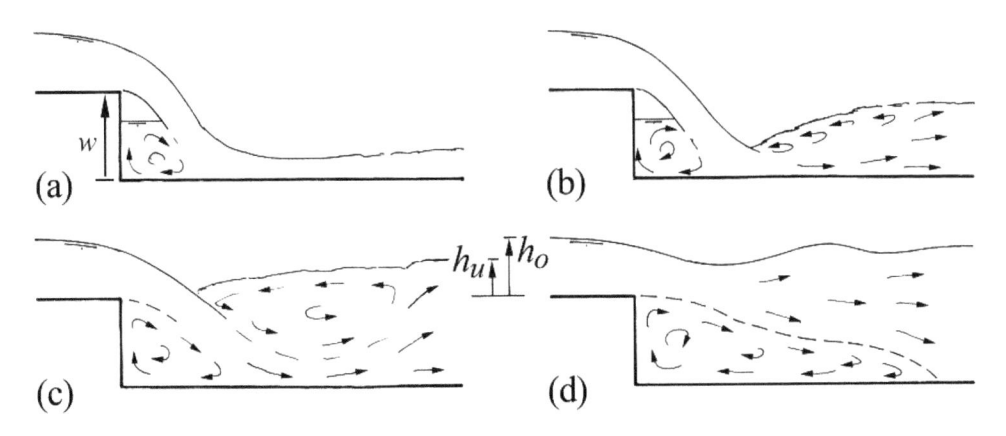

Figure 5.31 Drop structure in prismatic rectangular channel, four basic flow types (a) free-falling jet, (b) hydraulic jump onto tailwater bottom, (c) plunging jet flow, (d) undular surface jet

in 1954, and the Straight Drop Spillway Stilling Basin of Donnelly and Blaisdell in 1965, of which Vischer and Hager (1995) provide a review.

For free jet flow the so-called end overflow results, involving a prismatic approach flow channel. The hydraulic problems involved with this configuration are complex, given the transition from bottom-bounded free surface flow to jet flow through the atmosphere, for which the standard hydraulic equations do not apply. The problem has received considerable attention, given the basic hydraulic problem requiring for an adequate solution. Of relevance is the effect of streamline curvature, resulting in flows in which non-hydrostatic pressure and non-uniform velocity distributions apply. Given the various texts treating this aspect, only the pertinent references are given (Montes, 1992; Marchi, 1993; Khan and Steffler, 1996; Ferro, 1999; Dey, 2002; Ramamurthy *et al.*, 2005; Hager, 2010; Castro-Orgaz and Hager, 2010, 2011, 2017). A state of the art until the early 1980s is provided by Whittacker and Schleiss (1984). Rock scour for high-head drop structures is detailed in Chapter 6.

5.4.2 Drop impact structures

A drop impact structure receives a nearly plane and vertical jet in which the energy is dissipated by jet diffusion and jet deflection (Figure 5.32). Usually, the structures have a large water cushion of thickness t_L, with a jet of thickness t_j and an impact velocity V_1 (Figure 5.38a). From a review on the vertical jet, Vischer and Hager (1995) found for the maximum pressure p_M, the transverse pressure distribution $\bar{p}(x)$ and the root-mean-square (rms) pressure fluctuation p'

$$p_M / (\rho V_1^2 / 2) = 7.4(t_j / t_L), \tag{5.72}$$

$$\bar{p}(x) / p_M = \exp[-0.023(x / t_L)^2], \tag{5.73}$$

$$\overline{(p'^2)^{1/2}} / (\rho V_1^2 / 2) = \alpha_p. \tag{5.74}$$

Extreme values are $\alpha_p = +0.28$ and $\alpha_p = -0.04$. Results on the round turbulent jet were also presented by Vischer and Hager (1995).

5.4.3 Scour characteristics at unlined drop structures

The following refers to free jets from drop structures discharging into plunge pools, involving a loose, granular sediment bed. The scour features typically issued by jets of ski jumps are detailed in Chapter 6. The study of Bormann and Julien (1991) includes both drops and free jets (Figure 5.33). With h_u as tailwater depth, z_e scour depth, z_j drop height, q unit discharge, V_o approach flow velocity, d_n representative grain diameter, and α_j impact jet angle, the test data indicate

$$\frac{z_e + z_j}{h_o} = 0.61 \left[\frac{V_o^2}{g h_o} \right]^{0.8} \left[\frac{h_o}{d_n} \right]^{0.4} \frac{\sin \alpha_j}{[\sin(25^{\circ} + \alpha_j)]^{0.8}}. \tag{5.75}$$

The impact angle α_j (in rad) for *submerged jets* is (Figure 5.33a)

$$\alpha_j = 0.32 \sin \alpha_b + 0.15 \ln \left(1 + \frac{z_j}{h_o} \right) + 0.13 \ln \left(\frac{h_u}{h_o} \right) - 0.05 \ln[V_o / (g h_o)^{1/2}]. \tag{5.76}$$

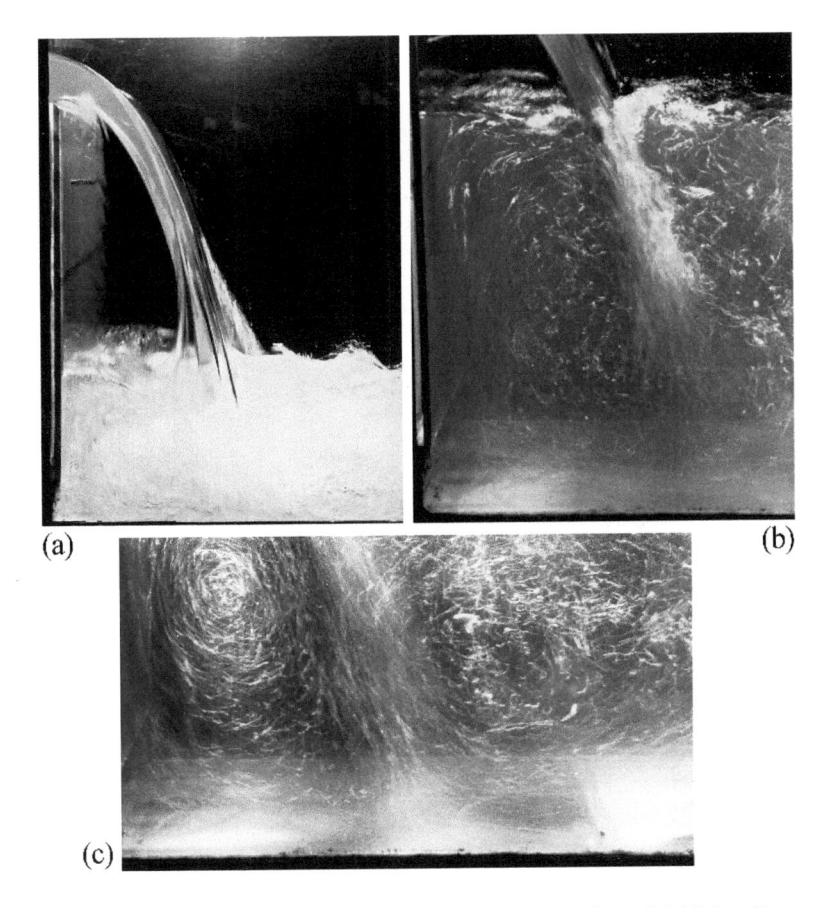

Figure 5.32 Drop impact structure with plunging jet for (a) low, (b) high tailwater levels, (c) vorticity of flow close to bottom (Courtesy Willi H. Hager)

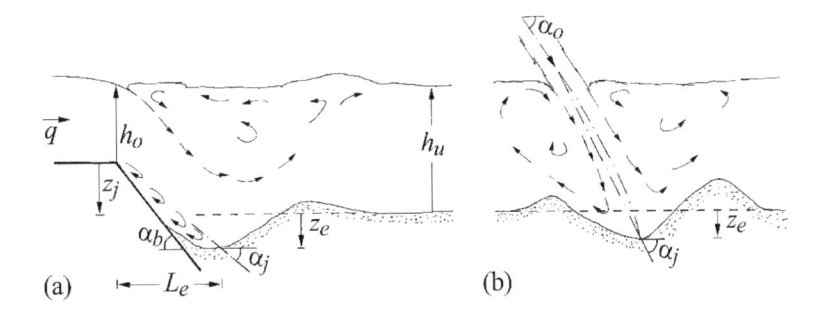

Figure 5.33 Scour in plunge pools for (a) submerged, (b) free jet flow issued from drop structures

For *free jets* the result is simply (Figure 5.33b)

$$\alpha_j = \alpha_o.$$ (5.77)

The location of maximum scour depth is with d_{90} as the dominant grain diameter

$$L_e / h_o = 0.61 \left[\frac{V_o^2 / (gh_o)}{\sin(25° + \alpha_j)} \right]^{0.8} \left[\frac{h_o}{d_{90}} \right]^{0.4}.$$ (5.78)

Scour depths downstream of PKWs founded on a loose bed are detailed in 3.5.

5.5 Free fall outlets

5.5.1 Introduction

Among various types of spillway outlet or transport structures, the free fall, spillway chute, and spillway cascade are the most prominent. These are dealt with herein and in Chapter 4 with regard to the hydraulic approach. Figure 5.34 sketches typical outlet structures.

The free fall outlet is an uncommon type of spillway outlet structure, mainly due to concerns with impact pressure and scour just beneath it. It applies exclusively at locations with excellent geologic underground conditions. Also, a stilling basin is often provided to create a water cushion by which the impact action of the falling jet is reduced. Thomas (1976) mentions that a unit discharge of 80 m²s⁻¹ should not be exceeded and that fall heights over 100 m have to receive a particular attention. He suggests a fully aerated and steady nappe to counter impact pulsations. Figure 5.34a refers to a dam with a free falling jet as outlet structure, involving an adversely sloping stilling basin to provide a minimum submergence effect. Figure 5.34b shows a tailwater view of Gebidem Dam, Switzerland, at discharge of 30 m³/s. Note the excellent rock characteristics at the dam toe on which the falling jet impacts, so that no additional action was considered necessary in this case.

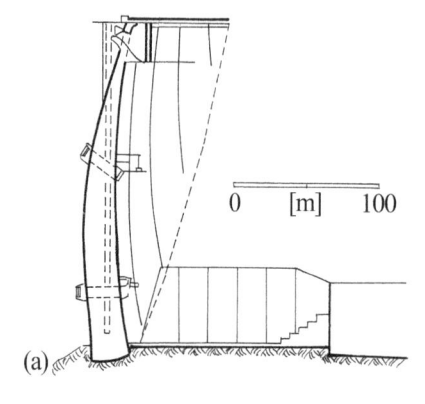

Figure 5.34 (a) Free fall outlet structure at Mratinje Dam, former Yugoslavia, (b) Gebidem Dam, Valais, Switzerland (*Schweizerischer Wasserwirtschafts-Verband* 1970, 42: 53)

Free fall outlets are provided in relatively wide valleys with excellent geological characteristics. The hydraulic design involves four main features: (a) overflow, (b) nappe flow, (c) impact action, and (d) stilling basin. Below, items (b) and (c) are dealt with. The overflow structure is analogous to standard spillway overflow, except for a small deflector discharging the jet into the atmosphere instead of a chute. The jet trajectory is then fully determined by the effects of gravity, viscosity, surface tension, and air entrainment. The intensity of the jet impact action depends on the relative jet thickness, and the degrees of turbulence and jet dispersion. The stilling basin, i.e. the plunge pool, corresponds to a separate hydraulic structure whose main characteristics are dealt with above. Selected features of the impact basin are presented below.

5.5.2 Jet trajectory

Water jets are complex to analyze due to the complicated interaction of gravity, viscosity, and capillarity effects. A further challenge stems from its two free surfaces in the atmosphere. Currently, these jets are numerically treated in 2D for inviscid flow (e.g. Dias and Tuck, 1991, Vanden Broeck and Keller, 1986). A large body of experimental data on spillway jets however defines the exact lower boundary geometry of the standard spillway. It is thus appropriate to consider the *substitute* sharp-crested weir as the origin of the coordinate system rather than the crest of the standard spillway. Accordingly, for a given overflow structure according to Figure 2.14b, consider the corresponding weir design as shown in Figure 2.14a.

Figure 5.35a defines the variables of sharp-crested, fully aerated weir flow (notations overbarred) for which the overflow jet is guided with two lateral walls, corresponding to confined trajectory flow. The notation is $\bar{w}$ for weir height, $\bar{H}$ for the head on the weir, $(\bar{x}; \bar{z})$ define the Cartesian coordinate system with origin at the weir crest, and $(\bar{e}; \bar{f})$ are the coordinates of the

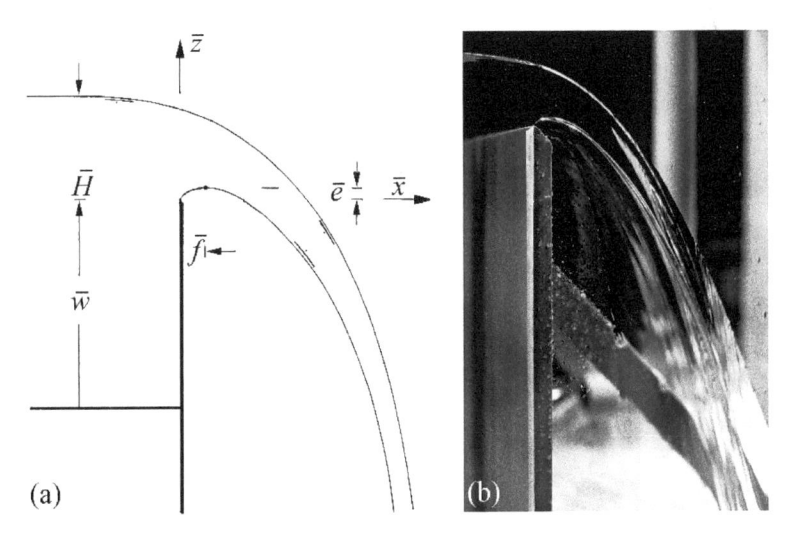

Figure 5.35 (a) Definition of nappe geometry, origin of coordinates $(\bar{x}; \bar{z})$ at weir crest, (b) overflow nappe of sharp-crested laboratory weir (Courtesy Willi H. Hager)

lower nappe maximum. Based on the data of USBR (1948), the data analysis of Rajaratnam *et al.* (1968) and with $\mu = V_o^2 / (2g\bar{H})$ as the approach flow index results in

$$\begin{aligned}
\bar{f} / \bar{H} &= 0.250 - 0.40\mu, \\
\bar{e} / \bar{H} &= 0.112 - 0.40\mu.
\end{aligned} \tag{5.79}$$

With $\bar{X} = 1.5(\bar{x}/\bar{H})$ and $\bar{Z} = 3.5(\bar{z}/\bar{H})$ as non-dimensional coordinates, the lower nappe profile reads as extension of Eq. (2.2)

$$\bar{Z} = \bar{X} \ln \bar{X} \left[1 + \frac{1}{6} \bar{X} \right]. \tag{5.80}$$

The upper nappe geometry has also been investigated. The vertical jet (subscript j) thickness t_j beyond the lower nappe maximum remains nearly constant at $\bar{t}_j / \bar{H}; = 0.70$, whereas the upper jet profile is approximately determined by adding $\bar{t}_j$ to the lower nappe elevation. The impact angle is thus determined at any elevation below the crest.

To counter jet vibration, the offtake portion of the overflow structure, especially at gate overflow, is equipped with *nappe splitters* (Figure 5.36) located at its most downstream point and projecting through the nappe at design discharge. These splitters should be extended over the lip to burst the nappe. The aeration of the compact nappe also enhances jet dispersion, so that less impact action is to be expected at the stilling basin as compared with a compact unaerated jet. According to Schwartz (1964), the maximum splitter spacing should be about two thirds of the fall height. An extensive account on flow-induced gate vibrations is given by Naudascher and Rockwell (1994).

Whereas additional aeration of a gated overflow nappe is essential to avoid nappe oscillation, it is not always needed for a massive free overfall considered here. Because the air has no access below the overflow nappe between gate piers, there is virtually no air restriction for free fall nappes. Both the significant fall heights and large water quantities result in spray

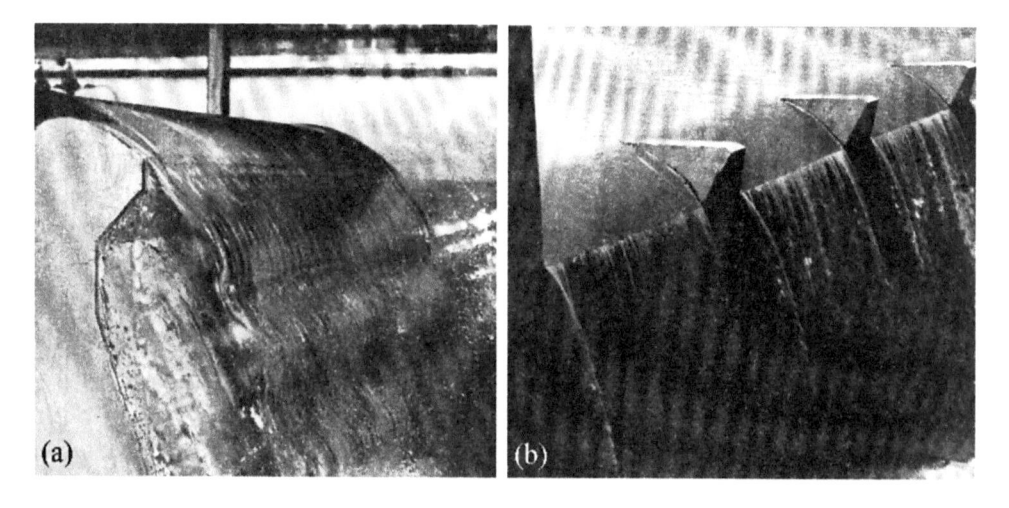

(a)   (b)

Figure 5.36 (a) Oscillating nappe, (b) nappe splitters (Schwartz, 1964)

generation, however. This is undesirable in cold regions and for power generation (switch-yards) in winter due to ice formation.

5.5.3 Jet impact

A jet issued from an overfall structure interacts with the surrounding air and generates an aerated turbulent jet (Figure 5.37). Depending on the relative jet thickness, its turbulence degree, and the fall height, the average air concentration may be small thereby resulting in a significant impact effect, or the jet is nearly dispersed in the atmosphere prior to impact. The latter case is certainly unusual for the design flow, so that free fall nappes have typically a high impact load. Because the overflow jet is not directed away from the dam as trajectory jets, a bottom impact protection is needed. Examples including the Kariba Dam involving orifice jets, i.e. directed away from the dam section, have resulted in excessive scour (Chapter 6). For free fall jets impacting close to the dam foundation, it is highly unadvisable to use an unlined plunge pool basin instead of a lined impact basin (5.4).

The impact jet characteristics are similar to these of the wall jet, involving a turbulent free zone of the approach flow jet, a deflection zone at the jet impact region, and a wall jet zone further downstream. A minimum water cushion must be provided to submerge the impinging jet. Details are given by Vischer and Hager (1995).

The time-averaged maximum (subscript M) bottom pressure p_{Mb} depends on the ratio of jet thickness t_j to the water cushion height t_L (Figure 5.38a). It is approximated with the jet impact velocity V_1 as

$$p_{Mb} / (\rho V_1^2 / 2) = 7.4(t_j / t_L).$$ (5.81)

The local pressure distribution is of Gaussian type, and the rms pressure fluctuations are correlated with the upper and lower extremes of +0.28 and −0.04, respectively, as

$$(\overline{p'^2})^{1/2} / (\rho V_1^2 / 2) = \alpha_p.$$ (5.82)

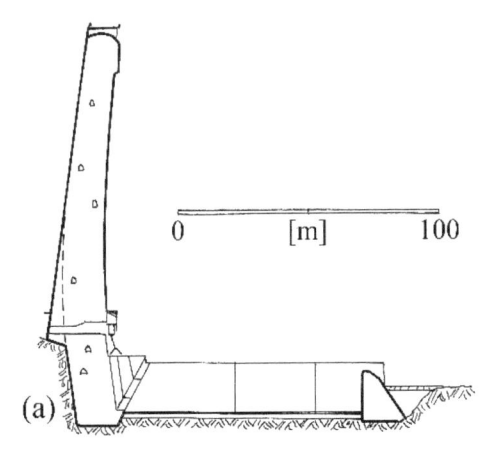

Figure 5.37 Morrow Point Dam, upper Colorado (USA), (a) section, (b) tailwater view (Koolgaard and Chadwick, 1988)

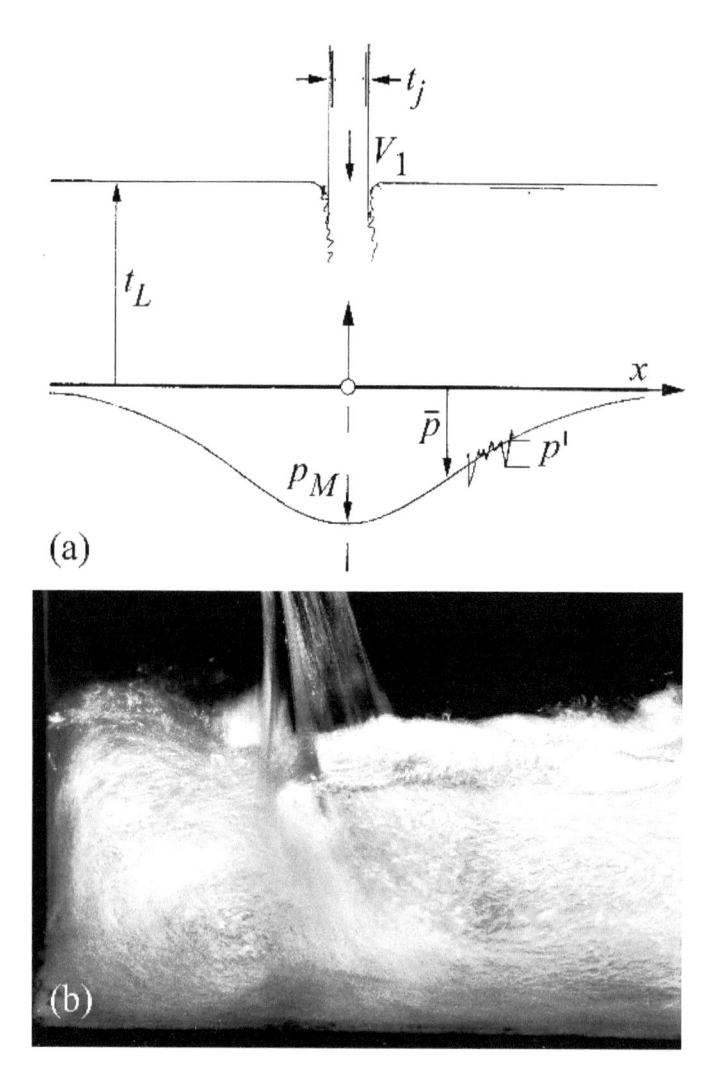

Figure 5.38 Impact jet action (a) notation, (b) laboratory impact jet with visible vortex action (Courtesy Willi H. Hager)

Stilling basin should have a standard lining as described in 5.3.6. General applicable information on impact jet basins is currently not available, so that detailed hydraulic modeling is recommended if a significant damage has to be expected. It should also be kept in mind that the impact jet divides at the basin bottom and that a significant flow portion can be directed upstream toward the dam (Figure 5.38b). Abutments have to be protected by walls inhibiting scour action. Chapter 5 Frontispiece (a) shows Schiffenen Dam (Switzerland) including three jets impacting the tailwater.

Notation

b	Channel width (m)
b_B	Block width (m)
b_s	Sill width (m)
b_1	Approach flow width of expanding stilling basin (m)
b_2	Tailwater width of expanding stilling basin (m)
d_n	Representative grain size (m)
e	Block spacing (m)
$\bar{e}$	Elevation of maximum lower nappe from weir crest (m)
F_B	Force on block (N)
F_1	Approach flow Froude number (-)
F_{1m}	Minimum approach flow required for sill action (-)
F_W	Wave Froude number (-)
$\bar{f}$	Distance of maximum lower nappe elevation from weir crest (m)
f_o	Friction factor (-)
g	Gravity acceleration (m/s^2)
H	Energy head (m)
$\bar{H}$	Head on weir crest (m)
H_1	Approach flow energy head (m)
H_2^*	Tailwater energy head of classical hydraulic jump (m)
h	Flow depth (m)
h_c	Critical flow depth (m)
h_M	Maximum flow depth (m)
h_m	Minimum flow depth (m)
h_o	Upstream flow depth (m)
h_u	Downstream flow depth (m)
h_W	Wave height (m)
h_1	Approach flow depth (m)
h_2^*	Sequent depth of classical hydraulic jump (m)
L_1, L_2	Wave lengths (m)
L_B	Basin length (m)
L_e	Location of maximum scour depth (m)
L_h	Horizontal stilling basin length (m)
L_j	Hydraulic jump length (m)
L_j^*	Length of classical hydraulic jump (m)
L_M	Location of maximum scour depth (m)
L_r^*	Length of roller of classical hydraulic jump (m)
L_s	Location of sill (m)
L_{sc}	Scour hole length (m)
M_z	Vertical momentum flux (m^2)
N	Coefficient and exponent (-)
N_1	Power-law exponent (-); also flow depth measured perpendicularly to chute bottom (m)
P	Dimensionless pressure component (-)
P_m	Maximum dimensionless pressure (-)
p	Pressure (N/m^2)
$\bar{p}$	Time-averaged pressure (N/m^2)

p_f	RMS-pressure (N/m^2)
p_M	Maximum pressure (N/m^2)
p_{Mb}	Maximum bottom impact pressure (N/m^2)
p'	Fluctuating pressure component (N/m^2)
Q	Discharge (m^3/s)
q	Unit discharge (m^2/s)
R_b	Bucket radius (m)
R_{bm}	Minimum bucket radius (m)
R	Reynolds number (-)
R$_1$	Approach flow Reynolds number (-)
r	Roller thickness (m)
S	Specific momentum (m^2)
S_e	Energy line slope (-)
S_M	Maximum relative sill height (-)
S_{opt}	Optimum relative sill or block height (-)
S_s	Relative sill height (-)
s	Sill height (m)
s_{opt}	Optimum sill or block height (m)
T	Vertical shear force (N)
T_W	Dimensionless wave period (-)
t_b	Minimum tailwater elevation (m)
t_j	Jet thickness (m)
t_L	Water cushion thickness (m)
t_s	Sill width (m)
t_W	Wave period (s)
U	Cross-sectional average velocity (m/s)
u	Horizontal velocity component (m/s)
u_m	Maximum cross-sectional forward velocity in hydraulic jump (m/s)
V	Average cross-sectional velocity (m/s)
V_1	Approach flow velocity (m/s)
V_2^*	Tailwater velocity of classical hydraulic jump (m/s)
v	Vertical velocity component (m/s)
w	Drop height (m)
$\bar{w}$	Weir height (m)
X	$= x/L_r^*$ Dimensionless location (-)
$\bar{X}$	Relative horizontal coordinate from weir crest (-)
X_a	$= (x - x_{1M})/(x_{2M} - x_{1M})$ (-)
X_c	$= (x - x_c)/L_r$ (-)
X_s	$= x_s/L_r^*$ (-)
X_w	$= (x - x_{1M})/(x_{3M} - x_{1M})$ (-)
X_1	$= x_1/L_r^*$ (-)
x	Streamwise coordinate (m)
$\bar{x}$	Horizontal coordinate from weir crest (m)
x_B	Basin length (m)
x_c	Location of critical point (m)
x_j	End location of hydraulic jump length (m)
x_s	Location of sill within expanding stilling basin (m)

x_1	Toe location of expanding hydraulic jump (m)
x_2	End location of expanding hydraulic jump roller (m)
Y	$= (h + r)/h_1$ (-)
Y^*	Sequent depth ratio of classical hydraulic jump (-)
Y_B	Sequent flow depths with block presence (-)
Y_M	Maximum relative total flow depth (-)
Y_s	Sequent depth ratio with sill presence (-)
Y_W	Sequent depth ratio under wave action (-)
ΔY_s	Sequent depth reduction due to sill presence (-)
y	Coordinate perpendicular to channel bottom (m)
Z	Vertical specific force (m^2)
$\overline{Z}$	Relative vertical coordinate from weir crest (-)
z	Vertical coordinate (m)
$\overline{z}$	Vertical coordinate from weir crest (m)
z_e	End scour depth (m)
z_j	Drop height (m)
z_M	Maximum scour depth (m)
z_o	Height of bed just beyond stilling basin elevation (m)
z_t	Tailwater bed elevation above stilling basin bottom (m)
α	Bottom inclination angle (-)
α_a	Apron angle (-)
α_j	Submerged jet impact angle (-)
α_k	Kinetic energy correction coefficient (-)
α_o	Free jet impact angle (-)
α_p	Pressure coefficient (-)
β	Width expansion ratio (-)
β_S	Momentum correction coefficient (-)
γ	$= \rho_w g$ (N/m^3)
γ_a	$= (h_a - h_1)/(h_{1M} - h_1)$ (-)
γ_w	$= (h_w - h_1)/(h_{3M} - h_1)$ (-)
χ	$= x/L_r$ (-)
χ_c	Relative location of critical point (-)
δ_0	Boundary layer thickness (m)
δ_1	Elevation of half-velocity above bottom (m)
η	$= z/h$ (-)
η^*	Efficiency of classical hydraulic jump (-)
Φ	Block force coefficient (-)
κ	von Karman coefficient (-)
Λ	Sequent depth ratio (-)
Λ_s	Relative sill location (-)
λ	$= \Lambda(\Lambda + 1)$ (-)
λ_1, λ_2	Coefficients varying with N (-)
μ	Approach flow index (-)
ν	Kinematic fluid viscosity (m^2/s)
ν_t	Turbulent eddy viscosity (m^2/s)
Θ	Inclination of stilling basin relative to horizontal (-)
ρ	Fluid density (kg/m^3)

ρ_b Relative bucket radius (-)
σ $= r/h_1$ (-)
σ_M Maximum relative roller thickness (-)
σ_W Wave steepness (-)
σ_x Normal streamwise turbulent stress (N/m^2)
ς $= h/h_1$ (-)
ς_c $= h_c/h_1$ (-)
τ_b Bed-shear stress (N/m^2)
τ_{xz} Reynolds stress (N/m^2)
τ_{yx} Tangential Reynolds stress (N/m^2)

References

Abou-Seida, M.M. (1963) Wave action below spillways. *Journal of the Hydraulics Division*, ASCE, 89(HY3), 133–152.

Barjastehmaleki, S., Fiorotto, V. & Caroni, E. (2016a) Spillway stilling basins lining design via Taylor hypothesis. *Journal of Hydraulic Engineering*, 10.1061/(ASCE)HY.1943–7900.0001133, 04016010.

Barjastehmaleki, S., Fiorotto, V. & Caroni, E. (2016b) Stability analysis of plunge pool linings. *Journal of Hydraulic Engineering*, 10.1061/(ASCE)HY.1943–7900.0001175.

Barjastehmaleki, S., Fiorotto, V. & Caroni, E. (2016c) Design of stilling basin linings with sealed and unsealed joints. *Journal of Hydraulic Engineering*, 10.1061/(ASCE)HY.1943–7900.0001218.

Basco, D.R. (1971) Optimized geometry for baffle blocks in hydraulic jumps. Proceedings of 14th *IAHR Congress* Paris, 2(B18), 1–8. IAHR, Delft.

Beichley, G.L. & Peterka, A.J. (1959) The hydraulic design of slotted spillway buckets. *Journal of Hydraulics Division ASCE*, 85(HY10), 1–36.

Berryhill, R.H. (1963) Experience with prototype energy dissipators. *Journal of the Hydraulics Division* ASCE, 89(HY3), 181–201; 90(HY1), 293–298; 90(HY4), 235.

Bormann, N.E. & Julien, P.Y. (1991) Scour downstream of grade-control structures. *Journal of Hydraulic Engineering*, 117(5), 579–594; 118(7), 1066–1073.

Bose, S.K. & Dey, S. (2007) Curvilinear flow profiles based on Reynolds averaging. *Journal of Hydraulic Engineering*, 133(9), 1074–1079.

Bowers, C.E. & Toso, J. (1988) Karnafuli Project: model studies of spillway damage. *Journal of Hydraulic Engineering*, 114(5), 469–483; 116(6), 850–855.

Bremen, R. & Hager, W.H. (1994) Expanding stilling basin. *Proceedings of ICE, Water Maritime & Energy*, 106, 215–228.

Cassidy, J.J., Locher, F.A., Lee, W. & Nakato, T. (1994) Hydraulic design for replacement of floor blocks for Pit 6 stilling basin. Proceedings of 18th *ICOLD Congress*, Durban, Q71(R40), 599–621.

Castro-Orgaz, O. & Hager, W.H. (2009) Classical hydraulic jump: basic flow features. *Journal of Hydraulic Research*, 47(6), 744–754.

Castro-Orgaz, O. & Hager, W.H. (2010) Moment of momentum equation for curvilinear free surface flow. *Journal of Hydraulic Research*, 48(5), 620–631; 49(3), 415–419.

Castro-Orgaz, O. & Hager, W.H. (2011) Vorticity equation for the streamline and the velocity profile. *Journal of Hydraulic Research*, 49(6), 775–783.

Castro-Orgaz, O. & Hager, W.H. (2017) *Non-hydrostatic Free Surface Flows*. Springer, Berlin.

Castro-Orgaz, O., Hager, W.H. & Dey, S. (2015) Depth-averaged model for undular hydraulic jump. *Journal of Hydraulic Research*, 53(3), 351–363.

Chanson, H. & Montes, J.S. (1995) Characteristics of undular hydraulic jumps: experimental apparatus and flow patterns. *Journal of Hydraulic Engineering*, 121(2), 129–144; 123(2), 161–164.

Dey, S. (2002) Free overfall in open channels: State-of-the-art review. *Flow Measurement and Instrumentation*, 13(5–6), 247–265.

Dias,F. & Tuck, E.O. (1991) Weir flows and waterfalls. *Journal of Fluid Mechanics*, 230, 525–539.

Elevatorsky, E.A. (1959) *Hydraulic Energy Dissipators*. McGraw Hill, New York.

Ferro, V. (1992) Theoretical end discharge relationship for free overfall. *Journal of Irrigation and Drainage Engineering*, 125(1), 40–44; 126(2), 133–138.

Gotoh, H., Yasuda, Y. & Ohtsu, I. (2005) Effect of channel slope on flow characteristics of undular hydraulic jumps. *Transactions of Ecology and Environment*, 83(1), 33–42.

Grillhofer, W. & Schneider, W. (2003) The undular hydraulic jump in turbulent open channel flow at large Reynolds numbers. *Physics of Fluids*, 15(3), 730–735.

Hager, W.H. (1985) Critical flow condition in open channel hydraulics. *Acta Mechanica*, 54(3–4), 157–179.

Hager, W.H. (1992) *Energy Dissipators and Hydraulic Jump*. Kluwer, Dordrecht.

Hager, W.H. (2010) *Wastewater Hydraulics: Theory and Practice*, 2nd ed. Springer, Berlin.

Hager, W.H. & Bremen, R. (1989) Classical hydraulic jump: sequent depths. *Journal of Hydraulic Research*, 27(5), 565–585.

Hager, W.H. & Hutter, K. (1983) Approximate treatment of plane hydraulic jump with separation zone above the flow zone. *Journal Hydraulic Research*, 21(3), 195–204.

Hager, W.H. & Hutter, K. (1984a) Approximate treatment of plane channel flow. *Acta Mechanica*, 51(1/2), 31–48.

Hager, W.H. & Hutter, K. (1984b) On pseudo-uniform flow in open channel hydraulics. *Acta Mechanica*, 53(3–4), 183–200.

ICOLD (1987) Spillways for dams. *ICOLD Bulletin* 58. International Commission for Large Dams, Paris.

Jurisits, R. & Schneider, W. (2012) Undular hydraulic jumps arising in non-developed turbulent flows. *Acta Mechanica*, 223(8), 1723–1738.

Khan, A.A. & Steffler, P. (1996) Physically-based hydraulic jump model for depth-averaged computations. *Journal of Hydraulic Engineering*, 122(10), 540–548.

Koolgaard, E.B. & Chadwick, W.L. (1988) *Development of Dam Engineering in the United States*. Pergamon, New York.

Liu, P.Q. & Li, A.H. (2007) Model discussion on pressure fluctuation propagation within lining slab joints in stilling basins. *Journal of Hydraulic Engineering*, 10.1061/(ASCE)0733–9429(2007)133, 6(618), 618–624.

Mandrup-Andersen, V. (1978) Undular hydraulic jump. *Journal of the Hydraulics Division* ASCE, 104(HY8), 1185–1188; 105(HY9), 1208–1211.

Marchi, E. (1963) Contributo allo studio del risalto ondulato [Contribution to the undular hydraulic jump]. *Giornale del Genio Civile,* 101(9), 466–476 (in Italian).

Mason, P.J. (1982) The choice of hydraulic energy dissipator for dam outlet works based on a survey of prototype usage. *Proceedings of the Institution of Civil Engineers,* 72(1), 209–219; 74(1), 123–126.

McCorquodale, J.A. (1986) Hydraulic jump and internal flows. In: Cheremisinoff, N.P. (ed) *Encyclopedia of Fluid Mechanics* 2, 122–173. Gulf Publishing, Houston.

McCorquodale, J.A. & Khalifa, A. (1983) Internal flow in hydraulic jumps. *Journal of Hydraulic Engineering*, 109(5), 684–701; 110(9), 1508–1509.

Montes, J.S. (1992) A potential flow solution for the free overfall. Proc. *ICE* 96(6), 259–266; 112(1), 85–87.

Murzyn, F. & Chanson, H. (2007) Free surface, bubbly flow and turbulence measurements in hydraulic jumps. *Report* CH63/07, Div. Civ. Engng. The University of Queensland, Brisbane, Australia.

Naudascher, E. & Rockwell, D. (1994) Flow-induced vibrations. *IAHR Hydraulic Structures Design Manual* 7. Balkema, Rotterdam.

Novak, P. (1955) Study of stilling basins with special regard to their end sill. Proceedings of 6th *IAHR Congress*, The Hague, C(15), 1–14. IAHR, Delft.

Ohtsu, I., Yasuda, Y. & Gotoh, H. (2001) Hydraulic condition for undular jump formations. *Journal of Hydraulic Research*, 39(2), 203–209; 40(3), 379–384.

Ohtsu, I., Yasuda, Y. & Gotoh, H. (2003) Flow conditions of undular hydraulic jumps in horizontal rectangular channels. *Journal of Hydraulic Engineering*, 129(12), 948–955.

Peterka, A.J. (1958) *Hydraulic Design of Stilling Basins and Energy Dissipators*. Engineering Monograph 25. US Bureau of Reclamation, Denver.

Rajaratnam, N. (1965) The hydraulic jump as a wall jet. *Journal of the Hydraulics Division* ASCE, 91(HY5), 107–132; 92(HY3), 110–123; 93(HY1), 74–76.

Rajaratnam, N. (1967) Hydraulic jumps. *Advances in Hydroscience*, 4, 197–280.

Rajaratnam, N., Subramanya, K. & Muralidhar, D. (1968) Flow profiles over sharp-crested weirs. *Journal of the Hydraulics Division* ASCE, 94(HY3), 843–847.

Ramamurthy, A.S., Qu, J. & Vo, D. (2005) Volume of fluid model for an open channel flow problem. *Canadian Journal of Civil Engineering*, 32(5), 996–1001.

Reinauer, R. & Hager, W.H. (1995) Non-breaking undular hydraulic jump. *Journal of Hydraulic Research*, 33(5), 683–698; 34(2), 279–287; 34(4), 567–573.

Resch, F.J., Leutheusser, H.J. & Coantic, M. (1976) Etude de la structure cinématique et dynamique du ressaut hydraulique [Study of kinematic and dynamic structures of hydraulic jump]. *Journal of Hydraulic Research*, 14(4), 293–319 (in French).

Rodi, W. (1993) *Turbulence Models and Their Application in Hydraulics*, 3rd ed. IAHR Monograph. Balkema, Rotterdam.

Rouse, H., Siao, T.T. & Nagaratnam, S. (1959) Turbulence characteristics of the hydraulic jump. *Transactions of ASCE*, 124, 926–966.

Sanchez Bribiesca, J.S. & Capella Viscaino, A.C. (1973) Turbulent effects on the lining of stilling basin. Proceedings of 11th *ICOLD Congress*, Madrid, Q41(R2), 1575–1592. International Commission on Large Dams, Paris.

Schneider, W., Jurisits, R. & Bae, Y.S. (2010) An asymptotic iteration method for the numerical analysis of near-critical free-surface flows. *Journal of Heat & Fluid Flow*, 31(6), 1119–1124.

Schröder, R. (1963) Die turbulente Strömung im freien Wechselsprung [Turbulent flow in free hydraulic jump]. In: Press, H. (ed) *Mitteilung* 59, Institut für Wasserbau und Wasserwirtschaft. TU Berlin, Berlin (in German).

Schwartz, H.I. (1964) Projected nappes subject to harmonic pressures. *Proceedings of ICE*, 28(3), 313–325.

Serre, F. (1953) Contribution à l'étude des écoulements permanents et variables dans les canaux [Contribution to steady and unsteady open channel flows]. *La Houille Blanche*, 8(6–7), 374–388; 8(12), 830–887 (in French).

Steffler, P.M. & Jin, Y.C. (1993) Depth averaged and moment equations for moderately shallow free surface flow. *Journal of Hydraulic Research*, 31(1), 5–17.

Thomas, H.H. (1976) *The Engineering of Large Dams*. Wiley, London, New York.

USBR (1948) Studies of crests for overfall dams. Boulder Canyon Projects, Final Reports Part VI: hydraulic investigations. *Bulletin* 3. US Bureau of Reclamation, Department of the Interior, Denver.

Valiani, A. (1997) Linear and angular momentum conservation in hydraulic jump. *Journal of Hydraulic Research*, 35(3), 323–354.

Vanden-Broeck, J.-M. & Keller, J.B. (1986) Pouring flows. *Physics of Fluids*, 29(12), 3958–3961.

Vischer, D.L. & Hager, W.H. (1995) *Energy Dissipators*. IAHR Hydraulic Structures Design Manual 9. Balkema, Rotterdam.

Vischer, D.L. & Hager, W.H. (1998) *Dam Hydraulics*. Wiley, Chichester.

Wang, H. & Chanson, H. (2015) Air entrainment and turbulent fluctuations in hydraulic jumps. *Urban Water Journal*, 12(6), 502–518.

White, F.M. (1991) *Viscous Fluid Flow*. McGraw-Hill, New York.

Whittaker, J.G. & Schleiss, A. (1984) Scour related to energy dissipaters for high head structures. *Mitteilung 73*. Versuchsanstalt für Wasserbau, Hydrologie und Glaziologie VAW, ETH Zurich, Zürich.

Woycicki, K. (1931) Wassersprung, Deckwalze und Ausfluss unter einer Schütze [Hydraulic jump, roller and outflow from a gate]. *Dissertation* 639. ETH Zurich, Zürich, Switzerland (in German).

Bibliography

Aeration of drop flow

Chanson, H. (1995) Air entrainment in two-dimensional turbulent shear flows with partially developed inflow conditions. *International Journal of Multiphase Flow*, 21(6), 1107–1121.

Chanson, H. & Toombes, L. (1998) Air-water flow structures at an abrupt drop with supercritical flow. Proceedings of 3rd International Conference *Multiphase Flow*, Lyon, 2(2.5), 1–6.

Ervine, D.A. (1976) The entrainment of air in water. *Water Power & Dam Construction*, 28(12), 27–30.

Ervine, D.A. (1998) Air entrainment in hydraulic structures: a review. *Proceedings of ICE, Water, Maritime & Energy*, 130(3), 142–153.

Ervine, D.A., McKeogh, E.J. & Elsawy, E.M. (1981) Model-prototype conformity in hydraulic structures involving aeration. Proceedings of 19th *IAHR Congress*, New Delhi, 5(8), 65–72. IAHR, Delft.

Falvey, H.T. (1980) Air-water flow in hydraulic structures. *Engineering Monograph*, 41. US Department of the Interior, Washington DC.

Haindl, K. (1984) Aeration at hydraulic structures. In: Novak, P. (ed) *Developments in Hydraulic Engineering* 2, 113–158. Elsevier, London.

Kalinske, A.A. (1938) The hydraulics and pneumatics of the plumbing drainage system. *Bulletin*, 10, 1–8. Associated State Engineering Societies: Washington DC.

Kobus, H. (1984) Local air entrainment and detrainment. Proceedings of Symposium *Scale Effects in Modelling Hydraulic Structures*, Esslingen, 4(10), 1–10.

Laushey, L.M. & Mavis, F.T. (1953) Air entrained by water flowing down vertical shafts. Proceedings of 5th *IAHR Congress*, Minneapolis. IAHR, Delft, pp. 483–487.

Marquenet, G. (1953) Eintrainement d'air par un écoulement en conduit vertical. Application aux puits d'adduction secondaires [Air entrainment in vertical conduit flow: application to shafts]. Proceedings of 5th *IAHR Congress*, Minneapolis. IAHR, Delft, pp. 489–506 (in French).

Rajaratnam, N. & Kwan, A.Y.P. (1996) Air entrainment at drops. *Journal of Hydraulic Research*, 34(5), 579–588.

Drop structures

Blaisdell, F.W. & Donnelly, C.A. (1954) The Box Inlet Drop Spillway and its outlet. *Proceedings of ASCE*, 80(534), 1–41; 81(841), 1–2; also *Transactions of ASCE*, 121, 955–994.

Chamani, M.R. & Beirami, M.K. (2002) Flow characteristics at drops. *Journal of the Hydraulic Engineering*, 128(8), 788–791.

D'Agostino, V. & Ferro, V. (2004) Scour on alluvial bed downstream of grade-control structures. *Journal of Hydraulic Engineering*, 130(1), 24–37.

Donnelly, C.A. & Blaisdell, F.W. (1965) Straight Drop Spillway stilling basin. *Journal of the Hydraulics Division* ASCE, 91(HY3), 101–131; 82(HY1), 102–107; 82(HY4), 140–145.

Ervine, D.A. & Falvey, H.T. (1987) Behaviour of turbulent water jets in the atmosphere and in plunging pools. *Proceedings of the Institution of Civil Engineers*, 83, 295–314; 85, 359–363.

Ferreri, G.B. & Ferro, V. (1989) Esame sperimentale del comportamento idraulico di una sistemazione a salti di fondo: primi resultati (Experimental study of the hydraulic behavior of drops: first results). *Idrotecnica*, 16(3), 121–134 (in Italian).

Keutner, C. (1937a) Die Regelung kleinerer Wasserläufe durch Errichtung von Gefällstufen (The control of smaller conveyances by drops). *Die Bautechnik*, 15(13/14), 173–188 (in German).

Keutner, C. (1937b) Die Ausbildung der Gefällsbrechpunkte geregelter kleinerer Wasserläufe (The control of slope breaks of smaller conveyances). *Die Bautechnik*, 15(40/41), 518–533 (in German).

Little, W.C. & Murphey, J.B. (1982) Model study of low drop grade control structures. *Journal of the Hydraulics Division* ASCE, 108(HY10), 1132–1146.

Morris, B.T. & Johnson, D.C. (1943) Hydraulic design of drop structures for gully control. *Transactions of ASCE*, 108, 887–940.

Müller, R. (1946) Über die hydraulische Dimensionierung von Absturzbauwerken (On the hydraulic design of drop structures). *Zeitschrift für Schweizerisches Forstwesen*, 97(1), 41–48 (in German).

Noutsopoulos, G.C. (1984) *Channels and Channel Control Structures*, Smith, K.V.H. (ed). Springer, Berlin, pp. 167–181.

Ohtsu, I. & Yasuda, Y. (1997) Characteristics of flow passing over drop-structures. Proceedings of International Conference *Management of Landscapes Disturbed by Channel Incision*. University of Mississippi, Oxford, MS, pp. 1–6.

Paderi, F. & Benedini, M. (1964) Misuratore di portata a chiamata da gradino di fondo in un canale [Discharge measurement at a drop structure]. *L'Acqua*, 42(2), 3–15 (in Italian).

Pagliara, S. & Viti, C. (1991) Sul comportamento idraulico di un salto di fondo rigurgitato [On the hydraulic behavior of submerged drops]. *Idrotecnica*, 18(5), 343–348 (in Italian).

Rajaratnam, N. & Chamani, M.R. (1995) Energy loss at drops. *Journal of Hydraulic Research*, 33(3), 373–384; 34(2), 273–278.

Ramshorn, A. (1932) Die Energievernichtung bei Abstürzen und Schusstrecken in offenen Abwasser-kanälen [Energy dissipation at drops and chutes in open sewers]. *Die Bautechnik*, 10(12), 139–160 (in German).

Rand, W. (1955) Flow geometry at straight drop spillways. *Proceedings of ASCE*, 81(791), 1–13; 82(HY1, 881), 57–62; 82(HY3, 1010), 7–9.

Vickery, L. (1968) The rise of water level behind the nappe of a rectangular weir. *Journal of Mechanical Engineering Science*, 10(1), 81–84; 10(3), 297–298.

Hydraulic jump

Allen, J. & Hamid, H.I. (1968) The hydraulic jump and other phenomena associated with flow under rectangular sluice-gates. *Proceedings of ICE*, 40(3), 345–362; 42(4), 529–533.

Avery, S. & Novak, P. (1975) Oxygen uptake in hydraulic jumps and at overfalls. Proceedings of 16th *IAHR Congress*, Sao Paulo, 3(C38), 329–337; 6, 339–341.

Babb, A.F. & Aus, H.C. (1981) Measurement of air in flowing water. *Journal of the Hydraulics Division* ASCE, 107(HY12), 1615–1630.

Bischoff, H. & Gieseler, O. (1977) A method to determine the equivalent sand roughness by means of the circular hydraulic jump. Proceedings of 17th *IAHR Congress*, Baden-Baden, 6(S3), 613–616.

Chanson, H. (1996) *Air Bubble Entrainment in Free-surface Turbulent Shear Flows*. Academic Press, San Diego.

Garg, S.P. & Sharma, H.R. (1971) Efficiency of hydraulic jump. *Journal of the Hydraulics Division* ASCE, 97(HY3), 409–420; 97(HY9), 1570–1573; 97(HY10), 1790–1795; 97(HY11), 1923; 97(HY12), 2107–2110; 98(HY1), 278–284; 99(HY3), 527–529.

Herbrand, K. (1969) Der Wechselsprung unter dem Einfluss der Luftbeimischung (The hydraulic jump affected by air entrainment). *Wasserwirtschaft*, 59(9), 254–260 (in German).

Hornung, H.G., Willert, C. & Turner, S. (1995) The flow field downstream of a hydraulic jump. *Journal of Fluid Mechanics*, 287, 299–316.

Hoyt, J.W. & Sellin, R.H.J. (1989) Hydraulic jump as 'mixing layer'. *Journal of Hydraulic Engineering*, 115(12), 1607–1614.

Kobus, H.E. (1985) An introduction to air-water flows in hydraulics. *Mitteilung* 61. Institut für Wasserbau, Universität Stuttgart, Stuttgart.

Leutheusser, H.J. & Alemu, S. (1979) Flow separation under hydraulic jump. *Journal of Hydraulic Research*, 17(3), 193–206.

Leutheusser, H.J. & Kartha, V.C. (1972) Effects of inflow condition on hydraulic jump. *Journal of the Hydraulics Division* ASCE, 98(8), 1367–1385; 99(HY3), 550–551; 99(HY4), 698–699; 99(HY5), 859–860; 99(HY11), 2130–2131.

Long, D., Steffler, P.M., Rajaratnam, N. & Smy, P.R. (1991) Structure of flow in hydraulic jumps. *Journal of Hydraulic Research*, 29(2), 207–218.

Madsen, P.A. & Svendsen, I.A. (1983) Turbulent bores and hydraulic jumps. *Journal of Fluid Mechanics*, 129, 1–25.

Ohtsu, I. & Yasuda, Y. (1994) Characteristics of supercritical flow below sluice gate. *Journal of Hydraulic Engineering*, 120(3), 332–346.

Orlins, J.J. & Gulliver, J.S. (2000) Dissolved gas supersaturation downstream of a spillway. *Journal of Hydraulic Research*, 38(2), 151–159.

Quingchao, L. & Drewes, U. (1994) Turbulence characteristics in free and forced hydraulic jumps. *Journal of Hydraulic Research*, 32(6), 877–898.

Rajaratnam, N. (1961a) An experimental study of the air entrainment characteristics of hydraulic jump. *Journal of Institution of Engineers* India, 42(3), 247–273.

Rajaratnam, N. (1961b) The pre-entrained jump. *Civil Engineering and Public Works Review*, 56(10), 1349–1351; 56(11), 1469–1471.

Rajaratnam, N. (1962) Effect of air-entrainment on stilling basin performance. *Irrigation and Power*, 19(5), 334–343.

Rajaratnam, N. & Subramanya, K. (1969) Profile of the hydraulic jump. *Journal of the Hydraulics Division* ASCE, 94(HY3), 663–673; 95(HY1), 546–557; 95(HY2), 725–727; 96(HY2), 579–580.

Renner, J. (1974) Lufteintrag bei Deckwalzen (Air entrainment at rollers). *Wasserwirtschaft*, 54(11), 329–334 (in German).

Renner, J. & Naudascher, E. (1975) Air entrainment in surface rollers. *Journal of the Hydraulics Division* ASCE, 101(HY2), 325–327.

Resch, F.J. & Leutheusser, J.J. (1971) Mesures de turbulence dans le ressaut hydraulique [Turbulence measurements in hydraulic jumps]. *La Houille Blanche*, 26(1), 17–31 (in French).

Resch, F.J. & Leutheusser, H.J. (1972) Le ressaut hydraulique: mesures de turbulence dans la région diphasique [Hydraulic jump: measurements in the air-water region]. *La Houille Blanche*, 27(4), 279–293 (in French).

Resch, F.J., Leutheusser, H.J. & Alemu, S. (1974) Bubbly two-phase flow in hydraulic jump. *Journal of the Hydraulics Division* ASCE, 100(HY1), 137–149.

Rouse, H. (1970) Work-energy equation for the streamline. *Journal of the Hydraulics Division* ASCE, 96(HY5), 1179–1190; 96(HY10), 2159–2163; 96(HY12), 2668–2674; 97(HY7), 1138.

Wilson, E.G. & Turner, A.A. (1972) Boundary layer effects on hydraulic jump location. *Journal of the Hydraulics Division* ASCE, 98(HY7), 1127–1142; 99(HY7), 1170–1173; 100(HY2), 320.

Wu, S. & Rajaratnam, N. (1995) Free jumps, submerged jumps and wall jets. *Journal of Hydraulic Research*, 33(2), 197–212.

Yeh, H.H. (1991) Vorticity-generation mechanisms in bores. *Proceedings of the Royal Society*, London, Series A, 432, 215–231.

Pressure fluctuations

Abdul Khader, M.H. & Elango, K. (1974) Turbulent pressure field beneath a hydraulic jump. *Journal of Hydraulic Research*, 12(4), 469–489.

Akbari, M.E., Mittal, M.K. & Pande, P.K. (1982) Pressure fluctuations on the floor of free and forced hydraulic jumps. Proceedings of the International Conference *Hydraulic Modelling in Civil Engineering Structures*, Coventry, C(1). BHRA, Cranfield UK, pp. 87–96.

Anastasi, G. (1981) Ancrage du radier d'un basin amortisseur renforcé: Essais sur modèle hydroélastique [Anchorage of stilling basin floor: experiments on a hydro-elastic model]. Proceedings of 19th *IAHR Congress*, New Delhi, D, 251–258 (in French).

Bellin, A. & Fiorotto, V. (1995) Direct dynamic force measurement on slabs in spillway stilling basins. *Journal of Hydraulic Engineering*, 121(10), 686–693.

Bowers, C.E. & Tsai, F.Y. (1969) Fluctuating pressures in spillway stilling basins. *Journal of the Hydraulics Division* ASCE, 95(HY6), 2071–2079; 96(HY6), 1369–1370; 96(HY7), 1640–1643; 96(HY54–1758; 97(HY3), 454–455.

Fiorotto, V. & Rinaldo, A. (1988) Sul dimensionamento delle protezioni di fondo in bacini di dissipazione: Nuovi risultati teorici e sperimentali [On the design of stilling basin floor protections: new theoretical and experimental results]. *Giornale del Genio Civile*, 126(7–9), 179–201 (in Italian).

Fiorotto, V. & Rinaldo, A. (1992) Fluctuating uplift and lining design in spillway stilling basins. *Journal of Hydraulic Engineering*, 118(4), 578–596; 119(4), 526–528.

Fiorotto, V. & Salandin, P. (2000) Design of anchored slabs in spillway stilling basins. *Journal of Hydraulic Engineering*, 126(7), 502–512.

Gunko, F.G. (1967) Macroturbulences of flows below spillways of medium head dams and their protection against undermining. Proceedings of 12th *IAHR Congress*, Fort Collins, 2(B16), 135–143.

Khatsuria, R.M. & Deolalikar, P.B. (1987) Various approaches for determining thickness of apron lining of stilling basins: an evaluation in respect of adequacy. Proceedings of International Symposium *New Technologies in Model Testing in Hydraulic Research*, Pune, 1, 121–125.

Lopardo, R.A., De Lio, J.C. & Henning, R.E. (1987) Modelling techniques for preventing cavitation in structures submerged in hydraulic jumps. Proceedings of 25th *IAHR Congress*, Lausanne B, 177–182.

Narayanan, R. (1980) Cavitation induced by turbulence in stilling basins. *Journal of the Hydraulics Division* ASCE, 106(HY4), 616–619; 107(HY2), 244245; 107(HY10), 1271–1272.

Narayanan, R. & Reynolds, A.J. (1968) Pressure fluctuations in reattaching flows. *Journal of the Hydraulics Division* ASCE, 94(HY6), 1383–1398; 95(HY5), 1746–1750; 96(HY4), 1041–1042.

Petrikat, K., Abdul Khader, M.H. & Knoll, M. (1969) Vibration due to pressure fluctuations on baffle piers in cavitating-supercavitating flows. Proceedings of 13th *IAHR Congress*, Kyoto, 5(1), 77–81.

Rahman, M.A. (1972) Damage to Karnafuli Dam spillway. *Journal of the Hydraulics Division* ASCE, 98(HY12), 2155–2170; 99(HY11), 2148–2154; 100(HY11), 1720–1721.

Rinaldo, A. (1985) Un criterio di dimensionamento delle protezioni di fondo in bacini di smorzamento (A design criterion for stilling basin protection). *Giornale del Genio Civile*, 123(4–6), 165–186 (in Italian).

Sanchez Bribiesca, J.L. & Fuentes Mariles, O.A. (1979) Experimental analysis of macroturbulence effects on the lining of stilling basins. Proceedings of 13th *ICOLD Congress*, New Delhi, Q50(R6), 85–103.

Toso, J.W. & Bowers, C.E. (1988) Extreme pressures in hydraulic-jump stilling basins. *Journal of Hydraulic Engineering*, 114(8), 829–843.

Uppal, H.L., Gulati, T.D. & Sharma, B.A.D. (1967) A study of causes of damage to the central training wall Bhakra Dam spillway. *Journal of Hydraulic Research*, 5(3), 209–224.

Vasiliev, O.F. & Bukreyev, V.I. (1967) Statistical characteristics of pressure fluctuations in the region of hydraulic jump. Proceedings of 12th *IAHR Congress*, Fort Collins, 2(B1), 1–8; 5, 223.

Rough channel jump

Carollo, F.G., Ferro, V. & Pampalone, V. (2007) Hydraulic jumps on rough beds. *Journal of Hydraulic Engineering*, 133(9), 989–999.

Ead, S.A. & Rajaratnam, N. (2002) Hydraulic jumps on corrugated beds. *Journal of Hydraulic Engineering*, 128(7), 656–663.

Hardwick, J.D., Grant, J.C. & Donald, A.S. (1997) Design of a roughened apron to dissipate energy. *Proceedings of ICE, Water, Maritime & Energy*, 124(2), 86–94.

Hughes, W.C. & Flack, J.E. (1984) Hydraulic jump properties over a rough bed. *Journal of Hydraulic Engineering*, 110(12), 1755–1771.

Kawagoshi, N. (1993) Hydraulic jump and wave at the bed protection structures of abrupt drop provided with the artificial roughness. *Transactions of JSIDRE*, 168, 19–30 (in Japanese with English summary).

Khapaeva, A.K. (1970) Hydraulic jump over a smooth or rough bottom regarded as a wall jet. *Hydrotechnical Construction*, 4(5), 248–259.

Leutheusser, H.J. & Schiller, E.J. (1975) Hydraulic jump in a rough channel. *Water Power & Dam Construction*, 27(5), 186–191.

Mohamed Ali, H.S. (1991) Effect of roughened-bed stilling basin on length of rectangular hydraulic jump. *Journal of Hydraulic Engineering*, 117(1), 83–93.

Mohamed Ali, H.S. (1994) Bed roughened stilling basins. *Proceedings of ICE, Water, Maritime & Energy*, 106(1), 33–41.

Rajaratnam, N. (1968) Hydraulic jumps on rough beds. *Transactions of Engineering Institute of Canada*, 11(A2), 1–8.

Scour

Ali, K.H.M. & Lim, S.Y. (1986) Local scour caused by submerged wall jets. *Proceedings of ICE*, 81(4), 607–645; 83(5), 875–886.

Ali, K.H.M. & Salehi Neyshaboury, A.A. (1991) Localized scour downstream of a deeply submerged horizontal jet. *Proceedings of ICE*, 91(1), 1–18;

Altinbilek, H.D. & Basmici, Y. (1973) Localized scour at the downstream of outlet structures. Proceedings of 11th *ICOLD Congress*, Madrid, Q41(R7), 105–122.

Anonymous (1948) Model studies of Imperial Dam, desilting works, All-American Canal Structures. Boulder Canyon Project: Final Reports 6: Hydraulic investigations. *Bulletin* 4. USBR, Washington DC.

Balachandar, R. & Kells, J.A. (1997) Local channel scour in uniformly graded sediments: the timescale problem. *Canadian Journal of Civil Engineers*, 24(5), 799–807.

Breusers, H.N.C. & Raudkivi, A.J., eds. (1991) *Scouring*. IAHR Hydraulic Structures Design Manual 2. Balkema, Rotterdam.

Carstens, M.R. (1966) Similarity laws for localized scour. *Journal of the Hydraulics Division* ASCE, 92(HY3), 13–36; 93(HY2), 67–71; 94(HY1), 303–306.

Chatterjee, S.S. & Ghosh, S.N. (1980) Submerged horizontal jet over erodible bed. *Journal of the Hydraulics Division* ASCE, 106(HY11), 1765–1782; 107(HY4), 530–532; 108(HY6), 797–798.

Colaric, J., Pichon, E. & Sananes, F. (1967) Etude des affouillements à l'aval d'un seuil déversant (Scour downstream of an overflow sill). Proceedings of 12th *IAHR Congress*, Fort Collins, 3(C37), 1–8 (in French).

Dargahi, B. (2003) Scour development downstream of a spillway. *Journal of Hydraulic Research*, 41(4), 417–426.

de Groot, M.B., Bliek, A.J. & van Rossum, H. (1988) Critical scour: new bed protection design method. *Journal of Hydraulic Engineering*, 114(10), 1227–1240.

Dietz, J.W. (1972) Modellversuche über die Kolkbildung (Model tests on scour development). *Die Bautechnik*, 49(5), 162–168; 49(7), 240–245 (in German).

Di Stefano, C. & Ferro, V. (1998) Calculating average filling rock diameter for gabion-mattress channel design. *Journal of Hydraulic Engineering*, 124(9), 975–978.

Farhoudi, J. & Smith, K.V.H. (1982) Time scale for scour downstream of hydraulic jump. *Journal of the Hydraulics Division* ASCE, 108(HY10), 1147–1162; 109(8), 1182–1183.

Froehlich, D.C. & Benson, C.A. (1996) Sizing dumped rock riprap. *Journal of Hydraulic Engineering*, 122(7), 389–396; 124(6), 652–653.

Govinda Rao, N.S. & Sarma, K.V.N. (1967) Scour function. *Journal of Institution Civil Engineers*, India, 47(5), 260–286; 48(7), 1199.

Hoffmans, G.J.C.M. & Pilarczyk, K.W. (1995) Local scour downstream of hydraulic structures. *Journal of Hydraulic Engineering*, 121(4), 326–340; 122(7), 419–420.

Hogg, A.J., Huppert, H.E. & Dade, W.B. (1997) Erosion by planar turbulent wall jets. *Journal of Fluid Mechanics*, 338, 317–340.

Keutner, C. (1936) Massnahmen zur Bekämpfung der Kolkbildung stromab von Stauanlagen mit Wehrboden [Means to fight scour downstream of dams with stilling basin]. *Der Bauingenieur*, 17(27/28), 279–289 (in German).

Knauss, J. (1980) Special experiences and resultant design procedure from model tests with spillway structures at German dams. *Wasserwirtschaft*, 70(3), 84–88 (in German).

Koloseus, H.J. (1984) Scour due to riprap and improper filters. *Journal of Hydraulic Engineering*, 110(10), 1315–1324.

Lane, E.W. & Bingham, W.F. (1935) Protection against scour below overfall dams. *Engineering News-Record*, 114(March 14), 373–378.

Mason, P.J. (1989) Effects of air entrainment on plunge pool scour. *Journal of Hydraulic Engineering*, 115(3), 385–399; 117(2), 256–265.

Mason, P.J. (1993) Practical guidelines for the design of flip buckets and plunge pools. *Water Power & Dam Construction*, 45(9/10), 40–45.

Maynord, S.T., Ruff, J.F. & Abt, S.R. (1989) Riprap design. *Journal of Hydraulic Engineering*, 115(7), 937–949; 116(4), 609; 117(4), 540–544.

Meyer-Peter, E. (1927) Die hydraulischen Modellversuche für das Limmatkraftwerk Wettingen der Stadt Zürich [The hydraulic model tests for the hydropower plant on Limmat River at Wettingen of Zurich City]. *Schweizerische Bauzeitung*, 89(21), 275–279; 89(22), 291–297; 90(4), 51–52 (in German).

Meyer-Peter, E. & Müller, R. (1948) Affouillements en aval des barrages [Scour downstream of dams]. Proceedings of 2nd *IAHR Congress*, Stockholm, (29), 1–16 (in French).

Mohamed, M.S. & McCorquodale, J.A. (1992) Short-term local scour. *Journal of Hydraulic Research*, 30(5), 685–699.

Nik Hassan, N.M.K. & Narayanan, R. (1985) Local scour downstream of an apron. *Journal of Hydraulic Engineering*, 111(HY11), 1371–1385.

Orhon, M. & Bilgi, V. (1994) Deterioration of spillway of Seyhan Dam. Proceedings of 18th *ICOLD Congress*, Durban, Q71(R10), 137–147.

Prins, J.E. (1963) The time scale in scour model research. *La Houille Blanche*, 18(2), 183–188 (in French, with English summary).

Rajaratnam, N. (1980) Erosion by circular wall jets in cross flow. *Journal of the Hydraulics Division* ASCE, 106(HY11), 1867–1883; 107(HY11), 1571.

Rajaratnam, N. & Berry, B. (1977) Erosion by circular turbulent wall jets. *Journal of Hydraulic Research*, 15(3), 277–289.

Rehbock, T. (1926) Die Bekämpfung der Sohlen-Auskolkung bei Wehren durch Zahnschwellen (The fight against scour at weirs using teeth sills). *Schweizerische Bauzeitung*, 87(3), 27–31; 87(4), 44–46; 87(7), 85–86 (in German).

Rehbock, T. (1928) Die Verhütung schädlicher Kolke bei Sturzbetten (The avoidance of harmful scour at stilling basins). *Schweizerische Wasserwirtschaft*, 20(3), 35–40; 20(4), 53–58 (in German).

Shafai-Bajestan, M. & Albertson, M.L. (1993) Riprap criteria below pipe outlet. *Journal of Hydraulic Engineering*, 119(2), 181–200.

Stevens, M.A., Simmons, D.B. & Lewis, G.L. (1976) Safety factors for riprap protection. *Journal of the Hydraulics Division* ASCE, 102(HY5), 637–655; 102(HY12), 1791; 103(HY4), 457–458; 103(HY7), 810; 104(HY2), 299.

Tödten, H. (1976) Zur Beurteilung der Energiedissipation von Tosbecken (Toward the criteria of energy dissipation in stilling basins). *Der Bauingenieur*, 51(11), 429–433 (in German).

Uyumaz, A. (1988) Scour downstream of vertical gate. *Journal of Hydraulic Engineering*, 114(7), 811–816.

Wang, S.-y. & Shen, H.W. (1985) Incipient sediment motion and riprap design. *Journal of Hydraulic Engineering*, 111(3), 520–538.

Sills and blocks

Blasidell, F.W. (1948) Development and hydraulic design, Saint Anthony Falls stilling basin. *Transactions of ASCE*, 113, 483–561.

Bradley, J.N. & Peterka, A.J. (1957) Hydraulic design of stilling basins: high dams, earth dams, and large canal structures. *Journal of the Hydraulics Division* ASCE, 83(HY5) Paper 1402, 1–14; 84(HY5) Paper 1616, 25–91; 84(HY5) Paper 1832, 61–84.

Forster, J.W. & Skrinde, R.A. (1950) Control of the hydraulic jump by sills. *Transactions of ASCE*, 115, 973–1022.

Francis, J.R.D. & Phelps, H.O. (1967) Open channel flow-force analysis. *Proceedings of the Institution of Civil Engineers*, 39(1), 95–101; 40(3), 393–396.

Hager, W.H. & Li, D. (1992) Sill-controlled energy dissipator. *Journal of Hydraulic Research*, 30(2), 165–181; 31(2), 282–283.

Karki, K.S. & Mishra, S.K. (1987) Force on baffle wall under submerged hydraulic jump. Proceedings of 15th National Conference *Fluid Mechanics and Fluid Power*, Srinagar, pp. 73–77.

Locher, F.A. & Hsu, S.T. (1984) Energy dissipation at high dams. *Developments in Hydraulic Engineering*, 2, 183–238.

Narayanan, R. & Schizas, L.S. (1980) Force fluctuations on sill of hydraulic jump. *Journal of the Hydraulics Division* ASCE, 106(HY4), 589–599; 107(HY3), 383.

Narayanan, R. & Schizas, L.S. (1980) Force on sill of forced jump. *Journal of the Hydraulics Division* ASCE, 106(HY7), 1159–1172; 107(HY7), 949–951; 108(HY2), 285.

Novak, P. & Cabelka, J. (1981) *Models in Hydraulic Engineering*. Pitman, Boston.

Ohtsu, I. (1981) Forced hydraulic jump by a vertical sill. *Transactions of JSCE, Hydraulics and Sanitary Division*, 13(7), 165–168.

Ohtsu, I., Yasuda, Y. & Hashiba, H. (1996) Incipient jump conditions for flows over a vertical sill. *Journal of Hydraulic Engineering*, 122(8), 465–469.

Ohtsu, I., Yasuda, Y. & Yamanaka, Y. (1991) Drag on vertical sill of forced jump. *Journal of Hydraulic Research*, 29(1), 29–47; 30(2), 277–288.

Rai, S.P. (1987) Wall-wakes on moderate adverse pressure gradients. *Journal of Hydraulic Research*, 24(5), 377–390.

Rajaratnam, N. (1964) The forced hydraulic jump. *Water Power*, 16(1), 14–19; 16(2), 61–65.

Rajaratnam, N. & Murahari, V. (1971) A contribution to forced hydraulic jumps. *Journal of Hydraulic Research*, 9(2), 217–240.

Rajaratnam, N. & Rai, S.P. (1979) Plane turbulent wall wakes. *Journal of Engineering Mechanics Division* ASCE, 105(EM5), 779–794.

Rand, W. (1957) An approach to generalized design of stilling basins. *Trans. New York Academy of Science*, 20(2), 173–191.

Rand, W. (1965) Flow over a vertical sill in an open channel. *Journal of the Hydraulics Division* ASCE, 91(HY4), 97–121.

Rand, W. (1966) Flow over a dentated sill in an open channel. *Journal of the Hydraulics Division* ASCE, 92(HY5), 135–153; 93(HY3), 229–230; 94(HY1), 307–308.

Rand, W. (1967) Efficiency and stability of forced hydraulic jump. *Journal of the Hydraulics Division* ASCE, 93(HY4), 117–127; 94(HY3), 774; 94(HY6), 1529.

Rand, W. (1970) Sill-controlled flow transitions and extent of erosion. *Journal of the Hydraulics Division* ASCE, 96(HY4), 927–939; 97(HY2), 359–360.

Shukry, A. (1957) The efficacy of floor sills under drowned hydraulic jumps. *Journal of the Hydraulics Division* ASCE, 83(HY3) Paper 1260, 1–18; 83(HY5) Paper 1417, 31; 83(HY6) Paper 1456, 15–24; 84(HY1) Paper 1558, 33–37; 84(HY5) Paper 1832, 35–38.

Stanley, C.M. (1934) Study of stilling-basin design. *Transactions of* ASCE, 99, 490–523.

Steele, I.C. & Monroe, R.A. (1929) Baffle-pier experiments on models of Pit River Dams. *Transactions of* ASCE, 94, 452–546.

Sumi, T. (1988) Forced hydraulic jump type. Proceedings of International Symp. *Hydraulics for High Dams*, Beijing, 106–113.

Tyagi, D.M., Pande, P.K. & Mittal, M.K. (1978) Drag on baffle walls in hydraulic jump. *Journal of the Hydraulics Division* ASCE, 104(HY4), 515–525.

Stilling basin

Alam, S. (1988) A new conception of energy dissipation for medium head overflow spillways discharging into erodible river beds. Proceedings of 16th *ICOLD Congress*, San Francisco, C(12), 1109–1125.

Anonymous (1964) Energy dissipators for spillways and outlet works. *Journal of the Hydraulics Division* ASCE, 90(HY1), 121–147; 90(HY4), 359–363; 90(HY5), 201–222; 91(HY2), 292–300.

Arthur, H.G. & Jabara, M.A. (1967) Problems involved in operation and maintenance of spillways and outlets at Bureau of Reclamation Dams. Proceedings of 9th *ICOLD Congress*, Istamboul, Q33(R5), 73–93.

Berryhill, R.H. (1957) Stilling basin experiences of the Corps of Engineers. *Journal of the Hydraulics Division* ASCE, 83(HY3, Paper 1264), 1–36; 83(HY6, Paper 1456), 25–26; 84(HY5, Paper 1832), 39–40.

Bradley, J.N. & Peterka, A.J. (1957) The hydraulic design of stilling basins. *Journal of the Hydraulics Division* ASCE, 83(HY5), 1–32; 83(HY5), 1–24; 84(HY2, 1616), 25–91; 84(HY5, 1832), 61–84.

Cassidy, J.J. (1990) Fluid mechanics and design of hydraulic structures. *Journal of Hydraulic Engineering*, 116(8), 961–977.

Center, G.W. & Rhone, T.J. (1973) Emergency redesign of Silver Jack Spillway. *Journal of the Power Division* ASCE, 99(PO2), 265–279.

Cochrane, R.B. (1959) Operation of spillways in Northwest Projects. *Journal of the Hydraulics Division* ASCE, 85(HY8), 7–38.

Dietz, J.W. & Pulina, B. (1975) Tosbecken und Sohlensicherung am Rheinwehr Iffezheim [Stilling basin and bottom protection of Rhine Weir Iffezheim]. *Wasserwirtschaft*, 65(9), 226–233 (in German).

Garbrecht, G. (1959) Über die Berechnung von Sturzbetten [On the computation of stilling basins]. *Wasserwirtschaft*, 49(2), 66–73; 49(3), 95–105 (in German).

Gedney, R.H. (1961) Stilling basin damage at Chief Joseph Dam. *Journal of the Hydraulics Division* ASCE, 87(HY2), 97–120.

Gong, Z., Liu, S., Xie, S. & Lin, B. (1987) Flaring gate piers. In: Kia, A.R. & Albertson, M.L. (eds) *Design of Hydraulic Structures*. Colorado State University, Fort Collins, pp. 139–146.

Hager, W.H. (1994) Central counter-current stilling basin. *Dam Engineering*, 5(2), 15–28.

Hartung, F. (1972) Gestaltung von Hochwasserentlastungsanlagen bei Talsperrendämmen [Design of spillways of dams]. *Wasserwirtschaft*, 62(1/2), 39–51 (in German).

Keim, S.R. (1962) The Contra Costa energy dissipator. *Journal of the Hydraulics Division* ASCE, 88(HY2), 109–122; 88(HY5), 335–336; 88(HY6), 193–195; 89(HY5), 121–123.

Knauss, J. (1971) Hydraulische Probleme beim Entwurf von Hochwasserentlastungsanlagen an grossen und kleinen Dämmen [Hydraulic design problems of spillways of large and small dams]. *Bericht* 22. Versuchsanstalt für Wasserbau, TU München, München (in German).

Lowe III, J., Bangash, H.D. & Chao, P.C. (1979) Some experiences with high velocity flow at Tarbela Dam Project. Proceedings of 13th *ICOLD Congress*, New Delhi, Q50(R13), 215–247,

Novak, P., Moffat, A.I.B., Nalluri, C. & Narayanan, R. (1989) *Hydraulic Structures*. Unwin Hyman, London.

Pillai, N.N., Goel, A. & Dubey, A.K. (1989) Hydraulic jump type stilling basin for low Froude numbers. *Journal of Hydraulic Engineering*, 115(7), 989–994.

Rath, G.B., Vittal, N. & Ranga Raju, K.G. (1983) Performance of stilling basins under submerged flow conditions. Proceedings of 20th *IAHR Congress*, Moscow, 7, 443–450.

Regan, R.P., Munch, A.V. & Schrader, E.K. (1979) Cavitation and erosion damage of sluices and stilling basins at two high-head dams. Proceedings of 13th *ICOLD Congress*, New Delhi, Q50(R11), 177–198.

Rudavsky, A.B. (1976) Selection of spillways and energy dissipators in preliminary planning of dam developments. Proceedings of 12th *ICOLD Congress*, Mexico, Q46(R9), 153–180.

Scott, K.F., Reeve, W.T.N. & Germond, J.P. (1968) Farahnaz Pahlavi Dam at Latiyan. *Proceedings of ICE*, 39(3), 353–395; 41(3), 585–595.

Simmons, Jr., W.P. (1964) Hydraulic design of transitions for small canals. Water Resources Technical Publication, *Engineering Monograph* 33. USBR, Denver.

Smith, C.D. (1955) Hydraulic design of outlet basin for single culvert. *The Engineering Journal*, 38(8), 1063–1069.

Smith, C.D. (1988) Outlet structure design for conduits and tunnels. *Journal of Waterway, Port, Coastal, and Ocean Engineering*, 114(4), 503–515.

Smith, C.D. & Korolischuk, E.M. (1973) Modified USBR impact basin. *Journal of the Hydraulics Division* ASCE, 99(HY1), 283–287.

Strassburger, A.G. (1973) Spillway energy dissipator problems. Proceedings of 11th *ICOLD Congress*, Madrid, Q41(R16), 249–268.

Tung, Y.-K. & Mays, L.W. (1982) Optimal design of stilling basins for overflow spillways. *Journal of the Hydraulics Division* ASCE, 108(HY10), 1163–1178; 110(1), 79–82.

USBR (1987) *Design of Small Dams*, 4th ed. US Bureau of Reclamation, Denver CO.

Zai-Chao, L. (1987) Influence area of atomization-motion downstream of dam. Proceedings of 22nd *IAHR Congress*, Lausanne, B, pp. 203–208.

Zhenlin, D., Lizhong, N. & Longde, M. (1988) Some hydraulic problems of slit-type buckets. Proceedings of the International Symposium *Hydraulics for Large Dams*, Beijing, pp. 287–294.

Zizhong, G. (1988) Energy dissipation of high-velocity flow. Proceedings of the International Symp. *Hydraulics for High Dams*, Beijing. IAHR, Delft, pp. 17–32.

Submerged hydraulic jump

Albertson, M.L., Dai, Y.B., Jensen, R.A. & Rouse, H. (1950) Diffusion of submerged jets. *Transactions of* ASCE, 115, 639–697.

Cola, R. & Fioratti, M. (1986) Esame del comportamento di vasche di dissipazione a risalto annegato [Study of behavior of submerged stilling basin]. Proceedings of the 20th *Convegno di Idraulica e Costruzioni Idrauliche*, Padova, 731–761 (in Italian).

Einwachter, J. (1933) Der Wechselsprung mit gestauter Deckwalze [The submerged hydraulic jump]. *Wasserkraft und Wasserwirtschaft*, 28(17), 200–202 (in German).

Govinda Rao, N.S. & Rajaratnam, N. (1963) The submerged hydraulic jump. *Journal of the Hydraulics Division* ASCE, 89(HY1), 139–162; 89(HY4), 277–279; 89(HY5), 147–152; 90(HY3), 313–316.

Gunal, M. & Narayanan, R. (1998) k-ε turbulence modelling of submerged hydraulic jump using boundary-fitted coordinates. *Proceedings of ICE, Water, Maritime & Energy*, 130(2), 104–114.

Long, D., Rajaratnam, N. & Steffler, P.M. (1990) LDA study of flow structure in submerged hydraulic jump. *Journal of Hydraulic Research*, 28(4), 437–460.

Lopardo, R.A. & Sauma Haddad, J.C. (1993) Estimation of fatigue aspects on structures submerged in macroturbulent flows. Proceedings of *Advances of Hydrosciene and Engineering*, Washington DC, 1, 1006–1012.

Nagaratnam, S. & Mura Hari, V. (1969) The submerged hydraulic jump in trapezoidal channels. *Irrigation and Power*, 27(1), 41–50.

Narasimhan, S. & Bhargava, V.P. (1976) Pressure fluctuations in submerged jump. *Journal of the Hydraulics Division* ASCE, 102(HY3), 339–350; 102(HY12), 1785–1787; 103(HY7), 810.

Narayanan, R. (1978) Pressure fluctuations beneath submerged jump. *Journal of the Hydraulics Division* ASCE, 104(HY9), 1331–1342; 105(HY7), 917–919; 106(HY5), 938.

Ohtsu, I., Yasuda, Y. & Awazu, S. (1990) Free and submerged hydraulic jumps in rectangular channels. *Report* 35. Research Institute of Science and Technology, Nihon University, Tokyo.

Rajaratnam, N. (1965) Submerged hydraulic jump. *Journal of the Hydraulics Division* ASCE, 91(4), 71–96; 92(HY1), 146–155; 92(HY2), 420–421; 92(HY4), 154–156; 92(HY6), 207; 93(HY3), 179.

Smetana, J. (1934) Experimental study on the submerged hydraulic jump. *Prace e Studie*, 13, 1–40. Institute of Hydrology and Hydrotechnics T.G. Masaryk, Prague (in Czech, with French Summary).

Wu, S. & Rajaratnam, N. (1995) Effect of baffles on submerged flows. *Journal of Hydraulic Engineering*, 121(9), 644–652; 123(5), 479–481.

Undular hydraulic jump

Abbott, M.B. & Rodenhuis, G.S. (1972) A numerical simulation of the undular hydraulic jump. *Journal of Hydraulic Research*, 10(3), 239–257; 12(1), 141–157.

Benet, F. & Cunge, J.A. (1971) Analysis of experiments on secondary undulations caused by surge waves in trapezoidal channels. *Journal of Hydraulic Research*, 9(1), 11–33.

Benjamin, T.B. & Lighthill, M.J. (1954) On cnoidal waves and bores. *Proceedings of the Royal Society*, Series A, 224, 448–460.

Einwachter, J. (1935) Wassersprung- und Deckwalzenlänge [Lengths of hydraulic jump and roller]. *Wasserkraft und Wasserwirtschaft*, 30(8), 85–88 (in German).

Holtorff, G. (1967) Der gewellte Wassersprung [The undular hydraulic jump]. *Die Wasserwirtschaft*, 57(12), 427–432 (in German).

Iwasa, Y. (1955) Undular jump and its limiting condition for existence. Proceedings of 5th *Japan National Congress for Applied Mechanics*, pp. 315–319.

Jones, L.E. (1964) Some observations on the undular jump. *Journal of the Hydraulics Division* ASCE, 90(HY3), 69–82; 90(HY6), 351–354; 91(HY1), 185–192; 92(HY4), 103–110.

Meyer, R.E. (1967) Note on the undular jump. *Journal of Fluid Mechanics*, 28(2), 209–221.

Montes, J.S. (1986) A study of the undular jump profile. Proceedings of 9th *Australasian Fluid Mechanics Conf.*, Auckland, 148–151.

Montes, J.S. & Chanson, H. (1998) Characteristics of undular hydraulic jumps: experiments and analysis. *Journal of Hydraulic Engineering*, 124(2), 192–205.

Ohtsu, I., Yasuda, Y. & Gotoh, H. (2001) Hydraulic condition for undular-jump formations. *Journal of Hydraulic Research*, 39(2), 203–209; 40(3), 379–384.

Ohtsu, I., Yasuda, Y. & Gotoh, H. (2003) Flow conditions of undular hydraulic jumps in horizontal rectangular channels. *Journal of Hydraulic Engineering*, 129(12), 948–955.

Peregrine, D.H. (1966) Calculations of the development of an undular bore. *Journal of Fluid Mechanics*, 25(2), 321–330.

Preissmann, A. & Cunge, J.A. (1967) Low-amplitude undulating hydraulic jump in trapezoidal canals. *Journal of Hydraulic Research*, 5(4), 263–279.

Sandover, J.A. & Zienkiewicz, O.C. (1957) Experiments on surge waves. *Water Power*, 10(11), 418–424.

Sandover, J.A. & Taylor, C. (1962) Cnoidal waves and bores. *La Houille Blanche*, 17, 443–455.

Zienkiewicz, O.C. & Sandover, J.A. (1957) The undular surge wave. Proceedings of the 7th *IAHR Congress*, Lisbon, D(25), 1–12.

Chapter 6 Frontispiece (a) Spillway operation of Kariba Dam in 2010: Double curvature arch dam 128 m high with 617 m long crest. Orifice spillway contains six caterpillar gates of 8.80 m height and 9.15 m width, providing total spilling capacity of 9500 m³/s (Courtesy Bernard Goguel), (b) Deriner Dam, Turkey, impacting jets from two outlets thereby greatly reducing scour action (Courtesy Marc Balissat, Baden, Switzerland)

Chapter 6

Ski jump and plunge pool

6.1 Introduction

Besides the stilling basin (Chapter 5), the ski jump, often referred to as flip bucket or terminal jet deflector, and plunge pools are among the important terminal structures of a dam spillway. Whereas ski jumps are mainly employed to direct a high-speed chute flow away from the dam, the plunge pool allows for the adequate energy dissipation so that the tailwater of the dam is free from turbulence concentrations, high-speed flow, surface waves, and excessive air concentrations, similar to the tailwater of the stilling basin (Chapter 5).

Given that standard stilling basins are limited in the approach flow velocity to some 20 m/s because of cavitation problems, tailwater requirements, and extreme pressure fluctuations, alternatives were developed as dam heights increased. Starting in the 1930s, the ski jump was introduced in which the flow at the downstream chute end is deflected into the air to direct the air-water flow away from the immediate dam structure, impacting at a location where the flow has no more harm to the dam. This feature was found important to avoid damages of the dam foundation, but it was also realized that concerns previously mentioned with stilling basins were practically removed. In turn, due to high-speed jet impact, a so-called scour hole develops, which can become deep, thereby endangering the stability of the valley slopes and indirectly dam abutments, as was experienced for decades at Kariba Dam in Africa, among others. This example will be considered below in detail, based on recent laboratory tests, to detail procedures to avoid damages as experienced from the 1960s until now.

Ski jumps or trajectory basins are usually arranged straight in plan view, because minor plan curvature results in excessive shock waves (Chapter 4). For similar reasons, the widths of the chute and the ski jump are identical to suppress shock waves due to width reductions or expansions, as discussed in Chapter 4. In the longitudinal section, the ski jump includes a rounded base at the chute end with a takeoff angle of the order of the approach flow angle, so that the high-speed flow is directed into the atmosphere as a free jet aerated all over its surface. In contrast, the approach flow portion of stilling basins includes only a change of direction from the chute angle to the horizontal, and the flow is not aerated from below. The selection of either of these basic hydraulic structures depends on site conditions and the approach flow velocity.

Terminal jet deflectors are essentially a generalized form of ski jumps, with the particularity of transverse jet deflection in addition to the streamwise take-off into the atmosphere. These poorly studied structures are often employed as terminal elements of bottom outlets (Chapter 8), including a jet deflection of some 45° to realign the flow to the tailwater. A particular problem relies on flow choking, given that the high-speed flow may break down if the deflector geometry is too abrupt, so that a hydraulic jump may develop at the takeoff location, resulting in a hydraulic drop rather than in a trajectory jet. As for the ski jump, the scour hole is then too close to the dam structure, undermining its foundation by the drop flow. Two main jet deflectors are presented below, one for small and the other for higher supercritical approach flows.

The main issue of ski jumps and trajectory jet deflectors is the generation of a scour hole sufficiently away from the dam. Regarding the physical processes involved and the scour potential assessment methods, a distinction is made between scour in loose, granular material, and scour in partially fissured rock. Despite various research efforts in the past, relatively few definite results are so far available. A scour hole has implications on dam hydraulics, namely the formation of spray, destabilization of adjacent slopes and danger of slides into the scour hole, and the deposition of scour material in the tailwater, resulting in a backwater effect which may reach the takeoff elevation of outlet structures thereby reducing the gross head and consequently power generation. These issues are to be assessed carefully so that the combination of ski jump or trajectory deflector with the plunge pool scour hole represents an adequate design. The final goal is to avoid damage to the dam from these outlet elements.

The current spillway design has a tendency to increase the unit discharge. In gated chute ski jumps, i.e. flip buckets, unit discharges ranging from 200 to 300 m^3/sm are not rare anymore since the cavitation risk of chutes is mitigated by bottom aerators (Schleiss, 2002). This trend is also confirmed by many high dam projects in China and built in narrow valleys subjected with large discharges. Special experiences on high gravity, high arch, and high rockfill dams are reported by Gao et al. (2011).

This chapter aims to highlight the main features encountered with ski jumps or similar flip buckets as the origins of high-speed jets up to jet impact at the plunge pool sufficiently far away from any dam structures or embankments which may be endangered by the enormous energy release due to air-water mixture jet impact. As to plunge pools, both the granular and the rock scour types are considered based mainly on recent hydraulic laboratory techniques, by which the complexities of jet impact can be understood from the engineering perspective. The chapter both aims to present the underlying flow features and the main hydraulic relations by which the main questions can be solved, and to provide a keen outlook to a fascinating field in hydraulics that has to be considered in the future, given the many still unsolved problems.

6.2 Ski jump

6.2.1 Description of structure and takeoff

In a trajectory basin or ski jump, i.e. flip buckets, the energy is mainly dissipated by jet dispersion and flow turbulence. To become hydraulically effective, the approach flow energy head should at least be 30 m, and the discharge in $[m^3s^{-1}]$ be smaller than $250(H - H_t)$ with $H_t = 8$ m (Coyne, 1951; Maitre and Obolensky, 1954; Mason, 1982, 1993). Ski-jump spillways are currently widely used in dam engineering for they appear to be the only hydraulic element allowing for the technically sound control of large quantities of excess energy usually during the flood season. Further details are provided by e.g. Marcano et al. (1988).

To design all features of the flip bucket and the plunge pool for the probable maximum flood (PMF) condition is often not economical (Mason, 1983; Schleiss, 2002). More frequent floods as the 500-year flood for the ski-jump geometry, and the 100- to 200-year flood for plunge pool pre-excavation are recommended. However, significant damages endangering the dam should not occur even for the PMF.

From a hydraulic perspective, the ski-jump spillway is composed of (Figure 6.1):

① Approach flow chute,
② Deflection and takeoff at flip bucket,
③ Dispersion of water jet in air,
④ Jet impact and scour hole, and
⑤ Tailwater zone.

The basic ski-jump spillway is composed of a rectangular constant width chute and a circular-shaped take-off bucket. Special shapes involve buckets at orifice outlets in arch dams, and flip buckets or terminal jet deflectors at bottom outlets, with a curved jet trajectory in the plan view (Chapter 8). Below, the standard structure is considered.

The optimum take-off elevation of a ski jump from the dam toe relative to the dam height is 30% to 50%. If it is too high, the velocity is not large enough for sufficient jet dispersion, whereas the trajectory length is too short otherwise. The elevation of the bucket lip has to be above the maximum tailwater level to (1) prevent material entering the bucket causing abrasion and (2) counter cavitation damage due to submergence fluctuations. An exception is the slotted-bucket stilling basin (Chapter 5).

The ski-jump bucket has to deflect the water flow into the air and guide it to the proper impact location. It should operate properly under all discharges and be designed for both the static and the dynamic pressure loads. The bucket shape is normally circular of bucket (subscript b) radius R_b. The approach flow slope should be smaller than 4:1 whereas the take-off angle α_j ranges between 20 and 40° (Figure 6.2).

Geometrical ski-jump deflection angles α_j between 15° and 35° are recommended by Peterka (1964), up to 30° are proposed by USBR (1978), Shivashankara Rao (1982) mentions values between 30° and 40°, whereas the typical spectrum of 20° to 40° is proposed by Vischer and Hager (1998), while Mason (1993) refers to the widely adopted range from 30° to 35°. The angle between the lower jet surface and the concrete structure downstream of the lip should exceed ~40° to avoid local sub-pressure on the concrete surface (Mason, 1993).

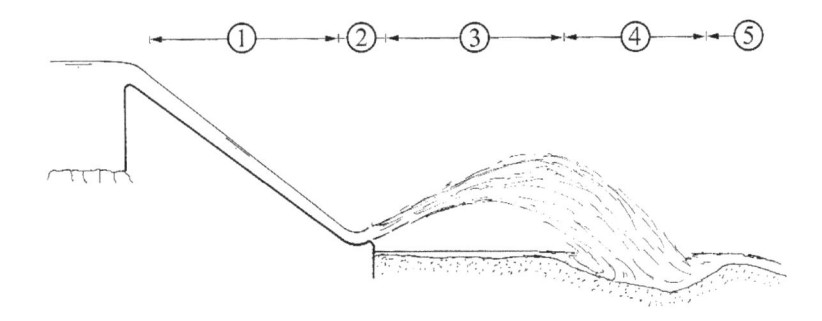

Figure 6.1 Portions of ski-jump spillway, for details see main text

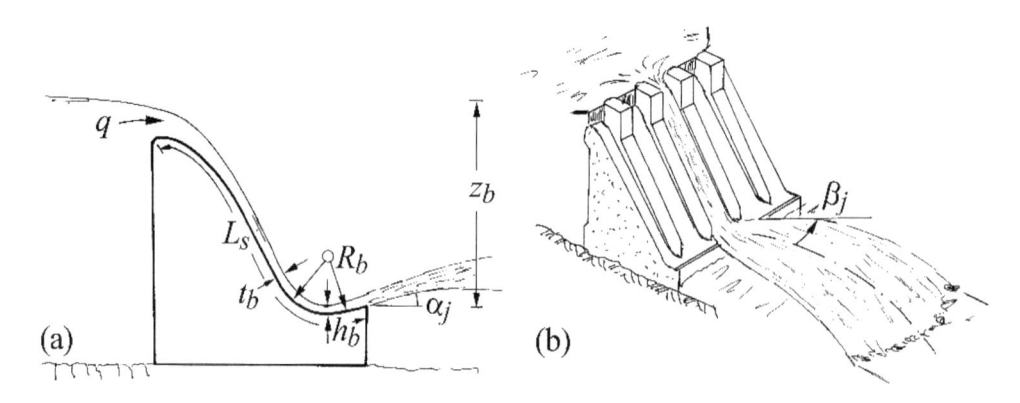

Figure 6.2 Ski jump (a) section with notation, (b) transverse jet spread

Assuming concentric streamlined flow of approach flow depth t_b and Froude number $F_o = q / \left(g t_b^3 \right)^{1/2}$, the maximum (subscript M) pressure head due to flow deflection is (Hager and Schleiss, 2009)

$$\frac{p_M}{\rho g t_b} = \frac{t_b}{R_b} F_o^2. \tag{6.1}$$

This maximum pressure head is assumed to apply along the entire bucket length in a preliminary design, independent of the deflection angle (Prasad, 1984; Rajan and Shivashankara Rao, 1980; Shivashankara Rao, 1982; Varshney and Bajaj, 1970; Lenau and Cassidy, 1969; Chen and Yu, 1965; Henderson and Tierney, 1962; Balloffet, 1961; Elevatorski, 1958).

Vischer and Hager (1995) reviewed various approaches for the bucket radius R_b. With $H_o = V_o^2 / 2g$ as the approach flow energy head and p_M the maximum pressure, Damle *et al.* (1966) and the USBR (1987) proposed, respectively,

$$R_b / t_b = \left(H_o / t_b \right)^{1/2}, \tag{6.2}$$

$$R_b / t_b = V_o^2 / (2 p_M / \rho). \tag{6.3}$$

The *flaring gate pier* is a recent development of the ski jump to enhance jet dispersion (Figure 6.3). These are located upstream from the bucket inducing air pockets at the pier wake. An air-water mixture is thus deflected into the air, instead of a compact water flow (Gao, 2011). No practical experience of this Chinese design is currently available, however.

The *slit-type flip bucket* was introduced by Zhenlin *et al.* (1988) mainly for bottom outlets. The bucket contracts the flow, promotes air entrainment and jet dispersion, therefore (Figure 6.4). The contraction ratio must be carefully selected to inhibit flow choking and the formation of a hydraulic jump on the bucket. Information on chute contractions (Chapter 4) serves as a preliminary guide in addition to Gao's (2011) approach. This design looks promising but experience is currently also not available.

A principal disadvantage of ski jumps along with jet generation is spray formation, leading to saturation and possible slides of adjacent valley flanks, and freezing on roads and

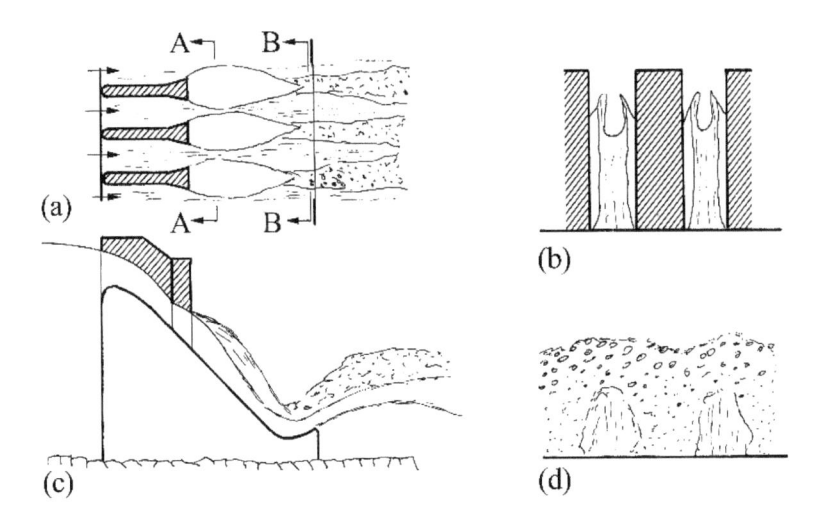

Figure 6.3 Flaring gate pier (a) plan, (b) side view, (c) section A-A, (d) section B-B (Gong *et al.*, 1987)

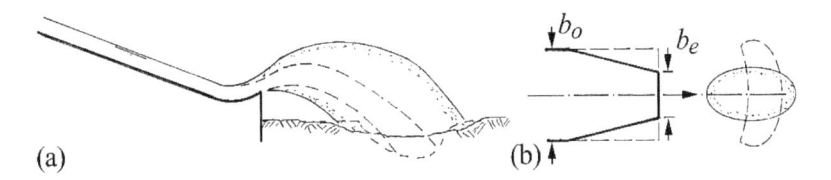

Figure 6.4 Slit-type flip bucket (a) section, (b) plan and jet section. (- - -) Jet caused by conventional bucket (Zhenlin *et al.*, 1988)

switchyards in cold regions. Further, residual jet impact energy can produce scour of the mobile riverbed or the adjacent rock. Chutes combined with ski jumps should only be applied for sediment free flows to avoid abrasion on the chute and flip bucket. The energy dissipation and scour action at jet impact is reduced by impact of two jets, as highlighted in the frontispiece of this chapter.

6.2.2 Jet trajectory and disintegration

Basic flow features

A vast number of papers is available on the disintegration of liquid jets in air. Most of these involve means to improve the compactness of the jet, such as in fire nozzles or irrigation sprinklers. In contrast, the formation of a highly disintegrated jet is required for energy dissipation. In the extreme, the more or less compact approach flow is deflected into the air by the trajectory bucket so that a highly concentrated spray falls back onto the tailwater, thereby reducing scour. Such a high degree of disintegration is not feasible, however. Spray flow by ski jumps was studied by Zai-Chao (1987). Figure 6.5 shows various zones to be distinguished.

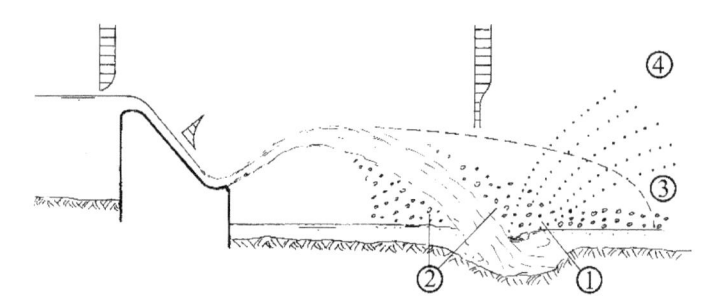

Figure 6.5 Spray flow induced by ski jump with ① splash drop, ② rainstorm, ③ atomization by rain, ④ atomization by wind (Zai-Chao, 1987)

The disintegration of a water jet in air is enhanced by:

- Approach flow turbulence
- Approach flow swirl
- Approach flow geometry
- Counter-current wind
- Fluid properties.

Because the jets studied under laboratory conditions normally have a small diameter, effects of surface tension and viscosity are significant. The number of parameters that influence a liquid jet in air is so large that few general results are available. From a literature study, the disintegration process is enhanced by (Vischer and Hager, 1995):

- Non-circular cross section to counter jet compactness
- 'Jet roughening' to increase turbulence level (beware of cavitation damage)
- Abrupt transition from bucket to air
- Jet aeration thereby creating an air-water mixture at take-off zone.

All these measures should be in a relation between adding to jet dispersion, cavitation damage as well as economy. Actually, few general guidelines are available, mainly because of lack of prototype observations. Jet dispersion as occurs at ski jumps is also governed by complex hydraulic phenomena including the interaction of fluid viscosity, surface tension, air entrainment, turbulence, and gravity.

Air entrainment in water jets was studied by Ervine and Falvey (1987). The lateral jet spreading was related to the turbulence number $T = u'/V_o$ with u' as the rms value of the instantaneous axial velocity and V_o as the approach flow velocity (Figure 6.6). For a typical turbulence intensity of 5% to 8%, the jet spread is $\alpha_d = 3\%$ to 4%, and the inner core has a decay angle of $\alpha_c = 0.5\%$ to 1%. A turbulent jet begins to break up if the inner core has completely disappeared, i.e. a relative breakup distance of $L_b/D_o = 50$ to 100. Based on experimental data and a literature survey, jet issuance parameters as the turbulence number T relevant for engineering practice are stated by Manso *et al.* (2008).

$$z = \tan \alpha_j x - \frac{g}{2V_j^2 \cos^2 \alpha_j} x^2. \tag{6.4}$$

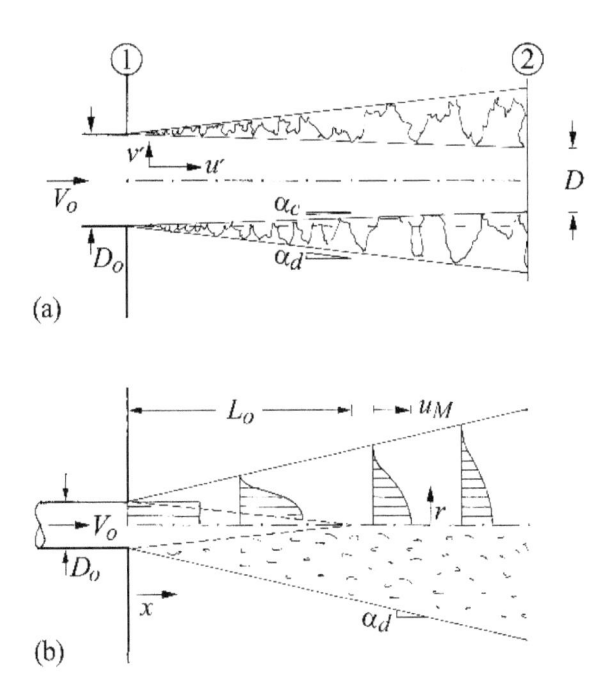

Figure 6.6 Disintegration of turbulent water jet in air (a) flow geometry, (b) jet spreading

With $H_j = V_j^2 / (2g)$ as the take-off velocity head and V_j as take-off velocity, the location x_M and the maximum jet elevation z_M are, respectively,

$$x_M / H_j = 2 \sin \alpha_j \cos \alpha_j; \quad z_M / H_j = \sin^2 \alpha_j. \tag{6.5}$$

Further, the local trajectory angle α_t is

$$\tan \alpha_t = \tan \alpha_j - (x / H_j) / (2 \cos^2 \alpha_j). \tag{6.6}$$

The trajectory length L_t is with the impact height z_i (Figure 6.7)

$$L_t / H_j = 2 \sin \alpha_j \cos \alpha_j \left[1 + \left(1 + \frac{z_i}{H_j \sin^2 \alpha_j} \right)^{1/2} \right]. \tag{6.7}$$

The above ballistic jet equation neglects jet disintegration as well as aerodynamic interaction by friction. Thus, effective trajectories observed on hydraulic models or especially on prototypes considerably differ from the simplified ballistic description. The reason for this difference is based on uncertainties concerning the effective take-off velocity V_j, the effective flow depth t at jet issuance, the jet take-off angle α_j, and the disintegration processes combined with the interaction of surrounding air. The observed prototype jet length is therefore shorter than from Eq. (6.7). To consider this reduced jet travel length, USBR (1978) proposes

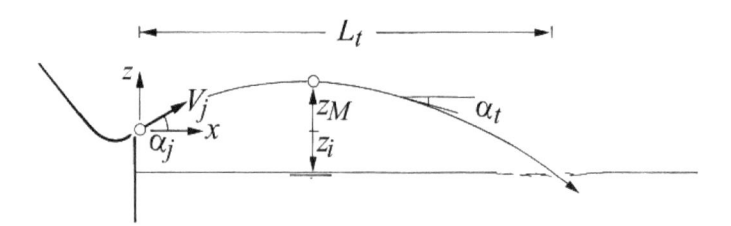

Figure 6.7 Definition of one-dimensional jet trajectory $z(x)$

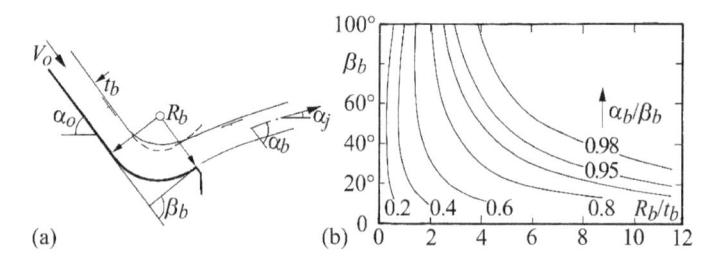

Figure 6.8 Takeoff angle $\alpha_j = \alpha_b - \alpha_o$ (a) flow geometry, (b) bucket angle ratio α_b/β_b versus relative bucket radius R_b/t_b for various deflection angles β_b

a factor of 1.1 to multiply with the negative term of Eq. (6.7). Gunko *et al.* (1965) observed a reduction of the trajectory length for $F > 6$. Kawakami (1973), comparing prototype data with the theoretical jet trajectories, observed an effect if $V_j > 13$ m/s.

The bucket take-off angle α_j does not coincide with the terminal bucket angle α_b because of the particular pressure distribution. Orlov (1974) presented a diagram to determine $\alpha_j = \alpha_b - \alpha_o$ versus the deflection angle β_b and the relative bucket radius R_b/t_b (Figure 6.8). Orlov used the conformal mapping method thereby limiting his results to $F_o > 5(\sin\alpha_o)^{1/2}$. Dhillon *et al.* (1981) or Pfister (2012) indicate that the effective jet take-off angle measured just downstream of the ski-jump lip differs from the geometrical (concrete surface) take-off angle α_j. Accordingly, using α_j for trajectory computation results in non-adequate, typically too long jets. In reality, the jet impact location on the plunge pool surface is closer at the dam toe than estimated with α_j.

To derive reliable jet trajectories, particularly in the jet far field, trajectories were fitted by Pfister *et al.* (2014) based on the data of Heller (2005), Schmocker (2008), and Balestra (2012). 'Virtual' upper (subscript U) and lower (subscript L) jet take-off angles α_U and α_L were thus determined. These are often significantly smaller than the geometrical value α_j. These take-off angles are based on the dimensionless parameter

$$\Lambda = \tan\alpha\left(1 - \frac{h_o}{R}\right)^{1/3}. \tag{6.8}$$

The virtual take-off angles relative to the horizontal are given by linear functions (Pfister *et al.*, 2016)

$$\tan\alpha_U = 0.84\,\Lambda - 0.04, \tag{6.9}$$

$$\tan\alpha_L = 0.80\,\Lambda - 0.07. \tag{6.10}$$

The entire upper and lower jet trajectories are now determined from Eq. (6.4) using either α_U or α_L instead of α_j.

The transverse jet expansion angle β_j (Figure 6.2b) was reanalyzed by Vischer and Hager (1995). It depends mainly on the bucket flow depth relative to the fall height z_b, and the relative discharge $\bar{q} = q / \left(gL_S^3 \right)^{1/2}$ with L_S as spillway length (Fig. 6.2a), resulting in

$$\tan \beta_j = \frac{1.05(h_b / z_b)^{1/2}}{\tanh(6\bar{q}^{1/3})}. \tag{6.11}$$

The angle ranges from $\beta_j = 5°$ to $10°$, i.e. important in defining the jet impact zone. The streamwise trajectory analysis based on virtual take-off angles indicates similar results.

Overview on experimental approaches

A number of laboratory studies on ski jumps were conducted, including the works of Heller *et al.* (2005), Steiner *et al.* (2008), Schmocker *et al.* (2008), Balestra (2012) and Pfister *et al.* (2014, 2016). Their main results are presented below.

Heller *et al.* (2005) determine three issues: (1) Definition of jet trajectories for black-water approach flow, (2) energy dissipation across a ski jump, and (3) choking flow features of the circular-shaped bucket. The bucket pressure distribution was studied along with the jet impact characteristics onto a rigid tailwater channel.

Steiner *et al.* (2008) studied whether the ski jump of triangular bucket geometry is an attractive alternative to the standard circular-shaped ski-jump bucket for two reasons: (1) ease in construction and (2) basis of 3D flip bucket design. Based on Heller *et al.* (2005) relating to the circular-shaped bucket geometry, additional data for the triangular bucket were collected to expand the previous research.

The main purpose of Schmocker *et al.*'s (2008) research was to analyze the air entrainment characteristics of plane jets downstream of a ski jump, both for pure water *and* pre-aerated approach flow conditions. The results relate to the development of the air concentration of highly turbulent water, and air-water jets in the atmosphere, and therefore emphasize the jet disintegration characteristics (Bin, 1993; Ervine, 1998) and the resulting plunge pool scour (Mason, 1993; Pagliara *et al.*, 2006).

As Steiner *et al.* (2008), Balestra (2012) conducted air concentration measurements along various jets issued by a classical ski jump. Pfister *et al.* (2014, 2016) analyzed all ski-jump data previously mentioned to derive general equations for the jet trajectories, the take-off angles, and air features along the jets. The following describes mainly experimental observations relative to ski jumps.

6.2.3 Bucket pressure, energy dissipation and choking features

Motivation and experimentation

The experiments conducted in a rectangular channel aimed to systematically vary both the bucket (subscript b) radius R_b and the deflection angle β_b. The supercritical approach (subscript o) flow depth h_o and average approach flow velocity $V_o = Q/(bh_o)$ both were varied, with Q as discharge and b as channel width. This study involved a horizontal approach flow channel to simplify the test setup (Heller, *et al.*, 2005).

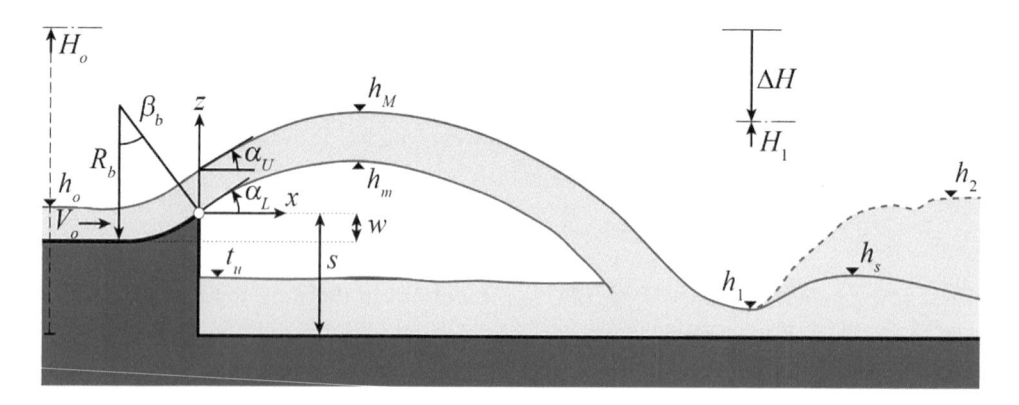

Figure 6.9 Definition sketch for plane ski-jump flow (Heller *et al.*, 2005)

Figure 6.9 shows a definition sketch. The approach flow is described by the Froude number $\mathsf{F}_o = V_o/(gh_o)^{1/2}$ and relative bucket radius h_o/R_b. Bucket height $w = R_b(1 - \cos\beta_b)$ generates two distinctly different flow types, namely choked and free bucket flows, as presented below. The jet generated by the ski jump is described with the lower (subscript L) and the upper (subscript U) jet trajectories whose origins are located at the bucket takeoff section $x = 0$, and at the bucket crest $z_L = 0$ for the lower, and $z_U = h_o$ for the upper trajectory, respectively. The corresponding takeoff angles are α_U and α_L, respectively. The jet trajectories attain a maximum elevation of h_M for the upper, and h_m for the lower boundary, respectively. The elevation difference $s-w$ between the approach flow and tailwater channels was kept constant, yet the parameter $S = s/h_o$ varied with the approach flow depth. The flow depth h_1 beyond jet impact determines the energy head $H_1 = h_1 + V_1^2/2g$. Given the strong development of spray and shocks, h_1 was indirectly determined by measuring the sequent flow depth h_2 of the hydraulic jump using the tailwater flap gate. The relative head loss across the ski jump is $\eta_E = \Delta H/H_o$. The height h_s of the shock (subscript s) waves on the channel sides was measured for unsubmerged tailwater flow. The recirculation (subscript u) flow depth t_u below the jet was measured by assuring atmospheric pressure in the air cavity.

Three bucket radii $R_b = 0.10$, 0.25, and 0.40 m were considered along with bucket deflection angles of $\beta_b = 10°$, 15°, 20°, 25°, 30°, and 40°. The approach flow depth h_o varied between 0.036 m and 0.095 m; scale effects are small for flow depths in excess of 0.040 m for black-water approach flow.

Bucket pressure characteristics

The theoretical (subscript T) dynamic pressure head $h_{PT} = p_M/(\rho g)$ exerted by potential (subscript P) flow with concentric streamlines of boundary radius R_b is from Eq. (6.1) (Juon and Hager, 2000)

$$h_{PT}/h_o = \mathsf{B}_o^2. \tag{6.12}$$

Here $\mathsf{B}_o = (h_o/R_b)^{1/2}\mathsf{F}_o$ is the so-called approach flow bend number as the product of the square root of relative bucket radius times the approach flow Froude number. The pressure head line was plotted for each run (Figure 6.10d) to detect the maximum h_{PM} and its location x_{PM} relative to the coordinate origin $x = 0$. Figure 6.10a shows the ratio h_{PM}/h_{PT} versus the relative bucket curvature (h_o/R_b) times $(40°/\beta_b)$. Symbols for the various radii are black for $R_b = 0.10$ m, light gray for $R_b = 0.25$ m, and dark gray for $R_b = 0.40$ m. From Figure 6.10a the theoretical value h_{PT} according to Eq. (6.12) applies to values $[(h_o/R_b)(40°/\beta_b)] < 0.20$, whereas otherwise

$$h_{PM} / h_{PT} = (1/5)[(h_o / R_b)(40° / \beta_b)]^{-1}, \quad (h_o / R_b)(40° / \beta_b)] \geq 0.20 .$$ (6.13)

Equation (6.13) differs considerably from Eq. (6.12). Inserting into Eq. (6.13) yields

$$h_{PM} / h_o = (1/5)(\beta_b / 40°)\mathsf{F}_o^2, \quad [(h_o / R_b)(40° / \beta_b)] \geq 0.20.$$ (6.14)

Equation (6.14) demonstrates the significance of both the approach flow depth h_o and particularly of F_o on the maximum dynamic bucket pressure head h_{PM}; it also states that h_{PM} is linearly related to the deflection angle β_b. Small deflection angles produce a smaller maximum pressure head than do large. However, for large h_o/R_b, the latter effect is insignificant on h_{PM}/h_o. These results apply for horizontal approach flows but are subject to variation particularly for small deflection angles for a nearly horizontal chute. A limitation of the results is presented below.

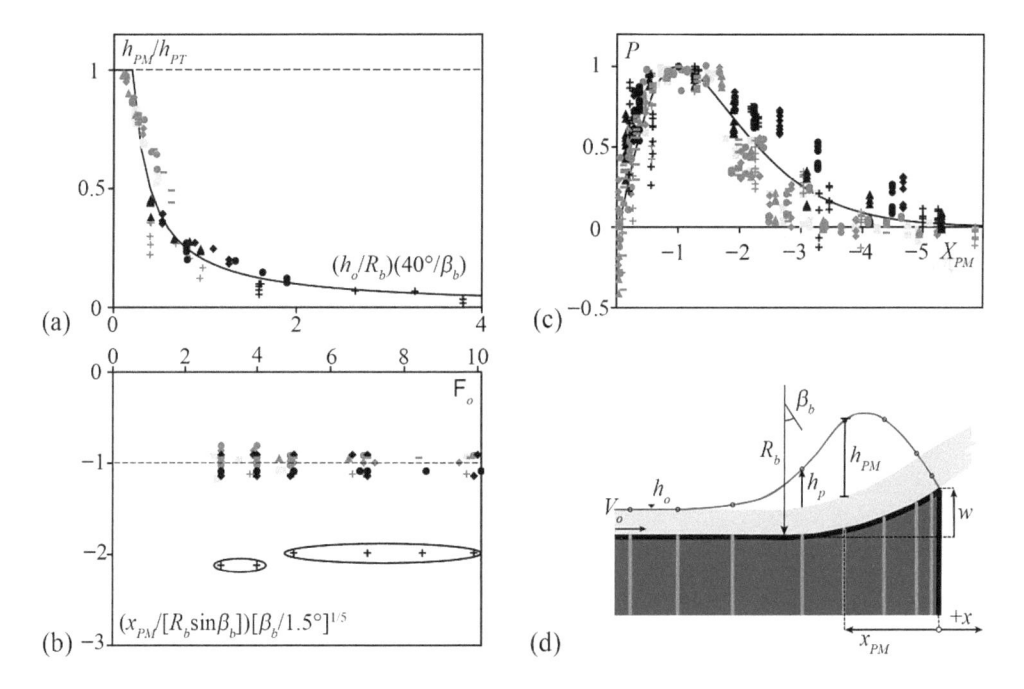

Figure 6.10 Dynamic pressure head characteristics (a) ratio between measured and theoretical pressure heads with (−−) Eqs. (6.13), (6.14), (b) location of maximum pressure head with (−−) Eq. (6.15), (c) local pressure head distribution with (−−) Eq. (6.16), (d) definition sketch (Heller *et al.*, 2005)

The location x_{PM} of the maximum dynamic pressure head is shown in Figure 6.10b as

$$x_{PM} / (R_b \sin\beta_b) = -(1.5° / \beta_b)^{1/5}, \quad \beta_b \geq 15°. \tag{6.15}$$

The data for $\beta_b = 10°$ deviate significantly from Eq. (6.15) because the bucket length is too short to attain the maximum pressure head. This restriction limits the experimental setup of the present tests. The effect of the approach flow channel slope is small if the bucket deflection is larger than 15°, therefore. This value is considered a minimum in hydraulic engineering, as follows from a literature review of existing bucket angles.

The dynamic pressure head distribution $h_p(x)$ along the bucket was analyzed using the maximum pressure head characteristics $(x_{PM}; h_{PM})$. Figure 6.10c shows the relative pressure head $P = h_p/h_{PM}$ versus $X_{PM} = x/x_{PM}$. The data scatter about the trend line

$$P = [X_{PM} \times \exp(1 - X_{PM})]^{1.5}. \tag{6.16}$$

The effect of non-hydrostatic pressure disappears if $x/(R_b \sin\beta_b) \leq -3$, corresponding to $X_{PM} \leq -5$ (Figure 6.10c). Note that the data for a bucket radius of $R_b = 0.10$ m are slightly above Eq. (6.16), whereas those with $R_b = 0.25$ m and $R_b = 0.40$ m are rather below it.

Jet trajectory characteristics

Figure 6.11 shows side views of jet trajectories for a bucket radius of $R_b = 0.10$ m, $\beta_b = 40°$, and $h_o = 0.05$ m, relating to $F_o = 3, 4, 5, 7$, and 10. Note that the takeoff angles α_U and α_L of both the upper and the lower jet trajectories are always smaller than the bucket deflection angle β_b. Further, the lower takeoff angle appears to be smaller than the takeoff angle of the upper jet trajectory. A jet from a ski jump thus increases its width as it travels across the atmosphere. How can this phenomenon be explained? What are the factors influencing the basic jet trajectory characteristics? These issues are dealt with in 6.2.2.

Figure 6.12 shows two jets for bucket radii of 0.10 m and 0.40 m, respectively, and otherwise identical approach flow parameters, namely $F_o = 5$, $h_o = 0.05$ m, and $\beta_b = 40°$. Note that (a) for a small bucket radius produces a takeoff angle much smaller than (b). A slight error in α_j has a considerable effect on the jet trajectory, particularly on the impact location resulting in scour at a jet impact location not planned.

Shock wave and jet recirculation zone

A plane water flow falling over a vertical drop produces shock waves in the tailwater channel due to jet deflection. Drop structures enhance significant turbulence beyond the impact location. The shock wave height is difficult to assess due to spray presence and high turbulence. The relative height of shock waves $Y_s = h_s/h_o$ increases with the approach flow Froude number. Because shock waves are a response of the drop structure, the impact features have to be considered in addition. A jet impacting almost vertically onto the tailwater produces more height than jets of small impact angle. Accordingly, the bucket deflection angle has also an effect. This phenomenon is observed in channels but hardly noted in prototype plunge pools. The following does therefore relate to observations conducted in the laboratory environment.

Figure 6.13a shows the relative shock wave height $(Y_s - 1)/(\sin\beta_b)^{1/2}$ versus $(F_o - 1)$. The data follow the straight

$$(Y_s - 1) / (\sin\beta_b)^{1/2} = 0.85(F_o - 1), \quad 2 \leq (F_o - 1) \leq 10. \tag{6.17}$$

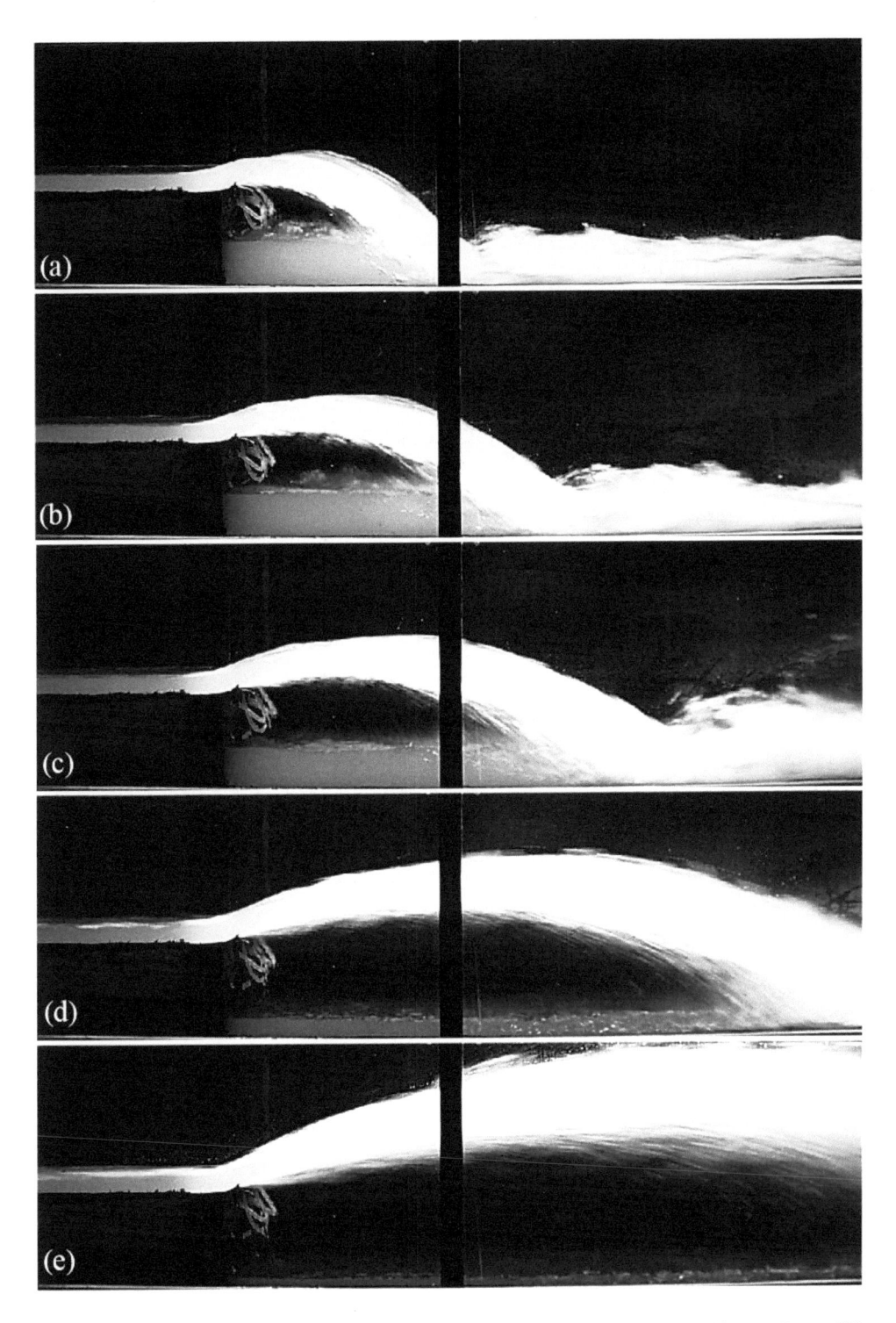

Figure 6.11 Comparison between various ski-jump jets with $R_b = 0.10$ m, $\beta_b = 40°$, $h_o = 0.05$ m and $F_o = $ (a) 3, (b) 4, (c) 5, (d) 7, (e) 10 (Heller *et al.*, 2005)

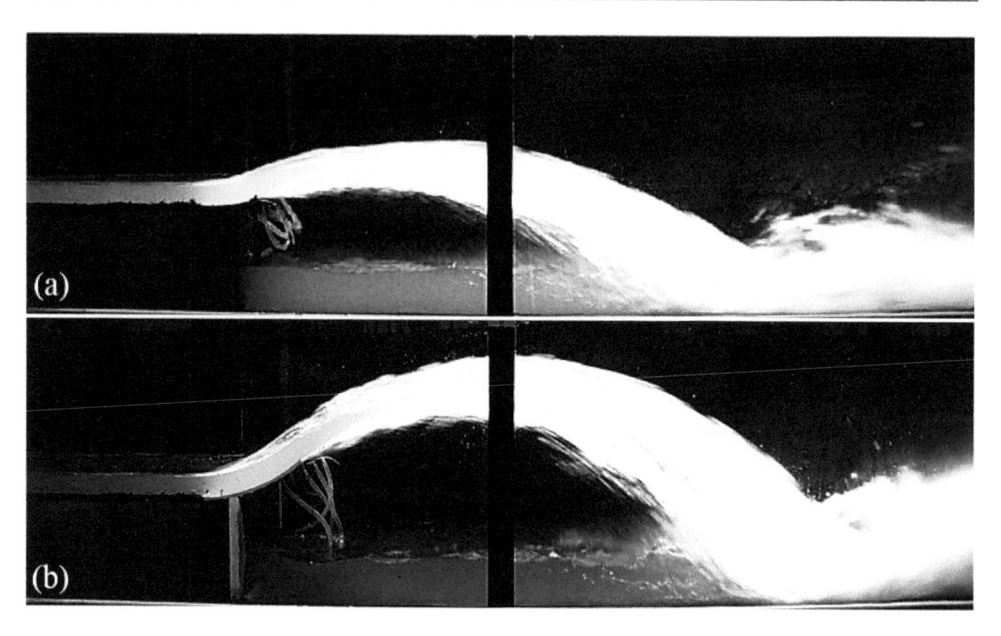

(a)

(b)

Figure 6.12 Effect of bucket radius on jet trajectories for $\beta_b = 40°$, $F_o = 5$, $h_o = 0.05$ m and $R_b = $ (a) 0.10 m, (b) 0.40 m (Heller *et al.*, 2005)

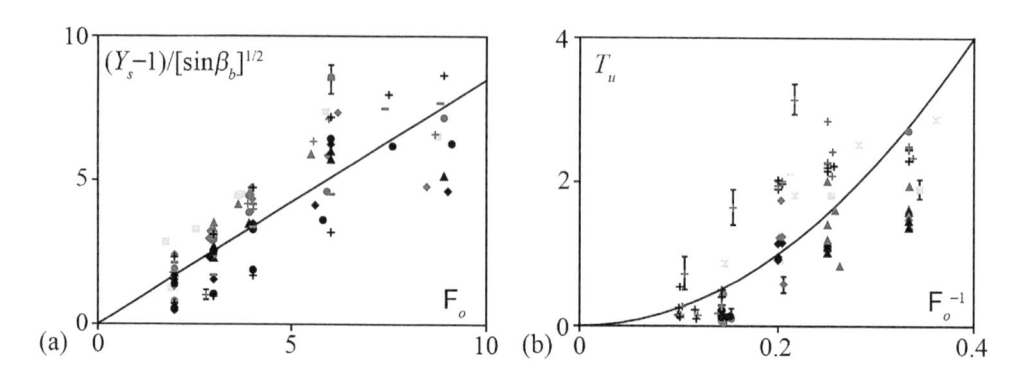

Figure 6.13 (a) Relative height of shock waves with (−−) Eq. (6.17), (b) relative recirculation flow depth with (−−) Eq. (6.18) (Adapted from Heller *et al.*, 2005)

Equation (6.17) reflects the trends mentioned above allowing for the estimation of the shock wave height in a prismatic channel. This height reduces in prototypes because the tailwater is much wider than the approach flow channel.

The recirculation flow zone below the jet is influenced by the jet impact onto the tailwater (Figure 6.9). Its relative height $T_u = t_u/h_o$ varies strongly with F_o as (Figure 6.13b)

$$T_u = \left(5F_o^{-1}\right)^2, \quad 3 < F_o < 10. \tag{6.18}$$

The vertical lines indicate the observational accuracy (±1 cm) of selected data.

Energy dissipation

The relative energy dissipation of a trajectory jet is $\eta_E = \Delta H/H_o$, from the horizontal approach flow channel beyond to impact onto the tailwater, with $H_o = s - w + h_o + V_o^2/2g$ as energy head based on the tailwater channel elevation. The value $E = \eta_E - (\beta_b - A_b)/\beta_R$ varies with the relative drop height $S = s/h_o$. The energy dissipation rate increases as the elevation difference between the takeoff point and the tailwater channel increases. This appears logic because the jet impact angle onto the tailwater channel increases with S, so that a significant portion of the jet energy content is dissipated by impact. Note that both coefficients A_b and β_R reduce as the bucket radius increases. With $A_b = 60°[\tanh(h_o/R_b)]^{1/2}$ and $\beta_R = 10A_b$, Heller *et al.* (2005) find

$$\eta_E = (\beta_b - A_b)/10A_b + 0.060S, \ 0 < S < 9. \tag{6.19}$$

The energy dissipation across a ski jump increases with the relative drop height, the deflection angle and smaller relative bucket curvature, therefore. Typically, energy dissipation rates of 40% are attained. In prototypes, η_E may be higher because of additional air entrainment into the jet, tailwater depth, as well as the pool topography.

Chocking flow characteristics

A bucket is a local element of negative bottom slope: Flows of too small velocity are submerged by the bucket crest becoming locally subcritical (Figure 6.15d). Each ski jump is designed for supercritical flow across the entire structure, to deflect any discharge away from the dam toe. The water otherwise falls close to the bucket end onto the foundation causing scour. Each trajectory bucket must be checked for the choking flow condition; these flows must be avoided by a suitable discharge regime. Choking usually occurs for small discharges avoided by a suitable spillway regime.

A distinction between the increasing and the decreasing discharge regimes applies for the analysis of choking flow. The application of the momentum and the energy equations is useful to predict the basic choking phenomenon. Consider Figure 6.14a for the formulation of the momentum theorem, with D as the dynamic and St as the static pressure distributions. If a partial hydraulic jump is located in the bucket, the flow depth at its end may be assumed critical, i.e. $h_c = (Q^2/gb^2)^{1/3}$. The pressure distribution on the spillway bucket is complicated. Assuming hydrostatic pressure distribution both upstream of the bucket and at the brink section allows for an estimate. The momentum equation in the horizontal direction reads with $q = Q/b$ as unit discharge

$$(1/2)h_o^2 + q^2/(gh_o) = (1/2)(h_c + w)^2 + q^2/(gh_c). \tag{6.20}$$

With $h_c/h_o = \mathsf{F}_o^{2/3}$, Eq. (6.19) may be expressed for the lower choking bound as

$$W = w/h_o = [1 - 2\mathsf{F}_o^{4/3} + 2\mathsf{F}_o^2]^{1/2} - \mathsf{F}_o^{2/3}. \tag{6.21}$$

The momentum equation based on Figure 6.14b applies to obtain an upper choking bound. The flow depth at the bucket start is assumed to be $(h_c + w)$ whereas the flow leaves the bucket end again with the critical flow depth. Equating forces thus gives

$$(1/2)h_o^2 + q^2/(gh_o) = (1/2)(h_c + w)^2 + q^2/[g(h_c + w)]. \tag{6.22}$$

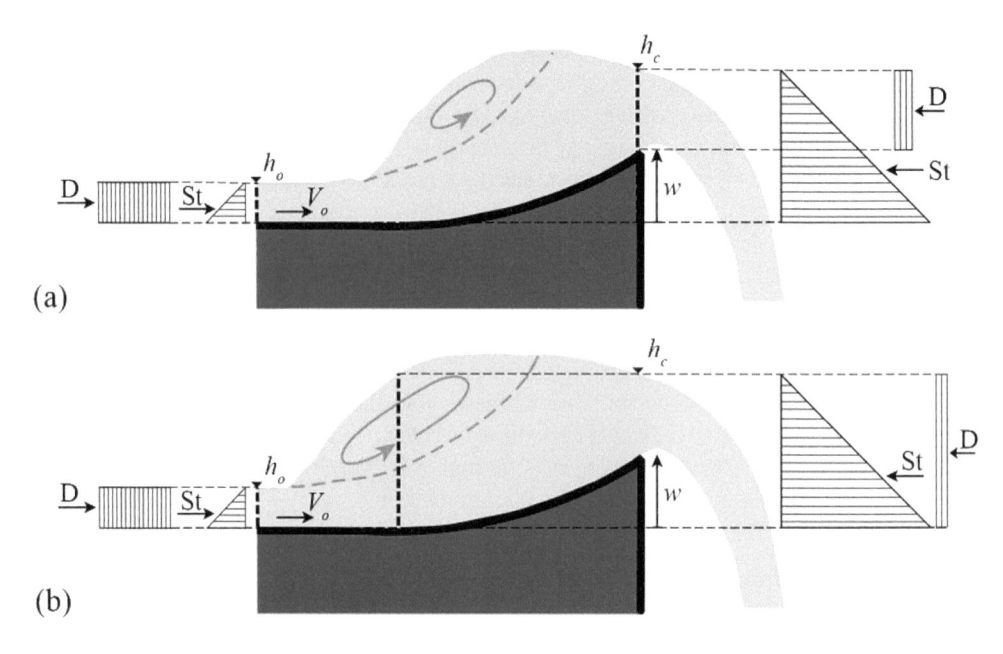

Figure 6.14 Control volume and forces to predict flow choking (a) upper bound, (b) lower bound (Adapted from Heller *et al.*, 2005)

Similar to Eq. (6.21), the result is (Heller *et al.*, 2005)

$$W = (1/2)[(1+8\mathsf{F}_o^2)^{1/2} - 1] - \mathsf{F}_o^{2/3}. \tag{6.23}$$

Figure 6.15 (a, b) refers to the test data obtained for the increasing (sign +), and the decreasing (sign −) discharge regimes, respectively, along with (Heller *et al.*, 2005)

$$W_+ = 0.60(\mathsf{F}_o - 1)^{1.2}, \quad 1 < \mathsf{F}_o < 4, \tag{6.24}$$

$$W_- = 0.90(\mathsf{F}_o - 1)^{0.9}, \quad 1 < \mathsf{F}_o < 4. \tag{6.25}$$

Figure 6.15c compares the choking flow condition for the rising discharge regime with Eqs. (6.21) and (6.23), noting reasonable agreement with the simplified model. According to Eqs. (6.24, 6.25), choking flow is suppressed by a small relative bucket height W. For a typical deflection angle of $\beta_b = 30°$ and a relative bucket curvature of $h_o/R_b = 0.10$, $W = 10(1 - \cos 30°) = 1.34$, resulting in $\mathsf{F}_{o+} > 2.95$ to exclude choking.

6.2.4 Ski jump with triangular bucket

Motivation

Steiner *et al.* (2008) studied whether ski jumps with a triangular bucket are an attractive alternative to the standard circular-shaped ski jump bucket for two reasons: (1) ease of

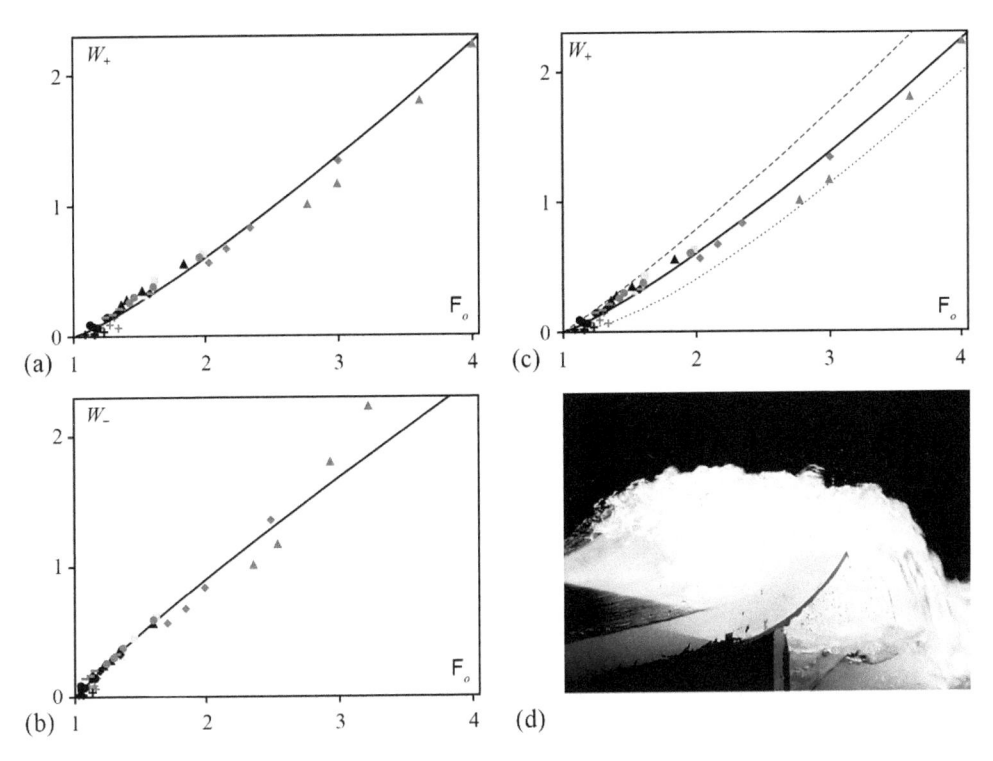

Figure 6.15 Choking flow characteristics for β_b = (+) 10°, (-) 15°, (•) 20°, (■) 25°, (♦) 30°, (▲) 40° (a) increasing discharge regime $W_+(F_o)$ with (--) Eq. (6.24), (b) decreasing discharge regime $W_-(F_o)$ with (--) Eq. (6.25), (c) comparison of Eqs. (6.24) and (6.21, 6.23), (d) typical flow for Q_- and R_b = 0.40 m, β_b = 40° (Adapted from Heller *et al.*, 2005)

construction and (2) basis of 3D flip bucket design. Ski jumps have been and still are widely used but few general guidelines are currently available. Based on Heller *et al.* (2005) relating to the circular-shaped bucket geometry, tests for the triangular bucket were conducted to expand the research.

A triangular wedge-shaped bucket was inserted in the channel used by Heller *et al.* (2005) to answer three questions: (1) What are differences between the circular- and triangular-shaped bucket geometries in terms of hydraulic performance? (2) How large is the energy dissipation across this setup? (3) What are its choking flow features? The bucket pressure distribution and the jet impact features were also investigated.

The study was limited to a horizontal approach flow channel. The approach flow was non-aerated 'black-water', whereas the cavity below the jet was aerated to assure atmospheric pressure. The approach flow was characterized by the approach flow Froude number $F_o = V_o/(gh_o)^{1/2}$, the relative bucket height w/h_o and the deflector angle γ (Figure 6.16a). The jet generated by the ski jump is described by the lower (subscript L) and the upper (subscript U) jet trajectories whose origins are at the bucket takeoff section $x = 0$, and at the bucket crest $z_L = 0$ for the lower and $z_U = h_o$ for the upper trajectory, respectively. The jet takeoff angles are α_U and α_L, and the jet trajectories attain a maximum (subscripts M and m) elevation h_M for the

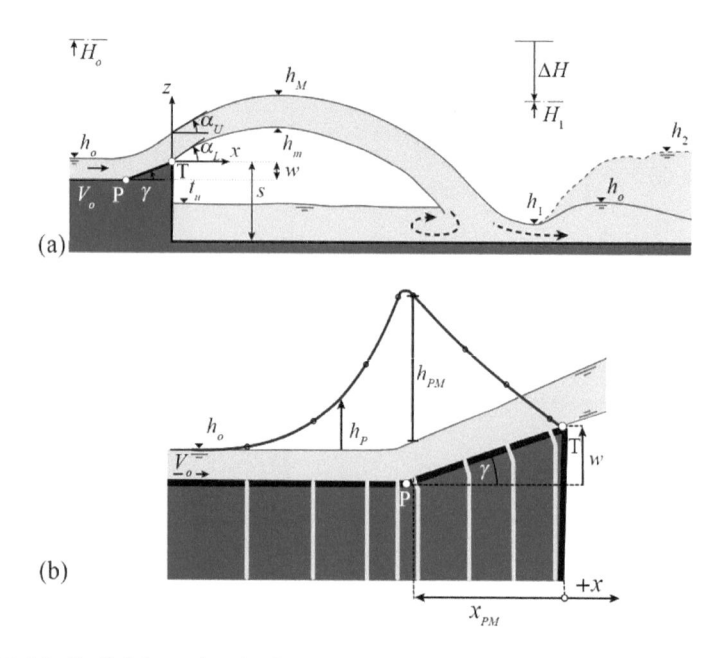

Figure 6.16 (a) Definition sketch for plane ski-jump flow using triangular deflector, (b) streamwise section along deflector with typical bottom pressure distribution (Steiner *et al.*, 2008)

upper, and h_m for the lower jet trajectories, respectively. Although the elevation difference $s - w$ between the approach flow and the tailwater channels was kept constant, parameter $S = s/h_o$ varied with the approach flow depth. The flow depth h_1 beyond jet impact (subscript 1) determines the energy head $H_1 = h_1 + V_1^2/2g$. It was indirectly determined by measuring the sequent flow depth h_2 of the hydraulic jump generated with the tailwater flap gate. The relative head loss across the ski jump is $\eta_E = \Delta H/H_o$. The maximum height h_s of the shock (subscript s) waves on the channel sides for unsubmerged tailwater, and the recirculation (subscript u) flow depth t_u below the jet cavity were also measured.

The deflector angles were $\gamma = 8.2°$, $11.3°$, $18.3°$, $26.2°$, and $33.2°$, whereas bucket heights w varied from 0.014 m to 0.075 m. The approach flow depth h_o varied from 0.045 m to 0.070 m to exclude scale effects. Approach flow Froude numbers $\mathsf{F}_o = 3$, 5, and 8 were used, and an additional series with $\mathsf{F}_o = 7$ involved a large deflector angle.

Bucket pressure characteristics

Figure 6.16b shows a streamwise deflector section with the typical pressure distribution. Upstream from the deflection point P, the pressure is hydrostatic because of parallel horizontal streamlines. In its vicinity, the dynamic pressure head (subscript P) rises to the maximum h_{PM} then reducing to atmospheric pressure $h_p = -h_o$ at the take-off point T. The maximum dynamic pressure head relative to flow depth h_o follows from (Figure 6.17a)

$$h_{PM}/h_o = 0.30[\mathsf{F}_o(\tan\gamma)^{0.20}]^2 . \tag{6.26}$$

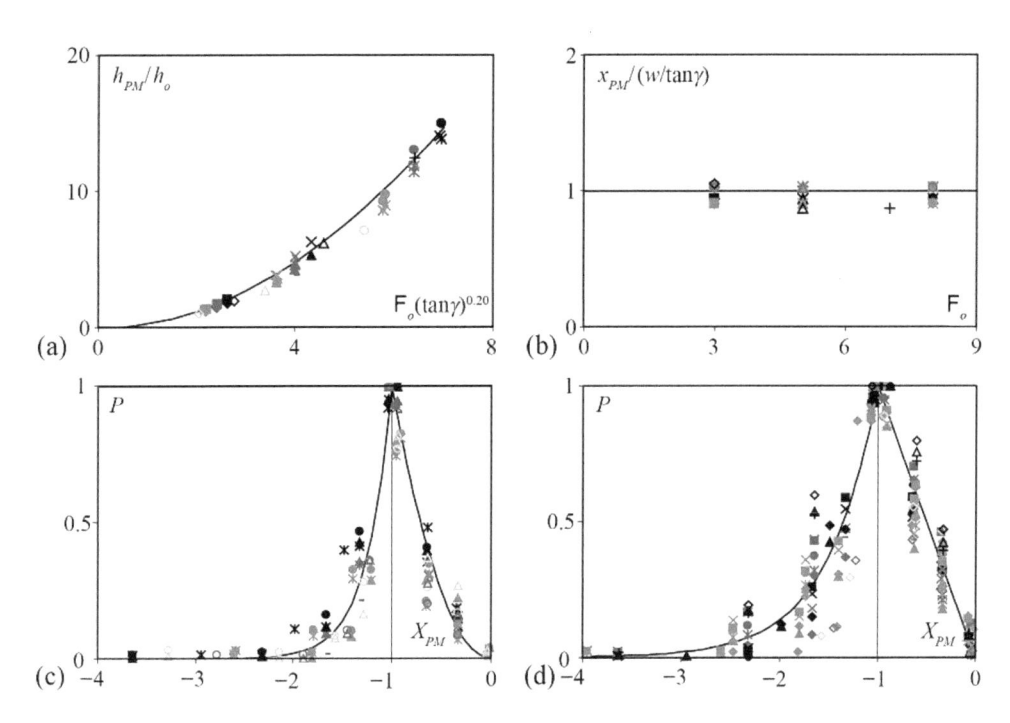

Figure 6.17 Dynamic pressure head features (a) maximum pressure head h_{PM}/h_o versus $F_o(\tan\gamma)^{0.20}$, (b) location of maximum $x_{PM}/(w/\tan\gamma)$ versus F_o, pressure head distributions $P(X_{PM})$ for (c) $\Gamma \leq 0.057$, (d) $\Gamma > 0.057$. Note different vertical scales of P for $X_{PM} \leq -1$, and $-1 < X_{PM} < 0$ (Steiner *et al.*, 2008)

Accordingly, the approach flow Froude number has a quadratic effect on h_{PM} and the deflector angle a relatively small, but still noticeable influence.

The pressure headline includes location x_{PM} of the maximum pressure head (Figure 6.16b), reaching its maximum $x_{PM} = w/\tan\gamma$ at the deflector point P (Figure 6.17b). The dynamic pressure head distribution $h_p(x)$ along the deflector was analyzed using the maximum pressure head characteristics $(x_{PM}; h_{PM})$. The data analysis resulted in two distributions, depending on parameter $\Gamma = (h_o/w)(\sin\gamma/F_o)$. For $\Gamma \leq 0.057$, i.e. for large F_o and w, and a small γ, the dynamic pressure head distribution is sharp-peaked, whereas it is fuller if $\Gamma > 0.057$. The distribution is described with two different coordinates to account for the two asymptotes $h_p(x \to -\infty) = 0$, and $h_p(x = 0) = -h_o$. With $X_{PM} = x/x_{PM} \leq 0$, non-dimensional coordinates are $P = (h_p + h_o)/(h_{PM} + h_o)$ for $-1 \leq X_{PM} \leq 0$, and $P = (h_p/h_{PM})$ for $X_{PM} < -1$. Figure 6.17(c, d) shows the relative pressure head distributions versus $X_{PM} = x/x_{PM}$ with $\varepsilon = 2$ for $\Gamma \leq 0.057$ and $\varepsilon = 1$ for $\Gamma > 0.057$ as

$$\left(h_p + h_o\right)/\left(h_{PM} + h_o\right)|X_{PM}{}^\varepsilon|, \quad \text{for} -1 \leq X_{PM} \leq 0, \tag{6.27}$$

$$\left(h_p / h_{PM}\right) = \exp[2\varepsilon\left(1 + X_{PM}\right)], \quad \text{for } X_{PM} < -1. \tag{6.28}$$

These results are compared below with the circular-shaped bucket characteristics.

Jet trajectory characteristics

Figure 6.18 shows side views of jet trajectories for $F_o = 3, 5$, and 7, $\gamma = 33.2°$, $w = 0.05$ m, and $h_o = 0.05$ m; the significant effect of F_o is evident. The takeoff angles α_U and α_L of both the upper and lower jet trajectories are usually smaller than the bucket deflector angle γ (Dhillon et al., 1981, Heller et al., 2005), with $\alpha_L < \alpha_U$. The jet thickness issued from a ski jump thus increases as it travels through the atmosphere. The flow depth along a deflector increases, from h_o at the upstream section to $h_o + \Delta h$ at the takeoff section because of kinetic energy losses and pressure rearrangement.

The upper and lower jet trajectories $z(x)$ issued from a ski jump follow mass point dynamics based on Eqs. (6.4) to (6.7). With $h_M/h_o = z_o/h_o + (1/2)F_o^2\sin^2\alpha_j$ as maximum jet elevation, jets of arbitrary takeoff angle α_j are described with the generalized coordinates $X_j = (2x/h_o) \times 1/[F_o^2\sin(2\alpha_j)]$ and $Z_j = [(z - z_o)/(h_M - z_o)]$ (Figure 6.16a). Both the upper and lower jet trajectories are thus expressed as

$$Z_j = 2X_j - X_j^2, \quad X_j \geq 0. \tag{6.29}$$

Figure 6.19 (a, c) relates to the generalized upper and lower jet trajectories for $0 \leq X_j \leq 8$. All data follow well Eq. (6.29). The experimental jet trajectories $z_j(x)$ were compared with

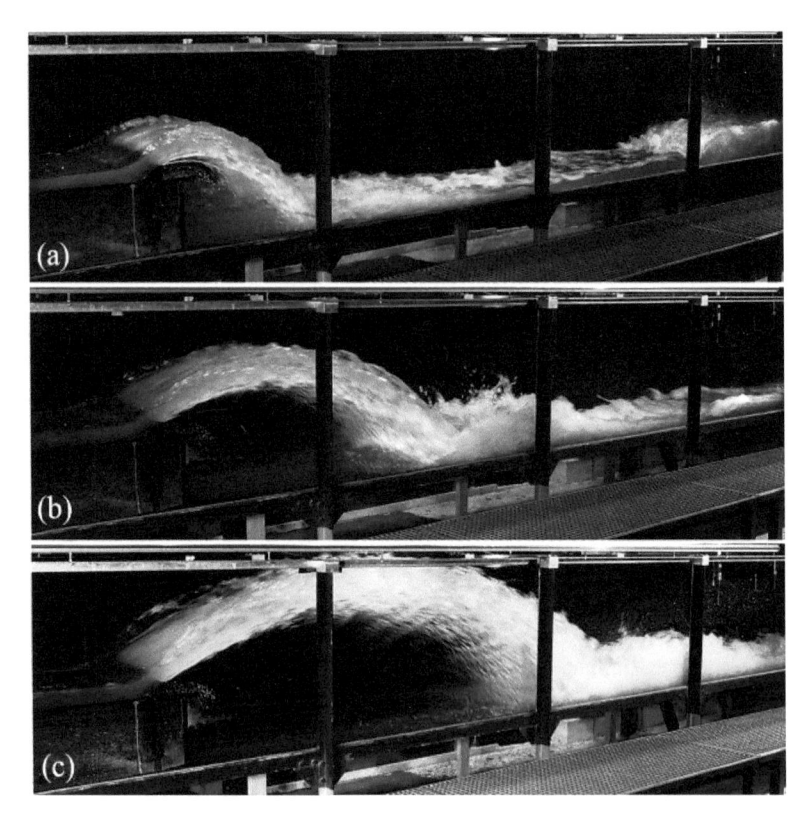

Figure 6.18 Ski-jump jets for $w = 0.05$ m, $\gamma = 33.2°$, $h_o = 0.05$ m, F_o = (a) 3, (b) 5, (c) 7 (Steiner *et al.*, 2008)

Eq. (6.4) by optimizing the angles α_U and α_L. The result α_j/γ was plotted versus $(w/h_o)(\tan\gamma)/$ F_o. Figure 6.19 (b, d) compares the test data with

$$(\alpha_j / \gamma) = \psi_j[1.05 - (w/h_o)(\tan\gamma)/F_o], \quad 0.01 \le (w/h_o)(\tan\gamma)/F_o \le 0.30. \tag{6.30}$$

Here, $\psi_j = 1$ for the upper and $\psi_j = 1/1.12 = 0.89$ for the lower takeoff angle. Both α_U and α_L follow the same expression except for the multiplier ψ_j. The jet expansion angle is thus close to 7° for jets issued by the circular-shaped deflector (Heller et al., 2005). For small values of $(w/h_o)(\tan\gamma)/F_o$, $\alpha_j/\gamma > 1$ resulting from the constant flow depth assumption across the deflector. Note the larger data scatter for the lower than for the upper jet trajectory because of access difficulties.

Heights of shock wave and recirculation zone

Shock waves in the tailwater channel result from jet impact (Vischer and Hager, 1998). Drop structures produce significant turbulence beyond the impact location. In prototype applications with a wide tailwater valley and a deep plunge pool after scour action, shock waves are smaller but still of concern for adjacent structures.

The relative shock wave height h_s/h_o (Figure 6.9) increases with F_o and γ because a large jet impact angle produces a higher shock wave as compared with a small. Figure 6.20a relates $(h_s/h_o - 1)$ to $(F_o - 1)(\tan\gamma)^{0.40}$; the test data follow (Heller et al., 2005)

$$(h_s / h_o - 1) = (F_o - 1)(\tan\gamma)^{0.40}, \quad \text{for } 1 \le (F_o - 1)(\tan\gamma)^{0.40} \le 6. \tag{6.31}$$

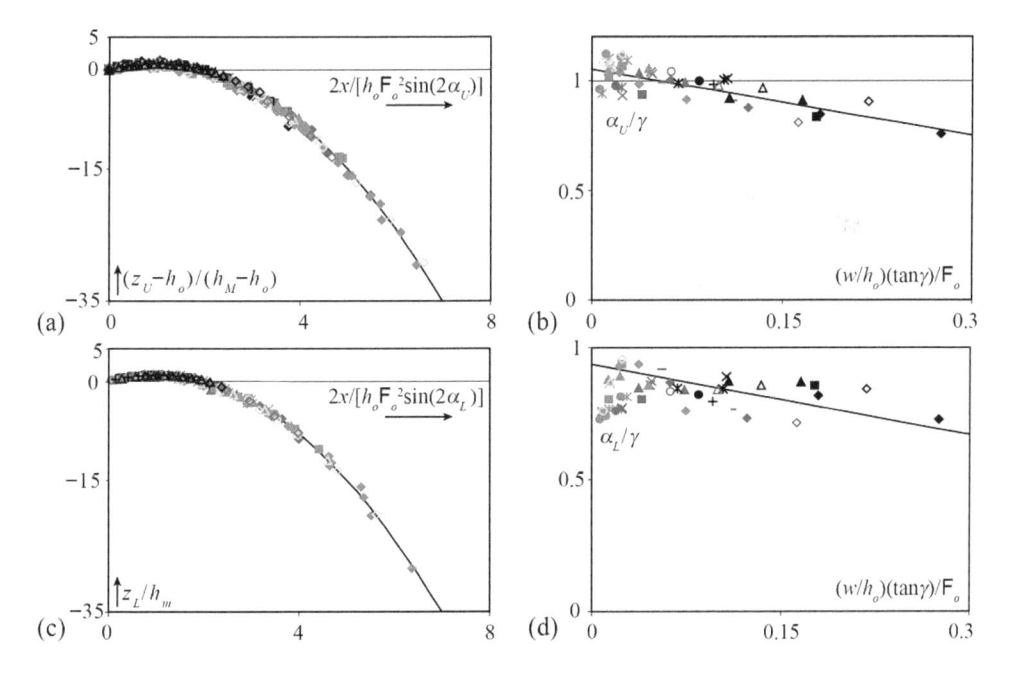

Figure 6.19 (a) Upper, (c) lower jet trajectories as compared with (—) Eq. (6.29); takeoff angle ratio (b) α_U/γ, (d) α_L/γ versus $[(w/h_o)(\tan\gamma)/F_o]$, (—) Eq. (6.30) (Steiner et al., 2008)

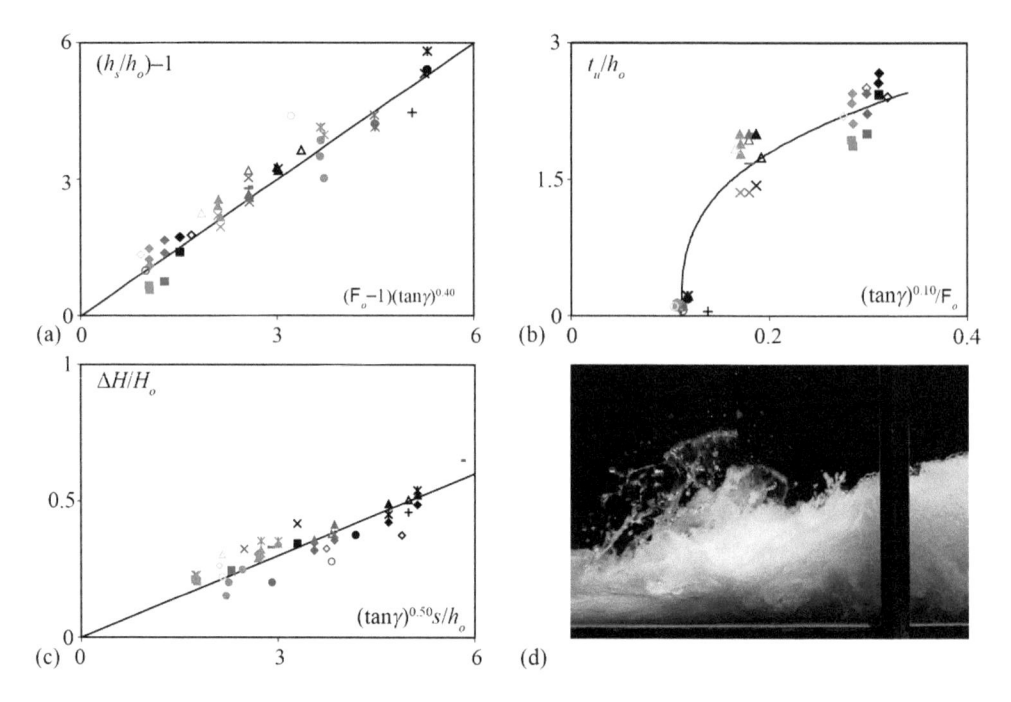

Figure 6.20 (a) Relative shock wave height $[(h_s/h_o)-1]$ versus $[(F_o-1)(\tan\gamma)^{0.40}]$ with (—) Eq. (6.31), (b) relative recirculation flow depth t_u/h_o versus $[(\tan\gamma)^{0.10}/F_o]$ with (—) Eq. (6.32), (c) energy dissipation rate $\Delta H/H_o$ versus $[(\tan\gamma)^{0.50}(s/h_o)]$ with (—) Eq. (6.33), (d) hydraulic jump to determine indirectly energy dissipation (Steiner *et al.*, 2008)

Jet impact affects the recirculation flow zone in the tailwater channel (Figure 6.16a). Its relative height t_u/h_o again varies with F_o and slightly with γ as (Figure 6.20b)

$$t_u / h_o = 4\{[(\tan\gamma)^{0.10}/F_o)]-0.11\}^{0.33}, \quad \text{for } 0.11 < (\tan\gamma)^{0.10}/F_o < 0.35. \tag{6.32}$$

Figure 6.21 compares the heights of the recirculation zone for otherwise identical flow conditions except for an increase of F_o. Note the large depth t_u for a steep impact angle at $F_o = 3$, and a small value of t_u at $F_o = 7$.

Energy dissipation

Ski jumps are mainly applied to divert a high-speed jet away from a dam structure (Rajan and Shivashankara Rao, 1980). The energy dissipation is accomplished in two steps, namely (1) jet travel in the air and jet impact and (2) jet recirculation upstream of jet impact point including turbulence production (Figure 6.17a). In contrast to real-world applications, this study does not account for tailwater scour, given additional complications in the test setup. The following therefore refers to jet impact onto a horizontal fixed channel bottom. Plunge pool scour is dealt with in 6.4 and 6.5.

The total relative energy (subscript E) dissipation of a trajectory jet, from the horizontal approach flow channel to beyond impact onto the unsubmerged tailwater is expressed as

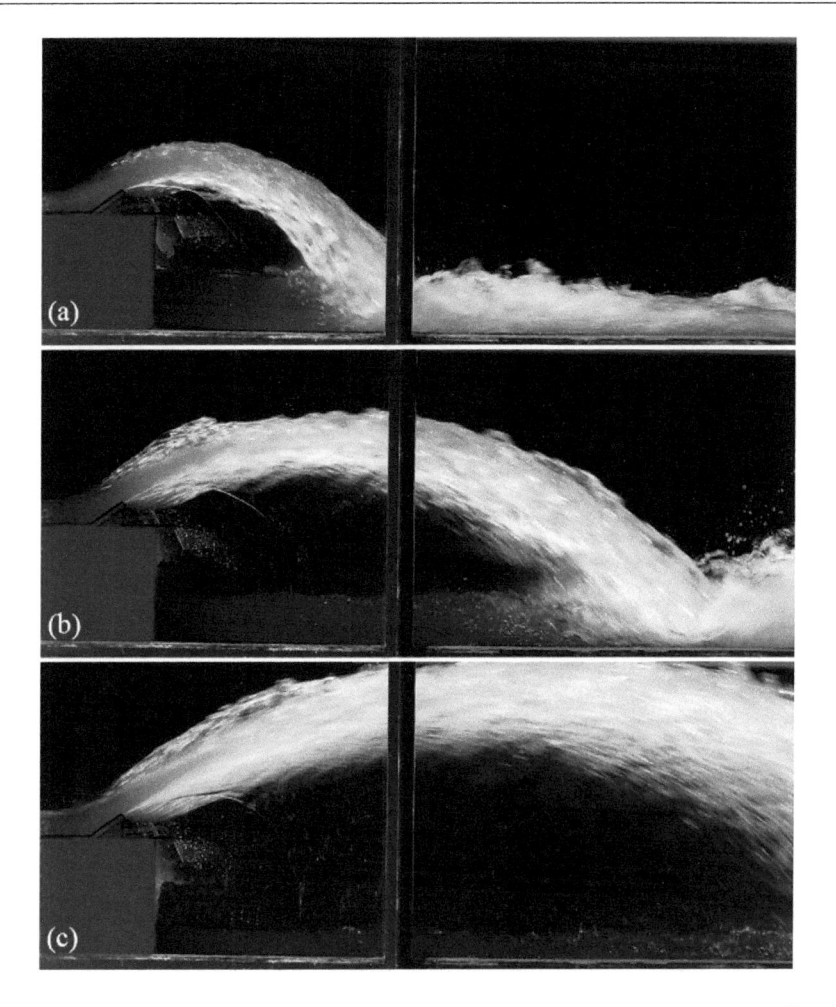

Figure 6.21 Effect of approach flow Froude number on recirculation depth for F_o = (a) 3,
(b) 5, (c) 7 (Steiner *et al.*, 2008)

$\eta_E = \Delta H/H_o$, where $H_o = s - w + h_o + V_o^2/2g$ is the approach flow energy head and $\Delta H = H_o - H_1$, with $H_1 = h_1 + V_1^2/2g$ as the tailwater energy head. Figure 6.20c shows $\Delta H/H_o$ versus the combined parameter $(\tan\gamma)^{0.50}(s/h_o)$ indicating an increase of $\Delta H/H_o$ both with s/h_o and γ. The jet impact angle therefore has a significant effect on energy dissipation since it tends to increase with γ and s/h_o. The test data follow

$$\eta_E = 0.10(\tan\gamma)^{0.50}(s/h_o), \quad \text{for } 1.7 < (\tan\gamma)^{0.50}(s/h_o) < 5. \tag{6.33}$$

The energy dissipation to beyond the jet impact point may become large. Typically, energy dissipation rates of 40% are attained. In prototypes, this may be higher because of the additional jet air entrainment and increased turbulence production. The rectangular tailwater channel has a small effect on energy dissipation accomplished by the jet flow. No information is available on the energy dissipation beyond jet impact, however, due to the experimental facility used.

Choking flow characteristics

Because the bucket of a ski jump has a negative bottom slope, flows of small velocity may become submerged and then are locally subcritical. Ski jumps are designed for supercritical flow to deflect any discharge far away from the dam toe. For choking flow conditions, the water drops close to the bucket end onto the foundation causing scour (Figure 6.22). Trajectory buckets must therefore be checked for the breakdown of supercritical flow. Choking occurs under small discharges; these should be excluded except for the starting and the shutdown phases of spillway flow.

The analysis of choking flow differs for the increasing (starting phase; sign $_+$) and decreasing (shutdown phase; sign $_-$) discharge regimes. Heller $et\ al.$ (2005) applied the momentum and energy equations to predict basic choking relations, resulting in Eqs. (6.24, 6.25) between the choking Froude number F_o and the relative deflector heights w_+/h_o and w_-/h_o. The data of Steiner $et\ al.$ (2008) are shown in Figure 6.23 along with two straight lines as the limit between the two flow regimes, given by

$$w_\pm / h_o = C_\pm(F_o - 1). \tag{6.34}$$

Herein, $C_+ = 0.70$ for the increasing, and $C_- = 0.925$ for the decreasing discharge regimes, respectively. For a certain w/h_o, F_o must therefore be larger for the increasing than for the

Figure 6.22 Choking flow development of decreasing discharge regime, (a) formation of oblique shock waves, (b) incipient hydraulic jump on deflector, (c) hydraulic jump on approach flow channel (Steiner $et\ al.$, 2008)

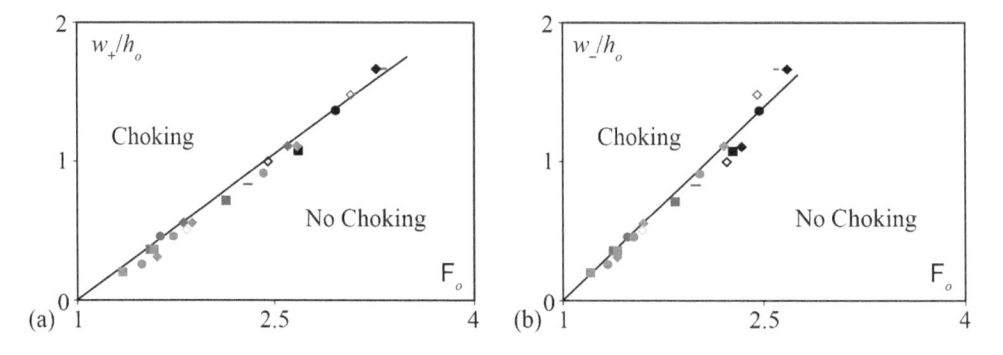

Figure 6.23 Choking characteristics for (a) increasing discharge regime w_+/h_o versus F_o, (b) decreasing discharge regime w_-/h_o versus F_o with (−−) Eq. (6.34) (Steiner $et\ al.$, 2008)

decreasing regime. The circular and the triangular bucket shapes result in nearly identical choking relations, therefore.

Comparison of results

The purpose of Steiner *et al.*'s research was to introduce an alternative – and simpler – geometry for spillway trajectory buckets. Figure 6.24 shows the extreme cases allowing for a comparison of the two designs: (a) buckets for which the takeoff angle is nearly identical, whereas (b) with buckets of identical base length. Note that the deflector angle in (a) is $\gamma = \beta_b$, whereas $\gamma = \beta_b/2$ in (b). Further, the deflector height is related to the circular-shaped bucket as $w = R_b(1 - \cos\beta_b)$, in which β_b is the deflection angle and R_b the bucket radius (Heller *et al.*, 2005).

The comparison of results is straightforward. The parameters of the circular-shaped (subscript S) bucket are h_o, F_o, β_b, and R_b, whereas h_o, F_o, w, and γ are these of the triangular (subscript T) deflector. All previous results may therefore be compared with each other for the two limit cases using the ratio $\eta = $ [Parameter T/Parameter S].

Steiner *et al.* (2008) note that the maximum dynamic pressure head of the triangular deflector is always larger than that of the circular-shaped bucket because of abrupt flow deflection by the deflector. Given that the dynamic pressure distribution resulting from the triangular deflector is concentrated at the flow deflection zone, that of a circular-shaped bucket has a longer streamwise extension.

Both α_U and α_L are compared with the same formulation, because they differ only by the constant $(1/1.12)$. Except for extremely low deflectors as $w/h_o = 0.36$, the takeoff angles are similar, yet with a higher value of the triangular than of the circular-shaped bucket, almost independent of F_o. For otherwise identical base parameters, the triangular deflector gives larger throwing distances than the circular-shaped bucket, which may be advantageous in applications.

The shock wave heights h_s resulting from the triangular deflector and the circular-shaped bucket are nearly identical for $\gamma = \beta_b/2$, whereas it is 40% larger for $\gamma = \beta_b$.

The recirculation depth ratio between the two designs indicates nearly no effect of γ if $F_o < 7$. The absolute values of t_u for large F_o are of the order of h_o. Therefore, a ski-jump bucket is hardly submerged from the tailwater, except for deposits downstream from the scour hole.

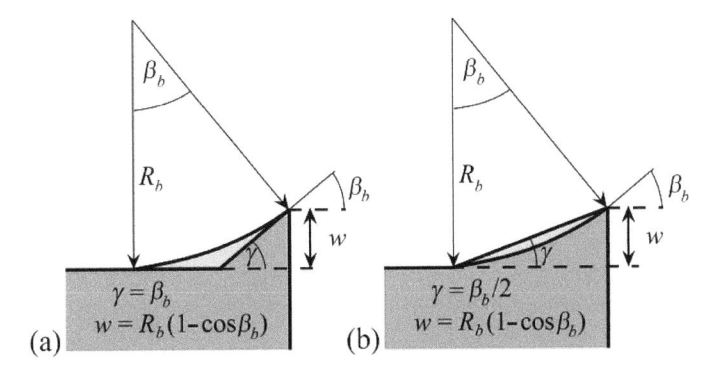

Figure 6.24 Sketch of extreme deflector designs (a) identical deflector angle, (b) identical base length

The jet energy dissipation features of the triangular deflector and the circular-shaped bucket are comparable. The choking flow characteristics for both increasing ($_+$) and decreasing ($_-$) discharge regimes is also similar.

Ski-jump buckets are currently mostly circular-shaped. Steiner *et al.* (2008) assessed the performance of the triangular deflector as an alternative. From visual observation, no obvious disadvantages were noted as compared with jet deflection from the standard bucket. Hydraulically, there are definite differences, however. The maximum dynamic pressure is significantly larger for the triangular deflector as for the corresponding circular-shaped bucket, yet the total dynamic pressure force is larger for the latter because of the long dynamic pressure load extension. Due to the large overpressure at the deflection point, the cavitation potential is negligibly small.

The takeoff angles of the two designs are comparable for usual relative bucket heights of the order of $w/h_o \cong 2$, but they are larger for the triangular deflector of small value w/h_o. No definite pros or cons result from this fact, because the jet trajectories can be directed to the optimum impact location with the two independent parameters deflector angle γ and deflector height w. This statement applies also to the circular-shaped bucket with parameters β_b and R_b. The remaining flow features were also compared for the two corresponding designs without stating a major drawback of either type. The triangular-shaped deflector is thus considered a definite element for ski jumps, and of interest in terms of simple shape and as a basis for 3D flip buckets. Limitations are (1) $8.2° \leq \gamma \leq 33.2°$ for the deflector angle; (2) $3 \leq F_o \leq 8$ for the approach flow Froude number; and (3) $0.20 \leq w/h_o \leq 1.67$ for the relative deflector height.

The air concentration features of deflector-generated jets were provided by Pfister and Hager (2009), focusing on the black-water jet core as reference length, and fitting to the latter the average and minimum streamwise air development of the jet.

6.2.5 Air entrainment in ski-jump jets

Motivation and experimentation

Despite ski jumps are incorporated in many hydraulic schemes, relatively few works on their basic hydraulic features are available. Most observations are site specific, so that generally applicable comprehensive design guidelines of ski jumps cannot be specified. Most of these hydraulic structures are therefore model-tested prior to the final design stage. The loss of a ski jump or damages caused by scour or cavitation in the approach flow chute may result in severe consequences, as for Karun I Dam in Iran. The entire spillway in general – and the dissipator in particular – therefore require a detailed hydraulic consideration (Vischer and Hager, 1998; Khatsuria, 2005).

The purpose of Schmocker *et al.*'s (2008) study was to analyze the air entrainment characteristics of a plane jet downstream of a ski jump, both for pure water *and* pre-aerated approach flow conditions. The results thus relate to the development of the air concentration of highly turbulent pure water, and air-water jets in the atmosphere, emphasizing the jet disintegration characteristics (Bin, 1993; Ervine, 1998) and the resulting plunge pool scour (Mason, 1993; Pagliara *et al.*, 2006). The present results are also relevant to analyze water jets for fire fighters, yet with the opposite purpose. Whereas water jets should remain highly compact for fire fighting, ski jumps should produce an air-water flow with a minimum potential for plunge pool scour. If the jet thickness is too large, the throwing distance too short or the impact angle to steep, then plunge pools tend to become excessively deep. The addition of special elements to disperse artificially a water jet issued from a ski jump was not tested.

The hydraulic model is shown in Figure 6.25 with (from left) the approach flow pipe of 250 mm internal diameter, the jet box allowing for variable h_o, take-off elevation 0.304 m above channel bottom, the tailwater channel and the instrumentation used. A fixed bucket geometry was employed involving a take-off angle of $\beta_b = 30°$, a bucket radius of $R_b = 0.40$ m and discharge $Q = 150$ l/s in the 0.50 m wide channel. Air was supplied to the pre-aerated jets, resulting in an approach flow air concentration $C_o = Q_A/(Q_A + Q_W)$, with subscripts A and W relating to air and water, respectively. The air concentration profile was measured shortly upstream from the bucket, resulting in the typical chute distribution (4.2.3). At jet take-off, the air concentration distribution is affected by streamline curvature.

An RBI twin fiber-optical probe was employed to record local air concentrations $C > 0.1\%$; the local air-water mixture velocity was recorded for local air concentrations $C > 1\%$ (Boes and Hager, 2003; Kramer et $al.$, 2006). The data correlation for pre-aerated jet flow was generally poorer than of non-aerated approach flows, similar to the deflector tests (Chapter 4).

Concentration profiles

The inset of Figure 6.26 sketches the air concentration profile $C(Z)$, with the vertical dimensionless coordinate $Z = (z - z_L)/(z_U - z_L)$ relative to the upper (subscript U) and the lower (subscript L) jet trajectories. Close to take-off, the profile is nearly rectangular, with a black-water core spanning over 70% of the jet (subscript j) thickness h_j. Further downstream, air is entrained along the upper and lower jet trajectories by turbulence production along the chute and by interfacial mixing along the air-water boundary layer. As noted from Figure 6.26, the concentration profiles are asymmetrical about the jet center line: Location z_m of the minimum (subscript m) air concentration is above $0.50Z$. Whereas water drops ejected from the upper jet trajectory eventually return onto it, these ejected along the lower jet trajectory fall onto the channel bottom.

Figure 6.25 Ski-jump model with jet box on the left (Heller *et al.*, 2005)

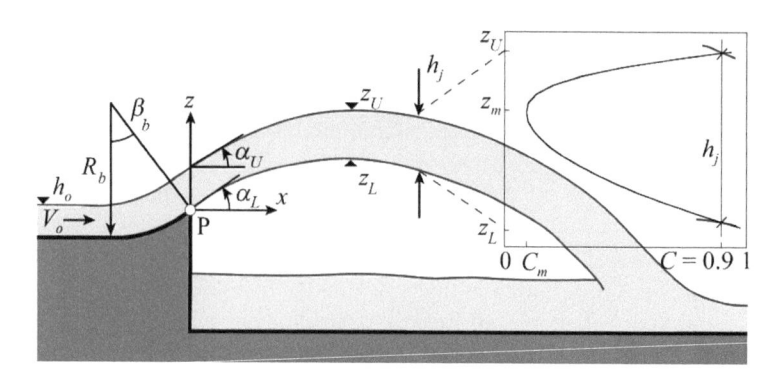

Figure 6.26 Sketch of ski-jump flow, with notation (Schmocker *et al.*, 2008)

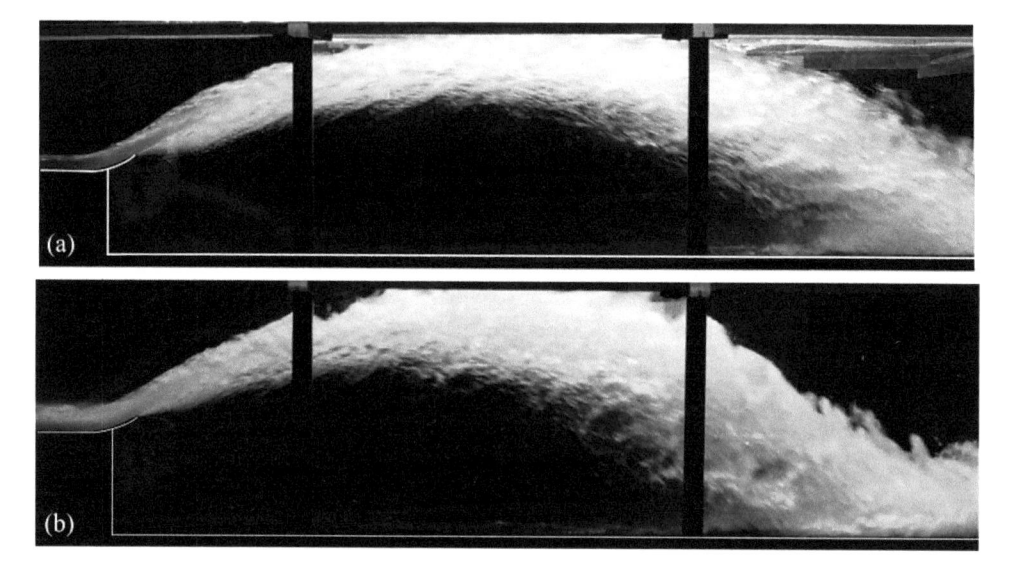

Figure 6.27 Side view of jets with F_o = 8, h_o = 0.045 m and C_o = (a) 0, (b) 0.20. Compare with Figure 6.28 for jet air concentrations (Schmocker *et al.*, 2008)

Of particular relevance are the upper and the lower jet trajectories $z_U(x)$ and $z_L(x)$ plus the location z_m and the amount C_m of minimum air concentration. Figure 6.27 shows the jet dispersion effect for two tests. Upstream from the bucket the flow is slightly aerated at the free surface for C_o = 0, followed by a dark gray jet core from the take-off section to almost maximum jet elevation; further downstream the jet contains lots of air impacting onto the channel bottom as a fully disintegrated air-water flow. For C_o = 0.20, in turn, the jet thickness is larger and the impact location closer to the take-off section because of increased turbulence production and energy dissipation.

Streamwise air concentration profiles

The air concentration profiles $C(x, z)$ are shown as contour plots of jets horizontally stretched. Figure 6.28 relates to $C_o = 0$ and $C_o = 0.20$, with the streamwise development of the jet air content. Two particular angles of disintegrating jets were considered, namely the jet core angle γ_i as the limit of the jet core, and the transverse jet growth angle γ_j describing the jet surface where the air concentration has the standard value $C = 0.90$ (Vischer and Hager, 1998). The jet core angle was defined by isolines of jet air concentrations of $C = 0.10$ and 0.20.

Figure 6.29a shows a definition sketch for a jet issued from a nozzle of height h_o into a stagnant fluid, in contrast to the present free surface jet flow. The jet thickness at a specific location is $h_j = z_U - z_L$, whereas the core thickness is h_i. Both angles γ_i and γ_j are known not to vary with F_o for turbulent flow. Figure 6.29b shows h_j/h_o versus $x/[h_o(1 - C_o)]$, and the trend line as

$$h_j/h_o = 1.26 + 0.051(x/[h_o(1-C_o)]), \quad \text{for } (x/h_o) < 60. \tag{6.35}$$

From Eq. (6.35) the jet spread angle is $\gamma_j = (1/2)\arctan[0.051/(1 - C_o)]$, i.e. $\gamma_j = 1.5°$, equivalent to a full-spread angle of $2\gamma_j = 2.9°$ for $C_o = 0$ and $2\gamma_j = 3.6°$ for $C_o = 0.20$, respectively. Similar values are reported for turbulent jets (Rajaratnam, 1976; Chanson, 1991; Khatsuria, 2005; Annandale, 2006).

Figure 6.29 (c, d) shows the jet core thickness h_i/h_o again versus $x/[h_o(1 - C_o)]$ along the isolines of $C = 0.10$ and $C = 0.20$, respectively. Note that the two air concentrations considered for the jet core development result in similar findings except for the exact numerical values. The jet core decay angles are $\gamma_i(C = 0.10) = 1.6°$, $\gamma_i(C = 0.20) = 1.2°$ for $C_o = 0$, and $\gamma_i(C = 0.10) = 2.0°$, $\gamma_i(C = 0.20) = 1.5°$ for $C_o = 0.20$, respectively. The angles for pre-aerated flows are larger than these of non-aerated flows. The data relating to a test with $h_o = 0.070$ m have a different trend in the far jet field; they do not start at (0; 1) due to jet curvature across the ski jump and a slight surface aeration of the approach flow. Their trend lines are expressed with the jet core coefficient $D_{\text{core}} = 0.056$ for $C = 0.10$, and $D_{\text{core}} = 0.041$ for $C = 0.20$ as (Schmocker et al., 2008)

$$h_i/h_o = 1.05\,[1 - D_{\text{core}}(x/[h_o(1-C_o)])]. \tag{6.36}$$

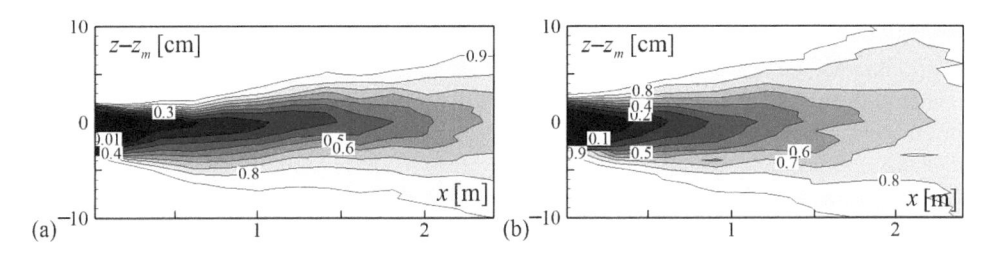

Figure 6.28 Contour plots of streamwise jet air concentration $C(x; z)$ for $C_o =$ (a) 0, (b) 0.20, $\mathsf{F}_o = 8$ (Schmocker et al., 2008)

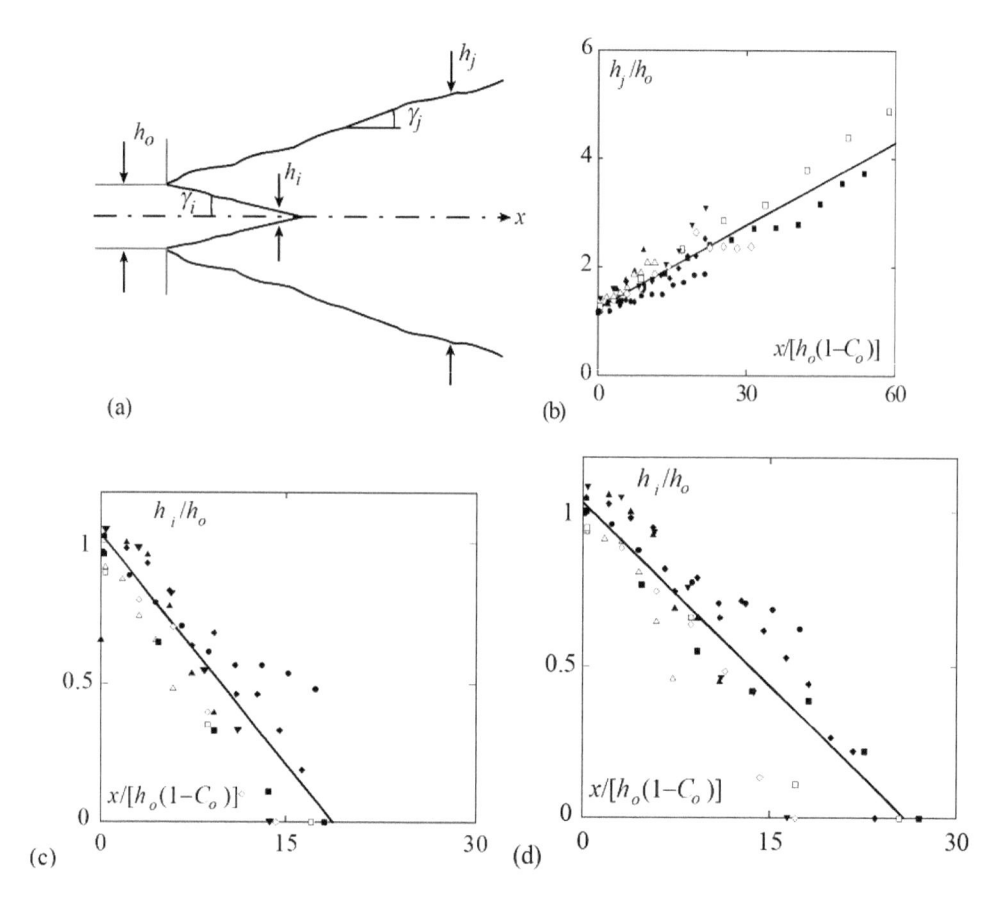

Figure 6.29 (a) Definition sketch of nozzle-issued one-phase flow jet, (b) jet thickness h_j/h_o versus $x/[h_o(1-C_o)]$ with (−) Eq. (6.35), jet core thickness h_i/h_o versus $x/[h_o(1-C_o)]$ for $C =$ (c) 0.10, (d) 0.20 with (−) Eq. (6.36) (Schmocker *et al.*, 2008)

Jet trajectories

The jet trajectories $z_U(x)$ and $z_L(x)$ of a high-speed air-water flow are usually defined by a jet air content $C = 0.90$. The present results based on the air concentration probe were compared with these of Heller *et al.* (2005) measured with a trajectory point gage (6.2.3). Figure 6.30 relates to a (a) non-aerated and (b) pre-aerated jet. The data of the two samples essentially collapse, mainly in the near flow field up to the trajectory maxima. Further downstream, $h_j(x)$ is larger for the concentration probe data than the point gage data. Jet pre-aeration thus increases the jet thickness, mainly along the upper jet trajectory due to increased turbulence and air addition.

Cross-sectional air entrainment characteristics

Figure 6.31a shows the streamwise development of the cross-sectional average (subscript a) air concentrations $C_a(x)$ defined as

$$C_a = \frac{1}{h_j} \int_{z_L}^{z_U} C(z) \cdot \mathrm{d}z. \tag{6.37}$$

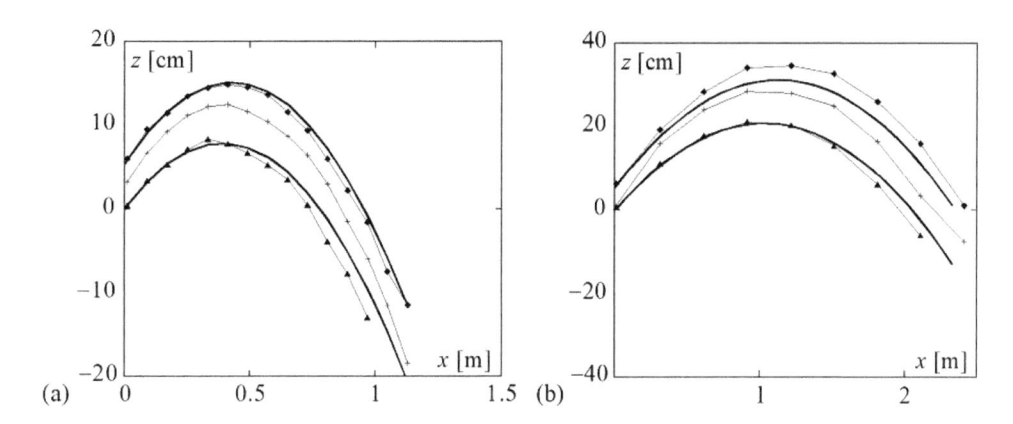

Figure 6.30 Comparison of jet trajectories ($\blacklozenge$) $z_U(x)$, (+) $z_m(x)$, ($\blacktriangle$) $z_L(x)$ according to (–) Eq. (6.4) for h_o = 0.045 m and (a) F_o = 5, C_o = 0, (b) F_o = 8.2, C_o = 0.20 (Schmocker *et al.*, 2008)

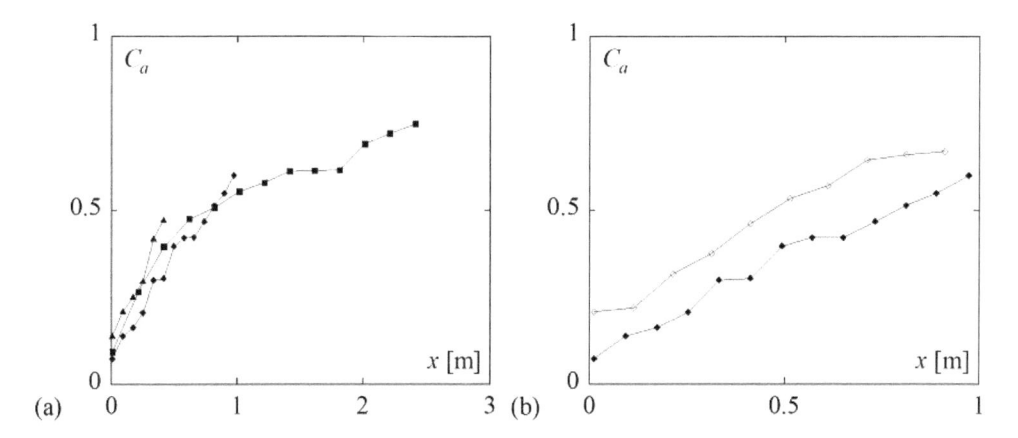

Figure 6.31 Cross-sectional average air concentration $C_a(x)$ for (a) C_o = 0 and F_o = ($\blacktriangle$) 3, ($\blacklozenge$) 5, ($\blacksquare$) 8, (b) comparison of C_a between C_o = ($\blacklozenge$) 0 and ($\diamond$) 0.21 for F_o = 5 and h_o = 0.045 m (Schmocker *et al.*, 2008)

All data follow a trend, with an initially steep increase reducing for larger values of x to finally tend toward the asymptotic value $C_a \rightarrow 1$. Figure 6.31b compares the data for C_o = 0 and 0.21, for F_o = 5. The difference $\Delta C_a(x)$ between the two remains almost constant with $\Delta C_a \cong 0.15 < C_o$ = 0.21 because of air detrainment in the ski jump reach. Note that $C_a(0) > 0$ because of the slight surface air entrainment prior to jet take-off.

The effect of the approach flow depth h_o on $C_a(x)$ is included by the dimensionless coordinate x/h_o. Further effects originate from the relative bucket curvature h_o/R_b and F_o (Heller *et al.*, 2005), so that $X = (x/h_o)(R_b/h_o)(1/F_o) = x \cdot R_b/(F_o h_o^2)$ is the governing dimensionless coordinate. The approach flow Reynolds and Weber numbers are confined to $7.1 \times 10^4 \leq R_o \leq 2.6 \times 10^5$ and $49 \leq W_o \leq 132$, respectively. Here $R_o = V_o h_o/\nu$ and $W_o = (\rho_W V_o^2 h_o/\sigma)^{1/2}$ with ν as kinematic viscosity, ρ_W as fluid density, and σ as surface tension. The test data are analyzed and generalized below.

The minimum (subscript m) cross-sectional air concentration C_m increases with x. The test data are plotted as $C_m(X)$ in Figure 6.32a. Except for one test, all data collapse on a single curve both for non-aerated and pre-aerated jets (Schmocker et al., 2008).

The velocity data relate to air-water jets with $C > 1\%$. Figure 6.32b shows the relative maximum (subscript M) cross-sectional velocity of pre-aerated jets as $V_M(X)/V_o$ indicating a significant velocity decay along the jet trajectory following

$$V_M / V_o = 1.12(1 - 0.0125X), \quad \text{for } 0 < X < 50. \tag{6.38}$$

Air concentration profiles

An air concentration profile is made up of the upper and the lower jet trajectories $z_U(x)$ and $z_L(x)$, respectively, plus the location $z_m(x)$ and the value C_m (Figure 6.26). A data analysis indicated that both the upper and lower concentration portions are represented by power functions. The data were normalized as $c = (C - C_m)/(0.90 - C_m)$ providing boundary values $c(C = C_m) = 0$ and $c(C = 0.90) = 1$; and $Z = (z - z_L)/(z_U - z_L)$ with $0 \leq Z \leq 1$. Constant values of $Z_m = 0.63$ (±0.05) for non-aerated and $Z_m = 0.54$ (±0.07) for pre-aerated flows were used. The data are expressed as (Schmocker et al., 2008)

$$c = \zeta^{1/m}. \tag{6.39}$$

Here $(1/m)$ is an exponent, $\zeta = [(Z - Z_m)/(1 - Z_m)]$ and $\zeta = [(Z - Z_m)/(-Z_m)]$ for the upper and the lower jet portions, respectively. The data trend follows with $m = 0.176$ for $Z_m \leq Z \leq 1$, whereas $m = 0.224$ for $0 \leq Z \leq Z_m$. Figure 6.33 compares the data with Eq. (6.39) at the dimensionless location X; the data quality of non-aerated tests is again better than for pre-aerated flow. The data scatter appears to increase with F_o, given the highly turbulent flow pattern. The data at small distances X tend to a more triangular and those for large X to a more rectangular shape than predicted from Eq. (6.39). More data under a wider test program would have to be collected for a generalized data analysis.

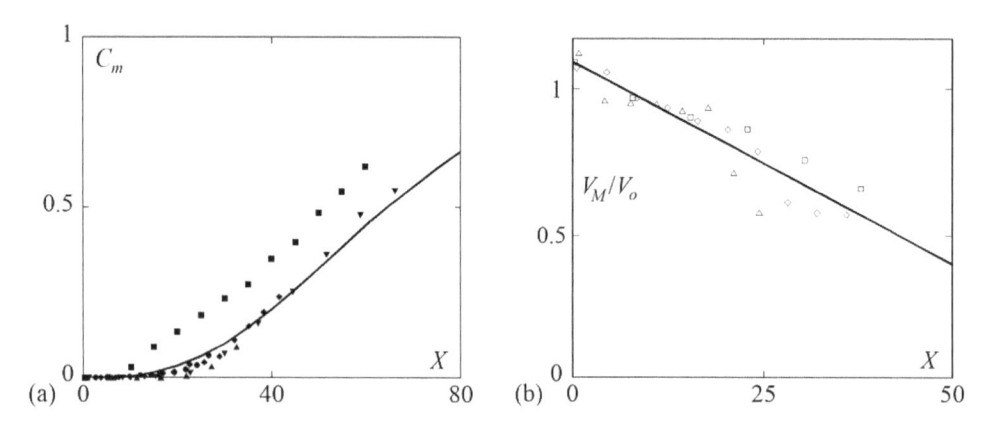

Figure 6.32 (a) Minimum air concentration $C_m(X)$ for non-aerated jets, (b) cross-sectional maximum relative velocity $V_M/V_o(X)$ for pre-aerated jets with (−) Eq. (6.38) (Schmocker et al., 2008)

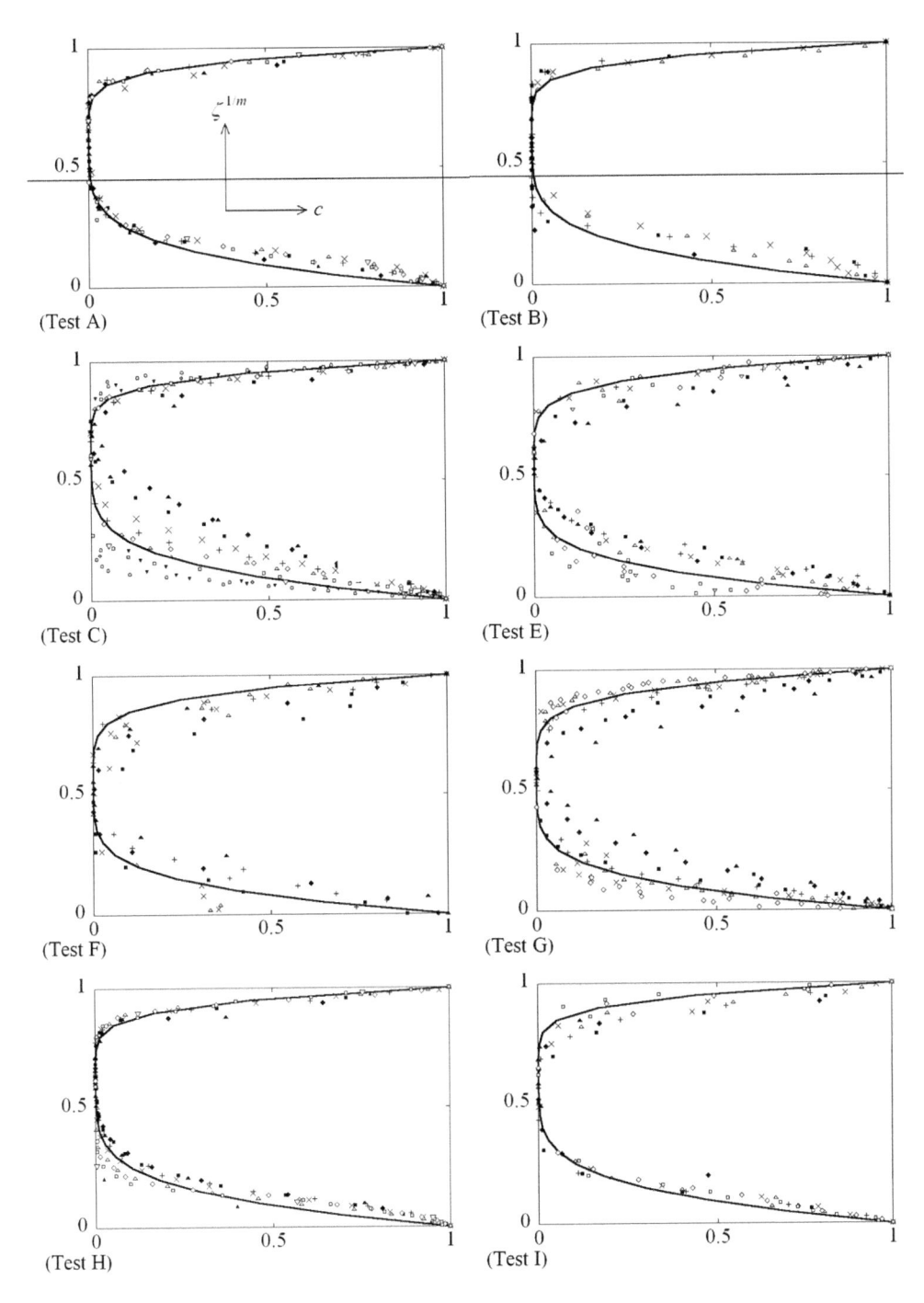

Figure 6.33 Generalized air concentration profile $c(\zeta)$ with (−) Eq. (6.39) (Schmocker *et al.*, 2008)

Discussion of results

The previous results allow for an appreciation of the jet air concentration prior to jet impact. If the impact (subscript s) elevation is assumed to be identical to the jet take-off elevation, the horizontal distance between the two is $l_s = \sin(2\alpha) \times [V_o^2/g]$ from Eq. (6.7). Further, the relative bucket radius under design conditions is of the order of $R_b/h_o = 10$ (Vischer and Hager, 1998). The corresponding dimensionless distance thus is $L_s = (l_s/h_o)(R_b/h_o)F_o^{-1} = 10F_o\sin(2\alpha)$. The average impact air concentration depends on $C_{as} = C_{as}(C_o, F_o, \alpha)$. The minimum value of C_{as} results for $C_o = 0$. Assuming a typical take-off angle $\alpha = 30°$ results in a relation between C_{as} and F_o, with $C_{as}(F_o = 4) = 0.56$, $C_{as}(F_o = 6) = 0.74$ and $C_{as}(F_o = 8) = 0.86$. The values for a jet with $C_o = 0.20$ are some 10% higher. For prototype condition with $F_o > 5$, the average impact jet air concentration is thus always larger than 50%. For small α values, the air entrainment is considerably reduced. The take-off angle is the relevant design variable affecting the disintegration process, therefore.

A similar analysis applies for the minimum jet air concentration C_{ms} at jet impact. With the same basic parameters as above for the average jet air concentration results $C_{ms}(F_o = 4) = 0.14$, $C_{ms}(F_o = 6) = 0.34$ and $C_{ms}(F_o = 8) = 0.55$ (Schmocker *et al.*, 2008). For pre-aerated flow, these numbers are significantly increased. Therefore, typical ski-jump jets hardly have a black-water core, and are considerably aerated at the impact location. The test procedure of e.g. Pagliara *et al.* (2006) for scour holes with an air-water mixture pipe flow to be dealt with below are considered realistic, therefore. As to scale effects of aerated jets, few information is available. Comparing laboratory jets with typical prototype jets indicates significant differences, mainly in the amount of air contained in prototype jets. Given the difficulties in dealing with these in the field, their aeration characteristics need to be considered in hydraulic laboratories.

6.2.6 Generalized jet air concentration features

The streamwise jet air concentration characteristics are described by Pfister *et al.* (2014) uniquely by the black-water (subscript bw) core length L_{bw}, based on Schmocker (2006) and Balestra (2012). Parameter L_{bw} is defined based on the minimum (subscript m) air concentration C_m measured within a certain air concentration profile. The black-water core is located between the jet take-off section at $x = 0$ and the profile at location x where $C_m = 0.01$. Values $C_m \leq 0.01$ along L_{bw}, whereas further downstream $C_m > 0.01$. Note that coordinate x is parallel to the chute bottom upstream of the flip bucket, and that the origin is at the take-off section. The black-water core length is given as (Pfister *et al.*, 2014, Figure 6.34a)

$$\frac{L_{bw}}{h_o} = 76F_o^{-1}(1+\tan\delta)^{-4}(1+\sin\varphi). \tag{6.40}$$

With h_o as the approach flow depth, F_o as the approach flow Froude number, δ as the equivalent deflector angle, and φ as the chute bottom angle relative to the horizontal, predictions follow Eq. (4.86), so that the following remains also for $\varphi = 0$. With L_{bw} and the inclined coordinate x, the streamwise normalization for the airflow features is represented by the relative black-water core length $\chi = x/L_{bw}$. Accordingly, the reach $0 < \chi \leq 1$ corresponds to the black-water core portion, whereas $\chi > 1$ to the fully aerated jet portion. Straub and Anderson (1960) defined the average air concentration C_a as the integral of the cross-sectional

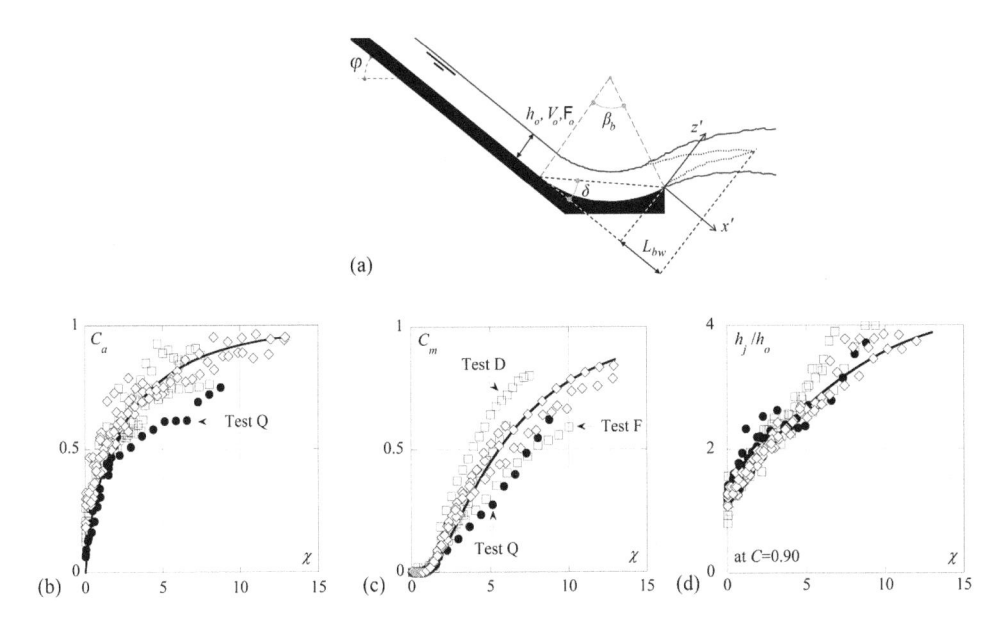

Figure 6.34 Takeoff jet features (a) definition sketch, (b) average air concentration $C_a(\chi)$ with (−) Eq. (6.34), (c) minimum air concentration $C_m(\chi)$ with (−) Eq. (6.35), (d) normalized jet thickness $h_j/h_o(\chi)$ with (−) Eq. (6.36) for ϕ = (•) 0, (□) 12°, (◊) 30° based on Pfister *et al.* (2014)

distribution up to the aerated flow depth, herein equivalent to the jet thickness, with the boundaries z_U and z_L at the upper and lower jet surfaces, respectively, following Eqs. (6.4) and (6.6).

The average air concentrations C_a are directly related to the individual and relative black-water core length χ. Values of C_a increase particularly along the first jet reach, tending to $C_a \to 1$. At $\chi = 1.7$, for instance, $C_a \cong 0.50$, whereas $C_a = 0.90$ at $\chi \cong 8.8$. Along the jet, the average air concentrations can be estimated from (Figure 6.34b)

$$C_a = \tanh(0.4\chi^{0.6}). \tag{6.41}$$

The minimum air concentration C_m within a profile also is a function of χ only. For non-aerated black-water approach flow, as tested by Pfister *et al.* (2014), the initial value at $\chi = 0$ corresponds to $C_m \cong 0$ as observed at the chute bottom, and $C_m = 0.01$ at $\chi = 1$ by definition. Further downstream, C_m values significantly increase tending to $C_m \to 1$ (Figure 6.34c). Note besides the pronounced increase along $1 < \chi < 5$ the much smaller slope of the trend line for $\chi > 10$, resulting in the generalized estimation

$$C_m = \left[\tanh(0.4(\chi-1)^{0.6})\right]^{2.5}. \tag{6.42}$$

The normalized jet thickness h_j/h_o versus χ is given from the streamwise development of the average air concentration as (Figure 6.34d)

$$\frac{h_j}{h_o} = 1.3 + 3\tanh(0.1\chi). \tag{6.43}$$

Using the constant initial jet thickness $h_j = 1.3h_o$ at $\chi = 0$ provides similar results as that with the geometrically corrected value $h_j = [1/\cos\beta_b]h_o$, assuming a constant flow depth across the bucket. Pfister *et al.* (2014) suggest general streamwise air concentration profiles along disintegrating jets.

6.3 Flip bucket

6.3.1 Types of bucket geometries

The flip bucket is a generalized hydraulic element by which a high-speed water flow is deflected in plan view, and the jet impact location is controlled with a ski-jump type element by which the flow is transformed into an air-water jet, so that the advantages of this design are also included. Flip buckets are applied in hydraulic engineering if a high-speed flow needs readjustment of direction, so as with a bottom outlet whose streamwise axis is typically by 45° away from the axis of the tailwater valley. Flip buckets thus result in the action of the ski jump plus in a deflection of the approach flow. The flip bucket thus combines a jet deflection both in the streamwise section (jet lifting) and in the plan view (jet deflection) without generating excessive shock waves upstream from the control element (Figures 6.35, 6.36).

Figure 6.35 Flip bucket spillway of Shuibuya Dam (233 m) in China (Courtesy Martin Wieland)

Flip buckets were often applied in the past, but it appears that there are even less generalized guidelines than for the basic ski jump. Mason (1993) provides comments including a state of the art. Rhone and Peterka (1959) were among the first to propose the flip bucket to 'deflect or flip tunnel spillway discharges'. The US Bureau of Reclamation (USBR) developed various bucket designs for discharges ranging from 650 to 27,000 m³/s for approach flow heads from 130 to 200 m. The main hydraulic problems relating to flip buckets are (1) complex flow features demanding laboratory tests for a thorough appreciation of their effect, (2) transition from normally circular-shaped tunnel invert to rectangular approach flow cross section, (3) cavitation damage by the generation of low-pressure zones of a poorly designed bucket or drainage slits, and (4) addition of appurtenances to improve jet dispersion which may have adverse effects again with regard to cavitation damage. The finally retained design appears to be complicated; it has hardly been improved over the past 60 years, so that little information is available. Experiences on spillway operation with a particular regard to flip buckets were reported by Del Campo et al. (1967), resulting in operational guidelines. Depending on the jet impact location, deflectors spread or elongate a jet (Şentürk, 1994; Vischer and Hager, 1998). USBR (1978) proposed various designs, yet most of these are site specific (Khatsuria, 2005) involving complex geometry (Novak et al., 2010), requiring hydraulic model tests. Damage of flip buckets at Karun I Dam in Iran (Figure 6.36) are addressed by Jalalzadeh et al. (1994).

An appropriate bucket geometry has to be defined depending on the aspired jet trajectory and its footprint features (Pfister and Schleiss, 2015). Some of the most frequent ski-jump types are shown in Figure 6.37, in which 'bucket type' refers to the longitudinal and transversal bucket arrangements, whereas 'chute end layout' (Figure 6.38) describes the sidewall geometry in plan view. The sidewalls may not only vary at the bucket, but already upstream to avoid hydraulic phenomena as flow separation or shock waves at expansions or contractions.

Optimizing a flip bucket relative to jet energy dissipation only may result in high dynamic forces on the bucket, flow choking up to relatively large discharges, and generation of pronounced shock waves. The optimum combination of bucket type and chute end layout allows for defining an appropriate jet impact location in the plunge pool along with a high degree of jet disintegration. Bucket types include (Figure 6.37):

- Bucket Type 1 with transversally horizontal take-off lip, representing the standard configuration resulting in a straight and relatively compact jet.
- Bucket Type 2 with transversally inclined take-off lip generating a relatively long jet trajectory at the steep side (small bucket radius) and a relatively short at the flat side (large bucket radius), next to a slight lateral jet deflection.
- Bucket Type 3 includes a transversally rounded take-off lip, so that the jet take-off angle is steep near the sidewalls (for a concave lip curvature) but flat in the bucket center. Consequently, relatively long lateral trajectories are produced near the sidewalls, reducing toward the center. The vertical jet spread is reduced and the jet footprint more concentrated. A convex curvature reduces the lateral take-off angles and thereby shortens these trajectories, resulting in a rounded footprint area.
- Bucket Type 4 with a drainage channel improves its performance under small discharges by avoiding choking. No hydraulic jump in the bucket occurs due to the counter slope for relatively small discharges, so that a free jet emerges. This type is mainly an option for unregulated spillways.

Figure 6.36 (a) Karun III Dam, Iran. Double-arch dam of 205 m total height, with flood release structures comprising two chutes followed by flip buckets, two low-level outlets and three ogee crest overfall sections of 18,000 m³/s combined discharge capacity. Outlet jets plunge into concrete-lined plunge pool, 400 m long and 50 m wide (Photo: Iran Water & Power Resources Development Co. 2005), (b) Karun I flip bucket showing both originally strongly sloping take-off, and improved horizontal take-off located above it (Jalalzadeh *et al.*, 1994)

- Bucket Type 5 with insert(s) enhances jet disintegration and the footprint area. The insert(s) also produces additional jet turbulence and lift the central jet portion.
- Bucket Type 6 with baffles roughens the lower jet trajectory surface enhancing the flow turbulence in the lower jet part, and thus the jet disintegration process. The footprint area is enlarged thereby reducing the density of the impact energy. In addition, the footprint is not coherent but consists of several individual patches, at least up to certain discharges.

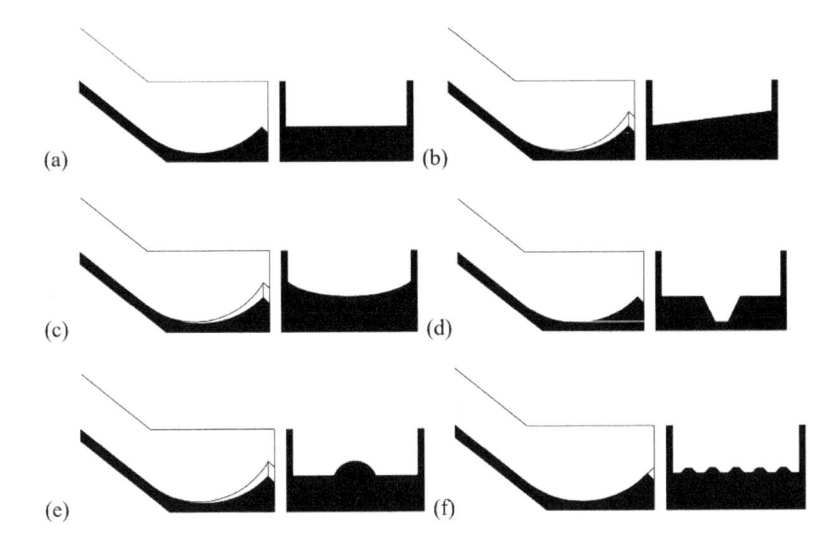

Figure 6.37 Sketches of bucket Types (a) 1, (b) 2, (c) 3, (d) 4, (e) 5, and (f) 6, showing longitudinal section along final chute end as well as transversally along bucket take-off lip in downstream direction (Pfister and Schleiss, 2015)

In addition to the bucket types, the chute end layout with its sidewall configuration strongly affect the shape and the surface area of the jet and its footprint. The standard configuration with parallel sidewalls is shown in Figure 6.38, including bucket Type 1. The jet spreads transversally due to sidewall-induced turbulence (Ervine and Falvey, 1987) highlighting a slightly rounded footprint at jet impact. This shape is due to the flow velocity distribution at the take-off lip. Near the bottom and at the sidewalls, the local flow velocity is some 80% of the average value (Chow, 1959). Accordingly, the resulting lateral jet trajectories are shorter as compared with those in the center, where the local velocity corresponds to about 120% of the average value.

The footprint shape is affected by varying the bucket take-off lip and its angle, whereas the velocity distribution is determined by the chute and cannot be altered. An example is shown in Figure 6.38b, combing the chute end Layout 1 with bucket Type 2. A small geometrical take-off angle is noted at left (large bucket radius with low take-off elevation), and a high take-off angle at right. A shorter jet trajectory thus emerges at left than at right thereby rotating the footprint. The jet is also slightly deviated to the left due to the lateral acceleration generated by the bucket. The slightly rounded footprint remains due to the particular chute velocity distribution. Note that a similar footprint results with an oblique take-off lip in plan view (Berchtold and Pfister, 2011, Figure 6.38a). Combining bucket Type 6 with the chute end Layout 1 is shown in Figure 6.38c. The baffles add turbulence to the jet, enhancing its disintegration process. The footprint is spread over a larger area and partially is discontinuous.

All bucket Types from Figure 6.37 can be combined with the chute end Layouts 1–4 shown in Figure 6.38. Contracting or expanding sidewalls are often employed. They either reduce the jet width thereby increasing its vertical thickness (Figure 6.38d), or transversally widen the jet thereby expanding its footprint (Figure 6.38e). The widening angle η_W per sidewall is limited to $\tan\eta_W - 1/(3F_o)$ to avoid flow separation from the wall (USBR, 1987). Widening

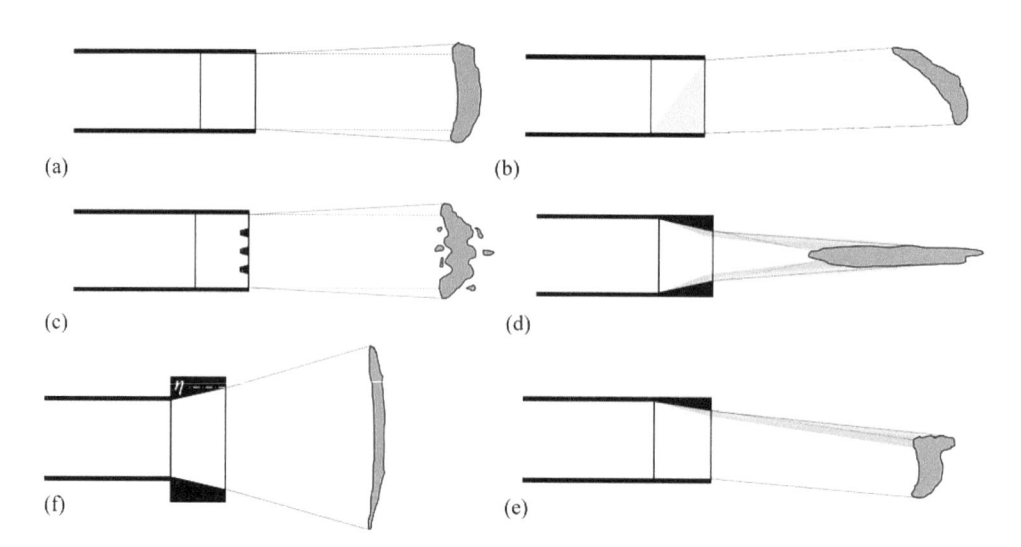

Figure 6.38 Sketches of chute end layouts with their resulting jet footprints in plan view. Layout 1 with parallel sidewalls combined with (a) bucket Type 1, (b) bucket Type 2, (c) bucket Type 6 (Mason, 1983), (d) Layout 2, contraction or slit-type bucket (Zhenlin *et al.*, 1988; Dai *et al.*, 1988), (e) Layout 3, expansion, (f) Layout 4, vertical triangular-shaped deflector (Juon and Hager, 2000; Lucas *et al.*, 2013, 6.3.2) (Pfister and Schleiss, 2015)

thus often starts upstream from the bucket (Berchtold and Pfister, 2011). For contractions (Figure 6.38d), shock waves are generated at the start of the lateral deflectors (Vischer and Hager, 1998). Given that deflectors represent a positive abrupt wall deflection, positive shocks with increased flow depth result. These represent zones of high unit discharge, continuing or spreading along the jet trajectory and producing a flow concentration at the footprint. The contraction must carefully be designed to avoid flow choking in the bucket.

In extreme cases, as for the slit-type bucket, the left- and right-side shock waves merge, the jet is straightened up (Figures 6.38d, 6.3), enhancing vertical jet dispersion. This increases the upper jet take-off angle resulting in a long and thin footprint area. For small discharges, these buckets concentrate the flow and create a relatively long jet, thereby reducing the spectrum of jet travel length due to varying discharges.

6.3.2 Horizontal triangular-shaped flip bucket

Motivation and experimentation

Given the relevance of high-speed jet deflection and the few studies conducted in the past, a triangular-shaped flip bucket deflector was developed based on systematic model observations (Lucas *et al.*, 2013). The design applies for a relatively small approach flow Froude number.

A jet box issued a rectangular supercritical approach (subscript o) flow of depth h_o and average velocity V_o resulting in the approach flow Froude number $F_o = V_o/(gh_o)^{1/2}$. The 0.50 m wide channel was reduced to 0.30 m to allow for lateral jet deflection up to the deflector section (Figure 6.39). The origin of the coordinate system (x, y, z) is at the tailwater channel bottom below the deflector end. The maximum (subscript M) elevation z_M of the upper

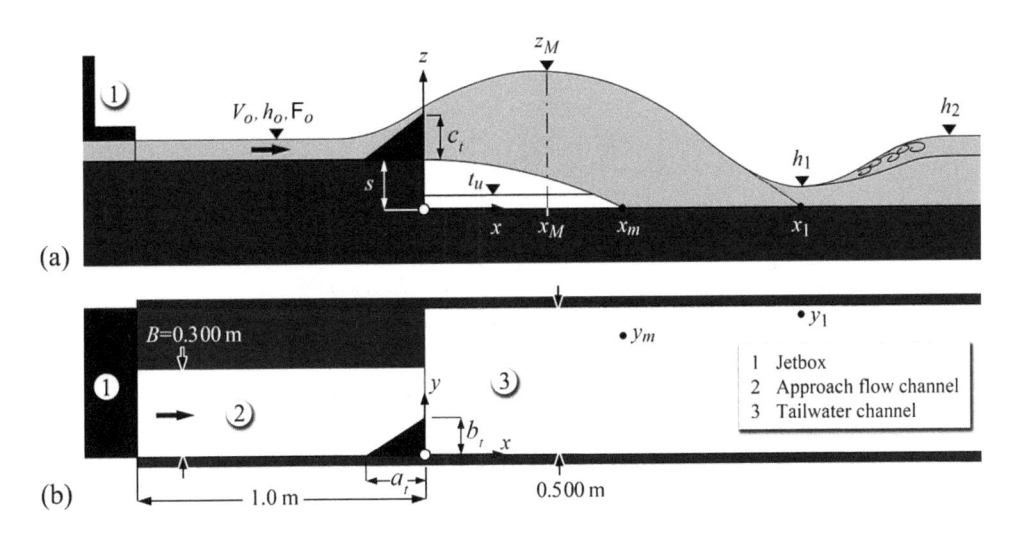

Figure 6.39 Definition sketch of test setup for triangular flip deflector (a) longitudinal section, (b) plan view (see also Figure 6.42 b, 6.42c) (Lucas *et al.*, 2013)

jet trajectory and its position x_M were measured. The impact points of both the upper x_1 and lower x_m jet trajectories onto the tailwater channel were determined visually from the channel side. The jet deflection $\tan\delta_j = y_1/x_1$ of the upper trajectory was measured between the origin and the impact point y_1.

The test program included eight right-angled tetrahedrons, and two vertical plate deflectors located with their ends at $x = 0$; they are characterized by the base length a_t in the streamwise, b_t in the transverse directions, and height c_t (Figure 6.39). The asymmetric tetrahedrons were turned to all three possible positions. Two vertical plate deflectors with $a_t = 0$ were compared with the tetrahedrons. The drop height from the approach flow to the tailwater channel was fixed to $s = 250$ mm. Tests for $F_o = 3$ to 10 using $h_o = 50$ mm as base approach flow depth were conducted. Selected tests involved $h_o = 25$ mm and 75 mm to analyze scale effects (Heller, 2011). It was found that the data pertaining to the relative maximum elevation of the upper jet trajectory for $h_o = 25$ mm are significantly lower than these for $h_o = 50$ mm and 75 mm, for which the data nearly collapse, so that only the latter tests are considered.

Jet trajectory

Figure 6.40 shows images for the 3 deflectors of the model family for $h_o =$ (a) 25 mm, (b) 50 mm and (c) 75 mm with $F_o = 3$, 5, and 6.6. Note on the left the 0.30 m wide approach flow channel, containing clear-water appearing in black, while the air-water mixture jet flow on the right is white. A shock wave is generated at jet impact onto the tailwater bottom. The images with $F_o = 3$ indicate flow de-aeration beyond jet impact.

The relative maximum elevation $Z_M = (z_M - s)/h_o$ of the upper jet trajectory and its relative distance $X_M = x_M/h_o$ increase both with F_o, the deflector angle $\alpha_D = \arctan(b_t/a_t)$ and the relative deflector height c_t/h_o as (Lucas *et al.*, 2013)

$$Z_M = 1 + 0.14\left[F_o\sin\alpha_D\left(\frac{c_t}{h_o}\right)^{1/3}\right]^2, \quad \text{if } 1.5 < F_o\sin\alpha_D\left(\frac{c_t}{h_o}\right)^{1/3} < 9; \tag{6.44a}$$

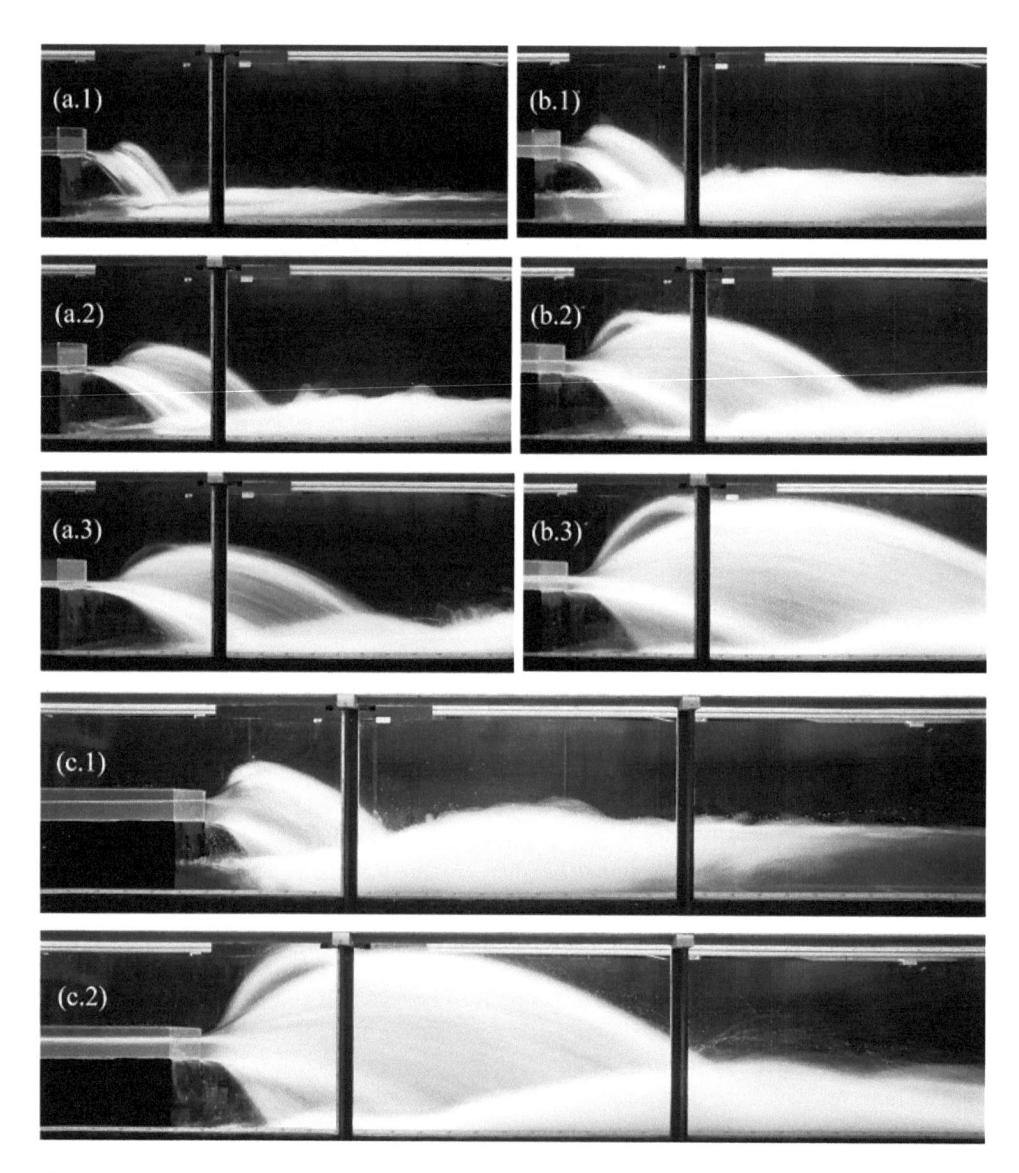

Figure 6.40 Images of (a.1) to (a.3) Deflector 6, h_o [mm] = 25, F_o = 3, 5, 6.6; (b.1) to (b.3) Deflector 1, h_o [mm] = 50, F_o = 3, 5, 6.6; (c.1) to (c.2) Deflector 3, h_o [mm] = 75, F_o = 3, 5. Channel height is 0.70 m for length scale (Lucas *et al.*, 2013)

$$X_M = 0.30\left[F_o\left(\frac{c_t}{h_o}\right)^{0.1}\right]^2, \qquad \text{if } 2 < F_o(c_t/h_o)^{0.1} < 9. \tag{6.44b}$$

Note that X_M and Z_M mainly depend on F_o, whereas the effect of c_t/h_o is small.

Equation (6.44a) is similar to $h_M/h_o = 1 + 0.5F_o^2\sin^2\alpha$ of Steiner *et al.* (2008) (6.2.3) for the maximum elevation of the upper jet trajectory h_M relating to plane deflectors with α as the jet take-off angle, whereas this research is concerned with 3D deflectors spanning partially over the approach flow channel with α as streamwise deflector angle. The relevant Froude number at deflector impact for the upper jet trajectory is smaller than the axial F_o due to wall friction. Given the transverse deflector position, the ratio of reduced to axial approach flow velocities is ≈ 0.80. Inserting this reduction into the above equation improves the data fit. Both involve F_o and the deflector angle α to power 2. Steiner *et al.*'s (2008) modified equation is $h_M / h_o = 1 + 0.32 F_o^2 \sin^2 \alpha$.

The relative upper jet length from $x = 0$ to jet impact $X_1 = x_1 / h_o$ again varies with F_o, along with a minor effect of the relative deflector height, as (Figure 6.41a)

$$X_1 = \left[F_o \left(\frac{c_t}{h_o} \right)^{0.1} \right]^2, \qquad \text{if } 2 < F_o \left(\frac{c_t}{h_o} \right)^{0.1} < 9. \tag{6.45}$$

The lower jet trajectory corresponds essentially to the jet issued by an end overfall (Hager, 2010), depending only on F_o (Figure 6.41b). The deviations of $X_m = x_m / h_o$ for small F_o are due to the fact that Hager's equation applies for a constant drop height s, whereas the present drop flow was affected by the recirculation depth t_u (Figure 6.39). For small F_o, t_u increases so that the effective drop height reduces to $s - t_u$. The relative distance X_m to jet impact increases linearly with F_o as (Figure 6.41b)

$$X_m = 3F_o - 4, \qquad \text{if } 2 < F_o < 7, \, s/h_o \cong 5. \tag{6.46}$$

Jet deflection

The jet deflection $\tan \delta_j = y_1 / x_1$ (Figures 6.38, 6.42) reduces with increasing F_o but increases with relative deflector height c/h_o. Angle δ_j varied between $3°$ and $29°$. Under standard

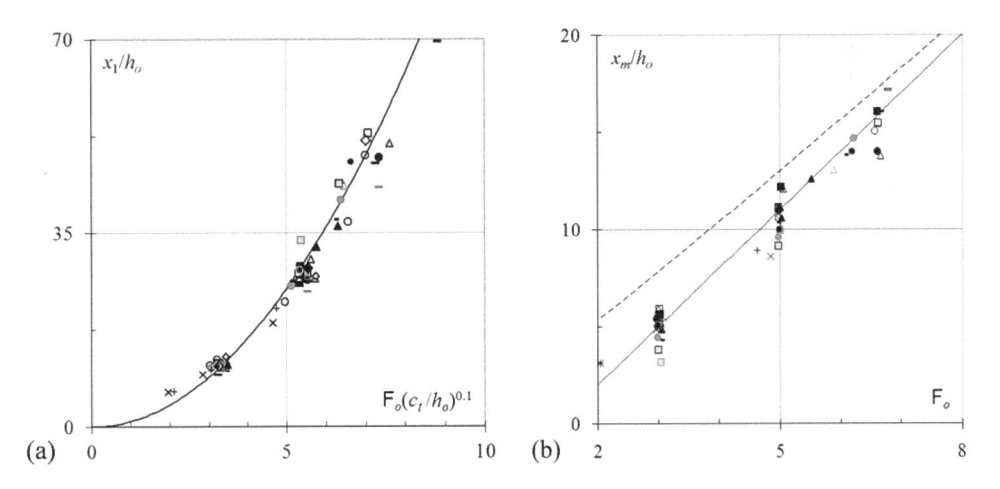

Figure 6.41 Relative distance from take-off (a) $X_1 = x_1 / h_o$ of upper jet trajectory to jet impact versus $F_o (c/h_o)^{0.1}$, (b) (—) $X_m = x_m / h_o$ of lower jet trajectory to jet impact versus F_o compared with (---) end overfall trajectory (Lucas *et al.*, 2013)

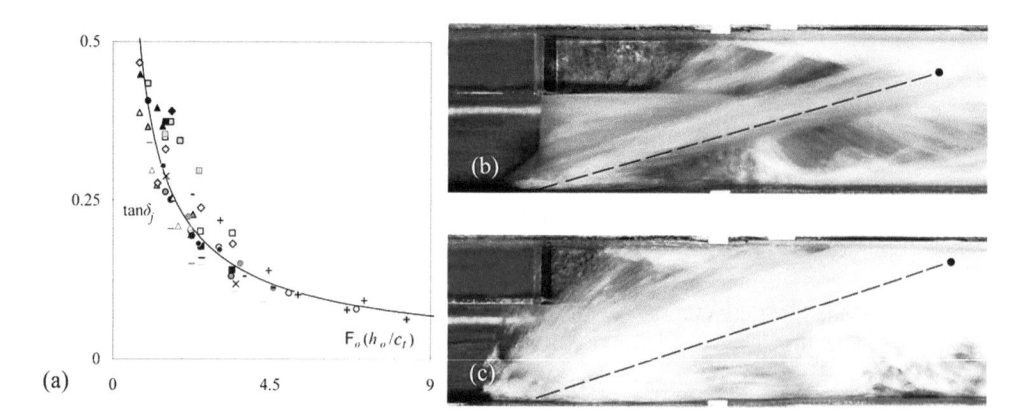

Figure 6.42 (a) Jet deflection $\tan\delta_j$ versus $F_o(h_o/c_t)$; top views with ($\bullet$) impact point of upper jet trajectory onto tailwater bottom under identical $F_o = 5.0$ and $h_o = 50$ mm for deflector with $b_t =$ (b) 100 mm, (c) 200 mm; (---) jet axis (Lucas *et al.*, 2013)

conditions ($F_o = 5.0$, $h_o = 50$ mm), its maximum was $\delta_j = 27°$. For small c/h_o, the deflector effect on the jet is small. The test data follow (Figure 6.42a)

$$\tan\delta_j = \frac{2}{5}\left[F_o\frac{h_o}{c_t}\right]^{-4/5}, \qquad \text{if } 0.75 < F_o(h_o/c_t) < 8.5. \tag{6.47}$$

Energy dissipation

The total energy dissipation of a hydraulic element is defined as $\eta_E = (H_o - H_1)/H_o$, with $H_o = s + h_o + (V_o^2)/(2g)$ as the approach flow energy head, and $H_1 = h_1 + (V_1^2)/(2g)$ as the tailwater energy head, respectively. Most energy is dissipated at jet impact (Chaudhry, 1993). Assuming a classical hydraulic jump beyond jet impact for tests with $t_u = 0$, the sequent flow depth h_1 follows from the measured flow depth h_2, resulting in H_1 and η_E. This method neither applies for roughly $F_o < 4$ because then $t_u > 0$, nor for $F_o > 5.5$ and $h_o = 75$ mm, because then h_2 cannot be measured.

The relative energy dissipation varied between 40% and 70%, similar to Steiner *et al.* (2008). It depends weakly both on F_o and on $C_t = (b_t c_t)/(B h_o)$ describing the relative deflector height c/h_o and width b/B (Figure 6.39). Increasing b/B results in enhanced jet dispersion and so in higher energy dissipation. The test data follow

$$\eta_E = \frac{1}{3}\left[F_o C_t^{1/2}\right]^{1/3}, \qquad \text{if } 2 < F_o C_t^{1/2} < 8. \tag{6.48}$$

Flow choking

In supercritical chute flow, any obstruction may cause a flow breakdown (6.2.3). As to the triangular-shaped deflectors it was found that flow choking depends on F_o, the ratio b_t/a_t and the relative flow depth h_o/B. The experimental data indicate

$$F_o^2 \geq \frac{3}{4}\frac{b_t}{a_t}\left(\frac{B}{h_o}\right), \qquad \text{if } 0.75 < F_o\left(\frac{h_o}{B}\right)^{0.5} < 1.25. \tag{6.49}$$

For typical values of $B/h_o = 10$ and $b_i/a_t = 1$, the minimum Froude number to avoid flow choking is 2.74, i.e. of the order of the lower limit of F_o investigated.

The effect of simple-shaped deflectors on chute flow thus includes a main interest on maximum jet deflection in terms of deflector geometry and the approach flow conditions. Scale effects occur for approach flow depths below 50 mm. No prototype data are so far available given the absence of the novel structure proposed.

The maximum elevation of the upper jet trajectory and its streamwise position are determined. The impact of the upper jet trajectory onto the tailwater bottom compares well with that of flip buckets with similar take-off angles as the streamwise angles studied here. The lower jet trajectory is only marginally influenced by the deflector following the trajectory of the end overfall. The deflector geometry affects the jet shape and therefore energy dissipation. The jet features of vertical plate deflectors are similar to tetrahedron deflectors, except for the formation of shock waves upstream of the deflector due to the abrupt cross-sectional change. The proposed equations allow for the determination of the main jet parameters.

Jet deflection angles of the proposed elements range from 20° to 25° (Lucas et al., 2013). Higher deflections result only for approach flow Froude numbers below 3, irrelevant in the design, however, provoking in addition conditions close to flow choking. Overall, the simple geometry and the straightforward design of the tested deflectors are considered advantageous as compared with conventional flip buckets for almost horizontal approach flow, if the above limitations are respected.

6.4 Granular scour

6.4.1 Granular scour and assessment methods

The adequate performance of a ski jump is mainly related to the quality of the impact zone: If it behaves differently than assumed, the entire dam can be damaged with significant consequences to dam safety. In the past, damage to plunge pools and stilling basins of dams including the Kariba (Zambia and Zimbabwe) or the Tarbela (Pakistan) power plants have led to serious concerns and considerable rehabilitation works after dam completion, to guarantee dam safety. The actual scour holes were often much larger than predicted.

Scour – in contrast to erosion – involves large-scale vortices and jet turbulence by which rock or a granular matrix undergo a fracturing process. Erosion mainly involves shear forces by which the granular matrix considered is superficially attacked, resulting in a planar reduction of matrix elevation, whereas scour rather results in holes in this matrix by which its internal stability may be lost. Note that both erosion and scour may be not visible after e.g. a flood, because portions of the attacked matrix area are covered with loose material. It would however be a big mistake to think that these matrix portions are undamaged, given that scour holes often are only partially covered with granular material.

The scour process is made up of two stages: (1) the disintegration phase, during which the matrix is fractured due to dynamic pressure action; and (2) the transportation phase, during which lifted rock fragments or granular material are entrained by the flow and deposited at the rims of the scour area.

The temporal progress of scour should be modeled physically for each ski jump. Appropriate predictions of prototype scour result from scale models only for granular bed material or strongly fractured rock. For compact, massive rock, scour has to be assessed by combined physical and numerical modeling (6.5). The scour material may be reproduced based on Froude similitude using a filler to simulate material cohesion, reproducing scour geometries closer to prototypes. Incohesive material is employed to study the ultimate scour area, or the scour process to be modeled under the least additional effects offering the maximum scour hole dimensions, as in 6.5.2. A narrow scour hole involving steep slopes as observed only in rock may lead to even deeper ultimate scour even for good-quality rock, as confirmed by the experience of Kariba Dam (6.5.6). The physical processes involved in rock scour are described in 6.5.3.

A state of the art of empirical scour assessment methods until the early 1980s is provided by Whittaker and Schleiss (1984). Some formulas include both model and prototype observations. Figure 6.43 relates to scour below overfall and ski jumps.

Based on tests with gravel beds and drop heights of up to 2 m, Mason (1989) observed a significant effect of the approach flow jet aeration on the scour depth. With $V_e \cong 1.1$ m/s as

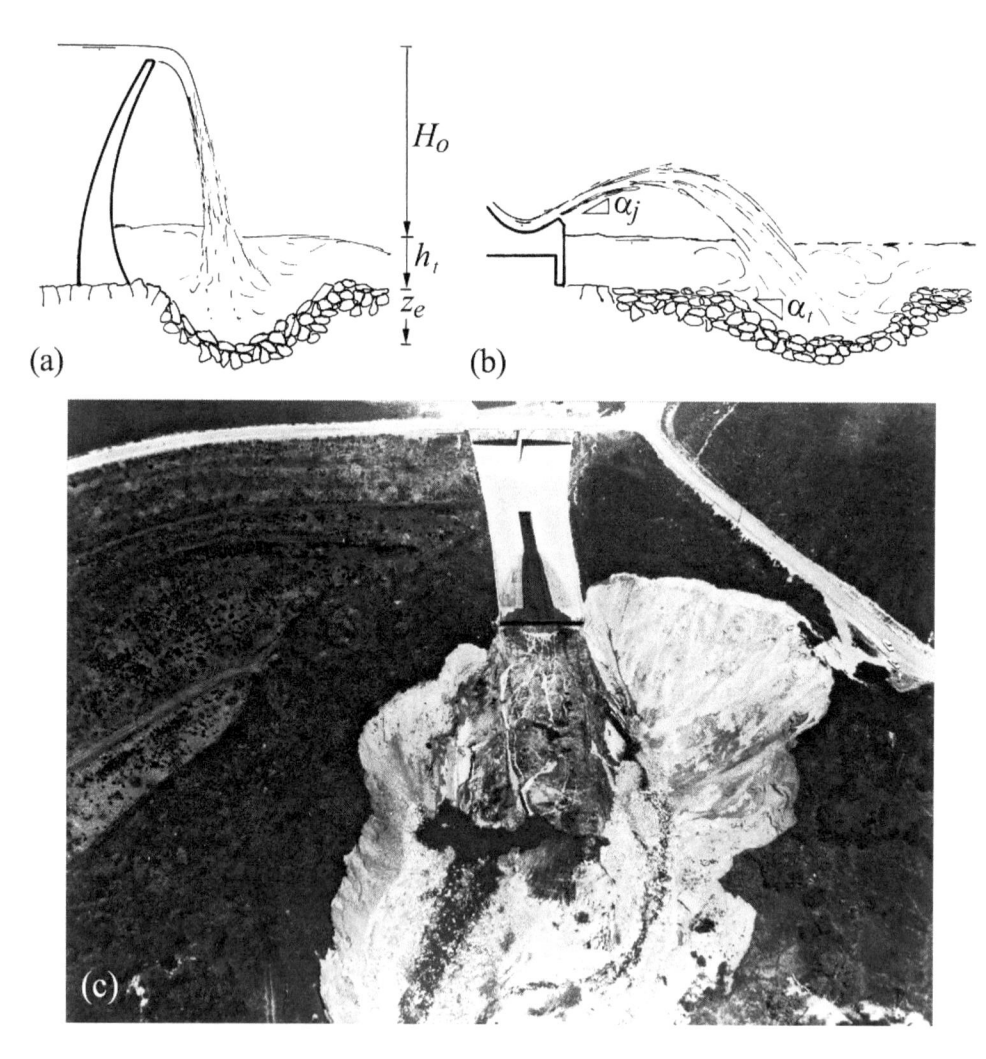

Figure 6.43 Scour below (a) overfall, (b) ski jump, (c) photo of ski jump with scour hole at Nacimiento Dam after 1969 flood (Mason, 1984)

entrainment velocity, V_i as impact velocity, H_o as fall height, and t_i as impact jet thickness, the aeration ratio $\beta_A = q_A/q$ between the air (subscript A) and water unit discharges is

$$\beta_A = 0.13\left[1-\frac{V_e}{V_i}\right]\left[\frac{H_o}{t_i}\right]^{0.45}.$$ (6.50)

Another approach to estimate the jet air concentration is provided in 6.2.2. The *ultimate* scour depth $z_{e\infty}$ relative to the tailwater depth h_t in granular material depends mainly on the tailwater Froude number $F_t = q/\left(gh_t^3\right)^{1/2}$, the ratio β_A, but is nearly independent of relative particle size d_m/h_t. For $\beta_A < 2$ follows (Mason, 1989)

$$\frac{z_{e\infty}}{h_t} = 1 + 3.4F_t^{0.6}(1+\beta_A)^{0.3}(h_t/d_m)^{0.06}.$$ (6.51)

The effect of F_t is thus significant, so that the ultimate scour depth is reduced by an adequate scour hole submergence. A typical mean particle size is $d_m = 0.25$ m. Note that Eq. (6.51) does not depend on the fall height H_o, since Mason (1989) used a relation between β_A and H_o developed by Ervine (1976). The empirical formula is accurate for model data and appears to give a reasonable upper bound of scour depth for prototype conditions. Additional information is provided by Mason (1993, 2011). Figure 6.43c shows a massive prototype scour highlighting both the large damage and the danger and risk for a dam.

6.4.2 Effect of jet air content

Motivation

Plunge pool scour depends on factors including the jet velocity, jet air content, jet thickness, tailwater depth above the original sediment bed, sediment size, and its granulometry. However, few information is available for the maximum deposition height due to free jet scour, and the scour width to be expected. The temporal scour development has also not received general attention.

Many scale model scour studies were conducted employing granular material, yet most relating to site-specific conditions, thus not allowing for a generalized analysis. A reproduction of the entire spillway including the ogee, the chute, take-off, and the free jet in the air impacting onto a sediment surface is often out of experimental possibilities, and the data would result in significant scale effects due to a too large scaling factor. The hydraulic conditions of a jet having traveled through the air for many meters can hardly be characterized in terms of average jet surface, average jet velocity and average air concentration. Therefore, the prototype configuration was simplified by accounting exclusively for a pre-aerated jet whose characteristics are known shortly upstream from impact onto the water body contained in the scour hole. Its cross-sectional shape was circular, as discussed below, in contrast to Mason (1989) who considered a rectangular turbulent jet (6.4.1). The test jet impacted some five diameters vertically above the water surface, allowing for sufficient space to observe the scour progress at the impact location. The main results are provided below.

Experimental setup and procedure

The experiments were conducted in a rectangular channel 0.500 m wide and 7 m long (Figure 6.44). A flap gate at the downstream channel end maintained the desired water elevation. During the tests, both water and air were independently supplied to a pipe of internal diameter $D = 0.10$ m. It was mounted axially above the channel axis, with an elbow at its end to produce a free jet of $-30°$ deflection from the horizontal. Air (subscript A) was discharged directly into the pipe, 6 m upstream from its efflux to allow for sufficient mixing with water flow. Both water jets and air-water mixtures were studied, with a maximum of $\beta_A = Q_A/Q \cong 3$.

Three almost uniform crushed sediments were used: Sediment A with $d_{50} = 0.041$ m, $d_{90} = 0.056$ m and a sediment non-uniformity coefficient $\sigma_s = (d_{84}/d_{16})^{1/2} = 1.38$, Sediment B with $d_{50} = 0.013$ m, $d_{90} = 0.019$ m and $\sigma_s = 1.46$, and Sediment C with $d_{50} = 0.0072$ m, $d_{90} = 0.0096$ m, and $\sigma_s = 1.28$. In total, 38 tests were conducted, of which 17 were water experiments, and 21 air-water mixture tests. The effect of jet diameter was analyzed with a hose of internal diameter 0.019 m fixed onto the pipe to assure identical direction and distance from the water surface. The water surface profile $h(x)$ with x as the streamwise coordinate was read with a conventional point gage. The sediment surface profile $z(x)$ was recorded with a point gage extended with a circular metal ring of 0.04 m diameter, of reading accuracy $0.50d_{50}$.

Once sediment was horizontally placed in the test reach, and the water elevation was adjusted, pre-selected air was discharged into the pipe. Water was then added within few seconds, and time t set to zero. The sediment surface profile was measured at $t = 1, 5, 20$ and 120 minutes, with an appropriate spacing, depending on the scour hole length. The water surface profile was measured only once because it hardly changed with time. Compared

Figure 6.44 Test setup looking in flow direction with ① mixture approach flow pipe, ② sediment bed, ③ point gage, ④ sediment gage, ⑤ tailwater gate (Canepa and Hager, 2003)

with pier scour, the sediment surface profile develops much faster for plunge pool scour to the end scour profile, mainly for large sediment size. Once a test was completed, the water was abruptly stopped to keep the final dynamic scour hole shape. The flap gate was then slowly lowered to measure the scour profile under 'dry' conditions. Large differences mainly in the scour hole were observed between the 'wet' and 'dry' conditions, the latter producing always less scour depth. The 'dry' scour profiles were not further processed, therefore.

Main parameters

Figure 6.45 shows a definition plot for plunge pool scour. A fully mixed jet made up of water discharge Q and air discharge Q_A of diameter D with an angle α relative to the horizontal impacts a water body of thickness h_t. The sediment bed of average grain size d_{50} and non-uniformity σ_s is originally horizontal. Once the sediment surface approaches the end scour condition, it is characterized with the ultimate scour depth z_m at location x_m relative to the scour origin at x_o. The ultimate aggradation height z_M is located at x_M. Coordinates x_a and x_u

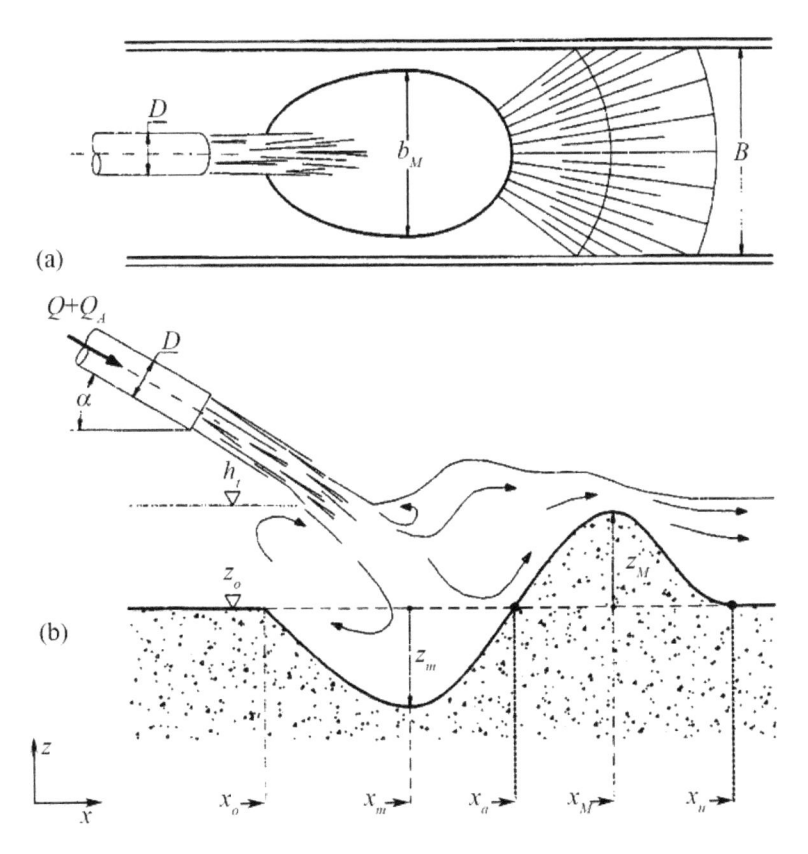

Figure 6.45 Definition sketch for plane plunge pool scour (a) plan, (b) streamwise section (Canepa and Hager, 2003)

define the locations where the scour profile $z(x)$ crosses the original sediment surface both up- and downstream of the aggradation zone. The ultimate surface width of the scour hole is b_M (Figure 6.45a).

Plunge pool scour is a three-phase process involving sediment, fluid, and gas flows. Usual scour features are characterized by the densimetric (subscript d) particle Froude number $F_d = V_o/(g'd_i)^{1/2}$, with V_o as the approach flow velocity, $g' = [(\rho_s - \rho)/\rho]g$ the reduced gravity acceleration, with ρ_s and ρ as densities of sediment and water, and d_i as the determining sediment size. The number F_d is relevant for processes involving solid and liquid phases for which the liquid surface remains almost horizontal, as for density currents, with the common Froude number close to zero and thus irrelevant. For the three-phase flows considered here, a modified Froude number F_β was defined in which $F_\beta \to F_d$ for a jet air content $\beta_A = Q_A/Q = 0$. What is the effect of β_A on plunge pool scour?

Consider the scour depth z_m close to end scour for an almost uniform sediment. The value z_m is supposed to increase as the jet velocity V, the jet diameter D and the jet angle α increase, and as the density difference $(\rho_s - \rho)/\rho$, and the sediment size d decrease. As to the air content β_A, a distinction must be made between water velocity V, and the air-water mixture (subscript AW) velocity V_{AW}. Using V would suggest that the scour depth increases with increasing air content, because of the increased mixture velocity $V_{AW} = V(1 + \beta_A)$. The test data for water flow alone $(\beta_A = 0)$ were considered thus first, to add the effect of air in a second step. The influence of scour duration is investigated separately below; the following refers to close to end scour data, for which no substantial increase of scour occurred even after a long time.

Extreme ultimate scour elevations

Figure 6.46a refers to the dimensionless scour depth $Z_m = z_m/D$ versus $F_\beta = F_d$ for jets issued with both the hose and the pipe. F_β varied between 2.5 and 15 for the three sediments used; the linear data trend follows the relation (Canepa and Hager, 2003)

$$Z_m = 0.37F_\beta. \tag{6.52}$$

The relevance of F_β for the present problem is thus verified. The ultimate aggradation height was treated accordingly, with the identical result (Figure 6.46b)

$$Z_M = 0.37F_\beta. \tag{6.53}$$

To detect the effect of air content on the scour depth, the white-water jet data were plotted as $Z_m/(0.37F_{AW})$ versus β_A, as suggested by Eqs. (6.52, 6.53). Figure 6.47a indicates that for both Z_m and Z_M the identical expression $(1 + \beta_A)^{-0.75}$ fits the data well, so that the governing expression for three-phase scour flow involving $F_\beta < 10$ to 15 is

$$Z_m = Z_M = 0.37F_\beta = 0.37V(1 + \beta_A)^{0.25}/(g'd_{90})^{1/2}. \tag{6.54}$$

The sediment size d_{90} instead of d_{50} was selected because it is easier to determine in applications. Equation (6.54) thus describes the extreme ultimate scour elevations for both the black-water and white-water jet configurations.

The data scatter of the functions $Z_m(F_\beta)$ and $Z_M(F_\beta)$ was considered small enough to propose the simple design Eq. (6.54) involving equal scour depth and aggradation height. It states that

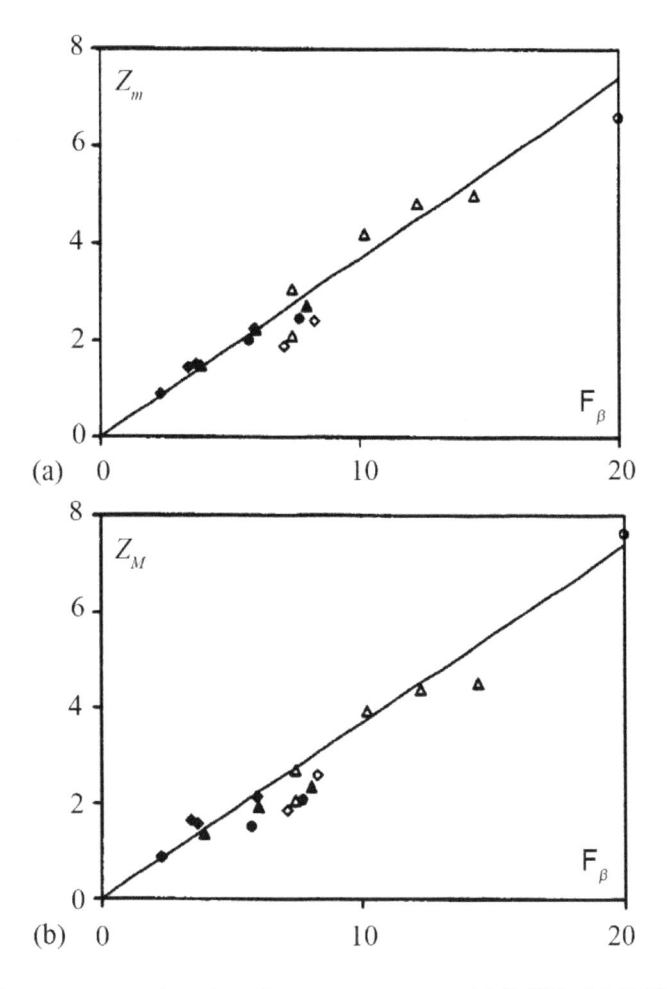

Figure 6.46 Extreme scour elevations for pure-water tests (a) $Z_m(F_\beta)$, (b) $Z_M(F_\beta)$ for d_{90} = (◇)
0.056 m, (△) 0.019 m, (○) 0.0096 m for pipe (closed) and hose (open symbols)
tests (Canepa and Hager, 2003)

the jet velocity has the dominant effect on plunge pool scour. The effect of air content must
be discussed in terms of the jet velocity selected: (1) using the pure-water flow velocity V,
the scour depth increases as air is added because of the increased jet velocity $V_{AW} = V(1 + \beta_A)$,
whereas (2) scour depth reduces when increasing the air addition for a constant mixture velocity V_{AW} of which the jet density is smaller than 1. Depending on the velocity
selected, the scour depth increases, or decreases with the jet air content β_A. Note that $F_\beta = V(1 + \beta_A)^{0.25}/(g'd_{90})^{1/2}$ is based on water velocity V.

Manso *et al.* (2004) stated that adding air to a certain water discharge generates on the one
hand jets of increased kinetic energy resulting in higher scour depth. On the other hand, jets
of the identical total discharge under increasing aeration have lower kinetic energy, explaining the reduction of scour depth. The latter procedure is preferred to transfer the results to

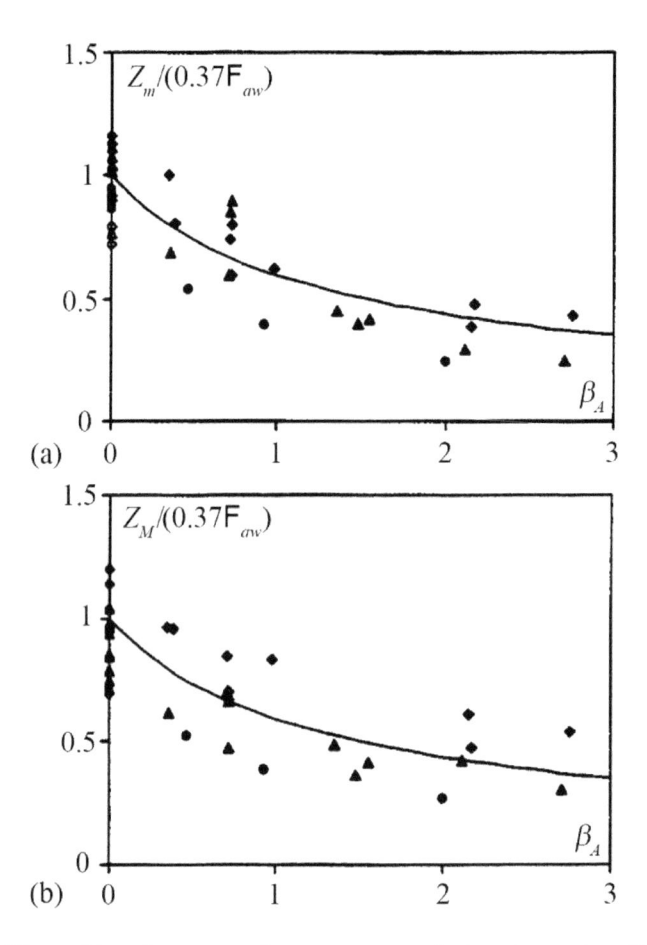

Figure 6.47 Effect of jet air content β_A on extreme relative scour elevations $Z/(0.37F_{aw})$. Notation Figure 6.45 (Canepa and Hager, 2003)

prototype cases where the jet velocity remains equal to the water velocity as far as the core of the jet persists at the impact point.

Longitudinal and transverse scour characteristics

The length parameters of a scour hole are defined as $L_m = x_m - x_o$, $L_a = x_a - x_o$, $L_M = x_M - x_o$, and $L_u = x_u - x_o$ (Figure 6.45). As for the extreme ultimate scour elevations, the scaling length corresponds to the jet diameter D, so that the relative lengths L_m/D, L_a/D, L_M/D, and L_u/D are plotted versus F_β. Figure 6.48 (a–d) demonstrates agreement of the data with the linear fit (Canepa and Hager, 2003)

$$L_i / D = C_L F_\beta.$$ (6.55)

The constants are $C_L(i = m) = 1.00$, $C_L(i = a) = 1.65$, $C_L(i = M) = 2.30$, and $C_L(i = u) = 3.15$. The scour hole width b_M was found to follow (Figure 6.50e)

$$b_M / D = 1.55 \mathsf{F}_\beta. \tag{6.56}$$

The scour plan shape is elliptic of length $x_a - x_o$ and width b_M (Figure 6.45a).

Scour hole profile

As mentioned, the axial scour hole profile was measured at various times t after test initiation. For the sediments employed, close to 'end' scour occurred after some 20 min. These profiles were plotted non-dimensionally as $Z = (z - z_m)/(z_M - z_m)$ versus the streamwise coordinate $X = (x - x_o)/(x_M - x_o)$ imposing scour start at $(X; Z) = (0; 0.5)$, scour maximum at $Z = 0$, and the aggradation maximum at $(X; Z) = (1; 1)$. Figure 6.48f shows that the data define an asymmetric sinusoidal curve, whose trend line $Z(X)$ is fitted with the two curves (Canepa and Hager, 2003)

$$Z = 0.50[1 + \sin(3.49(X - 0.90)], \ 0 \le X \le 0.4; \tag{6.57}$$

$$Z = 0.50[1 + \sin(5.71(X - 0.725)], \ 0.45 \le X \le 1.275. \tag{6.58}$$

For tests of small overflow depth $(h_t - z_M)$, the aggradation was eroded due to overflow velocities larger than the sediment entrainment velocity so that the scour hole profile followed a flatter curve (Figure 6.48f). The remainder of the data agrees with the above similitude procedure. Due to the small jet impact angle $\alpha = -30°$, the profile is steeper in the rising than in the falling portion. For $\alpha = -90°$, two banks on either scour hole side establish, causing a symmetrical profile (Rajaratnam, 1982a, b).

Comparison with other approaches

The sum $(z_m + h_t)$ was often considered as the variable scour depth, instead of z_m as above. The total length composed of maximum scour depth plus flow depth was then normalized with the jet (subscript j) energy head $H_j = V_j^2 / 2g$. Mason (1989) proposed Eq. (6.50) involving three separate effects: (1) entrainment velocity V_e [= 1.1 m/s], (2) jet velocity V_j, and (3) relative fall height H_o/t_j. The latter effect is comparatively small; further h_t is contained on both sides of Eq. (6.51). Note that Mason (1989) used d_{50} instead of d_{90}. Another feat is the inclusion of the aeration effect $(1 + \beta_A)$ in Eq. (6.51) almost to the same power as in Eq. (6.50).

Upon comparing Mason's approach for a plane jet of width a with the present data, overall agreement is noted (Canepa and Hager, 2003). As to F_β, velocity V represents the water velocity, and $a = (\pi/4)^{1/2}D$ corresponds to the equivalent jet width. Mason's original formula relates to the 45° jet impact angle, yet the effect of impact angle is not included. Based on Canepa and Hager (2003), the effect of the still water depth h_t is small. In applications, h_t is difficult to assess because the flow over and around the sediment aggradation involves complexities. Fortunately, the present formulation does not include h_t explicitly. This effect, among others, needs additional research.

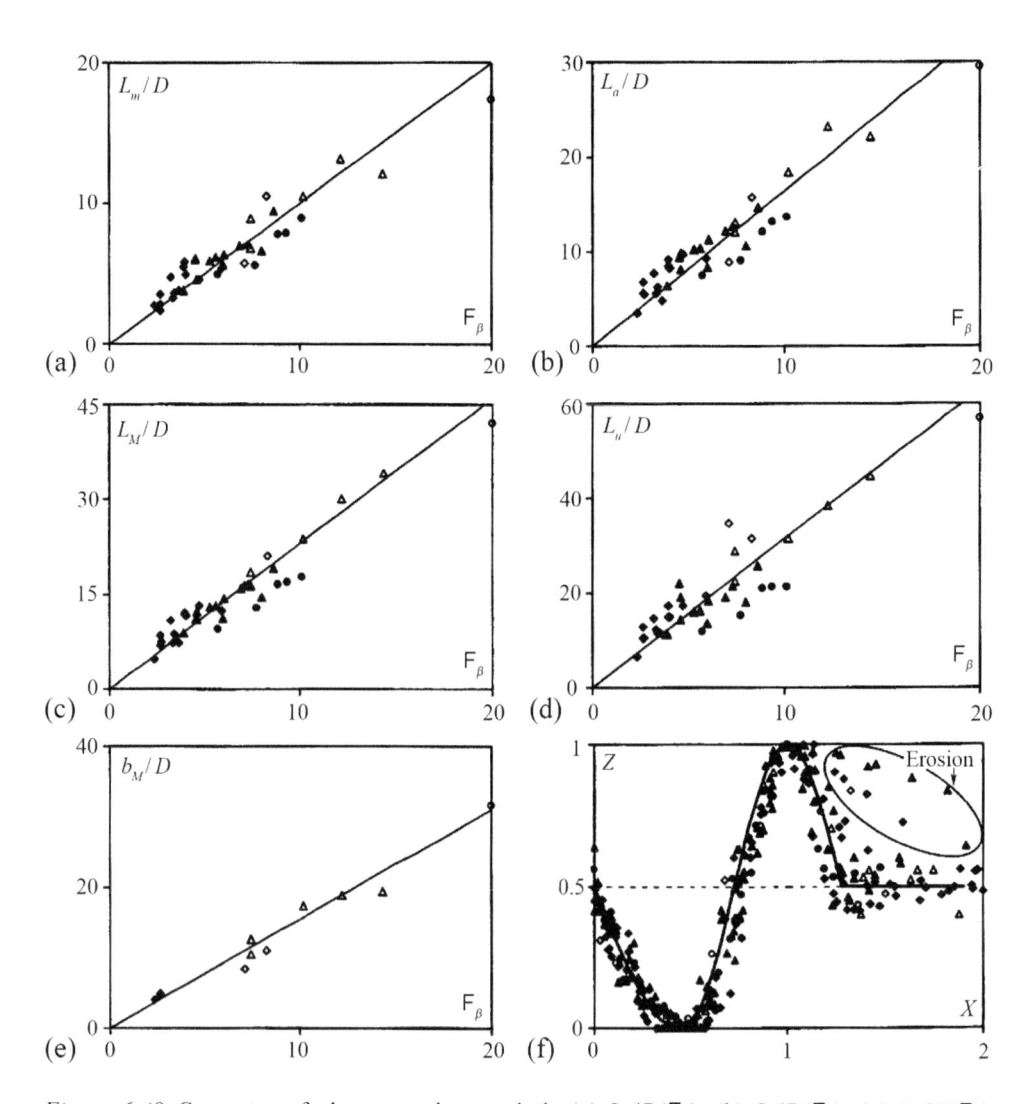

Figure 6.48 Geometry of plunge pool scour hole (a) $L_m/D(\mathsf{F}_\beta)$, (b) $L_a/D(\mathsf{F}_\beta)$, (c) $L_M/D(\mathsf{F}_\beta)$, (d) $L_u/D(\mathsf{F}_\beta)$, (e) $b_M/D(\mathsf{F}_\beta)$ with (−) Eqs. (6.55, 6.56); (f) generalized axial scour hole profile $Z(X)$, (−) Eqs. (6.57, 6.58) (Canepa and Hager, 2003)

Blaisdell and Anderson (1991) defined the densimetric particle Froude number as $\mathsf{F}_{d50} = V/(g'd_{50})^{1/2}$ relating to sediment size d_{50}. For pipe exit elevations larger than $1D$ above the water surface, the ultimate total scour depth $\underline{Z}_m = (z_m + h_t)/D$ is

$$\underline{Z}_m = 10.5[1 - \exp(-0.35(\mathsf{F}_{d50} - 2))]. \tag{6.59}$$

Equation (6.59) has two features: (1) no scour at all for $\mathsf{F}_{d50} < 2$, and (2) exponential decay to ultimate scour depth $\underline{Z}_m = 10.5$ for large F_{d50}, typically >10. Feature (1) was not noticed by Canepa and Hager because the pipe was supplied with a minimum discharge to avoid free surface pipe flow. In addition, the curves shown in Figure 6.46 may be extrapolated to the origin, whereas Blaisdell and Anderson's plots start at point $F_{d50} = 2$. Because the tests of Canepa and Hager (2003) involved nearly uniform material, the ratio d_{50}/d_{90} is almost equal to 0.70. Inserting in Eq. (6.59) would slightly modify the constants. Equations (6.54) and (6.59) may not be directly compared because there is no explicit effect of flow depth h_t in Eq. (6.54).

6.4.3 Hydraulics of plane plunge pool scour

Motivation and test setup

Whereas Canepa and Hager (2003) proposed a suitable three-phase flow Froude number and studied the effect of jet air content, Pagliara *et al.* (2006) investigated additional effects, namely these of jet shape, submerged or unsubmerged jet impact, jet velocity, jet impact angle, and the jet air concentration. All data relate to conditions with an insignificant erosion of the sediment aggradation (ridge) downstream of the scour hole. These effects were considered outside the scope given the large number of remaining parameters. The tests refer to a quasi-2D scour formation whose surface is completely fragmented thus retaining no additional safety against scour.

The tests were conducted in the channel used by Canepa and Hager (2003). The air-water mixture jet was generated with a circular pipe of internal diameter $D = 0.070$ m. Water discharges were up to $Q = 0.025$ m³/s, whereas the air discharge was limited to $Q_A = 0.045$ m³/s. The arrangement used had distinctive advantages as compared with a standard plunge pool scour setup: The air-water (subscript *AW*) mixture velocity V_{AW} was defined within ±5%, the model was short, so that all hydraulic parameters prior to jet impact onto the tailwater were adequately defined. Figure 6.49a shows the test setup.

Four jet impact angles $\alpha = 30°$, 45°, 60°, and 90° were retained. The jet shape was circular, except for a series involving a rectangular slit of 0.029 m, with the slit axis positioned both horizontally and vertically. Plunge pool scour normally involves rock (6.5). A completely disintegrated rock bed simulated with sediment of grain sizes d_{50} or d_{90}, and a sediment non-uniformity parameter $\sigma_s = (d_{84}/d_{16})^{1/2}$, offers the least resistance against scour. Canepa and Hager (2003) accounted for the effect of grain size for nearly uniform sediment. Pagliara *et al.* (2006) considered in turn a sediment made up by $d_{16} = 5.2$ mm, $d_{50} = 6.5$ mm, $d_{84} = 7.8$ mm, and $d_{90} = 8.0$ mm, so that $\sigma_s = 1.22$.

The air-water mixture jet was studied both for submerged (S) and unsubmerged (U) tailwater conditions. The latter corresponds to the usual arrangement with a jet traveling through the atmosphere and impacting onto an almost stagnant water body. For submerged flow, the pipe length was increased so that the mixture flow issued below the water surface. Both configurations have practical relevance, so that the differences between these two flow types were investigated.

A typical test lasted for 1 h, including preparation of a horizontal sediment surface, air supply to pipe, setting time to zero once water discharge was added, observation of axial scour profiles at various times, and sediment surface recording for 'dry' bed conditions, i.e. after water had been drained from the channel at test end (Figure 6.49b). The sediment surface was recorded under dynamic conditions, i.e. during test progress, except for 1 min after test initiation when measuring only the maximum scour depth. This contrasts other works employing a static sediment surface record, i.e. the jet being stopped or deflected away from

Figure 6.49 (a) Generation of 2D plunge pool scour using pressurized air-water mixture flow, (b) plunge pool scour for 'dry' conditions after test completion (Pagliara *et al.*, 2006)

the scour hole and measurements proceeding below still water. The tests were continued then until conditions close to end scour resulted, for which the scour hole undergoes practically no more temporal changes.

Effect of jet shape

Figure 6.50 shows a definition sketch of the main parameters. An air-water jet of diameter D and mean cross-sectional velocity $V_{AW} = (Q_A + Q_W)/[\pi D^2/4]$ impacts a water body of water depth h_t under the angle α. This height either submerges the conduit outlet resulting in submerged (S) flow, or produces unsubmerged (U) outflow. The discharges are Q for water and Q_A for air. The original elevation of the sediment bed is z_o. The high-speed jet generates an almost plane scour hole of maximum depth z_m at location x_m from the scour origin x_o. The maximum deposition (ridge) height is z_M at location x_M. The scour hole profile $z(x)$ intersects the original bed at locations x_a and x_u, respectively. The axial lengths defining the scour hole geometry are $l_m = x_m - x_o$, $l_a = x_a - x_o$, $l_M = x_M - x_o$, and $l_u = x_u - x_o$.

For two-phase jet flow involving air and water, the densimetric particle Froude number is $F_{di} = V_W/(g'd_i)^{1/2}$ with $V_W = Q_W/[\pi D_W^2/4]$ as water velocity, $g' = [(\rho_s - \rho)/\rho]g$ as the reduced gravity acceleration with ρ_s and ρ as densities of sediment and water, respectively, and d_i the determining grain size. The significant length scales either are the pipe diameter D, or the black-water jet diameter D_W. The jet air content is $\beta_A = Q_A/Q_W$, resulting in a black-water jet for $\beta_A = 0$, or in a white-water jet for $\beta_A > 0$.

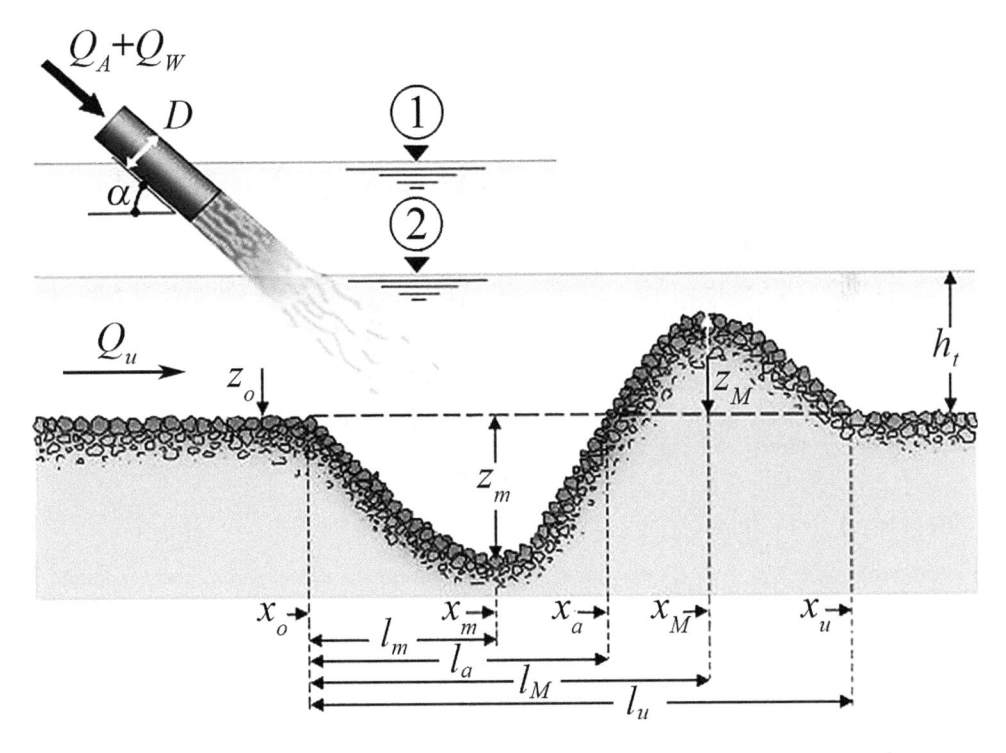

Figure 6.50 Definition sketch for 2D plunge pool scour under ① submerged, ② unsubmerged flow

The effect of jet shape on the scour hole depth is important, because flip buckets may generate a jet geometry deviating largely from the circular jet shape as used in the tests. Therefore, a special test series was conducted using four different jet shapes: circular conduits of internal diameters (1) $D = 0.100$ m (Canepa and Hager, 2003), (2) $D = 0.070$ m; and rectangular jet (3) of width $b_j = 0.100$ m and height $h_j = 0.029$ m, and (4) inverted with $b_j = 0.029$ m and $h_j = 0.100$ m. Only black-water tests were considered; the effect of air supply is discussed below. The equivalent (subscript e) diameter of configurations (3) and (4) is $D_e = (4b_j h_j/\pi)^{1/2} = 0.061$ m.

The relative scour depth $Z_m = z_m/D$ versus $F_{d90} = V/(g'd_{90})^{1/2}$ for all data available and both S and U flow conditions indicates that there is practically no jet shape effect if the identical cross-sectional average jet velocity $V = Q/(\pi D_e^2/4)$ is considered (Pagliara et al., 2006). This result simplifies the application of the present research to practice.

Effect of jet impact angle

The maximum scour depth Z_m depends on four independent functions f_1 to f_4 as

$$Z_m = f_1(F_{d90}) \times f_2(\alpha) \times f_3(\beta_A) \times f_4(S \text{ or } U). \tag{6.60}$$

The black-water data were expressed as $Z_{m+} = Z_m/f_2(\alpha)$ with $f_2 = 0.38\sin(\alpha + 22.5°)$ for both submerged (S) and unsubmerged (U) flows ($f_4 = 1$) for the four jet impact angles $\alpha = 30°$, $45°$, $60°$, and $90°$. The curve $Z_{m+}(F_d)$ is a straight line. The difference between the S and U data is small. For unsubmerged jet flow, the distance from the pipe outlet to the tailwater surface was close to $2D$. Therefore, for black-water flow, scour depths are practically identical whether the jet impacts onto a water body, or is issued below it. This important result applies for a tailwater level of at least $3D$ above the original sediment bed. The maximum scour depth for $\beta_A = 0$ thus is from Eq. (6.60)

$$Z_m = -0.38\sin(\alpha + 22.5°)F_{d90}, \quad 2 < F_{d90} < 20, \quad 30° \le \alpha \le 90°. \tag{6.61}$$

It was originally thought that the effect of the jet impact angle would follow the $\sin\alpha$ function. However, the scour depth is larger for 60° than for 90°. This may be explained with: (1) deposition height Z_M (see below) is significantly larger for $\alpha = 60°$ than for 90° so that less sediment is suspended in the more confined scour hole of a vertical jet; and (2) erosion of the deposition crest is larger for 60° than for the vertical jet because of increased streamwise velocity under otherwise equal conditions.

Effect of jet air content

The jet air content $\beta_A = Q_A/Q$ plays a significant role on the scour parameters, as stated in 6.4.2. This effect is either described by the black-water velocity $V = Q/(\pi D^2/4)$ with D_W as black-water jet diameter, or air-water mixture velocity $V_{AW} = Q(1 + \beta_A)/(\pi D^2/4)$. The related Froude numbers are $F_{d90} = V/(g'd_{90})^{1/2}$ and $F'_{d90} = V_{AW}/(g'd_{90})^{1/2}$, respectively. The function $f_{3m}(\beta_A)$ in Eq. (6.60) involves $Z_{m++} = Z_{m+}/F'_{d90} = Z_m/[F'_{d90} \times f_2(\alpha)]$ versus β_A. For $\beta_A < 12$ with $m = 0.75$ for the unsubmerged, and $m = 0.50$ for the submerged jet configuration, respectively, the data follow (Pagliara et al., 2006)

$$f_{3m}(\beta_A) = (1+\beta_A)^{-m}. \tag{6.62}$$

The effect of β_A is thus slightly larger for unsubmerged than for submerged jets.

The final equation for the maximum scour hole depth Z_m generated by a turbulent air-water mixture jet onto nearly uniform sediment thus reads (Pagliara *et al.*, 2006)

$$Z_m = z_m / D_e = -0.38\sin(\alpha + 22.5°)\mathsf{F}_{d90} \, (1+\beta_A)^{-m},$$
$$2 < \mathsf{F}_{d90} < 20, \, 30° \le \alpha \le 90°. \tag{6.63}$$

It includes the effects of equivalent jet diameter D_e for non-circular jets, the jet impact angle α, the mixture jet velocity V_{AW}, relative density times the gravity acceleration $g' = [(\rho_s - \rho)/\rho]g$ between the sediment and fluid phases, grain size d_{90}, and the jet air content β_A. Predictions for all 200 data sets for both S and U flows follow Eq. (6.63) with a correlation coefficient of 0.91. Note that white-water jets were almost uniformly aerated, in contrast to typical prototype jets with a black-water core and increasing air concentration from the jet axis toward the jet surface, an effect considered small. Both the externally aerated high-speed jet and jet impact topographies deviating strongly from the present setup must be studied separately to confirm the above procedure.

Maximum ridge height

The maximum height of deposition $Z_M = z_M/D$ is of interest for predicting the tailwater phenomena. The data were processed as previously for black-water tests. In contrast to the maximum scour hole depth, a differentiation between unsubmerged (U) and submerged (S) flow conditions is significant here.

The maximum ridge height increases linearly with F_d but it is almost double for unsubmerged than for submerged flow because the additional air entrained by the unsubmerged plunging jet lifts the sediment particles close to the free surface from where they are transported to the ridge (subscript R). Note that even for black-water approach flow, large air quantities are entrained for U conditions. The data further indicate for all S flows a decrease of Z_M with α. For submerged flows, this decrease is even larger for small α. The data are expressed with $A_R = 0$, $B_R = 16$ and $n = 1$ for U-flows, and $A_R = 0.3$, $B_R = 438$ and $n = 2$ for S-flows, respectively, as (Pagliara *et al.*, 2006)

$$Z_M = A_R + B_R\alpha^{1.2n}\mathsf{F}_{d90}, \quad 2 < \mathsf{F}_{d90} < 20, \, 30° \le \alpha \le 90°. \tag{6.64}$$

The ridge thus remains small for either a large jet impact angle α or small F_{d90}.

The effect of jet air content β_A on Z_M was analyzed as previously for the maximum scour depth, thereby accounting for both U- and S-flow conditions. The function $Z_{M+} = Z_M/f_2(\alpha)$ versus β_A for U-flow follows Eq. (6.62) with $F = 1$ as (Pagliara *et al.*, 2006)

$$f_{3M}(\beta_A) = F(1+\beta_A)^{-0.50}, \quad \beta_A < 12. \tag{6.65}$$

For S-flow, the data follow a different trend, with an initial increase to a maximum, and then a decrease similar to U-flow. For $\beta_A > 2.5$, these data follow Eq. (6.65) with $F = 1.5$. Note a large scatter in both formulations compared with the scour hole depth attributed to erosion of the ridge crest for small overflow depths. Equation (6.65) represents a logic extension of Eq. (6.62) but is subject to further analysis.

Scour hole geometry

The relative scour hole length $L_m = l_m/D$ from start to its maximum depth (Figure 6.50) versus F_{d90} for $\alpha = 30°$ and unsubmerged black-water flow was studied. The data of Canepa and Hager (2003) and these of Pagliara *et al.* (2006) follow a straight line. The remaining data were fitted accordingly. These relating to U- and S-flows have a similar trend as previously, with the S-data slightly below the U-data. Their trends follow with $C_s = 0.90$ for the submerged, and $C_s = 1.0$ for the unsubmerged jet flows

$$L_m = 1.5 + (C_s - 0.01\alpha)F_{d90}, \qquad 2 < F_{d90} < 20, \ 30° \leq \alpha \leq 90°. \qquad (6.66)$$

Length $l_a = x_a - x_o$ from the origin x_o to point x_a where the scour profile intersects the original bed elevation varies also with both α and F_{d90}. The data are expressed with $\mu_a = 1.05$ for U-, and $\mu_a = 0.85$ for S-jets, respectively, as (Pagliara *et al.*, 2006)

$$L_a = 2.5[1 + \mu_a \exp(-0.030\alpha)F_{d90}], \qquad 2 < F_{d90} < 20, \ 30° \leq \alpha \leq 90°. \qquad (6.67)$$

The distance of the maximum sediment deposition from the origin l_M was treated accordingly. The non-dimensional length $L_M = l_M/D$ varies with F_{d90} and α with $\mu_M = 1.5$ for U- and $\mu_M = 1$ for S-jet conditions, respectively, as

$$L_M = 2.5[1 + 30\mu_M \alpha^{-1.25}F_{d90}], \qquad 2 < F_{d90} < 20, \ 30° \leq \alpha \leq 90°. \qquad (6.68)$$

The relative total length $L_u = l_u/D$ from the scour hole start to the ridge end follows a similar trend with $\mu_u = 1.5$ for U- and $\mu_u = 1$ for S-jet conditions, respectively, as

$$L_u = 2.5[1 + 38.5\mu_u \alpha^{-1.25}F_{d90}], \qquad 2 < F_{d90} < 20, \ 30° \leq \alpha \leq 90°. \qquad (6.69)$$

Lengths L_M and L_u thus differ by a factor of $\mu_M/\mu_u = 1.30$. These lengths are the basis of the generalized scour profile. The previously established relations for the four length scales across a scour hole were processed by comparing observations with predictions according to Eqs. (6.66) to (6.69), to define the generalized scour hole profile.

Scour hole profile

Figure 6.51a shows the non-dimensional plot of Z/Z_m versus $X_m = (x - x_m)/(x_a - x_o)$ for $\alpha = 45°$, with all data passing through the profile minimum $(0; -1)$. Note that the falling profile limb is slightly higher for the submerged than for the unsubmerged jet data. The curves start at $X_m = -0.62$ and -0.69, respectively, passing through $X_m = +0.38$ and $+0.31$ at the other side. Due to the selected data normalization, the ridge profiles are poorly reproduced; a different treatment is presented below. Note also the larger ridge heights for the white-water than for the black-water jet data, as previously stated.

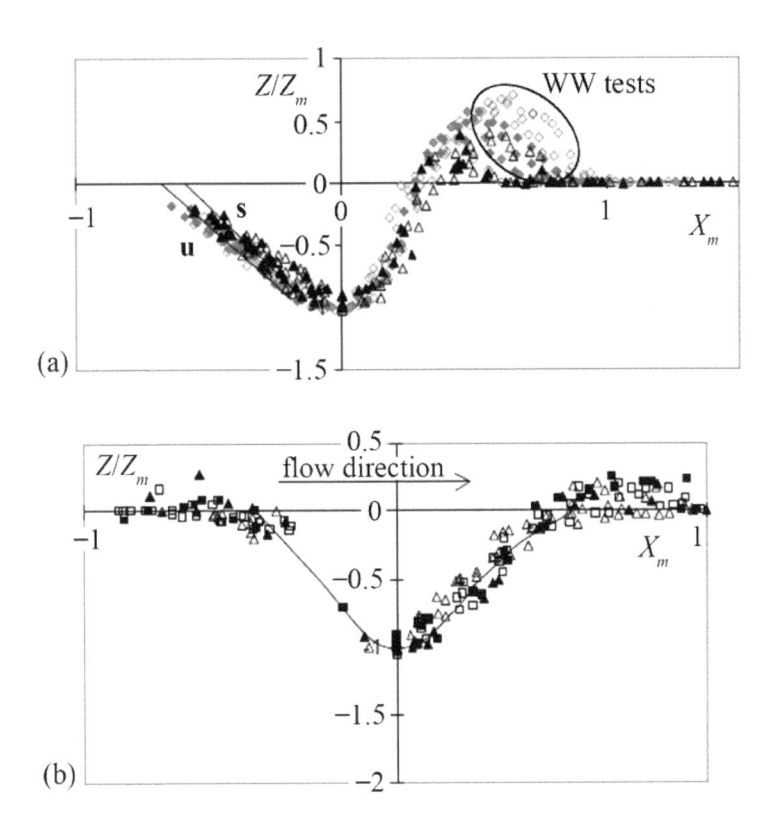

Figure 6.51 Scour hole profile $Z/Z_m(X_m)$ for white- and black-water jets, and **u** unsubmerged and **s** submerged jet flow data for (a) $\alpha = 45°$, (b) $\alpha = 90°$ (Pagliara *et al.*, 2006)

A generalized data analysis for the scour hole profiles and jet angles $30° \leq \alpha \leq 60°$ indicates the typical scatter of Figure 6.51a. All profiles follow (Pagliara *et al.*, 2006)

$$Z/Z_m = -1 + (4/9)X_m + 6.08X_m^2 + 4.75X_m^3 \qquad \text{for } Z \leq 0, \ 30° \leq \alpha \leq 60°. \qquad (6.70)$$

The normalization selected thus allows for a simple data analysis independent of jet impact angle α, jet air content β_A and the tailwater conditions (U; S), because these effects are contained in the previously established scalings. Figure 6.52a compares all data for $\alpha = 30°$, $45°$, and $60°$ with Eq. (6.70), resulting in an acceptable overall fit.

Figure 6.51b relates to $\alpha = 90°$ and shows an almost symmetrical scour hole profile for vertical jet impact. This feature is accounted for with the symmetry (subscript s) parameter $X_s = 2(x_o - x_m)/(x_a - x_o)$ to result in $X_s = -1$ for perfect symmetry. Figure 6.52b relates to the ridge profile $Z/Z_M(X_M)$ with $X_M = (x - x_a)/(x_M - x_a)$ for $30° \leq \alpha \leq 90°$. All profiles pass thus through the points (0; 0) and (1; 1) and tend to disperse further downstream. The profile equation follows the parabola (Pagliara *et al.*, 2006)

$$Z/Z_M = 2X_M - X_M^2, \qquad 0 \leq X_M \leq 2, \quad 30° \leq \alpha \leq 90°. \qquad (6.71)$$

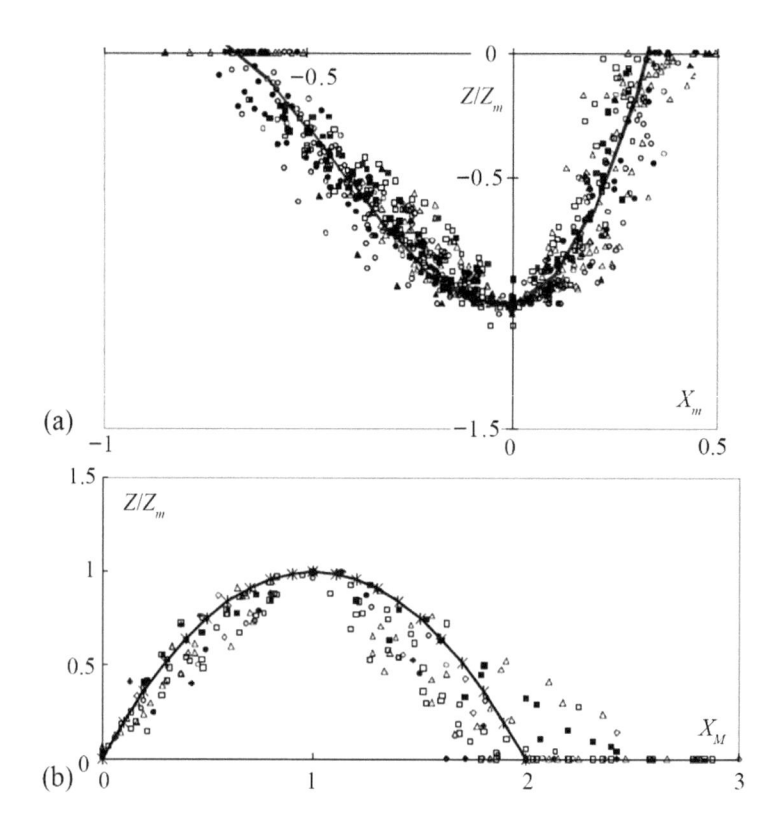

Figure 6.52 General scour hole profiles for (a) scour hole $Z/Z_m(X_m)$ with (−) Eq.(6.70) for $30°\leq\alpha\leq60°$, (b) ridge profile $Z/Z_M(X_M)$ for $\alpha = 30°$, (−) Eq.(6.71) (Pagliara *et al.*, 2006)

Equations (6.70) and (6.70) predict the general scour hole and ridge profiles provided that the tailwater elevation is so high that no appreciable erosion occurs on the ridge crest. Its erosion pattern needs attention because of combined deposition and erosion.

Effect of tailwater depth

The effect of tailwater depth $T_s = h/D$ was studied using only black-water test data. The relative scour depth $Z_m = z_m/D$ was plotted versus $F_{d90} = V/(g'd_{90})^{1/2}$ for $0.7 \leq T_s \leq 10$. Two features are noted: (1) relative scour depth Z_m increases linearly with F_{d90}, and (2) Z_m decreases as T_s increases, if $4 \leq F_{d90} \leq 20$; practically no scour occurs for $F_{d90} < 3$. The maximum scour depth is $Z_m = \tau F_{d90}$, in which (Pagliara *et al.*, 2006)

$$\tau = 0.12\ln(1/T_s) + 0.45, \qquad \text{for } 0.7 \leq T_s \leq 10. \tag{6.72}$$

Accordingly, there is a significant tailwater effect on the scour hole geometry (Figure 6.53). For $T_s > 5$, the maximum scour depth (1) is comparatively small because of a relatively high ridge. The ridge height increases for $3 < T_s < 4.5$ to its absolute maximum (see below),

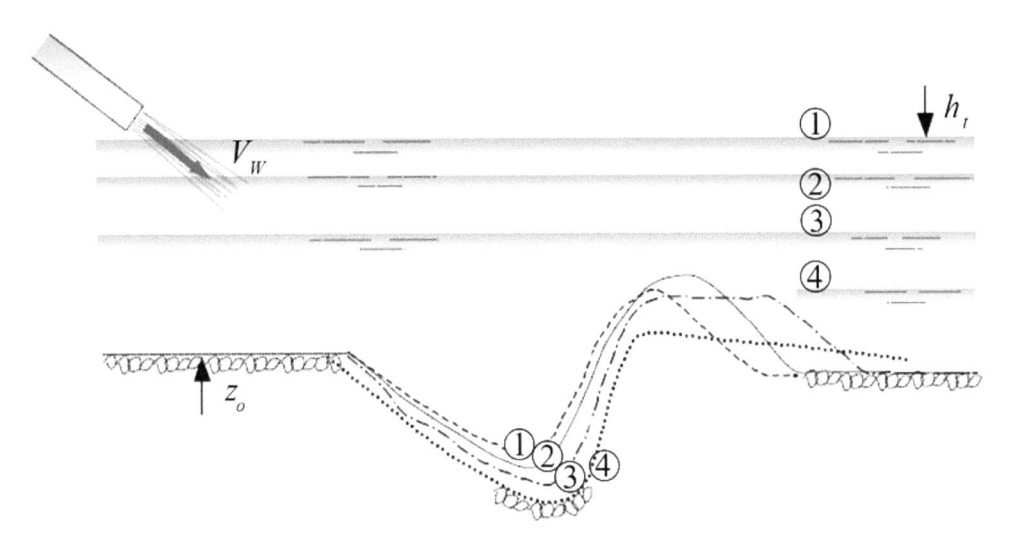

Figure 6.53 Scheme of tailwater elevation ① to ④ on maximum scour hole depth and ridge height

associated with a small increase of the scour depth (2), because of the shorter distance between jet impact and the scour hole. Further reducing T_s to 2 erodes the ridge crest, depending on the grain size and the water velocity in the ridge region, resulting in a deeper scour hole (3). If $T_s \cong 1$, the ridge is eroded and the scour hole remains unprotected (4). Ridge erosion may not thoroughly be analyzed with the parameters introduced because sediment transport depends on additional quantities.

Effect of granulometry

To study the effect of granulometry, the tailwater effect was eliminated by using the reduced scour depth $Z'_m = Z_m / \tau$. The scour hole depth increases with the non-uniformity parameter $\sigma_s = (d_{84}/d_{16})^{1/2}$. Scour holes consisting of non-uniform particles tend to become rougher in the region of jet impact, with the fine material deposited on the ridge and at the scour hole contour. This effect is smaller for scour holes made up of almost uniform sediment. The data fit the relations (Pagliara *et al.*, 2006)

$$Z'_m = \Sigma \cdot \mathsf{F}_{d90},$$ (6.73)

$$\Sigma = -(0.33 + 0.57\sigma_s), \quad \text{for } 1 \le \sigma_s \le 3.$$ (6.74)

Effect of upstream velocity

The effect of upstream (subscript u) velocity V_u on the scour features was established with 40 separate tests by a systematic variation of the upstream Froude number $\mathsf{F}_u = V_u/(gh_t)^{1/2}$

(Figure 6.50). As F_u is increases, suspended sediment in the scour hole is transported into the tailwater, resulting in a deeper scour hole and usually a smaller ridge because of ridge erosion.

The effects of tailwater $f_4(T_s)$ and sediment non-uniformity $f_5(\sigma_s)$ are accounted for with Eqs. (6.72) and (6.74). The effect of F_u on the relative maximum scour depth $Z_m'' = Z_m / [f_1(F_d) \times f_4(T_s) \times f_5(\sigma_s)]$ is

$$Z_m'' = 1 + F_u^{0.50}, \qquad \text{for } F_u < 0.30 . \tag{6.75}$$

The data scatter is large, yet the effect evident. Note the increase of scour depth of more than 50% for $F_u = 0.30$, so that the upstream Froude number must be limited unless an artificial ridge removal is a concern.

Ridge removal

In special tests, the sediment ridge (subscript r) was constantly removed during an experiment to provide a horizontal sediment surface downstream of the scour hole, to explore whether the scour depth increases. As mentioned, a ridge protects a scour hole from further deepening, yet this may be undesirable under certain conditions. The maximum scour hole depth Z_m was plotted versus F_{d90} for various relative tailwater depths T_s under ridge removal. Some increase of the scour hole depth was noted as compared with conditions when the ridge is not removed. The slope τ of the curve $Z_m(F_{d90})$ increases independent of all other parameters, so that Eq. (6.72) with $C_r = 0.45$ for ridge presence and $C_r = 0.52$ for artificial ridge removal is generalized to

$$\tau = 0.12 \ln(1/T_s) + C_r, \qquad \text{for } T_s^{-1} > 0.05 . \tag{6.76}$$

General equation for maximum scour hole depth

Based on Canepa and Hager (2003) and Pagliara *et al.* (2006), the effects of all independent parameters affecting the maximum scour depth $Z_m = z_m/D$ are assessed as

Densimetric Froude number $\quad F_{d90} = V_W / (g'd_{90})^{1/2}$

$$f_1(F_{d90}) = -F_{d90}. \tag{6.77}$$

Jet impact angle α

$$f_2(\alpha) = -[0.38\sin(\alpha + 22.5°)]. \tag{6.78}$$

Jet air entrainment β_A

$$f_3(\beta_A) = (1 + \beta_A)^{-m}. \tag{6.79}$$

Tailwater effect T_s

$$f_4(T) = [0.12\ln(1/T_s) + C_r]/0.30. \tag{6.80}$$

Sediment non-uniformity σ_s

$$f_5(\sigma_s) = -[0.33 + 0.57\sigma_s]. \tag{6.81}$$

Upstream flow effect F_u

$$f_6(F_u) = 1 + F_u^{0.50}. \tag{6.82}$$

The final expression for the maximum scour depth thus is (Pagliara *et al.*, 2006)

$$Z_m = f_1(F_{d90}) \cdot f_2(\alpha) \cdot f_3(\beta_A) \cdot f_4(T_s) \cdot f_5(\sigma_s) \cdot f_6(F_u). \tag{6.83}$$

All individual effects are independently listed from each other. The exponent in Eq. (6.79) is $m = 0.75$ for unsubmerged and $m = 0.50$ for submerged jet flow, respectively. The coefficient $1/(0.3)$ in Eq. (6.80) was introduced to be coherent with Eq. (6.63) established for $T_s \cong 3.5$, i.e. $f_4(T_s = 3.5) = [0.12\ln(1/3.5) + 0.45]/0.3 = 1$. The agreement of all test data with Eq. (6.83) is good for the entire range of experiments and similar to these of Canepa and Hager (2003) for a total of 435 tests.

Further observations

The effect of sediment non-uniformity σ_s on the dimensional scour hole geometry $z(x)$ is shown in Figure 6.54a. All tests refer to black-water jets with $\sigma_s = 1.22$ (almost uniform), 1.73, and 2.66: a scour hole deepens as granulate becomes uniform. In all three cases a ridge was formed whose height increases as the scour hole becomes deeper. Also included are the static scour holes, demonstrating a strong difference with the dynamic scour hole patterns (Pagliara *et al.*, 2004a).

The effect of the upstream Froude number F_u on the dimensional scour hole geometry is shown in Figure 6.54b. Both the scour depth and ridge height are relatively small for $F_u \cong 0$. Increasing F_u under otherwise identical conditions, the scour hole continuously deepens. This is not true for the ridge, however. As F_u is increased, the deposition height first increases to a maximum (second largest discharge in this example), then reduces because of the increased transport capacity of the combined jet and upstream discharges. As mentioned, this aspect needs further attention.

Comparison with existing procedure

Pagliara *et al.* (2004b) compared the above data sets with existing formulae for the maximum scour depth z_m (Hager, 1998). In chronological order the experimental studies of Schoklitsch (1932), Veronese (1937), Kotoulas (1967), Martins (1975, 1977), Mason and Arumugam (1985), Mason (1989), D'Agostino (1994) and D'Agostino and Ferro (2004) were accounted for. Two types of scour formula exist, namely those of Schoklitsch to Martins involving the

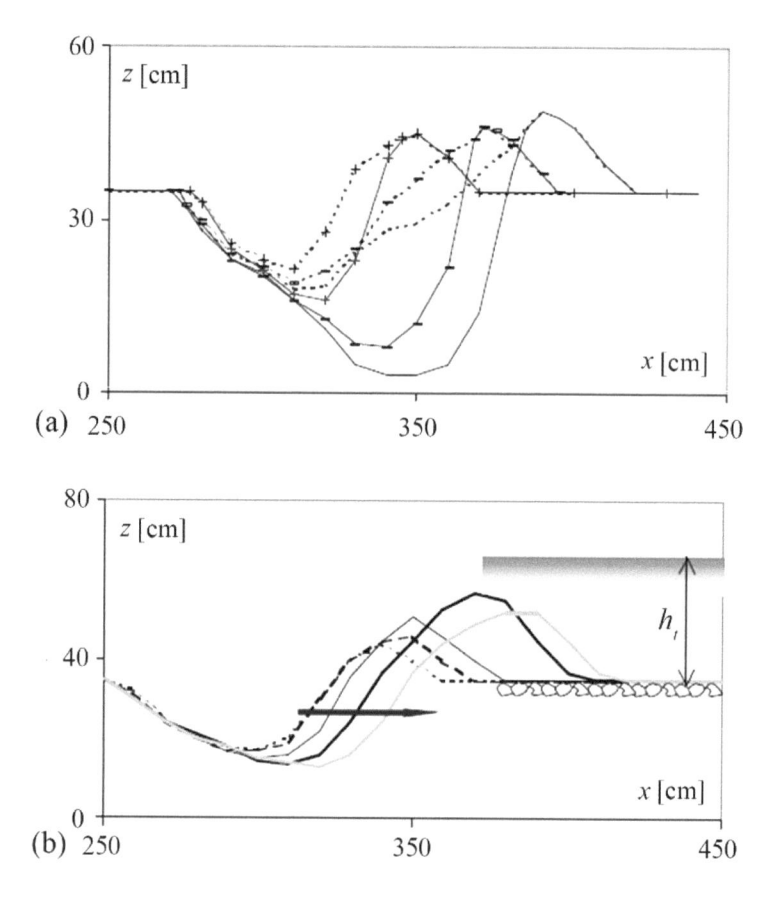

Figure 6.54 Effect of (a) sediment non-uniformity σ_s on scour hole geometry for σ_s = (—)
1.22, (- -) 1.73, (++) 2.66 under (−) dynamic and (. . .) corresponding static
scour conditions, (b) upstream Froude number F_u = (. . .) 0.02, (---) 0.04, (− −)
0.06, (—) 0.12, (−) 0.17 (Pagliara *et al.*, 2006)

sum of maximum scour depth and tailwater height $(z_m + h_t)$ against discharge per unit width q,
energy head H and sediment size d_{50} or d_{90}; and the recent formulae involving more complex
parameters. Notably, all formulae except for Mason (1989) are dimensionally incorrect. Several effects were not accounted for including the jet impact angle, sediment non-uniformity,
and the tailwater submergence. Pagliara *et al.* (2004b) found a reasonable performance of the
Mason (1989) formula, whereas the remainder except for Schoklitsch, Veronese, and partly
Kotoulas produced poor results. Currently, Mason's formula mainly applies for predicting
the maximum scour depth thereby accounting for the effect of jet aeration.

Canepa and Hager (2003) investigated whether the Froude similitude applies for the three-
phase jet flow configuration, and whether an extended version of the densimetric Froude
number accounts correctly for the flow conditions. Their results were slightly modified by
Pagliara *et al.* (2006) using the additional data. All except for Canepa and Hager's (2003)
'hose data' are included in the updated data analysis.

Limitations

The above research involves 435 tests including these of Canepa and Hager (2003). The main parameters influencing plunge pool scour are:

1 *Densimetric particle Froude number* F_{d90} as ratio between the approach flow water jet velocity V_W relative to the scaling velocity computed as the square root of reduced gravity acceleration g' times sediment size d_{90}. F_{d90} varied between 4 and 20. Scale effects are absent if $V_W > 1$ m/s and sediment size $d_{50} > 1$ mm.
2 *Jet impact angle* α between $30°$ and $90°$; its effect on plunge pool scour is not as large as would be expected; scour hole depths increase with the jet angle, whereas the ridge height decreases for otherwise identical conditions.
3 *Jet air content* β_A as ratio between air and water discharges has a strong effect on scour. The larger β_A, the smaller is the scour hole. Values of $\beta_A < 12$ were tested, for which the fluid phase was only a small portion of the gas phase. According to Manso *et al.* (2004), the effect of jet air content can only be noted by comparing jets of identical kinetic energy.
4 *Relative tailwater height* T_s as ratio between tailwater depth h_t and jet diameter D ranging between 0.7 and 10. Its effect on the maximum scour depth is strong. A scour hole is deep for small T_s reducing for larger T_s, under otherwise identical conditions. The ridge height has a maximum for $T_s \cong 3.5$ due to the combined effect of jet diffusion and ridge erosion.
5 *Sediment non-uniformity* σ_s up to 2.7. Scour hole is deeper for uniform sediment than for non-uniform granular bed under otherwise identical conditions. This is similar to fluvial erosion due to armoring of the sediment bed.
6 *Upstream Froude number* F_u up to 0.44 with a limited effect on the scour process. Scour hole increases with F_u whereas the ridge may become either smaller or larger as compared with $F_u = 0$.

A number of other parameters was also systematically accounted for, including the continuous ridge removal or jets that are either free or submerged from the tailwater. A major conclusion is that all results apply only under Froude similitude. Further, the scour surface was made up with incohesive granulate in which an almost plane plunge pool scour developed. The maximum scour hole depth follows Eqs. (6.77) to (6.83) within an accuracy of $\pm20\%$.

6.4.4 Hydraulics of spatial plunge pool scour

Motivation

An extension to the three-dimensional (3D) plunge pool scour was conducted by Pagliara *et al.* (2008a). Given the large number of parameters, some of the previously studied effects were excluded. These include the systematic variation of the scoured sediment, the effect of air entrainment into the high-speed jet, the addition of an approach flow to the plunge pool zone, and the submerged jet flow configuration. Therefore, all data refer to one sediment size, black-water jets, without the addition of an approach flow for unsubmerged jet flows. The equations describing 3D plunge pool scour account for the previously established findings by expanding the equations of Pagliara *et al.* (2006) for the additional effects.

The main parameter of 3D plunge pool scour is the scour width ratio $\lambda = B_m/B$ with B_m as the maximum scour hole width and B as width of the jet impact zone, equivalent to the channel width. The parameter λ solely describes whether a scour hole may be considered 2D or 3D. The latter has the particular plan shape shown in Figure 6.55a, with an elliptic base and a half-moon shaped ridge surrounding the scour hole. The 2D scour hole in contrast has both a cylindrical scour hole and ridge. The ridge presence is significant, both for the 2D and 3D plunge pool arrangements, because of flow recirculation developed in the scour hole. Ridge removal results in a deeper scour hole associated with a smaller tailwater (Pagliara *et al.*, 2006).

Spatial plunge pool scour has received small research attention. The main activities were directed to selected site conditions not allowing for a generalized approach. The works of Prof. Rajaratnam, University of Alberta, Canada are a notable exception. Rajaratnam and Berry (1977) used air jet nozzle diameters of 0.925 and 0.25 in. studying the scour hole features of sand and polystyrene beds. Rajaratnam (1980) then studied the scour hole characteristics under presence of a cross-flow, a setup not further investigated hereafter. The studies of Mih (1982), and Mih and Kabir (1983) explored the possibility of fine sediment removal from a granular bed with a high-speed nozzle. Anderson and Blaisdell's (1982) work is excluded here because of the relatively small jet impact velocity that led to a curved jet trajectory, in contrast to the almost straight trajectory considered. A similar work of Bormann and Julien (1991) involves a jet trajectory deviating appreciably from the straight, given the relatively small jet Froude numbers. The minimum densimetric particle Froude number herein was 4, providing an almost linear jet trajectory, as is typical for ski jumps close to the jet impact zone. All works involving a vertical circular jet were also excluded because of almost perfect radial symmetry of the scour hole. The main results of a hydraulic model study are thus highlighted along with a discussion and the limitations of the results to be accounted for their application in practice.

Experiments

The model tests involved a channel of 25 m total length of which a test reach was confined to 6 m. The rectangular channel was 0.80 m wide and 0.90 m high; reduced widths of 0.40 and 0.20 m were also used. The sediment bed was graded prior to test start and the initial tailwater water depth was h_t.

The test procedure followed Pagliara *et al.* (2006) involving circular pipes of internal diameters $D = 16, 27, 35,$ and 51 mm. Only black-water jets were considered because the

Figure 6.55 3D plunge pool scour for (a) free ($\alpha = 30°$), (b) submerged jet flow ($\alpha = 45°$) (Pagliara *et al.*, 2008a)

effect of air addition is analyzed in 6.4.3. The water (subscript W) discharge varied from $Q_W = 0.35$ to 10.0 l/s. The relative tailwater elevation $T_s = h/D$ ranged between 0.80 and 10, i.e. similar to Pagliara *et al.* (2006). Jet impact angles were $\alpha = 30°$, 45°, and 60°. One nearly uniform sediment was considered with $d_{10} = 8.6$ mm, $d_{16} = 9.0$ mm, $d_{60} = 10.6$ mm, $d_{84} = 11.4$ mm, and $d_{90} = 11.6$ mm resulting in $\sigma_s = (d_{84}/d_{16})^{1/2} = 1.13$. This research involved a total of 291 tests of which 198 were made in the 0.80 m wide, 29 in the 0.40 m wide channel, and 64 aimed to compare the findings relative to the plane flow conditions with the 0.20 m wide test reach. As to the jet impact angle, 127 tests involved $\alpha = 30°$, 99 tests $\alpha = 45°$, and 65 tests $\alpha = 60°$.

Figure 6.55b shows a side view of the test setup for 45° jet flow. For each test, the axial bed profile was measured as described by Pagliara *et al.* (2006) at times $t = 1, 5, 20$ min and at test end typically 40 min from test initiation at $t = 0$. The transverse scour hole geometry was taken at test end prior to water stop. Once the channel was drained from water, the so-called static scour hole was also recorded.

Classification of spatial scour holes

The spatial features of plunge pool scour are more complex than these of the plane configuration. The jet impact onto the sediment bed involves essentially a radial flow from the impact point with a main momentum component directed from the approach flow direction. Secondary currents are set up so far hardly described. The following aims at a short account of the main flow features.

The ridge downstream from the scour hole has a significant effect on the flow pattern in the scour hole. According to Pagliara *et al.* (2006), a ridge removal increases the scour hole depth because of the surface recirculation set up by ridge presence. The features in the main flow direction are: upstream from the jet impact zone, sediment is not scoured but slides by avalanches into the scour hole, from where it is entrained by the jet and scoured up on the opposite scour hole zone. Depending on the jet impact angle α, two zones are distinguished: (1) active cup-shaped scour zone close to jet impact with a direct sediment entrainment ending with a small rim, and (2) passive scour hole zone from the rim up to the sediment ridge, where material is deposited by suspended material. The angles of the latter zone and of the material deposited beyond the ridge tend to the angle of natural repose below water.

Lowering the tailwater increases the 3D flow pattern by an increase of the velocity components in the transverse directions. For a tailwater depth smaller than 3D, the ridge may be higher than the water elevation and the flux is fully deviated laterally. A similar phenomenon occurs as the jet impact angle α increases associated with a portion of the flow deflected into the upstream direction.

For a high (denoted H) tailwater (subscript Tw) and a low (denoted L) jet impact angle, the scour hole has a large streamwise extension, referred to as Type $H_{Tw}L_\alpha$ (Figure 6.56). As α increases (denoted I for intermediate) under a similar tailwater elevation, a vortex develops upstream from the jet impact zone resulting in a less elongated scour hole, referred to as Type $H_{Tw}I_\alpha$. A dune develops along the tailwater flanks of the scour hole. Further increasing α, Type $H_{Tw}H_\alpha$ develops with an increase of the ridge and a further change to an almost circular plunge pool. As the tailwater elevation is reduced, the effect of α reduces and a large portion of the water flow is deflected by the ridge presence (Pagliara *et al.*, 2008a).

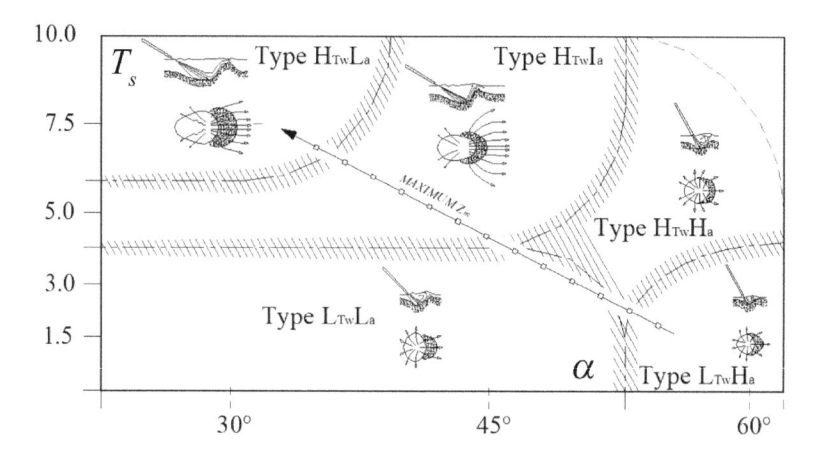

Figure 6.56 Types of 3D plunge pool scour versus impact angle α and relative tailwater

Scour depth features

Figure 6.57 shows a sketch of the spatial (subscript T for 3D) scour hole. It includes in the side view the still water depth h_t, the maximum (subscript m) scour depth z_{mT} and the maximum (subscript M) ridge elevation z_{MT}. The main streamwise scour hole lengths are l_{iT} as the distance from the geometrical point of jet impact onto the original sediment bed to scour hole start, l_{mT} as the distance from the scour hole origin to the point of maximum scour depth and l_{aT} as scour hole length. Further lengths are those relative to the ridge maximum l_{MT} and to the end of the sediment deposition l_{uT}. The parameters in plan view are the maximum scour hole width B_M, the intended angle ω of the ridge extension measured from the point of maximum scour depth, and s_{tot} as half ridge length with s as the coordinate measured from the maximum ridge elevation to the ridge start (Figure 6.57). All tests involved unsubmerged black-water jet flow.

Pagliara *et al.* (2006) observed that the maximum scour hole depth z_m of the 2D flow configuration is linearly related to the densimetric particle Froude number $\mathsf{F}_{d90} = V/(g'd_{90})^{1/2}$, with $g' = [(\rho_s-\rho)/\rho]g$ as reduced gravity acceleration involving densities ρ_s and ρ of sediment and water. This finding applies equally for spatial jet flow. The relative scour hole depth $Z_m = z_m/D(\mathsf{F}_{d90})$, e.g. for $T_s \cong 10$, is significantly larger for the 3D configuration as compared with the 2D setup. This follows from the momentum concentration in the spherically shaped lower scour hole portion for the 3D setting, i.e. the concentration of jet momentum. However, for $\alpha = 45°$ and $T_s \cong 5$, the data almost collapse with the 2D scour relation. For $\alpha = 60°$ and $T_s \cong 1$ the 3D scour hole depth is even smaller than in the 2D case. Accordingly, depending on these two and possibly other parameters, the scour hole depth may be larger, equal or smaller than for the 2D configuration. It was found that Z_m increases with T_s as compared with the 2D setup.

The data for the maximum scour depth $Z_m(\mathsf{F}_{d90})$ for $\alpha = 30°$ and a variable relative tailwater submergence T_s confirm that as larger T_s, as deeper is Z_m. The effect of T_s on Z_m is inversely contained in the scour depth relation of Pagliara *et al.* (2006). Note that the relation between Z_m and F_{d90} is practically linear. As to the jet impact angle, the smaller α, the deeper is the

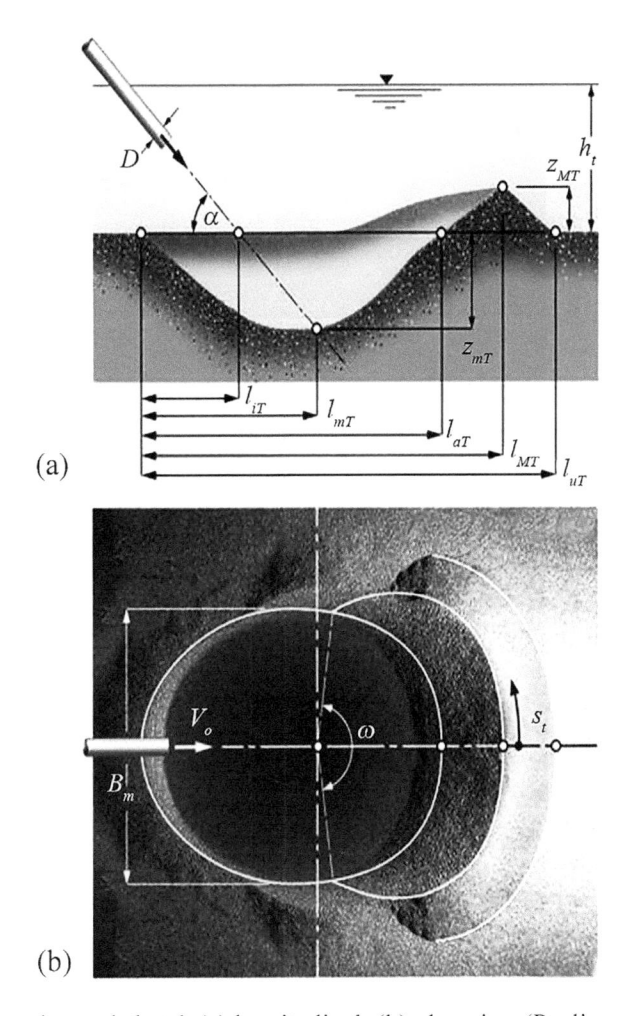

(a)

(b)

Figure 6.57 Experimental sketch (a) longitudinal, (b) plan view (Pagliara *et al.*, 2008a)

scour hole because of the modified ridge geometry. Whereas the ridge in the 2D setup results in deep scour holes for large α, the partial ridge geometry in the 3D scour hole retains less suspended sediment. Figure 6.58 indicates a larger scour surface for a small jet angle as compared with large α.

Equation for scour depth

Pagliara *et al.* (2006) proposed Eqs. (6.77) to (6.83) for the maximum 2D scour hole depth. To expand their validities for spatial scour holes, a number of additional effects have to be accounted for. 3D plunge pool scour is analyzed with the width parameter $\lambda = B_m/B$ in which B_m is the maximum scour hole width and B the channel width. For $\lambda < 1.5$ the scour plan shape is spatial, whereas it is nearly plane for $\lambda > 3$. The relation $Z_{mT}/Z_m(T)$ for $1.5 < \lambda < 2$ and the three jet impact angles indicates that $Z_{mT}/Z_m \cong 1$.

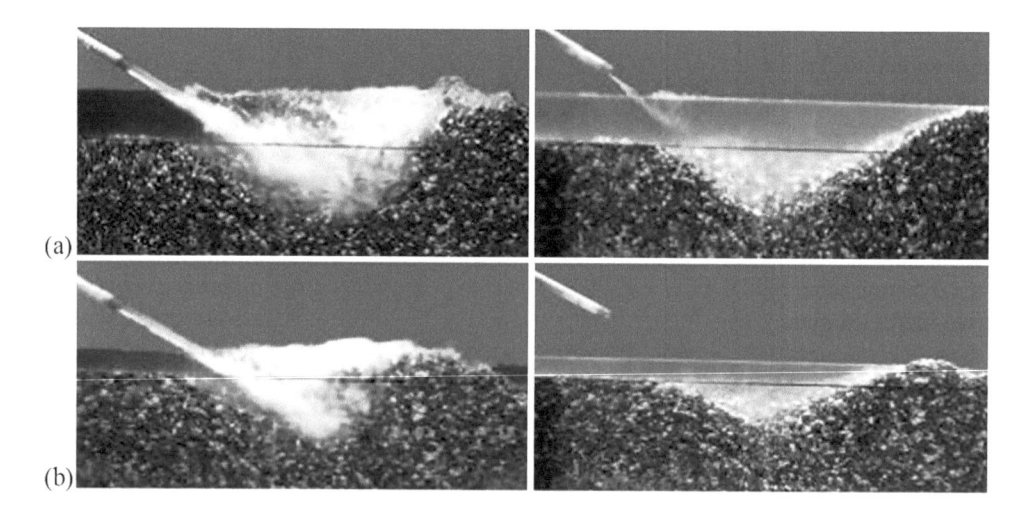

Figure 6.59 Comparison between dynamic (left) and static (right) scour holes for $\alpha = 30°$ and T_s = (a) 5, (b) 3. Note large amount of suspended sediment in dynamic scour hole (Pagliara *et al.*, 2008a)

Scour hole length features

The essential parameters of the plane scour hole geometry include l_m, l_a, l_M and l_u. All dimensionless lengths $L_m = l_m/D$, $L_a = l_a/D$, $L_M = l_M/D$, and $L_U = l_u/D$ increase with F_{d90} and vary with the jet impact angle α. In addition, the relative tailwater elevation T_s has an effect. Pagliara *et al.* (2006) propose Eqs. (6.66) and (6.67) for unsubmerged 2D scour holes. The effect of spatial flow is accounted for with parameter ζ_{LaT} equal to $\zeta_{LaT} = 1$ for $\lambda \geq 3$, whereas Pagliara *et al.* (2008a) suggest for $\lambda \leq 1.50$

$$\zeta_{LaT} = [0.168 + 0.062T_s + 0.020\alpha°] \cdot [1.217 - 0.013F_{d90}] \cdot [0.851 + 0.345\lambda] . \tag{6.92}$$

The modified Eq. (6.67) reads then

$$L_{aT} = 2.5[1 + 1.05\exp(-0.030\alpha°)F_{d90}] \cdot \zeta_{LaT} . \tag{6.93}$$

Likewise, the relative length L_m from Eq. (6.66) is extended to the 3D case as

$$L_{mT} = [1.5 + (1 - 0.01\alpha°)F_{d90}] \cdot \zeta_{LmT} \tag{6.94}$$

with $\zeta_{LmT} = 1$ for $\lambda > 3$; whereas for $\lambda < 1.5$

$$\zeta_{LmT} = [1.705 - 0.012\alpha° - 0.010F_{d90}] \cdot [0.719 + 0.061T_s] . \tag{6.95}$$

In the transition range $1.5 < \lambda < 3$ linear interpolation applies between the limit values $\lambda = 1.50$ and $\lambda = 3$. The validities of Eqs. (6.92) to (6.95) are as for the previous fits.

Scour hole plan features

The spatial scour hole characteristics depend solely on the relative scour hole width $\lambda = B_m/B$. The maximum scour hole width B_m increases with the parameters F_{d90}, α, and T_s. Plots $B_m(F_{d90})$ for various tailwater elevations T_s and $\alpha = 45°$ indicate that large values of T_s produce a large value of B_m. A scour hole tends to be of circular shape for large jet angles, whereas it is elongated otherwise. Accordingly, the maximum scour hole width $B_M = B_m/D$ is related to length L_M. The relation among the governing parameters states that B_M increases with α as (Pagliara et al., 2008a)

$$B_M = [0.500 + 0.008\alpha°] \cdot L_{aT}. \tag{6.96}$$

The ridge profile has the plan shape of a half moon. The ridge start is described with the intended angle ω (Figure 6.57) originating at the point of maximum scour depth (Figure 6.57). For the sediments used and impact angles ranging between 30° and 60°, $\omega = 160°$ ($\pm 40°$), independent of T_s and F_{d90}.

The non-dimensional axial scour hole profile is described with the normalized coordinates $X_m = (x - x_m)/l_{aT}$ und $Z_m = z/z_{mT}$ (Pagliara et al., 2006), thereby accounting for the scalings in Eqs. (6.87) and (6.93). The streamwise degree of scour hole symmetry (subscript S) is described with parameter $X_S = 2(x_o - x_m)/l_{aT}$ (Pagliara et al., 2006). For $X_S = -1$, the point of maximum scour is at $0.50l_{aT}$. The symmetry parameter X_S of 3D scour holes increases from about $X_S(\alpha = 30°) = -1.4$ to $X_S(\alpha = 45°) = -1.25$ parallel to 2D scour holes. For larger jet impact angles, the two curves diverge pointing at the differences between the 2D and 3D scour hole patterns.

Length of impact point from scour origin

The dimensionless length $L_{iT} = l_{iT}/D$ relates the scour hole origin to the fictitious jet impact point on the scour surface (Figure 6.57). The relation $L_{iT}(F_{d90})$ versus tailwater elevation T_s and $\alpha = 30°$ is (Pagliara et al., 2008a)

$$L_{iT} = [(0.035 + 0.034T_s - 0.001\alpha°)F_{d90} + (7.667 - 0.100\alpha°)](1.285 - 0.361\lambda). \tag{6.97}$$

Parameter L_{iT} essentially increases with T_s and decreases with the jet impact angle α. The validity of Eq. (6.97) is as for the other parameters.

Scour hole and suspended sediment volumes

The volume of the scour hole w_m under dynamic jet flow conditions is of interest. Its dimensionless value is $W_m = (g'd_{90})^{1/2}[w_m/(Q_WD)]$; it depends on F_{d90}. Further, W_m increases with T_s, because the ridge is an obstacle for large values of T_s deflecting the flow, thereby increasing the amount of suspended sediment over the scour hole. From 3D tests, the dimensionless scour volume W_m is (Pagliara et al., 2008a)

$$W_m = 1.124 \cdot [-0.929 + 0.014\alpha°)T_s - 0.500]F_{d90}. \tag{6.98}$$

The suspended sediment volume w_s is equal to the difference of the scour hole and ridge (subscript r) volumes ($w_m - w_r$). The dimensionless suspended sediment volume $W_s = (W_m - W_r)/W_m$ increases with the jet impact angle α, whereas the effect of F_{d90} is relatively small. It was also found that W_s increases as T_s decreases. The data follow under the usual limitations (Pagliara *et al.*, 2008a)

$$W_s = [0.015F_{d90} + (-0.078 + 0.001\alpha°)T_s + (0.140 + 0.007\alpha°)](0.966 - 0.012\lambda). \quad (6.99)$$

Maximum ridge height

Pagliara *et al.* (2006) describe the importance of the ridge in 2D plunge pool scour. Experiments were conducted in which the ridge was constantly removed resulting in an increase of the scour hole depth, because the ridge protects a scour hole from excessive depth, associated with a promotion of suspended flow. A ridge enhances flow recirculation above the scour hole, both in the 2D and 3D jet arrangements. The dimensionless ridge height was defined as for the 2D scour hole as $Z_M = z_M/D$ (Figure 6.57). A data plot of $Z_M(F_{d90})$ for a range of T_s and $\alpha = 30°$ indicates a strong effect of both F_{d90} and T_s. The data follow under the usual limitations (Pagliara *et al.*, 2008a)

$$Z_{MT} = ([0.010 + (0.044 - 0.001\alpha°)T_s] \cdot F_{d90} + 0.800) \cdot (0.815 + 0.507\lambda). \quad (6.100)$$

The effect of T_s on Z_{MT} is larger under small than under large jet angles α.

The longitudinal ridge profile above the original sediment bed was also studied. The local ridge or dune (subscript d) elevation z_d is defined as $Z_d = z_d/z_M$ versus the streamwise variable $S_{tot} = s_T/s_{tot}$, where s_{tot} is the total length between the dune origin and the axial ridge maximum. The longitudinal dimensionless profile $Z_M(S_{tot})$ for all jet impact angles investigated, for the sediment used, follows the linear trend

$$Z_d = 1 - S_{tot}. \quad (6.101)$$

Discussion and limitations

Spatial plunge pool scour is governed by: (1) jet hydraulics including jet approach flow velocity V, jet diameter D, jet impact angle α and tailwater depth h_t; (2) granulometry including sediment diameter d_{90}, sediment density ρ_s and sediment non-uniformity parameter σ_s; and (3) scour geometry involving extreme scour and ridge elevations z_m and z_M, streamwise and transverse scour extensions l_m, l_a, l_M, l_u, and b_M, plus the scour hole topography. Most of these parameters were systematically varied and experimentally determined.

Froude similitude was admitted, implying the limitation of members involved in the densimetric Froude number $F_{d90} = V/(g'd_{90})^{1/2}$: jet flows should be fully turbulent of minimum jet Reynolds number $R_o = V_oD/v = 10^5$. Jet impact angle are confined to $30° \le \alpha \le 60°$. Tailwater depths are $h \ge 0.05$ m, and tailwater ratios $T_s = h/D \ge 0.80$. Sediment of median grain size below 1 mm involves viscous effects. Other limitations relate to the relative scour width $\lambda = B_m/B$ from 0.20 to 1.5. The value of $\lambda = 3$ is considered the minimum for a practically plane scour pattern, whereas the transitional regime applies for $1.5 < \lambda < 3$. The tailwater ratio T_s varied from 0.80 to 10. Densimetric Froude numbers F_{d90} ranged from 4 to 30, resulting in nearly straight jet trajectories from the pipe outlet section to the scour hole impact zone.

An important question relating to plunge pool scour is: Are 3D scour holes deeper than 2D holes? This may not easily be answered, because the ratio between the two may be smaller or larger than 1, depending on the parameter constitution. 3D scour holes are generally deep for large tailwater submergence $T_s > 5$ and jet impact angles $\alpha < 45°$. Their ratio is smaller than 1 if $T_s < 3$ and $\alpha \cong 60°$. For intermediate values of the controlling parameters T_s and α, differences between the 3D and the 2D scour hole depths are small. The third important parameter of plunge pool scour is F_{d90}: both the maximum scour hole depth and ridge height are directly proportional to it.

Further results apply to the difference between the static and the dynamic scour holes. The effect is important for larger jet impact angles α but negligible for $\alpha = 30°$. The 3D scour hole involves two portions, namely the cup-shaped lower portion of the active scour zone and the surrounding scour hole slopes close to the angle of repose. The ridge plan geometry is half-moon-shaped bordering the downstream scour hole for $\alpha = 30°$. For $\alpha = 60°$, the complete scour hole is surrounded by the ridge whose maximum elevation is located opposite of the jet approach flow. The scour volume data in terms of the governing flow parameters indicate that the suspended sediment volume may be larger than 50% of the scour hole volume for $\alpha = 60°$, decreasing with α. Ridge presence is an important scour holes feature protecting it from excessive depth, as it deflects the jet for small relative tailwater elevation. A classification of the effects of tailwater elevation and jet impact angle on the scour hole features is provided. This research thus adds to the work on the plane scour hole characteristics constituting a preliminary design guideline.

6.4.5 3D Flow features in plunge pool

Experimentation

Based on previous experimentation on plunge pool scour, Unger and Hager (2007) addressed the flow field generated by a water jet directed onto a sediment bed. The flow field was studied by a non-intrusive and instantaneous approach using large-scale non-laser 2D particle image velocimetry (PIV). A test illustrates the relevant flow processes and the basic hydraulic features of plunge pool scour are discussed.

Tests were conducted in a scour channel (Hager *et al.*, 2002) involving a nearly uniform sediment of median grain size $d_{50} = 1.15$ mm along with a sediment non-uniformity parameter $\sigma_s = (d_{84}/d_{16})^{0.5} = 1.15$. The water jet was generated by a high-precision dosing-pump attached to a flexible piping system mounted on a streamwise and transversely adjustable carriage to vary the jet impact position, jet diameter and jet impact angle. This water supply system was part of a closed recirculation system to ensure a constant tailwater depth during a test, irrespective of jet discharge Q.

To allow for excellent flow visualization and to record the scour topography, the jet was placed parallel to the glassed channel sidewall distant by 3 mm. Pagliara *et al.* (2008a) noted a small boundary layer effect along the glass wall. Therefore, this half-model setup is equivalent to the arrangement with a centered jet. The jet diameter was $D_{test} = 21.4$ mm and jet impact angles were $\alpha = 30°$, 45°, and 60°. The equivalent jet diameter is $D = 2^{0.5}D_{test} = 30.3$ mm. The jet was deflected from the horizontal sediment bed until reaching the test discharge. The time was set to $t = 0$ once the jet impacted the sediment bed. The flow field and the scour topography were measured at $t = 12,600$ s, i.e. after the end scour hole had practically developed (Pagliara *et al.*, 2008a).

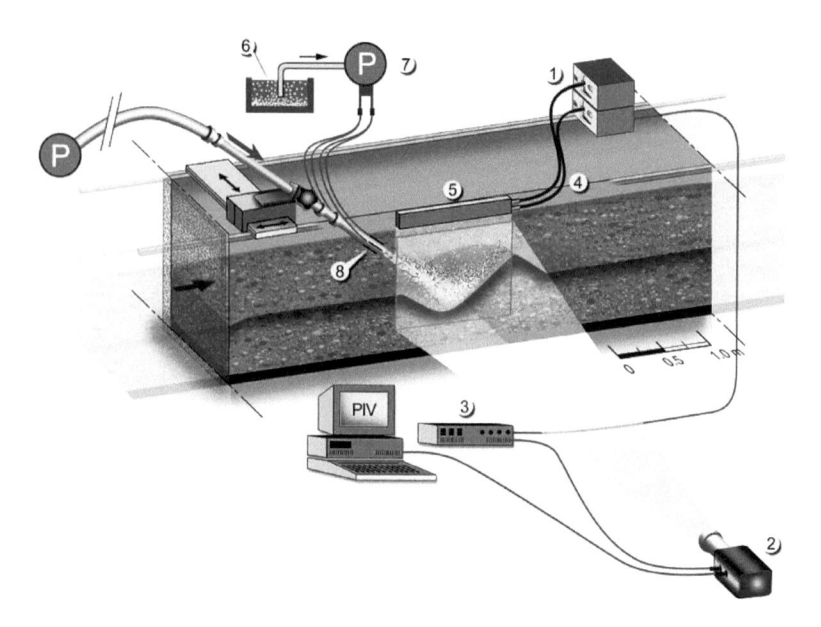

Figure 6.60 Sketch of experimental setup including PIV system with ① flashlight strobos-
copes, ② CCD camera, ③ Programmable Timing Unit, ④ Fiberoptic light
guide, ⑤ Line converter, ⑥ particle reservoir, ⑦ hose pump, ⑧ particle injection nozzles (Unger and Hager, 2007)

The scour topography was measured: (1) during the test at $t = 12,600$ s using the optical
PIV recordings of the CCD camera similar to the method proposed by Pagliara *et al.* (2008a),
and (2) after jet discharge was stopped. For the latter, a conventional shoe gage mounted on
a moveable carriage was used. Due to optical inaccessibility, the sediment deposition zone
far from the model symmetry axis was partly hidden for the camera; at $t = 12,600$ s only
measurements in planes $y/D \leq 2.50$ were possible, with y as the transverse coordinate. For a
jet impact angle of $\alpha = 30°$, no essential differences occurred between the two tests. Due to
the large amount of suspended sediment, a significant difference occurred between the (1)
so-called dynamic scour topography during a test, and the (2) static counterpart after the jet
was stopped for impact angles $\alpha > 35°$ (Pagliara *et al.*, 2004a). The relevant scour topography
relates to the dynamic condition, because the static scour hole is much less deep due to 'avalanche flow' of the suspended sediment. The dynamic scour topography was thus considered.

A planar large-scale PIV setup was employed to determine the velocity vector fields
simultaneously and non-intrusively. Let z be the vertical coordinate above the original sediment
bed. The flow field was determined both at various elevated horizontal layers z/D, and in
vertical layers parallel to the jet flow direction y/D, allowing for a quasi-spatial visualization.
As shown in Figure 6.60 for vertical layers, the camera was placed sideways of the channel
and the line converter illuminated the flow field of interest from the top across the free water
surface. For the horizontal flows fields, the CCD camera position and the illumination were
inverted. The water surface was covered with a thin Plexiglas-plate to inhibit optical problems
with surface waves, thereby not affecting the flow. Unger and Hager (2007) provide further
details on experimentation. Figure 6.60 shows a sketch of the PIV system.

Vertical flow fields

The test conditions for the test selected were: Jet impact angle $\alpha = 30°$, jet diameter $D_{test} = 21.4$ mm with $D = 30.3$ mm, relative tailwater elevation $T_s = h_t/D = 7.92$ with $h_t = 0.24$ m, jet discharge $Q = 0.50$ l/s resulting in the densimetric jet Froude number $F_{d90} = V_o'/(g'd_{90})^{1/2} = 9.40$ with the approach flow jet velocity of $V_o = 1.39$ m/s, $g' = [(\rho - \rho_s)/\rho]g$ as the reduced gravity acceleration and $\tau_j = (g'd_{90})^{0.5}\cdot t/D = 6.07 \times 10^4$ as non-dimensional scour time (Unger and Hager, 2007). The approach flow jet Reynolds number was $R = V_o D/v = 4.17 \times 10^5$ with v as kinematic fluid viscosity, resulting in fully turbulent jet flow. Figure 6.61 sketches the scour hole arrangement including notation.

The vertical flow fields shown in Figure 6.62 refer to different layers in the transverse direction parallel to the channel symmetry axis, namely at $y/D = 0.50$ as longitudinal jet symmetry axis; at $y/D = 0.83$ along the rear jet boundary in the transverse direction, and in farther rearward layers at $y/D = 1.65, 2.48, 3.30, 4.13, 4.95$, and 6.61. No PIV records were available directly at the pipe outlet due to the high seeding density. From the plane $y/D = 0.50$ the jet expands under an angle of some 5°. The maximum jet velocity decreases along the longitudinal jet axis from $V/V_o = 0.71$ at $(x/D; z/D) = (-4.5; +5.5)$ close to the pipe outlet to $V/V_o = 0.37$ at $(+7.6; -1.9)$ near the sediment surface. Parallel to the increasing sediment surface, starting from the maximum scour depth at $(+7.1; -2.7)$ to the sediment deposition maximum at $(+18.3; +2.9)$, relatively high bottom velocities are generated with $V/V_o = 0.24$ at $(+13.2; -0.2)$ for $y/D = 0.50$ and $V/V_o = 0.22$ at $(+14.5; +0.4)$ for $y/D = 0.83$, respectively. Outside from the longitudinal jet symmetry axis, these bottom velocities decrease continuously from $V/V_o = 0.19$ at $(+11.9; -0.2)$ for $y/D = 1.65$ to $V/V_o = 0.11$ at $(+17.1; +2.7)$ for $y/D = 4.95$. In the plane $y/D = 1.65$ only the upper jet portion close to the sediment surface and the deflected portion inside the scour hole remain visible, whereas the flow velocities are too small outside this region. For the planes $y/D \geq 2.48$, notable velocities develop exclusively inside the scour

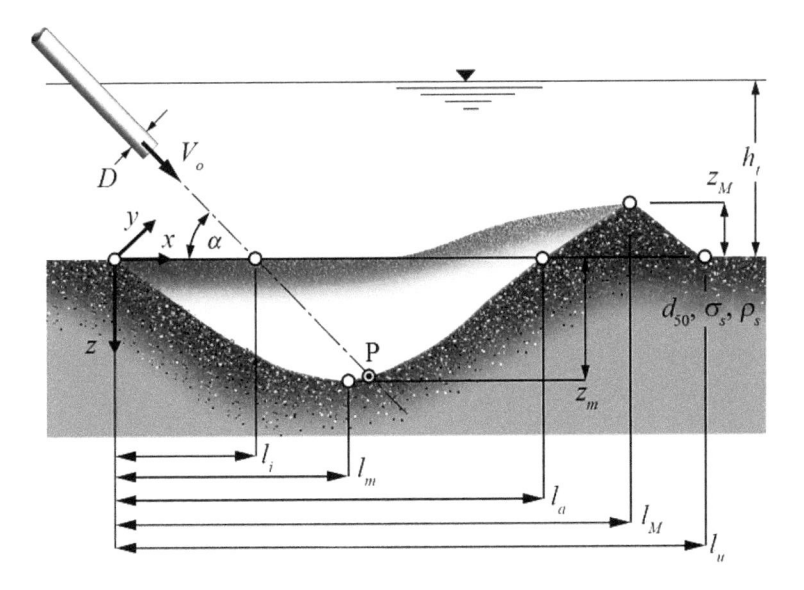

Figure 6.61 Typical scour hole topography, with notation

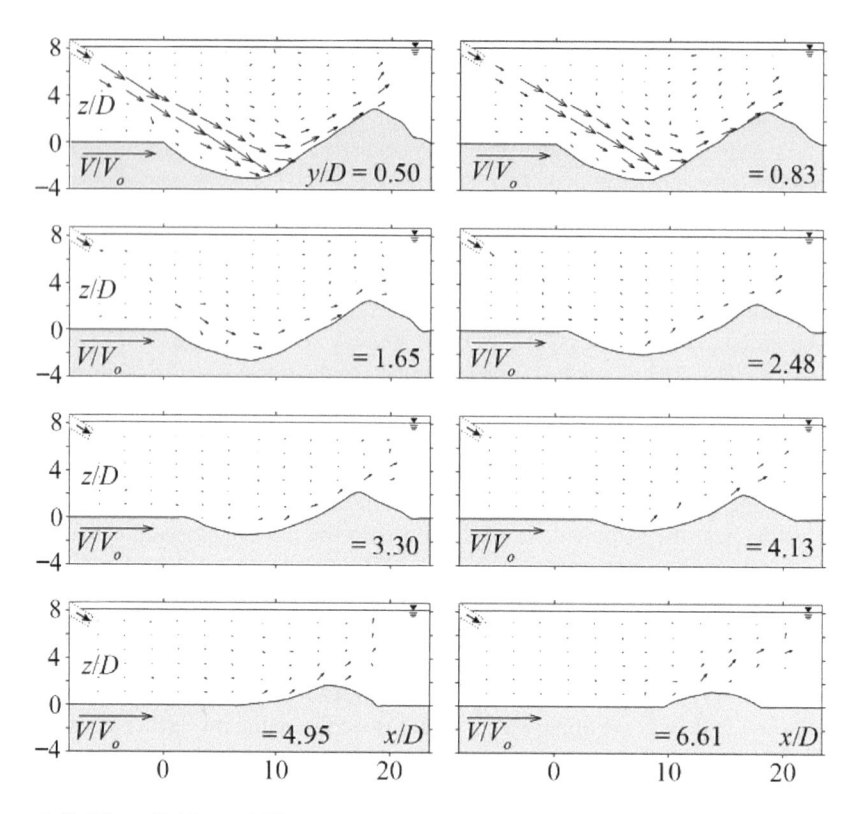

Figure 6.62 Flow fields at different vertical planes $0.50 \leq y/D \leq 6.61$ (Unger and Hager, 2007)

zone due to jet deflection. From test limitations, no velocities were recorded in the flow recirculation beyond the sediment ridge.

To illustrate the main flow structures, streamline plots were extrapolated from the velocity vector fields. Figure 6.63 shows these for conditions considered previously. In the plane $y/D = 0.50$, the linear expanding jet was detected from the pipe outlet up to its impact slightly downstream from the maximum scour depth. The flow outside the jet core is entrained by the jet due to pressure reduction. Upstream of the impact point the jet is deflected toward the water surface along the scour hole ridge under an angle of ~30°. At (+13.5; +5.5) a source is generated resulting in a small recirculation at (+10.9; +2.9). However, no recirculating flow is noted to develop inside the scour region. In the plane $y/D = 0.83$ the flow upstream of the jet impact point is similar to $y/D = 0.50$ except for a source flow at (+1.4; +6.4). The flow above the deflected jet upstream from its impact point is first directed against the main jet flow direction and then redirected oppositely. For $y/D \geq 1.65$, the region upstream of the maximum scour depth includes numerous highly unsteady singularities and small vortices due to turbulent diffusion processes. Downstream of the maximum scour depth, a deflected flow develops along the scour boundary in all planes. Its deflection angle increases from 30° at $y/D = 1.65$ to ~45° for $y/D \geq 6.61$. Above the jet core, clockwise rotating vortices are generated, whose position and size are subjected to turbulent fluctuations.

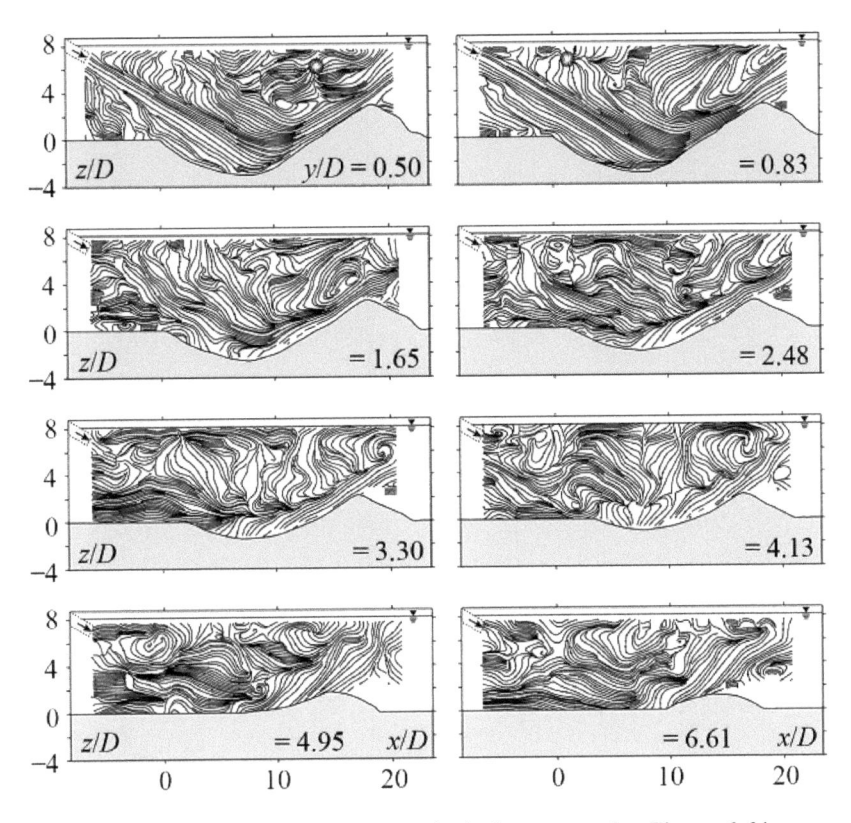

Figure 6.63 Streamline plots in different vertical planes, see also Figure 6.64

Figure 6.64 refers to the identical hydraulic conditions as Figures 6.62 and 6.63. The velocity vector fields are shown in various elevated planes, namely at z/D = 7.07, 4.95, 3.30, 2.48, 1.65, 0.83, 0, -0.83, and -1.65; the scour hole topography is plotted with the relative scour depth $Z_s = z/D$. In the plane z/D = 7.07, corresponding to the elevation of the pipe outlet, no significant velocities are generated except for the jet intake region (-5.5; $+0.6$) with maximum values of V/V_o = 0.13 and V/V_o = 0.11 in the region ($+20$; $+8.0$) due to jet deflection at the scour hole. Jet velocity decreases continuously from V/V_o = 0.49 at z/D = 4.95 to V/V_o = 0.35 at z/D = -1.65. Its position simultaneously migrates downstream due to the jet slope of α = 30°. At the downstream end of the flow field, i.e. at the deposition maximum, a transverse deflected flow occurs in all planes. Its velocities increase with decreasing depth from V/V_o = 0.11 at z/D = 7.07 to V/V_o = 0.19 at z/D = -0.83 with the deflection angle ranging from 30° to 40°.

Figure 6.65 shows the corresponding streamlines for the different horizontal layers. The closely spaced streamlines parallel to the channel sidewall starting from the plane z/D = 4.95 demonstrate the geometrically unaffected approach flow jet. A small flow toward the jet develops in all planes due to pressure reduction inside the jet. Outside of it, low insensitive singularities and unsteady clockwise rotating vortices are set up.

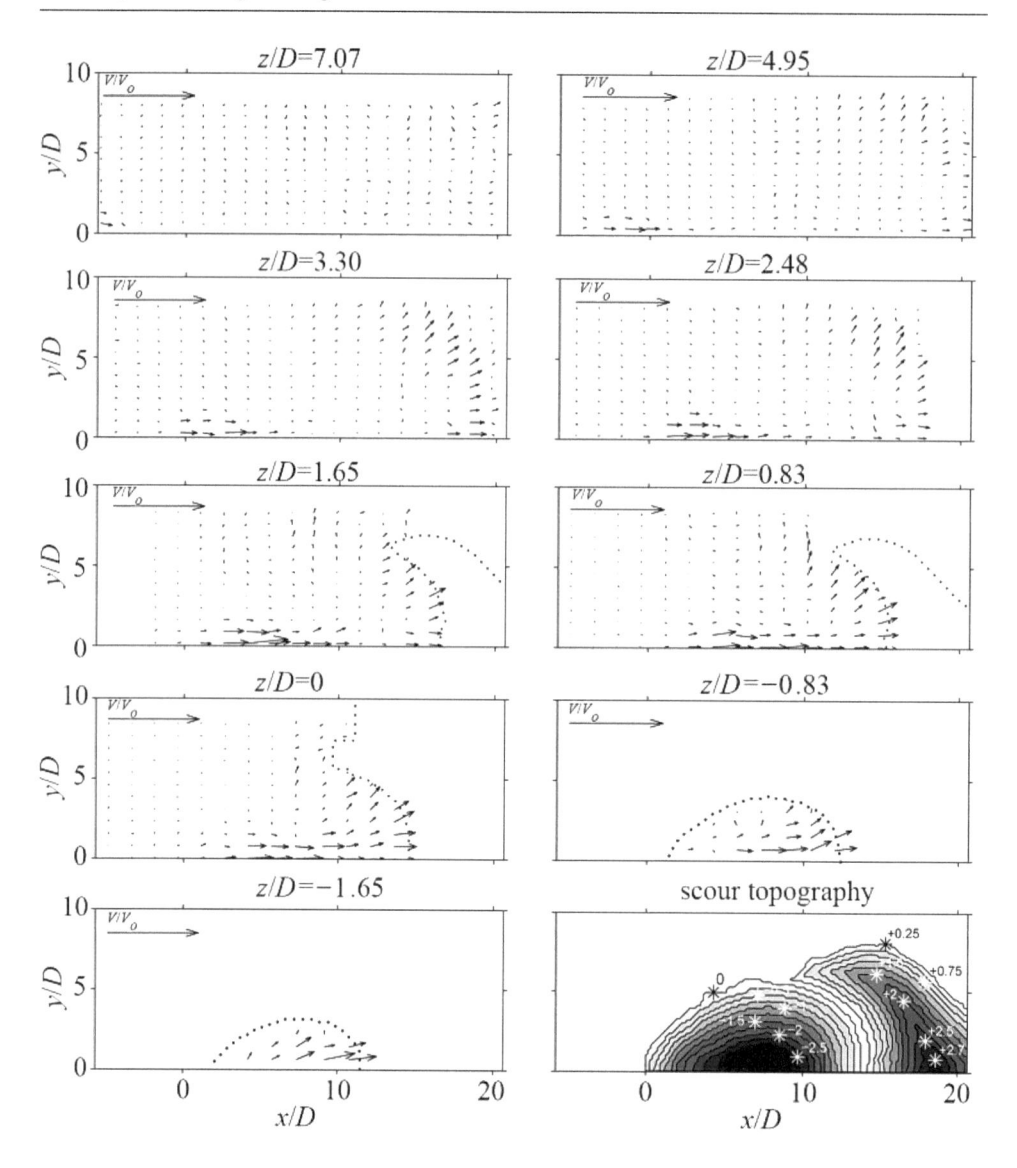

Figure 6.64 Flow field at various elevated horizontal layers $-1.65 \leq z/D \leq 7.07$ including scour hole topography, ($\bullet \bullet \bullet$) limit curve from where PIV data are not available because of scour topography

The resulting scour hole is essentially parabola-shaped. It starts in the jet axis at location $x/D = 0$, maximum scour depth is $z/D = 3.0$ at $x/D = 7.5$, and its end is at $x/D = 13.0$. Beyond the scour hole forms a sediment ridge of maximum height $z/D = 2.8$ above the original bed level; the ridge end is at $x/D = 25.5$ outside of the observational area. The final scour hole shape close to the jet impact point is cup-shaped, whereas the angles close to the scour hole start and the ridge maximum are equal to the natural slope angle of repose below water of the sediment involved.

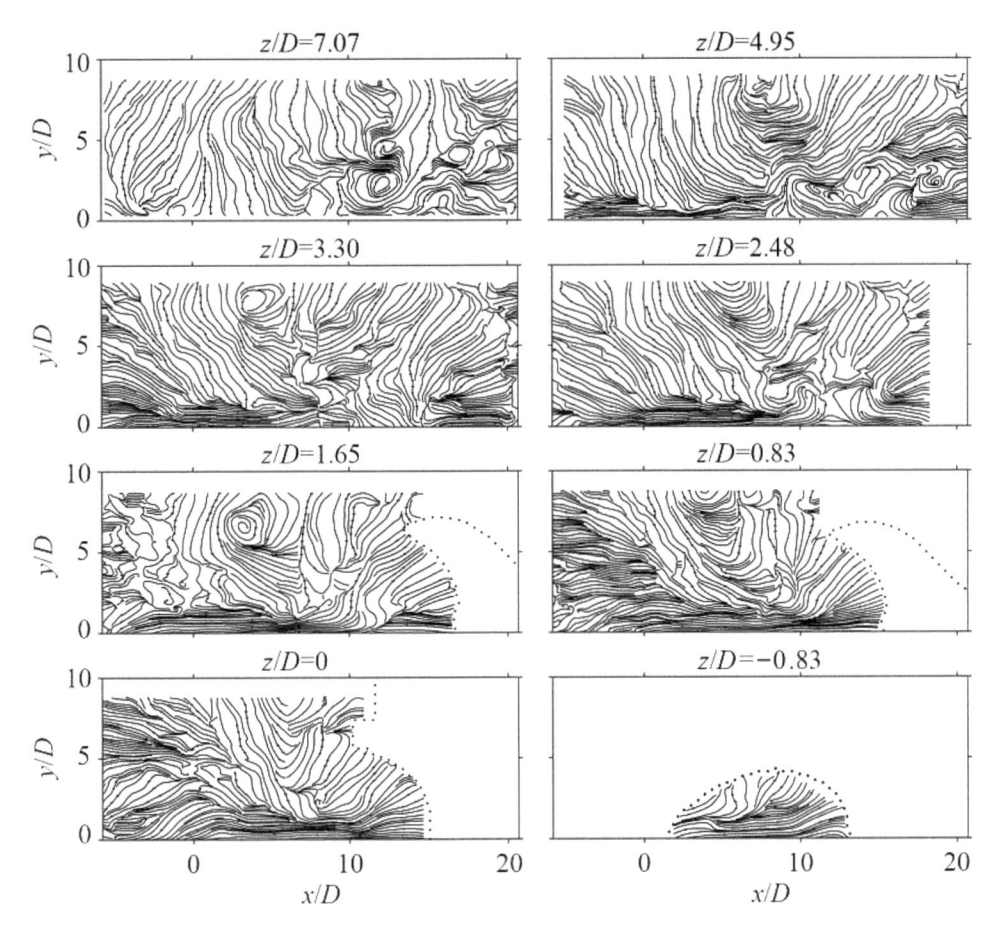

Figure 6.65 Streamline plots at various horizontal elevated layers $-1.65 \leq z/D \leq 7.07$ (details in Figure 6.64)

Discussion of results

The current knowledge on the flow features in plunge pool scour holes indicates that a jet toward a mobile sediment bed generates relatively simple flow structures of which only the flow inside the scour hole is jet-affected. According to Figure 6.66 the turbulent jet expands under an angle of some 5°. From its impact point slightly downstream of the maximum scour depth, the jet is deflected toward the scour hole ridge along the sediment surface. Simultaneously the jet expands radially downstream of its impact point in the lateral direction under an angle of 30° to 40°. In the longitudinal jet symmetry axis, relatively high velocities are generated along the rising scour topography, starting slightly downstream of the maximum scour depth. As a result, the sediment is entrained only in this region, as confirmed visually. Inside the scour hole, no major flow recirculation is generated because the jet impact angle is close to the natural angle of repose below water.

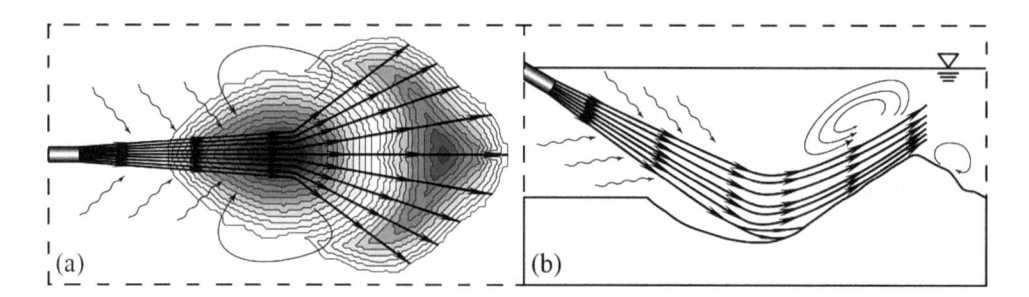

Figure 6.66 Principal flow features inside developed scour hole generated by plunging turbulent jet in (a) plan, (b) section (Unger and Hager, 2007)

As shown in Figure 6.66, only minor flow activity occurs outside from the main jet flow. The fluid of the approach flow region is entrained by the jet due to the negative pressure gradient, resulting in two recirculation zones laterally from the jet impact point (Figure 6.66a). Additionally, above the maximum ridge height, a small but highly unsteady anti-clockwise rotating recirculation was observed (Figure 6.66b). Downstream of the maximum ridge height close to the decreasing ridge surface occurs a clockwise rotating recirculation. Due to the limited observational window for the 30° run, it was not detected but observed for all runs with $\alpha > 30°$. All other singularities generated by these flows appear to be unstable resulting from turbulent dissipation at the level of micro-vortices. The resulting scour hole topography is parabola-shaped including a maximum scour depth slightly upstream of the jet impact point and a maximum ridge height beyond the scour hole, confirming the observations of Pagliara *et al.* (2008a) and Rajaratnam and Berry (1977).

6.4.6 Temporal evolution of spatial plunge pool scour

Experimentation

The question of how a plunge pool scour hole temporally develops was addressed by Pagliara *et al.* (2008b), among others, based on detailed experimentation (Hager *et al.*, 2002). Because only the temporal advance of plunge pool scour was studied, the test program involved one nearly uniform sediment of median diameter $d_{50} = 1.15$ mm, $d_{90} = 1.35$ mm, and $\sigma_s = (d_{84}/d_{16})^{1/2} = 1.15$. Black-water jets were generated with a pump attached (Figure 6.60). Jet diameters were $D_{test} = 21.7$ mm and 35.0 mm, and jet impact angles $\alpha = 30°$, 45°, and 60° from the horizontal.

The sediment surface was horizontal at test start with the tailwater level set to ±1 mm. The pump was directly supplied from the test channel so that its water level remained constant during a test. Time was set to $t = 0$ once the jet impacted the sediment bed. To allow for scour visualization as discussed in 6.4.5, the half-model arrangement was again applied. The equivalent jet diameter is $D = 2^{1/2}D_{test}$, thereby requiring equal jet velocity as for the full model setup (Unger and Hager, 2007).

The longitudinal scour hole profiles were recorded with a monochrome progressive scan CCD camera (Figure 6.60). It is important that neither air bubbles nor suspended sediment inhibit optical access to the flow field. These conditions were not always realized due to sediment resuspension and air entrained by the jet. If the sediment surface was only partially visible at one instant, conditions improved shortly thereafter allowing for an accurate definition of the entire scour hole profile over time.

A total of 54 tests were conducted with test durations up to 1 day involving both 2D and 3D scour holes including Submerged (S-jets) and Unsubmerged (U-jets). For the latter, the jet impinged onto the water surface, whereas the tailwater elevation was higher than the pipe outlet for S-jets (Pagliara *et al.*, 2008b). The densimetric particle Froude number based on sediment size d_{90} is $\mathsf{F}_{d90} = V/(g'd_{90})^{1/2}$ with $V = Q/(\pi D^2/4)$ as the median jet velocity, $g' = [(\rho_s - \rho)/\rho]g$ as the reduced gravity acceleration with ρ_s and ρ as the densities of sediment and water, respectively.

Scour hole features

Figures 6.67 and 6.68 show photographs of scour holes at various times t and $\alpha = 30°$ and $60°$, i.e. for a small and a large jet impact angle. For $\alpha < 35°$, the scour hole is essentially parabola-shaped. Its origin is at location $x = 0$, the maximum (subscript m) scour depth is at $x = l_m$ and its end at $x = l_a$ (Figure 6.61). Beyond the scour hole forms a sediment ridge of maximum (subscript M) height z_M above the original bed level at $x = l_M$ and the ridge ends at $x = l_u$. The final scour hole geometry close to the jet impact point P is cup-shaped, whereas its angles close to the scour hole origin and the ridge are equal to the natural angle of repose ϕ below water of the sediment involved.

Figure 6.67 shows images for $\mathsf{F}_{d90} = 33.4$, $\alpha = 30°$ for an S-jet issued from the upper left of the photos. At test start $t = 0$, the sediment surface is horizontal. At $t = 1$ s, surface turbulence is noted due to jet impact onto the sediment surface. At $t = 4$ s, an initial, almost symmetrical scour hole about the point of maximum scour depth forms, with sediment transport both in the longitudinal and lateral directions. At $t = 8$ s, the scour hole shape becomes asymmetric, with dominant sediment transport in the streamwise direction. From $t = 12$ s, the sediment transport capacity is too small to transport all scoured material, resulting in sediment deposition beyond the flow recirculation above the scour hole, thereby forming the sediment ridge as an important feature of the entire scour hole. At $t = 51$ s, the upstream ridge portion is shaped, whereas the tailwater portion develops much slower because of limited sediment supply over the ridge. Note the comparatively large distances between the ridge maximum and scour hole origin at $t = 51$ and 86 s as compared with earlier time, and the strong growth of scour depth within this period. The *developing* scour phase is by now completed.

During the *developed* scour phase, the scour hole origin does hardly move until test end at $t = 3400$ s, because its slope remains constant (Figure 6.67). The upstream scour hole portion therefore only advances as the maximum scour hole depth increases, whereas the ridge undergoes a substantial variation during this phase. At $t = 125$ s, the scour hole base is cup-shaped, because of the too steep tailwater scour hole portion. During this phase, its length from the maximum scour depth to the maximum ridge elevation ($l_M - l_m$) is made up of two parts, namely the (1) active portion close to the maximum scour depth due to jet impact, and (2) passive scour hole portion further downstream, whose angle is essentially equal to ϕ. At $t = 298$ s, the cup-shaped scour hole base is still visible, together with an intense sediment transport, pointing at the significant spatial scour hole expansion. At $t = 1200$ s, the ridge has reached its final shape, but the entire formation still moves into the tailwater, as observed at $t = 3400$ s. The last photo relates to the static scour hole condition, i.e. after jet flow is

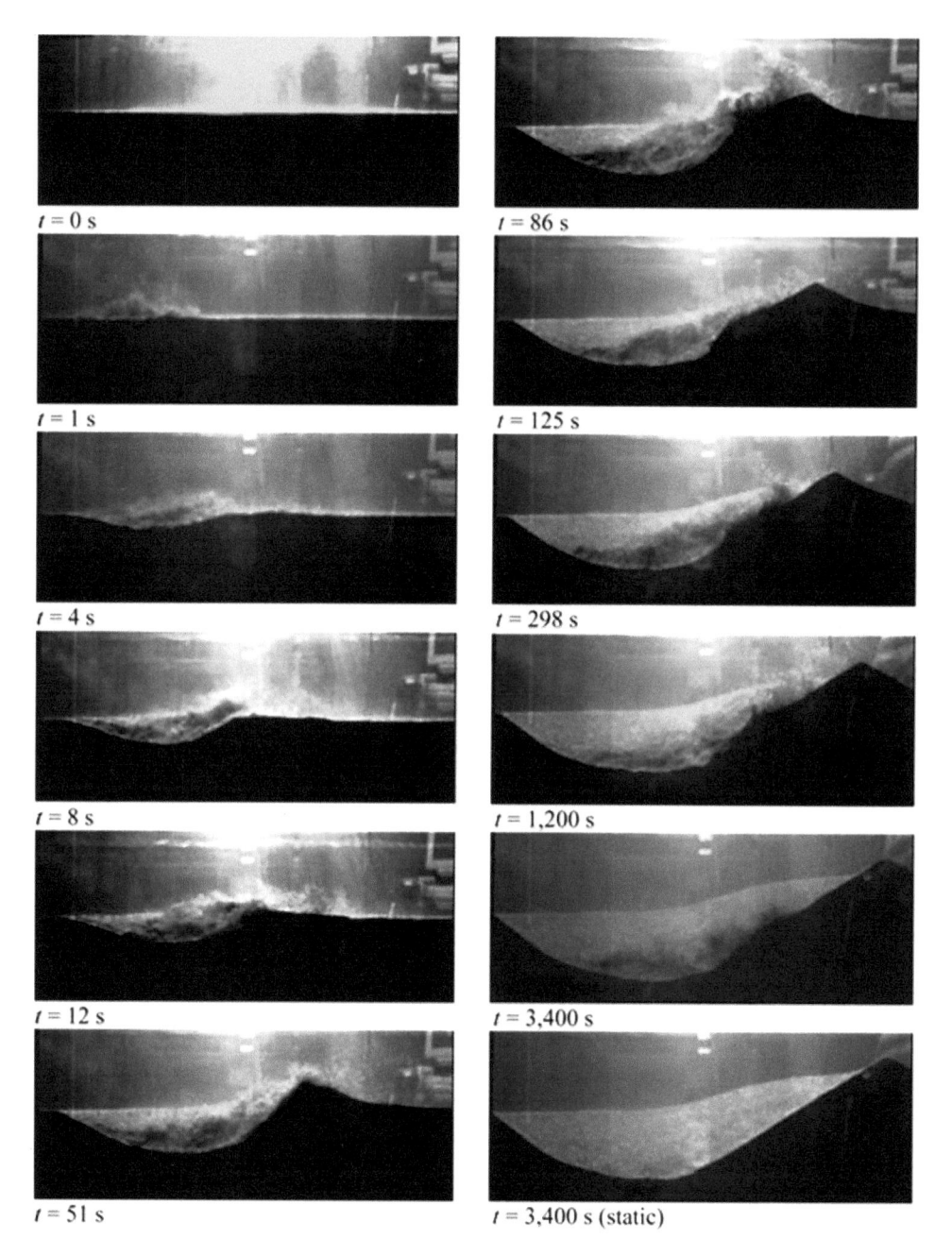

$t = 0$ s

$t = 1$ s

$t = 4$ s

$t = 8$ s

$t = 12$ s

$t = 51$ s

$t = 86$ s

$t = 125$ s

$t = 298$ s

$t = 1,200$ s

$t = 3,400$ s

$t = 3,400$ s (static)

Figure 6.67 Temporal scour hole development, test with $F_{d90} = 33.4$, $\alpha = 30°$, S-jet, sequence of photos from $t = 0$ to 3400 s (Pagliara *et al.*, 2008b)

stopped. A notable difference occurs mainly at the scour hole base, where the 'cup' portion has disappeared and the scour hole surface has slopes always equal or smaller than ϕ.

Figure 6.68 relates to a test with $F_{d90} = 27.8$, $\alpha = 60°$, and a U-jet, i.e. with a similar value of F_{d90} as in Figure 6.67 but a large jet impact angle α. At time $t = 3$ s, a sediment cloud limits visual access to the scour hole, which eventually moves in the tailwater direction. From $t = 7$ s, ridges both upstream and downstream of the jet impact point evolve, with the upstream ridge still advancing at $t = 4500$ s. The observational quality of the scour hole surface for $\alpha = 60°$ is complicated due to suspended sediment, so that no tests were conducted for vertical jet impact. The geometries of the dynamic and the static scour holes at time $t = 6060$ s are completely different given the cup-shaped and much deeper scour hole under dynamic test conditions. Pagliara et al. (2004b) noted that the latter is the relevant scour hole geometry in hydraulic engineering, because the static scour hole has a significantly smaller depth due to the deposition of the previously suspended sediment into the scour hole once the jet is stopped.

The non-dimensional scour hole depth $Z_m = z_m/D$ versus the non-dimensional time $\tau_j = (g'd_{90})^{1/2} \cdot t/D$ (Oliveto and Hager, 2002) is plotted semi-logarithmically in Figure 6.69

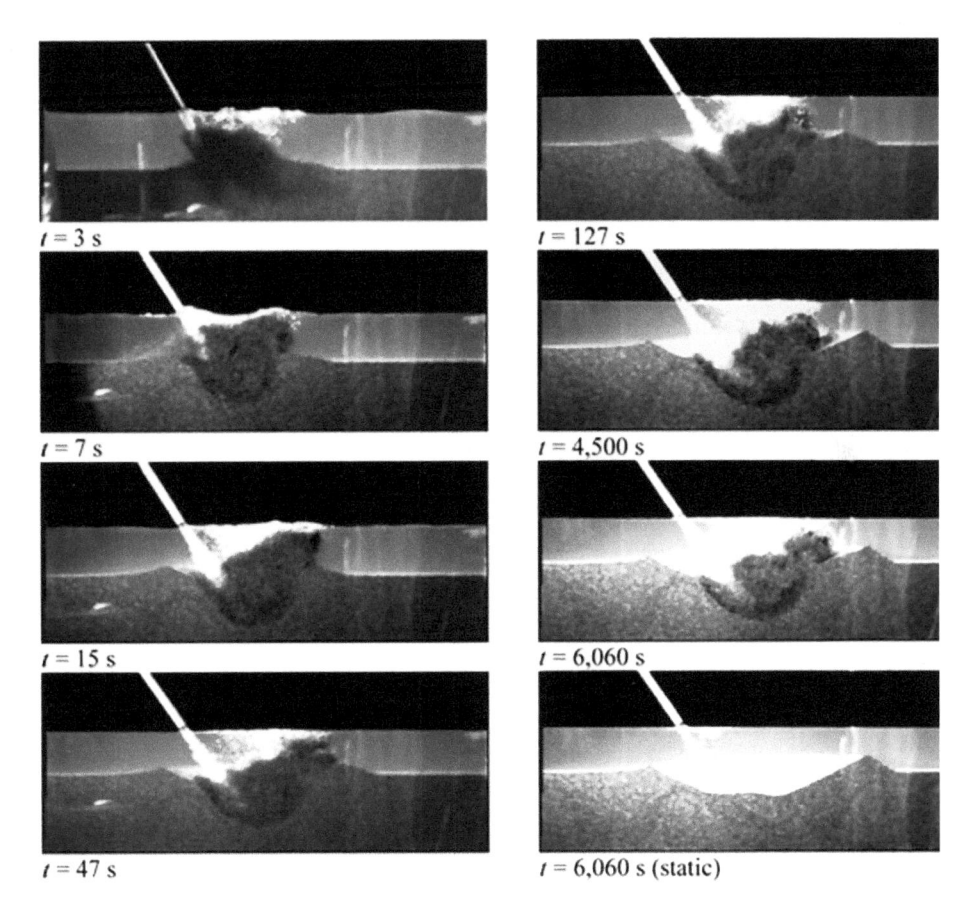

$t = 3$ s

$t = 127$ s

$t = 7$ s

$t = 4,500$ s

$t = 15$ s

$t = 6,060$ s

$t = 47$ s

$t = 6,060$ s (static)

Figure 6.68 Test with $F_{d90} = 27.8$, $\alpha = 60°$, U-jet, photo sequence from $t = 0$ to 6060 s (Pagliara et al., 2008b)

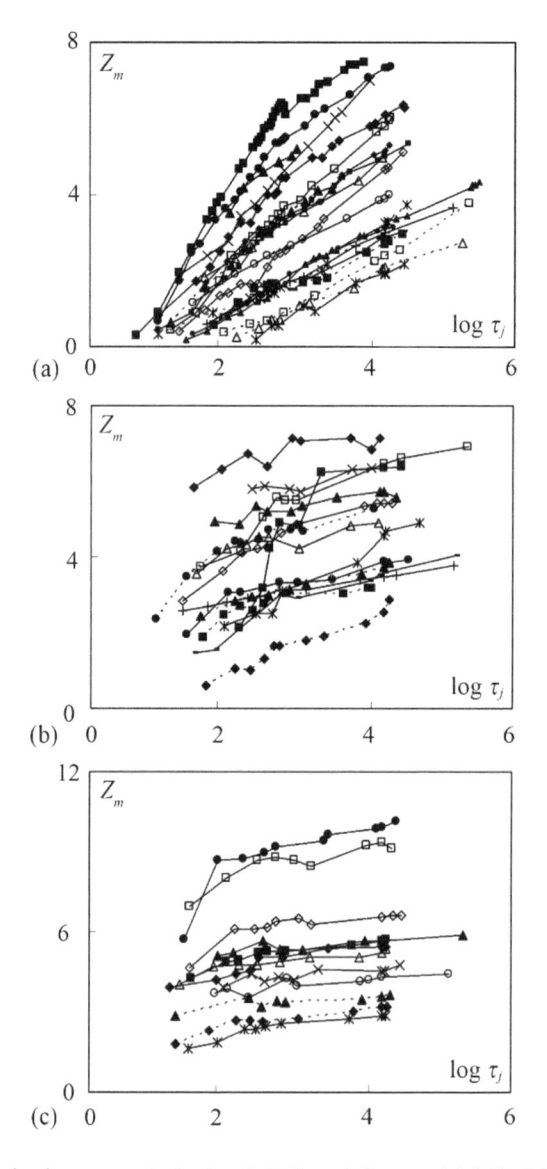

Figure 6.69 Dimensionless scour hole depth $Z_m(\log\tau_j)$ for $\alpha =$ (a) 30°, (b) 45°, (c) 60°. Light symbols U-jets, full symbols S-jets (Pagliara *et al.*, 2008b)

for $\alpha = 30°$, 45°, and 60°. The slopes of these curves for a specific α remain nearly constant after the developing time, for both S-jets and U-jets, so that $Z_m(\log\tau_j)$ is essentially straight. Each line is defined by the slope and an arbitrarily selected origin. The developed scour hole regime is well defined at $\tau_j = 15,000$ ($\log\tau_j = 4.18$), i.e. at $t \cong 0.5$ h for the sediment and jet diameters employed. Similar plots result for the dimensionless maximum ridge location $L_M = l_M/D(\log\tau_j)$ and height $Z_M = z_M/D(\log\tau_j)$. Note that $Z < 0$ for elevations above the original sediment bed (Figure 6.61).

Experimental results

The temporal effect of two-phase flows following Froude similitude depends on the densimetric particle Froude number $\mathbf{F}_d = V/V_R$ as ratio of the effective to the reference (subscript R) velocities. With the jet diameter D as reference length L_R and $V_R = (g'd)^{1/2}$ as reference velocity, the reference time is $t_R = L_R/V_R$. For pier and plunge pool scour, the pier and jet diameters are the reference lengths resulting in $t_R = D/(g'd)^{1/2}$ so that the relative time is $\tau_j = t/t_R = (g'd)^{1/2} \cdot t/D$, with $d = d_{90}$ (Pagliara *et al.*, 2006).

The reduced scour hole depth $\Delta Z_m = Z_m - Z_m(\tau_j = 15{,}000)$ for $\alpha = 30°$ versus $\log\tau_j$ with $Z_m(\tau_j = 15{,}000)$ was computed from Pagliara *et al.* (2008b). The data of the S-jets were noted to follow a slightly different trend than these of U-jets. In both cases, however, the transition (subscript *t*) from the developing to the developed scour phase occurs at $150 < \tau_j < 850$ ($2.2 < \log\tau_j < 3$). For $\tau_j > \tau_t$, the data follow $\Delta Z_m = -C_{1m} + C_{2m} \cdot \log\tau_j$, implying that $\Delta Z_m = C_{2m} \cdot \log(\tau_j/15{,}000)$. The parameter C_{2m} retains all end scour effects, whereas the temporal effect is reflected by the log-term. The data for $\alpha = 45°$ and $60°$ were treated similarly. The terms C_{1m} and C_{2m} were plotted versus α for both the S- and U-jets. The trend lines almost collapse and are described with $A_1 = 1$ for S-jets and $A_1 = 1.12$ for U-jets with α [deg.] during the developed scour phase ($\tau_j > \tau_t$) as

$$\Delta Z_m = 2{,}000 A_1 \alpha^{-2.2} \cdot \log(\tau_j / 15{,}000). \tag{6.102}$$

Accordingly, unsubmerged jets produce a slightly larger scour hole depth under otherwise identical flow conditions. To develop a final equation for $Z_m(\tau_j)$, the value $Z_m(\tau_j = 15{,}000)$ and the transition time τ_t must be known, to limit Eq. (6.102) against the lower end. The data indicate for $30° \leq \alpha \leq 60°$ (Pagliara *et al.*, 2008b)

$$\tau_t = A_1^2(-0.78\alpha^2 + 51.68\alpha - 150). \tag{6.103}$$

Parameter $Z_m(\tau_j = 15{,}000)$ is plotted for the developed scour phase $\tau_j > \tau_t$ in Figures 6.70 and 6.71 for $\alpha = 30°$ and $60°$, respectively. The origin ($\log\tau_j; Z_m$) = (0; 0) was linearly connected with $Z_m(\tau_j = \tau_t)$ to describe the developing scour phase $Z_{mt} = Z_m(\tau_t)$ as

$$Z_m / Z_{mt} = \log\tau_j / \log\tau_t, \quad 0 < \tau_j < \tau_t. \tag{6.104}$$

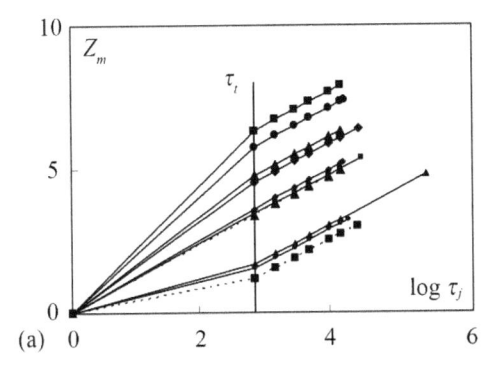 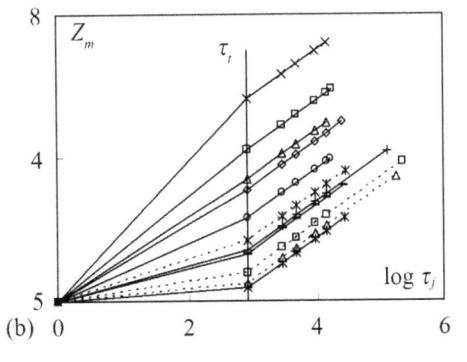

Figure 6.70 Maximum dimensionless scour hole depth $Z_m(\log\tau_j)$ for $\alpha = 30°$ and (a) S-jets, (b) U-jets (Pagliara *et al.*, 2008b)

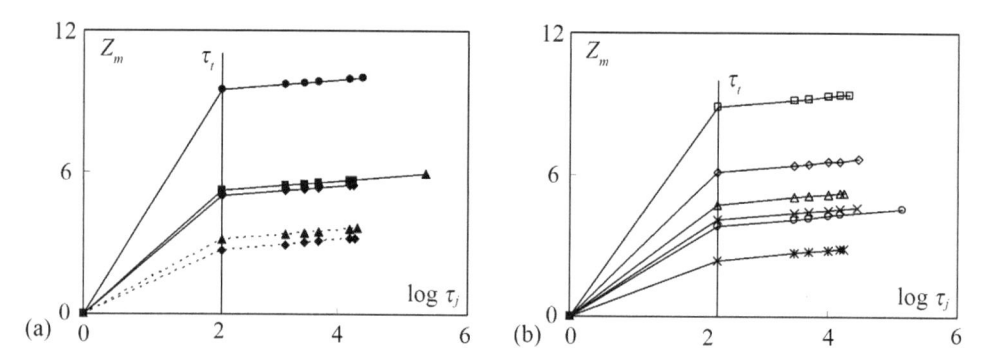

Figure 6.71 Effect of jet type on $Z_m(\log\tau_j)$ for $\alpha = 60°$ and (a) S-jets, (b) U-jets (Pagliara et al., 2008b)

The equation for the developed scour phase is with τ_t from Eq. (6.103) versus both α and the jet flow regime

$$Z_m(\tau_j) = 2,000 A_1 \alpha^{-2.2} \cdot \log(\tau_j / 15,000) + Z_m(\tau_j = 15,000), \quad \tau_j \geq \tau_t. \tag{6.105}$$

A comparison of the test data with Eq. (6.104) for the developing, and Eq. (6.105) for the developed scour phases indicates that the data scatter about the line of perfect fit is larger in Eq. (6.104).

Additional scour parameters

The above procedure was applied to the location of the maximum ridge height $L_M(\tau_j)$, its elevation $Z_M(\tau_j)$, the locations of maximum scour depth $L_m(\tau_j)$ and the scour hole intersection with the original bed level $L_a(\tau_j)$ (Figure 6.61), allowing for the description of the entire scour hole geometry versus time, and the governing base parameters.

The location of the maximum ridge height $L_M = l_M/D$ was treated as $Z_m(\tau_j)$ in Eqs. (6.104) and (6.105). With $A_1 = 1.00$ for S-jets and $A_1 = 1.12$ for U-jets as previously, the equation of the maximum ridge height location for $30° \leq \alpha \leq 60°$ reads with $\mu_M = 1.5$ for U-jets and $\mu_M = 1$ for S-jets during the developing scour phase (Pagliara et al., 2008b)

$$L_M / L_{Mt} = \log\tau_j / \log\tau_t, \quad 0 < \tau < \tau_t. \tag{6.106}$$

The corresponding equation for the developed scour phase reads with $L_{Mt} = L_M(\tau_t)$

$$L_M = 8,800 A_1 \alpha^{-2.2} \log(\tau_j / 15,000) + L_M(\tau_j = 15,000), \text{ for } \tau_j \geq \tau_t. \tag{6.107}$$

Equations (6.106) and (6.107) compare with the test data better than the data for Z_m.

Quantity $\Delta Z_M(\log\tau_j)$ of the maximum ridge elevation was investigated for U-jets, resulting with $A_1 = 1$ for S-jets and $A_1 = 1.12$ for U-jets, respectively, in (Pagliara et al., 2008b)

$$\Delta Z_M = -C_{2M}\log\tau_j + C_{1M} = -2,300(1/A_1)\alpha^{-2.2} \cdot \log(\tau_j / 15,000) . \tag{6.108}$$

Both jet types follow again the identical relation, except for the numerical constant A_1.

The relation $Z_M(\log\tau_j)$ during the developing scour phase was also assumed to increase linearly from $Z_M(\tau_j = 0)$ to $Z_M(\tau_t)$, with the transition time τ_t given in Eq. (6.103). The temporal advance of the maximum ridge height $Z_M(\tau_j)$ is then (Pagliara et al., 2008b)

$$Z_M / Z_{Mt} = \log\tau_j / \log\tau_t, \qquad 0 < \tau_j < \tau_t; \tag{6.109}$$

$$Z_M = 2,300(1/A_1)\alpha^{-2.2}\cdot\log(\tau_j/15,000) - Z_M(\tau_j = 15,000), \; \tau_j \geq \tau_t. \tag{6.110}$$

The locations of the dimensionless maximum scour depth $L_m = l_m/D$ and of the scour profile intersection with the original bed $L_a = l_a/D$ (Figure 6.61) were investigated for the three jet impact angles. In contrast to parameters Z_m, L_M, and Z_M previously addressed, the relation $\Delta L_a = L_a(\tau_j) - L_a(\tau_j = 15,000)$ versus $(\log\tau_j)$ involves only small differences between the S- and U-jets, so that only one relation was retained as

$$L_a / L_{at} = \log\tau_j / \log\tau_t, \qquad 0 < \tau_j < \tau_t; \tag{6.111}$$

$$L_a = 5,600\alpha^{-2.2}\cdot\log(\tau_j/15,000) + L_a(\tau_j = 15,000), \qquad \tau_j \geq \tau_t. \tag{6.112}$$

Following the previous procedure with $L_a(\tau_j = 15,000)$ results for $30° < \alpha < 60°$ in

$$L_{at} = 5,600\alpha^{-2.2}\cdot\log(\tau_t/15,000) + L_a(\tau_j = 15,000). \tag{6.113}$$

The test data indicate an excellent agreement with Eq. (6.113) for the developed and reasonable agreement with Eq. (6.112) for the developing scour phase.

The location of maximum scour hole depth L_m versus $(\log\tau_j)$ for the two scour phases follows (Pagliara et al., 2008b)

$$L_m / L_{mt} = \log\tau_j / \log\tau_t, \qquad 0 < \tau < \tau_t; \tag{6.114}$$

$$L_m = (1.39 - 0.017\alpha)[\log\tau_j - 4] + L_m(\tau_j = 15,000), \qquad \tau_j > \tau_t. \tag{6.115}$$

The origin of the scour hole profile $L_t(\tau_j = 15,000) = l_t(\tau_j = 15,000)/D$ as distance to the jet impact point P (Figure 6.61) was previously determined by Pagliara et al. (2008a).

In summary, plunge pool scour is relevant both in hydraulic engineering as also from conceptual considerations. This research deals with the temporal advance of plunge pool scour for incohesive sediment subjected with a circular jet under impact angles from 30° to 60°. Based on previous work, in which the effects of the basic hydraulic parameters were studied for end scour, and a time scale derived from Froude similitude, the time history of the major quantities describing the scour hole and the ridge geometry was studied. These quantities include the maximum scour hole depth and the ridge height, plus their locations relative to the scour hole origin.

A distinction is made between submerged and unsubmerged jets; the latter result in slightly larger values, based on coefficient A_1. The effect of the jet impact angle is almost inversely quadratic, i.e. smaller angles produce a deeper and longer scour hole as do larger jet impact angles. The data analysis leads to a distinction between the developing and the developed scour phases, separated at transition time τ_t. The quantities investigated advance during both phases logarithmically with time. Because all relations are derived from the test data, they are limited to the test conditions.

6.5 Rock scour

6.5.1 Introduction and challenges

Scour of rock is of concern when assessing the effect of overtopping jets plunging onto rock foundations downstream of dams or designing plunge pools. The main questions arising in view of safety against scour formation downstream of spillways and bottom outlets are discussed. First, the physical processes involved in the scour formation are analyzed based on the state-of-the-art of knowledge and understanding. Various scour evaluation methods from simple empirical formulas to complex scour models are then discussed. The approach developed at Ecole Polytechnique Fédérale de Lausanne (EPFL) is then highlighted. The so-called comprehensive scour method (CSM) is based on physics and entails the analysis of fluctuating turbulent pressures and their interaction with rock formations. The potential and extent of rock scour is determined by employing the principles of fracture mechanics and Newton's second law. Continuing research into various aspects of turbulent flow and its interaction with rock formations refine this procedure.

The main difficulties encountered when estimating scour depths are discussed, including the choice of the appropriate theory, interpretation of hydraulic model tests and prototype observations, the scour rate and the prevailing discharge. The selection of the flood return period is discussed, for which the scour formation and the control measures have to be evaluated, as also options of measures for scour control.

In today's spillway design of dams there is a tendency of increasing the unit discharge of high-velocity jets leaving the appurtenant structures. For gated chute flip bucket (ski jump) spillways, unit discharges between 200 and 300 $m^3/s/m$ are not rare anymore, since the cavitation risk is mitigated by chute bottom aerators. Uncontrolled crest spillways for arch dams are currently designed for specific discharges from 70 m^2/s to 120 m^2/s by installing crest gates. With the latest high-pressure gate technology, low-level orifice spillways evacuate unit discharges of up to 400 m^2/s. Therefore, it is challenging for dam designers to answer the questions (Schleiss, 2002):

- What will be the evolution and extent of scour at jet impact zone?
- Are the stability of valley slopes and the foundation at the downstream dam end endangered?
- Is a tail pond dam required to create a water cushion, and how does it affect the scour depth?
- Is a pre-excavation of the rocky riverbed required and/or is plunge pool lining necessary?
- Is the powerhouse operation influenced by scour formation?

Rock scour is a complex problem studied extensively but often in a simplified way by using granular material. As illustrated in Figure 6.72, rock scour is described by a series of physical processes involving (Bollaert, 2002):

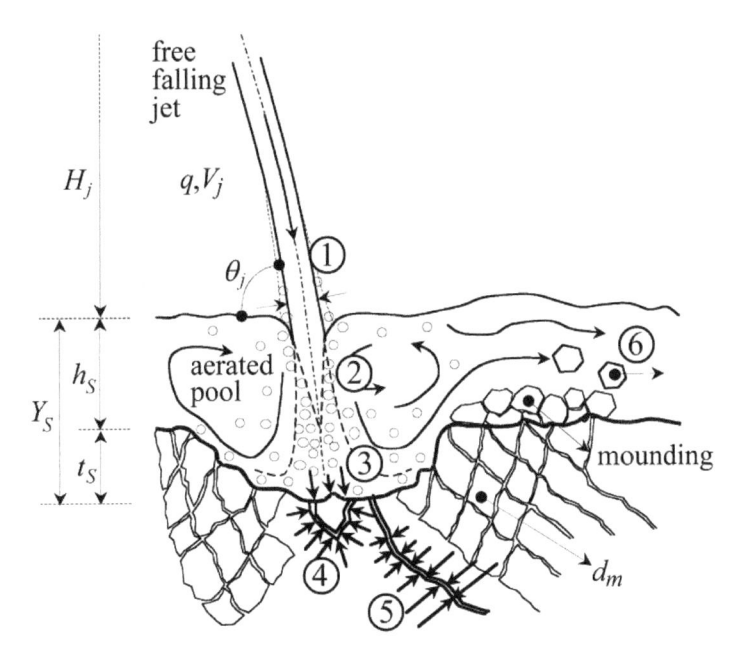

Figure 6.72 Main parameters and physical-mechanical processes involved in scour phenomena with ① plunging jet impact, ② diffusive shear layer, ③ bottom pressure fluctuations, ④ hydrodynamic fracturing, ⑤ hydrodynamic uplift, ⑥ downstream transport (Bollaert and Schleiss, 2005)

① Free-falling jet behavior in air and aerated jet impingement
② Plunging jet behavior and turbulent flow in plunge pool
③ Pressure fluctuations at water-rock interface
④ Propagation of dynamic water pressures into rock joints
⑤ Hydrodynamic fracturing of closed ended rock joints and rock splitting
⑥ Ejection of rock blocks by dynamic uplift into plunge pool, including removal of fractured rock block elements
⑦ Break-up of rock blocks by ball milling effect of turbulent flow in plunge pool
⑧ Formation of downstream mound and displacement of scoured material by sediment transport.

The existing scour evaluation methods are grouped as (Bollaert and Schleiss, 2005):

- Empirical approaches based on laboratory and field observations
- Analytical-empirical methods combining laboratory and field observations with physics
- Approaches based on extreme fluctuating pressures at plunge pool bottom
- Techniques based on temporal-mean and instantaneous pressure differences thereby accounting for rock characteristics
- Scour model based on fully transient water pressures in rock joints.

The most common methods used for the (temporal) scour evaluation due to falling, high-speed jets are shown in Figure 6.73 along with the considered physical parameters related to the three phases water, rock, and air involved in the scour process.

Besides empirical and semi-empirical formulae of limited applicability to practice, the recent research on rock scour resulted in pragmatic methods to quantify its occurrence, extent, and rate (Bollaert, 2002; Annandale, 2006). Annandale (1995, 2006) developed a threshold relation defining incipient conditions for rock scour by applying a geo-mechanical index, known as the Erodibility Index, to quantify the relative ability of rock to resist the erosive capacity of water. He quantifies stream power to estimate the relative magnitude of the erosive capacity of water jets. The threshold defining incipient conditions for rock scour relates the Erodibility Index to stream power. To predict the potential for and the maximum extent of rock scour, use is made of the Erodibility Index Method. Large-scale model test are often used which allow for a more appropriate assessment of scour in rock. An example is the spillway of the Three Gorges Dam, where experimental investigations of fluctuation uplift on rock blocks at the bottom of the scour pool have been conducted (Liu *et al.*, 1998; Liu, 1999).

Bollaert (2002) systematically investigated rock scour processes by employing a comprehensive physics-based approach. Using both fracture mechanics and basic Newtonian force relations, he developed techniques that apply to directly quantify the erosive capacity of water and use this information to determine whether rock is scoured by brittle fracture, in fatigue, or by means of block removal. Using these methods, both the extent and rate of rock scour are quantified. Systematic experimental investigations of plunge pools with different lateral

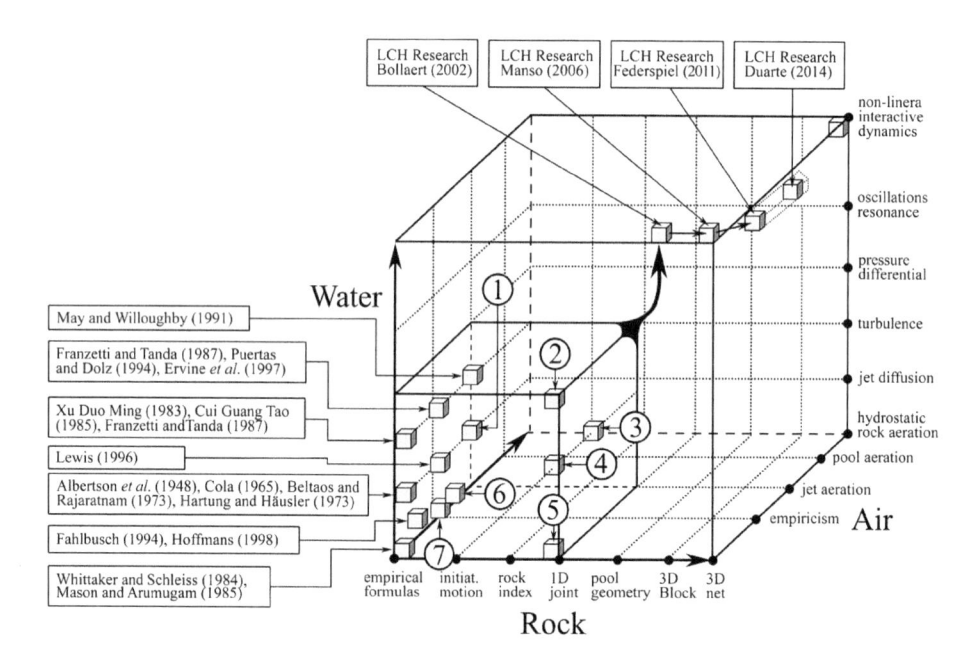

Figure 6.73 Graphical representation of existing scour evaluation methods involving parameters rock, water, air, and time with ① McKoegh & Elsawy (1980), Bin (1984), Ervine (1998), Borer *et al.* (1998), ② Fiorotto & Rinaldo (1992), Liu *et al.* (1998), Fiorotto & Saladin (2000), ③ Annandale (1995), Annandale *et al.* (1998), ④ Spurr (1985), ⑤ Yuditskii (1963), Gunko *et al.* (1965), Reinius (1986), Otto (1989), ⑥ Poreh & Hafez (1967), Stein *et al.* (1993), Chee & Yuen (1985), Bormann & Julien (1991), ⑦ Mason (1989) (Completed by Duarte 2014, from Bollaert and Schleiss, 2005)

confinements indicated that the pool geometry influences not only the plunging jet diffusion and the air entrainment in the pool, but also the impact pressures at the water-rock interface and inside the fissured rock mass (Manso, 2006; Manso *et al.*, 2009). Favorable plunge pool geometries may limit the pressure fluctuations at the water-rock interface so that the scour potential is reduced (Bollaert *et al.*, 2012; Federspiel *et al.*, 2011). Recent research investigated the effect of jet aeration at issuance (Duarte *et al.*, 2012; Duarte, 2014; Duarte *et al.*, 2015, 2016a, b).

6.5.2 Comprehensive scour method

Motivation

The impact of high-velocity plunging jets produces strong dynamic pressures at the pool bottom, which are transferred into rock joints by transient flow governed by the propagation of pressure waves. The latter is strongly influenced by the air content in the plunge pool and the underlying rock fissures. For closed-end rock joints, as encountered in a partially jointed rock mass, the reflection and superposition of pressure waves including resonance phenomena generate a hydrodynamic loading at the joint tip. If the corresponding stresses there exceed the fracture toughness and the initial compressive stresses of the rock, it will crack and the joint grows further. For open-end rock joints in a fully jointed rock mass, the pressure waves inside the joints create a significant dynamic uplift force on the rock blocks. This force breaks up the remaining rock bridges in the joints by fatigue and, if high enough, ejects the so formed rock blocks from the rock mass into the macro-turbulent plunge pool flow. To reproduce and understand these complex interactions between high-turbulent flow in plunge pools and the underlying rock, an experimental setup was designed, allowing for the observation of the relevant phenomena in near-prototype scale.

Experimental setup

A 300 mm diameter water supply conduit with a cylindrical jet outlet models the jet (Figure 6.74). Due to constructive limitations, the supply conduit has a 90° bend just upstream

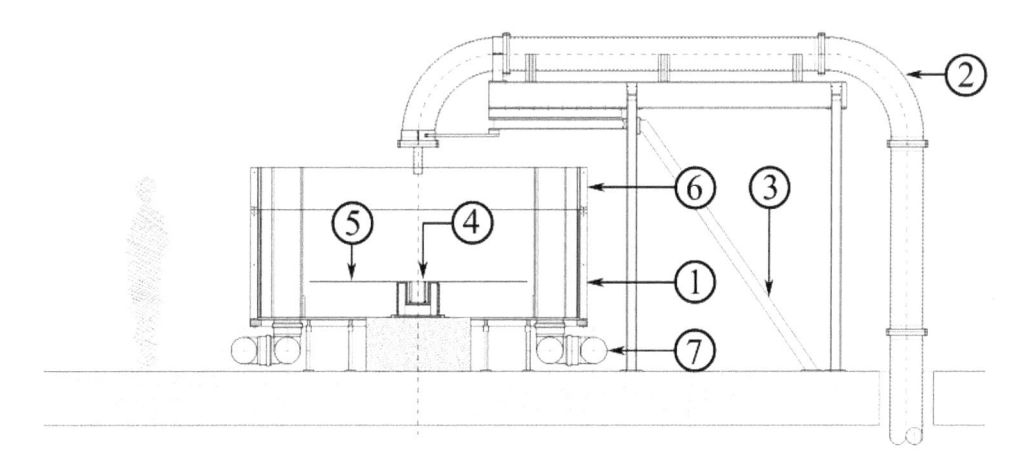

Figure 6.74 General view of experimental facility (transverse section) with ① plunge pool, ② water supply, ③ support of water supply, ④ new plunge pool bottom (measurement box and highly instrumented block), ⑤ new level of plunge pool bottom, ⑥ new plunge pool maximum water level, ⑦ water restitution (Federspiel *et al.*, 2011) (see also Figure 6.76)

of the jet outlet system of diameter 72 mm. A 3 m diameter and 1.4 m high cylindrical basin in steel-reinforced Lucite simulates the plunge pool. Its bottom is made of rigid steel frames, covered by a 10 mm opaque Lucite plate. Inside the basin, two rectangular boxes made of Lucite adjust the water level by a flat plate. The water flowing over these plates runs downstream into four restitution conduits.

A restitution system consisting of four 220 mm diameter conduits simulates the downstream conditions. These conduits are connected to overflow boxes discharging the water into the main laboratory reservoir. The 63 m head pump is located in the main reservoir, from where the water is pumped into the supply conduit. After restitution, the water returns to the main reservoir resulting in a closed water circulation system. The maximum discharge is 250 l/s, resulting in maximum jet outlet velocities of 30 m/s, similar to prototype values.

The measurement box into which the highly instrumented Intelligent Block (IB) is inserted, is a structure composed of steel plates (Figure 6.75) 402 mm in length, 402 mm in width, and 340 mm in height. The thickness of the steel plates is 20 mm. A large series of interconnected cavities inside this box allows for inserting pressure and displacement transducers. To allow for manipulation inside these cavities (i.e. modify transducers' position), all lateral walls have a 250 mm movable lid. The walls near the IB are pre-perforated to change the transducer positions. This allows for measuring water pressure generated by the jet inside the

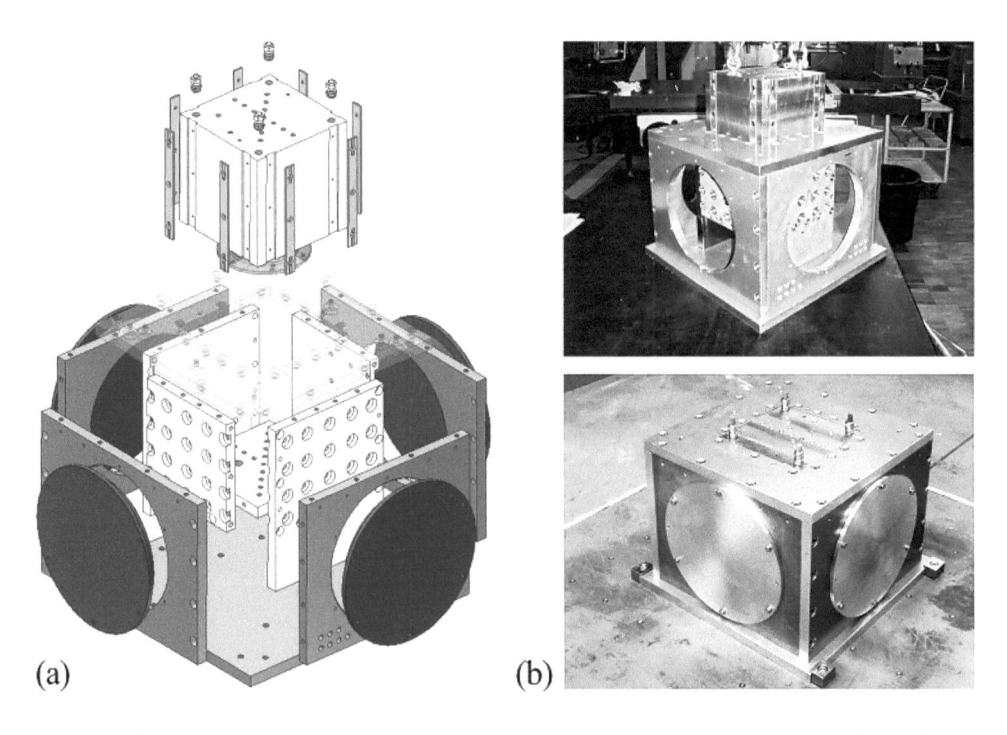

(a)   (b)

Figure 6.75 (a) Axonometric view of measurement box (surrounding rock mass) and highly instrumented Intelligent Block, (b) inserting block in measurement box (top) and closed measurement box after sensor installation (bottom) (Federspiel *et al.*, 2011)

Figure 6.76 (a) Experimental facility with measurement box and Intelligent Block. Note absence of final plunge pool bottom (Figure 6.74 ⑤), (b) jet impact for 0.1 m water level in plunge pool (Federspiel *et al.*, 2011)

joints between the IB and the surrounding box, and to measure its corresponding displacement. The box is impermeable to protect the electrical equipment.

In the measurement box center, a large cavity allows for inserting the IB. This cavity is 202 mm long, 202 mm wide and 201 mm high. The IB has cubic shape of 200 mm side length. The thickness of the steel plates was optimized to have a density similar to prototype rock of 2400–2500 kg/m^3. On top of the IB, holes were pre-perforated to fix the pressure transducers. In total, 95 different positions allow for pressure transducer insertion. For the acceleration transducer and the displacement transducer, however, the positions are fixed. In the space between the measurement box and the IB, a 3D fissure of 1 mm width is so created. Pressure and vibration transducers inserted in the IB measure the pressure at the pool bottom under high-speed jet impact and block vibration. Along the IB outside walls, eight vertically oriented guide plates limit the degree of block freedom to vertical movements. Both the IB and the measurement box have been placed inside the existing 3 m diameter cylindrical basin simulating the plunge pool (Figures 6.75, 6.76).

Features of new rock scour model

Using this test setup, a new, completely physically based engineering model was developed to predict the ultimate scour depth of jointed rock (Bollaert, 2002; Bollaert *et al.*, 2003 a, b; Bollaert *et al.*, 2005; Manso *et al.*, 2009; Federspiel *et al.*, 2011; Duarte *et al.*, 2013). The initial scour model of Bollaert (2002) for 2D joints and rigid rock blocks incorporates two sub-models expressing two failure criteria of the rock mass. The Comprehensive Fracture Mechanics (CFM) model determines the ultimate scour depth by expressing the instantaneous or time-dependent crack propagation due to hydrodynamic pressures. The Dynamic Impulsion (DI) model describes the ejection of rock blocks from their mass. Each sub-model consists of a physical approach to describe the destruction of rock mass (Figure 6.72, ④ and ⑤).

The most appropriate depends on the geo-mechanical rock features, mainly on its break-up degree.

The scour model includes three modules: falling jet, plunge pool, and rock mass. The latter contains the two mentioned sub-models describing the rock failure criteria.

Falling jet module

This module describes the main geometrical and hydraulic characteristics of a falling jet issued from a dam outlet structure. The determining jet parameters are the mean jet velocity V_i and diameter D_i or width B_i, as well as its turbulence intensity T_i defined as the rms value of the velocity fluctuations divided by the mean velocity of the jet

$$\mathsf{T} = u'/U . \tag{6.116}$$

The jet trajectory is assessed by the ballistic theory, Eq. (6.4). Experience indicates, however, that prototype high-velocity jets are influenced by air friction resulting typically in jet trajectories reaching only 85–90% of the theoretical length. Knowing the jet trajectory starting at its issuance (subscript i), the jet impact zone (subscript j) is defined as well as the jet contraction (diameter D_j at impact) due to gravity forces as

$$D_j / D_i = \left(V_i / V_j \right)^{1/2} . \tag{6.117}$$

The gravity-influenced jet geometry affects the mean impact jet velocity V_j depending on the jet velocity V_i at issuance and the free fall height H_j as (Figure 6.77)

$$V_j = \left(V_i^2 + 2gH_j \right)^{1/2} . \tag{6.118}$$

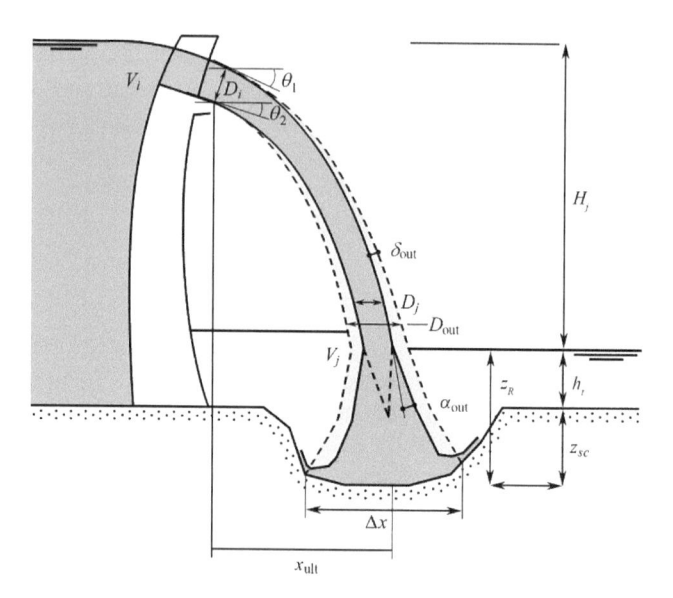

Figure 6.77 Definition sketch of free overfall jet parameters

Rectangular jets issued from gated orifices are deformed in the air becoming either elliptical or circular. To be on the safe side, circular jets at the impact location should be assumed, given their highest scour potential.

The turbulence intensity T determines the lateral diffusion angle δ_{out} of a plunging jet; it depends on distance x along the jet axis, starting at the dam outlet as (Ervine and Falvey, 1997)

$$\delta_{out} / x = 0.38T . \tag{6.119}$$

Typically, the lateral diffusion angle of turbulent jets reaches values in the range of 3% to 4%. For prototype jets, the turbulence intensity is in the range of 4% to 5%, but may reach 7% to 8% for jets issued from high-pressure orifices with high diffusion. Ervine and Falvey (1997) describe this jet diffusion versus T and the jet Froude number. Accounting for the lateral jet diffusion and the initial jet diameter D_i at the outlet, the jet diameter D_{out} at the impact zone located by distance L_j away from jet issuance is

$$D_{out} = D_i + 2\delta_{out}L_j . \tag{6.120}$$

The effect of the impact jet angle is small for large values of 70° to 90°. For smaller angles, these parameters apply as for a vertical impact, but the underwater travel jet distance Y_j has to be determined along the inclined jet direction (Figure 6.77).

Plunge pool module

This module describes the main hydraulic and geometrical characteristics of the jet plunging into the scour hole or plunge pool, and derives the hydrodynamic pressures at the rock surface due to jet impact. The knowledge of the scour hole water depth Y_S is essential; it is defined as the distance from the tailwater surface to the scour hole bottom. It increases during the scour process. Its temporal evolution is determined by the scour depth h_S already formed and the mound thickness t_S of material deposited downstream of the scour hole, i.e. $Y_S(t) = h_S + t_S$.

Jet diffusion in the scour hole depends on the parameter Y_S/D_j. The mean dynamic pressure coefficient C_p is defined with p_{mean} as mean dynamic pressure head as

$$C_p = (p_{mean} - Y_S) / \left(V_j^2 / 2g\right). \tag{6.121}$$

Let the mean impact jet velocity be V_j, and the impact air concentration α_j. The time-averaged pressure coefficient C_p varies with Y_S/D_j as (Ervine *et al.*, 1997; Kraatz, 1965)

$$C_p = 38.4\left(1-\alpha_j\right)^{1.345} \cdot \left(\frac{D_j}{Y_S}\right)^2, \quad \text{for } Y_S / D_j \geq 4 \text{ to } 6; \tag{6.122}$$

$$C_p = 0.85, \quad \text{for } Y_S / D_j < 4-6. \tag{6.123}$$

The mean air concentration C_a is defined as the volumetric air-to-water ratio β_A as

$$C_a = \frac{\beta_A}{1+\beta_A} . \tag{6.124}$$

For prototype jets, β_A ranges between 2 and 3, corresponding to C_a = 65% to 75%.

The dynamic pressure fluctuations are characterized by the pressure coefficient C'_p, defined with σ_p as the RMS-value of the pressure fluctuation in [m] as

$$C'_p = \sigma_p / \left(V_j^2 / 2g\right). \tag{6.125}$$

From experiments, C'_p varies with $Y_J = Y_S/D_J$ and a_J defining the jet turbulence intensity as (Bollaert, 2002)

$$C'_p = a_J + 0.0716Y_J - 0.0079Y_J^2 + 0.000215Y_J^3. \tag{6.126}$$

For T = 3% to 5%, a_J = 0.10, whereas a_J = 0.15 for higher T. Equation (6.126) is valid for $Y_J < 20$. For $Y_J \geq 20$, Eq. (6.126) should be used by setting Y_J = 20.

The mean and fluctuating pressures due to jet impact onto a rock surface create pressure waves traveling inside the rock fissures, which follow

$$\left.\begin{array}{lll} \Gamma^+ = 4 + 2Y_J & \text{for } Y_J < 8 \\ \Gamma^+ = 20 & \text{for } 8 \leq Y_J \leq 10 \\ \Gamma^+ = 40 - 2Y_J & \text{for } 10 < Y_J \end{array}\right\} \text{minimum value;} \tag{6.127}$$

$$\left.\begin{array}{lll} \Gamma^+ = -8 + 2Y_J & \text{for } Y_J < 8 \\ \Gamma^+ = 8 & \text{for } 8 \leq Y_J \leq 10 \\ \Gamma^+ = 28 - 2Y_J & \text{for } 10 < Y_J \end{array}\right\} \text{maximum value.} \tag{6.128}$$

The amplification factor Γ^+ between the maximum pressures inside the joints and the rms pressures at the water-rock interface according to Eqs. (6.127) and (6.128) applies only for fractured rock with wide fissures of about 1 mm.

Rock mass module

Two sub-models are used to express the failure criteria of the rock mass, namely the Comprehensive Fracture Mechanics (CFM) model determining the ultimate scour depth by expressing the instantaneous or time-dependent crack propagation due to hydrodynamic pressures, and the Dynamic Impulsion (DI) model describing the ejection of rock blocks from their mass.

COMPREHENSIVE FRACTURE MECHANICS (CFM) MODEL

This module compares the maximum pressure peak inside the rock fissures with its resistance to cracking or fatigue. The maximum pressure peak results from (Bollaert, 2002, 2004)

$$P_{max}[\text{Pa}] = \gamma_W C_{pmax}\left(V_j^2 / 2g\right) = \gamma_W (C_p + \Gamma^+ C'_P)\left(V_j^2 / 2g\right). \tag{6.129}$$

Pressure coefficients C_p and C'_p are defined in Eqs. (6.121) and (6.125), respectively, and γ_W is the specific weight of water in [N/m³]. The pressure peaks inside fissures have a cyclic character. Their amplitude Δp_c and frequency f_c are important in view of brittle or fatigue crack propagation in the rock mass. The amplitude is assumed equal to the maximum pressure peak. The frequency f_c depends on the fissure length L_f as

$$f_c = c_j / (4L_f) \text{ for closed end fissures;} \tag{6.130a}$$

$$f_c = c_j / (2L_f) \text{ for open end fissures.} \tag{6.130b}$$

Here c_j is the pressure wave celerity in water, strongly affected by jet air concentration. Low values of $c_j = 100$ to 200 m/s apply under prototype conditions.

According to the fracture potential, the occurrence of rock fractures is assessed by a stress intensity factor K_I as

$$K_I = P_{max} F_p (\pi L_f)^{1/2}. \tag{6.131}$$

Factor F_p represents a correction depending on the boundary conditions of the initial fissure geometry (Figure 6.78a). Figure 6.78b shows the variation of F_p versus the initial fissure geometry for a semi-elliptical joint, and a single-edged joint versus its persistence EL. Elliptical joints are present in fairly fractured rock masses, whereas single-edged joints rather in strongly fractured rock masses.

The local stresses at the fracture ends K_I describe the fracture toughness of rock masses K_{Ic} depending on its Ultimate Compression Strength UCS [MPa], or Tensile Strength TS [MPa], involving σ_c as the initial confining stress in the rock mass as

$$K_{I,TS} = (0.105 \text{ to } 0.132)TS + 0.054\sigma_c + 0.5276; \tag{6.132a}$$

$$K_{I,UCS} = (0.008 \text{ to } 0.010)UCS + 0.054\sigma_c + 0.42. \tag{6.132b}$$

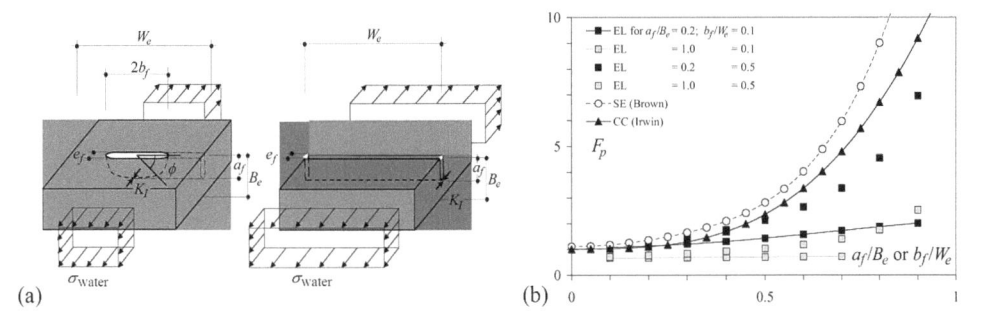

Figure 6.78 (a) Geometrical definitions of semi-elliptical and single-edged joints, (b) correction factor F_p and its relation to initial fissure geometry (Bollaert, 2002, 2004). Water pressures are applied from outside the joints having a length of $2b_f$, a depth of a_f and a width of e_f, $EF = b/W_e$ is the persistence of a joint in an element of length W_e

Brittle fracture and the resulting explosive crack propagation occurs if $K_I > K_{I,TS}$ or $K_{I,UCS}$. Crack propagation by fatigue is time-dependent and expressed with N as the numbers of peak pressure cycles and ΔK_I as its amplitudes as (Bollaert, 2002, 2004)

$$\frac{\mathrm{d}L_f}{\mathrm{d}N} = C_r \left(\Delta K_I / K_{Ic}\right)^{m_r}. \tag{6.133}$$

Further, C_r and m_r are rock parameters obtained by experiments (Table 6.1).

DYNAMIC IMPULSION (DI) MODEL

Once the cracks are completely formed in the rock mass due to jet impact, rock blocks are ejected by dynamic uplift pressures into the scour hole. Block ejection is caused by the impulse $I_{\Delta p}$ resulting from dynamic uplift pressure during the pulse time Δt_{pulse} of the acting net forces with m_b as the mass of rock as

$$I_{\Delta p} = \int_0^{\Delta t_{\text{pulse}}} \left(F_u - F_o - G_b - F_{sh}\right)\mathrm{d}t = m_b V_{up}. \tag{6.134}$$

Here F_u and F_o are the forces acting on the up- and downward directions, G_b is the immerged block weight, and F_{sh} the shear forces along the joint. The impulse $I_{\Delta p}$ acting on a rock block of mass m_b will result in an upward block movement with a displacement velocity of V_{up}. The forces over the block result from the dynamic pressures at the scour hole bottom and the block size. The uplift forces underneath result from pressure waves traveling inside the rock fissures. These are defined by determining the net impulsion $I_{\max}$ by multiplying the maximum net force with its persistence. The equivalent net pressure is non-dimensionalized by dividing it by the incoming kinetic velocity head $V_j^2 / 2g$. The non-dimensional corresponding duration results from division with the characteristic period T_p for pressure waves inside rock joints, i.e. $T_p = 2L/c_j$ for an open end joint, with L_f as the fissure length. Coefficient C_I of the net impulsion is given by $C_{up} \cdot T_{up} = V_j^2 L_f / \left(gc_j\right)$. The experimentally determined net impulsion coefficient C_I for an open-end rock joint is

$$C_I = 1.22 - 0.119Y_j + 0.0035Y_j^2. \tag{6.135}$$

Table 6.1 Fatigue exponent m_r and coefficient C_r for different types of rock (Bollaert, 2002)

Type of rock	Exponent m_r	Coefficient C_r
Arkansas Novaculite	8.5	1.0×10^{-8}
Mojave Quartzite	10.2–12.9	3.0×10^{-10}
Tennessee Sandstone	4.8	4.0×10^{-7}
Solenhofen Limestone	8.8–9.5	1.1×10^{-8}
Falerans Micrite	8.8	1.1×10^{-8}
Tennessee Marble	3.1	2.0×10^{-6}
Westerley Granite	11.8–11.9	8.0×10^{-10}
Yugawara Andesite	8.8	1.1×10^{-8}
Black Gabbro	9.9–12.2	$4.0 \times 10^{-9} - 5.0 \times 10^{-10}$
Ralston Basalt	8.2	1.8×10^{-8}
Whin Sill Dolerite	9.9	4.0×10^{-9}

Block ejection is determined by solving Eq. (6.134) for the displacement velocity V_{up} and transferring this ejection velocity to a vertical displacement h_{up} using

$$V_{up} = \left(2gh_{up}\right)^{1/2}. \tag{6.136}$$

The understanding of the physical processes was further enhanced by considering the plunge pool geometry and the lateral confinement of the plunging jet (Manso, 2006), as well as the dynamic response of a rock block due to the fluid-structure interaction (Federspiel *et al.*, 2011; Asadollahi *et al.*, 2011; Duarte, 2014).

The Comprehensive Scour Model (CSM) allows for predicting the scour depth in the jet impact region characterized by the turbulent shear layer, where normally the deepest scour is expected. At the stagnation point, the jet is deflected along the pool bottom partially in the up- and downstream directions, depending on the jet impact angle θ_j. The water flowing along the pool bottom in this wall jet region creates quasi-steady uplift forces over and under rock blocks, depending on their orientation and protrusion (Reinius, 1986). Rock blocks may be ejected by these uplift forces or peeled off by high-velocity wall jets. This process determines the 3D evolution and geometry of rock scour. Bollaert (2012) enhanced CSM with a so-called quasi-steady impulsion model, to predict the scour evolution along the wall jet region.

6.5.3 CSM with active jet air entrainment

Plunge pool module

The plunge pool module represents jet diffusion throughout the pool depth. This process dissipates a fraction of the jet energy. The jet entrains large air quantities into the water pool at the plunging section, thereby strongly influencing the diffusion properties. The jet aeration or air-to-water ratio is defined as $\beta_A = Q_A/Q_W$, with Q_A and Q_W as the air and water discharges, respectively. The proposal of Ervine and Falvey (1997) is considered for practical purposes

$$\beta_A = K_1 \left(1 - \frac{V_e}{V_i}\right) \left(\frac{L_j}{D_i}\right)^{1/2}. \tag{6.137}$$

Here K_1 is a parameter ranging from 0.20 for smooth turbulent jets to 0.40 for rough jets, L_j is the distance from jet issuance to jet impact, and V_e is the jet onset velocity at the plunging section initiating air entrainment, normally taken as ~1 m/s. The mean density ρ_{AW} of the air-water jet inside the pool is with ρ_A and ρ_W as air and water densities, respectively,

$$\rho_{AW} = \frac{1}{1+\beta_A}\rho_W + \frac{\beta_A}{1+\beta_A}\rho_A. \tag{6.138}$$

The energy input to the process is determined by the kinetic energy per unit volume of the air-water jet at the plunge section as

$$E_k = \frac{1}{2}\rho_{AW}V_i^2. \tag{6.139}$$

After plunging into the pool with aeration β_A, mean density ρ_{AW}, and kinetic energy E_k, the dissipation process of the jet starts. The inner jet core progressively disintegrates from its borders toward the jet axis, where the flow remains approximately at identical velocity as at the plunge section. The jet core reduces with y_{core} as the length of core decay and v as kinematic fluid viscosity according to Duarte (2014) as

$$\frac{y_{core}}{D_i} = 7.74 \times 10^{-6} \frac{V_i D_i}{v}, \quad \text{if } 7.74 \times 10^{-6} \frac{V_i D_i}{v} \leq A';$$

$$\frac{y_{core}}{D_i} = A', \quad \quad \text{if } 7.74 \times 10^{-6} \frac{V_i D_i}{v} \geq A'. \tag{6.140}$$

For submerged jets $A' = 3.5$, whereas it is 7.8 for plunging jets (Duarte, 2014; Duarte *et al.*, 2016a, b, c). The term $V_i D_i/v$ corresponds to the jet Reynolds number at the plunge section. Once the jet core is disintegrated, the jet velocity decays linearly across the pool depth for both submerged and plunging jets.

The remaining kinetic jet energy is converted into dynamic pressures acting on the plunge pool bottom. The time-averaged pressures p_{mean} are highest at the intersection of the jet axis with the water-rock interface, referred to as stagnation. The time-averaged pressure coefficient is $C_p = (p_{mean} - \rho_w g Y_j)/E_k$. For non-aerated jets at stagnation, C_p is reproduced by (Duarte, 2014; Duarte *et al.*, 2016a, b, c)

$$C_p = \psi \left(0.926 - 0.0779 \frac{Y_j - y_c}{D_i} \right)^2, \quad \text{if } Y_j > y_{core}; \tag{6.141}$$

$$\psi^{-1} = 1 + \exp\left[-5.37 \times 10^{-6} \left(\frac{V_i D_i}{v} - 6.63 \times 10^5 \right) \right]. \tag{6.142}$$

If $Y < y_{core}$, the jet core impacts directly onto the rock bottom so that $C_p = 0.86$. The parameter ψ describes the energy loss due to the impact region formed near the intersection of the jet centerline with the pool bottom (Beltaos and Rajaratnam, 1973). Parameter ψ is a logistic function of the jet velocity, reaching asymptotically the value 1 for high jet velocities.

Aerated jets have a lower momentum as compared with otherwise identical clear-water jets, due to a reduced mean density of the air-water mixture (Ervine and Falvey, 1987; Manso *et al.*, 2004). Air bubbles reduce the shear stresses with the surrounding water in the pool, resulting in lower velocity decay rates and higher C_p values for the aerated jets at the bottom (Duarte, 2014). The effect of jet aeration β_A on C_p was assessed by Duarte (2014) for tests with submerged jets, as the full air entrainment discharge was provided at the nozzle. A linear increase was observed with the C_p^A value taking into account jet aeration at issuance as

$$\frac{C_p^A}{C_p} = 1 + 0.4\beta_A. \tag{6.143}$$

Dynamic Impulsion Method

The Dynamic Impulsion Method has the objective to assess the scour potential of plunging jets by means of their capacity to remove mobilized rock blocks from the pool bottom

(Bollaert, 2002; Bollaert and Schleiss, 2005). In contrast to CFM, the DI method does not consider the temporal evolution of the scour hole, but computes the equilibrium or ultimate scour depth. It is based on the non-dimensional maximum dynamic impulsion coefficient C_I^{max}, defined as the non-dimensional uplift force acting on the rock block during a given time period. The impulse $I_{\Delta p}$ due to dynamic uplift pressures is defined as time integration of the forces applied on the block

$$I_{\Delta p} = \int_0^{\Delta p} \left(F_v - F_u - W_t - F_{hf} - F_{sf} \right) dt = (m_b + m_{add}) V_b. \tag{6.144}$$

Here Δp_{pulse} is the pulse time, F_v the sum of the vertical forces around the block due to the impinging jet, F_u the fluid resistance inside the fissures to a change in volume, W_i the immerged block weight, F_{hf} and F_{sf} are the hydraulic and solid friction forces on the vertical fissures, m_b is the block mass, m_{add} the added mass of the block and V_b the block displacement velocity. Equation (6.144) corresponds to a complete formulation of the impulse. For practical engineering applications, the stabilizing forces may be neglected, resulting in (Bollaert and Schleiss, 2005)

$$I_{\Delta p} = \int_0^{\Delta p} \left(F_v - W_t \right) dt = m_b V_b. \tag{6.145}$$

The impulse $I_{\Delta p}$ is considered whenever net uplift forces exist. The maximum impulse of a test run is I^{max}. Time is non-dimensionalized by dividing it by the period of the pressure waves inside the joints $T_p = 2L_f/c_j$, with L_f as the fissure length and c_j the wave celerity (Bollaert and Schleiss, 2005). For simplicity, the rock blocks are considered to have a square base of side length x_b and height z_b. Hence, $L_f = 2z_b + x_b$. The forces are non-dimensionalized by transforming them into a pressure acting on a block side (i.e. top or bottom of the block of area x_b^2) and dividing the result by the kinetic energy per unit volume E_k. The maximum dynamic impulsion coefficient C_I^{max} thus becomes

$$C_I^{max} = \frac{I^{max} c_j}{x_b^2 \rho_{AW} V_i^2 L_f}. \tag{6.146}$$

According to Duarte (2014), the extreme values of C_I^{max} range roughly between 0.35 and 0.15. The effect of the issued jet aeration β_1 is relatively small and the values decrease for high β_1 toward approximately 0.20. As to the effect of the issued jet velocity, C_I^{max} decays smoothly toward 0.2 for high jet velocities; based on the test data, the use of $C_I^{max} = 0.2$ is proposed (Duarte, 2014). This corresponds particularly to the strongly aerated high-velocity jets as occur under prototype conditions.

The maximum dynamic impulsion on dislodged rock blocks on the plunge pool bottom depends on the kinetic energy dissipation of the jet in the water pool and on the maximum impulsion coefficient acting on a block. The former is represented by the time-averaged pressure coefficient C_p^A accounting for the aeration effect. The maximum dynamic impulsion is

$$I^{max} = C_I^{max} C_p^A \frac{\rho_{AW} V_i^2}{2} x_b^2 \Delta p. \tag{6.147}$$

The vertical block displacement is then

$$h_{up} = \frac{V_b^2}{2g}.$$ (6.148)

This approach differs from Bollaert (2002) and Bollaert and Schleiss (2005), who suggest a maximum impulsion depending exclusively on C_I^{max}, instead of both C_I^{max} and C_p^A as in Eq. (6.147). They propose an empirical relation for C_I^{max} as decreasing function of the relative pool depth based on their experimental data (Eq. 6.135) as

$$C_I^{max} = 0.0035 \left(\frac{Y_j}{D_i} \right)^2 - 0.119 \left(\frac{Y_j}{D_i} \right) + 1.22 .$$ (6.149)

In the proposed modification for taking into account the active jet aeration, C_p^A is a decreasing function of the relative pool depth, according to Eqs. (6.141) and (6.143), reflecting jet dissipation along the pool. The pressure rise due to lower velocity decay caused by air entrainment is represented in the formulation of C_p^A in Eq. (6.143). The pressure reduction due to a lower apparent density of aerated jets is reproduced by a lower kinetic energy of the jet as stated in Eq. (6.147).

6.5.4 Difficulties in estimating scour depth

Engineers are confronted in practice by issues which are challenging the task of scour assessment. These issues are described in the following according to Schleiss (2002).

Which is the appropriate formula or theory?

In the feasibility design stage of a dam project, easily applicable empirical and semi-empirical formulae furnish a first estimate of the expected ultimate scour depth downstream of spillway structures during their lifetime. Most of the formulae were developed for a specific case such as a ski jump, a free crest overfall, or an orifice spillway. Therefore, a careful selection of the appropriate equations should be made for each project. However, even after careful selection, the results of the remaining formulas often indicate a large scatter. Then, the engineering decision is based on a statistical analysis of the results by using, for example, the average of all formulae, or the positive standard deviation in a more conservative way. This analysis accounts for a certain spillway discharge with the corresponding tailwater level by varying the characteristic rock block size according to the expected geological conditions. Prototype scour data of an existing dam are occasionally available, located at a similar geological site as the new project. Then, the formula predicting best this measured scour depth can be identified and applied to the new project.

How model tests be performed and interpreted?

If free jets impinging on rock underlying a plunge pool have to be modeled in a laboratory, three main difficulties arise (Whittaker and Schleiss, 1984):

- Appropriate choice of material behaving dynamically in the model as fissured rock as does in prototype

- Grain size effects
- Aeration effects.

In most models the disintegration process, i.e. the hydrodynamic fracturing of closed-end rock joints and splitting of rock into rock blocks, is assumed to have taken place. Thus only the ejection of the rock blocks into the macro-turbulent plunge pool flow and its transport from the scour hole is modeled. Reasonable results are obtained if fissured rock is modeled by appropriately shaped concrete elements (Martins, 1973), but their regular pattern and size is not fully representing a rock mass with several intersecting fracture sets. Nevertheless, when modeling rock, crashed gravel should be used, having at least a grain size distribution as the expected rock blocks, instead of round river gravel. In any case, model tests cannot simulate the breakup of the rock blocks by the ball milling effect of the turbulent flow in the plunge pool. Therefore, a mound is formed in the model which is higher and more stable than in the prototype. As a result the prototype scour depth is underestimated. This is compensated to some extent by selecting the material carefully including down-scaling. Normally good predictive results for scour depth result by using non-cohesive material, but the scour extent may not be correct because steep and near vertical slopes are not reproduced. Therefore the use of slightly cohesive material by adding cement, clay caulk, chalk powder, or paraffin wax to the crushed gravel is proposed (Johnson, 1977; Gerodetti, 1982; Quintela and Da Cruz, 1982; Mason, 1984).

It is known that for grain sizes smaller than 2–5 mm, the ultimate scour depth becomes constant (Veronese, 1937; Mirtskhulava et al., 1967; Machado, 1982). For an acceptable scale of a comprehensive dam model of scale 1:50–1:70, the smallest reproducable prototype rock blocks are in the range of 0.1–0.35 m.

Air entrainment cannot be scaled appropriately in comprehensive models unless using an unpractically large model scale of the order of 1:10. Air entrainment has a high random character, influencing a scour process considerably (4.2). Mason (1989) studied systematically the effect of jet aeration on scour in a mobile gravel bed with a specially designed laboratory apparatus for head drops of up to 2 m.

Hydraulic models with rigid bottoms are equipped with high sensitivity pressure transmitters to measure dynamic pool pressures generated by high-velocity plunging jets. Different measures to reduce these dynamic pressures may be studied (Berchtold et al., 2011).

Hybrid modeling appears to correctly represent the behavior of rock scour, as done successfully for defining mitigation measures for the scour of Kariba Dam (Noret et al., 2013; Bollaert et al., 2012). Historical scour geometries were reproduced in a physical model with a fixed mortar surface equipped with pressure sensors. Dynamic pressures were measured at the water-rock interface under various spillway modes. The recorded dynamic pressures were employed in the comprehensive scour assessment method of Bollaert and Schleiss (2005) to predict at which depth rock blocks can be formed by fatigue fissure propagation and ejected into the plunge pool by dynamic pressures.

How to analyze prototype observations properly?

When analyzing prototype observations on the scour depth to derive an equation for similar conditions, the following questions have to be answered:

1   What was the operational spilling duration for different specific discharges (discharge-duration curve)? An example is shown in Figure 6.79.

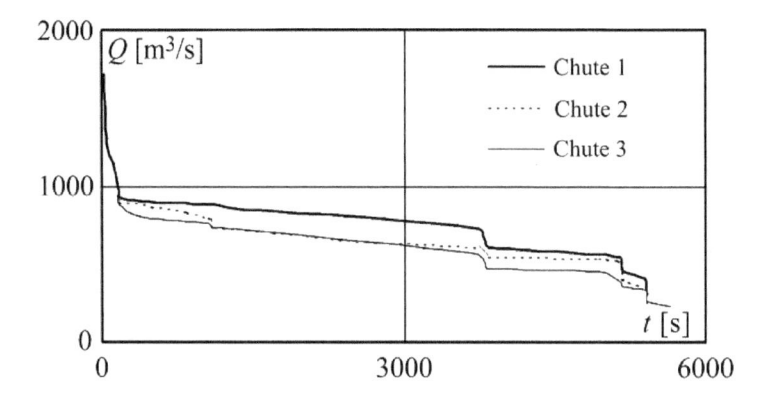

Figure 6.79 Discharge-duration curve of chute spillway, Karun I Dam in Iran, from March 1980 to July 1988 (width per chute is 15 m) (Schleiss, 2002)

2   Which was the prevailing, specific discharge having formed the scour depth?
3   Was the duration of this discharge long enough to create ultimate scour?

Since it is often difficult to answer these questions, probably significant uncertainties have been introduced in the existing formulas derived from prototype observations. This explains the large scatter when predicting scour features for other prototypes.

Can ultimate scour depth form during operation and what is scour rate?

The scour depth estimated by using empirical and semi-empirical formulas occurs in principle only after a long duration of spillway operation, once steady flow conditions in the scour hole occur. This will happen only after a minimum duration of spillway operation, depending mainly on the quality and jointing of the rock mass. Therefore, the specific discharge having a sufficiently long duration to form the ultimate scour during the technical life of the dam has to be known. Higher and therefore rare discharges are not able to create ultimate scour.

Since plunge pool scour depth $(t_s + h_t)$ is known to develop at an exponential rate with time t, the scour rate follows with a_s as site-specific constant and h_t as the tailwater depth the relation (Spurr, 1985)

$$\left(t_s + h_t\right)(t) = \left(t_s + h_t\right)_{\text{end}} \left(1 - e^{-a_s{}^{t/t}e}\right).$$
(6.150)

The evaluation of t_e as the time at which scour equilibrium is attained depends on how rapidly hydro-fracturing and material wash out from the scour hole occur, taking into account the primary and secondary rock characteristics. The primary rock characteristics comprise RQD, joint spacing, uniaxial compressive strength, and the angle of jet impact compared with the main faults or bedding planes; the secondary characteristics are the hardness and degree of weathering (Spurr, 1985). Knowing the scour depth which occurred during a certain period of operation, and estimating the maximum scour depth by one of the formulae, the site-specific constant a_s/t_e is determined. As a rough estimation based on prototype data, the ultimate scour is normally attained only after $t_e = 100$ to 300 h of spillway operation for a certain discharge considered. It is concluded that the ultimate scour depth for a certain specific discharge occurs only if its duration is long enough. Scour for a smaller duration is estimated with an exponential rate relation.

Which is the prevailing discharge for scour formation?

A flood event and the corresponding discharge curve of the spillway is characterized by a hydrograph (Figure 6.80). For all discharges of the hydrograph with a duration T shorter than time t_e at which scour equilibrium is attained, the ultimate scour will not be reached. Knowing the scour rate relation (Eq. 6.150), the prevailing discharge producing maximum scour depth during the flood is determined. The scour is estimated successively for discharges $q_u(t = t_e = t_u)$, $q_1(t_1 < t_u)$, $q_2(t_2 < t_1 < t_u)$,. ....., $q_{peak}(t_{peak} < t_i < t_u)$. The discharge resulting in the deepest scour is prevailing.

Note that these considerations are valid only for ungated free surface spillways. For gated free surface spillways, the discharge may not be directly related to the reservoir inflow but be prescribed by operational rules. When lowering the reservoir level during floods, outflow discharges are higher than the reservoir inflow. This applies also for pressurized orifice spillways.

Spillways are designed for the so-called design or project flood, typically a 1000-year flood, and checked for the so-called safety check flood, typically between the 10,000-year flood and the Probable Maximum Flood PMF. The question arises for what flood the scour depth has to be evaluated, and on which structural scour control measures they have to be based (Schleiss, 2002).

As discussed above, the ultimate scour depth occurs only once steady conditions in the scour hole occur, after a certain duration of spillway operation. Therefore, it is conservative to base the estimation of the scour depth or the design of mitigation measures on low frequency floods (PMF or 1000-year flood). It appears to be unlikely that during the technical life of the dam these rare floods produce ultimate scour depth. Therefore, for each flood with a certain return period, the prevailing discharge and the maximum scour depth have to be

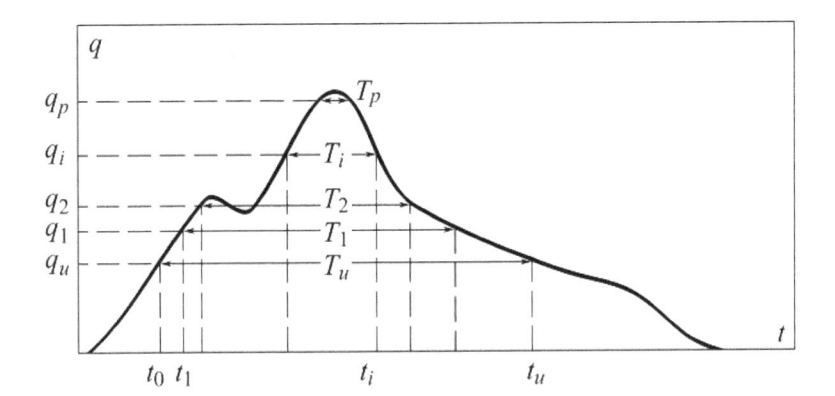

Figure 6.80 Flood event and corresponding discharge curve $q(t)$ of spillway showing discharges with duration shorter than at instant $t_e = t_u$, at which equilibrium scour and its ultimate depth are attained to determine prevailing discharge (Schleiss, 2002)

determined according to 6.5.4. Furthermore, it has to be decided under which flood return period the maximum scour depth during the technical life of the dam shall be estimated. The probability of the occurrence of a flood with a given return period during the useful life of a dam is

$$r = 1 - (1 - 1/n)^m. \tag{6.151}$$

Here r is the risk or probability of occurrence, n the flood return period in years, and m the useful life of the dam. According the Eq. (6.151) the probability of occurrence of a 200-year flood during 200 years of operation is 63%, whereas for a 1000-year flood it is only 20%.

It appears reasonable to select a design discharge with a probability of occurrence of about 50% during the useful lifetime of a dam for the scour evaluation and the protection measures. Higher design discharges with a lower probability of occurrence are considered too conservative. Note that for gated free surface and orifice spillways, high discharges can be released at any time by gate opening. Furthermore, for low-level outlet spillways, the core impact jet velocity is nearly independent of discharge.

6.5.5 Measures for scour control

Overview

To avoid scour damage, three active options are considered (Schleiss, 2002):

- Avoid scour formation completely
- Design water release structures so that scour occurs far from dam foundation and abutments
- Limit scour extent.

Since structures for scour control are expensive, only the two latter items appear economically feasible (Ramos, 1982). Besides elongating as much as possible the jet impact zone by an appropriate design of the water release structures, the extent of scour is influenced by the following measures:

- Limitation of specific spillway discharge
- Forced aeration and splitting of jets leaving spillway structures
- Increasing tailwater depth by tail pond dam
- Pre-excavation of plunge pool.

The scour location depends on the selected type of spillway and its design.

To avoid scour completely, structural measures as lined plunge pools are required. Besides active options, scour damage is also prevented by passive measures, i.e. by protecting dam abutments with anchors against instability due to scour formation.

Limitation of specific spillway discharge

This measure is mainly important for arch dams and free ogee crest spillways with jet impact close to the dam. The jet is guided by an appropriate crest lip design for a given specific discharge at a certain distance from the dam toe. If the dam foundation were endangered by scour, the discharge per unit width of the ogee crest has to be limited. But by reducing the specific discharge, the available velocity at the crest lip and therefore the travel distance of the jet is also reduced.

For gated ski-jump spillways and low-level outlets, the specific discharge depends on the size of the outlet openings. Since the available outlet velocity is high enough to divert the jet far away from the dam and its foundation, the limitation of the specific discharge is normally less important than for free crest spillways.

Forced aeration and jet splitting

To split and aerate jets leaving flip buckets and crest lips, they are often equipped with baffle blocks, splitters or deflectors. Further, high-velocity flows in chutes are normally aerated by aeration ramps and slots along the chute. All these measures increase the air entrainment, thereby reducing the scour capacity of plunging jets. Nevertheless, the amount of air entrained is difficult to estimate. Because of scale effects, the efficiency of these measures can only be checked qualitatively by hydraulic modeling.

Martins (1973) suggested a reduction of 25% of the calculated scour depth under high air entrainment, and 10% under intermediate air entrainment. Mason (1989) proposed an empirical expression considering the volumetric air-to-water ratio β_A. The proposed empirical equation based on spillway models does not depend on the fall height H_j, since a direct relationship between β_A and H_j developed by Ervine (1976) was used. The empirical formula is accurate for model data and seems to give a reasonable upper bound of scour depth under prototype conditions.

For prototype scour in fractured rock, air strongly influences the water hammer velocity and consequently resonance effects of pressure waves in rock joints. Recent research reveals that forced aeration may reduce pressure fluctuations at the water-rock interface, but may increase resonance effects (Duarte et al., 2012, 2013). Forced aeration thus increases under certain conditions the risk of rock block break-up and ejection into the plunge pool.

Increasing tailwater depth by tail pond dam

Another way to control jet scour is to increase the tailwater depth by a tail pond dam downstream of the jet impact zone. The efficiency of a water cushion is often overestimated (Häusler, 1980). For plunge pool depths smaller than 4–6 times the jet diameter, core jet impact is normally observed at the plunge pool bottom (Bollaert, 2002). The jet core is characterized by a constant velocity and is not influenced by the outer two-phase shear layer conditions of the impinging jet. The pressures are also constant with low fluctuations, having significant spectral energy at high frequencies of up to several 100 Hz. For tailwater depths larger than 4–6 times the jet diameter (or thickness), developed jet impact occurs at the plunge pool bottom involving a different pressure pattern produced by a turbulent two-phase shear layer. Significant fluctuations are produced with high spectral content at frequencies of up to 100 Hz. Maximum fluctuations have been observed for tailwater depths between 5 and 8 jet diameters (Bollaert and Schleiss, 2003a, b). Substantial high values persist up to tailwater depths of 10–11 jet diameters. From these observations it is concluded that water cushions in the range of 5–11 jet diameters can generate even more severe dynamic pressures relative to an interaction with rock joints at the plunge pool bottom than lower tailwater. Thus only water cushions deeper than 11 jet diameters at impact have a reducing effect on the scour formation in rock.

This tendency is confirmed by the empirical scour equation of Martins (1973), producing a maximum scour depth for a certain tailwater depth. Johnson (1967) found that too small water cushions are even worse than no cushion, since the material is transported more easily out of the plunge pool. Note that an increased tailwater level by a tail pond dam may interfere with bottom outlets.

Pre-excavation of plunge pool

Pre-excavation increases in principle the tailwater depth and in view of scour control the same remarks are valid as given above. Pre-excavation of the expected scour may be also appropriate if the eroded and by the river transported material can form dangerous tailwater deposits, as near powerhouse outlets. These deposits tend to increase the tailwater level thereby reducing power production. These problems are absent if the scour is formed about 200 m upstream of the powerhouse outlet.

Pre-excavation of the scour hole is also often considered when instabilities of the valley slopes have to be feared. The excavation has then to be stabilized by anchors and other measures. As an alternative, even under these conditions, pre-excavation is omitted if the valley slopes are stabilized by appropriate measures in such a way that the slopes be stable even after ultimate scour formation.

The selection of the design discharge for an excavated geometry of the plunge pool is based on considerations similar to these discussed in 6.5.4. It is often economically uninteresting to excavate deeper than the scour depth, which would be formed by a 50-year to 100-year flood. If instabilities of the valley slopes, for a deeper scour due to higher spillway discharges, have to be feared, rock anchors and pre-stressed tendons apply (Figure 6.81).

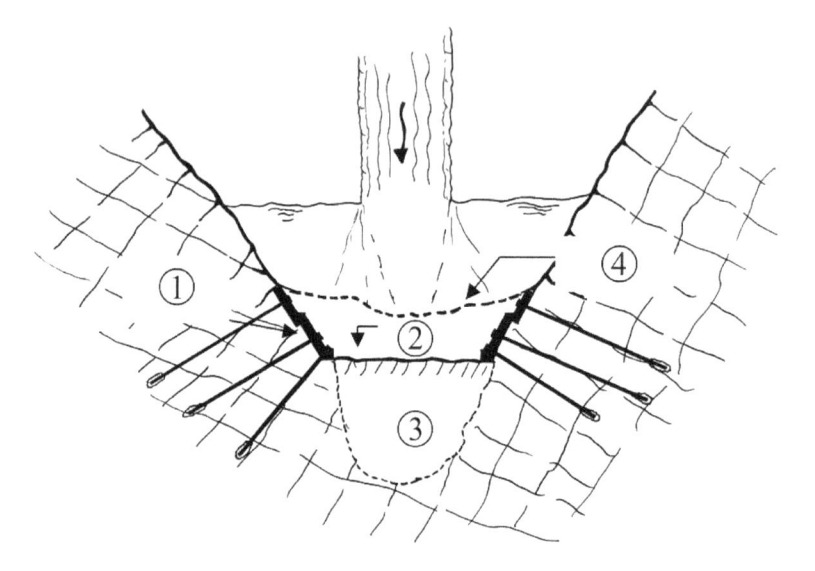

Figure 6.81 ① Pre-excavation of ② initial rock plunge pool for scour depth forming by 50- to 100-years flood and ③ slope stabilization measures for ④ ultimate scour formation to avoid abutment instabilities; slope stabilization measures (Schleiss, 2002)

Instead of full excavation, pre-splitting of plunge pool rock appears adequate under certain conditions, rendering the rock more readily erodible at certain places. Pre-splitting is therefore used to affect the eventual geometry of the plunge pool without the expense of a complete excavation. If scour affects power production, complete excavation may be more appropriate, however.

The pre-excavated plunge pool geometry has to be based on the expected natural scour geometry. Empirical formulas were proposed to estimate its length (Martins, 1973; Kotoulas, 1967) and width (Martins, 1973). Amanian and Urroz (1993) made tests on a model scale flip bucket spillway with a gravel bed plunge pool to develop equations describing the scour hole geometry created by jet impact into the plunge pool. They observed that the performance of the pre-excavated scour holes are best if those are close to the self-excavated hole for the same flow parameters.

For the well-known scour hole case at Kariba Dam on Zambezi River, having reached a depth of over 80 m below the tailwater level, an innovative solution was developed by reshaping the existing scour hole with an excavation mainly in the downstream direction (Noret *et al.*, 2013), significantly reducing dynamic pressures at the water-rock interface due to jet impact. It was proved with hybrid modeling that the reshaped scour hole will no more progress in the future (Bollaert *et al.*, 2012).

Concrete-lined plunge pools

If absolutely no scour formation in the rock downstream of a dam is accepted, a plunge pool has to be reinforced and tightened by concrete lining. Since the thickness of the lining is limited by structural and economic reasons, normally high-tension or pre-stressed rock anchors are required to ensure the lining stability in view of the high dynamic loading. Furthermore, the lining surface has to be protected with reinforced (wire mesh, steel fibers) and high tensile concrete of high resistance against abrasion. Construction joints have to be sealed with efficient water stops. In addition, the lining stability against static uplift pressure during dewatering of the plunge pool has to be guaranteed by a drainage system, limiting also the dynamic uplift pressures under limited cracking of the lining.

The design of plunge pools linings may be based on the following sequence of events as for stilling basins (Fiorotto and Rinaldo, 1992; Fiorotto and Salandin, 2000):

- Pulsating pressures can damage joint seals between slabs (construction joints)
- Extreme pressures may propagate through these joint seals, from the upper to the lower slab surfaces
- Instantaneous pressure differentials between the two can reach high values
- Resulting force due to the pressure differential may exceed slab weight and anchor's resistance.

The propagation of dynamic pressures through fissures in the lining reveals also the presence of water hammer effects, amplifying pulsating uplift pressures underneath concrete slabs (Bollaert, 2002; Melo *et al.*, 2006; Liu, 2007; Fiorotto and Caroni, 2007).

Since cracking of the plunge pool lining cannot be excluded, the assumptions of an absolutely tight lining and neglecting dynamic uplift are on the unsafe side. If the high dynamic pressures can propagate through a small, local fissure from the upper to the lower concrete slab surface, dynamic pressures from underneath will locally lift the slab, finally resulting in a progressive failure of the entire plunge pool lining. Thus, the concept of a tight plunge pool lining is as risky as the chain concept: the system's resistance is given by the weakest link, i.e. the local permeability of the concrete slab. Furthermore, fluctuating pressures due to plunging jets are high compared with the proper slab weight and thus result in vibrations of the concrete slab and ultimately in crack development in the concrete and in possible fatigue failure of the anchors.

Cracking of a concrete slab before operation cannot be fully ruled out even when using sophisticated construction joints and water stops, because of temperature effects when filling the plunge pool with water, and of underground deformation. Therefore, the design criteria of the plunge pool liner have to take into account:

- Load case of dynamic uplift pressures during operation. Reinforcement of concrete slab has to be designed for crack width limitation under possible dynamic vibration modes (depending on anchor pattern and stiffness);
- Grouted and pre-stressed rock anchors have to be designed for fatigue.

The drainage system under a plunge pool lining is significant since it increases the safety against dynamic uplift pressures. However, since limited cracking of the lining cannot be excluded, as mentioned, pressure waves are transferred through the cracks into the drainage system. Its response in view of these dynamic pressures with a wide range of frequencies at the entrance of a possible crack has to be controlled by a transient analysis, to exclude pressure amplification in the drainage system (Mahzari *et al.*, 2002). These amplifications in the drainage system or any dynamic pressure underneath the slabs determine the anchor design (Mahzari and Schleiss, 2010).

Since plunge pool linings are a risky concept, a 'belt-and-braces' approach for prototype design should be used with the following recommendations:

- Slabs are sized generously so that localized instantaneous fluctuations average over a larger area, and slabs are generally reinforced;
- Water bars are provided between slabs to prevent dynamic pressure exchange with the slab underside;
- Anchor bars as a further precaution are used, based on mean pressure differentials across different parts of the apron assuming that some water bars may have failed;
- Generously sized under-drainage is provided so that any pressure fluctuations feeding through small joints between the slabs are dissipated in the generously sized drainage zones.

Although the physical understanding of scour processes has considerably improved during the past two decades, the scour evaluation still remains a challenge for dam designers. Scour models are now available taking into account pressure fluctuations in the plunge pool and the propagation of transient water pressures into the joints of the underlying rock mass as the Comprehensive Scour Method. However, further research is needed especially regarding the physical process of rock block ejection due to dynamic uplift, and the influence of air bubbles in the plunge pool.

Remaining challenges

Despite the development of complex theoretical models including all physical processes, there still remain open questions, which have to be answered based on engineering judgment. Above all, rock parameters as number, spacing, direction, and persistency of fracture sets, *in-situ* stresses in the rock mass, fracture toughness, or unconfined compressive strength, have to be collected during the geotechnical dam reconnaissance at the location of the expected scour hole. These parameters will also influence the temporal scour evolution together with the hydrological conditions during the considered lifetime of the dam, involving further uncertainties. To check and calibrate complex scour models, more detailed prototype data on the scour evolution with fully documented discharge records are needed. In principal, these observations are essential for a continuous safety assessment of a dam and allow for predicting the scour evolution.

6.5.6 Case study: Kariba Dam scour hole

Description of hydraulic scheme

Kariba Dam is located on the Zambezi River between Zambia and Zimbabwe, where it creates one of the largest man-made reservoirs in the world. The hydropower plant has a capacity of 1266 MW with current refurbishment works to increase the total capacity to 1450 MW (Noret *et al.*, 2013). The flood release devices comprise six middle outlet gates of 1500 m³/s discharge capacity each. No additional structure is located outside the dam with the purpose of flood release (see chapter frontispiece).

As a result of long spilling periods since operational start in 1959, a deep and steep-sided scour hole was formed at the bottom of the plunge pool downstream of the dam. Bathymetry campaigns indicate that the pool bottom was at elevation 306 m a.s.l. in 1981 (Figure 6.82), corresponding also to the elevation in 2001. This is still considered to be the current shape

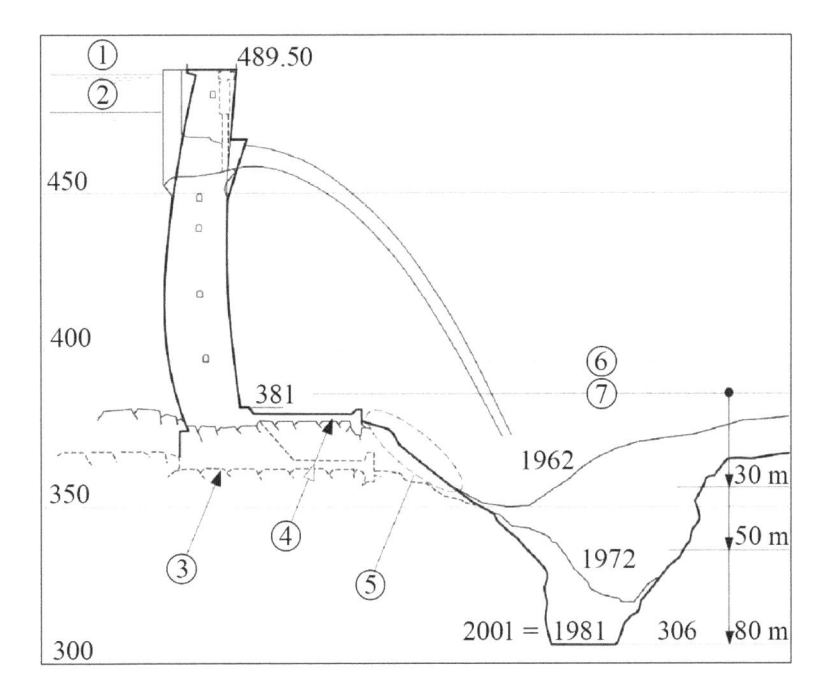

Figure 6.82 Longitudinal section of Kariba plunge pool with temporal evolution of scour hole between 1962 and 2001 including ① maximum flood level, ② minimum operational level, ③ deepest section, ④ apron, ⑤ maintenance works, level of power stations and ⑥ 396 m plus 3 gates, ⑦ 386 m plus full load (Noret *et al.*, 2013)

These results are shown and compared with the longitudinal profiles measured in 1972 and 1981 in Figure 6.83.

The differences between the two failure criteria $h_{up}/z_c = 1$ and 0.25 is small. The results for the ultimate scour depth are close to the deepest point of the pool bottom as measured in 1981 and 2001. A failure criterion of $h_{up}/z_c = 0.25$ resulted in exactly the same pool bottom elevation of 306 m a.s.l. It indicates that the scour hole has attained its scour potential considering the capacity of the impinging jet to eject blocks from the rock mass.

The original DI method based on the failure criterion $h_{up}/z_c = 0.25$ gives a bottom elevation of 289.5 m a.s.l. Note that the DI method takes only into account the erosion capacity of the turbulent shear layer of the jet. It thus considers the effect of a direct jet impact but does not include the influence of rollers formed by jet deflection against the scour hole. This effect is accounted for by the enhanced CSM with a so-called quasi-steady impulsion model (Bollaert, 2012), which predicts the scour evolution along the wall jet region by the process of pealing up rock layers. As highlighted by Manso *et al.* (2009), the geometry of the plunge pool bottom generates induced flow patterns and has a strong influence on the way the jet dissipates and is deflected at the pool bottom. Subsequently to jet impact at the intersection of the jet axis with the water-rock interface, the jet is deflected and forms a wall jet parallel to the bottom.

As to Kariba Dam, jet deflection towards upstream of the impingement point is a specific concern, which may cause erosion near the dam foundations. The current reshaping efforts

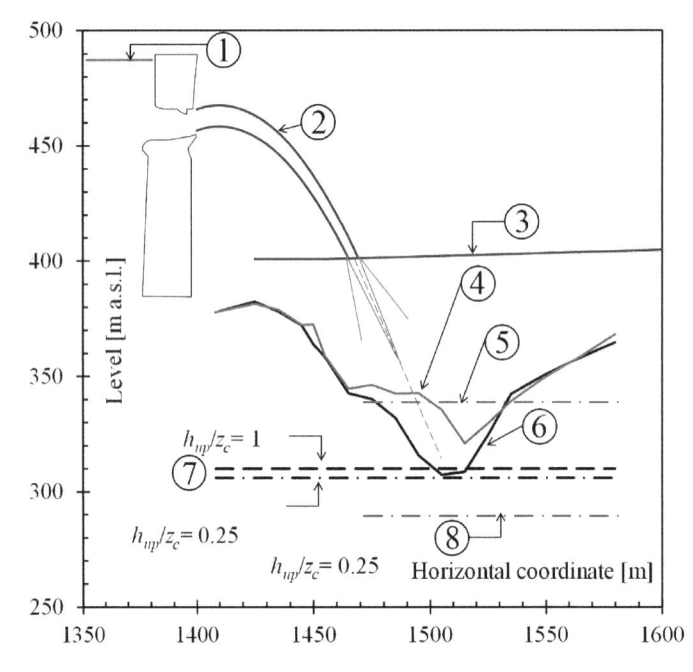

Figure 6.83 Results of adapted Dynamic Impulsion (DI) method for Kariba Dam scour hole with pool bottom in (——) 1972, (——) 1981 and 2001; (---) adapted DI method for either $h_{up}/z_c = 1$ or 0.25; (---) original DI method with $h_{up}/z_c = 0.25$. ① Reservoir level, ② jet issued, ③ tailwater level, ④ scour hole in 1972, ⑤ Mason and Arumugam (1985), ⑥ scour hole in 1981, ⑦ adapted DI method for $h_{up}/z_c = 1$ (upper) and 0.25 (lower straight), ⑧ original DI method (Bollaert, 2002)

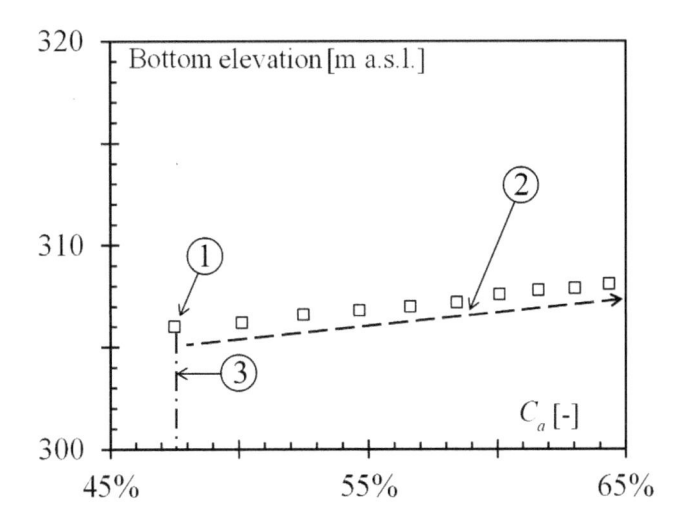

Figure 6.84 Bottom elevation of Kariba Dam scour hole at ultimate scour depth in [m a.s.l.] versus entrained jet air concentration C_a, (□) model results, (---) tendency of results; computational results with adapted DI method for $h_{up}/z_c = 0.25$ with ① estimated bottom level at ultimate scour depth, ② effect of hypothetic jet aeration increase, ③ estimated entrained air concentration (Duarte *et al.*, 2016)

have the objective of leading the deflection of the impinging jet towards downstream, thus avoiding further scour toward the dam. With the pre-excavation works, the scour is expected not to deepen any further (Noret *et al.*, 2013; Bollaert *et al.*, 2012, 2013).

Other uncertainties arise from model parameters. For example, it highly depends on the wave celerity considered. Although the celerity used in this case is a robust result of the experimental campaign, the study might be extended in the future to assess the effect of block geometry relative to the jet and pool geometric scales and interlocking of the joints on the wave celerity. Moreover the dynamic impulse applied on a rock block was computed neglecting the stabilizing forces of the moving block as a simplifying assumption for engineering practice.

The adapted DI method allows for simulating changes in the jet air entrainment rates. Although purely hypothetical for Kariba Dam, the effect of adding air to the jets at the orifices on the ultimate scour depth are represented in Figure 6.83. The failure criterion used was $h_{up}/z_c = 0.25$. Increasing entrained air concentrations were simulated, starting with the estimated value of 48% of the existing jets, up to 65%.

Figure 6.84 shows that hypothetical air addition to the jets reduces the scour depth. The bottom elevation increases steadily with the entrained air concentration, yet this increase is small. The simulated air concentration increase of 17% resulted in an ultimate scour depth only 2.1 m smaller, raising the bottom from 306 to 308.1 m a.s.l.

Notation

A   Coefficient (-)

A_b Bucket coefficient (-)

A_R	Ridge constant (-)
A'	Jet type parameter (-)
A_1	Jet type coefficient (-)
a	Equivalent jet width (m)
a_f	Fracture depth (m)
a_j	Jet turbulence intensity (-)
a_s	Site-specific scour rate decay (-)
a_t	Element base length (m)
B	Jet width at impact zone (m); also coefficient (-)
B_i	mean jet width (m)
B_M	$= B_m/D$ (-)
B_m	Maximum scour hole width (m)
B_o	Approach flow bend number (-)
B_R	Ridge coefficient (-)
b	Channel width (m)
b_e	End width of slit-type flip bucket (m)
b_f	Fracture length (m)
b_j	Jet width (m)
b_M	Surface width of scour hole (m)
b_o	Approach flow width of slit-type flip bucket (m)
b_t	Element width (m)
C	Air concentration (-)
C_a	Cross-sectional average air concentration (-)
C_{as}	Average impact air concentration (-)
C_I	Dimensionless dynamic impulsion coefficient (-)
C_I^{max}	Dimensionless maximum dynamic impulsion coefficient (-)
C_L	Scour hole length coefficient (-)
C_m	Minimum air concentration (-)
C_{ms}	Minimum jet air concentration (-)
C_o	Approach flow air concentration (-)
C_p	Mean dynamic pressure coefficient (-)
C'_p	Dynamic pressure coefficient (-)
$C'_{p\ max}$	Maximum pressure coefficient (-)
C_r	Ridge presence or absence coefficient (-); also rock parameter (-)
C_s	Scour hole coefficient (-)
C_t	Deflector aspect ratio (-)
C_{1m}	Coefficient (-)
C_{2m}	Coefficient (-)
C_+	Increasing discharge coefficient (-)
C_-	Decreasing discharge coefficient (-)
c	Relative air concentration (-)
c_a	Relative average air concentration (-)
c_j	Pressure wave celerity (m/s)
c_t	Element height (m)
D	Internal pipe diameter (m); also variable jet diameter (m)
D_{core}	Jet core coefficient (-)
D_e	Equivalent jet diameter (m)

D_i	Impact jet diameter (m)
D_j	Mean jet diameter at jet impact (m)
D_o	Jet approach flow diameter (m)
D_{out}	Jet diameter at impact location (m)
D_{test}	Equivalent jet diameter (m)
D_W	Black-water jet diameter (m)
d_i	Determining sediment size (m)
d_m	Mean sediment size (m); also equivalent rock block diameter (m)
d_{50}	Median sediment size (m)
E	Relative energy dissipation (-)
E_k	Kinetic energy per unit volume (J)
EL	Joint persistence (-)
e_f	Fracture width (m)
F	Ridge coefficient (-)
F_p	Pressure correction coefficient (-)
F_{hf}	Hydraulic friction force on vertical fissures (N)
F_o	Force acting on downstream side of block (N)
F_{sf}	Solid friction force on vertical fissures (N)
F_u	Force acting on upstream side of block (N)
F_v	Sum of vertical forces around block (N)
F_d	Densimetric Froude number (-)
F_{d50}	Densimetric Froude number based on d_{50} (-)
F_{di}	Densimetric Froude number based on particular sediment size $= V_W/(g'd_i)^{1/2}$ (-)
F_o	Approach flow Froude number (-)
F_{sh}	Shear forcers along block joint (N)
F_u	Fluid resistance inside fissures (N)
F_t	Tailwater Froude number (-)
F_u	Upstream velocity related Froude number (-)
F_v	Sum of vertical forces around block (N)
F_β	Densimetric Froude number for air-water mixture flow (-)
f_c	Pressure frequency (1/s)
G_b	Immersed block weight (N)
g	Gravity acceleration (m/s^2)
g'	Reduced gravity acceleration (m/s^2)
H	Energy head (m)
H	High tailwater condition (-)
H_j	Take-off velocity head (m); also free-falling jet height (m)
H_o	Approach flow energy head (m); also fall height (m)
H_t	Threshold energy head (m)
H_1	Energy head beyond jet impact (m)
ΔH	Energy head loss (m)
h_b	Flow depth over bucket (m)
h_c	Critical flow depth (m)
h_f	Fall height of jet (m)
h_i	Jet core thickness (m)
h_j	Jet thickness (m)
h_M	Maximum elevation of upper jet trajectory (m)

h_m	Maximum elevation of lower jet trajectory (m)
h_o	Approach flow depth (m)
h_P	Pressure head (m)
h_{PM}	Maximum pressure head (m)
h_{PT}	Potential flow pressure head (m)
h_s	Shock wave height (m)
h_t	Tailwater depth (m)
h_{up}	Vertical rock displacement (m)
h_1	Flow depth beyond jet impact (m)
h_2	Sequent flow depth to h_1 (m)
Δh	Flow depth increase across deflector (m)
I_{max}	Net maximum impulsion (N)
$I_{\Delta p}$	Impulsion (N)
K_1	Jet air aeration coefficient (-)
K_I	Stress intensity factor (-)
K_{Ic}	Fracture toughness of rock mass depending on UCS or TS (MPa)
ΔK_I	Amplitude of dynamic local stresses at fracture end (MPa)
L	Low tailwater condition (-)
L_a	Scour hole length (m)
L_{aT}	3D Scour hole length (m)
L_b	Jet break-up length (m)
L_{bw}	Black-water jet core length (m)
L_f	Fissure length (m)
L_{iT}	$= l_{iT}/D$ (-)
L_j	Distance of jet impact from jet issuance (m)
L_M	Distance of maximum ridge location from scour hole origin (m)
L_m	Distance of maximum scour depth location from scour origin (m)
L_R	Reference length (m)
L_S	Spillway length (m)
L_s	Dimensionless jet travel distance (-)
L_t	Jet trajectory length (m)
L_u	Combined length of scour hole and ridge (m)
l_{aT}	3D scour hole length (m)
l_{iT}	3D scour hole length up to maximum depth (m)
l_{MT}	Distance to 3D maximum ridge elevation (m)
l_{mT}	Distance from 3D scour hole start to its deepest point (m)
l_s	Horizontal jet travel distance (m)
l_{uT}	Overall length of 3D scour hole and ridge (m)
m	Exponent (-)
m_{add}	Added block mass (g)
m_b	Rock or block mass (g)
m_r	Rock parameter (-)
N	Number of peak pressure cycles (-)
n	Exponent (-); also flood return period in years (9)
P	Relative pressure head (-)
P_a	Pre-aeration coefficient (-)
P_{max}	Maximum pressure peak (Pa)

$p_M/\rho g$	Maximum pressure head (m)
p_{mean}	Mean dynamic pressure head (m)
Δp	Pulse time (s)
Δp_c	Pressure amplitude (N/m^2)
Q	Discharge (m^3/s)
Q_A	Air discharge (m^3/s)
Q_u	Upstream discharge to scour hole (m^3/s)
Q_W	Water discharge (m^3/s)
q	Unit discharge (m^2/s)
$\bar{q}$	Relative unit discharge (-)
q_A	Unit air discharge (m^2/s)
q_P	Unit peak discharge (m^2/s)
R_b	Bucket radius (m)
R_o	Approach flow Reynolds number (-)
r	Radial coordinate (m); also risk of probability of occurrence (-)
S	$= s/h_o$ (-)
S_{tot}	$= s_T/s_{tot}$ (-)
s	Approach flow elevation above tailwater bottom (m)
s_T	Coordinate measured from maximum 3D ridge elevation to ridge start (m)
s_{tot}	Half 3D ridge length (m)
T_p	Pressure wave period (s)
T_s	$= h_o/D$ relative tailwater depth (-)
TS	Tensile strength (Pa)
T_u	$= t_u/h_o$ (-)
Tw	Tailwater
T	Turbulence number (-)
t	Time (s)
t_b	Bucket approach flow depth (m)
t_e	Equilibrium time (s)
t_i	Jet impact thickness (m)
t_{peak}	Time to peak (s)
t_R	Reference time (s)
t_s	Plunge pool scour depth (m)
t_u	Recirculation flow depth (m); also ultimate scour depth time (s)
U	Mean jet velocity (m/s)
UCS	Compression strength (Pa)
u'	rms-value of velocity (m/s)
u_M	Maximum cross-sectional jet velocity (m/s)
V	Velocity (m/s)
V_{AW}	Mean cross-sectional velocity (m/s)
V_{aw}	Air-water mixture velocity (m/s)
V_b	Block displacement velocity (m/s)
V_e	Entrainment velocity (m/s)
V_i	Jet impact velocity (m/s); also mean jet velocity (m/s)
V_j	Jet take-off velocity (m/s); also total jet energy head (m)
V_M	Maximum velocity of pre-aerated jet (m/s)
V_o	Approach flow velocity (m/s)

V_R	Reference velocity (m/s)
V_u	Upstream velocity (m/s)
V_{up}	Ejection velocity (m/s)
V_w	Jet water velocity (m/s)
V_1	Velocity beyond jet impact (m/s)
v	Horizontal velocity component (m/s)
W	Relative bucket height (-)
W_e	Element length (m)
W_i	Immersed block weight (N)
W_m	$= (g'd_{90})^{1/2}[w_m/(Q_W D)]$ (-)
W_r	$= (g'd_{90})^{1/2}[w_r/(Q_W D)]$ (-)
W_s	$= (W_s - W_r)/W_m$ (-)
W_t	Immersed block weight (N)
W_o	Approach flow Weber number (-)
w	Bucket height (m)
w_m	Scour hole volume (m³)
w_r	Ridge sediment volume (m³)
w_s	Suspended sediment volume in scour hole (m³)
X	Dimensionless horizontal coordinate (-)
X_j	Generalized streamwise jet coordinate (-)
X_M	$= (x - x_a)/(x_M - x_a)$ Relative location of maximum jet elevation (-)
X_m	x_m/h_o (-); also $(x - x_m)/(x_a - x_o)$ (-); also $(x - x_m)/l_{aT}$ (-)
X_{PM}	$= x/x_{PM}$ (-)
X_S	$2(x_o - x_m)/l_{aT}$ (-)
X_s	$= 2(x_o - x_m)/(x_a - x_o)$ (-)
X_1	$= x_1/h_o$ (-)
x	Streamwise coordinate (m)
x_a	Location of upstream crossing of scour hole and original sediment bed (m)
x_b	Block square base length (m)
x_M	Location of maximum jet elevation (m); also location of maximum ridge height (m)
x_m	Location of lower jet trajectory impact onto tailwater bottom (m); also location of maximum scour hole depth (m)
x_o	Location of scour hole origin (m)
x_{PM}	Location of maximum pressure (m)
x_u	Location of downstream crossing of scour hole and original sediment bed (m)
x_{ult}	Distance of maximum ultimate rock scour depth from jet issuance
x_1	Jet impact location of upper trajectory on tailwater bottom (m)
Y	Sum of tailwater and scour hole depths (m)
Y_J	$= Y_J/D_j$ Relative scour hole water depth (-)
Y_S	Scour hole water depth (m)
Y_s	$= h_s/h_o$ (-)
y	Transverse coordinate (m)
y_{core}	Jet core thickness (m)
y_1	Transverse jet impact location (m)
Z	Dimensionless vertical coordinate (-)
Z_d	$= z_d/z_M$ (-)
Z_j	Generalized vertical jet coordinate (-)

Z_M Relative maximum jet elevation (-); also maximum relative ridge elevation (-)

Z_{MT} $= z_{MT}/D$ (-)

Z_m $= z_m/h_j$ (-); also dimensionless scour hole depth (-)

$\underline{Z}_m$ Ultimate dimensionless scour hole depth (-)

Z_{mT} $= z_{mT}/D$ (-)

Z_{mt} Developing scour phase parameter (-)

Z_s $= z_s/D$ (-)

Z_{sT} $= z_{sT}/D$ (-)

z Vertical coordinate (m)

z_b Fall height (m); also block height (m)

z_c Cavity depth (m)

z_d Local ridge elevation (m)

$z_{e\infty}$ Ultimate scour depth (m)

z_i Elevation difference between jet take-off and tailwater (m)

z_L Lower jet trajectory elevation (m)

z_M Maximum jet elevation (m); also maximum ridge height (m)

z_{MT} Maximum 3D ridge elevation (m)

z_m Location of minimum air concentration (m); also maximum scour hole depth (m)

z_{mT} Maximum 3D scour hole depth (m)

z_o Jet takeoff elevation (m)

z_s Maximum 2D scour hole depth (m)

z_{sT} Maximum static 3D scour hole depth (m)

z_U Upper jet trajectory elevation (m)

α Jet take-off angle (-)

α_b Terminal bucket angle (-)

α_c Inner core jet angle (-)

α_D Deflector angle (-)

α_d Jet spread angle (-)

α_i Jet impact air concentration (-)

α_j Jet takeoff angle (-)

α_L Lower trajectory angle (-)

α_O Jet angle of upper jet trajectory (-)

α_o Upstream bucket angle (-)

α_t Local jet trajectory angle (-)

α_U Jet angle of lower jet trajectory (-)

$\alpha°$ 3D jet impact angle in degrees (-)

β Aeration rate (-)

β_a Aeration ratio (-)

β_b Bucket deflection angle (-)

β_j Transverse jet expansion angle (-)

β_R $= 10A_b$ (-)

χ Relative distance based on black-water core length (m)

Γ Deflector parameter (-)

Γ^+ Amplification factor (-)

γ Deflector angle (-)

γ_i Jet core angle (-)

γ_j Jet growth angle (-)

γ_W Specific water weight (N/m³)

δ Local streamline angle (-)

δ_j Jet deflection angle (-)

δ_{out} Lateral diffusion angle of plunging jet (-)

ε Deflector coefficient (-)

ϕ Angle of repose (-)

η_E Relative energy loss (-)

φ Chute bottom angle (-)

Λ Jet take-off parameter (-)

λ $= B_m/B$ scour width ratio (-)

μ_a Coefficient relating to L_a for either U or S type jet flows (-)

μ_M Coefficient for relating to L_M either U or S type jet flows (-)

μ_u Coefficient for relating to L_u either U or S type jet flows (-)

v Kinematic fluid viscosity (m²/s)

ψ Energy loss coefficient due to jet intersection (-)

ψ_j Jet angle coefficient (-)

ρ Density (kg/m³)

ρ_A Density of air (g/m³)

ρ_{AW} Density of air-water mixture (g/m³)

ρ_s Sediment density (kg/m³)

ρ_W Fluid density (kg/m³)

σ Fluid surface tension (N/m)

σ_c Initial confining stress (Pa/m²)

σ_p RMS pressure fluctuation coefficient (-)

σ_s Sediment non-uniformity coefficient (-)

σ_W Surface tension of water (N/m)

τ Maximum scour depth parameter (-)

τ_j $= (g'd_{90})^{1/2}(t/D)$ (-)

τ_t Transition time (-)

ζ Relative jet elevation (-)

ζ_{LaT} 3D scour hole a-length factor (-)

ζ_{LmT} 3D scour hole m-length factor (-)

ζ_{ZsT} 3D scour hole depth factor (-)

ω Intended angle of 3D ridge extension (-)

References

Amanian, N. & Urroz, G.E. (1993) Design of pre-excavated scour hole below flip bucket spillways. Proceedings of International ASCE Symposium *Hydraulic Engineering*, San Francisco, pp. 856–860.

Anderson, C.L. & Blaisdell, F.W. (1982) Plunge pool energy dissipators for pipe spillways. In: Smith, P.E. (ed) Proceedings of Conference *Applying Research to Hydraulic Practice*, Jackson MI. ASCE, New York, pp. 289–298.

Annandale, G.W. (1995) Erodibility. *Journal of Hydraulic Research*, 33(4), 471–494; 35(2), 280–284.

Annandale, G.W., Wittler, R.J., Ruff, J. & Lewis, T.M. (1998) Prototype validation of erodibility index for scour in fractured rock media. Proceedings of the 1998 *International Water Resources Engineering Conference*, Memphis, Tennessee, United States.

Annandale, G.W. (2006) *Scour Technology*. McGraw-Hill, New York.

Asadollahi, P., Tonon, F., Federspiel, M. & Schleiss, A.J. (2011) Prediction of rock block stability and scour depth in plunge pools. *Journal of Hydraulic Research*, 49(5), 750–756.

Balestra, A.A. (2012) Einfluss vom Zuflussgefälle auf Skisprünge [Effect of chute slope on ski jumps]. Master Thesis, VAW, ETH Zurich, Zürich (unpublished, in German).

Balloffet, A. (1961) Pressures on spillway flip buckets. *Journal of the Hydraulics Division* ASCE, 87(HY5), 87–98; 88(HY2), 175–178; 88(HY4), 279–281.

Beltaos, S. & Rajaratnam, N. (1973) Plane turbulent impinging jets. *Journal of Hydraulic Research*, 11(1), 29–59.

Berchtold, T. & Pfister, M. (2011) Measures to reduce dynamic plunge pool pressures generated by a free jet. In: Schleiss, A.J. & Boes, R.M. (eds) Proceedings of International Symposium *Dams and Reservoirs under Changing Challenges*. 79th Annual Meeting of ICOLD, Swiss Committee on Dams, Lucerne, pp. 639–700, ISBN 978-0-415-68267-1.

Bin, A.K. (1993) Gas entrainment by plunging liquid jets. *Chemical Engineering Science*, 48(21), 3585–3630.

Blaisdell, F.W. & Anderson, C.L. (1991) Pipe plunge pool energy dissipators. *Journal of Hydraulic Engineering*, 117(3), 303–323; 118(10), 1448–1453.

Boes, R. & Hager, W.H. (2003) Hydraulic design of stepped spillways. *Journal of Hydraulic Engineering*, 129(9), 671–679.

Bollaert, E. (2002) Transient water pressures in joints and formation of rock scour due to high-velocity jet impact. In: A. Schleiss (ed) *Communication* 13. Laboratory of Hydraulic Constructions (LCH). Ecole polytechnique fédérale de Lausanne, EPFL, Lausanne.

Bollaert, E. & Schleiss, A.J. (2003a) Scour of rock due to the impact of plunging high velocity jets 1: a state-of-the-art review. *Journal of Hydraulic Research*, 41(5), 451–464.

Bollaert, E. & Schleiss, A.J. (2003b) Scour of rock due to the impact of plunging high velocity jets 2: experimental results of dynamic pressures at pool bottoms and in one- and two-dimensional closed end rock joints. *Journal of Hydraulic Research*, 41(5), 465–480.

Bollaert, E.F.R. (2004) A comprehensive model to evaluate scour formation in plunge pools. *International Journal of Hydropower & Dams*, 11(4), 94–101.

Bollaert, E. & Schleiss, A.J. (2005) Physically based model for evaluation of rock scour due to high-velocity jet impact. *Journal of Hydraulic Engineering*, 131(4), 153–165.

Bollaert, E.F.R. (2012) Wall jet rock scour in plunge pools: a quasi-3D prediction model. *International Journal of Hydropower & Dams*, 19(4), 70–77.

Bollaert, E., Duarte, R., Pfister, M., Schleiss, A. & Mazvidza, D. (2012) Physical and numerical model study investigating plunge pool scour at Kariba Dam. Proceedings of 24th *ICOLD Congress*, Kyoto, Q94(R17), 241–248.

Bollaert, E.F.R., Munodawafa, M.C. & Mazvidza, D.Z. (2013) Kariba Dam plunge pool scour: quasi-3D numerical predictions. *La Houille Blanche*, 68(1), 42–49.

Bormann, N.E. & Julien, P.Y. (1991) Scour downstream of grade-control structures. *Journal of Hydraulic Engineering*, 117(5), 579–594; 118(7), 1066–1073.

Canepa, S. & Hager, W.H. (2003) Effect of jet air content on plunge pool scour. *Journal of Hydraulic Engineering*, 129(5), 358–365; 130(11), 1128–1130;

Chanson, H. (1991) Aeration of a free jet above a spillway. *Journal of Hydraulic Research*, 29(5), 655–667.

Chaudhry, M.H. (1993) *Open-channel Flow*. Prentice Hall, Englewood Cliffs NJ.

Chee, S.P. & Yuen, E.M. (1985) Erosion of unconsolidated gravel beds. *Canadian Journal of Civil Engineering*, 12(3), 559–566.

Chen, T.-C. & Yu, Y.-S. (1965) Pressure distributions on spillway flip buckets. *Journal of the Hydraulics Division* ASCE, 91(HY2), 51–63; 91(HY6), 171–172; 92(HY5), 193–195.

Chow, V.T. (1959) *Open Channel Hydraulics*. McGraw-Hill, New York.

Coyne, A. (1951) Observations sur les déversoirs en saut de ski [Observations on ski jumps]. Proceedings of 4th *ICOLD Congress*, New Delhi, Q12(R89), 737–756 (in French).

D'Agostino, V. (1994) Indagine sullo scavo a valle di opere trasversali mediante modello fisico a fondo mobile [Study on scour using physical modelling]. *L'Energia Elettrica*, 71(2), 37–51 (in Italian).

D'Agostino, V. & Ferro, V. (2004) Scour on alluvial bed downstream of grade control structures. *Journal of Hydraulic Engineering*, 130(1), 24–37.

Dai, Z., Ning, L. & Miao, L. (1988) Some hydraulic problems of slit-type flip buckets. Proceedings of International Symposium *Hydraulics for High Dams*, Beijing, 1–8.

Damle, P.M., Venkatraman, C.P. & Desai, S.C. (1966) Evaluation of scour below ski-jump buckets of spillways. *Golden Jubilee Symposia, Model and Prototype Conformity*, 1, 154–163.

Del Campo, A., Trincado, J. & Rossello, J.G. (1967) Experience obtained in the operation of spillways in 'Saltos del Sil' hydroelectric system. Proceedings of 9th *ICOLD Congress*, Istamboul, Q33(R21), 347–364.

Dhillon, G.S., Sakhuja, V.S. & Paul, T.C. (1981) Measures to contain throw of flip bucket jet in installed structures. *Journal of Irrigation and Power India*, 38(3), 237–245.

Duarte, R., Bollaert, E., Schleiss, A.J. & Pinheiro, A. (2012) Dynamic pressures around a confined block impacted by plunging aerated high-velocity jets. Proceedings of 2nd *European IAHR Congress*, Munich (CD-Rom).

Duarte, R, Schleiss, A.J. & Pinheiro, A. (2013) Dynamic pressure distribution around a fixed confined block impacted by plunging and aerated water jets. Proceedings of 35th *IAHR World Congress*, Chengdu (CD-Rom).

Duarte, R. (2014) Influence of air concentration on rock scour development and block stability in plunge pools. In: Schleiss, A.J. (ed) *Communication* 59, Laboratory of Hydraulic Constructions (LCH), Ecole polytechnique fédérale de Lausanne, EPFL, Lausanne.

Duarte, R., Schleiss, A.J. & Pinheiro, A. (2015) Influence of jet aeration on pressures around a block embedded in a plunge pool bottom. *Environmental Fluid Mechanics*, 15(3), 673–693.

Duarte, R., Schleiss, A.J. & Pinheiro, A. (2016a) Effect of pool confinement on pressures around a block impacted by plunging aerated jets. *Canadian Journal of Civil Engineering*, 43(3), 201–210.

Duarte, R., Pinheiro, A. & Schleiss, A.J. (2016b) Dynamic response of an embedded block impacted by aerated high-velocity jets. *Journal of Hydraulic Research*, 54(4), 399–409.

Duarte, R., Pinheiro, A. & Schleiss, A.J. (2016c) An enhanced physically based scour model for considering jet air entrainment. *Engineering*, 2(3), 294–301.

Elevatorski, E.A. (1958) Trajectory bucket-type energy dissipators. *Journal of the Power Division* ASCE, 84(PO1, 1553), 1–17.

Ervine, D.A. (1976) The entrainment of air in water. *Water Power and Dam Construction*, 28(12), 27–30.

Ervine, D.A. & Falvey, H.T. (1987) Behaviour of turbulent water jets in the atmosphere and in plunging pools. *Proceedings of the Institution of Civil Engineers*, 83(1), 295–314; 85(2), 359–363.

Ervine, D.A. & Falvey, H.T. (1997) Pressure fluctuations on plunge pool floors. *Journal of Hydraulic Research*, 35(2), 257–279; 37(2), 272–288.

Ervine, D.A. (1998) Air entrainment in hydraulic structures: a review. *Proceedings of ICE, Water, Maritime and Energy*, 130(3), 142–153.

Federspiel, M.P.E.A. (2014) Response of an embedded block impacted by high-velocity jets. In: Schleiss, A.J. (ed) *Communication*, 47, Laboratory of Hydraulic Constructions (LCH), Ecole polytechnique fédérale de Lausanne, EPFL, Lausanne.

Federspiel, M., Bollaert, E. & Schleiss, A.J. (2011) Dynamic response of a rock block in a plunge pool due to asymmetrical impact of a high-velocity jet. In: Proceedings of 34th *IAHR World Congress* Brisbane, Australia, 2404–2411 (CD-Rom).

Fiorotto, V. & Rinaldo, A. (1992) Fluctuating uplift and lining design in spillway stilling basins. *Journal of Hydraulic Engineering*, 118(4), 578–596.

Fiorotto, V. & Salandin, P. (2000) Design of anchored slabs in spillway stilling basins. *Journal of Hydraulic Engineering*, 126(7), 502–512.

Fiorotto, V. & Caroni, E. (2007) Discussion of Forces on plunge pool slabs: influence of joints location and width. *Journal of Hydraulic Engineering*, 133(10), 1182–1184.

Gao, J., Liu, Z. & Guo, J. (2011). Newly achievements on dam hydraulic research in China. In: Valentine, E.M., Apelt, C.J., Ball, J. & Chanson, H. (eds) Proceedings of 34th *IAHR World Congress*, Brisbane, 2436–2443 (CD-Rom).

Gerodetti, M. (1982) Auskolkung eines felsigen Flussbettes: Modellversuche mit bindigen Materialen zur Simulation des Felsens [Rock scour: model experiments involving cohesive material to simulate rock]. In: D. Vischer (ed) *Arbeitsheft* 5, VAW, ETHZ, Zürich (in German).

Gong, Z., Liu, S., Xie, S. & Lin, B. (1987) Flaring Gate Piers. In: Kia, A.R. & Albertson, M.L. (eds) *Design of Hydraulic Structures*, Colorado State University, Fort Collins, 139–146.

Gunko, F.G., Burkov, A.F., Isachenko, N.B., Rubinstein, G.L., Soloviova, A.G. & Yudiitsky, G.A. (1965) Research on the hydraulic regime and local scour of river bed below spillways of high-head dams. Proceedings of 11th *IAHR Congress*, Leningrad, 1(50), 1–14.

Hager, W.H. (1998) Plunge pool scour: early history and hydraulicians. *Journal of Hydraulic Engineering*, 124(12), 1185–1187.

Hager, W.H., Unger, J. & Oliveto, G. (2002) Entrainment criterion for bridge piers and abutments. *River Flow*, 2002, (2), 1053–1058. D. Bousmar, Y. Zech, eds. Balkema, Lisse.

Hager, W.H. & Schleiss, A.J. (2009) *Constructions hydrauliques: Ecoulements stationnaires* [Hydraulic structures: steady flows]. PPUR, Lausanne (in French).

Hager, W.H. (2010) *Wastewater Hydraulics: Theory and Practice*, 2nd ed. Springer, Berlin, Germany.

Häusler, E. (1980) Zur Kolkproblematik bei Hochwasser-Entlastungsanlagen an Talsperren mit freiem Überfall [On the scour problem at dam spillways with free overfall]. *Wasserwirtschaft*, 70(3), 97–99 (in German).

Heller, V., Hager, W.H. & Minor, H.-E. (2005) Ski jump hydraulics. *Journal of Hydraulic Engineering*, 131(5), 347–355; 132(10), 1115–1117.

Heller, V. (2011) Scale effects in physical hydraulic engineering models. *Journal of Hydraulic Research*, 49(3), 293–306; 50(2), 244–250.

Henderson, F.M. & Tierney, D.G. (1962) Flow at the toe of a spillway. *La Houille Blanche*, 17(6), 728–739; 18(1), 42–50.

Jalalzadeh, A.A., Fouladi, C., Mehinrad, A. & Bayatmakoo, K. (1994) Spillway damage and the construction of a new flip bucket at Karun I. *Hydropower and Dams*, 1(11), 66–69.

Johnson, G. (1967) The effect of entrained air in the scouring capacity of water jets. Proceedings of 12th *IAHR Congress,* Fort Collins, 3, 218–226.

Johnson, G. (1977) Use of a weakly cohesive material for scale model scour tests in flood spillway design. Proceedings of 17th *IAHR Congress,* Baden-Baden, 4(C63), 509–512.

Juon, R. & Hager, W.H. (2000) Flip bucket without and with deflectors. *Journal of Hydraulic Engineering*, 126(11), 837–845.

Kawakami, K. (1973) A study on the computation of horizontal distance of jet issued from ski-jump spillway. *Proceedings of Japanese Society of Civil Engineers*, 58(5), 37–44 [in Japanese].

Khatsuria, R.M. (2005) *Hydraulics of Spillways and Energy Dissipators*. Dekker, New York.

Kotoulas, D. (1967) Das Kolkproblem unter besonderer Berücksichtigung der Faktoren 'Zeit' und 'Geschiebemischung' im Rahmen der Wildbachverbauung [The scour problem by considering the effects of time and sediment mixture in torrent rehabilitation]. Schweizerische Anstalt für das forstliche Versuchswesen, *Mitteilung* 43(1), 1–67 (in German).

Kraatz, W. (1965) Flow characteristics of a free circular water jet. Proceedings of 11th *IAHR Congress*, Leningrad, 1(44), 1–13. IAHR, Delft.

Kramer, K., Hager, W.H. & Minor, H.-E. (2006) Development of air concentration on chute spillways. *Journal of Hydraulic Engineering*, 132(9), 908–915.

Lenau, C.W. & Cassidy, J.J. (1969) Flow through spillway flip bucket. *Journal of the Hydraulics Division* ASCE, 95(HY2), 633–648.

Liu, P.Q., Dong, J.R. & Yu, C. (1998) Experimental investigation of fluctuation uplift on rock blocks at the bottom of the scour pool downstream of Three-Gorges Spillway. *Journal of Hydraulic Research*, 36(1), 55–68.

Liu, P.Q. (1999) *Mechanism of Energy Dissipation and Hydraulic Design for Plunge Pools Downstream of Large Dams*. China Institute of Water Resources and Hydropower Research, Beijing PRC.

Liu, P.Q. & Li, A.H. (2007) Model discussion on pressure fluctuation propagation within lining slab joints in stilling basins. *Journal of Hydraulic Engineering*, 133(6), 618–624.

Lucas, J., Hager, W.H. & Boes, R.M. (2013) Deflector effect on chute flows. *Journal of Hydraulic Engineering*, 139(4), 444–449; 140(6), 07014008.

Machado, L.I. (1982) O sistema de dissipação de energía proposto para a Barragem de Xingo [On the system of energy dissipation for the Xingo Dam]. Transactions of International Symp. *Layout of Dams in Narrow Gorges* ICOLD, Brazil, 1, 253–262.

Mahzari, M., Arefi, F., Schleiss, A. (2002) Dynamic response of the drainage system of a cracked plunge pool liner due to free falling jet impact. Proc. Intl. Workshop *Rock scour due to falling high-velocity jets* Lausanne: 227-237, A. Schleiss, E. Bollaert, eds. Balkema: Rotterdam.

Mahzari, M. & Schleiss, A.J. (2010) Dynamic analysis of anchored concrete linings of plunge pools loaded by high velocity jet impacts issuing from dam spillways. *Dam Engineering*, 20(4), 307–327.

Maitre, R. & Obolensky, S. (1954) Study of some flow characteristics in the downstream part of spill-ways. *La Houille Blanche*, 9(7/8), 481–511.

Manso, P., Fiorotto, V., Bollaert, E. & Schleiss, A.J. (2004) Discussion of Effect of jet air content on plunge pool scour. *Journal of Hydraulic Engineering*, 130(11), 1128–1130.

Manso, P. (2006) The influence of pool geometry and induced flow patterns in rock scour by high-velocity plunging jets. In: Schleiss, A. (ed) *Communication* 25. Laboratory of Hydraulic Construction (LCH). Ecole polytechnique fédérale de Lausanne, EPFL, Lausanne.

Manso, P., Bollaert, E. & Schleiss, A.J. (2008) Evaluation of high-velocity plunging jet-issuing characteristics as a basis for plunge pool analysis. *Journal of Hydraulic Research*, 46(2), 147–157.

Manso, P.A., Bollaert, E.F.R. & Schleiss, A.J. (2009) Influence of plunge pool geometry on high-velocity jet impact pressures and pressure propagation inside fissured rock media. *Journal of Hydraulic Engineering*, 135(10), 783–792.

Marcano, A., Patiños, A. & Castro, C. (1988) Selection of energy dissipators for the spillways of Lower Caroni projects. Proceedings of the 16th *ICOLD Congress*, San Francisco, Q63(R45), 745–767.

Martins, R. (1973) Contribution to the knowledge on the scour action of free jets on rocky river-beds. Proceedings of 11th *ICOLD Congress*, Madrid, Q41(R44), 799–814.

Martins, R.B.F. (1975) Scouring of rocky riverbeds by free-jet spillways. *Water Power & Dam Construction*, 27(4), 152–153.

Martins, R. (1977) Cinemática do jacto livre no ámbito das estructuras hidráulicas [Free-jet kinematics in the scope of hydraulic structures]. *Memória* 486. LNEC, Lisboa.

Mason, P.J. (1982) The choice of hydraulic energy dissipator for dam outlet works based on a survey of prototype usage. *Proceedings of the Institution of Civil Engineers*, 72(1), 209–219; 74(1), 123–126.

Mason, P.J. (1983) Energy dissipating crest splitters for concrete dams. *Water Power & Dam Construction*, 35(11), 37–40.

Mason, P.J. (1984) Erosion of plunge pools downstream of dams due to the action of free-trajectory jets. *Proceedings of ICE*, 76(2), 523–537; 78(4), 991–999.

Mason, P.J. & Arumugam, K. (1985) Free jet scour below dams and flip buckets. *Journal of Hydraulic Engineering*, 111(2), 220–235; 113(9), 1192–1205.

Mason, P.J. (1989) Effects of air entrainment on plunge pool scour. *Journal of Hydraulic Engineering*, 115(3), 385–399; 117(2), 256–265.

Mason, P.J. (1993) Practical guidelines for the design of flip buckets and plunge pools. *Water Power & Dam Construction*, 45(9/10), 40–45.

Mason, P.J. (2011) Plunge pool scour: an update. *Hydropower & Dams*, 18(6), 123–124.

Melo, J.F., Pinheiro, A.N. & Ramos, C.M. (2006) Forces on plunge pool slabs: influence of joints location and width. *Journal of Hydraulic Engineering*, 132(1), 49–60.

Mih, W.C. (1982) Scouring effects of water jets impinging on non-uniform streambeds. In: Smith, P.E. (ed) Proceedings of Conference *Applying Research to Hydraulic Practice*, Jackson MI, ASCE, New York, 270–279.

Mih, W.C. & Kabir, J. (1983) Impingement of water jets on nonuniform streambeds. *Journal of Hydraulic Engineering*, 109(4), 536–548.

Mirtskhulava, T.E., Dolidze, I.V. & Magomeda, A.V. (1967) Mechanism and computation of local and general scour in non-cohesive, cohesive soils and rock beds. Proceedings of 12th *IAHR Congress*, Fort Collins, 3, 169–176.

Noret, C., Girard, J.-C., Munodawafa, M.C. & Mazvidza, D.Z. (2013) Kariba Dam on Zambezi River: stabilizing the natural plunge pool. *La Houille Blanche*, 68(1), 34–41.

Novak, P., Guinot, V., Jeffrey, A. & Reeve, D.E. (2010) *Hydraulic Modeling: An Introduction.* Spon, Abingdon UK.

Oliveto, G. & Hager, W.H. (2002) Temporal evolution of clear-water pier and abutment scour. *Journal of Hydraulic Engineering*, 128(9), 811–820.

Orlov, V. (1974) Die Bestimmung des Strahlsteigwinkels beim Abfluss über einen Sprungschanzen-Überfall (Determination of jet rise angle of flow discharging over a ski jump). *Wasserwirtschaft-Wassertechnik*, 24(9), 320–321 (in German).

Pagliara, S., Hager, W.H. & Minor, H.-E. (2004a) Plunge pool scour in prototype and laboratory. Proceedings of International Conference *Hydraulics of Dams and River Structures* Tehran: 165–172. Balkema: Lisse.

Pagliara, S., Hager, W.H. & Minor, H.-E. (2004b) Plunge pool scour formulae: an experimental verification. Proceedings of 29th *Convegno di Idraulica e Costruzioni Idrauliche*, Trento, 1, 1131–1138.

Pagliara, S., Hager, W.H. & Minor, H.-E. (2006) Hydraulics of plane plunge pool scour. *Journal of Hydraulic Engineering*, 132(5), 450–461.

Pagliara, S., Amidei, M. & Hager, W.H. (2008a) Hydraulics of 3D plunge pool scour. *Journal of Hydraulic Engineering*, 134(9), 1275–1284.

Pagliara, S., Hager, W.H. & Unger, J. (2008b) Temporal evolution of plunge pool scour. *Journal of Hydraulic Engineering*, 134(11), 1630–1638.

Peterka, A.J. (1964) Hydraulic design of stilling basins and energy dissipators. *USBR Engineering Monograph* 25, Denver, US.

Pfister, M. & Hager, W.H. (2009) Air concentration characteristics of drop- and deflector-generated jets. Proceedings of 33rd *IAHR Congress*, Vancouver. IAHR, Madrid, pp. 4909–4916.

Pfister, M. (2012) Jet impact angle on chute downstream of aerator. Proceedings of the 4th IAHR International Symposium on *Hydraulic Structures*, Porto (P), 7, 1-9 (CD-ROM).

Pfister, M., Hager, W.H. & Boes, R.M. (2014) Trajectories and air flow features of ski-jump-generated jets. *Journal of Hydraulic Research*, 52(3), 336–346.

Pfister, M. & Schleiss, A.J. (2015) Ski jumps, jets and plunge pools. In: Chanson, H. (ed) *IAHR Monograph Energy Dissipation in Hydraulic Structures*. CRC Press, Taylor & Francis Group, London.

Pfister, M., Hager, W.H. & Boes, R.M. (2016) Closure to Trajectories and air flow features of ski jump-generated jets. *Journal of Hydraulic Research*, 54(2), 247. DOI: 10.1080/00221686.2016. 1140092.

Prasad, R.K. (1984) Pressure distribution on ski-jump buckets. *Irrigation and Power*, 41(3), 311–317.

Quintela, A.C. & Da Cruz, A.A. (1982) Cabora-Bassa dam spillway: conception, hydraulic model studies and prototype behaviour. Transactions of International Symposium *Layout of Dams in Narrow Gorges*, ICOLD, Brazil, 1, 301–309.

Rajan, B.H. & Shivashankara Rao, K.N. (1980) Design of trajectory buckets. *Irrigation and Power* India, 37(1), 63–76.

Rajaratnam, N. (1976) *Turbulent Jets*. Elsevier, Amsterdam NL.

Rajaratnam, N. & Berry, B. (1977) Erosion by circular turbulent wall jets. *Journal of Hydraulic Research*, 15(3), 277–289.

Rajaratnam, N. (1980) Erosion by circular wall jets in cross flow. *Journal of the Hydraulics Division* ASCE, 106(HY11), 1867–1883.

Rajaratnam, N. (1982a) Erosion by submerged circular jets. *Journal of the Hydraulics Division* ASCE, 108(HY2), 262–267; 109(HY2), 324.

Rajaratnam, N. (1982b) Erosion by unsubmerged plane water jets. Proceedings of Conference *Applying Research to Hydraulic Practice*. ASCE, New York, 280–288.

Ramos, C.M. (1982) Energy dissipation on free jet spillways: bases for its study in hydraulic models. Transactions of International Symposium *Layout of Dams in Narrow Gorges*, ICOLD, Rio de Janeiro, Brazil, 1, 263–268.

Reinius, E. (1986) Rock erosion. *Water Power and Dam Construction*, 38(6), 43–48.

Rhone, T.J. & Peterka, A.J. (1959) Improved tunnel spillway flip buckets. *Journal of the Hydraulics Division* ASCE, 85(HY12), 53–76.

Schleiss, A.J. (2002) Scour evaluation in space and time: the challenge of dam designers. In: Schleiss, A.J. & Bollaert, E. (eds) Proceedings of International Workshop *Rock Scour Due to Falling High-velocity Jets* Lausanne. CRC Balkema, Rotterdam, 3–22.

Schmocker, L. (2006) Belüftungs-Eigenschaften von Skisprüngen [Aeration characteristics of ski jumps]. *Master Thesis*, VAW, ETH Zurich, Zürich (unpublished, in German).

Schmocker, L., Pfister, M., Hager, W.H. & Minor, H.-E. (2008) Aeration characteristics of ski jump jets. *Journal of Hydraulic Engineering*, 134(1), 90–97.

Schoklitsch, A. (1932) Kolkbildung unter Überfallstrahlen [Scour below overflow jets]. *Wasserwirtschaft*, 25(24), 341–343 (in German).

Şentürk, F. (1994) *Hydraulics of Dams and Reservoirs*. Water Resources Publications, Highlands Ranch CO.

Shivashankara Rao, K.N. (1982) Design of energy dissipators for large capacity spillways. Transactions of International Symposium *Dams in Narrow Gorges*, ICOLD, Rio de Janeiro, 1, 311–328.

Spurr, K.J.W. (1985) Energy approach to estimating scour downstream of a large dam. *Water Power & Dam Construction*, 37(7), 81–89; 38(2), 7–8.

Steiner, R., Heller, V., Hager, W.H. & Minor, H.-E. (2008) Deflector ski jump hydraulics. *Journal of Hydraulic Engineering*, 134(5), 562–571.

Straub, L.G. & Anderson, A.G. (1960) Self-aerated flow in open channels. *Transactions of ASCE*, 125, 456–481; 125, 485–486.

Unger, J. & Hager, W.H. (2007) Flow features in plunge pools. Proceeding of 32nd *IAHR Congress*, Venice, (407), 1–10. IAHR, Madrid.

US Bureau of Reclamation (1978) Hydraulic design of stilling basins and energy dissipators. *Engineering Monograph*, 25. USBR, Denver.

US Bureau of Reclamation (1987) *Design of Small Dams*. Water Resources Technical Publication. US Department of the Interior, Washington DC.

Varshney, R.S. & Bajaj, M.L. (1970) Ski-jump buckets on Indian dams. *Journal of Irrigation and Drainage*, 27(4), 383–393.

Veronese, A. (1937) Erosioni di fondo a valle di un scarico [Bottom erosion due to an outlet]. *Annali dei Lavori Pubblici*, 75(9), 717–726 (in Italian).

Vischer, D.L. & Hager, W.H. (1995) *Energy Dissipators*. IAHR Hydraulic Structures Design Manual 9. Balkema, Rotterdam.

Vischer, D.L. & Hager, W.H. (1998) *Dam Hydraulics*. Wiley, Chichester UK.

Whittaker, J.G. & Schleiss, A. (1984) Scour related to energy dissipaters for high head structures. *Mitteilung* 73. Versuchsanstalt für Wasserbau, Hydrologie und Glaziologie VAW, ETH, Zürich.

Yuditskii, G.A. (1963) Actual pressure on the channel bottom below ski-jump spillways. *Izvestiya Vsesoyuznogo Nauchno – Issledovatel – Skogo Instituta Gidrotekhniki*, 67, 231–240.

Zai-Chao, L. (1987) Influence area of atomization-motion downstream of dam. Proceedings of 22nd *IAHR Congress*, Lausanne, B, 203–208.

Zhenlin, D., Lizhong, N. & Longde, M. (1988) Some hydraulic problems of slit-type buckets. Proceedings of International Symposium *Hydraulics for Large Dams*, Beijing, 287–294.

Bibliography

Air entrainment

Adler, A. (1934) Momentaufnahmen von Flüssigkeitsstrahlen [Instant images of liquid jets]. *Z. Physikalischen und Chemischen Unterricht*, 47(2), 63–69 (in German).

Bin, A.K. (1984) Air entrainment by plunging liquid jets. In: Kobus, H. (ed) Proceedings of Symposium *Scale Effects in Modelling Hydraulic Structures*, Esslingen, 5(5), 1–6.

Bin, A.K. (1988) Minimum air entrainment velocity of vertical plunging liquid jets. *Chemical Engineering Science*, 43(2), 379–389.

Chen, T.-F. & Davis, J.R. (1964) Disintegration of a turbulent water jet. *Journal of the Hydraulics Division* ASCE, 90(HY1), 175–206; 90(HY6), 273.

Di Silvio, G. (1970) Indagine sperimentale sulle condizioni di similitudine di grossi getti liquidi liberamenti effluenti [Experimental study on the similitude conditions of large free jets]. *L'Energia Elettrica*, 47(5), 285–294 (in Italian).

Dodu, J. (1957) Etude de la couche limite d'air autour d'un jet d'eau à grande vitesse [Study of the boundary layer around a high-speed water jet]. Proceedings of 7th *IAHR Congress*, D(6), 1–12 (in French).

Dodu, J. (1964) Contribution à l'étude de la dispersion des jets liquides à grande vitesse [Contribution to the study of the dispersion of high-speed liquid jets]. *Publications Scientifiques et Techniques du Ministère de l'Air*, 407. Paris (in French).

Ervine, D.A. & Falvey, H.T. (1988) Aeration in jets and high velocity flows. In: Burgi, P. (ed) Proceedings of *Model-Proto Correlation of Hydraulic Structures*, ASCE, New York, pp. 22–55.

Ervine, D.A., McKeogh, E. & Elsawy, E.M. (1980) Effect of turbulence intensity on the rate of air entrainment by plunging water jets. *Proceedings of the Institution of Civil Engineers*, 69(2), 425–445.

Freeman, J.R. (1889) Experiments relating to hydraulics of fire streams. *Transactions of ASCE*, 21, 303–461.

Hänlein, A. (1931) Über den Zerfall eines Flüssigkeitsstrahles [On the disintegration of a liquid jet]. *Forschung auf dem Gebiete des Ingenieurwesens*, 2(4), 139–149 (in German).

Homma, M. & Sakamoto, T. (1959) Breaking up of a circular-shaped stream of falling water. Proceedings of 8th *IAHR Congress*, Montreal, 2(D), 1–14.

Horeni, P. (56). Desintegration of a free jet of water in air. *Prace a Studie* 93. VUV, Prague (in Czech, with English summary).

Howe, J.W. & Posey, C.J. (1940) Characteristics of high-velocity jets. In: Howe, J.W. & McNown, J.S. (eds) Proceedings of 3rd *Hydraulics Conference*. University of Iowa, Iowa City IA, 315–332.

Hoyt, J.W., Taylor, J.J. & Runge, C.D. (1974) The structure of jets of water and polymer solution in air. *Journal of Fluid Mechanics*, 63(4), 635–640.

Hoyt, J.W. & Taylor, J.J. (1977) Waves on water jets. *Journal of Fluid Mechanics*, 83(1), 119–127.

Isachenko, N.B. & Chanishvili, A.G. (1968) Distortion of a jet diverted by a flip bucket. *Izvestiya Vsesoyuznogo Nauchno*, 87, 253–262.

Kobus, H. (1985) An introduction to air-water flows in hydraulics. *Mitteilung* 61. Institut für Wasserbau, Universität, Stuttgart.

Littaye, G. (1942) Contribution à l'étude des jets liquides [Contribution to the study of liquid jets]. *Publications Scientifiques et Techniques du Secrétariat d'Etat à l'Aviation* 181. Blondel la Rougery, Paris (in French).

Marchetti, M. & delle Chiaje, M. (1961) Caratterizzazione di getti frazionati da lance antincendi: distribuzione della loro precipitazione su terreno orizzontale [Characterization of fractioned jets issued with fire nozzles: distribution of precipitation on a horizontal bottom]. *Antincendio e Protezione Civile*, 18(32), 3–17 (in Italian).

Mathieu, F. (1960) Contribution à l'étude de l'action d'un jet gazeux sur la surface libre d'un liquide [Contribution to the study of the action of a gazous jet on a liquid surface]. *Revue Universelle des Mines*, 103(7), 309–321 (in French).

Mathieu, F. (1962) Nouvelles recherches sur l'action d'un jet gazeux sur la surface libre d'un liquid au repos [New researches on the action of a gazous jet on the free surface of a still liquid]. *Revue Universelle des Mines*, 105(7), 482–499 (in French).

McKeogh, E.J. & Elsawy, E.M. (1980) Air retained in pool by plunging water jet. *Journal of the Hydraulics Division* ASCE, 106(HY10), 1577–1593.

McKeogh, E.J. & Ervine, D.A. (1981) Air entrainment rate and diffusion pattern of plunging liquid jets. *Chemical Engineering Science*, 36(7), 1161–1172.

Neményi, P. (1933) *Wasserbauliche Strömungslehre* [Hydraulic structures]. Barth, Leipzig (in German).

Oehler, T. (1930) Der Wasserstrahl und seine Auflösung in Tropfen [The water jet and its disintegration into drops]. *Technische Mechanik und Thermodynamik*, 1(9), 329–338 (in German).

Oguey, P., Mamin, M. & Baatard, F. (1951) La dispersion du jet d'eau à grande vitesse [The dispersion of a high-speed water jet]. *Publications* 16, 17. Ecole Polytechnique de l'Université de Lausanne, Lausanne (in French).

Ohnesorge von, W. (1936) Die Bildung von Tropfen an Düsen und die Auflösung flüssiger Strahlen [The generation of drops at nozzles and the disintegration of fluid jets]. *ZAMM*, 16(6), 355–358 (in German).

Ohnesorge von, W. (1937) Anwendung eines kinematographischen Hochfrequenzapparates mit mech-anischer Regelung der Belichtung zur Aufnahme der Tropfenbildung und des Zerfalls flüssiger Strahlen [Application of a cinemato-graphic high-frequency apparatus involving mechanical control to shoot liquid jet disintegration]. *Dissertation* TH Berlin. Triltsch, Würzburg (in German).

Rajaratnam, N., Steffler, P.M., Rizvi, S.A.H. & Smy, P.R. (1994) An experimental study of very high velocity circular water jets in air. *Journal of Hydraulic Research*, 32(3), 461–470.

Schweitzer, P.H. (1937) Mechanism of disintegration of liquid jets. *Journal of Applied Physics*, 8(8), 513–521.

Sene, K.J. (1988) Air entrainment by plunging jets. *Chemical Engineering Science*, 43(10), 2615–2623.

Taylor, J.J. & Hoyt, J.W. (1983) Water jet photography: techniques and methods. *Experiments in Fluids*, 1(3), 113–120.

Van de Sande, E. & Smith, J.M. (1973) Surface entrainment of air by high velocity water jets. *Chemical Engineering Science*, 28(5), 1161–1168.

Weber, C. (1931) Zum Zerfall eines Flüssigkeitsstrahles [On the disintegration of a liquid jet]. *ZAMM*, 11(2), 136–154 (in German).

Bucket flow

Auroy, F. (1951) Les évacuateurs de crues du barrage de Chastang [The spillway of Chastang Dam]. Proceedings of 4th *ICOLD Congress*, New Delhi, Q12(R82), 661–686 (in French).

Bourgin, F., Lebreton, J.-C., Pera, J.-A., Rueff, J., Vormeringer, F., Dauzier, C., Guilhot, S., Le May, Y. & Terrassa, X. (1967) Considérations sur la conception d'ensemble des ouvrages d'évacuation provisoires et définitives des barrages [Considerations on the overall concept of provisional and definite overflow structures]. Proceedings of 9th *ICOLD Congress*, Istamboul, Q33(R27), 459–493 (in French).

Corlin, B. & Larsen, P. (1979) Experience from some overflow and side spillways. Proceedings of 13th *ICOLD Congress*, New Delhi, Q50(R37), 627–647.

Engez, N. (1959) Betrachtungen über die Hochwasserüberfälle in Bauart Sprungschanze bei der neuen Elmali-Talsperre (Istanbul) [Considerations on the ski jumps at the new Elmali Dam]. *Die Bautechnik*, 36(10), 378–386 (in German).

Galindez, A., Guinea, P.M., Lucas, P. & Aspuru, J.J. (1967) Spillways in a peak flow river. Proceedings of the 9th *ICOLD Congress*, Q33(R22), 365–389.

Groupe de Travail (1979) Quelques problèmes particuliers posés par les déversoirs à grande capacité: Tapis de protection, dissipation d'énergie par déflecteurs, et aération et cavitation produite par les écoulements à grande vitesse [Some particular problems posed by large capacity spillways: protection carpets, energy dissipation by deflectors, and aeration and cavitation due to high-speed flows]. Proceedings of the 13th *ICOLD Congress*, New Delhi, Q50(R38), 649–673 (in French).

Istamboul, Erpicum, S., Archambeau, P., Dewals, B. & Pirotton, M. (2010) Experimental investigation of the effect of flip bucket splitters on plunge pool geometry. *Wasserwirtschaft*, 100(4), 110–112.

Jiang, S. (1988) Turbulent boundary layer in vertical curves. *Journal of Hydraulic Engineering*, 114(7), 783–797.

Orlov, V.G. (1969) Plotting of a free-surface curve and a pressure distribution curve to create favourable conditions immediately below a spillway dam. *Izvestiya VNIIG*, 87, 282–303 (In Russian, with English summary).

Orlov, V. (1975) Die Strömung über eine Sprungschanze mit Strahlablenker [The flow over a ski jump with jet deflection]. *Wissenschaftliche Zeitschrift der TU Dresden*, 24(1), 259–262 (in German).

Roose, K. & Gilg, B. (1973) Comparison of the hydraulic model tests carried out for the ski-jump shaped spillways of the Smokovo and Paliodherli Dams. Proceedings of 11th *ICOLD Congress*, Madrid, Q41(R37), 671–689.

Rouvé, G. (1971) Zur Dimensionierung von Sprungschanzen im Wasserbau [On the design of ski jump buckets]. *Wasserwirtschaft*, 61(9), 259–261 (in German).

Sentürk, F. (1967) Examples of ski jumps from Turkey. Proceedings of 9th *ICOLD Congress*, Istamboul, 6(Q33), 247–262.

Shuguang, L. & Zhengxiang, L. (1988) Gravity-affected potential flows past spillway flip buckets. *Journal of Hydraulic Engineering*, 114(4), 409–427.

Tinney, E.R., Barnes, W.E., Rechard, O.W. & Ingram, G.R. (1961) Free streamline theory for segmental jet deflections. *Journal of the Hydraulics Division* ASCE, 87(HY5), 135–148; 88(HY1), 159–161; 88(HY3), 215–216.

Yang, S.L. (1994) Dispersive-flow energy dissipator. *Journal of Hydraulic Engineering*, 120(12), 1401–1408.

General

Coyne, A. (1944) Prototypes modernes de barrages et d'usines hydroélectriques [Modern dam and hydro-electric prototypes]. *Travaux*, 28(2), 25–29 (in French).

Lin, B., Li, G. & Chen, H. (1986) Hydraulic research in China. *Journal of Hydraulic Engineering*, 113(1), 47–60.

Townson, J.M. (1991) *Free-surface Hydraulics*. Unwin Hyman, London.

Jet trajectory

Aspuru, J.J. & Blanco, J.L. (1979) Comparaison entre prototype et modèle réduit: Cas de l'évacuateur supérieur en saut de ski du Barrage d'Almendra (Tormes), Espagne [Comparison between prototype and scale model: superior ski jump of Almendra Dam, Spain]. Proceedings of 13th *ICOLD Congress*, New Delhi, C(15), 227–234 (in French).

Escande, L. (1953) Evacuateur de crues pour barrages de grande hauteur [Spillways for dams of large height]. *Nouveaux compléments d'hydraulique*, 1, 16–33. Publications Scientifiques et Techniques du Ministère de l'Air, 280. Paris (in French).

Fagerburg, T.L. (1979) Spillway velocity measurement and flip bucket trajectory, Raystone Dam, Juniata River, Pennsylvania. *Miscellaneous Paper* HL-79-3. US Army Corps of Engineers, Baltimore MD.

Hänlein, K. (1918) Über Flüssigkeitsstrahlen [On liquid jets]. *Zeitschrift Gesamte Turbinenwesen*, 15(20), 173–174; 15(21), 184–186; 15(22), 189–191; 15(23), 200–202 (in German).

Hatton, A.P. & Osborne, M.J. (1979) The trajectories of large fire fighting jets. *International Journal of Heat & Fluid Flow*, 1(1), 37–41.

Lencastre, A. (1985) State of the art on the dimensioning of spillways for dams. *La Houille Blanche*, 40(1), 19–52 (in French).

Puertas, J. & Dolz, J. (1994) Criterios hidráulicos para el diseño de cuencos de disipación de energía en presas bóveda con vertido libre por coronación [Hydraulic criteria for the design of energy dissipation basins for arch dams with overfall spillage]. *Ph.D. Thesis*. Polytechnical University of Catalunya, Barcelona, Spain (in Spanish).

Rouse, H., Howe, J.W. & Metzler, D.E. (1952) Experimental investigation of fire monitors and nozzles. *Transactions of ASCE*, 117, 1147–1188.

Satterly, J. & Gilmore, O.A. (1938) Further study of an inclined liquid jet. *Transactions of Royal Society Canada*, Series 3, 32, 17–27.

Scimemi, E. (1934) Sui getti liquidi [On liquid jets]. *Atti* del Reale Istituto Veneto di Scienze, *Lettere ed Arti*, 94(2), 579–586 (in Italian).

Zunker, F. (1928) Beachtenswertes bei Weitstrahlregnern [Noteworthies for long-range sprinklers]. *Kulturtechniker*, 31(5), 319–342 (in German).

Scour

Ade, F. & Rajaratnam, N. (1998) Generalized study of erosion by circular horizontal turbulent jets. *Journal of Hydraulic Research*, 36(4), 613–635.

Aderibigbe, O. & Rajaratnam, N. (1995) Reduction of scour below submerged impinging jets by screens. *Proceedings of ICE, Water, Maritime & Energy*, 112(3), 215–226.

Aderibigbe, O. & Rajaratnam, N. (1996) Erosion of loose beds by submerged circular impinging vertical turbulent jets. *Journal of Hydraulic Research*, 34(1), 19–33; 35(4), 567–574.

Aderibigbe, O. & Rajaratnam, N. (1998) Effect of sediment gradation on erosion by plane turbulent wall jet. *Journal of Hydraulic Engineering*, 124(10), 1034–1042.

Annandale, G.W. (1994) Taking the scour out of water power. *Water Power & Dam Construction*, 46(11), 46–49.

Atkinson, C.H. & Overbeeke, K. (1973) Design and operation of Lower Notch Spillway. Proceedings of 11th *ICOLD Congress*, Madrid, Q41(R53), 957–976.

Bohrer, J.G., Abt, S.R. & Wittler, R.J. (1998) Predicting plunge pool velocity decay of free falling, rectangular jet. *Journal of Hydraulic Engineering*, 124(10), 1043–1048.

Breusers, H.N.C. & Raudkivi, A.J., eds. (1991) *Scouring*. Hydraulic Structures Design Manual 2, Balkema, Rotterdam.

Fahlbusch, F.E. (1994) Scour in rock riverbeds downstream of large dams. *Hydropower & Dams*, 1(1), 30–32.

Guinea, P.M., Lucas, P. & Aspuru, J.J. (1973) Selection of spillways and energy dissipators. Proceedings of 11th *ICOLD Congress*, Madrid, Q41(R66), 1233–1254.

Hartung, F. & Häusler, E. (1973) Scours, stilling basins and downstream protection under free overfall jets at dams. Proceedings of 11th *ICOLD Congress*, Madrid, Q41(R3), 39–56.

Häusler, E. (1972) Der Kolk unterhalb der Kariba-Staumauer [The scour downstream of Kariba Dam]. *Tiefbau*, 14(10), 953–962 (in German).

Hemphill, R.W. & Bramley, M.E. (1989) *Protection of River and Canal Banks: A Guide to Selection and Design*. Butterworths, London.

Hoffmans, G.J.C.M. (1998) Jet scour in equilibrium phase. *Journal of Hydraulic Engineering*, 124(4), 430–437.

ICOLD (1987) Spillways for dams. *ICOLD Bulletin* 58. International Commission for Large Dams, Paris.

Kulkarni, V.N. & Patel, I.C. (1981) Ski-jump spillway for India's Ukai Dam. *Water Power & Dam Construction*, 33(9), 44–48.

Mason, P.J. & Arumugam, K. (1985) A review of 20 years of scour development at Kariba Dam. Proceedings of 2nd International Conference of the *Hydraulics of Floods & Flood Control*, Cambridge, 63–71. BHRA, Cranfield.

Novak, P., Moffat, A.I.B., Nalluri, C. & Narayanan, R. (1989) *Hydraulic structures*. Unwin Hyman, London.

Otto, B. (1989) Scour potential of highly stressed sheet-jointed rocks under obliquely impinging plane jets. *Ph.D. Thesis*. James Cook University of North Queensland, Townsville.

Poreh, M. & Hefez, E. (1967) Initial scour and sediment motion due to an impinging submerged jet. Proceedings of 12th *IAHR Congress*, Fort Collins, 3(C2), 9–16.

Quintela, A. de C., Fernandes, J. De S. & da Cruz, A.A. (1979) Barrage de Cahora-Bassa: problèmes posés par le passage des crues pendant et après la construction [Cahora-Bassa Dam: problems of floods during and after construction]. Proceedings of 13th *ICOLD Congress*, New Delhi, Q50(R41), 713–739; 5(Q50), 536–540 (in French).

Rajaratnam, N. (1981) Erosion by plane turbulent jets. *Journal of Hydraulic Research*, 19(4), 339–358.

Rajaratnam, N., Aderibigbe, O. & Pochylko, D. (1995) Erosion of sand beds by oblique plane water jets. *Proceedings of ICE, Water, Maritime & Energy*, 112(1), 31–38.

Rajaratnam, N. & Macdougall, R.K. (1983) Erosion by plane wall jets with minimum tailwater. *Journal of Hydraulic Engineering*, 109(7), 1061–1064.

Robinson, A.R. (1971) Model study of scour from cantilevered outlets. *Transactions of ASAE*, 14(3), 571–581.

Rouse, H. (1939) Criteria for similarity in the transportation of sediment. Proceedings of 1st Hydraulics Conference. *Studies in Engineering*, 20, 33–49. Iowa State University, Iowa City IA.

Stein, O.R., Alonso, C.V. & Julien, P.Y. (1993) Mechanics of jet scour downstream of a headcut. *Journal of Hydraulic Research*, 31(6), 723–738; 32(6), 951–957.

Stein, O.R. & Julien, P.Y. (1994) Sediment concentration below free overfall. *Journal of Hydraulic Engineering*, 120(9), 1043–1059.

Vega da Cunha, L. (1976) Time evolution of local scour. *Memoria*, 477. LNEC, Lisboa.

Wieland, H. & Mock, F.J. (1984) Felskolk als alternatives Tosbecken [Rock scour as alternative stilling basin]. *Wasser und Boden*, 36(9), 434–436 (in German).

Wood, I.R. (1960) The design of the spillway for the Upper Cotter Dam. *Journal of the Institution of Engineers* Australia, 32(6), 105–112.

Yildiz, D. & Üzücek, E. (1994) Prediction of scour depth from free falling flip bucket jets. *Water Power & Dam Construction*, 46(11), 50–56.

Turbulent jet impact

Ackermann, N.L. & Undan, R. (1970) Forces from submerged jets. *Journal of the Hydraulics Division* ASCE, 96(HY11), 2231–2240; 97(HY8), 1242–1243; 97(HY9), 1530–1533.

Albertson, M.L., Dai, Y.B., Jensen, R.A. & Rouse, H. (1948) Diffusion of submerged jets. *Proceedings of ASCE*, 74, 1571–1596.

Beltaos, S. (1976) Oblique impingement of plane turbulent jets. *Journal of the Hydraulics Division* ASCE, 102(HY9), 1177–1192.

Beltaos, S. & Rajaratnam, N. (1974) Impinging circular turbulent jets. *Journal of the Hydraulics Division* ASCE, 100(HY10), 1313–1328.

Beltaos, S. & Rajaratnam, N. (1978) Impingement of axisymmetric developing jets. *Journal of Hydraulic Research*, 15(4), 311–326.

Cola, R. (1965) Energy dissipation of a high velocity vertical jet entering a basin. Proceedings of 11th *IAHR Congress*, Leningrad, Soviet Union 1(52), 1–13.

Cola, R. (1966) Diffusione di un getto piano verticale in un bacino d'acqua d'altezza limitata [Diffusion of a plane vertical jet in a basin of limited depth]. *L'Energia Elettrica*, 43(11), 649–667 (in Italian).

Davanipour, T. & Sami, S. (1977) Short jet impingement. *Journal of the Hydraulics Division* ASCE, 103(HY5), 557–567; 103(HY12), 1504–1506; 104(HY7), 1105–1107.

Donaldson, C.P. & Snedeker, R.S. (1971) A study of free jet impingement. *Journal of Fluid Mechanics*, 45(2), 281–319, 45(3), 477–512.

Elsawy, E.M. & McKeogh, E.J. (1977) Study of self-aerated flow with regard to modelling criteria. Proceedings of 17th *IAHR Congress*, Baden-Baden, 1(A60), 475–482.

Franzetti, S. (1980) Pressioni idrodinamiche sul fondo di una vasca di smorzamento [Hydrodynamic pressures on the bottom of an energy dissipator]. *L'Energia Elettrica*, 57(6), 280–285 (in Italian).

Franzetti, S. & Tanda, M.G. (1987) Analysis of turbulent pressure fluctuation caused by a circular impinging jet. International Symp. *New Technology in Model Testing in Hydraulic Research*, India, 85–91.

Gill, M.A. (1979) Hydraulics of rectangular vertical drop structures. *Journal of Hydraulic Research*, 17(4), 289–302.

Gutmark, E., Wolfshtein, M. & Wygnanski, I. (1978) The plane turbulent impinging jet. *Journal of Fluid Mechanics*, 88(4), 737–756.

Häusler, E. (1966) Dynamische Wasserdrücke auf Tosbeckenplatten infolge freier Überfallstrahlen bei Talsperren [Dynamic water pressures on stilling basin floors due to free overflow jets at dams]. *Wasserwirtschaft*, 56(2), 42–49 (in German).

Hom-ma, M. (1953) An experimental study on water fall. Proceedings of the 5th *IAHR Congress*, Minneapolis, 477–481.

Kobus, H., Leister, P. & Westrich, B. (1979) Flow field and scouring effects of steady and pulsating jets impinging on a movable bed. *Journal of Hydraulic Research*, 17(3), 175–192.

Leske, W. (1963) Die Umlenkung eines frei herabfallenden, ebenen Strahls an einer horizontalen Sohle [Deflection of a freely falling, plane jet at a horizontal bottom]. *Wissenschaftliche Zeitschrift der TU Dresden*, 12(6), 1749–1765 (in German).

Lewis, T. (1996) Predicting impact velocities of developed jets. *Dam Foundation Erosion Study*, Internal Report, Colorado State University, Fort Collins CO.

Looney, M.K. & Walsh, J.J. (1984) Mean-flow and turbulent characteristics of free and impinging jet flows. *Journal of Fluid Mechanics*, 147, 397–429.

May, R.W.P. & Willoughby, I.R. (1991) Impact pressures in plunge pool basins due to vertical falling jets. *Report* SR 242. HR Wallingford, UK.

Oesterlen, F. (1926) Zur Ausbildung von Turbinensaugrohren [On the design of turbine draft tubes]. *Hydraulische Probleme*, 111–132. VDI-Verlag, Berlin (in German).

Petrillo, A. (1985) Analisi delle fluctuazioni di pressione sul fondo di una vasca di smorzamento di getti circolari [Analyses of pressure fluctuations due to circular jets on the bottom of a stilling basin]. *Idrotecnica*, 12(3), 153–163 (in Italian).

Petrillo, A. & Ranieri, M. (1985) Azioni dinamiche di un getto circolare [Dynamic action of a circular jet]. *Idrotecnica*, 12(2), 71–83 (in Italian).

Ramos, C.M. (1979) Statistical characteristics of the pressure field of crossed flows in energy dissipation structures. Proceedings of 13th *ICOLD Congress*, New Delhi, Q50(R24), 403–415.

Reich, F. (1927) Umlenkung eines freien Flüssigkeitsstrahles an einer ebenen Platte [Deflection of a free fluid jet at a plane plate]. *Zeitschrift VDI*, 71(8), 261–264 (in German).

Schach, W. (1934) Umlenkung eines freien Flüssigkeitsstrahles an einer ebenen Platte [Deflection of a free fluid jet at a plane plate]. *Ingenieur-Archiv*, 5(4), 245–265 (in German).

Tao, C.G., Ji Yong, L. & Xingrong, L. (1985) *Efeito do impacto, no leito do rio, da lâmina descarregada sobre uma barragem-abóbada* [Effect of the impact, on the riverbed, of the jet spilled over an arch dam]. Translation from Chinese by Pinto de Campos, J.A. LNEC, Lisboa (in Portuguese).

Taylor, G. (1966) Oblique impact of a jet on a plane surface. *Philosophical Transactions* of the Royal Society, London, Series A 260: 96–100.

Werner, W. (1966) Der schräge Stoss eines freien Flüssigkeitsstrahles auf eine ebene Platte [The oblique impact of a free fluid jet on a plane plate]. *Wissenschaftliche Zeitschrift der TU Dresden*, 15(1), 45–52 (in German).

Xerez, A.C., Granger Pinto, H. & Cunha Ferreira, A. (1967) Problèmes hydrauliques dans les barrages voutes et voutes multiples [Hydraulic problems at dams with single and multiples arches]. Proceedings of 9th *ICOLD Congress*, Istamboul, Q33(R18), 297–313 (in French).

Xu-Duo-Ming (1983) *Pressão no fundo de um canal devido ao choque de um jacto plano, e suas características de fluctuação* [Pressure on the bottom of a channel due to the impact of a plane jet, and its fluctuation characteristics]. Translation from Chinese by J.A. Pinto de Campos, Lisboa.

Chapter 7 Frontispiece Seujet free-surface diversion channel on Rhone River, Geneva (a) weir and ship lock during construction of first phase, (b) overview with cofferdam during first phase (Courtesy for both photos Division des Ponts et des Eaux, Genève, Switzerland). Note that these two photos complement each other, allowing for a view both at the approach flow reach, and into the cofferdam

River diversion structures

7.1 Introduction

A dam is a major engineering structure closing the lower portion of a valley. Both during the construction and later in service, a dam should not create any risk during floods for the downstream portion of the dam. For example, floods may arrive during construction and are then particularly dangerous because the structure is not finalized and thus prone to damage, including the failure of cofferdams. The design discharge for dam diversions is discussed in Chapter 1. Although the diversion structures under discussion are often provisional during construction, they have to be correctly and economically designed because any failure may have catastrophic consequences in the tailwater. A substitute waterway, the so-called dam diversion, must be operational for the river to bypass the dam site so that floods do neither harm the construction area nor pose an increased risk to the tailwater (Figure 7.1).

In 7.2 and 7.3, two types of diversions are presented, namely the *diversion tunnel* guiding the river integrally around the construction site, and the *diversion channel* guiding the river across the construction site with a local river constriction. A main concern in diversion tunnel hydraulics is the potential transition from free surface to pressurized flow, particularly in regard to the design discharge and the definition of the tunnel roughness. Also, moving hydraulic jumps and low pressure due to limited aeration may produce unstable tunnel flow. Accordingly, choking tunnel flow is of relevance, so that conditions in terms of the choking number are stated. Diversion tunnels are often transferred after dam construction into permanent hydraulic structures as intakes, spillways or bottom outlets.

A *diversion channel* has particular flow features, mainly as regards the flow type. If the constriction rate is too strong, transitional flow occurs, so that the structure is hydraulically comparable with a Venturi flume. For entirely subcritical flow across this constriction, standing waves may be a nuisance causing erosion in the tailwater. The upstream head-discharge curve is an important design basis for the cofferdams and consequently for the economic and safe construction of the dam itself.

A *culvert* is a transition or diversion structure across the dam body mainly during construction for small upstream heads. It may be transferred after construction to a permanent low-level or a bottom outlet. Of particular relevance in culvert design are the various flow types, as described in 7.4. The safe selection and determination of the governing flow type is essential for the correct hydraulic design of culverts.

As mentioned, diversion tunnels bypass a river around a dam site, whereas culverts cross the site, and diversion channels are led merely through the construction site. The principles of all three transition structures are outlined, their ranges of application are indicated and a

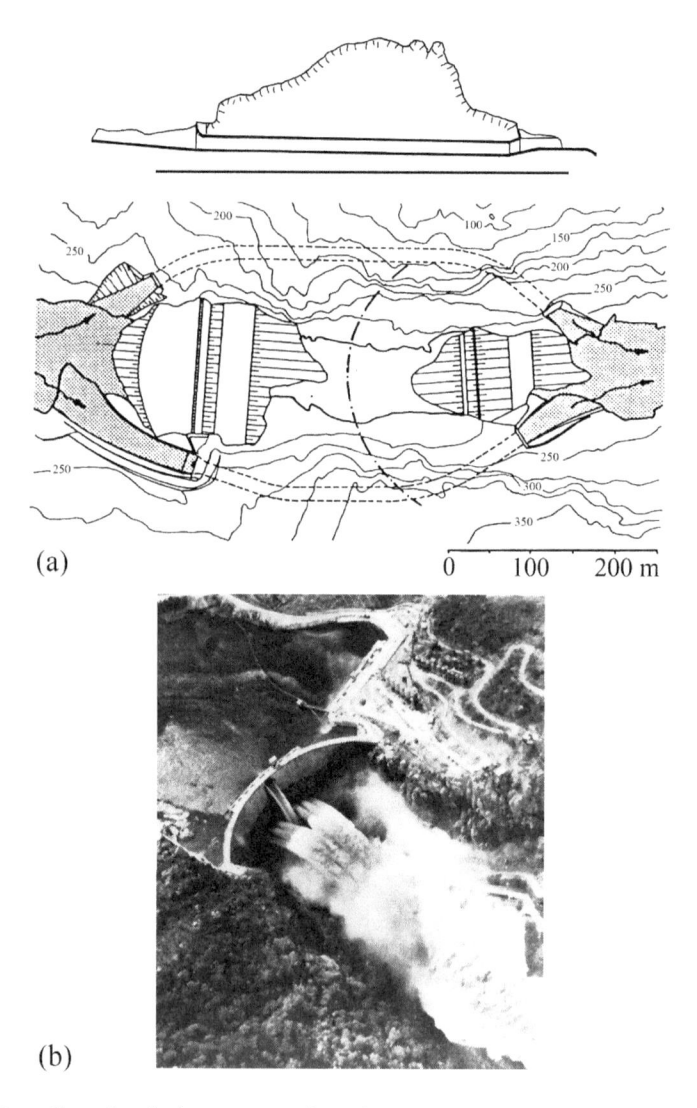

(a)

0   100   200 m

(b)

Figure 7.1 Dam diversion during construction of Cabora Bassa Dam on Zambezi River (Africa)
(a) Plan and section of diversion tunnel, (b) final dam (*ICOLD* Q.50, R.41)

simplified hydraulic design guideline is provided. There are a number of additional points to be considered for a particular *in-situ* structure, mainly with regard to aspects of both erosion and sedimentation, and the outlet flow features. These items are site-specific, however, so that a generalized design can hardly be provided. An innovative concept of erosion protection by concrete prisms at the diversion tunnel outlet is presented in 7.2. Means for surface protection of cofferdams to resist potential overtopping are also mentioned. A number of relevant references is listed for further reading.

In addition to the above problems, scour at bridge elements are added in 7.5, given the relevance in dam hydraulics. A distinction between abutment and pier scour has to be made.

It is demonstrated, however, that both issues follow the identical approach, except for a coefficient accounting for the geometrical shape of the two elements. The approach is based on the densimetric particle Froude number, by which the effects of both the sediment and fluid densities are included. This approach applies for flows with a comparatively small free-surface effect, as is typical for low subcritical flows as occur in relation with bridge scour.

7.2 Diversion tunnel

7.2.1 Introduction

Figure 7.2 shows the photograph of a typical diversion tunnel intake. The construction site of the dam is protected by two cofferdams: The upstream dam guides the water away from the valley into the diversion tunnel whereas the downstream cofferdam returns the discharge back to the river. For larger design floods, several diversion tunnels may be arranged (Figure 7.2). Diversion tunnels can also run around both valley sides (Figure 7.1a). These tunnels should operate for the construction design flood under *free-surface flow* conditions for reasons of flow stability, safety, and to divert float. The cross-sectional geometry and the layout are designed for their final purpose: Either, the tunnel is closed after construction, or it will be connected to a bottom outlet (Chapter 8), a spillway structure as a morning glory spillway (Chapter 3), or a water intake structure (Chapter 8).

The layout of the diversion tunnel depends on the available bottom slope. For small slopes generating subcritical flow, there is practically no margin except for no deposition of material, whereas for larger slopes with a supercritical flow, the tunnel intake zone includes an acceleration reach followed by an equilibrium reach limited by the erosion potential of the

Figure 7.2 Intake to diversion tunnel of Karakaya Dam (Turkey) during construction (*Hydraulic Engineering Works* Italstrade: Milan 1990)

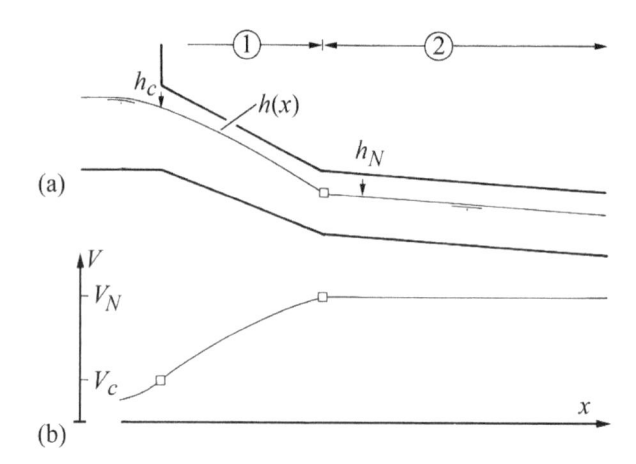

Figure 7.3 Intake of diversion tunnel (a) section, (b) streamwise increase of average velocity with ① Acceleration reach, ② Equilibrium reach

tunnel bottom portion (Figure 7.3). The flow has to be accelerated from the intake to the critical velocity V_c at the grade break to the uniform (subscript N) velocity reach of average velocity V_N along the equilibrium reach. If the acceleration reach is practically a chute, then the tunnel bottom slope can be relatively large. For excavated tunnels, the maximum slope is nearly 10% due to constructional reasons. The slope in the equilibrium reach should satisfy the following requirements:

- Reduction of cross-sectional area for economic reasons
- Generation of stable supercritical flow, i.e. with a Froude number larger than 1.5–2 to inhibit undular flow
- Allowance for sediment transport, if not retained upstream from the cofferdam or intake
- Prevention of tunnel abrasion (Chapter 9; Vischer *et al.*, 1997).

For concrete-lined diversion tunnels, flow velocities should not exceed some 10 m/s to avoid excessive abrasion.

The diversion tunnel consists of three major portions, namely:

- Inlet for flow acceleration equipped with stop logs for closure
- Tunnel
- Outlet with energy dissipation measures.

Each of these three portions has hydraulic particularities summarized below.

7.2.2 Inlet flow

The purpose of the inlet structure is to accelerate the flow to the tunnel velocity, resulting in a smooth transition from the river or the reservoir to the tunnel flow, and to provide sufficient air for ensuring atmospheric pressure in the tunnel. Because the cross-sectional shape of the channel guiding the river to the diversion tunnel is nearly trapezoidal, and the tunnel is either

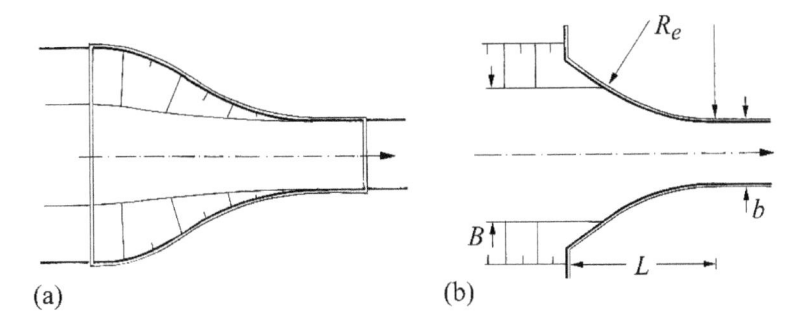

Figure 7.4 Intake structures according to (a) Hinds (1928), (b) Smith (1967)

of horseshoe or circular shape, a complicated transition shape is needed for acceptable flow conditions. Furthermore, the width of the upstream inlet crest has to be wide enough that river closure is practically feasible. In the case of one cofferdam, the required head increase to force the river flow into the diversion tunnel should remain below 2 m to allow for successful river closure. The critical flow depth on the inlet crest should thus not exceed 1.3 m for the prevailing discharge during river closure (Schleiss, 2018).

Hinds (1928) proposed a warped wall inlet (Figure 7.4a). For small and medium discharges, a simpler structure was proposed by Smith (1967), whose length L is only $1.25(B-b)$, with B and b as the upstream bottom width, and tunnel width, respectively (Figure 7.4b). The intake (subscript e) radius is $R_e = 1.65(B-b)$. The energy loss ΔH_e across the intake is expressed in terms of the tailwater (subscript t) velocity $V_t = Q/(bh_t)$ and the width ratio $\beta_e = b/B$ as

$$\Delta H_e = 0.06(1-\beta_e)\frac{V_t^2}{2g}. \tag{7.1}$$

To prevent a standing wave pattern, the Froude number $\mathsf{F}_t = V_t/(gh_t)^{1/2}$ based on the tailwater (subscript t) depth should not exceed 0.67.

For a transition from subcritical to supercritical flow at the intake, the location of critical flow is determined using the theorem of Jaeger (Jaeger, 1949; Hager and Castro-Orgaz, 2016). The location of critical flow for a certain discharge is where the energy head H has a minimum (Figure 7.5). For any inlet geometry this corresponds to the section for which the energy head of the critical flow has its maximum. The free-surface profile is determined with conventional backwater curves, starting at the critical point with the critical depth $h_c = (Q^2/gb^2)^{1/3}$, moving in the upstream direction for subcritical, and in the downstream direction for supercritical flows (Chow, 1959; Hager and Schleiss, 2009; Hager, 2010). Based on the application of the generalized Bernoulli equation, a certain level difference between the inlet crest and the tunnel invert is required to accelerate the flow along the inlet to reach uniform flow at the tunnel entrance. A detailed procedure for its design is provided by Schleiss (2018).

7.2.3 Tunnel flow

As mentioned, uniform flow conditions along the diversion tunnel should be reached for the design flood. For lower discharges, free surface flow in a tunnel is determined as for open channels involving backwater curves. Figure 7.6 shows a definition sketch with x as the longitudinal coordinate, z as invert height, h as flow depth, V as average cross-sectional velocity,

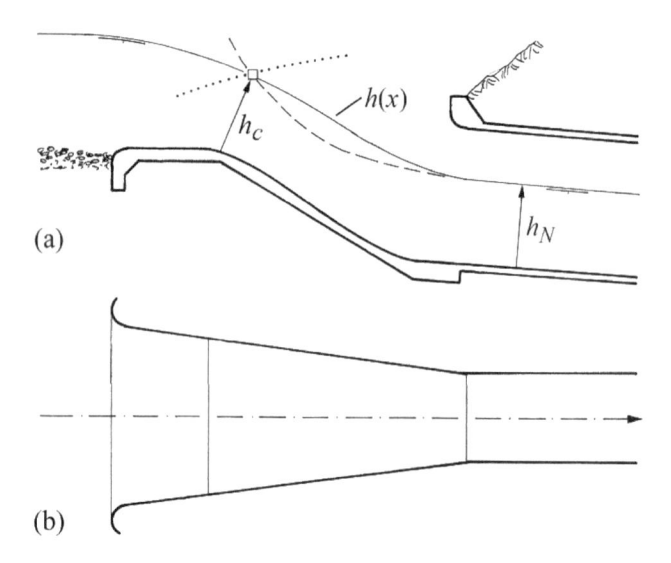

Figure 7.5 Transition from sub- to supercritical flow at diversion intake with bottom slope increase and width decrease with (a) (- - -) uniform flow depth profile $h_N(x)$, (. . .) critical flow depth profile $h_c(x)$, (—) effective free-surface profile $h(x)$, (b) plan

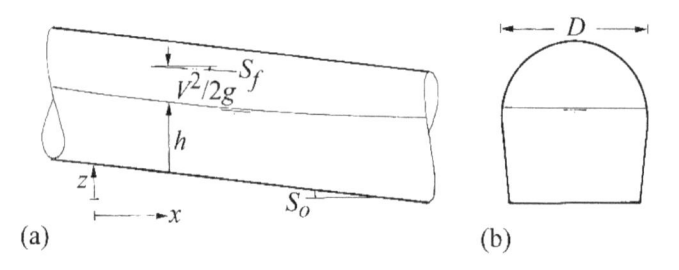

Figure 7.6 Backwater curves in diversion tunnel (a) longitudinal section, (b) transverse section

S_o as the bottom slope, and S_f as friction slope. According to the generalized Bernoulli equation, the change of energy head $H = h + V^2/2g$ is equal to the energy increase due to the bottom slope $S_o = -dz/dx$ minus the friction slope S_f, i.e. (Chow, 1959)

$$\frac{d(h+V^2/2g)}{dx} = S_o - S_f. \tag{7.2}$$

Equation (7.2) is based on both uniform velocity and hydrostatic pressure distributions, applying to most tunnel diversions. The *Froude number* is defined as ratio of average flow velocity V to the propagation velocity c of a shallow water wave

$$F = V / c. \tag{7.3}$$

The latter is equal to the square root of the cross-sectional area A divided by the free-surface width $B_s = \partial A / \partial h$ times the gravity acceleration g, i.e.

$$c = [gA / (\partial A / \partial h)]^{1/2}. \tag{7.4}$$

Equating Eqs. (7.3, 7.4) yields for the backwater curve in *prismatic* channels

$$\frac{\mathrm{d}h}{\mathrm{d}x} = \frac{S_o - S_f}{1 - \mathsf{F}^2}. \tag{7.5}$$

Backwater curves are dominated by two particular cases (Hager, 2010):

1 $S_o = S_f$, i.e. the *uniform flow* condition, for which the flow depth h does not change along the channel because $\mathrm{d}h/\mathrm{d}x = 0$.
2 $\mathsf{F}^2 = 1$, i.e. the *critical flow* condition, for which the surface profile is theoretically vertical because $\mathrm{d}x/\mathrm{d}h = 0$.

Uniform flow, on the one hand, describes the equilibrium of tractive and resisting forces, comparable to a condition of static flow. It is governed by the bottom slope, fluid viscosity, the boundary roughness pattern, velocity, and the channel geometry. Uniform flow follows the resistance law given below. It is also identified by the uniform flow velocity V_N or the uniform flow depth h_N.

Critical flow, on the other hand, corresponds to the condition of minimum energy head, and the *transition* between subcritical and supercritical flows. The critical flow is also characterized by the critical flow depth h_c. For subcritical flow, velocity V is smaller than the wave propagation velocity c, or the flow depth h is larger than the critical flow depth h_c so that flow perturbations propagate both in the upstream and downstream directions. Subcritical flow is described as smooth, relatively slow, and nearly one-dimensional for which backwater curves are a good approximation.

For supercritical flow, in contrast, the propagation velocity c is smaller than the tunnel velocity V, or $h < h_c$ in terms of flow depths, and perturbations propagate only in the flow direction. These flows are characterized as rough, relatively fast, and two-dimensional, given the oblique surface pattern behind any flow perturbation (Chapter 4). Figure 7.7 shows

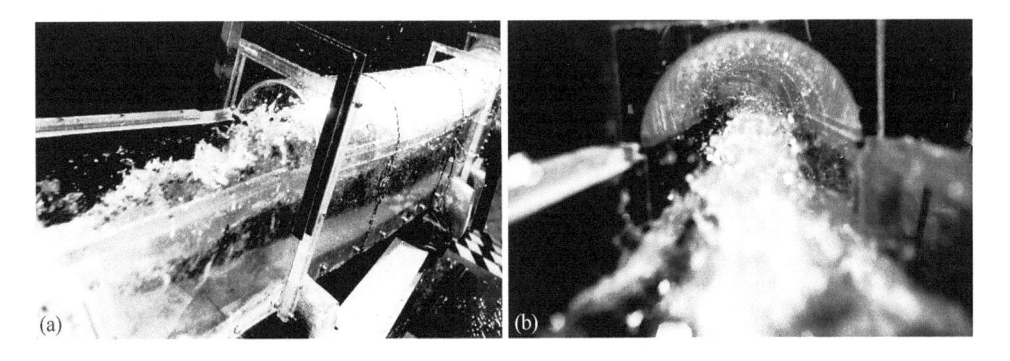

(a)   (b)

Figure 7.7 Flow in curved tunnel with formation of shock waves and air entrainment. Views from (a) upstream, (b) downstream (Courtesy Willi H. Hager)

curved tunnel flow with so-called shock waves along which all perturbations are propagated. These waves tend to flow concentrations so that a supercritical flow is improved by rendering the surface pattern more uniform (Chapter 4).

The *friction slope* S_f for continuous flows is computed according to the resistance laws of hydraulics. The classic resistance equation according to Darcy-Weisbach reads with f as the friction factor, D_h the hydraulic diameter defined as four times the hydraulic radius $R_h = A/P_h$ with P_h as wetted perimeter, and $V^2/2g$ as velocity head

$$S_f = \frac{f}{D_h}\frac{V^2}{2g}. \qquad (7.6)$$

The *friction factor f* depends on:

- Resistance characteristics of the fluid, i.e. kinematic viscosity
- Resistance characteristics of boundary material.

According to the concept of *equivalent roughness*, a uniform sand roughness height k_s [m] causes the resistance effect identical to an arbitrary roughness pattern. The scaling length of the roughness height is the hydraulic diameter D_h, with $\varepsilon = k_s/D_h$ as relative roughness. With $R = VD_h/v$ as the Reynolds number, Colebrook and White (1937) proposed the *universal friction law*

$$f^{-1/2} = -2\log\left[\frac{\varepsilon}{3.7} + \frac{2.51}{Rf^{1/2}}\right]. \qquad (7.7)$$

A flow is turbulent if $R > 2300$. In civil engineering applications, flows are commonly turbulent. For $[\varepsilon/3.7] \ll [2.51/(Rf^{1/2})]$, wall roughness is insignificant and the flow is said to be in the hydraulic *smooth* regime. In contrast, the hydraulic *rough* regime applies if $[\varepsilon/3.7] \gg [2.51/(Rf^{1/2})]$. If both wall roughness and viscosity effects are relevant, the flow is in the transitional regime.

Because the exact definition of the equivalent roughness height k_s is often difficult, and due to the small effect of viscosity for flows in hydraulic structures, the usual approach is to assume these flows as turbulent rough. Then, f depends exclusively on the relative roughness as

$$f^{-1/2} = -2\log[\varepsilon/3.7]. \qquad (7.8)$$

The relevant roughness domain includes $5 \times 10^{-4} < \varepsilon < 5 \times 10^{-2}$. For usual Reynolds numbers R from 10^5 to 10^7, Eq. (7.8) is approximated with the power function

$$V = KS_o^{1/2}R_h^{2/3}. \qquad (7.9)$$

This is the *GMS formula* of the Frenchman G-P. Gauckler (1826–1905), redeveloped by the Irishman Robert Manning (1816–1897), and tested with additional data by the Swiss Albert Strickler (1887–1963). The formula is simple in application, and thus popular. In the USA, it is referred to as the Manning formula. Either of the quantities V, S_o, or R_h are explicitly computed; the Strickler roughness value K, or the Manning friction factor $1/n$, are well-known dimensional quantities to the experienced hydraulic engineer. Table 7.1 provides a summary

Table 7.1 Roughness coefficient *n* according to Manning, and $K = 1/n \ [m^{1/3}/s]$ according to Strickler (Adapted from Chow, 1959)

Type			Minimum	Normal	Maximum
A CLOSED-CONDUIT FLOWING PARTLY FULL					
Metal	Brass, smooth		0.009	0.010	0.013
	Steel,	lockbar and welded	0.010	0.012	0.014
		reveted and spiral	0.013	0.016	0.017
	Cast iron, coated		0.010	0.013	0.014
	Corrugated metal, subdrain		0.017	0.019	0.021
Nonmetal	Cement,	neat	0.010	0.011	0.013
		mortar	0.011	0.013	0.015
	Concrete,	straight and clean	0.010	0.011	0.013
		with bends, some debris	0.011	0.013	0.014
		unfinished, steel form	0.012	0.013	0.014
		unfinished, rough	0.015	0.017	0.020
B LINED OR BUILT-UP CHANNELS					
Metal	smooth surface		0.011	0.012	0.014
	corrugated		0.021	0.025	0.030
Nonmetal	Cement, smooth		0.010	0.011	0.013
Concrete	Trowel finish		0.011	0.013	0.015
	Float finish		0.013	0.015	0.016
	unfinished		0.014	0.017	0.020
	Gunite, good section		0.016	0.019	0.023
	good excavated rock		0.017	0.020	–
	irregular rock		0.022	0.027	–
	vegetal lining		0.030	. . . . .	0.500
C EXCAVATED OR DREDGED					
Earth	straight and uniform and clean		0.016	0.018	0.020
	winding and sluggish	no vegetation	0.023	0.025	0.030
		weeds	0.030	0.035	0.040
		stony bottom	0.025	0.035	0.040
	rock cuts	smooth and uniform	0.025	0.035	0.040
		jagged and irregular	0.035	0.040	0.050
	not maintained	dense weeds	0.050	0.080	0.120
		dense bush, high stage	0.080	0.100	0.140
D NATURAL STREAMS					
Minor streams clean and straight			0.025	0.030	0.033
	clean, winding, with pools		0.033	0.040	0.045
	clean, but with cobbles		0.045	0.050	0.060
	weedy, deep pools, underbrush		0.075	0.100	0.150
	mountain stream	gravel	0.030	0.040	0.050
		boulders	0.040	0.050	0.070
Flood plains no brush, short grass			0.025	0.030	0.035
	cultivated areas, no crop		0.020	0.030	0.040
	brush	scattered	0.035	0.050	0.070
		dense	0.070	0.100	0.160
Major streams regular			0.025	. . . . .	0.060
	irregular		0.035	. . . . .	0.100

of n-values, with more detail given, e.g. by Chow (1959) or Schröder (1990). The actual value of the roughness coefficient is a matter of guess work, finally, and should be carefully selected based on site conditions, experience, and the importance of the structure considered.

Gravel bed rivers, which are typical for the up- and downstream portions of diversion structures, are characterized by the gravel diameter d_n, with n as the percentage of grains by weight smaller than indicated by the value n. For bottom slopes $4 \times 10^{-3} < S_o < 2.5 \times 10^{-2}$ and hydraulic radii of 10^{-1} m $< R_h < 10^1$ m, Strickler (1923) proposed the empirical relation (Hager, 2015)

$$K = 1/n = 21.1d_{50}^{-1/6}. \tag{7.10}$$

Increasing d_{50} gives an increase in roughness. Meyer-Peter and Müller (1948) related their considerations to the characteristic grain diameter d_{90} instead of d_{50}, i.e. to almost the largest sediment size of a riverbed, and proposed in dimensionless form

$$\frac{(1/n)d_{90}^{-1/6}}{g^{1/2}} = 8.2. \tag{7.11}$$

Neill, a discusser of Bray and Davar (1987), proposed the simple dimensional relation

$$1/n = 10S_o^{-1/6}. \tag{7.12}$$

He recommended this check equation for all calculations. Clearly, the grain diameter varies with the river bottom slope and so does the roughness parameter.

Jarrett (1984) proposed for streams of average bottom slope $S_o > 0.2\%$ the formula

$$V = 3.81S_o^{0.21}R_h^{0.83}. \tag{7.13}$$

It was calibrated with data for gravel rivers of $S_o < 4\%$ and hydraulic radii $R_h < 2.1$ m.

Rock tunnels of river diversions have received particular attention, because of:

- Largeness for which the universal friction law was not tested
- Particular surface roughness patterns
- Effects of deposits and organic growths.

The contributions of Colebrook (1958), the ASCE Task Force (1965), and Barr (1973) revealed that their roughness is sometimes considerably larger than anticipated. Algae growth on the concrete lining surfaces increases the flow resistance, causing it to vary seasonally. The design of large tunnels should thus at least include an assessment of the minimum and maximum friction coefficients. Also, an initially smooth surface may become rough due to sediment deposits or abrasion.

7.2.4 Choking flow

Bend flow chokes in a tunnel if its capacity is too small to generate free-surface flow. These flows were studied by Gisonni and Hager (1999) for a 45° deflection angle. The results apply approximately also for other deflection angles, given the particular flow pattern. Figure 7.8

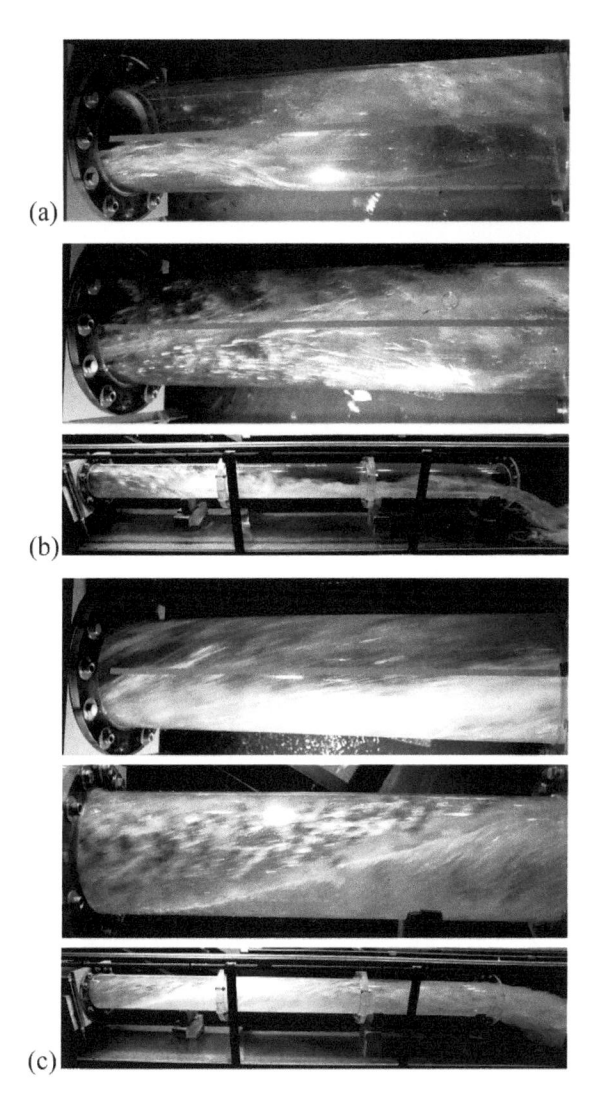

Figure 7.8 Flow downstream of 45°-tunnel bend for y_o = 0.24 and F_o = (a) 5.2, (b) 5.8, (c) 11.8 with general view (top), plan view on bend outflow (center or bottom), and side view (bottom of c) (Gisonni and Hager, 1999)

relates to the straight tailwater reach downstream of the circular tunnel, whose approach flow filling ratio is $y_o = h_o/D$ = 0.24 for a range of approach flow pipe Froude numbers F_o = $Q/(gh_oD^4)^{1/2}$ (Chapter 4), with D as the tunnel diameter, h_o as the approach flow depth for a relative tunnel radius of R/D = 3. For F_o = 5.2 (Figure 7.8a), so-called stratified tunnel flow is generated, for which the air phase is above the water phase. As F_o = 5.8 is reached (Figure 7.8b), the transition from stratified to annular flow occurs just downstream of the bend. Upon further increasing to F_o = 11.8, the entire tailwater reach falls into the annular flow regime (Hager, 2010).

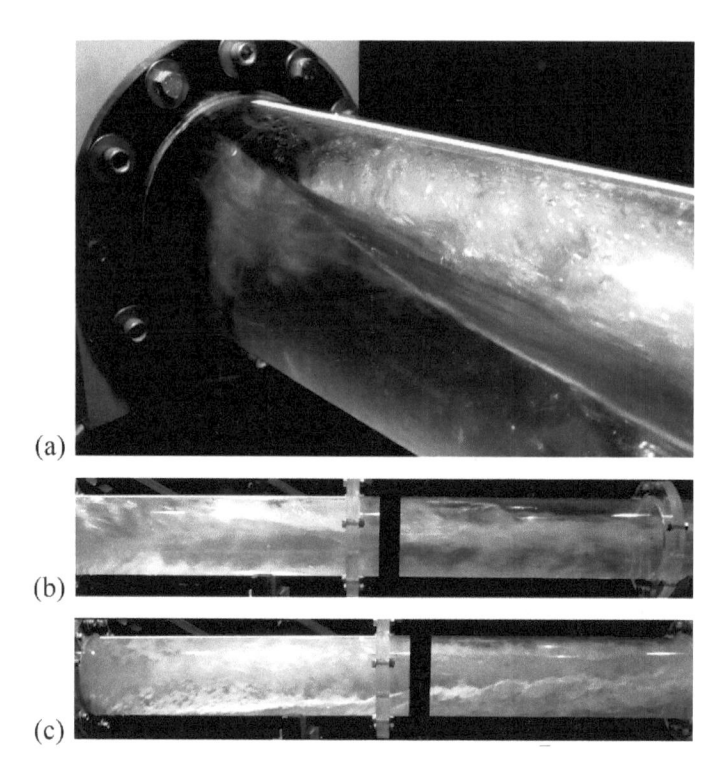

Figure 7.9 Tunnel flow downstream of 45°-deflection for y_o = 0.61 and F_o = (a) 2.0, (b) 2.75, (c) 4.2 (Gisonni and Hager, 1999)

Figure 7.9 relates to a higher filling ratio of y_o = 0.61, for which the flow structure is similar as for y_o = 0.24, but a reduced transition Froude number between stratified and annular flows. Figure 7.9a shows the transverse surface profile consisting mainly of black-water along the outer, and a white-water front along the inner tailwater sides. If either y_o or F_o are increased then the tunnel flow chokes resulting in pressurized two-phase flow. The transition occurs at $F_o \cong 2.7$ (Figure 7.9b, c) indicating annular flow for F_o = 4.2 including a vortex spanning from the bottom of the tunnel outlet toward the top of the bend end. The following relates to the main hydraulic features.

Figure 7.10 shows a definition plot for the flow considered, including the approach flow water discharge Q and the approach flow air discharge Q_A, the distance d_1 from the bend end to the first (subscript 1) wave maximum h_M and height h_2 of the second (subscript 2) wave maximum. As demonstrated by Hager (2010), the sequent depth ratio of the flow in a straight circular tunnel is $Y = h_2/h_1 = F_1^{0.90}$, with subscript 1 relating here to the supercritical approach flow. Choking from free surface to pressurized tunnel flow occurs nominally for $h_1 = D$. Inserting this condition in the previous sequent depth relation leads to the so-called choking number

$$C = F_1/(D/h_1) = Q/(gD^3h_1^2)^{1/2}. \tag{7.14}$$

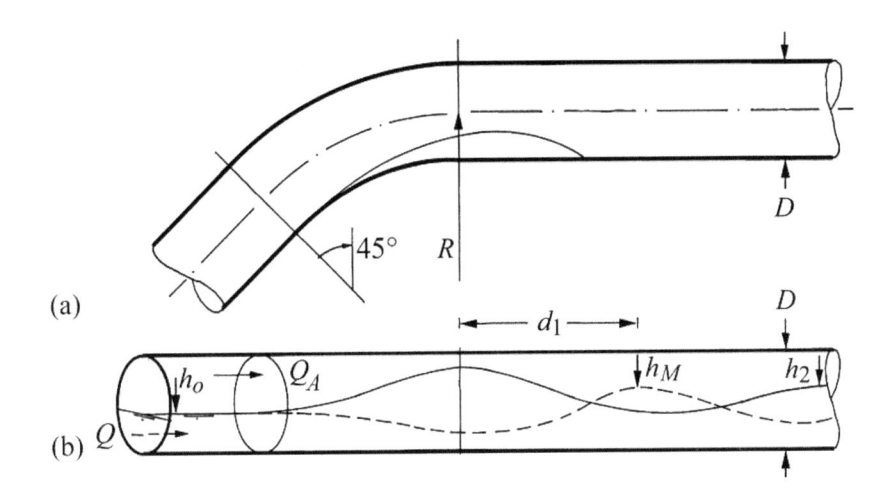

Figure 7.10 Definition plot for supercritical tunnel bend flow with (a) plan, (b) streamwise section

Note that all three govern-ing parameters discharge Q, tunnel diameter D, and approach flow depth h_1 have an almost linear effect on C. The choking number so defined includes both the filling ratio of the approach flow and its dynamics as expressed by the approach flow pipe Froude number.

A tunnel bend generates shock waves for supercritical flow reducing the capacity as compared with a straight tunnel reach. The choking number of tunnel bend flow is from Eq. (7.14) $C_o = F_o/(D/h_o)$ a relevant parameter. The relative wave maximum $Y_M = (h_M/h_o)(h_o/D)^{2/3}$ in the tailwater reach (Figure 7.10b) is (Gisonni and Hager, 1999)

$$Y_M = 1.10 \tanh(1.4C_o), \quad C_o > 0.80. \tag{7.15}$$

For $C_o < 0.80$, the tunnel flow remains subcritical because a hydraulic jump forms at the bend inlet. For $C_o > 1$, the maximum relative wave height remains almost constant at $Y_M = 1.1$ indicating $h_M = 1.1(h_o D^2)^{1/3}$, corresponding to a maximum filling ratio of $h_M/D = 1.1(h_o/D)^{1/3}$. The distance d_1 from bend end to wave 1 follows the expression $d_1/D = 4.8(C_o - 0.8)^{1/2}$. At incipient choking, the first wave is located some 2 tunnel diameters downstream from the tunnel end. Wave 2 downstream of the maximum wave is always smaller, i.e. $h_2 < h_M$, and further waves were hardly visible. No effect of a small streamwise bottom slope on this and the following relations was noted.

Under annular flow downstream of the bend tunnel results a transition to stratified flow at length d_a once the rotational component is insufficient. The relative distance is

$$d_a/D = 0.85(C_o - 1)^2, \quad C_o > 1. \tag{7.16}$$

The tunnel thus issues stratified flow at $C_o = 1$, corresponding to stratified tunnel bend flow. The distance of the first wave crest from the tunnel outlet is

$$d_1/D = 4.8(C_o - 0.8)^{1/2}, \quad C_o > 0.8. \tag{7.17}$$

For incipient choking, the first wave is located about 2 tunnel diameters downstream from the bend end, with a corresponding filling ratio of $h_M/D = 0.97(h_o/D)^{1/3}$.

The air discharge Q_A across the tunnel bend increases with C_o but decreases with the filling ratio, because the cross-sectional area for airflow reduces. Accordingly, the usual air-water discharge ratio as presented by Hager (2010) was generalized to $\beta_{A+} = (Q_A/Q)(1 - y_o)^2$. The tests indicated that β_{A+} increases essentially with C_o if $C_o > 1$. For lower values of C_o there is also airflow but mainly due to free-surface entrainment, as for open-channel flow. The data follow (Gisonni and Hager, 1999)

$$\beta_{A+} = 3\tanh[3(C_o - 1)], \; C_o > 1. \tag{7.18}$$

This indicates a maximum of $Q_A/Q = 3(1-y_o)^2$ for $C_o > 1$, providing values of $Q_A/Q = 1$ to 2. This is a rather large air discharge ratio asking for a separate tunnel air supply.

7.2.5 Outlet structure

The outlet of a diversion tunnel is normally directly connected to the river. For a mobile riverbed, the erosion potential has to be limited by protections as described in 7.2.6. If the diversion tunnel is integrated into a permanent structure, a stilling basin is provided to dissipate excess kinetic energy (Figure 7.11). The latter structure is required either if the tunnel flow is supercritical, or the elevation of the tunnel outlet relative to the river bottom is larger

Figure 7.11 Outlet of bypass for Adda River (Valtellina, Italy) due to large rock slide in 1987, tunnel diameters of 4.2 and 6.0 m (*Italstrade*, Milan 1990)

than needed for subcritical flow to establish. The design of the junction structure involving outlet and river depends on site conditions, so that no general guidelines are available. There are two aspects of relevance, however:

1   Tailwater level should be so low that tunnel flow is *not* submerged, because the tunnel may seal otherwise and yield air-water flows associated with sub-pressure.
2   If the approach flow Froude number to the tailwater stilling basin is too small, problems with the stability of energy dissipation arise, resulting in the formation of weak jumps, asymmetric flow, blowout, and tailwater waves.

If the diversion tunnel is subsequently used as a tunnel spillway, as for a morning glory spillway (Chapter 3), provision for the final design should be made. A ski jump (Chapter 6) is a common design for locations with space limitations.

7.2.6 Erosion protection at tunnel outlet

Overview of protection measures

Water released from diversion tunnels into rivers should not result in riverbed scour, endangering the stability of the tunnel outlet itself or cause failure of any hydraulic structures near the scour zone such as cofferdams. Outlet structures are therefore required to reduce flow velocity and to ensure energy dissipation without dangerous scour (Emami and Schleiss, 2006a, b).

Knowing the expected scour depth and extension, the stability of the outlet structure (or tunnel portal) may be guaranteed by cutoff walls. These reinforced-concrete walls have to be designed deep enough and founded on sound rock. Another possibility is to protect the mobile riverbed by a concrete slab or even a stilling basin. As non-permanent structures, these solutions are expensive and difficult to build in the presence of water, however.

The existence of deep alluvium at the diversion tunnel outlets of the Seymareh and Karun IV Dams in Iran revealed execution problems for these traditional erosion protection structures at the tunnel outlet resulting in high construction cost. Thus, as an innovative protection measure against scour, large unreinforced-concrete prisms were placed in the riverbed close to the outlet of the diversion tunnels (Figure 7.12). These prisms result from dividing concrete cubes diagonally. This method was used successfully as bank and bed erosion protection measures in steep mountain rivers in Switzerland (Schleiss 1998; Meile *et al.*, 2004).

The applications at Karun IV and Seymareh Dams have demonstrated that the protection of the alluvial riverbed with concrete prisms is a promising solution for limiting scour from the points of view safety and economy (Emami and Schleiss, 2006b). Figure 7.13 illustrates the use of concrete prisms as an erosion protection at the outlets. To proof the innovative concept and to establish general design criteria, systematic physical model tests have been performed (Emami, 2004).

For the cases with and without protection, a total of 24 tests were performed in the wide parameter range varying tailwater level, prism size, discharge intensity, length of protected area, and densimetic Froude number involving the riverbed material (Emami, 2004). The test setup is shown in Figure 7.14 with the scour geometry at the end of two tests.

In a first step, the variation of the maximum scour depth $y_{TW} = h_{TW}/D$ downstream of diversion tunnels without any protection was analyzed versus the densimetric Froude number

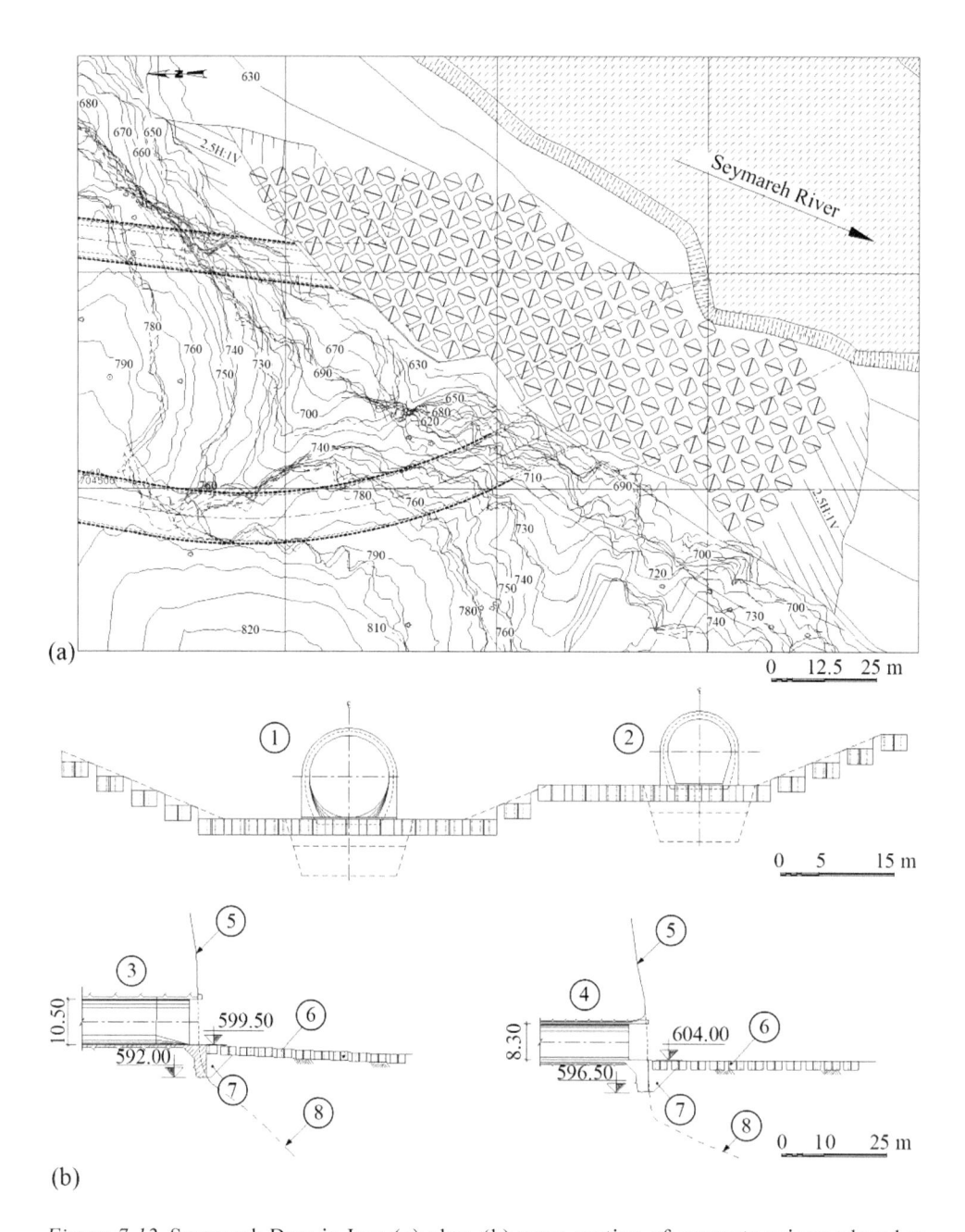

Figure 7.12 Seymareh Dam in Iran (a) plan, (b) cross section of concrete prisms placed as erosion protection at diversion tunnel outlets with ① Tunnel 1, ② Tunnel 2, ③ outlet of Tunnel 1, ④ outlet of Tunnel 2, ⑤ original ground line, ⑥ concrete blocks, ⑦ compacted rockfill, ⑧ assumed rock line (Emami and Schleiss, 2006a)

Figure 7.13 Diagonally divided concrete cubes placed as erosion protection at Diversion Tunnel 1 of Seymareh Dam in Iran with ① lost formwork (Emami and Schleiss, 2006a) (Courtesy Soleyman Emami)

Figure 7.14 Experimental setup showing scour hole versus tailwater level h_{TW} relative to tunnel diameter D for Q = 12.5 l/s and tailwater ratio h_{TW}/D = (a) 1.1 (high), (b) 0.2 (low) (Emami, 2004)

$\mathsf{F}_o = V_o/[(\rho_s/\rho-1)gd_{50}]^{1/2}$ and the tailwater depth h_{TW}. Here, V_o is the velocity at the tunnel outlet of diameter D, ρ_s is sediment density with grain size d_{50} and ρ is fluid density. The ratio h_{TW}/D ranged from 0.1 to 1.1 as is typical at diversion tunnels of dams. The test results were fitted semi-logarithmically as

$$y_{TW} = a_{TW} \cdot \ln\mathsf{F}_o + b_{TW}. \tag{7.19}$$

Table 7.2 Coefficients a_{TW} and b_{TW} of Eq. (7.19) versus relative tailwater depth for $8.5 < F_o$ <14.5 and $0.10 < h_{TW}/D < 1.10$ (Emami, 2004)

Scour hole characteristics	y_{TW}	a_{TW}	b_{TW}
Maximum scour depth	d_{sc}/D	$-0.60(h_{TW}/D) + 1.80$	$1.23(h_{TW}/D) - 2.25$
Maximum scour length	L_{TW}/D	$-0.38(h_{TW}/D) + 13.20$	$6.08(h_{TW}/D) - 21.95$
Distance of d_{sc} from pipe outlet	x_{TW}/D	$+0.86(h_{TW}/D) + 4.49$	$1.00(h_{TW}/D) - 7.97$
Maximum scour width	w_{TW}/D	$-0.42(h_{TW}/D) + 3.53$	$-3.33(h_{TW}/D) + 0.78$

According to the tests, the tailwater level significantly influences the scour hole geometry. The scour hole is moved downstream from the outlet with increasing tailwater level. Consequently, the coefficients a_{TW} and b_{TW} in Eq. (7.19) are affected by h_{TW}. Within the tested parameter ranges, coefficients a_{TW} and b_{TW} depend linearly on h_{TW}/D, as stated in Table 7.2 for the various scour characteristics including the maximum scour depth d_{sc}/D, maximum scour length L_{TW}/D, the distance of maximum scour depth from the outlet x_{TW}/D, and the maximum scour hole width w_{TW}/D.

Empirical scour relations

To develop dimensionless relations for the protection of scour holes by *concrete prisms*, the so-called prism or block Froude number F_b was introduced based on the concept of the densimetric Froude number as

$$F_b = V_o / \sqrt{(\rho_b/\rho - 1)g V_{cube}^{1/3}}. \tag{7.20}$$

Here ρ_b the block density whereas the equivalent cube size is $V_{cube}^{1/3} = (a_b^3/2)^{1/3}$, with a_b as the side length of a diagonally sliced cube.

The scour data in terms of F_b follow again a linear fit of the form $y_b = a_b \cdot F_b + b_b$. The coefficients a_b and b_b depend almost linearly on the tailwater depth divided by the equivalent cube dimension $h_{TW}/V_{cube}^{1/3}$ (Emami, 2004). Based on the test data, the coefficients of the following scour characteristics are specified (Figure 7.15, Table 7.3):

- Maximum scour depth d_{sc}/D
- Scour depth at pipe or tunnel outlet d_{toe}/D
- Maximum scour width w_{sc}/D
- Up- and downstream limits of scour hole x_1/L_P, x_3/L_P
- Distance of deepest scour point from pipe or tunnel outlet x_2/L_P
- Required length of protected area L_{Req}/D.

Note from Table 7.3 that only the maximum scour width is independent of h_{TW}.

Failure criteria

Any failure of the area protected by concrete prisms occurs in principle if either the scour extends over the protected area and/or if the concrete prisms are swept away by the flow

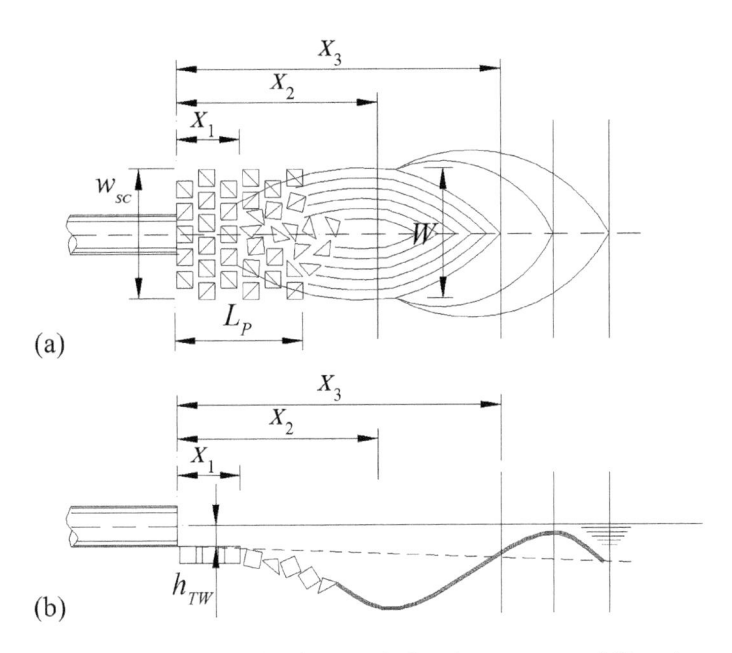

(a)

(b)

Figure 7.15 Definition of scour hole characteristics downstream of diversion tunnel outlet (a) plan, (b) streamwise section

Table 7.3 Coefficients a_b and b_b of equation $y_b = a_b \cdot F_b + b_b$ for scour hole characteristics with protection by concrete prisms (Figure 7.15, Emami and Schleiss, 2006a)

Scour hole characteristics	y_b	a_b	b_b
Maximum scour depth	d_{sc}/D	$-0.01(h_{TW}/V_{cube}^{1/3}) + 0.87$	$0.38(h_{TW}/V_{cube}^{1/3}) - 1.00$
Scour depth at pipe outlet	d_{toe}/D	$-0.11(h_{TW}/V_{cube}^{1/3}) + 0.38$	$0.09(h_{TW}/V_{cube}^{1/3}) - 0.37$
Maximum scour width	w_{sc}/D	2.00	1.50
Upstream boundary of scour	x_1/L_P	$-0.27(h_{TW}/V_{cube}^{1/3}) + 0.09$	$0.88(h_{TW}/V_{cube}^{1/3}) - 0.29$
Distance of d_{sc} from pipe outlet	x_2/L_P	$-0.07(h_{TW}/V_{cube}^{1/3}) + 0.36$	$0.62(h_{TW}/V_{cube}^{1/3}) - 0.50$
Downstream boundary of scour	x_3/L_P	$-0.25(h_{TW}/V_{cube}^{1/3}) + 1.13$	$1.00(h_{TW}/V_{cube}^{1/3}) - 1.45$
Required length of protected area	$L_{P,Req}/D$	$-0.37(h_{TW}/V_{cube}^{1/3}) + 3.63$	$0.39(h_{TW}/V_{cube}^{1/3}) + 0.38$

(Emami, 2004; Emami and Schleiss, 2006a). The experimental study revealed that the failure of prisms is affected by the velocity at the pipe outlet V_o, the mass densities of the prisms and water, the prism size $V_{cube}^{1/3}$, the tailwater depth h_{TW} and the length of the protected area L_P. Based on observations, the failure of the protected area occurred if one or more of the following conditions were met (Figure 7.12):

- Scour depth at tunnel outlet higher than 50% of tunnel diameter
- Maximum scour depth higher than 2 times tunnel diameter
- Maximum scour width higher than width of protected area.

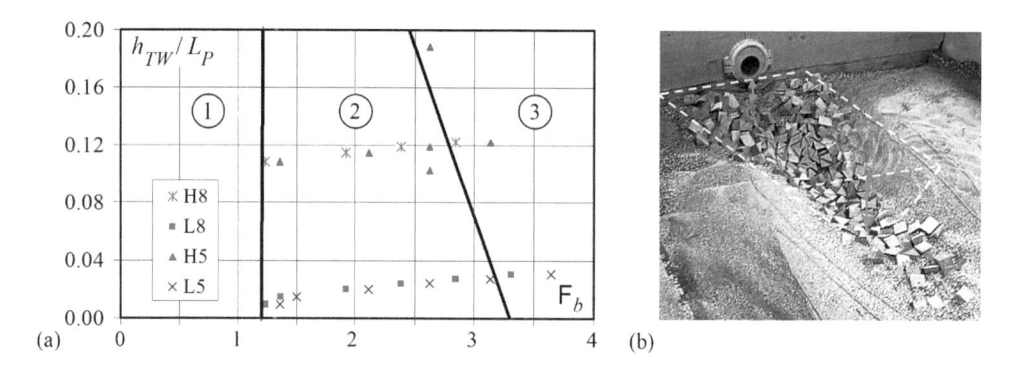

Figure 7.16 (a) Failure diagram for prism protection h_{TW}/L_P versus prism Froude number F_b, (−) Eq. (7.21) with ① No movement, ② Acceptable prism movement, ③ Prism failure, (b) example of failure of protected area (Emami, 2004)

It was noted from the tests that the failure of the protected area strongly depends on the protection length L_P and the tailwater level h_{TW} for a given prism size expressed by the prism Froude number F_b. To define a dimensionless failure diagram to protect prisms, the relation between the prism Froude number F_b and parameter h_{TW}/L_P was plotted (Figure 7.16). Based on the above criteria, the failure diagram was divided into the three regions 'No movement', 'Acceptable movement' and 'Failure'.

Figure 7.16b shows the successive failures of the area protected by concrete prisms with increasing discharge. The minimum required protection length $L_{P,\mathrm{Req}}$ downstream of the outlet to avoid any failure in the protected area is

$$\mathsf{F}_b = 3.32 - 4.26(h_{TW}/L_{P,\mathrm{Req}}). \tag{7.21}$$

The location of the scour hole was analyzed for protection lengths $6 < L_p/D < 11$. The required protection length was then defined if the following conditions were satisfied:

- Prisms are not eroded from downstream due to high tailwater
- Maximum scour is pushed by (1 to 2)D downstream of outlet (low tailwater).

The latter ensures the stability of the tunnel outlet. The minimum required protection length $L_{P,\mathrm{Req}}$ follows the empirical formula given in Table 7.3 resulting from the tests. If a protection length $L_p > L_{P,\mathrm{Req}}$ is selected, the prism size can be reduced according to the failure diagram (Figure 7.16). The width of the protection area was always $w_p/D = 7.5$, selected according to the flow diffusion angle for maximum discharges released from the tunnel. It can be considered as design value.

Design recommendations

The design discharge for checking the prism stability results from an analysis of the diversion system in view of construction cost and damages during floods at the construction site. The required prism size should then be determined by considering a safety factor. For the design discharge, a safety factor of $\beta_s = 1.3$ is recommended for prisms obtained by dividing cubes diagonally. The safety factor is applied on the prism number when using the failure diagram $(\beta_s \cdot F_b)$. The prism stability should be checked for the safety discharge $(\beta_s \geq 1)$ corresponding to the maximum capacity of the diversion system under extreme conditions.

Within the application range of the developed scour formulas $(0.10 < h_{TW}/V_{cube}^{1/3} < 2.90$ and $8.5 < F_o < 14.5)$, the minimum required prism size $a_{b,min}$ should be close to 45% of the tailwater depth $(a_{b,min} \geq 0.45h_{TW})$ for the design discharge. The required prism dimension is then checked by using the failure diagram (Figure 7.16). According to the tests, the maximum acceptable prism spacing should not exceed 40% of the prism size $(s_b \leq 0.40a_b)$. A minimum prism spacing of 0.50 m is recommended for construction reasons (excavation and formwork). The length of the protected area should correspond at least to the required length as defined in Table 7.3. The width of the protected area should be at least $7.5D$.

Concrete prisms should be used by dividing cubes diagonally as they are the most resistant and the design criteria have been developed for them. As long as they are not too flat, prisms obtained from square blocks are also acceptable. Since the prism Froude number is based on the equivalent cube size, the design criteria and design formulae still apply. If rather flat prisms are used (prism height 60% to 80% of its side length), the safety factor should be increased to $\beta_s = 1.5$.

The concrete prisms can be cast in place after excavation of the cube using a diagonal lost formwork to create them. The alternate solution is to precast a reinforced-concrete formwork and fill it on site with mass concrete. A cement content of 250 kg/m³ should be used for the concrete to ensure a sufficient shear and tensile strength of the prisms. Since erosion of material between the prisms can be accepted, no filter is required. Furthermore, in prototype applications, such a filter is practically not feasible. Once the material between the prisms is eroded, the latter tilts and forms a structure which protects against further erosion.

7.2.7 Surface protection of cofferdams

Since there is a practical limit of the discharge capacity of diversion tunnels regarding their maximum feasible excavation diameter (currently up to 16 m in excellent rock), the cofferdams may be overtopped during critical construction phases. To avoid the complete failure of cofferdams with catastrophic concerns in view of construction cost and time, their surface can be reinforced by adequate overlays to resist controlled overtopping. Widely used surface protection measures are:

- Riprap
- Reinforced slopes by anchored wire mesh
- Gabion boxes or mattresses
- Pre-casted concrete slabs and blocks.

A *riprap protection* is based on the critical surface flow velocity V_{cr} (Isbash, 1935)

$$V_{cr} = 1.2\left[\frac{\rho_b - \rho}{\rho}2gd_b\cos\phi_d\right]^{1/2}. \tag{7.22}$$

Here, d_b is the equivalent block diameter, ρ_b block density (typically 2650 kg/m³), ρ water density and ϕ_d the upstream angle of the cofferdam. Schematic examples of reinforced slopes by anchored wire mesh and gabions are shown in Figures 7.17 and 7.18. The surface protection of the downstream cofferdam of Cabora Bassa Dam is seen in Figure 7.19 (Quintela et al., 1979; GT-CFGB, 1973.

Concrete slabs have the disadvantage to accelerate the flow along the downstream cofferdam side and thus enhance the risk of toe erosion. This risk can be mitigated by using a macro-roughness lining system as protection consisting of precast concrete elements (Manso

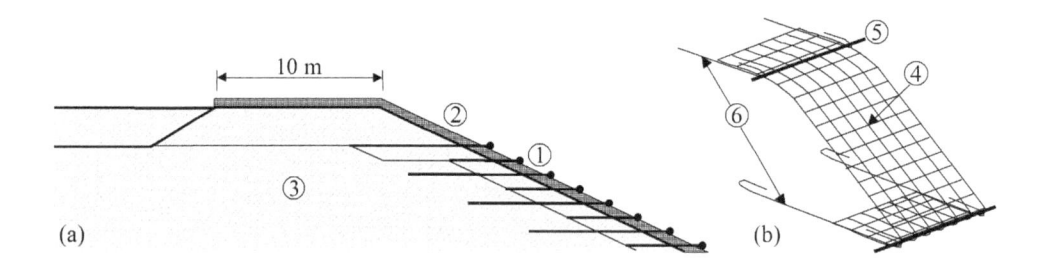

Figure 7.17 Protection by reinforcing slopes with anchored wire mesh including (a) ① steel bars 4 m and 10 m long, ② wire mesh, ③ riprap, (b) details of ④ wire mesh ⑤ thick steel bar ⑥ anchor bars (Schleiss, 2015)

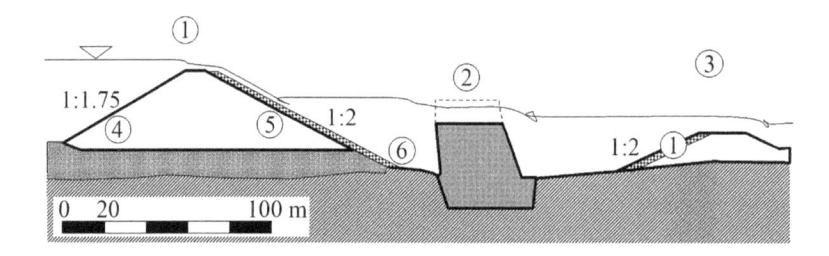

Figure 7.18 Surface protection using gabions with ① upstream cofferdam, ② dam under construction, ③ downstream cofferdam, ④ riprap, ⑤ gabion mattresses, ⑥ toe of cofferdam (Schleiss, 2015)

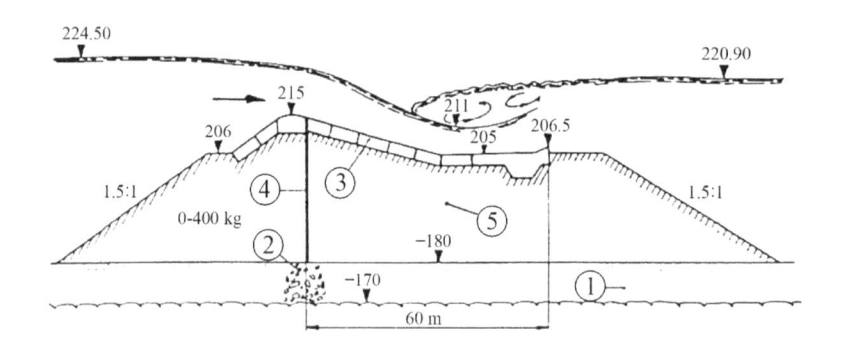

Figure 7.19 Protection of Cabora Bassa cofferdam with concrete slabs 3 m thick and size of 7×7 m² with ① alluvial riverbed, ② grouting, ③ concrete slabs, ④ sheet pile, ⑤ riprap (GT-CFGB, 1973)

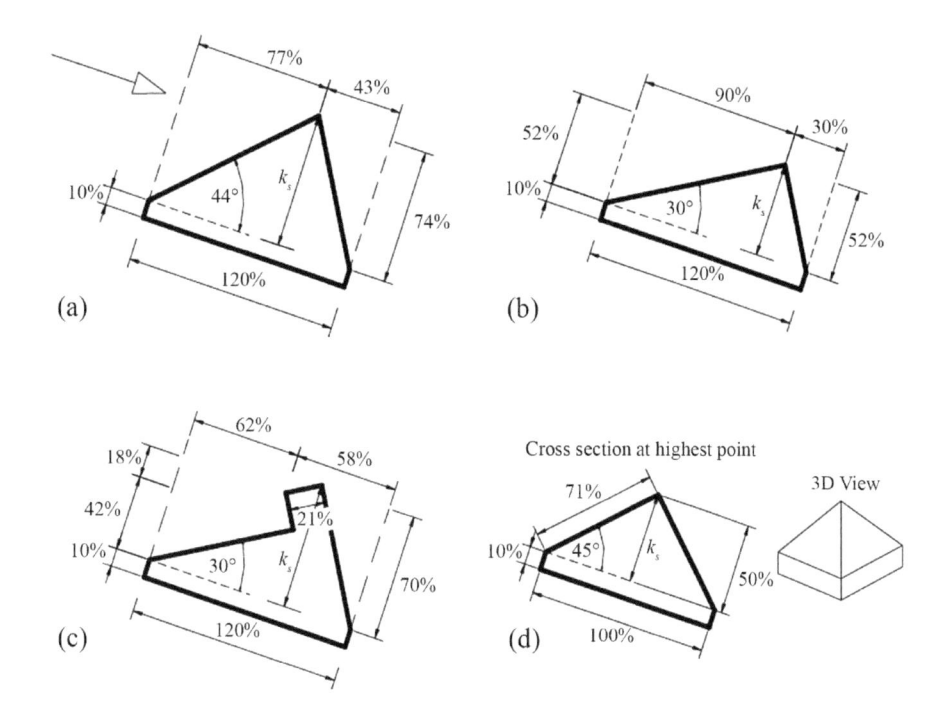

Figure 7.20 Dimensionless geometry of tested types of concrete lining blocks on a 1V:3H slope (a) Type 1, (b) Type 2, (c) Type 2+ES, (d) Type 3. Roughness height k_s and block geometry are defined as multiple of element width $b_E = 100\%$ (Manso and Schleiss, 2002)

and Schleiss, 2002). The main difference between this system and other existing concrete element systems is the stability concept, based on the self-weight of the blocks. Several types of elements were developed and tested in a physical model for a typical dam slope of 1V:3H (Figure 7.20) (Manso, 2002). Failure conditions were identified after submitting the elements to increasing discharges.

The largest element, the 44° negative sloped step (Type 1), does not present any advantage in terms of stability compared with other stepped-like elements, or at least not within the range of discharges tested. The 30° negative step element (Type 2) is clearly the solution withstanding highest discharges. By addition of an end sill (Type 2+ES), higher energy-dissipation efficiency seems likely to occur by increasing the momentum exchanges at the pseudo-bottom interface. The pyramid element (Type 3) is a good alternative for moderate discharges, creating a highly complex flow pattern and presenting the lowest velocities for the observed range of discharges.

For the preliminary design of the concrete elements, the design charts of Figures 7.21–7.24 apply. These exclude any drainage layer below the concrete elements, the recommended conservative design assumption. For a detailed stability analysis, the procedure of Manso and Schleiss (2002) should be considered. This surface protection system is also envisaged as overflow structure for low-risk embankment dams.

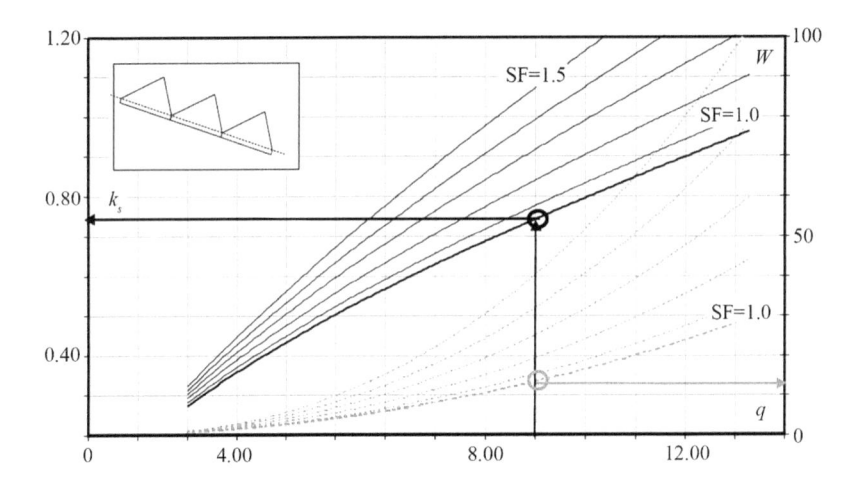

Figure 7.21 Design chart for 44° negative slope step (Type 1). Safety Factor values SF between 1 and 1.5, without drainage, versus roughness height k_s [m], unit discharge q [m²/s], and weight W [kN] for concrete density of 2400 kg/m³; dam slope 1V:3H. (○) Experimental observation (Manso and Schleiss, 2002). Roughness height k_s of elements according to Figure 7.20

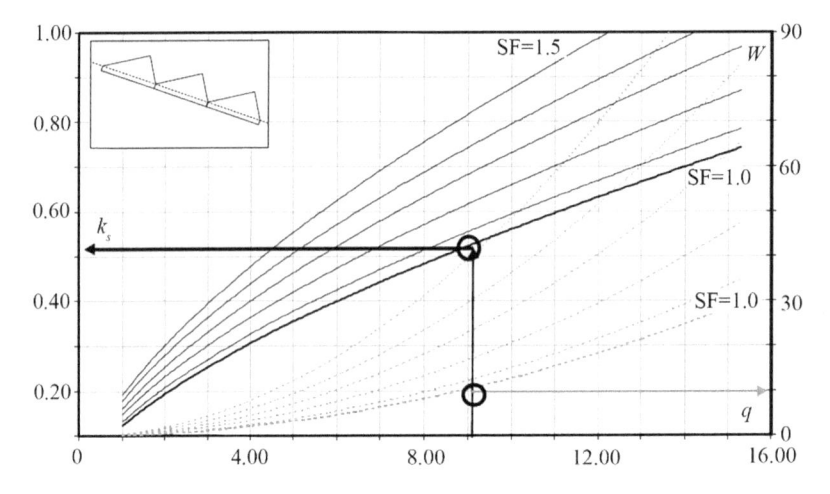

Figure 7.22 Design chart for 30° negative slope step (Type 2). Safety Factor values SF between 1 and 1.5, without drainage, versus roughness height k_s [m], unit discharge q [m²/s], and weight W [kN] for concrete density of 2400 kg/m³; dam slope 1V:3H. (○) Experimental observation (Manso and Schleiss, 2002). Roughness height k_s of elements according to Figure 7.20

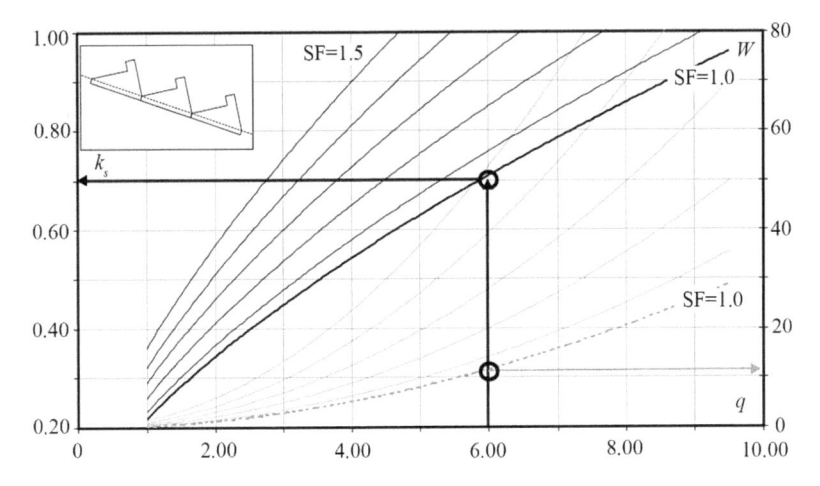

Figure 7.23 Design chart for 30° negative slope step (Type 2+ES). Safety Factor values SF between 1 and 1.5, without drainage, versus roughness height k_s [m], unit discharge q [m²/s], and weight W [kN] for concrete density of 2400 kg/m³; dam slope 1V:3H. (O) Experimental observation (Manso and Schleiss, 2002). Roughness height k_s of elements according to Figure 7.20

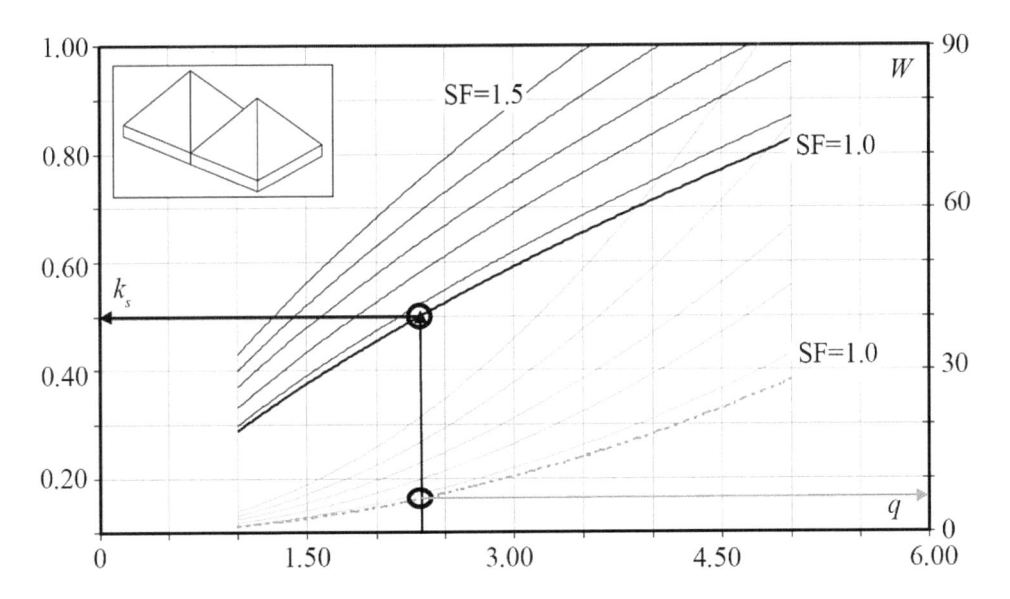

Figure 7.24 Design chart for 30° negative slope step (Type 3). Safety Factor values SF between 1 and 1.5, without drainage, versus roughness height k_s [m], unit discharge q [m²/s], and weight W [kN] for concrete density of 2400 kg/m³; dam slope 1V:3H. (O) Experimental observation (Manso and Schleiss, 2002). Roughness height k_s of elements according to Figure 7.20

7.3 River diversion

7.3.1 Effect of constriction

Instead of diverting a river into a tunnel or a culvert using cofferdams, as previously discussed, the river may also laterally be constricted so that a portion of construction is initiated in the contracted river, provided site conditions allow for this procedure. During a second stage, the already finished portion is returned to the river so that the construction proceeds to the adjacent portion. Often, several construction stages are necessary. Figure 7.25 shows the diversion channel of Clyde Dam, NZ, designed for a discharge of 1800 m^3s^{-1}, corresponding to the flood with a 15-year return period. The freeboard is 1 m below the upstream cofferdam crest. In November 1984 a flood with discharge of 2200 m^3s^{-1} was diverted without problems (Vischer, 1987).

Figure 7.26 refers to the Al-Massira multipurpose scheme, Morocco. The buttress dam is 79 m high and the spillway design discharge is 6000 m^3s^{-1}, with 1380 m^3s^{-1} discharge

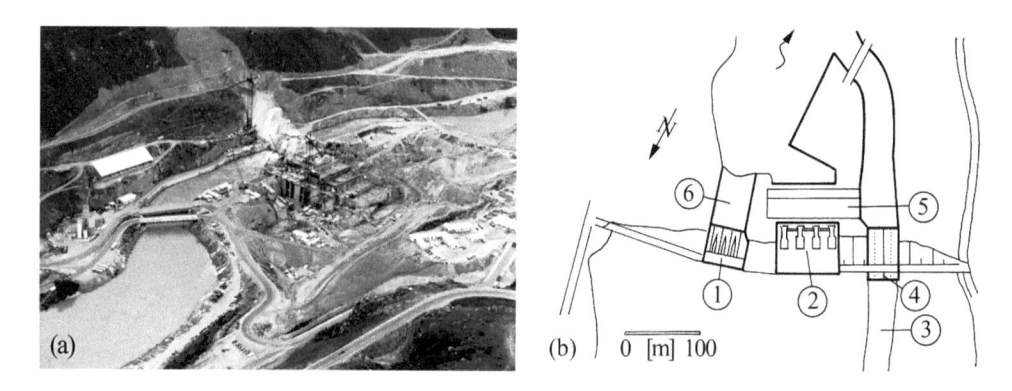

Figure 7.25 Clyde Hydropower Station (a) river diversion to allow for construction, (b) final design with ① spillway, ② penstocks, ③ diversion channel, ④ diversion sluices, ⑤ powerhouse, ⑥ stilling basin (Vischer, 1987)

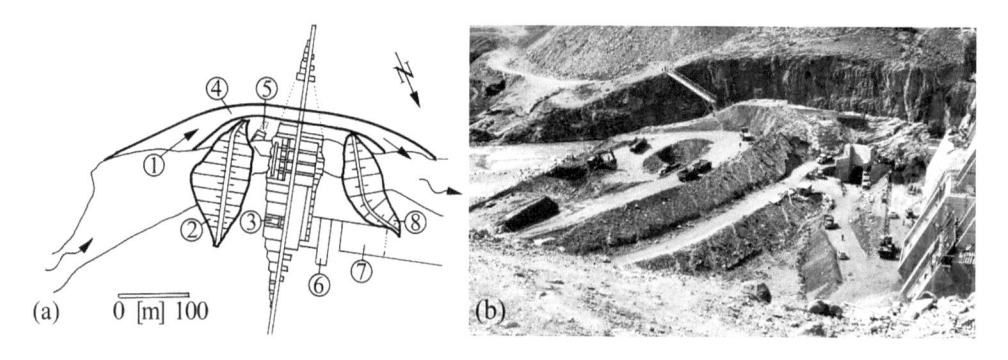

Figure 7.26 Al-Massira Dam, Morocco, (a) plan with ① diversion channel, ② upstream cofferdam, ③ intakes, ④ spillway and bottom outlet, ⑤ diversion tunnel, ⑥ powerhouse, ⑦ tailrace channel, ⑧ downstream cofferdam, (b) intake and dam under construction (Vischer, 1987)

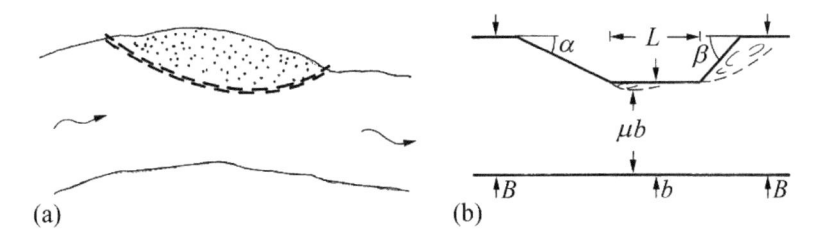

Figure 7.27 River diversion (a) typical configuration with a cofferdam, (b) hydraulic abstraction

capacity of the bottom outlet. The diversion channel of 2600 m³s⁻¹ discharge capacity is located on the left river side. In the center portion it consists of a rectangular channel 18 m wide, and the total length is 400 m (Figure 7.26a). Due to sound rock quality, nearly vertical walls were excavated.

The effect of a *constriction* on river flow may become relevant, so that a hydraulic design has to account for the submergence effect, allowing for works under flood water. Often, a river at the dam site has an almost rectangular cross section, with an approach flow width B and a constricted width b (Figure 7.27). The transition from the approach flow river to the constriction depends on the particular site conditions, but a polygonal geometry involving the upstream angle α and the downstream angle β is quite general. Rounded corners improve the discharge capacity, as described below. For diversions of more complex geometry, scale modeling is recommended. The effect of a loose boundary may be important in connection with sediment deposits and erosion. The bottom roughness is specified in 7.2.3.

Depending on the discharge Q, the width ratio $\psi = b/B \leq 1$, the length ratio $\lambda = L/b$ and the angles α and β, two basic flow types for subcritical approach flow apply:

- Transition from subcritical approach to supercritical tailwater flow for a strong constriction
- Subcritical flow throughout for a weak constriction.

The second case is typical in applications, because a strong constriction causes a significant submergence into the approach flow river, to be considered in the design of the required height of the cofferdam. Further, scour problems in the tailwater are relevant. Both cases are discussed subsequently.

7.3.2 Transitional flow

For a transition from sub- to supercritical flow, *critical flow* is known to occur at the location of minimum flow width (Hager, 2010). For a polygonal constriction, the minimum or critical width occurs slightly downstream from the leading edge at the entrance of the constriction, and the corresponding contraction coefficient is μ (Figure 7.27b). Based on the momentum equation, Hager and Dupraz (1985) found that

$$\mu = \hat{\mu} + \frac{1}{3}(1 - \hat{\mu})(\psi + 2\psi^4),$$
(7.23)

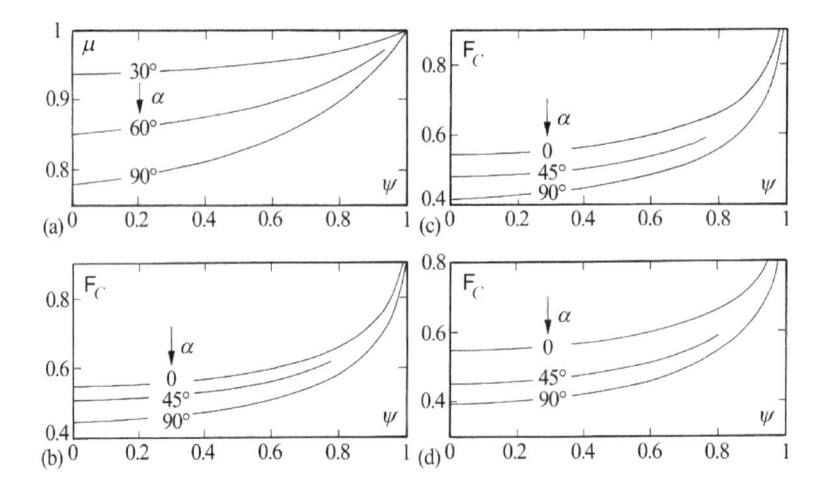

Figure 7.28 Polygonal constriction under transitional flow (a) contraction coefficient μ
for $\lambda = 1$, relative discharge $F_C = Q/(gb^2h_o^3)^{1/2}$ for $\lambda = $ (b) 0, (c) 1, (d) ∞ (Hager
and Dupraz, 1985)

$$\hat{\mu} = (1+\lambda)^{-1}\left[\frac{1+\sin(\alpha/2)}{(1+\sin(\alpha/2))^2} + \lambda\left(1-\frac{3}{8}\sin^{0.8}(\alpha/2)\right)\right]. \tag{7.24}$$

Clearly, the tailwater angle β has no effect on the contraction coefficient μ. Figure 7.28a
shows $\mu(\psi)$ for various angles α and $\lambda = 1$. For $\psi = 1$ or $\alpha = 0$ results simply $\mu = 1$.

The *discharge-head relation* across the constriction follows from the energy equation by
assuming that the friction slope is nearly compensated for by the bottom slope. The relation
for the upstream and the critical energy heads H_o and H_c thus reads

$$H_o = h_o + \frac{Q^2}{2gB^2h_o^2} = \frac{3}{2}\left(\frac{Q^2}{g\mu^2b^2}\right)^{1/3}. \tag{7.25}$$

With $F_C = Q/(gb^2h_o^3)^{1/2}$ as constriction Froude number follows

$$\mu = \left(\frac{3}{2+\psi^2F_C^2}\right)^{3/2}. \tag{7.26}$$

The relation between F_C, ψ and α is plotted in Figure 7.28(b–d) for various values of λ. For
any geometry assumed, the parameter F_C, and thus Q, or h_o are determined. Also, the effect
of a modification of any parameter on the other may be deduced. The effect of the upstream
angle on F_C for constriction rates below $\psi = 0.6$ is relatively small.

7.3.3 Subcritical flow

For a relatively large tailwater (subscript t) depth h_t, or for a larger value of ψ, the flow is no
more transitional but becomes submerged. Figure 7.29 refers to typical sectional views of

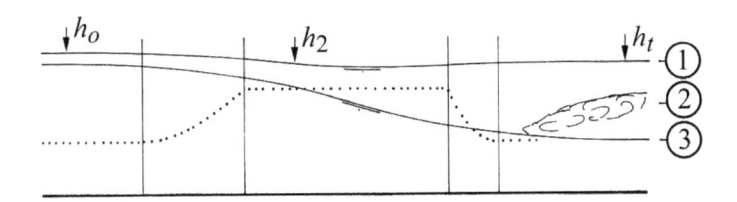

Figure 7.29 Surface profiles across river constriction with subcritical approach flow and ① subcritical tailwater, ② local transition to supercritical flow followed by hydraulic jump, ③ supercritical tailwater. (. . .) Critical depth profile

both cases. Also plotted are the cases with a transition from sub- to supercritical flow, and with the presence of a hydraulic jump in the tailwater.

The contraction coefficient μ for subcritical flow across a constriction depends not only on the approach flow angle α, the constriction rate ψ, and the constriction length λ but in addition on the ratio of flow depths in the contracted and upstream sections $Z_C = h_o/h_2$. An expression for μ was determined by Hager and Schleiss (2009) as

$$\mu = \mu_0 + (1-\mu_0)\left[\frac{\psi + 3\psi^4}{4}\right],$$ (7.27)

$$\mu_0 = \frac{3-2Z_C}{1+\sin(\alpha/2)} + 2(Z_C - 1)\left[1 - \frac{3}{8}\sin^{0.8}(\alpha/2)\right].$$ (7.28)

By accounting for the local head losses in the expanding tailwater portion, a relation for the depth ratio $T = h_o/h_t \geq 1$ versus the tailwater Froude number $\mathsf{F}_t = Q/(gB^2h_t^3)^{1/2}$ results. Figure 7.30 refers to constriction rates between $\psi = 1/3$ and $5/6$. Computations were checked with model observations. The dotted lines refer to the transition from submerged to free flow. According to Figure 7.30, the effects of both α and β on T are relatively small, whereas F_t and ψ have a significant effect.

The effect of a *rounded* constriction geometry in plan was considered by Hager and Schleiss (2009). Separation of flow due to contracted width μb is inhibited (i.e. $\mu = 1$) if the curvature radius upstream from the contracted cross section (subscript C) is larger than $R_o/B = \mathsf{F}_C^2$ in which $\mathsf{F}_C = Q/(gb^2h_C^3)^{1/2}$. For critical flow, i.e. $\mathsf{F}_C = 1$, the minimum curvature radius is thus $R_o = b$. The effect of flow contraction is therefore avoided with a comparably small effort.

7.4 Culvert

7.4.1 Introduction

A culvert is a temporary opening left in a concrete dam or an artificial hydraulic tunnel structure placed in the rock foundation of an embankment dam. A culvert thus transfers flood water during construction either by free surface or pressurized flow. The culvert design is closely related to the definition of the *flow type* for any relevant discharge and tailwater level.

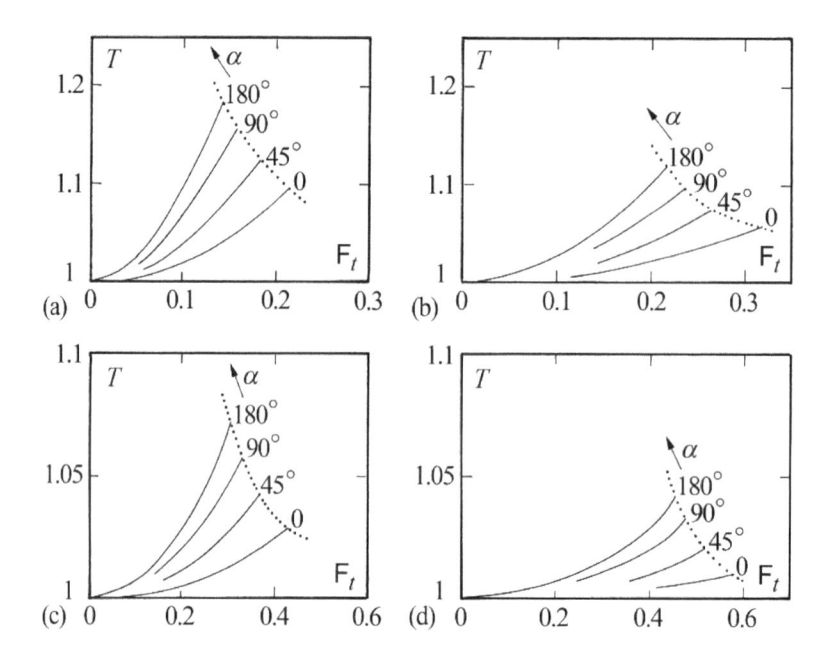

Figure 7.30 Depth ratio $T = h_o/h_t$ versus $\mathsf{F}_t = Q/(gB^2h^3_t)^{1/2}$ for various approach flow angles α and constriction rates ψ = (a) 1/3, (b) 1/2, (c) 2/3, (d) 5/6 (Hager, 1987)

It involves the computation of the head-discharge relation. Culverts are occasionally also used in parallel. At construction end, they are either closed with stop logs and/or a concrete plug, or integrated into permanent water release structures as bottom outlets and intakes (Chapter 8). The following presents a generalized approach to obtain preliminary design information. Culverts should be checked with regard to vortex setup in the approach flow domain, and the scour development at the culvert outlet. These aspects are mentioned below. Figure 7.31 shows a typical culvert as used for a dam.

Culverts have a wide variety of cross-sectional shapes including rectangular, circular, and horseshoe. Temporary openings in concrete dams have normally a rectangular cross section combined with a triangular roof, to allow for tight plugging with concrete (Figure 7.32). The circular culvert has a diameter D and length L_d. The bottom slope is S_o, the approach flow energy head is H_o relative to the inlet invert, and the outlet energy head is H_t relative to the outlet invert. The inlet is often curved with a curvature radius of r_d. Figure 7.33 refers to this simplified culvert structure. Culverts are considered long if $L_d/D > 10$ to 20, for which the flow patterns become not as complex as for the short culvert, to be excluded here.

Chow (1959) summarized the main culvert flow types as (Figure 7.34):

1   For $H_o < 1.2D$ and $S_o > S_c$ with S_c as the critical bottom slope (i.e. the slope for which the uniform flow is equal to the critical flow), *critical flow* develops at the inlet.
2   For $H_o > (1.2$ to $1.5)D$ and for free outflow results in so-called gate flow. The culvert inlet corresponds hydraulically to a gate structure, with unlimited air access from the tailwater into the culvert.

Figure 7.31 Culvert structure at Al-Massira Dam (Morocco) during construction, employed as bottom outlets after construction (Courtesy VAW 51/6/3, ETH Zurich)

Figure 7.32 Openings in arch dam during construction left as culverts (Courtesy Richard Sinniger)

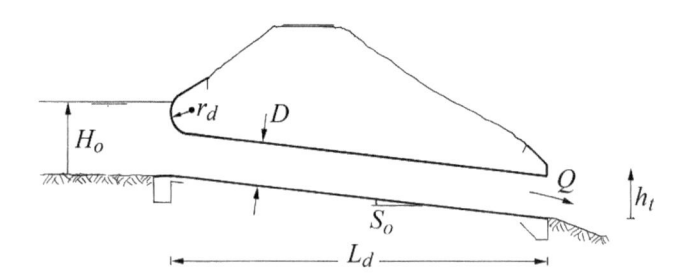

Figure 7.33 Standard culvert structure, notation

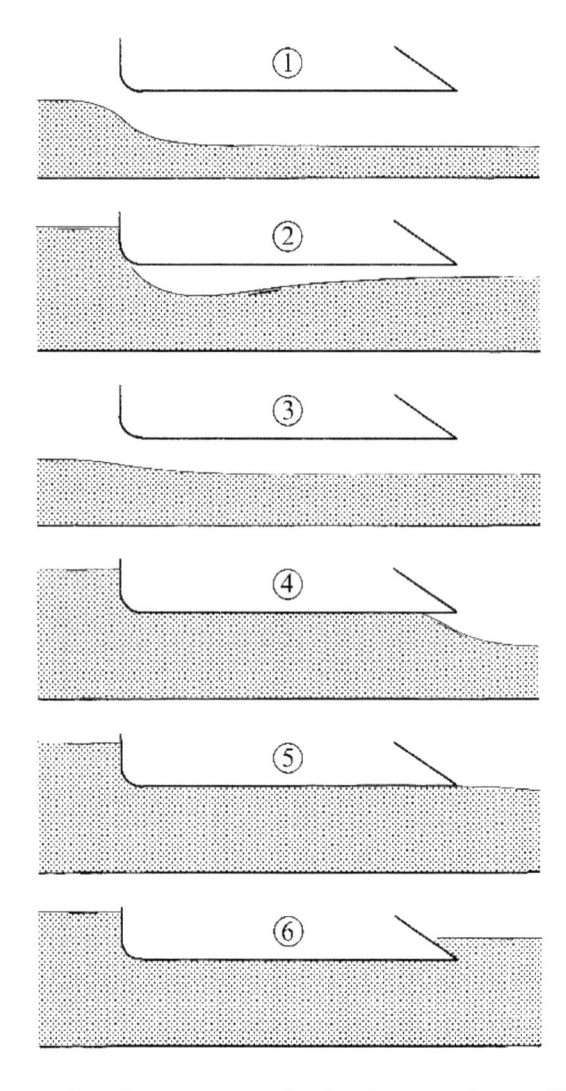

Figure 7.34 Flow types in culvert structure, for details see main text (Hager, 2010)

3   For $H_o < 1.2D$ and $S_o < S_c$ subcritical flow develops, controlled either at the culvert out-
     let, or in the tailwater channel. In general, backwater curves dominate this flow type, and
     uniform flow may be assumed for long culverts.
4   For a tailwater submergence $h_c < H_t < D$ and small bottom slope $(S_o < S_c)$, the culvert is
     partially submerged.
5   For *pressurized flow* with the outflow depth equal to the culvert height, the tailwater may
     become supercritical.
6   For a completely *submerged outlet* $(H_t > D)$ and $S_o < S_c$ pressurized culvert flow occurs
     with a subcritical outflow.

7.4.2 Hydraulic design

A culvert structure may be designed based on 4 fundamental flow regimes including:
   Critical flow (subscript c) in circular pipes follows (Hager, 2010)

$$\frac{H_{oc}}{D} = \frac{5}{3}\left[\frac{Q}{(gD^5)^{1/2}}\right]^{3/5}. \tag{7.29}$$

Uniform flow (subscript N) depends on the so-called culvert roughness characteristics
$\chi = KS_o^{1/2}D^{1/6}g^{-1/2}$, with $K = 1/n$, as (Hager, 2010)

$$\frac{H_{oN}}{D} = \frac{2}{\sqrt{3}}\left(\frac{Q}{KS_o^{1/2}D^{8/3}}\right)^{1/2}\left[1+\left(\frac{9\chi}{16}\right)^2\right]. \tag{7.30}$$

For uniform flow to occur, the uniform flow depth h_N is limited to 90% of the culvert diameter. For $\chi > 2$, the uniform flow becomes supercritical, so that Eq. (7.29) applies.
   Gate flow (subscript g) is influenced by the discharge coefficient C_d. For a curved inlet
geometry of relative curvature $\eta_d = r_d/D$ (Hager, 2010)

$$C_d = 0.96[1+0.50\exp(-15\eta_d)]^{-1}. \tag{7.31}$$

According to the Bernoulli equation, the discharge obtains

$$Q_g = C_d(\pi/4)D^2[2g(H_o - C_dD)]^{1/2}. \tag{7.32}$$

The discharge increase versus C_d is limited to $\eta_d \leq 0.15$. Rounding the inlet by radius
$r_d = 0.15D$ yields an inflow without separation from the vertex. Note that the pressure may
drop below atmospheric pressure, however, so that vibrations due to boundary layer detachment may result.
   Pressurized flow (subscript p) is influenced by the head losses across the culvert. If $\Sigma\xi$
represents the sum of all losses, the discharge equation with $H_d = H_o + S_oL_d{-}H_t$ as the head
on the culvert reads

$$Q_p = (\pi/4)D^2[2gH_d/(1+\Sigma\xi)]^{1/2}. \tag{7.33}$$

For a reasonably rounded inlet and a straight prismatic tunnel, the resistive forces are made up
by friction forces, i.e. $\Sigma\xi = 2.4^{4/3}gL_d/(K^2D^{4/3})$. Figure 7.35 shows a generalized discharge-head

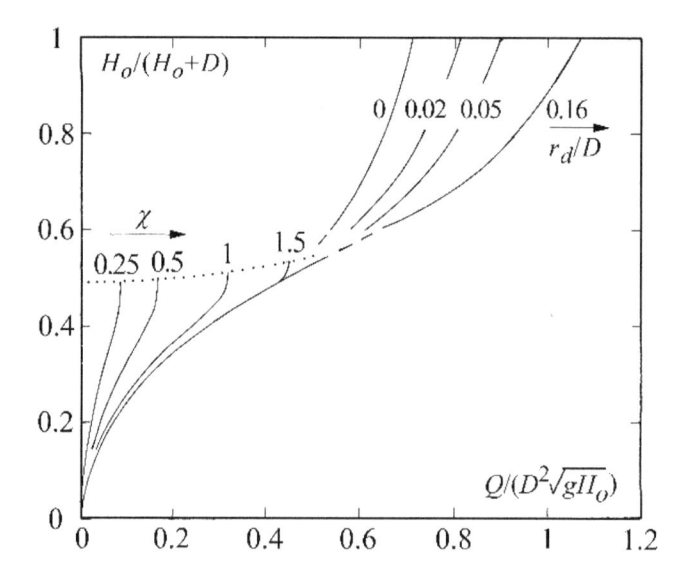

Figure 7.35 Culvert design diagram $Q/[D^2(gH_o)^{1/2}][H_o/(H_o + D)]$ versus roughness characteristics χ from free surface to pressurized flows, and relative inlet curvature r_d/D (Hager and Schleiss, 2009)

diagram including the critical, the uniform, and the gate type flows. A simple design is thus possible, which correspondingly applies to other cross-sectional shapes. The computational approach was verified by model observations of Hager and Del Giudice (1998) resulting in acceptable agreement.

7.5 Pier and abutment scour

7.5.1 Introduction

Abutment scour is a significant problem along cofferdams built for river constrictions during dam construction as well as at temporary or permanent bridges close to the site. Despite a significant research activity in the past decades starting mainly after World War II, pier and abutment scour in rectangular straight rivers remains a problem of engineering concern. Oliveto and Hager (2002, 2005) and Kothyari *et al.* (2007) have developed a semi-empirical procedure allowing for a straightforward determination of the most important features of bridge scour, so that their approach is highlighted. Their work introduces a new engineering tool based on similitude and detailed hydraulic experimentation. Tests under controlled laboratory conditions were conducted for three uniform and three non-uniform sediments. Their research refers to the temporal evolution of scour depth versus the approach flow depth and approach flow velocity, for relative densities between sediment and water from 0.42 to 1.65, the latter relating to sand or gravel. Based on the concept of the densimetric particle Froude number, the effects of granulometry are incorporated. Some two hundred tests lasting from several hours up to more than one month were conducted, representing the basis of this work. In addition, the available literature data including those of Chabert and Engeldinger (1956) were considered to validate the design proposal. The approach presented satisfies most of these data sets, provided the model limitations are satisfied. These include in particular Froude similitude, i.e. grain sizes larger than about 0.9 mm to exclude viscous effects, flow conditions either in the clear water or in the transition to live-bed scour, as also flow depths larger than 20 mm to exclude macro-roughness effects. The present results are based on fundamental hydraulic parameters allowing for a straightforward application to engineering design.

The literature review refers to scour studies describing the *temporal scour progress* only for mainly clear-water conditions with an essentially plane sediment bed. In the 1990s, the scour evolution was not yet a standard in hydraulic engineering because of lack of reliable data, so that the maximum scour depth was determined under the so-called equilibrium state (Breusers and Raudkivi, 1991; Hoffmans and Verheij, 1997).

The oldest and most complete experimental study on pier scour was conducted by Chabert and Engeldinger (1956), involving two test channels, various circular-shaped piers, five uniform sediments of grain sizes between 0.26 and 3.2 mm, and three approach flow depths of 0.10, 0.20, and 0.35 m. Further, six other pier shapes were tested and additional experiments aimed to find optimum arrangements for scour protection. The study involved some 300 tests lasting from a few hours to days. Maximum scour resulted at the transition between the clear-water and the live-bed regimes. Despite the large data set, no general scour relation was established.

Colorado State University CSU (1962) published a report with temporal scour data for fine sand to be considered below. Dietz (1972) investigated pier scour for artificial uniform sediment of density 1045 t/m³. His two tests are an important basis to verify the density effect on pier scour. Cunha (1975) concluded from a literature survey that most of the literature tests lasted for less than 3 h, and that the proposed equations were based on isolated data series. He introduced for his tests a power law up to equilibrium conditions, which was reached for only three of his tests, however.

Franzetti *et al.* (1989) conducted a longtime test to check whether end pier scour existed, which occurred only after 100 h. Ballio *et al.* (2000) noted no equilibrium state for abutment scour lasting over 1000 hours. In both test programs, plastic material was employed. Yanmaz and Altinbilek (1991) conducted almost 50 clear-water bridge pier experiments for nearly uniform sediment of 0.84 and 1.07 mm grain sizes. Circular and square shaped piers were tested. A differential equation for the progress of the scour hole depth was solved numerically. The non-dimensional representation of time involving parameters of geometry and granulometry is notable.

Kothyari *et al.* (1992) conducted tests in a 1 m wide rectangular channel using pier diameters from 0.065 to 0.17 m. Uniform sediment between 0.40 and 4.0 mm, and sediment mixtures of medium grain size 0.50 and 0.71 mm were employed. A computational scheme relating a number of experimental constants was provided to predict the scour depth for both steady state and unsteady approach flow conditions. An effect of sediment mixture was noted for $\sigma > 1.124$, with $\sigma = (d_{84}/d_{16})^{1/2}$ as sediment non-uniformity parameter. Here d_i is the sediment size $<i\%$. An effective mixture size diameter was determined that increases with $\sigma^{2/3}$, so that a sediment mixture undergoes less scour than the corresponding uniform sediment due to the armoring process in the scour hole for otherwise equal conditions.

Kohli (1998) investigated out-of-the-river building scour as a type of abutment scour, employing abutment widths between 0.05 and 0.20 m. The data of his tests will be considered below. Melville and Chiew (1999) did not detect a viscous effect and introduced an equilibrium time at 95% of the end scour. Similar to Cunha (1975), the normalized scour depth involving equilibrium depth was plotted against normalized time, and an effect of relative approach flow velocity was observed. Their analysis was based on the data of Ettema (1980) and these collected at the University of Auckland and the Nanyang Technological University. A procedure for temporal scour advance under well-defined flow conditions was proposed.

Cardoso and Bettess (1999) conducted experiments in a 2.44 m wide channel including a lateral flood plain. Six abutments ranging between 0.147 and 0.80 m, and an almost uniform sediment of 0.835 mm grain size were considered. The interaction between the flood plain and the main channel was relatively small, so that scour in the flood plain was considered as in a rectangular channel.

This review of available data in 2002 demonstrates that most of the works toward the understanding of scour evolution were conducted from the 1980s. A trend away from the design based on the equilibrium scour concept is noted, because floods often last much shorter than equilibrium time, and the resulting design appears uneconomic. An account for both the hydraulic and hydrologic features is required, therefore.

7.5.2 Experimental setup

The experiments were conducted in two rectangular channels, of widths $B = 1.00$ m, and 0.50 m. The data collected from the latter resulted in no width effect (Figure 7.36). The 1 m wide main scour channel had a total length of 11 m, with a working section of 5 m, a glass wall on the left side and a smooth steel wall on the right. Given that scour is a local phenomenon, the sediment was placed horizontally, and the free surface measured along the flow. The determining approach flow depth h_o was taken at $(B-b_p)/2$ or $(B-D_p)/2$ upstream from the

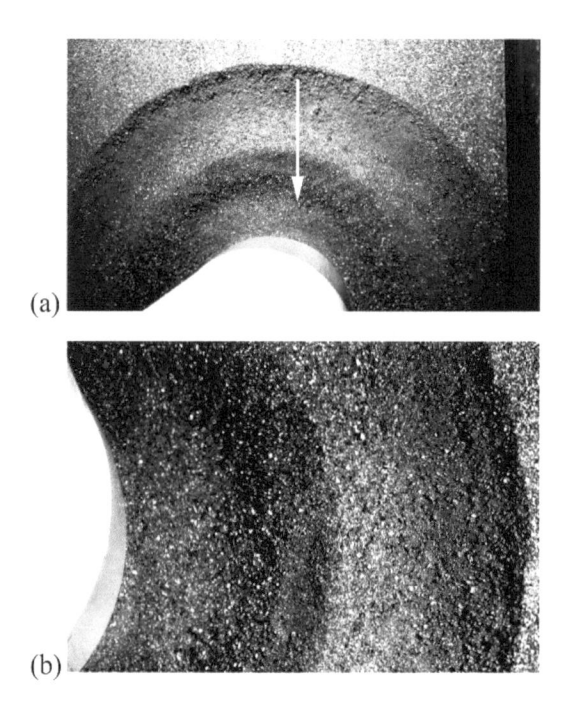

(a)

(b)

Figure 7.36 Structure of pier scour after scour hole drainage (a), top view, (b) detail (Hager *et al.*, 2004)

scour element face, with b_p and D_p as the width and diameter of abutment and pier, respectively. A flow straightener placed at the upstream channel end inhibited flow concentrations, suppressed local turbulence, and reduced wavy surface patterns. All tests were run under plane bed conditions, and the bed elevation was estimated to some 15% of the median grain size d_{50}. The sediment surface was measured with a so-called shoe-gage having a 4 mm by 2 mm wide horizontal plate at its base, whereas the water surface was read with a point gage, typically to ± 1 mm. The bed upstream of the scour hole remained horizontal under the clear-water scour regime, even after weeks of experimentation.

Six different sediments were used, three of which were uniform with grain sizes of 0.55, 3.3 and 4.8 mm, and three mixtures, with $d_{50} = 5.3$, 1.2, 3.1 mm, and $\sigma = 1.43$, 1.80, and 2.15, respectively. The uniform sediment (subscript s) with $d_{50} = 3.3$ mm had a mass density of $\rho_s = 1.42$ t/m^3, whereas the remainder had $\rho_s = 2.65$ t/m^3. These tests are summarized by Oliveto and Hager (2002). Next to d_{50}, σ and the relative density $\rho' = (\rho_s - \rho)/\rho$ with ρ as the fluid density, the test duration, the element geometry D_p or b_p, the approach flow depth h_o, the densimetric particle Froude number F_d, the threshold Froude number F_t both specified below, and the approach flow Reynolds number $R_o = 4h_o V_o/v$ with v as the kinematic viscosity were considered.

The abutments used were always rectangular in plan, of widths $b_p = 0.05$, 0.10, 0.20, 0.40, and 0.60 m. According to Kohli (1998), the effect of the streamwise abutment length is small. The circular cylinders had diameters of $D_p = 0.011$, 0.022, 0.050, 0.064, 0.110, 0.257, 0.400, and 0.500 m. All elements were manufactured of transparent Plexiglas to allow for visual determination of scour depths to ± 1 mm during test progress. Visualization across an element simplifies the processing of the scour data, mainly during the initial scour phase when data are collected in a higher sequence than later when the scour process has slowed down.

The maximum discharge was $Q = 0.130$ m^3/s. The experiments were started once the bed and the elements were properly placed by submerging the working section using the tailwater flap gate, which adjusted the downstream flow depth to the next 1 mm. The flap gate was lowered within 10–20 s to the pre-selected flow depth. The temporal test start ($t = 0$) was set at scour inception. Subsequent data were taken at times $t = 1$, 3, 6, 10, (15), 20 minutes up to several days or even weeks in selected runs.

For abutments, scour always started downstream of the corner region protruding into the upstream flow, whereas the initial movement was on both pier sides, at ~70° downstream from the channel axis (Figure 7.37). Scour in the channel axis for piers, or at the sidewall for abutments, started always late as compared with the temporal origin. However, after sufficiently long test duration, the maximum scour depth migrated from the element side toward the pier axis or the channel sidewall, respectively. The maximum scour depth $z(t)$ thus is either along the element side, or at the pier axis (or channel wall for abutments), respectively (Oliveto and Hager, 2002).

7.5.3 Scour depth equation

Turbulent two-phase flows of water and sediment exhibit a complicated flow pattern. In scour holes, flows are complex due to a spatial vorticity system including the (1) horseshoe vortex as driving mechanism along the element upstream base, (2) surface vortex system due to stagnation flow, (3) tornado vortices along element sides, and (4) rear vortices induced by von Karman vortex streets (Kohli, 1998). Few successful numerical models for local scour processes are available, given the 3D computation of a turbulent two-phase flow. A simplified approach was thus considered based on the analogy between local scour and resistance of a rigid body in inviscid fluid flow.

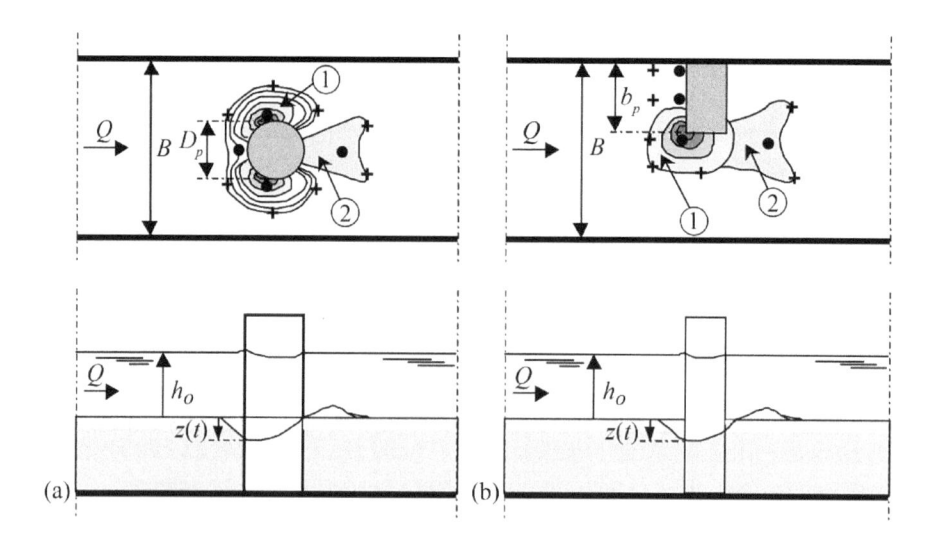

Figure 7.37 Definition sketch and measurement points for (a) pier, (b) abutment. Points defining (●) scour or aggradation depths, (+) scour or aggradation area with ① scour hole, ② deposition area

The resistance force F of a body in fluid flow is from Newton $F = \rho A V_o^2/2$, with ρ as fluid density, A as cross-sectional area perpendicular to the approach flow direction, and V_o as the approach flow velocity. The effect of the angle of attack is not considered here because all elements were placed perpendicularly to the channel axis. Sediment scour in fluid flow occurs because the hydrodynamic forces on the grains are larger than these retaining them from movement. Here, the gravity force only applies, so that the model follows this concept, thus accounting for inertia forces.

The scour depth z at a certain time t increases with a reference (subscript R) length L_R, involving the abutment width b_p or the pier diameter D_p, as also the approach flow depth h_o, i.e. the approach flow section A on the scour element, as for the resistance problem. From purely geometrical considerations, $L_R = b_p^\alpha h_o^\beta$ for the abutment, and $L_R = D_p^\alpha h_o^\beta$ for the pier, with $\alpha + \beta = 1$. The scour depth further increases with the approach flow velocity V_o, approximately following the square relation. In addition, the scour depth decreases with increasing density difference $(\rho_s - \rho)/\rho$ between the two fluids involved. For a two-phase gravity flow, the densimetric (subscript d) particle Froude number $\mathsf{F}_d = V_o/(g'd_{50})^{1/2}$ is the determining parameter, thus (Rajaratnam and Nwachukwu, 1983), with $g' = \{(\rho_s - \rho)/\rho\}g$ as the reduced gravity acceleration. The scour depth thus depends essentially on F_d, which includes both the hydraulics and the granulometric patterns. Note that F_d is the ratio of the actual approach flow velocity V_o and the reference velocity $V_R = (g'd_{50})^{1/2}$, so that the reference time is $t_R = L_R/V_R$. Accordingly, the time scale involves the element geometry, the approach flow depth, and sediment density and granulometry. Scour thus is controlled by the dimensionless scour time $T = t/t_R = \{(g'd_{50})^{1/2}/L_R\}t$, to be verified with the experimental data.

Scour of any element starts under well-defined conditions, referred to as inception condition (Hager and Oliveto, 2002). If no element is present, the inception (subscript i) conditions follow Shields' diagram. Accordingly, a scour equation would contain a difference between the actual densimetric Froude number F_d and the inception Froude number F_{di}, so that scour develops only if $F_d > F_{di}$. The term F_{di} is normally much smaller than F_d, however, and may often be neglected.

The dimensionless scour depth $Z = z(t)/L_R$, or any other parameter including scour width or the lateral scour extent, vary essentially as $Z \sim F_d{}^\gamma \times f(T)$, where $f(T)$ increases with time. From past research, the function $f(T)$ is known to be logarithmic. Based on data analyses, the scour depth evolution follows (Oliveto and Hager, 2002)

$$Z = 0.068N\sigma^{-1/2}F_d^{1.5}\log(T),\ F_d > F_{di}.\tag{7.34}$$

Here $Z = z/(b_p{}^2 h_o)^{1/3}$ or $Z = z/(D_p{}^2 h_o)^{1/3}$ for the abutment and pier, respectively, N as shape number equal to $N = 1$ for the circular pier, and $N = 1.25$ for the rectangular abutment (or pier). Sediment non-uniformity σ has a definite effect on scour as also for the entrainment of bed load transport (Hager and Oliveto, 2002). Equation (7.34) was validated with all available data (7.5.1). With $D_* = (g'/v^2)^{1/3}d_{50}$ as dimensionless grain size, the inception Froude number for the hydraulic rough regime $(D_* \geq 150)$ is

$$F_{di} = 1.65\left(R_{ho}/d_{50}\right)^{1/6}.\tag{7.35}$$

Oliveto and Hager (2002) provide also approximations for $D_* < 150$, i.e. for the turbulent smooth and transition regimes. Further, the threshold (subscript t) densimetric Froude number $F_t = V_o/V_t$ with $V_t = (g'd_{50})^{1/2}F_t$, and $F_t = F_{di}$ according to Eq. (7.35) without any scour element present upstream of pier or abutment is for $D_* \geq 150$

$$F_t = \frac{V_o}{1.65\ (g'd_{50})^{1/2}\ \sigma^{1/3}\left(R_{ho}/d_{50}\right)^{1/6}}.\tag{7.36}$$

Subcritical open channel flows are generally influenced by two distinctly different effects, namely relative roughness and viscosity. The latter occurs also for scour if the approach flow grain Reynolds number $R_{o*} = du_*/v < 200$, for which the flow is in the transition regime where both relative roughness and viscosity have an effect on scour. Shields introduced this fact and many approaches have been forwarded since both for the entrainment condition and sediment transport. However, no specific viscous effect for scour appears to have been made, although it undergoes a similar effect. Using the definition of the shear velocity $u_* = (gSh_o)^{1/2}$, and for conditions close to uniform flow in a wide river, the effect of local hydraulic gradient S is small, so that it follows the Manning-Strickler equation $V = (1/n)S^{1/2}h^{2/3}$, with $(1/n) = 6.75g^{1/2}d_{50}{}^{-1/6}$ as roughness coefficient (Hager and Oliveto, 2002). Accordingly, $u_* = (V/6.75)(d/h)^{1/6}$, and the grain Reynolds number in terms of basic parameters obtains $R_{o*} = (1/6.75)(dV_o/v)(d/h_o)^{1/6}$. For flows in the fully turbulent regime, the relative grain size is $(g'/v^2)^{1/3}d > 150$. Eliminating v gives $R_{o*} = 272F_d(d/h_o)^{1/6}$. The value of F_d in the experiments of Oliveto and Hager (2002) was $\cong 2$, indicating with $R_{o*} > 200$ the criterion for the lower limit of sediment size $d/h_o > (200/544)^6 = 0.002$.

The second criterion for the transitional flow regime is $(g'/v^2)^{1/3}d_{50} > 10$ to 20. For sand and water flow with $g' = 16.2$ m/s² and $v = 10^{-6}$ m/s², the result obtains $d_{50} > 0.40$ to 0.80 mm. Accordingly, two conditions for viscosity have to be satisfied: (1) minimum grain size relative to undisturbed flow depth, and (2) minimum absolute grain size.

7.5.4 Limitations and further results

Equation (7.34) to predict the scour depth advance is valid if (Figure 7.38):

- Channel is almost rectangular in section and straight in plan.
- Distribution of roughness pattern is nearly uniform.
- Element scoured is either circular cylinder or rectangular abutment, with approach flow perpendicular to front side. The abutment is located either along the channel side, or at the channel axis as during river diversion.
- Fluid is water, but might be different if viscosity is similar. The sediment is sand or gravel, or even different, if its density is well in excess of fluid density.
- Approach flow conditions remain constant with time, i.e. both flow depth and cross-sectional average flow velocity are not subject to significant variations.
- Medium sediment diameter $d_{50} < (5$ to $10)h_o$ to exclude effects of macro roughness.
- Medium sediment size $d_{50} > 0.8$ mm, i.e. minimum size due to viscosity (Figure 7.39).
- Equation (7.34) applies only for sediment entrainment close to scour element.
- Threshold Froude number in the upstream sediment bed is $F_t < 1.2$, resulting in no significant approach flow sediment transport.
- Flow conditions are such that no significant bed forms develop.

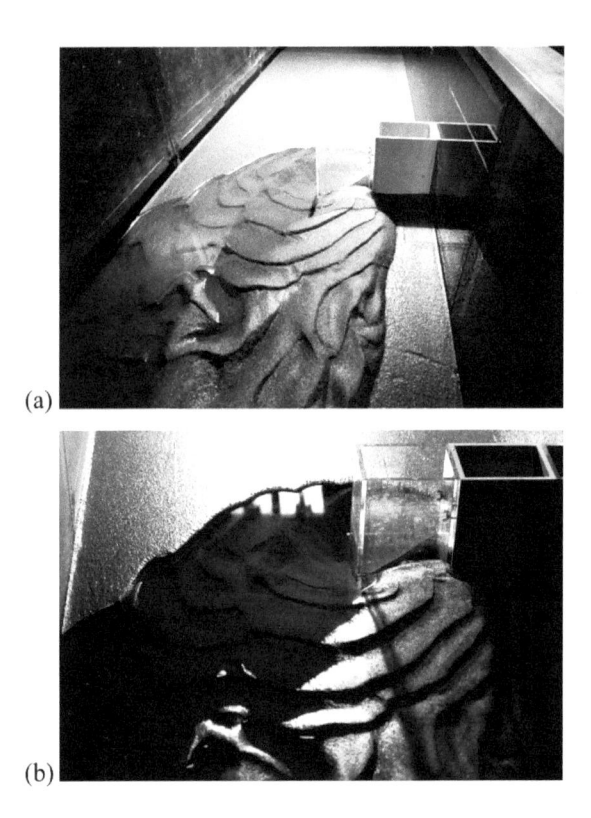

(a)

(b)

Figure 7.38 Scour around relatively (a) narrow, (b) wide abutments (Hager *et al.*, 2004)

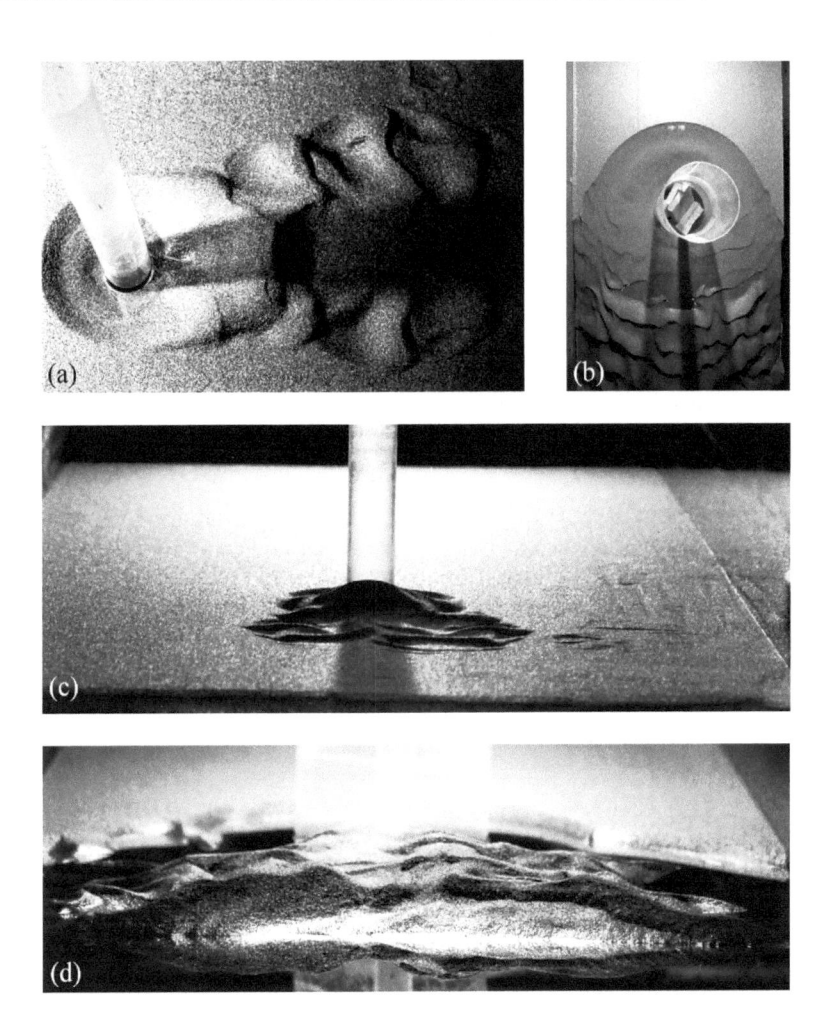

Figure 7.39 Pier scour in (a, b) plan, (c, d) views from downstream for small and large hydraulic loads, and fine sediment (Hager *et al.*, 2004)

Therefore, both the temporal evolution of pier and abutment scour under clear-water conditions are experimentally considered based on a similarity approach and extended laboratory tests under well-controlled flow conditions. For flows in the non-viscous regime, Eq. (7.34) accounts for the temporal advance of scour depth. The effect of element shape, i.e. either circular pier or sharp-crested rectangular abutment, involves only the shape coefficient N. The temporal scour evolution $z(t)$ depends on three main parameters, namely:

- Reference length ($D_p^{2/3}h_o^{1/3}$) for the pier, and ($b_p^{2/3}h_o^{1/3}$) for the abutment
- Densimetric mixture Froude number $F_{dm} = \sigma^{-1/3}V_o/(g'd_{50})^{1/2}$, with $\sigma = 1$ for uniform sediment
- Relative time T involving element geometry and sediment characteristics.

This approach was tested with extended laboratory observations, and verified with all existing literature data sets. The present prediction, based also on the extension of Oliveto and Hager (2005), is in agreement with all these sets, provided the limitations stated are satisfied, involving both minimum grain size and minimum flow depth to exclude mainly viscous effects, and several geometrical requirements.

The approach of Kothyari et al. (2007) is essentially based on the above data set involving a total of more than 650 temporal scour tests. In contrast to the previous approach, the entrainment condition is explicitly contained in the scour equation, so that the final result is improved from the physical perspective. However, given that the effective densimetric particle Froude number F_d is normally much larger than F_{di}, computational differences remain small. More importantly, the logarithmic progress of the scour depth evolution was investigated. It tends to an end for sediment mixtures because these are likely to form a stable armor layer over the scoured surface due to partial removal of fines from the parent bed material around the obstruction. This phenomenon is referred to as 'end scour' for which the scour hole does not undergo further temporal changes. It is governed by the characteristics of the bed material, the approach flow features, and the element shape. If the armor layer of the scoured surface consists of coarse (subscript c) material of size $d_c = \sigma d_{50}$, the corresponding densimetric particle Froude number is $F_{dc} = \sigma^{-1/2}F_d$. For a rough armor layer, F_{di} is related to (R_{ho}/d_{50}) according to Eq. (7.35). The data set was analyzed in terms of relative time $T_e = t_e/t_R$ at which the end (subscript e) scour is established. It was found (Kothyari et al., 2007)

$$\log T_e = 4.8 F_d^{1/5}. \tag{7.37}$$

The temporal scour advance is thus logarithmic for $t < t_e$, from when it stops due to the armoring effect.

The present approach is based on fundamental hydraulic quantities including cross-sectional velocity, approach flow depth, medium grain size, and transverse extension of the bridge element. Therefore, Eq. (7.34) to predict the temporal scour evolution is straightforward, and no additional site analyses are needed. A simple relation to determine the threshold condition in the approach flow channel is proposed.

7.5.5 Effect of flood wave

Introduction

The previous treatment of bridge scour involved a constant approach flow discharge Q, by which a scour is logarithmically generated until the end scour time is reached, from when the scour hole does no more undergo significant changes. In applications, the assumption of the constant discharge scenario is often unrealistic, given the long times required to establish the end scour. The effect of unsteady approach flow as a simple flood wave is thus considered, based on Hager and Unger (2010).

Equation (7.34) rewritten in dimensional quantities reads with $T_M = t/t_M$ as the time t relative to the time to peak t_M of the flood hydrograph, and $\gamma = t_M/t_R$ so that $T = \gamma T_M$

$$z/(D_p^2 h_o)^{1/3} = 0.068 N \sigma^{-1/2} \left(\frac{V_o}{(g' d_{50})^{1/2}} \right)^{3/2} \log(\gamma T_M). \tag{7.38}$$

Except for the flood approach flow parameters h_o and V_o, all quantities in Eq. (7.38) remain constant with time, i.e. auxiliary parameter $\alpha_a = 0.068 N \sigma^{-1/2} = \text{const}$. The ratio $\gamma = [\sigma^{1/3}(g'd_{50})^{1/2}t_M]/(h_o D_p^2)^{1/3}$ varies as $10^2 < \gamma < 10^5$ for extreme prototype values $1 < \sigma < 3$, $10^{-3} < d_{50}$ [m] $< 10^{-1}$, $1 < h_o$ [m] < 10, and $0.5 < D_p$ [m] < 5, allowing for simplifications.

The maximum scour depth depends essentially on V_o and slightly on h_o, so that the latter is approximated with the uniform flow formula. In a wide rectangular river, the unit discharge $q = Q/B$ is from Manning-Strickler's equation with K [m$^{1/3}$/s] as the Strickler roughness coefficient, and S_o as bottom slope

$$q = KS_o^{1/2} h_o^{5/3}. \tag{7.39}$$

Eliminating h_o, $V_o = q/h_o$ in Eq. (7.38), and using $z_M = [q_M^{0.8} D_p^{2/3} \alpha_a (KS_o^{1/2})^{0.7}]/(g'd_{50})^{0.75}$ with q_M as maximum unit flood discharge gives, with $Q_M = q/q_M$ and $Z_M = z/z_M$,

$$z/z_M = (q/q_M)^{4/5} \log(\gamma t/t_M), \quad \text{or} \quad Z_M = Q_M^{4/5} \log(\gamma T_M). \tag{7.40}$$

A typical flood wave is single-peaked, starts at time $t = 0$ with the base discharge q_0 to reach the maximum (subscript M) unit flood discharge q_M at time to peak $t = t_M$. Often, $q_0 \ll q_M$, so that $q_0/q_M \to 0$. Whether the base discharge results in scour or not is analyzed with Eq. (7.34) (Hager and Oliveto, 2002). Single-peaked flood waves follow with the hydrograph shape parameter $n > 1$ the function

$$Q_M = [T_M \exp(1 - T_M)]^n. \tag{7.41}$$

Below the values $n = 2$, 5, and 10 are considered, based on a best fit with the design hydrograph. With $Z_M = z/z_M$ and normalized time $T_M = t/t_M$ relative to time of peak $T_M = 1$, the governing Eq. (7.40) for scour evolution is

$$Z_M(T_M) = [Q_M(T_M)]^{0.80n} \log(\gamma T_M). \tag{7.42}$$

It relates the scour depth increase $Z_M(T_M)$ to the flood hydrograph $Q_M(T_M)$. Note from Eq. (7.40) that Q_M varies almost linearly with the maximum scour depth Z_M.

Computational approach

The computation of the scour advance involves a step-method of time step ΔT_M (Hager and Unger, 2010). Figure 7.40a shows the scour evolution $Z_M(T_M)$ for various values of γ and $n = 5$. All curves start at $T_M \cong 0.50$ increasing to the end value $Z_{Me} = z_e/z_M$. The latter is reached faster for large than for small γ. Figure 7.40b shows normalized curves Z_M/Z_{Me} versus T_M for $10^2 \leq \gamma \leq 10^6$ indicating a small effect of γ. The end scour is practically reached at $T_M = 1.1$, i.e. 10% after time to peak. The essential scour advance thus is confined to $0.5 \leq T_M \leq 1.1$. The three average curves $Z_M/Z_{Me}(T_M)$ for $n = 2$, 5, and 10 as shown in Figure 7.40c follow for $2 \leq n \leq 10$

$$Z/Z_{Me} = 0.5 \left\{ 1 + \tanh\left[(1.1 + 0.98n) \cdot (T_M - (0.92 - 0.66 \cdot 0.70^n)) \right] \right\}. \tag{7.43}$$

Scour initiation (subscript i) is defined at time $Z_M/Z_{Me} = 1\%$, whereas the scour is considered completed (subscript C) at $Z_M/Z_{Me} = 99\%$. The relative time of scour initiation $T_i = t/t_M$

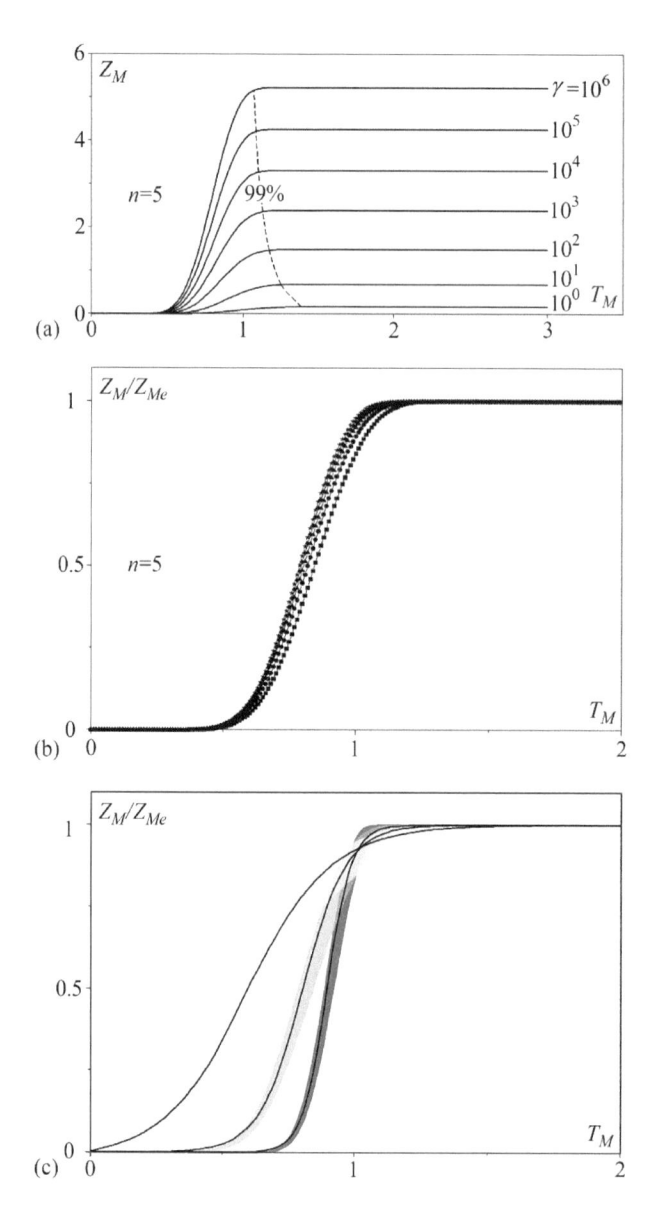

Figure 7.40 Scour depth advance (a) $Z_M(T_M)$, (b) $Z_M/Z_{Me}(T_M)$ for various values of γ, (c) ranges of curves $Z_M/Z_{Me}(T_M)$ for $2 \le n \le 10$ and (—) Eq. (7.43) for $n = $ (––) 2, (■■) 5, (■) 10

increases with n from $T_i(n = 2) = 0.08$, to $T_i(n = 5) = 0.43$ and $T_i(n = 10) = 0.70$ (Figure 7.40c). The time of completed scour $T_C = t_c/t_M$ decreases as n increases from 1.34 for $n = 2$, to 1.20 for $n = 5$, and to 1.11 for $n = 10$. The scour duration (subscript d) $T_d = T_C - T_i$ then is $T_d(n = 2) = 1.26$, $T_d(n = 5) = 0.77$, and $T_d(n = 10) = 0.41$.

The end scour depth Z_{Me} versus γ was investigated in terms of n. Figure 7.41a shows that the curves $\log(Z_{Me})$ versus $\log(\gamma)$ increase almost linearly for $10^2 \le \gamma \le 10^5$. The smaller n,

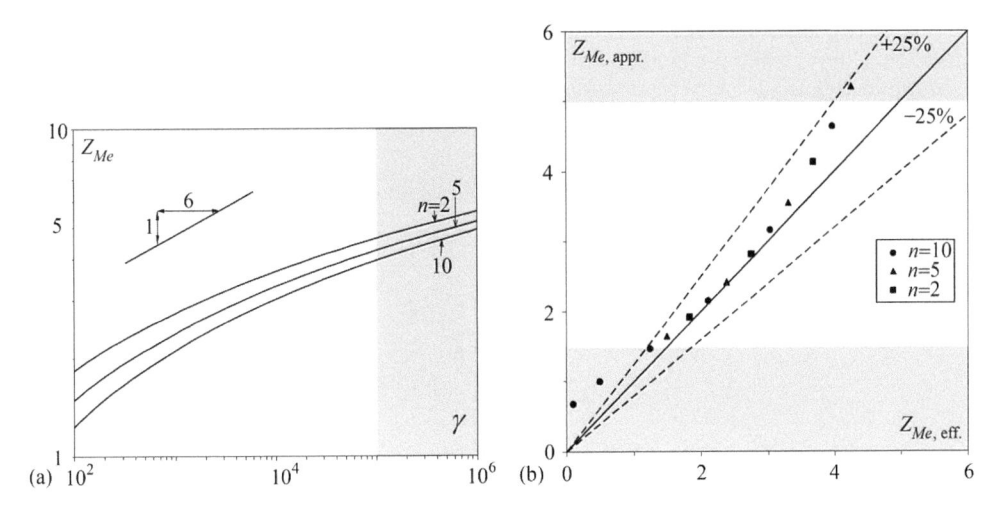

Figure 7.41 End scour depth (a) logarithmic plot of $Z_{Me}(\gamma)$ for $n = 2, 5, 10$, (b) comparison of end scour depth with (−) Eq. (7.44)

the higher is $Z_{Me}(\gamma)$, whose increase follows nearly $n^{-1/6}$. The relative end scour depth then simply is

$$Z_{Me} = (\gamma / n)^{1/6}. \tag{7.44}$$

Figure 7.41b compares the approximated (subscript appr.) Eq. (7.44) with the effective (subscript eff.) mathematical solution. The approximation applies for $1.5 \leq Z_{Me} \leq 5$ within ±25%. Defining the maximum velocity as $V_M = q_M/h_M = KS_o^{1/2}h_M^{2/3}$ gives with the corresponding densimetric particle Froude number $F_{dM} = V_M/(g'd_{50})^{1/2}$

$$\frac{z_e}{(h_M^4 D_p^5)^{1/9}} = 0.068 N\sigma^{-1/2}F_{dM}^{3/2}\left(\frac{\sigma^{1/3}(g'd_{50})^{1/2}t_M}{h_M n}\right)^{1/6}. \tag{7.45}$$

Thus z_e relative to $(h_M^4 D_p^5)^{1/9}$ varies essentially with F_{dM}, the element shape parameter N but only slightly with $\sigma, \rho_s/\rho, d_{50}, t_M, h_M$, and n.

Experiments

Three sediment mixtures with $d_{50} = 3.1$ mm, 3.9 mm, and 1.1 mm and $\sigma = 2.15, 2.10$, and 1.15, respectively, were considered. All flood waves involved $n = 10$ with t_M varied systematically between 300 and 3600 s. The peak discharges were between 0.05 and 0.11 m³/s, resulting in $h_M \cong 0.15$ m; two pier diameters $D_p = 0.11$ and 0.20 m were used, resulting in $810 \leq \gamma \leq 8200$ (Hager *et al.*, 2002). Both $F_{dM} = V_M/(g'd_{50})^{1/2}$ and $F_{tM} = V_M/V_t$ as threshold (subscript *t*) Froude number were related to the peak discharge, with V_t = threshold velocity (Hager and Oliveto, 2002), with $1.96 < F_{dM} < 3.68$ and $0.55 < F_{tM} < 1.18$. Two tests ranged between the clear-water and live-bed scour regimes.

Flood waves were generated with an externally controlled pump thereby starting with a pre-set initial flow depth. Readings were made at time increments $\Delta t \cong t_M/10$ both at the pier side and at the pier front; the larger of the two was retained for the data analysis. Shortly after time to peak had passed, the scour stopped indicating that end scour was reached. Scour duration is thus essentially confined to the rising limb.

Figure 7.42a shows $Z_M/Z_{Me}(T_M)$ for $n = 10$ for tests with $F_t < 0.90$ and $F_t \geq 0.90$. The latter data are systematically too high in the initial scour phase, whereas the data with $F_t < 0.90$ follow Eq. (7.43). This is due to the limitation of Eq. (7.38) for conditions close to live-bed scour. Note that Z_{Me} was determined with Eq. (7.45). Figure 7.42b examines the relation $q(h_o)$

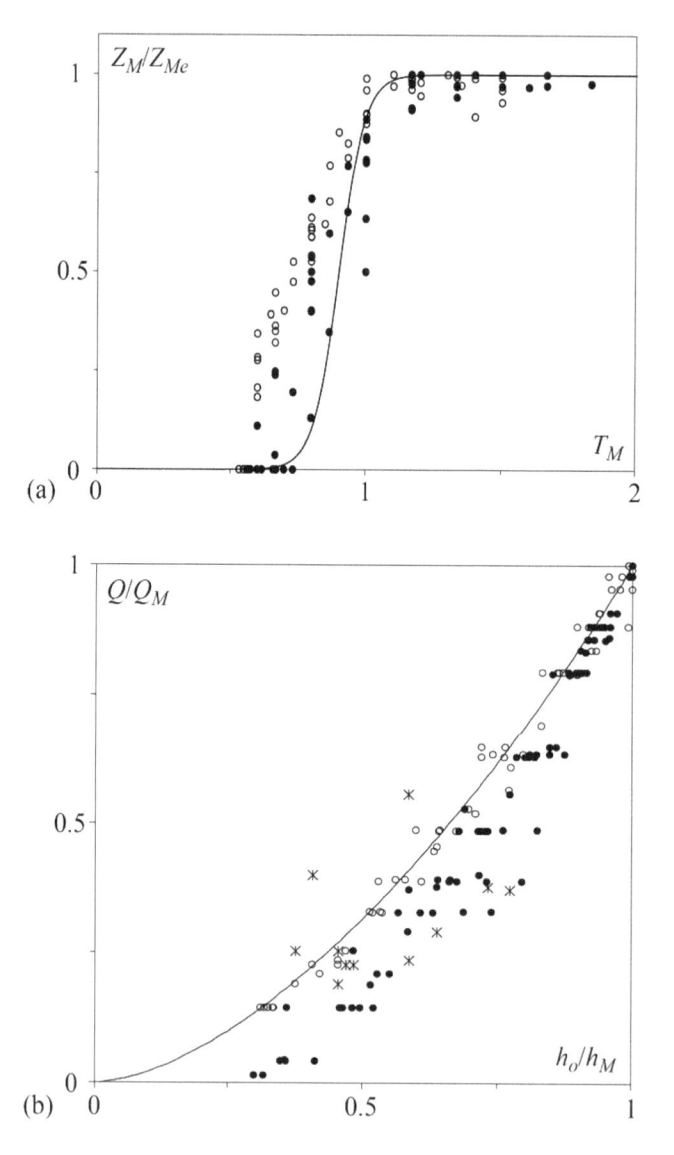

Figure 7.42 (a) Comparison of $Z_M/Z_{Me}(T_M)$ between data and (—) Eq. (7.43), $n = 10$, (b) relation $Q/Q_M(h_o/h_M)$ from data and (– –) Eq. (7.46) for (•) $F_{tM} < 0.90$, (∘) $F_{tM} \geq 0.90$, (*) scour inception. For symbols see Hager and Unger (2010)

given by Eq. (7.39). At peak discharge, q is expressed as $q_M = KS_o^{1/2}h_M^{5/3}$. Dividing Eq. (7.39) by the latter equation results in

$$Q/Q_M = (h_o/h_M)^{5/3}. \tag{7.46}$$

Figure 7.42b shows that the deviation of data from Eq. (7.46) reduces as $h_o/h_M \to 1$. This effect stems from the downstream boundary condition imposed by the tailwater flap gate, involving submerged flow for relatively small discharges instead of uniform flow from Eq. (7.46). Note that the approach flow depth increases under pier presence.

End scour depth

A data analysis indicates that all test data are within $\pm 25\%$ of Eq. (7.45), and that the correlation improves as F_{dM} increases, due to the relatively small threshold Froude numbers tested by Oliveto and Hager (2005). Equation (7.45) is expressed as

$$z_e = (0.068N\sigma^{-4/9}n^{-1/6}\frac{V_M^{3/2}t_M^{1/6}}{(g'd_{50})^{2/3}}(h_M D_p^2)^{5/18}. \tag{7.47}$$

The dominant term for z_e is V_M, followed by the term involving sediment size. Large values of V_M or small d_{50} result in a large end scour depth. The latter depends also on the scour element shape, as expressed by N. As sediment non-uniformity σ increases, e.g. from $\sigma = 1$ to $\sigma = 3$, z_e reduces by 39%. A pier diameter D_p increase by a factor of 2 results in 47% additional end scour depth. The effects of h_M and t_M are small, however. Increasing t_M by a factor of 10 results in a modest increase of z_e to only 1.47. Note that z_e in Eq. (7.47) is governed by a total of 11 parameters describing the flood wave (n, q_M, h_M, t_M, ρ), the scour element (N, D_p) and the sediment (d_{16}, d_{50}, d_{84}, ρ_s).

Kothyari et al. (1992) and Chang et al. (2004) reported test data with a wider range of the time ratio $1.4 \times 10^5 < \gamma < 2.4 \times 10^5$ than tested herein, resulting in a satisfactory agreement (Figure 7.42b). Chang et al. (2004) initiated a flood wave with conditions above sediment entrainment resulting in a rapid scour depth rise close to scour start (Figure 7.42a). From time $T_M = 0.5$, the flood hydrograph was similar to Eq. (7.41). Predictions compare favorably with their observations from that time, and their end scour depth practically collapses with Eq. (7.45). Kothyari et al. (1992) initiated the scour process from a finite discharge $q_0/q_M = 0.75$ so that the initial scour advance was much higher than according to Eq. (7.41). Their flood hydrograph was stopped at $T_M = 1$ so that the true end scour depth was not attained. Despite z_e was not fully reached, Kothyari et al.'s (1992) data practically collapse with Eq. (7.43) except for a time shift of 0.20 due to their particular scour initiation process.

The flood wave effect on the end scour depth results by dividing the unsteady by the steady scour depths from Eqs. (7.43) and (7.37), respectively, by assuming that the maximum scour depth occurs at $t = t_M$. Steady approach flow overestimates the scour depth under flood waves by about $+30\%$, with a correlation coefficient of 0.91.

The end scour depth due to a single-peaked flood wave in a plane rectangular sediment bed containing a single pier (or abutment) was thus considered under clear-water conditions. Based on the Oliveto-Hager scour equation for steady approach flow, and a model flood hydrograph, the evolution of the maximum scour depth is predicted. The results indicate that the effect of the time ratio γ on temporal scour increase is insignificant, allowing for a simplified representation of numerical results according to Eq. (7.43). Equation (7.45) describes the end scour depth versus the densimetric particle Froude number at peak discharge and normalized time to peak. The end scour depth depends essentially on the peak approach flow velocity, neutrally on the element shape parameter, the pier diameter, the sediment non-uniformity, the fluid and sediment densities, and the sediment size, but only slightly on the additional variables. The time to sediment entrainment and the scour time relative to time to peak are also specified, resulting in a simple application to design.

7.5.6 Protection against scour using riprap

Introduction

Piers and abutments are essential structural components of bridges. Their failure may have significant concerns on these infrastructural element. They may be subjected to scour by severe hydraulic conditions, or if the bed sediment size is relatively small, resulting in a large scour hole exposing the entire bridge foundation. The addition of riprap to these bridge elements increases their hydraulic resistance against scour, so it is extensively used in hydraulic engineering. Given that riprap elements are relatively expensive, their optimum placement is essential for an economical and safe design.

Although bridges failed through all times, and despite riprap was applied from ancient times, the systematic hydraulic study to improve bridge foundations started only recently. A first indication on the required riprap diameter was given by Breusers *et al.* (1977). Another classical reference to riprap sizing is Izbash (1935) as referred to by both Neill (1973) and Breusers *et al.* (1977). Based on these indications, Wörman (1989) proposed an equation for the riprap thickness required, thereby involving the characteristic sizes of riprap and base sediment: a Froude-type number expresses the ratio of average approach flow velocity and the square root of gravity acceleration times riprap thickness. Parola (1993) summarized proposals relative to the riprap size required thereby introducing a riprap stability number N_c involving the approach flow velocity and riprap size. He conducted experiments with circular and rectangular piers protected by three layers of uniform riprap. Riprap failure was defined when the second riprap layer was partially exposed after a test duration of 0.5 hour. A design equation was proposed involving the parameters previously mentioned.

Chiew (1995) presented a thorough analysis of riprap failure mechanisms at piers. His experiments involved a pier diameter of $D_p = 0.070$ m inserted in almost uniform sand of size $d = 1$ mm. The riprap sizes were $d_R = (2.6, 4,$ and $4.85)d$. The riprap extent was defined by the diameter and riprap thickness. All tests involved an approach flow depth of 0.20 m under clear-water conditions. Three failure modes were identified:

1   *Shear failure*, riprap is unable to withstand downflow and horseshoe vortex associated with pier scour mechanism.
2   *Winnowing failure*, underlying fine bed material escapes through riprap voids.
3   *Edge failure*, riprap at interface to bed material slides into scoured surface.

Chiew noted no scour for a ratio of approach flow and critical sediment entrainment velocities <0.30. An expansion of this finding led to a design equation similar to Izbash's equation, in which the relative riprap size d_R/h_o depends only on the approach (subscript o) flow Froude number $F_o = V_o/(g'h_o)^{1/2}$ with $g' = [(\rho_s - \rho)/\rho]g$ = reduced gravity acceleration, ρ_s and ρ as sediment and fluid density, V_o = approach flow velocity, and h_o = approach flow depth. Lauchlan and Melville (2001), Chiew (2000) and Lim and Chiew (2001) also investigated riprap characteristics under live-bed scour, a topic not considered here. Work on riprap usage in fluvial hydraulics originates from Lauchlan and Melville (2001). The failure mechanisms defined by Chiew (1995) were confirmed. Their work mainly investigated the effect of riprap elevation relative to the undisturbed bed sediment. A design equation for the riprap diameter was proposed depending on parameters such as placement depth, pier diameter to sediment size, riprap coverage area, riprap layer thickness, pier shape and alignment effects. A summary of the design equations is available (Melville and Coleman, 2000).

The following presents a novel design basis for the minimum riprap diameter required under steady flow for clear-water conditions. The results relate to a singular circular-shaped pier placed in bed sediment and protected by riprap of uniform size. The results of Unger and Hager (2006) are only concerned with the riprap failure condition thus neglecting the temporal scour advance once riprap failure has occurred.

Conceptual approach

Erosion of a sediment surface consisting of uniform grains by a steady discharge was first described with the Shields diagram relating the Shields parameter to the grain Reynolds number. The Shields diagram includes three domains, depending on the dimensionless sediment size $D_* = (g'/v^2)^{1/3}d_{50}$, where v is kinematic viscosity and d_{50} the median bed sediment size. For $D_* \leq 10$, flows are in the turbulent smooth regime, these for $10 < D_* < 150$ in the turbulent transition regime whereas those with $D_* \geq 150$ are in the turbulent rough regime. For the latter, viscous effects are absent and flows follow Froude similitude. Viscous effects are small for $D_* > 80$. Sediment entrainment (subscript i) in the rough turbulent regime follows with $\sigma = (d_{84}/d_{16})^{1/2}$ as bed sediment non-uniformity parameter and $R_{ho} = Bh_o/(B + 2h_o)$ as the hydraulic radius with B as channel width (Hager and Oliveto, 2002)

$$\sigma^{-1/3}V_i/(g'd_{50})^{1/2} = 1.65(R_{ho}/d_{50})^{1/6}. \tag{7.48}$$

According to Eq. (7.48), the entrainment velocity V_i increases in a wide river both as the sediment size and flow depth increase. Due to sediment hiding, V_i is larger for sediment mixtures than for uniform sediment of equal median size d_{50}. Equation (7.48) is also true for a sediment bed in which the pier diameter tends to $D_p \to 0$. As it increases in size, an additional effect of relative pier size $\beta_p = D_p/B < 1$ results whereas all other parameters are included in Eq. (7.48) and do not reappear, therefore (Hager and Oliveto, 2002).

Adding a riprap protection to the pier yields additional parameters to the sediment entrainment condition made up only by the relative riprap (subscript R) diameter $\delta = d_R/d_{50}$, and the aerial riprap extent expressed with the number n_R of riprap rows placed around a pier. The entrainment condition thus reads functionally

$$\sigma^{-1/3}V_{iR}/(g'd_{50})^{1/2} = f[R_{ho}/d_{50}, D_p/B, d_R/d_{50}, \text{ and } n_R]. \tag{7.49}$$

The entrainment function f of a pier in a sediment bed is different from that of the combined system pier-riprap. The function f for the latter was determined with systematic hydraulic modeling.

Experiments

PARAMETERS AND PROCEDURE

The tests were conducted in the scour channel described by Hager *et al.* (2002). A total of 35 main experiments involved two bed sediments of median sizes d_{50} = 5.0 mm and 1.1 mm, and $\sigma = (d_{84}/d_{16})^{1/2}$ = 2.29 and 1.18, respectively. The riprap sediment was almost uniform and involved diameters d_R = 8.4, 12, 13, 19, 23.6, 30, and 50 mm, resulting in diameter ratios $\delta = d_R/d_{50}$ from 2.4 to 45. Conditions with n_R = 1 to 10 riprap rows were studied, pier diameters D_p ranged from 110 to 457 mm resulting in riprap to pier diameters of 1–4.8. Rounded riprap elements instead of crushed aggregate was used to study critical failure conditions, because more angular material interlocks and strengthens to a certain extent the riprap. A total of 104 tests were conducted.

The main tests were conducted with half-cylinders positioned along the glass wall. Pier diameters were large enough to suppress effects of the wall boundary layer on the failure process (Oliveto and Hager, 2002). For selected runs, full cylinders were placed in the channel axis and the riprap failure compared with the previously described arrangement. Differences were found to be small, and the advantages for accurate failure observations considered superior as compared with tests at the channel axis. Accordingly, the relative pier width $\beta_p = D_p/B$ ranged from 0.055 to 0.23 for the half-width setup, and from 0.11 to 0.20 for the full-width setup (Unger and Hager, 2006).

Figure 7.43 shows a definition sketch. A coordinate system (x; y) was placed at the center of the circular pier, with x and y as streamwise and transverse coordinates, and ϕ as angle measured from the approach flow direction. The sediment (subscript s) bed is described with the parameters ρ_s, d_{50}, and σ, whereas the riprap layer with d_R, n_R, and the riprap diameter $D_R = D_p + 2n_R d_R$. All tests were conducted with one riprap layer whose surface was exactly flush with the bed sediment surface. This was considered critical for riprap failure, because the

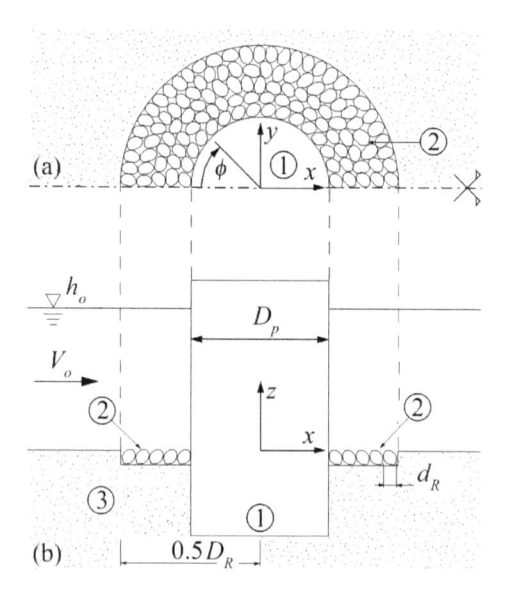

Figure 7.43 Definition sketch for riprap arrangement around pier (a) plan, (b) section with ① pier, ② riprap (n_R = 5), ③ sediment bed with ρ_s, σ, d_{50}

addition of a filter or a riprap sublayer causes failure at more severe hydraulic conditions. One riprap layer is a conservative design in practice: A second or even third riprap layer improves the interlocking of the riprap elements and the riprap performance. These conditions were considered too difficult to simulate in model tests, so that the simplest setup was retained.

To initiate tests, the bed sediment and riprap area were first placed (Hager *et al.* (2002), thereby submerging the entire test reach to facilitate positioning. Discharge was then set to the test value thereby suppressing sediment entrainment by sufficient submergence. The tailwater level was lowered with an externally controlled flap gate to a height where entrainment of the bed sediment in the pier rear was observed.

Failure of riprap was attained by progressive tailwater lowering, thereby keeping all the other parameters constant. The tailwater level and thus the approach flow depth were lowered by $\Delta h \approx 2$ mm in a time increment of $\Delta t = 60$ s. With a typical approach flow velocity of $V_o \approx 0.3$ m/s, the relative drawdown speed is $(\Delta h/\Delta t)/V_o \cong 10^{-4}$. To analyze this effect on riprap failure, tests lasting for up to one day were conducted prior to the last drawdown step before riprap failure. It was found that riprap remained stable independent of Δt except for the Undermining failure mode (see below).

FAILURE DEFINITION AND MECHANISMS

A riprap is considered failed if the first riprap element along the pier perimeter has dislodged. Therefore, failure has not yet occurred as a riprap element in the third row had dislodged, and e.g. others follow in the second row. This failure condition refers to only the dislodgement of the first riprap stone along the pier circumference, because scour initiates rapidly then. A test was usually stopped after riprap failure.

The progress of sediment transport from bed sediment entrainment to riprap failure was typically as follows (Figure 7.44): (a) Bed sediment is entrained at $\phi \cong 140°$. Decreasing the tailwater level creates a small scour hole in this reach eventually expanding as the tailwater is further lowered (Figure 7.44b). The scour migrates then rapidly upstream along the interface between the bed sediment and riprap with a maximum scour depth at $\phi \approx 75°$, associated with a sediment deposition in the rear of the pier (Figure 7.44c). Further reducing the tailwater level increases both the maximum scour depth and its extent around the pier, except for pier nose and rear (Figure 7.44d). This interface scour of small depth extends from $\phi \approx 60°$ to $150°$.

The riprap failure mode depends on various factors. Figure 7.45 shows a typical temporal progress of scour with (a) the original test arrangement, (b) 'small scour' in the rear region of the pier, (c) upstream migration of scour surface yet without any riprap element being dislodged at the pier, and (d) failure as the first riprap element at the pier circumference is dislodged. Note that the pier front was never scoured for all tests conducted. Some riprap locations require thus less protection than others.

Although Chiew (1995) had defined three riprap failure modes, Unger and Hager (2006) proposed alternatively:

1 *Rolling* according to Shields involving a small sediment size ratio $\delta = d_R/d_{50}$, i.e. a riprap sized close to the bed sediment. A riprap element of any except of the most peripheral row rolls out from the riprap unit causing sliding or rolling of neighbor elements (Figure 7.46).

2 *Undermining* as failure mode for $d_R \gg d_{50}$. Instead of being transported, riprap elements sink into the bed sediment. Because a riprap filter was absent, undermining occurred at the location with the largest surface deformations $\phi \approx 75°$ (Figure 7.47). The riprap at the pier rear is covered by bed sediment being eroded upstream.

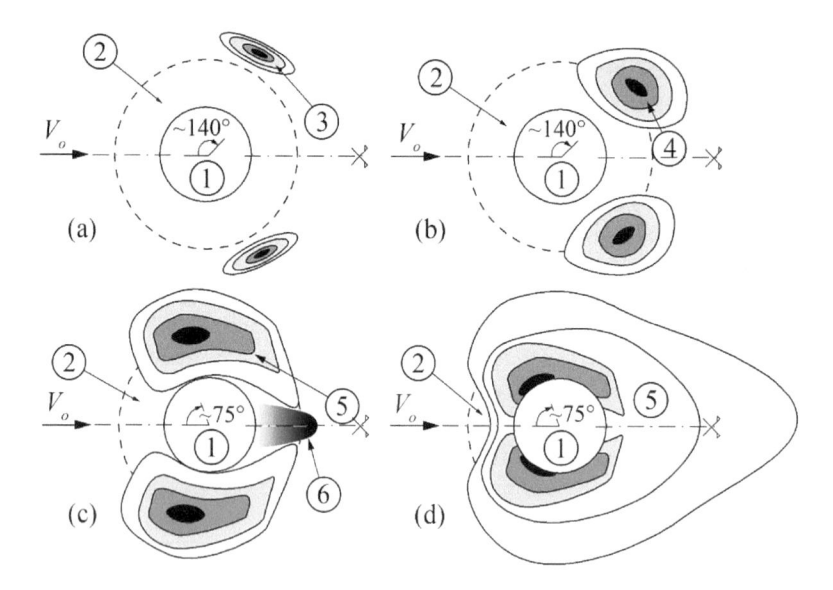

Figure 7.44 Scheme of riprap failure progress, from (a) sediment entrainment to (d) dislodgement of first riprap member along pier perimeter, with ① pier, ② riprap, ③ initial erosion, ④ small scour, ⑤ definite scour, ⑥ deposition (Unger and Hager, 2006)

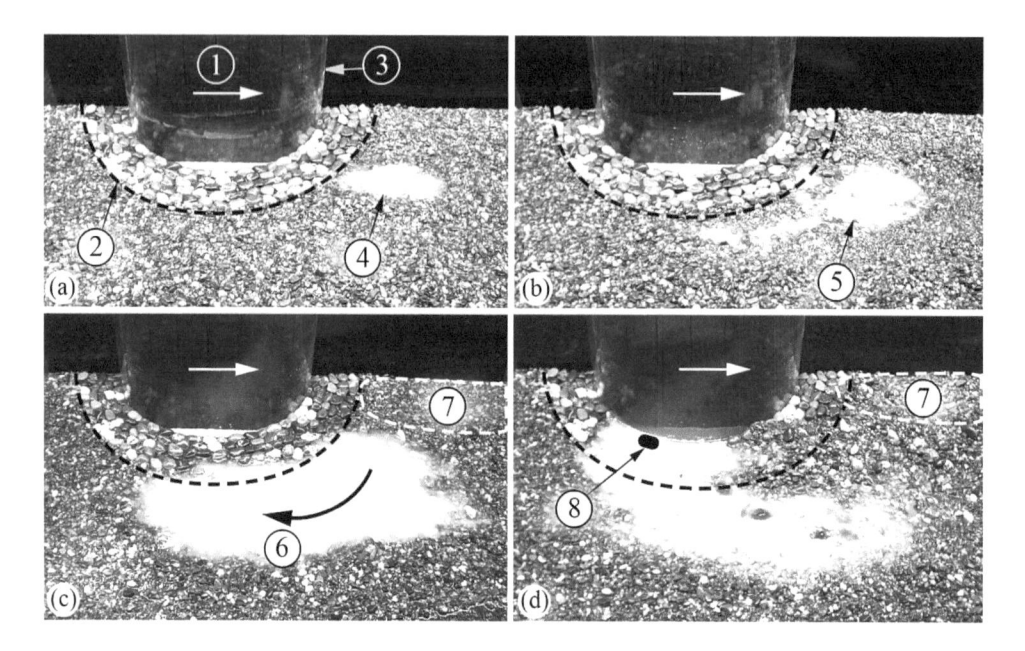

Figure 7.45 Test view of riprap failure progress with ① flow direction, ② original riprap, ③ pier, ④ erosion, ⑤ small scour, ⑥ migration, ⑦ deposition, ⑧ failure (Unger and Hager, 2006)

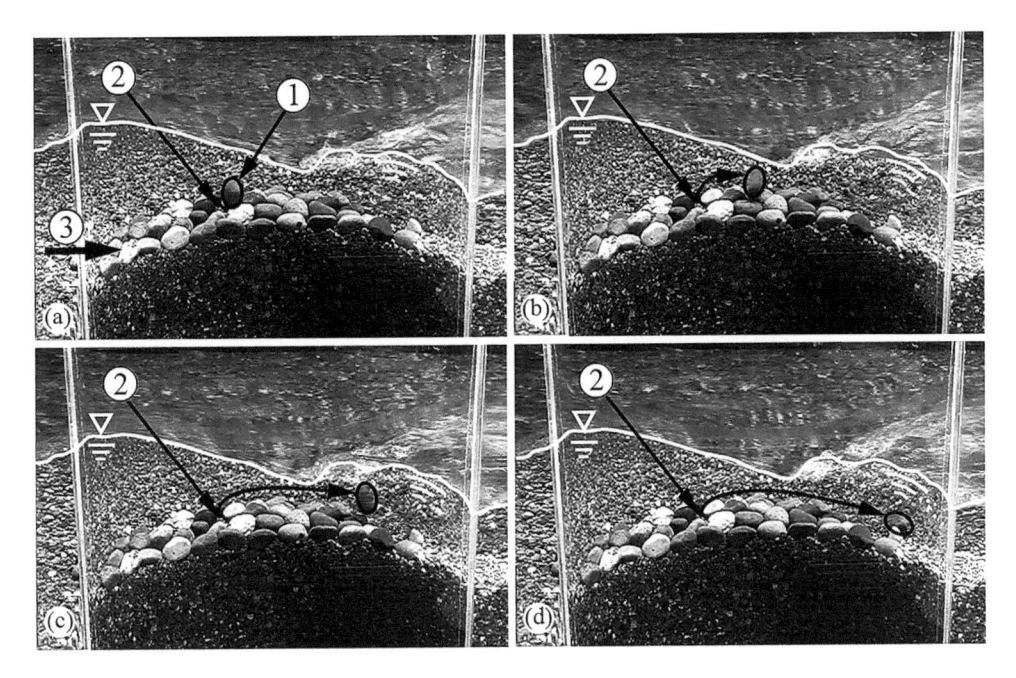

Figure 7.46 Failure of riprap by Rolling for $n_R = 5$, $D_p = 457$ mm, $d_{50} = 5.0$ mm, $d_R = 30$ mm, $Q = 0.060$ m^3/s at (a) to (d) successive times with ① riprap element, ② gap, ③ flow (Unger and Hager, 2006)

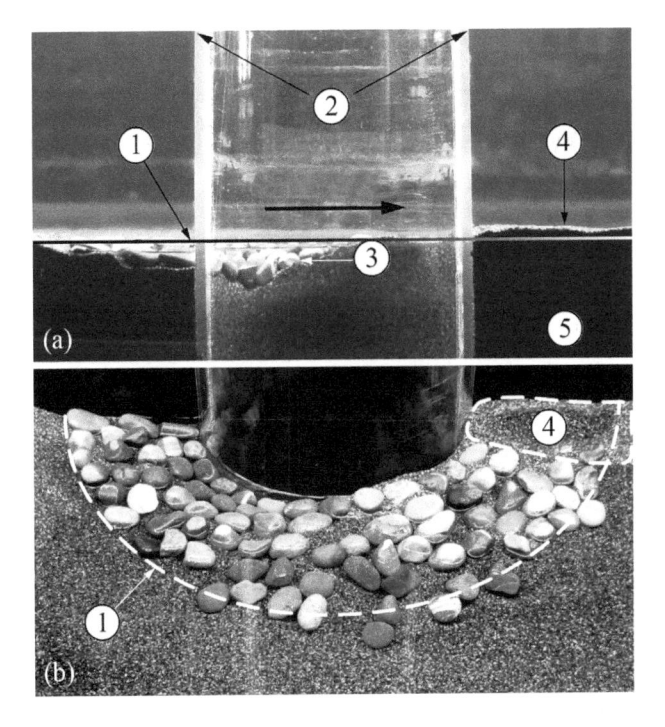

Figure 7.47 Failure of riprap by Undermining with $n_R = 5$, $D_p = 260$ mm, $d_{50} = 1.14$ mm, $d_R = 23.6$ mm, $Q = 0.080$ m^3/s (a) side view, (b) top view with ① original riprap level, ② pier, ③ riprap, ④ deposition, ⑤ sediment bed (Unger and Hager, 2006)

3 *Sliding* once an interface scour has developed along the riprap periphery; the first riprap member fails at $\phi \approx 75°$ (Chiew, 1995). As the tailwater level decreases, more elements at almost the same location slide sideways until the process has reached the pier perimeter. Figure 7.48a shows a flow with a relatively wavy surface containing an initial erosion surface at $\approx 120°$, for which the pier is still protected by at least one riprap row. In Figure 7.48b the interface scour has expanded upstream in which riprap elements progressively slid so that failure was reached, with the first riprap element being dislodged at an angle of 75°.

These failure modes may be subject to variation if more than one riprap layer and a filter are provided. This effect was not investigated here for reasons stated previously.

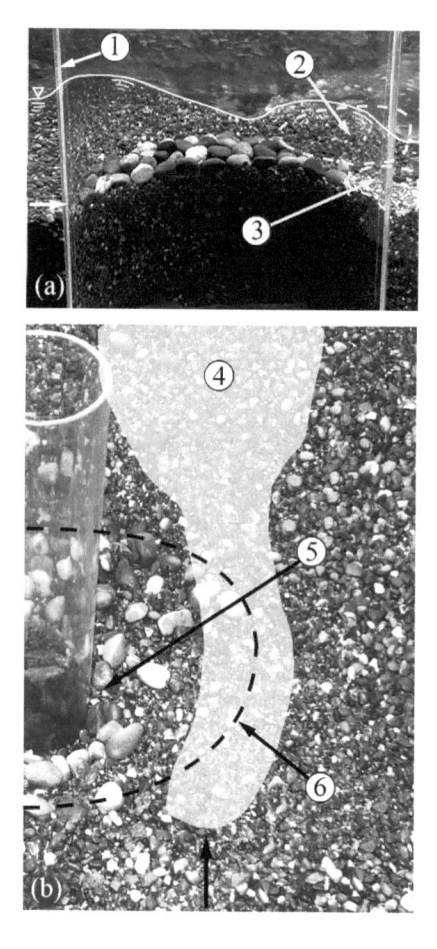

Figure 7.48 Failure of riprap by Sliding for $n_R = 5$ (a) initial phase with $D_p = 457$ mm, $d_{50} = 5.0$ mm, $d_R = 30$ mm, $Q = 0.080$ m³/s, (b) final phase with $D_p = 110$ mm, $d_{50} = 5.0$ mm, $d_R = 19$ mm, $Q = 0.100$ m³/s ① pier, ② initial erosion, ③ deposition, ④ interface scour, ⑤ riprap failure, ⑥ original riprap (Unger and Hager, 2006)

Experimental results

The test program aimed at a definition of riprap failure. For each test, the approach flow failure velocity V_{iR} was determined to compute the failure densimetric Froude number $F_{iR} = V_{iR}/(g'd_{50})^{1/2}$. Based on Eq. (7.49), the reduction factor Φ_R of pier *and* riprap presence was determined to (Figure 7.49, Unger and Hager, 2006)

$$\Phi_R = 1 - \beta_p^{1/4}, \qquad 0 < \beta_p < 0.25. \tag{7.50}$$

Note that $\Phi_R = 1$ for $\beta_p = D_p/B = 0$ and that Φ_R decreases rapidly as the relative pier size increases. Hager and Oliveto (2002) found for the pier without riprap presence the corresponding factor $\Phi_\beta = 1-(2/3)\beta_p^{1/4}$ (Figure 7.49).

Because all observations were essentially conducted in the turbulent rough regime, viscous effects on riprap failure were considered small. Based on a modified Shields' condition (7.48), the riprap load parameter $R_L = [\sigma^{-1/3}V_{iR}/(g'd_{50})^{1/2}]/[1.65(R_{ho}/d_{50})^{1/6}\Phi_R]$ includes all but the effects of riprap placement around a singular pier. Figure 7.50 shows R_L plotted against relative riprap size $\delta = d_R/d_{50}$ in semi-logarithmic scales, so that the number of riprap rows n_R plots as straights from the origin (1; 1). The latter corresponds to the Shields' condition in the turbulent rough regime, because $R_L = 1$ if riprap tends asymptotically to the bed sediment size ($\delta = 1$). Based on the observations of the failure mechanism, a distinction between the three failure modes was added, namely: (1) Rolling for relatively small riprap up to $\delta = 15$, (2) Undermining for relatively large riprap typically in excess of $\delta = 15$, and (3) Sliding as the intermediate failure mode for relatively few riprap rows and $5 < \delta < 50$. Figure 7.50 is limited to clear-water approach flow. The data may be expressed as

$$R_L = \delta^{0.20 n_R^{0.40}}, \text{ for } \delta<50 \text{ and } n\leq10. \tag{7.51}$$

For novel riprap placement, the riprap load parameter is given so that the combination of riprap size and riprap rows may be computed; for an existing riprap arrangement, the hydraulic failure conditions may be predicted, therefore.

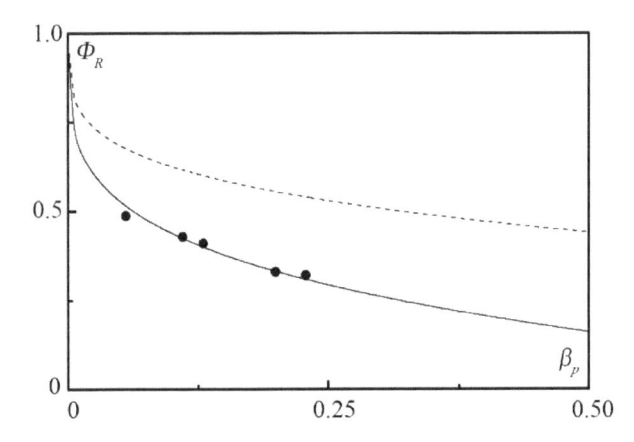

Figure 7.49 Effect Φ_R of relative pier width β_p, (——) $\Phi_\beta = 1-(2/3)\beta_p^{1/4}$ with, and (- - -) Eq. (7.50) without riprap placement

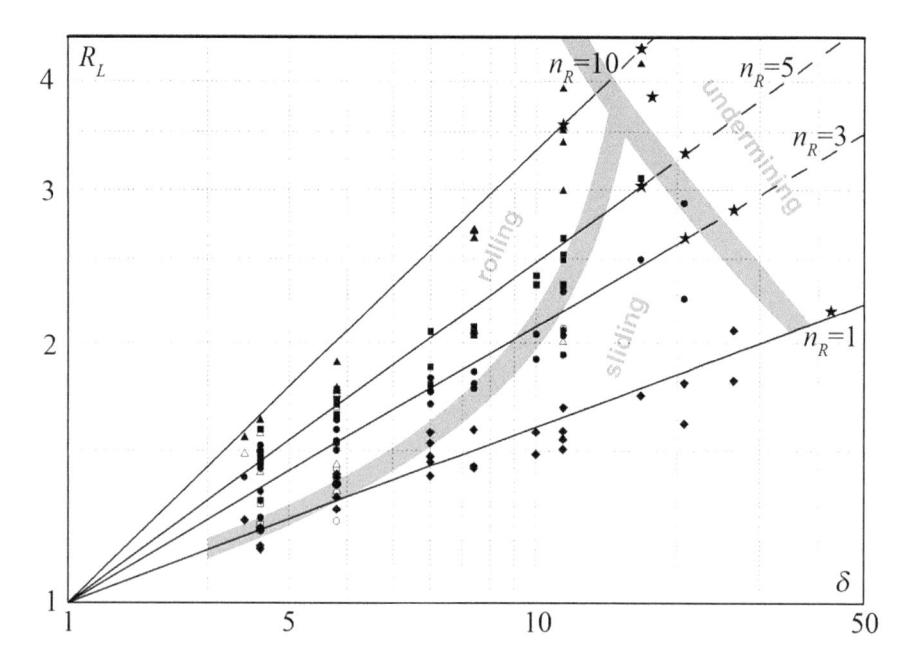

Figure 7.50 Generalized riprap failure diagram, riprap load parameter R_L versus relative riprap size δ for riprap rows $n_R = (\blacklozenge)$ 1, $(\bullet)$ 3, $(\blacksquare)$ 5, $(\blacktriangle)$ 10, $(\bigstar)$ Undermining, $(-)$ Eq. (7.51), including predominant failure modes (in gray)

Design equation and limitations

The previous test data apply for riprap design by assigning a safety factor, depending on the knowledge of the individual parameters and the relevance of the bridge to be protected. In the following, no safety factor is introduced, therefore.

Combining Eqs. (7.48), (7.50) and (7.51) gives for the riprap failure condition

$$\frac{V_{iR}}{\sigma^{1/3}(g'd_{50})^{1/2}} = 1.65(R_{ho}/d_{50})^{1/6}\left[1-\beta_p^{1/4}\right]\delta^{0.20n^{0.40}}. \tag{7.52}$$

This statement is subject to limitations including:

1. Froude similitude; effects of viscosity are small, corresponding to relative sediment size $D_* > 80$, approach flow depths $h_o > 50$ mm and $\mathsf{R}_{o*} = V_o d_{50}/v > 300$.
2. Granular material with d_{50}, σ and ρ_s for bed sediment, and d_R and n_R for almost uniform single-layered riprap, with $2 < \delta < 50$, $1 < \sigma < 3$ and $1 \leq n_R \leq 10$.
3. Relative pier diameter $0.05 < \beta_p < 0.25$.
4. Riprap arranged concentrically around pier with n_R rows of size d_R.
5. Approach flow threshold Froude number $0.60 < \mathsf{F}_{ot} = V_o/V_t < 1.2$ (Oliveto and Hager, 2002).

The threshold (subscript t) velocity is defined as $V_t = [1.65(g'd_{50})^{1/2}\sigma^{1/3}(R_{ho}/d_{50})^{1/6}]$ and $\mathsf{F}_{ot} \geq 1$ indicates entrainment of bed sediment in the approach flow region. According to Eq. (7.52) the entrainment velocity increases as the (1) difference between sediment and fluid densities, (2) riprap grain size, (3) relative approach flow depth, (4) sediment non-uniformity, and (5) number of riprap rows increase, but as the relative pier size decreases. These parameters define the design equation consisting of products of individual effects, so that it may be simply solved for each parameter.

The results were further compared with the computational approach. A comparison between the data and the predictions for the failure modes Sliding and Undermining demonstrates that all data are contained within ±15%. As to the failure mode Rolling, the accuracy of Eq. (7.52) is lower than for the other two modes. Note that Rolling involves a more dynamic effect than Undermining. Turbulence determines Rolling, and once a riprap element has left the arrangement, the progress depends strongly on details such as the exact riprap placement, turbulence of the approach flow, and speed of tailwater lowering. This larger uncertainty in the prediction of Rolling is mirrored by a poorer data correlation. A safety factor previously mentioned should thus be somewhat larger for the Rolling failure mode than for Sliding and Undermining.

Discussion of results

The findings of Melville and Coleman (2000) may be expressed as

$$d_R/h_o = A_R \cdot F_o^{\alpha_R} \cdot g^{\gamma_R}. \tag{7.53}$$

Here, A_R is a constant, $F_o = V_o/(gh_o)^{1/2}$ is the approach flow Froude number, and α_R, γ_R are exponents. Note that riprap failure was defined differently among the various authors, so that a comparison of results may only be an appreciation of the order of magnitude. A comparison of the present results with these listed by Melville and Coleman (2000) as $d_R/h_o(F_o)$ yields: the suggestions of (1) Breusers et al. (1977), (2) Austrodas, and (3) Croad (Melville and Coleman, 2000) are much higher than the remainder. Riprap failure was defined in (1) according to the Izbash critical velocity involving only d_R; the failure definition in (2) depends on a velocity factor ranging from 0.89 to 2.289; no further comments exist to (3). Equation (7.52) was inserted in a plot using a riprap extension $d_R n_R = D_p$ assuming a wide river, $R_{ho}/h_o \cong 1$. Further parameters varied as 1.1 mm $\leq d_{50} \leq$ 5 mm, 1.1 $\leq \sigma \leq$ 2.3, and 0.05 $\leq \beta_p \leq$ 0.23, with B = 1 m.

Except for Chiew (1995) and Parola (1993), all proposals involve a direct relation between d_R/h_o and F_o, independent of relative pier diameter D_p/B, relative sediment size d_{50}/h_o and the number of riprap rows n_R. For the conditions adopted, the suggestions of Quazi and Peterson, Farraday and Charlton, Parola and Jones, Richardson and Davis (Melville and Coleman, 2000), Parola (1993), and Chiew (1995) are in the domain covered by the present proposal. The result of Lauchlan and Melville (2001) is slightly higher, whereas that of Breusers and Raudkivi (1991) is much lower, based on a completely different riprap failure definition.

The recommendations of Parola (1993) and Chiew (1995) were further investigated because their data include an effect of relative riprap diameter D_p/d_R. A comparison between the previous two proposals with Eq. (7.52) for a range of relative riprap diameters 4 $\leq D_p/d_R \leq$ 33 as investigated by Parola (1993) cover practically the same domain in the plot d_R/h_o versus F_o. Closer inspection reveals that the effect of relative riprap size is inversely accounted for by Chiew (1995). Both Parola (1993) and Unger and Hager (2006) find that riprap size increases with the pier diameter, whereas Chiew states the opposite. The effect of grain size as retained by Chiew (1995) and Unger and Hager (2006) reveals for the example adopted

($B = 1$ m, $d_R n_R = D_p$, $R_{ho}/h_o = 1$ and $\beta_p = 0.13$) that Chiew's riprap diameter decreases as the size of the bed sediment increases, whereas d_{50} has practically no effect on d_R for the case considered.

The present work may also be employed to determine practical issues of riprap placement. As mentioned, riprap always failed at an angle $\phi \cong 75°$, i.e. almost sideways of the pier, although sediment entrainment started at the pier rear at $\cong 150°$. Notably, riprap elements at the pier front were never dislodged. Therefore, the largest riprap elements should be placed at the pier sides. It was also observed that the riprap elevation relative to the original bed sediment is important. If riprap blocks are 'just' dropped at a pier, the effective pier diameter increases and scour starts earlier than without any riprap exposure. Except for urgency cases when a pier has to be saved, riprap must be carefully placed for optimum protection and its economic use.

Temporal scour advance after riprap failure

Some tests were conducted beyond riprap failure to study the temporal development of bridge pier scour, as compared with pier scour without any riprap placement. Their purpose was to define differences between riprap-protected and unprotected piers. In the latter case, pier scour for originally riprap-protected piers that had failed would develop according to standard procedures (Melville and Coleman, 2000; Oliveto and Hager, 2002), whereas significant research is needed otherwise to determine the added safety of a failed riprap around a bridge pier.

The experiments were directed exclusively to the riprap failure modes Sliding and Undermining (Unger and Hager, 2006). From Figure 7.51, the scour progress differs between the

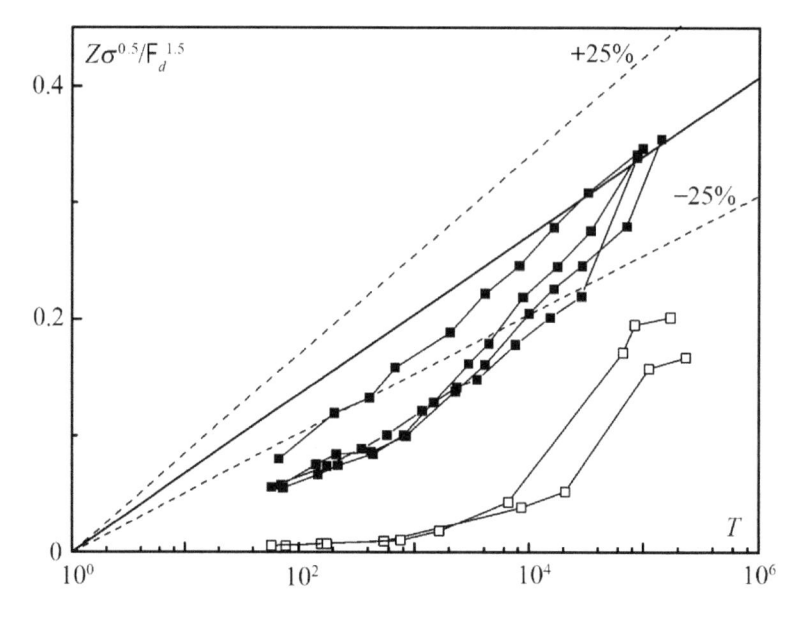

Figure 7.51 Temporal progress of pier scour $Z\sigma^{1/2}/F_d^{1.5}[T]$ after riprap failure as compared with (——) Eq. (7.34) for pier scour without riprap placement for failure mode (■) Sliding, (□) Undermining (Oliveto and Hager, 2002)

two failure modes: Whereas the scour depth is initially slow for both modes, the equation of Oliveto and Hager (2002) is relatively fast reached for failure mode Sliding, whereas scour of the mode Undermining progresses much slower not reaching the scour depth of the unprotected pier. Therefore, a riprap-protected pier has added safety against scour for failure mode Undermining ($\delta > 20$); yet additional cost is needed for the proper filter design. However, for $\delta < 20$, the scour advances almost independently from riprap presence for failure mode Sliding. Scour progress is then fast, and the scour depth of the unprotected pier is reached within a short time.

Given that the usual riprap size $\delta = d_R/d_{50}$ is between 10 and 20 (Breusers and Raudkivi, 1991), no additional safety is available for piers protected by a riprap once failure has occurred. The results of this research are thus important for bridge pier protection because knowledge on riprap failure adds to the safety of bridge structures.

Notation

A	Cross-sectional area (m²); also factor (-)
a_b	Side length of block (m)
$a_{b,min}$	Minimum required prism size (m)
a_{TW}	Coefficient (-)
B	Approach flow channel width (m)
B_s	Surface width (m)
b	Tunnel width (m); also abutment width
b_b	Block dissipator coefficient (-)
b_{TW}	Coefficient (-)
C_d	Discharge coefficient (-)
C	Choking number (-)
C_o	Approach flow choking number (-)
c	Shallow-water wave celerity (m/s)
D	Diameter (m)
D_*	Dimensionless grain size (-)
D_h	Hydraulic diameter (m)
D_p	Pier diameter (m)
D_R	riprap extension
d	Sediment diameter (m)
d_a	Length from tunnel end to reach stratified flow (m)
d_b	Equivalent block diameter (m)
d_c	Sediment size of armor layer (m)
d_{min}	Minimum sediment size to avoid significant scale effects (m)
d_R	riprap size (m)
d_{sc}	Maximum scour depth at tunnel outlet (m)
d_{toe}	Scour depth at tunnel outlet (m)
d_1	Distance from bend end to first wave maximum (m)
d_{50}	Median sediment size (m)
F	Resistance force (N)
F	Froude number (-)
F_b	Densimetric block Froude number (-)
F_C	Constriction Froude number (-)

F_d	Densimetric Froude number (-)
F_{dc}	Densimetric Froude number of armor layer (-)
F_{di}	Inception densimetric Froude number (-)
F_{dm}	Sediment mixture densimetric Froude number (-)
F_{dM}	Maximum densimentric Froude number (-)
F_o	Approach flow Froude number (-)
F_t	$= V_t/(gh_t)^{1/2}$ tailwater Froude number (-); also threshold Froude number (-)
F_{tM}	$= V_M/V_t$ (-)
F_1	$= Q/(gh_1D^4)^{1/2}$ (-)
f	Friction factor (-)
g	Gravity acceleration (m/s²)
g'	Reduced gravity acceleration (m/s²)
H	Energy head (m)
H_c	Critical energy head (m)
H_o	Approach flow energy head (m)
H_{oc}	Critical approach flow energy head (m)
H_{oN}	Uniform approach flow energy head (m)
H_d	Net head on culvert (m)
H_t	Tailwater energy head (m)
ΔH_e	Energy head loss (m)
h	Flow depth (m)
h_C	Flow depth at constricted section (m)
h_c	Critical flow depth (m)
h_M	Maximum wave height (m)
h_N	Uniform flow depth (m)
h_o	Approach flow depth (m)
h_{TW}	Tailwater flow depth at tunnel outlet (m)
h_t	Tailwater flow depth (m)
h_2	Second wave height (m)
K	Strickler's roughness coefficient (m$^{1/3}$/s)
k_s	Equivalent sand roughness height (m)
L	Transition length (m); also constriction length (m)
L_d	Culvert length (m)
L_P	Length of protection (m)
L_R	Reference length (m)
$L_{P,Req}$	Required protection length (m)
L_{TW}	Maximum scour hole length (m)
N	Shape number (-)
n	Percentage value of sediment mixture smaller than indicated (-); also hydrograph shape parameter (-); also number of riprap rows (-)
$1/n$	Manning's roughness value (s/m$^{1/3}$)
P_h	Wetted hydraulic perimeter (m)
Q	Discharge (m³/s)
Q_A	Airflow discharge (m³/s)
Q_g	Discharge under gated flow regime (m³/s)
Q_M	$= q/q_M$ (-)
Q_p	Pressurized flow discharge (m³/s)
q	Unit width discharge (m²/s)

q_M	Maximum unit discharge (m²/s)
q_0	Base unit discharge (m²/s)
R	Radius of curvature (m)
R_e	Intake radius (m)
R_h	Hydraulic radius (m)
R_{ho}	hydraulic radius of approach flow (m)
R_L	riprap load parameter (-)
R_o	Approach flow curvature radius (m)
R	Reynolds number (-)
R_o	Approach flow Reynolds number (-)
R_{o*}	Grain Reynolds number (-)
r_d	Culvert intake radius (m)
S	Hydraulic gradient (-)
S_c	Critical slope (-)
S_f	Friction slope (-)
S_o	Bottom slope (-)
s_b	Prism spacing (m)
T	$= h_o/h_t$ (-); also relative time $= t/t_R$ (-)
T_C	$= t_C/t_M$ (-)
T_d	$= t_d/t_M$ (-)
T_e	$= t_e/t_R$ (-)
T_i	$= t_i/t_M$ (-)
T_M	Relative maximum time to peak (-)
t	Time (s)
t_C	Time to scour completion (s)
t_d	Duration of scour activity (s)
t_e	Time at which end scour is formed (s)
t_i	Time to scour initiation (s)
t_M	Time to peak (s)
t_R	Reference time (s)
$u*$	Shear velocity (m/s)
V	Average cross-sectional flow velocity (m/s)
V_c	Critical flow velocity (m/s)
V_{cr}	Critical surface velocity (m/s)
V_{cube}	Equivalent cube volume (m³)
V_M	Maximum velocity during unsteady approach flow scour (m/s)
V_N	Uniform velocity (m/s)
V_o	Approach flow velocity (m/s)
V_R	Reference velocity (m/s)
V_t	Tailwater velocity (m/s); also threshold velocity (m/s)
W	Block weight (kN)
w_P	Protected scour hole width (m)
w_{sc}	Maximum scour width (m)
w_{TW}	Maximum scour hole width (m)
x	Streamwise coordinate (m)
x_{TW}	Distance of maximum scour depth from outlet (m)
x_1	Upstream location of scour hole (m)
x_2	Location of scour hole minimum elevation (m)

x_3	Downstream location of scour hole (m)
Y	Sequent depth ratio (-); also relative wave height $= h_2/h_1$ (-)
Y_M	Relative maximum wave height (-)
y_b	Variable of block energy dissipator (-)
y_o	Approach flow filling ratio (-)
y_{TW}	Variable of tunnel outlet scour hole (-)
Z	Maximum dimensionless scour depth (-)
Z_C	Relative flow depth in constriction (-)
Z_M	$= z/z_M$ (-)
Z_{Me}	End value of Z_M (-)
z	Invert elevation (m); also maximum scour depth (m)
z_e	End scour depth (m)
z_M	Absolute maximum scour depth (m)
α	Upstream angle of constriction (-); also exponent (-)
α_a	Constant in flood wave effect (-)
β	Downstream angle of constriction (-); also exponent (-)
β_{A+}	Air-water discharge ratio (-)
β_e	$= b/B$ contraction width ratio (-)
β_p	Relative pier size
β_s	Safety factor (-)
γ	$= t_M/t_R$ (-); also exponent (-)
χ	Roughness characteristics (-)
δ	Ratio between riprap and bed sediment sizes (-)
ε	Relative roughness (-)
η_d	Relative curvature radius (-)
Φ	Effect of relative pier size β_p (-)
ϕ	Angular coordinate (-)
ϕ_d	Downstream angle of cofferdam (-)
λ	Length ratio of constriction (-)
μ	Contraction coefficient (-)
μ_0	Base contraction ratio (-)
ξ	Head-loss coefficient (-)
v	Kinematic viscosity (m²/s)
ψ	Width ratio of constriction (-)
ρ	Fluid mass density (g/m³)
ρ'	Reduced density (-)
ρ_b	Block mass density (g/m³)
ρ_s	Sediment mass density (g/m³)
σ	Sediment non-uniformity parameter (-)

References

ASCE Task Force (1965) Factors influencing flow in large conduits. *Journal of the Hydraulics Division* ASCE, 91(HY6), 123–152; 92(HY4), 168–218; 93(HY3), 181–187.

Ballio, F., Crippa, S., Fioroni, M. & Franzetti, S. (2000) Effetto del restringimento di sezione sui processi erosivi in prossimità delle spalle dei ponti (Effect of contraction on the scour process near a bridge pier). Proceedings of the 27th *Convegno di Idraulica e Costruzioni Idrauliche*, Genova, 1, 195–203 (in Italian).

Barr, D.I.H. (1973) Resistance laws for large conduits. *Water Power*, 24(8), 290–304.

Bray, D.I. & Davar, K.S. (1987) Resistance to flow in gravel bed rivers. *Canadian Journal of Civil Engineering*, 14, 77–86; 14, 857–858.

Breusers, H.N.C., Nicollet, G. & Shen, H.W. (1977) Local scour around cylindrical piers. *Journal of Hydraulic Research*, 15(3), 211–252; 16(3), 259–260.

Breusers, H.N.C. & Raudkivi, A.J. (1991) *Scouring*. IAHR Hydraulic Structures Design Manual 2, Balkema, Rotterdam.

Cardoso, A.H. & Bettess, R. (1999) Effects of time and channel geometry on scour at bridge abutments. *Journal of Hydraulic Engineering*, 125(4), 388–399.

Chabert, J. & Engeldinger, P. (1956) Étude des affouillements autour des piles de ponts [Study of scour at bridge piers] *Série* A. Laboratoire National d'Hydraulique, Chatou (in French).

Chang, W.Y., Lai, J.S. & Yen, C.L. (2004) Evolution of scour depth at circular bridge piers. *Journal of Hydraulic Engineering*, 130(9), 905–913.

Chiew, Y.-M. (1995) Mechanics of riprap failure at bridge piers. *Journal of Hydraulic Engineering*, 121(9), 635–643; 123(5), 481–482.

Chiew, Y.-M. (2000) Failure behavior of riprap layer at bridge piers under live-bed conditions. *Journal of Hydraulic Engineering*, 126(1), 43–55.

Chow, V.T. (1959) *Open Channel Hydraulics*. McGraw-Hill, New York.

Colebrook, C.F. (1958) The flow of water in unlined, lined, and partly lined rock tunnels. *Proceedings of the Institution of Civil Engineers*, 11, 103–132; 12, 523–562.

Colebrook, C.F. & White, C.M. (1937) Experiments with fluid friction in roughened pipes. *Proceedings of the Royal Society* London, Series A, 161, 367–381.

Colorado State University (1962) Analytical study of local scour. *Report*. Department of Civil Engineering, Colorado State University, Fort Collins, CO.

Cunha, L.V. (1975) Time evolution of local scour. Proceedings of the 14th *IAHR Congress*, Paris B36, 285–299.

Dietz, J.W. (1972) Systematische Modellversuche über die Pfeilerkolkbildung [Systematic tests on the development of pier scour]. BAW *Mitteilungsblatt* 31. Bundesanstalt für Wasserbau, Karlsruhe (in German).

Ettema, R. (1980). Scour at bridge piers. *Report* 216. School of Engineering, University of Auckland, Auckland, NZ.

Emami, S. (2004) Erosion protection downstream of diversion tunnels using concrete prisms: design criteria based on a systematic physical model study. In: Schleiss, A. (ed) *Communication* 18. Laboratory of Hydraulic Constructions (LCH). LCH-EPFL, Lausanne.

Emami, S. & Schleiss, A.J. (2006a) Design of erosion protection at diversion tunnel outlets with concrete prisms. *Canadian Journal of Civil Engineering*, 33(1), 81–92.

Emami, S. & Schleiss, A.J. (2006b) Performance of large concrete prisms as erosion protection at Seymareh diversion tunnel outlets during a large flood. Proceedings of the 21st *ICOLD Congress*, Barcelona, 5(C12), 99–813.

Ettema, R. (1980) Scour at bridge piers. *Report* 216. School of Engineering, University of Auckland, Auckland NZ.

Franzetti, S., Larcan, E. & Mignosa, P. (1989) Erosione alla base di pile circolari di ponti: Verifica sperimentale dell'ipotesi di esistenza di una situazione finale di equilibrio [Scour at the base of circular bridge piers: experimental verification of the equilibrium scour hypothesis]. *Idrotecnica*, 16(3), 135–141 (in Italian).

Gisonni, C. & Hager, W.H. (1999) Studying flow at tunnel bends. *Hydropower and Dams*, 6(2), 76–79.

GT-CFGB (1973) Groupe de travail du Comité Français des Grands Barrages: Méthodes de dérivation pendant la construction. Proceedings of the 11th *ICOLD Congress*, Madrid, Q36(R16), 565–590 (in French).

Hager, W.H. (1987) Discharge characteristics of local, discontinuous contractions II. *Journal of Hydraulic Research*, 25(2), 197–214.

Hager, W.H. & Del Giudice, G. (1998) Generalized culvert design. *Journal of Irrigation and Drainage Engineering*, 124(5), 271–274.

Hager, W.H. & Dupraz, P.A. (1985) Discharge characteristics of local, discontinuous contractions I. *Journal of Hydraulic Research*, 23(5), 421–433.

Hager, W.H. & Oliveto, G. (2002) Shields' entrainment criterion in bridge hydraulics. *Journal of Hydraulic Engineering*, 128(5), 538–542.

Hager, W.H., Unger, J. & Oliveto, G. (2002) Entrainment criterion for bridge piers and abutments. In: Bousmar, D. & Zech, Y. (eds) Proceedings of *River Flow*, 2002 (2), 1053–1058.

Hager, W.H., Oliveto, G., Pagliara, S. & Unger, J. (2004) Recent advances in scour hydraulics. In: Greco, M. (ed) Proceedings of *River Flow*, 1, 3–12. Balkema, Leiden, NL.

Hager, W.H. & Schleiss, A.J. (2009) *Constructions hydrauliques* (Hydraulic structures) Presses Polytechniques et Universitaires Romandes, Lausanne (in French).

Hager, W.H. & Unger, J. (2010) Bridge pier scour under flood waves. *Journal of Hydraulic Engineering*, 136(10), 842–847.

Hager, W.H. (2010) *Wastewater Hydraulics*. Springer, Berlin.

Hager, W.H. (2015) Albert Strickler: his life and work. *Journal of Hydraulic Engineering*, 141(7), 02515002, 1–5.

Hager, W.H., Castro-Orgaz, O. (2016) Transcritical flow in open channel hydraulics: from Böss to De Marchi. *Journal of Hydraulic Engineering*, 142(1), 02515003, 1–9.

Hinds, J. (1928) The hydraulic design of flume and syphon transitions. *Transactions of ASCE*, 92, 1423–1459.

Hoffmans, G.J.C.M. & Verheij, H.J. (1997) *Scour Manual*. Balkema, Rotterdam.

Izbash, S.V. (1935) Construction of dams and other structures by dumping stones into flowing water. *Izvestiya VNIIG*, 17, 12–66 (in Russian, with English summary).

Jaeger, C. (1949) *Technische Hydraulik*. Birkhäuser, Basel (in German).

Jarrett, R.D. (1984) Hydraulics of high-gradient-streams. *Journal of Hydraulic Engineering*, 110(11), 1519–1539; 113(7), 918–929.

Kohli, A. (1998) Kolk an Gebäuden in Überschwemmungsebenen [Building scour in flood plains]. *Ph.D. thesis* 12592. ETH Zurich, Zürich, Switzerland (in German).

Kothyari, U.C., Garde, R.J. & Ranga Raju, K.G. (1992) Temporal variation of scour around circular bridge piers. *Journal of Hydraulic Engineering*, 118(8), 1091–1106.

Kothyari, U.C., Hager, W.H. & Oliveto, G. (2007) Generalized approach for clear-water scour at bridge elements. *Journal of Hydraulic Engineering*, 133(11), 1229–1240; 135(3), 240–242.

Lauchlan, C.S. & Melville, B.W. (2001) Riprap protection at bridge piers. *Journal of Hydraulic Engineering*, 127(5), 412–418.

Lim, F.-H. & Chiew, Y.-M. (2001) Parametric study of riprap failure around bridge piers. *Journal of Hydraulic Research*, 39(1), 61–72.

Manso, P.A. (2002) Stability of linings by concrete elements for surface protection of overflow earthfill dams. In: Schleiss, A. (ed) *Communication* 12, Laboratory of Hydraulic Constructions, LCH-EPFL, Lausanne, Switzerland.

Manso, P.A. & Schleiss, A.J. (2002) Stability of concrete macro-roughness linings for overflow protection of earth embankment dams. *Canadian Journal of Civil Engineering*, 29(5), 762–776.

Meile, T., Bodenmann, M., Schleiss, A. & Boillat, J.-L. (2004) Flood protection concept for the river Gamsa in Canton Wallis. Intl. Symp. *Interpraevent*, Riva/Trient, Italy, Theme VII, 219–230 (in German).

Melville, B.W. & Chiew, Y.-M. (1999) Time scale for local scour at bridge piers. *Journal of Hydraulic Engineering*, 125(1), 559–665.

Melville, B.W. & Coleman, S.E. (2000) *Bridge Scour*. Water Resources Publications, Highlands Ranch CO.

Meyer-Peter, E. & Müller, R. (1948) Formulas for bed-load transport. Proceedings of the 2nd *IAHR Congress*, Stockholm, 2, 1–26.

Neill, C. (1973) *Guide to Bridge Hydraulics*. Road and Transportation Association of Canada, University of Toronto, University Press, Toronto.

Oliveto, G. & Hager, W.H. (2002) Temporal evolution of clear-water pier and abutment scour. *Journal of Hydraulic Engineering*, 128(9), 811–820.

Oliveto, G. & Hager, W.H. (2005) Further results to time-dependent local scour at bridge elements. *Journal of Hydraulic Engineering*, 131(2), 97–105; 132(9), 997.

Quintela, A.C., Fernandez, J.S. & Cruz, A.A. (1997) Le barrage de Cahora-Bassa: Problèmes posés par le passage des crues pendant et après la construction. Proceedings of the 13th *ICOLD Congress*, New Delhi, 3, Q50(R41), 713–730 (in French).

Parola, A.C. (1993) Stability of riprap at bridge piers. *Journal of Hydraulic Engineering*, 119(10), 1080–1093.

Rajaratnam, N. & Nwachukwu, B.A. (1983) Erosion near groyne-like structures. *Journal of Hydraulic Research*, 21(4), 277–287.

Schleiss, A., Aemmer, M., Philipp, E. & Weber, H. (1998) Erosion protection at mountain rivers with buried concrete blocks. *Wasser Energie Luft*, 90(3/4), 45–52 (in German).

Schleiss, A. (2015) *Aménagements hydrauliques*. Polycopié EPFL, Section de Génie Civil (484 p.) EPFL, Lausanne (in French).

Schleiss, A.J. (2018) The safety challenge of river diversion during construction of dams. Proceedings of 5th International Symposium on *Dam Safety*, Istanbul, 1, 1–19, Hasan Tosun et al. (eds).

Schröder, R.C.M. (1990) Hydraulische Methoden zur Erfassung von Rauheiten [Hydraulic methods to determine roughnesses]. *DVWK Schrift*, 92. Parey, Hamburg and Berlin (in German).

Smith, C.D. (1967) Simplified design for flume inlets. *Journal of the Hydraulics Division* ASCE, 93(HY6), 25–34; 94(HY3), 813–815; 94(HY4), 1152–1153; 94(HY6), 1544–1545; 95(HY4), 1456–1457.

Strickler, A. (1923) Beiträge zur Frage der Geschwindigkeitsformel und der Rauhigkeitszahlen für Ströme, Kanäle und geschlossene Leitungen [Contributions to the question of velocity formula and roughness numbers for rivers, channels and closed conduits]. *Mitteilung* 16. Amt für Wasserwirtschaft, Bern (in German).

Unger, J. & Hager, W.H. (2006) Riprap failure at circular bridge piers. *Journal of Hydraulic Engineering*, 132(4), 354–362.

Vischer, D. (1987) Das Wasserkraftwerk Clyde am Clutha in Neuseeland [The hydro power plant Clyde on Clutha River in New Zealand]. *Wasserwirtschaft*, 77(2), 63–68 (in German).

Vischer, D.L., Hager, W.H., Casanova, C., Joos, B., Lier, P. & Martini, O. (1997) Bypass tunnels to prevent reservoir sedimentation. Proceedings of the 19th *ICOLD Congress*, Florence, Q74(R37), 605–624.

Wörman, A. (1989) Riprap protection without filter layers. *Journal of Hydraulic Engineering*, 115(12), 1615–1630.

Yanmaz, A.M. & Altinbilek, H.D. (1991) Study of time-dependent local scour around bridge piers. *Journal of Hydraulic Engineering*, 117(10), 1247–1268; 118(11), 1593–1597.

Bibliography

Backwater curves

Chow, V.T. (1981) Hydraulic exponents. *Journal of the Hydraulics Division* ASCE, 107(HY11), 1489–1499.

Hager, W.H. (1987) Abfluss im U-Profil [Flow in U-shaped channel]. *Korrespondenz Abwasser*, 34(5), 468–482 (in German).

Kouwen, N., Harrington, R.A. & Solomon, S.I. (1977) Principles of graphical gradually varied flow model. *Journal of the Hydraulics Division* ASCE, 103(HY5), 531–541; 104(HY5), 803–805; 104(HY12), 1677.

Meselhe, E.A., Sotiropoulos, F. & Holly, Jr., F.M. (1997) Numerical simulation of transcritical flow in open channels. *Journal of Hydraulic Engineering*, 123(9), 774–783.

Minton, P. & Sobey, R.J. (1973) Unified nondimensional formulation for open channel flow. *Journal of the Hydraulics Division* ASCE, 99(HY1), 1–12; 100(HY1), 238–239; 100(HY3), 482–485; 101(HY1), 191–193.

Müller, R. (1972) Geschlossene Berechnung von Stau- und Senkungslinien [Closed form computation of backwater curves]. *Mitteilung*, 9, 43–91. Institut für Hydraulik und Gewässerkunde, TU München, München (in German).

Peruginelli, A. & Viti, C. (1977) Ricerca sperimentale su un canale a pendenza critica unica [Experimental research on the flow in a critical-sloped channel]. *Giornale del Genio Civile*, 115(10/12), 403–414 (in Italian).

Rajaratnam, N., Van der Vinne, G. & Katopodis, C. (1986) Hydraulics of vertical slot fishways. *Journal of Hydraulic Engineering*, 112(10), 909–927.

Rhodes, D.G. (1995) Newton-Raphson solution for gradually varied flow. *Journal of Hydraulic Research*, 33(2), 213–218; 33(5), 731–735.

Bend flow in open channels

Allen, J. & Chee, S.P. (1962) The resistance to the flow of water round a smooth circular bend in an open channel. *Proceedings of the ICE*, 23(3), 423–434; 26(2), 335–344.

Anonymous (1930) Hydraulic characteristics of flow of water around bends. *Engineering News-Record*, 105(Sep. 04), 376–377; 105(Nov. 06), 739–740; 105(Nov. 27), 859; 105(Dec. 11), 939–940; 111(Nov. 09), 570.

Apmann, R.P. (1964) A case history in theory and experiment: fluid flow in bends. *Isis*, 55(4), 427–434.

Apmann, R.P. (1972) Flow processes in open channel bends. *Journal of the Hydraulics Division* ASCE, 98(HY5), 795–810; 99(HY4), 687–691.

Blanckaert, K. & Graf, W.H. (2004) Momentum transport in sharp open-channel bends. *Journal of Hydraulic Engineering*, 130(3), 186–198.

Böss, P. (1934) Anwendung der Potentialtheorie auf die Bewegung des Wassers in gekrümmten Kanal- oder Flussstrecken [Application of potential flow theory on the flow of water in curved conveyances]. *Der Bauingenieur*, 15(25/26), 251–253 (in German).

Bridge, J.S. & Jarvis, J. (1982) The dynamics of a river bend: a study in flow and sedimentary processes. *Journal of Sedimentology*, 29(4), 499–541.

Chang, H.H. (1984) Variation of flow resistance through curved channels. *Journal of Hydraulic Engineering*, 110(12), 1772–1782.

Damaskinidou-Georgiadou, A., Smith, K.V.H. (1986) Flow in curved converging channel. *Journal of Hydraulic Engineering*, 112(6), 476–496.

de Vriend, H.J. (1977) A mathematical model of steady flow in curved shallow channels. *Journal of Hydraulic Research*, 15(1), 37–53.

Einstein, H.A. & Harder, J.A. (1954) Velocity distribution and the boundary layer at channel bends. *Transactions of AGU*, 35(1), 114–120.

Engelund, F. (1974) Flow and bed topography in channel bends. *Journal of the Hydraulics Division* ASCE, 100(HY11), 1631–1648; 101(HY9), 1290–1291; 101(HY10), 1367–1369.

Fox, J.A. & Ball, D.J. (1968) The analysis of secondary flow in bends in open channels. *Proceedings of the ICE*, 39(3), 467–475; 40(4), 581–584.

Francis, J.R.D. & Asfari, A.F. (1971) Velocity distribution in wide, curved open-channel flows. *Journal of Hydraulic Research*, 9(1), 73–89; 9(3), 485–489.

Hersberger, D., Franca, M. & Schleiss, A.J. (2016) Wall-roughness effects on flow and scouring in curved channels with gravel beds. *Journal of Hydraulic Engineering*, 10.1061/(ASCE)HY.1943–7900.0001039, 04015032.

Ippen, A.T. & Drinker, P.A. (1962) Boundary shear stresses in curved trapezoidal channels. *Journal of the Hydraulics Division* ASCE, 88(HY5), 143–180; 89(HY2), 189–191; 89(HY3), 327–345; 90(HY2), 351–356.

Jin, Y.-C., Steffler, P.M. & Hicks, F.E. (1990) Roughness effects on flow and shear stress near outside bank of curved channel. *Journal of Hydraulic Engineering*, 116(4), 563–577.

Jobes, J.G. & Douma, J.H. (1942) Testing theoretical losses in open channel flow. *Civil Engineering*, 12(11), 613–615.

Leschziner, M.A. & Rodi, W. (1979) Calculation of strongly curved open channel flow. *Journal of the Hydraulics Division* ASCE, 105(HY10), 1297–1314; 106(HY10), 1713–1714; 107(HY1), 142–143; 107(HY9), 1111–1112.

Lien, H.C., Yang, J.C. & Yeh, K.C. (1999) Bend-flow simulation using 2D depth-averaged model. *Journal of Hydraulic Engineering*, 125(10), 1097–1108; 127(2), 167–170.

Mockmore, C.A. (1944) Flow around bends in stable channels. *Transactions of ASCE*, 109, 593–628.

Pacheco-Ceballos, R. (1983) Energy losses and shear stresses in channel bends. *Journal of Hydraulic Engineering*, 109(6), 881–896; 110(9), 1281–1286.

Prus-Chacinski, T.M. (1956) Patterns of motion in open-channel bends. *Journal of the Institution of Water Engineers*, 10, 420–426.

Ramponi, F. (1940) Sul moto dell'acqua nei canali aperti ad asse curvilinea (On water flow on open channel curves) *L'Energia Elettrica*, 17(4), 194–205 (in Italian).

Rozovskii, I.L. (1957) *Flow of water in bends of open channels*. Academy of Sciences of the Ukrainian SSR, Kiev.

Sellin, R.H.J., Ervine, D.A. & Willetts, B.B. (1993) Behaviour of meandering two-stage channels. Proceedings of the ICE, *Water, Maritime & Energy*, 101(2), 99–111; 112(2), 176–178.

Shukry, A. (1949) Flow around bends in an open flume. *Transactions of ASCE*, 115, 751–779.

Soliman, M.M. & Tinney, E.R. (1968) Flow around 180° bends in open rectangular channels. *Journal of the Hydraulics Division* ASCE, 94(HY4), 893–908; 95(HY2), 729–731; 95(HY3), 1064; 96(HY1), 2258.

Steffler, P.M., Rajaratnam, N. & Peterson, A.W. (1985. Water surface at change of channel curvature. *Journal of Hydraulic Engineering*, 111(5), 866–870.

Varshney, D.V. & Garde, R.J. (1975) Shear distribution in bends in rectangular channels. *Journal of the Hydraulics Division* ASCE, 101(HY8), 1053–1066; 102(HY5), 688–689; 102(HY12), 1771.

Varshney, D.V. (1977) Scour around bends in alluvial channels. *Transactions of the Institution of Engineers* India, 58(9/11), 91–98.

Yen, C.H. & Howe, J.W. (1942) Effects of channel shape on losses in a canal bend. *Civil Engineering*, 12(1), 28–29.

Yen, C.-l. & Yen, B.C. (1971) Water surface configuration in channel bends. *Journal of the Hydraulics Division* ASCE, 97(HY2), 303–321; 98(HY5), 953.

Bridge hydraulics and scour protection measures

Ahmed, F. & Rajaratnam, N. (1998) Flow around bridge piers. *Journal of Hydraulic Engineering*, 124(3), 288–300.

Cardoso, A.H., Simarro, G., Fael, C., Le Doucen, O. & Schleiss, A.J. (2010) Toe protection for spill-through and vertical-wall abutments. *Journal of Hydraulic Research*, 48(4), 491–498.

Cardoso, A.H., Simarro, G., Le Doucen, O. & Schleiss, A.J. (2010) Sizing of riprap for spill-through abutments. Proceedings ICE – *Water Management*, 163(10), 499–507.

Carstens, M.R. (1966) Similarity laws for localized scour. *Journal of the Hydraulics Division* ASCE, 92(HY3), 13–36; 92(HY6), 271–278; 93(HY2), 67–71; 94(HY1), 303–306.

Chang, F.M. & Yevdjevich, V.M. (1962) Analytical study of local scour. *Report* CER62FMC26. Civil Engineering Section, Colorado State University, Fort Collins, CO.

Chiew, Y.-M. & Melville, B.W. (1987) Local scour around bridge piers. *Journal of Hydraulic Research*, 25(1), 15–26.

Dargahi, B. (1990) Controlling mechanism of local scouring. *Journal of Hydraulic Engineering*, 116(10), 1197–1214; 118(2), 504–505.

Dey, S., Bose, S.K. & Sastry, G.L.N. (1995) Clear water scour at circular piers: a model. *Journal of Hydraulic Engineering*, 121(12), 869–876.

Graf, W.H. & Yulistiyanto, B. (1998) Experiments on flow around a cylinder: the velocity and vorticity fields. *Journal of Hydraulic Research*, 36(4), 637–653.

Graf, W.H. & Altinakar, M.S. (1998) *Fluvial hydraulics*: flow and transport processes in channels of simple geometry. Wiley, Chichester.

Graf, W.H. & Istiarto, I. (2002) Flow pattern in the scour hole around a cylinder. *Journal of Hydraulic Research*, 40(1), 13–20; 41(4), 443–446.

Hager, W.H. & Kohli, A. (1997) Kolk an Quaderelementen [Scour at quadratic elements]. *Österreichische Wasser- und Abfallwirtschaft*, 49(7/8), 145–153 (in German).

Hamill, L. (1993) A guide to the hydraulic analysis of single-span arch bridges. Proceedings of the ICE, *Municipal Engineer*, 98, 1–11.

Hamill, L. (1997) Improved flow through bridge waterways by entrance rounding. Proceedings of the ICE, *Municipal Engineer*, 121, 7–21.

Hamill, L. (1999) *Bridge Hydraulics*. Spon, London.

Kohli, A. & Hager, W.H. (1999) Building scour in floodplains. Proceedings of the ICE, *Water and Maritime Engineering*, 148(2), 61–80.

Kothyari, U.C., Garde, R.J. & Ranga Raju, K.G. (1992) Live-bed scour around cylindrical bridge piers. *Journal of Hydraulic Research*, 30(5), 701–715; 31(2), 279–281.

Kuhnle, R.A., Alonso, C.V. & Shields, Jr., F.D. (1999) Geometry of scour holes associated with 90° spur dikes. *Journal of Hydraulic Engineering*, 125(9), 972–978.

Lim, S.-Y. (1997) Equilibrium clear-water scour around an abutment. *Journal of Hydraulic Engineering*, 123(3), 237–243; 124(10), 1069–1073.

Lim, S.-Y. & Cheng, N.-S. (1998) Prediction of live-bed scour at bridge abutments. *Journal of Hydraulic Engineering*, 124(6), 635–638; 125(9), 985–986.

Marchi, E. (1994) Il rigurgito dovuto alle pile di ponte di forma circolare [The head-losses due to presence of circular bridge piers]. *Idrotecnica*, 21(5), 263–271 (in Italian).

Melville, B.W. & Sutherland, A.J. (1988) Design method for local scour at bridge piers. *Journal of Hydraulic Engineering*, 114(10), 1210–1226; 116(10), 1290–1293.

Melville, B.W. (1992) Local scour at bridge abutments. *Journal of Hydraulic Engineering*, 118(4), 615–631; 119(9), 1064–1073.

Melville, B.W. (1995) Bridge abutment scour in compound channels. *Journal of Hydraulic Engineering*, 121(12), 863–868.

Melville, B.W. & Raudkivi, A.J. (1996) Effects of foundation geometry on bridge pier scour. *Journal of Hydraulic Engineering*, 122(4), 203–209.

Molinas, A., Kheireldin, K. & Wu, B. (1998) Shear stress around vertical wall abutments. *Journal of Hydraulic Engineering*, 124(8), 822–830.

Murillo, J.A. (1987) The scourge of scour. *Civil Engineering*, 57(7), 66–69.

Rajaratnam, N. & Nwachukwu, B.A. (1983) Flow near groin-like structures. *Journal of Hydraulic Engineering*, 109(3), 463–480.

Richardson, E.V. (1998) History of bridge scour research and evaluations in the United States. In: Richardson, E.V. & Lagasse, P.F. (eds) *Stream Stability and Scour at Highway Bridges*. ASCE, Reston VA, pp. 15–40.

Shen, H.W., Schneider, V.R. & Karaki, S.S. (1966) *Mechanics of Local Scour*. Bureau of Public Roads. Civil Engineering Dept., Engineering Research Center, Colorado State University, Fort Collins, CO.

Sturm, T.W. & Janjua, N.S. (1994) Clear-water scour around abutments in floodplains. *Journal of Hydraulic Engineering*, 120(8), 956–972.

Umbrell, E.R., Young, G.K., Stein, S.M. & Sterling Jones, J. (1998) Clear-water contraction scour under bridges in pressure flow. *Journal of Hydraulic Engineering*, 124(2), 236–240; 125(7), 785.

Yulistiyanto, B., Zech, Y. & Graf, W.H. (1998) Flow around a cylinder: shallow-water modeling with diffusion-dispersion. *Journal of Hydraulic Engineering*, 124(4), 419–429.

Culverts

Abt, S.R., Kloberdanz, R.L. & Mendoza, C. (1984) Unified culvert scour determination. *Journal of Hydraulic Engineering*, 110(10), 1475–1479.

Abt, S.R., Ruff, J.F. & Doehring, F.K. (1985) Culvert slope effect on outlet scour. *Journal of Hydraulic Engineering*, 111(10), 1363–1367.

Abt, S.R., Ruff, J.F., Doehring, F.K. & Donnell, C.A. (1987) Influence of culvert shape on outlet scour. *Journal of Hydraulic Engineering*, 113(3), 393–400.

Argue, J.R. (1960) New structure for roadway pipe culverts. *Journal of Institution of Engineers, Australia*, 32(6), 123–129.

Aronson, H.G. (1961) Prefabricated reducers as entrances for pipe culverts. *Journal of the Highway Division* ASCE, 87(HW1), 1–20; 87(HW3), 35–37; 89(HW1), 115–120.

Bauer, W.J. (1959) Improved culvert performance through design and research studies. *Civil Engineering*, 29(3), 167–169; 29(7), 492; 29(11), 650.

Blaisdell, F.W. (1954) Hydraulic fundamentals of closed conduit spillways. *Proceedings of the ASCE*, 79(Sep.354), 1–13; 80(Sep. 491), 1–6; 80(Sep. 538), 11–13.

Blaisdell, F.W. & Donnelly, C.A. (1956) Hood inlet for closed conduit spillways. *Agricultural Engineering*, 37(10), 670–672.

Blaisdell, F.W. (1958) Hydraulics of closed conduit spillways. St. Anthony Falls (SAF) Hydraulic Laboratory, *Technical Paper* 19, Series B. University of Minnesota, Minneapolis.

Blaisdell, F.W. (1960) Hood inlet for closed conduit spillways. *Journal of the Hydraulics Division* ASCE, 86(HY5), 7–31; 86(HY8), 75–77; 86(HY9), 179–183; 87(HY1), 181–186; 88(HY4), 197–207.

Blaisdell, F.W. (1966) Hydraulic efficiency in culvert design. *Journal of the Highway Division* ASCE, 92(HW1), 11–22; 93(HW1), 84; 93(HW2), 192–194.

Blaisdell, F.W. (1967) Flow in culverts and related design philosophies. *Journal of the Hydraulics Division* ASCE, 92(HY2), 19–31; 92(HY5), 261–271; 93(HY1), 85–91; 93(HY3), 188–190; 94(HY2), 531–540.

Bodhaine, G.L. (1982) Measurement of peak discharge at culverts by indirect methods. *Techniques of Water-Resources Investigations*, A3, 3–60.

Carstens, M.R. & Holt, A.R. (1956) Demonstration of possible flow conditions in a culvert. Culvert-flow characteristics, *Bulletin*, 126, 1–23. Highway Research Board, Washington DC.

Dasika, B. (1995) New approach to design of culverts. *Journal of Irrigation and Drainage Engineering*, 121(3), 261–264; 121(5), 367; 123(1), 71–72.

French, J.L. (1964) Tapered inlets for pipe culverts. *Journal of the Hydraulics Division* ASCE, 90(HY2), 255–299; 90(HY6), 315–330; 91(HY3), 287–296.

French, J.L. (1969) Nonenlarged box culvert inlets. *Journal of the Hydraulics Division* ASCE, 95(HY6), 2115–2137.

Garg, S.P. (1966) Distribution of head at a rectangular conduit outlet. *Journal of the Hydraulics Division* ASCE, 92(HY4), 11–31; 93(HY2), 79–81; 94(HY2), 540–541.

Li, W.-H. & Patterson, C.C. (1956) Free outlets and self-priming action of culverts. *Journal of the Hydraulics Division* ASCE, 82(HY3, Paper 1009), 1–22; 82(HY6, Paper 1131), 5–8; 83(HY1, Paper 1177), 23–40; 83(HY4, Paper 1348), 3–5.

Lim, S.-Y. (1995) Scour below unsubmerged full-flowing culvert outlets. Proceedings of the ICE, *Water Maritime & Energy*, 112(3), 136–149.

Mavis, F.T. (1934) Capacity of creosoted-wood culverts studies. *Engineering News-Record*, 113(Oct. 18), 486–487.

Mavis, F.T. (1942) The hydraulics of culverts. *Bulletin*, 56. Engineering Experiment Station, Pennsylvania State College, University Park PA.

Metzler, D.E. & Rouse, H. (1959) Hydraulics of box culverts. *Bulletin*, 38. Studies in Engineering, State University of Iowa, Iowa City IA.

Milano, V. (1978) Ricerca sperimentale sulla copertura di un tronco di un corso d'acqua a forte pendenza [Experimental research on the cover of a steeply-sloping watercourse]. *Idrotecnica*, 5(6), 221–236 (in Italian).

Neill, C.R. (1962) Hydraulic capacity of large corrugated metal culverts. *The Engineering Journal, Canada*, 45(2), 33–38.

Rice, C.E. (1967) Effect of pipe boundary on hood inlet performance. *Journal of the Hydraulics Division* ASCE, 93(HY4), 149–167.

Robin, R.C. (1936) The discharge of rectangular culverts. *Transactions of Institution of Engineers, Australia*, 8(2), 83–93.

Simmons, W.P. (1964) Transitions for canals and culverts. *Journal of the Hydraulics Division* ASCE, 90(HY3), 115–153

Smith, C.D. & Oak, A.G. (1995) Culvert inlet efficiency. *Canadian Journal of Civil Engineers*, 22(3), 611–616.

Straub, L.G. & Morris, H.M. (1951a) Hydraulic data comparison of concrete and corrugated metal culvert pipes. *Technical Paper* 3, Series B. St. Anthony Falls (SAF) Hydraulic Laboratory, University of Minnesota, Minneapolis.

Straub, L.G. & Morris, H.M. (1951b) Hydraulic tests on concrete culvert pipes. *Technical Paper* 4, Series B. St. Anthony Falls (SAF) Hydraulic Laboratory, University of Minnesota, Minneapolis.

Yarnell, D.L. (1924) The flow of water through pipe culverts. *Public Roads*, 5(1), 19–33; 7(3), 149–151.

Open channel inlet

Ashino, I. (1969) On the theory of the additional loss at pipe entrance in viscous fluid. *Bulletin JSME*, 12(51), 522–529.

Chen, C.-l. & Etemadi, B. (1984) Convergent flow pattern at an abrupt contraction. *Journal of Engineering Mechanics*, 110(6), 894–910.

Dubois, R.H. (1948) Effect of end flares on capacity of irrigation siphon tubes. *Agricultural Engineering*, 29(8), 355–356.

Escande, L. (1940) Recherches sur l'écoulement de l'eau à l'entrée d'un canal découvert [Researches on water flow at the entry of an open channel]. *Le Génie Civil*, 116(9), 152–154; 116(10), 164–166 (in French).

Jegorow, S.A. (1935) Überfall über breite Wehrkrone mit Kreisöffnung [Overflow over broad-crested weir with circular opening]. *Wasserkraft und Wasserwirtschaft*, 30(3), 31–34 (in German).

Levi, L. & Clermont, F. (1970) Etude des pertes de charges singulières dans les convergents coniques [Study of head losses in conical converging pipes]. *Le Génie Civil*, 147(10), 463–470 (in French).

Liong, S.-Y. (1984) Channel design and flow operation without choke. *Journal of Irrigation and Drainage Engineering*, 110(4), 403–407.

Marchetti, M. (1953) Considerazioni sulle perdite di carico dovuto a bocchelli e diaframmi di misura [Considerations on the head losses due to openings and diaphragms]. *L'Energia Elettrica*, 30(1), 6–11 (in Italian).

Mazumder, S.K., Ahuja, K.C. (1978) Optimum length of contracting transition in open channel subcritical flow. Institution of Engineers, India, *Journal CI*, 58(3), 218–223.

Rajaratnam, N. (1963) The circular broad-crested weir. *Water and Water Engineering*, 67(9), 361–363.

Rajaratnam, N., Katopodis, C. & Sabur, M.A. (1991) Entrance region of circular pipes flowing partly full. *Journal of Hydraulic Research*, 29(5), 685–698.

Rossmiller, R.L. & Dougal, M.D. (1982) Tapered inlet design using specific energy curves. *Journal of the Hydraulics Division*, ASCE, 108(HY1), 127–135.

Schlag, A. (1927) Note sur le coefficient de debit des tuyères [Note on the discharge coefficient of inlets]. *Revue Universelle des Mines*, 70(1), 64–74 (in French).

Seely, F.B. (1917) The effect of mouthpieces on the flow of water through submerged short pipe. *Engineering Survey*, 39(11), 957–958.

Smith, C.D. (1966) Head losses at open channel inlets. *Transactions of Engineering Institute of Canada*, 9(A2), 1–8.

Stewart, C.B. (1907) Experiments with submerged tubes 4 ft. square. *Engineering Record*, 56(13), 352–354.

Stewart, C.B. (1908) The flow of water through submerged tubes: results of experiments at the University of Wisconsin. *Engineering News*, 59(2), 35–38.

Subcritical contraction flow

Biery, P.F. & Delleur, J.W. (1962) Hydraulics of single span arch bridge constrictions. *Journal of the Hydraulics Division* ASCE, 88(HY2), 75–108; 88(HY5), 327–333; 89(HY3), 291–293.

Cottman, N.H. & McKay, G.R. (1990) Bridges and culverts reduced in size and cost by use of critical flow transitions. *Proceedings of the ICE*, 88(3), 421–437; 90(3), 643–645.

Escande, L. (1939) Recherches sur l'écoulement de l'eau entre les piles de ponts [Researches on water flow through bridge piers]. *Le Génie Civil*, 115(6), 113–117; 115(7), 138–140; 115(13), 259–260 (in French).

Fiuzat, A.A. & Skogerboe, G.V. (1983) Comparison of open channel constriction rates. *Journal of Hydraulic Engineering*, 109(12), 1589–1602.

Formica, G. (1955) Esperienze preliminary sulle perdite di carico nei canali, dovute a cambiamenti di sezione [Preliminary experiments on the head losses due to cross-sectional changes]. *L'Energia Elettrica*, 32(7), 554–567 (in Italian).

Frantz, H.-R. (1963) Einfluss von Profileinengungen auf den Rückstau in offenen Gerinnen [Effect of channel constriction on backwater in open channels]. *Dissertation*. TH Carolo-Wilhelmina, Braunschweig (in German).

Kazemipour, A.K. & Apelt, C.J. (1983) Effects of irregularity of form on energy losses in open channel flow. *Transactions of Institution of Engineers*, Australia, 25, 294–300.

Kindsvater, C.E., Carter, R.W. & Tracy, H.J. (1953) Computation of peak discharge at contractions. *Geological Survey Circular, Transactions of Institution of Engineers*, 284. US Dept. of the Interior, Washington DC.

Kindsvater, C.E. & Carter, R.W. (1955) Tranquil flow through open-channel constrictions. *Transactions of ASCE*, 120, 955–992.

Laco, V. (1964) Die Einlaufflügelform eines niedrigen Wehres [The inlet shape of a low-crested weir]. *Vodohospodarsky Casopis*, 12, 201–218 (in Czech, with German Abstract).

Lane, E.W. (1919) Experiments on the flow of water through contractions in an open channel. *Transactions of ASCE*, 83, 1149–1219.

Lane, R.G.T. (1967) Temporary dam construction under water and overtopped by floods. Proceedings of the 9th *ICOLD Congress*, Istamboul, Q35(R4), 59–83.

Nicollet, G. (1982) Hydraulique des ouvrages de franchissement des vallées fluviales [The hydraulics of structures for crossing river valleys]. *La Houille Blanche*, 37(4), 289–308 (in French).

Römer, H. (1973) Diversion methods during the construction of dams on rivers. Proceedings of the 11th *ICOLD Congress*, Madrid, Q41(R41), 727–764.

Saladin, P. (1995) Indagine sperimentale sulla localizzazione del risalto a mezzo di quinte [Experimental study on the localization of the hydraulic jump in a contraction]. *Giornale del Genio Civile*, 131(1/3), 47–71 (in Italian).

Smith, F.T. & Duck, P.W. (1980) On the severe non-symmetric constriction, curving or cornering of channel flows. *Journal of Fluid Mechanics*, 90(4), 727–753.

Stefan, H. & Lunow, H.-J. (1968) Einstau und Fliesswechsel an Engstellen offener Gerinne [Backwater and flow change at constrictions of open channels]. *Die Bautechnik*, 45(6), 205–214 (in German).

Vallentine, H.-R. (1958) Flow in rectangular channels with lateral constriction plates. *La Houille Blanche*, 13(1), 75–84.

Vittal, N. & Chiranjeevi, V.V. (1983) Open channel transitions: rational method of design. *Journal of Hydraulic Engineering*, 109(1), 99–115; 109(12), 1778.

Wu, B. & Molinas, A. (2001) Choked flows through short contractions. *Journal of Hydraulic Engineering*, 127(8), 657–662.

Subcritical expansion flow

Adams, E.W. & Stamou, A.I. (1989) Bistable flow patterns in a free surface water channel. *Transactions of ASME*, 111(12), 408–413.

Austin, L.H., Skogerboe, G.V. & Bennett, R.S. (1971) Outlet transitions with triangular-shaped baffles. *Journal of the Irrigation and Drainage Division* ASCE, 97(IR3), 433–448.

Camichel, C., Parmentier, J. & Escande, L. (1935) Les indéterminations et les solutions multiples dans leurs rapports avec l'hydraulique fluviale [The instabilities and the multiple solutions in relation with fluvial hydraulics]. *La Technique Moderne*, 27(17), 579–584 (in French).

Chaturvedi, R.S. (1962) Expansive sub-critical flow in open channel transitions. *Journal of Institution of Engineers* India, 43, 447–487; 44, 279–288; 44, 804–805.

Garde, R.J., Ranga Raju, K.G. & Mishra, R.C. (1979) Subcritical flow in open channel expansions. *Irrigation and Power*, 36(1), 45–54.

Ishihara, T. & Shikata, T. (1966) Flow behaviours at sudden expansions of open channels. *Transactions of JSCE*, 128, 12–28 (in Japanese).

Kells, J.A. & Smith, C.D. (1988) Head recovers at submerged abrupt conduit outlets. *Canadian Journal of Civil Engineers*, 15, 272–274; 16, 206–207.

Michels, V. (1980) Efficient pipe-culvert exit transitions. Proceedings of the 7th Conference *Australasian Hydraulics and Fluid Mechanics* Brisbane, 317–321.

Naib, S.K.A. (1966) Mixing of a subcritical stream in a rectangular channel expansion. *Journal of the Institution of Water Engineers*, 20, 199–206.

Nashta, C.F. & Garde, R.J. (1988) Subcritical flow in rigid-bed open channel expansions. *Journal of Hydraulic Research*, 26(1), 49–65; 27(4), 556–558.

Ramamurthy, A.S., Basak, S. & Rama Rao, P. (1970) Open channel expansions fitted with local hump. *Journal of the Hydraulics Division* ASCE, 96(HY5), 1105–1113; 97(HY1), 202–204; 97(HY2), 362–365; 97(HY12), 2081–2082.

Smith, C.D. & Yu, J.N.G. (1966) Use of baffles in open channel expansions. *Journal of the Hydraulics Division* ASCE, 92(HY2), 1–17; 92(HY5), 255–261; 92(HY6), 212–215; 93(HY1), 78–85; 93(HY4), 273–275.

Sobey, I.J. & Drazin, P.G. (1986) Bifurcations of two-dimensional channel flows. *Journal of Fluid Mechanics*, 171, 263–287.

Chapter 8 Frontispiece (a) Operating spillway and closed bottom outlets, total discharge capacity of 1000 m³/s at Schiffenen Dam, Switzerland, (b) bottom outlet of Rossens Dam, Switzerland (Both photos Courtesy Michael Pfister)

Chapter 8

Intakes and outlets

8.1 Introduction

This chapter deals with both intakes and outlets of dams. Sections 8.1–8.4 deal with intakes, including high submergence intakes in 8.2, low submergence intakes in 8.3, and practical aspects in 8.4. As to outlets, gate flow is considered in 8.5, including the standard gate and the hinged flap gate, followed by the hydraulics of standard gate flow. Low-level outlets are detailed in 8.6.

The purposes of submerged intake structures are twofold:

1. Water withdrawal through intakes for power production, irrigation or drinking-water supply
2. Intake of outlet structures as e.g. for orifice spillways or bottom outlets.

In the first case, the flow velocity is dictated by the user and controlled by the hydrome-chanical equipment as turbines or valves. Typical design velocities in pressure tunnels and shafts or penstocks are some m/s. The water is thus accelerated to this velocity at the intake structure. The main design goal for water withdrawal normally is to keep the intake losses as low as economically possible by limiting the flow velocities at the intake. A trash-rack is provided to retain float and debris.

In the second case, the flow velocity depends on the available pressure head. Intake structures located deeply under the upstream water level thus have a high velocity, whereas structures located close to the water level have velocities as in the first case. For both cases, therefore, similar and different design principles apply. The common design requirements for all submerged intake structures include:

- Minor setup of vortices under relatively small intake head
- Economical limitation of intake losses and velocities under water withdrawal
- No flow separation at inlet structure
- Vibration control
- Emergency closure.

Reservoir intakes for power production, irrigation works or drinking-water supply are located at the reservoir banks (Figure 8.1a), at the reservoir bottom (Figure 8.1b) or combined with the dam structure for concrete dams. *Withdrawal structures* have to be protected against floating debris by trash-racks. *Bottom outlets* have both a service and an emergency gate for safety reasons, and are often designed for sediment flushing. Both types are dealt with in this chapter.

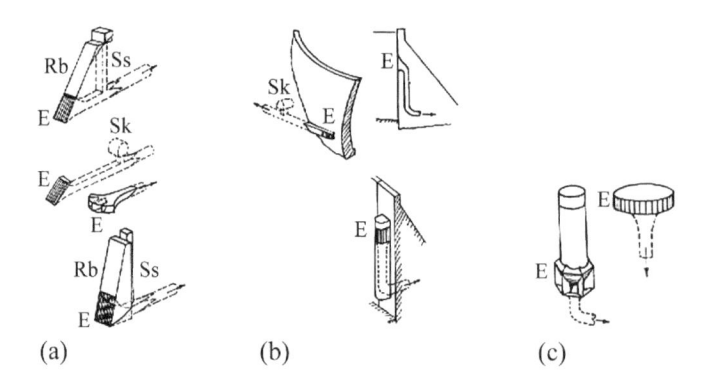

Figure 8.1 Arrangement of reservoir intakes (a) at reservoir banks, (b) at reservoir bottom, (c) or combined with concrete dams. E: intake entrance; Sk: gate chamber, Ss: gate shaft; Rb: trash-rack and stop lock rails

Intakes are somewhat contrary to outlets in view of their hydraulic features. Whereas the latter issue a concentrated jet into the atmosphere or into a tailwater, and energy dissipation is a main concern, intakes behave nearly as a potential flow with the characteristic decrease of pressure due to velocity increase. Depending on whether the head on the intake structure is low or high, significantly different flows are generated, and the associated hydraulic problems are correspondingly different.

For *low-head intakes* slightly submerged, the intake pressure may be so low as to set up an air-water mixture. Further, especially for axisymmetric approach flow conditions, vortices are generated and a significant swirl component is set up at the intake. A hydraulic design thus involves guidelines to inhibit both flows with a too strong swirl, and with too large air content, because the conduit flow conditions are poor otherwise, resulting in significant damage.

For *high-head intakes*, the pressure along the wetted structural boundary may locally drop below vapor pressure. The intake geometry must therefore be designed so as the pressure is always above vapor pressure, and cavitation damage is no concern. Based on the trajectory of an orifice jet, the boundary geometry is adopted to inhibit these damages.

Figure 8.2 shows a typical withdrawal structure located at the reservoir banks. The bellmouth-shaped inlet is connected to a circular-shaped penstock of 6.15 m diameter and 115 m³/s discharge capacity. The tunnel is lined with 700 mm reinforced concrete. Downstream, 76 m from the intake, is located the gate shaft with a slide and a roll gate of 4 × 5.5 m² section each. Both gates are operated from the top of the gate shaft. The transition from the circular penstock to the rectangular gate location is steel lined. The trash-rack is connected with the cleaning machine operating also from the gate shaft. Note both the anti-vibration elements added to the racks, and the complicated transition geometry of the intake. The maximum head on the intake structure is 35 m.

The following highlights the hydraulic problems related to an intake structure. The hydraulics of high-head intakes are reviewed in 8.2, and those referring to low-head intakes in 8.3. Design relations for cavitation control, and vortex flow are particularly addressed. Further, the air entrainment rate at horizontal circular conduits is detailed, based on recent research.

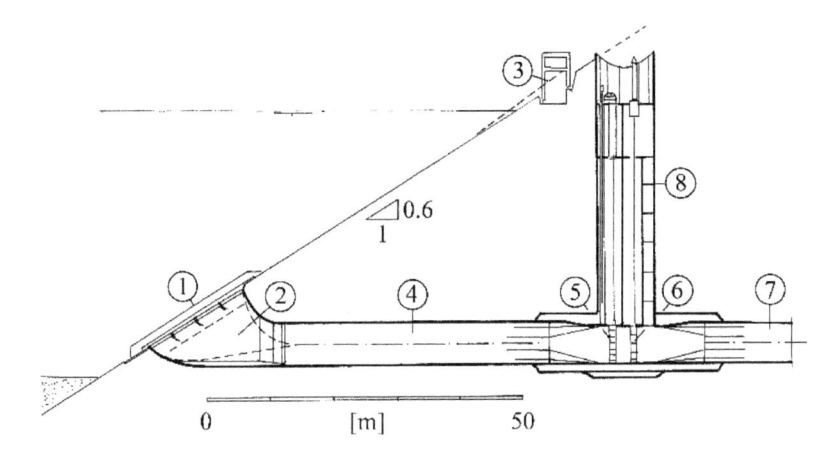

Figure 8.2 Withdrawal structure of Hydroprado Power Plant (Colombia) with ① intake trash-rack, ② intake transition, ③ trash-rack cleaner, ④ penstock, ⑤ slide gate, ⑥ wheel gate, ⑦ penstock to power plant, ⑧ gate shaft (Tröndle, 1974)

8.2 High submergence intakes

8.2.1 Design principles

Highly submerged inlet structures as used for bottom outlets and low-level orifice spillways have to be safe under all hydraulic conditions, including (1) cavitation, (2) vibration, and (3) flow stability. In addition, their design should be simple and economic, along with a high discharge capacity. Because the head on an inlet structure at the bottom of a dam easily is in excess of 100 m, the resulting velocities are typically 50 m/s or even more, giving rise to low boundary pressure and potential danger of *cavitation damage*.

Bottom and low-level outlets comprise several elements, as shown in Figure 8.3. Depending on the tunnel length and the gate position relative to the conduit diameter, a distinction

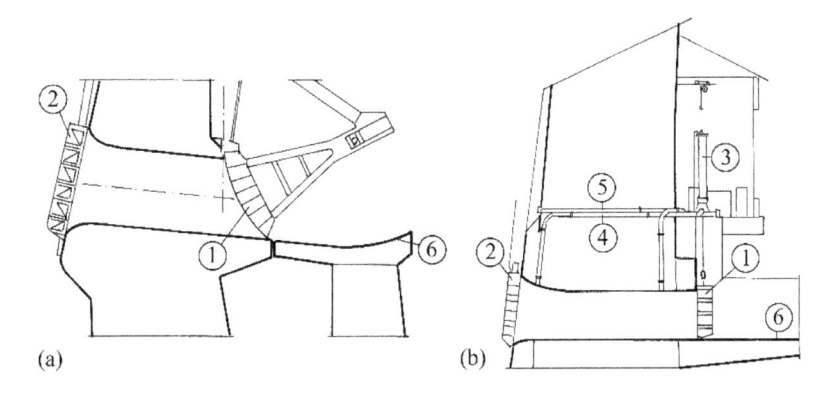

Figure 8.3 Intake portions of Japanese bottom outlets (a) Ayakita Dam, (b) Muromaki Dam with ① main gate, ② emergency gate, ③ oil-pressure cylinder, ④ pressurized air supply pipe, ⑤ fill pipe, ⑥ chute (Fujimoto and Takasu, 1979)

Figure 8.4 Upstream view of Karakaya Intake during construction (Italstrade S.p.A., Milan)

is made between the bellmouth-shaped inlet, the pressure tunnel or *penstock*, the gates, and the tailwater tunnel. In particular cases, a pressure tunnel may be absent so that the gates are located just downstream of the inlet structure. In the following, attention is paid to the inlet of the intake structure for high-head flow. The gate chamber and the tailwater tunnel are described in 8.6.

The geometry of the intake portions of withdrawal and bottom outlet structures is similar. Figure 8.4 relates to Karakaya Dam, Turkey (Özis and Özel, 1989). In total, six intakes are connected to separate penstocks, of which each has a trash-rack of 18×18 m^2 surface area. Together with the bulkhead gates placed with a mobile crane located at the dam crest, a total of 1668 tons of steel was used for the intake structure.

8.2.2 Orifice flow

The definition of the geometry of a high submergence intake for bottom and low-level outlets is based on jet flow issued from a sharp-crested orifice. At design discharge, the boundary pressure is atmospheric as for the corresponding liquid jet. This principle corresponds to the definition of the crest geometry of the standard overflow structure, therefore. For corresponding heads, the surface pressures of both the free liquid jet and the pressurized flow are both atmospheric, so that cavitation is no concern. Based on the vertical jet flow from an orifice

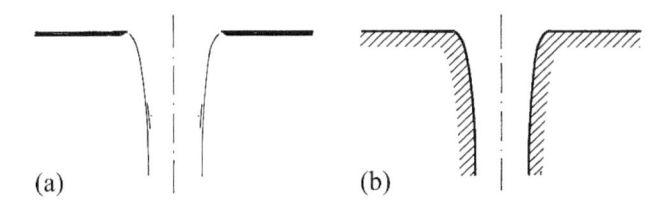

Figure 8.5 Geometry of (a) axisymmetric liquid jet with vertical axis, (b) corresponding intake

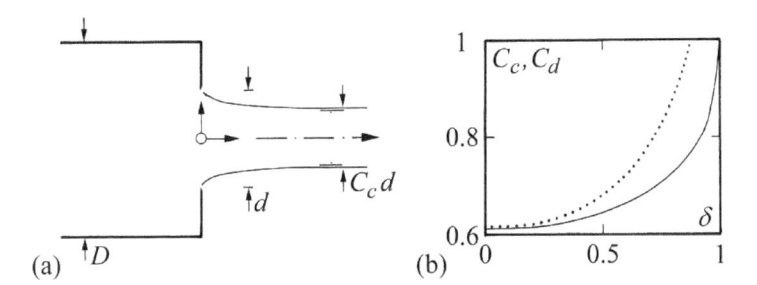

Figure 8.6 Hydraulic characteristics of orifice flow (a) definition of geometry, (b) (———) contraction C_c and ($\cdots$) discharge C_d coefficients versus diameter ratio $\delta = d/D$

with a horizontal plane and with a standard sharp crest (Figure 8.5), Liskovec (1955) drew the conclusions:

- For orifice diameters in excess of 60–70 mm, effects of viscosity and surface tension are absent, and scaling follows Froude similitude
- For a head larger than six orifice diameters, gravity plays a small role
- Generalized equation for the jet geometry is available both in terms of orifice or contracted diameters.

Based on the irrotational flow theory, Kirchhoff and later von Mises defined the jet geometry issued from a tank into space by neglecting the gravity effect (Figure 8.6a). Rouse and Abul-Fetouh (1950) defined the geometry of these jets experimentally. For a cylindrical tank of diameter D with an orifice of diameter d, the discharge Q under a head h is with C_d as discharge coefficient and g as gravity acceleration

$$Q = C_d A (2gh)^{1/2}. \tag{8.1}$$

With C_c as the contraction coefficient, i.e. the ratio of the asymptotic jet diameter to the orifice diameter, and with a and A as the areas of the orifice and the tank sections, respectively, the potential flow theory yields

$$C_d = \frac{C_c}{[1 - (C_c a / A)^2]^{1/2}}. \tag{8.2}$$

The coefficient C_c depends exclusively on the diameter ratio $\delta = d/D$. For $\delta \to 0$, i.e. for a small orifice in a large tank, the asymptotic value tends to $C_c = \pi/(\pi+2) = 0.611$. For the other extreme $\delta \to 1$ follows $C_c = 1$. The relation $C_c(\delta)$ is implicitly given by

$$C_c^{-1} = 1 + \frac{2}{\pi}\left[\frac{1}{C_c\delta} - C_c\delta\right]\arctan(C_c\delta). \tag{8.3}$$

Figure 8.6b shows that both C_c and C_d as computed from Eqs. (8.2) and (8.3) are nearly constant if $\delta < 0.20$. For high-head intakes, this condition is always satisfied, so that the effect of δ is negligible.

Rouse and Abul-Fetouh (1950) determined the jet profiles by using the electrical analogy, from which jet contraction is completed within a distance of nearly $x = d$. The pressure distribution on the orifice plate was also determined.

For an orifice with a non-horizontal outflow plane, the effect of gravity has to be accounted for. However, due to the short reach of jet contraction, this effect can often be neglected, if the jet Froude number is large, as for high-head inlets. The detailed jet geometry and the intake shape for these inlets are presented subsequently.

8.2.3 Inlet geometry

The inlet portion of an intake structure has to satisfy various conditions:

- Positive boundary pressures
- Absence of cavitation zones
- Continuous reduction of pressure headline
- Minor head losses, i.e. good hydraulic performance
- Control of vibration
- Economic design.

Withdrawal structures of relatively small submergence must be protected by a trash-rack against floating debris and equipped with a cleaning device. Because floating debris is rare in large reservoir depths, bottom outlets are only equipped with a large spaced trash-rack made of concrete beams, whose spacing must be just below the smallest gate dimension. Effects of trash-rack vibrations are discussed in 8.4. Attention is paid to the transition curve from the upstream dam face or reservoir bank slope to the intake penstock. The intake height is a, and b is the intake width. According to USACE (1957) ellipses or compound ellipses approximate the inlet geometry. Figure 8.7 shows a definition plot and defines the origin of the coordinate system $(x; y)$. For a sloping dam face, slightly modified curves apply.

According to USACE, the boundary pressure along the transition may be negative if a single elliptical curve is adopted. Experiments of McCormmach (1968) proved this finding so that the compound elliptical transition curve is recommended. With the dimensionless coordinates $X = x/a$, or $X = x/b$ depending on whether the vertical or the horizontal transition curve is considered, and scalings $Y = y/a$ or $Y = y/b$, respectively, the transition geometry for the *four-sided intake* configuration is defined as

$$(X-1)^2 + \left(\frac{Y}{0.32}+1\right)^2 = 1, \quad 0 < X < 0.33; \tag{8.4}$$

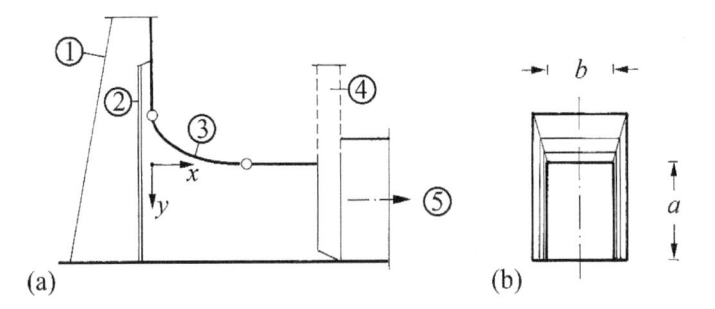

(a) (b)

Figure 8.7 Intake geometry of high-head intake (a) longitudinal section, (b) cross section with ① upstream pier face, ② trash-rack, ③ transition curve, ④ gate with gate slot, ⑤ penstock

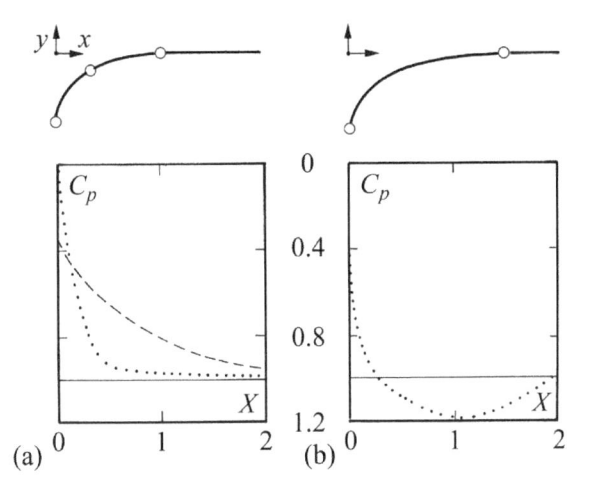

Figure 8.8 Intake transition curve (a) four-sided, (b) roof curve only. Transition geometry (top) and pressure coefficient $C_p(X)$ ($\cdots$) along vertical plane of symmetry and top corner, (- - -) horizontal plane of symmetry (USACE, 1957)

$$(X-1.2)^2 + \left(\frac{Y}{0.16}+1\right)^2 = 1, \quad 0.33 \le X \le 1. \tag{8.5}$$

Figure 8.8 shows the transition geometry along with the pressure coefficient

$$C_p = \Delta H / (V^2 / 2g). \tag{8.6}$$

Here $\Delta H(x)$ is the pressure head drop from the reservoir into the penstock or pressure tunnel. Note that it is continuous without any local minima provoking cavitation. The coefficient tends asymptotically to $C_p = 1$, so that the head loss across the transition (subscript t) is $\Delta H_t = V^2/2g$ with V as penstock or pressure tunnel velocity (Figure 8.8a).

Figure 8.9 Four-sided intake configuration of two low-level outlets of Karun III Dam in Iran (during construction in 2002) used as bottom outlet and for flood evacuation. Two small intakes for releasing residual irrigation flow are also seen in the center (Courtesy Anton J. Schleiss)

For intakes involving only a *roof transition* with $X = x/a$ and $Y = y/a$ as shown in Figure 8.7a, USACE (1957) recommends the single elliptical curve

$$\left(\frac{2}{3}X - 1\right)^2 + \left(\frac{3}{2}Y + 1\right)^2 = 1, \quad \text{for } 0 \le X \le 1.5. \tag{8.7}$$

The minimum pressure coefficient $C_p = 1.2$ occurs at location $X = 1.2$ (Figure 8.8b) which is considered acceptable.

Figure 8.9 shows the four-sided intake configuration of the two low-level outlets of Karun III Dam in Iran used as bottom outlet and for flood evacuation. Note also the two small intakes for releasing residual irrigation flow.

The guidelines for inlet geometries apply also for low submergence intakes (8.3). For intakes used as inflow and outflow structures of pumped-storage power plants, the outflow jet has to be guided laterally over a certain distance to avoid jet oscillations, by which head losses are strongly increased. An example is shown in Figure 8.10. The improvement of the pressure features of entirely steel-lined intakes by bulkhead slots involving slot templates during high discharges is considered by Arnold *et al.* (2018).

8.3 Low submergence intakes

8.3.1 Vortex flow

For intakes serving power production, irrigation or drinking-water supply, low submergence may occur during reservoir drawdown. Under these conditions, those withdrawal structures

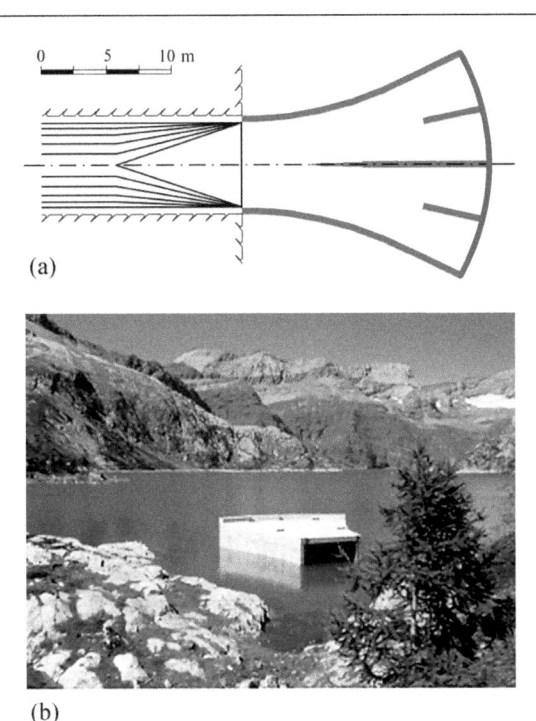

(a)

(b)

Figure 8.10 (a) Elevation of in- and outflow structures of pumped-storage power plant *Nant de Drance*, Emosson Dam reservoir, Switzerland, of design discharge 120 m³/s, (b) prefabricated intake structure at reservoir shore to avoid reservoir emptying, then lowered to foundation platform at reservoir bottom, and finally connected to pressure tunnel (Courtesy Anton J. Schleiss)

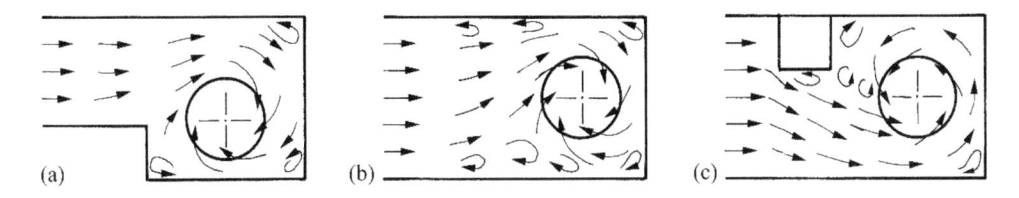

Figure 8.11 Vorticity due to (a) offset, (b) velocity gradient, (c) obstruction (Knauss, 1987)

are thus prone to vortex formation. A *vortex* is a coherent structure of rotational flow mainly caused by the eccentricity of the approach flow to a hydraulic sink: asymmetric approach flow conditions and obstruction effects among other reasons can also set up intake vortices. Figure 8.11 shows major sources of vorticity presence.

Vortices in hydraulic structures typically occur at intakes and at pump sumps, of which the latter are excluded. The intake direction varies from vertical to horizontal. The simplest vortex occurs at the vertical intake, discussed below mainly to set out the flow principle, although its practical relevance in dam hydraulics is small. More relevant is the vortex formation at nearly horizontal takeoff conduits, so that their characteristics are also discussed.

Vortices have four main disadvantages in hydraulic designs:

- *Air entrainment* with effects on hydraulic machinery or hydropneumatics, and pressure surges;
- *Swirl entrainment* associated with an increase of head loss and efficiency reduction in hydraulic machinery;
- *Enhancement of cavitation and vibration* with a reduced longevity of important mechanical parts;
- *Float entrainment* including wood or ice, and blockage of screens, or damage of coatings.

Ideally, the transition from free surface to pressurized flow should be uniform, steady, and of single phase. Air and float entrainment is the worst consequence of an intake structure, so that an acceptable hydraulic design avoids this condition in particular.

8.3.2 Vertical intake vortex

The basic type of vortex flow occurs in a large cylindrical containment with a much smaller circular sharp-crested orifice at its center. Even for this simple configuration, no physical solution is currently available. Effects including streamline curvature and turbulence are significant, not to mention viscosity and surface tension for scale models, on which most of the available data were obtained.

The Rankine combined vortex is a simplified model of typical vortices occurring in nature (Figure 8.12). It is composed of a forced vortex in the core zone and a free vortex region away from the core. In the core zone, viscosity affects the flow resulting in zero central velocity, whereas potential flow dominates the free vortex zone. The two flow regions are matched at radius r_o as a fraction of the orifice radius $R = D/2$.

In the vortex core, the tangential (subscript t) velocity v_t is assumed to vary linearly with the radial coordinate r and the angular velocity ω as

$$v_t = \omega r, \quad 0 < r \le r_o. \tag{8.8}$$

In the free vortex region, the circulation Γ is constant, so that

$$v_t = \frac{\Gamma}{2\pi r}, \quad r_o < r. \tag{8.9}$$

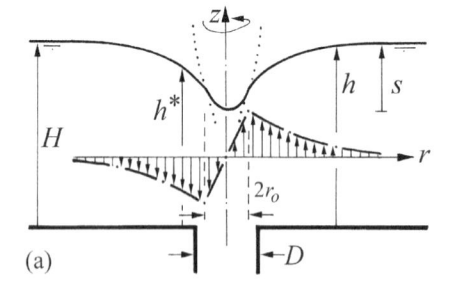

Figure 8.12 Rankine combined vortex (a) free-surface profile $h(r)$ and distribution of tangential velocity $v_t(r)$, (b) typical vortex flow pattern

This simplified model accounts neither for vertical velocity components nor for vertical variations of the horizontal velocities.

According to Anwar (1966), the depth s of the free-surface depression is with the maximum tangential velocity $v_{tM} = v_t(r = r_o)$ approximated as (Figure 8.12a)

$$s = 0.60[v_{tM}^2 / (2g)]. \tag{8.10}$$

The location of maximum velocity is nearly $r_o = (3/4)R$. The relation between ω and Γ thus is $\Gamma/\omega = 2\pi r_o^2 = 3.5R^2$.

Vortices are classified according to their strength intensity. A visual classification fits best for defining the circulation due to difficulties in measuring a swirl parameter. Figure 8.13 is based on this concept in which six vortex strengths are distinguished:

1   Coherent surface swirl
2   Surface dimple, with coherent surface swirl
3   Dye core to intake, with coherent swirl along column
4   Vortex pulling floating trash, but no air entrainment
5   Vortex pulling air bubbles to intake
6   Vortex with full air core.

A fundamental vortex, referred to as the bathtub vortex, depends on the dimensionless parameters:

$$\text{submergence Froude number} \quad \mathsf{F} = \frac{Q}{(\pi/4)D^2(gh)^{1/2}}$$

$$\text{circulation number} \quad \mathsf{N}_\Gamma = \frac{\Gamma D}{Q}$$

$$\text{radial Reynolds number} \quad \mathsf{R} = \frac{Q}{vh}$$

$$\text{Weber number} \quad \mathsf{W} = \frac{Q}{(\pi/4)D^2(\rho D/\sigma)^{1/2}}.$$

Here, Q is discharge, D intake diameter, h intake submergence, v kinematic viscosity, ρ mass density, and σ surface tension. The relative submergence $y = h/D$ is also relevant. For $\mathsf{W}^2 > 10^4$, the effect of surface tension is insignificant (Hecker, 1987). Viscous effects are dominant if

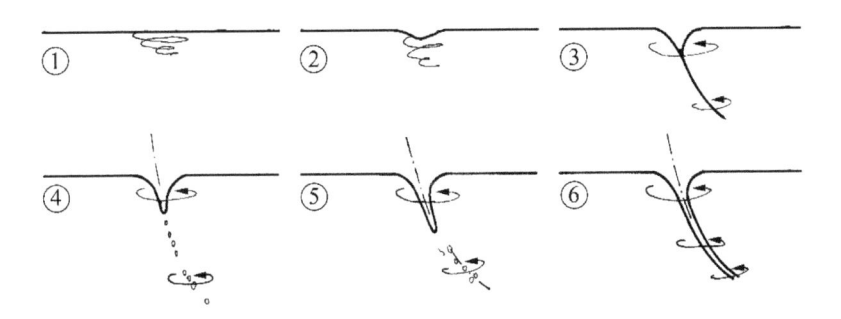

Figure 8.13 Vortex type classification, details in text (Hecker, 1984)

$R < 3 \times 10^4$. For prototype structures, the vortex characteristics are thus determined by the parameters F, N_f and y. The difficult parameter to estimate is the circulation Γ because it may even be interrelated with the other parameters. Usually, Γ is estimated only on scale models or prototype structures if available, or from numerical simulations (Hecker, 1987). Scale effects, which are particularly important for the prediction of air entrainment in prototype structures, are addressed by e.g. Chang and Prosser (1987), and are further considered below.

8.3.3 Limit or critical intake submergence

To avoid air entrainment and to counter swirl entrainment, an adequate intake submergence is required (Jain *et al.*, 1978). Figure 8.14 refers to the three basic intake configurations characterized by the orientation Φ relative to the vertical direction, and the distance of the intake center from the invert. The orientations shown in Figure 8.14 include $\Phi = 0$, $\pi/2$, and π, respectively.

Knauss (1987) defined the limit (subscript L) submergence height h_L as the height under which vortex type 5 is generated. For $h \geq h_L$ air bubbles are entrained continuously but intermittently, and there is no air core into the intake structure. The effects of circulation and Froude number are combined in the *swirl number* $C = \Gamma/(gD^3)^{1/2}$ with D as the pipe diameter. The limit submergence varies thus with

$$h_L/D = f(\Phi, C). \tag{8.11}$$

The limit submergence was evaluated for various type structures, including these shown in Figure 8.14. It was found that the relation between the limit submergence and C is nearly linear, so that with C as a constant of proportionality

$$h_L/D = C\frac{\Gamma}{(gD^3)^{1/2}}. \tag{8.12}$$

For identical swirl number, the effect of structural geometry is thus described by the constant C. Further, a simple relation exists between radius r_o as defined in Figure 8.12a and the

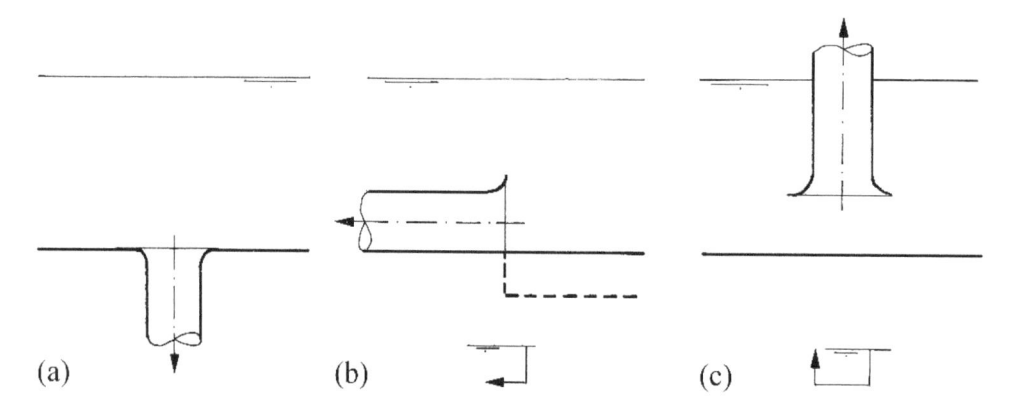

Figure 8.14 Basic intake types and notation (a) vertical intake, (b) horizontal intake, (c) pump intake, based on intake orientation Φ

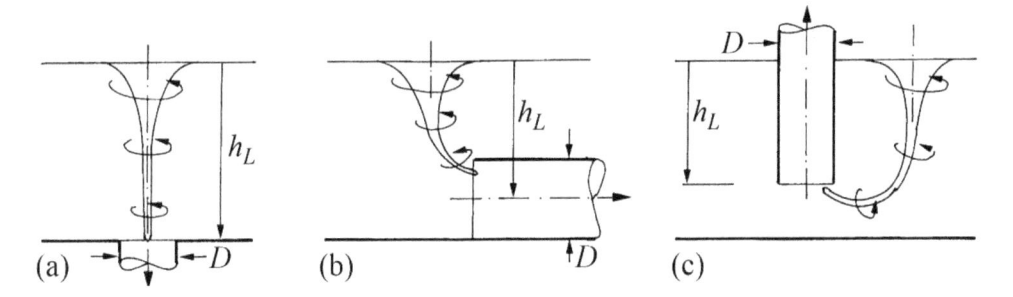

Figure 8.15 Limit submergence height h_L for (a) vertical intake, (b) horizontal intake, (c) pump intake

maximum tangential velocity. With $C = C(\Phi)$, i.e. as a function of only the intake orientation, the limit submergence height is

$$h_L/D = [C(r_o/D)]^2. \tag{8.13}$$

This equation provides a means to determine the constant C.

Figure 8.15 refers to typical structures shown in Figure 8.14. Because of the intake shape, the flow has to contract at the inlet section. If a vortex has a central air core extending beyond the inlet section, air is continuously entrained so that the intake submergence is insufficient. The indicators of intake direction and swirl number are the *surface vortex diameter* and the *vortex length* up to the inlet section. Both parameters increase with the intake direction and the swirl number (Figure 8.15).

Knauss (1987) determines the main features of horizontal intake flow under limit submergence. From the data of Amphlett (1976), and Anwar *et al.* (1978), he deduced a simple relation for the radius of maximum tangential velocity as

$$r_o/D = 0.0109[(h_L/D) + 1.45]^{1/2}. \tag{8.14}$$

From this and analogous findings, parameter C is determined as

Vertically downward intake	$C = 110$	(8.15)
Horizontal intake	$C = 90$	(8.16)
Vertically upward intake	$C = 75$.	(8.17)

The vortex orientation effect is thus considerable and yields an increase or decrease of 20%, respectively, as compared with the horizontal intake arrangement. For oblique intake orientation, linearly interpolate between the numbers previously specified.

Based on model tests of Gordon (1970) and Hecker (1981), and with $F = V/(gD)^{1/2}$ as the pipe Froude number, Knauss (1987) proposes as *minimum* intake submergence under normal approach flow conditions

$$(h_L/D) = 1 \text{ to } 1.5, \quad \text{for } F \leq 0.25; \tag{8.18}$$
$$(h_L/D) = (1/2) + 2F; \text{ for } F > 0.25. \tag{8.19}$$

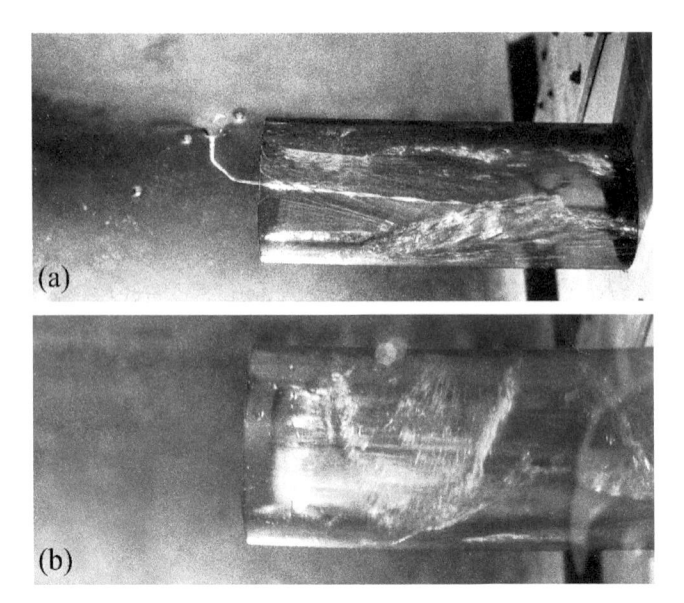

Figure 8.16 Inflow to horizontal pipe (a) vortex intensity 6, (b) swirling pipe flow due to intake vortex (Courtesy Willi H. Hager)

For $F < 0.25$ the intake structure is large and a typical pipe velocity is 2 m/s. For $F > 2$, a small or medium intake structure prevails, with a typical pipe velocity of 4 m/s. These recommendations do not include special features of vortex suppression. Figure 8.16 shows inflows to a horizontal intake pipe. In Figure 8.16a, the full air-core vortex is seen, whereas the vorticity setup in the pipe flow is noted in Figure 8.16b. Based on vortex observations at hydropower intakes, Nagarkar (1986) suggested the empirical formula involving the intake velocity V [m/s] and the conduit diameter D [m] as

$$h_L \geq 4.4(VD^{1/2})^{0.54}. \tag{8.20}$$

Practical experience has demonstrated that this formula gives a conservative value of the minimum required submergence.

The performance of large intakes at hydropower plants is optimized by hydraulic model tests. With structural modifications of the intake the optimum arrangement is successively attempted, avoiding any permanent vortex higher than type 2, and any intermittent vortex higher than type 3 (Figure 8.13) for the design discharge. Below, the critical intake submergence is further detailed for a limited air entrainment rate.

8.3.4 Air entrainment

Introduction

Air entraining vortices are a typical phenomenon at intakes of hydroelectric power plants and other pressurized systems. Their occurrence and the resulting effects on the pressure system have to be considered during both the design and operation stages. A common design criterion of intakes is to avoid air entrainment by vortices as these involve a major source of air entrainment into the pressurized tailwater system. The knowledge on air entrainment, especially the correlation to a flow pattern, is still poorly understood. The relative air entrainment rate $\beta = Q_A/Q_W$ expresses the relation between the volumetric air (subscript A) discharge and the volumetric water (subscript W) discharge. The average (subscript a) air concentration $C_a = Q_A/(Q_A + Q_W) = \beta/(\beta + 1)$ is also used instead of β (e.g. Padmanabhan, 1984). For small Q_A, both values β and C_a are nearly identical, because at $\beta = C_a/(1-C_a)$ = [1, 10]% the relative difference $(\beta-C_a)/C_a$ is only [0.99, 4.8]%. The research of Möller et al. (2015) attempted to quantify β due to intake vortices in terms of the hydraulic intake characteristics.

Air in pressurized systems changes the flow properties from single-phase to two-phase flow, along with strong implications on the operation and safety of pressurized systems. These effects include: (1) efficiency reduction of turbines and pumps, (2) unsteady flow behavior, i.e. pulsations and pressure surges in the system with corresponding mechanical loads on the involved components, and (3) steady effects as discharge reduction due to the presence of air bubbles and local corrosion damage, respectively. The first group refers to direct effects of air presence in hydraulic machinery. Here, even minute β values cause a reduction in efficiency of around 1%. An air entrainment rate of 1.5% leads to an efficiency reduction of up to 16% (Denny and Young, 1957; Papillon et al., 2000). For $\beta = 4\%$ occurs a further exponential efficiency reduction. Depending on the type of pump (axial pumps are more sensitive than are centrifugal), a sudden efficiency drop down to a total flow interruption may occur at air entrainment rates between $\beta = 7$ to 20% (Chang, 1977; Poullikkas, 2000).

A literature review concerning intake vortices indicates that only few publications deal with the air entrainment as to its quantification, e.g. Iversen (1953), Denny and Young (1957), Hattersley (1965), and Padmanabhan (1984). No valuable information appears to quantify the air entrainment versus the common flow features as the pipe or submergence Froude numbers. Furthermore, a large number of researches using the term 'air entraining vortices' provides no detailed information on entrainment rates or void fractions (e.g. Haindl, 1959; Jain et al., 1978, Chang and Lee, 1995). Suerich-Gulick et al. (2014) present a model to estimate the key vortex characteristics of free-surface vortices at low-head hydropower intakes. However, a link to the amount of entrained air is missing. First steps to quantify β were presented by Iversen (1953) involving a pump sump with a vertically upward oriented bell-mouth intake. The air entrainment measurement involved a so-called air separator. The pipe was thereby interrupted by a closed cylindrical tank in which the air separated from the water and accumulated at the top section. According to Denny and Young (1957) the air entrainment rate at pump sumps is typically $\beta = 5\%$ reaching $\beta = 10\%$ in extremes. Studies of Hattersley (1965) indicate that β ranges between 0.06 and 0.73%, thus one to two magnitudes smaller than the former value.

For horizontal intakes, Padmanabhan (1984) states maximum air concentration rates versus the submergence Froude number $F_{co} = V_D/(gh)^{0.5}$, with V_D as the average velocity at the characteristic intake cross section, h as intake submergence relative to the pipe axis and g as gravity acceleration. Hydraulic model data with two horizontal intake pipes and a basin representing the pump sump indicate that $C_{a,max} = 15\%$, so that the maximum relative air entrainment rate $\beta_{max} = C_a/(1-C_a) = 18\%$. The majority of all data are below 1% of the void fraction, with a considerable data scatter. Instead of correlating C_a to the pertinent flow parameters, an envelope line of 'maximum air entrainment' was proposed.

Experimentation

Laboratory experiments were conducted at a large-scale tank-pipe system (Figure 8.17) consisting of a 50 m³ test steel tank, a 14 m long pressurized-pipe system, and a 130 m³ underground reservoir. The maximum discharge was $Q_W = 0.510$ m³/s in a closed loop. The tank was symmetrically arranged to the main flow direction. The water flow from the main tank into the pressurized-pipe system was discharged through a horizontal intake pipe, with diameter $D = 0.39$ m of the sharp-crested inlet. The intake and combined Froude numbers, $\mathsf{F}_D = V/(gD)^{0.5}$ and $\mathsf{F}_{co} = V/(gh)^{0.5}$, varied up to 2, and 2.8, respectively, whereas the relative intake submergence h/D was up to 4.

The model employed exceeds the generally accepted limits regarding the similitude criteria of intake-vortex studies: Viscous effects remain small for intake Reynolds numbers $\mathsf{R}_D = V_D D/\nu > 2.4 \times 10^4$ (Daggett and Keulegan, 1974), with ν as kinematic viscosity. Intake Weber numbers of $\mathsf{W}_D{}^2 = V_D(\rho D/\sigma)^{0.5} > 121$ avoid effects of surface tension in model vortex formation (Ranga Raju and Garde, 1987).

The approach flow velocities to the intake pipe ranged from ~0.1 to ≥3–4 m/s and amount to ~10 m/s for the entrained air in the air tube. Scale family tests were conducted to proof the feasibility of the model investigation, with respect to the data transferability to prototype scale (Möller et al., 2012). Based on Möller (2013), the scale similarity was proven by scale family tests for $0.2 \leq \mathsf{F}_D \leq 1.5$, $0.75 \leq h/D \leq 3.0$, $D = [200, 300, 400, 500]$ mm, with $h/D = 0$ referring to a water level at the pipe axis. The study of vortex-induced air entrainment based on Froude scaling applies to intake diameters of $D \geq 400$ mm with regard to scale effects in the air-core diameter for the selected test setup. Viscosity and surface tension effects are considered negligible if $\mathsf{R}_D \geq 6 \times 10^5$, and $\mathsf{W}_D{}^2 > 3200$, respectively. These numbers are one

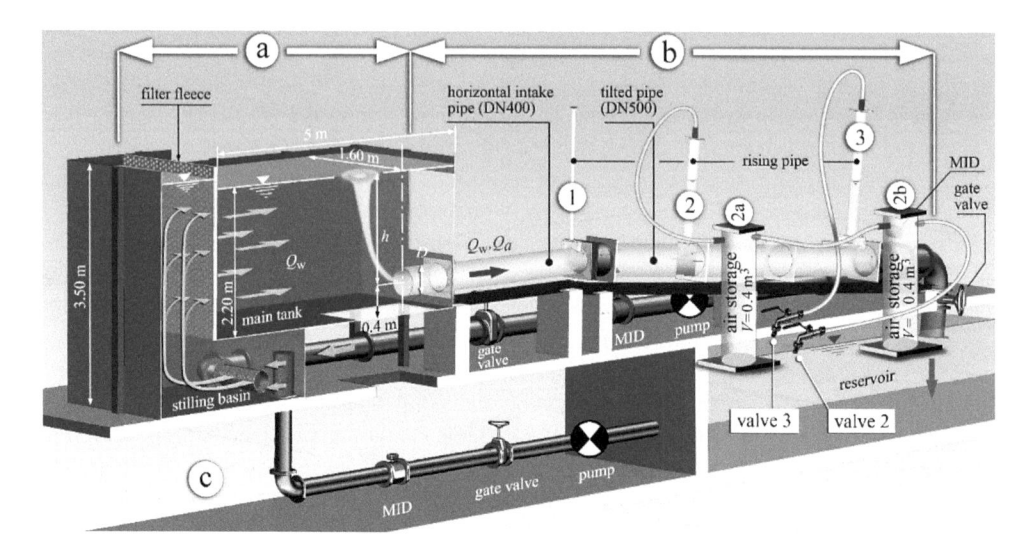

Figure 8.17 Experimental test facility (a) 50 m³ experimental tank with stilling basin and filter fleece for flow uniformization, and intake pipe, (b) 14 m long pressurized-pipe system with de-aeration device consisting of three riser pipes and two air storages, (c) 130 m³ reservoir and pumps (Möller *et al.*, 2015)

order of magnitude higher than those of Daggett and Keulegan (1974), or Ranga Raju and Garde (1987).

Air entrainment rate

Experiments were conducted at a conduit diameter of $D = 400$ mm for $0.4 \leq \mathsf{F}_D \leq 1.2$ and $1.25 \leq h/D \leq 2.5$, corresponding to $0.36 \leq \mathsf{F}_{co} \leq 0.80$. For $[\mathsf{F}_D, \mathsf{F}_{co}] = [0.36, 0.50]$ the air entrainment was far below the measurable limit of $\beta = 10^{-5}$, although the presence of a stable vortex during two hours. Figure 8.18a shows a typical air-entraining intake vortex at $h/D = 1.5$ and $\mathsf{F}_D = 0.8$. The resulting bubbles in the de-aeration pipe are shown in Figure 8.18b. The water level in the riser pipes equals the piezometric head at atmospheric ambient pressure. If the pipes are closed on top, the pressure increases with each accumulated bubble rising in the pipe, resulting in a water level drop measured by both an ultrasonic sensor and a pressure sensor. At each further time step, the mass of collected air was determined using the ideal gas law. Subsequently, the difference of the actual and the initial atmospheric pressures allows for the computation of the entrained air volume, and thus indirectly for the air discharge Q_A, ranging from 1.9×10^{-5} to 3.6×10^{-4} m³/s at different h/D and F_D values.

A broad analysis of all test data was conducted by employing two average values characterizing the measured air entrainment, namely the mean (subscript mean) value over an entire test duration Q_{amean}, and the characteristic maximum value related to vortex type $VT6$ phase (Figure 8.13). The latter does not equal the absolute maximum during a run, but corresponds to a two-times averaged value, first within a single $VT6$ phase, and second over all $VT6$ phases related to the vortex phases during which high air discharge was measured and therefore referred to as the characteristic maximum. Figure 8.19 shows that β strongly correlates with F_{co}. A regression analysis with β_{mean} as the mean air entrainment rate, β_{VT6} the

Figure 8.18 Air discharge Q_A at $h/D = 1.5$, $\mathsf{F}_D = 0.8$ (a) air-entraining vortex, (b) de-aeration at riser pipe

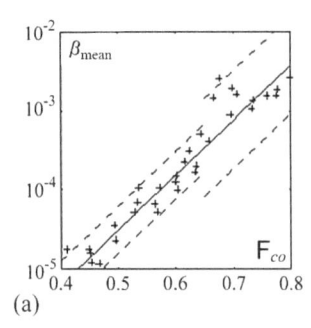 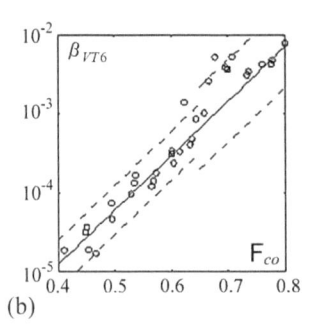

(a)   (b)

Figure 8.19 Air entrainment rates versus $F_{co} = V_D/(gh)^{0.5}$ with ±95% prediction bands. Subdivision at $F_{co} = 0.66$ accounts for varying prediction widths (a) mean air entrainment rate $\beta_{mean}(F_{co})$ compared with Eq. (8.21), (b) $\beta_{VT6}(F_{co})$ compared with Eq. (8.22)

mean air entrainment rate of *VT6* phases, and $F_{co} = V_D/(gh)^{0.5}$ as combined Froude number results in (Möller *et al.*, 2015)

$$\beta_{mean} = 1.04 \times 10^{-8} \exp(16F_{co}), \tag{8.21}$$
$$\beta_{VT6} = 2.00 \times 10^{-8} \exp(16F_{co}). \tag{8.22}$$

The coefficients of determination $R^2 \geq 0.9$ are high for both relations. Test limitations are $F_{co} = [0.36–0.80]$, given that the air entrainment rate at $F_{co} = 0.36$ was below the measurement limit of $\beta = 1 \times 10^{-5}$. A simple rule of thumb is $\beta_{VT6} \approx 2\beta_{mean}$.

Padmanabhan (1984) appears to provide the best data set concerning measured air entrainment rates. Möller *et al.*'s (2015) β_{VT6} values lie within his data cloud. His inception point of air entrainment matches well at $F_{co} \approx 0.3$, whereas Möller *et al.* (2014) determined 0.4. The 'maximum air entrainment' line of Padmanabhan (1984) allows for a distinction between the two data sets. With $\beta = C_a/(1-C_a)$, the separation line follows $\beta_{VT6} = 0.12F_{co} - 0.044$. Padmanabhan's envelope curve is clearly above the corresponding values β_{VT6} of Möller *et al.* (2015). Their air entrainment rates never reach Padmanabhan's level. Consequently, Padmanabhan's (1984) line of maximum air entrainment conservatively overestimates the incipient point of air entrainment.

Critical intake submergence

The critical (subscript *cr*) intake submergence h_{cr} is widely used to determine the incipient state at which no air is entrained due to intake vortices. This parameter follows from a visualization of the vortex formation. Thus, both h_{cr} and the vortex appearance concerning air entrainment were visually observed. The data enabled for a novel approach, given that h_{cr} was determined in direct correlation with the minimum air entrainment rate of $\beta = 1 \times 10^{-5}$. The critical submergence corresponds to the air inception level at which air is entrained. Therefore, the relative critical intake submergence was reanalyzed once the smallest rate β_{cr} became measurable. Figure 8.20 shows Möller *et al.*'s (2015) data for β_{VT6} as compared with these of Knauss (1987) and Gordon (1970). The two hydraulic design parameters $F_{co,cr} = V_D/(gh_{cr})^{0.5}$ and $F_D = V_D/(gD)^{0.5}$ are recast to $(h/D)_{cr} = F_D^2/F_{co,cr}^2$, with $F_{co,cr}(\beta_{VT6} = 10^{-5}) = 0.39$

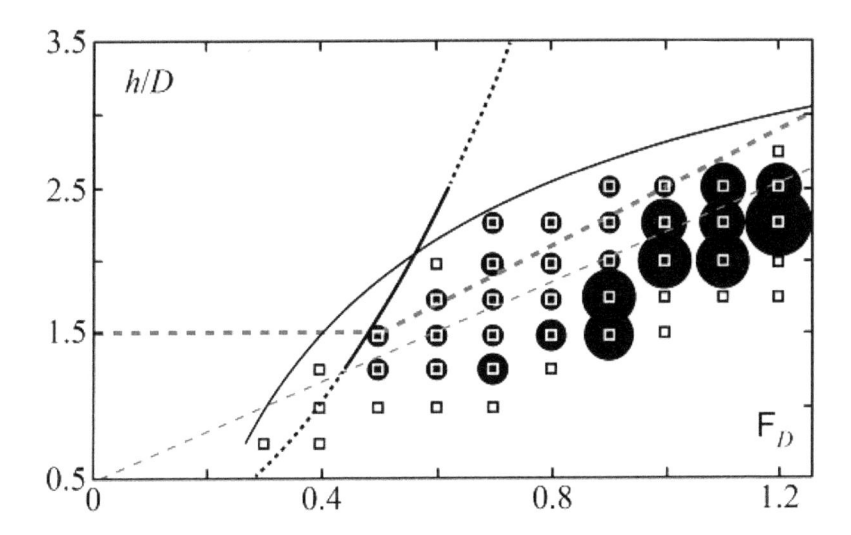

Figure 8.20 Critical relative intake submergence defined at $\beta = 1 \times 10^{-5}$ and critical submergences compared with (– –) Knauss (1987), (– –) Gordon (1970) and (□) Meyer (2011). (●) Measured β_{VT6}, whose size is linearly correlated with measured air entrainment rate, Eq. (8.23)

from Eq. (8.22) and $F_{co,cr}(\beta_{mean} = 10^{-5}) = 0.43$ from Eq. (8.21), resulting in the critical submergence in the test range $1.25 \leq (h/D)_{cr} \leq 2.5$. Statements outside this range are uncertain, e.g. for $h/D > 2.5$ the curve increases over-proportionally as compared with Knauss (1987). To include this shortcoming, the range $0.75 \leq (h/D)_{cr} \leq 3.0$ was also considered, resulting in (Möller *et al.*, 2015)

$$(h/D)_{cr} = -2.5F_D^{-0.45} + 5.3, \qquad 0.75 \leq (h/D) \leq 3.0. \tag{8.23}$$

Equation (8.23) is also included in Figure 8.20. Note that it is in contrast to the linear relation suggested by Gordon (1970) or Knauss (1987). However, the latter two and all other published criteria for intake submergence are based on observations of presumed air entrainment due to vortex presence. Vice versa, Möller *et al.*'s (2015) approach is based on real air entrainment data, for which the occurrence, frequency, and the vortex type are of minor importance. From the model tests ranging within $0.5 \leq F_D \leq 1.0$, both approaches of Knauss (1987) and Gordon (1970) underestimate the occurrence of air entrainment and the subsequent dangers to the pressurized systems.

8.3.5 Design recommendations

The designer of an intake structure has various procedures to improve the vortex features. These refer to the approach flow geometry, and special vortex suppression devices. According to Rutschmann *et al.* (1987) the approach flow is modified by:

- Uniform approach flow by appropriate appurtenances,
- Elements directing flow to intake,

- Elimination of secondary flow regions by isolating structures or adding injectors,
- Streamlining of boundaries and piers,
- Partial gate closure, and
- Acceleration of approach flow by tapering section.

In addition, improvement results by streamline elongation toward the intake:

- Higher tailwater submergence,
- Lower intake level,
- Modifying inflow direction,
- Horizontal roof above intake, and
- Reduction of approach flow velocity by widening cross section.

Anti-vortex devices frequently used for vortex suppression include:

- Vertical rows of walls,
- Horizontal beams and concrete grids,
- Floating rafts, and
- Flow straighteners.

All devices, among others, aim at breaking up a significant swirl portion (Gulliver *et al.*, 1986). Note that the devices induce a self-perturbation, so that they should be used only after detailed model observation to inhibit over-forcing. Because the mechanics of vortex flows is complex, physical modeling is recommended for all cases in which a performance failure leads to a significant head loss and damage potential due to vibrations. This is normally the case for large intakes at hydropower plants.

Figure 8.21 reviews various successful techniques adopted in existing intake structures (Rutschmann *et al.*, 1987). In Figure 8.21a, the vortex is suppressed with buoys and a horizontal beam. The maximum intake head is nearly 5 m, whereas the minimum intake section is 4.5 m × 11.4 m, and the design discharge amounts to 92 m^3/s. The intake of the Mt. Elbert Powerplant (USA) is protected against vortices by a horizontal rack (Figure 8.21b). The intake head varies between 10 and 20 m, the intake diameter is 4.75 m, and the design discharge 102 m^3/s.

Bremgarten Powerplant (Switzerland) has an injector favoring parallel flow to the intake pipe, so that stagnant surface water is suppressed (Figure 8.21c). Also, floating debris is forced to move toward the trash-rack. The intake head is 15.5 m, the intake section 8.6 m^2, and the design discharge 100 m^3/s.

For El Cajon hydropower plant (Honduras), an intake tower stabilizes the vortex formation and reduces rotation by sucking in water from the dead-water zone close to the dam (Figure 8.21d). Intake heads vary between 9 m and 41.5 m, the intake diameter is 12 m, and the design discharge 1780 m^3/s.

After this review of the current design practice, the question remains whether vortices should be prevented, or not. The design policy is toward an optimum between the hydraulic performance, safety, and economy. Consequently, many vortices have no adverse effect on the performance and safety of hydraulic structures; their removal would result in a highly uneconomic design, however. If an air core of, say 10%, of the intake diameter is considered typical, its cross section is only 1% of the pipe section so that a typical air ingestion is

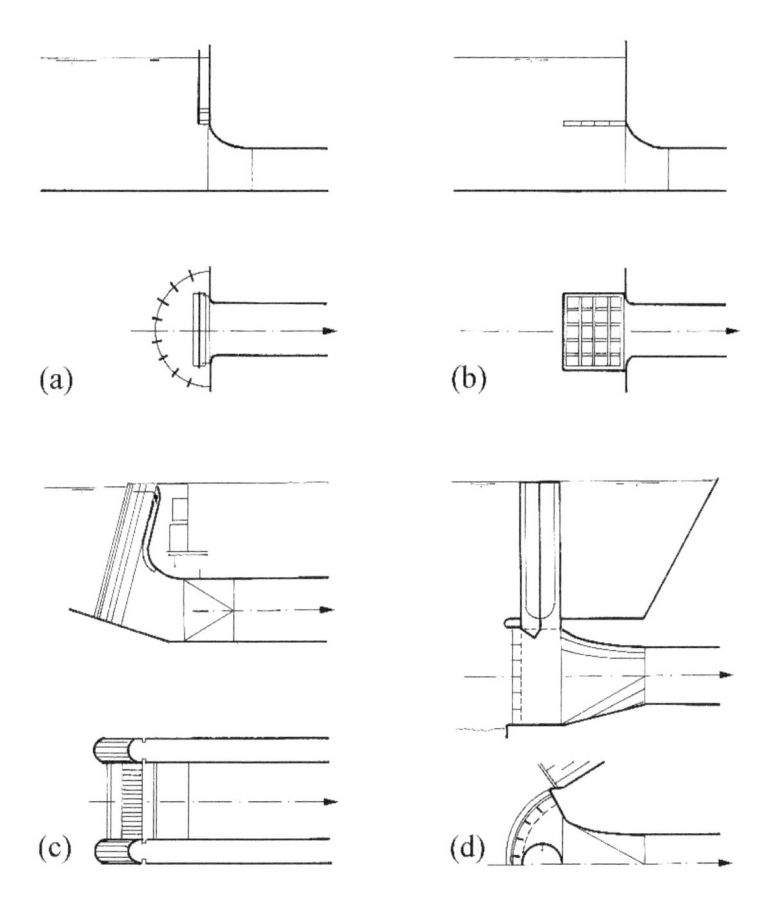

Figure 8.21 Intake structures in section (top) and plan (bottom) of hydropower plants (a) Gevelinghausen (Germany), (b) Mt. Elbert, (c) Bremgarten, (d) El Cajon

lower than 2% even for strong air-core vortices. Unless the air is locally concentrated in the pipe system, small difficulty should result from this low air entrainment. Pipe bends located downstream of the inlet section may cause as much swirl and flow asymmetry as these of intake vortices. Therefore, one should evaluate whether the flow characteristics are adverse for the specific application rather than adopting generally that vortices are to be avoided as a matter of principle (Knauss, 1987).

8.4 Practical aspects

8.4.1 Floating debris and trash-rack vibrations

Intake structures are covered with water, so that float is not entrained easily into the pipe inlet. However, suspended load as roots of trees or ingestion of floating matter by surface vortices may give significant operational problems, as mentioned in 8.2. Intake structures are therefore protected with a trash-rack. The bar distance e depends on the degree of protection.

As a rule of thumb, $e = (2/3)E$ where E corresponds to the minimum plant width (minimum flow spaces at turbines or gates). The bar distance is also often selected to avoid the entrainment of large fish.

Trash-racks should be obviously avoided for outlet structures used for flood diversion and sediment flushing. Further, gates without gate slots should be used as sector gates. Accordingly, the intake dimension has to be selected so that obstruction is impossible even from the largest float. For bottom outlets, a large spacing trash-rack with bars often made by rounded concrete beams is normally provided to retain logs of wood contained in the sediment, thus preventing clogging of the outlet gates.

For withdrawal structures, the mean velocity ranges between 0.5 and 1 m/s at the bellmouth entry section, over periods of hours or days. Local velocities through intakes often reach the double of the mean value. Trash-racks have thus to be designed against flow-induced vibrations, producing fatigue failure of the trash-rack bars (Figure 8.22). Vibrations are enhanced by flow separation from the bars by vortex shedding. Figure 8.23 relates to the *Strouhal number* $S = fd/V$ for various bar profiles, from which the frequencies of vortex detachment is determined as

$$f = S(V/d). \tag{8.24}$$

For bar diameters or widths of $d = 0.1$ m and $V = 1$ m/s velocity, these range from 0.3 to 1 Hz.

Figure 8.22 Fatigue failure at trash-racks due to flow-induced vibrations by vortex shedding. Failure occurs at intersections of vertical trash-rack bars and horizontal or inclined circular bars ensuring the distance between vertical bars (Courtesy: Anton J. Schleiss)

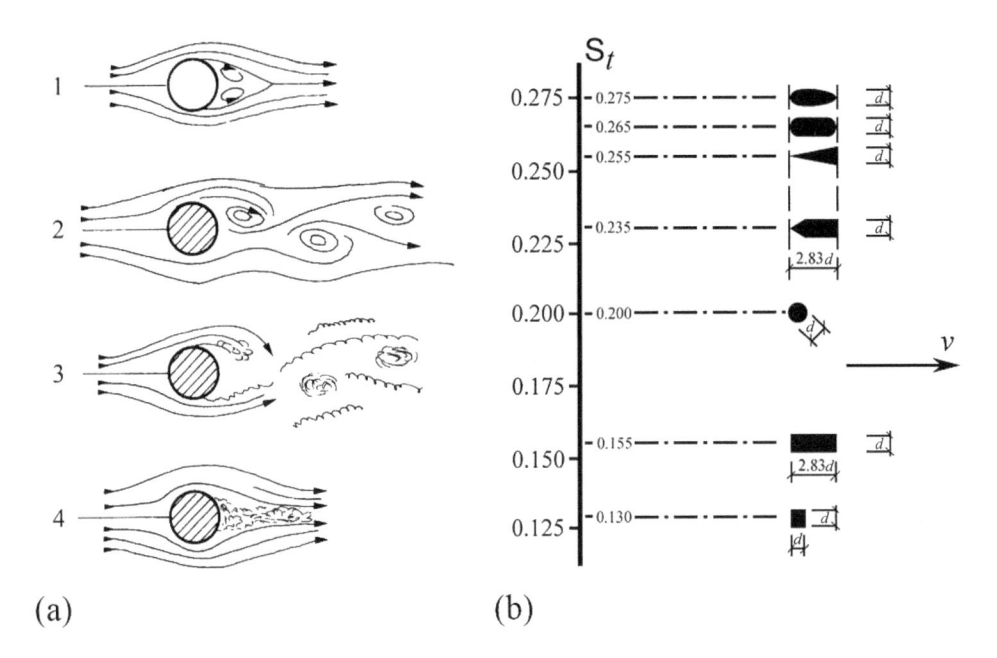

(a)   (b)

Figure 8.23 Natural frequencies of typical bar profiles (a) flow conditions and vortex shedding for circular-shaped cylinder with $R = Vd/v =$ (1) 20, (2) 10^2, (3) 10^4, (4) 10^8 (Levin, 1967), (b) Strouhal numbers S_t for various rack bar shapes of width d and length l, or $2.83d$

The frequency of vortex shedding is compared with the natural frequencies of the trash-rack bars. The first natural frequency is (Schleiss, 1985; Schleiss and Fust, 1992)

$$ f_E = \frac{M}{h_b^2} \sqrt{\frac{E_s g I}{A_b \left(\rho_s + \rho \left(a_{eff} / d \right) \right)}} . \tag{8.25} $$

Here h_b is the spacing of the horizontal circular bars keeping the distance between the vertical trash-rack bars, M is a fixation coefficient between the trash-rack bars and the horizontal circular bars ($M = 1.57$ for free rotation, $M = 3.57$ for rigid fixation). Further, E_s is the modulus of elasticity of steel, I the lateral inertial moment of the steel bars perpendicular to the flow direction, A_b the area of the trash-rack bars, ρ and ρ_s are the densities of water and steel, respectively. In addition, a_{eff} considers the water mass vibrating with the bars corresponding to 70% of the bar length l but in the maximum the spacing between the bars.

The next three higher natural frequencies result by multiplying the first frequency by 4, 9, and 16 for free rotation with $M = 1.57$ and 2.75, 5.43, and 8.93 for rigid fixation $M = 3.57$. To avoid resonance vibrations, the frequency of vortex shedding should be $f < (0.60 \div 0.65) f_E$ for the first natural frequency. For higher natural frequencies, the difference to the frequency of the vortex shedding has to be at least 20% to 30%. Besides increasing the cross section of the steel bars, which is not economical, reducing the spacing between the horizontal circular bars is often an alternative. Besides the single trash-rack bars, check also whether the entire rack tables are excited by flow-induced vortex shedding at the bars. This is avoided by fixing the rack tables (panels) laterally at the intake sidewalls or by adding inclined circular bars as panel stiffeners. Rectangular trash-rack bars are less excited by vortex shedding than hydrodynamic-shaped bars. Even if the frequency of vortex shedding and the natural frequency of the trash-rack bars are far enough away from each other, small vibrations still occur and the structural elements, especially the weldings of the trash-rack panels have to be designed for fatigue resistance. More design details are discussed by Schleiss and Fust (1992), or Naudascher and Rockwell (1994).

The design of a trash-rack includes means for rack cleaning. For trash-racks close to the free surface of the reservoir, a rack cleaner is recommended. For trash-racks of intakes located deeply below the reservoir free surface, a rack pedestal should at least be considered in the design that is accessible during drawdown of the reservoir level. Cleaning is then possible with racks or other suitable tools. Clearly, trash-racks with a small bar distance clog faster than if the bar distance is wide. Therefore, and because of a reduction of head losses, the bar distance is not reduced beyond the minimum value previously mentioned.

8.4.2 Emergency gate closure

The possibility of placing emergency gates in front of a service gate is discussed in Chapter 2 and in 8.2. Below, emergency gates of withdrawal structures are considered. Their discharge is regulated close to the user, but not at the intake structure, by means of a pump at the pumping station, a turbine at the powerhouse, or valves for irrigation works. Two cases are distinguished:

1   Pressure conduits connected to a single aggregate
2   Pressure conduits discharging into a set of parallel aggregates and thus bifurcating in a manifold pipe.

For both cases, an emergence gate, for high-pressure heads typically a spherical valve, is placed just upstream of each aggregate for safety and maintenance purposes. For a long waterway system including pressure tunnels and shafts, also an emergency gate, typically a roller gate or a butterfly valve, is required directly downstream of the intake for maintenance purposes and to limit outflowing water in case of its failure. For very long pressure tunnels, a further emergence gate may be placed at the upper end of the steel-lined pressure shaft or penstock but downstream of the surge tank for safety and maintenance reasons. For that purpose, normally butterfly valves are used. Further, for the maintenance of intake structures including the trash-racks, stop-locks are recommended. This specially applies if a reservoir drawdown is unacceptable. Bottom outlets as a safety structure of dams and reservoirs are always equipped with both an emergency and a service gate. For the maintenance of the intake structure, stop-locks are also recommended.

8.5 Gate flow

8.5.1 Introduction

Gates constitute a fundamental hydraulic structure allowing for the control of an upstream water elevation. Among various gate designs, the so-called standard gate with a vertical gate structure containing a standard crest positioned in an almost horizontal, smooth, and rectangular channel is particularly relevant in low-head applications. Surface roughness of both the channel and the gate is small and thus negligible. These gates are used in hydraulic laboratories, irrigation canals, large sewers, or in hydraulic structures.

Compared with overflow structures, or in particular the standard sharp-crested weir, standard gates have received scarce attention. The knowledge is particularly poor regarding basic hydraulics, whereas studies on their vibration features are available. Roth and Hager (1999) describe new findings on standard gate flow, including (1) scale effects, (2) coefficient of discharge, (3) surface ridge, (4) shock wave features, (5) velocity field, (6) bottom and gate pressure distributions, (7) corner vortices, and (8) vortex intensities. A novel device to reduce shock waves in the tailwater is also proposed.

The present knowledge on gates is summarized by Lewin (1995), including a short chapter on vertical gates containing information on both discharge and contraction coefficients, with a relatively large scatter of data, so that gate flow was by then far from being understood from this point of view.

Historical studies on underflow gates are available; it is currently a common belief that the discharge characteristics of vertical gates have been detailed in the 19th century. This is not the case, however, because of the accuracy of discharge measurement, and the small hydraulic models often used. Well-known approaches include those of Boileau (1848), Bornemann (1871, 1880), containing summaries of the experiments of Lesbros, Haberstroh (1890), Gibson (1920), Hurst and Watt (1925), Keutner (1932, 1935), Fawer (1937), Escande (1938), Gentilini (1941a,b), and Smetana (1948). The exact geometrical configurations in these experimental studies are often poorly defined, and the data are not always available. Details of gate fixation are also not described.

The first modern study relating to free gate flow was conducted by Rajaratnam and Subramanya (1967). The coefficient of discharge C_d was related to the difference of flow depths in the up- and downstream sections $(h_o-C_c a)$, with h_o as the approach flow depth, C_c the coefficient of contraction and a the gate opening. According to observations for both free and submerged flow, C_d is exclusively a function of the relative gate opening a/h_o, increasing slightly with a/h_o, starting from $C_d = 0.595$. The effect of skin friction was attributed for deviations between computations based on the potential flow theory and observations.

Rajaratnam (1977) conducted a second study on vertical gates in a rectangular channel 0.311 m wide, with gate openings between 0.026 and 0.101 m. The axial free-surface profile downstream of the gate section was shown to be self-similar. Noutsopoulos and Fanariotis (1978) pointed at the significant data scatter relating to both coefficients of contraction and discharge. Deviations between theory and observations were attributed to the spatial flow characteristics, and the too narrow channels often employed.

Nago (1978) conducted observations in a 0.400 m wide rectangular channel for $a = 0.060$ m. The coefficient C_d was found to decrease with increasing relative gate opening, from 0.595 for $a/h_o \rightarrow 0$ to 0.52 for $a/h_o = 0.50$. Rajaratnam and Humphries (1982) considered the free-flow characteristics upstream of a vertical gate, for gate openings of $a = 0.025$ and 0.050 m. Their data refer to the upstream recirculation zone, the bottom pressure distribution, and the velocity field.

Montes (1997) furnished a solution to the 2D gate outflow using conformal mapping, compared the coefficient of contraction with experiments, and identified deviations due to viscous effects. The surface profiles up- and downstream from the gate section were studied exclusively in terms of the gate opening. Energy losses across a gate were related to the boundary layer development and the spatial flow features upstream from the gate.

The purpose of the research of Roth and Hager (1999) was to clarify several points of standard gate flow under free, i.e. unsubmerged flow conditions, including the discharge coefficient, the ridge position, the velocity and pressure distributions, and the shock wave development in the tailwater previously not considered. These results may attract and guide numerical flow modelers, and add to the current design guidelines in hydraulic engineering.

8.5.2 Vertical planar gate flow

Experimentation

Systematic tests were conducted in a 0.500 m wide rectangular channel. The width of the approach flow channel was also reduced to $b = 0.245$ and 0.350 m. An aluminum gate 0.499 m wide, 0.600 m high and 0.010 m thick was used, of which the crest was of standard geometry, i.e. 2 mm thick with a 45° bevel on the downstream side. The gate was mounted with variable openings from the channel bottom. No gate slots were provided and water tightness was assured with a tape. Only free gate flow was tested. The gate opening varied from $a = 10$ to 120 mm.

The experimental program aimed at analyzing scale effects, free-surface profiles, the development of corner eddies, the determination and reduction of shock waves, and the velocity and pressure characteristics in gate vicinity (Roth and Hager, 1999). Figure 8.24 shows a definition sketch of the standard gate. The main parameters of gate flow are discharge Q, approach flow depth h_o, and gate opening a. Further, x is streamwise coordinate measured from the gate section, z the vertical coordinate measured form the channel bottom, x_R the ridge position (see below), h_p piezometric head on the channel bottom, x_s the position of maximum shock wave height h_s, and h_u as the axial downstream depth.

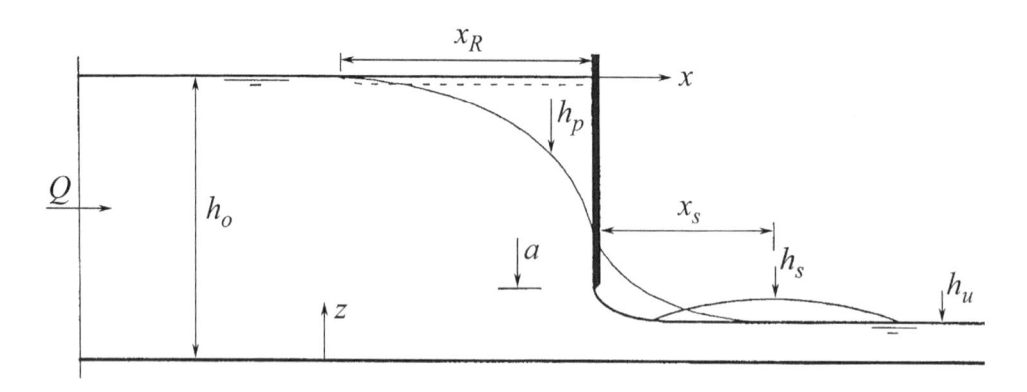

Figure 8.24 Definition plot of standard gate, with notation

Discharge coefficient

The discharge below a gate structure is expressed for free flow as

$$Q = C_d ab(2gh_o)^{1/2}. \tag{8.26}$$

Figure 8.25a shows $C_d(A)$ with $A = a/h_o$ as the relative gate opening. For $a \geq 0.050$ m all data follow a single curve based on Froude similitude. For $a < 0.05$ m, the curves $C_d(A)$ split, however, because of effects of viscosity, pointing at scale effects. For extremely small values of A, surface tension dominates the flow, so that C_d decreases sharply to zero. This domain was not investigated.

All curves $C_d(A)$ start close to 0.60, decrease to a minimum (subscript m) value C_{dm} with the corresponding relative gate opening A_m, and increase again. Both A_m and C_{dm} vary with the gate Reynolds number $R_a = a(2ga)^{1/2}\nu^{-1}$, with ν as kinematic viscosity. For $R_a < 5 \times 10^4$ the data follow the curves (Figure 8.25b, c)

$$A_m = 0.05 + 0.40 \cdot \log(R_a/1000); \tag{8.27}$$
$$C_{dm} = 0.60 - (1/18) \cdot \log(R_a/1000). \tag{8.28}$$

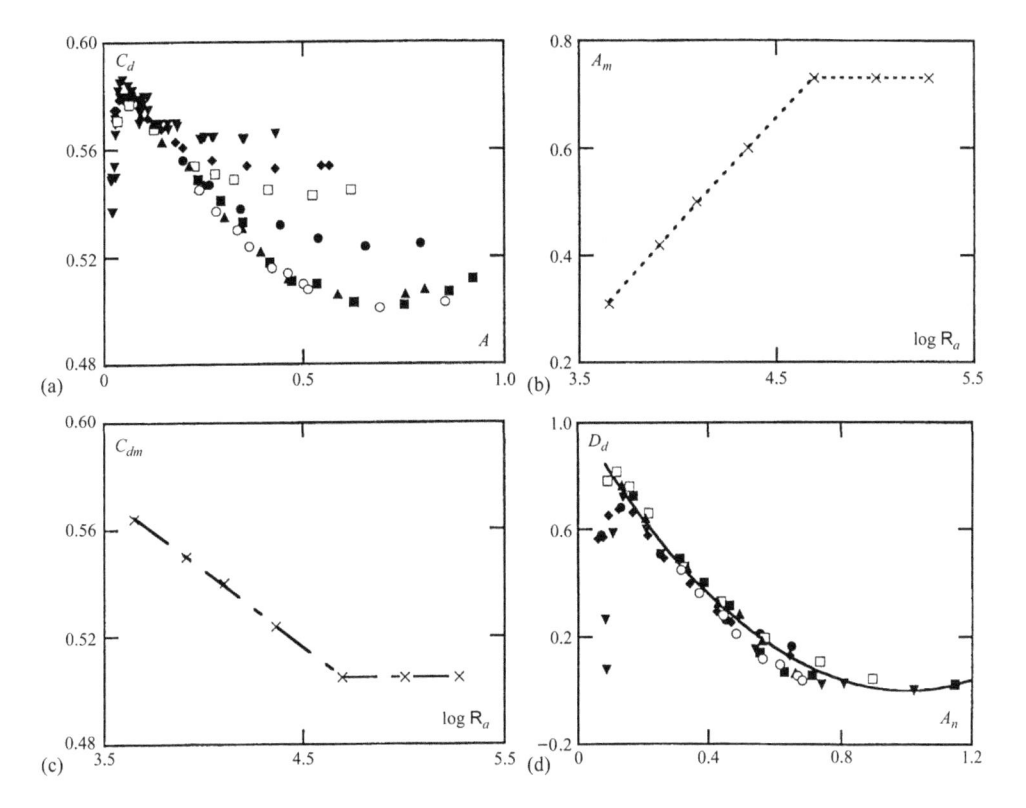

Figure 8.25 Discharge coefficient of standard gate flow (a) $C_d(A)$, (b) $A_m(R_a)$ with (···) Eq. (8.27), (c) $C_{dm}(R_a)$ with (---) Eq. (8.28), (d) relative value of $D_d(A_n)$ with (—) Eq. (8.29) for a (mm) = (▼) 10, (◆) 15, (□) 20, (●) 30, (▲) 50, (■) 80, (○) 120 (Roth and Hager, 1999)

Let $D_d = (C_d - C_{dm})/(C_{do} - C_{dm})$ be a measure of scale effects in the discharge coefficient, then all C_d data are solely expressed versus $A_n = A/A_m$, with $C_{do} = 0.594$ as the base value for small A as (Figure 8.25d)

$$D_d = (1 - A_n)^2. \tag{8.29}$$

This relation describes the discharge coefficient including viscous effects.

The discharge coefficient was also determined for channel widths of $b = 0.245$ and 0.350 m. The limit (subscript L) gate openings for Froude similitude to apply are $a_L = 0.090$ and 0.070 m, respectively. Accordingly, the channel width has a significant effect on the limit gate opening. The reason for this scale effect is fluid viscosity (Montes, 1997). The head losses then exceed a certain limit value due to an extremely small downstream flow depth. To advance a hydraulic approach, the viscosity effect was computed for a simplified flow configuration with a flow depth $h = h_u = C_c a$ along the contraction length $2a$ downstream from the gate. The friction slope S_f times the contraction length divided by the contracted velocity $u_u = Q/(C_c ab)$ was postulated an index for Reynolds effects. The parameter $\Phi = S_f (2a)/[u_u^2/(2g)]$ is almost a constant for the three limit conditions previously determined. Knowing $\Phi = \Phi_L$ allows to predict a_L for channel widths smaller or larger than tested. Based on Roth and Hager (1999), the limit condition $A = 0.25$ yields for all three channel widths tested the value $\Phi_L = 0.020$. For $b > 1$ m, the minimum value tends to $a_L = 0.045$ m. For any channel width, a minimum gate opening of $\cong 0.050$ m is thus required for inviscid flow to occur. For narrower channels, the minimum gate opening increases, as detailed above.

Reynolds ridge and shock waves

Harber and Gulliver (1992) analyzed the so-called Reynolds ridge visible in front of gates and corresponding to the plunging point of stagnation flow (Figure 8.26b). The Reynolds ridge depends on surface tension and was also determined by Rajaratnam and Humphries (1982). It is a significant feature related to water quality because floating matter is retained from downflow, except for entrainment by corner vortices. For common laboratory water, the so-called surface tension height is $[\sigma/(\rho g)]^{1/2} = 2.7 \times 10^{-3}$ m, with σ as surface tension, and

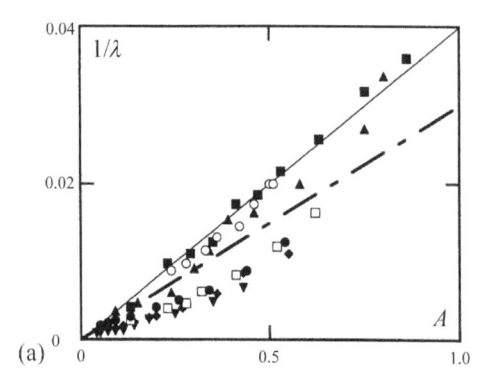
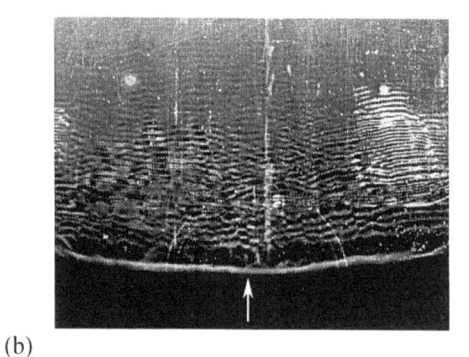

Figure 8.26 (a) Inverse relative ridge position $\lambda^{-1}(A)$ with $(-\cdot-)$ Rajaratnam and Humphries (1982) for viscous flow and (—) Eq. (8.30), (b) Reynolds ridge upstream of standard gate (Roth and Hager, 1999)

ρ as fluid density. The relative ridge position $\lambda = x_R/[\sigma/(\rho g)]^{1/2}$ upstream from the gate (Figure 8.24) varies inversely with the relative gate opening. The data of Roth and Hager (1999), and Rajaratnam and Humphries (1982) follow the trend (Figure 8.26a)

$$\lambda^{-1} = 0.04A. \tag{8.30}$$

Accordingly, the ridge position increases as the relative gate opening decreases.

Shocks downstream of a gate are due to corner vortices characterized by the maximum wave height h_s, the corresponding location x_s and the shock angle θ (Figure 8.24). Their features vary with the contracted Froude number $\mathsf{F}_c = Q/[bh_u(2gh_u)^{1/2}] = 2^{1/2}C_d/(C_c^{3/2}A^{1/2})$. For inviscid gate flow C_c and C_d are nearly identical, so that the governing shock wave parameter is $T = (2/A)^{1/2}-1$. The experimental data indicate for the relative height of shock waves $Y_s = h_s/h_u$, the location of maximum height $X_s = x_s/h_u$, and the shock angle θ (Roth and Hager, 1999)

$$Y_s = 0.30T^{2/3}, \tag{8.31}$$
$$X_s = 2T, \tag{8.32}$$
$$\cot\theta = (20/3)T. \tag{8.33}$$

Velocity and pressure distributions

Figure 8.27a shows the axial horizontal velocity component U for a gate opening $a = 0.080$ m. The normalizing velocity $(2gh_o)^{1/2}$ refers to downstream conditions, where the velocities are properly scaled. Upstream from the gate the velocity scale would be $Q/(bh_o)$. For both $X = x/a < -2$ and $X > +2$, the velocity distribution is almost uniform, except for the bottom boundary layer. Close to the gate section, the velocity increases significantly with depth $Z = z/a$. Figure 8.27b is based on video observations in which 0.2 mm VESTIRON particles were illuminated with a light sheet.

The bottom pressure head $h_p(x)$ varies along the channel, from $h_p = h_o$ upstream of the gate to $h_p = h_u$ sufficiently downstream. Figure 8.28a refers to the axial bottom pressure head $H_p = (h_p-h_u)/(h_o-h_u)$ versus dimensionless location $X = x/a$. All data refer to the inviscid flow condition, for which the data follow (Roth and Hager (1999)

$$H_p = 1 - \exp[-(1/3)(X-1.7)^2]. \tag{8.34}$$

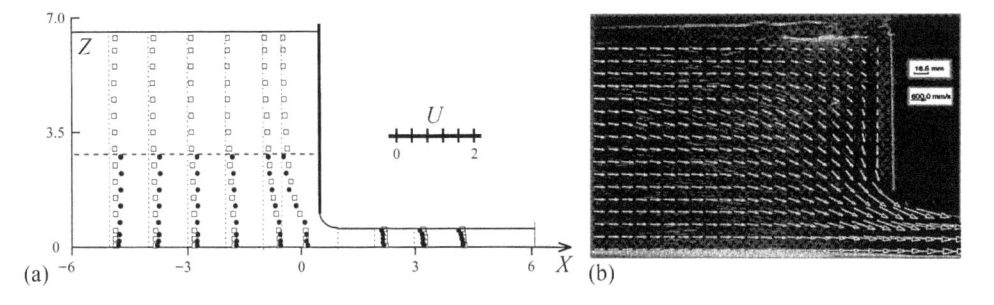

(a)   (b)

Figure 8.27 (a) Axial streamwise velocity component $U(X, Z)$ for h_o (mm) = ($\bullet$) 235, ($\square$) 530, (b) velocity vector field for $h_o = 0.300$ m, $a = 0.080$ m, $b = 0.500$ m (Roth and Hager, 1999)

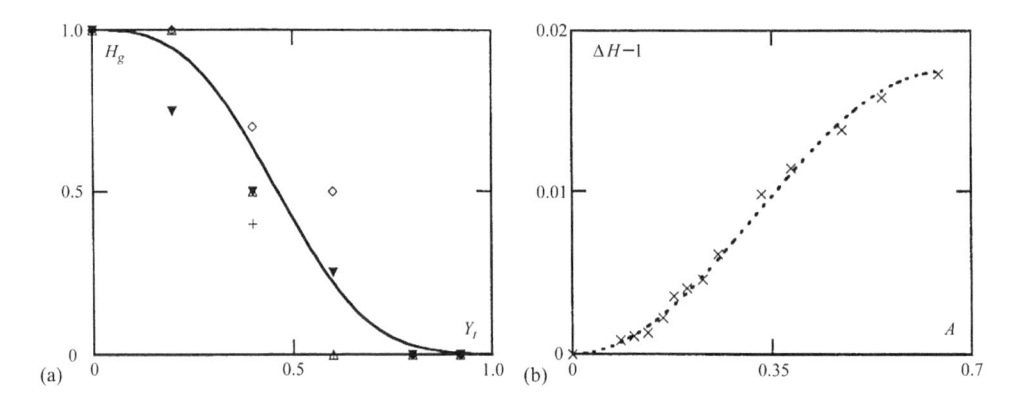

Figure 8.30 Transverse surface profile upstream of standard gate (a) $H_g(Y_t)$ for h_o (mm) = ($\blacktriangledown$) 156, ($\diamond$) 218, ($\triangle$) 280, (+) 379, (—) Eq. (8.38), (b) $\Delta H(A)$ with ($\times$) observations, ($\cdots$) Eq. (8.39) for $a = 0.08$ m and $b = 0.50$ m (Roth and Hager, 1999)

the gate, i.e. the stagnation (subscript s) surface profile $H_s(y_t)$, with y_t as the transverse coordinate measured from the channel axis. Let H_o and H_w be the stagnation depths in the channel axis (subscript o) and close to the wall (subscript w), respectively. With $H_g = (H_s - H_w)/(H_o - H_w)$ and $Y_t = y/(b/2)$ the dimensionless profile $H_g(Y_t)$ at the gate (subscript g) varies as shown in Figure 8.30a. The data fit follows

$$H_g = \exp(-7Y_t^3). \tag{8.38}$$

The temporally averaged free-surface profile has extremes close to the walls; their differences $\Delta H = (H_o - H_w)/H_o$ increases with A (in rad.) as (Figure 8.30b)

$$\Delta H = 0.0175[1 - (\cos 2.4A)^2]. \tag{8.39}$$

The maximum value is close to 2%, indicating that this small difference is able to develop large intake vortices due to the stagnation phenomenon.

Figure 8.31 is a definition sketch for corner vortices generated at the standard gate. The vortex (subscript v) core distance from the gate is x_v, and the distance from the channel wall is y_v. With $X_v = x_v/a$ and $Y_v = y_v/a$ as dimensionless coordinates of the vortex centers, the observations yield with $\alpha = a/b$ (Figure 8.32)

$$X_v = (1/2)A^{-0.75}; \tag{8.40}$$
$$Y_v = (2/3)\exp(-3\alpha). \tag{8.41}$$

The streamwise distance of the vortex center therefore increases as the relative gate opening A decreases due to the increased distance between the upstream flow surface to the outlet section. The vortex center is also closer to the wall for a wide than a narrow channel. Figure 8.32c shows the flow pattern with the presence of an intake vortex, whereas Figure 8.32d shows a surface vortex of intensity 4 (8.3.2), with the Reynolds ridge on the left,

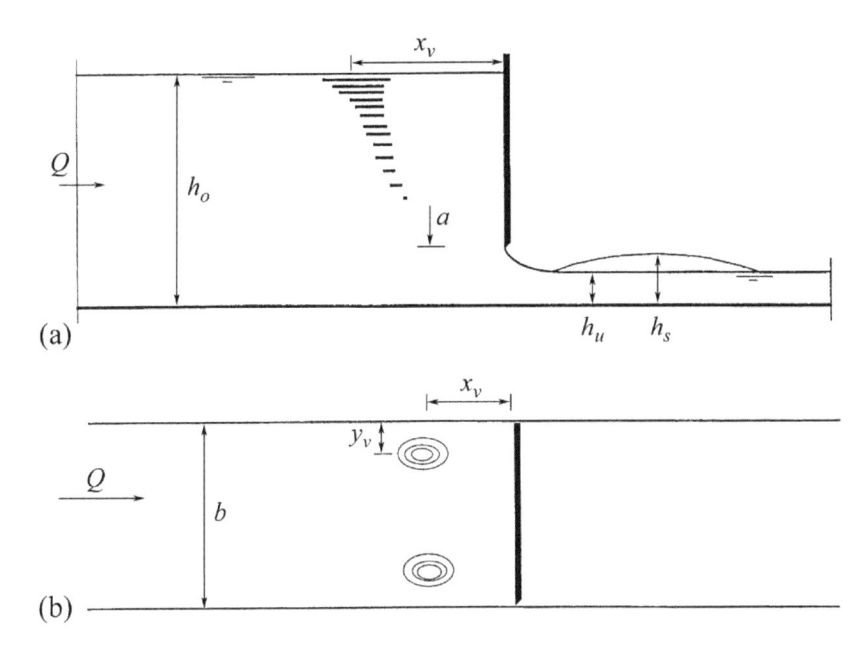

Figure 8.31 Definition sketch for corner vortices (a) section, (b) plan

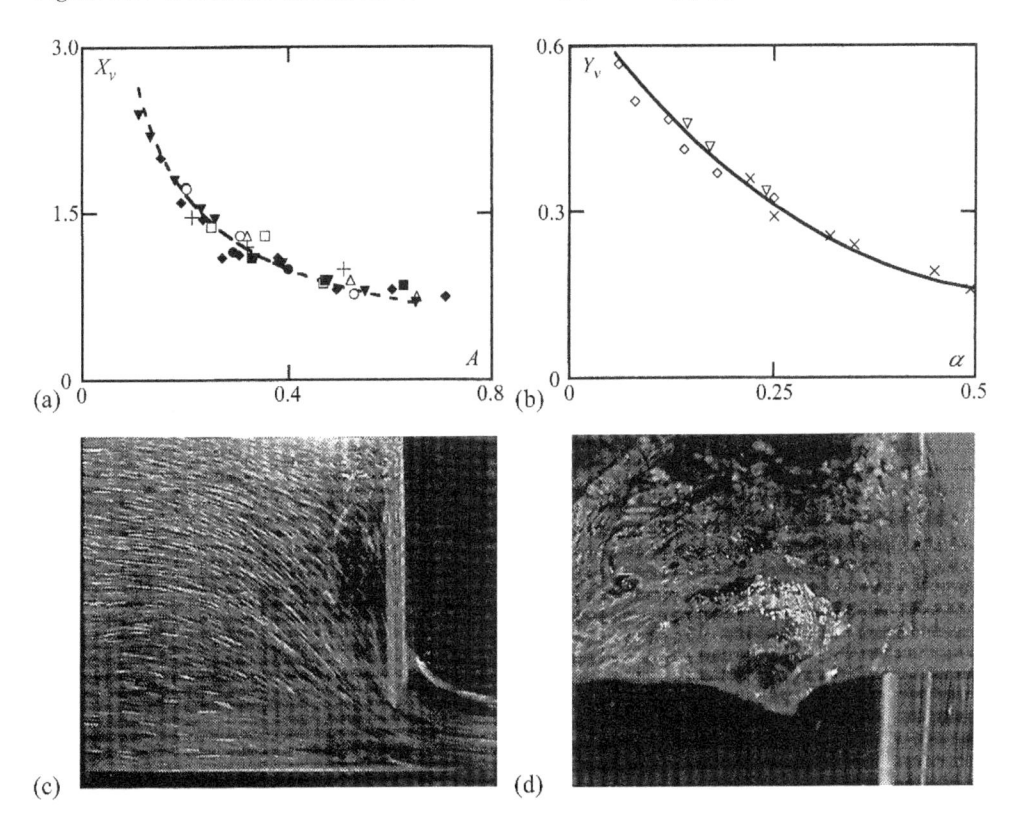

Figure 8.32 Intake vortex at standard gate (a) location $X_v(A)$ for $b = 0.50$ m and a (mm) = (▼) 50, (●) 80; $b = 0.35$ m and a (mm) = (○) 60, (+) 80; $b = 0.245$ m and a (mm) = (△) 80, (□) 100, (◆) 120, (···) Eq. 8.40), (b) vortex distance from wall $Y_v(\alpha)$ for b (m) = (×) 0.245, (▽) 350, (◇) 0.500, (—) Eq. (8.41), (c) vortex flow toward gate opening, (d) surface vortex with Reynolds ridge (Roth and Hager, 1999)

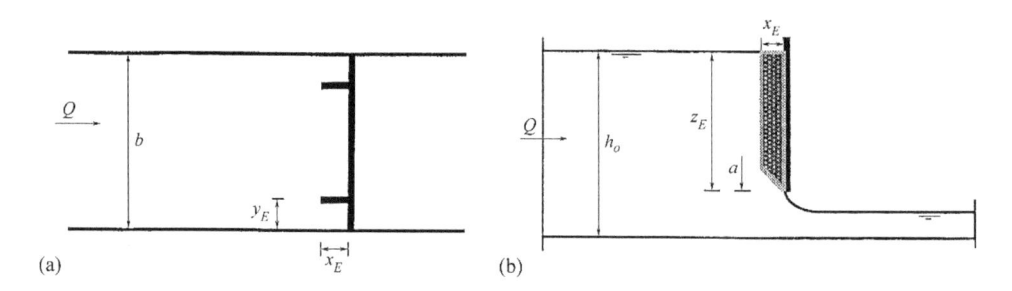

Figure 8.33 Anti-vortex element at standard gate (a) plan, (b) side view (Roth and Hager, 1999)

and the gate at right. Based on Hecker (1984), the vortex intensity for inviscid standard gate flow is typically 5 for $A > 0.15$, whereas it decreases toward 0 (no visible surface swirl) as $A < 0.15$.

Given the problems with vortex presence at intakes, and with shock waves in the tailwater reach of a gate or particularly in bottom outlets downstream of high-head gates, means to significantly reduce vortex action were sought based on the previous results. Shock waves include the following disadvantages: (1) Increased freeboard in tunnels and chutes, (2) local air entrainment and thus bulkage of air-water high-speed flow, and (3) asymmetry of tailwater flow. Two configurations to reduce shock waves due to stagnation flow at gates were tested, namely (1) vertical plates located close to sidewall to cut the transverse stagnation currents from the channel axis to the walls, and (2) horizontal triangular corner plates to cut the vertical vortex paths. Because (2) was ineffective, design (1) was further considered. Figure 8.33 shows the so-called anti-vortex elements (subscript E) attached at the upstream gate wall, consisting of two vertical plates of length x_E away from either wall by y_E and of height z_E. The performance criteria include:

*   Reduction of shock waves,
*   Removal of corner vortices, and
*   Simplicity of design and economy.

The lateral element position was optimum when located at the vortex core, i.e. $y_E = y_v$. The performance increases as the relative element height $Z_E = z_E/(h_o - a)$ tends to 1, so that the element should span over the entire gate height. The relative element length $X_E = x_E/y_v$ varied up to 2; no effect resulted if $X_E \geq 1$, whereas the performance increased strongly as X_E increased from 0 to 1, so that $X_E = 1$ was considered the optimum. Therefore, the optimum element is simply described with $x_E = y_E = y_v$, i.e. the corner portion between the element and the wall has square shape in plan (Figure 8.33a). The cross-sectional element geometry was varied as shown in Fig 8.34b with rectangular, slice-shape, and triangular bottom shape. Given that no effect was observed, the latter was selected because of ease in design.

The shock wave reduction by the anti-vortex element was determined based on the parameter $D_m = (h_{rm} - h_u)/(h_s - h_u)$ with h_{rm} as the reduced shock wave height, and h_s as the shock wave height without element presence (Figure 8.34). Figure 8.34a shows $D_m(a)$ for both viscous and inviscid flows, indicating for the latter a typical value of $D_m = 0.5$, which is considered relevant both from the hydraulic aspect as also from cost-benefit.

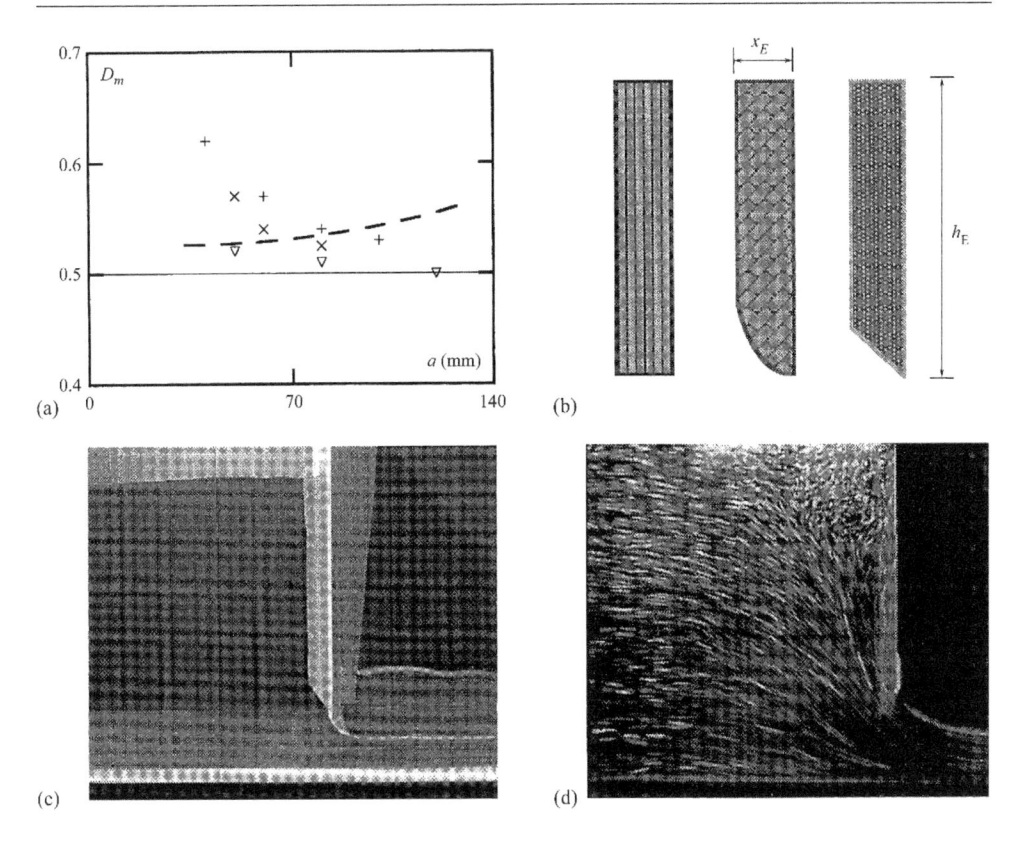

Figure 8.34 Anti-vortex element for standard gate (a) reduction ratio D_m versus a (mm) with (---) limit between flows with and without scale effects, (b) designs of bottom shapes, (c) flow with optimum element mounted on front side, but none at back side with resulting downstream flow surfaces along walls, (d) corner vortex as origin of shock waves (Roth and Hager, 1999)

8.5.3 Hinged sloping flap gate

Introduction

Any control of flow involves two main options:

- Discharge under variable head so that discharge remains constant, and
- Head under variable discharge so that the head remains constant.

With Q as discharge, C_d as discharge coefficient, F as gate outflow area, g as gravity acceleration and h_o as approach flow depth, both types have widespread applications in hydraulic engineering. They are described for free gate flow with

$$Q = C_d F (2gh_o)^{1/2}. \tag{8.42}$$

For constant discharge control any of the three variables C_d, F, or h_o is a governing parameter so that their combined effect assures Q = const. Typical hydraulic elements include the so-called throttle hose (Vischer, 1979) in which fluid pressure on a hose in a basin is applied to control discharge. In turn, Brombach and Horlacher (1996) proposed an elastic steel sheet to control discharge thereby influencing C_d. Levi (1991) among others recommended a floating siphon to keep the variable h_o constant.

The ratio $Q/(C_dF)$ needs to be kept constant if the approach flow head in a canal has to remain constant under variable discharge. The Neyrpic Gate serves this purpose involving a balanced gate and a float (Bos, 1976). The research of Raemy and Hager (1998) is based on a device previously suggested by Pethick and Harrison (1981) and Kay and Ashton (1983). The novelty of the so-called Hinged Flap Gate is a counterweight attached non-perpendicularly to the gate, so that problems with instability of the former device are reduced under all discharges up to the design discharge. The purpose of the following is: (1) Presentation of pressure distributions for the inclined sharp-crested gate, and (2) application of moment of momentum equation to one of the rare problems in hydraulic engineering excluding rotatory hydraulic machinery.

Experimentation

Tests were conducted in a 0.50 m wide and 0.70 m deep rectangular, horizontal channel. Discharges of up to 100 l/s were considered; all flows were non-submerged from the tailwater. The flap gate was 0.415 m long and 0.499 m wide. Standard gate crests were fitted along both the top and bottom surfaces; pressure tapings were drilled through the aluminum gate for pressure readings. The gate was suspended on a pivot located 15 mm below the top crest, so that the resulting frictional moment remained extremely small. Water tightness between the gate and the channel sidewalls was secured with conventional adhesive tapes, so that the gate was both absolutely watertight and completely free in the movement along the walls. Figure 8.35 shows the Hinged Flap Gate and explains its hydraulic features.

The pivot elevation was so adjusted that the lower gate crest touched the bottom at vertical gate position. The gate remained thus vertical as $Q = 0$. It starts to open if the upstream

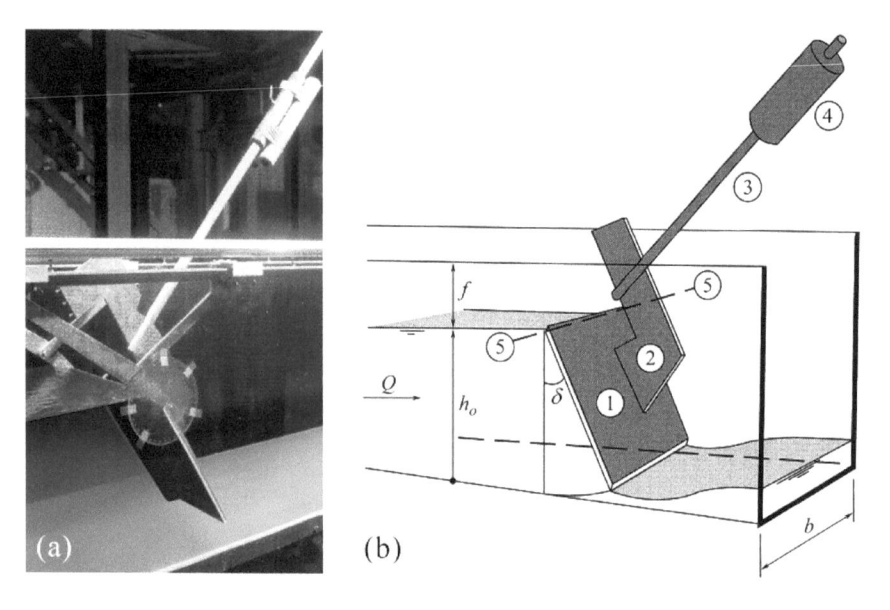

Figure 8.35 Hinged Flap Gate (a) side view, (b) gate elements with ① flap gate, ② weight suspension, ③ rod, ④ counterweight, ⑤ gate suspension (Raemy and Hager, 1998)

head equals the gate height with gate overflow about to start. Free-surface elevations were measured with a point gage and pressure heads with a manometer.

To apply the momentum, and the moment of momentum equations, the pressure distribution on the gate must be known. This knowledge is only partially available from Fawer (1937) and Gentilini (1941a), or Southwell and Vaisey (1946) and Anwar (1964). The pressure distribution on sloping gates of shell-type geometry was given by Nago (1977), including force ratios between dynamic and hydrostatic conditions, and the discharge coefficients. Han and Chow (1981) or Naghdi and Vongsarnpigoon (1986) furnished the corresponding numerical data, yet without general results.

Free gate flow is characterized by free surfaces both at the top and lower gate crests (Figure 8.36). Rajaratnam and Humphries (1982) describe a surface eddy with small velocities extending into the approach flow and a main contracting flow zone below it. The pressure distribution along the gate is thus nearly hydrostatic from the top crest, reducing as the lower crest is approached due to increasing flow velocity, becoming atmospheric at the lower crest. The pressure force on a sloping gate depends mainly on the gate angle δ and slightly on the approach flow Froude number $\mathsf{F}_o = V_o/(gh_o)^{1/2}$, whose effect was demonstrated to be small. Figure 8.37 shows typical pressure distributions on the gate for three gate angles δ, from

Figure 8.36 Hinged Flap Gate with pressure tapings and connections to manometer for δ = (a) 29.7°, (b) 26.1° (Raemy and Hager, 1998)

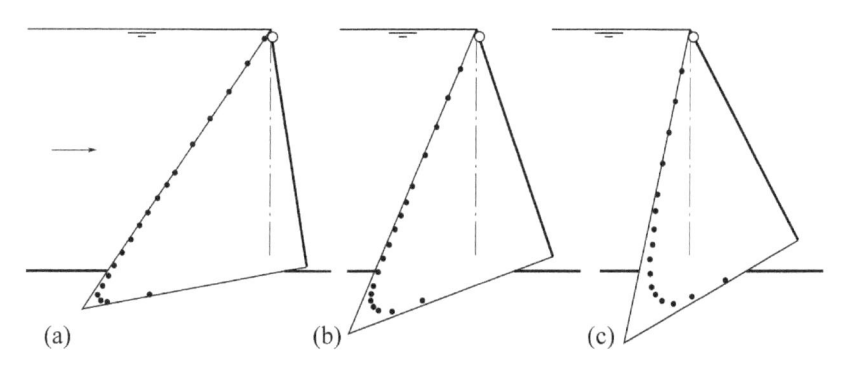

Figure 8.37 Pressure distribution of sloping gate with (●) dynamic and (–) static pressure for δ = (a) 10°, (b) 20°, (c) 30° (Raemy and Hager, 1998)

where the pressure deviation from the hydrostatic pressure distribution increases as the lower gate crest is approached; deviations increase also with δ. The hydrostatic (subscript s for static) pressure force S_s and dynamic (subscript d) pressure force S_d result from integration of pressure over the gate length. The two are identical for $\delta = 0$ given that then $Q = 0$.

Control of hinged sloping gate

Let L_g be the gate (subscript g) length and $D = L_g/h_o$ relative gate length. Figure 8.38a shows the reducing pressure (subscript p) force ratio $\sigma_p = S_d/S_s$ with D following the expression (Raemy and Hager, 1998)

$$\sigma_p = 1 - (1/7)D\tan\delta. \tag{8.43}$$

The moment exerted by the pressure force on the flap gate is determined with the moment ratio $\mu = M_d/M_s$ as (Figure 8.38b)

$$\mu = 1 - (1/4)D^{1/2}\tan\delta. \tag{8.44}$$

The deviation of dynamic pressure as compared with the static pressure thus increases with both D and δ.

A Hinged Flap Gate has to balance two moments, to be in equilibrium for upstream level control (Figure 8.39):

- Dynamic moment M_d due to hydrodynamic pressure as disturbing moment,
- Sum of moments $M_g + M_w$ due to gate weight G and counterweight W as restoring moment.

The equilibrium condition for $h_o = L_g$ reads with e as the distance from the pivot to the center of gravity of the counterweight system, and ε as the angle of the counterweight relative to the normal of the gate

$$M_d = (1/2)GL_g\sin\delta + eW\cos(\varepsilon+\delta). \tag{8.45}$$

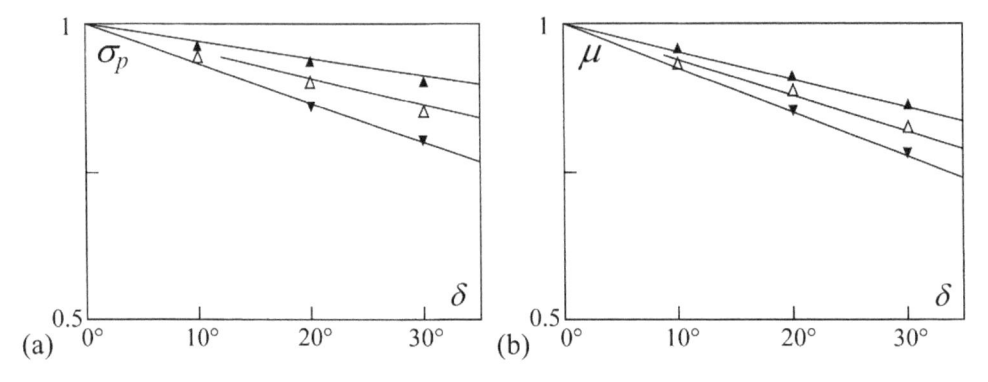

Figure 8.38 Hinged Flap Gate (a) pressure force ratio $\sigma_p(\delta)$, (b) moment ratio $\mu(\delta)$ for $D = (\blacktriangle)$ 1, $(\triangle)$ 1.56, $(\blacktriangledown)$ 2.38 (Raemy and Hager, 1998)

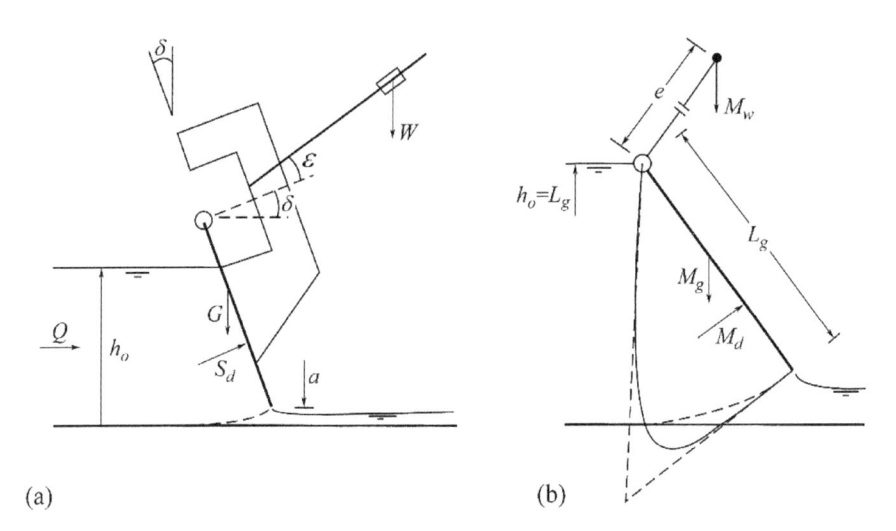

Figure 8.39 (a) Forces, (b) moments governing Hinged Flap Gate flow (Raemy and Hager, 1998)

The static moment of the gate is $M_s = (1/2)\rho gbL_g^2(2L_g/3)\cos\delta$. With $M_d = \mu M_s$ and for $D = 1$ follows from Eqs. (8.44) and (8.45)

$$(1/3)\rho gbL_g^3 - eW\cos\varepsilon = [(1/2)GL_g - eW\sin\varepsilon + (1/12)\rho gbL_g^3]\tan\delta. \tag{8.46}$$

This relation has to be satisfied for any angle δ, so that two conditions result. First, for the restoring moment

$$eW/(\rho gbL_g^3) = (3\cos\varepsilon)^{-1}. \tag{8.47}$$

Second, for the angle ε,

$$\sin\varepsilon = (eW)^{-1}[(1/2)GL_g + (1/12)\rho gbL_g^3]. \tag{8.48}$$

Eliminating the term (eW) from these two conditions gives

$$\tan\varepsilon = \frac{(3/2)G}{\rho gbL_g^2} + \frac{1}{4}. \tag{8.49}$$

This yields a minimum angle of $\varepsilon = 14°$ ($\tan\varepsilon = 1/4$) for a gate of weight much smaller than the corresponding water weight ρgbL_g^2. The leading term in Eq. (8.49) is (1/4).

With the trigonometric relation $(\cos\varepsilon)^{-1} = (1 + \tan^2\varepsilon)^{1/2}$, the dimensionless restoring moment becomes (Raemy and Hager, 1998)

$$\frac{eW}{\rho gbL_g^3} \simeq 0.344\left[1 + \frac{(1+3)G}{\rho gbL_g^2}\right]. \tag{8.50}$$

For $G/(\rho g b L_g^2) \ll 1$, the restoring moment required from the counterweight is $eW/(\rho g b L_g^3) = (e/L_g)[W/(\rho g b L_g^2)] = 0.35$ implying that the product of length ratio (e/L_g) and the weight ratio should be close to 1/3. In applications, typically one has $e/L_g = 3$, so that $W = (1/10)(\rho g b L_g^2) \cong 1/10$, i.e. the counterweight is a small fraction of the reference water weight, of the order of 1%. Therefore, the effect of non-hydrostatic pressure distribution on the gate is essential.

As to the performance of the Hinged Flap Gate, experiments were conducted with a gate of weight $G = 46$ N, a supporting construction of weight 27.7 N, whereas that of the counterbalance was 106.6 N, i.e. $W = 134.3$ N. The distance between the pivot and gravity center was $e = 0.860$ m. The approach flow depth was adjusted to $h_o = 0.415$ m, and the counterweight had an angle of $\varepsilon = 18.5°$ for $Q(\delta = 0) = 0$. There was practically no effect on the approach flow depth when increasing discharge Q, with extreme level variations of only ±1% for $0 < \delta < 45°$. The gate was manually also moved out of the equilibrium position to study its stability. After either increasing or decreasing the gate angle δ by say ±20°, it immediately returned to the equilibrium position by an aperiodic movement. The gate stability increases with higher discharges, with longer oscillations occurring only for $\delta < 10°$. An addition of arrest devices to inhibit gate rotation for $\delta < 0°$ or $\delta > 45°$ is thus required, limiting the work domain to $0 < \delta < 45°$.

The discharge capacity of the Hinged Flap Gate results from Eq. (8.42) with $F = ab$ and $C_d = 0.64$ (±2%). For $h_o = L_g$ follows

$$Q = 0.64bh_o(2gh_o)^{1/2}(1 - \cos\delta).\tag{8.51}$$

For a maximum gate angle of 45°, $Q/(gb^2L_g^3)^{1/2} = 0.27$ as relative discharge capacity, varying linearly with b and over-proportionally with L_g. For small values of δ this simplifies to $Q/(gb^2L_g^3)^{1/2} = 0.45\delta^2$, i.e. quadratically with δ. The hydraulic (subscript H) freeboard f_H required for the completely blocked Hinged Flap Gate ($\delta = 0$) follows with C_W as overflow discharge coefficient and h_W as the head on the 'weir' measured from the channel bottom as (Figure 8.35)

$$Q_W = C_W b[2g(h_W - L_g)^3]^{1/2}.\tag{8.52}$$

Combining Eqs. (8.51) and (8.52) results in

$$\frac{h_W}{L_g} \simeq 1 + \left[(1 - \cos\delta)\frac{C_d}{C_W}\right]^{2/3}.\tag{8.53}$$

For the maximum angle $\delta = 45°$ follows $h_W/L_g = 1.6$, so that the channel height must be at least 60% higher than the gate length L_g for free surface flow to occur, i.e. $f_H > (h_W - L_g)$. For the typical angle $\delta = 30°$, this height reduces to $h_W = 1.35L_g$.

Figure 8.40 shows side views of the Hinged Flap Gate for $Q = 10, 50, 100$ l/s, corresponding to $\delta = 13°, 31°, 45°$. For $\delta = 13°$ the relative gate opening amounts to only $A = 1 - \cos\delta = 0.026$, so that $C_c a = 0.007$ m as contracted flow depth. The flow features are typical for $\delta = 31°$ involving well-developed contracted flow and a stable gate position, whereas the effect of approach flow velocity is significant for $\delta = 45°$ and instability effects start to appear so that this value was selected as the upper limit.

Figure 8.41 relates to $\delta = 31°$ including an approach flow view, side view of the contracted flow, upstream view with the typical capillary surface waves, and a downstream view. No

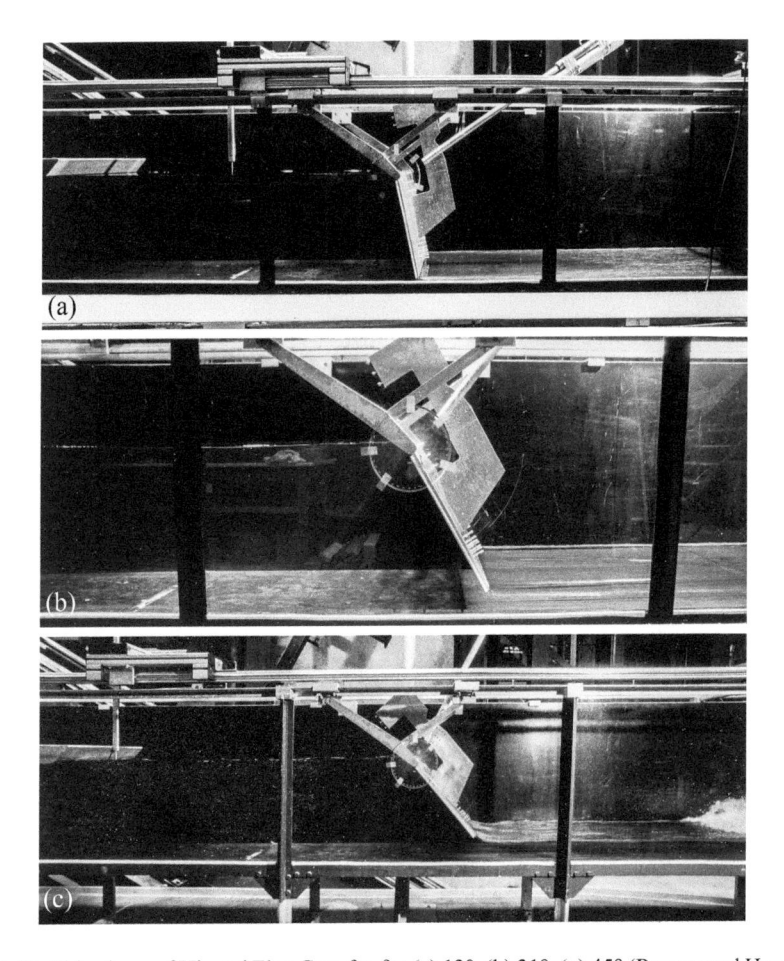

Figure 8.40 Side views of Hinged Flap Gate for δ = (a) 13°, (b) 31°, (c) 45° (Raemy and Hager, 1998)

Figure 8.41 Hinged Flap Gate with δ = 31° (a), tailwater view (b) side view, (c) approach flow view, (d) transition from pressurized to free-surface flow (Raemy and Hager, 1998)

shock waves are noted in the tailwater due to gate presence. Summarizing, the Hinged Flap Gate is able to control an upstream head by purely hydraulic means, without floats or pressure differentials. Its features are a counterweight with an adjusted lever-arm length, the angle ε relative to the gate plane, and a well-defined weight allowing for the simple, effective, and economic control of relatively small discharges.

8.5.4 Hydraulics of standard vertical gate

Introduction

The standard sluice gate relates to a planar vertical gate with a sharp-crested opening placed in a smooth and prismatic, horizontal, rectangular channel (Figure 8.42a). Whereas the flow features away from the gate section are detailed in 8.5.2, this section deals with the hydraulic features of gate flow, thereby particularly addressing the effects of streamline curvature and slope. Experimental results on its basic flow features provide Rajaratnam and Humphries (1982), Othsu and Yasuda (1994), and Roth and Hager (1999). The flow is made up by the approach flow portion up to the gate section involving an internal jet along the channel bottom with a vortex flow zone above it, and the free jet portion in the tailwater (Figure 8.42b). Computational simulations based on the ideal fluid flow theory are also available (Fangmeier and Strelkoff, 1968; Montes, 1997). In spite of these results originating from physical and 2D mathematical modeling, the knowledge of standard gate flow is considered incomplete for one-dimensional (1D) modeling. Free gate flow is often treated assuming conservation of energy between the approach flow and the tailwater portions (Rouse, 1950; Montes, 1998). Flow features of engineering interest including the free-surface profile and the bottom pressure distribution are thereby overlooked, so that they remain unexplained in the context of 1D hydraulics.

Standard sluice gate flow includes hydraulic features incompatible with parallel-streamlined flow considerations due to the non-hydrostatic pressure distribution. The Boussinesq-type approximation introduces the vertical acceleration effect in the 1D flow equations (Boussinesq, 1877; Matthew, 1991; Bose and Dey, 2007, 2009; Castro-Orgaz and Hager, 2009), thereby overcoming the standard 1D limitation. Boussinesq's theory was considered by Serre (1953) for the free jet portion yet without checking the theoretical results with neither

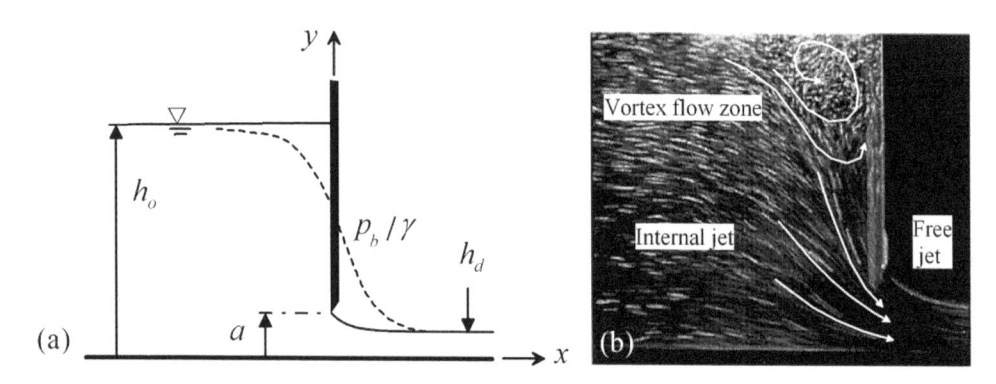

Figure 8.42 Standard sluice gate flow (a) definition sketch with (---) bottom pressure head distribution, (b) typical internal flow structure of approach flow (Photo VAW)

experimental nor 2D computations. Benjamin (1956) in vain fitted the 2D free-surface jet profile to a standard solitary wave profile. Both Serre (1953) and Benjamin (1956) overlooked the effects of the upstream free-surface profile and the bottom pressure features. The contraction coefficient C_c was assumed to be known, rather than determined by theory.

The purpose of the research of Castro-Orgaz and Hager (2014) was to fill in this gap by studying the free surface and bottom pressure features using the 1D approach, including the contraction and discharge coefficients, and the pressure distributions both on the gate and on the bottom. A Boussinesq-type approximation was used by neglecting frictional effects because the equations of inviscid flow apply for this flow type (Montes, 1997; Roth and Hager, 1999). This theoretical approach provides a 1D estimation of all features of engineering interest of standard free gate flow.

Free-surface profile

The prediction of the free jet portion in the tailwater of a standard sluice gate was considered using the full 2D potential flow model. A solution of the Laplace equation was sought (Fangmeier and Strelkoff, 1968, or Montes, 1997), the latter of which reviewed the relevant 2D numerical solutions from 1860. In contrast, the computation of free jets using a 1D model received almost no attention. The first attempt was due to Fawer (1937) based on an extended Boussinesq-type energy equation. His model was only valid for inclined gates, serving to estimate the contraction coefficient. Serre (1953) proposed a theoretical model for the free jet surface profile, yet again without verifying the results with experimental data or 2D computations. Using a 1D model, Benjamin (1956) tried to fit a standard solitary wave to the 2D free-surface profile from Southwell and Vaisey (1946). The fit was stated to be valid only far from the gate lip. A hybrid solution method was then proposed using a 2D solution near the gate section. The 1D modeling of free jets is thus reconsidered using the Boussinesq-type equations (Serre, 1953; Benjamin and Lighthill, 1954; Hager and Hutter, 1984)

$$E = h + \frac{q^2}{2gh^2}\left(1 + \frac{2hh_{xx} - h_x^2}{3}\right), \quad S = \frac{h^2}{2} + \frac{q^2}{gh}\left(1 + \frac{hh_{xx} - h_x^2}{3}\right). \tag{8.54}$$

Here h is flow depth, E specific energy head, S specific momentum, q unit discharge and subscripts indicate ordinary differentiation with respect to the horizontal coordinate x. The first Eq. (8.54) is also expressed as

$$\frac{q^2}{6g}\frac{d}{dh}\left(\frac{h_x^2}{h}\right) = E - h - \frac{q^2}{2gh^2}. \tag{8.55}$$

It straightforwardly integrates to (Serre, 1953)

$$\frac{q^2}{6g}h_x^2 = Eh^2 - \frac{h^3}{2} + \frac{q^2}{2g} - Sh. \tag{8.56}$$

For inviscid flow the jet invariants are $E = h_d + q^2/(2gh_d^2)$ and $S = h_d^2/2 + q^2/(gh_d)$. Imposing the asymptotic free jet condition $h_x(h \rightarrow h_d) \rightarrow 0$ gives with $h_d = C_c a$, $\mathsf{F}_d = q/(gh_d^3)^{1/2}$ as tailwater Froude number, a as gate opening, and C_c as contraction coefficient

$$h_x^2 = \frac{3}{F_d^2}\left(\frac{h}{h_d}-1\right)^2\left(F_d^2-\frac{h}{h_d}\right).$$

(8.57)

Its general integral is with ω as a constant (Serre, 1953)

$$\frac{h}{h_d}=1+4(F_d^2-1)\frac{\omega\exp\chi}{(1+\omega\exp\chi)^2}, \qquad \chi=\frac{(3F_d^2-3)^{1/2}}{F_d}\frac{x}{h_d}, \qquad F_d^2=2\left(\frac{E}{C_c a}-1\right).$$

(8.58)

Imposing the boundary condition $h(x = 0) = a$ at the gate section results in the quadratic equation

$$1+4(F_d^2-1)\frac{\omega}{(1+\omega)^2}-\frac{1}{C_c}=0.$$

(8.59)

For $\omega = 1$, Eq. (8.59) reduces to the standard solitary wave profile (Benjamin, 1956)

$$\frac{h}{h_d}=1+4(F_d^2-1)\frac{\exp\chi}{(1+\exp\chi)^2}=1+(F_d^2-1)\,\mathrm{sech}^2\left(\frac{\chi}{2}\right).$$

(8.60)

Equation (8.58) is compared in Figure 8.43a with 2D computations of Montes (1997) using $C_c = 0.61$, the mean value supported by 2D computations. The agreement of Eq. (8.58)$_1$ with 2D results is good for $E/a \geq 2$, except for slight deviations near $x = 0$ for $E/a = 2$ and 3.

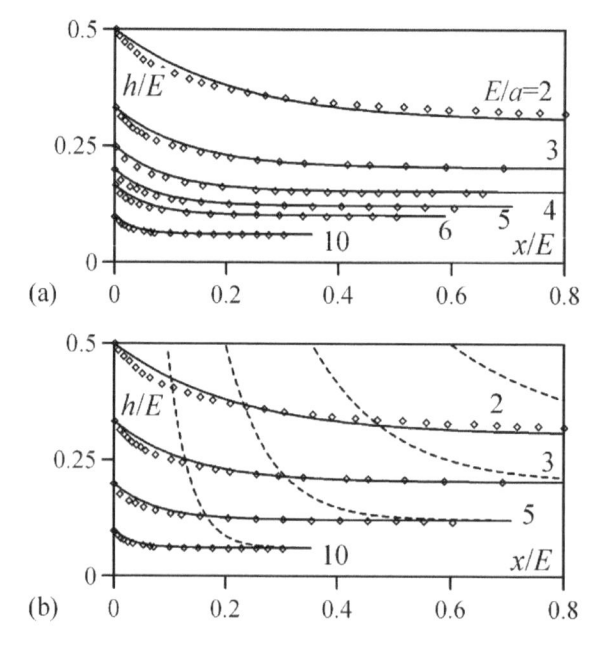

Figure 8.43 Free jet surface profile $h/E(x/E)$ (a) comparison of (—) Boussinesq-type solution Eq. (8.58) with ($\Diamond$) 2D potential flow computations (Montes, 1997), (b) idem including (- - -) standard solitary wave (Benjamin, 1956)

A limitation of 1D models as compared with the full 2D solution relates to the exact boundary conditions at $x = 0$. From Eq. (8.57), $h_x(0) \cong -1$, whereas the surface profile is vertical at the gate lip from 2D results. Given this local effect, its importance is considered small for larger x.

Equation (8.60) is compared in Figure 8.43b for selected tests with Eq. (8.58)$_1$ and the 2D results. Note that the agreement is generally poor and limited to the tailwater solitary wave portion for large E/a. This explains the failure of Benjamin's approach, given that his solitary wave does not account for the correct boundary conditions at $x = 0$, whereas Eq. (8.58)$_1$ correctly describes the generalized solitary wave-like profile.

An additional limitation of Serre (1953) and Benjamin (1956) is that their values C_c were assumed as model input, given that use of the invariant $S = (C_c a)^2/2 + q^2/(gC_c a)$ was made for a known value of E/a. However, numerical computations first use Eq. (8.54), with C_c then found as part of the solution. The discharge q was first estimated using $C_c = 0.61$ as indicated by 2D potential flow computations. Equation (8.54)$_1$ for $h(x)$ was solved using the fourth-order Runge-Kutta method, subject to the boundary conditions $h(0) = a$ and $h_x(x \to \infty) \to 0$. The value of $h_x(0)$ must be iteratively determined until constant jet height results in the tailwater, with C_c determined numerically. The results for C_c are plotted in Figure 8.44a and compared with these of Montes (1997). The agreement is good, supporting the 1D model as approximate solution of a 2D potential flow problem. The experimental data of Roth and Hager (1999) are also included, indicating $C_c \cong 0.59$, i.e., 3% lower than

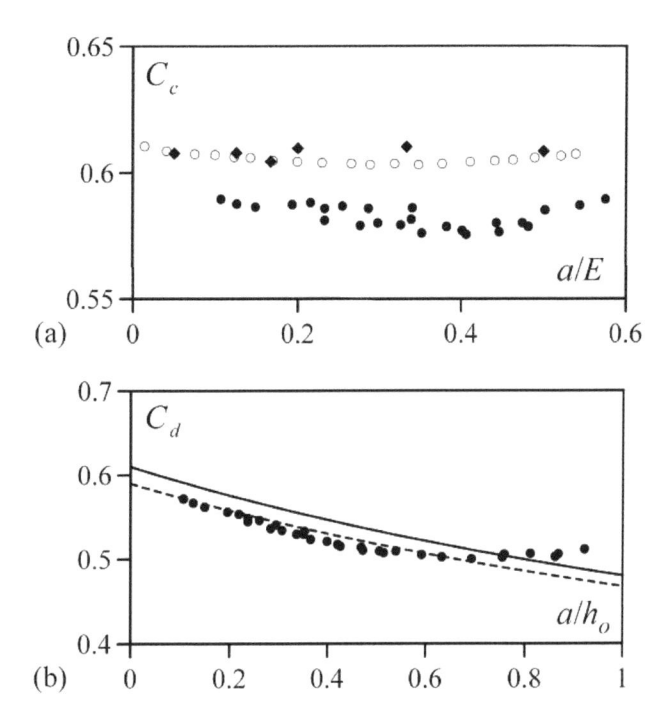

Figure 8.44 Coefficient of (a) contraction $C_c(a/E)$, with (○) 2D computations (Montes, 1997), (◆) numerical solution of Eq. (8.54), (b) discharge $C_d(a/h_o)$ for $C_c = $ (—) 0.61 and (---) 0.59; (●) experimental data (Roth and Hager, 1999)

0.61 from the ideal fluid flow theory. The discharge coefficient C_d is given by e.g. Montes (1998), or Roth and Hager (1999), as $C_d = q/[a(2gh_o)^{1/2}] = C_c(1 + C_c a/h_o)^{-1/2}$, and compared in Figure 8.44b using $C_c = 0.61$ with the experimental data of Roth and Hager (1999) for $a > 50$ mm. The agreement is good except for $a/h_o > 0.70$. An additional prediction is included for $C_c = 0.59$ as indicated by experiments, leading to closer agreement. Deviations between both predictions are less than 3% for $a/h_o > 0.3$, so that $C_c = 0.61$ is a good estimate for engineering purposes.

All previous 1D computations assumed a horizontal upstream free surface. Benjamin and Lighthill (1954) and Hager and Hutter (1984) demonstrated that the only possible steady-state water wave in a subcritical flow ($F < 1$) conserving E, S, and q is the cnoidal wave. If the approach flow to the gate is assumed horizontal, the water surface profile approaching the sluice gate and the stagnation point is violated, so that the approach flow to the gate is physically better represented by a cnoidal-wave train. Figure 8.45 compares the 2D surface profiles of Montes (1997) with the cnoidal-wave solution of Eq. (8.56), thereby imposing the flow depth h_o as boundary condition at $x = -5a$, and $h_x = -0.05$ as arbitrary value adopted there to deviate the flow from uniformity. Results indicate that the upstream approach flow profile indeed follows the cnoidal-wave solution of small wave amplitude and large wave length.

The transition from supercritical ($F > 1$) to subcritical ($F < 1$) flows close to the critical depth appears as an undular hydraulic jump. This transition was characterized as solitary wave connected with a cnoidal wave (Iwasa, 1955; Marchi, 1963; Hager and Hutter, 1984). The solitary wave provides transitional flow from the supercritical approach flow in the form of a wave emerging in subcritical flow. A local loss of energy permits then to connect this solitary wave portion with a cnoidal-wave train in the subcritical reach. Standard sluice gate flow provokes a transition from sub- ($F < 1$) to supercritical ($F > 1$) flows, with the upstream portion as a cnoidal wave. A drop in momentum S provoked by the gate presence permits to emerge below the gate a solitary wave, asymptotic to the tailwater flow. Thus, the transitional flow at a standard sluice gate follows Benjamin and Lighthill's (1954) theory, providing with the undular jump a complete image of flow transitions in straight-bottomed channels.

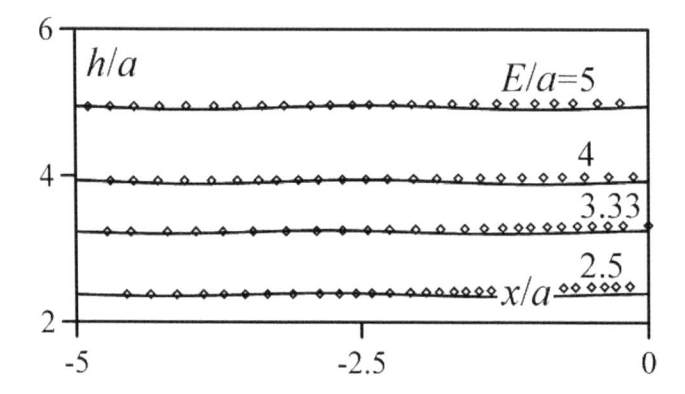

Figure 8.45 Approach flow surface profile $h/a(x/a)$ with (—) 1D, ($\diamond$) 2D computation (Montes, 1997)

Gate section

The pressure at the gate section is non-hydrostatic (Montes, 1997). From the ideal fluid flow theory, the horizontal velocity is zero at the vertical gate plane, so that the vertical velocity is large. A Boussinesq-type development using the Cartesian system of reference, taking streamline curvatures and slopes with reference to the *x*-coordinate is inadequate, because of $h_x \rightarrow \infty$ at the gate, so that a special solution must be developed for the gate plane. The exact 2D boundary condition at the gate lip is with v_e as the corresponding vertical velocity (Castro-Orgaz and Hager, 2014)

$$\frac{v_e^2}{2g} = E - a. \tag{8.61}$$

Following the standard Boussinesq equations, the vertical velocity is assumed to vary linearly along the gate plane as

$$v = v_e \left(1 - \frac{y-a}{E-a} \right). \tag{8.62}$$

For strongly vertical flows, Fawer's (1937) theory applies, thereby generalizing Eq. (8.62) with m_v as Fawer type exponent and $\eta = (y-a)/(E-a)$ to

$$v = v_e \left(1 - \eta^{m_v} \right). \tag{8.63}$$

The pressure distribution on the gate is then given by the Bernoulli equation as

$$\frac{P}{\gamma(E-a)} = 2\eta^{m_v} - \eta^{2m_v} - \eta . \tag{8.64}$$

The pressure force F_p at the gate section follows by integrating Eq. (8.64) as

$$\frac{F_p}{(E-a)^2} = \int_0^1 \frac{P}{\gamma(E-a)} d\eta = \frac{2}{m_v + 1} - \frac{1}{2m_v + 1} - \frac{1}{2}. \tag{8.65}$$

This constitutes the first relationship $F_p(m_v)$. A second identity to be satisfied by F_p originates from the momentum balance at the boundary sections as

$$F_P = \left(\frac{h_o^2}{2} + \frac{q^2}{gh_o} \right) - \left(\frac{C_c^2 a^2}{2} + \frac{q^2}{gC_c a} \right). \tag{8.66}$$

In addition, for given *E*, the discharge *q* follows from conservation of energy

$$q = C_c a \left[2g(E - C_c a) \right]^{1/2}. \tag{8.67}$$

The unit discharge *q* was computed for given values of *E* and *a* from Eq. (8.67) with $C_c = 0.61$. The upstream flow depth h_o was computed using *E* and *q*, and the force F_p from Eq. (8.66).

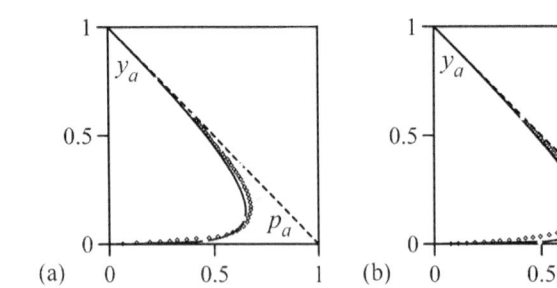
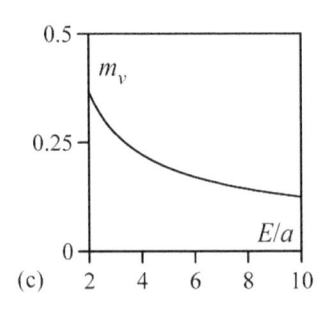

Figure 8.46 Pressure distribution $y_a = (y-a)/(E-a)$ versus $p_a = p/\gamma(E-a)$ on vertical gate for $E/a =$ (a) 2.5 ($m_v = 0.31$) with (—) Eq. (8.64), ($\diamond$) 2D computation (Montes, 1997), ($\circ$) experimental data (Finnie and Jeppson, 1991), (b) 3.33 ($m_v = 0.25$), with (—) Eq. (8.64), ($\diamond$) 2D computation (Montes, 1997), ($\circ$) 2D data (Cheng *et al.*, 1981), (c) $m_v = m_v(E/a)$ (Adapted from Castro-Orgaz and Hager, 2014)

With given F_p, parameter m_v results from Eq. (8.65), and the pressure distribution follows from Eq. (8.64). A test case for $E/a = 2.5$ is shown in Figure 8.46a, where the Boussinesq-type pressure distribution agrees well with 2D data (Montes, 1997) and experiments (Finnie and Jeppson, 1991). The 1D results agree well with 2D data of Montes (1997) and Cheng *et al.* (1981) for $E/a = 3.33$ (Figure 8.46b). The variation of m_v with E/a is shown in Figure 8.46c (Castro-Orgaz and Hager, 2014).

Bottom pressure profile

A gate provokes a drop in specific momentum S permitting the approaching cnoidal wave to pass below the gate, thereby transforming into a solitary wave, with E kept constant. The flow depth at the gate is discontinuous, therefore, whereas the bottom (subscript b) pressure profile $p_b(x)$ remains continuous, approaching asymptotically the up- and downstream flow depths (Figure 8.42a). The bottom pressure associated with Eq. (8.56) is (Matthew, 1991; Castro-Orgaz and Hager, 2009)

$$\frac{p_b}{\gamma} = h + \frac{q^2}{gh^2}\left(\frac{hh_{xx} - h_x^2}{2}\right). \tag{8.68}$$

From Eq. (8.58)$_1$, the derivatives h_x and h_{xx} and thus p_b are determined. The results are shown in Figure 8.47a for $E/a = 5$, and compared with 2D results. Note that the agreement is only good for $x/a > 1$ because of the limitation of Eq. (8.68) to weakly curved flow. Let K be a curvature distribution parameter, with Eq. (8.68) based on $K = 1$. Both the specific energy E and bottom pressure p_b are expressed versus K as (Fawer, 1937)

$$E = h + \frac{q^2}{2gh^2}\left(1 + \frac{2hh_{xx}}{K+2} - \frac{h_x^2}{3}\right), \quad \frac{p_b}{\gamma} = h + \frac{q^2}{gh^2}\left(\frac{hh_{xx}}{K+1} - \frac{h_x^2}{2}\right). \tag{8.69}$$

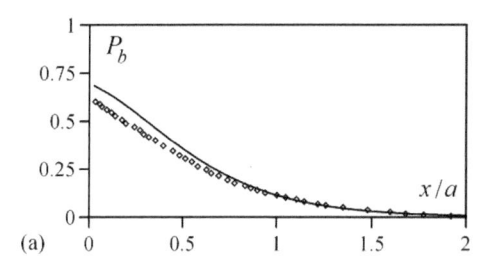 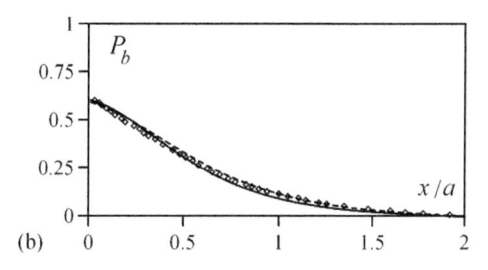

Figure 8.47 Normalized free jet bottom pressure profile $P_b = (p_b/\gamma a - C_c)/(E/a - C_c)$ versus x/a, comparison of Boussinesq-type solution with 2D potential flow results for $K =$ (a) 1, (b) 1.4 with 1D computation for $E/a =$ (—) 5, (- - -) 2, (◇) 2D computation (Montes, 1997)

Equation $(8.69)_1$ for h was solved subject to $h(0) = a$ and $h_x(x \to +\infty) \to 0$ (Castro-Orgaz and Hager, 2014). A value of $h_x(0)$ is first assumed, so that the system of equations is solved numerically and the tailwater conditions at $x/a = +6$ are revised. If $h_x(6)$ was not close to zero then computations restarted with a new value of $h_x(0)$. Once $h(x)$ determined, the computation of $p_b(x)$ is straightforward. Results are shown in Figure 8.47b for $K = 1.4$, indicating an excellent reproduction of the bottom pressure features. The numerical free-surface profile remains almost unaffected as compared with the results of Eq. (8.58).

The cnoidal-wave profile of the approach flow produces a slight deviation of bottom pressure from the free-surface profile, indicating that the quadratic and linear profiles for the horizontal and vertical velocity components (u, v) of potential flow (Matthew, 1991) do not reproduce the internal jet-like features. Keutner (1935) stated that non-hydrostatic bottom pressure of the upstream flow is associated with an internal jet originating at the upstream approach flow section transforming at the gate flow into a free jet. To explain the bottom pressure features of the approach flow, Keutner's method is developed. The flow is divided in two layers, an internal jet transporting discharge q and a recirculating fluid above it. With p_s as pressure at the jet surface and s as jet surface profile, the vertical Euler equation yields the approach flow pressure distribution p as

$$\frac{p}{\gamma} = \frac{p_s}{\gamma} + s - y + \int_y^s \frac{u^2}{g} \frac{\partial}{\partial x}\left(\frac{v}{u}\right) dy. \tag{8.70}$$

Approximating u by the depth-averaged velocity q/s, and using the resulting vertical velocity profile obtained with the 2D continuity equation, S and p_b are given by

$$S = \frac{(s + p_s/\gamma)^2}{2} + \frac{q^2}{gs}\left(1 + \frac{ss_{xx} - s_x^2}{3}\right), \quad \frac{p_b}{\gamma} = \frac{p_s}{\gamma} + s + \frac{q^2}{gs^2}\left(\frac{ss_{xx} - s_x^2}{2}\right). \tag{8.71}$$

Equation (8.71) generalizes the advance of Valiani (1997) for hydrostatic submerged jets. Experiments indicate that the vertical velocity component is significant (Roth and Hager, 1999), so that this effect needs to be retained. In the single-layer Boussinesq model leading to Eq. (8.54), the development considers the non-hydrostatic pressure distribution and the non-uniform velocity profile (Matthew, 1991; Castro-Orgaz and Hager, 2014). The latter is considered in the mainstream layer $s = s(x)$ for the two-layer model leading to Eq. (8.71). However, recirculating flow is excluded in the roller layer $h(x)-s(x)$. The equations are further improved for a nonlinear curvature distribution, resulting as in Eq. (8.69) in (Castro-Orgaz and Hager, 2014)

$$S = \frac{(s + p_s/\gamma)^2}{2} + \frac{q^2}{gs}\left(1 + \frac{ss_{xx}}{K+2} - \frac{s_x^2}{3}\right), \quad \frac{p_b}{\gamma} = \frac{p_s}{\gamma} + s + \frac{q^2}{gs^2}\left(\frac{ss_{xx}}{K+1} - \frac{s_x^2}{3}\right). \qquad (8.72)$$

The sum $p_s/\gamma + s = h_p$ is the effective pressure head. Note that $p_s(x = 0) = 0$ despite the local water column $(E-a)$. Assuming that p_s equals the local flow depth is not reliable, given the significant vertical velocities close to the gate, and the corresponding non-hydrostatic pressure. Given the complex flow pattern as the flow approaches the gate, the simplest approximation is to assume h_p = constant for the internal jet. Its value will be determined using as boundary condition the bottom pressure at $x = 0$. Computations with Eq. (8.72)$_1$ used the fourth-order Runge-Kutta method with $s(0) = a$ and $K = 1.4$, as previously for the free jet portion. The value of S is taken as an invariant, determined from the known values of h_o and q. The value $s_x(0)$ was thereby iteratively adjusted until reaching the asymptotic condition $p_b/\gamma(x \to -1.5a) \to h_o$. Results for $E/a = 2.5$ are shown in Figure 8.48a. Note that the agreement of the 2D bottom pressure and the two-layer approximation of Keutner (1935) is excellent, justifying that the internal flow features close to the gate are provoked by an internal jet, and not by the free-surface cnoidal-wave-like configuration. The theoretical shape of the internal jet is shown in Figure 8.48b; note its similarity with the Benjamin-Cola cavity bubble (Hager, 1999).

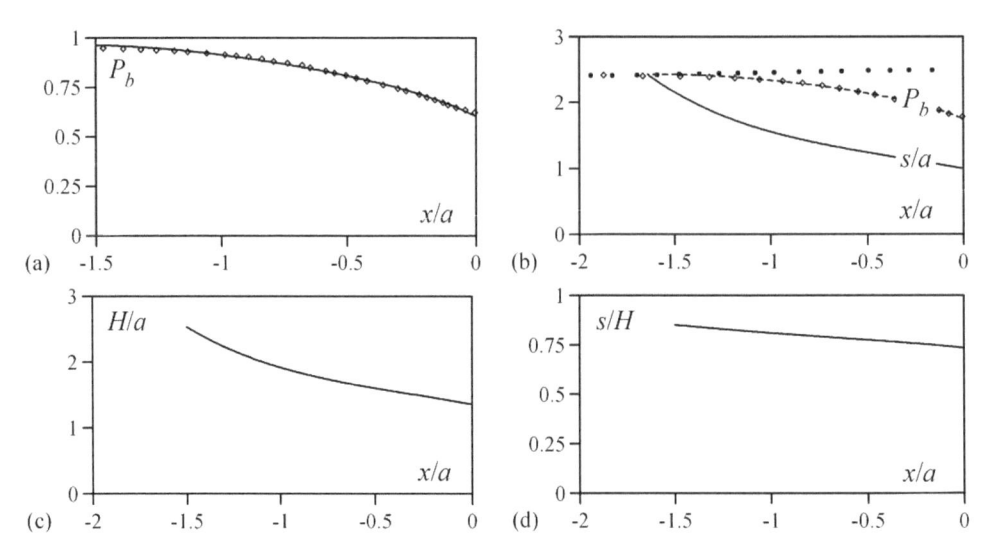

Figure 8.48 Approach flow portion (a) bottom pressure distribution $P_b = [(p_b/\gamma a - C_c)/(E/a - C_c)](x/a)$ with (—) 1D computation, (◇) 2D computation (Montes, 1997), (b) internal jet features (—) $s/a(x/a)$, (---) $P_b = p_b/\gamma a(x/a)$, (◇) 2D bottom pressure computation (Montes, 1997), (•) 2D free-surface computation (Montes, 1997), (c) pseudo-specific energy head $H/a(x/a)$, (d) relative jet thickness $s/H(x/a)$ (Castro-Orgaz and Hager, 2014)

The free-surface profile $h = h(x)$ is discontinuous at the gate section and does not explain the transition from sub- to supercritical flow across the critical flow section. In contrast, the bottom pressure profile $p_b = p_b(x)$ is continuous, with $p_b(0)/E = 0.63$ at the gate section, i.e. close to 2/3 as the ratio of depth to specific energy at critical flow for parallel-streamlined flow. For the latter $h = p_b/\gamma$ so that $p_b/\gamma E = 2/3$. The discontinuity in $h(x)$ provokes a drop in the depth profile that does not allow for $h/E = 2/3$, whereas the bottom pressure profile permits a continuous transitional flow. It is therefore relevant to investigate transitional flow in terms of p_b, which is related to critical flow by the internal jet profile $s(x)$. The latter controls the bottom features of the approach flow, and defines the pseudo-specific energy head H as

$$H = s + \frac{q^2}{2gs^2}\left(1 + \frac{2ss_{xx}}{K+2} - \frac{s_x^2}{3}\right). \tag{8.73}$$

The profile $H(x)$ was determined from the jet thickness $s(x)$ (Figure 8.48c). Note that the minimum of H occurs at $x = 0$. Therefore, pseudo-critical flow as section of minimum specific energy occurs at $x = 0$, representing a continuous bottom pressure profile as the flow transition. In addition, the ratio of jet thickness to pseudo-specific energy head s/H is plotted in Figure 8.48d. At $x = 0$, $s/H = 0.737$, i.e. close to the minimum specific energy head for streamlined critical flow (Montes, 1998). Note the discontinuity of s_x at $x = 0$ from the two-layer and single-layer models used for modeling the upstream and free jet portions, respectively. The function H has thus not a minimum in the sense of the extreme $\partial H/\partial s = 0$ of a smooth function $H = H(s)$. Computations indicate that the two-layer Boussinesq model predicts the smallest value of H at the gate.

Effect of tailwater submergence

As the tailwater level increases, the jet becomes submerged, resulting in a submerged hydraulic jump (Henry, 1950; Montes, 1998). The submergence effect on the drowned jet is investigated by applying Eq. $(8.72)_1$ to the data of Henry (1950). With h_d as tailwater depth, the momentum is conserved along a submerged jump (Henry, 1950; Montes, 1998). Assuming $K = 1$, Eq. $(8.72)_1$ reads (Castro-Orgaz and Hager, 2014)

$$S = \frac{h^2}{2} + \frac{q^2}{gs}\left(1 + \frac{ss_{xx} - s_x^2}{3}\right) = \frac{h_d^2}{2} + \frac{q^2}{gh_d}. \tag{8.74}$$

The free-surface profile $h(x)$ must be prescribed to compute the profile $s(x)$ of the submerged jet. The discharge q was estimated from conservation of energy between the upstream section and the vena contracta. Integration of Eq. (8.74) with $s(0) = a$ for $E/a = 13$ and $h_d/a = 7.9$ results in Figure 8.49a. The free-surface profile was assumed to vary linearly, starting at the gate section with $h/a = 7.14$. The theoretical jet profile agrees well with the data of Henry (1950). The theoretical Eq. (8.74) predicts the rapid jet contraction near the gate at $x/a \approx 1$ followed by a long expansion portion toward the roller end. A less submerged jump of $E/a = 7.95$ and $h_d/a = 4$ is considered in Figure 8.49b. The theoretical prediction fairly agrees with observations. Deviations are attributed to the assumed linear free-surface profile. These computations support the extended Boussinesq equations for submerged jets.

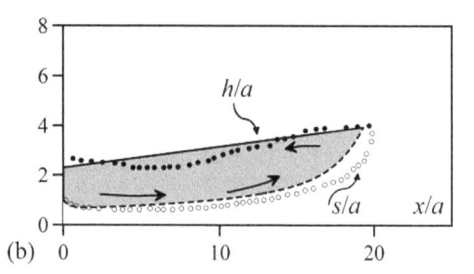

Figure 8.49 Submerged hydraulic jump in tailwater gate portion for $(E/a, h_t/a)$ = (a) (13, 7.9), (b) (7.95, 4.0), with (—) assumed free-surface profile $h/a(x/a)$, (---) computed submerged jet $s/a(x/a)$, (●) measured free-surface profile (Henry, 1950), (○) measured jet profile (Henry, 1950) (grey area: roller)

In summary, Castro-Orgaz and Hager (2014) developed a 1D approach based on the Boussinesq-type approximation, limited to 2D potential flow, thereby ignoring e.g. vortices. The study of the free-surface profile indicates that the upstream portion is a cnoidal wave, undergoing a momentum drop provoked by the gate presence, thereby emerging below the gate opening similar to a solitary wave of equal specific energy head. This explains the free-surface behavior in the context of the steady water wave theory. The bottom pressure profile is a continuous gate flow feature indicating that the curvature distribution exhibits a non-linear behavior. Both the contraction and discharge coefficients determined from the model agree well with literature data. A novel Boussinesq-type approximation was developed at the gate plane, resulting in the pressure distribution of the gate. Good agreement was found between bottom pressure simulations on the jet portion and 2D data. A two-layer approach was developed for bottom pressure of the approach flow, in agreement again with 2D data. The numerical results of the internal jet flow were used to investigate critical flow, from where the pseudo-specific energy head of the jet has a minimum at the gate section. The two-layer model was also used to simulate the drowned jet features in submerged hydraulic jumps. These flows have not received much attention in the past, however, mainly as to stability aspects of the submerged jet flow.

8.6 Low-level outlet

8.6.1 Design principles

According to 2.1, a bottom outlet serves various purposes including:

- Controlled reservoir filling,
- Urgency reservoir drawdown,
- Sediment flushing, and
- Flood and residual discharge release.

Because the velocity V at the bottom outlet is nearly as large as given by the Torricelli formula $V = (2gH_o)^{1/2}$, with H_o as head on the outlet and g as the gravity acceleration, concerns

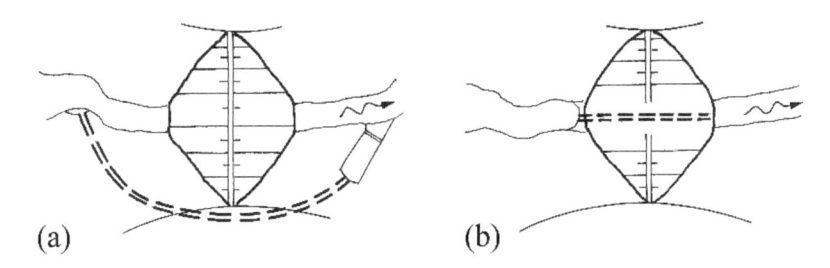

Figure 8.50 Arrangement of bottom outlet, (a) combination with diversion tunnel, (b) culvert type

with cavitation, abrasion, and aerated flow are particular hydraulic problems. Additional problems include:

- Sediment transport due to reservoir sedimentation,
- Gate blockage due to floating debris or sediment deposits,
- Gate vibration due to high-speed flow, and
- Choking of tunnel flow due to limited air supply.

Gate vibration is a particular problem of bottom outlets considered by Naudascher (1991), and Naudascher and Rockwell (1994). Problems with sediment are treated in Chapter 9 in relation to sediment bypass tunnels. Choking of the bottom outlet is considered in connection with air-water flow (8.6.7 and 8.6.8).

A low-level or bottom outlet has to be designed so that it may be operated under all conditions for which it was planned. Usually two outlet gates are provided, namely the (1) safety gate or maintenance gate either open or closed, and (2) service gate or regulating gate with a variable opening. According to Blind (1985), a bottom outlet should be provided at each dam of a certain size, particularly for emergency repair. A useful design involves the combination of diversion tunnel and bottom outlet (Figure 8.50a). For smaller embankment dams, particularly flood retention dams that are impounded only temporarily, or for concrete dams, a culvert type bottom outlet may be considered because of the simple design (Figure 8.50b).

Blind (1985) describes four arrangements for the bottom outlet (Figure 8.51):

(a) Diversion tunnel used as bottom outlet, with access through shaft.
(b) Bottom outlet culvert, not accessible except for minimum reservoir level, with two gates close to inlet to shorten pressurized outlet portion.
(c) Diversion tunnel as combined spillway and bottom outlet for morning glory spillways, with a pipe in upstream sealed tunnel portion to form bottom outlet.
(d) Gravity dam with bottom outlet, much shorter than for an embankment dam.

Since bottom outlets should not be integrated into high and permanently impounded earth- and rockfill dams due to the risks of differential settlements, the development of preferential seepage paths and internal erosion, they have to be combined with river diversion structures such as diversion tunnels. A typical example of the bottom outlet in the former diversion tunnel of Alicura Dam, Argentina, is shown in Figure 8.52a (Minor, 1988).

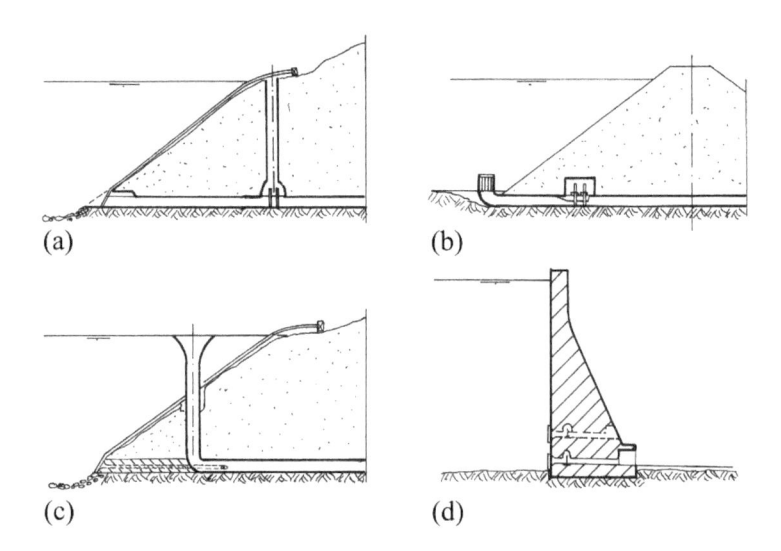

(a)   (b)

(c)   (d)

Figure 8.51 Basic arrangements of bottom outlet (adopted from Blind, 1985), for details see main text

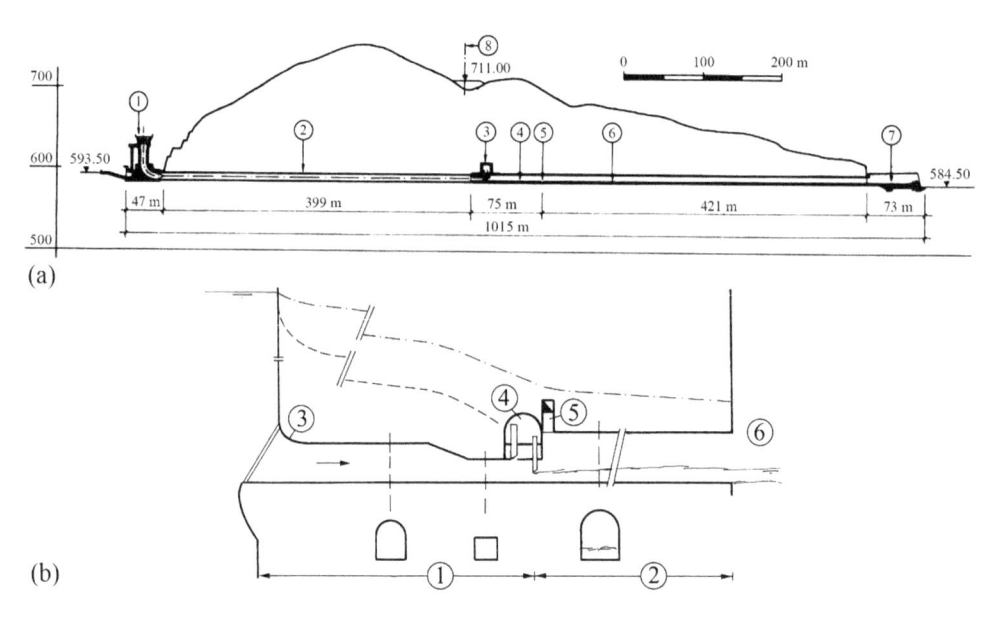

(a)

(b)

Figure 8.52 (a) Bottom outlet combined with diversion tunnel at Alicura Dam with ① intake and largely spaced concrete beams as trash-rack, ② upstream tunnel of diameter 9.0 m, ③ gate chamber downstream of dam grout curtain, ④ 75 m steel-lined portion of downstream tunnel, ⑤ slot bottom aerator, ⑥ concrete-lined downstream tunnel, ⑦ ski jump as outlet structure, ⑧ axis of dam crest. Details of gate chamber see Figure 8.55c (Minor, 1988). (b) Hydraulic scheme of bottom outlet with (- - -) pressure headline, (-.-) energy headline: ① Pressurized and ② free-surface flow portions, ③ tunnel inlet, ④ gate chamber, ⑤ air supply, ⑥ outlet

The technical requirements for a bottom outlet include (Giesecke, 1982):

- Smooth flow for completely opened outlet structure,
- Excellent performance for all discharges under partial opening,
- Effective energy dissipation at terminal outlet,
- Structure without leakage,
- Simple and immediate application,
- Easy access for maintenance and service,
- Economic and useful design, and
- Longevity.

A bottom outlet is not a structure for permanent use due to severe limitations in terms of cavitation, hydrodynamic forces, abrasion, and vibrations. However, it should allow for complete reservoir emptying, as previously mentioned.

Figure 8.52b sketches the hydraulic configuration of a bottom outlet. Note the pressurized flow upstream, and the free-surface flow downstream from the gate. At the tunnel inlet typically provided with a trash-rack, the water is accelerated to the tunnel velocity normally made up of a horseshoe profile. Shortly upstream from the gate chamber, the section contracts to the rectangular cross section to cause sufficient backpressure and to accommodate the gates. Downstream from the gate chamber, the tunnel is expanded mainly at the tunnel ceiling and sometimes also laterally. For short tunnels, no additional aeration is required. For long tunnels relative to the tunnel diameter, an air supply discharging behind the gate chamber provides sufficient air for free-surface flow under practically atmospheric pressure. The air supply conduit has to be designed so that the gate chamber is safe against submergence from the tunnel.

It is imperative that submergence of the bottom outlet is inhibited. The transition from pressurized to free-surface flow has to be located exactly downstream of the gate, therefore. This condition requires sufficient aeration and a tunnel roof high enough so that surging flow may not develop in the downstream tunnel portion. The discharge is then fully controlled with the gates.

The hydraulics of a bottom outlet across an embankment dam is sketched in Figure 8.53. The tunnel is of constant diameter, without flow aeration (Fig 8.53a). *Pressurized flow* thus

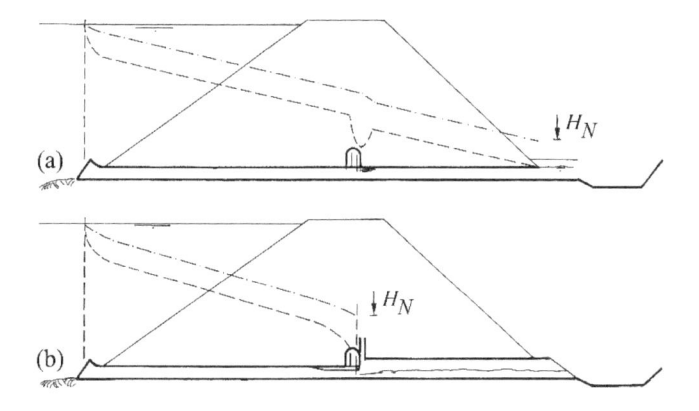

Figure 8.53 Bottom outlet across embankment dam (a) fully pressurized flow, (b) free-surface flow downstream of gate chamber

results for a large discharge. Therefore, the discharge depends mainly on the tunnel section F, but not on the gate section f. The latter adds only to the loss of head, involving an additional energy loss. In contrast, for *free-surface flow*, the gate section f controls the discharge linearly downstream from the gate chamber. In practice, this feature of bottom outlets has sometimes been overlooked, particularly for smaller structures, causing costly adaptions if rehabilitated due to safety reasons.

This Section is organized as follows: In 8.6.2 the various gate types and the corresponding design requirements are highlighted. Sub-Section 8.6.3 deals with gate vibrations, whereas 8.6.4 highlights the hydraulics of bottom outlets including the aeration features. Sub-Section 8.6.5 deals with cavitation aspects in general, and particular aspects in relation with bottom outlets.

8.6.2 Gate types

Figure 8.54 shows the most frequent types of outlet structures used (Task Committee, 1973; Giesecke, 1982; Sagar, 1995). These include (a) Wedge gate moved vertically, with complete gate sealing only at fully closed position. Instead of a wedge, two interconnected and displacing plates are also in use. A disadvantage of this gate type are the gate slots as for roller gates where complex hydraulic currents occur and sediment may enter. Flush conduits should be provided to ensure complete closure; (b) Slide gate as the common bottom outlet gate type. Wheel gates are considered for large sections under moderate heads known to be less prone to vibration; (c) Radial or sector gate does not use gate slots, and sealing is simple; the forces are concentrated to the gate trunnions, so that the abutments are highly stressed. This is one of the favorite arrangements for bottom outlets, especially in combination with sediment flushing, and of all applications of gate flow with large discharges in general. Further, for relatively low discharges and constant release as for example for irrigation outlets (d) Hollow jet valve comparable to the ring valve apply, but with an aeration device to break the compactness of the outflow jet; (e) Ring valve as an element that can be displaced axially of excellent hydraulic performance. The valve may be regarded as a pipe extension, which

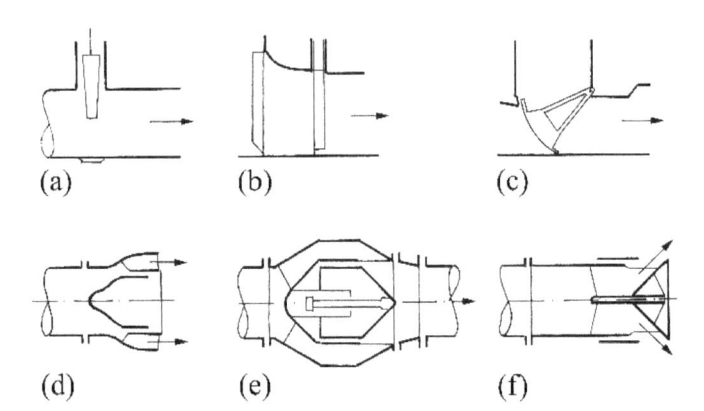

(a) (b) (c)

(d) (e) (f)

Figure 8.54 Types of valves and gates for outlet structures with (a) wedge gate, (b) slide gate, (c) sector gate, (d) hollow jet valve, (e) ring valve, (f) conic valve, details see text

can be displaced axially. Accordingly, a ring jet develops with small losses due to streamlining. It applies as either a regulating or a safety device for high pressures without leakage problems. The ring valve applies as terminal structure of a bottom outlet; (f) Conic valve as a simple and effective device with a displacing element. The hollow jet is dispersed into the atmosphere so that problems with cavitation are absent due to flow aeration. Also, vibrations are not a concern for all degrees of valve opening. Maintenance is simple because all parts are located outside from the bottom outlet. Often, a stilling basin is not required but the outlet should be covered due to massive spray action.

According to Sagar (1979), vibration, non-closure, cavitation, and abrasion are the main reasons for failure of high head gates. In the following, some features of modern gate experiences are reviewed.

Slide gates should have a lip sloping under $45°$ to inhibit vibration and to reduce the downpull forces. The offsets due to the gate slot should be minimized. Depending on the velocity and the sediment concentration, a bottom outlet should be lined. The tolerances should be closely checked. All parts should not only be carefully welded to avoid distortion, but often need stress-relieving prior to machining. Slide gates are suitable for heads up to 200 m, normally in a tandem arrangement. They are typically arranged at the downstream outlet end or within the conduit up to sizes of 10 m². The gate area should be slightly smaller than the approach flow conduit to ensure positive water pressure if the gate is fully opened. These gates are unsuitable for outlets where self-closing is required for quick shut-off in an emergency.

Fixed wheel gates are suitable as self-closing emergency gates. The slots are slightly wider than of slide gates rendering concerns with hydraulic disturbances. They apply also as regulating gates for heads of less than 100 m. To overcome the seal, the wheel and the guide frictional forces, the submerged gate weight should be at least 30% larger. The bottom shape, the gate slot, and the air vents should carefully be designed to avoid vibration. According to Sagar (1979), a vertical lip should be used extended 0.5 times the depth of the bottom horizontal beam. The downstream corners of the gate slots should be offset relative to the upstream corners. Self-lubricating bronze brushings are satisfactory for the wheel assemblies, and the grease has to be compatible with the bearing. Rubber seals with Teflon-coated contact surfaces are still a popular design to ensure water tightness.

Radial gates have no gate slots and thus are valuable, provided the top sealing is designed as to avoid undesirable water jets during partial gate operation. A standard seal ensures water tightness for the fully closed gate, and the anti-jet seal fixed to the frame gives a watertight contact at all gate positions. If the gates are operated with a hydraulic hoist, the cylinder must be properly hinged to ensure smooth gate operation without undue stresses in the stems and lifting mechanism. The gate trunnion has to be protected against corrosion and debris.

Based on failures of high-head gates, Sagar (1979) recommends:

- Bonnet covers be bolted down to bonnets or gate frames to distribute the water load on the concrete surroundings, and direct load transfer is avoided,
- Adequate aeration be ensured downstream of the gate to inhibit problems with cavitation and vibration. The inlets for air vents are located so as to provide an air supply uninterrupted by tailwater, waves or debris,
- Open gate shafts never be submerged due to significant vertical flow, and
- Top and side seals be effective under all positions of regulating gate.

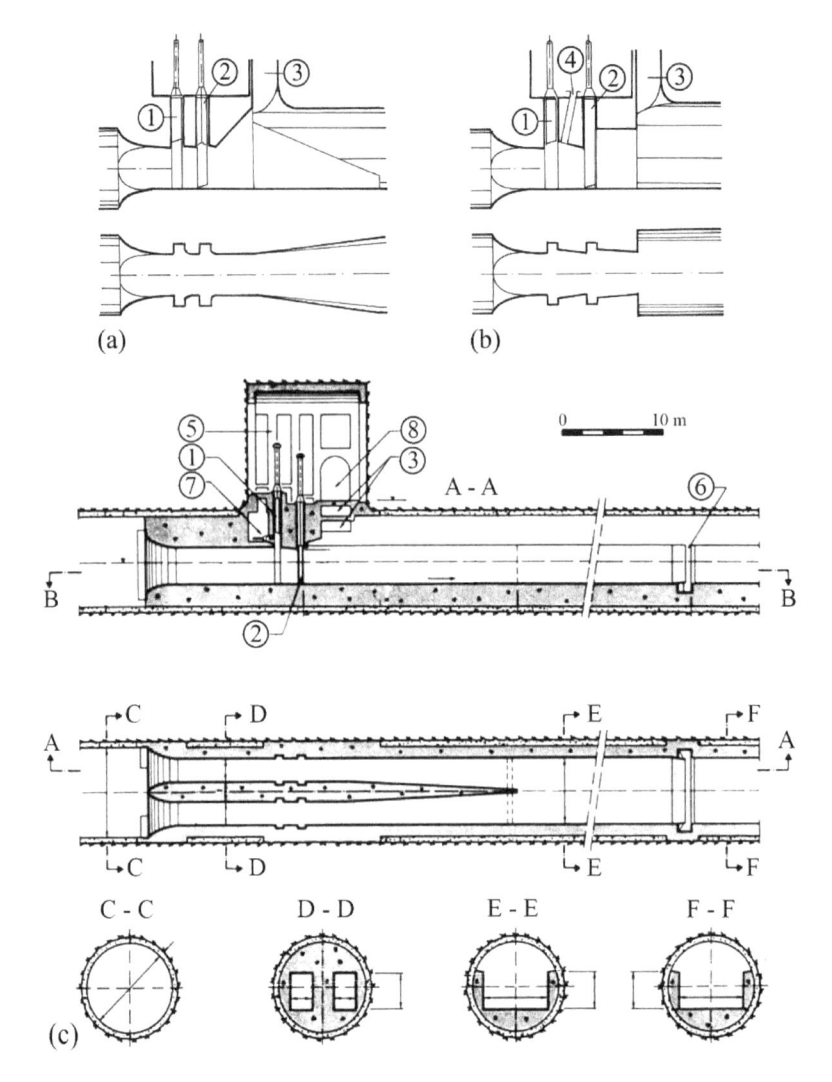

Figure 8.55 Bottom outlets in section and plan view (a) original, (b) final design (Dammel 1977). (c) Bottom outlet of Alicura Dam installed in diversion tunnel, with details of gate chamber with aeration and ① safety gate, ② service gate, ③ tailwater aeration conduit, ④ intermediate aeration ⑤ gate chamber, ⑥ aeration slot, ⑦ service flap gate, ⑧ tunnel access (Minor, 1988)

As an example of gate flow optimization, consider Figure 8.55 with the (a) original and (b) final design. It refers to a project for the drinking-water supply of Athens, Greece. The head on the wheel gate is 125 m, the gate area is 3.1 m × 3.1 m and the maximum discharge 125 m^3s^{-1}. Hydraulic modeling was used to answer questions of:

• Shape of gate geometry,
• Hydrodynamic forces on gate,

- Flow conditions in gate slots, and
- Requirements of aeration.

The inlet geometry, as defined by the US Army Corps of Engineers (1957), was found satisfactory for a transition from the 6 m pipe to the 3.1 m square section. The transition from the gate slots to the tailwater tunnel of horseshoe-shape induced low pressures; linear transitions were found to perform better than ellipses. The cross section abruptly expanded to 4.5 m behind the service gate instead of a linear height increase. Accordingly, the jet would spring clear from the gate crest, and unique wall pressure conditions developed.

If the regulating gate should be inhibited for any reason, the safety gate must close without problems. Vibrations are then of particular concern due to the interrelation between the two gates (Naudascher and Rockwell, 1994). An appropriate aeration between the tandem gates was found essential.

Figure 8.55c shows the gate chamber of the bottom outlets of the diversion tunnel of Alicura Dam with the maintenance (emergency) and service sliding gates (Minor, 1988). Downstream of the gates the high-speed jets are guided by an intermediate wall to avoid rooster tails and choking of the tunnel flow. The design discharge of this facility is 3000 m³/s, and the design head 135 m.

Glen Canyon Dam, the largest and most important feature of the Colorado River project, has a height of 216 m. The diversion tunnel of 12.5 m internal diameter was plugged after use and serves now as the bottom outlet with three parallel conduits extending through the plug (Figure 8.56). The flow is regulated by tandem slide gates of 2.1 m × 3.2 m, and the bottom

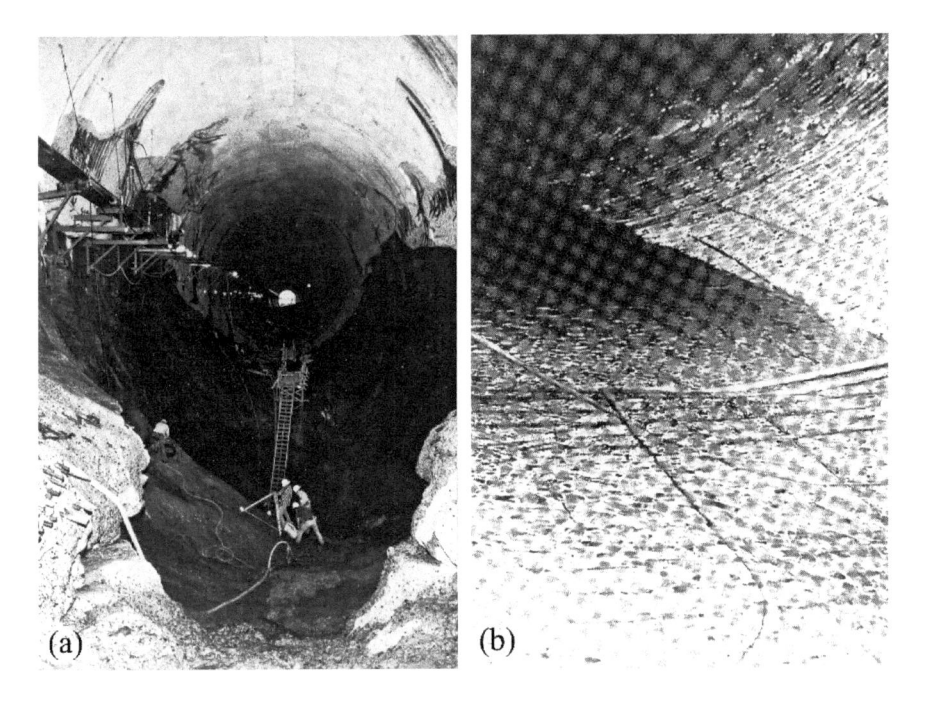

Figure 8.56 Glen Canyon Dam (USA) (a) damages of tunnel invert (Falvey, 1990), (b) view of tailwater invert

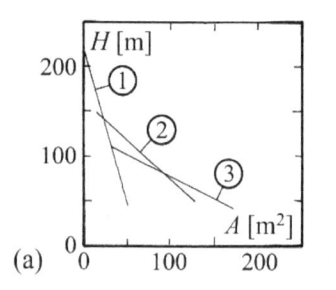 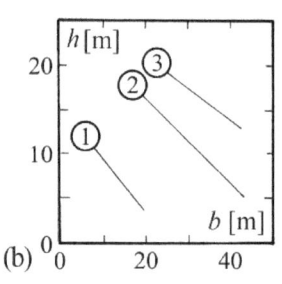

Figure 8.57 (a) Cross-sectional area A (m²) of high-head pressure rates versus head H (m) for ① slide, ② caterpillar, and ③ Tainter/fixed-wheel gates, (b) height of underflow gates h (m) versus width b (m) for ① slide, ② fixed-wheel, and ③ Tainter gates (Adapted from Erbisti, 1981)

and sides of the downstream conduits are lined with 19 mm steel. Air to each regulating gate is supplied through the space above the free water surface in the downstream conduits.

After two years of operation, the tunnels were inspected and considerable erosion and cavitation damage was found due to depressions in the invert profile and along the steel lining (Wagner, 1967; Falvey, 1990). Other damage was also noted at the top between the service and the safety gates. The major damage to the gates and the conduit liner was attributed to irregularities or misalignment of the fluidway surfaces: Irregularities must be controlled by rigid manufacturing and installation tolerance. It was concluded that the prototype paints must be carefully applied to avoid surface-roughness enhancing cavitation. It should also be considered that the roughness may be initiated by abrasion due to sediment transport in the bottom outlet. The abrasive action may be increased at locations of a hydraulic jump presence where the material circulates as in a ball-mill, so that all foreign material should be removed periodically and kept from entering from a hydraulic jump basin.

Any gate selection should account for past experience with successful designs and manufacturing capacities. According to Erbisti (1981), gate designers appear to be rather conservative in their approach, preferring whenever possible to adopt a type of gate already proven in practice, unless a new proposed type has demonstrated greater safety and economy. The continuous development of hydroelectric projects demands equipment of increasing size.

Based on a review of existing gates including more than 4000 references, Erbisti related various types of gates to the gate dimensions. Figure 8.57a refers to high-pressure gates indicating that slide gates of restricted area of 20 m² are exclusively used for heads larger than 150 m. Fixed wheel and Tainter gates are nearly applied to the same extent whereas caterpillar gates are more suitable than fixed-wheel gates for heads in excess of 100 m. The relation between height and width for underflow gates is highlighted in Figure 8.57b. Currently, the Tainter gate is widely used. Due to the reduction of the friction forces, the gate width was progressively increased in the past.

8.6.3 Gate vibrations

Naudascher and Rockwell (1994) distinguish between predominantly (1) extraneously induced and (2) instability-induced exiting. Examples of (1) are the skin plate of sluice gates, the tandem gate during emergency closure, and the Tainter gate for two-phase flow. Exiting type (2) occurs for high-head leaf gates with an unstable jet, a cylinder gate of bistable underflow, or leaf gates involving impinging shear layers (Figure 8.58).

To protect a gate against flow-induced vibrations in the *vertical* direction, either sufficient damping or a bottom shape from which the flow remains unseparated or stably reattached must be provided. Naudascher and Rockwell (1994) pointed to the effect of gate crest geometry on the latter condition. Figure 8.59 lists crest shapes where vertical vibration may, or may not occur. The crest height ratio a/e should be much larger than unity. Other examples of vortex-induced vibrations relate to a gate withdrawn into the gate chamber (Figure 8.58a), pressure fluctuations in a rectangular gate slot (Figure 8.58b), and a tandem gate arrangement with the service gate stuck (Figure 8.58c).

Horizontal gate vibrations associated with underflow are not as dangerous as vertical excitations. Stable lip shapes are comparable to that of Figure 8.59e, if the lip thickness is small and the gate passes quickly through small degrees of openings.

The most dangerous mechanisms for movement-induced excitation of gates and gate seals are those involving coupling with fluid flow pulsations (Naudascher and Rockwell, 1994). These occur if a seal arrangement gives rise to flow through a leak channel with an upstream constriction, to be eliminated or moved toward the downstream end of the flow passage.

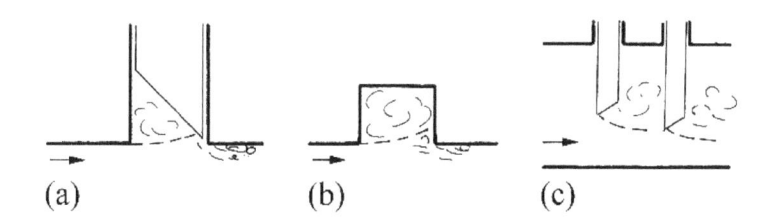

Figure 8.58 Sources of gate vibration with impinging shear-layer instability, details see text (Adapted from Naudascher and Rockwell, 1994)

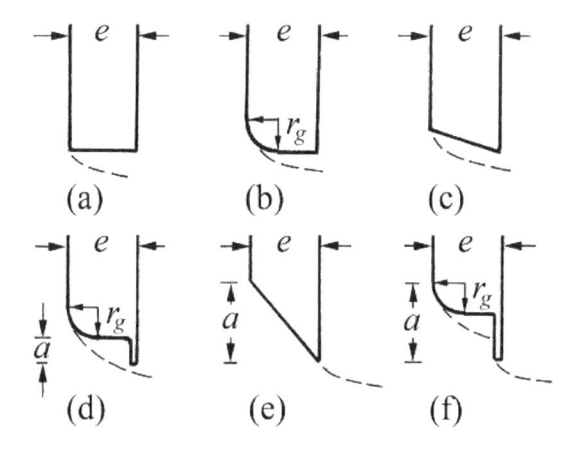

Figure 8.59 Crest shapes for bottom outlets (a) to (d) unstable, (e) and (f) stable under vertical gate vibrations (Naudascher and Rockwell, 1994)

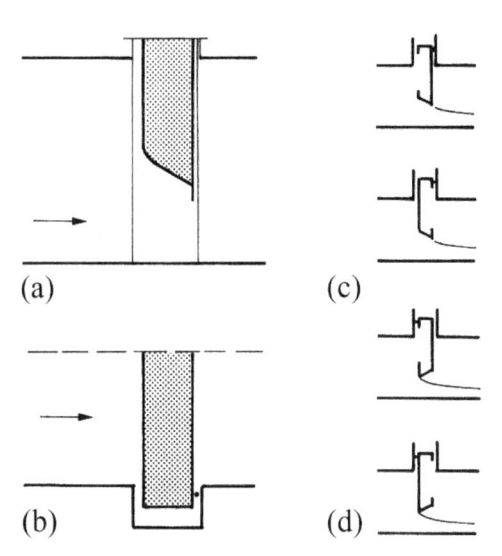

Figure 8.60 Common geometry of high-head leaf gate (a) side view, (b) plan view, (c), (d) possible top seal and skin plate arrangements (Naudascher, 1991)

The leaf gate is the most widely used high-head gate. Of all gate types, it causes the greatest problems with hydrodynamic loading and downpull. The pressure along the crest of the leaf gate is dramatically reduced due to the high efflux velocities. The downpull thus often exceeds the gate weight. Figure 8.60 shows side and plan views of a common leaf gate, together with typical seal arrangements. The most feasible arrangement is shown in Figure 8.60c top, with seal and skin plate at the downstream side. Despite the advantage of downpull reduction, gates with upstream seals apply only for low heads due to vortex action in the gate slots.

8.6.4 Hydraulics of high-head gates

Discharge equation

Gate flow may be either free or submerged. For *free gate flow*, the space behind the gate is filled with air of pressure head h_a. If the efflux is into the atmosphere, $h_a = 0$. Submerged gate flow in bottom outlets should be avoided, as discussed in 8.6.1. Based on energy considerations, the gate underflow discharge Q is (Naudascher, 1991)

$$Q = C_c ab[2g(H - H_e - C_c a - h_a)]^{1/2}. \tag{8.75}$$

Here C_c is the contraction coefficient, a the gate opening, b the gate width, and $H-H_e$ the head on the gate with H_e as the energy head loss from the entrance to the gate section (Figure 8.61). With C_o as a discharge coefficient for the seal system, a_3 the width between the slot and the gate, H_{ce} the head loss between the approach flow and the upstream gate face, and h_s

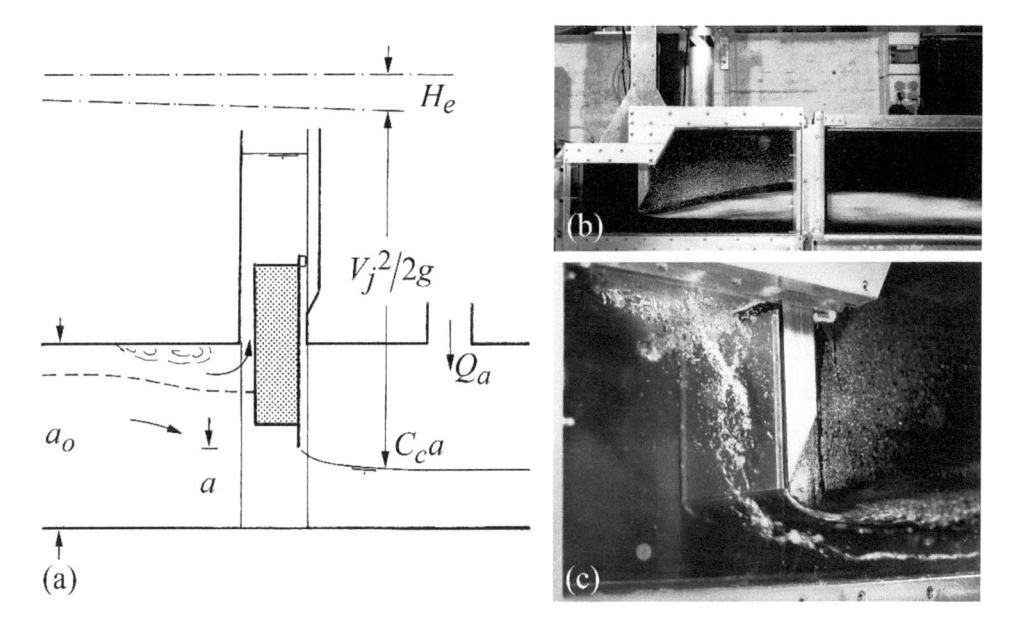

Figure 8.61 Free gate flow (a) definition sketch, (b) overall view, (c) detail of gate flow (Photo VAW 45/72/12)

as the vertical distance between the datum and the section of jet issue, the overflow discharge Q_o of a high-head gate is

$$Q = C_o a_3 b [2g(H - H_e - H_{ce} - h_s - h_a)]^{1/2} .$$ (8.76)

The discharge coefficient C_d of gate flow represents with ΔH as difference head from Eq. (8.75) the ratio

$$C_d = Q / (2g\Delta H)^{1/2}.$$ (8.77)

Parameter C_d depends on a number of variables including the relative gate opening $\eta = C_c ab/A_o$, with $A_o = a_o b_o$ as the approach flow section, the loss factor, the aspect ratio and the approach flow velocity distribution. Accordingly, one should rather use the contraction coefficient C_c than the lump parameter C_d.

Provided $F_j < 4$, the contraction coefficient C_c depends on the efflux Froude number $F_j = V_j/(gC_c a)^{1/2}$. For large Froude numbers as typically encountered in prototype high-head gates, neither this free-surface effect nor the effect of viscosity have to be accounted for because potential flow with the gravity effect absent governs these flows, i.e. the gate geometry has the significant effect on C_c. Figure 8.62 shows contraction coefficients for the typical leaf gate geometry of relative crest radius $R_g = r_g/e = 0.4$ and $a_o/e = 6$.

Gate overflow is complicated by corner eddies. Naudascher (1991) accounted for the detailed flow structure between the gate body and the gate chamber. The downpull of a gate is a specialized domain of hydrodynamic forces and not discussed here because of

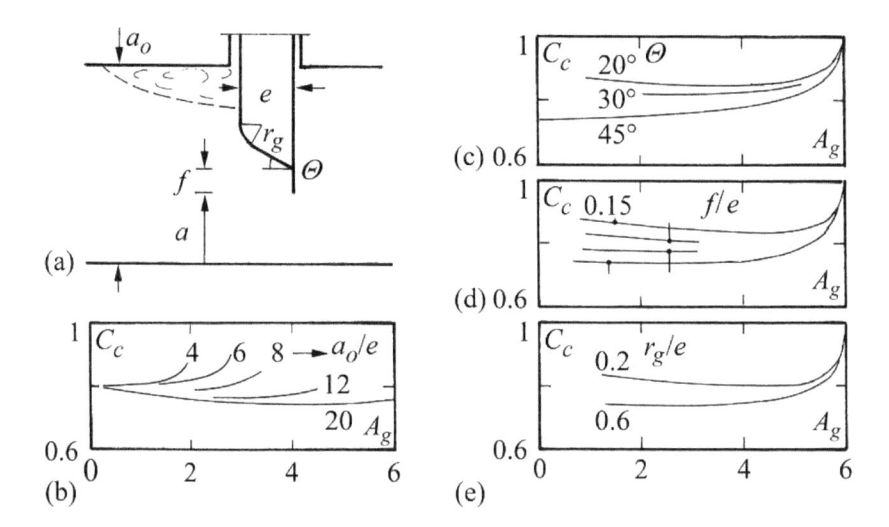

Figure 8.62 Contraction coefficient C_c (a) Definition sketch, effects of (b) gate thickness $A_g = a/e$ ($\theta = 30°$, $R = 0.4$, $f/e = 0$), (c) crest angle $\theta(A_g)$, (d) relative off-set length $F = f/e$ and (e) relative gate radius $R_g = r_g/e$ versus gate opening $A_g = a_g/e$ for $r_g/e = 0.4$, $a_o/e = 6$ (Adapted from Naudascher, 1991)

Naudascher's exhaustive treatment. For important gates of bottom outlets, model experiments are necessary and recommended.

Gate slots

The performance of slide and wheel gates depends highly on the supporting gate slots. Ball (1959) has presented basic information, so that damages relating to both the gate and the outlet structure should not occur. The tests were made with the gate fully open and for pressurized flow; it was demonstrated that this corresponds to the relevant flow configuration for gate slots. Among various slot geometries, the configuration with abrupt up- and downstream corners was retained provided the downstream corner was recessed to obtain a gradual convergence toward the tailwater wall (Figure 8.63a). A final design of the US Army Corps of Engineers (1968) involves a rounded slot geometry 58% deep of the gate length, as shown in Figure 8.63b.

Figure 8.64 refers to photographs of gate slots with and without the presence of a guide rail. The system of vortices is seen to be significantly different for two otherwise identical configurations.

8.6.5 Cavitation and cavitation damage

Description

Cavitation is defined as the formation of a bubble or void in liquid (4.4). Cavitation occurs by decreasing the local pressure under constant temperature. The local pressure reduction

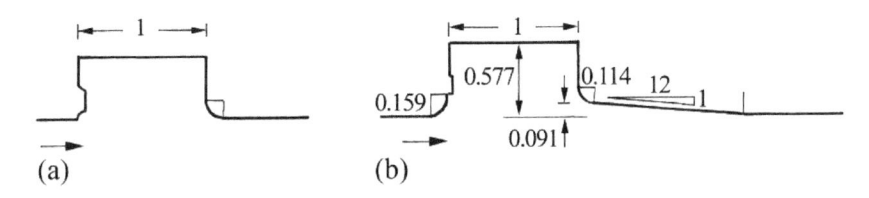

Figure 8.63 Gate slot (a) without, (b) with downstream offset (recommended). All lengths multiplied by slot length

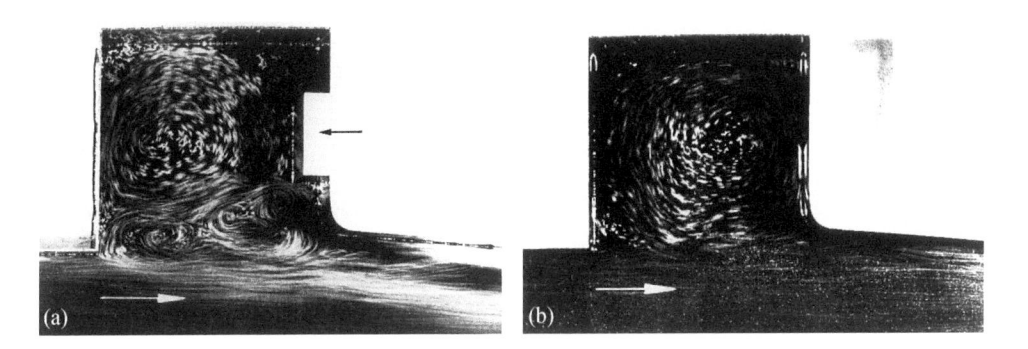

Figure 8.64 Effect of guide rail in gate slot (a) with, (b) without rail (*Proc. Institution of Civil Engineers* 1985, 79(2): 755)

can be caused in fluid flow by a reduction of the total energy head because of increase in elevation, by a local increase of velocity (e.g. due to contraction) or by turbulence, vortices or large-scale flow separation. The water flowing in hydraulic structures contains air bubbles of various sizes and with impurities. These conditions are necessary to initiate cavitation, and determine the potentials of damage and noise.

If the pressure in a fluid flow is continuously reduced due to velocity increase such as over a spillway crest, a critical point is reached at which cavitation starts, referred to as *incipient cavitation*. Similarly, if cavitation exists and the velocity is reduced, a critical condition is reached where cavitation disappears, referred to as desinent cavitation. The incipient and desinent states do not occur at identical flow conditions.

The *cavitation index* is the hydraulic parameter describing the cavitation process as

$$\sigma = \frac{p - p_0}{\rho V_0^2 / 2}. \tag{8.78}$$

Herein, p is the local pressure, and p_0 and V_0 are the reference (Subscript 0) pressure and velocity, typically of the upstream flow. To avoid ambiguities, both vapor and reference pressures are referenced to absolute zero pressure. Although one parameter such as the cavitation index is not able to describe the various complex flow features, it is a useful quantity to indicate the state of cavitation. For example, for flow past a sudden into-the-flow offset, cavitation will not occur if $\sigma > 1.8$. For $\sigma \leq 1.8$, however, more and more cavitation bubbles form, detected visually as a fussy white cloud. For an even lower cavitation index, the

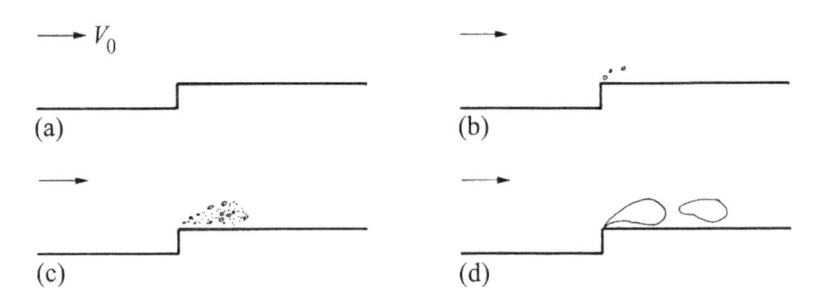

Figure 8.65 Development of cavitation for flow past sudden into-the-flow offset with
$\sigma = $ (a) 3, (b) 1.8 with incipient cavitation where cavitation bubbles occasionally appear, (c) 0.3 < σ < 1.8 with developed cavitation, (d) σ < 0.3 super-
cavitation (Falvey, 1990). Reference quantities referred to upstream flow

cloud consisting of individual bubbles suddenly forms long super-cavitating vapor pockets
(Figure 8.65).

Bubble collapse dynamics

To simulate the bubble collapse dynamics, water compressibility has to be considered. A bubble collapse consists of phases in which the bubble diameter decreases from the original size
as the pressure is increased, reaching a minimum and then grows or rebounds, as shown in
Figure 8.66a. The process is repeated for several cycles with the bubble diameter decreasing
during each cycle until finally becoming microscopic in size. During the rebound phase, a
shock wave forms with the shock celerity equal to the speed of sound in water. At a distance
of two times the initial bubble radius from the collapse center, the pressure intensity is about
200 times the ambient pressure at the collapse site. The time for a bubble to collapse depends
mainly on the initial bubble radius and is of the order of some micro-seconds. If the bubble

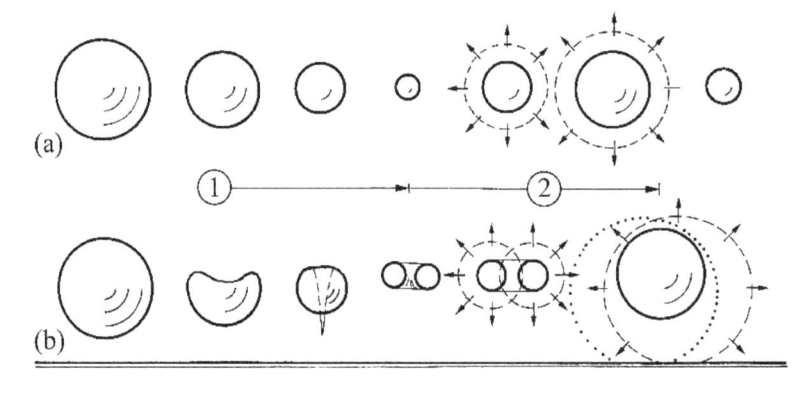

Figure 8.66 Collapse of individual bubble (a) in quiescent fluid, (b) near boundary with ①
collapse, ② rebound (Adapted from Falvey, 1990)

collapses near a solid boundary, the boundary reflects the wave toward the bubble causing an asymmetric collapse. As shown in Figure 8.66b this causes one side of the bubble to deform into a micro-jet penetrating at the opposite bubble side. The velocity of this micro-jet is large and the shock wave generates pressure so high that it may cause *cavitation damage* to the solid boundary.

If more than one bubble is present, a collapse of the first will produce shock waves radiating to other bubbles. These shocks cause the unsymmetrical collapse of neighboring bubbles. The ultra-jets have velocities in the order of 1.5 times the sonic water velocity generating higher pressure intensities than either spherical shock waves or micro-jets. If a bubble in a swarm collapses, the resulting shock wave will cause other bubbles to collapse in its vicinity, and the collapse process continues in the form of a chain reaction until the remainder of the swarm collapses simultaneously (Falvey, 1990). The synchronous collapse of a bubble swarm produces higher pressure intensity than the random collapse of individual bubbles in a swarm. A theory for the prediction of the pressure magnitude generated by a bubble swarm does not yet exist.

Damage caused by a group of bubbles trapped by a vortex can be many times larger than caused by the collapse of an individual bubble. Shear flows as occur typically along a boundary thus collect bubbles on their axes.

Cavitation characteristics

An irregularity of a flow surface either may be an isolated roughness or uniformly distributed roughness. Typical isolated roughness includes (Figure 8.67):

- Offset into-the-flow,
- Offset away-from-the-flow.
- Grooves, and
- Protruding joints.

Cavitation is formed by turbulence along a shear zone due to the sudden change of flow direction. Depending on the roughness shape, cavitation bubbles collapse either within the flow or near the flow boundary.

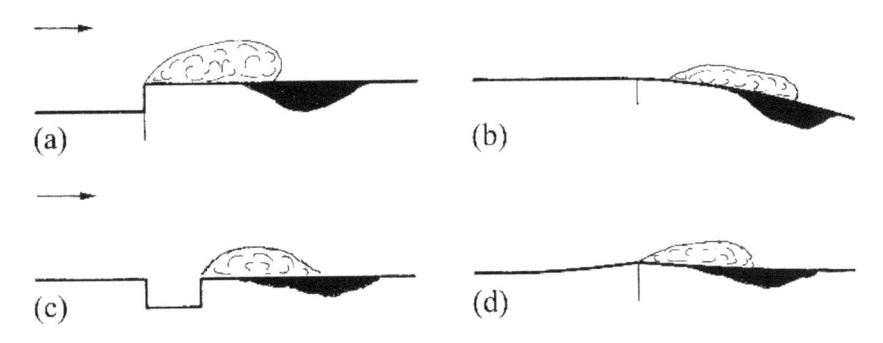

Figure 8.67 Isolated roughness elements with vapor cavities and damage zones, details see text

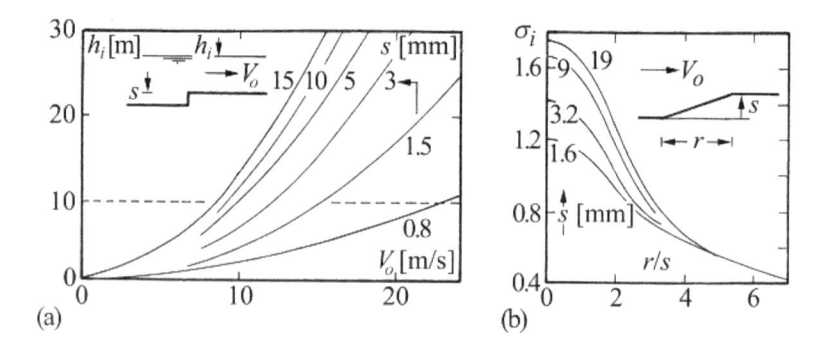

Figure 8.68 Incipient cavitation for (a) abrupt offset $h_i(V_o)$, (b) cavitation index $\sigma_i(r/s)$ for chamfered offset (Falvey, 1990)

Figure 8.68a refers to the cavitation characteristics of the into-the-flow offset. For an approach velocity of, say 20 m/s, and an offset of only 0.8 mm, the pressure head for incipient cavitation to occur is 7.7 m above vacuum. Cavitation is thus no problem for pressure over atmospheric pressure (+10 m). However, if the offset has double height (1.6 mm), then the pressure head has to be at least 17.8 m above vacuum.

Figure 8.68b refers to a generalized plot of the *cavitation index* $\sigma_i = (h-h_v)/(V_o^2/2g)$ for incipient cavitation, with h as the absolute pressure head, h_v as the vapor (subscript v) pressure head, and $V_o^2/2g$ as the approach flow velocity head. Falvey (1990) presented a comprehensive data set for other surface irregularities, including holes, grooves, uniform roughness, and combinations of uniform and isolated roughness. Note that the superposition principle is applicable, at least to lowest order.

Cavitation damage

Cavitation as such is no danger for any hydraulic structure. If cavitation occurs close to flow boundaries, then cavitation damage may occur, however, with the potential to seriously damaging a structure. The surface damage starts at the downstream end of the cloud of collapsing cavitation bubbles. Later, an elongated hole is formed in the boundary surface. As time progresses, the hole deepens due to high velocity jet flow impacting onto the downstream end of the hole. In contrast, this flow is able to create high pressures within minute cracks around individual pieces of aggregate. Pressure differences between the impact zone and the surrounding area are able to break away parts of the aggregate swept away by the flow. This process resembles erosion yet the loss of material due to cavitation is different.

Figure 8.69 shows a hole 11 m deep in the spillway invert of the left tunnel of Glen Canyon Dam, USA. Concrete lumps were found attached to the end of the reinforcing steel. At this stage, a high-speed flow acting on the lumps rip reinforcement bars from the concrete even though the steel was imbedded as deep as 150 mm. After the tunnel lining was penetrated, erosion continued into the underlying rock. Once damage penetrated the liner, the integrity of the structure became a major concern.

Based on the typical value of $\sigma = 2$ for the cavitation index and for barometric pressure conditions, the resulting velocity for cavitation inception is about 10 m/s. It is thus prudent to investigate the possibility of cavitation whenever the velocity of flow exceeds this limit value.

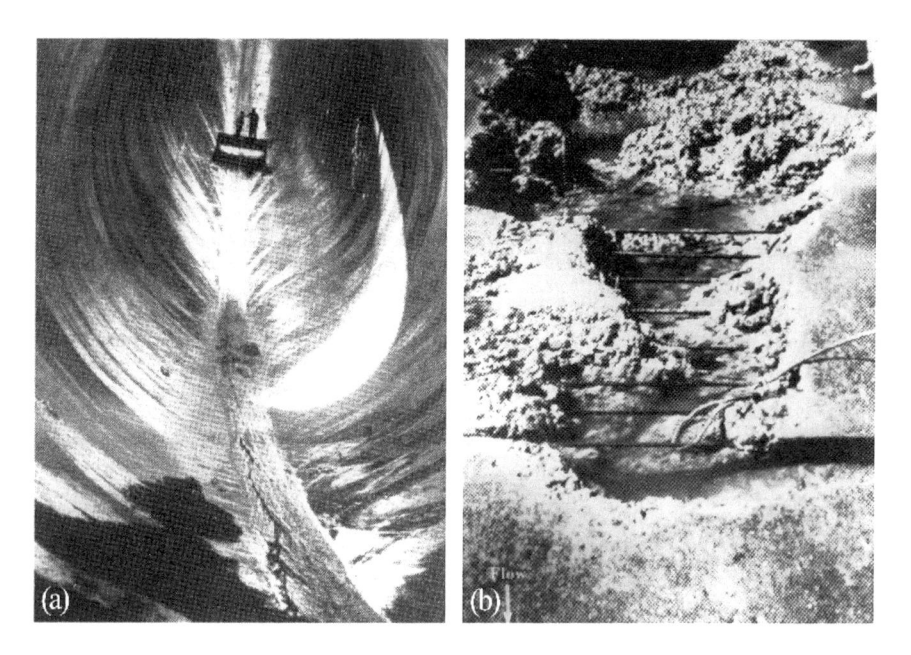

Figure 8.69 Cavitation damage to Glen Canyon Dam tunnel spillway (Falvey, 1990)

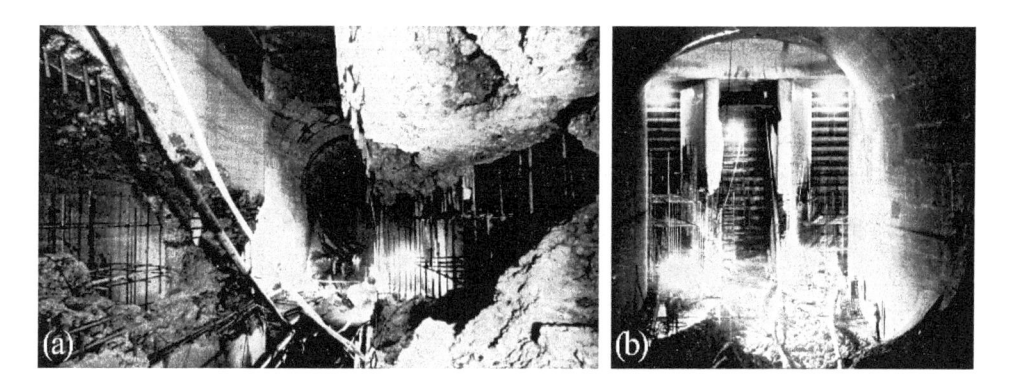

Figure 8.70 Cavitation damage to Tarbela Tunnel 2, Pakistan, with views from (a) upstream,
(b) downstream (Kenn and Garrod, 1981)

The resistance of a certain surface to cavitation damage depends on factors such as the ultimate material strength, the ductility, and the homogeneity. Strength and ductility combine into the parameter *resilience*, i.e. the area under the stress-strain curve of a material. A comparative cavitation resistance of various materials indicates a relatively poor resistance of concrete and polymer impregnated concrete, followed by aluminum or copper, and finally by stainless steel. There exist no materials, however, not prone to cavitation damage. The effect of exposure time is discussed by Falvey (1990). Figure 8.70 shows typical damages on Tunnel 2 of Tarbela Dam.

Cavitation control

Cavitation damage is controlled by two methods:

- Control of cavitation index by geometry, if possible, and
- Control of cavitation damage by appropriate flow aeration.

The irregularities of a bottom outlet have to be estimated case specific, based on accuracy specification, quality of material, and site considerations. The concept of incipient cavitation has to be applied; it must be determined whether or not cavitation damage occurs. These computations involve a prediction of the surface profiles based on backwater curves, the determination of the pressure head curves for all relevant discharges, and the computation of the cavitation index curves. If there are locations where cavitation is predicted, then the spillway geometry or the boundary smoothness have to be improved. If both approaches fail, the flow has to be aerated, as described below. Note that the capacity of aerated spillway flow has to be larger because of the increase of air-water discharge. Details on spillway aerators are given in Chapter 4.

8.6.6 Passive and active air entrainment

Studies on required air demand

Whereas the flow is pressurized upstream from gates of a bottom outlet, free-surface flow develops in the tailwater, depending on the flow characteristics in the outlet tunnel. Because free gate outflow greatly reduces the potentials of gate vibration and cavitation damage, a bottom outlet has to be designed for free-surface flow. Flow aeration originates from three different sources (Figure 8.71):

① Tunnel outlet in countercurrent airflow along outlet roof,
② Air supply conduit by which under pressure of surface airflow is reduced, and
③ Bottom aerator countering problems with cavitation damage.

Due to limited air availability, the airflow features in a bottom outlet are more complex as compared with chute aerators. The hydraulics of air-water flows in bottom outlets is currently not yet fully understood, so that hydraulic modeling based on sufficiently large-scale

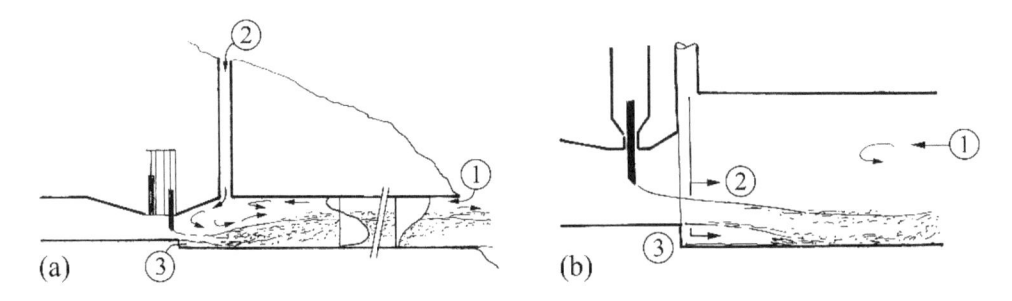

Figure 8.71 Air entrainment by gated outflow in bottom outlet tunnel (a) overall view, (b) gate chamber with ① counter-current air flow in tunnel, ② air supply conduit, ③ bottom aerator

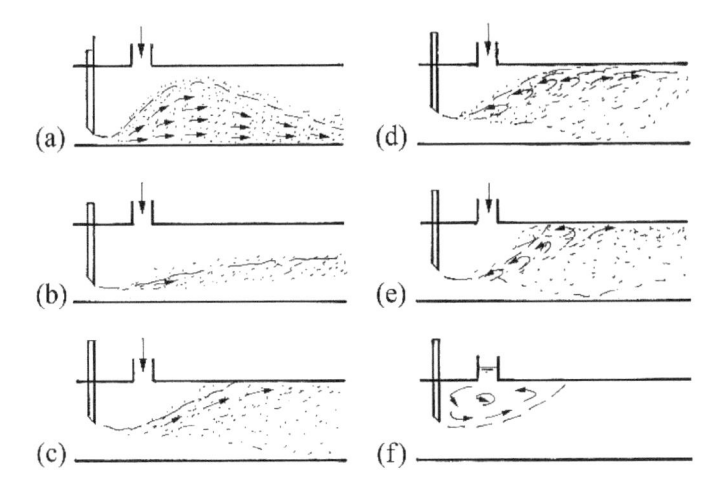

Figure 8.72 Classification of flow types *without* bottom aerator, for details see text (Adapted from Sharma, 1976)

models is recommended. The Froude similitude governs such flow, and successful studies have been conducted in the past (Giezendanner and Henry, 1980; Anastasi, 1983). These and other studies stressed the effect of flow type on the air entrainment characteristics.

Sharma (1976) classified the flow types without bottom aerator (Figure 8.72):

a   *Spray flow* for relative gate opening below 10%, with high air entrainment
b   *Free supercritical flow*, including shock waves and two-phase flow
c   *Foamy flow* for tunnel almost full with air-water flow
d   *Hydraulic jump* with free-surface tailwater flow due to tailwater submergence
e   *Hydraulic jump* with transition to pressurized tailwater flow
f   *Fully pressurized flow* caused by deep tailwater submergence.

Only types (a) to (c) are relevant because of dangerous surging conditions otherwise.

Rabben and Rouvé (1985) presented a preliminary design for air entrainment. A fictitious cross-sectional air conduit section $A_a^* = A_a(1 + \Sigma\xi_i)^{-1/2}$ was defined with $\Sigma\xi_i$ as the sum of all head losses from the atmosphere to the gate chamber. For *spray flow* (Figure 8.72a) of relative gate opening <6% and $\mathsf{F}_c \geq 20$, the entrained air ratio $\beta = Q_A/Q$ is

$$\beta = (A_a^* / A_d)\mathsf{F}_c. \tag{8.79}$$

Here, the tailwater tunnel section is A_d, and $\mathsf{F}_c = q/[g(C_c a)^3]^{1/2}$ is the Froude number relative to the contracted (subscript c) gate flow depth. The air discharge is Q_A, and the water discharge $Q = qb_g$ with b_g as gate width, g as the gravity acceleration, and a as the gate opening.

For *free gate flow* (Figure 8.72b) of relative gate opening >12%, and $\mathsf{F}_c \leq 40$, a similar expression for the entrained air ratio is

$$\beta = 0.94(A_a^* / A_d)^{0.90}\mathsf{F}_c^{0.62}. \tag{8.80}$$

For tunnel flow with a *hydraulic jump*, the tailwater depth or the corresponding pressure head for conduit flow is computed with a backwater curve based on the outflow submergence. For conditions analogous to free flow (Figure 8.72d) results

$$\beta = 0.019(A_a^* / A_d)^{0.099} \mathsf{F}_c^{0.969}. \tag{8.81}$$

This is much lower than for spray flow. Alternative approaches are due to Bollrich (1963) and Sharma (1976). In all approaches, the effect of tunnel length is excluded, so that these relations apply for short tunnels. For longer tunnels, less air is entrained because of de-aeration processes.

Bottom outlets with *aerator presence* are governed by complex flow phenomena. Figure 8.73a shows a definition plot. Rabben and Rouvé (1985) proposed a preliminary design procedure, yet additional experiments have to be conducted before design application. A summary of the preliminary design procedure is given by Sinniger and Hager (1989) or Hager and Schleiss (2009).

The effect of relative tunnel length and a standardization of the jet aeration have not yet been investigated. The actual tendency is to aerate the jet not only from the bottom, but also from the sides just beyond the gate section. A design procedure is not available so that model observations on a sufficiently large-scale model are required. Information on recent prototype observations are due to Volkart and Speerli (1994), and Lier and Volkart (1994).

Two-phase flow in bottom outlets

Bottom outlets are relatively short for arch or gravity dams but considerably longer for embankment dams and for diversions through dam abutments. To reduce the length of the pressurized reach, long bottom outlets are divided into a pressurized portion controlled with a high-head gate and an outlet tunnel discharging supercritical flow into the atmosphere. The gate is located slightly upstream from the grout curtain so that the lengths of pressurized and open tunnel reaches are roughly equal (Figure 8.52). The free-surface portion is of particular hydraulic interest due to air-water mixture flow, air entrainment and detrainment, air supply system, as well as potential cavitation damage and abrasion with flows containing sediment.

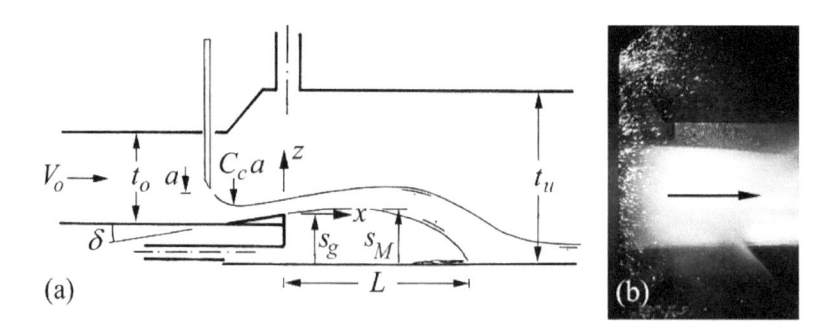

Figure 8.73 Bottom aerator of bottom outlet (a) schematic flow structure and air supply, (b) photo of slide gate at Panix Dam, Switzerland, for gate opening of 10 mm, and a pressure head of 50 m. Note shock waves due to gate slots and spray development due to small opening (Photo VAW 43/62/28)

Speerli (1998) presented new data on two-phase high-speed tunnel flows based on laboratory observations. Various flow patterns are described, the development of air concentration along the tunnel, and the maximum air concentration are analyzed. The mixture-flow profile is computed using approximations for supercritical flow, based on a modified uniform flow depth concept. Design recommendations are proposed based on the results of past studies. A computational procedure is also described with a typical example. Note that the design of air vents was excluded, however.

The current knowledge of two-phase flows in bottom outlets is limited. Because of challenges for laboratory and prototype observations, only one important contribution per decade was roughly published. Falvey (1980) summarized the main findings. Kalinske and Robertson (1943) studied air entrainment in pipes involving a hydraulic jump. Their 0.15 m diameter pipe was sloped up to 30%, and the air entrainment was found to be mainly a function of the Froude number upstream of the hydraulic jump. Based on U.S. prototype observations, Campbell and Guyton (1953) observed larger air entrainment than Kalinske and Robertson due to effects of turbulence and viscosity. Their relation for the air ratio involves the Froude number at the contracted section downstream of the outlet gate. A similar estimate for relative air discharge based on prototype observations was proposed by the U.S. Army Corps of Engineers (1964). Levin (1965) and Wisner (1967) obtained similar results, with coefficients accounting for the bottom outlet geometry.

Ghetti and Di Silvio (1967) were the first to account for the interactions between air supply and air entrainment for flows in bottom outlets. Their laboratory tests verified prototype observations on Italian dams. The effect of a limited air section above the mixture section of two-phase flow was discussed. Sharma (1976) classified the three flow types as spray, free surface, and hydraulic jump flows recommending individual equations for the corresponding air entrainment ratios. Rabben and Rouvé (1984) conducted laboratory tests by systematically varying the loss characteristics of the air supply system. The air ratio coefficient was considered a function of both the Froude number at the contracted section and the reduced cross-sectional area of the air supply conduit in which the hydraulic loss characteristic was integrated.

Speerli (1998) stated that the effect of bottom outlet geometry, i.e., its length, width, and height, as well as the governing flow mechanisms was unknown. The purpose of his study was to obtain insight into these phenomena. Given the large installation and the involved instrumentation, the effects of tunnel length and air supply were determined for a rectangular tunnel of fixed bottom slope.

8.6.7 Interaction of water flow and air entrainment

Test facility

Tunnel flow downstream (subscript u) of a bottom outlet depends on the gate opening a, the approach flow energy head H_E, the gate geometry, the loss characteristics of the air supply system, and the tunnel geometry. For a rectangular tunnel section as abstraction of a partly filled horseshoe section, the geometry is given by the bottom slope S_u, tunnel width B_u, tunnel height H_u, and tunnel length L_u. The widths of the approach flow duct, the gate, and the tunnel were equal; the bottom was straight, without aerators downstream from the gate section. The contracted (subscript c) Froude number $F_c = V_c/(gh_c)^{1/2}$ relates to the section downstream from the gate, and F_c includes the contraction coefficient C_c, in which $V_c = Q/(B_u h_c)$, with $h_c = C_c a$, and g as gravity acceleration. The fluid characteristics are density ρ of air (subscript a)

and water (subscript w), the corresponding kinematic viscosities v, surface tensions σ, and temperatures T. In addition, the tunnel boundary roughness and the air supply system need to be considered. The air supply discharge Q_{Ao} from the supply system (subscript o), for instance, depends on all these parameters.

To reduce the number of variables, simplifications were introduced (Speerli, 1998):

- No presence of hydraulic jumps along tunnel;
- Effect of roughness variation is small;
- Effect of gate slots is small, except for spray flow;
- Variation of turbulence level is small for single gate arrangement;
- Turbulence level is so high that viscous effects are absent;
- Ratios of fluid densities (ρ_a/ρ_w) and viscosities (v_a/v_w) are almost constant;
- Effect of Weber number remains constant for the two fluids involved, thereby accepting scale effects when upscaling from lab to prototype results; and
- Effect of fluid temperature variation is small.

The air supply discharge thus depends on the tunnel geometry (B_u, H_u, L_u), the loss coefficient ξ_o of the air supply system, relative gate opening $A = a/a_0$ (0 to 100%) and the approach flow energy head H_E. The tunnel section remained constant $H_u/B_u = 1.5$.

Figure 8.74 sketches the test facility used consisting of pumps, an approach flow conduit, a gate chamber, a tunnel, and a return flow system. The total pump capacity was 360 l/s for

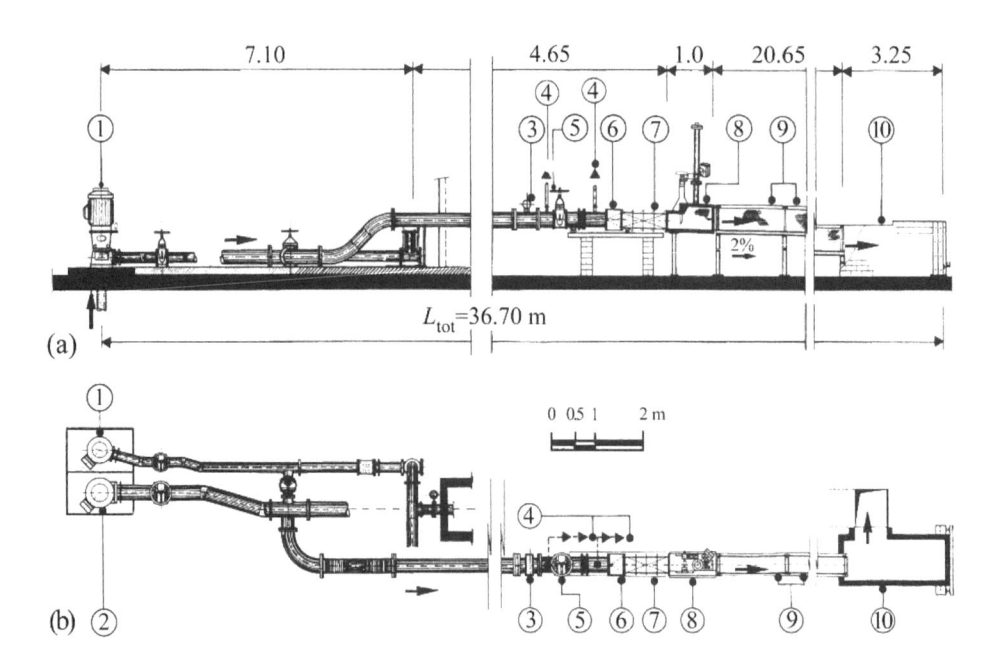

Figure 8.74 Test facility (a) section, (b) plan with (① pump 1, ② pump 2, ③ Inductive Discharge Measurement, ④ de-aeration, ⑤ valve, ⑥ transition element, ⑦ flow straightener, ⑧ gate chamber with slide gate and air supply conduit, ⑨ bottom outlet tunnel, ⑩ recirculation (Speerli, 1998)

approach flow heads of up to 25 m. The tunnel was 0.30 m wide and 0.45 m high. Except for along the tunnel sides, where shock waves were generated due to gate presence (Speerli and Hager, 1999), the free-surface downstream from the gate was smooth and non-aerated up to the air inception point. The gate was vertical and moved from the fully closed to the fully open positions. Downstream from the gate, the tunnel height increased at 45° to the tunnel height H_u. The tunnel was 21 m long and of 2% bottom slope. A rubber sheet was provided at the end of the air-proof section to inhibit air access from the tailwater.

The air supply conduit of 0.10 m diameter ran vertically into the upstream tunnel end (Figure 8.74). The air supply was controlled by orifices of throttling degrees ranging from 0% to 100%. The mixture velocity of tunnel flow ranging between 6 and 25 m/s was measured with capacity probes (Volkart, 1978). Five probes were set in parallel across the tunnel. Their distances from the gate section were 4.5, 8.9, 13.35, and 17.75 m, respectively. The local air concentration was measured by accounting for the difference in electric conductivity between air and water (Volkart, 1988). With h as flow depth, effects of viscosity and surface tension remained small if the Reynolds number $R = 4VR_h/v > 10^5$, and the Weber number $W = V^2 h(\rho_w/\sigma_{aw}) > 110$ (Speerli, 1998).

Flow patterns

Figure 8.75 shows side views and a downstream view of laterally expanding shock waves, starting from the vena contracta section and eventually crossing the axis of the bottom outlet. Their origins are due to stagnation flow upstream the gate, resulting in air entrainment and flow non-uniformity (Figure 8.61c; Speerli and Hager, 1999).

Figure 8.76 shows typical mixture flows for a relative gate opening of 17%. For $H_E = 10$ m, the airflow above the air-water mixture flow is small, and the threads suspended at the tunnel roof remain nearly vertical. Increasing H_E to 20 or 25 m increases the counter airflow because of sub-pressure generation downstream of the outlet gate. Depending on the discharge, the air vent, relative tunnel filling, and tunnel length, strong countercurrent airflow

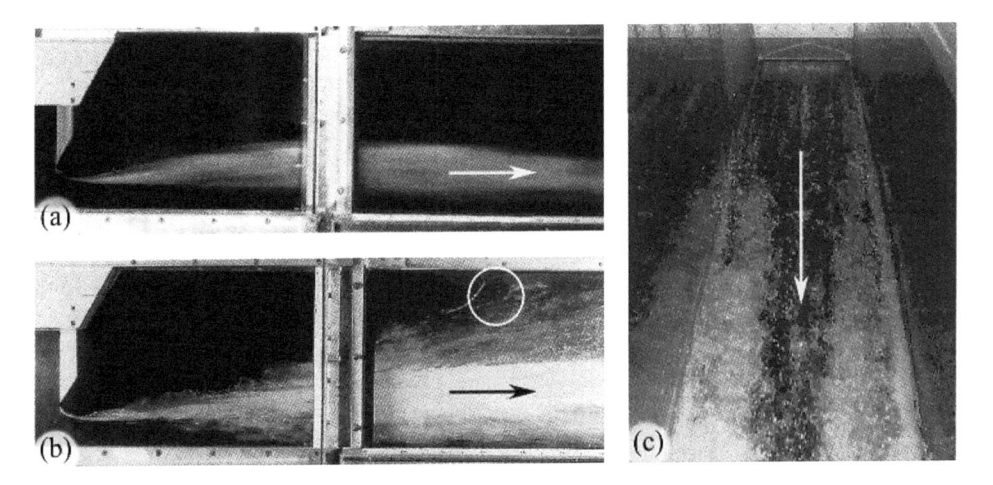

Figure 8.75 Shock waves in bottom outlet for (a) $H_E = 5$ m, (b) $H_E = 15$ m, (c) view from tailwater for $H_E = 10$ m and 33% relative gate opening (Speerli and Hager, 2000)

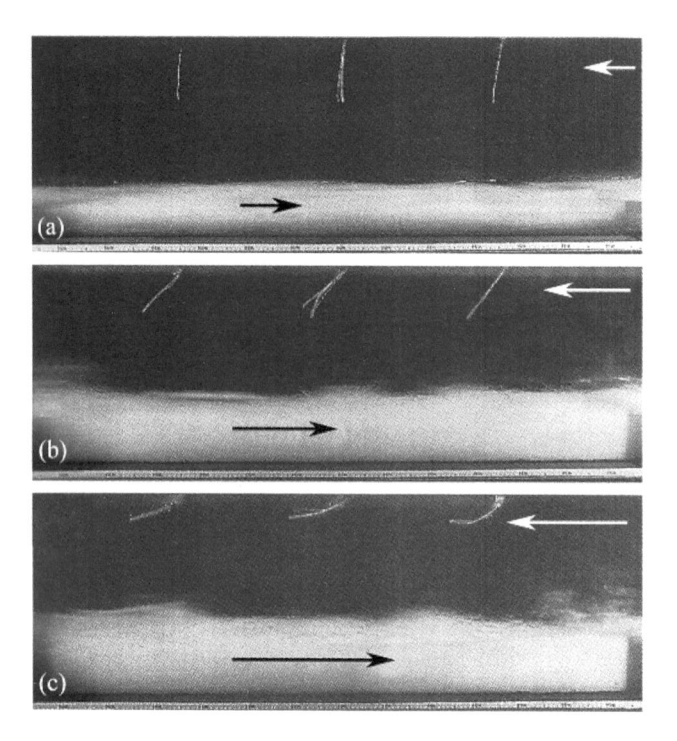

Figure 8.76 Tunnel flow with air-water mixture along bottom below countercurrent airflow for $A = 16.7\%$ and $H_E =$ (a) 10 m, (b) 20 m, (c) 25 m. ($\leftarrow$) countercurrent airflow (Speerli and Hager, 2000)

results. Air-water mixture flow below airflow was considered only; the selected approach flow conditions excluded slug flow.

Air entrainment features

Due to high flow velocity, air is entrained downstream from the air inception point generating highly turbulent flow with a rough free surface. Depending on the bottom slope, the mixture flow accelerates or decelerates. For $S_u = 2\%$ as selected for the tests, the flow decelerated, but supercritical flow was maintained, so that hydraulic jumps were absent. Typically, the flow was aerated beyond the vena contracta section, along the sidewalls due to shock wave presence and at the center tunnel portion due to the thickening of the turbulent boundary layer beyond the flow depth (Wood, 1991). The flow then decelerated along with air detrainment. The tunnel length has a significant effect on the aeration pattern of bottom outlet flow.

The transverse air distribution profiles were found to be practically uniform, so that only the axial air concentration distribution is considered. Figure 8.77 relates to the axial air concentration profiles C versus relative mixture (subscript g) depth $H_g = z/h_{99}$, with z as the vertical coordinate measured from the tunnel invert and h_{99} as the mixture depth with 99% air concentration. The curves $C(H_g)$ depend on the location downstream from the gate for small relative gate opening A, while this effect disappears with increasing A value. The air

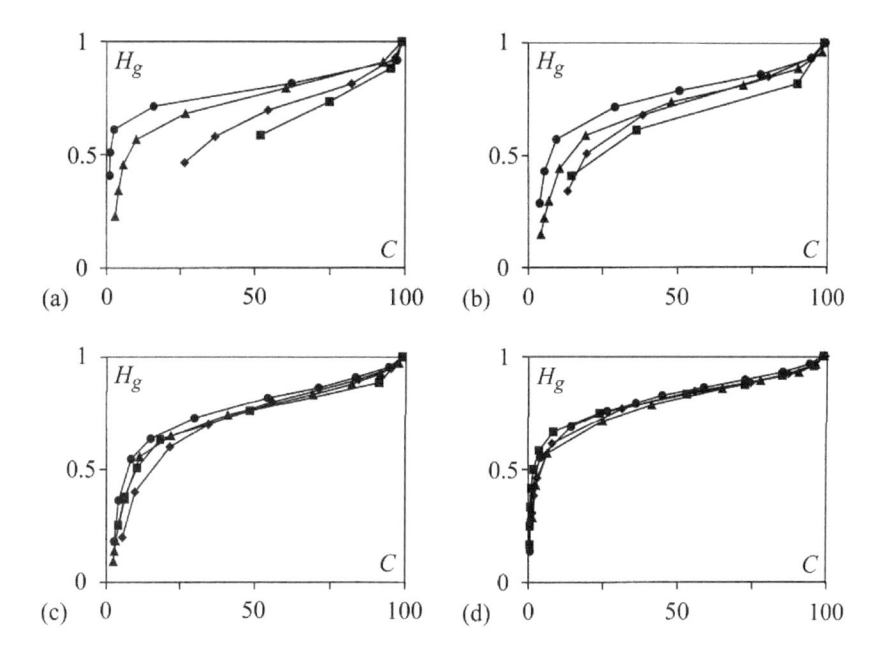

Figure 8.77 Air concentration distribution $C(H_g)$ at locations (■) A, (◆) B, (▲) C,(●) D for $H_E = 10$ m and relative gate openings $A =$ (a) 6.7%, (b) 13.3%, (c) 26.7%, (d) 40% (Speerli and Hager, 2000)

concentration is close to zero near the tunnel invert increasing in an S-shape toward the mixture surface. These distributions are comparable with those of spillway flows (Wood, 1991, Chapter 4).

The average air concentration $\bar{C}$ at a specific cross section is of concern. Figure 8.78 shows $\bar{C}(X_L)$ with $X_L = x/L_u \leq 1$ as relative tunnel length, measured from the gate section. Data were collected from section A ($x = 4.5$ m) to section D ($x = 17.75$ m) and completed close to the vena contracta section. The average air concentration increases sharply from the vena contracta section to section A, then decreasing due to air detrainment. Figure 8.78 also shows the average cross-sectional mixture velocities $\bar{v}_g$ versus X_L. Water discharge Q_W was related to the mixture velocity as $Q_W = (1-\bar{C})A_g\bar{v}_g$, with A_g as the cross-sectional area of mixture flow.

Of relevance is the maximum (subscript M) air concentration $\bar{C}_M$ downstream of the vena contracta section relative to the Froude number $\mathsf{F}_c = V_c/(gh_c)^{1/2}$ as (Figure 8.79a)

$$\bar{C}_M = 0.70[1-\exp(-0.05(\mathsf{F}_c-\mathsf{F}_0))]. \tag{8.82}$$

The threshold Froude number was fitted to $\mathsf{F}_0 = 6$ (Volkart, 1978). For large F_c result therefore average maximum air concentrations of up to 70%.

Reinauer and Hager (1996) derived an equation for the drawdown profile of hypercritical flow with Froude numbers $\mathsf{F} > 3$. Assuming shallow 1D flow in a rectangular tunnel of

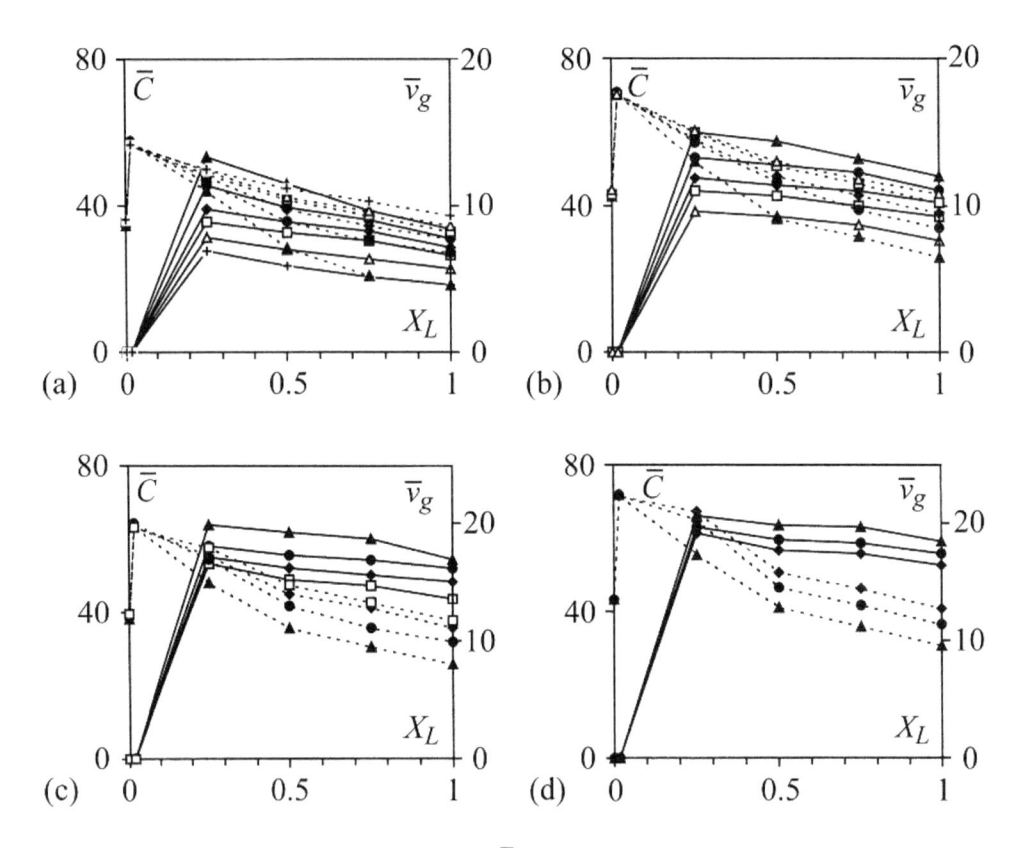

Figure 8.78 Average air concentration (—) $\bar{C}$ and ($\cdots$) mixture velocity $\bar{v}_g$ [m/s] versus relative tunnel length X_L for $H_E =$ (a) 10 m, (b) 15 m, (c) 20 m, (d) 25 m and relative gate openings $A =$ (▲) 6.7%, (●) 13.3%, (◆) 20%, (□) 26.7%, (△) 33.3%, (+) 43.3% (Speerli and Hager, 2000)

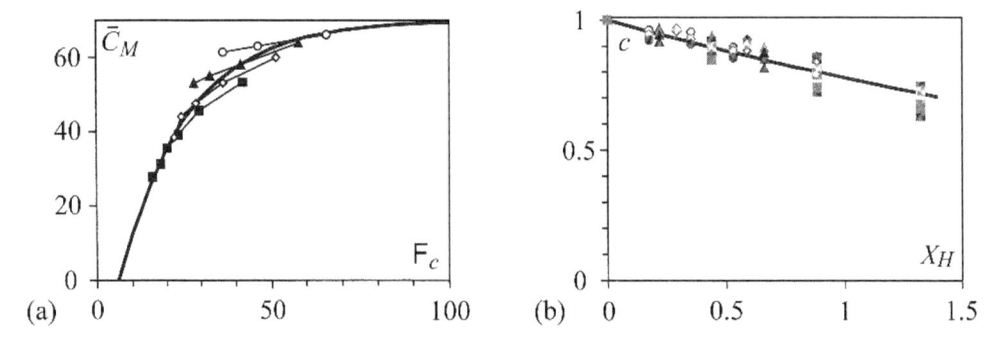

Figure 8.79 (a) Maximum average air concentration $\bar{C}_M(F_c)$ with (—) Eq. (8.82), (b) decay of relative air concentration $c = \bar{C}/\bar{C}_M$ versus streamwise coordinate X_H for $H_E =$ (■) 10 m, (◇) 15 m, (▲) 20 m, (○) 25 m for A between 6.7% and 43.3%, (—) Eq. (8.91)

small bottom slope S_u, for which the flow depth is more representative than the hydraulic radius, the drawdown equation for the mixture-flow depth $h_g(x)$ follows

$$\frac{\mathrm{d}h_g}{\mathrm{d}x} = \frac{S_u - S_f}{-\mathsf{F}_g^2}.$$ (8.83)

The friction slope S_f according to Darcy-Weisbach with f as friction coefficient is with subscript 0 relating to approach flow, and subscript N to uniform flow conditions

$$S_f = (1/8)f_0\mathsf{F}_0^2(h_0/h_N)^3.$$ (8.84)

Let $X_N = S_u x/h_N$ be relative location and $Y_g = h_g/h_N$ relative flow depth, so that Eq. (8.83) is expressed following Reinauer and Hager (1996) with $h_k = [Q_g^2/(gB_u^2)]^{1/3}$ as critical mixture-flow depth and $\chi = (h_N/h_k)^3(S_u x/h_N)$ as relative location

$$\frac{\mathrm{d}Y_g}{\mathrm{d}\chi} = 1 - Y_g^3.$$ (8.85)

Its solution is subject to the boundary condition $Y_g(\chi = 0) = 0$. A simplification involving the tangent-hyperbolic function is (Speerli and Hager, 2000)

$$Y_g = \tanh(1.1\chi).$$ (8.86)

For a given boundary (subscript b) mixture depth $Y_g = Y_{gb}$, the corresponding location χ_b is given by Eq. (8.86), thereby defining the downstream mixture flow depth (Figure 8.80a).

The previous analysis accounts for cross-sectional simplifications and neglects air entrainment and detrainment. Figure 8.80b shows a plot of all measured mixture-flow profiles $Y_g(\chi)$ indicating that the data follow with the previously determined parameters and a fit value $\chi_A = 0.1$ the expression (Speerli and Hager, 2000)

$$Y_g = \tanh(\chi - \chi_A)^{1/3}.$$ (8.87)

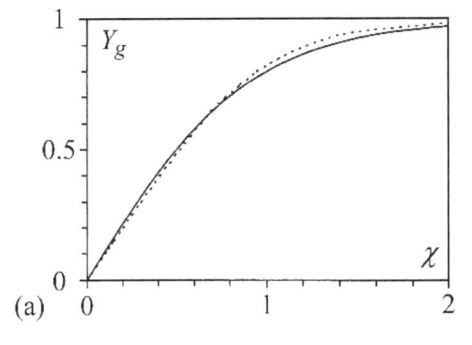
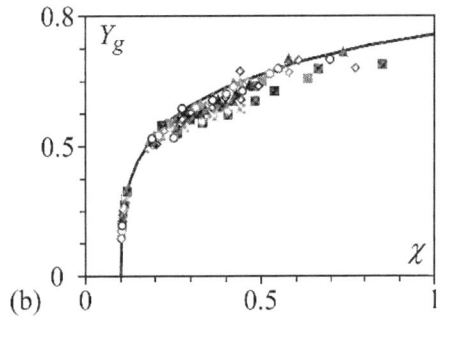

Figure 8.80 Relative mixture-flow depth profile $Y_g(\chi)$ (a) according to Eq. (8.86), (b) data compared with Eq. (8.87) for $H_E = $ (■) 10 m, (◇) 15 m, (▲) 20 m, (○) 25 m and for A between 6.7% and 43.3% (Speerli and Hager, 2000)

The mixture-flow depth $Y_g(\chi)$ from Eq. (8.87) increases more rapidly than from Eq. (8.86) because of air entrainment for small χ and less rapidly than that from Eq. (8.86) because of air detrainment for larger χ. Maximum values of $Y_g = 0.70$ were attained. For a sufficiently long tunnel, uniform flow results as $Y_g(\chi \to \infty) = 1$.

For a bottom slope of $S_u = 2\%$, the mixture-flow depth always increased. From the continuity equation $Q_W = (1 - \overline{C}) B_u h_g v_g$, the average air concentration $\overline{C}(x)$ is thus smaller than the corresponding decay of the function $\overline{v}(x)$. Accordingly, for a constant discharge per unit width Q_W/B_u, the mixture-flow depth must increase to finally attain the uniform flow depth. The tunnel model was too short for complete air detrainment containing uniform pure water flow far downstream from the gate. The uniform flow depth thus corresponds to the largest possible tunnel flow depth and, therefore, is of design relevance for long tunnels. This important finding is valid only for flows with a Froude number of uniform flow larger than about 2, thereby excluding undular flows in the transcritical regime (Chapter 5).

Both previously determined scalings χ and Y_g depend on the uniform flow depth h_N. For pure-water flow, it follows the Darcy-Weisbach equation as

$$S_u = S_f = \frac{v^2}{2g} \frac{f}{4R_h}. \tag{8.88}$$

Here $v = Q/(Bh_N)$ is the cross-sectional average velocity, f the friction coefficient, and R_h the hydraulic radius. The laboratory experiments were conducted in the turbulent smooth regime, for which f varies only with the Reynolds number R. For $10^5 < \mathrm{R} < 10^7$, the Colebrook-White equation simplifies to $f = 0.20\mathrm{R}^{-0.2}$. Inserting into Eq. (8.88), the dimensionless discharge $q\mathrm{R} = Qv^{1/9}/[B^{8/3}(gS_u)^{5/9}]$ is related to the relative tunnel flow depth $m = h_N/B$ as (Speerli and Hager, 2000)

$$q\mathrm{R} = 4.4m^{4/3}, \quad \text{if } 0.2 < m < 1.5. \tag{8.89}$$

The uniform pure-water flow depth, i.e., the scaling introduced previously, varies mainly with discharge per unit width Q/B, and slightly with the bottom slope S_u as

$$h_N = \frac{Q^{3/4}v^{1/12}}{3B(gS_u)^{5/12}}. \tag{8.90}$$

The streamwise coordinate $\chi = (h_N/h_k)^3(S_u x/h_N)$ is thus also described as $\chi = x/x_0$ with the scaling length $x_0 = 8Q^{1/2}(vgS_u)^{-1/6}$ when inserting h_N and h_k. This length scale is relevant for the longitudinal development of the mixture-flow profile.

For a given air concentration $\overline{C}(x)$ and maximum air concentration $\overline{C}_M$ from Eq. (8.82) follows the decay of the average air concentration. Let $c = \overline{C}/\overline{C}_M$ be the reference air concentration, and the streamwise coordinate be $X_H = (x - x_A)/H_E$, with x_A as the distance from the gate to section A. Figure 8.79b indicates that (Speerli and Hager, 2000)

$$c = \exp(-0.25X_H). \tag{8.91}$$

The relative concentration decreases exponentially with the downstream location. For small X_H, the concentration decay is $d\overline{C}/dx = -0.25\overline{C}_M/H_E$, or $d\overline{C}/d(x/C_c a) = -0.50\mathrm{F}_c^{-2}$ when accounting for the energy equation. It asymptotically tends to $d\bar{C}/d(x/C_c a) = -0.35F_c^{-2}$, i.e. varies inversely with the square of the contracted Froude number.

Design recommendations

Bottom outlets are a significant component of each modern dam. Velocities reach 30–40 m/s with maxima up to 60 m/s. The failure of a bottom outlet has serious consequences for the entire dam structure. Based on present and past experiments, the proposed design guidelines include:

- Except for spray flow conditions due to small gate opening, the tunnel should be designed for supercritical mixture flow,
- Design Froude number of water flow should be at least 2, or better even 3, to inhibit surface undulations,
- Submergence from tunnel outlet is to be avoided to exclude hydraulic jumps inducing slug flow,
- Tunnel height should be significantly larger than gate height to allow for free surface air-water mixture tunnel flow,
- Tunnel filling should be less than 70% to inhibit slug flow,
- Maximum airflow velocities for which effects of air compressibility are absent should be 50 m/s; higher velocities may induce undesirable vibrations,
- Maximum air sub-pressure should be around −1.5 m of water head, and
- Aerators are required if bottom air concentration falls below minimum for cavitation protection (Falvey, 1980).

These indications demonstrate that further research is needed mainly in terms of cavitation protection. Few results are available regarding abrasion of bottom outlets. In addition, the choking flow features, i.e. the breakdown of free surface air-mixture flow and the transition to pressurized flow have also not been investigated so far.

Computational example

Development of mixture-flow depth

1. Water discharge $Q_W(H_E)$ $Q_W = C_c a B_u [2g(H_E - C_c a)]^{1/2}$
2. Uniform flow depth from Eq. (8.90) $h_N = Q_W^{3/4} v_W^{1/12}/[3B_u(gS_u)^{5/12}]$
3. Scaling length $x_0 = 8Q_W^{1/2}/(v_W g S_u)^{1/6}$
4. Relative position in tunnel $\chi = x/x_0$
5. Corresponding mixture-flow depth from Eq. (8.87) with $\chi_A = 0.10$, $Y_g = \tanh(\chi - \chi_A)^{1/3}$
6. Mixture-flow depth is then $h_g = Y_g h_N$, so that mixture profile $h_g(\chi)$ is determined

Air concentration

7. Maximum air concentration from Eq. (8.82) $\bar{C}_M = 0.70[1 - \exp(-0.05(F_c - 6))]$
8. Streamwise coordinate $X_H = (x - x_A)/H_E$
9. Decay of air concentration from Eq. (8.91) $c = \exp(-0.25X_H)$
10. Air concentration then follows $\bar{C} = c\bar{C}_M$

Consider a tunnel length of $L_u = 200$ m, tunnel width of $B_u = 2$ m, approach flow energy head of $H_E = 100$ m, gate opening of $a = 0.50$ m, a contraction coefficient of $C_c = 0.67$, a bottom slope of $S_u = 2\%$, and water viscosity of $v_W = 1 \times 10^{-6}$ m²/s. Results:

1. $Q_W = 29.6$ m³/s
2. $h_N = 1.32$ m
3. $x_0 = 571$ m
4. Since $Y_g(0) = h_c/h_N = 0.335/1.32$, $\chi(0) = 0.117$ so that $x_A(0) = x_0\chi(0) = 67$ m. At tunnel outlet $\chi(267$ m$) = 0.468$
5. $Y_g = 0.615$
6. Mixture-flow depth at tunnel outlet is $h_g = 0.81$ m
7. With $\mathsf{F}_c = 24.4$, Eq. (8.82) gives for $\bar{C}_M = 0.42$
8. With $x_A/L_u = 0.25$ (Figure 8.79a), $X_H = (x-0.25L_u)/H_E = (X_L-0.25)(L_u/H_E) = 2X_L-0.5$. At tunnel outlet, $X_L = 1$, and $X_H = 1.5$
9. $c(X_H = 1.5) = 0.69$
10. At tunnel outlet, mean air concentration is $\bar{C} = 0.29$.

8.6.8 Recent experimentation on air demand

Introduction

Bottom outlets constitute a key safety element of high-head dams. Their main purposes are the control and fast drawdown of the reservoir water level during floods, imminent danger, and structural damage of the dam or maintenance works. Additional purposes include residual discharge release, flood diversion or sediment flushing. The transition from pressurized to free-surface flow at the gate section results in a high-speed water jet in the tailwater tunnel. For the large energy heads encountered, this can result in flow velocities of up to 50 m/s. These high velocities associated with an extreme turbulence level lead to considerable air entrainment into the flow, with a potential to create negative air pressures in the tunnel. These are known to aggravate problems with gate vibrations, cavitation damage and slug flow formation (Dettmers, 1953; Falvey, 1980; Hohermuth, 2017). These problems can be mitigated by sufficient aeration via air vents. The question of how much air supply is sufficient to prevent the above stated problems has not yet been answered, however. Despite the relevance of bottom outlets, design recommendations and research on high-speed tunnel flow is surprisingly scarce. Available recommendations for bottom outlet design are mostly qualitative (e.g. Douma, 1955; USACE, 1964; Falvey, 1980; Vischer and Hager, 1998). An overview of existing air demand formulae is given in Figure 8.81 and Table 8.1.

The air demand of closed-conduit flow was first studied in small-scale experiments by Kalinske and Robertson (1943). They found that the air demand $\beta = Q_{Ao}/Q_W$, i.e., the ratio of air discharge through the air vent Q_{Ao} to the water discharge Q_W increases with increasing Froude number F_o upstream of the hydraulic jump (Table 8.1). Based on prototype data from five dams in the U.S., Campbell and Guyton (1953) observed a higher air demand, attributed to turbulence and viscous effects. Wunderlich (1963) deduced β from air velocity profiles above the mixture surface. A similar attempt was made by Campbell and Guyton (1953), but then linking β to the Froude number at the vena contracta (subscript c) F_c (Table 8.1). The U.S. Army Corps of Engineers (USACE, 1964) proposed a design equation similar to Campbell and Guyton (1953) based on additional prototype data from Norfork and Pine Flat

Table 8.1 Overview of existing air demand design equations for bottom outlets, with Spray = spray flow, FSF = free-surface flow and HJ = hydraulic jump (Hohermuth et al., 2020)

Authors	Equation	Flow pattern	Data
Kalinske and Robertson (1943)	$\beta = 0.0066(F_o - 1)^{1.4}$	HJ	Model tests: $L_t = 9.75$ m, $h_t = 0.15$ m (circular)
Campbell and Guyton (1953)	$\beta = 0.04(F_c - 1)^{0.85}$	FSF, HJ, large A_M	Prototype data: Denison, Hulah, Norfork, Pine Flat, and Tygart dam (USACE, 1964)
Wunderlich (1963)	$\beta = r\left(\dfrac{A_t}{A_c} - 1\right),\quad r = \left(\dfrac{A_c}{A_t}\right)^2$	FSF	Prototype data: Lumiei (Dettmers, 1953), Mauvoisin (Schilling, 1963), Norfork Dams (USACE, 1954)
USACE (1964)	$\beta = 0.03(F_c - 1)^{1.06}$	FSF, HJ, large A_M	Prototype data (USACE, 1964)
Levin (1965)	$\beta = K(F_c - 1),\quad K = 0.025 - 0.12$	FSF	Prototype data (not specified)
Wisner (1967)	$\beta = 0.033(F_c - 1)^{1.4}$	Spray	Model tests (not specified), prototype data (Campbell and Guyton, 1953; Guyton, 1958; Mura et al., 1959)
	$\beta = 0.024(F_c - 1)^{1.4}$	FSF	
	$\beta = 0.014(F_c - 1)^{1.4}$	HJ ($F_c > 8$)	
	$\beta = 0.04(F_c - 1)^{0.85}$	HJ ($F_c < 8$)	
Lynse and Guttormsen (1971)	$\beta = 1.2\left(\dfrac{A_t}{A_c}\right)^{0.2} - 1$	FSF, HJ, large A_M	Prototype data (own data, USACE, 1964)
Sharma (1973)	$\beta = 0.2\mathsf{F}_c$	Spray	Model tests: $L_t = 8$ m, $w_t = 0.1$ m, $h_t = 0.15$ m, Prototype data of above authors
	$\beta = 0.09\mathsf{F}_c$	FSF	
	$\beta = 0.0066(F_c - 1)^{0.85}$	HJ	
Rabben (1984)	$\beta = \left(\dfrac{A_v^*}{A_t}\right)\mathsf{F}_c,\quad A_v^* = \dfrac{A_v}{(\zeta + 1)^{0.5}}$	Spray	Model tests: $L_t = 4$ m, $w_t = 0.15$ m, $h_t = 0.32$ m
	$\beta = 0.94\left(\dfrac{A_v}{A_t}\right)^{0.9}\mathsf{F}_c^{0.62}$	FSF	
	$\beta = 0.019\left(\dfrac{A_v}{A_t}\right)^{0.099}\mathsf{F}_c^{0.969}$	HJ	
Speerli (1998)	$Q_{Ao} = 0.022 H_E\left(\dfrac{L_t}{h_t}\right)^{0.167} A_M^{0.5}\left(gw_t^3\right)^{0.5}\zeta^{-0.43}$	FSF	Model tests: $L_t = 20$ m, $w_t = 0.3$ m, $h_t = 0.45$ m

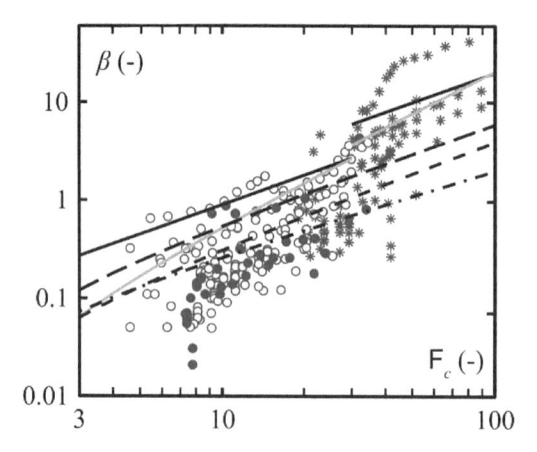

Figure 8.81 Comparison of prototype air demand β versus the Froude number at the vena contracta F_c with available design equations. Prototype data and design equations are summarized in Table 8.1. (●) Spray, (O) FSF, (●) HJ, (—) Sharma (1973), (——) Wisner (1967), (— — —) Levin (1965), (- - -) USACE (1964), (-··-) Campbell and Guyton (1953)

Dams. The main American investigations focused on bottom outlets where the gate area A_g is equal to the cross-sectional area of the tunnel A_t. For these, the hydraulic jumps in the tunnel lead to smaller β as compared with free-surface flow if the gate opening is sufficiently large. Levin (1965) included a shape factor K to account for the effects of profile transition, i.e. a distinction between $A_g = A_t$ (Type A), $A_g < A_t$ with a smooth transition (Type B), and $A_g < A_t$ with an abrupt transition (Type C) to the outlet tunnel. Wisner (1967) distinguished between the flow patterns spray, free-surface flow and hydraulic jump. Ghetti and Di Silvio (1967) recognized the effect of the air vent loss coefficient ζ on β, yet without presenting a design equation. Lynse and Guttormsen (1971) developed a design equation for β including only the cross-sectional flow areas A_c and A_t (Table 8.1) based on data of USACE (1964) and prototype data of two Norwegian dams. Sharma (1973) defined six possible tunnel flow patterns; (i) spray flow, (ii) free-surface flow, (iii) foamy flow (full flowing tunnel), (iv) hydraulic jump followed by free flow, (v) hydraulic jump followed by pressurized flow and (vi) pressurized flow (submerged jump). Air demand design equations for flow patterns (i) to (iii) were presented. Rabben (1984) systematically varied the loss coefficient of the air vent ζ including its effect on β via a reduced cross-sectional area of the air vent A_v^* (Table 8.1). Speerli (1999) was the first to include the effect of tunnel length L_t in Q_{Ao}. However, this is limited to relative gate openings below 40%.

The air demand β is typically expressed as a function of the Froude number at the vena contracta F_c. Prototype data and existing design equations scatter over a wide range, even if the flow pattern is accounted for (Figure 8.81). Therefore, it is admitted that the scatter is due to effects of the air vent and tunnel geometries. The main objectives of Hohermuth *et al.*'s (2020) study was to determine the governing air demand parameters of bottom outlets and to improve general process understanding. Hydraulic model tests were conducted to allow for a systematic variation of hydraulic conditions, air vent loss characteristics, and both tunnel length and slope. The resulting design equations and guidelines for the air demand include the effects of all relevant parameters. However, the effects of the currently available prototype data are not yet clearly assessed, so that additional research is required.

Experimental Setup

A physical model of a typical low-level outlet was built at the Laboratory of Hydraulics, Hydrology and Glaciology (VAW), ETH Zurich, Switzerland (Figure 8.82a). Two high-head pumps provided a maximum energy head at the gate of $H_E = 30$ m at a water discharge of $Q_W \approx 600$ l/s. A rectangular sharp-crested gate without gate slots was installed to control the discharge (Figure 8.82b). The gate had a maximum opening of $a_{max} = 0.25$ m and was operated by a stepper motor allowing for a continuous variation of the gate opening a. The rectangular tunnel had a constant width of $w_t = 0.2$ m and an approach flow height h_o equaling the maximum gate height h_g, i.e., $h_o = h_g = a_{max} = 0.25$ m upstream of the gate, that increased to the tunnel height (invert to crown) of $h_t = 0.3$ m downstream of the gate (Figure 8.82b).

The outlet tunnel had a maximum downstream tunnel length of $L_t = 20.6$ m that was varied using detachable elements. All channel walls were of PVC or of acrylic glass with a hydraulic roughness of $k_s \approx 0.01$ mm. The gate chamber was connected to a circular air vent. Two air vent diameters d were tested and orifice plates were used to vary the loss coefficient ζ of the air-supply system, leading to different air vent parameter values $A_v^* = A_v/[A_t(\zeta+1)^{0.5}]$ given in Table 8.2, with A_t, A_v as the tunnel and air vent cross-sections, respectively. A similar air vent was located at the downstream tunnel end to measure the air flow Q_{Au} into or out of the tunnel (Figure 8.83). The

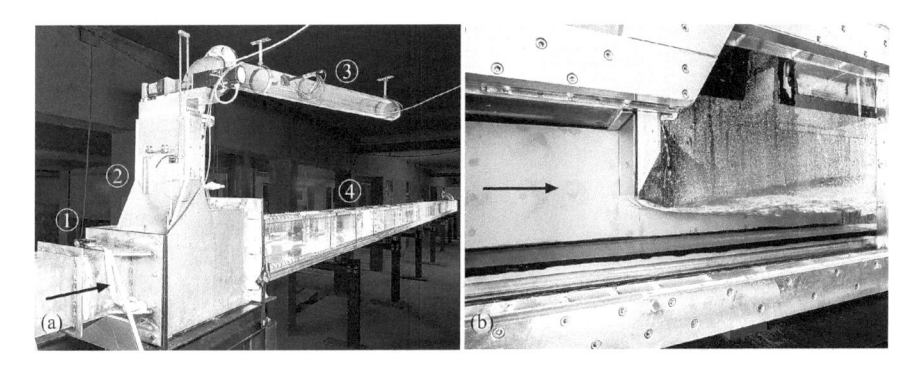

Figure 8.82 (a) Bottom outlet model with ① pressurized inflow, ② gate chamber, ③ air vent, ④ bottom outlet tunnel, (b) gate chamber for relative gate opening of $a/a_{max} = 0.4$ (Hohermuth *et al.*, 2020)

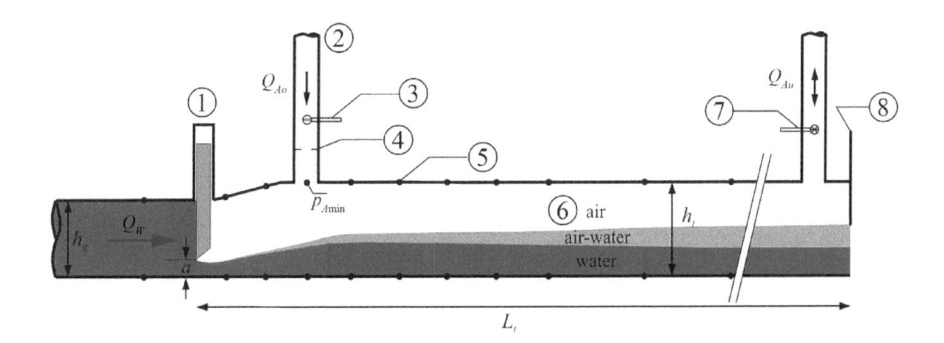

Figure 8.83 Sketch of bottom outlet scale model, with notation and ① gate chamber, ② air vent, ③ thermal anemometer, ④ orifice plate, ⑤ pressure transducer, ⑥ air-water flow, ⑦ vane anemometer, ⑧ adjustable tailwater gate (Hohermuth *et al.*, 2020)

Table 8.2 Overview of used air vent diameters d, orifice openings, loss coefficient ζ, and air vent parameter A_v^*

d (m)	A_v/A_t (-)	orifice opening (%)	ζ (-)	A_v^* (-)
0.1	0.13	100	0.7	0.10
		60	2.7	0.07
		40	9.3	0.04
		30	20	0.03
		26	28	0.02
0.15	0.29	100	0.1	0.28
		60	1.9	0.17
		40	8.3	0.10
		30	18.2	0.07

downstream air vent had a diameter of $d = 0.15$ m and a loss coefficient of $\zeta \sim 0.1$. Comparative measurements with and without the downstream air vent showed no significant differences in Q_{Ao} and the air pressure inside the tunnel. The tunnel cross-section above the air-water mixture was blocked with a gate at the downstream tunnel end (Figure 8.83). The gate had a 5 cm long rubber lip, which was adjusted so that the rubber lip would graze over the air-water flow mixture. Consequently, air had to flow through the second air vent thereby enabling a measurement of Q_{Au} with ±20% accuracy. Depending on the flow conditions, the air flow through the downstream air vent was directed either into or out of the tunnel. Choosing w_t as reference length results in a scale factor of $\lambda \approx 5$–20 for typical reservoir dam low-level outlets.

The water discharge Q_w was measured with an inductive flow meter of ±0.5% accuracy of the Measurement Value (MV) and ±0.4 l/s Absolute Error (AE). The centerline air flow velocity in the air vent V_{Ao} was measured with a thermal anemometer (Höntzsch TA10–185GE 140/p16 ZG1b) while the centerline air flow velocity at the downstream tunnel end V_{au} was measured with a bidirectional vane anemometer (Höntzsch ZSR25GFE-mn120/100/p6). The thermal anemometer had a MV accuracy of ±2.5% whereas the vane anemometer had a MV accuracy of ±1.5% and an AE value ±0.2 m/s. The air discharges through the air vent Q_{Ao} and through the downstream tunnel end Q_{Au} were computed iteratively from the measured air velocity assuming a logarithmic velocity profile. The minimum air pressure $p_{A,\min}$ was recorded at the end of the air vent (Figure 8.83). The water pressure measured upstream of the gate $p_{W,o}$ was used to compute $H_E = p_{W,o}/(\rho_W g) + [Q_W/(w_t h_g)]^2/(2g)$ with ρ_W = water density and g = gravity acceleration. The contraction coefficient C_c of the gate was determined from ortho-rectified images of the gate outflow resulting in values from 0.56 to 0.68, consistent with the approach of Montes (1997). The Froude number at the vena contracta (subscript c) F_c was calculated as $F_c = [Q_W/(aw_t C_c)]/(gaC_c)^{0.5}$. All measurements were averaged over 180 s to ensure a stable mean value for all tested conditions.

Test Program and Procedure

The following parameters were systematically varied to investigate their effects on the air demand. For the horizontal tunnel ($S_t = 0.00$), the relative gate opening a/a_{max} was increased from 0.1 to 0.8 in steps of 0.1 and H_E was varied from 5 to 30 m in steps of 5 m. Five orifice plates were applied resulting in ζ-values of $\zeta = 2p_{Amin}/(0.5\rho_A V_{Ao}^{2})-1 = 0.7, 2.7, 9.3, 20$, and 28 for $d = 0.1$ m, leading to the A_v^*-values given in Table 8.2. All parameter combinations were tested for three tunnel lengths $L_t = 20.6, 12.6$, and 6.6 m resulting in 368 test runs with free-surface flow conditions.

For a tunnel slope of $S_t = 0.04$, only tunnel lengths of $L_t = 20.6$ and 6.6 m were tested and a/a_{max} was increased from 0.2 to 1 in steps of 0.2. Four different orifice plates were tested for the larger air vent $d = 0.15$ m (Table 8.2), leading to another 176 test runs with free-surface flow. The tunnel width w_t and height h_t were not varied during these tests. The water discharge Q_W, the energy head at the gate H_E, the Froude number at the vena contracta F_c, air discharges Q_{Ao} and Q_{Au}, and the air pressure in the tunnel were determined for each test run. Hydraulic jumps and pressurized flow in the downstream tunnel were not investigated in this study.

Scale effects have to be considered, especially if two-phase flow is involved. Froude scaling was applied leading to smaller Reynolds R and Weber W numbers than in prototypes. It is generally accepted that scale effects are negligible if R and/or W are above a certain threshold (Heller, 2011). The threshold values are established by a comparison with prototype data. The only available scale series for bottom outlets was done by Rabben (1984). A final conclusion for free-surface flow was not possible because the measurement errors for the smaller models were too large. Consequently, Rabben (1984) compared his model-derived formula with prototype data for the Norfork (USACE, 1954) and the Mauvoisin (Schilling, 1963) Dams. He states that a 'quite good' agreement was achieved for the Norfork dam, whereas his equation underestimates the air demand at Norfork Dam by up to 40% and by a factor of more than 3 at Mauvoisin Dam. As no final conclusion can be drawn, the present findings will be compared to prototype data in the discussion below.

Minimum threshold numbers are available for two related physical phenomena, namely the air entrainment in chute flow and the disintegration of water jets in air. The strictest criteria for air entrainment in chute flows is due to Skripalle (1994) with $W_c = (V_c^2 \rho h_c/\sigma)^{0.5} > 170$, where ρ = liquid density, and σ = surface tension coefficient (Pfister and Chanson, 2012). No significant scale effects regarding air entrainment are expected in all present tests, therefore (Hohermuth et al., 2020).

Model effects are not caused by the model scale but are attributed to an incorrect representation of the prototype, e.g., geometric details or inflow conditions. The geometry of the gate and the gate slots have a decisive effect on the spray formation (Sharma, 1973; Rabben, 1984; Speerli, 1998). Its formation is underestimated herein as the gate slots were not modeled. Wall roughness has no significant effect on β (Sharma, 1973). Thus, no effect of the small tunnel roughness (corresponding to steel or smooth concrete) is expected. Based on these concerns, spray flow is excluded from the present data analysis due to scale and model effects, while free-surface flow data is not expected to be influenced to a significant degree.

Effects of flow pattern, gate opening and energy head

The flow pattern has a strong effect on air demand (Figure 8.84). Three regions occur: (I) spray flow for small relative gate openings, (II) free-surface flow characterized by an

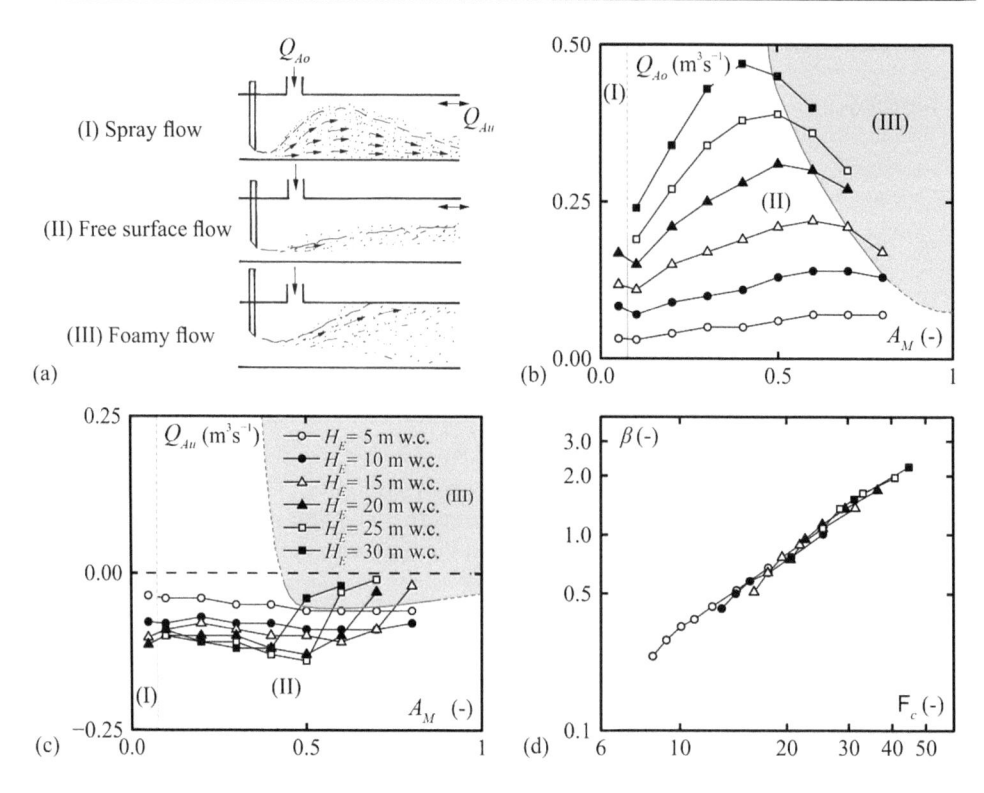

Figure 8.84 (a) Flow patterns as defined by Sharma (1973), effect of flow patterns, H_E (see c) and $A_M = a/a_{max}$ on (b) Q_{Ao}, and (c) Q_{Au}, (d) air demand $\beta(F_c)$, data for $S_t = 0$, $L_t = 20.6$ m, and $\zeta = 0.7$ (Hohermuth *et al.*, 2019)

almost continuous increase in Q_{Ao} with increasing A_M, and (III) foamy flow marked by a distinct drop in Q_{Ao} and an increase in Q_{Au}. For foamy flow, for which the tunnel is flowing full, the decrease in Q_{Ao} is explained by the decreasing cross section available for air transport above the mixture flow and a successive increase in Q_{Au}. Small A_M lead to $Q_{Au} < 0$, i.e. an airflow out of the tunnel. This is attributed to the rough air-water mixture surface dragging air out of the tunnel. However, for foamy flow at large A_M, Q_{Au} tends to zero indicating no airflow into or out of the tunnel end. As airflow above the air-water mixture is blocked for foamy flow conditions, Q_{Ao} originates only from air entrained into the flow. This leads to smaller Q_{Ao} as compared with free-surface flow, where air is also dragged out of the tunnel above the air-water mixture, adding to Q_{Ao}.

In the free-surface flow region, Q_{Ao} increases with increasing H_E (Figure 8.84b). For small H_E air discharge increases with increasing A_M, leading to maximum values for the largest gate opening. The maximum in Q_{Ao} is shifted towards smaller gate openings of $A_M = 0.4$ to 0.6 as H_E increases. Smaller gate openings have a different behavior, because Q_{Ao} from air entrainment is smaller than for larger H_E and foamy flow is not reached for large A_M. An increase in Q_{Ao} does not necessarily indicate an increase in $\beta = Q_{Ao}/Q_W$, as Q_W also increases with both increasing H_E and A_M.

For one specific bottom outlet configuration, the air demand β follows the power law $\beta = c_1 \mathsf{F}_c^{c_2}$, with c_1, c_2 as constants for each bottom outlet configuration (Figure 8.84d). For tests with constant ζ, L/h_t and S_t, the data for β collapse, demonstrating that the effects of H_E and A_M are incorporated in F_c. However, c_1 and c_2 vary as the air vent loss coefficient ζ, relative tunnel length L/h_t or tunnel slope S_t are modified.

Effects of air vent resistance, tunnel length and bottom slope

The air vent loss coefficient ζ has a strong effect on β (Figure 8.85a). An increase in ζ causes a significant decrease in β, as less air is supplied across the air vent. The almost parallel shift of the trend lines indicates a change in c_1, while c_2 (the exponent in $\beta = c_1 \mathsf{F}_c^{c_2}$) remains almost constant. Increasing ζ further leads to a decrease in p_{Amin}, as the tunnel is not sufficiently supplied with air. (Figure 8.85b). The minimum air pressure follows $p_{Amin} = (\zeta + 1)\rho V_{Ao}^2$. In analogy with a water pump, where the bottom outlet has the ability to transport a certain Q_{Ao} for a given pressure difference, different operating points as illustrated by the arrows in Figure 8.85b apply. The operation point is controlled by changing ζ; a small ζ leads to a large Q_{Ao} and a moderate p_{Amin}, while a large ζ leads to a smaller Q_{Ao} at a significantly smaller p_{Amin}. Excessively small p_{Amin} may aggravate problems with cavitation and gate vibrations. Furthermore, the airflow Q_{Au} from the downstream tunnel end increases with decreasing p_{Amin}, creating a countercurrent airflow in the tunnel. Strong countercurrent airflows set up instabilities at the air-water interface eventually leading to slug flow (Hohermuth, 2017). The minimum tolerable p_{Amin} has to be determined site specifically; general recommendations range from -0.5 to -2 m. (Dettmers, 1953; Douma, 1955; Wunderlich, 1963). Note that the magnitude of pressure fluctuations increases with decreasing mean pressure, possibly leading to gate vibrations even prior to the minimum tolerable mean pressure is reached. Air vent loss coefficients of $\zeta > 20$ are rarely observed in prototypes; they are generally not recommended, and thus excluded from the final data analysis.

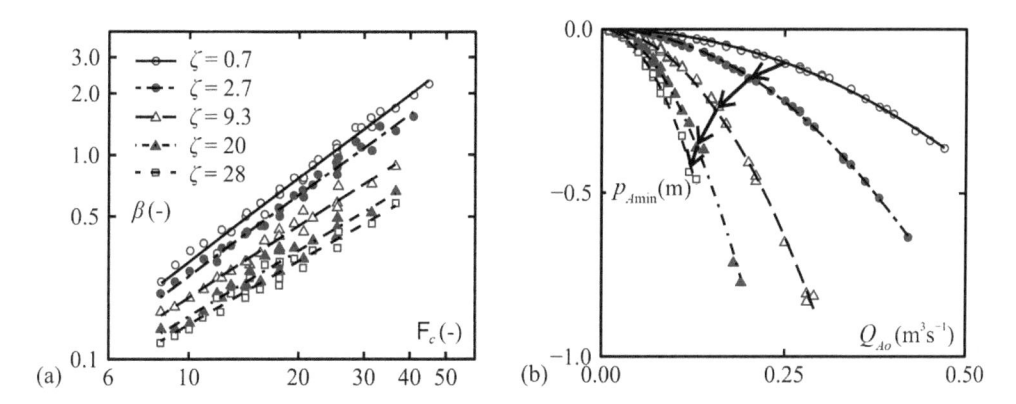

Figure 8.85 (a) Air demand $\beta(\mathsf{F}_c)$ for different ζ-values, (b) air pressure p_{Amin} at the air vent end versus Q_{Ao}. Arrows indicate whether ζ increases for a specific bottom outlet. Data for $S_t = 0$ and $L_t = 20.6$ m (Hohermuth *et al.*, 2020)

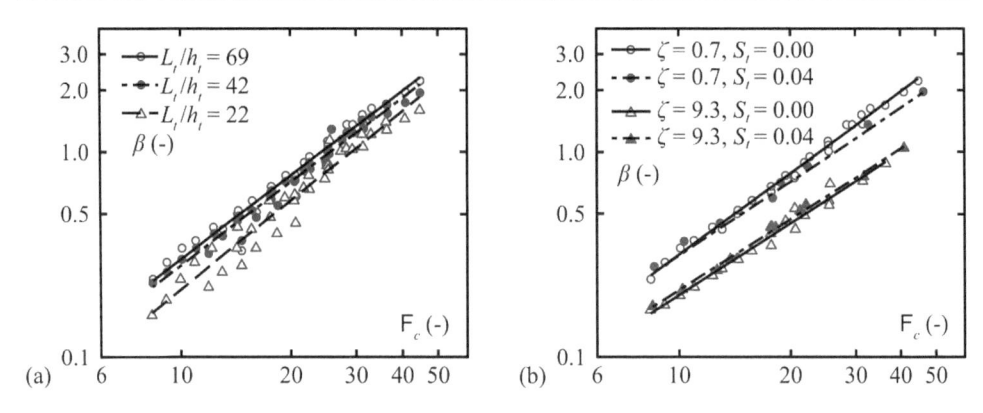

Figure 8.86 Effects of (a) relative tunnel length L_t/h_t, and (b) tunnel slope S_t on $\beta[\mathsf{F}_c]$. Data in (a) for $S_t = 0$ and $\zeta = 0.7$, data in (b) for $L_t = 20.6$ m and different ζ and S_t as indicated (Hohermuth *et al.*, 2020)

A decrease in relative tunnel length L_t/h_t decreases β (Figure 8.86a), as air is able to enter from the downstream tunnel end more easily, leading to an increase in Q_{Au} with decreasing L_t/h_t. Further, Q_{Au} increases for small L_t/h_t because the apparent loss coefficient of the outlet tunnel for air flow decreases with decreasing L_t/h_t. The minimum air pressure p_{Amin} also increases with L_t/h_t due to a decrease in β (i.e. Q_{Ao}).

Increasing S_t from 0.00 to 0.04 leads to slightly smaller mixture-flow depths, due to the accelerating gravity effect. This effect is apparent especially further downstream of the gate, as the flow depth close to the gate is mainly governed by H_E for a given gate opening. Consequently, the free cross-sectional area above the mixture flow slightly increases, thereby allowing for more air to enter from the downstream tunnel end. This increase in Q_{Au} leads to a slight but noticeable decrease in β, especially for small ζ. However, no significant effect was measurable for larger ζ (Figure 8.86b).

Governing flow patterns

The main flow processes encountered in bottom outlets were photographed. As noted from Figure 8.82b, the flow detaches at the sharp gate lip and shock waves are formed along both tunnel sides. These shocks are induced by corner vortices upstream of the gate (Roth and Hager, 1999) and their height increases with decreasing relative gate opening. For a small relative gate opening and high energy heads, the shocks generate intense spray. The shock height decays some 5 m downstream of the gate section. A backwater curve develops for the air-water mixture beyond the vena contracta section along with flow aeration due to the high flow velocity and the turbulence level.

Spray flow occurs for $A_M < 0.2$; however, due to the absence of gate slots and smaller absolute velocities, the spray is less pronounced than in prototypes. For $0.2 \le A_M < 0.6$, free-surface flow was noted for all tested parameter configurations. At higher relative gate openings $0.6 \le A_M \le 0.8$, the tunnel close to its downstream end is almost flowing full, corresponding to foamy flow (Sharma, 1973). A hydraulic jump with subsequent fully pressurized flow is generated once the mixture surface reaches the tunnel soffit for $A_M > 0.8$. The hydraulic jump travels upstream resulting in pressurized flow with a drowned aeration portion. The tunnel outlet remained unsubmerged for all tests, so that a hydraulic jump with subsequent free-surface flow was not observed.

Throttling the aeration chamber associated with a large loss coefficient ζ leads to flow pulsations, especially for large energy heads. Kelvin-Helmholtz instabilities are set up at the air-water interface so that the mixture surface is sucked toward the tunnel soffit leading to foamy slugs, completely filling the tunnel cross section (Figure 8.87a). Only one slug is formed at a certain time. These slugs travel downstream along the entire tunnel and completely block the air supply from the tunnel end (Figure 8.87b–d). Stratified free-surface flow was observed in the tailwater of these slugs (Figure 8.87e, f). The slugs lead to a large negative pressure in the tunnel section upstream of the slug. For high energy heads 20 m $\leq H_E \leq 30$ m, pressure drops are below the measurement limit of -1 m in the model. In

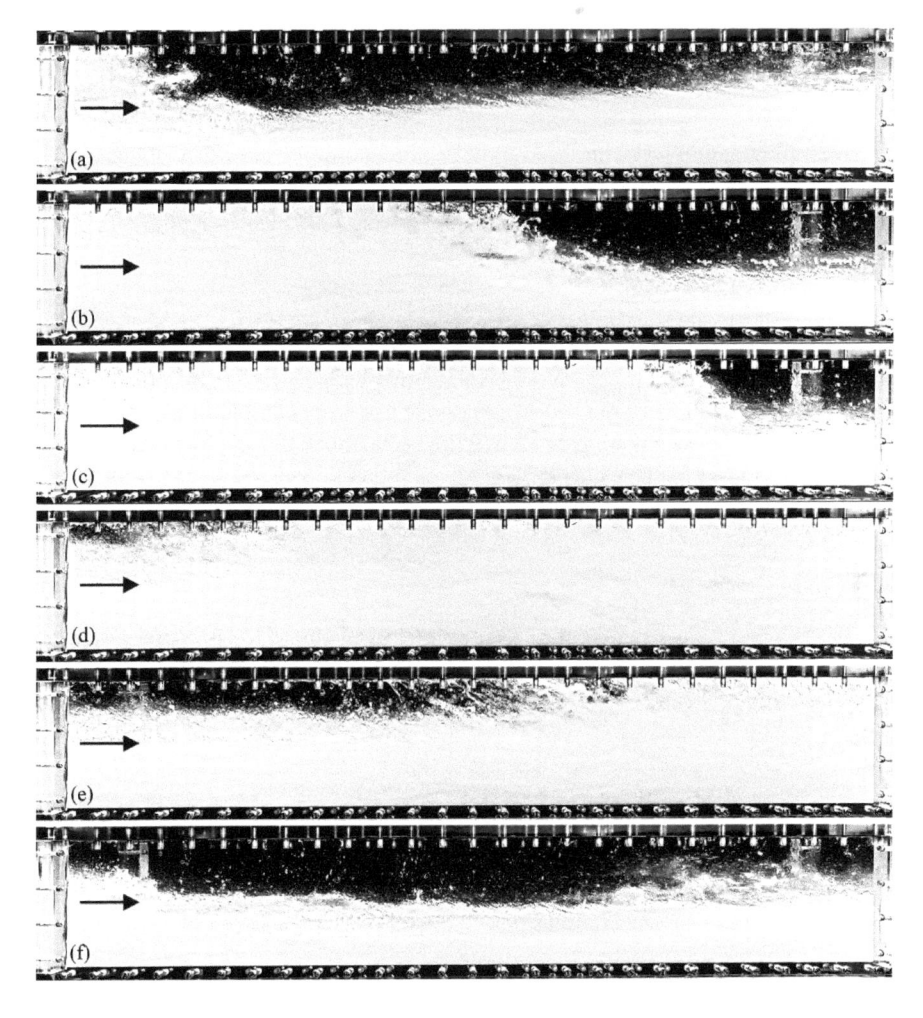

Figure 8.87 Photo sequence (a) to (f) of foamy slug flowing from left to right for $H_E = 25$ m, $L = 20.6$ m, $A_M = 0.4$, $\zeta = 24$. Photos taken 10 m downstream of gate as side views (Hohermuth, 2017)

prototypes with higher H_E, slug flow can lead to severe cavitation damage and flow chocking, to be avoided.

Design equation for air demand

Based on a dimensional analysis, the following governing parameters for relative air demand were identified: $\beta = f$ (flow pattern, F_c, A_v^*, L/h_t, S_t). Free-surface flow is the desired (design) flow pattern, whereas spray flow is affected by scale and model effects. Thus, the regression was limited to free-surface flow, resulting for the *model data* in (Hohermuth *et al.* 2020)

$$\beta = 0.037\,F_c^{1.3}\left(\frac{A_v}{A_t\left(\zeta+1\right)^{0.5}}\right)^{0.8}\left(\frac{L_t}{h_t}\right)^{0.25}\left(1+S_t\right)^{-1.5}. \tag{8.92}$$

Equation (8.92) indicates a dominant effect of F_c on β with an exponent of 1.3, followed by the effect of the air vent parameter, implicitly containing the effect of ζ with an exponent of −0.4. An increase in L/h_t increases β with an exponent of 0.25, whereas the decreasing effect of $(1+S_t)$ follows a power function with an exponent of −1.5. However, as the range tested in S_t was small (0.00 and 0.04), similar to most prototype low-level outlets, the effect is small yet statistically significant. Excluding S_t from the regression has no significant effect on the pre-factor and the exponents if rounded off to the nearest 0.01, as given in Eq. (8.92). Therefore, S_t is kept in the regression. Equation (8.92) is valid within the following parameter ranges $8 \le F_c \le 45$, $0.03 \le A_v^* \le 0.28$, $22 \le L/h_t \le 69$, $0.00 \le S_t \le 0.04$, and tunnels without lateral profile transitions ($A_g/A_t \approx 0.83$).

Most predicted (subscript pred) β-values are within ±20% of the observation data (Figure 8.88). However, for $\beta_{pred} < 0.7$, the values are overpredicted for some of the data with $L/h_t = 22$. This deviation is on the conservative side and such small β-values for short tunnels are not critical from a design point of view. There is no evident trend in the relative error for larger β-values (Figure 8.88b), thus indicating that all effects of the investigated parameters F_c, ζ, L/h_t, and S_t are adequately included in Eq. (8.92).

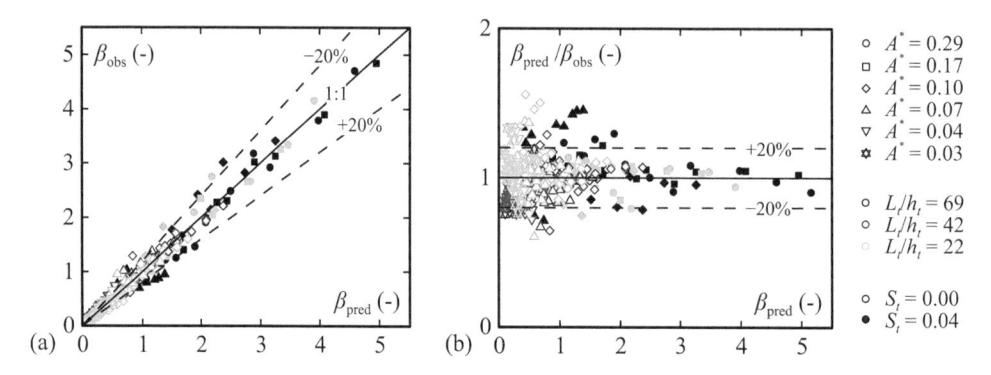

Figure 8.88 Comparison of relative air demand (a) observed versus predicted β-values based on (—) Eq. (8.92), (b) relative error versus predicted β-value, model data (Hohermuth *et al.* 2020)

Contrary to the existing air demand design equations (Table 8.1) based on a constant pre-factor and an exponent B (i.e., $\beta = $ const.$\times F_c^{B}$), the new Eq. (8.92) incorporates a coefficient based on A_v^*, L/h_t, and S_t. This renders Eq. (8.92) applicable to low-level outlets of different air vent size, air vent loss coefficient, tunnel length, and tunnel slope.

Comparison to prototype data

The comparison of model-based results with prototype data aims to assess scale and model effects and to extend the parameter range investigated in the model study. However, only a limited number of available prototype data include all parameters in Eq. (8.92), since especially ζ was often not assessed. The four data sets from the literature including all parameters are Norfork Dam (USACE 1954a), Mauvoisin Dam (Schilling 1963), Curnera Dam (Lier and Volkart 1994) and Panix Dam (Volkart and Speerli 1994). These four data sets cover a wide range of parameters (Table 8.3).

A sufficiently high prototype data quality is necessary for a meaningful comparison of air demand data to model results. The quality of prototype data is mainly affected by: (i) accuracy of the measurement instrumentation and possible interpolation methods, (ii) the intrusiveness of the instruments, and (iii) the measurement location. As to effect (i), it is straightforward to assess the largest uncertainty in Q_{Ao}, which stems from the interpolation of point velocity measurements. The effects of intrusiveness (ii) can be more severe; if orifice flow meters (as at Norfork Dam) or Venturi meters (as at Mauvoisin Dam) are used, the loss coefficient of the air vent can be significantly increased due to partial flow area obstruction compared to the normal, unrestricted air vent. Highly intrusive flow meters thus do not allow for measuring the correct air demand of a low-level outlet. The measurement location (iii) can also considerably affect the measured air discharge. Measuring close to the gate chamber (as at Curnera Dam) leads to a larger air discharge as opposed to close to the air vent portal (as at Mauvoisin and Norfork Dams) due to air expansion caused by the pressure drop along the air vent and possible altitude difference. Additional

Table 8.3 Parameter ranges covered in model tests and prototype data sets (Hohermuth *et al.* 2020)

	Model tests	Curnera	Mauvoisin	Norfork	Panix
F_c (-)	8–45	18–30	32–36	8–21	9–27
a/a_{max} (-)	0.2–0.8	0.17–0.55	0.14–0.17	0.17–0.92	0.12–0.83
L/h_t (-)	69, 42, 22	73	110	21	16
S_t (-)	0.00, 0.04	0.04	0.03	0.15	0.05
A_v^* (-)	0.02–0.28	0.058, 0.035	0.13	0.026, 0.025, 0.023, 0.015	0.19, 0.17, 0.13
R (-)	$3.3 \cdot 10^5$– $2.5 \cdot 10^6$	$7.7 \cdot 10^6$– $3.2 \cdot 10^7$	1.3–$1.7 \cdot 10^7$	$5.0 \cdot 10^6$– $3.0 \cdot 10^7$	$3.1 \cdot 10^6$– $2.4 \cdot 10^7$
W (-)	$1.9 \cdot 10^2$– $7.5 \cdot 10^2$	$2.2 \cdot 10^3$– $4.9 \cdot 10^3$	$3.4 \cdot 10^3$– $3.9 \cdot 10^3$	$1.5 \cdot 10^3$– $3.6 \cdot 10^3$	$1.2 \cdot 10^3$– $3.2 \cdot 10^3$
Lateral profile transition	no	smooth	smooth	no	abrupt
A_g/A_t (-)	0.83	0.21	0.44	1	0.46
Gate slots	no	yes	yes	yes	yes
Gate lip	sharp crest	45°lip	45°lip	45°lip	45°lip

uncertainties arise if velocity meters are placed inconsiderately; at Curnera Dam, for instance, only one vane anemometer was placed in the centerline of the cross-section just downstream of a junction in the air vent, where an asymmetric velocity profile is expected. Given that the effects of (ii) and (iii) cannot be quantified for all prototype data, the following relative standard uncertainties μ_r were assumed for the comparison: $\mu_r(V_{Ao})/V_{Ao} = 0.15$, $\mu_r(p_{Amin})/p_{Amin} = 0.10$, $\mu_r(Q_W)/Q_W = 0.05$, $\mu_r(F_c)/F_c = 0.05$, $\mu_r(\text{length})/\text{length} = 0.02$. Therefore, the highest experimental uncertainty of prototype measurements originates from the velocity data, whereas the smallest occurs for the length measurements, i.e. the tunnel length or the tunnel width.

In Figure 8.89a, the observed relative air demand β_{obs} is compared with the predicted values using the model-based prediction, Eq. (8.92). The model-based results underestimate the prototype air demand by roughly a factor of $\beta_{obs}/\beta_{pred} \approx 2.2$, which is similar to the ratio of model predictions and prototype data for chute aerators ($\beta_{obs}/\beta_{pred} \approx 1.5$ to 1.9, Gardarsson et al. 2015). Nevertheless, the good collapse between the different data sets indicates that the relative effects of F_c, A_v^*, and L_t/h_t are adequately considered. Note that the available prototype data are not spread out over a wide enough range to enable an estimation of the pre-factor and the exponents in Eq. (8.92) directly from the prototype data. Therefore, only the pre-factor was adjusted to match the *prototype data*, resulting in

$$\beta = 0.08 \, F_c^{1.3} \left(\frac{A_v}{A_t (\zeta + 1)^{0.5}} \right)^{0.8} \left(\frac{L_t}{h_t} \right)^{0.25}. \tag{8.93}$$

Since the effect of S_t in the regression did not lead to a statistically significant result for the prototype data, it was dropped from the regression analysis. The resulting Eq. (8.93) captures the overall trend of all prototypes within ±30% of the predicted values (Figure 8.89b). This is

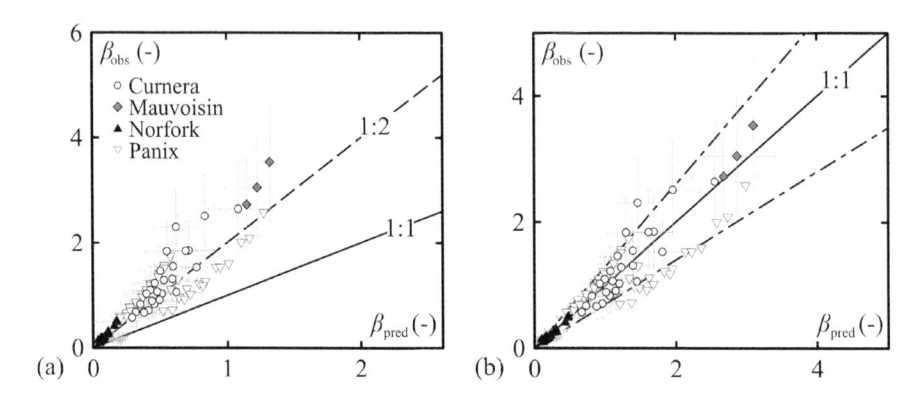

Figure 8.89 Comparison of relative air demand (a) observed prototype data versus predicted β based on (−−) Eq. (8.92), (b) observed prototype data versus adjusted prediction based on (−−) Eq. (8.93), (−·−·) = ±30%. Vertical error bars = 95% confidence interval based on the assumed standard uncertainties, horizontal error bars = 95% prediction interval including effects of input parameter uncertainty and uncertainty of regression

a significant improvement as compared to the scatter observed for prediction equations only considering F_c as an explanatory variable (Figure 8.81).

The significant underestimation of prototype air demand by the model-based Eq. (8.92) is due to both scale and model effects (as outlined in the section Experimental setup). Therefore, more prototype data are needed to quantify the contributions of different effects. For practical applications, Eq. (8.93) is proposed along with the Design recommendations given below.

Design Recommendations

Low-level outlets are key safety elements of high-head dams; their failure can have serious consequences for the entire dam structure and reservoir operation. Based on the presented experiments, prototype data, and past experiences, the following design recommendations are given (Hohermuth *et al.* 2020):

- Low-level outlets should be designed for free-surface flow in the downstream tunnel section. However, spray formation for small gate openings cannot be avoided. The transition to foamy or pressurized flow should be avoided by limiting the ratio of the mixture flow depth to the tunnel height to 70% to 80% (Speerli and Hager 2000, Hohermuth 2019).
- The relative air demand β for free-surface flow can be estimated using Eq. (8.93) within the limitations given in Table 8.3. Note that intense spray formation, strong profile transitions and bottom and/or sidewall aerators may significantly increase the relative air demand β compared to Eq. (8.93).
- The negative air pressures in the outlet tunnel can be assessed based on β, the air vent area A_v and the loss coefficient of the air vent ζ, yielding $p_{Amin} = (\zeta+1)\rho_A\beta^2 Q_W^2/(2A_v^2)$. The minimum allowable air pressure has to be determined on a case-by-case basis, with general recommendations ranging from -0.5 to -2 m (Dettmers 1953, Douma 1955, Wunderlich 1963, USACE 1980). Excessive negative pressures induce a counter-current air flow in the tunnel, which can trigger slug flow. As no detailed criteria are yet available, air vents with $A_v^* < 0.05$ should be avoided as a general recommendation.

The above recommendations are in agreement with and extend upon the Hydraulic Design of Reservoir Outlet Works manual by UASCE (1980). While the Hydraulic Design Criteria 050–1 endorses the use of the USACE (1964) equation (Table 8.1) based on limited prototype data, the newly proposed Eq. (8.93) enables quantifying the effects of the air vent loss coefficient, the air vent size, and the tunnel length. Additionally, the use of the air vent parameter A_v^* allows to meet criteria regarding maximum air velocity and minimum air pressure. The qualitative recommendation by USACE (1980) to use generous bend radii and gradual transitions can now be quantified with $A_v^* < 0.05$.

Notation

A	Relative gate opening $= a/h_o$ (-); also Cross-sectional tank area (m^2)
A_b	Area of bar (m^2)
A_c	Cross-sectional area of vena contracta (m^2)

A_g Cross-sectional area of mixture flow (m²); also cross-sectional area of gate (m²)
A_m Value of A with minimum C_d (-)
A_M = a/a_{max}, relative gate opening (-)
A_n Normalized A value = A/A_m (-)
A_t Cross-sectional area of tunnel (m²)
A_v^* Air vent parameter (-)
A_v Air vent cross-section (m²)
a Gate opening (m); also Cross-sectional orifice area (m²)
a_{eff} Vibrating water mass (m³)
a_{max} Maximum gate opening (m)
B Tunnel width (m)
b Channel width (m)
C Swirl number (-)
C Air concentration (-); also Constant (-)
C_A Average air concentration (-)
C_c Contraction coefficient (-)
C_d Discharge coefficient (-)
C_{do} Basic discharge coefficient (-)
$\overline{C}$ Average cross-sectional air concentration (-)
$\overline{C}_M$ Maximum air concentration (-)
C_p Pressure coefficient (-)
C_W Weir discharge coefficient (-)
c Relative air concentration (-)
c_d Drag coefficient (-)
D Relative gate opening (-); also Tank diameter (m); also Orifice diameter (m)
D_d Normalized value of C_d (-)
D_m Shock wave height parameter (-)
d Orifice diameter (m); also bar diameter or width (m)
d_D Drop diameter (m)
E Specific energy head (m); also Minimum flow space (m)
E_s Modulus of elasticity (N/m²)
e Distance of pivot to counterweight center (m); also Bar distance (m)
F Froude number (-); also Submergence Froude number (-); also pipe Froude number (-)
F_c Froude number at vena contracta (-)
F_{co} Submergence Froude number (-)
$F_{co,cr}$ Critical submergence Froude number (-)
F_D Combined Froude number (-)
F_o Froude number upstream of hydraulic jump (-)
F Gate outflow section (m²)
F_p Gate force (N)
f Friction coefficient (-); also Frequency (1/s)
f_H Hydraulic freeboard (m)
G Gate weight (N)
g Gravity acceleration (m/s²)
H Pseudo-specific energy head (m)
H_E Approach flow energy head (m); also energy head at gate section (m)

H_g	Relative mixture-flow depth (-)
H_o	Approach flow energy head to gate flow (m)
H_p	Relative bottom pressure (-)
H_{pg}	Normalized gate pressure head (-)
H_s	Stagnation flow depth (m)
H_u	Tunnel height (m)
ΔH	Difference of extreme stagnation depths (-); also Pressure head drop (m)
h	Flow depth (m); also Head on orifice (m); also Intake submergence (m)
h_b	Bar spacing (m)
h_c	Flow depth at vena contracta (m)
h_{cr}	Critical intake submergence (m)
h_d	Downstream flow depth (m)
h_g	Mixture-flow depth (m); also gate height (m)
h_k	Critical flow depth (m)
h_L	Submergence height (m)
h_o	Approach flow depth (m)
h_p	Pressure head (m)
h_{pg}	Gate pressure head (m)
h_{rm}	Reduced shock wave height (m)
h_t	Tunnel height (m)
h_{99}	Mixture depth with 99% air concentration (m)
I	Inertial moment (m^4)
K	Curvature distribution parameter (-); also profile factor (-)
k_s	Roughness height (m)
L	Tunnel length (m)
L_g	Gate length (m)
L_t	Tunnel length of bottom outlet (m)
l	Bar length (m)
l_R	Reference length (m)
M	Moment on gate (Nm); also Fixation coefficient (-)
m	Relative tunnel flow depth (-)
m_v	Vertical velocity distribution parameter (-)
N_Γ	Circulation number (-)
P	Profile effect number (-)
P_h	Hydrostatic gate pressure force (N)
p	Pressure (N/m^2)
p_A	Air pressure (N/m^2)
p_{Amin}	Minimum air pressure (N/m^2)
p_b	Bottom pressure (N/m^2)
p_s	Interface pressure (N/m^2)
p_W	Water pressure (N/m^2)
Q	Discharge (m^3/s)
Q_A	Air discharge (m^3/s)
Q_{Amean}	Mean air discharge (m^3/s)
Q_{Ao}	Air supply discharge (m^3/s)

Q_{Au}	Tunnel air discharge (m³/s)
Q_W	Water discharge (m³/s)
q	Unit discharge (m²/s)
q_R	Dimensionless discharge (-)
q_W	Unit water discharge (m²/s)
R	Reynolds number (-); also Radial Reynolds number (-)
R_a	Gate Reynolds number $= a(2ga)^{1/2}v^{-1}$ (-)
R_D	Intake Reynolds number (-)
R_c	Reynolds number at vena contracta section (-)
R	Orifice radius (m)
R_h	Hydraulic radius (m)
r	Radial coordinate (m)
r_o	Matching radius (m)
S	Strouhal number (-)
S	Specific momentum (m²); also bottom slope (-)
S_d	Dynamic pressure force (N)
SF	Safety factor (-)
S_f	Friction slope (-)
S_g	Relative gate opening (-)
S_p	Pressure force (N)
S_s	Static pressure force (N)
S_t	Bottom slope of tunnel (-)
S_u	Tunnel bottom slope (-)
s	Internal jet profile depth (m); also Depth of free-surface depression (m)
s_g	Gate opening (m)
s_0	Maximum gate opening (m)
T	Shock wave parameter $= (2/A)^{1/2}-1$ (-)
U	Relative streamwise velocity (-)
u	Streamwise velocity component (m/s)
u_D	Drop velocity (m/s)
u_R	Reference velocity (m/s)
V	Average cross-sectional velocity (m/s)
V_{Ao}	Air velocity in air vent (m/s)
V_{Au}	Air velocity at downstream tunnel end (m/s)
V_c	Water velocity at vena contracta section (m/s)
V_D	Average intake velocity (m/s)
v	Velocity (m/s)
v_e	Vertical velocity at gate lip (m/s)
$\overline{v}_g$	Average mixture velocity (m/s)
V_r	Relative velocity between water drops and air (m/s)
v_t	Tangential velocity (m/s)
v_{tM}	Maximum tangential velocity (m/s)
W	Weber number (-)
W_c	Weber number at vena contracta section (-)
W_D	Intake Weber number (-)
W	Weight of counterweight (N)

w_g	Gate width (m)
w_t	Tunnel width (m)
X	Relative streamwise coordinate $= x/a$ (-); also $= x/b$ (-)
X_L	Relative tunnel length (-)
X_N	Relative location (-)
X_s	Relative position of maximum shock wave height $= x_s/h_u$ (-)
X_v	Dimensionless location of vortex center $= x_v/a$ (-)
x	Streamwise coordinate (m)
x_v	Location of vortex center (m)
Y	Normalized transverse coordinate $=$ (-); also $= y/a, = y/b$ (-)
Y_g	Relative mixture depth (-)
Y_s	Relative shock wave height $= h_s/h_u$ (-)
Y_t	Normalized transverse coordinate $= y/(b/2)$ (-)
Y_v	Dimensionless location of vortex center $= y_v/a$ (-)
y	Vertical coordinate (m), also Relative submergence (-)
y_t	Transverse coordinate (m)
y_v	Distance of vortex center from wall (m)
Z	Relative position above channel bottom $= z/a$ (-)
$\overline{Z}$	Dimensionless depth $= (z-a)/(H_o-a)$ (-)
z	Vertical coordinate (m)
α	Aspect ratio (-)
β	Relative air entrainment rate (-)
β_{cr}	Critical relative air entrainment rate (-)
β_{mean}	Mean relative air entrainment rate (-)
β_{VT6}	Relative air entrainment rate relating to vortex type 6 (-)
Γ	Circulation (m²/s)
γ	Specific weight of water (N/m³)
γ_p	Relative gate pressure (-)
δ	Gate angle (-); also Ratio of diameters $= d/D$ (-)
ε	Counterweight angle (-)
η	Dimensionless vertical distance along gate plane (-)
Φ	Intake orientation (-); also Friction gradient parameter (-)
θ	Shock angle (-)
λ	Relative ridge length (-)
μ	Moment ratio $= M_d/M_s$ (-)
Π	Force ratio (-)
χ	Normalized distance (-)
ν	Kinematic viscosity (m²/s)
ρ	Fluid density (kg/m³)
ρ_A	Air density (kg/m³)
ρ_s	Steele density (kg/m³)
σ	Surface tension (N/m)
σ^p	Pressure force ratio (-)
ξ	Loss coefficient (-)
ζ	Air vent loss coefficient (-)
ω	Constant parameter (-); also Angular velocity (m/s)

References

Intakes

Amphlett, M.B. (1976) Air-entraining vortices at a horizontal intake. *Report* 7. Hydraulics Research Station, Wallingford.

Anwar, H.O. (1966) Formation of a weak vortex. *Journal of Hydraulic Research*, 4(1), 1–16.

Anwar, H.O., Weller, J.A. & Amphlett, M.B. (1978) Similarity of free vortex at horizontal intake. *Journal of Hydraulic Research*, 16(2), 95–105.

Arnold, R., Bezzi, A., Lais, A. & Boes, R.M. (2018) Intake structure design of entirely steel-lined pressure conduits crossing an RCC dam. Proceedings of the Conference *Hydro 2018*, Paper 04.02, 1–9. Gdansk, Poland.

Chang, E. (1977) Review of literature on the formation and modelling of vortices in rectangular pump sumps. *Technical Report* TN1414, British Hydromechanics Research Association (BHRA), Cranfield.

Chang, E. & Prosser, M.J. (1987) Basic results of theoretical and experimental results. Swirling flow problems at intakes. In: Knauss, J. (ed) *IAHR Hydraulic Structures Design Manual* 1, 39–55. Balkema, Rotterdam.

Chang, K.S. & Lee, D.J. (1995) An experimental investigation of the air entrainment in the shutdown cooling system during mid-loop operation. *Annals of Nuclear Energy*, 22(9), 611–619.

Daggett, L.L. & Keulegan, G.H. (1974) Similitude in free-surface vortex formations. *Journal of the Hydraulics Division* ASCE, 100(HY11), 1565–1581; 101(HY9), 1287–1289; 101(HY11), 1449–1453.

Denny, D.F. & Young, G.A.J. (1957) The prevention of vortices and swirl at intakes. Proceedings of the 7th *IAHR Congress*, Lisbon, C1, 1–18.

Fujimoto, S. & Takasu, S. (1979) Historical development of large capacity outlets for flood control in Japan. Proceedings of the 13th *ICOLD Congress*, New Delhi, Q50(R25), 417–438.

Gordon, J.L. (1970) Vortices at intakes. *Water Power*, 22(4), 137–138.

Gulliver, J.S., Rindels, A.J. & Lindblom, K.C. (1986) Designing intakes to avoid free-surface vortices. *Water Power & Dam Construction*, 38(9), 24–28.

Haindl, K. (1959) Contribution to air entrainment by a vortex. Proceedings of the 8th *IAHR Congress*, Montreal, D(16), 11–17.

Hattersley, R.T. (1965) Hydraulic design of pump intakes. *Journal of the Hydraulics Division* ASCE, 91(HY2), 223–249.

Hecker, G.E. (1981) Model-prototype comparison of free surface vortices. *Journal of the Hydraulics Division* ASCE, 107(HY10), 1243–1259; 108(HY11), 1409–1420; 109(3), 487–489.

Hecker, G.E. (1984) Scale effects in modelling vortices. Proceedings of the Symposium *Scale Effects in Modelling Hydraulic Structures*, Esslingen, 6(1), 1–9.

Hecker, G.E. (1987) Fundamentals in vortex intake flow. Swirling flow problems at intakes. *IAHR Hydraulic Structures Design Manual* 1, 13–38. Balkema, Rotterdam.

Hohermuth, B., Schmocker, L., & Boes, R.M. (2020). Air demand of low-level outlets for large dams. *Journal of Hydraulic Engineering*, provisionally accepted.

Iversen, H.W. (1953) Studies of submergence requirements of high-specific-speed pumps. *Transactions of ASME*, 75(4), 635–641.

Jain, A.K., Ranga Raju, K.G. & Garde, R.J. (1978) Vortex formation at vertical pipe inlet. *Journal of the Hydraulics Division* ASCE, 104(HY10), 1429–1445; 105(HY10), 1328–1336; 106(HY1), 211–213; 106(HY9), 1528–1530.

Knauss, J. (1987) Swirling flow problems at intakes. *IAHR Hydraulic Structures Design Manual* 1. Balkema, Rotterdam.

Levin, L. (1967) Problèmes de perte de charge et stabilité des grilles de prises d'eau [Problems of head-loss and stability of intake racks]. *La Houille Blanche*, 22(3), 271–278 (in French).

Liskovec, L. (1955) Suitable inlet form of pressure conduits. Proceedings of the 6th *IAHR Congress*, The Hague, C(14), 1–11.

McCormmach, A.L. (1968) Dworshak dam spillway and outlets hydraulic design. *Journal of the Hydraulics Division* ASCE, 94(HY4), 1051–1072.

Meyer, A. (2012) Modellfamilie zum Lufteintrag durch Einlaufwirbel [Model family on air entrainment by intake vortices]. *Master Thesis*, VAW, ETH Zurich, Zürich, Switzerland (unpublished, in German).

Minor, H.-E. (1988) Der Grundablass der Wasserkraftanlage Alicura in Argentinien [The bottom outlet of the hydropower plant Alicura in Argentina]. *Wasserwirtschaft*, 77(6), 1–4 (in German).

Möller, G., Detert, M., & Boes, R.M. (2012) Air entrainment due to vortices: State-of-the-art. Proceedings of the 2nd *IAHR Europe Congress*, Paper B16 (CD-Rom).

Möller, G. (2013) Vortex-induced air entrainment rate at intakes. In: Boes, R.M. (ed) *Mitteilung* 220. Laboratory of Hydraulics, Hydrology and Glaciology (VAW), ETH Zurich, Zürich, Switzerland.

Möller, G., Detert, M. & Boes, R.M. (2015) Vortex-induced air entrainment rates at intakes. *Journal of Hydraulic Engineering*, 141(11)1943–7900.0001036

Nagarkar, P.K. (1986) *Submergence criteria for hydroelectric intakes*, India (Unpublished report).

Naudascher, E. (1992) *Hydraulik der Gerinne und Gerinnebauwerke* [Hydraulics of channels and hydraulic structures], 2nd ed. Springer, Wien (in German).

Naudascher, E. & Rockwell, D. (1994) Flow-induced vibrations. *IAHR Hydraulic Structures Design Manual* 7. Balkema, Rotterdam.

Özis, Ü. & Özel, I. (1989) Karakaya dam and powerplant. *Water Power & Dam Construction*, 41(7), 20–24.

Padmanabhan, M. (1984) Air ingestion due to free-surface vortices. *Journal of Hydraulic Engineering*, 110(12), 1855–1859.

Papillon, B., Kirejczyk, J. & Sabourin, M. (2000) Atmospheric air admission in hydroturbines. Proceedings of *HydroVision*, Charlotte, NC, 3C, 1–11.

Poullikkas, A. (2000) Effects of entrained air on the performance of nuclear reactor cooling pumps. Proceedings of the *Melecon 2000: Information Technology and Electrotechnology for the Mediterranean Countries*, pp. 1028–1031.

Ranga Raju, K.G. & Garde, R.J. (1987) Modelling of vortices and swirling flows. Swirling flow problems at intakes. In: Knauss, J. (ed) *IAHR Hydraulic Structures Design Manual* 1, 77–90. Balkema, Rotterdam.

Rouse, H. & Abul-Fetouh, A.-H. (1950) Characteristics of irrotational flow through axially symmetric orifices. *Journal of Applied Mechanics*, 17(12), 421–426.

Rutschmann, P., Volkart, P. & Vischer, D.L. (1987) Design recommendations. Swirling flow problems at intakes. In: Knauss, J. (ed) *IAHR Hydraulic Structures Design Manual* 1, 91–100. Balkema, Rotterdam.

Schleiss, A. (1985) Trash-rack vibrations in hydroelectric power-plants. *Wasser, Energie, Luft*, 77(10), 299–303 (in German).

Schleiss, A. & Fust, A. (1992) Analysis of trash-rack vibrations at the hydroelectric power plant of Laufenburg at River Rhine. Proceedings of the Symposium *Operation, maintenance and refurbishment of hydroelectric power plants. Bericht* 73, 225–237. Lehrstuhl für Wasserbau und Wasserwirtschaft, TU München (in German).

Suerich-Gulick, F., Gaskin, S., Villeneuve, M. & Parkinson, É. (2014) Characteristics of free surface vortices at low-head hydropower intakes. *Journal of Hydraulic Engineering*, 140(3), 291–299.

Tröndle, E. (1974) Hidroprado Columbien: Wasserkraftanlage [Hidroprada Colombia: hydropower station]. *Wasserwirtschaft*, 64(2), 33–41 (in German).

Outlets

Anastasi, G. (1983) Besondere Aspekte der Gestaltung von Grundablässen in Stollen [Particular design aspects for bottom outlets]. *Wasserwirtschaft*, 73(12), 501–509 (in German).

Anwar, H.O. (1964) Discharge coefficients for control gates. *Water Power*, 16(4), 152–159.

Ball, J.W. (1959) Hydraulic characteristics of gate slots. *Journal of the Hydraulics Division* ASCE, 85(HY10), 81–114; 86(HY4), 121–126; 86(HY5), 133–143; 86(HY6), 87–89; 87(HY1), 155–163.

Benjamin, T.B. (1956) On the flow in channels when rigid obstacles are placed in the stream. *Journal of Fluid Mechanics*, 1, 227–248.

Benjamin, T.B. Lighthill, M.J. (1954) On cnoidal waves and bores. *Proc. Roy. Soc.* London A 224, 448–460.

Blind, H. (1985) Design criteria for reservoir bottom outlets. *Water Power & Dam Construction*, 37(7), 30–33.

Boileau, P. (1848) Mémoire sur le jaugeage des cours d'eau à faible ou à moyenne section [Memoir on the discharge measurement for small or medium cross-sections]. *Journal de l'Ecole Polytechnique*, Paris, 33, 129–234 (in French).

Bollrich, G. (1963) Belüftung von Grundablassverschlüssen [Aeration of bottom outlets]. *Wissenschaftliche Zeitschrift*, Dresden, 12(6), 1709–1717 (in German).

Bornemann, K.R. (1871) Versuche über den Ausfluss unter Wasser bei Schützen [Experiments on flow below gates]. *Civilingenieur*, 17, 45–60 (in German).

Bornemann, K.R. (1880) Über den Ausfluss bei Schützen und schützenartigen Mündungen [On outflow at gates and gate-type orifices]. *Civilingenieur*, 26, 297–376 (in German).

Bos, M.G. (1976) Discharge measurement structures. *Report* 4. Laboratorium voor Hydraulica en afvoerhydrologie, Landbouwhogeschool, Wageningen NL.

Bose, S.K. Dey, S. (2007) Curvilinear flow profiles based on Reynolds averaging. *J. Hydr. Engng.* 133(9), 1074–1079.

Bose, S.K. Dey, S. (2009) Reynolds averaged theory of turbulent shear flows over undulating beds and formation of sand waves. *Physical Review* E 80(036304), 1–9.

Boussinesq, J. (1877) Essai sur la théorie des eaux courantes [Memoir on the theory of flowing water]. *Mémoires* présentés par divers savants à l'Académie des Sciences, Paris, 23, 1-680 (in French).

Brombach, H. & Horlacher, H.-B. (1996) Selbstregulierender Auslaufschlitz für Regenüberlaufbecken [Self-regulating overflow slit for rain overflow basins]. *Wasserwirtschaft*, 86(3), 128–132 (in German).

Campbell, F.B. & Guyton, B. (1953) Air demand in gated outlet works. Proceedings of the 5th *IAHR Congress,* Minnesota, 529–533. IAHR, Delft.

Castro-Orgaz, O. & Hager, W.H. (2014) Transitional flow at standard sluice gate. *Journal of Hydraulic Research*, 52(2), 264–273.

Cheng, A.H.-D., Liggett, J.A. & Liu, P.L.-F. (1981) Boundary calculations of sluice and spillway flows. *Journal of the Hydraulics Division* ASCE, 107(HY10), 1163–1178.

Dammel, W. (1977) Einige Aspekte zur Konstruktion von Tiefschützen [Some aspects for the design of high-head gates]. *Bauingenieur*, 52(9), 353–355 (in German).

Dettmers, D. (1953) Beitrag zur Frage der Belüftung von Tiefschützen (A contribution to the aeration of high-head gates) *Mitteilung*, 4, 22–68. Hannoversche Versuchsanstalt für Grund- und Wasserbau, Franzius-Institut, Technische Hochschule Hannover, Hannover (in German).

Douma, H. (1955) Hydraulic design criteria for reservoir outlets. Proceedings of the 6th *IAHR Congress,* The Hague, C(10), 1–20.

Erbisti, P.C. (1981) Hydraulic gates: the state-of-the-art. *Water Power & Dam Construction*, 33(4), 43–48.

Escande, L. (1938) Etude théorique et expérimentale de l'écoulement par vanne de fond [Theoretical and experimental study on flow below gates]. *Revue Générale de l'Hydraulique*, 4(19), 25–29; 4(20), 120–128 (in French).

Falvey, H.T. (1980) Air-water flow in hydraulic structures. *Engineering Monograph*, 41. U.S. Department of the Interior, Water and Power Resources Service: Denver CO.

Falvey, H.T. (1990) Cavitation in chutes and spillways. *Engineering Monograph*, 42. US Bureau of Reclamation: Denver.

Fangmeier, D.D. & Strelkoff, T.S. (1968) Solution for gravity flow under a sluice gate. *Journal of the Engineering Mechanics Division* ASCE, 94(EM1), 153–176; 94(EM4), 1009–1010; 94(EM6), 1585–1588.

Fawer, C. (1937) Calcul de la contraction causée par une vanne plane dans le cas d'un écoulement dénoyé [Computation of contraction due to flow below a free gate]. *Bulletin Technique de la Suisse Romande*, 63, 192–198; 63, 217–219; 63, 245–252; 63, 283–287 (in French).

Finnie, J.I. Jeppson, R.W. (1991) Solving turbulent flows using finite elements. *J. Hydr. Engng.* 117(11), 1513–1530.

Gardarsson, S. M., Gunnarsson, A., Tomasson, G. G. & Pfister, M. (2015) Karahnjukar dam spillway: Comparison of operational data and results from hydraulic modelling. Proceedings of the *HYDRO 2015*, Paper 22.05, Bordeaux, France.

Gentilini, B. (1941a) Efflusso dalle luci soggiacenti alle paratoie piane inclinate e a settore [Flow below plane and sector gates]. *L'Energia Elettrica*, 18(6), 361–380 (in Italian).

Gentilini, B. (1941b) Sui processi di efflusso piano [On processes of plane outflow]. *L'Energia Elettrica*, 18(4), 213–233; 18(6), 361–380 (in Italian).

Ghetti, A. & Di Silvio, G. (1967) Investigation on the running of deep gated outlet works from reservoirs. Proceedings of the 9th *ICOLD Congress*, Istanbul, Q33(R48), 837–852.

Gibson, A.H. (1920) Experiments on the coefficients of discharge under rectangular sluice-gates. *Institution of Civil Engineers*, Selected Paper, 207, 427–434.

Giesecke, J. (1982) Verschlüsse in Grundablässen: Funktion und Ausführung [Gates of bottom outlets: performance and design]. *Wasserwirtschaft*, 72(3), 97–104 (in German).

Giezendanner, W. & Henry, P. (1980) Vidange de fond à grande vitesse [Bottom outlets of high velocity flows]. *Ingénieurs et Architectes Suisses*, 106(25), 387–394 (in French).

Haberstroh, C.E. (1890) Experiments on flow of water through large gates and over wide crests. *Journal of Assoc. Engng. Societies*, 5, 1–11.

Hager, W.H. (1999) Cavity flow from a nearly horizontal pipe. *Intl. J. Multiphase Flow*, 25(2), 349–364.

Hager, W.H. Hutter, K. (1984) On pseudo-uniform flow in open channel hydraulics. *Acta Mech.* 53(3–4), 183–200.

Hager, W.H. & Schleiss, A.J. (2009) *Constructions Hydrauliques: Ecoulements Stationnaires* [Hydraulic structures: steady flows]. Presses Polytechniques et Universitaires Romandes, Lausanne (in French).

Han, T.Y. & Chow, W.L. (1981) The study of sluice gate and sharp-crested weir through hodograph transformations. *Journal of Applied Mechanics*, 48(6), 229–238.

Harber, C.D. & Gulliver, J.S. (1992) Surface films in laboratory flumes. *Journal of Hydraulic Research*, 30(6), 801–815.

Hecker, G.E. (1984) Scale effects in modelling vortices. In: Kobus, H. (ed), Proceedings of the Symposium on *Scale Effects in Modelling Hydraulic Structures*, 6, 1–9. Technische Akademie, Esslingen.

Heller, V. (2011) Scale effects in physical hydraulic engineering models. *Journal of Hydraulic Research*, 49(3), 293–306, DOI:10.1080/00221686.2011.578914.

Henry, H.R. (1950) *A study of flow from a submerged sluice gate*. M.S. Thesis. Dept. of Mechanics and Hydraulics, State University of Iowa, Iowa City IA.

Hohermuth, B. (2017) Air demand of high-head bottom outlets. Proceedings of the 37th *IAHR World Congress*, Kuala Lumpur, Malaysia, pp. 2956–2965.

Hohermuth, B. (2019) Aeration and two-phase flow characteristics of low-level outlets. *VAW-Mitteilung* 253, Versuchsanstalt für Wasserbau, Hydrologie und Glaziologie (VAW), (R.M. Boes, ed.), ETH Zurich, Zürich, Switzerland.

Hohermuth, B., Schmocker, L. & Boes, R.M. (2017) Bottom outlet test Malvaglia. *Internal Report* (commissioned work).

Hohermuth, B., Schmocker, L. & Boes, R.M. (2018) Performance of middle and bottom outlet during a flushing event in Luzzone. *Internal Report* (commissioned work).

Hohermuth, B., Schmocker, L., Boes, R.M. (2020) Air demand of low-level outlets for large dams. *J. of Hydraulic Engineering*, in print.

Hurst, H.E. & Watt, D.A.F. (1925) The similarity of motion of water through sluices of the Assuan dam. *Proceedings of the Institution of Civil Engineers*, 218, 72–180.

Iwasa, Y. (1955) Undular jump and its limiting conditions for existence. Proc. 5th *Japan Natl. Congress Applied Mech.* II-14, 315–319.

Kalinske, A.A. & Robertson, J.W. (1943) Closed conduit flow. *Transactions of ASCE*, 108, 1435–1447.

Kay, M.G. & Ashton, D.A. (1983) A laboratory investigation of counterbalanced flap gates for water level control. *Journal of the Institution of Water Engineers*, 37, 506–512.

Kenn, M.J. & Garrod, A.D. (1981) Cavitation damage and the Tarbela Tunnel collapse of 1974. *Proceedings of the Institution of Civil Engineers*, 70(1), 65–89; 70(1), 779–810.

Keutner, C. (1932) Wasserabführungsvermögen von scharfkantigen und abgerundeten Planschützen [Discharge capacity of sharp-crested and rounded plan gates]. *Bautechnik*, 10(21), 266–269; 10(24), 303–305 (in German).

Keutner, C. (1935) Die Strömungsvorgänge an unterströmten Schütztafeln mit scharfen und abgerundeten Unterkanten [Flow processes at gates with sharp and rounded crests]. *Wasserkraft und Wasserwirtschaft*, 30(1), 5–8; 30(2), 16–21 (in German).

Levi, E. (1991) The floating siphon: an alternative to canal level control. Proceedings of the 24th *IAHR Congress*, Madrid D, 313–320. IAHR, Delft.

Levin, L. (1965) Calcul hydraulique des conduites d'aération des vidanges de fond et dispositifs déversants [Hydraulic computation of aeration conduits at bottom outlets]. *La Houille Blanche*, 20(2), 121–126 (in French).

Lewin, J. (1995) *Hydraulic gates and valves*. Thomas Telford, London.

Lier, P. & Volkart, P.U. (1994) Prototype investigation on aeration and operation of the Curnera high head bottom outlet. Proceedings of the 18th *ICOLD Congress*, Durban, Q71(R36), 535–553.

Lynse, D.K. & Guttormsen, O. (1971) Air demand in high head regulated outlet works. Proceedings of the 14th *IAHR Congress*, Paris, 5, 77–80.

Marchi, E. (1963) Contributo allo studio del risalto ondulato [Contribution to the study of the undular hydraulic jump]. *Giornale del Genio Civile* 101(9), 466–476 (in Italian).

Matthew, G.D. (1991) Higher order one-dimensional equations of potential flow in open channels. *Proc. ICE* 91(3), 187–201.

Montes, J.S. (1997) Irrotational flow and real fluid effects under planar sluice gates. *Journal of Hydraulic Engineering*, 123(3), 219–232; 125(2), 208–213.

Montes, J.S. (1998) *Hydraulics of open channel flow*. ASCE Press, Reston VA.

Naghdi, P.M. & Vongsarnpigoon, L. (1986) Steady flow past a sluice gate. *Physics of Fluids*, 29(12), 3962–3970.

Nago, H. (1977) Hydraulic pressure acting on shell-type gates. *Transactions of JSCE*, 9(267), 170–172.

Nago, H. (1978) Influence of gate-shapes on discharge coefficients. *Transactions of Japanese Society of Civil Engineers*, 10(270), 116–119.

Naudascher, E. (1991) Hydrodynamic forces. *IAHR Hydraulic Structures Design Manual* 3. Balkema, Rotterdam.

Naudascher, E. & Rockwell, D. (1994) Flow induced vibrations. *IAHR Hydraulic Structures Design Manual* 7. Balkema, Rotterdam.

Noutsopoulos, G.K. & Fanariotis, S. (1978) Discussion of Free flow immediately below sluice gates. *Journal of the Hydraulics Division* ASCE, 104(HY3), 451–454.

Ohtsu, I. Yasuda, Y. (1994) Characteristics of supercritical flow below sluice gate. *J. Hydr. Engng.* 120(3), 332–346.

Pethick, R.W. & Harrison, A.J.M. (1981) The theoretical treatment of the hydraulics of rectangular flap gates. Proceedings of the 19th *IAHR Congress*, New Delhi, B(C12), 247–254.

Pfister, M. & Chanson, H. (2012) Discussion of Scale effects in physical hydraulic engineering models by V. Heller. *Journal of Hydraulic Research*, 50(2), 244–246.

Rabben, S.L. (1984) Untersuchung der Belüftung an Tiefschützen unter besonderer Berücksichtigung von Massstabseffekten [Investigation of aeration of high-head gates under special consideration of scale effects]. *Mitteilung* 53. Institut für Wasserbau und Wasserwirtschaft, RWTH Aachen, Aachen, (in German).

Rabben, S.L. & Rouvé, G. (1984) Air demand of high head gates: model family studies to quantify scale effects. Proceedings of the Symposium *Scale Effects in Modelling Hydraulic Structures*, Esslingen, Germany, 4(9), 1–3.

Rabben, S.L. & Rouvé, G. (1985) Belüftung von Grundablässen [Aeration of bottom outlets]. *Wasserwirtschaft*, 75(9), 393–399 (in German).

Raemy, F. & Hager, W.H. (1998) Hydraulic level control by Hinged Flap Gate. Proceedings of the ICE, *Water, Maritime & Energy*, 130, 95–103.

Rajaratnam, N. (1977) Free flow immediately below sluice gates. *Journal of the Hydraulics Division* ASCE, 103(HY4), 345–351; 103(HY11), 1371–1373; 104(HY3), 451–454; 104(HY10), 1462–1463.

Rajaratnam, N. & Humphries, J.A. (1982) Free flow upstream of vertical sluice gates. *Journal of Hydraulic Research*, 20(5), 427–437.

Rajaratnam, N. & Subramanya, K. (1967) Flow equation for the sluice gate. *Journal of the Irrigation and Drainage Division* ASCE, 93(ID3), 167–186.

Reinauer, R. & Hager, W.H. (1996) Generalized drawdown curve for chutes. Proceedings of the Institution of Civil Engineers, *Water, Maritime and Energy*, 118(4), 196–198.

Roth, A. & Hager, W.H. (1999) Underflow of standard sluice gate. *Experiments in Fluids*, 27(4), 339–350.

Rouse, H. (1950) *Engineering hydraulics*. Wiley, New York.

Sagar, B.T.A. (1979) Safe practices for high head outlet gates. Proceedings of the 13th *ICOLD Congress*, New Delhi, Q50(R27), 459–467.

Sagar, B.T.A. (1995) ASCE hydrogates task committee design guidelines for high-head gates. *Journal of Hydraulic Engineering*, 121(12), 845–852.

Serre, F. (1953) Contribution à l'étude des écoulements permanents et variables dans les canaux [Contribution to the study of steady and varied flows in channels]. *La Houille Blanche* 8(3), 830–872 (in French).

Schilling, H. (1963) Luftmengenmessungen in Grundablässen von Stauanlagen [Air discharge measurements in bottom outlets of dams]. *Escher Wyss Mitteilungen*, 36(2/3), 41–47 (in German).

Sharma, H.R. (1973) Air demand for high head gated conduits. *Ph.D. thesis*. University of Trondheim, The Norwegian Institute of Technology, Trondheim.

Sharma, H.R. (1976) Air entrainment in high head gated conduits. *Journal of the Hydraulics Division* ASCE, 102(HY11), 1629–1646; 103(HY10), 1254–1255; 103(HY11), 1365–1366; 103(HY12), 1486–1493; 104(HY8), 1200–1202.

Sinniger, R.O. & Hager, W.H. (1989) *Constructions hydrauliques* [Hydraulic structures]. Presses Polytechniques Universitaires Romandes, Lausanne (in French).

Skripalle, J. (1994) Zwangsbelüftung von Hochgeschwindigkeitsströmungen an zurückspringenden Stufen im Wasserbau [Forced aeration of high-velocity jets at backward facing steps in hydraulic structures]. *Mitteilung* 124. Institut für Wasserbau und Wasserwirtschaft, TU Berlin, Berlin (in German).

Smetana, J. (1948) Ecoulement de l'eau au-dessous d'une vanne et forme rationelle de la surface d'appui de la vanne [Flow of water below gates and rational free surface profile]. *La Houille Blanche*, 4(1/2), 41–53; 4(3/4), 126–146 (in French).

Southwell, R.V. & Vaisey, G. (1946) Relaxation method applied to engineering problems XII: Fluid motions characterized by free streamlines. *Philosophical Transactions of Royal Society London,* Series A, 240, 117–161.

Speerli, J. (1998) *Strömungsprozesse in Grundablassstollen* [Flow processes in bottom outlets]. *Dissertation* ETHZ 12583, Zürich, Switzerland; also *VAW-Mitteilung* 163, H.E. Minor, ed. VAW, ETH-Zurich, Zürich (in German).

Speerli, J. & Hager, W.H. (1999) Discussion of Irrotational flow and real fluid effects under planar sluice gates. *Journal of Hydraulic Engineering*, 125(2), 208–210.

Speerli, J. & Hager, W.H. (2000) Air-water flow in bottom outlets. *Canadian Journal of Civil Engineering*, 27(3), 454–462.

Task Committee (1973) High head gates and valves in the United States. *Journal of the Hydraulics Division* ASCE, 99(HY10), 1727–1775.

US Army Corps of Engineers (1954a) Slide gate tests, Norfork Dam, North Fork River, Arkansas. *Technical Memorandum* 2–389. Waterways Experiment Station, Vicksburg, Mississippi, USA.

US Army Corps of Engineers (1954b) Tainter gate tests, Norfork dam, North Fork River, Arkansas. *Technical Memorandum* 2–387. Waterways Experiment Station, Vicksburg, Mississippi, USA.

US Army Corps of Engineers (1957) Sluice entrances flared on four sides. *Hydraulic Design Criteria*, Hydraulic design chart 211. US Army Engineer Waterways Experiment Station, Vicksburg, MS.

US Army Corps of Engineers (1964) Air demand-regulated outlet works. *Hydraulic Design Criteria*, Sheet 050–1/2/3, 211–1/2, 212–1/2, 225–1. U.S. Corps of Engineers, Vicksburg, MS.

US Army Corps of Engineers (1968) Gate slots: pressure coefficients. *Hydraulic Design Criteria* Hydraulic Design Chart, 212(1/1). US Army Engineer Waterways Experiment Station, Vicksburg MS.

US Army Corps of Engineers (USACE) (1980) Hydraulic design of reservoir outlet works. *Engineering Manual* 1110–2–1602. Department of the Army, U.S. Army Corps of Engineers, Washington, DC.

USBR (1948) Studies of crests for overfall dams. Boulder Canyon Projects, Final Reports Part VI: Hydraulic investigations. *Bulletin* 3. US Bureau of Reclamation, Dept. of the Interior: Denver.

Valiani, A. (1997) Linear and angular momentum conservation in hydraulic jump. *J. Hydr. Res.* 35(3), 323–354.

Vischer, D. (1979) Die selbsttätige Schlauchdrossel zur Gewährleistung konstanter Beckenausflüsse [The self-controlled hose throttle allowing for constant reservoir outflow]. *Wasserwirtschaft*, 69(12), 371–371 (in German).

Vischer, D. & Hager, W.H. (1998) *Dam hydraulics*. Wiley, Chichester.

Volkart, P. (1978) Hydraulische Bemessung steiler Kanalisationsleitungen unter Berücksichtigung der Luftaufnahme [Hydraulic design of steep sewers considering air entrainment]. *VAW Mitteilung* 30. ETH, Zürich, Switzerland (in German).

Volkart, P. (1988) Instrumentation for measuring local air concentration in high-velocity free-surface flow. Proceedings of the International Symposium *Hydraulics for High Dams*, Beijing, China, pp. 1088–1096.

Volkart, P.U. & Speerli, J. (1994) Prototype investigation on the high velocity flow in the high head tunnel outlet of the Panix dam. Proceedings of the 18th *ICOLD Congress*, Durban, Q71(R6), 55–78.

Wagner, W.E. (1967) Glen Canyon diversion tunnel outlets. *Journal of the Hydraulics Division* ASCE, 93(HY6), 113–134.

Wisner, P. (1967) Hydraulic design for flood control by high head gated outlets. Proceedings of the 9th *ICOLD Congress*, Istanbul, 5(C12), 495–507.

Wood, I.R. (1991) Air entrainment in free-surface flows. *IAHR Hydraulic Structures Design Manual* 4. Balkema, Rotterdam.

Wunderlich, W. (1963) Die Grundablässe an Talsperren, Teil 2 [Bottom outlets of dams, Part 2]. *Wasserwirtschaft*, 53(4), 106–114 (in German).

Bibliography

Intakes

Bloomer, N.T., Markland, E. & Power, G. (1955) Design of two-dimensional entrances to hydraulic channels. *The Engineer*, 199(June 3), 765–767.

Guyton, B. (1958) Field investigations of spillways and outlet works. *Journal of the Hydraulics Division* ASCE, 84(HY1, Paper 1532), 1–21.

Huan-Wen, H., Fu-Yi, C. & Sheng, C. (1979) Studies on the configuration of short intakes for free-flow spillway tunnels. Proceedings of the 13th *ICOLD Congress*, New Delhi, Q50(R57), 1013–1034.

Joglekar, D.V. & Damle, P.M. (1957) Cavitation free sluice outlet design. Proceedings of the 7th *IAHR Congress*, Lisboa, B(6), 1–10.

Liskovec, L. (1951) A study of the inlet shape of reservoir outlets. Proceedings of the 4th *ICOLD Congress*, New Delhi, Q12(R61), 544–558.

Liskovec, L. (1961) Research on inlets for pressure conduits. *Prace a Studie*, 102. VUV, Prague (in Czech, with English summary).

Margaritora, G. (1968) Contributo sperimentale allo studio dei raccordi di imbocco per condotte quadrate [Experimental study on the transition between intake and quadratic conduit]. *L'Acqua*, 46(1), 3–14 (in Italian).

McNown, J.S. (1947) Pressure distribution and cavitation on submerged boundaries. In: Howe, J.W. & McNown, J.S. (eds), Proceedings of the 3rd *Hydraulics Conference*, Iowa. University of Iowa, Iowa, pp. 192–208.

Novak, P. & Cabelka, J. (1984) *Models in hydraulic engineering: Physical principles and design applications*. Pitman, Boston.

Rao, P.V. (1968) Boundary-layer development at curved conduit entrances. *Journal of the Hydraulics Division* ASCE, 94(HY1), 195–217; 95(HY1), 479–480; 95(HY5), 1720–1721.

Rao, P.V. (1973) Hydraulic design of sluice entrances in dams. *Journal of Institution of Engineers*, India, 53(3), 181–186.

Rouse, H. & Hassan, M.M. (1949) Cavitation-free inlets and contractions. *Mechanical Engineering*, 71(3), 213–216.

Thomas, H.A. & Schuleen, E.P. (1942) Cavitation in outlet conduits of high dams. *Transactions of ASCE*, 107, 421–456.

VAW (2010) Kraftwerk Handeck 2: Physikalische Modellversuche zum Einlaufbauwerk [Hydropower Station Handeck 2: Physical scale model tests for intake optimization]. *Technical Report* 4280. Laboratory of Hydraulics, Hydrology and Glaciology (VAW), ETH Zurich, Zürich, Switzerland (unpublished, in German).

Intake vortices

Anwar, H.O. (1965a) Flow in a free vortex. *Water Power*, 17(4), 153–161.

Anwar, H.O. (1965b) Coefficients of discharge for gravity flow into vertical pipes. *Journal of Hydraulic Research*, 3(1), 1–19.

Anwar, H.O. (1967) Vortices at low-head intakes. *Water Power*, 19(11), 455–457.

Anwar, H.O. (1968a) Vortices in a viscous fluid. *Journal of Hydraulic Research*, 6(1), 1–14.

Anwar, H.O. (1968b) Prevention of vortices at intakes. *Water Power*, 19(10), 393–401.

Anwar, H.O. & Amphlett, M.B. (1980) Vortices at vertically inverted intake. *Journal of Hydraulic Research*, 18(2), 123–134.

Anwar, H.O. (1983) The non-dimensional parameters of free-surface vortices measured for horizontal and vertically inverted intakes. *La Houille Blanche*, 38(1), 11–25.

Broadbent, E.G. & Moore, D.W. (1982) The two-dimensional oscillations of straight filaments with a power-law distribution of vorticity. *Proceedings of the Royal Society London*, Series A, 384, 1–29.

Chen, Y.N. (1979) From bath-tub vortex to pump-intake vortex. *Schweizer Ingenieur und Architekt*, 97(42), 845–852.

Cola, R. & Trivellato, F. (1988) Il vortice a superficie libera e ad asse verticale in un campo di moto indefinite: La distribuzione della velocità [Free surface vortices with vertical axes in a field of indefinite motion: Velocity distribution]. *Idrotecnica*, 15(6), 457–465 (in Italian).

Dallwig, H.-J., Schröder, R.C.M. & Zäschke, E. (1979) Wirbelbildung an Entnahmebauwerken [Vortex generation at intakes]. *Wasser und Boden*, 30(4), 101–104 (in German).

Dhillon, G.S., Sakhuja, V.S. & Paul, T.C. (1981) Modelling criteria for vortex formation at pipe intakes. Proceedings of the 19th *IAHR Congress*, New Delhi, 5, 223–234; 6, 332–335.

Einstein, H.A. & Li, H. (1955) Le vortex permanent dans un fluide réel [The steady vortex in a real fluid]. *La Houille Blanche*, 10(4), 483–496 (in French).

Farrell, C. & Cuomo, A.R. (1984) Characteristics and modeling of intake vortices. *Journal of Engineering Mechanics*, 110(5), 723–742.

Gardner, G.C. (1990) Deep swirling flow down a plughole with air entrainment. *Intl. Journal of Multiphase Flow*, 16(6), 1003–1021.

Green, S.I., ed. (1995) *Fluid vortices*. Kluwer, Dordrecht.

Gulliver, J.S., Rindels, A. & Lindblom, K.C. (1983) Guidelines for intake design without free surface vortices. Proceedings of *Waterpower '83*(3), 1472–1483.

Gulliver, J.S. & Rindels, A.J. (1987) Weak vortices at vertical intakes. *Journal of Hydraulic Engineering*, 113(9), 1101–1116; 115(5), 703–707.

Hite, Jr., J.E. & Mih, W.C. (1994) Velocity of air-core vortices at hydraulic intakes. *Journal of Hydraulic Engineering*, 120(3), 284–297; 121(5), 440–441; 121(8), 631.

Holtorff, G. (1964) La surface libre et les conditions de similitude du vortex [The free surface and similitude conditions of the vortex]. *La Houille Blanche*, 19(3), 377–384 (in French).

Kaufmann, W. (1963) *Technische Hydro- und Aeromechanik* [Technical hydro- and aeromechanics]. Springer, Berlin (in German).

Levi, E. (1967) I vortici e l'idraulica [The vortices and hydraulics]. *L'Acqua*, 45(6), 163–173 (in Italian).

Levi, E. (1972) Experiments on unstable vortices. *Journal of the Engineering Mechanics Division* ASCE, 98(EM3), 539–559; 99(EM1), 227–228; 99(EM5), 1089–1094.

Levi, E. (1991) Vortices in hydraulics. *Journal of Hydraulic Engineering*, 117(4), 399–413.

Marris, A.W. (1967) Theory of the bathtub vortex. *Journal of Applied Mechanics*, 34(1), 11–15.

McCorquodale, J.A. (1968) Scale effects in swirling flow. *Journal of the Hydraulics Division* ASCE, 94(HY1), 285–300; 95(HY1), 487–495; 95(HY5), 1722–1723.

Odgaard, A.J. (1986) Free-surface air core vortex. *Journal of Hydraulic Engineering*, 112(7), 610–620; 114(4), 447–452.

Padmanabhan, M. & Hecker, G.E. (1979) Flow characteristics of reactor recirculation sumps. Proceedings of the 18th *IAHR Congress*, Cagliari, 4(D6), 453–460.

Paul, T.C., Sakhuja, V.S. & Dhillon, G.S. (1987) Analytical models for free-surface air-core vortex. *Indian Journal of Power & River Development*, 37(8/9), 161–170.

Pugh, C.A. (1981) Intakes and outlets for low-head hydropower. *Journal of the Hydraulics Division* ASCE, 107(HY9), 1029–1045; 108(HY11), 1406–1408.

Reddy, Y.R. & Pickford, J. (1974) Vortex suppression in stilling pond overflow. *Journal of the Hydraulics Division* ASCE, 100(HY11), 1685–1698; 101(HY9), 1291–1295; 101(HY12), 1543; 102(HY6), 780–781.

Rohan, K. (1966) Determination of air core cross-section of the drain vortex. *Vodohospodarsky*, 14(1), 20–31 (in Czech, with English summary).

Rouse, H. (1963) On the role of eddies in fluid motion. *American Scientist*, 51(9), 285–314.

Sharma, H.D. & Sharma, H.R. (1979) Air-entrainment problems at intakes. Proceedings of the 3rd *World Congress on Water Resources*, Mexico City, 7, 3081–3093.

Stevens, J.C. & Kolf, R.C. (1959) Vortex flow through horizontal orifices. *Transactions of ASCE*, 124, 871–893.

Wickenhäuser, M. (2008) Zweiphasenströmung in Entlüftungssystemen von Druckstollen [Two-phase flow in de-aeration systems of pressurized tunnels]. In: Minor, H.-E. (ed) *Mitteilung* 205, Laboratory of Hydraulics, Hydrology and Glaciology (VAW), ETH Zurich, Zürich (in German).

Yildirim, N. & Kocabas, F. (1995) Critical submergence for intakes in open channel flow. *Journal of Hydraulic Engineering*, 121(12), 900–905; 123(6), 588–590.

Yildirim, N. & Kocabas, F. (1998) Critical submergence for intakes in still-water reservoir. *Journal of Hydraulic Engineering*, 124(1), 103–104.

Yildirim, N., Kocabas, F. & Gülcan, S.C. (2000) Flow-boundary effects on critical submergence of intake pipe. *Journal of Hydraulic Engineering*, 126(4), 288–297; 133(4), 461.

Yildirim, N. & Kocabas, F. (2002) Prediction of critical submergence for an intake pipe. *Journal of Hydraulic Research*, 40(4), 507–518.

Zielinski, P.B. & Villemonte, J.R. (1968) Effect of viscosity on vortex-orifice flow. *Journal of the Hydraulics Division* ASCE, 94(HY3), 745–752; 95(HY1), 567–570; 95(HY5), 1736–1737.

Orifice flow

Chanson, H., Aoki, S.-i. & Maruyama, M. (2002) Unsteady two-dimensional orifice flow: a large-size experimental investigation. *Journal of Hydraulic Research*, 40(1), 63–71.

Cozzo, G. (1973) Azione di una piastra piana sull'efflusso da un boccaglio [Effects of a flat plate on the outflow of a nozzle]. *Memorie e Studi* 259. Istituto di Idraulica e Costruzioni Idrauliche, Politecnico di Milano: Milano. Istituto di Idraulica dell'Università di Padova: Padova (in Italian).

D'Alpaos, L. (1975) Effetto della tensione superficiale sulla traiettoria di un getto liquido sottile: Esame sperimentale [Effect of surface tension on a thin liquid jet: experimental study]. *Studi e Ricerche*, 305. Istituto di Idraulica dell'Università di Padova: Padova (in Italian).

De Marchi, G. (1925) Esperienze sulla contrazione delle vene liquide [Experiments on the contraction of liquid jets]. *Annali dei Lavori Pubblici*, 63(8), 689–731 (in Italian).

De Martino, G. & Ragone, A. (1975) L'impiego delle luci a feritoia nelle vasche di sedimentazione [The use of orifices in sedimentation basins]. *Ingegneria Sanitaria*, 23(3), 113–131 (in Italian).

De Martino, G. & Ragone, A. (1984) Effects of viscosity and surface tension on slot weirs flow. *Journal of Hydraulic Research*, 22(5), 327–341.

Dong, W.-g. & Lienhard, J.H. (1986) Contraction coefficients for Borda mouthpieces. *Journal of Fluids Engineering*, 108(3), 377–379.

Gardner, G.C. (1983) Flooded countercurrent two-phase flow in horizontal tubes and channels. *Intl. Journal of Multiphase Flow*, 9(4), 367–382.

Geer, J.F. & Strikwerda, J.C. (1980) Vertical slender jets. *Journal of Fluid Mechanics*, 101, 53–63.

Geer, J.F. & Strikwerda, J.C. (1983) Vertical slender jets with surface tension. *Journal of Fluid Mechanics*, 135, 155–169.

Gibson, A.H. (1945) *Hydraulics and its applications*. Constable, London.

Goh, K.H.M. & Tuck, E.O. (1985) Thick waterfalls from horizontal slots. *Journal of Engineering Mathematics*, 19(4), 341–349.

Grose, R.D. (1985) Orifice contraction coefficient for inviscid incompressible flow. *Journal of Fluids Engineering*, 107(1), 36–43.

Hansen, M. (1949) Über das Ausflussproblem [On the orifice problem]. *VDI-Forschungsheft*, 428. Deutscher Ingenieur-Verlag, Düsseldorf (in German).

Jeler, V. (1971) La cavitation dans les conduites de vidange très courtes de type ajoutage extérieur [Cavitation in short emptying conduits with external jet nozzle]. Proceedings of the 14th *IAHR Congress,* Paris, 5(216), 103–108 (in French).

Keutner, C. (1934) Einfluss der Querschnittsform einer Wandöffnung auf Wasserabführung und auf Querschnitt des ausfliessenden Strahles [Effect of cross-sectional geometry of wall opening on discharge and on cross-sectional jet shape]. *Die Bautechnik*, 12(19), 243–245; 12(21), 270–273 (in German).

Kubie, J. (1986) Hydrodynamics of outflows from vessels. Encyclopedia of fluid mechanics 2 In: Cheremisinoff, N.P. (ed) *Dynamics of Single Fluid Flows and Mixing*, 1187–1207, N.P. Cheremisinoff, ed. Gulf, Houston.

Larock, B.E. (1969) Jets from two-dimensional symmetrical nozzles of arbitrary shape. *Journal of Fluid Mechanics*, 37, 479–489.

Lauffer, H. (1934) Einfluss der Oberflächenspannung auf den Ausfluss aus Poncelet-Öffnungen [Effect of surface tension on the outflow from Poncelet orifices]. *Forschung auf dem Gebiet des Ingenieurwesens*, 5(6), 266–274 (in German).

Lienhard V, J.H. & Lienhard IV, J.H. (1984) Velocity coefficients for free jets from sharp-crested orifices. *Journal of Fluids Engineering*, 106(1), 13–17.

Marchetti, M. (1963) Caratteristice geometriche dei getti da bocce di diversa forma aderte in superficie piana verticale [Geometrical characteristics of orifice jets of various outlet shapes in a vertical plane]. *Memorie e Studi*, 216. Istituto di Idraulica e Costruzioni Idrauliche, Politecnico, Milano (in Italian).

Margaritora, G. (1966) Ricerca sperimentale sull'efflusso da tubi addizionale a sezione quadrata e rettangolare [Experimental study on the outflow from tubes of quadratic and rectangular cross-sections]. *Pubblicazione*, 93. Istituto di Costruzioni Idrauliche, Università, Roma (in Italian).

McNown, J.S. & Ling, S.C. (1955) Inlets for square conduits. *La Houille Blanche*, 10(5), 775–781.

McNown, J.S. (1993) Contraction of jets from conduits. *Journal of Hydraulic Research*, 31(5), 579–586; 32(4), 636–638.

O'Brien, M.P. & Folsom, R.G. (1937) Modified I.S.A. orifice with free discharge. *Transactions of ASME*, 59(1), 61–64; 59(6), 756–757.

Pai, S.-I. (1954) *Fluid dynamics of jets*. Van Nostrand, New York.

Patta, F. (1971) Sull'efflusso da luci munite di spina centrale cilindrica [On orifice efflux guided with an axial thorn]. *L'Acqua*, 49(6), 95–111 (in Italian).

Rouse, H. (1934) On the use of dimensionless numbers. *Civil Engineering*, 4(11), 562–568.

Schoder, E.W. & Dawson, F.M. (1927) *Hydraulics*. McGraw-Hill, New York.

Smith, J.F.D. & Steele, S. (1935) Rounded-approach orifices. *Mechanical Engineering*, 57(12), 760–780; 58(4), 256–257.

Tison, L.-J. & Heyndrickx, G. (1943) La détermination des débits au moyen d'orifices en mince paroi [The determination of discharge using sharp-crested orifices]. *Revue Générale de l'Hydraulique*, 9(36), 219–225 (in French).

Tuck, E.O. (1987) Efflux from a slit in a vertical wall. *Journal of Fluid Mechanics*, 176, 253–264.

Vanden-Broeck, J.-M. (1984) The effect of surface tension on the shape of the Kirchhoff jet. *Physics of Fluids*, 27(8), 1933–1936.

von Mises, R. (1917) Berechnung von Ausfluss- und Überfallzahlen [Computation of outflow and overflow coefficients]. *Zeitschrift des Vereines Deutscher Ingenieure*, 61(21), 447–452; 61(22), 469–474; 61(23), 493–498 (in German).

Zhang, Z. & Cai, J. (1999) Compromise orifice geometry to minimize pressure drop. *Journal of Hydraulic Engineering*, 125(11), 1150–1153.

Outflow vortex

Benfratello, G. (1962) Ricerca sperimentale su scarichi a pozzo aventi imbocchi di forme speciali [Experimental research on shaft inflow of special shapes]. *L'Acqua*, 40(3), 49–74 (in Italian).

Binnie, A.M. (1938) The use of a vertical pipe as an overflow for a large tank. *Proceedings of the Royal Society* A, 168, 219–237.

Binnie, A.M. & Hookings, G.A. (1948) Laboratory experiments on whirlpools. *Proceedings of the Royal Society* A, 194, 398–415.

Binnie, A.M. & Davidson, J.F. (1949) The flow under gravity of a swirling liquid through an orifice-plate. *Proceedings of the Royal Society* A, 199, 443–457.

Binnie, A.M. & Teare, J.D. (1956) Experiments on the flow of swirling water through a pressure nozzle and an open trumpet. *Proceedings of the Royal Society* A, 235, 78–89.

Binnie, A.M., Hookings, G.A. & Kamel, M.Y.M. (1957) The flow of swirling water through a convergent-divergent nozzle. *Journal of Fluid Mechanics*, 3, 261–274.

Binnie, A.M. & Sims, G.P. (1969) Air entrainment by flowing water under reduced atmospheric pressure. *Journal of Hydraulic Research*, 7(3), 279–299.

Crump, E.S. (1955) A vortex siphon spillway for maintaining a constant water level upstream of a structure. *Proceedings of the ICE*, 5, Part 3, (2), 139–154.

Kubie, J. & Oates, H.S. (1980) Aspects of outflow from large vessels. *Journal of Fluids Engineering*, 102(3), 324–329.

Schlag, A. (1969) Recherches expérimentales sur l'écoulement par déversoir-puits [Experimental research on flow in shaft-spillways]. *La Houille Blanche*, 14(2), 127–136 (in French).

Outlets

Addison, H. (1931) The flow of water through groups of sluices: Experiments on scale models. The Institution of Civil Engineers *Selected Engineering Paper*, 105, 3–48.

Addison, H. (1937) Supplementary notes on flow through model sluices. *Journal of Institution of Civil Engineers*, 8, 53–72; 9, 447–449.

Allen, J. & Hamid, H.I. (1968) The hydraulic jump and other phenomena associated with flow under rectangular sluice-gate. *Proceedings of the ICE*, 40(3), 345–362; 42(4), 529–533.

Angus, R.W. Bryce, J.B. (1937) Model tests on spillways in the power dam at Abitibi Canyon. *Bulletin* 150: 1-21. Faculty of Applied Science and Engineering, School of Engineering Research, University of Toronto: Toronto.

Anonymous (1969) Theory and practice of model experiments. Annual Report. Central Board of Irrigation and Power, *Publication* 75. Government of India: New Delhi.

Apjohn, J.H. (1881) Determine the true coefficient of the discharge of the head sluice of the Midnapore Canal. *Transactions of Institution of Civil Engineers,* Ireland, 13, 249–262.

Auterio, M. (1963) Distribuzione delle pressioni sulle traverse a tetto [Pressure distributions at roof weirs]. Proceedings of the 8th *Convegno di Idraulica*, Pisa, C(7), 1–12 (in Italian).

Binnie, A.M. (1952) The flow of water under a sluice-gate. *Journal of Mechanics and Applied Mathematics*, 5(4), 395–407.

Binnie, A.M. (1979) Unstable flow under a sluice-gate. *Proceedings of the Royal Society*, London, Ser. A, 367, 311–319.

Birkhoff, G. (1961) Calculation of potential flows with free streamlines. *Journal of the Hydraulics Division* ASCE, 87(HY6), 17–22; 88(HY2), 187–189; 88(HY3), 223–224; 88(HY4), 293–299.

Blaisdell, F.W. (1937) Comparison of sluice-gate discharge in model and prototype. *Transactions of ASCE*, 102, 544–560.

Bradley, J.N. (1954) Rating curves for flow over drum gates. *Transactions of ASCE*, 119, 403–433.

Burt, C.M., Angold, R., Lehmkuhl, M. & Styles, S. (2001) Flap gate design for automatic upstream canal water level control. *Journal of Irrigation and Drainage Engineering*, 127(2), 84–91; 128(4), 264–265.

Castro-Orgaz, O., Mateos, L. & Dey, S. (2013) Revisiting the energy-momentum method for rating vertical sluice gates under submerged flow conditions. *Journal of Irrigation and Drainage Engineering*, 139(4), 325–335.

Caulk, D.A. (1976) On the problem of fluid flow under a sluice gate. *Intl. Journal of Engineering Science*, 14(12), 1115–1125.

Chow, W.L. & Chan, P.C.T. (1981) The effect of gravity on certain curved channel potential flows. *Journal of Fluids Engineering*, 103(12), 639–643.

Chung, Y.K. (1972) Solution of flow under sluice gate. *Journal of Engineering Mechanics Division* ASCE, 98(EM1), 121–140.

Clemmens, A.J., Strelkoff, T.S. & Replogle, J.A. (2003) Calibration of submerged radial gates. *Journal of Hydraulic Engineering*, 129(9), 680–687.

Collins, D.L. (1976) Discharge computations at river control structures. *Journal of the Hydraulics Division* ASCE, 102(HY7), 845–863; 103(HY12), 1481–1484; 104(HY8), 1199.

Cozzo, G. (1978) Una formula per il calcolo del coefficiente d'efflusso delle luci sotto paratoie [A formula to determine the discharge coefficient of gate flow]. *L'Energia Elettrica*, 55(11/12), 504–513 (in Italian).

Einwachter, J. (1932) Berechnung der Deckwalzenbreite des freien Wechselsprunges [Computation of roller length of free hydraulic jump]. *Wasserkraft und Wasserwirtschaft*, 27(21), 245–249 (in German).

Erbisti, P.C.F. (1999) The historical development of hydraulic gates. *Hydropower & Dams*, 6(2), 49–54.

Fellenius, W. & Lindquist, E. (1933) Researches concerning the discharge conditions at the Vargön Weir for weekly regulation of Göta Älv. *Meddelande 7*. Vattenbyggnadsinstitutionen Kungl. Tekniska Högskolan, Stockholm (in German, with English Summary).

Ferro, V. (2000) Simultaneous flow over and under a gate. *Journal of Irrigation and Drainage Engineering*, 126(3), 190–193; 127(9/10), 325–328.

Franke, P.-G. & Valentin, F. (1969) The determination of discharge below gates in case of variable tailwater conditions. *Journal of Hydraulic Research*, 7(4), 433–447.

Fröhlich, E. (1921) Bestimmung der Durchfluss-Koeffizienten für das Stauwehr Augst-Wyhlen [Determination of discharge coefficient for Augst-Wyhlen Dam]. *Schweizerische Bauzeitung*, 78(20), 233–238 (in German).

Gill, M.A. (1982) Discharge characteristics of radial gates. *Water Power & Dam Construction*, 33(3), 39–41.

Godwin, J. (1993) Trends and recent developments in the use of gates. *Water Power & Dam Construction*, 45(2), 40–42.

Govinda Rao, N.S. & Rajaratnam, N. (1963) The submerged hydraulic jump. *Journal of the Hydraulics Division* ASCE, 89(HY1), 139–162; 89(HY4), 277–279; 89(HY5), 147–152; 90(HY3), 313–316.

Gryzwienski, A. (1930) Neueste Fortschritte auf dem Gebiete der Wehrkonstruktionen [Latest developments in weir designs]. *Wasserwirtschaft*, 23(18/19), 416–421 (in German).

Hager, W.H. (1994) Impact hydraulic jump. *Journal of Hydraulic Engineering*, 120(5), 633–637.

Hartung, F. (1954) Wehranlagen mit Absenkschützen [Weir structures with lowering gates]. *Der Bauingenieur*, 29(10), 392–397 (in German).

Hartung, F.K. (1956) Die Entwicklungstendenzen des Grosswehrbaus bei den Flusskraftwerken in Mitteleuropa [Developing tendencies of large weir structures in rivers of Central Europe]. *Österreichische Bauzeitschrift*, 11(7/8), 137–155 (in German).

Hartung, F. (1970) Die strömungstechnische Entwicklung in Konstruktion und Gestaltung der Staustufen [The hydraulic development of construction and design of dam structures]. *Tiefbau*, 12(3), 201–230 (in German).

Hartung, F. (1973) Gates in spillways of large dams. Proceedings of the 11th *ICOLD Congress*, Madrid, Q41(R72), 1361–1374.

Hartung, F. (1975) Gedanken zur Gestaltung von Klappenwehren [Thoughts on the design of flap gates]. *Wasserwirtschaft*, 65(9), 238–243 (in German).

Heinemann, E. (1979) Beitrag zur Vermeidung der Wirbelbildung vor Tauchwänden [Contribution to the avoidance of vortex flow ahead of diving walls]. *Mitteilung* no, 27. Institut für Wasserbau und Wasserwirtschaft, RWTH Aachen, Aachen (in German).

Helmy, S. & Lemoine, R. (1959) Flow under a control gate in a tunnel: distribution of pressures for different gate openings derived from the theoretical flow net. Proceedings of the 18th *IAHR Congress*, Montreal, 18(A), 1–9.

Hilgard, K.E. (1904) Über Walzenwehre [On roller weirs]. *Schweizerische Bauzeitung*, 43(6), 65–69; 43(7), 86–88; 43(8), 99 (in German).

Horton, R.E. (1934) Discharge coefficients for Tainter gates. *Engineering News-Record*, 112(1), 10–12; 112(14), 483; 112(26), 846.

Isaacs, L.T. (1977) Numerical solution for flow under sluice gates. *Journal of the Hydraulics Division* ASCE, 103(HY5), 473–481.

Isaacs, L.T. & Allen, P.H. (1994) Contraction coefficients for radial sluice gates. *International Conference of Hydraulics in Civil Engineering*, Brisbane. Institution of Engineers, Australia, Brisbane, pp. 261–265.

Jones, J.H. (1929) The sluice-discharge measurements at Assouan. *Engineering*, 128, 413–414; 128, 845–846; 129, 23.

Josserand, A., Milan, D. & Berthollon, G. (1980) Vibrations des vannes des aménagements hydroélectriques: Connaissances actuelles, exemples industriels [Gate vibrations in hydraulic engineering: current knowledge, industrial examples]. *La Houille Blanche*, 35(7/8), 485–491 (in French).

Kapur, A.D. & Reynolds, A.J. (1967) Reattachment downstream of a control gate. *Journal of Hydraulic Research*, 5(1), 1–14.

Kirkpatrick, K.W. (1957) Discharge coefficients for spillways at TVA Dams. *Transactions of ASCE*, 122, 190–210.

Klassen, V.J. (1967) Flow from a sluice gate under gravity. *Journal of Mathematical Analysis and Applications*, 19(2), 253–262.

Krzeczkowski, S.A. (1980) Measurement of liquid droplet disintegration mechanisms. *Intl. Journal of Multiphase Flow*, 6(3), 227–239.

Larock, B.E. (1969) Gravity-affected flow from planar sluice gates. *Journal of the Hydraulics Division* ASCE, 95(HY4), 1211–1226; 96(HY4), 1050–1052; 96(HY10), 2121–2122.

Larock, B.E. (1970) A theory for free flow beneath radial gates. *Journal of Fluid Mechanics*, 41(4), 851–864.

Lemos, F.O., Martins, H.F., Peixeiro, L.C. & Leite, D.O. (1973) An accident with the big Tainter gate of a spillway. Proceedings of the 11th *ICOLD Congress*, Madrid, C15, 993–1005.

Levin, L. (1971) Phénomènes vibratoires observés sur des vannes-segment déversantes [Vibrational phenomena observed on overflowing segment gates]. *La Houille Blanche*, 26(6), 553–562 (in French).

Lewin, J. (1980) Hydraulic gates. *Journal of Institution of Water Engineers*, 34(5), 237–255.

Lewin, J. (1983) Vibration of hydraulic gates. *Journal of Institution of Water Engineers*, 37, 165–179; 37, 420.

Maitre, R. (1952) Study of the working conditions of surface spillways with partial opening of gates. *La Houille Blanche*, 7(4), 232–244 (in French with Summary in English).

Marchi, E. (1954) Efflusso da una luce soggiacente ad una paratoia piana a spigolo vivo [Free outflow from a plane gate]. *Rendiconti Accademia Scienze,* Bologna, 1–14 (in Italian).

Marchi, E. (1958) Ancora sull'efflusso da una luce di fondo [Again on the outflow from a bottom opening]. *Atti Accademia Scienze*, Bologna, 5, 122–123 (in Italian).

Martin, W.W., Naudascher, E. & Padmanabhan, M. (1975) Fluid-dynamic excitation involving flow instability. *Journal of the Hydraulics Division* ASCE, 101(HY6), 681–698.

Mayer, P.G. (1965) Discharge characteristics of partially open Tainter gates. *Water Power*, 17(9), 366–370.

McGee, R.G. (1988) Prototype evaluation of Libby Dam aeration system. In: Burgi, P.H. (ed), Proceedings of the International Symposium on *Model-Prototype Correlation of Hydraulic Structures*, Colorado Springs. ASCE, New York, pp. 138–147.

McNown, J.S., Hsu, E.-Y. & Yih, C.-S. (1955) Applications of the relaxation technique in fluid mechanics. *Transactions of ASCE*, 120, 650–686.

Mevorach, J. & Zanker, K. (1977) System of slightly immersed gates. *Journal of Hydraulic Research*, 16(1), 45–53; 17(1), 71–74.

Müller, H. (1935) Rechnerische Ermittlung der Strömungsvorgänge an scharfkantigen Planschützen [Computational procedure of flow processes at sharp-crested plan gates]. *Wasserkraft und Wasserwirtschaft*, 30(24), 281–284 (in German).

Müller, O. (1933) Schwingungsuntersuchungen an unterströmten Wehren [Vibration tests of flow below weirs]. *Mitteilung* 13. Preussische Versuchsanstalt für Wasserbau und Schiffbau, Berlin (in German).

Müller, O. (1937) Neuere Schwingungsuntersuchungen an unterströmten Wehren [Recent tests of gate vibrations]. *Bautechnik*, 15(6), 65–69 (in German).

Naudascher, E. (1962) Vibration of gates during overflow and underflow. *Transactions of ASCE*, 127, 384–407.

Naudascher, E., Rao, P.V., Richter, A., Vargas, P. & Wonik, G. (1986) Prediction and control of downpull on tunnel gates. *Journal of hydraulic Engineering*, 112(5), 392–416.

Novak, P. & Cabelka, J. (1981) *Models in hydraulic engineering*. Pitman, Boston.

Ogihara, K. (1967) A fundamental study of the vibration phenomena in a sluice gate. *Transactions of JSCE*, 141, 31–41.

Pajer, G. (1937) Über den Strömungsvorgang an einer unterströmten scharfkantigen Planschütze [On the flow process at a sharp-crested plan gate]. *Zeitschrift für Angewandte Mathematik und Mechanik*, 17(5), 259–269 (in German).

Pani, B.S., Satish, M.G. & Ramamurthy, A.S. (1979) Sluice gates with cylindrical lips. *Water Power & Dam Construction*, 30(2), 38–42.

Paul, T.C. & Dhillon, G.S. (1986) Dimensioning vertical lift gates. *Water Power & Dam Construction*, 38(11), 45–47.

Powley, R.L. & Haid, B.H. (1991) Flow through gated conduits at partial and full gate openings. *Canadian Journal of Civil Engineering*, 18(1), 43–52.

Rajaratnam, N. & Subramanya, K. (1967) Flow immediately below submerged sluice gate. *Journal of the Hydraulics Division* ASCE, 93(HY4), 57–77; 94(HY1), 340–341; 94(HY2), 601–603; 94(HY6), 1528.

Ramamurthy, A.S., Subramanya, K. & Pani, B.S. (1978) Sluice gates with high discharge coefficients. *Journal of Irrigation and Drainage Division* ASCE, 104(IR4), 437–441.

Ramos, C.M. (1982) Crossed jet stilling basins: Design criteria. *Memoria*, 577. Laboratorio Nacional de Engenharia Civil LNEC: Lisboa (in Portuguese, with Abstract in English).

Ransford, G.D. (1983) Pressures on spillway aprons downstream from partly raised crest gates. *Journal of Hydraulic Research*, 21(4), 303–314.

Rhone, T.J. (1959) Problems concerning the use of low head radial gates. *Journal of the Hydraulics Division* ASCE, 85(HY2), 35–65; 85(HY7), 151–154; 85(HY9), 113–117; 86(HY3), 31–36.

Rogala, R. & Winter, J. (1985) Hydrodynamic pressures acting upon hinged-arc gates. *Journal of Hydraulic Engineering*, 111(4), 584–599.

Rouvé, G. & Abdul Khader, M.H. (1969) Transition from conduit to free surface flow. *Journal of Hydraulic Research*, 7(3), 375–404.

Rouvé, G. & Abdul Khader, M.H. (1969) Lösungsverfahren für Potential-Strömungen mit freier Oberfläche [Solution procedures for potential flows with a free surface]. *Wasserwirtschaft*, 59(4), 95–102 (in German).

Sakhuja, V.S., Paul, T.C. & Singh, S. (1984) Air entrainment distortion in free surface flows. Proceedings of the International Symposium on *Scale Effects in Modeling Hydraulic Structures*, Esslingen, Germany, 4(8), 1–4.

Sehgal, C.K. (1996) Design guidelines for spillway gates. *Journal of Hydraulic Engineering*, 122(3), 155–165.

Shammaa, Y., Zhu, D.Z. & Rajaratnam, N. (2005) Flow upstream of orifices and sluice gates. *Journal of Hydraulic Engineering*, 131(2), 127–133.

Simmons, W.P. (1965) Experiences with flow-induced vibrations. *Journal of the Hydraulics Division* ASCE, 91(HY4), 185–204; 92(HY2), 432–439; 93(HY2), 45.

Stoney, F.G. (1876) Construction of large sluices for irrigation and navigation. *Transactions of Institution of Civil Engineers*, Ireland, 11, 24–39.

Swamee, P.K., Pathak, S.K., Mansoor, T. & Ojha, C.S.P. (2000) Discharge characteristics of skew sluice gates. *Journal of Irrigation and Drainage Engineering*, 126(9/10), 328–334.

Toch, A. (1955) Discharge characteristics of Tainter gates. *Transactions of ASCE*, 120, 290–300.

Vanden-Broeck, J.-M. (1986) Flow under a gate. *Physics of Fluids*, 29(10), 3148–3151.

Vanden-Broeck, J.-M. (1997) Numerical calculations of the free-surface flow under a sluice gate. *Journal of Fluid Mechanics*, 330, 339–347.

Warncke, A., Gharib, M. & Roesgen, T. (1996) Flow measurements near a Reynolds ridge. *Journal of Fluids Engineering*, 118(9), 621–624.

Chapter 9 Frontispiece (a, b) Gries Reservoir, Switzerland, prior and after reservoir drawdown (Courtesy Dr. Daniel Ehrbar, ETH Zurich), (c) sediment bypass tunnel of Palagnedra Dam, Switzerland (Courtesy Dr. Christian Auel, ETH Zurich)

Chapter 9

Reservoir sedimentation

9.1 Involved processes and sustainable reservoir use

Sedimentation is known as the process of filling natural lakes and man-made reservoirs by sediment transforming finally into land again. Its main reason is the sediment yield transported by rivers as suspended or bed load into the reservoirs (Groupe de Travail, 1976; Annandale, 1987; Morris and Fan, 1998). Both *bed* and *suspended* sediment loads originate from soil and rock erosion in the catchment area of the reservoir, also referred to as *denudation*. Suspended fine sediments also result from surface erosion as well as crashing and abrasion of coarser sediments transported by rivers.

The amount of suspended load is often much larger than the bed load, representing typically 80–90% of the total sediment yield. At the river mouth into the reservoir, the bed load settles and contributes to the delta development (Figure 9.1). The larger components of the suspended load settle close to the delta, whereas finer particles are taken into the reservoir settling mainly at stagnation-water zones. Extremely fine particles remain suspended and may leave the reservoir with the outflow.

The inflow to a reservoir is a mixture of water and sediment. Due to the density difference, so-called density currents develop from the mixture of water and fine sediment. Density currents in general are two-phase flows involving a small density difference, similar to cold and warm water, or water and sludge. In contrast, air-water flows have a density ratio of nearly 1000. Also, the two fluids should be miscible and the density difference be a function of differences in the temperature, the salt content, or the sediment content, independent of pressure and elasticity of the fluids involved.

The sediment-laden flow has a larger density than the reservoir water thus moving along the reservoir bottom toward the dam. Figure 9.1 shows a typical configuration with A river inflow, B clear-water reservoir, and C recirculation zone. The density current is seen to move along the thalweg toward the dam outlet. Upstream, the coarse material is deposited as a delta and from the plunge point, the current dives along the reservoir bottom to reach the outlet zone. Upstream from the outlet crest, fine material deposits in the upstream direction. Details on density currents are provided in 9.6.

Due to local sediment deposition, the density current is diluted along its trajectory. In the delta, where the velocity of the density current is already reduced, particles larger than, say, 0.05 mm are deposited. The finest silt particles of diameter less than 0.01 mm are carried up to the deepest reservoir zone in front of the dam and normally near the outlet, while the intermediate fraction is deposited along zone B.

Although the reasons and the involved processes of reservoir sedimentation are well-known, sustainable and preventive measures are rarely considered in the design of new reservoirs. To

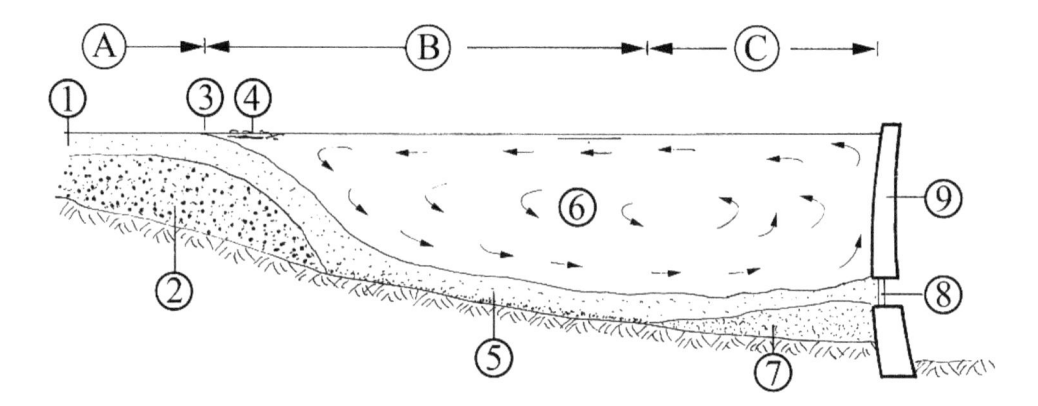

Figure 9.1 Density current in reservoir with ① approach flow, ② delta, ③ plunging point, ④ floating debris, ⑤ density current, ⑥ clear water, ⑦ sediment deposition, ⑧ outlet, and ⑨ dam

avoid operational problems of powerhouses and water supply facilities, sedimentation is often treated for existing reservoirs with measures, which are efficient during a limited period only. Since most of the measures will lose their effect, a sustainable reservoir operation and thus the water supply, as well as the production of valuable peak energy are endangered. Today's worldwide yearly mean loss of reservoir storage capacity due to sedimentation is already higher than the increase of capacity by the addition of new reservoirs for irrigation, drinking water, and hydropower. Depending on the region, it is commonly accepted that about 1–2% of the worldwide capacity is annually lost (Jacobsen, 1998). In Asia, for example, 80% of the useful storage capacity for hydropower production will be lost already in 2035 (Basson, 2009). In the European Alps, the loss-rate of reservoir capacity is significantly below world average; the main process in narrow and long reservoirs is the formation of turbidity currents, transporting fine sediment regularly toward the dam, thereby increasing sediment levels up to 1 m/a close to the dam toe. The outlet devices including intakes and bottom outlets are therefore often already affected by sediment presence after 40–50 years of operation, even in catchments with moderate surface erosion rates. The effects of climate change will increase the future sediment yield of reservoirs mainly in Alpine regions due to glacier retreat and melting of permafrost grounds. To counter the reduction of reservoir volume due to sedimentation processes, a total investment of 13–19 billion US$ would be required for storage replacement, equaling some 20–30% of the yearly global budget for reservoir operation and maintenance (Schleiss *et al*, 2010). The measures effectively taken to counter reservoir sedimentation are by far smaller than required for a sustainable long-term reservoir management.

9.2 Sedimentation rate and sediment distribution

The currently worldwide installed water capacity of all reservoirs is about 7000 km³, from which some 4000 km³ are used for energy production, irrigation, and water supply (Basson, 2009). Their mean age is between 30 and 40 years. It is estimated that about 1% of the worldwide water storage capacity is lost annually by sedimentation. The highest average sedimentation rate occurs in arid regions, as in the Middle East, Australia and Oceania, as well as in

Africa (Table 9.1). Batuca and Jordaan (2000) provide a detailed collection of sedimentation rates in regions all over the world.

Table 9.2 details for different world regions the dates when 80% of the useful reservoir volumes for hydropower production and 70% of the reservoirs for other usage are lost by sedimentation, respectively (Basson, 2009). Note that in Asia, for example, 80% of the useful storage capacity for hydropower production is estimated to be lost already in 2035. In view of the increasing energy and food demands, this issue addresses a serious problem: 70% of the storage volumes used for irrigation will be filled up by sediment already in 2025 in Asia, in 2030 in the Middle East, and in 2040 in Central America. This fact underlines that reservoir sedimentations endangers the sustainable energy and food production in many regions of the world.

The sedimentation rate of each particular reservoir is highly variable. It depends particularly on the climatic situation, the geomorphology, and the conception of the reservoir including its outlet works. For Alpine reservoirs, Beyer Portner and Schleiss (2000) describe the annual erosion volume per surface by an empirical equation based on a detailed analysis of 19 Swiss reservoirs. On the average, only about 0.2% of the storage capacity is annually lost in Swiss reservoirs. This relatively low sedimentation rate in the Alps is due to the geologic characteristics, involving mainly rock, and the catchment areas located at high altitudes (Oehy, 2003).

Based on the observation of existing reservoirs, empirical methods were developed to characterize the sediment distribution along a reservoir under so-called equilibrium conditions

Table 9.1 Average sedimentation rates in different regions (Basson, 2009)

Region	Average sedimentation rate (%/a)
Africa	0.85
Asia	0.79
Australia & Oceania	0.94
Central America	0.74
Europe	0.73
Middle East	1.02
North America	0.68
South America	0.75

Table 9.2 Dates when 80% of useful reservoir volumes for hydropower production, and 70% of reservoirs for other uses are lost by sedimentation, respectively (Basson, 2009)

Region	Hydropower dams: Date 80% filled with sediment	Non-hydropower dams: Date 70% filled with sediment
Africa	2100	2090
Asia	**2035**	**2025**
Australasia	2070	2080
Central America	2060	**2040**
Europe & Russia	2080	2060
Middle East	2060	**2030**
North America	2060	2070
South America	2080	2060

(Borland and Miller, 1958). For fine sediments transported and distributed over the entire reservoir length, the principle of minimum stream power applies (Chang, 1979; Rooseboom and Annandale, 1983; Annandale, 1984, 1996) to predict the equilibrium sedimentation pattern. Based on the analysis of Iranian reservoirs, Rahmanian and Banihashemid (2012) demonstrated that accounting for the effects of both wetted perimeter variations on sedimentation behavior through the reservoir, and the hydraulic radius of the reservoir cross section, the predicted quality relative to the cumulative sediment deposition along the reservoir increases.

The coarse sediment transported mainly as bed load into the reservoir is deposited in the delta. Due to the reservoir backwater into the approach flow river, the reservoir delta grows so that aggradation of the river develops (9.7.1). The accumulating sediments successively reduce the water storage capacity of the reservoir (Fan and Morris, 1992a, b). At long term the reservoir operates thus only at reduced functional efficiency. Declining the storage volume reduces and eventually eliminates the capacity for flow regulation along with water supply, energy, and flood control benefits (Graf, 1983; ICOLD, 1989, 2012). Reservoir sedimentation can even lead to a perturbation of the operating intake, and to sediment entrainment in waterway systems and hydropower schemes (Schleiss *et al.*, 1996; De Cesare, 1998; Lai *et al.*, 2015). Depending on the degree of sediment accumulation, the outlet works may be clogged by sediments. Blockage of intake and bottom outlet structures or damage to gates not designed for sediment passage present also severe security problems (Schleiss *et al.*, 1996). Other consequences of sediment reaching intakes are abrasion of hydraulic machinery, reducing their efficiency and increasing maintenance cost (Boillat and Delley, 1992; Boes *et al.*, 2013; Felix *et al.*, 2013; Abgottspon *et al.*, 2014).

9.3 Evolution of knowledge and management competence

Based on the analysis of more than 370 cites, De Cesare and Lafitte (2007) concluded the following. Field observations on reservoir sedimentation and the related physical processes appeared as early as in 1936, but some 80% of all publications date from the 1980s. Reservoir sedimentation surveys were reported only from 1960, whereas overall reservoir sedimentation studies were published from the 1970s, so that the few sources published before 1950 remain exceptional. Most methods and techniques for achieving the sustainable use of reservoirs are found in the late 1980s and the 1990s. The subject is currently of wide interest including state-of-the-art contributions from field studies, but the direct application on the ongoing planning of new reservoirs is often still insufficient.

The effective knowledge on reservoir sedimentation, its related processes, and the sustainable management and the awareness in the planning phases of new hydropower and water supply schemes were simply not available in a consistent and commonly accepted manner before 1950. Outstanding publications on reservoir sedimentation and examples of sediment management of hydropower schemes appeared prior to 1950, but the broad understanding of the importance of reservoir sedimentation was not yet generally developed. This statement is underlined by the fact that even recent design books on reservoir and dam engineering rapidly pass over the subject topic, underestimating its importance or simply neglecting it. Studies from the late 1980s describe in a much broader view the transportation and sedimentation processes. Both experimental and numerical simulation techniques allowed for an insight into the structure of turbidity currents; since the end of the 1990s the major task for avoiding sediments in large reservoirs was identified to support the long-term operation of bottom outlets and water intakes. Note that most of the existing artificial lakes are still not operated sustainably. Based on the current knowledge and technology, future hydropower and water supply projects with storage reservoirs are operated more efficiently and productively, ensuring their long-term sustainability.

It is currently a must to consider all aspects relating to reservoir sedimentation for any new hydropower and reservoir project. The necessity of sustainable sediment management was neglected for a long time, including knowledge of the decisive erosion, transportation, and deposition processes, which led to massive reservoir siltation. Overlooking the fundamentals of reservoir sedimentation management methods and techniques is hardly understandable and not sustainable for any new reservoir project. It is thus difficult to understand why some engineers, in spite of available knowledge, still plan, design, build and operate hydropower and water supply schemes with such a deliberate loss of storage capacity, resulting in future problems of sediment entrainment at intake structures, turbines, and water supply facilities, respectively, and the decline of the downstream river.

Schleiss *et al.* (2016) present the state-of-the-art and the main scientific advances in terms of prevention and mitigation measures against reservoir sedimentation. Furthermore, the main research challenges regarding open questions as well as the emerging research methodologies to study sedimentation processes are discussed.

9.4 Measures against reservoir sedimentation

9.4.1 Overview

Over the past years, various measures against reservoir sedimentation have been proposed (Annandale, 1987; Morris and Fan, 1998; Schleiss and Oehy, 2002; Sumi, 2005; Boes and Hagmann, 2015), of which not all are sustainable, efficient, and affordable, however. For example, dam heightening or increasing the intake level of outlet works does not provide a long term and sustainable solution.

One way to classify the currently known desilting measures is to subdivide into measures taken in the catchment area, in the reservoir, or at the dam itself (Figure 9.2). These measures can be implemented before or during reservoir construction, either as preventive measure or only after reservoir sedimentation has occurred. The latter retroactive measures are mainly used to restore the reservoir volume by sediment removal. In general, preventive measures are preferable, as retroactive desiltation measures for existing dams with a significant sediment aggradation are normally more difficult and costly. The selection of feasible retroactive measures tends to be much more restricted as compared with preventives measures (Müller and De Cesare, 2009; Boes, 2011).

Alternatively, reservoir sedimentation mitigation measures can be classified according to the desilting concept (Sumi *et al.*, 2004; Boes and Hagmann, 2015). Each concept is further divided into a number of desilting types and methods (Figure 9.3). The careful selection of reservoir location and layout may greatly reduce its service longevity regarding reservoir sedimentation on the one hand. On the other hand, particularly for existing reservoirs with severe sedimentation problems, sediment yield reduction, sediment routing, and sediment removal are distinguished. While the two former are mainly preventive to avoid sediment supply to the reservoir, the two latter are retroactive, demanding the handling of sediment already transported into or even settled in the reservoir.

Despite this classification, guidelines on how to tackle sedimentation issues are not yet detailed. As the environmental, technical, and economic reservoir boundary

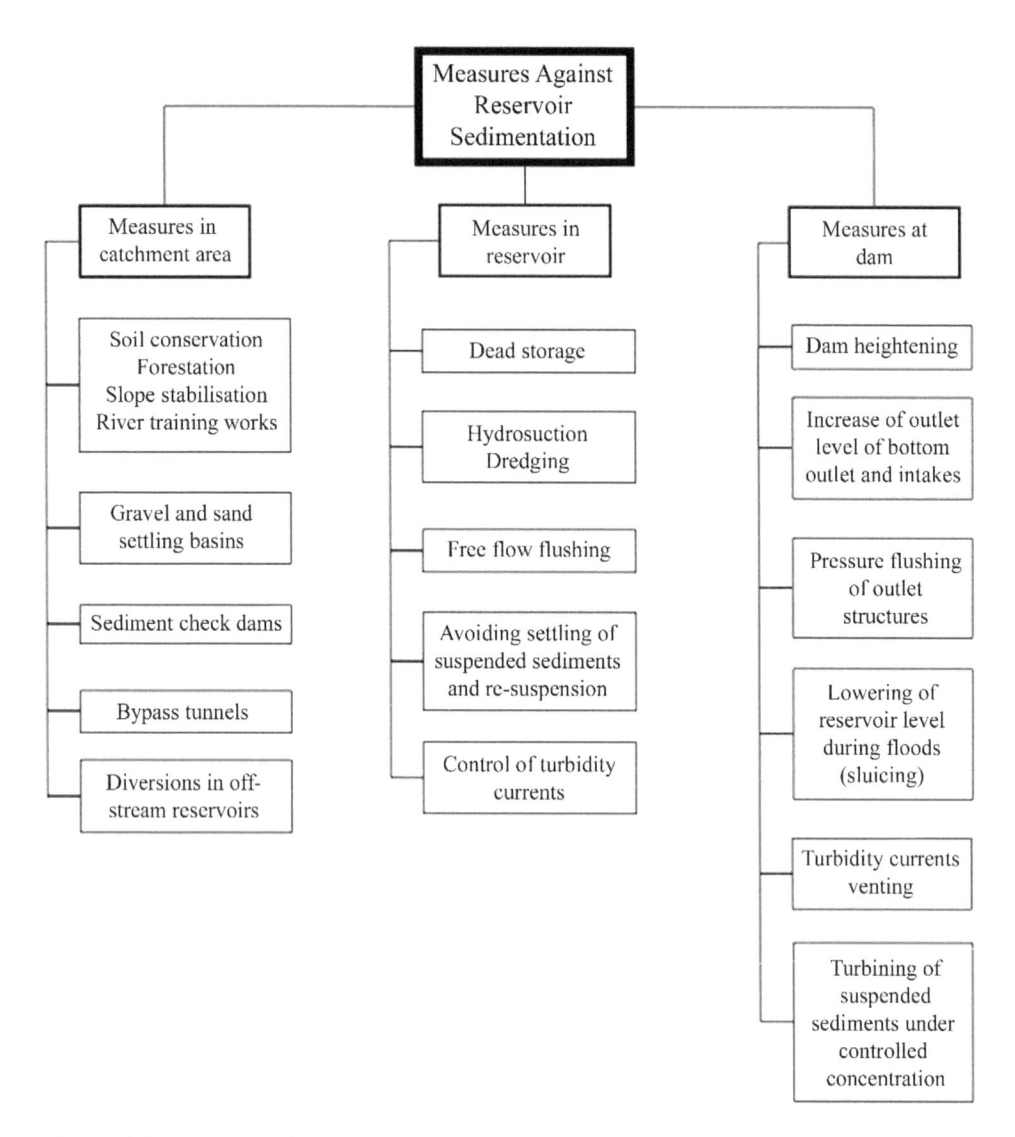

Figure 9.2 Inventory of possible measures for sediment management in the catchment area, the reservoir and at the dam (Adapted from Schleiss and Oehy, 2002)

conditions are highly site-specific, each case has to be individually treated and opti-mized. A first decision on the concept to be selected should be based on the following parameters:

- Mean Annual Sediment inflow volume (MAS) (see Figure 9.18 below);
- Reservoir life value, i.e. ratio between reservoir volume (so-called CAPacity, CAP) and MAS;

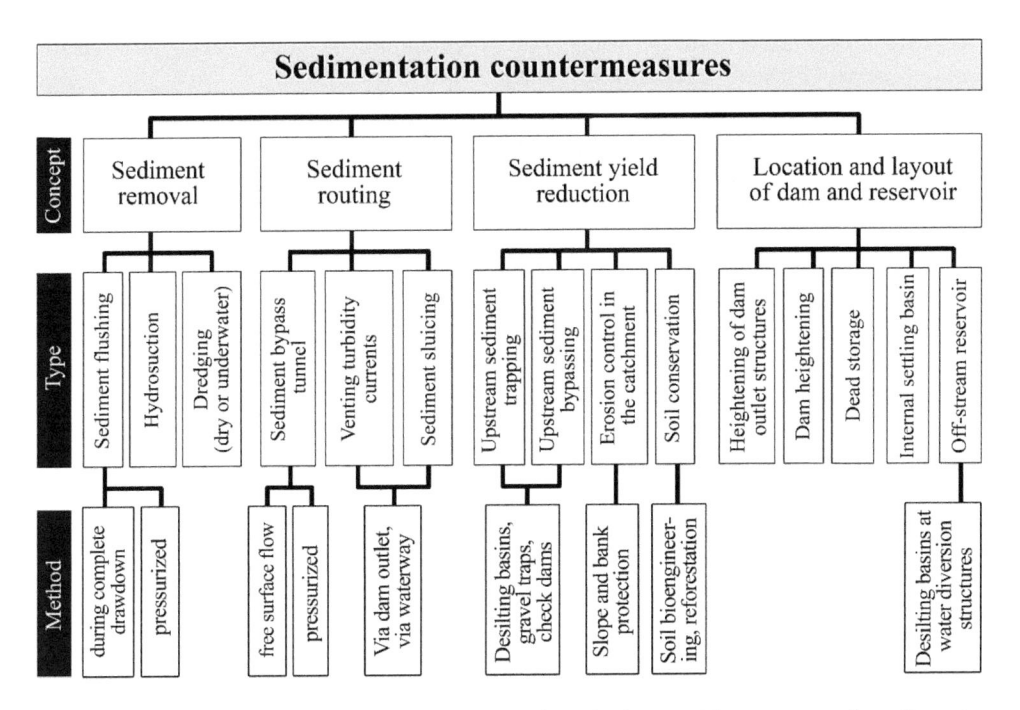

Figure 9.3 Classification of concepts, types and methods to mitigate reservoir sedimentation (Adapted from Boes and Hagmann, 2015)

- Capacity-inflow ratio (CIR = CAP/MAR), i.e. the ratio between the reservoir storage volume and the Mean Annual Runoff (MAR), as measure of the mean residence time of suspended sediment particles in the reservoir.

A sustainable sediment management usually combines different measures. Whenever possible, the extent of sedimentation should be minimized from the beginning by carefully selecting the location and layout of reservoirs and dams. Off-stream reservoirs (Figures 9.2 and 9.3), mainly filled with water from water intakes with desilting facilities, may be viable options for areas with high sediment yields. In the framework of the Interreg III B European research project ALPRESERV, detailed recommendations have been stated for sustainable reservoir use regarding sedimentation by considering both ecological and economical aspects (Jenzer Althaus and De Cesare, 2006).

9.4.2 Measures in catchment area

The retention of sediments in a catchment area deals with the root of the problem. In a large catchment, this is hardly possible due to economic reasons, whereas in a small or medium catchment area, the erosion process is reduced by two methods:

- Soil conservation, and
- Sediment retention measures.

Methods involving soil conservation are lengthy and costly, becoming effective only after decades. Soil conservation needs to be supported by agricultural development. In non-vegetated areas these activities are practically impossible.

Methods involving measures in the catchment are based on the reduction of sediment input into rivers by slope and bank protection. Accordingly, slide areas and bank instabilities along rivers are prevented by river training works. In steep rivers, bed incision and depth erosion have to be limited by transverse sills and check dams.

Sediment retention and settling basins created by check dams may be arranged at particular locations. They retain mainly bed load as soon as they are submerged during floods. However, bed load or gravel retention basins influence the sediment transport only locally. If they are located just upstream from the reservoir, they are designed as a storage zone to be normally cleared annually or after each large flood. There results a gravel deficit in the downstream river reach, however, along with riverbed degradation. This erosion process, in turn, may reduce the long-term stability of the gravel retention basin (Figure 9.4). Therefore, the latter is effective only if the:

- Tailwater river is resistant against large scale-erosion,
- Erosion is reduced by suitable protection, or
- Gravel retention basin is located immediately upstream of reservoir.

If gravel retention basins are relatively small, they do not significantly reduce the sediment input into large reservoirs. Therefore, larger sediment check dams are placed directly at the entrance of the large reservoirs upstream of the delta. The question relating to the retention of various sediment grain sizes depends on the retention volume created by check dams. As a secondary dam is fully submerged under maximum reservoir level, gravel as well as sand are effectively settled. The silt fraction is mostly carried into the reservoir depositing there (Figure 9.5). The trap efficiency of secondary sediment check dams depends on the CIR, i.e. the ratio between the storage volume and the annual inflow as a measure of the mean residence time of suspended sediment particles in the reservoir. The larger this CIR, the better is the reservoir retention potential to trap a significant amount of sediments. For a small ratio, the reservoir degrades to a storage basin for gravel as previously mentioned under the measures against erosion. These check dams can retain at least nearly all bed load and also coarser suspended sediment as sand fractions. Nevertheless, their volume has to correspond to an annual sediment yield designed for a certain probability. The volume of the bed load retaining dams should be such that they store at least the bed load yield of one 50-year flood, or

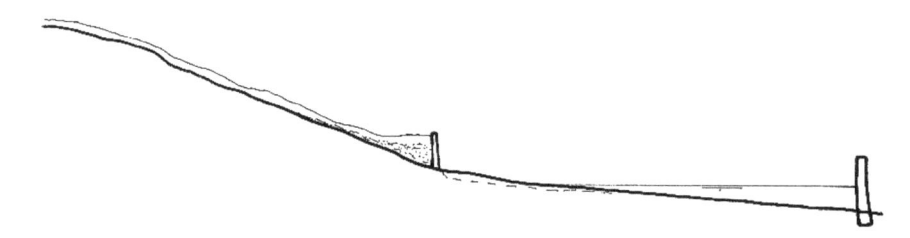

Figure 9.4 Gravel retention basin at grade break upstream from reservoir, with (- - -) possible erosion of tailwater river due to sediment deficit

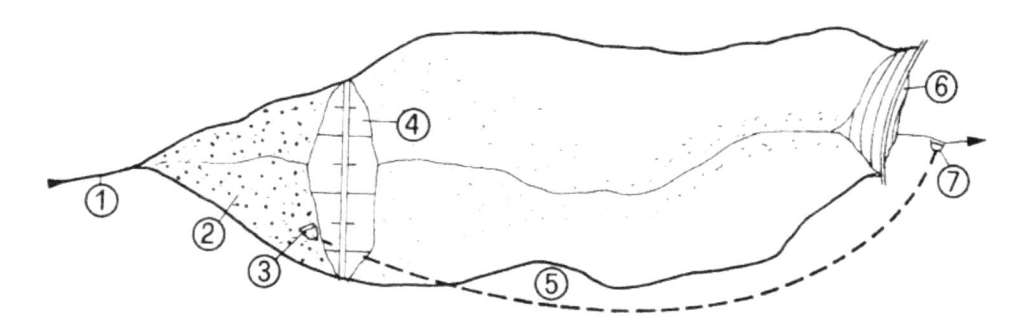

Figure 9.5 Gravel storage dam to counter reservoir sedimentation ① approach flow river, ②
pre-reservoir, ③ bypass tunnel inlet, ④ sediment check dam, ⑤ sediment bypass
tunnel, ⑥ main dam, ⑦ bypass outlet

two 10-year floods. Trapping fine sediment settled by the check dams requires a sufficiently large retention volume. For example, a trap efficiency of 90% of fine sediment is reached only if the check dam creates a storage volume of at least 1/10 of the yearly inflow.

As mentioned, sediment check dams at the inlet of large reservoirs have to be emptied regularly to remove sediments. They can be combined with Sediment Bypass Tunnels (SBTs) and structures (Figure 9.5). Up to a certain flood event, typically a 5- to 10-year flood, these bypass structures flush coarse (bed load) and fine suspended sediment around the main reservoir. The SBT is in operation only during these floods and its tractive forces must be large enough to prevent clogging by larger deposits of granular material. The SBT must further resist abrasion. The inlet to the SBT must satisfy the conditions stated in 9.4.3. Detailed information on SBT are provided in 9.5.

The risk of reservoir sedimentation scales directly with the size of the directly controlled catchment area. Therefore, in river catchment areas with high sediment transport, it is wise not to build the reservoir in the main valley, i.e. in-stream along the main watercourse, but to foresee an off-stream reservoir, which directly controls a smaller catchment area less affected by sedimentation. The water is then diverted from the main river by an intake structure into the off-stream reservoir. Nevertheless, the intake and diversion structures have to be equipped with efficient sediment management structures including flushing devices and sand traps. Mainly suspended sediment-laden water up to a certain design discharge is diverted into the off-stream reservoir. The *Längental* Reservoir in the Austrian Alps is an example of such an off-stream reservoir with a direct catchment of only 15% (Boes and Reindl, 2006).

9.4.3 Measures in reservoir

Dead storage

A still widely used method to conserve the storage capacity is to oversize reservoirs, i.e. to keep some of the impoundment available for sedimentation. If this volume is not available for reservoir operation, it is referred to as dead storage. Its volume is designed so that the sediment deposits do not reach the level of the intake structures during a certain period. Often a period of 50 years applies for which the sediments entering the reservoir should not fill up this dead storage. Since turbidity currents re-suspend and then transport fine sediments which have been settled in the upper reservoir portion (Figure 9.1) during yearly floods directly in front of the dam, the dead storage may be filled up much faster than designed, as indicated by many examples worldwide (e.g. Schleiss *et al.*, 1996; Boes and Hagmann, 2015). In view of the high investment for dam construction, oversizing a reservoir with a significant dead storage is neither an economical nor a sustainable measure.

Flushing

Hydraulic flushing was practiced in Spain already in the 16th century (Brown, 1944). Since then arguments about whether flushing is more suitable for large or small reservoirs have been forwarded. Several reasons prevent its use, because fish may be harmed (Salih, 1994) or the released sediments are heavily polluted (Scheuerlein, 1995). Sediment flushing involves a technique whereby previously accumulated and deposited sediments in a reservoir are hydraulically eroded and removed by accelerated flows created when the bottom or low-level outlets of the dam are opened. Sediment flushing is discussed by Shen and Lai (1996) or Morris and Fan (1998). Practical criteria for successful flushing are given by White (2001).

A distinction between flushing under pressure and free-flow flushing is made. Flushing under pressure may release only fine sediments near the flushing outlets. If at the same time the reservoir water level is drawn-down, the reservoir flow velocities are increased so that fine suspended sediment does not settle. After a full reservoir drawdown, free-flow flushing occurs during which also coarse sediment along the reservoir bottom is mobilized due to high bed-shear stress. The features of both flushing under pressure, and free-flow flushing are detailed in the following.

During flushing under pressure, water is released through the bottom outlets while the water level in the reservoir is kept high. The bottom outlet of a dam is located normally at the front portion of the reservoir sediments. Therefore, the outlet is used to clear the material settled just upstream. Flushing material under pressure leaves a cone with side slopes of up to 1:1, depending on the natural angle of repose under submergence of the deposited sediment. The effect of flushing under pressure is thus limited to sediment deposits located close to the flushing devices (Figure 9.6), so that only a small portion of the entire reservoir is flushed.

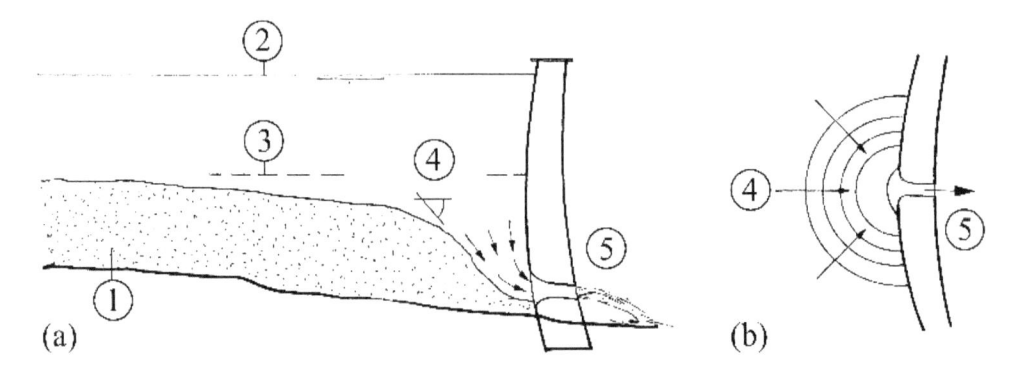

Figure 9.6 Flushing under pressure of sediment front by bottom outlet, (a) section, (b) plan with ① reservoir deposits, ② maximum and ③ minimum reservoir levels, ④ flushing cone, ⑤ bottom outlet

Nevertheless, by continuously lowering the reservoir level during flushing at flood events, the flow velocities along the reservoir increase and fine sediments may be hindered to settle. This requires a sufficiently large discharge capacity of the flushing outlet, however.

Because a bottom outlet is not permanently in operation, it may clog due to the advancement of the sediment front, or due to underwater slides. Clogging may rarely occur due to logs sunk to the reservoir bottom. An upper sediment concentration of 5% is reported to yield no problems with transport in bottom outlets (Dawans *et al.*, 1982). Figure 9.7 shows the regularly flushed bottom outlet of the Gebidem Storage Scheme, Switzerland (Meile *et al.*, 2014). If the outlet gets clogged as both the flap and the radial gates are opened, the injector pit is operated. Due to the enormous hydraulic grade at the sediment front, packets of material pass as a water-sediment flow. If the sediment concentration stays within the mentioned limits, this two-phase flow continues until the inlet portion to the bottom outlet is cleared. Otherwise, the injector has to be operated again to clean the outlet.

The design of the injector pit is based on the following requirements:

- Inlet to pit should not self-clog,
- Inlet level is under minimum reservoir level,
- Efficiency is sufficient to clear bottom outlet, and
- Structure is as small as possible not interfering with normal outlet.

Note that clogging is most probable for bottom outlets without a special anti-clogging device as injector pits (Krumdieck and Chamot, 1979). The latter allow for a dilution and blowout of the consolidated sediments, as shown in Figure 9.8.

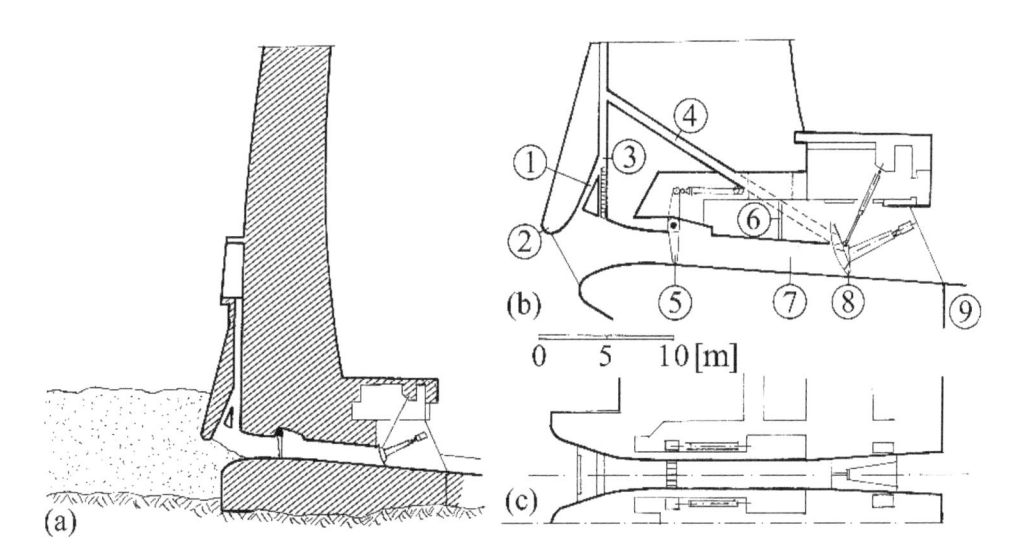

Figure 9.7 Gebidem Dam bottom outlet with flushing gates (a) overall section, (b) detailed section, (c) plan with ① jetting pipe, ② guard nose, ③ bulkhead gate, ④ compensation water pipe, ⑤ flap gate, ⑥ bypass, ⑦ steel-blinded bottom outlet, ⑧ radial gate, ⑨ steel-lined tailwater invert (Adapted from Dawans *et al.*, 1982)

Figure 9.8 Flushing under pressure with bottom outlet at Jiroft Dam, Iran (Courtesy Soleyman Emami)

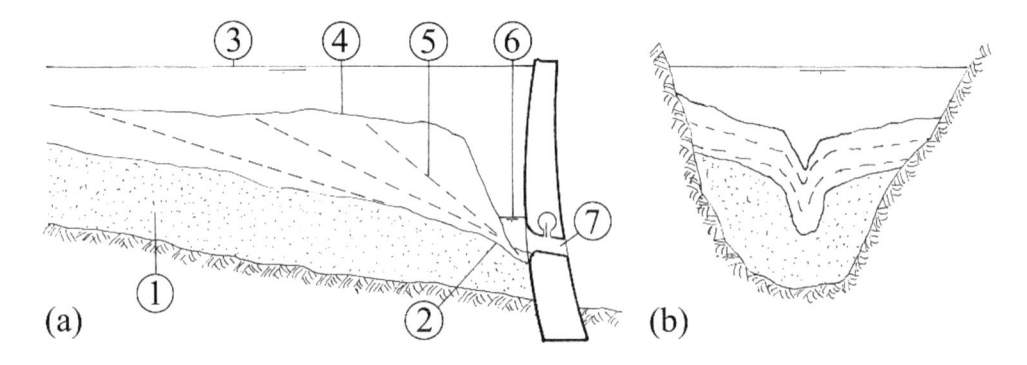

Figure 9.9 Reservoir flushing (a) streamwise section, (b) cross section with ① remaining sedimentation, ② erosion cone, ③ maximum reservoir level, ④ reservoir bottom prior to flushing, and ⑤ after some flushing time, ⑥ minimum reservoir level, ⑦ bottom outlet

As mentioned, the effect of flushing under pressure is local, and counters reservoir sedimentation only for small basins. Under *free-flow flushing*, the reservoir is lowered so that free-surface flow results along the reservoir thalweg, resembling natural riverine conditions. Figure 9.9 shows the temporal development of free-flow flushing and the resulting reservoir erosion down to the un-erodible bottom. Erosion starts at the cone close to the outlet developing upstream along the thalweg. The erosion process ends when the tractive forces become smaller than the resistive forces, and an equilibrium bottom topography has developed. Note the canyon-like erosion along the thalweg, and the relatively small erosion along the reservoir flanks due to bank slides (see also Figure 9.54).

Free-flow flushing can erode formerly deposited sediments transporting these along the reservoir bottom through the flushing devices. Compared with pressurized flushing, a much greater sediment load and even consolidated sediments are thus evacuated. Free-flow flushing by lowering the reservoir is the only effective measure to erode and transport coarse sediments deposited in the delta region. For run-of-river power plants with a limited storage capacity, free-flow flushing after each flood season is mandatory to ensure the operation safety of intakes.

If the water level is drawn down during flushing, the sediment removal occurs in several phases. Flushing is most effective during the first hours after the stored water in the reservoir has been released. A free-flow phase then begins with high rates of sediment removal in the first few days or weeks, but once a stream has re-established its original gradient through the reservoir basin, the amount of sediment picked up and transported greatly decreases (Brown, 1944) and the reservoir turbidity level remains stable. The flushing timing is important, both for economic and ecological reasons. In regions with pronounced flood seasons, flushing has to be executed before the start of the yearly flood. If the prime purpose of a dam is irrigation water storage, flushing is done at the end of the irrigation season, coinciding with the period of least water demand. Kereselidze et al. (1986) recommend that flushing should be performed immediately before fish spawn and after they rear the fry. Shen and Lai (1996) state that flushing should be performed regularly, especially for cohesive clay deposits, before deposits consolidate. Flushing becomes more effective the longer it lasts, the narrower, straighter, and shorter the reservoir is, the higher the discharge of the flushing stream is, the larger the dimensions and the lower the location of the flushing outlet are, the finer and the rounder sediment particles are, the younger and the less consolidated the sediments are, and the steeper the original stream gradient through the reservoir is (Orth, 1934, see Brown, 1944).

Flushing is often associated with adverse environmental impacts. The methods must be selected with the aim of limiting the impact of the downstream river reaches (OFEFP, 1994). Legal requirements may also restrict or prohibit the practice of removing solids from surface waters along with reintroduction into the flow at a later time (Suter, 1998; Boillat and Pougatsch, 2000). An optimum flushing program is determined for each reservoir with a maximum reservoir effect and a minimum tailwater impact. Accordingly, the transport capacity of the downstream river should be so large that deposits remain small (Figure 9.10). Environmental concerns regarding fish and water flora have to be accounted for to respect the ecological balance. Typical maximum volumetric sediment concentrations to be respected in the tailwater amount to 10 ml/l in Switzerland (Boes and Hagmann, 2015).

Hydro-suction

Different systems with fixed pipes intended for routing sediment through or around a dam were suggested. Hotchkiss and Huang (1995) state that a shock to the downstream reach associated with flushing is avoided by using a hydro-suction sediment-removal system, because it is continuous and of longer duration. These techniques try to return the system to more natural pre-dam conditions by releasing sediments in accordance with the downstream transport capacity.

Hydro-suction systems remove deposited sediments by using the available energy head due to the difference between water levels up- and downstream of a dam, or driven by engines (Figure 9.11). Hotchkiss and Huang (1995) describe their design and presented field

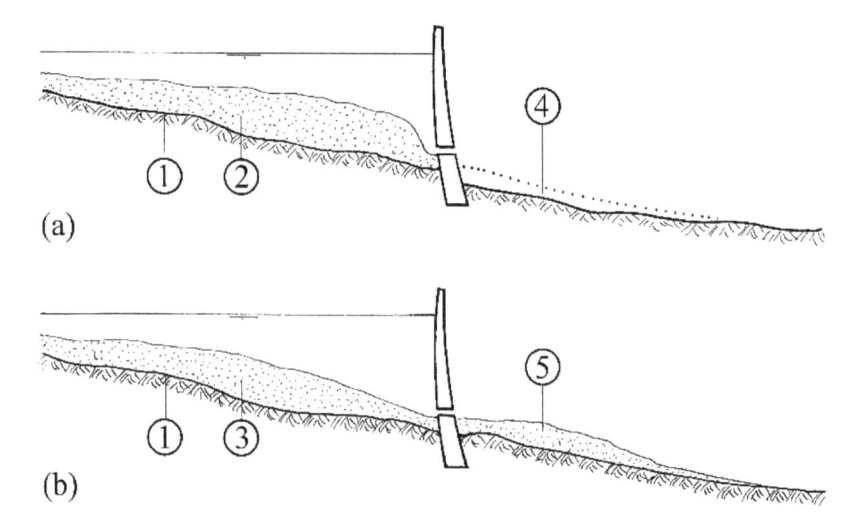

Figure 9.10 Effect of (a) reservoir sedimentation, (b) reservoir flushing on tailwater river with ① original riverbed, ② sedimentation zone, ③ reduced zone after flushing, ④ degradation, ⑤ tendency of tailwater aggradation

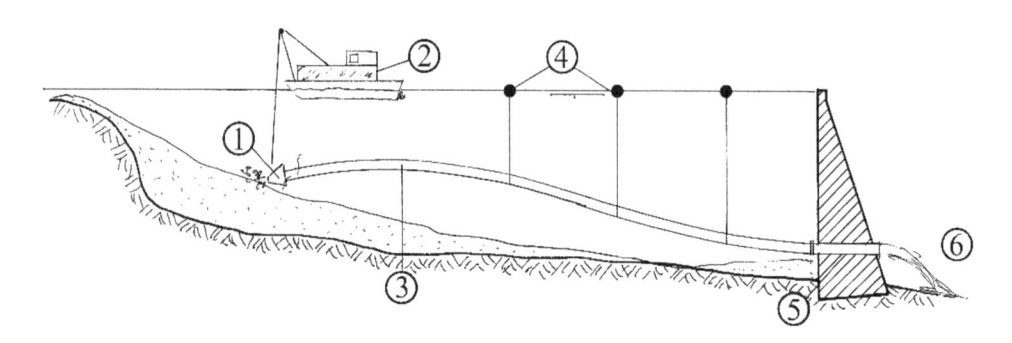

Figure 9.11 Sediment removal with dredger connected to syphon pipe with ① head of pipe, ② dredging platform, ③ connecting hose, ④ buoys for pipeline suspension, ⑤ bottom outlet, ⑥ tailwater

tests. The pumping device is equipped with a drill head to facilitate the disintegration of the deposits. The volume of the pumped mixture is normally discharged in decantation basins. It can also be diverted into the downstream river with a controlled discharge. This type of operation was applied at Luzzone Reservoir, Switzerland, to clear the water intake opening. An extraction of 17,000 m³ of sediments in 1983 and 25,000 m³ in 1984 were thereby achieved (Boillat and Pougatsch, 2000).

A main problem with a gravity suction system is that the approach flow velocities drop quickly as the radial distance from the suction point increases, and sediments are no longer entrained. To be efficient, the pipe suction end has to be repositioned almost continuously. But with a long pipe and high sediment volume, the head loss increases, potentially choking the pipeline (Alam, 1999). A special suction system described by Jacobsen (1998) and Jacobsen and Jiménez (2015) involved a slotted pipe laid on the reservoir bottom. An alternative of continuous and controlled sediment transfer system by hydro-suction using swimming automated or manual vessels is described by Schüttrumpf and Detering (2011). Sediment is then evacuated by suction either through the bottom outlets, over the spillway or across other outlets. Direct sediment transfer is limited to sediment particle sizes of up to 150 mm which in most cases represents the major fraction of all sediment involved. Sediment dredged by hydro-suction may also be added directly at powerwater intakes once the coarser fractions are reduced (De Cesare *et al.*, 2009; Schüttrumpf and Detering, 2011; Felix *et al.*, 2016, 2017), see 9.4.4.

Dredging

One obvious alternative to flushing or routing sediments through a reservoir is underwater dredging or dry excavation of the deposited material (Figure 9.12). Its drawback is the high cost for sediment removal, but as reservoir level drawdown for flushing and sluicing may not solve all sediment-related problems, the impounded reach will need to be dredged due to continuous gravel accumulation (Morris and Fan, 1998). Dredging requires less drawdown of the reservoir water level than flushing.

Another concern is the deposition of sediments after dredging. Depending on the legal framework, a deposit having been stored over years in the reservoir and thus potentially contaminated is often not permitted, or should be avoided due to ecological constraints. Returning it to the downstream river portion is delicate since polluted sediments are not appropriate for the ecological system and its concentration is difficult to control. Furthermore, muddy lake deposits in the lower reservoir portion, mainly composed of silt and clay, are not easy to dredge because of their high water content. Together with possible organic

Figure 9.12 (a) Dredging by hydro-suction at Margaritze Reservoir, Austria, (b) composition analysis of sediments at Pieve di Cadore Reservoir, Italy (Courtesy ALPRESERV)

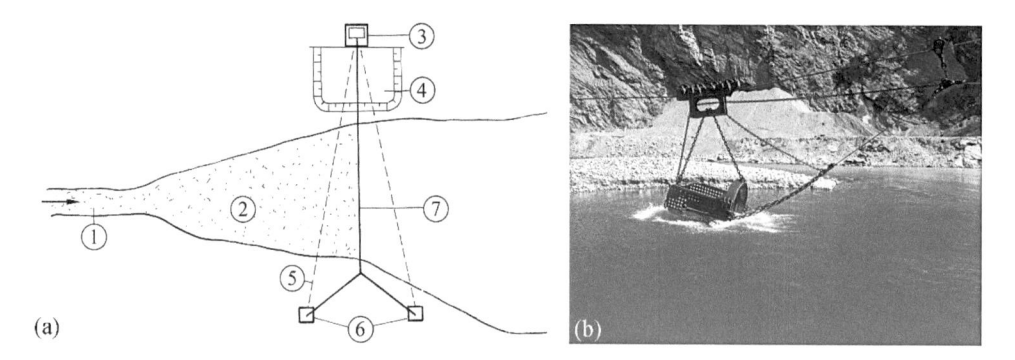

Figure 9.13 (a) Permanent dragline dredging station at reservoir inlet zone ① approach flow river, ② delta, ③ dredging tower, ④ sediment deposit zone, ⑤ stabilization cable, ⑥ cable fixation, ⑦ main cable, (b) example of Solis Reservoir, Switzerland (Copyright VAW)

matter contents they are also difficult to deposit or to re-use. Gravel and coarse sand fractions may be used for construction purposes, however.

Depending on the local water depth, the quality of sediments, and the transport systems, sediments may be removed either from the shore or from a boat. Because the cohesion of the sediments settled is large, simple suction devices cannot be used. Therefore, their head has to be extended with water jet nozzles or with a rotating head to loosen the material for easy dredging.

A particular solution proposed for small lakes in Italy (Roveri, 1981) involves a dredger mounted on a swimming platform connected to a pipeline discharging at the dam outlet. The hydraulic sediment transport, also referred to as hydro-transport, is affected by the siphon alone so that the effect of the conventional sediment outlet is extended into the reservoir. The dredger has thus no pumps because the head between the reservoir level and the outlet is sufficient for hydro-transport (Figure 9.12a).

If the reservoir inlet zone is relatively narrow and the gravel delta well-defined, then a dredging system may be effective if the reservoir level does not strongly vary (Figure 9.13). The removal system is either mobile or operated from the shore. Such a dredging system has proven effective and simple. The material removed applies as dam filler or for concrete. For clean delta material it also applies for ecological sediment replenishment downstream of the dam. Further details state Breusers *et al.* (1982). Note that dredged sludge is not easily deposited because of dewatering and quality. The material is thus often returned to the tailwater river if not contaminated.

Turbidity current control by obstacles and water jet screens

As mentioned, turbidity currents are the main transportation process of fine sediments along the reservoir down to the dam in narrow and long reservoirs. To stop turbidity currents in the upstream reservoir portion and to force them to settle, Oehy (2003) systematically investigated the effects of obstacles, screens, water jets, and bubble curtains by means of physical experiments and numerical simulations.

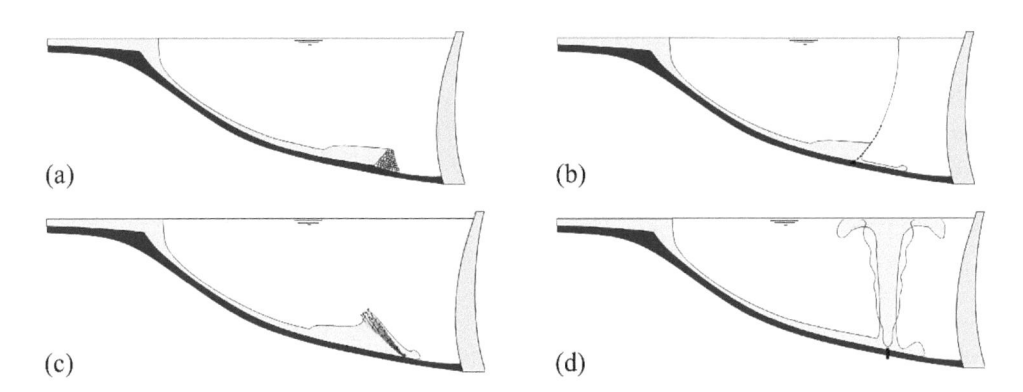

Figure 9.14 Technical measures to control turbidity currents in reservoir (a) embankment dam as obstacle, (b) floating geotextile screen, (c) upstream-directed water jets, (d) bubble screen (Adapted from Oehy, 2003)

Obstacles are created by embankment dams placed across the reservoir thalweg and erected with excavation material before impounding. During operation they are always submerged. They can stop turbidity currents and cause the fine sediments to settle (Figure 9.14a). These currents are also stopped by energy-dissipating elements as a geotextile screen, spanned by steel ropes over the lower portion of a reservoir cross section, or suspended by floats or booms to follow reservoir level variations. The screen does not have to be absolutely impermeable but be a mesh of certain porosity. The turbidity current will then be partially stopped, slowed down and diluted in such a way that it dies out (Figure 9.14b). The turbidity current is further diverted into specific reservoir zones by means of obstacles and screens, where sediment deposits do not affect the intake operation.

Obstacles such as embankment dams effectively block subcritical turbidity currents (Oehy and Schleiss, 2004, 2007). The obstacle has to be at least twice as high as the approach turbidity current, resulting in obstacle heights from 20 to 30 m for typically 10 m high turbidity currents in Alpine reservoirs. The storage volume upstream of the obstacle should be as large as possible. Therefore, obstacles should be placed at the downstream end of mild reservoir bottom slopes or, if present, at the end of negative bottom slopes (Oehy and Schleiss, 2001). During each flood resulting in a turbidity current, the storage volume upstream of the obstacle is somewhat reduced. The case study of Grimsel Reservoir (Oehy and Schleiss, 2001) indicates that obstacles 20 m high lose their retaining capacity after some 50 years. A more sustainable solution involves the construction of several obstacles along the reservoir. For Grande Dixence Dam, Switzerland, a submerged dam in the reservoir erected prior to today's main dam also has the function of an obstacle. Depending on the sedimentation level, adaptive measures as the orifice elevation increase of the obstacle have to be envisaged (Bretz and Barras, 2012; Boes and Hagmann, 2015).

An obstacle also applies to divert turbidity currents to reservoir areas in which the deposits do not affect the operation of the intake and outlet structures. If the intake of the bottom outlet is sufficiently far away from the power intake, turbidity currents are directed to the bottom outlet by means of obstacles. This increases the efficiency of regular flushing operations by means of the bottom outlet, especially if they are conducted during floods.

The efficiency of permeable screens or geotextile curtains for slowing down turbidity currents depends mainly on their porosity and mesh size (Oehy and Schleiss, 2004, 2007). To retain a significant amount of sediments, the porosity should be lower than 30%. Simulations demonstrated that permeable screens are insensitive to the approach flow conditions, i.e. they also perform well for supercritical turbidity currents. The screen should be roughly three times higher than the height of the approaching current. For typical practical applications this would lead to curtains of about 30 m height. The mesh openings should be in the range of 50–100 mm to avoid clogging by suspended particles.

The highest pressure force on the curtain occurs if the turbidity current strikes against it. The pressure is composed of a hydrostatic and a dynamic portion. The hydrostatic pressure remains moderate since the density difference of the turbidity current and clear reservoir water is small. The total pressure on the geotextile curtain reaches 70–150 N/m² if the approach flow velocity of the turbidity current ranges between 0.2 and 0.35 m/s, and average sediment concentrations are 0.5 g/l.

Screens or geotextile curtains have similar applications in reservoirs as obstacles. In comparison to embankment dams, the installation of geotextile screens is less time-consuming and allows also for reservoir operations as a short and limited water level drawdown. Installing several screens in series increases sediment retention efficiency. The screens are made of fiber-reinforced geotextile bands attached to horizontal steel cables spanned over the reservoir valley and/or suspended from floats at the free reservoir surface.

Numerical simulations and physical experiments of a turbidity current across an inclined water jet screen indicate that turbidity currents are considerably slowed down by a jet screen and that most of the sediment is retained (Oehy et al., 2010). The jet screen is adapted to sub- and supercritical approach flow conditions, yet it is more effective for subcritical currents (Figure 9.14c). The deposits downstream of the screen were reduced optimally to 50% as compared with free flowing turbidity currents.

In practice, the local conditions of the reservoir morphology determine the most successful measure for blocking turbidity currents, either using solid obstacles, geotextile screens, or water jets. Numerical simulations allow for determining the optimum location and the governing parameters of these measures. For a preliminary assessment of the effect of solid obstacles, geotextile screens and water jets on the turbidity current flow, analytical approaches based on free-surface flows and shallow-water approximation are given in 9.6.3–9.6.5.

Avoiding settling of suspended sediments and re-suspension

If turbidity currents cannot be stopped in the reservoir as mentioned, they reach the dam and create a muddy lake in front of it eventually settling down. A new idea is to whirl up these fine sediments of mean diameter below 60 μm near the dam and the intakes keeping them in suspension, allowing for a continuous release through the turbines (Figure 9.14d, see 9.4.4). This is realized by special water jets arrangements as developed by Jenzer Althaus et al. (2011, 2012). It was found that a configuration of four jets arranged in a circle on a horizontal plane in front of the intake maintains the fine sediments in suspension (Jenzer Althaus, 2011; Jenzer Althaus et al., 2015). Based on a preliminary case study for Mauvoisin Dam, Switzerland, it was demonstrated that a significant amount of sediment may be released continuously through the intake during powerhouse operation with this circular jet arrangement, resulting in an acceptable sediment concentration regarding turbine abrasion and downstream ecology. Alpine reservoirs are often fed by transfer tunnels bringing water to the reservoir from neighboring catchment areas. If the transfer tunnels enter the reservoirs with a considerable height above the maximum reservoir level, they have good potential of feeding these jet arrangements, so that no pumps with power supply are required.

9.4.4 Measures at dam

Heightening of dam, intake and bottom outlet structures

The heightening of a dam and its outlet structures is an alternative for compensating the loss of reservoir capacity due to sedimentation. Although it might be cost-effective in the mid-term, dam heightening does not provide a sustainable solution of the sedimentation problem. Construction of new dams to solve the sedimentation problem in the future leads to the same problems. With the heightening of both the intake and the bottom outlet structures, the dead storage volume is increased and sediment entrainment into the intakes is prevented for a certain period. The heightened intakes may be combined with flushing tunnels so that the sediment cone can be evacuated, as designed at the new Mauvoisin intake (Figure 9.15)

Sluicing

During sluicing of sediment, the reservoir water level is drawn down partially to allow for sediment-laden inflow to pass the reservoir with a minimum of deposition. Typical of

Figure 9.15 New power intake at Mauvoisin Dam equipped with flushing devices (Courtesy LCH, EPFL Lausanne)

sluicing in a flood-detention reservoir is that during a rising flood-water level the outflowing sediment discharge is always smaller than that of the inflow. During drawdown, the outflowing sediment discharge is larger than the inflow, due to erosion in the reservoir (Fan, 1985). Since the inflowing sediment concentration during a flood tends to be highest during the rising hydrograph limb, the reservoir is filled with less turbid water following the flood peak (Fan and Morris, 1992b). Sluicing operations should be timed to accommodate the higher sediment concentrations brought in by flood flows. By opening bottom outlets fast enough, the rate of the outflow increase can become equal to the rate of increase of an incoming flood. The detention effect and the least alteration of the hydrograph of the sediment are then minimized. Thus, the sediment outflow approaches the natural flow condition. Sluicing is successful if the capacity of the outflow structures is adequate, the operation is done judiciously, the river is transporting mainly suspended sediments, and the flow hydrograph is predictable with confidence at the dam site (Basson and Rooseboom, 1997). The capacity of the outflow structures has to be such that during lowering of the reservoir water level sufficiently high flow velocities occur so that suspended sediment cannot settle anymore, but pass through the reservoir. This operation is successful if the reservoir volume is relatively small compared with the inflowing flood volume.

Turbidity current venting

Venting of density currents means that the incoming sediment-laden flow is routed under the stored water and through the bottom or low-level outlets, or alternatively through power intakes, if located deep enough, and routed through waterways and powerhouses (Lai *et al.*, 2015). Since the reservoir may stay impounded during the release of density currents, this method is widely used in arid regions where water is in shortage. Fan and Morris (1992b) suggested that density-current venting is well-suited at large reservoirs with a multiyear storage capacity where drawdown is avoided. Venting operations have a much better chance of accomplishing their purpose if timed to intercept gravity underflows as they reach the dam. The correct timing of gate opening and closing is important. Either a too late sluice operation or a too small gate opening result in a smaller amount of sediment discharged from the reservoir. In contrast, if the gate is opened too early or the opening is too large, valuable water is lost (Chen and Zhao, 1992). The capacity of the bottom outlets has to be sufficient to allow for venting turbidity currents, i.e. at least for the discharge of turbidity current reaching the dam. Swiss examples of turbidity venting is described by Müller and De Cesare (2009) or Boes and Hagmann (2015). Besides examples from Switzerland, numerous worldwide applications of turbidity current venting are described by Chamoun *et al.* (2016a, b) and Chamoun (2017). Recommendations on how to perform efficiently turbidity current venting are given in 9.7.2.

Turbining of suspended sediments under controlled concentration

To mitigate reservoir sedimentation and to re-establish sediment continuity without causing high Suspended Sediment Concentrations (SSC) in the river downstream of the dam, fine sediment from reservoirs can be conveyed through power waterways and hence turbines to downstream river reaches (Jenzer Althaus, 2011; Jenzer Althaus *et al.*, 2015; Felix *et al.*, 2017). The SSC has to be controlled not only for environmental reasons in the tailwater, but also to avoid enhanced turbine abrasion. Sediment particles from reservoirs can be transported into power waterways by (Figures 9.2, 9.3):

- Sluicing of sediment-laden water, possibly with the aid of jet-induced turbulence to avoid settling, as mentioned in 9.4.3 (Jenzer Althaus, 2011; Jenzer Althaus *et al.*, 2015);
- Venting of turbidity currents, under an adequate intake position and routing through waterway and powerhouse, see above (Schlegel and Dietler, 2010; Lai *et al.*, 2015);
- Re-mobilisation of settled fine (often cohesive) sediment by hydro-suction or air lift (DWA, 2006; Schüttrumpf and Detering, 2011; Jacobsen and Jiménez, 2015), thereby controlling the area of sediment removal, the sediment transport rate, and to exclude larger particles, see 9.4.3.

With sediment conveyance from reservoirs through power waterways, the SSC in the river downstream of the hydropower plant's water restitution point is relatively low, because the sediment is transported during long periods in a large volume of water. In the river reach between the dam and the water restitution point, i.e. in the residual flow reach, SSC is not increased. Other advantages are that no further transport, dewatering and land-based sediment disposal are required, and no flushing water is lost for electricity generation.

To limit turbine erosion, the following factors are favorable for this option of sediment handling:

- Low to medium head hydropower plants, thus smaller flow velocities in turbines;
- Relatively small particles (clay, fine and medium silt, ≤ 20 μm, DWA, 2006);
- Low particle hardness (typically in catchment areas with sedimentary rocks);
- Suitable turbine design (e.g. seals, trunnions, cooling water system adapted to high SSC).

Sediment deposits close to intakes and dams are generally to be removed. In medium and large reservoirs, deposits in these zones typically consist of fine particles, because coarser sediment particles settle more upstream in delta zones close to the reservoir inflow. Because small particles cause less turbine erosion than larger particles at a given SSC (Nozaki, 1990; Winkler *et al.*, 2011), less erosion damage is expected from such fine sediment.

Sediment conveyance from reservoirs through power waterways was practiced e.g. in the three hydropower plants equipped with Francis turbines listed in Table 9.3. These hydropower plants have moderate heads and rather small reservoirs. Sediment particles (mainly silt) were transported by hydro-suction from the bottom of reservoirs in front of the intakes, where they were entrained into the power waterway and passed through the turbines. The admissible increases of SSC in the downstream river reaches were defined by environmental authorities as a function of the season and of the SSC in the rivers upstream (subscript us) of the restitution points (SSC_{us}). The hydro-suction units were controlled in terms of the measured SSC_{us} or corresponding turbidity. In the second example in Table 9.3, a more dynamic sediment mobilization regime was selected for environmental reasons. When the hydro-suction installation was operating at full capacity, i.e. removing 200 m^3 of sediment deposits per hour, the SSC in the turbine water was 1.4 g/l at design discharge. This is considerably lower than typical SSC during reservoir flushing operations. The resulting downstream SSC are similar to natural conditions before the construction of the hydropower plants on this river.

Table 9.3 Examples of HydroPower Plants (HPP) with sediment removal from reservoirs by hydro-suction and transfer of fine sediment through power waterway to downstream river reach (Adapted from Felix *et al.*, 2016).

HPP (river) and reservoir names, country	Gross head	Design discharge of turbines Q_d	Admissible increase of SSC in the downstream river reach	Admissible sediment return at Q_d
HPP Kubel (Sitter), Gübsensee, Switzerland (De Cesare *et al.* 2009)	97 m	16 m³/s	0 in winter, 0.2 g/1 otherwise	270 t/day
HPP Walgauwerk (Ill), Compensation basins Rodund, Austria (Sollerer and Matt 2013)	162 m	68 m³/s	0 if $SSC_{us} \leq 0.05$ g/1, 0.2 g/1 if $0.05 < SSC_{us} < 0.2$ g/1, $1.5SSC_{us}$ if $SSC_{us} \geq 0.2$ g/1	1170 t/day ≤ 4800 t/day
HPP Langenegg, Reservoir Bolgenach, Austria (DWA 2006)	280 m	approx. 30 m³/s	practiced restitutions of 0.02 g/1 to 0.2 g/1 (mainly at high natural Q) were below admissible SSC	Practiced ≤ 500 t/day (80 000 t/year)

To assess the trade-off between reservoir desilting and increased turbine wear, the hydro-abrasive turbine erosion needs to be estimated. A state-of-the-art turbine erosion model has been calibrated with comprehensive data sets from field studies at high-head hydropower plants. More information on the effects of suspended sediments on turbine erosion and turbine efficiency as well as suspended sediment monitoring techniques are presented by Felix *et al.* (2018).

9.5 Sediment bypass tunnel

9.5.1 General

Sediment Bypass Tunnels (SBTs) are considered a measure in the catchment area (Figure 9.2), or in the reservoir, depending on the intake location (Section 9.5.3); they belong to the sediment routing concept (Figure 9.3). During floods, the sediment-laden water is conveyed through the tunnel to the dam tailwater. SBTs feature several advantages over other desilting countermeasures. They have positive effects in eco-morphology, because sediment conveyance may significantly decelerate or even stop riverbed erosion and increase the morphological variability downstream of a dam. Mainly sediments provided from the upstream river reach are conveyed through SBTs since remobilization of accumulated sediments in the reservoir hardly occurs. The sediment concentration in the tailwater of a dam is thus not affected by the reservoir itself and is of natural character. An SBT can thus greatly improve the sediment continuity despite river impoundment by dams. Second, SBTs have demonstrated to be an effective countermeasure against reservoir sedimentation. Asahi SBT in

Japan has e.g. greatly reduced the severe aggradation in terms of accumulated sedimentation volume since its commissioning in 1998 (Figure 9.16). During an exceptionally large flood due to a typhoon in 2011, the routing of sediments around the dam greatly helped to limit the sediment inflow into the reservoir. The mean annual bypass efficiencies amount to 77% at Asahi and to 94% at Nunobiki SBT, extending the respective so-called reservoir life values by 450 and 1200 years, respectively (Auel *et al.*, 2016). Reservoir life is defined as the ratio of reservoir capacity (CAP) and the Mean Annual Sediment (MAS) load, representing the theoretical duration until the reservoir is completely filled with sediment. Bypass efficiencies of Swiss SBTs up to 83% have been reported, depending on the intake position (9.5.3) and operational aspects (9.5.9) (Boes *et al.*, 2018).

Whereas bed load deposition may be completely prevented with an SBT, the deposition of fines in the reservoir depends on the design discharge of the tunnel. The higher the SBT design capacity and thus the flood recurrence interval, the higher is the share of the incoming suspended load to be conveyed through the tunnel and the smaller the amount of fines entering the reservoir. The main drawback of an SBT is related to economic considerations. The implementation of an SBT is not only costly from an investment perspective, but also requires regular maintenance. Due to high flow velocities of up to 20 m/s (Auel and Boes, 2011) and high sediment loads, invert abrasion is generally a severe problem, requiring costly repair works and maintenance (Figure 9.17) and limiting the number of SBTs currently in service. An SBT should therefore be considered as an adequate desilting measure rather for small to medium-sized reservoirs with Capacity-Inflow Ratios of about CIR = 0.003 to 0.2 and with typical reservoir life values of up to some hundred years (Sumi and Kantoush, 2011). The prototype SBTs currently in service mainly in Switzerland and Japan underline this application range (Figure 9.18) for which SBTs are most effective. To date, SBTs in Switzerland were only constructed at reservoirs with relatively small volumes varying between 0.06×10^6 and 4.3×10^6 m³ with a mean value of 1.4×10^6 m³. In contrast, SBTs in Japan have been placed at larger reservoirs with volumes from 0.76 to 58×10^6 m³,

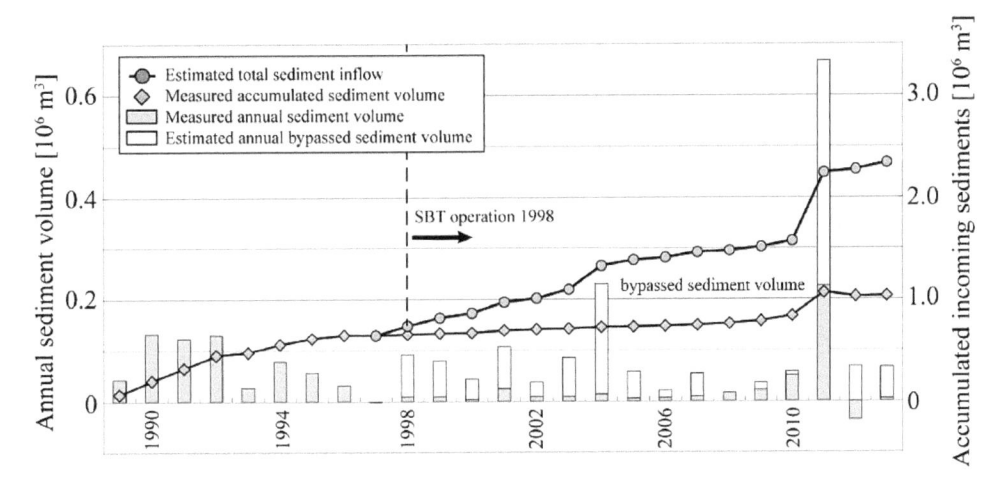

Figure 9.16 Development of reservoir sedimentation volume at *Asahi* Reservoir, Japan, prior to and after commissioning of SBT in 1998 (Auel *et al.*, 2016, Adapted from Fukuroi, 2012)

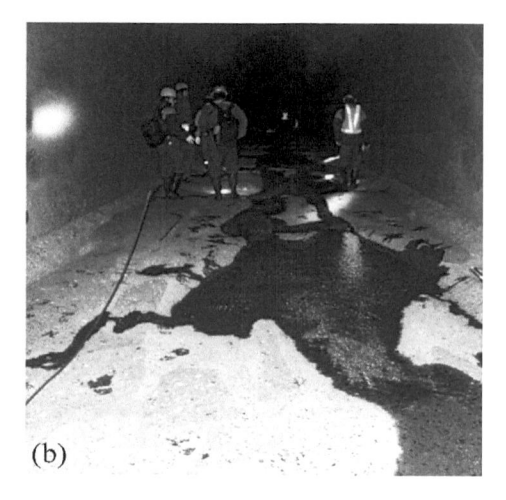

Figure 9.17 Invert abrasion at (a) Palagnedra SBT, Switzerland (Courtesy Dr. Christian Auel), (b) Asahi SBT, Japan (Adapted from photo by T. Koshiba)

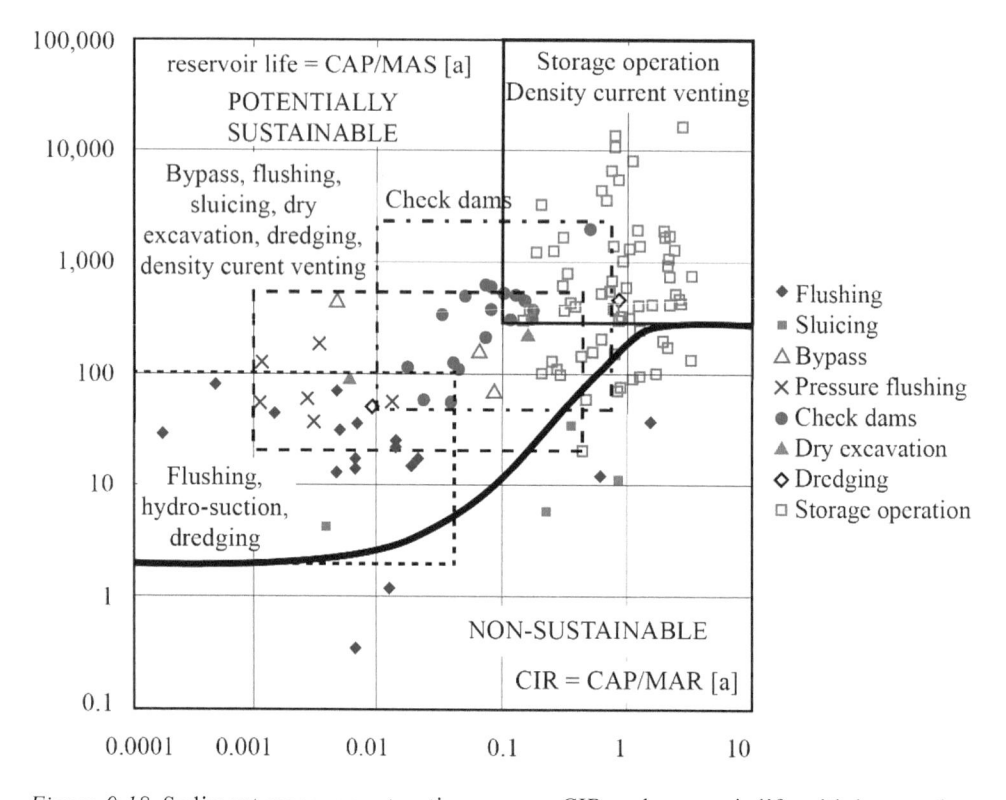

Figure 9.18 Sediment management options versus CIR and reservoir life with bypass data from Japanese and Swiss reservoirs (Adapted from Annandale, 2013)

with a mean value of 22.3×10^6 m³ (Auel, 2014). For these small to medium-sized reservoirs, bed load aggradation and delta formation are more critical concerns than the sedimentation of fines, since a large amount of incoming fines stays in suspension due to the relatively short residence time and are discharged via the outlet works.

However, as floods up to the SBT discharge capacity are conveyed around the dam without being damped in the reservoir, SBTs are most appropriate in regions of high water availability, while the flood runoff is of prime importance in arid environment with pronounced dry seasons storage. Moreover, mainly tunneling cost favors the use of SBTs at rather small reservoirs due to short tunnel length. SBTs can substitute reservoir flushing, or represent a complementary measure, e.g. in combination with drawdown flushing (De Cesare et al., 2015; Beck et al., 2016). If SBTs are operated during natural flood events, the sediments are discharged to the downstream valley in an eco-friendly way. An overview of worldwide SBTs in operation, under construction or planned as well as on state-of-the-art SBT research is given by Boes (2015) and Sumi (2017). To reduce negative effects of SBT hydro-abrasion, it is recommended to (i) optimize the hydraulic design to limit the particle impact (9.5.5), and (ii) select sustainable and optimally abrasion-resistant invert lining material (9.5.8).

9.5.2 Suitable bypassing discharge and target sediment granulometry

The design of the SBT discharge capacity depends above all on an economic tunnel cross-sectional area and on the hydrological catchment conditions. According to Vischer et al. (1997) and Sumi et al. (2004), SBT design discharges typically correspond to between a one- and a ten-year flood event. However, particularly for reservoirs with small catchments impounded by embankment dams, a higher recurrence interval of up to 100 years may by preferable to complement the service spillway capacity (Boes and Reindl, 2006). Absolute values of SBT design discharges currently in service or under construction range from 40 to 400 m³/s (Auel and Boes, 2011; Auel, 2014; De Cesare et al., 2015). It should thereby be kept in mind that the surplus flow exceeding the discharge capacity has to be conveyed to the downstream reservoir reach. Because routing of all incoming sediments is achieved only up to the SBT design discharge, a rather high flood return period should be selected. The duration of operation ranges from a few days/yr. to more than 100 days/yr., depending on local hydrology and reservoir size (Hagmann et al., 2016). The majority of existing SBTs is operated during less than 30 days/yr. (De Cesare et al., 2015).

The target sediment grain size distribution and volumes of both bed load and suspended load to be bypassed through an SBT have to be evaluated site-specifically with regard to the total sediment yield, acceptable reservoir storage loss, and sediment flushing and venting capacity of other hydraulic structures including bottom outlets and power intakes (De Cesare et al., 2015). Most SBTs are designed and operated to convey as much incoming sediment as possible, from coarse bed load to suspended fines. However, especially for long SBTs as at Miwa Dam, Japan, the coarse material is trapped and dredged upstream, and only the suspended load is bypassed around the dam (Hagmann et al., 2016). For inflow discharges exceeding the SBT design capacity, suspended sediments may however still enter the reservoir and accumulate to some extent in the downstream reservoir section, while the conveyance of all bed load toward the SBT intake should be aimed at by means of a guiding structure (Auel and Boes, 2011; Section 9.5.3). Model studies of Chespí-Palma Real SBT, Ecuador, evidenced a sediment bypass efficiency of almost 100% for frequent floods under upstream drawdown, reducing to 20% under larger flood events above the SBT discharge capacity (De Cesare *et al.*, 2015).

9.5.3 Hydraulic design

General

A typical SBT layout includes a *radial* gate (Chapter 8) as regulating device at the tunnel inlet, followed by a mostly short and steep flow accelerating reach and a more gently sloped standard cross section up to the outlet with a subsequent energy dissipator (Figure 9.19, Chapter 7). Free-surface flow is preferred, but pressurized flow for the design discharge may also occur, depending on the gate location determining the control section. SBTs as Patrind, Pakistan (Beck *et al.*, 2016), or Rizzanese, France (Carlioz and Peloutier, 2014; Laperrousaz and Carlioz, 2015), are gate-controlled from the downstream end, due to site constraints disabling a gate installation at the inlet. In the latter cases, free-surface, mixed, and pressurized flows may occur, depending on the discharge scenario. Regarding flow velocity, there is a trade-off: While it must be sufficiently high to avoid sediment deposition, the maximum velocities should be as small as possible to control the extent of invert abrasion (Hagmann *et al.*, 2016). In most SBTs, therefore, the flow is accelerated at the inlet either by pressurized flow upstream of the gate or by a steep section downstream to quickly reach quasi-uniform supercritical flow.

Whenever possible, bends in plan view should be avoided to reduce shock waves and secondary currents, causing locally high specific sediment transport rates. In the Solis SBT, Switzerland, for instance, the sediment transport at the tunnel outlet is clearly concentrated on the orographic right side due to a right-hand side bend further upstream (Albayrak *et al.*, 2015, Figure 9.31). In a first design step, the design discharge (9.5.2) and the tunnel intake location have to be defined. Thereafter, the proper SBT is designed in streamwise direction as given hereafter.

Intake location

Two different locations are generally adopted for the SBT intake, both affecting the entire SBT design and the reservoir operation during sediment routing. Its most common location, applied for the majority of SBTs in Switzerland and Japan, is at the reservoir head (position A,

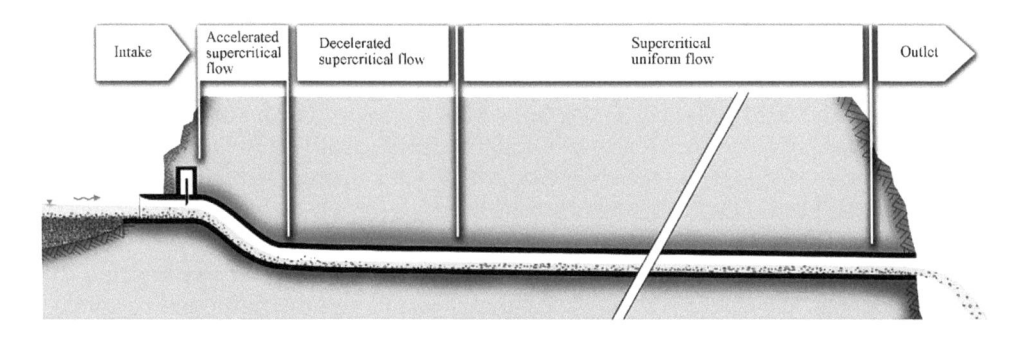

Figure 9.19 Sketch of SBT (Auel, 2014)

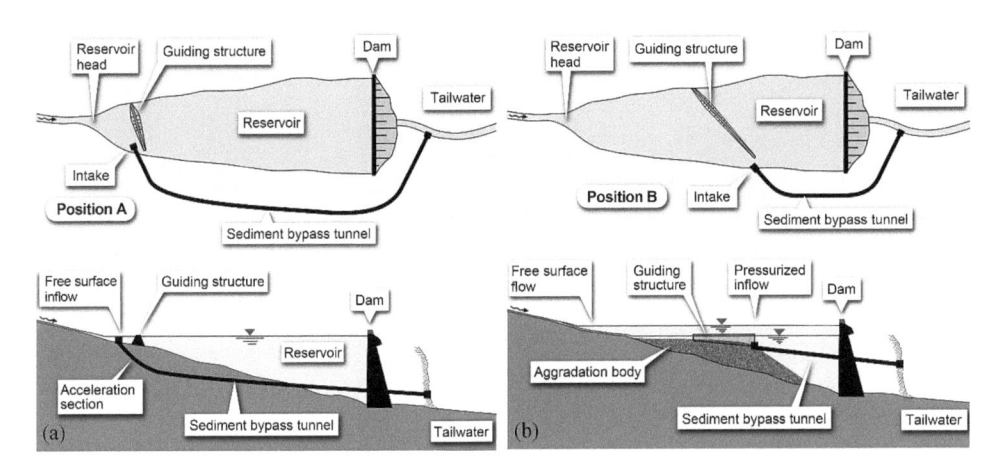

Figure 9.20 Sketches of two different SBT systems relative to location of tunnel intake (a) free-surface inflow at reservoir head (position A), (b) pressurized inflow downstream of reservoir head (position B) (Auel and Boes, 2011)

Figure 9.20a). Another suitable location is downstream of the reservoir head close to the dam, for retrofitting existing, partially silted reservoirs with an SBT preferably just downstream of the pivot point of the aggradation body (position B, Figure 9.20b). The advantages of position A include: (i) Complete reservoir is kept free from sediments, and (ii) reservoir level during bypass operation is independent from the upstream river reach and kept at full supply level. Depending on topography, disadvantages are the long distance of the reservoir head to the tailwater causing high construction cost, and the free-surface flow conditions at the intake requiring a steep acceleration section potentially provoking significant abrasion damage at the tunnel invert due to high flow velocities (9.5.4).

For SBTs of position B type, the intake inflow is pressurized so that an acceleration section can be waived. As a major drawback only the reservoir section downstream from the intake is kept free from sediment accumulation, and the reservoir level has to be temporarily lowered to a certain level to sustain a sufficiently high sediment transport capacity in the upper reservoir reach upstream from the SBT intake (9.5.4). The effects of the two intake locations on both SBT operations and each individual tunnel element are discussed below.

Guiding structure

The guiding structure, also referred to as diversion weir or sill, sediment check dam or partition dam, ensures that the incoming sediment-laden flow is conveyed to the SBT intake. According to Vischer *et al.* (1997) and Sumi *et al.* (2004), the guiding structure is located next to the intake structure in most existing SBTs, crossing the reservoir from the intake to the opposite reservoir bank.

The guiding structure should preferably not be overtopped during SBT operation to avoid sediment input and accumulation in the reservoir. However, if the flood event exceeds the SBT design capacity, the guiding structure has to be securely overtopped or openings in the guiding structure are to be designed to lead the surplus flow to the dam outlet structures

Figure 9.21 (a) Tunnel intake with skimming wall for floating debris detention in November 2012, view from left bank, reservoir level at approx. 820 m a.s.l. (b) detail of guiding structure during bypass operation on May 3, 2013, reservoir level at approx. 816 m a.s.l. (Courtesy ewz) (Adapted from Oertli and Auel, 2015)

(Auel and Boes, 2011). Depending essentially on the requested height, the guiding structure can be designed as a small concrete or reinforced cofferdam or a vertical sheet pile wall (Figure 9.21b). Detailed hydraulic studies and designs of diversion facilities are given by Kashiwai *et al.* (1997), Auel *et al.* (2010), De Cesare *et al.* (2015), and Oertli and Auel (2015).

Intake

The SBT intake typically consists of an intake trumpet followed by a sluice or a radial gate. During normal reservoir operation, the gate is closed. During flood events, the gate is opened, however, and the sediment-laden discharge is routed through the SBT. The design of the intake depends directly on the selection of the intake location given above.

  If the intake is located at the reservoir head (position A), the discharge is conveyed into the SBT under free-surface flow (Figure 9.22a). The tunnel invert level at the intake is flush to the riverbed. Downstream of the gate, the discharge has to be accelerated to generate supercritical flow. This is achieved by a short and steep acceleration reach (see above). For position B type intakes, the tunnel invert level can be lower than the riverbed and the surrounding aggradation body, respectively. A certain energy head is thus available and the discharge is conveyed into the intake trumpet under pressurized flow (Figure 9.22b). However,

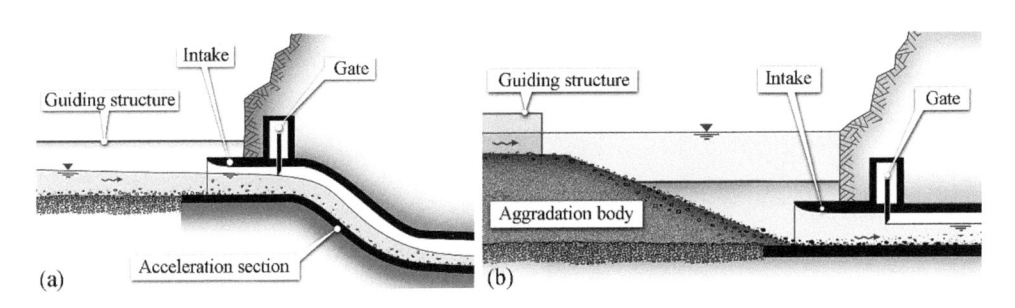

Figure 9.22 Sketches of SBT intakes at position (a) A with free-surface inflow, (b) B with pressurized inflow (Auel and Boes, 2011)

downstream of the gate, the discharge is routed through the SBT under free-surface flow. Due to the significant energy head, the flow velocity downstream of the gate is high, so that an acceleration reach is waived (see above). An SBT intake of type B with 1000 m^3/s design capacity is under construction in Taiwan to sluice fine sediment (Kung *et al.*, 2015). Some type B intakes may be operated both under free-surface flow (reservoir drawn down), and under pressurized flow at high reservoir levels. Examples of such a bimodal intake design are given by Carlioz and Peloutier (2014), De Cesare *et al.* (2015), Grimaldi *et al.* (2015) or Beck *et al.* (2016).

Keeping floating debris away from an SBT intake is advisable to prevent clogging by logs or rootstocks at the control gate or elsewhere in the tunnel. Racks consisting of vertical poles or bars should be applied with care as sediment transport may become interrupted if wooden debris is retained at these structures (Hagmann *et al.*, 2015). From experience with driftwood retention racks involving vertical poles, the sediment transport capacity is considerably reduced once a significant amount of floating debris is trapped (Schmocker and Weitbrecht, 2013). However, one central pillar may act as a driftwood alignment straightening the incoming driftwood as e.g. at the flood bypass tunnel Campo Vallemaggia, Switzerland (Figure 9.23). According to Lange and Bezzola (2006), driftwood may still be blocked at this pillar. Note that this tunnel is operated under free-surface flow conditions at the intake, thus the tunnel may still operate during blockage because the water freely overflows the blockage barrier. Furthermore, a potential blockage can be removed more easily for type A than type B intake. Pillars or obstacles in general should be avoided in front of the intake unless blockage of the tunnel is accepted for rare events.

If driftwood enters the tunnel, different hazard scenarios are possible if the gate is not completely opened (Figure 9.24). A trunk or rootstock can be blocked at the Tainter gate and cause accumulation by further incoming wood. Thus, the gate operation is endangered and the gate cannot be closed. These hazards are reduced by a total gate opening to unblock the accumulated wood. To avoid driftwood being sucked into the tunnel intake by a vortex or rotating surface flow, a skimming wall is preferable to keep floating debris away from the intake (Figure 9.21a). This wall may safely guide the driftwood-laden surface flow toward the dam if a partial flow of about 20% of the incoming flood is not bypassed through the SBT. In addition, a skimming wall requires less effort to remove retained wood and debris after a flood as compared with racks and fences. Examples of driftwood retention at SBT intakes are given by Auel *et al.* (2010), Kashiwai and Kimura (2015), and Sakurai and Kobayashi (2015).

Figure 9.23 ③ Pillar for driftwood alignment at flood bypass tunnel Campo Vallemaggia, Switzerland, with ① dividing pillar, ② tunnel entrance (Adapted from Lange and Bezzola, 2006)

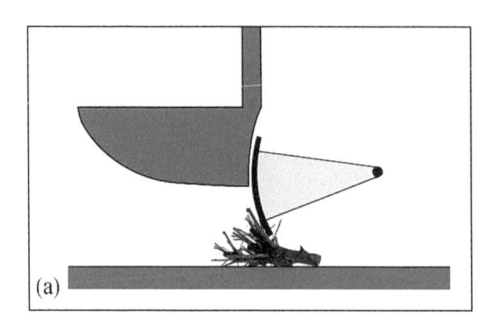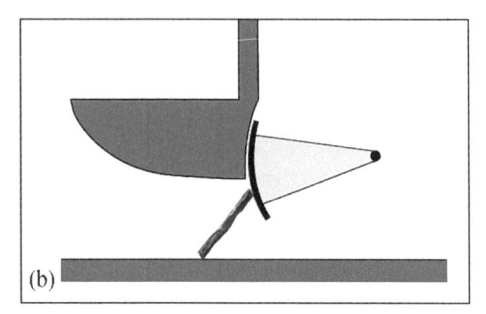

Figure 9.24 Possible driftwood problems at tunnel gates with blockage by (a) rootstock, (b) single log

Tunnel section

The SBT connects the upstream tunnel intake with the downstream outlet structure located at the dam tailwater. Typical tunnel lengths of existing SBTs vary between 250 and 4300 m, and typical quasi-uniform design flow velocities range between 7 and 15 m/s, while hydraulic sections range from 20 to 30 m², resulting in Froude numbers between 1.4 and 3.1.

The cross section of most SBTs is of horseshoe or archway shape (the latter is also referred to as hood shape). The circular shape should be avoided as the sediment transport is concentrated at the lowest invert point causing severe abrasion problems (9.5.4). Another drawback of circular versus archway or horseshow shaped cross section is the challenging trafficability during tunnel construction and maintenance.

Steeply sloping acceleration reach

According to Chervet and Vischer (1996) and Harada *et al.* (1997), most SBTs in Switzerland and Japan include a short and steeply sloping acceleration reach because of their intake location at the reservoir head (Type A). The aim of this steep reach is to accelerate the discharge to supercritical uniform flow. The maximum velocities at the end of the accelerating section are between 7 and 20 m/s. These high velocities are required to ensure the requested sediment transport capacity in the upper tunnel reach. Supercritical flow is promoted to keep the tunnel cross section in an economical range (compare next section). Typical bottom slopes of the acceleration reach range from 15 to 35% (Auel, 2014).

Standard smoothly sloping reach

Bottom slopes of the constantly sloping main tunnel reaches vary between 1 and 6.9% (Vischer *et al.*, 1997; Sumi *et al.*, 2004; Auel and Boes, 2011; Auel, 2014; Carlioz and Peloutier, 2014). The discharge is conveyed under supercritical flow to ensure both a sufficient sediment transport capacity and an economic tunnel cross section.

To ensure free-surface flow in the SBT, choking should be avoided to achieve safe operating conditions of sufficient cross section for aeration and air circulation. The maximum filling ratio should thus not exceed $Y_N = 75$ to 80%, with $Y_N = h_{out}/D_t$ or $Y_N = h_{out}/h_t$, with h_{out} as clear-water flow depth at the tunnel outlet for horseshoe- or archway-shaped sections, respectively, and D_t as tunnel diameter and h_t as tunnel height. The selection of the invert slope has to satisfy the two conflicting challenges to prevent sediment aggradation while limiting hydro-abrasion, as mentioned above.

Outlet structure

The sediment-laden flow is discharged at the outlet structure into the tailwater downstream of the dam. The following aspects regarding its design have to be respected: (i) Sufficient transport capacity in the downstream river reach to avoid sedimentation in the outlet vicinity and further downstream; this should typically be no problem because the sediment transport process in the entire river system is revitalized to its original condition before dam construction; (ii) tunnel outlet should release sediments sufficiently far away from the dam outlet structures to avoid sedimentation and backwater effects in their vicinity; (iii) design a drop from the tunnel outlet to the river reach to avoid backward aggradation in the SBT (Figure 9.25); (iv) angle between centerline of the tunnel outlet and the river thalweg should be kept small to reduce erosion impact on the opposite river bank; (v) scour due to jet impact from the tunnel outlet has to be monitored and countermeasures have to be taken, if necessary (Auel and Boes, 2011).

9.5.4 Hydro-abrasion processes

Abrasion is a serious concern for most SBTs. The extent of damages along the wetted perimeter by coarse sediment-laden flow, i.e. mainly on the invert and the lower parts of

Figure 9.25 Cantilevering tunnel outlet of Solis SBT near Alvaschein, Canton of Grisons, Switzerland, during operation on May 23, 2014 (Oertli and Auel, 2015)

the tunnel walls (Figure 9.17), typically increases with sediment load, particle hardness and size, and flow velocities or shear stress. The particle hardness is mainly characterized by the quartz content of the sediment mineralogical composition (Boes *et al.*, 2014). Abrasion is a wear phenomenon involving progressive material loss due to hard particles forced against and moving along a solid surface. In bedrock rivers, abrasion is the driving process for bed incision (Whipple *et al.*, 2000; Sklar and Dietrich, 2004, or Turowski, 2012), while at hydraulic structures such as spillways, weirs, flushing channels and SBTs, abrasion causes severe damage of the invert surface mostly made of concrete, natural stone or steel (Ishibashi, 1983; Jacobs *et al.*, 2001; Helbig *et al.*, 2012; Mechtcherine *et al.*, 2012). There are a number of models to predict abrasion (9.5.6). While those for the prediction of bedrock incision rate (Sklar and Dietrich, 2004; Lamb *et al.*, 2008, or Beer and Turowski, 2015) focus on typical flow conditions in river systems in the sub- and low supercritical flow regimes, these for abrasion prediction on hydraulic structure surfaces (Ishibashi, 1983; Helbig and Horlacher, 2007; Auel *et al.*, 2017a, b; 2018a) account for highly supercritical flows.

Hydro-abrasive damage on an invert of a hydraulic structure occurs if the (i) flow-induced bed-shear stress exceeds a critical value so that numerous particles start being transported and impacting, and (ii) impacting forces exceed the material resistance. Depending on the flow conditions and particle size and shape, sediment particles are transported in sliding, rolling or saltation modes causing grinding, rolling or saltating impact stress and thus wear on the bed (Figure 9.26, Boes *et al.*, 2014).

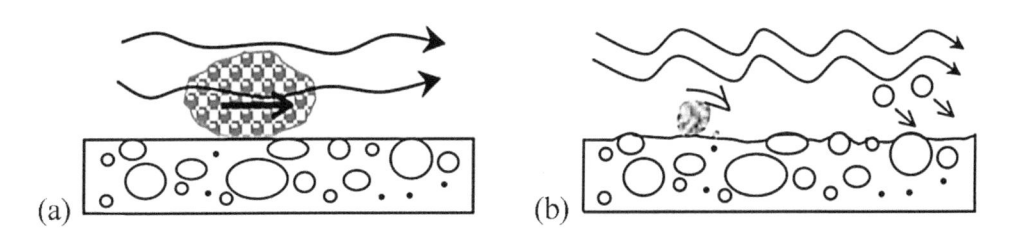

Figure 9.26 Abrasion processes for flow-sediment-structure interactions (a) grinding, (b) combination of grinding and impingement (Jacobs *et al.*, 2001)

According to Sklar and Dietrich (2001, 2004), the relevant process causing abrasion is saltation, whereas sliding and rolling do not cause significant wear. Therefore, for an optimum SBT design in terms of minimizing hydro-abrasion, expected hydraulic conditions, particle size and distribution, particle transport modes, particle impact velocities and particle trajectories should be determined. The rolling, saltation, and suspension probabilities of different particle sizes are important since the transport mode directly affects the particle impact energy on the bed. Moreover, the particle trajectories determine the number of impacts per unit length. Estimates of these hydraulic characteristics from recent findings on bed load particle motion are given in 9.5.5. These computations allow for an iterative optimization of the hydraulic conditions as basis for the selection of an appropriate SBT invert material.

9.5.5 Bed load particle motion dynamics

Transport mode

Auel *et al.* (2017a) performed hydraulic model tests to study bed load particle motion in supercritical open-channel flow over a planar bed of low relative roughness height simulating flows in high-gradient mountain streams and hydraulic structures such as SBTs. Under these conditions, particles are dominantly transported in saltation with minor parts in rolling and a few small particles in the suspension mode. The change from rolling to saltation is estimated as

$$P_R = 5.1 \times 10^{-3} \left(T^* \left(\frac{k_s}{d} \right)^2 \right)^{-0.64} , \quad R^2 = 0.85, \quad 0 < P_R \leq 1; \quad k_s/d \leq 0.06. \tag{9.1}$$

Here P_R is the rolling probability, $T^* = (\theta_S/\theta_{Sc})-1$ the excess transport stage, with θ_S as Shields' (subscript S) parameter and θ_{Sc} as critical (subscript c) Shields parameter for incipient motion, k_s as equivalent sand-roughness height, and d as sediment particle diameter. The Shields parameter $\theta_S = \tau/[(\rho_s-\rho)gd]$ is a non-dimensional measure for bed-shear stress τ, with ρ_s as sediment density, ρ as fluid density, and g as gravity acceleration. Auel *et al.* (2017a) determined from the experimental data $\theta_{Sc} = 0.005$ for supercritical open-channel flow over a planar bed of low relative roughness height $k_s \ll d$. In the experimental setup of Auel *et al.* (2017a), $k_s = 0.2$ mm, i.e. an order of magnitude smaller than d, and the maximum ratio was $k_s/d \approx 0.06$.

Particle trajectories

The data of Auel *et al.* (2017a) evidence that particle trajectories on planar beds are flat and long compared with alluvial bed data. Their experimental data follow with L_p as particle hop length

$$\frac{L_p}{d} = 2.3\left(T^*\right)^{0.8}, \quad R^2 = 0.73. \tag{9.2}$$

With h as flow depth, Auel *et al.* (2017a) propose as relation for the normalized hop height H_p/d on planar beds of low relative roughness heights ($k_s/h < 0.01$)

$$H_p/d = 0.27T^{*0.5}, \quad R^2 = 0.80, \qquad \text{for } H_p + d/2 \leq h. \tag{9.3}$$

For movable beds with $k_s/h \gg 0.01$, the relation by Sklar and Dietrich (2004) describes the relative hop height adequately as

$$H_p/d = 1.44T^{*0.5}. \tag{9.4}$$

Bed impact

Auel *et al.* (2017b) describe the characteristics of oblique particle impacts. The ratio of mean particle impact velocity V_{im} to the mean particle velocity V_p is

$$V_{im} = 0.98V_p, \quad R^2 = 0.995, \qquad k_s/h \ll 0.1. \tag{9.5}$$

Therefore, $V_{im} \approx V_p$ for planar beds. The vertical impact velocity W_{im} was low as compared with the resultant value, namely $W_{im} = (0.054 \pm 0.014)V_{im}$ due to the flat and long trajectories, as noted above. The best data fit follows with $s = \rho_s/\rho$ as relative sediment density from (Auel *et al.*, 2017b)

$$\frac{W_{im}}{\left[(s-1)gd\right]^{0.5}} = 0.1T^{*0.39}, \quad R^2 = 0.69. \tag{9.6}$$

9.5.6 Mechanistic abrasion model

A mechanistic river incision model based on abrasion due to saltating particles was proposed by Sklar and Dietrich (2004) encompassing the analysis of particle motion characteristics for a wide data range. The magnitude of abrasion expressed as vertical abrasion rate A_r [m/s] follows

$$A_r = \frac{Y_M}{k_v f_t^2} \cdot \frac{W_{im}^2}{L_p} \cdot q_s \cdot \left(1 - \frac{q_s}{q_s^*}\right). \tag{9.7}$$

Here Y_M [Pa] is Young's modulus of elasticity of the bed material, f_t [Pa] the splitting tensile strength of the bed material, k_v [-] the bed material resistance coefficient encompassing both the particle and bed material characteristics, q_s [kg/(sm)] the specific gravimetric bed load rate, and q_s^* [kg/(sm)] the specific gravimetric bed load transport capacity. The last term on the right-hand side of Eq. (9.7) is related to the cover effect accounting for bed load partly covering the bed, resulting in reduced impact energy (Sklar and Dietrich, 1998, or Turowski, 2009). Applying their findings for the vertical impact velocity W_{im} (Eq. 9.6) and hop length L_p (Eq. 9.2) for supercritical open-channel flow over a planar bed of low relative roughness height $k_s \ll D$, Auel et al. (2017b) adapted Eq. (9.7) as

$$A_r = \frac{Y_M}{k_v f_t^2} \cdot \frac{\left(0.1 T^{*0.39}\left[(s-1)gd\right]^{0.5}\right)^2}{2.3 T^{*0.8} d} q_s \left(1 - \frac{q_s}{q_s^*}\right) \approx \frac{Y_M}{k_v f_t^2} \cdot \frac{(s-1)g}{230} q_s \left(1 - \frac{q_s}{q_s^*}\right). \qquad (9.8)$$

Here $T^{*0.78} \approx T^{*0.8}$ is applied, so that A_r only scales with the specific gravimetric bed load rate q_s. Equation (9.8) does not account for the mode shift from saltation to suspension included in Sklar's original model by a nonlinear function additionally increasing the hop length at high transport stages. This extension is not considered here, but might have to be included in Eq. (9.8) for dominant suspended load share in the total sediment transport.

To apply the saltation-abrasion model for hydraulic structures where concrete is the dominant lining material, the material property parameters have to be adapted. The decisive parameter describing the concrete material strength is compression, as concrete bears only little tension without reinforcement (Arioglu et al., 2006). Hence, for concrete abrasion, the compression strength is typically used (Jacobs et al., 2001; Helbig and Horlacher, 2007; Mechtcherine et al., 2012; Helbig et al., 2012). A correlation between the splitting tensile and the cylindrical compression strength f_c gives with both f_t and f_c in [MPa] (Arioglu et al., 2006)

$$f_t = 0.387 f_c^{0.63}, \qquad 4 < f_c < 120 \text{ [MPa]}. \qquad (9.9)$$

Young's modulus Y_M for concrete is determined with ρ_c [kg/m³] as concrete density as (Noguchi et al., 2009)

$$Y_M = k_1 k_2 \cdot 3.35 \times 10^4 \left(\frac{f_c}{60}\right)^{1/3} \left(\frac{\rho_c}{2,400}\right)^2, \qquad 40 < f_c < 160 \text{ [MPa]}. \qquad (9.10)$$

The correction factors k_1 and k_2 vary from 0.95 to 1.20 and account for the type of coarse aggregate and admixtures, respectively. Assuming $k_1 = k_2 = 1$ for the sake of simplicity, and $\rho_c = 2400$ kg/m³ as a common value (Noguchi et al., 2009), Eq. (9.8) reduces for abrasion of concrete to

$$A_r \approx c_a \frac{(s-1)g}{k_v f_c^{0.93}} q_s \left(1 - \frac{q_s}{q_s^*}\right) \text{ [m/s]}. \qquad (9.11)$$

Here f_c in [Pa] and $c_a = 94.4$ [Pa$^{-0.07}$] as abrasion coefficient because Eqs. (9.9, 9.10) are not unit compliant. A decisive parameter affecting the abrasion rate is the abrasion coefficient k_v assumed to depend on the material properties. For typical conditions at hydraulic structures and SBTs with $f_c \geq 30$ [MPa] and $f_t \geq 3.3$ [MPa], $k_v \approx 2 \times 10^5$ for concrete (Auel et al., 2017b, Müller-Hagmann et al., 2020) and $k_v \approx 2.4 \times 10^6$ for granite as invert material (Müller-Hagmann et al., 2020).

9.5.7 Lining material

Although there is a large variety of materials employed in SBTs, including particularly high priced supplies as epoxy resin mortar and rubber plates on steel, medium- and high-strength concrete is still the most widely used. While the former concrete standard EN 206–1 defined high-strength concrete with a compressive strength class higher than C50/60 (in the current standard EN 206 there is no more definition for high-strength concrete) and the American Concrete Institute (ACI) defines high-strength concrete as having a 28 day compressive strength of at least f_c = 55 MPa (ACI, 2013), so-called Ultra-High Performance Concrete (UHPC) features 150 MPa $\leq f_c \leq$ 250 MPa (Fehling et al., 2005). High Performance Concrete (HPC) and UHPC are defined as concrete meeting special combinations of performance and uniformity requirements not always achieved routinely using conventional constituents and normal mixing, placing, and curing practices (ACI, 2013).

Based on a long-term field study at Runcahez SBT between 1995 and 2016 (Jacobs et al., 2001; Jacobs and Hagmann, 2015; Müller-Hagmann, 2017), the decisive material characteristics of concrete as to its abrasion resistance are not completely known. The splitting tensile strength, and the fracture energy show a moderate correlation to abrasion. The compressive strength shows a weaker correlation. However, due to the fact that in most cases only compressive strength data are available, compressive strength is still often used as the 'characteristic' parameter. To guarantee sufficient resistance to hydro-abrasion excluding extremely severe conditions, the compressive strengths of SBT linings should be f_c > 60 [MPa], while the fracture energy should be >200 [J/m^2] at 28 days. From Eq. (9.9) the splitting tensile strength is thus f_t > 5.1 [MPa]. Natural stone material such as cast basalt or granite can also be used (Müller-Hagmann, 2017; Auel et al., 2018b), and steel armoring in reaches of high wear, e.g. in the acceleration reach near the intake gate, have been applied successfully. For the selection of adequate material, not only the initial investment, but the total life-cycle cost including maintenance and repair should be considered and weighed (Müller-Hagmann, 2017; Boes et al., 2018). For this purpose, more research is needed to better predict abrasion depths and service life of different materials (Hagmann et al., 2016; Müller-Hagmann, 2017).

Results from in-situ SBT tests at Runcahez, Pfaffensprung and Solis, Switzerland, indicate that the mean abrasion rates tend to increase with decreasing compressive and splitting tensile strengths of the invert concrete (Boes et al., 2014; Jacobs and Hagmann, 2015; Müller-Hagmann, 2017; Müller-Hagmann et al., 2020). One of the important findings from the Pfaffensprung SBT abrasion study is the significantly higher hydro-abrasion resistance of granite compared with the implemented high-strength concrete invert ($f_c \approx$ 110 [MPa]) under very severe conditions. The results show that the mean abrasion rate of the granite plates is considerably smaller (by a factor of 6–7) than for the concrete invert (Boes et al., 2014; Hagmann et al., 2015; Müller-Hagmann, 2017; Boes et al., 2018) due to the considerably higher abrasion coefficient of the former (9.5.6). This suggests that the used type of granite is a better choice as invert material over high-strength concrete for very severe abrasion conditions as at Pfaffensprung SBT in terms of life-cycle cost (Müller-Hagmann, 2017). For low to medium severe abrasion conditions, invert materials with a lower abrasion resistance (e.g. HPC) are sufficient, however. By considering these results, a granite pavement was selected for the rehabilitation of the formerly steel-lined Mud Mountain bypass tunnel in the USA (Auel et al., 2018b).

The results of the Pfaffensprung SBT study show that damages typically occur in the form of grooves along the joints of basalt and granite plates (Figure 9.27a), while a wavy pattern

Figure 9.27 Abrasion patterns at Pfaffensprung SBT (a) grooves forming along joints of granite plates, (b) undular invert at steel-fibre high-strength concrete test field with f_c > 70 [MPa] (Boes *et al.*, 2014)

of abrasion occurs on high-strength concrete (Figure 9.27b). To further reduce the abrasion on granite, this result suggests that plates should not be placed in parallel to the main flow direction and a jointless tight installation between the plates should be achieved. It should be noted that the latter is challenging requiring special knowledge and skills of the construction team, which are not always available. The joints between the granite plates and the SBT sidewalls should be as small as possible and filled with a high-abrasion resistant material such as a special mortar with basalt aggregates.

Whereas natural stone material, e.g. cast basalt plates, is supposed to have a high-abrasion resistance against the pure particle grinding action, their brittleness may favor fracturing by impinging particles for saltating sediments. The risk of fracturing largely depends on both the particle size determining the impact energy and the thickness of the invert liner. For particularly large saltating sediment particles in the multi-decimeter range, either steel or cementitious material such as high-strength concrete may show improved resistance. As steel linings are often too costly for abrasion protection of large areas such as in SBTs, high-strength concrete becomes an interesting and economical alternative (Boes *et al.*, 2014).

9.5.8 Design of tunnel invert lining

General

Large particles possess a large particle mass, thus higher impact energy is transferred by these on the surface than by small particles (Auel *et al.*, 2014). A combination of large particle size and high flow velocities thus results in high mean invert abrasion. Depending on the invert material, sediment properties and sediment transport intensity, the abrasion rate varies; in general, typical mean abrasion rates range from a few microns to some hundredth of millimeters per hour of operating time, i.e. A_r ≈ 1 to 10 μm/h for Swiss and Japanese SBTs (Jacobs *et al.*, 2001; Kataoka, 2003; Sumi *et al.*, 2004; Fukuroi, 2012; Hagmann *et al.*, 2015; Jacobs and Hagmann, 2015; Auel *et al.*, 2018b). However, under extreme floods or severe conditions, considerable local abrasion rates and depths may be attained, as for the Palagnedra SBT which was in continuous operation for 10 months following the extraordinary 1978 Flood, resulting in invert abrasion of up to 2.7 m depth into the underlying bedrock (Hagmann *et al.*, 2015, Figure 9.17a). Abrasion rates may then increase from $A_r \approx$ 10^2 to 10^3 µm/h, i.e. up to some mm/h.

The areas particularly prone to hydro-abrasion are those exposed to bed load transport, i.e. the invert and the walls up to a height of the expected bed load layer thickness, typically of 0.5 m. An appropriate selection and correct placement of sustainable and optimum abrasion-resistant lining material at both the invert and the mentioned wall sections requires the following:

- If most of the particles are transported in the rolling or sliding motion with only minor saltation (Eq. 9.1), abrasion processes are expected to be mainly grinding and only weakly impinging. Hence using natural stones such as granite and cast basalt as invert lining material or high-strength concrete are viable options. As to the former, the use of quadratic plates placed diagonally to the flow direction or of hexagonal plates is recommended (Figure 9.28), to avoid joints parallel to the flow which are more easily abraded. However, abrasion also starts at upstream edges perpendicular to the flow (Figure 9.27a), so that the placement of type (a) is preferred over (b) in Figure 9.28. The plates should be embedded into a special mortar.
- If saltation is expected to be the main particle transport mode and/or the sediment is rather coarse combined with high flow velocities, high-strength concrete with $f_c >$ 60 [MPa] (i.e. C60/85 and higher) and $f_t > 5$ [MPa] is preferable. Concrete curing is critical and should be performed carefully (Jacobs *et al.*, 2001).

Spatial abrasion distribution and consequences for invert placement

In practice, the SBT invert will be non-uniformly abraded depending on the flow conditions. The tunnel width to water depth aspect ratio, B/h, determines if the flow is three (3D) or two dimensional (2D). The critical aspect ratio $(B/h)_{cr}$ is around 4–5. For lower values, the flow is 3D, so that strong secondary currents across the tunnel width cause higher bed-shear stress, sediment transport and hence abrasion at the tunnel side corners as compared to the center. The results of Auel *et al.* (2014) confirm the strong effect of aspect ratio on the abrasion pattern (Figures 9.29, 9.30), showing continuous abrasion with two deep lateral incision channels developing along the walls. The abrasion depths and widths are

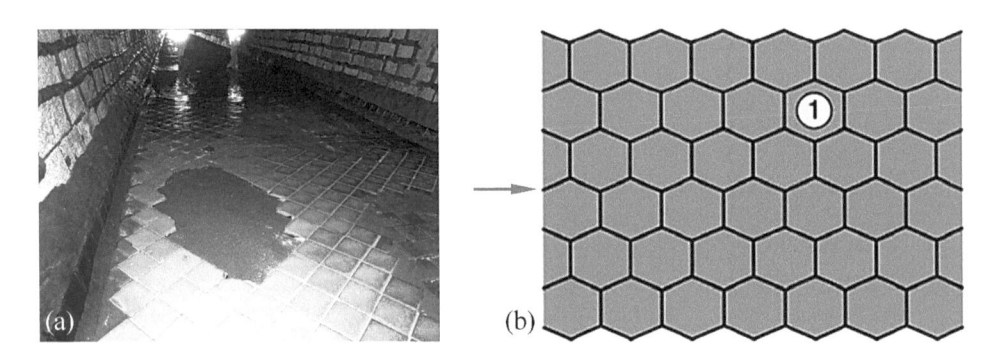

Figure 9.28 (a) Diagonally checkered cast basalt plates with local damages at SBT Paffensprung, Switzerland (Courtesy Michelle Müller-Hagmann), (b) schematic plan view of thin pavement of hexagonal natural stone plates (Jacobs *et al.*, 2001)

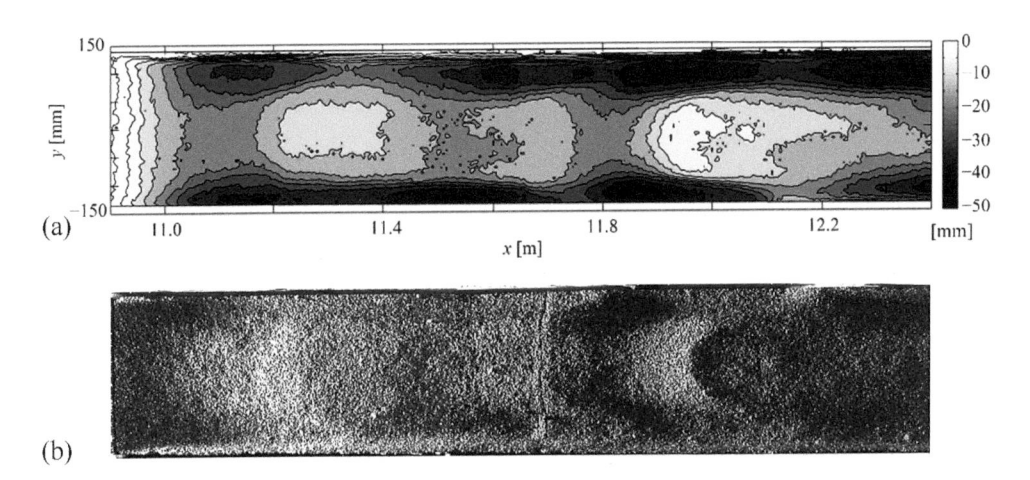

(a)

(b)

Figure 9.29 Topography (a) scan and (b) photograph after 1.5 h of abrasion test for aspect ratio of 2.6 and Froude number F = 4 (Auel *et al.*, 2014)

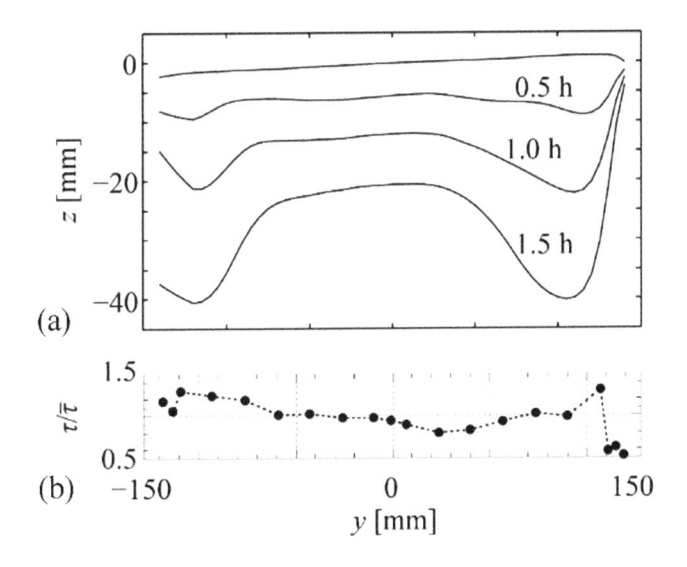

Figure 9.30 (a) Spatially averaged cross sections after various runtimes for conditions as in Figure 9.29, (b) corresponding normalized bed-shear stress (Auel *et al.*, 2014)

expanding toward the flume center with time. The abrasion pattern matches well with the spanwise bed-shear stress distribution across the flume (Figure 9.30). These lateral incision channels have also been observed e.g. at Runcahez prototype SBT (Jacobs and Hagmann, 2015).

The aspect ratio of most SBTs for the relevant load cases is $B/h < 4$, i.e. the flow is 3D. The abrasion pattern with two lateral incision channels along the walls as shown in Figures 9.29

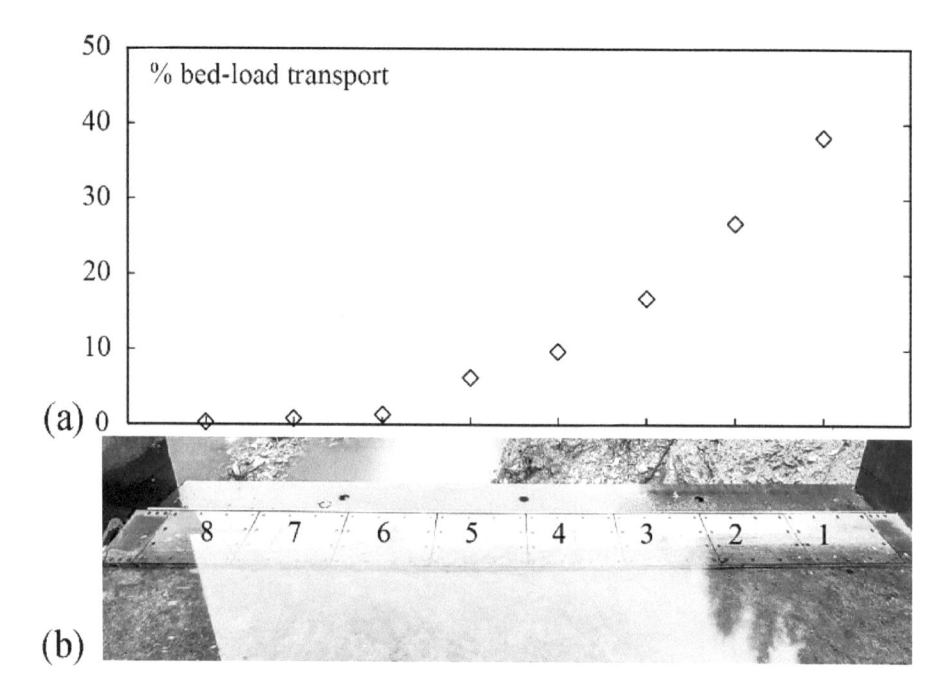

Figure 9.31 (a) Downstream view of transverse bed load transport distribution during flood on August 13th, 2014, registered by Swiss plate geophone system (Adapted from Albayrak *et al.*, 2015) consisting of (b) eight units across tunnel width installed at outlet of Solis SBT (Adapted from Hagmann *et al.*, 2015)

and 9.30 is expected to form. For straight tunnels it is thus an option to place high-strength concrete or natural stones across the outer quarter of the SBT width, while a conventional and more economical concrete may be used in the center half. For curved flow due to bends in plan view, however, the sediment particles are transported mainly along the inner SBT sidewall due to secondary currents (Figure 9.31, Albayrak *et al.*, 2015), inducing higher abrasion rates there, as observed for instance in Asahi SBT (Nakajima *et al.*, 2015) and Solis SBT. Then, a high-strength concrete or natural stones applies solely in this inner half across the invert width.

The choice whether to use different material across the tunnel width is a site-specific economic optimization on the one hand, and a question of practicability on the other, as the additional construction joints may themselves initiate local abrasion. So-called hot joints should therefore be favored for concrete usage, i.e. the adjacent concrete field in the transverse direction should be poured before setting of the previous. To check for the correctness of the assumptions made in the invert design procedure, it is recommended to visually inspect the tunnel invert and to measure abrasion depths, if applicable, after each flushing or flood event in the first years of operation until experience on the pattern and magnitude of abrasion has been gained (9.5.9).

Requirements on invert lining thickness and implementation

Equation (9.8) allows for the prediction of concrete or natural stone abrasion under super-critical free-surface flow over a planar bed of low relative roughness height $k_s \ll d$, as typically found in SBTs. In general, the invert slab thickness should be designed so that a major refurbishment should not become necessary for at least some years. As a worst case scenario, even at the design flood with design duration, there should be no local abrasion in excess of the slab thickness. Often, a wearing surface top layer of high-strength concrete or natural stone material is used, while the tunnel concrete underneath features a conventional concrete some 0.3 m thick.

Typical wearing surface invert concrete characteristics and their layer thicknesses used in Swiss SBTs are given in Tables 9.4 and 9.5, while Table 9.6 features natural stone and steel invert characteristics. Regarding the SBT invert placement, the proper implementation is relevant, since small damages can rapidly spread out. The flatness of the concrete surface is also important to avoid locally pronounced abrasion or cavitation damages due to surface irregularities leading to larger secondary damages.

9.5.9 Tunnel operation, maintenance, and rehabilitation

Depending on the SBT intake location, reservoir operation during sediment routing varies. For position A (Figure 9.20a), the gate is opened during floods and the water-sediment mixture is routed through the tunnel under free-surface flow. The reservoir level can be kept at full supply level. The incoming flow and sediment are conveyed into the dam tailwater, independent of the reservoir level.

Table 9.4 Invert concrete characteristics used in sediment bypass tunnels in Switzerland (part I)

Layer thickness [m]		concrete aggregate		cement		water	
		Grain size distribution	mass [kg/m³]	type	mass [kg/m³]	w/c ratio [-]	mass [kg/m³]
Normal invert (High strength concrete with steel fibers)	0.3	0/4: 40% 4/8: 24% 8/16: 35%	1900	CEM II/A-D 52.5N	536	0.33	177
Concrete with steel fibers	0.3	0/4: 41% 4/8: 22% 8/16: 36%	1740	CEM II/A-D 52.5R	450	0.41	185
Concrete with shrinkage reduction	0.3	0/4: 40% 4/8: 24%	1900	CEM II/A-D 52.5R	390	0.44	172
high alumina cement	0.15	Alag fine 0/4: 50% Alag coarse 4/10: 50%	2060	high alumina cement concrete	515	0.4	206
Ultra high performance concrete	0.08	quartz sand	870	CEM II/B-M	1100	0.17	187

Table 9.5 Invert concrete characteristics used in sediment bypass tunnels in Switzerland (part II)

		fibers		admixture		material properties		
	Layer thickness [m]	Art	mass [kg/m³]	type	mass [kg/m³]	$f_{c,28d}$ [MPa]	$f_{bt,28d}$ [MPa]	Euro-code
Normal invert (High strength concrete with steel fibers)	0.3	Steel Dramix RC 80/30-BP	45	Water-reducing admixture: Glenium Sky 587	2.5	101		C70/85 with 45 kg/m³ steel fibers
Concrete with steel fibers	0.3	Steel Dramix RC 80/3 0-BP	60	Water-reducing admixture: Glenium Sky 587	2.5	78.9	12.4	C55/67 with 60 kg/m³ steel fibers
Concrete with shrinkage reduction	0.3	Polymer fibers: Superfiber 40/8	3	Shrinkage reduction: D-Zero 32	11.7	84.7	10.8	C50/60 with 3 kg/m³ polymer fibers & shrinkage reduction
high alumina cement	0.15				0	86.3	11.5	
Ultra high performance concrete	0.08	Steel 13/0.16	236	Water-reducing admixture: Sika P5	35	132.5	25.5	

Table 9.6 Natural stone and steel invert characteristics used in sediment bypass tunnels in Switzerland

Material	$f_{c,28d}$ [MPa]	$f_{t,28d}$ [MPa]	Geometry / layer thickness [m]
Cast basalt plates with mortar-filled joints	300–450	45	$0.2 \times 0.2 \times 0.05$
Granite plates without joints (flush)	260	>10°	$1.0 \times 1.0 \times 0.3$
Steel S235 JR on self-compacting concrete SCC C35/45	$f_y = 225$ * 66.9	9.1	0.02 0.28

* f_y = yield strength
° independent of sample age
$f_{c,28d}$ = f_c value after 28 days
$f_{bt,28d}$ = bending tensile strength after 28 days

At position B (Figure 9.20b), the reservoir level has to be lowered prior to a flood event depending on the distance of the reservoir head to the tunnel intake. The reservoir reach upstream of the intake has to be subjected to free-surface flow to ensure high transport capacity so that incoming sediment is transported toward the intake. The reservoir level has to be kept at a certain lower elevation to avoid the interruption of sediment transport (Auel and Boes, 2011). Albayrak *et al.* (2015) and Hagmann *et al.* (2015) demonstrated the sensitivity of the sediment transport capacity on the reservoir level based on measurements at Solis SBT, Switzerland. Once the reservoir level exceeded a critical threshold, the bed load transport dropped. Figure 9.32 shows time series of the volumetric suspended and bed load transport rates in Solis SBT and the reservoir level during a large flood event. Monitoring data indicate that the lower the reservoir level, the coarser sediment material is transported through the SBT. The data from eight geophone sensors installed across the tunnel invert at the outlet (Figure 9.31) revealed that during reservoir drawdown, three peaks in bed load transport occurred at certain reservoir elevations (arrows in Figure 9.32). Turbidity data revealed a similar behavior for the suspended sediment transport rate, but the peaks occurred before those of the bed load (Figure 9.32). In addition, suspended sediment transport through the SBT occurred prior to drawdown of the reservoir level.

For position B type SBT, an adequate operation mode of SBT and reservoir is thus a prerequisite to obtain high sediment bypassing efficiencies. By drawing down the reservoir level during flood events, the armoring layer at the reservoir head can be set into motion and the transport capacity in the entire reservoir upstream of the SBT intake increases. This keeps the rise of bed level low at the reservoir head and leads to a faster delta progression and lower sedimentation levels, resulting in a larger active reservoir storage volume. As soon as the delta front reaches the SBT inlet, the large sediment inflow can be sluiced through the SBT. SBT operation experiences are given by Jacobs and Hagmann (2015), Laperrousaz and Carlioz (2015), Müller and Walker (2015), Oertli and Auel (2015), and Sakurai and Kobayashi (2015).

Maintaining an SBT in a secure state may require significant efforts and be costly. It should be considered from the start by estimating the expected abrasion depths (9.5.8) and by

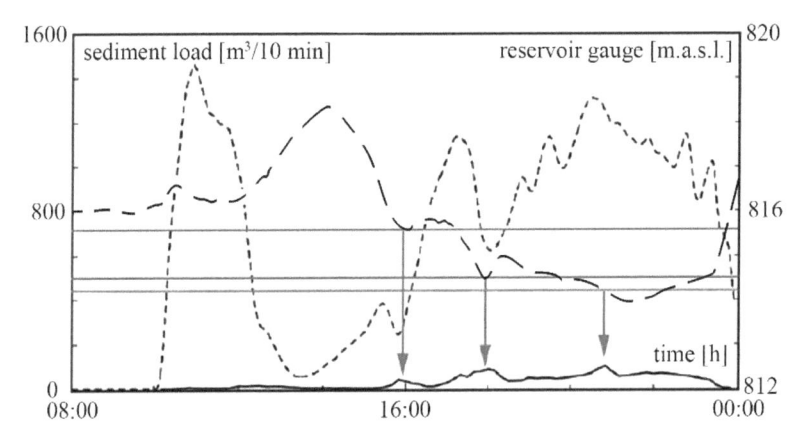

Figure 9.32 Time series of (---) suspended and (—) bed load transport at Solis SBT with (– –) reservoir level during flood event of August 13, 2014 (Albayrak *et al.*, 2015)

assessing the life-cycle cost of the invert material selected. Depending on the extent of invert abrasion and the endangerment of the tunnel stability as well as consequences of a failure, rehabilitation works at existing SBTs are performed on a regular schedule, e.g. during each low-flow season, or at irregular intervals, e.g. after decades only (Hagmann *et al.*, 2016). Full or partial maintenance work should be carried out once local abrasion depths reach about half to 2/3 of the wearing surface slab thickness. Therefore, it is strongly recommended to visually inspect the tunnel after the first and each consecutive operation to be aware of the current situation. An access route for trucks and large machinery to the tunnel inside should be well designed from the beginning. Field experience with maintenance and rehabilitation of SBTs is reported by Jacobs and Hagmann (2015), Baumer and Radogna (2015), Nakajima *et al*. (2015), and Müller and Walker (2015).

9.5.10 Instrumentation and monitoring techniques

Modern SBTs should be equipped with instrumentation to monitor and follow-up their operation and behavior, e.g. regarding flow velocities, sediment transport and the state of invert abrasion. This allows for the collection of operational data, bypass efficiency rating, abrasion and damage observation, and an analysis of flushing events to optimize efficiencies and to minimize abrasion (De Cesare *et al.*, 2015; Hagmann *et al.*, 2016). If real-time monitoring techniques are applied, the operation may even be adapted during the event, e.g. to maximize the flushed sediment volume. The main parameters to be monitored in real-time are the reservoir water level, the suspended sediment concentration (e.g. by means of turbidimeters), intake gate position (e.g. by displacement transducers), tunnel flow velocity (e.g. by radar), tunnel flow depth (e.g. by pressure sensors) and, optimally, the bed load transport rate (e.g. by means of geophones or hydrophones). 3D laser scanning can be effectively utilized for periodical monitoring of abrasion patterns. More information on the Swiss plate geophone system in particular is reported by Wyss *et al.* (2014, 2016) and on SBT instrumentation and monitoring and its setup in general by Albayrak *et al.* (2015) and Hagmann *et al.* (2015).

9.5.11 Ecological impacts of SBT operation

Re-establishing the sediment continuum in a river is one of the main purposes of any SBT. They allow for conveyance of sediments, both bed and suspended loads, from the upstream river and reservoir to the downstream reach having the potential to enhance river morphology of the often sediment-depleted river stretches downstream of dams. The routing of large sediment loads around the reservoir during flood events resembles the natural riverine conditions leading to a reduced erosion and river incision potential in the downstream reach. This in turn may affect the river ecosystem, altering the habitat conditions for the biota, particularly fish and macroinvertebrates (Facchini *et al.*, 2015). Therefore, the downstream effects and the bypassing efficiency of SBT operations become more and more important. The downstream morphology and ecology should be monitored to optimize the SBT operation both in terms of the downstream effects on fauna and flora, and regarding the bypassing efficiency, i.e. the ratio of bypassed to incoming sediment volumes. The downstream situation should be compared with undisturbed river reaches upstream of the reservoir to assess the ecological effect of SBT operation.

Facchini *et al.* (2015) examined the eco-morphological effects in the downstream river system by monitoring the riverbed elevation and its grain size distribution at different cross sections downstream of the Solis SBT outlet. Riverbed levels were measured in several river cross sections along some 6 km from the SBT outlet in 2012 and 2013, and after the August 2014 Flood, while the grain size distribution was measured in the same years in the middle and downstream portions, some 3 and 6 km from the outlet, respectively. The results indicate that flood events with SBT operation cause considerable changes in the river morphology and in the grain size distribution of the bed material (Figure 9.33). Martin *et al.*'s (2015) investigations show that the habitat quality of the river is affected by both the Solis Dam spillway and SBT operation, and that the extent of disturbance increases with discharge. Compared to a multiyear frequency of these events, the recovery duration is relatively short of the order of several weeks to months.

Auel *et al.* (2017c) monitored four reservoirs with SBTs in Japan (Asahi, Koshibu) and Switzerland (Solis, Pfaffensprung) to analyze their effects in terms of up- to downstream morphological and biotic changes. Sediment grain size distributions, local bed characteristics, microhabitat abundance and invertebrate richness were analyzed. It was found that GSD at reservoirs with newly established SBTs are fine in the up- (US) and coarse in their downstream (DS) reaches due to lack of conveyed sediments in the past. An analysis of biotic data directly below the dams, but upstream of the SBT outlet (DS-D, Figure 9.34d) revealed that microhabitat richness is low whereas lentic species abundance is high compared to the upstream reach, while these differences decrease further downstream (Figure 9.34a–c). Microhabitat and invertebrate richness in the DS adjust to the US values

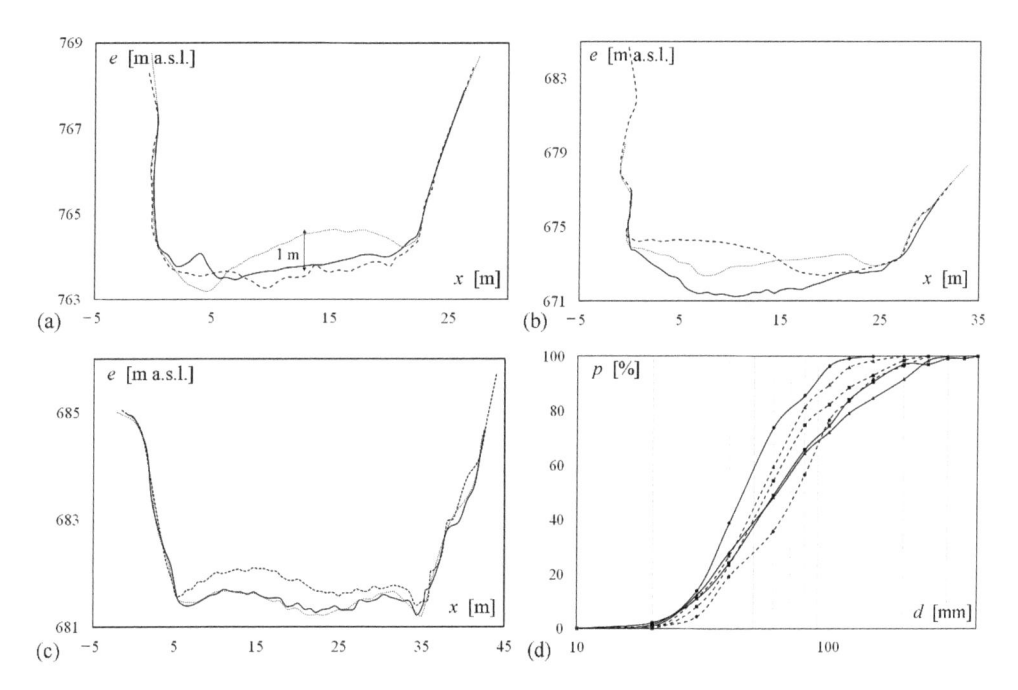

Figure 9.33 Temporal development of river cross section elevation $e(x)$ (a) #16 ca. 0.3 km, (b) #13 ca. 3 km and (c) #10 ca. 6 km downstream of Solis SBT (Courtesy Meisser Vermessungen AG), (d) grain size distribution comparison $p(d)$ in (—) 2012, (– –) 2013 and (····) 2014 (Courtesy Hunziker, Zarn & Partner AG) (Facchini *et al.*, 2015)

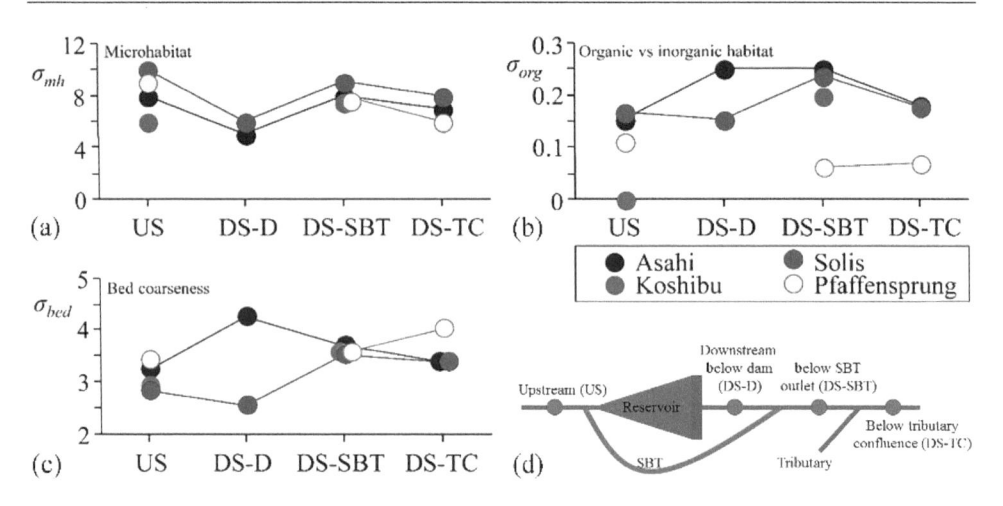

Figure 9.34 (a) Microhabitat richness σ_{mh}, (b) relative abundance of organic versus inorganic habitats σ_{org} versus stream location, (c) bed coarseness index σ_{bed}, (d) sketch of sample locations (Auel *et al.*, 2017c)

with increasing number of operation years, clearly highlighting the positive effects of long-term SBT operation.

9.6 Turbidity currents

9.6.1 Definition

A density current is a layer of fluid flowing in an ambient fluid of different density. The density difference may result from changes in temperature, salinity, dissolved solid concentration, or due to different fluids involved. For a mixture of suspended sediments and ambient water, density currents are more specifically named turbidity currents, corresponding to flows driven by density differences caused by suspended fine solid material in an ambient fluid. They appear in different forms, depending on the density of the mixed (water/sediment, subscript *m*) fluid. If its density ρ_m is above that of the ambient (subscript *w* for water) fluid ρ_w, the turbidity current is formed along the bottom (Figure 9.35a). If the density of the mixed fluid is less than that of the ambient fluid, the current is lifted to the surface (Figure 9.35b). Sometimes, the reservoir is stratified so that the mixed fluid flows as an intrusion (Figure 9.35c). Generally, it is the first type of current transporting the largest quantities of sediments downstream.

To travel long distances, the velocity of a turbidity current must be sufficiently large to generate the turbulence required to maintain its sediment load in suspension, thereby maintaining the density difference between the gravity-induced current and the ambient water. During passage the turbidity current may unload or re-suspend fine granular material.

The sediment exchange at the reservoir bottom involves a flux between the bed and current, separated into sediment entrainment and sediment deposition terms evaluated at a reference height slightly above the real bed level (Parker *et al.*, 1986; De Cesare,

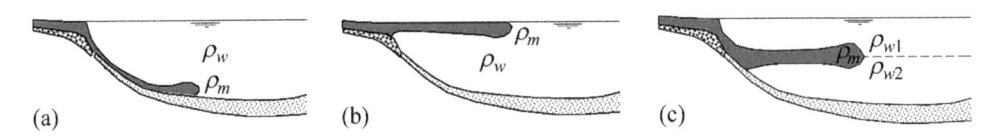

Figure 9.35 Types of turbidity currents in reservoir for (a) $\rho_m > \rho_w$ along reservoir bottom, (b) $\rho_m < \rho_w$ along reservoir surface, (c) $\rho_{w1} < \rho_m < \rho_{w2}$ intrusive due to density stratification (Adapted from Oehy, 2003)

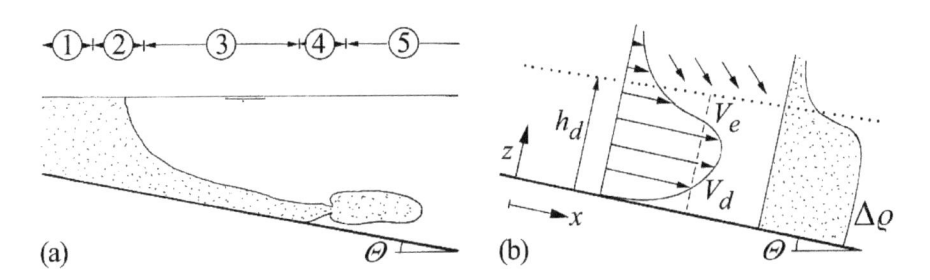

Figure 9.36 Density flow into stratified reservoir (a) flow zones, (b) notation with (. . .) interface

1998). Suspended sediment is constantly falling out of the current at a rate given by the sediment fall velocity and the mean volumetric concentration of suspended particles near the bed.

The turbidity current exerts a stress on the reservoir bed capable of entraining sediments from the bed into suspension. If the entrainment rate is less than the depositional rate, then the turbidity current experiences a net loss of granular material, so that the sediment concentration in the current decreases. The driving force acting on the current decreases, therefore, causing its deceleration and eventually to vanish. In contrast, a higher flow velocity of the turbidity current produces a rate of sediment entrainment from the bed that is higher than the depositional rate. The current density increases and the turbidity current accelerates. As the bed-stress increases further and more sediment is entrained, a self-reinforcing cycle is created allowing for the development of a self-sustaining turbidity current gradually reaching high speeds. The availability of bed sediment for entrainment, the reservoir geometry, or eventually the damping of turbulence at high concentrations limit the growth of gravity-driven flow.

Turbidity currents have features similar to open-channel flows, differing essentially because the *buoyancy* of the surrounding fluid reduces the gravity force. Figure 9.36 shows the zones involved in a density current. The approach flow ① of density ρ_o is analyzed based on conventional hydraulics. At the plunge point ②, equilibrium between the density difference and the baroclinic pressure exists (see below). Beyond the plunge point, a two-layer flow ③ develops associated with mixing across the *interface*. If the reservoir is stratified, the density current may reach a depth where it becomes neutrally buoyant. Accordingly, separation ④ from the inclined bottom occurs to form an intrusion ⑤.

The flow of density currents depends mainly on the reduced gravity $g' = (\Delta\rho/\rho_o)g$, with g as the gravity acceleration, ρ_o the approach flow density, and $\Delta\rho$ the density difference between the inflow and the ambient fluid. If the Reynolds number $R = uh/v > 2 \times 10^3$, with u and h as the representative streamwise velocity and flow depth, and v as kinematic viscosity, the current is turbulent with mixing along the interface. The stability of the interface depends on the *Richardson number* (Alavian *et al.*, 1992)

$$\mathsf{Ri} = g' \frac{h\cos\theta}{u^2} .$$

(9.12)

Here, θ is the slope of the incline, and Ri expresses the balance between the effective gravity perpendicular to the interface and the shear across the interface. Decreasing Ri increases the amount of mixing. The buoyancy thus either accelerates and destabilizes, or damps and stabilizes a density current.

9.6.2 Plunge point and equilibrium flow

With $\mathsf{F}_p = V_o/(gh_p)^{1/2}$ as the Froude number based on the approach flow velocity V_o and depth h_p at the plunge point (subscript p), Q_o as the approach flow discharge, and b as the approach flow width, then for $0.1 < \mathsf{F}_p < 0.7$ in terms of 1D hydraulics (Savage and Brimberg, 1975)

$$h_p = \left(\frac{Q_o^2}{g'b^2} \right)^{1/3} \mathsf{F}_p^{-2/3}.$$

(9.13)

For 3D inflow, i.e. sloping bottom and expanding sides, the plunging characteristics are more complex, involving stalled and no-stalled flows. For a channel of constant width and lateral slopes of the trapezoidal section m between 0.2 and 0.8, the term F_p is empirically defined with S_o as bottom slope, and C_d as the total friction coefficient of the underflow with $0.01 < C_d < 0.09$ as (Savage and Brimberg, 1975)

$$\mathsf{F}_p = \frac{2.05}{1+m} \left(\frac{S_o}{C_d} \right)^{0.478} .$$

(9.14)

Generalized relations are provided by Graf (1983) and Alavian *et al.* (1992).

After plunging, turbidity currents can normally be separated into two portions: The front or head, which has as driving forces essentially the pressure gradient due to the density difference between the front and the ambient fluid ahead of it, and the body, driven by the gravity force of the heavier fluid (mixed water/sediment). The front flow portion is unsteady while the body core flow is considered steady. The quantity of suspended sediment is not conserved and is free to exchange with the bed sediment by bed erosion and deposition causing self-acceleration of the current by entrainment of bed sediment. Figure 9.37 illustrates a typical turbidity current.

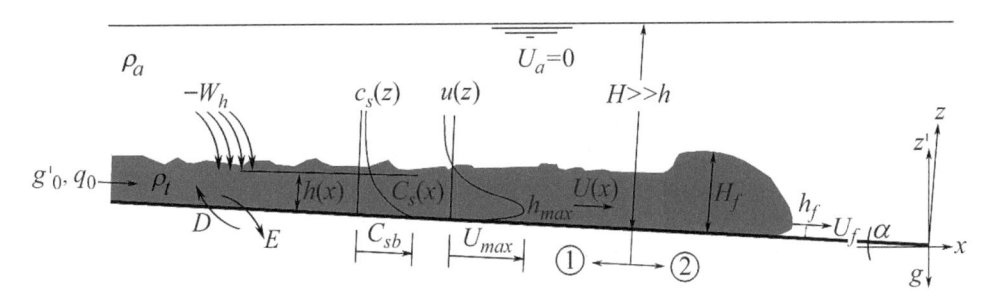

Figure 9.37 Characteristics of turbidity current flowing along bottom (Adapted from Graf and Altinakar, 1995)

The *uniform flow* of a density current (upstream zone in Figure 9.37) depends on the bottom topography and the stratification characteristics. The Richardson number $\mathrm{Ri} = g'h_d\cos\theta/V_d^2$ of the density (subscript d) current is also important. Typical velocity and density distributions of a plane density current are shown in Figure 9.36b. The interface is defined at a location where the density gradient has its maximum. Let the entrainment (subscript e) constant be $E = V_e/V_d$ with V_e as the entrainment velocity, $C_d = \tau_d/(\rho_d V_d^2)$ the bottom drag coefficient, and D_1 and D_2 density profile correction coefficients. The profile change of the density current then follows (Alavian *et al.*, 1992)

$$\frac{h_d}{3\mathrm{Ri}}\frac{d\mathrm{Ri}}{dx} = \frac{\left(1+\frac{1}{2}D_1\mathrm{Ri}\right)E - D_2\mathrm{Ri}\tan\theta + C_d}{1 - D_1\mathrm{Ri}}. \tag{9.15}$$

For uniform density distribution both $D_1 = D_2 = 1$. From experiments with turbulent density currents, $D_1 = 0.6$ to 1.0, and $D_2 = 0.95$ to 1.1. Equation (9.15) corresponds to a generalized backwater equation for the unknown $h_d(x)$.

For equilibrium (subscript u) flow, the Richardson number remains constant and

$$\mathrm{Ri}_u = \frac{E_u + C_d}{D_2\tan\theta - \frac{1}{2}D_1 E_u}. \tag{9.16}$$

Also, $dh_u/dx = E_u$, so that with the boundary condition $h_d(x = x_o) = h_o$ close to the plunging point,

$$\frac{h_u - h_o}{x - x_o} = E_u. \tag{9.17}$$

The *uniform velocity* is equal to

$$V_u = \left(\frac{V_d h_d g'_d \cos\theta}{\mathrm{Ri}_u}\right)^{1/3}. \tag{9.18}$$

The term $V_d h_d g'_d$ is referred to as the buoyancy flux, which may remain constant. The entrainment constant is related to the Richardson number as

$$E / E_0 = \mathrm{Ri}^{-\gamma}. \tag{9.19}$$

According to Alavian *et al.* (1992), the constants are $E_0 = 1.5 \times 10^{-3}$, $\gamma = 1$ for Ri>2×10^{-1}, whereas $E_0 = 2.8 \times 10^{-3}$, $\gamma = 1.2$ for turbidity currents. The previous system of equations was also established for flow in a triangular channel and on a laterally unbounded surface. An alternative approach is given by Graf (1983), and Graf and Altinakar (1993). They have also examined the front of the current. Note the marked similarity between density currents and dambreak waves (Chapter 11). Altinakar *et al.* (1996) compare the body of the current with a wall jet involving two regions, namely the wall and the jet regions (Figure 9.38). The height h_{max} is attained where the velocity has its maximum, $u = U_{max}$, separating these two regions into the:

- Wall region $z < h_{max}$, turbulence created at wall, entrainment of sediments
- Free region $z > h_{max}$, turbulence due to friction and entrainment of ambient fluid.

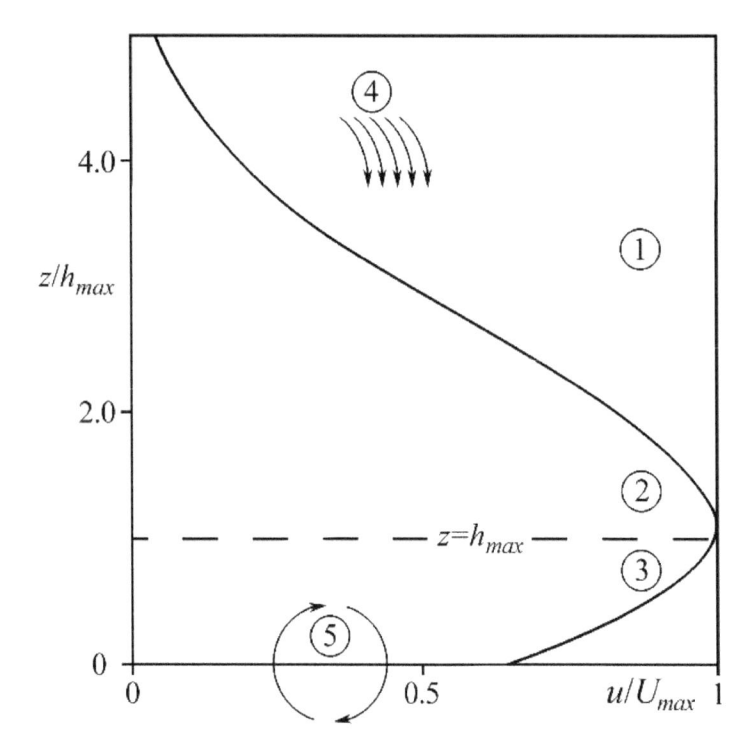

Figure 9.38 Schematic dimensionless velocity profile $u/U_{max}(z/h_{max})$ for turbidity currents with ① free jet region, ② maximum velocity, ③ wall region, ④ water entrainment, ⑤ erosion/deposition (Adapted from De Cesare, 1998)

Altinakar *et al.* (1996) experimentally found that

$$h_{max}/h = 0.30; \; U_{max}/U = 1.3; \; h_t/h = 1.3. \tag{9.20}$$

Here, h is the average current height, and h_t the current height at which $u \equiv 0$. Based on a Chézy-type relationship, a simple relation between the front (subscript f) velocity U_f and the front height H_f for a large range of slopes and roughness is (Turner (1973)

$$U_f = l_1(g'H_f)^{1/2}. \tag{9.21}$$

Here, the constant l_1 ranges between 0.63 (Altinakar *et al.*, 1990) and 0.75 (Middleton, 1966; Turner, 1973), with 0.63 more appropriate for small slopes; g' is the reduced gravity acceleration expressed as

$$g' = g(\frac{\rho_m - \rho_w}{\rho_w}). \tag{9.22}$$

After Britter and Linden (1980), the front velocity U_f can also be expressed with l_2 as a constant depending on the bottom slope and the Reynolds number R as

$$U_f = l_2 B_0^{1/3}. \tag{9.23}$$

For bottom slopes <5%, Altinakar *et al.* (1990) found values varying linearly between 0.7 and 1.0. Choi and Garcia (1995) proposed $l_2 = 1$, based on numerical experiments and the laboratory data of Altinakar *et al.* (1990). With h_0 as initial flow depth and U_0 initial velocity, B_0 is the initial (subscript 0) buoyancy flux per unit width defined as

$$B_0 = g_0'h_0U_0 = g_0'q_0. \tag{9.24}$$

The body of a turbidity current is considered 2D, and the flow turbulent and incompressible. The body velocity is normally larger than the velocity of the head. This difference increases with the bottom slope; to maintain flow continuity, the front height is always larger than the body height. The front height increases with the entrainment of ambient fluid, depending on the distance traveled.

9.6.3 Flow over obstacle

As a turbidity current meets an obstacle, a portion of the denser fluid may flow over it along with a hydraulic jump or bore traveling upstream (Figures 9.39, 9.40). The current is partially or even totally blocked by the obstacle. Consider the fluid traveling with constant flow depth h_1 and velocity U_1 and a moving hydraulic jump propagating upstream with velocity U_j as well as the sequent depth h_2 with a smaller velocity U_2. The obstacle height is h_m.

The flow over an obstacle is described by two independent variables, namely the densimetric (subscript d) approach flow Froude number F_{d1} and the ratio $H_m = h_m/h_1$ between the obstacle height h_m and the approach flow depth h_1. If the ambient fluid motion is ignored ($U_a \cong 0$), the density current has common features with free-surface flows, if the shallow-water approximation is valid (Rottman *et al.*, 1985). Neglecting fluid entrainment and

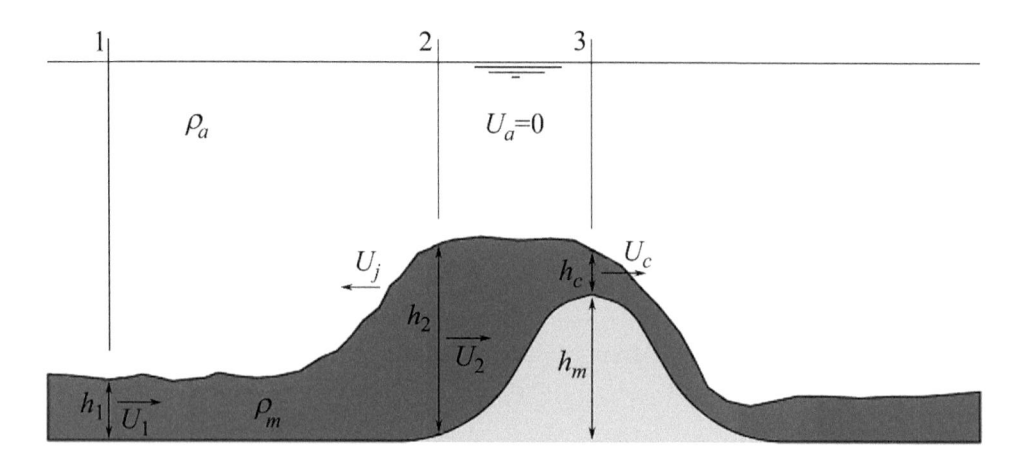

Figure 9.39 Flow of density current over obstacle (Adapted from Oehy, 2003)

friction ($E_w = 0$, $u_{*b} = 0$), as well as erosion and deposition, the steady-state inflow is described with U_1 and h_1 as upstream values of U and h as

$$Uh = U_1 h_1, \tag{9.25}$$

$$h + h_m + \frac{U^2}{2g'} = h_1 + \frac{U_1^2}{2g'}. \tag{9.26}$$

For each value of F_{d1} exists a limit $H_m = H_{mc}$ at which the flow on the crest is critical (subscript c). Applying Bernoulli's equation between sections 1 and 3 (Figure 9.39),

$$H_{mc} = 1 + \frac{\mathsf{F}_{d1}^2}{2} - \frac{3}{2}\mathsf{F}_{d1}^{2/3}. \tag{9.27}$$

To define limits for flows over an obstacle, Eq. (9.27) is solved for F_{d1} as (Figure 9.39)

$$\mathsf{F}_{d1}^2 = (H_m - 1)^2 \left(\frac{H_m + 1}{2H_m} \right). \tag{9.28}$$

Based on Baines (1995) and considering that $U_j = 0$ (Long, 1970)

$$H_m = \frac{(1 + 8\mathsf{F}_{d1}^2)^{3/2} + 1}{16\mathsf{F}_{d1}^2} - \frac{1}{4} - \frac{3}{2}\mathsf{F}_{d1}^{2/3}. \tag{9.29}$$

Steady flow over the obstacle only occurs left of the curve from Eq. (9.27) combined with Eqs. (9.28, 9.29) (full line in Figure 9.40). The flow is supercritical in regions B and C, and subcritical in region A. In region D, the turbidity current is partially blocked as $H_m > H_{mc}$. An internal bore, i.e. a moving hydraulic jump, is then formed propagating upstream, dissipating energy to match the steady solution to the upstream flow condition. The sequent depths of the

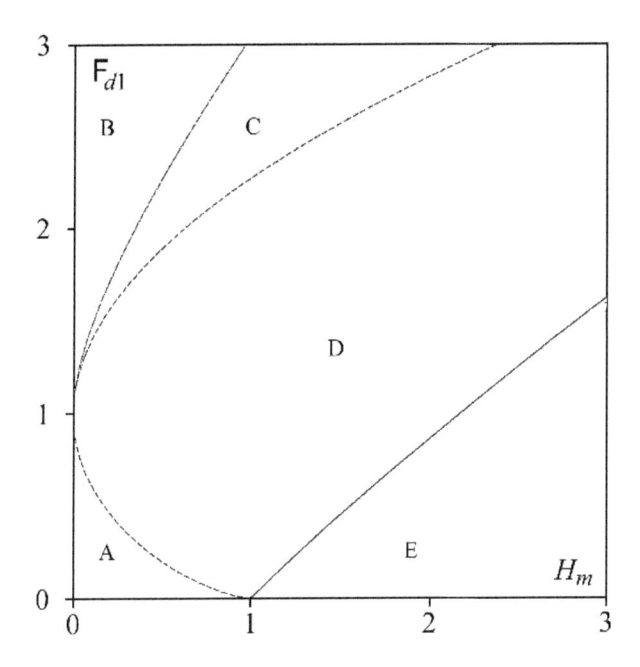

Figure 9.40 Flow regimes of shallow-layer flow over obstacle in terms of $H_m = h_m/h_1$ versus
F_{d1} (Adapted from Oehy, 2003)

hydraulic jump follow Bélanger's equation for single-phase open-channel flow based on the densimetric Froude number F_{d1}

$$\frac{h_2}{h_1} = \frac{1}{2}\left((1+8F_{d1}^2)^{1/2} - 1\right). \tag{9.30}$$

In region E of Figure 9.40, the obstacle is high enough to completely block the approach turbidity current. Figure 9.41 shows the corresponding flow features in a laboratory test, resembling supercritical water flow over a sill (Chapter 5).

9.6.4 Flow across screen

The interaction of a steady gravity current with a screen is solved by a similar method as used for the obstacle (Figures 9.42, 9.43), depending mainly on the screen porosity. Since the screen is permeable, the current does not climb as high as at an obstacle (Rottman *et al.*, 1985).

The hydraulics of the turbidity current across a screen is presented below following Oehy (2003), thereby excluding deposition. In Figure 9.42, h_1 and U_1 are the upstream (subscript 1) flow depth and velocity U_1, U_j is the upstream propagation velocity of a hydraulic jump, h_2 and U_2 are the corresponding current flow depth and velocity at screen upstream vicinity, and h_3 and U_3 the tailwater depth and velocity. Further, f represents screen porosity.

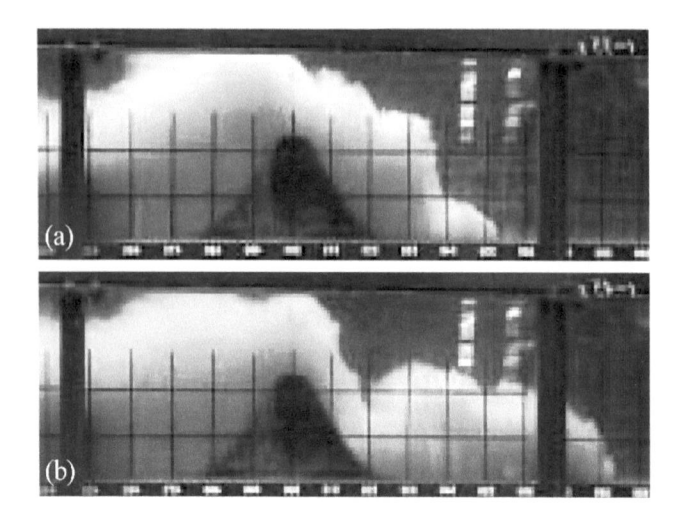

Figure 9.41 Turbidity current over rigid obstacle in laboratory flume (a) 20 s, (b) 40 s after impacting obstacle toe (Adapted from Oehy and Schleiss, 2007)

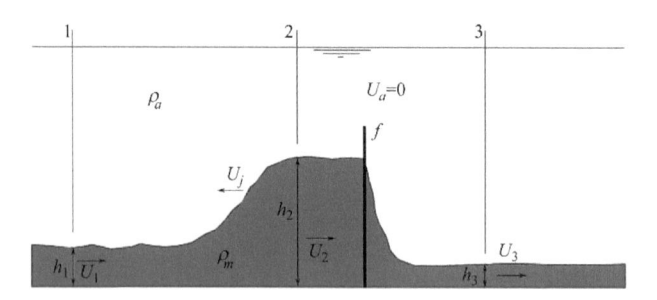

Figure 9.42 Flow across screen, notation (Adapted from Oehy, 2003)

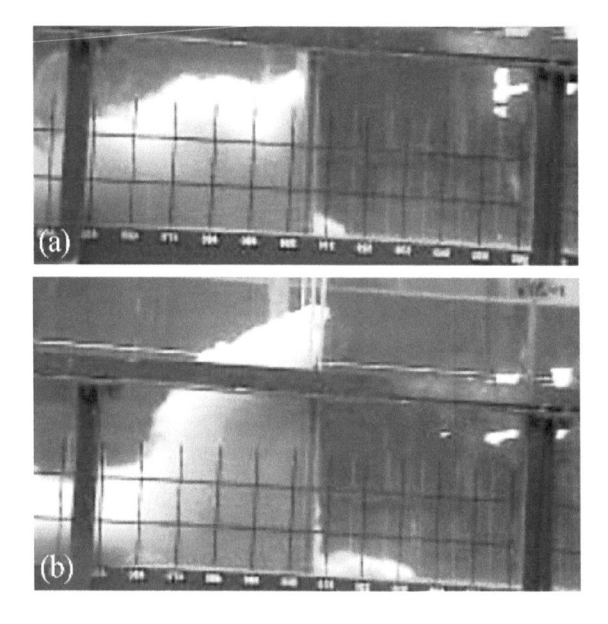

Figure 9.43 Turbidity current across porous laboratory screen (a) 10 s, (b) 20 s after impacting the obstacle toe (Adapted from Oehy *et al.*, 2007)

The ratio $H_p = h_3/h_2$ between the down- and upstream flow depths varies with the porosity f as shown in Figure 9.44. Note the almost linear relation between H_p and f. The relation $H_j = h_2/h_1$ in terms of screen porosity and F_{d1} follows from Figure 9.45.

The ratio η of the turbidity current crossing the screen to the approach flow is

$$\eta = \frac{U_2 h_2}{U_1 h_1}.$$ (9.31)

Figure 9.46 shows the relation η (f, F_{d1}) from which a relatively small effect of the upstream densimetric Froude number is noted if $F_{d1} < 0.7$.

The unit force F exerted on the screen versus its porosity f and the upstream turbidity current depth h_2 is (Oehy, 2003)

$$F = \left[(\frac{1}{2} - \frac{f}{1+f})(1-f^2) \right] \rho g' h_2^2.$$ (9.32)

Note that the relative force $F/(\rho g' h_2^2)$ depends only on the screen porosity f.

9.6.5 Control by opposing jets

Oehy et al. (2010) explored the backing-up of turbidity currents in reservoirs by means of opposing submerged jets. Their simulations and experiments indicated that these jets stop turbidity currents forcing sedimentation upstream of it. Bühler et al. (2013) assumed that the: (1) slope and sedimentation over the control section are negligible, (2) jets emerge from a number of equidistant nozzles arranged in the spanwise direction, and (3) they are inclined at an angle θ relative to the bed (Figure 9.47). The jets are associated with a unit discharge q, and a momentum flux m per unit width, regardless the number of nozzles, so that the jets are replaced by an equivalent line source (from a slot). The upper fluid layer is assumed to be at rest ($u_a = 0$).

The conservation equations for the fluxes of volume, momentum, and buoyancy are

$$\gamma h_1 u_1 + q = \gamma h_2 u_2,$$ (9.33)

$$\frac{1}{2} g'_1 h_1^2 + \beta \gamma u_1^2 h_1 - m\cos\theta = \frac{1}{2} g'_2 h_2^2 + \beta \gamma u_2^2 h_2,$$ (9.34)

$$g'_1 h_1 u_1 = g'_2 h_2 u_2 = B.$$ (9.35)

The momentum coefficient β and the factor modifying the current depth γ are assumed to be identical on both sides of the injection (subscript i). With $q_i = \gamma u_i h_i$ follows from mass conservation $q_1 + q = q_2$. Let the injection Richardson number be $Ri_i = Ri_i^*/(\beta\gamma)$, so that $Ri_i = 1$ for critical flow. The flux $\gamma u_i^2 h_i$ is expressed as $\{B/(\beta\gamma Ri_i)\}^{1/3} q_i$, from which (Bühler et al., 2013)

$$\frac{1}{Ri_1^{1/3}} (1 + \frac{1}{2} Ri_1) = \frac{q_2}{q_1} \frac{1}{Ri_2^{1/3}} (1 + \frac{1}{2} Ri_2) + \frac{\gamma^{1/3} m\cos\theta}{q_1 (\beta^2 B)^{1/3}}.$$ (9.36)

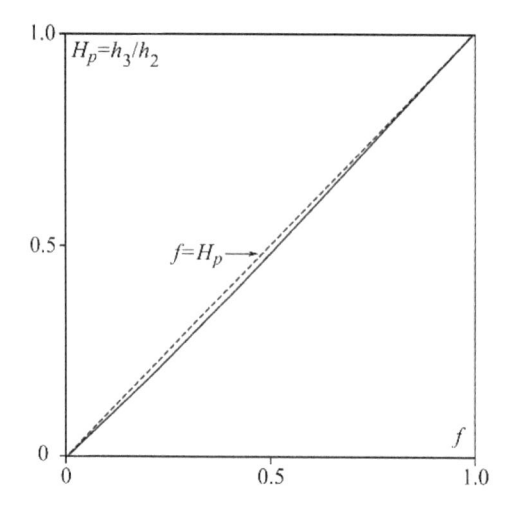

Figure 9.44 Ratio of density current heights down- and upstream of screen $H_p = h_3/h_2$ versus effective porosity f with $(-)$ solution of Oehy (Adapted from Oehy, 2003)

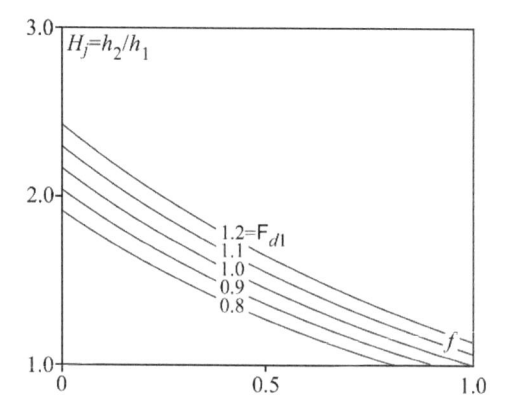

Figure 9.45 $H_j = h_2/h_1$ versus screen porosity f and F_{d1} (Adapted from Oehy, 2003)

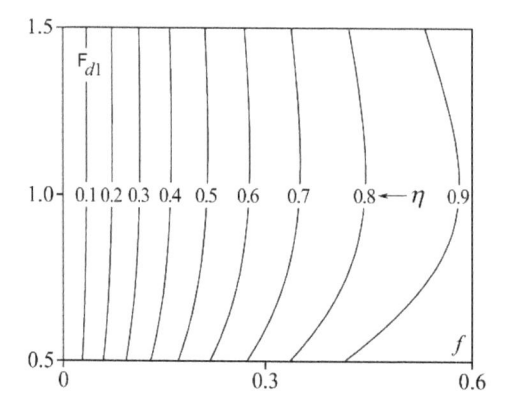

Figure 9.46 Ratio of incoming turbidity current crossing screen versus effective porosity f and upstream densimetric Froude number F_{d1} (Adapted from Oehy, 2003)

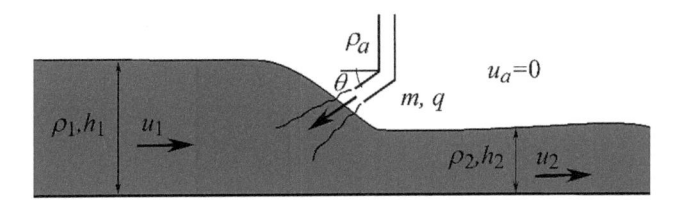

Figure 9.47 Definition sketch of jet opposing a turbidity current (Adapted from Bühler et al., 2013)

For free-surface flow $g' = g$, $\gamma = 1$, so that $u_i^2 h_i$ is expressed as $\{gq_i/(\beta \mathrm{Ri}_i)\}^{1/3} q_i$, and

$$\frac{1}{\mathrm{Ri}_1^{1/3}}(1+\frac{1}{2}\mathrm{Ri}_1) = \left(\frac{q_2}{q_1}\right)^{4/3} \frac{1}{\mathrm{Ri}_2^{1/3}}(1+\frac{1}{2}\mathrm{Ri}_2) + \frac{m\cos\theta}{(\beta^2 gq_1^4)^{1/3}}.$$ (9.37)

Solutions result by substituting $\mathrm{Ri}_1 = z_s^3$, and by denoting the right-hand side of Eq. (9.36) or (9.37) by r, leaving $z_s^3 - 2rz_s + 2 = 0$. A further substitution $z_s = 2(2r/3)^{1/2}\cos\omega$ leads to solutions of the form $\cos(3\omega) = -[3/(2r)]^{3/2}$, so that $\omega = \arccos[-3/(2r)]^{3/2}/3$. Solutions $z_s = \mathrm{Ri}_1^{1/3} = 2(2r/3)^{1/2}\cos\omega$ represent the subcritical branch in Figure 9.48, whereas $z_s = \mathrm{Ri}_1^{1/3} = 2(2r/3)^{1/2}\cos(\omega + 4\pi/3)$ applies to supercritical flow (Figure 9.49). An analogous procedure applies to cases for which the downstream flow conditions, and Ri_2 are known. From Eq. (9.36) follows for gravity currents

$$\frac{1}{\mathrm{Ri}_2^{1/3}}(1+\frac{1}{2}\mathrm{Ri}_2) = \frac{q_1}{q_2}\frac{1}{\mathrm{Ri}_1^{1/3}}(1+\frac{1}{2}\mathrm{Ri}_1) - \frac{\gamma^{1/3}m\cos\theta}{q_2(\beta^2 B)^{1/3}}.$$ (9.38)

Computational example

Consider an undisturbed turbidity current in a reservoir flowing along the thalweg as sketched in Figure 9.47 with velocity $u_0 = 0.6$ m/s and depth $h_0 = 2$ m just upstream of the location of opposing jet nozzles (Bühler *et al.*, 2013). The sediment concentration according to typical turbidity currents is assumed 1%, corresponding to $g'_0 = 0.16$ m/s^2, and $B_0 = 0.19$ m^3/s^3. With $\beta = 1.07$ and $\gamma = 1.4$ this results in $\mathrm{Ri}_0 = 0.59$, whereas the unit width flux is $q_0 = 1.68$ m^2/s. A unit injection stream of $q = 0.05$ m^2/s is available releasing a momentum of $m = 0.5$ m^3/s^2 at a nozzle inclination of $\theta = 10°$. To what value Ri_1 is the upstream flow backed up?

The buoyancy loss in the backed up flow region is estimated to 50%, so that $B_1 = B_2 = 0.096$ m^3/s^3; neglecting entrainment gives $q_0 = q_1$. The downstream Richardson number is assumed equal to that of the undisturbed current on the same slope, i.e. $\mathrm{Ri}_2 = \mathrm{Ri}_0$. The factor r according to Eq. (9.37) then is $r = 2.27$. As the upstream flow is less supercritical than that downstream, the upper branch of Figure 9.27 applies, with $\mathrm{Ri}_1^{1/3} = 1.85$, or $\mathrm{Ri}_1 = 6.33$. Therefore, the value of $u_1 = [B_1/(\mathrm{Ri}_1 \beta \gamma)]^{1/3} = 0.22$ m/s along with $h_1 = q_1/(\gamma u_1) = 5.53$ m. The total flow depth thus increases from $\gamma h_0 = 2.8$ m to about $\gamma h_1 = 7.74$ m due to the back-up effect. Determining the relevant flow parameters is obviously more difficult for internal flows than for flows in a stilling basin. The buoyancy loss

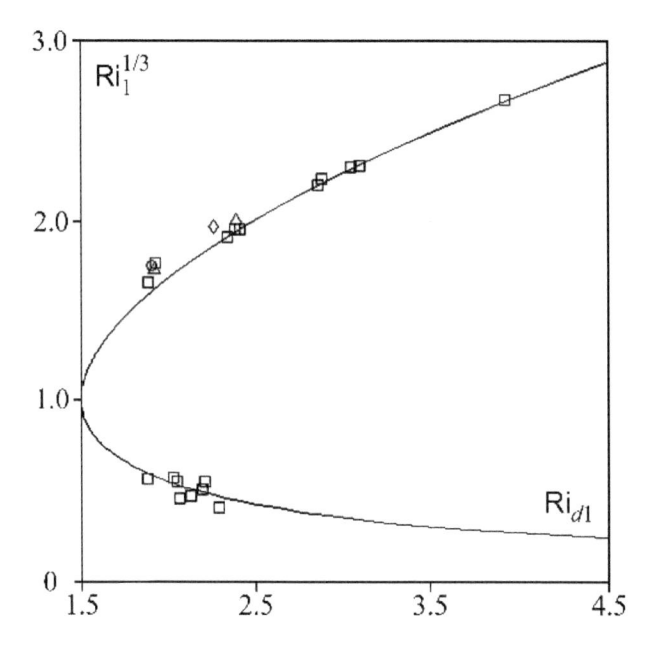

Figure 9.48 Relation between downstream condition $\mathrm{Ri}_{d1} = (q_2/q_1)^{4/3}\mathrm{Ri}_2^{-1/3}(1 + \mathrm{Ri}_2/2) + m\cos\theta(\beta^2 gq_1^4)^{1/3}$ and upstream Richardson number Ri_1 from (−) Eq. (9.37), comparison with experiments using (□) eight 5 mm pipes, (○) four 7.5 mm pipes, (◇) eight 10 mm pipes, and (△) two 10 mm pipes (Adapted from Bühler *et al.*, 2013)

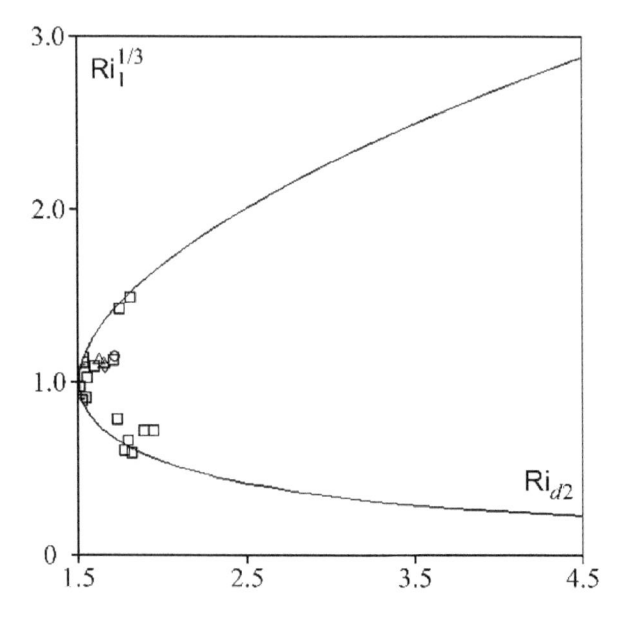

Figure 9.49 Relation between upstream condition $\mathrm{Ri}_{d2} = (q_1/q_2)^{4/3}\mathrm{Ri}_1^{-1/3}(1 + \mathrm{Ri}_1/2) - m\cos\theta(\beta^2 gq_2^4)^{1/3}$ and downstream Richardson number Ri_2. Symbols as in Figure 9.48 (Adapted from Bühler *et al.*, 2013)

in particular follows from an iterative procedure by using numerical models, including ANSYS-CFX12 or FLOW3D. Note, however, that the jets also help to keep fine sediments in suspension after a turbidity current has died out, or during normal operation in turbid reservoirs.

9.6.6 Intrusion

If the densities of the current and the stratified ambient become equal, the density current leaves the bottom and propagates horizontally into the reservoir (Figure 9.36a). Let F_i be the densimetric Froude number at the intrusion (subscript i) point, and G the *Grashof number* involving buoyancy frequency, reservoir length, and the average vertical eddy viscosity. Depending on the dimensionless number

$$I = F_i G^{1/3} , \tag{9.39}$$

three flow regimes occur (Alavian *et al.*, 1992):

1 $I>1$, inertial and buoyancy forces are in equilibrium, propagation speed is $c_i = 0.194(g_i' h_i)^{1/2}$, and the inflow thickness $h_i = 3(q_i^2/g_i')^{1/3}$,
2 $P^{-5/6}<I<1$, with P as *Prandtl number*, viscosity and buoyancy forces dominate, and
3 $I<P^{-5/6}$, viscosity and diffusion dominate.

9.7 Sedimentation control

9.7.1 Turbulent suspension

For a given reservoir, a definite relation between average sediment concentration and average stream power exists, defined as velocity times gradient upstream from the cross section considered. As the sediment capacity decreases within the reservoir backwater zone, the sediment load decreases rapidly with distance, so that material in colloidal suspension reaches the dam (ICOLD, 1989). Figure 9.50 shows that the vertical distribution of sediment concentration across the Verwoerd Reservoir (South Africa) is nearly uniform whereas for other reservoirs a typical density current occurs. Figure 9.51 shows observations on the Sautet Reservoir (France).

Fine particles in reservoirs are mainly distributed due to turbulent suspension. The variation of concentration C in the vertical direction z under equilibrium conditions depends on the turbulent diffusion D_d and is expressed as (Rouse, 1937)

$$\frac{C}{C_a} = \left(\frac{D_d - z}{z} \frac{z_a}{D_d - z_a} \right)^{m_d} . \tag{9.40}$$

Here C_a is the concentration at the reference location $z = z_a$ above the bed, with the exponent m_d equal to

$$m_d = \frac{2.1 V_s}{(gD_d S_e)^{1/2}} = 2.1N. \tag{9.41}$$

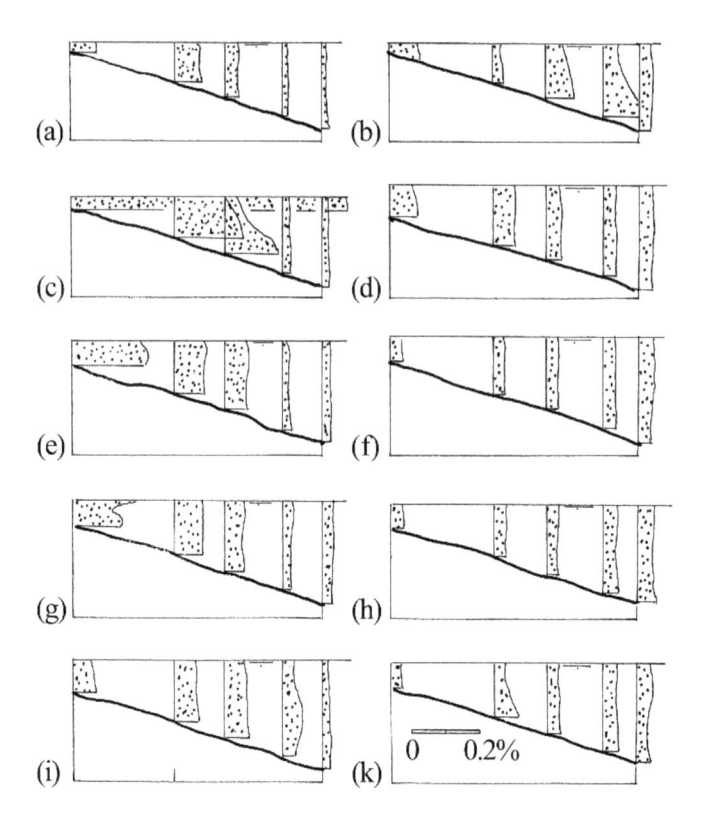

Figure 9.50 Distribution of sediment concentration in Verwoerd Reservoir (SA) for various reservoir in- and outflows, (a) to (k) refer to various reservoir in- and outflow configurations in 1974 (Adapted from ICOLD, 1989)

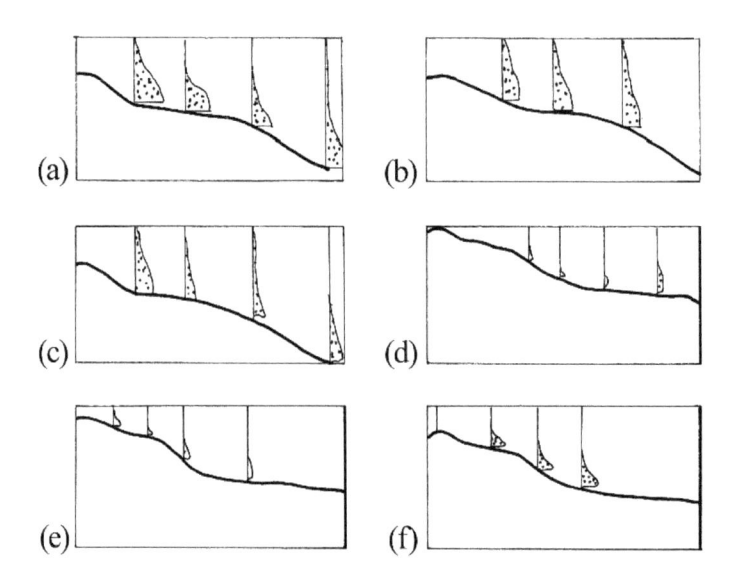

Figure 9.51 Variations of sediment concentrations at Sautet Reservoir (France) in time and space (Adapted from ICOLD, 1989)

Here V_s is the settling velocity of particles considered, D the flow depth, and S_e the energy line slope. With $V^* = (gD_dS_e)^{1/2}$ as shear velocity, m_d may be regarded as a ratio of two velocities. For $m_d \to 0$ the variation of sediment concentration is negligible, resulting in a situation as shown in Figure 9.50 d, f, h. The limit value for the number $N = V_s/(gD_dS_e)^{1/2}$ is 8.3, i.e. $m_d = 17.4$. For $N > 8.3$ the movement of sediment through the turbulent suspension ceases; the sediment concentration along the bed is much higher than close to the reservoir surface.

9.7.2 Recommendations on turbidity current venting

A turbidity current develops if a layer of sediment-laden fluid moves in ambient fluid of slightly less density. The two layers are separated by a sharp discontinuity. These currents typically occur if large quantities of sediment are transported into a reservoir, as during floods where the currents are able to penetrate deeply into the reservoir.

The driving force of turbidity currents is the density difference $\Delta\rho$ associated with the sediment concentration times the bed slope S_o; the resisting force is equal to density ρ times the slope S_f of the energy line. The *density number*

$$D = \frac{\Delta\rho S_o}{\rho S_f} \tag{9.42}$$

corresponds to an indicator of the relative importance of the two effects. Therefore, turbidity currents are enhanced provided:

- Density differences are large,
- Flow depth is large,
- Bed slope is large, and
- Velocity is low.

Favorable conditions for turbidity currents are thus heavily charged floods entering a deep reservoir of steep bed slope. Hydraulic criteria include (ICOLD, 1989)

$$F_d < 1, \tag{9.43}$$

$$\frac{S_o C^2 D_d^5}{Q^2} > 10^4. \tag{9.44}$$

Equation (9.43) involves the *densimetric Froude number* $F_d = V/(g'D_d)^{1/2}$ with $g' = (\Delta\rho/\rho)g$ as reduced gravity acceleration. Accordingly, the turbidity current should remain subcritical. Based on D from Eq. (9.31), Eq. (9.44) determines whether a constant density ratio $\Delta\rho/\rho$ is assumed if the Chézy friction formula with C_{Ch} as friction coefficient is adopted.

For many reservoirs designed for a, say, 50-years sediment dead storage, turbidity currents are not that significant regarding volume losses because the sediment is mainly moving along the foreset slope, i.e. shifting of material from up- to downstream locations takes place within the reservoir. However the problem is that at each turbidity current event a high amount of sediment is deposited just upstream of the dam, affecting only after some decades the operation of bottom outlets and power intakes. The control of turbidity currents within a reservoir by obstacles and water jet screens is described in 9.6.3 to 9.6.5, whereas venting of turbidity current has also been outlined.

The effect of the outlet discharge or the capacity of the low-level outlets on the efficiency of turbidity current venting has been studied systematically by Chamoun (2017) both with flume experiments and numerical simulations. The influence of the venting degree, defined as the ratio between outflow and turbidity current discharges, on the efficiency of venting, was highlighted. The optimum venting degree depends on the reservoir slope at the outlet vicinity. For turbidity currents reaching the outlet on a horizontal bed in the experimental configuration, a venting degree of nearly 100% resulted in the highest venting efficiency (Chamoun et al., 2016b). For steeper reservoir bed slopes (i.e., 2.4% and 5.0%), the optimum efficiency is obtained with a venting degree of about 135%. Within the range of slopes tested, it was noted that the steeper the reservoir bed slope at the outlet vicinity, the more efficient is venting. The main reason is that the upstream reflection of the current by the dam decreases under higher slopes. Thus, venting of turbidity currents should be started from the very start of dam operation. This helps in maintaining a cone at the dam vicinity avoiding the filling of the dead storage and thus the development of a flatter bed.

The following recommendations are given for achieving high venting efficiencies (Chamoun, 2017):

- The timing or start of venting should be synchronized with the arrival of the turbidity current at the dam. In any case, an early venting is more efficient than a late venting. Venting should be started latest when the turbidity current reaches the aspiration cone of the bottom outlet having a length of typically five times the aspiration height of the outlet. In practice, one of the most important conditions for a successful venting operation is to have measurements which allow to detect turbidity currents and estimate their arrival time. Based on the required time to open the gate and reach a stable flow ($\sim$ 5–10 minutes) and for typical velocities of turbidity currents (0.3 to 1 m/s), it is highly recommended to measure velocities and/or concentrations close to the reservoir bed, at a location of about 200–300 m upstream of the outlet.
- Venting should last as long as there is a turbidity current approaching the dam. Furthermore, it should be maintained after the end of the flood event entering the reservoir to avoid clogging of the riverbed downstream by fine sediment in high concentrations. Thus, venting should be continued until outflow concentrations become low (< 0.5 g/l) to 'rinse' the downstream river with clear water.
- The bottom outlet should be positioned in a way to minimize the dead storage since a low position of the bottom outlet results in higher venting efficiencies.
- The dimensions of the bottom outlet(s) should be such that the aspiration zone created by the opening(s) englobe the tail of the turbidity behind its head.

Venting of turbidity currents can be combined when appropriate with drawdown flushing for delta mobilization. Venting releases mainly fine sediments such as silt and clay, while the downstream river needs also coarse sediments for a healthy ecosystem. Fine material alone can harm fish habitats in the downstream river. Therefore, replenishment techniques that supply coarse sediments to the downstream environment should be considered along with venting (Battisacco et al., 2016).

In reservoirs where turbidity currents form during floods, it is recommended to first open bottom outlets before operating the spillway. This leads to the evacuation of possible turbidity currents reaching the dam or at least unblocks the outlets, freeing a cone upstream which might have been filled with sediments in past events.

9.7.3 Sediment flushing

Sediments are flushed out of a reservoir if sufficient transport capacity is available. In the absence of density currents, the typical concentrations during flushing are of the same order as these encountered during river floods. The transport capacity depends on the grain (subscript g) Reynolds number $R_g = (gD_dS_f)^{1/2}(d_{50}/v)$, with d_{50} as the median particle size and v as kinematic viscosity. For $R_g<13$, a *laminar* boundary layer develops in the reservoir and sediment transport occurs if (ICOLD, 1989)

$$N < 0.63R_g, \quad \text{for } R_g < 13.$$ (9.45)

For a *turbulent* boundary layer as generally developed in practice, the condition for incipient sediment transport is

$$N < 8.3, \text{ for } R_g \geq 13.$$ (9.46)

These equations are valid for unconsolidated particles with $N = V_s/(gD_dS_e)^{1/2}$ as defined above. Velocities larger than 1 m/s are required to re-entrain particles after deposition, whereas deposition occurs if the velocity is smaller than 0.5 m/s. The mechanism of sediment flushing across the dam is described in 9.4.3.

9.7.4 Selection of reservoir geometry and locations of inlets and outlets

The geometry of shallow reservoirs has a strong effect on the flow pattern and sedimentation process of suspended sediments (Camnasio *et al.*, 2011, 2013). The trap efficiency and the sedimentation index can be expressed with empirical relations as a function of a geometrical shape factor defined by Kantoush (2008). Furthermore, the flushing efficiency for free flow and drawdown flushing can be also linked with geometrical shape factors (Kantoush and Schleiss, 2009; Kantoush *et al.*, 2008a, b). Thus, if there is freedom to select an optimum geometry of shallow reservoirs in the design procedure, as it is often the case for compensation basins, then the sedimentation process can be reduced significantly by using appropriate shapes.

Furthermore, the location of outlets and inlets in shallow but also in deep reservoirs can have an influence on the reservoir hydrodynamics and consequently can influence strongly sedimentation (Guillen *et al.*, 2017). For example, if a reservoir is completed with additional intakes or inlets for a pumped-storage power plant, their location should be chosen, based on hydrodynamic modeling, so that a kind of short-circuit between inlet and outlet occurs.

9.8 Secondary hydraulic effects

9.8.1 Upstream river

Due to the reservoir backwater into the approach flow river, the reservoir delta grows and river aggradation develops. The prediction of delta growth is important thus for taking measures against sedimentation. The US Bureau of Reclamation (Strand and Pemberton, 1982) proposed a semi-empirical method involving the original bottom slope and the topset bed slope. The point of intersection P between the topset and the foreset slopes is established as median operating reservoir level (Figure 9.52).

A consequence of delta formation is the growth of grass and bushes. This anticipated dense growth of vegetation has an obstructive effect on floods and should be included in the roughness estimation for backwater prediction. Figure 9.53 shows reservoir sedimentation of Lake Ksob (Algeria), Lake Mead (USA) and the Bill Williams River entering Lake Havasu (USA). The light color represents the delta intrusion into the reservoir with the prominent phreatophytic growth.

9.8.2 Downstream river

The release of clear water with a capacity to pick-up a sediment load changes the stable downstream river by degradation. This process moves progressively downstream until reaching a new equilibrium condition. The degradation has undesirable consequences (ICOLD, 1989):

- In-channel structures including bridges or culverts are subject to erosion,
- Endangerment of valuable agricultural, industrial, and residential properties within channel banks,
- Coarsening of bottom material and change of vegetation along channel banks with effect on flora, and
- Erosion of tributary rivers due to drop of hydraulic control.

USBR recommends two methods to counter a degradation process. The *Armoring Method* involves the application of large and coarse material not transported by normal river discharge. Under this process the finer material is sorted out and vertical degradation slows down until the armor is sufficiently deep. The armoring depth is roughly three armoring particle diameters, or 0.15 m, whichever is smaller.

The *Stable Slope Method* involves a stream slope for which the bed material is no longer transported. The resulting slope results from the Meyer-Peter and Müller (MPM) sediment transport formula, or Shield's diagram for no motion (Simons and Senturk, 1992; Julien, 2010). The design discharge is either the bank-full flow, or the 2-years flood peak discharge.

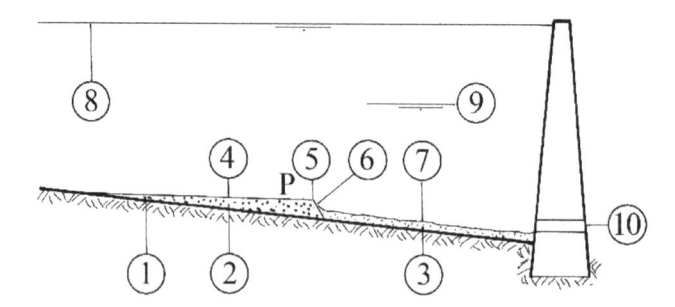

Figure 9.52 Delta formation in reservoir due to sedimentation ① coarse sediment, ② original bed slope, ③ fine sediment, ④ topset slope, ⑤ pivot point P, ⑥ foreset slope, ⑦ bottomset slope, ⑧ maximum reservoir level, ⑨ normal reservoir level, ⑩ bottom outlet (Adapted from Strand and Pemberton, 1982)

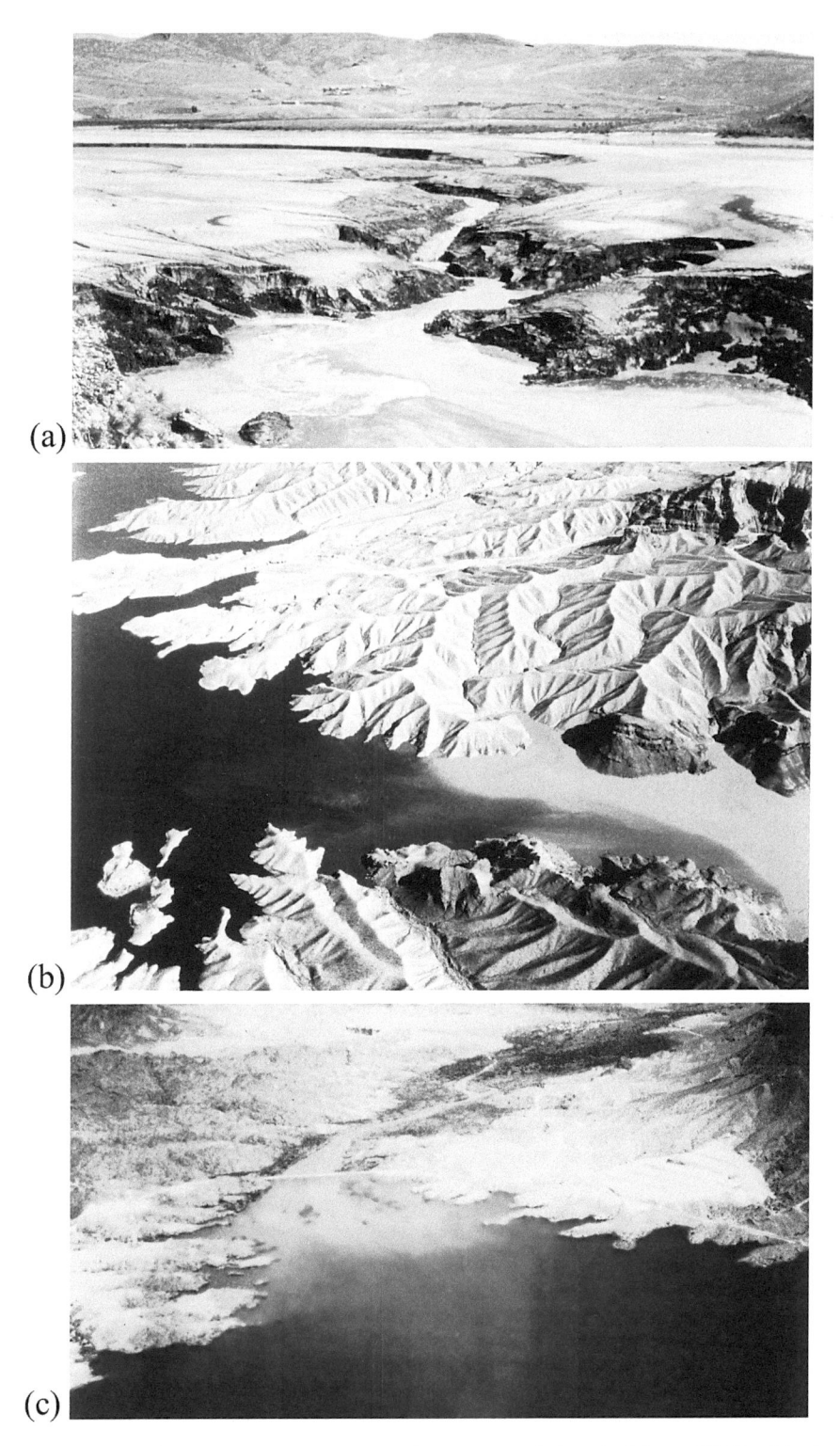

Figure 9.53 (a) Lake Ksob in Algeria, (b) Lake Mead on Colorado River (USA) with plunging front on right side, (c) delta of Lake Havasu at mouth of Bill Williams River (Adapted from ICOLD, 1989)

(a)

(b)

Figure 9.54 Gebidem Dam, Switzerland (a) upstream end of free-surface flushing (Courtesy VAW 5213, ETH Zurich), (b) intake cone at dam (Courtesy VAW 3686, ETH Zurich)

9.8.3 Replenishment or disposal of sediments

Dredged sediments can be released to the river below the dam in the framework of an eco-morphological replenishment mitigation measure, or may be discharged to the closest disposal area. Discharge to the river without prior removal of the fine sediment portion can result in high turbidity flows detrimental to ecology and recreation. Land disposal can result in spoil piles and drainage from the spoiled material affecting the surface and the groundwater.

A replenishment of adequate gravel material downstream of large dams combined with an artificial flood release for mobilizing sediment can restore the morphology of rivers downstream of dams to some extent. Battissacco *et al.* (2016) performed systematic physical flume experiments to study which arrangements of sediment deposits downstream of a dam are suitable for recreating morphological variability. Recommendations were given regarding the placement of the artificial gravel deposits and the required discharge to mobilize them, with the purpose that deposits recreate morphological bed forms when transported in the river reach downstream. The concept was tested at Rossens Dam in Switzerland (Stähly et al., 2019).

Water quality regulations may prohibit a return of dredged material to the river especially out of flood periods. In the USA, schemes of total containment of the water and sediment mixture have been developed (ICOLD, 1989). The concept consists of a diked pond in which the water evaporates and where the solids are then used as fill material. A common practice is to establish a vegetation to enhance the water loss through evapotranspiration. Sometimes it is difficult to find a suitable location for a pond due to topographic reasons or public acceptance. The *ultima ratio* then leads to:

*   Increase reservoir dead storage by building new bottom outlet or even new intakes at higher elevation, or
*   Detach a bay from the reservoir by auxiliary dam to create a disposal area.

Both solutions result of course in a loss of usable reservoir storage.

Notation

A_r	Vertical abrasion rate (m/s)
B	Tunnel width (m)
B_0	Initial buoyancy flux (m^3/s^3)
b	Approach flow width (m)
C	Concentration (-)
C_a	Reference concentration (-)
C_{Ch}	Chézy friction coefficient (m$^{1/2}$/s)
C_d	Total friction coefficient (-)
c_a	Abrasion coefficient (Pa$^{-0.07}$)
c_i	Current propagation speed (m/s)
D	Density profile correction coefficients (-)
D	Density current number (-)
D_d	Current flow depth (m); also turbulent diffusion (m)
d	Sediment particle diameter (m)
d_{50}	Median grain size (m)
E	Entrainment constant $= V_e/V_d$ (-)
E_0	Alavian's entrainment constant (-)
E_u	Equilibrium entrainment constant (-)
E_w	Water entrainment coefficient (-)
e	Elevation of river cross section (m)

F	Unit force on screen (N/m²)
F_d	Densimetric Froude number $= V/(g'h)^{1/2}$ (-)
F_i	Densimetric Froude number at intrusion point (-)
F_p	Plunge point Froude number $= V_o/(gh_p)^{1/2}$ (-)
f	Screen porosity (-)
f_c	Cylindrical compression strength (Pa)
f_t	Splitting tensile strength of bed material (Pa)
G	Grashof number (-)
g	Gravity acceleration (m/s²)
g'	Reduced gravity acceleration $= (\Delta\rho/\rho_o)g$ (m/s²)
g'_0	Initial reduced current gravity acceleration (m/s²)
H_f	Current front height (m)
H_j	Relative current jump height $= h_2/h_1$ (-)
H_m	Relative obstacle height $= h_m/h_1$ (-)
H_{mc}	Relative obstacle height under critical flow $= h_{mc}/h_1$ (-)
H_p	Ratio between down- and upstream current flow depths $= h_3/h_2$ (-)
h	Flow or current depth (m)
h_0	Initial current depth (m)
h_1	Approach flow depth to density jump (m)
h_2	Sequent depth of density jump (m)
h_3	Tailwater current depth (m)
h_d	Depth of density current (m)
h_i	Current inflow thickness (m)
h_m	Obstacle height (m)
h_{max}	Current depth at location of current maximum velocity (m)
h_p	Flow depth at plunge point (m)
h_t	Current height (m)
h_u	Equilibrium flow depth (m)
I	Dimensionless number composed of F_i and G (-)
k_1, k_2	Correction factors for Young's modulus (-)
k_s	Equivalent sand-roughness height (m)
k_v	Bed material resistance coefficient (-)
L_p	Particle hop length (m)
l_1, l_2	Current constant (-)
m	Lateral slope of trapezoidal section (-); also momentum flux (m³/s²)
m_d	Exponent in diffusion relation (-)
N	Settling process number $= V_s/(gD_dS_e)^{1/2}$ (-)
P	Prandtl number (-)
P_R	Rolling probability (-)
p	Passage of sediment (%)
Q_o	Approach flow discharge (m³/s)
q_0	Initial unit discharge $= h_0U_0$ (m²/s)
q_i	Injection unit discharge (m²/s)
q_s	Specific gravimetric bed load rate (kg/sm)
q_s^*	Specific gravimetric bed load transport capacity (kg/sm)
R	Reynolds number $= uh/v$ (-)
R_g	Grain Reynolds number $= (gD_dS_f)^{1/2}(d_{50}/v)$ (-)
Ri	Richardson number $= g'h_d\cos\theta/V_d^2$ (-)
Ri_0	Initial current Richardson number (-)
Ri_u	Equilibrium Richardson number (-)
r	Right-hand side expression of Eqs. (9.25, 9.26) (-)

S_e	Energy line slope (-)
S_f	Friction slope (-)
S_o	Bottom slope (-)
s	Relative sediment density (-)
T^*	Excess transport stage (-)
U_0	Initial current velocity (m/s)
U_1	Approach flow velocity to density jump (m/s)
U_2	Sequent velocity of density jump (m/s)
U_3	Tailwater current velocity (m/s)
U_a	Ambient fluid velocity (m/s)
U_f	Current front velocity (m/s)
U_j	Speed of moving density jump (m/s)
U_{max}	Maximum velocity of current (m/s)
u	Streamwise velocity (m/s)
u_0	Initial current velocity (m/s)
u_{*b}	Bottom friction velocity (m/s)
V^*	Shear velocity (m/s)
V_d	Velocity of density current (m/s)
V_e	Entrainment velocity (m/s)
V_{im}	Mean particle impact velocity (m/s)
V_o	Approach flow velocity (m/s)
V_p	Mean particle velocity (m/s)
V_s	Settling velocity (m/s)
V_u	Equilibrium flow velocity (m/s)
W_{im}	Vertical impact velocity (m/s)
x	Streamwise coordinate (m)
Y_M	Young's modulus of elasticity (Pa)
z	Vertical coordinate (m)
z_a	Reference elevation above bed (m)
z_s	Substitution = $\mathrm{Ri}_1^{1/3}$ (-)
β	Momentum coefficient (m^3/s^2)
γ	Modifying current depth factor (-)
η	Ratio of relative turbidity current crossing screen (-)
v	Kinematic viscosity (m^2/s)
$\Delta\rho$	Density difference between inflow and ambient fluid (t/m^3)
ρ_1	Approach flow water-sediment density (t/m^3)
ρ_2	Tailwater water-sediment density (t/m^3)
ρ_a	Ambient water density (t/m^3)
ρ_c	Concrete density (t/m^3)
ρ_m	Water-sediment density (t/m^3)
ρ_o	Approach flow water-sediment density (t/m^3)
ρ_w	Water density (t/m^3)
θ	Slope of incline (-)
θ_S	Shields' parameter (-)
θ_{Sc}	Critical Shields' parameter (-)
σ_{bed}	Bed coarseness index (-)
σ_{mh}	Microhabitat richness index (-)
σ_{org}	Organic versus inorganic habitat richness (-)
τ	Bed-shear stress (N/m^2)
ω	Substitution = $\arccos[-3/(2r)]^{3/2}/3$ (-)

References

Abgottspon, A., Staubli, T., Felix, D., Albayrak, I. & Boes, R.M. (2014) Monitoring suspended sediment and turbine efficiency. *Hydro Review Worldwide*, 22(4), 28–36.

ACI, American Concrete Institute (2013) *ACI Concrete Terminology*, An ACI Standard. Report ACI CT-13, Farmington Hills, MI, USA.

Alam, S. (1999) The influence and management of sediment at hydro projects. *Hydropower & Dams*, 6(3), 54–57.

Alavian, V., Jirka, G.H., Denton, R.A., Johnson, M.C. & Stefan, H.G. (1992) Density currents entering lakes and reservoirs. *Journal of Hydraulic Engineering*, 118(11), 1464–1489.

Albayrak, I., Felix, D., Hagmann, M., & Boes, R.M. (2015) Suspended sediment and bed load transport monitoring techniques. Proc. 38th *Dresdner Wasserbaukolloquium*, Dresdner Wasserbauliche Mitteilungen, Heft 53, 405–414.

Altinakar, M.S., Graf, W.H. & Hopfinger, E.J. (1990) Weakly depositing turbidity currents on a small slope. *Journal of Hydraulic Research*, 28(1), 55–80.

Altinakar, M.S., Graf, W.H. & Hopfinger, E.J. (1996) Flow structure in turbidity currents. *Journal of Hydraulic Research*, 34(5), 713–718.

Annandale, G.W. (1984) Predicting the distribution of deposited sediment in southern African reservoirs. Proceedings of the International Symposium on *Challenges in African Hydrology and Water Resources*, Harare, Zimbabwe. *IAHS Publication*, 144, 549–558.

Annandale, G.W. (1987) *Reservoir sedimentation*. Elsevier, Amsterdam.

Annandale, G.W. (1996) Spatial distribution of deposited sediment in reservoirs. Proceedings of the International Conference on *Reservoir Sedimentation*. Colorado State University, Fort Collins.

Annandale, G.W. (2013) Quenching the thirst: Sustainable water supply and climate change, *Createspace*, ISBN-10: 1480265152.

Arioglu, N., Canan Girin, Z., & Arioglu, E. (2006) Evaluation of ratio between splitting tensile strength and compressive strength for concretes up to 120 MPa and its application in strength criterion. *ACI Materials Journal*, 103(1), 18–24.

Auel, C. (2014) Flow characteristics, particle motion and invert abrasion in sediment bypass tunnels. *VAW Mitteilung* 229, R.M. Boes, ed. Versuchsanstalt für Wasserbau, Hydrologie und Glaziologie, ETH Zurich, Zürich.

Auel, C., Albayrak, I., & Boes, R.M. (2014) Turbulence characteristics in supercritical open-channel flows: Effects of Froude number and aspect ratio. *Journal of Hydraulic* Engineering, 140(4), 04014004, 16 pages.

Auel, C., Berchtold, T., & Boes, R. (2010) Sediment management in the Solis reservoir using a bypass tunnel. Proc. of 8th *ICOLD European Club Symposium*, Innsbruck, Austria, 455–460 (hard copy), 438–443 (CD).

Auel, C., & Boes, R.M. (2011) Sediment bypass tunnel design – review and outlook. Proc. ICOLD Symposium *Dams under changing challenges* (A.J. Schleiss & R.M. Boes, eds.). Proc. 79th Annual Meeting, Lucerne. Taylor & Francis, London, 403–412.

Auel, C., Albayrak, I., Sumi, T., & Boes, R.M. (2017a) Sediment transport in high-speed flows over a fixed bed: 1. Particle dynamics. *Earth Surface Processes and Landforms*, 42(9), 1365–1383, DOI: 10.1002/esp.4128.

Auel, C., Albayrak, I., Sumi, T., & Boes, R.M. (2017b) Sediment transport in high-speed flows over a fixed bed: 2. Particle impacts and abrasion prediction. *Earth Surface Processes and Landforms*, 42(9), 1384–1396, DOI: 10.1002/esp.4132.

Auel, C., Boes, R.M. & Sumi. T. (2018a) Invert abrasion prediction at Ashai sediment bypass tunnel based on Ishibashi's formula. *Journal of Applied Water Engineering and Research*, 6(2), 125–138, https://doi.org/10.1080/23249676.2016.1265470.

Auel, C., Kobayashi, S., Takemon, Y., & Sumi. T. (2017c) Effects of sediment bypass tunnels on grain size distribution and benthic habitats in regulated rivers. *International Journal of River Basin Management* 15(4), 433–444. https://doi./10.1080/15715124.2017.1360320.

Auel, C., Kantoush, S.A. & Sumi, T. (2016) Positive effects of reservoir sedimentation management on reservoir life: Examples from Japan. Proceedings of the *84th Annual Meeting of ICOLD*, Johannesburg, South Africa, 4.11–4.20.

Auel, C., Thene, J.R., Carroll, J., Holmes, C. & Boes, R.M. (2018b) Rehabilitation of the Mud Mountain bypass tunnel invert. Proceedings of the 26th *ICOLD Congress*, Vienna, Austria Q100(R4), 51–71.

Baines, P.G. (1995) *Topographic effects in stratified flow*. Cambridge University Press, Cambridge.

Baumer, A., & Radogna, R. (2015) Rehabilitation of the Palagnedra sediment bypass tunnel (2011–2013). Proceedings of the 1st international Workshop on Sediment Bypass Tunnels, Zürich. *VAW Mitteilung* 232, 235–245, R.M. Boes, ed. Versuchsanstalt für Wasserbau, Hydrologie und Glaziologie, ETH Zurich, Zürich.

Basson, G.R. (2009) Management of siltation in existing and new reservoirs. Proceedings of the 23rd *ICOLD Congress*, Brasilia, 2, General Report Q89 (on CD).

Basson, G.R. & Rooseboom, A. (1997) Dealing with reservoir sedimentation. *Report* 779/971998 (A comprehensive survey report). South African Water Res. Comm. WRC, Pretoria.

Battisacco, E., Franca, M.J. & Schleiss, A.J. (2016) Sediment replenishment: Influence of the geometrical configuration on the morphological evolution of channel-bed. *Water Resources Research*, 52(11), 8879–8894.

Batuca, G.D. & Jordaan, M.J. (2000) *Silting and desilting of reservoirs*. Balkema, Rotterdam.

Beck, C., Lutz, N., Lais, A., Vetsch, D., & Boes, R.M. (2016) Patrind Hydropower Project, Pakistan: Physical model investigations on the optimization of the sediment management concept. Proceedings of Conference *Hydro2016*, Montreux, Switzerland, Paper 26.08.

Beer, A.R., & Turowski, J.M. (2015) Bedload transport controls bedrock erosion under sediment-starved conditions. *Earth Surface Dynamics,* 3: 291–309. DOI: 10.5194/esurf-3-291-2015.

Beyer Portner, N. & Schleiss, A. (2000) Bodenerosion in alpinen Einzugsgebieten in der Schweiz [Soil erosion of the Swiss Alpine regions]. *Wasserwirtschaft*, 90(2), 88–92 (in German).

Boes, R. & Reindl, R. (2006) Nachhaltige Massnahmen gegen Stauraumverlandung alpiner Speicher [Sustainable measures to counter sedimentation of Alpine reservoirs]. Proceedings of the Symposium *Stauhaltungen und Speicher: Von der Tradition zur Moderne. Report*, 46(1), 179–193. Institut für Wasserbau und Wasserwirtschaft, TU Graz, Austria (in German).

Boes, R.M. (2011) Nachhaltigkeit von Talsperren angesichts der Stauraumverlandung [Sustainability of dams in view of reservoir sedimentation]. In: Schüttrumpf, H. (ed) *Mitteilung*, 164, 161–174. Lehrstuhl und Institut für Wasserbau und Wasserwirtschaft, RWTH Aachen, Germany (in German).

Boes, R.M., Auel, C., Hagmann, M. & Albayrak, I. (2014) Sediment bypass tunnels to mitigate reservoir sedimentation and restore sediment continuity. Proceedings of the Conference *Reservoir Sedimentation* (Schleiss, A.J., De Cesare, G., Franca, M.J., Pfister, M., eds.). Taylor & Francis Group, London, UK, 221–228, ISBN 978-1-138-02675-9.

Boes, R.M., ed. (2015) Proc. 1st Intl. Workshop on Sediment Bypass Tunnels, Zürich. *VAW Mitteilung* 232. Versuchsanstalt für Wasserbau, Hydrologie und Glaziologie, ETH Zurich, Zürich, 258 pages.

Boes, R.M., & Hagmann, M. (2015) Sedimentation countermeasures: Example from Switzerland. Proc. 1st Intl. Workshop on Sediment Bypass Tunnels Zürich. *VAW Mitteilung* 232, 193–210, R.M. Boes, ed. Versuchsanstalt für Wasserbau, Hydrologie und Glaziologie, ETH Zurich, Zürich.

Boes, R.M., Felix, D. & Albayrak, I. (2013) Schwebstoffmonitoring zum verschleissoptimierten Betrieb von Hochdruck-Wasserkraftanlagen [Monitoring of suspended sediment to optimize wear at high-head dams]. *Wasser, Energie, Luft*, 105(1), 35–42 (in German).

Boes, R.M., Müller-Hagmann, M., Albayrak, I., Müller, B., Caspescha, L., Flepp, A., Jacobs, F. & Auel, C. (2018) Sediment bypass tunnels: Swiss experience with bypass efficiency and abrasion-resistant invert materials. Proceedings of the 26th *ICOLD Congress*, Vienna, Austria Q100(R40), 625–638.

Boillat, J.-L. & Pougatsch, H. (2000) State-of-the-art of sediment management in Switzerland. International Workshop and Symposium *Reservoir Sedimentation Management*, Tokyama, JP, 35–45.

Boillat, J.-L. & Delley, P. (1992) Transformation de la prise d'eau de Malvaglia: Etude sur modèle et réalisation [Transformation of Malvaglia intake: Model study and realisation]. *Wasser Energie Luft*, 84(7/8), 145–151 (in French).

Borland, W.M. & Miller, C.R. (1958) Distribution of sediment in large reservoirs. *Journal of the Hydraulics Division* ASCE, 84(HY2), 1–18.

Bretz, N.-V. & Barras, M. (2012) Sediment management at Grande Dixence Dam. Proceedings of the International Symposium *Dams for a changing world*, Kyoto, Paper 346, 6p.

Breusers, H.N.C., Klaassen, G.J., Barkel, J. & van Roode, F.C. (1982) Environmental impact and control of reservoir sedimentation. Proceedings of the 14th *ICOLD Congress*, Rio de Janeiro, Q54(R23), 353–372.

Britter, R.E. & Linden, P.F. (1980) The motion of the front of a gravity current travelling down an incline. *Journal of Fluid Mechanics*, 99(3), 531–543.

Brown, C.B. (1944) The control of reservoir silting. *Miscellaneous Publication*, 521. US Department of Agriculture, Washington DC.

Bühler, J., Oehy, C. & Schleiss, A.J. (2013) Jets opposing turbidity currents and open channel flows. *Journal of Hydraulic Engineering*, 139(1), 55–59.

Camnasio, E., Orsi, E. & Schleiss, A.J. (2011) Experimental study of velocity fields in rectangular shallow reservoirs. *Journal of Hydraulic Research*, 49(3), 352–358.

Camnasio, E., Erpicum, S., Orsi, E., Pirotton, M., Schleiss, A.J. & Dewals, B.J. (2013) Coupling between flow and sediment deposition in rectangular shallow reservoirs. *Journal of Hydraulic Research*, 51(5), 535–547.

Carlioz, P., & Peloutier, V. (2014) Implementing a sediment transit gate at Rizzanese Dam. Proc. Intl. ICOLD Symposium *Dams in a Global Environmental Challenges*, Bali, Indonesia, Abstract number 346.

Chamoun, S., De Cesare, G. & Schleiss, A.J. (2016a) Venting turbidity currents for the sustainable use of reservoirs. *International Journal of Hydropower & Dams*, 23(5), 64–69.

Chamoun, S., De Cesare, G. & Schleiss, A.J. (2016b) Managing reservoir sedimentation by venting turbidity currents: A review. *International Journal of Sediment Research*, 31(3), 195–204.

Chamoun, S. (2017) Influence of outlet discharge on the efficiency of turbidity current venting. *Ph.D. Thesis* 7736. Ecole Polytechnique Fédérale de Lausanne. In: Schleiss, A. (ed) *Communication*, 72, Laboratory of Hydraulic Constructions. EPFL: Lausanne, Switzerland.

Chang, H.H. (1979) Minimum stream power and river channel patterns. *Journal of Hydrology*, 41(3–4), 303–332.

Chen, J. & Zhao, K. (1992) Sediment management in Nanqin Reservoir. *International Journal of Sediment Research*, 7(3), 71–84.

Chervet, A., & Vischer, D. (1996) Geschiebeumleitstollen bei Stauseen: Möglichkeiten und Grenzen [Sediment bypass tunnels at reservoirs: Potential and limits]. Proc. Intl. Symposium: Verlandung von Stauseen und Stauhaltungen, Sedimentprobleme in Leitungen und Kanälen, *VAW Mitteilung* 143, 25–43, D. Vischer, ed. Versuchsanstalt für Wasserbau, Hydrologie und Glaziologie, ETH Zurich, Zürich (in German).

Choi, S-U. & Garcia, M.H. (1995) Modeling of one-dimensional turbidity currents with a dissipative-Galerkin finite element method. *Journal of Hydraulic Research*, 33(5), 623–648.

Dawans, P., Charpié, J., Giezendanner, W. & Rufenacht, H.P. (1982) Le dégravement de la retenue de Gebidem: Essais sur modèle et expériences sur prototype [The sediment reduction of Gebidem Dam: Model experiments and prototype experiences]. Proceedings of the 14th *ICOLD Congress*, Rio de Janeiro, Q54(R25), 383–407 (in French).

De Cesare, G. (1998) Alluvionnement des retenues par courants de turbidité [Reservoir sedimentation by density currents]. *Ph.D. Thesis* 1820. Ecole Polytechnique Fédérale de Lausanne. In: Schleiss,

A. (ed) *Communication* 7. Laboratory of Hydraulic Constructions. EPFL, Lausanne, Switzerland (in French).

De Cesare, G. & Lafitte, R. (2007) Outline of the historical development regarding reservoir sedimentation. Proceedings of the 32nd *IAHR Congress*, Venice, CD-Rom.

De Cesare, G., Baumann, R., Zuglian, R. & Binder, F. (2009) Sedimentausleitung aus dem Speicher Gübsensee über die Triebwasserleitung [Sediment diversion from Gübsensee Reservoir through the power waterway]. *Wasser, Energie, Luft*, 101(3), 203–206 (in German).

De Cesare, G., Manso, P., Daneshvari, M. & Schleiss, A.J. (2015) Laboratory research: Bed load guidance into sediment bypass tunnel inlet. Proc. 1st Intl. Workshop on Sediment Bypass Tunnels Zürich. *VAW Mitteilung* 232, 169–179, R.M. Boes, ed. Versuchsanstalt für Wasserbau, Hydrologie und Glaziologie, ETH Zurich, Zürich.

DWA (2006) *Entlandung von Stauräumen* [Removal of reservoir sediments]. Deutsche Vereinigung für Wasserwirtschaft, Abwasser und Abfall e.V., Hennef, Germany (in German).

Facchini, M., Siviglia, A. & Boes, R.M. (2015) Downstream morphological impact of a sediment bypass tunnel – preliminary results and forthcoming actions. Proc. 1st Intl. Workshop on Sediment Bypass Tunnels Zürich. *VAW Mitteilung* 232, 137–146, R.M. Boes, ed. Versuchsanstalt für Wasserbau, Hydrologie und Glaziologie, ETH Zurich, Zürich.

Fan, J. (1985) Methods of preserving reservoir capacity. In: Bruk, S. (ed) *Methods of Computing Sedimentation in Lakes and Reservoirs: A Contribution to the International Hydrological Programme*, IHP-II Project A. 2.6.1, 65–164. UNESCO, Paris.

Fan, J. & Morris, G.L. (1992a) Reservoir sedimentation I: Delta and density current deposits. *Journal of Hydraulic Engineering*, 118(3), 354–369.

Fan, J. & Morris, G.L. (1992b) Reservoir sedimentation II: reservoir desiltation and long-term storage capacity. *Journal of Hydraulic Engineering*, 118(3), 370–384.

Fehling, E., Schmidt, M., Teichmann, T., Bunje, K., Bornemann, R. & Middendorf, B. (2005) Entwicklung, Dauerhaftigkeit und Berechnung ultrahochfester Betone (UHPC) [Development, durability and computation of ultra high performance concrete (UHPC)]. *Structural Materials and Engineering*, Series 1 (Schmid, M., Fehling, E., eds.), Kassel University, Germany (in German).

Felix, D., Albayrak, I., Boes, R.M., Abgottspon, A., Deschwanden, F. & Gruber, P. (2013) Measuring suspended sediment: Results from the first year of the case study at HPP Fieschertal in the Swiss Alps. Proceedings of the Conference *Hydro 2013*, Innsbruck, Austria, 18(03) (Abstract book).

Felix, D., Albayrak, I., Abgottspon, A. & Boes, R.M. (2016) Optimization of hydropower plants with respect to fine sediment focusing on turbine switch-offs during floods. IOP Conf. Series, *Earth and Environmental Science*, 49, DOI:10.1088/1755-1315/49/12/122011.

Felix, D., Albayrak, I. & Boes, R.M. (2017) Weiterleitung von Feinsedimenten via Triebwasser als Massnahme gegen die Stauraumverlandung [Transfer of fine sediments via turbine water as a measure against reservoir sedimentation]. *Wasser, Energie, Luft*, 109(2), 85–90 (in German).

Felix, D., Albayrak, I., Boes, R.M., Abgottspon, A. & Staubli, T. (2018) Dealing with Pelton turbine erosion based on systematic monitoring. *International Journal of Hydropower & Dams*, 25(5), 84–92.

Fukuroi, H. (2012) Damage from Typhoon Talas to civil engineering structures for hydropower and the effect of the sediment bypass system at Asahi Dam. Proc. Intl. Symposium on *Dams for a changing World – Need for Knowledge Transfer across the Generations and the World*. Kyoto, Japan.

Graf, W.H. (1983) The hydraulics of reservoir sedimentation. *Water Power & Dam Construction*, 35(4), 45–52; 35(9), 33–38; 36(4), 37–40 (including abundant bibliography).

Graf, W.H. & Altinakar, M.S. (1993) *Hydraulique fluviale* [Fluvial hydraulics]. Presses Polytechniques Universitaires Romandes, Lausanne (in French).

Graf, W.H. & Altinakar, M.S. (1995) Courants de turbidité [Density currents]. *La Houille Blanche*, 50(7), 28–37 (in French).

Grimaldi, C., Micheli, F. & Bremen, R. (2015) Sediment management in Andean Region: Chespí-Palma Real project. Proc. 1st Intl. Workshop on Sediment Bypass Tunnels Zürich. *VAW Mitteilung* 232, 85–94, R.M. Boes, ed. Versuchsanstalt für Wasserbau, Hydrologie und Glaziologie, ETH Zurich, Zürich.

Meile, T., Bretz, N.-V., Imboden, B. & Boillat, J.-L. (2014) Reservoir sedimentation management at Gebidem Dam (Switzerland) In: Schleiss, A.J., De Cesare, G., Franca, M.J. & Pfister, M. (eds) *Reservoir Sedimentation*, 245–255. Taylor & Francis, London.

Middleton, G.V. (1966) Experiments on density and turbidity currents I: motion of the head. *Canadian Journal of Earth Sciences*, 3(3), 523–546.

Morris, G.L. & Fan, J. (1998) *Reservoir sedimentation handbook: Design and management of dams, reservoirs, and watersheds for sustainable use.* McGraw-Hill, New York.

Müller, P. & De Cesare, G. (2009) Sedimentation problems in the reservoirs of the Sarganserland Kraftwerke: Venting of turbidity currents as the essential part of the solution. Proceedings of the 23rd *ICOLD Congress*, Brasilia, Brazil, Q89(R21), 51–52 (Abstract book).

Müller, B., & Walker, M. (2015) The Pfaffensprung sediment bypass tunnel: 95 years of experience. Proc. 1ˢᵗ Intl. Workshop on Sediment Bypass Tunnels, Zürich. *VAW Mitteilung* 232, 247–258, R.M. Boes, ed. Versuchsanstalt für Wasserbau, Hydrologie und Glaziologie, ETH Zurich, Zürich.

Müller-Hagmann, M. (2017) Hydroabrasion in high speed flow at sediment bypass tunnels. In: Boes, R.M. (ed) *VAW-Mitteilung* 239. Versuchsanstalt für Wasserbau, Hydrologie und Glaziologie, Zürich.

Müller-Hagmann, M., Albayrak, I, Auel, C., & Boes, R.M. (2020) Field investigation on hydroabrasion in high-speed sediment-ladden flows at sediment bypass tunnels. *Water*, 12(2), 469, 1–27, https://www.mdpi.com/2073-4441/12/2/469.

Nakajima, H., Otsubo, Y., & Omoto, Y. (2015) Abrasion and corrective measures of a sediment bypass system at Asahi Dam. Proc. 1ˢᵗ Intl. Workshop on Sediment Bypass Tunnels Zürich. *VAW Mitteilung* 232, 21–32, R.M. Boes, ed. Versuchsanstalt für Wasserbau, Hydrologie und Glaziologie, ETH Zurich, Zürich.

Noguchi, T., Tomosawa, F., Nemati, K.M., Chiaia, B.M., & Fantilli, A.P. (2009) A practical equation for elastic modulus of concrete. *ACI Structural Journal*, 106(5), 690–696.

Nozaki, T. (1990) Estimation of repair cycle of turbine due to abrasion caused by suspended sand and determination of desilting basin capacity. *Report*, Japan International Cooperation Agency, Tokyo.

Oehy, C. (2003) Effects of obstacles and jets on reservoir sedimentation due to turbidity currents. *Ph.D. Thesis* 2684. Ecole Polytechnique Fédérale de Lausanne, and *Communication* 15, Laboratory of Hydraulic Constructions, A. Schleiss, ed. EPFL Lausanne, Switzerland.

Oehy, C. & Schleiss, A. (2001) Numerical modelling of a turbidity current passing over an obstacle: Practical application in Lake Grimsel, Switzerland. Proceedings of the International Symposium *Environmental Hydraulics*, Tempe AZ (CD-Rom).

Oehy, C. & Schleiss, A.J. (2004) Management of reservoir sedimentation due to turbidity currents by technical measures. In: Yazdandoost, F. & Attari, J. (eds) Proceedings of the International Conference *Hydraulics of Dams & River Structures*, Tehran, Iran. Balkema, Rotterdam, 263–270.

Oehy, C. & Schleiss, A.J. (2007) Control of turbidity currents in reservoirs by solid and permeable obstacles. *Journal of Hydraulic Engineering*, 133(6), 637–648.

Oehy, C., De Cesare, G. & Schleiss, A.J. (2010) Effect of inclined jet screen on turbidity current. *Journal of Hydraulic Research*, 48(1), 81–90.

Oertli, C., & Auel, C. (2015) Solis sediment bypass tunnel: First operation experiences. Proc. 1ˢᵗ Intl. Workshop on Sediment Bypass Tunnels Zürich. *VAW Mitteilung* 232, 223–233, R.M. Boes, ed. Versuchsanstalt für Wasserbau, Hydrologie und Glaziologie, ETH Zurich, Zürich.

OFEFP (1994) Conséquences écologiques des curages de bassins de retenue [Ecological consequences of reservoir cleanings]. *Cahier de l'environnement*, 219. Office fédéral de l'environnement, des forêts et du paysage, Berne (in French).

Parker, G., Fukushima, Y. & Pantin, H.M. (1986) Self-accelerating turbidity currents. *Journal of Fluid Mechanics*, 171, 145–181.

Rahmanian, M.R. & Banihashemid, M.A. (2012) Characterization of sedimentation pattern in reservoirs. *Canadian Journal of Civil Engng.* 39(8), 951–956.

Rooseboom, A. & Annandale, G.W. (1983) Reservoir sedimentation and stream power. Proceedings of the D.B. Simons Symposium *Erosion and Sedimentation*. Colorado State University, Fort Collins CO, 1–20.

Rottman, J.W., Simpson, J.E., Hunt, J.C.R. & Britter, R.E. (1985) Unsteady gravity current flows over obstacles: Some observations and analysis related to phase II trials. *Journal of Hazardous Materials*, 11(1–3), 325–340.

Rouse, H. (1937) Modern conceptions of the mechanics of turbulence. *Transactions of* ASCE, 102, 463–543.

Roveri, E. (1981) Conservazione della capacità utile nei laghi artificiali [Capacity conservation of artificial lakes]. *Wasser Energie Luft*, 73(9), 199–201 (in Italian).

Sakurai, T., & Kobayashi, K. (2015) Operations of the sediment bypass tunnel and examination of the auxiliary sedimentation measure facility at Miwa Dam. Proc. 1st Intl. Workshop on Sediment Bypass Tunnels, Zürich. *VAW Mitteilung* 232, 33–44, R.M. Boes, ed. Versuchsanstalt für Wasserbau, Hydrologie und Glaziologie, ETH Zurich, Zürich.

Salih, E.-T.H.M. (1994) The effects of flushing on the fish community in the Khashm el Girba Reservoir, Eastern Sudan. *M.Sc. Thesis*, Dept. Fisheries and Marine Biology, University of Bergen, Norway.

Savage, S.B. & Brimberg, J. (1975) Analyses of plunging phenomena in water reservoirs. *Journal of Hydraulic Research*, 13(2), 187–205.

Scheuerlein, H. (1995) Downstream effects of dam construction and reservoir operation. Proceedings of the 6th International Symposium *River Sedimentation, Management of Sediment: Philosophy, Aims and Techniques*, New Delhi. Balkema, Rotterdam, 1101–1108.

Schlegel, B. & Dietler, T. (2010) Dez Dam, Iran: High sedimentation rates within the reservoir requires construction of flushing tunnels. *Wasserwirtschaft*, 100(4), 90–92 (in German).

Schleiss, A., Feuz, B., Aemmer, M. & Zünd, B. (1996) Verlandungsprobleme im Stausee Mauvoisin: Ausmass, Auswirkungen und mögliche Massnahmen [Sedimentation problems at Mauvoisin Dam: Extent, effects and possible means]. International Symposium *Verlandung von Stauseen und Stauhaltungen, Sedimentprobleme in Leitungen und Kanälen, VAW Mitteilung*, 142(1), 37–58, ETH Zürich (in German).

Schleiss, A. & Oehy, C. (2002) Verlandung von Stauseen und Nachhaltigkeit [Reservoir sedimentation and sustainability]. *Wasser, Energie Luft*, 94(7/8), 227–234 (in German).

Schleiss, A., De Cesare, G. & Jenzer Althaus, J. (2010) Verlandung der Stauseen gefährdet die nachhaltige Nutzung der Wasserkraft [Reservoir sedimentation endangers sustainable use of hydropower]. *Wasser, Energie, Luft*, 102(1), 31–40 (in German).

Schleiss, A.J., Franca, M.J., Juez, C. & De Cesare, G. (2016) Reservoir sedimentation. *Journal of Hydraulic Research*, 54(6), 595–614.

Schmocker, L., & Weitbrecht, V. (2013) Driftwood: Risk analysis and engineering measures. *Journal of Hydraulic Engineering*, 139(7), 683–695.

Schüttrumpf, H. & Detering, M. (2011) Innovative sediment handling to restore reservoir capacity. In: Schleiss, A.J. & Boes, R.M. (eds) Proceedings of the ICOLD Symposium *Dams Under Changing Challenges*, 79th Annual Meeting, Lucerne. Taylor & Francis, London, 345–352.

Shen, H.W. & Lai, J.-S. (1996) Sustain reservoir useful life by flushing sediment. *International Journal of Sediment Research*, 11(3), 10–17.

Simons, D.B. & Senturk, F. (1992) *Sediment transport technology*. Water Resources Publications, Fort Collins CO.

Sklar, L.S., & Dietrich, W.E. (1998) River longitudinal profiles and bedrock incision models: Stream power and the influence of sediment supply. Rivers over rock: Fluvial processes in bedrock channels (eds. Winkler, K.J., Wohl, E.E.). *Geophysical Monograph Series*, 107, 237−260.

Sklar, L.S., & Dietrich, W.E. (2001) Sediment and rock strength controls on river incision into bedrock. *Geology*, 29(12), 1087−1090. DOI: 10.1130/0091-7613(2001)029<1087:SARSCO>2.0.CO;2.

Sklar, L.S., & Dietrich, W.E. (2004) A mechanistic model for river incision into bedrock by saltating bed load. *Water Resources Research* 40(W06301). DOI: 10.1029/2003WR002496.

Sollerer, F. & Matt, P. (2013) Sediment management of reservoirs: Sediment discharge in dependence on the suspended load concentration in the run-off water. Proceedings of the *Hydro Conference*, Innsbruck, Austria, Paper no. 18.08.

Stähly, S., Franca, M.J., Robinson, C.T. & Schleiss, A.J. (2019) Sediment replenishment combined with an artificial flood improves river habitats downstream of a dam. *Scientific Reports*, 9(1), 5176.

Strand, R.I. & Pemberton, E.L. (1982) *Reservoir sedimentation: Technical guideline*. US Bureau of Reclamation, Denver CO.

Sumi, T., & Kantoush, S.A. (2011) Comprehensive sediment management strategies in Japan: Sediment bypass tunnels. Proc. 34th *IAHR World Congress*, Brisbane, Australia, 1803–1810.

Sumi, T., Okano, M. & Takata, Y. (2004) Reservoir sedimentation management with bypass tunnels in Japan. Proceedings of the 9th International Symposium *River Sedimentation*, Yichang, China, 1036–1043.

Sumi, T. (2005) Sediment flushing efficiency and selection of environmentally compatible reservoir sediment management measures. Proceedings of the 2nd East Asia International ICOLD Symposium *Sediment Management and Dams*, pp. 9–22.

Sumi, T. (ed) (2017) Proc. 2nd Intl. *Workshop on Sediment Bypass Tunnels*. Kyoto University, Kyoto, Japan. http://ecohyd.dpri.kyoto-u.ac.jp/en/index/SBTworkshop.html

Sumi, T. & Kantoush, S. (2011) Sediment management strategies for sustainable reservoirs. In: Schleiss, A.J. & Boes, R.M. (eds) Proceedings of the ICOLD Symposium *Dams and Reservoirs Under Changing Challenges*. 79th Annual Meeting of ICOLD, Lucerne, Switzerland. Taylor and Francis, London, 353–362.

Suter, P. (1998) Verlandung und Spülung des Rempenbeckens der AG Kraftwerk Wägital [Reservoir sedimentation and flushing of Rempen basin, Wägital Power Plant]. *Wasser, Energie, Luft*, 90(5), 127–131 (in German).

Turner, J.S. (1973) *Buoyancy effects in fluids*. Cambridge University Press, Cambridge.

Turowski, J.M. (2009) Stochastic modeling of the cover effect and bedrock erosion. *Water Resources Research*, 45(W03422). DOI:10.1029/2008WR007262.

Turowski, J.M. (2012) Semi-alluvial channels and sediment-flux-driven bedrock erosion. *Gravel-Bed Rivers: Processes, Tools, Environments* (eds. M. Church, P.M. Biron and A.G. Roy). John Wiley & Sons, Ltd., Chichester, UK. DOI: 10.1002/9781119952497.ch29.

Vischer, D., Hager, W. H., Casanova, C., Joos, B., Lier, P., & Martini, O. (1997) Bypass tunnels to prevent reservoir sedimentation. Proc. 19th *ICOLD Congress*, Florence, Italy Q.74(R.37), 605–634. ICOLD, Paris.

Whipple, K.X., Hancock, G.S., & Anderson, R.S. (2000) River incision into bedrock: Mechanics and relative efficacy of plucking, abrasion and cavitation. *Geological Society of America Bulletin*, 112(3), 490−503. DOI: 10.1130/0016-7606(2000)112<0490:RIIBMA>2.3.CO;2.

White, R. (2001) *Evacuation of sediments from reservoirs*. Telford, London.

Winkler, K., Dekumbis, R., Rentschler, M., Parkinson, E. & Garcin, H. (2011) Understanding hydro-abrasive erosion. Proceedings of the Conference *Hydro*, Prague, Czech Republic. Aqua-Media Intl. Ltd., London.

Wyss, C.R., Rickenmann, D., Fritschi, B., Turowski, J.M., Weitbrecht, V., & Boes, R.M. (2014) Bedload grain size estimation from the indirect monitoring of bedload transport with Swiss plate geophones at the Erlenbach stream. Proc. Intl. *River Flow Conference*, 1907–1912 (Schleiss, A.J., De Cesare, G., Franca, M.J., & Pfister, M., eds.), ISBN 978-1-138-02674-2. Taylor & Francis Group, London, UK.

Wyss, C.R., Rickenmann, D., Fritschi, B., Turowski, J.M., Weitbrecht, V., & Boes, R.M. (2016) Measuring bedload transport rates by grain-size fraction using the Swiss plate geophone signal at the Erlenbach. *Journal of Hydraulic Engineering*, 140(4), 04016003-1, 11 pages.

Bibliography

Akiyama, J. & Stefan, H.G. (1984) Plunging flow into a reservoir: Theory. *Journal of Hydraulic Engineering*, 110(4), 484–499; 111(1), 175–177.

ASCE Task Committee (1973) Sediment control methods: reservoirs. *Journal of the Hydraulics Division* ASCE, 99(HY4), 617–635; 100(HY2), 332–335; 100(HY5), 696–697.

Baines, P.G. (1984) A unified description of two-layer flow over topography. *Journal of Fluid Mechanics*, 146, 127–167.

Bonnecaze, R.T., Huppert, H.E. & Lister, J.R. (1993) Particle-driven gravity currents. *Journal of Fluid Mechanics*, 250, 339–369.

Brooks, N.H. & Koh, R.C.Y. (1969) Selective withdrawal from density-stratified reservoirs. *Journal of the Hydraulics Division* ASCE, 95(HY4), 1369–1400.

Bühler, J. (1994) Simple internal waves and bores. *Journal of Hydraulic Engineering*, 120(5), 638–645; 121(7), 575; 121(9), 683.

Chen, Y.H., Lopez, J.L. & Richardson, E.V. (1978) Mathematical modeling of sediment deposition in reservoirs. *Journal of the Hydraulics Division* ASCE, 104(HY12), 1605–1616.

De Cesare, G., Schleiss, A. & Hermann, F. (2001) Impact of turbidity currents on reservoir sedimentation. *Journal of Hydraulic Engineering*, 127(1), 6–16; 128(6), 644–645.

Denton, R.A. (1990) Accounting for density front energy losses. *Journal of Hydraulic Engineering*, 116(2), 270–275; 117(9), 1222–1225.

Garcia, M.H. (1993) Hydraulic jumps in sediment-driven bottom currents. *Journal of Hydraulic Engineering*, 119(10), 1094–1117.

Garde, R.J. & Swamee, P.K. (1973) Analysis of aggradation upstream of a dam. Proceedings of the International IAHR Symp. *River Mechanics*, Bangkok, 1, 13–22.

Graf, W.H. (1971) *Hydraulics of sediment transport*. McGraw-Hill, New York.

Gröbelbauer, H.P., Fanneløp, T.K. & Britter, R.E. (1993) The propagation of intrusion fronts of high density ratios. *Journal of Fluid Mechanics*, 250, 669–687.

Hinwood, J.B. (1969) The study of density-stratified flows up to 1945. *La Houille Blanche*, 25(4), 347–359; 27(8), 709–722.

Kersey, D.G. & Hsü, K.J. (1976) Energy relations of density-current flows: an experimental investigation. *Journal of Sedimentology*, 23(6), 761–789.

Klemp, J.B., Rotunno, R. & Skamarock, W.C. (1994) On the dynamics of gravity currents in a channel. *Journal of Fluid Mechanics*, 269, 169–198.

Klemp, J.B., Rotunno, R. & Skamarock, W.C. (1997) On the propagation of internal bores. *Journal of Fluid Mechanics*, 331, 81–106.

Kordas, B. & Ratomski, J. (1976) Hydraulic method of sedimentation forecasting in reservoirs. Proceedings of the 12th *ICOLD Congress*, Mexico, Q47(R25), 1099–1107.

Lee, H.-Y. & Yu, W.-S. (1997) Experimental study of reservoir turbidity current. *Journal of Hydraulic Engineering*, 123(6), 520–528.

List, E.J. (1982) Turbulent jets and plumes. *Annual Review of Fluid Mechanics*, 14, 189–212.

Lüthi, S. (1980) Some new aspects of two-dimensional turbidity currents. *Journal of Sedimentology*, 28(1), 97–105.

Okada, T. & Baba, K. (1982) Sediment release plan at Sakuma Reservoir. Proceedings of the 14th *ICOLD Congress*, Rio de Janeiro, Q54(R4), 41–64.

Pedersen, F.B. (1972) Gradually varying two-layer stratified flow. *Journal of the Hydraulics Division* ASCE, 98(HY1), 257–268; 99(HY3), 534.

Rajaratnam, N. & Powley, R.L. (1990) Hydraulic jumps in two-layer flows. *Proceedings of the ICE*, 89(2), 127–142.

Rajaratnam, N., Tovell, D. & Loewen, M. (1991) Internal jumps in two moving layers. *Journal of Hydraulic Research*, 29(1), 91–106.

Rooseboom, A. (1976) Reservoir sediment deposition rates. Proceedings of the 12th *ICOLD Congress*, Mexico, Q47(R21), 1049–1060.

Rottman, J.W. & Simpson, J.E. (1983) Gravity currents produced by instantaneous releases of a heavy fluid in a rectangular channel. *Journal of Fluid Mechanics*, 135, 95–110.

Simpson, J.E. (1972) Effects of the lower boundary on the head of a gravity current. *Journal of Fluid Mechanics*, 53, 759–768.

Simpson, J.E. (1982) Gravity currents in the laboratory, atmosphere, and ocean. *Annual Review of Fluid Mechanics*, 14, 213–234.

Soni, J.P. (1977) Stability of sediment-laden underflows passing through reservoirs. *Indian Journal of Power & River Valley Development*, 27(5), 142–146.

Soni, J.P. (1979) Prototype observation in relation to reducing sedimentation in reservoirs. *Indian Journal of Power & River Valley Development*, 29(1), 11–16.

Szechowycz, R.W. & Qureshi, M.M. (1973) Sedimentation in Mangla Reservoir. *Journal of the Hydraulics Division* ASCE, 99(HY9), 1551–1572; 100(HY9), 1283–1285; 101(HY4), 406–407.

Wood, I.R. & Simpson, J.E. (1984) Jumps in layered miscible fluids. *Journal of Fluid Mechanics*, 140, 329–342.

Yalin, M.S. (1977) *Mechanics of sediment transport*, 2nd ed. Pergamon Press, Oxford.

Yücel, Ö. (1976) Model investigations of reservoir sedimentation. Proceedings of the 12th *ICOLD Congress*, Mexico, Q47(R11), 899–911.

Chapter 10 Frontispiece Impulse wave (a-d) generation and propagation, (e, f) slide impact in Alpine Lake at Grindelwald, Switzerland (all photos Courtesy H.-R. Burgener). Maximum wave height was estimated between 5 and 10 m, the maximum wave run-up in slide direction was between 10 and 15 m

Chapter 10

Impulse waves in reservoirs

10.1 Introduction

Impulse waves are generated by various causes, including rockfalls, landslides, ice falls, glacier calvings, or snow avalanches. These waves may endanger a dam and the reservoir due to wave run-up, or due to dam overtopping and the resulting floods in the tailwater. Impulse waves can also lead to significant shore erosion resulting in secondary events (Vischer *et al.*, 1991).

There exists a significant body of knowledge on the mechanics of wave generation, wave propagation along the reservoir and wave run-up or overtopping. The results refer mainly to plane flow conditions and these are normally expanded to spatial situations. The phenomenon is governed by Froude similitude if scale effects are excluded with hydraulic modeling currently as the approach to be considered for all projects where impulse waves are a concern.

The following discussion is mainly a summary of studies conducted at VAW, ETH Zurich, referring to both the plane and the spatial impulse waves. All three stages of flow, including wave generation, wave propagation, and wave run-up are accounted for. It should be stressed that impulse waves need to be considered mainly for safety reasons, given the enormous potential of destruction. Note further the notion Mega-tsunamis, given the extreme wave heights in excess of 100 m, as compared with normal tsunamis, whose wave heights amount to typically 10 m. Following the 2004 Southeast Asia Tsunami, and the 2011 Fukushima Disaster, among others, these extreme waves have received careful attention from the public. Notwithstanding, these incidents need to be considered carefully at all dam sites involving a potential of slides into the water body.

Chapter 10 includes a total of 8 sub-chapters on plane impulse waves, and a single sub-chapter on spatial impulse waves, expanding the former knowledge to the 3D environment. As to the 2D waves, the fundamentals of impulse waves are discussed first, followed by the main features of wave generation and propagation, in which the so-called impulse wave parameter is introduced. Expressions for the main parameters are provided exclusively in terms of the fundamental wave input parameters, allowing for a thorough engineering design. In 10.4, the impulse wave types are described, limitations are specified and the water wave theory is shortly exposed, to allow for a classification of these waves mainly when numerical methods are considered. In 10.5 wave transformation from the solitary wave to overland flow is detailed, considering a setup similar to what occurred during the Fukushima Disaster. In 10.6 the underwater deposition characteristics are specified based on the analysis of numerous laboratory observations, allowing to estimate the impulse wave input parameters from the slide deposition characteristics. Rigid dam overtopping due to impulse waves is considered in 10.7, allowing to determine the main overtopping features. Similarly, 10.8 deals with the erodible dam overtopping, also providing the main features under this scenario of engineering interest. In 10.9, a novel approach to observe and analyze spatial impulse waves is highlighted allowing for detailed studies of the main flow features as observed during laboratory experimentation. Although the experimental approaches for the 2D and 3D impulse waves differ slightly, the overall result allows for a good estimate of the main hydraulic wave parameters which currently is a requirement for the design of each larger dam. Although the current knowledge on impulse waves can be considered good, there remain lots of cases where additional insight into this highly fascinating feature is required, particularly from the side of numerical modeling not detailed here, however.

10.2 Fundamental approaches

10.2.1 Wave theories and impulse waves

Waves are amenable to mathematical treatment; there exists a large body of literature relating to water waves (Wiegel, 1964). Their main parameters are (Figure 10.1a):

- *Wave amplitude* a_M defined as crest elevation over undisturbed water line,
- *Wave length* L_w between two adjacent wave crests, and
- *Water depth* h_o as elevation of the still water level over the bottom.

Depending on the wave shallowness $\tau = a_M/h_o$ and the wave steepness $\epsilon = h_o/L_w$, various wave types are distinguished (Figure 10.1b):

1 *Linear waves* for which both $\tau = 1$ and $\epsilon^2 = 1$,
2 *Airy waves* with $\tau \gg \epsilon^2$; these waves are sinusoidal,
3 *Boussinesq waves* with $\tau \approx \epsilon^2$ for which curvature effects become significant; these waves are nonlinear, and
4 *Dispersive waves* for which $\tau \ll \epsilon^2$.

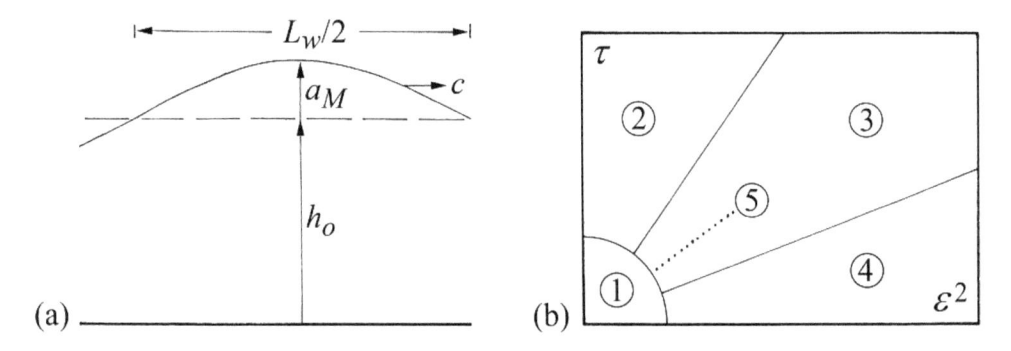

(a)   (b)

Figure 10.1 Water wave theories (a) schematic wave propagation c with notation, (b) effects of wave shallowness $\tau = a_M/h_o$ and wave steepness $\epsilon = h_o/L_w$ on wave theories (Sander and Hutter, 1992)

Usually, these wave theories account only for the gravity force thereby neglecting the effects of viscosity and surface tension. Impulse waves as treated in this chapter are typical phenomena where gravity is the dominant parameter. They are normally intermediate between the deep-water and shallow-water waves with $\tau \approx \epsilon^2$. The solitary wave is a particular wave type relevant to impulse waves. A general account of water waves is given, e.g. by Le Mehauté (1976).

Waves generated in a reservoir include three mechanisms (Figure 10.2):

1 Wave generation by mass impact onto water body,
2 Wave propagation in the reservoir, and
3 Wave run-up on topography, such as opposite a shore or a dam, with possible wave overtopping.

Impulse waves are generated by various sources, including rock, ice or snow avalanches, glacier calving, shore instabilities, and earthquakes, among others. The characteristics of *plane* impulse waves are summarized as follows:

- First wave height normally corresponding to the maximum elevation (Noda, 1970; Wiegel *et al.*, 1970). It contains more energy than all following waves;
- Waves height decay during propagation over nearly horizontal bottom (Hunt, 1988);
- Wave pattern depending only on reservoir bathymetry and reservoir volume once a wave has traveled over several of its wavelengths; and
- Dispersive waves generated by a horizontal mass displacement transform with increasing propagation length to solitary wave, followed by sinusoidal wave train (Miller, 1960a).

The main parameters by which impulse waves are governed include the Froude number based on the slide impact velocity onto the water surface, the slide volume, the impact angle with respect to the reservoir axis and the distance from the impact location. Figure 10.3 shows a typical event with the impact location P, and the wave generation in the reservoir. *Spatial* impulse waves are more complex and the results may not be generalized so easily as for the plane impulse waves.

The following deals with both plane and spatial impulse waves (10.9). The latter are currently in the research stage, and the results are limited to the main wave features, given the complex wave propagation characteristics. In turn, plane impulse waves have received

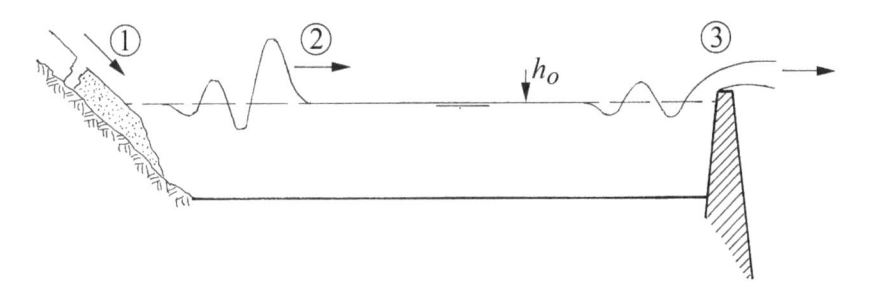

Figure 10.2 Impact wave mechanisms on reservoir, schematic

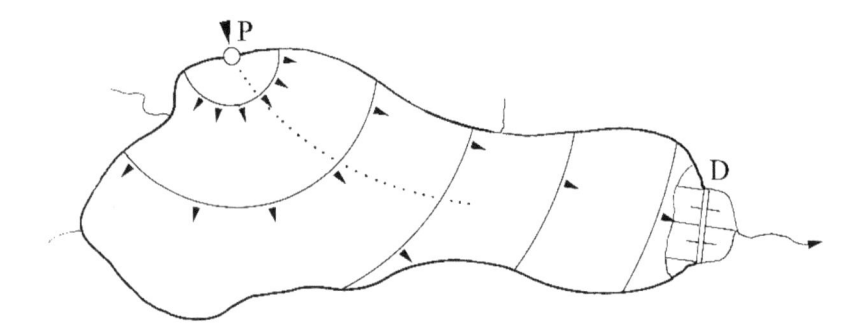

Figure 10.3 Impulse wave in reservoir generated at point P and propagating towards dam D

considerable attention in the past, so that the results allow for a detailed analysis of simplified arrangements. The main knowledge is presented in 10.3 relating to the wave generation and propagation, in 10.4 relating to landslide-generated impulse waves, in 10.5 dealing with the transformation of the solitary wave to overland flow, in 10.6 dealing with the underwater deposition features of impulse wave events, and in 10.7 and 10.8 to the overtopping features of impulse waves over rigid and granular dams. Most of these results are based on laboratory experimentation along with a systematic data analysis, so that the relevant features of impulse waves in reservoirs are presented.

10.2.2 Wave generation by moving wedge

The momentum transfer of a slide mass falling into the water body of a reservoir generating impulse waves is difficult to assess. The effects of mass fracturing and air entrainment complicate the process. These two latter aspects have hardly received attention, given the laboratory environment in which these processes cannot be studied adequately. Generally, slides into a water body therefore relate either to a rigid body, or to a granular, incohesive sediment by which the water quality is not subjected by turbid effects, and in which viscous effects remain absent.

Sander and Hutter (1992) have considered the following problem: What are the wave characteristics due to a moving body submerged in a rectangular channel? Both moving wedges (subscript W) of translation velocity V_W and rotating plates were studied. The determining wave parameters are the relative dislocation d/h_o, the wedge Froude number $F_W = V_W/(gh_o)^{1/2}$ and the wave period $T_W = t_W(g/h_o)^{1/2}$ with h_o as the still water depth (Figure 10.4).

For $0.1 < F_W < 1.1$ the waves generated have always the same characteristics, namely a steep front extending to a first wave maximum followed by a slightly slower decay in the first wave tail. Usually, a wave train was produced starting with a compact front wave and a wave tail ordered according to the wave amplitudes.

The maximum possible wave height h_∞ relative to the initial flow depth h_o, and the corresponding relative wave length L_∞/h_o depend on the wedge angle α displaced by d/h_o over the rectangular channel bottom. Both h_∞ and L_∞ serve as scaling values (Figure 10.5).

For a given wedge Froude number F_W, wedge angle α, relative dislocation d/h_o, and thus known scaling values h_∞ and L_∞, the maximum wave amplitude a_M/h_∞ and the corresponding wave length ratio L_M/L_∞ results from Figure 10.6, from which the governing parameter of

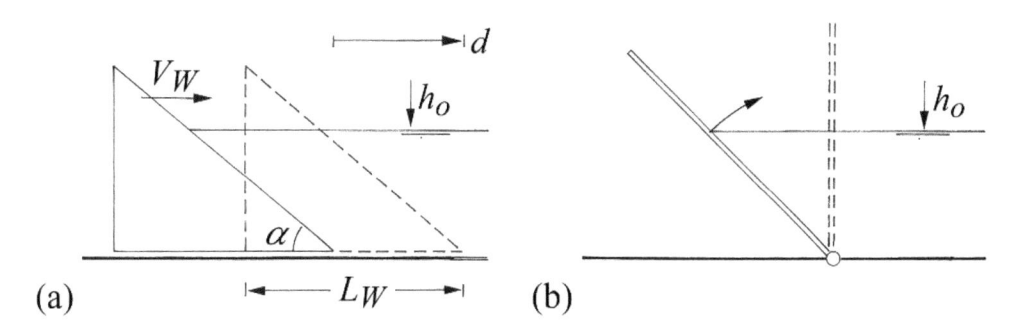

Figure 10.4 Experimentation of Sander and Hutter (1992) (a) moving wedge, (b) rotating plate

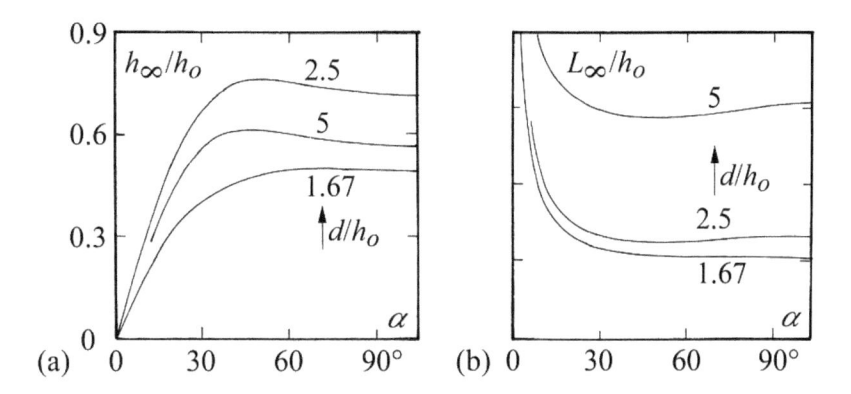

Figure 10.5 Scaling parameters for waves generated by moving wedge (a) maximum possible wave amplitude ratio h_∞/h_o, (b) corresponding wave length ratio L_∞/h_o versus wedge angle α and relative dislocation d/h_o (Sander and Hutter, 1992)

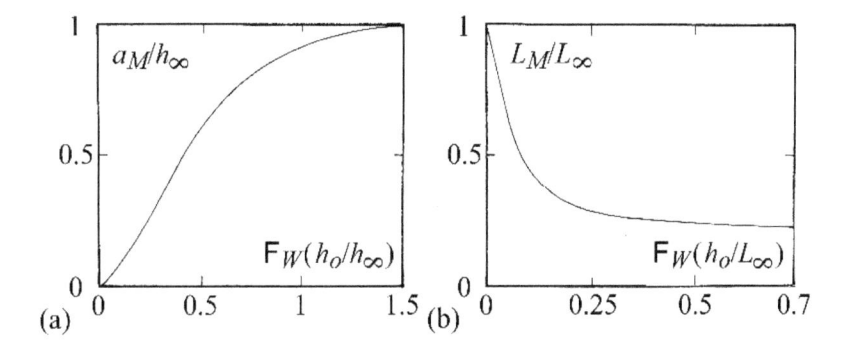

Figure 10.6 Impulse waves generated by moving wedge (a) Maximum wave amplitude a_M/h_∞ versus $F_W(h_o/h_\infty)$, (b) corresponding wave length ratio L_M/L_∞ versus $F_W(h_o/L_\infty)$ (Sander and Hutter, 1992)

wave propagation is seen to be $F_W(h_o/h_\infty)$, and $F_W(h_o/L_\infty)$. For given α, the wave amplitude ratio a_M/h_∞ increases with increasing wedge velocity and decreases as h_∞ increases. To the lowest order of approximation, the linear relation $a_M/h_\infty = F_W(h_o/h_\infty)$ holds, provided that $a_M/h_\infty < 1$. Thus, independent of the wedge angle α, the result is simply

$$a_M / h_o = F_W. \tag{10.1}$$

Accordingly, the wave amplitude increases proportionally with F_W and h_o.

10.2.3 Wave generation by falling mass

According to Huber (1982), Switzerland has been endangered almost during each decade of the 20th century by impulse waves occurring in natural lakes. A recent example in Lake of Lucerne has been reported by Fuchs and Boes (2010). The damage of property has been substantial, and people have lost their lives in two cases. Other occurrences in reservoirs, especially in Italy and China, resulted in large fatalities and catastrophes.

Rock avalanches or bank slides along steep reservoir shores include various problems, such as the form and extent of the slide, the mass transfer on to the water body and the main characteristics of the impulse waves generated (Vischer, 1986). Accordingly, these topics have to be addressed by experts in geology and wave mechanics. The latter part is dealt with hereafter, based on the knowledge of the main slide parameters. These include the slide volume V_s of density ρ_s, the still water depth h_o, the hillslope angle α at the impact site, and the distance x from the impact location (Figure 10.7). All points where wave heights have to be predicted should be visible from the impact site. They are described by the horizontal impact angle γ. Indirect waves are influenced by wave reflection and diffraction, and therefore are much smaller than direct waves, to be considered below (Figure 10.7).

The range of validity of the test results is (Huber, 1982):

- Relative wave height is smaller than wave breaking limit, i.e. $a_M/h_o < 0.78$;
- Relative propagation ranges are $5 < x/h_o < 100$ for 2D, and $5 < x/h_o < 30$ for 3D waves;

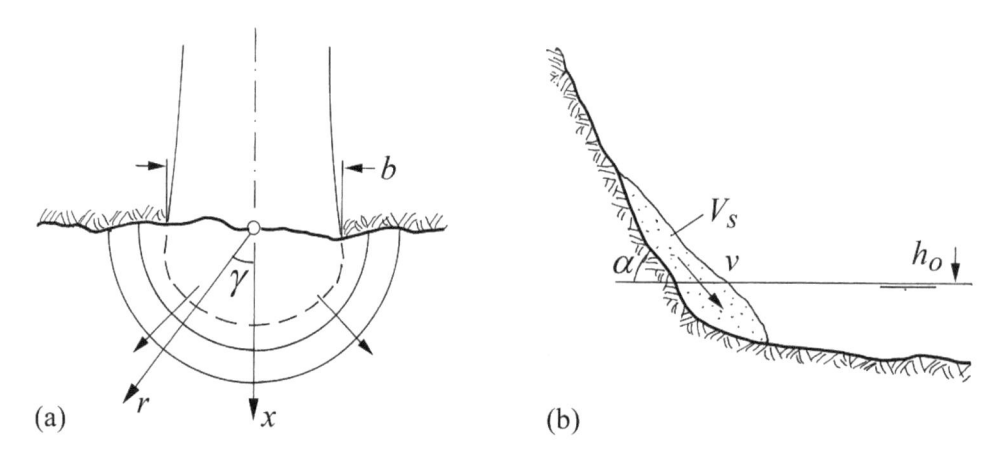

(a)   (b)

Figure 10.7 Definition plot for wave generation in reservoirs (a) plan, (b) section

- Slide velocity is larger than about 50% of wave celerity c;
- Slide angle α between 28° and 60°, because friction impedes granular slides for $\alpha < 25°$; and
- Slide mass represents a dense debris flow. Compact rock slides result in higher waves, whereas a disintegrated mass due to blasting yields much smaller waves.

These data were reanalyzed following within ±15% (Huber and Hager, 1997)

$$\frac{a_M}{h_o} = 0.88 \sin \alpha \left(\frac{\rho_s}{\rho_w}\right)^{1/4} \left(\frac{V_s}{bh_o^2}\right)^{1/2} \left(\frac{h_o}{x}\right)^{1/4}. \tag{10.2}$$

The following statements apply for waves in slide impact direction ($\gamma = 0$):

1 Slide angle α has the major effect on the relative wave height, followed by the relative volume $V_s/(bh_o^2)$ with b as slide width. The effects of density ratio (ρ_s/ρ_w) and relative location (x/h_o) are relatively small;
2 Effect of slide Froude number based on still water depth h_o is insignificant. Accordingly, a rapid slide in a deep reservoir produces nearly the same wave height as a slower slide in a correspondingly shallower reservoir.
3 Wave propagation velocity follows the expression for a surge or a solitary wave as

$$c^2 = g(h_o + a_M). \tag{10.3}$$

The impulse wave problem is thus approximated with the Boussinesq equations, given that the effects of streamline curvature are significant (Figure 10.8).

For *spatial* impulse waves, the following comments apply:

- Wave propagation is semi-radial, from impact location toward reservoir (Figure 10.3);
- Highest wave location is in extension of the slide direction; lateral waves at equal distance from impact site are considerably lower; and
- Decay of wave maximum follows 2D experiments. Huber (1982) allows for an estimation of the wave decay.

3D test data of Huber have been reanalyzed; it was found that the wave reduction in a reservoir depends mainly on the angle γ relative to the slide direction (Figure 10.7). Also, the decay of wave height in a reservoir is significantly larger than in the flume. The test results follow with ±20% accuracy (Huber and Hager, 1997)

$$\frac{a_M}{h_o} = 2 \cdot 0.88 \sin \gamma \cos^2 \left(\frac{2\alpha}{3}\right) \left(\frac{\rho_s}{\rho_w}\right)^{\frac{1}{4}} \left(\frac{V_s}{bh_o^2}\right)^{\frac{1}{2}} \cdot \left(\frac{r}{h_o}\right)^{-\frac{2}{3}}. \tag{10.4}$$

Accordingly, the decay of wave height follows $(h_o/r)^{2/3}$, with r as the radial coordinate from the impact site. For lateral angles $\gamma < ±20°$ nearly no wave height reduction occurs. The effect of variable reservoir depth h_o is incorporated in the analysis by accounting for the average reservoir depth. An alternative approach as well as computational examples are presented by Huber and Hager (1997).

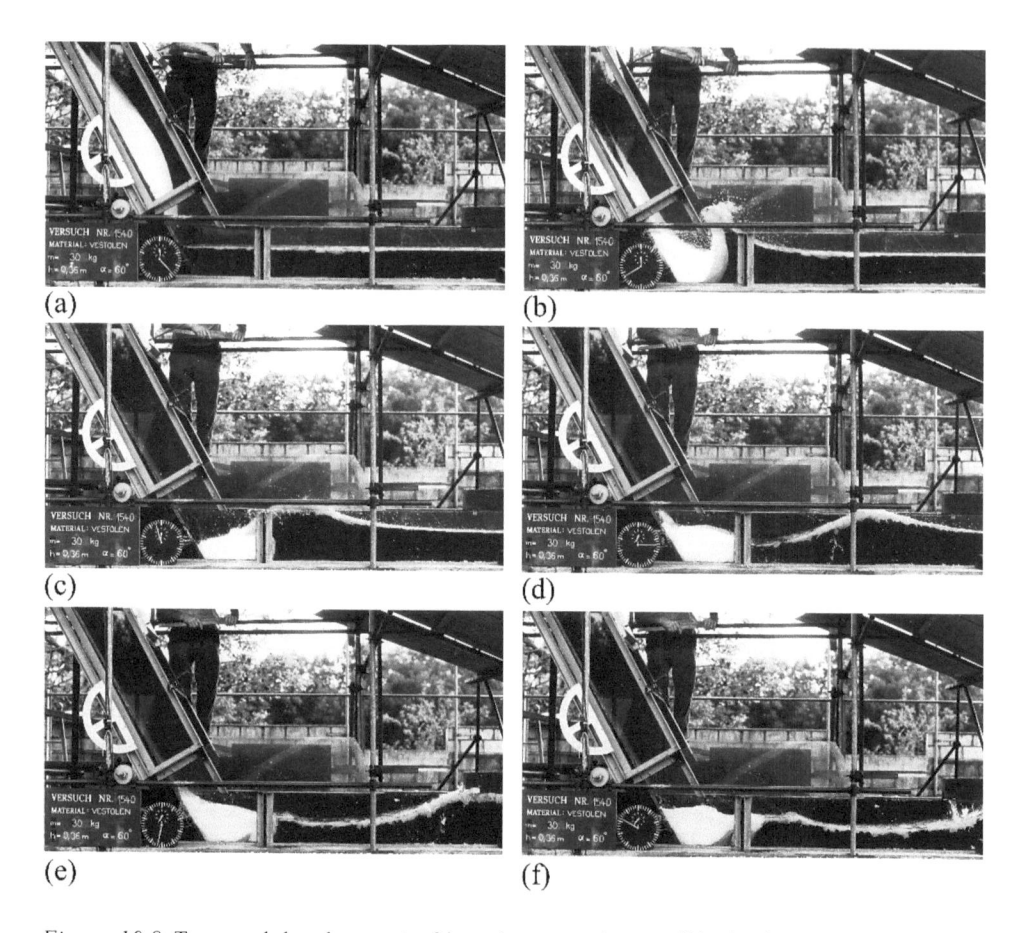

Figure 10.8 Temporal development of impulse wave due to slide (Huber, 1982)

Huber (1982) apparently was the first conducting systematic laboratory experiments on both 2D and 3D impulse waves. Although his PhD thesis (Huber 1980) includes numerous data and observations, its important drawback is the applied wave generation mechanism. Instead of systematically varying the slide impact velocity, this value resulted from the slide characteristics (granulate employed, its shape, and location above stillwater level). The effect of the slide Froude number was thereby small because it was hardly subject to variation in the model tests. As shown below, this effect is the most prominent as it directly affects the primary wave height. This particularity of Huber's (1982) test campaign resulted in a revival of the impulse wave research at VAW, ETH Zurich, during the past two decades.

10.2.4 Wave run-up and overtopping features

Figure 10.9 shows the *plane* impulse wave on a nearly horizontal bottom propagating toward a dam. The maximum (subscript M) wave height from the wave trough to the wave peak is h_M, the stillwater depth is h_o, L_w is the wave length, and the wave front propagation velocity is c_w.

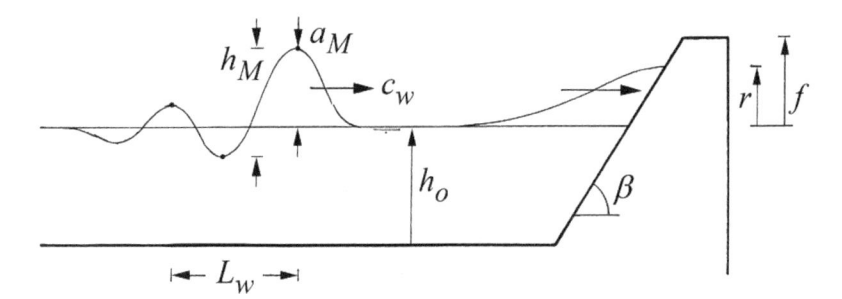

Figure 10.9 Wave propagation and run-up on dam, definition of variables

Due to the circular wave propagation, one may assume a nearly orthogonal wave run-up onto a dam. The quantities to be determined are the run-up height r, the overtopping volume V_d if the run-up height r is larger than the freeboard height f, and the time of overtopping t_d.

Based on a dimensional analysis, Müller (1995) demonstrated that both the relative run-up height $R = r/h_o$ and the overtopping volume V_d/h_o^3 depend on the relative wave height h_M/h_o, wave steepness $\epsilon = a_M/L_w$, wave period $\tau = T_w/(h_o/g)^{1/2}$ with $T_w = L_w/c_w$, and run-up angle β. For sufficiently large waves, Froude similitude governs the processes.

The wave run-up originates either from wind waves of short wave length, or from tsunamis of long-wave lengths. Whereas wind waves are periodical in deep water, tsunamis are long waves consisting typically of a solitary front wave followed by a periodic wave train. Herbich (1990) distinguishes between non-breaking and breaking tsunamis, of which the latter break far away or close to the coast. These waves are notorious for their destructive action.

The impulse wave considered belongs to waves in the transition regime with a wave shallowness h_o/L_w between 0.5 and 2, with properties close to shallow-water flows. Accordingly, the wave parameters are variable with location and time. On reaching the shore, the wave piles up and possibly breaks. Impulse waves are thus comparable with tsunamis. Figure 10.10 refers to run-up on a dam with an upstream slope of 1:3. Breaking is not seen to occur.

The relative run-up height $R = r/h_o$ was previously related to the relative wave height h_M/h_o, the wave steepness $\varepsilon_A = a_M/L_w$, the relative wave period $\tau = T_w(g/h_o)^{1/2}$, and the run-up angle β (Figure 10.9). The effect of wave period was found to be negligible from some 200 experiments, so that (Müller, 1995)

$$R = 1.25 \left(\frac{\pi}{2\beta} \right)^{0.2} \left(\frac{h_M}{h_o} \right)^{1.25} \left(\frac{h_M}{L_w} \right)^{-0.15}. \tag{10.5}$$

Increasing the run-up angle β reduces the run-up height R. In the test domain of $18° \le \beta < 90°$, this effect gives maximum variations of up to 40%. The relative wave height is the governing parameter; the tests refer to $0.01 \le h_M/h_o \le 0.51$. The wave steepness has a relatively small effect within $0.001 \le h_M/L_w \le 0.0135$. Waves with $B_w < 3$ break as they run up a shore, with $B_w = \tan\beta/(h_M/L_w)^{1/2}$ as the wave breaking index.

As impulse waves can reach several dozen meters of height, wavelengths of 100s of meters and velocities up to 30 m/s, they may endanger dams by overtopping or by impact action. The

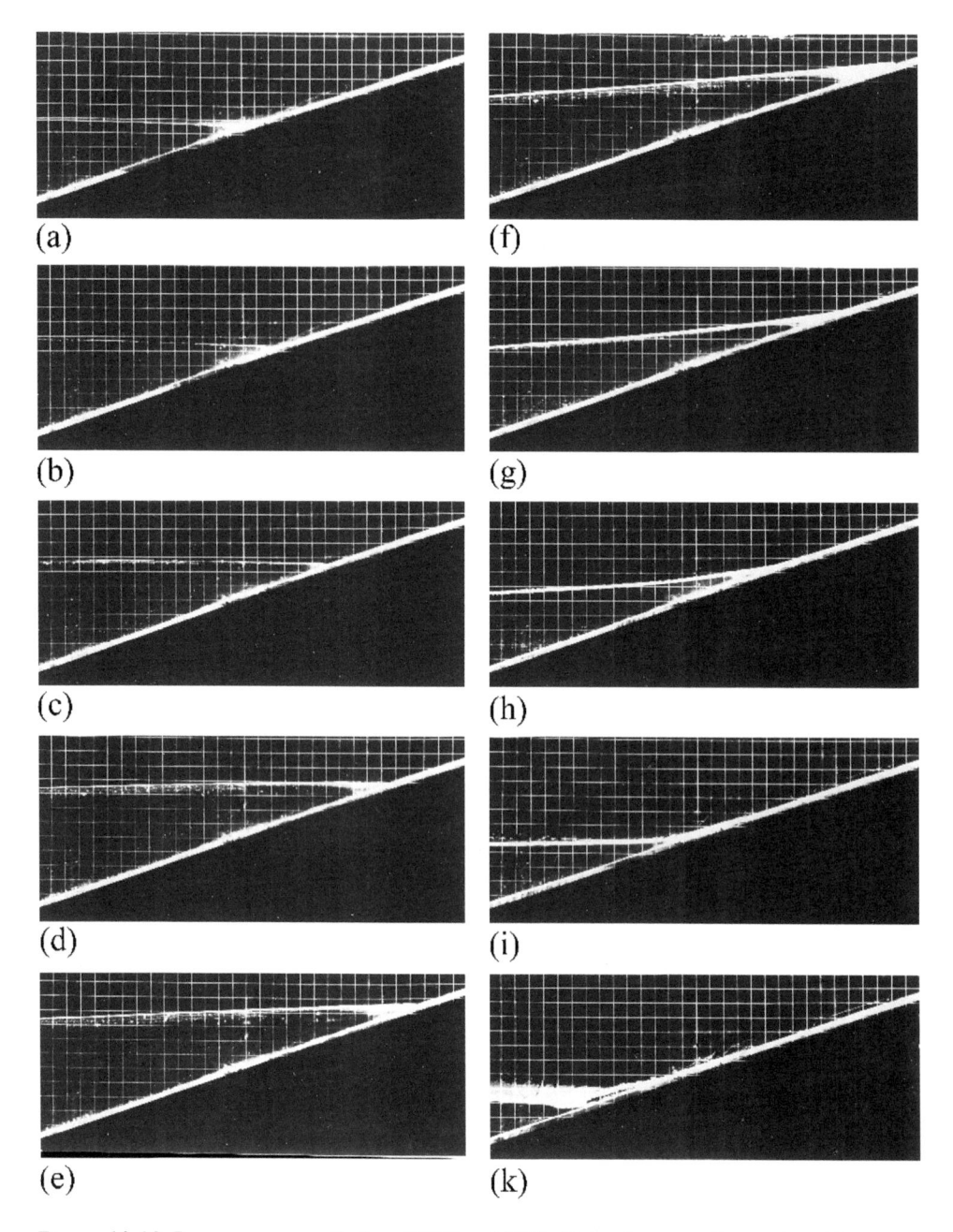

Figure 10.10 Run-up on smooth dam (Müller, 1995). Note absence of wave breaking

most tragic accident occurred at Vaiont Dam, Northeastern Italy, on Oct. 9, 1963. After the first reservoir filling up to a freeboard of 20 m and a subsequent rapid reservoir drawdown, a rock avalanche of some $300 \cdot 10^6$ m³ occurred over a width of 2 km. The arch dam was over-topped with a height of 100 m, slightly damaging the left crest zone (Figure 10.11). The dam remained stable, but a water mass of $40 \cdot 10^6$ m³ overtopped it reaching the Piave Valley a few kilometers downstream. Some 2000 persons died, and terrible devastation resulted. For the first time, the potential of impulse waves was demonstrated, asking for a detailed hydraulic safety check for all dams with a potential for slides into a reservoir.

Figure 10.12 refers to dam overtopping by an impulse wave. Wave breaking in front of the dam is infrequent due to the large bottom slope, and the overtopping jet is usually aerated both along its upper and lower trajectories. The overtopping process is considered as quasi-steady spillway flow. The differences are mainly in the modified approach flow conditions involving a rising surface toward the dam. From detailed observations, Müller (1995) found that:

- Wave front is less steep than wave tail; and
- Maximum overtopping level occurs after (2/3) of the overtopping period.

A generalized wave profile was also described.

The overtopping mechanism is so complex that a simplified approach for the reference volume is given. Müller (1995) related the overtopping volume V_d to the reference volume V_o which would occur under zero freeboard. Clearly, the relative overtopping volume V_d/V_o reduces from the maximum value for zero freeboard $f = 0$ to the run-up height r. Müller's result for the envelope of experimental data is

$$\frac{V_d}{V_o} = \left(1 - \frac{f}{r}\right)^2 . \tag{10.6}$$

Figure 10.11 Arch Dam of Vaiont (Italy): after overtopping resulted a 100 m high impulse wave (a) view on failure surface, (b) reservoir remnant (after Italian legislation, no copyright permission is needed for documents older than 20 years)

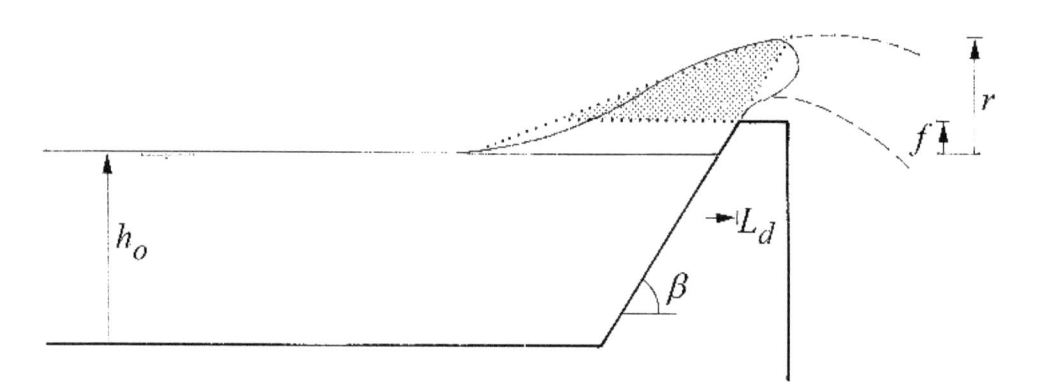

Figure 10.12 Overtopping of dam with (· · ·) overtopping volume (Müller, 1995)

This follows when assuming that the shape of the free surface just upstream of the dam is nearly a triangular wedge of height $r-f$, and the corresponding maximum value has a height equal to the run-up height r (Figure 10.12). Figure 10.13 shows a series of photographs analogous to Figure 10.10 for a broad-crested dam of impact slope 1:1.

The maximum overtopping volume V_o for waves over a dam without freeboard, i.e. if the reservoir is completely filled to the crest, depends on (Figure 10.12):

- Angle β of dam cross section on waterside;
- Crest geometry (i.e. sharp-, round-, or broad-crested);
- Overtopping time t_o;
- Maximum overtopping height h_d; and
- Shallowness parameter of approach flow.

According to Müller (1995), the following semi-empirical relations apply. The overtopping time t_o is mainly related to the wave period t_w of the first wave as

$$t_o(\frac{g}{h_o})^{1/2} = 4\left[t_w(\frac{g}{h_o})^{1/2}\right]^{4/9}. \tag{10.7}$$

The maximum overtopping height h_d at a vertical dam crest is equal to $h_d = 0.96h_M$, so that the heights h_d of overtopping and h_M of the approach wave are practically equal. With c_o as crest shape factor, the reference overtopping volume V_o per unit width is thus (Müller, 1995)

$$V_o = \sqrt{2}c_o(gh_M^6 h_o^2 t_w^2)^{2/9}. \tag{10.8}$$

The parameter V_o depends strongly on the maximum wave height h_M of the approach flow, whereas the effects of still water depth h_o, and wave period t_w are modest. The accuracy of

Eq. (10.8) is $\pm 10\%$. The crest shape coefficient $c_o = c_\beta \cdot c_\xi$ in Eq. (10.8) depends on the angle β, the crest width ratio $\xi_d = h_d/L_d$ and the base value $c_{90°}$ for the sharp-crested standard weir (Figure 10.12). The effect of β is approximated as

$$c_\beta = c_{90°} + 0.05\sin\left[\frac{2}{3}(90° - \beta)\right].$$ (10.9)

A maximum occurs for $\beta \approx 30°$, as was experimentally found (Hager, 1994).

The effect of relative crest width $\xi_d = h_d/L_d$ on the overtopping volume follows for $0.2 < \xi_d < 2.1$ as

$$c_\xi = 0.60 + 0.19\xi_d.$$ (10.10)

Here $c_\xi \to 1$ for larger ξ_d, corresponding to the sharp-crested weir. A standard value is $c_\xi = 1$.

For known characteristics of the approach flow wave, the dam geometry, and reservoir topography, estimates of the main overtopping parameters for a nearly plane flow may thus be predicted. For spatial flow configurations, recourse to sufficiently large hydraulic models were recommended, based on Froude similitude. Huber and Hager (1997) present a typical example.

The above indications are based mainly on laboratory experimentation conducted on a sufficiently large size so that scale effects are negligible. Impulse waves are a gravity-affected phenomenon in which Froude similitude applies. Typical wave periods are 1/2 to some minutes. The waves are in the transition wave zone or behave as shallow-water-waves. The wave type is in the spectrum of solitary and sinusoidal waves. The tests described were conducted in the late 20th century with limitations mainly in instrumentation and data analysis. From a modern perspective, improved observations could be collected, therefore.

The run-up of plane impulse waves on a dam or a shore is comparable with that of tsunamis. Wave breaking does hardly occur, because the wave is fully reflected, or it breaks before run-up. The run-up height depends on the approach flow amplitude and less on the run-up angle and the wave steepness. Ice in the reservoir up to a thickness of 0.50 m has no effect on the run-up process. Also, rough run-up boundaries have no effect on the run-up height (Müller, 1995). Instead, the reservoir bathymetry has a significant effect on the run-up of impulse waves. Further details are provided in 10.7.

Dam overtopping has the following features:

- *Maximum height* of overtopping wave over a dam crest is always smaller than maximum height of the corresponding run-up wave;
- *Overtopping volume* related to reference volume at zero freeboard increases quadratically with decreasing freeboard;
- *Impulse wave profile* normalized by wave amplitude and wave length is characteristic both during wave propagation in the reservoir and overtopping;
- *Reference overtopping volume* V_o depends on approach flow wave amplitude and less on reservoir depth and wave period; and
- *Spatial features* of overtopping waves are complex. Hydraulic modeling is strongly recommended, therefore.

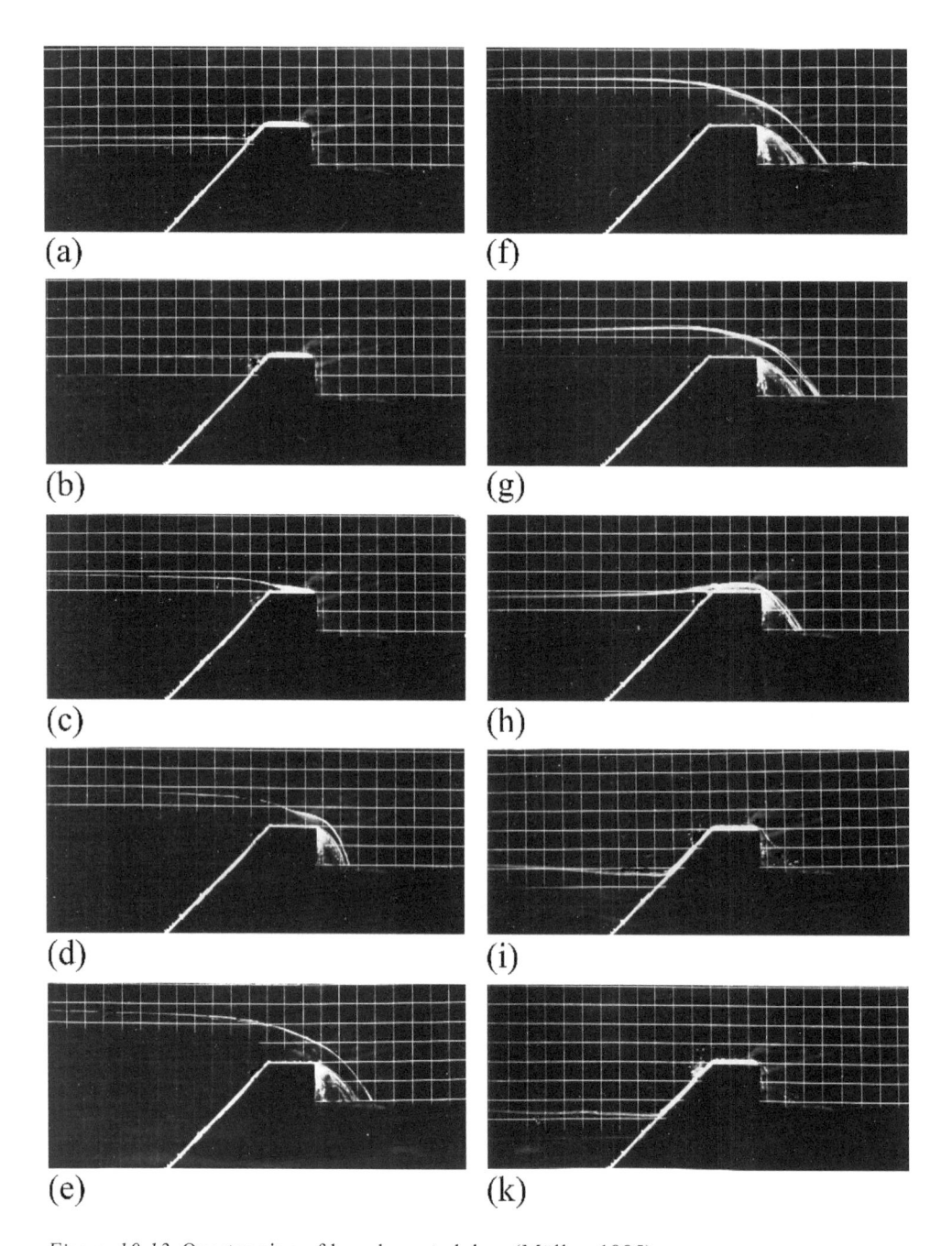

Figure 10.13 Overtopping of broad-crested dam (Müller, 1995)

Given the above limitations, VAW, ETH Zurich, started in 1998 a fundamental research program on impulse waves, in which the following issues were studied:

- 2D impulse wave generation and propagation;
- Effects of the governing parameters including slide density, slide speed, slide thickness, slide angle, among others;
- 2D wave run-up, overtopping, and overland flow;
- Velocity and vorticity fields at slide impact region;
- Underground deposition features of slides into reservoirs; and
- Expansion to 3D impulse waves.

The following summarizes the main findings, based on the previous results.

10.3 2D impulse wave generation and propagation

10.3.1 Review of research activities

Rockfalls, landslides, shore instabilities, snow avalanches, glacier calvings, or even asteroid impacts may generate large water waves in oceans, bays, lakes, or reservoirs. These tsunami-type waves, referred hereafter to as impulse waves, are relevant for the Alpine environment because of steep valley sides, both possible large slide masses and impact velocities, and the considerable number of artificial reservoirs. In contrast to seismically generated tsunamis with a potential global destructive effect, subaerial landslide-generated impulse waves are more local phenomena (Monserrat *et al.*, 2006) due to the considerable wave decay. Nevertheless, fatalities due to these documented catastrophes have exceeded 20,000 fatalities (Schnitter, 1964; Slingerland and Voight, 1979). Mainly passive methods are available to prevent damages due to these waves including evacuation, water level drawdown in artificial reservoirs, freeboard control, or blasting of possible slides. These methods require detailed knowledge of the wave features such as wave height or wave decay. Because these parameters are currently not satisfactorily predictable, the water level of many artificial reservoirs is kept well below the maximum possible level (Panizzo *et al.*, 2005).

Five strategies allow for knowledge expansion on impulse waves, namely (1) specific prototype studies (e.g. WCHL, 1970 for Mica Reservoir), (2) numerical simulations (e.g. Quecedo *et al.*, 2004; Liu *et al.*, 2005; Falappi and Gallati, 2007), (3) prediction based on field data (e.g. Ataie-Ashtiani and Malek Mohammadi, 2007), (4) analytical calculations (e.g. Di Risio and Sammarco, 2008), or (5) generalized model studies because of their ease and economic application to practice (Heller *et al.*, 2009; Evers *et al.*, 2019).

Several general 2D model studies were conducted in wave channels, where waves have been measured in two dimensions propagating along a channel, or wave basins (3D), where waves propagated freely and were measured in three dimensions. These studies were either based on block or granular slides (Heller, 2008; 2007). Important advances in subaerial landslide-generated impulse wave research include the 2D wave type classification of Noda (1970), the introduction of 2D governing dimensionless parameters of Kamphuis and Bowering (1972), and the state-of-the-art of Slingerland and Voight (1979). Data were provided with some 1100 2D and 3D experiments involving granular slide material (Huber, 1980), the independent variation of the governing parameters and the velocity vector fields in the channel axis using Particle Image Velocimetry (PIV) of Fritz (2002) and Fritz *et al.* (2003a), respectively, and the 3D study of Panizzo *et al.* (2005) based on a block slide model. General 2D studies were also conducted e.g. by Law and Brebner (1968), Wiegel *et al.* (1970), Monaghan and Kos (2000), Walder *et al.* (2003), and Panizzo (2004) based on block slides and by e.g. Huber and Hager (1997), Fritz *et al.* (2004), Zweifel (2004), and Zweifel *et al.* (2006) based on granular slide models. The 3D studies of Johnson and Bermel (1949) are based on a block slide model whereas Huber and Hager (1997) reanalyzed the granular slide data of Huber (1980). A computational method based on general model studies for reservoirs including the wave generation, propagation, run-up, dam overtopping, and the forces on dams was presented by Heller *et al.* (2009), and extended by Evers *et al.* (2019).

Although a considerable number of general model studies exists, the prediction of impulse wave features remains challenging. Most general model studies provide an empirical equation for the maximum wave features in the slide impact zone. However, the wave height, wave amplitude, wave period, and wave length at a specific location of the reservoir have to be available to determine the effects of impulse waves on a shore or dam (Müller, 1995; Heller *et al.*, 2009). Therefore, the wave transformation versus the distance from the slide impact zone was also analyzed. Heller *et al.* (2009) and Heller and Hager (2010) attempted to provide information on all aforementioned 2D wave features. Complete empirical design criteria for both the maximum wave features in the slide impact zone and the wave transformation in the wave propagation zone are presented versus the impulse product parameter P including the governing parameters in wide test ranges. These equations were compared with other empirical models for the 1958 Lituya Bay case.

10.3.2 Experimentation

Experimental setup

The experiments were conducted in a prismatic rectangular water wave channel 11 m long, 0.500 m wide, and 1 m deep consisting of glass and steel plates. The pneumatic landslide generator (Fritz and Moser, 2003) accelerated a slide box filled with granular slide material with up to 8 bar air pressure down a 3 m long hillslope ramp. As the slide box reached the maximum velocity its front flap opened. The slide left the box accelerating further down the hillslope ramp due to gravity acceleration g generating an impulse wave in the wave channel. The entire pneumatic landslide generator was adjustable to various still water depths h and slide impact angles α. It allowed for the independent variation of all test parameters as a basic requirement in general model studies for an isolated quantification of the effect of each parameter involved. The inner slide box lengths were 0.600 m, 0.300 m, or 0.150 m whereas the inner box heights were 0.236 m, 0.118 m, or 0.059 m. The inner box width was always 0.472 m corresponding to 94% of the channel width $b = 0.500$ m (Figure 10.14).

The granular (subscript g) slide consisted of four cylindrically shaped grains of diameters 2 mm $\leq d_g \leq$ 8 mm, bulk slide densities 573 kg/m³$\leq \rho_s \leq$ 1592 kg/m³, grain densities $\rho_g = \rho_s/(1-n)$ in the range of 955 kg/m³$\leq \rho_g \leq$ 2745 kg/m³ with bulk slide porosities 39% $\leq n \leq$ 45%, and dynamic bed friction angles 20° $\leq \delta \leq$ 27°. The slide material was made up of barium-sulfate (BaSO$_4$) compounded with polypropylene (PP) or pure PP (Heller, 2007).

All test data were recorded in the channel axis. The main measurement techniques included two Laser Distance Sensors (LDS) to scan slide profiles, seven Capacitance Wave Gages (CWG) to record the wave profiles, and Particle Image Velocimetry (PIV) for determining the velocity vector fields in the slide impact zone, as discussed by Fritz (2002). The

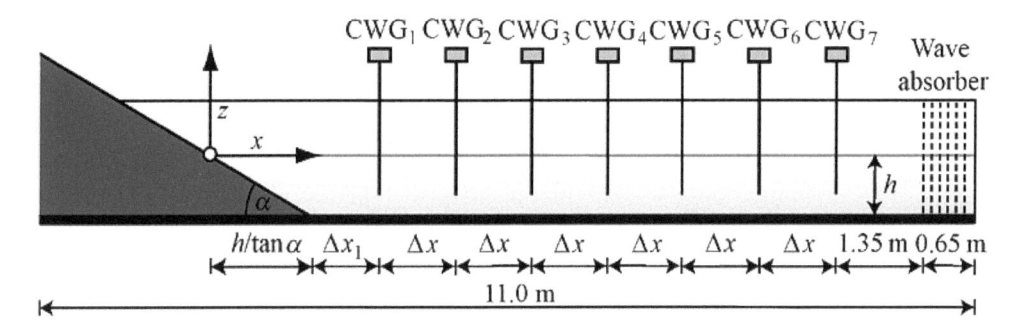

Figure 10.14 Side view of wave channel with seven Capacitance Wave Gages (CWG) along channel axis with spacing for $\alpha = 30°$: $\Delta x_1 = 0.71$ m, $\Delta x = 1.00$ m; $\alpha = 45°$: $\Delta x_1 = 1.13$ m, $\Delta x = 1.00$ m; $\alpha = 60°$: $\Delta x_1 = 1.27$ m, $\Delta x = 1.06$ m; $\alpha = 90°$: $\Delta x_1 = 1.31$ m, $\Delta x = 1.26$ m (Heller and Hager, 2010)

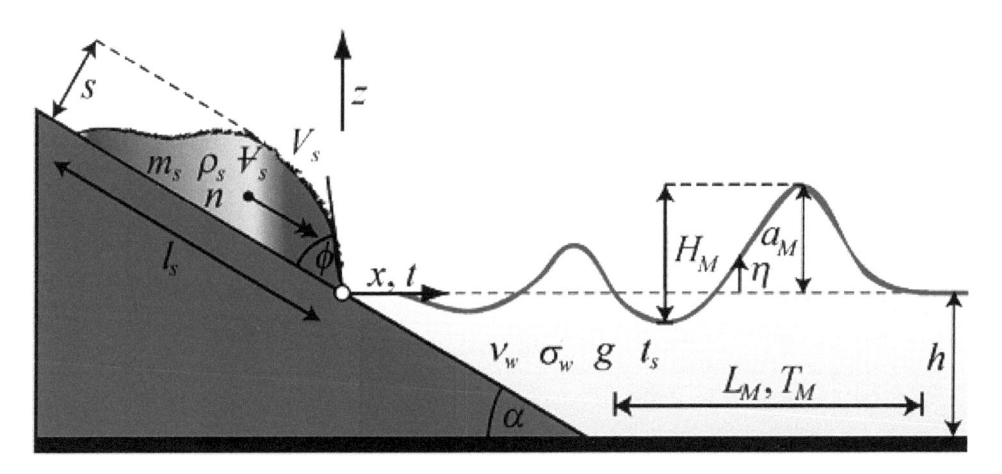

Figure 10.15 Definition sketch of slide impacting water body along with governing slide and water body parameters, and maximum impulse wave characteristics (Heller and Hager, 2010)

LDSs were arranged parallel to the hillslope ramp and their data allowed for the determination of the slide thickness s and the slide impact velocity V_s (Figure 10.15). The bulk slide volume Ψ_s was set identical to the inner box volume; the corresponding bulk slide density $\rho_s = m_s/\Psi_s$ was determined with the bulk slide volume Ψ_s and the slide mass m_s prior to box filling. All three measurement techniques were triggered simultaneously with the start button of the pneumatic landslide generator. The data from the LDSs and CWGs were digitized with an analog-to-digital converter and recorded during 10 s with the software LabVIEW® 6.1. The seven CWGs were arranged as shown in Figure 10.14 in the $(x; z)$ plane with a spacing Δx_1 between CWG_1 and the hillslope ramp end, and equi-distant spacing Δx between two neighboring CWGs which varied both with α. The horizontal distance between the hillslope ramp end and the coordinate origin was $h/\tan\alpha$. Note that CWGs normally underestimate wave heights consisting of a water-air mixture.

Governing parameters

Figure 10.15 shows a definition sketch of the slide, the water body, and the impulse wave features in the $(x; z)$ plane. The coordinate origin is defined at the intersection of the still water surface and the hillslope ramp. The 7 governing parameters affecting the maximum wave height H_M, maximum wave amplitude a_M of period T_M and length L_M include: still water depth h, slide impact velocity V_s, slide thickness s, bulk slide volume Ψ_s, bulk slide density ρ_s, slide impact angle α, and grain diameter d_g. The test ranges of the governing and dimensionless parameters are shown in Table 10.1 for the present and the two previous studies of Fritz (2002) and Zweifel (2004). The bulk slide porosity n, slide mass m_s, slide length l_s, slide front angle ϕ, and time of underwater landslide motion t_s defined as the duration between slide impact and slide stop were not, or only indirectly considered. The parameters n, m_s, ϕ, and t_s are dependent parameters and their consideration would violate the basic requirement of dimensional analysis (Buckingham, 1914). The parameter n is included in ρ_s, and that of $m_s = \rho_s \Psi_s$ in ρ_s and Ψ_s, respectively. The slide front angle ϕ depends on α and V_s, whereas t_s depends on several governing parameters (Panizzo et al., 2005). The parameter t_s was further not considered because it is undefined for $\rho_s < \rho_w$, with ρ_w as water density; slides with $\rho_s > \rho_w$ often impact the opposite shore resulting in a not fully developed t_s, in contrast to general model testing in horizontal channels or basins. The effect of slide length l_s was found negligible relative to wave period T (e.g. Kamphuis and Bowering, 1972).

Table 10.1 Overview of experimental parameters and dimensionless quantity ranges of present study, Fritz (2002), Zweifel (2004), and all VAW runs (Heller, 2007)

Symbol	Dimension	Present study	Fritz (2002)	Zweifel (2004)	All runs
		Experimental ranges			
h	(m)	0.15, 0.20, 0.30, 0.45, 0.60	0.30, 0.45, 0.675	0.15, 0.30, 0.60	0.150–0.675
s	(m)	0.053–0.249	0.050–0.199	0.075–0.189	0.050–0.249
d_g	(mm)	2, 4, 5, 8, and mixture	4	4, 5	2.0–8.0
x	(m)	0.00–8.90	0.00–7.81	0.00–7.73	0–8.90
V_s	(m/s)	2.06–8.34	2.76–8.20	2.60–8.77	2.06–8.77
Ψ_s	(m³)	0.017, 0.033, 0.067	0.017, 0.033, 0.067	0.017, 0.033, 0.067	0.0167–0.0668
ρ_s	(kg/m³)	≈ 590, ≈ 1720	1,720	590, 1010, 1340, 1720	590–1720
t	(s)	0–10	0–10	0–10	0–10
m_s	(kg)	19.69–113.30	27.03–108.12	10.09–108.87	10.09–113.30
α	(°)	30, 45, 60, 90	45	45	30–90
F	(–)	0.86–6.83	1.08–4.66	1.25–4.89	0.86–6.83
S	(–)	0.09–1.64	0.076–0.663	0.125–1.134	0.09–1.64
M	(–)	0.110–10.020	0.119–2.403	0.110–3.588	0.110–10.020
X	(–)	0–59	0–25	0–49	0–59
D_g	(–)	0.003–0.040	0.006, 0.008, 0.013	0.007–0.033	0.003–0.040
T_r	(–)	0–81	0–57	0–81	0–81
i	(–)	211	137	86	434

The following governing dimensionless parameters are relevant:

Slide Froude number $\mathsf{F} = V_s/(gh)^{1/2}$;
Relative slide thickness $S = s/h$;
Relative slide mass $M = m_s/(\rho_w bh^2)$;
Relative grain diameter $D_g = d_g/h$;
Slide impact angle α;
Relative streamwise distance $X = x/h$; and
Relative time $T_r = t(g/h)^{1/2}$.

The model considered is based on Froude similitude; scale effects relative to the maximum wave amplitude a_M are small for Reynolds numbers $\mathsf{R} = g^{1/2}h^{3/2}/v_w \geq 3 \times 10^5$ and Weber numbers $\mathsf{W} = \rho_w gh^2/\sigma_w \geq 5000$, where v_w is kinematic water viscosity and σ_w surface tension of water (Heller *et al.*, 2008). Still water depths of $h \geq 0.200$ m at the slide impact zone account roughly for these criteria. Yet, 25 test were conducted with $h = 0.150$ m which were not excluded from the data analysis because their maximum scale effect relative to the wave amplitude or height was estimated to be smaller than $\pm 15\%$ compared with runs with a negligible scale effect (Heller *et al.*, 2008).

The impulse product parameter is a dimensionless parameter combination resulting in optimized coefficients of determinations R^2 defined as

$$\mathsf{P} = \mathsf{F}S^{1/2}M^{1/4}\{\cos[(6/7)\alpha]\}^{1/2}. \tag{10.11}$$

It is related to the streamwise slide momentum flux component (Zweifel *et al.*, 2006) expressed as $(\rho_s Q_s V_s \cos\alpha)^{1/2} \cong (\rho_s sbV_s^2\cos\alpha)^{1/2} = \rho_s^{1/2}s^{1/2}b^{1/2}V_s(\cos\alpha)^{1/2}$ for a slide discharge Q_s. Because $\cos(90°) = 0$ would result in no impulse wave action, the term $(\cos\alpha)^{1/2}$ was replaced with $\{\cos[(6/7)\alpha]\}^{1/2}$ resulting from the data analysis. The relative effects of V_s and $s^{1/2}$ are correctly accounted for in Eq. (10.11). The relative effect of ρ_s, linearly included in M with $m_s = \rho_s \forall_s$, is with $\rho_s^{1/4}$ smaller than $\rho_s^{1/2}$ in the streamwise slide momentum flux component because the slide energy to impulse wave energy conversion of slides with $\rho_s < \rho_w$ is more efficient than for $\rho_s > \rho_w$, where a part of the slide energy is 'lost' due to impact onto the channel bottom (Heller, 2007). The impulse product parameter P excludes the relative grain diameter D_g because it was found to have a negligible effect on all impulse wave features in the data analysis. The parameters X and T_r are also not contained in P because they are relevant only for wave propagation. Note that P consists exclusively of basic slide parameters and water properties and therefore may be estimated prior to slide impact.

10.3.3 Experimental results

Overview

The subaerial landslide impulse wave generation features a high-speed granular slide impact, the impact crater and primary wave formation, the crater collapse, the phase mixing of water, granulate, and air, and the wave run-up (Fritz *et al.*, 2003b). A photo sequence of a test with $\mathsf{F} = 3.88$, $S = 0.78$, $M = 1.21$, $\alpha = 90°$, and $D_g = 0.013$ is shown in Figure 10.16 for both the slide impact zone, with a time interval of $\Delta t \approx 0.39$ s between two images, and the wave propagation zone. The slide material with $\rho_s = 1634$ kg/m^3 reaches in Figure 10.16a

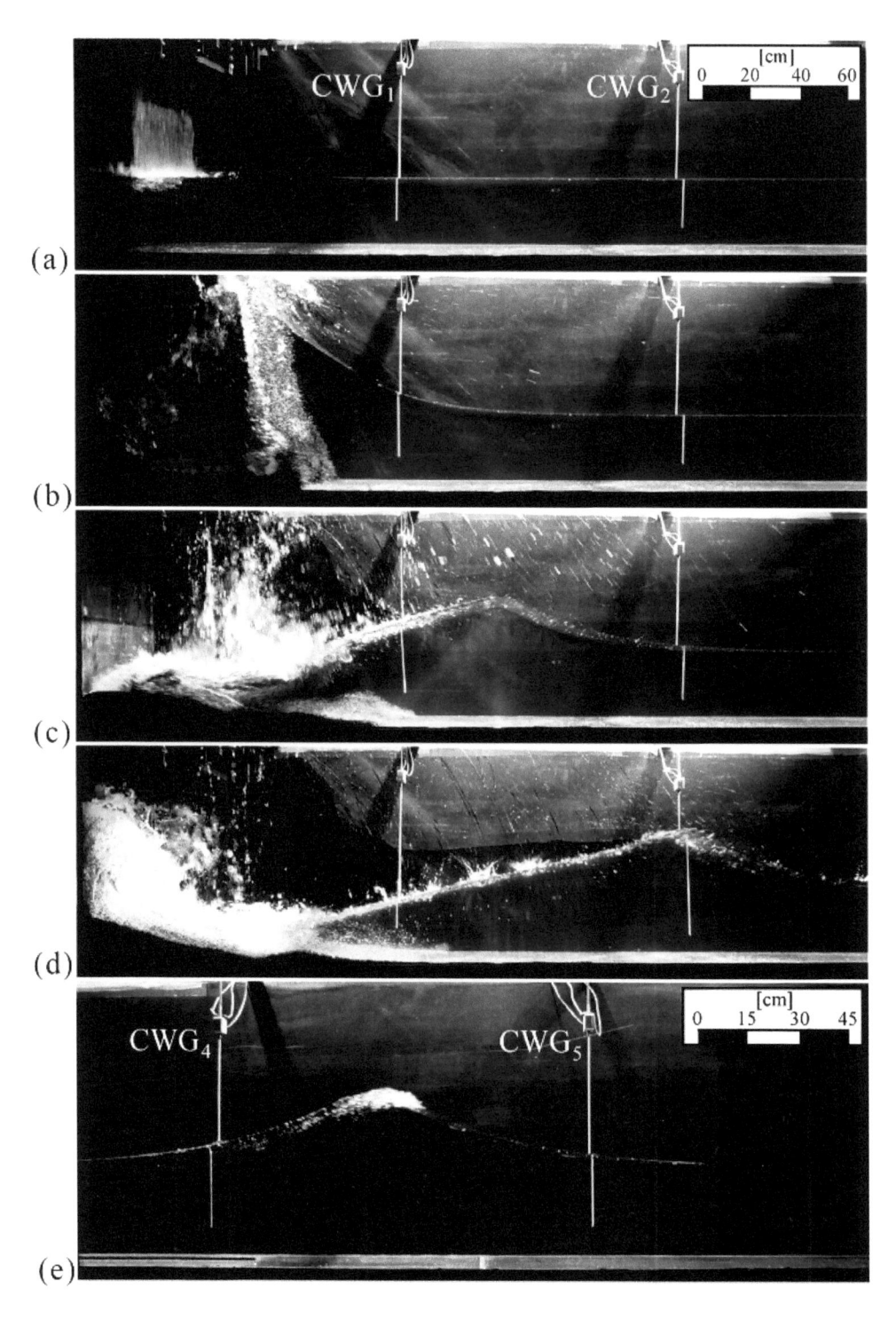

Figure 10.16 Photo sequence of granular slide impact, impact crater formation, primary impulse wave generation and propagation, crater collapse, and flow run-up for $h = 0.300$ m, $\rho_s = 1634$ kg/m³, $\alpha = 90°$, F = 3.88, $S = 0.78$, $M = 1.21$, $D_g = 0.013$ with $T_r \approx$ (a) 0.0, (b) 2.3 (c) 4.6, (d) 6.9 and (e) impulse wave of identical run in channel section at $X = 13.8$–25.6 with $T_r \approx 15.1$ (Heller and Hager, 2010)

the water surface $(t = 0)$ at $h = 0.300$ m. The granulate impacts the wave channel bottom in Figure 10.16b. A water splash larger than the 1 m high wave channel develops associated with an air cavity resulting in an almost vertically lifted water body. The air cavity collapses and the water body divides to form the primary impulse wave of maximum amplitude and a backward current in Figure 10.16c. The backward flow runs up the vertical ramp in Figure 10.16d, the crest of the primary impulse wave reaches CWG_2 while the granular slide comes to rest on the channel bottom (Figure 10.14).

The wave propagation zone is characterized by air detrainment, wave propagation, wave frequency dispersion, and wave decay. Figure 10.16e shows the impulse wave in the channel section between the relative distances $X = 13.8$ to 25.6. The crest of the primary impulse wave is located between CWG_4 and CWG_5 of spilling breaker type having a whitecap due to the large relative wave height.

Figure 10.17 relates to the relative wave profiles η/h, with η as the water surface displacement (Figure 10.15) versus relative time $T_r = t(g/h)^{1/2}$ of the same run as shown in Figure 10.16. Figure 10.17 shows from top (a) to bottom (g) the data at CWG_1 to CWG_7 with a constant spacing between two CWGs of $\Delta x = 1.26$ m (Figure 10.14). In Figure 10.17a at CWG_1 the primary impulse wave has a relative wave amplitude $A = a/h = 0.63$ which increases to the maximum $A_M = a_M/h = 0.82$ at CWG_3 where the maximum relative wave height is $Y_M = H_M/h = 1.04$ (Figure 10.17c). As found below, the maximum wave height is on the average indeed (5/4) times the maximum wave amplitude. The impulse wave undergoes a wave height decay in

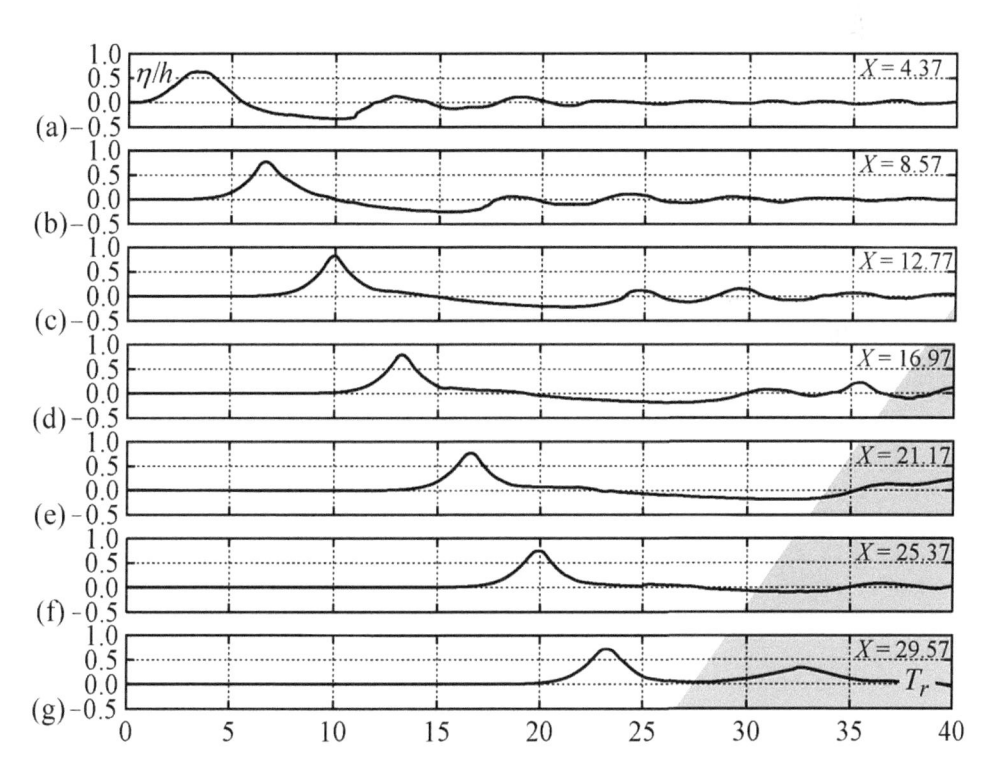

Figure 10.17 Relative wave profiles η/h versus relative time $T_r = t(g/h)^{1/2}$ of run in Figure 10.16 at (a) CWG_1, (b) CWG_2, (c) CWG_3, (d) CWG_4, (e) CWG_5, (f) CWG_6, and (g) CWG_7; gray area is influenced by wave reflection (Heller and Hager, 2010)

Figure 10.17(d to g). The gray zones in Figure 10.17(c to g) refer to wave data affected by wave reflection at the wave absorber located 1.35 m downstream of CWG_7 (Figure 10.14).

Slide impact zone

The extreme wave features applicable for narrow reservoirs ($X < X_M = x_M/h$ with the streamwise distance x_M at a_M) are presented based on CWG measurements. The relative maximum wave amplitude $A_M = a_M/h$ and the relative maximum wave height $Y_M = H_M/h$ are shown in Figure 10.18(a, b) versus P, respectively. The parameters a_M and H_M defined in Figure 10.15 are the largest measured values independent from position in the wave train. In about 95% of the experiments, the primary wave crest had the largest wave amplitude and height, respectively. The remaining 5% related to Stokes-like waves (Stokes, 1847; Heller et al., 2009). The 25 experiments (*) conducted with $h = 0.150$ m were estimated to have a scale effect of less than 15% as compared with runs of negligible scale effect. A correction would only slightly change the data fits yet they are applicable to a much wider parameter range. Maximum relative wave amplitudes and heights of up to $A_M = 2.5$ and $Y_M = 2.6$, respectively, were measured. These maximum values result from the crater rim or water splash (Figure 10.16b) and not of a well-defined wave profile. Multiple regression with a deviation of ±30% and a coefficient of determination $R^2 = 0.88$ for A_M in Figure 10.18a and $R^2 = 0.82$ for Y_M in Figure 10.18b, respectively, yield (Heller and Hager, 2010)

$$A_M = (4/9)P^{4/5}, \qquad \text{for } X < X_M; \text{ and} \qquad (10.12)$$

$$Y_M = (5/9) P^{4/5}, \qquad \text{for } X < X_M. \qquad (10.13)$$

Equations (10.12, 10.13) are identical expect for the pre-factors, resulting in $H_M = (5/4)a_M$. Therefore, the wave trough of landslide-generated impulse wave is on average 20% of the maximum wave height H_M. Note that this criterion is a good approximation for cnoidal-,

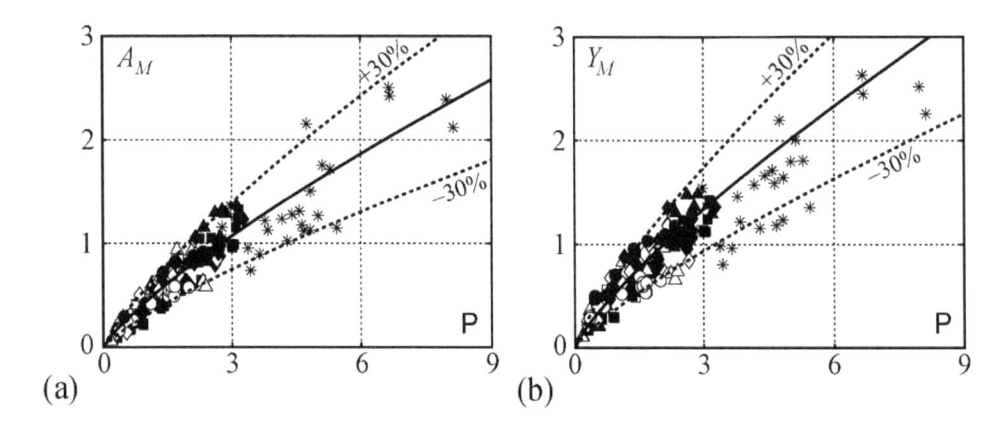

(a) (b)

Figure 10.18 (a) Relative maximum wave amplitude $A_M(P)$ and (−) Eq. (10.12) with (--) ±30% deviation ($R^2 = 0.88$), (b) relative maximum wave height $Y_M(P)$ and (−) Eq. (10.13) with (--) ±30% deviation ($R^2 = 0.82$); legend in Figure 10.20 (Heller and Hager, 2010)

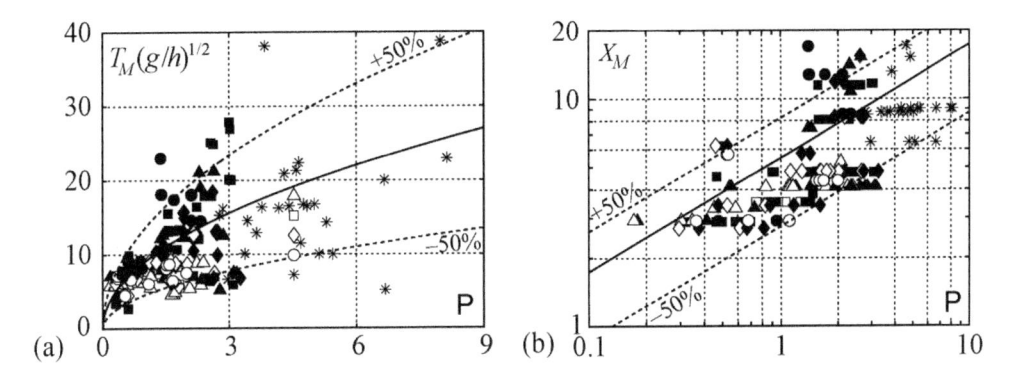

Figure 10.19 (a) Relative wave period $T_M(g/h)^{1/2}$ of a_M versus P and (–) Eq. (10.14) with (--) ±50% deviation ($R^2 = 0.33$), (b) relative distance X_M at a_M versus P and (–) Eq. (10.16) with (--) ±50% deviation ($R^2 = 0.23$); legend in Figure 10.20 (Heller and Hager, 2010)

solitary-, and bore-like wave types but only a rough estimate for Stokes-like wave types (Heller and Hager, 2010).

Figure 10.19a shows the relative wave period $T_M(g/h)^{1/2}$, defined in Figure 10.15 as duration between the first and second water surface up-crossing of maximum wave amplitude a_M, versus P. The definition of T_M is consistent since subaerial landslide-generated impulse waves result always in a leading elevation wave. Relative maximum wave periods ranged within $2.7 \leq T_M(g/h)^{1/2} \leq 38.8$. A multiple regression of deviation ±50% and $R^2 = 0.33$ in Figure 10.19a resulted in (Heller and Hager, 2010)

$$T_M(g/h)^{1/2} = 9P^{1/2} \qquad \text{for } X < X_M. \tag{10.14}$$

The correlation for T_M is less good as for a_M or H_M (Figure 10.18). To determine the effect of impulse waves on dams, this is less important than H_M (Heller *et al.*, 2009).

The last required parameter is the wave length $L_M = cT_M$ of periodic waves (Dean and Dalrymple, 2004), corresponding to an estimate since the impulse waves studied have translational character. The wave celerity c of landslide-generated impulse waves follows the solitary wave celerity as (e.g. Kamphuis and Bowering, 1972; Huber, 1980; Heller, 2007)

$$c = [g(h + a)]^{1/2}. \tag{10.15}$$

The relative distance $X_M = x_M/h$ is required to decide whether Eqs. (10.12) to (10.14) in the slide impact zone, or Eqs. (10.17) to (10.19) in the wave propagation zone apply. The relative distance X_M at a_M versus P is shown in Figure 10.19b. Because the location of a_M is for more than 90% of the runs identical to the location of H_M, the data in Figure 10.19b apply also as an estimate of the relative distance of H_M. Multiple regression with a deviation of ±50% and $R^2 = 0.23$ yields (Heller and Hager, 2010)

$$X_M = (11/2)P^{1/2}. \tag{10.16}$$

The relative distance is in the range of $2.67 \leq X_M \leq 17.13$. About 90% of all data are contained within ±50% of the prediction of Eq. (10.16) in Figure 10.19b.

Wave propagation zone

Subaerial landslide-generated impulse waves may reach a considerable wave height in the slide impact zone such as some 160 m in the below discussed 1958 Lituya Bay case (Fritz et al., 2001). However, they decay faster due to air entrainment, large turbulence production, amplitude dispersion, and frequency dispersion than do e.g. wind waves. An endangered shore or dam may be located farther away from the slide impact zone than X_M. The relative distance $X = x/h$ has also to be included in the data analysis, therefore. The relative distances of the CWGs are available from Figure 10.14 and were in the range of $2.69 \leq X \leq 59.15$. The X values of the CWGs depend besides h also on α because CWG locations were adjusted for each α.

The relative wave amplitude decay $A(X) = a(x)/h$ was evaluated from the data of all seven CWGs. For the wave height decay $Y(X) = H(x)/h$ only the first six CWGs were considered because the wave trough of CWG_7 was affected by wave reflection (Figure 10.17g). Figure 10.20a,b shows $A(X)$ and $Y(X)$, respectively, both versus the product $PX^{-1/3}$. Because the relative distance X has a negative exponent, the coordinate origin in Figure 10.20 corresponds to $X \rightarrow \infty$. The seven relative wave amplitudes and the six relative wave heights, respectively, for each run are connected with a solid line. The symbols with (*) relate to the 25 runs possibly affected by a scale effect. They were all bore-like waves (Shen and Meyer, 1963; Heller and Hager, 2010) and decay generally faster than do these with a negligible scale effect. The multiple regression with a deviation of $\pm 30\%$ and $R^2 = 0.81$ for $A(X)$ in Figure 10.20a and $R^2 = 0.80$ for $Y(X)$ in Figure 10.20b, respectively, yield (Heller and Hager, 2010)

$$A(X) = (3/5)(PX^{-1/3})^{4/5}, \qquad \text{for } X \geq X_M; \tag{10.17}$$

$$Y(X) = (3/4)(PX^{-1/3})^{4/5}, \qquad \text{for } X \geq X_M. \tag{10.18}$$

A comparison of Eqs. (10.17) and (10.18) results in $H = (5/4)a$, as already observed for the slide impact zone. Both the relative wave height and amplitude decay with $X^{-4/15}$, in agreement with other studies based on wave channels with values between $X^{-1/5}$ and $X^{-1/2}$. A nearly identical wave decay was observed by Wiegel et al. (1970) with $X^{-1/5}$ and Huber and Hager (1997) with $X^{-1/4}$ (Table 10.2). The wave decay depends on the wave type (Huber, 1980; Heller, 2007). Nevertheless, the wave decays for each individual wave type, including Stokes-, cnoidal- and solitary-, and bore-like wave types as noted by Heller (2007), result in worse coefficients of determination than $R^2 = 0.81$ based on Figure 10.20a and Eq. (10.17).

Figure 10.21 shows the relative wave period $T(x)(g/h)^{1/2}$ versus $PX^{5/4}$. The multiple regression with a deviation of $\pm 50\%$ and $R^2 = 0.66$ yields (Heller and Hager, 2010)

$$T(x)(g/h)^{1/2} = 9(PX^{5/4})^{1/4}, \qquad \text{for } X \geq X_M. \tag{10.19}$$

In contrast to the decay of $a(x)$ and $H(x)$ with $X^{-4/15}$, the wave period $T(x)$ increases with $X^{5/16}$. Scale effects alone do not explain the scatter in Figure 10.21 for the runs (*) with $h = 0.150$ m. All these runs resulted in an outward collapsing impact crater (Fritz et al., 2003b) and bore-like waves, for which T is obviously ill-defined. Nevertheless, most of the data in Figure 10.21 follow the prediction of Eq. (10.19). The wave length L was determined again from $L = c \cdot T$ with the wave celerity c available from Eq. (10.15).

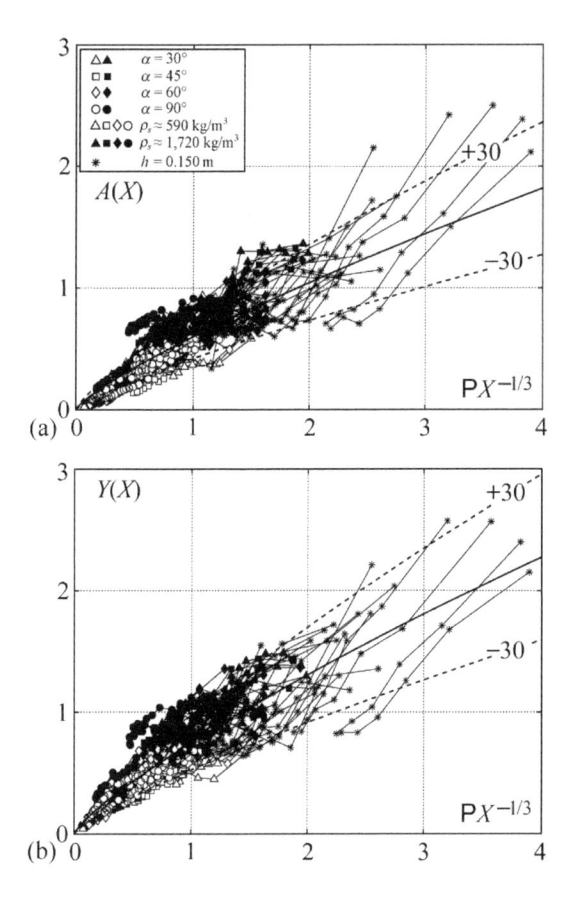

Figure 10.20 (a) Wave amplitude decay $A(X)$ versus $PX^{-1/3}$ and ($-$) Eq. (10.17) ($R^2 = 0.81$) with (--) $\pm30\%$ deviation, (b) wave height decay $Y(X)$ versus $PX^{-1/3}$ and ($-$) Eq. (10.18) ($R^2 = 0.80$) with (--) $\pm30\%$ deviation; data belonging to one run are connected with solid line (Heller and Hager, 2010)

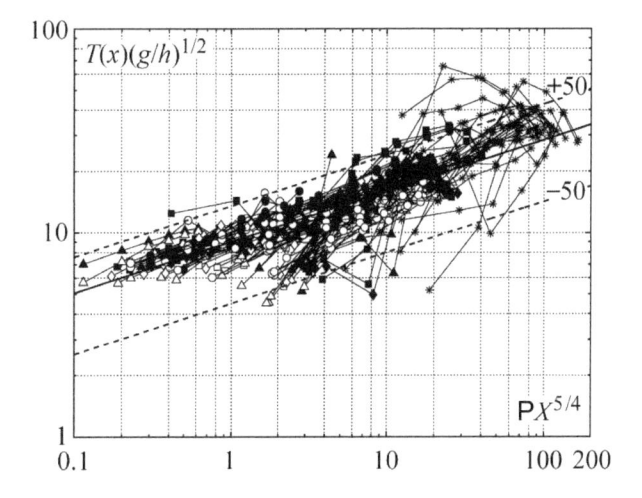

Figure 10.21 Wave period decay $T(X)$ versus $PX^{5/4}$ and ($-$) Eq. (10.19) ($R^2 = 0.66$) with (--) $\pm50\%$ deviation; legend Figure 10.20 (Heller and Hager, 2010)

Table 10.2 Experimental model comparison for parameters in Table 10.3. Top: 2D relative maximum wave amplitude A_M (or relative primary wave amplitude a_p/h) and wave height Y_M. Center: 2D relative wave amplitude $A(X)$ and wave height decay $Y(X)$ at $X = 10$. Bottom: 3D relative maximum wave amplitude $A_M(r/h, \gamma)$ and wave height decay $Y(r/h, \gamma)$ at $r/h = 10$; deviations Δa and ΔH relative to values from Eqs. (10.12) and (10.13), respectively (Heller and Hager, 2010)

Reference	Equation	a_M $a(x)$ $a(r,\gamma)$ (m)	Δa (%)	H_M $H(x)$ $H(r,\gamma)$ (m)	ΔH (%)
Eq. (10.12) (2D granular slide model)	$A_M = (4/9)P^{4/5}$	143	Basis		
Eq. (10.13)	$Y_M = (5/9)P^{4/5}$			179	Basis
(2D granular slide model) Noda (1970) (2D piston-type model)	$A_M = 1.32F$	428	+199		
Monaghan and Kos (2000) (2D block slide model)	$a_p/h = 3(m_s/(40\rho_w sbh))^{2/3}S^{2/3}$	111	−22		
Fritz *et al.* (2001) (2D case study with granulate)		152	+6	162	−9
Fritz (2002)/Fritz *et al.* (2004) (2D granular slide model)	$A_M = (1/4)F^{7/5}S^{4/5}$	95 83	−34 −42		
Walder *et al.* (2003) (2D block slide model)	$A_M = 1.32\{T_s/[\Psi_s/(bh^2)]\}^{-0.68}$ with $T_s = 4.8(l_s/h)^{2/5}$	151	+6		
Zweifel (2004)/Zweifel *et al.* (2006) (2D granular slide model)	$A_M = (1/3)FS^{1/2}M^{1/4}$				
Kamphuis and Bowering (1972) (2D block slide model)	$Y(X) = F^{0.7}[0.31 + 0.2\log(Sl_s/h)] + 0.35\exp(-0.08X)$			132 79	−26 −56
Huber and Hager (1997) (2D granular slide model)	$Y(X) = 0.88\sin\alpha(\rho_g/\rho_w)^{1/4} [\Psi_g/(bh^2)]^{1/2}X^{-1/4}$				
Slingerland and Voight (1979) (3D granular bag model)	$\log(A_M) = -1.25 + 0.71\log[(l/2)(\Psi_g/(h^3))(\rho_g/\rho_w)F^2]$	253	+77		
Huber and Hager (1997) (3D granular slide model)	$Y(r/h, \gamma) = 2 \cdot 0.88(\sin\alpha)\cos^2(2\gamma/3) \times (\rho_g/\rho_w)^{1/4}[\Psi_g/(bh^2)]^{1/2}(r/h)^{-2/3}$			60	−66
Panizzo *et al.* (2005) (3D block slide model)	$Y(r/h,\gamma) = 0.07[T_s/(bs/h^2)]^{-0.45} \times (\sin\alpha)^{-0.88}\exp(0.6\cos\gamma)(r/h)^{-0.44}$ with $T_s = 0.43(bs/h^2)^{-0.27}F^{-0.66} \times (\sin\alpha)^{-1.32}$			32	−82

Comparison with former VAW data

The empirical equations based on 211 runs were confirmed with the 137 runs of Fritz (2002) and 86 runs of Zweifel (2004) conducted in the same hydraulic model. Their parameter ranges are shown in Table 10.1. Note that Zweifel (2004) applied the largest slide impact

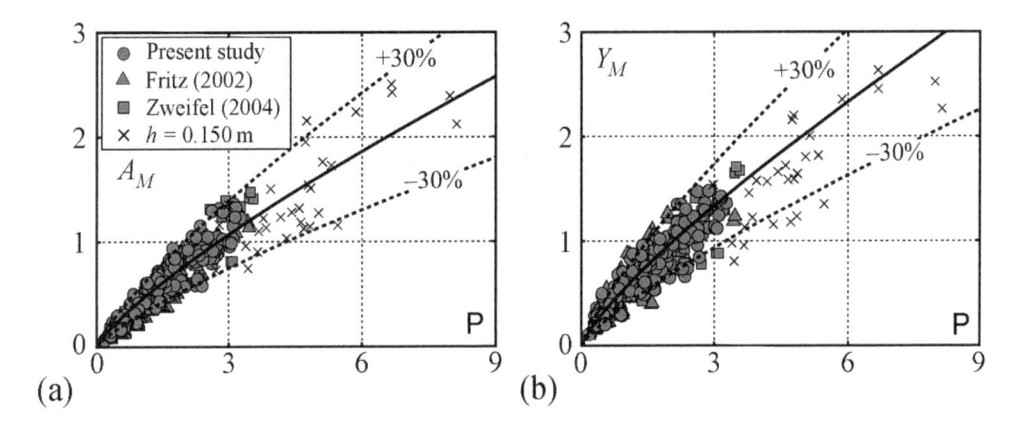

Figure 10.22 Data from all VAW runs (a) relative wave amplitude $A_M(\mathsf{P})$ and (−) Eq. (10.12) with (--) ±30% deviation ($R^2 = 0.89$), (b) relative wave height $Y_M(\mathsf{P})$ and (−) Eq. (10.13) with (--) ±30% deviation ($R^2 = 0.85$) with legend in (a) (Heller and Hager, 2010)

velocity $V_s = 8.77$ m/s, whereas Fritz (2002) conducted tests up to $h = 0.675$ m and used the smallest slide thickness of $s = 0.050$ m. Heller (2007) expanded the ranges of nearly all dimensionless parameters (Table 10.1).

Figure 10.22 replots Figure 10.18 with the relative maximum wave amplitudes A_M and the relative maximum wave heights Y_M versus P including the runs of Heller (2007), those of Fritz (2002) and Zweifel (2004), and the 31 tests with $h = 0.150$ m. Equations. (10.12) and (10.13) agree with the data of the two previous studies, and the goodness of fit is even improved in Figure 10.22a with $R^2 = 0.89$ as compared with $R^2 = 0.88$ in Figure 10.18a, and in Figure 10.22b with $R^2 = 0.85$ as compared with $R^2 = 0.82$ in Figure 10.18b. The goodness of fit of nearly all presented equations is improved if all VAW data are considered including the wave period T_M from Eq. (10.14), the amplitude decay $a(x)$ from Eq. (10.17), or the wave height decay $H(x)$ from Eq. (10.18).

Lituya Bay case

The 1958 Lituya Bay case is convenient for a comparison of results because it is well documented, its governing parameters match the present parameter ranges (Table 10.1), and the Lituya Glacier on the northwest side of the water body reduced 3D effects. This event was described by Miller (1960a), Slingerland and Voight (1979), Fritz *et al.* (2001), among others. Numerical simulations of this event are available from Mader and Gittings (2002), Quecedo *et al.* (2004), Schwaiger and Higmann (2007), Weiss and Wünnemann (2007), Weiss *et al.* (2009), among others. The T-shaped bay is located near the St. Elias Mountains in Alaska, where the main bay is about 12 km long and 1.2 km to 3.3 km wide, except for the exit to the Pacific Ocean being 300 m wide. On July 9, 1958, an 8.3 moment magnitude earthquake with the epicenter located 21 km southeast of Lituya Bay lasted for 1–4 min. Therefore, a rockslide of grain density $\rho_g = 2700$ kg/m^3 was triggered from up to 914 m above sea level on a slope of hillslope angle $\alpha = 40°$ (Figure 10.23). The governing parameters are listed in Table 10.3 together with the references. The mean slide width was $b = 823$ m, the maximum slide thickness $s = 92$ m, and the slide length $l_s = 970$ m. The slide centroid started at $\Delta z_{sc} = 609$ m above the still water surface. The slide dimensions resulted in a slide grain volume of $V_g = 30.6 \times 10^6$ m³. Heller (2007) assumed a bulk slide porosity of $n = 40\%$ during slide impact due to crushing of the grain volume, resulting in a bulk slide volume of $V_s = V_g/(1-n) = 51.0 \times 10^6$ m³ and a bulk slide density $\rho_s = \rho_g(1-n) = 1620$ kg/m³. Note that also l_s, b, and s may change for a granular slide from the initial slide position to the slide impact location. However, the original values of b and s were applied assuming that their increase due to the increases of n and V_s and the decrease due to the slide dispersion during slide movement compensate. The mean still water depth in the slide impact zone was $h = 122$ m. The slide impact velocity V_s followed from an energy balance between the slide release and impact location (Körner, 1976) using a dynamic bed friction angle of $\delta = 14°$ (Slingerland and Voight, 1979) as

$$V_s = \sqrt{2g\Delta z_{sc}(1 - \tan\delta\cot\alpha)} = 92\,\text{m/s}. \tag{10.20}$$

The rockslide in the Lituya Bay generated an impulse wave of maximum run-up height $R = 524$ m on the opposite shore at distance $x \approx 1350$ m and a run-up angle $\beta = 45°$. The wave height was computed inversely by Fritz et al. (2001) with the solitary wave run-up equation of Hall and Watts (1953) as $H = 162$ m.

If the present empirical equations are applied to the Lituya Bay case, the first step is to decide with the use of Eq. (10.16) whether the maximum wave height H_M from Eq. (10.13) or the wave height $H(x)$ from Eq. (10.18) is relevant. From Eq. (10.16) $x_M = 1232$ m, i.e. slightly larger than the horizontal distance from the intersection of the hillslope with the still water surface to the hillslope end of the opposite shore $x = 1350$ m − 122 m = 1228 m. Note that the calculated value $H = 162$ m is located at the hillslope end of the opposite shore and the distance 1350 m from Figure 10.23 has to be reduced by $h/\tan\beta = 122$ m. Equation (10.13) for $X < X_M$ is therefore relevant, resulting in a maximum wave height of $H_M = 179$ m.

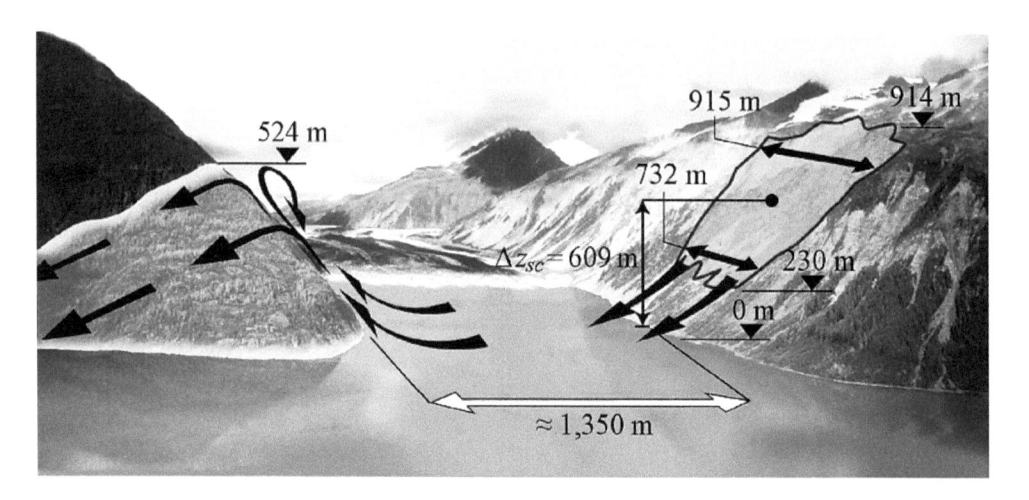

Figure 10.23 1958 Lituya Bay case with rockslide dimensions and maximum wave run-up height $R = 524$ m (after Fritz et al., 2001)

Table 10.3 Governing parameters of 1958 Lituya Bay case and associated references

Symbol	Dimension	Value	References and comments
h	(m)	122	Miller (1960); Slingerland and Voight (1979); Fritz *et al.* (2001); among others
s	(m)	92	Miller (1960); Slingerland and Voight (1979); Fritz *et al.* (2001); among others
b	(m)	823	Miller (1960); Slingerland and Voight (1979); Fritz *et al.* (2001); among others
l_s	(m)	970	Slingerland and Voight (1979); Fritz *et al.* (2001); among others
V_s	(m/s)	92	According to Eq. (10.20) with a dynamic bed friction angle of $\delta = 14°$ from Slingerland and Voight (1979) and $\Delta z_{sc} = 609$ m from Miller (1960a)
$\mathcal{V}_g$	(m³)	30.6×10^6	Miller (1960a); Slingerland and Voight (1979); Fritz *et al.* (2001); among others
$\mathcal{V}_s$	(m³)	51.0×10^6	With $n = 40\%$ and volume increase due to the crushing of the grain volume $\mathcal{V}_g$
ρ_g	(kg/m³)	2,700	Miller (1960a); Slingerland and Voight (1979); Fritz *et al.* (2001); among others
ρ_s	(kg/m³)	1,620	$\rho_s = \rho_g (1 - n)$ with $n = 40\%$
α	(°)	40	Slingerland and Voight (1979); among others
F	(-)	2.66	
S	(-)	0.75	
M	(-)	6.74	With $n = 40\%$
l_s/h	(-)	7.95	
P	(-)	3.37	

The northwest side of the Lituya Bay slide was barred by the Lituya Glacier whereas the southeast side was open. A part of the wave energy was therefore 'lost' on the southeast side, resulting in a slightly smaller wave height $H = 162$ m as compared with the prediction $H_M = 179$ m based on the 2D approach. It is often ignored in practice that empirical equations based on a prismatic wave channel or rectangular wave basin are primarily applicable to similar prototype geometries thereby following the fundamental principle of geometrical similarity (Hughes, 1993). Deviations from these idealized geometries may result in appreciable effects due to e.g. refraction or wave reflection (Camfield, 1980).

Comparison with other experimental data

The values $H_M = 179$ m from the previous section and $a_M = 143$ m resulting from Eq. (10.12) for the Lituya Bay case are shown in Table 10.2, serving as comparison basis with ten other empirical equations and a case study. The deviations Δa and ΔH, respectively, to this basis are expressed in percents on the right-hand side of Table 10.2. 2D formulas for a_M, H_M, or the primary wave amplitude a_P were presented by Noda (1970) for a piston-type model, by Monaghan and Kos (2000) and Walder *et al.* (2003) for block models, and by Fritz *et al.* (2001), Fritz (2002), and Zweifel (2004) for granular slide models.

The piston-type model of Noda (1970) strongly overestimates the maximum wave amplitude with $a_M = 428$ m (+199%). In this model, the impulse waves were generated with a horizontal moving plank much larger than h, not allowing water to flow over its back, whereas in the Lituya Bay case a part of the water flowed over the back of the landslide (Fritz et al., 2001). The Scott Russell wave generator of Monaghan and Kos (2000) consisting of a vertical dropping rigid box predicts the primary wave amplitude to $a_P = 111$ m (-22%). The parameter a_P is compared with a_M because the primary wave coincides with the maximum wave in 95% of the runs of Heller (2007). Fritz et al. (2001) built a specific model for the 1958 Lituya Bay case using a granular slide. They measured at $X = 7.25$ the value $a_M = 152$ m ($+6\%$) and $H_M = 162$ m (-9%), the latter perfectly matching the back-calculated value of $H = 162$ m from observation at $X = 10$. The maximum wave amplitude $a_M = 95$ m according to Fritz (2002) is 34% smaller and $a_M = 151$ m according to Zweifel (2004) is $+6\%$ larger than Heller's value $a_M = 143$ m although the identical hydraulic model was applied. The discrepancy of -34% is explained with the neglect of the relative slide mass M by Fritz (2002). Recall that the data of both Fritz (2002) and Zweifel (2004) perfectly match the prediction of Eq. (10.12) for a_M and even improve the goodness of fit. A small value of $a_M = 83$ m (-42%) yields the block model of Walder et al. (2003), who used the relative time of underwater landslide motion $T_s = t_s(g/h)^{1/2}$. Their experiments were conducted with still water depths of $h = 0.051$ m to 0.130 m, i.e. clearly in the range of non-negligible scale effects $h < 0.200$ m thereby resulting in smaller wave heights.

The center part of Table 10.2 includes equations by authors having offered only an equation for the wave amplitude $a(x)$ or the wave height decay $H(x)$, respectively, but without an indication on the maximum wave features. They were again applied to the Lituya Bay case at $X = 1228/122 = 10$ and compared with the present prediction.

2D wave height decay formulas were presented by Kamphuis and Bowering (1972) for a block model and by Huber and Hager (1997) for a granular slide model. Note that Zweifel (2004) also offered a formula for the wave amplitude decay $a(x)$. Because the basis in Table 10.2 was determined with the formula for the maximum wave amplitude, the formula for a_M of Zweifel (2004) was also considered. The solution $H(X = 10) = 132$ m of Kamphuis and Bowering (1972) underestimates the value $H_M = 179$ m of Heller and Hager (2010) by 26%, whereas the solution of Huber and Hager (1997) results with $H(X = 10) = 79$ m in a considerably smaller wave height (-56%). The 2D equation of Huber and Hager (1997) yields rather conservative values, as also noted by e.g. Fritz et al. (2004).

The lower portion of Table 10.2 shows the 3D wave amplitude $a(r, \gamma)$ and height $H(r, \gamma)$ decay formulas. Smaller values are expected as compared with the basis due to the radial and transverse spread of the wave energy in 3D. Surprisingly, the equation of Slingerland and Voight (1979) based on a granular bag model yields with $a_M(r/h \approx 4, \gamma = 0°) = 253$ m a larger value ($+77\%$) than the basis $a_M = 143$ m based on 2D. Note that their maximum wave amplitude is only applicable for $r/h \approx 4$ and a wave propagation angle $\gamma = 0°$. At $r/h \approx 4$ the wave height is indeed larger than at $r/h = 10$ according to Fritz et al. (2001). However, this does not explain the large difference of $+77\%$. The formula of Slingerland and Voight (1979) is based on two 3D case studies and is probably hardly applicable. The slide model of Huber and Hager (1997) results in $H(r/h = 10, \gamma = 0°) = 60$ m (-66%). The value $H(r/h = 10, \gamma = 0°) = 32$ m of Panizzo et al. (2005) is almost 2 times smaller than $H = 60$ m of Huber and Hager (1997) or -82% as compared with Heller and Hager (2010). It is even more surprising since Panizzo et al. (2005) applied a block model resulting generally in higher waves than granular slide models (Zweifel, 2004). No explanation for this effect was offered.

The comparison in Table 10.2 results partly in strong differences. An obvious reason is that equations based on 2D and 3D should not be compared because the wave decay differs considerably due to different geometries. However, also within the studies based on identical geometries, the deviations are up to a factor of 3. Possible reasons are not negligible scale effects, limitations of some studies in Table 10.2 not satisfying Lituya Bay parameters, slide modeling as a rigid block or with granulate, splash formation complicating the definition of wave features, application of different measurement systems, or identical measurement systems but at different locations.

Limitations of equations

The equations of Heller and Hager (2010) apply to subaerial but not to submarine, submerged, subaqueous, or underwater landslide-generated impulse waves as addressed by Najafi-Jilani and Ataie-Ashtiani (2008), Bardet *et al.* (2003), or Watts (2000). Although many identical parameters including the hillslope angle or the slide volume affect both phenomena, their relative importance is different and they occur in different ranges than investigated at VAW (e.g. underwater slides propagate already at slope angles around $1°$; Hampton *et al.*, 1996). Parameters not relevant for subaerial slides such as the initial still water depth at center point of sliding mass (Najafi-Jilani and Ataie-Ashtiani, 2008) affect in addition underwater landslide-generated impulse waves. Further, air can have a dominant effect in subaerial slides resulting in impact craters (Figure 10.16b) whereas mainly water and solid are involved in underwater slides.

Applying the above equations exceeding the limitations in Table 10.1 may result in larger deviations of the wave parameters than shown in Figures 10.18–10.22. The effect of the relative grain diameter D_g was found small in the range $0.003 \leq D_g \leq 0.040$. However, if the grain diameter is so small that water enters only slowly into the bulk pore volume, larger waves may occur and the effect of D_g may no more be negligible. The same effect is expected if the mass propagates not as a granular slide but as a block (e.g. as rockfall as specified by Cruden and Varnes, 1996), resulting in much larger maximum wave amplitudes depending on the slide Froude number (Zweifel, 2004). Effects such as refraction or reflections are relevant if the equations based on simple channel geometries are applied to real reservoirs (Heller *et al.*, 2009).

In summary, the 2D generation and propagation of subaerial landslide-generated impulse waves were studied based on Froude similitude and granular slide material. The seven independently tested governing parameters include the still water depth h, slide impact velocity V_s, slide thickness s, bulk slide volume $⩝$, bulk slide density ρ_s, slide impact angle α, and grain diameter d_g. The results are summarized as follows:

1. Governing dimensionless parameters are slide Froude number $F = V_s/(gh)^{1/2}$, relative slide thickness $S = s/h$, relative slide mass $M = m_s/(\rho_w bh^2)$, and slide impact angle α.
2. Impulse product parameter $P = FS^{1/2}M^{1/4}\{\cos[(6/7)\alpha]\}^{1/2}$ is relevant parameter combination for all presented empirical equations. It accounts for streamwise slide momentum flux component. The wave features increase with increasing P. Parameter F is dominant dimensionless parameter of impulse wave generation.
3. Slide impact angle α included as $\{\cos[(6/7)\alpha]\}^{1/2}$ affects the wave properties. The wave features increase with decreasing α if the remaining dimensionless parameters stay constant.
4. Two sets of empirical equations involve maximum wave features a_M and H_M, T_M in the slide impact zone ($X < X_M$), and the wave decay features $a(x)$, $H(x)$, and $T(x)$ in the wave propagation zone ($X \geq X_M$).

5　Wave amplitude is on average 80% of the wave height if both the maximum wave features a_M with H_M and the wave decay $a(x)$ with $H(x)$ are compared.

6　Decays of both the relative wave amplitude $A(X)$ and relative wave height $H(X)$ are proportional to relative streamwise distance $X^{-4/15}$, whereas the relative wave period $T(x)(g/h)^{1/2}$ increases with $X^{5/16}$.

7　Validation of equations with another 223 VAW runs results in improved goodness of fit.

8　Wave height was compared with eleven existing approaches for 1958 Lituya Bay case. Similar model studies result in considerable differences.

Limitations of the proposed results are highlighted. For their application to real-world geometries consider Heller *et al.* (2009). Note the importance of the slide velocity V_s or the slide Froude number F, given that these affect the impulse product parameter linearly, whereas the other parameters have only a secondary effect on the main wave features.

10.4 Impulse wave types

10.4.1 Motivation and experimentation

Subaerial landslide-generated impulse waves involve nonlinear and intermediate- to shallow-water waves of small to considerable fluid mass transport (Heller, 2007). They are dispersive (Kamphuis and Bowering, 1972) and cover a wide wave type spectrum. Depending on the wave type, the wave profile, the amount of fluid mass transport, run-up height, or wave force on a structure differ. Knowledge of the wave type is therefore a prerequisite to reliably predict the effects of impulse waves on a dam (Heller and Hager, 2011).

Figure 10.15 sketches an impulse wave in the (x, z) plane involving the basic wave parameters. These are also defined in Figure 10.34 for the sinusoidal wave. Wave type classifications of impulse waves exist for wave channel experiments. Four approaches are available to classify impulse waves:

(i)　Optical wave profile inspection
(ii)　Nonlinearity a/H
(iii)　Ursell parameter $U = (H/L)/(h/L)^3 = HL^2/h^3$
(iv)　Wavelet transform analysis.

A linear wave according to method (ii) requires $a/H = 0.5$ and a solitary wave $a/H = 1$. The Ursell parameter U in method (iii) is herein defined with the wave height H in analogy to Le Méhauté (1976), Huber (1980), or Panizzo *et al.* (2005). For a linear wave it tends to $U \to 0$. According to Ursell (1953) U is a more appropriate criterion to identify linear waves than simply $a/H = 0.5$ since these require not only identical crest and trough amplitudes but also small ratios of H/h and H/L.

Existing wave type classifications may fail if applied as predictive method because they do not include all governing parameters, are based on block and not on granular slides (Noda, 1970; Wiegel *et al.*, 1970; Panizzo *et al.*, 2002, 2005), resulting in design diagram regions including several wave types (Zweifel *et al.*, 2006), or include no design diagram to predict the governing wave type (Huber, 1980; Panizzo *et al.*, 2002). The following aims to classify the maximum impulse wave types in the near field due to granular slides based on optical wave profile inspection and the criteria defined in Table 10.4. The purpose is to

Table 10.4 Criteria for wave type classification (Heller and Hager 2011)

Stokes-like wave	cnoidal-like wave	solitary-like wave	bore-like wave
wave profile symmetric (to both axis)	wave profile symmetric (to vertical axis)	wave profile symmetric (to vertical axis)	irregular wave profile (to both axes)
trough identical long as crest	trough longer than crest		steep wave front, flat wave tail
multiple equivalent crests (at least two)	multiple crests (at least two)	one dominant crest	one dominant crest
$a \approx a_t$		trough nearly absent $(a_t < a/4)$	
no air transport $a_M < h/2$ ($\approx$ intermediate-water)	small air transport $a_M < h$ (intermediate-water)	small air transport intermediate-water	large air transport intermediate- to shallow-water

predict the wave types directly versus the governing parameters, so that the wave parameters addressed by Heller and Hager (2010) are not required. It concludes further the 2D impulse wave research of Fritz (2002), Zweifel (2004) and Heller (2007), including the effects of hillslope angle, and an extension of most dimensionless governing parameter ranges (Heller and Hager, 2010).

The tests were conducted in the prismatic, horizontal wave channel previously employed by Fritz (2002) and Zweifel (2004). The front sidewall was made of glass, whereas the back sidewall and the bottom consisted of continuous steel plates, and glass, respectively. The granular material was filled in the slide box and accelerated with the pneumatic landslide generator down a 3 m long hillslope ramp. Typically, 1.3 m above the still water surface, the box front flap opened and the slide moved free on the ramp before impacting the water body to generate impulse waves. The wave types were determined from the wave profiles and controlled with video recordings. Details on the physical model are given by Heller (2007) and are detailed in 10.3.2.

Figure 10.24 shows the relevant parameters. The pneumatic landslide generator allowed for a systematic variation of the 7 governing parameters in the test ranges: still water depth $0.150 \text{ m} \leq h \leq 0.675 \text{ m}$, slide impact velocity $2.06 \text{ m/s} \leq V_s \leq 8.77 \text{ m/s}$, slide thickness $0.050 \text{ m} \leq s \leq 0.249 \text{ m}$, bulk slide volume $0.017 \text{ m}^3 \leq V_s \leq 0.067 \text{ m}^3$, bulk slide density $590 \text{ kg/m}^3 \leq \rho_s \leq 1720 \text{ kg/m}^3$, slide impact angle $30° \leq \alpha \leq 90°$, and grain diameter $2.0 \text{ mm} \leq d_g \leq 8.0 \text{ mm}$. The governing parameters V_s and s were related to the slide impact location. The streamwise distance $0 \leq x \leq 8.90 \text{ m}$ and times $0 \leq t \leq 10 \text{ s}$ are relevant in the wave propagation zone.

The dimensional analysis of Heller *et al.* (2008) resulted in the dimensionless governing parameters: slide Froude number $F = V_s/(gh)^{1/2}$ varied within $0.86 \leq F \leq 6.83$, relative slide thickness $S = s/h$ $(0.09 \leq S \leq 1.64)$, relative slide mass $M = m_s/(\rho_w bh^2)$ $(0.11 \leq M \leq 10.02)$, relative grain diameter $D_g = d_g/h$ $(0.003 \leq D_g \leq 0.040)$, slide impact angle α $(30° \leq \alpha \leq 90°)$, relative streamwise distance $X = x/h$ $(0 \leq X \leq 59.2)$, and relative time $T_r = t(g/h)^{1/2}$ $(0 \leq T_r \leq 81)$. Negligible scale effects with regard to the relative maximum wave amplitude, due to surface tension or kinematic viscosity, for impulse waves involving a granular slide require by rule of thumb $h \geq 0.200 \text{ m}$ (Heller *et al.*, 2008). Some tests were conducted with $h = 0.150 \text{ m}$; scale effects in this range affect the wave height or wave decay, but hardly change the wave type.

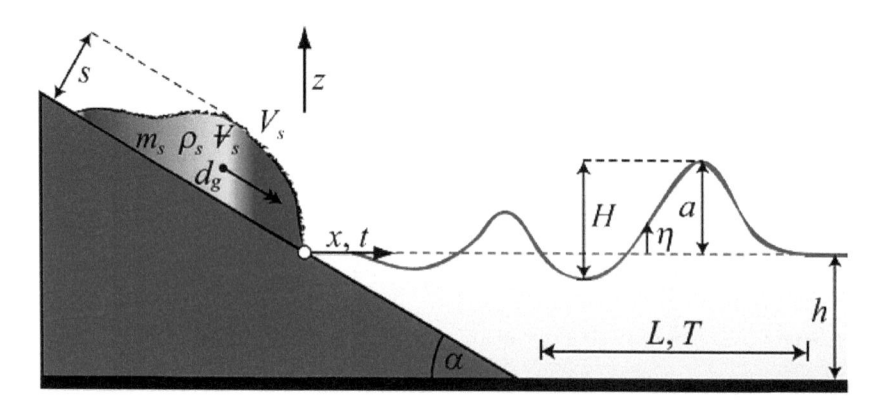

Figure 10.24 Definition sketch of governing parameters and wave features in (x, z) plane

10.4.2 Experimental results and discussion

Wave type classification

The wave length L was determined with the measured wave celerity c and wave period T as $L = cT$. Impulse waves measured at CWG_1 ranged within $5.1 \leq L/h \leq 82.7$, corresponding to intermediate- to shallow-water waves, therefore. The primary wave was in 95% equal to the maximum wave. About 90% of the generated maximum waves were in the intermediate-water wave range. Further, the Ursell parameter at CWG_1 was in the range $2.7 \leq U \leq 10{,}340$. Linear waves with $U \to 0$ were not observed. The impulse waves studied are therefore characterized as nonlinear, intermediate- to shallow-water waves with small to considerable fluid mass transport. The maximum waves were allocated to the four nonlinear types: (1) Stokes wave, (2) cnoidal wave, (3) solitary wave, and (4) bore (see 10.4.3).

The following wave type classification includes not only impulse waves with a 'perfect' but also with a transient profile because of their relevance in practice. The criteria for the classification defined in Table 10.4 are individually discussed hereafter for each wave type. Impulse waves may change their type over a short reach. Bore-like wave profiles in the slide impact zone for instance transform to a solitary- or cnoidal-like wave profile farther away from the impact zone due to energy dissipation and air detrainment. For the present classification, the most characteristic wave type of the maximum wave between the first and the last CWG $(2.7 \leq X \leq 59.2)$ was selected.

Heller *et al.* (2009) identified the wave type product $T = S^{1/3}M\cos[(6/7)\alpha]$ as the relevant number characterizing a landslide impacting a water body. Figure 10.25 relates T to the slide Froude number F for the 211 tests of Heller (Figure 10.25a) and for all tests including the data of Fritz (2002) with 137 runs and Zweifel (2004) with 86 runs (Figure 10.25b). The symbols refer to Stokes-like waves (o), cnoidal-like waves (□), solitary-like waves (◇), and bore-like waves (▽). Data shown with (*) were conducted in the range of possible scale effects, all of which resulted in bore-like waves. Open data in Figure 10.25a refer to runs with $\rho_s \approx 590$ kg/m³ and dark data to runs with $\rho_s \approx 1720$ kg/m³, whereas dark data

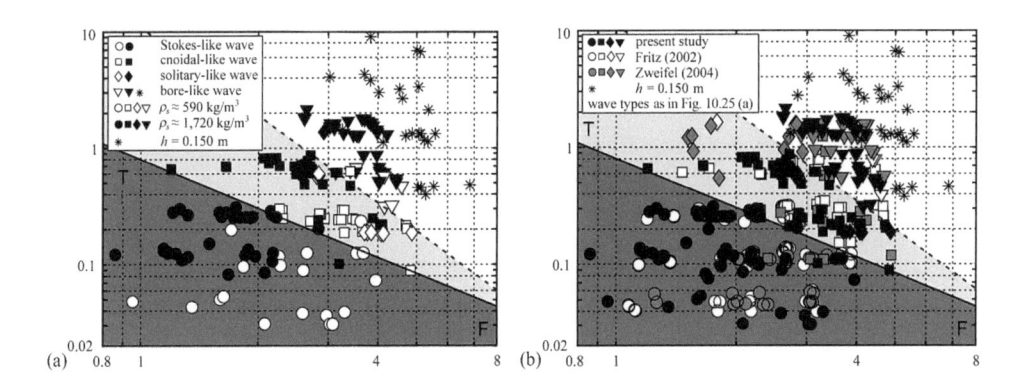

Figure 10.25 Wave type classification based on Table 10.4: wave type product
T = $S^{1/3}M\cos[(6/7)\alpha]$ and (–) Eq. (10.21) (= instead of <) and (--) Eq. (10.23)
(= instead of >) versus F for (a) data of Heller (2007), (b) combined data of
Heller (2007), Fritz (2002), and Zweifel (2004) (Heller and Hager, 2011)

in Figure 10.25b refer to Heller's (2007) data, open data to runs of Fritz (2002), and light
shaded data to Zweifel's (2004).

All four wave types occurred at all considered slide impact angles α. However, solitary-
like wave types are particularly well-developed for $\alpha = 90°$ originating from Russell's (1837)
experiments where solitary waves were generated with a vertically dropped block. Three
wave type zones may be identified in Figure 10.25, namely

$$T < \frac{4}{5}F^{-7/5} \qquad \text{Stokes-like waves} \tag{10.21}$$

$$\frac{4}{5}F^{-7/5} \leq T \leq 11F^{-5/2} \quad \text{cnoidal- or solitary-like waves} \tag{10.22}$$

$$T > 11F^{-5/2} \qquad \text{bore-like waves.} \tag{10.23}$$

The cnoidal- and solitary-like waves are not separable in Figure 10.25. A reason appears to
be the theoretical similarity between these two with the solitary wave as a special case of the
cnoidal wave (see 10.4.4). The number of outliers in Figure 10.25a and b is 8 (4% of 211)
and 37 (9% of 434), respectively. The slide Froude number F and the relative slide mass M
are the dominant parameters determining the impulse wave type. Runs with large values of
F, S, M, and small α tend to generate bore-like waves because of the large streamwise slide
momentum flux component represented by the impulse product parameter P = $FS^{1/2}M^{1/4}\{\cos$
$[(6/7)\alpha]\}^{1/2}$. In contrast, small F, S, M, and large α values tend to generate Stokes-like waves
in accordance with a small P. A wave type classification based only on P was not success-
ful even though it is the relevant parameter for most wave features. In the following, the
observed four wave types are discussed individually for the wave propagation zone, whereas
their effects on a dam are determined by Heller *et al.* (2009).

Stokes-like waves

Figure 10.26a–c shows a photo sequence of Stokes-like impulse waves generation involving $\rho_s = 608$ kg/m³, $\alpha = 60°$, and small relative values of F = 1.36, $S = 0.23$, and $M = 0.11$ to satisfy Eq. (10.21). The water splash generated at impact in Figure 10.26a is relatively small, entraining limited air that affects mainly the primary wave surface. The slide to wave energy conversion is energetically efficient since a slide with $\rho_s < \rho_w$ is totally damped by the surrounding water body and no energy is 'lost' during impact on the channel bottom (Heller, 2007). The maximum of the primary impulse wave reaches CWG_1 in Figure 10.26c. Several similar pure-water waves reach the wave propagation zone at relative distances $X = 6.1$ to 10.7 in Figure 10.26d–f. Note that the primary wave is not necessarily the wave of maximum height for Stokes-like waves.

Typical wave profiles of Stokes-like impulse waves are shown in Figure 10.27. The relative wave profiles η/h versus relative time T_r are shown from top to bottom at CWG_1 in Figure 10.27a to CWG_7 in Figure 10.27g. The gray zone is affected by wave reflection at the wave absorber. These wave profiles allow for a classification according to Table 10.4. Stokes-like waves feature a nearly symmetrical wave profile both in the vertical and the horizontal axes where multiple crests occur. Due to small governing dimensionless parameters and small wave amplitudes, i.e. much smaller than at wave breaking, negligible air transport occurs in the wave propagation zone. The maximum wave amplitude $a_M < h/2$ was small to assure the intermediate-water range to which the Stokes-wave theory applies (Le Méhauté, 1976). As compared with the theoretical Stokes-wave profile (Figure 10.34b), the observed profiles are not monochromatic, decay, and have no front trough. The following

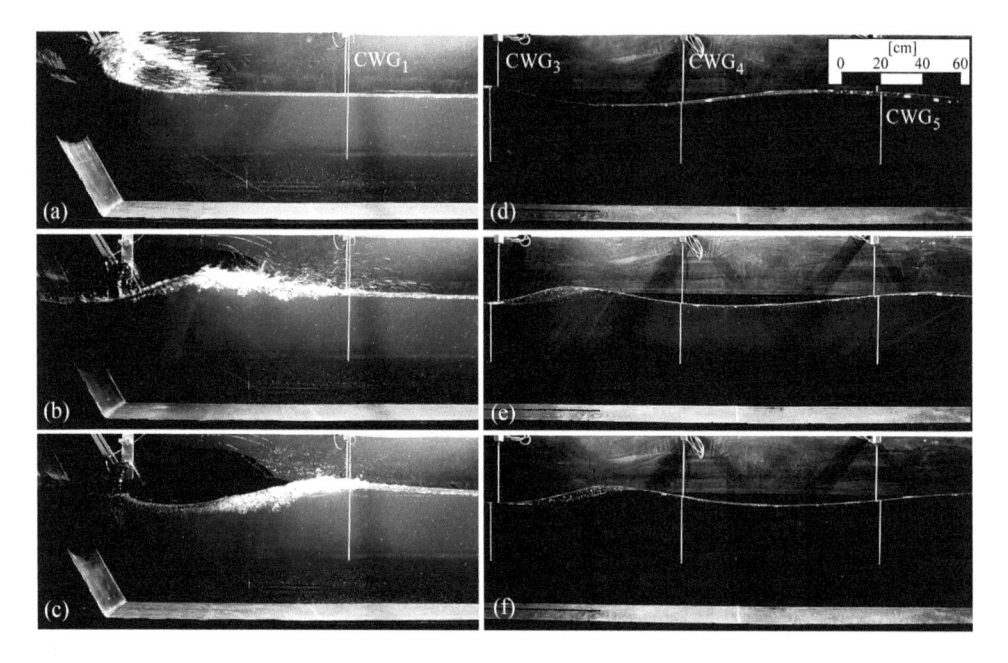

Figure 10.26 Stokes-like wave generation (a)-(c) and propagation (d)-(f) for $h = 0.600$ m, $\rho_s = 608$ kg/m³, $\alpha = 60°$, F = 1.36, $S = 0.23$, $M = 0.11$, $X = 6.1$–10.7, and $T_r \approx$ (a) 0.8, (b) 2.4, (c) 4.0, (d) 18.5, (e) 19.8, (f) 21.0 (Heller and Hager, 2011)

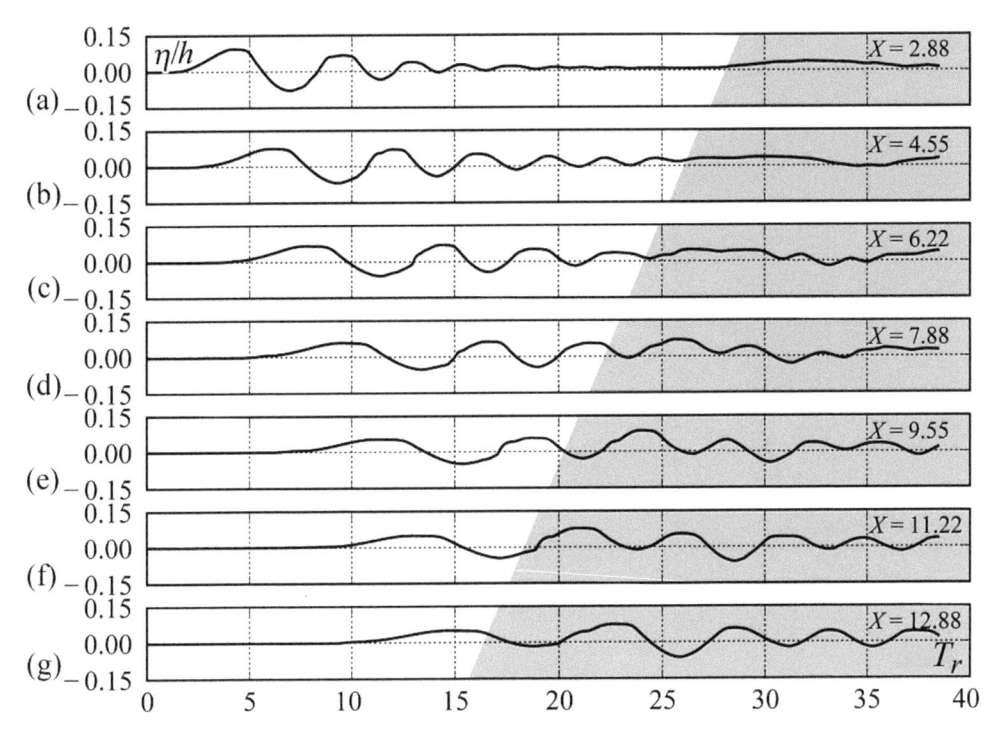

Figure 10.27 Stokes-like wave profiles for $h = 0.600$ m, $\rho_s = 608$ kg/m^3, $\alpha = 45°$, $F = 1.18$, $S = 0.17$, $M = 0.29$ at (a) CWG$_1$, (b) CWG$_2$, (c) CWG$_3$, (d) CWG$_4$, (e) CWG$_5$, (f) CWG$_6$, and (g) CWG$_7$; grey area is influenced by wave reflection (Heller and Hager, 2011)

studies are relevant to determine the effects of Stokes-like waves: Miche (1951) for run-up of oscillatory waves, Sainflou (1928), Minikin (1950), Wiegel (1964), Novak *et al.* (2001), or Dean and Dalrymple (2004) for computing the forces of a standing wave (clapotis) on a vertical wall, or Tanimoto *et al.* (1984) for predicting the force of a sinusoidal shallow-water wave.

Cnoidal-like wave

Figure 10.28 (a-c) shows a photo sequence of a cnoidal-like impulse wave generation involving $\rho_s = 610$ kg/m^3, $\alpha = 30°$, and medium to large relative values of $F = 2.27$, $S = 0.40$, and $M = 0.45$ to satisfy Eq. (10.22). The granular slide material impacts the water body in Figure 10.28a to generate a considerable water splash in Figure 10.28b. The amplitude envelope and the splash profile are visible on the wet dark zone of the channel back wall. The splash impacts the water surface in Figure 10.28 (b-c) entraining a large amount of air. The impulse wave travels to the wave propagation zone at relative distances $X = 10.3$ to 17.6 in Figure 10.28 (d-f). In Figure 10.28d the cnoidal-like impulse wave transports still a large amount of air from the slide impact process, escaping in Figure 10.28e so that the wave consists almost only of water in Figure 10.28f.

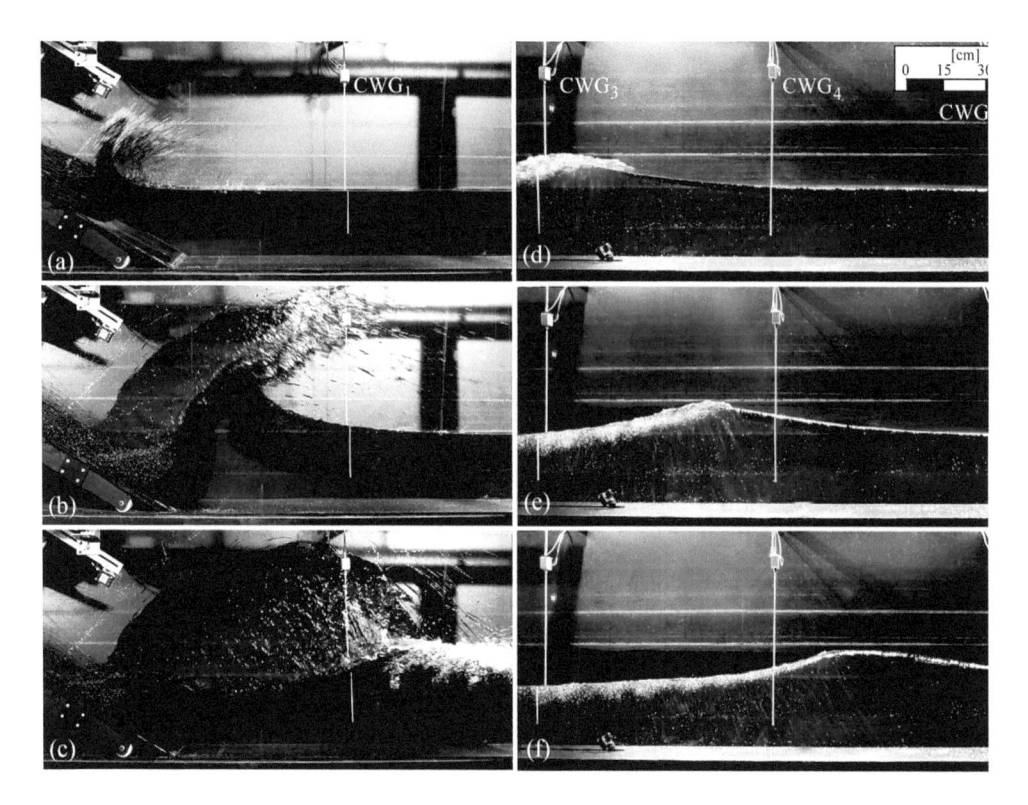

Figure 10.28 Cnoidal-like wave generation (a)-(c) and propagation (d)-(f) for $h = 0.30$ m, $\rho_s = 610$ kg/m^3, $\alpha = 30°$, $F = 2.27$, $S = 0.40$, $M = 0.45$, $X = 10.3$–17.6, and $T_r \approx$ (a) 1.1, (b) 3.4, (c) 5.7, (d) 11, (e) 13.1, (f) 15.1 (Heller and Hager, 2011)

Typical wave profiles of cnoidal-like impulse waves are shown in Figure 10.29. The relative wave profiles η/h versus relative time T_r are shown from top to bottom at CWG$_1$ in Figure 10.29a to CWG$_7$ in Figure 10.29g. These waves were selected according to the criteria of Table 10.4: Cnoidal-like waves feature a symmetrical wave profile to the vertical axis where multiple crests occur. The wave troughs are longer than the wave crests and the wave crest amplitudes are larger than the wave trough amplitudes. The maximum amplitude a_M is larger than of Stokes-like impulse waves yet limited to $a_M < h$. Cnoidal-like wave types may transport air as shown in Figure 10.28 due to impact crater formation. As compared with the theoretical cnoidal-wave profile described in 10.4.4 the observed profiles are not shallow water, but intermediate-water waves. Further, they are not periodic and the primary wave has no front trough.

The effects of cnoidal-like waves are determined with the results of Müller (1995). Relevant is the literature review below on solitary waves given the similar profiles of cnoidal- and solitary-like waves by comparing the maximum wave of Figure 10.29d with that of Figure 10.31c. Further an identical method has to be applied in practice for both cnoidal- and solitary-like waves because they are inseparable in Figure 10.25.

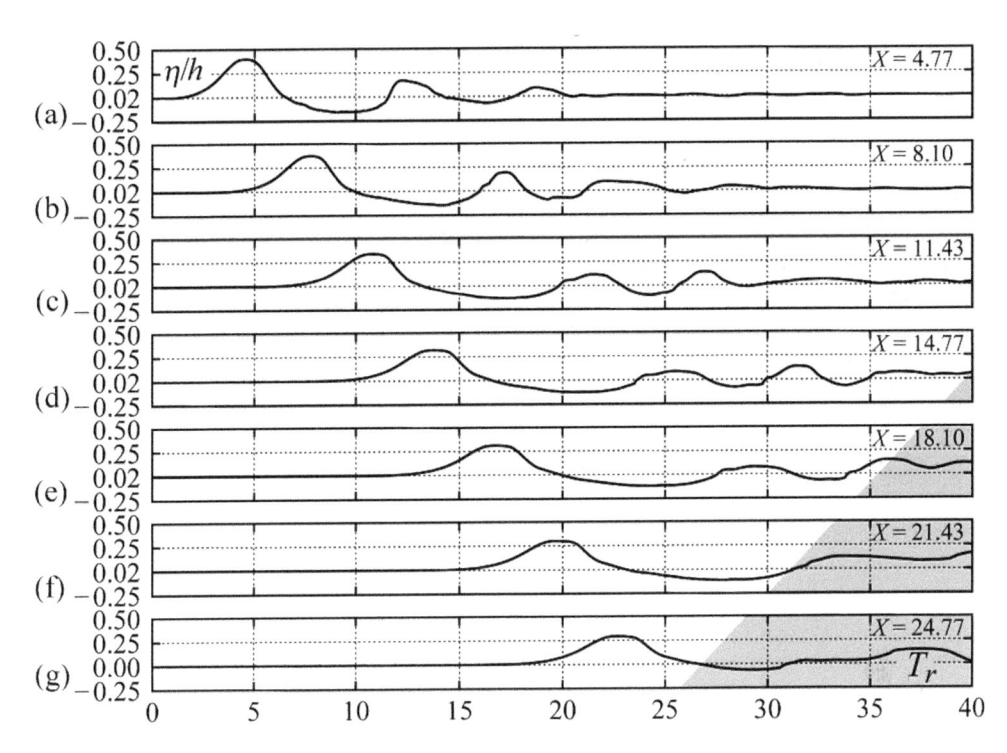

Figure 10.29 Cnoidal-like wave profiles for $h = 0.300$ m, $\rho_s = 610$ kg/m³, $\alpha = 45°$, F = 2.70, $S = 0.34$, $M = 1.11$ at (a) CWG_1, (b) CWG_2, (c) CWG_3, (d) CWG_4, (e) CWG_5, (f) CWG_6, (g) CWG_7 (Heller and Hager, 2011)

Solitary-like wave

Figure 10.30 (a-c) shows a photo sequence of solitary-like impulse wave generation generated using $\rho_s = 609$ kg/m³, $\alpha = 90°$, and large relative values of F = 3.77, $S = 0.81$, and $M = 0.90$, satisfying Eq. (10.22). The vertically impacting granular slide in Figure 10.30a generates a solitary-like wave in analogy to the vertically dropped box of Russell (1837). The generated water splash is relatively small resulting in comparably small air entrainment in Figure 10.30b and detrainment in Figure 10.30c. In the wave propagation zone in Figure 10.30 (d-f) at relative distances $X = 15.3$ to 24.3 the impulse wave consists only of the water phase. The solitary-like wave is well-developed in Figure 10.30e. Its large decay is noted on the wet dark zone along the channel back wall.

The wave profiles of Figure 10.30 are shown in Figure 10.31. The relative wave profiles η/h versus relative time T_r are shown from the top to the bottom at CWG_1 in Figure 10.31a to CWG_7 in Figure 10.31g. Solitary-like waves were selected with the criteria shown in Table 10.4. These waves feature a symmetrical wave profile to the vertical axis with only a singular or solitary dominant crest. The wave trough is nearly absent following the criterion $a_t < a/4$. Maximum relative wave amplitudes $0.49 \leq a_M/h \leq 0.94$ range close to the maximum stable wave height $H = 0.78h$ (McCowan, 1894). Solitary-like waves transport in the

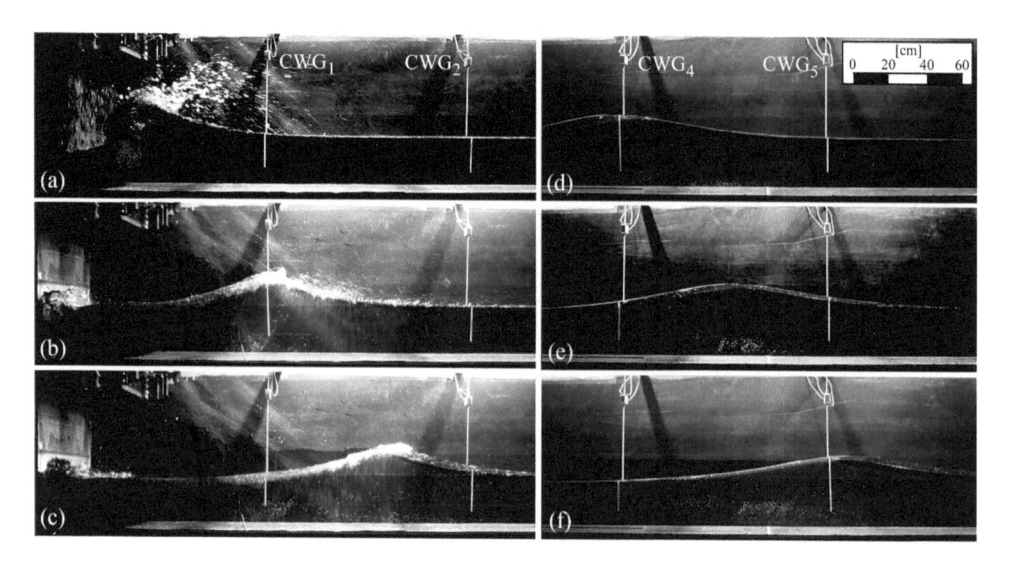

Figure 10.30 Solitary-like wave generation (a)-(c) and propagation (d)-(f) for $h = 0.30$ m, $\rho_s = 609$ kg/m^3, $\alpha = 90°$, $F = 3.77$, $S = 0.81$, $M = 0.90$, $X = 15.3 - 24.3$, and $T_r \approx$ (a) 2.1, (b) 4.1, (c) 6.1, (d) 15.5, (e) 17.5, (f) 19.5 (Heller and Hager, 2011)

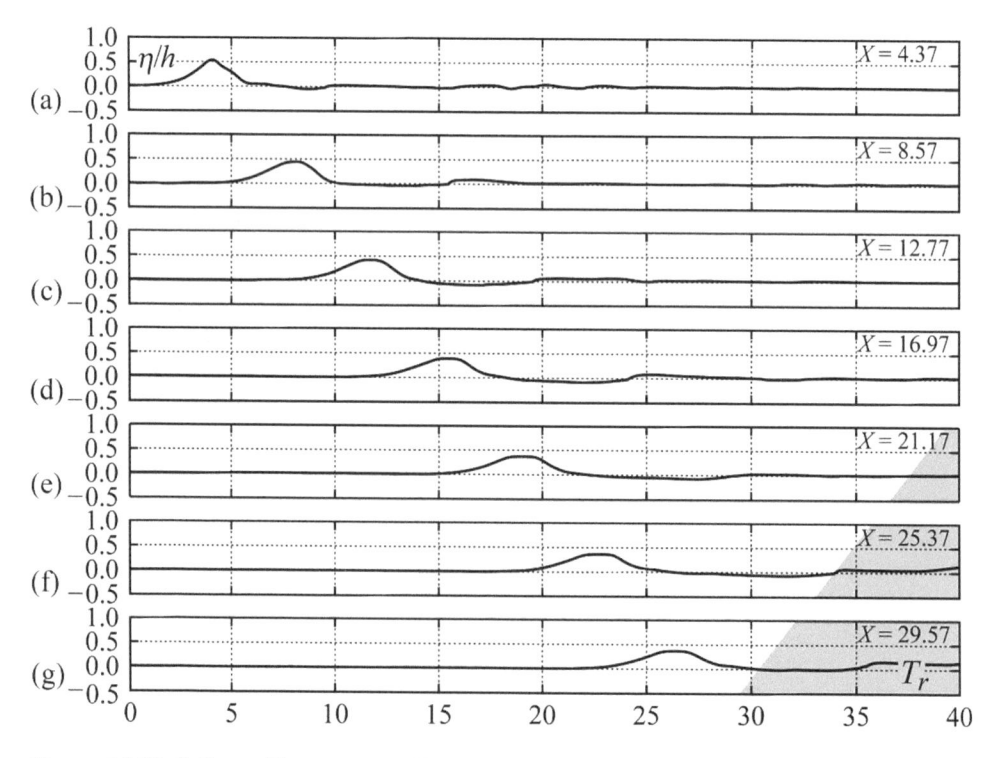

Figure 10.31 Solitary-like wave profiles from Figure 10.30 at (a) CWG$_1$, (b) CWG$_2$, (c) CWG$_3$, (d) CWG$_4$, (e) CWG$_5$, (f) CWG$_6$, (g) CWG$_7$; grey area is influenced by wave reflection (Heller and Hager, 2011)

wave propagation zone almost no air. These waves were in the intermediate-water range, in contrast to the theory which applies to the shallow-water regime (see 10.4.4). The wave length and period, respectively, are not infinite and the wave decays (Figure 10.30f) due to amplitude dispersion, frequency dispersion, and both turbulence in the slide impact zone and in the boundary layers. The observed maximum waves are followed by a small wave trough, in contrast to theory.

The effects of solitary waves were the purpose of extensive research: Hall and Watts (1953), Synolakis (1987), Liu *et al.* (1991), Zelt (1991), Teng *et al.* (2000), Li and Raichlen (2001), Gedik *et al.* (2005), and Hughes (2004), among others, studied solitary wave run-up; Müller (1995) focused on both the run-up and dam overtopping, and Ramsden (1996) and Cooker *et al.* (1997) investigated the solitary wave force on vertical walls.

Bore-like wave

Figure 10.32 (a-c) shows a photo sequence of a bore-like impulse wave generation using $\rho_s = 1664$ kg/m^3, $\alpha = 60°$, and very large relative values of $F = 4.22$, $S = 0.61$, and $M = 2.47$ to satisfy Eq. (10.23). The impacting slide mass in Figure 10.32a generates a considerable air cavity. A large amount of the slide energy is 'lost' during impact on the channel bottom and the slide-to-wave energy conversion is energetically inefficient (Heller, 2007). Nevertheless, a relatively large impulse wave is generated in Figure 10.32b and a large amount of air is entrained in Figure 10.32c. A bore-like wave reaches the wave propagation zone at relative distances $X = 14.9$ to 24.3 characterized by large energy dissipation and both air entrainment at the wave front and detrainment at the wave tail. The entire water column moves horizontally as noted from Figure 10.32f on the rising air bubbles accumulated prior to the experiment on the channel bottom.

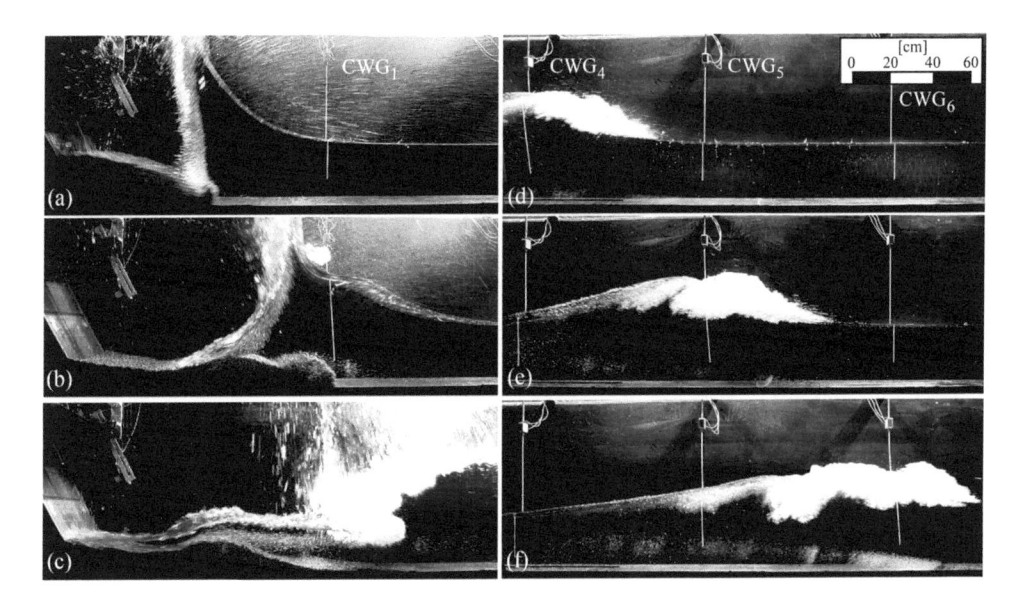

Figure 10.32 Bore-like wave generation (a)-(c) and propagation (d)-(f) for $h = 0.30$ m, $\rho_s = 1664$ kg/m^3, $\alpha = 60°$, $F = 4.22$, $S = 0.61$, $M = 2.47$, $X = 14.9–24.3$, and $T_r \approx$ (a) 2.3, (b) 4.6, (c) 6.7, (d) 10.6, (e) 12.9, (f) 15.0 (Heller and Hager, 2011)

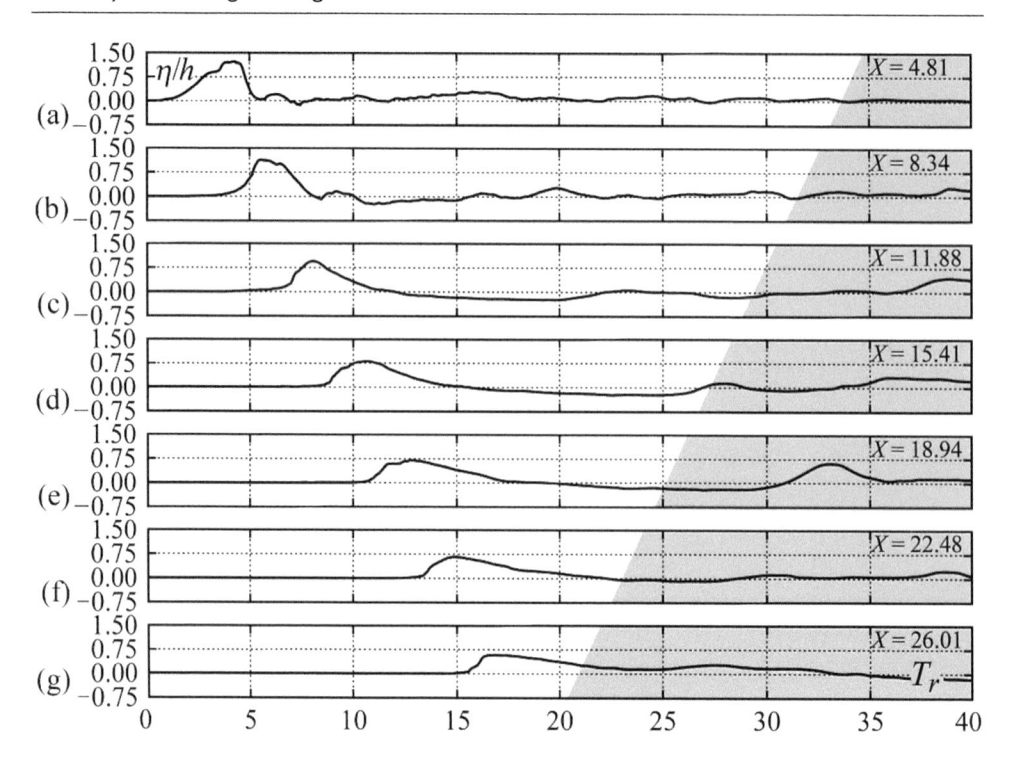

Figure 10.33 Bore-like wave profiles of Figure 10.32 at (a) CWG$_1$, (b) CWG$_2$, (c) CWG$_3$, (d) CWG$_4$, (e) CWG$_5$, (f) CWG$_6$, (g) CWG$_7$; grey area is influenced by wave reflection (Heller and Hager, 2011)

The wave profiles of Figure 10.32 are shown in Figure 10.33. The relative wave profiles η/h versus relative time T_r are shown from the top to the bottom at CWG$_1$ in Figure 10.33a to CWG$_7$ in Figure 10.33g. The criteria in Table 10.4 are again applied to classify these wave profiles. Bore-like waves with their typical dominant crest feature an unsymmetrical wave profile both relative to the vertical and the horizontal axes. The steep wave front, the flat wave tail, and the large air transport are typical for these waves. The maximum wave amplitude a_M prior to bore formation, consisting rather of a water sheet (Figure 10.32a) than a well-developed wave profile, was up to 2.5 times the still water depth h. The observed waves were in the intermediate- to shallow-water range, in contrast to the theoretical solitary wave profile which applies to shallow-water waves (see 10.4.4). The fluid flow is further highly turbulent and rotational.

The run-up of bores is discussed by Shen and Meyer (1963), Miller (1968), Yeh (1991), and Hughes (2004). Ikeno *et al.* (2001) investigated the bore wave force on a vertical wall.

10.4.3 Shortcut on nonlinear wave theories

Stokes-wave theory: The Stokes-wave profile is shown in Figure 10.34b with a_t as the distance from the still water surface to the wave trough and a as the wave amplitude. As compared with the sinusoidal wave shown in Figure 10.34a, the Stokes-wave profile has steeper

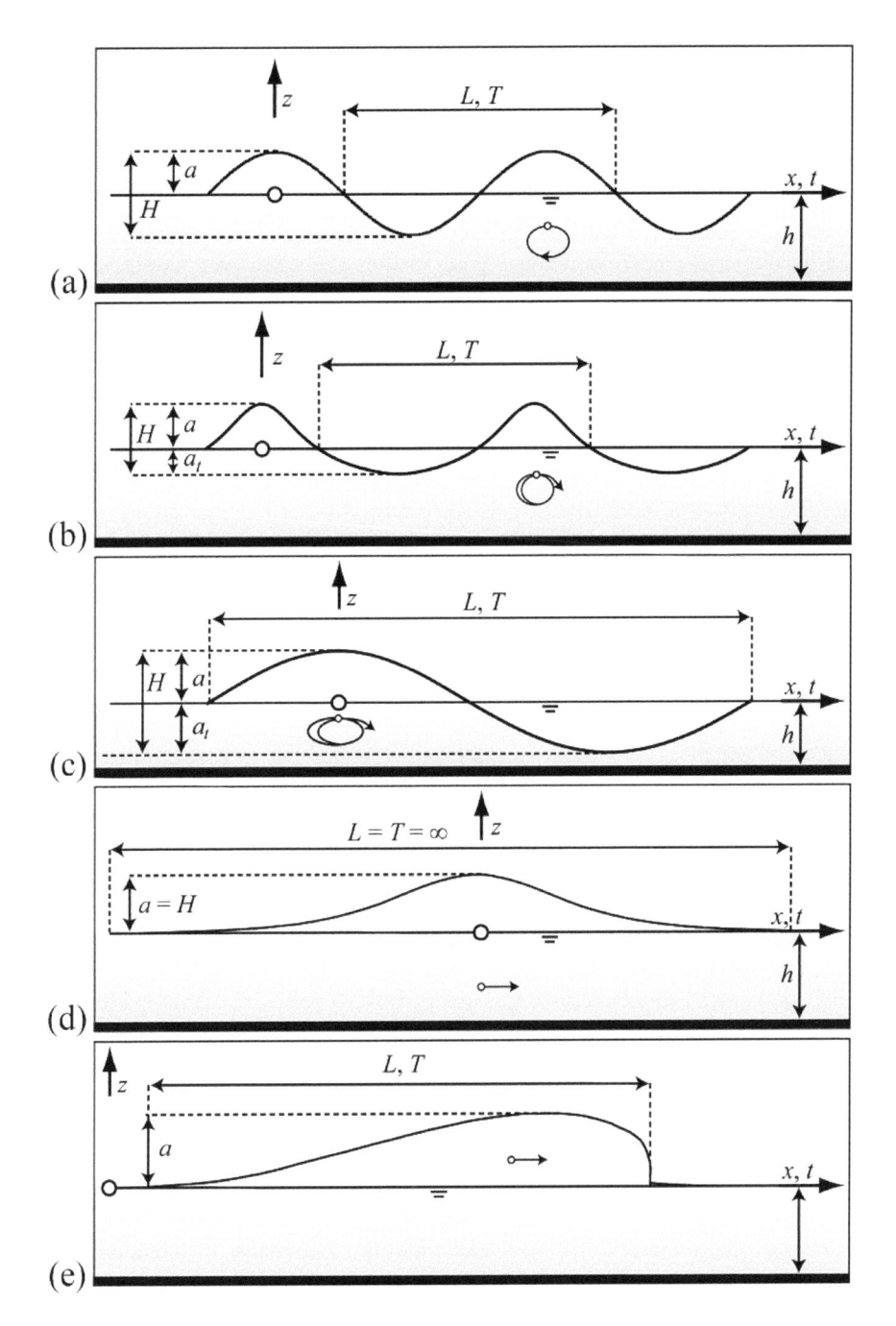

Figure 10.34 Wave profiles with basic parameters for (a) sinusoidal wave, (b) Stokes waves, (c) cnoidal wave, (d) solitary wave, (e) bore (Heller *et al.*, 2009)

crests and flatter troughs whereas it is still symmetrical about the vertical plane (Wiegel, 1964). This theory originates from Stokes (1847) and is based on potential (irrotational) flow with a non-hydrostatic pressure distribution. The analytical description of the wave profile involves a power series based on (H/L), commonly to the fifth order (Fenton, 1985). The water particles describe open orbits; the waves are oscillatory of intermediate character and a small fluid mass transport, therefore. The range of the Stokes waves is defined according to Le Méhauté (1976) as $2 \leq L/h \leq 20$. Keulegan (1950) suggested for Stokes waves of fifth order the limitation $L/h < 10$, corresponding to an Ursell parameter of roughly $U < 10$.

Cnoidal-wave theory: The cnoidal-wave theory was developed by Korteweg and de Vries (1895) with its wave profile shown in Figure 10.34c. The wave crest amplitude is larger than the wave trough and the wave trough portion is longer than both the wave crest and the sinusoidal wave trough portions (Figure 10.34a). The name cnoidal is derived from the Jacobian elliptic function 'cn' to describe its profile (Wiegel, 1960). This wave type is derived by assuming hydrostatic pressure distribution for the first order (Isobe, 1985), and non-hydrostatic pressure distribution for the second order approximations in combination with irrotational flow (Le Méhauté, 1976). This theory allows for the existence of periodic waves in shallow water (Dean and Dalrymple, 2004). The profile has mainly oscillatory character, although fluid mass transport occurs. This wave type is especially appropriate for $U > 25$ and according to Keulegan (1950) valid in the range of $L/h \geq 10$. The cnoidal-wave theory is spanning the range between the solitary wave theory on the one hand (for $T \rightarrow \infty$), and the linear wave theory on the other hand (Wiegel, 1960; Dean and Dalrymple, 2004).

Solitary wave theory: The solitary wave type consists of only one water surface elevation and no wave trough so that the wave length L is theoretically infinite (Figure 10.34d). Russell (1837) investigated solitary waves in a hydraulic model and derived its basic features. The theory of the solitary wave was independently derived by Boussinesq (1871) and Rayleigh (1876) by assuming both non-hydrostatic pressure distribution and rotational flow. This wave is translative involving a considerable fluid mass transport (Le Méhauté, 1976). For this wave type the nonlinearity, tending to steepen the wave front, balances dispersion, which tends to spread the wave front. Therefore, its profile remains constant without damping along the travel distance in a horizontal channel of constant width b, if boundary layer effects are neglected. The solitary wave celerity c is (Laitone, 1960)

$$c = [g(h + a)]^{1/2}. \tag{10.24}$$

This relation is accepted to describe the celerity of subaerial landslide-generated impulse waves (e.g. Kamphuis and Bowering, 1972; Huber, 1980; Fritz, 2002; Heller, 2007).

Bore theory: The tidal bore and roller wave theory describes e.g. plunging breakers after breaking in the shoaling region (Wiegel, 1964). A bore wave profile shown in Figure 10.34e consists of a steep front and a flat portion behind the crest. The water particles move in the bore propagation direction. The system of equations is based on hydrostatic pressure distribution and irrotational flow. A bore describes an extreme shallow-water wave with fluid mass transport. The wave profile is evaluated with the momentum, the continuity, and the characteristic equations (Le Méhauté, 1976).

In summary, the wave types of subaerial landslide-generated impulse waves are studied based on Froude similitude and a total of 434 granular slide tests. The effects of the 7 governing parameters, namely the still water depth h, slide impact velocity V_s, slide thickness s, bulk slide volume Ψ_s, bulk slide density ρ_s, slide impact angle α, and grain diameter d_g, are systematically considered. The wave types were determined by optical wave profile inspection. The main conclusions are:

1. Wave type classification is based on wave type product $T = S^{1/3}M\cos[(6/7)\alpha]$ and slide Froude number F.
2. Parameters for wave type classification are $F = V_s/(gh)^{1/2}$, slide thickness $S = s/h$, and relative slide mass $M = m_s/(\rho_w bh^2)$. The slide impact angle α affects the wave type with $\cos[(6/7)\alpha]$ whereas the effect of the relative grain diameter is small.
3. Four wave and transient types are described with four nonlinear wave theories, namely the Stokes wave, cnoidal wave, solitary wave, and bore theory.

The proposed wave type classification applies in practice since it depends only on basic slide parameters or the still water depth to be estimated prior to an event.

10.5 Transformation of solitary wave to overland flow

10.5.1 Motivation and experimentation

As stated, impulse waves are generated by an impulse transfer to a water body caused by e.g. avalanches, glacier calvings, landslides, or rockfalls. The long-wave behavior combined with the short propagation distance within reservoirs often leads to small wave attenuation and thus to a large damage potential for humans and the near shore infrastructure. Damages occur due to direct wave impact, impact of float and debris, and their deposits after water retreat. Dam overtopping can even result in reservoir failure and thus to catastrophic events. For a hazard assessment, the overland flow depth and flow velocity correspond to the governing parameters of interest. The entire impulse wave process may be separated into various independent regions, namely ① generation, ② propagation, ③ wave run-up, and ④ overland flow (Figure 10.35b).

The largest wave run-up caused by an impulse wave event occurred in 1958 at Lituya Bay, Alaska (see 10.3.3). The best-known impulse wave event occurred at Vaiont Reservoir, Italy (see 10.2.4). A recent event occurred in 2010 at Laguna 513, a glacial lake near Carhuaz, Peru. A snow-ice avalanche generated a wave overtopping a natural rock dam with a water volume of 10^6 m³. The overland flow continuously entrained material during propagation forming a debris flow causing damage to the infrastructure at Carhuaz, 15 km distant to the lake (Schneider *et al.*, 2014). The latter incident highlights the importance of global warming, since temperature rise leads to glacier retreat and thus to the formation of proglacial lakes.

The impulse wave generation process was investigated either using a simplified solid body (e.g. Russell, 1837; Wiegel, 1955, or Kamphuis and Bowering, 1972) or due to the advance of laboratory equipment, using granular slide material (e.g. Huber, 1980; Fritz, 2002; Zweifel, 2004; Heller, 2007; Mohammed, 2010, or Bregoli *et al.*, 2013). Investigations involved either 2D models modeling the process characteristics in the main impulse direction or even 3D investigations in a wave basin. As a result of past research and a literature review, an assessment guideline for landslide-generated impulse waves was published (Heller *et al.*, 2009; Evers *et al.*, 2019).

The solitary wave run-up was physically investigated by Hall and Watts (1953) deducing two different run-up relations for slopes <12° and >12°, with respect to the initial

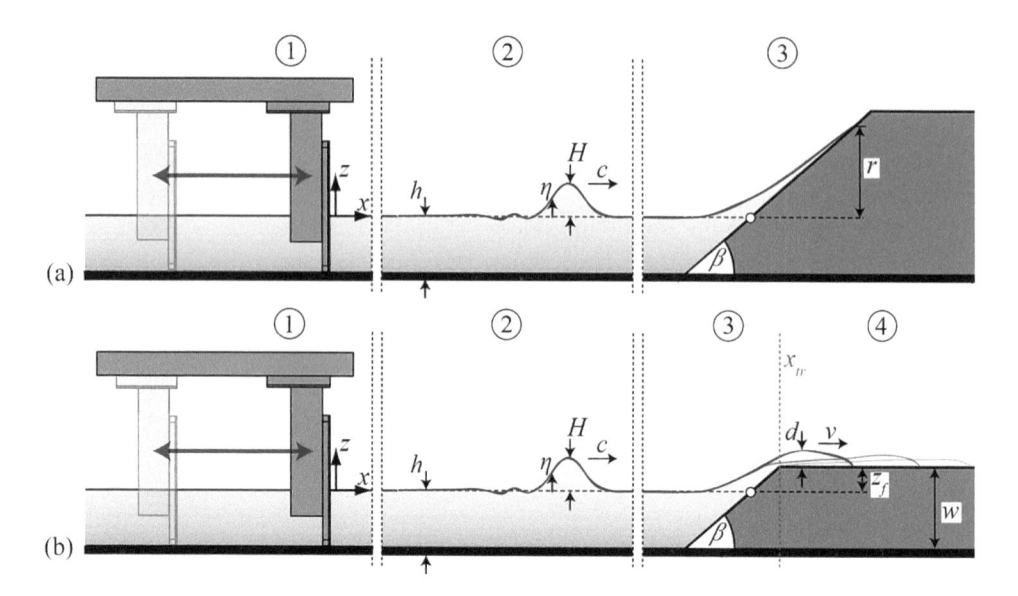

Figure 10.35 Definition scheme of test setup with ① wave generation, ② wave propagation, ③ wave run-up, and ④ overland flow regions for plane (a) wave run-up, (b) overland flow (Fuchs and Hager, 2015)

relative wave height and the shore slope. Su and Mirie (1980) analytically considered the collision of two solitary waves. They state that the maximum surface elevation is equal to a solitary wave interacting with a vertical wall for two identical waves and negligible viscous effects. An analytical consideration of solitary wave shoaling was presented by Pedersen and Gjevik (1983) including a comparison with experimental data. They stressed that no satisfactory results followed due to the errors introduced by approximations. Synolakis (1987) defined an analytically derived 'run-up law' supported by a comparison to test data for a gentle slope of 1:19.85 (V:H). He also proposed a breaking criterion stating that breaking wave run-up is significantly smaller as compared with the non-breaking run-up for a similar wave height. Shore roughness effects were addressed by Teng *et al.* (2000) who conducted experiments on both smooth shores and shores with attached gravel. Whereas on gentle ($\beta = 10°$) rough slopes the wave run-up was reduced by some 50%, this effect was smaller on the steep slope. Li and Raichlen (2001) improved the run-up relation of Synolakis (1987) by introducing a higher-order boundary condition. Some 10% higher run-up values resulted with a better agreement to the experimental data for solitary waves with $\varepsilon = H/h = 0.0185$–$0.34$ on steep 1:2.08 (V:H), and gentle 1:19.85 (V:H) slopes. Based on the approach of Iribarren and Nogales (1949) for periodic waves, Fuhrman and Madsen (2008) defined a surf similarity parameter ξ_s for solitary waves, describing the fundamental wave-shore interaction features, i.e. wave reflection, wave run-up, and wave transmission. Goseberg (2011) studied long-wave run-up on a 1:40 (V:H) 2D sloping beach with urban coastal structural macro roughness. The long surge waves of $\varepsilon = 0.05$–0.095 led to an increase in overland flow velocities for an obstructed cross section as compared with the plane beach. Baldock *et al.* (2012) experimentally studied plane wave run-up of non-breaking and breaking solitary waves on a gentle 1:9.35 (V:H) slope. They also removed a portion of the run-up slope to measure the corresponding overtopping volumes for various freeboard values. They noted higher overtopping volumes for solitary bores as compared with non-broken solitary waves, given the small relative freeboard. Vegetation effects were addressed in a numerical model by Tang *et al.* (2013), indicating that flow velocities and run-up heights are reduced for increasing plant height and blockage ratio of a shore vegetation.

Whereas 2D wave run-up was intensively studied in the past, wave-induced overland flow was rarely addressed. The majority of overland flow investigations is related to dam-break waves resulting from a much higher upstream reservoir depth as compared with an incoming sea wave. Goring (1978) experimentally analyzed the solitary and cnoidal-wave propagation from deep-ocean to a horizontal continental shelf with the water surface above the shelf level. Due to the small deep-water amplitudes of tsunamis, he suggested the linear non-dispersive wave theory to model prototype events. Wave-induced overland flow was investigated by Zelt and Raichlen (1991) both numerically and experimentally under zero freeboard, i.e. for the still water level corresponding to the shore level. On dry and wet shores, solitary waves of relative heights $\varepsilon = 0.14$ and 0.21 generated maximum overland flow depths of 3–5 cm. A similarity between wave-induced overland flow and a dambreak wave was noted, yet no general expression for the overland flow features was proposed. Schüttrumpf and Oumeraci (2005) studied both regular and irregular wave run-up on sea dikes. Flow velocities and the sheet thickness responsible for the landward erosion potential were quantified for the sea- and landward slopes, and the horizontal dike crest. Due to the limited dike crest width, the corresponding propagation distances were comparably small, however. Quiroga and Cheung (2013) modeled solitary wave transformation from the deep-ocean to a horizontal shelf similar to Goring (1978) in a 104 m long channel. Sælevik *et al.* (2013) studied solitary wave run-up for $\varepsilon = 0.2$–0.49 on a composite beach. They compared their data with a plane 10° shore, observing distinct differences in the free-surface elevation and flow velocities.

To improve solitary wave-induced overland flow feature predictions, the research of Fuchs and Hager (2015) involved a 2D physical model including a smooth and impermeable shore with a horizontal overland flow portion. The resulting overland flow depths and flow velocities were investigated for various incoming wave heights, shore angles, and the shore freeboard.

Laboratory tests were conducted in the 11 m long, 0.5 m wide and 1 m deep impulse wave channel, consisting of smooth steel plates except for its front wall and parts of the channel bottom made of glass to allow for optical access. The equivalent sand-roughness height was estimated to 0–0.003 mm for glass and 0.1 mm for steel, respectively (Montes, 1998; Reeve *et al.*, 2004). The channel was equipped at its front end with a pneumatically driven piston-type wave generator. A smooth shore made of polyvinyl chloride (PVC) was added at the channel rear end (Figure 10.35). The governing parameters are x and z as the horizontal and vertical coordinates, the still water depth h, the free water surface profile η, the wave height H, the wave celerity c, the shore angle β, the shore height w, and the freeboard z_f, all determining the target variables, namely the maximum run-up r for plane wave run-up, and the overland flow depth d and flow velocity v for a horizontal overland flow portion. The entire test setup was 2D; shore roughness, permeability, and vegetation effects were excluded.

As the fit equations presented below were empirically deduced from the test data, they are valid for the specified range of basic parameters. The test program included a variation of the still water depth $h = 0.16, 0.18, 0.20, 0.22$, and 0.24 m, resulting in relative freeboards of $Z_f = z/h = 0.042–0.563$. For each still water depth, waves with $\varepsilon = 0.1, 0.2, 0.3, 0.4, 0.5, 0.6$, and 0.7 were investigated. Three different shore slopes of $\tan\beta = 1/1.5, 1/2.5$, and 1/5.0 were tested with the steepest slope of $\approx 34°$ roughly corresponding to the natural angle of repose of sand and the gentle slope of $\approx 11°$ typically found at lake shores. Overland flow tests were repeated 4 times to deduce flow front velocities as the sample average of 5 identical tests.

High-quality images were taken for process documentation using a Nikon D3x camera with a sample frequency of some 5 Hz. For the run-up tests, r was measured optically using a standard digital video camera and run-up markers attached to the shore. The free water surface was measured using UNAM 3016 Ultrasonic Distance Sensors (UDSs) as a non-intrusive measurement method with a specified accuracy of <0.3 mm and a sample frequency of 60 Hz. To improve signal accuracy, a standard MATLAB© smoothing filter with a span width corresponding to <0.1 s was applied. Two UDS's were placed upstream of the shore inset to measure the incoming deep-water wave conditions, and six UDS's were positioned on the horizontal overland flow portion spaced by $2w$ to measure the free-flow surface (Figure 10.36). The data evaluation was based on maximum values (subscript max) of the local flow depth d_{max} at a certain position and the front velocity v_f determined as the average velocity over the distance between two neighboring sensors.

With UDS records as point measurements involving data black-out under a steep flow front, the acquired flow velocity information at the transition point (subscript tr) $x = x_{tr}$ are limited. A Particle Image Velocimetry (PIV, Figure 10.37) system was thus used to reliably determine the maximum depth-averaged horizontal flow velocity $\tilde{v}_{x,max}$ and the maximum unit discharge $q_0 = q_{max}$ at $x = x_{tr}$.

The double Nd:YAG laser with a double pulse of 225 mJ was operated at a repetition rate of 15 Hz. The pulse separation time was selected between $\Delta t = 1300–5600$ μs, to obtain a particle displacement corresponding to 1/4 of the correlation grid size of 32 pixels, i.e. $32/4 = 8$ pixels. The laser light sheet entered the channel through the glassed bottom and was reflected downstream by a 2″ mirror. The highly refractive spherical seeding particles of $\approx 1,000$ kg/m³ density had a mean diameter of 0.3–0.5 mm. Particle motion was captured by

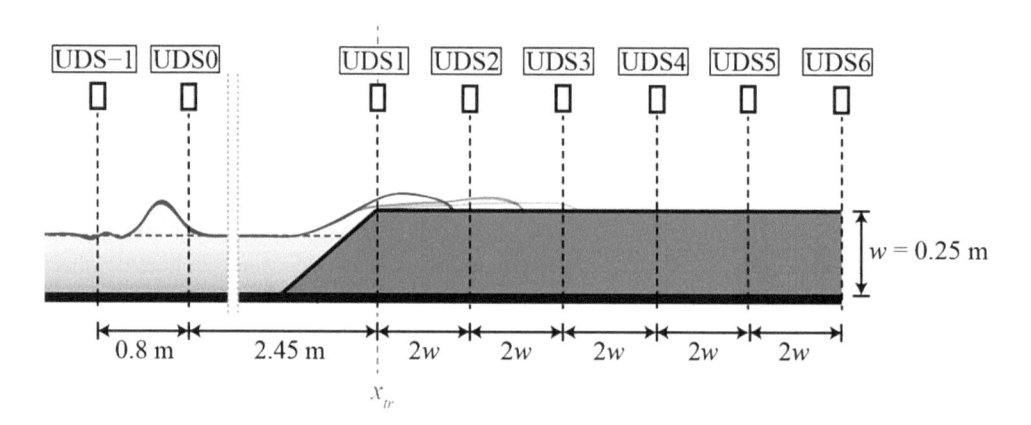

Figure 10.36 Definition sketch of UDS location (Fuchs and Hager, 2015)

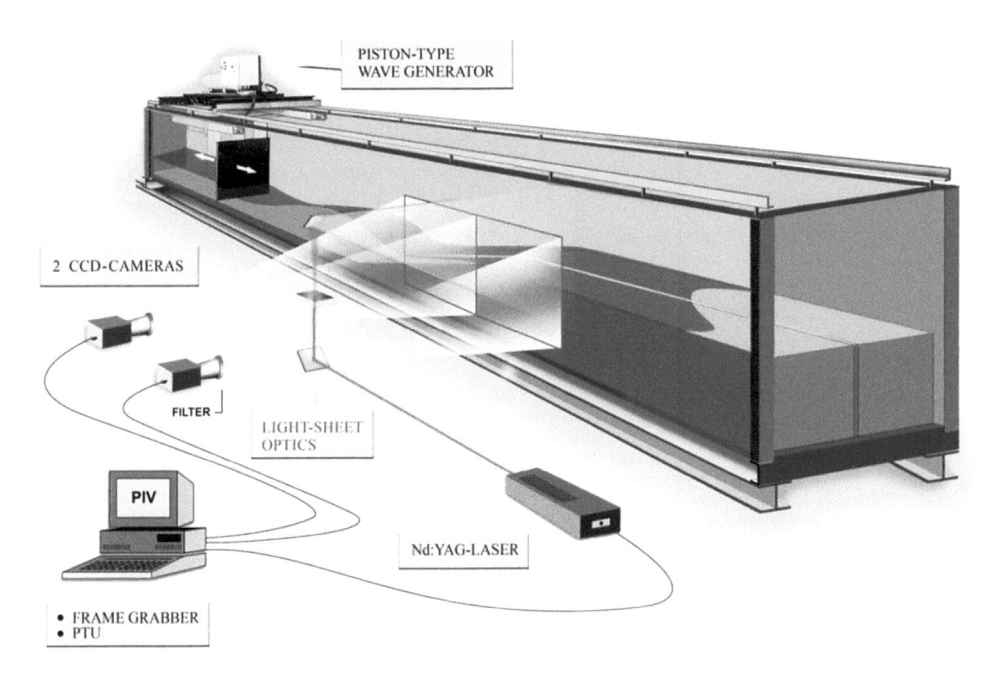

Figure 10.37 Definition sketch of PIV setup (Fuchs and Hager, 2015)

two 1MP CCD cameras mounted orthogonally to the light sheet plane. Two 532 nm line-pass filters attached to the camera lenses improved the image quality.

The quadratic Field Of View (FOV) of one camera was 1008(H) by 984(V) pixels, corresponding to 0.55 m $\times$ 0.55 m. Both camera images were combined to a 1.10 m wide FOV. The images were processed using the DaVis© 6.2 and 6.3 software of LaVision GmbH, Germany. Image pre-processing involved a high-pass filter of 10 pixels scale length and a 5-counts background subtraction filter to improve the image quality. A standard multi-pass Fast Fourier Transformation algorithm was applied to obtain velocity vectors with an interrogation window size recursively decreasing from 64 $\times$ 64 to 32 $\times$ 32 pixels, and a final overlap of 50%. The resulting spatial vector resolution was 8.8 mm $\times$ 8.8 mm. Post-processing involved a minimum vector peak ratio of 1.1 and a median filter (Raffel *et al.*, 1998).

Using a piston-type wave generator moving the entire water column from the channel bottom to the free water surface, all types of shallow-water waves can be generated. Fuchs and Hager (2015) focused on solitary waves, simplifying the test procedure: these waves are easy to generate, they are rapidly established after a short propagation distance from the generator paddle, and reflection effects are small. Due to absence of a wave trough, the solitary wave inherits a larger potential energy as periodic waves. Solitary waves are thus considered a conservative assumption. The waves were generated using Goring's (1978) method. The wave heights measured are affected by wave attenuation and thus are slightly smaller than targeted. Whereas targeted values of H were used for notation, the effectively measured values were employed for the data analysis. Due to minor gap losses at the generator paddle and wave attenuation during wave propagation, the incident wave heights are reduced to an average of some 98%. The standard error of wave height of 20 identical waves generated at $h = 0.20$ m and $\varepsilon = 0.5$ was 0.1 mm, stating excellent test repeatability.

For wave phenomena involving a free water surface Froude similitude applies as scaling law. Scale effects were investigated to ensure transferability of the test data to prototype conditions (Fuchs and Hager, 2013). For given overland flow front velocity v_f, maximum overland flow depth d_{max}, kinematic fluid viscosity $v = 10^{-6}$ m²/s, water surface tension $\sigma = 0.074$ kg/s², and water density $\rho = 1000$ kg/m³, the minimum Reynolds and Weber numbers for the overland flow phase are

$$R_f = \frac{v_f d_{max}}{v} > 6{,}300, \tag{10.25}$$

$$W_f = \frac{v_f}{\left[\sigma / \rho d_{max}\right]^{1/2}} > 10. \tag{10.26}$$

14% of all data were affected by scale effects and thus excluded from the data analysis.

10.5.2 Plane wave run-up

Wave run-up on a linearly inclined smooth slope was studied as a suitable condition to compare with overland flow results and literature data. Normalized run-up values $R = r/h$ of tests with $h = 0.20$ m plotted versus ε are compared with predictions from Hall and Watts (1953) [H&W] and Su and Mirie (1980) [S&M] in Figure 10.38. Maximum run-up values correspond to $r \approx 3H$, with a descending order from the gentle to the steep slopes for $\varepsilon \leq 0.5$,

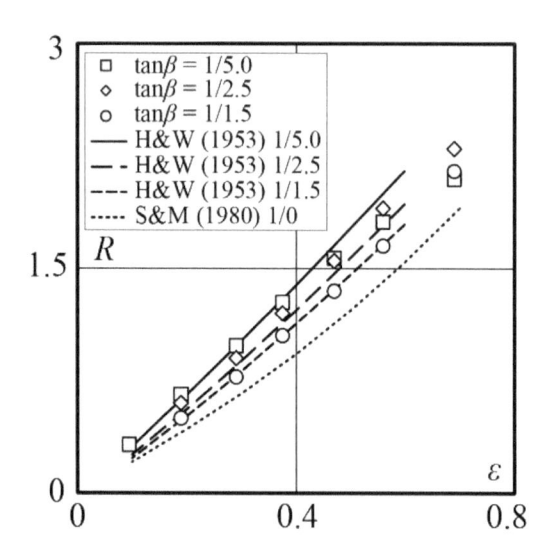

Figure 10.38 Relative run-up $R(\varepsilon)$ compared with predictions of Hall and Watts (1953) [H&W] and Su and Mirie (1980) [S&M] (Fuchs and Hager, 2015)

whereas for $\varepsilon \geq 0.6$ the largest run-up was measured for $\tan\beta = 1/2.5$. The agreement with the prediction of Hall and Watts (1953) is good for $\varepsilon \leq 0.5$, corresponding to their test range. For $\tan\beta = 1/1.5$ the agreement is excellent up to $\varepsilon = 0.7$. Although no experiments were conducted for a vertical wall ($\tan\beta = 1/0$), the prediction of S&M was added for completeness supporting the abovementioned trend of higher run-up values for decreasing shore steepness.

According to a least squares data analysis, the run-up data follow with $R^2 = 0.98$

$$R = \frac{r}{h} = 3(\tan\beta)^{-0.05}\varepsilon. \tag{10.27}$$

A comparison of measured run-up values and the run-up predicted (subscript pred) using Eq. (10.27) is shown in Figure 10.39a for the tests of Fuchs and Hager (2015), selected data of H&W and the data of Sælevik et al. (2013). Whereas prediction and measurements agree well for the present experiments and the data of Sælevik et al. (2013), several H&W data are slightly overestimated. The ratio between predicted and measured run-up is plotted versus $\xi_s = \tan\beta/\varepsilon$ in Figure 10.39b for the data of Fuchs and Hager (2015), the data of Sælevik et al. (2013) and these of H&W. Deviations are thus not related to a specific range of wave heights or shore angles but scatter equally for the entire parameter range. Whereas the VAW data scatter within $R_{meas}/R_{pred} = 0.86$–1.07, these of Sælevik et al. (2013) are within $R_{meas}/R_{pred} = 0.90$–1.05 and those of H&W scatter by $R_{meas}/R_{pred} = 0.63$–1.08. The large deviations of H&W are suspected to be caused by the outdated laboratory equipment and the wave generation mechanism using a moving mass. Results on wave reflection coefficients and free-surface profiles of plane wave run-up are provided by Fuchs and Hager (2013).

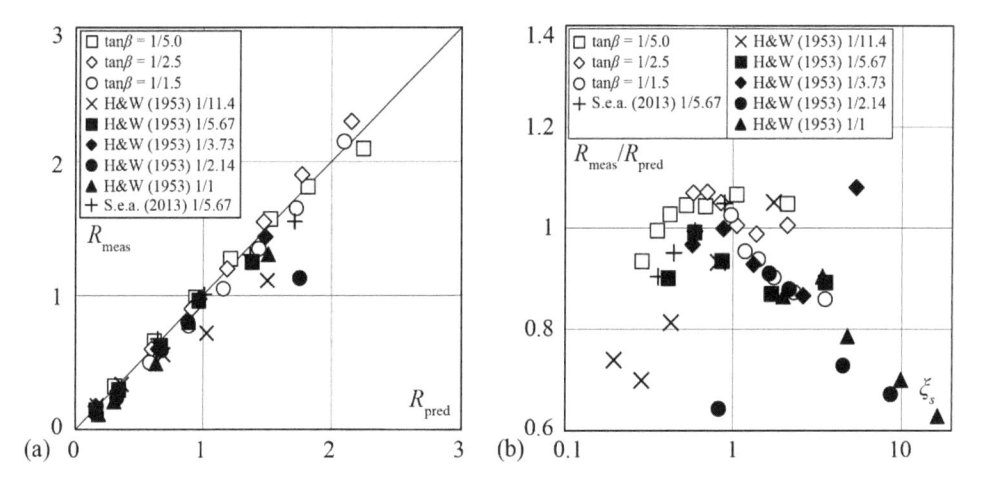

Figure 10.39 Relative run-up R (a) measured versus predicted values using Eq. (10.27), (b) deviations R_{meas}/R_{pred} versus solitary wave Iribarren number ξ_s for present study (open symbols), selected data of Hall and Watts (1953) [H&W] ($\times$ and closed symbols), and Sælevik et al. (2013) [S.e.a.] (+) (Fuchs and Hager, 2015)

Evers and Boes (2019) considered plane wave run-up of non-breaking impulse waves including solitary waves, thereby expanding the run-up angle to $10°\leq\beta\leq90°$. The data analysis led to a generalized Eq. (10.27), namely $R = 2e^{0.4\varepsilon}(90°/\beta)^{0.2}$. The relative wave amplitude ε has again a significant effect, whereas the effect of the run-up angle is found to be $(90°/\beta)^{0.2}$, i.e. more sensitive than expressed in Eq. (10.27).

10.5.3 Plane overland flow

In addition to the two governing parameters of plane wave run-up ε and β, freeboard $z_f = w - h$ determines the overland flow characteristics. Given that it is large, the wave shore interaction is similar to plane wave run-up. In contrast, for small values of z_f, a major portion of the incoming wave energy is transmitted to the connected horizontal portion so that even small waves lead to overland flow. Whereas for plane wave run-up the maximum water surface elevation as a static parameter is the target variable, flow depths d and flow velocities v along the entire horizontal portion are of interest for overland flow. The following deals with the governing parameters affecting overflow flow, considers the overland flow front propagation, the overland flow depth, and the overland flow velocity, allowing for a complete description of this phenomenon.

Figure 10.40 shows an image sequence of the overland flow process. It consists of three regions, namely the solitary wave approach (Figure 10.40a), the transformation of wave motion to overland flow (Figure 10.40b–c), and distinct overland flow without wave motion (Figure 10.40d–e). As long as no fluid reaches the horizontal overland flow portion, the incoming solitary wave is similar to plane wave run-up (Figure 10.40a). The wave becomes compressed and thus asymmetric, transforming into overland flow (Figure 10.40b) with

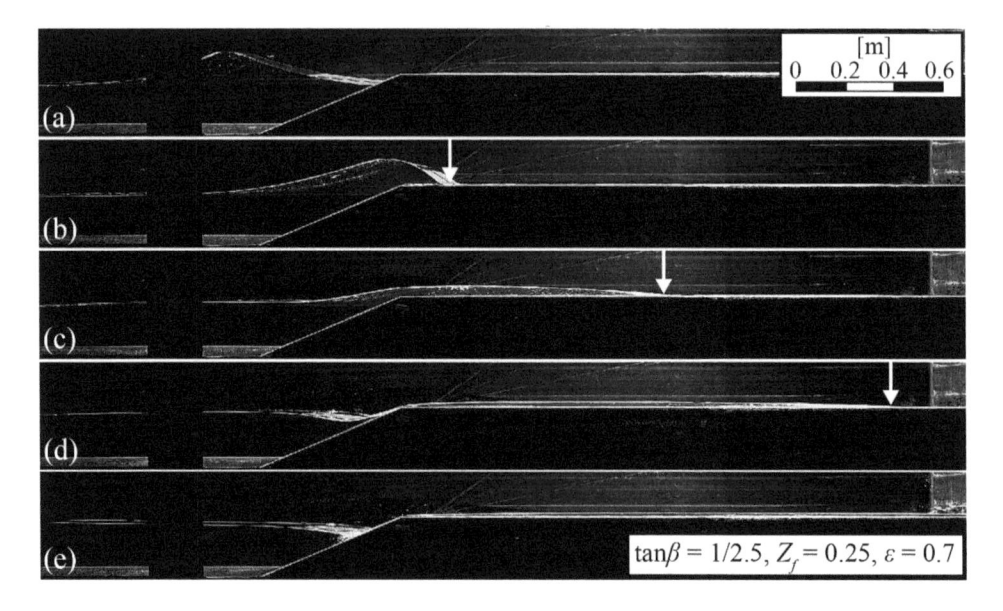

Figure 10.40 Overland flow for $\tan\beta = 1/2.5$, $Z_f = 0.25$ and $\varepsilon = 0.7$, time increment between images $\Delta t = 0.4$ s, arrows indicate flow front (Fuchs and Hager, 2015)

the wave peak located on the shore slope and not reaching the transition point x_{tr}. This conversion includes the transfer of potential energy to flow front acceleration and is completed in Figure 10.40c. The flow front has then reached almost midway of the overland flow portion. The overland flow characteristics are fully established involving a small flow depth and air entrained at the flow tip due to turbulence (Figure 10.40d). Flow reversal of the previous 'wave crest' occurs on the shore slope. Bore formation and air entrainment are small as compared with plane run-up since the down-rush velocity is comparably small (Figure 10.40e).

The three main parameters H, β and z_f were varied widely to study the overland flow characteristics. The relative wave height was found to have an increasing effect on the target variables d and v on the overland flow portion, similar to wave run-up. The free-surface elevation and thus the onshore flow depth increases with H. With g as the gravity acceleration, both the solitary wave celerity (Russell, 1837)

$$c = \left[g(h+H) \right]^{1/2} \tag{10.28}$$

and the maximum horizontal particle velocity under the wave crest (Laitone, 1959),

$$v_{x,\max} = c\left\{ -1 + \varepsilon\left[1 + \frac{\varepsilon}{4}\left(3\left(\frac{z}{h} \right)^2 - 5 \right) \right] \right\}, \tag{10.29}$$

depend on the wave height, so that the resulting overland flow velocity increases with H. The shore angle β only locally affects the target variables at the transition point as shown in Figure 10.41. Given a gentle value of $\tan\beta = 1/5$ the wave gradually transforms into overland flow (Figure 10.41a), with a small flow depth already at $x = x_{tr}$ and the maximum flow velocity located at the overland flow front. In contrast, for $\tan\beta = 1/1.5$, the wave is decelerated and compressed leading to a reduced flow front velocity and an increased free-surface

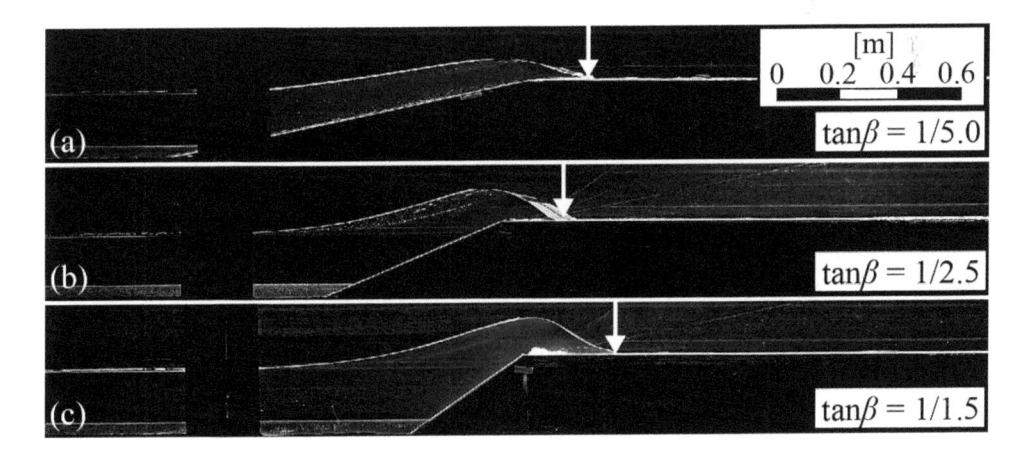

Figure 10.41 Overland flow for $\varepsilon = 0.7$, $Z_f = 0.25$, $\tan\beta = $ (a) 1/5.0, (b) 1/2.5, (c) 1/1.5, arrows indicate flow front (Fuchs and Hager, 2015)

elevation $\eta(x_{tr})$ (Figure 10.41c). Yet, the overland flow is accelerated due to hydrostatic pressure so that potential energy is transformed into kinetic energy. The shore slope thus has only a minor effect on $v_{x,max}$ in the 'far field'.

An image comparison of the overland flow characteristics depending on the relative freeboard Z_f is shown in Figure 10.42. Since the shore height was kept constant at $w = 0.25$ m and only the still water depth h was varied to study different $Z_f = (w–h)/h$, the subfigure size was increased (Figure 10.42a) and decreased (Figure 10.42c), respectively, to maintain visually constant h. The freeboard indicates a process similarity to plane wave run-up with large values of Z_f corresponding to large run-up heights required before the fluid reaches the overland flow portion (Figure 10.42a). In contrast, for small values of Z_f, a major portion of the incoming wave is transmitted to the overland flow portion (Figure 10.42a). Given $Z_f > R$, no overland flow is generated.

Rearranging Eq. (10.27), the minimum (subscript min) relative wave height required to induce overland flow is

$$\varepsilon_{min} = \frac{z_f \left(\tan \beta \right)^{0.05}}{3h}. \tag{10.30}$$

By subtracting ε_{min} from the incoming wave height, the wave portion effectively (subscript eff) determining the overland flow characteristics is

$$\varepsilon_{eff} = \varepsilon - \varepsilon_{min}. \tag{10.31}$$

Safety measures and damage estimations are required for an overland flow hazard assessment. Evacuation plans are primarily based on the flow front propagation, i.e. the duration of the flow front to reach a given location. The damage potential is then assessed based on local flow depth and flow velocity. The propagation of overland flow front $x^*(T)$ with $x^* = (x–x_{tr})/w$ and relative time $T = t(c/H)$ is shown for $\varepsilon = 0.5$ in Figure 10.43a. Whereas the overland flow front decelerates during propagation on the overland flow portion for a

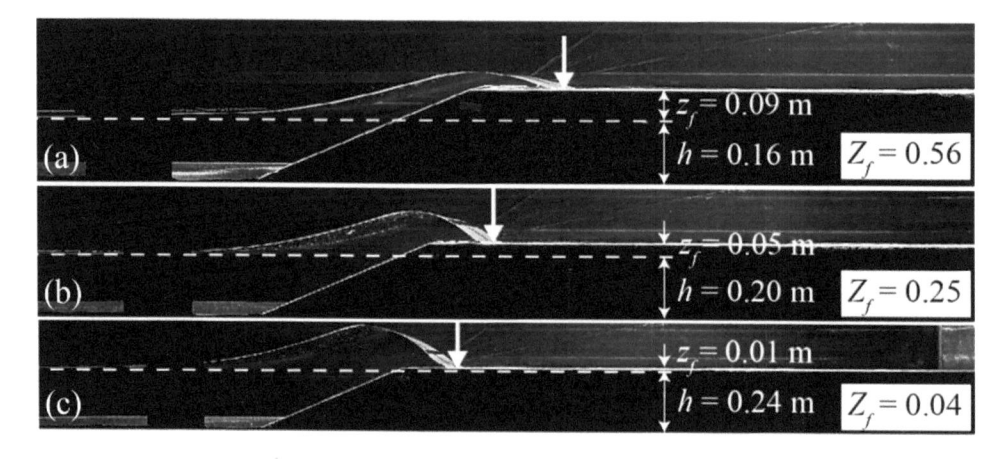

Figure 10.42 Overland flow for $\varepsilon = 0.7$, $\tan\beta = 1/2.5$, $Z_f =$ (a) 0.56, (b) 0.25, (c) 0.04, arrows indicate flow front, images are magnified to maintain similar visual still water depth (Fuchs and Hager, 2015)

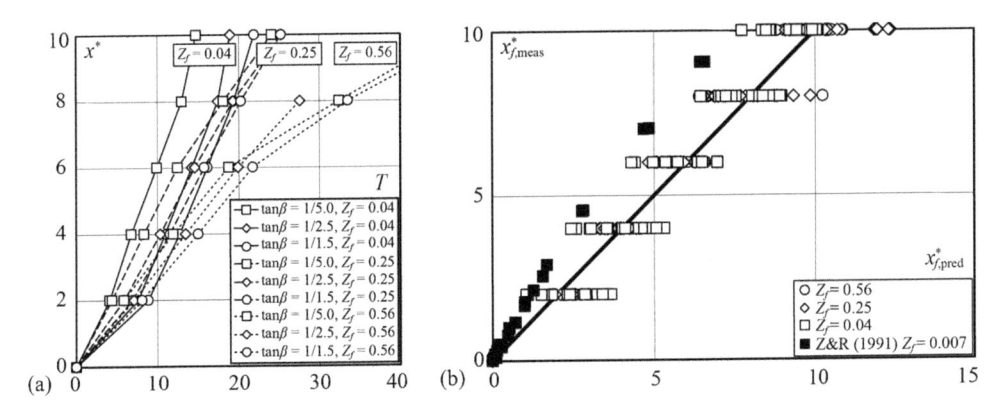

Figure 10.43 Overland flow (a) front position $x^*(T)$ for $\varepsilon = 0.5$ and various values of $\tan\beta$ and Z_f, (b) comparison of predicted and measured relative front position x_f^*, with full symbols as data of Zelt and Raichlen (1991) [Z&R] (Fuchs and Hager, 2015)

large value of $Z_f = 0.56$, it propagates almost linearly for $Z_f = 0.25$ and even accelerates if the freeboard is small ($Z_f = 0.04$). Given Z_f is large, the incoming wave is first transformed into a shore slope-parallel run-up flow before the fluid reaches the transition point where a second change of flow direction then forms the horizontal overland flow (Figure 10.42a). Therefore, a portion of the incoming wave energy is reflected or dissipated during the run-up process thereby not contributing to overland flow. For small Z_f, however, the incoming wave is significantly transmitted to the overland flow portion. The wave is then compressed at the shore and overland flow is subsequently accelerated, as mentioned. Whereas front propagation is almost equal for the different Z_f investigated for small x^*, it significantly differs for large x^* according to the above description. For an incident wave height of $\varepsilon = 0.5$, the flow front has reached $x^* = 10$ at $T \approx 45$ given that $Z_f = 0.56$, as compared with $T \approx 18$ for $Z_f = 0.04$, resulting in a factor of ≈ 2.5 as compared with ≈ 2.0 for $\varepsilon = 0.7$, and ≈ 3.5 for $\varepsilon = 0.3$. The spread of the flow front propagation with respect to the relative wave height $T(x^* = 8; \varepsilon = 0.3)/T(x^* = 8; \varepsilon = 0.7) \approx 7.5$ is large for $Z_f = 0.56$ as compared with $T(x^* = 10; \varepsilon = 0.3)/T(x^* = 10; \varepsilon = 0.7) \approx 3.4$ for $Z_f = 0.04$. In addition, the data spread is large for small propagation distances x^* and small at the end of the overland flow portion.

The overland flow propagation is generally fast for gentle slopes since the overland flow features are then fully established at the transition point, as compared with the steep slope. For a small $Z_f = 0.04$, the spread of the flow front propagation with respect to the shore slope $T(x^* = 4; \varepsilon = 0.5; \tan\beta = 1/1.5)/T(x^* = 4; \varepsilon = 0.5; \tan\beta = 1/5.0) \approx 1.9$ is large as compared with $T(x^* = 4; \varepsilon = 0.5; \tan\beta = 1/1.5)/T(x^* = 4; \varepsilon = 0.5; \tan\beta = 1/5.0) \approx 1.25$ for $Z_f = 0.56$.

The relative flow front position x_f^* is retained by parameter combination ($R^2 = 0.95$)

$$x_f^* = \left(1.7Z_f + 0.5\right)\left[\frac{T\varepsilon^{Z_f+1.45}}{\left(\tan\beta\right)^{0.25}}\right]^{1.35-1.15Z_f}. \tag{10.32}$$

Predicted and measured values of the flow front position x_f^* are compared in Figure 10.43b for various values of Z_f, including laboratory data of Zelt and Raichlen (1991) for $\tan\beta = 1/2.75$,

$Z_f = 0.007$, and $\varepsilon = 0.14, 0.21$. Whereas for small x_f^* a much faster overland flow propagation was observed than predicted by Eq. (10.32), their data become parallel but with an offset with increasing x_f^*. This difference is due to the rounded transition in their setup, in contrast to the sharp transition of Fuchs and Hager (2015). The distance of the flow front to propagate is therefore shorter, so that the overland flow portion is reached faster. Their relative shore height $Z_f = 0.007$ was also slightly beyond the specified limit of validity $Z_f = 0.04–0.56$ of Eq. (10.32).

The hazard assessment step below is based on local flow depths and flow velocities to determine forces on infrastructure. The flow depth assessment is divided into maximum flow depth at (1) transition point x_{tr}, and (2) a certain location along the overland flow portion. Values measured at the transition point $d_0 = d_{max}(x = x_{tr})$ are plotted versus the incident wave height H in Figure 10.44a for the three shore slopes and selected values of Z_f. As for the wave run-up heights, the flow depths are proportional to H. In addition, d_0 strongly depends on the freeboard since a major portion of the wave energy is reflected for large values of Z_f. Given the small freeboard of $Z_f = 0.04$, the maximum flow depths can be larger than the incident wave height. As described, a steep slope of $\tan\beta = 1/1.5$ leads to a more abrupt wave reflection and thus to larger flow depth at x_{tr} ($d_0/H = 1.04$ [$H = 0.168$ m] to 1.27 [$H = 0.024$ m]), as compared with $\tan\beta = 1/5.0$. This slope effect is prominent for large incident wave heights [$d_0(\varepsilon = 0.7; Z_f = 0.04; \tan\beta = 1/1.5)/d_0(\varepsilon = 0.7; Z_f = 0.04; \tan\beta = 1/5.0) = 173$ mm/137 mm = 1.26], whereas differences remain minute for small H [$d_0(\varepsilon = 0.1; Z_f = 0.04; \tan\beta = 1/1.5)/d_0(\varepsilon = 0.1; Z_f = 0.04; \tan\beta = 1/5.0) = 30.8$ mm/31.6 mm = 0.97]. The effect of shore slope also increases with increasing freeboard [$d_0(\varepsilon = 0.7; Z_f = 0.56; \tan\beta = 1/1.5)/d_0(\varepsilon = 0.7; Z_f = 0.56; \tan\beta = 1/5.0) = 62.7$ mm/25.7 mm = 2.44]. The shore slope $\tan\beta$ also affects the decreasing character of Z_f. For $\tan\beta = 1/1.5$, the flow depths of the small freeboard $Z_f = 0.04$ are by a factor ≈ 1.5 larger than these of the large freeboard $Z_f = 0.56$. For the gentle shore slope $\tan\beta = 1/5.0$, this ratio increases to ≈ 2.8.

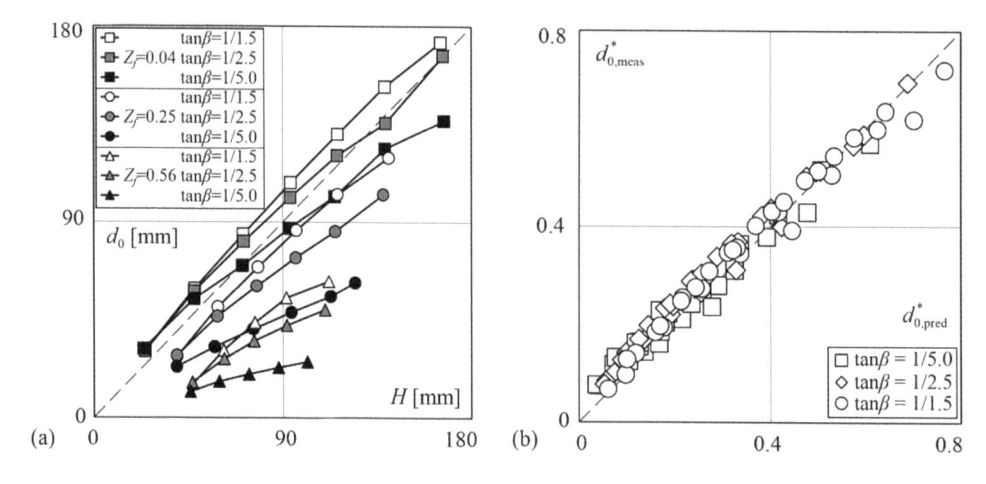

Figure 10.44 Overland flow depth at transition point (a) maximum values d_0 versus incoming wave height H for various values of Z_f and $\tan\beta$, (b) comparison of values predicted using Eq. (10.33) and these measured for various $\tan\beta$ (Fuchs and Hager, 2015)

Given the effective wave height in Eq. (10.31), the relative maximum flow depth d_0^* at x_{tr} follows the parameter combination ($R^2 = 0.96$)

$$d_0^* = \frac{d_0}{h} = \frac{(0.4\tan\beta + 0.9)\,\varepsilon_{eff}}{(Z_f + 1)^{0.45\cot\beta}}.\tag{10.33}$$

Predicted values of d_0^* agree well with these measured, as shown in Figure 10.44b.

The maximum flow depths on the connected overland flow portion $d_{max}(x)$ are evaluated based on Eq. (10.33). Figure 10.45a shows values of d_{max} measured for $\tan\beta = 1/2.5$ and various Z_f. Note that only data in accordance with Eqs. (10.25; 10.26) were considered for the data analysis. The maximum flow depths range between $d_{max}(x^* = 0) = 166.5$–30.5 mm and $d_{max}(x^* = 10) = 20.6$–7.9 mm, depending on ε and Z_f. Normalizing d_{max} with d_0 separates the overland flow characteristics from the incident wave features so that the relative wave height ε is excluded. The relative maximum overland flow depth d_{max}/d_0 is plotted versus the relative overland flow position for various shore slopes $\tan\beta$ in Figure 10.45b as

$$x_{max}^* = x^*\left(Z_f + 1\right)^{5.3\tan\beta - 1.4}.\tag{10.34}$$

The measured data are represented by the hyperbolic tangent fit as ($R^2 = 0.99$)

$$\frac{d_{max}}{d_0} = 1 - \tanh\left[0.54\left(x_{max}^*\right)^{0.39}\right].\tag{10.35}$$

The maximum flow depth along the shore therefore rapidly reduces within a short propagation distance. At a relative overland flow position $x^*_{max} = 5$, only $\approx 23\%$ of the initial flow depth d_0 remains. Data of Schüttrumpf and Oumeraci (2005) are included in Figure 10.45b

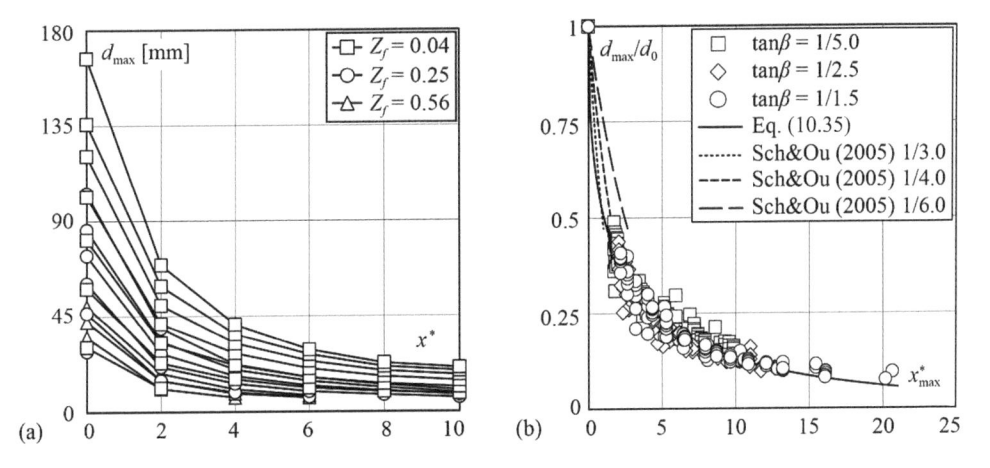

Figure 10.45 Maximum flow depth on overland flow portion with (a) $d_{max}(x^*)$ for $\tan\beta = 1/2.5$ and various Z_f, (b) $d_{max}/d_0(x^*_{max})$ for various $\tan\beta$ including data of Schüttrumpf and Oumeraci (2005) [Sch&Ou (2005)] (Fuchs and Hager, 2015)

for $Z_f = 0.33$ and $\tan\beta = 1/3$, $1/4$ and $1/6$. They proposed an exponential decay for the maximum flow depth on a dike crest (subscript cr) of width b_{cr} as

$$\frac{d_{max}}{d_0} = \exp\left(-0.75\frac{x}{b_{cr}}\right). \tag{10.36}$$

The dike overtopping features were investigated for $b_{cr} = 0.30$ m, leading to small relative overland flow positions in their experiments. However the agreement is good particularly for the steep slopes $\tan\beta = 1/3$ and $1/4$. Using both Eqs. (10.33) and (10.35) the maximum flow depth during the entire overland flow process is determined with respect to the initial parameters.

In addition to UDS free-surface measurements, PIV data focusing on the transition point were conducted to obtain velocity information. The flow field of overland flow which has just reached the horizontal overland flow portion is shown in Figure 10.46 for identical incident conditions $\varepsilon = 0.50$, $Z_f = 0.14$ and $\tan\beta =$ (a) $1/5$, (b) $1/2.5$, and (c) $1/1.5$. Fluid motion indicated by the grayscale background contour shows that the gradual bathymetry change of the gentler slopes $\tan\beta = 1/5$ and $1/2.5$ leads to a transformation from wave motion to overland flow, indicated by the vertical bounds of the particle velocity ranges with the maximum particle velocities located at the flow front. In contrast, for $\tan\beta = 1/1.5$, the velocity distribution is similar to the undisturbed solitary wave with the maximum particle velocity under the wave crest at $x^*\approx-1.5$. For $\tan\beta = 1/2.5$ the wave crest is located at $x^*\approx-1.2$ whereas for $\tan\beta = 1/5$ at $x^*\approx-0.8$, respectively. At the sharp transition from the shore slope to the horizontal plane, the flow is deflected upwards in the slope-parallel direction. This effect is pronounced at $x^* \approx 0.2$ for steeper slopes (Figure 10.46b, c) but almost absent for $\tan\beta = 1/5$ (Figure 10.46a).

Velocity profiles at the transition point $x = x_{tr}$ were extracted from PIV data. Depth-averaged maximum values of the horizontal velocity component $\tilde{v}_{x,max}$ are plotted in Figure 10.47a versus ε for various Z_f and $\tan\beta$. Flow velocities are larger for high incoming waves. For large values of $Z_f = 0.56$ the velocities are considerably lower as compared with the flow on a shore of smaller freeboard. The flow then has to adapt to two successive direction changes before reaching the horizontal portion at $x > x_{tr}$. A portion of the incoming wave energy is dissipated or reflected and does not contribute to the overland flow, as described. Measurements corresponding to $Z_f = 0.56$ are thus excluded from the data evaluation in Figure 10.47b. In addition, velocities are strongly affected by the shore slope. For a steep slope of $\tan\beta = 1/1.5$, flow characteristics at x_{tr} are determined by the incoming wave features (Figure 10.46c) as indicated by the lower flow velocities as compared with well-established overland flow features generated on the gentler slopes $\tan\beta = 1/2.5$ ($\tilde{v}_{x,max} \leq 1.32°$ m/s) and $1/5$ ($\tilde{v}_{x,max} \leq 2.01°$ m/s). For $\tan\beta = 1/1.5$ the curves are ordered by the values of the corresponding Z_f, in contrast to values of $\tilde{v}_{x,max}$ for the gentle slope $\tan\beta = 1/5$, because velocities are then affected by the overland flow behavior, i.e. the velocity concentration at the flow front.

The maximum depth-averaged horizontal flow velocity normalized by the incident wave celerity is plotted in Figure 10.47b versus the transition point velocity parameter as

$$A_{tr} = \varepsilon_{eff}\left(Z_f + 1\right)^{3\tan\beta+2.5}\left(\cot\beta\right)^{0.5}. \tag{10.37}$$

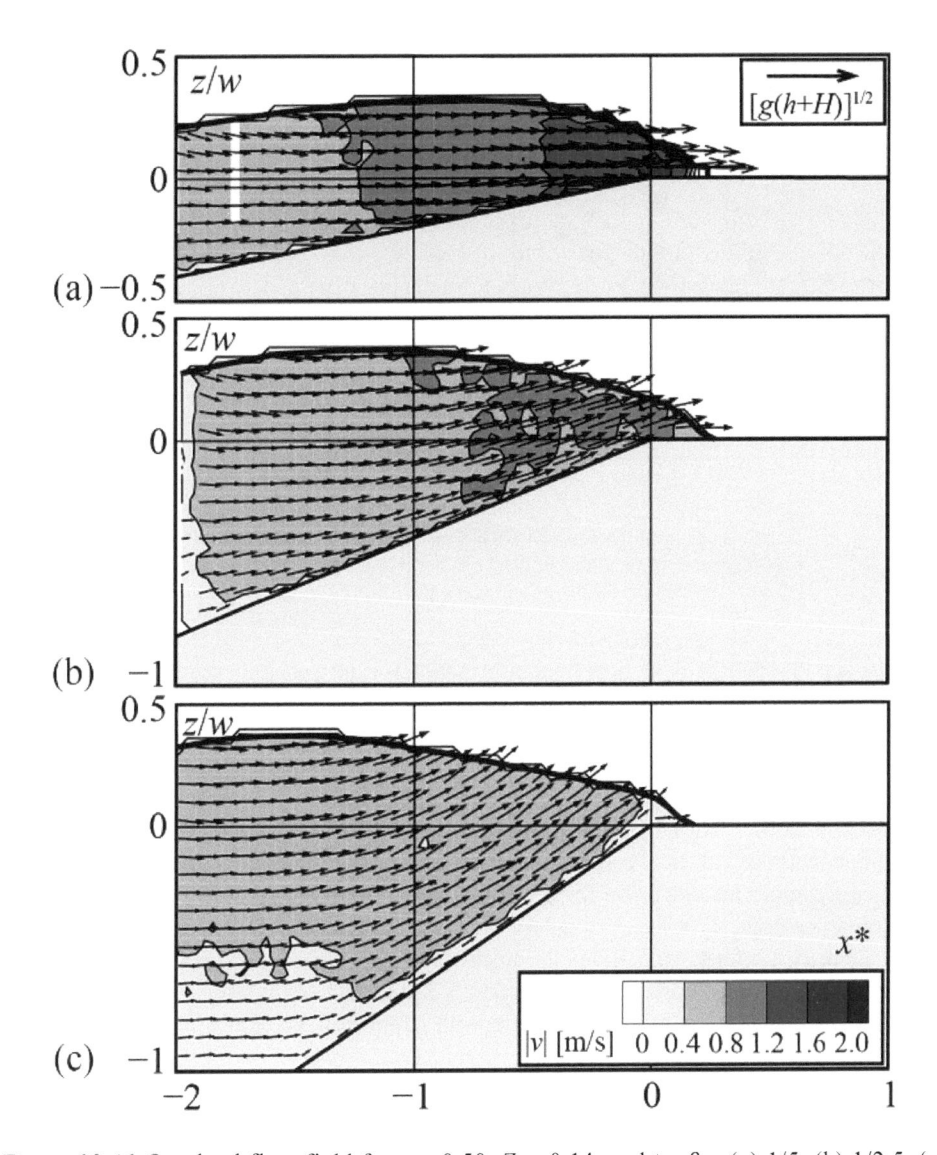

Figure 10.46 Overland flow field for $\varepsilon = 0.50$, $Z_f = 0.14$, and $\tan\beta$ = (a) 1/5, (b) 1/2.5, (c) 1/1.5 (Fuchs and Hager, 2015)

The experimental data follow the hyperbolic tangent function as ($R^2 = 0.98$)

$$\tilde{v}_{x,max}/c = 5\tanh\left(0.13A_{tr}^{0.5}\right).$$

(10.38)

Full symbols correspond to $Z_f = 0.56$ and were excluded from the data analysis.

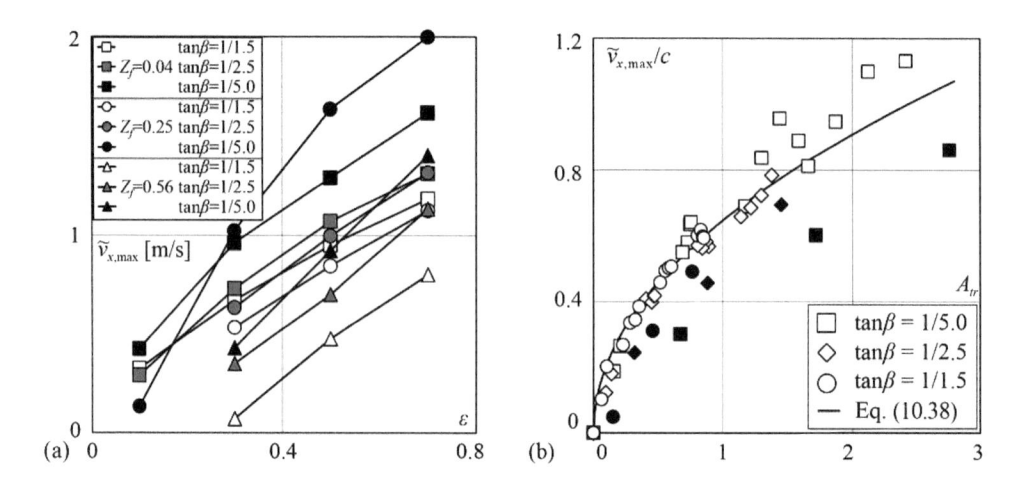

Figure 10.47 Depth-averaged horizontal flow velocity at transition point (a) maxima versus incoming relative wave height ε for various Z_f and $\tan\beta$, (b) normalized velocity versus transition point velocity parameter A_{tr} (Fuchs and Hager, 2015)

The above PIV data provide flow field information for the transition point, where the wave motion is transformed into overland flow, depending on the three basic parameters ε, Z_f and $\tan\beta$. However, for larger distances x^*, the overland flow features are fully established with the maximum flow velocities located at the flow front. The corresponding front velocity data therefore adequately represent the overland flow features. Flow front velocities on the overland flow portion normalized by incoming wave celerity are shown in Figure 10.48a for $Z_f = 0.14$ and various values of ε and $\tan\beta$. Similar to the maximum flow velocities at x_{tr}, front velocities are larger for larger incoming wave heights. The shore slope effect is concentrated at the transition point x_{tr}. Whereas the flow is strongly accelerated from $v/c \approx 0.7$ at $x^* = 1$ to $v/c \approx 1.25$ at $x^* = 5$ for $\varepsilon = 0.5$, given the steep slope $\tan\beta = 1/1.5$, the front velocity remains almost constant during propagation with $v/c \approx 1.3$ at $x^* = 1$–9 for $\tan\beta = 1/5$. For larger x^* the shore slope has only a minor effect on v_f with larger flow velocities for the gentle slope $\tan\beta = 1/5$.

Front velocities at $x^* = 5$ represent the maximum during overland flow propagation. For $x^* = 1$–5 the potential energy converted to kinetic energy may accelerate the flow, given a steep shore slope. For $x^* = 5$–9 the energy dissipation due to turbulence and viscous effects dominates leading to reduced front velocities. Therefore, the values $v_{f,max} = v_f(x^* = 5)$ were determined as the maximum onshore flow velocity. Figure 10.48b shows $v_{f,max}/c$ versus ε_{eff}/W involving the relative shore height $W = w/h = Z_f + 1$. The normalized maximum flow front velocity is represented by ($R^2 = 0.96$)

$$\frac{v_{f,max}}{c} = 1.6 \tanh\left[2.2\left(\varepsilon_{eff}/W\right)^{0.75}\right]. \tag{10.39}$$

The maximum 'far field' flow front velocity is therefore almost independent of the shore slope. Maximum front velocities are $v_f \approx 1.6c$ for large wave heights ε and small freeboard

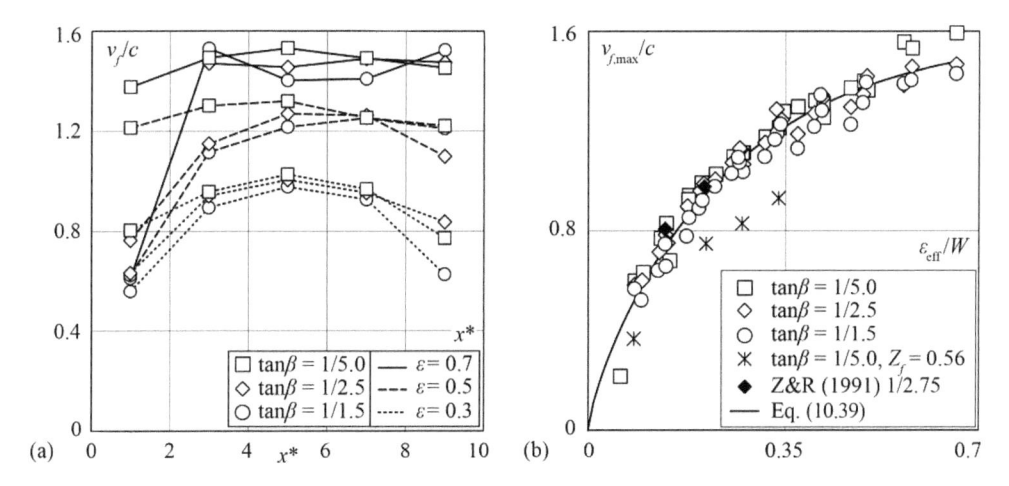

Figure 10.48 Relative overland flow front velocity (a) along overland flow portion for $Z_f = 0.14$ and various ε and $\tan\beta$, (b) maximum values versus $\varepsilon_{\text{eff}}/W$, with ($*$) for $\tan\beta = 1/5$ and $Z_f = 0.56$ excluded from analysis, full symbols for data of Zelt and Raichlen (1991) [Z&R] (Fuchs and Hager, 2015)

Z_f corresponding to a small relative shore height W. The maximum front velocity is therefore about 1/3 higher than the maximum particle velocities at the transition point from Eq. (10.38). The data ($\blacklozenge$) of Zelt and Raichlen (1991) agree well with the prediction. Their data correspond to $\tan\beta = 1/2.75$, $Z_f = 0.007$, and $\varepsilon = 0.14$ and 0.21, respectively. In contrast to Fuchs and Hager (2015) involving a sharp transition from the slope to the horizontal plane, their shore insets were rounded. However, the shore geometry has a negligible effect on the maximum 'far field' flow front velocity, as compared with the effects of ε and Z_f.

The wave-induced overland flow discharge per unit width at $x = x_{tr}$ was determined as the horizontal particle velocity v_x integrated over the flow depth d as

$$q = \int_{z=0}^{z=d} v_x dz \; . \tag{10.40}$$

The observed discharge characteristics are similar to unsteady flow, supporting the analogy to a dambreak or surge flow with a larger flow velocity and correspondingly smaller flow depth at the ascending stage, as compared with the descending stage. The maximum discharge at the transition point $q_0 = q_{\max}(x = x_{tr})$ according to Eq. (10.40) is plotted versus ε in Figure 10.49a for various Z_f and $\tan\beta$. As both particle velocity and flow depth at the transition point increase with larger ε and smaller Z_f, the value of q_0 follows this trend. No distinct shore slope effect is observed.

Maximum discharge normalized with the maximum mass flux under the incoming solitary wave crest is plotted in Figure 10.49b versus the discharge parameter B as

$$B = \frac{\varepsilon_{\text{eff}}}{\left(Z_f + 1\right)^{1.5\cot\beta}\left(\cot\beta\right)^{0.5}} \; . \tag{10.41}$$

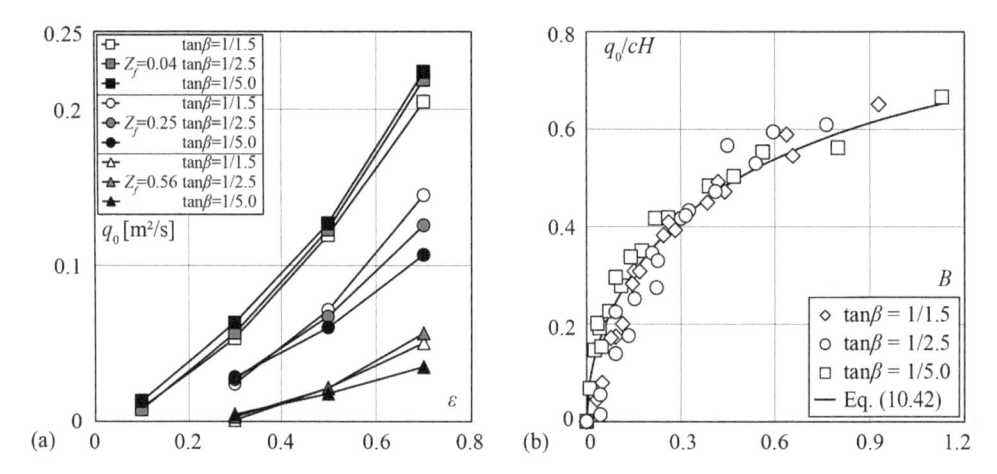

Figure 10.49 Specific discharge at transition point (a) maximum $q_0(\varepsilon)$ for various Z_f and tanβ, (b) normalized maximum discharge $q_0/(cH)$ versus parameter B (Fuchs and Hager, 2015)

The experimental data are represented by the hyperbolic tangent function ($R^2 = 0.95$)

$$q_0/(cH) = 0.83\tanh(B^{0.5}).$$ (10.42)

For large wave heights and small freeboards the discharge tends to 83% of the maximum mass flux within the solitary wave $\tilde{v}_x(h + H) = cH$.

In summary, physical model tests were conducted to describe the plane solitary wave run-up and solitary wave-induced overland flow. The study is based on the variation of three basic parameters: (1) relative wave height $H/h = 0.1$–0.7, (2) relative freeboard $z_f/h = 0.04$–0.56, and (3) shore slope tan$\beta = 1/1.5$–1/5.

Larger waves are demonstrated to lead to a stronger wave-shore interaction since the potential and kinetic energies are related to wave height. The freeboard indicates a similarity to plane wave run-up. Whereas for a small freeboard a major wave portion is transmitted onshore, the wave energy is reflected and dissipated during wave run-up reducing the subsequent overland flow features if the freeboard is large. The shore slope affects the overland flow features particularly at the transition point. On a gentle slope, the incoming wave is continuously transformed into overland flow. The flow features are then fully established at the transition point with large flow velocities located at the flow front and corresponding small flow depths. In contrast, for a steep shore, the incoming wave is abruptly decelerated leading to larger flow depth but small flow velocities. The potential energy of the increased flow depth is converted into kinetic energy and thus into flow acceleration. The shore slope therefore has only a minor effect on the maximum flow velocities in the 'far field'.

Prediction equations for both the plane wave run-up height and maximum flow depth, the maximum velocity and discharge at the transition point for a connected horizontal plane are given. Equations are also provided for the overland flow propagation, the maximum flow depth along the overland flow portion, and the maximum 'far field' flow velocity, all contributing to a detailed impulse wave hazard assessment.

10.6 Underwater deposition features

10.6.1 Motivation and data basis

The large damage potential due to run-up or dam overtopping results from the long-wave characteristics and the short propagation distances. Damages by impulse waves are caused by (1) wave impact on structures, (2) driftwood and float, and (3) their deposits after water retreat. Recent events were in 2007 at Lake Lucerne, Switzerland (Fuchs and Boes, 2010) or in 2010 at Lake Chehalis, Canada (Brideau et al., 2011).

Laboratory tests involving solid blocks or granular slide models in a 2D or even 3D test setup were conducted to study the wave generation process. A large test number conducted in a wide range of initial parameters contributed to the assessment guideline for landslide-generated impulse waves in reservoirs (Heller et al., 2009) including a literature review on impulse wave generation.

Heim (1932) summarized his observations on a large number of past Swiss rockfall events, thereby describing the main failure processes and differentiating between three characteristic reaches: (1) fall, (2) jump, and (3) runout of the slide material. The 'Fahrböschung' approach (an average slope α from the top of the potential slide down to the end of the deposition zone) was proposed to determine the maximum horizontal Δx and vertical Δz slide travel distances, also known as the runout relation by

$$\mu = \Delta z \, / \, \Delta x = \tan \alpha. \tag{10.43}$$

Here, $\mu \approx 0.6$ as the apparent friction coefficient for typical rock material. However, significantly lower values down to $\mu = 0.1$ resulted from case analyses, e.g. for clay containing soil or extremely large landslides with volumes in excess of 100,000 m^3 (Scheidegger, 1973). An inverse relation between $\tan\alpha$ and the failure rock volume was established by Hampton et al. (1996) based mainly on submarine slides and their deposition pattern, involving much more volume than subaerial slides. Runout lengths vary due to lateral confinement or obstacle presence. Predictions based on the analysis of past events are challenging, since reported slide lengths vary up to a factor of 8 for one event (Hampton et al., 1996). Savage and Hutter (1989) developed an Eulerian and Lagrangian mathematical model for dry granular slide propagation without water interaction. The slide spread of physical tests followed the Lagrangian scheme.

The simplest approach to determine the maximum slide velocity involves the energy transformation $V_s = (2g\Delta z)^{1/2}$ with g as gravity acceleration and Δz as elevation difference, thereby neglecting frictional effects. Slingerland and Voight (1982) expanded Eq. (10.43) to

$$V_s = \left\{ 2gx'_t \left[\sin \alpha - f_k \cos \alpha \right] \right\}^{1/2}. \tag{10.44}$$

Here x'_t = travel distance and the kinetic friction coefficient $f_k = 0.25$ is an average value based on slide event back analyses. Watts (2000) proposed an equation to estimate the underwater solid block motion for 'early times', i.e. until the block reaches its slide equilibrium velocity. Since primary wave generation corresponds to this first block motion the equilibrium slide motion was excluded. The center of mass of a granular slide behaves similar to a solid body. The motion of a deforming slide is separated thus into the center of mass motion and an additional deformation stretching the granular slide. However, the process of subaqueous slides significantly differs from subaerial landslides. Whereas underwater generated slides accelerate and spread, subaerial slides impacting a water body are abruptly decelerated and compressed. Underwater slides are thin with a ratio of mean thickness to length of $s/l_s \approx$ 0.01 (Watts and Grilli, 2003).

Subaerial landslide-generated impulse waves were originally investigated using solid block models to facilitate experimentation (Wiegel, 1955; Noda, 1970). As the measurement and test equipment improved, granular slides were used (Huber, 1980; Fritz et al., 2004; Zweifel et al., 2006; Heller et al., 2008; Mohammed, 2010), with a literature review on impulse wave generation by Heller et al. (2009). Scale effects for landslide-generated impulse waves resulted for tests with still water depths $h < 0.20$ m (Heller et al., 2008). During the slide impact phase air entrainment is then reduced due to surface tension affecting the impact crater formation thereby leading to a smaller amplitude of the primary wave and affecting the final slide deposition pattern.

More than 300 impulse wave tests in a wide range of basic parameters were conducted at VAW along with video recording. 41 tests covering a wide range of initial parameters were analyzed by Fuchs et al. (2013). An overview of the overall parameter ranges and the parameters covered by the selected tests is given in Figure 10.50. The upper portion contains initial basic parameters whereas dimensionless parameters are listed in the lower portion. The selected tests represent the entire parameter range, except for bulk slide density ρ_s. Slides with $\rho_s < 1400$ kg contained granular material with a grain density of $\rho_g = 955$ kg, i.e. floating

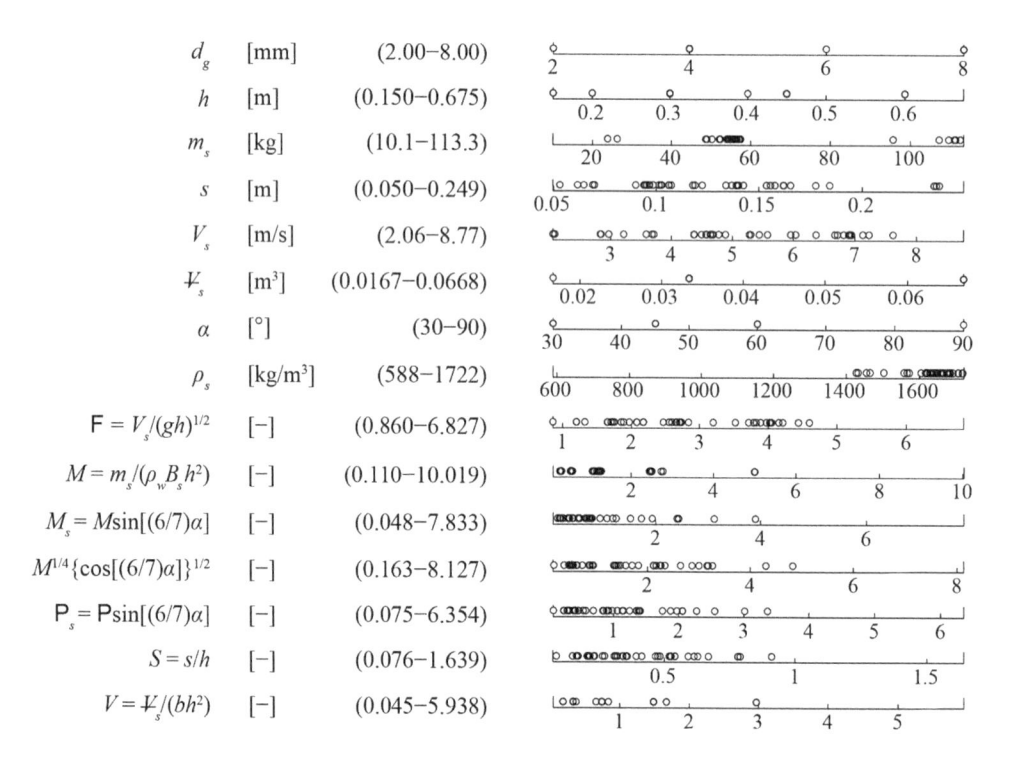

Figure 10.50 Parameter ranges of VAW tests and (○) parameters of selected tests (Fuchs et al., 2013)

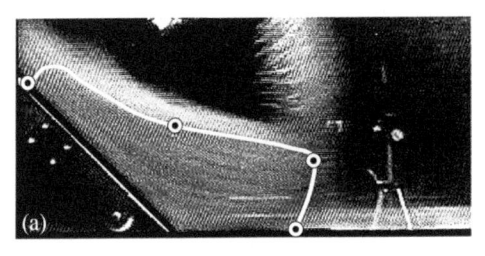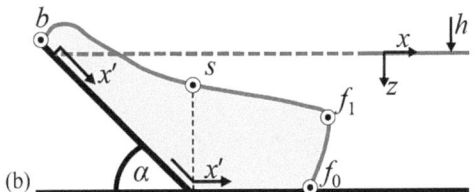

Figure 10.51 Definition scheme (a) slide impact onto water body, (b) notation (Fuchs *et al.*, 2013)

grains. These rapidly lose their compactness when impacting water and are not detectable from camera records so that these were not considered. Although scale effects are expected for tests with $h < 0.20$ m, they were included into the data analysis but marked (*) in all plots.

Figure 10.51 shows a definition scheme with x and z as the horizontal and vertical coordinates, respectively, and x' measured along the channel bottom. Four distinctive points describing the underwater slide geometry were extracted from each image by visual detection: f_0 as slide front position along the channel bottom, f_1 as maximum slide front position, s as the maximum slide thickness and b as the rear slide position. The temporal origin $t = 0$ was set at slide front impact onto the still water.

10.6.2 Test results

Image sequences of three typical slide impact processes having a time step between still images of $\Delta t \approx 3/25$ s, illustrate the different aspects of the slide impact process. The relating initial test parameters are specified in Table 10.5 and the corresponding filtered slide profiles $\xi(t)$ measured by LDSs shown in Figure 10.52. Test A shows an average-type shape of steep front followed by a gentle back. Test B starts with a small first leading 'foot' followed by the actual slide of steep front and back. This 'foot' is a typical feature also observed for dambreak waves (Lauber and Hager, 1996). The slide of Test C has a small thickness of round shape. Note the deposits of thickness $\xi \approx 0.02$ m on the slide ramp above the water level for $t \geq 0.9$ s (Figure 10.53x).

Figure 10.53a–h shows a 'standard test' of typical prototype slide impact angle of $\alpha = 45°$ for mountainous regions. The majority of the tests analyzed bared these slide dynamics and impact crater formation. The slide reaches the still water surface in Figure 10.53a ($t = 0$). Due to the high impact velocity of $V_s = 6.68$ m/s, large splash is displaced orthogonally to the slide impact direction. The still water surface remains undisturbed. The slide reaches the channel bottom and is deflected in Figure 10.53b. The slide back (b in Figure 10.51b) is still in motion on the slide ramp. Note the distinct slide 'nose' formation (f_1). The impact crater formation is pronounced but in progress and its water-air boundary difficult to detect. The slide front motion between Figure 10.53b–c is much reduced compared with Figure 10.53a-b pointing at significant slide deceleration. The slide back has left the slide ramp and impact crater formation is at its maximum. In Figure 10.53d–h the impact crater collapses in the outward direction thereby forming the primary impulse wave leaving the field of view in Figure 10.53g. This crater hardly affects the geometry of the slide back, which is considered at rest from Figure 10.53d, whereas the slide front is still protruding into the water body up to Figure 10.53h.

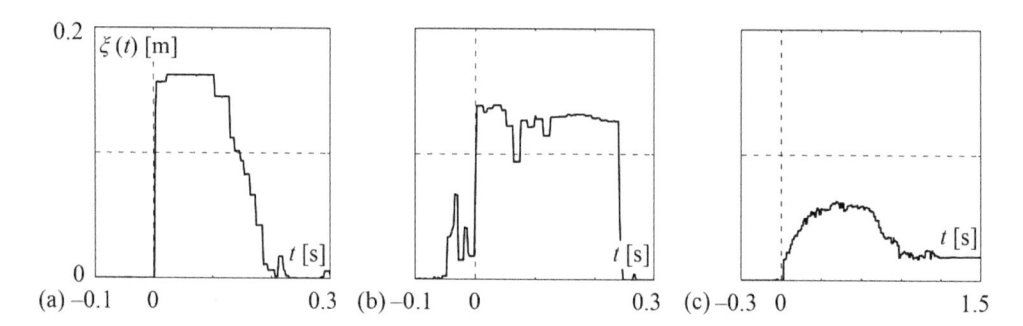

Figure 10.52 Filtered slide profiles of Tests (a) A, (b) B, (c) C (Fuchs *et al.*, 2013)

Table 10.5 Initial test parameters of Tests A, B, and C shown in Figure 10.53

	A	B	C
d_g [mm]	4	4	4
h [m]	0.30	0.45	0.30
M_s [kg]	109.59	57.40	55.11
s [m]	0.162	0.139	0.065
V_s [m/s]	6.68	4.37	2.06
Ψ_s [m³]	0.067	0.033	0.033
α [°]	45	90	30
ρ_s [kg/m³]	1640	1719	1650

Figure 10.53i–p corresponds to a test with a slide impact angle of $\alpha = 90°$. The slide reaches the still water surface in Figure 10.53i ($t = 0$). Due to lower slide impact velocity of $V_s = 4.37$ m/s, splash is much reduced as compared with Test A. The slide protrudes into the water body in Figure 10.53j, thereby separating both at the slide ramp and the water body. Separation at the slide ramp induces a slide deflection in the horizontal direction. In Figure 10.53k, the slide front is 'bent' and points at the longitudinal channel axis without having reached the channel bottom yet. Separation from the water body induces an inward collapsing water crater in Figure 10.53l–n, exerting large pressure forces on the slide back and pushing granular material to the slide ramp. The slide front protrudes the water body thereby increasing the crater size in the horizontal direction (Figure 10.53m) and the subsequent air detrainment after crater collapse. The water crater formation also generates the primary impulse wave (Figure 10.53k-m) which has not yet completely left the field of view in Figure 10.53p. A secondary impulse wave is formed at the channel front end by the collapsing impact crater.

Figure 10.53q–x describes Test C with a gentle slide impact angle of $\alpha = 30°$, an impact velocity of $V_s = 2.06$ m/s and a small slide thickness of $s = 0.065$ m. The slide reaches the still water surface in Figure 10.53q ($t = 0$). Only minor splash is generated by the small impact velocity (Figure 10.53r) and the water flow does not separate from the slide (Figure 10.53s). Fluid enters the slide pores leading to a less compact slide front (Figure 10.53s,t), reaching the channel bottom in Figure 10.53t. The momentum transfer from slide to water is almost completed leading to primary impulse wave generation. The slide front motion is observed up to Figure 10.53x. The slide back is deposited on the ramp. The primary impulse wave is still located in the center of the field of view. Note the different progress of the slide front motions of all three tests.

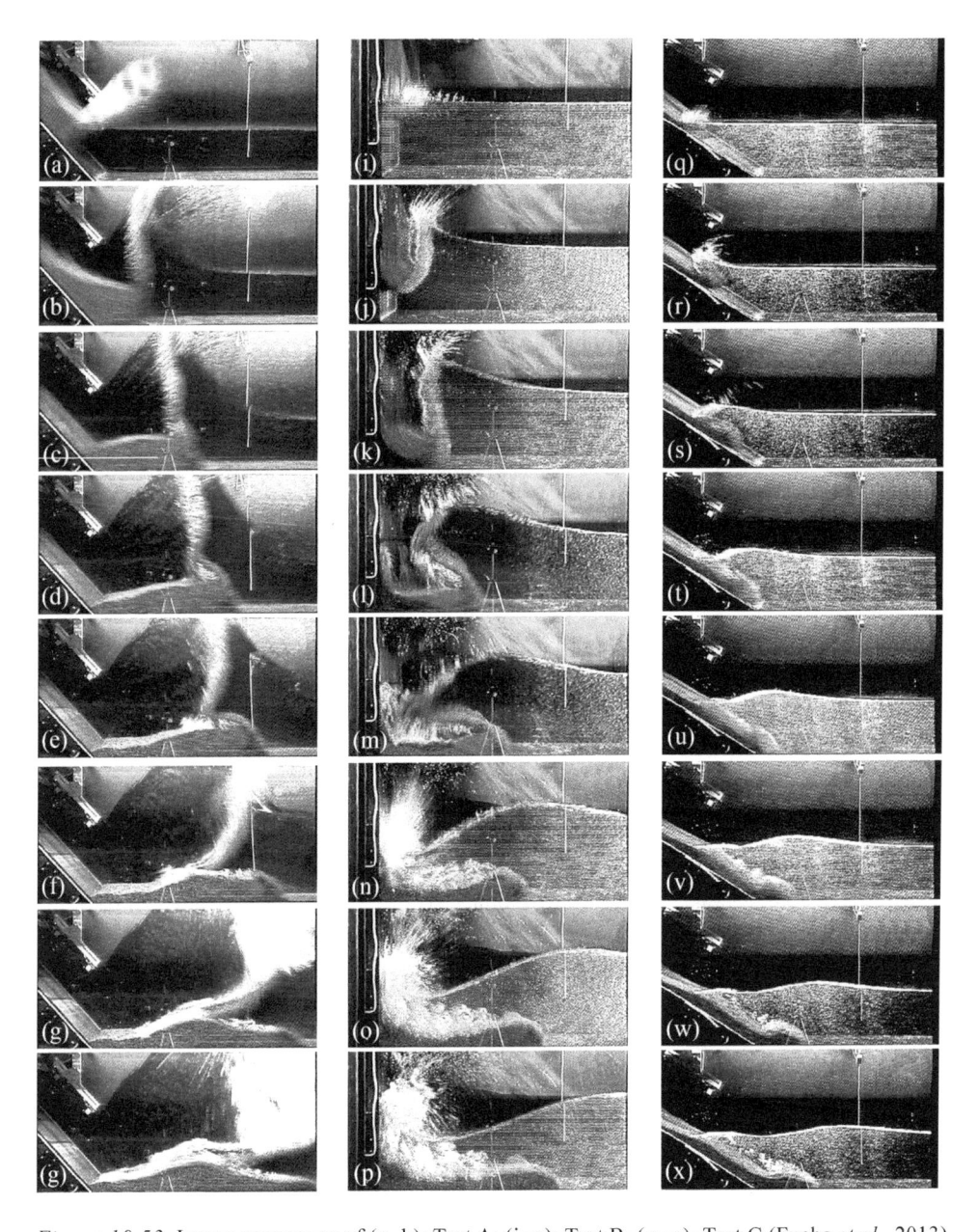

Figure 10.53 Image sequences of (a–h): Test A; (i–p): Test B; (q–x): Test C (Fuchs *et al.*, 2013)

The slide front position x'_{f0} was extracted from still images. Figure 10.54a shows the normalized slide front propagation $F_0 = x'_{f0}/x'_{f0, \, end}$ plotted versus the normalized time $T = t/t_{end}$. The data follow the fit ($R^2 = 0.96$)

$$F_0 = 1.03 \tanh(2T). \tag{10.45}$$

The slide motion is continuously decelerated during underwater propagation, marking a distinct difference to subaqueous slides, which first accelerate during the triggering slope failure process. The inclination of Eq. (10.45) indicates an over-proportionally fast slide motion for $T < 0.45$, whereas the motion is relatively slow for $T \geq 0.45$. Figure 10.54b shows the normalized slide front velocity determined as the sliding average of three consecutive images against normalized time T as

$$V_{F0,n} = \frac{\Delta x'_{F0}}{\Delta T} = \frac{x'_{F0,n+1} - x'_{F0,n-1}}{T_{n+1} - T_{n-1}}. \tag{10.46}$$

The first derivative of Eq. (10.45) directly gives ($R^2 = 0.64$)

$$V_{F0}(T) = \frac{\mathrm{d}F_0}{\mathrm{d}T} = 2.06 \cdot \cosh(2T)^{-2}, \qquad \text{if } 0.1 \leq T \leq 1. \tag{10.47}$$

For $T < 0.1$, Eq. (10.47) significantly underestimates the slide front velocity. The data scatter in Figure 10.54b arises due to measurement inaccuracy, particularly for fast slides and small time steps ΔT.

The underwater slide propagation dynamics are described with Eqs. (10.45) and (10.47), yet finite values for slide deposition characteristics are necessary to asses a particular slide event. The slide deposition geometry was analyzed to extract the final (subscript 'end') position of the slide front $x'_{f0, end}$, the maximum deposition thickness $z_{s, end}$, its position $x'_{s, end}$, and the relative slide propagation duration $T_{end} = t_{end}(g/l)^{0.5}$. Linear least-squares correlations between these and the initial slide parameters M_s and $P_s = P\sin(6\alpha/7)$ are shown in Figure 10.55. For the front position and thickness as main underwater landslide deposition parameters, the correlation is good. The parameter $x'_{f0, end}/h$ increases from 2 to 8 whereas $z_{s, end}/h$ decreases from 0.9 to 0.3 (Figure 10.55a-d), respectively, both for increasing P_s and M_s. Accordingly, the larger the slide impact (involving slide impact velocity as the governing basic parameter) the further the slide front penetrates into the water body and the thinner

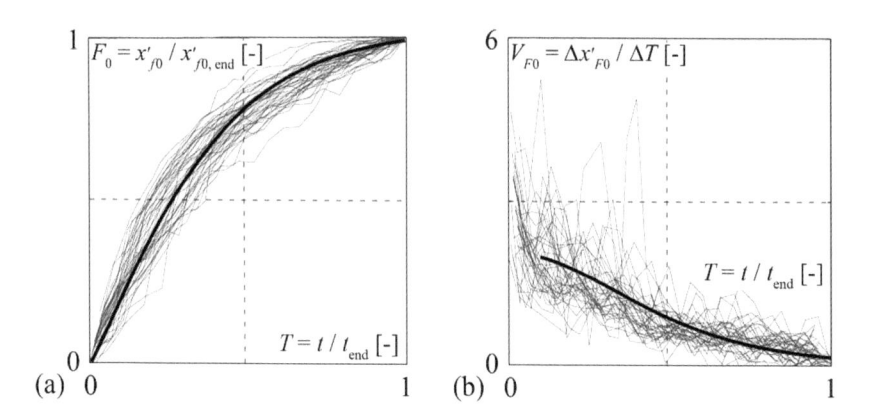

Figure 10.54 Normalized underwater slide dynamics (a) slide front propagation $F_0(T)$ with (—) Eq. (10.45), (b) slide front velocity $V_{F0}(T)$ with (—) Eq. (10.47); (—) individual tests (Fuchs *et al.*, 2013)

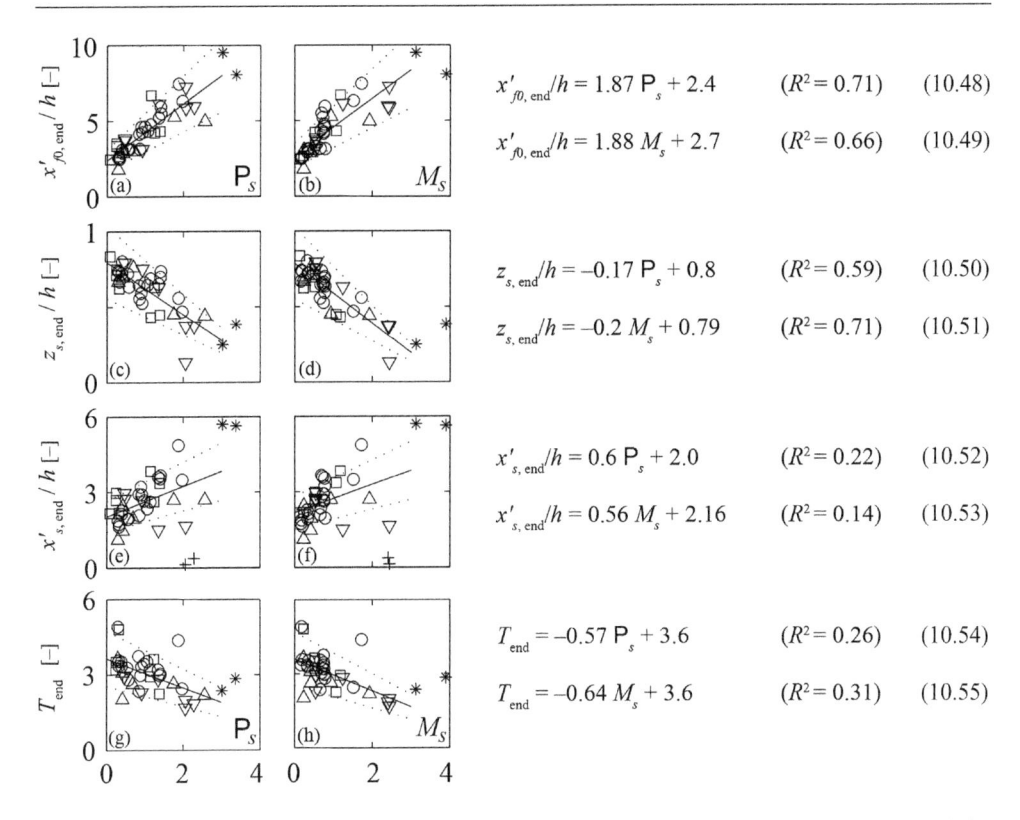

$$x'_{f0,\,end}/h = 1.87\,P_s + 2.4 \qquad (R^2 = 0.71) \qquad (10.48)$$

$$x'_{f0,\,end}/h = 1.88\,M_s + 2.7 \qquad (R^2 = 0.66) \qquad (10.49)$$

$$z_{s,\,end}/h = -0.17\,P_s + 0.8 \qquad (R^2 = 0.59) \qquad (10.50)$$

$$z_{s,\,end}/h = -0.2\,M_s + 0.79 \qquad (R^2 = 0.71) \qquad (10.51)$$

$$x'_{s,\,end}/h = 0.6\,P_s + 2.0 \qquad (R^2 = 0.22) \qquad (10.52)$$

$$x'_{s,\,end}/h = 0.56\,M_s + 2.16 \qquad (R^2 = 0.14) \qquad (10.53)$$

$$T_{end} = -0.57\,P_s + 3.6 \qquad (R^2 = 0.26) \qquad (10.54)$$

$$T_{end} = -0.64\,M_s + 3.6 \qquad (R^2 = 0.31) \qquad (10.55)$$

Figure 10.55 Characteristic deposition parameters versus P_s (left) and M_s (right) (a) and (b) $x'_{f0,\,end}$, (c) and (d) $z_{s,\,end}$, (e) and (f) $x'_{s,\,end}$, and (g) and (h) T_{end}, for $\alpha = (\square)\ 30°$, ($\circ$) 45°, ($\triangle$) 60°, ($\triangledown$) 90° with ($*$) tests potentially affected by scale effects, ($\cdots$) ±30% (Fuchs *et al.*, 2013)

becomes the slide deposit. The other deposit parameters $x'_{s,\,end}$ and T_{end} (Figure 10.55e-h) follow weakly the correlation equations with a tendency for decreasing duration and increasing protrusion distance of the deposition maximum ($x'_{s,\,end}$) as P_s and M_s increase.

The impulse product parameter P controls the impulse wave amplitude for the wave generation process (Heller *et al.*, 2009). The impact angle-corrected relative slide mass, in turn, determines the governing slide deposition characteristics thereby excluding the slide impact dynamics as

$$M_s = m_s / (\rho_w B_s h^2)\sin[(6/7)\,\alpha]. \qquad (10.56)$$

Therefore, the data are compared with the 'Fahrböschung' method of Heim (1932). The theoretically required travel distance x'_t to obtain the slide impact velocity V_s was back-calculated for the given slopes and friction coefficients using Eq. (10.44). The run-out relation given by Eq. (10.43) was determined using the theoretically traveled elevation difference Δz between the slide top in its initial position down to the channel bottom $z = -h$, and the ordinate for the streamwise slide displacement Δx from the back end of the initial slide position to the slide deposit front $x'_{f0,\,end}$. The value of $\Delta z/\Delta x$ therefore increases with the slide impact angle α except for $\alpha = 90°$ for which the run-out length significantly scatters, indicating a prominent effect of the slide-water interaction (Fuchs *et al.*, 2013). The impact crater formation and collapse then lead to additional deposit pattern deformation. If $\Delta z/\Delta x$ is plotted versus $\tan\alpha$ according to Scheidegger (1973), an improved correlation results as ($R^2 = 0.89$, Fuchs *et al.*, 2013)

$$\Delta z / \Delta x = 0.4746 \tan\alpha + 0.302, \qquad 30° \leq \alpha \leq 60°. \tag{10.57}$$

To assess impulse wave events, a conflict may arise due to the definition of the slide velocity. Whereas the slide front velocity V_{f0} is easier to determine especially for prototype events, the slide-centroid velocity V_s is an important parameter for the impulse wave generation process (Fritz *et al.*, 2004). Therefore, the two velocities prior to impact onto the still water are $V_s < V_{f0}$. This finding supports Watts (2000), according to whom the slide motion consists of two superposed mechanisms: the slide centroid moves with a certain velocity analogue to a moving solid body and the slide additionally spreads over the travel distance leading to an increased front velocity V_{f0} compared to V_s as (Fuchs *et al.*, 2013; $R^2 = 0.74$)

$$V_s[\text{m/s}] = 0.80 V_{f0}[\text{m/s}] + 0.86, \qquad 2 \leq V_{f0}[\text{m/s}] \leq 8.5. \tag{10.58}$$

The normalized slide front propagation shown in Figure 10.54a stipulates that the underwater slide characteristics are similar for all tests. The normalized slide length versus normalized time though indicates large deviations. Whereas the slide length from $t = 0$ to t_{end} was deduced from video records, the characteristic slide length l_s at $t = 0$ was determined using the slide-centroid velocity V_s. Depending on the initial slide characteristics, the slide length stretches by a factor of 1–7 compared with the slide box length l_{box} at test start $t = t_0$. The slide length rapidly reduces after water impact due to slide compaction with a minimum as the channel bottom is reached. From channel bottom contact to the complete still stand, the slide spreads again so that l_s increases by a factor of 1.3–5 compared with l_{box}.

This research thus analyzes landslide-generated impulse waves to determine both the underwater slide dynamics and the final deposition pattern. This knowledge applies for both the numerical modeling of a slide impact using the moving boundary approach, or the assessment of historical deposits resulting from slides into water bodies. The slide deposition length is of special concern if safety measures dictate the control of endangered infrastructure.

An interesting fact is the independence of the final deposition pattern from slide dynamics. This pattern in terms of front position and maximum thickness is defined solely by the relative impact angle-corrected slide mass M_s involving the slide mass, the still water depth and the slide impact angle. A method to determine the final slide deposition pattern is compared with a classical run-out relation overlooking water interaction effects. The relation between the maximum vertical to horizontal travel distances of subaqueous slides therefore mainly depends on the impact angle.

10.7 Rigid dam overtopping

10.7.1 Motivation and experimentation

Rapid mass wasting into reservoirs generates impulse waves propagating across the water body and either run up on the opposite shore, or overtop a dam (ICOLD, 2002), with a potential to damage a dam structure, or result in tailwater flooding. Because these damages have to be strictly avoided to satisfy the basic requirements of dam safety, and given the complex and involved physical characteristics of overtopping waves, a hydraulic laboratory research aimed to address the hydraulic features. In a first step, overtopping was studied without considering a possible erosion of the crest of earth and rock-fill dams, thus the notion rigid overtopping is used.

Overtopping of oscillatory waves at sea defense structures is an extensive research field within coastal engineering (Pullen *et al.*, 2007). While these findings apply to reservoir settings under wind waves (Yarde *et al.*, 1996), only few studies describe overtopping by impulse waves. In both physical and numerical models, solitary waves are applied to simulate the overtopping characteristics of impulse waves generated by landslides on dam structures, since this wave type represents the extreme (Heller and Hager, 2011). Dodd (1998), Hsiao and Lin (2010), Hunt-Raby *et al.* (2011), and Lin *et al.* (2011) placed a dam-like structure at the top end of a smoothly sloping beach, so that the generated solitary wave is subject to shoaling effects including wave breaking prior to overtopping. These test setups thus deal with conditions usually found along coastlines. In reservoirs, the dam structure is commonly located at lower regions of the bathymetry excluding a distinct shoaling phase prior to overtopping. Özhan and Yalçıner (1990), Müller (1995), Jervis and Peregrine (1996), Stansby (2003), and Xiao *et al.* (2008) placed an impounded dam structure on a flat bottom, allowing for undisturbed solitary wave propagation up to it. From an engineering point of view, the overtopping volume is a basic parameter for hazard assessment of the dam tailwater. Among the studies of undisturbed solitary waves, only Müller (1995) and Özhan and Yalçıner (1990) provide equations to predict the overtopping volume. A discussion of the approach by Müller (1995) is given by Heller *et al.* (2009), reviewed in 10.2.4: Computation of (1) run-up height at inclined plane; (2) overtopping volume without freeboard with empirical overfall equation; (3) actual overtopping unit length volume based on (1) and (2) including freeboard consideration. The overtopping duration equation of Müller (1995) is limited to cases without freeboard. The present one-step approach to predict overtopping volumes is compared with the data of Müller (1995).

The plane (2D) overtopping behavior of solitary waves at dams of different dam front face angles and crest widths was modeled in a rectangular wave channel 11 m long and $b = 0.50$ m wide. Its front sidewall and two thirds of the channel bottom were made of glass whereas the other sidewalls involve steel plates. The PVC dam with its origin at $x = 0$ m was placed at a horizontal distance of 4.85 m from the center position of the wave paddle (Figure 10.56). Solitary waves were generated using a pneumatic piston-type wave generator. Fuchs (2013) describes both wave channel and generator.

The three governing variables include the wave overtopping volume V, maximum wave overtopping depth a_M, defined as the maximum water elevation at $x = 0$, and the wave overtopping duration t_O, as duration between the initial and final wave crossing of the horizontal line at elevation $w = 0.30$ m and $x = 0$. All relevant wave parameters are shown for solitary, and for solitary-like waves in Figure 10.57, concerning all waves generated in the present study, and these studied by Müller (1995), respectively. The following parameters were varied: Still water depth h, wave amplitude a, dam front face angle β, and dam crest width b_K. The sharp-crested overtopping geometry was considered; the space below the tailwater and the dam face remained unaerated due to the short overtopping duration. The dam height was kept constant at $w = 0.30$ m. The parameters derived from the data analysis include the relative wave amplitude $\varepsilon = a/h$, wave celerity $c = [g(h + a)]^{1/2}$ with g as gravity acceleration, and the freeboard $f = w - h$.

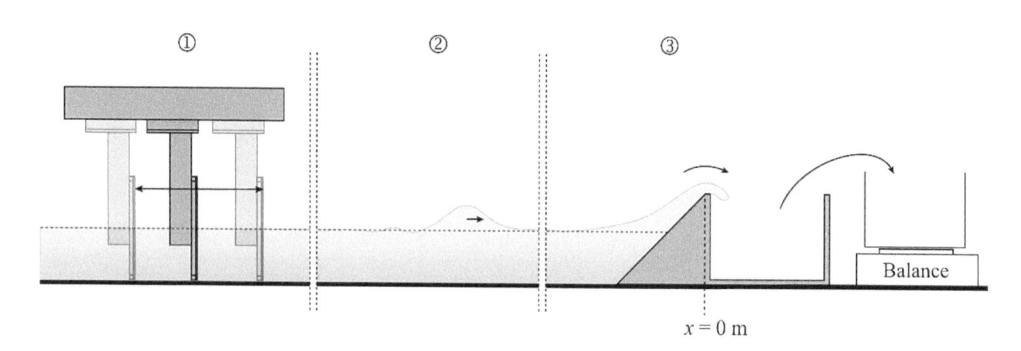

Figure 10.56 Test setup with ① Wave generation, ② Wave propagation, ③ Wave overtopping (Kobel *et al.*, 2017)

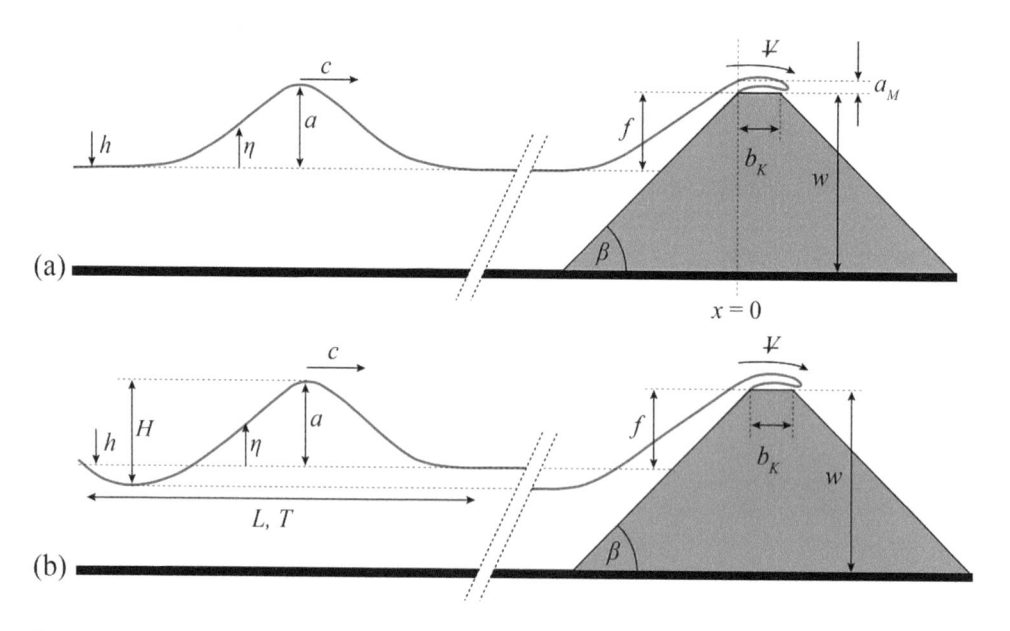

Figure 10.57 Definition sketch for overtopping of (a) solitary wave of amplitude a, (b) solitary-like wave of height H (Kobel *et al.*, 2017)

Parameter combinations of the still water depth h ranging from 0.10 to 0.30 m and relative wave amplitudes ε between 0.10 and 0.70 leading to dam overtopping were incorporated into the test program (Kobel *et al.*, 2017). Data for which the maximum overtopping depth was below $a_M = 0.05$ m were excluded from the fit equations due to possible scale effects (Le Méhauté, 1976; Sarginson, 1972; Heller, 2011).

The standard crest width was $b_K = 0.02$ m. Its effect was considered by repeating the test program for $b_K = 0.06$ m and 0.16 m at $\beta = 90°$. Similarly, the effect of the dam front face angle β was reconsidered for $\beta = 18.4°$ and 45° at $b_K = 0.02$ m. A total of 75 tests was conducted.

A high-speed camera operating at 40 Hz recorded the overtopping process. The maximum overtopping depth a_M at $x = 0$ m was manually extracted from the images. The ratio between pixel and effective length was determined from the known crest width of 0.020 m resulting in 0.378 mm for one pixel. The overtopping volume was weighed with a calibrated balance to ±1 g. The still water depth h was adjusted to the test elevation prior to each test using a point gage.

10.7.2 Overtopping processes

Figures 10.58 to 10.61 describe the overtopping features by qualitative assessment of the recorded high-speed images. The effect of relative wave amplitude for the vertical dam face and a narrow crest width is highlighted in Figure 10.58. The overflowing surface profile is similar to standard weir overflow for small ε, except for the sloping free surface due to the approaching solitary wave (Figure 10.58a). Given the relative small maximum overflow depth of 0.12 m, effects of surface tension result in the non-sharp lower nappe profile along the glass wall. Increasing ε to 0.50 results in a more peaked free-surface profile whose crest is slightly upstream of the dam crest. The approach flow surface profile becomes steeper than for $\varepsilon = 0.30$; in addition, the overflowing nappe is almost circularly shaped due to the unaerated space below the lower nappe. These effects increase for the highest ε value tested (Figure 10.58c). Given the extreme free-surface curvatures along with small overflow depths at overtopping start and end, hydraulic experimentation appears to be the only means by which this physical process is adequately explored. For smaller values of β, the effect of wave steeping and free-surface curvature at the dam crest vicinity is less pronounced.

The effect of dam front face angle on the overtopping process is shown in Figure 10.59. All images relate to conditions prior to jet impact onto the tailwater. For $\beta = 18.4°$ the upstream free-surface profile nearly follows the channel bottom, with a slight wave crest curvature at the dam crest. The overflowing jet resembles partial weir flow with strong turbulence generation at its front (Figure 10.59a). As the angle increases to $\beta = 45°$, the overflow process becomes more horizontally compacted, with a strong surface curvature at the dam crest, a circularly shaped lower nappe profile and a less turbulent jet front (Figure 10.59b). For $\beta = 90°$ these effects further increase leading to a sharp-peaked wave crest and a steep upstream free-surface slope (Figure 10.59c). Note the smooth free surfaces except for the jet front. The process thus does not greatly depend on airflow except once the jet has impacted the tailwater.

Figure 10.60 relates to the effect of still water depth. Note the increasing slope of the upstream water surface elevation $\eta(x)$ as the difference $f = w - h$ reduces. The maximum wave

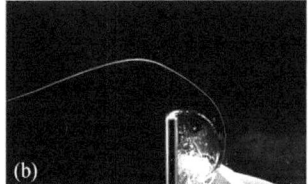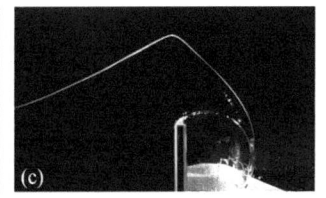

Figure 10.58 Effect of relative wave amplitude on overtopping process for $w = 0.30$ m, $h = 0.25$ m, $b_K = 0.02$ m, $\beta = 90°$ and ε [-] = (a) 0.30, (b) 0.50, (c) 0.70 (Kobel et al., 2017)

Figure 10.59 Effect of dam front face angle on overtopping process for $w = 0.30$ m, $h = 0.20$ m, $b_K = 0.02$ m, $\varepsilon = 0.70$ and β [°] = (a) 18.4, (b) 45, (c) 90 (Kobel *et al.*, 2017)

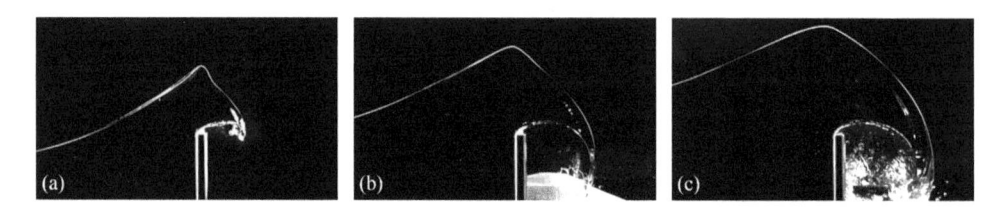

Figure 10.60 Effect of still water depth on overtopping process for $w = 0.30$ m, $\beta = 90°$, $\varepsilon = 0.70$, h [m] = (a) 0.20, (b) 0.25, (c) 0.30 (Kobel *et al.*, 2017)

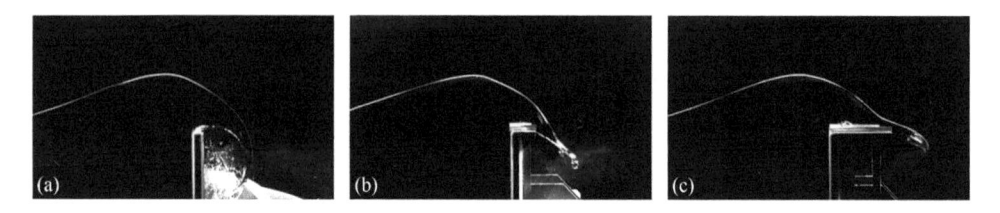

Figure 10.61 Effect of crest width on overtopping process for $w = 0.30$ m, $\beta = 90°$, $\varepsilon = 0.50$, $h = 0.25$ m and b_K [m] = (a) 0.02, (b) 0.06, (c) 0.16 (Kobel *et al.*, 2017)

crest curvature also increases with f, whereas the jet geometry in the tailwater looks similar as in Figures 10.58 and 10.59.

The effect of dam crest width b_K is reproduced in Figure 10.61. Increasing it causes an additional convex curvature on the tailwater wave surface portion due to the longer crest width (Figure 10.61c), thereby affecting the overtopping features. This effect is particularly pronounced at small freeboards f.

10.7.3 Experimental results

Overtopping volume

The test results were systematically analyzed to obtain an empirical fit between the over-topping volume $⩊$ and the governing parameters. These include the still water depth h, the relative wave amplitude ε, the dam front face angle β, and dam crest width b_K. Two approaches are presented below: (1) an approximation excluding the effect of β, and (2) a more accurate approach including β.

The relative overtopping volume $V/(bh^2)$ increases significantly with the relative still water depth h/w including the effect of the freeboard, moderately with the relative wave amplitude ε, and slightly with the relative maximum wave height a_w/b_K with $a_w = h + a - w$ as the effective wave amplitude as ($R^2 = 0.96$, Kobel et al., 2017)

$$\frac{V}{bh^2}=1.42\left[\varepsilon\left(\frac{h}{w}\right)^{2.5}\left(\frac{a_w}{b_K}\right)^{0.105}\right]^{0.8}=1.42W_1^{0.8}, \quad \begin{array}{l} 0.35 < W_1^{0.8} < 0.95, \\ 0.07 < b_K/w < 0.53 \ . \end{array} \tag{10.59}$$

Figure 10.62a shows the relation $V/(bh^2)$ versus $W_1 = \varepsilon(h/w)^{2.5}(a_w/b_K)^{0.105}$ as the wave overtopping volume parameter. Note the excellent fit for $W_1^{0.8} > 0.35$, whereas extreme deviations are $\pm 30\%$ otherwise.

An increase of the dam front face angle β from $18.4°$ to $90°$ causes a reduction in the overtopping volume $V/(bh^2)$. This effect increases with the relative still water depth h/w and the relative wave amplitude ε. Note that no effect of macro roughness was studied, so that the overtopping volume continuously increases with decreasing dam front face angle β. Without freeboard ($h = w$), the reduction of $V/(bh^2)$ due to the decrease of the dam front face angle is almost constant at 7%. To include the effect of dam front face angle, the exponent of h/w depends on both β and ε as ($R^2 = 0.99$)

$$\frac{V}{bh^2}=1.35\left[\varepsilon\left(\frac{h}{w}\right)^{2\left(\frac{\beta}{\varepsilon\,90°}\right)^{0.25}}\left(\frac{a_w}{b_K}\right)^{0.12}\right]^{0.7}=1.35W_2^{0.7}, \quad \begin{array}{l} 0.15 < W_2^{0.7} < 0.95, \\ 0.07 < b_K/w < 0.53 \ . \end{array} \tag{10.60}$$

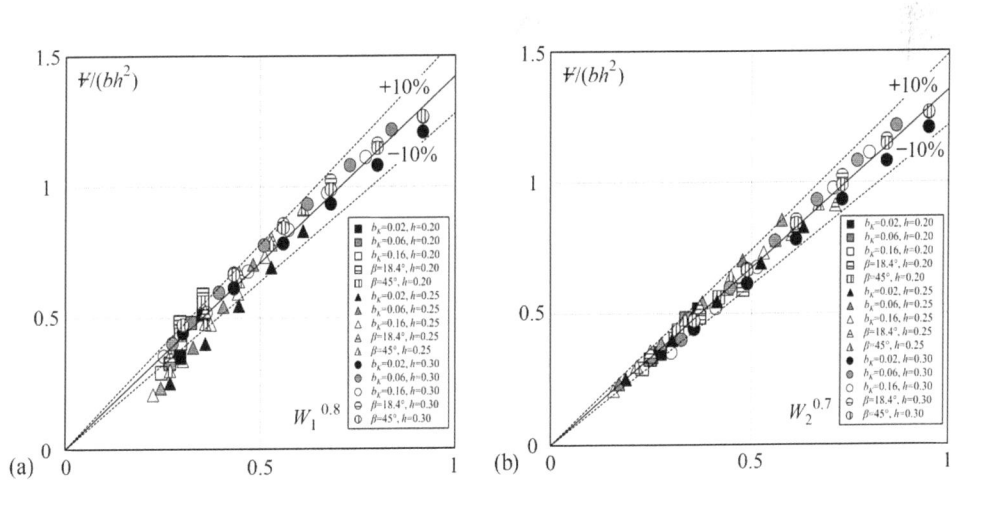

Figure 10.62 Relative overtopping volume $V/(bh^2)$ versus (a) $W_1^{0.8}$ according to (—) Eq. (10.59), (---) $\pm 10\%$ deviation, (b) $W_2^{0.7}$ according to (—) Eq. (10.60), (---) $\pm 10\%$ deviation. Symbols: b_K [m], h [m], β [°]; if not stated, then $\beta = 90°$, $b_K = 0.02$ m (Kobel et al., 2017)

Maximum deviations of Eq. (10.60) from the test data are -12% for $W_2^{0.7} = 0.3$ and $+10\%$ for $W_2^{0.7} = 0.6$ (Figure 10.62b). Equations (10.59, 10.60) are similar in the terms except for the exponent of h/w; whereas it remains constant at 2.5 in Eq. (10.59), it varies from 1.92 to 10 in the extreme in Eq. (10.60). Note its slight reduction by a smaller overall exponent of 0.7 in Eq. (10.60) as compared with 0.8 in Eq. (10.59).

Maximum overtopping depth

The second key variable is the maximum wave overtopping depth a_M at the dam crest ($x = 0$, Figure 10.57a). The normalized value a_M/w versus the relative wave amplitude ε, the relative still water depth h/w and the dam front face angle β follows the empirical fit ($R^2 = 0.95$, Kobel *et al.*, 2017)

$$\frac{a_M}{w} = 1.34\left[\varepsilon\left(\frac{h}{w}\right)^{1.7}\left(\frac{\beta}{90°}\right)^{0.25}\right] = 1.34\,E_1; \quad 0.40 < E_1 < 0.70. \tag{10.61}$$

Figure 10.63a shows the relation $a_M/w[E_1]$ with $E_1 = \varepsilon(h/w)^{1.7}(\beta/90°)^{0.25}$ as maximum wave overtopping depth parameter. Note the excellent fit for $E_1 > 0.40$, whereas maximum deviations are $\pm 30\%$ otherwise. The data scatter is strongly reduced for $E_2 < 0.2$ (Figure 10.63b) with the extended fit equation ($R^2 = 0.99$)

$$\frac{a_M}{w} = 1.32\left[\varepsilon\left(\frac{h}{w}\right)^{4\left[\left(\frac{\beta}{90°}\right)^{-0.21}-\varepsilon\right]}\left(\frac{\beta}{90°}\right)^{0.16}\right] = 1.32\,E_2; \; 0.10 < E_2 < 0.75. \tag{10.62}$$

As to the exponent of the term h/w in Eq. (10.62), its maximum is 4.78, whereas the minimum is 1.20. Note the slight change of the exponent of the term $(\beta/90°)$, whereas the effect of the dam crest width b_K does not appear neither in Eqs. (10.61) nor (10.62).

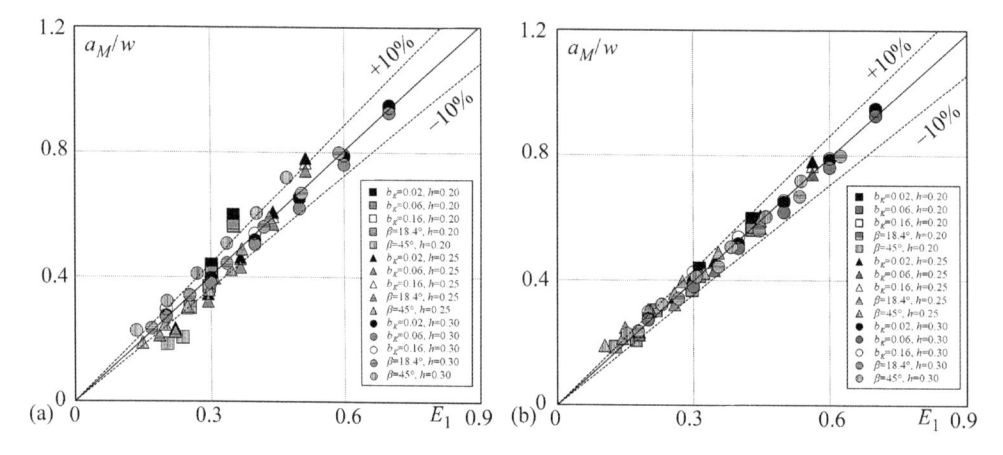

Figure 10.63 Relative maximum overtopping depth a_M/w versus (a) E_1 according to (—) Eq. (10.61), (---) $\pm 10\%$ deviation, (b) E_2 according to (—) Eq. (10.62), (---) $\pm 10\%$ deviation (Kobel *et al.*, 2017)

Overtopping duration

The wave overtopping duration t_O is relevant to estimate the flood hydrograph after a dam overtopping scenario. As opposed to Müller's approach, Eq. (10.63) applies for cases with a freeboard $f > 0$ (Heller et al., 2009). Figure 10.64 shows the inverse relative overtopping duration $T_O = w^{1/2}/(g^{1/2}t_O)$ versus the dimensionless wave overtopping duration parameter $F = \varepsilon^{0.2}(h/w)^{(-0.9/\varepsilon)(\beta/90°)^{0.4}}$ as ($R^2 = 0.97$, Kobel et al., 2017)

$$\frac{w^{0.5}}{t_O g^{0.5}} = 0.15\left[\varepsilon^{0.2}\left(\frac{h}{w}\right)^{\frac{0.9}{\varepsilon}\left(\frac{\beta}{90°}\right)^{0.4}}\right]^{1.9} = 0.15F^{1.9}, \quad 0.50 < F^{1.9} < 2.60. \tag{10.63}$$

The minimum exponent of the term (h/w) is -4.5, whereas the maximum amounts to -0.70 generating the significant effect on T_O. Note the comparatively small effect of ε in Eq. (10.63), and again the total absence of the effect of b_K. To obtain an estimate of resulting parameters, consider as example $h = 0.25$ m, $w = 0.30$ m, $\varepsilon = 0.50$ and $\beta = 90°$. First $F = 1.21$, from where $T_O = 0.22$ based on Eq. (10.63), so that $t_O = 0.81$ s. Note the extremely short time of only $t_O = 8.1$ s for a prototype scale of 100.

Individual parameter effects

Although Eqs. (10.59, 10.60) do not predict any overtopping volume for waves of which $a_w = a + h - w < 0$, i.e. if the approach flow wave amplitude is located below the dam crest elevation, overtopping actually occurs due to effects of wave compression (Figure 10.65). Therefore, the proposed equations do not provide information on incipient overtopping. To

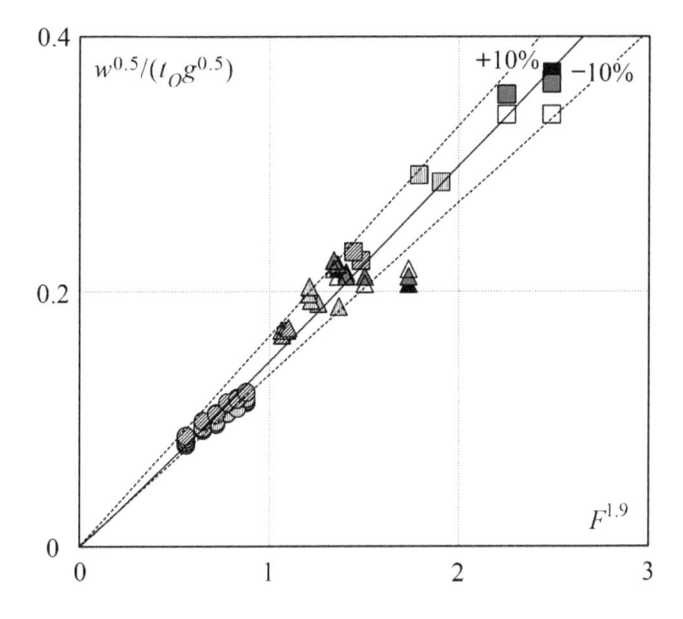

Figure 10.64 Inverse relative overtopping duration $w^{0.5}/(t_O g^{0.5})$ versus $F^{1.9}$ according to (—) Eq. (10.63), (---) ±10% deviation. Symbols as in Figure 10.62 (Kobel et al., 2017)

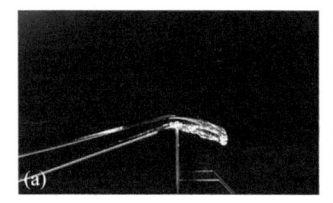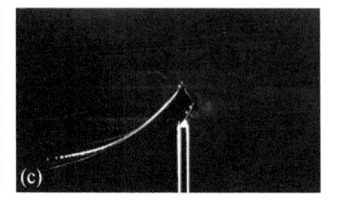

Figure 10.65 Occurrence of wave overtopping despite $a_w < 0$ for $h = 0.15$m, $\varepsilon = 0.70$, $w = 0.30$ m, β [°] = (a) 18.4, (b) 45, (c) 90 (Kobel *et al.*, 2017)

determine the minimum freeboard required to prevent it, the approach of Fuchs and Hager (2015) applies, except for $\beta = 90°$, for which no data were there collected.

Solving Eqs. (10.59) and (10.61) explicitly for the unknowns V and a_M reveals that the effect of still water depth h is dominant, followed by that of dam height w, whereas that of the wave amplitude a is almost linear, and the effects of both a_w and b_K are negligible in the first case. In the second case, the wave amplitude effect is dominant, whereas the relative still water depth h/w has a medium effect, and $\beta/90°$ has an only minor influence.

Comparison with literature data

Equation (10.60) for the overtopping volume V was expanded with the approach of Müller (1995) by comparing his complete data set, except for cases for which $a_w/b_K < 0$, with the present parameters. Figure 10.57b sketches Müller's setup, including the relative wave height H/h, wave non-linearity a/H, wave steepness H/L, relative wave period $T(g/H)^{1/2}$, relative wave celerity $c/(gh)^{1/2}$, relative wave length L/h and relative angle $\beta/90°$. In contrast to Kobel *et al.*'s (2017) research, Müller (1995) considered solitary-like waves limited by $0.01 < L^2H/h^3 < 4\pi^2/3 = 13.16$ and $3 < h/H < 48$. A parameter range check indicates that Eq. (10.60) does not apply to his data. Kobel *et al.*'s research is generalized by including Müller's data for the wave overtopping volume.

Müller (1995) studied solitary-like, cnoidal, Stokes, and Airy waves limited by $2 < h/H < 100$ and $0.03 < h/L < 0.15$. Whereas he proposed an equation for the unit overtopping volume V/b in terms of the above listed parameters plus coefficients detailing the dam crest shape, Kobel *et al.*'s research is solely based on the additional effect of the relative wave height $A = a/H$, with $A = 1$ for the solitary wave, whereas all other above addressed waves satisfy $A < 1$. To apply the results of the present research, Eq. (10.60) thus was retained and the effect of A added. Figure 10.66a includes all of Müller's data relating to solitary-like waves as previously defined. A data reanalysis indicates that ($R^2 = 0.88$, Kobel *et al.*, 2017)

$$\frac{V}{bh^2} = 1.35\left[\frac{a}{H}\right]^{1.5}\left[\varepsilon\left(\frac{h}{w}\right)^{\frac{2}{\varepsilon}\left(\frac{\beta}{90°}\right)^{0.25}}\left(\frac{a_w}{b_K}\right)^{0.12}\right]^{0.7} = 1.35A^{1.5}W_2^{0.7}, \quad \begin{array}{l} 0.02 < W_2^{0.7} < 0.59, \\ 0.63 < A < 0.93, \\ 0.07 < b_K/w < 0.53. \end{array} \quad (10.64)$$

Further, Eq. (10.64) was also applied to all the remaining 318 data of Müller (1995), resulting in a similar trend ($R^2 = 0.89$). Figure 10.66b indicates a larger data scatter as Figure 10.66a, yet Eq. (10.64) approximately applies also to this complex case. The deviation of Eq. (10.64) for small values of $A^{1.5}W_2^{0.7}$ in Figure 10.66a is attributed to overtopping conditions for which $a_w < 0.04$ m, i.e. to data possibly affected by scale effects (Le Méhauté, 1976; Sarginson, 1972; Heller, 2011). Note that Eq. (10.64) requires much less detailed knowledge of fundamental wave and dam data as compared with Müller, whose final fit equation provides results of similar data scatter. Also note that solitary waves produce the maximum overtopping volume if all other parameters remain identical. The generalized approach includes thus also solitary-like waves. A simple prediction of the fundamental overtopping characteristics is thus amenable.

10.8 Erodable dam overtopping

10.8.1 Motivation and literature review

Dams impounding water are either human made or natural. Artificial embankment dams can suppress floods, but also provide hydropower. Natural dams can emerge through landslides, blocking a valley, or forming moraines shaped by glacial activity. They can alter and impound rivers or create lakes after glacier retreat. For landslide dams, the most common failure cause is overtopping (Costa and Schuster, 1988). Other common failure modes include geological instability or internal erosion. A common failure mechanism for moraine-dammed lakes in steep mountain areas is overtopping and breaching by waves generated in the lake (Costa and Schuster, 1988; Klimeš et al., 2016). Moraine dams pose a significant hazard because the dam faces feature steep slope angles of more than 40° (Costa and Schuster, 1988) and the lakes impounded behind them are directly downstream from crevassed glaciers and steep rock slopes. The waves are generated by landslides, snow avalanches or other mass wasting into a large water body (Heller et al., 2009), referred to as impulse waves. Dam overtopping by impulse waves can have catastrophic consequences (Kobel et al., 2017), given that the tailwater valley is flooded and the dam itself breaches in extreme scenarios, causing an outburst flood.

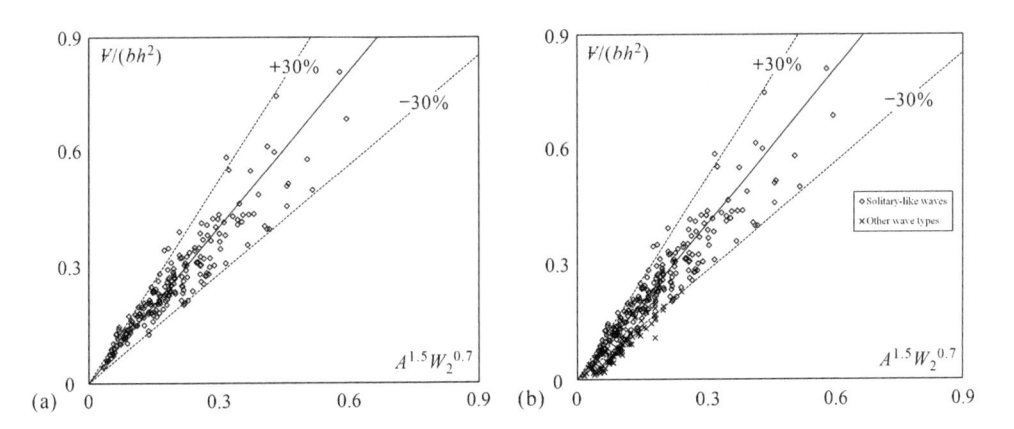

Figure 10.66 Relative wave overtopping volume $V/(bh^2)$ versus $A^{1.5}W_2^{0.7}$ according to (—) Eq. (10.64), (---) ±30% deviation for (a) solitary-like waves, (b) all compared wave overtopping data of Müller (1995) (Kobel et al., 2017)

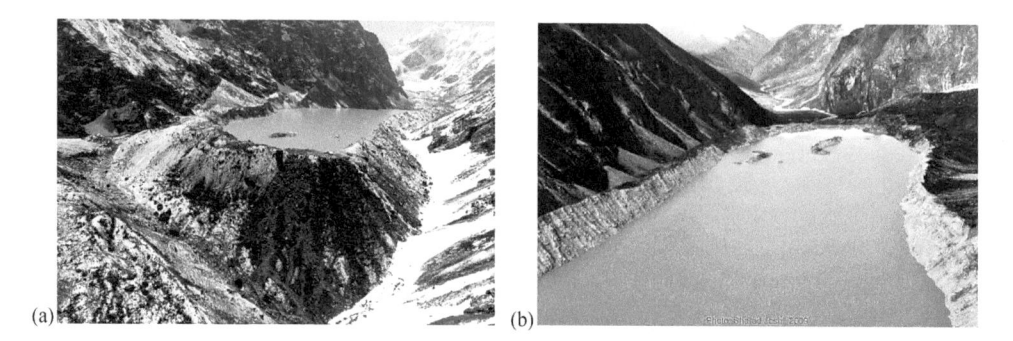

Figure 10.67 Lake Tsho Rolpa in Himalayans impounded by 216 m high end moraine dam with retreating Trakarding Glacier in background (photo Brian Collins/ USGS (CC-BY 2.0)), views from (a) downstream, (b) upstream of lake with freeboard and downslope Rolwalling Valley (photo Sharad Joshi/ICIMOD) (Huber *et al.*, 2017)

Carey (2008) provides an eyewitness report from a Glacial Lake Outburst Flood event in Peru: 'On December 13, 1941, the blinding force of nature unleashed its power, making the mountains shake; the avalanche came, killing and swallowing, destroying and demolishing everything in its path. It cut short the lives of thousands of innocent people who just happened to live or be located in this picturesque and beautiful land. [The flood] converted the city and its precious lands into a heaping pile of inert material'. This event, leading to the destruction of large parts of the city of Huaraz and nearly 5000 fatalities, was triggered by a glacier break off into Lake Palcacocha. The ice masses generated an impulse wave with a height up to 15 m (Carey, 2010) overtopping and eroding the top of the lake's moraine dam thereby presumably creating an incision for a gradually advancing dam breach. The water flooded downstream Lake Jiracocha and made its moraine dam fail as well. An estimated 8×10^6 m³ of water and debris eventually impacted the city of Huaraz, 23 km downstream of Lake Palcacocha (Carey, 2008). In 2003, a landslide into Lake Palcacocha generated waves of 8 m height. The released flood affected the drinking-water infrastructure in the valley downstream and cut off the citizens of Huaraz from water supply for more than a week (Klimeš *et al.*, 2016). Further events triggered by impulse wave impact on natural dams include outburst floods e.g. at Lake Dig Tsho, Nepal, in 1985 (Vuichard and Zimmermann, 1987), and at Nostetuko Lake in 1983 (Blown and Church, 1985; Clague and Evans, 2000) as well as Queen Bess Lake in 1997 (Clague and Evans, 2000), both Canada.

Figure 10.67 shows the glacial Lake Tsho Rolpa, Nepal. Its end moraine has a vertical height of 216 m and impounds a 3.45 km long and 500 m wide water body of 86×10^6 m³ (ICIMOD, 2011). Due to its steep moraine dam and small freeboard of 5 m it is considered the most unstable moraine dam in Nepal (ICIMOD, 2011). Therefore, the dam is equipped with an artificial drainage installation to avoid further water level elevation due to normal inflow (Rana *et al.*, 2000). However, under impulse waves generated by ice falls from Trakarding Glacier, the dam would be without protection. These outburst flood events and the example of Tsho Rolpa highlight the relevance of gaining knowledge of the interaction processes between impulse waves and dams composed of granular material, since it represents the triggering mechanism for dam failure and subsequent outburst floods. Schaub *et al.* (2013) predict a worldwide increase of these events at newly created lakes due to deglaciation.

Among the various wave types, the solitary wave is the most extreme of impulse waves (Synolakis, 1988; Heller and Hager, 2011). Given its relatively simple definition and the wave of highest energy, it is suitable to study its effect on a granular dam. The goal of Huber *et al.* (2017) was to study overtopping processes of solitary waves by hydraulic experimentation. Dikes in contrast to dams are also referred to as levees and typically placed along rivers to protect a valley from flooding. Embankment dams are normally made up of larger elements and usually have a core and surface protection. This research is directed to dam overtopping of which the dam structures are simplified for modeling reasons, containing therefore neither dam core nor surface protection. Given that these two dam additions improve their stability and seepage characteristics, the test setup used corresponds to the structure of minimum safety against damage.

From the engineering point of view, the overtopping volume is a basic parameter for hazard assessment of the dam tailwater, whereas the overtopping duration is important to determine the flood hydrograph. To evaluate the dam safety after wave overtopping or even its vulnerability to a complete failure, it is further relevant to assess the erosion features due to overtopping. To achieve this goal, granular dams in a 2D setup were considered. A systematic variation of the still-water depth, wave amplitude, dam face slope, sediment size and non-uniformity, and dam crest width was conducted. The results of this research allow for the prediction of these quantities based on fit equations.

Schmocker (2011) and Schmocker and Hager (2012) conducted laboratory tests on 2D dike breaches under constant overflow. Their goals included model limitations, understanding of processes and fit equations describing the overtopping features. Both the dike and sediment dimensions were systematically varied to provide information on test repeatability, sidewall effects, scale effects, the breach process, seepage effect, breach discharge, and sediment transport. The seepage effect was noted to be absent if an adequate dike drainage is provided. The relevant model limitations include (Chapter 11):

- Minimum grain size $d = 1$ mm to avoid cohesive effects, to apply Froude similitude
- Maximum grain size $d = 5.5$ mm to prevent dike sliding due to seepage
- Minimum dike width $b = 0.20$ m to avoid sidewall effects
- Minimum dike height $w = 0.20$ m to inhibit errors due to too small dike dimensions

These works detail the effects of dam face slope angles, grain size and sediment non-uniformity on the breach process (Schmocker *et al.*, 2014). The dike erosion advances fast for a steep dike slope because of high flow velocities. During the initial breach phase, erosion is fast for a large sediment diameter d, in contrast to common sediment transport. Similar effects are also expected in this research.

Müller *et al.* (2016) expanded the aforementioned study considering 2D granular dike breach scenarios involving (1) constant approach flow discharge, and (2) constant headwater elevation. The effects of crest width (referred to as crest length by Schmocker and Hager, 2012) and up- and downstream dike slopes were accounted for by the so-called dike shape parameter μ. Furthermore, the dike height and the sediment diameter were varied. The results include expressions for the temporal advance of dike height and breach discharge.

Kobel *et al.* (2017) studied the overtopping processes of solitary and solitary-like waves on a rigid laboratory dam. The still-water depth h, the relative wave amplitude $\varepsilon = a/h$, the dam front face angle β, and the dam crest width b_K were varied. Based on a data analysis, empirical fits for the overtopping depth a_M, the overtopping duration t_O, and the overtopping water volume Ψ were found as stated in Eqs. (10.59), (10.61), and (10.63). It was noted that the still-water depth, dam height w and wave amplitude are the governing parameters, whereas the effects of the dam front face angle and crest width are small.

Awal *et al.* (2010) conducted experiments on wave overtopping of sand dams. The waves were generated by impact of a rigid block into the water. A total of 32 tests was conducted, involving various block shapes, dam shapes, dam material and still-water levels. The temporal advance of reservoir water level, overtopping discharge and eroded sediment were measured. Their research demonstrates that the amount of drainage water is independent of the crest width for a freeboard $f = 0$, whereas more water was drained for a triangular than a trapezoidal dam if $f > 0$. The peak discharge occurred earlier for the triangular than the trapezoidal dam. A steeper front face slope led to a higher peak discharge, larger erosion and a higher total volume of drained water.

Balmforth *et al.* (2008, 2009) theoretically studied whether a large displacement wave leads to a catastrophic erosional incision of a moraine lake. Their laboratory experiments as well as numerical simulations indicate that a single wave is unable to break the dam, whereas an almost-filled reservoir can ultimately fail under repeated dam overtopping due to wave reflection. Their conceptual model expresses the dam breach process as a competition between erosion, lake drainage, and wave damping, so that a threshold based on initial wave amplitude and erosion rate is proposed.

These indications reveal that the dam overtopping process is currently not fully understood, even for the simplified case of the plane granular dam under the typical approach flow scenario of a solitary wave. Huber *et al.* (2017) expanded the current knowledge by a detailed process description and fit equations allowing for the assessment of the dam breach failures due to overtopping.

10.8.2 Experimental program

Wave channel

Granular dams overtopped by a solitary wave were investigated with a hydraulic scale model. Given the 2D problem, tests were run in a rectangular wave channel. The dam was placed in the channel center portion, including a bottom drainage below it. The water depths were measured with an Ultrasonic Distance Sensor (UDS). A high-speed camera was installed at the channel side to capture the overtopping process; the overtopping volume was determined with a balance. The 2D wave test channel was 11.0 m long, 1.0 m deep and 0.50 m wide (± 2.0 mm) (Figure 10.68). The channel bottom was made of polyvinyl chloride (PVC), the front wall of glass and the other walls of steel. The solitary impulse waves were generated by a pneumatically driven piston-type wave generator, displacing the entire water column by a vertical plate reaching down to the channel bottom (Fuchs, 2013).

The dams were placed into the channel so that the origin of the streamwise coordinate $x = 0$ was at the dam crest and its distance from the wave paddle nearly 5 m (± 0.10 m). The channel was filled with tap water up to the test still-water depth h. Because of water drainage before test start, the value of h refers to time when the wave generator was activated. The wave amplitude reduction along the approach flow reach was smaller than 1% of w so that it was neglected. A dam drainage was installed to ensure dam stability and seepage control during a test. The dams were placed on a 51 mm high platform, so that the water drained through the gap between the platform and the channel bottom (Figure 10.68). A sieve of fine wire mesh prevented sediment washout. The overtopping discharge was pumped into a

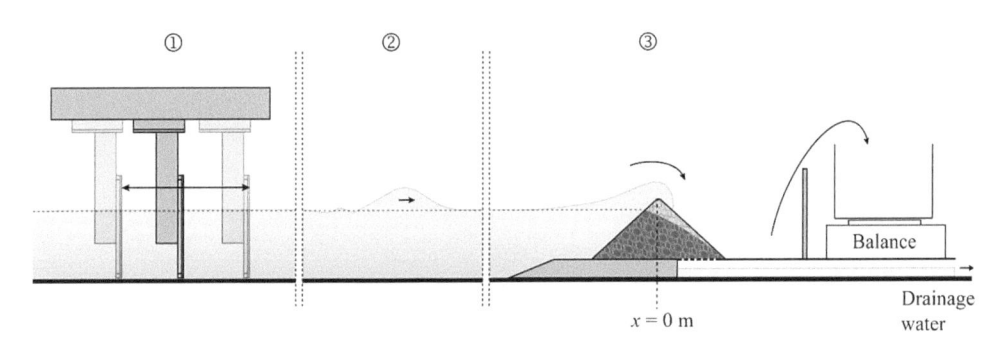

Figure 10.68 Definition sketch of test setup with ① wave generator, ② wave propagation, ③ wave run-up and overtopping

Figure 10.69 Photograph of dam setup before test start (Huber *et al.*, 2017)

container located on a balance to measure the overtopping water volume. The platform was 1.50 m long with a 27° wedge at its front to avoid wave modification. The dam height w was measured from the channel bottom including the platform thickness. The dam face slopes of both the up- and downstream faces were always identical (Figure 10.69).

Figure 10.70 shows a definition sketch for the overtopping tests. Herein x and z are the horizontal and vertical coordinates, respectively, η the free water surface elevation, w the dam height, and h the still-water depth. Further a is the initial wave amplitude, β the dam face angles, b_K the crest width and d the grain size. Six variables were examined, namely the:

- Maximum wave overtopping depth a_M measured from the original dam crest elevation
- Wave overtopping volume Ψ
- Wave overtopping duration t_O between initial and final wave crossing of initial dam height at $x = 0$ (Figure 10.71)
- Eroded crest depth h_e as minimum vertical distance from initial dam crest to highest dam point after a test (Figure 10.70)

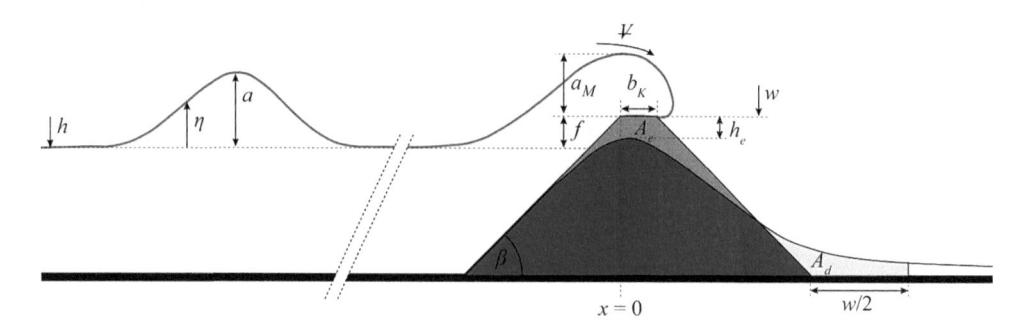

Figure 10.70 Definition sketch (Huber *et al.*, 2017)

- Eroded sediment area A_e removed during overtopping process
- Deposited sediment area A_d during overtopping measured up to $x = b_K + Sw + w/2$ (Figure 10.70).

The following parameters were varied:

- Relative still-water depth $h/w = 0.7, 0.8, 0.9, 0.95$
- Relative wave height $\varepsilon = a/h = 0.25, 0.30, \ldots 0.75$
- Dam face slope $S = 1/\tan\beta = 1.5, 2.0, 2.5, 3.0$
- Relative crest width $b_K/w = 0, 0.1, 0.5$
- Sediment grain size $d = 1.23, 1.65, 2.68$ mm.

Water depths were measured with an UDS UNAM 30®, Baumer electric, Switzerland, of 60 Hz frequency and an accuracy of $<\pm0.30$ mm. The minimum distance between the water surface and the UDS was 70 mm to avoid inaccuracies due to tilted surfaces (Fuchs, 2013).

Sediments were characterized by the median diameter d and the non-uniformity coefficient $\sigma_g = (d_{84}/d_{16})^{0.5}$. Three uniform, homogenous sediment sizes of density $\rho_S \approx 2550$ kg/m³ and $\sigma_g < 1.2$ were used (Figure 10.72a-c): $d = 1.23, 1.65, 2.68$ mm. The sediment was always dry at test start. Out of these three sediments, a total of four mixtures was prepared: Mix1–2 as compound of sediments with $d = 1.23$ and $d = 1.65$ mm; Mix1–3 of $d = 1.23$ and $d = 2.68$ mm; Mix2–3 of $d = 1.65$ and $d = 2.68$ mm; and Mix1–2–3 of $d = 1.23$, $d = 1.65$ and $d = 2.68$ mm. Figure 10.72d shows the grain size distributions of the four mixtures with $d_{50} = 1.39$ mm and $\sigma_g = 1.4$ for Mix1–2; $d_{50} = 2.04$ mm and $\sigma_g = 1.6$ for Mix1–3; $d_{50} = 2.38$ mm and $\sigma_g = 1.4$ for Mix2–3; and $d_{50} = 1.84$ mm and $\sigma_g = 1.5$ for Mix1–2–3, respectively.

A high-speed camera was placed perpendicularly to the dam at $x = 1$ m from the glass-sided channel wall. The 5.5 megapixel pco.edge® scientificCMOS camera was combined with a Zeiss® Distagon T* 2/28 lens. With a maximum resolution of 2560×2160 pixels, the camera records up to 100 fps. The view frame was reduced to 2560×1728 pixels to control the data storage: The camera was set to 50 fps and an exposure time of 0.15 ms. A measuring tape was fixed at the channel bottom wall to determine the image reference scale. The camera setup was surrounded with black cloth to avoid reflection from the glass wall. The illumination of the water surface was placed at the dam crest zone. The overtopping time t_O was evaluated by the number of frames between $t = 0$ and overtopping end. The overtopping

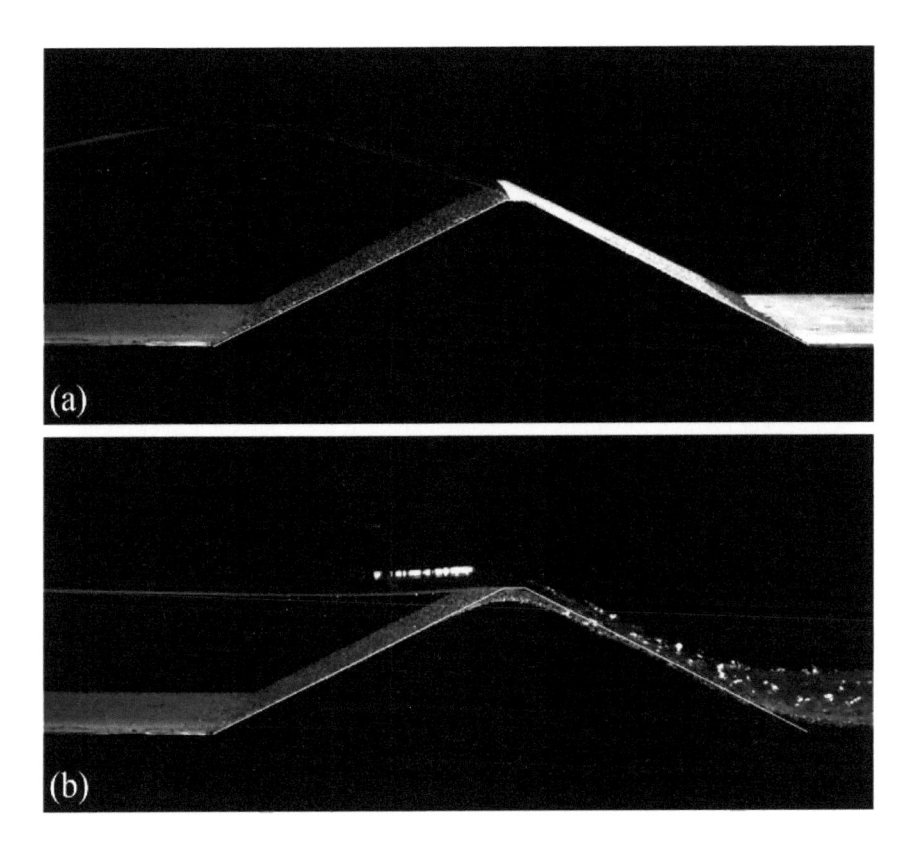

Figure 10.71 Dam overtopping process at (a) $t = 0$ (wave toe reaches dam crest), (b) $t = t_O$ (wave drops below initial crest elevation) (Huber *et al.*, 2017)

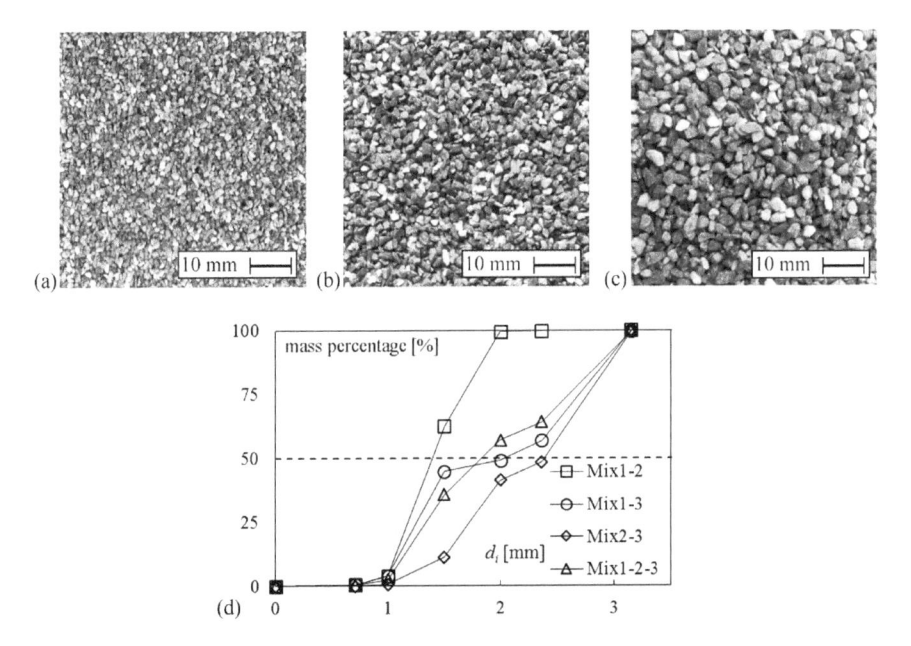

Figure 10.72 Granular sediment with d [mm] = (a) 1.23, (b) 1.65, (c) 2.68; (d) grain size distributions of sediment mixtures with (---) median sediment size (Huber *et al.*, 2017)

depth a_M, eroded height h_e, eroded dam area A_e and deposited dam area A_d were evaluated in AutoCAD 2016. The initial and end dam surfaces were plotted so that their difference allowed for the analysis of h_e, A_e, and A_d. The frame with the highest water surface elevation at $x = 0$ determined the overtopping depth a_M.

Scale effect tests

The test setup followed the procedure of Schmocker (2011). Potential scale effects were studied using scale families. According to Yalin (1971), the ratio of a variable in the model (subscript M) to the corresponding prototype variable (subscript P) is defined by the scale factor $\lambda = L_M/L_P$. The perfect similitude between the two is attained if geometric, kinematic, and dynamic similitudes are reached. Given the free-surface flow problem, Froude similitude was considered, so that effects of Reynolds and Weber numbers had to be studied. According to Schmocker (2011), Reynolds effects are expected for small dam overtopping depths, small sediment size, whereas Weber effects typically occur when surface ripples result, or soap fluid would be used. Given the limitations including $a_M \geq 0.05$ m, $d > 1$ mm and the usage of clean water, scale effects due to both small Reynolds and Weber numbers should not be expected.

Ten tests were conducted to detect significant scale effects. According to Schmocker (2011), a dam height of $w = 0.20$ m does not lead to scale effects, so that this value was taken here as reference ($\lambda = 1$). To investigate the effect of dam height w, four dams with $w = 0.30$ m ($\lambda = 1.5$), four dams with $w = 0.20$ m ($\lambda = 1$), and two with $w = 0.225$ m ($\lambda = 1.125$) were tested. The corresponding sediment sizes were $d = 1.23$ mm for both $\lambda = 1$ and 1.125, and $d = 1.65$ mm for $\lambda = 1.5$. Note that the ratio of sediment sizes $(1.65/1.23 = 1.34)$ is exact for the cases $\lambda = 1.125$ and $\lambda = 1.5$, but slightly too small for $\lambda = 1$. For all three λ, tests were run with relative wave amplitudes $\varepsilon = 0.3$ and 0.7, thereby keeping all other parameters constant (Tests 2 to 11, Table 10.6). Note that Tests (2, 3), (4, 5), etc. served also as repeatability tests (see below).

Table 10.6 Experimental test program. Numbers in bold refer to main test variable (Huber *et al.*, 2017)

	Test	Symbol	w [m]	S [-]	d [mm]	h [m]	b_K [m]	ε [-]
Repeatability	13, 26, 27	⊗	0.20	2	1.23	0.18	0.02	0.5
	2, 3	●	0.30	2	1.65	0.27	0.03	**0.3**
	4, 5	○						**0.7**
Scale families	8, 9	◑	0.20	2	1.23	0.18	0.02	**0.3**
	6, 7	⊜						**0.7**
	11	◍	0.225		1.23	0.20	0.02	**0.3**
	10	◓						**0.7**
	15	◆	0.20	2	**1.65**	0.18	0.02	0.5
	18, 19	◇			**2.68**			
	21	◁	0.20	2	1.23	**0.14**	0.02	0.6
	20	◈				**0.16**		
	22	⊕				**0.18**		
	12	▲	0.20	2	1.23	0.18	**0.00**	0.5
	14	△					**0.10**	

	Test	Symbol	w	S	d	h	b_K	ε
			[m]	[-]	[mm]	[m]	[m]	[-]
Main tests	35	⚠	0.20	2.5	1.23	0.18	**0.00**	0.5
	23	⚠					**0.02**	
	34	⚠					**0.10**	
	32	⚠	0.20	3	1.23	0.18	**0.00**	0.5
	24	⚠					**0.02**	
	33	⚠					**0.10**	
	16	■	0.20	2	1.23	0.18	0.02	**0.25**
	17	☐						**0.75**
	31	▥	0.20	1.5	1.23	0.18	0.02	**0.35**
	30	▤						**0.7**
	29	▨	0.20	3	1.23	0.18	0.02	**0.35**
	28	▨						**0.7**
	36	▦	0.30	2	1.23	0.21	0.03	**0.35**
	37	▩						**0.7**
	38	▣	0.30	2	1.23	0.285	0.03	**0.4**
	39	▨						**0.6**
	1	◈	0.20	2	1.65	0.18	0.02	0.6
	25	◈	0.20	2	2.68	0.14	0.02	0.6
	40	◈	0.20	2	1.23	0.16	0.00	0.6
	41	◈	0.20	1.5	1.23	0.15	0.02	0.7
Mixtures	42	✕	0.20	2	**1.18**	0.18	0.02	0.5
	43	✚			**1.80**			
	44	✳			**2.20**			
	45	▬			**1.59**			

Main tests

An overall number of 28 main tests (Table 10.6) was conducted to investigate the effect of each test parameter on the dam overtopping features, thereby keeping all but one parameter constant, and varying only this last parameter. Tests (42–45) involve sediment mixtures.

The dimensions of the governing test parameters were selected based on the reference dam height of $w = 0.20$ m. If the ratio between reference model and prototype dam heights is assumed to be 1:100, then the prototype dam height would be $w_p = 20$ m. The corresponding prototype sizes would thus amount to 0.123 m $\leq d_p \leq 0.268$ m, 14 m $\leq h_p \leq 18$ m, $0 \leq b_{KP} \leq 10$ m. Further, dam face slopes S were varied between 1.5 and 3, as is typical in applications, and the relative wave amplitude ε ranged from 0.25 to 0.75. Smaller values produced extremely small waves in conflict with scale effects, whereas larger led to wave breaking. Few additional tests were conducted to fill in gaps in the results. Test 19, in which the drainage was not capable of discharging all water, and Test 36, in which the overtopping depth was only 38 mm instead of the minimum of 50 mm, were not further analyzed.

10.8.3 Experimental results

Overview

Dam overtopping and erosion by a solitary wave were studied to answer the questions: How much water is expected to overtop?; How deep is the dam eroded at the dam crest zone, and how large is the eroded volume?; and How long is the overtopping duration?

Figure 10.73 shows the overtopping process of Test 26 of which $w = 0.20$ m, $S = 2$, $b_K = 0.02$ m, $h = 0.18$ m, $d = 1.23$ mm and $\varepsilon = 0.5$ (Table 10.6), a standard test used also for test repeatability. Note that the first and the last image have a time lag of >1 s, while the process captured in frames (b) to (e) takes only 0.34 s. At $t = -1.24$ s, the initial conditions are set, as stated above. The solitary wave peak is seen to approach the dam crest at $t = -0.04$ s, with its front located slightly beyond the dam crest at $t = 0.04$ s, so that the temporal origin is correctly selected at $t = 0$ when the wave front is at $x = 0$. At $t = 0.14$ s, the wave front has moved to half height of the tailwater dam face. Note the slight erosion along both the up- and downstream crest reaches, and the small reduction of dam height. At $t = 0.30$ s, this process has developed resulting in additional erosion mainly along the crest zone and sediment deposition in the tailwater reach. The process is complicated by air bubbles originally contained in the dry dam material entrained by fluid overtopping. At $t = 1.30$ s, the overtopping process is completed revealing an asymmetric dam body. Note that the erosion process is mainly confined to the originally dry dam reach, whereas the originally wet reach undergoes practically no erosion.

Figure 10.74 shows the overtopping process of a wave of relative amplitude $\varepsilon = 0.35$ and a dam face slope of $S = 3$ (Test 29), whereas Figure 10.75 highlights the process for $\varepsilon = 0.6$ and $b_K = 0$ (Test 40). The other dimensions are identical as in Figure 10.73. Given the smaller values of ε and $1/S$ in Test 29, the erosive action is reduced and the free-surface curvature is lower as in Figure 10.75. Note also the smaller amount of air entrained into the flow. In contrast, Figure 10.75 relates to values by which the overtopping action is enhanced. Note the steep wave front at $t = 0.06$ s, and the flow separation from the dam crest at $t = 0.18$ s,

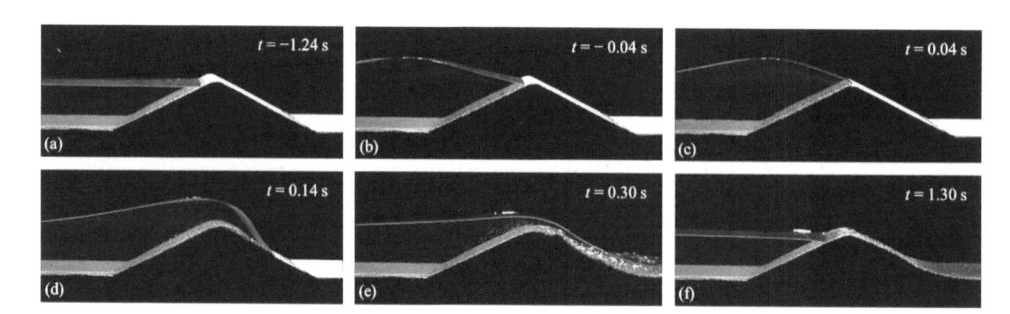

Figure 10.73 Overtopping process of solitary wave with $w = 0.20$ m, $S = 2$, $b_K = 0.02$ m, $h = 0.18$ m, $d = 1.23$ mm and $\varepsilon = 0.5$ (Test 26) (Huber *et al.*, 2017)

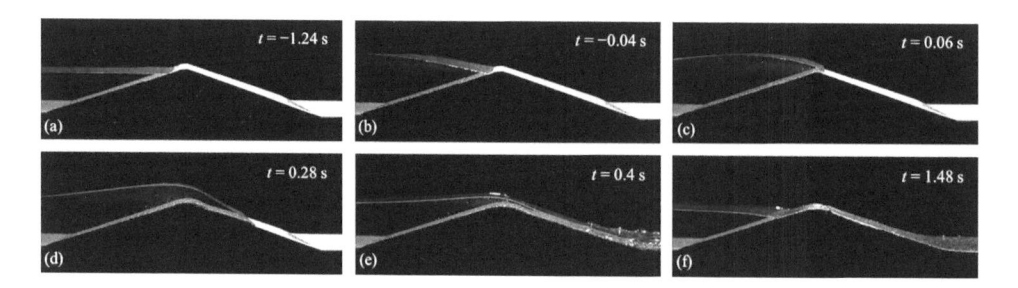

Figure 10.74 Overtopping process of solitary wave with $w = 0.20$ m, $S = 3$, $b_K = 0.02$ m, $h = 0.18$ m, $d = 1.23$ mm and $\varepsilon = 0.35$ (Test 29) (Huber *et al.*, 2017)

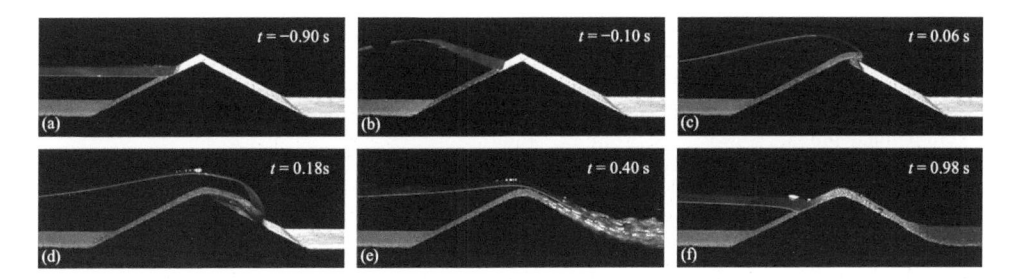

Figure 10.75 Overtopping process of solitary wave with w = 0.20 m, S = 2, b_K = 0.00 m, h = 0.16 m, d = 1.23 mm and ε = 0.6 (Test 40) (Huber *et al.*, 2017)

leading to increased erosive action due to flow impact onto the tailwater dam face. This increase is also evidenced by the additional air entrainment into the flow. Given the enormously short overtopping duration of the order of 1 s, the final reduction of dam crest height is comparatively small, as are the deposited dam areas (see below). The sediment deposition zone starts slightly upstream from the tailwater dam toe, from where it is distributed along the tailwater channel.

Test repeatability

The dam erosion process is affected by various factors. To check test repeatability, three tests with identical parameters were analyzed. The maximum deviation Δ_{max} for all measured parameters was determined as Δ_{max} = max($|\bar{p}-p_i|/\bar{p}$), with $\bar{p}$ as average among the individual test data p_i. Table 10.7 shows a maximum deviation of below 5% for the overtopping depth and time. The eroded dam area and the overtopping volume have an accuracy of below ±10%, whereas the eroded crest depth has a deviation of 17%, corresponding to an absolute height deviation of 1.8 mm, i.e. close to the measurement accuracy. Given these modest deviations from the target values, the present test setup was considered to satisfy quality requirements.

Scale effects

Potential scale effects were analyzed by comparing tests for ε = 0.30, and two for ε = 0.70 for dam heights w = 300 mm (λ = 1.5) and w = 200 mm (λ = 1). The grain sizes were d = 1.65 mm (λ = 1.5) and d = 1.23 mm (λ = 1.125). Because sediment of d = 1.11 mm producing a scale factor of λ = 1 was unavailable, two tests with ε = 0.70 and ε = 0.30 were conducted with w = 0.225 m (λ = 1.125). The other parameters were S = 2, h/w = 0.90 and b_K = $w/10$ (Tests 2–11 in Table 10.6). Table 10.8 compares the test data for w = 0.20 m and w = 0.30 m (Tests 2–9). Note the extremely small deviations in overtopping depth and time of the order of 6% and 2%, and medium deviations of the eroded crest depth, eroded dam area and deposited dam area of up to 22%, respectively, whereas the overtopping volume exhibits large deviations of up to 60%. The two numbers relating to the eroded crest depth shed light on additional problems mainly due to the granular matrix observed with image analysis. Based on e.g. Schmocker (2011), deviations below one sediment grain size cannot be visualized, so that the percentage deviation should be replaced by the absolute length deviation.

Table 10.7 Results from repeatability tests 13, 26, 27 (Huber *et al.*, 2017)

			Average	Δ_{max}
				[%]
Overtopping depth	a_M	[mm]	80	3.8
Overtopping volume	Ψ	[l]	11.5	7.8
Overtopping time	t_O	[s]	0.85	3.1
Eroded crest depth	h_e	[mm]	10.8	16.9
Eroded dam area	A_e	[mm²]	2,901	7.8
Deposited dam area	A_d	[mm²]	1,153	15.6

Table 10.8 Comparison of scale effect tests with $w = 0.30$ m and $w = 0.20$ m, maximum deviations in bold

		$\varepsilon = 0.7$			$\varepsilon = 0.3$		
		$w = 0.30$ m	$w = 0.20$ m	Δ_{max}	$w = 0.30$ m	$w = 0.20$ m	Δ_{max}
		Down-scaled	*Measured*	[%]	*Down-scaled*	*Measured*	[%]
Overtopping depth [mm]	d_0	109	114	**4.9**	47	50	**6.4**
Overtopping volume [l]	Ψ	9.8	15.9	**62.1**	3.6	4.5	**26.6**
Overtopping time [s]	t_O	0.81	0.83	**2.3**	0.85	0.86	**1.5**
Eroded crest depth [mm]	h_e	11.3	13.3	**18.0**	8.0	8.0	**0.4**
Eroded dam area [mm²]	A_e	3,414	2,664	**22.0**	2,867	2,761	**3.7**
Deposited dam area [mm²]	A_d	1,122	1,341	**19.5**	1,092	1,298.5	**18.9**

The second scale effect series (Tests 10–11) with $w = 0.225$ m instead of 0.20 m was compared with the data pertaining from $w = 0.30$ m. Differences in a_M and t_O were similar as previously, these in h_e, A_e, and A_d are again markedly larger yet with much less deviation between the two tests for h_e. As to Ψ, deviations for both tests are now of the order of 30%. Given the extremely unsteady overtopping features lasting 1 s only, involving small sediment sizes determined with image analysis, no significant scale effects were detected. The larger deviations in the two volumetric measurements are also reflected below in the data analysis.

Overtopping depth

The overtopping depth a_M is a key variable. An empirical relation between the still-water depth h, the relative wave amplitude ε, the dam front face angle β, and the dam crest width b_K was found by Kobel *et al.* (2017) for the rigid dam setup (9.7). Given that a_M increases with the relative wave amplitude ε, the relative still-water depth h/w, and the dam face angle ($\beta/90°$), their parameter E_1 defined in Eq. (10.61), was also applied herein. The present data follow Eq. (10.61) entirely except for the constant 1.34, now reduced to 1.25. Figure 10.76a shows an excellent agreement with the test data and ($R^2 = 0.90$)

$$\frac{d_0}{w} = 1.25\left[\varepsilon\left(\frac{h}{w}\right)^{1.7}\left(\frac{\beta}{90°}\right)^{0.25}\right] = 1.25E_1, \qquad 0.16 < E_1 < 0.46. \qquad (10.65)$$

The following limitations apply: $0.25 < \varepsilon < 0.75$, $0.70 < h/w < 0.95$, and $0.20 < \beta/90° < 0.37$. Note that neither the grain size nor the sediment non-uniformity affect the overtopping depth. Only two data points are located outside the $\pm20\%$ margin, both pertaining to the minimum still-water depth $h = 0.14$ m, resulting in relatively small values of $a_M \cong 40$ mm. According Kobel et al. (2017), the fit quality drops as $E_1 < 0.40$, an observation noted here as well.

Overtopping volume

The effects of the dam face slope and the crest width are described by either the *dike* shape parameter $\mu = [(1/2)(S_o + S_d) + L_K/w]/2.5 = [S + (L_K/w)]/2.5$ (Müller et al., 2016), or $\gamma = 0.1S + (b_K/w)$ as the *dam* shape parameter. For the values of V, t_o, and h_e, parameter γ better describes the effects of dam face slope and crest width, whereas parameter μ results in an improved description of the eroded and deposited dam areas. Note that the effect of S in μ is much higher than in γ because the crest region is immediately rounded for granular dams at overtopping start, whereas a rigid crest guides the oncoming wave above the crest during the entire overtopping process.

A data analysis based on Kobel et al. (2017) failed, given the significant differences between the two setups. The overtopping volume increases similarly as does a_M/w. Let $W = \varepsilon(h/w)^{2.5}\gamma^{-1/9}$, then the best data fit limited to the identical constraints as Eq. (10.65) is provided by (Figure 10.76b, $R^2 = 0.93$)

$$\frac{V}{bh^2} = 1.53\left[\varepsilon\left(\frac{h}{w}\right)^{2.5}\gamma^{-1/9}\right] = 1.53W, \qquad 0.23 < W < 0.65. \tag{10.66}$$

Equation (10.66) depends mainly on the relative still-water depth h/w, linearly on the relative wave amplitude ε, and slightly on the dam shape parameter γ. For $W > 0.45$, maximum

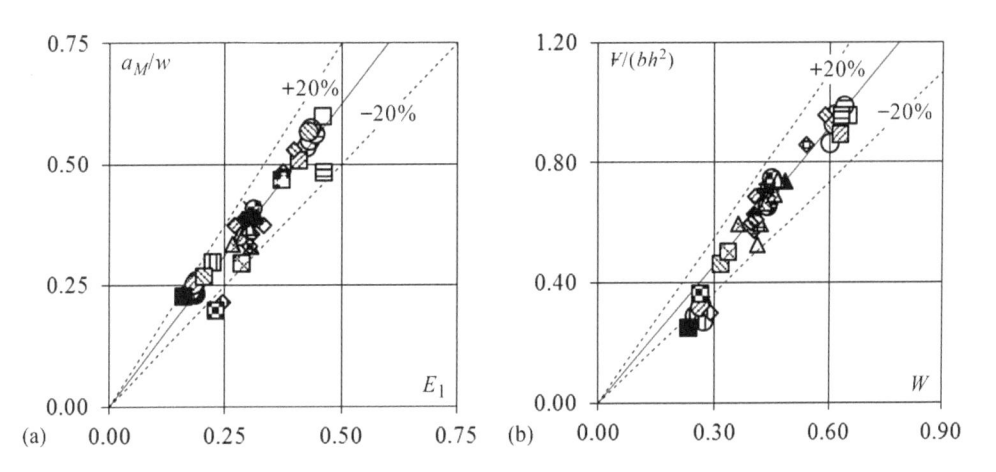

Figure 10.76 (a) (——) Relative maximum overtopping depth $a_M/w[E_1]$ according to Eq. (10.65), (b) (——) relative overtopping volume $V/(bh^2)[W]$ according to Eq. (10.66), (---) $\pm$ 20% deviation. Symbols as in Table 10.6 (Huber et al., 2017)

deviations are less than ±10%. For smaller values, the overtopping volume is overestimated by up to ±35%. Overall, the fit is good, given that only 4 test have deviations >20%. Given problems with the dam drainage particularly for large sediment size, Tests 1, 2, 18, 19, 31 and 44 had to be excluded for this analysis. Note also that the sediment characteristics have no effect on the overtopping volume.

Wave overtopping duration

The wave overtopping duration is important to estimate a flood hydrograph in the tailwater valley. Figure 10.77a shows the inverse relative overtopping duration $w^{0.5}/(t_o g^{0.5})$ versus the dimensionless wave overtopping duration parameter $F = (h/w)^{-5/3}\gamma^{-0.2}$ as ($R^2 = 0.90$)

$$\frac{w^{0.5}}{t_o g^{0.5}} = 0.11\left[\left(\frac{h}{w}\right)^{-\frac{5}{3}}\gamma^{-0.2}\right] = 0.11F, \qquad 1.21 < F < 2.41. \tag{10.67}$$

The limitations include: $0.2 < \gamma < 0.8$, $0.7 < h/w < 0.95$. The equation depends slightly on the dam shape parameter, and strongly on the relative still-water depth h/w. However, neither the wave amplitude, the sediment grain size nor the non-uniformity affect the overtopping duration. Its maximum deviation from Eq. (10.67) is less than ±10%.

Eroded crest depth

The assessment of the eroded crest depth allows to determine the minimum vertical distance from the initial dam crest to the highest dam elevation after a test. The normalized value h_e/w versus the relative wave amplitude ε, the relative still-water depth h/w and the dam

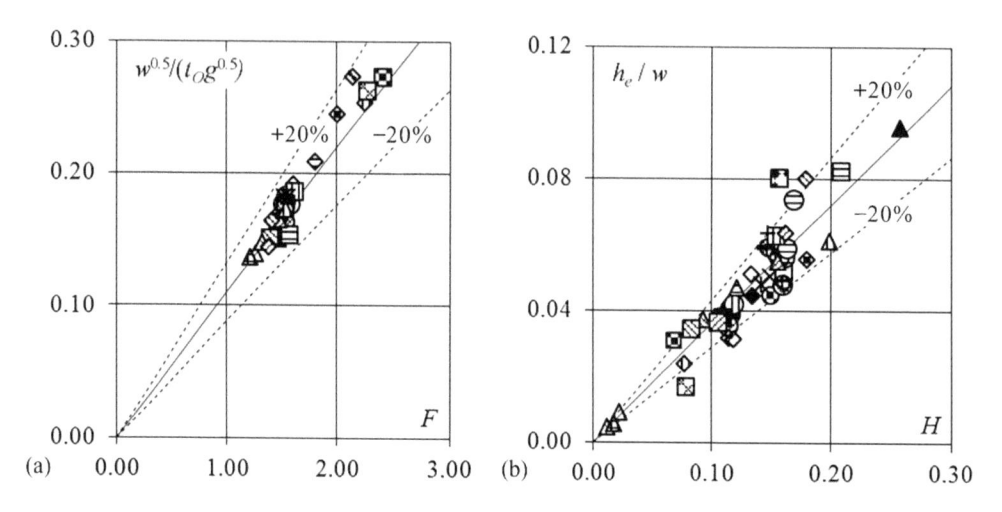

Figure 10.77 (a) (—) Inverse relative overtopping duration $w^{0.5}/(t_o g^{0.5})[F]$ according to Eq. (10.67), (b) (—) normalized eroded crest depth $h_e/w[H]$ according to Eq. (10.68), (---) ±20% deviation. Symbols as in Table 10.6 (Huber *et al.*, 2017)

shape parameter γ, combined to $H = \tanh(2.8\varepsilon)(h/w)^3\exp(-4.9\gamma)$, follows the empirical fit (Figure 10.77b, $R^2 = 0.86$)

$$\frac{h_e}{w} = 0.36\left[\tanh(2.8\varepsilon)\left(\frac{h}{w}\right)^3\exp(-4.9\gamma)\right] = 0.36H, \quad 0.01 < H < 0.26. \tag{10.68}$$

Limitations include: $0.25 < \varepsilon < 0.75$, $0.7 < h/w < 0.95$, $0.2 < \gamma < 0.8$. Deviations are less than $\pm20\%$ for 85% of all test data, yet are $\pm40\%$ for Tests 1 and 37 for unknown reasons. The eroded crest depth depends again significantly on the relative still-water depth, and nearly linearly on the relative wave amplitude and dam shape parameter. As previously, the sediment characteristics do not affect these results.

Eroded dam area

The eroded dam area A_e was normalized with the original dam area $A_0 = w(b_K + Sw)$. It is approximated with the combined parameter $A = \mu^{-1.2}(h/w)^3(d/w)^{2/3}$ by (Figure 10.78a, $R^2 = 0.77$)

$$\frac{A_e}{A_0} = 1.2\left[\mu^{-1.2}\left(\frac{h}{w}\right)^3\left(\frac{d}{w}\right)^{2/3}\right] = 1.2A, \quad 0.011 < A < 0.045. \tag{10.69}$$

Limitations are: $0.64 < \mu < 1.4$, $0.7 < h/w < 0.95$, $0.004 < d/w < 0.013$. A_e/A_0 depends strongly on the relative still-water depth h/w, almost linearly on the dam shape parameter μ, and slightly on the sediment diameter d. Note that sediment non-uniformity has no effect, and A_e/A_0 does not depend on the relative wave amplitude ε. A deviation $> \pm20\%$ applies to 17% of the data, with a maximum of 37%.

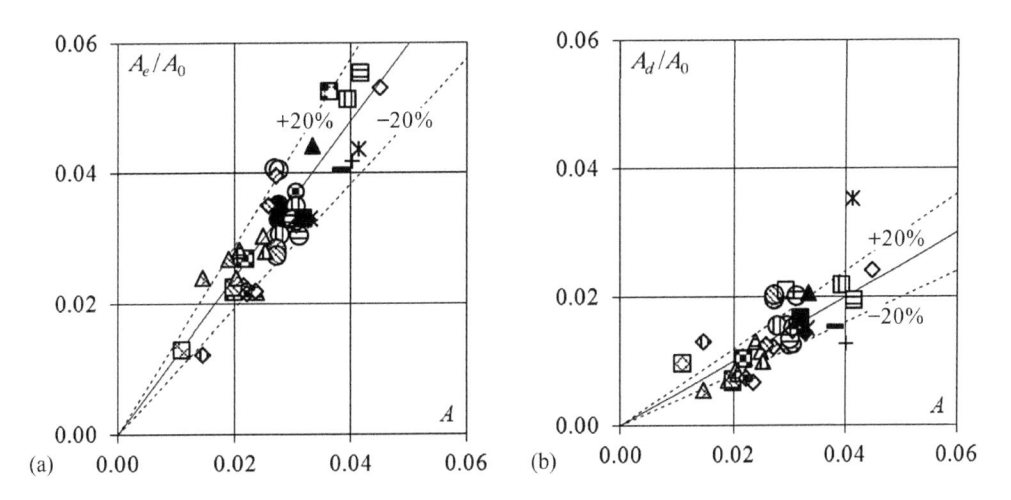

Figure 10.78 (a) (——) Normalized eroded dam area $A_e/A_0[A]$ according to Eq. (10.69), (b) (——) normalized deposited dam area $A_d/A_0[A]$ according to Eq. (10.70), (---) $\pm20\%$ deviation. Symbols as in Table 10.6 (Huber *et al.*, 2017)

Deposited dam area

The deposited dam area is measured from the point at which the sediment is deposited on the original dam up to distance $w/2$ from the tailwater dam toe, namely $x = b_K + w(S + 1/2)$, as it could not be observed further downstream. Therefore, these data scatter noticeably. The best estimate for the relative deposited dam area A_d/A_0 follows the data fit (Figure 10.78b, $R^2 = 0.54$)

$$\frac{A_d}{A_0} = 0.5\left[\mu^{-1.2}\left(\frac{h}{w}\right)^3\left(\frac{d}{w}\right)^{2/3}\right] = 0.5A, \qquad 0.011 < A < 0.045. \tag{10.70}$$

Limitations are as for Eq. (10.69). Except for the coefficient 0.5, Eq. (10.70) is identical with Eq. (10.69), depending also on μ, h/w and d, whereas the effects of relative wave amplitude ε and sediment non-uniformity are again absent. Deviations are up to ±60% mainly due to the reasons previously explained.

10.8.4 Discussion of results

The scale effect tests evidence a deviation of the overtopping volume by about ±30% from the fit equations. For both $\lambda = 1$ and $\lambda = 1.12$, the measured volume is larger than expected by down-scaling the results from $w = 0.30$ m. Both Schmocker (2011) and Fuchs (2013) demonstrate that no significant scale effects apply for the model dimensions used in their setups. For the test dimensions employed herein, the data fit is excellent, excluding thereby scale effects. The (slight) data scatter is attributed to measurement inaccuracies and the small absolute values of the governing test parameters.

The overtopping depth does neither depend on the sediment diameter nor on the crest width. The effect of the dam face angle is small, resulting in slightly larger values of a_M/w with increasing β, as also noted by Kobel *et al.* (2017). The difference between the two leading factors, 1.25 in Eq. (10.65), and 1.34 in (Eq. 10.61), results from the two different settings, once involving a granular matrix, and second a rigid dam. From this perspective, the present results relating to the *hydraulic* parameters remain independent from both sediment size *and* non-uniformity, given the close agreement with Kobel *et al.* (2017).

The equations for the eroded and the overtopping depths, Eqs. (10.68) and (10.65), look similar in terms of governing parameters. Both h_e/w and d_0/w increase with the relative wave amplitude, relative still-water depth and dam face angle. The extra effect of b_K/w contained in Eq. (10.68) is not contained in its hydraulic counterpart for reasons of flow separation at the crest region. The eroded crest depth increases with increasing dam face angle both according to Schmocker (2011) and Eq. (10.68).

It is obvious that the eroded and deposited integral areas must be identical. The purpose of Eq. (10.70) is rather to obtain an idea how much of the total eroded dam area is deposited within the dam vicinity, namely within $w/2$ from the tailwater dam toe. Equations (10.69) and (10.70) are identical except for the leading terms 1.2 and 0.5, respectively. This suggests that about 40% is deposited within $w/2$ for the present setting. The finding according to which both A_e and A_d increase with the relative sediment diameter is surprising from the perspective of standard sediment transport, for which the opposite applies. This issue is also discussed by Schmocker (2011); the reason for it appears to be related to increased turbulence and streamline curvature along the dam crest.

The question: How can dam overtopping be controlled for a set of governing parameters? cannot be answered. In applications, dam engineers would ask for the limit (subscript L) still-water depth h_L for a given wave characteristic ε and dam parameters w, S, b_K, d, σ_g. Given that tests with an overtopping depth $a_M \leq 0.05$ m were excluded for reasons of scale effects, this aspect has not further been studied. An alternative simplified procedure involves the condition according to which the eroded crest depth should be smaller than the original dam freeboard, $h_e \leq w-h$. This task is accomplished by applying Eq. (10.68).

As to upscaling to prototype scale, the following issues should be considered for a 100 m high dam:

- Smallest used relative sediment size is $d/w = 0.004$, corresponding to boulders of 0.4 m diameter. Such a dam would evidently consist of a granular mixture retaining water under massive seepage.
- Both single-grained and mixtures of granular sediment were used, yet without any filter or surface protection. These two additions are important in dam engineering, but are hardly model tested, given the small sediment size involved.
- Maximum tested relative still-water depths h/w were 0.95. For dams of 100 m height, the freeboard can be smaller than 5 m, so that the fit equations do no more apply. The 95% freeboard limit avoided wetting and erosion of the dam crest prior to test start.
- Overtopping time is narrowly limited to $F > 1.21$, requiring additional tests involving larger crest widths and gentler dam slopes to extend this limit.
- Tests were stopped once the solitary wave had overtopped the dam, excluding a complete dam failure. Extending the test duration by adding also reflected wave overtopping, the dam crest would continuously be eroded until failure (Balmforth *et al.*, 2008, 2009).

Based on Froude similitude, the minimum values of the dam height, sediment size, still-water depth and wave amplitude were fixed to avoid scale effects. A systematic test program involving the above parameters plus various dam face slopes, dam crest widths and sediment non-uniformities allowed for the assessment of the overtopping depth and volume, wave overtopping duration, and eroded crest depth, eroded dam area, and deposited dam area. The test program involved the systematic variation of the six governing parameters influencing the overtopping process. Based on a data analysis, fit equations are provided, and the results are critically assessed to highlight the leading effects. Remarkably, the present results agree with these for the overtopping depth of rigid dams (10.7).

For the overtopping volume, eroded crest depth and overtopping duration, the *dam* shape parameter was introduced. For the eroded and deposited dam areas, the *dike* shape parameter was retained, however. Only these two values depend on the sediment diameter, yet not on the sediment non-uniformity. According to all fit equations, the relative still-water depth has the leading effect on the overtopping features, whereas the wave amplitude is the second main factor except for both the eroded and the deposited dam areas. This research highlights the advantages of hydraulic model tests when highly unsteady two-phase flows are considered.

10.9 Spatial impulse waves

10.9.1 Motivation

Massive waves may be generated if a landslide, an avalanche, or glacier calving interacts with a large body of water. The wave generation mechanism involves a momentum transfer from the slide mass to the water column. Waves generated by extremely rapid, gravity-driven subaerial mass movements are commonly referred to as impulse waves (Heller *et al.*, 2009); they may be observed in coastal areas as well as in inland waters (Roberts *et al.*, 2014). An impulse wave event involves three phases as shown in Figure 10.02. During the wave generation phase, the subaerial slide mass transfers its momentum to the water column, thereby generating a wave train with several crests and troughs. The amplitudes and wavelengths of the wave train are transformed in the wave propagation stage. Depending on the bathymetry of the water body, these transformation effects include refraction, diffraction, and shoaling (Heller *et al.*, 2009). In addition, the waves are subject to frequency and amplitude dispersion (Heller and Hager, 2010). Wave run-up (e.g. Müller, 1995; McFall and Fritz, 2017), overland flow (e.g. Fuchs and Hager, 2013), or wave overtopping (e.g. Müller, 1995; Kobel *et al.*, 2017) characterize the wave impact phase. The first two phases are in the scope of this study.

A multitude of impulse wave events have been reported. One of the most prominent cases is the event in Lituya Bay, USA, which was triggered by an earthquake in 1958 and caused run-up heights of more than 500 m (Miller, 1960b). Bornhold *et al.* (2007) describe a historical example in the 16th century, when impulse waves generated by a rockslide at Knight Inlet, Canada, presumably destroyed an indigenous settlement. Impulse wave events of the 21st century involving subaerial landslides in coastal areas include Paatuut, Greenland, in 2000 (Dahl-Jensen *et al.*, 2004), Aysén Fjord, Chile, in 2007 (Sepúlveda *et al.*, 2010), Taan Fjord, USA, in 2015 (George *et al.*, 2017), and Karrat Fjord, Greenland, in 2017 (Poli, 2017). A well-documented event in inland waters occurred in 2007 when a three million cubic meters landslide impacted Chehalis Lake, Canada, causing run-up heights of 38 m at the opposite shore (Roberts *et al.*, 2013; Wang *et al.*, 2015). In a water body impounded behind an artificial or natural dam structure, impulse wave overtopping may have a devastating impact on downstream areas. The wave overtopping event at Vajont Dam, Italy, in 1963 claimed nearly 2000 fatalities (Genevois and Ghirotti, 2005). Risley *et al.* (2006) conducted a hazard analysis of potential impulse wave overtopping at the Usoi landslide dam, Tajikistan. A global overview with 254 events covering both inland and coastal waters was compiled by Roberts *et al.* (2014).

Based on the slide properties and the bathymetry, a prediction of the generated wave magnitude may be conducted. Heller *et al.* (2009) describe five different prediction methods within the context of hazard assessment: (1) generally applicable equations developed from model tests, (2) prototype-specific model tests, (3) numerical simulations, (4) empirical equations derived from field data, and (5) analytical investigations. Heller *et al.* (2009) and Evers *et al.* (2018) discuss the capabilities and limitations of these methods. Item (1) yields estimated wave magnitudes at low cost and time requirements. Therefore, this method is advantageous when an event is imminent or as a first assessment to determine whether additional more accurate investigations are needed.

Two types of impulse wave propagation patterns are commonly investigated in model tests: unidirectional and omnidirectional. While the former is investigated in wave flumes (2D), e.g. by Fritz *et al.* (2004), Zweifel *et al.* (2006), and Di Risio and Sammarco (2008), the latter is studied in wave basins (3D). Generally applicable equations developed from model tests accounting for omnidirectional impulse wave propagation were presented by Panizzo *et al.* (2005), Heller *et al.* (2009), Mohammed and Fritz (2012), Heller and Spinneken (2015), and McFall and Fritz (2016). Evers and Hager (2016a) applied the equations of the former four 3D studies to selected experiments to assess their applicability on an extended range of test parameters. They found limitations regarding slide model replication, i.e. rigid block vs. deformable granular slides, constant test parameters, e.g. a single slide impact angle or slide width, or a lack of experimental validation. In addition, all studies were based on wave gage data from a limited number of measurement locations. Bregoli *et al.* (2017) demonstrated the advantages of the videometric measurement approach: First, it allows for a continuous representation of the water surface and therefore an adaptive tracking of specific wave conditions in the impact region. Second, in contrast to capacitance wave gages, the measurement devices may not be damaged by the impacting slide mass. Note that the measurement approach by Bregoli *et al.* (2017) was limited to a single wave propagation direction along the slide axis, although their experiments were conducted in a wave basin.

The main objective of this section is to advance the understanding of spatial propagation of landslide-generated impulse waves by applying a videometric measurement technique to track arbitrary water surfaces in 3D. For designing protection measures against wave impact, engineers require quantitative information on expected wave magnitudes at a certain location depending on its position relative to the wave source. Therefore, generally applicable empirical prediction equations are derived from a broad set of slide parameters for key wave characteristics to improve the quality of hazard assessments.

10.9.2 Experimental setup

The experiments were conducted in a 4.5 m × 8 m wave basin 0.75 m high (Figure 10.79). The impulse waves were generated with a movable chute featuring an inclinable sliding plane, a release box, and lateral guidance walls. The sliding plane allowed for impact angles of 30°, 45°, 60°, and 90°. Mesh-packed slides were applied, allowing for a simplified handling similar to rigid slides while maintaining a granular matrix and deformability. Evers and Hager (2015a) demonstrated that the impulse wave characteristics generated by this type of slide are predictable with empirical equations derived from experiments with free granular slides within a similar scatter range, e.g. ±30% for the wave height, provided that the slide-centroid velocity is the reference velocity in both cases. The mesh-packed slides were composed of granular material loosely filled into bags made of sifting media with a mesh opening of 500 μm and 47% porosity. The granular material was made of 87% BaSO4 and 13% PP of grain (subscript g) diameter $d_g = 8$ mm, grain density $\rho_g = 2420$ kg/m^3, and a bulk slide density of $\rho_s = 1338$ kg/m^3 (Heller, 2008). The release box allowed for predefining the slide geometry including the slide width between 0.25 m and 1.00 m. The slides were accelerated by gravity from various drop heights after manual release. During acceleration, the slide deformation was found negligible and the slide-centroid velocity was consequently equal to the slide front velocity. The latter was measured with two Laser Light Barriers mounted perpendicularly to the sliding plane. The sensors' short response time <0.5 ms resulted in a measurement accuracy of approximately ±5% for the maximum measured slide impact velocity of 4.76 m/s over a measuring section of 0.1 m. A porous filter foam was placed in the basin acting as wave damper to reduce the waiting time between experimental runs (Evers *et al.*, 2019).

The application of a videometric measurement system required an opaque fluid allowing for the projection of a grid pattern on the free surface. Therefore, deionized water was mixed with 4 kg/m³ titanium dioxide (TiO2) pigments, as demonstrated by Przadka *et al.* (2012) and Evers and Hager (2015b). The addition of TiO2 has negligible effects on both water viscosity and surface tension (Przadka *et al.*, 2012). A regular grid pattern with 79×79 intersections was projected onto the water surface with a projector, thereby defining the measurement zone. As the general wave propagation pattern is considered symmetrical, the measurement zone covered only half of the full wave pattern (Figure 10.79). The commercially available ProSurf-system AICON 3D was applied to track the spatial positions of the grid intersections. It comprised four cameras positioned around the wave basin (Figure 10.79), synchronized by a control box to an acquisition rate of 24 Hz. The cameras' spatial positions and orientations were calibrated, enabling the tracking of the grid intersection by triangulation (Evers, 2017).

Figure 10.80 shows the water surface profiles of five experimental runs repeated with identical preset parameters at times $t = 0.5$ s, 1.5 s, and 2.5 s from slide impact onto the water surface. Within the vicinity of the slide impact location at $r = 0$ m, the measured profiles feature larger deviations, while wave crests and troughs are captured sufficiently consistent and detailed for larger propagation distances. However, the deviations are not necessarily related to inaccuracies of the measurement system but are also caused by the test repeatability, e.g. the slide impact velocities range within the measurement accuracy of the laser

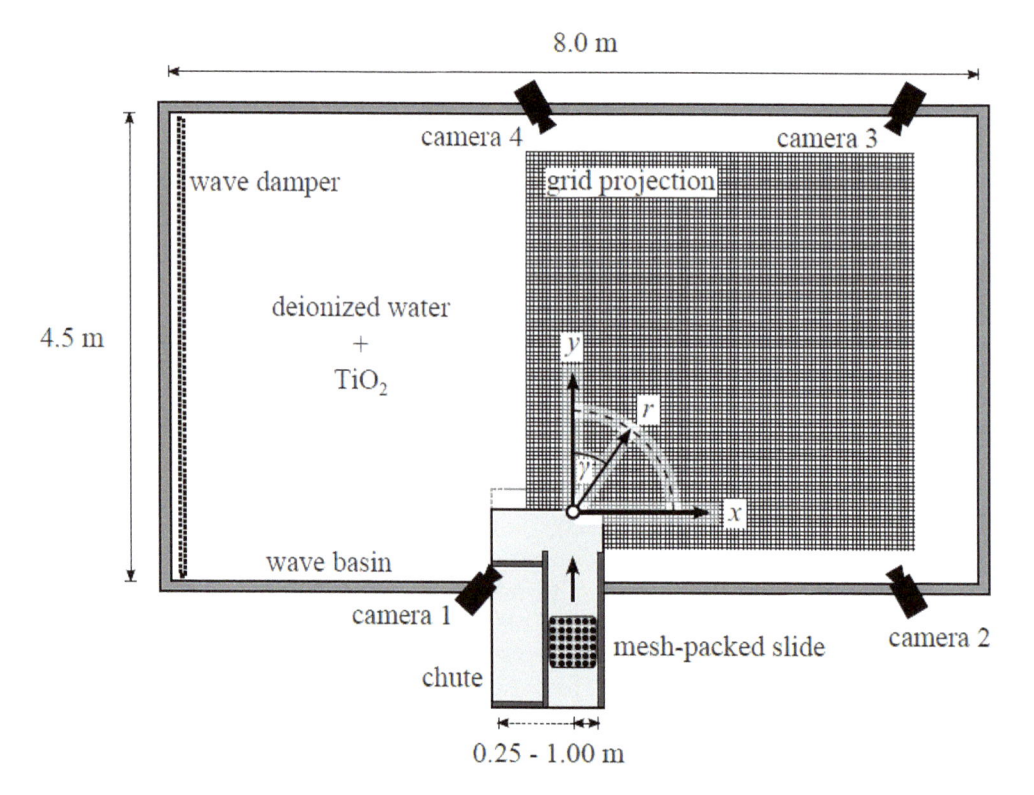

Figure 10.79 Experimental setup of 3D wave basin (Adapted from Evers, 2017)

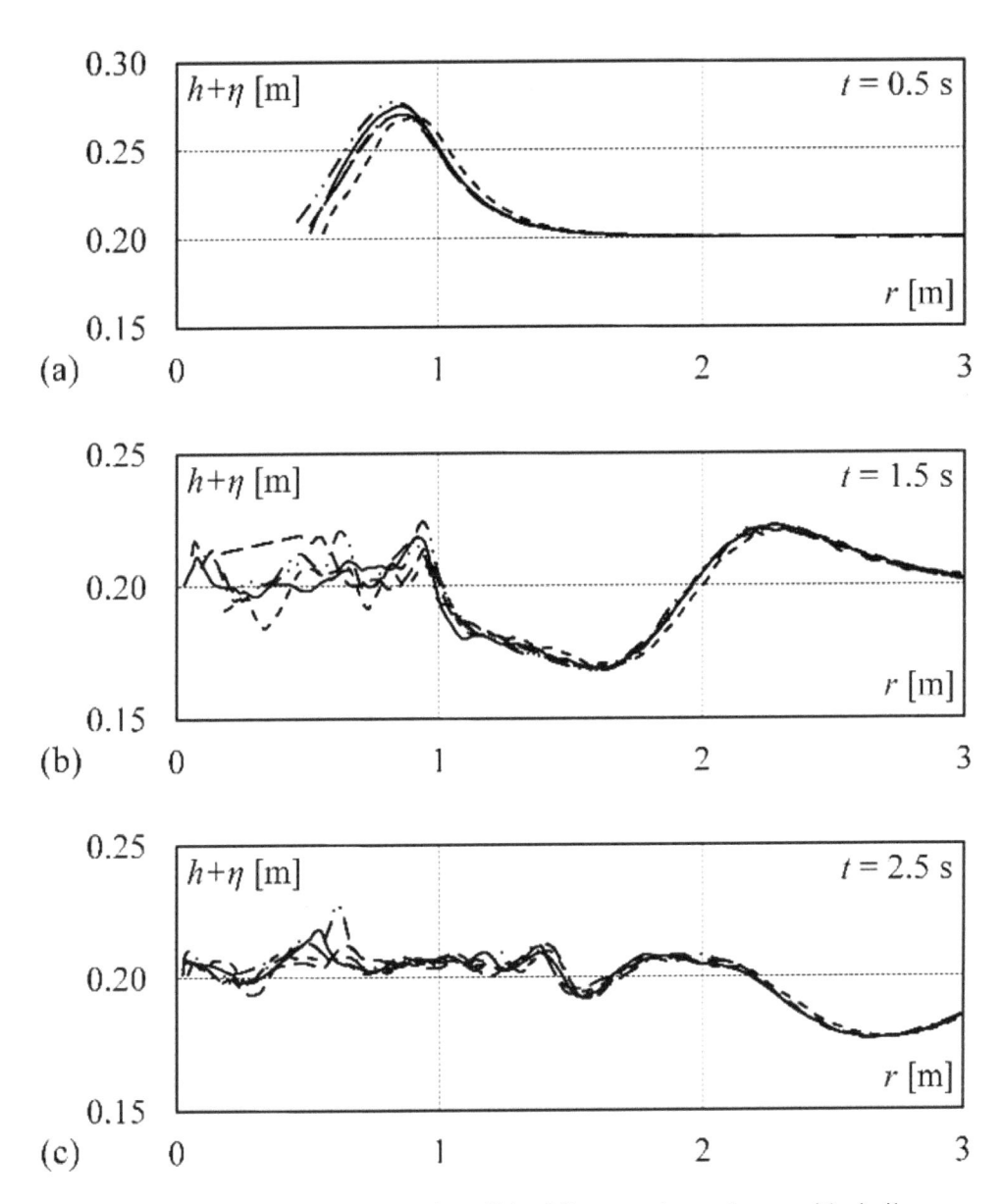

Figure 10.80 Water surface profiles $[h + \eta](r)$ of five experimental runs with similar governing parameters for wave propagation angle $\gamma = 0°$ (Evers *et al.*, 2019)

light barriers between 4.35 m/s and 4.55 m/s. The measurement accuracy of the videometric system was discussed by Frank and Hager (2014). They found absolute deviations of <2 mm to point gage measurements of solid surfaces. Evers and Hager (2015b) compared the videometrically tracked water level displacements with those measured with capacitance wave gages and found good overall agreement. Based on these findings, the test accuracy for water surface tracking was estimated to ±1 mm.

10.9.3 Process description

Governing parameters

Both, the governing wave generation parameters as well as the targeted wave characteristics are shown in Figure 10.81. The wave generation parameters include the slide impact velocity V_s, the slide mass m_s, the slide thickness s, the slide width b, the slide impact angle α, and the stillwater depth h. The wave characteristics included in the analysis are the first wave crest amplitude a_{c1}, the first wave trough amplitude a_{t1}, the first wave height H_1, the first wave period T_1, the second wave crest amplitude a_{c2}, as well as the celerities of the first and the second wave crests c_{c1} and c_{c2}. The positions of the respective wave features were defined within a polar coordinate system with the pole representing the slide impact location and the angular and radial coordinates the wave propagation angle γ and distance r, respectively.

The dimensionless quantities derived from the governing wave generation parameters include as for the plane impulse waves the slide Froude number $\mathsf{F} = V_s/(gh)^{0.5}$, the relative slide mass $M = m_s/(\rho_w bh^2)$, the relative slide thickness $S = s/h$, and the relative slide width $B = b/h$. Most existing empirical equations for the prediction of wave characteristics include these parameters (Di Risio *et al.*, 2011; Evers and Hager, 2016a). The first three parameters F, M, and S were included by Heller and Hager (2010) in the impulse product parameter $\mathsf{P} = \mathsf{F}S^{0.5}M^{0.25}\{\cos([6/7]\alpha)\}^{0.5}$, describing impulse wave propagation in 2D wave channels.

Table 10.9 gives an overview of the experimental parameter ranges. The complete data set including the experimental parameters and the definition of test numbers is provided by Evers (2018). A total of 74 experiments were conducted and analyzed.

The propagation distance of an outgoing impulse wave train within the basin was limited by vertical sidewalls. Wave reflection was accounted for by extrapolating the position of the first initial uplifting of the stillwater level closest to one of the basin walls both in space and time. The extrapolation involved the propagation celerity of the first initial

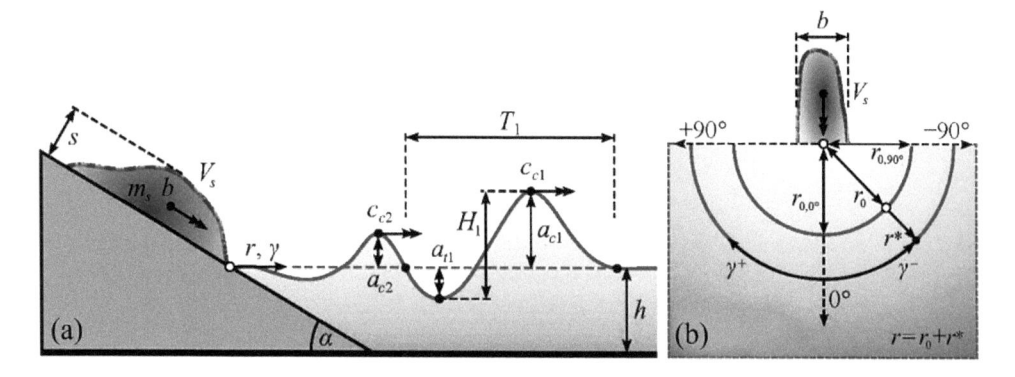

Figure 10.81 Definition plot for spatial impulse wave propagation with governing parameters and wave characteristics with (a) side view, (b) plan view (Adapted from Evers, 2017)

Table 10.9 Overview of experimental quantities (Evers *et al.*, 2018)

Parameter	Test range	Dimensionless parameter	Test range
V_s [m/s]	0.72–4.76	F	0.40–3.40
m_s [kg]	10–40	M	0.25–1.00
s [m]	0.06–0.12	S	0.15–0.6
b [m]	0.25–1.00	B	0.83–5.00
α [°]	30–90	α	30–90
h [m]	0.2–0.4	-	-
-	-	P	0.13–2.08

uplifting before reflection as well as the residence time at the wall (Chen *et al.*, 2015). Wave features affected by reflection from the sidewalls were excluded from the analysis. To avoid significant scale effects, the tests were conducted with stillwater depths $h \geq$ 0.20 m (Heller *et al.*, 2008).

Water surface evolution

Applying the videometric system yields a quasi-continuous contour representation of the water surface. Figure 10.82 highlights oblique views of the grid projection of Test 043 (Evers, 2018). At $t = 0.179$ s, the splash screen of the impact crater has attained its highest elevation level. The impact crater then collapses up to 0.750 s. While the first wave crest is then hardly visible, the first wave trough has clearly formed at $t = 1.125$ s. The water surface close to the slide impact location is irregularly distorted at this stage. At $t = 1.500$ s, the second wave crest forms. Note the spilling second wave crest for small wave propagation angles γ. The second wave crest has further propagated at $t = 1.875$ s along with reduced wave spilling.

Figure 10.83 shows the contour plots of Test 043 (Figure 10.82). These plots were produced by interpolating between the projected grid intersections tracked by the videometric system. Note that Figure 10.83 starts at $t = 0.375$ s, while Figure 10.82 includes $t = 0.179$ s as the first time step. The slide impacts the water surface at $t = 0$ s at the origin of the y-axis. At $t = 0.375$ s, the first wave crest has just emerged; it has a large amplitude for $\gamma = 0°$, decreasing toward $\gamma = 90°$ (x-axis). The blank area close to the impact location arises from the simultaneous collapse of the impact crater. The continuous tracking of the water surface is limited by the strong deformation of the water surface. At $t = 0.750$ s, the first wave crest propagated to $r \approx 1.1$ m is subjected to amplitude decay. The first trough of the wave train emerges at the same time. At $t = 1.125$ s, the first wave trough is fully captured by the videometric measurement system. The water surface around the impact location is strongly distorted and consequently the blank area increases. The second wave crest is fully developed at $t = 1.500$ s and has further propagated at $t = 1.875$ s, while the first wave crest has partially left the measurement area. A second wave trough has formed at $t = 2.250$ s. A qualitative visual comparison with the first wave shown at full extent at $t = 1.125$ s reveals that the wave length of the second wave is substantially shorter, as was also observed by Mohammed and Fritz (2012) and McFall and Fritz (2016). However, note that the water surface for propagation distances $r > 1.5$ m at $t = 2.250$ s is potentially

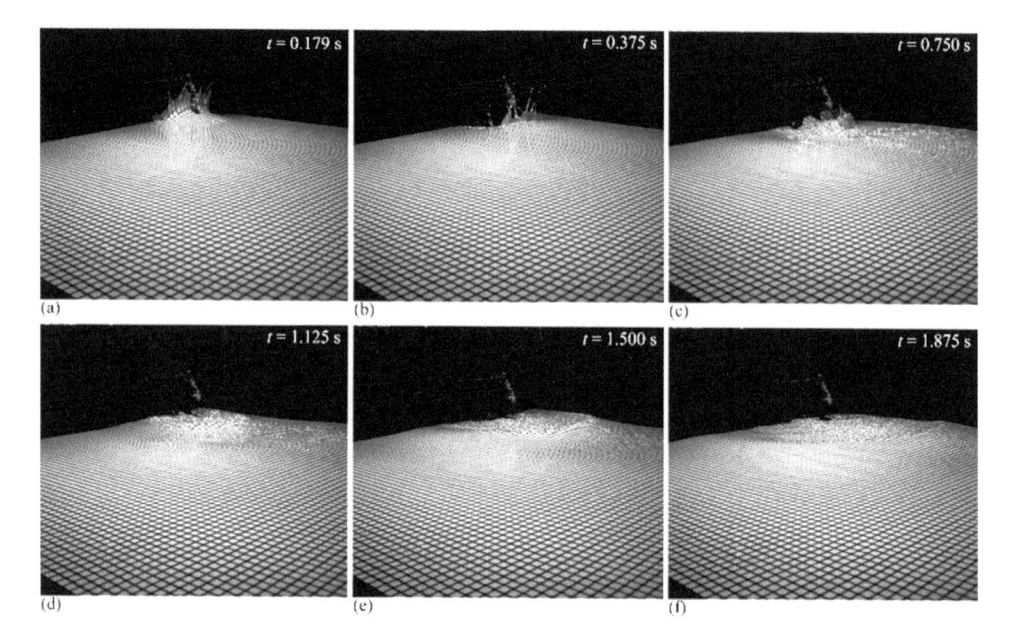

Figure 10.82 Oblique view of Test 043 (Evers, 2018) with $V_s = 2.38$ m/s, $m_s = 20$ kg, $s = 0.06$ m, $b = 0.50$ m, $\alpha = 60°$, $h = 0.40$ m, and $P = 0.26$ (Adapted from Evers, 2017)

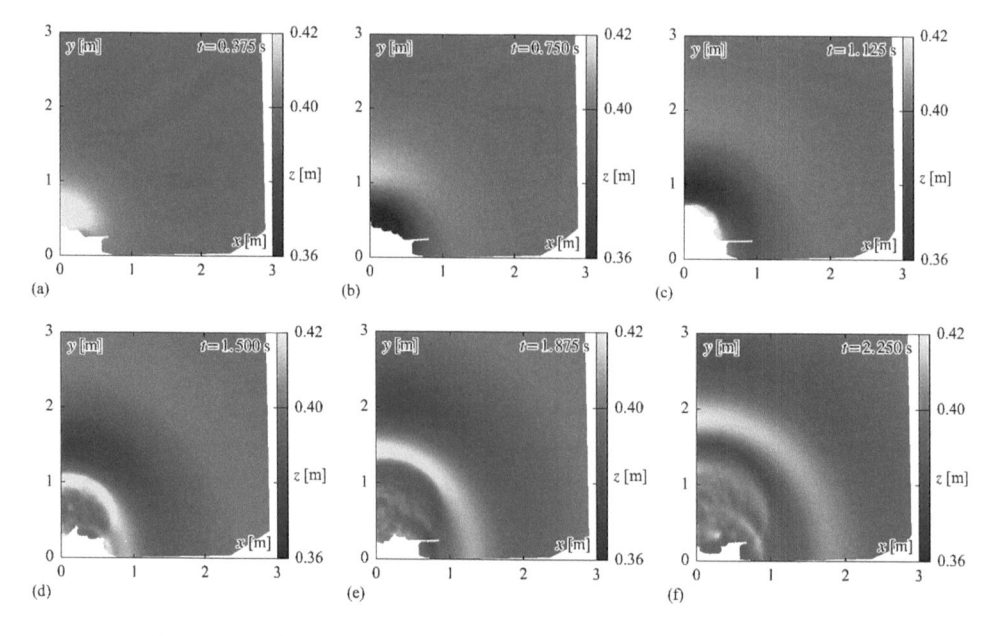

Figure 10.83 Contour plots $z(x, y)$ of Test 043 (Evers, 2018)

affected by reflections from the basin sidewalls; the second wave trough amplitude was not included in the quantitative analysis, therefore.

The effect of slide width b on spatial impulse wave propagation is shown in Figure 10.84. All plots refer to the identical instant at $t = 1.50$ s for three different test runs. While their impulse product parameters P were similar, the slide widths were $b = 0.25$ m, 0.50 m, and 1.00 m, respectively. The first wave crest amplitudes a_{cl} ($y \approx 2$ m) increase significantly with increasing slide width for spatial, i.e. 3D, wave propagation. In contrast, P was derived from tests in a wave flume. In this 2D type of setup, a doubling of the flume width along with slide width b and the slide mass m_s has no effect on the generated wave magnitudes (Evers and Hager, 2016b). Figure 10.84 demonstrates that the effect of b has to be accounted for in 3D water bodies to allow for the prediction of spatial impulse wave propagation.

Figure 10.85 shows the water surface profiles $[h + \eta](r)$ for $\gamma = 0°$, 45°, and 90° during the wave generation process. The water surface is undisturbed at $t = 0.000$ s. At $t = 0.292$ s, the impacting slide creates an impact crater displacing the water surface. The crater collapses and the first wave crest starts to emerge from the slide impact zone at $t = 0.417$ s. The crest is fully developed at $t = 0.500$ s. The wave propagation distance r of the initial first wave crest amplitude $a_{0,cl}$ is defined as the impact radius r_0. At $t = 1.00$ s and 1.292 s, the respective water surface profiles indicate that also the initial first wave trough and the second wave crest amplitudes are formed at the impact radius r_0.

Wave generation and propagation is divided into the: (I) slide impact zone and the (II) wave propagation zone (Figure 10.86), separated by the impact radius r_0. Zone I is affected by the generation of the impact crater and its collapse, involving strong turbulence and air entrainment. Distinct wave characteristics, e.g. wave amplitudes, become quantifiable first in Zone II by applying the videometric system (Figure 10.85). In addition to the wave propagation angle γ, the surrogate radial wave propagation distance $r^* = r - r_0$ describes the outgoing impulse wave train beyond r_0 within the polar coordinate system (Figures 10.81b and 10.86).

The first wave crest amplitudes a_{cl}, i.e. the shape of the first wave crest, is plotted versus the wave propagation angle γ in Figure 10.87a, for time steps between $t = 0.38$ s and 2.17 s. During wave propagation, a_{cl} decreases for all γ, the decay rate being higher for $\gamma = 0°$ than for 90° (Figure 10.87). While $a_{cl,90°}$ accounts for 8% of $a_{cl,0°}$ at $t = 0.38$ s, this ratio increases to 43% at $t = 2.17$ s (Figure 10.87b), indicating a lateral spreading effect on the wave crest normal to its propagation direction.

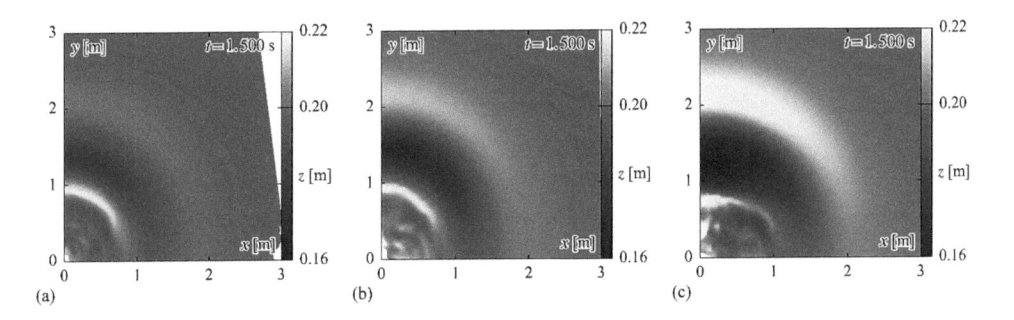

(a)   (b)   (c)

Figure 10.84 Contour plots $z(x, y)$ of Tests 072, 004, and 059 (Evers, 2018) with $[V_s; m_s; s; b; \alpha; h, P]$ (a) [3.03 m/s; 10 kg; 0.06 m; 0.25 m; 60°; 0.20 m; 0.93], (b) [2.70 m/s; 20 kg; 0.06 m; 0.50 m; 60°; 0.20 m; 0.83], (c) [2.78 m/s; 40 kg; 0.06 m; 1.00 m; 60°; 0.20 m; 0.86]

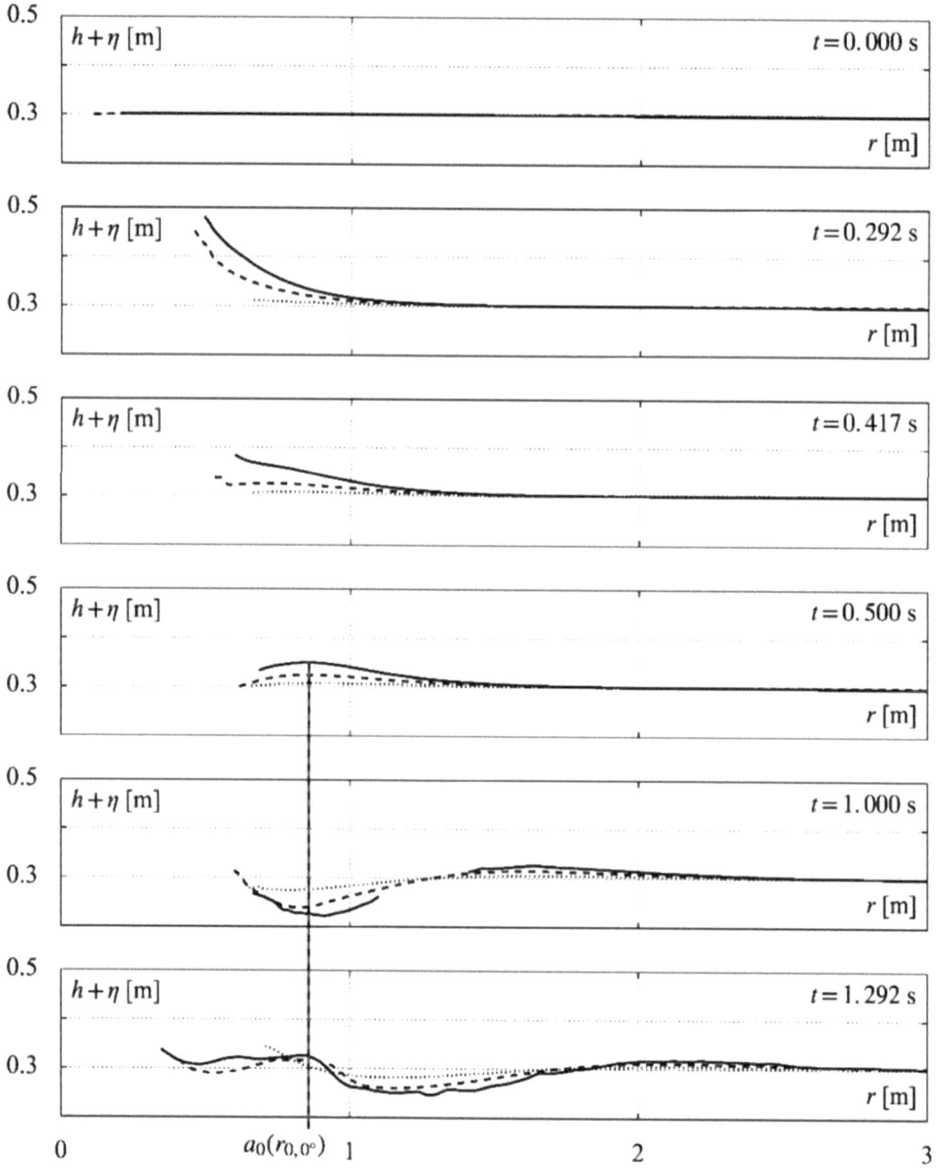

Figure 10.85 Profile plots $[h + \eta](r)$ of Test 033 (Evers, 2018) including impact radius r_0 and initial first wave crest amplitude $a_{0,c1}$ with $V_s = 4.55$ m/s, $m_s = 20$ kg, $s = 0.12$ m, $b = 0.50$ m, $\alpha = 60°$, $h = 0.30$ m, and P $= 1.08$ at $\gamma = (—)\ 0°, (--)$ 45°, $(\cdots)$ 90° (Evers, 2017)

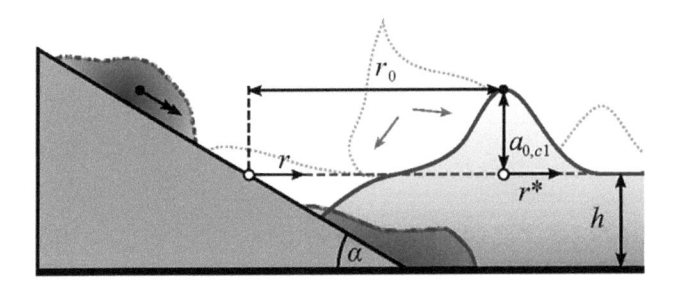

Figure 10.86 Definition plot of impact radius r_0 and the initial amplitude of the first wave crest $a_{0,c1}$

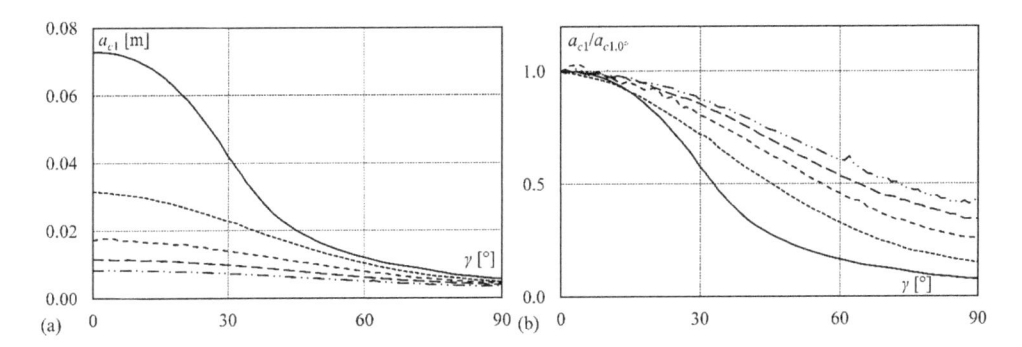

Figure 10.87 First wave crest elevation profiles $a_{c1}(\gamma)$ of Test 004 (Evers, 2018) at $t = 0.38$ s, 0.83 s, 1.29 s, 1.75 s, and 2.17 s for (a) a_{c1}, (b) $a_{c1}/a_{c1,0°}$ versus γ (Evers *et al.*, 2019)

10.9.4 Experimental results

Slide impact zone

The definition plots (Figures 10.81b, 10.86) show the impact radius r_0, representing the boundary between the slide impact and the wave propagation zones; it defines the location at which the impulse wave characteristics become first measurable. The relative impact radius is defined as $R_0 = r_0/h$, varying for spatial wave propagation with the wave propagation angle γ. It is governed by the impulse product parameter P, the relative slide width B, and the effective slide impact angle $\alpha_{\text{eff}} = (6/7)\alpha$. For $\gamma = 0°$ and $90°$, the measured values R_0 are approximated by (Evers, 2017)

$$R_{0,0°} = r_{0,0°} / h = 2.5(PB\cos\alpha_{\text{eff}})^{0.25}, \tag{10.71}$$

$$R_{0,90°} = r_{0,90°} / h = (B/2) + 1.5(P\cos\alpha_{\text{eff}})^{0.25}. \tag{10.72}$$

The cosine of α_{eff} is included twice, since it is also part of P. As the center of the slide width represents the origin of $R_{0,90°}$, its minimum value is $B/2$ (Figure 10.81b); i.e. very wide slides create large $R_{0,90°}$ regardless the magnitude of the other slide parameters. For γ between $0°$ and $90°$, R_0 is approximated by an ellipse in polar form as

$$R_0(\gamma) = \sqrt{\frac{R_{0,0°}^2 R_{0,90°}^2}{R_{0,0°}^2 \sin^2 \gamma + R_{0,90°}^2 \cos^2 \gamma}} \,. \tag{10.73}$$

The measured impact radii $R_{0,meas}$ for all γ versus that predicted $R_{0,pred}$ includes 65 of the total 74 experiments. For 9 experiments, the water surface close to the slide impact location was not sufficiently captured by the videometric system, but these experiments are still included in the results of the wave propagation zone. The coefficient of determination is $R^2 = 0.71$. For $R_0 > 3$, the predicted values tend to underestimate the measured values. However, most measured values scatter within a $\pm 30\%$ range.

The relative initial first wave crest amplitude $A_{0,c1} = a_{0,c1}/h$ represents the maximum wave crest elevation at R_0 for $\gamma = 0°$. It is governed by the impulse product parameter P, the relative slide width B, and the effective slide impact angle α_{eff} following the expression (Evers, 2017)

$$A_{0,c1} = a_{0,c1}/h = 0.2P^{0.5}B^{0.75}(\cos\alpha_{eff})^{0.25}. \tag{10.74}$$

For γ between $0°$ and $90°$, the relative first wave crest amplitude A_{c1} at R_0 follows

$$A_{c1}(R_0) = A_{0,c1}\left[\text{sech}\left(3.2\tfrac{\gamma}{90°}\right)\right]^{\cos\alpha_{eff}}. \tag{10.75}$$

The hyperbolic secant function describes the initial wave crest shape over γ, featuring an inflection point, which may also be observed in the measurements shown in Figure 10.87a. The effect of α on the initial wave crest shape is accounted for by the exponent. While the ratio of $A_{c1}(R_{0,0°})$ to $A_{c1}(R_{0,90°})$ is $\cong 1.75$ for vertical slide impact ($\alpha = 90°$), it increases to approximately 9.6 for $\alpha = 30°$. The measured initial first wave amplitudes $A_{c1,meas}$ at R_0 for all γ of the 65 experiments scatter by $\pm 30\%$ versus the predicted $A_{c1,pred}$ ($R^2 = 0.91$).

Wave propagation zone

The following wave characteristics were tracked and analyzed in the wave propagation zone: first wave crest amplitude a_{c1}, first wave trough amplitude a_{t1}, second wave crest amplitude a_{c2}, first wave height H_1, first wave period T_1, as well as the first and the second wave celerities c_{c1} and c_{c2} (Figure 10.81). These wave features allow for assessing the magnitude and consequently the impact of the outgoing wave train front at a specific location relative to the wave source.

The definition plots (Figures 10.81b, 10.86) show the surrogate radial wave propagation distance r^*. It extends the impact radius r_0 and is the radial coordinate for describing the wave characteristics in the wave propagation zone, i.e. outside the slide impact zone. The relative wave propagation distance r/h is defined as the sum of the relative impact radius R_0 and the relative surrogate radial wave propagation distance R^*

$$r/h = r_0/h + r^*/h = R_0 + R^*. \tag{10.76}$$

The crest elevation above the stillwater surface of the first outgoing wave is defined as the first wave crest amplitude a_{c1} (Figure 10.81). The measured relative first wave crest amplitudes $A_{c1} = a_{c1}/h$ for all γ are approximated by (Evers, 2017)

$$A_{c1}(R^*,\gamma) = A_{0,c1} \exp\left(-0.4 A_{0,c1}^{-0.3} \sqrt{R^*}\right)\left[\operatorname{sech}\left(3.2\tfrac{\gamma}{90°}\right)\right]^{\cos \alpha_{\mathrm{eff}} \exp\left(-0.15\sqrt{R^*}\right)}. \tag{10.77}$$

Equation (10.77) expands Eq. (10.75) by including R^* to describe the wave decay effect. The first exp-function accounts for the amplitude decay in the propagation direction $\gamma = 0°$. The second exp-function in the exponent of the hyperbolic secant function includes the effect of lateral wave crest spreading, i.e. the decreasing ratio of the amplitudes at $\gamma = 0°$ to $90°$ during wave propagation as shown in Figure 10.87. Based on a histogram highlighting the distribution of the measured wave amplitudes versus those predicted for 208,685 data points extracted in $1°$ steps between $\gamma = 0°$ and $90°$ from the interpolated water surface measurements are included in the analysis. A data analysis of Evers (2017) indicates that 90% of the measured data scatter within a range of -45% and $+25\%$ around Eq. (10.77) ($R^2 = 0.89$).

The depression below the stillwater surface following a_{c1} is defined as the first wave trough amplitude a_{t1} (Figure 10.81). Its relative value $A_{0,t1}$ at R_0 and $\gamma = 0°$ follows (Evers, 2017)

$$A_{0,t1} = a_{0,t1}/h = 0.35(PB\cos\alpha_{\mathrm{eff}})^{0.5}. \tag{10.78}$$

The measured relative first wave crest trough $A_{t1} = a_{t1}/h$ is approximated by (Evers, 2017)

$$A_{t1}(R^*,\gamma) = A_{0,t1} \exp\left(-0.4 A_{0,t1}^{-0.3} \sqrt{R^*}\right)\left[\operatorname{sech}\left(3.6\tfrac{\gamma}{90°}\right)\right]^{\cos \alpha_{\mathrm{eff}} \exp\left(-0.15\sqrt{R^*}\right)}. \tag{10.79}$$

A total of 90% of the measured data scatter between -20% and $+40\%$ ($R^2 = 0.91$). The underestimation is highest close to the slide impact location and slightly increases for $r/h > 7$. Furthermore, the measured values are slightly underestimated for γ between $20°$ and $80°$.

The second wave crest amplitude a_{c2} is defined as the crest elevation above the water surface of the second outgoing wave (Figure 10.81). The relative initial second wave crest amplitude $A_{0,c2}$ at R_0 and $\gamma = 0°$ follows (Evers, 2017)

$$A_{0,c2} = a_{0,c2}/h = 0.14(PB\cos\alpha_{\mathrm{eff}})^{0.25}. \tag{10.80}$$

The measured relative second wave crest amplitudes $A_{c2} = a_{c2}/h$ for all γ are approximated by

$$A_{c2}(R^*,\gamma) = A_{0,c2} \exp\left(-0.1 A_{0,c2}^{-0.3} \sqrt{R^*}\right)\left[\operatorname{sech}\left(3\tfrac{\gamma}{90°}\right)\right]^{\cos \alpha_{\mathrm{eff}} \exp\left(-0.15\sqrt{R^*}\right)}. \tag{10.81}$$

Compared with A_{c1} (Eq. 10.77), the scatter is substantially larger with 90% of the data points within -60% and $+50\%$ ($R^2 = 0.69$). The largest scatter occurs for $r/h < 5$, i.e. close to the slide impact location.

The first wave height H_1 is defined as the distance between the water surface elevation of the first wave crest and trough amplitudes at the same location (Figure 10.81). The measured relative values $Y_1 = H_1/h$ are therefore given as

$$Y_1(R^*, \gamma) = A_{c1} + A_{t1}. \tag{10.82}$$

90% of the data scatter evenly between -30% and $+25\%$ with the median close to zero ($R^2 = 0.93$).

The first wave period T_1 is defined as the time difference between the initial uplifting of the first wave crest and the stillwater level intersection following the first wave trough at the same location (Figure 10.81). The initial uplifting was tracked at the location where the water level exceeded an absolute threshold equal to the estimated measurement accuracy of 1 mm above stillwater level. The measured relative first wave period $T_1(g/h)^{0.5}$ is approximated by

$$T_1(R^*, \gamma) \cdot (g/h)^{0.5} = 10(Y_1)^{0.2} + (R^*/2). \tag{10.83}$$

The data have a narrow scatter range with 90% between -10% and $+15\%$ ($R^2 = 0.86$). The value of $T_1(g/h)^{0.5}$ only depends on the relative wave height Y_1 and the relative surrogate radial wave propagation distance R^*.

The first and the second wave celerities c_{c1} and c_{c2} are defined as the propagation celerities of the first and second wave crests, respectively. The measured relative first wave celerities $c_{c1}/(gh)^{0.5}$ are approximated by (Evers, 2017)

$$c_{c1}/(gh)^{0.5} = 0.95(1 + A_{c1})^{0.5}. \tag{10.84}$$

The data scatter with 90% between -15% and $+10\%$. The measured relative second wave celerities $c_{c2}/(gh)^{0.5}$ are approximated by (Evers, 2017)

$$c_{c2}/(gh)^{0.5} = 0.7(1 + A_{c2})^{0.5}. \tag{10.85}$$

The data scatter with 90% between -25% and $+15\%$. The measured values deviate acceptably from these predicted.

10.9.5 Discussion of results

The governing parameters given in Table 10.9 represent the limitations of the empirical equations proposed in the preceding section. The slide impact velocities V_s cover both sub- and supercritical ranges of the slide Froude number F. Although the slide mass m_s was varied in the experiments, the bulk slide density $\rho_s = 1338$ kg/m³ remained constant. However, all equations include the impulse product parameter P, representing the momentum transfer per unit slide width, derived from 2D experiments with ρ_s between 590 kg/m³ and 1720 kg/m³ (Heller and Hager, 2010). Therefore, the application range of the equations presented may be potentially extended to bulk slide densities lower than the water density although these were not covered in the present experiments. The limitations of the wave propagation angle γ have to be assessed based on the water body geometry. Mohammed and Fritz (2012) and McFall and Fritz (2016, 2017) studied the edge wave propagation at a straight beach extending the sliding plane. At Chehalis Lake, these edge waves had a lower but nonetheless significant run-up compared to the waves for $\gamma = 0°$ (Roberts *et al.*, 2013). The experimental setup of this study resembles the tip of a headland, since it is not bounded by an inclined shoreline, and edge waves were not accounted for. Heller *et al.* (2012) conducted experiments in a wave basin with confining sidewalls, adjusted to a basin side angle θ. For θ between 30° and 90°, with $\theta = 0°$ representing a 2D flume, no significant effect on the maximum wave features was found at $\gamma = 0°$. Edge waves strongly depend on bathymetry and topography of the water body requiring a case-by-case assessment. The maximum relative wave propagation distance reached in the experiments was $r/h \approx 16$. While this upper limitation is insufficient to cover long distances in oceans, it is suitable for smaller confined basins, e.g. bays or fjords. The lower limit of r/h is directly defined by the impact radius R_0. The approach by Heller *et al.* (2009) states $r/h = 5$ as global lower limit. Therefore, the present approach involving R_0 allows for the prediction of impulse wave characteristics closer to the impact location.

Equations (10.74), (10.78), and (10.80) describe the initial wave amplitudes $A_{0,c1}, A_{0,t1}, A_{0,c2}$ at the impact radius R_0 for $\gamma = 0°$. Parameter P, introduced by Heller and Hager (2010) based on 2D experiments, is included in all three equations. The results of this study indicate that P also applies to the generation and propagation of impulse waves in 3D. However, the relative slide width B has to be added as additional parameter, since P solely accounts for the unit momentum transfer from the slide to the water column. The effective slide impact angle α_{eff} is also added as additional parameter, although it is already included in P; it has a strong influence in spatial environments. The features of 3D impulse waves therefore are based on these of 2D impulse waves and extended by typical 3D wave parameters as described.

Bregoli *et al.* (2017) proposed for the maximum wave amplitude a_{max} and its distance x_{max} from the impact location

$$a_{\text{max}} / h = 0.118 S^{0.459} \left(\frac{l_s}{h} \right)^{0.463} \mathsf{F}^{0.554} \triangleq A_{0,c1}, \tag{10.86}$$

$$x_{\text{max}} / h = 3.97 S^{0.198} \left(\frac{m_s V_s^2}{\rho_w gbl_s h^2} \right)^{0.277} \triangleq R_{0,0°}. \tag{10.87}$$

Equations (10.86) and (10.87) correspond to the relative initial wave amplitude $A_{0,c1}$ (Eq. 10.74) and the relative impact radius $R_{0,0°}$ (Eq. 10.71) for a single wave propagation angle $\gamma = 0°$, respectively. In Figure 10.88, Eqs. (10.86) and (10.87) by Bregoli *et al.* (2017) are applied to predict the data set of the present study. The slide length is substituted with $l_s = m_s/(sb\rho_s)$. While the measured relative impact radii $R_{0,0°}$ for $\gamma = 0°$ agree well with the prediction, the predicted relative initial wave amplitudes $A_{0,c1}$ largely underestimate the measured values. Compared to $R_{0,0°}$, the equation for x_{max}/h by Bregoli *et al.* (2017) does not include the effect of the slide impact angle α. While Kamphuis and Bowering (1972) found an only minor effect of α on the generated wave magnitudes, Heller and Hager (2010) included its effect into the impulse product parameter P. In addition, the equation for a_{max}/h does not include the effect of the slide mass m_s, the slide width b, and again α, which may explain the observed deviations.

Equations (10.77), (10.79), and (10.81) include the initial amplitudes describing the spatial wave propagation process with the relative surrogate wave propagation radius R^* and the wave propagation angle γ as polar coordinates. $R^* = 0$ is equivalent to the impact radius R_0 and the equations are defined for $R^* \geq 0$. The wave amplitude decay processes are described by exponentials, also contained in Hughes (1993) for different wave attenuation equations as well as by Bregoli *et al.* (2017) in their relations describing impulse wave propagation for $\gamma = 0°$. While the exponential function in Eq. (10.77) includes a factor of -0.4, Eq. (10.81) has a lower factor of -0.1, indicating that a_{c2} is subject to a smaller wave decay rate than a_{c1}. However, the initial first wave crest amplitude $A_{0,c1}$ (Eq. 10.74) is generally larger than the second $A_{0,c2}$ (Eq. 10.80). In combination with the different decay rates, the ratio of A_{c1} to A_{c2} is subject to permanent variation during wave propagation. To describe the wave crest and trough shapes along γ, hyperbolic secant functions were selected based on experimental data (Figure 10.87). Their exponents include α_{eff} and R^* as governing parameters accounting for the lateral spreading during wave propagation. A slightly different exponent to describe this effect was originally introduced by Heller *et al.* (2015).

In the context of hazard assessment, the predicted wave characteristics represent input parameters to estimate the impact at the shoreline. Besides wave crest amplitudes, also the wave height, the wave period, and the wave length are accounted for, e.g. by Müller (1995). Among various water wave types (Heller, 2008), the solitary wave causes run-up heights with the highest elevation (Synolakis, 1988). Impulse waves exhibit features of wave types within the range of intermediate-water to shallow-water waves depending on the generation process (Fritz *et al.*, 2004; Heller and Hager, 2011). The measured wave crest celerities of this study are presented in relation to solitary waves (Eqs. 10.84, 10.86). While the first wave crest propagates at approximately 95% of the solitary wave celerity, the second wave crest reaches 70%. The celerity reduction from the first to the second wave crest is on average 26%. This is a slightly higher reduction than 18% to 23% observed by Mohammed and Fritz (2012) and 23% by McFall and Fritz (2016).

The videometric measurement system allows for a quasi-continuous representation of the water surface and consequently a high number of data. This measurement technique allowed for adaptively tracking the newly introduced impact radius r_0. Another advantage of the high

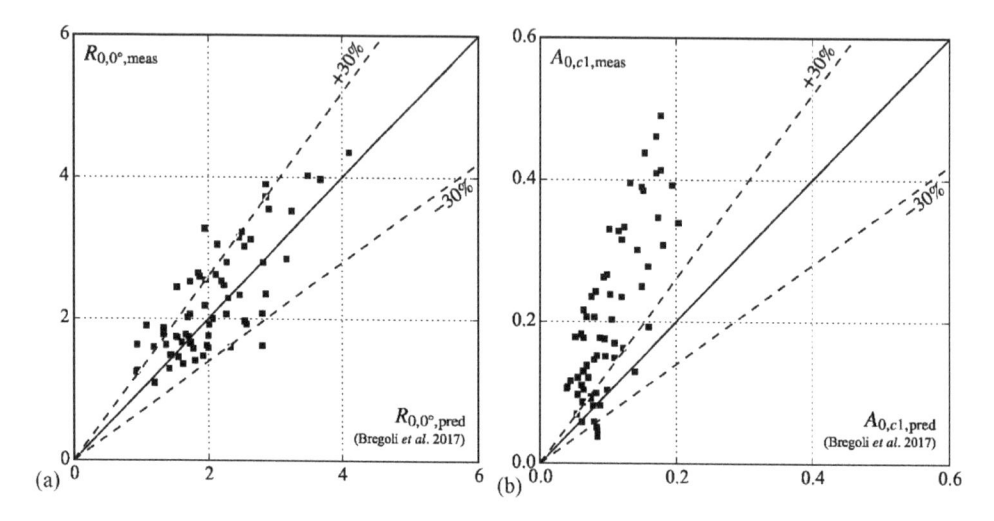

Figure 10.88 (a) Measured impact radii $R_{0,0°,meas}$ versus predicted $R_{0,0°,pred}$ (Eq. 10.73), (b) measured initial first wave crest amplitudes $A_{0,c1,meas}$ versus predicted $A_{0,c1,pred}$ (Eq. 10.74); values predicted with equations by Bregoli *et al.* (2017)

number of data points is highlighted by box plots (Evers, 2017). These plots allow for assessing the evolution of the impulse wave features as well as their respective scatter along the polar coordinates r/h and γ.

10.9.6 Relevance for practice

Summarizing the above findings relating to spatial impulse waves, one may state:

- The impulse product parameter, originally derived from 2D wave flume experiments, in combination with the slide width and the additional effect of the slide impact angle, is suitable to quantify the spatial impulse wave generation and propagation features.
- The impact zone close to the slide impact location is influenced by the collapse of the impact crater, strong turbulence, and arbitrary distortion of the water surface, while distinct characteristics of the outgoing wave train, e.g. amplitudes, are quantifiable in the wave propagation zone.
- The transition from the impact zone to the wave propagation zone is confined by the impact radius r_0, which is governed by the slide characteristics; its spatial extent is approximated by an ellipse.
- The wave crest and trough shapes of the spatially propagating wave train are approximated by a hyperbolic secant function featuring an exponent accounting for the lateral spreading of the wave crest during propagation.
- The wave decay is approximated by an exponential function; the first wave crest amplitude is thereby subject to a higher wave decay rate than the second wave crest amplitude.
- While the first wave crest propagates at 95% of the solitary wave celerity, the second does at only 70%.
- Equations (10.77) and (10.79) define the first wave crest and trough amplitudes, Eq. (10.83) relates to the first wave period, whereas Eqs. (10.84) and (10.85) refer to the first and the second wave celerities.

The equations derived allow for predicting impulse wave magnitudes at a specific location relative to the wave source based on the governing slide parameters. In combination with literature equations describing the wave impact as run-up heights, a preliminary hazard assessment for potential landslide-generated impulse wave events at prototype scale may be conducted with the present background information.

Notation

A = Relative wave amplitude $A = a/h$, also $= a/H$ (-); also dimensionless eroded dam area parameter (-)

A_0 = Original dam area (m^2)

$A_{0,c1}$ = Relative initial first wave crest amplitude for $\gamma = 0°$ (-)

$A_{0,c2}$ = Relative initial second wave crest amplitude for $\gamma = 0°$ (-)

$A_{0,t1}$ = Relative initial first wave trough amplitude for $\gamma = 0°$ (-)

A_{c1} = Relative first wave crest amplitude (-)

A_{c2} = Relative second wave crest amplitude (-)

A_d = Eeposited dam area (m^2)

A_e = Eroded dam area (m^2)

A_M = Relative maximum wave amplitude $A_M = a_M/h$ (-)
A_{t1} = Relative first wave trough amplitude (-)
A_{tr} = Transition point velocity parameter (-)
a = Wave amplitude (m)
$a_{0,c1}$ = Initial first wave crest amplitude for $\gamma = 0°$ (m)
$a_{0,c2}$ = Initial second wave crest amplitude for $\gamma = 0°$ (m)
$a_{0,t1}$ = First wave trough amplitude for $\gamma = 0°$ (m)
a_{c1} = First wave crest amplitude (m)
a_{c2} = Second wave crest amplitude (m)
a_M = Maximum wave amplitude (m)
a_{max} = Maximum wave amplitude for $\gamma = 0°$ by Bregoli et al. (2017) (m)
a_P = Primary wave amplitude (m)
a_{t1} = First wave trough amplitude (m)
a_w = Effective wave amplitude, $a_w = h + a - w$ (m)
B = Discharge parameter (-); also relative slide width = b/h (-)
B_s = Slide width (m)
b = Channel width (m); also slide width (m)
b_{cr} = Dike crest width (m)
b_K = Dam crest width (m)
c = Wave celerity, or solitary wave celerity $c = [g(h + H)]^{1/2}$ (ms^{-1})
c_{c1} = First wave crest celerity (m/s)
c_{c2} = Second wave crest celerity (ms)
D_g = Relative grain diameter $D_g = d_g/h$ (-)
d = Flow depth (m), also sediment diameter (m)
d_g = Grain diameter (m)
d_{max} = Maximum flow depth at certain position (m)
d_{50} = Median sediment size (m)
d_i = Sediment size with i% passing sieve (m)
d_0 = Maximum flow depth at transition point (m)
$d_0{}^*$ = Relative maximum flow depth at transition point $d_0{}^* = d_0/h$ (-)
E = Maximum wave overtopping depth parameter (-)
E_1 = Maximum wave overtopping depth parameter (-)
F = Wave overtopping duration parameter (-)
F = Slide Froude number $\mathsf{F} = V_s/(gh)^{1/2}$ (-)
F_0 = Relative slide front position (-)
f = Freeboard, $f = w - h$ (m)
f_k = Kinetic friction coefficient (-)
G = Parameter after Huber and Hager (1997), $G = 0.88(\sin\alpha)(\rho_g/\rho_w)^{1/4}[V_g/(bh^2)]^{1/2}$ (-)
g = Gravity acceleration (ms^{-2})
H = Wave height (m), also eroded depth parameter (-)
H_1 = First wave height (m)
H_M = Maximum wave height (m)
h = Still-water depth (m)
h_e = Eroded crest depth (m)
h_o = Stillwater depth (m)
L = Wave length (m)
L_M = Wave length of a_M (m)
L_w = Wave length (m)

l_{box} = Length of slide box (m)

l_s = Slide length (m)

M = Relative slide mass $M = m_s/(\rho_w bh^2)$ (-)

M_s = Impact angle corrected relative slide mass (-)

m_s = Slide mass (kg)

n = Bulk slide porosity (%)

P = Impulse product parameter $\mathsf{P} = \mathsf{F}S^{1/2}M^{1/4}\{\cos[(6/7)\alpha]\}^{1/2}$ (-)

P_s = Impact angle corrected impulse product parameter (-)

$\bar{p}$ = Average data value (-)

p_i = Individual data value (-)

Q_s = Slide discharge (m³s⁻¹)

q = Discharge per unit width (m²s⁻¹)

q_0 = Maximum discharge at transition point (m²s⁻¹)

R = Relative run-up height $R = r/h$ (-)

R_0 = Relative impact radius (-)

R^2 = Coefficient of determination (-)

R^* = Relative surrogate radial wave propagation distance (-)

R = Reynolds number $\mathsf{R} = g^{1/2}h^{3/2}/v_w$ (-)

R_f = Overland flow Reynolds number $\mathsf{R}_f = (v_f d_{max})/v$ (-)

r = Run-up height, radial distance (m)

r^* = Surrogate radial wave propagation distance (m)

r_0 = Impact radius (m)

S = Relative slide thickness $S = s/h$ (-), also dam face slope (-)

s = Slide thickness (m)

T = Dimensionless wave type product $\mathsf{T} = S^{1/3}M\cos[(6/7)\alpha]$ (-)

T = Relative time $T = t(c/H)$, wave period (-)

T_1 = First wave period (s)

T_M = Wave period of a_M (s)

T_r = Relative time $T_r = t(g/h)^{1/2}$ (-)

T_s = Relative time of underwater landslide motion $T_s = t_s(g/h)^{1/2}$ (-)

T_w = Relative wave period $T_w = t_w(g/h_o)^{1/2}$ (-)

t = Time (s)

t_{end} = Slide propagation duration (s)

t_0 = Time of test start (s)

t_O = Wave overtopping duration (s)

t_s = Time of underwater landslide motion (s)

t_w = Wave period (s)

U = Ursell parameter $\mathsf{U} = (H/L)/(h/L)^3 = HL^2/h^3$ (-)

V = Relative slide volume (-)

V_{f0} = Slide front velocity (ms⁻¹)

V_s = Slide impact velocity (ms⁻¹)

$\forall_O$ = Overtopping volume (m³)

$\forall_g$ = Slide grain volume (m³)

$\forall_s$ = Bulk slide volume identical to inner box volume (m³)

v = Flow velocity (ms⁻¹)

v_f = Flow front velocity (ms⁻¹)

v_x = Horizontal velocity component (ms⁻¹)

$\tilde{v}_{x,max}$ = Depth-averaged horizontal velocity (ms⁻¹)

W	= Relative shore height $W = w/h$ (-), also wave overtopping volume parameter (-)
W_1, W_2	= Wave overtopping volume parameter (-)
W	= Weber number $\mathsf{W} = \rho_w gh^2/\sigma_w$ (-)
W_f	= Overland flow Weber number $\mathsf{W}_f = v_f/[\sigma/(\rho d_{max})]^{1/2}$ (-)
w	= Shore height, dam height (m)
X	= Relative streamwise distance $X = x/h$ (-)
X_M	= Relative streamwise distance at a_M, $X_M = x_M/h$ (-)
x	= Horizontal streamwise coordinate (m)
x'	= Streamwise coordinate along channel bottom (m)
x'_{f0}	= Streamwise coordinate of slide front position (m)
x'_s	= Streamwise coordinate of maximum deposition thickness (m)
x'_t	= Streamwise slide travel distance (m)
x_M	= Streamwise distance at a_M (m)
x_{max}	= Distance from impact location of a_{max} by Bregoli $et\ al.$ (2017) (m)
x^*	= Relative overland flow propagation distance $x^* = (x-x_{tr})/w$ (-)
x_{tr}	= Location of transition from shore slope to overland flow portion (m)
Y	= Relative wave height $Y = H/h$ (-)
Y_1	= Relative first wave height (-)
Y_M	= Relative maximum wave height $Y_M = H_M/h$ (-)
Z_f	= Relative freeboard $Z_f = z_f/h$ (-)
z	= Vertical coordinate (m), also dam surface elevation (m)
z_f	= Freeboard (m)
z_s	= Vertical coordinate of maximum deposition thickness (m)
α	= Slide impact angle, i.e. hillslope angle (°)
α_{eff}	= Effective slide impact angle, $\alpha_{eff} = (6/7)\alpha$ (°)
β	= Shore angle, run-up angle (°), also dam front face angle (°)
γ	= Wave propagation angle (°), also dam shape parameter (-)
Δa	= Deviation of wave amplitude in terms of a basis (%)
ΔH	= Deviation of wave height in terms of a basis (%)
Δt	= Time interval (s)
Δx	= Spacing between two CWGs (m)
Δx_1	= Spacing between CWG_1 and hillslope ramp end (m)
Δ_{max}	= Maximum deviation (%)
Δz_{sc}	= Centroid drop height (m)
β	= Dam front face angle (°)
δ	= Dynamic bed friction angle (°)
ε	= Relative wave height $\varepsilon = H/h$, also relative wave amplitude $= a/h$ (-)
ϵ	= Wave steepness $\epsilon = h_o/L_w$ (m)
η	= Free water surface elevation (m)
ϕ	= Slide front angle (°)
φ'	= Internal friction angle (°)
λ	= Scale factor (-)
ρ_g	= Grain density (kgm^{-3})
ρ_s	= Bulk slide density (kg/m^3), also sediment density (kg/m^3)
ρ_w	= Water density $\rho = 1000$ kg/m^3 (kgm^{-3})
σ	= Water surface tension $\sigma = 0.07275$ kg/s^2 (kgs^{-2})
σ_g	= Sediment non-uniformity coefficient (-)

μ = Apparent coefficient of friction (-), also dike shape parameter (-)
ν = Kinematic viscosity of water $\nu = 10^{-6}$ (m²s⁻¹)
θ = Basin side angle (°)
τ = Wave shallowness $\tau = a_M/h_o$ (-)
ξ = Slide profile (m)
ξ_s = Solitary wave Iribarren number $\xi_s = \tan\beta/\varepsilon$ (-)

References

Ataie-Ashtiani, B. & Malek Mohammadi, S. (2007) Near field amplitude of subaerial landslide generated waves in dam reservoirs. *Dam Engineering*, 17(4), 197–222.

Awal, R., Nakagawa, H., Fujita, M., Kawaike, K., Baba, Y. & Zhang, H. (2010) Experimental study on glacial lake outburst floods due to waves overtopping and erosion of Moraine Dam. *Annuals of Disaster Prev. Res. Inst.*, No. 53 B, Kyoto University, Kyoto.

Baldock, T.E., Peiris, D. & Hogg, A.J. (2012) Overtopping of solitary waves and solitary bores on a plane beach. *Proceedings of the Royal Society* A, 468, 3494–3516.

Balmforth, N.J., von Hardenberg, J., Provenzale, A. & Zammett, R. (2008) Dam breaking by wave-induced erosional incision. *Journal of Geophysical Research*, 113(1), F01020, DOI:10.1029/2007JF000756.

Balmforth, N.J., von Hardenberg, J. & Zammett, R.J. (2009) Dam-breaking seiches. *Journal of Fluid Mechanics*, 628, 1–21. DOI:10.1017/S0022112009005825

Bardet, J.-P., Synolakis, C.E., Davies, H.L., Imamura, F. & Okal, E.A. (2003) Landslide tsunamis: Recent findings and research directions. *Pure and Applied Geophysics*, 160, 1793–1809.

Blown, I. & Church, M. (1985) Catastrophic lake drainage within the Homathko River basin, British Colombia. *Canadian Geotechnical Journal*, 22(4), 551–563. DOI:10.1139/t85-075

Bornhold, B.D., Harper, J.R., McLaren, D. & Thomson, R.E. (2007) Destruction of the first nation's village of Kwalate by a rock avalanche-generated tsunami. *Atmosphere-Ocean*, 45(2), 123–128.

Boussinesq, J. (1871) Théorie de l'intumescence liquide appelée onde solitaire ou de translation, se propageant dans un canal rectangulaire [Theory of liquid disturbance referred to as solitary wave propagating in a rectangular channel]. *Comptes Rendus* de l'Académie des Science, Paris, 72, 755–759 (in French).

Bregoli, F., Bateman Pinzón, A., Medina Iglesias, V. & Gómez Cortéz, D.A. (2013) Experimental studies on 3D impulse waves generated by rapid landslides and debris flow. *Italian Journal of Engineering Geology and Environment*, Book Series, 6, 115–122.

Bregoli, F., Bateman, A. & Medina, V. (2017) Tsunamis generated by fast granular landslides: 3D experiments and empirical predictors. *Journal of Hydraulic Research*, 55(6), 743–758.

Brideau, M.-A., Sturzenegger, M., Stead, D., Jaboyedoff, M., Lawrence, M., Roberts, N., Ward, B., Millard, T. & Clague, J. (2011) Stability analysis of the 2007 Chehalis lake landslide based on long-range terrestrial photogrammetry and airborne LiDAR data. *Landslides*, 9(1), 75–91.

Buckingham, E. (1914) On physically similar systems: Illustrations of the use of dimensional equations. *Physical Review*, 4, 345–376.

Camfield, F.E. (1980) Tsunami engineering. *Special Report* 6, Department of the Army, Coastal Engineering Research Center, Virginia.

Carey, M. (2008) Disasters, development, and glacial lake control in twentieth-century Peru. *Mountains: Sources of Water, Sources of Knowledge*. Springer Netherlands, pp. 181–196. DOI:10.1007/978-1-4020-6748-8_11

Carey, M. (2010) *In the shadow of melting glaciers: Climate change and Andean Society*. Oxford University, Oxford.

Chen, Y. Y., Kharif, C., Yang, J. H., Hsu, H. C., Touboul, J. & Chambarel, J. (2015). An experimetal study of steep solitary wave reflection at a vertical wall. *European Journal of Mechanics*-B/Fluids, 49, 20–28. https://doi.org/10.1016/j.euromechflu.2014.07.003

Clague, J.J. & Evans, S.G. (2000) A review of catastrophic drainage of moraine-dammed lakes in British Columbia. *Quaternary Science Reviews*, 19(17–18), 1763–1783. DOI:10.1016/S0277-3791(00)00090-1

Cooker, M.J., Weidman, P.D. & Bale, D.S. (1997) Reflection of a high-amplitude solitary wave at a vertical wall. *Journal of Fluid Mechanics*, 342, 141–158.

Costa, J.E. & Schuster, R.L. (1988) The formation and failure of natural dams. *Bulletin Geological Society of America*, 100(7), 1054–1068. DOI:10.1130/0016-7606(1988)100<1054:TFAFON>2.3.CO;2.

Cruden, D.M. & Varnes, D.J. (1996) Landslide types and processes. In: Turner, A.K. & Schuster, R.L. (eds) *Landslides*, Special Report, 247, 36–75.

Dahl-Jensen, T., Larsen, L.M., Pedersen, S.A.S., Pedersen, J., Jepsen, H.F., Pedersen, G., Nielsen, T., Pedersen, A.K., von Platen-Hallermund, F. & Weng, W. (2004) Landslide and tsunami 21 November 2000 in Paatuut, West Greenland. *Natural Hazards*, 31(1), 277–287.

Dean, R.G. & Dalrymple, R.A. (2004) *Water wave mechanics for engineers and scientists*. Advanced series on ocean engineering 2. World Scientific, Singapore.

Di Risio, M. & Sammarco, P. (2008) Analytical modeling of landslide-generated waves. *Journal of Waterway, Port, Coastal, Ocean Engineering*, 134(1), 53–60.

Di Risio, M., De Girolamo, P. & Beltrani, G.M. (2011) Forecasting landslide generated tsunamis: a review. In: Mörner, N.-A. (ed) *The Tsunami Threat: Research and Technology*.

Dodd, N. (1998) Numerical model of wave run-up, overtopping, and regeneration. *Journal of Waterway, Port, Coastal, and Ocean Engineering*, 124(2), 73–81.

Evers, F.M. (2017) Spatial propagation of landslide generated impulse waves. In: Boes, R. (ed) *Doctoral Dissertation* No 24650 and *VAW-Mitteilung*, 244. ETH Zurich, Zürich. https://doi.org/10.3929/ethz-b-000209471

Evers, F.M. (2018) Hydraulic scale model experiments on spatial propagation of landslide generated impulse waves [Dataset]. *Zenodo*. https://doi.org/10.5281/zenodo.1069077

Evers, F.M., & Boes, R.M. (2019) Impulse wave runup on steep to vertical slopes. *Journal of Marine Science and Engineering*, 7(1), 8. https://doi.org.10.3390/jmse7010008.

Evers, F.M. & Hager, W.H. (2015a) Impulse wave generation: Comparison of free granular with mesh-packed slides. *Journal of Marine Science and Engineering*, 3, 100–110. https://doi.org/10.3390/jmse3010100

Evers, F.M. & Hager, W.H. (2015b) Videometric water surface tracking: Towards investigating spatial impulse waves. Proceedings of the 35th *IAHR Congress*, The Hague. https://doi.org/10.3929/ethz-a-010630706

Evers, F.M. & Hager, W.H. (2016a) Spatial impulse waves: Wave height decay experiments at laboratory scale. *Landslides*, 13(6), 1395–1403. https://doi.org/10.1007/s10346-016-0719-1

Evers, F.M. & Hager, W.H. (2016b) Generation and spatial propagation of landslide generated impulse waves. *Coastal Engineering Proceedings*, 35(1) https://doi.org/10.9753/icce.v35.currents.13

Evers, F.M., Schmocker, L., Fuchs, H., Schwegler, B., Fankhauser, A.U. & Boes, R.M. (2018) Landslide generated impulse waves: assessment and mitigation of hydraulic hazards. Proceedings of the *ICOLD Congress*, Vienna, Q102(R40), 679–694.

Evers, F.M., Hager, W.H. & Boes, R.M. (2019) Spatial impulse wave generation and propagation. *Journal of Waterway, Port, Coastal, and Ocean Engineering*, 145(3), 04019011. https://doi.org/10.1061/(ASCE)WW.1943-5460.0000514.

Evers, F.M., Heller, V., Fuchs, H., Hager, W.H. & Boes, R.M. (2019) Landslide-generated impulse waves in reservoirs: Basics and computation. *VAW-Mitteilung* 254 (R. Boes, ed.). VAW, ETH Zurich, Zürich.

Falappi, S. & Gallati, M. (2007) SPH simulation of water waves generated by granular landslides. Proceedings of the 32nd *IAHR Congress*, Venice, 933, 1–10. IAHR, Madrid.

Fenton, J.D. (1985) A fifth-order Stokes theory for steady waves. *Journal of Waterway, Port, Coastal, and Ocean Engineering*, 111(2), 216–234.

Frank, P.-J. & Hager, W.H. (2014) Spatial dike breach: Accuracy of photogrammetric measurement system. Proceedings of Conference *River Flow 2014*. https://doi.org/10.1201/b17133-219

Fritz, H.M., Hager, W.H. & Minor, H.-E. (2001) Lituya Bay case: rockslide impact and wave run-up. *Science of Tsunami Hazards*, 19(1), 3–22.

Fritz, H.M. (2002) Initial phase of landslide generated impulse waves. *Ph.D. Thesis* 14871. ETH Zurich, Zürich; also *VAW-Mitteilung* 178. VAW, ETH Zurich, Zürich.

Fritz, H.M. & Moser, P. (2003) Pneumatic landslide generator. *International Journal of Fluid Power*, 4(1), 49–57.

Fritz, H.M., Hager, W.H. & Minor, H.-E. (2003a) Landslide generated impulse waves. 1. Instantaneous flow fields. *Experiments in Fluids*, 35, 505–519.

Fritz, H.M., Hager, W.H. & Minor, H.-E. (2003b) Landslide generated impulse waves. 2. Hydrodynamic impact craters. *Experiments in Fluids*, 35, 520–532.

Fritz, H.M., Hager, W.H. & Minor, H.-E. (2004) Near field characteristics of landslide generated impulse waves. *Journal of Waterway, Port, Coastal, Ocean Engineering*, 130(6), 287–302.

Fuchs, H. & Boes, R. (2010) Berechnung felsrutschinduzierter Impulswellen im Vierwaldstättersee [Computation of rockfall-induced impulse waves in the Lake of Lucerne]. *Wasser Energie Luft*, 102(3), 215–221 (in German).

Fuchs, H. (2013) Solitary impulse wave run-up and overland flow. *Ph.D. thesis* Nr. 21174. ETH Zurich, Zürich; also *VAW-Mitteilung* 221, R. Boes, ed. VAW, ETH Zurich, Zürich.

Fuchs, H. & Hager, W.H. (2013) Solitary impulse wave run-up characteristics. Proceedings of the 35th *IAHR World Congress*, Chengdu. Tsinghua University Press (CD-Rom).

Fuchs, H., Winz, E. & Hager, W.H. (2013) Underwater landslide characteristics from 2D laboratory modelling. *Journal of Waterway, Port, Coastal, Ocean Engineering*, 139(6), 480–488; 140(6), 08214001.

Fuchs, H. & Hager, W.H. (2015) Solitary impulse wave transformation to overland flow. *Journal of Waterway, Port, Coastal, Ocean Engineering*, 10.1061/(ASCE)WW.1943–5460.0000294, 04015004.

Fuhrman, D.R. & Madsen, P.A. (2008) Surf similarity and solitary wave runup. *Journal of Waterway, Port, Coastal, and Ocean Engineering*, 134(3), 195–198.

Gedik, N., Irtem, E. & Kabdasli, S. (2005) Laboratory investigation on tsunami run-up. *Ocean Engineering*, 32(5–6), 513–528.

Genevois, R. & Ghirotti, M. (2005) The 1963 Vaiont landslide. *Giornale di Geologia Applicata*, 1(1), 41–52. https://doi.org/10.1474/GGA.2005-01.0-05.0005

George, D.L., Iverson, R.M. & Cannon, C.M. (2017) New methodology for computing tsunami generation by subaerial landslides: application to the 2015 Tyndall Glacier landslide, Alaska. *Geophysical Research Letters*, 44, 7276–7284. https://doi.org/10.1002/2017GL074341

Goring, D.G. (1978) Tsunamis: The propagation of long waves onto a shelf. *Ph.D. Thesis*. California Institute of Technology, Pasadena, USA.

Goseberg, N. (2011) The run-up of long waves: Laboratory-scaled geophysical reproduction and onshore interaction with macro-roughness elements. *Ph.D. Thesis*. Leibniz Universität, Hannover, Germany.

Hager, W.H. (1994) Dammüberfälle [Dam overfalls]. *Wasser und Boden*, 46(2), 33–36 (in German).

Hall, J.V. & Watts, G.M. (1953) Laboratory investigation of the vertical rise of solitary wave on impermeable slopes. *Technical Memo*. 33. U.S. Army Corps of Engineers, Beach Erosion Board.

Hampton, M.A., Lee, H.J. & Locat, J. (1996) Submarine landslides. *Reviews of Geophysics*, 34(1), 33–59.

Heim, A. (1932) Bergsturz und Menschenleben [Rockfall and human life]. *Beiblatt zur Vierteljahresschrift der Natf. Ges. Zürich*, 20, 1–218 (in German).

Heller, V. (2007) Landslide generated impulse waves: Prediction of near field characteristics. *Ph.D. Thesis* 17531. ETH Zurich, Zürich; also *VAW-Mitteilung* 204. VAW, ETH Zurich, Zürich.

Heller, V. (2008) Landslide generated impulse waves: experimental results. In: Smith, J.M. (ed) Proc. 31st International Conference *Coastal Engineering*, Hamburg, 2, 1313–1325.

Heller, V. (2011) Scale effects in physical hydraulic engineering models. *Journal of Hydraulic Research*, 49(3), 293–306; 50(2), 246–250.

Heller, V. & Hager, W.H. (2010) Impulse product parameter in landslide generated impulse waves. *Journal of Waterway, Port, Coastal, Ocean Engineering*, 136(3), 145–155.

Heller, V. & Hager, W.H. (2011) Wave types of landslide generated impulse waves. *Ocean Engng.*, 38(4), 630–640.

Heller, V., Hager, W.H. & Minor, H.-E. (2008) Scale effects in subaerial landslide generated impulse waves. *Experiments in Fluids*, 44, 691–703.

Heller, V., Hager, W.H. & Minor, H.-E. (2009) Landslide generated impulse waves in reservoirs: Basics and computation. In: Boes, R. (ed) *VAW-Mitteilung*, 211. VAW, ETH Zurich, Zürich.

Heller, V., Moalemi, M., Kinnear, R.D. & Adams, R.A. (2012) Geometrical effects on landslide-generated tsunamis. *Journal of Waterway, Port, Coastal, and Ocean Engineering*, 138(4), 286–298. https://doi.org/10.1061/(ASCE)WW.1943-5460.0000130

Heller, V. & Spinneken, J. (2015) On the effect of the water body geometry on landslide–tsunamis: Physical insight from laboratory tests and 2D to 3D wave parameter transformation. *Coastal Engineering*, 104, 113–134. https://doi.org/10.1016/j.coastaleng.2015.06.006

Herbich, J.B. (1990) *Handbook of coastal and ocean engineering*, 1. Gulf Publishing, Houston.

Hsiao, S.-C. & Lin, T.-C. (2010) Tsunami-like solitary waves impinging and overtopping an impermeable seawall: Experiment and RANS modeling. *Coastal Engineering*, 57(1), 1–18.

Huber, A. (1980) Schwallwellen in Seen als Folge von Felsstürzen [Surge waves due to rock falls]. In: Vischer, D. (ed) *VAW-Mitteilung*, 180. ETH Zurich, Zürich (in German).

Huber, A. (1982) Impulse waves in Swiss lakes as a result of rock avalanches and bank slides. Proceedings of the 14th *ICOLD Congress*, Rio de Janeiro, Q54(R29), 455–476.

Huber, A. & Hager, W.H. (1997) Forecasting impulse waves in reservoirs. Proceedings of the 18th *ICOLD Congress*, Florence, C31, 993–1005.

Huber, L.E., Evers, F.M. & Hager, W.H. (2017) Solitary wave overtopping at granular dams. *Journal of Hydraulic Research*, 55(6), 799–812.

Hughes, S.A. (1993) *Physical models and laboratory techniques in coastal engineering*. Advanced Series on Ocean Engineering 7. World Scientific, London.

Hughes, S.A. (2004) Estimation of wave run-up on smooth, impermeable slopes using the wave momentum flux parameter. *Coastal Engineering*, 51(11–12), 1085–1104.

Hunt, B. (1988) Water waves generated by distant landslides. *Journal of Hydraulic Research*, 26(3), 307–322.

Hunt-Raby, A.C., Borthwick, A.G.L., Stansby, P.K. & Taylor, P.H. (2011) Experimental measurement of focused wave group and solitary wave overtopping. *Journal of Hydraulic Research*, 49(4), 450–464.

ICOLD (2002) Reservoir landslides: Investigation and management – Guidelines and case histories. *Bulletin*, 124. ICOLD, Paris.

Ikeno, M., Mori, N. & Tanaka, H. (2001) Experimental study on tsunami force and impulsive force by a drifter under breaking bore like tsunamis. Proceedings of the 48th Japanese Conf. *Coastal Engineering* JSCE, 50, 721–725.

International Centre for Integrated Mountain Development (ICIMOD) (2011) *Glacial lakes and glacial lake outburst floods in Nepal*. ICIMOD, Kathmandu. ISBN 978-9-291-15193-6.

Iribarren, C.R. & Nogales, C. (1949) Protection des ports [Harbour protection]. Proc. 17th International *Navigation Congress*, Lisbon, Section 2, Comm. 4, 31–80 (in French).

Isobe, M. (1985) Calculation and application of first-order cnoidal wave theory. *Coastal Engineering*, 9(4), 309–325.

Jervis, M. & Peregrine, D.H. (1996) Overtopping of waves at a wall: A theoretical approach. Proceedings of the 25th International Conference *Coastal Engineering*, 1(25), 2192–2205. ASCE, Reston, VA.

Johnson, J.W. & Bermel, K.J. (1949) Impulsive waves in shallow water as generated by falling weights. *Transactions of American Geophysics Union*, 30(2), 223–230.

Kamphuis, J.W. & Bowering, R.J. (1972) Impulse waves generated by landslides. Proceedings of the 12th *Coastal Engineering Conference,* Washington, DC, 1, 575–588.

Keulegan, G.H. (1950) Wave motion. In: Rouse, H. (ed) *Engineering hydraulics.* Wiley, New York, pp. 711–768.

Klimeš, J., Novotný, J., Novotná, I., de Urries, B.J., Vilímek, V., Emmer, A., Strozzi, T., Kusák, M., Rapre, A.C., Hartvich F. & Frey, H. (2016) Landslides in moraines as triggers of glacial lake outburst floods: Example from Palcacocha Lake (Cordillera Blanca, Peru). *Landslides,* 1–17. DOI:10.1007/s10346-016-0724-4

Kobel, J., Evers, F.M. & Hager, W.H. (2017) Impulse wave overtopping at rigid dam structures. *Journal of Hydraulic Engineering,* 143(6), 0401 7002.

Körner, H.J. (1976) Reichweite und Geschwindigkeit von Bergstürzen und Fliessschneelawinen [Distance and speed of rock falls and avalanches]. *Rock Mechanics,* 8(4), 225–256 (in German).

Korteweg, D.J. & de Vries, G. (1895) On the change of form of long waves advancing in a rectangular canal, and on a new type of long stationary waves. *Philosophical Magazine* Ser. 5, 39, 422–443.

Laitone, E.V. (1959) Water waves. IV, Shallow water waves. *Technical Report* 82–11. University of California, Berkeley.

Laitone, E.V. (1960) The second approximation to cnoidal and solitary waves. *Journal of Fluid Mechanics,* 9, 430–444.

Lauber, G. & Hager, W.H. (1996) Experiments to dambreak wave: Horizontal channel. *Journal of Hydraulic Research,* 36(3), 291–307.

Law, L. & Brebner, A. (1968) On water waves generated by landslides. Proceedings of the 3rd Australasian Conference on *Hydraulics and Fluid Mechanics,* Sydney, pp. 155–159.

Le Mehauté, B. (1976) *An introduction to hydrodynamics and water waves.* Springer, New York.

Li, Y. & Raichlen, F. (2001) Solitary wave runup on plane slopes. *Journal of Waterway, Port, Coastal, and Ocean Engineering,* 127(1), 33–44.

Lin, T.-C., Hwang, K.-S., Hsiao, S.-C. & Yang, R.-Y. (2011) An experimental observation of a solitary wave impingement, run-up and overtopping on a seawall. *Journal of Hydrodynamics,* Ser. B, 24(1), 76–85.

Liu, P.L.-F., Synolakis, C.E. & Yeh, H.H. (1991) Report on the International Workshop on long-wave run-up. *Journal of Fluid Mechanics,* 229, 675–688.

Liu, P.L.-F., Wu, T.-R., Raichlen, F., Synolakis, C.E. & Borrero, J.C. (2005) Runup and rundown generated by three-dimensional sliding masses. *Journal of Fluid Mechanics,* 536, 107–144.

Mader, C.L. & Gittings, M.L. (2002) Modeling the 1958 Lituya Bay mega-tsunami, II. *Science of Tsunami Hazards,* 20(5), 241–245.

McCowan, J. (1894) On the highest wave of permanent type. *Philosophical Magazine* Ser. 5, 38, 351–358.

McFall, B.C. & Fritz, H.M. (2016) Physical modelling of tsunamis generated by three-dimensional deformable granular landslides on planar and conical island slopes. *Proceedings of the Royal Society* A, 472(2188), 20160052. https://doi.org/10.1098/rspa.2016.0052

McFall, B.C. & Fritz, H.M. (2017) Runup of granular landslide-generated tsunamis on planar coasts and conical islands. *Journal of Geophys. Research Oceans,* 122(8), 6901–6922. https://doi.org/10.1002/2017JC012832

Miche, R. (1951) Le pouvoir réfléchissant des ouvrages maritimes [The force reflected by maritime structures]. *Annales des Ponts et Chaussées,* 121(May/June), 285–319 (in French).

Miller, D.J. (1960a) Giant waves in Lituya Bay. *Geological Survey, Professional Paper,* 354(C), 51–83.

Miller, D.J. (1960b) The Alaska earthquake of July 10, 1958: Giant wave in Lituya Bay. *Bulletin of the Seismological Society of America,* 50(2), 253–266.

Miller, R.L. (1968) Experimental determination of run-up of undular and fully developed bores. *Journal of Geophysical Research,* 73(14), 4497–4510.

Minikin, R.R. (1950) *Winds, waves and maritime structures.* C. Griffin, London.

Mohammed, F. (2010) Physical modeling of tsunamis generated by three-dimensional deformable granular landslides. *Ph.D. Thesis.* Georgia Institute of Technology, Atlanta GA.

Mohammed, F. & Fritz, H.M. (2012) Physical modeling of tsunamis generated by three-dimensional deformable granular landslides. *Journal of Geophysical Research*, 117, C11015. https://doi.org/10.1029/2011JC007850

Monaghan, J.J. & Kos, A. (2000) Scott Russell's wave generator. *Physics of Fluids*, 12(3), 622–630.

Monserrat, S., Vilibić, I. & Rabinovich, A.B. (2006) Meteotsunamis: atmospherically induced destructive ocean waves in the tsunami frequency band. *Natural Hazards and Earth System Sciences*, 6, 1035–1051.

Montes, S. (1998) *Hydraulics of open channel flow.* ASCE Press, Reston, VA.

Müller, C., Frank, P.-J. & Hager, W.H. (2016) Dyke overtopping: Effects of shape and headwater elevation. *Journal of Hydraulic Research*, 54(4), 410–422. DOI:10.1080/00221686.2016.1170072

Müller, D.R. (1995) Auflaufen und Überschwappen von Impulswellen an Talsperren [Run-up and overtopping of impulse waves at dams]. *Dissertation* 11113. Eidg. Technische Hochschule ETH, Zürich; also *VAW-Mitteilung* 137. VAW, ETH Zurich, Zürich. (in German).

Najafi-Jilani, A. & Ataie-Ashtiani, B. (2008) Estimation of near-field characteristics of tsunami generation by submarine landslide. *Ocean Engineering*, 35(5–6), 545–557.

Noda, E. (1970) Water waves generated by landslides. *Journal of the Waterways, Harbors and Coastal Engineering Division* ASCE, 96(WW4), 835–855.

Novak, P., Moffat, A.I.B., Nalluri, C. & Narayanan, R. (2001) *Hydraulic structures.* Spon, London.

Özhan, E. & Yalçıner, A.C. (1990) Overtopping of solitary waves at model sea dikes. Proceedings of the 22nd International Conference *Coastal Engineering*, 1(22), 1487–1498. ASCE, Reston, VA.

Panizzo, A. (2004) Physical and numerical modelling of subaerial landslide generated waves. *Ph.D. Thesis.* Università degli studi, L'Aquila, Italy.

Panizzo, A., Bellotti, G. & De Girolamo, P. (2002) Application of wavelet transform analysis to landslide generated waves. *Coastal Engineering*, 44(4), 321–338.

Panizzo, A., De Girolamo, P. & Petaccia, A. (2005) Forecasting impulse waves generated by subaerial landslides. *Journal of Geophysics Research*, 110 C12025, 1–23.

Pedersen, G. & Gjevik, B. (1983) Run-up of solitary waves. *Journal of Fluid Mechanics*, 135, 283–299.

Poli, P. (2017) Creep and slip: Seismic precursors to the Nuugaatsiaq landslide (Greenland). *Geophysical Research Letters*, 44, 8832–8836. https://doi.org/10.1002/2017GL075039

Przadka, A., Cabane, B., Pagneux, V., Maurel, A. & Petitjeans, P. (2012) Fourier transform profilometry for water waves: How to achieve clean water attenuation with diffusive reflection at the water surface? *Experiments in Fluids*, 52(2), 519–527. https://doi.org/10.1007/s00348-011-1240-x

Pullen, T., Allsop, N.W.H., Bruce, T., Kortenhaus, A., Schüttrumpf, H. & van der Meer, J.W. (2007) EurOtop – Wave overtopping of sea defences and related structures: assessment manual. In: German Coastal Engineering Research Council (KFKI) (ed) *Die Küste*, 73. Boyens Medien GmbH & Co. KG, Heide i. Holstein.

Quecedo, M., Pastor, M. & Herreros, M.I. (2004) Numerical modelling of impulse wave generated by fast landslides. *International Journal of Numerical Methods of Engineering*, 59(12), 1633–1656.

Quiroga, P.D. & Cheung, K.F. (2013) Laboratory study of solitary-wave transformation over bed-form roughness on fringing reefs. *Coastal Engineering*, 80(10), 35–48.

Raffel, M., Willert, C.E. & Kompenhans, J. (1998) *Particle Image Velocimetry: A practical guide.* Springer, Berlin.

Ramsden, J.D. (1996) Forces on a vertical wall due to long waves, bores, and dry-bed surges. *Journal of Waterway, Port, Coastal, and Ocean Engineering*, 122(3), 134–141.

Rana, B., Shrestha, A.B., Reynolds, J.M. & Aryal, R. (2000) Hazard assessment of the Tsho Rolpa Glacier Lake and ongoing remediation measures. *Journal of Nepal Geological Society*, 22, 563–570.

Rayleigh, L. (1876) On waves. *Philosophical Magazine* Ser. 5, 1, 257–279.

Reeve, D., Chadwick, A. & Fleming, C. (2004) *Coastal engineering: Processes, theory and design practice.* Taylor & Francis, London.

Risley, J.C., Walder, J.S. & Denlinger, R.P. (2006) Usoi dam wave overtopping and flood routing in the Bartang and Panj rivers, Tajikistan. *Natural Hazards*, 38(3), 375–390. https://doi.org/10.1007/s11069-005-1923-9

Roberts, N.J., McKillop, R.J., Lawrence, M.S., Psutka, J.F., Clague, J.J., Brideau, M.-A. & Ward, B.C. (2013) Impacts of the 2007 landslide-generated tsunami in Chehalis Lake, Canada. In: Margottini, C., Canuti, P. & Sassa, K. (eds) *Landslide Science and Practice*, 6, 133–140. https://doi.org/10.1007/978-3-642-31319-6_19

Roberts, N.J., McKillop, R., Hermanns, R.L., Clague, J.J. & Oppikofer, T. (2014) Preliminary global catalogue of displacement waves from subaerial landslides. In: Sassa, K., Canuti, P. & Yin, Y (eds) *Landslide Science for a Safer Geoenvironment*, 3, 687–692. https://doi.org/10.1007/978-3-319-04996-0_104

Russell, J.S. (1837) *Report* of the Committee on Waves. Report of the 7th Meeting of the British Association for the Advancement of Science Liverpool. Murray, London, pp. 417–496.

Saelevik, G., Jensen, A. & Pedersen, G. (2013) Runup of solitary waves on a straight and a composite beach. *Coastal Engineering*, 77(7), 40–48.

Sainflou, G. (1928) Essai sur les digues maritimes verticales [Essay on vertical maritime dikes]. *Annales des Ponts et Chaussées*, 98, 5–48 (in French).

Sander, J. & Hutter, K. (1992) Evolution of weakly non-linear shallow water waves generated by a moving boundary. *Acta Mechanica*, 91, 119–155.

Sarginson, E.J. (1972) The influence of surface tension in weir flow. *Journal of Hydraulic Research*, 10(4), 431–446.

Savage, S.B. & Hutter, K. (1989) The motion of a finite mass of granular material down a rough incline. *Journal of Fluid Mechanics*, 199, 177–214.

Schaub, Y., Haeberli, W., Huggel, C., Künzler, M. & Bründl, M. (2013) Landslides and new lakes in deglaciating areas: A risk management framework. In: Margottini, C. *et al.* (eds) *Landslide Science and Practice*, 7, Springer, Berlin Heidelberg, pp. 31–38. DOI:10.1007/978-3-642-31313-4_5

Scheidegger, A.E. (1973) On the prediction of the reach and velocity of catastrophic landslides. *Rock Mech.* 5, 231–236.

Schmocker, L. (2011) Hydraulics of dike breaching. *Ph.D. Thesis* No. 19983. ETH Zurich, Zürich, Switzerland. DOI:10.3929/ethz-a-006716949; also *VAW-Mitteilung* 218. VAW, ETH Zurich, Zürich.

Schmocker, L. & Hager, W.H. (2012) Plane dike-breach due to overtopping: Effects of sediment, dike height and discharge. *Journal of Hydraulic Research*, 50(6), 576–587. DOI:10.1080/00221686.2012.713034

Schmocker, L., Frank, P.-J. & Hager, W.H. (2014) Overtopping dike-breach: Effect of grain size distribution. *Journal of Hydraulic Research* 52(4), 559–564. DOI:10.1080/00221686.2013.878403

Schneider, D., Huggel, C., Cochachin, A., Guillén, S. & García, J. (2014) Mapping hazards from glacier lake outburst floods based on modelling of process cascades at Lake 513, Carhuaz, Peru. *Advances in Geosciences*, 35, 145–155.

Schnitter, G. (1964) Die Katastrophe von Vaiont in Oberitalien [The catastrophe of Vaiont in Upper Italy]. *Wasser- und Energiewirtschaft*, 56(2/3), 61–69 (in German).

Schüttrumpf, H.F.R. & Oumeraci, H. (2005) Layer thickness and velocities of wave overtopping flow at seadikes. *Coastal Engineering*, 52(6), 473–495.

Schwaiger, H.F. & Higman, B. (2007) Lagrangian hydrocode simulations of the 1958 Lituya Bay tsunamigenic rockslide. *Geochemistry, Geophysics, Geosystems*, 8(7), Q07006.

Sepúlveda, S.A., Serey, A., Lara, M., Pavez, A. & Rebolledo, S. (2010) Landslides induced by the April 2007 Aysén Fjord earthquake, Chilean Patagonia. *Landslides*, 7(4), 483–492. https://doi.org/10.1007/s10346-010-0203-2

Shen, M.C. & Meyer, R.E. (1963) Climb of a bore on a beach. *Journal of Fluid Mechanics*, 16, 113–125.

Slingerland, R.L. & Voight, B. (1979) Occurrences, properties and predictive models of landslide-generated impulse waves. In: Voight, B. (ed) *Rockslides and Avalanches*, 2, 317–397.

Slingerland, R.L. & Voight, B. (1982) Evaluating hazard of landslide-induced water waves. *Journal of Waterway, Port, Coastal, Ocean Engineering*, 108(4), 504–512; 110(1), 111–113.

Stansby, P.K. (2003) Solitary wave run up and overtopping by a semi-implicit finite-volume shallow-water Boussinesq model. *Journal of Hydraulic Research*, 41(6), 639–647.

Stokes, G.G. (1847) On the theory of oscillatory waves. *Transactions of Cambridge Philosophical Society*, 8, 441–455.

Su, C.H. & Mirie, R.M. (1980) On head-on collisions between two solitary waves. *Journal of Fluid Mechanics*, 98, 509–525.

Synolakis, C.E. (1987) The runup of solitary waves. *Journal of Fluid Mechanics*, 185, 523–545.

Synolakis, C.E. (1988) Are solitary waves the limiting waves in long wave runup? *Coastal Engineering Proceedings*, 21, 219–233. https://doi.org/10.1061/9780872626874.015

Tang, J., Causon, D., Mingham, C. & Qian, L. (2013) Numerical study of vegetation damping effects on solitary wave run-up using the nonlinear shallow water equations. *Coastal Engineering*, 75(1), 21–28.

Tanimoto, K., Tsuruya, K. & Nakano, S. (1984) Tsunami force of Nihonkai-Chubu Earthquake in 1983 and cause of revetment damage. Proceedings of the 31st Japanese Conference *Coastal Engineering* JSCE, pp. 257–261.

Teng, M.H., Feng, K. & Liao, T.I. (2000) Experimental study on long wave run-up on plane beaches. In: Chung, J.S., Olagnon, M. & Kim, C.H. (eds) Proceedings of the 10th International Conference *Offshore and Polar Engineering*, Seattle, 3, 660–664. ISOPE, California, USA.

Ursell, F. (1953) The long-wave paradox in the theory of gravity waves. *Proceedings of the Cambridge Philosophical Society*, 49(4), 685–694.

Vischer, D.L. (1986) Rockfall-induced waves in reservoirs. *Water Power and Dam Construction*, 38(9), 45–48.

Vischer, D., Funk, M. & Müller, D. (1991) Interaction between a reservoir and a partially flooded glacier: problems during the design stage. Proceedings of the 18th *ICOLD Congress*, Vienna, Q64(R8), 113–135.

Vuichard, D. & Zimmermann, M. (1987) The 1985 catastrophic drainage of a moraine-dammed lake, Khumbu Himal, Nepal: Cause and consequences. *Mountain Research and Development*, 7(2), 91–110. DOI:10.2307/3673305

Walder, J.S., Watts, P., Sorensen, O.E. & Janssen, K. (2003) Tsunamis generated by subaerial mass flows. *Journal of Geophysical Research*, 108(B5), 2,236 2–1–2–19.

Wang, J., Ward, S.N. & Xiao, L. (2015) Numerical simulation of the December 4, 2007 landslide-generated tsunami in Chehalis Lake, Canada. *Geophysical Journal International*, 201(1), 372–376. https://doi.org/10.1093/gji/ggv026

Watts, P. (2000) Tsunami features of solid block underwater landslides. *Journal of Waterway, Port, Coastal, and Ocean Engineering*, 126(3), 144–152.

Watts, P. & Grilli, S.T. (2003) Underwater landslide shape, motion, deformation, and tsunami generation. Proceedings of the 13th *International Offshore and Polar Engineering Conference*, Honolulu, pp. 364–371.

WCHL (1970) Hydraulic model studies: Wave action generated by slides into Mica Reservoir, British Columbia. *Report*. Western Canada Hydraulic Laboratories, Vancouver.

Weiss, R., Fritz, H.M. & Wünnemann, K. (2009) Hybrid modeling of the mega-tsunami runup in Lituya Bay after half a century. *Geophysics Research Letters*, 36 L09602, 1–6.

Weiss, R. & Wünnemann, K. (2007) Understanding tsunami by landslides as the next challenge for hazard, risk and mitigation: Insight from multi-material hydrocode modeling. *Eos Transactions of American Geophysical Union*, 88(52), S51C-06. San Francisco CA.

Wiegel, R.L. (1955) Laboratory studies of gravity waves generated by the movement of a submerged body. *Transactions of American Geophysical Union*, 36(5), 759–774.

Wiegel, R.L. (1960) A presentation of cnoidal wave theory for practical application. *Journal of Fluid Mechanics*, 7, 273–286.

Wiegel, R.L. (1964) *Oceanographical engineering.* Prentice-Hall, London.

Wiegel, R.L., Noda, E.K., Kuba, E.M., Gee, D.M. & Tornberg, G.F. (1970) Water waves generated by landslides in reservoirs. *Journal of Waterways, Harbors and Coastal Engineering Division* ASCE, 96(WW2), 307–333; 97(WW2), 417–423; 98(WW1), 72–74.

Xiao, H., Huang, W. & Tao, J. (2008) Numerical modeling of wave overtopping a levee during Hurricane Katrina. *Computers & Fluids*, 38(5), 991–996.

Yalin, M.S. (1971) *Theory of hydraulic models.* Macmillan, London UK.

Yarde, A.J., Banyard, L.S. & Allsop, W. (1996) Reservoir dams: Wave conditions, wave overtopping and slab protection. *Technical Report*, HR Wallingford.

Yeh, H.H. (1991) Tsunami bore runup. *Natural Hazards*, 4(2–3), 209–220.

Zelt, J.A. (1991) The run-up of nonbreaking and breaking solitary waves. *Coastal Engineering*, 15(3), 205–246.

Zelt, J.A. & Raichlen, F. (1991) Overland flow from solitary waves. *Journal of Waterway, Port, Coastal, and Ocean Engineering*, 117(3), 247–263.

Zweifel, A. (2004) Impulswellen: Effekte der Rutschdichte und der Wassertiefe [Impulse waves: Effects of slide density and still water depth]. *Ph.D. Thesis* 15596. ETH Zurich, Zürich; also *VAW-Mitteilung* 186. VAW, ETH Zurich, Zürich (in German).

Zweifel, A., Hager, W.H. & Minor, H.-E. (2006) Plane impulse waves in reservoirs. *Journal of Waterway, Port, Coastal, Ocean Engineering*, 132(5), 358–368.

Bibliography

Abdul Khader, M.H., Rai, S.P. & Yong, D.M. (1991) An experimental study of wave runup on steep curvilinear slopes. *Journal of Hydraulic Research*, 29(3), 403–415; 30(3), 423–427.

Bertacchi, P., Fanelli, M. & Maione, U. (1989) Une vue d'ensemble des problèmes concernant l'alerte hydraulique causée par le barrage naturel et par le lac qui s'est formé par suite de l'éboulement de Val Pola [An overall view of the problems concerning the hydraulic alert caused by the natural dam and the lake formed after the landslide of Val Pola]. *La Houille Blanche*, 44(5), 376–386 (in French).

Chang, K.-A. & Liu, P.L.-F. (1999) Experimental investigation of turbulence generated by breaking waves in water of intermediate depth. *Physics of Fluids*, 11(11), 3390–3400.

Chaudhry, M.H., Mercer, A.G. & Cass, D. (1983) Modeling of slide-generated waves in a reservoir. *Journal of Hydraulic Engineering*, 109(11), 1505–1520.

Das, M.M. & Wiegel, R.L. (1972) Waves generated by horizontal motion of a wall. *Journal of Waterways, Harbors and Coastal Engineering Division* ASCE, 98(WW1), 49–65.

Davidson, D.D. & McCartney, B.L. (1975) Water waves generated by landslides in reservoirs. *Journal of the Hydraulics Division* ASCE, 101(HY12), 1484–1501.

Dean, R.G. & Dalrymple, R.A. (1991) *Water wave mechanics for engineers and scientists.* World Scientific Publishing Company.

Debnath, L. (1994) *Nonlinear water waves.* Academic Press, Boston.

Eie, J., Solberg, G., Tvinnereim, K. & Tørum, A. (1971) Waves generated by landslide under arctic conditions. Proceedings of the 1st International Conference *Port and Ocean Engineering*, 1, 489–513.

Harbitz, C.B., Pedersen, G. & Gjevik, B. (1993) Numerical simulation of large water waves due to landslides. *Journal of Hydraulic Engineering*, 119(12), 1325–1342.

Hu, S.-L. & McCauley, J.L. (1997) Estimation of wave overtopping rates for irregular waves. *Journal of Waterway, Port, Coastal, and Ocean Engineering*, 123(5), 266–273.

Huber, A. (1980) Felsstürze in Seen und hierdurch erzeugte Schwallwellen: Erkenntnisse über die Entstehung und Ausbreitung dieser Wellen [Landslides into lakes and resultant impulse waves: Knowledge about the development and propagation of these waves]. Proceedings of *Interpraevent* Bad Ischl, 2, 57–68 (in German).

Huber, A. (1992) Auswirkungen von Massenstürzen und Lawinenniedergängen auf Stauhaltungen [Effects of landslides and avalanches on dams]. *Wasser, Energie, Luft*, 79(11/12), 309–313 (in German).

Huber, A. (1992) Der Val Pola Bergsturz im oberen Veltlin vom 28. Juli 1987 [The Val Pola Slide in Upper Valtelline of July 28, 1987]. *Eclogae Geologicae Helvetiae*, 85(2), 307–325 (in German).

Iversen, H.W. (1952) Laboratory study of breakers. *NBS Circular* 521 Gravity waves, pp. 9–32.

Jansen, P.C.M. (1986) Laboratory observations of the kinematics in the aerated region of breaking waves. *Ocean Engineering*, 9(5), 453–477.

Johns, B. (1980) The modeling of the approach of bores to a shoreline. *Coastal Engineering*, 3(3), 207–219.

Keller, H.B., Levine, D.A. & Whitham, G.B. (1960) Motion of a bore over a sloping beach. *Journal of Fluid Mechanics*, 7, 302–316.

Lemos, C.M. & Martins, M.L. (1996) A numerical study of the impact of solitary waves on obstacles of simple geometrical configuration. In: Rahman, M. & Brebbia, C.A. (eds) Proceedings of the 1st International Conference *Advances in Fluid Mechanics*. Computational Mechanics Publications, London, pp. 113–123.

Longuet-Higgins, M.S. (1973) A model of separation at the free surface. *Journal of Fluid Mechanics*, 57, 129–148.

Longuet-Higgins, M.S. & Turner, J.S. (1974) An entraining plume model of a spilling breaker. *Journal of Fluid Mechanics*, 63, 1–20.

Longuet-Higgins, M.S. (1983) Wave set-up, percolation and undertow in the surf zone. *Proceedings of the Royal Society* A, 390, 283–291.

Madsen, P.A. & Svendsen, I.A. (1983) Turbulent bores and hydraulic jump. *Journal of Fluid Mechanics*, 129, 1–25.

Mohapatra, P.K., Murty Bhallamudi, S. & Eswaran, V. (2000) Numerical simulation of impact by bores against inclined walls. *Journal of Hydraulic Engineering*, 126(12), 942–945.

Müller, D. & Huber, A. (1992) Auswirkungen von Schwallwellen auf Stauanlagen [Effects of impulse waves on dams]. *Wasser, Energie, Luft*, 84(5/6), 96–100 (in German).

Müller, D. & Vischer, D. (1996) Bemessungsansätze für das Auflaufen und Überschwappen von Impulswellen an Talsperren [Design guidelines for run-up and overtopping of impulse waves at dams]. *Wasserwirtschaft*, 86(11), 560–564 (in German).

Nakamura, M., Shiraishi, H. & Sasaki, Y. (1969) Hydraulic characteristics of tsunami acting on dikes. Proceedings of the 13th *IAHR Congress*, Kyoto, 3, 45–59. IAHR, Delft.

Neuhauser, E. (1979) Modellversuche über die Wirkung von Schwallwellen am Staudamm Gepatsch [Model tests to study the effect of surge waves on the Gepatsch rockfill dam]. *Oesterr. Wasserwirtschaft*, 31(5/6), 191–201 (in German).

Noda, E.K. (1971) Water waves generated by a local surface disturbance. *Journal of Geophysical Research*, 76(30), 7389–7400.

Nola, A. (1968) Spinte delle onde su pareti verticali e inclinate [Thrusts by waves on vertical and inclined walls]. *L'Acqua*, 46(3), 1–13 (in Italian).

Papanicolaou, P. & Raichlen, F. (1987) Wave characteristics in the surf zone. In: Dalrymple, R.A. (ed) Proceedings of the Conference *Coastal Hydrodynamics*, Newark. ASCE, New York, pp. 765–780.

Pouliquen, O. (1999a) Scaling laws in granular flows down rough inclined planes. *Physics of Fluids*, 11(3), 542–548.

Pouliquen, O. (1999b) On the shape of granular fronts down rough inclined planes. *Physics of Fluids*, 11(7), 1956–1958.

Prins, J.E. (1958) Characteristics of waves generated by a local disturbance. *Transactions of AGU*, 39(5), 865–874.

Pugh, C.A. & Harris, D.W. (1982) Prediction of landslide-generated water waves. Proceedings of the 14th *ICOLD Congress*, Rio de Janeiro, Q54(R20), 283–316.

Ramsden, J.D. & Raichlen, F. (1990) Forces on vertical wall caused by incident bores. *Journal of Waterway, Port, Coastal, and Ocean Engineering*, 116(5), 592–613.

Raney, D.C. & Butler, H.L. (1976) Landslide generated water wave model. *Journal of the Hydraulics Division* ASCE, 102(HY9), 1269–1282.

Sibul, O. (1955) Flow over reefs and structures by wave action. *Transactions of AGU*, 36(1), 61–63.

Stevanella, G. (1988) Evaluation of waves due to landslides. Proceedings of the 16th *ICOLD Congress*, San Francisco, Q63(R87), 1501–1513.

Stive, M.T.F. (1984) Energy dissipation in waves breaking on gentle slopes. *Coastal Engineering*, 8(2), 99–127.

Stoker, J.J. (1957) *Water waves.* Interscience, New York.

Svendsen, I.A. (1984) Wave heights and set-up in a surf zone. *Coastal Engineering*, 8(4), 303–329.

Synolakis, C.E. (1987) The runup and reflection of solitary waves. In: Dalrymple, R.A. (ed) Proceedings of the Conference *Coastal Hydrodynamics*, Newark. ASCE, New York, pp. 533–547.

Thomson, W. (Lord Kelvin) (1887) On the waves produced by a single impulse in water of any depth, or in a dispersive medium. *Proceedings of the Royal Society* A, 43, 80–83.

Thornton, E.R. (1979) Energetics of breaking waves within the surf zone. *Journal of Geophysical Research*, 84(C8), 4931–4938.

Titov, V.V. & Synolakis, C.E. (1995) Modeling of breaking and nonbreaking long-wave evolution and runup using VTCS-2. *Journal of Waterway, Port, Coastal, and Ocean Engineering*, 121(6), 308–316.

Volkart, P. (1974) Modellversuche über die durch Lawinen verursachten Wellenbewegungen im Ausgleichbecken Ferden im Lötschental [Model tests on the wave patterns at Ferden Basin in Lötschental due to avalanches]. *Wasser- und Energiewirtschaft*, 66(8/9), 286–292 (in German).

Volkart, P. (1975) Talsperren: Überschwappvorgänge infolge Lawinen- und Eisniedergängen in alpinen Staubecken [Dams: overtopping due to avalanches and ice falls at alpine dams]. Proceedings of *Interpraevent*, Innsbruck, pp. 255–270 (in German).

Watts, P. (1998) Wavemaker curves for tsunamis generated by underwater landslides. *Journal of Waterway, Port, Coastal, and Ocean Engineering*, 124(3), 127–137.

Wu, C.-S. (1987) The energy dissipation of breaking waves. In: Dalrymple, R.A. (ed) Proceedings of the Conference *Coastal Hydrodynamics*, Newark. ASCE, New York, pp. 740–750.

Zweifel, A., Zuccalà, D. & Gatti, D. (2007) Comparison between computed and experimentally generated impulse waves. *Journal of Hydraulic Engineering*, 133(2), 208–216.

Chapter 11 Frontispiece (a) Zeyzoun rockfill embankment Dam, Syria, 43 m high and nearly 5 km long. Its reservoir capacity was 71×10^6 m³. An 80 m wide breach formed on June 4, 2002 by overtopping due to flooding (published by T. Sakamoto and N. Yasuda 2009; courtesy © UNESCO-Encyclopedia of Life Support Systems (EOLSS) from Sakamoto, T., Yasuda, N. (2009). Monitoring and evaluating dams and reservoirs. *Water storage, transport and distribution*: 176–196, Y. Takahasi, ed. Eoloss Co. Ltd., Oxford, UK), (b) 21 Mile earthfill Dam failed on Feb. 8, 2017, near Montello, NV, USA releasing fast-moving water (courtesy Elko Daily Free Press, published on Feb. 10, 2017)

Chapter 11

Dam breach

11.1 Introduction

A dam failure may release large quantities of water creating major flood waves in the tail-water and causing serious damages. Singh (1996) noted 1000 dam failures since the 12th century, of which some 200 have occurred in the 20th century causing a loss of over 8000 lives and enormous damages. Even if the records are too limited for a statistical assessment, the annual probability of dam failure are theoretically estimated to 10^{-4}, so that the failure probability during a dam lifetime of 100 years is 10^{-2}.

Schnitter (1993) demonstrated by a data collection of ICOLD that the percentage of embankment dams failing in a certain year has dropped at least tenfold during the first half of the 20th century. Modern dams thus exhibit remarkable longevity, and have currently a good safety record compared with other hydraulic structures. Dam safety is still a large issue in dam engineering, however, so that the study of consequences of dam breaches is a basic requirement by authorities for various existing dams.

A dam and its reservoir may be threatened by factors including:

- Floods,
- Rockfalls and/or landslides,
- Earthquakes,
- Deterioration and instability of heterogeneous foundation,
- Poor quality of construction and construction materials,
- Differential settlements,
- Improper reservoir management, and
- Acts of war and terrorism.

The *causes* of dam failure include (Singh, 1996):

- 30% ($\pm$5%) floods exceeding spillway discharge capacity,
- 37% ($\pm$8%) geotechnical problems (seepage, piping, internal erosion, excess of pore pressures, fault movement, settlements),
- 10% ($\pm$5%) slides (earth, rock, glacier, avalanches), and
- 23% ($\pm$12%) improper design and construction, inferior material quality, acts of war, lack of operation and maintenance.

Foundation problems and overtopping count thus to the determining reasons for dam failures. Also, the failure probability is much greater for embankment than for concrete and masonry dams. After 1900, almost half of the failures were due to overtopping (ICOLD, 1973; Schnitter, 1993; Singh, 1996; Foster *et al.*, 2000). In 41% of those cases, the spillway was under-designed, whereas in 21% overtopping was caused by problems with spillway gate operation. Safety assessments of large dams are mainly based on deterministic considerations in which dam performances are evaluated for a number of scenarios of the aforementioned threats. Even if often limited by practical constraints, probabilistic approaches have been used more often recently since they have the potential to go further than deterministic methods by detecting weak parts of the dam system and accounting for the combination or cascading of events and threats (Matos *et al.*, 2018; Peter, 2017; Peter *et al.*, 2018a, b).

Each reservoir constitutes a potential hazard to the tailwater region. As witnessed in many instances, as on the Indus River valley in Pakistan, floods may be devastating for a densely populated region. The Kabul River with a basin of almost 100,000 km^2 experienced large-scale riverine and flash floods causing more than 1100 casualties. Given the number of dams erected in these valleys, additional potential for even higher flooding is present, so that the consequences of their failure must be assessed *prior* to dam erection based on a dam breach assessment and inundation mapping. Several basic methods are currently available, namely hydraulic laboratory studies to settle hydraulic questions, as presented below, as well as analytical, empirical, and numerical approaches, accounting for particular hydraulic, hydrologic, and geologic details of the dam environment, based on the validation of laboratory data. While both empirical and analytical models are strongly simplified requiring only few parameters, many physical processes are implied and numerous parameters need to be defined in numerical models, resulting in a high computational effort. Parameter models represent a trade-off between simplification and comprehensiveness, hence the computational cost and number of required parameters are at a corresponding level. The focus herein is on hydraulic laboratory studies.

Books summarizing unsteady flows including the dambreak problem and methods of solution were presented by Mahmood and Yevjevich (1975) in a series of three volumes, one of which is a bibliography including nearly 1900 citations with a short discussion of the approaches used (Miller and Yevjevich, 1975). Earlier accounts of note were written by Stoker (1957) counting to the early books on water waves in general, and dambreak flows in particular using computational approaches, and Roache (1972) with his book on computational fluid dynamics. A number of books were then published in the 1980s around Jean Cunge at Grenoble, France. These include Abbott (1979) on basic numerical approaches, Cunge *et al.*, (1980) on practical aspects in open-channel flows, Chaudhry (1993) on open-channel flow modeling, Toro and Clarke (1998) on numerical methods for wave propagation, and Toro (2001) on shock-capturing methods in hydraulic engineering. General accounts on dam breaches and dam safety include an overview of Novak *et al.*, (2001) also detailing instrumentation and surveillance, ICOLD (1998) with a state-of-the-art on the dambreak flood analysis, an account on dambreak floods by Fread (1996), a presentation of mainly American dams and the development of dam engineering in the United States by Kollgaard and Chadwick (1988), USBR (1983) with a safety evaluation of existing dams, or Jansen (1980) on Dams and public safety.

Given the risk potential of dams in terms of lives and infrastructure mainly in the tailwater, these hydraulic structures need a particular attention to inhibit any failure scenario under all circumstances. Although a dam failure or breach normally involves a local threatening, the consequences may be dramatic as witnessed in the past, including for instance at the 1976 Great Teton Dam Breach, or during the 2005 Katrina Disaster, both in the USA. It is therefore of key relevance to avoid a repetition of these and other similar scenarios, so that a special attention from the engineering community is required in the prediction and assessment of dam breaches. The purpose of this chapter is to highlight recent findings allowing to predict the formation of dam breaches and to propose adequate methods to counter them.

A distinction is made between the failures of concrete and embankment dams, involving instantaneous and progressive breaches, respectively. Whereas the former normally relates to arch or gravity dams of a certain height, the latter include a core sealing or a surface protection to retain water bodies. Given that embankments cannot be adequately modeled experimentally due to limitations in Froude scaling, as highlighted below, the purely granular dam composed of sediment of a certain limit size is used to represent its main features. The following relates to both the instantaneous and progressive dam breaches, given that both apply to hydraulic laboratory simulations if the scaling problems encountered are adequately considered and the simplifications critically accounted for.

11.2 Empirical breach data

11.2.1 Breach characteristics and examples

Dams fail either progressively or instantaneously. The type of failure depends on the cause of failure and the dam type. If a dam fails instantaneously, i.e. a large portion or even the entire dam is removed within a short time, then a sudden release of water generates a flood wave propagating into the tailwater. Because the baseflow can often be neglected, the dambreak wave propagates practically over a dry bed. A negative wave is created upstream propagating into the reservoir. Its topography controls the negative wave features. Since an instantaneous failure constitutes the most adverse hydraulic condition, it is commonly adopted for dambreak modeling and prescribes the upper bounds for the expected damages. Concrete dams failing by overturning or sliding are typical examples of *instantaneous failure*, to be dealt with in Section 11.5.

Earth or rockfill dam disasters are often due to *gradual failure* over a period of time. The failure duration lasts typically from minutes to hours, because the assumption of instantaneous failure is unrealistic for earth dams. Then no shock front is developed at the wave tip and the flow is gradually varied. This failure mechanism is dealt with in section 11.3.

An example of *instantaneous dam failure* occurred on Dec. 2, 1959, at *Malpasset Dam*, Southern France (James, 1988). The collapse after a foundation instability occurred as a sudden burst that destroyed practically the entire arch dam, resulting in the loss of several hundreds of lives and millions of dollars in property damage. The dam was located in a narrow gorge of Reyran River 12 km upstream of Fréjus, a resort community on the French Riviera. The extremely thin dam had a double-curvature arch 66 m high with a crest 222 m long. Its thickness varied from 6.8 m at the base to 1.5 m at the crest. The dam had an ungated spillway 30 m long; the 1.5 m diameter bottom outlet was controlled by a butterfly valve, and a reinforced-concrete apron was erected for downstream scour control.

Shears and faults are numerous in the dam foundation zone, orientated in the same directions as the jointing. The foundation was not grouted, except for the contact zone immediately below the concrete blocks. East of the river, the dam rested on a *wedge of gneiss* and it was not known that this wedge was detached from the underlying rock mass by two converging discontinuities. The geological investigations included site inspections by a professor of geology, eight grout holes 30 m deep, and examinations of the foundation excavation. This study was based on the supposition that a gravity dam would be constructed. It is unclear whether the geologist was duly informed when the design was changed to an arch dam.

Dam construction started in 1952 while reservoir filling began in late 1954. Exceptional floods in November 1959 filled the last 4 m of storage within three days. Some cracks were observed in the concrete apron a few weeks before dam failure. In the afternoon of the failure day, a group of engineers decided to open the bottom outlet to control the rising reservoir pool. At 6 p.m. the caretaker left the dam crest from a routine job. At 9.13 p.m. Electricité de France registered a power outage of the Malpasset 10 kV line. The first failure of a modern concrete arch dam had occurred catastrophically. The witness closest to the scene was about 1.6 km downstream reporting first feeling a ground trembling, followed by a loud, brief rumble, then a strong blast of air. Finally, the water arrived in two pulses, a wave that overtopped the stream banks and then a wall of water, which the witness barely escaped.

Figure 11.1a shows dam remnants. According to the dam designers *Coyne et Bellier*, the failure was due to secondary alteration of the rock at the left abutment. As the reservoir filled, the hydrostatic force increased against the 'underground dam' that had been created by compression of the rock abutment. Ultimately, a crack opened in the foundation along the upstream face of this barrier and the base of the rock wedge was exposed to full reservoir pressure. Unable to withstand this added force, the rock wedge slid outward and upward along the plane of the fault. The vertical component of this force raised the dam, causing a rotation intact as though hinged at its extremity at the right abutment (Figure 11.1a).

An example of a *progressive dam failure* occurred to Teton Dam (Idaho, USA), located in a steep-walled canyon at Rexburg Bench (Jansen, 1988). In the reservoir area consisting of volcanic rock, the permeability is high so that an appreciable percentage of the water stored is conveyed away, with significant recharge of the regional groundwater table. The dam has a compacted central core, zoned earth and gravel fill embankment. Its height above bedrock was 126 m. The dam crest was 950 m long, and nearly had a 3H:1V slope, with a total fill volume of 7.65×10^6 m^3.

Dam construction lasted from April 1972 to November 1975. Storage started in December 1975, with the auxiliary low-level outlet as tunnel under the right abutment in operation. The released discharge was 8.5 m^{3}s^{-1} until May 1976. Due to heavy spring runoff, the discharge was increased finally to 27 m^{3}s^{-1}, somewhat in excess of its rated capacity. The combined capacity of the outlet and the main river outlet works located under the left abutment was 120 m^{3}s^{-1}. The river outlet was incomplete, however, up to the time of dam failure, and hence was unavailable for participation in planned control of the reservoir filling schedule.

Teton Dam failed during the first filling at a reservoir depth of 84 m, 7 m below full supply level. Between 7 a.m. on June 5, 1976, when the initial damaging leaks were first seen in the right dam groin, and noon of that day, its total breach and failure occurred, starting with the appearance of muddy springs, followed quickly by piping through the embankment, and ending with the crest collapse into the rapidly enlarged 'pipe'. By 6 p.m. the reservoir was virtually empty, having released 3×10^8 m^3 water with a peak outflow of 28,000 m^{3}s^{-1}. It was concluded that the failure cause was the inadequate protection of parts of the impervious core material from internal erosion (Jansen, 1988).

A number of dam failures including details of causes and damages are summarized by Singh (1996). The above two examples may be considered extremes as relates to the dam type, but also to the scenarios encountered. Following these catastrophes, detailed investigations were undertaken to investigate the failure causes both in terms of damage compensation as also in terms of engineering knowledge for the future. Recent dam failures of

(a)

(b)

Figure 11.1 Remnants of (a) Malpasset Dam, France (Anonymous, 1959), (b) Teton Dam, USA (Anonymous, 1977)

embankment dams by overtopping or accidental leakage and processes for breach widening and deepening are described by Vogel *et al.*, (2014).

11.2.2 Breach characteristics and temporal breach development

Based on 52 historical dam failures, Singh (1996) summarizes the characteristics of dam breaches at embankment dams. The breach shape is approximated in all cases as trapezoidal, with a ratio between the top and the bottom widths of 1.29($\pm$0.18), and extreme values from 1.06 to 1.74. The ratio of top width B to breach depth d_b depends linearly on the ratio of H_s/H_d, where $H_s = V_s^{1/3}$ with V_s as the dam storage volume (Figure 11.2). For the dams mentioned $B/d_b = 0.40H_s/H_d$, provided the dam height $H_d > 8$ m (Figure 11.2a). The angle between the breach side slope and the vertical was typically between 40° and 50°. The failure time t_f was between 0.5 and 12 h. For most of the cases it was less than three hours, however. With a probability of 50%, the failure time was less than 1.5 h for the data of Singh (1996). For $1 < H_s/B_a < 10$, the failure time follows $(g/H_d)^{1/2} \cdot t_f = 1.5(H_s/B_a)$, with B_a as the average failure width. The relative peak discharge Q_p is related to the relative dam height as $Q_p/(gB_a^2H_d^3)^{1/2} = 1.25 \times 10^{-2}(H_s/H_d)$ for dams having failed after 1925 (Figure 11.2b). These relations apply for preliminary analysis of the dam breach characteristics, i.e. the breach geometry and the outflow hydrograph. Note that both the typical storage depth H_s and the dam height H_d are the significant parameters.

A dam breach is modeled as a two-phase water-sediment flow using the water-volume balance

$$A_s(H_H)\frac{\mathrm{d}H_H}{\mathrm{d}t} = Q_b \tag{11.1}$$

plus the erosion rate $\mathrm{d}z/\mathrm{d}t$ as a function of velocity across the breach

$$\frac{\mathrm{d}z}{\mathrm{d}t} = -\alpha_e u^\beta. \tag{11.2}$$

Here, A_s is the reservoir surface, H_H the headwater surface elevation from a reference datum, t is time and Q_b the breach discharge (Figure 11.3). Further $z = z(t)$ is the temporal

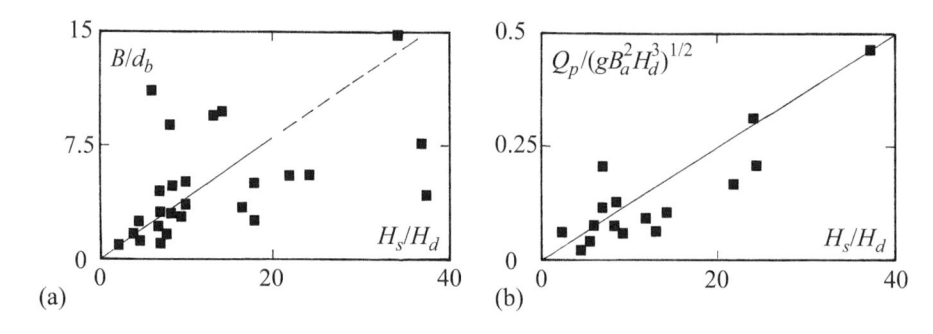

Figure 11.2 Empirical correlations for dam breach characteristics in terms of relative dam storage height H_s/H_d (a) relative top width B/d_b, (b) Relative peak discharge $Q_p/(gB_a^2H_d^3)^{1/2}$ (adopted from Singh, 1996)

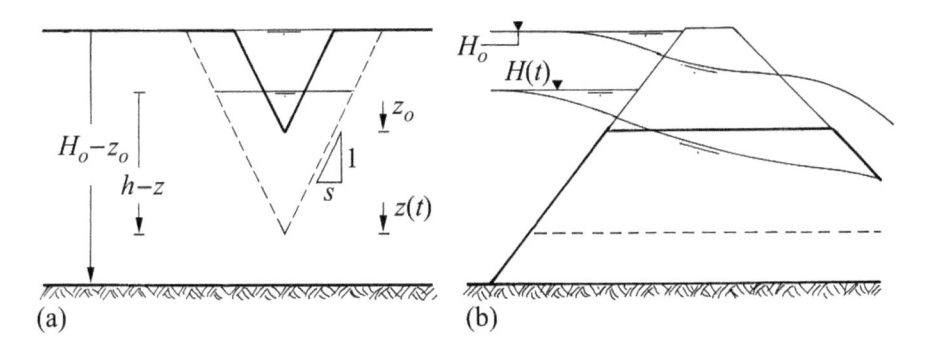

Figure 11.3 Definition of dam breach for embankment dams (a) front view, (b) streamwise section

development of the breach bottom, u the average breach velocity and α_e [ms^{-1}]$^{1-\beta}$ and β [–] are coefficients. Equation (11.1) holds if the difference between the reservoir in- and outflows over the spillway, the bottom outlet and the powerhouse are much smaller than the breach discharge. During the breach event, essentially all discharge flows across the breach section, therefore.

The breach discharge is equal to the breach velocity times the breach section, i.e. $Q_b = uA_b$, with $u = C_d[2g(H_H - z)]^{1/2}$, with C_d as the discharge coefficient. Eliminating the time differential gives the differential equation for $H_H(z)$ as

$$\frac{dH_H}{dz} = \alpha_e^{-1}(A_b / A_s)[2gC_d^2(H_H - z)]^{(1-\beta)/2}.$$ (11.3)

The breach shape is approximately trapezoidal as stated above, with a bottom breach width b and a side slope l(vertical):s(horizontal), i.e.

$$A_b = b(H_H - z) + s(H_H - z)^2.$$ (11.4)

For $s = 0$ the breach shape is rectangular, whereas for $b = 0$ it is triangular (Figure 11.3a). To reduce the number of parameters, and to simplify, only the latter case is considered here ($b = 0$). Inserting in Eq. (11.3) and using the transformation $h = H_H - z$ gives

$$1 + \frac{dh}{dz} = \frac{s}{\alpha_e A_s}(2gC_d^2)^{(1-\beta)/2} h^{(5-\beta)/2}.$$ (11.5)

This is solved subject to the two initial conditions $H_H = H_o$ and $z = z_o$ at $t = 0$. With the dimensionless parameters $H_H = h/h_o$ and $Z = z/z_o$, and the constants $\gamma = h_o/z_o > 1$ and $C^2 = \alpha_e A_s(2gC_d^2 h_o)^{(\beta-1)/2}/(sh_o^2)$, Eq. (11.5) reads

$$1 + \gamma\frac{dH_H}{dZ} = C^{-2}H_H^{(5-\beta)/2}.$$ (11.6)

For $H_H(Z = 1) = 1$, the solutions for the linear ($\beta = 1$) and the quadratic ($\beta = 2$) erosion rate laws, respectively, read with $\delta = C^{2/3}$ (Singh, 1996)

$$\beta = 1 \quad Z(H_H) = 1 - \frac{1}{2}\gamma C^2 \ln\left[\frac{C+H_H}{C-H_H} \cdot \frac{C-1}{C+1}\right] \tag{11.7}$$

$$\beta = 2 \quad Z(H_H) = 1 - 2\sqrt{3}\gamma \arctan\left[\frac{6\delta(H_H^{1/2}-1)}{3\delta^2 + (2H_H^{1/2}+\delta)(2+\delta)}\right] +$$

$$\frac{\gamma}{3\delta}\ln\left[\frac{H_H + \delta H_H^{1/2}+\delta^2}{(\delta - H_H^{1/2})^2} \cdot \frac{(\delta-1)^2}{1+\delta+\delta^2}\right]. \tag{11.8}$$

Figure 11.4a shows Eq. (11.7) for $C>1$. For $C>2$, the effect of C is small. The final drawdown level $Z(H_H = 0) = Z_\infty$ is plotted versus C^{-1} in Figure 11.4b.

The temporal evolution of the function $z(t)$ follows from Eq. (11.2) by eliminating h with Eq. (11.6), resulting in (Singh, 1996)

$$\alpha_e C_d (2g/h_o)^{1/2} t = -C^{1/2}\left[\frac{1}{2}\ln\left(\frac{C^{1/2}+H_H^{1/2}}{C^{1/2}-H_H^{1/2}} \cdot \frac{C^{1/2}-1}{C^{1/2}+1}\right)\right.$$

$$\left. + \arctan\left(\frac{(H_H^{1/2}-1)C^{1/2}}{H_H^{1/2}+C}\right)\right], \text{for } \beta = 1; \tag{11.9}$$

$$\text{and } 3\alpha_e gC_d^2 t = -\ln\left[\frac{H_H^{3/2}(1-C^2)}{H_H^{3/2}-C^2}\right], \text{for } \beta = 2. \tag{11.10}$$

To reach $H_H = 0$, infinite time is thus required. Also, the time increases as C reduces. For $C>2$, the effect of C is insignificant and $H_H = [1-(1/2)\alpha_e C_d(2g/h_o)^{1/2}t]^2$ for $\beta = 1$, whereas $H_H = \exp(-2\alpha_e gC_d^2 t)$ for $\beta = 2$, independent of both C and γ.

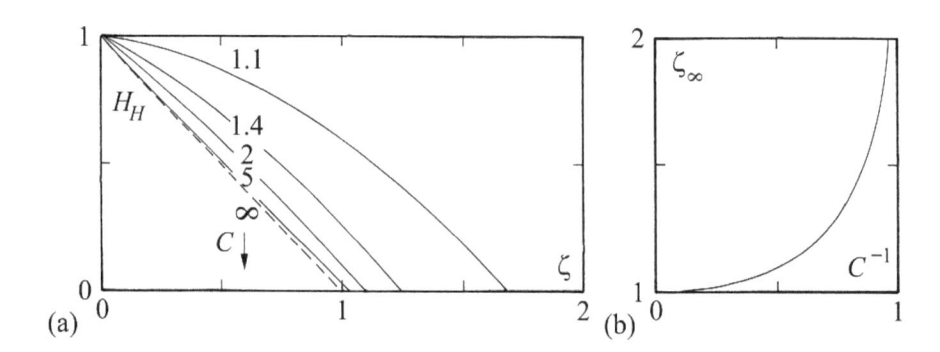

Figure 11.4 (a) Relation between $H_H(\zeta)$ with $\zeta = (Z-1)/(\gamma C)$ for various values of C according to Eq. (11.7), (b) final drawdown level $\zeta_\infty(C^{-1})$

This method was successfully applied by Singh (1996) to model the breach evolution of prototype dams. The reservoir surface was estimated as the reservoir volume divided by the reservoir filling height. The effect of the erosivity coefficient α_e is significant and both α_e and relative time vary linearly with C. It was deduced from a series of historical cases that α_e in the linear model ($\beta = 1$) is about one order of magnitude larger than for the quadratic erosion model ($\beta = 2$). The performance of the linear erosion rate was better than of the quadratic erosion model, but a definite data lack was stated to adequately describe the erosion process. From the available data, one may estimate $\alpha_e = 0.15(H_s/H_d)^{-2}$ for $\beta = 1$, and $\alpha_e = 4 \times 10^{-3}(H_s/H_d)^{-1}$ for $\beta = 2$, for dams having failed after 1920. The correlation for $\beta = 2$ is better than for $\beta = 1$.

According to Singh (1996), significant experimental and numerical work has to be completed in this field to understand the basic features of progressive breach modeling. His approach is considered the first semi-empirical model for the temporal evolution of dam breaches. A complete breach model would include knowledge of:

- Reservoir routing,
- Breach dynamics versus hydrodynamics and soil mechanics, and
- Downstream routing based on water and sediment hydrographs at dam site.

Downstream routing has to account for both the river and the floodplain, involving water depths, duration, and extent of inundation. Accordingly, the effects of a potential dam failure could then be predicted.

An alternative approach to that of Singh is proposed by Froehlich (2016). When assessing potential hazards of catastrophic flooding created by breached embankment dams, an appropriate inflow design flood needs to be selected to prepare emergency action plans. These breaches are often considered by assuming a trapezoidal-shape defined by its final height, base width or average width, and side slopes, along with the time needed for the breach to completely form. The data of 111 embankment dam failures were evaluated to obtain expressions for expected values of the final breach width, side slope, and formation time, along with expressions for the variances and prediction intervals of the parameters considered.

11.3 Progressive 2D breach

11.3.1 Introduction

An embankment of earth- or rockfill is designed and constructed to contain, control, or divert water for flows up to a certain return period. Embankments are designed for a defined discharge or a certain water level, for which no overtopping occurs, including freeboard due to wind waves, ice flow, excessive embankment settlings or discharge fluctuations. Following freeboard and discharge uncertainties, overtopping is a serious issue in terms of a breach. Although these risks are generally known, relatively few steps were so far taken from the hydraulic engineering community relating to research on the progressive breach formation. One of the reasons appears to be the complexity of the breach phenomenon, which is affected by both hydraulic and geotechnical processes. The breach process is influenced by seepage, overtopping erosion, sliding failure and various geotechnical and material characteristics. A systematic study of the plane breach processes due to overtopping using a hydraulic model is therefore of concern.

Despite the manifold construction types of embankment dams, they all share one common failure process, i.e. the development of a breach, i.e. the gradual removal of dam material induced by water flow involving mixed-regime flows, high sediment transport and fast morphological changes (ASCE/EWRI, 2011), hereafter referred to as breaching. The breach process can be subdivided into: (1) breach initiation, (2) breach formation, and (3) breach growth (Morris *et al.*, 2008). Various triggering events may cause different processes taking place during the breach initiation phase. The triggering events are categorized into (Peter, 2017):

Overtopping: Heavy rainfall or blockage of spilling structures may lead to rising the reservoir water level until the dam crest is reached and the dam body is overtopped. If the shear stress induced by the water flow exceeds a critical value dependent on the dam surface protection, a small channel across the dam crest may form.

Internal erosion: Seepage may lead to a pipe being formed across the dam body, also denoted as piping. If the pipe becomes sufficiently large, it collapses and the dam material above the pipe settles, often causing sink holes up to the dam crest.

Settling: Earthquakes, slope failure, or foundation problems may lead to a partial settling of the dam crest, similar to the piping event.

All three processes finally result in an incision (or 'notch') in the dam crest (Aufleger and López, 2016). The time needed to form this initial breach varies from seconds (in case of settlings) over hours (in case of overtopping) to days or even weeks (in case of piping, Peter, 2017). During breach initiation, discharge rates are still low and do not influence the hydrograph of a failing dam (Wahl, 2004). If the reservoir water level is still high enough to keep water flowing through the initial breach after the breach initiation phase, a further enlargement of the breach is possible. After developing a sufficiently large breach, the flow rate increases rapidly and the failure cannot be prevented. After the onset of the breach, the embankment starts to be eroded rapidly vertically by removal of dam material at the breach bottom and laterally by slumps of dam material falling into the breach. At some point the breach discharge reaches a maximum referred to as peak discharge. Although the discharge is now decreasing and the breach bottom has often reached the dam foundation, the breach can still grow laterally until the reservoir level has reached the level of the breach bottom (Vetsch and Boes, 2016).

Hydraulic model studies on progressive breaches due to overtopping differ with regard to the erosion process (2D or 3D), constant or falling reservoir levels, and the presence or absence of surface or core layers. All researches mainly determined the breach process, the breach profiles and the breach discharge. An overview on past hydraulic breach models due to overtopping was presented by Schmocker and Hager (2009) and Morris (2009). A summary of the most recent works is given below.

Chinnarasri *et al.*, (2003) observed four stages in plane erosion, namely: (1) Small crest erosion after initial overtopping, (2) slope sliding failure with ongoing erosion, (3) wavelike-shaped profile, and (4) large sediment wedge deposition with small slope at erosion end. Similar processes were also noted by Dupont *et al.*, (2007). They additionally observed sliding of the lower part of the downstream slope just before actual overtopping. The erosion progresses from the downstream face toward the dam crest, with the downstream face rotating around a pivot point. With ongoing erosion, anti-dunes are generated on the downstream face.

Coleman *et al.*, (2002) presented embankment-breach tests under constant reservoir level to describe the spatial breach process. Flow through a pilot channel located at the channel sidewall initially erodes a small breach channel on the embankment downstream face from the crest to the toe. The breach expands vertically and then laterally as the breach channel approaches the embankment foundation. Chinnarasri *et al.*, (2004) studied the breach geometry of homogenous embankments under falling reservoir level. They also observed primarily vertical erosion in the initial breach with subsequent lateral erosion after the breach had approached the fixed embankment foundation. Visser *et al.*, (2006) distinguished five stages in the process of overtopping breaches. During Stages I and II, the breach starts due to water entering the pilot channel gradually, increasing its width and height. In Stage III, the breach growth accelerates until the pilot channel is completely washed out at the breach section. The breach expands then mainly laterally in Stage IV and erosion decelerates in Stage V under decreasing backwater and hence decreasing breach discharge. Large-scale modeling was conducted by Hoeg *et al.*, (2004) and Løvoll (2006). Details on the dam construction, the experimental procedure and the dam material used are given first. Next, breaches by through flow, by overtopping, and by piping are detailed. It was found that the dams tested were found to be more resistant to breach initiation and failure than based on existing guidelines. However, once the erosion reaches the upstream corner of the dam crest, the breach formation was noted to be rapid. Løvoll (2006) assessed the governing breach mechanisms of up to 6 m high dams.

Schmocker and Hager (2009) focused on scale effects in laboratory breach tests along with model limitations. Gregoretti *et al.* (2010) conducted similar tests on the failure of homogenous dams built of coarse material on a sloping bed. Depending on the downstream dam face angle, they observed the three failure types overtopping, headcutting, and sliding of the downstream dam face. Pickert *et al.*, (2011) conducted spatial embankment-breach tests due to overtopping, dividing the failure of homogeneous embankments into two breach phases. Their erosion process depends on the sediment diameter and is especially influenced by apparent cohesion. The lateral erosion is described as a combination of continuous erosion and sudden collapse of the breach side slopes.

Schmocker and Hager (2012) investigated the effect of various parameters on the *plane* breach process due to overtopping. A simple model embankment of trapezoidal shape consisting of homogenous sand or gravel was considered to reduce the work effort and to limit the breach process to pure overtopping erosion. For incohesive sediment, the erodibility depends primarily on the grain size distribution, density and grain shape. Of particular interest is the temporal breach progress, the maximum breach height and the breach discharge. The main model parameters include the sediment diameter, the embankment height and the inflow discharge. The following aims to summarize the main findings based on systematic laboratory experimentation. Both 2D and 3D breach tests are described, thereby including the effects of additional parameters as compared with these listed above. The main features of 2D progressive breaches are more or less well understood, whereas these for the 3D breaches are more complex and deserve additional effort both from the experimental as also from the numerical sides. It should be stressed that all tests involved incohesive granular material by which upscaling to prototype conditions follows Froude similitude.

11.3.2 Hydraulic modeling

Model channel

The hydraulic test conditions of Schmocker and Hager (2012) include (Figure 11.5): (1) Trapezoidal model embankment, (2) homogenous, incohesive sediment, (3) homogenous

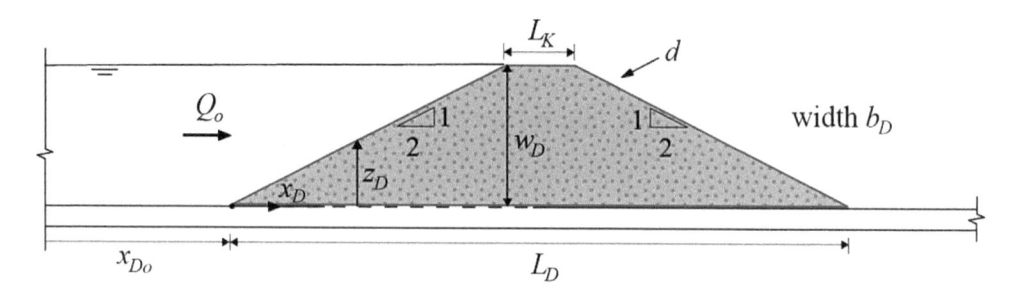

Figure 11.5 Streamwise view of 2D breach model, notation, with (– –) bottom drainage

embankment, (4) seepage control by bottom drainage, (5) steady inflow, falling reservoir level, and (6) optical recording across the lateral glass wall. The plane breach tests were conducted in a horizontal, glass-sided-breach channel. The intake of the 0.40 m wide channel was 0.66 m long, equipped with a flow straightener to generate undisturbed approach flow conditions. The eroded sediment was collected at the channel end.

Trapezoidal embankments (subscript D) were inserted at distances x_{Do}, with the origin of the coordinate system (x_D, y_D, z_D) at their upstream toe. The parameters investigated include: Inflow (subscript o) discharge $Q_o = 1$ to 64 l/s, height $w_D = 0.10$ to 0.40 m, width $b_D = 0.10$ to 0.40 m, crest (subscript K) length $L_K = 0.05$ to 0.20 m and uniform sediment of mean diameter $d = 0.31$ to 8.0 mm, of density $\rho_s \approx 2650$ kg/m^3 and uniformity coefficient $\sigma_s = (d_{84}/d_{16})^{0.5}$ < 1.2. Both the up- and downstream face slopes were $S_o = S_d = 1{:}2$ (V:H) resulting in a total embankment length of $L_D = 4w_D + L_K$.

To prevent seepage-induced failure prior to overtopping as e.g. sliding of the downstream face or erosion of the downstream toe, seepage through the embankment was reduced with a bottom drainage. A second PVC bottom was added to the original channel with drainage holes located along its upstream portion. Sediment washout was prevented using a fine wire net. As the upstream water level increased, a portion of the seepage discharge passed the drain so that sliding of the downstream slope was prevented. Seepage failure was not completely avoided for $d = 8.0$ mm due to the high sediment porosity. The drainage discharge was 3% of Q_o, reaching a maximum of 8%.

Test procedure and program

All tests were conducted using a constant approach flow discharge Q_o, referred to as the inflow. The dry material neither was neither compacted nor were surface or core layers added. The setting was accurate to one grain size. The inflow was added fast to attain steady state before overtopping started. The flow overtopped the embankment over its entire width thereby resulting in a 2D breach. The duration from erosion start to test end was between 500 and 1000 s, depending on d, w_D, and Q_o. The overflow was always free without any tailwater effect. Tests were stopped once the equilibrium erosion state was reached, i.e. erosion had stopped and the breach profile remained stable, or when the entire embankment was eroded. Given the short breach duration and the relatively high overflow discharges, no constant reservoir level was attained. The results are therefore limited to breach processes with a falling reservoir level.

The model used involved a frontal approach flow, mainly to reduce the laboratory effort and the test analysis. The parallel approach flow scenario as e.g. used by Roger *et al.* (2009) or Rifai *et al.* (2017) was not considered, therefore. Further, the constant inflow scenario was selected to obtain a basic and simple upstream boundary condition. In prototypes, overtopping occurs once the water level exceeds the crest elevation. The overtopping discharge depends on the breach progress and the inflow. The breach and inflow discharges interact so that there is hardly a constant inflow to the breach section. A precise upstream boundary condition, similar to Q_o, is therefore not attained in nature. However, the effect of inflow on the breach shape and the breach time is relevant in practice. The effect of a constant headwater elevation is discussed below.

A total of 31 tests were conducted to investigate the breach process and the breach discharge. The constant inflow discharge Q_o is represented by the critical flow depth $h_c = (Q_o^2/gb_D^2)^{1/3}$. Schmocker and Hager (2009) state the following limitations to prevent scale effects: (1) minimum height of 0.20 m and width for ease in laboratory embankment setting and to avoid errors due to too small lengths, (2) minimum grain size of $d = 1$ mm due to apparent viscosity, both (3) maximum grain size of $d = 5.5$ mm, and (4) minimum unit inflow of $q_o = 20$ l/(sm) due to viscous effects.

The breach process was recorded with a 30 Hz CCD camera, resulting in a length accuracy of ± 2 mm. The images were analyzed using a standard graphics program to determine the following parameters at various times t (Figure 11.6): Water surface $h_D(x,t)$, sediment surface $z_D(x,t)$, overflow depth $h_o(t)$, crest radius $R_D(t)$, up- and downstream face angles $\alpha_u(t)$ and $\alpha_d(t)$, and embankment volume $V_D(t)$ within $0 \leq x \leq L_D$. To compare the overall erosion process for different tests, dimensionless sediment surface profiles $Z_D(X_D)$ were considered with $X_D = x_D/L_D$ and $Z_D = z_D/w_D$. Test start $t = 0$ s was set when the upstream water level reached the embankment crest. The breach discharge Q was determined using the broad-crested weir formula (Chapter 2)

$$Q = C_d b_D \left(2gH_o^3\right)^{1/2}. \tag{11.11}$$

Here C_d = discharge coefficient, b_D = overflow width, g = gravity acceleration, and $H_o = h_o + Q^2/[2gb_D^2(h_o + w_D)^2]$ = approach flow energy head with h_o = overflow depth. Compared with the broad-crested weir, the breach profiles exhibit a rounded crest and sloping up- and

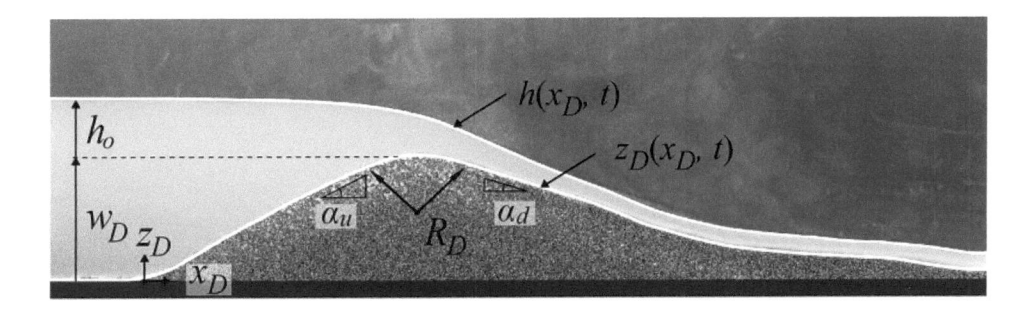

Figure 11.6 Measured data from camera images (Schmocker *et al.*, 2011)

downstream faces (Figure 11.6), of which both affect the discharge coefficient. The discharge coefficient of Schmocker *et al.*, (2011) is

$$C_d = \frac{2}{3\sqrt{3}}\left(1+\frac{3\rho_k'}{11+4.5\rho_k'}\right)$$

(11.12)

with

$$\rho_k' = \frac{H_o}{R_D}\left(\frac{\alpha_u+2\alpha_d}{270}\right)^{\frac{1}{3}}.$$

(11.13)

Effect of inflow discharge

The effect of discharge on the breach process is shown for Q_o = 4, 8 and 16 l/s with all other parameters kept constant, resulting in different longitudinal breach profiles (Figure 11.7). For Q_o < 4 l/s, the crest is only slightly rounded of almost triangular shape as compared with a wide round-crested breach shape for higher Q_o. For small Q_o, the eroded material deposits below the original embankment due to the small stream power. Therefore, the face slope downstream of the crest is almost constant. For higher Q_o, the eroded material is transported further downstream, resulting in a steeper downstream face angle and the typical rounded breach crest. Whereas the crest radius remains almost constant for Q_o = 4 l/s, it increases continuously with time for Q_o = 8 l/s and 16 l/s.

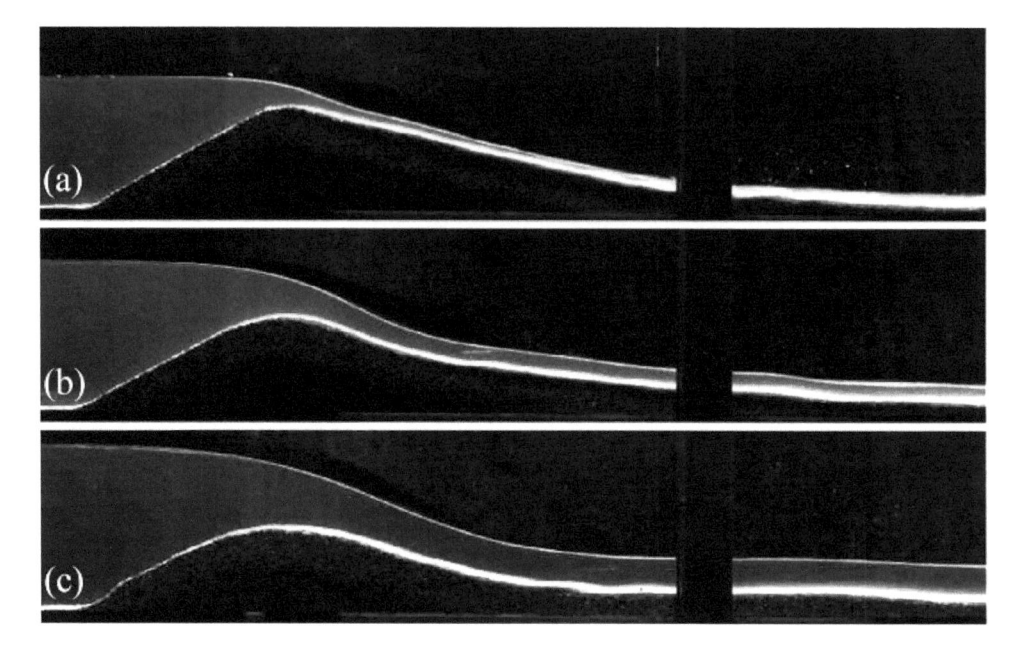

Figure 11.7 Effect of inflow discharge Q_o on streamwise breach profiles at t = 8 s (a) triangular breach for Q_o = 4 l/s, round-crested breach shapes for Q_o = (b) 8 l/s, (c) 16 l/s (Schmocker and Hager, 2012)

Effect of embankment height

The effect of the initial embankment height w_D on the breach process with all other parameters kept constant is evident. Due to the decreased volume, the erosion process for small w_D is faster. At $t = 6$ s, the erosion for $w_D = 0.15$ m and 0.20 m is advanced, whereas for $w_D = 0.30$ m it nearly keeps its original shape as the overtopping flow just causes small erosion of the downstream toe. All embankments then undergo continuous erosion with time. Although these with $w_D = 0.15$ m and 0.20 m are eroded faster than with $w_D = 0.30$ m, the relative height and slope of the deposited material is similar. At $t = 100$ s, the relative maximum height of all three breaches is nearly identical forming a tailwater wedge. The breach profiles therefore converge with time.

Effect of sediment diameter

Figure 11.8 shows the surface profiles $Z_D(X_D)$ at various times t for $d = 2$ mm, 4 mm and 5.5 mm with all other parameters kept constant. The breach process is divided into the initial phase up to $t \approx 10$ s and the secondary breach phase for $t > 10$ s. After $t = 2$ s, the breach profiles are identical as overtopping starts and the erosion of the downstream face is initiated. At $t = 4$ s and 6 s (Figure 11.8c, d), the breach profiles differ considerably, with the lowest elevation for $d = 5.5$ mm and the highest for 2 mm. During the initial breach phase, erosion is faster with increasing d. For $d = 2$ mm, the erosion proceeds uniformly as the grains are entrained and transported mostly by rolling. The downstream face is still stable despite water flow, whereas for $d = 5.5$ mm, the downstream face becomes unstable. Small sediment portions slide down the face with the coarser particles transported by rolling and sliding. The bottom drainage prevents sliding failure prior to overtopping but does not prevent seepage. As overtopping starts, the material saturation is higher for the coarser material. After $t = 4$ s, the downstream slope is thus smaller for $d = 5.5$ mm as compared with 2 mm.

At $t = 10$ s the breach profiles for the tested sediments start to converge and are nearly identical at $t = 20$ s (Figure 11.8e, f). The profiles then start to deviate again in the second phase, as the erosion is primarily governed by rolling sediment transport. The erosion proceeds slower for the coarse material given its higher erosive resistance. At $t = 50$ s and 100 s (Figure 11.8g, h), the breach profile for $d = 5.5$ mm has the highest elevation, whereas that for $d = 2$ mm has the lowest. All breach profiles remain stable, reaching the equilibrium stage after $t \approx 300$ s.

Effect of soil saturation

The uniform material with $d = 0.31$ mm (fine sand) behaves cohesively, resulting in a different breach profile compared with $d \geq 1$ mm. Figure 11.9 compares the temporal breach profiles $z_D(x_D)$ for (a) $d = 2$ mm and (b) $d = 0.31$ mm, a fine sand exhibiting cohesive pattern, with all other parameters kept constant. The erosion starts at the downstream crest for both sediment diameters. Then, the breach profiles for $d = 2$ mm show the typical round-shaped crest with the material deposited directly below the downstream toe. The downstream slope decreases with time. For $d = 0.31$ mm, no eroded material was deposited downstream of the embankment as the fine material was transported away from it. Thus, the downstream slope remains almost constant during the entire breach but is continuously eroded backward. Further, neither surface slips nor sliding on the downstream face were noted due to the soil-water content

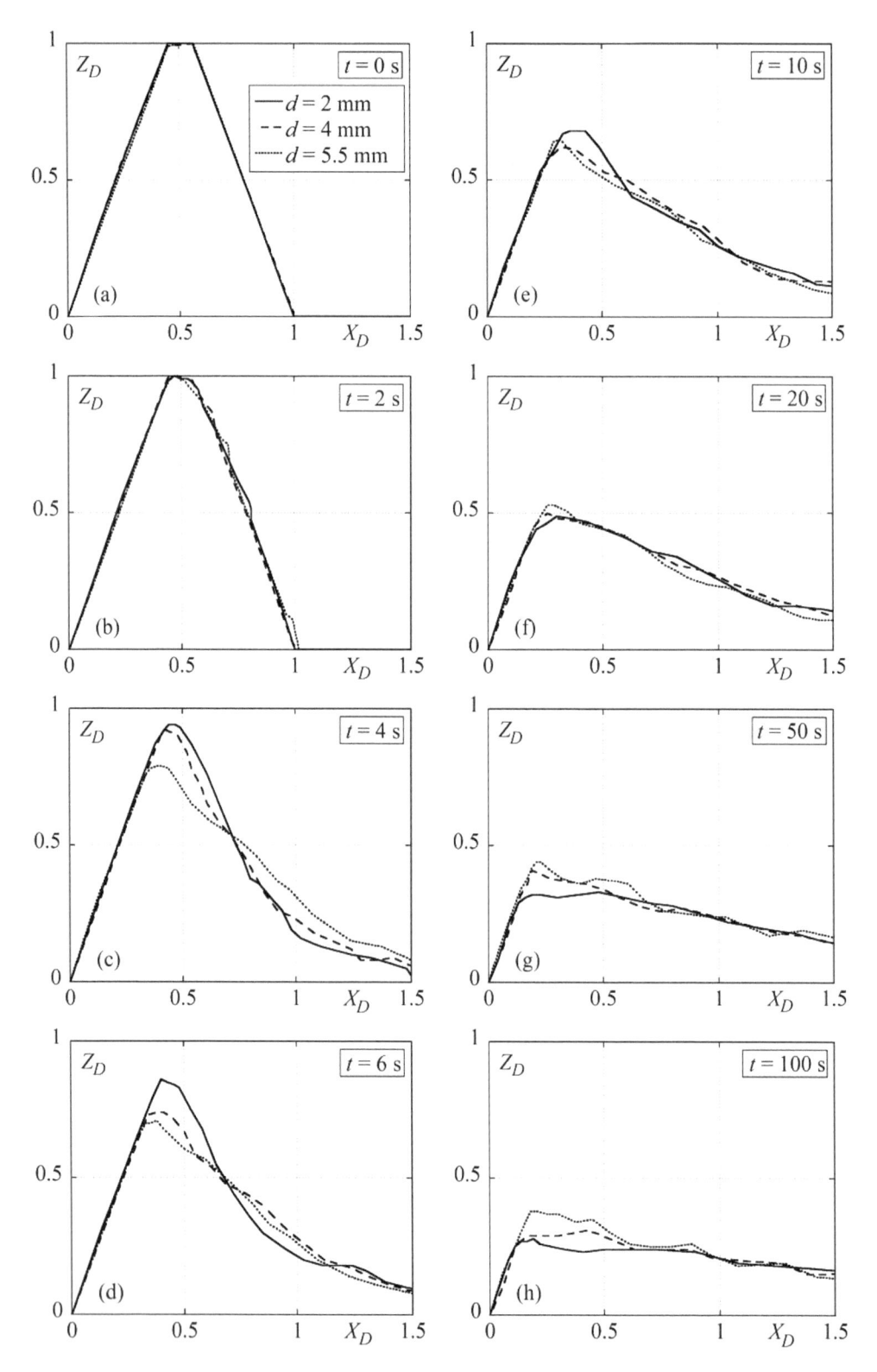

Figure 11.8 Effect of sediment diameter d on surface profiles $Z_D(X_D)$ at various times t for $d = 2$ mm, 4 mm and 5.5 mm (Schmocker and Hager, 2012)

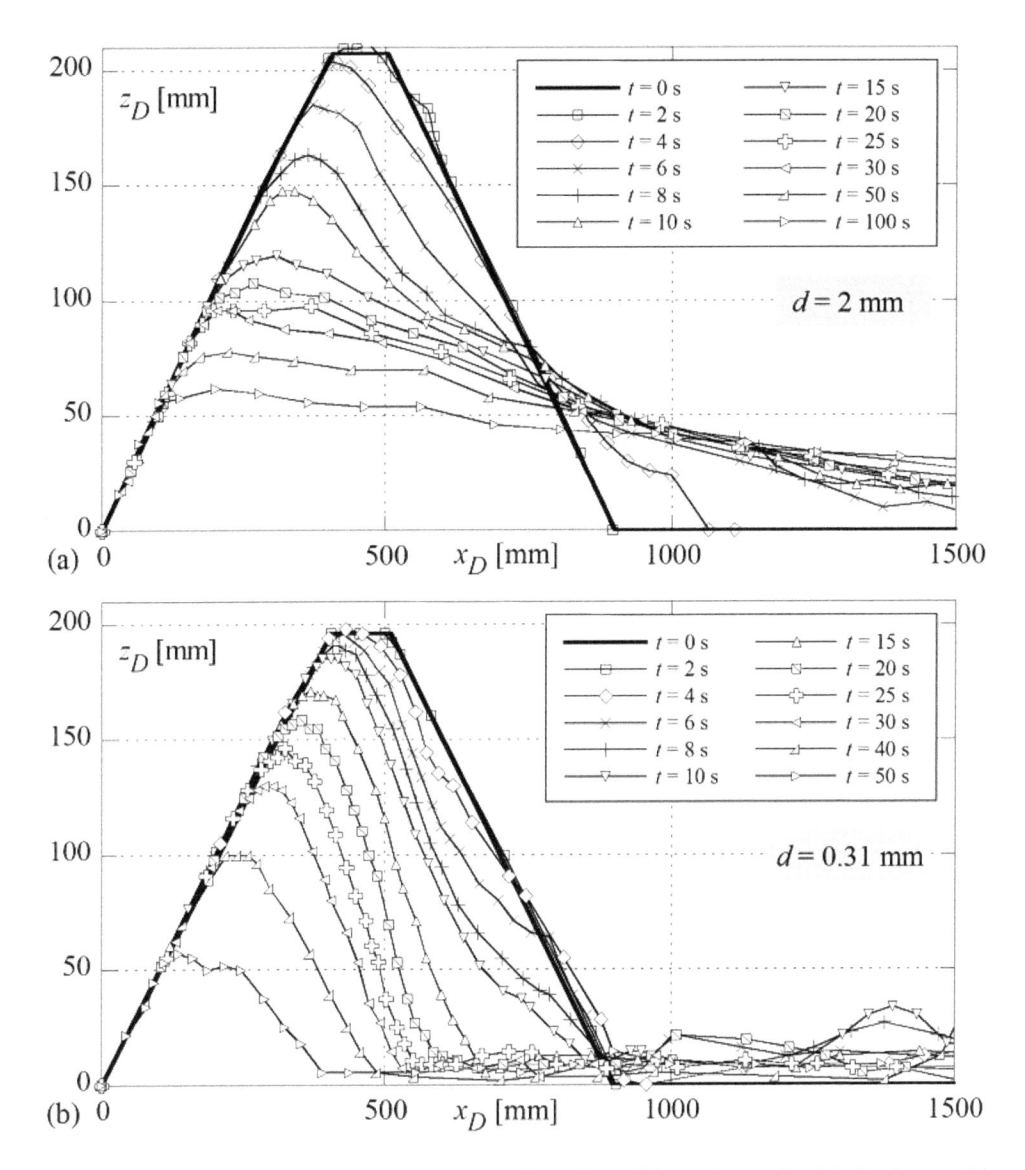

Figure 11.9 Effect of sediment diameter d on breach profiles $z_D(x_D, t)$ for (a) $d = 2$ mm, (b) $d = 0.31$ mm (Schmocker and Hager, 2012)

of the fine sand and the resulting effects of cohesion or tensile strength of which both vary with the degree of saturation. Its tensile strength stabilizes the downstream face.

During the breach with $d = 0.31$ mm, neither abrupt failure nor a spontaneous collapse of the sediment body resulted. The erosion advanced gradually with smooth erosion profiles. For cohesive material as clay, a partial collapse of the embankment must be expected due to shear failure and undercutting of the downstream slope (Powledge *et al.*, 1989). To avoid effects of apparent cohesion, the minimum sediment diameter required is $d \geq 1$ mm, as previously stated.

11.3.3 Normalized results

Governing parameters

To generalize the results, the governing parameters are described versus quantities accelerating the breach process in the numerator, and quantities decelerating the breach process in the denominator, i.e.

$$\text{Breach process} \sim \frac{h_c, x_D}{w_D, g', L_K, S_o, S_d, d, \sigma_s, \theta_s}. \tag{11.14}$$

The gravity acceleration and both the sediment and water densities are replaced by the submerged specific sediment gravity $g' = [(\rho_s - \rho)/\rho]g$. As fluid viscosity and surface tension were constant in all tests they are neglected, given that scale effects are excluded. Parameters L_K, g', S_o, S_d and x_D were not systematically varied during this first stage of experimentation, and are therefore dropped. As only uniform sediment was tested and the submerged angle of repose was similar for all sediments, both σ_s and θ_s are dropped as well. The effect of d on the breach process depends on the breach stage. For large sediment, the material saturation is higher due to increased seepage. Sliding sediment transport and small surface slips therefore increase for coarse material at test initiation. After this short initial phase, the erosion proceeds slower for coarse than for fine material given its higher erosive resistance. The governing parameters for the breach process considered here are, therefore,

$$\text{Breach process} \sim \frac{h_c}{w_D, d}. \tag{11.15}$$

The exponents of the three parameters were determined using regression analysis.

Maximum embankment height

For constant inflow, the maximum (subscript M) embankment height z_{DM} decreases fast after overtopping and then slows down with time as both the height and the energy slope decrease. The analysis of all test data indicates that z_{DM} decreases faster with: (1) increasing Q_o due to the high erosion potential, (2) decreasing embankment height due to the reduced volume to be eroded, and (3) decreasing sediment diameter due to increased erodibility. The maximum height reaches eventually the maximum equilibrium (subscript e) height z_{DMe} after time t_e. The maximum embankment height z_{DM} was normalized with its original height w_D and z_{DMe} as

$$Z_{DM} = \frac{z_{DM} - z_{DMe}}{w_D - z_{DMe}}. \tag{11.16}$$

All tests start therefore at $Z_{DM} = 1$ and tend to $Z_{DM} = 0$. A dimensionless time was introduced to account for the temporal erosion process. The effect of all breach parameters on the erosion process was assessed using regression analysis resulting in dimensionless time in terms of h_c, d, w_D as (Schmocker and Hager, 2012)

$$T = t(g')^{1/2} \frac{h_c}{d^{1/2} w_D}. \tag{11.17}$$

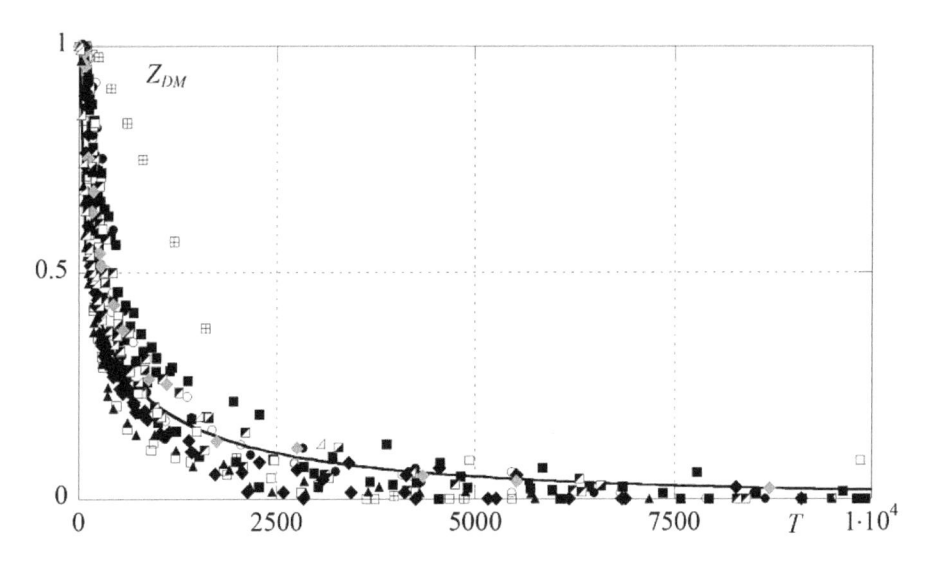

Figure 11.10 Dimensionless maximum embankment height $Z_{DM}(T)$, (–) Eq. (11.18), (⊞) $d = 0.31$ mm (Schmocker and Hager, 2012)

Figure 11.10 shows the dimensionless maximum height $Z_{DM}(T)$ along with the fit equation for $T \leq 10^4$ ($R^2 = 0.72$, Schmocker and Hager, 2012)

$$Z_{DM} = \exp\left(-0.10T^{0.4}\right). \tag{11.18}$$

For small Q_o, Z_{DM} remains constant after initial overtopping as the flow first passes along the crest before erosion starts at the downstream face. The larger L_K, the longer the maximum height remains constant after breach initiation. Therefore, the data trend is not exponential but rather follows a Gaussian, yet an exponential overall fit was selected as this effect applies only for small Q_o. Except for the described above test with $d = 0.31$ mm, involving apparent cohesion, all data follow well Eq. (11.18) when considering the numerous factors affecting the breach process. Note that the erosion process reaches the equilibrium stage and is completed ($Z_{DM} < 0.1$) at $T \cong 2500$, i.e. at $t > 2500(w_D/h_c)(d/g')^{1/2}$. The breach process is thus long for high embankments as compared with the critical depth, and, to a smaller extent, for large sediment size.

The final embankment height reaches the maximum equilibrium height z_{DMe}. Figure 11.11 shows the dimensionless expression $Z_{DMe} = z_{DMe}/w_D$ versus $h_c/(d^{2/3}w_D^{1/3})$. The fit equation for $2.5 \leq h_c/(d^{2/3}w_D^{1/3}) < 12$ is ($R^2 = 0.92$, Schmocker and Hager, 2012)

$$Z_{DMe} = \exp\left[-0.54\left(\frac{h_c}{d^{2/3}w_D^{1/3}}\right)^{2/3}\right]. \tag{11.19}$$

Besides inflow Q_o or h_c, the sediment diameter is a governing parameter, controlling the equilibrium stage by sediment transport. The maximum equilibrium height reduces with both decreasing d and w_D and increasing h_c. For small $h_c/(d^{2/3}w_D^{1/3})$, the embankment would remain in its original shape, i.e. $Z_{DMe} \to 1$, whereas it is completely eroded for very large

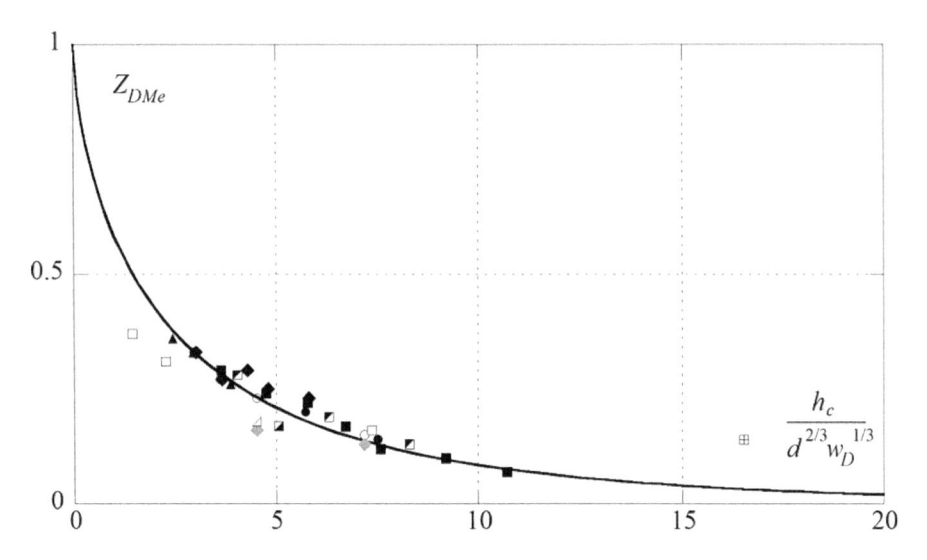

Figure 11.11 Dimensionless maximum equilibrium height $Z_{DMe} = z_{DMe}/w_D[h_c/(d^{2/3}w_D^{1/3})]$, (−) Eq. (11.19) (Schmocker and Hager, 2012)

$h_c/(d^{2/3}w_D^{1/3})$, resulting in $Z_{DMe} \to 0$. The data for the tests with scale effects follow the overall trend, whereas the test with $d = 0.31$ mm again deviates from the fit equation due to effects of apparent cohesion.

Embankment volume evolution

The temporal embankment volume V_D was determined between $0 \leq x \leq L_D$. Its initial (subscript 0) maximum volume is V_{D0} at $t = 0$ s. For all Q_o, V_D decreases fast soon after initial overtopping and then slows down with time as the erosive potential reduces. Similar to $z_{DM}(t)$, the embankment volume decreases fast with (1) increasing Q_o due to the high erosion potential, (2) decreasing embankment height due to the reduced volume to be eroded, and (3) decreasing d due to increased erodibility. For all tests, V_D eventually reached the equilibrium volume V_{De} after time t_e. All tests thus start at $(V_D - V_{De})/(V_{D0} - V_{De}) = 1$, tending to 0 as $t \to t_e$. The dimensionless time T is again given by Eq. (11.17). The test data for $T \leq 10^4$ follow ($R^2 = 0.89$, Schmocker and Hager, 2012)

$$\frac{V_D - V_{De}}{V_{D0} - V_{De}} = \exp\left(-0.01T^{0.7}\right). \tag{11.20}$$

All data including those with scale effects follow the overall trend. Although the maximum height is affected by apparent cohesion for $d = 0.31$ mm, the temporal eroded volume is similar to that of the other diameters. For large times, the volume reaches the equilibrium value V_{De}.

The dimensionless equilibrium volume V_{De}/V_{D0} versus $h_c/(d^{2/3}w_D^{1/3})$ follows for $2.5 \leq h_c/(d^{2/3}w_D^{1/3}) < 12$ the fit equation ($R^2 = 0.85$, Schmocker and Hager, 2012)

$$\frac{V_{De}}{V_{D0}} = \exp\left(-0.30\left(\frac{h_c}{d^{2/3}w_D^{1/3}}\right)^{0.85}\right).$$

(11.21)

Besides the inflow, the sediment diameter is again the governing parameter, as the equilibrium stage is mainly controlled by sediment transport and thus by incipient motion. The equilibrium volume reduces with both decreasing d and w_D and increasing h_c. The data for the tests with scale effects follow also the overall trend. The determination of the equilibrium volume was complicated again due to bed forms and 3D erosion patterns at large times.

Maximum breach discharge

The breach discharge increases fast after initial overtopping to attain the maximum breach discharge $Q_M > Q_o$. After reaching the maximum outflow, the breach discharge decreases slowly to $Q \rightarrow Q_o$ and then remains constant. Given the constant inflow scenario, Q_M depends on Q_o, in contrast to a constant headwater level, for which Q_M constantly increases to the final value (see below).

The dimensionless maximum breach discharge Q_M/Q_o was always attained within $t = 15$ s. The following applies: (1) For fixed height w_D and sediment diameter d, Q_M/Q_o decreases with increasing Q_o. The higher the inflow, the faster the erosion is initiated and the less important is the volume of the upstream reservoir. For high Q_o, $Q_M/Q_o \rightarrow 1$ as the embankment is then instantly washed away thereby not affecting the breach discharge, (2) for fixed height w_D and inflow Q_o, Q_M/Q_o decreases with increasing d. For large d, the erosion is initiated fast due to increased seepage and surface sliding. The reservoir level does thus not significantly increase as compared with an embankment made up of fine sediment, and (3) for constant d and Q_o, Q_M/Q_o increases with increasing w_D. The increased reservoir volume for higher embankments results in an increased breach discharge due to the stored water upstream of it.

All data were normalized and related to $(h_c^{2/3}d^{1/3}/w_D)$. As the embankment during the initial stage is eroded faster for coarse sediment, the sediment diameter d is now in the numerator. Figure 11.12 shows (Q_M/Q_o)–1 versus $(h_c^{2/3}d^{1/3}/w_D)$ along with the fit limited to $0.05 \leq (h_c^{2/3}d^{1/3}/w_D) \leq 0.16$ ($R^2 = 0.91$, Schmocker and Hager, 2012)

$$\frac{Q_M}{Q_o} - 1 = 1.7\exp\left(-20\frac{h_c^{2/3}d^{1/3}}{w_D}\right).$$

(11.22)

All data agree well with Eq. (11.22) except for those with a scale effect. In summary, Q_M/Q_o reduces with decreasing w_D and both increasing h_c and d. For small $(h_c^{2/3}d^{1/3}/w_D)$, $Q_M \cong 1.5Q_o$, whereas for $(h_c^{2/3}d^{1/3}/w_D) > 0.20$, $Q_M/Q_o \rightarrow 1$, indicating that an embankment is instantaneously washed away or the reservoir has no effect on the breach discharge.

A comparison of the maximum breach discharge with existing formulae is difficult, given the constant inflow scenario. Most empirical fits derived from embankment dam failures (e.g. Fröhlich, 1995) are based on the reservoir volume. Therefore, the estimated maximum breach discharge considerably exceeds the maximum breach discharge of the present tests.

Effects of sediment mixture and grain size

Given that both hydraulic and geotechnical aspects affect an embankment breach, complex phenomena resulting from seepage, overtopping erosion, sliding failure and various

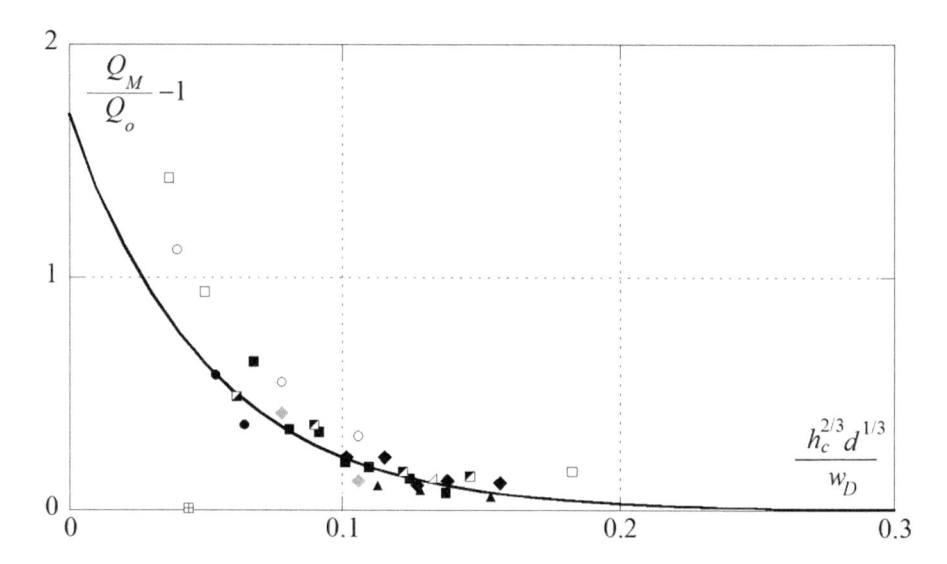

Figure 11.12 Dimensionless maximum breach discharge $(Q_M/Q_o)-1$ versus $(h_c^{2/3}d^{1/3}/w_D)$, $(-)$ Eq. (11.22) (Schmocker and Hager, 2012)

geotechnical characteristics of the embankment material have to be assessed. In addition to these noted under 11.3.3, additional work has been conducted by Ribi *et al.*, (2008), Cao *et al.*, (2011), Sun *et al.*, (2012), Kamalzare *et al.*, (2012), or Orendorff *et al.*, (2013). Van Emelen *et al.*, (2015) combined laboratory tests conducted under constant approach flow discharge with a numerical model analyzing the effect of various sediment transport equations on the breach process. Most research is limited to one specific test setup or test parameter, so that there is still a lack of laboratory data on the breach process including breach profiles, free-surface profiles, or breach discharge, to test and further improve existing numerical breach models.

Additional systematic 2D breach tests were therefore conducted by Schmocker *et al.* (2014), based on the same test setup as described above, to expand their findings relative to the effect of the grain size distribution. The temporal progress of the breach surface was used to quantify the effects on the overall breach process. As previously, a constant inflow discharge Q_o was added, represented by the critical flow depth $h_c = (Q_o^2/gb_D^2)^{1/3}$. Given the channel width of $b_D = 0.20$ m, the reservoir volume is comparatively small. A variation of the inlet distances x_D, and thus of the reservoir volume, had no effect on the overall breach process. The reservoir was filled until the flow overtopped over its entire width thereby resulting in a 2D breach, with the plane erosion pattern along the channel sidewall representative for the entire width. The overflow remained again unaffected of the tailwater and tests were stopped once the breach profiles remained constant. Given the short breach duration and the relatively high overflow, no constant but a failing reservoir level was attained. After a test, the water surface $h_D(x_D,t)$ and sediment surface $z_D(x_D,t)$ were determined from the images. To compare the overall erosion process for different tests, dimensionless sediment surface profiles $Z_D(X_D)$ were considered with $X_D = x_D/L_D$ and $Z_D = z_D/w_D$. The test start $t = 0$ s was set once the upstream water level reached the upper embankment crest.

The effect of sediment mixture on the breach process was studied using two uniform sediments with $d = 2.0$ and 4.0 mm both with a uniformity coefficient $\sigma_g < 1.2$ (Tests 1–6), and three sediment mixtures (Tests 7–15). Each uniform sediment and each grain size distribution was tested for three discharges Q_o. The mixtures were screened using 11 different sediment classes, the smallest of $d = 0.75$–1.02 mm and the largest of $d = 7.5$–8.3 mm, to exclude both effects of apparent cohesion ($d_{50} > 1$ mm) and failure of the bottom drainage for $d_{50} \geq 8$ mm (Schmocker and Hager, 2009, 2012). Figure 11.13 shows the grain size distributions, respectively, for $d_{50} = 1.8$ mm; $\sigma_g = 2.1$ (fine mixture), $d_{50} = 3.4$ mm; $\sigma_g = 2.2$ (medium mixture), and $d_{50} = 5.9$ mm; $\sigma_g = 1.6$ (coarse mixture). The medium mixture follows the Fuller distribution exhibiting a theoretical minimum pore volume. The uniform sediment has density $\rho_s \approx 2670$ kg/m³ in contrast to the three mixtures with $\rho_s \approx 2600$ kg/m³. The dry material was neither compacted nor were surface or core layers added. The compaction effect was found insignificant for the current laboratory setup and the materials used.

Figure 11.14 shows the surface profiles $Z_D(X_D)$ at various times t for different d_{50}, σ_g and $Q_o = 4$ l/s by keeping constant all other parameters. The ordinates are stretched to allow for a better comparison. No apparent effect of sediment mixture exists and all profiles $Z_D(X_D)$ exhibit almost equal shape. Deviations are noted at $t = 4$ s and 6 s (Figure 11.14b, c) in which the downstream face slopes of Tests 4 and 13 involving coarse sediment became unstable, resulting in a surface slip due to increased sediment saturation. The downstream face of all other tests remained stable despite water flow over the crest. This effect disappears once the profiles start to converge from $t = 10$ s (Figure 11.14d). Therefore, embankment erosion proceeds faster with increasing d_{50} within the first breach stage, almost independent of σ_g. Once the downstream face is stable again, erosion is governed by sediment transport, so that the erosion proceeds slower for large d_{50}. Again, no effect of σ_g is observed (Schmocker et al., 2014).

The erosion for Tests 6, 12 and 15 with $Q_o = 16$ l/s (Figure 11.15) is delayed shortly after overtopping, as the overtopping water seeps into the downstream face for the coarser

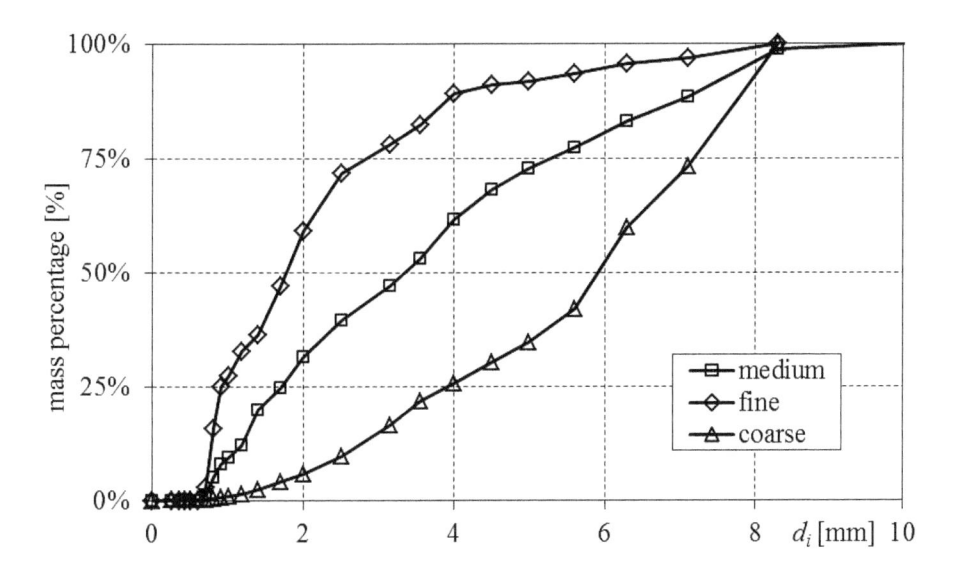

Figure 11.13 Grain size distributions of three sediment mixtures (Schmocker and Hager, 2012)

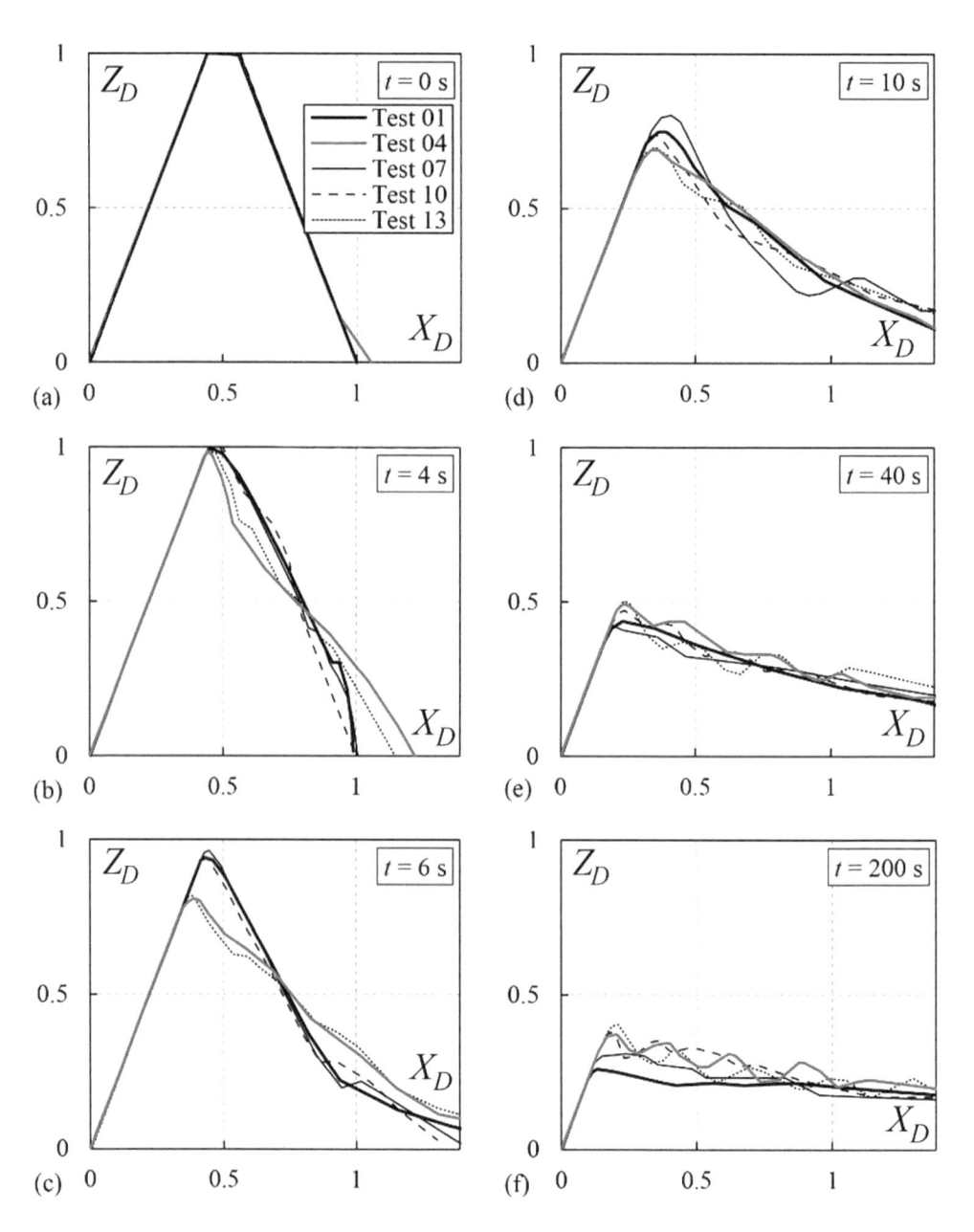

Figure 11.14 Breach surface profiles $Z_D(X_D)$ at various times t for $Q_o = 4$ l/s and various d_{50} and σ_g with all other parameters kept constant (Schmocker and Hager, 2012)

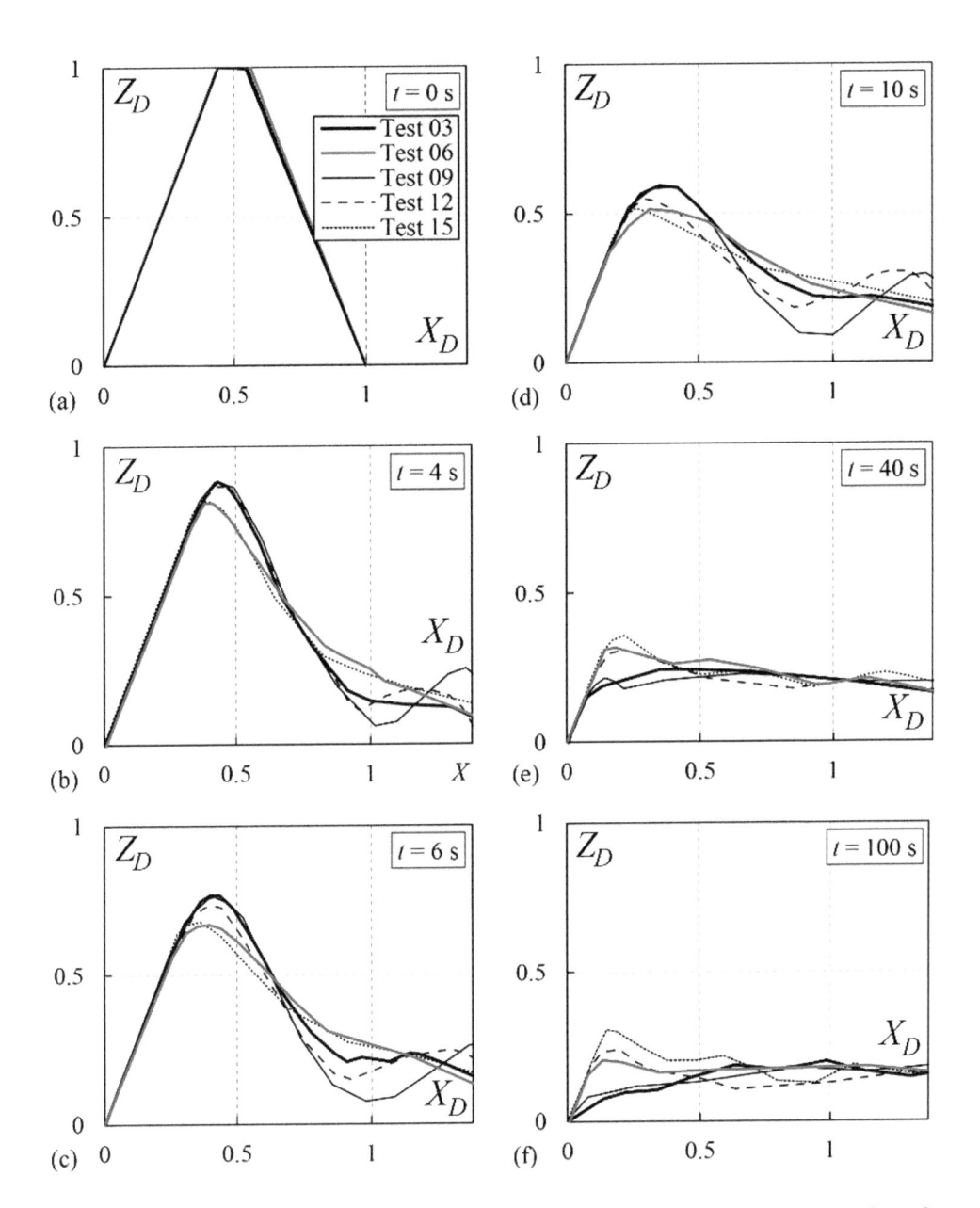

Figure 11.15 Breach surface profiles $Z_D(X_D)$ at various times t for $Q_o = 16$ l/s and various d_{50} and σ_g with all other parameters kept constant (Schmocker and Hager, 2012)

material. This infiltration is almost invisible for the finer sediment for which the erosion process starts immediately. This effect is only visible during the first seconds after overtopping. After $t = 6$ s (Figure 11.15c), Test 3 ($d_{50} = 2$ mm) and Test 9 ($d_{50} = 1.8$ mm) exhibit the highest crest elevation and both Test 6 ($d_{50} = 4$ mm) and Test 15 ($d_{50} = 5.9$ mm) the lowest. The fine mixture (Test 9) prevents sliding of the downstream face due to seepage almost completely. At $t = 10$ s (Figure 11.15d), the downstream faces of the uniform material $d_{50} = 2$ mm, the fine (Test 9) and the medium (Test 12) sediment mixtures are more stable and result therefore in a steeper downstream slope and a rounded breach profile compared with the three other tests. The breach profiles start to converge from $t = 20$ s (Figure 11.15e). Again a higher embankment height is attained for large d_{50} (Tests 6, 12, 15). Besides, no apparent effect of σ_g is visible so that the breach process depends mainly on d_{50}. Both Tests 3 and 9, and Tests 6 and 15 with similar d_{50} have an almost equal breach advance, although σ_g is different. For $Q_o = 8$ l/s, the results for the downstream face slope and breach profiles are between these of $Q_o = 4$ l/s and 16 l/s.

The effect of the grain size distribution on the overall erosion process is therefore small compared with the effect of d_{50}. Sediment mixtures have no significant effect on the breach process for the test range studied, given that the hydraulic conditions are much above the sediment entrainment condition. For laboratory research on embankment breaches due to overtopping, uniform material simulates the general characteristics of the embankment-breach process adequately. This is an important finding relating to both embankment design and laboratory studies. If the effect of sediment non-uniformity would be large, then their design would be much more involved. Fortunately, this effect is small as compared with others.

The temporal advance of maximum height z_{DM} was normalized with $h_c/(w_D d_{50})$ as breach parameter (Schmocker and Hager, 2012). The breach process is accelerated with h_c and decelerated with both w_D and d_{50}. In general, z_{DM} decreases fast after initial overtopping and then advances slow as both the height and the energy line slope reduce, reaching eventually the maximum equilibrium embankment height z_{DMe} after time t_e. As above, the maximum height z_{DM} was normalized with the original height w_D and z_{DMe} according to Eq. (11.16). The time scale followed Eq. (11.17) except for inclusion of the parameter $\sigma_g = (d_{84}/d_{16})^{1/2}$ as $T = [(g'\sigma_g)^{1/2}(h_c/d_{50}^{1/2}w_D)]t$. A comparison of the dimensionless height advance $Z_{DM}(T)$ with Eq. (11.18) indicates good agreement ($R^2 = 0.76$), as also to the equilibrium height Z_{DMe}, closely following Eq. (11.19). Both relations do not include the effect of sediment non-uniformity, given that it was not observed during the tests. The results of Schmocker and Hager (2012) thus include also breaches made up of non-uniform sediment.

11.3.4 Generalized approach

Introduction

The main limitations and simplifications in 2D embankment-breach experimentation allowing for the application of Froude similitude include (Schmocker *et al.*, 2014):

- Sufficiently large discharge and sediment diameter to exclude scale effects;
- Relatively narrow channel to generate 2D erosion;
- Loose, non-compacted embankment without surface layer and vegetation;
- Uniform sediment since effect of sediment mixture on erosion is small;

- Bottom drainage to exclude failure by seepage prior to overtopping;
- Frontal approach flow to embankment to reduce 3D flow features;
- Excellent water quality to analyze images taken through glass sidewall;
- No tailwater submergence to restrict analysis to free overtopping.

The following analyses breach data for two extreme breach scenarios: (1) Constant approach flow discharge (Scenario 1), (2) constant headwater elevation (Scenario 2). Scenario (1) typically occurs for an embankment discharging laterally under small approach flow Froude number, whereas Scenario (2) occurs if a large reservoir is controlled by the embankment. Combinations of these extremes occur in applications. The experiments were conducted in the breach channel described in 11.3.2.

The three parameters investigated by Müller *et al.*, (2016) include the up- and downstream face slopes 1:S_o and 1:S_d, retained below only by S_o and S_d, respectively, and the crest length L_K (11.2.5). Three test Series 1–3 were conducted, with Series 1 relating to the effect of the face slopes, Series 2 to the crest length, and Series 3 to the constant headwater elevation. Series 1 was subdivided into tests involving equal slopes $S_o = S_d$, constant slope in the tailwater but varied values upstream, and vice versa. Slopes included values of 1.5, 2, 2.5 and 3. The embankment height was always $w_D = 0.20$ m, the width $b_D = 0.20$ m, and the median sediment diameter $d = 1.75$ mm. The effect of crest length involved $S_o = S_d = 2$. The embankment length $L_D = (S_o + S_d)w_D + L_K$ was systematically varied with $L_K = 0, 0.10, 0.20$ and 0.40 m, as compared with the reference of $L_K = 0.10$ m and $S_o = S_d = 2$ (Schmocker and Hager, 2012). The third test series combines these three parameters. The test program is stated in Table 11.1.

Table 11.1 Test program of Series 1–3 (Müller *et al.*, 2016)

Item		Test	L_K (m)	S_o (-)	S_d (-)	Q_o (l/s)	h_H (m)	Symbol
Embankment slopes	$S_o = S_d$	1, 2, 3	0.1	1.5	1.5	4, 8, 16	–	□, △, ◇
		4, 5, 6	0.1	3	3	4, 8, 16	–	○, ■, ▲
		7, 8, 9	0.1	2	2	4, 8, 16	–	◆, ●, ▥
		10, 11, 12	0.1	2.5	2.5	4, 8, 16	–	△, ◈, ◍
	$S_o \neq S_d$	13, 14, 15	0.1	1.5, 3, 2.5	2	8	–	▨, ▲, ◆
		16, 17, 18	0.1	2	1.5, 3, 2.5	8	–	●, ▨, ▲
Crest length		19, 20, 21	0.0	2	2	4, 8, 16	–	◆, ●, ■
		22, 23, 24	0.4	2	2	4, 8, 16	–	▲, ◆, ●
		25, 26, 27	0.1	2	2	4, 8, 16	–	–, +, –
		28, 29, 30	0.2	2	2	4, 8, 16	–	X, +, X
Validation		31	0.0	1.5	1.5	8	–	∗
		32	0.4	3	3	8	–	∗
Constant reservoir level		33	0.0	1.5	1.5	–	0.245	□
		34	0.1	1.5	1.5	–	0.245	△
		35	0.1	2	2	–	0.245	◇
		36	0.1	3	3	–	0.245	○
		37	0.4	3	3	–	0.245	▥
		38	0.0	1.5	1.5	–	0.226	■
		39	0.1	1.5	1.5	–	0.226	▲
		40	0.1	2	2	–	0.226	◆
		41	0.1	3	3	–	0.226	●
		42	0.4	3	3	–	0.226	▨

The images recorded allow for visualizing both the water and sediment surfaces to analyze the breach process stages. Using DaVis® software, the coordinates $(x_D; z_D)$ of the water and sediment surfaces at times t were determined. The drainage (subscript dr) discharge Q_{dr} was based on the data of two ultrasonic probes of ±2% accuracy, one of which measured the approach flow elevation whereas the other the water table in the drainage water tank. These determined the breach (subscript b) discharge $Q_b = Q_o + Q_H - Q_{dr}$, with Q_H as headwater discharge due to the storage effect (Frank and Hager, 2015). The noise of the ultrasonic data was removed using a filter (Savitzky and Golay, 1964).

Images of the test setup with constant Q_o were taken at times $t = 0, 4, 6, 10, 20, 40, 100$ and 200 s. For Scenario 2, $t = 0, 4, 6, 10, 20, 30, 40$ and 50 s were analyzed due to the faster erosion advance. Normalized streamwise and vertical sediment surface profiles $X_D = x_D/L_D$ and $Z_D = z_D/w_D$ involved L_D and w_D based on the reference embankment to generalize the analysis of the breach advance in time and space.

Constant approach flow discharge scenario

The effects of embankment face slope and crest length were systematically considered by Müller *et al.*, (2016). Figure 11.16 shows images of a typical breach test. At test start ($t = 0$ s), the headwater level has exactly reached the upstream crest. Sediment transport is initiated

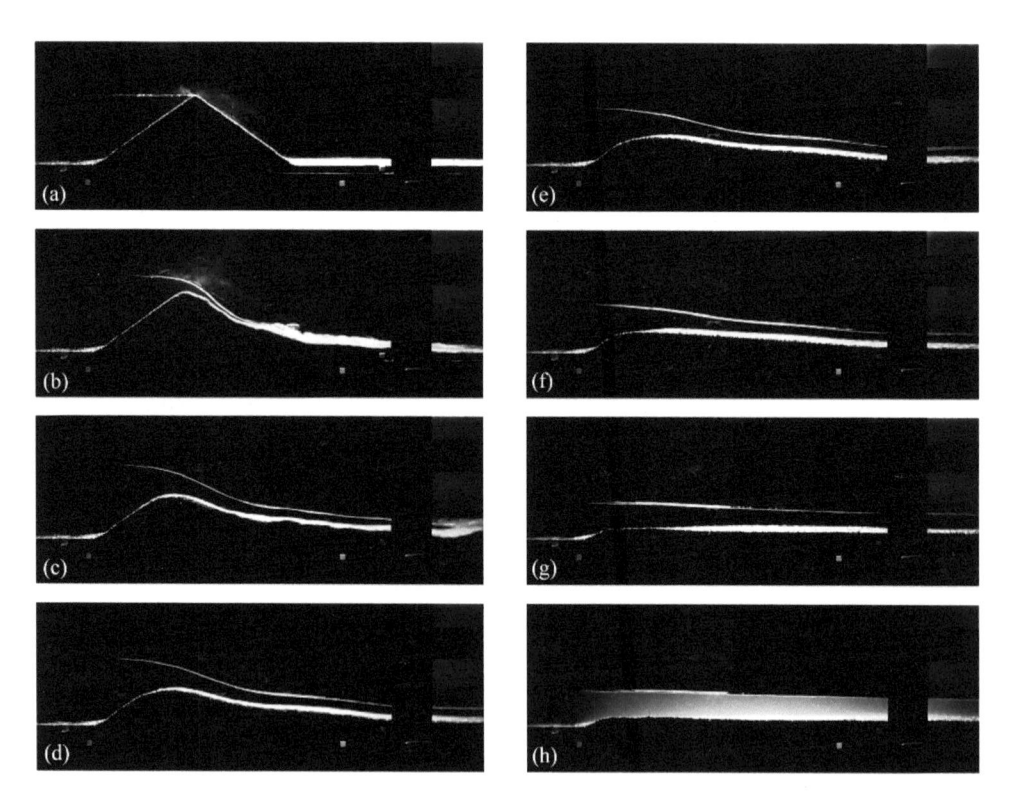

Figure 11.16 Breach advance of Test 31 with $Q_o = 8$ (l/s), $L_K = 0$ (m), $S_o = S_d = 1.5$ at times t (s) = (a) 0, (b) 2, (c) 4, (d) 6, (e) 10, (f) 20, (g) 100, (h) 200 (Müller *et al.*, 2016)

along the downstream face at $t = 2$ s resulting in a small deposit expanding the embankment length thereby creating a wavy downstream face. At $t = 4$ s, erosion has reduced the height and a large sediment wedge has formed downstream the original setting. Note that sediment transport along the downstream channel covers the entire bottom and that the originally triangular crest develops into the round-crested profile. At $t = 6$ s, the height has considerably reduced to $z_{DM} \cong w_D/2$. From around $t > 20$ s the erosion is strongly reduced along with a much less steep embankment, and the breach discharge tending to Q_o after having reached its peak at $t \cong 6.5$ s. For larger time t, an almost horizontal sediment deposit develops tending ultimately to the equilibrium sediment profile.

Figure 11.17 highlights the temporal evolution $Z_D(X_D, t)$ of Test 31 shown in Figure 11.16, demonstrating the rotational movement from the original triangular setting to the finally almost rectangular equilibrium deposit. Note the typical almost constant embankment height at $X_D = 1$ during the entire breach process.

Test series 1 was conducted to study the effects of the face slopes S_o and S_d. It was subdivided into (1) equal face slopes $S_o = S_d = 1.5, 2, 2.5, 3$ (Tests 1 to 12), and (2) constant downstream face slope $S_d = 2$ under varied S_o, and vice versa (Tests 13 to 18; note also Table 11.1).

Figure 11.18 refers to the normalized surface profiles $Z_D(X_D)$ based on the original length L_D and height w_D of the reference setting. Each plot shows the breach process at a certain time stated at the upper right-hand corner. The circle (o) at $t = 0$ s marks the common position of the upstream crests, so that the headwater storage volume changed only slightly with S_o. For small time t, the erosion progresses rapidly ($t <\cong 20$ s) but then slows down as time advances ($20 < t < 200$ s) tending as $t \rightarrow \infty$ to the equilibrium height z_{DMe}. The pattern of this process thus follows the logarithmic trend. The data analysis indicates that the steeper the face slopes, the faster is the erosion process and the lower becomes the equilibrium height. This effect is obvious at $t = 10$ s and 20 s. For large t the profiles $Z_D(X_D)$ look similar. The overall trends for $Q_o = 4$ and 16 l/s are similar as for $Q_o = 8$ l/s, indicating a higher erosion

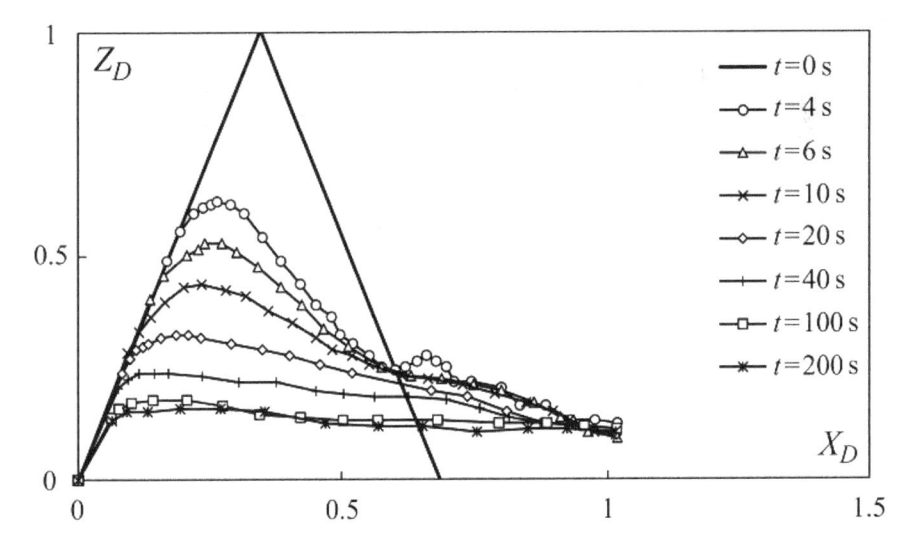

Figure 11.17 Temporal breach profiles $Z_D(X_D)$ of Test 31 at various times t (s) for $L_K = 0$ mm, $S_o = S_d = 1.5$, $Q_o = 8$ l/s (Müller *et al.*, 2016)

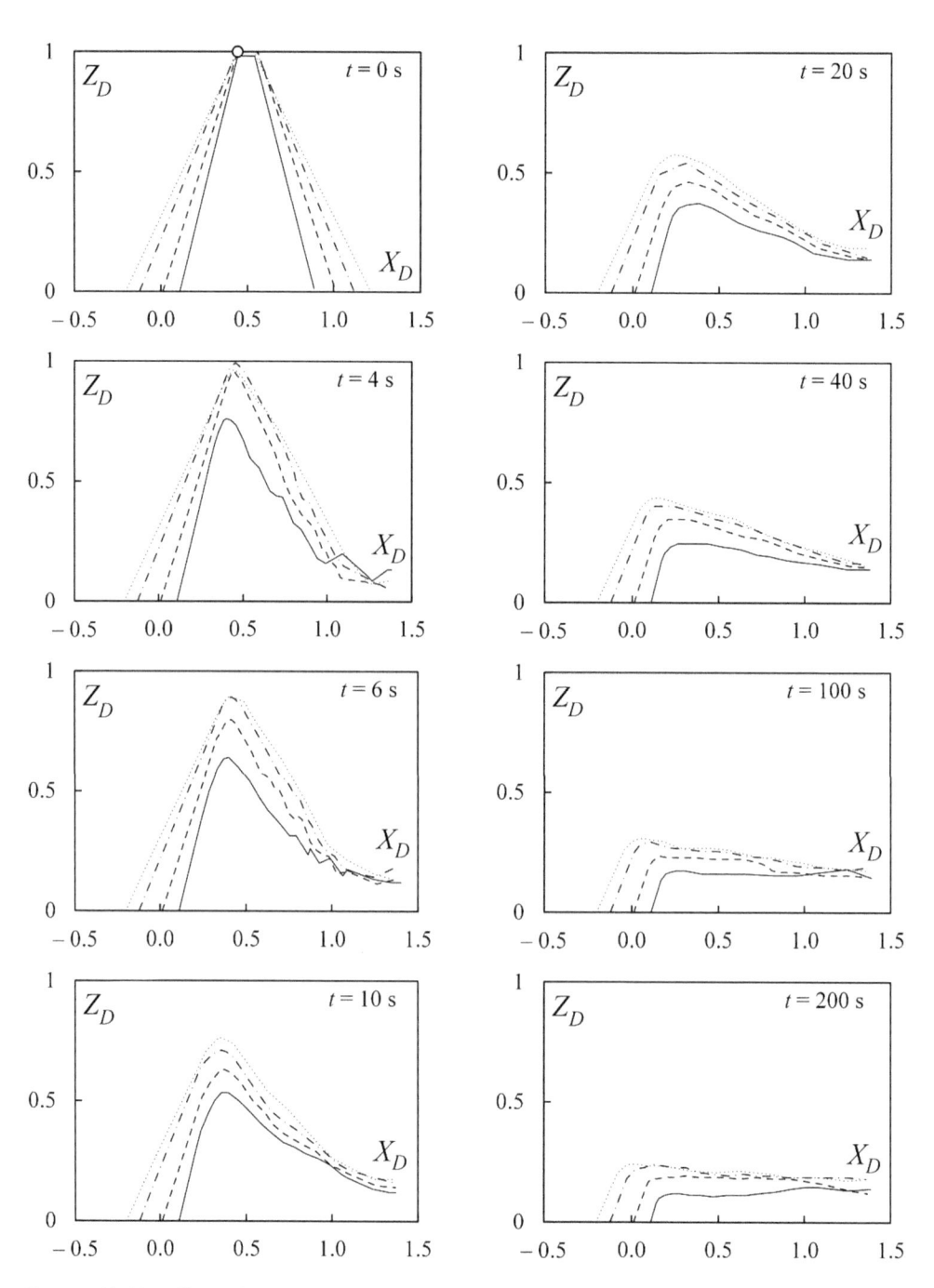

Figure 11.18 Effect of up- and downstream face slopes on erosion process profiles $Z_D(X_D)$ for $Q_o = 8$ l/s with $S_o = S_d =$ (−−) 1.5, (- - -) 2, (−·−) 2.5, (····) 3 (Tests 2, 5, 8, 11) (Müller *et al.*, 2016)

velocity as discharge Q_o increases, so that $z_{DM}(t)$ reduces faster. For the smallest discharge $Q_o = 4$ l/s, erosion started only at $t = 4$ s until when embankments with $S_o = S_d \geq 2$ retained their original height. For the largest $Q_o = 16$ l/s, however, the erosion process at $t = 4$ s was fully developed. The erosion was so strong that the upstream toe moved in the flow direction, typically as $t > 100$ s.

Figure 11.19 details the effect of the approach flow discharge for the reference setting. The smallest value of $Q_o = 4$ l/s produced a peak-shaped breach profile, whereas higher discharges tend to a more round-crested profile $Z_D(X_D)$. Whereas the crest radius remained nearly constant for $t > 20$ s and $Q_o = 4$ l/s, it continuously increased for higher discharges.

Tests 13–18 involved varied values of S_o and $S_d = 2$, and vice versa for $Q_o = 8$ l/s (Table 11.1), with the same trend as described above, namely a rapid erosion at small values of t and a trend toward the equilibrium height z_{DMe} as $t \rightarrow \infty$. The steeper the face slope, the faster is the erosion process, particularly at test start. For these tests, z_{DMe} remained nearly constant regardless of face slope.

As to the effect of crest length, Figure 11.20 shows the normalized profiles $Z_D(X_D)$ for $L_K = 0, 0.100, 0.200$ and 0.400 m and $Q_o = 8$ l/s, similar to Figures 11.18 and 11.19. Note the comparable processes for the effect of face slopes: The shorter the crest length, the faster is the erosion and the lower the equilibrium height z_{DMe}. The setting with $L_K = 0.40$ m retained its trapezoidal shape until $t = 6$ s.

Based on Schmocker et al., (2014) with $S_o = S_d = 2$ and $L_K = 0.10$ m (reference setting), their results were generalized by accounting for these three effects. Embankment shapes ranged from the triangular steep shape with $L_K = 0$ (Tests 19 to 21) to the wide trapezoidal shape with $L_K = 0.40$ m (Tests 22 to 24). Further tests involved other values of L_K (Tests 25 to 30). The combined effect of the three parameters S_o, S_d, and L_K is included by the unit embankment volume $[(1/2)(S_o + S_d)w^2 + L_K w]$, or non-dimensionally by $[(1/2)(S_o + S_d) + L_K/w]$, corresponding to $[(1/2)(2+2) + 0.10/0.20] = 2.5$ for the reference setting, so that the embankment shape parameter is $\mu = [(1/2)(S_o + S_d) + L_K/w]/2.5$. Based on Eq. (11.19), the data are generalized with $R^2 = 0.80$ to

$$Z_{DMe} = \exp\left(-0.54\left(\frac{h_c}{d^{2/3}\mu^{2/3}w_D^{1/3}}\right)^{2/3}\right), \quad 2.5 < \frac{h_c}{d^{2/3}\mu^{2/3}w_D^{1/3}} < 12. \quad (11.23)$$

Figure 11.21 relates to the dimensionless equilibrium height $Z_{DMe}[h_c/(d^2\mu^2 w_D)^{1/3}]$ including $\mu^{2/3}$ from a best-fit analysis, so that its effect is identical to that of d. The effect of embankment shape is thus simply integrated with μ, reducing to Eq. (11.19) for $\mu = 1$. The result reflects that an embankment with a comparatively large μ value behaves similar to that subjected with a small discharge, or large grain size, for which Z_{DMe} increases. In contrast, for embankments with a small μ value, the effects are reversed, resulting in a smaller Z_{DMe} value. The four independent parameters h_c, d, μ, and w_D form in Eq. (11.23) a single non-dimensional parameter, containing in addition S_o, S_d and L_K.

Figure 11.22 shows the dimensionless temporal height advance $Z_{DM}(T)$ accounting for μ in the dimensionless time based on Eq. (11.17) and the data analysis with $R^2 = 0.83$

$$T = \left[g^{1/2}\frac{h_c}{d^{1/2}\mu^{3/2}w_D}\right]t. \quad (11.24)$$

A comparison between the original plots of Schmocker and Hager (2012) with Figures 11.21 and 11.22 indicates even better fits for the present data set up to $T = 8000$. Up to $T = 900$,

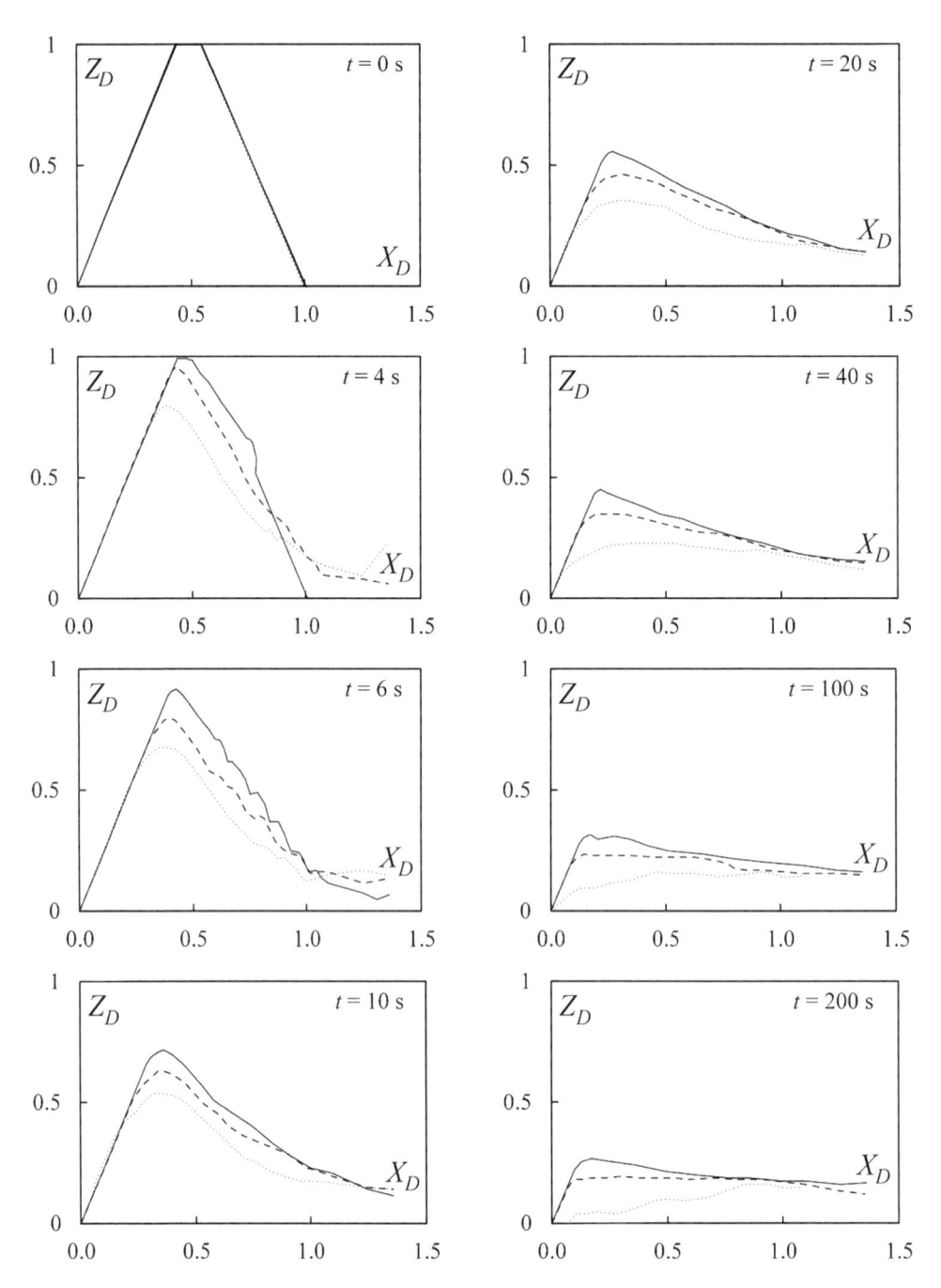

Figure 11.19 Effect of approach flow discharge on breach process profiles $Z_D(X_D)$ of reference setting, for Q_o (l/s) = (−−) 4, (- - -) 8, (····) 16 (Tests 7, 8, 9) (Müller *et al.*, 2016)

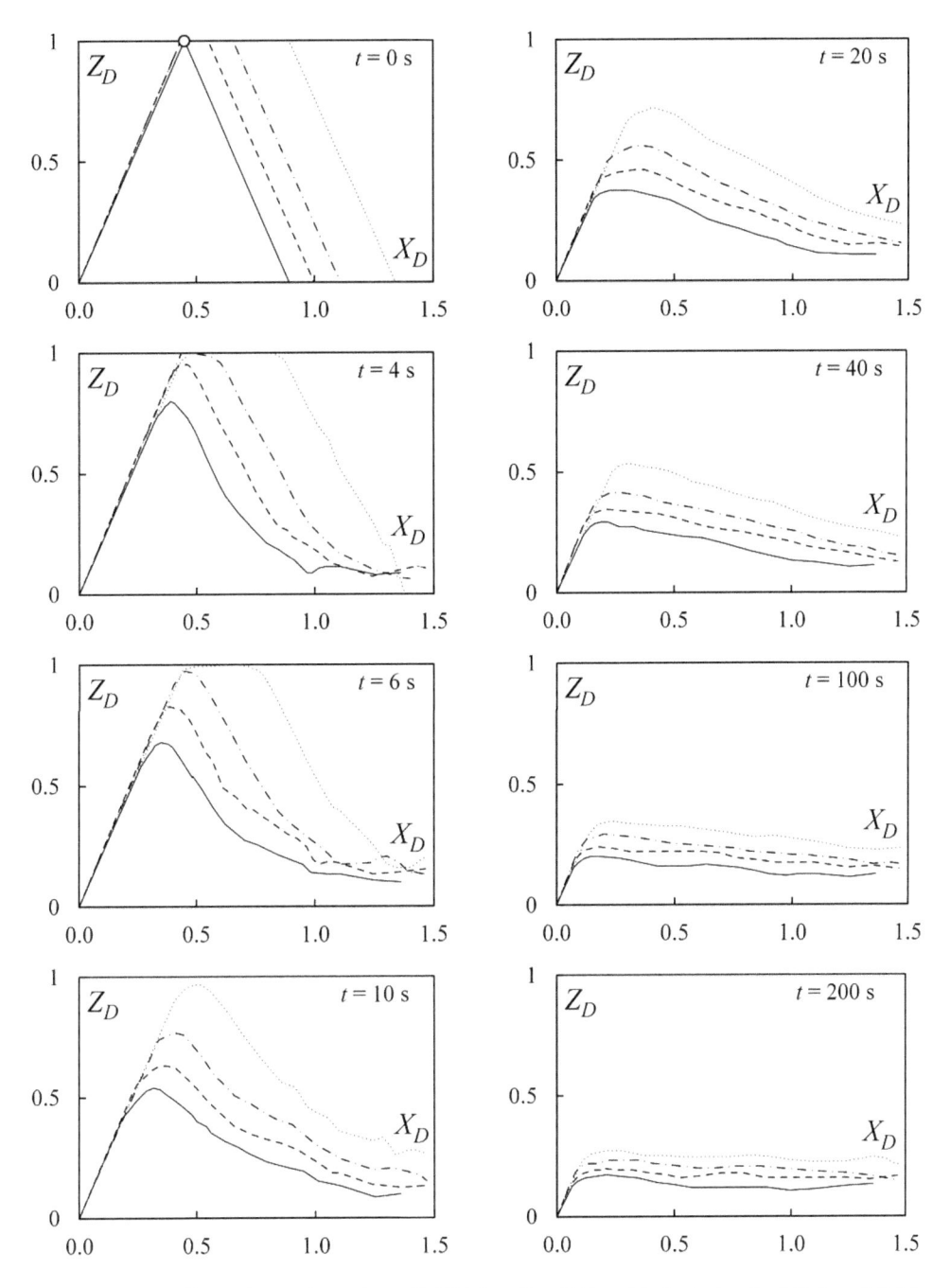

Figure 11.20 Effect of crest length on $Z_D(X_D)$ for Q_o = 8 l/s with L_K (m) = (——) 0, (- - -) 0.10, (—·—) 0.20, (····) 0.40 (Tests 20, 23, 26, 29) (Müller *et al.*, 2016)

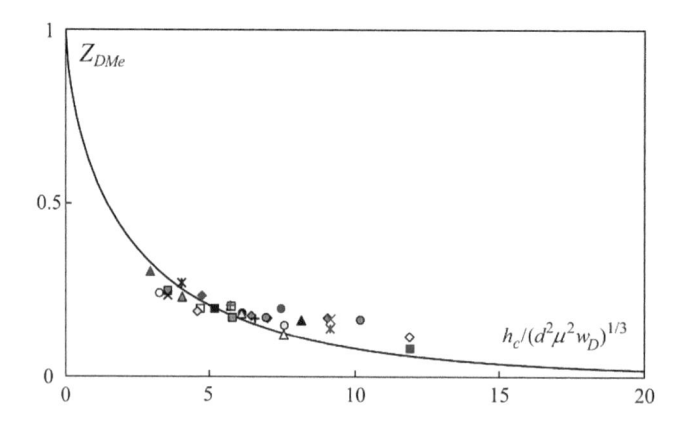

Figure 11.21 Dimensionless maximum equilibrium height $Z_{DMe}[h_c/(d^2\mu^2 w_D)^{1/3}]$ with (—) Eq. (11.23), Symbols in Table 11.1 (Müller *et al.*, 2016)

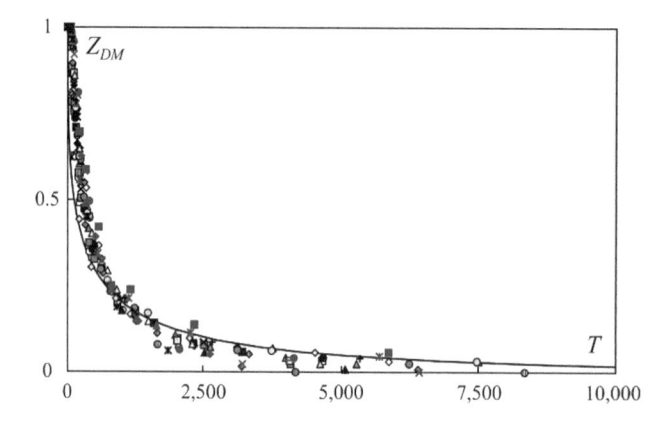

Figure 11.22 Dimensionless maximum height advance $Z_{DM}(T)$ with (—) Eq. (11.18) based on Eq. (11.24), symbols in Table 11.1 (Müller *et al.*, 2016)

the data lie rather above the curve whereas for $T > 900$ they are mainly below. Equilibrium time occurs at $T_e = t_e g'^{1/2} h_c/(d^{1/2}\mu^{3/2} w_D) \cong 5000$. The validation tests (*) follow the curve and the general data trend.

The maximum breach discharge Q_M is based on the standard reservoir storage equation, stating that $Q_b - Q_o + Q_{dr}$ equals the temporal in- or decrease of the headwater elevation times the headwater area $Q_H = -A_H dh_H/dt$. The headwater area A_H is made up of the approach flow length 1.66 m times width b_D from 11.2.5, plus $(1/2)b_D w_D S_o$ as upstream embankment portion. The data are plotted in Figure 11.23 based on Figure 11.12 including the effect of μ as $(Q_M/Q_o - 1)$ versus $(h_c^2\mu d)^{1/3}/w_D$ with $R^2 = 0.83$ as

$$\frac{Q_M}{Q_o} - 1 = 1.7 \exp\left(-20\frac{h_c^{2/3}\mu^{1/3}d^{1/3}}{w_D}\right).$$

(11.25)

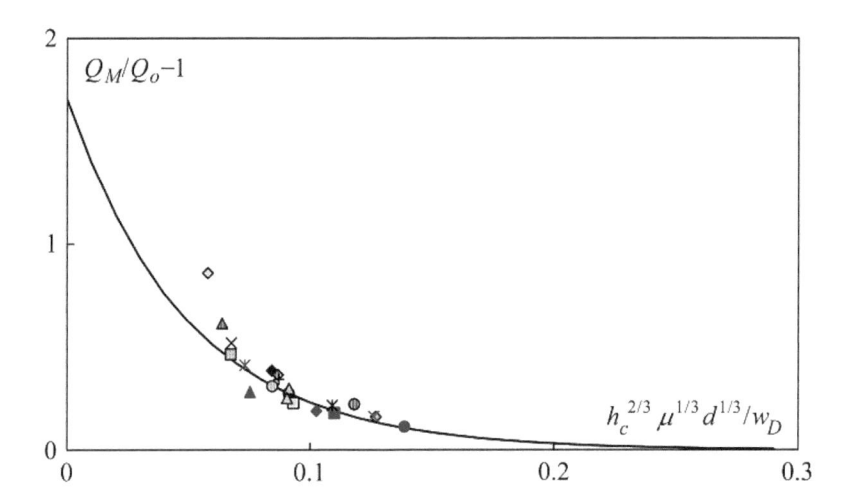

Figure 11.23 Relative breach discharge [$(Q_M/Q_o)-1$] versus $(h_c^2\mu d)^{1/3}/w_D$ for Scenario 1 with (—) Eq. (11.25), symbols in Table 11.1 (Müller *et al.*, 2016)

Because not all drainage discharge hydrographs were available, Figure 11.23 includes only 23 tests. The data point with the highest [$(Q_M/Q_o)-1$] representing Test 19 for which $\mu = 0.8$ and $Q_o = 4$ l/s, i.e. small values, deviates from the general trend line. Further, the maximum breach discharge Q_M corresponds almost to the constant approach flow discharge Q_o if $(h_c^2\mu d)^{1/3}/w_D > 0.10$, whereas its effect becomes significant otherwise. The effect of embankment height is dominant.

Constant headwater elevation scenario

As embankments breach much faster under this scenario than under the constant approach flow discharge, the times analyzed were reduced to $t = 0, 4, 6, 10, 20, 30, 40,$ and 50 s as compared with up to 200 s for Scenario 1 (Table 11.1). Once erosion starts, the maximum height decreases fast, slowing down after larger t. In all tests the embankments were completely removed so that $z_{Me} \rightarrow 0$, as compared with Eq. (11.23) for Scenario 1. Normalized results of Tests 33–37 are shown in Figure 11.24. Note the similar overall behavior of embankment shape during the breach process as for Scenario 1 (Figure 11.18). Differences between Scenarios 1 and 2 include: (i) smaller z_{DM} values for $t > 20$ s in Scenario 2 than in 1; and (ii) upstream face slope reducing earlier in Scenario 2 ($t > 40$ s) than in 1 ($t > 200$ s).

The data were analyzed as for Scenario 1 in terms of the maximum height advance $z_{DM}(t)$. Because $z_{DM}(t \rightarrow \infty) \rightarrow 0$, the relative maximum embankment elevation simplifies to $Z_{DM} = z_{DM}/w_D$. A data analysis indicated that the temporal advance defined in Eq. (11.24) for Scenario 1 equally applies to Scenario 2, so that the effect of the embankment shape parameter μ is accounted therein. The critical depth h_c, remaining constant in Scenario 1, satisfies in Scenario 2 the condition of constant headwater elevation $h_H = z_{DM} + (3/2)h_c$. Given that h_H is a prefixed design value and z_{DM} was measured, h_c is known so that the relation $Z_{DM}(T)$ is known. Figure 11.25 compares the test data (Table 11.1) with the best data fit ($R^2 = 0.97$)

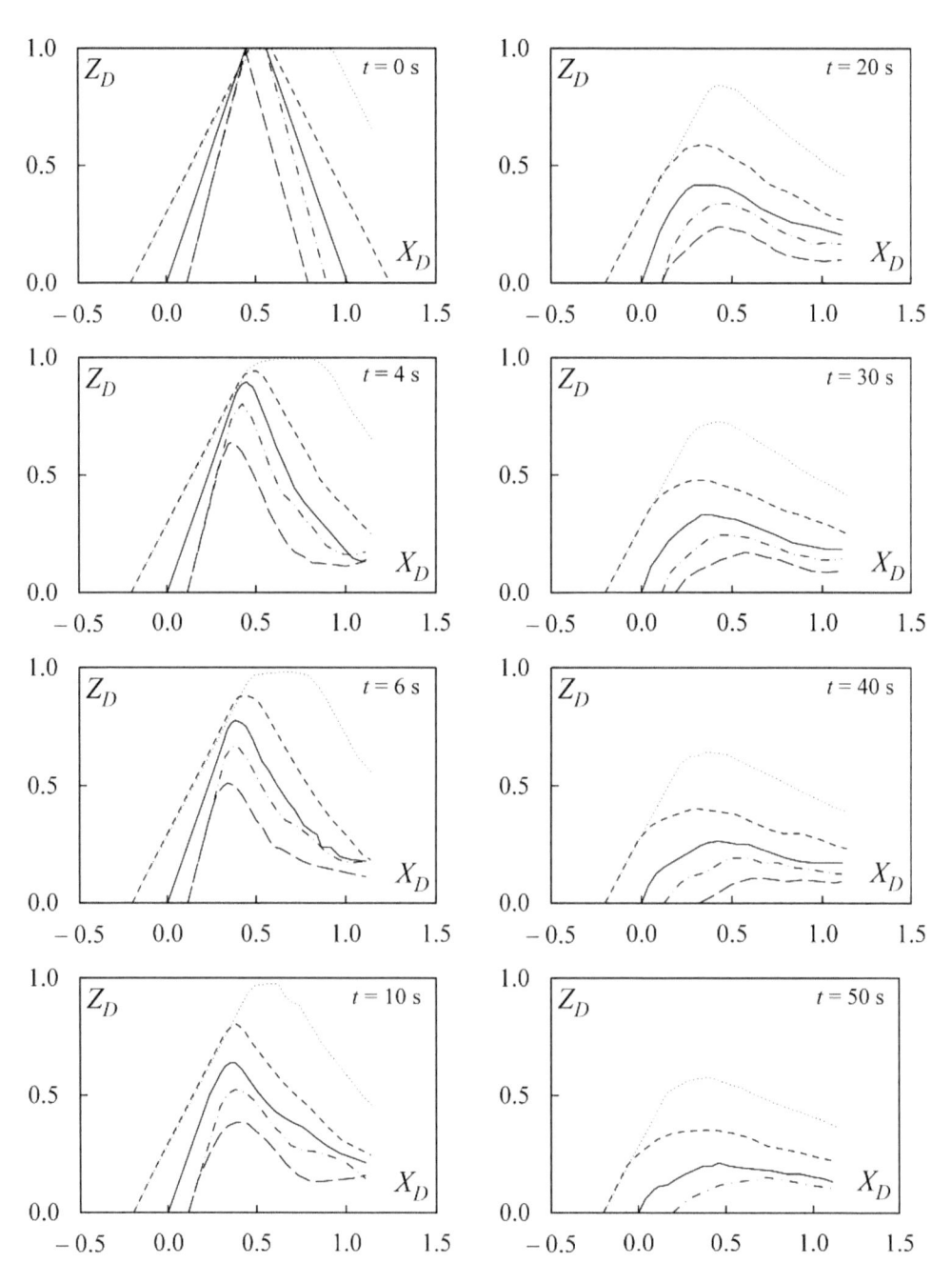

Figure 11.24 Effect of constant headwater elevation on profile advance $Z_D(X_D)$ for $h_H = 0.245$ m; Test (—) 33, (−·−) 34, (−−) 35, (- - -) 36, (····) 37 (Müller *et al.*, 2016)

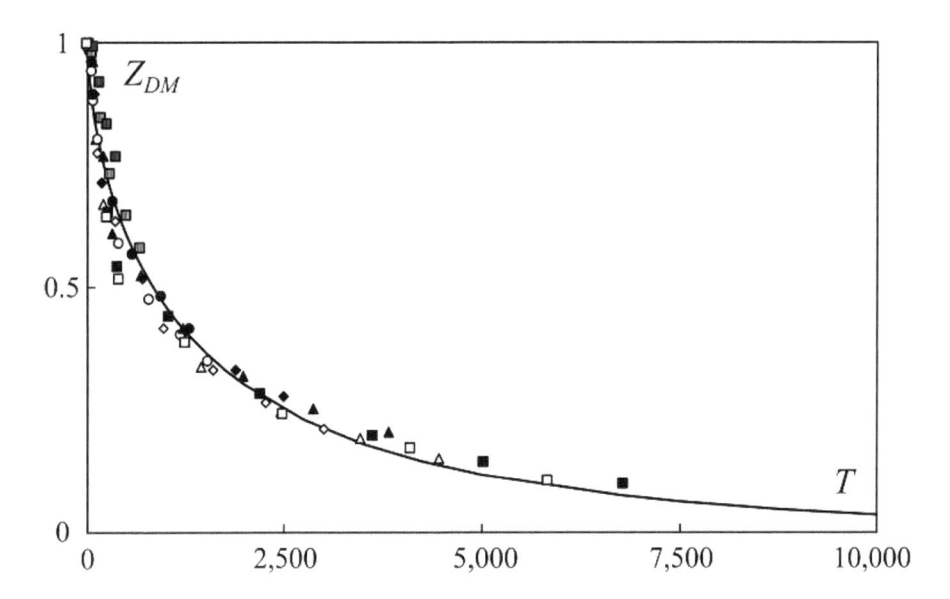

Figure 11.25 Dimensionless maximum embankment height $Z_{DM}(T)$ with (—) Eq. (11.26) based on Eq. (11.24), symbols in Table 11.1 (Müller *et al.*, 2016)

$$Z_{DM} = \exp\left(- 0.01T^{0.63}\right), T < 7 \times 10^3. \tag{11.26}$$

As to the maximum breach discharge, the headwater elevation h_H was considered, varying from $h_o = 0$ at overtopping start ($t = 0$) to the final target value h_H. Accordingly, the normalized headwater elevation is $H_H = h_o/h_H$ with $H_H(T = 0) = 0$ and $H_H(T \to \infty) = 1$. It was further noted that time T defined in Eq. (11.24) for Scenario 1 applies equally to Scenario 2. Figure 11.26 shows $H_H(T)$ following the best-fit relation ($R^2 = 0.96$)

$$H_H = 1.1\tanh(0.06T^{1/3}), \quad T < 7 \times 10^3. \tag{11.27}$$

The breach discharge increases massively up to time $T \cong 10^3$, from when it tends to the equilibrium value. Note further the differences between the two scenarios relating to the breach discharge: In Scenario 1, the approach flow discharge Q_o is reached for a large value of $(h_c^2\mu d)^{1/3}/w_D$, whereas the equilibrium headwater elevation (or the corresponding discharge) is only reached after a long time $T > 10^4$ in Scenario 2.

The three governing variables of the embankment shape parameter μ involve the two face slopes S_o and S_p, and the crest length L_K, varying within relatively small bounds from a minimum of 0.6 for the triangular-shaped steep to 2 for the wide trapezoidal-shaped setting. Its effect is only contained in the adjusted time scale T and the equilibrium height Z_{DMe}. In the latter parameter both the sediment diameter and μ have the identical exponent of 2/3, whereas the exponents of d and μ in T are 0.5 and 1.5, respectively. This feature suggests that the effect of μ contains a variation of the sediment diameter. A certain target grain diameter therefore results by either changing the grain size d, or by modifying parameter μ, given that the product of $(\mu d)^{2/3}$ for Z_{DMe}, and $(\mu^3 d)^{1/2}$ for T have to be constant if all other

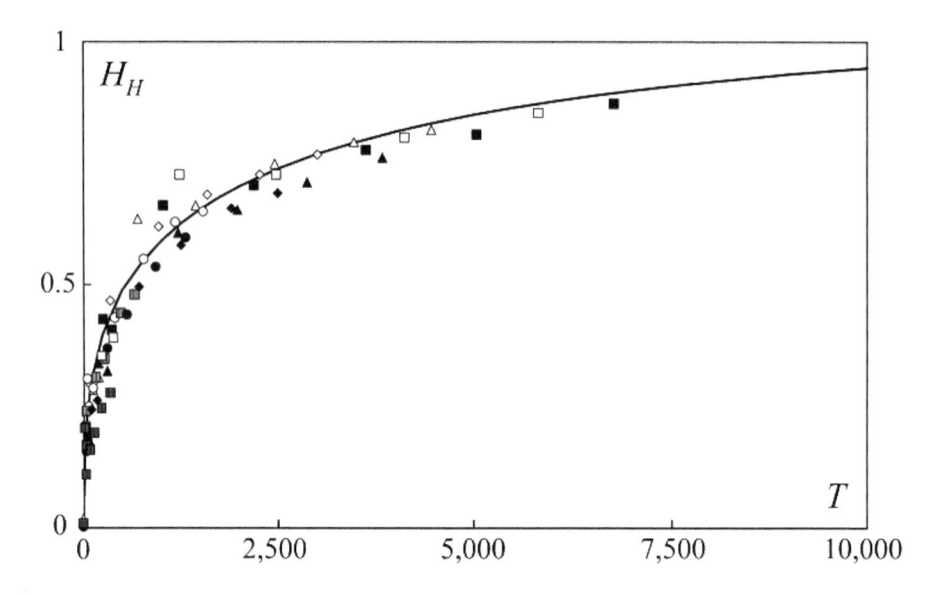

Figure 11.26 Dimensionless breach discharge $H_H(T)$ for Scenario 2 with (—) Eq. (11.27), symbols in Table 11.1 (Müller *et al.*, 2016)

parameters remain unchanged. Based on the current limitations in terms of the critical depth ($h_c > 0.05$ m), sediment size ($1 \le d$ [mm]≤ 8), or embankment height ($w_D \ge 0.20$ m), these are partially removed by adequately selecting any of the dimensional parameters (h_c, d, w_D, μ) thereby modifying the absolute physical dimensions. This research thus not only adds in terms of process understanding, but conceptionally by allowing for generalized modeling.

The 2D breach includes three stages, including (Figure 11.22) (1) $0.20 \le Z_{DM} \le 1$, (2) $0.05 \le Z_{DM} \le 0.20$, and (3) $Z_{DM} < 0.05$. The transition from (1) to (2) thus corresponds to $T_{12} \cong 1000$. Note the strong effects of μ and (h_c/w_D), whereas that of d is small. To increase time T_{12}, both the embankment has to be wide and trapezoidal, and its relative height w_D/h_c has to be large. If a breach is initiated, then the transition to the developed breach stage occurs within hours. 'Additional safety' as vegetation, riprap or compaction will delay a breach, yet the failure duration is not essentially increased.

11.4 Fuse plug

11.4.1 Main features

As mentioned, inadequate spillway capacity is responsible for about one third of all dam failures, so that all existing dams must be critically reviewed in terms of the spillway design discharge thereby ensuring that its capacity is adequate, or that any shortfall is compensated for with other means, including the increase of overflow depth, widening of existing spillways, or provision of additional overflow structures. The upgrading of existing spillways with these means may cause excessive cost so that options were proposed including the fuse plug, a safe and economical alternative if properly designed. Fuse plugs are commonly

only allowed for ensuring additional discharge capacity beyond the design flood. Further, the erosion has to be limited by a concrete crest and an appropriate tailwater scour control. In addition, the fuse gate spillway was found an alternative involving the advantages of an uncontrollable spillway while maximizing the reservoir storage capacity and providing additional dam safety. Below, both types of structures are described, based mainly on Khatsuria (2005), who has given a more extensive account on this recent addition to dams.

A fuse plug corresponds to an erodible, pre-determined separate section of an earthfill dam designed to wash out if the approach discharge exceeds the spillway capacity discharge, and the reservoir reaches the design crest elevation. Fuse plugs are progressively eroded within a certain duration thereby releasing a surplus discharge without endangering dam safety. An uncontrolled auxiliary spillway would serve the same purpose but its length would have to be excessive to keep the overflow depth small. In contrast, the fuse plug provides identical discharge capacity with a much reduced width because it is gradually washed away.

The selection criteria for a fuse plug include:

- *Topography* with a saddle close to the main dam so that spillway structure can be erected;
- *Geology* consisting of sound rock for its foundation, to withstand the erosion during fuse plug washout. It is often required to provide concrete cutoff walls beneath the fuse plug to limit the undermining of the foundation;
- *Tailwater* conveying the flow from the fuse plug into the main river so that no adjacent structures are endangered. The tailwater should also not be clogged by fuse plug material; and
- *Maintenance* should be considered in economic analyses even though these structures work only during a short period and very infrequently.

Currently only few documented cases of fuse plug prototype action are available. In addition there exists only scarce information on laboratory observations relative to their hydraulic performance. Fuse plugs tend to stabilize and to compact due to traffic, vegetal growth and armoring over sufficiently long time. It is therefore unclear whether they really are eroded as anticipated. Khatsuria (2005) reports of only some 20 sites where they were erected, involving maximum unit discharges of 83 $m^3/(sm)$, overflow heights of up to 10 m, maximum height from base elevation to maximum water level of up to 13.5 m, and breach lengths up to 1200 m.

A fuse plug includes the following elements (Khatsuria, 2005):

- *Pilot channel* initiating overflow once the reservoir level reaches the crest elevation. The embankment material below the pilot channel should be highly erodible to ensure the effective washout of the fuse plug;
- *Impervious core* as key element, corresponding to a thin core inclined in the downstream direction, preventing any washout for discharges below design discharge, but collapsing under its own weight once overtopping starts. This core may get dried out and thus crack because the reservoir level will be rarely at its elevation, so that it should be constantly maintained;
- *Filters* covering the core to prevent piping and premature washout;
- *Sand and gravel* involve the major portion of the embankment; its material gradation affect the rate of washout and erosion; and
- *Slope protection* consisting of riprap and coarse gravel on both fuse plug faces to protect it against wind, waves, rainfall, and snowmelt.

The present design guidelines are mainly based on empirical knowledge of Tinney and Hsu (1962), Pugh (1985), USBR (1987) and Wahl (1993). A fuse plug should be designed as a zoned earth- and rockfill dam but should washout as predicted once overtopped. Its washout should start at the pilot channel and general washout should then initiate rapidly. Long fuse plugs are subdivided into sections using splitter walls. By keeping the embankment crest at successively higher elevations along the length and providing pilot channels at each section, the washout process is matched to cater to successive infrequent floods, but the entire fuse plug would not washout unless the full capacity of the auxiliary spillway is exceeded.

The lateral erosion rate is a primary design factor. Generally, fuse plugs operate only under floods with a return period in excess of the design flood. Figure 11.27 shows a streamwise section of trapezoidal shape including the impervious core inclined toward downstream, the fall of water beyond it and the erosive action by the falling water jet. Note that remnants of the fuse plug are visible at the original fuse plug end, but that most material is transported further

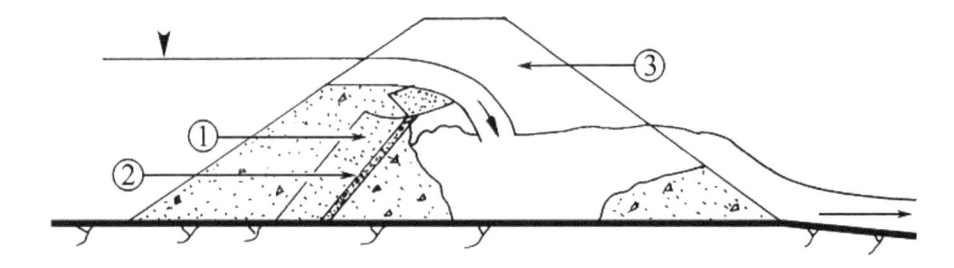

Figure 11.27 Streamwise schematic section of fuse plug with ① inclined impervious core, ② sand filter, ③ pilot channel (Adapted from Khatsuria, 2005)

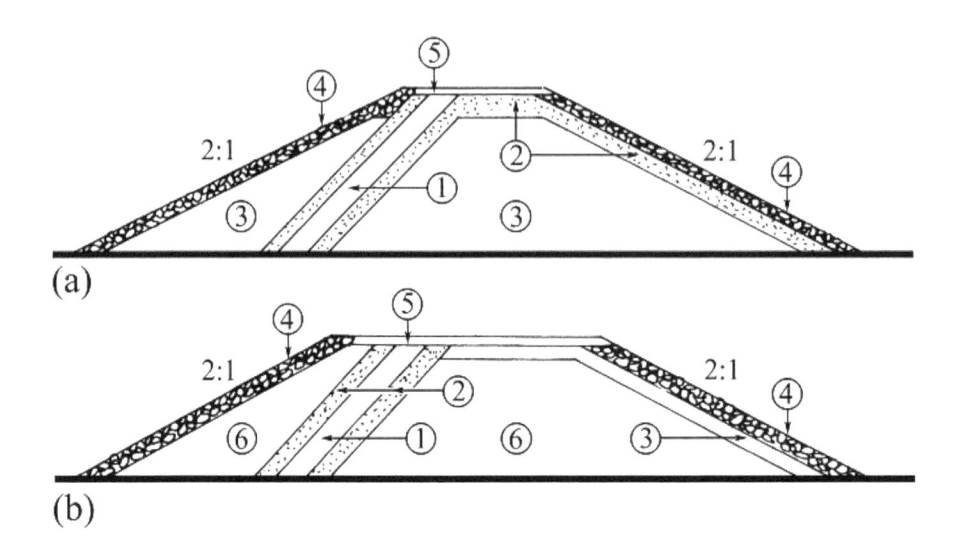

Figure 11.28 Sections across fuse plug (a) embankment, (b) pilot channel with ① core, ② sand filter, ③ sand and gravel, ④ slope protection, ⑤ gravel surface, ⑥ compacted rockfill (Adapted from Khatsuria, 2005)

downstream to assure removal within a short time. The overhanging core breaks continuously off both under its own weight and the overflowing water.

Figure 11.28 shows typical sections across the embankment and the pilot channel. Core material normally comprises silt and clay. Filter zones are provided both up- and downstream to prevent piping across cracks. The compacted sand and gravel zones in the main fuse

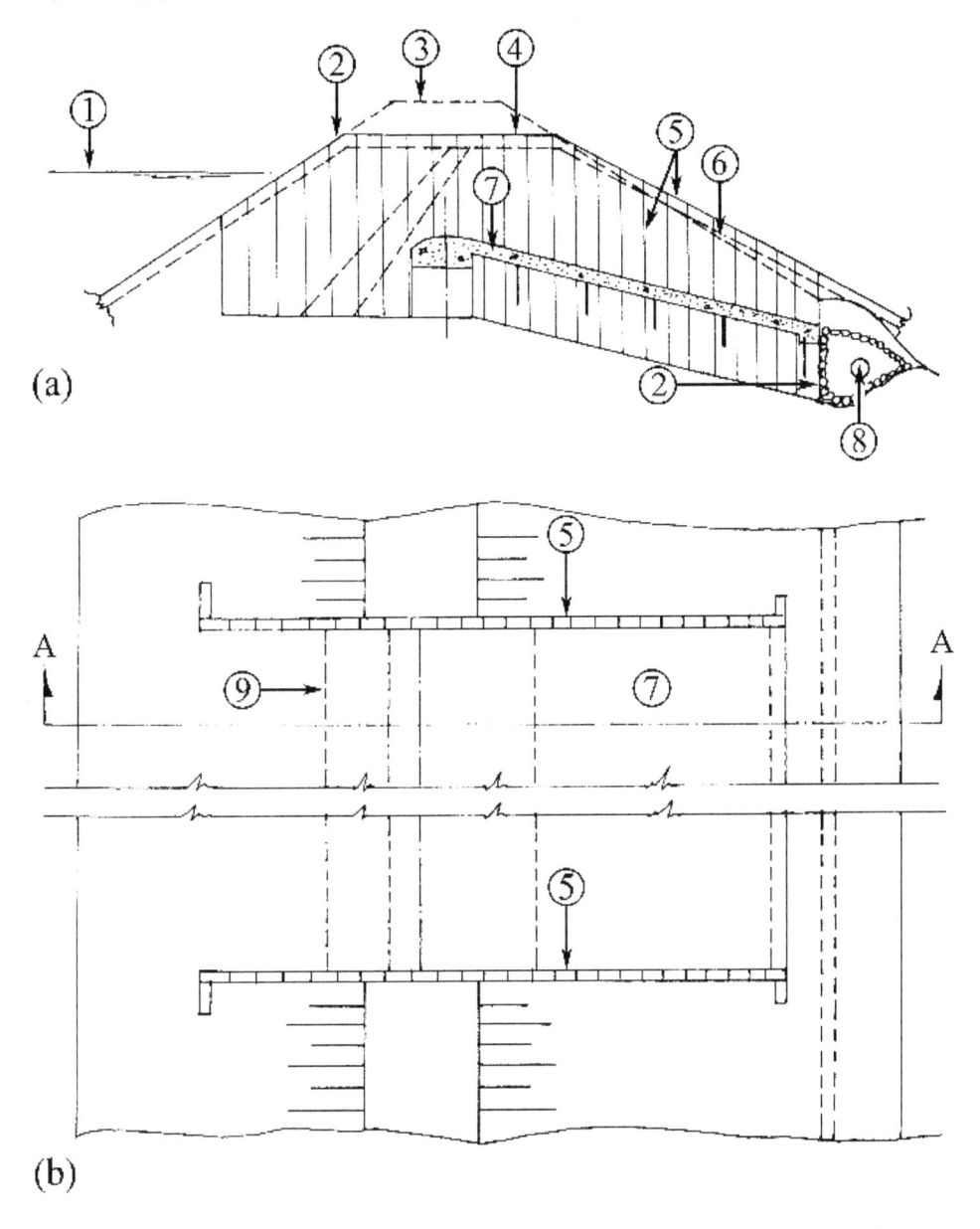

Figure 11.29 Fuse plug in existing dam (schematic) (a) section A-A, (b) plan with ① normal reservoir level, ② riprap, ③ top of existing dam, ④ lowered fuse plug crest, ⑤ sheet piling, ⑥ slope of earth dam, ⑦ chute slab, ⑧ drain pipe, ⑨ lowered fuse plug crest (Adapted from Khatsuria, 2005)

embankment and the compacted rockfill zone in the pilot channel are visible; these should be incohesive and easily erodible to initiate the washout process.

Based on Pugh (1985), material gradation suggested by Khatsuria (2005) includes for the core $d_{50} = 0.05$ mm, $\sigma_s = (d_{84}/d_{16})^{1/2} = 6$; for the filter $d_{50} = 0.5$ mm, $\sigma_s = 2.2$; for sand and gravel $d_{50} = 3$ mm, $\sigma_s = 3$; for the rock-fill $d_{50} = 15$ mm, $\sigma_s = 3$; and for the slope protection $d_{50} = 100$ mm, $\sigma_s = 1.1$. Figure 11.29 shows a typical setup of an existing dam. Note the segregation between the non-overtopping dam portion and the fuse plug, and the placement of the rigid overflow section into the fuse plug body so that the erosion is limited to the extent required for dam safety.

11.4.2 Case study

Introduction

Schmocker *et al.*, (2013) describe a series of laboratory tests conducted at ETH Zurich to examine the feasibility of a fuse plug design at Hagneck Canal along Aare River, Switzerland. Two designs were tested and conclusions given below indicate further design principles when applying fuse plugs to relatively small dams within the river engineering environment.

Hagneck Canal was erected in the late 19th century providing along 8 km a man-made diversion of Aare River to Lake Biel in western Switzerland. The 140 year-old canal needs to be upgraded because the current design discharge of 1200 m³/s cannot be discharged without major problems. The 2005 flood discharge of 1500 m³/s caused almost overtopping of the existing river levees, and both internal erosion and seepage were observed along the levees.

The improved project aims to increase the flooding safety by two measures: (1) existing levees are elevated to discharge the 1500 m³/s design flood including a freeboard of 1 m, and (2) placement of fuse plug at the improved section to prevent uncontrolled levee overtopping up to discharges of 1800 m³/s. Figure 11.30 shows the cross section across Aare river including the fuse plug on top of the existing levee based on a rigid weir sill. To guarantee sufficient discharge capacity the latter is 1.7 m below the elevated levee crest. The fuse plug is designed as 1.2 m high and 300 m long earth embankment. Under emergency flooding it should erode entirely within one hour and provide an additional discharge capacity of 300 m³/s, corresponding to 1 m²/s unit discharge to limit the Aare River discharge to 1500 m³/s. The surplus flow is discharged into a flood plain with a comparatively low damage potential.

The literature review of Schmocker *et al.*, (2013) indicates that current knowledge is limited to comparatively large dams, and that laboratory tests describe a specific application of

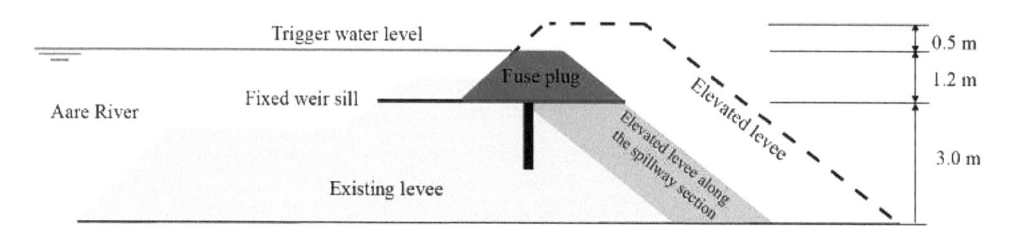

Figure 11.30 Embankment cross section along Aare River levee with existing levee, fuse plug and elevated levee (Schmocker *et al.*, 2013)

the concept so that few general design guidelines are available. Further, only few data exist on the operation or the long-term behavior of fuse plugs.

Experimentation

The fuse plug design and the performance were studied using hydraulic models of scale 1:5 and 1:2.5. The final design was tested at prototype dimensions using prototype material to minimize large-scale effects. In nature, the erosion process of a fuse plug is mainly 3D, including both vertical and lateral erosion. In the hydraulic model tests, the erosion process was investigated first 2D with a cross-sectional model to allow for a detailed optical analysis of the breach profiles through the channel sidewall. An additional small-scale test was conducted to study the 3D erosion process. The model tests further adopted a frontal approach flow to the fuse plug crest, given that the river flow has a small velocity not affecting the performance of the fuse plug.

A constant headwater level scenario was adopted once overtopping started. For the prototype test, a falling water level had to be accepted during the breach due to the limited pump discharge of 0.300 m³/s. The fuse plug was overtopped once the river water level exceeded its crest elevation with the overtopping discharge depending on the breach progress and the river discharge. The breach and river discharges interact so that there is hardly a constant inflow to the breach section. Given the relatively high flood discharge, an almost constant water level in Aare River may be assumed.

The tests were conducted in two rectangular, horizontal channels. The channel for the small-scale tests was 0.20 m wide, 0.70 m high and 8 m long with a discharge capacity of 0.070 m³/s. Its intake was equipped with a flow straightener to generate undisturbed inflow whereas the eroded sediment was collected at the channel end. For all tests, the existing flood levee was not modeled but the fuse plugs were directly erected on the channel bottom to guarantee complete erosion and prevent backwater effects. The prototype tests were conducted in a 15 m long, 2 m high and 1 m wide channel of 0.300 m³/s discharge capacity (Figure 11.31). This channel was also used for the 3D tests to investigate the lateral breach process. The erosion processes were observed across the channel sidewall and from above. Both the water and the sediment surfaces were recorded across the sidewall. The upstream

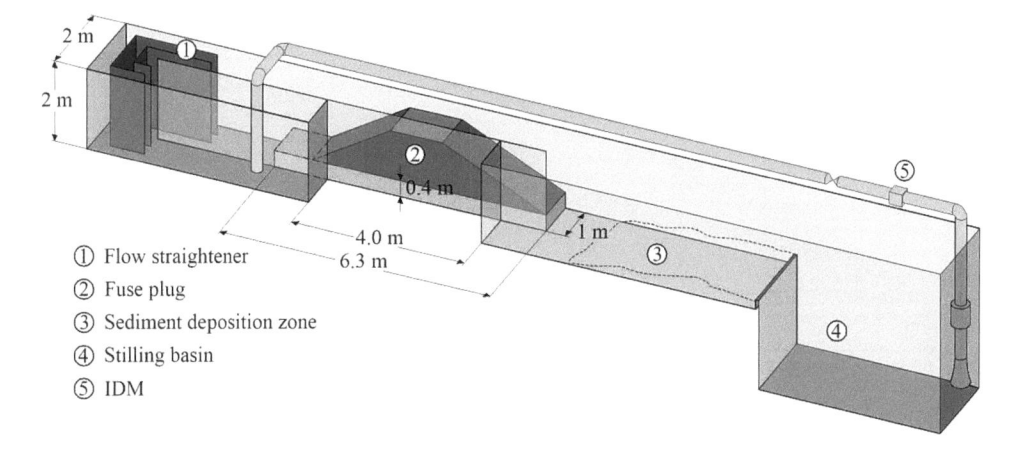

① Flow straightener
② Fuse plug
③ Sediment deposition zone
④ Stilling basin
⑤ IDM

Figure 11.31 Schematic side view of hydraulic model (Schmocker *et al.*, 2013)

water level was recorded with an Ultrasonic Distance Sensors, whereas the inflow to the upstream reservoir was recorded with an electromagnetic flow meter (IDM).

Two different fuse plug designs were tested. Design A involved a fuse plug with an inclined silt core adapted from Pugh (1985), whereas Design B corresponded to a fuse plug with a fine sand body. Figure 11.32 shows both designs in prototype dimensions and the material types according to the Unified Soil Classification System (USCS). Both fuse plugs were 1.2 m high, of up- and downstream slopes S_u = 2:3 and S_d = 1:2. The fuse plug width met the limited available space on the existing levee. The filter layer thickness was large enough to prevent segregation during construction and to provide a self-healing effect in case of cracks of the core material. A 40° core angle was selected to facilitate core collapse but prevent large shielding of the downstream material (Pugh, 1985). For Design B (Figure 11.32b), the core was dropped and the entire embankment body was made of sand. A seepage protection was placed to prevent piping along the fuse plug base, yet the stabilizing effect of vegetation was not tested.

The discharge capacity was determined based on the broad-crested weir formula

$$Q = C_d b \left(2gH_o^3 \right)^{0.5}. \tag{11.28}$$

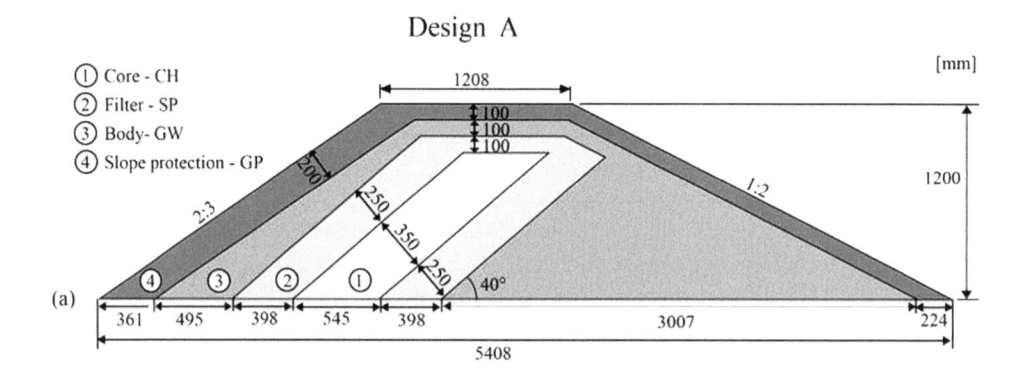

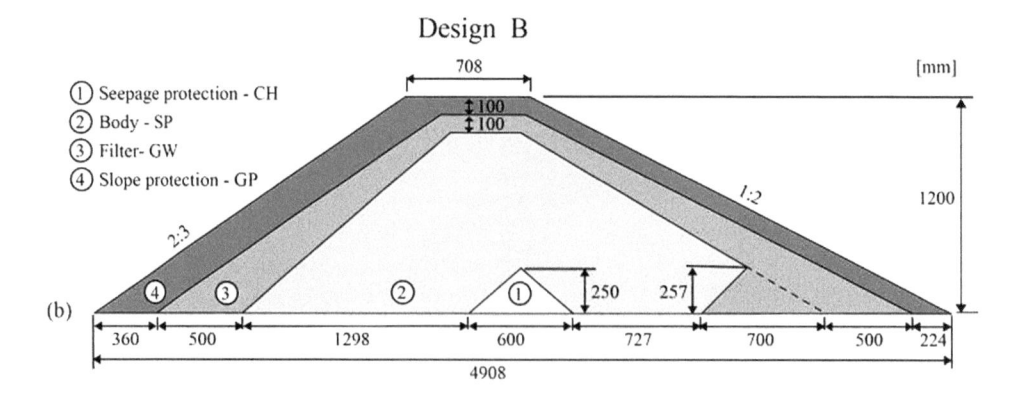

Figure 11.32 Fuse plug cross section for (a) Design A with inclined core, (b) Design B with sand body, CH = fat clay, SP = poorly graded sand, GW = well-graded gravel and GP = poorly graded gravel (Schmocker *et al.*, 2013)

Here C_d = discharge coefficient, b = overflow width, g = gravity acceleration and H_o = approach flow (subscript o) energy head. The discharge coefficient of broad-crested weirs of Fritz and Hager (1998) was used with $\xi = H_o/(H_o + L_K)$ = relative crest length

$$C_d = 0.43 + 0.06\sin\left[\pi(\xi - 0.55)\right]. \tag{11.29}$$

For a crest length of L_K = 5.408 m as base length of the fuse plug, b = 300 m and with an overflow depth of $H_o \approx h_o$ = 1.2 m, the unit discharge is $q \approx 2.2$ m²/s, so that the required discharge of $q \approx 1.0$ m²/s was available for complete fuse plug erosion.

In step 1, the core material was defined to meet the three criteria: (1) permeability $k \leq 10^{-7}$ m/s to prevent seepage, (2) limited stability to guarantee fast collapse, and (3) optimum shrinkage characteristics to reduce cracks as the core may dry out. Fat clay of high plasticity was selected. The filter covering the core as well as both the body and the slope protection was selected according to filter criteria (SN670 125a, 1983) to prevent internal erosion and sediment washout (Schmocker et al., 2013). The following materials met the filter criteria and thus were selected: Poorly graded sand for filter; well-graded gravel for embankment body material; and poorly graded gravel for surface protection. The grain size distributions used in the prototype test included d_{50} = 0.009 mm, $\sigma_s = (d_{84}/d_{16})^{1/2}$ = 3.7 for core; d_{50} = 0.15 mm, σ_s = 2.4 for filter; d_{50} = 2.7 mm, σ_s = 2.4 for embankment body; and d_{50} = 25 mm, σ_s = 1.8 for slope protection. Note the differences between these and the numbers given in 11.4.1. Further details on the soil characteristics are provided by Schmocker et al., (2013).

The test program for the fuse plug with the inclined core was tested 2D at scale 1:5 (Tests 1–3), 1:2.5 (Test 5) and 1:1 (Tests 6, 7) and 3D at scale 1:5 (Test 4). The fuse plug involving the sand body was tested at scales 1:5 (Test 8) and 1:1 (Test 9). The material was placed as obtained from the gravel plant but not dried. The water content was 30% for the core, 12% for the filter and 4% for the embankment body. Except for Tests 4 and 9, all fuse plugs were built and the breach tests conducted within one day. The fuse plug in Test 4 was kept dry for 2 months to evaluate the core behavior. In Test 9, the reservoir was kept constant below the fuse plug crest to evaluate the model seepage discharge. Given the limited space in the model channel, the fuse plug was always constructed and compacted manually.

The reservoir was filled slowly to observe seepage processes. Once overtopping started, the discharge was continuously adjusted to keep the reservoir level constant. This was achieved at the start, but failed at large times as the erosion process was too fast or the pump had reached its capacity. The tests were stopped once the entire fuse plug was eroded or if no more change occurred in the fuse plug body. For the 3D test, a 0.02 m wide and 0.02 m deep triangular pilot channel was provided by removing the surface protection of the fuse plug along the channel sidewall.

As the breach flow across a fuse plug is a gravity dominated free-surface process, Froude similitude was applied. However, as sediment transport is also involved, the model must further simulate the tractive shear stress. The shear stress on a sediment particle fluctuates due to turbulence and both the drag force and turbulence vary with viscous forces, i.e. the Reynolds number. Therefore, a Froude similitude model does not necessarily simulate both tractive forces and sediment transport accurately, as the Reynolds number may be too small in the model. The critical shear stress for incipient motion depends on the grain Reynolds number but is constant at 0.047 if the latter is in excess of 200. However, if the sediment is scaled according to Froude, the grain Reynolds number may fall below 200 and the unit sediment discharge rate for the model would be higher than for the prototype. The sediment size must therefore be adjusted to compensate for too small grain Reynolds number (Schmocker *et al.*, 2013).

Experimental results

Figure 11.33 shows the typical temporal advance in prototype time of the fuse plug breach for Test 1. As soon as the reservoir water level rises above the impermeable core, seepage through the downstream fuse plug body is initiated. The water level further increases until reaching the trigger water level. Seepage increases considerably as the water flows through the surface protection, yet the fuse plug is still stable. Overtopping starts with increasing reservoir level; the downstream slope protection becomes destabilised starting to slide, thereby initiating the breach process for the overflow depth of $h_o \cong 0.05$ m at $t = 0$ s. The embankment body is then exposed due to erosion and sliding failure of the slope protection ($t = 22$ s). The erodible sediment is removed fast due to the increased discharge at $t = 90$ s until the impermeable core partly collapses due the undercutting process and its own weight ($t = 134$ s). This process is repeated until the core is no more undercut after $t \cong 300$ s. The fuse plug was completely eroded after $t = 400$ s. Complete fuse plug erosion in 1:5 scale occurred only in Tests 1 and 4 as the last core part remained stable for Tests 2 and 3, see below.

Figure 11.34 shows the advance of the 2D breach with an inclined core at prototype scale. The reservoir level was raised until seepage of the fuse plug was initiated after the water level exceeded the core height. A major seepage through the downstream fuse plug body started once the water passes the surface protection. The water at the downstream toe of the fuse plug was clear, indicating no significant internal erosion. The surface protection

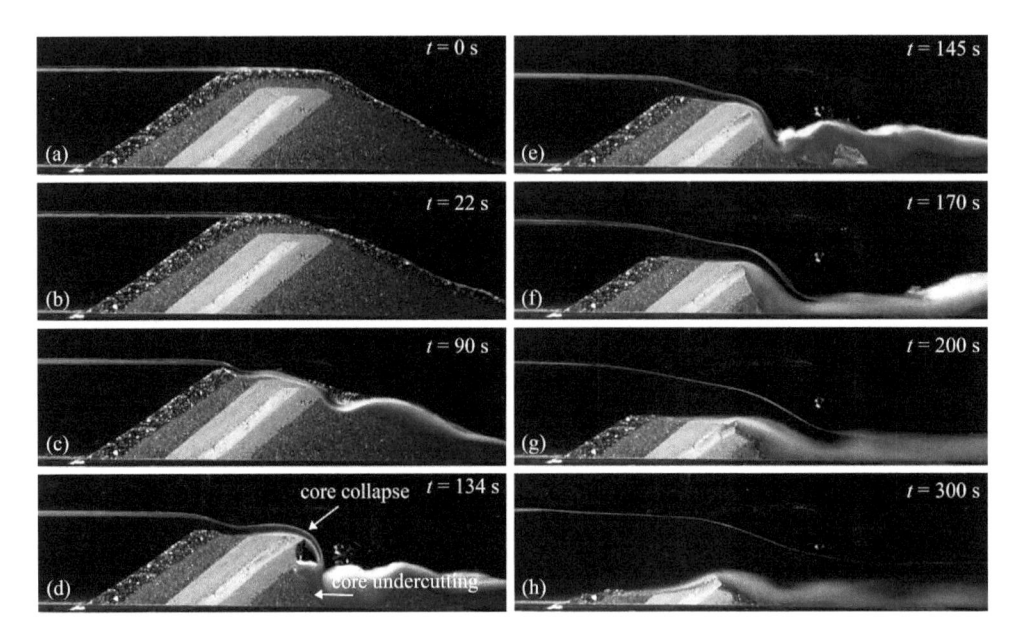

Figure 11.33 Temporal advance of fuse plug erosion at prototype times t for 1:5 scale test at time t [s] = (a) 0, (b) 22, (c) 90, (d) 134, (e) 145, (f) 170, (g) 200, (h) 300 (Schmocker *et al.*, 2013)

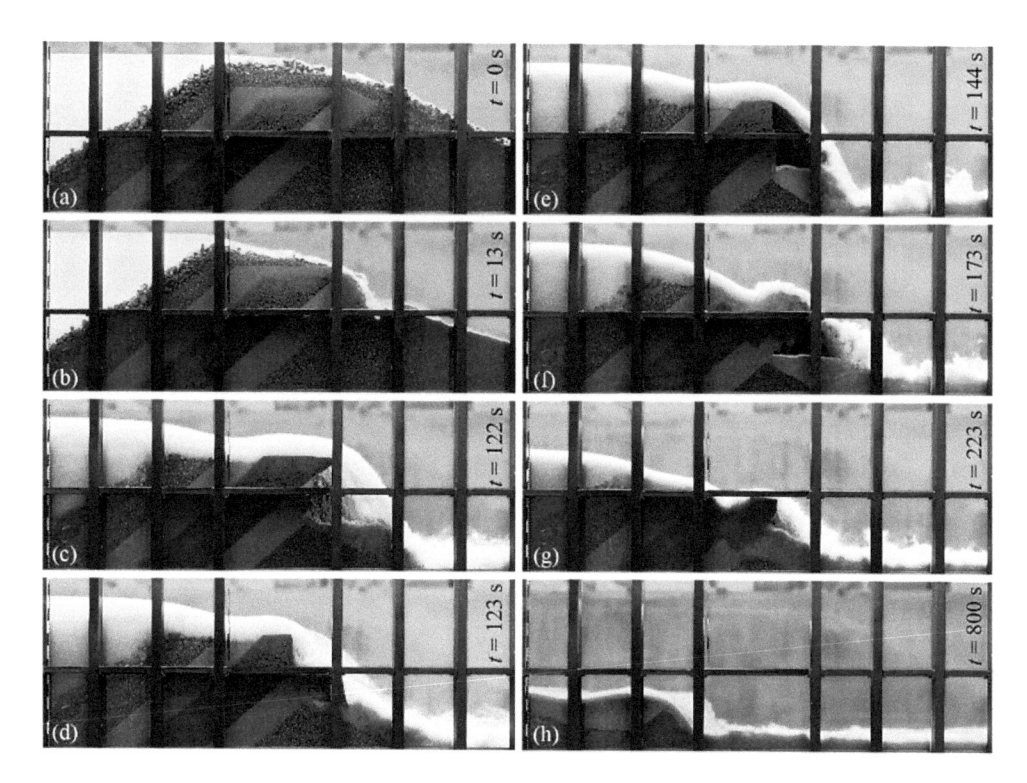

Figure 11.34 Temporal advance of fuse plug erosion with inclined core under prototype dimension at time t [s] = (a) 0, (b) 13, (c) 122, (d) 123, (e) 144, (f) 173, (g) 223, (h) 800 (Schmocker *et al.*, 2013)

remained stable to an overflow depth of $h_o = 0.06$ m above the crest. Then the erosion process was initiated due to a surface slip of the surface protection ($t = 0$ s). The breach advanced fast as the surface protection was removed and the body material was eroded. The core was undercut and collapsed due to its own weight and the water load. Equal to the small-scale tests, the core was no more undercut with ongoing erosion but remained stable at a height of 0.30 m at $t = 800$ s. The duration of erosion for the prototype fuse plug was around $t \approx$ 13 min.

Due to the limited unit inflow of $q \approx 0.3$ m²/s, the relative hydraulic load is smaller as compared with prototype conditions where a constant water level in River Aare is maintained. The core may therefore erode completely in a prototype event. However, despite the 0.30 m remaining fuse plug height, the required unit spillway capacity of $q \approx 1.0$ m²/s may be provided, as an overflow depth of $h_o = 0.90$ m and a corresponding unit discharge of $q \approx$ 1.4 m²/s are reached. The feasibility of the fuse plug design and the erosion advance was thus verified in the prototype as well.

Figure 11.35 shows the 3D erosion process across the sidewall and from above for prototype times using a half model at 1:5 scale in the 1.0 m wide channel. The breach started at $t = 0$ s with discharge entering the pilot channel. The erosion first advanced vertically because the pilot channel was located along the channel sidewall. After 63 s, an erosion channel

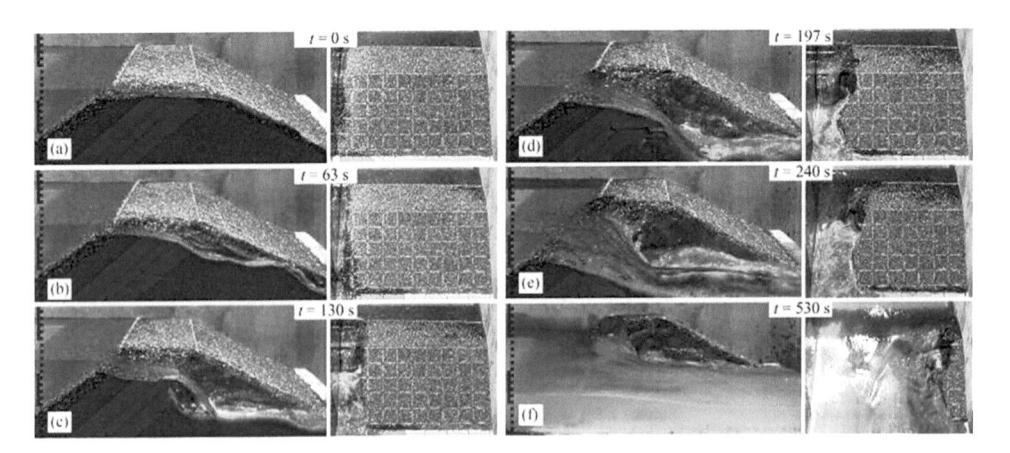

Figure 11.35 Temporal advance of 3D fuse plug erosion for 1:5 scale test at prototype times t [s] = (a) 0, (b) 63, (c) 130, (d) 197, (e) 240, (f) 530 (Schmocker *et al.*, 2013)

parallel to the downstream face slope formed along the sidewall. The surface protection was completely removed whereas the impermeable core partly collapsed similar to the 2D case. In phase 2, the breach developed laterally due to tractive shear stress and turbulence. Undercutting of the breach channel side slopes and the core caused large sediment volumes to collapse into the channel center from where it was transported downstream. The impermeable core had only a minor effect on the lateral erosion process as it was continuously undercut.

The discharge was continuously increased to maintain a constant reservoir level but prevent entire overtopping of the fuse plug. The breach developed both in vertical and lateral directions with time reaching the center of the 1.0 m wide channel after $t \approx 240$ s. Assuming that the channel wall represents the breach centerline and that the breach migrates in both directions, the lateral erosion rate (including initial breach in the pilot channel) was $v_e = 0.25$ m/min in 1:5 scale, and $v_e \cong 0.5$ m/min or $v_e \cong 30$ m/h in prototype scale, respectively. To guarantee the complete erosion of the 300 m long fuse plug within one hour, ten designated pilot channels have to be installed.

For Design B, only the prototype test is described. As the fuse plug with a sand body has no impervious layer, seepage may affect its stability. The reservoir upstream of the fuse plug was therefore filled up to 0.20 m below the crest and its level was kept constant. The seepage line reached the downstream toe after 4 h along with a continuous unit seepage discharge of $q = 6.5 \cdot 10^{-6}$ m²/s. Given the clear seepage flow, significant washout of the filter material was again excluded. The water level was then continuously raised until initiating erosion of the fuse plug. A small slip of the surface protection was observed once the overflow depth was $h_o = 0.03$ m, and the erosion started once $h_o = 0.05$ m at $t = 0$ s (Figure 11.36). The surface protection and the filter on the downstream slope were completely eroded within seconds, from when the fuse plug body was eroded. Compared to the fuse plug with the inclined core, the sand body was eroded continuously and no slope failures or collapses of large fuse plug parts were observed. The downstream slope remained nearly constant until reaching the seepage protection at $t = 100$ s, from when it acted as fixed point resulting in a continuous reduction of the fuse plug slope.

Figure 11.36 Temporal advance of fuse plug erosion with sand body, prototype times
t [s] = (a) 0, (b) 25, (c) 50, (d) 75, (e) 100, (f) 200 (Schmocker *et al.*, 2013)

The fuse plug eroded down to the seepage protection within t = 3 min. However, the seepage protection was not eroded in the hydraulic model and the final fuse plug height remained therefore stable at w = 0.25 m. A reservoir drop had to be accepted during the tests due to the limited discharge. Again, the overflow depth for a constant upstream water level would be h_o = 0.95 m resulting in a unit spillway discharge of $q \approx 1.5$ m²/s, thus achieving the required discharge. The seepage protection height may further be reduced as the risk of seepage erosion of the fuse plug was considered small. The fuse plug design with the sand body thus satisfies all stated requirements, providing an equivalent solution as the fuse plug with inclined core.

Discussion of results

To compare the various tests, the temporal evaluation of the maximum (subscript m) fuse plug height $w_m(t)$ was determined from the camera images and converted to prototype dimension. Figure 11.37a shows $w_m(t)$ for the 2D breach tests with inclined core (Tests 1–4, 6–8). As the surface protection was eroded during the initial phase, $w_m(t)$ is similar for all tests until the core height is reached. Then, $w_m(t)$ has a high scatter due to the random process of core collapse. Further, differences result due to: (1) Reservoir drop for 1:1 tests. The unit discharge in prototype scale was $q \approx 0.3$ m²/s compared with $q \approx 10$ m²/s for the 1:5 tests. Their erosion process should therefore progress comparatively faster; (2) although the core was properly scaled, the core stability may vary due to construction techniques. As the core was inserted in 0.10 m layers and compacted at 1:1 scale, the core could not be compacted at 1:5 scale but was prebuilt and inserted in one piece; and (3) effects of channel width resulted due to the adhesion of the core to the channel sidewall at the 1:5 and 1:2.5 scales. The core may further stabilize itself due to an arch effect.

However, the overall erosion process and erosion velocity were similar for all tests and the erosion end was always reached within some 500 s. The two 1:1 tests (Tests 7, 8) have an equal decrease in w_m including four main core breaks until the core remains stable once it is no more undercut at $w_m \approx 0.30$ m. The 1:2.5 scale test erodes slightly faster and the core remains stable at $w_m \approx 0.16$ m. Only Tests 1 and 4 of the 1:5 scale tests were eroded completely. Tests 2 and 3 eroded comparatively slower and remained stable at $w_m \approx 0.50$ m. The reason for these differences was attributed to the sidewall effect. As the fuse plug erosion depends mainly on the core stability, the setup procedure and material characteristics may further affect fuse plug erosion in small scale. Test 4, for which the fuse plug remained dry for 4.5 months, had a nearly equal erosion process as Test 1 of which the fuse plug was eroded immediately after setup.

Figure 11.37b shows $w_m(t)$ for the 2D breach tests with the sand body (Tests 9, 10). Both the 1:5 and 1:1 models have an identical decrease. Compared with the inclined core fuse plug, w_m reduces continuously with ≈ 0.35 m/min. vertical erosion velocity. The final stage is attained once the erosion reaches the seepage protection so that both fuse plugs remained stable at $w_m \approx 0.25$ m. Although the fine sand was not scaled for the 1:5 tests, no major scale effects on the erosion process occurred because the hydraulic conditions during the breach are much above the sediment entrainment condition.

The fuse plug erosion process with the inclined core is similar to these of Tinney and Hsu (1962) or Pugh (1985). The erosion times however deviate from the literature data. The initial breach section of the 4 m high fuse plug tested at Oxbow Dam eroded within 100 s to its base, much faster than in the 1:1 tests with the 1.2 m high fuse plug. The reason was stated to be the constant water level adapted at the Oxbow field test. The lateral erosion rate of $v_e = 0.25$ m/min for Test 5 with a fuse plug height of 0.24 m (1:5 scale) was also below literature values. The lateral erosion rates after the initial breach were $v_e = 0.46$ m/min for a 0.30 m high fuse plug (Pugh, 1985), and 1 m/min for a 0.20 m high fuse plug (Tinney and Hsu, 1962). However, the lateral erosion rate was determined after the breach in the pilot channel had reached the fuse plug base as the lateral erosion rate in the VAW tests was determined including the erosion in the pilot section, as the fuse plug was only 1.0 m long.

Pugh's (1985) empirical formula to predict the lateral erosion rate v_e in [ft./h] for fuse plugs of height w [ft.] with an inclined core between 3 and 9 m high is

$$v_e = 13.2w + 150. \tag{11.30}$$

This would result in $v_e = 62$ m/h for the present fuse plug, compared with the observed prototype value of 30 m/h in Test 5. The VAW small-scale tests therefore not directly compare with the large-scale field tests. Given the progressive breach of the inclined core, the material characteristics, and the different setups, a scatter in the lateral erosion rate is expected. However, the various models tested at VAW demonstrate the adaptability of small-scale tests to investigate the fuse plug erosion process.

In summary, extreme flood events may be controlled by a fuse plug embankment. Two fuse plug designs were studied and the general feasibility was proved using both small-scale and prototype model tests. The main fuse plug body consisted of highly erodible material covered with a coarse slope protection. For Design A, an inclined impermeable core prevented seepage and internal erosion for water levels below the trigger water level, whereas for Design B the fuse plug body consisted of fine sand providing sufficient impermeability to prevent seepage failure. Both designs meet the required criteria as they remained stable for

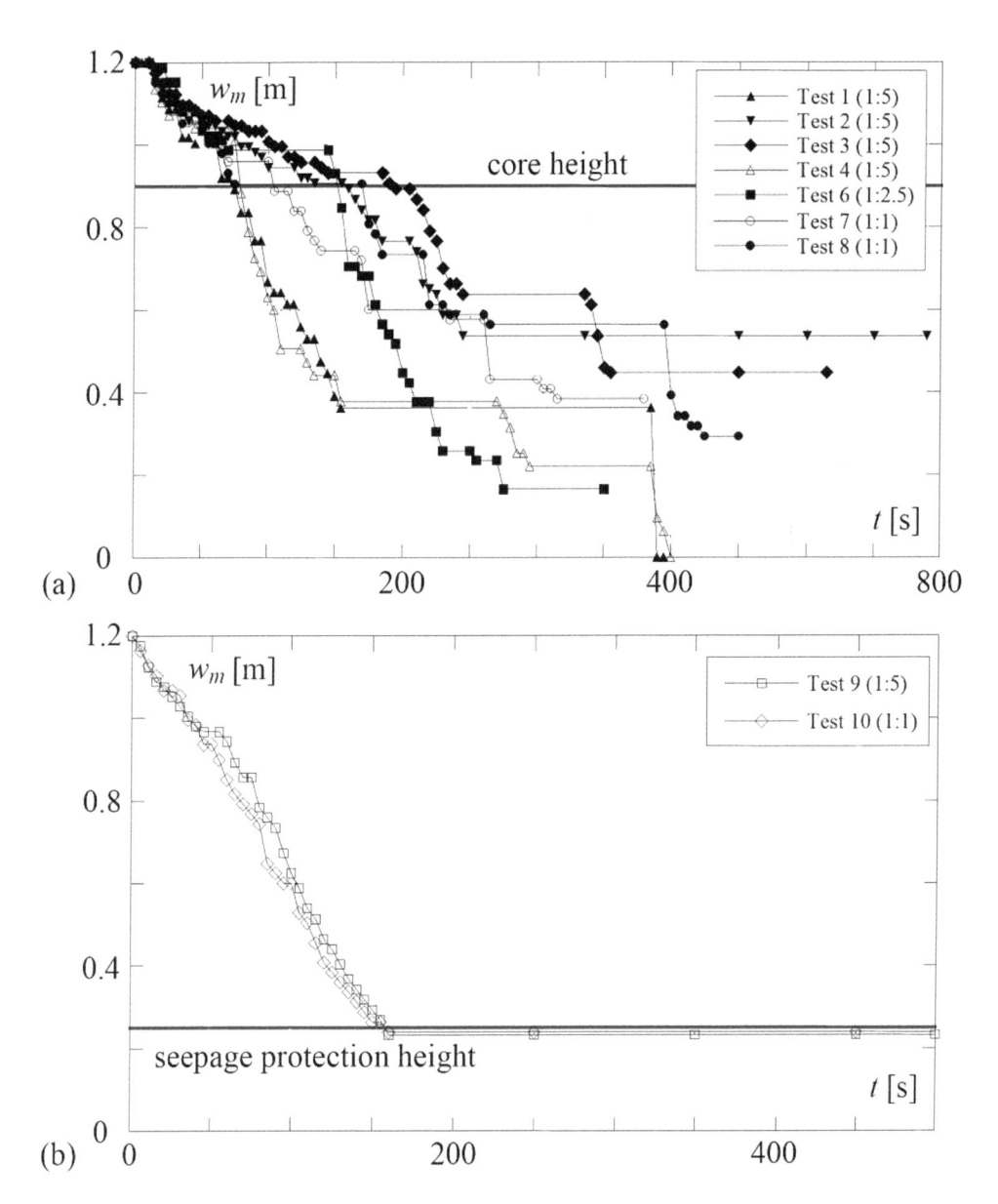

Figure 11.37 Temporal advance of maximum fuse plug height $w_m(t)$ in prototype dimension for 2D erosion with (a) inclined core, (b) sand body (Schmocker *et al.*, 2013)

water levels below the trigger water level and eroded fast once being overtopped. For Design A, the erosion was dominated by the highly erodible embankment body material followed by undercutting and collapse of the impermeable core. The fuse plug eroded stepwise due to a collapse of the core. Design B breached continuously as the material was entrained. In 1:1 scale, the fuse plug with the core reached its stable end stage after roughly twice the time of the fuse plug with the sand body. However, no complete fuse plug erosion occurred for both designs. Either the inclined core remained stable once it was no more undercut (Design A) or the tested seepage protection was too stable (Design B).

The fuse plug Design A was successfully adopted for any scale and especially for spillways at dam structures. The small fuse plug height adopted would result in a high constructional effort. Further, as the fuse plug operates only during extreme floods, a water level reaching the fuse plug is rare and complete impermeability is not required. The fuse plug Design B was thus selected for Hagneck Canal, mainly due to its simple structure and constructional ease. In general, the fuse plug concept with the fine sand body is thus preferable for small fuse plug heights as along rivers, whereas the fuse plug with the inclined core appears favorable as an addition for dam spillways.

Several points need to be kept in mind when transferring the results from model tests to prototype application. Both fuse plug stability and erosion process depend on the material characteristics, construction technique and compaction methods. The fuse plug performance may thus vary in prototype as the construction must be simplified compared with the model tests. Further, the long-term behavior was not tested and the scale models lack additional slope protection. The erosion will therefore initiate for higher overflow depths as compared with the model tests. A maintenance concept is therefore inevitable to guarantee a correct long-term operation of any fuse plug.

11.5 Instantaneous 2D breach

11.5.1 De Saint-Venant equations

The equations of the dambreak wave have been derived by De Saint-Venant (1871) assuming 1D flow for which the pressure distribution is hydrostatic and the velocity distribution uniform. With x as the location measured from the dam section in the downstream direction, t as time from break start, v as cross-sectional average velocity, h as the flow depth, A as the cross-sectional area, S_o as the bottom slope and S_f as the friction slope (Figure 11.38), these equations read (Liggett, 1994; Chaudhry, 1993)

$$\frac{\partial A}{\partial t} + \frac{\partial (vA)}{\partial x} = 0,$$
(11.31)

$$\frac{1}{g}\frac{\partial v}{\partial t} + \frac{\partial}{\partial x}\left(h + \frac{v^2}{2g}\right) = S_o - S_f.$$
(11.32)

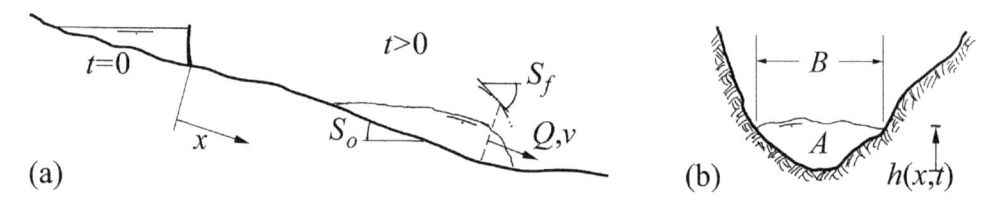

Figure 11.38 Definition of main variables in dambreak flow (a) streamwise, (b) transverse sections (Vischer and Hager, 1998)

The continuity Eq. (11.31) satisfies conservation of mass, i.e. balances the temporal change of cross-sectional area plus the spatial change of the discharge. The second, dynamic equation requires that the temporal change of velocity plus the spatial change of the energy head is equal to the bottom slope minus the friction slope.

With B as the free-surface width, the wave celerity c is

$$c = (gA / B)^{1/2}. \tag{11.33}$$

The characteristic equations of the hyperbolic system (11.31), (11.32) have interesting features and read (e.g. Abbott, 1966)

$$\frac{dv}{dt} \pm \frac{g}{c}\frac{dh}{dt} = g(S_o - S_f), \tag{11.34}$$

$$\frac{dx}{dt} = v \pm c. \tag{11.35}$$

First, the two sets of Eqs. (11.31, 11.32) and (11.34, 11.35) are completely equivalent. Whereas the first set is expressed as system of nonlinear partial differential equations, the second set comprises two ordinary differential equations (compatibility equations) valid along two characteristic curves defined by Eqs. (11.35). These plot two curves on the x–t plane, referred to as the positive (+) and the negative (−) characteristics. For subcritical flow ($v<c$) the slope of the negative characteristic is negative, whereas the slope of the positive characteristic is positive. For supercritical flow ($v>c$) the slope of both characteristics is positive (Figure 11.39). Solving the first set of equations involves rectilinear coordinates, currently a standard procedure for finite-difference methods (Chaudhry, 1993). The method of characteristics, in turn, has been popular until the 1970s. The computation involves nonlinear coordinates so that the locations of the unknowns $h(x, t)$ and $v(x, t)$ have to be computed by interpolation. If hydraulic shocks are generated, the second set of equations may be of particular interest.

The systems (11.31; 11.32) and (11.34; 11.35) are equations for the unknowns velocity v and flow depth h as functions of location x and time t. They are subject to appropriate boundary and initial conditions. Figure 11.40 shows the x–t plane on which the two unknowns $v(x, t)$ and $h(x, t)$ are sought in the method of characteristics. A distinction between subcritical flow $v<c$, and supercritical flow $v>c$ is required. From Cunge *et al.* (1980), and with x_0, x_1 as

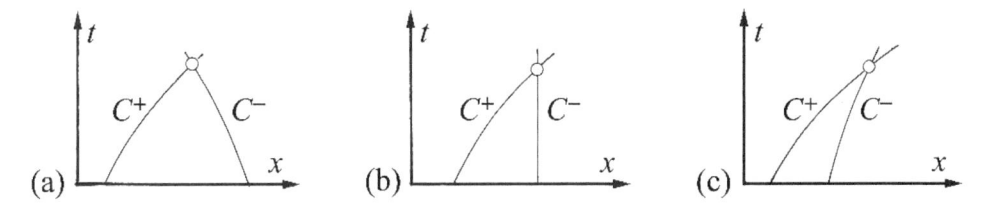

Figure 11.39 Characteristic curves in x–t plane for (a) subcritical, (b) critical, (c) supercritical flow (Vischer and Hager, 1998)

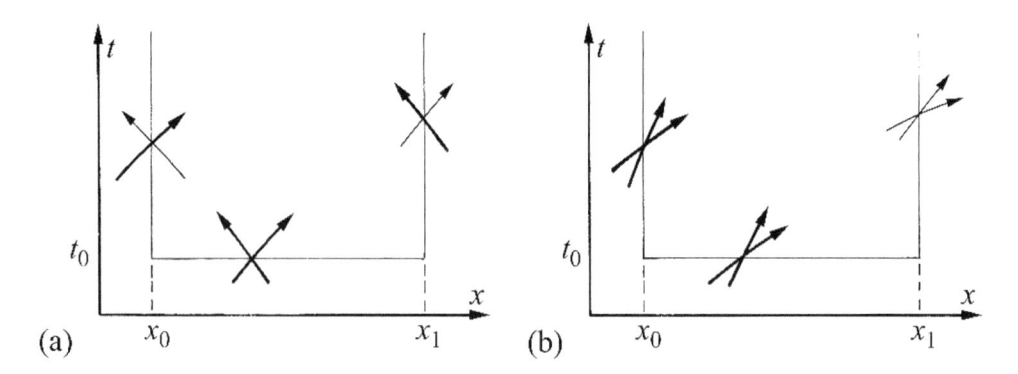

Figure 11.40 Characteristics at limits of computational domain, boundary or initial conditions in bold for (a) subcritical, (b) supercritical flow

up- and downstream ends of the computational domain, one boundary condition has to be specified for each characteristic entering the computational domain at its limits. The initial condition represents a special boundary condition in time. Therefore:

- Two conditions have to be imposed as initial conditions at $t = t_0$,
- One boundary condition is required both at the up- ($x = x_0$) and downstream ($x = x_1$) boundaries for *subcritical* flow, and
- Two conditions define the upstream boundary ($x = x_0$) for *supercritical* flow.

The finite-difference methods are discussed by e.g. Chaudhry (1993) or Liggett (1994). Particular aspects involve the computational stability of various schemes, their mathematical consistencies and the treatment of initial and boundary conditions. This is not further treated here, given the standard works in computational hydraulics (Castro-Orgaz and Hager, 2019).

11.5.2 Ritter's solution

Although de Saint-Venant (1871) solved his equations for the special case in which the friction slope is compensated for by the bottom slope, i.e. $S_o - S_f = 0$, for all x and t, Ritter in 1892 first used these equations for dambreak flows in rectangular channels (Hager and Chervet, 1996). The approach was extended by Su and Barnes (1970) to cross sections of triangular ($\lambda = 2$), parabolic ($\lambda = 3/2$) and rectangular ($\lambda = 1$) shape by introducing the shape factor $a^2 = A/(Bh) = \lambda^{-1/2}$, so that the wave celerity is $c = a(gh)^{1/2}$. The generalized characteristic equations then read with $w = (2/a) \cdot (gh)^{1/2}$

$$\frac{\mathrm{d}(v \pm w)}{\mathrm{d}t} = 0, \quad \text{or} \quad (v \pm w) = \text{const. along} \tag{11.36}$$

$$\frac{\mathrm{d}x}{\mathrm{d}t} = v \pm c . \tag{11.37}$$

Because the water is initially at rest, their solution reads

$$(gh)^{1/2} = \left(\frac{a}{2+a^2}\right)\left[\frac{2}{a}(gh_o)^{1/2} - \frac{x}{t}\right], \tag{11.38}$$

$$v = \left(\frac{2}{2+a^2}\right)\left[a(gh_o)^{1/2} + \frac{x}{t}\right]. \tag{11.39}$$

Obviously, this result satisfies Froude similitude.

Let the upstream (subscript o) stagnant flow depth be $h = h_o$ (Figure 11.41a). With the dimensionless quantities scaled to length h_o, time $(h_o/g)^{1/2}$, and velocity $(gh_o)^{1/2}$, the solution reads with $C = a^{-1}(h/h_o)^{1/2}$ and $V = v/(gh_o)^{1/2}$

$$C = \left(\frac{a}{2+a^2}\right)\left[2 - a\frac{X}{T}\right], \tag{11.40}$$

$$V = \left(\frac{2}{2+a^2}\right)\left[a + \frac{X}{T}\right]. \tag{11.41}$$

The functions C and V are shown in Figure 11.41 as $h/h_o = (C/a)^2$ and $v/(gh_o)^{1/2}$ versus $m = X/T$. The features of these plots are the positive propagation velocity $m_+(C = 0) = 2/a$, i.e. $m_+ = 2$, $6^{1/2} = 2.45$, $2^{3/2} = 2.83$ for the rectangular, parabolic, and triangular cross sections, respectively. In turn, the negative propagation velocity is $m_-(C = a) = -a$, i.e. $m_- = -1$, $-(2/3)^{1/2}$, $-(1/2)^{1/2}$, respectively. For equal depth h_o, a dambreak wave has the largest front velocity for the triangular section, but the largest upstream velocity for the rectangular section.

At the dam section (subscript 0) where $m(x = 0) = 0$, the flow depth remains constant at $h_0/h_o = [2/(2+a^2)]^2$, i.e. $h_0/h_o = (2/3)^2 = 0.44$, $(3/4)^2 = 0.56$, $(4/5)^2 = 0.64$, respectively. The free-surface shape is parabolic for all three cross sections, rather steep in the upstream portion, and tending to zero flow depth at the positive wave front. For the velocities, differences

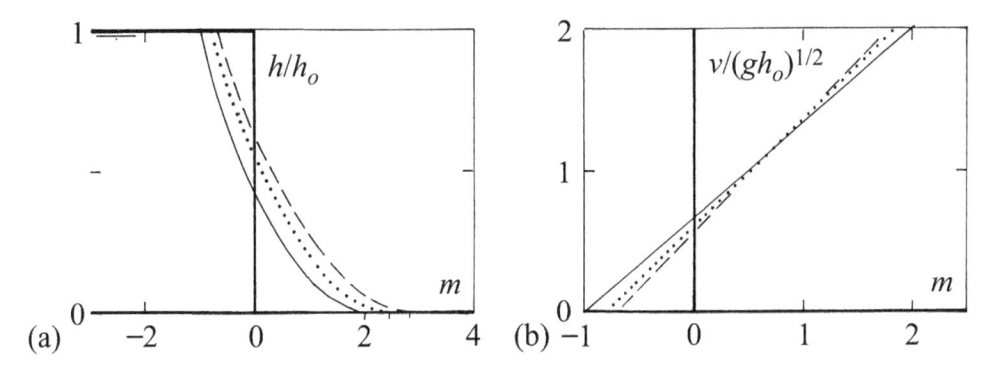

Figure 11.41 Dimensionless dambreak wave without friction for (—) rectangular, (. . .) parabolic, (- -) triangular horizontal channels (a) free-surface profiles $h/h_o(m)$, (b) velocity profiles $V = v/(gh_o)^{1/2}$ versus $m = X/T = x/[(gh_o)^{1/2}t]$ (Su and Barnes, 1970)

are even smaller than for the free-surface profile, as noted from Figure 11.41b. Note the linear velocity distribution between the two dambreak fronts. Most of the research on dambreak waves was conducted in rectangular channels. Note that the effect of cross-sectional shape is small, so that it may be often neglected.

Given the particular interest in the rectangular channel shape, Eqs. (11.40, 11.41) are restated here, corresponding to Ritter's dambreak wave solution, reading with $a = 1$

$$C = \left(\frac{1}{3}\right)\left[2 - \frac{X}{T}\right],$$

(11.42)

$$V = \left(\frac{2}{3}\right)\left[1 + \frac{X}{T}\right].$$

(11.43)

Note the extremely simple solution for a complicated phenomenon. Its main features are $C(X/T = 2) = 0$, so that $X/T = 2$ defines the positive wave front, whereas $C(X/T = -1) = 1$, corresponding to the location of the negative wave front, at which $h = h_o$. The relative velocities at these two sections are $V(X/T = -1) = 0$ and $V(X/T = 2) = 2$. Note that the energy head $H_e = h + V^2/2g$ at these two locations and at the dam section ($X = 0$) related to flow depth h_o is $H/h_o(X/T = -1) = 1$, $H/h_o(X/T = 0) = 2/3$, and $H/h_o(X/T = 2) = 2$. Further, the corresponding Froude numbers $\mathsf{F} = V/(gh)^{1/2}$ equal $\mathsf{F}(X/T = -1) = 0$, $\mathsf{F}(X/T = 0) = 1$, and $\mathsf{F}(X/T = 2) \to \infty$. These features evidence that the dambreak flow solution close to the positive wave front can hardly match physical reasoning, so that a special treatment is required, as presented below.

11.5.3 Dressler's asymptotic solution

Dressler (1952) studied the frictional effect on the dambreak wave in the rectangular, horizontal channel by setting for the friction slope according to Darcy-Weisbach

$$S_f = \frac{f}{4R_h}\frac{v^2}{2g}.$$

(11.44)

For a wide rectangular channel, the hydraulic radius is $R_h = h$, so that $S_f = fv^2/(8gh)$. The governing system of equations is, from Eqs. (11.31; 11.32) with $C = (h/h_o)^{1/2}$, $V = v/(gh_o)^{1/2}$, $X = x/h_o$, $T = (g/h_o)^{1/2}t$ and $R_f = f/8$

$$2\frac{\partial C}{\partial T} + C\frac{\partial V}{\partial X} + 2V\frac{\partial C}{\partial X} = 0,$$
$$\frac{\partial V}{\partial T} + V\frac{\partial V}{\partial X} + 2C\frac{\partial C}{\partial X} + R_f\frac{V^2}{C} = 0.$$

(11.45)

A perturbation solution for small values of TR_f was considered, expressed as

$$C = C^0 + C^1 TR_f + C^2(TR_f)^2 + ...,$$

(11.46)

$$V = V^0 + V^1 TR_f + V^2(TR_f)^2 + ...$$

(11.47)

Here C^0, V^0 are the zeroth-order solutions according to Ritter, as defined in Eqs. (11.42; 11.43). With the perturbation parameters $\sigma = TR_f$, $M = (1+m)/(2-m)$, and when accounting for expansions up to order 1, the approximate solution reads

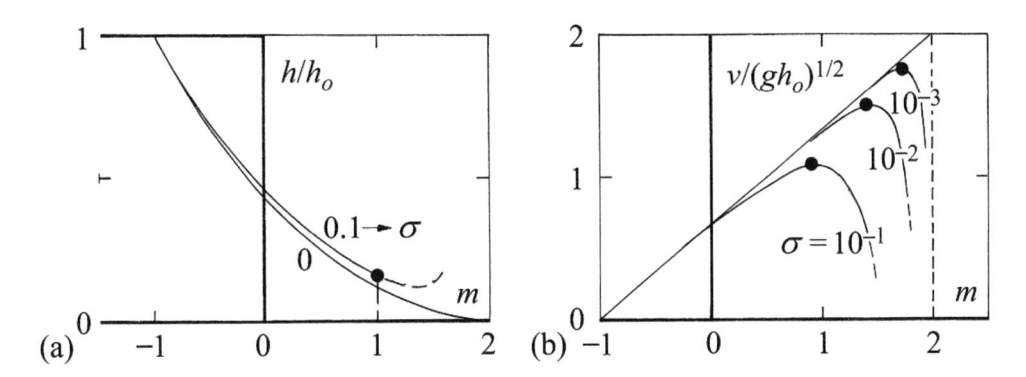

Figure 11.42 Dambreak wave according to Dressler (1952) (a) relative flow depth h/h_o, (b) relative velocity $v/(gh_o)^{1/2}$ versus m for various σ with ($\bullet$) wave tip location

$$C = \frac{1}{3}(2-m)\left[1-\frac{1}{2}M^2\sigma\right],$$
(11.48)

$$V = \frac{2}{3}(1+m)\left[1-\frac{2M}{2-m}\sigma\right].$$
(11.49)

For $\sigma = 0$ results the original Ritter solution which agrees with experiments up to about $m = 1$ (Dressler, 1952). At the wave front portion, the frictional effect is significant, so that the terms in the square brackets become relevant. Note that friction causes a much larger effect on the velocity V than on the celerity C.

Figure 11.42 shows the relative flow depth h/h_o and the relative velocity $v/(gh_o)^{1/2}$ versus $m = X/T$ for various σ. For $\sigma = 0$ the solutions are as for the rectangular channel (Figure 11.41). For $\sigma > 0$, there is nearly no velocity modification if $m<1$ (Figure 11.42b). For larger m, however, the curves split from the curve with $\sigma = 0$, similar to a boundary layer. Dressler (1952) suggested that the solution for $\sigma>0$ is valid up to the maximum (subscript M) velocity V_M, and introduced at the wave front (subscript F) the so-called wave tip of constant velocity $V_F = V_M$. From Eq. (11.48), the maximum of σ for any value of m is $\sigma_M = (2-m)^2/(12M)$. Inverting gives $m_M \cong 2-3\sigma^{1/3}$, so that

$$V_M = 2(1-\sigma^{1/3})\left[1-\frac{2}{3}\sigma^{1/3}(1-\sigma^{1/3})\right].$$
(11.50)

For $\sigma = 0.1$ (Figure 11.42a), $C_M = 0.90$ and with $m_M = 0.61$, the corresponding flow depth is from Eq. (11.48) $h/h_o = 0.19$. The tip location is thus at $m = 1$. Dressler (1954) checked these results against laboratory data noting agreement with his wave tip approach.

11.5.4 Pohle's 2D approach

The shallow-water equations are based on the assumptions of hydrostatic pressure and uniform velocity distributions. To be sure, these assumptions are invalid close to the wave fronts propagating in the up- and downstream directions. At small time, the domain close to the dam section is thus governed by streamline curvature effects.

By using the Lagrangian representation for a particle located at position x from the dam section at elevation z above the horizontal floor, Pohle (1952) set

$$x = x_o + x^{(1)}(x_o, z_o)t + x^{(2)}(x_o, z_o)t^2 + \ldots, \tag{11.51}$$

$$z = z_o + z^{(1)}(x_o, z_o)t + z^{(2)}(x_o, z_o)t^2 + \ldots. \tag{11.52}$$

The functions $x^{(1)}$; $z^{(1)}$ represent velocities whereas $x^{(2)}$; $z^{(2)}$ are particle accelerations. Because the water is initially at rest, $x^{(1)} = z^{(1)} = 0$. Using conformal mapping,

$$x^{(2)}(x_o, z_o) = \frac{g}{2\pi} \ln \left[\frac{\cos^2\left(\frac{\pi z_o}{4h_o}\right) + \sinh^2\left(\frac{\pi x_o}{4h_o}\right)}{\sin^2\left(\frac{\pi z_o}{4h_o}\right) + \sinh^2\left(\frac{\pi x_o}{4h_o}\right)} \right], \tag{11.53}$$

$$z^{(2)}(x_o, z_o) = -\frac{g}{\pi} \arctan \left[\frac{\sin\left(\frac{\pi z_o}{4h_o}\right)}{\sinh\left(\frac{\pi x_o}{2h_o}\right)} \right]. \tag{11.54}$$

If expansions are carried to second order, Eqs. (11.51) to (11.54) define the free-surface profile at small times t. For $z_o = h$ the result is $x^{(2)} = 0$ at the free surface, i.e. all particles move in the vertical direction immediately after dambreak initiation. Up- and downstream from the dam section, the free surface is defined by Eq. (11.54) as

$$\frac{h}{h_o} = 1 - \frac{gt^2}{\pi h_o} \arctan\left[\left(\sinh\frac{-\pi x}{2h_o}\right)^{-1}\right], \quad x \le 0, \tag{11.55}$$

$$\frac{h}{h_o} = \frac{4}{\pi} \operatorname{arccot}\left[\exp\left(\frac{\pi x}{gt^2}\right)\right] - \frac{gt^2}{2h_o}, \quad x \ge 0. \tag{11.56}$$

At the dam section ($x = 0$), one has simply

$$\frac{h}{h_o} = 1 - \frac{gt^2}{h_o}. \tag{11.57}$$

At the first instant, therefore, the particle located at the edge between the reservoir surface and the dam section moves like a free-falling body. Later, the free surface is curved, in the reservoir with the center of curvature below the free surface, and vice versa downstream from the dam section (Figure 11.43). Equations (11.55) and (11.56) are approximated with $X = x/h_o$, and $T = (g/h_o)^{1/2} \cdot t$ as

$$\frac{h}{h_o} = 1 - \frac{1}{2}T^2[1 + \tanh X], \quad X \le 0; \tag{11.58}$$

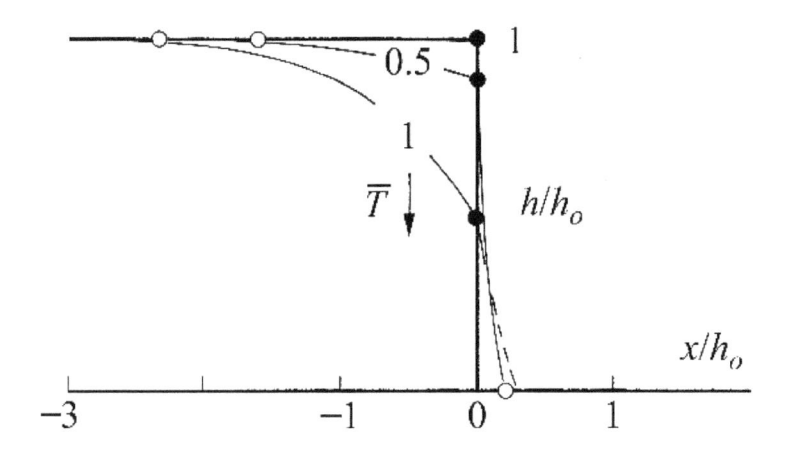

Figure 11.43 Solution of Pohle (1952) as modified by Martin (1990) for initial dambreak wave ($T\leq 0.7$) with (•) water depth at dambreak section, (○) positive and negative wave fronts

$$\frac{h}{h_o}=1-\tanh(2X/T^2)-\frac{1}{2}T^2,\quad X\geq 0\,. \tag{11.59}$$

Letting $h/h_o\rightarrow 0.99$ and 0, respectively, the wave front locations $X_F(T)$ follow

$$X_F=\mathrm{Arctanh}\left(\frac{0.02}{T^2}-1\right),\quad X_F<0, \tag{11.60}$$

$$X_F=\frac{1}{2}T^2\mathrm{Arctanh}\left(1-\frac{1}{2}T^2\right),\quad X_F\geq 0\,. \tag{11.61}$$

Note that $X_F\rightarrow 0$ at $T\rightarrow 2^{1/2}$. The validity of Eqs. (11.60) and (11.61) is thus limited to $T\ll 1$, as demonstrated by Martin (1990). For very small times T, Eq. (11.61) follows

$$X_F=\frac{1}{2}T^{3/2}\,. \tag{11.62}$$

This is much smaller than $X_F=2T$ according to Ritter.

Martin (1990) presented a complete model for the dambreak wave in the reservoir domain ($x\leq 0$) using Pohle's solution for $T<0.7$, and a modified Ritter solution for larger times. This solution was found accurate until wave reflections occur from the upstream wall of the horizontal channel. The discharge across the dambreak section is

$$Q_o/Q_R=T/0.7,\quad T<0.7\,. \tag{11.63}$$

It is identical to $Q_o = Q_R = (8/27)b(gh_o^3)^{1/2}$ given by Ritter (subscript R). The discharge decreases as soon as the negative wave has reached the upstream boundary.

11.5.5 Hunt's asymptotic solution

Another relevant solution to the dambreak wave is due to Hunt (1984). A plane of slope S_o contains a dam initially filled with water up to height h_o (Figure 11.44). If locations far downstream of the dam are considered, the leading terms of Eq. (11.32) are $S_o - S_f = 0$. A so-called outer solution exists by using the Darcy-Weisbach equation

$$S_f = \frac{fV^2}{8gh} \tag{11.64}$$

and the continuity equation for the rectangular channel

$$\frac{\partial h}{\partial t} + \frac{\partial(Vh)}{\partial x} = 0. \tag{11.65}$$

Eliminating the friction slope S_f, and with the scaling parameters (involving asterisk *) h^*, V^*, L^*, $t^* = L^*/V^*$ yields, with $U = V/V^*$, $\Psi = h/h^*$, $\chi = x/L^*$, and $\tau_H = t/t^*$,

$$1 - \frac{U^2}{\Psi} = 0. \tag{11.66}$$

This setting represents the kinematic wave approximation. Eliminating the velocity in Eq. (11.66) gives for $\Psi(\chi,\tau_H)$ the nonlinear partial differential equation of first order

$$\frac{3}{2}\Psi^{1/2}\frac{\partial \Psi}{\partial \chi} + \frac{\partial \Psi}{\partial \tau_H} = 0. \tag{11.67}$$

The characteristic equations corresponding to Eq. (11.67) require $d\Psi/d\tau_H = 0$ along $d\chi/d\tau_H = (3/2)^{1/2}$. Note that this system of equations involves a kinematic shock (subscript s) $\chi = \chi_s(\tau_H)$ of which $\Psi = 0$ in front of this shock. Upstream from the shock

$$\Psi = (2\chi/3\tau_H)^2, \quad 0 < \chi \leq \chi_s. \tag{11.68}$$

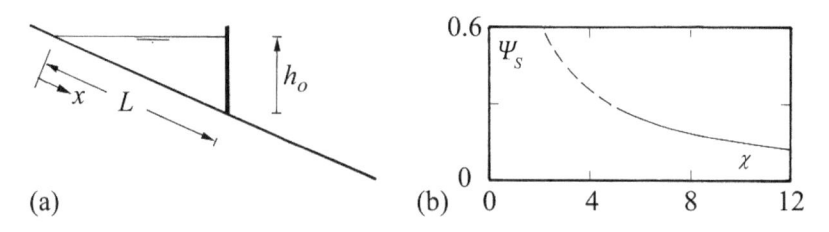

(a) (b)

Figure 11.44 Kinematic wave approximation for dambreak wave (Hunt, 1984) (a) parameter definition, (b) shock height Ψ_s versus location χ according to Eq. (11.69)

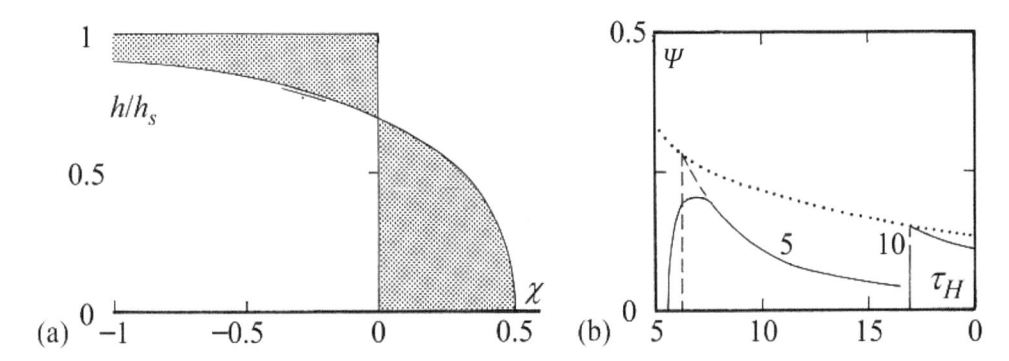

Figure 11.45 (a) Wave tip region $h/h_s(\chi)$ according to Eq. (11.71), (b) dambreak wave $\Psi(\tau_H)$
according to Hunt (1984) with (. . .) shock profile, (- - -) outer solution, (—)
composite solution

At the shock results with $A = (1/2)L/h_o$ as normalized reservoir volume

$$\Psi_s = \left(\frac{2A}{\tau_H}\right)^{2/3}, \quad \text{or} \quad \Psi_s = \frac{3}{2\chi}. \tag{11.69}$$

The solution is correct for locations $\chi = x/L \geq 5$ downstream of the dam (Hunt, 1984), in which
the origin of x is at the reservoir upstream end (Figure 11.44a).

The inner solution relates to the hydraulic shock mechanism. From an asymptotic analysis, Hunt (1984) demonstrates that the velocities close to the shock vary with τ but not with
Ψ. Conservation of mass gives an additional relation to solve the system of the governing
equations. The shock location χ_s and the free-surface profile Ψ_s are

$$\chi_s = \frac{3}{2}\tau_H^{2/3} + \frac{1}{2}\tau_H^{-2/3}, \tag{11.70}$$

$$\frac{\chi - \chi_s}{\Psi_s} = \Psi + \ln(1-\Psi) + \frac{1}{2}. \tag{11.71}$$

Figure 11.45a shows Eq. (11.71). Due to mass conservation the shaded areas on both sides of
$\chi = 0$ are equal. The composite solution is equal to the sum of Eqs. (11.70) and (11.71) minus
that of Eq. (11.69) for $\chi < \chi_s$, and equal to Eq. (11.71) at the tip region (Figure 11.45b). Hunt
stressed that his matched asymptotic expansion becomes asymptotically valid once the shock
has moved over five times the reservoir length.

11.5.6 Front treatment

Chanson (2009) attempted to describe the front portion of the dambreak problem using
analytical solutions based on the method of characteristics. His results apply to either the
horizontal or the constantly sloping rectangular channel initially completely dry downstream
from the dam section, and to the turbulent flow motion.

Figure 11.46 shows the dambreak wave in the horizontal, initially dry rectangular channel. Along the wave tip region ($x_1 < x < x_s$), the flow velocity remains at a certain time nearly constant ($dv/dt = \partial v/\partial t + v\partial v/\partial x = 0$ in Eq. 11.32); the governing dynamic equation reduces with $S_f = fv^2/(8g)$ for a wide rectangular channel with f as the friction factor assumed to be constant along the wave front to

$$\frac{\partial h}{\partial x} + \frac{fv^2}{8h} = 0. \tag{11.72}$$

Note that x_1 describes the transition from the Ritter wave profile to the wave front portion, and x_s the proper wave front. Integrating Eq. (11.72) gives the parabola

$$h = \frac{f^{1/2}v}{2}(x_s - x)^{1/2}. \tag{11.73}$$

Note the vertical free-surface slope dh/dx at the wave front required to balance the significant frictional effect.

At the transition between the two wave portions $x = x_1$, both the flow depth h and the velocity v must be identical to generate a continuous wave profile. Accordingly, two equations result from Eqs. (11.42) and (11.43)

$$C_1 = \left(\frac{1}{3}\right)\left[2 - \frac{X_1}{T}\right], \tag{11.74}$$

$$V_1 = \left(\frac{2}{3}\right)\left[1 + \frac{X_1}{T}\right]. \tag{11.75}$$

Further, mass conservation must be satisfied, so that the fluid mass in the wave tip region ($x_1 < x < x_s$) equals that with ($x_1 < x < 2(gh_o)^{1/2}t$). Equating yields (Chanson, 2009)

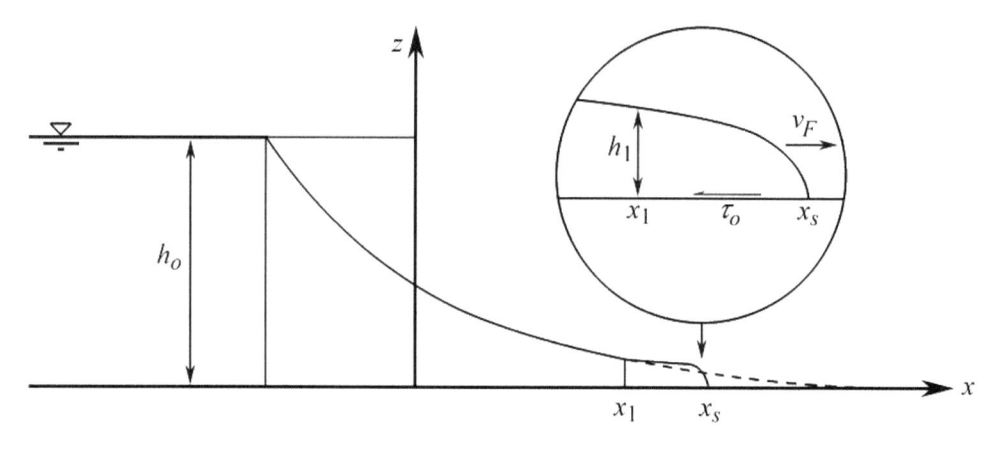

Figure 11.46 Wave front portion with wave tip region (right) and Ritter wave (left) (Adapted from Chanson, 2009)

$$T = \left(\frac{8}{3f}\right)\left[\frac{(1-(1/2)V)^3}{V^2}\right]. \tag{11.76}$$

The dimensionless wave front location then is

$$X_s = \left(\frac{3}{2}V - 1\right)T + \frac{4}{fV^2}\left(1-\frac{V}{2}\right)^4. \tag{11.77}$$

Further, the free-surface profile satisfies from Eqs. (11.42) and (11.73)

$$\frac{h}{h_o} = \left(\frac{1}{9}\right)\left[2-\frac{X}{T}\right]^2, -T \leq X \leq \left[\frac{3}{2}V-1\right]T, \tag{11.78}$$

$$\frac{h}{h_o} = \left(\frac{f^{1/2}V}{2}\right)[X_s - X]^{1/2}, \left[\frac{3}{2}V-1\right]T \leq X \leq X_s. \tag{11.79}$$

The solution of the dambreak wave is now complete, with the velocity versus time in Eq. (11.76), and the wave profile specified by Eqs. (11.78) and (11.79).

According to Chanson (2009), the assumption of a constant friction factor in the wave tip region is rough. For computational reasons, a simplified expression for f was therefore considered, in which the effects of relative roughness height and Reynolds number are contained, thereby accounting for wall friction and viscous effects. A similar procedure than for a constant friction factor was conducted, resulting in more complex expressions. Comparing with experimental data results in a good agreement both for the wave front, and the composite wave profile. Given that most of the experimental data are almost historical, a new research should be conducted involving modern instrumentation.

Chanson (2009) also considered the effect of a mild bottom slope for the dambreak wave tip region, based on Peregrine and Williams (2001) and Chanson (2005). The details of this

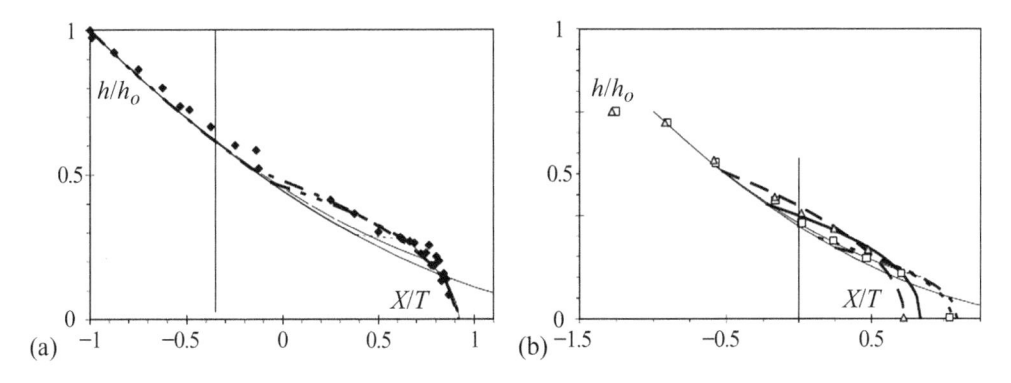

Figure 11.47 Comparison of analytical model and experimental data for dimensionless free-surface profiles $h/h_o(X/T)$ (Adapted from Chanson, 2009) (a) from (—) Eq. (11.74), (– –) Eqs. (11.78, 11.79) for $f = 0.03$, (– –) Dressler (1952), (···) Whitham (1955) and experimental data of Schoklitsch for $h_o = 0.074$ m, $t = 9.4$ s, (b) at various times

approach are not further presented here, however, and recourse to the mentioned works should be made. Figure 11.47 shows two typical plots highlighting Chanson's validation of his approach with experimental data. A more recent fully analytical solution of the dambreak wave tip region was conducted by Deng *et al.* (2018).

11.5.7 Experimental approach

Experimental observations

Lauber (1997) conducted experiments in a rectangular prismatic and hydraulically smooth channel. Based on preliminary experimentation, it was found that dambreak flows follow essentially Froude similitude if the initial flow depth is $h_o \geq 0.300$ m. Effects of surface tension and viscosity are then negligible, at least up to (30 to 50)h_o downstream of the dam section. Also, a dambreak may be considered instantaneous if the period of dam removal is smaller than $1.25(h_o/g)^{1/2}$. All tests were conducted in the initially dry channel, because the effect of initial tailwater depth was found to be significant. Figure 11.48 compares the wave fronts of the dry with the initially wetted downstream channels, noting that the wave front is smooth and continuous in the dry, but of bore-type in the initially wetted channel. The wetting effect may be compared with flows in hydraulically rough channels.

The data presented have been obtained with a high-speed video camera positioned successively at several locations along the 14 m long, 0.50 m wide and 0.70 m high test channel. Observations were taken with a temporal interval of 50 ms, and were repeated at the next section. A vertical gate without gate slots was used as the rupture mechanism resulting in a high degree of reproduction without generating any shock waves or other flow disturbances. Both flow depths and time-averaged velocities were measured to 1 mm, and to ±0.05 m/s. The bottom slope S_o was varied from horizontal to +50% (26.5°). This data set is unique allowing both for the formulation of a semi-empirical approach, and representing a data basis for advanced numerical methods. A summary of the results is provided by Lauber and Hager (1998 a, b).

Horizontal smooth channel

Figure 11.49a shows a definition sketch of the hydraulic configuration studied. A dam of reservoir length L_B is initially filled with water up to a height h_o. The origin of the coordinate

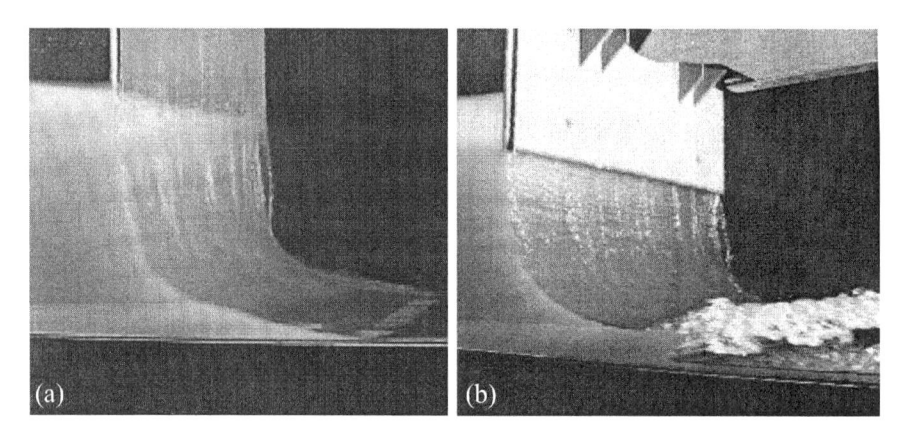

Figure 11.48 Wave fronts close to dam section of initially (a) dry, (b) wetted downstream channel (Lauber, 1997)

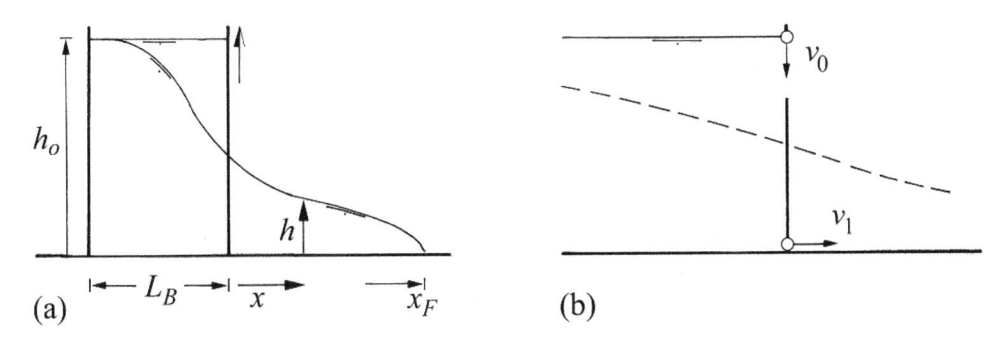

Figure 11.49 Definition of dambreak wave in (a) horizontal channel, (b) (——) initial condition $h(x, 0)$ and (- - -) free surface after long time (Adapted from Lauber, 1997)

system $(x; h)$ is located at the toe of the dam section. At time $t = 0$, the dam is suddenly removed and the questions to be answered refer to the distribution of flow depth $h(x, t)$, cross-sectional average velocity $v(x, t)$, and discharge $Q(x, t)$. Particular questions refer to the wave front (subscript F), the maximum (subscript M) cross-sectional flow depth, and the corresponding time of occurrence.

Consider the initial conditions (Figure 11.49b) with the typical velocities v_0 at the top and v_1 at the bottom of the dam section. Because all particles are accelerated from rest, $v_0 = gt$ and the depth $h_M = (4/9)h_o$ are reached within a short time. The velocity head of the bottom particle is $v_1^2/2g = (5/9)h_o$ along with the front velocity $v_1 = [(5/9)2gh_o]^{1/2}$. The developing phase is referred to as the initial dambreak wave, in which $v_F = v_1$. The initial wave phenomenon is mainly governed by orifice-type flow at the dam section. The so-called dynamic dambreak wave with origin at the free surface of the dam section has a velocity $v_0 = 2(gh_o)^{1/2}$ according to Ritter and starts at time $T_0 = 2^{1/2}$, after having reached the channel bottom. Up to time T_0 this particle behaves essentially as a free-falling body (Pohle, 1952). Because $v_0 > v_1$, the dynamic wave overtakes the initial wave at time $2(gh_o)^{1/2}(T-2^{1/2}) = [(5/9)2gh_o]^{1/2}$, i.e. at $T = 3.0$ upon assuming that the bottom slope compensates the frictional effect (Figure 11.50).

During the first instances, the flow in the vicinity of the dam section is governed by streamline curvature effects, and a 2D approach is appropriate (11.5.4). This phase is terminated once the dynamic dambreak wave starts at the dam section, i.e. at time $T_0 \cong 2^{1/2}$. Then, all variables except for these at the front vary gradually, so that the de Saint-Venant shallow-water-equations apply. Time T_0 is neither affected by friction nor by the bottom slope because of closeness to the dam section, and short duration.

POSITIVE WAVE FRONT

The positive dambreak wave front (subscript F) in a rectangular channel of bottom angle α is determined by Eqs. (11.34) and (11.35) as

$$\frac{d(v+2c)}{dt} = g(\sin \alpha - S_f) \text{ along } \frac{dx}{dt} = v + c. \tag{11.80}$$

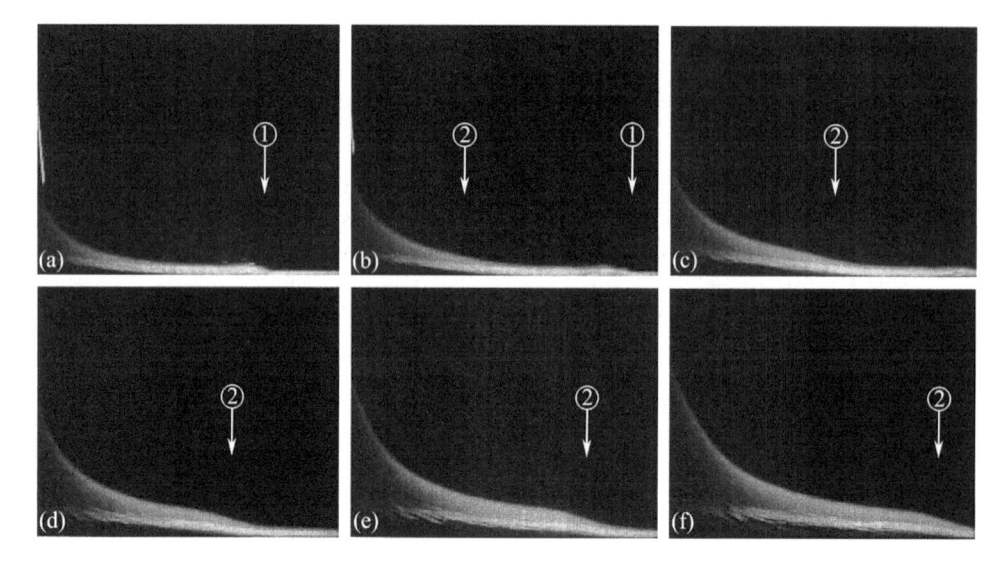

Figure 11.50 (a) to (c) Formation, (d) to (f) initial wave ① passed by dynamic wave ② (Lauber, 1997)

Integrating the dimensionless Eq. (11.80) subject to the initial conditions $V_F(T=2^{1/2})=2$ and $c=0$ at the wave front gives

$$V_F = 2 + (\sin\alpha - S_f)(T - 2^{1/2}). \tag{11.81}$$

If the hydraulic radius is replaced by the flow depth h, the friction slope at the wave tip region, i.e. close to the dambreak front, is

$$S_f = \frac{f_a}{4h_a}\frac{v^2}{2g}. \tag{11.82}$$

The friction coefficient f_a and the corresponding flow depth h_a are taken as averages $h_a = \sigma h_o$ and $f_a = 0.20 R_a^{-0.20}$ for turbulent smooth flow. The average Reynolds number is $R_a = 4q/v$ with $q = (8/27)(gh_o^3)^{1/2}$ as the maximum discharge per unit width. For the tests with $h_o = 0.30$ m, $q = 0.251$ m²s⁻¹ and thus $R_a = 0.61 \times 10^6$, the average computed friction factor is $f_a = 0.014$. Inserting the average friction slope $S_f = V_F^2 f_a (8\sigma)$ in Eq. (11.81) and solving the quadratic equation for V_F yields, with $\tau = T - 2^{1/2}$ and $j = S_o \sigma/f_a$,

$$V_F = \frac{4\sigma}{f_a\tau}\left[\left(1 + \frac{f_a}{\sigma}\tau\left(1 + \frac{1}{2}S_o\tau\right)\right)^{1/2} - 1\right]. \tag{11.83}$$

Two cases are relevant:

- $\tau \ll 1$, for which $V_F = 2 + O(\tau)$, i.e. no effect of neither bottom nor friction slopes. The front velocity V_F then is equal to Ritter's prediction, and
- $\tau \gg 1$, for which to order τ^{-1}

$$V_F = 2(2j)^{1/2}\left[1+\frac{1-(2j)^{1/2}}{S_o\tau}\right]. \tag{11.84}$$

Figure 11.51a compares data of Lauber (1997) with Eq. (11.84). All velocities start at the initial velocity $V_1 = 1.05$, jump to the dynamic (subscript d) wave velocity $V_d = 2$ following the prediction for larger time. For all flows both in horizontal and sloping channels, the coefficient is $\sigma = 0.06$ and the tip region has a height of about 6% of the initial water depth h_o. For $j = 0.5$, corresponding to $S_o = j/4 = 0.125$, the front velocity remains constant at $V_F = 2$. For large time, the front tends from Eq. (11.84) to the limit velocity $V_{F\infty} = 2(2j)^{1/2}$. Note that $V_{F\infty}$ corresponds to the uniform velocity of the tip region. Pseudo-uniform flow conditions are reached as $\tau > 40$, for $j > 0.25$.

The wave front location $X_F(\tau)$ is determined by $dX_F/d\tau = V_F$ thereby imposing the initial condition $X_F(\tau = 0) = 0$. Integrating Eq. (11.83) leads to a singularity. Therefore, the modified Eq. (11.84) was used by imposing $V_F(\tau = 0) = 2$, i.e. with $\overline{\tau} = (f_a/\sigma)\tau$

$$V_F = 2(2j)^{1/2}\left[1+\frac{1-(2j)^{1/2}}{S_o\tau+(2j)^{1/2}}\right], \quad j > 0.1; \tag{11.85}$$

$$V_F = \frac{4}{\overline{\tau}}[(1+\overline{\tau})^{1/2}-1], \quad j = 0. \tag{11.86}$$

Integration subject to $X_F(\tau = 0) = 0$ gives

$$X_F = 2(2j)^{1/2}\left[\tau + \frac{1-(2j)^{1/2}}{S_o}\ln\left(1+\frac{S_o\tau}{(2j)^{1/2}}\right)\right], j > 0.1, \tag{11.87}$$

$$X_F = \frac{4\sigma}{f_a}[2(1+\overline{\tau})^{1/2}-2+\ln\left[\left(4\frac{[1+\overline{\tau}]^{1/2}-1]}{\overline{\tau}[(1+\overline{\tau})^{1/2}+1]}\right)\right], j = 0. \tag{11.88}$$

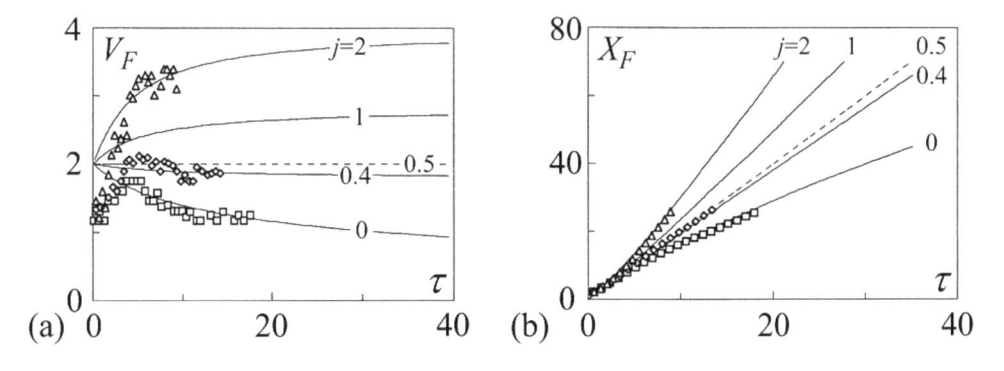

Figure 11.51 (a) Front velocity $V_F(\tau)$ for various bottom slopes $j = S_o\sigma/f_a$ with $f_a/\sigma = 0.25$, (b) Front location $X_F(\tau)$ with $\tau = T-2^{1/2}$, for $S_o = (\square)$ 0, $(\lozenge)$ 0.1, $(\triangle)$ 0.5, (- - -) equilibrium slope (Lauber and Hager, 1998a)

Figure 11.51b compares data with Eqs. (11.87) and (11.88). The positive wave front is thus determined by an analytical approach.

NEGATIVE WAVE FRONT

According to Ritter, the negative wave front (subscript nF) has a propagation velocity $V_{nF} = -1$. Figure 11.52a compares the data with this value, yet deviation is noted. According to Figure 11.52b showing surface profiles in the reservoir, Ritter's solution is correct below the surface, but a surface current moves faster than the wave body.

According to Lauber (1997), the negative front velocity is $v_{nF} = -(2gh_o)^{1/2}$, or

$$V_{nF} = -2^{1/2}.$$
(11.89)

Integrating subject to the initial condition $X_{nF}(T = 0) = 0$ gives for the front location

$$X_{nF} = -2^{1/2}T, \quad T < 7.$$
(11.90)

Note that the initial positive wave front velocity is smaller than predicted by Ritter, but the negative wave front velocity is significantly larger. Ritter's solution currently is used for the initiation of dambreak waves. As a result, the present initialization of dambreak waves is in contrast with observations.

MAXIMUM FLOW CHARACTERISTICS

At any location $x>0$ the flow depth is initially $h = 0$ up to the arrival of the positive wave front, increases to a maximum depth h_M and decreasing again. In the horizontal channel, the maximum (subscript M) flow depth is influenced by the relative location (x/h_o) and the relative reservoir length $\lambda_B = L_B/h_o$. Figure 11.53a shows $Y_M = h_M/h_o$ versus the combined parameter $\bar{X} = \lambda_B(x/h_o)^{-2/3}$, resulting in perfect agreement. The data are expressed, with notations barred, as (Lauber, 1997)

$$Y_M = \frac{4}{9}(1 + \bar{X}^{-1})^{-5/4}.$$
(11.91)

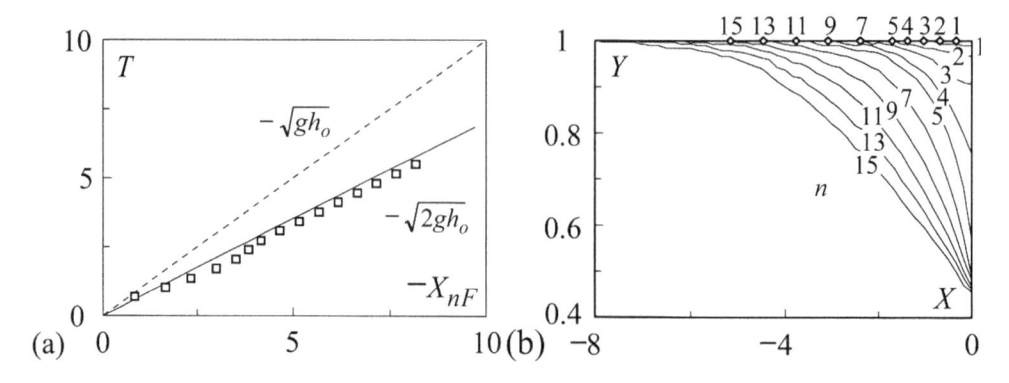

Figure 11.52 Negative dambreak wave front (a) wave location $X_{nF}(T)$ according to (- - -) Ritter, (—) Eq. (11.90), (□) observations, (b) surface profiles $Y(X)$ at successive times $T = n \times \Delta T$ with $T = (g/h_o)^{1/2} \times 60$ ms $= 0.343$, (◊) Eq. (11.90) (Lauber and Hager, 1998a)

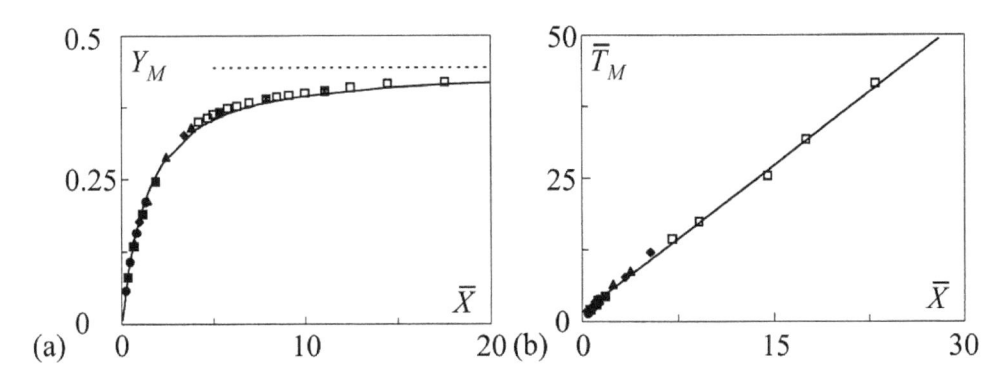

Figure 11.53 Dambreak wave maximum (a) maximum flow depth $Y_M = h_M/h_o$ with (—) Eq. (11.91) and (···) $Y_M = 4/9$, (b) time $\overline{T}_M$ versus non-dimensional location $\overline{X} = \lambda_B(x/h_o)^{-2/3}$ with (—) Eq. (11.93) (Lauber and Hager, 1998a)

Interestingly, the maximum flow depth increases linearly with the relative reservoir length and decreases with $X^{2/3}$. The controlling parameter is $\overline{X} = L_B/(h_o x^2)^{1/3}$. The absolute maximum $Y_M = 4/9$ agrees with Ritter's solution and is asymptotically reached as $X \to 0$. For $X \gg 1$, the asymptotic solution of Eq. (11.91) is

$$h_M / h_o = (4/9)[L_B^3 / (h_o x^2)]^{5/12} . \tag{11.92}$$

The maximum flow depth is thus related to L_B, and the relative distance x/h_o.

The time of maximum flow depth $\overline{T}_M = (g/h_o)^{1/2}(x/h_o)^{-2/3}t_M$ increases linearly with $\overline{X} = \lambda_B X^{-2/3}$ as shown in Figure 11.53b as (Lauber, 1997)

$$\overline{T}_M = 1.7(1+\overline{X}). \tag{11.93}$$

This time corresponds to the propagation times of the negative wave upstream to the reflection boundary plus that to section $\overline{X}$. At $\overline{X} = 0$, it is $\overline{T}_M = 1.7$ from Eq. (11.93).

The propagation velocity of the wave maximum is $d\overline{X}_M/d\overline{T}_M = 1.7^{-1}$ from Eq. (11.93). The corresponding Froude number $F_M = (dX_M/dT)Y_M^{-1/2}$ is thus

$$F_M = \frac{4}{3}(1+X^{-1})^{5/8} . \tag{11.94}$$

For large X, the so-defined Froude number tends asymptotically to 4/3.

The general dambreak wave profile is normalized (subscript N) as follows: The origin at the wave front is $Y_N(T_N = T_F) = 0$, and $Y_N(T_N = T_M) = Y_M$ at the wave maximum. From the data analysis, the time scales of the rising $(T_N < T_M)$ and the falling wave branches are different resulting in the wave profiles, respectively, (Lauber, 1997)

$$Y_N = T_N^{*1/3} \quad \text{with} \quad T_N^* = \frac{T-T_F}{T_M-T_F}, \quad 0 \le T_N^* < 1; \tag{11.95}$$

$$Y = 1.1T_N^{-1} \quad \text{with} \quad T_N = \frac{T}{T_M}, \quad T_N \ge 1. \tag{11.96}$$

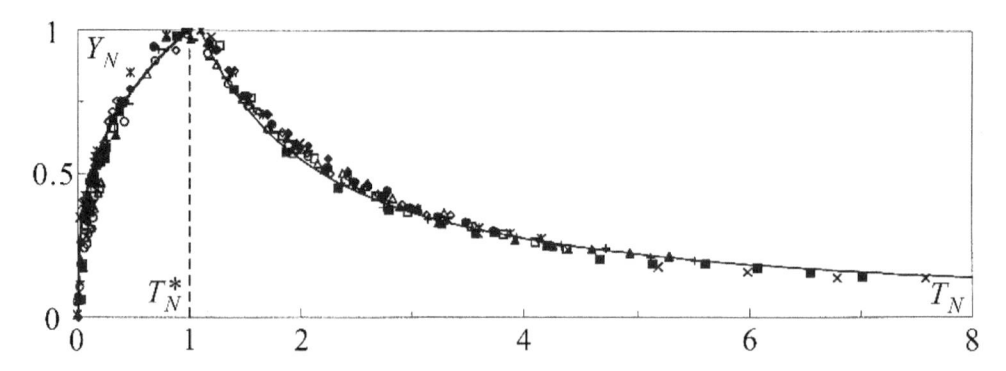

Figure 11.54 Generalized wave profile $Y_N(T_N^*)$ for $T_N^* < 1$ and $Y_N(T_N)$ for $T_N > 1$, (–) Eqs. (11.95), (11.96) (Lauber and Hager, 1998a)

These two equations are valid for $\lambda_B \geq 3$, and allow for determining the complete wave profile downstream of the dam section for $T < 50$ and locations $X > 0$ (Figure 11.54).

WAVE VELOCITY

The time-averaged velocity $v(x, t)$ was measured in turbulent smooth flow using PIV. At a certain time and location, the transverse velocity distribution is nearly uniform, except for a thin boundary layer close to the channel bottom. Reynolds numbers upstream from the tip region typically range from 10^5 to 10^6. Wüthrich *et al.*, (2018) confirmed the existence of a thin boundary layer using Ultrasonic Velocity Profilers. Their data indicate that a constant velocity profile neglecting the boundary layer is acceptable for waves propagating on a horizontal smooth channel.

Figure 11.55 shows the experimentally captured dimensionless velocity distribution $V(X)$ for $\lambda_B = 11.05$. Prior to wave reflection at the dam section ($T < 22$), the velocity distribution is nearly linear, starting at $X = -X_F$, crossing $V = 2/3$ according to Ritter at $X = 0$, and extending to the positive wave front according to Eq. (11.88). For larger times, the velocity profiles rotate about the end point $X = -\lambda_B$, are nearly straight up to a maximum and decrease gradually toward the front velocity. For small distances, the maximum velocity at a certain time is thus at the wave front, shifting progressively into the wave body for larger downstream locations. The velocity increase is thus almost linear from the negative to the positive wave fronts.

WAVE DISCHARGE

With given functions $Y(X, T)$ for flow depth, and $V(X, T)$ for velocity results in the discharge $Q(X, T) = YV$, in which $Q = q/(gh_o^3)^{1/2}$. Figure 11.56 shows the hydrographs $Q(T)$ at various locations X again for $\lambda_B = 11.05$. Note that the maximum discharge $Q_M = 8/27$ as predicted by Ritter is confirmed by the data at $X = 0$. For larger X, Q_M decreases considerably to, say, $Q_M(X = 27) = 0.2$ (−30%).

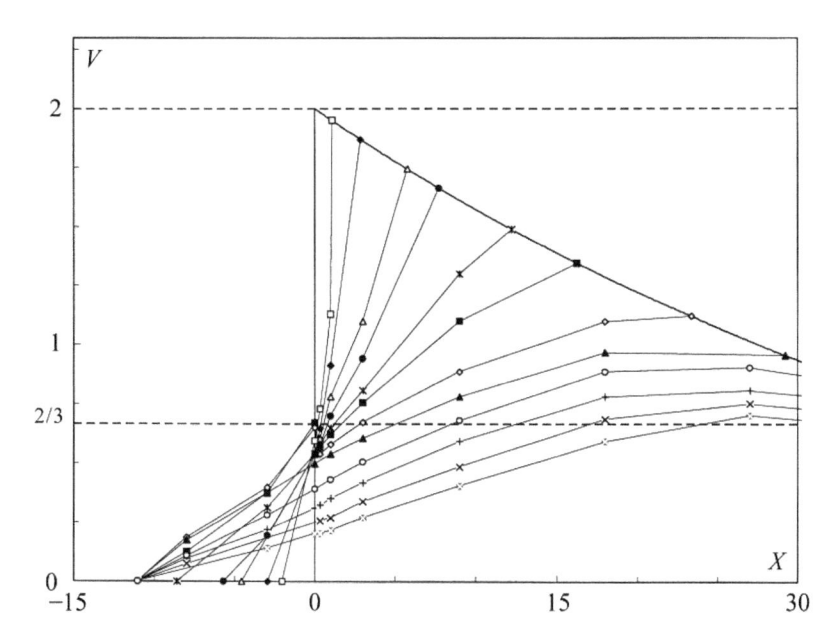

Figure 11.55 Velocity distribution $V(X)$ at times T = (□) 2.05, (♦) 3, (△) 4.57, (•) 5.7, (*) 8.6, (■) 11.45, (◊) 17.15, (▲) 22.9, (○) 28.6, (+) 34.3, (×) 40.0, (□) 45.75; (—) dynamic wave front, (- - -) Ritter velocity at dam section (Lauber and Hager, 1998a) copyright © International Association for Hydro-Environment Engineering and Research, reprinted by permission of Taylor & Francis Ltd, http://www.tandfonline.com on behalf of International Association for Hydro-Environment Engineering and Research

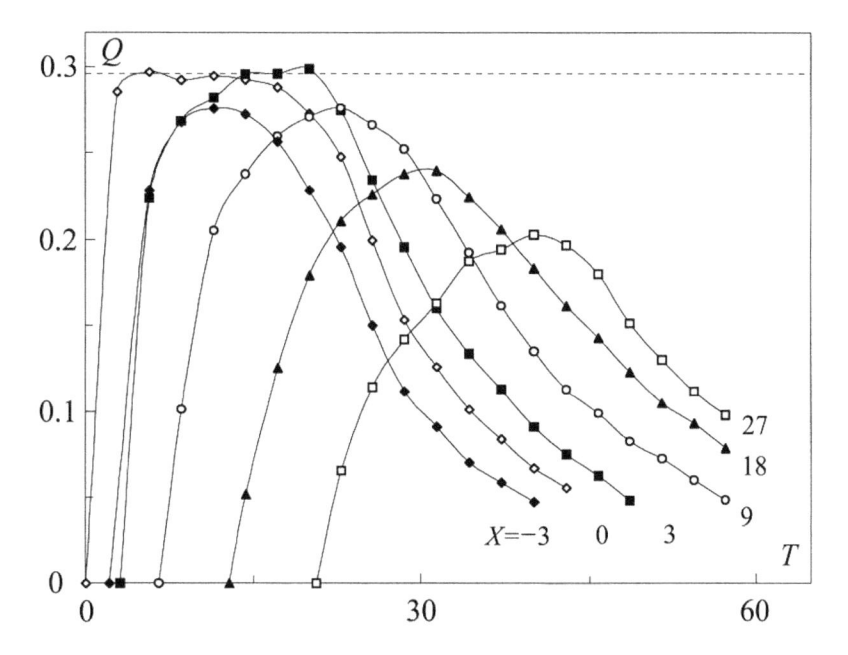

Figure 11.56 Discharge distribution $Q(T)$ at various locations X, with positive front locations and (- - -) Ritter's solution $Q_M = 8/27$ for $\lambda_B = 11.05$ (Lauber and Hager, 1998a)

Sloping smooth channel

WAVE PROFILES

Figure 11.57a shows a rectangular channel of bottom slope S_o. The flow depth h is measured perpendicularly to the bottom, and the origin of the streamwise coordinate x is at the bottom of the dam section. Note that the reservoir is limited to $L_B = h_o/S_o$. Non-dimensional coordinates are $X = x/h_o$ for location, $T = (g/h_o)^{1/2}\cdot t$ for time and $Y = h/(h_o\cos\alpha)$ for flow depth with α as the bottom inclination angle. Experiments were conducted for $S_o = 0.1$ and 0.5 to allow for the presentation of a generalized approach.

Figure 11.58 shows the wave profiles $Y(X)$ at various times T for $S_o = 0.10$ and 0.50. Their downstream portions flatten as the slope increases, whereas a maximum flow depth occurs upstream from the dam section. After long time, the reservoir is emptied, and a compact wave with moving positive and negative fronts runs down the channel.

The positive wave front was determined for both the horizontal and the sloping channels. Figure 11.59 shows the negative wave front in the sloping channel given by $dx/dt = -[g(h_o+x\sin\alpha)]^{1/2}$. Integrating subject to the initial condition $x(t = 0) = 0$ yields

$$T_R = \frac{2}{\sin\alpha}[1-(1-X\sin\alpha)^{1/2}].\tag{11.97}$$

Figure 11.59 compares the modified Ritter solution (subscript R) with observations, resulting in disagreement, as for the horizontal channel. Dividing T_R by $2^{1/2}$, i.e. $T_{nF} = T_R/2^{1/2}$, renders agreement as previously, so that (Lauber and Hager, 1998b)

$$T_{nF} = \frac{2^{1/2}}{\sin\alpha}\left[1-(1-X\sin\alpha)^{1/2}\right].\tag{11.98}$$

Solving for the wave front location gives

$$X_{nF} = -(\sin\alpha)^{-1}\left[1-\left(1-\frac{T\sin\alpha}{2^{1/2}}\right)^2\right].\tag{11.99}$$

Lauber (1997) confirmed the validity of Eqs. (11.98) and (11.99) also for $S_o = 50\%$.

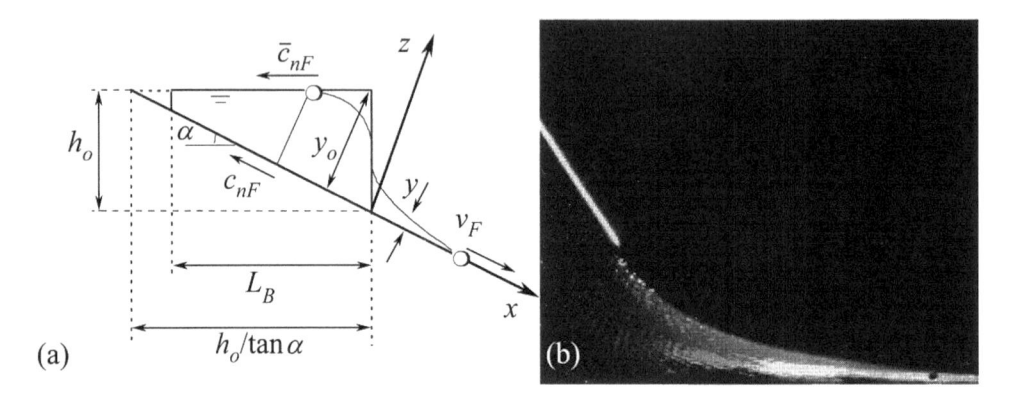

(a)　(b)

Figure 11.57 (a) Definition plot for dambreak wave in sloping channel, (b) initial wave for $S_o = 0.5$ with channel bottom parallel to lower border of photo (Lauber and Hager, 1998b) copyright © International Association for Hydro-Environment Engineering and Research, reprinted by permission of Taylor & Francis Ltd, http://www.tandfonline.com on behalf of International Association for Hydro-Environment Engineering and Research

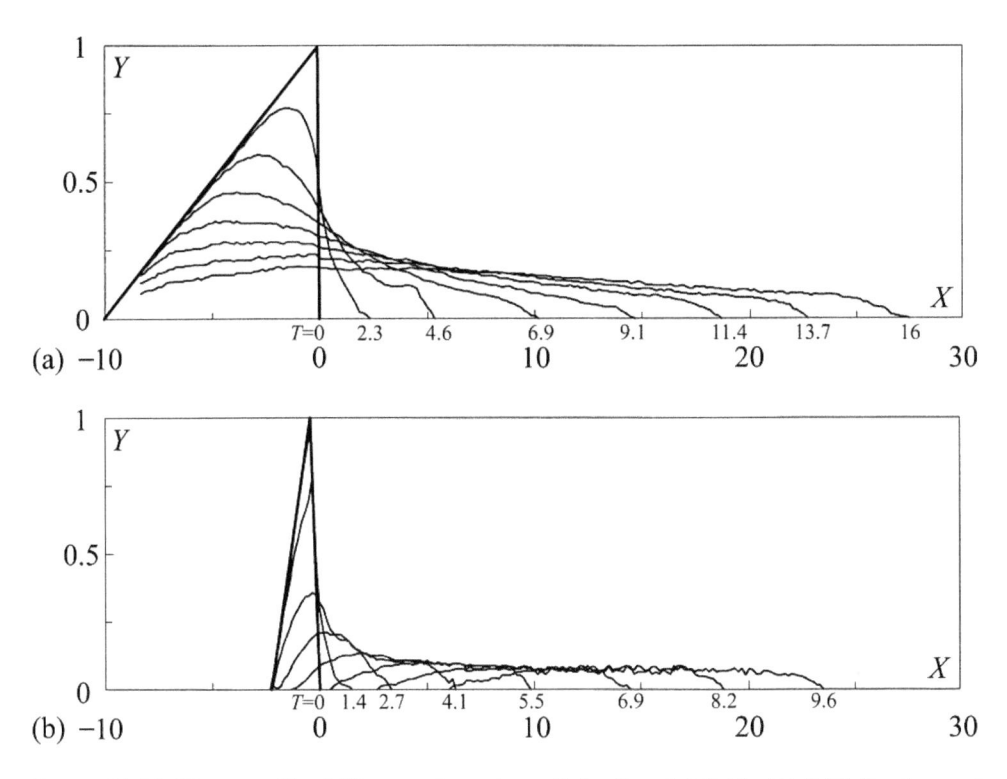

Figure 11.58 Wave profile $Y(X)$ at various times T for S_o = (a) 0.10, (b) 0.50 (Lauber and Hager, 1998b)

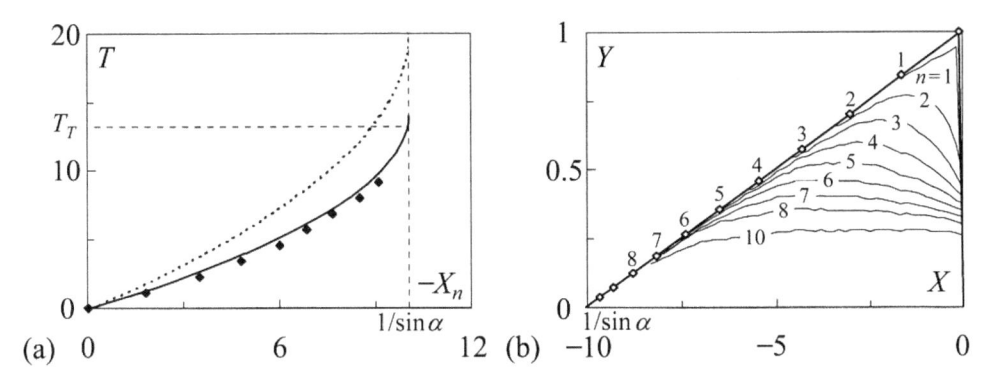

Figure 11.59 (a) Negative wave front for S_o = 10%, (- - -) Eq. (11.97), (♦) observations and (—) Eq. (11.98), (b) negative wave profiles $Y(X)$ at times $T = n \times \Delta T$ with $\Delta T = (g/h_o)^{1/2} \times 0.2$ s = 1.14, (◊) front locations according to Eq. (11.99) (Lauber and Hager, 1998b)

WAVE MAXIMA

The controlling wave maximum parameter in the horizontal channel is $\bar{X} = \lambda_B X^{-2/3}$. With $\varphi = (\sin\alpha)^{1/4}$ as the slope parameter, the dimensionless location is extended to $\bar{X} = \lambda_B^{1-\varphi} \cdot X^{-2/3+0.2\varphi}$. The maximum flow depth of sloping channels follows Eq. (11.91) as

$$Y_M = \frac{4}{9}(1+\bar{X}^{-1})^{-5/4} .$$
(11.100)

The agreement between observations and Eq. (11.100) is excellent if $\bar{X} > 0.2$. For large $\bar{X}$, i.e. large λ_B and small X, the maximum depth tends asymptotically to $Y_M = 4/9$, as predicted by Ritter. For small $\bar{X}$, i.e. large bottom slope, large distance or small reservoir length, the maximum flow depth tends to

$$Y_M = \frac{4}{9}\left[\frac{L_B}{h_o}\left(\frac{x}{h_o}\right)^{1/4}\right]^{\varphi}\left(\frac{h_o^5}{x^2 L_B^3}\right)^{5/12} .$$
(11.101)

If small bottom slopes $0 \leq S_o < 0.05$ are excluded, and with $L_B = h_o/S_o$, then the relevant parameter is $X^* = (h_o/x\cdot\sin\alpha)^{2/3}$. The maximum wave height is then simply (Figure 11.60)

$$Y_M = \frac{4}{9}[\tanh(0.067 X^*)]^{1/3} .$$
(11.102)

For small $X^* < 10$, $\tanh X^* \cong X^*$ so that $Y_M = 0.18(h_o/x\cdot\sin\alpha)^{2/9}$. The maximum flow depth thus decreases with decreasing location x/h_o but increasing angle α. The parameter X^* represents the ratio of bottom elevation difference between the dam section and point x, and the dam height. Accordingly, it does not matter whether x is large and $\sin\alpha$ small, but whether the product $(x\cdot\sin\alpha)$ is small or large compared with the initial flow depth h_o (Figure 11.60b).

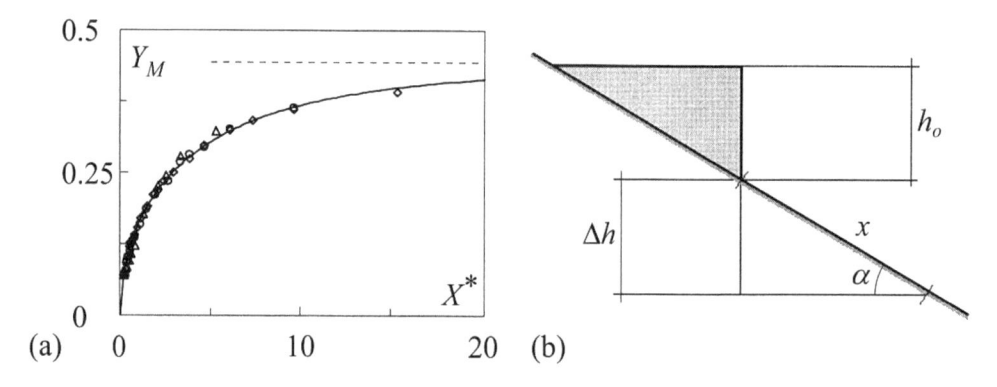

Figure 11.60 (a) Maximum flow depth $Y_M(X^*)$ with $X^* = [(h_o/x)\sin\alpha]^{2/3}$ for $S_o = (\lozenge)$ 0.1, ($\circ$) 0.2, (Δ) 0.5, (- - -) Ritter's solution, (b) sketch of ratio $\Delta h = x\cdot\sin\alpha$ and h_o (Lauber and Hager, 1998b)

The time of maximum flow depth occurrence T_M follows from Eq. (11.93). With $\bar{T}_M = X^{-2/3} \cdot T$ and $\bar{X} = \lambda_B^{1-\varphi} \cdot X^{-2/3+0.2\varphi}$, it follows up to any bottom slope $S_o \leq 50\%$

$$\bar{T}_M = 1.7(1+\bar{X}). \tag{11.103}$$

CHANNEL EMPTYING PROCESS

Whereas a horizontal channel empties only after a long time, depending mainly on the reservoir length, the emptying process of a sloping channel follows elementary hydraulics. Figure 11.61 shows the positive and negative wave fronts $X_F(T)$ and $X_{nF}(T)$. The overall wave length is $\Delta X = X_F - X_{nF}$ and increases with time, i.e. the wave length increases as it moves down a channel, so that the positive wave propagation is faster than wave recession. From Eq. (11.98), the upper reservoir end is reached as $X = -(\sin\alpha)^{-1}$, at the drying (subscript T) time $T_T = 2^{1/2}(\sin\alpha)^{-1}$. As the flow depth at the negative front is $h_{nF} = h_T = 0$, the corresponding velocity is $v_T = (\sin\alpha - S_f)gt$, or

$$V_T = (\sin\alpha - S_f)T . \tag{11.104}$$

Inserting for $S_f = f_m V_T^2/(8\sigma)$ and solving for V_T gives

$$V_T = \frac{4\sigma}{f_a T}\left[\left(1+\frac{f_a \sin\alpha}{2\sigma}T^2\right)^{1/2} - 1\right]. \tag{11.105}$$

With $dX_T/dT = V_T$ and the initial condition $X_T(T = T_T) = -(\sin\alpha)^{-1}$, the drying front-time relation $X_T(T)$ is

$$\frac{f_a}{4\sigma}\left(X_T + \frac{1}{\sin\alpha}\right) = \left(1+\frac{f_a \sin\alpha}{2\sigma}T^2\right)^{1/2} - \left(1+\frac{f_a}{\sigma \sin\alpha}\right)^{1/2}$$
$$- \ln\left(\frac{1+\left(1+\dfrac{f_a \sin\alpha}{2\sigma}T^2\right)^{1/2}}{1+\left(1+\dfrac{f_a}{\sigma \sin\alpha}\right)^{1/2}}\right). \tag{11.106}$$

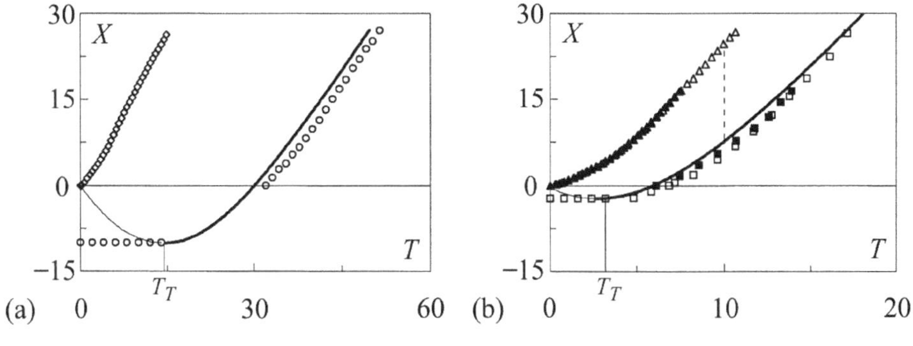

Figure 11.61 Positive and negative wave fronts for S_o = (a) 10%, (b) 50% and h_o [mm] = (Δ, $\blacktriangle$) (300, 500) mm, $L_B = h_o/S_o$, (- -) wave length (Lauber and Hager, 1998b)

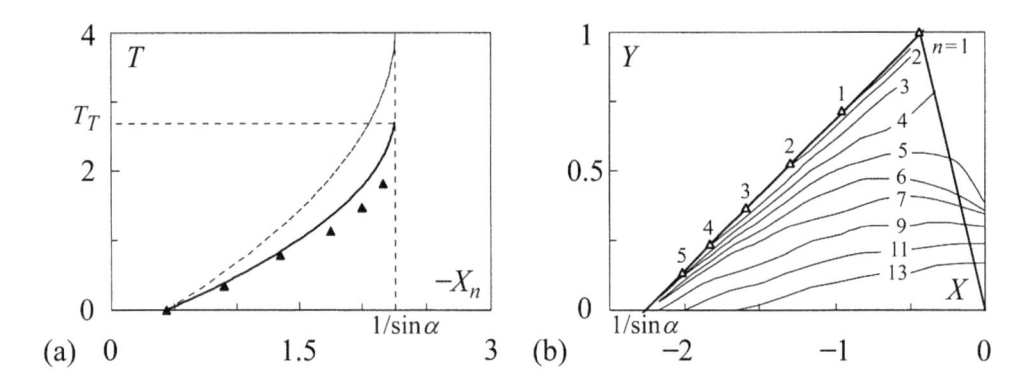

Figure 11.62 (a) Negative wave front, (b) negative wave profiles for S_o = 50% (Lauber and Hager, 1998b)

Figure 11.61 shows the negative front and the drying front, together with the positive wave front for S_o = 10% and 50%. For small slopes the wave expands with time, whereas it is of almost constant length for larger slopes. Note that the prediction according to Eq. (11.106) for the drying front agrees with the data (Figure 11.62).

WAVE PROFILES

Figure 11.63 shows the surface profiles $Y(T)$ at various locations X. For S_o = 0.1, the profiles are single-peaked with the rising limb much shorter than the falling. In contrast, waves in a steep channel have a plateau of maximum depth developing into a nearly trapezoidal wave profile. In both cases, the maximum wave height decreases considerably as the location increases, as given by Eq. (11.102).

For small slopes, a normalized wave profile results as for the horizontal channel. Using the time of positive wave front T_F, the time of the negative front T_{NF}, the time of the maximum flow depth T_M and the maximum flow depth h_M gives, with $Y_N = h/h_M$

$$Y_N = T_N^{*1/3}, \quad T_N^* = \frac{T - T_F}{T_M - T_F}, \quad \text{for } 0 < T_N < 1; \tag{11.107}$$

$$Y_N = \exp[3(T_N - 1)^2], \quad T_N^* = \frac{T - T_M}{T_{nF} - T_M}, \quad \text{for } 1 < T_N < 2. \tag{11.108}$$

Equation (11.107) for the rising wave profile is thus independent of the bottom slope (Figure 11.64).

WAVE VELOCITY

Velocities in the dambreak wave body were measured as in the horizontal channel. Experimental data are shown in Figure 11.65 as $V(X)$ at various times T. The envelope curves of the positive and negative wave fronts are also plotted. Each curve starts at the latter point $(X_{nF}; T)$, increases almost linearly to a maximum and tends then to the positive front $X_F(T)$.

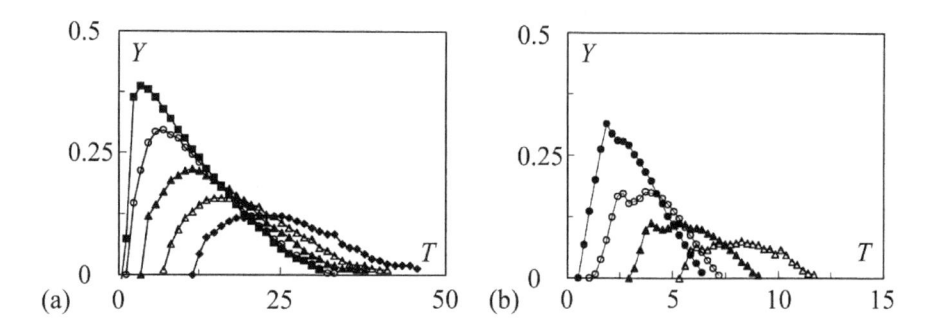

Figure 11.63 Wave profile $Y(T)$ for S_o = (a) 0.1, (b) 0.5 at locations X = ($\bullet$) 0.1, ($\blacksquare$) 0.167, ($\circ$) 1, ($\blacktriangle$) 4, ($\triangle$) 10 and ($\blacklozenge$) 20 (Lauber, 1997)

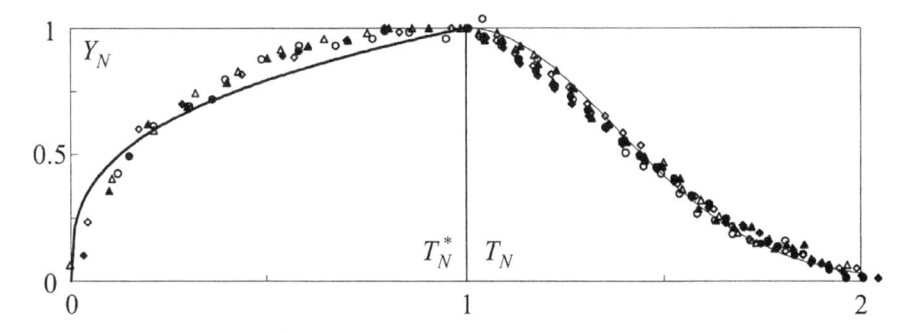

Figure 11.64 Generalized wave profiles $Y_N(T_N^*)$ and $Y_N(T_N)$ for S_o = 0.1 and various X between 0.5 and 20, (—) Eqs. (11.107), (11.108) (Lauber and Hager, 1998b)

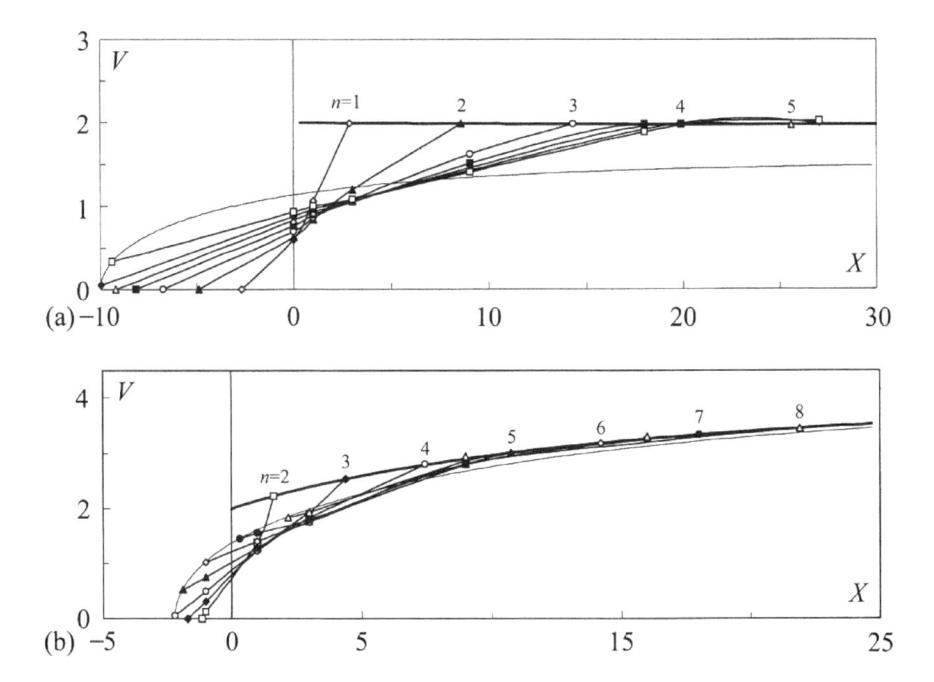

Figure 11.65 Wave velocity $V(X)$ at times $T = n \times \Delta T$ with $\Delta T = (g/h_o)^{1/2} \cdot 0.5$ s = 2.86 for $\lambda_B = (\sin\alpha)^{-1}$ and S_o = (a) 0.1, (b) 0.5, with (—) positive and (–) negative wave fronts (Lauber and Hager, 1998b)

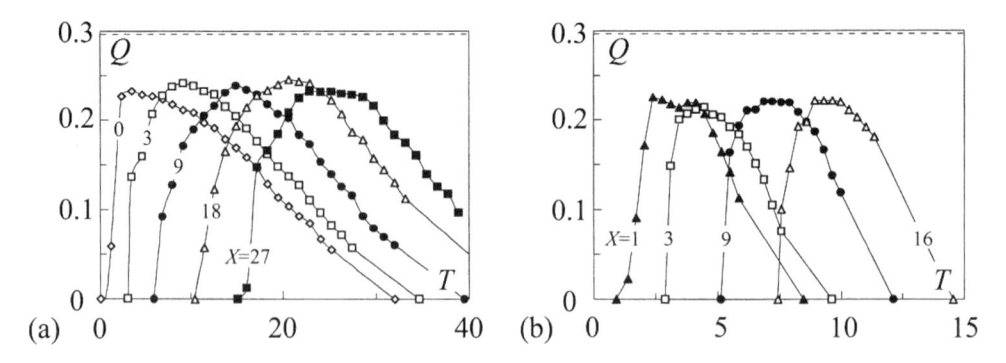

Figure 11.66 Discharge $Q(T)$ at various locations X and $\lambda_B = (\sin\alpha)^{-1}$ for S_o = (a) 0.1, (b) 0.5, (- - -) $Q = Q_M = 8/27$ (Lauber and Hager, 1998b)

Except for small times, the maximum velocity occurs slightly upstream of the positive wave front due to the wave tip region governed by a strong frictional effect. The velocity increases with the flow depth, according to the dynamic equation. At the wave rear of small depth, the velocity reduces again. The maximum wave velocity is nearly as large as the positive front velocity for both S_o = 0.1 and 0.5.

WAVE DISCHARGE

The discharge $Q = YV$ is determined for given flow depth $Y(X, T)$ and velocity $V(X, T)$. Figure 11.66 relates to S_o = 0.1 and 0.5 again, demonstrating that the wave body has a nearly constant discharge of $Q = q/(gh_o^3)^{1/2} = 8/27$, according to Ritter for $X = 0$. The present data indicate that Q_M = 0.24, nearly independent of location and bottom slope, if $X<30$. Starting with $Q = 0$ at the negative wave front, the hydrograph increases rapidly to Q_M, staying at the maximum discharge nearly all over the wave length. This result suggests that the effects of increasing flow depth and decreasing velocity in the wave body nearly compensate. The discharge in the wave body is thus nearly constant and equal to the maximum discharge Q_M.

11.5.8 Dam-break waves for silted-up reservoirs

Many dams face serious problems of reservoir sedimentation, which may significantly influence the behavior of dam-break waves. Created by silted-up reservoirs, these waves were investigated experimentally (Duarte *et al*, 2011). The effect of the silting degree of the reservoir and the sediment grain size distribution (sand mean diameter of 1 and 2 mm) on the general dam-break wave characteristics, final sediment deposition configuration, the position of the positive water wave front and its celerity, and the maximum wave depth, were studied. It was observed that the silting degree plays a more important role in the propagation of the positive front rather than the grain size distribution.

The empirical equation for the estimation of wave front celerity V_F, and that accounting for the silting degree of the reservoir ϕ, reads

$$V_F = \frac{\left(0.52 T e^{1-\frac{T}{2.2}}\right)^{1.95}}{\frac{0.7}{1.5^\phi}e^\phi}. \tag{11.109}$$

The silting degree is defined as the ratio h_o/h_{so} between the initial water depth h_o and the reservoir sediment depth h_{so}.

The results also reveal that for values $\phi < 0.5$, positive wave front celerities after $T = 1.5$ tend to the value $2^{1/2}$, as obtained in previous studies for clear water dam-break waves (Roche et al, 2008; Hogg and Pritchard, 2004). Regarding risk management, it was observed that a silted-up reservoir produces a lower and less rapid downstream dam-break wave than a clear water-filled reservoir. Nevertheless, sediments represent a higher risk for the near-field zone downstream of the dam, where they will create a deposition cone.

Notation

A	Cross-sectional area $= A/(Bh) = \lambda^{-1/2}$ (m²)
A_H	Headwater area (m²)
A_s	Reservoir surface (m²)
a	Dambreak shape parameter (-)
B	Breach top width (m)
B_a	Average dam failure width (m)
b	Bottom breach width (m)
b_D	Embankment width (m)
C	Relative flow depth of dambreak wave $h/h_o = (C/a)^2$ (-)
C_d	Discharge coefficient (-)
c	Wave celerity (m/s)
d	Average sediment diameter (m)
d_b	Breach depth (m)
dz/dt	Dam erosion rate (m/s)
F	Froude number $= V/(gh)^{1/2}$ (-)
F_M	Froude number of dambreak wave maximum $= (dX_M/dT)Y_M^{-1/2}$ (-)
f	Friction coefficient (-)
f_a	Average front friction coefficient $= 0.20R_a^{-0.20}$ (-)
g	Gravity acceleration (m/s²)
$H = h/h_o$	Relative breach flow depth (-)
H_d	Dam height (m)
H_e	Energy head $= h + V^2/2g$ (m)
H_H	Normalized headwater elevation (m)
H_o	Approach flow energy head $= h_o + Q^2/[2gb^2(h_o + w)^2]$ (m)
H_s	Dam storage volume length, $H_s = V_s^{1/3}$ (m)
h	Flow depth (m)
h_a	Average front flow depth $= \sigma h_o$ (m)
h_c	Critical flow depth $= (Q_o^2/gb^2)^{1/3}$ (m)
h_H	Target headwater elevation (m)
h_o	Approach flow depth (m)
j	Reduced bottom slope $= S_o \sigma/f_a$ (-)
k	Permeability coefficient (m/s)
L^*	Length scale of Hunt's dambreak solution (m)
L_B	Reservoir length (m)
L_D	Embankment length (m)
L_K	Crest length (m)
m	Non-dimensional dambreak wave location $= X/T$ (-)
n	Counter (-)
Q	Non-dimensional dambreak discharge $= q/(gh_o^3)^{1/2}$ (-)

Q_b	Breach discharge (m³/s)
Q_{DM}	Maximum breach discharge (m³/s)
Q_{dr}	Drainage discharge (m³/s)
Q_H	Headwater discharge (m³/s)
Q_M	Relative maximum dambreak discharge (-)
Q_o	Approach flow discharge (m³/s)
q	Unit discharge (m²/s)
q_o	Unit approach flow discharge (m²/s)
Q_P	Dam failure peak discharge (m³/s)
Q_R	Discharge according to Ritter = $(8/27)b(gh_o^3)^{1/2}$ (m³/s)
R^2	Coefficient of determination (-)
R_a	Average Reynolds number = $4q/v$ (-)
R_D	Crest radius (m)
R_f	Reduced friction factor = $f/8$ (-)
R_h	Hydraulic radius (m)
S_d	Downstream face slope (-)
S_f	Friction slope (-)
S_o	Bottom slope, upstream face slope (-)
s	Side slope (-)
T	Time (s), also dimensionless breach time (-)
$\overline{T}_M$	Relative time of maximum flow depth = $(g/h_o)^{1/2}(x/h_o)^{-2/3}t_M$ (-)
T_N	Normalized dambreak time (-)
T_{nF}	Relative time of negative wave front (-)
T_T	Relative dambreak drying time (-)
t	Time (s)
t^*	Time scale of Hunt's dambreak solution = L^*/V^* (s)
t_f	Dam failure duration (s)
t_M	Time of maximum flow depth (s)
U	Non-dimensional velocity of Hunt's dambreak solution = V/V^* (-)
u	Average breach velocity (m/s)
V	Relative velocity of dambreak wave = $v/(gh_o)^{1/2}$ (-)
V^*	Velocity scale for Hunt's dambreak solution (m/s)
V_0	Relative dynamic dambreak velocity = $2(gh_o)^{1/2}$ (-)
V_D	Embankment volume (m³)
V_{De}	Equilibrium embankment volume (m³)
V_{D0}	Initial embankment volume (m³)
V_F	Relative dambreak front velocity = $v_F/gh_o)^{1/2}$ (-)
$V_{F\infty}$	Asymptotic relative front velocity (-)
V_{nF}	Negative relative front velocity (-)
V_s	Dam storage volume (m³)
V_T	Relative dambreak drying velocity (-)
v	Cross-sectional average velocity (m/s)
v_e	Lateral erosion rate (m/s)
v_T	Dambreak drying velocity (m/s)
w	Dambreak wave celerity = $(2/a)\cdot(gh)^{1/2}$ (m/s)
w_D	Original embankment height (m)
w_m	Maximum fuse plug height (m)
X	Non-dimensional dambreak coordinate (-)
$\overline{X}$	Combined relative streamwise coordinate = $\lambda_B(x/h_o)^{-2/3}$ (-)

$\bar{X}$ Generalized parameter for sloping dambreak $= \lambda_B^{1-\varphi} \cdot X^{-2/3+0.2\varphi}$ (-)

X^* Pertinent parameter of sloping dambreak $= (h_o/x \cdot \sin\alpha)^{2/3}$ (-)

ΔX Overall dambreak wave length $= X_f - X_{nF}$ (-)

X_D Streamwise coordinate $= x_D/L_D$ (-)

X_{nF} Negative relative front location (-)

x Streamwise coordinate (m)

x_1 Transition location from Ritter wave profile to wave front portion (m)

x_D Streamwise embankment location (m)

x_{Do} Location of upstream embankment toe (m)

x_s Location of proper wave front (m)

Y Normalized flow depth of dambreak $= h/(h_o\cos\alpha)$ (-)

Y_M Maximum relative dambreak wave height $= h_M/h_o$ (-)

Y_N Normalized dambreak wave height (-)

y_D Transverse coordinate (m)

Z_∞ Non-dimensional final reservoir drawdown level (-)

$Z = z/z_o$ Relative breach elevation (-)

Z_D Non-dimensional vertical embankment surface elevation $= z_D/w_D$ (-)

Z_{DM} Dimensionless maximum embankment height (-)

Z_{DMe} Dimensionless equilibrium embankment height (-)

z Coordinate perpendicular to x, local embankment height (m)

z_D Embankment elevation above channel bottom (m)

z_{DM} Non-dimensional maximum embankment height (-)

z_{DMe} Non-dimensional equilibrium embankment height (-)

$z(t)$ Temporal development of breach bottom (m)

α Angle of sloping plane in dambreak problem (°)

α_d Downstream embankment face angle (-)

α_e Erosivity coefficient $(m/s)^{1-\beta}$

α_u Upstream embankment face angle (-)

β Coefficient (-)

$\gamma = h_o/z_o$ Breach constant of Singh (-)

φ Slope parameter in sloping dambreak wave $= (\sin\alpha)^{1/4}$ (-)

μ Embankment shape parameter (-)

λ Cross-sectional shape parameter (-)

λ_B Relative reservoir length $= L_B/h_o$ (-)

θ_s Submerged angle of repose (°)

ρ Fluid density (kg/m³)

ρ_k' Crest curvature parameter (-)

ρ_s Sediment density (kg/m³)

σ_s Sediment non-uniformity parameter $= (d_{84}/d_{16})^{0.5}$ (-)

τ Relative time of dynamic dambreak wave $= T-2^{1/2}$ (-)

$\bar{\tau}$ Reduced value of $\tau = (f_a/\sigma)\tau$ (-)

τ_H Relative time of Hunt's dambreak solution $= t/t^*$ (-)

Ψ Relative flow depth of Hunt's dambreak solution $= h/h^*$ (-)

Ψ_s Relative shock location of Hunt's dambreak wave (-)

χ Relative distance of Hunt's dambreak solution $= x/L^*$ (-)

χ_s Relative shock location of Hunt's dambreak wave (-)

ζ Dimensionless drawdown level (-)

ζ_∞ Final drawdown level (-)

References

Abbott, M.B. (1966) *An introduction to the method of characteristics*. Thames & Hudson, London.

Abbott, M.B. (1979) *Computational hydraulics: Elements of the theory of free surface flows*. Pitman, London.

Anonymous (1959) French dam collapse: Rock shift was probable cause. *Engineering News-Record*, 163(Dec. 10), 24–25; 163(Dec. 17), 44.

Anonymous (1977) Teton Dam failure. *Civil Engineering*, 47(8), 56–61.

ASCE/EWRI (2011) Earthen embankment breaching. *Journal of Hydraulic Engineering*, 137(12), 1549–1564. http://dx.doi.org/10.1061/(ASCE)HY. 1943–7900.0000498 ASCE/EWRI Task Committee on Dam/ Levee Breaching.

Aufleger, M. & López, D. (2016) Die Dammkrone als Indikator für die Talsperrensicherheit in Extremsituationen [Dam crest as indicator of dam safety in extreme situations]. *Wasserwirtschaft*, 106(6), 136–139 (in German).

Cao, Z., Yue, Z. & Pender, G. (2011) Landslide dam failure and flood hydraulics I: Experimental investigation. *Natural Hazards*, 59(2), 1003–1019.

Castro-Orgaz, O. & Hager, W.H. (2019) *Hydrostatic free surface flows*. Springer, Berlin.

Chanson, H. (2005) Applications of the Saint-Venant equations and method of characteristics to the dam break wave problem. *Report* CH55/05. Dept. of Civil Engineering, University of Queensland, Brisbane, Australia.

Chanson, H. (2009) Application of the method of characteristics to the dam break wave problem. *Journal of Hydraulic Research*, 47(1), 41–49.

Chaudhry, M.H. (1993) *Open-channel flow*. Prentice Hall, Englewood Cliffs, NJ.

Chinnarasri, C., Tingsanchali, T., Weesakul, S. & Wongwises, S. (2003) Flow patterns and damage of dike overtopping. *International Journal of Sediment Research*, 18(4), 301–309.

Chinnarasri, C., Jirakitlerd, S. & Wongwises, S. (2004) Embankment dam breach and its outflow characteristics. *Civil Engineering Environment Systems*, 21(4), 247–264.

Coleman, S.E., Andrews, D.P. & Webby, M.G. (2002) Overtopping breaching of noncohesive homogenous embankments. *Journal of Hydraulic Engineering*, 128(9), 829–838.

Cunge, J.A., Holly, F.M. & Verwey, A. (1980) *Practical aspects of computational river hydraulics*. Pitman, London.

Deng, X., Liu, H. & Lu, S. (2018) Analytical study of dam-break wave tip region. *Journal of Hydraulic Engineering*, 144(5), 04018015.

De Saint-Venant, B. (1871) Théorie du mouvement non permanent des eaux, avec application aux crues de rivières et à l'introduction des marées dans leur lits [Theory of unsteady water flow, with application to floods in rivers and introduction of tides in their beds]. *Comptes Rendues de l'Académie des Sciences*, Paris, 173, 147–154; 173, 237–240 (in French).

Dressler, R.F. (1952) Hydraulic resistance effects upon the dambreak functions. *Journal of Research*, 49(3), 217–225.

Dressler, R.F. (1954) Comparison of theories and experiments for the hydraulic dambreak wave. Proceedings of *Association Internationale d'Hydrologie*, Rome, 3, 319–328.

Duarte, R., Ribeiro, J., Boillat, J.-L. & Schleiss, A.J. (2011) Experimental study on dam-break waves for silted-up reservoirs. *Journal of Hydraulic Engineering*, 137(11), 1385–1393.

Dupont, E., Dewals, B.J., Archambeau, P., Erpicum, S. & Pirotton, M. (2007) Experimental and numerical study of the breaching of embankment dam. Proceedings of the 32nd *IAHR Congress*, Venice, 1(178), 1–10.

Foster, M., Fell, R. & Spannagle, M. (2000) The statistics of embankment dam failures and accidents. *Canadian Geotechnical Journal*, 37(5), 1000–1024.

Frank, P.-J. & Hager, W.H. (2015) Spatial dike breach: Sediment surface topography using photogrammetry. Proceedings of the 36th *IAHR World Congress*, The Hague, 1613–1622. IAHR, Madrid.

Fread, D.L. (1996) Dam-breach floods. In: Singh, V.P. (ed) *Hydrology of disasters*. Kluwer, Dordrecht, pp. 85–126.

Fritz, H.M. & Hager, W.H. (1998) Hydraulics of embankment weirs. *Journal of Hydraulic Engineering*, 124(9), 963–971.

Froehlich, D.C. (1995) Peak outflow from breached embankment dam. *Journal of Water Resources Planning and Management*, 121(1), 90–97; 122(7/8), 314–319.

Froehlich, D.C. (2016) Empirical model of embankment dam breaching. Proceedings of the 8th Conference *River Flow*, St. Louis MS, DOI:10.1201/9781315644479-285.

Gregoretti, C., Maltauro, A. & Lanzoni, S. (2010) Laboratory experiments on the failure of coarse homogeneous sediment natural dams on a sloping bed. *Journal of Hydraulic Engineering*, 136(11), 868–879.

Hager, W.H. & Chervet, A. (1996) Geschichte der Dammbruchwelle [History of dambreak wave]. *Wasser Energie Luft*, 88(3/4), 49–54 (in German).

Höeg, K., Løvoll, A. & Vaskinn, K.A. (2004) Stability and breaching of embankment dams: Field tests on 6 m high dams. *Hydropower & Dams*, 11(1), 88–92.

Hogg, A.J. & Pritchard, D. (2004) The effects of hydraulic resistance on dam-break and other shallow inertial flow. *Journal of Fluid Mechanics*, 501, 179–212.

Hunt, B. (1984) Perturbation solution for dam-break floods. *Journal of Hydraulic Engineering*, 110(8), 1058–1071.

ICOLD (1973) *Lessons from dam incidents.* International Commission on Large Dams, Paris.

ICOLD (1998) Dam-break flood analysis: Review and recommendations. *Bulletin* 111. International Commission on Large Dams, Paris.

James, L.B. (1988) The failure of Malpasset dam. In: Jansen, R.B. (ed) *Advanced dam engineering for design, construction and rehabilitation.* Van Nostrand Reinhold, New York, pp. 17–27.

Jansen, R.B. (1980) *Dams and public safety.* US Water and Power Resources Service, Denver, CO.

Jansen, R.B. (1988) *Advanced dam engineering for design, construction and rehabilitation.* Van Nostrand Reinhold, New York.

Kamalzare, M., Stuetzle, C., Chen, Z.X., Zimmie, T.F., Cutler, B. & Franklin, W.R. (2012) Validation of erosion modeling: Physical and numerical. Proceedings of Conference *GeoCongress*, 2012, Oakland CA, pp. 710–719, DOI:10.1061/9780784412121.074.

Khatsuria, R.M. (2005) *Hydraulics of spillways and energy dissipators.* Dekker, New York.

Kollgaard, E.B. & Chadwick, W.L., eds. (1988) *Development of dam engineering in the United States.* Pergamon Press, New York.

Lauber, G. (1997) Experimente zur Talsperrenbruchwelle im glatten geneigten Rechteckkanal [Experiments to dambreak wave in the smooth sloping rectangular channel]. *Dissertation* 12115. Eidgenössische Technische Hochschule ETH, Zürich (in German).

Lauber, G. & Hager, W.H. (1998a) Experiments to dambreak wave: Horizontal channel. *Journal of Hydraulic Research*, 36(3), 291–307.

Lauber, G. & Hager, W.H. (1998b) Experiments to dambreak wave: Sloping channel. *Journal of Hydraulic Research*, 36(5), 761–773.

Liggett, J.A. (1994) *Fluid mechanics.* McGraw-Hill, New York.

Løvoll, A. (2006) Breach formation in rockfill dams: Results from Norwegian field tests. Proceedings of the 22nd *ICOLD Congress*, Barcelona, Q86(R4), 35–51.

Mahmood, K. & Yevjevich, V., eds. (1975) *Unsteady flow in open channels.* Water Resources Publications, Fort Collins, CO.

Martin, H. (1990) Plötzlich veränderliche instationäre Strömungen in offenen Gerinnen [Abruptly unsteady open channel flows]. In: Bollrich, G. (ed) *Technische Hydromechanik.* Verlag für Bauwesen, Berlin, pp. 565–635 (in German).

Matos, J.P., Mignan, A. & Schleiss, A.J. (2018) The role of uncertainty in dam failure frequency estimates: A conceptual case study. Proceedings of the 26th *ICOLD Congress*, Vienna, Q101(R19), 48–50 (Abstract book).

Miller, Jr., W.A. & Yevjevich, V. (1975) *Unsteady flow in open channels 3: Bibliography.* Water Resources Publications, Fort Collins, CO.

Morris, M.W., Hanson, G. & Hassan, M. (2008) Improving the accuracy of breach modelling: Why are we not progressing faster? *Journal of Flood Risk Management*, 1(3), 150–161.

Morris, M.W. (2009) Breaching processes: A state of the art review. FLOODsite *Report* T06–06–03. FLOODsite, www.floodsite.net.

Müller, C., Frank P.-J. & Hager, W.H. (2016) Dyke overtopping: Effects of shape and headwater elevation. *Journal of Hydraulic Research*, 54(4), 410–422.

Novak, P., Moffat, A.I.B., Nalluri C. & Narayanan, R. (2001) *Hydraulic structures,* 3. Spon, London.

Orendorff, B., Al-Riffai, M., Nistor, I. & Rennie, C.D. (2013) Breach outflow characteristics of non-cohesive embankment dams subject to blast. *Canadian Journal of Civil Engineering,* 40(3), 243–253.

Peregrine, D.H. & Williams, S.M. (2001) Swash overtopping a truncated plane beach. *Journal of Fluid Mechanics,* 440, 391–399.

Peter, S.J., Siviglia, A., Nagel, J., Marelli, S, Boes, R.M., Vetsch, D. & Sudret, B. (2018a) Development of probabilistic dam breach model using Bayesian inference. *Water Resources Research,* 54(7), 4376–4400.

Peter, S.J., Vetsch, D.F., Siviglia, A. & Boes, R. (2018b) Probabilistische Dammbruchanalyse [Probabilistic dam-break analysis]. *Wasser, Energie, Luft,* 110(3), 179–185 (in German).

Pickert, G., Weitbrecht, V. & Bieberstein, A. (2011) Breaching of overtopped river embankments controlled by apparent cohesion. *Journal of Hydraulic Research,* 49(2), 143–156.

Pohle, F.V. (1952) Motion of water due to breaking of a dam, and related problems. *Circular,* 521, 47–53. Department of Commerce, National Bureau of Standards. Government Printing Office: Washington DC.

Powledge, G.R., Ralston, D.C., Miller, P., Chen, Y.H., Clopper, P.E. & Temple, D.M. (1989) Mechanics of overflow erosion on embankments 2: Hydraulic and design considerations. *Journal of Hydraulic Engineering,* 115(8), 1056–1075.

Pugh, C.A. (1985) Hydraulic model studies of fuse plug embankments. *Report* REC-ERC-85–7. US Bureau of Reclamation, Denver, CO.

Ribi, J.M, Pury, J. & Boillat, J.-L. (2008) Breach formation in a fuse plug lateral weir. In: Altinakar, M.S., Kokpinar, M.A., Aydin, I., Cokgor, S. & Kirkgoz, S. (eds) Proceedings of the International Conference *River Flow.* Çeşme, Izmir, Turkey, 539–545. IAHR, Madrid.

Rifai, I., Erpicum, S., Archambeau, P., Violeau, D., Pirotton, M., El Kadi Abderrezzak, K. & Dewals, B. (2017) Overtopping induced failure of noncohesive, homogeneous fluvial dikes. *Water Resources Research,* 53(4), 3373–3386.

Roache, P.J. (1972) *Computational fluid dynamics.* Hermosa, Albuquerque NM.

Roche, O., Montserrat, S., Niño, Y. & Tamburrino, A. (2008) Experimental observations of water-like behavior of initially fluidized, dam break granular flows and their relevance for the propagation of ash-rich pyroclastic flows. *Journal of Geophysical Research,* 113, B12203.

Roger, S., Dewals, B., Erpicum, S., Schwanenberg, D., Schüttrumpf, H., Köngeter, J. & Pirotton, M. (2009) Experimental and numerical investigations of dike-break induced flows. *Journal of Hydraulic Research,* 47(3), 349–359.

Savitzky, A. & Golay, M.J. (1964) Smoothing and differentiation of data by simplified least squares procedures. *Analytical Chemistry,* 36(8), 1627–1639.

Schmocker, L. & Hager, W.H. (2009) Modelling dike breaching due to overtopping. *Journal of Hydraulic Research,* 47(5), 585–597.

Schmocker, L. & Hager, W.H. (2012) Plane dike-breach due to overtopping: Effects of sediment, dike height and discharge. *Journal of Hydraulic Research,* 50(6), 576–586.

Schmocker, L., Halldórsdóttir, B.R. & Hager, W.H. (2011) Effect of weir face angles on circular-crested weir flow. *Journal of Hydraulic Engineering,* 137(6), 637–643.

Schmocker, L., Höck, E., Mayor, P. & Weitbrecht, V. (2013) Hydraulic model study of the fuse plug spillway at Hagneck Canal, Switzerland. *Journal of Hydraulic Engineering,* 139(7), 683–695.

Schmocker, L., Frank, P.-J. & Hager, W.H. (2014) Overtopping dike-breach: Effect of grain size distribution. *Journal of Hydraulic Research,* 52(4), 559–564.

Schnitter, N.J. (1993) Dam failures due to overtopping. Proceedings of the Workshop *Dam Safety Evaluation,* Grindelwald, 1, 13–19.

Singh, V.P. (1996) *Dam break modelling technology.* Kluwer, Dordrecht.

SN 670 125a (1983) *Filtermaterialien* [Filter materials]. Vereinigung Schweizerischer Strassenfachleute, Zürich (in German).

Stoker, J.J. (1957) *Water waves.* Interscience, New York.

Su, S.-T. & Barnes, A.H. (1970) Geometric and frictional effects on sudden releases. *Journal of the Hydraulics Divisions* ASCE, 96(HY11), 2185–2200.

Sun, E., Zhang, X., Li, Z. & Wang, Y. (2012) Tailings dam flood overtopping failure evolution pattern. Proceedings of the International Conference Modern Hydraulic Engineering. *Procedia Engineering,* 28, 356–362.

Tinney, E.R. & Hsu, H.Y. (1962) Mechanics of washout of an erodible fuse plug. *Transactions of ASCE*, 127, 31–59.

Toro, E.F. & Clarke, J.F., eds. (1998) *Numerical methods for wave propagation*. Kluwer, Dordrecht, NL.

Toro, E.F. (2001) *Shock-capturing methods for free-surface shallow flows*. Wiley, New York.

USBR (1983) *Safety evaluation of existing dams*. US Bureau of Reclamation, Denver, CO.

USBR (1987) *Guidelines for using fuse plug embankments in auxiliary spillways*. U.S. Department of the Interior, Bureau of Reclamation, Denver, CO.

Van Emelen, S., Zech, Y. & Soares-Frazão, S. (2015) Impact of sediment transport formulations on breaching modelling. *Journal of Hydraulic Research*, 53(1), 60–72.

Vetsch, D.F. & Boes, R.M. (2016) Vereinfachte Modellierung des progressiven Bruchs bei kleinen Erdschüttdämmen [Simplified modelling of progressive dam breaching at small embankment dams]. *WasserWirtschaft*, 106(6), 140–143 (in German).

Vischer, D.L. & Hager, W. H. (1998) *Dam hydraulics*, Wiley & Sons, Chichester.

Visser, P.J., Zhu, Y. & Vrijling, J.K. (2006) Breaching of dikes. Proceedings of the 30th International Conference *Coastal Engineering*, San Diego, CA, pp. 2893–2905.

Vogel, A., Courivaud, J.-R. & Jarecka, A. (2014) Dam failures of embankment dams caused by overtopping or accidental leakage: New aspects for breach widening and deepening processes. In: Toledo, M.Á., Morán, R. & Oñate, E. (eds) Proceedings of the 1st International Seminar on *Dam Protection Against Overtopping and Accidental Leakage*, Madrid, 59–71. Taylor & Francis, London.

Wahl, T.L. (1993) Hydraulic model study of horseshoe dam fuse plug auxiliary spillway. *Report*, 93–10. USBR, Denver CO.

Wahl, T.L. (2004) Uncertainty of predictions of embankment dam breach parameters. *Journal of Hydraulic Engineering*, 130(5), 389–397.

Whitham, G.B. (1955) The effects of hydraulic resistance in the dam-break problem. *Proceedings of the Royal Society London* A, 227, 399–407.

Wüthrich, D., Pfister, M., Nistor, I. & Schleiss, A.J. (2018) Experimental study of tsunami-like waves on dry and wet bed generated with a vertical release technique. *Journal of Waterway, Port, Coastal, and Ocean Engineering*, 144(4), 04018006. DOI:10.1061/(ASCE)WW.1943-5460.0000447.

Bibliography

Progressive 2D dam breach

Abt, S.R. & Johnson, T.L. (1991) Riprap design for overtopping flow. *Journal of Hydraulic Engineering*, 117(8), 959–972.

Albert, R. & Gautier, J. (1992) Evacuateurs fondés sur remblai [Stilling basins placed on a bank]. *La Houille Blanche*, 47(2/3), 147–157 (in French).

Das, B.P. (1972) Stability of rockfill in end-dump river closures. *Journal of the Hydraulics Division* ASCE, 98(HY11), 1947–1967.

Das, B.P. (1973) Bed scour at end-dump channel constrictions. *Journal of the Hydraulics Division* ASCE, 99(HY12), 2273–2291.

Johnson, F.A. & Illes, P. (1976) A classification of dam failures. *Water Power*, 28(12), 43–45.

Leopardi, A., Oliveri, E. & Greco, M. (2003) Numerical simulation of gradual earth-dam failure. *L'Acqua*, (2), 47–54.

Martins, R. (1981) Hydraulics of overflow rockfill dams. *Memória*, 559. LNEC: Lisboa.

Powledge, G.R., Ralston, D.C., Miller, P., Chen, Y.H., Clopper, P.E. & Temple, D.M. (1989) Mechanics of overflow erosion on embankments 1: Research activities. *Journal of Hydraulic Engineering*, 115(8), 1040–1055.

Sandover, J.A. & Tallis, J.A. (1966) Construction of loose-tip cofferdams. *Water Power*, 18(7), 273–279.

Sandover, J.A. (1969) Discharge coefficients of constrictions in open channels. *Water Power*, 21(7), 256–261.

Sandover, J.A. (1970) Backwater effects due to channel constrictions. *Water Power*, 22(1), 28–32; 22(4), 125.

Sarkaria, G.S. & Dworsky, B.H. (1968) Model studies of an armoured rockfill overflow dam. *Water Power*, 20(11), 455–462.

Wahl, T.L. (1997) Predicting embankment dam breach parameters: A needs assessment. Proceedings of the 27th *IAHR Congress*, San Francisco, 4, 48–53.

Progressive 3D dam breach

Boes, R., Frank, P.-J. & Hager, W.H. (2017) Spatial breach development of homogeneous non-cohesive levees and embankment dams due to overtopping. Proceedings of the 85th *ICOLD Annual Meeting*, Prague, Czech Republic, Paper, 344, 1–10.

Frank, P.-J. & Hager, W.H. (2014) Spatial dike breach: Accuracy of photogrammetric measurement system. In: Schleiss, A.J. *et al.*, (eds) Proceedings of Conference *River Flow 2014*, 1647–1654. Taylor & Francis Group, London (CD-ROM).

Frank, P.-J. & Hager, W.H. (2015) Spatial dike breach: Sediment surface topography using photogrammetry. E-Proceedings of the 36th *IAHR World Congress*, The Hague, The Netherlands, 1613–1622. IAHR, Madrid.

Frank, P.-J. & Hager, W.H. (2016) Challenges of dike breach hydraulics. In: Constantinescu, G., Garcia, M.H. & Hanes, D.R. (eds) Proceedings of Conference *River Flow 2016*, St. Louis, 428–435. Taylor & Francis Group, London.

Rifai, I., Erpicum, S., Archambeau, P., Benoit, M., Pirotton, M., Dewals, B., El kadi Abderrezzak, K. & Dewals, B. (2015) Physical modeling of lateral dike breaching due to overtopping. Proceedings of the 9th Symposium *River, Coastal and Estuarine Morphodynamics – RCEM 2015*. http://hdl.handle.net/2268/177595.

Rifai, I., Erpicum, S., Archambeau, P., Violeau, D., Pirotton, M., El kadi Abderrezzak, K. & Dewals, B. (2016) Monitoring topography of laboratory fluvial dike models subjected to breaching based on a laser profilometry technique. Proceedings of the International Symposium *River Sedimentation* Stuttgart. http://hdl.handle.net/2268/195875

Rifai, I., Erpicum, S., Archambeau, P., Violeau, D., Pirotton, M., El kadi Abderrezzak, K. & Dewals, B. (2017) Overtopping induced failure of noncohesive, homogeneous dikes. *Water Resources Research*, 53(4), 3373–3386.

Rifai, I. (2018) Rupture de digues fluviales par surverse [Overtopping induced fluvial dike failure]. *Ph.D. Thesis*, Université de Liège, Liège, Belgique; Université Paris-Est, Marne la vallée, France (in French).

Dambreak analysis

Abbott, M.R. (1956) A theory of the propagation of bores in channels and rivers. *Proceedings of the Cambridge Philosophy Society*, 52, 344–362.

Akanbi, A.A. & Katopodes, N.D. (1988) Model for flood propagation on initially dry land. *Journal of Hydraulic Engineering*, 114(7), 689–706; 116(2), 292–294.

Aureli, F., Mignosa, P. & Tomirotti, M. (2000) Numerical simulation and experimental verification of dam-break flows with shocks. *Journal of Hydraulic Research*, 38(3), 197–206.

Balloffet, A., Cole, E. & Balloffet, A.F. (1974) Dam collapse wave in a river. *Journal of the Hydraulics Division* ASCE, 100(HY5), 645–665.

Barr, D.I.H. & Das, M.M. (1980) Numerical simulation of dam-burst and reflections, with verification against laboratory data. *Proceedings of the ICE*, 69(2), 359–373; 71(1), 273–276.

Barr, D.I.H. & Das, M.M. (1981) Simulation of surges after removal of a separating barrier between shallower and deeper bodies of water. *Proceedings of the ICE*, 71(3), 911–919; 73(2), 533–537; 73(3), 698.

Basco, D.R. (1989) Limitations of the Saint-Venant equations in dam-break analysis. *Journal of Hydraulic Engineering*, 115(7), 950–965.

Bellos, C.V. & Sakkas, J.G. (1987) 1-D dam-break flood-wave propagation on dry bed. *Journal of Hydraulic Engineering*, 113(12), 1510–1524; 115(6), 859; 115(8), 1153–1157.

Benoist, G. (1989) Les études d'ondes de submersion des grands barrages d'EDF [The failure waves studies devoted to large dams of EDF]. *La Houille Blanche*, 44(1), 43–54 (in French).

Carballada, L. (1989) Dam break analysis on a river system. *La Houille Blanche*, 44(1), 55–64.

Cavaillé, Y. (1965) Contribution à l'étude de l'écoulement variable accompagnant la vidange brusque d'une retenue [Contribution to the study of unsteady flow due to abrupt emptying of a reservoir]. *Publications* Scientifiques et Techniques du Ministère de l'Air 410. Paris.

Četina, M. & Rajar, R. (1994) Two-dimensional dam-break flow simulation in a sudden enlargement. In: Molinaro, P. & Natale, L. (eds) Proceedings of *Modelling of Flood Propagation Over Initially Dry Areas*, 268–282. ASCE, New York.

Chen, C.-l. (1980) Laboratory verification of a dam-break flood model. *Journal of the Hydraulics Division* ASCE, 106(HY4), 535–556; 107(HY2), 246–249; 108(HY2), 275–284.

Chen, C.-l. & Armbruster, J.T. (1980) Dam-break wave model: Formulation and verification. *Journal of the Hydraulics Division* ASCE, 106(HY5), 747–767.

Chervet, A. & Dallèves, P. (1969) Calcul sur ordinateur d'écoulements instationnaires dans les canaux découverts [Numerical computation of unsteady open channel flows]. Proceedings of the 11th *IAHR Congress*, Kyoto 1(A29), 259–266 (in French).

Chervet, A. & Dallèves, P. (1970) Calcul de l'onde de submersion consécutive à la rupture d'un barrage [Computation of dambreak wave]. *Schweizerische Bauzeitung*, 88(19), 420–432 (in French).

Clavenad, C. (1885) De la propagation de la marée dans les canaux et dans les fleuves [On the propagation of tides in canals and rivers]. *Le Génie Civil*, 7(13), 196–199 (in French).

Cunge, J.A. (1970) Calcul de propagation des ondes de ruptures de barrage [Computation of dambreak wave propagation]. *La Houille Blanche*, 25(1), 25–33 (in French).

Dassie, G. (2010) Simulazione numerica di prove sperimentali relative a crolli diga [Numerical simulation of dam break experimental tests]. *L'Acqua*, (4), 49–62 (in Italian).

Delis, A.I., Skeels, C.P. & Ryrie, S.C. (2000) Evaluation of some approximate Riemann solvers for transient open channel flows. *Journal of Hydraulic Research*, 38(3), 217–231.

De Marchi, G. (1945) Onde di depressione provocato da apertura di paratoia in un canale indefinito [Positive dambreak wave in an indefinite channel]. *L'Energia Elettrica*, 22(1–2), 1–13 (in Italian).

Deymie, P. (1935) Note sur la propagation des intumescences allongées [Note on the propagation of long waves]. *Bulletin PIANC*, 10(20), 64–69 (in French).

Elliot, R.C. & Chaudhry, M.H. (1992) A wave propagation model for two-dimensional dam-break flows. *Journal of Hydraulic Research*, 30(4), 467–483.

Estrade, J., Gras, R. & Nahas, N. (1965) Etudes théoriques et experimentales relatives aux ondes et submersion [Theoretical and experimental studies on the waves and submergion]. Proceedings of the 11th *IAHR Congress*, Leningrad, 3(33), 1–5 (in French).

Fantoli, G. (1925) Sul passaggio dell'onda di piena nella supposta rotta di un serbatoio [On dam break wave propagation]. *Annali delle Utilizzazioni delle Acque*, 2(1), 11–27 (in Italian).

Faure, J. & Nahas, N. (1961) Etude numérique et expérimentale d'intumescences à forte courbure du front [A numerical and experimental study of steep-fronted waves]. *La Houille Blanche*, 16(5), 576–587 (in French).

Fennema, R.J. & Chaudhry, M.H. (1987) Simulation of one-dimensional dam-break flows. *Journal of Hydraulic Research*, 25(1), 41–51; 27(1), 175–177.

Fraccarollo, L. & Toro, E.F. (1995) Experimental and numerical assessment of the shallow water model for two-dimensional dam-break type problems. *Journal of Hydraulic Research*, 33(6), 843–864.

Frank, J. (1951) Betrachtungen über den Ausfluss beim Bruch von Stauwänden [Considerations on the outflow after break of dams]. *Schweizerische Bauzeitung*, 69(29), 401–406 (in German).

Fread, D.L. & Harbaugh, T.E. (1973) Transient hydraulic simulation of breached earth dams. *Journal of the Hydraulics Division* ASCE, 99(HY1), 139–154.

Freeman, J.C. & Le Méhauté, B. (1964) Wave breakers on a beach and surges on a dry bed. *Journal of the Hydraulics Division* ASCE, 90(HY2), 187–216.

Fujihara, M. & Borthwick, A.G.L. (2000) Godunov-type solution of curvilinear shallow-water equations. *Journal of Hydraulic Engineering*, 126(11), 827–836.

Gozali, S. & Hunt, B. (1993) Dam-break solutions for a partial breach. *Journal of Hydraulic Research*, 31(2), 205–214.

Greenspan, H.P. & Young, R.E. (1978) Flow over containment dyke. *Journal of Fluid Mechanics*, 87(1), 179–192.

Hicks, F.E., Steffler, P.M. & Yasmin, N. (1997) One-dimensional dam-break solutions for variable width channels. *Journal of Hydraulic Engineering*, 123(5), 464–468.

Hunt, B. (1982) Asymptotic solution for dam-break problem. *Journal of the Hydraulics Division* ASCE, 108(HY1), 115–126; 109(4), 633–640.

Hunt, B. (1983) Asymptotic solution for dam break on sloping channel. *Journal of Hydraulic Engineering*, 109(12), 1698–1706; 110(10), 1515–1516.

Hunt, B. (1984) Dam-break solution. *Journal of Hydraulic Engineering*, 110(6), 675–686; 111(6), 1040–1042.

Hunt, B. (1987) A perturbation solution of the flood-routing problem. *Journal of Hydraulic Research*, 25(2), 215–234; 26(3), 343–349.

Hunt, B. (1987) An inviscid dam-break solution. *Journal of the Hydraulic Research*, 25(3), 313–327.

Hunt, B. (1989) The level-reservoir approximation for unsteady flows on sloping channels. *Journal of Hydraulic Research*, 27(3), 347–354.

Jeyapalan, J.K., Duncan, J.M. & Seed, H.B. (1983) Analyses of flow failures of mine tailings dams. *Journal of Geotechnical Engineering*, 109(2), 150–189; 110(3), 454–458.

Jha, A.K., Akiyama, J. & Ura, M. (1995) First- and second-order flux difference splitting schemes for dam-break problem. *Journal of Hydraulic Engineering*, 121(12), 877–884.

Katopodes, N. & Strelkoff, T. (1978) Computing two-dimensional dam-break flood waves. *Journal of the Hydraulics Division* ASCE, 104(HY9), 1269–1288.

Katopodes, N.D. & Schamber, D.R. (1983) Applicability of dam-break flood wave models. *Journal of Hydraulic Engineering*, 109(5), 702–721.

Kordas, B. & Witkowska, H. (1978) Dam break flood in nonprismatic channel. *Archiwum Hydrotechniki*, 25(3), 309–323.

Martin, H. (1983) Dam-break wave in horizontal channels with parallel and divergent side walls. Proceedings of the 20th *IAHR Congress*, Moscow, 2, 494–505.

Martin, H. & Bollrich, G. (1989) Berechnungsgrundlagen für Schwall- und Sunkwellen sowie Dammbruchprobleme [Computational bases for surge and dambreak waves]. Proceedings of the Symposium *Hydraulik offener Gerinne. Wiener Mitteilungen*, 79, 139–164. Österr. Wasserwirtschaftsverband, Wien (in German).

Martin, H. (1990) Plötzlich veränderliche instationäre Strömungen in offenen Gerinnen [Abruptly-varied unsteady open channel flows]. In: Bollrich, G. (ed) *Technische Hydromechanik*, 2, 565–635. VEB Verlag für Bauwesen, Berlin (in German).

Martin, J.C. & Moyce, W.J. (1952) An experimental study of the collapse of liquid columns on a rigid horizontal plane. *Philosophical Transaction* A, 244, 312–324.

Miller, S. & Chaudhry, M.H. (1989) Dam-break flows in curved channel. *Journal of Hydraulic Engineering*, 115(11), 1465–1478.

Molinaro, P. (1991) Dam-break wave analysis: A state-of-the-art. In: Ouazar, D., Ben Sari, D. & Brebbia, C.A. (eds) Proceedings of the 2nd International Conference *Computer Methods in Water Resources* 2: *Computational Hydraulics and Hydrology*. Springer, Berlin, pp. 77–87.

Peviani, M.A. & Robles, C. (1994) Mathematical modelling of the overtopping and breaking of natural earth-dams. Proceedings of the International Workshop *Floods and Inundations Related to Large Earth Movements*, Trento B(9), 1–17.

Ponce, V.M. & Tsivoglou, A.J. (1981) Modeling gradual dam breaches. *Journal of the Hydraulics Division* ASCE, 107(HY7), 829–838.

Preissmann, A. (1960) Propagation des intumescences dans les canaux et les rivières [Wave propagation in channels and rivers]. Proceedings of the Premier *Congrès de l'Association Française de Calcul*, Grenoble, pp. 433–442 (in French).

Preissmann, A. & Cunge, J.A. (1961) Calcul du mascaret sur machine électronique [Tidal bore calculation on an electronic computer]. *La Houille Blanche*, 16(5), 588–596 (in French).

Rajar, R. (1973) Modèle mathématique et abaques sans dimensions pour la détermination de l'écoulement qui suit la rupture d'un barrage [Mathematical model and dimensionless plots to determine dambreak flow]. Proceedings of the 11th *ICOLD Congress*, Madrid, Q40(R34), 503–521 (in French).

Rajar, R. (1978) Mathematical simulation of dam-break flow. *Journal of the Hydraulics Division* ASCE, 104(HY7), 1011–1026; 105(HY8), 1043–1044; 106(HY3), 453–454.

Ré, R. (1946) Etude du lacher instantané d'une retenue d'eau dans un canal par la méthode graphique [Study of instantaneous dam breach in a channel by the graphical method]. *La Houille Blanche*, 2(5), 181–187 (in French).

Ritter, A. (1892) Die Fortpflanzung der Wasserwellen [Propagation of water waves]. *Zeitschrift des Vereines Deutscher Ingenieure*, 36(33), 947–954 (in German).

Sakkas, J.G. & Strelkoff, T. (1973) Dam-break flood in a prismatic dry channel. *Journal of the Hydraulics Division* ASCE, 99(HY12), 2195–2216; 100(HY10), 1501–1502.

Sakkas, J.G. & Strelkoff, T. (1976) Dimensionless solution of dam-break flood waves. *Journal of the Hydraulics Division* ASCE, 102(HY2), 171–184; 102(HY12), 1782–1784; 103(HY8), 932–933.

Savic, L.J. & Holly, Jr., F.M. (1993) Dambreak flood waves computed by modified Godunov method. *Journal of Hydraulic Research*, 31(2), 187–204.

Schamber, D.R. & Katopodes, N.D. (1984) One-dimensional models for partially breached dams. *Journal of Hydraulic Engineering*, 110(8), 1086–1102.

Singh, V.P. & Quiroga, C.A. (1987) A dam-breach erosion model. *Water Resources Management*, 1(3), 177–221.

Singh, V.P. & Quiroga, C.A. (1988) Dimensionless analytical solutions for dam-breach erosion. *Journal of Hydraulic Research*, 26(2), 179–197; 27(3), 447–454.

Singh, V.P. & Scarlatos, P.D. (1988) Analysis of gradual earth-dam failure. *Journal of Hydraulic Engineering*, 114(1), 21–42.

Tingsanchali, T. & Rattanapitikon, W. (1993) 2-D mathematical modelling for dam break wave propagation in supercritical and subcritical flows. Proceedings of the 25th *IAHR Congress*, Tokyo, A(1), 25–32.

Vasiliev, O.F. (1970) Numerical solution of the non-linear problems of unsteady flows in open channels. In: Ehlers, J. (ed) Proceedings of the 2nd International Conference Numerical Methods in Fluid Dynamics. *Lecture Notes in Physics*, 8, 410–421. Springer, Berlin.

Volz, C., Frank, P.-J., Vetsch, D.F., Hager, W.H. & Boes, R.M. (2017) Numerical embankment breach modelling including seepage flow effects. *Journal of Hydraulic Research*, 55(4), 480–490.

Wang, Z. & Shen, H.T. (1999) Lagrangian simulation of one-dimensional dam-break flow. *Journal of Hydraulic Engineering*, 125(11), 1217–1220.

Wasley, R.J. (1961) Hydrodynamics of flow into curb-opening inlets. *Journal of the Engineering Mechanics Division* ASCE, 87(EM4), 1–18; 87(EM6), 185–186; 88(EM2), 153–156; 88(EM3), 157–160.

Wu, C., Huang, G. & Zheng, Y. (1999) Theoretical solutions of dam-break shock wave. *Journal of Hydraulic Engineering*, 125(11), 1210–1215.

Wurbs, R.A. (1987) Dam-breach flood wave models. *Journal of Hydraulic Engineering*, 113(1), 29–46; 114(5), 565–569.

Yang, J.Y., Chang, S.H. & Hsu, C.A. (1993) Computations of free surface flows 1: One-dimensional dam-break flow. *Journal of Hydraulic Research*, 31(1), 19–34.

Dambreak experimentation

Ackerman, A.J. (1932) Models predict behaviour of failing dam. *Civil Engineering*, 2(7), 415–420.

Aguirre-Pé, J., Plachco, F.P. & Quisca, S. (1995) Tests and numerical one-dimensional modelling of a high-viscosity fluid dam-break wave. *Journal of Hydraulic Research*, 33(1), 17–26.

Bechteler, W. & Kulisch, H. (1994) Physical 3D-simulations of erosion-caused dam-breaks. International Workshop *Floods and inundations related to large earth movements*, Trento B(8), 1–11.

Bell, S.W., Chaudhry, M.H. & Elliot, R.C. (1992) Experimental results of two-dimensional dam-break flows. *Journal of Hydraulic Research*, 30(2), 225–252.

Bellos, C.V., Sakkas, J.G. & Soulis, J.V. (1992) Experimental investigation of two-dimensional dam-break induced flows. *Journal of Hydraulic Research*, 30(1), 47–63.

Betâmio de Almeida, A. & Bento Franco, A. (1994) Modeling of dam-break flow. In: Chaudhry, M.H. & Mays, L.W. (eds) *Computer modeling of free-surface and pressurized flows*. Kluwer, Dordrecht, pp. 343–373.

Blaser, F. & Hager, W.H. (1999) Positive front of dambreak wave on rough bottom. Proceedings of the 28th *IAHR Congress*, Graz, pp. 1–6 (CD-Rom).

Bon, Jr., W., Veiga Pinto, A., Maranha das Neves, E. & Martins, R. (1982) Rockfill deformations forecast and overflow rockfill dams. Proceedings of the 14th *ICOLD Congress*, Rio de Janeiro, Q55(R23), 391–419.

Cassidy, J.J. (1993) Flood data and the effect on dam safety. Proceedings of the Workshop *Dam Safety Evaluation*, Grindelwald, 4, 1–15.

Drobir, H. (1971) Der Ausfluss aus einem Speicher beim Bruch einer Talsperre [Reservoir outflow following a dambreak]. *Mitteilung* 17. Institut für Wasserwirtschaft und konstruktiven Wasserbau, TH Graz (in German).

Dussault, J.-G., Marche, C., Quach, T.T. & Carballada, L. (1982) L'étude du comportement des ondes de rupture de barrage: Une donnée essentielle pour les mesures de protection civil [Study of behavior of dambreak waves: An important basis for civil protection mesures]. Proceedings of the 14th *ICOLD Congress*, Rio de Janeiro, Q52(R12), 219–237 (in French).

Escande, L., Nougaro, J., Castex, L. & Barthet, H. (1961) Influence des quelques paramètres sur une onde de crue subite à l'aval d'un barrage [Effect of some parameters on the dambreak wave]. Proceedings of the 9th *IAHR Congress*, Dubrovnik, pp. 1198–1206, also published in *La Houille Blanche*, 16(5), 565–575 (in French).

Estrade, J. (1967) Contribution à l'étude de la suppression d'un barrage: Phase initiale de l'écoulement [Contribution to the study of dambreak flow: initial flow phase]. *Bulletin* de la Direction des Etudes et Recherches, Série A, 1, 1–90 (in French).

Estrade, J. & Gras, R. (1967) Ecoulement consécutif à la rupture d'un barrage [Flow following a dambreak]. Proceedings of the 10th *Convegno di Idraulica*, Cagliari, pp. 130–138 (in French).

Fiori, A. & Guercio, R. (1996) Modelling the overtopping of embankment dams. *Hydropower & Dams*, 3(2), 59–63.

Kenfaoui, M. & Marche, C. (1986) Comment évaluer l'érosion d'une vallée sous l'écoulement de rupture d'un barrage (How to evaluate the erosion in a valley under flow following a dambreak?) *Canadian Journal of Civil Engineering*, 13(4), 474–484 (in French).

Khan, A.A., Steffler, P.M. & Gerard, R. (2000) Dam-break surges with floating debris. *Journal of Hydraulic Engineering*, 126(5), 375–379.

Kosorin, K. (1986) Initial phase of dam break wave: Experimental part. *Vodohospodárstvy Časopis*, 34(1), 71–81 (in Czech, with English Summary).

Kubo, N. & Shimura, H. (1983) Reflection of negative surges from boundaries of various kinds. *Transactions of JSIDRE* (107), 9–19 (in Japanese with English Summary).

Lauber, G. & Hager, W.H. (1997) Ritter's dambreak wave revisited. Proceedings of the 27th *IAHR Congress*, San Francisco, 4, 258–262.

Levin, L. (1952) Mouvement non permanent sur les cours d'eau à la suite de rupture de barrage [Unsteady water flow following a dambreak]. *Revue Générale de l'Hydraulique*, 18(72), 297–315 (in French).

MacDonald, T.C. & Langridge-Monopolis, J. (1984) Breaching characteristics of dam failures. *Journal of Hydraulic Engineering*, 110(5), 567–586; 111(7), 1123–1132.

Martin, H. (1981) Dammbruchkurven in horizontalen Rechteckkanälen [Free surface profiles of dambreak flows in horizontal rectangular channels]. *Acta Hydrophysica*, 26(1), 45–65 (in German).

Menendez, A.N. & Navarro, F. (1990) An experimental study on the continuous breaking of a dam. *Journal of Hydraulic Research*, 28(6), 753–772.

Olivier, H. (1967) Through and overflow rockfill dams: New design techniques. Proceedings of the *ICE*, 36(3), 433–471; 37(4), 855–888.

Penman, A.D.M. (1990) The safety and rehabilitation of tailings dams. *Water Power & Dam Construction*, 42(5), 30–37.

Reinauer, R. & Lauber, G. (1996) Steile Kanäle im wasserbaulichen Versuchswesen [Steep channels in hydraulic experimentation]. *Schweizer Ingenieur und Architekt*, 114(8), 121–124 (in German).

Schoklitsch, A. (1917) Über Dammbruchwellen [On dambreak waves]. *Sitzungberichte* IIa, 126, 1489–1514. Kaiserliche Akademie der Wissenschaften, Wien, Mathematisch-Naturwissenschaftliche Klasse (in German).

Schween, W. (1936) Beitrag zum Ablauf von Schwallwellen in nichtbegehbaren Leitungen auf Grund von Messungen in Strassenkanälen der Stadt Dresden [Experimental contribution to wave propagation in sewers at Dresden City]. *Die Städtereinigung*, 34(20), 513–517; 34(21), 540–545; 34(22), 565–570; 34(23), 589–596; 34(24) 611–615; 35(1), 15–18; 35(2), 34–38 (in German).

Simmler, H. & Sametz, L. (1982) Dam failure from overtopping studied on a hydraulic model. Proceedings of the 14th *ICOLD Congress*, Rio de Janeiro, Q52(R26), 427–445.

Soulis, J.V. (1992) Computation of two-dimensional dam-break flood flows. *International Journal for Numerical Methods in Fluids*, 14(6), 631–664.

Stansby, P.K., Chegini, A. & Barnes, T.C.D. (1998) The initial stages of dam-break flows. *Journal of Fluid Mechanics*, 374, 407–424.

Townson, J.M. & Al-Salihi, A.H. (1989) Models of dam-break flow in R-T space. *Journal of Hydraulic Engineering*, 115(5), 561–575.

Trifonov, E.K. (1933) Experimental investigation of positive wave's propagation along dry bottom. *Izvestija VNIIG*, 10, 169–188 (in Russian, with English summary).

Trifonov, E.K. (1935) Etude expérimentale de la propagation d'une onde positive le long d'un fond sec [Experimental study on the propagation of a positive wave on dry bottom]. *Bulletin PIANC*, 10(19), 66–77 (in French).

Volkoff, N.M. (1956) Formation d'une onde lors d'un déversement dans un lit à sec [Wave formation following a breach into a dry channel]. *Traduction*, 574. Electricité de France, Paris (in French).

Wörman, A. & Olafsdottir, R. (1992) Erosion in a granular medium interface. *Journal of Hydraulic Research*, 30(5), 639–655.

Wörman, A. (1993) Seepage-induced mass wasting in coarse soil slopes. *Journal of Hydraulic Engineering*, 119(10), 1155–1168.

Historical case studies

Anonymous (1889) La rottura del serbatoio di Sonzier presso Montreux [Dam break at Sonzier near Montreux]. *L'Ingegneria Civile*, 15(1), 1–24 (in Italian).

Anonymous (1889) The Johnstone Disaster. *The Engineering and Building Record*, 20(June 8), 16–17; 20(June 15), 29–32.

Anonymous (1895) La rupture du barrage de Bouzey [Bouzey Dam failure]. *Revue Universelle* 1(May 20), 433–437 (in French).

Anonymous (1895) Failure of the Bouzey Dam. *Engineering*, 59(May 3), 583–586; 59(May 17), 645.

Anonymous (1897) La rupture du barrage du réservoir de Bouzey [The failure of Bouzey Dam]. *Le Génie Civil*, 27(2), 17–23 (in French).

Anonymous (1900) The failure of the Austin Dam. *Engineering News*, 43(16), 250–254; 43(17), 274–275; 43(18), 290–291; 43(19), 308–309; 43(26), 428; 44(23), 390–391; 45(22), 392–393; 46(10), 160–161; 47(4), 70; 47(9), 176–177.

Anonymous (1900) The disaster to the water power plant at Hannawa Falls NY. *Engineering News*, 43(17), 277–280; 43(19), 307–308.

Anonymous (1910) Designs and contract for rebuilding the Colorado River Dam at Austin, Texas. *Engineering News*, 63(15), 440–443.

Anonymous (1911a) Another Austin Dam Failure and its lessons. *Engineering News*, 66(14), 410–411; 66(14), 417–422; 66(15), 446–447; 66(18), 543; 66(20), 594–595; 66(22), 661.

Anonymous (1911b) The destruction of the Austin Dam. *Engineering Record*, 64(Oct. 7), 429–436.

Anonymous (1957) Diga del Vaiont [Vaiont Dam]. *Giornale del Genio Civile*, 95(8), 514–517 (in Italian).

Anonymous (1960) The Report on the failure of the Malpasset Dam. *Civil Engineering and Public Works Review*, 55(7), 921–922.

Anonymous (1963) Vaiont Dam survives immense overtopping. *Engineering News-Record*, 171(Oct. 17), 22–23.

Anonymous (1964a) Lessons learnt from dam disasters. *Engineering*, 197(May 15), 681–683.

Anonymous (1964b) Vajont. *Rivista Tecnica della Svizzera Italiana*, 55, 342–347 (in Italian).

Anonymous (1967) Dam designers focus on safety. *Engineering News-Record*, 179(Sep. 14), 15–17.

Anonymous (1976) Grout curtain failure may have triggered Teton Dam collapse. *Engineering News-Record*, 196(June 10), 10–11; 196(June 17), 9–10; 197(Sep. 23), 42; 198(Jan. 13), 8–9.

Anonymous (1980) Final Teton report faults designer. *Engineering News-Record*, 204(Feb. 14), 31.

Anonymous (1982) Bases techniques des plans d'alertes destines à faciliter la protection des populations en aval des barrages [Technical bases of evacuation plans to facilitate the protection of tailwater population]. Proceedings of the 14th *ICOLD Congress*, Rio de Janeiro Q52(R74), 1263–1288 (in French).

Anonymous (1985) Four major dam failures re-examined. *Water Power & Dam Construction*, 37(11), 33–46.

Anonymous (1993) Investigation work begins at Peruća Dam, Croatia. *Water Power & Dam Construction*, 45(4), 15–22.

Bernard, S. (1889) La catastrophe de Johnstown [The Johnstown disaster]. *Cosmos*, 13(232), 373–378; 13(233), 411–414 (in French).

Brou, W.C. (1954) La protection des barrages en temps de guerre [Protection of dams in war time]. *La Technique de l'Eau*, 8(7), 9–17 (in French).

Budweg, F.M.G. (1982) Safety improvements taught by dam incidents and accidents in Brazil. Proceedings of the 14th *ICOLD Congress*, Rio de Janeiro, Q52(R73), 1245–1262.

Cheng, S.-T. (1993) Statistics on dam failures. In: Yen, B.C. (ed) *Reliability and uncertainty analyses in hydraulic design*. ASCE, New York, pp. 97–105.

Coutinho Rodrigues, M.J.M. (1995) Dam failures: Statistical analysis. *ICOLD Bulletin*, 99. International Commission of Large Dams: Paris.

Davidson, J.R. (1914) Some dam failures. *Transactions of Liverpool Engineering Society*, 35, 177–212.

Davies, W.E. (1973) Buffalo Creek Dam disaster: Why it happened. *Civil Engineering*, 43(7), 69–72; 44(2), 107.

De Alba, P.A., Seed, H.B., Retamal, E. & Seed, R.B. (1988) Analyses of dam failures in 1985 Chilean Earthquake. *Journal of Geotechnical Engineering*, 114(12), 1414–1434.

De Marchi, G. (1945) Sull'onda di piena che seguirebbe al crollo della diga di Cancano [On the dambreak wave following the failure of Cancano Dam]. *L'Energia Elettrica*, 22(8–10), 157–169 (in Italian).

Frank, W. (1988) The cause of the Johnstown Flood. *Civil Engineering*, 58(5), 63–66.

Golzé, A.R. (1971) Model law to improve dam safety. *Civil Engineering*, 41(3), 53–56; 41(6), 40; 41(7), 56; 41(7), 85.

Goubet, A. (1979) Risques associés aux barrages [Risks associated with storage dams]. *La Houille Blanche*, 34(8), 475–490 (in French).

Gruner, E. (1963) Dam disasters. *Proceedings of the ICE*, 24(1), 47–60; 27(2), 343–376.

Grzywienski, A. (1971) Failure of conventional dams by overtopping. *Proceedings of the ICE*, 48(1), 35–50.

Hager, W.H. & Lauber, G. (1996) Hydraulische Experimente zum Talsperrenbruchproblem [Hydraulic experiments to dambreak problem]. *Schweizer Ingenieur und Architekt*, 114(24), 515–524 (in German).

Hartung, F. (1954) Der Wiederaufbau des 1941 zerstörten Stauwerkes Dnjeprostroj [Reconstruction of the 1941 destroyed Dnjeprostroj Dam]. *Schweizerische Bauzeitung*, 72(17), 239–244 (in German).

Hatton, T.C. (1912) The Austin Dam and its failure. *Engineering News*, 68(14), 605–606; 68(14), 635–636.

Hilgard, K.E. (1925) Zwei Expertenberichte über die Ursachen des Einsturzes der Gleno-Staumauer in Oberitalien [Two expert reports on causes of the Gleno Dam failure]. *Schweizerische Bauzeitung*, 85(22), 279–283 (in German).

Jaeger, C. (1963) The Malpasset Report. *Water Power*, 15(2), 55–61; 15(4), 137–138; 15(6), 228–230.

Jessup, W.E. (1964) Baldwin Hills Dam failure. *Civil Engineering*, 34(2), 62–64; 34(4), 66.

Kelen, N. (1924) Die norditalienische Dammbruchkatastrophe [The North Italian dam failure]. *Beton und Eisen*, 23(1), 6–8 (in German).

Kiersch, G.A. (1964) Vaiont Reservoir disaster. *Civil Engineering*, 34(4), 32–39; 34(7), 70; 35(3), 63; 35(9), 74.

Kirschmer, O. (1949) Zerstörung und Schutz von Talsperren und Dämmen [Destruction and protection of dams]. *Schweizerische Bauzeitung*, 67(20), 277–281; 67(21), 300–303 (in German).

Langlois, L. (1897) Rupture du barrage de Bouzey [Failure of Bouzey Dam]. *Bulletin Technologique*, 51, 1413–1595 (in French).

Lebreton, A. (1985) Ruptures and serious accidents on dams from 1964 to 1983. *La Houille Blanche*, 40(6/7), 529–544 (in French).

Lempérière, F. (1993) Dams that have failed by flooding: An analysis of 70 failures. *Water Power & Dam Construction*, 45(9/10), 19–24.

Ludin, A. (1924) Der Einsturz der Gleno-Talsperre [The failure of Gleno Dam]. *Deutsche Wasserwirtschaft*, 19(2), 33–48; 19(8), 188–189 (in German).

Macchione, F. & Sirangelo, B. (1990) Floods resulting from progressively breached dams. Proceedings of Conference *Hydrology in Mountainous Regions*, pp. 325–332, also *IAHS Publication* 194.

Noetzli, F.A. (1928) Der Bruch der St. Francis-Staumauer in Kalifornien [The failure of St. Francis Dam in California]. *Schweizerische Bauzeitung*, 91(16), 193–196; 91(24), 295–297 (in German).

Quast, H. (1969) Der Wiederaufbau der kriegszerstörten Möhnetal- und Edertalsperrenmauer im Jahre 1943 [Reconstruction of the war-affected Möhnetal and Edertal Dams in 1943]. *Tiefbau*, 11(9), 733–742 (in German).

Ricoy de Oliveira, A. & De Moraes Leme, C.R. (1985) Adding 1000 m^3/s to Euclides da Cinha Dam outflow. Proceedings of the 15th *ICOLD Congress*, Lausanne, Q59(R7), 93–106.

Robert, B. & Paré, J.-J. (1995) Rupture du barrage du lac Beloeil: Causes et conséquences [Dam break at Lake Beloeil: Causes and consequences]. *Canadian Journal of Civil Engineering*, 22(3), 506–513 (in French).

Schnitter, G. (1964) Die Katastrophe von Vaiont in Oberitalien [The Vaiont Disaster in Upper Italy]. *Wasser- und Energiewirtschaft*, 56(2/3), 61–69 (in German).

Schnitter, G. (1972) Das Unglück am Vajont [Disaster at Vaiont]. *Schweizerische Bauzeitung*, 90(39), 948–954 (in German).

Schnitter, N. (1976) Statistische Sicherheit der Talsperren [Statistical safety of dams]. *Wasser, Energie, Luft*, 68(5), 126–129 (in German).

Serafim, J.L. (1981) Safety of dams judged from failures. *Water Power & Dam Construction*, 33(12), 32–35.

Smith, N.A.F. (1994) The failure of the Bouzey Dam in 1895. *Construction History*, 10(1), 47–65.

Stucky, A. (1924) La rupture du barrage du Gleno [The failure of Gleno Dam]. *Bulletin Technique de la Suisse Romande*, 50(6), 65–71; 50(7), 79–82; 50(9), 107–108 (in French).

Stucky, A. (1924) Der Talsperrenbruch im Val Gleno [Dam failure in Gleno Valley]. *Schweizerische Bauzeitung*, 83(6), 63–67; 83(7), 74–76; 83(8), 92; 83(25), 295–296 (in German).

Taylor, T.U. (1900) The Austin Dam. *Water Supply and Irrigation Paper* 40. US Geological Survey, Washington, DC.

Travaglini, G. (2010) Il Vajont: I primi interventi dopo il disastro [Vajont: The measures after the disaster]. *L'Acqua*, 4, 9–16 (in Italian).

Weber-Ebenhof, A.R. von (1895) Der Bruch der Staumauer von Bouzey und die daraus für die Jaispitzbach-Reservoire zu ziehenden Lehren [Failure of Bouzey Dam and lessons to be drawn for the Jaispitzbach Reservoirs]. *Österr. Monatsschrift für den öffentlichen Baudienst*, 1, 161–167 (in German).

Wüstemann, G. (1960) The breaching of the Oros earth Dam in the State of Ceara, North-east Brazil. *Water and Water Engineering*, 64(8), 351–355.

Subject Index

Note: Page numbers in *italic* indicate a figure and page numbers in **bold** indicate a table on the corresponding page.

Authors' Index